Texts in Applied Mathematics 48

Texts in Applied Mathematics

1. *Sirovich*: Introduction to Applied Mathematics.
2. *Wiggins*: Introduction to Applied Nonlinear Dynamical Systems and Chaos.
3. *Hale/Koçak*: Dynamics and Bifurcations.
4. *Chorin/Marsden*: A Mathematical Introduction to Fluid Mechanics, 3rd ed.
5. *Hubbard/West*: Differential Equations: A Dynamical Systems Approach: Ordinary Differential Equations.
6. *Sontag*: Mathematical Control Theory: Deterministic Finite Dimensional Systems, 2nd ed.
7. *Perko*: Differential Equations and Dynamical Systems, 3rd ed.
8. *Seaborn*: Hypergeometric Functions and Their Applications.
9. *Pipkin*: A Course on Integral Equations.
10. *Hoppensteadt/Peskin*: Modeling and Simulation in Medicine and the Life Sciences, 2nd ed.
11. *Braun*: Differential Equations and Their Applications, 4th ed.
12. *Stoer/Bulirsch*: Introduction to Numerical Analysis, 3rd ed.
13. *Renardy/Rogers*: An Introduction to Partial Differential Equations.
14. *Banks*: Growth and Diffusion Phenomena: Mathematical Frameworks and Applications.
15. *Brenner/Scott*: The Mathematical Theory of Finite Element Methods, 2nd ed.
16. *Van de Velde*: Concurrent Scientific Computing.
17. *Marsden/Ratiu*: Introduction to Mechanics and Symmetry, 2nd ed.
18. *Hubbard/West*: Differential Equations: A Dynamical Systems Approach: Higher-Dimensional Systems.
19. *Kaplan/Glass*: Understanding Nonlinear Dynamics.
20. *Holmes*: Introduction to Perturbation Methods.
21. *Curtain/Zwart*: An Introduction to Infinite-Dimensional Linear Systems Theory.
22. *Thomas*: Numerical Partial Differential Equations: Finitc Difference Methods.
23. *Taylor*: Partial Differential Equations: Basic Theory.
24. *Merkin*: Introduction to the Theory of Stability of Motion.
25. *Naber*: Topology, Geometry, and Gauge Fields: Foundations.
26. *Polderman/Willems*: Introduction to Mathematical Systems Theory: A Behavioral Approach.
27. *Reddy*: Introductory Functional Analysis with Applications to Boundary-Value Problems and Finite Elements.
28. *Gustafson/Wilcox*: Analytical and Computational Methods of Advanced Engineering Mathematics.
29. *Tveito/Winther*: Introduction to Partial Differential Equations: A Computational Approach.
30. *Gasquet/Witomski*: Fourier Analysis and Applications: Filtering, Numerical Computation, Wavelets.

(continued after index)

Diederich Hinrichsen   Anthony J. Pritchard

Mathematical Systems Theory I

Modelling, State Space Analysis, Stability and Robustness

With 178 Figures

Springer

Diederich Hinrichsen
Fachbereich Mathematik
und Informatik
Universität Bremen
Bibliotheksstr. 1
28359 Bremen, Germany
dh@mathematik.uni-bremen.de

Anthony J. Pritchard
Institute of Mathematics
University of Warwick
CV4 7AL Coventry
United Kingdom
ajp@maths.warwick.ac.uk

Series Editors

J.E. Marsden
Control and Dynamical Systems, 107-81
California Institute of Technology
Pasadena, CA 91125
USA
marsden@cds.caltech.edu

L. Sirovich
Division of Applied Mathematics
Brown University
Providence, RI 02912
USA
chico@camelot.mssm.edu

S.S. Antman
Department of Mathematics
and
Institute for Physical Science
and Technology
University of Maryland
College Oark, MD 20742-4015
USA
ssa@math.umd.edu

ISBN 978-3-540-44125-0 e-ISBN 978-3-540-26410-1
DOI 10.1007/978-3-540-26410-1
Springer Heidelberg Dordrecht London New York

Mathematics Subject Classification (2000): 93XX, 34XX, 15XX, 47A55

Library of Congress Control Number: 2004115457

Cover design: deblik, Berlin

Printed on acid-free paper

Springer is part of Springer Science+Business Media (www.springer.com)

Für Malte und Moritz Hinrichsen

To Buddug, Sian, Catrin, Rhian and Ceri Pritchard

Preface

The origins of this book go back more than twenty years when, funded by small grants from the European Union, the control theory groups from the universities of Bremen and Warwick set out to develop a course in finite dimensional systems theory suitable for students with a mathematical background, who had taken courses in Analysis, Linear Algebra and Differential Equations. Various versions of the course were given to undergraduates at Bremen and Warwick and a set of lecture notes was produced entitled "Introduction to Mathematical Systems Theory". As well as ourselves, the main contributors to these notes were Peter Crouch and Dietmar Salamon. Some years later we decided to expand the lecture notes into a textbook on mathematical systems theory. When we made this decision we were not very realistic about how long it would take us to complete the project. Mathematical control theory is a rather young discipline and its foundations are not as settled as those of more mature mathematical fields. Its basic principles and what is considered to be its core are still changing under the influence of new problems, new approaches and new currents of research. This complicated our decisions about the basic outline and the orientation of the book. During the period of our writing, problems of uncertainty and robustness, which had been forgotten for some time in 'modern control', gradually re-emerged and came to the foreground of control theory. Convinced of their key importance we finally deemed it necessary to make them a central subject of the book. Indeed we had already worked on problems of uncertainty ourselves, trying to develop tools for their analysis in state space theory where they had been largely neglected in the aftermath of geometric control theory. Our endeavour to develop a mathematical framework for dealing with such problems, both in the analysis and in the synthesis of control systems, brought up new research problems, and this interaction between the work on the book and work on research further delayed its completion.

Our aim has been to give a rigorous and detailed mathematical treatment of the basic elements of systems theory which could serve as a reference. But we also wanted to do justice to the origins of the subject in engineering and illustrate its interdisciplinary character by many examples and discussions on aspects of application. With this in mind we decided at an early stage that the book should be focussed on finite dimensional time-invariant linear systems. There were two main reasons for this choice. Firstly, nearly all the main problems, concepts and approaches in the theories of nonlinear and infinite dimensional control have their origins in linear finite dimensional theory. Secondly, advanced theories require more sophisticated mathematics, and there is the risk that technical problems of mathematics obscure the system theoretic content. This was in conflict with our wish to write a book accessible to students of mathematics after two years of study and to concentrate on the main issues and fundamental concepts of systems theory. Nevertheless, in spite of the focus on finite dimensional linear systems we have made it a rule to develop the basic system theoretic notions in full generality. Throughout the book the presentation proceeds in a systematic way from the abstract to the concrete. The exposition is restricted to time-invariant linear systems only where a development for other classes of systems would require advanced mathematical tools beyond those outlined in the appendix. For instance, we do not touch on any topics of nonlinear systems and control theory which require the use of differential geometric tools, nor do we deal with infinite dimensional systems theory since then a substantial preparation in functional analysis would be necessary.

The first two chapters of this volume are of an introductory nature whereas the others are more demanding and prepare the reader for research. The rigorous mathematical treatment is complemented by many examples, illustrations and explanatory comments. Also computational issues are discussed. As such, we hope the volume will be useful for established researchers in systems theory as well as those just starting in the field. For teaching it can be used at two different levels. The material can be filtered to obtain undergraduate courses, and individual graduate courses can be based on single or pairs of chapters. Indeed we have based undergraduate courses on Chapter 3, graduate ones on Chapters 3, 4, and Chapters 4, 5 and a seminar on Chapter 1. It is our experience that a first course in mathematical systems theory in the third year of a mathematics curriculum is an excellent way of showing students the usefulness of what they have studied in their first two years. In control theory they can learn that methods from different mathematical fields, like analysis, linear algebra, differential equations, complex analysis, integral transformations and numerical analysis, which they have studied separately in their first years, must be combined to develop a successful theory for applications.

The book is divided into two volumes. The second one will be concerned with *control* aspects and contains chapters on controllability and observability, input-output systems, geometric control theory, the linear quadratic problem and H_∞ control theory. The present first volume consists of five chapters and is concerned mainly with *systems analysis*. At the end of this volume there is a detailed index preceded by a glossary and an extensive bibliography. Every chapter, with the exception of the first, has the same format. Each is divided into sections and subsections with exercises and notes and references at the end of each section. Sections are numbered consecutively within chapters and subsections are numbered consecutively within sections. For example, Section 5.3 is the third section in Chapter 5 and Subsection 5.3.1 is the first subsection in Section 5.3. Theorems, propositions, definitions etc. are numbered consecutively by chapter and section in a single list and are indexed with three numbers. Thus Theorem 5.1.8 refers to a theorem in Section 1 of Chapter 5 and is the eighth theorem or example etc. in the list of that section. Figures and tables are numbered consecutively, e.g. Figure 4.1.7 could be followed by Table 4.1.8. Equations are numbered by single numbers in each section, and are referenced by this number in the section where it occurs. For example (9) refers to the ninth equation in the same section. However, within say Chapter 3, the ninth equation in Section 2 is written (2.9) when cross-referenced in say Section 3, whereas, if the equation is referred to in any other chapter we give the triple (3.2.9). Exercises are referenced in a similar way, i.e. we write Ex. 9, Ex. 2.9 or Ex. 3.2.9. A survey of the material in each of the chapters can be obtained by looking at the table of contents. Below we give a brief overview.

The first chapter is of an illustrative and motivational character. It presents a series of dynamic models from six areas of application and explains by examples how dynamic phenomena in different fields of science and engineering can be translated into appropriate mathematical representations. It also shows how typical system theoretic problems and concepts arise in these fields. The descriptive style adopted in this chapter is rather different from the mathematical style of the ensuing chapters. Most of the sections just give a catalogue of examples from the corresponding field of application. The sections on *mechanics* and *electromagnetism* are different. These fields have their own well-established theories of dynamics. In fact control theory has emerged from mechanical and electrical engineering which are still the main areas of application. We therefore deemed it appropriate to explain some of the scientific principles behind the dynamic models in these areas and sketch some modelling techniques in use. Altogether, the chapter is meant as an introduction to dynamic models and an illustration of the diversity of dynamical phenomena to which system theoretic concepts may be applied. Some of the models described here are taken up later in the examples of the following chapters.

The introduction to mathematical systems theory begins with Chapter 2. Some readers may prefer to start directly with this chapter and go back to Chapter 1 for more details whenever an example from the first chapter is used for illustration. Chapter 2 provides an introduction to state space theory. We have chosen to use the input-state-output approach put forward by Kalman. The general concept of a dynamical system is developed and then it is specialized to the linear case. Continuous time and discrete time systems are treated in parallel and are interrelated by a discussion of sampling and approximations problems. Some preliminary elements of input-output theory are also introduced and the relationship between the analysis of input-output systems in time and in frequency domain is explained.

The next chapter deals with stability theory. Some elements of topological dynamics and Liapunov's stability theory are developed in a general setting and then specialized to different classes of systems. A notable feature of this chapter is that the sections on Liapunov's analytical approach are complemented by an extensive final section on classical algebraic stability theory.

One would expect to find some of the material of the previous chapters in a book on systems theory, but the inclusion of a chapter on perturbation theory (the subject of Chapter 4) might seem surprising. We felt it was necessary because many of the results we give permeate various branches of systems theory but are rarely explicitly stated and proved in books on systems and control. Moreover we wished to address the robustness question in a general setting and so needed to introduce some elements of μ-analysis.

The final chapter of this first volume reflects our joint research on uncertain systems. Our main objective is to develop a spectral theory for uncertain time-invariant linear systems. We do this via *spectral value sets* and *stability radii* and most of the chapter is devoted to deriving both qualitative and quantitative results for them.

However we also deal with the problem of transient deviations of trajectories from an equilibrium point and in a final section obtain results for stability radii of uncertain systems with respect to time-varying, nonlinear and dynamic perturbations.
Since the range of mathematics used in this volume is quite wide we have included some of the background mathematics in fairly substantial appendices.
We have tried our best to eliminate any errors in the book. However our experience has shown that this is a never ending process and we would be very grateful if readers could communicate to us any errors and inaccuracies they encounter in this volume.
In conclusion we would like to thank those colleagues who helped us, directly or indirectly, with the preparation of this book. As students of mathematics we did not come into contact with systems theory. We learnt it whilst lecturing at university and have been strongly influenced by friends and colleagues who at an early stage in our careers introduced us to their fields of research during periods when they were guest professors of our universities or when we were invited to their research centres. We benefited greatly from their knowledge and advice, and would like to express our special thanks to Roger Brockett, Chris Byrnes, Ruth Curtain, Paul Fuhrmann, Michiel Hazewinkel, Michael Heymann, Alan Laub, Larry Markus, Howard Rosenbrock, Jan Willems, Murray Wonham and Jerzy Zabczyk. We also owe thanks to our doctoral students and co-workers at that time, who are now friends and colleagues. Their enthusiasm and manifold contributions spurred our research and without them we would not have undertaken this project.
More recently, we have profited from the expertise of the many people who visited us in Bremen and Warwick. In particular we are indebted to Vladimir Kharitonov. His series of lectures on algebraic stability theory in Bremen helped us with the preparation of Section 3.4. Our doctoral students and colleagues Eduardo Gallestey, Michael Karow, Elmar Plischke and Fabian Wirth have collaborated with us in the research which led to the results presented in Chapter 5. Many of the examples and figures in this chapter are due to them. Fabian read some of the sections and made suggestions for their improvement. We also would like to thank Buddug Pritchard who helped us with the English. In the early days Bernd Kelb typed some of the sections, computed some of the examples, constructed some of the figures and helped us with LaTeX. More recently Elmar has taken on this role. Not only has he contributed in research to the development of the material on transient behaviour in Chapter 5, he has also computed many figures and read, and suggested improvements to many of the sections. Moreover he has been a rock for us with his technical knowledge of and expertise with the computer. Whenever we had problems with Unix, Linux, LaTeX, xfig, MATLAB he willingly gave us his assistance and always did so with a wry sense of humour. Finally we would like to thank the team at Springer, in particular Ruth Allewelt and Martin Peters who have been most helpful, patient and understanding.

Bremen
Warwick
October, 2004

Diederich Hinrichsen
Tony Pritchard

Contents

Preface vii

1 Mathematical Models 1
1.1 Population Dynamics . . . 2
1.1.1 Notes and References . . . 6
1.2 Economics . . . 8
1.2.1 Notes and References . . . 12
1.3 Mechanics . . . 13
1.3.1 Translational Mechanical Systems . . . 13
1.3.2 Mechanical Systems with Rotational Elements . . . 18
1.3.3 The Variational Method . . . 27
1.3.4 Notes and References . . . 38
1.4 Electromagnetism and Electrical Systems . . . 39
1.4.1 Maxwell's Equations and the Elements of Electrical Circuits . . . 39
1.4.2 Electrical Networks . . . 50
1.4.3 Notes and References . . . 55
1.5 Digital Systems . . . 56
1.5.1 Combinational Switching Networks . . . 59
1.5.2 Sequential Switching Networks . . . 62
1.5.3 Notes and References . . . 68
1.6 Heat Transfer . . . 70
1.6.1 Notes and References . . . 72

2 Introduction to State Space Theory 73
2.1 Dynamical Systems . . . 74
2.1.1 The General Concept of a Dynamical System . . . 74
2.1.2 Differentiable Dynamical Systems . . . 83
2.1.3 System Properties . . . 88
2.1.4 Linearization . . . 92
2.1.5 Exercises . . . 94
2.1.6 Notes and References . . . 98
2.2 Linear Systems . . . 100
2.2.1 General Linear Systems . . . 100
2.2.2 Free Motions of Time–Invariant Linear Differential Systems . . . 104
2.2.3 Free Motions of Time–Invariant Linear Difference Systems . . . 113
2.2.4 Infinite Dimensional Systems . . . 115
2.2.5 Exercises . . . 121
2.2.6 Notes and References . . . 123
2.3 Linear Systems: Input–Output Behaviour . . . 124

2.3.1 Input-Output Behaviour in Time Domain . . . 124
2.3.2 Transfer Functions . . . 138
2.3.3 Relationship Between Input–Output Operators and Transfer Matrices . . . 147
2.3.4 Exercises . . . 151
2.3.5 Notes and References . . . 153
2.4 Transformations and Interconnections . . . 154
2.4.1 Morphisms and Standard Constructions . . . 154
2.4.2 Composite Systems . . . 160
2.4.3 Exercises . . . 166
2.4.4 Notes and References . . . 167
2.5 Sampling and Approximation . . . 168
2.5.1 A/D- and D/A-Conversion of Signals . . . 169
2.5.2 The Sampling Theorem . . . 171
2.5.3 Sampling Continuous Time Systems . . . 175
2.5.4 Approximation of Continuous Systems by Discrete Systems . . . 177
2.5.5 Exercises . . . 189
2.5.6 Notes and References . . . 192

3 Stability Theory **193**
3.1 General Definitions . . . 194
3.1.1 Local Flows . . . 195
3.1.2 Stability Definitions . . . 199
3.1.3 Limit Sets . . . 202
3.1.4 Recurrence . . . 206
3.1.5 Attractors . . . 211
3.1.6 Exercises . . . 213
3.1.7 Notes and References . . . 215
3.2 Liapunov's Direct Method . . . 217
3.2.1 General Definitions and Results . . . 217
3.2.2 Time–Varying Finite Dimensional Systems . . . 229
3.2.3 Time–Invariant Systems . . . 235
3.2.4 Exercises . . . 248
3.2.5 Notes and References . . . 251
3.3 Linearization and Stability . . . 253
3.3.1 Stability Criteria for Time-Varying Linear Systems . . . 254
3.3.2 Time–Invariant Systems: Spectral Stability Criteria . . . 263
3.3.3 Numerical Stability of Discretization Methods . . . 268
3.3.4 Liapunov Functions for Time-Varying Linear Systems . . . 272
3.3.5 Liapunov Functions for Time-Invariant Linear Systems . . . 282
3.3.6 Exercises . . . 291
3.3.7 Notes and References . . . 295
3.4 Stability Criteria for Polynomials . . . 296
3.4.1 Stability Criteria and the Argument Principle . . . 297
3.4.2 Characterization of Stability via the Cauchy Index . . . 308
3.4.3 Hermite Forms and Bézoutiants . . . 313
3.4.4 Hankel Matrices and Rational Functions . . . 320
3.4.5 Applications to Stability . . . 334
3.4.6 Schur Polynomials . . . 340

3.4.7 Algebraic Stability Domains and Linear Matrix Equations 357
3.4.8 Exercises 361
3.4.9 Notes and References 366

4 Perturbation Theory 369
4.1 Perturbation of Polynomials 369
4.1.1 Dependence of the Roots on the Coefficient Vector 370
4.1.2 Polynomials with Holomorphic Coefficients 376
4.1.3 The Sets of Hurwitz and Schur Polynomials 384
4.1.4 Kharitonov's Theorem 389
4.1.5 Exercises 393
4.1.6 Notes and References 396
4.2 Perturbation of Matrices 398
4.2.1 Continuity and Analyticity of Eigenvalues 398
4.2.2 Estimates for Eigenvalues and Growth Rates 404
4.2.3 Smoothness of Eigenprojections and Eigenvectors 409
4.2.4 Exercises 426
4.2.5 Notes and References 429
4.3 The Singular Value Decomposition 431
4.3.1 Singular Values and Singular Vectors 431
4.3.2 Singular Value Decomposition 435
4.3.3 Matrices Depending on a Real Parameter 439
4.3.4 Relations between Eigenvalues and Singular Values 444
4.3.5 Exercises 446
4.3.6 Notes and References 448
4.4 Structured Perturbations 449
4.4.1 Elements of μ-Analysis 449
4.4.2 μ-Values for Real Full-Block Perturbations 465
4.4.3 Exercises 480
4.4.4 Notes and References 481
4.5 Computational Aspects 484
4.5.1 Condition Numbers 485
4.5.2 Matrix Transformations 492
4.5.3 Algorithms 501
4.5.4 Exercises 513
4.5.5 Notes and References 515

5 Uncertain Systems 517
5.1 Models of Uncertainty and Tools for their Analysis 520
5.1.1 General Definitions and Basic Properties 520
5.1.2 Perturbation Structures 530
5.1.3 Exercises 540
5.1.4 Notes and References 542
5.2 Spectral Value Sets 544
5.2.1 General Definitions and Results 544
5.2.2 Complex Full-Block Perturbations 556
5.2.3 Real Full-Block Perturbations 561
5.2.4 The Unstructured Case (Pseudospectra) 569
5.2.5 Exercises 580

5.2.6 Notes and References . . . 583
5.3 Stability Radii . . . 585
5.3.1 General Definitions and Results . . . 586
5.3.2 Complex Full-Block Perturbations . . . 591
5.3.3 Real Full-Block Perturbations . . . 596
5.3.4 Hamiltonian Characterization of the Complex Stability Radius . . . 602
5.3.5 The Unstructured Case . . . 609
5.3.6 Dependence on System Data . . . 614
5.3.7 Stability Radii and the Cayley Transformation . . . 617
5.3.8 Exercises . . . 621
5.3.9 Notes and References . . . 624
5.4 Root Sets and Stability Radii of Polynomials . . . 625
5.4.1 General Formulas . . . 625
5.4.2 Complex Perturbation Structures . . . 633
5.4.3 Real Perturbation Structures . . . 637
5.4.4 Exercises . . . 644
5.4.5 Notes and References . . . 646
5.5 Transient Behaviour . . . 648
5.5.1 Transient Bounds and Initial Growth Rate . . . 648
5.5.2 Contractions and Estimates of the Transient Bound . . . 658
5.5.3 Spectral Value Sets and Transient Behaviour . . . 669
5.5.4 Robustness of (M, β)-Stability . . . 675
5.5.5 Exercises . . . 680
5.5.6 Notes and References . . . 684
5.6 More General Perturbation Classes . . . 686
5.6.1 The Perturbation Classes . . . 687
5.6.2 Stability Radii . . . 696
5.6.3 The Aizerman Conjecture . . . 701
5.6.4 Exercises . . . 709
5.6.5 Notes and References . . . 711

Appendix **715**
A.1 Linear Algebra . . . 715
A.1.1 Norms of Vectors and Matrices . . . 715
A.1.2 Spectra and Determinants . . . 719
A.1.3 Real Representation of Complex Matrices . . . 720
A.1.4 Direct Sums and Kronecker Products . . . 720
A.1.5 Hermitian Matrices . . . 722
A.2 Complex Analysis . . . 724
A.2.1 Topological Preliminaries . . . 724
A.2.2 Path Integrals . . . 725
A.2.3 Holomorphic Functions . . . 727
A.2.4 Isolated Singularities . . . 729
A.2.5 Analytic Continuation . . . 732
A.2.6 Maximum Principle and Subharmonic Functions . . . 733
A.3 Convolutions and Transforms . . . 735
A.3.1 Sequences: Convolution and **z**-Transforms . . . 735
A.3.2 Lebesgue Spaces, Convolution of Functions, Laplace Transforms . . . 739
A.3.3 Fourier Series and Fourier Transforms . . . 744

A.3.4 Hardy Spaces 750
A.4 Linear Operators and Linear Forms 753
A.4.1 Summability and Generalized Fourier Series 753
A.4.2 Linear Operators on Banach Spaces 754
A.4.3 Linear Operators on Hilbert Spaces 757
A.4.4 Spectral Theory 759

References **763**

Glossary **789**

Index **795**

Chapter 1

Mathematical Models

In this chapter we present a range of dynamical systems from different areas of application and use them as examples to illustrate some typical problems from systems and control theory. Several of the mathematical models we introduce and discuss in the following sections will be taken up as examples in later chapters.
The development of mathematical systems theory starts in the next chapter. The readers who prefer to go directly to Chapter 2 can do so without any difficulty as the mathematical exposition in that chapter is self-contained and independent of following material. On encountering an example based on a dynamic model from Chapter 1, they may wish to look back to its origin here to find more details and get additional background information.
This chapter consists of six sections in which we present dynamical models from the following areas:

- Biology (Population Dynamics)
- Economics
- Mechanics
- Electromagnetism and Electrical Systems
- Digital Systems
- Heat Transfer

The mathematical models in the first three sections are described by *ordinary differential equations* and by *difference equations*. Also in Section 1.4, although the basic equations of electromagnetism are *partial differential equations*, we will only consider so-called *lumped models* of electromagnetic devices which again are described by ordinary differential equations. Different types of models are presented in the remaining two sections. In Section 1.5 we consider digital systems which have only a finite number of different states and are represented as finite automata. In the last section we deal with an example of a distributed parameter system described by partial differential equations.
In all these sections we will not only discuss the mathematical models but also point out some of the problems encountered in determining a mathematical model for a real process. While most of the sections just present a gallery of typical examples, some modelling methods will be sketched out in the sections on mechanical and electrical systems.

1.1 Population Dynamics

In order to predict or estimate the growth of a given population one needs a dynamical model. Such models may also be useful if one wants to control the development of a population. For example problems of control arise in fisheries management where one would like to keep fishing at a sustainable level and maximize the average catch over long time periods. In other applications interaction between different populations may be important and one may make use it for control purposes, e.g. in pest control where one introduces predators to reduce the pest. In this section we consider two classical models of population dynamics.

Example 1.1.1. (Logistic growth model). The simplest growth model is

$$\dot{x}(t) = ax(t). \tag{1}$$

Here $x(t)$ is the size, density or biomass of a given population at time t and the growth parameter a is the *intrinsic growth rate* (difference between the birth rate and the death rate) of the population. If the initial size of the population is $x(0) = x_0 > 0$ the development follows the exponential law $x(t) = e^{at}x_0$. Thus we have exponential growth if $a > 0$ (i.e. the birth rate is larger than the death rate) and exponential decay if $a < 0$. The idea that human populations when "unchecked by the difficulties of subsistence" have a positive constant natural growth rate goes back to Malthus. In his *Essay on Population (1798)* he contrasted the natural geometric growth of mankind with the linear growth of subsistence resources and drew far reaching conclusions from this which had a profound effect on political economics.

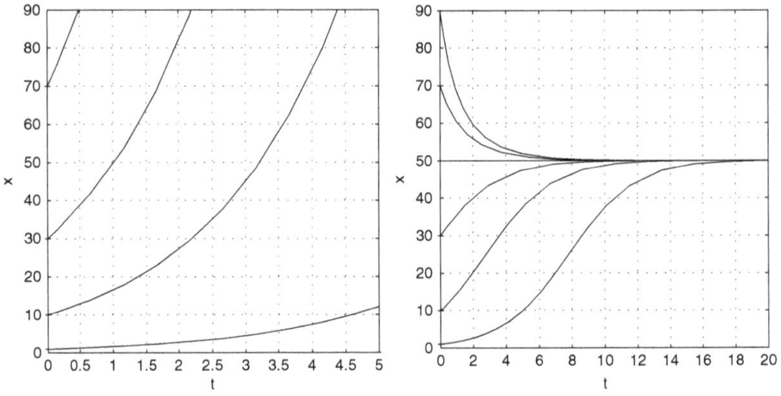

Figure 1.1.1: Exponential and logistic growth models

The exponential growth model, although adequate in many applications over a limited time span becomes unrealistic in the long run since $e^{at}x_0 \to \infty$ as $t \to \infty$. The growth rate $\dot{x}(t)/x(t)$ cannot be constant over arbitrarily long periods of time, since resources are limited. As the population becomes larger and larger, restraining factors will have an increasingly negative effect on population growth ("crowding"). In 1838 Verhulst proposed another growth model which incorporated the limiting factors and accounted for the fact that individuals compete for food, habitat, and other limited resources,

$$\dot{x}(t) = r(K - x(t))x(t). \tag{2}$$

According to this model a small population will initially grow at an exponential rate rK but as the population increases the growth rate will be diminished.
If the system is initially at $x_0 = K$ then it will remain at $x(t) = K$ for all time. Then the population is at an equilibrium $x(t) \equiv \bar{x} = K$, $t \geq 0$. If $0 < x_0 < K$ the population $x(t)$ will increase continuously and approximate K as $t \to \infty$. If $x_0 > K$, the population size $x(t)$ will converge towards K from above. In fact the following formula for the solution is easily obtained by separation of variables

$$x(t) = \frac{K}{1 + (K/x_0 - 1)\, e^{-rKt}}.$$

The graphs of these solutions are called *logistic curves* and Verhulst's model is also known as the *logistic growth model.* Figure 1.1.1 illustrates that $x(t) \equiv K$ is a stable equilibrium, i.e. all trajectories with initial state $x_0 > 0$ converge towards this equilibrium as $t \to \infty$. The saturation level K is interpreted as the *environmental carrying capacity* of the corresponding ecosystem. Now suppose that we want to describe the dynamics of a fish population under the influence of fishing. If $u(t) \geq 0$ is the catch rate and we assume the logistic growth model for the undisturbed fish population, we obtain *Schaefer's model*

$$\dot{x}(t) = r(K - x(t))x(t) - u(t). \tag{3}$$

Note that only non-negative solutions $x(t, u) \geq 0$ make sense. Given an initial state $x_0 > 0$ and a fixed time period $[t_0, t_1]$, a fishing policy $u(\cdot) : [t_0, t_1] \to \mathbb{R}_+$ may be called "admissible" if it leads to a non-negative solution $x(t, u)$ of (3) for $t \in [t_0, t_1]$ and "optimal" if it maximizes the overall catch during that period. Such an "optimal" fishing policy will, however, lead to depletion at time t_1. To prevent this one may wish to impose a "terminal constraint" $x(t_1) \geq x_1$ where $x_1 > 0$ is a lower bound to an acceptable fish population at the end of the period. Thus we end up with the following *optimal control problem*:

$$\text{Maximize} \int_{t_0}^{t_1} u(t)dt \;\; \text{subject to} \;\; u(t) \geq 0,\; x(t, u) \geq 0,\; t \in [t_0, t_1],\; x(t_1) \geq x_1.$$

If $u(t)$ is required to be constant, the problem is easily solved, see Ex. 2.1.15.
Another optimal control problem which can be solved by elementary means is the *optimal constant-effort harvesting problem.* Here the harvesting rate $u(t)$ is by definition proportional to $x(t)$, i.e. $u(t) = cx(t)$. This is a simple example of *feedback control* where the control variable $u(t)$ is determined as a given function of the instantaneous state $x(t)$ of the system. Following this control strategy one obtains a Verhulst model in which the parameters have changed

$$\dot{x}(t) = r(K - c/r - x(t))x(t).$$

If $c < rK$ there is an equilibrium solution $x(t) = \bar{x} = K - c/r$, $t \geq 0$ corresponding to the constant harvesting policy $u(t) = c\bar{x}$, $t \geq 0$. Again one can determine the optimal constant harvesting policy which yields the highest sustainable harvesting rate, see Ex. 2.1.15. □

Remark 1.1.2. Although the logistic model is a widely used and successful model which predicts quite well the growth of various laboratory populations (see *Notes and References*), it is a highly simplified model. It is based on a number of assumptions which are not usually satisfied when the growth of a species in a real ecosystem is considered, e.g.

(i) The influence of environmental factors on the growth of the species is assumed to be constant in time. But these factors and the behaviour of a species usually vary with the time of the year. Also there are often random variations in the environment.

(ii) The effects of limited resources are assumed to affect all individuals of the species in an equal manner. A more realistic model would take the spatial distribution of the species and its resources into account (partial differential equations).

(iii) It is assumed that the birth and death rates of the population respond instantly to the population size, whereas usually there is a delay between birth and the ability to give birth.

(iv) The age distribution of the population is assumed to be constant or that if it changes it does not influence the growth of the species.

Although the assumptions are not realistic, highly simplified models like that of Verhulst are often of great scientific value. Their purpose is not to give an accurate portrait of an underlying real process but to enhance the understanding of some of its internal mechanisms. As such they can be more important motors for scientific progress than complex "realistic" simulation models[1]. □

Often the dynamics of a population are strongly influenced by the interaction with other populations in the same ecosystem. Several species may compete for the same natural resources or a species may be predatory on some species while serving as prey for others. In the following example we describe a classical predator-prey model due to Lotka and Volterra[2].

Example 1.1.3. (Predator-prey system). Suppose that an island is populated by goats and wolves. The goats survive by eating the island's vegetation and the wolves survive by eating the goats. Often oscillations are observed in the development of such predator-prey populations. If, initially, there are only a few wolves but many goats, the wolves have a lot to eat and the number of goats will be diminished while the number of wolves will increase until there are not enough goats to feed them. Then the number of wolves will be reduced so that the goats will be able to recover and this closes the cycle. The classical Lotka-Volterra model for such a predator-prey system is

$$\begin{aligned} \dot{x}_1 &= ax_1 - bx_1x_2 \\ \dot{x}_2 &= -cx_2 + dx_1x_2, \end{aligned} \tag{4}$$

where x_1 and x_2 are the densities (number per unit area) of the prey and predator populations respectively, and a, b, c, d are positive constants. The model mirrors a qualitative feature which has been observed in many real predator-prey systems, the persistence of periodic fluctuations. This is illustrated in Figure 1.1.2. $\bar{x} = (c/d,\ a/b)$ is an equilibrium point of (4) and any initial state $x^0 \neq \bar{x},\ x_1^0 > 0, x_2^0 > 0$ leads to a periodic trajectory cycling around this equilibrium point in the positive orthant.

Clearly, this is a simplistic model and does not aim at simulating or predicting a real process. The model is based on the following assumptions.

[1] "This work seeks to gain general ecological insights with the help of general mathematical models. That is to say the models aim not at realism in detail, but rather at providing mathematical metaphors for broad classes of phenomena. Such models can be useful in suggesting interesting experiments or data collecting enterprises, or just in sharpening discussion." (R. M. May, Preface of "Stability and Complexity in Model Ecosystems").

[2] The story of how Volterra came to design the model (independently of Lotka) is interesting. For many years fishermen had observed periodic fluctuations between sharks and their prey populations in the Adriatic Sea. During World War I, commercial fishing was greatly reduced and so it was expected that there would be plentiful fish stocks for harvesting after the war was over. Instead the catches of commercially valuable fish declined after the war while the number of sharks increased.

(i) In the absence of predators the prey population grows exponentially with rate a.

(ii) In the absence of prey the predator population decreases at the death rate c.

(iii) The growth of the predator population depends affinely on the food intake, i.e. on predation.

(iv) Predation depends on the likelihood that a victim is encountered by a predator and this likelihood is proportional to the product $x_1 x_2$ of the two populations' densities.

An assumption similar to (iv) is made in chemical kinetics where, according to the so-called law of mass action, the rate of molecular collisions of two substances in a given solution is assumed to be proportional to the product of their concentrations.

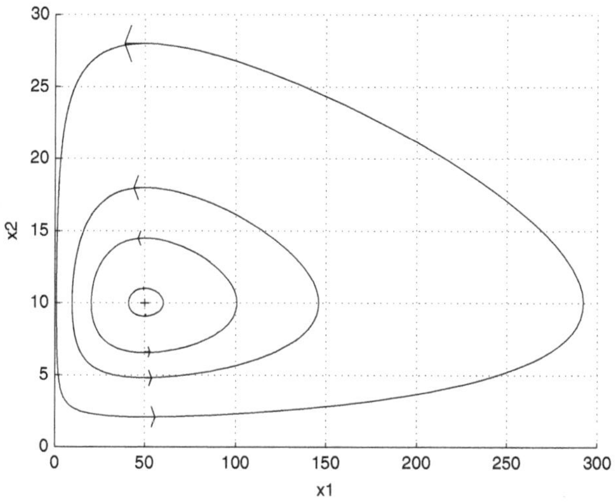

Figure 1.1.2: Predator-prey trajectories

Many "more realistic" models have been obtained from (4) by modifying the predator-free prey growth term ax_1 to include crowding effects or by allowing for saturation effects and lags in the predators' response to increasing prey densities. For instance, in order to eliminate the assumption that the prey grows exponentially in the absence of predators one could introduce a term $-ex_1^2$ in the first equation of (4) which accounts for the effect of crowding on the growth of the prey (see Example 1.1.1).

$$\begin{aligned} \dot{x}_1 &= ax_1 - bx_1 x_2 - ex_1^2 = e(a/e - x_1)x_1 - bx_1 x_2 \\ \dot{x}_2 &= -cx_2 + dx_1 x_2. \end{aligned} \tag{5}$$

This drastically alters the qualitative behaviour of the predator-prey system. In the absence of predators the prey now evolves according to a logistic growth model with carrying capacity a/e. Moreover, the new system does not always have an equilibrium with positive coordinates. In fact the equilibrium equations are

$$(a - bx_2 - ex_1)x_1 = 0, \quad (-c + dx_1)x_2 = 0$$

and these equations have a (unique) positive solution $\bar{x} = (c/d, (da - ec)/bd)$ if and only if $a/e > c/d$. Figure 1.1.3 illustrates the changed behaviour of the modified predator-prey system (5). In particular, it has no non-constant periodic solutions and its only

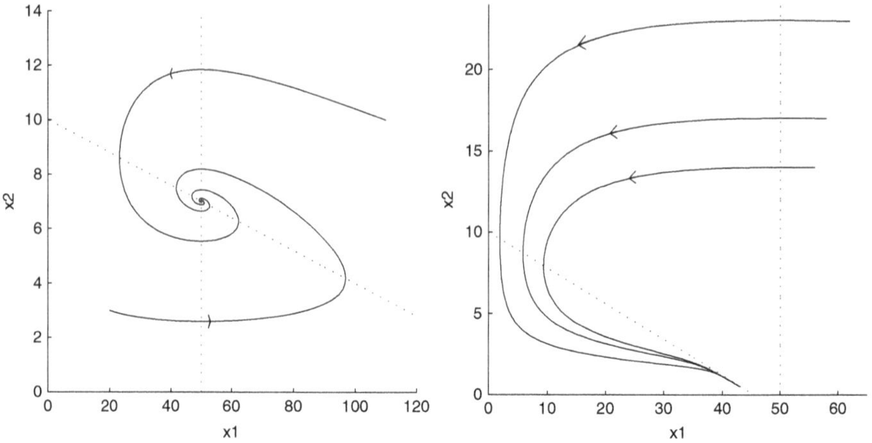

Figure 1.1.3: The effect of crowding

positive equilibrium point $\bar{x} = (c/d,\ (ad-ec)/bd)$ is now asymptotically stable. It attracts all trajectories starting at initial states close to it, but not necessarily those starting further away, see Figure 1.1.3. In Chapter 3 we will show how the stability or instability of an equilibrium point can be examined for a given set of parameters. Using these results it is possible to show that the other two equilibrium points $(0,0)$ and $(a/e,0)$ are unstable. The qualitative changes between (4) and (5) do not depend on the size of $e > 0$ which can be arbitrarily small. This shows that the classical predator-prey system is not structurally stable in the sense that a perturbation of the model, however small, may exhibit a qualitatively different behaviour.

In spite of their simplicity predator-prey systems and other models of two species are used in a number of control applications, e.g. in the management of renewable resources or in pest control where predators are introduced to control pests feeding on agricultural crops. Consider for instance a predator-prey system of salmon and herring in marine fishing. Choosing a suitable predator-prey model and adding control terms to both equations (catch rates) one may ask what are the optimal sustainable harvesting rates given the prices for salmon and herring, and what is the corresponding equilibrium point of the system, i.e. the stocks of salmon and herring which allow one to realize the optimal rates. If this optimal equilibrium point is found, The problem then arises of how the optimal equilibrium can be attained from a given initial population of salmon and herring by applying a suitable fishing policy. This is a controllability problem which we consider in Vol. II. In order to be of any practical value, the optimal equilibrium must be asymptotically stable since otherwise unavoidable small deviations of the populations from their optimal sizes may lead to large deviations from the optimal equilibrium solution. But asymptotic stability is not enough. It is important that this property is preserved under perturbations which reflect the uncertainties about the model and its parameters. This is a problem of robust stability which we will analyze in Chapter 5. □

1.1.1 Notes and References

Modelling in general

There are a number of introductions to dynamical systems which emphasize modelling. In particular, we recommend the book by *Luenberger* (1979) [349]. Many elementary

examples of control systems can be found in a collection of case studies by *MacClamroch* (1980) [369]. Additional information about modelling and a large number of examples can be found in textbooks on the analysis, modelling and design of dynamic systems, see for example *Ogata* (1992) [397], *Burton* (1994) [84], *Close and Frederick* (1995) [105].
The reader who is interested in *modelling techniques* for a variety of physical systems is referred to *Wellstead* (1979) [516], *Shearer et al.* (1967) [462], *MacFarlane* (1970) [355].

Population Dynamics

A comprehensive textbook discussing dynamic models in various areas of biology and the life sciences is *Murray* (1993) [385], see also *Edelstein-Keshet* (1988) [146], *Hoppensteadt and Peskin* (1992) [263]. The book by Murray also contains an extensive bibliography.
Population Dynamics is one of the core subjects of mathematical biology and was amongst the first areas in life sciences which attracted mathematical methods. Classical references are *Malthus* (1798) [358], *Verhulst* (1938) [506], [347], *Volterra* (1927) [509] [510], *Kostitzin* (1934) [315] and *Kolmogoroff* (1936) [313]. English translations of some of their papers and brief discussions of their work can be found in [489].
Various empirical investigations have shown a good fit between the logistic model and the growth of actual laboratory populations, see e.g. *Lotka* (1924) [347] (Drosophila) and *Gause* (1959) [185] (Paramecium caudatum). A detailed discussion of the Verhulst model can be found in *May* (1981) [368]. The behaviour of the *discrete time logistic equation* $x(t+1) = rx(t)(1-x(t))$ has been analyzed by means of cobweb diagrams in *Edelstein-Keshet* (1988) [146]. The qualitative features of the model change drastically at certain critical parameter values (bifurcation values) and for certain values of r chaotic behaviour is observed, see Ex. 3.1.15. A discussion of this model in the context of Population Dynamics can be found in *May* (1976) [367].
The controlled Verhulst equation (3) was used by *Schaefer* (1954) [449] to study the tuna fisheries of the tropical Pacific. It is probably the simplest dynamical model in Bio-economics (an interdisciplinary field which combines Mathematics, Biology and Economics), and has been used to study the effect of harvesting on growing populations. A standard reference on this subject is *Clark* (1976) [101], see also [102]. Control aspects are also important in bio-technology. A book on modelling bio-reactions and bio-reactors is *Nielsen and Villadsen* (1994) [392].
There is a large variety of models for interacting populations and some of them can already be found in the classical references above. These models play an important role in theoretical Ecology, see *Pielou* (1977) [411], *May* (1981) [368] and [366]. Important areas to be analyzed are the existence and stability of equilibria, the existence and stability of periodic solutions, their dependency on parameters, the effect of lags, the relationship between stability and complexity, the effect of competition, age structure and migration on growth rates, the extinction of species etc. An interesting mathematical discussion of various two species models is given in *Hirsch and Smale* (1974) [258]. The problem of robust stability or "resilience" is of particular interest in Ecology, for a discussion in the context of "complexity versus stability", see *May* (1974) [366].
Supplemented with a control term population models are also used in the management of renewable resources, see *Clark* (1985) [102]. Other areas of application include Epidemiology *Bailey* (1975) [31]), theories of evolution *Hofbauer and Sigmund* (1988) [259], and pest control (rabies, weed dispersal, foot and mouth, etc.), see e.g. *Evans and Pritchard* (2001) [153] and the references therein.

1.2 Economics

In contrast to the previous examples we will now consider dynamic models which evolve in discrete time $t = 0, 1, 2, \ldots$. The time axis is sub-divided in periods of equal length and $x(t)$ denotes the value of x in the period t. Usually economic data is not available in a continuous way, but is given as a time series accumulated over certain periods (days, months, years,...). So discrete time models are particularly appropriate here.

Example 1.2.1. (Cobweb model). Supply and demand of a given commodity depend upon its price. With an increasing price p the supply $S(p)$ increases whilst the demand $D(p)$ decreases. Given the supply and demand curves of a commodity its equilibrium price will be that value $\bar{p}$ which clears the market, i.e. the supply matches the demand. Thus $\bar{p}$ is the abscissa of the intersection point of the supply and the demand curves, see Figure 1.2.1. This is a *static* supply and demand model for determining the price of a single commodity in a market. It remains unclear how this equilibrium price is actually realized by the interaction of sellers and buyers in the market place. But the model is not unreasonable if we assume that the commodity is not stored (and will perish if it is not sold). Let us now consider an economy where pork for example is produced for immediate

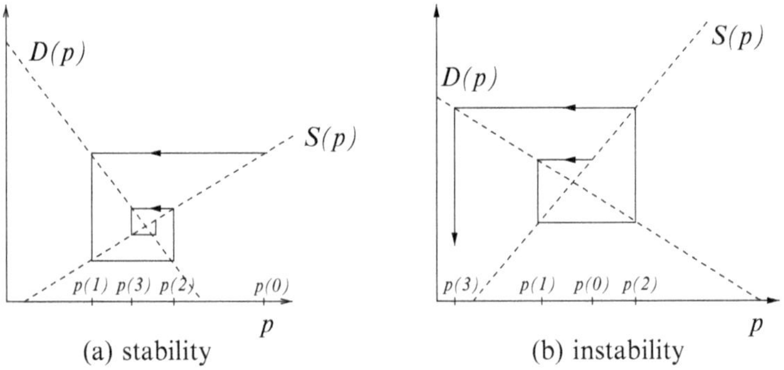

Figure 1.2.1: Cobweb Diagram

consumption and let us take into account the fact that the production (raising pigs) takes time. Choosing the production time as the basic period, the supply of pigs at time $t \in \mathbb{N}$ will depend on the price $p(t-1)$ valid at the time $t-1$ when the decision was taken to produce the pigs for consumption in period t. On the other hand the actual demand for pork at time t depends on the current price $p(t)$. Let us assume – according to the above static model – that the price $p(t)$ is determined in such a way that the complete supply is sold at time t. Assuming that the demand curve is strictly decreasing and its range contains the range of the supply function there will exist a unique value of $p = p(t)$ for which this happens, $p(t) = D^{-1}(S(p(t-1)))$. Thus, starting with an initial price $p(0) = p_0$ the prices $p(t)$ will develop according to the difference equation

$$p(t+1) = D^{-1}(S(p(t))), \quad t \in \mathbb{N}, \quad p(0) = p_0. \tag{1}$$

Using the given supply and demand curves the solution $p(\cdot)$ of this initial value problem can easily be constructed, see Figure 1.2.1. The initial price $p(0)$ determines the supply

$S(p(0))$ at period 1 via the supply curve and this supply determines the equilibrium price $p(1)$ which clears the market at period 1 as the unique solution p of $D(p) = S(p(0))$. Going through the same cycle with $p(1)$ instead of $p(0)$ and continuing the process we obtain a sequence $(p(t))_{t\in\mathbb{N}}$ of prices. The corresponding time series of purchases/sales is given by $D(p(t)) = S(p(t-1))$, $t \in \mathbb{N}$. The cobweb-like picture generated in this way led to the naming of the model.

$\bar{p}$ is an equilibrium of the above system i.e. a solution starting in $\bar{p}$ will always remain there, if and only if, it is a fixed point of the function on the RHS[1] of the difference equation. Equivalently, $S(\bar{p}) = D(\bar{p})$. So $(\bar{p}, D(\bar{p}))$ is just the intersection point of the demand and supply curves. In the situation depicted in Figure 1.2.1(a) the prices $p(t)$ and purchases/sales $D(p(t)) = S(p(t-1))$ converge towards the equilibrium values $\bar{p}$ and $D(\bar{p})$ as $t \to \infty$ (asymptotic stability). The second picture shows that a different configuration of the two curves can lead to a diverging spiral around the equilibrium point (instability). This means that a small initial deviation of $p(0)$ from $\bar{p}$ will lead to ever larger oscillations of prices and purchases/sales around their equilibrium values. Here a weakness of the model becomes apparent since in this case prices will eventually become negative.

We now analyze the conditions under which stability and instability may occur. For this let us suppose that the supply and demand curves are linear,

$$D(p) = D_0 - ap, \qquad S(p) = S_0 + bp$$

where $D_0 \geq 0, a, b > 0$ and $S_0 \in \mathbb{R}$ are constants. Then the price p clearing the market with supply $S > 0$ is determined by the equation $D_0 - ap = S$, i.e. $p = (D_0 - S)/a$. Hence the difference equation (1) reads

$$p(t+1) = (D_0 - S_0 - bp(t))/a = (D_0 - S_0)/a - (b/a)p(t), \quad t \in \mathbb{N}, \quad p(0) = p_0. \qquad (2)$$

The corresponding equilibrium price is $\bar{p} = (D_0 - S_0)/(a+b)$. An easy calculation shows that the deviations from the equilibrium $x(t) = p(t) - \bar{p}$ satisfy the difference equation

$$x(t+1) = -(b/a)x(t), \quad t \in \mathbb{N}, \quad x(0) = p_0 - \bar{p}.$$

The solution of this equation is $x(t) = (-b/a)^t x(0)$ and so we have damped oscillations (asymptotic stability) if and only if $b < a$ i.e. the demand curve is steeper than the supply curve. In economic terms this means that the consumers react more sensitively to changes in the price than the suppliers. Similarly we have instability if and only if $b > a$. Equality between the two parameters leads to periodic oscillations around the equilibrium.

The cobweb model assumes that the suppliers do not learn from past experience - in making their production decision they always expect the price in the next period to be equal to the present one. This is rather unrealistic. The following model, due to Goodwin, assumes that price expectations which guide the production decisions are modified by past experiences according to the rule

$$\hat{p}(t) = p(t-1) + \rho[p(t-1) - p(t-2)]$$

where $\rho \in \mathbb{R}$ is a constant. The case $\rho = 0$ corresponds to the cobweb model. Usually the value of ρ is chosen between -1 and 0, in which case the price is expected to move in the opposite direction to that of the previous period, i.e. suppliers expect oscillations in the price. If $\rho > 0$ the price is expected to move in the same direction as in the previous

[1]RHS: right hand side, LHS: left hand side.

period. Assuming the same linear demand and supply curves as before we are led to the following difference equation for the price clearing the market at period $t+1$

$$p(t+1) = (D_0 - S(\hat{p}(t)))/a = (D_0 - S_0)/a - (b/a)\,\{p(t-1) + \rho[p(t-1) - p(t-2)]\}.$$

Equivalently

$$p(t+1) = (1+\rho)(b/a)\,p(t-1) - (b/a)\,\rho p(t-2) + (D_0 - S_0)/a. \tag{3}$$

This is a difference equation with two time lags and hence *two* initial values, say $p(0)$ and $p(1)$, have to be specified to start up an iterative solution process. In the next chapter we will derive explicit formulas for the solutions of such equations. (3) has the same equilibrium solution $p(t) \equiv \bar{p} = (D_0 - S_0)/(a+b)$ as (1), but now it is no longer immediately obvious how the asymptotic stability of this equilibrium depends on the parameters (a, b, ρ) of the system. In Chapter 3 we will develop methods for analyzing stability properties of equilibria of discrete time systems, see Ex. 3.3.16. □

The cobweb model is concerned with a micro-economic dynamical problem – the price dynamics in a single product market. In contrast we will now consider a model for the dynamics of a whole national economy. One would expect such a model to involve an enormous number of difference equations representing the production, pricing and consumption of a large variety of goods, incomes, saving and investment activities, tax flows and public expenditures etc. In fact such large, "realistic" models have been built in econometrics and have been used for economic forecasting and policy making. On the other hand highly aggregated models are used in theoretical macro-economics in order to gain insight into basic economic mechanisms. The next example deals with a classical model of the business cycle.

Example 1.2.2. (Samuelson-Hicks multiplier-accelerator model). We begin with a nonlinear version of the model. The basic variables are

$Y(t)$	the total national income (= national product) in year t
$C(t)$	the total consumer expenditure in year t
$I(t)$	the total (net) investment in year t
$G(t)$	the total government expenditure in year t.

We make the following assumptions:

(i) the total national product is the sum of consumer, investment and government expenditure,
$$Y(t) = C(t) + I(t) + G(t), \quad t \in \mathbb{N}, \tag{4}$$

(ii) the consumer expenditure in year $t+1$ depends only on the income in the previous two years t and $t-1$,
$$C(t+1) = f(Y(t), Y(t-1)), \tag{5}$$

(iii) the investment in year $t+1$ only depends on the increase of national income from year $t-1$ to year t,
$$I(t+1) = h(Y(t) - Y(t-1)). \tag{6}$$

Substituting (6) and (5) in (4) gives

$$Y(t+1) = f(Y(t), Y(t-1)) + h(Y(t) - Y(t-1)) + G(t+1). \tag{7}$$

(7) is an example of a nonlinear second order difference equation. Given future government expenditure $G(t),\ t = 2, 3, \ldots$ and the national income $Y(0)$, $Y(1)$ in the initial two years one can solve (7) recursively to determine the future national income $Y(t),\ t = 2, 3, \ldots$. Since the government is free to determine its expenditures (within certain constraints) $G(t)$ represents a control variable.
Let us now suppose that $Y(t) \equiv \overline{Y}$ is some given equilibrium solution of (7) corresponding to constant government expenditure $G(t) = \overline{G}$, i.e.

$$\overline{Y} = f(\overline{Y}, \overline{Y}) + h(0) + \overline{G}. \tag{8}$$

In order to analyze the system's behaviour close to this equilibrium solution let $y(t) = Y(t) - \overline{Y}$, $u(t) = G(t) - \overline{G}$ and assume that up to first order we have

$$f(\overline{Y} + y_1, \overline{Y} + y_2) \sim f(\overline{Y}, \overline{Y}) + c_1 y_1 + c_2 y_2, \quad h(y) \sim h(0) + ay.$$

The constant a is called the *acceleration coefficient*, $c = c_1 + c_2$ the *marginal propensity to consume* and $s = 1 - c$ the *marginal propensity to save* (it is always assumed that $0 < c < 1$). Subtracting (8) from (7) we obtain to first order

$$y(t+1) = cy(t) + k(y(t) - y(t-1)) + u(t), \qquad k = a - c_2. \tag{9}$$

This is the Samuelson-Hicks multiplier-accelerator model. It describes how the deviations $y(t) = Y(t) - \overline{Y}$ of the actual national product from an equilibrium $\overline{Y}$ evolves given the initial deviations $y(0)$, $y(1)$ and the deviation $u(t) = G(t) - \overline{G}$ of the government expenditure from the constant value $\overline{G}$.
In the fifties considerable attention was paid to the possibility of "progressive expansion" of an economy in the presence of constant government expenditures. For the above linear model, this question is easily analyzed. $y(t) = (1+r)^t y_0$ with $r \in \mathbb{R}$, $y_0 \neq 0$ solves (9) with $u(t) \equiv 0$ for all $t \in \mathbb{N}$ if and only if $(1+r)^2 = c(1+r) + k(1+r-1)$, i.e.

$$r^2 - (k - s - 1)r + s = 0, \qquad (s = 1 - c).$$

This equation has a positive solution r (and hence (9) has a solution with constant growth rate $r > 0$) if and only if

$$k - s - 1 > 0 \text{ and } (k - s - 1)^2 \geq 4s, \quad \text{i.e.} \quad k \geq (1 + \sqrt{s})^2.$$

It was concluded from this result that, even with fixed government expenditure, an acceleration coefficient of moderate size could produce enough investment to make a constant growth rate of the national income possible. Although this result seems to be quite satisfactory at first sight, it must be regarded with some scepticism. The linear multiplier-accelerator model (9) is at best an appropriate model for small deviations $y(t)$ from the equilibrium solution $\overline{Y}$. Therefore the solution $y(t) = (1+r)^t y_0$ will, in the long run, move out of the neighbourhood of the origin where the model is meaningful. Adequate models for long term economic growth cannot be expected to be linear. Assuming the validity of the nonlinear model the significance of the above analysis for the long term behaviour of (7) is that the equilibrium solution $Y(t) \equiv \overline{Y}$ is unstable if the parameters of the linearization (9) satisfy the inequality $k = a - c_2 \geq (1 + \sqrt{s})^2$. We will illustrate this in Chapter 3. Another question which is of obvious importance for the theory of the business cycle is to determine those values of the parameters a, c_1, c_2 for which the solutions of the linear model are oscillatory. This question can be answered by applying the formulas for solutions derived in the next chapter or via the spectral analysis of Chapter 3. □

1.2.1 Notes and References

Standard references for dynamic models in Economics are *Allen* (1959) [9], *Gandolfo* (1980) [181]. The cobweb model can be found in these books and they also discuss models *with stocks* or *inventories* where supply and demand may be different. Goodwin's model which allows for the influence of past price changes on the suppliers' price expectations is described in [200]. In the econometric literature there are reports on single markets of a particular commodity where prices show an oscillatory behaviour similar to that generated by an undamped cobweb model.

The multiplier-accelerator model is discussed in most textbooks on Mathematical Economics and Macro-economics. The model was first described by *Samuelson* (1939) [446] and later elaborated by *Hicks* (1950) [227]. Various stabilization policies for these type of models have been suggested and analyzed by *Phillips* (1954) [409]. As in classical control engineering Phillips distinguishes between *proportional, derivative* and *integral* stabilization policies and analyzes their effects on the national income in the presence of constant external disturbances.

1.3 Mechanics

In this section we describe some mathematical models of simple mechanical systems. The modelling of such systems is based on the laws of classical mechanics and various techniques have been developed for this over the centuries. These methods have been corroborated by experiments and as a consequence reliable models are available for a great number of mechanical devices. We begin by describing a modelling technique which builds up an approximate *lumped model* of a mechanical system by representing it as an interconnection of ideal translational and rotational *elements* characterized by simple *constitutive laws*. To understand the interaction of these elements within the system, the forces and torques generated by the connection of one element with another must be considered. In a final subsection we briefly describe the variational (Lagrangian or Hamiltonian) approach to modelling which is based on energy considerations. Here the interconnecting forces and torques do not play a role. For this approach an elegant and powerful coordinate free framework has been developed in the general setting of symplectic manifolds, see *Notes and References*. However, an exposition of this framework is beyond the scope of this section. Instead we limit ourselves to a description of the variational method based on local (generalized) coordinates. We emphasize that the purpose of this section is not to give an introduction to classical mechanics, but to present some models of technical mechanical devices and sketch a few modelling techniques.

1.3.1 Translational Mechanical Systems

The dynamic behaviour of a mechanical system is described by vectors of displacements, velocities, forces and torques. A common modelling technique is to represent a mechanical system approximately as an interconnection of a finite number of idealized elements (masses,[1] springs, dampers, transformers and their rotational counterparts). The behaviour of each element is governed by a simple law relating the external force to the displacement, velocity or acceleration associated with the element. This law is called the *constitutive* relation or equation of the element.[2] Table 1.3.1 summarizes the constitutive laws for a pure mass, a linear spring and a linear damper. In the table arrows are associated with the forces. This does not mean that the forces are actually in these directions since the magnitude of $F(t)$ may be negative. For example if for the spring $y_{12}(t) > \overline{y}_{12}$ then the force required to produce the extension is in the direction shown. However if $y_{12}(t) < \overline{y}_{12}$, then one needs to compress the spring, so $F(t) < 0$ and the force is actually in the opposite direction to the one shown.

For a single particle, *Newton's Second Law of Motion* states that *the sum of the forces acting on the particle is equal to the time rate of change of its linear momentum.* Therefore the constitutive law of a mass element is given by $\frac{d}{dt}(mv(t)) = F(t)$. Here

[1]It may seem strange to some readers that mass is regarded as a constitutive element of a mechanical system in parallel with springs and dashpots. This is, however, common practice in the modelling of engineering systems. The reader should distinguish between the fundamental concept of *mass* in theoretical mechanics and the notion of a *mass element* as a building block ("pure mass") in the modelling of a mechanical system.

[2]Throughout the present and the following section the predicate *constitutive* will only be used in this terminological sense, see [84], [105].

Symbol	Constitutive Law	Variables
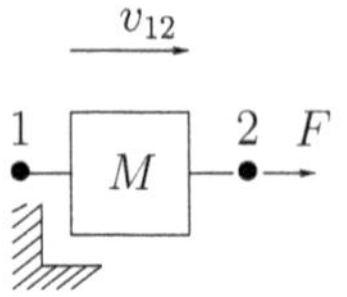	$\frac{d}{dt}(Mv(t)) = F(t)$	$v(t) = v_{12}(t)$ velocity of mass $F(t)$ force applied to mass
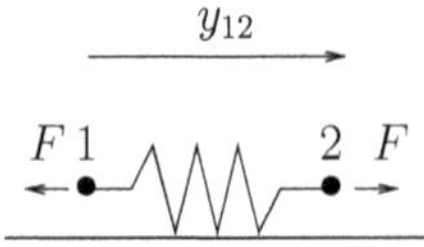	$ky(t) = F(t)$	$y(t) = y_{12}(t) - \overline{y}_{12}$ net elongation $F(t)$ force applied to spring
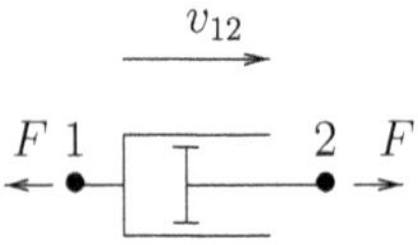	$cv(t) = F(t)$	$v(t) = v_{12}(t)$ relative velocity of piston $F(t)$ force applied to damper

Table 1.3.1: Symbols and constitutive laws of mass, spring, and damper

the velocity and acceleration must be measured with respect to an inertial reference frame (in classical mechanics this is usually fixed at the centre of the Sun).

The constitutive law of the linear spring is given by Hooke's law. In reality this linear relation between force and elongation will only be approximately valid within certain bounds on the elongation. Hence the use of a linear spring element in a mechanical model imposes constraints on the variables involved.

Similarly for models involving a damper. A physical realization of a linear damper is a dashpot where a piston moves through an oil-filled cylinder and there are holes in the face of the piston through which the oil passes. If the rates of flow are kept within certain bounds viscous damping results in a linear relation between the force and the relative velocity of the piston with respect to the cylinder. At higher velocities such a dashpot will show nonlinear characteristics.

The spring, damper and mass in the above table are also idealized objects from another point of view. Any real spring has some (albeit comparatively small) inertia and damping. Similarly any damper has some mass and exhibits small spring effects. We may account for the difference between the real devices and idealized objects by lumping all inertias of a given mechanical system together in the masses, all stiffness effects in the springs and all frictional forces in the dashpots ("lumped parameter model"). This lumped parameter approach to modelling a mechanical system is not limited to linear models. Nonlinear relations between stresses and deflections in a mechanical system can be modelled by nonlinear springs, and nonlinear viscous frictions between adjacent bodies can be modelled by nonlinear dampers.

If we describe a mechanical system as an interconnection of a finite number of masses, springs and dampers, a model of the overall system is obtained by combining the constitutive relations of its elements with the *interconnection laws* governing the interaction between them. Throughout this section we will assume that the forces between mechanical elements obey Newton's third law of action and reaction: *Any force of one element on another is accompanied by a reaction force on the first*

element of equal magnitude and opposite direction along the line joining them, see Table 1.3.1 where the forces on the left of the spring and damper symbols are the reaction forces to those on the right. There are various methods of obtaining the equations for the overall mechanical system from the constitutive relations of its elements and the *interconnection laws* e.g. bond graph methods and network methods, see Section 1.4. For more detailed information about this mass-spring-damper modelling approach, see *Notes and References*.
We now give a few examples of mechanical systems.

Example 1.3.1. (Trolley). Consider a trolley of mass M moving on rails under the influence of a force $\beta u(t)$ as in Figure 1.3.2. Here β is a constant which converts the control variable u (e.g. a voltage) into a force. We neglect all frictions present in the system –

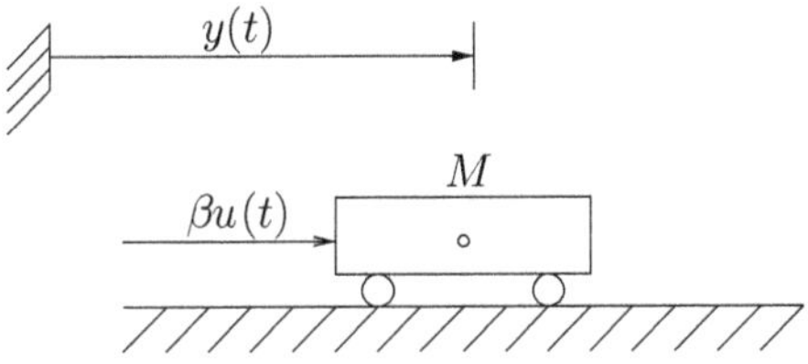

Figure 1.3.2: Pure mass: trolley

friction between wheels and rails, friction in the wheel bearings, drag friction of the trolley moving through the atmosphere. We also neglect the masses of the wheels and assume that the trolley behaves like a rigid body. Finally we assume that the line of action of the force is through the trolley's centre of mass, parallel to the rails. So the trolley does not rotate and, under the influence of gravity, does not loose contact with the rails. Since the mass of the trolley is constant Newton's second law yields the following scalar equation of motion,

$$M\ddot{y}(t) = \beta u(t) \tag{1}$$

where $y(t)$ is the displacement of the centre of mass of the trolley from a fixed point in an inertial reference frame. In order to determine the motion of the system for $t \geq 0$ it is necessary to know the initial position $y(0)$ and the initial velocity $\dot{y}(0)$ of the trolley. Moreover the exterior force $\beta u(t)$ must be known as a function of time $t \geq 0$. If we consider the force as a control variable and fix a rest position at $y = 0$ as the set point of the trolley, a typical control problem is to find a feedback control law $u(t) = f(y(t), \dot{y}(t))$ which brings the trolley back or approximately back to the prescribed rest position from any given pair of initial values $(y(0), \dot{y}(0))$. If we assume that the control values are limited by $|u(t)| \leq c$, $t \geq 0$ where $c > 0$ is a given constant, a typical optimal control problem is: given the initial conditions $(y(0), \dot{y}(0))$, find a control $u(\cdot) : [0, t_1] \to [-c, c]$ which steers the trolley back to the rest position $(y(t_1), \dot{y}(t_1)) = (0, 0)$ in minimal time t_1. Additionally constraints may be imposed on the trajectory of the trolley (e.g. $|y(t)| \leq d, d > 0$) and this leads to an *optimal control problem with state constraints*. □

In the next example we consider interconnections of mechanical elements. The harmonic oscillator is used as a highly simplified model for many technical systems. We illustrate this by a mass-spring-damper model for an automobile suspension system.

Example 1.3.2. (Linear oscillator). Consider the vertical motion of a mass M sliding in some bearing and suspended to a support by a spring as in the left hand figure in Figure 1.3.3. Besides the exterior forces (gravity and an additional time-depending force

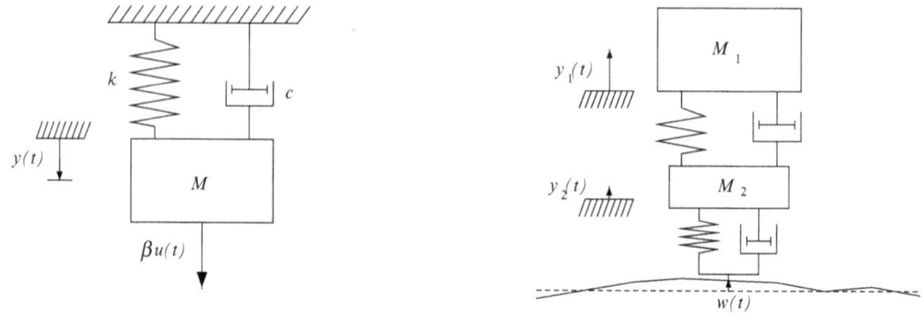

Figure 1.3.3: Mass-spring-damper systems

$\beta u(t)$) two types of interior forces act on the mass. These are modelled by a linear spring and a linear damper with coefficients k and c, respectively. Let us determine the equation of motion of the above mass-spring-damper system. The behaviour of the system is completely described by the vertical position and velocity of the mass. In order to eliminate the gravitational force we introduce the displacement y of the centre of mass from its equilibrium position under the influence of gravity. By Newton's second law the sum of the forces acting on M must equal $M\ddot{y}$. Note that the force exerted by the spring and the damper on the mass is opposite to the direction of the displacement and velocity respectively. The resulting equation of motion is

$$M\ddot{y}(t) + c\dot{y}(t) + ky(t) = \beta u(t). \tag{2}$$

We will now construct a simple mass-spring-damper model for an automobile suspension system. The purpose of such a suspension system is to smooth the response of the car body to the irregular ups and downs of the road. We will only consider the vertical movements of the car body and axles and make the non-realistic assumptions that both axles move in the same way so that they can be lumped together and the rotational motion of the car body can be ignored. We first assume that the road is flat. Since the car body and the axle can move independently, we need two position variables y_1 and y_2. As reference points for these positions we choose the rest positions of the car body (mass M_1) and of the axle (mass M_2) over the nominal road level under the influence of gravity. The tyres are modelled as springs with comparatively high stiffness k_2 coupled in parallel with a dashpot accounting for the energy dissipation through the tyres. The main suspension mechanism consisting of coil springs, leaf springs and shock absorbers, is modelled in a lumped manner by a linear spring and a linear damper connecting the axle with the car body, see the right hand figure in Figure 1.3.3. Let $w(t)$ be the displacement of the point of contact between tyre and road from the nominal road level. $w(t)$ is determined by the profile of the road and the position of the car. The tyre spring force (in an upward direction) corresponding to the deviation of the tyre from the nominal road level is $k_2 w$, the corresponding frictional force is $c_2\dot{w}$. Applying Newton's second law to each of the two masses and Newton's third law to the interaction between the two masses we obtain the equations of motion

$$\begin{aligned} M_1\ddot{y}_1 + c_1(\dot{y}_1 - \dot{y}_2) + k_1(y_1 - y_2) &= 0 \\ M_2\ddot{y}_2 + c_1(\dot{y}_2 - \dot{y}_1) + k_1(y_2 - y_1) + c_2\dot{y}_2 + k_2 y_2 &= c_2\dot{w} + k_2 w. \end{aligned} \tag{3}$$

Here $w(t)$ may be considered as a perturbation and an important design objective would be to ensure that the road conditions which the car is likely to encounter, will not generate vibrations of the car body i.e. values of $y_1(t)$ and $\dot{y}_1(t)$ which are not acceptable from the point of view of passenger comfort. If the suspension mechanism can be controlled, a typical problem would be to design a feedback control which decouples the vertical velocity of the car body $\dot{y}_1(t)$ as much as possible from the (largely unknown) perturbations $w(t)$ generated by the irregular road surface (*disturbance attenuation problem*). □

The previous two examples deal with translational mechanical systems whose movements are restricted to one direction. Arbitrary motions of a mass in three dimensional space are governed by a vector version of Newton's second law. Here and in the next section all vectors in $\mathbb{R}^3$ or families of such vectors are written in bold face and we use vector analysis definitions and notations as found, for example, in [362]. We assume that the positions are determined with respect to a cartesian coordinate system which is fixed in an inertial frame. If the position vector of a particle of mass m at time t is denoted by $\mathbf{r}(t)$ and $\mathbf{F}(t)$ is the vector sum of all individual forces applied to the mass at time t, then

$$\dot{\mathbf{p}}(t) = (m\ddot{\mathbf{r}})(t) = \mathbf{F}(t), \tag{4}$$

where $\mathbf{p} = m\dot{\mathbf{r}}$ is the *linear momentum* of the mass point.
Now consider a system of N particles with masses m_i at positions $\mathbf{r}_i$, $i \in \underline{N}$. The linear momentum of such a system is by definition the sum of the linear momenta of each particle,

$$\mathbf{p}(t) = \sum_{i=1}^{N} \mathbf{p}_i(t) = \sum_{i=1}^{N} m_i\dot{\mathbf{r}}_i(t). \tag{5}$$

Applying Newton's second law to each of the particles we must distinguish between the external forces $\mathbf{F}_i^e(t)$ and the interactive forces $\mathbf{F}_{ij}(t)$ between the particles of the system. Summing over all particles we obtain from (4)

$$\dot{\mathbf{p}}(t) = \sum_{i=1}^{N} m_i\ddot{\mathbf{r}}_i(t) = \sum_{i=1}^{N} \mathbf{F}_i^e(t) + \sum_{i,j=1,\, i\neq j}^{N} \mathbf{F}_{ij}(t). \tag{6}$$

By Newton's third law of action and reaction $\mathbf{F}_{ij}(t) + \mathbf{F}_{ji}(t)$ is zero for all t and $i,j \in \underline{N}$, $i \neq j$ and so the second term on the right vanishes. Hence, if we define the *total external force* and the *centre of mass* of the system at time t by

$$\mathbf{F}^e(t) = \sum_{i=1}^{N} \mathbf{F}_i^e(t), \qquad \bar{\mathbf{r}}(t) = \sum_{i=1}^{N} \frac{m_i\mathbf{r}_i(t)}{M} \quad \text{where } M = \sum_{i=1}^{N} m_i \tag{7}$$

equations (5) and (6) can be written in the form

$$\mathbf{p}(t) = M\dot{\bar{\mathbf{r}}}(t) \quad \text{and} \quad \dot{\mathbf{p}}(t) = M\,\ddot{\bar{\mathbf{r}}}(t) = \mathbf{F}^e(t). \tag{8}$$

In particular, *the centre of mass of the system moves as if the total external force were acting on the entire mass of the system concentrated at the centre of mass.*
In order to describe a rigid body in three-dimensional space, the position of its centre of mass and the orientation of the rigid body with respect to an inertial reference frame must be specified. We therefore need a counterpart of Newton's second law for rotational motions.

1.3.2 Mechanical Systems with Rotational Elements

Consider a fixed point O in an inertial reference frame with origin O^*. The *angular momentum* $\mathbf{H}(t)$ of a particle of mass m about the reference point O is defined by the vector product

$$\mathbf{H}(t) = \mathbf{r}(t) \times \mathbf{p}(t) = \mathbf{r}(t) \times m\dot{\mathbf{r}}(t)$$

where $\mathbf{r}(t)$ is the "moment arm", i.e. the vector from the point O to the position of the particle, and $\mathbf{p}(t) = m\dot{\mathbf{r}}(t)$ is the linear momentum of the mass (with respect to the inertial reference frame). The corresponding *moment of force* or *torque* $\mathbf{N}(t)$ due to the force $\mathbf{F}(\mathbf{t})$ is defined by $\mathbf{N}(t) = \mathbf{r}(t) \times \mathbf{F}(t)$. As a consequence of (4) one obtains the following relation between the net torque applied to the particle and the rate of change of its angular momentum

$$\dot{\mathbf{H}}(t) = \frac{d}{dt}(\mathbf{r}(t) \times m\dot{\mathbf{r}}(t)) = \dot{\mathbf{r}}(t) \times m\dot{\mathbf{r}}(t) + \mathbf{r}(t) \times m\ddot{\mathbf{r}}(t) = \mathbf{r}(t) \times \mathbf{F}(t) = \mathbf{N}(t). \quad (9)$$

Note that the angular momentum and the torque both depend upon the point O about which moments are taken.

Let us now consider a system of N particles with the same setup as that which led to (8). The *total angular momentum* of such a system about O is obtained by summing up the angular momenta of all the particles, i.e.

$$\mathbf{H}(t) = \sum_{i=1}^{N} \mathbf{r}_i(t) \times m_i\dot{\mathbf{r}}_i(t)\,,$$

so that by (6) and (9)

$$\dot{\mathbf{H}}(t) = \sum_{i=1}^{N} \mathbf{r}_i(t) \times m_i\ddot{\mathbf{r}}_i(t) = \sum_{i=1}^{N} \mathbf{r}_i(t) \times \left(\mathbf{F}_i^e(t) + \sum_{j=1,\, j\neq i}^{N} \mathbf{F}_{ij}(t)\right).$$

Now $\mathbf{F}_{ij}(t) = -\mathbf{F}_{ji}(t)$ by Newton's third law, and the same law implies that the vectors $\mathbf{r}_i(t) - \mathbf{r}_j(t)$ and $\mathbf{F}_{ij}(t)$ are linearly dependent. Hence, if

$$\mathbf{N}^e(t) = \sum_{i=1}^{N} \mathbf{r}_i(t) \times \mathbf{F}_i^e(t) \quad (10)$$

is the *total external torque*, then

$$\dot{\mathbf{H}}(t) = \mathbf{N}^e(t) + \sum_{i=1}^{N} \sum_{j=i+1}^{N} (\mathbf{r}_i(t) - \mathbf{r}_j(t)) \times \mathbf{F}_{ij}(t) = \mathbf{N}^e(t). \quad (11)$$

So *the rate of change of the total angular momentum of a system of particles about a fixed point O is equal to the sum of the moments of the external forces about O.*

By (8) the total linear momentum of a system of N particles is the same as if the entire mass were concentrated at the centre of mass and moving with it. We now develop a counterpart of this result for the angular momentum which includes the possibility that the point about which we take moments is moving. Let us denote this moving point by O_t and suppose the vector from O^* to O_t is $\mathbf{r}^*(t)$, the vector from O_t to the centre of mass is $\overline{\mathbf{r}}(t)$, the vector from the centre of mass to the i-th particle is $\mathbf{r}_i'(t)$ and $\mathbf{v}^*(t) = \dot{\mathbf{r}}^*(t)$, $\overline{\mathbf{v}}(t) = \mathbf{v}^*(t) + \dot{\overline{\mathbf{r}}}(t)$, $\mathbf{v}_i(t) = \overline{\mathbf{v}}(t) + \dot{\mathbf{r}}_i'(t)$ are the velocity vectors of O_t, of the centre of mass and of the i-th particle (with respect to the inertial frame). The angular momentum about O_t takes the form[3]

[3] In order to compactify the equations we drop the time variable t where necessary.

$$\mathbf{H} = \sum_{i=1}^{N} (\overline{\mathbf{r}} + \mathbf{r}_i') \times m_i(\overline{\mathbf{v}} + \dot{\mathbf{r}}_i')$$
$$= \sum_{i=1}^{N} \overline{\mathbf{r}} \times m_i \overline{\mathbf{v}} + \sum_{i=1}^{N} \mathbf{r}_i' \times m_i \dot{\mathbf{r}}_i' + \left(\sum_{i=1}^{N} m_i \mathbf{r}_i' \right) \times \overline{\mathbf{v}} + \overline{\mathbf{r}} \times \frac{d}{dt} \sum_{i=1}^{N} m_i \mathbf{r}_i'.$$

Since $\sum_{i=1}^{N} m_i \mathbf{r}_i'(t) = 0$ by the definition of the centre of mass (7), the last two terms on the right hand side vanish and we obtain

$$\mathbf{H}(t) = \overline{\mathbf{r}}(t) \times M\,\overline{\mathbf{v}}(t) + \sum_{i=1}^{N} \mathbf{r}_i'(t) \times m_i\, \dot{\mathbf{r}}_i'(t). \tag{12}$$

Note that by the above argument $\mathbf{H}'(t) = \sum_{i=1}^{N} \mathbf{r}_i'(t) \times m_i\, \dot{\mathbf{r}}_i'(t) = \sum_{i=1}^{N} \mathbf{r}_i'(t) \times m_i \mathbf{v}_i(t)$ is the angular momentum of the system about the centre of mass. *Thus the total angular momentum of the system about O_t is the angular momentum of its total mass concentrated at the centre of mass, plus the angular momentum of the system about the centre of mass.* Only if the centre of mass is at rest (i.e. $\overline{\mathbf{v}} = 0$) will the angular momentum be independent of the reference point O_t and its velocity. In this case $\mathbf{H}(t)$ reduces to the angular momentum taken about the centre of mass. Differentiating $\mathbf{H}(t) - \mathbf{H}'(t) = \overline{\mathbf{r}}(t) \times M\,\overline{\mathbf{v}}(t)$ we obtain

$$\dot{\mathbf{H}} - \dot{\mathbf{H}}' = \dot{\overline{\mathbf{r}}} \times M\overline{\mathbf{v}} + \overline{\mathbf{r}} \times \mathbf{M}\dot{\overline{\mathbf{v}}} = (\overline{\mathbf{v}} - \mathbf{v}^*) \times M\overline{\mathbf{v}} + \overline{\mathbf{r}} \times \mathbf{M}\dot{\overline{\mathbf{v}}} = -\mathbf{v}^* \times M\overline{\mathbf{v}} + \overline{\mathbf{r}} \times \mathbf{M}\dot{\overline{\mathbf{v}}}.$$

In particular if O_t is the moving centre of mass we have $\dot{\mathbf{H}}(t) = \dot{\mathbf{H}}'(t)$. So *in calculating the rate of change of angular momentum of a particle system about its centre of mass, we may treat the centre of mass as if it were at rest.*
Let $\mathbf{F}_i^e(t)$ be the external forces, $\mathbf{F}^e(t) = \sum_{i=1}^{N} \mathbf{F}_i^e(t)$ the total external force and define the total torque about the moving reference point O_t by (see (10))

$$\mathbf{N}^e(t) = \sum_{i=1}^{N} (\overline{\mathbf{r}}(t) + \mathbf{r}_i'(t)) \times \mathbf{F}_i^e(t).$$

Then, if $\mathbf{N}^{e*}(t)$ is the total torque about O^*, we get

$$\mathbf{N}^{e*}(t) = \sum_{i=1}^{N} (\mathbf{r}^*(t) + \overline{\mathbf{r}}(t) + \mathbf{r}_i'(t)) \times \mathbf{F}_i^e(t) = \mathbf{r}^*(t) \times \mathbf{F}^e(t) + \mathbf{N}^e(t) = \mathbf{r}^*(t) \times M\dot{\overline{\mathbf{v}}}(t) + \mathbf{N}^e(t).$$

since $\mathbf{F}^e(t) = M\dot{\overline{\mathbf{v}}}(t)$, see (8). The total angular momentum $\mathbf{H}^*(t)$ about O^* satisfies

$$\mathbf{H}^*(t) = \sum_{i=1}^{N} (\mathbf{r}^*(t) + \overline{\mathbf{r}}(t) + \mathbf{r}_i'(t)) \times m_i \mathbf{v}_i(t) = \mathbf{H}(t) + \mathbf{r}^*(t) \times M\overline{\mathbf{v}}(t).$$

Since we have $\dot{\mathbf{H}}^*(t) = \mathbf{r}^*(t) \times M\dot{\overline{\mathbf{v}}}(t) + \mathbf{N}^e(t)$ by (11) we get

$$\dot{\mathbf{H}} = \mathbf{r}^* \times M\dot{\overline{\mathbf{v}}} + \mathbf{N}^e - \dot{\mathbf{r}}^* \times M\overline{\mathbf{v}} - \mathbf{r}^* \times M\dot{\overline{\mathbf{v}}} = \mathbf{N}^e - \mathbf{v}^* \times M\overline{\mathbf{v}}. \tag{13}$$

In particular if O_t is the moving centre of mass we have $\dot{\mathbf{H}}(t) = \mathbf{N}^e(t)$. Therefore *the rate of change of the angular momentum of a particle system about its centre of mass is the sum of the moments about the centre of mass of all the external forces, irrespective of whether the centre of mass is moving or at rest.*

There is an *angular momentum law* for rigid bodies which complements Newton's second law. However we will not develop this for general rotational motions in $\mathbb{R}^3$, since in the following examples we only consider *plane* rotational systems. This means, in particular, that all the elements are rotating around axes which are parallel to each other and all forces are restricted to the plane. This assumption greatly simplifies the analysis. If we describe the motion of a system in an inertial reference frame where the z axis is parallel to the axes of rotation, then all vector products of vectors in the x, y plane are parallel to the z axis. As a consequence only the z-coordinates of these vector products are nontrivial. Now consider any particle of mass m rotating about an axis parallel to the z axis through a fixed point $O = (x_0, y_0, 0)$ in the x, y plane and let $(x_0, y_0, 0) + \mathbf{r}(t) = (x_0, y_0, 0) + (x(t), y(t), 0)$ be the coordinates of the particle at time t. Since by assumption the distance $\|\mathbf{r}(t)\| = r = (x(t)^2 + y(t)^2)^{1/2}$ between the particle and the point O is constant we obtain by differentiation

$$x(t)\dot{x}(t) + y(t)\dot{y}(t) = 0.$$

Hence there exists a real number $\omega(t)$ satisfying

$$\dot{\mathbf{r}}(t) = (\dot{x}(t), \dot{y}(t), 0) = \omega(t)(-y(t), x(t), 0).$$

Let $\boldsymbol{\omega}(t) = (0, 0, \omega(t))$, then $\boldsymbol{\omega}$ is called the *angular velocity* of the particle about O and we obtain $\dot{\mathbf{r}}(t) = \boldsymbol{\omega}(t) \times \mathbf{r}(t)$. The angular momentum of the particle about O is $\mathbf{r} \times m\dot{\mathbf{r}} = m\mathbf{r} \times (\boldsymbol{\omega} \times \mathbf{r}) = m(0, 0, \omega(x^2 + y^2))$. Hence, for plane rotations, the equation of motion (9) is reduced to the scalar differential equation

$$\frac{d}{dt}\left[m\omega(t)(x(t)^2 + y(t)^2)\right] = mr^2\dot{\omega}(t) = N(t). \tag{14}$$

Here

$$N(t) = x(t)F_2(t) - y(t)F_1(t) \tag{15}$$

is the z-component of the torque generated by a given force $\mathbf{F}(t) = (F_1(t), F_2(t), 0)$ applied to the particle.

Now consider a two dimensional rigid body $B \subset \mathbb{R}^2$ rotating in the x, y plane about a perpendicular axis through a fixed point O with angular velocity ω. Suppose that the rigid body has mass density $\rho(x, y), (x, y) \in B$. Then for this rigid body the angular momentum law takes the form

$$\frac{d}{dt}(J\omega)(t) = N(t) \quad \text{where} \quad J = \int_B \rho(x, y)(x^2 + y^2)dxdy \tag{16}$$

and J is the *moment of inertia* of the body about O. Moreover (16) also holds if O is a moving centre of mass. For many rigid bodies with uniform mass distribution the moments of inertia about given axes can be found in textbooks on analytic mechanics. The centre of mass $(\overline{x}, \overline{y})$ and total mass M of a body B with mass distribution $\rho(x, y)$ are given by

$$\overline{x} = \frac{1}{M}\int_B x\rho(x, y)dx\,dy, \quad \overline{y} = \frac{1}{M}\int_B y\rho(x, y)dx\,dy, \quad M = \int_B \rho(x, y)dx\,dy.$$

There is a close relationship between plane rotations and one-dimensional translational

Symbol	Constitutive Law	Variables
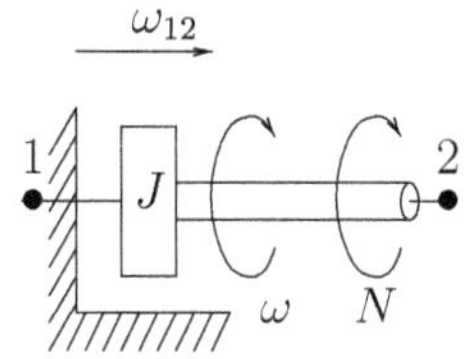	$\frac{d}{dt}(J\omega(t)) = N(t)$	$\omega(t) = \omega_{12}(t)$ angular velocity $N(t)$ torque applied about the axis
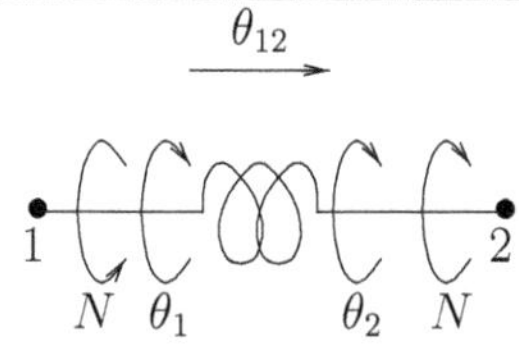	$k\theta(t) = N(t)$	$\theta(t) = \theta_{12}(t) = \theta_2(t) - \theta_1(t)$ relative angular displacement of torsional spring $N(t)$ torque applied to spring
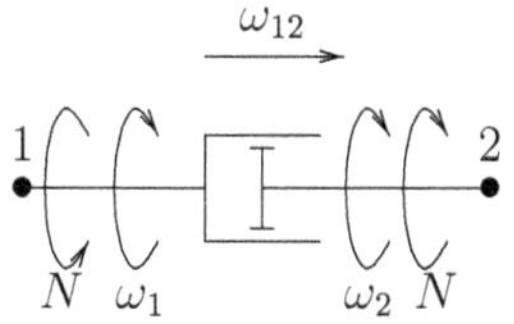	$c\,\omega(t) = N(t)$	$\omega(t) = \omega_{12}(t) = \omega_2(t) - \omega_1(t)$ relative angular velocity $N(t)$ torque applied to damper

Table 1.3.4: Symbols and constitutive laws of rotational elements

motions. The rotational counterparts of *displacements*, *velocities* and *forces* are *angles*, *angular velocities* and *torques*. The rotational counterpart of mass is, as we have seen, the moment of inertia. Table 1.3.4 summarizes the rotational counterparts of masses, springs and dampers (again the directions indicated by the arrows are arbitrary since the values of the functions may be positive or negative).
Physical devices which may be modelled as rotational springs are, for example, the mainspring of a clock or an elastic rod joining two masses rotating about the same axis. Rotational viscous damping occurs for example if two concentric cylinders separated by an oil film rotate with different angular velocities about a common axis.
The interconnection laws for rotational elements are strictly analogous to those for translational systems *if the interacting elements rotate about the same axis.* Then the torque exerted by one element on another is accompanied by a reaction torque of the same magnitude but of opposite direction on the first element. This holds, in general, but is no longer true if the elements rotate about different (albeit parallel) axes. For instance, the contact forces by which two gears act on one another are of equal magnitude and opposite direction, but the corresponding torques will be different if the radii of the gears are different.
In order to decide whether a rotation in the plane is positive or negative we have to fix an orientation of the plane (clockwise or anticlockwise)[4]. In the following examples we will always specify such an orientation. A directed angle (the direction being indicated by an arrowhead) is positive if it coincides with the given orientation of the plane, otherwise it is negative. The next example is a purely rotational

[4]Equivalently we could impose a direction to the axis of rotation and define the orientation of the plane by the right hand screw law.

mechanical system.

Example 1.3.3. (Pendulum). Consider a pendulum of length l suspended from a fixed point O as shown in Figure 1.3.5. We first model the pendulum as a point mass m

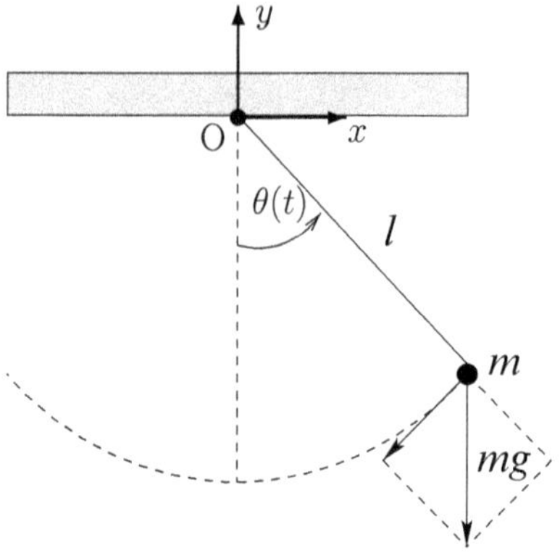

Figure 1.3.5: Pendulum

attached to a mass-less rigid rod of length l which rotates in the plane without any friction about O. Suppose that the directed angle from the downward vertical to the rod measured with respect to the anti-clockwise orientation is θ. Then the motion of the pendulum is completely described by the angle $\theta(t)$ as a function of time. Taking moments about O we obtain from (14) the following equation of motion

$$ml^2\ddot{\theta} = -mgl\sin\theta \tag{17}$$

where g is the gravitational constant.
Let us now abandon the assumption that the rod is mass-less and the rotation is without friction. Instead we assume that the pendulum is a plane rigid body of total mass m and there is viscous rotational friction with coefficient c at the pivot. Since the horizontal component of the gravitational force is zero, the torque about O exerted by the uniform gravitational field on the rigid body $B(t)$ at time t is by (15)

$$\int_{B(t)} x\rho(x,y)g\,dx\,dy = mg\overline{x}(t),$$

i.e. the torque is equal to the torque about O exerted by the gravitational force on a particle of mass m located at the centre of mass $(\overline{x}(t), \overline{y}(t))$ of the rigid body at time t, see Figure 1.3.6 (b). We therefore obtain from (16) the equation of motion

$$\frac{d}{dt}(J\dot{\theta})(t) = -c\dot{\theta}(t) - mg\overline{x}(t) = -c\dot{\theta}(t) - mg\bar{l}\sin\theta(t) \tag{18}$$

where J is the moment of inertia of the rigid body rotating about O and $\bar{l}$ is the distance of the centre of mass from O. Note that this equation specializes to (17) if the rigid body is a particle and no friction is present. If, for example, the pendulum consists of a slender bar of length l and mass m uniformly distributed along the bar then the moment of inertia about O would be $J = (1/3)ml^2$ and $\bar{l} = (1/2)l$.
Equations (17) and (18) are nonlinear time-invariant equations of second order. Given an initial angle $\theta(0) = \theta_0$ and an initial angular velocity $\dot{\theta}(0) = \dot{\theta}_0$ there exist unique solutions of (17) and (18) for all $t \in \mathbb{R}$ satisfying these initial conditions. The angle θ is treated

here as a real variable although only its values modulo 2π matter. Both systems (17) and (18) have the same equilibrium solutions corresponding to the pendulum being in a vertical position (either downward or upright) with zero angular velocity: If either of the two systems satisfies the initial conditions $(\theta(0), \dot{\theta}(0)) = (0, 0)$ or $(\theta(0), \dot{\theta}(0)) = (\pi, 0)$ it will remain in this position indefinitely. However, the two equilibria exhibit very different behaviour when the initial conditions are slightly perturbed. If the pendulum is initially in the downward rest position a slight perturbation will only lead to small deviations from the equilibrium (see Example 2.1.10), whereas an arbitrarily small initial perturbation of the upper rest position will produce large deviations, because the pendulum will fall down. Hence the first equilibrium position is stable and the other is unstable. Whilst these statements hold for both systems considered here, there is an important difference between them with regard to their behaviour in a neighbourhood of the stable equilibrium point. In the presence of friction the pendulum will gradually return to the downward equilibrium position whereas it will swing with constant amplitude about the equilibrium in the absence of friction. To determine the stability properties of an equilibrium point for a given system is a basic problem in control theory. Since in most applications there are no simple analytic formulas for the solutions of the equation of motion one needs to find a method which allows one to determine the stability or instability of an equilibrium without solving the differential equations. Such a method has been developed by Liapunov whose central idea was to use the energy or an energy-like real valued function for this purpose. This method will be studied in detail in Chapter 3. □

The unstable upward position of a pendulum can be stabilized by a control mechanism which applies a torque $N(t)$ to the pendulum depending on the deviation $\theta(t) - \pi$ from the equilibrium position. A more interesting problem is to stabilize the inverted pendulum by moving its base e.g. in a horizontal or vertical way. This leads to a mechanical system which combines translational and rotational movements.

Example 1.3.4. (Cart-pendulum system). Consider a pendulum which rotates about a pivot which is mounted on a cart. The cart has mass M and is driven on a horizontal rail by a force $\beta u(t)$ in the same way as the trolley in Example 1.3.1. However, here we allow for viscous friction between the cart and the rail. The centre of mass of the pendulum lies at a distance l from the pivot and the moment of inertia of the pendulum (modelled as a rigid body) *about its centre of mass* is J. We allow for viscous friction at the pivot point. The position of the cart is measured by the horizontal displacement r of its centre of mass from the origin of an inertial coordinate system. We assume that the centre of mass of the cart is moving along the x-axis of this coordinate system. The position of the pendulum is measured by the angular displacement θ of the line joining its centre of mass with the pivot from the downward vertical (measured in an anti-clockwise direction). Although we view the cart as a rigid body we assume that its motion is one-dimensional, i.e. the torques generated by the totality of forces acting on the cart are in balance. This means that we can neglect the moments about its centre of mass and treat the cart as a point mass. To simplify the notation we assume that the pivot point coincides with the centre of mass of the cart.

In order to obtain a model of the system we will use free-body diagrams for each element, representing all external and interactive forces between the elements by symbols together with arrows which define their "positive senses", see Figure 1.3.6: The forces are positive if they operate in the directions shown, they are negative if they operate in the opposite direction. For instance, if the system is at rest in the downward position, the force $F_2(t)$ acting on the pendulum at the hinge will be directed upwards and hence it will be positive with respect to the direction indicated in Figure 1.3.6. In general, all forces are vectors but since the cart's motion is restricted to one dimension we decompose the forces into their horizontal and vertical components.

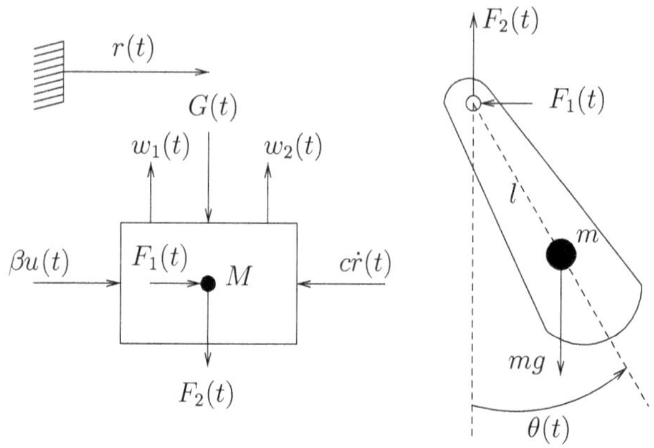

Figure 1.3.6: Free-body diagrams of cart and pendulum

The horizontal forces on the cart are the driving force $\beta u(t)$, the viscous friction force $-c\dot{r}(t)$ and the horizontal component of the (unknown) contact force, $F_1(t)$ at the pivot. The vertical forces on the cart are the contact forces $w_1(t)$, $w_2(t)$ through the wheels supporting the cart on the rail, the gravitational force $G(t)$ and the vertical component of the (unknown) contact force, $F_2(t)$ at the pivot. Since we assume that the cart is constrained to the one-dimensional motion along the rails, the vertical forces on the cart are in balance. The horizontal motion of the cart is governed by the equation

$$M\ddot{r}(t) = \beta u(t) - c\dot{r}(t) + F_1(t). \tag{19}$$

In order to describe the planar motion of the pendulum it is sufficient to consider the motion of its centre of mass and its rotation about its centre of mass. If $(x(t), y(t))$ denotes the coordinates of the centre of mass at time t with respect to the given inertial coordinate system, then (see (8)) the horizontal and vertical motions are determined by

$$\begin{aligned} m\ddot{x}(t) =& m\frac{d^2}{dt^2}\left[\ddot{r}(t) + l\left(\sin\,\theta(t)\right)\right] = -F_1(t) \\ m\ddot{y}(t) =& ml\frac{d^2}{dt^2}\left(-\cos\,\theta(t)\right) = -mg + F_2(t). \end{aligned}$$

Calculating the double derivatives we obtain

$$m\left[\ddot{r}(t) + l\ddot{\theta}(t)\cos\,\theta(t) - l\dot{\theta}(t)^2\sin\,\theta(t)\right] = -F_1(t) \tag{20}$$

$$ml\left[\ddot{\theta}(t)\sin\,\theta(t) + \dot{\theta}(t)^2\cos\,\theta(t)\right] = -mg + F_2(t). \tag{21}$$

Since the cart-pendulum system is described by two independent variables, $r(t)$ and $\theta(t)$, and since the two contact forces $F_1(t)$, $F_2(t)$ are unknown we need one more equation of motion. It remains to determine the rotation of the pendulum about its centre of mass.

The gravitational force does not produce any torque on the pendulum about its centre of mass, G. So the rotation of the pendulum is determined by the torque of the force $\mathbf{F}(t) = (-F_1(t), F_2(t))$ about $(x(t), y(t))$. Now the vector from $(x(t), y(t))$ to the pivot where the force $\mathbf{F}(t)$ is applied is given by $(-l\sin\theta(t), l\cos\theta(t))$ and so the torque of the force $\mathbf{F}(t)$ about $(x(t), y(t))$ is $-F_2(t)l\sin\theta(t) + F_1(t)l\cos\theta(t)$, see (15). The force of rotational friction $c_P\dot{\theta}(t)$ acts to oppose the motion. Therefore we obtain from (16) with O the centre of mass

$$J\ddot{\theta}(t) = -F_2(t)l\sin\theta(t) + F_1(t)l\cos\theta(t) - c_P\dot{\theta}(t). \tag{22}$$

Using (20) and (21) to express the unknown reaction forces $F_1(t), F_2(t)$ between the cart and the pendulum and replacing $F_1(t), F_2(t)$ by these expressions in (19), (22) we obtain the following two equations which describe the dynamic behaviour of the cart-pendulum system (we drop the time variable)

$$\begin{aligned} (M+m)\ddot{r} + (ml\cos\theta)\ddot{\theta} + c\dot{r} - ml\dot{\theta}^2\sin\theta &= \beta u \\ (ml\cos\theta)\ddot{r} + (J+ml^2)\ddot{\theta} + c_P\dot{\theta} + mgl\sin\theta &= 0. \end{aligned} \tag{23}$$

Subtracting suitable multiples of these equations from each other in order to eliminate firstly $\ddot{\theta}$ and then $\ddot{r}$ one obtains the equivalent equations

$$\begin{aligned} M(\theta)\ddot{r} &= (J+ml^2)(\beta u - c\dot{r} + ml\dot{\theta}^2\sin\theta) + ml\cos\theta\,(mgl\sin\theta + c_P\dot{\theta}) \\ M(\theta)\ddot{\theta} &= -ml\cos\theta\,(\beta u - c\dot{r} + ml\dot{\theta}^2\sin\theta) - (M+m)(c_P\dot{\theta} + mgl\sin\theta) \end{aligned} \tag{24}$$

where

$$M(\theta) = (M+m)J + ml^2M + m^2l^2\sin^2\theta.$$

Setting

$$x_1(t) = r(t), \quad x_2(t) = \theta(t), \quad x_3(t) = \dot{r}(t), \quad x_4(t) = \dot{\theta}(t)$$

yields the following system of nonlinear first order differential equations for the cart pendulum system

$$\begin{aligned} \dot{x}_1 &= x_3, \qquad \dot{x}_2 = x_4 \\ \dot{x}_3 &= \frac{1}{M(x_2)}\left[(J+ml^2)(\beta u - cx_3 + mlx_4^2\sin x_2) + ml\cos x_2\,(mgl\sin x_2 + c_Px_4)\right] \\ \dot{x}_4 &= \frac{-ml\cos x_2}{M(x_2)}\left(\beta u - cx_3 + mlx_4^2\sin x_2\right) - \frac{(M+m)}{M(x_2)}\left[c_Px_4 + mgl\sin x_2\right]. \end{aligned} \tag{25}$$

If $u(t) \equiv 0$ the system will remain at rest provided that the initial velocities $x_3(0) = \dot{r}(0)$, $x_4(0) = \dot{\theta}(0)$ are zero and the initial angular displacement $x_2(0) = \theta(0)$ is either zero or π. Cart pendulum systems which are required to operate close to these equilibrium positions occur in practice. For instance, consider a loading plant (see Figure 1.3.7(a)) where a grab is suspended from a cart rolling on horizontal rails. These plants operate around the downward position of the pendulum and are required to be close to this equilibrium before putting down the load. On the other hand consider the balancing problem illustrated by the inverse pendulum in Figure 1.3.7(b). Such inverse pendulum systems are used in university laboratories for experimentation with controllers which stabilize the system at the upward position. A more practical example of a three dimensional balancing problem is that of the control of a rocket in an upright position in preparation for launch. Another (not so obvious) example is that of maintaining a satellite in a prescribed orbit (see Example 2.1.27). For the inverted pendulum shown in Figure 1.3.7(b) it is usual to

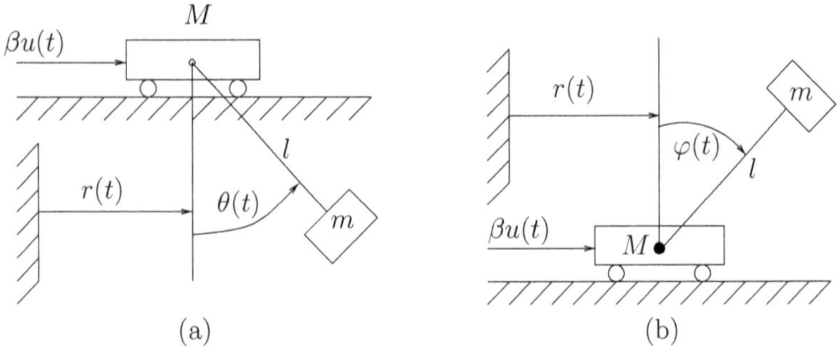

Figure 1.3.7: (a) Loading plant and (b) Inverted pendulum

express the equations of motion in terms of the angle[5] $\varphi = \theta - \pi$ (the deviation of θ from the equilibrium value π). Setting $\theta = \pi + \varphi$ in (24) yields

$$\begin{aligned} M(\varphi)\ddot{r} &= (J + ml^2)(\beta u - c\dot{r} - ml\dot{\varphi}^2 \sin\varphi) - ml\cos\varphi\,(-mgl\sin\varphi + c_P\dot{\varphi}) \\ M(\varphi)\ddot{\varphi} &= ml\cos\varphi\,(\beta u - c\dot{r} - ml\dot{\varphi}^2\sin\varphi) - (M+m)(c_P\dot{\varphi} - mgl\sin\varphi). \end{aligned} \tag{26}$$

Let

$$x_1(t) = r(t), \quad x_2(t) = \varphi(t), \quad x_3(t) = \dot{r}(t), \quad x_4(t) = \dot{\varphi}(t)$$

then one obtains a system of nonlinear first order equations similar to (25), but with a different sign pattern.

$$\begin{aligned} \dot{x}_1 &= x_3, \qquad \dot{x}_2 = x_4 \\ \dot{x}_3 &= \frac{1}{M(x_2)}\left[(J+ml^2)(\beta u - cx_3 - mlx_4^2 \sin x_2) - ml\cos x_2\,(-mgl\sin x_2 + c_P x_4)\right] \\ \dot{x}_4 &= \frac{ml\cos x_2}{M(x_2)}\left(\beta u - cx_3 - mlx_4^2\sin x_2\right) - \frac{(M+m)}{M(x_2)}\left[c_P x_4 - mgl\sin x_2\right]. \end{aligned} \tag{27}$$

Now assume that for the loading plant $|x_2(t)| = |\theta(t)|$ and $|x_4(t)| = |\dot{\theta}(t)|$ remain sufficiently small so that

$$\sin x_2(t) \approx x_2(t), \quad \cos x_2(t) \approx 1, \quad x_4(t)^2 \sin x_2(t) \approx 0, \quad \sin^2 x_2(t) \approx 0. \tag{28}$$

Then $M(x_2) \approx M_0 = (M+m)J + ml^2M$ is approximately constant and we obtain the following approximate linear equation of motion for the loading plant (pendulum down)

$$\dot{x} = Ax + bu, \tag{29}$$

where $x(t) = [r(t), \theta(t), \dot{r}(t), \dot{\theta}(t)]^\top$ and

$$A = \begin{bmatrix} 0 & 0 & 1 & 0 \\ 0 & 0 & 0 & 1 \\ 0 & a_{32} & a_{33} & a_{34} \\ 0 & a_{42} & a_{43} & a_{44} \end{bmatrix}, \quad b = \begin{bmatrix} 0 \\ 0 \\ b_3 \\ b_4 \end{bmatrix}. \tag{30}$$

With constant entries

$$\begin{aligned} a_{32} &= M_0^{-1}m^2l^2g, & a_{33} &= -M_0^{-1}(J+ml^2)c, & a_{34} &= M_0^{-1}mlc_P, \\ a_{42} &= -M_0^{-1}(M+m)mgl, & a_{43} &= M_0^{-1}mlc, \\ a_{44} &= -M_0^{-1}(M+m)c_P, & b_3 &= M_0^{-1}(J+ml^2)\beta, & b_4 &= -M_0^{-1}ml\beta. \end{aligned} \tag{31}$$

[5]Note that the angle $\varphi(t)$ as depicted in Figure 1.3.7 (b) is negative.

For the inverted pendulum if $|\varphi(t)|$ and $|\dot{\varphi}(t)|$ remain small, then (28) again holds but now $x_2(t) = \varphi(t)$ and $x_4(t) = \dot{\varphi}(t)$. The approximate linear model has matrices of the same form as (30), however some of the matrix entries have different signs

$$\begin{aligned} a_{32} &= M_0^{-1} m^2 l^2 g, & a_{33} &= -M_0^{-1}(J + ml^2)c, & a_{34} &= -M_0^{-1} mlc_P, \\ a_{42} &= M_0^{-1}(M+m)mgl, & a_{43} &= -M_0^{-1} mlc, & & \\ a_{44} &= -M_0^{-1}(M+m)c_P, & b_3 &= M_0^{-1}(J+ml^2)\beta, & b_4 &= M_0^{-1} ml\beta. \end{aligned} \tag{32}$$

For the purpose of automatic control, sensors are required which provide continuous information about the current state of the system. Let us consider the balancing problem for the inverted pendulum and suppose that we can measure the current values of $r(t)$, $\varphi(t)$. These measurements ("outputs") are related to the "state" $x(t) = [x_1(t), x_2(t), x_3(t), x_4(t)]^\top$ by the *output* or *measurement* equation

$$y = \begin{bmatrix} 1 & 0 & 0 & 0 \\ 0 & 1 & 0 & 0 \end{bmatrix} x. \tag{33}$$

The balancing problem consists in designing a *regulator* which keeps the pendulum in an upright position at a fixed value of r, say 0. The regulator accepts the values $y(t) \in \mathbb{R}^2$ as input values and produces the values $u(t) \in \mathbb{R}^1$ as output values. This must be done in such a way that the inherently unstable equilibrium $x_e = [0,\, 0,\, 0,\, 0]^\top$ becomes a stable equilibrium of the feedback system. This *stabilization problem* can be solved using the linearized equations about the equilibrium state x_e. Linear models are often sufficient in order to design stabilizing controllers even if the underlying system is nonlinear. By keeping the system close to the equilibrium position the controller ensures that the linearized model yields a good approximation of the nonlinear dynamics. This partially explains the surprising success of linear models in feedback control.
Another control problem is best explained for the loading plant. If it is required to position a load accurately at a certain point the question arises whether there exists a control function $u(\cdot)$ which steers the system from any given initial position to the desired final position in finite time. Additionally, it will be required that the load is at rest at the final position. This is a *controllability* problem. Note that for this problem the use of a linear model is questionable since this is a global problem and its solution requires a model which is accurate for a wide range of values of the system's state. □

1.3.3 The Variational Method

The previous example illustrates that even for an apparently simple mechanical device it is by no means trivial to find the equations of motion by analyzing the system as an interconnection of masses, springs and dampers. The interconnection of translational and rotational elements poses particular problems. The main difficulty in the modelling process is that the interaction between the elements must be described by introducing "contact forces" or "forces of constraint", which are not given a priori. They are among the unknowns of the problem and must be eliminated in order to get the system equations. Often the interconnective constraints are quite complicated and if there is a large number of them the above modelling procedure becomes cumbersome. For such cases an alternative procedure is available which is based on energy considerations. As a preparation we need some formulas for the energies of translational masses, springs and dampers and their rotational counterparts which we present in the next example. Additionally we discuss the kinetic and potential energy of a rigid body moving in three-dimensional space.

Example 1.3.5. The kinetic energy associated with a point mass m moving with velocity $\mathbf{v}(t)$ at time t is

$$\mathcal{T}(t) = (m/2)\,\|\mathbf{v}(t)\|^2 = (m/2)\,\langle \mathbf{v}(t), \mathbf{v}(t)\rangle.$$

For arbitrary motions of a rigid body in three dimensional space the determination of the kinetic energy is more complicated. First consider a rigid body composed of N point masses m_i. Let $\bar{\mathbf{r}}(t)$ be the position of the centre of mass of the body at time t (with respect to some inertial coordinate system) and fix a coordinate system in the body (moving with the body) whose origin is at the centre of mass. Suppose the body is rotating about an axis through the centre of mass with angular velocity $\boldsymbol{\omega}(t)$. So $\boldsymbol{\omega}(t) \in \mathbb{R}^3$ points in the *direction of the instantaneous axis of rotation* of the body about its centre of mass given by the right hand screw law. If $\tilde{\mathbf{r}}_i$ is the (constant) coordinate vector of the point mass m_i of the rigid body with respect to the body coordinates then the position vector of this point with respect to the inertial reference system is $\mathbf{r}_i(t) = \bar{\mathbf{r}}(t) + \tilde{\mathbf{r}}_i$ and the velocity vector of the point (with respect to the given inertial coordinate system) is

$$\mathbf{v}_i(t) = \dot{\mathbf{r}}_i(t) = \dot{\bar{\mathbf{r}}}(t) + \boldsymbol{\omega}(t) \times \tilde{\mathbf{r}}_i.$$

Hence the kinetic energy is

$$\begin{aligned} \mathcal{T}(t) &= \sum_{i=1}^{N} (m_i/2)\|\mathbf{v}_i(t)\|^2 = \sum_{i=1}^{N} (m_i/2)\langle \dot{\bar{\mathbf{r}}}(t) + \boldsymbol{\omega}(t) \times \tilde{\mathbf{r}}_i, \dot{\bar{\mathbf{r}}}(t) + \boldsymbol{\omega}(t) \times \tilde{\mathbf{r}}_i\rangle \\ &= (M/2)\,\|\dot{\bar{\mathbf{r}}}(t)\|^2 + \left\langle \dot{\bar{\mathbf{r}}}(t), \boldsymbol{\omega}(t) \times \sum_{i=1}^{N} m_i \tilde{\mathbf{r}}_i \right\rangle + \sum_{i=1}^{N} (m_i/2)\|\boldsymbol{\omega}(t) \times \tilde{\mathbf{r}}_i\|^2, \end{aligned}$$

where $M = \sum_{i=1}^{N} m_i$ is the total mass. The middle term in the above expression is zero since $\tilde{\mathbf{r}}_i$ is the vector from the centre of mass to the i-th point mass. The last term is a quadratic form in $\boldsymbol{\omega}(t)$, so we may write

$$\mathcal{T}(t) = (M/2)\,\|\dot{\bar{\mathbf{r}}}(t)\|^2 + (1/2)\,\langle \boldsymbol{\omega}(t), \mathbf{J}\boldsymbol{\omega}(t)\rangle,$$

where $\mathbf{J} = \mathbf{J}^\top \in \mathbb{R}^{3\times 3}$ is called the *moment of inertia matrix* of the rigid body. This analysis can be extended to continuous distributions and hence the above formula for the kinetic energy holds for arbitrary rigid bodies. So the kinetic energy of a rigid body is the kinetic energy obtained if all the mass of the body were concentrated at the centre of mass, plus the kinetic energy of its motion about the centre of mass.

We now consider *potential energy.* The potential energy stored in a translational or rotational spring displaced from its equilibrium state is equal to the work done to achieve this displacement. If the spring is translational and linear its potential energy at a displacement y is given by $(k/2)\,y^2$. Similarly the potential energy of a linear torsional spring (where the torque is $k\theta$) at an angular displacement θ is $(k/2)\,\theta^2$. Note that an ideal spring does not have kinetic energy since its mass is zero.

The potential energy of any point mass is defined relative to a given conservative field of force to which it is subjected, e.g. the gravitational field of the Earth. If $\mathbf{F} : \mathbb{R}^3 \to \mathbb{R}^3$ is a *conservative* field of force then the work done by moving a point mass from $\mathbf{a} \in \mathbb{R}^3$ to $\mathbf{b} \in \mathbb{R}^3$ only depends upon the points $\mathbf{a}, \mathbf{b} \in \mathbb{R}^3$ and not on the path along which the mass has been moved. Fixing a reference point O the potential energy of a particle positioned at a point P is, by definition, equal to the work needed in order to move the particle within the force field from O to P. The potential energy of a system of N point masses at positions $\mathbf{r}_1, \dots, \mathbf{r}_N$ is simply the sum of the individual potential energies. Approximating a rigid body of mass M by a system of point masses we see that at an altitude h above

the Earth (h not too large) its potential energy with respect to the gravitational field of the Earth is approximately Mgh. Note that the potential energy of a body relative to a conservative force field is only determined up to a constant depending on the reference point. For any system of point masses moving in a conservative field, if no energy dissipation occurs, then the sum of the kinetic and potential energies is constant in time.
Usually, dissipation of energy occurs because kinetic energy is transformed to thermal energy by friction. Frictional forces have to be overcome whenever bodies in contact have a relative velocity. A pure dissipator (damper) is an idealized object in which there is no kinetic or potential energy storage. However there is a dissipation of energy, for example the power absorbed at time t by a linear translational damper (where the force is cv) is $cv(t)^2$. Similarly the power absorbed by a linear rotational damper (where the torque is $c\omega$) is $c\omega(t)^2$. More generally, suppose that there is a system of N particles moving with velocities $\mathbf{v}_i(t) \in \mathbb{R}^3$, $i \in \underline{N}$ and that the particles are subjected to frictional forces which depend linearly on the velocities, $\mathbf{F}_i(t) = c_i\, \mathbf{v}_i(t)$, then the total energy dissipated is $\sum_{i=1}^{N} c_i \|\mathbf{v}_i(t)\|^2$. □

The variational method has been developed in the context of classical mechanics by, amongst many, Lagrange and Hamilton. We will not explain the derivation of the method here, nor discuss it in detail, but just sketch the essential steps to be followed. For a careful mathematical treatment, see *Notes and References.*
As the previous examples illustrate, the position vectors $\mathbf{r}_i(t)$ of the point masses of a mechanical system are usually not free to vary independently of each other. The constraints which limit their movements may be classified in various ways. If they can be expressed by equations of the form $f(\mathbf{r}_1, \ldots, \mathbf{r}_N, t) = 0$ they are called *holonomic*. A typical example is given by a rigid body where all the distances between its mass points are constant in time. Another example is given by a particle which moves along a curve (a bead sliding on a wire) or on a surface. Nonholonomic constraints are obtained if not only position but also velocity coordinates enter the constraint equation or if the constraint takes the form of an inequality (for example, gas molecules within a container). We will only consider holonomic constraints. All the constraints in our mechanical examples are of this type.
Now suppose that a system of N particles is given, together with a number of holonomic constraints of the form

$$f_j(\mathbf{r}_1, \ldots, \mathbf{r}_N, t) = 0, \quad j \in \underline{m} \tag{34}$$

where $\mathbf{r}_i \in \mathbb{R}^3$ denotes the position of the i-th particle and the f_j are real-valued smooth functions on $(\mathbb{R}^3)^N \times \mathbb{R}$. The set $\mathcal{M}(t)$ of all possible configurations of the system at time t, i.e. the set of all the vectors $\mathbf{r} = (\mathbf{r}_1, \ldots, \mathbf{r}_N) \in (\mathbb{R}^3)^N$ satisfying the constraints $f_1(\mathbf{r}, t) = 0, \ldots, f_m(\mathbf{r}, t) = 0$, is called the *configuration space* of the constrained mechanical system at time t. Let us fix the time t for a moment and consider the configuration space $\mathcal{M}(t)$. If the gradients of the functions $f_j(\cdot, t)$ are linearly independent at every point in $\mathcal{M}(t)$, the configuration space (at time t) carries the structure of an ℓ-dimensional differentiable manifold where $\ell = 3N - m$.[6] This implies that $\mathcal{M}(t)$ is provided with a finite or countable collection of *charts*, so that every point is represented in at least one chart. A chart is an open set U in $\mathbb{R}^\ell$

[6] In this case the constrained mechanical system is said to be a system with ℓ *degrees of freedom.*

with a diffeomorphic mapping ϕ from U onto some open subset $V = \phi(U)$ of $\mathcal{M}(t)$ which we write in the following way.

$$\phi : q \mapsto \mathbf{r}(q,t) = (\mathbf{r}_1(q,t), \ldots, \mathbf{r}_N(q,t)) \text{ where } \mathbf{r}_i(q,t) = \mathbf{r}_i(q_1, \ldots, q_\ell, t),\ q \in U. \quad (35)$$

The coordinates q_j, $j \in \underline{\ell}$ of the vector q are called the *generalized coordinates* of the configuration $\mathbf{r} = \mathbf{r}(q,t)$ (at time t). They yield a parametrization of the open subset $V = \phi(U)$ of the configuration space. The inverse mapping $\phi^{-1} : V \to U$ maps every configuration $\mathbf{r} \in V$ onto its *generalized coordinate vector* $q(\mathbf{r},t)$ (at time t). Note that the coordinates of the generalized coordinate vector $q \in U$ can be varied independently (provided q remains in the open set U) whereas the coordinates of a position vector $\mathbf{r} = (\mathbf{r}_1, \ldots, \mathbf{r}_N) \in V$ cannot be varied independently without violating some of the holonomic constraints (34) (at time t). In practice the generalized coordinates are usually obtained by applying the implicit function theorem to the constraint equations. This technique is sometimes called "elimination of the dependent coordinates". The position vectors $\mathbf{r}(q,t)$ corresponding to the generalized coordinate vectors $q \in U$ at time t automatically satisfy the constraints. The advantage of "getting rid of the constraints" by introducing the generalized coordinates is, however, not obtained without any cost. Firstly, the generalized coordinates describe, in general, only a part of the configuration space. Secondly, they need not have an immediate physical interpretation similar to the original position variables. And often they cannot be related to single elements of the system but are mathematical constructions for the description of the system as a whole.

Now suppose for a moment that the constraints (34) do not depend upon t (an assumption which is often satisfied in applications) so that the configuration manifold $\mathcal{M}(t) = \mathcal{M}$, the charts (U, ϕ) and the corresponding open subsets $V = \phi(U)$ of $\mathcal{M}$ are independent of time. Then, as the system moves, the position vector $\mathbf{r}(t)$ describes a curve in the configuration manifold $\mathcal{M}$. If $\mathbf{r}(t)$ belongs to the scope V of some chart (U, ϕ) for $t \in [t_1, t_2]$, the motion of the system during this time interval can alternatively be described in terms of the position vectors $\mathbf{r}_i(t)$, $i \in \underline{N}$ (satisfying the constraints) or in terms of the generalized coordinates $q_j(t)$, $j \in \underline{\ell}$.

We will now briefly point out how velocities, forces, kinetic and potential energies are expressed in terms of the generalized coordinates. With every family of velocity vectors $\mathbf{v}_i = \dot{\mathbf{r}}_i = \frac{d}{dt}\mathbf{r}_i$, $i \in \underline{N}$ which is consistent with the given constraints there is an associated generalized velocity vector $\dot{q} = \frac{d}{dt}q(\mathbf{r},t) = (\dot{q}_j)_{j \in \underline{\ell}}$ which can be determined by solving the system of $3N$ linear equations

$$\mathbf{v}_i = \dot{\mathbf{r}}_i = \sum_{j=1}^{\ell} \frac{\partial \mathbf{r}_i}{\partial q_j}\dot{q}_j + \frac{\partial \mathbf{r}_i}{\partial t}, \quad i \in \underline{N}. \quad (36)$$

Similarly, for any family of external forces $\mathbf{f}_i = (f_{i,1}, f_{i,2}, f_{i,3}) \in \mathbb{R}^3$, $i \in \underline{N}$ applied to the i-th particle at time t there is an associated vector $(F_1, \ldots, F_\ell) \in \mathbb{R}^\ell$ called the *generalized force* at time t defined by

$$F_j = \sum_{i=1}^{N} \left\langle \mathbf{f}_i, \frac{\partial \mathbf{r}_i}{\partial q_j} \right\rangle_{\mathbb{R}^3} = \sum_{i=1}^{N} \left(f_{i,1}\frac{\partial x_i(q,t)}{\partial q_j} + f_{i,2}\frac{\partial y_i(q,t)}{\partial q_j} + f_{i,3}\frac{\partial z_i(q,t)}{\partial q_j} \right), \quad j \in \underline{\ell}.$$

where $\mathbf{r}_i(q,t) = (x_i(q,t), y_i(q,t), z_i(q,t))$. If the i-th particle of the system has mass m_i and is moving with velocity $\mathbf{v}_i$, the associated kinetic energy of the system is

$\mathcal{T} = \sum_{i=1}^{N}(m_i/2)\|\mathbf{v}_i\|^2$. By means of (36) the kinetic energy can be expressed in terms of the generalized coordinates and velocities,

$$\mathcal{T}(q,\dot{q},t) = \sum_{i=1}^{N}(m_i/2)\|\mathbf{v}_i(q,t)\|^2 = \sum_{i=1}^{N}(m_i/2)\left\|\sum_{j=1}^{\ell}\frac{\partial \mathbf{r}_i}{\partial q_j}(q,t)\dot{q}_j + \frac{\partial \mathbf{r}_i}{\partial t}(q,t)\right\|^2 .$$

We see therefore that if the constraints are independent of time, then $\mathcal{T}$ is a homogeneous quadratic form in the generalized velocities. Now assume for a moment that the mechanical system is *conservative*, i.e. there exists a real valued function $\mathcal{W}(\mathbf{r}_1,\ldots,\mathbf{r}_N,t)$ such that the force $\mathbf{f}_i$ applied to the i-th particle is given by the i-th partial gradient of $\mathcal{W}$ (i.e. with respect to the coordinates x_i, y_i, z_i of $\mathbf{r}_i$)

$$\mathbf{f}_i(\mathbf{r}_1,\ldots,\mathbf{r}_N,t) = -\nabla_i\mathcal{W}(\mathbf{r}_1,..,\mathbf{r}_N,t).$$

In this case the generalized force is precisely the negative gradient of $\mathcal{W}$ viewed as a function of the generalized coordinates:

$$F_j(q,t)=\sum_{i=1}^{N}\left\langle \mathbf{f}_i,\frac{\partial \mathbf{r}_i}{\partial q_j}\right\rangle(q,t)=-\sum_{i=1}^{N}\left\langle \nabla_i\mathcal{W}(\mathbf{r}_1,\ldots,\mathbf{r}_N,t),\frac{\partial \mathbf{r}_i(q)}{\partial q_j}\right\rangle(q,t)=-\frac{\partial \mathcal{W}(q,t)}{\partial q_j}$$

where $\mathcal{W}(q,t) = \mathcal{W}(q_1,\ldots,q_\ell,t) := \mathcal{W}(\mathbf{r}_1(q,t),\ldots,\mathbf{r}_N(q,t),t)$ is called the *generalized potential energy*.

In 1788 Lagrange published in Paris his celebrated *Mécanique Analytique* [325] in which he set out a method for determining the equations of a mechanical system from a knowledge of the kinetic and potential energies. His ideas were developed further by Boltzmann in 1802 and Hamel in 1804 and the form in which we state the equations are essentially due to them, although they are widely referred to as Lagrange's equations. Lagrange [325] introduced what is now known as a *Lagrangian*:

$$L(q,\dot{q},t) = \mathcal{T}(q,\dot{q},t) - \mathcal{W}(q,t). \tag{37}$$

Then *Lagrange's equations* of motion take the form

$$\frac{d}{dt}\left(\frac{\partial L}{\partial \dot{q}_j}(q(t),\dot{q}(t),t)\right) - \frac{\partial L}{\partial q_j}(q(t),\dot{q}(t),t) = 0, \quad j=1,\ldots,\ell. \tag{38}$$

In practice most mechanical systems are not conservative, since, they either have significant internal frictions, or external forces are applied which are not derived from a potential. If F_j are the generalized forces which are not taken into account by the potential energy and $\mathcal{D}(\dot{q}) = \mathcal{D}(\dot{q}_1,\ldots,\dot{q}_\ell)$ is the total energy dissipated by linear dissipators (e.g. dampers), then the equations of motion take the form

$$\frac{d}{dt}\left(\frac{\partial L}{\partial \dot{q}_j}(q(t),\dot{q}(t),t)\right) - \frac{\partial L}{\partial q_j}(q(t),\dot{q}(t),t) + \frac{1}{2}\frac{\partial \mathcal{D}}{\partial \dot{q}_j}(\dot{q}(t)) = F_j(q(t),t), \; j=1,..,\ell. \tag{39}$$

Note that if the generalized external forces F_j do not depend on the generalized coordinates q they can easily be accounted for by modifying the potential energy

$$\mathcal{W} \rightsquigarrow \mathcal{W} - \sum_{j=1}^{\ell} F_j q_j. \tag{40}$$

By suitably modifying the Lagrangian it is also possible to include other generalized forces in Lagrange's equations (38), see *Notes and References.*
If, for a given mechanical system, generalized coordinates can be found, Lagrange's method is a very convenient way to eliminate the forces of constraint from the equations of motion. By this elimination the modelling procedure is greatly simplified. In fact, in order to model a complicated multi-body mechanical system by the free-body diagram approach illustrated in the previous examples, many vector forces and velocities must be handled, whereas whenever a Lagrangian formulation is applicable there is – in principle – a straight forward procedure for deriving the equations of motion. One "only" has to write three *scalar* functions $\mathcal{T}$, $\mathcal{W}$, $\mathcal{D}$ in generalized coordinates (which may not be so easy), form L, determine the generalized forces and substitute in (39). Sometimes, of course, one would like to know the contact forces and then it is necessary to resort to free-body diagrams. However, assuming Lagrange's equations have been solved for the generalized coordinates $q_i(t)$ as functions of time t and consequently the vector functions $\mathbf{r}_i(\cdot)$ are known, the equations for the contact forces obtained via free-body diagrams can often be easily resolved.

Remark 1.3.6. Lagrange's equations have the following interesting interpretation. Consider any given trajectory $\mathbf{r}(t)$ of a conservative mechanical system in configuration space from time t_0 to time t_1 and suppose that the trajectory remains inside the scope of a chart so that it can equivalently be described by a curve $t \to q(t) = (q_1(t), \ldots, q_\ell(t))$, $t \in [t_0, t_1]$ in $\mathbb{R}^\ell$ (satisfying $\mathbf{r}(t) = \mathbf{r}(q(t), t)$). Hamilton's Principle says: *The motion of a conservative system from time t_0 to time t_1 is such that the action integral*

$$\mathbf{I}(z(\cdot)) = \int_{t_0}^{t_1} L(z(t), \dot{z}(t), t)\, dt$$

is an extremum for the actual path of motion $q(\cdot)$ amongst all other curves $z(\cdot) : [t_0, t_1] \to \mathbb{R}^\ell$ connecting $q(t_0)$ with $q(t_1)$. There are global, coordinate free formulations of this principle which avoid the restriction to parts of the configuration manifold parametrized by a chart, see e.g. [1], [18].
It is shown in the calculus of variations that Lagrange's equations are exactly the necessary and sufficient conditions for the functional

$$\mathbf{I} : \{z(\cdot) \in C^1([t_0, t_1], \mathbb{R}^\ell);\; z(t_0) = q(t_0),\; z(t_1) = q(t_1)\} \to \mathbb{R}$$

to have an extremum at $z(\cdot) = q(\cdot)$. In 1766 Lagrange joined Euler as a court mathematician in Berlin under the patronage of Frederick the Great. Euler also developed necessary and sufficient conditions which are equivalent to those of Lagrange and it is usual, at least in the field of the calculus of variations, to refer to the equations as the *Euler-Lagrange equations.* The variational approach is of great importance since variational principles can be used in many fields of physics to express the equations of motion. This makes it possible to transfer the Lagrangian method to other fields and uncover structural analogies between them. □

Before we consider some examples we briefly outline the *Hamiltonian approach* to classical mechanics which yields another method for deriving the equations of motion of a conservative mechanical system. The result is a transformation of Lagrange's equations (38) which are second order into an equivalent system of *Hamiltonian equations* which are first order. This is accomplished by applying a Legendre transformation to the Lagrangian, see *Notes and References.* For arbitrary given q, t this transforms $L(q, \dot{q}, t)$ viewed as a function of $\dot{q}$ into a function of the new variable p where $\dot{q}$ and p are related via the formula $p = \partial L/\partial \dot{q}$,

$$H(q,p,t) = \langle p, \dot{q}\,(q,p,t)\rangle - L(q, \dot{q}\,(q,p,t), t), \quad (q,p,t) \in \mathbb{R}^\ell \times \mathbb{R}^\ell \times \mathbb{R}. \tag{41}$$

Here the function $\dot{q} = \dot{q}(q,p,t)$ is defined implicitly by the equation

$$p = \frac{\partial L}{\partial \dot{q}}\,(q, \dot{q}, t) \tag{42}$$

which is assumed to have a unique solution $\dot{q}$ for every $(q,p,t) \in \mathbb{R}^\ell \times \mathbb{R}^\ell \times \mathbb{R}$. H is called the *Hamiltonian* and $p = (p_1, p_2, \ldots, p_\ell)$ the *generalized momentum* of the conservative mechanical system. Now the total differential of the Hamiltonian

$$dH = \frac{\partial H}{\partial p}dp + \frac{\partial H}{\partial q}dq + \frac{\partial H}{\partial t}dt$$

is equal to the total differential of $\langle p, \dot{q}\rangle - L(q, \dot{q}, t)$,

$$dH = \langle \dot{q}, dp\rangle + \langle p, d\dot{q}\rangle - \left\langle \frac{\partial L}{\partial q}, dq\right\rangle - \left\langle \frac{\partial L}{\partial \dot{q}}, d\dot{q}\right\rangle - \frac{\partial L}{\partial t}dt.$$

where $\langle\cdot,\cdot\rangle$ denotes the usual inner product in $\mathbb{R}^\ell$. The second and fourth terms cancel because of (42), hence

$$\frac{\partial H}{\partial p} = \dot{q}, \quad \frac{\partial H}{\partial q} = -\frac{\partial L}{\partial q}, \quad \frac{\partial H}{\partial t} = -\frac{\partial L}{\partial t}.$$

Applying Lagrange's equations (38) we obtain *Hamilton's equations*

$$\dot{q} = \frac{\partial H}{\partial p}, \qquad \dot{p} = -\frac{\partial H}{\partial q}. \tag{43}$$

We now illustrate the Lagrangian and Hamiltonian approaches by deriving the equations of motion for the cart-pendulum system studied in Example 1.3.4.

Example 1.3.7. (Cart-pendulum system). In order to derive the equations of motion for the cart-pendulum system via Lagrange's equations we must determine the kinetic energy $\mathcal{T}$, the potential energy $\mathcal{W}$ and the dissipated energy $\mathcal{D}$ of this system in terms of its generalized coordinates r, θ and the corresponding velocities $\dot{r}$, $\dot{\theta}$. The kinetic energy $\mathcal{T}$ of the system is the sum of the kinetic energies of the cart and of the pendulum, and the latter is the sum of the kinetic energy of the centre of mass plus the energy of the pendulum rotating about its centre of mass, see Figure 1.3.6. Hence

$$\begin{aligned}\mathcal{T} &= (M/2)\,\dot{r}^2 + (m/2)\left[\left[\frac{d}{dt}(r + l\sin\theta)\right]^2 + \left[\frac{d}{dt}l\cos\theta\right]^2\right] + (J/2)\,\dot{\theta}^2 \\ &= (M/2)\,\dot{r}^2 + (J/2)\,\dot{\theta}^2 + (m/2)\,\left[(\dot{r} + l\dot{\theta}\cos\theta)^2 + (-l\dot{\theta}\sin\theta)^2\right].\end{aligned}$$

The potential energy, $\mathcal{W}$ is the same as that of a single mass m located at the centre of mass of the pendulum in a gravitational field, i.e $-mgl\cos\theta$ (modulo an additive constant).

Since the time varying external force $\beta u(t)$ does not depend on the generalized coordinates, we can take it into account by modifying the potential energy as in (40).

$$\mathcal{W} = -mgl\cos\theta - \beta u r.$$

$\mathcal{D}$ is the sum of the dissipated energies due to viscous friction $c\dot{r}$ between cart and rails and due to viscous friction $c_P\dot{\theta}$ at the pivot,

$$\mathcal{D} = c\dot{r}^2 + c_P\dot{\theta}^2.$$

The generalized Lagrange equations (39) in terms of the generalized coordinates and associated velocities r, θ, $\dot{r}$, $\dot{\theta}$ are

$$\begin{aligned}\frac{d}{dt}\left(\frac{\partial\mathcal{T}}{\partial\dot{r}}\right) - \frac{\partial\mathcal{T}}{\partial r} + \frac{\partial\mathcal{W}}{\partial r} + \frac{1}{2}\frac{\partial\mathcal{D}}{\partial\dot{r}} &=0\\ \frac{d}{dt}\left(\frac{\partial\mathcal{T}}{\partial\dot{\theta}}\right) - \frac{\partial\mathcal{T}}{\partial\theta} + \frac{\partial\mathcal{W}}{\partial\theta} + \frac{1}{2}\frac{\partial\mathcal{D}}{\partial\dot{\theta}} &=0.\end{aligned}$$

Or

$$\begin{aligned}&\frac{d}{dt}\left[M\dot{r} + m(\dot{r} + l\dot{\theta}\cos\theta)\right] + c\dot{r} = \beta u\\ &\frac{d}{dt}\left[J\dot{\theta} + m(\dot{r} + l\dot{\theta}\cos\theta)l\cos\theta + ml^2\dot{\theta}(\sin\theta)^2\right]\\ &\qquad + m(\dot{r} + l\dot{\theta}\cos\theta)l\dot{\theta}\sin\theta - ml^2\dot{\theta}^2\sin\theta\cos\theta + mgl\sin\theta + c_P\dot{\theta} = 0.\end{aligned}$$

A simple calculation yields the nonlinear differential equations

$$\begin{aligned}(M+m)\ddot{r} + ml\ddot{\theta}\cos\theta - ml\dot{\theta}^2\sin\theta + c\dot{r} &= \beta u\\ (J+ml^2)\ddot{\theta} + m\ddot{r}l\cos\theta + mgl\sin\theta + c_P\dot{\theta} &= 0.\end{aligned}$$

Thus the Lagrangian approach leads to the same equations of motion as the approach via free-body diagrams in Example 1.3.4, see (23).
Assuming that frictions can be neglected and the pendulum behaves like a point mass connected to a light rod of length l (i.e. $c = c_P = J = 0$), the nonlinear equations of motion reduce to

$$(M+m)\ddot{r} + ml\ddot{\theta}\cos\theta - ml\dot{\theta}^2\sin\theta = \beta u \tag{44}$$

$$l\ddot{\theta} + g\sin\theta + \ddot{r}\cos\theta = 0. \tag{45}$$

Setting $x_1 = r$, $x_2 = \theta$, $x_3 = \dot{r}$, $x_4 = \dot{\theta}$ (resp. $x_1 = r$, $x_2 = \varphi$, $x_3 = \dot{r}$, $x_4 = \dot{\varphi}$, see Example 1.3.4) the linearized models of the loading plant and the inverted pendulum, respectively, reduce to

$$\dot{x} = Ax + bu, \tag{46}$$

where (see (31), (32))

$$A = \begin{bmatrix} 0 & 0 & 1 & 0\\ 0 & 0 & 0 & 1\\ 0 & a_{32} & 0 & 0\\ 0 & a_{42} & 0 & 0\end{bmatrix},\quad b = \begin{bmatrix}0\\0\\b_3\\b_4\end{bmatrix},\qquad \begin{array}{ll} a_{32} = mg/M, & a_{42} = \mp(M+m)g/(Ml),\\ b_3 = \beta/M, & b_4 = \mp\beta/(Ml).\end{array}$$

Let us now consider Hamilton's equations for the frictionless case. By (42), the generalized momentum has components

$$p_1 = \frac{\partial L}{\partial\dot{r}} = (M+m)\dot{r} + ml\cos\theta\,\dot{\theta} \tag{47}$$

$$p_2 = \frac{\partial L}{\partial\dot{\theta}} = ml(\dot{r}\cos\theta + l\dot{\theta}). \tag{48}$$

Hence

$$\begin{aligned}\dot{r} &= (ml^2 p_1 - ml\cos\theta\, p_2)/(ml^2(M + m\sin^2\theta)) \quad &(49)\\ \dot{\theta} &= (-ml\cos\theta\, p_1 + (M+m)p_2)/(ml^2(M + m\sin^2\theta)). \quad &(50)\end{aligned}$$

So by (41) the Hamiltonian is

$$\begin{aligned}H(r,\, \theta,\, p_1,\, p_2) &= (ml^2 p_1^2 - 2ml\cos\theta\, p_1 p_2 + (M+m)p_2^2)/(2ml^2(M + m\sin^2\theta))\\ &\quad - mgl\cos\theta - \beta r u.\end{aligned}$$

Hamilton's equations are, therefore, (49), (50) augmented with

$$\begin{aligned}\dot{p_1} &= -\frac{\partial H}{\partial r} = \beta u\\ \dot{p_2} &= -\frac{\partial H}{\partial \theta} = -\, p_1 p_2 \sin\theta/(l(M + m\sin^2\theta)) - mgl\sin\theta\\ &\quad + (ml^2 p_1^2 - 2ml\cos\theta\, p_1 p_2 + (M+m)p_2^2)\sin\theta\cos\theta/(l^2(M + m\sin^2\theta)^2).\end{aligned} \qquad (51)$$

The linearization of equations (49), (50), (51) yields

$$\dot{\tilde{x}} = \tilde{A}\tilde{x} + \tilde{b}u, \qquad (52)$$

where $\tilde{x} = [r,\, \theta,\, p_1,\, p_2]^\top$, $\tilde{b} = [0,\, 0,\, \beta,\, 0]^\top$ and $\tilde{A}$ is the matrix

$$\tilde{A} = \begin{bmatrix} 0 & 0 & \tilde{a}_{13} & \tilde{a}_{14}\\ 0 & 0 & \tilde{a}_{23} & \tilde{a}_{24}\\ 0 & 0 & 0 & 0\\ 0 & \tilde{a}_{42} & 0 & 0 \end{bmatrix}, \qquad \begin{aligned}&\tilde{a}_{13} = 1/M, \quad \tilde{a}_{14} = \tilde{a}_{23} = -1/(Ml),\\ &\tilde{a}_{24} = (M+m)/(mMl^2), \quad \tilde{a}_{42} = -mgl.\end{aligned}$$

Equations (46) and (52) for the loading plant are two different mathematical models of the linearized system which are related by the transformations

$$\tilde{x} = Tx, \quad \tilde{A} = TAT^{-1}, \quad \tilde{b} = Tb,$$

where

$$T = \begin{bmatrix} 1 & 0 & 0 & 0\\ 0 & 1 & 0 & 0\\ 0 & 0 & t_{33} & t_{34}\\ 0 & 0 & t_{43} & t_{44} \end{bmatrix}, \quad t_{33} = M + m,\ t_{34} = t_{43} = ml,\ t_{44} = ml^2.$$

Such systems are called *similar* and we will discuss this concept in Section 2.4. □

We conclude this section by using Lagrange's equations to derive the equations of motion of an inverted double pendulum.

Example 1.3.8. (Inverted double pendulum). Consider a double pendulum which is mounted on a cart as illustrated in Figure 1.3.8. In a similar way to Example 1.3.4 we assume that the motion of the system is restricted to the vertical plane, the cart is moving on a horizontal rail with viscous friction and the two pendulums behave like rigid bodies with viscous friction at the pivots. Let m_i, l_i, J_i, c_i $(i = 1, 2)$ denote the mass, the distance between the centre of gravity and the lower hinge, the moment of inertia about the centre of mass and the friction coefficient for the lower $(i = 1)$ and the upper $(i = 2)$ pendulums. L is the total length of the lower pendulum and M, c_0 denote the mass and the friction coefficient of the cart. As generalized coordinates we choose the distance r of the cart

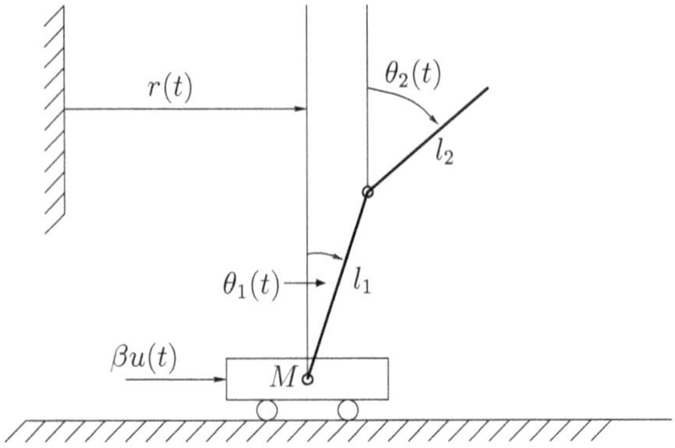

Figure 1.3.8: Double pendulum

from an inertial reference position, and the angles θ_i, $i = 1, 2$ of the two pendulums to the vertical, with *clockwise* orientation. We apply the method of Lagrange in order to find the equations of motion of the system. The *potential energy* of the system is equal to the sum of the potential energies of the masses m_i located at the centres of mass of the two pendulums, together with an adjustment for the external force. In terms of the chosen generalized coordinates it is given by

$$\mathcal{W} = m_1 g l_1 \cos\theta_1 + m_2 g (L\cos\theta_1 + l_2\cos\theta_2) - \beta u r. \tag{53}$$

The energy dissipated by the translational viscous friction between cart and rails and the rotational friction at the two hinges is given by

$$\mathcal{D} = c_0 \dot{r}^2 + c_1 \dot{\theta}_1^2 + c_2(\dot{\theta}_2 - \dot{\theta}_1)^2. \tag{54}$$

The kinetic energy is the sum of the kinetic energies of the cart plus the kinetic energies of the two pendulums:

$$\begin{aligned}
\mathcal{T}_0 &= (M/2)\,\dot{r}^2 \\
\mathcal{T}_1 &= (J_1/2)\,\dot{\theta}_1^2 + (m_1/2)\left\{\left[\frac{d}{dt}(r + l_1\sin\theta_1)\right]^2 + \left[\frac{d}{dt}(l_1\cos\theta_1)\right]^2\right\} \\
\mathcal{T}_2 &= (J_2/2)\,\dot{\theta}_2^2 + (m_2/2)\left\{\left[\frac{d}{dt}(r + L\sin\theta_1 + l_2\sin\theta_2)\right]^2 + \left[\frac{d}{dt}(L\cos\theta_1 + l_2\cos\theta_2)\right]^2\right\}.
\end{aligned}$$

A simple calculation yields the *total kinetic energy* of the system

$$\begin{aligned}
\mathcal{T} = {} & (M/2)\,\dot{r}^2 + (J_1/2)\,\dot{\theta}_1^2 + (J_2/2)\,\dot{\theta}_2^2 + (m_1/2)\left\{\dot{r}^2 + 2l_1\dot{r}\dot{\theta}_1\cos\theta_1 + l_1^2\dot{\theta}_1^2\right\} + \\
& (m_2/2)\left\{\dot{r}^2 + L^2\dot{\theta}_1^2 + l_2^2\dot{\theta}_2^2 + 2\dot{r}\left[L\dot{\theta}_1\cos\theta_1 + l_2\dot{\theta}_2\cos\theta_2\right] + 2Ll_2\dot{\theta}_1\dot{\theta}_2\cos(\theta_1 - \theta_2)\right\}.
\end{aligned} \tag{55}$$

Using (53), (54) and (55) we write down Lagrange's equations and obtain by elementary calculations the following equations of motion

$$\frac{d}{dt}\left(\frac{\partial \mathcal{T}}{\partial \dot{r}}\right) - \frac{\partial \mathcal{T}}{\partial r} + \frac{\partial \mathcal{W}}{\partial r} + \frac{1}{2}\frac{\partial \mathcal{D}}{\partial \dot{r}} = 0,$$

$$(M + m_1 + m_2)\ddot{r} + [(m_1 l_1 + m_2 L)\cos\theta_1]\ddot{\theta}_1 + (m_2 l_2 \cos\theta_2)\ddot{\theta}_2 \\ - (m_1 l_1 + m_2 L)\dot{\theta}_1^2 \sin\theta_1 - (m_2 l_2 \sin\theta_2)\dot{\theta}_2^2 + c_0\dot{r} = \beta u.$$

$$\frac{d}{dt}\left(\frac{\partial \mathcal{T}}{\partial \dot{\theta}_1}\right) - \frac{\partial \mathcal{T}}{\partial \theta_1} + \frac{\partial \mathcal{W}}{\partial \theta_1} + \frac{1}{2}\frac{\partial \mathcal{D}}{\partial \dot{\theta}_1} = 0,$$

$$[(m_1 l_1 + m_2 L)\cos\theta_1]\ddot{r} + (m_1 l_1^2 + m_2 L^2 + J_1)\ddot{\theta}_1 + [m_2 L l_2 \cos(\theta_1 - \theta_2)]\ddot{\theta}_2 \\ + m_2 L l_2 \dot{\theta}_2^2 \sin(\theta_1 - \theta_2) - (m_1 l_1 + m_2 L) g \sin\theta_1 + c_1\dot{\theta}_1 + c_2(\dot{\theta}_1 - \dot{\theta}_2) = 0.$$

$$\frac{d}{dt}\left(\frac{\partial \mathcal{T}}{\partial \dot{\theta}_2}\right) - \frac{\partial \mathcal{T}}{\partial \theta_2} + \frac{\partial \mathcal{W}}{\partial \theta_2} + \frac{1}{2}\frac{\partial \mathcal{D}}{\partial \dot{\theta}_2} = 0,$$

$$(m_2 l_2 \cos\theta_1)\ddot{r} + [m_2 L l_2 \cos(\theta_1 - \theta_2)]\ddot{\theta}_1 + (m_2 l_2^2 + J_2)\ddot{\theta}_2 + \\ - m_2 L l_2 \dot{\theta}_1^2 \sin(\theta_1 - \theta_2) - m_2 g l_2 \sin\theta_2 + c_2(\dot{\theta}_2 - \dot{\theta}_1) = 0.$$

Introducing the vector $z = [r, \theta_1, \theta_2]^\top \in \mathbb{R}^3$, the equations of motion can be written in the following concise form

$$K_1\ddot{z} = K_2\dot{z} + K_3 + k_4 u$$

where

$$K_1 = \begin{bmatrix} m_1 + m_2 + M & (m_1 l_1 + m_2 L)\cos\theta_1 & m_2 l_2 \cos\theta_2 \\ (m_1 l_1 + m_2 L)\cos\theta_1 & J_1 + m_1 l_1^2 + m_2 L^2 & m_2 l_2 L \cos(\theta_1 - \theta_2) \\ m_2 l_2 \cos\theta_2 & m_2 l_2 L \cos(\theta_1 - \theta_2) & J_2 + m_2 l_2^2 \end{bmatrix}$$

$$K_2 = \begin{bmatrix} -c_0 & (m_1 l_1 + m_2 L)\dot{\theta}_1 \sin\theta_1 & m_2 l_2 \dot{\theta}_2 \sin\theta_2 \\ 0 & -c_1 - c_2 & -m_2 l_2 L \dot{\theta}_2 \sin(\theta_1 - \theta_2) + c_2 \\ 0 & m_2 l_2 L \dot{\theta}_1 \sin(\theta_1 - \theta_2) + c_2 & -c_2 \end{bmatrix}$$

$$K_3 = \begin{bmatrix} 0 \\ (m_1 l_1 + m_2 L) g \sin\theta_1 \\ m_2 l_2 g \sin\theta_2 \end{bmatrix}, \quad k_4 = \begin{bmatrix} \beta \\ 0 \\ 0 \end{bmatrix}.$$

An equivalent system of first order equation is obtained by setting $x = \begin{bmatrix} z \\ \dot{z} \end{bmatrix} \in \mathbb{R}^6$

$$\dot{x} = \begin{bmatrix} \dot{z} \\ K_1^{-1}(K_2\dot{z} + K_3 + k_4 u) \end{bmatrix}. \tag{56}$$

Is it possible to stabilize the double pendulum in the upright position? Most readers will find it difficult to decide this question relying only on their physical intuition. In Vol. II we show that if the deviations from the upright position are small it is possible to find a control which restores this position in finite time. Then we prove that this implies the existence of a regulator which makes the upright position a stable equilibrium point of the feedback system. This regulator accepts as input values $r(t)$, $\dot{r}(t)$, $\theta_1(t)$, $\dot{\theta}_1(t)$, $\theta_2(t)$, $\dot{\theta}_2(t)$ and so sensors must determine these values for all $t \geq 0$. Since sensors are expensive one is interested in reducing the number. In particular the question arises whether or not it is possible to design a regulator which accepts as values, say $r(t)$, $\theta_1(t)$. This means that within the regulator it is necessary to reconstruct the angle $\theta_2(t)$ and the velocities $\dot{r}(t)$, $\dot{\theta}_1(t)$, $\dot{\theta}_2(t)$ from the measurements $r(t)$, $\theta_1(t)$. This is a typical *observability* problem, see Vol. II. □

1.3.4 Notes and References

Many books on modelling and dynamics contain chapters on the modelling of mechanical systems, see *Ogata* (1992) [397], *Burton* (1994) [84], *Close and Frederick* (1995) [105]. More realistic automobile suspension systems separating, for example, the motions of the front and the rear axles (see Example 1.3.2) can be found in [84] and [103]. In the presence of rotational friction at the hinge *Antman* (1998) [15] has shown that the derivation of the equations of motion of a compound pendulum may be flawed by the assumption that the reactive force at the hinge acts along the pendulum. Modelling the cart-pendulum system of Example 1.3.4 is discussed in more detail in *Clark* (1995) [103]. *Ackermann* (1977) [2] has analyzed the feedback control of the linearized loading plant. The inverted pendulum has been a favourite example for the illustration of modern control methods in textbooks since the sixties, see *Elgerd* (1967) [150]. The balancing problem was solved by various methods on the basis of the linearized model. However, the swinging up problem, i.e. moving the system from the downward to the upright rest position and keeping it there, requires the use of the nonlinear model. This problem has been studied in *Mori et al.* (1976) [382]. The stabilization of double and multiple pendulum systems has been investigated in *Furuta et al.* (1980) [177], *Maletinsky et al.* (1982) [357] and *Kwakernaak and Westdijk* (1995) [323].
An excellent introduction to Newtonian mechanics is contained in the first volume of the *Feynman Lectures on Physics* (1975) [161]. Brief introductions to Lagrangian and Hamiltonian modelling techniques from an engineering point of view are given in *MacFarlane* (1970) [355], *Wellstead* (1979) [516], *Burton* (1994) [84]. More information concerning variational principles, Lagrangian and Hamiltonian mechanics, can be found in standard textbooks on classical mechanics, *Whittaker* (1970) [520], *Gantmacher* (1975) [184], *Landau and Lifshitz* (1976) [329], *Goldstein* (1980) [194], *Chorlton* (1983), [99]. Lagrange's *Mécanique Analytique* is now available in English [325]. It is shown in [194] that by introducing velocity dependent potentials it is possible, under certain conditions, to include non-conservative generalized forces in Lagrange's equations (38). In particular the Lorentz force (4.12), which in general is not conservative, satisfies these conditions.
The Hamiltonian formulation was proposed by Hamilton in a British Association Report in 1834, although in part it had been anticipated by Lagrange and Poisson in 1809/1810. The Legendre transformation maps functions on a vector space to functions on the dual space: Let $f(x)$ be a convex function of $x \in \mathbb{R}^n$, then the Legendre transform is the function $g : \mathbb{R}^{n*} \to \mathbb{R}$ defined by

$$g(y) = F(y, x(y)) = \max_x F(y,x), \quad F(y,x) = \langle y, x \rangle - f(x), \quad y = \partial f / \partial x.$$

Details can be found in most references on mathematical physics, see e.g. *Courant and Hilbert* (1953) [112]. Modern advanced mathematical treatments of classical mechanics are given in *Arnold* (1978) [18], *Abraham and Marsden* (1978) [1] and *Marsden and Ratiu* (1999) [361]. These standard references develop the mathematical framework for a coordinate free treatment of the configuration space in the general setting of (symplectic) manifolds. [1] and [361] also contain many instructive historical remarks and comments on the literature.

1.4 Electromagnetism and Electrical Systems

This section is divided into two subsections. In the first we give a brief review of some of the historical developments of electromagnetism and describe the basic building blocks of circuits. Then in the second we show how to obtain the equations governing the current flows in networks of circuits.

1.4.1 Maxwell's Equations and the Elements of Electrical Circuits

Some of the most outstanding discoveries of the 19th century were connected with electricity and magnetism and their interaction. As a reference we quote *Richard Feynman* (1975) [161]: "From a long view of the history of mankind–seen from, say, ten thousand years from now–there can be little doubt that the most significant event of the 19th century will be judged as Maxwell's discovery of the laws of electrodynamics". In this subsection we recall some of the electromagnetic experiments that were carried out and indicate how the conclusions drawn from them can be formulated in a mathematical way, i.e. can be cast in the form of equations. We then use these equations in a number of different examples to obtain mathematical models of various electrical circuits and systems.

Maxwell's Equations

If a piece of amber is rubbed with a cloth and then the amber and cloth are separated they are found to attract each other. Such forces are called *electrical forces* and the amber and cloth are said to be electrified or charged with electricity. In 1729 *Gray* [201] discovered that some materials could convey electricity from one place to another. He carried out an experiment with a glass rod connected by a hemp cord of length 400 feet to an ivory ball and was able to electrify the ball by rubbing the glass tube. Further experiments were carried out by *Desaguliers* (1739) [128] who introduced the word *conductors* for those materials which transport electricity easily. *Cavendish* (1776) [93] anticipated Ohm's law, although much of his work was not published until 100 years later in a collection of his papers put together by *Maxwell* (1879) [365]. In 1821 *Ampère* put forward a workable definition of current and invented a galvanometer to measure it. He thought of voltage as the cause and current as the effect and although he knew that there was a relationship between them, he did not realize that, across a resistor, they are directly proportional. This discovery was made by *Ohm* (1826) [398]. He used as a source a thermoelectric battery with strips of copper and bismuth joined at their two ends. He kept one point of contact in boiling water and the other in ice and thereby obtained a stable current in an external circuit C which he connected across the two points of contact. Working with this rather deficient apparatus, Ohm performed a series of carefully devised experiments which established, for this circuit, the law of conduction (now known as *Ohm's law*):

> If I is the current in the circuit, C and V the voltage drop between the two points of contact, then $V = IR$, where R is a constant called the *resistance* which varies with the wire which is used to close the circuit, but does not depend on V or I.

That electric charges exert forces on each other with a magnitude inversely proportional to the square of the distance between them was suspected early in the 18th century. *Benjamin Franklin* (1755) [106] carried out experiments to determine this law and *Robinson* (1769) observed experimentally that the force was proportional to $r^{-2.06}$ where r is the distance between the charges. Although, as in the case of Cavendish (who gave the law as between $r^{-1.98}$ and $r^{-2.02}$), the results were not available universally and were published posthumously in 1822. Unaware of their results *Coulomb* (1785) [124] carried out completely different experiments and put forward the inverse square law and now the discovery is usually attributed to him. In 1936 *Plimpton and Lawton* [414] showed, experimentally, that the force deviated from an inverse square law by less than two parts in one billion.
Coulomb's experiments led to the formulation:

> If a charge of magnitude q is placed at the origin $\mathrm{O} \in \mathbb{R}^3$, then the force on a positive unit charge at a point $\mathbf{r}$ is proportional to $qr^{-2}\hat{\mathbf{r}}$, where $\hat{\mathbf{r}} = \mathbf{r}/\|\mathbf{r}\|$ and $r = \|\mathbf{r}\|$. Moreover the force is one of attraction if $q < 0$ and repulsion if $q > 0$.

Hence if we denote this *electrical force* at the point $\mathbf{r}$ by $\mathbf{E}(\mathbf{r})$ and write the constant of proportionality in the form $(4\pi\epsilon_0)^{-1}$, we have

$$\mathbf{E}(\mathbf{r}) = \frac{q\hat{\mathbf{r}}}{4\pi\epsilon_0 r^2} = -\operatorname{grad}\Phi(\mathbf{r}), \text{ where } \Phi : \mathbb{R}^3 \setminus \{\mathrm{O}\} \to \mathbb{R}, \quad \Phi(\mathbf{r}) = \frac{q}{4\pi\epsilon_0 r}. \tag{1}$$

$\Phi(\mathbf{r})$ is called the *electrostatic potential* at the point $\mathbf{r}$ and the constant of proportionality ϵ_0 the *permittivity* of free space.
The mapping $\mathbf{r} \to \mathbf{E}(\mathbf{r}) = q\,(4\pi\epsilon_0 r^2)^{-1}\,\hat{\mathbf{r}}$ defines a vector field on $\mathbb{R}^3 \setminus \{\mathrm{O}\}$ which assigns to each $\mathbf{r} \in \mathbb{R}^3 \setminus \{\mathrm{O}\}$ the electrical force exerted on a positive unit charge at that point[1]. This vector field is called the electric field of the charge q placed at O.
In the following most of the vector fields we consider will depend upon time.
There are two important quantities associated with a vector field which are used to describe the results of electromagnetic experiments, namely the *flux* and the *circulation*. These terms have their origins in fluid dynamics where the "flux of velocity" through a surface is the net amount of fluid going through the surface per unit time and the "circulation" around some loop is the net rotational motion around it. More generally for any vector field $\mathbf{F}$ the flux of $\mathbf{F}$ through a bounded oriented piecewise smooth surface S is defined by the surface integral[2]

$$\text{Flux of } \mathbf{F} \text{ through the surface } S = \int_S \langle \mathbf{F}, \mathbf{n} \rangle \, dS,$$

where $\mathbf{n}$ is the unit normal defining the orientation of S and $\langle \cdot, \cdot \rangle$ is the standard inner product on $\mathbb{R}^3$. The circulation of $\mathbf{F}$ around an orientated piecewise smooth closed curve C is

$$\text{Circulation of } \mathbf{F} \text{ around the curve } C = \oint_C \langle \mathbf{F}, \mathbf{t} \rangle \, ds,$$

[1]More generally, a vector field on some region $\mathcal{D} \subset \mathbb{R}^3$ is a map $\mathbf{F} : \mathcal{D} \to \mathbb{R}^3$ that assigns to each point $\mathbf{r}$ in its domain a vector $\mathbf{F}(\mathbf{r})$.

[2]In this section we often suppress the dependency of a vector field on space and time to simplify notation. Where necessary, we use either $\mathbf{r}$ or x as a space variable.

where $\mathbf{t}$ is the unit tangent to the curve. For the definition of the above integrals, see [362]. The line integral of a conservative electric field $\mathbf{E}$ along an arbitrary piecewise smooth curve C connecting $A \in \mathbb{R}^3$ to $B \in \mathbb{R}^3$, $\mathcal{V}_{AB} = \int_C \langle \mathbf{E}, \mathbf{t} \rangle \, ds$, is called the *voltage* (or *potential difference*) between the points A and B. We sometimes talk of the voltage of a point and by this we mean the difference in the potential of that point and the potential of an arbitrary established reference point called the *ground state.*

Now suppose that S is an orientated piecewise smooth closed surface in $\mathbb{R}^3$. Then it can be shown that the flux of the electric field $\mathbf{E}$ (given by (1)) through S is

$$\int_S \langle \mathbf{E}, \mathbf{n} \rangle \, dS = \frac{q\omega}{4\pi\epsilon_0}$$

where $\omega = 4\pi$ if O is inside S and $\omega = 0$ if O is outside S.

Let us now consider a continuous distribution of charge with volume charge density $\rho(x)$ in a bounded region $\Omega \subset \mathbb{R}^3$ and suppose that S encloses Ω, then the electric field $\mathbf{E}$ generated by this charge satisfies

$$\int_S \langle \mathbf{E}\,, \mathbf{n} \rangle \, dS = \frac{1}{\epsilon_0} \int_\Omega \rho(x) \, dx = \frac{Q}{\epsilon_0}, \tag{2}$$

where Q is the total charge on Ω and dx is the Lebesgue measure in $\mathbb{R}^3$. By the divergence theorem

$$\int_S \langle \mathbf{E}\,, \mathbf{n} \rangle \, dS = \int_\Omega \operatorname{div} \mathbf{E} \, dx \quad \text{and so} \quad \int_\Omega (\operatorname{div} \mathbf{E} - \rho/\epsilon_0) \, dx = 0.$$

Since this holds for any closed surface S it follows that $\operatorname{div} \mathbf{E} = \rho/\epsilon_0$. If additionally we suppose the electric field is derived from a potential Φ, then

$$\operatorname{div} \mathbf{E} = \rho/\epsilon_0 \quad \text{and we have Poisson's equation} \quad \triangle \Phi = -\rho/\epsilon_0, \tag{3}$$

where $\triangle$ is the Laplacian defined (in Cartesian coordinates) by

$$\triangle = \frac{\partial^2}{\partial x_1^2} + \frac{\partial^2}{\partial x_2^2} + \frac{\partial^2}{\partial x_3^2}. \tag{4}$$

In the case of free space where $\rho = 0$, $\operatorname{div} \mathbf{E} = 0$ and Φ satisfies the Laplace equation $\triangle \Phi = 0$.

An awareness of the existence of magnetized materials can be traced back to the Greeks who were familiar with loadstone and its power to attract iron. Indeed the term magnet came into use because loadstone pieces were found near the ancient Greek city called *Magnesia*[3]. Experiments with magnetic materials were of a much older vintage than those with electricity and the first application of magnetism, the compass, was used in Europe at the end of the twelfth century. *Newton* in *Principia* speculated that the law of force between two magnetic poles was proportional to the inverse cube of the distance between them, and *Michell* (1750) [373] was the first to give the correct law as being an inverse square. Thus if there is a magnetic pole of

[3]Plato in the dialogue *Ion* gives Socrates the words "impelling you like the power in the stone Euripides called the magnet....This stone does not simply attract iron rings, just by themselves; it also imparts to the rings a force enabling them to do the same thing as the stone itself".

strength m at the origin O, the force on a magnetic pole of positive unit strength at a point $\mathbf{r}$ is proportional to $mr^{-2}\hat{\mathbf{r}}$. Hence if we denote this *magnetic force* at the point $\mathbf{r}$ by $\mathbf{B}(\mathbf{r})$, we have

$$\mathbf{B}(\mathbf{r}) = \frac{m\hat{\mathbf{r}}}{4\pi\mu_0^{-1}r^2} = -\operatorname{grad}\Psi(\mathbf{r}), \qquad \Psi(\mathbf{r}) = \frac{m}{4\pi\mu_0^{-1}r}. \tag{5}$$

$\Psi(\mathbf{r})$ is called the *magnetostatic potential* at the point $\mathbf{r}$ and the constant of proportionality μ_0 the *permeability* of free space. The vector field $\mathbf{B} : \mathbf{r} \to \mathbf{B}(\mathbf{r})$ on the domain $\mathcal{D} = \mathbb{R}^3 \setminus \{O\}$ is called the *magnetic field* of the pole of strength m at O. The equations in (5) have the same form as those given in (1) and hence one can develop a theory of magnetostatics in parallel with that of electrostatics, see [151]. However there is an important difference. Whereas positive and negative electric charges can exist separately from each other, magnetic poles cannot. In any volume (no matter how small) the density of North poles is always the same as the density of South poles. So the net volume density must be zero and in analogy with the electric case the corresponding equations to (3) are

$$\operatorname{div}\mathbf{B} = 0 \quad \text{and} \quad \triangle\Psi = 0. \tag{6}$$

Now we leave the static case and consider the dynamic case where charges move and hence generate electric currents. In 1820 *Ørsted* conducted some experiments which showed that a magnetic field can be generated by an electric current flowing in a wire. *Faraday* (1821) [159] also discovered this and the precise relation as enunciated by *Ampère* takes the form:

> The circulation of a magnetic field in a non-magnetic medium around a closed path is equal to μ_0 times the total current flowing through a surface bounded by the path.

Suppose that at a point P with position vector $\mathbf{r}$ the volume charge density of electrons is $\rho(\mathbf{r})$ and their velocity[4] is $\mathbf{v}(\mathbf{r})$, then $\mathbf{j}(\mathbf{r}) = \rho(\mathbf{r})\mathbf{v}(\mathbf{r})$ is defined to be the *current density* at the point $\mathbf{r}$. So if S is an orientated piecewise smooth surface and I is the total current through S, we have

$$I = \int_S \rho\,\langle \mathbf{v}\,,\mathbf{n}\rangle\,dS = \int_S \langle \mathbf{j}\,,\mathbf{n}\rangle\,dS.$$

Hence if C is a closed orientated piecewise smooth curve Ampère's law takes the form

$$\oint_C \langle \mathbf{B}\,,\mathbf{t}\rangle\,ds = \mu_0 I = \mu_0 \int_S \langle \mathbf{j}\,,\mathbf{n}\rangle\,dS,$$

where the the surface S is such that $\partial S = C$. By Stokes' Theorem

$$\oint_C \langle \mathbf{B}\,,\mathbf{t}\rangle\,ds = \int_S \langle \operatorname{curl}\mathbf{B}\,,\mathbf{n}\rangle\,dS \quad \text{and so} \quad \int_S \langle \operatorname{curl}\mathbf{B} - \mu_0\mathbf{j}\,,\mathbf{n}\rangle\,dS = 0.$$

And since this holds for any surface S, we have

$$\operatorname{curl}\mathbf{B} = \mu_0\mathbf{j}. \tag{7}$$

This is the differential form of *Ampère's law*. Note that since div curl $= 0$, the above equation implies $\operatorname{div}\mathbf{j} = 0$ which we will see later is not in general true.

Another major advance was made in 1831 when *Faraday* [159] discovered, experimentally, that a current was induced in a conducting loop when the magnetic field changed. Faraday found that:

[4] Strictly speaking, $\mathbf{v}(\mathbf{r})$ is the *average* velocity of the electrons in a small volume containing P.

> The circulation of the electric field vector around a closed path is equal to the rate of decrease of the magnetic flux flowing through a surface bounded by the path.

The mathematical articulation of this law, now known as *Faraday's law* was first given by *Maxwell.*

$$\mathcal{V} := \oint_C \langle \mathbf{E}\,, \mathbf{t} \rangle \, ds = -\frac{d}{dt} \int_S \langle \mathbf{B}\,, \mathbf{n} \rangle \, dS, \tag{8}$$

where C and S are as above with $\partial S = C$. The circulation $\mathcal{V}$ of the electric field $\mathbf{E}$ around C is called the *induced voltage.* Using Stokes' Theorem, the differential form of Faraday's law is

$$\operatorname{curl} \mathbf{E} = -\frac{\partial \mathbf{B}}{\partial t}. \tag{9}$$

Maxwell, when only 24, set out to put Faraday's experimental work on a firm mathematical footing. The work, including a correction of *Ampère's law* (7) (which allowed for the possibility that $\operatorname{div} \mathbf{j} \neq 0$), culminated in his paper "A dynamical theory of the electromagnetic field" published in (1865) [363]. If ρ is the volume charge density of electrons and $\mathbf{v}$ their velocity, then given an orientated piecewise smooth closed surface S enclosing a volume Ω, conservation requires that the flux of electrons through S must be balanced by their rate of decrease in Ω, i.e.

$$\int_S \langle \rho \mathbf{v}\,, \mathbf{n} \rangle \, dS = -\frac{d}{dt} \int_\Omega \rho \, dx.$$

By the divergence theorem we get

$$\frac{\partial \rho}{\partial t} + \operatorname{div}(\rho \mathbf{v}) = 0.$$

This equation is called the *continuity equation.* Using the first equation in (3) and the fact that $\mathbf{j} = \rho \mathbf{v}$ yields

$$\operatorname{div} \left(\varepsilon_0 \frac{\partial \mathbf{E}}{\partial t} + \mathbf{j} \right) = 0.$$

We have seen that (7) implies $\operatorname{div} \mathbf{j} = 0$ which, in general, contradicts this equation. Maxwell saw that if, however, $\mu_0 \mathbf{j}$ was replaced with $\mu_0 (\varepsilon_0 \frac{\partial \mathbf{E}}{\partial t} + \mathbf{j})$ in (7), then there would be no contradiction. Therefore his equations consist of the first equations of (3) and (6) together with (9) and the adjustment to (7). Hence they take the form

$$\begin{aligned} \operatorname{div} \mathbf{E} &= \rho / \epsilon_0, \\ \operatorname{curl} \mathbf{E} &= -\frac{\partial \mathbf{B}}{\partial t}, \\ \operatorname{div} \mathbf{B} &= 0, \\ \operatorname{curl} \mathbf{B} &= \mu_0 \left(\varepsilon_0 \frac{\partial \mathbf{E}}{\partial t} + \mathbf{j} \right). \end{aligned} \tag{10}$$

Maxwell's hypotheses, together with confirmation of the correction term were substantiated experimentally by *Hertz* (1885) eight years after Maxwell's death.

There is a different version of Faraday's law for the case where the magnetic field $\mathbf{B}$ is constant in time but the wire circuit C is moving with a velocity $\mathbf{v}$. Then the induced voltage, $\mathcal{V}$, is given by

$$\mathcal{V} = \oint_C \langle \mathbf{v} \times \mathbf{B} \,, \mathbf{t} \rangle \, ds. \tag{11}$$

One can interpret the induced voltage $\mathcal{V}$ as being caused by an electric field $\mathbf{E}' = \mathbf{v} \times \mathbf{B}$, so that $\mathcal{V} = \oint_C \langle \mathbf{E}' \,, \mathbf{t} \rangle \, ds$. This suggests that if a charge of magnitude q is moving with velocity $\mathbf{v}$ in both an electrostatic field $\mathbf{E}$ and a magnetic field $\mathbf{B}$, the total force on it, $\mathbf{F}$, will be $q(\mathbf{E} + \mathbf{E}')$, i.e.

$$\mathbf{F} = q(\mathbf{E} + \mathbf{v} \times \mathbf{B}). \tag{12}$$

This is known as *Lorentz's force law* and its validity has been unquestionably established by experiments.

The Elements of Electric Circuits

In electrical engineering an important role is played by *circuits* in which power in the form of currents and fields is channelled by slender conductors (wires) connecting discrete elements. To understand the fine detail of the behaviour of these elements it is necessary to solve Maxwell's partial differential equations. Fortunately most elements are amenable to an adequately accurate, approximate treatment which simplifies the situation enormously. This is called the *lumped parameter* approximation and we now illustrate this with a number of examples.

Example 1.4.1. (Resistor). Consider a conductor made of homogeneous material in the form of a cylinder of length ℓ and cross section area S. It is assumed that the current density $\mathbf{j}$ and the electric field $\mathbf{E}$ within the conducting material are both constant and in the direction of the axis of the cylinder, $\hat{\mathbf{z}}$. A more general version of Ohm's law is $\mathbf{j} = \sigma \mathbf{E}$, where σ is called the *conductivity* of the material, see *Notes and References.* The voltage $\mathcal{V}$ between the ends of the cylinder and the total current I are

$$\mathcal{V} = \int_0^\ell \langle \mathbf{E} \,, \hat{\mathbf{z}} \rangle \, ds = \|\mathbf{E}\| \ell, \qquad I = \int_S \langle \mathbf{j} \,, \mathbf{n} \rangle \, dS = \|\mathbf{j}\| S.$$

Now since $\|\mathbf{j}\| = \sigma \|\mathbf{E}\|$, we have $\mathcal{V} = \left(\frac{\ell}{\sigma S}\right) I = RI$, where $R = \left(\frac{\ell}{\sigma S}\right)$ is the resistance. For example the resistance of a silver wire of length 1.265 m with a circular cross section of radius .048 cm is .0281 ohms.
Joule (1841) [281] reasoned, and then confirmed experimentally, that the energy dissipated as heat when a current I flows in a metallic conductor of resistance R is RI^2. □

Example 1.4.2. (Capacitor). Consider two parallel plates charged with constant charges of equal magnitude but opposite sign. If the distance between the plates is small compared with the size of the plates, the charge will reside almost entirely on the inner surfaces of the plates, the electric field will be zero in the interior of the plates and away from the edges of the plates the electric field between the plates is approximately normal to them. Hence in this region between the plates the potential will only change in a direction x_1 perpendicular to the plates. So Poisson's equation for the potential in Cartesian coordinates reduces to $\Phi_{x_1 x_1} = 0$, where $(\;)_{x_1} = \frac{\partial}{\partial x_1}$. The solution of this equation has the form $\Phi(x_1) = \alpha x_1 + \beta$, where α and β are constants. Suppose the plates are at $x_1 = a$ and $x_1 = b$ and the potentials are constant on each plate and are $\Phi(a) = V_a$ and $\Phi(b) = V_b$, then

$$\Phi(x_1) = \frac{(V_b - V_a)x_1 + bV_a - aV_b}{b - a}.$$

Now consider a closed cylindrical surface S where the axis of the cylinder is in the x_1 direction and the plane ends of area A are at $x_1 = a - \varepsilon$ and $x_1 = a + \varepsilon$ with $\varepsilon << b - a$. If q is the constant surface density of charge (positive on the one at $x_1 = a$ and negative on the other), then the charge enclosed in S is qA. Hence by (2)

$$qA/\epsilon_0 = \int_S \langle \mathbf{E}, \mathbf{n} \rangle \, dS = -\int_{S_1} \langle \operatorname{grad} \Phi, \mathbf{n} \rangle \, dS = \frac{V_a - V_b}{b - a} \int_{S_1} dS = \frac{(V_a - V_b)A}{b - a},$$

where S_1 is the plane surface at $x_1 = a + \varepsilon$. Thus

$$V_a - V_b = Q/C, \qquad C = \frac{A\epsilon_0}{b - a}$$

where Q is the total charge on the plate at $x_1 = a$. So the potential (or voltage) change across the plates is proportional to the charge. The proportionality constant C is called the *capacitance* and such a configuration is called a *capacitor* or *condenser*. The above result neglects fringing of the electric field at the edges of the plates, for a more accurate expression for the capacitance see *Notes and References.*

Now let us consider the electric energy stored in the capacitor. The capacitor is charged by connecting the plates in a circuit with a battery which has the effect of transferring charge from one plate to the other. If a small charge dQ is brought from a position $x_1 = b$ where the potential is $\Phi(b) = V_b$ to a position $x_1 = a$ where potential is $\Phi(a) = V_a$, then the work done is $dW = (V_a - V_b)dQ$. Hence $dW = (Q/C)dQ$ and so the the total work done in charging the capacitor is $W = Q^2/(2C)$, where $\pm Q$ are the final charges on the plates. □

Example 1.4.3. (Inductor). Consider a coil consisting of n turns of wire which are tightly wound on a toroidal frame of rectangular cross section and permeability μ_0. The inner and outer radii of the frame are r_1 and r_2, respectively, the height of the frame is h and there is a current of magnitude $I(t)$, $t \geq 0$ in the conducting wire.

Suppose that cylindrical coordinates are such that the z-axis is the axis of symmetry

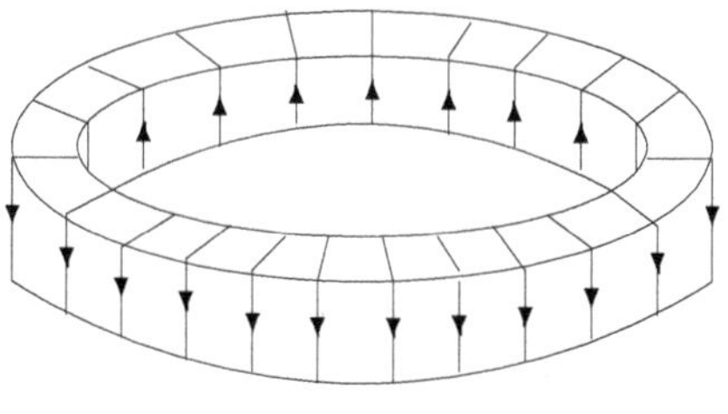

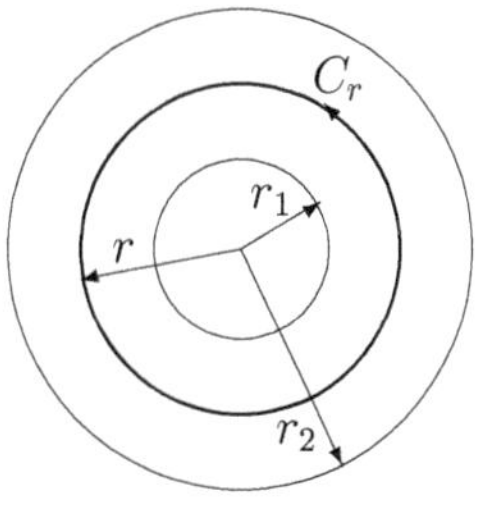

Figure 1.4.1: Toroidal inductor

and the frame is located between $r = r_1$ and $r = r_2$. Let C_r be a circular path within the toroidal frame of radius r with $r_1 < r < r_2$, We assume axial symmetry so that the magnetic field $\mathbf{B} = \mathbf{B}(r, z, t)$ only depends on r, z and t. By Ampère's law applied to the surface of the disk bounded by C_r, we have

$$\mu_0 \, n \, I(t) = \oint_{C_r} \langle \mathbf{B}, \mathbf{t} \rangle \, ds$$

where $\mathbf{t}$ is the unit tangent to C_r. If B_2 is the magnitude of the magnetic field in the direction $\mathbf{t}$, then

$$\oint_{C_r} \langle \mathbf{B}, \mathbf{t} \rangle \, ds = \int_0^{2\pi} B_2(r,z,t) \, ds = 2\pi \, r B_2(r,z,t).$$

So $B_2(r,z,t) = \mu_0 (2\pi\, r)^{-1} n I(t)$. Since the coil is tightly wound around the toroidal frame, every loop approximately traces out the perimeter of a surface which is a rectangular cross section of the frame. Let us apply *Faraday's law* to one such surface S. Then the normal to this surface $\mathbf{n} = \mathbf{t}$. So if $\mathcal{V}(t)$ is the induced voltage

$$\begin{aligned} \mathcal{V}(t) &= -\frac{d}{dt}\int_S \langle \mathbf{B}, \mathbf{n} \rangle \, dS = -\frac{d}{dt}\int_0^h \int_{r_1}^{r_2} B_2(r,z,t) dr dz = -\frac{\mu_0 n}{2\pi}\frac{dI}{dt}(t) \int_0^h \int_{r_1}^{r_2} r^{-1} dr dz \\ &= -\frac{\mu_0 n h}{2\pi} \ln{(r_2/r_1)} \frac{dI}{dt}(t). \end{aligned}$$

Since there are n such coils, if V is the total voltage dropped, we have

$$V(t) = L\dot{I}(t), \qquad L = \frac{\mu_0 n^2 h}{2\pi} \ln{(r_2/r_1)}.$$

The constant L is called the *inductance* and such a configuration is called an *inductor.* Now let us consider the magnetic energy stored in the inductor. Each charge in the wire is receiving energy at a rate $\langle \mathbf{E}, \mathbf{v} \rangle$ where $\mathbf{E}$ is the force on it and $\mathbf{v}$ is its velocity. So that if ρ is the density of charge per unit length the rate of doing work on the coil is

$$\frac{dW}{dt} = \oint_{coil} \langle \mathbf{E}, \mathbf{v} \rangle \rho \, ds = \oint_{coil} \langle \mathbf{E}, \mathbf{j} \rangle \, ds = I \oint_{coil} \langle \mathbf{E}, \mathbf{t} \rangle \, ds = VI = LI\frac{dI}{dt},$$

by (8). So we see that the energy required to build up the current I in the inductor is $W = (L/2)I^2$. □

Symbol	Constitutive Law	Variables
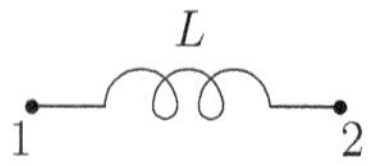	$V_1 - V_2 = L\dot{I}$	voltage change across an inductor of inductance L with a current I
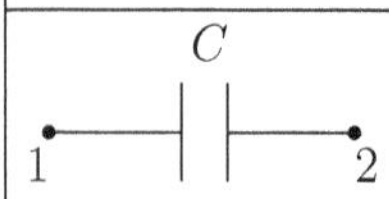	$V_1 - V_2 = Q/C$	voltage change across a capacitor of capacitance C with a charge Q on one plate and $-Q$ on the other
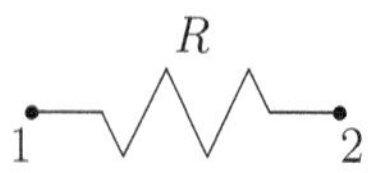	$V_1 - V_2 = IR$	voltage change across a resistor of resistance R with a current I

Table 1.4.2: Symbols and constitutive laws of a resistor, capacitor and inductor

Inductors, capacitors and resistors are the classical elements of electric circuits. Their symbols and constitutive laws are shown in Table 1.4.2. They are, respectively, the counterparts of masses, springs and dampers in mechanical systems. This correspondence is shown in Table 1.4.3 where M, k, c are the mass, spring constants and damping coefficient respectively, F is force, y displacement, v velocity and $\mathcal{T}, \mathcal{W}, \mathcal{D}$ are the kinetic energy of the mass, the potential energy of the spring and the energy

mass	inductor	spring	capacitor	damper	resistor
M	L	k	$1/C$	c	R
F	$V_1 - V_2$	F	$V_1 - V_2$	F	$V_1 - V_2$
v	I	y	Q	v	I
$F = M\dot{v}$	$V_1 - V_2 = L\dot{I}$	$F = ky$	$V_1 - V_2 = Q/C$	$F = cv$	$V_1 - V_2 = IR$
$\mathcal{T}=(M/2)v^2$	$W=(L/2)I^2$	$\mathcal{W}=(k/2)y^2$	$W=Q^2/(2C)$	$\mathcal{D}=cv^2$	$W=RI^2$

Table 1.4.3: Table of corresponding quantities

dissipated by the damper. For example in the first column mass corresponds to inductance, the force on the mass corresponds to the voltage change across the inductor and the velocity of the mass corresponds to the current in the inductor. The last two rows gives the corresponding constitutive laws and energies of the elements. The correspondence given in Table 1.4.3 is called the Force-Voltage analogy. There is also a Force-Current analogy, see [397]. These analogies suggest that the variational method described for mechanical systems in the previous section can also be applied to electrical systems, and indeed this is the case, see *Notes and References* and the following example.

Example 1.4.4. (Linear RLC circuit). Consider the circuit driven by a voltage source $e(t)$ as in Figure 1.4.4. The corresponding mechanical system is given on the left hand side of Figure 1.3.3. A systematic way of determining the laws of motion for the circuit will be explained in the next subsection. There we will see that by Kirchhoff's

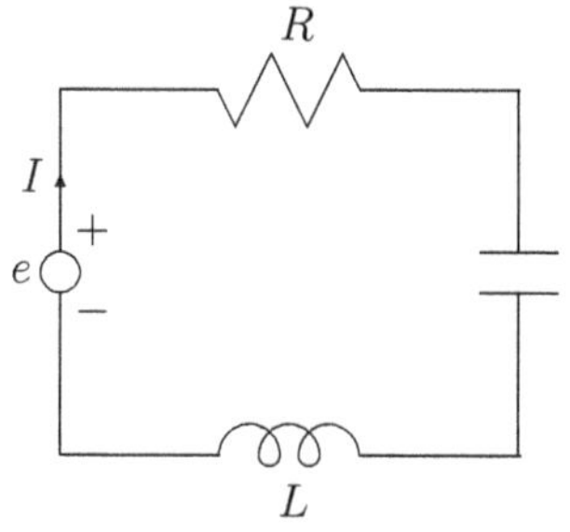

Figure 1.4.4: *RLC* circuit

voltage law the sum of the voltages around the closed circuit is zero. To be more precise if the current is in the direction indicated in Figure 1.4.4, then there will be a drop in the voltage across each of the elements. And Kirchhoff's law states that the total voltage drop across these elements must be balanced by that supplied by the voltage source. So if the voltage across the resistor, capacitor and inductor at time t are $V_R(t)$, $V_C(t)$, $V_L(t)$, respectively, we have

$$e(t) - V_R(t) - V_C(t) - V_L(t) = 0, \quad t \geq 0.$$

But if the current around the circuit at time t is $I(t)$ and the charge on the capacitor is $Q(t)$, then

$$V_R(t) = I(t)R, \quad V_C(t) = Q(t)/C, \quad V_L(t) = L\dot{I}(t), \quad I(t) = \dot{Q}(t), \quad t \geq 0.$$

Hence

$$e(t) = L\ddot{Q}(t) + R\dot{Q}(t) + Q(t)/C, \quad t \geq 0.$$

If $\mathcal{T} = (L/2)\dot{Q}^2$ is the magnetic energy of the inductor, $\mathcal{W} = Q^2/(2C)$ the electric energy of the capacitor, $\mathcal{D} = R\dot{Q}^2$ the energy dissipated by the resistor, $F(t) = e(t)$ and $L = \mathcal{T} - \mathcal{W}$, then the above equation can be obtained directly via the variational method by writing down Lagrange's equation (3.39).

$$\frac{d}{dt}\left(\frac{\partial L}{\partial \dot{Q}}(Q(t), \dot{Q}(t))\right) - \frac{\partial L}{\partial Q}(Q(t), \dot{Q}(t)) + \frac{1}{2}\frac{\partial \mathcal{D}}{\partial \dot{Q}}(\dot{Q}(t)) = F(t).$$

Setting $x_1 = Q$, $x_2 = \dot{Q}$, we can re-write the equation of motion as a system of first order equations, namely

$$\begin{bmatrix} \dot{x}_1 \\ \dot{x}_2 \end{bmatrix} = \begin{bmatrix} 0 & 1 \\ -1/LC & -R/L \end{bmatrix} \begin{bmatrix} x_1 \\ x_2 \end{bmatrix} + \begin{bmatrix} 0 \\ 1/L \end{bmatrix} e.$$

Suppose we are interested in determining the charge on the capacitor. It is difficult to measure the charge directly, so we may ask whether or not it is possible to determine the charge by measuring the current $I = \dot{Q} = x_2$. Setting $y = x_2 = [0\ 1]x$, this is an observability problem: given the observation $y(\cdot)$ and the input $e(\cdot)$ on some time interval, is it possible to determine the state $x(\cdot)$? □

In the following example we illustrate how the Lorentz force law (12) can be used to describe the interaction between electromagnetic forces and mechanical motion.

Example 1.4.5. (Loudspeaker). A loudspeaker is an electromechanical system in which the mechanical part is a loudspeaker diaphragm. Electromagnetic forces are used to make the diaphragm move and the consequent motion generates sound which is then transmitted through the air to the ear. Basically a signal from a tape, record, or disk generates an input voltage $e(t)$ in a circuit. Part of this circuit is in the form of a coil within a fixed permanent magnet. The motion of the electric charges in the coil interacts with the magnetic field generated by the magnet to produce a Lorentz force as given by (12). Since the speaker diaphragm is rigidly attached to the coil this force on the coil causes the diaphragm to move. The whole idea is that the diaphragm motion which produces the sound should be proportional to the original input signal. An idealized model is given in Figure 1.4.5.

The magnet is cylindrical with an inner solid cylindrical core which is the South pole and an outer concentric cylindrical shell which is the the North pole. It is assumed that this configuration results in a radial magnetic field in the air gap between the North and South poles directed to the axis of the magnet. In the figure the magnet is shown as dotted rectangles with small dots. The diaphragm is on the right of the figure and is shown as a rectangle with small circles, whereas the coil, which is rigidly connected to the diaphragm, is situated in the air gap between the North and South poles and is represented by small black circles inside other circles. It consists of n turns of wire each of which is at a distance a from the axis of the magnet. The motion of the diaphragm is modelled as an oscillator with mass m, damping c and stiffness k whereas the electric circuit is modelled as one which contains a resistor with resistance R and an inductor with inductance L. Suppose (r, θ, z) are cylindrical coordinates where the z-axis is along the central axis of the magnet directed from the magnet to the diaphragm. It is assumed that the diaphragm and coil are constrained so that only motion in the z direction is allowed. Then if $F(t)$ is the

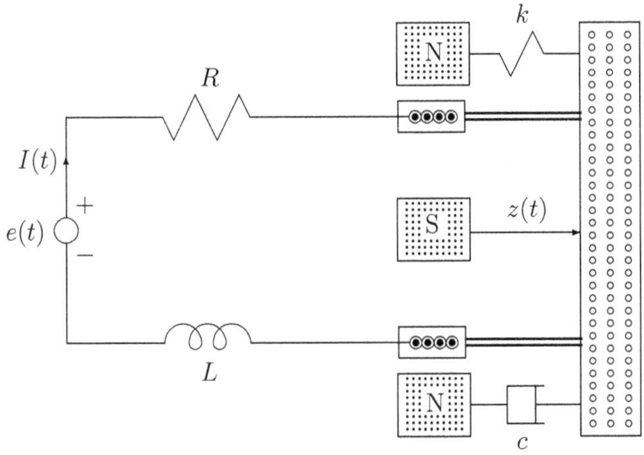

Figure 1.4.5: Loudspeaker

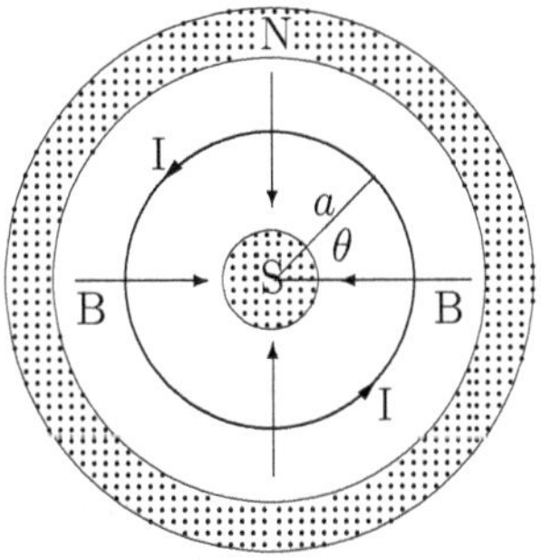

Figure 1.4.6: Magnet-coil geometry

component of the Lorentz force on the coil in this direction, since the coil and diaphragm are rigidly connected, the mechanical equation of motion of the diaphragm is

$$m\ddot{z}(t) + c\dot{z}(t) + kz(t) = F(t).$$

And if $V(t)$ is the voltage induced as a consequence of the motion of the coil in the magnetic field, then just as in Example 1.4.4, one obtains the following equation of motion for the current $I(t)$ in the circuit

$$L\dot{I}(t) + RI(t) = e(t) + V(t).$$

In order to complete the picture we find expressions for the terms $F(t)$ and $V(t)$. The magnitude of the magnetic field $\mathbf{B}$ at $r = a$ is denoted by B and it is assumed to be independent of z, θ and t. If at a point in the coil parametrized by an angle θ, $\theta \in [0, 2n\pi)$ there is a charge $q(t,\theta)$ which has a velocity $\mathbf{v}(t,\theta)$ the Lorentz force on it is $q\mathbf{v} \times \mathbf{B}$. The velocity $\mathbf{v}$ has a component $\mathbf{v_1}$ in the direction of the z-axis, but since $\mathbf{v_1} \times \mathbf{B}$ is perpendicular to the z-axis it will not make any contribution to $F(t)$. The other component of the charge's velocity is due to the movement of the charge around the coil. If its magnitude at (t,θ) is $v_2(t,\theta)$, then the magnitude of the Lorentz force F_2 in the direction of the z-axis is

$$F_2(t,\theta) = q(t,\theta)v_2(t,\theta)B.$$

Hence

$$\frac{d}{d\theta}F_2(t,\theta) = B\frac{d}{d\theta}(q(t,\theta)v_2(t,\theta)) = aBI(t).$$

So the total force in the direction of the z-axis is $F(t) = 2n\pi aBI(t)$. We see therefore that our mathematical model for the motion of the diaphragm is that of an oscillator driven by a force proportional to the current $I(t)$ in the coil.
The induced voltage $V(t)$ in the circuit is due to the motion of the coil in the z direction. In order to determine it we apply Faraday's law (11) with C being one loop of the coil in the magnet. If $\mathbf{t}$ is the unit tangent to C, we have

$$\mathcal{V} = \oint_C \langle \mathbf{v_1} \times \mathbf{B}\,, \mathbf{t}\rangle\, ds = -aB\dot{z}\int_0^{2\pi} d\theta = -2\pi aB\dot{z}.$$

And since the coil consists of n turns of wire the total induced voltage is given by $V(t) = -2n\pi aB\dot{z}(t)$. Setting $x_1 = z$, $x_2 = \dot{z}$, $x_3 = I$, we obtain the following state space system

$$\begin{bmatrix} \dot{x}_1 \\ \dot{x}_2 \\ \dot{x}_3 \end{bmatrix} = \begin{bmatrix} 0 & 1 & 0 \\ -k/m & -c/m & 2n\pi aB/m \\ 0 & -2n\pi aB/L & -R/L \end{bmatrix} \begin{bmatrix} x_1 \\ x_2 \\ x_3 \end{bmatrix} + \begin{bmatrix} 0 \\ 0 \\ 1 \end{bmatrix} e.$$

Suppose $y(t) = x_1(t) = z(t)$, then the design problem is that of choosing some or all of the parameters k, c, L, R, n, a, B so that $y(t)$ approximates the input $e(t)$ for all $t \geq 0$. □

1.4.2 Electrical Networks

In this subsection we give a brief account of how graph theoretical methods are used to obtain models of interconnected electrical systems. We will only consider electrical networks consisting of voltage sources, current sources, resistors, inductors and capacitors. For the modelling of more general networks and more detail of network methods, see *Notes and References.*
Determining the differential equations which govern a complicated network can be quite difficult. Nowadays it is common to use computer aided modelling procedures. These are based on a graph theoretical representation of the electrical network. Before describing the details we recall some basic facts from graph theory.
A directed graph $G = (V, E, \varphi)$ consists of a finite vertex set V, a finite edge set E and an *incidence map*

$$\varphi : E \to V^2, \qquad e \to \varphi(e) = (\varphi_1(e), \varphi_2(e)).$$

If $\varphi(e) = (v_1, v_2)$, then one calls v_1 the *initial* vertex and v_2 the *terminal* vertex of the edge e. Equivalently, edge e is said to be directed from v_1 to v_2. For a vertex $v \in V$ it is useful to define the sets of edges with initial and terminal vertex $v \in V$, viz.

$$\begin{aligned} E(v,\cdot) &= \varphi_1^{-1}(v) = \{e \in E;\, \varphi_1(e) = v\} \\ E(\cdot,v) &= \varphi_2^{-1}(v) = \{e \in E;\, \varphi_2(e) = v\}. \end{aligned} \tag{13}$$

The cardinalities of these sets are called the *out-degree*, $d_{out}(v)$, and *in-degree*, $d_{in}(v)$, of v, respectively. Then $d(v) = d_{out}(v) + d_{in}(v)$ is the total number of edges incident

on v and is called the *degree* of v.
A *path* of length $r \geq 1$ is a sequence $\underline{e} = (e_1, e_2, ..., e_r) \in E^r$ with the property

$$\varphi_1(e_{i+1}) = \varphi_2(e_i) =: v_{k_i}, \quad 1 \leq i \leq r-1.$$

$v_{k_0} := \varphi_1(e_1)$ is called the initial vertex and $v_{k_r} := \varphi_2(e_r)$ is called the terminal vertex. So one may equally think of $\underline{e}$ as a path from v_{k_0} to v_{k_r}. An *elementary* path is one in which $v_{k_1}, ..., v_{k_r}$ are distinct and a *cycle* is an elementary path with $v_{k_0} = v_{k_r}$. The directed graph G is said to be *strongly connected*[5] if for any two distinct vertices $v, v' \in V$ there exists a path from v to v'.
In our application to networks we do not always want the direction associated with an edge to play a role. A succinct way of achieving this with the above set up is to define for every edge $e \in E$ an additional edge $-e$ which is directed from the terminal vertex of e to the initial vertex of e. Let $-E$ denote the set of these additional edges. Then $E \cap -E = \emptyset$. φ is extended to an incident map $\tilde{\varphi}$ on $\tilde{E} = E \dot{\cup} -E$ by setting $\tilde{\varphi}(-e) = (\varphi_2(e), \varphi_1(e))$ for $e \in E$. This results in the graph $\tilde{G} = (V, \tilde{E}, \tilde{\varphi})$.
A graph $G' = (V', E', \varphi')$ is called a subgraph of $G = (V, E, \varphi)$ if $V' \subset V$, $E' \subset E$ and $\varphi' = \varphi\,|_{E'}$. A *spanning subgraph* of $G = (V, E, \varphi)$ is a subgraph $G' = (V, E', \varphi\,|_{E'})$ with the same vertex set as G. It is a *proper* spanning subgraph of G if $E' \neq E$.
Finally a *cut-set* C of G is a set of edges in E such that if all the edges $c, -c$ with $c \in C$ are removed from the graph $\tilde{G}$, the resulting graph decomposes into two strongly connected graphs (one of these may consist of a single vertex).
For the graph $\tilde{G}$ we need the concept of a subgraph which inherits the "undirected" structure of $\tilde{G}$. We say that $G' \subset \tilde{G}$ is a *symmetric subgraph* if it is a subgraph whose edge set E' has the following property: $e \in E' \Leftrightarrow -e \in E'$ for all $e \in E$. A tree in $\tilde{G}$ is a minimal symmetric strongly connected subgraph of $\tilde{G}$ (or equivalently, a symmetric strongly connected subgraph without non-trivial cycles). One can show that a symmetric subgraph of $\tilde{G}$ is a tree with n vertices if and only if it has $2(n-1)$ edges. Moreover, if one adds one edge $\tilde{e} \in \tilde{E}$ to a tree this creates exactly one cycle. A spanning tree in $\tilde{G}$ is a spanning subgraph which is a tree in $\tilde{G}$. One can show that $\tilde{G}$ always contains a spanning tree if it is strongly connected.
Now suppose that $\tilde{G}$ is strongly connected. Given a spanning tree T of $\tilde{G}$, any cycle obtained by adding to T an edge of the graph $\tilde{G}$ which is not an edge of the tree is called a *fundamental cycle* of $\tilde{G}$ (with respect to the given spanning tree).
In electrical networks there are no self-loops, i.e. there are no edges with the property that $\varphi_1(e) = \varphi_2(e)$. The constitutive laws of the elements in the network are assumed to be the ones given in Table 1.4.2 and the resistances, capacitances and inductances of the connecting wires are neglected. A directed graph of the network is defined by replacing every element (resistor, inductor, capacitor, voltage and current sources) by an edge and the junction points of the wires (where the elements are connected together) by a vertex. Let E be the corresponding set of edges (network elements) and V the set of vertices (junction points). If a network element e joins the junction points v and v', we may choose the direction of the edge e arbitrarily

[5] The directed graph G is called *connected* if the extended graph $\tilde{G}$ is strongly connected. Note that a directed graph consisting of just two vertices and one edge between them, is connected but not strongly connected.

by setting either $\varphi(e) := (v, v')$ or $\varphi(e) := (v', v)$. The incidence map $\varphi : E \to V^2$ is defined by choosing one of these two possibilities for each edge $e \in E$. With these specifications we obtain a directed graph $G = (V, E, \varphi)$ representing the electrical network. Associated with each edge $e \in E$ are two time-varying weighting functions $I_e(\cdot), V_e(\cdot) : [0, \infty) \to \mathbb{R}$; the current and voltage across the element which is represented by the edge. The direction of each edge is taken as reference direction for the current and the voltage drop. This is not a restriction, since negative values of I_e and V_e are allowed. However it does mean that I_e and V_e have the same sign. Since we also consider the graph $\tilde{G}$ we have to associate with each edge $-e \in -E$ a current and voltage and it is natural to set $I_{-e} = -I_e$ and $V_{-e} = -V_e$, respectively. In assembling an electrical network by interconnecting various elements there are constraints on the currents and voltages given by *Kirchhoff laws.* The *current law* can be expressed in terms of the graph G whereas we need the extended graph $\tilde{G}$ in order to state the *voltage law.*

Kirchhoff's current law states that the net current flow in and out of every vertex at the time t is zero, i.e.

$$\sum_{e \in E(\cdot, v)} I_e(t) - \sum_{e \in E(v, \cdot)} I_e(t) = 0, \quad t \geq 0, \quad v \in V. \tag{14}$$

Here $E(v, \cdot)$ and $E(\cdot, v)$ are defined by (13).
Since the current in the edge $-e \in -E$ is by definition $-I_e$ we could also have expressed the current law for the graph $\tilde{G}$ with the result that the LHS of (14) would have doubled.

Kirchhoff's voltage law states that the total voltage drop around every cycle in $\tilde{G}$ must be zero, i.e. if $\underline{e} = (e_1, e_2, ..., e_r)$ is a cycle in $\tilde{E}$, then

$$\sum_{j=1}^{r} V_{e_j}(t) = 0, \quad t \geq 0. \tag{15}$$

Suppose Kirchhoff's current law is written down for each vertex of G and we are given a cut-set for this graph. If we sum up the equations for all the vertices in either of the two subgraphs of G defined by the cut-set, only those currents entering or leaving the subgraph remain since the others cancel. So for the currents in the edges of the cut-set we have:

Cut-set condition: The sum of the currents entering one of the two subgraphs of G defined by a cut-set must equal the sum of the currents leaving it.

This version of Kirchhoff's current law is applied to each cut-set in E. Then Kirchhoff's voltage law is applied to each cycle in $\tilde{G}$. The resulting equations together with the constitutive laws of the elements are used to obtain a dynamical model for the electrical network. However there is a certain amount of redundancy if Kirchhoff's laws are applied to every cut-set and every cycle, in the sense that some of the equations are linearly dependent. Moreover it is not clear which variables should be eliminated and which ones retained in order to get a dynamical system model. Engineers have devised methods for overcoming these problems by means of a judicious choice of cut-sets, cycles and state space variables, see *Notes and References.* They recommend the following:

(C1) Select a spanning tree of the graph $\tilde{G}$ so that it contains all resistors, no current sources, and has as many capacitors and as few inductors as possible. In general these last two aims may be contradictory and a compromise must be made. For each edge in $e \in E$ of the tree, find a cut-set[6] (a subset of E) which contains the edge but no other edge in E of the spanning tree. Then for each such cut-set write down the equation determined by the corresponding *cut-set condition.*

(C2) For every fundamental cycle obtained by adding to the spanning tree any edge of $\tilde{G}$ write down Kirchhoff's voltage law (15).

(C3) For every edge of the graph write down the constitutive law of the corresponding element of the network.

(C4) Choose the charges on the capacitors and the currents through inductors which appear in the equations obtained by (C1), (C2) and (C3) as state space variables and eliminate all the others.

Example 1.4.6. Consider the network shown in Figure 1.4.7. The vertices of the asso-

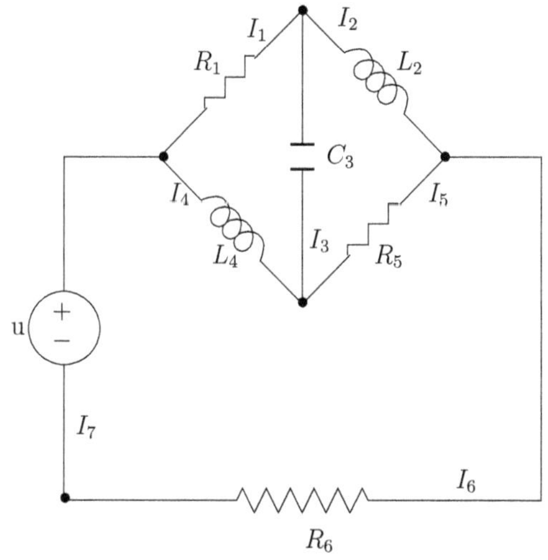

Figure 1.4.7: RLC Network

ciated graph correspond to the junction points marked with a • in Figure 1.4.8 and the edges correspond to the network elements (1 capacitor, 2 inductors, 3 resistances and a voltage source). Directions for the edges are chosen arbitrarily and the choice we have made is shown in the directed graph on the left of Figure 1.4.8. The extended graph $\tilde{G}$ is obtained from G by eliminating the arrowheads on the edges in G. Thus every line segment in the right hand graph of Figure 1.4.8 stands for a pair of edges $\{e_i, -e_i\}$ of $\tilde{G}$. There are many spanning trees of the graph $\tilde{G}$, e.g. $\{\pm e_1, \pm e_3, \pm e_5, \pm e_6\}$, $\{\pm e_3, \pm e_5, \pm e_6, \pm e_7\}$, and $\{\pm e_6, \pm e_7, \pm e_4, \pm e_3\}$. Guided by (C1) we choose to work with $\{\pm e_1, \pm e_3, \pm e_5, \pm e_6\}$ since

[6]The cut-set is uniquely determined. It consists of e together with all those edges in E which connect a vertex of one of the subgraphs with a vertex of the other.

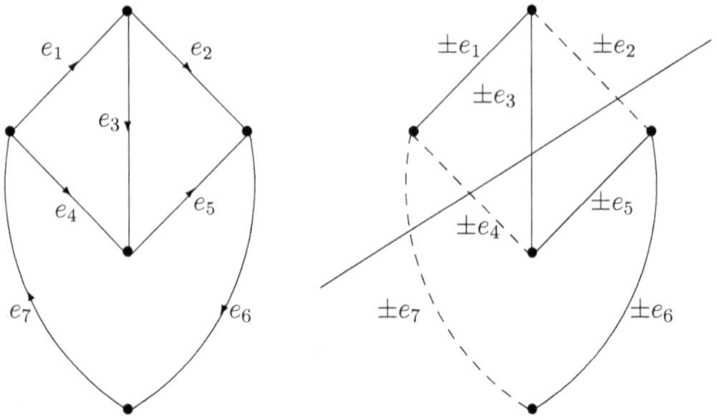

Figure 1.4.8: Directed graph G and spanning tree of $\tilde{G}$

it contains all resistors, the capacitor and no inductor. This tree is drawn with continuous edges in the right hand figure of Figure 1.4.8 whereas all other edges of the graph are dashed. The cut-set containing edge e_3, is $\{e_3, e_2, e_4, e_7\}$ and by the cut-set condition,

$$I_2 + I_3 + I_4 = I_7. \tag{16}$$

The cut-set containing edge e_1 is $\{e_1, e_4, e_7\}$, so

$$I_1 + I_4 = I_7, \tag{17}$$

The cut-set containing edge e_5 is $\{e_5, e_2, e_7\}$, so

$$I_5 + I_2 = I_7. \tag{18}$$

Finally the cut-set containing edge e_6 is $\{e_6, e_7\}$, so

$$I_6 = I_7. \tag{19}$$

Guided by (C2) we have to find the fundamental cycles in $\tilde{G}$ associated with the spanning tree $\{\pm e_1, \pm e_3, \pm e_5, \pm e_6\}$. These are $(-e_2, e_3, e_5)$ and the reverse cycle $(e_2, -e_5, -e_3)$, $(-e_4, e_1, e_3)$ and the reverse cycle $(e_4, -e_3, -e_1)$, $(e_1, e_3, e_5, e_6, e_7)$ and the reverse cycle $(-e_1, -e_7, -e_6, -e_5, -e_3)$. Applying Kirchhoff's voltage law to these cycles we have

$$\begin{aligned} -V_2 + V_5 + V_3 &= 0, & (20)\\ -V_4 + V_3 + V_1 &= 0, & (21)\\ V_1 + V_3 + V_5 + V_6 &= u. & (22) \end{aligned}$$

The equations (16)-(19) and (20)-(22) are augmented with the constitutive laws

$$V_1 = I_1 R_1, \;\; V_2 = L\dot{I}_2, \;\; V_3 = Q_3/C_3, \;\; V_4 = L\dot{I}_4, \;\; V_5 = I_5 R_5, \;\; V_6 = I_6 R_6. \tag{23}$$

We now follow (C4) and choose I_2, Q_3, I_4 as state variables and eliminate all the other variables in (16)-(23). To this end, from (16)–(19) we have

$$\begin{bmatrix} I_1 \\ I_5 \\ I_6 \end{bmatrix} = \begin{bmatrix} 1 & 1 & 0 \\ 0 & 1 & 1 \\ 1 & 1 & 1 \end{bmatrix} \begin{bmatrix} I_2 \\ I_3 \\ I_4 \end{bmatrix}$$

and substituting in (20)–(22) and using the expressions for V_1, V_5, V_6 in (23) yields

$$\begin{bmatrix} 1 & -R_5 & 0 \\ 0 & -R_1 & 1 \\ 0 & R & 0 \end{bmatrix} \begin{bmatrix} V_2 \\ I_3 \\ V_4 \end{bmatrix} = \begin{bmatrix} 0 & 1 & R_5 \\ R_1 & 1 & 0 \\ -(R_1+R_6) & -1 & -(R_5+R_6) \end{bmatrix} \begin{bmatrix} I_2 \\ V_3 \\ I_4 \end{bmatrix} + \begin{bmatrix} 0 \\ 0 \\ 1 \end{bmatrix} u\,,$$

where $R = R_1 + R_5 + R_6$. But

$$\begin{bmatrix} 1 & -R_5 & 0 \\ 0 & -R_1 & 1 \\ 0 & R & 0 \end{bmatrix}^{-1} = R^{-1} \begin{bmatrix} R & 0 & R_5 \\ 0 & 0 & 1 \\ 0 & R & R_1 \end{bmatrix}$$

and hence

$$R \begin{bmatrix} V_2 \\ I_3 \\ V_4 \end{bmatrix} = \begin{bmatrix} -R_5(R_1+R_6) & (R_1+R_6) & R_1R_5 \\ -(R_1+R_6) & -1 & -(R_5+R_6) \\ R_1R_5 & R_5+R_6 & -R_1(R_5+R_6) \end{bmatrix} \begin{bmatrix} I_2 \\ V_3 \\ I_4 \end{bmatrix} + \begin{bmatrix} R_5 \\ 1 \\ R_1 \end{bmatrix} u\,.$$

Then using (23) and $I_3 = \dot{Q}_3$ we get

$$R \begin{bmatrix} \dot{I}_2 \\ \dot{Q}_3 \\ \dot{I}_4 \end{bmatrix} = \begin{bmatrix} -R_5(R_1+R_6)/L_2 & (R_1+R_6)/L_2C_3 & R_1R_5/L_2 \\ -(R_1+R_6) & -1/C_3 & -(R_5+R_6) \\ R_1R_5/L_4 & (R_5+R_6)/L_4C_3 & -R_1(R_5+R_6)/L_4 \end{bmatrix} \begin{bmatrix} I_2 \\ Q_3 \\ I_4 \end{bmatrix} + \begin{bmatrix} R_5 \\ 1 \\ R_1 \end{bmatrix} u\,.$$

This is the dynamical model for the given RLC network obtained by following the guidelines (C1)-(C4). Note that

$$\begin{bmatrix} I_2 \\ Q_3 \\ I_4 \end{bmatrix} = \begin{bmatrix} 0 & -1 & 1 \\ -R_1C_3 & -R_5C_3 & -R_6C_3 \\ -1 & 0 & 1 \end{bmatrix} \begin{bmatrix} I_1 \\ I_5 \\ I_6 \end{bmatrix} + \begin{bmatrix} 0 \\ C_3 \\ 0 \end{bmatrix} u\,.$$

Using this transformation we could also write down differential equations for I_1, I_5, I_6 in violation of the guidelines. In this case both u and its derivative $\dot{u}$ will appear on the RHS. We will later restrict our analysis to dynamical models which do not contain derivatives of input variables. So without further modification the dynamical model in terms of I_1, I_5, I_6 will not fit this specification. □

1.4.3 Notes and References

A classical reference on electromagnetic theory is the book of *Elliott* (1966) which has been republished as an IEEE reprint, see [151]. Its main features are the historical material in each chapter and the development, via special relativity, of a complete electromagnetic theory. We also recommend the Lecture Notes on Physics by *Feynman* (1975) [161].
The more general version of Ohm's law and the effect of the fringing of the electric field on the capacitor considered in Example 1.4.2 can be found in [151]. A good book on vector fields developed through its application to engineering is *Shercliff* (1977) [463]. As an elementary mathematical introduction to Vector Analysis we recommend the textbook of *Marsden and Tromba* (1996) [362]. For a discussion of the modelling of electrical and electromechanical systems, see *Ogata* (1992) [397], *Burton* (1994) [84], *Close and Frederick* (1995) [105] and for references on electrical circuits see e.g. *Johnson et al.* (1992) [278] and *Wellstead* (1979) [516]. A comprehensive account of graph theory is contained in *Thulasiraman and Swamy* (1992) [495]. A concise description of how to use graph theoretical tools for the modelling of electrical networks can be found in *Zerz* (2000) [545], see also [278] and [516].

1.5 Digital Systems

In recent years, due mainly to the simultaneous dramatic improvement and reduction in cost of digital hardware, digital systems have become all pervasive in technology. They form a class of dynamical systems with quite distinctive features and therefore special engineering and mathematical disciplines have been developed for their analysis and design: "Theory of Switching Networks", "Automata Theory", "Logic Design", see *Notes and References.* Although these areas are not subjects of this book, it is appropriate to discuss some examples and special features of digital systems since they are not only an important class of dynamical systems in themselves but are also increasingly used in the control and measurement of analog signals and systems. Indeed many analog devices in signal processing, filtering and control have been replaced by digital counterparts which are often cheaper, more robust and more reliable.

The essential difference between analog and digital systems is that in the former ones input, output and internal state variables take on a continuous range of values whereas in the latter ones there are only a finite number of input, output and state values. Most digital systems are *binary*, i.e. their input, output and state variables take only two different values, "on" and "off". Physically these values may be encoded by different voltages (e.g. 5 volts versus 0 volts), by the flow or non-flow of an electrical current or by magnetic polarization (North and South). Mathematically the "on" and "off" values are usually represented by 1 and 0, the elements of the simplest nontrivial Boolean algebra $\mathbb{B} = \{0, 1\}$ or, alternatively, the binary field $\mathbb{Z}_2 = \mathbb{Z}/(2)$.

Because of its binary components a digital system is often viewed as a network of switches which operates in discrete time $t \in \mathbb{N}$ or $\mathbb{Z}$. There are two basic classes.

- *Combinational switching networks* are those whose current outputs depend only on the current inputs. Dynamical systems with this property are called *memoryless*, they transform the inputs directly into outputs without intermediate storage of energy or information. Physically, the output changes a short time after the input changes, but this short time delay is neglected in the mathematical description of the digital system. By convention the "current" input at time $t \in \mathbb{Z}$, $u(t)$, determines the "current" output, $y(t)$.[1] If such a combinational network has m input and p output channels its behaviour is completely described by a function F mapping the 2^m possible input vectors $u(t) \in \mathbb{B}^m$ into the corresponding output vectors $y(t) = F(u(t)) \in \mathbb{B}^p$.

- *Sequential switching networks* or *finite state machines* are those digital systems whose current outputs depend not only on the current inputs but also on the sequence of previous inputs. Such systems (for example a digital clock or a computer) contain memory elements in which information about the history of previous inputs is stored. The (binary) contents of all its, say n, memory elements form together a binary vector $x \in \mathbb{B}^n$ which is called the *state* of

[1] Alternatively, one could redefine the time dependence of the output function in such a way that the present input $u(t)$ determines the *next* output $y(t+1)$. In fact, this alternative convention is usually chosen in the mathematical description of *sequential* networks, see Example 1.5.2.

the system. The current state and the current input together determine the current output and the next state of the system. The behaviour of a sequential switching network is therefore described by two maps, which determine the current output and the next state as functions of the current input and the current state of the system. This is in contrast with combinational switching networks where there is no need to introduce the notion of state.

Before describing some elementary building blocks of these two types of digital systems we illustrate the difference between combinational and sequential switching networks by two examples.

Example 1.5.1. (Half and full adder). Suppose we want to add two binary digits A and B. A combinational switching network which performs this addition is called a *half adder*. It accepts two binary digits A and B (bits) as inputs and produces two binary digits as outputs, the "sum" $S = A \cdot (1 - B) + (1 - A) \cdot B$ in $\mathbb{Z}_2$ (i.e. $A + B$ mod 2) and the "carry" $C = A \cdot B$. The binary number CS formed by the two outputs is the dyadic representation of the sum of A and B in $\mathbb{Z}$, $A + B = C2^1 + S2^0$.

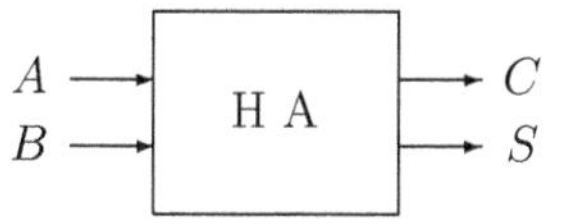

A	B	C	S
0	0	0	0
1	0	0	1
0	1	0	1
1	1	1	0

Figure 1.5.1: Block diagram and input-output table of half adder

When two binary numbers are added digit by digit, a third input must be considered, the carry-in from the next lower position. This yields the *full adder*. By a combination of half and full adders one can construct memoryless digital systems for the addition of arbitrary binary numbers of limited length. For instance, one can construct a machine for computing the sum of two binary numbers of 4 digits each by connecting in series one half adder and three full adders.

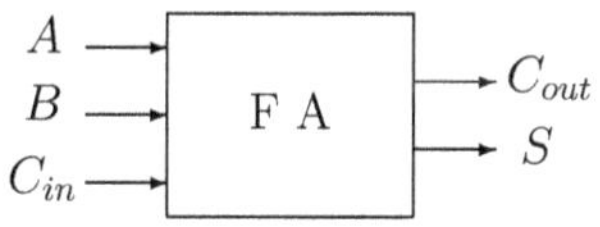

A	B	C_{in}	C_{out}	S
0	0	0	0	0
1	0	0	0	1
0	1	0	0	1
0	0	1	0	1
1	1	0	1	0
1	0	1	1	0
0	1	1	1	0
1	1	1	1	1

Figure 1.5.2: Block diagram and input-output table of full adder

In the above descriptions of the half and the full adder, time does not play a role since these digital systems are both *memoryless* and *time–invariant*. The current output $y(t) = (C_{out}, S)$ is completely determined by the *current* input $u(t) = (A, B)$ (resp. $u(t) = (A, B, C_{in})$); it does not depend upon the previous inputs (the system has no memory). Moreover, identical input vectors always determine the same output vector (independent of the time t at which they are applied), the input-output relationship does not change with time, it is time–invariant. □

Example 1.5.2. (Parity check machine). Whenever digital systems are used in computing or communication it is necessary to convert numbers and letters into strings of 1's and 0's. A map $F : U \to \mathbb{B}^p$ which maps a finite input alphabet (set of characters) U injectively into the set $\mathbb{B}^p$ of p-bit strings (code words) is called a block code of size p. An arbitrary string of p bits may or may not be a code word for the code F. An encoding device can be described as a memoryless time–invariant digital system which accepts inputs u from the finite input alphabet U and transforms these into outputs $y = F(u) \in \mathbb{B}^p$. A widely used alpha-numerical code is the ASCII code. This is a seven-bit code for the 10 decimal numbers, the 26 lower-case and 26 upper-case characters of the English language and a large number of special characters, such as "+", ")", "%" etc. With seven bits it is possible to encode at most 2^7 characters.
When information is encoded and transmitted some bits may be changed due to electrical noise or other transient failures. The change of a single bit can be detected by adding one bit to each code word in such a way that after this addition each valid code word has an even number of 1's, e.g. the ASCII code word for a is 1100001. This word has odd parity and so a 1 would be prefixed to the code word in order to achieve even parity. Thus the enlarged code word permitting error detection would be 11100001. If now one bit is changed in the code word, say by a transmission failure, the error would be detected by examining the parity of the transmitted word.[2] This can be done by a *parity checker*, a device which responds to a finite binary sequence $(u(0), u(1), \ldots, u(t))$ with the output $y(t + 1) = 0$ (in the next time unit) if the number of 1's in the sequence is even (no error), and with a 1 if not (error). The next output of a parity checker clearly depends not only on the current but also on the past inputs. If the number of ones in the past input sequence $(u(0), u(1), \ldots, u(t - 1))$ is even the next output $y(t + 1)$ is equal to the current input $u(t)$. If, however, the number of ones in the past input sequence is odd, the next output is the complement of the current input, $y(t+1) = \overline{u(t)} = 1 - u(t)$. These two cases lead to the idea of constructing a parity checker as a machine with two states, *Even* and *Odd*, which "remember" the parity of the past output sequence and are encoded by 0 and 1. The state transition of the parity checker under the influence of the present input is represented by its *state transition graph* and is explicitly described in the "next state table", see Figure 1.5.3.

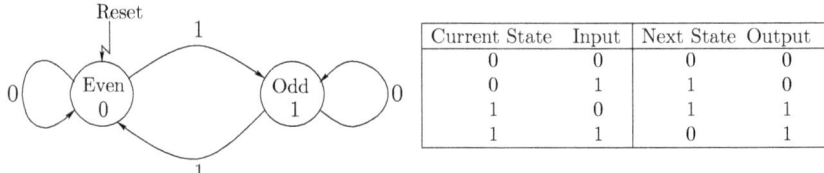

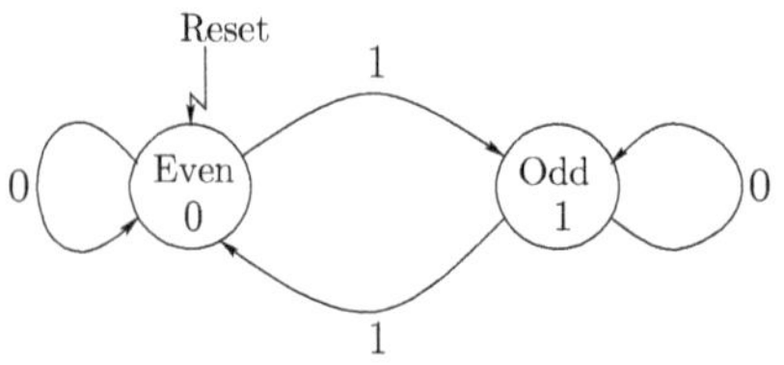

Current State	Input	Next State	Output
0	0	0	0
0	1	1	0
1	0	1	1
1	1	0	1

Figure 1.5.3: State transition graph and next state table of a parity checker

The system equations of the parity checking machine are

$$x(t + 1) = x(t) + u(t), \quad t \in \mathbb{N}, \quad x(0) = 0$$
$$y(t) = x(t)$$

[2]Note that if two bits are simultaneously changed in a code word this will *not* be discovered by a parity checker. However, the occurrence of a double error is much less probable than the occurrence of a single error (p^2 instead of p if p is the probability of a single error, assuming independence of the transmission errors).

where $u(t)$, $x(t)$, $y(t) \in \mathbb{Z}_2$ denote the current input, state and output, and $x(t+1) \in \mathbb{Z}_2$ is the next state. (The addition on the RHS of the first equation is taken in the binary field $\mathbb{Z}_2$). In order that this machine can be used for detecting errors in code words, a reset mechanism is needed which allows one to reset the state of the machine to 0 after the examination of each code word. □

1.5.1 Combinational Switching Networks

In this brief subsection we describe some of the elementary building blocks of combinational networks, the logic gates, and illustrate how simple arithmetic units, like the half and the full adder, can be built from these gates. We also explain by means of an example how the digital input–output behaviour of a gate can be approximately realized by a continuous nonlinearity.

Example 1.5.3. (Logic gates and half adder). A logic gate is an electronic device with two (or more) binary inputs and one binary output which performs simple logical operations. Its input–output behaviour can be described by a truth table or in terms of the three basic Boolean operations $\wedge$, $\vee$ and complementation. The three logic gates AND, OR, NOT which perform these operations are described in the following table together with a NOR gate which is a cascade connection of an OR-gate and the "inverter" NOT.

$y = \overline{u}$ $[= 1 - u]$

NOT

u	$y = \overline{u}$
0	1
1	0

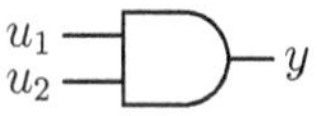

$y = u_1 \wedge u_2$ $[= u_1 \cdot u_2]$

AND

u_1	u_2	$y = u_1 \wedge u_2$
0	0	0
0	1	0
1	0	0
1	1	1

$y = u_1 \vee u_2$ $[= 1 - (1 - u_1) \cdot (1 - u_2)]$

OR

u_1	u_2	$y = u_1 \vee u_2$
0	0	0
0	1	1
1	0	1
1	1	1

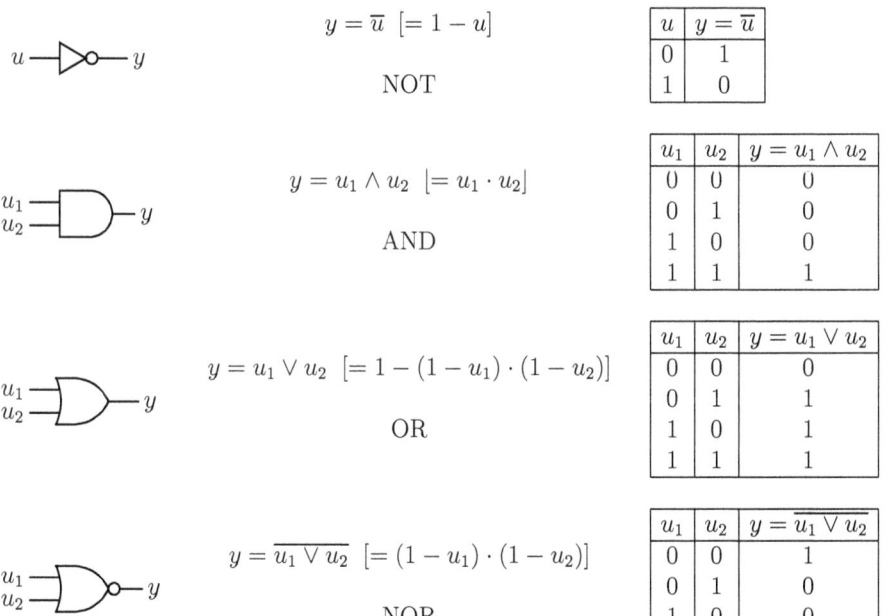

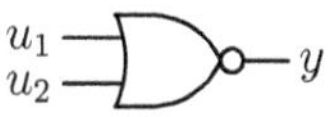

$y = \overline{u_1 \vee u_2}$ $[= (1 - u_1) \cdot (1 - u_2)]$

NOR

u_1	u_2	$y = \overline{u_1 \vee u_2}$
0	0	1
0	1	0
1	0	0
1	1	0

Table 1.5.4: Logic Gates

The table shows the standard symbols of these gates, their truth tables and the expression of their outputs in terms of their inputs (in the Boolean algebra $\mathbb{B}$). Note that these gates can also be described by arithmetic expressions in the binary field $\mathbb{Z}_2$, but while the AND gate corresponds to multiplication the OR gate *does not* correspond to addition in $\mathbb{Z}_2$ (although $\vee$ is often replaced by $+$ in textbooks on logic design). The four gates NOT,

AND, OR, NOR correspond, respectively, to the following four operations in $\mathbb{Z}_2$: $X \to \overline{X} = 1 - X$ (*complementation*), $(X, Y) \to X \cdot Y$ (*multiplication*), $(X, Y) \to 1 - (1 - X) \cdot (1 - Y)$ and $(X, Y) \to (1 - X) \cdot (1 - Y)$.
These gates can be combined to produce digital networks which perform more complicated logic or arithmetic functions. As an example we show in Figure 1.5.5 the realization of a

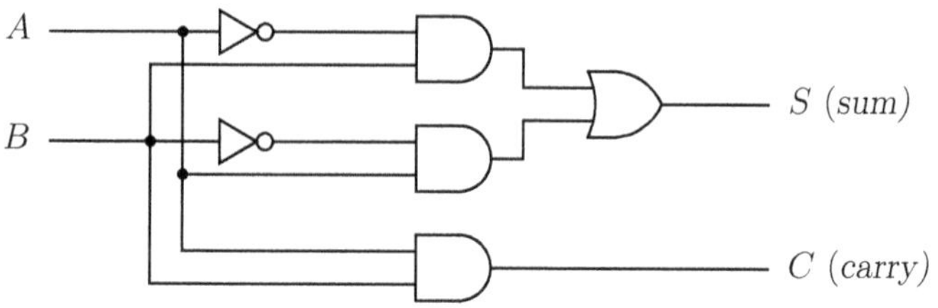

Figure 1.5.5: Realization of a half adder

half adder by a network of gates. The half adder is the simplest arithmetic circuit. The full adder (see Example 1.5.1) can be constructed from two half adders and an OR gate.

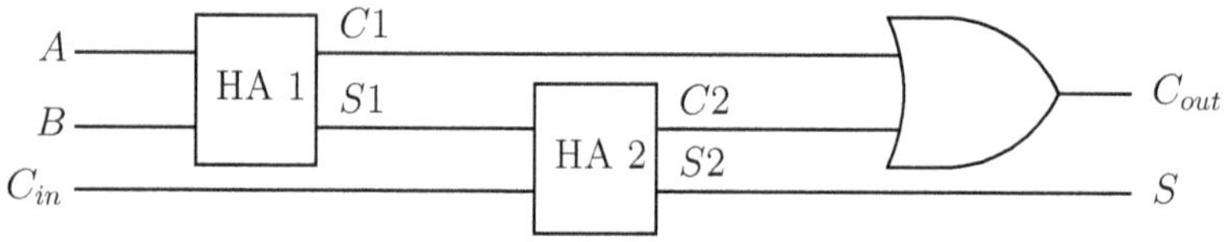

Figure 1.5.6: Realization of a full adder

Half and full adders are simple examples of *composite systems* i.e. systems composed of a number of interconnected subsystems. Very complex systems can be built in this way. In fact any function $F : \mathbb{Z}_2^m \to \mathbb{Z}_2$ or, equivalently, any logical/Boolean operation can be realized by the three gates NOT, AND, OR[3]. □

Usually a given Boolean operation can be realized by an interconnection of gates in many different ways. A basic problem in logic design is that of "minimal realization": Realize a given Boolean function by a network in which there is a minimum

- number of gates,
- number of gate inputs–this determines the amount of wiring within the network,
- number of cascaded levels of gates (i.e. the number of gates in the largest path from any input to any output). This number determines the overall time delay between inputs and outputs of the network.

Usually the above numbers cannot be minimized simultaneously so one must find a suitable compromise. Problems of minimal realization also arise in systems theory when one wants to realize a given input-output behaviour by a continuous or discrete time linear system with a minimal number of state variables, see Vol. II.
Before going on to discuss the notion of a *sequential network (finite state machine)*

[3]For a variety of reasons, NAND and NOR gates (i.e. the inverted AND and OR gates) are preferred in practice to AND and OR gates for realizing logic circuits.

it is useful to comment on some important points related to the physical realization of both kinds of switching networks (combinational and sequential). The physical components of a digital system are constructed from electronic building blocks (resistors, diodes, transistors) – in other words *a digital system is built with analog building blocks* operating in continuous time with real valued input, output and state coordinates. *Natura non facit saltus.* So the physical quantities within the system (voltages, currents) when they move from one of their two values to the other, will vary over a continuous range of transitional values. One must, therefore, distinguish between the digital system as a mathematical model and its physical realization by an electronic circuit. A precise modelling of the latter would be based on ordinary or partial differential equations with a continuous time domain, and these differential equations would describe not only the transition from the current steady state of the circuit to the next one but the whole continuous trajectories of its state and output vectors. In the above, what has been written about digital systems is concerned with their ideal mathematical behaviour and does not exactly apply to their physical realization. In the following example we illustrate how a simple digital system can be *approximately* realized by an analog device.

Example 1.5.4. (Inverter circuit). The logic inverter NOT is a digital system whose inputs and outputs are binary digits. It transforms the input 0 into the output 1 and the input 1 into the output 0. However, the circuit which realizes this ideal digital behaviour operates over electrical voltages rather than digits. It accepts arbitrary input voltages in the range of say, 0 to 5 volts and produces output voltages over the same range. The essential property which makes it a good realization of the logic inverter is that it transforms voltages which are "not too far" from 0 volts into voltages very close to +5 volts (representing a logical 1) and voltages which are "not too far" from +5 volts into voltages very close to 0 volts (representing a logical 0). A typical input-output behaviour of

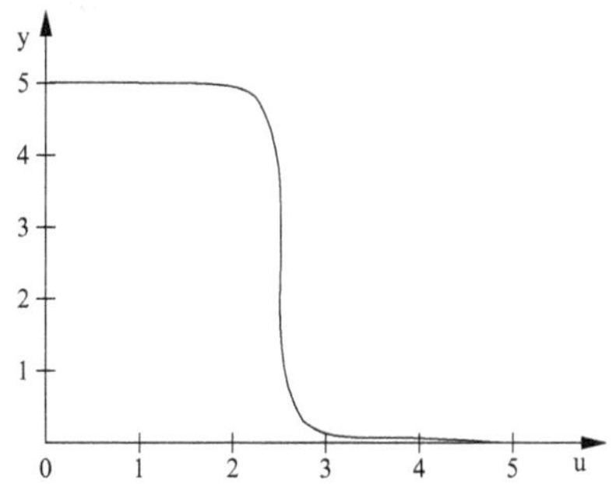

Figure 1.5.7: Input-output behaviour of an inverter circuit

such a circuit is shown in Figure 1.5.7. Here an input in the range of 0 to 2 volts produces an output of approximately 5 volts and an input in the range of 3 to 5 volts produces an output of approximately 0 volts. Thus we may say that the circuit "interprets" an input in the range of 0 to 2 volts as an input 0 producing an output 1 and if the input is in the range of 3 to 5 volts it interprets it as an input 1 and produces the output 0. So minor fluctuations in voltage levels are not misinterpreted by the circuit and have practically no influence on its output. A similar nonlinear behaviour which produces only two different output values in response to a relatively wide range of inputs values is also exhibited by other building blocks of digital systems. □

Remark 1.5.5. In general, due to unavoidable variations in the manufacturing process the input–output behaviour of an electrical device will differ from its prescribed performance. Furthermore, every interconnection of electrical components is subject to noise and signal degradation along the wires. Hence it is important that the components of a switching network are sufficiently tolerant with respect to input variations. The tolerance of a digital device to deviations of the input signal from the reference voltages is called its *noise margin.* A good noise margin of the components is fundamental for the accuracy and reliability of a digital system. Cascaded digital circuits with a good noise margin (such as the above inverter) can correct signal degradations. □

1.5.2 Sequential Switching Networks

Sequential networks are required if data is to be stored in a network for future use. In this subsection we describe how data can be stored by latches and flip–flops and we discuss the use of clocks in order to synchronize the network elements and thereby enhance the reliability of the network. We outline the main steps in the design of a finite state machine and illustrate this by constructing a three bit counter.
In the next example we describe some basic memory elements of sequential networks.

Example 1.5.6. (R–S latch and J–K latch). Broadly speaking a digital system consists of a memory part that stores past data and a combinational part by which new outputs are generated from the stored data and the current inputs. The basic memory elements of a digital system are constructed by feedback interconnection between a (small) number of gates. The most primitive memory devices are *latches*, these are circuits which "latch" onto one bit (0 or 1) and remember it. As an example we consider the R–S latch which is obtained by feedback coupling of two NOR gates. It follows immediately from the definition that a NOR gate acts as an inverter if one of the inputs is set to 0. If one of the inputs is set to 1 its output is always 0. Now consider the cross–coupled NOR

Figure 1.5.8: R–S latch: block diagram and realization by feedback of NOR gates

gates as depicted in Figure 1.5.8. It is required that the outputs of the two NOR gates have complementary values Q and $\overline{Q}$, respectively. The output Q of the lower NOR gate Q is said to be the state of the latch. If $Q = 1$ it is said to be in the *set* state and if $Q = 0$ it is said to be in the *reset* or *clear* state. Suppose that both inputs R and S are set to zero. Since each of the two NOR gates acts as inverter to the signal received from the other, the output values and hence the state remain unchanged (i.e. are stored) as long as both inputs are kept to zero. If $R = 0$ and $S = 1$ then the state (output of the lower NOR gate) is set to 1 whereas the output of the upper NOR gate is set to 0. If

$R = 1$ and $S = 0$ then Q is reset to 0. Therefore S is called the *set* input and R the *reset* input. What happens if both inputs are set to 1, i.e. if the latch is simultaneously set and cleared? In that case the outputs of both NOR gates would necessarily take the value 0 so that the complementarity assumption of the two outputs would be violated. Moreover, if afterwards both inputs were to be simultaneously changed from 1 to 0 at time t the resulting (next) state and output value $Q(t+\tau)$ would become unpredictable. If the upper gate switches first, its output $\overline{Q}$ would switch to 1 and so the next state would be set to $Q(t+\tau) = 0$. If the lower gate switches first, then its output would

$u_1(t)$	$u_2(t)$	$x(t)$	$x(t+\tau)$	Comment
0	0	0	0	HOLD
0	0	1	1	
1	0	0	1	SET
1	0	1	1	
0	1	0	0	RESET
0	1	1	0	
1	1	0	?	NOT ALLOWED
1	1	1	?	

Table 1.5.9: Next state table of the R–S latch

switch to $Q(t+\tau) = 1$ whilst $\overline{Q}$ would switch to 0. Thus the next state $Q(t+\tau)$ of the latch would depend upon which gate happens to be faster. Such a situation is referred to as a *race condition.* This unpleasant phenomenon is excluded if the two outputs never have the same value and this is secured if the input pair $(u_1, u_2) = (1, 1)$ is not allowed. For admissible input pairs the behaviour of the R–S latch is described by the *output map* $(y_1, y_2) = (x, 1-x)$ and the *next state map* $x(t+\tau) = (1-x(t))u_1(t) + x(t)(1-u_2(t))$, see the next state Table 1.5.9. Here τ is the propagation delay of the R–S latch, i.e. the time lag before the new steady state (output) is achieved in response to a change in the inputs.

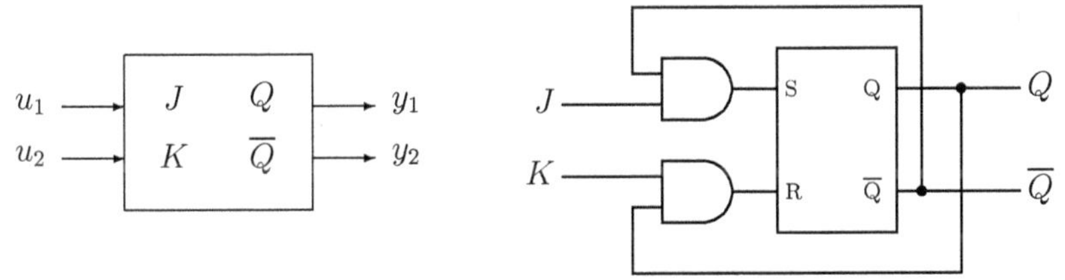

Figure 1.5.10: J–K Latch: Block diagram and circuit

In order to avoid the possibility of inadmissible inputs, the R–S latch can be connected with two additional AND gates as in Figure 1.5.10. By feeding back the outputs in the described manner it is guaranteed that the inputs R and S to the R–S latch are never simultaneously 1. The resulting circuit is called a *J-K latch* and is represented by the block diagram shown on the left in Figure 1.5.10. In addition to avoiding the forbidden input combination $(R, S) = (1, 1)$ at the internal R-S latch the configuration shows a new capability, *toggling.* If $J = K = 1$ then the current state $Q(t) = 0$ will toggle to $Q(t+\tau) = 1$ and the current state $Q(t) = 1$ will toggle to $Q(t+\tau) = 0$. Thus all possible input combinations lead to useful functions for the J-K latch: hold, reset, set, and toggle, see the next state Table 1.5.11. The behaviour of a J–K latch is described by the output

$u_1(t)$	$u_2(t)$	$x(t)$	$x(t+\tau)$	Comment
0	0	0	0	HOLD
0	0	1	1	
1	0	0	1	SET
1	0	1	1	
0	1	0	0	RESET
0	1	1	0	
1	1	0	1	TOGGLE
1	1	1	0	

Table 1.5.11: J–K Latch: Next state table

function $(y_1, y_2) = (x, 1-x)$ and the next state equation

$$x(t+\tau) = (1-x(t))u_1(t) + x(t)(1-u_2(t)) \quad x \in \mathbb{Z}_2,\ u \in \mathbb{Z}_2^2. \tag{1}$$

Note, however, that this equation is not to be understood in discrete time. Both the R–S and the J–K latches are *asynchronous* (or *unclocked*), i.e. they may change their state and outputs at any time in response to changes in the inputs.[4] This leads to a problem which becomes evident when these memory elements are realized by a circuit. For an asynchronous circuit to work properly, the inputs must be (approximately) constant for a sufficiently long time to allow the circuit to reach the corresponding next steady state. Moreover, only one external input should be effective (different from zero) at any given time. The reason for this is that if the two inputs $u_1(t) = u_2(t) = 1$ for a time interval longer than the propagation delay through the latch, the outputs will toggle an unknown number of times, determined by the length of the interval and the time lag with which a change in the output signal travels, via the feedback loop, through the circuit back to the output. The phenomenon of "oscillating outputs" caused by identical inputs $u_1(t) = u_2(t) = 1$ is illustrated in the timing diagram shown in Figure 1.5.12. So, although

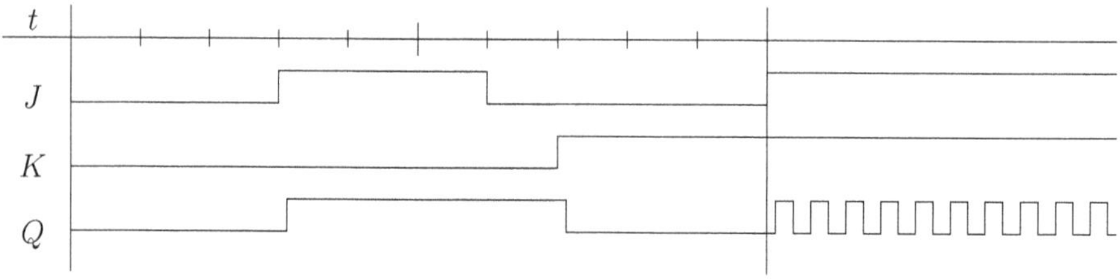

Figure 1.5.12: Timing behaviour of the J–K latch

all input combinations are allowable for the J–K latch, the problem of forbidden inputs reappears in a different form when the memory element is actually realized by a circuit. The input combination $u_1(t) = u_2(t) = 1$ causes the J–K latch to produce oscillating outputs in continuous time. □

We have seen in Example 1.5.4 that high reliability can be achieved in spite of unavoidable signal degradation and noise within a network if the network elements

[4] This is why we have denoted the next state by $x(t+\tau)$ in the above next state tables.

produce signals with only a finite number of steady state/output values ("quantization of signal values"). In order that only these values (representing 0 and 1) determine the behaviour of the network and that the transitional signal values have no effect, time must be discretized as well ("quantization of time"). This is performed by synchronizing the functioning of the network elements. A periodic signal (*clock*) is distributed throughout the circuit in order to ensure that all memory elements change state and output at approximately the same instant. The clock usually generates a square-wave pulse train. By adding for example the clock signal to the inputs of a J–K latch as in Figure 1.5.13 the output and state of this latch will be updated only if the clock is asserted (takes its upper value). When the clock is low,

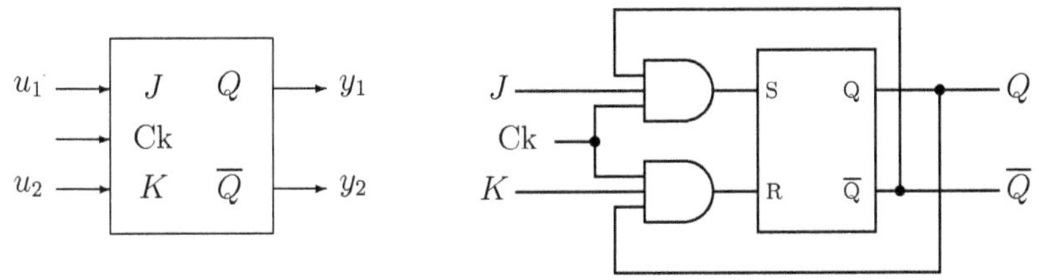

Figure 1.5.13: Clocked J–K Latch: Block diagram and circuit

the steering AND gates are disabled, and the output of the latch remains unaffected by the data inputs J and K. Such a method of synchronization is called *level triggering*, and level triggered storage devices are called *clocked latches.* If the inputs to the network elements do not change during the time when the clock is high and the corresponding steady state and output values are reached within one clock cycle the level triggered network behaves approximately like a digital system. However, level triggering cannot always handle *asynchronous* inputs, i.e. inputs which are changing whilst the clock is high. This may lead to racing problems and unpredictable outputs.

Flip–flops differ from latches in that their outputs change only with respect to the

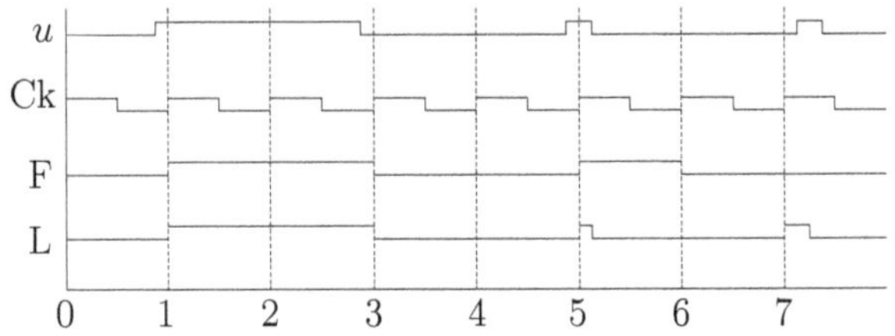

Figure 1.5.14: Time behaviour of a positive edge-triggered flip-flop (F) and a clocked latch (L)

clock whereas clocked latches change output if their inputs change (and the clock is high). *Edge–triggered* flip–flops respond to a rising or falling edge of the clock signal. This is of a very short duration so that racing and oscillating outputs are avoided. A positive (negative) edge–triggered flip–flop samples its inputs on the low–to–high (resp. high-to-low) transition of the clock and, after a short propagation delay, produces the next state corresponding to the current input and the current state. After this the input may change but the flip–flop will not respond until the next signal from the clock. This is in contrast to the behaviour of a clocked latch as illustrated in Figure 1.5.14. We see that the outputs differ if the input changes when the clock is high. This difference is particularly noticeable between the times 5 and 6 where the clocked latch responds to the decreasing input, but the output of the flip-flop remains at 1.

For reliable operation of flip–flops, the inputs must be "stable" (approximately constant) for a time interval from a *setup time* before, to a *hold time* after the clocking event, see Figure 1.5.15. Proper operation of the circuit requires that the steady

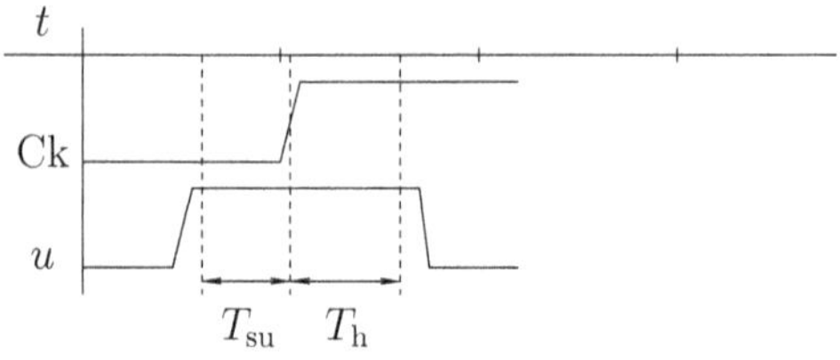

Figure 1.5.15: Setup T_{su} and hold T_{h} times

state value changes only once per clock cycle. In order to guarantee that the correct next state is achieved in spite of varying propagation delays of the input signals the period of the clock should be longer than the worst case propagation delay through the combinational network. If a network is designed in such a way that these constraints are respected, the resulting circuit behaves like a discrete time finite state machine. A careful timing methodology is fundamental for designing reliable sequential networks.

Different types of edge-triggered flip-flops can be created by an interconnection of latches, i.e. by an interconnection of feedback coupled gates. An edge-triggered J–K flip–flop for example, which is one of the most versatile and reliable flip–flops, can be built from 8–10 gates using suitable interconnections and feedback couplings. We do not go into details and refer the interested reader to the literature, see *Notes and References*. Edge triggered flip–flops are represented by block diagrams with a triangle in front of the clock input, see Figure 1.5.16.

In order to illustrate how the above memory elements are used to build a sequential network we conclude this section with an example of a finite state machine design. The main steps in such a design process are listed below.

1. ABSTRACT REPRESENTATION OF THE MACHINE. Identify the inputs, outputs, and introduce internal states of the machine which permit an easy description of the desired input–output behaviour. Draw a state diagram, i.e. a graph with vertices representing the states and directed arcs which represent the possible transitions from one state to the next one under the influence of the available inputs. Additionally, a next state table can be established. Describe the outputs associated with given input and state combinations.
2. STATE MINIMIZATION. Sometimes the first step results in a description that has a number of redundant states. These states can be eliminated without

affecting the input–output behaviour of the finite state machine. A reduction in the number of states usually reduces the number of logic gates and flip-flops which are needed for the realization of the finite state machine[5]. There are formal procedures and computational algorithms for state minimization, see *Notes and References.*

3. CHOICE OF FLIP–FLOPS for implementing the states.
4. IMPLEMENTATION OF THE FINITE STATE MACHINE. Realize the next state and output mappings by a combinational network connecting inputs, states and outputs.

As an illustration, let us design a synchronous binary counter. Counters are used in many digital systems (e.g. in digital clocks) to count events. They are amongst the simplest possible finite state machines. They typically have only one input (e.g. a square wave signal–the clock) and their outputs are identical with their current state. Their state transition graph consists of a single cycle joining the finitely many binary numbers through which the counter runs successively on each clock pulse.

Example 1.5.7. (Three bit counter). We construct a synchronous modulo-8 counter which is driven by a clock. Following the above procedure we begin with an abstract description of the digital system (Step 1). The clock is the only input to the counter. There are three binary output channels corresponding to the three bits Q_1, Q_2, Q_3 which are needed to represent the numbers $0, \ldots, 7$ in the dyadic system. We introduce 8 different states of the counter corresponding to the eight different output combinations and encode the states by the output combination they generate. On each clock pulse the counter advances successively through its 8 states in the following cycle

$$000 \to 001 \to 010 \to 011 \to 100 \to 101 \to 110 \to 111 \to 000.$$

In this simple case we may omit the state transition table. The output vector corresponding to the current state $x(t) = Q_3Q_2Q_1$ is (Q_1, Q_2, Q_3). If we want the present output of the counter to be a function of the present state alone, the number of states we have introduced is clearly minimal and we may skip Step 2.
To store the three binary digits Q_3, Q_2, Q_1 three flip–flops are needed. From the state transition graph we see that the digit Q_1 toggles at every clock pulse, the digit Q_2 toggles on every second clock pulse and the digit Q_3 on every fourth clock pulse. This suggests that a toggle flip–flop (T flip–flop) may be most suitable for the implementation of the counter. The T flip–flop has a single input that causes the stored state to remain un-

Figure 1.5.16: Edge-triggered T Flip–Flop: Block diagram and construction from an edge-triggered J–K flip–flop

[5]To realize a machine with n states at least m flip–flops are needed where $2^{m-1} < n \leq 2^m$.

changed if the input is zero and to be complemented when the input is asserted ($u = 1$). A toggle flip–flop can be constructed from a J–K flip–flop by tying its two inputs together (see Figure 1.5.16). If the input is 0, both J and K are 0 and the flip–flop holds its state; if the input is 1, both J and K are 1 and the flip–flop complements its state, see Table 1.5.11. The state transition of the positive edge triggered T flip–flop takes place on the rising clock edge after the toggle input is set ($u = 1$).
In the final step (Step 4) we express each bit of the next–state[6] $x(t+1) = Q_3^+ Q_2^+ Q_1^+$

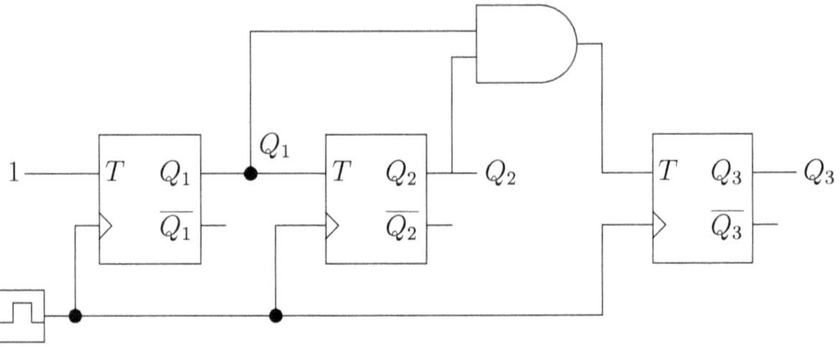

Figure 1.5.17: Three-bit counter circuit

as a combinational logic function of the current state bits and the clock signal. In this simple case the combinational logic for each of the three flip–flops can easily be determined by examining the state transition graph. The flip–flop storing Q_1 toggles on each clock pulse, the flip–flop storing Q_2 toggles at a clock pulse whenever Q_1 is asserted ($Q_1 = 1$) and the flip–flop storing Q_3 toggles at a clock pulse whenever Q_2 and Q_3 are asserted ($Q_1 = Q_2 = 1$). This leads to the circuit shown in Figure 1.5.17. □

Another simple and important class of sequential networks are shift registers, which play a key role in many finite state machines and communication systems. A simple example of a four bit shift register will be described in Example 2.1.7. For examples of more complicated finite state machines we refer to the literature, see *Notes and References*.

1.5.3 Notes and References

The realization of Boolean functions by *combinational* switching networks is based on Boolean algebra and discussed in all textbooks on switching theory and logic design. Important historical references are *Boole* (1849) [66] and *Huntington* (1904) [271].
Shannon (1938) [459] was the first to show how Boolean algebra could be applied to digital design. The digital designer wishes to realize a given Boolean function with the minimum number of gates and wires in order to reduce the size, power dissipation and cost of a digital circuit. There are many techniques (and CAD tools) for achieving minimal realizations of a given Boolean function, see *Katz* (1994) [296], *Fabricius* (1992) [154], *Wakerley* (1990) [513] and *Roth* (1985) [437].
The fundamental building blocks of *sequential* switching networks, latches and flip-flops

[6]That is the state in the next clock cycle.

are extensively discussed in most textbooks on digital systems, see the above references. The same holds for registers, memories and counters which can be built from such building blocks, see Example 2.1.7. A more complicated issue is the design of finite state machines for control and decision-making logic in digital systems. The central problem is the realization of a prescribed input-output behaviour by a finite state machine with minimal number of states (*minimal realization*) and efficient state encoding. Good references are *Roth* (1985) [437], *Green* (1986) [202], *Prosser and Winkel* (1987) [422] and *Katz* (1994) [296]. The latter is especially recommended since it contains many case studies (and two chapters describing how digital design techniques are applied to stored program computers).

The capabilities and behaviours of finite state machines are the subject of Automata Theory which had a strong influence on the early development of mathematical systems theory. Automata Theory studies finite machines as mathematical models of switching and encoding networks in abstraction from specific hardware realizations. For references and further comments on Automata Theory see the *Notes and References* in Section 2.1.

In using *discrete* time domains for modelling digital systems one should not overlook the fact that these systems are implemented by electronic circuits in *continuous* time. The resulting timing and synchronization problems are of fundamental importance in the practical design of digital systems. Detailed discussions of these issues can be found in *Katz* (1994) [296], see also *Mead and Conway* (1980) [370].

1.6 Heat Transfer

Heat transfer is the term used for the exchange of thermal energy. Here we only consider the transfer accomplished by conduction in a solid body and ignore convection and radiation effects. Nowadays it is usual to think of thermal energy or heat as the kinetic energy of the elementary particles of a solid, liquid or gas. The energy levels are a function of temperature with hot regions corresponding to high levels of energy. If a solid is a good electrical conductor there will be a large number of electrons which move freely through the lattice and thermal conduction is a consequence of this motion. In impure metals or in disordered alloys there is also a transfer of energy via lattice vibrations which may be comparable in magnitude to the electronic contribution. For gases the main mechanism is the exchange of kinetic energy from fast moving molecules to slow moving ones caused by collisions amongst themselves. In all cases (including liquids where a variety of mechanisms may be present) there is a flow of energy from regions of high temperature to ones of low temperature.
Given some initial temperature profile within a compact body $B \subset \mathbb{R}^3$, our objective is to describe the evolution of the profile in time. We will use the law of conservation of energy to obtain the corresponding differential equations. Let $V \subset B \subset \mathbb{R}^3$ be an arbitrary fixed, open, connected set with closure in the interior of the body B. We assume that its boundary S is orientated and piecewise smooth. If there are no sources of heat in V the conservation law states that:

> The rate of change of the thermal energy in V with respect to time is equal to the net flow of energy across the surface S of V.

We will now translate this law into mathematical formulas and make the statement more precise. Let $e(x,t)$ be the specific thermal energy (i.e. the energy per unit mass) at position $x = (x_1, x_2, x_3) \in \mathbb{R}^3$ and time t. We assume that the density $\rho = \rho(x)$ of the solid body is independent of time and temperature, then the total thermal energy in V is

$$\int_V \rho(x) e(x,t) dx$$

where dx denotes the Lebesgue measure in $\mathbb{R}^3$. Assuming that all the functions are continuously differentiable, the time rate of change of the thermal energy in V is

$$\frac{d}{dt} \int_V \rho(x) e(x,t)\, dx = \int_V \rho(x) e_t(x,t) dx, \quad e_t(x,t) = \frac{\partial e}{\partial t}(x,t).$$

Let $\mathbf{q} : B \times \mathbb{R} \to \mathbb{R}^3$ be the time-dependent vector field which describes the flow of thermal energy in the body B. The vector $\mathbf{q}(x,t)$ is called the *heat flux vector* at $x \in B$ at time t. Let $\mathbf{n}(x)$ denote the unit outward normal to the surface S at the point $x \in S$. By the conservation law,

$$\int_V \rho(x)\, e_t(x,t)\, dx = -\int_S \langle \mathbf{q}(x,t), \mathbf{n}(x) \rangle\, dS(x).$$

Applying the divergence theorem to the surface integral over S we obtain

$$\int_V (\rho(x)\, e_t(x,t) + \operatorname{div} \mathbf{q}(x,t))\, dx = 0.$$

Since the open set V is arbitrary, we have

$$\rho(x)\, e_t(x,t) = -\operatorname{div} \mathbf{q}(x,t), \quad x \in \operatorname{int} B,\ t \in \mathbb{R}. \tag{1}$$

In order to get an equation for the temperature $\Theta = \Theta(x,t)$, additional information of an empirical nature is required. For many materials the function e is linear in the temperature Θ over quite large temperature ranges. That is $e = c\,\Theta$, where $c = c(x)$ is time–invariant and is called the *specific heat* at $x \in \operatorname{int} B$ (the amount of heat absorbed by the body at the point x per unit mass per unit rise in temperature). Now the heat energy flows from hot to cold, so the heat flow in any direction $\mathbf{d} \in \mathbb{R}^3$ will be negative (i.e. $\langle \mathbf{d}, \mathbf{q} \rangle < 0$) if the temperature is rising in that direction (i.e. $\langle \mathbf{d}, \operatorname{grad}\Theta \rangle > 0$), and conversely, if $\langle \mathbf{d}, \operatorname{grad}\Theta \rangle < 0$ then $\langle \mathbf{d}, \mathbf{q} \rangle > 0$. As a consequence there will exist a positive scalar function k such that $\mathbf{q} = -k \operatorname{grad}\Theta$. k is called the *conductivity* and in general will vary with the medium itself, the position in the medium, the temperature and time. However if the temperature variations are not large, a first approximation which agrees with experiments is to assume that, for a given medium, $k = k(x)$ is only a function of position. This relationship was postulated by *Fourier* in 1822 and is now known as *Fourier's Law.* With these assumptions (1) becomes

$$c(x)\rho(x)\,\Theta_t(x,t) = \operatorname{div}\left(k(x) \operatorname{grad}\Theta(x,t)\right), \quad (x,t) \in \operatorname{int} B \times (0,\infty). \tag{2}$$

This is the general three-dimensional *heat equation.* If k does not depend on position, then one obtains the classical form of the heat conduction equation

$$\Theta_t(x,t) = \alpha(x) \triangle \Theta(x,t) \tag{3}$$

where $\alpha(x) = (c(x)\rho(x))^{-1}k$ is called the *thermometric conductivity* and $\triangle$ denotes the Laplacian. In order to solve it, an initial temperature distribution must be stipulated and boundary conditions must be specified which describe the way the body interacts with its surroundings. We illustrate this in the following example.

Example 1.6.1. Consider a metal rod heated in a furnace. The rod is assumed to be a cylinder of uniform cross sectional radius a which is heated by jets along its length. The heat from the jets affects the temperature distribution at the surface of the rod which in turn results in changes of the temperature within the rod. Suppose (r,ϕ,z) are cylindrical polar coordinates with the z-axis along the axis of the cylinder. We will assume that the heat supplied by the jets at point z along the rod and time t is the same for all values of ϕ and is given by $v(z,t)$. We will also assume the thermometric conductivity α is constant throughout the rod and the initial temperature distribution at time $t = 0$ is independent of ϕ. So it is reasonable to seek solutions Θ of (3) which have axial symmetry (i.e. are independent of ϕ), in which case (3) takes the form

$$\Theta_t(r,z,t) = \alpha \triangle \Theta(r,z,t) = \alpha\Theta_{zz}(r,z,t) + \alpha r^{-1}(r\Theta_r(r,z,t))_r. \tag{4}$$

Let

$$\overline{\Theta}(z,t) = (\pi a^2)^{-1} \int_0^{2\pi}\int_0^a \Theta(r,z,t)\, r\, dr\, d\phi = 2a^{-2}\int_0^a \Theta(r,z,t)\, r\, dr$$

be the average cross sectional area temperature, then integrating (4) over the cross section at z we get

$$\overline{\Theta}_t(z,t) = \alpha\overline{\Theta}_{zz}(z,t) + 2\alpha a^{-2}\left[r\Theta_r(r,z,t)\right]_0^a = \alpha\overline{\Theta}_{zz}(z,t) + 2\alpha a^{-1}\Theta_r(a,z,t).$$

But $v(z,t) = -k\Theta_r(a,z,t)$ and hence

$$\overline{\Theta}_t(z,t) = \alpha\overline{\Theta}_{zz}(z,t) + \beta v(z,t),$$

where $\beta = -2\alpha(ak)^{-1}$. Let us further assume that the distribution $b(\cdot)$ of the jets along the rod is fixed, but the magnitude can be varied in time by a control u, so that $v(z,t) = b(z)u(t)$, then

$$\overline{\Theta}_t(z,t) = \alpha\overline{\Theta}_{zz}(z,t) + \beta b(z)u(t) \tag{5}$$

Suppose the temperature at each end of the rod is kept at a constant value C, and the initial value of $\overline{\Theta}$ at $z \in [0,\ell]$ is $\overline{\Theta}_0(z)$, so that

$$\overline{\Theta}(0,t) = \overline{\Theta}(\ell,t) = C, \ \ t \geq 0, \qquad \overline{\Theta}(z,0) = \overline{\Theta}_0(z), \ \ z \in [0,\ell], \tag{6}$$

where ℓ is the length of the rod. Note that if the initial temperature profile is constant with $\overline{\Theta}_0(z) \equiv C$, then the corresponding solution of (5) and (6) with $u(t) = 0$, $t \geq 0$ is given by the equilibrium solution $\overline{\Theta}_0(z,t) = C$, $z \in [0,\ell], t \geq 0$. For any given solution $\overline{\Theta}(z,t)$ of the partial differential equation (5) let us denote by $\theta(z,t)$ the deviation of $\overline{\Theta}(z,t)$ from the equilibrium solution, i.e.

$$\overline{\Theta}(z,t) = \theta(z,t) + C, \quad \overline{\Theta}_0(z) = \theta_0(z) + C, \ (z,t) \in [0,\ \ell] \times \mathbb{R}_+ .$$

Then we obtain from (5) and (6) the *one-dimensional controlled heat equation*

$$\begin{aligned} \theta_t(z,t) &= \alpha\theta_{zz}(z,t) + \beta b(z)u(t) \\ \theta(0,t) = \theta(\ell,t) = 0, \quad \theta(z,0) &= \theta_0(z), \ (z,t) \in [0,\ \ell] \times \mathbb{R}_+. \end{aligned} \tag{7}$$

Finally suppose we sense the temperature at a given point $z_1 \in (0,\ell)$. In reality the sensor measures a weighted average of the temperature at nearby points. Let us assume that the measurement $Y(t)$ can be expressed in terms of the average temperature $\overline{\Theta}(z,t)$ in the form $Y(t) = \pi a^2 \int_0^\ell c(z)\overline{\Theta}(z,t)dz$, where the support of the continuous density $c(\cdot)$ is a small interval around z_1. If we denote by $y(t)$ the deviation of $Y(t)$ from the steady state output $Y_0(t) = C\pi a^2 \int_0^\ell c(z)\,dz$ (corresponding to the equilibrium solution $\overline{\Theta}_0(z,t) = C$), then

$$y(t) = \pi a^2 \int_0^\ell c(z)\,\theta(z,t)\,dz, \quad t \geq 0. \tag{8}$$

Equations (7) and (8) represent a single input single output system. The state of this system at each time t is given by the temperature profile $\theta(\cdot,t)$ which is an infinite dimensional object varying in a function space. Such systems are called *infinite dimensional.*
In applications the above model may be used to determine control laws which drive an initial temperature distribution to some desired final distribution in a given time interval (a controllability problem), see Subsection 2.2.4. Another possible application is to use the model to obtain an estimate of the whole temperature profile $\theta(\cdot,t)$ from the knowledge of the input and output functions $u(\cdot)$, $y(\cdot)$ on a given time interval $[0,\ T]$, $T > 0$ (an observability problem). □

1.6.1 Notes and References

J. B. Fourier's treatise on heat, "Théorie Analytique de la Chaleur", was published in 1822 and an English translation can be found in [171]. There are, of course, whole sections of libraries devoted to heat transfer. One book on the subject is *Ozisik* (1993) [401]. A similar statement is true for books on partial differential equations. We quote *Sobolev* (1964) [469] because some of the material in this section was based on it and because of the influence that Sobolev has had on the mathematical development. A standard reference for the control theory of infinite dimensional systems is *Curtain and Zwart* (1995) [116].

Chapter 2

Introduction to State Space Theory

State space theory deals with dynamical models describing both the internal dynamics of a given physical process and the interaction of the process with the outside world. In this chapter we introduce the general notion of a *dynamical system* and set the basis for the study of various important system classes.
We emphasize that for us a *dynamical system is a mathematical model* and hence should be carefully distinguished from the physical process for which it is a model. Dynamical systems of different types may be used as models of one and the same physical process. Nevertheless it will sometimes be convenient to use the word "system" for the real physical process described by the dynamical model and in this case we shall add the epithet "real" or "physical" whenever this is necessary for a clear distinction.
In Section 2.1 we begin with a description of the components which constitute the mathematical concept of a dynamical system and then give a very general definition. This definition incorporates the basic common structure of most dynamic state space models in current use and in particular comprises all the state space models described in Chapter 1. Its scope will be further illustrated by subsequent sections of this chapter. In the second and third section we focus on the class of *linear* systems and discuss in some detail the dynamics of linear models described by differential or difference equations with constant coefficients. The study of these models represents the core of dynamical systems theory and has strongly influenced the development of other branches. Section 2.2 is concerned with their free motions and Section 2.3 with their forced motions. We also describe some elements of input-output theory and explain the relationship between their representations in time and frequency domain. In Section 2.4 we introduce structure preserving mappings ("morphisms") between linear systems. We show how new systems of this class can be obtained via standard constructions and describe various interconnection schemes for building complex systems. Finally in the last section we analyze the problem of converting continuous time signals and systems into discrete time versions and vice versa. This is a problem of increasing importance due to the replacement of analog devices by digital ones in the control and measurement of processes which evolve continuously in time. Numerical Analysis offers many techniques for the discretization of

differential equations. We will describe some basic numerical schemes and indicate the difficulties which can occur in their use for approximating differentiable dynamical systems by discrete ones.

2.1 Dynamical Systems

In this section we introduce the general mathematical concept of a dynamical system in state space. This concept has evolved as a unification of a variety of notions which have been used in, for example, the classical theory of differentiable dynamical systems, circuit theory and automata theory. We will illustrate the scope of the general definition by different examples taken from these fields. In order to obtain additional structure we also introduce some basic properties which lead to a broad classification of dynamical systems. Since the section has mainly conceptual objectives the presentation is descriptive and contains just a few mathematical results.

2.1.1 The General Concept of a Dynamical System

Before presenting the formal definition we consider the main terms and relations which need to be specified in order to define a dynamical system.

Time domain. A dynamical system evolves in time and so the variables which describe the behaviour of the system are functions of time. With every dynamical system there is an associated *time domain* $T \subset \mathbb{R}$ which contains all the times t at which the system variables may be evaluated. The time domain may be continuous, i.e. an interval as in Example 1.1.1 where $T = [0, \infty)$ or discrete, i.e. T consists of isolated points in $\mathbb{R}$ e.g. $T = \mathbb{Z}$ or $T = \mathbb{N}$, see Example 1.2.1. For notational convenience we will write $[t_0, t_1)$ rather than $T \cap [t_0, t_1)$ in order to denote the interval $\{t \in T; t_0 \leq t < t_1\}$ in T whenever the underlying time domain is clear.

External variables. These are the variables which describe the interactions of the system with the exterior world. Since a complete description of all the interactions is never possible, the modeller must select a set of variables which are thought to be the most important for the problem in hand. In Example 1.1.1 ecological factors such as pollution may well affect the population dynamics but have not been taken into account in the model.
It is usual to divide the external variables into a family $u = (u_i)$ of *inputs* and a family $y = (y_i)$ of *outputs*. By "inputs" we mean those variables which model the influence of the exterior world on the physical system. These can be of different types — either *controlled* inputs or *uncontrolled* inputs (for instance, disturbances). By "outputs" we mean those variables with which the system acts on the exterior world. Sometimes the outputs are divided into two (not necessarily mutually disjoint) sets of variables. Those which are actually measured will be called *measurements* and those which must be controlled in order to meet specified requirements will be called *regulated*. In certain contexts it is important to distinguish between modelled inputs and outputs and the actual inputs and outputs of the physical system. In this book

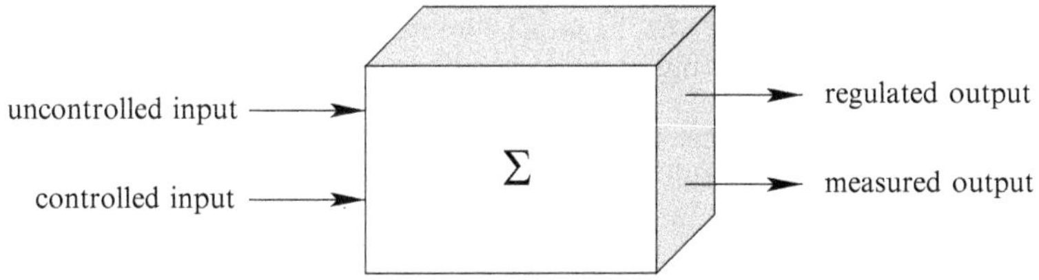

Figure 2.1.1: External variables

the external variables are to be understood as variables of the model and not as quantities of the underlying physical system.
It should be noted that it may not be a priori clear which external variables are to be considered as inputs and which as outputs. For instance, in the electrical circuit problem of Example 1.4.6 it is not obvious that the current should be taken as input and the voltage as output, or vice-versa. A general definition of dynamical systems which does not classify a priori the external variables into inputs and outputs has been developed by *J.C. Willems* (see *Notes and References*). We will not pursue this "behavioural approach" here, but presuppose that a distinction between inputs and outputs has already been made.
A dynamical system must specify the set U of input values (input alphabet) and the set Y of output values (output alphabet), for instance in Example 1.3.4 $U = \mathbb{R}$ and $Y = \mathbb{R}^2$ and in Example 1.1.1 $U = [0, \infty)$, $Y = [0, \infty)$. Throughout the text we assume that the set U of admissible input values does not change with time and does not depend on the values of any other system variables.
Let U^T denote the set of all functions $u(\cdot) : T \to U$. In general it is not possible to admit arbitrary functions $u(\cdot) \in U^T$ as input signals. For instance, in the controlled differential equations of Example 1.3.2 it would not be possible to allow for non-measurable controls since the equations could not be integrated. Therefore, in addition to the set of input values, we must specify a set $\mathcal{U} \subset U^T$ of admissible input functions. By an appropriate choice of $\mathcal{U}$, measurability or smoothness properties of the control functions can be imposed as well as time-varying constraints on the control values. Whenever there is a risk of confusion we distinguish in our notation between input *values* $u \in U$ and input *functions* $u(\cdot) \in \mathcal{U}$.
We do not include a space $\mathcal{Y}$ of admissible output functions in our general definition since this space is only occasionally needed in the context of state space theory. However, when we consider input-output systems the space $\mathcal{Y}$ of output signals will become important.

Internal state. The notion of state plays a central role in the definition of a dynamical system. Unlike external variables, the *internal* or *state* variables describe processes in the interior of the system. Not every set of internal variables of a system can be accepted as a state vector. Three basic conditions are required.

(I) The present state and the chosen control function together determine the future states of the system. More precisely, given the state $x(t_0) = x^0$ of the system at some time $t_0 \in T$ and a control $u(\cdot) \in \mathcal{U}$, the evolution of the system's state $x(t)$ is uniquely determined for all t in a suitable time interval

$T_{t_0,x^0,u(\cdot)}$ of T starting at t_0. $T_{t_0,x^0,u(\cdot)}$ may be considered as the "life span" of the trajectory $x(\cdot)$ starting at x^0 at time t_0 under the control $u(\cdot)$.

(II) Given $x(t_0) = x^0$ at some time $t_0 \in T$, the state $x(t)$ at any later time $t \in T$, $t \geq t_0$ only depends on the input values $u(s)$ for $s \in [t_0, t)$. Thus, at time t, the present state $x(t)$ is not influenced by the present and future values $u(s), s \geq t$ of the control. Moreover, knowledge of the state $x(t_0)$ at some time $t_0 < t$ supersedes the information about all previous input and state values.

(III) The output value at time t is completely determined by the simultaneous input and state values $u(t)$ and $x(t)$. In other words, the past inputs act on the present output only via accumulated effects on the system's present state.

These requirements ensure that the principle of causality is built into the concept of state. If we regard the output $y(t)$ as the "effect" of past and present "causes" (= inputs), then $u(t)$ represents the instantaneous cause and the state $x(t)$ incorporates the totality of past causes.

The choice of an adequate state vector is usually a much more difficult problem than the specification of the external variables. There are no general prescriptions. However, in physical systems state variables are often associated with the important energy stores of the system. For example, in mechanical systems the position and velocity of each mass point, or of each rigid body, are possible internal variables which together represent the state of the system at a given time. Similarly in electrical LRC circuits the charge on each capacitor and the current through each inductor may be chosen as the components of the state vector. Again, depending essentially on the objectives of the modeller the system may be roughly characterized by a few aggregated internal variables or may be more closely modelled by using a state vector with a large number or even infinitely many components. Since the state mediates the influence of past inputs on the output a rough characterization will in general yield only a rough approximation of the input-output behaviour of the physical plant.

The state variables need not represent physical quantities of the system. Indeed, from an information processing point of view the system's state may be regarded as a kind of continually updated memory or information storage. In this respect, the set X of possible states of the system can be substituted by any other set $\tilde{X}$ which is in one-to-one correspondence with X and hence can carry the same amount of information. So there is more scope for the definition of the state than for the definition of the external variables which usually refer to measurable or physically meaningful quantities. The arbitrariness can be reduced by requiring that the state of the system represents the *minimal* amount of information needed to describe the effect of past history on the future development of the system.

Conditions (I), (II) and (III) lead to the introduction of two maps which must be specified in the definition of every dynamical system.

State transition map. According to (I) and (II), the evolution of the state of a system (*trajectory*) can be described by a map φ called the *state transition map* as follows

$$x(t) = \varphi(t; t_0, x^0, u(\cdot)), \quad t \in T_{t_0,x^0,u(\cdot)}. \tag{1}$$

Actually, $\varphi(t;t_0,x^0,u(\cdot))$ only depends upon the restriction $u(\cdot)|[t_0,t)$ because of (II). In most applications this map is implicitly defined by the equations of motion of the system. If these are differential or difference equations in $x(\cdot)$ as in Example 1.3.4 (1.3.25) and Example 1.5.6 (1.5.1) an initial value problem with $x(t_0)=x^0$ must be solved for a given control function $u(\cdot)$ in order to obtain $\varphi(t;t_0,x^0,u(\cdot))=x(t)$, $t\in T_{t_0,x^0,u(\cdot)}$.

Output map. By requirement (III), the output of the system at time t is completely determined by the state and input values at time t,

$$y(t)=\eta(t,x(t),u(t)). \tag{2}$$

η is called the *output map*.

Differential equations can be solved forwards and backwards in time. Hence, if the state transition map is defined by a differential equation, the present state $x(t_0)=x^0$ has a life span $T_{t_0,x^0,u(\cdot)}$ which encompasses both past and future moments of time and the state trajectory $x(t)=\varphi(t;t_0,x^0,u(\cdot))$ is defined for $t<t_0$ as well. The following definition allows for this possibility.

Definition 2.1.1 (Dynamical system). A structure $\Sigma=(T,U,\mathcal{U},X,Y,\varphi,\eta)$ is said to be a *dynamical system* or *state space system* with *time domain* T, *input value space* U, *input function space* $\mathcal{U}$, *state space* X, *output value space* Y, *state transition map* φ and *output map* η, if T, U, $\mathcal{U}$, X, Y are non void sets, $T\subset\mathbb{R}$, $\mathcal{U}\subset U^T$, and $\eta: T\times X\times U\to Y$, $\varphi:\mathcal{D}_\varphi\to X$ (where $\mathcal{D}_\varphi\subset T^2\times X\times\mathcal{U}$) are functions such that the following axioms hold.

Interval Axiom: For every $t_0\in T$, $x^0\in X$, $u(\cdot)\in\mathcal{U}$ the *life span* of $\varphi(\cdot;t_0,x^0,u(\cdot))$

$$T_{t_0,x^0,u(\cdot)}=\{t\in T;(t;t_0,x^0,u(\cdot))\in\mathcal{D}_\varphi\} \tag{3}$$

is an interval in T containing t_0.

Consistency Axiom: For every $t_0\in T$, $x^0\in X$, $u(\cdot)\in\mathcal{U}$

$$\varphi(t_0;t_0,x^0,u(\cdot))=x^0. \tag{4}$$

Causality Axiom: For all $t_0\in T$, $x^0\in X$, $u(\cdot)$, $v(\cdot)\in\mathcal{U}$, $t_1\in T_{t_0,x^0,u(\cdot)}\cap T_{t_0,x^0,v(\cdot)}$

$$(\forall\, t\in[t_0,t_1):u(t)=v(t)) \quad\Rightarrow\quad \varphi(t_1;t_0,x^0,u(\cdot))=\varphi(t_1;t_0,x^0,v(\cdot)). \tag{5}$$

Cocycle property: If $t_1\in T_{t_0,x^0,u(\cdot)}$ and $x^1=\varphi(t_1;t_0,x^0,u(\cdot))$ for some $t_0\in T$, $x^0\in X$, $u(\cdot)\in\mathcal{U}$ then $T_{t_1,x^1,u(\cdot)}\subset T_{t_0,x^0,u(\cdot)}$ and

$$\varphi(t;t_0,x^0,u(\cdot))=\varphi(t;t_1,x^1,u(\cdot)),\quad t\in T_{t_1,x^1,u(\cdot)}. \tag{6}$$

The product space $T\times X$ is sometimes called the *event space* of Σ. We shall say that a control $u(\cdot)\in\mathcal{U}$ *transfers* an event (t_0,x^0) to (t_1,x^1) (notation: $(t_0,x^0)\overset{u(\cdot)}{\rightsquigarrow}(t_1,x^1)$) if $x^1=\varphi(t_1;t_0,x^0,u(\cdot))$. Although this expression intuitively only makes sense if $t_1\geq t_0$ it is convenient to use it also if $t_1<t_0$. The cocycle property says that if

a control $u(\cdot)$ transfers the event (t_0, x^0) to the event (t_1, x^1) and (t_1, x^1) to (t_2, x^2) then it also transfers (t_0, x^0) to (t_2, x^2). Without this assumption it would be impossible to interpret $\varphi(t; t_0, x^0, u(\cdot))$ as the state of Σ at time t when Σ is initialized at (t_0, x^0) and controlled by $u(\cdot)$. The axiom of consistency then implies that the argument x^0 of φ is in fact the initial state $x(t_0)$ of the system.
The interval axiom, the axiom of consistency and the axiom of causality together guarantee that the state of the system satisfies requirements (I) and (II). Requirement (III) is automatically satisfied if we interpret $y(t) = \eta(t, x, u)$ to be the output of Σ at time t when x is its state and u the instantaneous input value at time t.
For any $t_0 \in T$, $x^0 \in X$, $u(\cdot) \in \mathcal{U}$ the function

$$t \mapsto x(t) = \varphi(t; t_0, x^0, u(\cdot)), \quad t \in T_{t_0, x^0, u(\cdot)}$$

describes the evolution of the system's state and is called the (*state*) *trajectory* of Σ determined by the initial condition $x(t_0) = x^0$ and the control function $u(\cdot)$. Its domain of definition, $T_{t_0,x^0,u(\cdot)}$, is the *life span* of the trajectory. Its image $\{\varphi(t; t_0, x^0, u(\cdot));\ t \in T_{t_0,x^0,u(\cdot)}\}$ is said to be an *orbit* of Σ. The corresponding *output trajectory* or *output signal* is

$$y(\cdot) = y(\cdot; t_0, x^0, u(\cdot)) : \ t \mapsto y(t) = \eta(t, x(t), u(t)), \quad t \in T_{t_0,x^0,u(\cdot)}. \tag{7}$$

Definition 2.1.1 allows for the possibility that the state trajectory of a system starting at $x(t_0) = x^0$ under the control $u(\cdot) \in \mathcal{U}$ does not exist for all future times $t \geq t_0$. This may reflect a situation where the system "blows up" or the trajectory "leaves the state space" X under the influence of the control $u(\cdot)$. As an extreme case, Definition 2.1.1 allows for the possibility that $\varphi(t; t_0, x^0, u(\cdot))$ is not defined for any $t > t_0$ and we will express this by saying that the control $u(\cdot)$ *is not applicable* to Σ initialized at (t_0, x^0).

Remark 2.1.2. For some dynamical systems control aspects do not play a role. This can be expressed in the framework of Definition 2.1.1 by choosing for the input space U a singleton $\{u^*\}$ and for $\mathcal{U}$ the singleton which only consists of the constant input function $u(t) = u^*$, $t \in T$. Such a system will be called *uncontrolled* or *free*. In order to avoid dependency on the specific singleton it is convenient to use the standard singleton $\{\emptyset\}$ for U. In other situations measurement aspects may not be important. This can be expressed in the framework of Definition 2.1.1 by choosing for the output space the standard singleton so that there is only one constant output signal. Such a dynamic model will be called a *system without outputs*. □

Definition 2.1.3. A dynamical system Σ is said to be *complete* if, for all $(t_0,x^0,u(\cdot)) \in T \times X \times \mathcal{U}$,

$$T_{t_0,x^0,u(\cdot)} \supset T_{t_0} = \{t \in T;\ t \geq t_0\}.$$

Thus Σ is complete if and only if $\mathcal{D}_\varphi \supset T^2_\geq \times X \times \mathcal{U}$ where $T^2_\geq = \{(t, t_0) \in T^2; t \geq t_0\}$. Now suppose that Σ is complete and the system is initialized at (t_0, x^0), i.e. the initial state $x(t_0) = x^0$ is fixed. Then the output signal (7) is defined on T_{t_0} and the restriction $y(\cdot)\,|T_{t_0}$ of $y(\cdot) = y(\cdot; t_0, x^0, u(\cdot))$ only depends upon the restriction $v(\cdot) = u(\cdot)\,|T_{t_0} \in \mathcal{U}_{t_0} = \{u(\cdot)\,|T_{t_0};\ u(\cdot) \in \mathcal{U}\}$ by the causality axiom. By a slight abuse of notation we may therefore write $y(\cdot; t_0, x^0, u(\cdot)\,|T_{t_0})$ instead of $y(\cdot; t_0, x^0, u(\cdot))\,|T_{t_0}$. The input-output behaviour of Σ is then described by the following operator.

Definition 2.1.4. Given a complete system Σ and $(t_0, x^0) \in T \times X$ the *input-output operator* of Σ initialized at (t_0, x^0) is defined by

$$G_{t_0,x^0} : \mathcal{U}_{t_0} \to Y^{T_{t_0}}, \quad v(\cdot) \mapsto y(\cdot; t_0, x^0, v(\cdot)). \tag{8}$$

A complete dynamical system is called reversible if it is also a dynamical system for reverse time.

Definition 2.1.5. A complete dynamical system Σ is said to be *reversible* if

$$\mathcal{D}_\varphi = T^2 \times X \times \mathcal{U},$$

i.e. $T_{t_0,x^0,u(\cdot)} = T$ for all $(t_0, x^0, u(\cdot)) \in T \times X \times \mathcal{U}$.

Hence all state trajectories of a reversible system are defined on the whole time domain T. Given any event (t_1, x^1) and any $t_0 \in T$, $t_0 < t_1$, $u(\cdot) \in \mathcal{U}$, there exists a unique $x^0 \in X$ such that $u(\cdot)$ transfers (t_0, x^0) into (t_1, x^1). In fact this state is given by $x^0 = \varphi(t_0; t_1, x^1, u(\cdot))$. It is the only state with this property since, for every other $\hat{x}_0 \in X$ satisfying $(t_0, \hat{x}^0) \overset{u(\cdot)}{\rightsquigarrow} (t_1, x^1)$ it follows from the cocycle property and $(t_1, x^1) \overset{u(\cdot)}{\rightsquigarrow} (t_0, x^0)$ that $(t_0, \hat{x}^0) \overset{u(\cdot)}{\rightsquigarrow} (t_0, x^0)$, hence $\hat{x}^0 = x^0$ by the consistency axiom. Definition 2.1.1 of a dynamical system is far too general a definition on which to build a substantial mathematical theory. However we feel that it is useful

- for showing the unity of similar developments in different fields,
- for establishing bridges for the transfer of ideas from one area of application to another,
- for recognizing more clearly the additional structures of the objects in a particular field.

We will illustrate the definition with a simple example of a digital system (see Example 1.5.6). Digital systems have only finitely many states and are automata in the following sense.

Definition 2.1.6 (Automaton). A five tuple $\mathcal{A} = (U, X, Y, \psi, \eta)$ where U, X, Y are non-void sets and $\psi : X \times U \to X$, $\eta : X \times U \to Y$ are maps, is called an *automaton* with *input space* U, *state space* X, *output space* Y, *next-state function* ψ and *output function* η.

The dynamics of an automaton are described by the following state and output equations

$$\begin{aligned} x(t+1) &= \psi(x(t), u(t)), \quad t \in \mathbb{N} \\ y(t) &= \eta(x(t), u(t)) \end{aligned} \tag{9}$$

It follows that any automaton can be viewed as a dynamical system by setting $T = \mathbb{N}$, $\mathcal{U} = U^{\mathbb{N}}$ and defining $\varphi : T^2_{\geq} \times X \times \mathcal{U} \to X$ recursively by

$$\begin{aligned} \varphi(t_0 + k + 1; t_0, x^0, u(\cdot)) &= \psi(\varphi(t_0 + k; t_0, x^0, u(\cdot)), u(t_0 + k)), \quad k \in \mathbb{N} \\ \varphi(t_0; t_0, x^0, u(\cdot)) &= x^0. \end{aligned}$$

Example 2.1.7 (Switching networks). A (binary) switching network is an automaton whose input, state and output variables admit only two different values (symbolized by 0 and 1), so

$$U = \mathbb{Z}_2^m, \quad X = \mathbb{Z}_2^n, \quad Y = \mathbb{Z}_2^p$$

where $\mathbb{Z}_2 = \mathbb{Z}/2$ is the binary field (see Section 1.5). Physically, these two values may, for example, be realized by two different voltage levels. If $n = 0$ (so that the switching network has the trivial state space $\{0\}$ and trivial state transition map $\varphi \equiv 0$) the output at time t is completely determined by the input in time t, $y(t) = \eta(0, u(t))$. Dynamical systems with this property are called *memoryless* . They represent physical devices which directly transform inputs into outputs without intermediate storage of energy or information. Simple examples of memoryless switching networks are the logic gates described in Chapter 1. Their output map is given by truth tables.
Switching networks with memory are called *sequential* because a sequence of inputs must be specified in order to determine the output. The basic memory elements used in sequential networks are flip-flops. The "J–K flip-flop" described in Example 1.5.6 has input space $U = \mathbb{Z}_2^2$, output space $Y = \mathbb{Z}_2^2$, state space $X = \mathbb{Z}_2$, output function $y = [x, 1-x]^\top$ and next state function

$$\psi(x, u) = x(1 - u_2) + (1 - x)u_1 \quad x \in \mathbb{Z}_2,\ u \in \mathbb{Z}_2^2.$$

In large sequential networks it is common to synchronize the operation of all the flip-flops by a common clock or pulse generator emitting pulses at each time $k\tau$, $k \in \mathbb{N}$ where $\tau > 0$ is fixed. A synchronized sequential circuit changes state only after the occurrence of a clock pulse and the inputs and states of each of the flip-flops are not allowed to change at other times. It is natural to choose $T = \mathbb{Z}\tau$ as the time domain of such a system.
An important and simple example of a sequential network containing several flip-flops is the shift register. This is used in many digital systems to store and shift binary numbers arriving from a serial source. Figure 2.1.2 illustrates a four bit right shift regis-

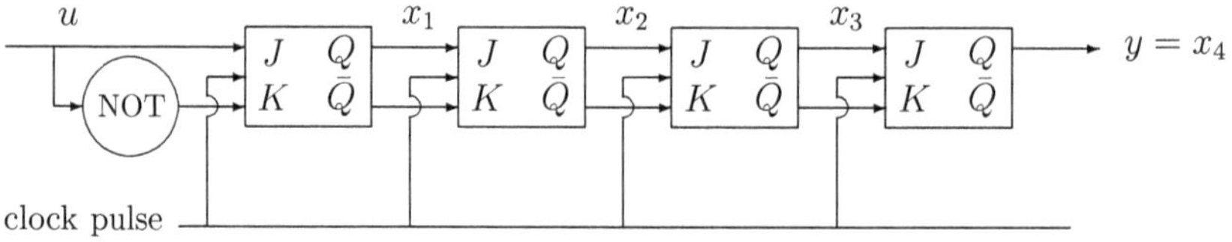

Figure 2.1.2: Shift register

ter constructed from clocked $J - K$ flip-flops. If the initial contents of the register is $x(0) = [x_1, x_2, x_3, x_4]^\top = [1, 0, 1, 1]^\top$ and the input sequence is $u(0) = 1$, $u(\tau) = 0$, $u(2\tau) = 0$, $u(3\tau) = 1$, then its successive states are $x(\tau) = [1, 1, 0, 1]^\top$, $x(2\tau) = [0, 1, 1, 0]^\top$, $x(3\tau) = [0, 0, 1, 1]^\top$, $x(4\tau) = [1, 0, 0, 1]^\top$. The bit shifted out of the right hand end is lost. We can construct a dynamical system modelling the register by setting $U = \mathbb{Z}_2$, $\mathcal{U} = U^T$, $X = \mathbb{Z}_2^4$, $Y = \mathbb{Z}_2$,

$$\begin{aligned} x_1((k+1)\tau) &= u(k\tau), & x_2((k+1)\tau) &= x_1(k\tau) \\ x_3((k+1)\tau) &= x_2(k\tau), & x_4((k+1)\tau) &= x_3(k\tau) \end{aligned}$$

and choosing $y(k\tau) = x_4(k\tau)$. Obviously the output at time $k\tau$, $k \geq 4$ is equal to the delayed input $u((k-4)\tau)$ and the state vector $x(k\tau)$ stores exactly the four preceding input values $u((k-i)\tau)$, $i = 1, 2, 3, 4$. Thus the shift register is an example which highlights the interpretation of the state as a continually updated memory of the system. □

Traditionally the concept of a dynamical system was more or less synonymous with "a system described by differential equations". The classical theory of dynamical systems was motivated by problems in mechanics particularly celestial mechanics. Then it was natural to assume that the external forces were given and not subject to human manipulation. This explains why input and output aspects are absent in the classical view of a dynamical system. The following concept of a *differentiable flow* can be regarded as the classical equivalent of a reversible dynamical system, in fact together with the concept of an automaton it motivated the more general Definition 2.1.1.

Definition 2.1.8 (Differentiable flow). A triple (T, X, ψ) is called a *differentiable flow* or *dynamical system in the classical sense* if $T \subset \mathbb{R}$ is an open interval, X an open subset of $\mathbb{K}^n$, $\mathbb{K} = \mathbb{R}$ or $\mathbb{C}$ (or, more generally a differentiable manifold) and ψ is a continuously differentiable map from $T^2 \times X$ into X, such that

$$\begin{aligned} \psi(t;t,x) &= x\,, \qquad t \in T,\ x \in X \\ \psi(t;t_1,\psi(t_1;t_0,x)) &= \psi(t;t_0,x)\,, \qquad t_0,t_1 \in T,\ x \in X\,. \end{aligned}$$

Local differentiable flows which avoid the completeness assumption are defined similarly by introducing initial time and state depending life spans T_{t_0,x^0} of the trajectories $\psi(\cdot;t_0,x^0)$. More general local flows will be considered later in the context of stability theory, see Chapter 3.
Local differentiable flows are usually generated via the solution of differential equations. Consider

$$\dot{x}(t) = g(t, x(t)) \tag{10}$$

where $g : T \times X \to \mathbb{K}^n$ is continuous, with T an open interval and X an open subset of $\mathbb{K}^n$. We say that $x(\cdot)$ is a solution of (10) on an open interval $I \subset T$ if $x(\cdot)$ is continuously differentiable on I, $(t, x(t)) \in T \times X$ for all $t \in I$ and $x(\cdot)$ satisfies (10) on I. We have the following theorem, see *Notes and References*.

Theorem 2.1.9. *Let $T \subset \mathbb{R}$ be an open interval, X an open subset of $\mathbb{K}^n$ and suppose that $g : T \times X \to \mathbb{K}^n$ is continuous and continuously differentiable with respect to x on $T \times X$. Then for any $(t_0, x_0) \in T \times X$, there exists a unique solution, $x(\cdot) = \psi(\cdot;t_0,x_0)$ of* (10) *on some maximal open interval $T_{t_0,x^0} \subset T$ containing t_0 such that $x(t_0) = x^0$. Moreover the set*

$$\mathcal{D}_\psi = \{(t, t_0, x_0);\ t \in T_{t_0,x^0}, (t_0, x_0) \in T \times X\},$$

is open in $T^2 \times X$ and $\psi : \mathcal{D}_\psi \to \mathbb{K}^n$ is continuously differentiable (ψ is said to be the general solution of (10)*).*

We see that under the conditions of the above theorem the differential equation (10) generates a *local* differentiable flow, (T, X, ψ). It will be shown later that if $X = \mathbb{K}^n$ and g is linearly bounded as in (22), then $T_{t_0,x^0} = T$ and hence in this case (T, X, ψ) is a differentiable flow in the sense of Definition 2.1.1.
To subsume a flow (T, X, ψ) under the general definition of a dynamical system we have to endow it with trivial inputs and outputs as described in Remark 2.1.2. Comparing Definitions 2.1.8 and 2.1.1 (under the completeness assumption) we note the following differences:

- only the evolution of the *state* is described,
- a smoothness condition is imposed on the state transition map,
- reversibility is built into the definition of a differentiable flow.

The following example illustrates the concept of a differential flow.

Example 2.1.10 (Pendulum). In Example 1.3.3 we saw that the equation of motion of a simple swinging pendulum of length l and mass m suspended from a fixed point is

$$ml^2\ddot{\theta} = -mgl\sin\theta \tag{11}$$

where g is the gravitational constant. Let $x = [x_1, x_2]^\top = [\theta, \dot{\theta}]^\top$, then

$$\dot{x}(t) = \begin{bmatrix} \dot{x}_1(t) \\ \dot{x}_2(t) \end{bmatrix} = \begin{bmatrix} x_2(t) \\ -gl^{-1}\sin x_1(t) \end{bmatrix} =: g(t, x(t)).$$

Suppose $T = \mathbb{R}$, $X = \mathbb{R}^2$, then it is easy to see that $g(\cdot,\cdot)$ satisfies the conditions of Theorem 2.1.9 and is linearly bounded. Hence there exists a unique solution $x(t) = \psi(t; t_0, x^0)$ on T satisfying $x(t_0) = x^0$, and (T, X, ψ) is a differentiable flow.
To obtain a graphical representation of the flow ψ the corresponding orbits $\{\psi(t; t_0, x^0);\ t \in T\}$ are provided with an orientation indicating the direction of motion as time increases. For a given $t_0 \in T$ the collection of oriented orbits corresponding to various initial conditions $x(t_0) = x^0$ in a given region of the state space form a so-called *phase-portrait* of the flow at time t_0, see Figure 2.1.3. The different character of these trajectories correspond

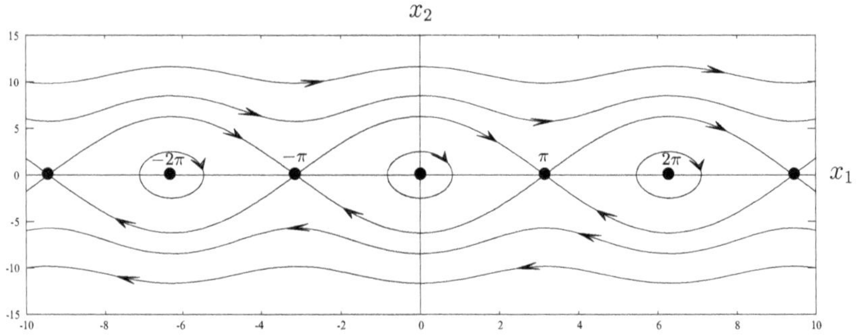

Figure 2.1.3: Phase portrait for the pendulum

to different motions of the pendulum. For example the pendulum stays at rest if it starts at $x^0 = [0, 0]^\top$ (vertically downwards with zero velocity) or at $x^0 = [\pi, 0]^\top$ (vertically upwards with zero velocity). It swings periodically backwards and forwards if it starts at $x^0 = [\theta_0, 0]^\top$ where $0 < \theta_0 < 2\pi$, $\theta_0 \neq \pi$ and rotates continuously around 0 if it starts at $x^0 = [0, \omega_0]^\top$ where $|\omega_0|$ is large enough. □

Remark 2.1.11. The equations of motion of a physical system are often described by higher order differential equations. In these cases a state vector must be found which enables the equations of motion to be transformed into an equivalent system of the form (10), see Ex. 8 and Ex. 9. □

Both examples considered in this subsection are complete systems. An example of a differentiable system which is not complete will be given in Example 2.1.16.

2.1.2 Differentiable Dynamical Systems

Let us now consider differentiable systems which are controlled and measured. Since we do not intend to develop a systematic theory of nonlinear control systems in this book, we will only deal with differentiable systems on open subsets $X \subset \mathbb{K}^n$ and not on general differential manifolds. In the following we suppose that every space $\mathbb{K}^\ell$, $\ell \in \mathbb{N}^*$ is provided with an arbitrary norm denoted by $\|\cdot\|$.

Definition 2.1.12 (Differentiable system). A dynamical system $\Sigma = (T, U, \mathcal{U}, X, Y, \varphi, \eta)$ is called *differentiable* if the following conditions are satisfied.

(i) $T \subset \mathbb{R}$ is an open interval.

(ii) U, Y are subsets of $\mathbb{K}^m$ and $\mathbb{K}^p$, X is an open subset of $\mathbb{K}^n$.

(iii) There exists a function $f : T \times X \times U \to \mathbb{K}^n$ such that for all $t_0 \in T$, $x^0 \in X$, $u(\cdot) \in \mathcal{U}$ the initial value problem

$$\begin{aligned} \dot{x}(t) &= f(t, x(t), u(t)), \quad t \geq t_0,\ t \in T \\ x(t_0) &= x^0 \end{aligned} \tag{12}$$

has a unique solution $x(\cdot)$ on a maximal open time interval I satisfying $I = T_{t_0, x^0, u(\cdot)}$ and $x(t) = \varphi(t; t_0, x^0, u(\cdot))$, $t \in I$.

(iv) $\eta : T \times X \times U \to Y$ is continuous.

Remark 2.1.13. A continuous time system $\Sigma = (T, U, \mathcal{U}, X, Y, \varphi, \eta)$ whose time interval $T \subset \mathbb{R}$ is not open will be called *differentiable* if it is obtained by restriction of the time domain from a differentiable system in the sense of the previous definition. □

Some remarks concerning the choice of $\mathcal{U}$ and the underlying solution concept for (12) are in order. Often it is necessary to consider jumps in the input functions. For example, if $u(\cdot)$ is a set point control a switch from one set point $u(t) = u_1$, $t \leq t_1$ to another $u(t) = u_2$, $t > t_1$ should be allowed. This leads to choices for $\mathcal{U}$ as the space of piecewise constant functions from T to U or the space of piecewise continuous functions $PC(T; U)$. Sometimes it will be necessary to extend the set of input signals to arbitrary Lebesgue measurable functions u which are *locally integrable* on T (i.e. $\int_a^b \|u(t)\| dt < \infty$ for all $a, b \in T$, $a < b$). Then $f(t, x, u(t))$ will not, in general, depend continuously on t for each fixed x, hence the solution concept used in Theorem 2.1.9 is not applicable. Instead we will call $x(\cdot) : I \to X$ a *solution of* (12) on an interval $I \subset T$ if it is *absolutely continuous* and satisfies (12) "almost everywhere" on I (that is "except on a set of Lebesgue measure zero"). Here "absolutely continuous" means that $x(\cdot)$ is continuous, differentiable almost everywhere (a.e.) with locally integrable derivative and can be reconstructed from its derivative by integration (see Definition A.3.12)

$$\int_{t_0}^{t} \dot{x}(s) ds = x(t) - x(t_0), \qquad t_0,\ t \in T,\ t \geq t_0.$$

For later use we formulate two basic results concerning the existence and uniqueness of solutions of differential equations with measurable RHS[1]

$$\dot{x}(t) = g(t, x(t)) \tag{13}$$

[1]RHS: right hand side, LHS: left hand side

where $g : T \times X \to \mathbb{K}^n$, $T \subset \mathbb{R}$ an interval and X an open subset of $\mathbb{K}^n$. We say that $g : T \times X \to \mathbb{K}^n$ satisfies the *Carathéodory conditions* if

(Car 1) $g(\cdot, x) : T \to \mathbb{K}^n$ is measurable for each fixed $x \in X$;

(Car 2) $g(t, \cdot) : X \to \mathbb{K}^n$ is continuous for each fixed $t \in T$;

(Car 3) $\|g(\cdot, \tilde{x})\|$ is locally integrable on T for some $\tilde{x} \in X$;

(Car 4) for each compact set $C = I \times K \subset T \times X$ there exists an integrable function $L_C(\cdot) : I \to \mathbb{R}_+$ such that

$$\|g(t,x) - g(t,y)\| \le L_C(t)\|x - y\|, \quad (t,x),(t,y) \in C. \tag{14}$$

Recall that in any metric space (X, d) the *distance* between a point $x \in X$ and a subset $S \subset X$ is defined by

$$\operatorname{dist}(x, S) = \inf\{d(x,y);\ y \in S\}. \tag{15}$$

Theorem 2.1.14 (Carathéodory). *If T is an open interval, X is an open subset of $\mathbb{K}^n$ and $g : T \times X \to \mathbb{K}^n$ satisfies the Carathéodory conditions on $T \times X$, then for any $(t_0, x^0) \in T \times X$ there exists a unique solution $x(\cdot) = \psi(\cdot; t_0, x_0)$ of (13) on some maximal open interval $T_{t_0,x^0} \subset T$ containing t_0, such that $x(t_0) = x^0$. Moreover*

(i) if $t_+(t_0, x^0) := \sup T_{t_0,x^0} < \sup T$ then $x(t)$ is unbounded as $t \nearrow t_+(t_0, x^0)$ or the boundary ∂X of X is not empty and $\operatorname{dist}(x(t), \partial X) \to 0$ as $t \nearrow t_+(t_0, x^0)$. An analogous statement holds for $t \searrow t_-(t_0, x^0) := \inf T_{t_0,x^0}$ if $t_-(t_0, x^0) > \inf T$.

(ii) If $\mathcal{D}_\psi$ is the domain of definition of the general solution ψ,

$$\mathcal{D}_\psi = \{(t, t_0, x_0);\ t \in T_{t_0,x^0}, (t_0, x_0) \in T \times X\},$$

then $\mathcal{D}_\psi$ is open in $T^2 \times \mathbb{K}^n$ and $\psi : \mathcal{D}_\psi \to \mathbb{K}^n$ is continuous.

If $t_+ = t_+(t_0, x^0) < \sup T$ as in (i) then t_+ is called a *finite escape time* of the solution $\psi(\cdot; t_0, x_0)$. If additionally $\psi(\cdot; t_0, x_0)$ is unbounded on $[t_0, t_+)$ we say that it "blows up" or "explodes" in finite time.
In order to establish that differentiable equations of the form (12) define a differentiable dynamical system one must verify that $g(t,x) = f(t, x, u(t))$ satisfies the Carathéodory conditions for all $u(\cdot) \in \mathcal{U}$. The following corollary gives a sufficient condition.

Corollary 2.1.15. *Suppose $T, U, \mathcal{U}, X, Y$ are sets as in Definition 2.1.12, $\eta : T \times X \times U \to Y$ is continuous and $f : T \times X \times U \to \mathbb{K}^n$ is jointly measurable in $(t, u) \in T \times U$ for every $x \in X$ and continuous in $x \in X$ for each fixed $(t, u) \in T \times U$. If $\mathcal{U} \subset U^T$ consists of locally L^p-integrable functions ($1 \le p < \infty$) on T and for each compact set $C = I \times K \subset T \times X$ there exist constants m_C, l_C such that*

$$\|f(t,x,u)\| \le m_C(\|u\|^p+1), \quad t \in I,\ u \in U \text{ for some } x \in X, \tag{16}$$

$$\|f(t,x,u)-f(t,y,u)\| \le l_C(\|u\|^p+1)\|x-y\|, \quad (t,x),(t,y) \in C,\ u \in U, \tag{17}$$

then the initial value problem

$$\begin{aligned} \dot{x}(t) &= f(t,x(t),u(t)), \quad t \in T \\ x(t_0) &= x^0 \end{aligned}$$

has a unique solution $x(\cdot) = x(\cdot\,; t_0, x^0, u(\cdot))$ *on a maximal interval of existence* $T_{t_0,x^0,u(\cdot)}$ *for all* $(t_0, x^0, u(\cdot)) \in T \times X \times \mathcal{U}$*. Moreover, if we define the state transition map* $\varphi : \mathcal{D}_\varphi \to X$ *by*

$$\varphi(t;t_0,x^0,u(\cdot)) = x(t;t_0,x^0,u(\cdot)), \ \mathcal{D}_\varphi = \{(t;t_0,x^0,u(\cdot)) \in T^2 \times X \times \mathcal{U};\ t \in T_{t_0,x^0,u(\cdot)}\}$$

then $\Sigma = (T, U, \mathcal{U}, X, Y, \varphi, \eta)$ *is a differentiable dynamical system.*

In general, differentiable systems are not complete. The following example illustrates this fact and shows that the maximal intervals of existence $T_{t_0,x^0,u(\cdot)}$ will in general depend on both x^0 and $u(\cdot) \in \mathcal{U}$.

Example 2.1.16 (Exploding solutions). Consider the initial value problem

$$\dot{x}(t) = x(t)^2 + u(t), \quad x(0) = x^0 \tag{18}$$

where $t \in T := \mathbb{R}$, $x^0 \in X := \mathbb{R}$. For the constant control $u(t) \equiv 1$, $t \ge 0$ we obtain the solution

$$x(t) = \tan(t + c(x^0)), \quad t \ge 0\,, \quad c(x^0) = \arctan x^0 \in (-\pi/2, \pi/2)$$

which "explodes" at the times $t_\pm(x^0) = \pm\pi/2 - c(x^0)$. Hence in this case the interval of existence is $(-\pi/2 - c(x^0), +\pi/2 - c(x^0))$. For the constant control $u(t) \equiv 0$ it is easily seen that $x(t) = x^0/(1 - x^0 t)$ is a solution of (18) on $(1/x^0, \infty)$ if $x^0 < 0$. For $x^0 = 0$ the solution is zero for all $t \in \mathbb{R}$ and for $x^0 > 0$ the interval of existence is $(-\infty, 1/x^0)$. □

We will now determine conditions under which a differentiable dynamical system with state space $X = \mathbb{K}^n$ is complete. The existence of solutions *in the large* (i.e. for all $\inf T < t < \sup T$) can be derived from Theorem 2.1.14 (i). Indeed, if $X = \mathbb{K}^n$ then $\sup T_{t_0,x^0} < \sup T$ (resp. $\inf T < \inf T_{t_0,x^0}$) can only occur when $x(t)$ is unbounded as $t \to \sup T_{t_0,x^0}$ (resp. $t \to \inf T_{t_0,x^0}$) . Thus we need criteria to ensure that a given solution will not escape to infinity at some time $\inf T < t_1 < \sup T$. Gronwall's lemma is fundamental for estimating the growth of solutions of differential equations. We give two versions of the lemma. The first one is important in this chapter, the more standard second version (which cannot be deduced from the first one) will be used in later chapters.

Lemma 2.1.17 (Generalized Gronwall inequality). *Suppose that* T *is an interval,* $a \in T$*,* $\beta(\cdot)$ *is a locally integrable non-negative function on* T *and* $\alpha(\cdot)$*,* $\xi(\cdot)$ *are non-negative continuous functions on* T *such that*

$$\xi(t) \le \alpha(t) + \left| \int_a^t \beta(r)\xi(r)dr \right|, \quad t \in T. \tag{19}$$

Then

$$\xi(t) \le \alpha(t) + \left| \int_a^t \alpha(r)\beta(r)\exp\left(\left| \int_r^t \beta(s)ds \right| \right) dr \right|, \quad t \in T. \tag{20}$$

Lemma 2.1.18 (Gronwall). *Suppose that T is an interval, $a \in T$ $\alpha \in \mathbb{R}$, $\beta(\cdot)$ is a locally integrable non-negative function on T and $\xi(\cdot)$ is a continuous function on T satisfying*

$$\xi(t) \leq \alpha + \int_a^t \beta(r)\xi(r)dr, \quad t \in T,\, t \geq a.$$

Then

$$\xi(t) \leq \alpha \exp\left(\int_a^t \beta(s)ds\right), \quad t \in T,\, t \geq a. \tag{21}$$

Proposition 2.1.19. *Suppose $T \subset \mathbb{R}$ is an open interval, $X \subset \mathbb{K}^n$ is open and $g : T \times X \to \mathbb{K}^n$ is affinely bounded, that is*

$$\|g(t,x)\| \leq M(t)\|x\| + m(t), \quad (t,x) \in T \times X. \tag{22}$$

where $M(\cdot)$, $m(\cdot)$ are locally integrable non-negative functions on T. Then every solution of (13) is bounded on every finite interval (t_1, t_2), $t_1, t_2 \in T, t_1 < t_2$ on which it is defined. If moreover $X = \mathbb{K}^n$ then every solution of (13) can be continued to all of T.

Proof: Let $x(\cdot)$ be a solution of (13) on $(t_1, t_2) \subset T$, $t_1, t_2 \in T$ and let $t_0 \in (t_1, t_2)$. It suffices to show that $x(\cdot)$ is bounded on $[t_0, t_2)$. The proof for $(t_1, t_0]$ is similar. Now

$$\begin{aligned}\|x(t)\| &\leq \|x(t_0)\| + \int_{t_0}^t \|g(r, x(r))\|dr \\ &\leq \left[\|x(t_0)\| + \int_{t_0}^t m(r)dr\right] + \int_{t_0}^t M(r)\|x(r)\|dr, \quad t_0 \leq t \leq t_2.\end{aligned}$$

Applying the generalized Gronwall inequality with $\alpha(t) = \|x(t_0)\| + \int_{t_0}^t m(r)dr$, $\xi(t) = \|x(t)\|$ and $\beta(t) = M(t)$ we see that $\|x(t)\|$ is bounded on $[t_0, t_2)$.
To conclude the proof, suppose that $X = \mathbb{K}^n$. It suffices to show that every *maximal* solution[2] $x(\cdot) : (t_-, t_+) \to X$ of (13) is defined on T. But if $t_+ < \sup T$ then $x(\cdot)$ would be bounded on $[t_0, t_+)$ for any $t_0 \in (t_-, t_+)$ and this would contradict Theorem 2.1.14 (i) since $\partial X = \emptyset$. $t_- = \inf T$ is shown similarly. □

As a corollary we obtain the following sufficient criterion for the completeness of a differentiable system.

Corollary 2.1.20. *Under the conditions of Corollary 2.1.15 with $X = \mathbb{K}^n$, if for every compact subinterval $I \subset T$ there exist constants C_I and c_I such that*

$$\|f(t,x,u)\| \leq C_I(\|u\|^p + 1)\|x\| + c_I(\|u\|^p + 1), \; (t,x,u) \in I \times X \times U \tag{23}$$

then the differentiable system $\Sigma = (T, U, \mathcal{U}, X, Y, \varphi, \eta)$ is complete and reversible.

Now let Σ be a differentiable system as in Definition 2.1.12 and $u(\cdot) \in \mathcal{U}$. The input function $u(\cdot)$ defines at every time $t \in T$ a *vector field* $x \mapsto f(t, x, u(t))$ on X. Of

[2] A solution of (13) which cannot be continued to a solution of (13) on a larger interval is called *maximal*.

particular importance are those states $\bar{x}$ at which the vector fields $x \mapsto f(t, x, u(t))$ vanish for all times $t \in T$

$$f(t, \bar{x}, u(t)) = 0, \quad t \in T. \tag{24}$$

These states are *singular points* for all the vector fields $x \mapsto f(t, x, u(t))$, $t \in T$. They represent equilibria of the system in the sense that if the state at an arbitrary initial time $t_0 \in T$ is $\bar{x}$ and Σ is controlled by $u(\cdot)$ then it remains in this state for all $t \in T_{t_0}$. The following definition applies to arbitrary dynamical systems.

Definition 2.1.21 (Equilibrium state). Let Σ be a dynamical system and $u(\cdot) \in \mathcal{U}$, then $\bar{x} \in X$ is said to be an *equilibrium state* of Σ under the control $u(\cdot)$ if

$$\varphi(t; t_0, \bar{x}, u(\cdot)) = \bar{x}, \qquad t_0, t \in T, \ \ t \geq t_0.$$

Systems which arise from technical processes are often designed to operate at a variety of equilibrium states. These different states are obtained by altering the input signal $u(\cdot)$. The next example describes a simple differentiable system in the sense of Definition 2.1.12 which has this property.

Example 2.1.22. Consider a tank of infinite height with constant cross sectional area a to which an incompressible fluid is supplied by a pipe with flow rate $u(t)$. The fluid leaves the tank via an orifice of cross sectional area a_0 (see Figure 2.1.4). Neglecting all inertia

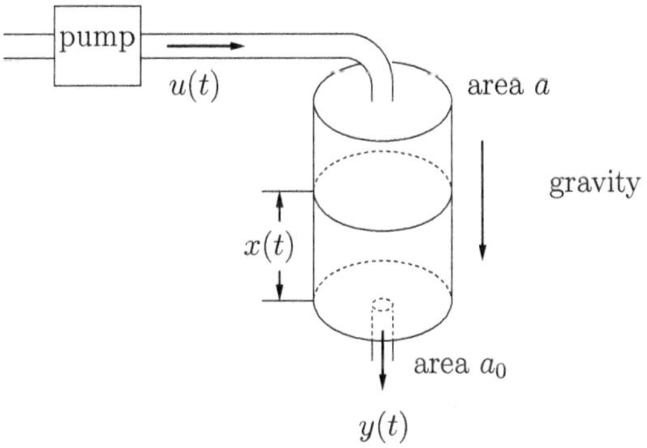

Figure 2.1.4: Fluid level control in a tank

effects of the fluid in the tank, the outlet flow rate y is related to the height x of the liquid level in the tank by the equation

$$y = a_0 \gamma \sqrt{2gx} \tag{25}$$

where γ is a "discharge coefficient" (0.62 for a sharp-edged orifice) and g is the gravitational constant. The principle of conservation of mass yields the following differential equation

$$\dot{x} = -\frac{a_0}{a}\gamma\sqrt{2gx(t)} + \frac{u(t)}{a}. \tag{26}$$

Clearly, (25) and (26) make sense only for $x > 0$. We regard x as the state of our system and choose $X = (0, \infty)$ to be the state space and $\mathcal{U}$ to be the set of all piecewise continuous non-negative functions $u(\cdot) : \mathbb{R} \to U = \mathbb{R}_+$. Applying Corollary 2.1.15 we see that for each

initial height $x^0 > 0$ of the liquid level in the tank and any control $u(\cdot) \in \mathcal{U}$ (26) admits a unique solution $x(t) > 0$ with $x(0) = x^0$ on some maximal time interval (t_1, t_2). Thus (25) and (26) define a differentiable dynamical system Σ with the above specification of $U, \mathcal{U}, X$ and $Y = (0, \infty)$. Since the RHS of (26) is affinely bounded $x(t)$ does not explode in finite time by Proposition 2.1.19. But $x(t)$ may leave the state space $X = (0, \infty)$ in finite time (e.g. for $u(\cdot) = 0$) so that Σ is not complete. If, however, $u(t) \geq \varepsilon > 0$ for all t then $x(t)$ cannot tend to 0 in finite time. Hence Σ is complete for all controls which are bounded away from zero.
Now suppose that the control is kept constant $u(t) \equiv \bar{u} > 0$, then for each value of $\bar{u}$ there is exactly one equilibrium state $\bar{x}$ namely

$$\bar{x}(\bar{u}) = \frac{1}{2g}\left(\frac{\bar{u}}{a_0\gamma}\right)^2.$$

The corresponding equilibrium output value is, as it should be, $\bar{y} = \bar{u}$. □

2.1.3 System Properties

In the previous subsections we introduced some special properties of dynamical systems - complete, reversible, differentiable. We shall now define further classifying properties which will play an important role later.
With respect to the time domain we distinguish between *continuous time* systems where T is a bounded or unbounded interval and *discrete time* systems where T is a discrete subset of $\mathbb{R}$. Typical discrete time domains are $T = \mathbb{Z}$, $T = \mathbb{N}$ or some corresponding equidistant time sequences $\mathbb{Z}_\mathcal{T} = \{k\tau;\ k \in \mathbb{Z}\}$, $\mathbb{N}_\mathcal{T} = \{k\tau;\ k \in \mathbb{N}\}$ where $\tau > 0$. Discrete time counterparts to differentiable systems are systems described by *difference equations.* Unlike differentiable systems they can be defined in a purely set theoretic framework.

Example 2.1.23 (Recursive system). Let U, X, Y be non-empty sets, $T = \mathbb{N}$ or $\mathbb{Z}$ and

$$f: T \times X \times U \to X, \quad \eta: T \times X \times U \to Y$$

be two arbitrary mappings. For any $u(\cdot) \in \mathcal{U} = U^T$, $t_0 \in T$, $x^0 \in X$ let $\varphi(t; t_0, x^0, u(\cdot))$, $t \in T$, $t \geq t_0$ be the unique solution of the recursive (or difference) equation

$$x(t+1) = f(t, x(t), u(t)) \tag{27}$$

with initial value $x(t_0) = x^0$. Then $\Sigma = (T, U, \mathcal{U}, X, Y, \varphi, \eta)$ is a discrete-time dynamical system. Every discrete time system (with the above time domains) can be described in this way (with possible restriction of $\mathcal{U}$) and, in particular, automata may be regarded as special recursive systems. □

Although the state of a dynamical system evolves in time, the system itself may be *time–invariant* in the sense that the state transition map is invariant with respect to time shifts and the output map does not depend explicitly on time. These systems are more easily analyzed than time–varying ones and so time invariance is often assumed although in reality the system dynamics may change slowly by the effect of growth, ageing, wear and tear, etc.. If $T \subset \mathbb{R}$, U is any non-empty set and $\tau \in \mathbb{R}$ we denote by S_τ the *shift operator* on U^T defined by

$$(S_\tau u)(t) = \begin{cases} u(t-\tau) & \text{if } t-\tau \in T \\ 0 & \text{otherwise.} \end{cases} \tag{28}$$

Definition (28) does not make sense if $t - \tau \notin T$ for some $t \in T$ and $0 \notin U$ (for instance, if U is not a subset of a vector space). Whenever we make use of the shift operator S_τ, it is implicitely assumed that either $t - \tau \in T$ for all $t \in T$ or U is a subset of a vector space V and contains the zero element of V. S_τ is called the *right* or *forward* shift if $\tau > 0$ and the *left* or *backward* shift if $\tau < 0$.

Definition 2.1.24 (Time–invariant system). A dynamical system Σ is said to be *time–invariant* if it satisfies the following axioms

(i) $T \subset \mathbb{R}$ contains 0 and is closed under addition, i.e. $T + T \subset T$.

(ii) $\mathcal{U}$ is invariant under the right shift, i.e. $S_\tau \mathcal{U} \subset \mathcal{U}$ for all $\tau \in T$, $\tau \geq 0$.

(iii) For every t_0, t, $\tau \in T$, $t \geq t_0$, $\tau \geq 0$ and every $x^0 \in X$, $u(\cdot) \in \mathcal{U}$

$$\varphi(t+\tau; t_0+\tau, x^0, S_\tau u(\cdot)) = \varphi(t; t_0, x^0, u(\cdot)).$$

(iv) The output map η does not depend on time, i.e. $\eta(t, x, u) = \eta(x, u)$, $t \in T$.

From (iii) we see that if $x(\cdot)$ is the state response to $u(\cdot)$ starting at (t_0, x^0), the state response $\tilde{x}(\cdot)$ to the control $\tilde{u} = S_\tau u$ starting at $(t_0 + \tau, x^0)$ is given by $\tilde{x}(t) = (S_\tau x)(t)$, $t \geq t_0 + \tau$ (see Figure 2.1.5).

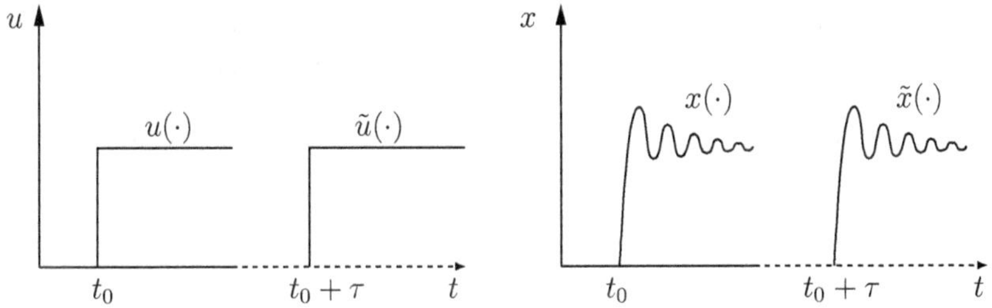

Figure 2.1.5: Time invariance.

The state transition map of a time–invariant system is completely determined by its state transition map at the fixed initial time $t_0 = 0$

$$\varphi(t; x^0, u(\cdot)) := \varphi(t; 0, x^0, u(\cdot)), \quad (t, x^0, u(\cdot)) \in T \times X \times \mathcal{U}.$$

A differentiable or recursive system

$$\begin{aligned} \dot{x}(t) &= f(t, x(t), u(t)), \ t \in \mathbb{R}_+ \\ y(t) &= \eta(t, x(t), u(t)) \end{aligned} \qquad \begin{aligned} x(t+1) &= f(t, x(t), u(t)), \ t \in \mathbb{N} \\ y(t) &= \eta(t, x(t), u(t)) \end{aligned}$$

is time–invariant if f and η do not depend explicitly on time. In particular every automaton is a time-invariant dynamical system.

Sometimes it is mathematically convenient to convert a time–varying differentiable or difference system into a time–invariant one by introducing time as a new state variable $x_{n+1}(t) = t$. Then the system equations become

$$\begin{aligned} \dot{x}(t) &= f(x_{n+1}(t), x(t), u(t)) \\ \dot{x}_{n+1}(t) &= 1 \\ y(t) &= \eta(x_{n+1}(t), x(t), u(t)), \end{aligned} \qquad \begin{aligned} x(t+1) &= f(x_{n+1}(t), x(t), u(t)) \\ x_{n+1}(t+1) &= x_{n+1}(t) + 1 \\ y(t) &= \eta(x_{n+1}(t), x(t), u(t)). \end{aligned}$$

Note that this method increases the dimension of the state space by one.

A system is called *finite*, *finite dimensional* or *infinite dimensional* depending on whether its state space X is a finite set, or a finite or infinite dimensional vector space. The system described in Examples 2.1.7 and 1.5.6 are finite, those in Examples 2.1.10, 2.1.22 are finite dimensional, and the heat equation of Section 1.6 describes an infinite dimensional system. Another infinite dimensional system is presented in the next example which illustrates very clearly the relationship between the state space and the memory of a system.

Example 2.1.25 (Delay system). Consider the system described by

$$\begin{aligned} \dot{x}(t) &= A_0 x(t) + A_1 x(t-h) + Bu(t) \\ y(t) &= Cx(t) \end{aligned} \tag{29}$$

where A_0, $A_1 \in \mathbb{R}^{n\times n}$, $B \in \mathbb{R}^{n\times m}$, $C \in \mathbb{R}^{p\times n}$ and $h > 0$ are given. Here the velocity $\dot{x}(t)$ depends not only on the present state $x(t)$ and control value $u(t)$ but also on the past value $x(t-h)$. Thus the system has a memory of positive length h whereas differentiable systems, in the sense of Definition 2.1.12, have only a memory of infinitesimal duration.

Mathematical models of the above type (both linear and nonlinear) play an important role whenever an action produces an effect with some delay. For example in engineering, feedback control systems sometimes contain long transmission lines which induce non-negligible time lags in the response of the plant regulator. In biology, the growth of a species is influenced by the time lag between birth and procreation. Also in economics, there is a time delay between an investment decision and its effect on productive capacity–the so-called "period of realization" of an investment.

In order to obtain a suitable state space for the system we have to find the amount of initial data required at any time $t_0 \in T = \mathbb{R}$ to determine the future evolution of $x(\cdot)$ on $[t_0, \infty)$. Obviously we need to know the values of $x(s)$ for $t_0 - h \le s \le t_0$. In fact we will show that for an arbitrary continuous initial function $z(\cdot) \in X := \mathcal{C}([-h, 0], \mathbb{R}^n)$ and piecewise continuous control $u(\cdot) : [t_0, \infty) \to \mathbb{R}^m$ there exists a unique continuous function $x(\cdot) : [t_0 - h, \infty) \to \mathbb{R}^n$ which coincides with $S_{t_0} z(\cdot)$ on $[t_0 - h, t_0]$ and satisfies (29) for $t \ge t_0$. We construct this solution by the *method of steps*. On the interval $[t_0, t_0 + h]$, $x(t)$ is uniquely determined by the variation-of-parameters formula for ordinary differential equations (see Example 2.2.1)

$$x(t) = e^{A_0(t-t_0)} z(0) + \int_{t_0}^{t} e^{A_0(t-s)} [A_1 z(s - t_0 - h) + Bu(s)] ds, \quad t \in [t_0, t_0 + h]. \tag{30}$$

If we set $x(t_0 + s) = z(s)$ for $s \in [-h, 0]$, then obviously $x(\cdot)$ is continuous on $[t_0 - h, t_0 + h]$. Now knowledge of $x(\cdot)$ on $[t_0, t_0 + h]$ enables us via (30) to construct $x(\cdot)$ on $[t_0 + h, t_0 + 2h]$ (replace t_0 by $t_0 + h$, $z(0)$ by $x(t_0 + h)$ and $z(s - t_0 - h)$ by $x(s - t_0 - h)$). Continuing this process we see that there is a unique continuous solution $x(\cdot)$ of (29) on $[t_0, \infty)$ with $x(t_0 + s) = z(s)$ for $s \in [-h, 0]$. Since we need to know the whole function segment

$$x_t : s \mapsto x(t+s) \quad s \in [-h, 0]$$

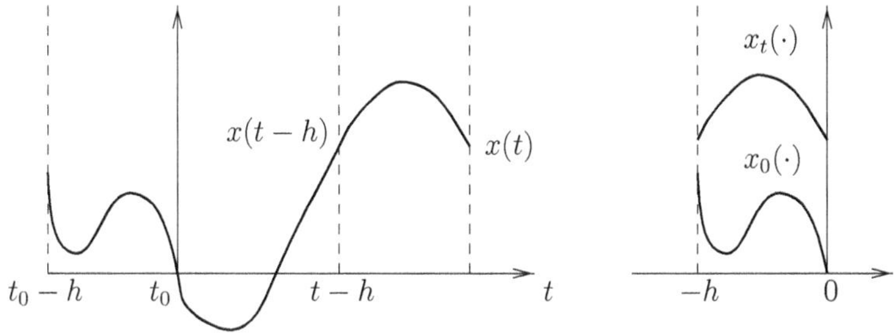

Figure 2.1.6: State of the delay system

in order to determine the system's future evolution under a given control $u(\cdot)$, we regard x_t as the state of our system at time t (see Figure 2.1.6) and take a suitable function space, e.g. $X = \mathcal{C}([-h, 0], \mathbb{R}^n)$, as state space. $x_t(\cdot)$ is simply the trajectory $x(\cdot)$ seen through a window of width h moving with time. The corresponding state transition map φ is given by

$$\varphi(t; t_0, z(\cdot), u(\cdot)) = x_t(\cdot)$$

where $t \in [t_0, \infty)$, $z(\cdot) \in X$ and $x(\cdot)$ is the corresponding solution of (29) with initial state $x_{t_0}(\cdot) = z(\cdot)$. If we apply a time shift $\tau \geq 0$, the solution of (29) on $[t_0 + \tau, \infty)$ with initial state $x_{t_0+\tau}(\cdot) = z(\cdot)$ and shifted input function $S_\tau u(\cdot)$ will be $S_\tau x(\cdot)$. Hence if the output map is given by

$$\eta(t, z(\cdot), u) = Cz(0), \quad z(\cdot) \in X, \ u \in \mathbb{R}^m$$

we have an example of a time–invariant infinite dimensional system.
Note that any solution of (29) must be absolutely continuous on its domain of definition. So if $z(\cdot) \in X$ is not absolutely continuous, then there cannot exist a solution of (29) on all of $\mathbb{R}$ which coincides with $z(\cdot)$ on $[-h, 0]$. Hence the system is not reversible. □

Besides *time–invariance* another system property which will play a central role throughout this book is that of *linearity*.

Definition 2.1.26 (Linear system). Let $\mathbb{K}$ be an arbitrary field. A dynamical system Σ is said to be $\mathbb{K}$-*linear* if

(i) U, $\mathcal{U}$, X, Y are vector spaces over $\mathbb{K}$,

(ii) the maps

$$\varphi(t; t_0, \cdot, \cdot) : \ X \times \mathcal{U} \to X \ \text{ and } \ \eta(t, \cdot, \cdot) : \ X \times U \to Y$$

are $\mathbb{K}$-linear for all t, $t_0 \in T$, $t \geq t_0$.

Condition (ii) implies that

$$\varphi(t; t_0, 0_X, 0_{\mathcal{U}}) = 0_X, \quad t, \ t_0 \in T, \ t \geq t_0$$

where 0_X is the origin in X and $0_{\mathcal{U}}$ the origin in $\mathcal{U}$ (zero function). This means that 0_X is an equilibrium state of Σ under the control $0_{\mathcal{U}}$ whenever Σ is linear. Example 2.1.25 is a linear system as is the system described by equations (1.3.29), (1.3.33) in Example 1.3.2.

2.1.4 Linearization

We conclude this section with some remarks on how linear models can be used to approximate the behaviour of a nonlinear differentiable system close to a given trajectory or equilibrium point. Let Σ be a differentiable dynamical system with state equation

$$\dot{x}(t) = f(t, x(t), u(t)), \quad t \in T \tag{31}$$

and output equation

$$y(t) = \eta(t, x(t), u(t)) \tag{32}$$

where $T \subset \mathbb{R}$ is an open interval, $U \subset \mathbb{R}^m$ and $X \subset \mathbb{R}^n$ are open, $Y = \mathbb{R}^p$, $\mathcal{U} = \mathcal{C}(T, U)$. Let $\tilde{x}(\cdot)$ be the trajectory corresponding to a given control $\tilde{u}(\cdot) \in \mathcal{U}$ and initial condition $(t_0, \tilde{x}^0) \in T \times X$, so that

$$\begin{aligned} \dot{\tilde{x}}(t) &= f(t, \tilde{x}(t), \tilde{u}(t)), \quad t \geq t_0,\ t \in T \\ \tilde{x}(t_0) &= \tilde{x}^0. \end{aligned}$$

We assume that the functions $f : T \times X \times U \to \mathbb{R}^n$ and $\eta : T \times X \times U \to Y$ are continuous and continuously differentiable with respect to (x, u) on $T \times X \times U$. Consider the Fréchet derivatives (Jacobians)

$$\begin{aligned} A(t) &= D_x f(t, \tilde{x}(t), \tilde{u}(t)) = \left[\frac{\partial f_i}{\partial x_j}(t, \tilde{x}(t), \tilde{u}(t))\right]_{n \times n} \\ B(t) &= D_u f(t, \tilde{x}(t), \tilde{u}(t)) = \left[\frac{\partial f_i}{\partial u_k}(t, \tilde{x}(t), \tilde{u}(t))\right]_{n \times m} \\ C(t) &= D_x \eta(t, \tilde{x}(t), \tilde{u}(t)) = \left[\frac{\partial \eta_i}{\partial x_j}(t, \tilde{x}(t), \tilde{u}(t))\right]_{p \times n} \\ D(t) &= D_u \eta(t, \tilde{x}(t), \tilde{u}(t)) = \left[\frac{\partial \eta_i}{\partial u_k}(t, \tilde{x}(t), \tilde{u}(t))\right]_{p \times m}. \end{aligned} \tag{33}$$

The linear differentiable system described by

$$\begin{aligned} \dot{x}(t) &= A(t)x(t) + B(t)u(t) \\ y(t) &= C(t)x(t) + D(t)u(t) \end{aligned} \tag{34}$$

is said to be the *linearization* of (31) and (32) along the pair $(\tilde{x}(\cdot), \tilde{u}(\cdot))$.
Let $\xi^0 \in \mathbb{R}^n$, $u(\cdot) \in \mathcal{U}$ and for all small $\varepsilon > 0$ denote by $x(t, \varepsilon)$ the solution of (31) corresponding to the control $u(t, \varepsilon) = \tilde{u}(t) + \varepsilon u(t)$ and the initial condition $x(t_0, \varepsilon) = \tilde{x}^0 + \varepsilon \xi^0$. It follows from basic results concerning the dependence of solutions on parameters and initial conditions that $x(t, \varepsilon)$ is differentiable with respect to ε at $\varepsilon = 0$ and the derivative $\xi(t) = \frac{\partial x}{\partial \varepsilon}(t, 0)$ satisfies

$$\dot{\xi}(t) = A(t)\xi(t) + B(t)u(t), \quad t \in T,\ t \geq t_0$$

(see *Notes and References*). Hence, if $\xi(\cdot)$ is a solution of (34) corresponding to a control $u(\cdot)$ and initial state ξ^0 then, for small $\varepsilon > 0$, $\tilde{x}(t) + \varepsilon \xi(t)$ is a first order approximation to the solution of (31) corresponding to the control $\tilde{u}(t) + \varepsilon u(t)$ and

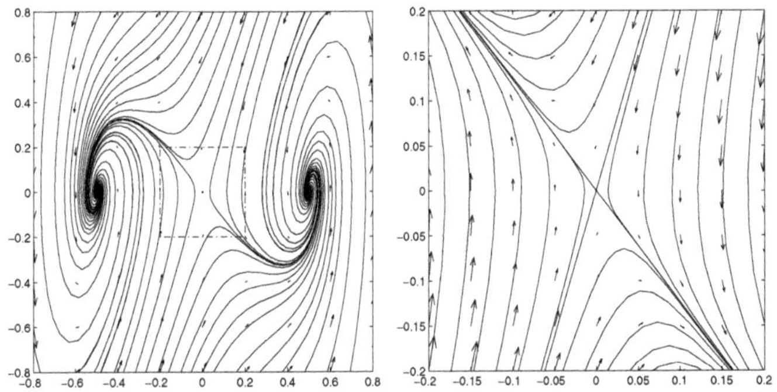

A nonlinear system   Its linearization at $\bar{x} = [0,0]^\top$

Figure 2.1.7: Phase portraits near an equilibrium point

initial state $\tilde{x}^0 + \varepsilon\xi^0$. Note however, that as $\varepsilon \to 0$ this approximation is, in general, only uniform in t on compact intervals. Nevertheless, the behaviour of the linear system (34) near the origin gives an approximate picture of the behaviour of the nonlinear system (31), (32) in a sufficiently small neighbourhood of the trajectory $\tilde{x}(t)$. The phase portraits in Figure 2.1.7 illustrate this for a time–invariant free system near an equilibrium solution $\tilde{x}(t) \equiv \bar{x}$ (saddle point). Note that the global properties of the nonlinear and linear systems are quite different.

As (33) shows, the linearized model is, in general, time varying *even if the nonlinear system is time–invariant*, and this is one of the main reasons for the importance of time–varying linear systems in control theory. However, if we linearize a time–invariant system at an equilibrium point corresponding to some constant control the linearized model will again be time–invariant.

Example 2.1.27 (Satellite). The motion of a satellite of mass $m = 1$ in a 2-dimensional central gravitational field of the form $k(r) = -\gamma r^{-2}$, $r \neq 0$ can be described by the following equations

$$\begin{aligned} \ddot{r}(t) &= r(t)\dot{\theta}^2(t) - \gamma r^{-2}(t) + u_1(t) && (35)\\ r(t)\ddot{\theta}(t) &= -2\dot{r}(t)\dot{\theta}(t) + u_2(t)\,. && (36) \end{aligned}$$

Here $r(t)$ is the distance of the satellite from the centre of gravitation at time t, $\dot{\theta}(t)$ is the angular velocity of the radius vector from the centre of gravitation to the satellite at time t, and $u_1(t)$, $u_2(t)$ are radial and tangential thrusts which we take to be control inputs.

If $u_1(\cdot) = u_2(\cdot) = 0$, the circular motion $r(t) = 1$, $\theta(t) = \sqrt{\gamma}\,t$ solves (35). Introducing the state variables $x_1 = r$, $x_2 = \dot{r}$, $x_3 = \theta$, $x_4 = \dot{\theta}$, we see that (35), (36) can be written in the form (31) where the coordinates of $f(t,x,u)$ are given by

$$\begin{aligned} f_1(t,x,u) &= x_2, & f_2(t,x,u) &= x_1x_4^2 - \gamma x_1^{-2} + u_1\\ f_3(t,x,u) &= x_4, & f_4(t,x,u) &= -2x_2x_4x_1^{-1} + u_2x_1^{-1}. \end{aligned}$$

If $\tilde{x}(t) = [1, 0, \sqrt{\gamma}\,t, \sqrt{\gamma}]^\top$, $\tilde{u}(t) = [0,0]^\top$ the linearized equation about this trajectory is

$$\dot{x} = Ax + Bu$$

where

$$A = \begin{bmatrix} 0 & 1 & 0 & 0 \\ 3\omega^2 & 0 & 0 & 2\omega \\ 0 & 0 & 0 & 1 \\ 0 & -2\omega & 0 & 0 \end{bmatrix} \qquad B = \begin{bmatrix} 0 & 0 \\ 1 & 0 \\ 0 & 0 \\ 0 & 1 \end{bmatrix} \tag{37}$$

and $\omega = +\sqrt{\gamma}$. If the distance $x_1(t)$ and the angle $x_3(t)$ are measured then the (linear) output equation is

$$y = Cx\,, \quad C = \begin{bmatrix} 1 & 0 & 0 & 0 \\ 0 & 0 & 1 & 0 \end{bmatrix}. \tag{38}$$

□

2.1.5 Exercises

1. *(RC network)* Introduce a suitable state vector and determine the state and the output equations of the electrical circuit represented in Figure 2.1.8. Choose the driving voltage

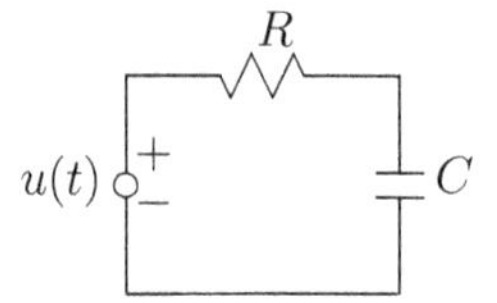

Figure 2.1.8: *RC* circuit

$u(\cdot)$ (piecewise continuous) as input and the current through the resistor R as output. Specify all the components of a differentiable dynamical system $\Sigma = (T, U, \mathcal{U}, X, Y, \varphi, \eta)$ modelling this circuit.

2. *(Tank system)* Determine the equation of motion and the output map of the fluid system shown in Figure 2.1.9. The cross sectional areas of the tanks are a_1, $a_2 > 0$. The

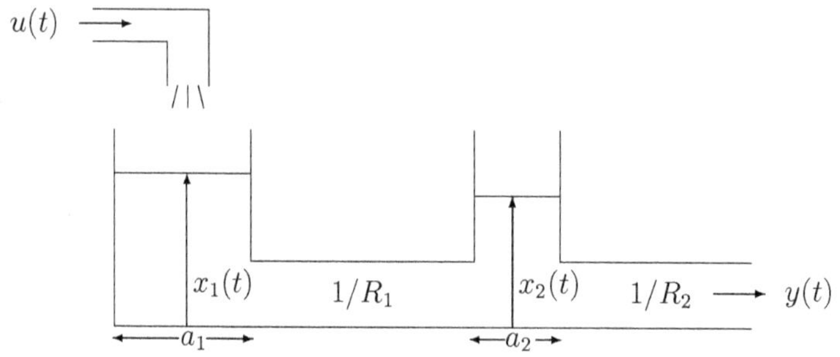

Figure 2.1.9: Fluid system

flow through the first orifice is proportional (with constant $1/R_1$) to $x_1(t) - x_2(t)$, and the flow through the second orifice is proportional with constant $1/R_2$ to $x_2(t)$. Specify all items of the corresponding dynamical system $\Sigma = (T, U, \mathcal{U}, X, Y, \varphi, \eta)$.

3. *(Mass-spring system)* Consider the mechanical system illustrated in Figure 2.1.10. Two masses m_1, m_2 are suspended on ideal springs with stiffness coefficients k_1, k_2, hanging from a fixed support. The outputs are the displacements $y_1(t)$, $y_2(t)$ of the two masses from their equilibrium positions. The input is a piecewise continuous force $u(t)$ applied

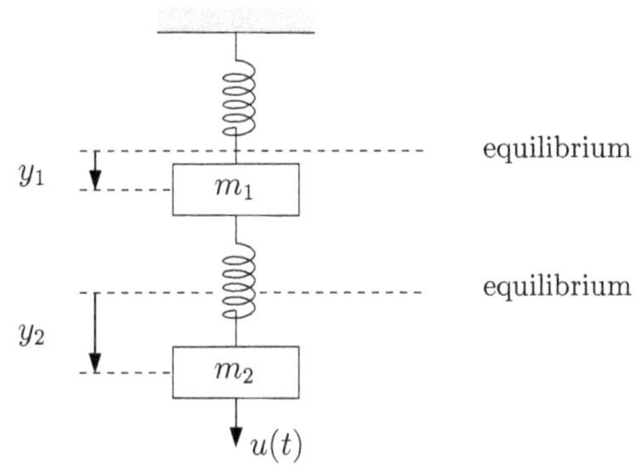

Figure 2.1.10: Mass-spring system

to the second mass. Assuming that the frictional resistances for the two masses in the surrounding medium are c_1 and c_2, we obtain the following equations of motion

$$\begin{aligned} m_1\ddot{y}_1 + c_1\dot{y}_1 + k_1 y_1 &= k_2(y_2 - y_1) \\ m_2\ddot{y}_2 + c_2\dot{y}_2 + k_2(y_2 - y_1) &= u \end{aligned} \tag{39}$$

Introduce a suitable state vector and specify a differentiable dynamical system Σ describing the above mechanical system. If $u(t) = \bar{u} \in \mathbb{R}$ is constant determine the corresponding set of equilibrium states. (see *Driver* (1977), [138, pp.173-74])

4. *(RLC circuit)* Consider the electrical circuit illustrated in Figure 2.1.11. If y_1, y_2 are the currents across the resistors R_1, R_2 show that the equations of motion are of the same form as (39) but with $\dot{u}(t)$ instead of $u(t)$ on the right hand side of the second equation. Introduce the currents through the inductor and the charge of the capacitor

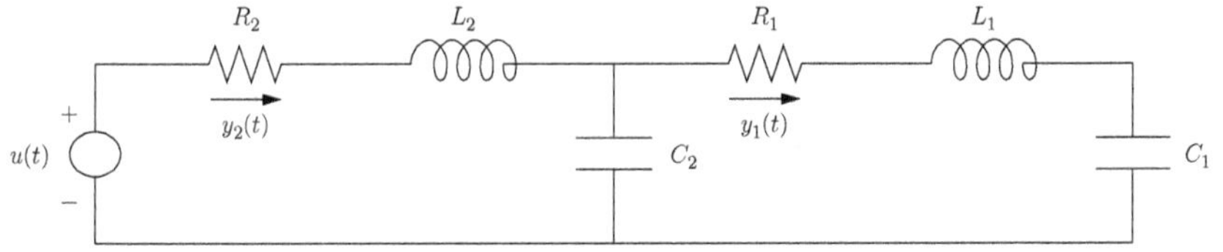

Figure 2.1.11: *RLC*-network

as state variables and specify the corresponding dynamical system Σ. Compare the state equation obtained with that of the previous exercise. Determine the equilibrium states corresponding to constant voltage $u(t) \equiv \bar{u}$. (see *Driver* (1977) [138, pp.177-178]).

5. *(Parity checker)* Construct an automaton $\mathcal{A} = (U, X, Y, \psi, \eta)$ with $U = Y = \mathbb{Z}_2$ and an initial state $x^0 \in X$ such that $\mathcal{A}$ acts as a "parity check machine" when initialized at x^0. This means it responds to an arbitrary finite sequence of zeros and ones with 0 if the number of ones is even and with 1 if the number of ones is odd (see *Birkhoff and Bartee* (1970) [60, pp.69-70]).

6. *(Coin operated dispenser)* Specify an automaton $\mathcal{A} = (U, X, Y, \psi, \eta)$ which models a candy machine that accepts two sorts of coins (nickels $N = 5$c and dimes $D = 10$c). The

price of a candy is 15c. According to the amount of money inserted, it returns nothing or a piece of candy or a piece of candy plus change. Assume that the candy store within the dispenser is infinite (see *MacClamroch* (1980) [369, pp.192]).

7. *(Communication system)* Consider a communication system of the structure as shown in Figure 2.1.12. A binary message (sequence of zeros and ones) is firstly encoded, then transmitted by a communication channel and then decoded to obtain the received message.

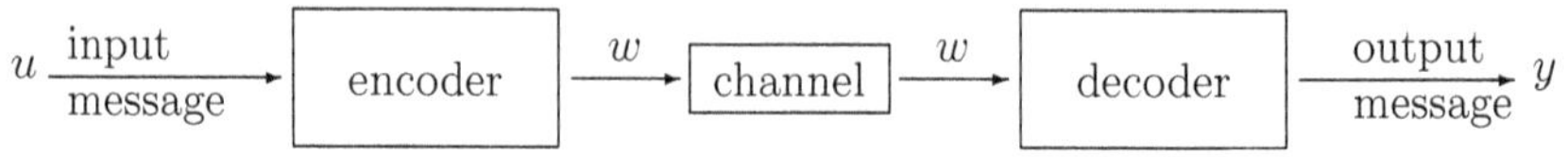

Figure 2.1.12: Communication system

Assume that the channel transmits the signal w without noise and that the encoder algorithm is described by

$$w(t) = u(t) + u(t-1) + u(t-2), \quad t \in \mathbb{N} \tag{40}$$

where we set $u(t) = 0$ for $t < 0$.

(i) Specify a dynamical system Σ_1 and an initial state $x(0) = x^0$ such that the corresponding input-output relation is identical with (40). Verify this for the input sequence (0 1 1 0 1 1 0 0 1 0 1 0 1).

(ii) Find a decoder algorithm which reconstructs the input message (such that $y(\cdot)$ is a shifted version of $u(\cdot)$).

(iii) Describe this decoder by a dynamical system Σ_2 and an appropriate initial state $\bar{x}(0) = \bar{x}^0$.

(iv) Specify a dynamical system Σ which describes the complete communication system, consisting of the encoder and the decoder coupled in series.

8. Sometimes the equations of motion of a differentiable system contains higher order derivatives of an internal variable z and/or derivatives of the control function u. The output equation may also contain derivatives of the internal variable z. In these cases one must seek a state vector x which enables these equations to be transformed into state and output equations of the prescribed form (see Definition 2.1.12). Find an appropriate state vector and determine the state and output equations corresponding to the following equations where a_i, $b_j \in \mathbb{R}$, $n \in \mathbb{N}$, $n \geq 1$ are given parameters

(i) $\ddot{z}(t) + a_1\dot{z}(t) + a_0 z(t) = u(t), \quad y(t) = \dot{z}(t),$

(ii) $y^{(n)}(t) + a_{n-1}y^{(n-1)}(t) + \ldots + a_1 y^{(1)}(t) + a_0 y(t) = u(t), \quad y^{(i)} = \dfrac{d^i y}{dt^i},$

(iii) $\dddot{y}(t) + a_2\ddot{y}(t) + a_1\dot{y}(t) = b_1\dot{u}(t) + b_0 u(t).$

9. Find a suitable state vector and determine state and output equations for the discrete time systems described by the following higher order difference equations on $T = \mathbb{N}$.

(i) $z(t+2) + a_1 z(t+1) + a_0 z(t) = u(t), \quad y(t) = z(t+1),$

(ii) $y(t+n) + a_{n-1}y(t+n-1) + \ldots + a_1 y(t+1) + a_0 y(t) = u(t),$

(iii) $y(t+n)+a_{n-1}y(t+n-1)+\ldots+a_0y(t)=b_{n-1}u(t+n-1)+\ldots+b_0u(t).$

10. Examine which of the following equations describe a complete dynamical system Σ. Determine the dimension of Σ and determine whether Σ is linear and/or time–invariant.

(i) $z(t)=u(t), \quad y(t)=\dot{z}(t), \quad t\in\mathbb{R},$

(ii) $\ddot{z}(t)=u(t), \quad y(t)=\dot{z}(t), \quad t\in\mathbb{R},$

(iii) $z(t)=\dot{u}(t), \quad y(t)=z(t), \quad t\in\mathbb{R},$

(iv) $\dot{z}(t)=z(t)u(t), \quad y(t)=\dot{z}(t), \quad t\in\mathbb{R},$

(v) $z(t)=(z(t-1)+z(t+1))/2, \quad y(t)=z(t), \quad t\in\mathbb{Z},$

(vi) $z(t)=(u(t-1)+u(t+1))/2, \quad y(t)=z(t-1), \quad t\in\mathbb{Z},$

(vii) $z(t+1)=z(t)+u(t-1), \quad y(t)=z(t-1), \quad t\in\mathbb{Z},$

(viii) $z(t+1)=-z(t)+u(t+1), \quad y(t)=z(t)^2, \quad t\in\mathbb{Z},$

(ix) $\dot{z}(t)=z(t)^{2/3}, \quad y(t)=z(t)^3, \quad t\in[0,\infty),$

(x) $\ddot{z}(t)+t\dot{z}(t)+t^2z(t)=u(t), \quad y(t)=\ddot{z}(t), \quad t\in[0,\infty),$

(xi) $z(t)=z(t-1)+z(t-2)+u(t), \quad y(t)=z(t), \quad t\in\mathbb{R},$

(xii) $\dot{z}(t)=z(t-1)+\dot{z}(t-1)+u(t), \quad y(t)=z(t-1/2), \quad t\in\mathbb{R}.$

11. Consider the scalar delay system

$$\dot{x}(t)=x(t-1)+u(t), \quad y(t)=x(t), \quad t\geq 0. \tag{41}$$

(i) Specify a dynamical system Σ described by these equations (see Example 2.1.25).

(ii) Determine the solution $x(\cdot)$ corresponding to the control function $u(t)\equiv 0$ and the initial condition $x(t)=1, \ t\in[-1,0]$ (give an explicit formula for $x(\cdot)$ on the intervals $[k,k+1], \ k\in\mathbb{N}$). Show that (41) cannot be solved backwards in time.

(iii) Solve (41) when $u(t)\equiv 1$ and $x(t)=1, \ t\in[-1,0]$.

(cf. *Bellman and Cooke* (1963))

12. Prove Gronwall's Lemma 2.1.18.

13. In Example 2.1.16 we saw that the equation $\dot{x}(t)=x(t)^2+u(t)$ does not have a common interval of existence for all initial states $x^0\in\mathbb{R}$ and constant controls $u(\cdot)$. Now introduce a small delay $\varepsilon>0$

$$\dot{x}(t)=x(t-\varepsilon)^2+u(t), \quad t\geq 0.$$

Specify T, X, U, $\mathcal{U}$, φ for a dynamical system Σ (without outputs) described by this equation of motion. Determine whether Σ is complete and/or reversible.

14. *(Euler's equations)* The rotation of a rigid body around its centre of mass is described by the equations

$$\begin{aligned} I_1\,\dot{\omega}_1(t) &= (I_2-I_3)\,\omega_2(t)\omega_3(t)+u_1(t)\\ I_2\,\dot{\omega}_2(t) &= (I_3-I_1)\,\omega_1(t)\omega_3(t)+u_2(t)\\ I_3\,\dot{\omega}_3(t) &= (I_1-I_2)\,\omega_1(t)\omega_2(t)+u_3(t)\end{aligned}$$

where ω is the angular velocity in a coordinate system coinciding with the principal axes of the rigid body. I_1, I_2, I_3 are the principal moments of inertia and $u = [u_1, u_2, u_3]^\top$ is the applied torque, see *Goldstein* (1980) [194, pp.158]. Assume $I_1 = I_2$ (symmetry) and consider the free motion ($\tilde{u} \equiv 0$)

$$\begin{aligned}\tilde{\omega}_1(t) &= \cos\left[\omega(I_2 - I_3)t/I_2\right]\\ \tilde{\omega}_2(t) &= \sin\left[\omega(I_3 - I_2)t/I_2\right]\\ \tilde{\omega}_3(t) &= \omega_0\end{aligned}$$

where $\omega_0 > 0$ is given. Linearize the above system about $(\tilde{\omega}(\cdot), \tilde{u}(\cdot))$.

15. **Case study: Fisheries model.** Consider a simple Verhulst model (see Example 1.1.1) for the dynamics of a fish population $x(t)$ in a pond

$$\begin{aligned}\dot{x}(t) &= \alpha x(t)(K - x(t)) - u(t)), \quad t \geq 0\\ x(0) &= x^0 > 0\end{aligned} \tag{42}$$

where $u(t)$ is the harvesting rate, $K > 0$ the saturation level of the population and $\alpha > 0$ a constant. We require that $x(t) \geq 0$, $u(t) \geq 0$ for all t.

(i) No harvesting. For $u(t) = 0$ determine an explicit formula for the solution of (42) on $[0, \infty)$ by separation of variables. Show that for all initial states $x^0 > 0$ the solution tends to the equilibrium state $\bar{x} = K$.

(ii) Constant harvesting with over-exploitation. Show that a constant harvesting rate $u(t) \equiv \bar{u} > \alpha K^2/4$ leads to depletion in finite time for all initial conditions.

(iii) Constant harvesting without over-exploitation. Assume that the harvesting rate $u(t) \equiv \bar{u}$ is constant but $\bar{u} < \alpha K^2/4$. Show that there exist two equilibrium states $\bar{x}$, $\hat{x}$, $0 < \bar{x} < \hat{x}$, and that for $x^0 > \bar{x}$ (42) admits a (unique) solution $x(t)$ on $[0, \infty)$ which tends to $\hat{x}$ as $t \to \infty$. However, if $x^0 < \bar{x}$, show that depletion occurs in finite time.

(iv) What happens if $u(t) \equiv \alpha K^2/4$, $t \geq 0$?

(v) Constant effort harvesting. If a constant effort for harvesting is made the harvesting rate $u(t)$ will be proportional to $x(t)$ so $u(t) = e\,x(t)$. Show that depletion occurs in finite time if $e > \alpha K$. Thus assume $e < \alpha K$. Prove that there exist two equilibrium states 0 and $\bar{x}$ and that for *any* initial state $x^0 > 0$ the corresponding solution of (42) tends to $\bar{x}$ as $t \to \infty$. Determine the effort coefficient $e \in [0, \alpha K]$ which leads to an equilibrium state with maximal harvesting rate $e\,\bar{x}$. Discuss the result in comparison with the result obtained in (iv).

(vi) Let $\mathcal{U}$ be the set of all piecewise continuous control functions $u(\cdot) : [0, \infty) \to [0, u_1]$ where $0 < u_1 < \alpha K^2/4$. Determine $x_1 > 0$ such that for any initial state $x^0 \in X := [x_1, \infty)$ and any control function $u(\cdot)$ there exists a (unique) solution of (42) on $[0, \infty)$.

2.1.6 Notes and References

A general concept of a dynamical system was first formulated by *Kalman* (1963) [287] (see also the introduction of *Kalman et al.* (1969) [290]). A concept of equal generality from an input-output point of view was defined by *Zadeh and Desoer* (1963) [542]. *Sontag* (1998)

has given a state space oriented general definition allowing for local existence of solutions in [472]. A novel comprehensive framework for the theory of dynamical systems has been developed by *J. C. Willems* in his *behavioural approach* to dynamical systems, see [529] and *Polderman and Willems* (1997) [416]. Its most salient feature is that it does not make an a priori distinction between inputs and outputs.
The relationship between the theory of dynamical systems and the theory of automata has been emphasized by *Arbib* (1968) [16]. The emergence of automata theory which was stimulated by the development of information-processing technology dates back to the fifties. An interesting early reference is the volume on Automata Theory in the *Annals of Mathematics Studies* series edited by *Shannon and McCarthy* (1956) [461]. This volume contains contributions by some of the pioneers of the field, Shannon, v. Neumann, Kleene, and Moore. Other important early contributions which led to the concept of *finite state machine* (Definition 2.1.6 with finite input, state and output sets) were *Huffman* (1954) [270], *Mealy* (1955) [371] and the papers collected in [381]. Finite state machines were studied as mathematical models of switching and encoding networks in abstraction from hardware considerations. The theory was strongly influenced by earlier developments in logic and the theory of computability of recursive functions, in particular by the concept of a Turing machine (invented by *Turing* in 1935). A comprehensive treatise on automata is *Eilenberg* (1974) [147], a more recent reference is *Khoussainov and Nerode* (2001) [307].
Many elementary examples of differentiable control systems are described in *MacClamroch* (1980) [369] and in the excellent introduction of *Luenberger* (1979) [349]. The satellite model (Example 2.1.27) is discussed in *Brockett* (1970) [77].
For the existence and uniqueness results from the theory of differential equations, see *Dieudonné* (1970) [132], *Hale* (1980) [214] and *Amann* (1990) [11]. The generalized Gronwall inequality can be found in [11] under the additional assumption that $\beta(\cdot)$ is continuous on T. However it is easy to see that local integrability suffices. For the more standard version of Gronwall's Lemma see e.g. [541]. Differential equations where the RHS depends measurably on t are carefully dealt with in *Aulbach and Wanner* (1996) [27]. For the continuous dependence of solutions on parameters, see e.g. [11], [214]. Differentiability theorems on which Subsection 2.1.4 on Linearization is based can be found in [11].
Some results on nonlinear control systems can be found in *Lee and Markus* (1967) [336]. A differential geometric approach has been developed by *Isidori* (1989) [275] and *Nijmeyer and van der Schaft* (1990) [393]. A comprehensive recent textbook which presents different methods for the study and design of nonlinear control systems both in state space and in an input-output framework is *Sastry* (1999) [448]. Advanced treatments of the classical theory of dynamical systems are *Palis and de Melo* (1982) [404], *Arnold* (1983) [20], *Devaney* (1989) [131] and *Katok and Hasselblatt* (1995) [295].
An elementary introduction to delay equations is given in *Driver* (1977) [138]. Many examples and interesting results on delay equations can be found in *Bellman and Cooke* (1963) [45]. Another standard reference for functional differential equations is *Hale* (1977) [213].

2.2 Linear Systems

The assumption of linearity allows every state and output trajectory to be represented as a linear combination ("superposition") of a fixed set of simpler trajectories. The use of this superposition principle for the analysis of linear systems is the central topic of this section.
We begin by considering *general* linear systems and show that every trajectory can be decomposed into a *free* motion which depends only on the initial state and a *forced* motion starting at zero which depends only on the control function. Then we specialize to the class of linear time–invariant finite dimensional systems described by differential and difference equations. Their free motions are analyzed in detail and we show that every free trajectory can be decomposed into generalized eigenmotions. The forced motions will be considered more thoroughly in the next section. We conclude with the study of a particular infinite dimensional system described by partial differential equations.

2.2.1 General Linear Systems

Let $\mathbb{K}$ be an arbitrary field and $\Sigma = (T, U, \mathcal{U}, X, Y, \varphi, \eta)$ a $\mathbb{K}$-linear system, then for every t_0, $t \in T$, $t \geq t_0$ and $\lambda_i \in \mathbb{K}$, $x_i \in X$, $u_i \in U$, $u_i(\cdot) \in \mathcal{U}$, $i = 1, \ldots, k$ we have

$$\varphi(t; t_0, \sum_{i=1}^{k} \lambda_i x_i, \sum_{i=1}^{k} \lambda_i u_i(\cdot)) = \sum_{i=1}^{k} \lambda_i \varphi(t; t_0, x_i, u_i(\cdot)) \tag{1}$$

$$\eta(t, \sum_{i=1}^{k} \lambda_i x_i, \sum_{i=1}^{k} \lambda_i u_i) = \sum_{i=1}^{k} \lambda_i \eta(t, x_i, u_i). \tag{2}$$

These equations express the *superposition principle* for the state and the output. As a special case we obtain the so-called *decomposition principle*

$$\varphi(t; t_0, x^0, u(\cdot)) = \varphi(t; t_0, x^0, 0_{\mathcal{U}}) + \varphi(t; t_0, 0_X, u(\cdot)). \tag{3}$$

This shows that every trajectory of a linear system can be decomposed into the sum of a *free motion* $t \mapsto \varphi(t; t_0, x^0, 0_{\mathcal{U}})$ which depends only on the initial state x^0 and a *forced motion* $t \mapsto \varphi(t; t_0, 0_X, u(\cdot))$ which depends only on the control $u(\cdot)$.
Again as a special case of (1), by setting $u_i(\cdot) = 0_{\mathcal{U}}$, we get the *superposition law of free motions*

$$\varphi(t; t_0, \sum_{i=1}^{k} \lambda_i x_i, 0_{\mathcal{U}}) = \sum_{i=1}^{k} \lambda_i \varphi(t; t_0, x_i, 0_{\mathcal{U}}) \tag{4}$$

and, by setting $x_i = 0_X$, the *superposition law of forced motions*

$$\varphi(t; t_0, 0_X, \sum_{i=1}^{k} \lambda_i u_i(\cdot)) = \sum_{i=1}^{k} \lambda_i \varphi(t; t_0, 0_X, u_i(\cdot)). \tag{5}$$

It is easy to see that the general superposition principle for state trajectories (1) is equivalent to the decomposition principle (3) together with the superposition laws (4) and (5).

The decomposition law leads us to introduce the following two families of linear maps. For any pair of times $(t, t_0) \in T^2_{\geq}$ we define the *evolution operator* $\Phi(t, t_0) : X \to X$ by

$$\Phi(t, t_0)x = \varphi(t; t_0, x, 0_{\mathcal{U}}), \quad x \in X \tag{6}$$

and the *input-to-state map* $\Theta(t, t_0) : \mathcal{U} \to X$ by

$$\Theta(t, t_0)u(\cdot) = \varphi(t; t_0, 0_X, u(\cdot)), \quad u(\cdot) \in \mathcal{U}. \tag{7}$$

The two maps are linear because of (4), (5). $\Phi(t, t_0)$ associates with any state x the state $x(t)$ at time t resulting from the free motion of Σ starting at $x(t_0) = x$. $\Theta(t, t_0)$ maps any control function $u(\cdot)$ onto the state $x(t)$ to which Σ is steered at time t by $u(\cdot)$ from the initial state $x(t_0) = 0$. By (3) all trajectories $t \mapsto \varphi(t; t_0, x^0, u(\cdot))$ of Σ are completely determined by these two families of linear operators

$$\varphi(t; t_0, x^0, u(\cdot)) = \Phi(t, t_0)x^0 + \Theta(t, t_0)u(\cdot), \quad (t, t_0) \in T^2_{\geq}.$$

We shall see later that for all linear systems of practical importance the linear map $\Theta(t, t_0)$ can be expressed with the aid of the operators $\Phi(t, s)_{(t,s) \in T^2_{\geq}}$. Therefore it is particularly important to study the properties of the family $(\Phi(t, s))_{(t,s) \in T^2_{\geq}}$. The axioms (1.4) and (1.6) of a state transition map imply the following basic equations

$$\begin{aligned} \Phi(t, t) &= I_X, \quad t \in T && (8) \\ \Phi(t_2, t_1) \circ \Phi(t_1, t_0) &= \Phi(t_2, t_0), \quad t_0, t_1, t_2 \in T, \quad t_0 \leq t_1 \leq t_2. && (9) \end{aligned}$$

A family $(\Phi(t, s))_{(t,s) \in T^2_{\geq}}$ of linear operators on X with these properties is called a *family of evolution operators* on X. If Σ is time–invariant we may fix $t_0 = 0$ and obtain a one-parameter family $(\Phi(t))_{t \in T_0}$ of linear operators $\Phi(t) : X \to X$ defined by

$$\Phi(t)x = \Phi(t, 0)x = \varphi(t; 0, x, 0_{\mathcal{U}}), \quad t \in T_0 = \{t \in T; t \geq 0\}. \tag{10}$$

Equations (8) and (9) then imply

$$\begin{aligned} \Phi(0) &= I_X && (11) \\ \Phi(t) \circ \Phi(s) &= \Phi(t + s), \quad s, t \in T, \quad t, s \geq 0. && (12) \end{aligned}$$

A family $(\Phi(t))_{t \in T_0}$ of linear operators on X with these properties is called a *semigroup of linear operators* on X. The theory of operator semigroups on normed spaces provides a mathematical basis for the study of infinite dimensional time–invariant linear systems (see *Notes and References* and Section 1.6). The following finite-dimensional example relates the abstract notions introduced above to familiar concepts from the theory of linear differential equations.

Example 2.2.1. (Linear differentiable systems). Let $T \subset \mathbb{R}$ be an interval, $X = \mathbb{K}^n$, $U = \mathbb{K}^m$, $Y = \mathbb{K}^p$, $\mathcal{U}$ any linear subspace of $L^1_{\text{loc}}(T, \mathbb{K}^m)$, e.g. $\mathcal{U} = PC(T, \mathbb{K}^m)$ and $A(\cdot) \in PC(T, \mathbb{K}^{n \times n})$, $B(\cdot) \in PC(T, \mathbb{K}^{n \times m})$, $C(\cdot) \in PC(T, \mathbb{K}^{p \times n})$, $D(\cdot) \in PC(T, \mathbb{K}^{p \times m})$. By Corollary 2.1.20, there exists a unique solution of the initial value problem

$$\begin{aligned} \dot{x}(t) &= A(t)x(t) + B(t)u(t), \quad t \in T && (13) \\ x(t_0) &= x^0 \end{aligned}$$

for every $u(\cdot) \in \mathcal{U}$, $t_0 \in T$, $x^0 \in X$. Recall that the fundamental matrix $X(t, t_0)$ associated with (13) is, by definition, the solution of the matrix differential equation

$$\begin{aligned} \dot{X}(t) &= A(t)X(t), \quad t \in T \\ X(t_0) &= I_n. \end{aligned} \tag{14}$$

This means that the columns $x^j(t, t_0)$ of $X(t, t_0)$ solve the initial value problems

$$\begin{aligned} \dot{x}(t) &= A(t)x(t), \quad t \in T \\ x(t_0) &= e^j \end{aligned}$$

where e^j is the j-th column of I_n, $j \in \underline{n}$. The variation-of-constants formula gives the following explicit representation for the solution $x(t) = \varphi(t; t_0, x^0, u(\cdot))$ of (13)

$$\varphi(t; t_0, x^0, u(\cdot)) = X(t, t_0)\, x^0 + \int_{t_0}^{t} X(t, s)B(s)u(s)\, ds, \quad t \in T. \tag{15}$$

If we set

$$\eta(t, x, u) = C(t)x + D(t)u$$

then $\Sigma = (T, U, \mathcal{U}, X, Y, \varphi, \eta)$ is a linear differentiable system (defined by the matrix-valued functions $A(\cdot)$, $B(\cdot)$, $C(\cdot)$, $D(\cdot)$). Since φ is defined on $T^2 \times X \times \mathcal{U}$, Σ is complete and reversible. Let us now determine the linear operators $\Phi(t, t_0)$, $\Theta(t, t_0)$ associated with Σ. As an immediate consequence of (6) and (15) we obtain for all t_0, $t \in T$

$$\Phi(t, t_0)x = X(t, t_0)x.$$

So the fundamental matrix of (13) is just the matrix representation of the evolution operator $\Phi(t, t_0)$ with respect to the standard basis of $\mathbb{K}^n$. In the sequel we shall use the same notation $\Phi(t, t_0)$ for both the linear operators and their matrix representations. Since $\Phi(t, t_0)\Phi(t_0, t) = I_n$, the operators $\Phi(t, t_0)$, $t, t_0 \in T$ are all invertible. From (7) and (15) we obtain for all t_0, $t \in T$

$$\Theta\,(t, t_0)u(\cdot) = \int_{t_0}^{t} \Phi(t, s)B(s)u(s)ds, \quad u(\cdot) \in \mathcal{U}. \tag{16}$$

This specifies the relation between the input-to-state operators $\Theta(t, t_0)$ and the evolution operators $\Phi(t, s)$ for the system Σ.
We conclude this example with a few words about the important special case where $T = \mathbb{R}$ and $A(t) \equiv A$, $B(t) \equiv B$, $C(t) \equiv C$, $D(t) \equiv D$ are independent of time. In this case the system equations are

$$\begin{aligned} \dot{x}(t) &= Ax(t) + Bu(t), \quad t \in \mathbb{R} \\ y(t) &= Cx(t) + Du(t). \end{aligned} \tag{17}$$

Let e^{At} denote the matrix exponential defined by the absolutely converging series

$$e^{At} = \sum_{k=0}^{\infty} \frac{t^k}{k!} A^k, \qquad t \in \mathbb{R}. \tag{18}$$

Then the fundamental matrix has the form $X(t, t_0) = e^{A(t-t_0)}$ and so the state transition map is given by

$$\varphi(t; t_0, x^0, u(\cdot)) = e^{A(t-t_0)}x^0 + \int_{t_0}^{t} e^{A(t-s)}Bu(s)\, ds, \quad t \in \mathbb{R}. \tag{19}$$

This formula shows that the system Σ is time–invariant, with associated semigroup of linear operators $\Phi(t) = e^{At}$. Since Σ is reversible this semigroup can actually be extended to a one-parameter *group* $\Phi = (e^{At})_{t\in\mathbb{R}}$ of linear operators on $\mathbb{K}^n$. □

In the following example we briefly discuss the discrete time counterpart of the previous example.

Example 2.2.2. (Linear difference systems). Let $\mathbb{K}$ be an arbitrary field, $U = \mathbb{K}^m$, $X = \mathbb{K}^n$, $Y = \mathbb{K}^p$, $T \subset \mathbb{Z}$ a time-domain satisfying $t \in T \Rightarrow t+1 \in T$, $\mathcal{U} = U^T$, and $A(\cdot) = (A(t))_{t\in T}$, $B(\cdot) = (B(t))_{t\in T}$, $C(\cdot) = (C(t))_{t\in T}$, $D(\cdot) = (D(t))_{t\in T}$ sequences of $n\times n$, $n\times m$, $p\times n$, $p\times m$ matrices over $\mathbb{K}$. Consider the discrete time counterpart of the system equations (13)

$$\begin{aligned} x(t+1) &= A(t)x(t) + B(t)u(t), \quad t\in T \\ y(t) &= C(t)x(t) + D(t)u(t). \end{aligned} \tag{20}$$

It is easily verified that for every $u(\cdot)\in\mathcal{U}$, $t_0\in T$, $x^0\in X$ the difference equation in (20) admits a unique solution $x(t) = \varphi(t; t_0, x^0, u(\cdot))$ with $x(t_0) = x^0$, namely

$$\varphi(t; t_0, x^0, u(\cdot)) = \Phi(t,t_0)x^0 + \sum_{s=t_0}^{t-1}\Phi(t,s+1)B(s)u(s), \quad t\in T_{t_0} \tag{21}$$

where $\Phi(t,s) = I_n$ for $s = t\in T$ and

$$\Phi(t,s) = A(t-1)A(t-2)\dots A(s), \quad s,t\in T,\ s<t.$$

If we set $\eta(t,x,u) = C(t)x + D(t)u$, then $\Sigma = (T, U, \mathcal{U}, X, Y, \varphi, \eta)$ is a discrete time linear time–varying dynamical system. The associated input-to-state operator $\Theta(t,t_0)$ can again be expressed in terms of the evolution operator $\Phi(t,s)$,

$$\Theta(t,t_0)(u(\cdot)) = \sum_{s=t_0}^{t-1}\Phi(t,s+1)B(s)u(s), \quad u(\cdot)\in\mathcal{U},\ t_0,t\in T,\ t_0<t.$$

In the time–invariant case where $A(t)$, $B(t)$, $C(t)$, $D(t)$ are constant matrices the system equations are

$$\begin{aligned} x(t+1) &= Ax(t) + Bu(t), \quad t\in\mathbb{Z} \\ y(t) &= Cx(t) + Du(t), \end{aligned} \tag{22}$$

so that the state transition map is given by

$$\varphi(t; t_0, x^0, u(\cdot)) = A^{(t-t_0)}x^0 + \sum_{s=t_0}^{t-1}A^{(t-1-s)}Bu(s).$$

It follows from this formula that Σ is time–invariant with an associated discrete semigroup of linear operators $(\Phi(t))_{t\in\mathbb{N}}$ given by

$$\Phi(t) = \Phi(t,0) = A^t, \quad t\in\mathbb{N}. \tag{23}$$

In contrast with the differentiable system of Example 2.2.1, Σ is not necessarily reversible. It is reversible if and only if A is nonsingular. □

In the next two subsections we will study the free motions of the time-invariant linear systems (17) and (22) in more detail.

2.2.2 Free Motions of Time–Invariant Linear Differential Systems

Let $A \in \mathbb{K}^{n \times n}$ be a given matrix. In this subsection we will study the state trajectories of the free system without output (see Remark 2.1.2) given by

$$\dot{x}(t) = Ax(t), \quad t \in \mathbb{R}. \tag{24}$$

First note that the origin $\bar{x} = 0$ is always a singular point of the vector field $x \mapsto Ax$ on $\mathbb{K}^n$ and is, therefore, an equilibrium point of (24). More generally, $\bar{x} \in \mathbb{K}^n$ is an equilibrium point if and only if $A\bar{x} = 0$, i.e. $\ker A$ is the set of equilibria of (24). We have seen in Example 2.2.1 that the trajectories of (24) (i.e. the free motions of (17)) are described by the group of linear operators

$$\Phi(t) = e^{At} = \sum_{k=0}^{\infty} \frac{t^k}{k!} A^k, \quad t \in \mathbb{R}. \tag{25}$$

This group has the following basic properties.

Lemma 2.2.3. *If $A \in \mathbb{K}^{n \times n}$, then for every $s,\, t \in \mathbb{R}$ we have*

(i) $\frac{d}{dt} e^{At} = Ae^{At} = e^{At}A$

(ii) $e^{A(t+s)} = e^{At}e^{As}$

(iii) $(e^{At})^{-1} = e^{-At}$

(iv) $e^{S^{-1}ASt} = S^{-1}e^{At}S, \quad S \in \mathbf{Gl}_n(\mathbb{K}).$

Proof: Properties (ii) and (iii) express the fact that $(e^{At})_{t \in \mathbb{R}}$ is a group of linear operators, (i) follows because e^{At} is the fundamental matrix of (24) at $t_0 = 0$ and (iv) follows from the series representation (25) since $(S^{-1}AS)^k = S^{-1}A^kS$, $k \in \mathbb{N}$ and the similarity action $A \mapsto S^{-1}AS$ is continuous on $\mathbb{K}^{n \times n}$. □

Our aim is to show that every trajectory of (24) can be represented as a superposition of a finite number of relatively simple trajectories, the (generalized) *eigenmotions*. These eigenmotions are easily determined once a basis of generalized eigenvectors of A has been found. Before we make this more precise we recall some spectral results from Linear Algebra. Suppose $\mathbb{K} = \mathbb{C}$ so that $\mathbb{C}^n$ is the state space and $A \in \mathbb{C}^{n \times n}$. Let $\sigma(A)$ denote the *spectrum of* A, i.e. the set of eigenvalues

$$\sigma(A) = \{\lambda \in \mathbb{C};\ \det(\lambda I_n - A) = 0\}.$$

$\sigma(A)$ is the set of roots of the characteristic polynomial of A

$$\chi_A(s) = \det(sI_n - A) = s^n + a_{n-1}s^{n-1} + \cdots + a_1 s + a_0. \tag{26}$$

Factorizing $\chi_A(s) \in \mathbb{C}[s]$ according to the Fundamental Theorem of Algebra we obtain

$$\chi_A(s) = \prod_{j=1}^{\ell} (s - \lambda_j)^{m(\lambda_j)}, \quad \lambda_i \neq \lambda_j \text{ for } i \neq j. \tag{27}$$

$m(\lambda_j)$ is said to be the *algebraic multiplicity* of the eigenvalue λ_j while $\dim\ker(\lambda_j I_n - A) \le m(\lambda_j)$ is its *geometric multiplicity.* The following well-known decomposition result is basic for our analysis.

Lemma 2.2.4 (Spectral Decomposition Lemma). *If* $\lambda_1, \ldots, \lambda_\ell$ *are the distinct eigenvalues of* $A \in \mathbb{C}^{n\times n}$ *with algebraic multiplicities* $m(\lambda_1), \ldots, m(\lambda_\ell)$ *then*

$$\mathbb{C}^n = \ker(\lambda_1 I_n - A)^{m(\lambda_1)} \oplus \cdots \oplus \ker(\lambda_\ell I_n - A)^{m(\lambda_\ell)} \tag{28}$$

i.e. $\mathbb{C}^n$ *is the direct sum (see Definition A.1.19) of the generalized eigenspaces* $\ker(\lambda_j I_n - A)^{m(\lambda_j)}$, $j \in \underline{\ell}$. *Moreover* $\dim\ker(\lambda_j I_n - A)^{m(\lambda_j)} = m(\lambda_j)$ *for each* $j \in \underline{\ell}$.

$z \in \mathbb{C}^n$ is said to be a *generalized eigenvector of order* $m \geq 1$ of A if

$$(\lambda I_n - A)^m z = 0 \quad \text{and} \quad (\lambda I_n - A)^{m-1} z \neq 0\,. \tag{29}$$

Hence the non-zero elements of $\ker(\lambda_j I_n - A)^{m(\lambda_j)}$ are the generalized eigenvectors of order $\leq m(\lambda_j)$. The projections corresponding to the decomposition (28)

$$\begin{aligned} P_j \;:\; \mathbb{C}^n &\longrightarrow \ker(\lambda_j I_n - A)^{m(\lambda_j)}\,, \quad j \in \underline{\ell} \\ x = x^1 \oplus \cdots \oplus x^\ell &\mapsto x^j \end{aligned}$$

are called *eigenprojections* of A. The following properties of the P_j, $j \in \underline{\ell}$ are obvious from the definition

$$P_j^2 = P_j\,, \quad P_j P_k = 0 \text{ if } j \neq k\,, \quad \sum_{j=1}^{\ell} P_j = I_n\,. \tag{30}$$

Moreover

$$A P_j = P_j A = \lambda_j P_j + N_j\,, \quad j \in \underline{\ell} \tag{31}$$

where $N_j = (A - \lambda_j I_n) P_j$ is nilpotent. N_j is called the *eigennilpotent* corresponding to the eigenvalue λ_j of A. Adding up these equalities and making use of (30) we obtain the *spectral representation* of A

$$A = A \sum_{j=1}^{\ell} P_j = \sum_{j=1}^{\ell} (\lambda_j P_j + N_j). \tag{32}$$

If $N_j = 0$, λ_j is said to be *semi-simple.* A is diagonalizable if and only if every eigenvalue is semi-simple and in this case

$$A = A \sum_{j=1}^{\ell} P_j = \sum_{j=1}^{\ell} \lambda_j P_j. \tag{33}$$

We now return to the free motions of (24). An initial state $z \in \ker(\lambda I_n - A)^m$ gives rise to the following *generalized eigenmotion* of (24)

$$e^{At} z = e^{\lambda t} e^{(A - \lambda I)t} z = e^{\lambda t} \sum_{j=0}^{m-1} \frac{t^j}{j!} (A - \lambda I)^j z\,, \quad t \in \mathbb{R}\,. \tag{34}$$

The trajectory remains in the linear subspace spanned by $z, Az, \ldots, A^{m-1}z$ for all $t \geq 0$. In particular, if z is an eigenvector, $Az = \lambda z$, then

$$e^{At} z = e^{\lambda t} z\,, \qquad t \in \mathbb{R} \tag{35}$$

remains always in the one-dimensional complex subspace through z. These trajectories are called (complex) *eigenmotions* of the system (24).
As functions of time, any generalized eigenmotion of order m (i.e. starting at a generalized eigenvector of order m) is the product of an *exponential* $e^{\lambda t}$ and a vector polynomial $\sum_{j=0}^{m-1} \frac{t^j}{j!}(A-\lambda I_n)^j z \in \mathbb{C}^n[t]$ of degree $m-1$. If $\operatorname{Re}\lambda \neq 0$ the exponential part determines the long term behaviour of the trajectory. $\|e^{At}z\|$ tends to zero or infinity depending on whether $\operatorname{Re}\lambda < 0$ or $\operatorname{Re}\lambda > 0$.
Remark 2.2.5. If $\lambda_0 = 0 \in \sigma(A)$ the associated eigenvectors are equilibrium points of (24). If z is an associated generalized eigenvector of order m then the corresponding generalized eigenmotion depends polynomially on time, $z(t) = e^{At}z = \sum_{j=0}^{m-1}(1/j!)A^j z\, t^j$. □

Since by Lemma 2.2.4 every initial state can be represented as a sum of generalized eigenvectors we obtain the following corollary.

Corollary 2.2.6. *Every trajectory of the free system* (24) *is a superposition of the generalized eigenmotions. More precisely, if $P_1, \dots, P_\ell$ are the eigenprojections of $A \in \mathbb{C}^{n\times n}$ corresponding to the distinct eigenvalues $\lambda_1, \dots, \lambda_\ell$ with algebraic multiplicities $m(\lambda_1), \dots, m(\lambda_\ell)$ then*

$$e^{At}x^0 = \sum_{j=1}^{\ell} e^{\lambda_j t} \sum_{k=0}^{m(\lambda_j)-1} \frac{t^k}{k!}(A-\lambda_j I_n)^k P_j x^0\,, \quad t \geq 0,\, x^0 \in \mathbb{C}^n\,. \tag{36}$$

In particular, if A is diagonalizable then

$$e^{At}x^0 = \sum_{j=1}^{\ell} e^{\lambda_j t} P_j x^0\,, \quad t \geq 0,\, x^0 \in \mathbb{C}^n\,. \tag{37}$$

The latter formula gives us a method for computing e^{At} in the diagonalizable case. If $(z^1, \dots, z^n)$ is a basis of eigenvectors of A, $Az^i = \lambda_i z^i$ then $S = [z^1, \dots, z^n] \in \mathbf{Gl}_n(\mathbb{C})$ satisfies

$$e^{At} = S\,\mathrm{diag}\,(e^{\lambda_1 t}, \dots, e^{\lambda_n t})\, S^{-1}\,. \tag{38}$$

For the general case, recall that if $J(\lambda, m)$ is a Jordan block of order $m \in \mathbb{N}^*$, i.e.

$$J(\lambda,m) = \begin{bmatrix} \lambda & 1 & & 0 \\ & \ddots & \ddots & \\ & & \ddots & 1 \\ 0 & & & \lambda \end{bmatrix} \in \mathbb{C}^{m\times m}, \quad \lambda \in \mathbb{C}, \ m \in \mathbb{N}, \tag{39}$$

then

$$e^{J(\lambda,m)t} = e^{\lambda t} \begin{bmatrix} 1 & t/1! & t^2/2! & \cdots & t^{m-1}/(m-1)! \\ & \ddots & \ddots & \ddots & \vdots \\ & & \ddots & \ddots & t^2/2! \\ 0 & & & \ddots & t/1! \\ & & & & 1 \end{bmatrix}, \quad t \in \mathbb{R}, \ \lambda \in \mathbb{C}, \ m \in \mathbb{N}. \tag{40}$$

Now suppose that $S^{-1}AS$ is in Jordan canonical form

$$S^{-1}AS = \oplus_{j=1}^{\ell} \oplus_{k=1}^{k_j} J(\lambda_j, m_{jk}) \tag{41}$$

where $m(\lambda_j) = \sum_{k=1}^{k_j} m_{jk}$ is the algebraic multiplicity of the eigenvalue $\lambda_j \in \sigma(A)$ and $\oplus$ denotes the direct sum of matrices, see Definition A.1.20. Then

$$e^{At} = S \left[\oplus_{j=1}^{\ell} \oplus_{k=1}^{k_j} e^{J(\lambda_j, m_{jk})t} \right] S^{-1}, \quad t \in \mathbb{R}. \tag{42}$$

Whilst these formulas are useful for analytical purposes (see Chapter 3) they should not be used for the numerical computation of e^{At}, see *Notes and References*.
In most applications where A is real, one is only interested in real state trajectories. What, then, is the significance of the above analysis?

Remark 2.2.7. From an operator theoretic point of view one has to distinguish between a linear map $L : \mathbb{R}^m \to \mathbb{R}^n$ and its *complexification* $L^{\mathbb{C}} : \mathbb{C}^m \to \mathbb{C}^n$ defined by

$$L^{\mathbb{C}}(x + \imath y) = Lx + \imath Ly, \qquad x, y \in \mathbb{R}^m. \tag{43}$$

However, if there is no risk of confusion, we use the same symbol for a matrix $L \in \mathbb{R}^{n \times m}$, the corresponding linear map $v \mapsto Lv$ from $\mathbb{R}^m$ to $\mathbb{R}^n$ and its complexification as a linear map from $\mathbb{C}^m$ to $\mathbb{C}^n$. Where necessary, we distinguish between the kernels (resp. ranges) of L and $L^{\mathbb{C}}$ by using the notations $\ker_{\mathbb{K}} L$ (resp. $\operatorname{im}_{\mathbb{K}} L$). □

For the rest of this subsection we suppose that $A \in \mathbb{R}^{n \times n}$. The *real eigenmotions* or *modes* of the system (24) are obtained by taking the real and imaginary parts of the complex eigenmotions. If $\lambda \in \sigma(A)$ is real and z a real eigenvector for λ, then the associated eigenmotion (35) is real. Whereas if $\lambda = \gamma + \imath\omega \in \sigma(A)$ is non-real $(\omega \neq 0)$ and $z = (z_i) \in \mathbb{C}^n$ is an associated eigenvector then $\overline{\lambda} = \gamma - \imath\omega \in \sigma(A)$ and the conjugate complex vector $\overline{z} = (\overline{z_i}) \in \mathbb{C}^n$ is an eigenvector of A for $\overline{\lambda}$. Choosing $\operatorname{Re} z = (1/2)(z + \overline{z}) \in \mathbb{R}^n$ and $\operatorname{Im} z = 1/(2\imath)(z - \overline{z}) \in \mathbb{R}^n$ as initial states we obtain the following *real eigenmotions* of (24)

$$e^{At}(\operatorname{Re} z) = \operatorname{Re}(e^{At}z) = \operatorname{Re}(e^{\lambda t}z) = e^{\gamma t}\left[(\cos\omega t)\operatorname{Re} z - (\sin\omega t)\operatorname{Im} z\right] \tag{44}$$
$$e^{At}(\operatorname{Im} z) = \operatorname{Im}(e^{At}z) = \operatorname{Im}(e^{\lambda t}z) = e^{\gamma t}\left[(\sin\omega t)\operatorname{Re} z + (\cos\omega t)\operatorname{Im} z\right]. \tag{45}$$

There is a qualitative difference between the modes corresponding to real and to non-real eigenvalues. In the real case we have a 'one-dimensional' trajectory along the real line $\mathbb{R}z$ which is contractive if $\lambda < 0$, constant if $\lambda = 0$ and expansive if $\lambda > 0$. In the complex case we have a 'two-dimensional' oscillatory motion in the plane spanned by $\operatorname{Re} z, \operatorname{Im} z \in \mathbb{R}^n$. This motion is contractive if $\operatorname{Re}\lambda < 0$, and expansive if $\operatorname{Re}\lambda > 0$. Some typical eigenmotions are shown in Figure 2.2.3.
Generalized real eigenmotions are obtained by taking the real and imaginary parts of generalized complex eigenmotions (34). If $\lambda \in \sigma(A)$ is real and z is a generalized real eigenvector of order m for λ then (34) is a generalized real eigenmotion, remaining for all $t \geq 0$ in the m-dimensional linear subspace

$$\operatorname{span}_{\mathbb{R}}\{z, Az, \ldots, A^{m-1}z\} \subset \mathbb{R}^n.$$

If $\lambda = \gamma + \imath\omega \in \sigma(A)$, $\omega \neq 0$ and $z \in \mathbb{C}^n$ is an associated generalized eigenvector of order m then we obtain two *generalized real eigenmotions* associated with the pair $\lambda, \overline{\lambda} \in \sigma(A)$ and the generalized eigenvector z

$$e^{At}\operatorname{Re} z = \operatorname{Re}(e^{At}z) = \operatorname{Re} e^{\lambda t}\sum_{j=0}^{m-1}\frac{t^j}{j!}(A-\lambda I_n)^j z \tag{46}$$
$$= e^{\gamma t}\sum_{j=0}^{m-1}\frac{t^j}{j!}\left[(\cos\omega t)\operatorname{Re}(A-\lambda I_n)^j z - (\sin\omega t)\operatorname{Im}(A-\lambda I_n)^j z\right],\quad t\geq 0,$$

$$e^{At}\operatorname{Im} z = \operatorname{Im}(e^{At}z) = \operatorname{Im} e^{\lambda t}\sum_{j=0}^{m-1}\frac{t^j}{j!}(A-\lambda I_n)^j z \tag{47}$$
$$= e^{\gamma t}\sum_{j=0}^{m-1}\frac{t^j}{j!}\left[(\cos\omega t)\operatorname{Im}(A-\lambda I_n)^j z + (\sin\omega t)\operatorname{Re}(A-\lambda I_n)^j z\right],\quad t\geq 0.$$

Both trajectories remain for all $t \geq 0$ in the $2m$-dimensional linear subspace

$$\operatorname{span}_{\mathbb{R}}\{\operatorname{Re}(A-\lambda I_n)^j z\,,\ \operatorname{Im}(A-\lambda I_n)^j z\,;\ \ j=0,\ldots,m-1\}\,.$$

Since A is real, its spectrum can be written in the form

$$\sigma(A) = \{\rho_1,\ldots,\rho_r,\lambda_1,\ldots,\lambda_c,\overline{\lambda}_1,\ldots,\overline{\lambda}_c\} \tag{48}$$

where $\rho_i \in \mathbb{R}$, $i \in \underline{r}$ and $\lambda_i \in \mathbb{C}\backslash\mathbb{R}$, $i \in \underline{c}$. The algebraic multiplicities of λ_i and $\overline{\lambda}_i$ are the same. Moreover, if $z^1,\ldots,z^{m(\lambda_i)}$ is a basis of $\ker(\lambda_i I_n - A)^{m(\lambda_i)}$ then $\overline{z^1},\ldots,\overline{z^{m(\lambda_i)}}$ is a basis of $\ker(\overline{\lambda}_i I_n - A)^{m(\overline{\lambda}_i)}$ and $\operatorname{Re} z^1,\ldots,\operatorname{Re} z^{m(\lambda_i)}$, $\operatorname{Im} z^1,\ldots,\operatorname{Im} z^{m(\lambda_i)} \in \mathbb{R}^n$ is a basis (over $\mathbb{R}$) of the $2m(\lambda_i)$-dimensional real linear subspace

$$\mathbb{R}^n \cap \left[\ker(\lambda_i I_n - A)^{m(\lambda_i)} \oplus \ker(\overline{\lambda}_i I_n - A)^{m(\overline{\lambda}_i)}\right] = \ker_{\mathbb{R}}(|\lambda_i|^2 I_n - 2(\operatorname{Re}\lambda_i)A + A^2)^{m(\lambda_i)}.$$

The real version of the spectral decomposition (28) is

$$\mathbb{R}^n = \oplus_{i=1}^r \ker_{\mathbb{R}}(\rho_i I_n - A)^{m(\rho_i)} \oplus \oplus_{i=1}^c \ker_{\mathbb{R}}(|\lambda_i|^2 I_n - 2(\operatorname{Re}\lambda_i)A + A^2)^{m(\lambda_i)}\,.$$

As a consequence we obtain the following real version of Corollary 2.2.6.

Corollary 2.2.8. *Let $A \in \mathbb{R}^{n\times n}$, then every real trajectory $e^{At}x^0$, $x^0 \in \mathbb{R}^n$ of* (24) *is a superposition of generalized real eigenmotions (modes).*

In the following example we determine the real modes of a linear oscillator.

Example 2.2.9. (Oscillator). Consider the motion of a unit mass connected to a support by a spring immersed in a homogeneous medium (see Figure 2.2.1). The spring constant is taken to be ν^2 (where $\nu \geq 0$) and the friction forces are proportional to the velocity with the constant 2α. If ξ measures the displacement of the mass from equilibrium then (see Example 1.3.2)

$$\ddot{\xi}(t) + 2\alpha\dot{\xi}(t) + \nu^2\xi(t) = 0. \tag{49}$$

Introducing the state vector $x = [x_1, x_2]^\top = [\xi, \dot{\xi}]^\top$ we obtain the state space model

$$\dot{x}(t) = Ax(t),\quad A = \begin{bmatrix} 0 & 1 \\ -\nu^2 & -2\alpha \end{bmatrix}.$$

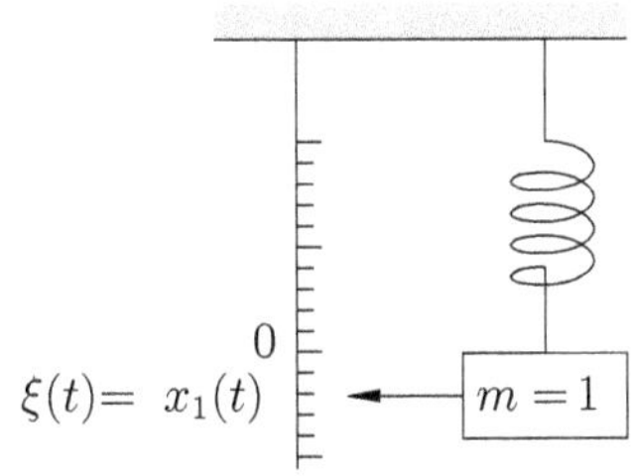

Figure 2.2.1: Mass-spring system

The eigenvalues of A are given by $\lambda_{1,2} = -\alpha \pm \sqrt{\alpha^2 - \nu^2}$ with corresponding eigenvectors

$$z^1 = \begin{bmatrix} 1 \\ \lambda_1 \end{bmatrix}, \quad z^2 = \begin{bmatrix} 1 \\ \lambda_2 \end{bmatrix}. \tag{50}$$

Clearly λ_1, λ_2 are real if and only if $|\alpha| \geq \nu$. In this case the eigenvalues have negative real parts (i.e. produce contracting eigenmotions) if and only if $\alpha > 0$. For $|\alpha| > \nu$ the real eigenmotions $x^1(t)$, $x^2(t)$ starting at z^1 resp. z^2 are

$$x^1(t) = e^{At}z^1 = e^{(-\alpha+\sqrt{\alpha^2-\nu^2})\,t}\begin{bmatrix} 1 \\ \lambda_1 \end{bmatrix}, \quad x^2(t) = e^{At}z^2 = e^{(-\alpha-\sqrt{\alpha^2-\nu^2})\,t}\begin{bmatrix} 1 \\ \lambda_2 \end{bmatrix}$$

or in terms of ξ (the first coordinate)

$$\xi^{1,2}(t) = e^{(-\alpha\pm\sqrt{\alpha^2-\nu^2})\,t}.$$

If $|\alpha| = \nu$, then $\lambda_1 = \lambda_2 = -\alpha$ and the corresponding generalized eigenspace is spanned by

$$z^1 = \begin{bmatrix} 1 \\ -\alpha \end{bmatrix}, \quad z^2 = \begin{bmatrix} 0 \\ 1 \end{bmatrix}$$

where z^1 is a real eigenvector and z^2 is a generalized eigenvector of second order. By (34) the corresponding (generalized) real eigenmotions are

$$x^1(t) = e^{-\alpha t}\begin{bmatrix} 1 \\ -\alpha \end{bmatrix}, \quad x^2(t) = e^{-\alpha t}\begin{bmatrix} 0 \\ 1 \end{bmatrix} + te^{-\alpha t}\begin{bmatrix} 1 \\ -\alpha \end{bmatrix}.$$

If $|\alpha| < \nu$ (hence $z^2 = \overline{z^1}$) then $\lambda_{1,2} = -\alpha \pm \imath\sqrt{\nu^2 - \alpha^2}$ and the corresponding real modes are by (44), (45) and (50)

$$x^1(t) = e^{-\alpha t}\left(\cos(\sqrt{\nu^2-\alpha^2}\,t)\begin{bmatrix} 1 \\ -\alpha \end{bmatrix} - \sin(\sqrt{\nu^2-\alpha^2}\,t)\begin{bmatrix} 0 \\ \sqrt{\nu^2-\alpha^2} \end{bmatrix}\right)$$
$$x^2(t) = e^{-\alpha t}\left(\sin(\sqrt{\nu^2-\alpha^2}\,t)\begin{bmatrix} 1 \\ -\alpha \end{bmatrix} + \cos(\sqrt{\nu^2-\alpha^2}\,t)\begin{bmatrix} 0 \\ \sqrt{\nu^2-\alpha^2} \end{bmatrix}\right).$$

In terms of the first coordinates $\xi_i(t)$ of $x^i(t)$, $i = 1, 2$ this yields the following oscillatory eigenmotions corresponding to initial conditions $\xi_1(0) = 1$, $\dot{\xi}_1(0) = -\alpha$ and $\xi_2(0) = 0$, $\dot{\xi}_2(0) = \sqrt{\nu^2 - \alpha^2}$

$$\xi_1(t) = e^{-\alpha t}\cos(\sqrt{\nu^2-\alpha^2}\,t), \quad \text{and} \quad \xi_2(t) = e^{-\alpha t}\sin(\sqrt{\nu^2-\alpha^2}\,t).$$

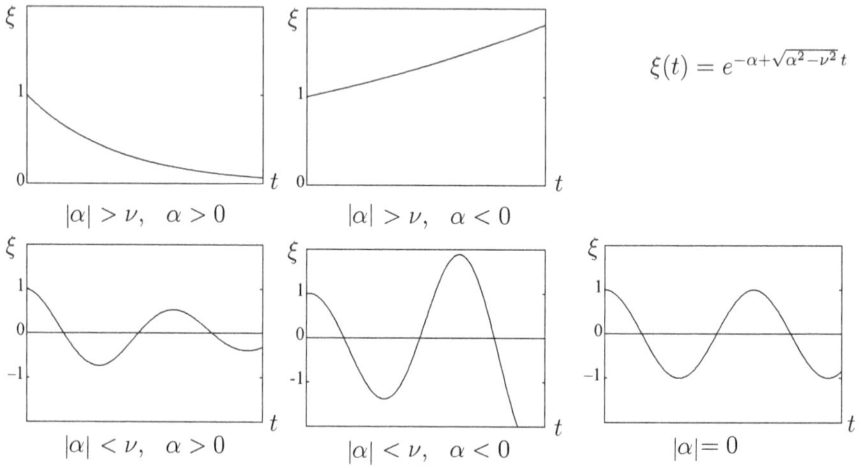

$$\xi(t) = e^{-\alpha t}\cos(\sqrt{\nu^2 - \alpha^2}\, t)$$

Figure 2.2.2: Eigenmotions of an oscillator

So if there is no damping ($\alpha = 0$) then the $\xi_i(\cdot)$ are periodic with period $2\pi/\nu$. If $\alpha > 0$ the oscillations die out whereas they are intensified if $\alpha < 0$. Some typical eigenmotions $\xi(\cdot)$ are shown in Figure 2.2.2. If we fix the coefficient $\nu > 0$ of the restoring force we observe a qualitative change as the damping $2\alpha > 0$ decreases from large values to zero. For $\alpha \geq \nu$ the eigenmotions converge monotonically to zero along the lines $\mathbb{R}z^i$ as $t \to \infty$ whereas, for $\alpha \in [0, \nu)$, the eigenmotions are oscillatory. □

Example 2.2.10. (Inverted pendulum). Consider the cart pendulum system of Example 1.3.4 with no damping. The matrix A of the linearization about the upright position is of the form

$$A = \begin{bmatrix} 0 & 0 & 1 & 0 \\ 0 & 0 & 0 & 1 \\ 0 & a_{32} & 0 & 0 \\ 0 & a_{42} & 0 & 0 \end{bmatrix} \tag{51}$$

where $a_{32} = M_0^{-1} m^2 l^2 g$, $a_{42} = M_0^{-1}(M+m)mgl$, see (1.3.32). Now

$$\det(\lambda I_n - A) = \lambda^2(\lambda - \sqrt{a_{42}})(\lambda + \sqrt{a_{42}}). \tag{52}$$

So A has the following eigenvalues and eigenvectors

$$\begin{array}{rclcrcl} \lambda_1 & = & 0 & , & z^1 & = & [1, 0, 0, 0]^\top \\ \lambda_2 & = & \sqrt{a_{42}} & , & z^2 & = & [1, a_{42}/a_{32}, \sqrt{a_{42}}, \sqrt{a_{42}}\, a_{42}/a_{32}]^\top \\ \lambda_3 & = & -\sqrt{a_{42}} & , & z^3 & = & [1, a_{42}/a_{32}, -\sqrt{a_{42}}, -\sqrt{a_{42}}\, a_{42}/a_{32}]^\top. \end{array}$$

By (52) we have $m(\lambda_1) = 2$, but the eigenspace of λ_1 is only one dimensional, so there is a generalized eigenvector $z^{1,2}$ of second order, for example $z^{1,2} = [0, 0, 1, 0]^\top$. All the eigenvalues are real and the corresponding eigenmotions are

$$e^{At} z^1 \equiv z^1, \quad e^{At} z^2 = e^{\sqrt{a_{42}}t} z^2, \quad e^{At} z^3 = e^{-\sqrt{a_{42}}t} z^3.$$

The first is an equilibrium state (upright position of the pendulum), the second is an expanding motion whilst the last is a contracting motion. The generalized eigenmode corresponding to the initial state $z^{1,2}$ is (see (34))

$$e^{At}z^{1,2} = z^{1,2} + tAz^{1,2} = z^{1,2} + tz^1.$$

As $t \to \infty$ we see that one eigenmode tends to zero, one is stationary and the other two both go to infinity (one exponentially and the other linearly). These results for the linearized model indicate (not unexpectedly) that the inverted pendulum has a relatively complicated dynamics which will probably be difficult to control. □

By superposition of the eigenmotions of a system (24), different patterns of free motions around the origin can be generated depending on the numbers of contracting, expanding, stationary or periodic eigenmotions of the system. For a two dimensional linear system it is possible to give a complete classification of the flow patterns around the equilibrium state $\bar{x} = 0$. Let (z^1, z^2) be a basis of $\mathbb{R}^2$ such that the matrix representation of $A : \mathbb{R}^2 \to \mathbb{R}^2$ with respect to this basis is in real Jordan canonical form. There are three types of 2×2 matrices in real Jordan form

$$\text{(i)} \quad \begin{bmatrix} \lambda_1 & 0 \\ 0 & \lambda_2 \end{bmatrix} \qquad \text{(ii)} \quad \begin{bmatrix} \lambda & 1 \\ 0 & \lambda \end{bmatrix} \qquad \text{(iii)} \quad \begin{bmatrix} \alpha & -\beta \\ \beta & \alpha \end{bmatrix}$$

where $\lambda_1, \lambda_2, \lambda, \alpha, \beta \in \mathbb{R}$. The corresponding free motions through any initial point $a \in \mathbb{R}^2$ are

$$\text{(i)} \quad \begin{bmatrix} a_1 e^{\lambda_1 t} \\ a_2 e^{\lambda_2 t} \end{bmatrix}, \quad \text{(ii)} \quad \begin{bmatrix} a_1 e^{\lambda t} + a_2 t e^{\lambda t} \\ a_2 e^{\lambda t} \end{bmatrix}, \quad \text{(iii)} \quad \begin{bmatrix} a_1 e^{\alpha t} \cos \beta t - a_2 e^{\alpha t} \sin \beta t \\ a_1 e^{\alpha t} \sin \beta t + a_2 e^{\alpha t} \cos \beta t \end{bmatrix}.$$

As a result we obtain for any $A \in \mathbb{R}^2$ a phase portrait of $\dot{x} = Ax$ around the origin which coincides qualitatively with exactly one of the patterns shown in Figure 2.2.3. The first six pictures correspond to the case (i), the next three pictures to case (ii) and the last six pictures to case (iii). Clearly the particular phase portrait will depend on the vectors z^1, z^2 and the magnitude of λ_1 and λ_2, λ, α and β.
In order to obtain a picture of the flow of a *nonlinear* time–invariant differentiable system with state space $\mathbb{R}^2$, an important first step is to determine the phase portraits of its linearizations around each of its equilibrium states. In a second step global features have to be specified, such as limit cycles, connections between saddle points (separatrices), connections to infinity (unbounded orbits), periodic orbits etc. There is a rich qualitative theory of differential systems in the plane (see *Notes and References*). We will not develop this. Instead we conclude this subsection with an illustrative example showing the phase portrait of a simple nonlinear system with two distinct equilibria (a saddle point and a stable focus, see Figure 2.2.4).

Example 2.2.11. (Nonlinear oscillator). Consider an oscillator with nonlinear restoring force $\ddot{\xi} + 2\dot{\xi} + 5\xi + \xi^2 = 0$ or

$$\dot{x}(t) = \begin{bmatrix} 0 & 1 \\ -5 & -2 \end{bmatrix} x(t) + \begin{bmatrix} 0 \\ -x_1^2(t) \end{bmatrix}. \tag{53}$$

Contrary to the linear oscillator (see Example 2.2.9), we have *two* equilibrium states $[0,0]^\top$

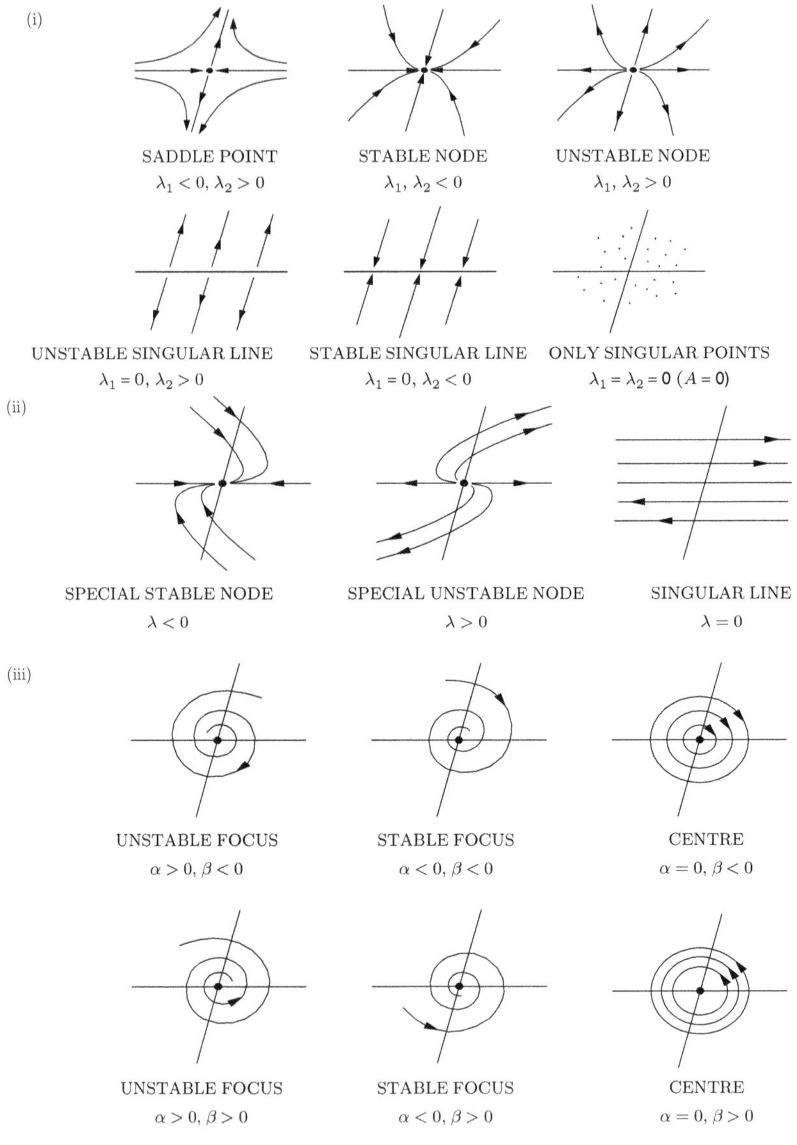

Figure 2.2.3: Phase portraits of two dimensional linear systems $\dot{x} = Ax$

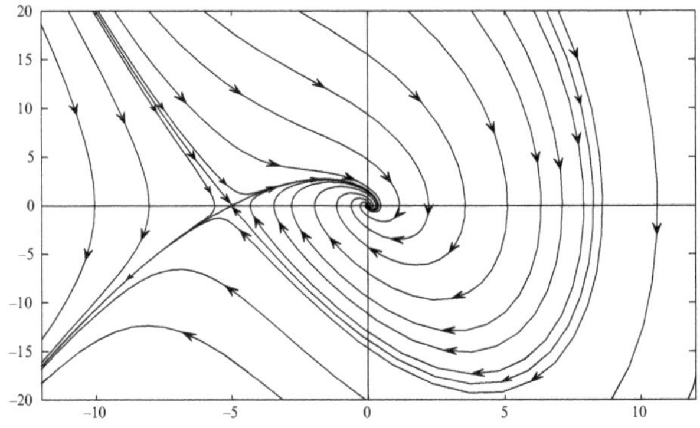

Figure 2.2.4: Phase portrait of the nonlinear oscillator (53) with two equilibria

and $[-5,0]^\top$. The matrix of the linearization about $[0,0]^\top$ is $A = \begin{bmatrix} 0 & 1 \\ -5 & -2 \end{bmatrix}$ which has eigenvalues $-1 \pm 2i$. Thus the local phase portrait of (53) around the origin has the form of a (nonlinear) *stable focus*. The matrix of the linearized system about $[-5,0]^\top$ is $A = \begin{bmatrix} 0 & 1 \\ 5 & -2 \end{bmatrix}$ and this matrix has real eigenvalues $-1 \pm \sqrt{6}$ with corresponding eigenvectors $z^1 = [1, -1+\sqrt{6}]^\top$, $z^2 = [1, -1-\sqrt{6}]^\top$. Thus $[-5,0]^\top$ is a *saddle point* of the nonlinear system (53). Figure 2.2.4 shows the actual trajectories of the nonlinear oscillator. □

2.2.3 Free Motions of Time–Invariant Linear Difference Systems

As in the differentiable case, every free motion of a discrete time system (22) can be represented as a superposition of generalized eigenmotions. Therefore the analysis of the free system

$$x(t+1) = Ax(t) \tag{54}$$

reduces more or less to the spectral analysis of the operator A. If $z \in \ker(A - \lambda I)^m$ is a generalized eigenvector of $A \in \mathbb{C}^{n\times n}$ corresponding to the eigenvalue $\lambda \in \sigma(A)$, the associated generalized eigenmotion of (54) in $\mathbb{C}^n$ is given by

$$\begin{aligned} A^t z &= [(A-\lambda I) + \lambda I]^t z = \sum_{\nu=0}^{t} \binom{t}{\nu} \lambda^{t-\nu} (A-\lambda I)^\nu z \\ &= \lambda^{t+1-m} \sum_{\nu=0}^{m-1} \binom{t}{\nu} \lambda^{m-1-\nu} (A-\lambda I)^\nu z, \quad t \geq m-1. \end{aligned} \tag{55}$$

If $\lambda = 0$ and z is an associated generalized eigenvector of order m then the free motions $A^t z$ ends at zero after m steps. This convergence to zero in finite time cannot occur in the continuous time case. However, if $\lambda \neq 0$ the generalized complex eigenmotion (55) is, as in the differentiable case, the product of an exponential term $(\lambda^t/\lambda^{m-1})$ in t and a vector polynomial $\sum_{\nu=0}^{m-1} \binom{t}{\nu} \lambda^{m-1-\nu}(A-\lambda I)^\nu z$ of degree $m-1$

in t.
In the case of an eigenvector ($m = 1$) we get the complex eigenmotion

$$A^t z = \lambda^t z, \quad t \in \mathbb{N}. \tag{56}$$

Now suppose $A \in \mathbb{R}^{n \times n}$. If $\lambda = r(\cos\theta + \imath \sin\theta) \in \sigma(A) \setminus \mathbb{R}$ and z is an associated eigenvector of A, the real eigenmotions corresponding to the pair of eigenvalues $\lambda, \overline{\lambda}$ and associated eigenvectors $z, \overline{z}$ are

$$\begin{aligned} A^t(\operatorname{Re} z) &= \operatorname{Re} A^t z = r^t[(\cos\theta t)\operatorname{Re} z - (\sin\theta t)\operatorname{Im} z], \quad t \in \mathbb{N} \\ A^t(\operatorname{Im} z) &= \operatorname{Im} A^t z = r^t[(\cos\theta t)\operatorname{Im} z + (\sin\theta t)\operatorname{Re} z], \quad t \in \mathbb{N}. \end{aligned} \tag{57}$$

If z is a generalized eigenvector of order m of A for λ the associated generalized real eigenmotions for $t \geq m-1$ are

$$\begin{aligned} A^t(\operatorname{Re} z) &= \sum_{\nu=0}^{m-1} \binom{t}{\nu} r^{t-\nu} \left[\cos(t-\nu)\theta \operatorname{Re}(A-\lambda I)^\nu z - \sin(t-\nu)\theta \operatorname{Im}(A-\lambda I)^\nu z\right] \\ A^t(\operatorname{Im} z) &= \sum_{\nu=0}^{m-1} \binom{t}{\nu} r^{t-\nu} \left[\cos(t-\nu)\theta \operatorname{Im}(A-\lambda I)^\nu z + \sin(t-\nu)\theta \operatorname{Re}(A-\lambda I)^\nu z\right]. \end{aligned} \tag{58}$$

As a discrete time counterpart to Corollaries 2.2.6 and 2.2.8 we obtain

Proposition 2.2.12. *Suppose $A \in \mathbb{K}^{n \times n}$, then very free motion of (54) in $\mathbb{K}^n$ can be represented as a sum of generalized eigenmotions of the form (55) if $\mathbb{K} = \mathbb{C}$ and (58) if $\mathbb{K} = \mathbb{R}$. If A is diagonalizable over $\mathbb{C}$ then all free motions of (54) in $\mathbb{K}^n$ are superpositions of eigenmotions of the form (56) if $\mathbb{K} = \mathbb{C}$ and (57) if $\mathbb{K} = \mathbb{R}$.*

Example 2.2.13. (Fibonacci's model). In 1202 the mathematician L. Fibonacci (1180 - 1240) introduced a model of a fictitious rabbit population which is a simple example of a population model with age structure. Assume that a single pair of rabbits starts the population. They reproduce twice, once at time 1 and once at time 2, then die. At each reproduction they produce a new pair of rabbits, one male and one female. These will go on to reproduce twice etc.. The resulting dynamics of the population is summarized in Table 2.2.5.

Time	0	1	2	3	4	5	6
Age group-1	1	1	2	3	5	8	13
Age group-2	0	1	1	2	3	5	8

Table 2.2.5: Number of rabbits pairs

If we denote by $x_1(t)$ and $x_2(t)$ the numbers of rabbit pairs in the first and second age group at time t and set $x(t) = [x_1(t), x_2(t)]^\top$ we obtain the following population model

$$x(t+1) = \begin{bmatrix} 1 & 1 \\ 1 & 0 \end{bmatrix} x(t), \qquad x(0) = \begin{bmatrix} 1 \\ 0 \end{bmatrix}. \tag{59}$$

Alternatively, the evolution of $\xi(\cdot) = x_1(\cdot)$ can be described by the second order difference equation (Fibonacci Renewal Equation)

$$\xi(t+2) = \xi(t+1) + \xi(t), \quad t \in \mathbb{N}$$

with initial conditions $\xi(0) = \xi(1) = 1$. The resulting sequence

$$1,\ 1,\ 2,\ 3,\ 5,\ 8,\ 13,\ 21,\ 34,\ 55,\ 89, \ldots$$

is called the *Fibonacci sequence* and plays a role in various fields of mathematics. The eigenvalues of the matrix in (59) are $\lambda_{1,2} = (1 \pm \sqrt{5})/2$ and the corresponding real modes are

$$z^1(t) = \left(\frac{1+\sqrt{5}}{2}\right)^t \begin{bmatrix} (1+\sqrt{5})/2 \\ 1 \end{bmatrix}, \quad z^2(t) = \left(\frac{1-\sqrt{5}}{2}\right)^t \begin{bmatrix} (1-\sqrt{5})/2 \\ 1 \end{bmatrix}.$$

The solution of the initial value problem (59) is of the form $x(t) = \alpha_1 z^1(t) + \alpha_2 z^2(t)$ for some $(\alpha_1, \alpha_2) \in \mathbb{R}^2$. Since $x(0) = [1, 0]^\top$ we obtain $\alpha_1 = 1/\sqrt{5}$, $\alpha_2 = -1/\sqrt{5}$ and hence the following analytic expression for the Fibonacci sequence

$$\xi(t) = \frac{1}{\sqrt{5}} \left[\left(\frac{1+\sqrt{5}}{2}\right)^{t+1} - \left(\frac{1-\sqrt{5}}{2}\right)^{t+1} \right].$$

□

2.2.4 Infinite Dimensional Systems

In Example 2.1.25 we showed that the state space for a delay equation is an infinite dimensional vector space. This is also the case for the partial differential equation described in Section 1.6. It is natural to ask therefore, whether there are results for infinite dimensional linear systems similar to those for time–invariant linear systems defined on finite dimensional vector spaces. The answer is often yes, but to develop these results in a general way is beyond the scope of this book. Instead we analyze the one-dimensional heat equation (taken from Section 1.6) in some detail and illustrate some of the main ideas and difficulties. As explained in Section 1.6 the evolution of the temperature in a heated metal bar of length ℓ with a fixed constant temperature at its ends can be determined via the equations

$$\begin{aligned} \frac{\partial \theta}{\partial t}(\xi, t) &= k\frac{\partial^2 \theta}{\partial \xi^2}(\xi, t) + b(\xi)u(t), \quad t \in (0, \infty), \quad \xi \in (0, \ell) && \text{(60a)} \\ \theta(0, t) &= \theta(\ell, t) = 0, \qquad t \in (0, \infty), && \text{(60b)} \\ \theta(\xi, 0) &= \theta_0(\xi), \qquad 0 \le \xi \le \ell. && \text{(60c)} \end{aligned}$$

The output equation is

$$y(t) = \int_0^\ell c(\xi)\theta(\xi, t)d\xi = \langle c(\cdot), \theta(\cdot, t)\rangle_{L^2} \tag{61}$$

where $\langle \cdot, \cdot \rangle_{L^2}$ is the inner product on $L^2(0, \ell; \mathbb{R})$. We will show that in a suitable setting we can write this controlled partial differential equation and the output equation in the form

$$\begin{aligned} \dot{x}(t) &= Ax(t) + Bu(t), \quad t \in (0, \infty), \qquad x(0) = x^0, && \text{(62)} \\ y(t) &= Cx(t), \end{aligned}$$

where the state space is an infinite dimensional Hilbert space. Let $X = L^2(0, \ell; \mathbb{R})$ and denote by $\mathcal{D}(A)$ the linear space

$$\mathcal{D}(A) = \{x\,;\, x(\cdot) \in \mathcal{C}^2([0, \ell], \mathbb{R}),\ x(0) = x(\ell) = 0\}$$

where $\mathcal{C}^2([0, \ell]; \mathbb{R})$ is the vector space of twice continuously differentiable functions on $[0, \ell]$ (at 0, and ℓ one-sided derivatives are considered). $\mathcal{D}(A)$ can be viewed as a linear subspace of X and will be endowed with the corresponding L^2 norm. We will assume

$$b(\cdot),\ \theta_0(\cdot) \in \mathcal{D}(A), \quad c(\cdot) \in X \quad \text{and} \quad \mathcal{U} = \mathcal{C}(\mathbb{R}_+, \mathbb{R}).$$

The operators A, B, C in (62) are defined by

$$\begin{aligned}
A &: \mathcal{D}(A) \to X, & (Az)(\xi) &= k\frac{d^2 z}{d\xi^2}(\xi), & \xi &\in (0, \ell), & z(\cdot) &\in \mathcal{D}(A) \\
B &: \mathbb{R} \to X, & (Bu)(\xi) &= b(\xi)u, & \xi &\in [0, \ell], & u &\in \mathbb{R} \\
C &: X \to \mathbb{R}, & Cz &= \langle c(\cdot), z(\cdot)\rangle_{L^2}, & & & z(\cdot) &\in X.
\end{aligned} \tag{63}$$

Before we can discuss the relationship between the operator differential equation (62) and the partial differential equation (60) we need to define what is meant by a solution of these equations.

Definition 2.2.14. A continuous function $\theta(\cdot,\cdot) : [0, \ell] \times \mathbb{R}_+ \to \mathbb{R}$ is said to be a solution of (60) if all the derivatives

$$\frac{\partial \theta}{\partial t}(\cdot,\cdot), \quad \frac{\partial \theta}{\partial \xi}(\cdot,\cdot), \quad \frac{\partial^2 \theta}{\partial \xi^2}(\cdot,\cdot)$$

exist and are continuous on $(0, \ell) \times (0, \infty)$ and the three equations in (60) are satisfied.

Definition 2.2.15. A continuous function $x(\cdot) : \mathbb{R}_+ \to X$ is a solution of (62) if it is Fréchet differentiable on $(0, \infty)$ and satisfies (62) in X.

Let us now construct the solution of (60). At the beginning of this section we saw that for time-invariant finite dimensional systems both the free and forced motions can be determined via a semigroup $\Phi(t)$ which is constructed from the (generalized) eigenmotions. This suggests that we should examine the eigenvalues and associated eigenvectors of the operator A,

$$A\psi = \lambda\psi, \quad \psi \in \mathcal{D}(A), \quad \psi \neq 0, \qquad \lambda \in \mathbb{C} \tag{64}$$

or, equivalently, the nontrivial solutions of

$$k\frac{d^2\psi}{d\xi^2}(\xi) = \lambda\psi(\xi), \quad \xi \in (0, \ell), \quad \psi(0) = \psi(\ell) = 0. \tag{65}$$

The differential equation in (65) has the general solution

$$\psi(\xi) = a e^{\sqrt{\lambda/k}\,\xi} + b e^{-\sqrt{\lambda/k}\,\xi}, \qquad a,\ b \in \mathbb{C}.$$

To satisfy the boundary condition $\psi(0) = 0$ we must have $a + b = 0$, hence $\psi(\xi) = 2a \sinh \sqrt{\lambda/k}\,\xi$. Now let $\sqrt{\lambda/k}\,\ell = \alpha + \imath\beta \quad \alpha,\ \beta \in \mathbb{R}$, then the second boundary condition $\psi(\ell) = 0$ implies

$$\sinh \sqrt{\lambda/k}\,\ell = \sinh(\alpha + \imath\beta) = \sinh\alpha\cos\beta + \imath\cosh\alpha\sin\beta = 0.$$

As a consequence we obtain $\alpha = 0$ and $\beta \in \mathbb{Z}\pi$, $\beta \neq 0$ (since $\beta = 0$ would yield the trivial solution). Hence $\sqrt{\lambda/k}\,\ell = \pm\imath n\pi$ for $n \in \mathbb{N}^*$. So there is an infinite sequence of eigenvalues and associated eigenvectors

$$\lambda_n = -k\frac{n^2\pi^2}{\ell^2}, \quad \psi_n(\xi) = \sqrt{\frac{2}{\ell}}\sin\frac{n\pi\xi}{\ell}, \quad n \in \mathbb{N}^* \tag{66}$$

with corresponding eigenmotions

$$\theta_n(\xi, t) = \sqrt{\frac{2}{\ell}}\exp\left(-k\frac{n^2\pi^2 t}{\ell^2}\right)\sin\frac{n\pi\xi}{\ell}, \quad t \geq 0, \qquad n \in \mathbb{N}^*.$$

It is easily verified that the eigenvectors $\psi_n(\cdot) \in \mathcal{D}(A) \subset L^2(0, \ell; \mathbb{R})$ are orthogonal to each other in the Hilbert space $L^2(0, \ell; \mathbb{R})$

$$\langle \psi_n, \psi_m \rangle_{L^2} = \int_0^\ell \psi_n(\xi)\psi_m(\xi)d\xi = \delta_{mn}. \tag{67}$$

In fact it is known from the theory of Fourier series that the functions $\{\psi_n\}_{n\in\mathbb{N}^*}$ form an orthonormal basis of the Hilbert space $L^2(0, \ell; \mathbb{R})$ in the sense that (67) holds and any $z(\cdot) \in L^2(0, \ell; \mathbb{R})$ can be expressed in a unique way as an infinite linear combination of the ψ_n's, viz $z(\xi) = \sum_{n=1}^{\infty} \alpha_n\psi_n(\xi)$. Here $\alpha_n = \langle z, \psi_n \rangle_{L^2}$, $n = 1, 2, \ldots$, and the equality is to be interpreted in the sense of $L^2(0, \ell; \mathbb{R})$, i.e. $\lim_{N\to\infty} \|z(\cdot) - \sum_{n=1}^{N} \alpha_n\psi_n(\cdot)\|_{L^2} = 0$ (see Section A.3). If, in particular, $z(\cdot) \in \mathcal{D}(A)$, then

$$\alpha_n = \int_0^\ell z(\xi)\psi_n(\xi)d\xi = -\frac{\ell^2}{n^2\pi^2}\int_0^\ell \frac{d^2 z}{d\xi^2}(\xi)\psi_n(\xi)d\xi, \quad n \in \mathbb{N}$$

on integrating by parts twice and using the fact that $z(0) = z(\ell) = 0$. Thus for $z(\cdot) \in \mathcal{D}(A)$, there exists a constant M such that

$$|\alpha_n| \leq \frac{M}{n^2}, \quad n \in \mathbb{N}^*. \tag{68}$$

Now, since the pre-Hilbert space $\mathcal{D}(A) \subset X$ has a basis consisting of eigenvectors of the operator A, if we mirror the development for time–invariant finite dimensional systems, we would expect the associated semigroup on $\mathcal{D}(A)$ to be given by the superposition of eigenmotions

$$(\Phi(t)z(\cdot))(\xi) = \sum_{n=1}^{\infty} e^{\lambda_n t}\langle z(\cdot), \psi_n(\cdot)\rangle_{L^2}\,\psi_n(\xi). \tag{69}$$

Then the solution of the controlled equation (60) would be

$$\theta(\xi,t) = (\Phi(t)\theta_0(\cdot))(\xi) + \int_0^t (\Phi(t-s)b(\cdot))(\xi)u(s)ds$$

(see (19)) or more explicitly

$$\theta(\xi,t) = \sum_{n=1}^{\infty} e^{\lambda_n t}\langle\theta_0(\cdot),\psi_n(\cdot)\rangle_{L^2}\,\psi_n(\xi) + \int_0^t \sum_{n=1}^{\infty} e^{\lambda_n(t-s)}\langle b(\cdot),\psi_n(\cdot)\rangle_{L^2}\,\psi_n(\xi)u(s)ds. \quad (70)$$

Using (68) with constants M_0, resp. M_b for $\theta_0(\cdot)$ and $b(\cdot) \in \mathcal{D}(A)$, it is easy to see that the series in (70) is uniformly absolutely convergent in $(\xi,t) \in [0,\ell] \times [0,t_1]$ for arbitrary $t_1 > 0$ and

$$|\theta(\xi,t)| \le \sqrt{\frac{2}{\ell}}M_0 \sum_{n=1}^{\infty} \frac{e^{\lambda_n t}}{n^2} + \frac{\sqrt{2\ell^3}}{k\pi^2}M_b \sup_{0\le s\le t}|u(s)| \sum_{n=1}^{\infty} \frac{1-e^{\lambda_n t}}{n^4}.$$

Therefore $\theta(\cdot,\cdot)$ is well defined and continuous on $[0,\ell] \times \mathbb{R}_+$.

Theorem 2.2.16. *Given $\theta_0 \in \mathcal{D}(A)$, $u(\cdot) \in \mathcal{U}$, then* (60) *has exactly one solution in the sense of Definition 2.2.14 and this solution is given by the function $\theta(\cdot,\cdot)$ defined by* (70).

Proof: It follows directly from the definition, from (70) and the above convergence result that $\theta(\cdot,\cdot)$ satisfies conditions (60b), (60c). In order to prove that $\theta(\cdot,\cdot)$ solves the partial differential equation (60a) one proceeds as follows. First it is shown that the partial derivatives $\frac{\partial\theta}{\partial t}(\xi,t)$ and $k\frac{\partial^2\theta}{\partial\xi^2}(\xi,t)$ can be calculated from (70) term by term. This can be done by proving that the resulting series are uniformly absolutely convergent on $[0,\ell] \times [0,t_1]$ for arbitrary $t_1 > 0$ (making use of the estimate (68) as above). Then comparing the two series for $\frac{\partial\theta}{\partial t}(\xi,t)$ and $k\frac{\partial^2\theta}{\partial\xi^2}(\xi,t)$ term by term it becomes clear that (60a) holds. We omit the details.
To prove uniqueness we assume that there is a second solution $\hat{\theta}(\cdot\,,\cdot)$ and set $e(\cdot\,,\cdot) = (\theta - \hat{\theta})(\cdot\,,\cdot)$. Then $e(\cdot\,,\cdot)$ must satisfy

$$\begin{aligned}
\frac{\partial e}{\partial t}(\xi,t) &= k\frac{\partial^2 e}{\partial\xi^2}(\xi,t), \quad t \in (0,\infty), \quad \xi \in (0,\ell) \\
e(0,t) &= e(\ell,t) = 0, \quad t \in \mathbb{R}_+ \\
e(\xi,0) &= 0, \qquad \xi \in [0,\ell].
\end{aligned}$$

Consider the function $E(t) = \int_0^\ell e^2(\xi,t)d\xi$, then

$$\frac{dE}{dt}(t) = 2\int_0^\ell e(\xi,t)\frac{\partial e}{\partial t}(\xi,t)d\xi = 2k\int_0^\ell e(\xi,t)\frac{\partial^2 e}{\partial\xi^2}(\xi,t)d\xi = -2k\int_0^\ell \left(\frac{\partial e}{\partial\xi}(\xi,t)\right)^2 d\xi$$

on integration by parts. So $E(t)$ is non-increasing, but we have $E(0) = 0$ and $E(t) \ge 0$. Hence $e(\cdot\,,\cdot) \equiv 0$ and the solution of (60) is unique. □

Up until now we have analyzed (60) on the pre-Hilbert space $\mathcal{D}(A)$. For technical reasons it is advantageous to associate with (60) a dynamical system with the whole Hilbert space X as state space and an extended space of control functions. To achieve this suppose that $\theta_0 \in X$ and $u(\cdot) \in \hat{\mathcal{U}} = L^2_{\mathrm{loc}}(\mathbb{R}_+, \mathbb{R})$. Then the series in (70) are still absolutely convergent for each $t \geq 0$ with respect to the norm of the Hilbert space $X = L^2(0, \ell; \mathbb{R})$. Hence the map

$$\varphi : \{(t, t_0) \in \mathbb{R}^2;\, t \geq t_0\} \times X \times \hat{\mathcal{U}} \to X, \quad (t; t_0, \theta_0, u(\cdot)) \mapsto \varphi(t; t_0, \theta_0, u(\cdot))(\cdot)$$

is well defined by $\varphi(t; t_0, \theta_0, u(\cdot))(\cdot) = z(\cdot)$ where for $\xi \in [0, \ell]$

$$z(\xi) = \sum_{n=1}^{\infty} e^{\lambda_n (t-t_0)} \langle \theta_0(\cdot), \psi_n(\cdot) \rangle_{L^2} \psi_n(\xi) + \int_{t_0}^{t} \sum_{n=1}^{\infty} e^{\lambda_n (t-s)} \langle b(\cdot), \psi_n(\cdot) \rangle_{L^2} \psi_n(\xi) u(s) ds. \quad (71)$$

This formula extends the solution formula (70) so that it is applicable for all $(\theta_0, u(\cdot)) \in X \times \hat{\mathcal{U}}$ instead of $(\theta_0, u(\cdot)) \in \mathcal{D}(A) \times \mathcal{U}$. The coordinates $\alpha_n(t)$, $n \in \mathbb{N}^*$ of $\varphi(t; t_0, \theta_0, u(\cdot))$ with respect to the basis $(\psi_n)_{n \in \mathbb{N}^*}$ of X are given by

$$\alpha_n(t) = e^{\lambda_n (t-t_0)} \langle \theta_0(\cdot), \psi_n(\cdot) \rangle_{L^2} + \int_{t_0}^{t} e^{\lambda_n (t-s)} \langle b(\cdot), \psi_n(\cdot) \rangle_{L^2} u(s)\, ds$$

and hence satisfy the differential equations

$$\dot{\alpha}_n(t) = \lambda_n \alpha_n(t) + \langle b(\cdot), \psi_n(\cdot) \rangle_{L^2} u(t).$$

Using this fact it is easy to see that φ satisfies the axioms of a state transition map (Definition 2.1.1). Therefore, defining the output map by $\eta(x, u) = \langle c, x \rangle_{L^2}$, we obtain a dynamical system $\Sigma = (\mathbb{R}, \mathbb{R}, \hat{\mathcal{U}}, X, \mathbb{R}, \varphi, \eta)$ with state space $X = L^2(0, \ell; \mathbb{R})$. This system is obviously linear and time-invariant. The associated operator semigroup (69) describing the free motions of the system is given by (69) where $z(\cdot)$ is now allowed to vary in X. The associated input-state map $\Theta(t, 0) : \hat{\mathcal{U}} \to X$ (see (7)) is given by

$$(\Theta(t, 0)u(\cdot))(\xi) = \int_0^t \sum_{n=1}^{\infty} e^{\lambda_n (t-s)} \langle b(\cdot), \psi_n(\cdot) \rangle_{L^2} \psi_n(\xi) u(s) ds, \quad \xi \in [0, \ell].$$

For initial states $x(0) = \theta_0 \in \mathcal{D}(A)$ and controls $u(\cdot) \in \mathcal{U}$, we have by Theorem 2.2.16

$$\varphi(t; 0, \theta_0, u(\cdot))(\cdot) = \theta(\cdot, t), \quad t \geq 0,$$

where $\theta(\cdot, \cdot)$ solves the partial differential equation (60). Hence $t \mapsto \varphi(t; 0, \theta_0, u(\cdot))(\cdot)$ describes the evolution of the temperature profile along the metal bar under the influence of the control $u(\cdot)$. A typical controlled temperature profile is shown in Figure 2.2.6. Here the initial temperature is zero along the bar and the object of the control is to steer the temperature to the profile $\theta(\xi, T) = \sin \pi\xi$, $\xi \in [0, \ell]$ in time $[0, 5]$. Since the process of heat propagation is not reversible we expect that the system Σ is not reversible. In fact the series in (71) do not converge in $X = L^2(0, \ell; \mathbb{R})$ for $t < t_0$. Moreover, it follows from an analysis of (71) that $\varphi(t; t_0, \theta_0, 0) \in \mathcal{D}(A)$ for arbitrary $t > 0$ and $\theta_0 \in X$, hence any free trajectory enters the dense subspace $\mathcal{D}(A) \subset X$ of smooth temperature profiles immediately after leaving the possibly discontinuous initial temperature profile θ_0.

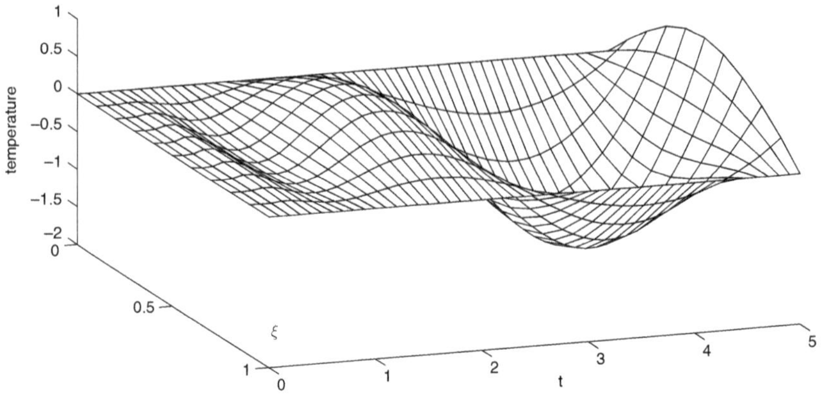

Figure 2.2.6: Evolution of the controlled temperature profile

Remark 2.2.17. We have used the abstract differential equation (62) only as a heuristic tool in order to obtain the expression (70) for a solution of the partial differential equation (60). In fact it is possible to show that, for every $x^0 \in X$, the semigroup $\Phi(t)$ on X yields a solution $\Phi(t)x^0$ of (62) for $u(t) \equiv 0$ in the sense of Definition 2.2.15, see *Notes and References*. Then using similar estimates as in the proof of Theorem 2.2.16 one can show that, for $\theta_0 \in \mathcal{D}(A)$, $u(\cdot) \in \mathcal{U}$, the solution $\theta(\cdot,\cdot)$ of (60) gives rise to a unique solution $t \mapsto x(t) = \theta(\cdot,t)$ of (62). For the more general initial condition $x(0) = \theta_0 \in X$ and arbitrary controls $u(\cdot) \in \hat{\mathcal{U}} = L^2_{\text{loc}}(\mathbb{R}_+;\mathbb{R})$, it can be shown that the solution $t \mapsto x(t) = \varphi(t;t_0,\theta_0,u(\cdot))$ is a *mild solution* of (62), i.e. satisfies the variation-of-constants formula

$$x(t) = \Phi(t)x^0 + \int_0^t \Phi(t-s)bu(s)ds$$

where the integral is a Bochner integral [538].
However, one should not be misled by the formal analogy with the finite dimensional situation. There are essential differences between the theories of finite and of infinite dimensional linear systems, not all of which are illustrated by the above example.

(i) As in the example the operator A is in general an *unbounded* linear operator and is not defined on the whole state space but only on a dense subspace $\mathcal{D}(A)$ of X.

(ii) The initial value problem (62) does not necessarily have a differentiable solution and the mild solution given by the variation-of-constants formula will in general only be a solution of (62) in the sense of Definition 2.2.15 if $x^0 \in \mathcal{D}(A)$ and $u(\cdot)$ is sufficiently smooth.

(iii) In contrast with the example the spectrum of the operator A will *not* in general consist of eigenvalues only (see Section A.4) and even if this is the case the generalized eigenvectors of A will *not* in general span the whole state space X.

(iv) The operators B and C need *not* be *bounded* (e.g. if the control is acting through the boundary conditions or if point measurements are taken).

As a consequence of (iii) the semigroup Φ cannot always be constructed via the eigenmotions by using a series representation as in (69). The relationship between semigroups of operators $(\Phi(t))_{t\in\mathbb{R}_+}$ and their "infinitesimal generators" A must be put on another footing and criteria must be found under which a given unbounded linear operator $A : \mathcal{D}(A) \to X$

"generates" a semigroup $(\Phi(t))_{t\in\mathbb{R}_+}$. This is done in the theory of operator semigroups, see Remark 5.5.44 and *Notes and References.* □

2.2.5 Exercises

1. Prove that a linear system with continuous coefficient matrices $A(t)$, $B(t)$, $C(t)$ and $D(t)$ is time–invariant if and only if the matrix functions are constant.

2. Determine the evolution operator $\Phi(t,t_0)$ of the time–varying scalar differential equation

$$\dot{x}(t) = a(t)x(t)$$

where $a(\cdot) : \mathbb{R} \to \mathbb{R}$ is continuous. Show that the result can be generalized to vector differential equations

$$\dot{x}(t) = A(t)x(t)$$

where $A(\cdot)$ is a continuous $n \times n$-matrix function such that $A(t)A(s) = A(s)A(t),\ s,t \in \mathbb{R}$. Apply this to $A(t) = a(t)A$ where $a(\cdot) : \mathbb{R} \to \mathbb{R}$ is continuous and $A \in \mathbb{R}^{n\times n}$.

3. If $A(\cdot) \in C([t_0,t_1];\mathbb{R}^{n\times n})$ is a continuous $n \times n$-matrix function show that the solution $\Phi(t,t_0)$ of the matrix differential equation

$$\dot{X}(t) = A(t)X(t), \quad t_0 \le t \le t_1, \quad X(t_0) = I_n$$

can be obtained as the uniform limit of the recursive sequence

$$\begin{aligned} \Phi_0(t,t_0) &\equiv I_n, \ t \in [t_0,t_1] \\ \Phi_k(t,t_0) &= I_n + \int_{t_0}^{t} A(s)\Phi_{k-1}(s,t_0)ds, \quad t \in [t_0,t_1] \end{aligned}$$

(cf. *Hale* (1980) [214, III.3]).

4. Find an appropriate state vector and determine matrices A,B,C,D for the discrete time linear system represented in the block diagram shown in Figure 2.2.7, where Δ is the unit time delay.

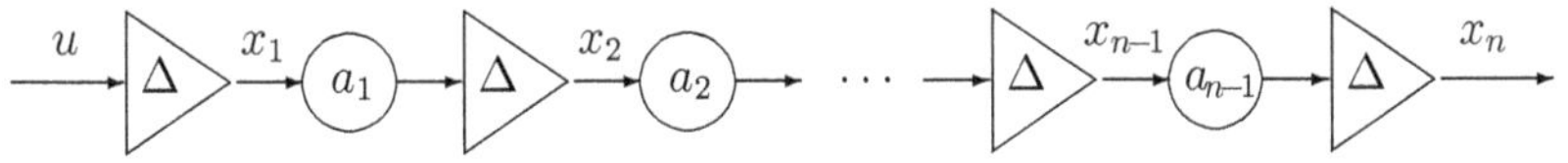

Figure 2.2.7: Block diagram of discrete time system

5. Consider the discrete time system (A,B,C,D) represented in Figure 2.2.7 (Ex. 4).

(i) Determine the spectrum of A.

(ii) Compute A^t, $t \in \mathbb{N}$.

(iii) Specify a basis of $\mathbb{R}^n$ consisting of generalized eigenvectors of A.

(iv) What can be said of the free motions of this system?

6. Compute $A^t, t \in \mathbb{N}$ and $e^{At}, t \in \mathbb{R}$ for the following matrices A where $\alpha,\beta > 0$

$$(i)\ \begin{bmatrix} 0 & \beta \\ \alpha & 0 \end{bmatrix}, \quad (ii)\ \begin{bmatrix} 0 & \beta \\ -\alpha & 0 \end{bmatrix}, \quad (iii)\ \begin{bmatrix} 0 & 1 & 0 \\ 0 & 0 & 1 \\ 0 & 0 & 0 \end{bmatrix}, \quad (iv)\ \begin{bmatrix} 0 & 1 & 0 \\ 0 & 0 & 1 \\ 1 & 0 & 0 \end{bmatrix}.$$

7. Compute the characteristic polynomial, the eigenvalues, the real modes and the free motions for the following systems and initial states

(i) $$x(t+1) = \begin{bmatrix} 0 & 1 & 0 \\ 0 & 0 & 1 \\ -6 & -11 & -6 \end{bmatrix} x(t), \quad x(0) = \begin{bmatrix} 1 \\ 0 \\ 0 \end{bmatrix},$$

(ii) $$x(t+1) = \begin{bmatrix} 0 & 1 & 0 \\ 0 & 0 & 1 \\ 0 & 0 & 0 \end{bmatrix} x(t), \quad x(0) = \begin{bmatrix} 1 \\ 1 \\ 1 \end{bmatrix},$$

(iii) $$\dot{x}(t) = \begin{bmatrix} 0 & 1 & 0 \\ 0 & -1 & 1 \\ -2 & 0 & -1 \end{bmatrix} x(t), \quad x(0) = \begin{bmatrix} 1 \\ 0 \\ -1 \end{bmatrix},$$

(iv) $$\dot{x}(t) = \begin{bmatrix} 0 & 1 \\ -1 & -2 \end{bmatrix} x(t), \quad x(0) = \begin{bmatrix} 1 \\ 1 \end{bmatrix}.$$

8. Let $\mathbb{Z}_3 = \{0, 1, 2\}$ be the field of integers modulo 3. Consider the finite linear machine over $\mathbb{K} = \mathbb{Z}_3$ represented in Figure 2.2.8 where Δ is the unit delay operator. The summer

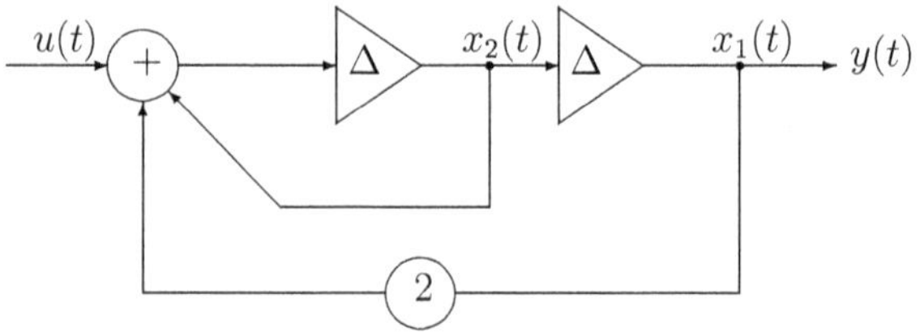

Figure 2.2.8: Finite linear machine

adds mod 3 and the gain multiplies mod 3.

(i) Determine the system equations
$$x(t+1) = Ax(t) + Bu(t), \; t \in \mathbb{N}; \quad y(t) = Cx(t) + Du(t).$$

(ii) Show that $A^2 - A + I = 0$.

(iii) Determine the fundamental matrix A^t, $t \in \mathbb{N}$.

(Note that all calculations have to be carried out in $\mathbb{Z}_3$)

9. Consider the electrical circuit represented in Figure 1.4.4. If the voltage source $e(t)$ is taken as input $u(t)$ and the charge on the capacitor is $q(t)$ it is shown in Example 1.4.4 that
$$L\ddot{q}(t) + R\dot{q}(t) + (1/C)q(t) = u(t).$$

(i) If the current $y(t)$ through the inductor together with the charge on the capacitor are state variables and the output is the current through the inductor specify the matrices A, B, C describing the system.

(ii) Determine the eigenvalues, eigenvectors and real eigenmodes of the system in terms of the parameters L, $R \geq 0$, $C > 0$.

(iii) Under which conditions do oscillating eigenmotions occur? Express the frequency in terms of R, C, L. What happens if $R = 0$? What happens if C approaches 0? What happens if $L = 0$?

(*Driver* (1977), pp.119-122 [138])

2.2.6 Notes and References

The analysis of free motions of linear differentiable systems is contained in most books on ordinary differential equations. For basic results we refer to the introductory textbooks *Driver* (1977) [138], *Polking et al.* (2001) [417] and the books recommended in the *Notes and References* of Section 2.1. For analogous results on difference equations see *Agarwal* (1992) [5] and *Kelley and Peterson* (2001) [298]. The results from Linear Algebra required for the spectral analysis of linear time-invariant systems can be found in most textbooks on Matrix Theory. Standard references which contain many results which are useful in different areas of Linear Systems Theory are *Gantmacher* (1959) [182] and *Horn and Johnson* (1985) [264].

For the numerical problems involved in computing the exponential of a matrix, see *Mohler and Van Loan* (1978) [378] and Section 2.5.

An excellent textbook on control theoretic aspects of linear differentiable systems is *Brockett* (1970) [77] which also contains many results on *time-varying* linear systems as does the book of *Rugh* (1993) [442]. For discrete time linear control systems see *Ogata* (1987) [396], *Franklin et al.* (1998) [169], and a comprehensive book which treats continuous time and discrete time systems in parallel is *Oppenheim et al.* (1997) [399].

Some results and a number of references concerning the qualitative theory of differential systems in the plane can be found in Section 3.1.

Example 2.2.13 (Fibonacci's model) is discussed in *Hoppenstead* (1982) [262]. Many interesting examples of a non-technological kind are described in the introductory text by *Luenberger* (1979) [349]. Linear models for real life mechanical, electrical and electromechanical control systems are discussed in *Franklin et al.* (1986) [168].

A classical reference for the theory of operator semigroups which also contains the necessary functional analytic foundations is *Hille and Phillips* (1957) [231]. Other references are *Pazy* (1983) [406] and the graduate textbook of *Engel and Nagel* (2000) [152], the latter one contains many interesting examples and applications. A comprehensive introduction to the theory of infinite dimensional linear systems is the excellent text book by *Curtain and Zwart* (1995) [116] and a concise introduction to infinite dimensional as well as finite dimensional control systems is given in *Zabczyk* (1992) [541].

2.3 Linear Systems: Input–Output Behaviour

In this section we study the forced motions of systems of the form (2.17) (resp. (2.22)) and begin by explaining how they can be represented as superpositions of trajectories generated by impulse controls. Then we analyze the input-output behaviour of time–invariant linear systems in time domain. Norms on the input and output function spaces are introduced and conditions are given under which the input-output map of a system is a *bounded linear operator* between these normed spaces. Systems with this property are called *input-output stable.* In the second subsection Laplace and Fourier transforms are used to obtain a frequency domain representation of the input-output behaviour in terms of transfer matrices. Under a diagonalizability assumption a dyadic decomposition of these matrices is constructed and it is shown how the transfer matrix is related to the response of the system to sinusoidal input signals. Finally the relationship between the time domain and frequency domain representations is discussed and leads in the case of input-output stable systems to a computable formula for the norm of the input-output operator.

2.3.1 Input-Output Behaviour in Time Domain

In many applications only the input-output behaviour of a system is of interest. We have seen that in state space theory a *pointwise* approach is possible because the instantaneous state of the system at time t contains all the necessary information needed to determine the effect of the past inputs upon the present output. In general the present output of a dynamical system is not determined by its present input alone, but by the whole control function on the preceding time interval. So if the concept of state is dropped, a *functional* viewpoint must be adopted which looks at the signal as a whole rather than its values at certain points in time. The analysis of the input-output behaviour of a system involves the investigation of the dependence of the output *function* ("output signal") on the input *function* ("input signal"). This is why functional analytical methods become important in this context (see Section A.3 and Section A.4). In this subsection we first initialize the linear state space systems (2.17) and (2.22) at $x(0) = 0$ and study the input-output behaviour on the time domain $T = \mathbb{R}_+$ (resp. $\mathbb{N}$). Then, under a stability assumption, we consider the input-output behaviour on the time domain $T = \mathbb{R}$ (resp. $\mathbb{Z}$). Throughout the section we assume that $U = \mathbb{K}^m$, $X = \mathbb{K}^n$ and $Y = \mathbb{K}^p$ for some given $m, n, p \in \mathbb{N}^*$.

Impulse Responses

For the systems (2.17) and (2.22) initialized at $x(0) = 0$ the dependence of the state and output trajectories on the control function is described by the *input-state map*

$$u(\cdot) \mapsto x(\cdot; u(\cdot)) = \varphi(\cdot\,; 0, 0, u(\cdot)), \quad u(\cdot) \in \mathcal{U} \tag{1}$$

and the *input-output map*

$$u(\cdot) \mapsto y(\cdot; u(\cdot)) = C\varphi(\cdot\,; 0, 0, u(\cdot)) + Du(\cdot), \quad u(\cdot) \in \mathcal{U}. \tag{2}$$

on the time domain $T = \mathbb{R}_+$ (resp. $\mathbb{N}$). We assume that in the discrete time case $\mathcal{U} = U^{\mathbb{N}}$ and in the continuous time case $\mathcal{U} = L^1_{\text{loc}}(\mathbb{R}_+; U)$. Obviously (1) is a special case of (2) with $Y = X$, $C = I_X$, $D = 0$ and thus it is sufficient to study (2). Let us first consider the discrete time case (2.22) with $T = \mathbb{N}$. Since $x(0) = 0$ we have

$$y(t; u(\cdot)) = Du(t) + \sum_{s=0}^{t-1} CA^{t-s-1}Bu(s), \quad u(\cdot) \in \mathcal{U}, \quad t \in \mathbb{N}. \tag{3}$$

(3) can be written in a more concise form using the *convolution of sequences*, see Section A.3.

Definition 2.3.1 (Convolution of sequences). If $g = (g(t))_{t\in\mathbb{N}}$, $v = (v(t))_{t\in\mathbb{N}} \in \mathbb{K}^{\mathbb{N}}$ are sequences over a field $\mathbb{K}$ the *convolution* $y = g*v$ is defined to be the sequence $y = (y(t))_{t\in\mathbb{N}} \in \mathbb{K}^{\mathbb{N}}$ given by

$$y(t) = \sum_{s=0}^{t} g(t-s)v(s), \quad t \in \mathbb{N}. \tag{4}$$

It is an easy exercise to prove that the set $\mathbb{K}^{\mathbb{N}}$ of scalar sequences is a commutative ring with respect to the operations $+$ and $*$. In fact, $\mathbb{K}^{\mathbb{N}}$ is even an integral domain (i.e. has no zero divisors) and has the sequence $(1, 0, 0, \ldots)$ as a unit element. Considering matrices and vectors with entries in this ring we define the convolution $\mathcal{G} * u$ of a sequence $\mathcal{G} = (\mathcal{G}(t))_{t\in\mathbb{N}}$ of $p \times m$-matrices and a sequence $u = (u(t))_{t\in\mathbb{N}}$ of m-vectors to be the sequence of p-vectors given by

$$(\mathcal{G} * u)(t) = \sum_{s=0}^{t} \mathcal{G}(t-s)u(s) = \sum_{s=0}^{t} \mathcal{G}(s)u(t-s), \quad t \in \mathbb{N}. \tag{5}$$

Using this notation (3) can be written in the form

$$y(t) = (\mathcal{G} * u)(t), \quad t \in \mathbb{N} \tag{6}$$

where

$$\mathcal{G} = (D, CB, CAB, \ldots, CA^{t-1}B, \ldots) \in (\mathbb{K}^{p\times m})^{\mathbb{N}}. \tag{7}$$

The matrices $CA^{t-1}B$, $t \geq 1$ occurring in this sequence are called the *Markov parameters* of the system (2.22). Together with the matrix D of direct input-output coupling they completely determine the input-output behaviour of (2.22) under the condition that the system is initialized at $x(0) = 0$. The sequence $\mathcal{G}$ can be viewed as a *weighting pattern* which determines the present output $y(t)$ as the weighted sum over the past and present inputs $u(s)$, $s \leq t$, see (5).
The sequence (7) can be obtained (at least theoretically) by a series of experiments as follows: Let $(e^1, \ldots, e^m)$ be the standard basis of $\mathbb{K}^m$ and suppose the following input signals are applied to the system

$$u^j(t) = \begin{cases} e^j & \text{if } t = 0, \ j \in \underline{m} \\ 0 & \text{if } t > 0 \end{cases}. \tag{8}$$

This particular test signal is called the jth *unit impulse* . By (6) the corresponding output sequences are

$$y(t; u^j(\cdot)) = (\mathcal{G} * u^j)(t) = \mathcal{G}(t)e^j = \mathcal{G}^j(t)$$

where $\mathcal{G}^j(t)$ is the jth column vector of $\mathcal{G}(t)$. Because of this property $\mathcal{G}$ is also called the *impulse response* of the system (2.22). We see, therefore, that if we were able to take exact measurements of the output signals (state trajectories for the special case $C = I_X$, $D = 0$) corresponding to the m test signals $u^1(\cdot), \ldots, u^m(\cdot)$, the input-output map (2) (resp. input-to-state map (1)) of the system would be completely determined.

Example 2.3.2. Consider the discrete time scalar system

$$\begin{aligned} x(t+1) &= ax(t) + bu(t), \quad x(0) = 0 \\ y(t) &= x(t) \end{aligned} \tag{9}$$

where a, $b \in \mathbb{R}$, $b \neq 0$ are given. Solving this equation for $u(\cdot) = (1, 0, 0, \ldots)$ we obtained the impulse response

$$g = (0,\, b,\, ab,\, a^2b, \ldots\,).$$

Let us compute the *step response* of (9) which is the output $\bar{y}(\cdot)$ corresponding to the constant input $\bar{u}(t) \equiv 1$. We get from (5)

$$\bar{y}(t; u(\cdot)) = \sum_{s=1}^{t} a^{s-1}b = \begin{cases} tb & \text{if } \ a = 1 \\ \dfrac{1-a^t}{1-a}b & \text{if } \ a \neq 1 \end{cases}, \qquad t \in \mathbb{N}.$$

Now $\bar{y}(t; u(\cdot)) - \bar{y}(t-1; u(\cdot)) = g(t)$ for $t \geq 1$. Furthermore $\lim_{t\to\infty} |\bar{y}(t; u(\cdot))| = \infty$ if $|a| > 1$ or $a = 1$, and $\lim_{t\to\infty} \bar{y}(t; u(\cdot)) = b(1-a)^{-1}$ if $|a| < 1$. Note that $\bar{x} = b(1-a)^{-1}$ is just the equilibrium state of the system (9) for the control $\bar{u}(t) \equiv 1$. □

A similar analysis can be carried out for continuous time systems with $T = \mathbb{R}_+$. For simplicity we assume that there is no direct input-output coupling so that $D = 0$, then the output function of (2.17) corresponding to an input $u(\cdot) \in L^1_{\text{loc}}(\mathbb{R}_+; \mathbb{K}^m)$ and zero initial condition is given by

$$y(t; u(\cdot)) = C\varphi(t; 0, 0, u(\cdot)) = \int_0^t Ce^{A(t-s)}Bu(s)ds, \quad t \in \mathbb{R}_+. \tag{10}$$

Recall the following definition, see Section A.3.

Definition 2.3.3 (Convolution of functions). If g, $v \in L^1_{\text{loc}}(\mathbb{R}_+; \mathbb{K})$ the convolution of g and v is defined almost everywhere by

$$(g * v)(t) = \int_0^t g(t-s)v(s)\, ds, \qquad a.e.\ t \in \mathbb{R}_+. \tag{11}$$

The integral in (11) exists almost everywhere and defines a locally integrable function. $L^1_{\text{loc}}(\mathbb{R}_+; \mathbb{K})$ is in fact a commutative ring over the field $\mathbb{K}$ with respect to the operations $+$ and $*$. Considering matrices and vectors with entries in this ring we define the convolution of a matrix function $\mathcal{G}(\cdot) \in L^1_{\text{loc}}(\mathbb{R}_+; \mathbb{K}^{p\times m})$ and a vector function $u(\cdot) \in L^1_{\text{loc}}(\mathbb{R}_+; \mathbb{K}^m)$ by

$$(\mathcal{G} * u)(t) = \int_0^t \mathcal{G}(t-s)u(s)\, ds = \int_0^t \mathcal{G}(s)u(t-s)\, ds, \qquad a.e.\ t \in \mathbb{R}_+. \tag{12}$$

Using this notation (10) can be written

$$y(t) = (\mathcal{G} * u)(t), \quad t \geq 0$$

where $\mathcal{G}$ is the continuous $p \times m$-matrix function (called the convolution kernel of (2.17) with $D = 0$) given by

$$\mathcal{G}(t) = Ce^{At}B, \quad t \geq 0. \tag{13}$$

Again $\mathcal{G}(t)$ can be viewed as a *weighting pattern.* However we run into some theoretical difficulties in attempting to mirror the interpretation of $\mathcal{G}(t)$ as an impulse response. In order to see why this is the case, let us consider a *test signal* of the form $u(\cdot) = v(\cdot)e^j$ where $v(\cdot) \in L^1_{\text{loc}}(\mathbb{R}_+; \mathbb{K})$. At time t the ith component of the corresponding output $y(t, v(\cdot)e^j)$ is

$$y_i(t, v(\cdot)e^j) = \int_0^t \mathcal{G}_{ij}(t-s)v(s)ds, \quad t \geq 0$$

and we would like this to equal $\mathcal{G}_{ij}(t)$. This means that $v(\cdot)$ must have the property

$$\mathcal{G}_{ij}(t) = \int_0^t \mathcal{G}_{ij}(t-s)v(s)ds, \quad t \geq 0. \tag{14}$$

However, there is no function $v(\cdot)$ satisfying (14) for a non-zero $\mathcal{G}_{ij}(t)$. In fact the commutative ring $L^1_{\text{loc}}(\mathbb{R}_+; \mathbb{K})$ does not have a unit element. This is an important difference between the discrete and continuous time cases. To get around this difficulty we show that it is possible to use piecewise continuous input signals which result in outputs *approximating* the components $\mathcal{G}_{ij}(t)$ of $\mathcal{G}(t)$.

Lemma 2.3.4. *Let $(u_k(\cdot))_{k\in\mathbb{N}}$ be a sequence of non-negative integrable functions on $\mathbb{R}_+$ such that*

$$\int_0^\infty u_k(s)ds = 1 \quad \text{and} \quad u_k|[\alpha_k, \infty) \equiv 0, \quad k \in \mathbb{N} \tag{15}$$

where $\alpha_k \searrow 0$ as $k \to \infty$. Then for every $f \in C(\mathbb{R}_+; \mathbb{R})$ we have

$$f(t) = \lim_{k\to\infty} (f * u_k)(t), \quad t \in \mathbb{R}_+ \tag{16}$$

uniformly on compact intervals in $\mathbb{R}_+$.

Proof: The statement follows from the uniform continuity of f on compact intervals $I \subset \mathbb{R}_+$ since

$$|(f * u_k)(t) - f(t)| \leq \int_0^t |f(t-s) - f(t)| u_k(s)ds \leq \sup_{0\leq s\leq\alpha_k} |f(t-s) - f(t)|, \quad t \in I.$$

□

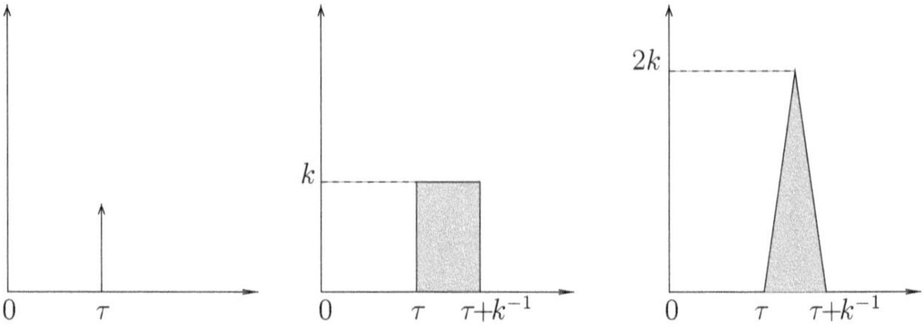

Figure 2.3.1: The Dirac impulse δ_τ and approximations

Now suppose that a sequence $(u_k(\cdot))_{k\in\mathbb{N}}$ of input signals is chosen to satisfy (15) where $\alpha_k \searrow 0$ as $k \to \infty$. Because of (16) such sequences are called *approximate identities*. If the input is $u_k(\cdot)e^j$, the ith component of the corresponding output will approximate $\mathcal{G}_{ij}(\cdot)$ uniformly on compact intervals.

$$\mathcal{G}_{ij}(t) = \lim_{k\to\infty} y_i(t; u_k(\cdot)e^j) = \lim_{k\to\infty} (\mathcal{G}_{ij} * u_k)(t), \text{ uniformly on compact intervals.} \quad (17)$$

Typical candidates for the test functions $u_k(\cdot)$, $k \in \mathbb{N}$ are shown on the right in Figure 2.3.1. They can be viewed as approximations of the so-called *Dirac impulse* δ_0 which is not a function but the unit point measure at 0. Because of (17) the $p\times m$-matrix function (13) is called the *impulse response* of the differentiable system (2.17) (with $D = 0$).

Remark 2.3.5. For every $\tau \in \mathbb{R}_+$, the *Dirac impulse* δ_τ (on $\mathbb{R}_+$) *at time* τ is the unit point measure at τ, defined by

$$\int f(s)\,\delta_\tau(d\,s) = f(\tau), \quad f \in \mathcal{C}(\mathbb{R}_+;\mathbb{K}). \quad (18)$$

Although δ_τ has no density with respect to the Lebesgue measure it is usual in the control literature to write

$$\int f(s)\delta_\tau(s)ds \;\text{ or }\; \int f(s)\delta(\tau - s)ds \;\text{ or }\; \int f(\tau - s)\delta(s)ds \quad \text{instead of} \quad \int f(s)\,\delta_\tau(d\,s).$$

Note, however, that this is only a suggestive notation and does not mean that δ_τ is a function. The general definition of convolution between a measure and a function (see Remark A.3.16) implies that for all $f \in \mathcal{C}(\mathbb{R}_+;\mathbb{K})$, $\tau \in \mathbb{R}_+$

$$(\delta_\tau * f)(t) = (f * \delta_\tau)(t) = \begin{cases} f(t-\tau), & t \in [\tau, \infty) \\ 0, & t \in [0, \tau) \end{cases}.$$

Hence δ_τ acts as a forward shift on f via the convolution. In particular, $\delta_0 * f = f * \delta_0 = f$ for $f \in \mathcal{C}(\mathbb{R}_+, \mathbb{K})$. This, together with Lemma 2.3.4 explains why we may regard δ_0 as the limit of the above sequences $(u_k(\cdot))$. Graphically the Dirac impulse δ_τ is represented by a vertical arrow of length 1 at τ (see Figure 2.3.1). □

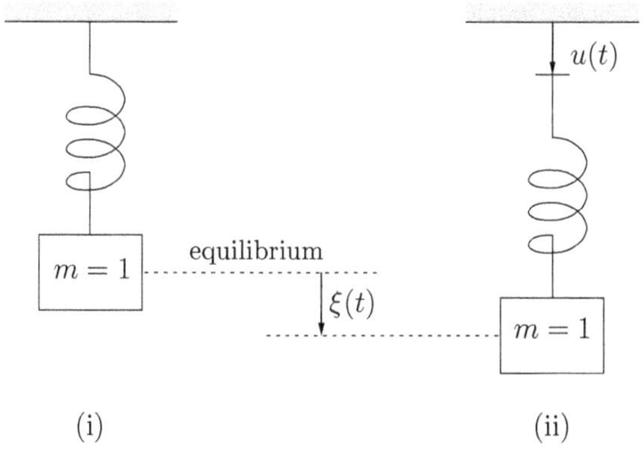

Figure 2.3.2: (i) Free linear oscillator at rest (ii) Forced linear oscillator in motion

Example 2.3.6. (Forced linear oscillator). Consider the mass-spring system described in Example 2.2.9 but suppose now that the support can be moved vertically. Let $u(t)$ denote the displacement of the support from some nominal reference point (see Figure 2.3.2). The displacement $\xi(t)$ of the mass from its equilibrium position (under the control $u(\cdot) = 0$) is taken as the output. Now since the restoring force is $\nu^2(\xi(t) - u(t))$ instead of $\nu^2\xi(t)$ we obtain the following equation of motion

$$\ddot{\xi}(t) + 2\alpha\dot{\xi}(t) + \nu^2\xi(t) = \nu^2 u(t), \quad y(t) = \xi(t), \ t \in \mathbb{R}_+.$$

The system matrices of the corresponding state space model are (see Example 2.2.9)

$$A = \begin{bmatrix} 0 & 1 \\ -\nu^2 & -2\alpha \end{bmatrix}, \quad B = \begin{bmatrix} 0 \\ \nu^2 \end{bmatrix}, \quad C = [1, \ 0].$$

The (scalar) impulse response of this system is $Ce^{At}B$, $t \geq 0$, and hence equal to the first coordinate of the free motion starting at $x^0 = [0, \ \nu^2]^\top$. To compute it we have to represent x^0 as a linear combination of the eigenvectors $z^1 = [1, \ \lambda_1]^\top$, $z^2 = [1, \ \lambda_2]^\top$ where $\lambda_{1,2} = -\alpha \pm \sqrt{\alpha^2 - \nu^2}$. A short calculation yields

$$x^0 = \frac{\nu^2}{2\sqrt{\alpha^2 - \nu^2}}(z^1 - z^2).$$

Hence the impulse response is

$$\mathcal{G}(t) = \frac{\nu^2}{2\sqrt{\alpha^2 - \nu^2}}(e^{\lambda_1 t} - e^{\lambda_2 t}), \quad t \geq 0.$$

For $|\alpha| < \nu$ this response is oscillating with frequency $\omega = \sqrt{\nu^2 - \alpha^2}$

$$\mathcal{G}(t) = \nu^2 e^{-\alpha t}\left[\frac{e^{\imath\omega t} - e^{-\imath\omega t}}{2\omega\imath}\right] = (\nu^2/\omega)e^{-\alpha t}\sin\omega t.$$

□

The fact that the impulse response completely determines the input-output behaviour (resp. forced motions) of the system is, as in the discrete time case, an

immediate consequence of the explicit formula for the output (10). However, it is instructive to explain this fact directly by the basic properties of time-invariance and linearity (superposition principle). For the sake of simplicity, we suppose that $m = p = 1$, i.e. Σ is a *single input single output* (siso) system.
For discrete time systems Σ of the form (2.22) the situation is simple. The output value $y(t)$ only depends on the input values $u(s)$, $s \in [0,t] \cap \mathbb{N}$. On the finite time set $[0,t] \cap \mathbb{N}$, $u(\cdot)$ can be represented as linear combination of shifted unit impulses (8). Hence, by linearity and time-invariance of Σ, $y(t)$ is completely determined if the system responses to the unit impulses are known.
For differentiable systems Σ of the form (2.17) (with $D = 0$) an additional property is used. For any $t_1 > 0$ the restriction of the output $y(\cdot; u(\cdot))|[0,t_1] \in \mathcal{C}([0,t_1];\mathbb{K})$ depends continuously on the restriction of the input $u(\cdot)|[0,t_1] \in L^1(0,t_1;\mathbb{K})$. Indeed, it follows immediately from (10) that, for arbitrary $u(\cdot)$, $v(\cdot) \in \mathcal{U}$ and any fixed $t_1 > 0$

$$\sup_{t\in[0,t_1]} |y(t;u(\cdot)) - y(t,v(\cdot))| \le K \int_0^{t_1} |u(s) - v(s)|ds, \tag{19}$$

where $K = \sup_{0\le s\le t_1} |Ce^{As}B|$.
Now consider for any *continuous* $u(\cdot) \in \mathcal{U}$ the step function approximation

$$v_k(t) = \sum_{j=0}^{k-1} u(jt_1/k)(1/k)S_{jt_1/k}w_k(t), \quad t \in [0,t_1] \tag{20}$$

where $S_{jt_1/k}$ is the forward shift by $\tau = jt_1/k$ and $w_k(\cdot)$ is the approximate Dirac impulse (see Figure 2.3.3)

$$w_k(t) = \begin{cases} k & \text{if } 0 \le t < t_1/k \\ 0 & \text{if } t \ge t_1/k. \end{cases}$$

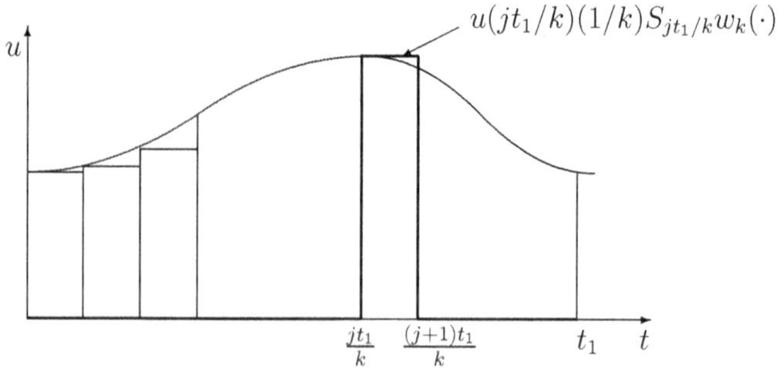

Figure 2.3.3: Step function approximation to an input

From the definition of the Riemann integral it is known that

$$\int_0^{t_1} |u(t) - v_k(t)|dt \to 0 \text{ as } k \to \infty$$

and hence by (19) $y(t; v_k(\cdot)) \to y(t; u(\cdot))$ as $k \to \infty$.
By linearity and time invariance of Σ the system's response to $v_k(\cdot)$ is the following superposition of shifted system responses to the approximate Dirac impulses $w_k(\cdot)$

$$y(t; v_k(\cdot)) = \sum_{j=0}^{k-1} u(jt_1/k)(1/k)S_{jt_1/k}y(t; w_k(\cdot)), \quad t \in [0, t_1].$$

Since by Lemma 2.3.4 $y(t; w_k(\cdot)) \to \mathcal{G}(t)$ uniformly on compact intervals as $k \to \infty$, we finally obtain, for any $t_1 > 0$,

$$\sup_{t\in[0,t_1]} |y(t; u(\cdot)) - \sum_{j=0}^{k-1} u(jt_1/k)(1/k)S_{jt_1/k}\mathcal{G}(t)| \to 0 \text{ as } k \to \infty. \tag{21}$$

Thus making use of only the three basic properties of linearity, time invariance and continuity, we see that the output signal $y(t; u(\cdot))$ corresponding to a continuous control function $u(\cdot) \in \mathcal{U}$ can be approximated by linear combinations of shifted impulse responses with coefficients determined by $u(\cdot)$.

Input-Output Operators in Time Domain

The input-output behaviour of the system (2.17)[1] with time domain $T = \mathbb{R}_+$, initialized at $x(0) = 0$, is described by the *input-output operator*

$$\begin{aligned} &\mathbb{L}_+ : L^1_{loc}(\mathbb{R}_+; \mathbb{K}^m) \to L^1_{loc}(\mathbb{R}_+; \mathbb{K}^p) \\ &(\mathbb{L}_+u)(t) = Du(t) + \int_0^t Ce^{A(t-\tau)}Bu(\tau)\, d\tau = Du(t) + (\mathcal{G} * u)(t), \quad t \in \mathbb{R}_+. \end{aligned} \tag{22}$$

Its counterpart for the discrete time system (2.22) on the time domain $T = \mathbb{N}$ is

$$\begin{aligned} &\mathbb{L}_+ : \ell^1_{loc}(\mathbb{N}; \mathbb{K}^m) = (\mathbb{K}^m)^{\mathbb{N}} \to \ell^1_{loc}(\mathbb{N}; \mathbb{K}^p) = (\mathbb{K}^p)^{\mathbb{N}} \\ &(\mathbb{L}_+u)(t) = Du(t) + \sum_{k=0}^{t-1} CA^{t-k-1}Bu(k) = (\mathcal{G} * u)(t), \quad t \in \mathbb{N}. \end{aligned} \tag{23}$$

Here the continuous and discrete time convolution kernels are given by

$$\mathcal{G}(t) = Ce^{At}B,\ t \in \mathbb{R}_+ \quad \text{and} \quad \mathcal{G}(t) = CA^{t-1}B,\ t \in \mathbb{N}^*,\ \mathcal{G}(0) = D. \tag{24}$$

Remark 2.3.7. Note that in the discrete time case the feedthrough matrix is determined by the convolution kernel $\mathcal{G}$ whereas we have seen that in the continuous time case it is not possible to express the direct input-output coupling via convolution with a (locally) integrable convolution kernel. However, if we allow for a Dirac impulse in the convolution kernel and set $\mathcal{G}(t) = \delta_0(t)D + Ce^{At}B$, $t \in \mathbb{R}_+$, then we may write the first equation in (40) as $y(t) = (\mathcal{G} * u)(t)$. More general convolution kernels involving *measures* or, more generally, *distributions* on $\mathbb{R}_+$ are considered in the literature, see *Notes and References*. □

[1] Now we allow for an arbitrary direct input-output coupling $D \in \mathbb{K}^{p\times m}$.

If S_τ, $\tau \in \mathbb{R}_+$ are the forward shift operators for the time domain $T = \mathbb{R}_+$ (see (1.28)), then

$$(\mathbb{L}_+ S_\tau u)(t) = Du(t-\tau) + \int_\tau^t Ce^{A(t-s)}Bu(s-\tau)\,ds = Du(t-\tau) + \int_0^{t-\tau} Ce^{A(t-\tau-s)}Bu(s)\,ds$$

and $(\mathbb{L}_+ S_\tau u)(t) = 0$ for $t < \tau$. Hence the continuous time input-output operator $\mathbb{L}_+$ is *time-invariant* in the sense that it commutes with the forward shift operators S_τ, $\tau \in \mathbb{R}_+$

$$(\mathbb{L}_+ S_\tau u)(\cdot) = (S_\tau \mathbb{L}_+ u)(\cdot), \quad u(\cdot) \in L^1_{\text{loc}}(\mathbb{R}_+; \mathbb{K}^m). \tag{25}$$

Similarly it can be shown that the discrete time input-output operator $\mathbb{L}_+$ commutes with the discrete time forward shift operators S_τ, $\tau \in \mathbb{N}$.

Up until now we have discussed the input-output behaviour of the linear systems (2.17) (resp. (2.22)) without comparing the sizes of the input and output signals. For such a comparison, which is of great importance in applications, we need to introduce *norms* on the signal spaces. Suppose that $\mathbb{K}^m$ and $\mathbb{K}^p$ are provided with arbitrary norms $\|\cdot\|_{\mathbb{K}^m}$ and $\|\cdot\|_{\mathbb{K}^p}$ and $\mathbb{K}^{p\times m}$ with the corresponding operator norm, see Definition A.1.4[2]. Then, for any fixed $q \in [1, \infty]$, the size of an input or output signals can be measured by their L^q-norm (resp. ℓ^q-norm), see Definitions A.3.10 and A.3.1). However, the system response $t \mapsto y(t; u(\cdot))$ is not necessarily in $L^q(\mathbb{R}_+; \mathbb{K}^p)$ (resp. $\ell^q(\mathbb{N}; \mathbb{K}^p)$) if $u(\cdot) \in L^q(\mathbb{R}_+; \mathbb{K}^m)$ (resp. $u(\cdot) \in \ell^q(\mathbb{N}; \mathbb{K}^m)$). Thus, in general, the input-output operator $\mathbb{L}_+$ may transform an input signal of finite L^q-norm (resp. ℓ^q-norm) into an output signal of infinite L^q-norm (resp. ℓ^q-norm). The system is called *input-output stable* or, more precisely, *L^q-stable* (resp. *ℓ^q-stable*) if this cannot happen. The following proposition gives a sufficient condition for L^q-stability (resp. ℓ^q-stability).[3]

Proposition 2.3.8. *Suppose*

$$\sigma(A) \subset \mathbb{C}_- \qquad (\textit{resp. } \sigma(A) \subset \mathbb{D}). \tag{26}$$

Then the continuous (resp. discrete) time input-output operator $\mathbb{L}_+$ given by (22) (resp. (23)) defines a bounded linear operator from $L^q(\mathbb{R}_+; \mathbb{K}^m)$ to $L^q(\mathbb{R}_+; \mathbb{K}^p)$ (resp. $\ell^q(\mathbb{N}; \mathbb{K}^m)$ to $\ell^q(\mathbb{N}; \mathbb{K}^p)$).

Proof: The linearity of $\mathbb{L}_+$ follows directly from the definition (see (22), (23)). By assumption there exists $\omega > 0$ such that $\operatorname{Re}\lambda < -\omega$ (resp. $|\lambda| < e^{-\omega}$) for all $\lambda \in \sigma(A)$. We will see later (Lemma 3.3.19) that this implies

$$\|e^{At}\| \le M_\omega e^{-\omega t},\ t \in \mathbb{R}_+, \quad \|A^t\| \le M_\omega e^{-\omega t},\ t \in \mathbb{N} \tag{27}$$

for a suitable constant $M_\omega \ge 1$. As a consequence, we have $\mathcal{G} \in L^1(\mathbb{R}_+; \mathbb{K}^{p\times m})$ (resp. $\mathcal{G} \in \ell^1(\mathbb{N}; \mathbb{K}^{p\times m})$). It then follows from (22) and the convolution inequality (A.3.24) that $y = \mathbb{L}_+ u \in L^q(\mathbb{R}_+; \mathbb{K}^p)$ for all $u \in L^q(\mathbb{R}_+; \mathbb{K}^m)$ and

$$\|\mathbb{L}_+ u\|_{L^q(\mathbb{R}_+;\mathbb{K}^p)} \le \left(\|D\|_{\mathcal{L}(\mathbb{K}^m,\mathbb{K}^p)} + \|\mathcal{G}\|_{L^1(\mathbb{R}_+;\mathbb{K}^{p\times m})}\right)\|u\|_{L^q(\mathbb{R}_+;\mathbb{K}^m)}, \quad u \in L^q(\mathbb{R}_+; \mathbb{K}^m).$$

[2]In the following we only make the norm explicit in the statements of theorems or propositions or where the particular norm used may be unclear

[3]We will see in the next chapter that condition (26) is equivalent to the asymptotic stability of the system (22) (resp. (23)), see Section 3.3.

Hence $\mathbb{L}_+$ is a bounded linear operator from $L^q(\mathbb{R}_+;\mathbb{K}^m)$ into $L^q(\mathbb{R}_+;\mathbb{K}^p)$ (see Section A.4). An analogous inequality holds in the discrete time case and this concludes the proof. □

In applications the square of the L^2-norm (resp. ℓ^2-norm) of a signal can often be interpreted as a measure of its energy, see Section 1.4. The previous proposition (with $q=2$) implies that, if (26) holds, the system (2.17) (resp. (2.22)) transforms finite energy input signals into finite energy output signals.
So far we have described the input-output behaviour on $\mathbb{R}_+$ (resp. $\mathbb{N}$) by setting the initial state to be zero at time $t_0=0$. We will now show that under assumption (26) we can do without fixing an initial condition and study the input-output behaviour of the systems (2.17) (resp. (2.22)) on the extended time domain $T=\mathbb{R}$ (resp. $\mathbb{Z}$). First we note that for any $t_0 \in \mathbb{R}$ the input-output behaviour of the continuous time system (2.17) with fixed initial state $x(t_0)=x^0$ at time $t_0\in\mathbb{R}$ and control functions $u(\cdot)\in L^q(t_0,\infty;\mathbb{K}^m)$ is given by

$$\begin{aligned} y(t) = y(t;t_0,x^0,u(\cdot)) &= Du(t) + Ce^{A(t-t_0)}x^0 + \int_{t_0}^t Ce^{A(t-s)}Bu(s)\,ds \\ &= y(t-t_0;0,x^0,S_{t_0}u(\cdot)) = Ce^{A(t-t_0)}x^0 + \mathbb{L}_+(S_{t_0}u(\cdot))|\mathbb{R}_+)(t-t_0), \quad t\ge t_0. \end{aligned} \tag{28}$$

It follows from (28) and the convolution inequality (A.3.24) that $y(\cdot;t_0,x^0,u(\cdot))\in L^2(t_0,\infty;\mathbb{K}^p)$ for every $u(\cdot)\in L^2(t_0,\infty;\mathbb{K}^m)$. Moreover if $D=0$ the output function $y(t)$ tends to zero as $t\to\infty$. To prove this last result we need the following lemma.

Lemma 2.3.9. *Suppose $t_0\in\mathbb{R}$ and $y(\cdot):[t_0,\infty)\to\mathbb{C}^p$ is absolutely continuous such that $y(\cdot)$, $\dot y(\cdot)\in L^2(t_0,\infty;\mathbb{C}^p)$, then $y(t)\to 0$ as $t\to\infty$.*

Proof: For the usual inner product on $\mathbb{C}^p$ and $t\ge t_1\ge t_0$, we have

$$\int_{t_1}^t [\langle \dot y(s), y(s)\rangle + \langle y(s), \dot y(s)\rangle]ds = \|y(t)\|_2^2 - \|y(t_1)\|_2^2.$$

Now since all norms on $\mathbb{C}^p$ are equivalent $y(\cdot),\dot y(\cdot)\in L^2(t_0,\infty;\mathbb{C}^p)$ where $\mathbb{C}^p$ is normed with the 2-norm. Hence $\langle \dot y(\cdot), y(\cdot)\rangle \in L^1(t_0,\infty;\mathbb{C})$ and so, for any given $\varepsilon>0$, there exists $t_\varepsilon\ge t_0$, such that for all $t_1\ge t_\varepsilon$

$$\left|\|y(t)\|_2^2 - \|y(t_1)\|_2^2\right| = \left|\int_{t_1}^t [\langle \dot y(s), y(s)\rangle + \langle y(s), \dot y(s)\rangle]ds\right| < \varepsilon, \quad t\ge t_1. \tag{29}$$

On the other hand since $y(\cdot)\in L^2(t_0,\infty;\mathbb{C}^p)$, there exists $t_1^\varepsilon\ge t_\varepsilon$ such that $\|y(t_1^\varepsilon)\|_2<\varepsilon$, hence choosing $t_1=t_1^\varepsilon$ in (29) we get $\|y(t)\|_2^2<2\varepsilon$ for $t\ge t_1^\varepsilon$. □

Proposition 2.3.10. *Suppose that $D=0$ and $\sigma(A)\subset\mathbb{C}_-$. Then there exists a constant K such that for all $t_0\in\mathbb{R}$, $x^0\in\mathbb{K}^n$, $u(\cdot)\in L^2(t_0,\infty;\mathbb{K}^m)$*

$$\|y(t;t_0,x^0,u(\cdot))\|_{\mathbb{K}^p} \le K\left[\|x^0\|_{\mathbb{K}^n} + \|u(\cdot)\|_{L^2(t_0,\infty;\mathbb{K}^m)}\right], \quad t\ge t_0, \tag{30}$$

where $y(\cdot)=y(\cdot;t_0,x^0,u(\cdot)):[t_0,\infty)\to\mathbb{K}^p$ is the associated output function of the system (2.17) defined by (28). Moreover $y(\cdot),\dot y(\cdot)\in L^2(t_0,\infty;\mathbb{K}^p)$ and $y(t)\to 0$ as $t\to\infty$.

Proof: Since $\|e^{At}\| \leq Me^{-\omega t}$ for some $M \geq 1$, $\omega > 0$ we obtain from (28)

$$\begin{aligned}\|y(t;t_0,x^0,u(\cdot))\| &\leq \|C\|Me^{-\omega(t-t_0)}\|x^0\| + \|C\|\|B\|M\int_{t_0}^t e^{-\omega(t-s)}\|u(s)\|\,ds \\ &\leq \|C\|M\|x^0\| + (M\|C\|\|B\|/\sqrt{2\omega})\|u(\cdot)\|_{L^2(t_0,\infty;\mathbb{K}^m)}, \quad t \geq t_0\end{aligned}$$

by the Cauchy-Schwarz inequality (A.3.20). This implies the inequality (30) for a suitably large constant $K > 0$. Now it follows from (28), (27) and the convolution inequality (A.3.24) that $y(\cdot) \in L^2(t_0,\infty;\mathbb{K}^p)$. Applying the same argument with $C = I_n$ we obtain $x(\cdot) \in L^2(t_0,\infty;\mathbb{K}^n)$. Moreover $y(\cdot) = Cx(\cdot)$ is absolutely continuous and since $\dot{y}(t) = C\dot{x}(t) = CAx(t) + CBu(t)$ for $t \geq 0$, we get $\dot{y}(\cdot) \in L^2(t_0,\infty;\mathbb{K}^p)$. Thus $y(t) \to 0$ as $t \to \infty$ by Lemma 2.3.9. □

Remark 2.3.11. Applying the above proposition with $C = I_n$ we see that if $\sigma(A) \subset \mathbb{C}_-$, then $\varphi(t;t_0,x^0,u(\cdot)) \to 0$ as $t \to \infty$, for all initial states $x^0 \in \mathbb{K}^n$ and input functions $u(\cdot) \in L^2(t_0,\infty;\mathbb{K}^m)$. □

Remark 2.3.12. The condition $D = 0$ is not needed in the discrete time case. The reason is that $\|u(t)\| \leq \|u(\cdot)\|_{\ell^2(t_0,\infty;\mathbb{K}^m)}$ for all $t \geq t_0$. We leave it to the reader (see Ex. 4) to prove the following discrete time counterpart of Proposition 2.3.10.
Suppose $\sigma(A) \subset \mathbb{D}$. Then there exists a constant K such that for all $t_0 \in \mathbb{Z}$, $x^0 \in \mathbb{K}^n$

$$\|y(t;t_0,x^0,u(\cdot))\| \leq K\left[\|x^0\| + \|u(\cdot)\|_{\ell^2(t_0,\infty;\mathbb{K}^m)}\right], \; u(\cdot) \in \ell^2(t_0,\infty;\mathbb{K}^m),\; t \in \mathbb{Z},\, t \geq t_0 \quad (31)$$

where $y(\cdot) = y(\cdot;t_0,x^0,u(\cdot))$ is the corresponding output function of the discrete time system (2.22) given by

$$y(t) = y(t;t_0,x^0,u(\cdot)) = CA^{t-t_0}x^0 + Du(t) + \sum_{k=t_0}^{t-1} CA^{t-k-1}Bu(k) \quad (32)$$

Moreover $y(\cdot) \in \ell^2(t_0,\infty;\mathbb{K}^p)$ and $y(t) \to 0$ as $t \to \infty$. □

Now let $t_0 \to -\infty$ in (28). Because of (27), we see that as t_0 goes back, the influence of the initial state on the output $y(t)$ gets less and less. This leads us to define the *input-output operator* of (2.17) with time domain $T = \mathbb{R}$ by

$$\begin{aligned}\mathbb{L} &: L^q(\mathbb{R};\mathbb{K}^m) \to L^q(\mathbb{R};\mathbb{K}^p) \\ (\mathbb{L}u)(t) &= Du(t) + \int_{-\infty}^t Ce^{A(t-s)}Bu(s)\,ds = Du(t) + (\mathcal{G} * u)(t) \quad t \in \mathbb{R}, \quad (33)\end{aligned}$$

where $\mathcal{G} * u$ is to be understood as a convolution of two functions defined on $\mathbb{R}$, see (A.3.23). In the discrete time case we define the *input-output operator* of (2.22) with the time domain $T = \mathbb{Z}$ by

$$\begin{aligned}\mathbb{L} &: \ell^q(\mathbb{Z};\mathbb{K}^m) \to \ell^q(\mathbb{Z};\mathbb{K}^p) \\ (\mathbb{L}u)(t) &= Du(t) + \sum_{s=-\infty}^{t-1} CA^{t-s-1}Bu(s) = (\mathcal{G} * u)(t), \quad t \in \mathbb{Z} \quad (34)\end{aligned}$$

where the convolution is as in (A.3.6). In both cases the convolution kernel $\mathcal{G}$ defined by (24) is trivially extended to $\mathbb{R}$ (resp. $\mathbb{Z}$) by setting $\mathcal{G}(t) = 0$ for $t < 0$. The assumption (26) then implies

$$\mathcal{G} \in L^1(\mathbb{R}; \mathbb{K}^{p\times m}), \qquad (\text{resp. } \mathcal{G} \in \ell^1(\mathbb{Z}; \mathbb{K}^{p\times m})). \tag{35}$$

As a consequence of Propositions A.3.14 and A.3.3 the input-output operator $\mathbb{L}$ is well defined, linear and bounded in both the continuous and the discrete time case, see Corollary 2.3.16.

Remark 2.3.13. Suppose $t_0 \in \mathbb{R}$ and $v(\cdot) \in L^q(-\infty, t_0; \mathbb{K}^m)$, then

$$x^0 = \int_{-\infty}^{t_0} e^{A(t_0-s)} Bv(s) ds \tag{36}$$

is well defined because of (27). In fact since $\|e^{At}\| \le Me^{-\omega t}$ for some $M \ge 1$, $\omega > 0$ the function $t \mapsto \|e^{At}\|$ is L^{q^*}-integrable on $\mathbb{R}_+$, where $q^* \in [1, \infty]$ is the conjugate exponent of q, and therefore $\|e^{A(t_0-s)}\|\, \|Bu(s)\|$ is integrable on $(-\infty, t_0]$ by the Hölder inequality (A.3.21) with $r = 1$. Now let $u(\cdot) \in L^q(t_0, \infty; \mathbb{K}^m)$ be arbitrary and denote by $u_v(\cdot) \in L^q(\mathbb{R}; \mathbb{K}^m)$ the extension of $u(\cdot)$ to $\mathbb{R}$ by $u_v(t) = v(t)$ for $t < t_0$. Then for $t \ge t_0$,

$$\int_{-\infty}^{t} Ce^{A(t-s)} Bu_v(s)\, ds = Ce^{A(t-t_0)} x_0 + \int_{t_0}^{t} Ce^{A(t-s)} Bu(s)\, ds, \quad u(\cdot) \in L^q(t_0, \infty; \mathbb{K}^m), \tag{37}$$

where x^0 is given by (36). Hence we recover from the input-output operator $\mathbb{L}$ on the time domain $\mathbb{R}$ the expression (28) for the input-output behaviour of the system (2.17) at the initial state $x(t_0) = x^0 \in \mathbb{K}^n$. An analogous result holds in the discrete time case.
In the first section we described how the internal state $x(t)$ at any time t incorporates the total effect of all past controls. This is again illustrated by (37) in combination with (36). These formulas show that, in order to predict the future output, once the state at time $x(t_0) = x^0$ (36) is known, one may forget about the previous control values $u(t)$, $t < t_0$. □

We have seen above that under the assumption (26) the input-output behaviour of a state space system with the time domain $\mathbb{R}$ (resp. $\mathbb{Z}$) can be described by a suitable convolution kernel. If one is only interested in the input-output behaviour of the system, it suffices to know the convolution kernel and one may forget about the state. Discarding the internal dynamics the state space system is reduced to a *black box model* or *input-output system.* An input-output system is basically just a map which associates with any input signal the corresponding output signal. In the remainder of this subsection we will make this concept more precise and consider especially those input-output systems whose behaviour is described by convolution kernels.

Definition 2.3.14. Let $T \subset \mathbb{R}$, $U, \mathcal{U} \subset U^T$, $Y, \mathcal{Y} \subset Y^T$ be non-empty sets and $\mathbb{G} : \mathcal{U} \to \mathcal{Y}$ a *causal* map, i.e. $\mathbb{G}$ satisfies

$$\forall t \in T \cap (-\infty, t_1] : u(t) = v(t) \quad \Rightarrow \quad (\mathbb{G}u)(t_1) = (\mathbb{G}v)(t_1) \tag{38}$$

for all $t_1 \in T$, $u(\cdot), v(\cdot) \in \mathcal{U}$. Then the sextuple $(T, U, \mathcal{U}, \mathbb{G}, Y, \mathcal{Y})$ is said to be an *input-output system* with time domain T, set of input values U, set of output values Y, set of input signals $\mathcal{U}$, set of output signals $\mathcal{Y}$ and input-output operator $\mathbb{G}$.

Input-output systems are represented by blockdiagrams as in Figure 2.3.4. If there is no risk of confusion, an input-output system is denoted by $(\mathcal{U}, \mathbb{G}, \mathcal{Y})$.
Every complete state space system Σ (see Definition 2.1.3) together with an initial

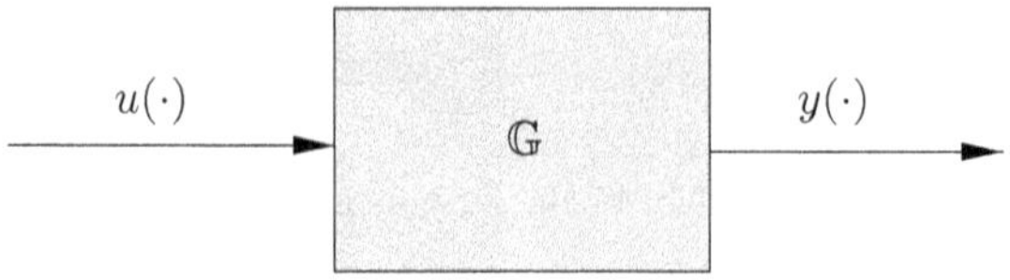

Figure 2.3.4: Input-output system with input-output operator $\mathbb{G}$

condition $x(t_0) = x^0$ where $(t_0, x^0) \in T \times X$ is fixed, defines an input-output system with time domain T_{t_0} and input-output operator (1.8).
An input-output system is said to be *linear* if $U, \mathcal{U}, Y, \mathcal{Y}$ are vector spaces over some field $\mathbb{K}$ and $\mathbb{G} : \mathcal{U} \to \mathcal{Y}$ is $\mathbb{K}$-linear. If Σ is a linear state space system with zero initial state $x(t_0) = 0$ then the corresponding input-output system with time domain T_{t_0} is linear.
An input-output system is said to be *time-invariant* if T contains 0 and is closed under addition, $\mathcal{U}, \mathcal{Y}$ are forward shift-invariant and $\mathbb{G}$ commutes with the forward shift operators S_τ, $\tau \in T$, $\tau \geq 0$, compare Definition 2.1.24.
A wide class of time-invariant linear input-output systems can be described by convolution kernels. In particular, most input-output systems which are given by *linear* ordinary, partial or delay differential equations with time-invariant parameters are *convolution systems.* An input-output system $(T, U, \mathcal{U}_+, \mathbb{G}_+, Y, \mathcal{Y}_+)$ with time domain $T = \mathbb{R}_+$ (resp. $T = \mathbb{N}$), input space $U = \mathbb{K}^m$, output space $Y = \mathbb{K}^p$ is called a convolution system if $\mathcal{U}_+$, $\mathcal{Y}_+$ are linear subspaces of $L^q_{\text{loc}}(\mathbb{R}_+, \mathbb{K}^m)$, $L^q_{\text{loc}}(\mathbb{R}_+, \mathbb{K}^p)$ for some $1 \leq q \leq \infty$ (resp. linear subspaces of $(\mathbb{K}^m)^{\mathbb{N}}$, $(\mathbb{K}^p)^{\mathbb{N}}$) and if the input-output operator $\mathbb{G}_+ : \mathcal{U}_+ \to \mathcal{Y}_+$ is of the form

$$\begin{aligned}
(\mathbb{G}_+ u)(t) &= Du(t) + \int_0^t \mathcal{G}(t-s)u(s)\,ds = Du(t) + (\mathcal{G} * u)(t), \ t \in \mathbb{R}_+, \ u(\cdot) \in \mathcal{U}_+ \\
(\mathbb{G}_+ u)(t) &= \mathcal{G}(0)u(t) + \sum_{s=0}^{t-1} \mathcal{G}(t-s)u(s) = (\mathcal{G} * u)(t), \quad t \in \mathbb{N}, \ u(\cdot) \in \mathcal{U}_+.
\end{aligned} \tag{39}$$

Here $D \in \mathbb{K}^{p\times m}$ is a given *feedthrough matrix* and $\mathcal{G} \in L^1_{\text{loc}}(\mathbb{R}_+; \mathbb{K}^{p\times m})$ (resp. $\mathcal{G} \in (\mathbb{K}^{p\times m})^{\mathbb{N}}$) is a given *convolution kernel.* Convolution systems with time domain $T = \mathbb{R}$ (resp. $T = \mathbb{Z}$) are defined in a similar way by equations of the form

$$\begin{aligned}
(\mathbb{G} u)(t) &= Du(t) + \int_{-\infty}^t \mathcal{G}(t-s)u(s)\,ds = Du(t) + (\mathcal{G} * u)(t), \ t \in \mathbb{R}, \ u(\cdot) \in \mathcal{U} \\
(\mathbb{G} u)(t) &= \mathcal{G}(0)u(t) + \sum_{s=-\infty}^{t-1} \mathcal{G}(t-s)u(s) = (\mathcal{G} * u)(t), \quad t \in \mathbb{Z}, \ u(\cdot) \in \mathcal{U}
\end{aligned} \tag{40}$$

where $\mathcal{U}$ is a linear subspace of $L^q(\mathbb{R}; \mathbb{K}^m)$ (resp. $\ell^q(\mathbb{Z}; \mathbb{K}^m)$). But for these equations to make sense one needs to assume that the convolution kernel is integrable (resp. summable), i.e. $\mathcal{G}$ satisfies assumption (35) where $\mathcal{G}(t) = 0$ for all $t < 0$ (resp. all $t \in \mathbb{Z}$, $t < 0$). In the next proposition S_τ denotes the shift operator (1.28) for the corresponding time domains $T = \mathbb{R}_+, \mathbb{N}, \mathbb{R}, \mathbb{Z}$. Note that, for all $\tau \in \mathbb{R}$ (resp. $\tau \in \mathbb{Z}$), the shift S_τ is a norm preserving automorphism of $L^q(\mathbb{R}; \mathbb{K}^k)$ (resp. $\ell^q(\mathbb{Z}; \mathbb{K}^k)$) and for all $\tau \in \mathbb{R}_+$ (resp. $\tau \in \mathbb{N}$) the forward shift S_τ is a norm preserving automorphism of $L^q(\mathbb{R}_+; \mathbb{K}^k)$ (resp. $\ell^q(\mathbb{N}; \mathbb{K}^k)$).

Proposition 2.3.15. *Suppose that $\mathcal{G} \in L^1(\mathbb{R}_+; \mathbb{K}^{p\times m})$ and $D \in \mathbb{K}^{p\times m}$ (resp. $\mathcal{G} \in \ell^1(\mathbb{N}; \mathbb{K}^{p\times m})$). Then, for arbitrary $1 \le q \le \infty$,*

(i) $\mathbb{G}_+$ defined by (39) yields a bounded linear operator from $\mathcal{U}_+ = L^q(\mathbb{R}_+; \mathbb{K}^m)$ to $\mathcal{Y}_+ = L^q(\mathbb{R}_+; \mathbb{K}^p)$ (resp. $\mathcal{U}_+ = \ell^q(\mathbb{N}; \mathbb{K}^m)$ to $\mathcal{Y}_+ = \ell^q(\mathbb{N}; \mathbb{K}^p)$) which is time-invariant, i.e. commutes with the shift operators S_τ, $\tau \in \mathbb{R}_+$ (resp. $\tau \in \mathbb{N}$).

(ii) $\mathbb{G}$ defined by (40) yields a bounded linear operator from $\mathcal{U} = L^q(\mathbb{R}; \mathbb{K}^m)$ to $\mathcal{Y} = L^q(\mathbb{R}; \mathbb{K}^p)$ (resp. $\mathcal{U} = \ell^q(\mathbb{Z}; \mathbb{K}^m)$ to $\mathcal{Y} = \ell^q(\mathbb{Z}; \mathbb{K}^p)$) which is time-invariant, i.e. commutes with the shift operators S_τ, $\tau \in \mathbb{R}$ (resp. $\tau \in \mathbb{Z}$).

$(\mathcal{U}_+, \mathbb{G}_+, \mathcal{Y}_+)$ and $(\mathcal{U}, \mathbb{G}, \mathcal{Y})$ are time-invariant linear L^q-stable (resp. ℓ^q-stable) input-output systems. Moreover,

$$\|\mathbb{G}_+\|_{\mathcal{L}(L^q(\mathbb{R}_+;\mathbb{K}^m), L^q(\mathbb{R}_+;\mathbb{K}^p))} = \|\mathbb{G}\|_{\mathcal{L}(L^q(\mathbb{R};\mathbb{K}^m), L^q(\mathbb{R};\mathbb{K}^p))} \le \|D\|_{\mathcal{L}(\mathbb{K}^m,\mathbb{K}^p)} + \|\mathcal{G}\|_{L^1(\mathbb{R}_+;\mathbb{K}^{p\times m})}$$

$$(\textit{resp. } \|\mathbb{G}_+\|_{\mathcal{L}(\ell^q(\mathbb{N};\mathbb{K}^m), \ell^q(\mathbb{N};\mathbb{K}^p))} = \|\mathbb{G}\|_{\mathcal{L}(\ell^q(\mathbb{Z};\mathbb{K}^m), \ell^q(\mathbb{Z};\mathbb{K}^p))} \le \|\mathcal{G}\|_{\ell^1(\mathbb{N};\mathbb{K}^{p\times m})}).$$

Proof: We only prove the statements for the continuous time case, the proof for the discrete time case is similar.
It follows from the integrability of the kernel and Proposition A.3.14 that the output signals $y(\cdot) = (\mathcal{G} * u)(\cdot)$ are q-integrable for all $u(\cdot) \in \mathcal{U}_+$ and all $u(\cdot) \in \mathcal{U}$. By the same proposition it follows that $\mathbb{G}_+ : \mathcal{U}_+ \to \mathcal{Y}_+$ and $\mathbb{G} : \mathcal{U} \to \mathcal{Y}$ are bounded linear operators satisfying the above inequality. The time-invariance of $\mathbb{G}_+, \mathbb{G}$ is shown in exactly the same way as the time-invariance of $\mathbb{L}_+$, see (25). This proves (i) and (ii) and the statement thereafter (which is equivalent to (i) and (ii)).
It only remains to prove that $\mathbb{G}$ and $\mathbb{G}_+$ have the same operator norm. Since the Banach spaces $L^q(\mathbb{R}_+; \mathbb{K}^k)$ can be embedded isometrically into the Banach spaces $L^q(\mathbb{R}; \mathbb{K}^k)$, $k = m, p$ (by trivial extension), we have $\|\mathbb{G}_+\| \le \|\mathbb{G}\|$. To prove the converse inequality, suppose $u(\cdot) \in L^q(\mathbb{R}_+; \mathbb{K}^m)$, let $\tau \le 0$ and define the shifted signal $u_\tau(\cdot) \in L^q(\mathbb{R}; \mathbb{K}^m)$ by $u_\tau(t) = u(t-\tau)$, $t \ge \tau$, $u_\tau(t) = 0$, $t < \tau$. In particular, $u_0(\cdot)$ is the trivial extension of $u(\cdot)$ to $\mathbb{R}$ and $u_\tau(\cdot) = (S_\tau u_0)(\cdot)$. The subspace $\mathcal{U}_{-\infty} \subset L^q(\mathbb{R}; \mathbb{K}^m)$ of all the shifted u_τ where $u(\cdot) \in L^q(\mathbb{R}_+; \mathbb{K}^m)$ and $\tau < 0$, is dense in $L^q(\mathbb{R}; \mathbb{K}^m)$. Since $\mathbb{G}$ commutes with the shift operator S_τ and $\mathbb{G}_+$ has the same operator norm as the restriction of $\mathbb{G}$ to $L^q(\mathbb{R}_+; \mathbb{K}^m) \subset L^q(\mathbb{R}; \mathbb{K}^m)$, the restriction of $\mathbb{G}$ to the normed subspace $\mathcal{U}_{-\infty}$ has the same operator norm as $\mathbb{G}_+$. Hence it follows from the continuity of $\mathbb{G}$ and the density of $\mathcal{U}_{-\infty}$ in $L^q(\mathbb{R}; \mathbb{K}^m)$ that $\mathbb{G}$ and $\mathbb{G}_+$ have the same norm. □

We have seen in the proof of Proposition 2.3.8 that the weighting pattern (13) and (7) of the state space systems (2.17) and (2.22) initialized at $x(0) = 0$ satisfies condition (35) if $\sigma(A) \subset \mathbb{C}_-$ and $\sigma(A) \subset \mathbb{D}$. Applying the previous proposition to these weighting patterns we obtain

Corollary 2.3.16. *Suppose $\sigma(A) \subset \mathbb{C}_-$ (resp. $\sigma(A) \subset \mathbb{D}$) then the input-output operator $\mathbb{L}$ of the state space system (2.17) (resp. (2.22)) defined by* (33) *(resp.* (34)*) is a time-invariant bounded linear operator from $L^q(\mathbb{R}; \mathbb{K}^m)$ to $L^q(\mathbb{R}; \mathbb{K}^p)$ (resp. from $\ell^q(\mathbb{Z}; \mathbb{K}^m)$ to $\ell^q(\mathbb{Z}; \mathbb{K}^p)$) and satisfies*

$$\|\mathbb{L}_+\|_{\mathcal{L}(L^q(\mathbb{R}_+;\mathbb{K}^m), L^q(\mathbb{R}_+;\mathbb{K}^p))} = \|\mathbb{L}\|_{\mathcal{L}(L^q(\mathbb{R};\mathbb{K}^m), L^q(\mathbb{R};\mathbb{K}^p))} \le \|D\|_{\mathcal{L}(\mathbb{K}^m,\mathbb{K}^p)} + \|\mathcal{G}\|_{L^1(\mathbb{R}_+;\mathbb{K}^{p\times m})}$$

$$(\textit{resp.} \quad \|\mathbb{L}_+\|_{\mathcal{L}(\ell^q(\mathbb{N};\mathbb{K}^m), \ell^q(\mathbb{N};\mathbb{K}^p))} = \|\mathbb{L}\|_{\mathcal{L}(\ell^q(\mathbb{Z};\mathbb{K}^m), \ell^q(\mathbb{Z};\mathbb{K}^p))} \le \|\mathcal{G}\|_{\ell^1(\mathbb{N};\mathbb{K}^{p\times m})}).$$

2.3.2 Transfer Functions

In the previous subsection we studied the input-output behaviour of linear state space systems in the *time domain*, i.e. state trajectories and input and output signals were considered as functions of time, and the system's input-output behaviour was modelled as a mapping between spaces of these time functions. We have seen that these mappings can be described by convolution kernels and this suggests that transform techniques may be a useful tool for their analysis (see Section A.3). In fact, Fourier and Laplace transforms have been used to describe the input-output behaviour of electrical circuits since the early decades of the past century. Via these transforms, convolution is converted into multiplication, and so the convolution operator of a system is transformed into a multiplication operator, determined by the Laplace transform of the convolution kernel, the so-called *transfer function* (or *transfer matrix*). A variety of graphical design methods has been developed in terms of these transfer functions, see *Notes and References.*

Transform techniques are based on the idea of representing continuous time signals as superpositions of harmonic oscillations. The variables of Fourier and Laplace transforms are interpreted as frequencies and therefore the analysis of the input-output behaviour of linear systems using these methods is called *frequency domain analysis.* In this subsection we give a brief introduction to some basic notions of this field. For a summary on transforms, see Section A.3.

Throughout the subsection it is assumed that all finite dimensional vector spaces $\mathbb{K}^m$, $\mathbb{K}^p$ are equipped with Euclidean norms and $\mathbb{K}^{p\times m}$ with the corresponding operator norm (spectral norm).

Signal Transforms

We first define the Fourier transform for signals defined on $T = \mathbb{R}$ and the Laplace transform for signals defined on $T = \mathbb{R}_+$. Then we define the discrete Fourier transform of discrete time signals on $T = \mathbb{Z}$ and the $\mathbf{z}$-transform of signals defined on $T = \mathbb{N}$.

The *Fourier transform* of a function $u(\cdot) \in L^1(\mathbb{R};\mathbb{K}^m)$ is defined by

$$\tilde{u}(\omega) = (\mathcal{F}u)(\omega) := \int_{-\infty}^{\infty} u(t)e^{-\imath\omega t}dt, \qquad \omega \in \mathbb{R}.$$

For every $u(\cdot) \in L^1(\mathbb{R};\mathbb{K}^m)$ the Fourier transform $\tilde{u}(\omega)$ is continuous in $\omega \in \mathbb{R}$ and tends to 0 as $|\omega| \to \pm\infty$ by Riemann's Lemma, see Proposition A.3.28. Note that if u takes its values in $\mathbb{R}^m$, then $\overline{\tilde{u}(\omega)} = \tilde{u}(-\omega)$.

We also need to consider the Fourier transform of signals $u(\cdot) \in L^2(\mathbb{R};\mathbb{K}^m)$. There is an initial difficulty to be overcome since $L^2(\mathbb{R};\mathbb{K}^m) \not\subset L^1(\mathbb{R};\mathbb{K}^m)$. However (see Plancherel's Theorem A.3.33) the Fourier transforms $\tilde{u}_N(\cdot)$ of the truncated functions $u_N(\cdot) = u(\cdot)1_{[-N,N]} \in L^1(\mathbb{R};\mathbb{K}^m)$ converge in $L^2(\mathbb{R};\mathbb{C}^m)$ to a limit $\tilde{u}(\cdot)$ called the *Fourier-Plancherel transform* of $u(\cdot)$,

$$\tilde{u}(\cdot) = \lim_{N\to\infty} \tilde{u}_N(\cdot) \text{ in } L^2(\mathbb{R};\mathbb{C}^m), \quad \tilde{u}_N(\omega) = \int_{-N}^{N} u(t)e^{-\imath\omega t}dt, \;\; \omega \in \mathbb{R}. \tag{41}$$

Note that for $u(\cdot) \in L^1(\mathbb{R};\mathbb{K}^m)$, the Fourier transform $\tilde{u}(\omega)$ is defined pointwise for every $\omega \in \mathbb{R}$, whereas for $u(\cdot) \in L^2(\mathbb{R};\mathbb{K}^m)$ the Fourier-Plancherel transform $\tilde{u}(\cdot)$ is

only determined as an element of $L^2(\mathbb{R};\mathbb{C}^m)$, i.e. almost everywhere.
For $u(\cdot) \in L^1_{loc}(\mathbb{R}_+;\mathbb{K}^m)$, $\alpha \in \mathbb{R}$ we set $u_\alpha(\cdot) : t \to e^{-\alpha t}u(t)$, $t \in \mathbb{R}_+$ and define

$$\mathcal{E}_\alpha(\mathbb{K}^m) = \{u(\cdot) \in L^1_{loc}(\mathbb{R}_+;\mathbb{K}^m);\ u_\alpha(\cdot) \in L^1(\mathbb{R}_+;\mathbb{K}^m)\}, \quad \alpha \in \mathbb{R}.$$

Functions belonging to $\mathcal{E}_\alpha(\mathbb{K}^m)$ for some $\alpha \in \mathbb{R}$ are called *Laplace transformable* and all signals occurring in control theory belong to this class. The *Laplace transform* (see Definition A.3.17) of $u(\cdot) \in \mathcal{E}_\alpha(\mathbb{K}^m)$ is defined by

$$\hat{u}(s) = (\mathcal{L}\,u)(s) := \int_0^\infty u(t)e^{-st}dt, \qquad \operatorname{Re} s \geq \alpha. \tag{42}$$

$\hat{u}(\cdot)$ is continuous and bounded on the closed set $\{s \in \mathbb{C}; \operatorname{Re} s \geq \alpha\}$. It is analytic on $\{s \in \mathbb{C}; \operatorname{Re} s > \alpha\}$ and will be identified with its analytic extensions to complex domains containing this set. Note that if u takes its values in $\mathbb{R}^m$, then $\overline{\hat{u}(s)} = \hat{u}(\overline{s})$.
Now suppose $u(\cdot) \in \mathcal{E}_\alpha(\mathbb{K}^m)$. Then for every $\beta \geq \alpha$ the Laplace transform $\hat{u}(\beta + \imath\omega)$ on the vertical line $\{\beta + \imath\omega; \omega \in \mathbb{R}\}$ can be expressed by the Fourier transform of the integrable function $u_\beta(\cdot)$ where $u_\beta(t) = u(t)e^{-\beta t}$, $t \geq 0$ and $u_\beta(t) = 0$, $t < 0$

$$(\mathcal{L}\,u)(\beta + \imath\omega) = \int_0^\infty (u(t)e^{-\beta t})e^{-\imath\omega t}dt = (\mathcal{F}\,u_\beta)(\omega), \quad \omega \in \mathbb{R}. \tag{43}$$

Hence $\hat{u}(\beta + \imath\omega) \to 0$ as $|\omega| \to \infty$ by Riemann's Lemma.
The counterpart of the Fourier transform for discrete time signals on $\mathbb{Z}$ (two-sided sequences) is the *discrete Fourier transform* . For an arbitrary summable two-sided sequence $u(\cdot) \in \ell^1(\mathbb{Z};\mathbb{K}^m)$ it is given by

$$\tilde{u}(\theta) = (\mathcal{F}_D\,u)(\theta) := \sum_{t=-\infty}^{\infty} u(t)e^{-\imath t\theta}, \quad \theta \in [-\pi,\pi]. \tag{44}$$

The series converges uniformly for $\theta \in [-\pi,\pi]$ and its limit $\tilde{u}(\cdot) : [-\pi,\pi] \to \mathbb{C}^m$ is continuous with $\tilde{u}(-\pi) = \tilde{u}(\pi)$. Note that if $u(\cdot)$ takes its values in $\mathbb{R}^m$, then $\overline{\tilde{u}(\theta)} = \tilde{u}(-\theta)$.
For signals $u(\cdot) \in \ell^2(\mathbb{Z};\mathbb{K}^m)$, we have a similar difficulty to that of the continuous time case since $\ell^2(\mathbb{Z};\mathbb{K}^m) \not\subset \ell^1(\mathbb{Z};\mathbb{K}^m)$. However since the functions $\psi_t(\theta) := (2\pi)^{-1/2}e^{-\imath t\theta}$, $t \in \mathbb{Z}$ form an orthonormal basis of the Hilbert space $L^2(-\pi,\pi;\mathbb{C}^m)$ (see Example A.4.6) the series in (44) converges in $L^2(-\pi,\pi;\mathbb{C}^m)$ to some function $\tilde{u}(\cdot) \in L^2(-\pi,\pi;\mathbb{C}^m)$, for every sequence $u(\cdot) \in \ell^2(\mathbb{Z};\mathbb{K}^m)$. Note again that for $u(\cdot) \in \ell^1(\mathbb{Z};\mathbb{K}^m)$, the discrete Fourier transform $\tilde{u}(\theta)$ is defined pointwise for every $\theta \in [-\pi,\pi]$, whereas for $u(\cdot) \in \ell^2(\mathbb{Z};\mathbb{K}^m)$ the transform $\tilde{u}(\cdot)$ is only determined as an element of $L^2(-\pi,\pi;\mathbb{C}^m)$, i.e. almost everywhere.
The counterpart of the Laplace transform for discrete time signals on $\mathbb{N}$ is the **z**-*transform* which associates with any one-sided sequence $u(\cdot) : \mathbb{N} \to \mathbb{K}^m$, the formal power series in z^{-1}

$$\hat{u}(z) = (\mathcal{Z}\,u)(z) = \sum_{t=0}^{\infty} u(t)z^{-t}, \tag{45}$$

see Definition A.3.5. If the sequence $(u(t))$ is exponentially bounded this formal power series defines a complex analytic function $\hat{u}(\cdot)$ on some neighbourhood of ∞.
For $u(\cdot) \in (\mathbb{K}^m)^{\mathbb{N}}$, $\gamma > 0$ we set $u_\gamma(t) = u(t)\gamma^{-t}$, $t \in \mathbb{N}$ and define

$$\mathcal{S}_\gamma(\mathbb{K}^m) = \{u(\cdot) \in (\mathbb{K}^m)^{\mathbb{N}};\ u_\gamma(\cdot) \in \ell^1(\mathbb{N};\mathbb{K}^m)\}, \quad \gamma > 0.$$

Then if $u(\cdot) \in \mathcal{S}_\gamma(\mathbb{K}^m)$ the series on the RHS of (45) is absolutely convergent for all $z \in \mathbb{C}$, $|z| \geq \gamma$ and defines a continuous function $\hat{u}(\cdot)$ on $\{z \in \mathbb{C}; |z| \geq \gamma\}$. This function is analytic on $\{z \in \mathbb{C}; |z| > \gamma\}$ and it will be identified with its analytic extensions to complex domains containing this set. We will use the same symbol $\hat{u}(z)$ to denote the *formal power series* (45) and the associated *complex analytic function*. Note that if $u(\cdot)$ takes its values in $\mathbb{R}^m$, then $\overline{\hat{u}(z)} = \hat{u}(\overline{z})$.

Now suppose $u(\cdot) \in \mathcal{S}_\gamma(\mathbb{K}^m)$ for some $\gamma > 0$. Then the holomorphic function given by the **z**-transform (45) may be viewed as the frequency domain representation of the discrete time signal $u(\cdot) \in \mathcal{S}_\gamma(\mathbb{K}^m)$. In fact, on any circle $\{re^{\imath\theta}; \theta \in [-\pi, \pi]\}$ with $r \geq \gamma$ the function $\hat{u}(z)$ defined by (45) can be expressed as the discrete Fourier transform of the summable sequence $u_r(\cdot)$ where $u_r(t) = u(t)r^{-t}$, $t \in \mathbb{N}$ and $u_r(t) = 0$, $t \in \mathbb{Z} \setminus \mathbb{N}$,

$$(\mathcal{Z}u)(re^{\imath\theta}) = \sum_{t=0}^{\infty}(u(t)r^{-t})e^{-\imath t\theta} = (\mathcal{F}_D u_r)(\theta), \quad \theta \in [-\pi, \pi]. \tag{46}$$

Transfer Matrices

We now turn from the representation of signals to the representation of input-output behaviours in frequency domain, and begin our discussion for convolution systems with time domain $T = \mathbb{R}_+$ (resp. $\mathbb{N}$) and input-output operator described by (39). Suppose that for some $\alpha \in \mathbb{R}$ (resp. $\gamma > 0$), $\mathcal{G}(\cdot) \in \mathcal{E}_\alpha(\mathbb{K}^{p\times m})$ and $u(\cdot) \in \mathcal{E}_\alpha(\mathbb{K}^m)$ (resp. $\mathcal{G}(\cdot) \in \mathcal{S}_\gamma(\mathbb{K}^{p\times m})$ and $u(\cdot) \in \mathcal{S}_\gamma(\mathbb{K}^m)$). Then it follows from (A.3.14) and (A.3.32) that the corresponding output signal $y(\cdot) = (\mathcal{G} * u)(\cdot)$ is in $\mathcal{E}_\alpha(\mathbb{K}^p)$ (resp. $\mathcal{S}_\gamma(\mathbb{K}^p)$). Taking the Laplace transform (resp. **z**-transform) of both sides of equation (39) we obtain (see Theorem A.3.21 (resp. Theorem A.3.7))[4]

$$\hat{y}(s) = G(s)\hat{u}(s), \quad \operatorname{Re} s \geq \alpha, \qquad \hat{y}(z) = G(z)\hat{u}(z), \quad |z| \geq \gamma \tag{47}$$

where $G(s)$ (resp. $G(z)$) is defined as follows.

Definition 2.3.17. Suppose that $D \in \mathbb{K}^{p\times m}$ and for some $\alpha \in \mathbb{R}$ (resp. $\gamma > 0$), $\mathcal{G}(\cdot) \in \mathcal{E}_\alpha(\mathbb{K}^{p\times m})$ (resp. $\mathcal{G}(\cdot) \in \mathcal{S}_\gamma(\mathbb{K}^{p\times m})$). Then the Laplace transform of $D\delta_0(t) + \mathcal{G}(\cdot)$ (resp. **z**-transform of $\mathcal{G}(\cdot)$)

$$G(s) = D + \int_0^\infty \mathcal{G}(t)e^{-st}dt, \ \operatorname{Re} s \geq \alpha, \quad G(z) = \mathcal{G}(0) + \sum_{t=1}^{\infty}\mathcal{G}(t)z^{-t}, \ |z| \geq \gamma, \tag{48}$$

is called the *transfer matrix* of the input-output system described by (39) or (40).

The convolution kernel $\mathcal{G}$ is uniquely determined in the continuous time case (almost everywhere) by its Laplace transform and in the discrete time case by its **z**-transform (see Theorems A.3.19 and A.3.8). So the transfer matrix completely determines the

[4]Note that in the discrete time the conditions $\mathcal{G} \in \mathcal{S}_\gamma(\mathbb{K}^{p\times m}), u \in \mathcal{S}_\gamma(\mathbb{K}^m))$ are not needed if an interpretation of the **z**-transform as a *function* on $\{z \in \mathbb{C}; |z| \geq \gamma\}$ is not required. The algebraic **z**-transform can be applied to arbitrary signals and convolution kernels on $\mathbb{N}$ and yields *formal power series* with matrix and vector coefficients, respectively, see Subsection A.3.1. The algebraic **z**-transform converts the convolution of time functions into the multiplication of formal power-series.

input-output operator of the convolution system described by (39) or (40).
Going back to the input-output behaviour of a state space system of the form (24) we can obtain an explicit expression for the associated transfer matrix by using the Laplace transform of $(e^{At})_{t\in\mathbb{R}_+}$ (resp. **z**-transform of $(A^t)_{t\in\mathbb{N}}$). However, it is also possible to obtain this expression directly from the system equations as we show in the following example.

Example 2.3.18. Consider the state space system of the form (2.17)

$$\begin{aligned} \dot{x}(t) &= Ax(t) + Bu(t), \quad t \in \mathbb{R}_+ \\ y(t) &= Cx(t) + Du(t). \end{aligned} \tag{49}$$

with initial state $x(0) = 0$, and let $u(\cdot)$ be a Laplace transformable input function. Since $\|e^{At}\| \le e^{\|A\|t}$, $t \ge 0$ the convolution kernel $t \mapsto e^{At}$ is exponentially bounded. Hence it follows from (A.3.32) that the state and output trajectories

$$t \mapsto x(t) = (e^{A\cdot} * Bu(\cdot))(t), \quad t \mapsto y(t) = Du(t) + Cx(t)$$

are Laplace transformable. Applying the Laplace transform to (49) we obtain

$$\begin{aligned} s\hat{x}(s) &= A\hat{x}(s) + B\hat{u}(s) \\ \hat{y}(s) &= C\hat{x}(s) + D\hat{u}(s), \quad \operatorname{Re} s \ge \alpha \end{aligned}$$

for some suitably large α. Therefore $\hat{y}(s) = (D + C(sI_n - A)^{-1}B)\hat{u}(s)$ and so the transfer matrix of the above system is given by

$$G(s) = D + C(sI_n - A)^{-1}B. \tag{50}$$

Note that this matrix-valued function is defined on $\rho(A) = \mathbb{C} \setminus \sigma(A)$. Since $(sI_n - A)^{-1} = \det(sI_n - A)^{-1} \operatorname{adj}(sI_n - A)$ where the adjugate $\operatorname{adj}(sI_n - A)$ is a polynomial matrix whose entries are of degree $\le n - 1$, we see that the transfer function of the time-invariant linear system (2.17) is a proper rational matrix, i.e. a matrix with entries $g_{ij}(s)$ satisfying

$$g_{ij} = \frac{p_{ij}}{q_{ij}}, \; p_{ij}, q_{ij} \in \mathbb{K}[s], \quad \deg p_{ij} \le \deg q_{ij}, \quad i = 1, ..., p, \; j = 1, ..., m.$$

If $D = 0$, $G(s)$ is strictly proper rational, i.e. its entries satisfy $\deg p_{ij} < \deg q_{ij}$. □

Remark 2.3.19. It is shown in realization theory that, conversely, for every proper rational matrix $G(s) \in \mathbb{K}^{p\times m}(s)$ there exists a time-invariant linear state space system of the form (49) whose transfer-matrix is $G(s)$. □

In the next example we present a system with an irrational transfer function.

Example 2.3.20. Consider the delay differential system

$$\dot{x}(t) = A_0x(t) + A_1x(t - h) + Bu(t), \quad y(t) = Cx(t), \quad t > 0,$$

where $(A_0, A_1, B, C) \in \mathbb{K}^{n\times n} \times \mathbb{K}^{n\times n} \times \mathbb{K}^{n\times m} \times \mathbb{K}^{p\times n}$, $h > 0$ are given and the initial state is zero: $x(\tau) = 0$, $\tau \in [-h, 0]$. A state space description of such a system has been presented in Example 2.1.25. Constructing the state trajectory by successive application of the variation-of-constant formula (see (1.30)) one can show that if the input u is exponentially bounded then so will $x(\cdot)$ and $y(\cdot)$ be exponentially bounded. Hence we may take the Laplace transform to obtain (see Proposition A.3.20)

$$s\hat{x}(s) = A_0\hat{x}(s) + A_1 e^{-hs}\hat{x}(s) + B\hat{u}(s), \quad \hat{y}(s) = C\hat{x}(s), \quad \operatorname{Re} s \geq \alpha$$

for some suitably large α. So the transfer matrix of the above system is given by

$$G(s) = C(sI_n - A_0 - e^{-hs}A_1)^{-1}B, \qquad \operatorname{Re} s \geq \alpha.$$

Note that this matrix function is no longer rational. Applying the inverse Laplace transform (see Theorem A.3.19) we conclude that the input-output behaviour of the above delay system can be described in time domain by a convolution kernel

$$y(t) = (\mathcal{G} * u)(t), \quad \text{where} \quad \mathcal{G} = \mathfrak{L}^{-1}(G).$$

The kernel $\mathcal{G}$ can be determined as follows. Let $t \mapsto \Phi(t) \in \mathbb{K}^{n\times n}$ be the fundamental solution of the delay equation, i.e. the matrix solution of the initial value problem

$$\dot{\Phi}(t) = A_0\Phi(t) + A_1\Phi(t-h),\ t \geq 0; \qquad \Phi(s) = 0,\ s \in [-h, 0),\ \ \Phi(0) = I_n$$

(Existence and uniqueness of the solution follow as in Example 2.1.25.) Then, see [213], the state trajectories with initial function zero are given by

$$x(t) = \int_0^t \Phi(t-\tau)Bu(\tau)d\tau, \quad t \geq 0.$$

Hence the input-output behaviour (starting at zero) will be described by a convolution operator with kernel $\mathcal{G}(t) = C\Phi(t)B$, $t \geq 0$. □

In the next example we derive a formula for the transfer matrix of the discrete time system (2.22).

Example 2.3.21. Consider the state space system of the form (2.22)

$$\begin{aligned} x(t+1) &= Ax(t) + Bu(t), \quad t \in \mathbb{N} \\ y(t) &= Cx(t) + Du(t) \end{aligned} \tag{51}$$

with initial state $x(0) = 0$. Applying the algebraic $\mathbf{z}$-transform we obtain for arbitrary input sequences $u(\cdot) : \mathbb{N} \to \mathbb{K}^m$

$$\begin{aligned} z\hat{x}(z) &= A\hat{x}(z) + B\hat{u}(z), \\ \hat{y}(z) &= C\hat{x}(z) + D\hat{u}(z) \end{aligned}$$

and hence

$$\hat{y}(z) = G(z)\hat{u}(z) \quad \text{where} \quad G(z) = D + C(zI_n - A)^{-1}B. \tag{52}$$

We know from Example 2.3.18 that $G(z)$ is proper rational and defined on $\rho(A)$. The corresponding formal power series in $\mathbb{K}^{p\times m}[[z^{-1}]]$ can be obtained by expressing the proper rational functions $g_{ij}(z) = q_{ij}(z)/p_{ij}(z)$ (via long division) in the form $g_{ij}(z) = \sum_{k=0}^{\infty} \gamma_k^{ij} z^{-k}$ (the Laurent expansion of $g_{ij}(z)$ at ∞, see Section A.2). □

Examples 2.3.18 and 2.3.21 show that the transfer functions of the continuous time system (2.17) and the discrete time system (2.22) coincide. This makes it possible to transfer results concerning the input-output behaviour from one class of systems to the other.

Dyadic Decomposition of Transfer Matrices

We now examine how the input and output signals of a state space system of the form (2.17) and (2.22) are coupled via the internal dynamics of the system. In order to simplify the analysis we will only consider the (generic) case where the system matrix A is diagonalizable. Then there exists a basis $v^1, ..., v^n$ of $\mathbb{C}^n$ consisting of eigenvectors of A. Let $V_j = \mathbb{C}v^j$, $j \in \underline{n}$ and let $\lambda_1, ..., \lambda_n$ be the corresponding (not necessarily distinct) eigenvalues of A. If $\tilde{P}_i : \mathbb{C}^n \to \mathbb{C}^n$ is the canonical projection from $\mathbb{C}^n = \oplus_{j=1}^n V_j$ onto V_i, $\sum_{j=1}^n \alpha_j v^j \mapsto \alpha_i v^i$, then these projections $\tilde{P}_i$, $i = 1, ..., n$ have the properties given in (2.30), (2.33) with $\ell = n$ and $\tilde{P}_i$ instead of P_i. In particular

$$(sI_n - A)^{-1} = \sum_{i=1}^n (s - \lambda_i)^{-1} \tilde{P}_i, \qquad s \in \rho(A).$$

If $(w^1, ..., w^n)$ is the biorthogonal basis of $(v^1, ..., v^n)$, i.e. $w^{j*}v^i = \delta_{ij}$ (Kronecker symbol) , then $\tilde{P}_i = v^i w^{i*}$. Hence the transfer matrix for (2.17) and (2.22) is

$$\begin{aligned} G(s) &= D + C(sI_n - A)^{-1}B = D + \sum_{i=1}^n (s-\lambda_i)^{-1} C\tilde{P}_i B = D + \sum_{i=1}^n (s-\lambda_i)^{-1}(Cv^i)(w^{i*}B) \\ &= D + \sum_{i=1}^n c^i (s - \lambda_i)^{-1} b^{i*} \end{aligned}$$

where $c^i = Cv^i \in \mathbb{C}^p$ and $b^i = B^* w^i \in \mathbb{C}^m$. This representation is called the *dyadic*

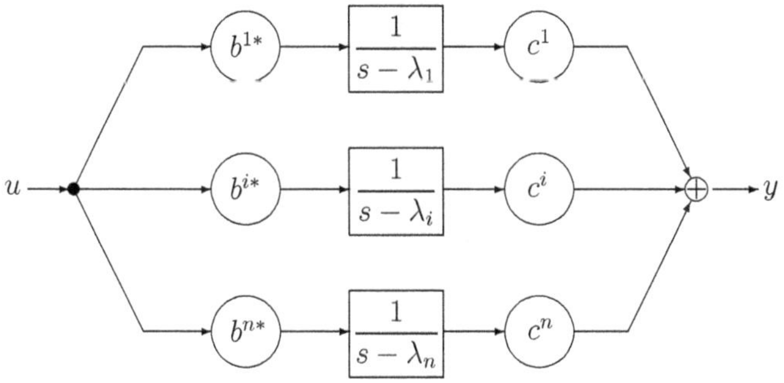

Figure 2.3.5: Dyadic decomposition of the transfer function with $D = 0$

decomposition of the transfer function for the system (A, B, C, D) and is illustrated in Figure 2.3.5. The components $b_1^i, ..., b_m^i$ of b^i specify the intensity by which the m inputs excite the i-th eigenmode of the system. Whereas the components $c_1^i, ..., c_p^i$ of c^i determine the intensity by which the i-th eigenmode influences the p outputs. If $b^i = 0$, no input will affect the i-th eigenmode and if $c^i = 0$, the i-th eigenmode will not affect the output. So, in both cases the i-th eigenmode is not important for the input-output behaviour of the system (A, B, C, D) initialized at $x(0) = 0$.

Interpretation of the Transfer Function: Response to Sinusoidal Inputs

In the previous subsection we showed that the weighting pattern $\mathcal{G}(t)$ of a continuous time convolution system can be approximately obtained by testing the input-output system with approximations to Dirac impulses. We will now show that the transfer function of such a system can be determined approximately by applying harmonic test signals of various frequencies to the system and interpolating the results. In contrast to the impulse approximations, this procedure is not only of theoretical but also of some practical importance, see *Notes and References.*

For simplicity we only consider the real scalar case with time domain $T = \mathbb{R}_+$ (resp. $T = \mathbb{N}$) and integrable (resp. summable) convolution kernel. Then the associated transfer function $g(s)$ is defined and continuous on the closed right half-plane $\overline{\mathbb{C}_+}$ (resp. on $\overline{\mathbb{D}_+}$) and hence also on the imaginary axis $\imath\mathbb{R}$ (resp. the unit circle $\partial\mathbb{D}$).

Proposition 2.3.22. *Let $g(\cdot)$ be the transfer function of a real scalar convolution system* (40) *with integrable kernel $\mathcal{G}(\cdot) \in L^1(\mathbb{R}_+;\mathbb{R})$ (resp. summable kernel $\mathcal{G}(\cdot) \in \ell^1(\mathbb{N};\mathbb{R})$) and a real feedthrough coefficient $D \in \mathbb{R}$. Then, for every $\omega \in \mathbb{R}$ (resp. $\theta \in [-\pi,\pi]$), the system response to the input signal $u(t) = \sin \omega t$, $t \in \mathbb{R}_+$ (resp. $u(t) = \sin \theta t$, $t \in \mathbb{N}$) approximates for large t the "steady state response"*

$$y^{ss}(t) = |g(\imath\omega)| \sin(\omega t + \varphi(\omega)), \quad t \geq 0 \quad (\textit{resp. } y^{ss}(t) = |g(e^{\imath\theta})| \sin(\theta t + \varphi(\theta)), \quad t \in \mathbb{N})$$

where $\varphi(\omega)$ (resp. $\varphi(\theta)$) is an argument function of $g(\imath\omega)$ (resp. $g(e^{\imath\theta})$).

Proof: We only prove the proposition for the continuous time case. The discrete time case is left to the reader, see Ex. 8. First note that

$$y^{ss}(t) = |g(\imath\omega)| \operatorname{Im} e^{\imath(\omega t + \varphi(\omega))} = |g(\imath\omega)| \operatorname{Im}(e^{\imath\varphi(\omega)} e^{\imath\omega t}) = \operatorname{Im}(g(\imath\omega)e^{\imath\omega t}), \quad t \geq 0.$$

The system response to $u(t) = \sin \omega t = \operatorname{Im} e^{\imath\omega t}$ is $y(t) = D \sin \omega t + (\mathcal{G} * u)(t)$ for $t \in \mathbb{R}_+$. On the other hand $\operatorname{Im}(g(\imath\omega)e^{\imath\omega t}) = \operatorname{Im}(De^{\imath\omega t} + \hat{\mathcal{G}}(\imath\omega)e^{\imath\omega t}) = D \sin \omega t + \operatorname{Im}(\hat{\mathcal{G}}(\imath\omega)e^{\imath\omega t})$. Therefore

$$\begin{aligned} |y(t) - \operatorname{Im}(g(\imath\omega)e^{\imath\omega t})| &= \Big| \int_0^t \mathcal{G}(\tau) \sin \omega(t-\tau)\, d\tau - \operatorname{Im} \int_0^\infty \mathcal{G}(\tau) e^{-\imath\omega\tau} d\tau\, e^{\imath\omega t} \Big| \\ &= \Big| \operatorname{Im} \int_t^\infty \mathcal{G}(\tau) e^{\imath\omega(t-\tau)} d\tau \Big|, \text{ as } \mathcal{G}(\cdot) \text{ is real} \qquad (53) \\ &\leq \Big| \int_t^\infty \mathcal{G}(\tau) e^{\imath\omega(t-\tau)} d\tau \Big| \leq \int_t^\infty |\mathcal{G}(\tau)| d\tau \to 0 \text{ as } t \to \infty. \end{aligned}$$

Since $y^{ss}(t) = \operatorname{Im}(g(\imath\omega)e^{\imath\omega t})$ this concludes the proof. □

In the continuous time case the steady state response $y^{ss}(t)$ is a harmonic oscillation of the same frequency as the harmonic input signal but amplified by $|g(\imath\omega)|$ and phase shifted by $\arg g(\imath\omega)$. Consequently, the values of the transfer function g on the imaginary axis can be determined experimentally (sometimes with great accuracy) by measuring for each interesting frequency ω the magnitude and phase shift (with respect to the phase of the input signal) of the steady state response. This procedure can, in principle, also be applied to multivariable convolution systems by feeding harmonic inputs successively into each of the m input channels (keeping the other input channels at zero) and measuring the amplifications and phase shifts of the system's responses on each of p output channels. Analogous results hold for the discrete time case.

Frequency Responses

We have just seen that the transfer function $g(s)$ of a scalar convolution system (40) with integrable kernel on $T = \mathbb{R}_+$ (resp. summable kernel on $T = \mathbb{N}$) is defined and continuous on the imaginary axis $\imath\mathbb{R}$ (resp. the unit circle $\partial\mathbb{D}$). $g(\cdot)$ is completely determined by its values on the imaginary axis (resp. the unit circle), see Proposition A.3.41 (resp. Proposition A.3.45). Thus the restriction of $g(s)$ to $\imath\mathbb{R}$ (resp. $\partial\mathbb{D}$) determines the input-output behaviour of the convolution system. This motivates the following definition.

Definition 2.3.23. Let $g(\cdot)$ be the transfer function of a real scalar convolution system with integrable kernel $\mathcal{G}$ on $\mathbb{R}_+$ (resp. summable kernel $\mathcal{G}$ on $\mathbb{N}$). Then the complex valued function $\omega \to g(\imath\omega)$ (resp. $\theta \to g(e^{\imath\theta})$) on $\mathbb{R}$ (resp. $[-\pi, \pi]$) is called the *complex frequency response*. The real valued function $\omega \to |g(\imath\omega)|$ (resp. $\theta \to |g(e^{\imath\theta})|$) is called the *amplitude (gain) response* and any (continuous) argument function $\omega \to \arg g(\imath\omega)$ (resp. $\theta \to \arg g(e^{\imath\theta})$) the *phase response* of the continuous (resp. discrete) time scalar convolution system (40).

The importance of these concepts for classical control theory, results from the fact that three of the four most prominent classical analysis and design techniques for linear siso systems (Nyquist, Bode, Nichols chart and root locus methods) are based on graphical representations of the frequency response. In particular, Nyquist's method proceeds from the *polar plot* $\{g(\imath\omega); \omega \in \mathbb{R}\}$ of the complex frequency response, and Bode's method proceeds from the graphs of the amplitude and the phase responses, see *Notes and References*.

Remark 2.3.24. Classical techniques have been developed for siso systems. Clearly they can be applied individually to represent graphically the influence of the j-th input channel on the i-th output of a multivariable system (2.17) or (2.22) described by the entry $g_{ij}(s)$ of the associated transfer matrix $G(s) = (g_{ij}(s)) \in \mathbb{K}^{p\times m}(s)$. However, these graphical methods are, in general, not suitable for analyzing the input-output behaviour of a multivariable system as a whole. □

Before illustrating the concept of *complex frequency response* by the *polar plots* of some simple siso systems let us make some general remarks concerning these plots.

(i) The transfer functions of stable real siso systems satisfy $\overline{g(\imath\omega)} = g(-\imath\omega)$ for all $\omega \in \mathbb{R}$, so that their polar plots are symmetric with respect to the real axis, and hence need only be computed for $\omega \geq 0$.

(ii) It is usual to indicate the orientation of the polar plot by an arrow showing the direction in which $g(\imath\omega)$ evolves as ω is increasing. If $g(s)$ and $h(s)$ are two transfer functions satisfying $h(s) = g(-s)$ then g and h have the same polar plots but they have reverse orientations.

(iii) For a continuous time siso convolution system with integrable kernel $\mathcal{G}$ and $D = 0$, the amplitude response $|g(\imath\omega)|$ tends to zero as $|\omega| \to \infty$ and so the harmonic inputs with large frequencies will be attenuated by the system, see (53). On the contrary, siso systems whose amplitude response is constant (i.e. $|g(\imath\omega)| = c > 0$ for all $\omega \in \mathbb{R}$) amplify/dampen harmonic inputs by the same factor c for all frequencies. They are called *all-pass* functions and in the rational case are characterized by the property of pole-zero symmetry with respect to the imaginary axis: if s_0 is a pole of g, then $-\overline{s_0}$ is a zero.

Example 2.3.25. The transfer function of any first order real siso system is of the form $d + b/(s+a)$. If $a \neq 0$ the corresponding polar plot is a circle since $|g(\imath\omega) - d - b/(2a)| = |b|/|2a|$ for all $\omega \in \mathbb{R}$. For $a = 0$ the polar plot is a vertical line since $g(\imath\omega) = d - \imath b/\omega$. In Figure 2.3.25 we illustrate, by their polar plots for $\omega \geq 0$, the frequency responses of five simple real siso systems (with and without delay). (The full polar plots are then obtained by adding their reflections about the real axis). Consider the transfer functions

$$g_1(s) = 1/(s^2+2s+5), \quad g_2(s) = 1/s(s^2+2s+5), \quad g_3(s) = (s+1)/(s^2+2s+5),$$

$$g_4(s) = e^{-s}/(1+s), \quad g_5(s) = 1/(s+1+e^{-s}).$$

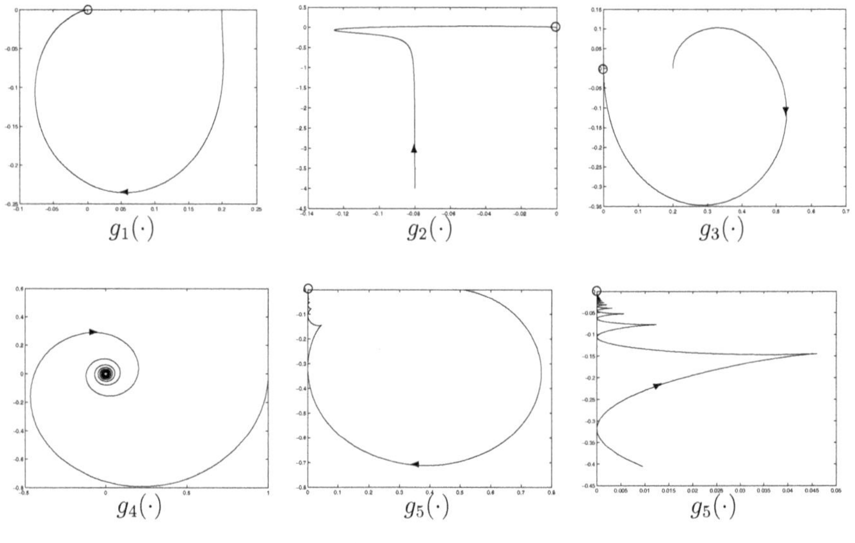

Figure 2.3.6: Polar plots of the transfer functions $g_1(\cdot), \ldots, g_5(\cdot)$

It is easy to construct state space systems of the form (49) whose transfer functions are $g_1(s)$, $g_2(s)$ and $g_3(s)$, see Ex. 11. $g_4(s)$ and $g_5(s)$ are the transfer functions of the delay systems

$$\dot{x}(t) = -x(t) + u(t-1),\ y(t) = x(t) \quad \text{and} \quad \dot{x}(t) = -x(t) - x(t-1) + u(t),\ y(t) = x(t).$$

In the six pictures of Figure 2.3.6 the origin in the complex plane is indicated by $\circ$. The pictures were produced via the `plot` command in MATLAB with a frequency band, in the main, from 0 to 100. The exceptions are the polar plot of g_2 (with frequency range $[0.5, 6]$) and the right hand plot of g_5 which is a zoom in (for higher frequencies) of the left hand one. g_1 has two complex poles at $-1 \pm 2\imath$ and its plot is a typical one for a second order system. g_2 is obtained from g_1 by adding a pole at the origin. Obviously the polar plot will be unbounded if there is a pole of the transfer function on the imaginary axis. Here $\operatorname{Re} g_2(\imath\omega) \to -.08$ and $\operatorname{Im} g(\imath\omega) \to -\infty$ as $\omega \to 0$ (see Ex. 10). g_3 is obtained from g_1

by adding a zero at $s = -1$. The polar plot is quite different to that of g_1. For small positive values of ω we have $\operatorname{Im} g_3(\imath\omega) \geq 0$ and for all ω we have $\operatorname{Re} g_3(\imath\omega) \geq 0$, whereas this is not the case for g_1. In fact g_3 is analytic on $\overline{\mathbb{C}_+}$ and maps this closed right half-plane into itself. Transfer functions with this property are called *positive real*. g_4 is the transfer function of a first order system where there is a delay of one unit in the input. The effect of the delay is to change the non-delayed plot of a circle to that of a spiral. Such behaviour is exhibited even if the delay is very small. This shows that in designing controls (or analyzing stability) by frequency domain methods one should be careful about neglecting delays. g_5 is another transfer function of a system with a delay, this time not in the control but in the state. For $\omega = 0$ we have $g_5(0) = 1/2$ which lies on the polar plot of $g(s) = 1/(s+2)$ which is a circle $\partial D((1/4, 0), 1/4)$ of radius $1/4$ around the centre $(1/4, 0)$. Then at $\omega = \pi$ it hits the imaginary axis $\imath\mathbb{R}$ for the first time. At $\omega = 2\pi$ it returns to the circle $\partial D((1/4, 0), 1/4)$ and at $\omega = 3\pi$ it hits $\imath\mathbb{R}$ again. The process is continued at multiples of π as seen in the right hand figure for g_5. Again this type of behaviour would also occur if the delay was small.
The transfer functions g_1, g_3, g_4 and g_5 are all in $H^\infty(\mathbb{C}_+;\mathbb{C})$, i.e. they are continuous and bounded on $\overline{\mathbb{C}_+}$ and analytic in $\mathbb{C}_+$, see Ex. 11. We will see in the next subsection that this implies that the corresponding convolution systems are L^2-stable. g_2 is not L^2-stable. □

2.3.3 Relationship Between Input–Output Operators and Transfer Matrices

In this subsection we consider convolution systems in both time domain and frequency domain and clarify the relationship between their input-output operators and transfer matrices. The technical development relies on Section A.3. Throughout the subsection it is assumed that all finite dimensional vector spaces $\mathbb{K}^p$, $\mathbb{K}^m$ are equipped with their standard Euclidean norms $\|\cdot\|_2$ and $\mathbb{K}^{p\times m}$ with the corresponding operator norm $\|\cdot\|_{2,2}$.
We suppose that an integrable convolution kernel $\mathcal{G}(\cdot) \in L^1(\mathbb{R}_+;\mathbb{C}^{p\times m})$ (resp. $\mathcal{G}(\cdot) \in \ell^1(\mathbb{N};\mathbb{C}^{p\times m})$) and a feedthrough matrix $D \in \mathbb{C}^{p\times m}$ are given and first consider the associated convolution system with the time domain $T = \mathbb{R}_+$ (resp. $\mathbb{N}$). This convolution system is described by $(\mathcal{U}_+, \mathbb{G}_+, \mathcal{Y}_+)$ where $\mathcal{U}_+ = L^q(\mathbb{R}_+;\mathbb{C}^m)$, $\mathcal{Y}_+ = L^q(\mathbb{R}_+;\mathbb{C}^p)$ (resp. $\mathcal{U}_+ = \ell^q(\mathbb{N};\mathbb{C}^m)$, $\mathcal{Y}_+ = \ell^q(\mathbb{N};\mathbb{C}^p)$) and the input-output operator $\mathbb{G}_+ : \mathcal{U}_+ \to \mathcal{Y}_+$ is of the form (39), see Proposition 2.3.15. In order to describe the relationship between $\mathbb{G}_+$ and the transfer matrix $G(\cdot)$ we introduce the Hardy spaces $H^q(\mathbb{C}_+;\mathbb{C}^n)$ (resp. $H^q(\mathbb{D}_+;\mathbb{C}^n)$).

Definition 2.3.26. For $1 \leq q \leq \infty$ denote by $H^q(\mathbb{C}_+;\mathbb{C}^n)$ the space of all analytic functions $v(\cdot)$ on $\mathbb{C}_+$ with values in $\mathbb{C}^n$ satisfying $\|v(\cdot)\|_{H^q(\mathbb{C}_+;\mathbb{C}^n)} < \infty$ where

$$\|v(\cdot)\|_{H^q(\mathbb{C}_+;\mathbb{C}^n)} = \begin{cases} \sup_{\alpha>0}\left(\int_{-\infty}^{\infty} \|v(\alpha+\imath\omega)\|_2^q d\omega\right)^{1/q}, & \text{if } 1 \leq q < \infty \\ \sup_{s\in\mathbb{C}_+} \|v(s)\|_2, & \text{if } q = \infty. \end{cases} \tag{54}$$

For some properties of these vector spaces see Subsection A.3.4. It is known that $H^q(\mathbb{C}_+;\mathbb{C}^n)$ provided with the norm (54) is a Banach space.

Definition 2.3.27. For $1 \le q \le \infty$ denote by $H^q(\mathbb{D}_+;\mathbb{C}^n)$ the space of all analytic functions $v(\cdot)$ on $\mathbb{D}_+$ with values in $\mathbb{C}^n$ satisfying $\|v(\cdot)\|_{H^q(\mathbb{D}_+;\mathbb{C}^n)} < \infty$ where

$$\|v(\cdot)\|_{H^q(\mathbb{D}_+;\mathbb{C}^n)} = \begin{cases} \sup_{r>1} \left(\int_{-\pi}^{\pi} \|v(re^{\imath\theta})\|_2^q d\theta\right)^{1/q}, & \text{if } 1 \le q < \infty \\ \sup_{z\in\mathbb{D}_+} \|v(z)\|_2, & \text{if } q = \infty. \end{cases} \tag{55}$$

Again it is known that $H^q(\mathbb{D}_+;\mathbb{C}^n)$ provided with this norm (55) is a Banach space. We now specialize to the case where $q = 2$ so that $\mathcal{U}_+ = L^2(\mathbb{R}_+;\mathbb{C}^m)$, (resp. $\mathcal{U}_+ = \ell^2(\mathbb{N};\mathbb{C}^m)$,) and $\mathcal{Y}_+ = L^2(\mathbb{R}_+;\mathbb{C}^p)$, (resp. $\mathcal{Y}_+ = \ell^2(\mathbb{N};\mathbb{C}^p)$). Let $G(s) = (D + \mathcal{L}\mathcal{G})(s)$ (resp. $G(z) = \mathcal{Z}\mathcal{G}(z)$) be the transfer matrix of the convolution system $(\mathcal{U}_+, \mathbb{G}_+, \mathcal{Y}_+)$, see Definition 2.3.17. Since $\mathcal{G}(\cdot) \in L^1(\mathbb{R}_+;\mathbb{C}^{p\times m})$, its Laplace transform is analytic on $\mathbb{C}_+$, bounded and continuous on $\overline{\mathbb{C}_+}$ so that $G(\cdot) \in H^\infty(\mathbb{C}_+;\mathbb{C}^{p\times m})$. Similarly, in the discrete time case, $G(z)$ is analytic on $\mathbb{D}_+$, bounded and continuous on $\overline{\mathbb{D}_+}$ so that $G(\cdot) \in H^\infty(\mathbb{D}_+;\mathbb{C}^{p\times m})$. Moreover by Proposition A.3.41 and Proposition A.3.45 we have

$$\|G(\cdot)\|_{H^\infty(\mathbb{C}_+;\mathbb{C}^{p\times m})} = \sup_{\omega\in\mathbb{R}} \|G(\imath\omega)\|_{2,2}, \quad \|G(\cdot)\|_{H^\infty(\mathbb{D}_+;\mathbb{C}^{p\times m})} = \max_{\theta\in[-\pi,\pi]} \|G(e^{\imath\theta})\|_{2,2}. \tag{56}$$

If $u(\cdot) \in L^2(\mathbb{R}_+;\mathbb{C}^m)$ (resp. $u(\cdot) \in \ell^2(\mathbb{N};\mathbb{C}^m)$), then $u(\cdot) \in \mathcal{E}_\alpha(\mathbb{C}^m)$ (resp. $\mathcal{S}_\gamma(\mathbb{C}^m)$) for every $\alpha > 0$ (resp. $\gamma > 1$). Therefore by (47)

$$\mathcal{L}((\mathbb{G}_+u)(\cdot))(s) = G(s)\hat{u}(s), \ \operatorname{Re} s > 0, \quad \mathcal{Z}((\mathbb{G}_+u)(\cdot))(z) = G(z)\hat{u}(z), \ |z| > 1 \tag{57}$$

for every $u(\cdot) \in L^2(\mathbb{R}_+;\mathbb{C}^m)$ (resp. $u(\cdot) \in \ell^2(\mathbb{N};\mathbb{C}^m)$). From (54) and (55) we get

$$\begin{aligned} \|G(\cdot)w(\cdot)\|_{H^2(\mathbb{C}_+;\mathbb{C}^p)} &\le \|G(\cdot)\|_{H^\infty(\mathbb{C}_+;\mathbb{C}^{p\times m})}\|w(\cdot)\|_{H^2(\mathbb{C}_+;\mathbb{C}^m)}, \quad w(\cdot)\in H^2(\mathbb{C}_+;\mathbb{C}^m) \\ \|G(\cdot)w(\cdot)\|_{H^2(\mathbb{D}_+;\mathbb{C}^p)} &\le \|G(\cdot)\|_{H^\infty(\mathbb{D}_+;\mathbb{C}^{p\times m})}\|w(\cdot)\|_{H^2(\mathbb{D}_+;\mathbb{C}^m)}, \quad w(\cdot)\in H^2(\mathbb{D}_+;\mathbb{C}^m). \end{aligned} \tag{58}$$

This shows that pointwise multiplication of an H^2-function by $G(s)$ yields an H^2-function. Let $M_G^+ : H^2(\mathbb{C}_+;\mathbb{C}^m) \to H^2(\mathbb{C}_+;\mathbb{C}^p)$ be the associated multiplication operator for the continuous time case defined by

$$(M_G^+ w)(s) = G(s)w(s), \quad w(\cdot) \in H^2(\mathbb{C}_+;\mathbb{C}^m), \ s \in \mathbb{C}_+, \tag{59}$$

and $M_G^+ : H^2(\mathbb{D}_+;\mathbb{C}^m) \to H^2(\mathbb{D}_+;\mathbb{C}^p)$ be its counterpart for the discrete time case defined by

$$(M_G^+ w)(z) = G(z)w(z), \qquad w(\cdot) \in H^2(\mathbb{D}_+;\mathbb{C}^m), \ z \in \mathbb{D}_+. \tag{60}$$

From (57) we see that the following diagrams commute

$$\begin{array}{ccc} L^2(\mathbb{R}_+;\mathbb{C}^m) & \xrightarrow{\mathbb{G}_+} & L^2(\mathbb{R}_+;\mathbb{C}^p) \\ \mathcal{L}\downarrow & & \downarrow\mathcal{L} \\ H^2(\mathbb{C}_+;\mathbb{C}^m) & \xrightarrow[M_G^+]{} & H^2(\mathbb{C}_+;\mathbb{C}^p) \end{array}, \qquad \begin{array}{ccc} \ell^2(\mathbb{N};\mathbb{C}^m) & \xrightarrow{\mathbb{G}_+} & \ell^2(\mathbb{N};\mathbb{C}^p) \\ \mathcal{Z}\downarrow & & \downarrow\mathcal{Z} \\ H^2(\mathbb{D}_+;\mathbb{C}^m) & \xrightarrow[M_G^+]{} & H^2(\mathbb{D}_+;\mathbb{C}^p) \end{array}.$$

By Theorem A.3.43 the normalized **z**-transform $(2\pi)^{-1/2}\mathcal{Z}$ is an isometry between the two ℓ^2 and H^2 spaces in the second diagram. Similarly by Theorem A.3.47 the normalized Laplace transform $(2\pi)^{-1/2}\mathcal{L}$ is an isometry between the two L^2 and H^2 spaces in the first diagram. Hence

$$\begin{aligned} \|\mathbb{G}_+\|_{\mathcal{L}(L^2(\mathbb{R}_+;\mathbb{C}^m),L^2(\mathbb{R}_+;\mathbb{C}^p))} &= \|M_G^+\|_{\mathcal{L}(H^2(\mathbb{C}_+;\mathbb{C}^m),H^2(\mathbb{C}_+;\mathbb{C}^p))}\,, \\ \|\mathbb{G}_+\|_{\mathcal{L}(\ell^2(\mathbb{N};\mathbb{C}^m),\ell^2(\mathbb{N};\mathbb{C}^p))} &= \|M_G^+\|_{\mathcal{L}(H^2(\mathbb{D}_+;\mathbb{C}^m),H^2(\mathbb{D}_+;\mathbb{C}^p))}\,. \end{aligned} \tag{61}$$

We will see later in Theorem 2.3.28 that the operator norms (61) of the multiplication operators M_G^+ can be computed by maximizing $\|G(s)\|_{2,2}$ on the imaginary axis and the unit circle, respectively.

We now turn to the case where $T = \mathbb{R}$ (resp. $\mathbb{Z}$) and consider the convolution system $(\mathcal{U},\mathbb{G},\mathcal{Y})$ where $\mathcal{U} = L^2(\mathbb{R};\mathbb{C}^m)$, $\mathcal{Y} = L^2(\mathbb{R};\mathbb{C}^p)$ (resp. $\mathcal{U} = \ell^2(\mathbb{Z};\mathbb{C}^m)$, $\mathcal{Y} = \ell^2(\mathbb{Z};\mathbb{C}^p)$) and the input-output operator $\mathbb{G}$ is given by (40). Again it follows from the convolution inequalities (A.3.24) and (A.3.7) that $\mathbb{G} : \mathcal{U} \to \mathcal{Y}$ is a bounded linear operator. Since the control functions may admit non-zero values on $(-\infty, 0)$ the Laplace transform $\mathcal{L}$ is no longer applicable to these signals. Instead we make use of the Fourier-Plancherel transform defined by (41) and the discrete Fourier transform defined by (44) where the limit of the series is to be understood in $L^2(-\pi,\pi;\mathbb{C}^m)$. We extend $\mathcal{G}(\cdot)$ trivially to $\mathbb{R}$ (resp. $\mathbb{Z}$) by setting $\mathcal{G}(t) = 0$ for $t < 0$. By (43) (with $\beta = 0$) and (46) (with $r = 1$) the Fourier transform (resp. discrete Fourier transform) of this extension is

$$\mathcal{F}(\mathcal{G})(\omega) = (\mathcal{L}\mathcal{G})(\imath\omega) = G(\imath\omega) - D, \quad \mathcal{F}_D(\mathcal{G})(\theta) = (\mathcal{Z}\mathcal{G})(e^{\imath\theta}) = G(e^{\imath\theta}).$$

Hence by Proposition A.3.35 (iii) (resp. Proposition A.3.39) the Fourier-Plancherel transform (resp. discrete Fourier transform) of the output signal $y(\cdot) = (\mathbb{G}u)(\cdot)$, $u(\cdot) \in \mathcal{U}$ is given by

$$\begin{aligned} \tilde{y}(\omega) &= (\mathcal{F}(Du + \mathcal{G} * u))(\omega) = G(\imath\omega)\tilde{u}(\omega), \quad \text{a.e.} \quad \omega \in \mathbb{R}, \\ \tilde{y}(\theta) &= (\mathcal{F}_D(\mathcal{G} * u))(\theta) = G(e^{\imath\theta})\tilde{u}(\theta), \quad \text{a.e.} \quad \theta \in [-\pi,\pi] \end{aligned} \tag{62}$$

where $\tilde{u}(\cdot) \in L^2(\mathbb{R};\mathbb{C}^m)$ (resp. $\tilde{u}(\cdot) \in L^2(-\pi,\pi;\mathbb{C}^m)$) denotes the Fourier-Plancherel transform (resp. discrete Fourier transform) of $u(\cdot)$. Now for $w(\cdot) \in L^2(\mathbb{R};\mathbb{C}^m)$ (resp. $w(\cdot) \in L^2(-\pi,\pi;\mathbb{C}^m)$) we have

$$\begin{aligned} \|G(\imath\cdot)w(\cdot)\|_{L^2(\mathbb{R};\mathbb{C}^p)} &\le \max_{\omega\in\mathbb{R}} \|G(\imath\omega)\|_{2,2}\|w(\cdot)\|_{L^2(\mathbb{R};\mathbb{C}^m)}, \\ \|G(e^{\imath\cdot})w(\cdot)\|_{L^2(-\pi,\pi;\mathbb{C}^p)} &\le \max_{\theta\in[-\pi,\pi]} \|G(e^{\imath\theta})\|_{2,2}\|w(\cdot)\|_{L^2(-\pi,\pi;\mathbb{C}^m)}. \end{aligned} \tag{63}$$

Hence the multiplication operators

$$\begin{aligned} M_G : L^2(\mathbb{R};\mathbb{C}^m) \to L^2(\mathbb{R};\mathbb{C}^p), \quad (M_G w)(\omega) = G(\imath\omega)w(\omega), \ \text{a.e.} \ \omega \in \mathbb{R}, \\ M_G : L^2(-\pi,\pi;\mathbb{C}^m) \to L^2(-\pi,\pi;\mathbb{C}^p), \quad (M_G w)(\theta) = G(e^{\imath\theta})w(\theta), \ \text{a.e.} \ \theta \in [-\pi,\pi] \end{aligned} \tag{64}$$

are well defined, linear and bounded. The equations in (62) imply that the following diagrams commute

$$\begin{array}{ccc} L^2(\mathbb{R};\mathbb{C}^m) & \xrightarrow{\mathbb{G}} & L^2(\mathbb{R};\mathbb{C}^p) \\ \mathcal{F}\big\downarrow & & \big\downarrow\mathcal{F} \\ L^2(\mathbb{R};\mathbb{C}^m) & \xrightarrow[M_G]{} & L^2(\mathbb{R};\mathbb{C}^p) \end{array} \qquad \begin{array}{ccc} \ell^2(\mathbb{Z};\mathbb{C}^m) & \xrightarrow{\mathbb{G}} & \ell^2(\mathbb{Z};\mathbb{C}^p) \\ \mathcal{F}_D\big\downarrow & & \big\downarrow\mathcal{F}_D \\ L^2(-\pi,\pi;\mathbb{C}^m) & \xrightarrow[M_G]{} & L^2(-\pi,\pi;\mathbb{C}^p) \end{array}$$

Since by Theorem A.3.33 the normalized Fourier transform $(2\pi)^{-1/2}\mathcal{F}$ is an isometry between the two L^2 spaces on the left diagram and by Remark A.3.38 the normalized discrete Fourier transform $(2\pi)^{-1/2}\mathcal{F}_D$ is an isometry between the two ℓ^2 and L^2 spaces of the right diagram, it follows that $\mathbb{G}$ and M_G have the same norm,

$$\begin{aligned} \|\mathbb{G}\|_{\mathcal{L}(L^2(\mathbb{R};\mathbb{C}^m),L^2(\mathbb{R};\mathbb{C}^p))} &= \|M_G\|_{\mathcal{L}(L^2(\mathbb{R};\mathbb{C}^m),L^2(\mathbb{R};\mathbb{C}^p))}, \\ \|\mathbb{G}\|_{\mathcal{L}(\ell^2(\mathbb{Z};\mathbb{C}^m),\ell^2(\mathbb{Z};\mathbb{C}^p))} &= \|M_G\|_{\mathcal{L}(L^2(-\pi,\pi;\mathbb{C}^m),L^2(-\pi,\pi;\mathbb{C}^p))}. \end{aligned} \tag{65}$$

The following theorem links these results with the corresponding ones for $\mathbb{G}_+$ given in (61). It will be used in Section 5.3 to characterize the complex stability radius.

Theorem 2.3.28. *Suppose* $\mathcal{G}(\cdot) \in L^1(\mathbb{R}_+;\mathbb{K}^{p\times m})$ *(resp.* $\mathcal{G}(\cdot) \in \ell^1(\mathbb{N};\mathbb{K}^{p\times m})$*) and* $D \in \mathbb{K}^{p\times m}$, $\mathbb{G} \in \mathcal{L}(L^2(\mathbb{R};\mathbb{K}^m), L^2(\mathbb{R};\mathbb{K}^p))$ *(resp.* $\mathbb{G} \in \mathcal{L}(\ell^2(\mathbb{Z};\mathbb{K}^m), \ell^2(\mathbb{Z};\mathbb{K}^p))$*),* $\mathbb{G}_+ \in \mathcal{L}(L^2(\mathbb{R}_+;\mathbb{K}^m), L^2(\mathbb{R}_+;\mathbb{K}^p))$ *(resp.* $\mathbb{G} \in \mathcal{L}(\ell^2(\mathbb{N};\mathbb{K}^m), \ell^2(\mathbb{N};\mathbb{K}^p))$ *are the input-output operators defined by* (39) *(resp.* (40)*) and* $G(\cdot)$ *is the associated transfer matrix* (48), *then*

$$\begin{aligned} \|G\|_{H^\infty} &= \sup_{\omega\in\mathbb{R}} \|G(\imath\omega)\|_{2,2} = \|\mathbb{G}_+\|_{\mathcal{L}(L^2(\mathbb{R}_+;\mathbb{K}^m),L^2(\mathbb{R}_+;\mathbb{K}^p))} = \|\mathbb{G}\|_{\mathcal{L}(L^2(\mathbb{R};\mathbb{K}^m),L^2(\mathbb{R};\mathbb{K}^p))}, \\ \|G\|_{H^\infty} &= \max_{\theta\in[-\pi,\pi]} \|G(e^{\imath\theta})\|_{2,2} = \|\mathbb{G}_+\|_{\mathcal{L}(\ell^2(\mathbb{N};\mathbb{K}^m),\ell^2(\mathbb{N};\mathbb{K}^p))} = \|\mathbb{G}\|_{\mathcal{L}(\ell^2(\mathbb{Z};\mathbb{K}^m),\ell^2(\mathbb{Z};\mathbb{K}^p))}. \end{aligned} \tag{66}$$

Proof: The proof is for the continuous time case. If $\mathbb{G}$ is a real convolution operator we have $\|\mathbb{G}\|_{\mathcal{L}(L^2(\mathbb{R};\mathbb{R}^m),L^2(\mathbb{R};\mathbb{R}^p))} = \|\mathbb{G}\|_{\mathcal{L}(L^2(\mathbb{R};\mathbb{C}^m),L^2(\mathbb{R};\mathbb{C}^p))}$, so we need only consider the case $\mathbb{K} = \mathbb{C}$. By (56)

$$\|G\|_{H^\infty(\mathbb{C}_+;\mathbb{C}^{p\times m})} = \sup_{\omega\in\mathbb{R}} \|G(\imath\omega)\|_{2,2}.$$

So because of (65) and Proposition 2.3.15 it remains to prove that

$$\|M_G\|_{\mathcal{L}(L^2(\mathbb{R};\mathbb{C}^m),L^2(\mathbb{R};\mathbb{C}^p))} = \sup_{\omega\in\mathbb{R}} \|G(\imath\omega)\|_{2,2}.$$

By (63) the inequality $\le$ holds. To prove the converse inequality let $\varepsilon > 0$ be arbitrary. Then there exist $\omega_0 \in \mathbb{R}$, $\delta > 0$ and $u \in \mathbb{C}^m$, $\|u\|_2 = 1$ such that

$$\|G(\imath\omega)u\|_2 \ge \sup_{\omega\in\mathbb{R}} \|G(\imath\omega)\|_{2,2} - \varepsilon, \quad \omega \in [\omega_0 - \delta, \omega_0 + \delta]. \tag{67}$$

This follows from the fact that $G(\cdot)$ is continuous and bounded on $\overline{\mathbb{C}_+}$: One first chooses ω_0 such that

$$\|G(\imath\omega_0)\|_{2,2} \ge \sup_{\omega\in\mathbb{R}} \|G(\imath\omega)\|_{2,2} - \varepsilon/2,$$

then $u \in \mathbb{C}^m$, $\|u\|_2 = 1$ such that $\|G(\imath\omega_0)u\|_2 = \|G(\imath\omega_0)\|_{2,2}$ (see Definition A.1.4) and finally $\delta > 0$ such that

$$\|G(\imath\omega)u - G(\imath\omega_0)u\|_2 \le \varepsilon/2, \quad \omega \in [\omega_0 - \delta, \omega_0 + \delta].$$

Define $\tilde{u}(\cdot) \in L^2(\mathbb{R};\mathbb{C}^m)$ by $\tilde{u}(\omega) = u/\sqrt{2\delta}$ for $\omega \in [\omega_0 - \delta, \omega_0 + \delta]$ and $\tilde{u}(\omega) = 0$ otherwise. Then $\|\tilde{u}\|_{L^2(\mathbb{R};\mathbb{C}^m)} = 1$ and by (67)

$$\int_{-\infty}^{\infty} \|(M_G\tilde{u})(\omega)\|_2^2 d\omega = \frac{1}{2\delta}\int_{\omega_0-\delta}^{\omega_0+\delta} \|G(\imath\omega)u\|_2^2 d\omega \ge \left(\sup_{\omega\in\mathbb{R}} \|G(\imath\omega)\|_{2,2} - \varepsilon\right)^2.$$

This completes the proof of the continuous time case. The proof for the discrete time case is similar, see Ex. 13. □

The following example illustrates how the norm of the input-output operator of a state space system (49) (resp. (51)) can be determined by applying (66).

Example 2.3.29. Consider the oscillator described in Example 2.3.6,

$$\ddot{\xi}(t) + 2\alpha\dot{\xi}(t) + \nu^2\xi(t) = \nu^2 u(t), \quad y(t) = \xi(t), \ \ t \in \mathbb{R}_+.$$

We assume $\alpha > 0$, $\nu \neq 0$ so that the corresponding state space system satisfies $\sigma(A) \subset \mathbb{C}_-$. The transfer function is $g(s) = \nu^2/(s^2 + 2\alpha s + \nu^2)$ and a simple calculation gives

$$\sup_{\omega\in\mathbb{R}} |g(\imath\omega)| = \begin{cases} 1 & \text{if} \quad \nu^2 \le 2\alpha^2 \\ \dfrac{\nu^2}{2\alpha\sqrt{\nu^2 - \alpha^2}} & \text{if} \quad \nu^2 > 2\alpha^2 . \end{cases}$$

The discrete time counterpart is

$$\xi(t+2) + 2\alpha\xi(t+1) + \nu^2\xi(t) = \nu^2 u(t), \quad y(t) = \xi(t), \quad t \in \mathbb{N}.$$

The corresponding state space system satisfies $\sigma(A) \subset \mathbb{D}$ if $1 > \nu^2 > |2\alpha| - 1$. The transfer function is as above and an easy calculation gives

$$\max_{\theta\in[-\pi,\pi]} |g(e^{\imath\theta})| = \begin{cases} \dfrac{|\nu|^3}{(1-\nu^2)[\nu^2 - \alpha^2]^{1/2}} & \text{if} \quad |\alpha|(1+\nu^2) < 2\nu^2 \\ \dfrac{\nu^2}{1 - 2|\alpha| + \nu^2} & \text{if} \quad |\alpha|(1+\nu^2) \ge 2\nu^2. \end{cases}$$ □

2.3.4 Exercises

1. Compute the impulse response $\mathcal{G} = (\mathcal{G}(t))_{t\in\mathbb{N}}$ and the *step response* (corresponding to the constant input $u(t) \equiv 1$, $t \in \mathbb{N}$) for the discrete time system (A, B, C) of Ex. 2.2.4 and the discrete time system (A, B, C, D) of Ex. 2.2.8.

2. Calculate the impulse response for the continuous time system (49) with

$$A = \begin{bmatrix} 0 & 1 \\ -1 & 0 \end{bmatrix}, \quad B = \begin{bmatrix} 0 \\ 1 \end{bmatrix}, \quad C = [1\,,\,0].$$

Find also the response $y^k(\cdot)$ to the input $u^k(\cdot)$ where

$$u^k(t) = \begin{cases} k & \text{if } t \in [0, 1/k) \\ 0 & \text{if } t \geq 1/k \end{cases}$$

and show that $y^k(\cdot)$ converges uniformly on compact intervals to the impulse response.

3. Consider the input-output relation of a siso convolution system of the form

$$y(t) = \int_0^t \mathcal{G}(t-\tau)u(\tau)d\tau, \quad t \geq 0,$$

where $\mathcal{G} : \mathbb{R}_+ \to \mathbb{R}$ is continuous. Let $\bar{y}(t)$, $t \geq 0$ be the *step response* (corresponding to $u(t) \equiv 1$, $t \geq 0$). Show that $\bar{y}(\cdot)$ is differentiable on $[0, \infty)$ and its derivative is the impulse response $\mathcal{G}(\cdot)$. Formulate and prove an analogous result for discrete time systems.

4. Prove the discrete time counterpart of Proposition 2.3.10 as stated in Remark 2.3.12.

5. Consider the scalar system

$$y^{(n)}(t) + a_{n-1}y^{(n-1)}(t) + \ldots + a_0 y(t) = b_{n-1}u^{(n-1)}(t) + \ldots + b_0 u(t), \quad y^{(i)} = \frac{d^i y}{dt^i}. \tag{68}$$

Find an appropriate state space model (A, B, C) for this system, see Ex. 2.1.8 and Ex. 2.1.9. Show that the impulse response $\mathcal{G}(t) = Ce^{At}B$ is a quasi-polynomial of the form $\mathcal{G}(t) = \sum_{i=1}^{\ell} p_i(t)e^{\lambda_i t}$ where the p_i are polynomials and $\lambda_1, \ldots, \lambda_\ell$ are the distinct roots of the characteristic polynomial $p(\lambda) = \lambda^n + a_{n-1}\lambda^{n-1} + \ldots + a_0$. Find an explicit expression for the transfer function and verify your answer by applying the Laplace transform to (68) with zero initial conditions.

6. Construct a dyadic decomposition of the system (A, B, C) where

$$A = \begin{bmatrix} 1 & 4 \\ 1 & -2 \end{bmatrix}, \quad B = \begin{bmatrix} 0 & 1 \\ 1 & 1 \end{bmatrix}, \quad C = I_2.$$

7. Calculate the impulse response $\mathcal{G}(t)$ and the transfer function $g(s)$ for the continuous time system (A, B, C, D) where

$$A = \begin{bmatrix} 0 & 1 \\ -1 & -1 \end{bmatrix}, \quad B = \begin{bmatrix} 0 \\ 1 \end{bmatrix}, \quad C = [1\,,\,0], \; D = 1.$$

Use MATLAB to plot the gain and frequency responses. Determine $\|g\|_{H^\infty(\mathbb{C}_+;\mathbb{C})}$.

8. Prove Proposition 2.3.22 for the discrete time case.

9. Find the convolution kernels $\mathcal{G}_i$ corresponding to the transfer functions $g_1, \ldots, g_5$ given in Example 2.3.25. In the case of g_5 you need only compute $\mathcal{G}_5(t)$ for $t \in [0, 3]$.

10. If, as in Example 2.3.25, $g_1(s) = 1/(s^2+2s+5)$ and $x(\omega) = \operatorname{Re} g_1(\imath\omega)$, $y(\omega) = \operatorname{Im} g_1(\omega)$ find an algebraic equation in x, y for the polar plot of $g_1(\cdot)$. Carry out the same programme for $g_2(s) = 1/s(s^2 + 2s + 5)$ and show that $\operatorname{Re} g_2(\imath\omega) \to -.08$ and $\operatorname{Im} g_2(\imath\omega) \to -\infty$ as $\omega \to 0$.

11. Prove that the transfer functions g_1, g_3, g_4 and g_5 given in Example 2.3.25 are in $H^\infty(\mathbb{C}_+; \mathbb{C})$ and calculate their H^∞-norms. Find corresponding state space models for the scalar transfer functions g_1, g_2 and g_3.

12. Consider the transfer function

$$G(s) = \frac{1}{s^2 + 2s + 2} \begin{bmatrix} s+1 & +1 \\ -1 & s+1 \end{bmatrix}.$$

Show that $\|G\|_{H^\infty(\mathbb{C}_+;\mathbb{C}^{2\times 2})} = 1$. For any $\varepsilon \in (0,1)$, construct a function $\tilde{u}(\cdot) \in L^2(\mathbb{R};\mathbb{C}^2)$ with $\|\tilde{u}\|_{L^2(\mathbb{R};\mathbb{C}^2)} = 1$ such that $\|M_G \tilde{u}\|^2_{L^2(\mathbb{R};\mathbb{C}^2)} \geq (1-\varepsilon)^2$, see the proof of Theorem 2.3.28.

13. Prove Theorem 2.3.28 for the discrete time case.

14. Consider the electrical circuit of Ex. 2.2.9 with input the driving voltage u and output the charge q on the capacitor. Determine the impulse response of the system. Let $u(t) = \sin \omega_0 t$ where $\omega_0 > 0$ is given. Specify conditions in terms of R, L, C under which the system admits a periodic trajectory with period $\omega_0 > 0$ (substitute $q(t) = a \cos \omega_0 t + b \sin \omega_0 t$ in the differential equation). Show that if these conditions are satisfied and $R > 0$ every trajectory of the system with initial state $x^0 \neq 0$ (under the control $u(t) = \sin \omega_0 t$) will approaches this periodic solution as $t \to \infty$.

2.3.5 Notes and References

The seminal monograph of *Desoer and Vidyasagar* (1975) [130] on input-output systems is still a standard reference. More recent textbooks which contain chapters on input-output systems are *Delchamps* (1988) [125], *Sontag* (1998) [472] and *Sastry* (1999) [448]. In [130] one can find details of a convolution algebra which allows for Dirac impulses in the convolution kernel, see also [116].
For background material on convolutions, **z** and Laplace transforms and Fourier transforms see the books recommended in Section A.3. An excellent introductory textbook which covers much of the material of this section, written from an engineering point of view and aimed at undergraduates is *Kwakernaak and Sivan* (1991) [322].
The Hardy space H^q play a role in systems theory in the context of robust control and H^∞ theory, see *Zhou et al.* (1996) [546].
Frequency response methods spread rapidly in the 1930's after the appearance of *Nyquist's* classical paper on feedback amplifier stability (1932) [395] which arose from problems of long distance telephony. By the early 1950's frequency domain methods dominated the analysis and design of automatic control systems. Nyquist's method proceeds from a modification of the *polar plot* of the complex frequency response. Bode's method proceeds from the graphs of amplitude and phase response and the Nichols chart combines the two Bode plots into a plot of the gain in decibels against phase shift in degrees, see e.g. *Macfarlane* (1979) [356]. Other standard references are [322], [168]. A recent book on system identification via frequency domain methods is *Pintelon and Schoukens* (2001) [413].

2.4 Transformations and Interconnections of Linear State Space Systems

In this section we only consider continuous time systems of the form (2.17) which we denote by the shorthand notation $\Sigma = (A, B, C, D)$. All the definitions and results can also be applied to discrete time systems of the form (2.22). The first subsection is concerned with showing that the systems (A, B, C, D) form a category and we specify some standard constructions for this category (subsystems, quotient systems, direct sum, ...). In the second subsection we introduce the basic coupling schemes for two systems $\Sigma_i = (A_i, B_i, C_i, D_i)$ $i = 1, 2$ and discuss the general form of a composite linear time–invariant system. As usual we do not distinguish notationally between a linear map and the matrix representing it with respect to a given basis. In Subsection 2.4.1 a coordinate free interpretation of A, B, C, D as linear maps between vector spaces will prevail. However in applications these maps will be described by matrices with respect to given bases of the input, state and output spaces.

2.4.1 Morphisms and Standard Constructions

Changes of bases in the state space X, the input space U and/or the output space Y of a system (A, B, C, D) lead to transformations of the matrices A, B, C, D. Hence a given physical system may be modelled by different quadruples of the form (A, B, C, D). This raises the question – which conditions render two systems to be isomorphic or similar? Two vector spaces (groups) are called isomorphic if there exists a linear isomorphism (resp. group isomorphism) between them. Using the terminology of category theory, isomorphisms are *invertible morphisms* and morphisms are *structure preserving* "maps" between structured objects of a given class. We do not intend to explore these generalities, but just to mention that the concepts of "isomorphism" and "(homo)morphism" are of fundamental importance in the construction of a mathematical theory.

Morphisms between dynamical systems can be defined in various ways. One possible definition would be:

If $\Sigma_i = (T, U_i, \mathcal{U}_i, X_i, Y_i, \varphi_i, \eta_i)$, $i = 1, 2$ *are two dynamical systems of a given class* $\mathcal{S}$ *then a morphism from* Σ_1 *to* Σ_2 *is a triple* (ρ, τ, σ) *consisting of maps* $\rho : U_1 \to U_2$, $\tau : X_1 \to X_2$, $\sigma : Y_1 \to Y_2$ *which have certain properties (such as smoothness, linearity etc.) depending on the specific class* $\mathcal{S}$. *Moreover the following three conditions must be satisfied for all* t, $t_0 \in T$, $u \in U_1$, $u(\cdot) \in \mathcal{U}_1$, $x \in X_1$.

$$u(\cdot) \in \mathcal{U}_1 \;\Rightarrow\; \rho \circ u(\cdot) \in \mathcal{U}_2 \quad \text{and} \quad T^{\Sigma_1}_{t_0,x,u(\cdot)} \subset T^{\Sigma_2}_{t_0,\tau(x),\rho\circ u(\cdot)} \tag{1}$$

$$\tau(\varphi_1(t; t_0, x, u(\cdot))) = \varphi_2(t; t_0, \tau(x), \rho \circ u(\cdot)), \quad t \in T^{\Sigma_1}_{t_0,x,u(\cdot)},\ u(\cdot) \in \mathcal{U}_1 \tag{2}$$

$$\sigma(\eta_1(t, x, u)) = \eta_2(t, \tau(x), \rho(u)). \tag{3}$$

Other definitions may allow for transformations of the time and for more general mappings between $\mathcal{U}_1$ and $\mathcal{U}_2$. We do not go into further details at this general level, but now give a precise definition for time–invariant linear finite dimensional systems.

Definition 2.4.1. (Linear system morphism). Consider two finite dimensional linear systems $\Sigma_i = (A_i, B_i, C_i, D_i)$ with input space U_i, state space X_i, output space Y_i $(i = 1, 2)$. A triple (R, T, S) of linear maps $R : U_1 \to U_2$, $T : X_1 \to X_2$, $S : Y_1 \to Y_2$ is called a *linear system morphism* from Σ_1 to Σ_2 (with the notation $(R, S, T) \in \mathrm{Mor}(\Sigma_1, \Sigma_2)$ or $(R, S, T) : \Sigma_1 \mapsto \Sigma_2$) if

$$A_2 T = T A_1, \quad B_2 R = T B_1, \quad C_2 T = S C_1, \quad D_2 R = S D_1, \tag{4}$$

i.e. the following diagrams commute

$$\begin{array}{ccccccc} U_1 & \xrightarrow{B_1} & X_1 & \xrightarrow{A_1} & X_1 & \xrightarrow{C_1} & Y_1 \\ {\scriptstyle R}\downarrow & & \downarrow{\scriptstyle T} & & \downarrow{\scriptstyle T} & & \downarrow{\scriptstyle S} \\ U_2 & \xrightarrow[B_2]{} & X_2 & \xrightarrow[A_2]{} & X_2 & \xrightarrow[C_2]{} & Y_2 \end{array} \qquad \begin{array}{ccc} U_1 & \xrightarrow{D_1} & Y_1 \\ {\scriptstyle R}\downarrow & & \downarrow{\scriptstyle S} \\ U_2 & \xrightarrow[D_2]{} & Y_2 \end{array}$$

If we represent the system (A, B, C, D) by the linear map

$$\Sigma : X \times U \to X \times Y, \quad \begin{bmatrix} x \\ u \end{bmatrix} \mapsto \begin{bmatrix} x \\ y \end{bmatrix} = \begin{bmatrix} A & B \\ C & D \end{bmatrix} \begin{bmatrix} x \\ u \end{bmatrix}, \tag{5}$$

then conditions (4) can be expressed equivalently by one compound equation

$$\begin{bmatrix} T & 0 \\ 0 & S \end{bmatrix} \Sigma_1 = \Sigma_2 \begin{bmatrix} T & 0 \\ 0 & R \end{bmatrix}. \tag{6}$$

In other words the matrices $T \oplus S = \mathrm{diag}\,(T, S)$ and $T \oplus R = \mathrm{diag}\,(T, R)$ intertwine the linear maps Σ_1 and Σ_2. Intertwining operators play an important role in system theory.

Remark 2.4.2. It is a simple matter to verify that the class of all time–invariant finite dimensional linear systems together with the morphisms defined above form a category if the composition of two morphisms is defined in the obvious way

$$(R_2, T_2, S_2) \circ (R_1, T_1, S_1) = (R_2 R_1, T_2 T_1, S_2 S_1).$$

□

A morphism $(R, T, S) \in \mathrm{Mor}(\Sigma_1, \Sigma_2)$ is called a *(linear system) isomorphism* if it admits a left and right inverse in the sense of the above composition, and this is the case if and only if $R : U_1 \to U_2$, $T : X_1 \to X_2$, $S : Y_1 \to Y_2$ are vector space isomorphisms.

In terms of matrix representations a linear system isomorphism describes changes of bases in the input, state and output spaces. In fact, consider system equations of the form (2.17)

$$\begin{aligned} \dot{x}(t) &= Ax(t) + Bu(t), \quad t \in \mathbb{R} \\ y(t) &= Cx(t) + Du(t), \end{aligned}$$

where $(A, B, C, D) \in \mathbf{L}_{n,m,p}(\mathbb{K}) := \mathbb{K}^{n\times n} \times \mathbb{K}^{n\times m} \times \mathbb{K}^{p\times n} \times \mathbb{K}^{p\times m}$, $n, m, p \geq 1$ and suppose we introduce new bases $(v^1, \ldots, v^m)$ in $U = \mathbb{K}^m$, $(z^1, \ldots, z^n)$ in $X = \mathbb{K}^n$, and $(w^1, \ldots, w^p)$ in $Y = \mathbb{K}^p$. The coordinate vectors of $u \in \mathbb{K}^m$, $x \in \mathbb{K}^n$, and $y \in \mathbb{K}^p$ with respect to the new bases are given by

$$\hat{u} = R^{-1}u, \quad \hat{x} = T^{-1}x, \quad \hat{y} = S^{-1}y \tag{7}$$

where

$$R = [v^1, \ldots, v^m] \in \mathbf{Gl}_m(\mathbb{K}), \; T = [z^1, \ldots, z^n] \in \mathbf{Gl}_n(\mathbb{K}), \; S = [w^1, \ldots, w^p] \in \mathbf{Gl}_p(\mathbb{K}).$$

In terms of the new coordinate vectors the system equations read

$$\begin{aligned} \dot{\hat{x}}(t) &= \hat{A}\hat{x}(t) + \hat{B}\hat{u}(t), \quad t \in \mathbb{R} \\ \hat{y}(t) &= \hat{C}\hat{x}(t) + \hat{D}\hat{u}(t) \end{aligned} \tag{8}$$

where

$$\hat{A} = TAT^{-1}, \quad \hat{B} = TBR^{-1}, \quad \hat{C} = SCT^{-1}, \quad \hat{D} = SDR^{-1}. \tag{9}$$

In applications the external variables often represent physical quantities such as current, velocity, temperature. Linear transformations of the input and output vectors would destroy this physical interpretation and so usually one does not consider coordinate transformations in the input and output spaces. If only linear coordinate transformations $\hat{x} = T^{-1}x$ in the state space are allowed, the system equations are transformed into

$$\begin{aligned} \dot{\hat{x}}(t) &= TAT^{-1}\hat{x}(t) + TBu(t) \\ y(t) &= CT^{-1}\hat{x}(t) + Du(t). \end{aligned} \tag{10}$$

This leads us to the more restrictive class of *similarity transformations*

$$T \cdot (A, B, C, D) = (TAT^{-1}, TB, CT^{-1}, D), \quad T \in \mathbf{Gl}_n(\mathbb{K}). \tag{11}$$

Definition 2.4.3. (Isomorphy, Similarity). Two finite dimensional linear systems $\Sigma_i = (A_i, B_i, C_i, D_i)$, $i = 1, 2$ are said to be

(i) *isomorphic* if there exists a linear system isomorphism $(R, S, T) : \Sigma_1 \mapsto \Sigma_2$,

(ii) *similar* if $U_1 = U_2$, $Y_1 = Y_2$, and there exists a linear isomorphism $T : X_1 \to X_2$ satisfying

$$A_2 = TA_1T^{-1}, \quad B_2 = TB_1, \quad C_2 = C_1T^{-1}, \quad D_1 = D_2. \tag{12}$$

The input–output operator of a linear system will in general change under arbitrary linear system isomorphisms, but not under similarity transformations.

Proposition 2.4.4. *The input–output operator and the transfer matrix of a linear system (2.17) are invariant under similarity transformations.*

Proof: It suffices to show the invariance of the transfer matrix. But this follows immediately from (12) since

$$C_2(sI - A_2)^{-1}B_2 + D_2 = C_1T^{-1}(sI - TA_1T^{-1})^{-1}TB_1 + D_1 = C_1(sI - A_1)^{-1}B_1 + D_1.$$

□

We now introduce the concepts of *subsystem* and *quotient system.*

Definition 2.4.5. (Subsystem). $\Sigma_1 = (A_1, B_1, C_1, D_1)$ is called a *subsystem* of $\Sigma_2 = (A_2, B_2, C_2, D_2)$ if there exist linear injections $R : U_1 \to U_2$, $T : X_1 \to X_2$, $S : Y_1 \to Y_2$ such that (4) holds. In this case (R, T, S) is called a *system embedding.*

Suppose a system (A, B, C, D) is given. If $U_1 \subset U$, $X_1 \subset X$, $Y_1 \subset Y$ are linear subspaces such that

$$BU_1 \subset X_1, \quad AX_1 \subset X_1, \quad CX_1 \subset Y_1, \quad DU_1 \subset Y_1 \tag{13}$$

then the system (A_1, B_1, C_1, D_1) obtained by restricting A, B, C, D to X_1, U_1, X_1, U_1 respectively is a subsystem of (A, B, C, D). The embedding is given by (R, T, S) where $R : U_1 \to U$, $T : X_1 \to X$, $S : Y_1 \to Y$ are the canonical injections.
Now let $\hat{R} : U \to U/U_1$, $\hat{T} : X \to X/X_1$, $\hat{S} : Y \to Y/Y_1$ be the natural projections. Then there exist linear maps $\hat{A}$, $\hat{B}$, $\hat{C}$, $\hat{D}$ which make the following diagrams commute, and the maps are uniquely determined by this property.

$$\begin{array}{ccccccc} U & \xrightarrow{B} & X & \xrightarrow{A} & X & \xrightarrow{C} & Y \\ \tilde{R}\big\downarrow & & \big\downarrow\tilde{T} & & \big\downarrow\tilde{T} & & \big\downarrow\tilde{S} \\ U/U_1 & \xrightarrow{\tilde{B}} & X/X_1 & \xrightarrow{\tilde{A}} & X/X_1 & \xrightarrow{\tilde{C}} & Y/Y_1 \end{array} \qquad \begin{array}{ccc} U & \xrightarrow{D} & Y \\ \tilde{R}\big\downarrow & & \big\downarrow\tilde{S} \\ U/U_1 & \xrightarrow{\tilde{D}} & Y/Y_1 \end{array} \tag{14}$$

Definition 2.4.6. (Quotient system). Suppose (A, B, C, D) is a linear system with input space U, state space X, output space Y and $U_1 \subset U$, $X_1 \subset X$, $Y_1 \subset Y$ are linear subspaces such that (13) holds. If Σ_1 denotes the corresponding subsystem of Σ then the linear system $\hat{\Sigma} = (\hat{A}, \hat{B}, \hat{C}, \hat{D})$ with input space $\hat{U} = U/U_1$, state space $\hat{X} = X/X_1$ and output space $\hat{Y} = Y/Y_1$ defined by (14) is called the *quotient system of* Σ *by* Σ_1 and is denoted by Σ/Σ_1.

If

$$\hat{R} : U \to \hat{U} = U/U_1, \; \hat{T} : X \to \hat{X} = X/X_1, \; \hat{S} : Y \to \hat{Y} = Y/Y_1$$

are the canonical projections then the quotient system $\hat{\Sigma} = \Sigma/\Sigma_1$ is uniquely determined by the property that $(\hat{R}, \hat{T}, \hat{S})$ is a linear system morphism from Σ to $\hat{\Sigma}$. This morphism is called the *canonical system projection* from Σ to $\hat{\Sigma}$.
Now suppose that U_2, X_2, Y_2 are algebraic complements of U_1, X_1, Y_1 in U, X, Y respectively, then A, B, C, D have the following representations with respect to the decompositions $U = U_1 \oplus U_2$, $X = X_1 \oplus X_2$, $Y = Y_1 \oplus Y_2$

$$A = \begin{bmatrix} A_{11} & A_{12} \\ 0 & A_{22} \end{bmatrix}, \; B = \begin{bmatrix} B_{11} & B_{12} \\ 0 & B_{22} \end{bmatrix}, \; C = \begin{bmatrix} C_{11} & C_{12} \\ 0 & C_{22} \end{bmatrix}, \; D = \begin{bmatrix} D_{11} & D_{12} \\ 0 & D_{22} \end{bmatrix}. \tag{15}$$

It is straightforward to verify that the isomorphisms

$$U_2 \cong U/U_1, \quad X_2 \cong X/X_1, \quad Y_2 \cong Y/Y_1$$

induced by the restriction of R, T, S to U_2, X_2, Y_2 respectively define a system isomorphism between $\Sigma_2 = (A_{22}, B_{22}, C_{22}, D_{22})$ and $(\hat{A}, \hat{B}, \hat{C}, \hat{D})$.

Example 2.4.7. Consider the mass-spring-damper system Σ shown in Figure 2.4.1. The masses m_1, m_2 slide on a horizontal surface without friction. The stiffness coefficients of the springs are k_1, k_2 and the damping coefficient is c. The outputs are the displacements $y_1(t)$, $y_2(t)$ of the two masses from some given equilibrium positions and the inputs $u_1(t)$,

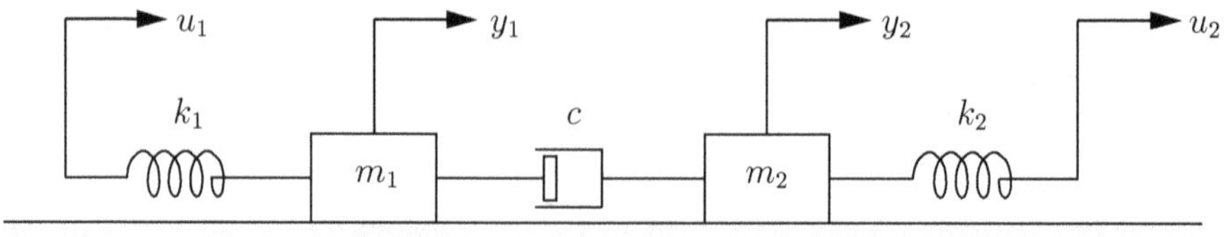

Figure 2.4.1: Mass-spring-damper system

$u_2(t)$ are the displacements of the outer ends of the springs from their corresponding rest positions. The equations of motion of this mechanical system are given by

$$\begin{aligned} m_1\ddot{y}_1 &= k_1(u_1 - y_1) + c(\dot{y}_2 - \dot{y}_1) \\ m_2\ddot{y}_2 &= k_2(u_2 - y_2) - c(\dot{y}_2 - \dot{y}_1). \end{aligned} \tag{16}$$

If we define the state variables to be $x_1 = y_1$, $x_2 = \dot{y}_1$, $x_3 = y_2$, $x_4 = \dot{y}_2$ we obtain the time–invariant linear system (A, B, C, D) where $C = \begin{bmatrix} 1 & 0 & 0 & 0 \\ 0 & 0 & 1 & 0 \end{bmatrix}$, $D = 0$ and

$$A = \begin{bmatrix} 0 & 1 & 0 & 0 \\ -k_1/m_1 & -c/m_1 & 0 & c/m_1 \\ 0 & 0 & 0 & 1 \\ 0 & c/m_2 & -k_2/m_2 & -c/m_2 \end{bmatrix}, \quad B = \begin{bmatrix} 0 & 0 \\ k_1/m_1 & 0 \\ 0 & 0 \\ 0 & k_2/m_2 \end{bmatrix}.$$

Now suppose $k_1/m_1 = k_2/m_2$ and define

$$X_1 = \{x \in \mathbb{R}^4;\ x_1 = x_3,\ x_2 = x_4\},\ U_1 = \{u \in \mathbb{R}^2;\ u_1 = u_2\},\ Y_1 = \{y \in \mathbb{R}^2;\ y_1 = y_2\}$$

then it is easy to verify the conditions (13) are satisfied. This means that if the initial state lies in X_1 and the same control is applied to both spring ends then the distance between the two masses remains constant ($y_1(t) = y_2(t)$), so there is no actual interaction between the masses via the damper. The subsystem $\Sigma_1 = (A_1, B_1, C_1, 0)$ obtained by the restriction of A, C, and B to X_1 and U_1, respectively, describes simultaneously the motions of two "decoupled" mass spring systems. To obtain a matrix representation of A_1, B_1, C_1 we choose the basis vectors $z^1 = [1,0,1,0]^\top$, $z^2 = [0,1,0,1]^\top$, $z^3 = [1,0,0,0]^\top$, $z^4 = [0,1,0,0]^\top$ in X, $v^1 = [1,1]^\top$, $v^2 = [1,0]^\top$ in U and $w^1 = [1,1]^\top$, $w^2 = [1,0]^\top$ in Y. With respect to these bases A, B, C have the following matrix representations (see (9))

$$A \sim \begin{bmatrix} 0 & 1 & 0 & 0 \\ -k_1/m_1 & 0 & 0 & c/m_2 \\ 0 & 0 & 0 & 1 \\ 0 & 0 & -k_1/m_1 & -(c/m_1 + c/m_2) \end{bmatrix}, \quad B \sim \begin{bmatrix} 0 & 0 \\ k_1/m_1 & 0 \\ 0 & 0 \\ 0 & k_1/m_1 \end{bmatrix},$$
$$C \sim \begin{bmatrix} 1 & 0 & 0 & 0 \\ 0 & 0 & 1 & 0 \end{bmatrix}. \tag{17}$$

X_1 is spanned by z^1, z^2, U_1 by v^1 and Y_1 by w^1. Hence

$$A_1 = \begin{bmatrix} 0 & 1 \\ -k_1/m_1 & 0 \end{bmatrix}, \quad B_1 = \begin{bmatrix} 0 \\ k_1/m_1 \end{bmatrix}, \quad C_1 = [1\ ,\ 0].$$

Equivalently the subsystem can be described by the following second order differential equation

$$\ddot{y}^1(t) + (k_1/m_1)y^1(t) = (k_1/m_1)\,u^1(t).$$

Since $k_1/m_1 = k_2/m_2$, the above equation is equivalent to

$$(m_1 + m_2)\ddot{y}^1(t) + (k_1 + k_2)(y^1(t) - u^1(t)) = 0$$

and hence describes the motion of the "aggregated" mass-spring system shown in Figure 2.4.2.

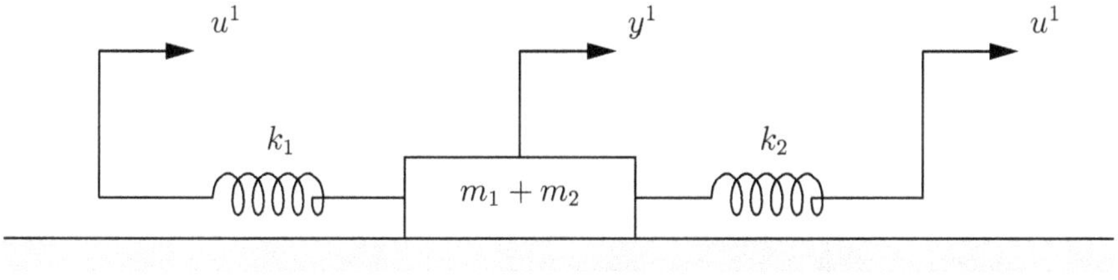

Figure 2.4.2: Aggregated mass-spring system

Now let us consider the quotient system $\hat{\Sigma} := \Sigma/\Sigma_1$. We choose as bases for the quotient spaces $\hat{X}$, $\hat{U}$, $\hat{Y}$ the vectors (equivalence classes) $z^3 + X_1$, $z^4 + X_1$ and $v^2 + U_1$, $w^2 + Y_1$ respectively. Then by (17) the matrix representation of the quotient system $\hat{\Sigma} = \Sigma/\Sigma_1$ is given by

$$\hat{A} = \begin{bmatrix} 0 & 1 \\ -k_1/m_1 & -(c/m_1 + c/m_2) \end{bmatrix}, \quad \hat{B} = \begin{bmatrix} 0 \\ k_1/m_1 \end{bmatrix}, \quad \hat{C} = [1\ ,\ 0].$$

This yields a second order differential equation

$$m_1 \frac{d^2\hat{y}}{dt^2} + c\frac{m_1 + m_2}{m_2}\frac{d\hat{y}}{dt} + k_1\hat{y} = k_1\hat{u}. \tag{18}$$

It can be interpreted as follows: Suppose that for given initial conditions $y_1(0) = \hat{y}(0)$, $\dot{y}_1(0) = \dot{\hat{y}}(0)$ and a control $u_1(\cdot) = \hat{u}(\cdot)$, the initial values $y_2(0)$, $\dot{y}_2(0)$ for the second mass and control $u_2(\cdot)$ are chosen in such a way that the centre of mass remains at equilibrium. Then $m_1y_1(t) + m_2y_2(t) \equiv 0$ and substituting for $y_2(t)$ in the first equation of (16) we see that $y_1(\cdot)$ satisfies (18). Hence (18) describes the equation of motion of the first mass under the assumption that the centre of mass of Σ remains at rest. $\hat{\Sigma}$ admits an analogous interpretation with respect to the second mass. □

Note that if a subsystem Σ_1 of Σ and the corresponding quotient system $\hat{\Sigma} = \Sigma/\Sigma_1$ are known, it is not, in general possible to reconstruct the complete system Σ from them. Whilst A_{11} and A_{22} can be reconstructed from Σ_1 and $\hat{\Sigma}$ respectively, this is not the case for A_{12}. Hence Σ is not, in general, the direct sum of Σ_1 and $\hat{\Sigma}$ in the sense of the following definition.

Definition 2.4.8. (Direct sum). Let $\Sigma_i = (A_i, B_i, C_i, D_i)$ be systems with state space X_i, input space U_i and output space Y_i, $i \in \underline{N}$. The direct sum is a system (A, B, C, D) with state space X, input space U and output space Y given by

$$X = \prod_{i=1}^{N} X_i, \quad U = \prod_{i=1}^{N} U_i, \quad Y = \prod_{i=1}^{N} Y_i,$$
$$A = \bigoplus_{i=1}^{N} A_i, \quad B = \bigoplus_{i=1}^{N} B_i, \quad C = \bigoplus_{i=1}^{N} C_i, \quad D = \bigoplus_{i=1}^{N} D_i\,. \tag{19}$$

2.4.2 Composite Systems

The direct sum is a trivial way of building a composite system from a collection of systems. In fact it is just a collection of uncoupled systems. Hence in a direct sum each subsystem can be studied independently of the other subsystems. This is not the case if the subsystems Σ_i, $i \in \underline{N}$ are *interconnected* within the composite system $\overline{\Sigma}$. In many areas of application one encounters large scale systems which are made up of complex arrays of many interconnected subsystems. The purpose of this subsection is to provide a general framework for describing such systems where $\Sigma_i = (A_i, B_i, C_i, D_i)$, $i \in \underline{N}$. In order to do this we introduce various interconnection schemes and also the general form of a composite time–invariant linear system.
The following four examples illustrate the most important ways that a system can be interconnected with other systems or with itself (feedback). We denote by $\overline{U}$, $\overline{X}$, $\overline{Y}$ the input, state and output spaces of the composite system. When $\overline{X}$ is not defined explicitly it is understood that $\overline{X} = X_1 \times X_2$.

Example 2.4.9. (Series connection). A *series connection* of Σ_1 and Σ_2 is obtained when the input of Σ_2 is connected to the output of Σ_1. Thus it requires $U_2 = Y_1$. The input and output of the composite system is u^1 and y^2 respectively and $\overline{U} = U_1$, $\overline{Y} = Y_2$. The composite system $\overline{\Sigma} = (\overline{A}, \overline{B}, \overline{C}, \overline{D})$ is given by

$$\overline{A} = \begin{bmatrix} A_1 & 0 \\ B_2C_1 & A_2 \end{bmatrix}, \quad \overline{B} = \begin{bmatrix} B_1 \\ 0 \end{bmatrix}, \quad \overline{C} = [0\ ,\ C_2], \quad \overline{D} = D_2D_1. \tag{20}$$

The transfer matrix of the series connection is simply the product of the individual transfer matrices

$$\overline{G}(s) = \overline{C}(sI - \overline{A})^{-1}\overline{B} + \overline{D} = G_2(s)G_1(s) \tag{21}$$

where $G_i(s)$, $i = 1, 2$ are the transfer function matrices of the subsystems connected in series. Thus multiplication of transfer matrices corresponds to series connection of the respective systems.

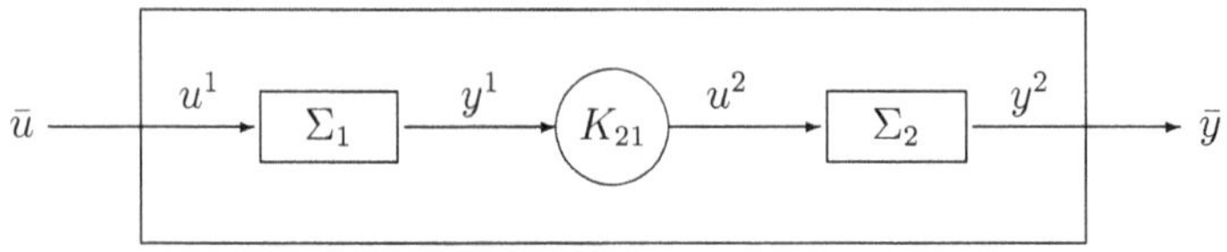

Figure 2.4.3: Series connection

If a direct coupling of the two subsystems in series is not possible because $Y_1 \neq U_2$ an *adapter* or *coupling matrix* $K_{21} : Y_1 \to U_2$ can be used to connect y^1 with u^2, viz. $u^2 = K_{21}y^1$, see Figure 2.4.3. The corresponding state space equation and transfer matrix are obtained by replacing C_1 by $K_{21}C_1$ in (20) and $G_1(s)$ by $K_{21}G_1(s)$ in (21). □

Example 2.4.10. (Parallel connection). A *parallel connection* of Σ_1, Σ_2 is obtained if both systems have the same input, and the output of the composite system is the sum of the individual outputs (see the left hand figure in Figure 2.4.4).
In this case $\overline{U} = U_1 = U_2$, $\overline{Y} = Y_1 = Y_2$, $u^1 = u^2 = \bar{u}$ and $\overline{\Sigma}$ is described by

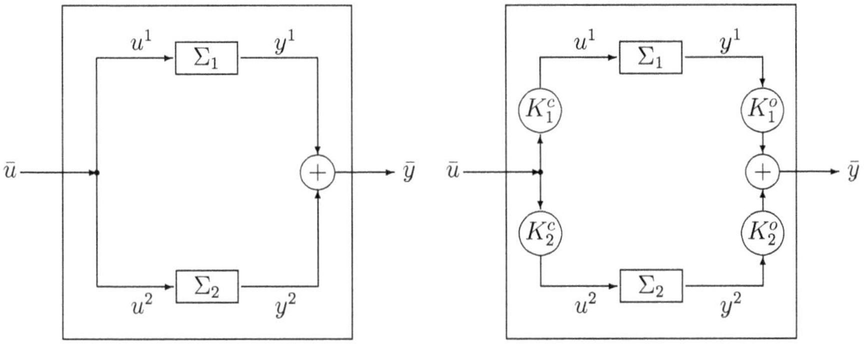

Figure 2.4.4: Parallel connection and extended parallel connection

$$\overline{A} = \begin{bmatrix} A_1 & 0 \\ 0 & A_2 \end{bmatrix}, \quad \overline{B} = \begin{bmatrix} B_1 \\ B_2 \end{bmatrix}, \quad \overline{C} = [C_1\,,\,C_2], \quad \overline{D} = D_1 + D_2. \tag{22}$$

The transfer matrix of Σ is given by $\overline{G}(s) = G_1(s) + G_2(s)$. Thus the addition of transfer matrices corresponds to the parallel connection of the respective systems. If Σ_1, Σ_2 do not have the same input and output spaces an extended parallel connection can be obtained by setting

$$u^1 = K_1^c\bar{u}, \quad u^2 = K_2^c\bar{u}, \quad \bar{y} = K_1^o y^1 + K_2^o y^2$$

where $K_i^c : \overline{U} \to U_i$ and $K_i^o : Y_i \to \overline{Y}$, $i = 1, 2$, are called *input and output coupling matrices* (see the right hand figure in Figure 2.4.4). The corresponding system equation and transfer matrix are easily determined. □

The third basic way of coupling two systems is via feedback. This configuration is of fundamental importance in control where a central problem is that of producing a desired input-output behaviour of a given system by feedback.

Example 2.4.11. (Dynamic output feedback). The feedback interconnection of

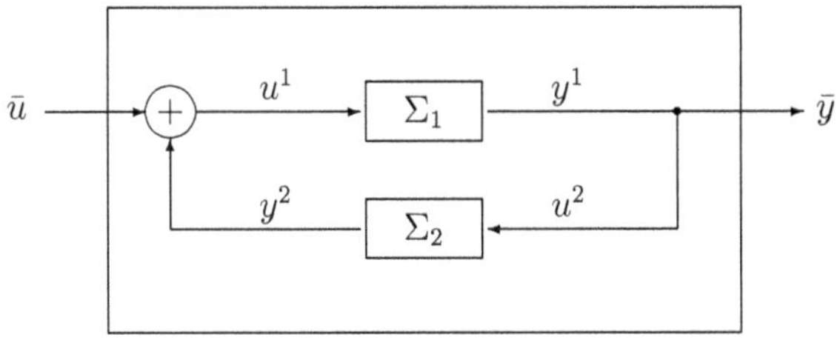

Figure 2.4.5: Dynamic output feedback

two systems Σ_1, Σ_2 connects the output of Σ_1 to the input of Σ_2 and the output of Σ_2 to the input of Σ_1 according to the formulas $u^2 = y^1$ and $u^1 = y^2 + \bar{u}$ (see Figure 2.4.5). $\bar{u}$ is considered as the input of the feedback system $\overline{\Sigma}$ and $\bar{y} = y^1$ as the output. Clearly, this interconnection presupposes that $Y_1 = U_2$ and $Y_2 = U_1$ (otherwise coupling matrices K_{ij}, $i, j = 1, 2$, $i \neq j$ must be used). But in contrast with the previous interconnections this compatibility condition is not sufficient for the feedback system to be well-defined. The couplings $u^2 = y^1$ and $u^1 = y^2 + \bar{u}$ lead to the feedback equations

$$u^1 = y^2 + \bar{u} = C_2 x^2 + D_2[C_1 x^1 + D_1 u^1] + \bar{u}$$

and similarly

$$u^2 = y^1 = C_1 x^1 + D_1[C_2 x^2 + D_2 u^2 + \bar{u}].$$

These equations can be solved for u^1 and u^2 if and only if *the matrices* $I_{U_1} - D_2 D_1$ *or, equivalently,* $I_{U_2} - D_1 D_2$ *are invertible.* This is the so-called *well-posedness condition* for the feedback configuration. If it is satisfied the feedback system is well defined and denoted by $\overline{\Sigma} = \Sigma_1 \Box \Sigma_2$. It has the input space $\overline{U} = U_1$, the output space $\overline{Y} = Y_1$, the state space $\overline{X} = X_1 \times X_2$ and its system equations are given by the data

$$\begin{aligned} \overline{A} &= \begin{bmatrix} A_1 + B_1\Pi_{21}D_2C_1 & B_1\Pi_{21}C_2 \\ B_2\Pi_{12}C_1 & A_2 + B_2\Pi_{12}D_1C_2 \end{bmatrix}, \quad \overline{B} = \begin{bmatrix} B_1\Pi_{21} \\ B_2\Pi_{12}D_1 \end{bmatrix}, \\ \overline{C} &= [C_1 + D_1\Pi_{21}D_2C_1 \quad D_1\Pi_{21}C_2], \qquad \overline{D} = D_1\Pi_{21} \end{aligned} \tag{23}$$

where $\Pi_{ij} = (I - D_i D_j)^{-1}$. The Laplace transform of the input and output signals are related by the feedback equations,

$$\hat{y}^1(s) = G_1(s)\hat{u}^1(s), \quad \hat{u}^1(s) = G_2(s)\hat{y}^1(s) + \hat{\bar{u}}$$

Assuming the well-posedness condition it is easily seen that $(I - G_2(s)G_1(s))$ is invertible (as a rational matrix). Hence the transfer matrix of the feedback system $\Sigma_1 \Box \Sigma_2$ is

$$\overline{G}(s) = G_1(s)(I - G_2(s)G_1(s))^{-1}. \tag{24}$$

Many variants of the above feedback configuration are used in control theory. For example it is sometimes necessary to distinguish between the to be controlled output variables of Σ_1 (which are taken as the output of $\overline{\Sigma}$) and the measured output variables of Σ_1 which can be used for feedback; or some of the input variables of Σ_1 cannot be controlled and only the remaining ones are available for feeding back the output of Σ_2. These configurations can be described by adding input and output coupling matrices to the above feedback configuration. □

A given system can also be coupled to itself by constant feedback couplings. Again, the possibilities of changing the system dynamics by constant linear feedback is a fundamental question in linear control theory.

Example 2.4.12. (Static state and output feedback). *Static state feedback* connects

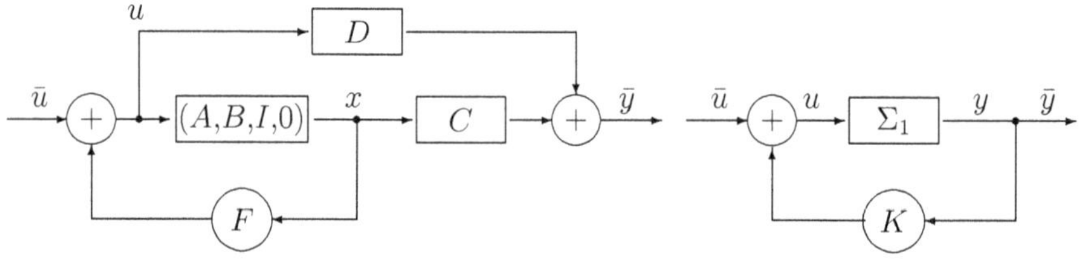

Figure 2.4.6: Static state and output feedback

the *state* of a system $\Sigma = (A, B, C, D)$ with the input of the same system via an affine transformation $u = Fx + \bar{u}$ (see Figure 2.4.6). In this case $\overline{U} = U$, $\overline{Y} = Y$, $\overline{X} = X$ and the feedback system is described by

$$\overline{A} = A + BF, \quad \overline{B} = B, \quad \overline{C} = C + DF, \quad \overline{D} = D. \tag{25a}$$

In the special case when $F = KC$, $D = 0$ we obtain *static output feedback* where the *output* of the system Σ is connected with the input of the same system via $u = Ky + \bar{u}$ (see Figure 2.4.6), so

$$\overline{A} = A + BKC, \quad \overline{B} = B, \quad \overline{C} = C. \tag{25b}$$

The respective transfer matrices of these feedback configurations are

$$\begin{aligned} \overline{G}(s) &= [C + DF](sI - A - BF)^{-1}B + D, \\ \overline{G}(s) &= C(sI - A - BKC)^{-1}B = G(s)(I - KG(s))^{-1}, \end{aligned} \tag{26}$$

where $G(s)$ is the transfer matrix of Σ. □

Let us now proceed to describe the general form of a composite system obtained by connecting finitely many subsystems $\Sigma_i = (A_i, B_i, C_i, D_i)$, $i \in \underline{N}$ via constant coupling matrices. All of the above examples (for $N = 2$) are special cases of the connection scheme shown in Figure 2.4.7.

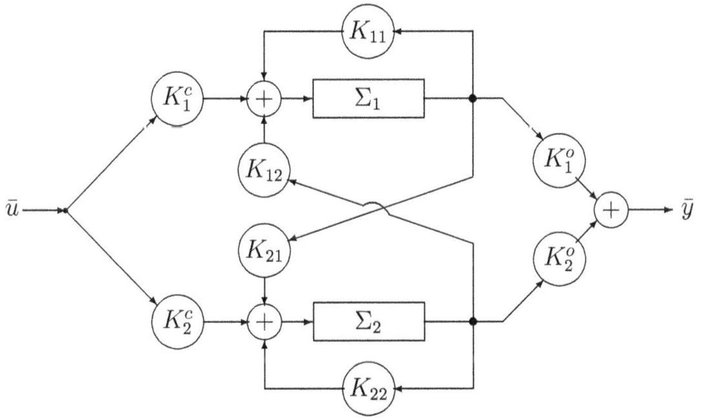

Figure 2.4.7: General composite system of two subsystems

For arbitrary $N \geq 1$ a general linear, time–invariant connection scheme for the systems $\Sigma_1, \Sigma_2, \ldots, \Sigma_N$ is given by the $(N+1)^2$ matrices

$$K_i^c : \overline{U} \to U_i, \quad K_{ij} : Y_j \to U_i, \quad K_i^o : Y_i \to \overline{Y}, \quad D_o : \overline{U} \to \overline{Y}$$

($i, j \in \underline{N}$) where $\overline{U} = \prod_{i=1}^N U_i$ is the input space, $\overline{Y} = \prod_{i=1}^N Y_i$ the output space and $\overline{X} = \prod_{i=1}^N X_i$ is the state space of the composite system.
The compound matrix $K = (K_{ij})_{i,j \in \underline{N}}$ is called the *matrix of interconnections* between the subsystems. If $K_{ij} \neq 0$ the output of Σ_j exercises an influence on the input of Σ_i. The matrices K_{ii} describe static feedback loops for the subsystems Σ_i. Since, in general, not all subsystems are connected with every other subsystem,

many of the matrices (K_{ij}) $i, j \in \underline{N}$ will be zero matrices.
The matrices

$$K^c = \begin{bmatrix} K_1^c \\ \vdots \\ K_N^c \end{bmatrix}, \quad K^o = [K_1^o \ldots K_N^o]$$

are called matrices of *input couplings* and *output couplings* respectively. K^o contains the matrix coefficients which specify those linear combinations of the subsystem's outputs which yield the output of the composite system. D^o describes the direct input-output coupling of $\overline{\Sigma}$,

$$\bar{y} = \sum_{i=1}^{N} K_i^o y^i + D^o \bar{u} = \sum_{i=1}^{N} K_i^o C_i x^i + \sum_{i=1}^{N} K_i^o D_i u^i + D_o \bar{u}. \tag{27}$$

The inputs u^i of Σ_i are obtained by adding the terms $K_{ij} y^j$ from each of the subsystems plus the terms $K_i^c \bar{u}$ from the external control $\bar{u}$,

$$u^i = \sum_{j=1}^{N} K_{ij} y^j + K_i^c \bar{u} = \sum_{j=1}^{N} K_{ij} C_j x^j + \sum_{j=1}^{N} K_{ij} D_j u^j + K_i^c \bar{u}. \tag{28}$$

In the general case when input-output couplings are present (28) is an implicit formula for the u^i's $i \in \underline{N}$ in terms of $x^1, \ldots, x^N$ and $\bar{u}$. So not every connection scheme is feasible and there will only exist unique solutions if the following *well-posedness condition* is satisfied by K

$$\det(I - K \operatorname{diag}(D_1, \ldots, D_N)) \neq 0. \tag{29}$$

Let (A, B, C, D) be the direct sum of the systems $\Sigma_1, \ldots, \Sigma_N$ as described by equation (19), then if (29) holds the unique solution of (28) can be expressed in the form

$$u = (I - KD)^{-1}(KCx + K^c \bar{u}) \tag{30}$$

where $u \in \overline{U}$, $x \in \overline{X}$ are the vectors with components u^i and x^i. Substituting (30) in the system equations of Σ_i and in (27) we see that the composite system is described by

$$\begin{aligned} \overline{A} &= A + B(I - KD)^{-1} KC, & \overline{B} &= BK^c \\ \overline{C} &= K^o C + K^o D (I - KD)^{-1} KC, & \overline{D} &= K^o D (I - KD)^{-1} K^c + D^o. \end{aligned} \tag{31}$$

If there are no direct input-output couplings in the subsystems, condition (29) is trivially satisfied and (31) simplifies to

$$\begin{aligned} \overline{A} &= \begin{bmatrix} A_1 + B_1 K_{11} C_1 & B_1 K_{12} C_2 & \ldots & B_1 K_{1N} C_N \\ B_2 K_{21} C_1 & A_2 + B_2 K_{22} C_2 & \ldots & B_2 K_{2N} C_N \\ \vdots & & & \vdots \\ B_N K_{N1} C_1 & & \ldots & A_N + B_N K_{NN} C_N \end{bmatrix} \\ \overline{B} &= BK^c, \quad \overline{C} = K^o C, \quad \overline{D} = K^o D K^c + D^o. \end{aligned} \tag{32}$$

The interconnection structure of a composite system $\overline{\Sigma}$ can be represented by a directed graph with $(N+2)$ nodes denoted by $\overline{U}$, $\overline{Y}$ and Σ_i $(i \in \underline{N})$, directed edges $\Sigma_j \to \Sigma_i$, $\overline{U} \to \Sigma_i$, $\Sigma_i \to \overline{Y}$ for those i, $j \in \underline{N}$ with $K_{ij} \neq 0$, $K_i^c \neq 0$, $K_i^o \neq 0$.

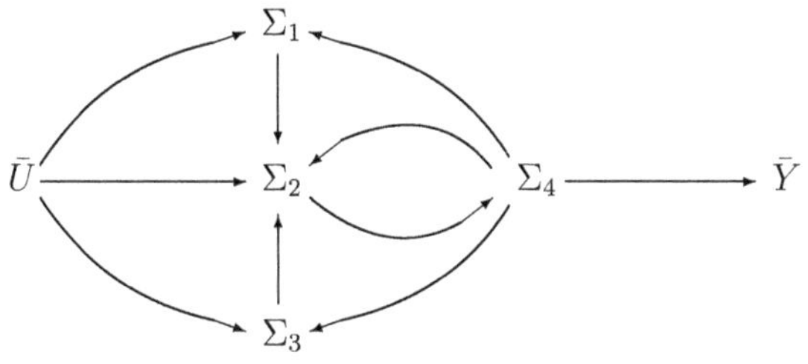

Figure 2.4.8: Directed graph

Example 2.4.13. If $N = 4$ and the interconnection structure is given by the matrices

$$K = \begin{bmatrix} 0 & 0 & 0 & K_{14} \\ K_{21} & 0 & K_{23} & K_{24} \\ 0 & 0 & 0 & K_{34} \\ 0 & K_{42} & 0 & 0 \end{bmatrix}, \quad K^c = \begin{bmatrix} K_1^c \\ K_2^c \\ K_3^c \\ 0 \end{bmatrix}, \quad K^o = \begin{bmatrix} 0 \\ 0 \\ 0 \\ K_4^o \end{bmatrix}$$

the representation as a directed graph is shown in Figure 2.4.8. □

A more detailed picture of the input and output couplings is obtained if the vertices $\bar{u}_1^i, \ldots, \bar{u}_m^i, \bar{y}_1^i, \ldots, \bar{y}_p^i$ are introduced. An edge $\bar{u}_j^i \to \Sigma_i$ or $\Sigma_i \to \bar{y}_k^i$ is drawn whenever the jth column of K_i^c or the kth row K_i^o is non-zero.

In particular any system $\Sigma = (A, B, C, D)$ with input space $\overline{U} = \mathbb{K}^m$, state space $\overline{X} = \mathbb{K}^n$ and output space $\overline{Y} = \mathbb{K}^p$ can always be represented as an interconnection of *integrators* (resp. *unit delays* in the discrete time case)

$$\Sigma_i : \dot{x}_i = u_i, \quad y_i = x_i \quad (i = 1, \ldots, n). \tag{33}$$

Indeed if we define

$$K^c = B, \quad K = A, \quad K_c^o = C, \quad D^o = D \tag{34}$$

it is easy to verify that the resultant composite system is identical with Σ. The graph (with separate representation of each input and output channel) associated with the interconnection scheme (34) is called the *system graph* of Σ.

Example 2.4.14. The matrices describing the overhead crane of Example 1.3.4 have the following structure

$$A = \begin{bmatrix} 0 & 0 & 1 & 0 \\ 0 & 0 & 0 & 1 \\ 0 & a_{32} & 0 & 0 \\ 0 & a_{42} & 0 & 0 \end{bmatrix}, \quad B = \begin{bmatrix} 0 \\ 0 \\ b_3 \\ b_4 \end{bmatrix}, \quad C = \begin{bmatrix} 1 & 0 & 0 & 0 \\ 0 & 1 & 0 & 0 \end{bmatrix}, \quad D = \begin{bmatrix} 0 \\ 0 \end{bmatrix}$$

where a_{32}, a_{42}, b_3, b_4 are determined by physical parameters. The corresponding system graph is shown in Figure 2.4.9.

□

Note that the system graph will in general be altered by a similarity transformation. Hence system graphs are only meaningful for the analysis of a system if the quantities and subsystems represented by the vertices correspond to real physical parts of the system, and if their interconnection is of importance in the overall analysis of the system. When this is the case, only system isomorphisms which do not destroy the structure of the graph can be allowed.

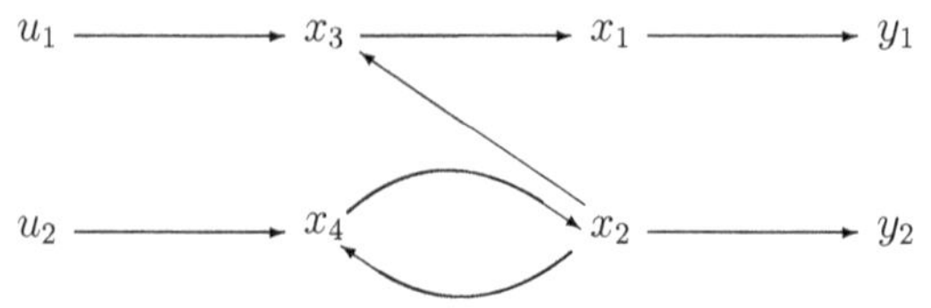

Figure 2.4.9: System graph for the overhead crane

2.4.3 Exercises

1. Consider the RLC-circuit described in Ex. 2.9

(i) Derive state space models of this circuit with state variables

(a) x_1 = current i through inductor, x_2 = charge q of capacitor

(b) $x_1 = i$, x_2 = voltage v across inductor

(c) x_1 = voltage across capacitor, $x_2 = i$.

(ii) Show that the linear systems (A, B, C) obtained in (i) are all similar.

(iii) Let $R = 3$, $L = 1$, $C = 0.5$. Define a state vector $x = [x_1, x_2]^\top$ for which the corresponding system matrix A is diagonal.

(iv) For which triples (R, L, C) are the resulting dynamical systems of (i) similar to the one defined in (iii).

2. Let

$$A = \begin{bmatrix} 0 & 1 \\ -1 & -2 \end{bmatrix}, \quad B = \begin{bmatrix} 0 \\ 1 \end{bmatrix}, \quad C = [0,\ 1].$$

Examine whether or not the triple (A, B, C) is similar or isomorphic to $(\hat{A}, \hat{B}, \hat{C})$ where

(i) $$\hat{A} = \begin{bmatrix} 1 & -2 \\ 0 & -1 \end{bmatrix}, \quad \hat{B} = \begin{bmatrix} 0 \\ 1 \end{bmatrix}, \quad \hat{C} = [1,\ 1].$$

(ii) $$\hat{A} = \begin{bmatrix} -1 & 0 \\ 0 & -1 \end{bmatrix}, \quad \hat{B} = \begin{bmatrix} -1 \\ 1 \end{bmatrix}, \quad \hat{C} = [-1,\ 0].$$

(iii) $$\hat{A} = \begin{bmatrix} -1 & 1 \\ 0 & -1 \end{bmatrix}, \quad \hat{B} = \begin{bmatrix} -1 \\ 1 \end{bmatrix}, \quad \hat{C} = [-1,\ 0].$$

(iv) $$\hat{A} = \begin{bmatrix} 0 & 1 \\ -1 & -2 \end{bmatrix}, \quad \hat{B} = \begin{bmatrix} 1 \\ 0 \end{bmatrix}, \quad \hat{C} = [2,\ 4].$$

3. Let (A, B, C) and $(\hat{A}, \hat{B}, \hat{C})$ be similar.

(i) Show by means of an example that $A = \hat{A}$, $B = \hat{B}$ does not imply $C = \hat{C}$.

(ii) Specify conditions for A, B under which $A = \hat{A}$, $B = \hat{B}$ does imply $C = \hat{C}$.

4. Show that the system $\Sigma_1 = (A_1, B_1, C_1)$ where

$$A_1 = \begin{bmatrix} 0 & 1 \\ 0 & 1 \end{bmatrix}, \quad B_1 = \begin{bmatrix} 0 & -2 \\ 0 & -3 \end{bmatrix}, \quad C_1 = [0,\ 1]$$

is a subsystem of $\Sigma = (A, B, C)$ where

$$A = \begin{bmatrix} 0 & 0 & 0 \\ -1 & 2 & 4 \\ 1 & -1 & -3 \end{bmatrix}, \quad B = \begin{bmatrix} 0 & 1 \\ 1 & -2 \\ -1 & 1 \end{bmatrix}, \quad C = \begin{bmatrix} 2 & -1 & 1 \\ -1 & 1 & 0 \end{bmatrix}.$$

Determine the quotient system Σ/Σ_1.

5. Let $\Sigma_i = (A_i, B_i, C_i, D_i)$ be two time-invariant linear systems. Prove

(i) if Σ_1 is a subsystem of Σ_2, then $\sigma(A_1) \subset \sigma(A_2)$,

(ii) if Σ_2 is a quotient system of Σ_1, then $\sigma(A_2) \subset \sigma(A_1)$.

6. Let (R, T, S) be a morphism from Σ to $\hat{\Sigma}$. Specify conditions under which, given Σ, the linear maps R, T, S uniquely determine the system $\hat{\Sigma} = (\hat{A}, \hat{B}, \hat{C}, \hat{D})$.

7. Extend the set $\mathrm{Mor}\,(\Sigma, \hat{\Sigma})$ by allowing in addition state feedback transformations $F : X \to \hat{U}$. Find a counterpart of (6) for these feedback morphisms $(R, T, S, F) : \Sigma \to \hat{\Sigma}$. Define a composition rule $(R, T, S, F) \circ (\overline{R}, \overline{T}, \overline{S}, \overline{F})$ and determine necessary and sufficient conditions for $(R, T, S, F) \in \mathrm{Mor}_{\mathrm{feedback}}(\Sigma, \hat{\Sigma})$ to be a *feedback isomorphism.*

8. Draw the graphs of the linear systems in

(i) Example 2.1.27,

(ii) Exercises 1.1, 1.2, 1.3, 1.4, 1.9(i) and 2.4

2.4.4 Notes and References

More details concerning categories of time–invariant linear systems can be found in *Prätzel-Wolters* (1983) [419]. System morphisms in the context of abstract realization theory are studied in *Sontag* (1990) [472].
The field of large scale systems and decentralized control has generated considerable interest amongst control theorists, see the special issue of *IEEE Transactions Automatic Control* (1978) [24], *Siljak* (1991)[465], the collection of papers edited by Leondes [339], [338] and the informative Control Handbook edited by *Levine* (1996) [342].

2.5 Sampling and Approximation: Relations Between Continuous and Discrete Time Systems

The practical implementation of a particular control scheme on a physical plant often involves both continuous and discrete time signals. Such systems are called *hybrid-time* or *sampled-data* systems and can arise, for example, when a digital computer is used to control a continuous time process. In these systems it is necessary to have interfaces which convert (continuous time) *analog* signals into (discrete time) *digital* signals (*A/D-converter*, *sampler*) and *digital* signals into *analog* ones (*D/A-converter, hold*).

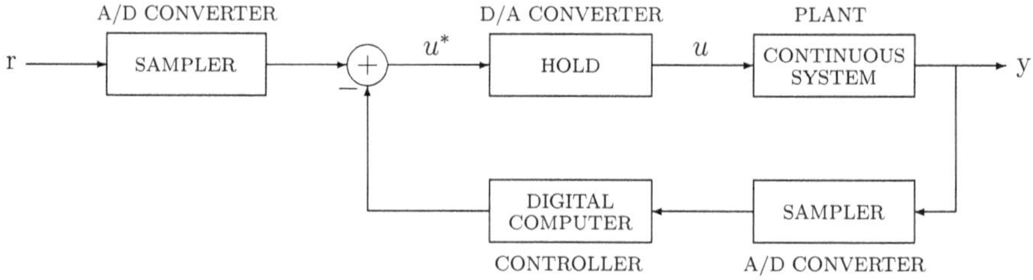

Figure 2.5.1: Digital control of a continuous time plant

The use of a digital computer as a controller implies two types of discretizations a) discretization of *time* and b) discretization of the system *parameters* and *variables* (quantization). Quantization effects arise because real numbers have to be stored and processed using a finite number of digits. The problem of how the various round off errors interact and propagate in a given feedback algorithm needs to be investigated by methods which have been developed in the field of Numerical Analysis and are outside the scope of this book (see *Notes and References*).

In this section we first examine the relationship between discrete and continuous time *signals* and prove a sampling theorem which specifies conditions under which a continuous time signal can be completely restored from its sampled values. Then we go on to examine the relationship between discrete and continuous time *systems*, neglecting quantization errors. We begin by describing the *sampling* of a differentiable system, i.e. the conversion of a continuous time system into a discrete time one by connecting it in series with a hold element and a sampler. We then discuss in some detail the use of numerical integration methods for the *approximation* of continuous time systems by discrete time systems. This is obviously of great importance for the digital simulation of continuous time processes and is relevant in many areas of control and communication where analog devices are replaced by "equivalent" digital devices. We use Euler's method as the simplest numerical integration scheme to explain some basic concepts. Some higher order methods (single and multistep) are also briefly described and their convergence properties are illustrated by an example with strong oscillations in the control. Whereas Numerical Analysis usually considers the approximation of single trajectories, for system theoretic purposes it is more important to consider the approximation of *differentiable systems* by *discrete time systems* with controls and initial states which are not fixed. At the end of the section we point out some specific difficulties related to this problem.

2.5.1 A/D- and D/A-Conversion of Signals

A *sampler* associates with each continuous time signal $f(\cdot)$ on $[0,\infty)$ a sequence $f^* = (f(t_k))_{k\in\mathbb{N}}$ of values of f at given sampling instants $t_k \in [0,\infty)$, $k \in \mathbb{N}$. We will find it useful to represent this discrete time signal by a series of impulses

$$f^*(t) = \sum_{k\in\mathbb{N}} f(t_k)\delta(t-t_k) = \sum_{k\in\mathbb{N}} f(t_k)\delta_{t_k}(t) \tag{1}$$

where $\delta(t-t_k) = \delta_{t_k}(t)$ is the Dirac impulse at t_k. Graphically (1) is represented by a sequence of vertical arrows of lengths $f(t_k)$ symbolizing the impulse $f(t_k)\delta(t-t_k)$ (see Figure 2.5.3). Usually equidistant sequences $t_k = k\tau$, $k \in \mathbb{N}$ are chosen. In this case

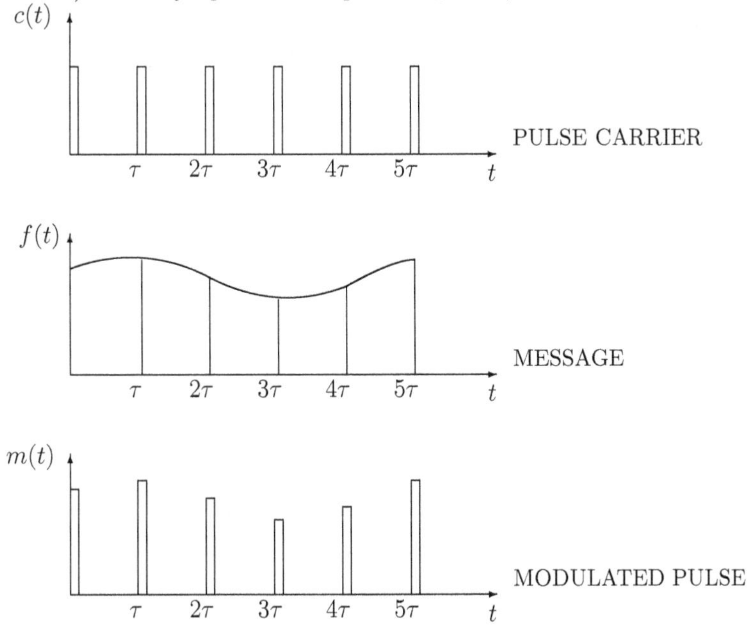

Figure 2.5.2: Pulse amplitude modulation

$\tau > 0$ is called the *sampling period*, $2\pi/\tau$ the *sample rate*. In communication theory frequent use is made of pulse amplitude modulation (see *Notes and References*). A continuous time signal ("message") $f(\cdot)$ modulates the amplitude of a unit pulse train $c(\cdot)$ of period τ ("the carrier") to give a modulated pulse $m(t) = f(t)c(t)$ (see Figure 2.5.2). The representation (1) corresponds to an *impulse modulation model* for the sampler where the pulses are idealized to have "infinitely small" width and the carrier signal is a train of impulses $\sum_{k\in\mathbb{N}} \delta_{k\tau}(\cdot)$ which are modulated by f to yield the sampled signal $f^*(\cdot) = \sum_{k\in\mathbb{N}} f(k\tau)\delta_{k\tau}(\cdot)$.

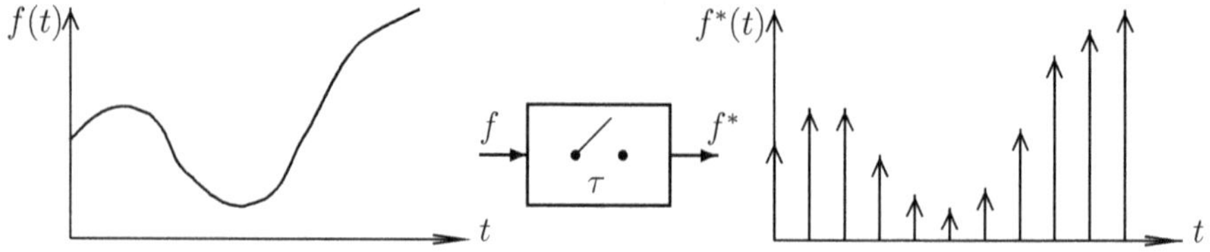

Figure 2.5.3: Ideal sampler with sampling period τ

A *hold* is a device which transforms a series of impulses into a continuous time signal $f(\cdot)$ by a given extrapolation formula. If a kth-order polynomial is used for the extrapolation it is called a kth*–order hold.* The simplest and most widely used is the zero-order hold H_τ^0 defined by $f(t) = f(k\tau)$ if $k\tau \le t < (k+1)\tau$ (see Figure 2.5.4). A first order hold H_τ^1 (see Figure 2.5.5) associates with f^* the piecewise linear signal

$$f(t) = f(k\tau) + [f(k\tau) - f((k-1)\tau)](t - k\tau)/\tau, \quad k\tau \le t < (k+1)\tau.$$

The continuous time signals produced by the zero and first order holds are both

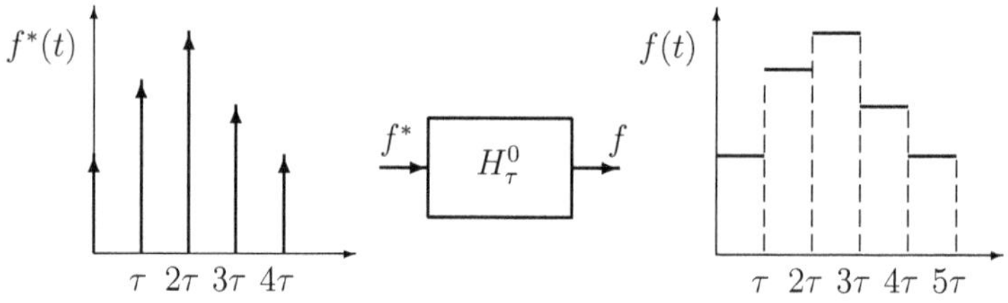

Figure 2.5.4: Zero order hold

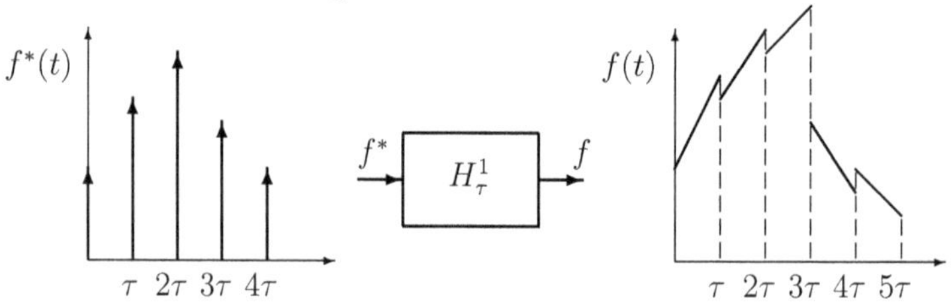

Figure 2.5.5: First order hold

discontinuous. A continuous signal may be produced by the following formula which yields a piecewise linear *interpolation* of the discrete time signal, i.e. a function $f(t)$ whose values at the sampling times $k\tau$ coincide with those of the discrete time signal,

$$f(t) = f(k\tau) + [f((k+1)\tau) - f(k\tau)](t - k\tau)/\tau, \quad k\tau \le t < (k+1)\tau. \qquad (2)$$

Unfortunately, this formula is not implementable since it requires the knowledge of $f((k+1)\tau)$ at time $t < (k+1)\tau$: the operator $f^* \mapsto f$ defined by (2) is not causal. To obtain causality it is necessary to introduce a delay τ, i.e.

$$f(t) = f((k-1)\tau) + [f(k\tau) - f((k-1)\tau)](t - k\tau)/\tau, \quad k\tau \le t < (k+1)\tau. \qquad (3)$$

This is a *delayed first order interpolator* since its value at the sampling time $k\tau$ coincides with the value of the discrete time signal at the previous sampling time $(k-1)\tau$.

Remark 2.5.1. Representing discrete time signals $(f(k\tau))_{k\in\mathbb{N}}$ as series of impulses $f^* = \sum_{k\in\mathbb{N}} f(k\tau)\delta_{k\tau}$ allows us to describe hold elements via convolutions. Let $h:\mathbb{R}\to\mathbb{R}$ be piecewise continuous and continuous on $\mathbb{R}\backslash\mathbb{N}\tau$. and zero on $(-\infty, 0)$. Then the convolution $f^* * h$ is almost everywhere defined and given by

$$(f^* * h)(t) = \sum_{k\in\mathbb{N}} f(k\tau)(\delta_{k\tau} * h)(t) = \sum_{k=0}^{n} f(k\tau)h(t-k\tau), \quad n\tau < t < (n+1)\tau.$$

It is easily verified (Ex. 1) that the zero order hold is described by the convolution kernel $h = 1_{[0,\tau)}$ and the first order hold is described by the piecewise linear convolution kernel

$$h:\ t \mapsto (1+t/\tau)1_{[0,\tau)}(t) + (1-t/\tau)1_{[\tau,2\tau)}(t).$$

□

An A/D-converter is a system which accepts continuous time signals (for example, voltages) and produces a sequence of binary numbers which represent the signal values at the sampling times. In practice it contains a sampler-and-hold element and the converter proper. The function of the sampler-and-hold is to sample the signal and hold its value long enough to allow the converter to code it by binary numbers and store it in a register. If we neglect the difference between the sampled signal value and its binary code the A/D-converter may be modelled as a sampler. Similarly the D/A-converter may be modelled as a hold.

2.5.2 The Sampling Theorem

When a continuous time signal is sampled there will in general be a loss of information and one would expect that the amount lost depends in some way on the rate of sampling. An analysis of this problem is obviously important in communication systems where continuous time signals are encoded, processed, transmitted and stored in a digital fashion. It is also important for the analysis and design of automatic control systems where a continuous time plant is regulated by digital controls. It is not a priori clear that a theoretically well behaved continuous time control law will actually perform well when implemented (digitally) on a computer. The same applies to the converse design methodology where the continuous time system is first discretized and then discrete time controls are designed for the discrete model. In both cases essential information may be lost either by sampling the output (sampled observations) or by discretizing the system.
In the following we determine conditions under which a continuous time signal $v(\cdot):\mathbb{R}\to\mathbb{C}$ can be completely reconstructed from its sampled values $v(k\tau)$, $k\in\mathbb{Z}$. Mathematically this is an interpolation problem.
A sampler with sampling period $\tau > 0$ cannot distinguish between a signal $v: t\mapsto v(t)$ and a signal $w: t\mapsto v(t)+\sin(2\pi t/\tau)$. This indicates that it might be useful to represent the signal as a superposition of harmonic oscillations. In order to explain how this can be done we will need to use some results on Fourier series and Fourier

transforms (see Sections A.3 and A.4).
It is well known that, for every $l > 0$, the functions $\psi_k : \theta \mapsto e^{\imath k\pi\theta/l}$, $k \in \mathbb{Z}$ form an orthonormal basis of the Hilbert space $L^2(-l, l; \mathbb{C})$ provided with the inner product

$$\langle u(\cdot), w(\cdot)\rangle = \frac{1}{2l}\int_{-l}^{l} u(\theta)\,\overline{w(\theta)}\,d\theta, \quad u(\cdot), w(\cdot) \in L^2(-l, l; \mathbb{C}). \tag{4}$$

In other chapters of this book the Hilbert space $L^2(-l, l; \mathbb{C})$ is provided with an inner product without the scalar $1/2l$. We have chosen not to do this here since the use of the inner product in (4) simplifies some of the formulas.
Every function $u(\cdot) \in L^2(-l, l; \mathbb{C})$, $l > 0$ is the sum of its Fourier series in $L^2(-l, l; \mathbb{C})$ (see Example A.4.6 and Theorem A.4.7)

$$u(\cdot) = \sum_{k\in\mathbb{Z}} c_k\psi_k(\cdot); \;\; \psi_k(\theta) = e^{\imath k\pi\theta/l}, \;\theta \in [-l, l]; \;\; c_k = \frac{1}{2l}\int_{-l}^{l} u(\theta)e^{-\imath k\pi\theta/l}d\theta, \; k \in \mathbb{Z} \tag{5}$$

where the two-sided sequence $(c_k)_{k\in\mathbb{Z}}$ of Fourier coefficients $c_k = \langle u(\cdot), e^{\imath k\pi(\cdot)/l}\rangle$ belongs to $\ell^2(\mathbb{Z}; \mathbb{C})$. Note that the sequence of harmonic oscillations $c_k\psi_k(\cdot)$, though summable in $L^2(-l, l; \mathbb{C})$, is not necessarily pointwise summable for all $\theta \in [-l, l]$.
It follows from (5) that the restriction of every signal $v(\cdot) \in L^2(\mathbb{R}; \mathbb{C})$ to any finite interval $[-l, l]$, $l > 0$ is almost everywhere equal, *on this interval*, to the sum of its Fourier series in $L^2(-l, l; \mathbb{C})$. But it is not possible, in general, to represent the signal $v(\cdot)$ on the *whole real axis* as a superposition of a *countable* set of harmonic oscillations $t \mapsto e^{\imath k\pi t/l}$. However, under an additional condition $v(\cdot) \in L^1(\mathbb{R}; \mathbb{C})$ can be represented as an integral over *all* harmonic oscillations $e^{\imath\omega t}$, $\omega \in \mathbb{R}$, with the Fourier transform $\tilde{v}(\cdot)$ as a density function. More precisely the Fourier transform $\tilde{v}(\cdot) : \mathbb{R} \to \mathbb{C}$ is defined by

$$\tilde{v}(\omega) = (\mathcal{F}v)(\omega) = \int_{-\infty}^{\infty} v(t)e^{-\imath\omega t}dt, \quad \omega \in \mathbb{R}. \tag{6}$$

Although $\tilde{v}(\cdot)$ is continuous it may not be integrable on $\mathbb{R}$. But when this is the case then

$$v(t) = \frac{1}{2\pi}\int_{-\infty}^{\infty} \tilde{v}(\omega)e^{\imath\omega t}d\omega\,, \quad \text{a. e. } t \in \mathbb{R}, \tag{7}$$

see Theorem A.3.29. If $v(\cdot) \in L^1(\mathbb{R}; \mathbb{C})$ is also continuous then equality holds in (7) for *all* $t \in \mathbb{R}$.
In the sampling theorem we will be concerned with signals $v(\cdot)$ of finite energy, i.e. $v(\cdot) \in L^2(\mathbb{R}; \mathbb{C})$. Since the Lebesgue measure of $\mathbb{R}$ is infinite, $L^2(\mathbb{R}; \mathbb{C})$ is not contained in $L^1(\mathbb{R}; \mathbb{C})$ and so the definition (6) of the Fourier transform is not directly applicable. However by Plancherel's Theorem A.3.33 for any given $v(\cdot) \in L^2(\mathbb{R}; \mathbb{C})$ the sequence of functions

$$\tilde{v}_N(\omega) = \int_{-N}^{N} v(t)e^{-\imath\omega t}dt, \quad \omega \in \mathbb{R}, \qquad N \in \mathbb{N}$$

converges in $L^2(\mathbb{R}; \mathbb{C})$. Its limit is again denoted by $\mathcal{F}v$ or $\tilde{v}$ and is called the Fourier-Plancherel transform of v. Let

$$v_N(t) = \frac{1}{2\pi}\int_{-N}^{N} \tilde{v}(\omega)e^{\imath\omega t}d\omega, \quad t \in \mathbb{R}, \qquad N \in \mathbb{N}$$

then by the inversion result in Plancherel's Theorem A.3.33 $v_N(\cdot)$ converges to $v(\cdot)$ in $L^2(\mathbb{R};\mathbb{C})$.
$v(\cdot) \in L^2(\mathbb{R};\mathbb{C})$ is said to be of *limited bandwidth* if there exists $\omega_0 < \infty$ such that

$$\tilde{v}(\omega) = 0 \quad \text{for all} \quad \omega \in \mathbb{R},\ |\omega| > \omega_0\,. \tag{8}$$

The smallest $\omega_0 \geq 0$ with this property is called the *bandwidth* of $v(\cdot)$ and is denoted by ω_v. The following theorem shows that it is possible to reconstruct the signal $v(\cdot)$ from its sampled values $v(k\tau)$, $k \in \mathbb{Z}$ *if $v(\cdot)$ is of limited bandwidth and the sampling frequency $2\pi/\tau$ is at least twice the bandwidth ω_v of the signal.* This reconstruction will be carried out via a series of sinc functions where the function sinc : $\mathbb{C} \to \mathbb{C}$ is defined by

$$\text{sinc} :\ z \mapsto \begin{cases} z^{-1}\sin z & , \quad z \neq 0 \\ 1 & , \quad z = 0 \end{cases}. \tag{9}$$

sinc$(\cdot)$ is an entire analytic function on $\mathbb{C}$ with the *globally* convergent power series

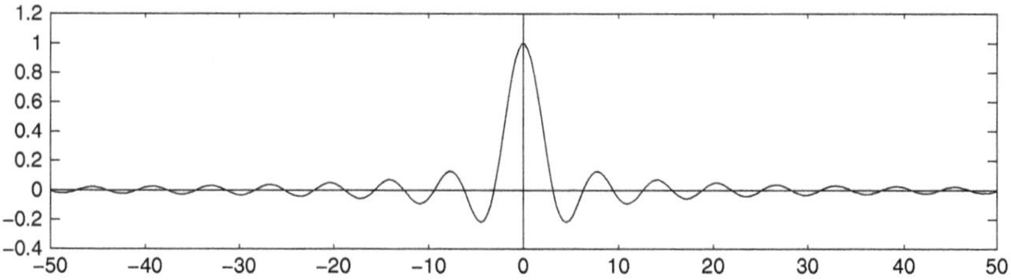

Figure 2.5.6: The sinc function

expansion

$$\text{sinc}(z) = \sum_{k=0}^{\infty} (-1)^k z^{2k}/(2k+1)!\,. \tag{10}$$

Moreover we note without proof that (see [479] and Ex. 3)

$$\sum_{k\in\mathbb{Z}} |\,\text{sinc}\,[t - k\pi]\,|^2 = 1, \quad t \in \mathbb{R}. \tag{11}$$

Theorem 2.5.2. (Sampling Theorem). *Suppose $v(\cdot) \in L^2(\mathbb{R};\mathbb{C})$ is a continuous function of limited bandwidth $\omega_v < \infty$ and τ is chosen such that*

$$0 < \tau < \pi/\omega_v. \tag{12}$$

Then the sequence of functions $(v(k\tau)\,\text{sinc}\,[(\pi/\tau)(\cdot) - k\pi])_{k\in\mathbb{Z}}$ is absolutely summable in $L^2(\mathbb{R};\mathbb{C})$ with sum $v(\cdot)$. Moreover, this sequence is pointwise absolutely summable, uniformly in $t \in \mathbb{R}$, and

$$v(t) = \sum_{k\in\mathbb{Z}} v(k\tau)\,\text{sinc}\,[(\pi/\tau)t - k\pi], \quad t \in \mathbb{R}. \tag{13}$$

Proof: Since v is of limited bandwidth we have $\tilde{v}(\cdot) \in L^1(\mathbb{R};\mathbb{C})$. Using the continuity of $v(\cdot)$ we obtain from (7) and (12)

$$v(t) = \frac{1}{2\pi}\int_{-\infty}^{\infty} \tilde{v}(\omega)e^{\imath\omega t}d\omega = \frac{1}{2\pi}\int_{-\pi/\tau}^{\pi/\tau} \tilde{v}(\omega)e^{\imath\omega t}d\omega, \quad t \in \mathbb{R}. \tag{14}$$

In particular

$$v(k\tau) = \frac{1}{2\pi}\int_{-\pi/\tau}^{\pi/\tau} \tilde{v}(\omega)e^{\imath\omega k\tau}d\omega, \quad k \in \mathbb{Z}. \tag{15}$$

The restriction $u(\cdot) = \tilde{v}(\cdot)|[-\pi/\tau, \pi/\tau]$ of $\tilde{v}(\cdot)$ to the interval $[-\pi/\tau, \pi/\tau]$ is square integrable. Hence, on this interval, it is the sum of its Fourier series in $L^2(-\pi/\tau, \pi/\tau; \mathbb{C})$ (see (5) with $l = \pi/\tau$

$$u(\cdot) = \sum_{k\in\mathbb{Z}} c_k\psi_k(\cdot),\ \psi_k(\theta) = e^{\imath k\tau\theta},\ \theta \in [-\pi/\tau, \pi/\tau],\ c_k = \frac{\tau}{2\pi}\int_{-\pi/\tau}^{\pi/\tau} \tilde{v}(\theta)e^{-\imath k\tau\theta}d\theta \tag{16}$$

where $(c_k)_{k\in\mathbb{Z}} \in \ell^2(\mathbb{Z};\mathbb{C})$. It follows from (15) that $c_k = \tau v(-k\tau)$ for $k \in \mathbb{Z}$ and so $\sum_{k\in\mathbb{Z}} |v(k\tau)|^2 < \infty$. Let ψ_k^e be the trivial extension of ψ_k to $\mathbb{R}$, i.e. $\psi_k^e(\omega) = e^{\imath k\tau\omega}1_{[-\pi/\tau,\pi/\tau]}(\omega)$. The two-sided sequence $(\psi_k^e(\cdot))_{k\in\mathbb{Z}}$ forms an orthogonal family in the Hilbert space $L^2(\mathbb{R};\mathbb{C})$ with $\|\psi_k^e(\cdot)\|^2_{L^2(\mathbb{R};\mathbb{C})} = 2\pi/\tau$. Hence (see Proposition A.4.3) $(c_k\psi_k^e(\cdot))_{k\in\mathbb{Z}} = (\tau v(-k\tau)\psi_k^e(\cdot))_{k\in\mathbb{Z}}$ is absolutely summable in $L^2(\mathbb{R};\mathbb{C})$ and since $\tilde{v}(\cdot)$ vanishes outside the interval $[-\pi/\tau, \pi/\tau]$, we obtain from (16) that

$$\tilde{v}(\cdot) = \sum_{k\in\mathbb{Z}} \tau v(-k\tau)\psi_k^e(\cdot) = \sum_{k\in\mathbb{Z}} \tau v(k\tau)e^{-\imath k\tau(\cdot)}\,1_{[-\pi/\tau,\pi/\tau]}(\cdot) \tag{17}$$

in $L^2(\mathbb{R};\mathbb{C})$. Applying the inverse Fourier-Plancherel transform to the function $1_{[-\pi/\tau,\pi/\tau]}(\cdot)$ we get, see (45)

$$\mathcal{F}^{-1}(1_{[-\pi/\tau,\pi/\tau]}(\cdot))(t) = (1/\tau)\,\mathrm{sinc}\,((\pi/\tau)\,t), \quad t \in \mathbb{R}.$$

and by the time shifting property of the Fourier-Plancherel transform we have

$$\mathcal{F}^{-1}(\tau e^{-\imath k\tau(\cdot)}1_{[-\pi/\tau,\pi/\tau]}(\cdot))(t) = \mathrm{sinc}\,[(\pi/\tau)t - k\pi], \quad t \in \mathbb{R},\ k \in \mathbb{Z}. \tag{18}$$

(see Proposition A.3.35). Since the inverse Fourier-Plancherel transform is a bounded linear operator on $L^2(\mathbb{R};\mathbb{C})$ it follows from (14), (17) and (18) that

$$v(\cdot) = \sum_{k\in\mathbb{Z}} v(k\tau)\,\mathrm{sinc}\,[(\pi/\tau)(\cdot) - k\pi] \tag{19}$$

where the sequence $(v(k\tau)\,\mathrm{sinc}\,[(\pi/\tau)(\cdot) - k\pi])_{k\in\mathbb{Z}} = (v(k\tau)\,\mathcal{F}^{-1}(\tau\psi_k^e(-(\cdot))))_{k\in\mathbb{Z}}$ is absolutely summable in $L^2(\mathbb{R};\mathbb{C})$ since we have shown above that the sequence $(\tau v(k\tau)\psi_k^e(-(\cdot)))_{k\in\mathbb{Z}}$ is absolutely summable in $L^2(\mathbb{R};\mathbb{C})$. This proves the first statement of the theorem.
Now $\sum_{k\in\mathbb{Z}} |v(k\tau)|^2 < \infty$ and it follows easily from (9) that the function $t \mapsto \sum_{k\in\mathbb{Z}} |\mathrm{sinc}\,[(\pi/\tau)t - k\pi]|^2$ is bounded on $\mathbb{R}$, see Ex. 3. Hence by the Cauchy-Schwarz inequality in $\ell^2(\mathbb{Z};\mathbb{C})$ (see (A.3.3)) the sequences $(v(k\tau)\,\mathrm{sinc}\,[(\pi/\tau)t - k\pi])_{k\in\mathbb{Z}}$, $t \in \mathbb{R}$ are absolutely summable, uniformly in $t \in \mathbb{R}$. As a consequence the sum of the series in (13) is continuous in $t \in \mathbb{R}$ and equals $v(t)$ almost everywhere (see Proposition A.3.11). Since $v(\cdot)$ is continuous by assumption, we finally obtain the equality in (13) for all $t \in \mathbb{R}$. □

The equation (13) is an explicit interpolation formula for the values of the time signal in between the sampling instants. Note, however that the sampled values $v(k\tau)$ of the past $k\tau < t$ as well as the future $k\tau > t$ are needed in order to compute the value of $v(\cdot)$ at time t. Hence the interpolation (13) *cannot* be implemented as a causal system with the sampled signal as input and the reconstructed signal as output.

Remark 2.5.3. (i) If the signal $v(t)$ is real then each term in the series on the RHS of (13) is also real.

(ii) If $t \in \mathbb{R}$ is replaced by $z \in \mathbb{C}$ in (14) then $v(\cdot)$ can be extended to a continuous function $v_{\mathbb{C}}(\cdot) : \mathbb{C} \to \mathbb{C}$ by

$$v_{\mathbb{C}}(z) = \frac{1}{2\pi} \int_{-\pi/\tau}^{\pi/\tau} \tilde{v}(\omega) e^{\imath\omega z} d\omega, \quad z \in \mathbb{C}.$$

$v_{\mathbb{C}}(\cdot)$ is analytic on $\mathbb{C}$ as can be shown by applying e.g. the theorems of Morera and Fubini. So we see that there would be no restriction in assuming that $v(\cdot) = v_{\mathbb{C}}(\cdot)\mid_{\mathbb{R}}$ is real analytic in the statement of the sampling theorem. Now the extension $v_{\mathbb{C}}(\cdot)$ satisfies the following exponential estimate

$$|v_{\mathbb{C}}(z)| \leq e^{\pi|z|/\tau} \frac{1}{2\pi} \int_{-\pi/\tau}^{\pi/\tau} |\tilde{v}(\omega)| d\omega = C e^{\pi|z|/\tau}, \quad z \in \mathbb{C}. \tag{20}$$

By a theorem of Paley-Wiener the fact that $v_{\mathbb{C}}(\cdot)$ is analytic on $\mathbb{C}$ and satisfies the inequality $|v_{\mathbb{C}}(z)| \leq C e^{\pi|z|/\tau}$, $z \in \mathbb{C}$ for some constant C is actually equivalent to $v(\cdot) \in L^2(\mathbb{R}, \mathbb{C})$ being of limited bandwidth $[-\pi/\tau, \pi/\tau]$, see *Notes and References*.

(iii) Let $l > 0$ and define the Paley-Wiener space

$$PW(l) = \{v(\cdot) \in L^2(\mathbb{R}; \mathbb{C})\,;\ \tilde{v}(\omega) = 0 \text{ for all } \ \omega \in \mathbb{R},\ |\omega| > l\}.$$

Then $PW(l)$ is a closed subspace of $L^2(\mathbb{R}; \mathbb{C})$. In the above proof (see (18)) we have seen that for $l = \pi/\tau$ the 2-sided sequence of sinc functions $(\operatorname{sinc}[l\,(\cdot) - k\pi])_{k\in\mathbb{Z}}$ form an orthogonal family in $L^2(\mathbb{R}, \mathbb{C})$. In fact it is a real orthogonal *basis* for the space $PW(l)$ and the formula (19) is just the expansion of v with respect to this basis. □

2.5.3 Sampling Continuous Time Systems

Consider a series connection of a zero-hold, a continuous time system Σ of the form (2.17) and a sampler. We write $(u^\tau(k))_{k\in\mathbb{N}}$ for the input sequence in order to indicate that the input value $u^\tau(k)$ is fed into the hold at time $k\tau$. The corresponding sampled states are given by

$$x^\tau(k) = e^{Ak\tau} x_0 + \int_0^{k\tau} e^{A(k\tau - s)} B u(s) ds.$$

Now $u(t) = u^\tau(k)$, $t \in [k\tau, (k+1)\tau)$, so the evolution of the sampled states is described by the difference equation

$$x^\tau(k+1) = e^{A\tau} x^\tau(k) + \left(\int_0^\tau e^{As} B ds \right) u^\tau(k).$$

The discrete time system $\Sigma^{(\tau)} = (e^{A\tau}, \int_0^\tau e^{As} B ds,\ C,\ D)$ is called the *sampled system* obtained from Σ by sampling at times $k\tau$, $k \in \mathbb{N}$. Note that the system matrix $e^{A\tau}$ of the sampled system is always nonsingular. This is a distinctive feature of discrete time systems obtained from sampling continuous time systems of the form (2.17).

Remark 2.5.4. If (i) only sampled times $k\tau$, $k \in \mathbb{N}$ are considered, (ii) only step inputs are used as controls, (iii) quantization errors are neglected, then the sampled system *exactly* reproduces the state and output values of the corresponding continuous time system at the times $t = k\tau$, $k \in \mathbb{N}$. □

Example 2.5.5. The linearized equations of motion of the inverted pendulum as described in Example 1.3.4 are

$$\dot{x}(t) = \begin{bmatrix} 0 & 0 & 1 & 0 \\ 0 & 0 & 0 & 1 \\ 0 & a_{32} & 0 & 0 \\ 0 & a_{42} & 0 & 0 \end{bmatrix} x(t) + \begin{bmatrix} 0 \\ 0 \\ b_3 \\ b_4 \end{bmatrix} u(t). \tag{21}$$

If the crane is controlled by a digital computer so that the force on the crane only changes at discrete instants $k\tau$, $k \in \mathbb{N}$, $\tau > 0$, then the sampled state trajectories $(x^\tau(k))_{k\in\mathbb{N}}$ are described by

$$x^\tau(k+1) = A_\tau x^\tau(k) + B_\tau u^\tau(k)$$

where

$$A_\tau = e^{A\tau} = \begin{bmatrix} 1 & a_{32}\,a_{42}^{-1}(\cosh(\sqrt{a_{42}}\tau) - 1) & \tau & a_{32}\,a_{42}^{-3/2}(\sinh(\sqrt{a_{42}}\tau) - \sqrt{a_{42}}\tau) \\ 0 & \cosh(\sqrt{a_{42}}\tau) & 0 & a_{42}^{-1/2}\sinh(\sqrt{a_{42}}\tau) \\ 0 & a_{32}a_{42}^{-1/2}\sinh(\sqrt{a_{42}}\tau) & 1 & a_{32}\,a_{42}^{-1}(\cosh(\sqrt{a_{42}}\tau) - 1) \\ 0 & \sqrt{a_{42}}\sinh(\sqrt{a_{42}}\tau) & 0 & \cosh(\sqrt{a_{42}}\tau) \end{bmatrix}$$

$$B_\tau = \int_0^\tau e^{As}B\,ds = \begin{bmatrix} b_4\,a_{32}\,a_{42}^{-2}(\cosh(\sqrt{a_{42}}\tau) - 1) + \tau^2(b_3 - b_4a_{32}\,a_{42}^{-1})/2 \\ b_4a_{42^{-1}}(\cosh(\sqrt{a_{42}}\tau) - 1) \\ b_4a_{32}\,a_{42}^{-3/2}\sinh(\sqrt{a_{42}}\tau) + \tau(b_3 - b_4a_{32}\,a_{42}^{-1}) \\ b_4\,a_{42}^{-1/2}\sinh(\sqrt{a_{42}}\tau) \end{bmatrix}$$

□

In general the sampled system will not yield information about the state values of the continuous time system between the sampling times. The dynamics of the continuous time system, particularly the location of the eigenvalues of A, and the frequency spectrum of the control and disturbance signals will determine which sampling rates are necessary to obtain sufficient information about the trajectories of the continuous plant from the sampled system. For example, if the feedback system shown in Figure 2.5.1 is required to track signals $r(t)$ having spectral content up to a frequency ω_0, then the sampling theorem enables us to specify an absolute lower bound on the sampling frequency $2\pi\tau^{-1} \geq 2\omega_0$. However in practice the sampling times have to be considerably higher (5-20 times this theoretical lower bound) depending upon dynamic characteristics of the closed loop system, such as its bandwidth, and additional performance requirements, e.g. reducing the delays between reference input and system response, see *Notes and References*.

2.5.4 Approximation of Continuous Systems by Discrete Systems

Control and communication systems are increasingly making use of digital rather than analog devices. Very often a feedback controller for a continuous plant is designed in continuous time and then a discrete time "equivalent" is implemented on the computer. In communication engineering the development of integrated circuit technology has generated a trend to replace analog by digital filters. A *filter* is a device which passes "desirable" frequency components of an input function (the useful signal) and rejects all others (noise). There are well established techniques for the design of analog filters, usually time invariant linear circuits, which meet prescribed performance specifications. Thus a common method in digital filter design is to first design a good analog filter and then approximate it by a digital filter. There are two aspects to the approximation problem:

(i) one must find a good discrete-time approximation $\Sigma^{(\tau)}$ of the continuous time system *at the sampling instants* and implement the discrete time system by a digital device,

(ii) the discrete time system $\Sigma^{(\tau)}$ must be converted into a continuous time system $\overline{\Sigma}_\tau$ by extrapolating the output values of $\Sigma^{(\tau)}$ between the sampling instants.

This latter conversion is usually carried out with a sampler and hold as illustrated in Figure 2.5.7. Note that if $\Sigma^{(\tau)}$ is a time-invariant linear system the resulting system

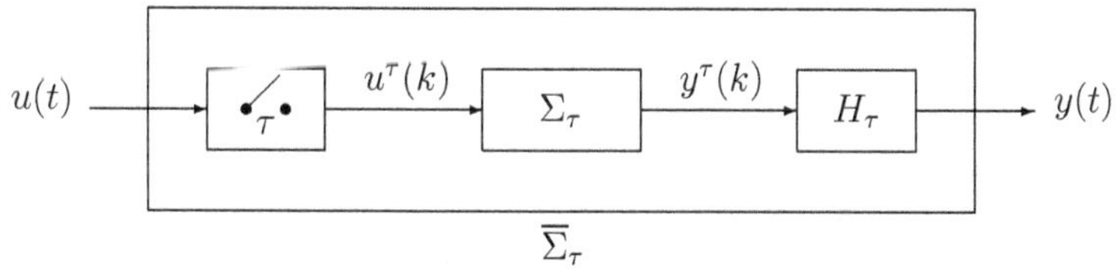

Figure 2.5.7: Conversion of a discrete time system into a continuous time system

$\overline{\Sigma}_\tau$ is linear but not time-invariant since $\overline{\Sigma}_\tau$ will only be invariant with respect to time shifts which are multiples of τ. If the sample period $\tau > 0$ cannot be made small, due to measurement costs or technical reasons, the choice of the extrapolating hold H_τ may be crucial for the performance of the continuous time system $\overline{\Sigma}_\tau$ as an approximation of the original system Σ.
In the following we will deal mainly with problem (i) and present various approximation schemes derived from numerical integration methods. However, we start with the ideal theoretical solution of the approximation problem for the system $\Sigma^{(\tau)}$.
Throughout this subsection $\|\cdot\|$ is an arbitrary norm on $\mathbb{R}^n$ and matrices are normed by the corresponding operator norm. We do not distinguish between these norms unless their use is unclear.

Sample and hold method

If Σ is given by the matrices (A, B, C, D), we have shown in the previous subsection, that under the assumption of Remark 2.5.4 approximation errors at the sampling instants can be avoided if we choose the sampled system

$$\Sigma^{(\tau)} = (e^{A\tau}, \int_0^\tau e^{As} B ds, C, D)$$

as the discrete time approximation of Σ. The resulting continuous time system $\overline{\Sigma}_\tau$ (Figure 2.5.7) is called the *hold equivalent* approximation of Σ. The approximation error $x(t) - \bar{x}^\tau(t)$ is zero for $t \in \tau\mathbb{N}$ and is reduced to the extrapolation error between the sampling instants. The extrapolation error depends on the sampling period and the chosen hold element.

Example 2.5.6. Let Σ be the scalar system

$$\begin{aligned} \dot{x}(t) &= ax(t) + bu(t), \quad a > 0 \\ y(t) &= x(t). \end{aligned} \tag{22}$$

Then $\Sigma^{(\tau)}$ has the form

$$\begin{aligned} x^\tau(k+1) &= e^{a\tau} x^\tau(k) + b\,a^{-1}(e^{a\tau} - 1)u^\tau(k) \\ y^\tau(k) &= x^\tau(k). \end{aligned}$$

Since $\Sigma^{(\tau)}$ reproduces exactly the state trajectory of the continuous time system Σ at the instants $k\tau$, the approximation error of the hold equivalent system only depends on the hold element in Figure 2.5.7. For the zero hold we have

$$x^\tau(t) = x^\tau(k) \quad \text{for} \quad t \in [k\tau, (k+1)\tau) = I_k(\tau)$$

and the corresponding approximation error is

$$\begin{aligned} |x(t) - x^\tau(t)| &= |(e^{at} - 1)x(k\tau) + \int_{k\tau}^t e^{a(t-s)} bu^\tau(k) ds\,|, \quad t \in I_k(\tau) \\ &\le |e^{a\tau} - 1||x(k\tau)| + |b\,a^{-1}(e^{a\tau} - 1)||u^\tau(k)|. \end{aligned}$$

□

It should be noted that the sampled system $\Sigma^{(\tau)}$ requires the computation of $e^{A\tau}$ and $\int_0^\tau e^{As} B ds$ and so can only be implemented approximately on the computer. This leads us to consider more direct approximating procedures.

Euler's method

Euler's method is the simplest integration method. Although it is numerically inefficient (see Table 2.5.12) we feel it is worthwhile discussing in some detail since it illustrates the concepts and problems which are typical for all finite difference methods. *An important advantage of these methods is that they are applicable to nonlinear as well as linear systems.* Since the output map of the approximating discrete time system is usually chosen to be the same as that of the continuous time system we disregard the output map in the following analysis. Consider a differentiable system Σ (as in Definition 2.1.12) with equation of motion

$$\dot{x}(t) = f(t, x(t), u(t)), \quad t \in T. \tag{23}$$

We assume that $U = \mathbb{R}^m$, $X = \mathbb{R}^n$, $T = [a,b] \subset \mathbb{R}$ is compact, $f : T \times X \times U \to X$ is continuous and satisfies for a given control $u(\cdot) \in \mathcal{U}$ and all $t \in T$ a Lipschitz condition

$$\|f(t,x,u(t)) - f(t,\bar{x},u(t))\| \le L\|x - \bar{x}\|, \quad x,\ \bar{x} \in \mathbb{R}^n \tag{24}$$

where L may depend upon the control $u(\cdot)$.
For a given step-size $\tau > 0$ and initial time $t_0 \in T$ we let $t_k = t_0 + k\tau$ and write N_τ for the largest natural number $N \le (b - t_0)/\tau$. The *approximate* value of $x(t_k)$ will be denoted by $x^\tau(k)$ and the control value at time $k\tau$ by $u^\tau(k)$.
Euler's method associates with the differential equation (23) (and initial time $t_0 \in T$) the difference equation

$$x^\tau(k+1) = x^\tau(k) + \tau f(t_k, x^\tau(k), u^\tau(k)), \tag{25}$$

with the domain $T_\tau = \{k \in \mathbb{N};\ 0 \le k \le N_\tau\}$. Equation (25) is obtained if the derivative in (23) is replaced by the difference quotient

$$[x(t+\tau) - x(t)]/\tau \ \text{ at } \ t = t_k\ .$$

Alternatively Euler's method can be interpreted in terms of *numerical integration.* In fact, integration of (23) over $[t_k, t_{k+1}]$ yields

$$x(t_{k+1}) = x(t_k) + \int_{t_k}^{t_{k+1}} f(t, x(t), u(t))dt. \tag{26}$$

Euler's method is obtained by approximating the above integral according to the "forward rectangular rule"

$$\int_{t_k}^{t_{k+1}} f(t,x(t),u(t))dt \approx \tau f(t_k, x(t_k), u(t_k)).$$

Example 2.5.7. For linear time invariant system (2.17), the discretized version (25) has the special form

$$x^\tau(k+1) = (I + \tau A)x^\tau(k) + \tau B u^\tau(k), \quad k \in \mathbb{N}. \tag{27}$$

Comparing this with the sampled system $\Sigma^{(\tau)}$ we see that the matrix exponential $e^{A\tau}$ is approximated by $I + A\tau$ while $\int_0^\tau e^{At}Bdt$ is approximated by τB.
Now let $u(t) \equiv 0$, $t_0 = 0$, $x(0) = x^\tau(0) = x^0$ and $\tau_N = t/N$ for a fixed $t > 0$, $N \in \mathbb{N}$. With respect to this step size the state $x(t) = e^{At}x^0$ of the continuous time system corresponds to the state $x^{\tau_N}(N)$ of the discrete time system (27). The approximation error is

$$\|x(t) - x^{\tau_N}(N)\| = \|e^{At}x^0 - (I + tA/N)^N x^0\|.$$

As in the scalar case it is easy to show

$$\|e^{At} - (I + tA/N)^N\| \to 0 \ \text{ as } \ N \to \infty.$$

Hence

$$\lim_{N\to\infty} \|x(t) - x^{\tau_N}(N)\| = 0.$$

So for any given initial state the corresponding free motion of (27) approximates (pointwise) the trajectory of the continuous time system at times $k\tau$ as $\tau \to 0$. □

The following theorem shows that the convergence result of the preceding example is true for arbitrary differential systems (23) if suitable smoothness assumptions are made.

Theorem 2.5.8. *Assume that $x(\cdot)$ is a twice continuously differentiable trajectory of (23) corresponding to some control function $u(\cdot)$ with $\|\ddot{x}(t)\| \le \gamma$ for $t \in [t_0, b)$. Let $x^\tau(\cdot)$ be the trajectory of (25) corresponding to the control function $u^\tau(\cdot) : k \mapsto u(t_k)$ and initial state $x^\tau(0) = x(t_0) + e^0$ where e^0 is an initial error. If f satisfies (24) then*

$$\|x(t_k) - x^\tau(k)\| \le e^{Lk\tau}\|e^0\| + \tau\gamma(e^{Lk\tau} - 1)/(2L), \quad 0 \le k \le N_\tau. \tag{28}$$

In particular if $\|x(t_0) - x^\tau(0)\| \le c_1\tau$ for some constant c_1, then

$$\max_{k \le N_\tau} \|x(t_k) - x^\tau(k)\| \le c\tau \tag{29}$$

for some constant $c \ge 0$.

Proof: Since $x(\cdot)$ is twice continuously differentiable, we have

$$x(t_{k+1}) = x(t_k) + \tau\dot{x}(t_k) + R(t_k). \tag{30}$$

where

$$\|R(t_k)\| \le (\tau^2/2) \max_{t_k \le t \le t_{k+1}} \|\ddot{x}(t)\| \le \tau^2\gamma/2. \tag{31}$$

Let $e_k = x(t_k) - x^\tau(k)$, $k \le N_\tau$, then from (30)

$$\begin{aligned} x(t_{k+1}) &= x(t_k) + \tau f(t_k, x(t_k), u(t_k)) + R(t_k) \\ x^\tau(k+1) &= x^\tau(k) + \tau f(t_k, x^\tau(k), u^\tau(k)), \quad u^\tau(k) = u(t_k). \end{aligned}$$

So

$$e_{k+1} = e_k + \tau[f(t_k, x(t_k), u(t_k)) - f(t_k, x^\tau(k), u^\tau(k))] + R(t_k).$$

Using (24) and (31) we obtain $\|e_{k+1}\| \le (1 + \tau L)\|e_k\| + \tau^2\gamma/2$. Therefore

$$\begin{aligned} \|e_k\| &\le (1 + \tau L)^k\|e^0\| + \sum_{i=0}^{k-1}(1 + \tau L)^i\tau^2\gamma/2 \\ &= (1 + \tau L)^k\|e^0\| + \tau\gamma[(1 + \tau L)^k - 1]/(2L) \end{aligned}$$

for $k \le N_\tau$. Since $(1 + \tau L)^k \le e^{k\tau L}$, the theorem is proved. □

The inequality (29) indicates that the approximation error should be halved when τ is halved (see Table 2.5.8). Although this characterizes the convergence rate of Euler's integration method, the explicit estimate (28) is much too conservative in most applications (see Table 2.5.8).
Note that the preceding theorem only describes the approximation of the trajectory $x(t)$ by the discrete time trajectory $x^\tau(k)$ at the instants $t_k = t_0 + k\tau$. In order to obtain an approximation on the whole interval $[t_0, t_0 + N_\tau]$ the integration method must be combined with an extrapolation procedure, as in Example 2.5.6. As a result the overall approximation error is a combination of integration errors and

extrapolation errors. For instance, if a zero hold is used for extrapolation, the overall error is

$$\|x(t) - \bar{x}^\tau(t)\| = \|x(t) - x^\tau(k)\| \leq \|x(t) - x(k\tau)\| + \|x(k\tau) - x^\tau(k)\| \leq c_2\tau + c\tau\,,$$

where $t_0 \leq t \leq t_0 + N_\tau$, c is any constant such that (29) is satisfied and

$$c_2 = \max\{\|\dot{x}(t)\|,\ t_0 \leq t \leq t_0 + N_\tau\}.$$

Example 2.5.9. Let us consider the same scalar equation as in Example 2.5.6 on a given time interval $[0, N_\tau]$. In order to apply Theorem 2.5.8 we require a bound on $|\ddot{x}(t)|$. If $u(\cdot)$ is differentiable, we have

$$\ddot{x}(t) = a\dot{x}(t) + b\dot{u}(t) = a^2x(t) + abu(t) + b\dot{u}(t).$$

Hence we can choose

	Exact solution	Error: $e(t) = x(t) - x^\tau(t/\tau)$			Error bound (28)
t	$x(t)$	$\tau = 0.2$	$\tau = 0.1$	$\tau = 0.05$	$\tau = 0.05$
2.0	2.631	-0.015	-0.007	-0.004	0.007
4.0	3.967	-0.024	-0.012	-0.006	0.017
8.0	5.956	-0.033	-0.016	-0.008	0.049
$u(t) \equiv 1$					
t	$x(t)$	$\tau = 0.2$	$\tau = 0.1$	$\tau = 0.05$	$\tau = 0.05$
2.0	2.624	-0.050	-0.013	-0.005	1.403
4.0	3.954	-0.089	-0.022	-0.009	3.118
8.0	5.934	-0.141	-0.034	-0.013	7.774
$u(t) = 1 + \sin 8\pi t$					

Table 2.5.8: Euler's method applied to $\dot{x} = -0.1x + u,\ \ x(0) = 1$

$$\gamma = a^2 \max_{0 \leq t \leq N_\tau} |x(t)| + |ab| \max_{0 \leq t \leq N_\tau} |u(t)| + |b| \max_{0 \leq t \leq N_\tau} |\dot{u}(t)|$$

where

$$\max_{0 \leq t \leq N_\tau} |x(t)| \leq e^{aN_\tau}|x^0| + |b\,a^{-1}(e^{aN_\tau} - 1)| \max_{0 \leq t \leq N_\tau} |u(t)|. \tag{32}$$

We see that the upper bound (28) for the integration error depends through γ not only on the control function $u(\cdot)|[0, N_\tau]$ but also on its rate of change $\dot{u}(\cdot)|[0, N_\tau]$. This dependence is illustrated in Table 2.5.8. In particular the integration error may become large if the control function changes rapidly even if τ is small. If a zero order hold is used for interpolation, the overall approximation error is

$$\begin{aligned} |x(t) - \bar{x}^\tau(t)| \ &\leq \ e^{aN_\tau}|e^0| + \gamma\,(2a)^{-1}(e^{aN_\tau} - 1)\tau \\ &\quad + [a \max_{0 \leq t \leq N_\tau} |x(t)| + |b| \max_{0 \leq t \leq N_\tau} |u(t)|]\tau. \end{aligned}$$

Here the third term (which represents an upper bound for the extrapolation error) can be estimated in terms of $|x^0|$ and $\max_{0 \leq t \leq N_\tau} |u(t)|$ using (32).
If $u(\cdot)$ is only piecewise differentiable with jumps at some $k\tau$, $k \leq N$, the preceding analysis must be applied successively to each interval on which $u(\cdot)$ is differentiable. □

Single and multi-step methods

Euler's method is a typical *single-step method.* These methods are characterized by the property that they only require knowledge of the present approximate state $x^\tau(k)$ in order to compute the next value $x^\tau(k+1)$. The general form of an explicit single-step method is

$$x^\tau(k+1) = x^\tau(k) + \tau F(t_k, x^\tau(k); \tau, f, u). \tag{33}$$

Here, for any given $t \in T$, $z \in \mathbb{R}^n$, $F(t, z; \tau, f, u)$ is a specific approximation of the difference quotient $\tau^{-1}[x(t+\tau) - z]$ where $x(\cdot)$ is the exact solution of (23) with $x(t) = z$ and control function $u(\cdot)$. In the special case of Euler's method we have $F(t, z; \tau, f, u) = f(t, z, u(t))$.
A $(\nu+1)$-step method requires the values $x^\tau(k), \ldots, x^\tau(k-\nu)$ in order to compute $x^\tau(k+1)$. The general form of such a multi-step method for $k \geq \nu$ is

$$x^\tau(k+1) = \sum_{j=0}^{\nu} a_j x^\tau(k-j) + \tau F(t_k, x^\tau(k+1), x^\tau(k), \ldots, x^\tau(k-\nu); \tau, f, u), \tag{34}$$

where $a_0, \ldots, a_\nu$ are given constants. The method is called *explicit* if F does *not* depend upon $x^\tau(k+1)$; otherwise it is called *implicit.* Hence in implicit methods it is necessary to solve (34) for $x^\tau(k+1)$ at each step. Nevertheless implicit methods are often more efficient than explicit ones.
Note that whereas single-step methods are self starting, multi-step methods need to be initialized by a single-step method.
In the following we will briefly describe some results for single-step methods and also *linear* multi-step methods of the following form

$$x^\tau(k+1) = \sum_{j=0}^{\nu} a_j x^\tau(k-j) + \tau \sum_{j=-1}^{\nu} b_j f(t_{k-j}, x^\tau(k-j), u(t_{k-j})), \quad k \geq \nu \tag{35}$$

where $a_0, \ldots, a_\nu$, $b_{-1}, \ldots, b_\nu$ are given constants. For the sake of simplicity we treat the scalar case. The expression (35) is a $(\nu+1)$-step method if $a_\nu \neq 0$ or $b_\nu \neq 0$. It is *explicit* if $b_{-1} = 0$ and *implicit* if $b_{-1} \neq 0$.

Example 2.5.10. (Trapezoidal and Heun's method). The trapezoidal method is defined by

$$x^\tau(k+1) = x^\tau(k) + (\tau/2)[f(t_k, x^\tau(k), u(t_k)) + f(t_{k+1}, x^\tau(k+1), u(t_{k+1}))]. \tag{36}$$

It is an *implicit* single-step method and is called the *trapezoidal method* since if f does not depend on x it reduces to the trapezoidal rule for numerical integration (see Figure 2.5.9). This implicit method can be converted into an explicit method by using Euler's method to predict $x^\tau(k+1)$ and then substituting this in the RHS of (36). As a result we obtain the so-called *Heun method*

$$x^\tau(k+1) = x^\tau(k) + (\tau/2)[f(t_k, x^\tau(k), u(t_k)) + f(t_{k+1}, x^\tau(k) + \tau f(t_k, x^\tau(k), u(t_k)), u(t_{k+1}))].$$

This is an explicit single-step method of the form (33) with

$$F(t, z; \tau, f, u) = (1/2)[f(t, z, u(t)) + f(t+\tau, z + \tau f(t, z, u(t)), u(t+\tau))].$$

It is a simple example of a *predictor-corrector algorithm* with Euler's method as predictor and the trapezoidal method as corrector. At each step it requires two evaluations of f. □

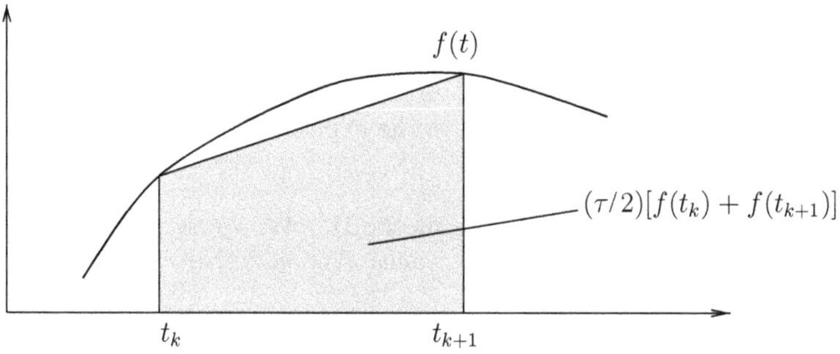

Figure 2.5.9: Trapezoidal method

Example 2.5.11. (Midpoint method). The *midpoint method* is defined by

$$x^\tau(k+1) = x^\tau(k-1) + 2\tau f(t_k, x^\tau(k), u(t_k)), \quad k \geq 1.$$

It is an explicit two-step method which requires one evaluation of f at each step and

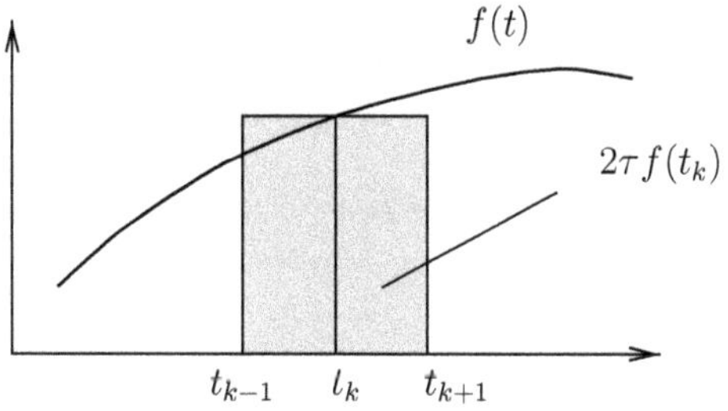

Figure 2.5.10: Midpoint rule

corresponds to the midpoint rule of numerical integration (see Figure 2.5.10). The value of $x^\tau(1)$ must be provided by a single step method. □

Example 2.5.12. (Classical Runge-Kutta method). The *Runge-Kutta method* is defined by

$$x^\tau(k+1) = x^\tau(k) + (\tau/6)[f_1^\tau(k) + 2f_2^\tau(k) + 2f_3^\tau(k) + f_4^\tau(k)]$$

where

$$\begin{aligned} f_1^\tau(k) &= f(t_k, x^\tau(k), u(t_k)), \\ f_2^\tau(k) &= f(t_k + \tau/2, x^\tau(k) + (\tau/2)f_1^\tau(k), u(t_k + \tau/2)), \\ f_3^\tau(k) &= f(t_k + \tau/2, x^\tau(k) + (\tau/2)f_2^\tau(k), u(t_k + \tau/2)), \\ f_4^\tau(k) &= f(t_k + \tau, x^\tau(k) + \tau f_3^\tau(k), u(t_{k+1})). \end{aligned}$$

It is an explicit single-step method requiring 4 evaluations of f per step. □

Euler's method, the trapezoidal and midpoint methods are special cases of numerical methods which are based on numerical integration procedures. The general idea of their construction is as follows:
Let $j \in \mathbb{N}$ be given and integrate (23) from t_{k-j} to t_{k+1}, $k \geq j$ to obtain

$$x(t_{k+1}) = x(t_{k-j}) + \int_{t_{k-j}}^{t_{k+1}} f(t, x(t), u(t))dt, \quad k \geq j. \tag{37}$$

Determine the polynomial $P(t)$ of a given degree $\nu \geq 0$ which coincides with the integrand $f(t, x(t), u(t))$ at the $(\nu+1)$ points t_i, $i \leq k+1$. The integral of $P(t)$ over the interval $[t_{k-j}, t_{k+1}]$ is then used as an approximation for the integral in (37). To illustrate the above scheme we derive the explicit method of Adams-Bashforth which is frequently used in practice.

Example 2.5.13. (Adams-Bashforth method). We choose $j = 0$ and replace the integrand in (37) by the interpolating polynomial $P_\nu(t)$ of degree ν in t satisfying

$$P_\nu(t_i) = f(t_i, x(t_i), u(t_i)), \quad i = k-\nu, \ldots, k.$$

Using the Lagrange's interpolation formula, we have

$$P_\nu(t) = \sum_{i=0}^{\nu} f\left(t_{k-i}, x(t_{k-i}), u(t_{k-i})\right) L_i(t)$$

where

$$L_i(t) = \prod_{\ell=0, \ell\neq i}^{\nu} \frac{t - t_{k-\ell}}{t_{k-i} - t_{k-\ell}}.$$

This yields the following *Adams-Bashforth integration formula*

$$x^\tau(k+1) = x^\tau(k) + \tau[\beta_{\nu 0} f_k + \beta_{\nu 1} f_{k-1} + \ldots + \beta_{\nu\nu} f_{k-\nu}]$$

where

$$\begin{aligned} \beta_{00} &= 1 \\ \beta_{\nu i} &= \frac{1}{\tau}\int_{t_k}^{t_{k+1}} L_i(t)dt = \int_0^1 \prod_{\ell=0,\ell\neq i}^{\nu} \frac{r+\ell}{-i+\ell} dr, \quad i = 0, \ldots, \nu, \\ f_\ell &= f(t_\ell, x^\tau(\ell), u(t_\ell)). \end{aligned}$$

Some values of $\beta_{\nu i}$ and the corresponding formulas are given in Table 2.5.11. □

i	0	1	2	3	$x^\tau(k+1) =$
β_{0i}	1				$x^\tau(k) + \tau f_k$
β_{1i}	$\frac{3}{2}$	$-\frac{1}{2}$			$x^\tau(k) + (\tau/2)\,[3f_k - f_{k-1}]$
β_{2i}	$\frac{23}{12}$	$-\frac{16}{12}$	$\frac{5}{12}$		$x^\tau(k) + (\tau/12)\,[23f_k - 16f_{k-1} + 5f_{k-2}]$
β_{3i}	$\frac{55}{24}$	$-\frac{59}{24}$	$\frac{37}{24}$	$-\frac{9}{24}$	$x^\tau(k) + (\tau/24)\,[55f_k - 59f_{k-1} + 37f_{k-2} - 9f_{k-3}]$

Table 2.5.11: Adams-Bashforth formulas

The above method can also be used to obtain implicit multi-step methods which are combined with a suitable explicit method (predictor) to yield efficient predictor-corrector algorithms. A popular example of this type is *Milne's method.*

Example 2.5.14. (Milne's method). Choose $j = 1$ in (37) and use a quadratic polynomial $P(t)$ which interpolates $f(t, x(t), u(t))$ on the nodes t_{k-1}, t_k, t_{k+1}. The result for $k \geq 1$ is

$$x^\tau(k+1) = x^\tau(k-1) + (\tau/3)[f_{k-1} + 4f_k + f_{k+1}], \text{ where } f_k = f(t_k, x^\tau(k), u(t_k)). \quad (38)$$

This is the corrector of Milne's method and if f does not depend on x it reduces to *Simpson's rule for numerical integration.* The predictor of Milne's method is obtained by using a quadratic interpolation to the integrand at t_{k-2}, t_{k-1}, t_k in (37)

$$x^\tau(k+1) = x^\tau(k-3) + (4\tau/3)[2f_{k-2} - f_{k-1} + 2f_k], \quad k \geq 3. \quad (39)$$

Thus in the kth step of Milne's method $x^\tau(k+1)$ is computed via (38) where $f_{k+1} = f(t_{k+1}, x^\tau(k+1), u(t_{k+1}))$ and $x^\tau(k+1)$ is evaluated by (39). □

Note that not all the methods we have presented define discrete time dynamical systems, in the sense of Definition 2.1.1. Indeed in the trapezoidal, Heun, Runge-Kutta and Milne methods (corrector plus predictor) the state $x^\tau(k+1)$ is influenced by the control value $u^\tau(k+1)$. This direct input-state coupling contradicts the axiom of causality. On the other hand if Euler's method, Milne's predictor, the midpoint rule or Adams-Bashforth methods are used for the discretization of a differentiable system (23), then discrete time dynamical systems are always obtained. However the discrete time systems obtained by the application of *multi-step* methods do not have the same state space as the corresponding continuous time system. If the method has a memory of length ν, i.e. if $x^\tau(k+1)$ is determined as function of $f_{k-\nu}, \ldots, f_k$, then since the past values of x^τ and u^τ have to be "stored" in the state vector the state space of the resulting discrete time system is $X^{\nu+1} \times U^\nu$. This may give rise to certain "instability phenomena" which are observed in multi-step methods such as the midpoint rule and Milne's method. If these are applied for example to linear systems of the form (2.17), rounding errors can incite *unstable parasitic* oscillations of the discretized model which do not correspond to any eigenmotion of the continuous time system (see Ex. 10, 11 and Subsection 3.3.3).
In Theorem 2.5.8 we showed that the approximation error for Euler's method can be bounded by $c\tau$ as $\tau \to 0$. This is expressed by saying that Euler's method is of order 1. More generally, a particular method is said to be *of order p* if it yields for all initial value problems

$$\begin{aligned} \dot{x}(t) &= f(t, x(t)), \quad t \in T = [a, b] \\ x(t_0) &= x^0 \end{aligned} \quad (40)$$

(where f has continuous and bounded derivatives up to order p on $T \times \mathbb{R}$) approximate solutions $x^\tau(\cdot)$ such that the global approximation error is order p. This means

$$\max_{0 \leq k \leq N_\tau} |x(t_k) - x^\tau(k)| = O(\tau^p) \text{ as } \tau \to 0,$$

whenever the initial errors tend to zero with order p

$$\max_{0 \leq i \leq \nu} |x(t_i) - x^\tau(i)| = O(\tau^p) \text{ as } \tau \to 0.$$

It can be shown that the Heun and midpoint methods are of order 2, the classical Runge-Kutta and Milne methods are of order 4. A linear multi-step method (35) is of order $p \geq 1$ if and only if $a_j \geq 0 \quad j = 0, \dots, \nu$ and

$$\sum_{j=0}^{\nu} a_j = 1 \quad \text{and} \quad \sum_{j=0}^{\nu} j a_j + \sum_{j=1}^{\nu} b_j = 1 \tag{41a}$$

$$\sum_{j=0}^{\nu} (-1)^j a_j + i \sum_{j=-1}^{\nu} (-j)^{i-1} b_j = 1 \quad \text{for} \quad i = 2, \dots, p \tag{41b}$$

(see *Atkinson* (1989) [26]). The following example illustrates how the order of a method is reflected in the reduction of the approximation error with diminishing stepsize.

Example 2.5.15. Table 2.5.12 displays the approximation errors $e_k = x(t_k) - x^{\tau}(k)$ for a variety of methods applied to the initial value problem

$$\dot{x}(t) = -0.1x(t) + 10\,(1 + \sin 10\pi t), \quad x(0) = 1.$$

All the multistep methods are initialized with "exact" values. Observe the behaviour of

			Euler (order 1)	Heun (order 2)	Milne (order 4)	Runge-K. (order 4)	Adams-B. (order 4)
τ	t_k	$x(t_k)$	e_k	e_k	e_k	e_k	e_k
0.2	5.0	39.829	1.218684	1.573321	1.541310	-0.689172	1.546893
	10.0	63.382	5.220485	5.649258	5.650283	-2.144811	5.631805
	20.0	86.342	17.319951	17.631950	17.645748	-6.221867	17.624505
0.1	5.0	39.829	-0.315121	-0.123519	-0.618127	0.005845	-0.798817
	10.0	63.382	-0.459733	-0.196185	-2.742330	0.009268	-0.983663
	20.0	86.342	-0.550889	-0.258111	-12.553442	0.012160	-1.422443
0.05	5.0	39.829	-0.121190	-0.026503	0.027124	0.000289	0.203423
	10.0	63.382	-0.172322	-0.042103	0.026776	0.000456	0.307603
	20.0	86.342	-0.200045	-0.055414	0.028282	0.000594	0.424276
0.001	5.0	39.829	-0.001891	-0.000010	0.000000	0.000007	0.000000
	10.0	63.382	-0.002604	-0.000016	0.000000	0.000008	0.000000
	20.0	86.342	-0.002896	-0.000021	0.000000	0.000006	0.000000
0.0005	5.0	39.829	-0.000943	-0.000002	0.000000	0.000007	0.000000
	10.0	63.382	-0.001298	-0.000004	0.000000	0.000008	0.000000
	20.0	86.342	-0.001442	-0.000005	0.000000	0.000006	0.000000

Table 2.5.12: Approximation errors and their dependence on the step size τ

the errors as τ is halved. In accordance with the sampling theorem for $\tau = 0.2 > \pi/\omega_u = \pi/10\pi$ no approximation is achieved. All the methods except those of Milne and 4-step Adams-Bashforth yield reasonable first approximation for $\tau = 0.1 = \pi/\omega_u$. The next halving ($\tau = 0.05$) yields an improvement of the approximation which is better than the orders of the various methods predict. For $\tau = 0.001 \to \tau = 0.0005$ the magnitude of the errors reduces more or less as the order predicts, with the exception of the Runge-Kutta method for which the error reduction rate deteriorates more and more. This is due to the increasing influence of the rounding errors (see Ex. 9). □

Small errors in the initial state and rounding errors may eventually lead to large errors in the solution if they incite unbounded eigenmotions of the discretized system. In Subsection 3.3.3 we will briefly discuss numerical stability properties of the above methods and we will see that (theoretically) very accurate methods such as Milne's may produce "unstable" discretized systems although the differential system itself is "stable". The selection of an adequate numerical method for a concrete initial value problem relies very much on experience and is still something of an art.
Additional problems arise when we apply theorems of Numerical Analysis, not to the problem of approximating a single solution of a differential equation, but to the much more complex problem of approximating a differential dynamical *system* by a discrete time one. We conclude this section by pointing out some specific difficulties in this context.

Dependence on the control functions

We have seen in Example 2.5.9 and Table 2.5.8 that the error bounds depend not only on the magnitude of the control function but also on the magnitude of its derivative. So we cannot expect that there exists a step-size τ which will yield good approximations for arbitrary control functions with values in a prescribed set. For example the Sampling Theorem suggests that the frequency spectrum of the input signals should be small outside $[-\pi/\tau, \pi/\tau]$.
The following example shows that bang-bang jumps of the control may cause considerable deviations and lead to oscillations around the exact solution which remain, even after the control $u(\cdot)$ has been switched off ($u(t) \equiv 0$, $t \geq t_1$).

Example 2.5.16. Consider the controlled harmonic oscillator without damping

$$\begin{bmatrix} \dot{x}_1 \\ \dot{x}_2 \end{bmatrix} = \begin{bmatrix} 0 & -1 \\ 1 & 0 \end{bmatrix} \begin{bmatrix} x_1 \\ x_2 \end{bmatrix} + \begin{bmatrix} 1 \\ 1 \end{bmatrix} u, \quad x(0) = \begin{bmatrix} 5/2 \\ 0 \end{bmatrix}. \tag{42}$$

Let $\tau = 0.05$ and choose

$$u(t) = \begin{cases} (-1)^k 20 & \text{if} \quad k \leq t < k+1; \ k = 0,1,2,3 \\ 0 & \text{if} \quad t \geq 4 \end{cases}.$$

For $t \geq 4$ the solution of (42) should coincide with the periodic free motion

$$\begin{bmatrix} \dot{x}_1 \\ \dot{x}_2 \end{bmatrix} = \begin{bmatrix} 0 & -1 \\ 1 & 0 \end{bmatrix} \begin{bmatrix} x_1 \\ x_2 \end{bmatrix}, \quad x(4) = \begin{bmatrix} -11.4 \\ -28.2 \end{bmatrix}. \tag{43}$$

Figure 2.5.13 shows the "exact" solution curve of (42) for $0 \leq t \leq 30$ (as obtained by a Runge-Kutta method with step size $\tau/10 = 0.005$). It also shows the approximate solution curves (computed with step size $\tau = 0.05$ over the same time interval) by *Euler's method*, the 4-*step method of Adams-Bashforth*, the *midpoint rule*, the *predictor of Milne's method* and the complete *Milne method* (predictor and corrector). All the multistep methods were initialized with accurate initial values (computed by Runge-Kutta's method with step size $\tau/10$).
Apart from the various deviations from the true solution (which are particularly large for the predictor of Milne's method) two facts are remarkable. If the same methods are applied to solve the initial value problem (43), all of them, with the exception of Euler's method, track the true circular solution very precisely. They do not show the dramatic

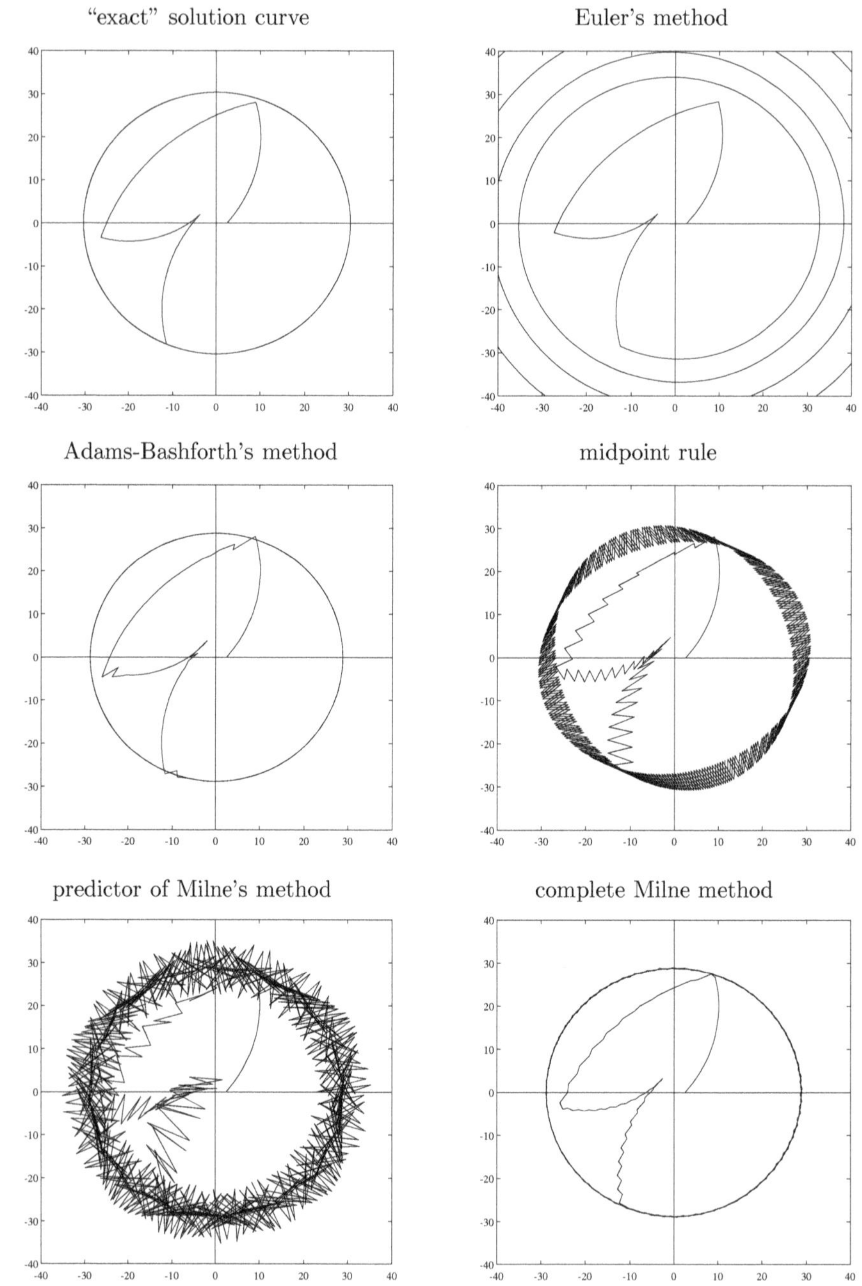

Figure 2.5.13: Approximation of a bang-bang controlled trajectory by various integration schemes

deviations which must, therefore, be caused by the large initial control oscillations. On the other hand Euler's method, which does not reproduce the periodic behaviour of the true solution, is very robust with respect to the effects of the control and shows for $t \geq 4$ the same qualitative behaviour when applied to (42) or (43). □

Advanced controls

We have seen above that several higher order methods (such as the methods of Heun, Runge-Kutta, all implicit multi-step methods) produce difference equations with direct input state couplings when applied to a controlled differential equation (23). To obtain a discrete-time dynamical system we have to introduce the *shifted* function $\bar{u}^\tau : \mathbb{N} \to \mathbb{R}$, $\bar{u}^\tau(k) = u(t_{k+1})$ instead of $u^\tau(k) = u(t_k)$ as input signal. The following example shows that it is essential to feed the discrete time systems generated by these methods with the anticipated control value $\bar{u}^\tau(k) = u(t_{k+1})$ at time t_k. Otherwise the order of convergence will not be preserved.

Example 2.5.17. Consider

$$\dot{x}(t) = -2x(t) + u(t), \quad x(0) = 1, \quad u(t) = t^2, \quad t \geq 0. \tag{44}$$

Application of Heun's method yields the following discrete time system

$$x^\tau(k+1) = (1 - 2\tau + 2\tau^2)x^\tau(k) + (\tau/2)u^\tau(k-1) + (\tau/2)(1-2\tau)u^\tau(k) \tag{45}$$

where $u^\tau(k) := u((k+1)\tau)$. Table 2.5.14 shows the difference between the solution of (44)

Step size		Exact solution	Heun	Heun (non-advanced control)
	t_k	$x(t_k)$	e_k	e_k
$\tau = 0.2$	1.000	0.352	-0.018	0.079
	5.000	10.250	-0.013	0.867
$\tau = 0.1$	1.000	0.352	-0.004	0.049
	5.000	10.250	-0.003	0.442
$\tau = 0.05$	1.000	0.352	-0.001	0.026
	5.000	10.250	-0.001	0.223

Table 2.5.14: Errors for advanced and non-advanced controls

and the solution of (45) with control $u^\tau(k) = (k+1)^2\tau^2$ and initial value $x^\tau(0) = 1$. The table also gives the approximation error when the non-advanced control $u^\tau(k) = k^2\tau^2$ is used in (45). A comparison of the results in this table shows that the use of non-advanced control not only deteriorates the approximation but changes the *order* of convergence. The same is true for the Runge-Kutta method. If an analogous time shift is applied to the control function in the difference equation obtained from (44) via the Runge-Kutta method, the resulting discretization approximates the solutions of (44) with order 1 instead of 4. □

2.5.5 Exercises

1. Prove the statements at the end of Remark 2.5.1.

2. Determine the convolution kernels which describe the first order interpolator (2) and the delayed first order interpolator (3).

3. (i) Prove that the function $\theta \mapsto \sum_{k\in\mathbb{Z}} |\text{sinc}\,[(\theta - k\pi]|^2$ is bounded on $\mathbb{R}$.

(ii) Prove (11).

4. Prove that under conditions of the Sampling Theorem $v(\cdot)$ is analytic on $\mathbb{R}$ and can be extended analytically to $\mathbb{C}$. Show that this extension satisfies the exponential estimate (20) (see Remark 2.5.3).

5. Let $t_0 < t_1 < t_2$ and f_0, f_1, f_2 be real numbers. Determine the quadratic polynomial $P(t)$ with $P(t_i) = f_i$, $i = 0, 1, 2$. Apply this to obtain a second order hold H^2_τ.
Suppose $u : \mathbb{R} \to \mathbb{R}$ is three times continuously differentiable, $u(t) = 0$ for $t \leq 0$ and $|u(t)| \leq M$ for $t \in \mathbb{R}$. Denote by $\bar{u}(\cdot)$ the function obtained by applying H^2_τ to the sampled signal $\sum_{k\in\mathbb{Z}} u(k\tau)\delta(t - k\tau)$. Show that $|u(t) - \bar{u}(t)| \leq M\tau^3$.
Find the impulse response and the step response of H^2_τ.

6. Determine the sampled system $\Sigma^{(\tau)}$ corresponding to the continuous time system

$$\dot{x}(t) = \begin{bmatrix} 0 & 1 \\ 0 & -1 \end{bmatrix} x(t) + \begin{bmatrix} 0 \\ 1 \end{bmatrix} u(t), \qquad y(t) = [0,\ 1]\, x(t).$$

7. Describe the sampling of a continuous time system (A, B, C, D) with a first order hold instead of a zero order hold by

(i) determining the equations of motion as higher order difference equations.

(ii) Finding a state space model of the sampled system (hint: it is a $(n+m)$-dimensional system).

(iii) Determining the class of controls for which the sampled system exactly reproduces the state values of the continuous time system at the sample times $k\tau$.

8. Derive an error estimation analogous to that given in Example 2.5.6 if a first order hold is used instead of a zero order hold for the second hold element (see Figure 2.5.5).

9. Suppose the conditions in Theorem 2.5.8 hold and Euler's method is applied to

$$\dot{x}(t) = f(t, x(t), u(t)), \quad x(t_0) = x^0.$$

Because of rounding errors the operations performed at each step are actually

$$\tilde{x}^\tau(k+1) = \tilde{x}^\tau(k) + \tau f(t_k, \tilde{x}^\tau(k), u(t_k)) + \rho_k$$

where ρ_k is the so-called *local rounding error.* Assume $\tilde{x}^\tau(t_0) = x^0$ and $|\rho_k| \leq \rho$ for all $k \geq 0$.

(i) Show the following estimate for the total error (approximation and quantization)

$$|x(t_k) - \tilde{x}^\tau(k)| \leq (\tau\gamma/2 + \rho/\tau)(e^{Lk\tau} - 1)/L, \quad 0 \leq k \leq N_\tau.$$

(ii) Discuss the behaviour of the above error bound as a function of τ.

(iii) Apply Euler's method to the initial value problem

$$\dot{x}(t) = -x^2(t), \quad x(0) = 1.$$

Find the exact solution, tabulate the errors $|x(3)-\tilde{x}^\tau(3/\tau)|$ for $\tau = 10^{-k}$, $k = 2, \ldots, 8$ and interpret the table with reference to (ii).

10. Case study: Instability of the midpoint rule.

(i) Write a computer program to solve the initial value problem

$$\dot{x}(t) = ax(t) + b, \quad x(0) = x^0$$

by the midpoint rule. Your program should be for arbitrary reals a, b, x^0, stepsize $\tau > 0$ and interval $[0, \bar{t}]$, $\bar{t} > 0$.

(ii) Choose $a = -2$, $b = 1$, $x^0 = 1$, $\tau = 0.02$ and $\bar{t} = 50$. Print the true solution values $x(k\tau)$ and the errors $x(k\tau) - x^\tau(k)$ for $k = m \cdot 100 + r$ where $m = 0,\ 1,\ 2,\ 4,\ 8,\ 16$ and $r = 1, \ldots, 10$.

(iii) Determine the discrete time system obtained by discretizing the equation from (i) via the midpoint rule. Find its eigenvalues and explain the results of (ii). (A systematic analysis will be given later in Subsection 3.3.3).

11. Case study: Impulse response.

(i) Write a computer program to solve $\dot{x} = Ax + bu$, $x(0) = x^0$ with $A \in \mathbb{R}^{2\times2}$, $b \in \mathbb{R}^2$ by Euler's method, Heun's method and the midpoint rule. Your program should be for an arbitrary matrix A, control of the form $\alpha \cdot 1_{[0,\beta]}$, $\alpha \in \mathbb{R}$, $\beta > 0$, initial states $x^0 \in \mathbb{R}^2$, stepsize τ and interval $[0, \bar{t}]$, $\bar{t} > 0$.

(ii) Use Euler's method to solve the above initial value problem with

$$A = \begin{bmatrix} 0 & 1 \\ -1 & 0 \end{bmatrix}, \ b = \begin{bmatrix} 0 \\ 1 \end{bmatrix}, \ x^0 = 0, \ u = \frac{1}{10\tau} 1_{[0,10\tau]}, \ \tau = 0.1,\ 0.01,\ 0.001, \ \bar{t} = 10$$

(print solution values at $t = 0.1, 0.2, \ldots, 1.0$ and $2.0, \ldots, 10.0$).

(iii) Repeat (ii) with $u = (1/10\tau^2)(1_{[0,10\tau]} - 1_{[10\tau,20\tau]})$.

(iv) Compare the solutions of (ii), (iii) with the solutions of the corresponding homogeneous equation $\dot{x} = Ax$ with initial values $x(0) = b = [0,1]^\top$ and $x(0) = Ab = [1,0]^\top$ (if available use a graphical display). Interpret the result of (ii) by means of Lemma 2.3.4 on the approximations of the Dirac impulse. Analyze the result of (iii) in the same way.

(v) Predict what will happen if the midpoint rule is used instead of Euler's method in (ii). Solve (ii) by means of Heun's method and by the midpoint rule (use a graphical display if available).

(vi) Determine the discrete time systems $x^\tau(k+1) = A_\tau x^\tau(k) + b_\tau u^\tau(k)$ obtained by each of the three methods in (i). Compare the spectra of A_τ with the spectrum of A. What happens if $\tau \to 0$? Determine the behaviour of the true solution $x(k\tau)$ and the approximate solutions $x^\tau(k)$ as $k \to \infty$. Try to explain the phenomena observed in (v).

2.5.6 Notes and References

There was very little work carried out on the analysis of sampled-data systems until 1950 when research was motivated by the first use of digital computers in control systems. Frequency domain analysis was used in the first generation of textbooks dedicated exclusively to sampled-data systems, *Franklin and Ragazzini* (1958) [170], *Jury* (1958) [283]. Later textbooks put much more emphasis on the state space approach, *Kuo* (1980) [321], *Ackermann* (1985) [3] and *Oppenheim et al.* (1997) [399]. Recently there has been an upsurge *in the analysis and design* of hybrid systems, formed when continuous time and discrete time systems are interconnected, in the context of H^∞ theory, see *Dullerud* (1996) [140] and the references therein.

Basic problems concerning the relationship between continuous time and discrete time signals and systems, e.g. sample rate selection, effects of quantization errors, approximation errors, are neglected in most control theoretic textbooks. Our guide in this important area has been the book by *Franklin and Powell* (1998) [169], which gives a good introduction from an engineering point of view, see also the final chapter in *Franklin et al.* (1986) [168], Chapters 9 and 10 in *Kwakernaak and Sivan* (1991) [322] and *Franklin et al.* (1998) [169]. These references also contain further information about the A/D and D/A conversion of signals.

In the context of interpolation theory *Whittaker* (1915) [519] proved that

$$f(z) = \sum_{k=-\infty}^{\infty} f(k\tau)\operatorname{sinc}[\pi(z - k\tau)/\tau], \quad z \in \mathbb{C}$$

for every analytic function $f : \mathbb{C} \to \mathbb{C}$ with $|f(z)| \leq Ce^{\pi|z|/\tau}$, see the monograph by *Stenger* (1993) [479]. By Theorem X in *Paley and Wiener* (1934) [403] we have seen in Remark 2.5.3 that these conditions are equivalent to the ones in Theorem 2.5.2. *Shannon* (1948) [460] was the first to state the sampling theorem in form we have given it and he also recognized its basic importance for communication theory. *Nyquist* (1928) [394] also made early contributions to the field and the sampling rate $1/\tau$ per second is known as the Nyquist sampling rate. For generalizations of the sampling theorem to irregular spaced samples see *Beutler* (1961) [53]. *Higgins* (1996) [228] contains many interesting historical remarks on the sampling theorem and shows how it plays a role in different areas of mathematics and engineering. Nowadays there are a great variety of sampling results available in the literature and a comprehensive theory is gradually evolving, see e.g. *Benedetto* (1992) [48].

Applications to signal processing and communication are discussed in the well-known introductory textbook of *Kwakernaak and Sivan* (1991) [322]. For further reading on communication systems we refer to *Benedetto et al.* (1987) [49] and *Carlson* (1986) [90].

The difficulties involved in computing the exponential of a matrix are discussed in the paper *Moler and van Loan* (1978) [378], and the update *Moler and van Loan* (2003) [379]. Numerical methods for solving ordinary differential equation are presented in most textbooks on numerical mathematics, see e.g. *Stoer and Bulirsch* (1993). A detailed study can be found in the two volumes *Hairer et al.* (1993)and (1996) [210], [211]. In the control engineering literature numerical integration methods are described as recipes for digital simulation of continuous time systems. However, one must be cautious since the approximation of continuous time *systems* by discrete time *systems* is complicated by the fact that instead of determining a fixed solution, it is necessary to consider the system behaviour for a variety of controls and initial states.

Chapter 3

Stability Theory

The Oxford English Dictionary's definition of stable is "not easily moved, changed or destroyed". Most of us have an intuitive notion of stability which corresponds more or less with this definition. However, in order to build a theory of stability it is necessary to be more precise about terms like "not easily moved or changed". We need to define the basic class of objects to which the notion of stability is applied and also specify the type of perturbations which are considered. In this chapter we study the stability of *state trajectories* under the influence of perturbations in the *initial state* and in the next two chapters we consider perturbations in the *system parameters*. The stability of output trajectories under the influence of perturbations in the input signal will be discussed in Volume II.

The development of modern stability theories was initiated by *Maxwell* (1868) [364] and *Vyshnegradskiy* (1876) [511] in their work on governors, but the importance of the concept of stability in many other scientific fields was soon recognized and now it is a cornerstone of applied mathematics. For example, the prediction of instabilities from a mathematical model has in many instances led to a confirmation that the model adequately represents the corresponding physical process. In 1923 G.I. Taylor [492], using the Navier-Stokes equations, showed that the flow of a viscous fluid between rotating cylinders would become unstable at a particular value of a parameter, now known as the Taylor number. He confirmed this experimentally and so increased confidence in modelling viscous fluid flows by the Navier-Stokes equations. Perhaps more relevant to this text is the fact that almost all control system designs are founded on a stability requirement and our treatment of the subject will be slanted in this direction.

In Section 3.1 we introduce the important concept of stability due to Liapunov, which is used to investigate variations in system trajectories with respect to perturbations in the initial conditions. We also discuss some basic notions from the qualitative theory of dynamical systems which enable us to describe more precisely the limiting behaviour of trajectories as time tends to infinity. The definitions in Section 3.1 will be given in terms of the system's trajectories, so that if the system is modelled by differential equations or difference equations, we must first solve these equations if we wish to check whether or not the conditions in the definitions hold. Liapunov had the ingenious idea of using generalized energy functions (Liapunov functions) to investigate stability properties *directly* from the differential or difference equations. In Section 3.2 we first introduce a very general concept of a Liapunov function for a flow on a metric space and determine conditions for (uniform) stability, asymptotic stability and instability. Again these conditions will be given in terms of the system's trajectories, however when we specialize the results to finite dimensional systems we will show that they can be verified directly. The theorems are illustrated by a number of examples from a variety of different fields. The main development in Section 3.3 is for linear finite dimensional systems. First we consider time-varying systems and relate stability and uniform asymptotic stability to the boundedness and exponential decay of the evolution operator Φ. Bohl and Liapunov exponents are introduced and it is shown that uniform asymptotic stability is equivalent to the fact that the upper Bohl exponent is negative, which is itself equivalent to the existence of a uniform estimate for the L^p-norm of Φ, $p > 0$. For time-invariant systems one does not need to compute Φ to see whether these conditions hold, instead all the stability properties are shown to be equivalent to constraints on the spectrum of the generator A. In order to prepare the ground for one of Liapunov's main results (*Liapunov's indirect method*) quadratic forms are used as Liapunov functions for the linear equations and stability and instability theorems deduced. These forms are then used for nonlinear systems to prove that uniform asymptotic stability of the linearized equation implies that the nonlinear system will also have this property. The section concludes with a theorem which shows that in the case of time-invariant nonlinear systems an equilibrium state is exponentially stable *if and only if* the linearization at the equilibrium state is exponentially stable.

The final section is a substantial one and contains many classical stability criteria for polynomials. *Routh* (1887), *Hurwitz* (1895) showed that instead of actually determining the spectrum of the system matrix, criteria for stability could be determined directly from the coefficients of its characteristic polynomial. In the section both real and complex polynomials are considered and we start by showing how the principle of the argument is used to prove the *Hermite-Biehler Theorem*, the test for Hurwitz stability by the *Routh Array*, and a characterization of Hurwitz stable polynomials by the *Cauchy index* of an associated real rational function. Then the *Hermite form*, *Bézoutiant* and *Hankel form* are introduced and these quadratic forms are used to derive algebraic stability tests for Hurwitz polynomials. In Subsection 3.4.6 the corresponding analytic and algebraic stability tests for the discrete time case are obtained. Finally, in the last subsection, stability criteria for other algebraic stability regions in the complex plane are given in terms of algebraic Liapunov-type equations.

3.1 General Definitions

In 1875 it was announced that the Adams Prize at the University of Cambridge for 1877 would be awarded for the best essay on "The Criteria of Dynamic Stability". The announcement went on to say (see *Fuller* (1976) [176]):

> *"To illustrate the meaning of the quotation imagine a particle to slide down inside a smooth inclined cylinder along the lowest generating line, or to slide*

down along the highest generating line. In the former case a slight derangement of the motion would merely cause the particle to oscillate about the generating line, while in the latter case the particle would depart from the generating line altogether. The motion in the former case would be in some sense stable, and in the latter case unstable ..."

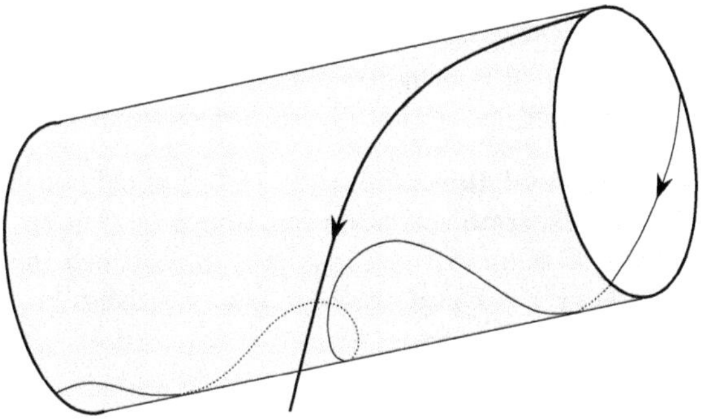

Figure 3.1.1: Stable and unstable motions on a cylinder

The winner of the prize was E. J. Routh of whom we will say more later (see Section 3.4). The stability criteria implied in the citation are not obvious. However, it is clear that they are concerned with the variation of a trajectory from a distinguished trajectory. This variation is achieved by a slight change in the state at a particular time, and the distinguished trajectory is said to be stable if the resulting perturbed trajectory is close to the distinguished one for all future times. We start with a mathematical formulation of this idea due to *Liapunov* (1893) [354], and give the definitions in their full generality so that they are applicable to a wide class of systems, discrete and continuous time, finite and infinite dimensional.
In the first subsection we introduce the concept of a local flow and illustrate its generation by difference and differential equations. Stability definitions for both trajectories and equilibrium states are given in the next subsection and we show how it is often possible to reduce the stability considerations of a trajectory to those of an equilibrium state. In the third subsection we restrict our considerations to time-invariant systems and introduce limit points and the limit set of a given trajectory. An important theorem is proved which gives conditions for the limit set to be non-empty and compact. As an example of the application of the theorem we study the flow generated by a nonlinear delay equation. Next, in a starred subsection we consider recurrency properties of time-invariant global flows; proving Birkhoff's and Poincaré's Recurrency Theorems and illustrating the results with a study of rational systems on the plane. Finally we extend the stability definitions to closed sets, introduce the basin of attraction and describe two examples – the Cayley Problem and the Lorenz attractor.

3.1.1 Local Flows

The purpose of this chapter is to study *the long term behaviour of trajectories under perturbations of the initial state*, so we will only consider time domains T which are unbounded to the right. We will also assume that the control $\bar{u}(\cdot) \in \mathcal{U}$ is fixed and neglect the output. In other words we study the local flow $\mathfrak{F} = (T, X, \varphi)$ determined by the fixed control $\bar{u}(\cdot)$

$$\varphi(t; t_0, x^0) = \varphi(t; t_0, x^0, \bar{u}(\cdot)) , \quad t \in T_{t_0}(x^0) := T_{t_0, x^0, \bar{u}(\cdot)} , \quad (t_0, x^0) \in T \times X .$$

The following concept of a *local flow* extends the notion of a differentiable local flow given in Definition 2.1.8. It does not require differentiability of the trajectories and is applicable to both continuous and discrete time systems.

Definition 3.1.1. (Local flow). $\mathfrak{F} = (T, X, \varphi)$ is said to be a *local flow* with time domain $T \subset \mathbb{R}$ and state transition function φ on a metric space (X, d), if for every $(t_0, x^0) \in T \times X$ there exists $t_+(t_0, x^0) \in (t_0, \infty]$ such that $\varphi(t; t_0, x^0) \in X$ is defined for all $t \in T_{t_0}(x^0) = T \cap \big[t_0, t_+(t_0, x^0)\big)$ and satisfies for all $(t_0, x^0) \in T \times X$, $t, t_1 \in T_{t_0}(x^0)$, $t \geq t_1$

(LF 1) $\varphi(t_0; t_0, x^0) = x^0$.

(LF 2) $T_{t_1}(\varphi(t_1; t_0, x^0)) = T_{t_1} \cap T_{t_0}(x^0)$ and $\varphi(t; t_0, x^0) = \varphi(t; t_1, \varphi(t_1; t_0, x^0))$.

(LF 3) The map $(t, s, x) \mapsto \varphi(t; s, x)$ is continuous in the sense: If (t_k, s_k, x^k) converges in $T \times T \times X$ to (t, s, x) where $t \in T_s(x)$, $t > s$, then $t_k \in T_{s_k}(x^k)$ for k sufficiently large and $\lim_{k \to \infty} \varphi(t_k; s_k, x^k) = \varphi(t; s, x)$.

$\mathfrak{F} = (T, X, \varphi)$ is called a *global flow* if additionally $t_+(t_0, x^0) = \infty$ for all $(t_0, x^0) \in T \times X$.

Note that in the discrete time case the above definition includes the possibility that there is no flow starting at (t_0, x^0), i.e. $T_{t_0}(x^0) = \{t_0\}$, whereas this cannot happen in the continuous time case.
(LF 1) and (LF 2) are counterparts of the consistency axiom and the cocycle property in the Definition 2.1.1 of a dynamical system. (LF 3) is a continuity condition which implies that the trajectories $t \mapsto \varphi(t, t_0, x^0)$ are continuous on $T_{t_0}(x^0)$ and the current state $x(t) = \varphi(t, t_0, x^0)$ depends continuously on the initial condition $x(t_0) = x^0$. By (LF 3)

$$\mathcal{D}_\varphi = \{(t; t_0, x^0) \in T \times T \times X; t > t_0, t \in T_{t_0}(x^0)\}$$

is an open subset of $T \times T \times X$ and $\varphi : \mathcal{D}_\varphi \to X$ is continuous.
$T_{t_0}(x^0)$ is called the *life span* of the trajectory $\varphi(\cdot\,; t_0, x^0)$, and it is required to be a (semi-open) *interval in* T. For every t_0, the set of initial states $x^0 \in X$ which generate, at time t_0, trajectories with *infinite life span* is denoted by

$$X_\infty(t_0) = \big\{x^0 \in X \,; \quad t_+(t_0, x^0) = \infty\big\} . \tag{1}$$

By (LF3), if $x^0 \in X_\infty(t_0)$ then, for every $t \in T$, $t > t_0$, there exists a positive $\delta = \delta(t, t_0, x^0) > 0$ such that $t \in T_\tau(x)$ for all $\tau \in [t_0 - \delta, t_0 + \delta] \cap T$ and $x \in B(x^0, \delta)$. But, in general, there will *not* exist $\delta > 0$ such that $t_+(t_0, x) = \infty$ for all $x \in B(x^0, \delta)$, i.e. $X_\infty(t_0)$ will in general not be open (see Example 3.1.5).
A local flow $\mathfrak{F}$ with time domain $T = \mathbb{R}_+$ or $\mathbb{N}$ is called *time–invariant*, if $T_{t_0}(x^0) = t_0 + T_0(x^0)$ and $\varphi(t_0 + t; t_0, x^0) = \varphi(t; 0, x^0)$ for all $t_0 \in T$, $t \in T_0(x^0)$. Then

$t_+(t_0, x^0) = t_0 + t_+(0, x^0)$ and $X_\infty(t_0) = X_\infty(0)$, for all $t_0 \in T$. For time–invariant local flows we simplify the notation and write $t_+(x^0)$, $T(x^0)$, $\varphi(t; x^0)$ and X_∞ instead of $t_+(0, x^0)$, $T_0(x^0)$, $\varphi(t; 0, x^0)$ and $X_\infty(0)$. Note that $\mathfrak{F}$ induces a global flow on X_∞.

Example 3.1.2. Consider the difference equation

$$x(t+1) = f(t, x(t)) \tag{2}$$

where $f : T \times X \to \tilde{X}$ is continuous, $T = \mathbb{N}$ or $\mathbb{Z}$, and X is an open subset of a metric space $(\tilde{X}, d)$. For every $(t_0, x^0) \in T \times X$ let $x(\cdot) = \varphi(\cdot; t_0, x^0)$ denote the solution of (2) with $x(t_0) = x^0$ on a maximal T-interval $T_{t_0}(x^0) = [t_0, t_1 + 1) \cap T$, $t_1 \in T$, i.e. $x(t) \in X$ for $t \in T_{t_0}(x^0)$, $x(\cdot)$ is a solution of (2) for all $t = t_0, \ldots, t_1$ and, if $t_1 < \infty$, then $f(x(t_1), t_1) \notin X$. Note that if $f(t_0, x^0) \notin X$ there is no flow starting (t_0, x^0).
If $t \in T_{t_0}(x^0)$, $t > t_0$ it is easily shown by induction that, for every $\varepsilon > 0$, there exists $\delta > 0$ such that $t \in T_{t_0}(x)$ and $d(\varphi(t; t_0, x), \varphi(t; t_0, x^0)) < \varepsilon$ for all $x \in B(x^0, \delta)$. Hence $\mathcal{F} = (T, X, \varphi)$ is a local flow on X. If the RHS of (2) does not depend upon t the corresponding local flow is time–invariant. In this case

$$X_\infty = \bigcap_{k \in \mathbb{N}} f^{-k}(X)$$

where $f^{-k}(X)$ is the preimage of $f^{-(k-1)}(X)$ by f, $k \in \mathbb{N}^*$, $f^0 = I_X$. □

In an analogous fashion a differential equation $\dot{x}(t) = f(t, x(t))$ on $\mathbb{K}^n$ defines a local flow if the right hand side satisfies the Carathéodory conditions. This is a consequence of the continuous dependence of solutions on initial conditions as stated in the following theorem, see *Notes and References*.

Theorem 3.1.3. *Let T be an interval, X an open subset of $\mathbb{K}^n$ and assume $f : T \times X \to \mathbb{K}^n$ satisfies the Carathéodory conditions. Suppose $[a, b] \subset T$, $t_0 \in [a, b]$, $x^0 \in X$ and $x(\cdot) = \varphi(\cdot; t_0, x^0)$ is a solution of*

$$\dot{x}(t) = f(t, x(t)), \tag{3}$$

on $[a, b]$ satisfying $x(t_0) = x^0$. Then there exists $\delta > 0$ such that for all $(\tau, x) \in T \times X$ satisfying $|\tau - t_0| < \delta$, $\|x - x^0\|_{\mathbb{K}^n} < \delta$ the (unique) solution $\varphi(\cdot; \tau, x)$ of (3) with $x(\tau) = x$ exists on $[a, b]$. Moreover,

$$\varphi(t; \tau, x) \to \varphi(t; t_0, x^0) \quad \text{as} \quad (\tau, x) \to (t_0, x^0)$$

uniformly in $t \in [a, b]$.

The study of local flows on metric spaces is called *Topological Dynamics.* One of its basic concepts is the notion of an *invariant set.*

Definition 3.1.4 (Invariant set). A non-empty subset $S \subset X$ is said to be *weakly invariant* for $\mathfrak{F}$ if $(t_0, x) \in T \times S$ implies $\varphi(t; t_0, x) \in S$ for all $t \in T_{t_0}(x)$. It is said to be *invariant* for $\mathfrak{F}$ if in addition $S \subset X_\infty(t_0)$ for all $t_0 \in T$.

X is weakly invariant for any local flow on X, whereas it is invariant if and only if $\mathfrak{F}$ is a global flow on X. $\mathfrak{F}$ induces a global flow on every invariant subset S of X. Now suppose that a local flow $\mathfrak{F}$ is time–invariant, then a subset $S \subset X$ is weakly

invariant for $\mathfrak{F}$ if and only if it is a union of the orbits $\mathcal{O}(x^0) = \{\varphi(t; x^0),\ t \in T(x^0)\}$, $x^0 \in S$ of $\mathfrak{F}$. By time–invariance, if $x^0 \in X_\infty$ then $\varphi(t; x^0) \in X_\infty$ for all $t \in T$. Hence X_∞ is the largest invariant set in X for $\mathfrak{F}$.
In the next example we present two local flows defined by differential equations, one time–invariant and the other time–varying. The example illustrates that $X_\infty(t_0)$ will in general depend upon t_0 in the latter case.

Example 3.1.5. Consider the scalar time–invariant system

$$\dot{x} = -x(1 - x^2)\,, \quad x(t_0) = x_0\,. \tag{4}$$

The solution is given by

$$\varphi(t; t_0, x_0) = x_0 / \big(x_0^2 - (x_0^2 - 1)e^{2(t-t_0)}\big)^{1/2}\,.$$

By Theorem 3.1.3 $\mathfrak{F} = (\mathbb{R}_+, \mathbb{R}, \varphi)$ is a time–invariant local flow and it is easy to see that

$$t_+(x_0) = \ln\{|x_0|/(x_0^2 - 1)^{1/2}\}\,,\ |x_0| > 1\,, \quad X_\infty = \{x \in \mathbb{R}\,; \quad |x| \le 1\}.$$

Any interval of the form $[a, b]$ where $-1 \le a \le 0 \le b \le 1$ is invariant for $\mathfrak{F}$. An interval $[a, b]$ with $b > 1$ is neither invariant nor weakly invariant for $\mathfrak{F}$. Intervals of the form $[a, \infty)$ with $a \ge 1$ are weakly invariant, but not invariant.

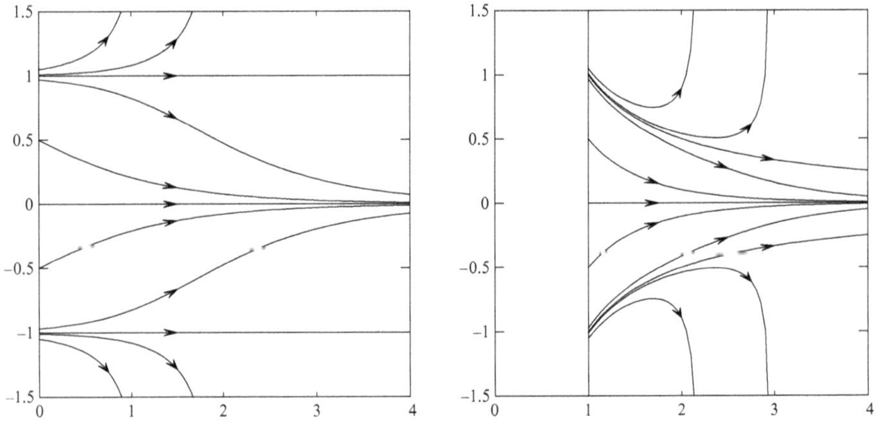

Figure 3.1.2: Local flows for (4) and for (5) with $t_0 = 1$

Now consider

$$\dot{\tilde{x}} = -(1 + 1/t)\tilde{x} + \tilde{x}^3 t^2\,, \quad t > t_0 > 0\,. \tag{5}$$

This equation is related to (4) in the sense that $\tilde{x}(t)$ solves (5) if and only if $x(t) = t\tilde{x}(t)$ solves (4). Hence the solution is given by

$$\tilde{\varphi}(t; t_0, x_0) = x_0 t_0 / [t\,(x_0^2 t_0^2 - (x_0^2 t_0^2 - 1)e^{2(t-t_0)})^{1/2}]\,.$$

So $\tilde{\mathfrak{F}} = \big((0, \infty), \mathbb{R}, \tilde{\varphi}\big)$ is a local flow and

$$t_+(t_0, x_0) = t_0 + \ln\{|x_0| t_0 / (x_0^2 t_0^2 - 1)^{1/2}\}\,,\ |x_0| t_0 > 1\,, \quad X_\infty(t_0) = \{x \in \mathbb{R}\,; \quad |x| \le t_0^{-1}\}.$$

The only invariant subset for $\tilde{\mathfrak{F}}$ is $\{0\}$. The intervals $[0, \infty)$ and $(0, \infty)$ are weakly invariant for $\tilde{\mathfrak{F}}$. □

3.1.2 Stability Definitions

In this subsection we introduce the basic notions of stability for local flows $\mathfrak{F} = (T, X, \varphi)$ on a metric space (X, d). We assume that T is unbounded to the right and is either an interval or a discrete additive semigroup of $\mathbb{R}$ (typically $T = \mathbb{R}_+$ or $T = \mathbb{N}$).

In order to prove that a given trajectory $\varphi(\cdot; t_0, x^0) : T_{t_0} \to X$ of a differential or difference equation is stable we have to show that all the solutions starting sufficiently close to x^0 at time t_0 have infinite life span, i.e. do not escape from the domain of definition in finite time (see Theorem 2.1.14). This is a natural requirement of stability; a trajectory (with infinite life span) should certainly not be called stable if a small deviation from the initial state leads to an explosion (finite escape time).

Definition 3.1.6. (Stability of a trajectory). A trajectory $t \mapsto \varphi(t; t_0, \overline{x})$, $\overline{x} \in X_\infty(t_0)$ of a local flow $\mathfrak{F}$ is said to be *stable* at time $t_0 \in T$ if for all $\varepsilon > 0$, there exists $\delta = \delta(\varepsilon, t_0) > 0$ such that $B(\overline{x}, \delta) \subset X_\infty(t_0)$ and for all $x^0 \in B(\overline{x}, \delta)$

$$d(\varphi(t; t_0, x^0), \varphi(t; t_0, \overline{x})) < \varepsilon \,, \quad t \in T_{t_0} \,. \tag{6}$$

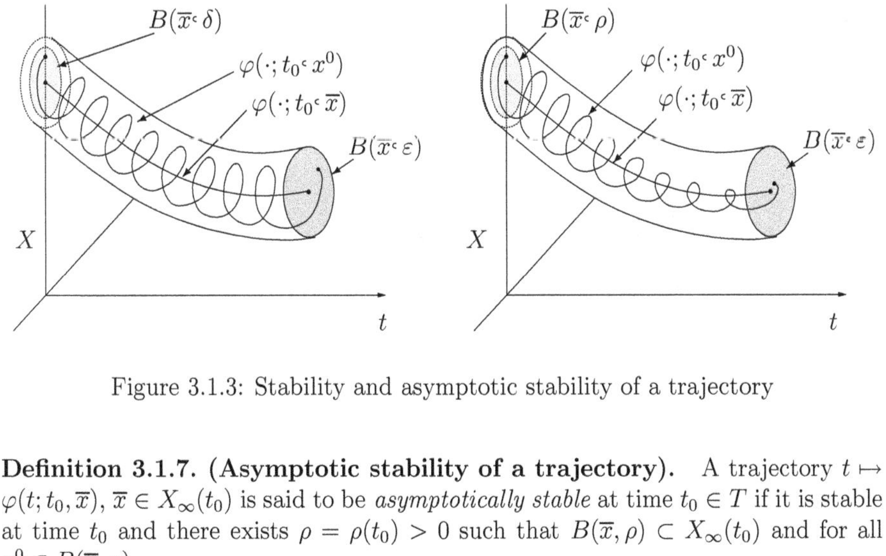

Figure 3.1.3: Stability and asymptotic stability of a trajectory

Definition 3.1.7. (Asymptotic stability of a trajectory). A trajectory $t \mapsto \varphi(t; t_0, \overline{x})$, $\overline{x} \in X_\infty(t_0)$ is said to be *asymptotically stable* at time $t_0 \in T$ if it is stable at time t_0 and there exists $\rho = \rho(t_0) > 0$ such that $B(\overline{x}, \rho) \subset X_\infty(t_0)$ and for all $x^0 \in B(\overline{x}, \rho)$

$$\lim_{t \to \infty} d(\varphi(t; t_0, x^0), \varphi(t; t_0, \overline{x})) = 0. \tag{7}$$

Here and in the following we omit obvious restrictions like "$t \in T_{t_0}$" under the limit. The trajectory $\varphi(t; t_0, x^0)$ is called *uniformly stable* or *uniformly asymptotically stable* if in the previous definitions δ, ρ do not depend on t_0 and the limit in (7) is uniform in t_0. If $\mathfrak{F}$ is time–invariant then (asymptotic) stability implies uniform (asymptotic) stability.

Now suppose that $\overline{x}$ is an equilibrium state of a local flow $\mathfrak{F}$, i.e. $\varphi(t; t_0, \overline{x}) = \overline{x}$ for all $t \in T_{t_0}$, $t_0 \in T$. Then the above definitions specialize to the following.

Definition 3.1.8. (Stability of an equilibrium state). An equilibrium state $\overline{x}$ of a local flow $\mathfrak{F}$ is *stable* at time $t_0 \in T$ if for all $\varepsilon > 0$ there exists $\delta = \delta(\varepsilon, t_0)$ such that $B(\overline{x}, \delta) \subset X_\infty(t_0)$ and

$$x^0 \in B(\overline{x}, \delta) \;\Rightarrow\; \varphi(t; t_0, x^0) \in B(\overline{x}, \varepsilon), \quad t \in T_{t_0}. \tag{8}$$

Definition 3.1.9. (Asymptotic stability of an equilibrium state).

(i) An equilibrium state $\overline{x}$ of a local flow $\mathfrak{F}$ is called *attractive* at time t_0 if there exists $\rho = \rho(t_0) > 0$ such that $B(\overline{x}, \rho) \subset X_\infty(t_0)$ and

$$x^0 \in B(\overline{x}, \rho) \Rightarrow \lim_{t \to \infty} \varphi(t; t_0, x^0) = \overline{x}. \tag{9}$$

(ii) $\overline{x}$ is said to be *asymptotically stable* at time t_0, if it is stable and attractive at time t_0.

(iii) If $\overline{x}$ is attractive the *basin of attraction* of $\overline{x}$ at time t_0 is given by

$$\mathcal{A}(t_0, \overline{x}) = \{x^0 \in X_\infty(t_0);\ \lim_{t \to \infty} \varphi(t; t_0, x^0) = \overline{x}\}. \tag{10}$$

$\overline{x}$ is said to be *globally attractive* at time t_0 if $\mathcal{A}(t_0, \overline{x}) = X$.

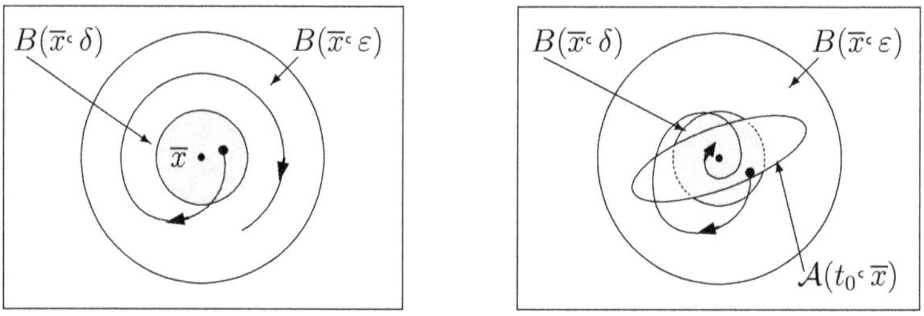

stable — asymptotically stable

Figure 3.1.4: Stability and asymptotic stability of an equilibrium point

An equilibrium point $\overline{x}$ will be *unstable* at time t_0 if *either* there does not exist $\delta > 0$ such that $B(\overline{x}, \delta) \subset X_\infty(t_0)$ *or* there exists $\varepsilon > 0$ and for every $\delta > 0$, no matter how small, there exist $x^0 \in B(\overline{x}, \delta)$ and $t \in T_{t_0}$ such that $\varphi(t; t_0, x^0) \notin B(\overline{x}, \varepsilon)$.

Example 3.1.10. (i) Consider again the time-invariant flow defined by (4). There are equilibrium points at $0, +1, -1$. If $x_0^2 > 1$, then $t_+(x^0) < \infty$ for all $t_0 \in \mathbb{R}_+$ and hence both $+1$ and -1 are unstable. Whereas if $x_0^2 < 1$ we have $|\varphi(t; t_0, x_0)| \le |x_0| e^{-(t-t_0)}/(1 - x_0^2)^{1/2}$ and hence the origin is uniformly asymptotically stable with basin of attraction $\mathcal{A}(0) = \{x \in \mathbb{R};\ |x| < 1\}$.
For the flow $\tilde{\mathfrak{F}}$ defined by (5) the origin is asymptotically stable at time $t_0 \in (0, \infty)$ with basin of attraction $\mathcal{A}(t_0, 0) = \{x \in \mathbb{R}\ ;\ \ |x| < t_0^{-1}\}$. Note that both the sets $\mathcal{A}(t_0, 0)$ and $X_\infty(t_0)$ contract to $\{0\}$ as $t_0 \to \infty$. Hence the origin is not uniformly attractive.
(ii) The linear scalar system

$$\dot{x} = \frac{\dot{g}(t)}{g(t)} x\,, \quad g(t) = t^{-2+\cos t}$$

defines a flow $\mathcal{F} = ((0,\infty), \mathbb{R}, \varphi)$ with $\varphi(t; t_0, x_0) = g(t)x_0/g(t_0)$, $t \in T_{t_0}$. The origin is stable and attractive at any time $t_0 > 0$. But it is not uniformly stable since if $t_k = 2k\pi$, $t_{0k} = (2k-1)\pi$, $k = 1, 2, \ldots$ then

$$\lim_{k\to\infty} \varphi(t_k; t_{0k}, x_0) = \lim_{k\to\infty} (2k\pi)^{-1} x_0/((2k-1)\pi)^{-3} = \infty\,.$$

(iii) Consider the time–invariant reversible flow on $\mathbb{R}^2 \setminus \{(0,0)\}$ described in polar coordinates by the differential equations

$$\dot{r} = r(1-r), \quad \dot{\theta} = \sin^2(\theta/2).$$

The flow has one equilibrium state which in Cartesian coordinates is given by $\bar{x} = (1,0)$. It is unstable, yet all trajectories $\varphi(t; x^0)$ in $\mathbb{R}^2 \setminus \{(0,0)\}$ are attracted by $\bar{x}$. So $\bar{x}$ is an attractive but unstable equilibrium state. Note further that any trajectory $\varphi(t; x^0)$ starting on the unit circle remains on the unit circle and $\varphi(t; x^0)$ converges towards $\bar{x}$ for $t \to \infty$ and for $t \to -\infty$. □

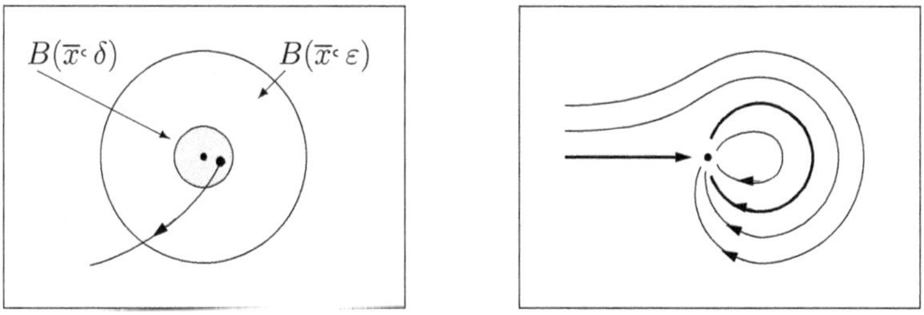

Unstable   Unstable and attractive

Figure 3.1.5: Unstable equilibrium points

For many local flows it is possible to reduce stability considerations of a trajectory to the corresponding considerations for an equilibrium state. To see how this can be carried out consider a dynamical system where the state evolution is determined via the differential equation

$$\dot{z}(t) = \bar{f}(t, z(t), u(t))\,, \quad z(t_0) = z^0\,. \tag{11}$$

For fixed $\bar{u}(\cdot) \in \mathcal{U}$ let the trajectory under consideration be

$$\bar{z}(t) = \varphi(t; t_0, \bar{z}) = \varphi(t; t_0, \bar{z}, \bar{u}(\cdot))\,, \quad t \in T_{t_0}.$$

Setting $z(t) = \bar{z}(t) + x(t)$ and $u(t) = \bar{u}(t)$ in (11), we obtain

$$\dot{x}(t) = \bar{f}(t, x(t) + \bar{z}(t), \bar{u}(t)) - \bar{f}(t, \bar{z}(t), \bar{u}(t)), \quad x(t_0) = z^0 - \bar{z}. \tag{12}$$

If we define

$$f(t,x) = \bar{f}(t, x + \bar{z}(t), \bar{u}(t)) - \bar{f}(t, \bar{z}(t), \bar{u}(t)) \quad \text{and} \quad x^0 = z^0 - \bar{z},$$

then (12) becomes

$$\dot{x}(t) = f(t, x(t)), \quad x(t_0) = x^0 \tag{13}$$

where $f(t,0) = 0$ for all $t \in T_{t_0}$. So the origin is an equilibrium state of (13) and the deviations of the solutions of (11) from the trajectory $\bar{z}(\cdot)$ are identical to the deviations of the solutions of (13) from the origin. In particular, the trajectory $\bar{z}(\cdot)$ of the local flow described by (11) is stable (asymptotically stable, unstable) if and only if the equilibrium point $\overline{x} = 0$ of (13) is stable (resp. asymptotically stable, unstable). Now suppose $\bar{f}$ is linear in z and u, $\bar{f}(t,z,u) = A(t)z + B(t)u$ then (13) becomes

$$\dot{x}(t) = A(t)x(t) \ . \tag{14}$$

Hence the (asymptotic) stability of the origin with respect to the free system (14) (at time t_0) implies the (asymptotic) stability of every trajectory

$$t \mapsto \varphi(t; t_0, z^0, \bar{u}(\cdot)) \,, \quad z^0 \in X \,,\ \bar{u}(\cdot) \in \mathcal{U}$$

of the original system $\dot{z}(t) = A(t)z(t) + B(t)\bar{u}(t)$. A similar statement can easily be proved for discrete time linear systems. So, by abuse of terminology we will call, *a linear system (asymptotically) stable if the origin is an (asymptotically) stable equilibrium state with respect to* (14). Note, however, that for nonlinear systems this phraseology is not permissible.

3.1.3 Limit Sets

In this subsection we suppose that $\mathcal{F} = (T, X, \varphi)$ is a *time–invariant* local flow on a metric space (X, d) with time domain $T = \mathbb{R}_+$ or $T = \mathbb{N}$. For such flows we introduce the concept of a limit point and limit set, prove a theorem which gives conditions for the limit set to be non-empty and compact, and illustrate the result by a differential delay example.

Definition 3.1.11 (Limit point and limit set). Let $x \in X_\infty$. A point $y \in X$ is said to be a *limit point* of the trajectory $\varphi(\cdot; x)$ if there exists a sequence (t_k) in T satisfying

$$y = \lim_{k\to\infty} \varphi(t_k; x) \quad \text{and} \quad \lim_{k\to\infty} t_k = \infty.$$

The set $\omega(x)$ of all limit points of $\varphi(\cdot; x)$ is called the *limit set* of $\varphi(\cdot; x)$.

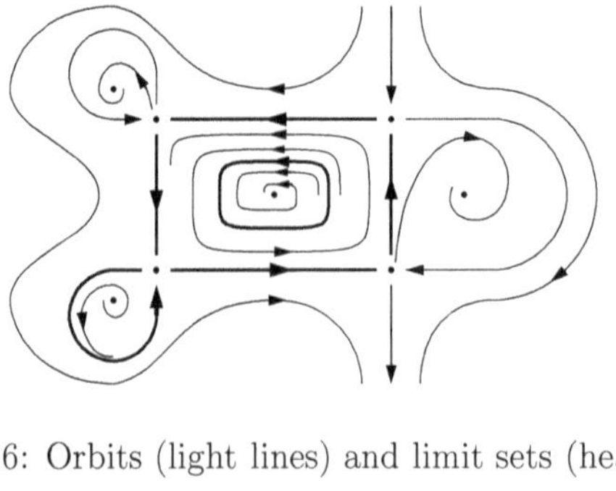

Figure 3.1.6: Orbits (light lines) and limit sets (heavy lines)

For every $x^0 \in X$ we denote by $\mathcal{O}(x^0)$ the (positive) orbit: $\mathcal{O}(x^0) = \{\varphi(t; x^0);\ t \in T(x^0)\}$. Then it follows from the continuity condition (LF 3) (see Ex. 4) that

$$\overline{\mathcal{O}(x^0)} = \mathcal{O}(x^0) \cup \omega(x^0), \quad x^0 \in X_\infty. \tag{15}$$

An equilibrium point $\bar{x}$ of $\mathcal{F}$ is attractive if and only if it belongs to the interior of X_∞ and is the unique limit point of all trajectories starting in a small neighbourhood of $\bar{x}$.

Remark 3.1.12. In the qualitative theory of differentiable systems *backward* limit points and *backward* limit sets $\alpha(x^0)$ where $t_k \to -\infty$ are also introduced. Intuitively $\alpha(x^0)$ is where the orbit of x^0 is "born" and $\omega(x^0)$ is where it "dies". Since we do not suppose $\mathcal{F}$ to be reversible, we will only consider *forward* limit points and omit this qualification in the sequel. □

Definition 3.1.13 (Minimal set). A subset $M \subset X$ is called *minimal* for a local flow if it is non-empty, closed and invariant and does not contain a proper subset with the same properties.

The simplest minimal sets consist of just one closed orbit. If x^0 is an equilibrium point or a periodic point (i.e. $x^0 = \varphi(\tau; x^0)$ for some $\tau > 0$) then $\mathcal{O}(x^0)$ is clearly minimal. All the minimal sets of the flow shown in Figure 3.1.6 are of this kind. In fact, equilibrium points and periodic orbits are the only bounded minimal sets of a differentiable system on $\mathbb{R}^2$. This follows from the theorem of Poincaré-Bendixson which we state here without proof, see *Notes and References.*

Theorem 3.1.14 (Poincaré-Bendixson Theorem). *The limit set of every bounded trajectory of a differentiable flow on the plane $\mathbb{R}^2$ either contains an equilibrium point or is a periodic orbit.*

Every closed (i.e. periodic) orbit of a differentiable planar flow is a Jordan curve and hence encircles a domain which is called its interior. It can be proved that the interior of every such orbit contains an equilibrium point.

Example 3.1.15. (Limit cycle). Consider the nonlinear differential equation

$$\dot{x}_1 = x_1 - x_2 - x_1(x_1^2 + x_2^2), \quad \dot{x}_2 = x_1 + x_2 - x_2(x_1^2 + x_2^2). \tag{16}$$

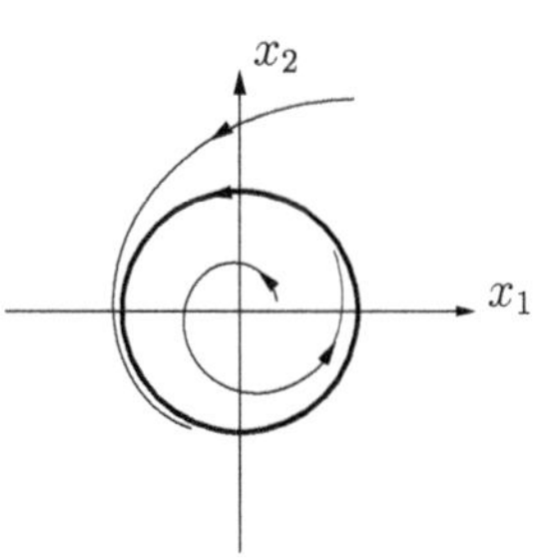

Figure 3.1.7: Stable limit cycle

If we introduce polar coordinates $x_1 = r\cos\theta$, $x_2 = r\sin\theta$, these equations become

$$\dot{r} = r(1 - r^2), \qquad \dot{\theta} = 1,$$

with solutions

$$r(t) = \frac{r_0\, e^t}{\sqrt{1 - r_0^2(1 - e^{2t})}}, \qquad \theta(t) = t + \theta_0.$$

Thus $r(t) \equiv 1$ if $r_0 = 1$, and $\lim_{t\to\infty} r(t) = 1$ if $r_0 \neq 0$. This shows that the unit circle $S^1 = \{x \in \mathbb{R}^2; x_1^2 + x_2^2 = 1\}$ is a periodic orbit of (16) and is the limit set $\omega(x^0)$ of any trajectory $\varphi(t;x^0)$ with $x^0 \neq 0$. Closed orbits which attract all trajectories starting in a suitable neighbourhood of the orbit are called *stable limit cycles.* The phase portrait of (16) is shown in Figure 3.1.7. □

Contrary to the two dimensional case, limit sets of a differentiable system in $\mathbb{R}^n$, $n \geq 3$ may be of a very complex structure (see Example 3.1.27). Nevertheless we will prove that all the limit sets of *bounded differentiable* trajectories share two nice properties of the limit cycle in Figure 3.1.7. They are invariant and connected. In the following theorem we call a subset $S \subset X$ *relatively compact in* X if its closure $\overline{S}$ in X is compact. We say that $\varphi(t;x^0)$, $x^0 \in X_\infty$, *approaches* a set S as $t \to \infty$ if

$$\lim_{t\to\infty} \operatorname{dist}(\varphi(t;x^0), S) = 0, \tag{17}$$

where dist is defined by (2.1.15). For any $\varepsilon > 0$, an ε-neighbourhood of S is denoted by $B(S,\varepsilon) = \{x \in X;\ \operatorname{dist}(x,S) < \varepsilon\}$.

Theorem 3.1.16. *If an orbit $\mathcal{O}(x^0)$, $x^0 \in X_\infty$ of a time-invariant local flow $\mathfrak{F}$ is relatively compact in X, then the corresponding limit set $\omega(x^0)$ is non-empty, compact, weakly invariant and is the smallest closed subset that $\varphi(t;x^0)$ approaches as $t \to \infty$. In the continuous time case $\omega(x^0)$ is connected.*

Proof: Note first that $\omega(x^0) = \bigcap_{\tau\in T} \overline{\Gamma}_\tau$ (see Ex. 4) where

$$\Gamma_\tau = \{\varphi(t;x^0);\ t \in T_\tau\}, \quad \tau \in T.$$

So $\omega(x^0)$ is the intersection of a decreasing family of non-empty compact sets and hence $\omega(x^0)$ itself is non-empty and compact.
Now let $y \in \omega(x^0)$ and $t \in T(y)$, then there exists a sequence (t_k) in T, such that $t_k \to \infty$ and $\varphi(t_k;x^0) \to y$ as $k \to \infty$. By (LF 2), $t \in T(\varphi(t_k;x^0))$ for k sufficiently large and $\varphi(t;\varphi(t_k;x^0)) \to \varphi(t;y)$ as $k \to \infty$. But $\varphi(t;\varphi(t_k;x^0)) = \varphi(t+t_k;x^0)$ and hence $\varphi(t,y) = \lim_{k\to\infty} \varphi(t+t_k;x^0) \in \omega(x^0)$. This proves that $\omega(x^0)$ is weakly invariant.
If S is any closed set in X satisfying (17) and $y = \lim_{k\to\infty} \varphi(t_k;x^0)$ is any limit point then $y \in S$; hence $\omega(x^0) \subset S$. To prove

$$\lim_{t\to\infty} \operatorname{dist}(\varphi(t;x^0), \omega(x^0)) = 0,$$

assume - by way of contradiction - that there exist an $\varepsilon > 0$ and a sequence (t_k) in T such that $t_k \to \infty$ as $k \to \infty$ and $\operatorname{dist}(\varphi(t_k;x^0), \omega(x^0)) \geq \varepsilon$ for all $k \in \mathbb{N}$. Since the sequence $(\varphi(t_k;x^0))_{k\in\mathbb{N}}$ is relatively compact in X, it contains a subsequence converging to some $x^* \in X$. But then $x^* \in \omega(x^0)$ by definition although we have by assumption $\operatorname{dist}(x^*, \omega(x^0)) \geq \varepsilon$. This contradiction proves the above equality.
It remains to prove the connectedness result. Suppose $T = \mathbb{R}_+$ and assume that $\omega(x^0)$ is the union of two disjoint closed non-empty sets Ω_1, Ω_2. Since Ω_1, Ω_2 are compact there exists $\varepsilon > 0$ such that the two open ε-neighbourhoods $U_1 = B(\Omega_1,\varepsilon)$ and $U_2 = B(\Omega_2,\varepsilon)$ are disjoint. Since $\varphi(t;x^0)$ approaches $\omega(x^0) = \Omega_1 \dot\cup \Omega_2$ as $t \to \infty$ there exists $\tau \in T$ such that

$$\varphi(t;x^0) \in B(\omega(x^0),\varepsilon) = U_1 \dot{\cup} U_2, \quad t \in (\tau,\infty).$$

On the other hand, $\omega(x^0)$ is the smallest closed set that $\varphi(t;x^0)$ approaches as $t \to \infty$. Hence $I_i := \{t \in (\tau,\infty);\ \varphi(t;x^0) \in U_i\}$, $i = 1,2$ are both non-empty. It follows that $(\tau,\infty) = I_1 \dot{\cup} I_2$ is a disjoint union of non-empty open subsets. This is a contradiction since the interval (τ,∞) is connected. □

Note that the assumption of compactness is indispensable here. This is illustrated in Figure 3.1.8 where the unboundedness of the trajectory prevents the two limit lines a, b from joining.

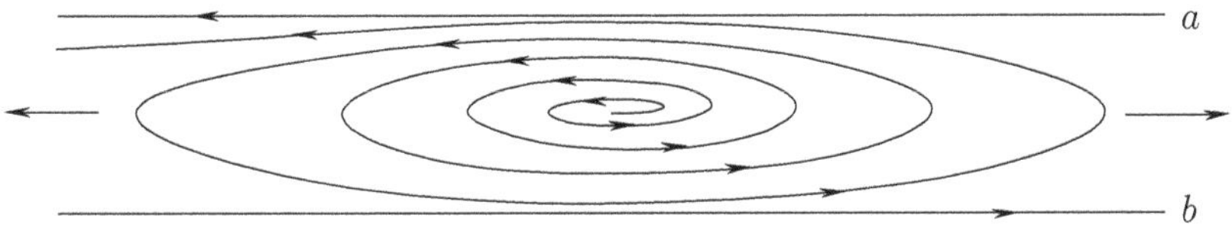

Figure 3.1.8: Unconnected limit set

Remark 3.1.17. If $\mathcal{F}$ is a local flow defined by a time–invariant differential equation satisfying the conditions of Theorem 2.1.14 on X, X open in $\mathbb{K}^n$, the above theorem is applicable to any bounded orbit $\mathcal{O}(x^0)$ whose closure is in X. Then $\omega(x^0)$ is not only weakly invariant but is in fact invariant. To see this note that if $\sup T(y) = t_+(y) < \infty$, $y \in \omega(x^0)$, then by Theorem 2.1.14 either $\lim_{t\to t_+(y)} \|\varphi(t;y)\|_{\mathbb{K}^n} = \infty$ or there is a sequence (t_k) in $T(y)$ such that $t_k \to t_+(y)$ and $\varphi(t_k;y) \to x^* \in \partial X$ as $k \to \infty$. Since $\varphi(t_k;y) \in \omega(x^0)$ by Theorem 3.1.16 both consequences contradict the fact that $\omega(x^0) \subset \overline{\mathcal{O}(x^0)}$ is compact in X (see also Lemma 3.2.14). □

The comments made in the above remark also hold true for certain classes of infinite dimensional systems as the following example illustrates.

Example 3.1.18. Let $X \subset \mathcal{C}$ be an open subset of the Banach space $\mathcal{C} = \mathcal{C}([-h,0],\mathbb{R}^n)$ provided with the sup-norm $\|\psi\| = \max_{t\in[-h,0]} \|\psi(t)\|_{\mathbb{R}^n}$, and let $f : X \to \mathbb{R}^n$ be continuously differentiable. Consider the time-invariant retarded differential equation

$$\dot{y}(t) = f(y_t)\,, \quad y_{t_0} = \psi \tag{18}$$

where $y_t \in \mathcal{C}$, $t \geq 0$ denotes the segment $s \mapsto y(t+s)$, $s \in [-h,0]$ and $\psi \in X$ is a given initial function (cf. Example 2.1.25). For any $t_0 \in \mathbb{R}$, a continuous function

$$y(\cdot) : [t_0 - h, t_1) \to \mathbb{R}^n, \quad t_1 > t_0$$

is called a solution of (18) if $y(t_0+t) = \psi(t)$ for $t \in [-h,0]$, $y(\cdot)$ is differentiable on (t_0,t_1) and satisfies $\dot{y}(t) = f(y_t)$ for $t \in (t_0,t_1)$. It is well known that (18) admits a unique solution $y(\cdot;t_0,\psi)$ on some maximal time interval $[t_0 - h, t_+(t_0,\psi))$. Define

$$\varphi(t;t_0,\psi) = y_t(t_0,\psi), \quad t \in T_{t_0}(\psi) := [t_0, t_+(t_0,\psi)), \quad (t_0,\psi) \in \mathbb{R} \times X\,. \tag{19}$$

Clearly $y(\cdot) : [t_0-h,t_1) \to \mathbb{R}^n$ is a solution of (18) if and only if $y(t_0+\cdot) : [-h,t_1-t_0) \to \mathbb{R}^n$ is a solution of (18) with 0 instead of t_0. Therefore

$$\varphi(t;t_0,\psi) = \varphi(t-t_0;0,\psi) =: \varphi(t-t_0;\psi), \quad t \in T_{t_0}(\psi) = t_0 + T_0(\psi).$$

Moreover, just as for ordinary differential equations one can prove the following basic properties [214].

(i) If $y(t) = y(t; 0, \psi)$ is a bounded solution on $[0, t_+(\psi))$ (where $t_+(\psi) := t_+(0, \psi)$) and the closure of the orbit $\mathcal{O}(\psi)$ in $\mathcal{C}$ is contained in X then $t_+(\psi) = \infty$.

(ii) If $\psi \in X$, $t < t_+(\psi)$ then $(t_k, \psi_k) \to (t, \psi)$ implies $t_k < t_+(\psi_k)$ for all k sufficiently large and $\varphi(t_k; \psi_k) \to \varphi(t; \psi)$ as $k \to \infty$.

As a consequence we see that $\mathfrak{F} = (\mathbb{R}_+, X, \varphi)$ is a time–invariant local flow.
Now let us assume that f maps closed bounded subsets of X into bounded sets of $\mathbb{R}^n$ and consider a bounded trajectory $y(t) = y(t; 0, \psi)$ satisfying $\overline{\mathcal{O}(\psi)} \subset X$. Then by property (i) $\psi \in X_\infty$. In order to apply Theorem 3.1.16 we have to show that $\mathcal{O}(\psi)$ is relatively compact in X, i.e. $\overline{\mathcal{O}(\psi)} \subset X$ is compact in $\mathcal{C}$. Let c be an upper bound of $\|f(y_t)\|_{\mathbb{R}^n}$, $t \geq 0$, then

$$\|y(t') - y(t)\| = \|\int_t^{t'} f(y_s)\, ds\| \leq c|t' - t|\,, \quad t, t' \geq 0\,.$$

Since $y_0 = \psi$ is uniformly continuous on $[-h, 0]$ it follows that $\mathcal{O}(\psi) = \{y_t;\ t \geq 0\} \subset \mathcal{C}$ is equicontinuous at each point $s \in [-h, 0]$: For any $\varepsilon > 0$ there exists $\delta > 0$ such that for all $s' \in [-h, 0]$, $|s' - s| < \delta$, we have

$$\|y_t(s') - y_t(s)\| = \|y(t + s') - y(t + s)\| < \varepsilon\,, \quad t \in \mathbb{R}_+\,.$$

By the theorem of Arzela–Ascoli any bounded and equicontinuous subset of $\mathcal{C}$ is relatively compact. So the conditions of Theorem 3.1.16 are satisfied. Applying this theorem, for any $x \in \omega(\psi)$ we have $\varphi(t; x) \in \omega(\psi) \subset \overline{\mathcal{O}(\psi)}$ for all $t \in T(x)$, and so $T(x) = \mathbb{R}_+$ by property (i). Concluding we have: if $y(\cdot, \psi)$ is a bounded trajectory satisfying $\overline{\mathcal{O}(\psi)} \subset X$ then the limit set $\omega(\psi)$ of $t \mapsto y_t(\psi)$ is a non-empty, compact, connected, invariant subset of $\mathcal{C}$ such that $y_t(\psi) \to \omega(\psi)$ as $t \to \infty$. □

3.1.4 Recurrence

We continue our development of time-invariant flows but now we make the further assumption that we have a *global* flow $\mathfrak{F} = (T, X, \varphi)$ so that $T(x) = T = \mathbb{R}_+$ or $\mathbb{N}$ for all $x \in X$. An important classification of orbits is based on the relation between the initial point x^0 and the corresponding limit set $\omega(x^0)$. Various concepts of "recurrence" or "almost periodicity" are used in the literature to describe the (approximate) return of a trajectory to its initial state as $t \to \infty$. In this subsection we first define the concept of a *wandering point* and prove a theorem due to Poincaré which gives conditions for the set of non-wandering points to be X. Then we introduce the notion of a *recurrent point* and prove Birkhoff's Recurrence Theorem which characterizes compact minimal sets in terms of recurrent points. Finally we give an example of a discrete time rational system on the plane which illustrates some of the ideas.
A point x^0 is called wandering if the trajectories which start near x^0 move uniformly away from their origins as $t \to \infty$. More precisely

Definition 3.1.19 (Wandering, non–wandering). A point $x^0 \in X$ is said to be a *wandering point* for $\mathfrak{F}$ if there exists a neighbourhood U of x^0 and $\tau \in T$ such that $U \cap U_t = \emptyset$ for all $t \in T_\tau$, where

$$U_t = \varphi(t; U) = \{\varphi(t; x);\ x \in U\}. \tag{20}$$

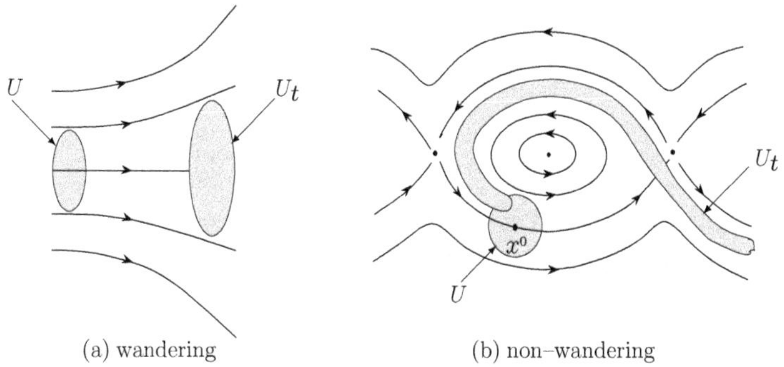

Figure 3.1.9: Wandering and non–wandering sets

Otherwise x^0 is called *non–wandering*. The set of non–wandering points is denoted by $\Omega(\mathfrak{F})$.

If x^0 is a wandering point in a state space of bounded volume, it is intuitively clear that the volumes of the moving neighbourhoods U_t must contract as $t \to \infty$. To make this precise, let X be an open set in $\mathbb{R}^n$ and λ be the Lebesgue measure on $\mathbb{R}^n$. A flow $\mathfrak{F}$ on X is said to be *volume preserving* if $\lambda(U_t) = \lambda(U)$ for all open subsets U in X and $t \in T$.

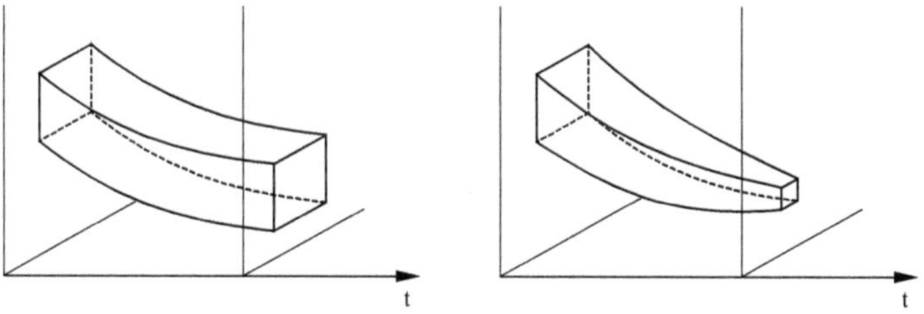

Figure 3.1.10: Volume preserving and volume contracting flows

If $\mathfrak{F}$ is described by a difference equation of the time–invariant form of (2), this is equivalent to

$$\lambda(f(U)) = \lambda(U), \quad U \subset X \text{ open.} \tag{21}$$

A differentiable system of the time–invariant form of (3) is volume preserving if and only if

$$\operatorname{div} f(x) = \sum_{i=1}^{n} \frac{\partial f_i}{\partial x_i}(x) = 0\,, \quad x \in X. \tag{22}$$

In fact, by Liouville's Theorem (see *Arnold* (1978) [18]), div f determines the rate of volume expansion due to the corresponding flow

$$\frac{d}{dt}(\lambda(U_t)) = \int_{U_t} \operatorname{div} f(x)\, dx. \tag{23}$$

Theorem 3.1.20 (Poincaré's Recurrence Theorem). *Suppose that* $\mathcal{F} = (T, X, \varphi)$ *is a global, time–invariant, volume preserving flow on an open set* $X \subset \mathbb{R}^n$ *of finite measure* $\lambda(X) < \infty$. *Then* $\Omega(\mathcal{F}) = X$.

Proof: Let x^0 be any point in X, V an open neighbourhood of x^0 and $\tau \in \mathbb{N}^*$. Then there exist t_1, $t_2 \in \mathbb{N}$, $t_2 - t_1 \geq \tau$ such that $\lambda(V_{t_1} \cap V_{t_2}) > 0$ since otherwise $\lambda(X) \geq \sum_{k \in \mathbb{N}} \lambda(V_{k\tau}) = \infty$. Let $W = V_{t_1} \cap V_{t_2}$ and $W_i = \{x \in X;\ \varphi(t_i; x) \in W\}$. Then $W \supset (W_i)_{t_i} \supset (W_i \cap V)_{t_i} \supset W$ and hence $\lambda(W_i) = \lambda(W_i \cap V) = \lambda(W) > 0$, $i = 1, 2$. To prove that x^0 is non–wandering it suffices to show that $V \cap V_{t_2 - t_1} \neq \emptyset$. Now suppose $V \cap V_{t_2 - t_1} = \emptyset$. Then $(V \cap W_1) \cap (W_2 \cap V)_{t_2 - t_1} = \emptyset$. So

$$\begin{aligned} \lambda(W_1) + \lambda(W_2) &= \lambda(W_1 \cap V) + \lambda\left((W_2 \cap V)_{t_2 - t_1}\right) = \lambda\left((W_1 \cap V) \cup (W_2 \cap V)_{t_2 - t_1}\right) \\ &\leq \lambda(W_1), \end{aligned}$$

since $(W_2)_{t_2 - t_1} \subset W_1$. Hence we have a contradiction. □

The following proposition shows that the dynamics of a system is ultimately concentrated on the set of all its non–wandering points; all limit sets and in particular all equilibrium points, periodic and recurrent orbits lie in $\Omega(\mathcal{F})$.

Proposition 3.1.21. *Suppose that* $\mathcal{F} = (T, X, \varphi)$ *is a global, time–invariant flow. Then the set* $\Omega(\mathcal{F})$ *of non–wandering points of* $\mathcal{F}$ *is closed, invariant and contains all the limit sets* $\omega(x)$, $x \in X$ *of trajectories of* $\mathcal{F}$.

Proof: $x \in \Omega(\mathcal{F})$ if and only if for every neighbourhood U of x and every $\tau \in T$ there exists $t \in T_\tau$, such that $U_t \cap U \neq \emptyset$. Hence $\Omega(\mathcal{F})$ is closed. In order to prove $\varphi(t; x^0) \in \Omega(\mathcal{F})$ for all $x^0 \in \Omega(\mathcal{F})$, $t \in T$, let V be a neighbourhood of $\varphi(t; x^0)$ and $\tau \in T$. By continuity of $\varphi(t; \cdot)$ there exists a neighbourhood U of x^0 such that $U_t \subset V$. Since x^0 is non–wandering there is $t_1 \in T_\tau$ satisfying $U_{t_1} \cap U \neq \emptyset$, hence

$$V_{t_1} \cap V \supset U_{t+t_1} \cap U_t \supset (U_{t_1} \cap U)_t \neq \emptyset$$

and so $\varphi(t; x^0) \in \Omega(\mathcal{F})$. Finally, let $z \in \omega(x)$, $x \in X$ and $t_k \to \infty$, $t_{k+1} \geq t_k$ such that $\varphi(t_k; x) \to z$ as $k \to \infty$. If U is any neighbourhood of z then there exists $k_0 \in \mathbb{N}$ satisfying $\varphi(t_k; x) \in U$, for $k \geq k_0$, hence $\varphi(t_k; x) = \varphi(t_k - t_{k_0}; \varphi(t_{k_0}; x)) \in U \cap U_{t_k - t_{k_0}} \neq \emptyset$ for $k \geq k_0$. □

"Non–wandering" is a very weak concept of recurrence (compare Figure 3.1.9(b)). A stronger recurrence concept requires that the trajectory $\varphi(t; x^0)$ itself - not only the moving neighbourhoods U_t (20) - returns arbitrarily close to x^0 as $t \to \infty$. Then x^0 is called (positively) *Poisson stable*. This recurrence condition can be expressed in terms of the limit set $\omega(x^0)$ in three equivalent ways (see Ex. 12)

$$x^0 \in \omega(x^0) \quad \text{or} \quad \varphi(t; x^0) \in \omega(x^0),\ t \in T \quad \text{or} \quad \overline{\mathcal{O}(x^0)} = \omega(x^0). \tag{24}$$

An even stronger condition is formulated in the next definition where it is required that the trajectory $\varphi(t; x^0)$ returns to any neighbourhood U of x^0 *at least once in every period of length* $\tau(U)$, where $\tau(U)$ is chosen suitably large.

Definition 3.1.22 (Recurrent point). A point $x^0 \in X$ is said to be *recurrent* for a time–invariant global flow $\mathfrak{F}$ if for any neighbourhood U of x^0 there exist $\tau(U) > 0$ and an increasing sequence $(t_k)_{k\in\mathbb{N}}$ in T such that

$$\varphi(t_k, x^0) \in U \ \text{ and } \ t_{k+1} - t_k < \tau(U),\ k \in \mathbb{N}, \qquad \lim_{k\to\infty} t_k = \infty. \tag{25}$$

A recurrent point x^0 clearly satisfies condition (24). In Figure 3.1.9(b) the non–wandering point x^0 does not satisfy (24) hence it is not recurrent. Another example of a non–wandering point which is not recurrent can be found in Example 3.1.24. The next theorem, due to Birkhoff, characterizes compact minimal sets as the limit sets of recurrent points with relatively compact trajectories.

Theorem 3.1.23 (Birkhoff's Recurrence Theorem). *Suppose that $\mathfrak{F} = (T, X, \varphi)$ is a global, time–invariant flow. Then*

(i) if $M \subset X$ is a compact minimal set for $\mathfrak{F}$ and $x \in M$, then x is recurrent and $\omega(x) = M$.

(ii) If x is recurrent and $\omega(x)$ is compact, then $M = \omega(x)$ is minimal.

Proof: (i) Suppose that M is compact and minimal for $\mathfrak{F}$. If $x \in M$, we have $\omega(x) \subset M$ is non–empty, invariant and compact by Theorem 3.1.16, and so $\omega(x) = M$ follows from Definition 3.1.13. If $x \in M$ were not recurrent there would exist an open neighbourhood U of x and an increasing sequence (t_k) in T such that

$$\varphi(t; x) \notin U \quad \text{for all} \quad t \in T \cap [t_k, t_k + k]. \tag{26}$$

Let x^0 be the limit of a convergent subsequence of $\varphi(t_k; x)$. Then, for any $t \in T$, $\varphi(t; x^0)$ is the limit of a convergent subsequence of $\varphi(t + t_k; x)$. But by (26) $\varphi(t + t_k; x) = \varphi(t; \varphi(t_k, x)) \notin U$ for all $k \in \mathbb{N}$ sufficiently large and since U is open this implies $\varphi(t; x^0) \notin U$. Hence $\mathcal{O}(x^0) \cap U = \emptyset$, but this would contradict $\omega(x^0) = M$. Thus x is recurrent.
(ii) Now suppose that x is recurrent and $\omega(x)$ is compact. By Theorem 3.1.16, $\omega(x)$ is invariant. Let $M \subset \omega(x)$ be a non-empty closed invariant set. We have to show $M = \omega(x)$. This is clear if $x \in M$. If $x \notin M$ there exist open ε-neighbourhoods U of x and V of M in X such that $U \cap V = \emptyset$. Choose any $x^0 \in M$ and (t_k) in T such that $t_k \to \infty$ and $\varphi(t_k; x) \to x^0$ as $k \to \infty$. By continuity, $\varphi(t; \varphi(t_k; x)) \to \varphi(t, x^0)$ uniformly in $t \in [0, \tau] \cap T$, for all $\tau \in \mathbb{N}$. Since $\varphi(t; x^0) \in M$ for all $t \in T$ there exists for each $\tau \in \mathbb{N}$ a $N_\tau \in \mathbb{N}$ such that $\varphi(t; \varphi(t_k; x)) = \varphi(t + t_k; x) \notin U$ for all $t \in T \cap [0, \tau]$, $k \geq N_\tau$. But this contradicts the assumption that x is recurrent. □

The following example illustrates the dichotomy of wandering and non–wandering points for a seemingly simple class of two-dimensional discrete time systems.

Example 3.1.24. (Rational systems in the plane). Let $f(z) = p(z)/q(z) \in \mathbb{C}(z)$ be any complex rational function with $\deg p + \deg q \geq 2$. The nonlinear difference equation

$$z(t+1) = f(z(t))\,, \quad t \in \mathbb{N} \tag{27}$$

defines a time–invariant, discrete time, global flow $\mathfrak{F}$ on the compactified complex plane $\overline{\mathbb{C}} = \mathbb{C} \cup \{\infty\}$ which we endow with the metric of the Riemannian sphere $S^2 \cong \overline{\mathbb{C}}$. f is said to be *normal* at $z^0 \in \overline{\mathbb{C}}$ if there exists a neighbourhood U of z^0 such that the family of maps

$$z \mapsto \varphi(t; z) = f^t(z),\ t \in \mathbb{N}, \quad \text{is equicontinuous on } U. \tag{28}$$

The set J of all points in $\overline{\mathbb{C}}$ where f is not normal is called the *Julia set* of f, in honour of *G. Julia* (1918) [282] who established the following properties of J

(i) J is a non-empty closed invariant set and for any neighbourhood U of a point $z \in J$ there exists a number $k \in \mathbb{N}$ such that $f^k(U \cap J) = J$.

(ii) If $z^0 \in \overline{\mathbb{C}}$ is an attractive equilibrium point of $\mathcal{F}$ then the boundary of the corresponding domain of attraction $\partial\mathcal{A}(z^0)$ is equal to J.

(iii) The periodic points in J are dense in J.

It follows from (i) that every point in J is non–wandering, i.e. $J \subset \Omega(\mathcal{F})$. The motion in J is "turbulent–like" in the following sense: If the initial state z^0 is only known up to finite precision then the trajectory $\varphi(t; z^0) = f^t(z^0)$ becomes unpredictable in finite time, i.e. $\varphi(t; z^0)$ may be anywhere in J for $t \geq k \gg 1$.
For illustrative purposes we shall now analyze one of the simplest systems of this kind

$$z(t+1) = z(t)^2, \quad t \in \mathbb{N}. \tag{29}$$

In particular, we will verify the above facts (i) – (iii) for this example. The system has three equilibrium points in $\overline{\mathbb{C}}$, $z_1 = 0$, $z_2 = \infty$ and $z_3 = 1$. Obviously z_1 and z_2 are asymptotically stable with domains of attraction

$$\mathcal{A}(0) = \{z \in \mathbb{C};\ |z| < 1\} \quad \text{and} \quad \mathcal{A}(\infty) = \{z \in \overline{\mathbb{C}};\ |z| > 1\}.$$

The boundaries of these basins are in fact identical, in accordance with (ii). Moreover it is not difficult to prove that $f(z) = z^2$ is normal at $z \in \overline{\mathbb{C}}$ if and only if $|z| \neq 1$. Hence $J = \{z \in \mathbb{C};\ |z| = 1\} = \partial\mathcal{A}(0) = \partial\mathcal{A}(\infty)$ and this concludes the verification of (ii). All the points in $\mathcal{A}(0) \setminus \{0\}$ and in $\mathcal{A}(\infty) \setminus \{\infty\}$ are wandering (converging to 0 and ∞ respectively) and so $\Omega(\mathcal{F}) = J \cup \{0, \infty\}$.
Whereas the asymptotic dynamics of $\mathcal{F}$ on the invariant open sets $\mathcal{A}(0)$ and $\mathcal{A}(\infty)$ are simple (convergence towards an attracting fixed point) the dynamics on J are quite complicated. To see this we parametrize the points in J by $z = e^{2\pi i\theta}$, $0 \leq \theta < 1$ and identify $\theta = \sum_{k=1}^{\infty} \theta_k 2^{-k}$ (dyadic expansion) with $(\theta_1, \theta_2, \theta_3, \ldots)$. Then the flow $\mathcal{F}$ on J can equivalently be described by the equation

$$\theta(t+1) = S\theta(t),$$

where $S : (\theta_1, \theta_2, \theta_3, \ldots) \mapsto (\theta_2, \theta_3, \theta_4, \ldots)$ is the left shift on the space Θ of all 0-1-sequences with the identification $(0, 0, 0, \ldots) = (1, 1, 1, \ldots)$. It is now easy to verify (i). Let $z = e^{2\pi i\theta}$ be any point in J and U an arbitrary neighbourhood of z. There exists $k \in \mathbb{N}$ such that U contains all $z' = e^{2\pi i\theta'}$ for which the first k digits of θ' and θ coincide. As a consequence we obtain $f^k(U \cap J) = J$.
Finally let us verify (iii). $z = e^{2\pi i\theta} \in J$ is periodic with period $\tau \in \mathbb{N}^*$ if and only if $z^{2\tau} = z$, i.e. $z^{2\tau - 1} = 1$ or, equivalently, the dyadic expansion of θ is periodic after $2\tau - 1$ digits. It follows that the set of periodic points in J is dense in J. Note that not all the points in J are recurrent. In fact, $z = e^{2\pi i\theta}$ is recurrent if and only if for every $k \in \mathbb{N}^*$ there exists $\tau(k) \in \mathbb{N}$ such that the block of the first k binary digits of θ recurs at least once every $\tau(k)$ digits in the dyadic expansion of θ. Hence $z = e^{2\pi i\theta}$ with $\theta = \sum_{k=1}^{\infty} 2^{-k!}$

is not recurrent. However, it can be shown that almost all points $z \in J$ (in the sense of the Lebesgue measure on the unit circle) are recurrent. Moreover it can be shown that $\mathfrak{F}$ is *ergodic* on J in the sense that - modulo zero sets - the only non-empty invariant subset of J is J.
By continuity the chaotic dynamics on J communicates itself to a close neighbourhood of J although the turbulent–like behaviour *near to* J is transient and dies out (with respect to the standard metric of $\overline{\mathbb{C}} \cong S^2$) as $t \to \infty$. In fact we know that for $|z^0| \neq 1$ and t sufficiently large $\varphi(t; z^0) = f^t(z^0)$ is either close to 0 or ∞ on $\overline{\mathbb{C}} \cong S^2$. □

3.1.5 Attractors

To describe the asymptotic behaviour of more complicated systems the stability concepts introduced in Definitions 3.1.8 and 3.1.9 for equilibrium states have to be extended to arbitrary closed invariant subsets of the state space. In keeping with our usual philosophy we give these definitions for general (possibly) time–varying local flows $\mathfrak{F} = (T, X, \varphi)$.

Definition 3.1.25 (Stable attractor). (i) A closed subset $\Omega \subset X$ is said to be *stable* at time t_0 if for every neighbourhood W of Ω there exists a neighbourhood V of Ω such that $V \subset X_\infty(t_0)$ and, for each $x^0 \in V$, $\varphi(t; t_0, x^0) \in W$ for all $t \in T_{t_0}$.

(ii) A closed subset $\Omega \subset X$ is called an *attractor* at time t_0 if there exists a neighbourhood V of Ω such that $V \subset X_\infty(t_0)$ and $\varphi(t; t_0, x^0) \to \Omega$ as $t \to \infty$ for every $x^0 \in V$.

(iii) The *basin of attraction* of an attractor Ω at time t_0 is given by

$$\mathcal{A}(t_0, \Omega) = \{x \in X_\infty(t_0);\ \varphi(t; t_0, x) \to \Omega \ \text{ as } \ t \to \infty\}.$$

Example 3.1.15 presents a 2-dimensional differential system whose asymptotic behaviour is determined by a limit cycle Ω encircling an unstable equilibrium point. The corresponding basin of attraction of Ω is the whole plane punctured at $\overline{x} = 0$. The following example shows that, for two-dimensional discrete time systems, the simplest attractors may have extremely complicated basins of attraction.

Example 3.1.26. (Cayley's problem). If Newton's method is applied to find the complex roots of a polynomial $p(z) = (z - \lambda_1) \cdots (z - \lambda_n)$, one obtains a discrete time rational system on the complex plane described by

$$z(t+1) = z(t) - \frac{p(z(t))}{p'(z(t))}\,, \quad t \in \mathbb{N}\,. \tag{30}$$

For a starting value $z(0)$ sufficiently close to a root λ_i the corresponding sequence $z(t) = \varphi(t; z(0))$ converges to λ_i. Hence the roots $\lambda_1, \ldots, \lambda_n$ of $p(z)$ are attractors of the system (30). Cayley raised the problem of determining the basins of attraction of these roots. He treated the quadratic case, but noted that the calculations for the cubic case appeared to be much more complicated. That this is indeed the case follows from the discoveries of *Julia* (1918) reported in Example 3.1.24. In the cubic case the system (30) has in general three attractors λ_1, λ_2, λ_3 and the corresponding three basins of attraction must all have the same boundary J in $\overline{\mathbb{C}}$. It is not easy to construct three sets in $\overline{\mathbb{C}}$ with this property

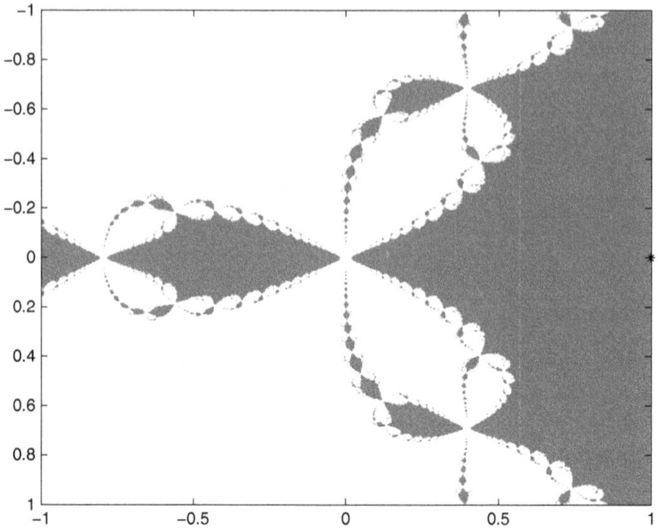

Figure 3.1.11: Basin of the attractor $\lambda_1 = 1$ of (30) with $p(z) = z^3 - 1$ in the region $|\operatorname{Re} z| \leq 1$, $|\operatorname{Im} z| \leq 1$

(try to colour the plane such that wherever two colours meet the third is present as well). The basin of attraction corresponding to the root 1 of $p(z) = z^3 - 1$ is shown in black in Figure 3.1.11. The basin consists of infinitely many connected components. □

The Poincaré-Bendixson Theorem ensures that in two dimensions the only compact minimal attractors of a differentiable system are equilibrium points and limit cycles. In higher dimensions much more is possible. In 1963 *Lorenz* [346] found what is probably the first example of a "strange attractor".

Example 3.1.27. (Lorenz attractor). If a fluid cell is heated from below and cooled from above the resulting convective motion in cross section can be modelled by partial differential equations. *Lorenz* expanded the solutions into an infinite number of modes. Setting all but three of them equal to zero he obtained the following system of ordinary differential equations

$$\begin{aligned} \dot{x}(t) &= \sigma(y(t) - x(t)) \\ \dot{y}(t) &= rx(t) - y(t) - x(t)z(t) \\ \dot{z}(t) &= x(t)y(t) - bz(t) \end{aligned} \tag{31}$$

where σ, r and b are (positive) physical parameters. Roughly speaking, x measures the rate of convective overturning and y, z the horizontal and vertical temperature variations. Figure 3.1.12 shows the projection onto the x, z plane of one computed solution of (31) when $\sigma = 10$, $b = 8/3$, $r = 28$. (Crossings are the result of projection). It is noteworthy that the general form of the figure does not depend upon the choice of initial conditions (provided that initial transient sections of the trajectory are ignored). On the other hand, the details of the figure, e.g. the exact sequence of loops which the trajectory performs, depend crucially on the initial condition (and the integration procedure chosen). As a consequence it is not possible to predict how an individual trajectory will develop over any longer time interval. Extensive numerical experiments seem to indicate that the "final

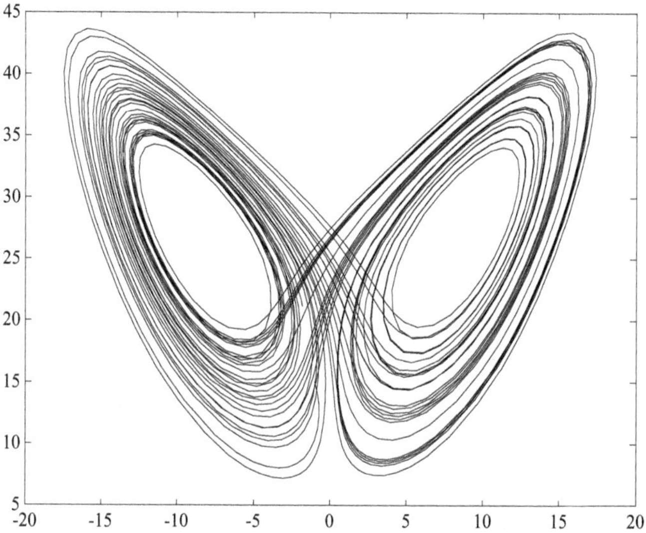

Figure 3.1.12: Trajectory of Lorenz equation

motions" of the Lorenz system are governed by a strange attractor whose projection is roughly the shape shown in Figure 3.1.12, the *Lorenz attractor*. *Lorenz* (1963) [346] proved that there is a bounded invariant ellipsoid $E_0 \subset \mathbb{R}^3$ into which all trajectories eventually enter. Moreover, the divergence of the vector field $(x, y, z) \mapsto (\sigma(y-x), rx-y-xz, xy-bz)$ is a negative constant $-(\sigma + b + 1)$ on $\mathbb{R}^3$. Thus the Lorenz flow is volume contracting. We conclude that the set $\Omega(\mathcal{F})$ of non–wandering points of (31) is a compact invariant set of Lebesgue measure zero. □

One of the most useful tools for determining (approximately) the locus of an attractor and its basin of attraction are Liapunov functions. An introduction to this method will be given in the next section. In particular we will see how the ellipsoid E_0 in the last example can be obtained via a Liapunov function (see Example 3.2.33).

3.1.6 Exercises

1. Let $\overline{x}$ be an equilibrium state of $\mathcal{F}$. Suppose that for every $K > 0$, $\varepsilon > 0$ there exists a point x^0 and some time $t \in T$ such that $d(x^0, \overline{x}) < \varepsilon$ and $d(\varphi(t; x^0), \overline{x}) > K d(x^0, \overline{x})$. Does this imply that $\overline{x}$ is unstable?

2. A function $V : G \to \mathbb{R}$ is said to be a *first integral* of the differential equation $\dot{x}(t) = f(t, x(t))$ if V is constant along every solution, i.e. $V(\varphi(t; x^0)) \equiv V(x^0)$. Find a first integral of the differential system $\ddot{y} + y^3 = 0$. Sketch its phase portrait and analyze the stability of its equilibrium state.

3. (Properties of Invariant Sets) Given a time–invariant global flow $\mathcal{F} = (\mathbb{R}_+, X, \varphi)$, prove

(i) the union and intersection of invariant sets are invariant.

(ii) Each subset $E \subset X$ contains a largest invariant subset M. M is the union of all orbits $\mathcal{O}(x^0)$ remaining in E.

(iii) The closure of an invariant set is invariant. If $\mathcal{F}$ is reversible (i.e. φ is defined and satisfies (LF1) – (LF2) on $T^2 \times X$) then the interior of an invariant set is invariant (if non-empty).

(iv) A set $S \subset X$ is invariant if and only if each of its connected components is invariant.

(v) If $K \subset X$ is compact and invariant then K contains a minimal set.

(vi) Every minimal set is connected.

4. (Properties of Limit Sets) Given a time–invariant global flow $\mathcal{F} = (\mathbb{R}_+, X, \varphi)$, prove

(i) $\omega(x^0) = \bigcap_{\tau \in T_0} \overline{\{\varphi(t; x^0);\ t \in T,\ t \geq \tau\}}$,

(ii) $\omega(x^0)$ is closed and invariant,

(iii) $\overline{\mathcal{O}(x^0)} = \mathcal{O}(x^0) \cup \omega(x^0)$ is invariant.

5. Prove the following discrete time counterpart of the connectedness statement in Theorem 3.1.16: Given a global flow $(\mathbb{N}, X, \varphi)$ on a metric space X and an $x^0 \in X$ which has a relatively compact orbit in X, then the (non-empty, compact, invariant) limit set $\omega(x^0)$ is invariantly connected, i.e. it is not the disjoint union of two non-empty closed invariant sets.

6. Prove: A non-empty set $S \subset X_\infty$ is minimal (with respect to a given time–invariant local flow $\mathcal{F}$) if and only if $\overline{\mathcal{O}(x)} = S$ for all $x \in S$.

7. (Equilibrium States) Given a local flow $\mathcal{F}$ on X. Prove

(i) if $\varphi(t; x^0) \neq x^0$ then there exists a neighbourhood U of x^0 such that $U \cap U_t = \emptyset$.

(ii) The set of all equilibrium points in X is closed in X_∞.

(iii) If $\varphi(t; x) \to y$ as $t \to \infty$ then y is an equilibrium point.

(*Bhatia and Szegö* (1970), pp. 16 [54]).

8. Consider the flow described in Example 3.1.10 (iii) and prove the assertions made there. In particular, show that $\overline{x}$ is an attractive but unstable equilibrium point. Compute a phase portrait of the flow on the rectangle $[-2, 2] \times [-2, 2]$.

9. Consider the system on $\mathbb{R}^2 \setminus \{0\}$ described in polar coordinates by

$$\dot{\theta} = \sin^2 \theta + (1 - r)^2, \quad \dot{r} = r(1 - r).$$

If $\omega(r_0, \theta_0)$ is the limit set of a solution initialized at (r_0, θ_0) show that $\omega(r_0, \theta_0) = \{r : r = 1\}$ if $r_0 \neq 1$. Prove that the orbits on $r = 1$ consist of the equilibrium points $\{\theta = 0\}$, $\{\theta = \pi\}$ and the arcs of the circle $\{\theta :\ 0 < \theta < \pi\}$, $\{\theta :\ \pi < \theta < 2\pi\}$ and show that the minimal sets on this circle are $\{\theta = 0\}$, $\{\theta = \pi\}$. (*Hale* (1969), pp. 48 [212]).

10. (Periodic Points) Given a time–invariant global flow $\mathcal{F}$, prove

(i) if $x \in X$ is periodic (i.e. generates a periodic orbit) then $\mathcal{O}(x)$ is a minimal set.

(ii) If $x \in X$ is periodic but not an equilibrium point then there is $\tau \in T$ such that τ is the smallest period of x (fundamental period). Moreover, if $\varphi(t; x) = x$ then $t = k\tau$ for some $k \in \mathbb{N}$.

(iii) If (x^k) is a sequence of periodic points with periods $\tau_k \to 0$ and $x^k \to x$ as $k \to \infty$ then x is an equilibrium point.

(iv) Given any $\rho > 0$ then the set of all $x \in X_\infty$ which are periodic with period $\tau \leq \rho$ is closed.

(*Bhatia and Szegö* (1970), pp. 18 [54]).

11. Prove that $x \in X_\infty$ is a periodic point for a time-invariant local flow $\mathcal{F} = (T, X, \varphi)$ if and only if $\mathcal{O}(x) = \omega(x)$. (*Bhatia and Szegö* (1970), pp. 32 [54]).

12. Prove the equivalence of the three conditions for Poisson stability in (24).

13.* Prove that if $\mathcal{F}$ is a global flow on a *complete* metric space X then the Poisson stable points are dense in the set $\Omega(\mathcal{F})$ of non–wandering points of $\mathcal{F}$ (*Bhatia and Szegö* (1970), pp. 36 [54]).

14. Consider the discrete time system $z(t+1) = f(z(t))$ where $f(z) = z^2$ on $\overline{\mathbb{C}}$, see Example 3.1.24. Prove that if $\tilde{z} \in J$ then $J = \overline{\{z \in \mathbb{C};\ f^t(z) = \tilde{z} \text{ for some } t \in \mathbb{N}\}}$.

15. Case study: Stability to Chaos. The population growth of a single species can sometimes be modelled (see Example 1.1.1) by a discrete time logistic equation of the form

$$N(t+1) = N(t)[1 + r(1 - N(t)/K)]\,, \quad t \in \mathbb{N} \tag{32}$$

where N denotes the population size of the species, r is the growth rate and K is the carrying capacity for the population. Show that if $N = K(1+r)x/r$, then equation (32) is transformed into

$$x(t+1) = ax(t)(1 - x(t))\,, \quad \text{where} \quad a = 1 + r. \tag{33}$$

(i) Prove that if $a \in [0, 4]$ and $x(0) \in I := [0, 1]$, then $x(t) \in I$ for all $t \in \mathbb{N}$. Show that there are two equilibrium states $x = 0$ and $x = 1 - 1/a$ and that for $a \in (1, 3)$ the former is unstable whilst the latter is asymptotically stable.

(ii) If $a > 3$ show that there is a periodic orbit of period 2. Take a value of $a \in (1, 1+\sqrt{6})$ and compute the solution of (33) for $t \in [0, 100]$ and a variety of initial states x_0. Does this suggest that the periodic orbit is the limit set $\omega(x_0)$ for all $x_0 \in (0, 1)$?

(iii) Show that if $a > 1 + \sqrt{6}$ there is a periodic orbit of period 4. By computational studies determine a value of a for which this periodic orbit is not the limit set $\omega(x_0)$ for a variety of initial states $x_0 \in (0, 1)$.

(iv) Take a value of $a > 3.57$ and compute the solution for $t \in [0, 1000]$ and a variety of initial states $x_0 \in (0, 1)$.

(*Li and Yorke* (1975) [343], *May* (1976) [367]).

3.1.7 Notes and References

Liapunov (1893) [354] was the first to define the concept of stability in a precise way. His stability theory dealt with the local properties of finite dimensional differentiable systems. Poincaré was mainly interested in global properties of planar systems. Many of the basic concepts and ideas of the qualitative theory of differential equations originate from his pioneering work (*Poincaré* (1892/99) [415]). Another landmark in the history of classical dynamical systems theory was the monograph of *Birkhoff* (1927) [61] which strongly influenced research in the 1930's and 1940's. A standard reference for the major developments in the qualitative theory up to the mid 1940's is *Nemytskii and Stepanov* (English translation, 1960) [389].

A proof of Theorem 3.1.3 can be found in standard references on ordinary differential equations such as *Hale* (1980) [214]. As a first introduction to geometric or qualitative aspects of differential equations we recommend *Arnold* (1978) [19]. Excellent advanced textbooks are *Arnold* (1983) [20] and *Palis and De Melo* (1982) [404]. A comprehensive compendium emphasizing Hamiltonian systems and applications to classical mechanics is *Abraham and Marsden* (1978) [1]. Topological aspects of dynamical systems on metric spaces are studied in *Bhatia and Szegö* (1970) [54].

Formal definitions of the notions of recurrence, limit point and minimal set go back to Birkhoff and have been central concepts of topological dynamics ever since. For further information, see *Sell* (1971) [457]. Details of Poincaré-Bendixson theory may be found in the classical texts of *Hirsch and Smale* (1974) [258], *Hartman* (1982) [217], *Hale* (1980) [214], *Palis and De Melo* (1982) [404]. A good more recent reference text is *Perko* (2001) [408].

Under certain conditions the non-wandering set of a differentiable system can be decomposed into a finite number of closed, connected invariant sets each of which contains a dense orbit (Spectral Decomposition Theorem, *Smale* (1967) [467]). These basic sets play the same fundamental role in differentiable dynamics as minimal sets play in topological dynamics, see *Abraham and Marsden* (1978) [1].

The iteration of rational maps on the plane was first analyzed systematically by *Julia* (1918) [282] and *Fatou* (1919/20) [160]. A nicely illustrated tutorial exposition of their findings can be found in *Peitgen et al.* (1984) [407]. Cayley's problem (Example 3.1.26) was first formulated in *Cayley* (1879) [94] and treated for $p(z) = z^2 - 1$ in *Cayley* (1890) [95]. For details of the Lorenz attractor (Example 3.1.27) see *Sparrow* (1982) [475]. A first introduction to "chaotic systems" is *Devaney* (1989) [131], and more comprehensive treatments are *Arrowsmith and Place* (1990) [22], *Katok and Hasselblatt* (1995) [295] and *Alligood et al.* (1997) [10].

3.2 Liapunov's Direct Method

The stability concepts we introduced in the first section have been defined in terms of the state transition map φ of a local flow $\mathfrak{F}$. However, to determine $\varphi(t;t_0,x^0)$ one must solve the equations of motion of the system. Hence a direct verification of stability properties is only possible if an explicit formula for the solution is available for every initial pair (t_0,x^0). In practice this will rarely be the case unless the system is *linear*. Liapunov developed two methods to cope with this dilemma. One natural idea is to use *linearizations* for the stability analysis of nonlinear systems (Liapunov's *indirect* method). This is a practical and efficient procedure which we describe in detail in Section 3.3. However, a serious drawback is that it only yields local information and does not give estimates for the basin of attraction. In this section we describe Liapunov's second method, often referred to as the *direct method*. It is applicable in situations of "marginal stability" where the linearization method does not work and in addition it also enables one to obtain estimates for the basin of attraction. Some of the results obtained by Liapunov's direct method will be used in Section 3.3 to develop the indirect method.

We begin by describing the basic idea of the direct method. A general concept of a Liapunov function is defined for any flow on a metric space and used to prove a very general stability criterion. This stability result is complemented by an instability theorem of similar generality. In the second subsection we characterize Liapunov functions for time-varying finite dimensional differential or difference systems by local properties which can be checked directly, without solving the system equations. This allows us to derive verifiable sufficient criteria for stability, uniform asymptotic stability, exponential stability and instability. Counterparts to these results for time-invariant systems are presented in the third subsection. Moreover we prove LaSalle's Invariance Principle and describe how Liapunov functions can be used to obtain estimates for basin of attraction. It is shown that these results are powerful tools for the stability analysis not only of equilibrium points but also of periodic orbits and more complicated compact invariant sets.

Throughout the section discrete and continuous time systems are considered simultaneously and the results are illustrated by examples of dynamical models from a range of areas of application.

3.2.1 General Definitions and Results

Liapunov's direct method was inspired by the use of energy functions in analyzing dynamical systems of classical mechanics. In order to illustrate the idea let us consider a time-invariant conservative mechanical system having n degrees of freedom. Its state is described by n generalized position coordinates $q_1,q_2,\dots,q_n$ and n generalized momentum coordinates $p_1,p_2,\dots,p_n$.

We assume that the kinetic energy of the system $\mathcal{T}$ is a positive definite quadratic form in $p=(p_1,\dots,p_n)$, whereas the potential energy $\mathcal{W}$ depends only on the generalized position vector $q=(q_1,\dots,q_n)$. The Hamiltonian $H(q,p)$ (see Subsection 1.3.3) is given by

$$H(q,p)=\mathcal{W}(q)+\mathcal{T}(p)\,,\quad (q,p)\in\mathbb{R}^{2n} \tag{1}$$

and represents for every state $(q,p) \in \mathbb{R}^{2n}$ the corresponding total energy of the system. Assuming that $q \mapsto \mathcal{W}(q)$ is continuously differentiable, the equations of motion are

$$\dot{q}(t) = \frac{\partial H}{\partial p}(q(t),p(t)) \,, \quad \dot{p}(t) = -\frac{\partial H}{\partial q}(q(t),p(t)). \tag{2}$$

Thus a state $(\bar{q},\bar{p})$ is an equilibrium state if and only if all the partial derivatives of H vanish at $(\bar{q},\bar{p})$, i.e. $\bar{q}$ is a critical point of $\mathcal{W}$ and $\bar{p}=0$.

Now suppose that the potential energy has a strict local minimum at $\bar{q}$. Then there exists a small open neighbourhood D of the equilibrium point $(\bar{q},0)$ in $\mathbb{R}^{2n}$ such that the sublevel sets

$$D_\varepsilon = \{(q,p) \in D;\ \ H(q,p) < H(\bar{q},0) + \varepsilon\}, \quad \varepsilon > 0$$

contract to $(\bar{q},0)$ as $\varepsilon \to 0$. We choose $\varepsilon > 0$ sufficiently small so that $\overline{D}_\varepsilon \subset D$. If the system is slightly perturbed from the equilibrium $(\bar{q},0)$ at time $t=0$, so that $(q(0),p(0)) \neq (\bar{q},0)$, then $H(q(0),p(0)) > H(\bar{q},0)$, i.e. the initial total energy is above the total energy of the system at the equilibrium point. But

$$\frac{dH}{dt}(q(t),p(t)) = \frac{\partial H}{\partial q}(q(t),p(t))\dot{q}(t) + \frac{\partial H}{\partial p}(q(t),p(t))\dot{p}(t) \equiv 0. \tag{3}$$

So the total energy is conserved and the state $(q(t),p(t))$ remains in the set of constant energy $H(q(t),p(t)) = H(q(0),p(0))$. In particular $(q(t),p(t))$ does not return to the equilibrium point $(\bar{q},0)$ as $t \to \infty$. So $(\bar{q},0)$ is not asymptotically stable. On the other hand, the sublevel sets $D_\varepsilon \subset \overline{D}_\varepsilon \subset D$ are invariant because $H(q(t),p(t)) \equiv H(q(0),p(0))$, $t \geq 0$ and a trajectory $(q(t),p(t))$ starting in D_ε cannot jump out of D, by continuity. Since D_ε contracts to $(\bar{q},0)$ as $\varepsilon \to 0$, we conclude that the equilibrium point $(\bar{q},0)$ of the conservative mechanical system is (marginally) stable.

In 1893 Liapunov realized that this method of analyzing the stability of mechanical systems could also be applied to arbitrary differential systems. This proved to be an extremely productive idea. Let $\overline{x}$ be an equilibrium point of a differential system. He showed in his PhD thesis [354] that any continuous real valued function V defined on some neighbourhood D of $\overline{x}$ could be used as an energy-type function to deduce stability, provided that it enjoys the following two properties

- $V(x)$ has a unique minimum in D at $\overline{x}$.
- $V(\varphi(t))$ decreases monotonically along every system trajectory $\varphi(t)$ contained in D.

The first property implies that the sublevel sets $D_\varepsilon = \{x \in D; V(x) < \varepsilon\}$ contract to $\overline{x}$ as $\varepsilon \to 0$. The second property is used to show (Figure 3.2.1) that the sublevel sets are invariant under the system's flow. From these two implications stability is deduced.

Throughout this subsection we assume that $\mathcal{F} = (T,X,\varphi)$ is a local flow on a metric state space (X,d) with time domain T, an interval in $\mathbb{R}$ or $\mathbb{Z}$ which is unbounded to the right. For every $(t_0,x^0) \in T \times X$, $t \mapsto \varphi(t;t_0,x^0)$, $t \in T_{t_0}(x^0)$ is a state trajectory of $\mathcal{F}$ (see Definition 3.1.1). In order to explain the simple logic behind Liapunov's idea and to emphasize its wide applicability we will give a general definition of a

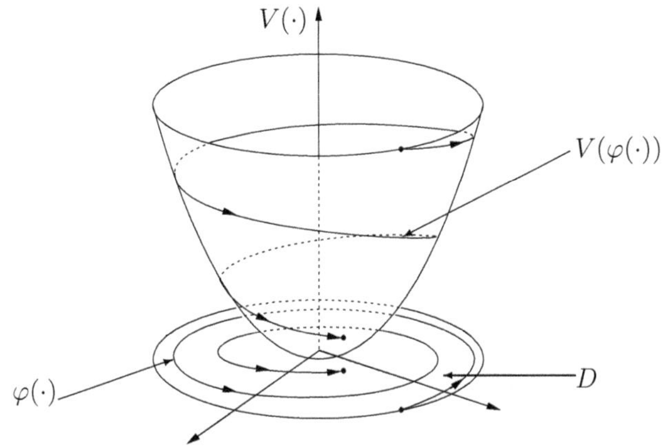

Figure 3.2.1: Liapunov function

Liapunov function which applies to discrete as well as to continuous time systems, to infinite dimensional systems as well as to systems on manifolds. But before we give the formal definition we introduce the following classes of *comparison functions* which are often used for obtaining estimates of solutions in stability analysis.

Definition 3.2.1. Let $0 < r_1 \leq \infty$. A function $\alpha : [0, r_1) \to \mathbb{R}_+$ is said to be

(i) *of class* $\mathcal{K}$ if $\alpha(\cdot)$ is monotonically increasing on $[0, r_1)$, $\alpha(r) > 0$ for $r \in (0, r_1)$ and $\lim_{r \searrow 0} \alpha(r) = 0$,

(ii) *of class* $\mathcal{K}_\infty$ if additionally $r_1 = \infty$ and $\lim_{r \to \infty} \alpha(r) = \infty$.

A function $\beta : \mathbb{R}_+ \times [0, r_1) \to \mathbb{R}_+$ is said to be *of class* $\mathcal{LK}$ if for each $t \in \mathbb{R}_+$, $\beta(t, \cdot) : [0, r_1) \to \mathbb{R}_+$ is of class $\mathcal{K}$ and, for each $r \in [0, r_1)$, $\beta(\cdot, r) : \mathbb{R}_+ \to \mathbb{R}_+$ is monotonically decreasing with $\lim_{t \to \infty} \beta(t, r) = 0$.

Some elementary properties of these functions are given in Ex. 1. In Ex. 2 and Ex. 3 the reader is asked to show how uniform stability and uniform asymptotic stability of an equilibrium point $\overline{x}$ of $\mathcal{F}$ can be characterized in terms of class $\mathcal{K}$ and class $\mathcal{LK}$ functions.

The following definition formalizes the first basic property of a Liapunov function. It applies to equilibrium points $\overline{x}$ as well as, more generally, to closed invariant sets Ω. For time-varying systems we have to consider Liapunov functions depending on time.

Definition 3.2.2. Let D be a subset of X and $\emptyset \neq \Omega = \overline{\Omega} \subset D$. A function $V : T \times D \to \mathbb{R}_+$ is said to be *positive definite* away from Ω on D if $V(T \times \Omega) = \{0\}$ and there exists $\alpha_1 \in \mathcal{K}$ on $[0, r_1)$ such that

$$V(t, x) \geq \alpha_1(d(x, \Omega)) \, , \quad (t, x) \in T \times D. \tag{4}$$

If additionally, there exists $\alpha_2 \in \mathcal{K}$ on $[0, r_2)$ such that

$$V(t, x) \leq \alpha_2(d(x, \Omega)) \, , \quad (t, x) \in T \times D \, , \tag{5}$$

then V is said to be *bounded by a class $\mathcal{K}$ function* on D.

Remark 3.2.3. (i) If there is a sequence (x^k) in D such that $d(x^k,\Omega) \to \infty$ as $k \to \infty$ then necessarily $r_1 = r_2 = \infty$ and this may seem at first sight to be a restriction. However since stability is a local property it will usually be possible to reduce D in order to examine whether or not a closed set Ω is stable with respect to a flow $\mathcal{F}$. Where the set D becomes very important is in determining global stability properties or in obtaining estimates for the basin of attraction. In this latter case we will see that in order to get good estimates we have to examine the interplay between the choice of the set D and upper and lower bounds for V.
(ii) It is often convenient to restrict the class $\mathcal{K}$ to those functions $\alpha(\cdot) : [0,r_1) \to \mathbb{R}_+$ which are continuous and strictly increasing with $\alpha(0) = 0$, see *Hahn* (1967), [209], *Khalil* (1996) [299]. We do not need these additional properties in our analysis. However, the reader is asked in Ex. 4 to show that there is no restriction in requiring that the class $\mathcal{K}$ functions α_1 and α_2 in (4) and (5) be continuous and strictly increasing. □

If V is positive definite away from a *point* $\overline{x}$, then all the functions $V(t,\cdot)$, $t \in T$ have a strict minimum on D at $\overline{x}$ and are bounded below *uniformly in* $t \in T$ by a positive definite function $x \mapsto \alpha_1(d(x,\overline{x}))$ which is independent of t. In applications positive definiteness of a time-varying function $V : T \times D \to \mathbb{R}_+$ is often proved by constructing a time-invariant lower bound $W_1(x) \leq V(t,x)$, $(t,x) \in T \times D$ which is itself positive definite. Note that a function $V : T \times D \to \mathbb{R}_+$ vanishing on $T \times \Omega$ is positive definite away from Ω if and only if for all $\varepsilon > 0$

$$\inf\{V(t,x);\ t \in T,\ x \in D \text{ and } d(x,\Omega) \geq \varepsilon\} > 0. \tag{6}$$

If $\Omega \subset D$ is compact and $U \subset D$ is a neighbourhood of Ω (i.e. $\Omega \subset \operatorname{int} U$) then $X \setminus U$ has a positive distance $d(X \setminus U, \Omega) = \min_{x\in\Omega} d(X \setminus U, x) > 0$ from Ω. So for every neighbourhood U of Ω there exists $\varepsilon > 0$ such that the ε-neighbourhood of Ω, $B(\Omega,\varepsilon) = \{x \in X;\ d(x,\Omega) < \varepsilon\}$, is contained in U.
Now let $W_1 : D \to \mathbb{R}_+$ be a time-invariant positive definite (away from Ω) lower bound for V, i.e. $W_1(x) \leq V(t,x)$ for all $(t,x) \in T \times D$. For any $t \in T$, $\rho > 0$ we denote by $D_\rho(t)$ the time-varying sublevel set of $V(t,\cdot)$ and by $D_\rho^{W_1}$ the corresponding time-invariant sublevel set of $W_1(\cdot)$

$$D_\rho(t) = \{x \in D;\ V(t,x) < \rho\}, \qquad D_\rho^{W_1} = \{x \in D;\ W_1(x) < \rho\}.$$

Then it follows from Definition 3.2.2 applied to $W_1 \leq V$, that

$$\Omega \subset D_\rho(t) \subset D_\rho^{W_1} \text{ and } \Omega = \bigcap_{\rho>0} D_\rho(t) = \bigcap_{\rho>0} D_\rho^{W_1},\ \ t \in T, \tag{7}$$

i.e the sublevel sets $D_\rho(t)$, $D_\rho^{W_1}$ shrink to Ω as $\rho \downarrow 0$. Let $V(t,x) \geq \alpha_1(d(x,\Omega))$ (resp. $W_1(x) \geq \alpha_1(d(x,\Omega))$) for all $(t,x) \in T \times D$ where $\alpha_1 \in \mathcal{K}$ on $[0,r_1)$, then for every $\varepsilon \in (0,r_1)$,

$$\bigcup_{t\in T} D_\rho(t) \subset B(\Omega,\varepsilon) \quad (\text{resp. } D_\rho^{W_1} \subset B(\Omega,\varepsilon)), \quad 0 < \rho \leq \alpha_1(\varepsilon). \tag{8}$$

So, for $\rho > 0$ sufficiently small, every ε-neighbourhood of Ω contains all sublevel sets $D_\rho(t)$, $t \in T$ which are themselves neighbourhoods of Ω if $V(t,\cdot)$ is continuous

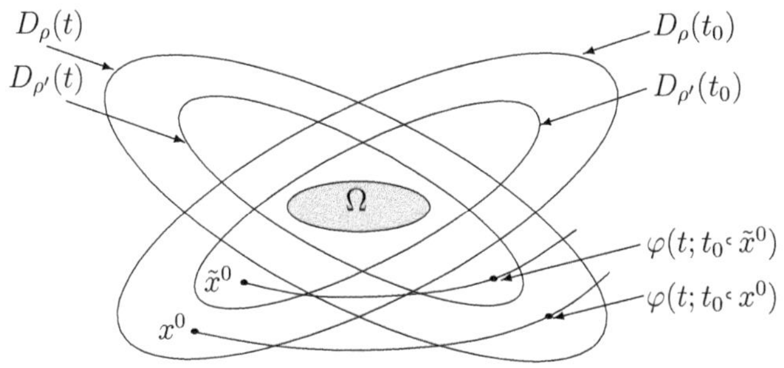

Figure 3.2.2: Time-varying sublevel sets of V $(\rho > \rho' > 0)$

(see Figure 3.2.2). On the other hand if V satisfies the boundedness condition (5) then for any $\rho > 0$ the intersection of all the time-varying sublevel sets $D_\rho(t)$ is a neighbourhood of Ω (see Figure 3.2.3).

$$\delta \in (0, r_2) \text{ and } \alpha_2(\delta) < \rho \Rightarrow B(\Omega, \delta) \cap D \subset D_\rho(t) \text{ for all } t \in T. \tag{9}$$

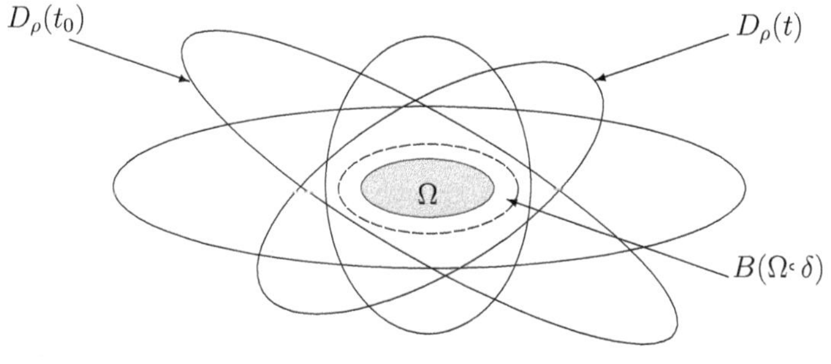

Figure 3.2.3: Intersection of time-varying sublevel sets

We now make precise the second main property of a Liapunov function which requires that V decreases along the trajectories of the flow. Whilst this property is sufficient to obtain stability results we need a stronger property in order to establish *asymptotic* stability. Roughly speaking V must decrease with "positive velocity" along all parts of trajectories which are outside any given ε-neighbourhood of Ω.

Definition 3.2.4. Let $\mathfrak{F} = (T, X, \varphi)$ be a local flow and D a neighbourhood of Ω in X.

(i) A function $V : T \times D \to \mathbb{R}$ is said to be *decreasing along the trajectories of* $\mathfrak{F}$ (or for short $\mathfrak{F}$*-decreasing*) if for all $(t_0, x^0) \in T \times D$ and all $t \in T_{t_0}(x^0)$

$$\varphi([t_0, t] \cap T; t_0, x^0) \subset D \;\Rightarrow\; V(t, \varphi(t; t_0, x^0)) \le V(t_0, x^0). \tag{10}$$

(ii) V is said to be *strictly* $\mathfrak{F}$*-decreasing away from* Ω if additionally, for all $(t_0, x^0) \in T \times D$, $t \in T_{t_0}(x^0)$ and every $\varepsilon > 0$,

$$\varphi([t_0, t] \cap T; t_0, x^0) \subset D \setminus B(\Omega, \varepsilon) \;\Rightarrow\; V(t, \varphi(t; t_0, x^0)) \le V(t_0, x^0) - \gamma(t - t_0) \tag{11}$$

for some function $\gamma(\cdot) = \gamma(\cdot, \varepsilon) : \mathbb{R}_+ \to \mathbb{R}_+$ satisfying $\lim_{\tau \to \infty} \gamma(\tau) = \infty$.

If V is strictly decreasing away from Ω then $V(t, \varphi(t; t_0, x^0))$ decreases by an amount $\gamma(t-t_0)$ which becomes arbitrarily large, provided the trajectory $\varphi(\cdot; t_0, x^0)$ remains outside a given ε–ball of Ω for a sufficiently large time $t - t_0$. Later, in applications to differential and difference equations we will always have $\gamma(\tau) > 0$ for $\tau > 0$ (which explains the terminology "strictly decreasing"). But this property is not really important whereas the asymptotic behaviour of $\gamma(\cdot)$ is essential. Note that the decrease of V in condition (11) only depends on the time difference $t - t_0$ and not on the initial time t_0. This is needed in order to establish *uniform* asymptotic stability in Theorem 3.2.7.

Remark 3.2.5. Suppose V is $\mathcal{F}$-decreasing. Then V is strictly decreasing away from Ω if, for every $\varepsilon > 0$ there exist $\tau \in T, \tau > 0$ and $\overline{\gamma} > 0$ such that for all $t_0 \in T$

$$\varphi([t_0, t_0 + \tau] \cap T; t_0, x^0) \subset D \setminus B(\Omega, \varepsilon) \ \Rightarrow\ V(t_0 + \tau, \varphi(t_0 + \tau; t_0, x^0)) \le V(t_0, x^0) - \overline{\gamma}.$$

In fact, let $t \in T,\ t_0 + k\tau \le t < t_0 + (k+1)\tau$ for some $k \in \mathbb{N}$ and $\varphi([t_0, t] \cap T; t_0, x^0) \subset D \setminus B(\Omega, \varepsilon)$. Set

$$t_j = t_0 + j\tau, \quad x^j = \varphi(t_j; t_0, x^0), \qquad j = 0, ..., k.$$

Then by (LF 2) we obtain $x^{j+1} = \varphi(t_j + \tau; t_j, x^j)$ for $j = 0, ...k-1$ and hence by induction

$$V(t, \varphi(t; t_0, x^0)) \le V(t_k, x^k) \le V(t_{k-1}, x^{k-1}) - \overline{\gamma} \le ... \le V(t_0, x^0) - k\overline{\gamma}.$$

Thus (11) is satisfied with $\gamma(\cdot)$ defined by $\gamma(t) = k\overline{\gamma},\ t \in [k\tau, (k+1)\tau),\ k \in \mathbb{N}$.
If V decreases with a guaranteed positive average velocity along the trajectories of the flow, i.e. given $\varepsilon > 0$ there exist $\tau > 0$ and $v = v(\varepsilon) > 0$ such that

$$\varphi([t_0, t] \cap T; t_0, x^0) \subset D \setminus B(\Omega, \varepsilon) \ \Rightarrow\ \frac{V(t_0, x^0) - V(t_0 + \tau, \varphi(t_0 + \tau; t_0, x^0))}{\tau} \ge v, \quad t_0 \in T,$$

then V is strictly $\mathcal{F}$-decreasing. □

Definition 3.2.6 (Generalized Liapunov function). Let $\mathcal{F} = (T, X, \varphi)$ be a local flow, $\Omega \subset X$ a closed set and D a neighbourhood of Ω. A continuous function $V : T \times D \to \mathbb{R}_+$ is said to be a *(strict) generalized Liapunov function* for $\mathcal{F}$ at Ω on $T \times D$ if

(i) V is positive definite away from Ω on $T \times D$,

(ii) V is (strictly) $\mathcal{F}$-decreasing on $T \times D$ away from Ω.

The mechanism of a proof of stability via Liapunov functions is illustrated in Figure 3.2.4. Suppose V is a generalized Liapunov function for $\mathcal{F}$ at Ω and

$$\alpha_1(d(x, \Omega)) \le V(t, x) \le \alpha_2(d(x, \Omega)), \qquad (t, x) \in T \times D$$

where $\alpha_1, \alpha_2 \in \mathcal{K}$. Given any $\varepsilon > 0$ sufficiently small, choose $\delta > 0$, such that $\alpha_2(\delta) < \alpha_1(\varepsilon)$. Then every trajectory $\varphi(t; t_0, x^0)$ starting at any time $t_0 \in T$ at $x^0 \in B(\Omega, \delta) \subset D_{\alpha_1(\varepsilon)}(t_0)$ (see (9)) remains in $D_{\alpha_1(\varepsilon)}(t) \subset B(\Omega, \varepsilon)$ for all $t \in T_{t_0}$.
For time-varying flows the existence of a (strict) generalized Liapunov function in a neighbourhood D of Ω is not sufficient to prove (asymptotic) stability. We need a weak additional assumption in order to ensure, that in the discrete time case, trajectories which start sufficiently close to Ω do not leap out of the domain D in one

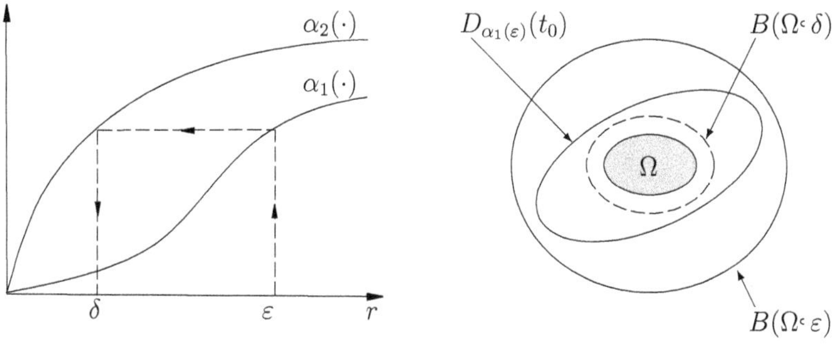

Figure 3.2.4: How to choose δ for a given $\varepsilon > 0$

step. In the continuous time case we must exclude the possibility that every neighbourhood of Ω, no matter how small, contains a trajectory of finite life time which never leaves that neighbourhood. The following theorem is our main Liapunov–type stability result for local flows.

Theorem 3.2.7 (Stability Theorem). *Let $\mathfrak{F} = (T, X, \varphi)$ be a local flow, $\Omega \subset X$ a compact set and D a neighbourhood of Ω. Suppose there is a neighbourhood $U \subset D$ of Ω in X such that, in the continuous time case, $\overline{U} \subset \operatorname{int} D$ and*

$$(t_0, x^0) \in T \times U \text{ and } t_+(t_0, x^0) < \infty \Rightarrow \exists t \in T_{t_0}(x^0) : \varphi(t; t_0, x^0) \notin U, \tag{12}$$

whilst in the discrete time case

$$(t_0, x^0) \in T \times U \Rightarrow t_0 + 1 \in T_{t_0}(x^0) \text{ and } \varphi(t_0 + 1; t_0, x^0) \in D. \tag{13}$$

If V is a generalized Liapunov function for $\mathfrak{F}$ at Ω on $T \times D$ with a positive definite lower bound $W_1(x) \le V(t, x)$, $(t, x) \in T \times D$, then

(i) Ω is $\mathfrak{F}$–invariant and stable at any time $t_0 \in T$. If $\rho > 0$ is such that $D_\rho^{W_1} \subset U$, then

$$x^0 \in D_\rho(t_0) \Longrightarrow t_+(t_0, x^0) = \infty \wedge \forall t \in T_{t_0} : \varphi(t; t_0, x^0) \in D_\rho^{W_1}. \tag{14}$$

(ii) If V is bounded in the sense of (5), then Ω is uniformly stable.

(iii) If V is strict generalized Liapunov function which is bounded in the sense of (5), then Ω is uniformly asymptotically stable.

(iv) If under the conditions of (iii) we have $W_1(x) \le V(t, x) \le W_2(x)$ for all $(t, x) \in T \times D$ and $W_1, W_2 : D \to \mathbb{R}_+$ are positive definite away from Ω, then $D_\rho^{W_2}$ is in the basin of attraction of Ω for all $\rho > 0$ such that $D_\rho^{W_1} \subset U$.

Proof: (i) Suppose $\rho > 0$ is such that $D_\rho^{W_1} \subset U$, and let $t_0 \in T$. We will show that every trajectory starting in $D_\rho(t_0)$ has infinite life span and does not leave $D_\rho^{W_1} \supset D_\rho(t_0)$. The first assertion follows from the second, since $\varphi(T_{t_0}(x^0); t_0, x^0) \subset D_\rho^{W_1} \subset U$ implies $t_+(t_0, x^0) = \infty$ by (12) (resp. (13)). Now assume by way of contradiction that the second assertion does not hold for some $x^0 \in D_\rho(t_0)$ and set

$t_1 = \inf\{t \in T_{t_0}(x^0); \varphi(t; t_0, x^0) \notin D_\rho^{W_1}\}$. In the discrete time case it follows that $\varphi(t_1; t_0, x^0) \notin D_\rho^{W_1}$, but by (13) we have $\varphi(t_1; t_0, x^0) \in D$ and so by (10)

$$W_1(\varphi(t_1; t_0, x^0)) \leq V(t_1, \varphi(t_1; t_0, x^0)) \leq V(t_0, x^0) < \rho.$$

Then $\varphi(t_1; t_0, x^0) \in D_\rho^{W_1}$ and so we have a contradiction. In the continuous time case $\varphi(t_1; t_0, x^0) \in \overline{D_\rho^{W_1}} \subset \overline{U} \subset \operatorname{int} D$ by the continuity of $\varphi(\cdot; t_0, x^0)$. Again by continuity there exists a time $t_2 > t_1$ in $T_{t_0}(x^0)$ such that $\varphi(t; t_0, x^0) \in D$ for all $t \in [t_0, t_2]$. But then (11) implies

$$W_1(\varphi(t; t_0, x^0)) \leq V(t, \varphi(t; t_0, x^0)) \leq V(t_0, x^0) < \rho, \quad t \in [t_0, t_2],$$

which contradicts the definition of t_1. This concludes the proof of (14). Now the stability of Ω at time t_0 follows from the fact that every ε-neighbourhood of Ω, $\varepsilon > 0$, contains $D_\rho^{W_1}$ for sufficiently small $\rho > 0$ and the corresponding $D_\rho(t_0)$ is a neighbourhood of Ω. Finally, since $\Omega \subset D_\rho(t_0)$ for every $\rho > 0$ and $\bigcap_{\rho>0} D_\rho(t) = \Omega$, $t \in T_{t_0}$, no trajectory starting in Ω at t_0 can leave it. So Ω is invariant and the proof of (i) is complete.

(ii) Suppose $W_1(x) \leq V(t, x) \leq W_2(x)$ for all $(t, x) \in T \times D$ where $W_1, W_2 : D \to \mathbb{R}_+$ are positive definite away from Ω (e.g. $W_i(x) = \alpha_i(d(x, \Omega))$, $i = 1, 2$ where $\alpha_1, \alpha_2 \in \mathcal{K}$). Given any $\varepsilon > 0$, choose $\rho > 0$ such that $D_\rho^{W_1} \subset U \cap B(\Omega, \varepsilon)$. Then $V(t, x) \leq W_2(x)$, $(t, x) \in T \times D$ implies that $D_\rho^{W_2} \subset D_\rho(t)$ for all $t \in T$ and so it

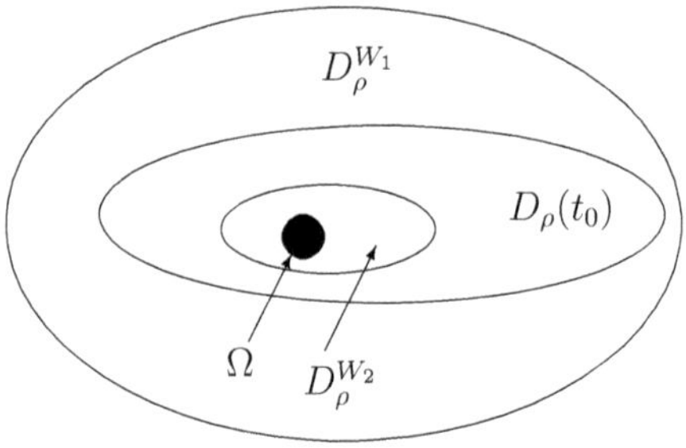

Figure 3.2.5: The sets $\Omega \subset D_\rho^{W_2} \subset D_\rho(t_0) \subset D_\rho^{W_1}$

follows from (14) that

$$x^0 \in D_\rho^{W_2} \Longrightarrow \forall t_0 \in T \,:\, t_+(t_0, x^0) = \infty \wedge \forall t \in T_{t_0} \,:\, \varphi(t; t_0, x^0) \in D_\rho^{W_1}. \tag{15}$$

But $D_\rho^{W_2}$ is a neighbourhood of Ω in X, so Ω is uniformly stable.

(iii) and (iv) Now assume additionally that V is a *strict* Liapunov function, and let $\rho > 0$ be such that $D_\rho^{W_1} \subset U$. Then again (15) holds. It remains to prove for every $x^0 \in D_\rho^{W_2}$ that $d(\varphi(t; t_0, x^0), \Omega) \to 0$ uniformly in t_0 as $t \to \infty$. By (8) it suffices to show that for every $\rho' \in (0, \rho)$ there exists a $\tau > 0$ such that $\varphi(t; t_0, x^0) \in D_{\rho'}^{W_1}$ for all $t \in T_{t_0+\tau}$ and all $t_0 \in T$. For this we only need to show that $\varphi([t_0, t_0 + \tau] \cap T; t_0, x^0) \cap D_{\rho'}^{W_2} \neq \emptyset$ for all $t_0 \in T$, since then, by (15)

$$\varphi(t; t_0, x^0) = \varphi(t; t_0 + \tau, \varphi(t_0 + \tau; t_0, x^0)) \in D_{\rho'}^{W_1}$$

for all $t \in T_{t_0+\tau}$. Now choose $\delta' > 0$ such that $B(\Omega, \delta') \subset D_{\rho'}^{W_2}$. By assumption there exists a $\gamma(\cdot) = \gamma(\cdot, \delta') : \mathbb{R}_+ \to \mathbb{R}_+$ satisfying $\lim_{r\to\infty} \gamma(r) = \infty$ and for every $t \in T_{t_0}$

$$\varphi([t_0, t] \cap T; t_0, x^0) \subset D \setminus B(\Omega, \delta') \;\Rightarrow\; V(t, \varphi(t; t_0, x^0)) \le V(t_0, x^0) - \gamma(t - t_0).$$

We choose $\tau > 0$ such that $\gamma(\tau) > \rho$ and suppose that for some $t_0 \in T$, $x^0 \in D_{\rho'}^{W_2}$ we have $\varphi(t; t_0, x^0) \notin B(\Omega, \delta')$ for all $t \in [t_0, t_0 + \tau] \cap T$. Then by (11)

$$V(t_0 + \tau, \varphi(t_0 + \tau; t_0, x^0)) \le V(t_0, x^0) - \gamma(\tau) \le \rho - \gamma(\tau) < 0.$$

and this contradiction concludes the proof. □

Remark 3.2.8. (i) If the flow $\mathfrak{F}$ in Theorem 3.2.7 is time–invariant and Ω is invariant the conditions (12) and (13) are always satisfied. This can be seen as follows. Since $\Omega \subset D$ is invariant every trajectory starting in Ω has an infinite life span. By the continuity assumption (LF3) of a local flow, there exists, for every $\tau > 0$ and each $x^0 \in \Omega$ a neighbourhood $U_{x^0} \subset D$ such that every trajectory $\varphi(t; x)$, $x \in U_{x^0}$ has a life span $> \tau$ and remains in D for $t \in [0, \tau] \cap T$. Hence $U = \bigcup_{x\in\Omega} U_x$ is a neighbourhood of Ω such that every trajectory starting in U, has a life span $> \tau$. Thus (13) follows in the discrete time case by setting $\tau = 2$. In the continuous time case U satisfies (12) whatever $\tau > 0$ we have chosen. Otherwise there would exist $x^0 \in U$ such that $t_+(x^0) < \infty$ and $\varphi(t; x^0) \in U$ for all $t \in T_{t_0}(x^0)$. But then $x_1 = \varphi(t_+(x^0) - \tau/2; x^0) \in U$ and so $\varphi(\cdot, x_1)$ would have a life span $< \tau/2$ which is a contradiction.
(ii) Even in the time-varying case condition (12) is automatically satisfied for local flows $\mathfrak{F}$ defined by differential equations on an open set X in a finite dimensional space $\mathbb{K}^n$. In fact, it follows from Theorem 2.1.14 that in this case (12) holds for every bounded neighbourhood U of Ω whose closure (in $\mathbb{K}^n$) is contained in X. In the discrete time case, if the local flow is defined by a difference equation as in Example 3.1.2, condition (13) need not be satisfied, e.g. if $x(t+1) = tx(t)$, $t \in \mathbb{N}$, $\Omega = \{0\}$ and D is bounded. However it will be satisfied if the sequence $f(t, \cdot)$, $t \in T_0$ is defined and equicontinuous at each point of Ω.
(iii) The proof of Theorem 3.2.7 shows that for all $\varepsilon > 0$ satisfying $\overline{B(\Omega, \varepsilon)} \subset U$ (where U is a neighbourhood of Ω satisfying (12) (resp. (13)) and every $t_0 \in T$, $x^0 \in D_\rho(t_0)$, $\rho = \alpha(\varepsilon)$

$$T_{t_0}(x^0) = T_{t_0} \quad \text{and} \quad \varphi(t; t_0, x^0) \in D_\rho(t) \subset B(\Omega, \varepsilon), \quad t \in T_{t_0}. \tag{16}$$

Moreover if V is also bounded in the sense of (5), then by (9)

$$\varphi(t; t_0, x^0) \in D_{\alpha(\varepsilon)}(t) \subset B(\Omega, \varepsilon), \quad (t_0, x^0) \in T \times B(\Omega, \delta), \quad t \in T_{t_0} \tag{17}$$

for any $\delta > 0$ such that $\beta(\delta) < \alpha(\varepsilon)$ and $B(\Omega, \delta) \subset D$. Finally, if V is a bounded strict generalized Liapunov function then we obtain the following lower bound for the basin of attraction of Ω at time t_0

$$\mathcal{A}(\Omega, t_0) \supset D_{\alpha(\varepsilon)}(t_0), \quad t_0 \in T. \tag{18}$$

(iv) If we replace "uniform asymptotic stability" in (iii) of the above theorem by "asymptotic stability", the assumption that V is strictly $\mathfrak{F}$-decreasing may be weakened so that γ may depend upon the initial time t_0; i.e. $\gamma(t) = \gamma(t; t_0, \varepsilon)$ where $\lim_{t\to\infty} \gamma(t; t_0, \varepsilon) \to \infty$. □

Many systems are not specified completely but contain parameters which can take a range of values. Sometimes these systems have a fixed equilibrium state for all possible values of the parameters. For such systems a stability analysis may require us to determine the set of parameter values for which the fixed equilibrium state is asymptotically stable. Since the use of the stability theorem yields only sufficient conditions for asymptotic stability, complementary information is needed to determine parameter values for which the equilibrium state is unstable. This information is provided by instability theorems. To obtain Liapunov–type instability theorems one uses real valued functions $V : T \times D \to \mathbb{R}$ which admit negative values in any neighbourhood of Ω and are $\mathcal{F}$-decreasing on the subset of $T \times D$ where $V < 0$.

Theorem 3.2.9 (Instability Theorem). *Let $\mathcal{F} = (T, X, \varphi)$ be a local flow. Suppose that Ω is a compact invariant set, $\varepsilon > 0$, $t_0 \in T$ and there exists a continuous function $V : T_{t_0} \times B(\Omega, \varepsilon) \mapsto \mathbb{R}$ satisfying $V(T_{t_0} \times \Omega) = \{0\}$ with the following properties.*

(i) For any $r \in (0, \varepsilon)$ there exists $x^0 \in B(\Omega, r)$ such that $V(t_0, x^0) < 0$.

(ii) There exists a class $\mathcal{K}$ function $\alpha : [0, \varepsilon) \to \mathbb{R}_+$ such that

$$(t, x) \in T_{t_0} \times B(\Omega, \varepsilon) \wedge V(t, x) < 0 \Longrightarrow V(t, x) \geq -\alpha(d(x, \Omega)).$$

(iii) For every $r \in (0, \varepsilon)$, there exists a function $\gamma(\cdot) = \gamma(\cdot, r) : \mathbb{R}_+ \to \mathbb{R}_+$ such that $\gamma(\tau) \to \infty$ as $\tau \to \infty$ and for all $x^0 \in B(\Omega, \varepsilon)$, $t \in T_{t_0}(x^0)$

$$\begin{aligned}&(\forall s \in [t_0, t) \cap T; \varphi(s; t_0, x^0) \in B(\Omega, \varepsilon) \setminus B(\Omega, r) \wedge V(s, \varphi(s; t_0, x^0)) < 0) \\ &\wedge (\varphi(t; t_0, x^0) \in B(\Omega, \varepsilon)) \Longrightarrow V(t, \varphi(t; t_0, x^0)) \leq V(t_0, x^0) - \gamma(t - t_0).\end{aligned} \tag{19}$$

Then Ω is unstable at the time t_0. More precisely, every point $x^0 \in B(\Omega, \varepsilon)$ such that $V(t_0, x^0) < 0$ generates a trajectory $\varphi(\cdot; t_0, x^0)$ which either has finite life span or leaves $B(\Omega, \varepsilon)$ at some time $t \in T_{t_0}$.

Proof: Let $x^0 \in B(\Omega, \varepsilon)$, $V(t_0, x^0) < 0$ and suppose by way of contradiction that $T_{t_0}(x^0) = T_{t_0}$ and $\varphi(t; t_0, x^0) \in B(\Omega, \varepsilon)$ for all $t \in T_{t_0}$. Choose $r \in (0, \varepsilon)$ such that $V(t_0, x^0) < -\alpha(r) < 0$. Then

$$(t, \varphi(t; t_0, x^0)) \in S = \{(s, x) \in T_{t_0} \times B(\Omega, \varepsilon); V(s, x) < 0 \wedge x \notin \overline{B(\Omega, r)}\}, \ t \in T_{t_0}. \tag{20}$$

In fact if this were not the case there would exist a smallest number $t_1 \in T_{t_0}$ such that $(t_1, \varphi(t_1; t_0, x^0)) \notin S$. So

$$\forall s \in [t_0, t_1) \cap T : \varphi(s; t_0, x^0) \in B(\Omega, \varepsilon) \setminus \overline{B(\Omega, r)} \wedge V(s, \varphi(s; t_0, x^0)) < 0.$$

It follows from (ii) and (iii) that

$$-\alpha(d(\varphi(t_1; t_0, x^0), \Omega)) \leq V(t_1, \varphi(t_1; t_0, x^0)) \leq V(t_0, x^0) < -\alpha(r) < 0.$$

Hence $d(\varphi(t_1; t_0, x^0), \Omega)) > r$ and so $(t_1, \varphi(t_1; t_0, x^0)) \in S$, a contradiction. Thus (20) holds for all $t \in T_{t_0}$ and by (iii) we obtain

$$V(t, \varphi(t; t_0, x^0)) \leq V(t_0, x^0) - \gamma(t - t_0), \quad t \in T_{t_0}.$$

But this implies that $V(t, \varphi(t; t_0, x^0))$ is not bounded below on T_{t_0}, although by (ii) $V(t, \varphi(t; t_0, x^0)) \geq -\alpha(\varepsilon)$ for all $t \in T_{t_0}$. Hence we again obtain a contradiction and it follows that either $T_{t_0}(x^0)$ is bounded or there exists a $t \in T_{t_0}$ such that $\varphi(t; t_0, x^0) \notin B(\Omega, \varepsilon)$. Thus Ω is unstable at t_0. □

Sometimes it is difficult to construct a *strict* Liapunov function in order to establish the asymptotic stability of an attractor Ω. For *time-invariant* local flows $\mathcal{F} = (T, X, \varphi)$ we now present an alternative way of proving asymptotic stability. It is based on the fact that every *relatively compact* trajectory $t \mapsto \varphi(t; x^0)$ with infinite life span is attracted by its limit set $\omega(x^0)$ (Theorem 3.1.16). LaSalle's Invariance Principle says that if $x^0 \in X_\infty$ and there is a time-invariant function $V(x)$ which is $\mathfrak{F}$-decreasing, then $\omega(x^0)$ is contained in a suitable level set $V^{-1}(c)$. Thus if, for example, $\overline{x}$ is an equilibrium point and $V^{-1}(c)$ does not contain an invariant subset $\neq \{\overline{x}\}$ for any $c \in \mathbb{R}$ then $\overline{x}$ is an attractor for $\varphi(t; x^0)$. Neither positive definiteness nor strict $\mathfrak{F}$- decreasing are required to draw this conclusion. LaSalle's Principle is based on the following simple lemma.

Lemma 3.2.10. *Let $\mathfrak{F} = (T, X, \varphi)$ be a time-invariant flow, $D \subset X$ and $V : D \to \mathbb{R}$ a continuous function. If $\overline{\mathcal{O}(x^0)} \subset D$ for some $x^0 \in X_\infty$ and $V(\cdot)$ decreases along the trajectory $\varphi(\cdot, x^0)$ then $\omega(x^0) \subset V^{-1}(c)$ for some $c \leq V(x^0)$.*

Proof: Let (t_k), (s_k) be two sequences in T such that $t_k \to \infty$, $s_k \to \infty$ and $\varphi(t_k; x^0) \to y$, $\varphi(s_k; x^0) \to z$ as $k \to \infty$. By taking subsequences if necessary we may suppose $s_k < t_k < s_{k+1}$ for all $k \in \mathbb{N}$. Then $V(\varphi(s_k; x^0)) \geq V(\varphi(t_k; x^0)) \geq V(\varphi(s_{k+1}; x^0))$, $k \in \mathbb{N}$ and hence, by continuity of $V(\cdot)$, $V(z) \geq V(y) \geq V(z)$. It follows that $\omega(x^0) \subset V^{-1}(c)$ for some $c \leq V(x^0)$. □

Theorem 3.2.11 (LaSalle's invariance principle for flows). *Suppose that $\mathfrak{F} = (T, X, \varphi)$ is a time-invariant local flow, $D \subset X$, $x^0 \in X_\infty$ and $V : D \to \mathbb{R}$ is a continuous function which is decreasing along the trajectory $\varphi(\cdot, x^0)$. If $\mathcal{O}(x^0)$, $x^0 \in X_\infty$, is a relatively compact orbit whose closure is contained in D then, for some $c \in \mathbb{R}$, $c \leq V(x^0)$,*

$$\varphi(t; x^0) \to M_c \text{ as } t \to \infty, \tag{21}$$

where $M_c \subset V^{-1}(c)$ is the largest weakly invariant subset of $V^{-1}(c)$.

Proof: By Theorem 3.1.16, $\omega(x^0)$ is a non-empty compact weakly invariant subset of D and $\varphi(t; x^0) \to \omega(x^0)$ as $t \to \infty$. It follows from the previous lemma that $\omega(x^0) \subset V^{-1}(c)$ for some $c \leq V(x^0)$, hence $\omega(x^0) \subset M_c$. □

The following direct consequence of the previous theorem provides a sufficient criterion for global asymptotic stability.

Corollary 3.2.12. *Suppose that $\mathfrak{F} = (T, X, \varphi)$ is a global flow, (i.e. $X_\infty = X$) and every orbit is relatively compact. If $V : X \to \mathbb{R}$ is continuous and $\mathfrak{F}$-decreasing on X, then $\bigcup_{c \in \mathbb{R}} M_c$ is a global attractor for $\mathfrak{F}$. In particular if V is a Liapunov function for $\mathfrak{F}$ at $\Omega = V^{-1}(0)$ and $M_c = \emptyset$ for every $c > 0$ then Ω is a globally asymptotically stable attractor for $\mathfrak{F}$.*

We now give an example illustrating how Liapunov's method (Theorem 3.2.7) and LaSalle's principle (Theorem 3.2.11) are applied to a nonlinear infinite dimensional system.

Example 3.2.13. Consider the one-dimensional time-invariant delay equation

$$\begin{aligned} \dot{x}(t) &= -x(t)^3 + bx(t-h)^3, \quad t \geq 0 \\ x(s) &= \psi(s), \quad -h \leq s \leq 0 \end{aligned} \tag{22}$$

where h is the length of memory of the system. (22) defines a time–invariant local flow $\mathcal{F} = (\mathbb{R}_+, X, \varphi)$ on the Banach space $X = \mathcal{C}([-h, 0]; \mathbb{R})$ (see Example 3.1.18). The zero function $\overline{x}(s) \equiv 0$, $s \in [-h, 0]$ is an equilibrium state. Consider the function

$$V(\psi) = (1/2)\psi(0)^4 + \int_{-h}^{0} \psi(s)^6 ds, \quad \psi \in X.$$

Note that this function is not positive definite away from $\overline{x}(s) \equiv 0$ if we provide X with the supremum norm $\|\cdot\|_\infty$. However, we are free to introduce another norm on X, e.g.

$$\|\psi\| := |\psi(0)| + \left[\int_{-h}^{0} |\psi(t)|^6 dt\right]^{1/6}. \tag{23}$$

With respect to the corresponding metric on X the function V is positive definite away from $\overline{x}(\cdot) \equiv 0$. We will now prove that V is decreasing along the trajectories if $|b| < 1$. For every solution $x(\cdot) = \varphi(\cdot, \psi)$ of the initial value problem (22),

$$V(x_t) = (1/2)x(t)^4 + \int_{t-h}^{t} x(s)^6 ds\,, \quad t \in [0, t_+(\psi)).$$

Multiplying (22) by $2x(t)^3$ and integrating from 0 to $t \in [0, t_+(\psi))$, we have

$$(1/2)(x(t)^4 - x(0)^4) = \int_0^t \big[- 2x(s)^6 + 2bx(s)^3 x(s-h)^3\big] ds.$$

Hence

$$V(x_t) - V(\psi) = \int_0^t \big[- 2x(s)^6 + 2bx(s)^3 x(s-h)^3\big] ds + \int_{t-h}^{t} x(s)^6 ds - \int_{-h}^{0} \psi(s)^6 ds.$$

But

$$\int_{t-h}^{t} x(s)^6 ds - \int_{-h}^{0} \psi(s)^6 ds = \int_0^t \big[x(s)^6 - x(s-h)^6\big] ds.$$

So

$$\begin{aligned} V(x_t) - V(\psi) &= \int_0^t \big[- x(s)^6 + 2bx(s)^3 x(s-h)^3 - x(s-h)^6\big] ds \\ &= - \int_0^t \Big[\big[x(s)^3 - bx(s-h)^3\big]^2 + (1-b^2)x(s-h)^6\Big] ds. \end{aligned} \tag{24}$$

Let us now assume $|b| < 1$, then we see that V is $\mathcal{F}$-decreasing and hence V is a time-invariant generalized Liapunov function for $\mathcal{F}$ on $(X, \|\cdot\|)$. Applying the Stability Theorem we obtain that the equilibrium state $\overline{x}(\cdot) \equiv 0$ is stable for (22) (with respect to the norm (23)).

One can in fact prove that $\overline{x}(\cdot) \equiv 0$ is asymptotically stable by showing that V is strictly $\mathcal{F}$-decreasing. But we find it easier to apply LaSalle's Invariance Theorem. For this, note that every solution $x(t) = \varphi(t, \psi)$, $\psi \in X$ is bounded (with respect to the supremum norm and the norm (23)) because

$$|x(t)|^4 \leq 2V(x_t) \leq 2V(\psi)\,, \quad t \in [0, t_+(\psi)). \tag{25}$$

Hence $t_+(\psi) = \infty$ for all $\psi \in X$ and the orbit $\mathcal{O}(\psi)$ is relatively compact in $(X, \|\cdot\|_\infty)$ (cf. Example 3.1.18). We may therefore apply Theorem 3.2.11. Now from (24) $V(\varphi(t;\psi)) < V(\psi)$, $t > 0$ if $\psi \not\equiv 0$. So $V^{-1}(c)$ does not contain a weakly invariant subset (i.e. $M_c = \emptyset$) if $c > 0$. But $V(\psi) > 0$ for non-zero $\psi \in X$ and so we conclude from Corollary 3.2.12 that

$$\varphi(t;\psi) \to 0 \text{ as } t \to \infty\,, \quad \psi \in X \tag{26}$$

in $(X, \|\cdot\|)$. It is easily deduced from (25) and $V(\psi) \leq \frac{1}{2}\|\psi\|_\infty^4 + h\|\psi\|_\infty^6$, that $\overline{x}(\cdot) \equiv 0$ is also stable for (22) with respect to the supremum norm. Therefore the origin in X is globally asymptotically stable for (22). □

Note that in the above example the function V was used to obtain information on the set X_∞. This idea is taken up in a more systematic way in the next subsection. Whilst the results of this subsection are very general, their application seems to suffer from the drawback that in order to verify the properties of a Liapunov function it is necessary to determine the trajectories $\varphi(t; t_0, x^0)$ of $\mathfrak{F}$, i.e. solve the equations of motion which define the flow. In the next subsection, we will show that for differentiable and discrete time systems in $\mathbb{K}^n$, Liapunov functions can be characterized by local properties which are directly verifiable without solving the system equations. The success of Liapunov's direct method relies on this fact.

3.2.2 Time–Varying Finite Dimensional Systems

In this subsection we show how Liapunov's direct method may be applied to study the stability properties of equilibrium points for *time-varying finite dimensional* systems. Since all norms on $\mathbb{K}^n$ are equivalent, any stability or instability statement is independent of the specific norm chosen. So in the rest of the section we provide $\mathbb{K}^n$ with the usual Euclidean norm $\|\cdot\|$ induced by the inner product $\langle\cdot,\cdot\rangle$. We suppose that $\overline{x}$ is an equilibrium point of the differential equation

$$\dot{x}(t) = f(t, x(t))\,, \quad t \in T = \mathbb{R}_+ \tag{27a}$$

or the difference equation

$$x(t+1) = f(t, x(t))\,, \quad t \in T = \mathbb{N} \tag{27b}$$

and make the following assumptions on f.

(A1) In the continuous time case $X \subset \mathbb{K}^n$ is an open set, $f : T \times X \to \mathbb{K}^n$ satisfies the Carathéodory conditions.

(A2) In the discrete time case $X \subset \mathbb{K}^n$ is an open set, $f(t,\cdot) : X \to \mathbb{K}^n$ is continuous for every $t \in T$ and $\{f(t,\cdot); t \in T\}$ is equicontinuous at $\overline{x}$, i.e. for every $\varepsilon > 0$ there exists $\delta > 0$ such that $\|f(t,x) - \overline{x}\| < \varepsilon$ for all $t \in T$ and $x \in \mathbb{K}^n$, $\|x - \overline{x}\| < \delta$.

We do not suppose that the equations (27) have global solutions $\varphi(t; t_0, x^0) \in X$ for all $(t_0, x^0) \in T \times X$. So we only have a local flow with state transition function $\varphi(\cdot; t_0, x^0)$ defined on maximal time intervals $T_{t_0}(x^0)$. The concept of stability (Definition 3.1.8) requires that trajectories starting near to $\overline{x}$ have an infinite life span.

Thus a first stability requirement for systems of the form (27) is that for $t_0 \in T$, if x^0 is near $\overline{x}$, the trajectory $\varphi(t; t_0, x^0)$ does not blow up as $t \to \infty$ or leave X in finite time. As an immediate consequence of Theorem 2.1.14 we obtain

Lemma 3.2.14. *Suppose that* $x(t) = \varphi(t; t_0, x^0)$ *is a bounded trajectory of* (27a) *such that the orbit closure* $\overline{\{x(t)\,; t \in T_{t_0}(x^0)\}} \subset X$, *then* $t_+(t_0, x^0) = \infty$.

We will now replace the condition of (strictly) $\mathfrak{F}$-decreasing in the definition of a generalized Liapunov function by an easily verifiable local condition. For this we suppose, throughout this subsection, that

(A3) $D \subset X$ is a neighbourhood of $\overline{x}$ and $V : T \times D \to \mathbb{R}$ is a continuous function. For the continuous time case we additionally assume that V is continuously differentiable on $T \times \operatorname{int} D$ and for $\mathcal{D}(\dot{V}) = T \times \operatorname{int} D$, we set

$$\dot{V}(t,x) = \langle \operatorname{grad} V(t,x), f(t,x)\rangle + \frac{\partial V}{\partial t}(t,x)\,,\ (t,x) \in \mathcal{D}(\dot{V}). \tag{28a}$$

In the discrete time case for $\mathcal{D}(\dot{V}) = \{(t,x) \in T \times D; f(t,x) \in D\}$, we set

$$\dot{V}(t,x) = V(t+1, f(t,x)) - V(t,x),\ (t,x) \in \mathcal{D}(\dot{V}). \tag{28b}$$

The functions $\dot{V}$ are called *derivatives of* V *along the solutions of* (27). In fact, if $(t, \varphi(t; t_0, x^0))$ remains in $\mathcal{D}(\dot{V})$ for $t \in [t_0, t_1) \cap T$, $t_1 \in T_{t_0}(x^0)$, then in the continuous time case[1]

$$\begin{aligned} \dot{V}(t, \varphi(t; t_0, x^0)) &= \langle \operatorname{grad} V(t, \varphi(t; t_0, x^0)), f(t, \varphi(t; t_0, x^0))\rangle + \frac{\partial V}{\partial t}(t, \varphi(t; t_0, x^0)) \\ &= \frac{dV}{dt}(t, \varphi(t; t_0, x^0))\,, \quad a.e.\ t \in [t_0, t_1). \end{aligned}$$

In the discrete time case

$$\dot{V}(t, \varphi(t; t_0, x^0)) = V(t+1, \varphi(t+1; t_0, x^0)) - V(t, \varphi(t; t_0, x^0))\,,\ \ t_0 \le t \le t_1 - 1\,.$$

So for both cases $\dot{V}(t, \varphi(t; t_0, x^0))$ is the *rate of change of* V *along the trajectory* $\varphi(\cdot; t_0, x^0)$. *Thus* V *is decreasing along the trajectories of* (27) *in* $\operatorname{int} D$ (cf. Definition 3.2.4) *if and only if* $\dot{V} \le 0$ *on* $\mathcal{D}(\dot{V})$. This condition can easily be checked if V and f are given.

We will now show that the other defining properties of a Liapunov function can be replaced by more easily verifiable criteria in the present setting. The verifications of these properties require the construction of (non-decreasing) real functions α_1, α_2 and γ satisfying (4), (5) and (11). The proof of the following lemma shows how any one of these functions may be constructed from a continuous function $W : D \to \mathbb{R}_+$ satisfying for some $\delta > 0$ with $\overline{B(\overline{x}, \delta)} \subset D$

$$W(\overline{x}) = 0, \quad W(x) > 0, \ \text{ for } x \in D \setminus \{\overline{x}\}, \quad \text{and} \quad \inf_{x \in D,\, \|x - \overline{x}\| \ge \delta} W(x) > 0. \tag{29}$$

It is easily seen that a continuous function $W: D \to \mathbb{R}_+$ satisfies (29) if and only if W is *positive definite away from* $\overline{x}$ *on* D in the sense of Definition 3.2.2.

[1] Note that $\varphi(\cdot; t_0, x^0)$ is absolutely continuous and hence differentiable almost everywhere.

Lemma 3.2.15. *Suppose D is a neighbourhood of $\overline{x}$, $W : D \to \mathbb{R}_+$ is continuous and $W(\overline{x}) = 0$.*

(i) If W satisfies (29) (resp. is bounded on $B(\overline{x}, r) \cap D$ for all $r > 0$) and $V : T \times D \to \mathbb{R}_+$ satisfies

$$V(T, \overline{x}) = \{0\}, \quad V(t,x) \geq W(x) \quad (\textit{resp. } V(t,x) \leq W(x)), \quad (t,x) \in T \times D,$$

then V is positive definite away from $\overline{x}$ (resp. bounded in the sense of (5)) on D.

(ii) If W satisfies (29) and $\dot{V}$ satisfies

$$\dot{V}(t,x) \leq -W(x)\,, \qquad (t,x) \in \mathcal{D}(\dot{V})$$

then V is strictly $\mathcal{F}$-decreasing away from $\overline{x}$.

Proof: (i) For $x \in D$, $\|x - \overline{x}\| = r$, we have

$$\inf_{r \leq \|y - \overline{x}\|,\, y \in D} W(y) \leq W(x) \leq \sup_{\|y - \overline{x}\| \leq r,\, y \in D} W(y)\,. \tag{30}$$

Let us denote the LHS by $\alpha_1(r)$ and the RHS by $\alpha_2(r)$ and set $r_1 = \sup_{y \in D} \|y - \overline{x}\|$. If W satisfies (29), then α_1 is of class $\mathcal{K}$ on $[0, r_1)$ and even on $[0, r_1]$ if $r_1 < \infty$ and there exists a $y \in D$ such that $\|y - \overline{x}\| = r_1$. If W is bounded on all bounded subset $B(\overline{x}, r) \cap D$, $r \in [0, r_1]$ then $\alpha_2 : [0, r_1] \to \mathbb{R}_+$ is of class $\mathcal{K}$. Hence (i) follows from (30) and Definition 3.2.2.

(ii) Suppose that for some $r > 0$ and $t \in T_{t_0}(x^0)$ we have

$$\varphi([t_0, t] \cap T; t_0, x^0) \subset D \setminus B(\overline{x}, r),$$

then, in the continuous time case[2]

$$V(t, \varphi(t; t_0, x^0)) - V(t_0, x^0) \;\leq\; -\int_{t_0}^{t} W(\varphi(\tau; t_0, x^0))d\tau \leq -\alpha_1(r)(t - t_0)$$

and a similar result holds in the discrete time case. We may therefore choose $\gamma(\tau, r) = \alpha_1(r)\tau$ in Definition 3.2.4. □

The previous lemma motivates the following definition of a Liapunov function for systems of the form (27). The defining properties of a Liapunov function are slightly stronger than those of a generalized Liapunov function, but they can be directly verified from the system equations without knowledge of their solutions.

Definition 3.2.16 (Liapunov function). Let D be a neighbourhood of $\overline{x}$ and $V :\; T \times D \to \mathbb{R}_+$ satisfy (A3). Then V is called a *Liapunov function* for (27) at $\overline{x}$ on $T \times D$ if $V(t, \overline{x}) = 0$ for all $t \in T$ and

(i) there exists a function $W_1 :\; D \to \mathbb{R}_+$ positive definite away from $\overline{x}$ on D such that

$$W_1(x) \leq V(t,x)\,, \quad (t,x) \in T \times D\,, \tag{31}$$

[2]Note that $v(t) = V(t, \varphi(t; t_0, x^0))$ is absolutely continuous so that $v(t) - v(t_0) = \int_{t_0}^{t} \dot{v}(s)\, ds$.

(ii) $\dot{V} \leq 0$ on $\mathcal{D}(\dot{V})$.

V is said to be a *strict Liapunov function* for (27) at $\overline{x}$ on $T \times D$ if in addition to (i) and (ii) there exists a function $W_3: D \to \mathbb{R}_+$ positive definite away from $\overline{x}$ on D, such that

$$\dot{V}(t,x) \leq -W_3(x), \quad (t,x) \in \mathcal{D}(\dot{V}). \tag{32}$$

Note that W_3 is independent of time. This enables us to derive sufficient Liapunov type conditions for *uniform* asymptotic stability. Strict Liapunov functions are not only useful tools for establishing asymptotic stability of an equilibrium point but also for estimating its basin of attraction. This is illustrated in the next theorem.

Theorem 3.2.17 (Liapunov Stability Theorem). *Let V be a Liapunov function on $T \times D$ for (27) at $\overline{x}$, then*

(i) $\overline{x}$ is stable at any time $t_0 \in T$.

(ii) If $V(t,x) \leq W_2(x)$ for all $(t,x) \in T \times D$ where $W_2: D \to \mathbb{R}_+$ is a continuous function with $W_2(\overline{x}) = 0$, then $\overline{x}$ is uniformly stable.

(iii) If additionally V is a strict Liapunov function then $\overline{x}$ is uniformly asymptotically stable.

(iv) Suppose in the case of (iii) that

$$W_1(x) \leq V(t,x) \leq W_2(x), \qquad (t,x) \in T \times D, \tag{33}$$

where $W_1, W_2 : D \to \mathbb{R}_+$ are positive definite away from $\overline{x}$. Let $\rho > 0$ be such that in the continuous (resp. discrete) time case

$$\overline{D_\rho^{W_1}} = \overline{\{x \in D; W_1(x) < \rho\}} \subset \operatorname{int} D \text{ is compact (resp. } f(T \times D_\rho^{W_1}) \subset D). \tag{34}$$

Then $D_\rho^{W_2}$ is in the basin of attraction of $\overline{x}$.

Proof: We apply Theorem 3.2.7 to the flow $\mathcal{F}$ defined by (27), with $\Omega = \{\overline{x}\}$. In the continuous time case let U be any bounded neighbourhood of $\overline{x}$ with closure $\overline{U} \subset \operatorname{int} D$. In the discrete time case, let U be any bounded neighbourhood containing $\overline{x}$ such that $f(t,U) \subset D$ for all $t \in T$ (such a neighbourhood of $\overline{x}$ exists by assumption (A2)). Then $T \times U \subset \mathcal{D}(\dot{V})$ and the conditions (12), (13) in Theorem 3.2.7 are satisfied. In the discrete time case this follows from the definition of U and in the continuous time case it is a consequence of Carathéodory's Theorem 2.1.14. Moreover, by Lemma 3.2.15, if V is a (strict) Liapunov function for (27) at $\overline{x}$ on $D \times T$ (in the sense of Definition 3.2.16), then V is a (strict) generalized Liapunov function for the flow generated by (27) at $\overline{x}$ on $T \times D$ (in the sense of Definition 3.2.6). Finally, if V is bounded above by a continuous function $W_2: D \to \mathbb{R}_+$ with $W_2(\overline{x}) = 0$, then, by Lemma 3.2.15, it is bounded in the sense of (5) on any compact neighbourhood of $\overline{x}$ in D. Therefore (i) – (iii) follow from Theorem 3.2.7.
(iv) Finally assume that V is a strict Liapunov function and (33), (34) are satisfied. Then $U := D_\rho^{W_1}$ is a neighbourhood of $\overline{x}$ with the properties specified above for both the continuous and discrete time cases. Hence the assertion (iv) follows from (iv) of Theorem 3.2.7. □

The following global version of the previous theorem is an easy consequence of the assertion (iv).

Corollary 3.2.18. *Let $X = \mathbb{K}^n$ and suppose that V is a strict Liapunov function on $T \times \mathbb{K}^n$ for* (27) *at $\overline{x}$. If* (33) *holds on $T \times \mathbb{K}^n$ and $W_1(x) \to \infty$ for $\|x\| \to \infty$, then $\overline{x}$ is globally uniformly asymptotically stable.*

Proof: Setting $D = \mathbb{K}^n$ the sublevel sets $D_\rho^{W_1}$ are bounded for all $\rho > 0$ by assumption and so $\overline{D_\rho^{W_1}}$ is compact. Hence all the assumptions of Theorem 3.2.17 are satisfied for all $\rho > 0$. Since the sublevel sets $D_\rho^{W_2}$, $\rho > 0$ cover $\mathbb{K}^n$ the result follows. □

Definition 3.2.19. An equilibrium point $\overline{x}$ of the nonlinear system (27) is said to be *exponentially stable at time* t_0 if it is stable and exponentially attractive at time t_0, i.e. there are $\delta = \delta(t_0) > 0$, $M = M(t_0) > 0$, $\omega = \omega(t_0) < 0$, such that $\varphi(t; t_0, x^0)$ exists for all $t \in T_{t_0}$ and

$$\|x^0 - \overline{x}\| < \delta \quad \Rightarrow \quad \|\varphi(t; t_0, x^0) - \overline{x}\| \le M e^{\omega(t-t_0)}, \quad t \in T_{t_0}. \tag{35}$$

If $\overline{x}$ is uniformly stable and (35) holds with constants δ, M, ω independent of t_0 then $\overline{x}$ is said to be *uniformly exponentially stable.*

The following corollary gives a sufficient condition for an even stronger version of exponential stability.

Corollary 3.2.20. *Suppose V is Liapunov function for* (27) *at $\overline{x}$ on $T \times D$ satisfying*

$$\alpha_1 \|x - \overline{x}\|^p \le V(t,x) \le \alpha_2 \|x - \overline{x}\|^p, \ \dot{V}(t,x) \le -\alpha_3 \|x - \overline{x}\|^p, \ (t,x) \in \mathcal{D}(\dot{V}) \tag{36}$$

for some positive constants α_1, α_2, α_3, p. Then there are constants $\delta > 0$, $M' > 0$ and $\omega < 0$ such that for all $t_0 \in T$, $x^0 \in B(\overline{x}, \delta)$, the solution $\varphi(\cdot; t_0, x^0)$ exists on T_{t_0} and

$$\|\varphi(t; t_0, x^0) - \overline{x}\| \le M' e^{\omega(t-t_0)} \|x^0 - \overline{x}\|, \quad t \in T_{t_0}. \tag{37}$$

In particular $\overline{x}$ is uniformly exponentially stable.
If $X = \mathbb{K}^n$, V is a global Liapunov function for (27) *and* (36) *holds for all $(t,x) \in T \times \mathbb{K}^n$, then $\overline{x}$ is globally uniformly exponentially stable, i.e* (37) *holds for all $(t_0, x^0) \in T \times \mathbb{K}^n$.*

Proof: By (A1), resp. (A2) we can choose $\varepsilon > 0$ such that $T \times B(\overline{x}, \varepsilon) \subset \mathcal{D}(\dot{V})$ and applying Theorem 3.2.17 we see that $\overline{x}$ is uniformly stable. Hence there exists $\delta > 0$ such that $x^0 \in B(\overline{x}, \delta)$ implies $\varphi(t; t_0, x^0) \in B(\overline{x}, \varepsilon)$ and so $(t, \varphi(t; t_0, x^0)) \in \mathcal{D}(\dot{V})$ for all $t \in T_{t_0}$, $t_0 \in T$. Now let $x^0 \in B(\overline{x}, \delta)$ then $v(t) = V(t, \varphi(t; t_0, x^0))$ satisfies

$$\dot{v}(t) \le -\alpha_3 \|\varphi(t; t_0, x^0) - \overline{x}\|^p \le -(\alpha_3/\alpha_2) v(t), \quad t \in T_{t_0}. \tag{38}$$

In the continuous time case $\frac{d}{dt}\left(e^{(\alpha_3/\alpha_2)t} v(t)\right) \le 0$ and so $v(t) \le v(t_0) e^{-(\alpha_3/\alpha_2)(t-t_0)}$. Hence for $t \in T_{t_0}$,

$$\alpha_1 \|\varphi(t; t_0, x^0) - \overline{x}\|^p \le v(t) \le v(t_0) e^{-(\alpha_3/\alpha_2)(t-t_0)} \le \alpha_2 \|x^0 - \overline{x}\|^p e^{-(\alpha_3/\alpha_2)(t-t_0)}.$$

So (37) holds with $M' = (\alpha_2/\alpha_1)^{1/p}$ and $\omega = -\alpha_3/\alpha_2 p$.
In the discrete time case (36) implies $v(t+1) - v(t) \le -(\alpha_3/\alpha_2)v(t)$ for $t \in T_{t_0}$ and so $\alpha_3 \le \alpha_2$. If $\alpha_3 = \alpha_2$ then necessarily $\varphi(t; t_0, x^0) = \overline{x}$ for $t \ge t_0 + 1$. Hence (37) holds for any $(M' \ge 1, \omega < 0)$. Now assume $\alpha_3 < \alpha_2$, then for $t \in T_{t_0}$

$$\alpha_1 \|\varphi(t; t_0, x^0) - \overline{x}\|^p \le v(t) \le (1 - \alpha_3/\alpha_2)^{t-t_0} v(t_0) \le (1 - \alpha_3/\alpha_2)^{t-t_0} \alpha_2 \|x^0 - \overline{x}\|^p.$$

So (37) holds with $\omega = (1/p)\ln(1 - \alpha_3/\alpha_2)$ and $M' = (\alpha_2/\alpha_1)^{1/p}$.
In both cases it follows from (37) that (35) holds with the same ω and $M = M'\delta$. In the global case, $\overline{x}$ is globally uniformly asymptotically stable by Corollary 3.2.18 and the above estimates are obtained for arbitrary $x^0 \in \mathbb{K}^n$. □

Example 3.2.21. Consider the discrete time system on $\mathbb{R}^2$ with time domain $T = \mathbb{Z}$

$$\begin{aligned} x_1(t+1) &= a(t)x_2(t)/(1 + x_1^2(t)) \\ x_2(t+1) &= b(t)x_1(t)/(1 + x_2^2(t)) \end{aligned} \tag{39}$$

where $|a(t)|, |b(t)| < 1$, $t \in \mathbb{Z}$. $\overline{x} = (0,0)$ is an equilibrium point of (39) on $D = \mathbb{R}^2$. Let us try the time-invariant function $V(t,x) = x_1^2 + x_2^2$ as a possible Liapunov function for (39) at $(0,0)$. V is positive definite on $\mathbb{R}^2$ away from $(0,0)$ and

$$\begin{aligned} \dot{V}(t,x) &= \frac{a^2(t)x_2^2}{(1+x_1^2)^2} + \frac{b^2(t)x_1^2}{(1+x_2^2)^2} - (x_1^2 + x_2^2) = \Big[\frac{a^2(t)}{(1+x_1^2)^2} - 1\Big]x_2^2 + \Big[\frac{b^2(t)}{(1+x_2^2)^2} - 1\Big]x_1^2 \\ &\le (a^2(t) - 1)x_2^2 + (b^2(t) - 1)x_1^2 \le 0, \quad x \in \mathbb{R}^2,\ t \in \mathbb{Z}. \end{aligned}$$

From Theorem 3.2.17 we conclude that $\overline{x} = (0,0)$ is uniformly stable. If there exist constants a, b such that $|a(t)| \le a < 1$, $|b(t)| \le b < 1$, $t \in \mathbb{Z}$, then by Corollary 3.2.20, $\overline{x} = (0,0)$ is uniformly exponentially stable. □

We now specialize the Instability Theorem 3.2.9 to the present setting.

Theorem 3.2.22 (Instability Theorem). *Let $t_0 \in T$ and D be a neighbourhood of $\overline{x}$. Suppose that $V : T \times D \to \mathbb{R}$ satisfies (A3), $V(T, \overline{x}) = \{0\}$ and has the following properties:*

(i) For any $r > 0$ there exists $x^0 \in B(\overline{x}, r)$ such that $V(t_0, x^0) < 0$.

(ii) $|V(t,x)| \le W(x)$ for all $(t,x) \in S = \{(s,x) \in T_{t_0} \times B(\overline{x}, \varepsilon);\ V(s,x) < 0\}$ where $\varepsilon > 0$ is such that $B(\overline{x}, \varepsilon) \subset \mathcal{D}(\dot{V})$ and $W : D \to \mathbb{R}_+$ is a continuous function which is bounded on $B(\overline{x}, r) \cap D$ for all $r > 0$ with $W(\overline{x}) = 0$.

(iii) In addition $\dot{V}(t,x) \le -W_3(x)$ for all $(t,x) \in S$ where $W_3 : D \to \mathbb{R}_+$ is positive definite away from $\overline{x}$.

Then $\overline{x}$ is unstable at t_0.

Proof: We apply Theorem 3.2.9 to the flow $\mathcal{F}$ defined by (27) with $\Omega = \{\overline{x}\}$. Then the restriction of V to $T \times B(\overline{x}, \varepsilon)$ satisfies the conditions (i)–(iii) of that theorem. In fact condition (i) of Theorem 3.2.9 follows directly from the above assumption (i). Setting $\alpha(r) = \sup\{W(y); y \in D, \|y - \overline{x}\| \le r\}$, then just as in the proof of Lemma 3.2.15 we see that $\alpha : [0, \varepsilon) \to \mathbb{R}_+$ is a class $\mathcal{K}$ function satisfying

$W(x) \le \alpha(\|x-\overline{x}\|)$, $x \in B(\overline{x},\varepsilon)$. So $V(t,x) \ge -\alpha(\|x-\overline{x}\|)$ for all $(t,x) \in S$ and hence condition (ii) of Theorem 3.2.9 holds. By assumption there exists a class $\mathcal{K}$ function $\alpha_3 : [0,\varepsilon) \to \mathbb{R}_+$ such that $W_3(x) \ge \alpha_3(\|x-\overline{x}\|)$ on $B(\overline{x},\varepsilon)$. Now if for any $r \in (0,\varepsilon)$

$$\forall s \in [t_0,t) \cap T : \ (s,\varphi(s;t_0,x^0)) \in S \text{ and } \varphi(s;t_0,x^0) \notin B(\overline{x},r),$$

then again just as in the proof of Lemma 3.2.15 we see that in the continuous (resp. discrete) time case (iii) implies

$$\begin{aligned} V(t,\varphi(t;t_0,x^0)) - V(t_0,x^0) &\le -\int_{t_0}^{t} W_3(\varphi(s;t_0,x^0))ds \ \ (\text{resp. } -\sum_{k=t_0}^{t-1} W_3(\varphi(k;t_0,x^0))) \\ &\le -\alpha_3(r)(t-t_0). \end{aligned}$$

Thus (iii) of Theorem 3.2.9 holds with $\gamma(t-t_0,r) = \alpha_3(r)(t-t_0)$. Therefore $\overline{x}$ is unstable at t_0. □

Example 3.2.23. Consider the differentiable system on $\mathbb{R}^2$ with time domain $T = \mathbb{R}$

$$\dot{x}_1 = tx_1 + x_2, \qquad \dot{x}_2 = x_1 - tx_2 + t^2 \sin x_2, \quad t \in \mathbb{R}.$$

Then $\overline{x} = (0,0)$ is an equilibrium point. Now if $V(x) = -x_1x_2$, we have $\dot{V}(x) = -x_2^2 - x_1^2 - t^2 x_1 \sin x_2$ for $x \in \mathbb{R}^2$, $t \in \mathbb{R}$. Let $t_0 \in \mathbb{R}$ and set

$$\varepsilon = \pi/2, \quad S = T_{t_0} \times \{x \in \mathbb{R}^2;\ x_1x_2 > 0,\ x_1^2 + x_2^2 < \pi^2/4\}$$

(see (ii) in the above theorem). Then $|V(x)| \le \frac{1}{2}(x_1^2 + x_2^2)$ and $\dot{V}(x) \le -(x_1^2 + x_2^2)$ on S. Hence we may apply Theorem 3.2.22 to conclude that $(0,0)$ is an unstable equilibrium state for the above system at any time $t_0 \in T$. □

3.2.3 Time–Invariant Systems

We now specialize the previous results to time-invariant versions of (27) and we will see that the stability and instability theorems take on simpler forms. In part this is due to the fact that the definiteness properties of V and $\dot{V}$ need not be expressed in terms of positive definite functions W_i. Instead these properties can be stated directly in terms of V and $\dot{V}$. The results in this subsection will be illustrated by some examples of classical stability problems.
The equations of motion are assumed to be of the form

$$\dot{x}(t) = f(x(t)), \ \ t \in T = \mathbb{R}_+; \qquad x(t+1) = f(x(t)), \ \ t \in T = \mathbb{N} \tag{40}$$

where f is Lipschitz continuous[3] (resp. continuous) on an open set $X \subset \mathbb{K}^n$. We assume that $\overline{x}$ is an equilibrium point of (40), i.e. $f(\overline{x}) = 0$ (resp. $f(\overline{x}) = \overline{x}$). By time-invariance the equilibrium point $\overline{x}$ is (asymptotically) stable if and only if it is uniformly (asymptotically) stable.

[3] A function $f : X \to \mathbb{K}^n$ is called *Lipschitz continuous* or *locally Lipschitz* if for every compact set $C \subset X$ there exists a constant L_C such that $\|f(x) - f(y)\|_{\mathbb{K}^n} \le L_C\|x-y\|_{\mathbb{K}^n}$ for all $x, y \in C$. In the time–invariant case the Carathéodory conditions reduce to Lipschitz continuity.

For time–invariant systems we only consider *time–invariant* Liapunov functions. In this case the derivative along the solutions of (40) (see (28)) takes a simpler form. Throughout the subsection we make the following assumption.

(A) $D \subset X$ is a neighbourhood of $\overline{x}$ and $V : D \to \mathbb{R}$ is continuous, and in the continuous time case continuously differentiable on int D. Then $\dot{V}$ is defined to be

$$\dot{V}(x) = \langle \operatorname{grad} V(x), f(x) \rangle, \quad x \in \mathcal{D}(\dot{V}) = \operatorname{int} D \tag{41a}$$

for the continuous time case and in the discrete time case:

$$\dot{V}(x) = V(f(x)) - V(x), \quad x \in \mathcal{D}(\dot{V}) = \{x \in D;\ f(x) \in D\}. \tag{41b}$$

Remark 3.2.24. Let $X_0 \subset X$ be an arbitrary weakly invariant set for (40). If $\dot{V}(x^0) \le 0$ for all $x^0 \in X_0 \cap \mathcal{D}(\dot{V})$ then V decreases along the flow defined by (40) on X_0 in the sense of Definition 3.2.4:

$$V(\varphi(t, x^0)) \le V(x^0) \quad \text{if} \quad x^0 \in X_0,\ t \in T(x^0) \text{ and } \varphi([t_0, t] \cap T; x^0) \subset D. \tag{42}$$

Making use of this fact most of the following results can be directly applied to flows induced by (40) on weakly invariant subsets X_0 (replace X by X_0 and interpret all topological statements relative to the induced topology of X_0). □

Theorem 3.2.25 (Stability Theorem). *Let $\overline{x}$ be an equilibrium point of* (40)*, D a neighbourhood of $\overline{x}$, $V : D \to \mathbb{R}_+$ a Liapunov function for* (40) *at $\overline{x}$, i.e. V satisfies (A) and*

$$V(\overline{x}) = 0, \quad V(x) > 0,\ x \in D \setminus \{\overline{x}\}, \quad \textit{and} \quad \dot{V}(x) \le 0,\ x \in \mathcal{D}(\dot{V}). \tag{43}$$

Then $\overline{x}$ is stable If V is a strict Liapunov function, i.e. additionally

$$\dot{V}(\overline{x}) = 0 \quad \textit{and} \quad \dot{V}(x) < 0, \quad x \in \mathcal{D}(\dot{V}) \setminus \{\overline{x}\} \tag{44}$$

then $\overline{x}$ is asymptotically stable

Note that V is not necessarily a Liapunov function *on* D in the sense of Definition 3.2.16 since the positivity assumption (43) does not necessarily imply positive definiteness away from $\overline{x}$ on the (possibly unbounded) set D. However the restriction of V to, for example, any compact neighbourhood $\tilde{D} \subset D$ of $\overline{x}$ yields a Liapunov function on $T \times \tilde{D}$ in the sense of Definition 3.2.16 (see Lemma 3.2.15). Therefore the above theorem is an immediate consequence of Theorem 3.2.17.
The condition $\dot{V}(x) \le 0$ is illustrated in Figure 3.2.6 for the continuous time case. It means that the "distance" of the point x from the given equilibrium point, as measured by $V(x)$, is decreased when x is moved in the direction of $f(x)$.
Before we illustrate the above results by some examples we first show how estimates for the basin of attraction of an equilibrium state can be obtained directly via the sublevel sets of Liapunov functions. For this the following lemma is useful and is of independent interest.

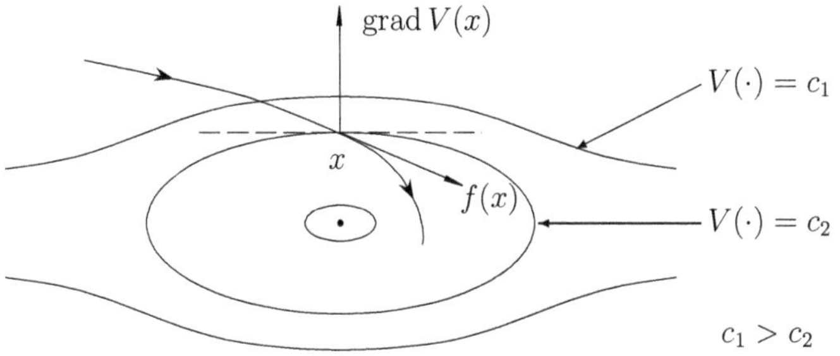

Figure 3.2.6: $\langle \operatorname{grad} V(x), f(x) \rangle \le 0$

Lemma 3.2.26. *Suppose D is any subset of X, $V : D \to \mathbb{R}$ is continuous and decreases along the flow defined by* (40) *(see* (42)*). If, for some $\rho \in \mathbb{R}$, the sublevel set $D_\rho = \{x \in D;\ V(x) < \rho\}$ satisfies*

$$\overline{D_\rho} \subset \operatorname{int} D \qquad (\textit{resp. } f(D_\rho) \subset D) \tag{45}$$

then D_ρ is weakly invariant for (40)*. In the discrete time case D_ρ is invariant.*

Proof: In the continuous time case, it follows from (45) that the boundary of D_ρ in $\mathbb{K}^n$ is given by

$$\partial D_\rho = \overline{D_\rho} \setminus D_\rho = \{x \in D; V(x) = \rho\} \subset \operatorname{int} D.$$

Hence, if $x \in D_\rho$ the solution $\varphi(\cdot; x)$ cannot leave D_ρ since otherwise there would exist by continuity a time $\bar{t} < t_+(x)$ such that $V(\varphi(\bar{t}; x)) = \rho > V(x)$ which contradicts the assumption that V decreases along the flow defined by (40). Hence D_ρ is weakly invariant.
For the discrete time case, if $x \in D_\rho$, then by (45) $f(x) \in D$ and $V(f(x)) \le V(x) < \rho$. Hence $f(D_\rho) \subset D_\rho$ and every $x \in D_\rho$ generates a trajectory with infinite life span which remains in D_ρ. □

The condition (45) is illustrated in Figure 3.2.7 for the continuous time case. If D_ρ is bounded, D_ρ is in fact invariant (see the proof of the next proposition). If D_ρ is unbounded, it may contain orbits with finite life span.

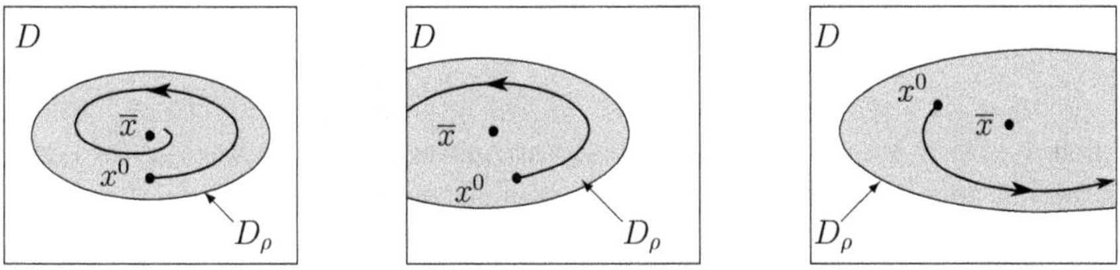

$\overline{D_\rho} \subset \operatorname{int} D$, D_ρ bounded $\overline{D_\rho} \not\subset \operatorname{int} D$ $\overline{D_\rho} \subset \operatorname{int} D$, D_ρ unbounded

Figure 3.2.7: The condition $\overline{D_\rho} \subset \operatorname{int} D$

Remark 3.2.27. Using the fact that positive definiteness of a Liapunov function ensures that the sublevel sets shrink to $\overline{x}$, the previous lemma can be used to give a simple direct proof of the stability statement in Theorem 3.2.25. (The reader is asked to do this in Ex. 12). □

Combining Lemma 3.2.26 with LaSalle's Invariance Principle (Theorem 3.2.11) we obtain the following result. It does not assume positive definiteness of V and is also applicable in situations where the domain of definition D contains more than one equilibrium point (see Example 3.2.34).

Theorem 3.2.28 (LaSalle's invariance principle). *Suppose $D \subset X$, $V : D \to \mathbb{R}$ satisfies (A) and $\dot{V}(x) \le 0$ for all $x \in \mathcal{D}(\dot{V})$. Let M be the largest invariant subset of $E = \{x \in \mathcal{D}(\dot{V});\ \dot{V}(x) = 0\}$ with respect to (40) and suppose that, for a given $\rho \in \mathbb{R}$, the sublevel set D_ρ of V satisfies*

(i) $\overline{D}_\rho \subset \operatorname{int} D$ (resp. $f(D_\rho) \subset D$).

(ii) Every orbit $\mathcal{O}(x^0) \subset D_\rho$ is bounded (in the continuous time case).

Then D_ρ is invariant and $\varphi(t, x^0) \to M \cap D_\rho$ as $t \to \infty$, for all $x^0 \in D_\rho$.

Proof: We prove the proposition for the continuous time case. By (i) $D_\rho = \{x \in \operatorname{int} D; V(x) < \rho\}$ is open and by the previous lemma it is weakly invariant. Let $x^0 \in D_\rho$. It follows from (i) and (ii) that the orbit closure $\overline{\mathcal{O}(x^0)}$ is compact and contained in $\operatorname{int} D$. Hence $t_+(x^0) = \infty$ by Lemma 3.2.14 and we see that D_ρ is in fact invariant. Applying Theorem 3.2.11 to the time–invariant flow $\mathcal{F} := (T, D_\rho, \varphi)$ defined by (40) on D_ρ we obtain $\varphi(t, x^0) \to M_c$ as $t \to \infty$, for some $c < \rho$, where M_c is the largest invariant set in $V^{-1}(c) \subset D_\rho$. Since $M_c \subset M \cap D_\rho$, this concludes the proof. □

If V is a Liapunov function the previous result implies the following estimate for the basin of attraction of $\overline{x}$.

Corollary 3.2.29. *Let $\overline{x}$ be an equilibrium point of (40), $D \subset X$ a neighbourhood of $\overline{x}$ and $V : D \to \mathbb{R}_+$ a Liapunov function for (40) at $\overline{x}$. Suppose that the sublevel set D_ρ is bounded and $\overline{D}_\rho \subset \operatorname{int} D$ (resp. $f(D_\rho) \subset D$) for a given $\rho > 0$. If the largest invariant set M in $E = \{x \in \mathcal{D}(\dot{V}), \dot{V}(x) = 0\}$ satisfies*

$$M \cap V^{-1}(c) = \emptyset, \quad c \in (0, \rho),$$

then $\overline{x}$ is asymptotically stable and D_ρ is contained in the basin of attraction of $\overline{x}$. In particular this holds if V is a strict Liapunov function.

Proof: The assumptions of LaSalle's Invariance Principle Theorem 3.2.28 are satisfied and $M \cap D_\rho = \{\overline{x}\}$. □

Recall that an equilibrium point $\overline{x}$ is said to be *globally asymptotically* stable for a flow on X if it is stable and globally attractive in the sense that its basin of attraction equals X. In the following corollary we assume that V is defined on $D = X$ and is therefore open in $\mathbb{K}^n$. We use the notation $\lim_{x \to \partial X} V(x) = \infty$ to mean that for every $r > 0$ there exists a compact set $K \subset X$ such that $V(x) > r$ for all $x \in X \setminus K$. In particular, if $X = \mathbb{K}^n$ then $\lim_{x \to \partial X} V(x) = \infty$ is equivalent to $\lim_{\|x\| \to \infty} V(x) = \infty$.

Corollary 3.2.30. *Suppose $\overline{x} \in X$ is an equilibrium point of (40) and $V : X \to \mathbb{R}_+$ satisfies (A) and*

$$V(\overline{x}) = 0; \quad V(x) > 0 \text{ and } \dot{V}(x) < 0, \quad x \in X \setminus \{\overline{x}\}. \tag{46}$$

If $\lim_{x\to\partial X} V(x) = \infty$ and in the discrete time case X is invariant, then $\overline{x}$ is globally asymptotically stable.

Proof: Since $\lim_{x\to\partial X} V(x) = \infty$, the sublevel set D_ρ of V is bounded with closure in D for every $\rho > 0$. In the discrete time case the invariance of X implies that $f(D_\rho) \subset X$. Hence the conditions of the previous corollary are satisfied for every D_ρ, $\rho > 0$. Now $X = \bigcup_{\rho>0} D_\rho$ and so the result follows. □

Example 3.2.31. Consider again the discrete time system of Example 3.2.21 where now $a(t) \equiv a$, $b(t) \equiv b$, $t \in \mathbb{N}$ are constant and suppose $|a|$, $|b| \leq 1$. We have seen that $V(x) = x_1^2 + x_2^2$, $x \in D = X = \mathbb{R}^2$ is a Liapunov function for the system at $\overline{x} = (0,0)$. Therefore, $\overline{x} = (0,0)$ is stable. Moreover,

$$\dot{V}(x) = \left[\frac{a^2}{(1+x_1^2)^2} - 1\right] x_2^2 + \left[\frac{b^2}{(1+x_2^2)^2} - 1\right] x_1^2 \leq (a^2-1)x_2^2 + (b^2-1)x_1^2 .$$

In Example 3.2.21 it was proved that the origin is exponentially stable if $|a| < 1$ and $|b| < 1$. We now use Theorem 3.2.28 and the notations therein to discuss the remaining possibilities.

(i) $a^2 < 1$, $b^2 = 1$. The set E is given by $E = \{x \in \mathbb{R}^2;\, x_2 = 0\}$. However, $f(x_1, 0) = (0, bx_1)$ and so the only invariant set in E is $M = \{(0,0)\}$. The assumptions of Corollary 3.2.29 are satisfied for all $\rho > 0$. Since $\mathbb{R}^2 = \bigcup_{\rho>0} D_\rho$, we conclude that the origin is globally asymptotically stable.

(ii) $a^2 = b^2 = 1$. In this case $E = \{x \in \mathbb{R}^2 :\ x_1 = 0\} \cup \{x \in \mathbb{R}^2 :\ x_2 = 0\}$. This set is

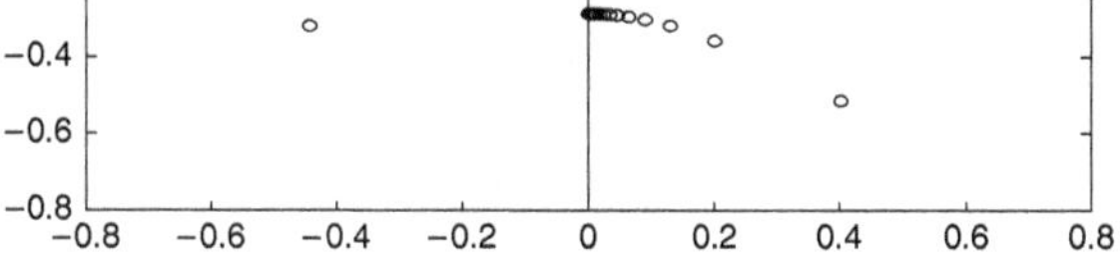

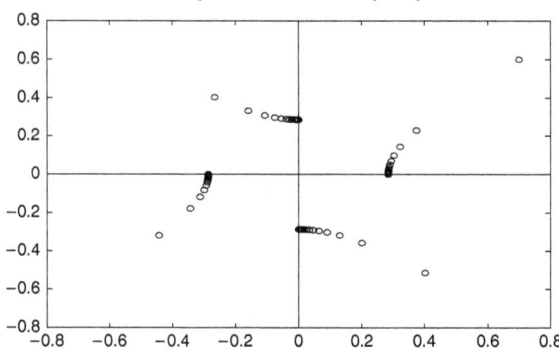

Figure 3.2.8: System (39) with $a = 1$ and $b = -1$

invariant and hence $M = E$. The hypotheses of Theorem 3.2.28 are satisfied for arbitrary $\rho > 0$. Now let $x^0 \in \mathbb{R}^2$, $x^0 \neq (0,0)$ and $c = \lim_{t\to\infty} V(\varphi(t;x^0))$, then $c \geq 0$. If $c = 0$ then $M \cap V^{-1}(c) = \{(0,0)\}$ and so $\varphi(t;x^0) \to (0,0)$ as $t \to \infty$. It is easily verified that $c = 0$ if x^0 belongs to the invariant set $\{x \in \mathbb{R}^2;\, x_1 = x_2\}$. If $c > 0$ then $M \cap V^{-1}(c)$ consists of the points $(\sqrt{c}, 0)$, $(0, \sqrt{c})$, $(-\sqrt{c}, 0)$ and $(0, -\sqrt{c})$ which are the intersections of E with the circle $x_1^2 + x_2^2 = c$. In this case there are two possibilities.

<u>If $ab = 1$</u>, then

$$f(\pm\sqrt{c},0) = (0, \pm b\sqrt{c}) \text{ and } f(0, \pm b\sqrt{c}) = (\pm ab\sqrt{c}, 0) = (\pm\sqrt{c}, 0).$$

Analogously $f^2(0, \pm\sqrt{c}) = f \circ f(0, \pm\sqrt{c}) = (0, \pm\sqrt{c})$. In this case the solution $\varphi(t; x^0)$ approaches a periodic motion of period 2 (which will depend on x^0).
If $ab = -1$, one verifies that

$$f^4(\pm\sqrt{c}, 0) = (\pm\sqrt{c}, 0) \text{ and } f^4(0, \pm\sqrt{c}) = (0, \pm\sqrt{c}).$$

In this case the solution approaches a periodic motion of period 4 (depending on x^0), see Figure 3.2.8. □

If the asymptotic behaviour of a system is too complicated to determine its limit sets exactly one may try to locate them approximately in attractive sublevel sets $[V \le c] = \{x \in D;\ V(x) \le c\}$.

Proposition 3.2.32. *Suppose $D \subset X$, $V : D \to \mathbb{R}$ satisfies (A), $c \in \mathbb{R}$, $\rho > c$ and $D_\rho = \{x \in D;\ V(x) < \rho\}$. If*

(i) D_ρ is bounded and $\overline{D}_\rho \subset \operatorname{int} D$ (resp. $f(D_\rho) \subset D$).

(ii) $V(x) > c \Rightarrow \dot{V}(x) < 0$ (resp. $V(f(x)) > c \Rightarrow \dot{V}(x) < 0$) for all $x \in D_\rho$.

Then the largest invariant set in $[V \le c]$ (and hence $[V \le c]$ itself) attracts all trajectories of (40) starting in D_ρ.
In particular, if $D = X = \mathbb{K}^n$, $\lim_{\|x\| \to \infty} V(x) = \infty$ and (ii) holds for all $x \in \mathbb{K}^n$ then $[V \le c]$ is a global attractor.

Proof: The proof is for the continuous time case. Let $\tilde{V} : D \to \mathbb{R}$, $\tilde{V}(x) = \max\{V(x), c\}$, then $\tilde{V}$ is continuous on D and coincides with V on $[V > c]$. We will show that $\tilde{V}$ is decreasing along the flow generated by (40). If this were not the case there would exist $x^0 \in D$ and $t_1 > 0$ such that $\tilde{V}(\varphi(t_1; x^0)) > \tilde{V}(x^0)$, which can only happen if $\tilde{V}(\varphi(t_1; x^0)) = V(\varphi(t_1; x^0)) > c$. Let $t_0 = \inf\{\tau > 0;\ V(\varphi(t; x^0)) > c \text{ for all } t \in [\tau, t_1]\}$. Then $t_0 < t_1$, $\dot{V}(\varphi(t; x^0)) < 0$ for all $t \in (t_0, t_1]$ and either $V(\varphi(t_0; x^0)) = c$ or $t_0 = 0$ and $V(x^0) > c$. Both possibilities lead to a contradiction since $V(\varphi(t; x^0))$ is decreasing on $[t_0, t_1]$. Thus $\tilde{V}$ is decreasing along the flow defined by (40).
It follows from (i) and Lemma 3.2.26 that D_ρ is weakly invariant under the flow of (40). Moreover, by (i) every orbit $\mathcal{O}(x^0)$, $x^0 \in D_\rho$ has compact closure in D. Hence all $x^0 \in D_\rho$ have an infinite life span and so D_ρ is invariant. By LaSalle's Invariance Principle (Theorem 3.2.11) there exists $\alpha \in \mathbb{R}$ such that $\varphi(t; x^0) \to M_\alpha$ as $t \to \infty$ where M_α is the largest invariant set in $\tilde{V}^{-1}(\alpha)$. But by (ii) $M_\alpha = \emptyset$ for $\alpha \ne c$ and this proves the first assertion since $\tilde{V}^{-1}(c) = [V \le c]$. Now assume $D = X = \mathbb{K}^n$ and $\lim_{\|x\| \to \infty} V(x) = \infty$. Then every sublevel set $D_\rho = \{x \in \mathbb{K}^n;\ V(x) < \rho\}$ is bounded and condition (i) is satisfied for all $\rho > c$. This proves the second assertion of the proposition. □

Example 3.2.33. Consider the Lorenz equation (1.31) with parameters $r, \sigma, b > 0$ and choose

$$V(x, y, z) = rx^2 + \sigma y^2 + \sigma(z - 2r)^2,\ (x, y, z) \in \mathbb{R}^3.$$

Then $V(x, y, z) \to \infty$ as $\|(x, y, z)\| \to \infty$ and

$$\dot{V}(x, y, z) = -2\sigma(rx^2 + y^2 + bz^2 - 2brz).$$

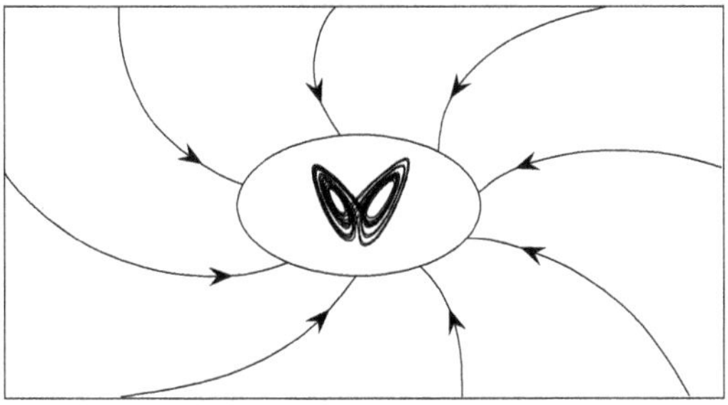

Figure 3.2.9: Attractive sublevel set containing an attractor

Let $C = \{(x,y,z);\ \ rx^2 + y^2 + bz^2 - 2brz \le 0\}$ and $c = \max\{V(x,y,z);\ (x,y,z) \in C\}$. Then V and c satisfy Proposition 3.2.32. Therefore $[V \le c]$ is a global attractor. In particular, the Lorenz attractor (see Example 3.1.27) is contained in the ellipsoid $[V \le c]$. All trajectories of the Lorenz system eventually enter the ellipsoid $[V \le c+\varepsilon]$ where $\varepsilon > 0$ is arbitrary. □

The next three extended examples illustrate the application of the previous results. We begin with some comments on the motivation of Liapunov's direct method by the use of energy functions in mechanics as mentioned in the introduction. The total energy $H(q,p)$ of an autonomous mechanical system (2) with n degrees of freedom yields a Liapunov function at each equilibrium point $(\bar{q},0)$ where $H(\cdot)$ admits a strict local minimum. If the system is conservative the Hamiltonian H is a *first integral* of the system, i.e. every trajectory remains within a surface of constant energy $\{(q,p) \in \mathbb{R}^n \times \mathbb{R}^n;\ H(q,p) = \text{const}\}$. In this case the local minima $(\bar{q},0)$ are stable but not asymptotically stable. If however the mechanical system dissipates energy e.g. by generating heat, one would expect the local minima to be asymptotically stable. Our first example illustrates this.

Example 3.2.34. (Pendulum with and without friction). Consider a simple swinging pendulum of unit length and mass m (see Examples 1.3.4, 2.1.10). Allowing for friction in the bearing, the equations of motion in state space form are

$$\begin{aligned} \dot{\theta}(t) &= \omega(t) \\ \dot{\omega}(t) &= -g \sin\theta(t) - (c/m)\omega(t) \end{aligned} \qquad (47)$$

where $\theta(t)$ is the angle, $\omega(t)$ the angular velocity and c the friction coefficient. $(\bar{\theta},\bar{\omega}) \in \mathbb{R}^2$ is an equilibrium point of (47) if and only if $\bar{\theta} \in \mathbb{Z}\pi$, $\bar{\omega} = 0$. It is intuitively clear that the state $(\bar{\theta},\bar{\omega}) = (0,0)$ (vertically downwards with zero velocity) is a stable equilibrium point of (47), which is asymptotically stable if energy is dissipated by friction, i.e. $c > 0$. The mechanical energy of the system (kinetic plus potential energy) is

$$V(\theta,\omega) = (1/2)m\omega^2 + mg(1 - \cos\theta). \qquad (48)$$

V defines a Liapunov function for (47) on $D = (-2\pi, 2\pi) \times \mathbb{R}$ at $(\bar{\theta},\bar{\omega}) = (0,0)$ (in the sense given in the Stability Theorem 3.2.25). In fact, V is continuously differentiable, positive definite on D and satisfies

$$\dot{V}(\theta,\omega) = mg\omega\sin\theta - mg\omega\sin\theta - c\omega^2 = -c\omega^2 \leq 0.$$

Given any $\varepsilon > 0$, V is positive definite away from $(0,0)$ on $\tilde{D} = [-2\pi+\varepsilon, 2\pi-\varepsilon] \times \mathbb{R}$. So $(\bar{\theta},\bar{\omega}) = (0,0)$ is stable for all values of $c \geq 0$. If $c = 0$, the energy V is a *first integral* of (47), i.e. the trajectories of (47) remain in the level sets of V. For small $\alpha > 0$ the level sets $V^{-1}(\alpha)$ are closed curves around $(0,0)$ (see Figure 2.1.3). Since these curves do not contain an equilibrium point, they are closed orbits of (47). This can easily be verified by elementary means but can also be inferred directly from the Poincaré-Bendixson Theorem 3.1.14. Thus all trajectories in a small neighbourhood of $(0,0)$ are periodic and $(\bar{\theta},\bar{\omega}) = (0,0)$ is a *centre* (see Figure 2.2.3).

In the presence of friction, i.e. $c > 0$, we expect the origin to be asymptotically stable (Figure 3.2.10) and we are interested in obtaining a good estimate for its basin of attraction. We will apply Corollary 3.2.29 and we will see that estimate depends crucially on the choice of D. In the present case

$$D_\rho = \{(\theta,\omega) \in D;\ (1/2)m\omega^2 + mg(1-\cos\theta) < \rho\},$$

and so D_ρ is bounded. We would like to choose ρ such that $\overline{D}_\rho \subset \operatorname{int} D = D$. However since $V(0,\theta) \to 0$ as $\theta \to 2\pi$, no such ρ exists. We must therefore take a smaller D. Let $D = [-\frac{3}{2}\pi, \frac{3}{2}\pi] \times \mathbb{R}$, then $\overline{D}_{mg} \subset \operatorname{int} D$ (and this will not be the case for any D_ρ with $\rho > mg$). The largest invariant set M in $E = \{(\theta,\omega) \in D;\ \dot{V}(\theta,\omega) = 0\} = [-\frac{3}{2}\pi, \frac{3}{2}\pi] \times \{0\}$ is $M = \{(0,0), (-\pi,0), (\pi,0)\}$. But the equilibrium points $(\pm\pi,0) \notin V^{-1}(\alpha)$ for any $\alpha < mg$, so by Corollary 3.2.29, $(0,0)$ is asymptotically stable and D_{mg} is contained in the basin of attraction of $(0,0)$. The question is can we do better by further reduction of D. Let $D = (-\pi,\pi) \times \mathbb{R}$, then $\overline{D}_\rho \subset \operatorname{int} D = D$ provided we choose $\rho < 2mg$ (and this will not be the case for any $\rho \geq 2mg$). The largest invariant set M in $E = \{(\theta,\omega) \in D;\ \dot{V}(\theta,\omega) = 0\} = (-\pi,\pi) \times \{0\}$ is $M = \{(0,0)\}$. By Corollary 3.2.29, $(0,0)$ is asymptotically stable and $D_{2mg} = \bigcup\{D_{\rho;\ \rho<2mg}\}$ is contained in the basin of attraction of $(0,0)$. Note that D_{2mg} consists of all those states (θ,ω) in D which generate periodic orbits around $(0,0)$ *if no friction is present* (swinging motions without rotations, see Figure 2.1.3).

By an analogous analysis on $(k2\pi,0) + D$ one verifies that the equilibrium points $(k2\pi,0)$, $k \in \mathbb{Z}$ are stable if $c = 0$, asymptotically stable if $c > 0$ and in this case the basin of attraction of $(k2\pi,0)$, $k \in \mathbb{Z}$ contains the translates $(k2\pi,0) + D_{2mg}$.

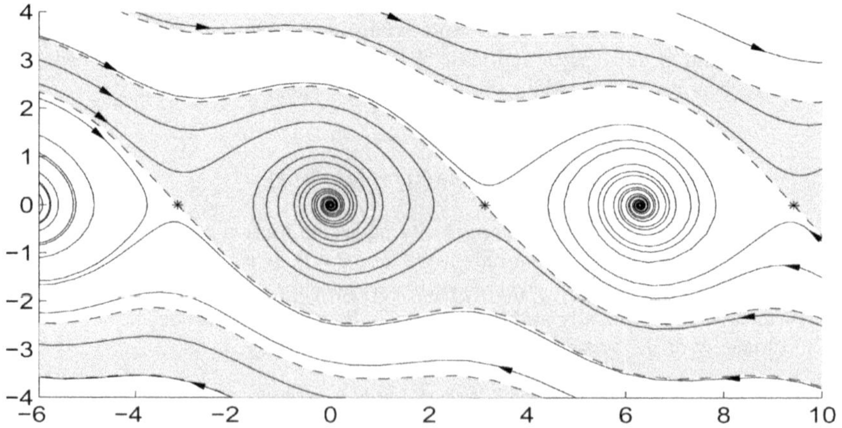

Figure 3.2.10: Phase portrait of pendulum with friction

To obtain a global result we have to consider the above V on $D = \mathbb{R}^2$ and apply Theorem 3.2.28. It is not difficult to prove that all trajectories of (47) are bounded if $c > 0$. Since $M = \mathbb{Z}2\pi \times \{0\}$ is the largest invariant subset of $E = \{(\theta,\omega) \in \mathbb{R}^2;\ \dot{V}(\theta,\omega) = 0\} = \mathbb{R} \times \{0\}$, every trajectory of (47) converges to some equilibrium point $(k2\pi,\ 0)$. This is in accordance with our physical intuition. Because of friction, the pendulum rotates only a finite number k times and then continues swinging with ever decreasing amplitudes gradually approaching the equilibrium point $(k2\pi, 0)$. □

In the following example we study a "conservative" ecological system.

Example 3.2.35. Consider the Lotka-Volterra model for predator-prey system of two interacting populations (cf. Example 1.1.3)

$$\begin{aligned} \dot{x}_1(t) &= ax_1(t) - bx_1(t)x_2(t) \\ \dot{x}_2(t) &= -cx_2(t) + dx_1(t)x_2(t) \end{aligned} \tag{49}$$

where x_1 and x_2 are the densities (number per area) of the prey and predator populations, respectively, and a, b, c, d are positive constants. Note that no trajectory of (49) starting in $D = (0,\infty) \times (0,\infty)$ can leave D, since $\mathbb{R}_+ \times \{0\}$ and $\{0\} \times \mathbb{R}_+$ are invariant under (49). The system has two equilibrium points $(0,0)$ and $(c/d, a/b)$. $(0,0)$ is unstable since the restriction of (49) to the invariant set $\mathbb{R} \times \{0\}$ yields the unstable linear system $\dot{x}_1 = ax_1$. To study the second equilibrium point it is convenient to normalize the variables. Setting $z_1 = dx_1/c$, $z_2 = bx_2/a$ we obtain the transformed equations of motion

$$\begin{aligned} \dot{z}_1(t) &= az_1(t)(1 - z_2(t)) \\ \dot{z}_2(t) &= -cz_2(t)(1 - z_1(t)) \end{aligned} \tag{50}$$

with equilibrium point $\overline{z} = (1,1)$. In order to find a first integral of (50) note that

$$\frac{\dot{z}_2(t)}{\dot{z}_1(t)} = -\frac{cz_2(t)(1 - z_1(t))}{az_1(t)(1 - z_2(t))}$$

and hence

$$c\dot{z}_1(t) - c\frac{\dot{z}_1(t)}{z_1(t)} + a\dot{z}_2(t) - a\frac{\dot{z}_2(t)}{z_2(t)} = 0.$$

Integrating with respect to t, we obtain

$$W(z_1, z_2) := cz_1 - c\log z_1 + az_2 - a\log z_2, \quad z \in D \tag{51}$$

is constant along the trajectories of (50). It is easily verified that $V(z_1, z_2) = W(z_1, z_2) - W(1,1)$ is positive definite away from $\bar{z} = (1,1)$. So V is a Liapunov function for (50) at $(1,1)$ and this equilibrium point is stable. In fact, the level sets $\{z \in D;\ V(z) = \gamma\}$ are closed curves for all $\gamma > 0$ (see Figure 1.1.2) and none of these curves contains an equilibrium point. Hence by the Poincaré-Bendixson Theorem 3.1.14 these level curves are periodic orbits of (50). Thus $\overline{z} = (1,1)$ is a centre and all the trajectories of (50) in D are periodic.
The effect of "friction" in a Lotka-Volterra model (due to overcrowding, see Figure 1.1.3), is considered in Ex. 14. □

We now study an extended example which illustrates how to use strict Liapunov functions to obtain estimates for the basin of attraction of an asymptotically stable equilibrium point. Often *various* Liapunov functions have to be constructed in order to get an approximate picture of this region. The construction of "good" Liapunov functions (i.e. those which yield good estimates of the basin of attraction) is still something of an art although some constructive methods have been suggested in the literature, see *Notes and References* and the Exercises.

Example 3.2.36. (Van der Pol equation). Van der Pol suggested the following second order differential equation as a possible model for the study of oscillations in vacuum tubes

$$\ddot{z} + \varepsilon(1 - z^2)\dot{z} + z = 0. \tag{52}$$

In a paper published in 1926 he discovered that if $\varepsilon < 0$, the trajectories converge to a stable limit cycle for all non-zero initial states. Figure 3.2.11 shows this limit cycle and some additional trajectories in the $(z, \dot{z})$ plane. Reversing time in (52) leads to the same

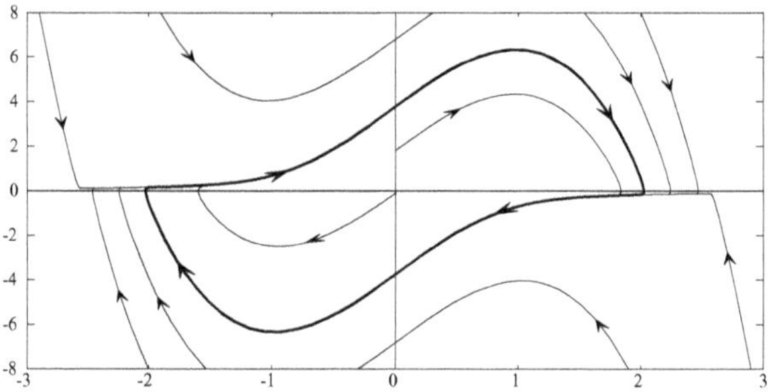

Figure 3.2.11: Stable limit cycle for $\varepsilon = -4.0$

equation but with changed sign of ε: if $z(\cdot)$ is a solution of (52) then $y(t) := z(-t)$ is a solution of (52) with ε replaced by $-\varepsilon$. Hence if $\varepsilon > 0$ the equilibrium point at the origin is asymptotically stable for (52) and its basin of attraction is simply the reflection of the interior of the Van der Pol's limit cycle about the z–axis (because $\dot{y}(t) = -\dot{z}(-t)$). As an example for more complicated situations where the basin of attraction may be unknown, we now show how to estimate the basin for the Van der Pol equation with positive ε via a family of Liapunov functions. Our knowledge in the present case allows us to judge the tightness of these estimates.
Let $x_1 = z$, $x_2 = \dot{z}$, then (52) becomes

$$\dot{x}_1 = x_2, \quad \dot{x}_2 = -\varepsilon(1 - x_1^2)x_2 - x_1. \tag{53}$$

For (53) with $\varepsilon > 0$ consider the functions $V_r : \mathbb{R}^2 \to \mathbb{R}$

$$V_r(x) = x_2^2 + r\varepsilon(x_1 - x_1^3/3)x_2 + (r\varepsilon^2/2)(x_1 - x_1^3/3)^2 + x_1^2$$

where $0 \le r \le 2$. Then

$$\dot{V}_r(x) = -\varepsilon[(2 - r)(1 - x_1^2)x_2^2 + r(1 - x_1^2/3)x_1^2]. \tag{54}$$

Note that

$$V_r(x) = [x_2 + (r\varepsilon/2)(x_1 - x_1^3/3)]^2 + x_1^2 + (r\varepsilon^2/4)(2 - r)x_1^2(1 - x_1^2/3)^2.$$

Hence if $0 \le r \le 2$, $V_r(x) > 0$ for any $x \ne 0$ and $\lim_{\|x\| \to \infty} V_r(x) = \infty$. In order to

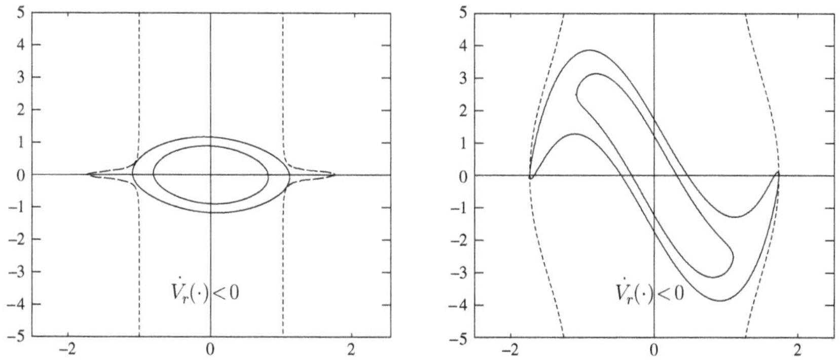

Figure 3.2.12: The curves $\dot{V}_r(\cdot) = 0$ (dashed lines) and $V_r(\cdot) = \rho$ (solid lines) for two values of ρ with $r = 0.05$ in the left figure and $r = 1.9$ in the right figure

apply Corollary 3.2.29 we have to choose a neighbourhood $D(r)$ of $\overline{x} = (0,0)$ on which V_r is a strict Liapunov function. Then choose ρ as large as possible so that the set $D_\rho(r) = \{x \in D(r);\ V_r(x) < \rho\}$ is bounded and $\overline{D_\rho(r)} \subset \operatorname{int} D(r)$. We denote this choice of ρ by $\rho(r)$ and illustrate the procedure in Figure 3.2.12 for $\varepsilon = 4.0$.

For $r = 0.05$ we find that $\rho(r)$ may be chosen to be approximately 1.37. Note that $D_{1.37}(.05)$ does not contain the point $(0.5, -2.0)$. However for $r = 1.9$ we may choose $\rho(r) \approx 3.0$ and $D_3(1.9)$ does contain the point $(0.5, -2.0)$.

Then $\cup_{r\in[0,2]} D_{\rho(r)}(r)$ will yield an estimate for the basin of attraction and this is illustrated in Figure 3.2.13 (shaded area) together with the exact boundary obtained by reflecting the limit cycle at the z axis.

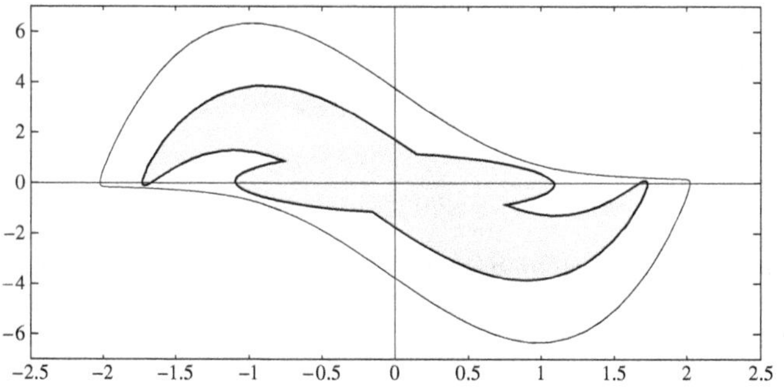

Figure 3.2.13: Estimate for the region of asymptotic stability of the origin, $\varepsilon = 4.0$

□

Our final result in this subsection strengthens the Instability Theorem 3.2.22 for the time–invariant case.

Theorem 3.2.37 (Instability Theorem). *Let $\overline{x}$ be an equilibrium point of (40), D a compact subset of X such that $\overline{x} \in D$ and assume that in the discrete time case $f(D) \subset X$. Suppose there exists a continuously differentiable[4] (resp. continuous) real-valued function V defined on D with the following properties.*

[4]By definition, V is continuously differentiable on a subset $D \subset \mathbb{K}^n$ if V is defined and continuously differentiable on some open set $U \supset D$.

(i) $V(\overline{x}) = 0$ *and for every* $r > 0$ *there exists* $x^0 \in D \cap B(\overline{x}, r)$ *such that* $V(x^0) < 0$.

(ii) $\dot{V}(x) \leq 0$ *for all* $x \in S = \{x \in D;\ V(x) < 0\}$.

(iii) There are no invariant sets in $E = \{x \in S;\ \dot{V}(x) = 0\}$.

Then for every $x^0 \in D$ *such that* $V(x^0) < 0$, *the trajectory* $\varphi(\cdot; x^0)$ *leaves* D *at some time* $t > 0$. *In particular, if* D *is a neighbourhood of* $\overline{x}$ *then* $\overline{x}$ *is unstable.*

Proof: The proof is for the continuous time case. Because of (i), it suffices to prove the first assertion of the theorem. Suppose there exists $x^0 \in D$ such that $V(x^0) < 0$ and $\varphi(t; x^0) \in D$ for all $t \in T(x^0)$. Then this trajectory has infinite life span by Lemma 3.2.14 since $D \subset X$ is compact. By Theorem 3.1.16 the limit set $\omega(x^0)$ is nonempty and weakly invariant. Again by Lemma 3.2.14, $\omega(x^0) \subset D$ is even invariant. By (ii)

$$s \geq t \geq 0 \quad \Rightarrow \quad V(\varphi(s; x^0)) \leq V(\varphi(t; x^0)) \leq V(x^0) < 0,$$

and so, by Lemma 3.2.10, there exists $c < 0$ such that $\omega(x^0) \subset V^{-1}(c) \subset S$. Since $\omega(x^0)$ is invariant and $V(\cdot)$ is constant on $\omega(x^0)$ it follows that $\omega(x^0) \subset E$. But this contradicts (iii). □

Note that assumptions (ii) and (iii) are automatically satisfied if

$$\dot{V}(x) < 0, \quad x \in D \setminus \{\overline{x}\}. \tag{55}$$

Example 3.2.38. Consider the conservative mechanical system described in the introduction to this section. The equations of motion are

$$\dot{q} = \frac{\partial H}{\partial p}, \quad \dot{p} = -\frac{\partial H}{\partial q} \tag{56}$$

where the Hamiltonian is given by $H(q, p) = \mathcal{T}(p) + \mathcal{W}(q)$. In this example we assume that $(\bar{q}, \bar{p}) = (0, 0)$ is an equilibrium point of (56), the kinetic energy is a positive quadratic form in p and $\mathcal{W}$ is continuously differentiable function which is strictly concave on some ball $B = B(0, r)$ around $\bar{q} = 0$ and admits a local maximum at $\bar{q} = 0$. By strict concavity

$$\mathcal{W}(q) + \langle \mathcal{W}'(q), (0 - q) \rangle > \mathcal{W}(0) \quad q \in B,\ q \neq 0$$

and hence $\langle \mathcal{W}'(q), q \rangle < 0$ for all $q \in B$, $q \neq 0$. Consider the function $V(q, p) = -\langle q, p \rangle$ on a compact neighbourhood $D \subset B \times \mathbb{R}^n$ of $(0, 0)$. Then the derivative of V along the trajectories of (56) is

$$\dot{V}(q, p) = -\langle \mathcal{T}'(p), p \rangle + \langle \mathcal{W}'(q), q \rangle = -2\mathcal{T}(p) + \langle \mathcal{W}'(q), q \rangle < 0, \quad (q, p) \in D \setminus \{0, 0\}.$$

Thus $\dot{V}$ satisfies (55) on D. Now for any $\rho > 0$ we can always choose an initial state (q_0, p_0) of norm less than ρ such that $-\langle q_0, p_0 \rangle < 0$. Hence $(0, 0)$ is unstable by Theorem 3.2.37. □

A finite-dimensional system is called *positive* it its state space is the positive orthant $X = \mathbb{R}^n_+$, $n \in \mathbb{N}^*$. Equations of the form (40) describe positive systems if every trajectory $\varphi(t; x^0)$ starting in the positive orthant remains in it for all $t \in T(x^0)$, i.e. if $\mathbb{R}^n_+$ is weakly invariant. For example, if the flow of the Lotka-Volterra model (49) is restricted to $\mathbb{R}^n_+$, we obtain a positive system.
If the stability of an equilibrium point $\overline{x} \in \mathbb{R}^n_+$ of a positive system is investigated, the analysis must be restricted to the state space of the system, so only points and trajectories in $X = \mathbb{R}^n_+$ are considered and the neighbourhoods of any $x \in \mathbb{R}^n_+$ must be understood as neighbourhoods of x in $X = \mathbb{R}^n_+$. More generally, suppose that X_0 is a weakly invariant subset for the flow defined by (40). In order to analyze the stability of an equilibrium point $\overline{x} \in X_0$ with respect to the induced flow on X_0 we combine the results of Subsection 3.2.1 with those of the present subsection. For instance, in order to establish stability of $\overline{x}$ we construct a continuously differentiable (resp. continuous) function $V : D \to \mathbb{R}$ on a neighbourhood D of $\overline{x}$ in $\mathbb{K}^n$, which satisfies the properties of a Liapunov function *on* X_0:

$$V(\overline{x}) = 0; \quad V(x) > 0, \ \ x \in (X_0 \cap D) \setminus \{\overline{x}\}; \quad \dot{V}(x) \le 0, \ \ x \in X_0 \cap \mathcal{D}(\dot{V}). \quad (57)$$

In fact, the restriction of such a function to $D_0 := X_0 \cap D$ is then a generalized Liapunov function for the induced flow on X_0 (see Remark 3.2.24) and the stability results of Subsection 3.2.1 are applicable. In the following discrete time example the natural state space is a compact subset of $\mathbb{R}^2_+$.

Example 3.2.39. We consider a discrete time epidemic model for the spread of gonorrhoea. Let us assume that the male and female populations are constant in time and the disease is transmitted by sexual contact of infected members of one population with members of the other. Recovery from the disease is possible, but no one is immune from infection. Let $x_1(t)$, resp. $x_2(t)$ be the fraction of males, resp. females infected at time t. A possible model is

$$\begin{aligned} x_1(t+1) &= \alpha x_2(t)(1 - x_1(t)) + (1-\beta)x_1(t) \\ x_2(t+1) &= \gamma x_1(t)(1 - x_2(t)) + (1-\delta)x_2(t) \end{aligned} \quad (58)$$

where the first terms represent the fractions which are newly infected by contact with infected members and the second terms represent the fractions that have not recovered within one time unit. We assume $0 < \alpha, \beta, \gamma, \delta < 1$. Since solutions of (58) are only meaningful as long as $0 \le x_i(t) \le 1$, $i = 1, 2$ we restrict (58) to the compact set

$$X_0 = \{x \in \mathbb{R}^2;\ 0 \le x_1 \le 1,\ 0 \le x_2 \le 1\}$$

which is invariant under the global flow defined by (58). Let us analyze the stability of the equilibrium state $\overline{x} = (0,0) \in X_0$. Consider

$$V(x) = (\gamma + \delta)x_1 + (\alpha + \beta)x_2, \quad x \in D = \mathbb{R}^2$$

which is positive on $X_0 \setminus \{\overline{x}\}$ and satisfies

$$\begin{aligned} \dot{V}(x) &= (\gamma+\delta)[\alpha x_2(1-x_1) + (1-\beta)x_1] - (\gamma+\delta)x_1 \\ &\quad + (\alpha+\beta)[\gamma x_1(1-x_2) + (1-\delta)x_2] - (\alpha+\beta)x_2 \\ &= (\alpha\gamma - \beta\delta)(x_1 + x_2) - [\alpha(\gamma+\delta) + \gamma(\alpha+\beta)]x_1 x_2. \end{aligned}$$

If $\alpha\gamma \leq \beta\delta$, then V is a (generalized) Liapunov function at $\{\overline{x}\} = \{(0,0)\}$ for the system (58) on $D_0 = X_0$. Since the only invariant set in $\dot{V}^{-1}(0)$ is $\overline{x} = (0,0)$, the largest invariant subset M_c of $V^{-1}(c)$ is empty for every $c > 0$. Hence it follows from Corollary 3.2.12 that $\overline{x} = (0,0)$ is a globally asymptotically stable equilibrium point of the system defined by (58) on X_0. So for any initial distributions of the disease the number of infected people will gradually decay to zero as $t \to \infty$.
Now suppose $\alpha\gamma - \beta\delta > 0$. Then $\dot{V}(x) > 0$ for $x \in X_0 \setminus \{(0,0)\}$ if

$$(\alpha\gamma - \beta\delta)(x_1 + x_2) > [\alpha(\gamma + \delta) + \gamma(\alpha + \beta)]x_1x_2.$$

It is easy to show that this will be the case if $x \in \tilde{D}$ where $\tilde{D}$ is the set of all $(x_1, x_2) \in X_0$ satisfying

$$x_1 + x_2 < \eta := \frac{4(\alpha\gamma - \beta\delta)}{\alpha(\gamma + \delta) + \gamma(\alpha + \beta)}.$$

Hence the function $\tilde{V} = -V$ satisfies (55) on $\tilde{D}$ and Theorem 3.2.37 can be applied to $\tilde{V}$ and any compact neighbourhood C of $(0,0)$ in $\tilde{D}$. So in this case $\overline{x}$ is an unstable equilibrium point of the flow defined by (58) on X_0, and every trajectory no matter how close to $\overline{x} = (0,0)$ in X_0 will eventually leave C. In other words, the total fraction of infected members of the population will eventually be greater than or equal to η. □

3.2.4 Exercises

1. Let α_1 and α_2 be class $\mathcal{K}$ functions on $[0, r_1)$, α_3 and α_4 be class $\mathcal{K}_\infty$ functions and β be a class $\mathcal{LK}$ function on $\mathbb{R}_+ \times [0, r_1)$. Prove

(i) $\alpha_1 \circ \alpha_2$ belongs to class $\mathcal{K}$ on $[0, r_1)$.
(ii) $\alpha_3 \circ \alpha_4$ belongs to class $\mathcal{K}_\infty$.
(iii) If $\sigma(s, r) = \alpha_1(\beta(\alpha_2(s), r))$, $(s, r) \in \mathbb{R}_+ \times [0, r_1)$, then σ belongs to class $\mathcal{LK}$ on $\mathbb{R}_+ \times [0, r_1)$.

2. Let $\mathcal{F} = (T, X, \varphi)$ be a local flow on a metric space X and $\overline{x} \in X$ an equilibrium point of this flow. Suppose there exists a neighbourhood U of $\overline{x}$ such that $t_+(t_0, x^0) = \infty$ for all $(t_0, x^0) \in T \times U$. Prove the following.

(i) If there exists a class $\mathcal{K}$ function α such that for all $x^0 \in B(\overline{x}, \rho) \subset U$, $\rho > 0$ sufficiently small we have

$$d(\varphi(t; t_0, x^0), \overline{x}) \leq \alpha(d(x^0, \overline{x})), \quad t \in T_{t_0},\ t_0 \in T,$$

then $\overline{x}$ is uniformly stable.

(ii) If there exists a class $\mathcal{LK}$ function β such that for all $x^0 \in B(\overline{x}, \rho) \subset U$, $\rho > 0$ sufficiently small we have

$$d(\varphi(t; t_0, x^0), \overline{x}) \leq \beta(t - t_0, d(x^0, \overline{x})), \quad t \in T_{t_0},\ t_0 \in T,$$

then $\overline{x}$ is uniformly asymptotically stable.

3. Show that the sufficient stability conditions specified in the previous exercise are also necessary. (See Khalil (1996) [299] Lemma 3.3 and Section A.3).

4. Let $\alpha : [0, r_1) \to \mathbb{R}_+$ be a class $\mathcal{K}$ function . Prove that there exist continuous and strictly increasing functions $\alpha_1, \alpha_2 : [0, r_1) \to \mathbb{R}_+$ of class $\mathcal{K}$ such that

$$\alpha_1(r) \leq \alpha(r) \leq \alpha_2(r), \quad r \in [0, r_1).$$

(Hint: To construct α_1, define $\alpha_1(r) = \int_0^r \alpha(\rho)d\rho$ for $r \in [0,1] \cap [0, r_1)$ and if $r > 1$ define

$$\alpha_1(r) = \alpha_1(1) + \int_1^r (\alpha(\rho) - \alpha_1(1))d\rho \quad \text{for} \quad r \in [1,2] \cap [0, r_1),$$

etc. To construct α_2, define for any $r \in (0, r_1)$, $\tilde{\alpha}(r) = (\tilde{r} - r)^{-1} \int_r^{\tilde{r}} \alpha(\rho)d\rho$ where $\tilde{r} = \min\{2r, (r + r_1)/2\}$ and $\tilde{\alpha}(0) = 0$. Then $\tilde{\alpha} : [0, r_1) \to \mathbb{R}_+$ is continuous and satisfies $\alpha(r) \leq \tilde{\alpha}(r) \leq \alpha(\tilde{r})$. Now set $\alpha_2(r) = \max_{0 \leq \rho \leq r} \tilde{\alpha}(\rho) + r$).

5. Use Liapunov's direct method to show that the origin for each of the following systems is stable
(i) $\dot{x}_1 = -x_1^3 - x_2^2$, $\dot{x}_2 = x_1 x_2 - x_2^3$.
(ii) $\dot{x}_1 = -6x_1/(1 + x_1^2)^2 + 2x_2$, $\dot{x}_2 = -2x_1 - 2x_2/(1 + x_1^2)^2$.
(iii) $\dot{x}_1 = x_2(1 - x_1)$, $\dot{x}_2 = -x_1(1 - x_2)$, use $V(x) = -x_1 - \log(1 - x_1) - x_2 - \log(1 - x_2)$.
(iv) $\dot{x}_1 = -2x_1 + 2x_2^4$, $\dot{x}_2 = -x_2$, use $V(x) = 6x_1^2 + 12x_2^2 + 4x_1 x_2^4 + x_2^8$.
(v) $\dot{x}_1 = x_2$, $\dot{x}_2 = \sin\alpha - \sin(\alpha + x_1)$, where $0 < \alpha < \pi/2$, use $V(x) = \frac{1}{2}x_2^2 + \cos\alpha - \cos(\alpha + x_1) - x_1 \sin\alpha$.

6. Analyze the stability properties of the equilibrium points of the following system by Liapunov's direct method

$$x_1(t+1) = x_1(t)^2 - x_2(t)^2, \quad x_2(t+1) = 2x_1(t)x_2(t).$$

7. If $c > 0$ prove that for some suitable D

$$V(\theta, \omega) = (1/2)m\omega^2 + mg(1 - \cos\theta) + (c/2)\,\theta\omega + (c^2/4m^2)\theta^2$$

is a strict Liapunov function at $(0,0)$ on D for the system given in Example 3.2.34.

8. Consider the time-varying second order system $y(t+2) + a(t)y(t) = 0$ or, equivalently,

$$x_1(t+1) = x_2(t), \quad x_2(t+1) = -a(t)x_1(t),$$

where $a(t) \in \mathbb{R}$, $t \in \mathbb{N}$ is given such that $a(t)^2 \leq \alpha < 1$ for all t. Prove that the equilibrium point $\overline{x} = 0$ is globally asymptotically stable.

9. Consider the time-varying system (5) in Example 3.1.5, $\dot{x} = -(1 + 1/t)x + x^3 t^2$. Show that if $V(t, x) = t^2 x^2$, then $\dot{V} = -2V(1 - V)$. Hence prove that the origin is asymptotically stable with basin of attraction $\mathcal{A}(t_0, 0) \supset \{x \in \mathbb{R};\ V(x, t_0) < 1\}$.

10. Consider the linear system $\ddot{x} + \dot{x} + e^{-t}x = 0$ and the function $V(t, x) = x^2 + e^t \dot{x}^2$. Prove that the origin is stable at any time $t_0 \geq 0$. Is it uniformly stable?

11. Use the function $V(x, y) = (x^2 + y^2)$ for the system

$$\dot{x} = a(t)y + b(t)x(x^2 + y^2), \quad \dot{y} = -a(t)x + b(t)y(x^2 + y^2),$$

to determine conditions on $b(t)$ for the zero state to be (i) stable, (ii) uniformly asymptotically stable.

12. Prove Theorem 3.2.25 making use of Lemma 3.2.26.

13. A model for the oscillations of a damped hard spring is Duffing's equation

$$\ddot{y} + \dot{y} + ay + y^3 = 0.$$

If $a \geq 0$ use the Liapunov function $V(y, \dot{y}) = ay^2 + (1/2)y^4 + \dot{y}^2$ to prove that $(0,0)$ is globally asymptotically stable. If $a < 0$ show that there are three equilibrium states, two of which are asymptotically stable, and one of which is unstable. Find estimates for the basins of attraction for the stable equilibrium states.

14. Consider the following predator prey model which accounts for the effects of crowding on the prey

$$\dot{x}_1 = ax_1 - bx_1x_2 - ex_1^2, \quad \dot{x}_2 = -cx_2 + dx_1x_2.$$

(cf. Example 3.2.35). Note that in the absence of predators ($x_2 \equiv 0$) the prey grows according to a logistic law of growth. Assume $a/e > c/d$ so that the system has a positive equilibrium point $\overline{x}_1 = c/d$, $\overline{x}_2 = (da - ec)/bd$. Proceed as in Example 3.2.35 to analyze the stability of this equilibrium point via the "ecological Liapunov function" (51).

15. A gradient system on an open set $X \subset \mathbb{R}^n$ is given by a differential equation

$$\dot{x}(t) = -\operatorname{grad} V(x(t))$$

where $V: X \to \mathbb{R}$ is a C^2-function. Prove:

(i) $\dot{V}(x) = -\|\operatorname{grad} V(x)\|^2$, $x \in X$.

(ii) If $\overline{x}$ is an isolated minimum of V then $\overline{x}$ is an asymptotically stable equilibrium point.

(iii) Find the asymptotically stable equilibria when $V: \mathbb{R}^2 \to \mathbb{R}$ is given by $V(x, y) = x^2(x-1)^2 + y^2$.

16. *Steepest descent.* Suppose that a function $f: \mathbb{R}^n \to \mathbb{R}$ is to be minimized and assume

(i) f has a unique minimum at the point $\overline{x}$,

(ii) f is continuously differentiable and $\operatorname{grad} f(x)$ vanishes only at $\overline{x}$,

(iii) $f(x) \to \infty$ as $\|x\| \to \infty$.

Let $\alpha: \mathbb{R}^n \to \mathbb{R}_+$ be a continuous stepsize function such that for each $x \in \mathbb{R}^n$

$$f(x - \alpha(x) \operatorname{grad} f(x)) < f(x), \quad x \neq \overline{x}.$$

Prove that the method of steepest descent

$$x_{k+1} = x_k - \alpha(x_k) \operatorname{grad} f(x_k),$$

converges from any initial value x^0 to the minimum $\overline{x}$.

17. Consider the nonlinear oscillator (Liénard's equation) $\ddot{y} + f(y)\dot{y} + g(y) = 0$ or, equivalently, $\dot{x}_1 = x_2$, $\dot{x}_2 = -f(x_1)x_2 - g(x_1)$, where the damping $f(y)$ and the restoring force $g(y)$ are given real valued continuous functions on $\mathbb{R}$ with $g(0) = 0$. Use the total energy

$$V(x) = (1/2)x_2^2 + G(x_1), \quad G(x_1) = \int_0^{x_1} g(\xi)\, d\xi$$

as Liapunov function to prove that the origin is a globally asymptotically stable equilibrium point under the assumptions

$$yg(y) > 0 \text{ for } y \neq 0, \quad f(y) > 0 \text{ for } y \neq 0, \quad G(y) \to \infty \text{ as } |y| \to \infty.$$

18. In the situation of Ex. 17 consider

$$V(x) = x_2^2 + F(x_1)x_2 + (1/2)F(x_1)^2 + 2G(x_1), \quad F(x_1) = \int_0^{x_1} f(\xi)\, d\xi$$

Determine conditions on f, g for the origin to be a stable equilibrium state. If $f(x) = 1-x^2$, $g(x) = x + x^3$, determine estimates for the basin of attraction of the origin.

19. *Krasovskii's method* Consider the system $\dot{x} = f(x)$ where $f : X \to \mathbb{R}^n$ is continuously differentiable and $f(\overline{x}) = 0$ if and only if $\overline{x} = 0$. Krasovskii's method is to try $V(x) = \langle f(x), f(x)\rangle$ as a possible Liapunov function. Prove that V is a Liapunov function for $\dot{x} = f(x)$ at $\overline{x} = 0$ if

$$\forall x \in \mathbb{K}^n : \langle x, Fx\rangle \le 0 \quad \text{where} \quad F = \left(\frac{\partial f_i}{\partial x_j}(0)\right) + \left(\frac{\partial f_i}{\partial x_j}(0)\right)^\top .$$

Apply Krasovskii's method to prove that the equilibrium state $\overline{x} = (0,0)$ of the system $\dot{x}_1 = -3x_1 + x_2, \quad \dot{x}_2 = x_1 - x_2 - x_2^3$ is globally asymptotically stable.

3.2.5 Notes and References

Good introductory texts on Liapunov stability are *LaSalle and Lefschetz* (1961) [332] and *J.L. Willems* (1970) [530]. Comprehensive works on Liapunov stability and related properties of differential systems are *Hahn* (1967) [209], *Krasovskii* (1963) [316] and *Yoshizawa* (1966) [537]. There are of course many different versions of the stability and instability theorems. For example one may dispense with the boundedness condition in Theorem 3.2.17 (ii) if one assumes that the right hand side of (27a) is bounded, see *Hahn* (1967) [209].
A Llapunov theory for general dynamical systems on metric spaces has been developed by various authors, see *Zubov* (1964) [547] and *Bhatia and Szegö* (1970) [54]. More instability criteria can be found in *LaSalle and Lefschetz* (1961) [332], *Hahn* (1967) [209], *Yoshizawa* (1963) [536], *Chetaev* (1961) [98]. In *Krasovskii* (1963) [316], *Yoshizawa* (1966) [537] and *Hahn* (1967) [209] converse theorems are proved which give conditions under which certain stability properties imply the existence of Liapunov functions of corresponding type. Converse theorems for flows on metric spaces can be found in *Bhatia and Szegö* (1970) [54]. Constructional methods have been suggested by *Zubov* (1964) [547], *Schultz and Gibson* (1962) [452], *Krasovskii* (1963) [316] and *Brockett* (1966) [76]. The construction of a Liapunov function for dissipative input-output systems is discussed in *Willems* (1971) [528].
One of the first uses of Liapunov functions in control theory was by *Lur'e* (1951) [352]. He considered the stability problem for the system

$$\dot{x} = Ax + bu, \quad u = f(\sigma), \quad \dot{\sigma} = c^\top x - ru, \tag{59}$$

where b, $c \in \mathbb{R}^n$ and r is a scalar. For this system Lur'e introduced a Liapunov function containing an integral term

$$V(x,\sigma) = x^\top Px + \int_0^\sigma f(s)ds\,, \tag{60}$$

see Ex. 3.24.
A concise exposition of LaSalle's invariance principle and its applications to discrete and continuous time systems is given in *LaSalle and Artstein* (1976) [331] and Example 3.2.35 (spread of gonorrhoea) is based on this reference. A standard reference for the mathematical theory of epidemics is *Bailey* (1957) [30].

Stability results for *time-varying systems* are presented in most of the books cited here, see also, *Miller and Michel* (1982) [375] and *Khalil* (1996) [299]. For results on the stability of periodic trajectories see e.g. *Rouché and Mawhin* (1980) [438].
Liapunov functions for differentiable systems on smooth manifolds have been considered by *Shub* (1973) [464]. For infinite dimensional systems stability results see the survey by *Pritchard and Zabczyk* (1983) [421].
Liapunov's direct method for discrete time systems has been described in *Hahn* (1958) [207] and *Bertram and Kalman* (1960) [52].
In recent years a good deal of research activity has been devoted to the stability analysis of *sets of systems* (systems with time-varying perturbations and controls, systems described by differential or difference inclusions). Here again Liapunov techniques play an important role, see *Lin et al.* (1996) [345], *Camilli et al.* (2002) [88], *Wirth* (2002) [532]. For a survey of some of these results see *Sontag* (2000) [473]. A research monograph in this area is *Grüne* (2002) [203].
In many areas such as economics, population dynamics, process control etc., components of the state vector are restricted to be positive quantities see *Luenberger* (1979) [349]. Liapunov functions for such systems are analyzed in *Arrow and Hahn* (1971) [21].

3.3 Linearization and Stability

Linearization of a nonlinear differential or difference equation of the form (2.27) about a given trajectory (in particular an equilibrium state) yields a linear system which, in general, will be time-varying. Since stability is a local property one might expect that the linearization provides sufficient information to determine whether or not the trajectory is stable. This is the idea behind an approach adopted by Liapunov which is now known as *Liapunov's indirect method.* In order to prepare the ground for the development of this method we need to consider stability problems for linear systems. Actually linear models are often used in areas of application and especially in control, so linear stability analysis is important in its own right. Moreover, as we shall see, the theory is well developed and yields a number of specific stability criteria which lead to computable tests. By using Liapunov's indirect method these linear stability tests can be applied to nonlinear systems as well.
Our stability analysis will be for both time-varying and time–invariant linear systems and we will also include methods related to Liapunov functions since they will enable us to prove the validity of the indirect method. In the last section of this chapter the stability analysis of linear systems will be continued with a derivation of the classical algebraic stability criteria for *time–invariant* linear systems.
In the first subsection we characterize the asymptotic and exponential stability of time–varying linear systems via their evolution operator Φ and associated Liapunov and Bohl coefficients. Whilst these results are quite satisfactory they suffer from the drawback that one needs to compute Φ in order to check them. In Subsection 3.3.2 time-invariant systems are considered and we show that the conditions are equivalent to constraints on the spectrum of A. We illustrate the results with some examples and also carry out an extended case study where the numerical stability of linear multi-step discretization methods, described in Section 2.5 is analyzed. In Subsection 3.3.4 we examine the possibility of using time–dependent quadratic forms as Liapunov functions for time-varying linear systems. Then these quadratic forms are used for nonlinear systems (2.27) to derive stability properties of a given trajectory from stability properties of the associated linearized model (Liapunov's indirect method). In fact we will see that the properties of asymptotic stability and of instability can be tested via the linearized model. However this is not possible for (marginal) stability, since the stability of a solution which is not asymptotically stable can be destroyed by arbitrary small perturbations of the system equation (see Subsection 3.3.2). Finally, in the last subsection we consider time-invariant systems and *time–invariant* quadratic Liapunov functions. For their construction a linear matrix equation, the *algebraic Liapunov equation*, must be solved. We analyze this equation in some detail and characterize the asymptotic stability and instability of a time-invariant linear system via the solutions of the associated algebraic Liapunov equation. This in turn allows us to conclude that an equilibrium point $\overline{x}$ of a nonlinear system is exponentially stable if and only if the spectrum $\sigma(A)$ of the matrix obtained from the linearization at $\overline{x}$ satisfies $\sigma(A) \in \mathbb{C}_-$ (resp. $\sigma(A) \in \mathbb{D}$). In addition, if $\sigma(A) \notin \overline{\mathbb{C}_-}$ (resp. $\sigma(A) \notin \overline{\mathbb{D}}$), then $\overline{x}$ is an unstable equilibrium point of the nonlinear system.

3.3.1 Stability Criteria for Time-Varying Linear Systems

In this subsection we analyze the stability of finite dimensional time-varying linear systems described by differential and difference equations. Recall that for these systems the (asymptotic) stability of the equilibrium solution at the origin is equivalent to the (asymptotic) stability of any other solution (see Subsection 3.1.2). Hence we may attribute the stability properties to the system itself instead of the solutions. We first express the stability properties of a linear system in terms of the associated evolution operator $\Phi(t, t_0)$. We then introduce Liapunov and Bohl exponents which measure the (uniform) exponential growth (or decrease) of the trajectories. The main theorem of the subsection is Theorem 3.3.15 where uniform exponential stability is shown to be equivalent to the Bohl exponent being negative and also to the existence of a uniform estimate on the L^p-norm of the evolution operator Φ.
We consider the following system equations

$$\dot{x}(t) = A(t)x(t), \quad t \in T \subset \mathbb{R} \tag{1a}$$
$$x(t+1) = A(t)x(t), \quad t \in T \subset \mathbb{Z} \tag{1b}$$

where the time domain T is either an interval in $\mathbb{R}$ or in $\mathbb{Z}$ which is unbounded to the right. By assumption $A(\cdot) \in PC(T; \mathbb{K}^{n\times n})$ in (1a) and $A(t) \in \mathbb{K}^{n\times n}$, $t \in T$ in (1b). In both cases $A(\cdot)$ generates an evolution operator $\Phi(\cdot, \cdot)$ (see Section 2.2), and the solution of (1) satisfying $x(t_0) = x^0$ is given by $x(t) = \Phi(t, t_0)x^0$, $t \in T_{t_0}$ for all $x^0 \in \mathbb{K}^n$. Linear systems of the form (1) induce a global flow $\mathfrak{F} = (T, X, \varphi)$ on $X = \mathbb{K}^n$ given by $\varphi(t; t_0, x^0) = \Phi(t, t_0)x^0$, $(t_0, x^0) \in T \times \mathbb{K}^n$, $t \in T_{t_0}$.
As in the previous section we provide $\mathbb{K}^n$ with the Euclidean norm and $\mathbb{K}^{n\times n}$ with the corresponding operator norm (spectral norm). The first two propositions are immediate consequences of Definitions 3.1.8, 3.1.9.

Proposition 3.3.1. *Let $(z^1, \ldots, z^n)$ be any basis of $\mathbb{K}^n$. Then the following statements are equivalent.*

(i) The system (1) is stable at time t_0 (resp. uniformly stable).

(ii) There exists a constant M which may depend on t_0 (resp. independent of t_0) such that $\|\Phi(t, t_0)\| \le M$ for all $t \in T_{t_0}$.

(iii) There exists a constant M which may depend on t_0 (resp. independent of t_0) such that $\|\Phi(t, t_0)z^i\| \le M$ for all $t \in T_{t_0}$, $i \in \underline{n}$.

Proof: (i) $\Rightarrow$ (ii). Suppose that (1) is stable at time t_0 (resp. uniformly stable). Then for $\varepsilon = 1$, there exists $\delta > 0$ depending on t_0 (resp. independent of t_0) such that

$$\|x^0\| \le \delta \quad \Rightarrow \quad \|\Phi(t, t_0)x^0\| \le 1, \quad t \in T_{t_0}.$$

Hence $\|\Phi(t, t_0)\| \le \delta^{-1}$ for all $t \in T_{t_0}$.
As (ii) $\Rightarrow$ (iii) is trivial it only remains to prove (iii) $\Rightarrow$ (i). Suppose (iii). Since there exist $a, b > 0$ such that

$$a \max_{i\in\underline{n}} |\xi_i| \le \|\sum_{i=1}^{n} \xi_i z^i\| \le b \max_{i\in\underline{n}} |\xi_i|, \quad \xi \in \mathbb{K}^n. \tag{2}$$

we have for all $x^0 = \sum_{i=1}^n \xi_i z^i \in \mathbb{K}^n$,

$$\|\Phi(t,t_0)x^0\| = \|\Phi(t,t_0)\sum_{i=1}^n \xi_i z^i\| \le \max_{i\in\underline{n}} |\xi_i| \sum_{i=1}^n \|\Phi(t,t_0)z^i\| \le a^{-1}nM\|x^0\|.$$

This proves (i). □

Proposition 3.3.2. *Let $(z^1,\ldots,z^n)$ be any basis of $\mathbb{K}^n$. Then the following statements are equivalent.*

(i) The system (1) *is asymptotically stable at time t_0 (resp. uniformly asymptotically stable).*

(ii) The system (1) *is globally asymptotically stable at time t_0 (resp. globally uniformly asymptotically stable).*

(iii) $\|\Phi(t,t_0)\| \to 0$ as $t\to\infty$ (resp. uniformly in t_0).

(iv) For $i\in\underline{n}$, $\|\Phi(t,t_0)z^i\| \to 0$ as $t\to\infty$ (resp. uniformly in t_0).

Proof: (i) ⇒ (ii) follows directly from linearity and (ii) ⇒ (iv) and (iii) ⇒ (i) are trivial.
(iv) ⇒ (iii). Suppose (iv) holds, then for every $\varepsilon > 0$ there exists a time $\tau(\varepsilon)$ depending on t_0 (independent of t_0) such that $\|\Phi(t,t_0)z^i\| < \varepsilon$ for all $t \in T_{t_0+\tau(\varepsilon)}$, $i \in \underline{n}$. But then, for every $x^0 = \sum_{i=1}^n \xi_i z^i$, $\|x^0\| = 1$ we have $\max_{i\in\underline{n}} |\xi_i| \le a^{-1}$ where $a > 0$ satisfies (2), and thus

$$\|\Phi(t,t_0)x^0\| = \|\sum_{i=1}^n \xi_i \Phi(t,t_0)z^i\| \le a^{-1}n\varepsilon\,, \quad t\in T_{t_0+\tau(\varepsilon)}\,,$$

hence (iii) holds. □

Remark 3.3.3. In the discrete time case

$$\Phi(t,t_0) = A(t-1)A(t-2)\ldots A(t_0)\,, \quad t\in T_{t_0}\,. \tag{3}$$

So if (1b) is (asymptotically) stable at time $t_0 \in T$ it will also be (asymptotically) stable at time $\tau \in T$ for all $\tau < t_0$. A similar statement also holds for $\tau \in T$, $\tau > t_0$ provided that $\det A(k) \neq 0$ for $k = t_0,\ldots,\tau-1$. Furthermore, taking norms in (3) we obtain

$$(\forall t\in T : \|A(t)\| \le \gamma) \quad \Rightarrow \quad \|\Phi(t,t_0)\| \le \gamma^{t-t_0}\,, \quad t_0\in T\,, \quad t\in T_{t_0}\,. \tag{4}$$

Hence the zero state of (1b) will be uniformly stable if $\|A(t)\| \le 1$, $t \in T$ and it will be uniformly asymptotically stable if $\|A(t)\| \le \gamma < 1$ for all $t \in T$. These conditions, however, are far from being necessary. □

An estimate of the spectral norm $\|\Phi(t,t_0)\|$ for the continuous time case is provided by the next lemma.

Lemma 3.3.4. *If $\Phi(t,t_0)$ is the evolution operator of* (1a), *then*

$$e^{-\int_{t_0}^t \|A(s)\|ds} \le \sigma_{\min}(\Phi(t,t_0)) = \|\Phi(t_0,t)\|^{-1}, \;\; \|\Phi(t,t_0)\| \le e^{\int_{t_0}^t \|A(s)\|ds}\,, \;\; t\ge t_0\,. \tag{5}$$

Proof: Since $\Phi(t_0,t) = \Phi(t,t_0)^{-1}$, we have

$$\begin{aligned}\frac{\partial}{\partial t}\Phi(t,t_0) &= A(t)\Phi(t,t_0),\\ \frac{\partial}{\partial t}\Phi(t_0,t) &= -\Phi(t,t_0)^{-1}A(t)\Phi(t,t_0)\Phi(t,t_0)^{-1} = -\Phi(t,t_0)^{-1}A(t) = -\Phi(t_0,t)A(t)\end{aligned}$$

for a.e. $t > t_0$. Integrating yields

$$\Phi(t,t_0) - I = \int_{t_0}^{t} A(s)\Phi(s,t_0)ds, \qquad \Phi(t_0,t) - I = -\int_{t_0}^{t} \Phi(t_0,s)A(s)ds, \quad t \geq t_0.$$

Hence for $t \geq t_0$

$$\|\Phi(t,t_0)\| \leq 1 + \int_{t_0}^{t} \|A(s)\|\|\Phi(s,t_0)\|ds, \quad \|\Phi(t_0,t)\| \leq 1 + \int_{t_0}^{t} \|\Phi(t_0,s)\|\|A(s)\|ds.$$

By Gronwall's Lemma 2.1.18, we have

$$\|\Phi(t,t_0)\| \leq e^{\int_{t_0}^{t}\|A(s)\|ds}, \qquad \|\Phi(t_0,t)\| \leq e^{\int_{t_0}^{t}\|A(s)\|ds}, \quad t \geq t_0.$$

So the second inequality holds and the first is a consequence of $\sigma_{\min}(\Phi(t,t_0)) = \|\Phi(t_0,t)\|^{-1}$ for $t \geq t_0$. □

As a corollary of this lemma and Propositions 3.3.1, 3.3.2 we obtain

Corollary 3.3.5. *The continuous time system* (1a) *with time-domain* $T = [t_0, \infty)$ *is uniformly stable if* $\int_{t_0}^{\infty} \|A(s)\|ds < \infty$. *It is (asymptotically) stable at time* $t_1 \in T$ *if and only if it is (asymptotically) stable at time* $t_0 \in T$.

Proof: The first part is clear from the previous Proposition 3.3.1 and the above lemma. The second follows from the estimates

$$\|\Phi(t,t_0)\| \leq \|\Phi(t,t_1)\|\|\Phi(t_1,t_0)\|, \ \|\Phi(t,t_1)\| \leq \|\Phi(t,t_0)\|\|\Phi(t_0,t_1)\|, \ t \geq t_1 \geq t_0 \quad (6)$$

and Proposition 3.3.2 and (5). □

The following proposition shows that for periodic systems (1) the stability properties can be characterized via those of an associated time–invariant linear system. We will see in the next subsection that efficient stability tests are available for such systems.

Proposition 3.3.6. *Suppose the generators* $A(\cdot)$ *of* (1) *are periodic with period* $\tau \in T$, $T = \mathbb{R}$ *or* $\mathbb{Z}$, $\tau > 0$: $A(t+\tau) = A(t)$, $t \in T$. *Then* (1) *is uniformly stable (uniformly asymptotically stable) if and only if the time–invariant discrete time system*

$$\hat{x}(k+1) = \Phi(\tau, 0)\hat{x}(k), \quad k \in \mathbb{N} \qquad (7)$$

is stable (asymptotically stable) where Φ *is the evolution operator generated by* (1).

Proof: By periodicity $\Phi(t,t_0)=\Phi(t+\tau,t_0+\tau)$, $t\in T_{t_0}$, $t_0\in T$. Hence if $t\in T_{t_0}$ and

$$t_0=k_0\tau+t_0', \qquad t=k\tau+t', \quad 0\le t_0',\, t'<\tau, \quad k,\, k_0\in\mathbb{N}, \tag{8}$$

then

$$\begin{aligned}\Phi(t,t_0) &= \Phi(t,k\tau)\Phi(k\tau,(k-1)\tau)\cdots\Phi((k_0+1)\tau,t_0) \\ &= \Phi(t',0)\Phi(\tau,0)^{k-k_0-1}\Phi(\tau,t_0').\end{aligned} \tag{9}$$

By (3), (5) there exists $c>0$ such that $\|\Phi(t',0)\|$, $\|\Phi(\tau,t_0')\|\le c$ for all t_0', $t'\in[0,\tau]\cap T$. Therefore (9) implies

$$\|\Phi(t,t_0)\|\le c^2\|\Phi(\tau,0)^{k-k_0-1}\|, \quad t\in T_{t_0} \text{ as in } (8).$$

Applying Proposition 3.3.1 (Proposition 3.3.2) we see that (1) is uniformly (asymptotically) stable if (7) has this property. The converse implication is obvious since $\Phi(\tau,0)^k=\Phi(k\tau,0)$. □

The next example shows that a system (1) may be unstable even though every time–invariant system $\dot{x}(t)=A(\tau)x(t)$ (resp. $x(t+1)=A(\tau)x(t)$) frozen at time $\tau\in T$ is asymptotically stable. It is also possible that every frozen system is unstable yet (1) is stable, see Ex. 7.

Example 3.3.7. Consider the two dimensional periodic system of period 2π, where

$$A(t)=\begin{bmatrix}\cos t & -\sin t\\ \sin t & \cos t\end{bmatrix}\begin{bmatrix}-1 & -5\\ 0 & -1\end{bmatrix}\begin{bmatrix}\cos t & \sin t\\ -\sin t & \cos t\end{bmatrix}. \tag{10}$$

Then $\sigma(A(\tau))=\{-1\}$, $\tau\in\mathbb{R}_+$ and we will see in the next subsection that time–invariant continuous time systems with spectrum in the open left half plane are asymptotically stable. However it is easily verified that the evolution operator generated by $A(\cdot)$ is such that

$$\Phi(t,0)=\begin{bmatrix}e^t(\cos t+\frac{1}{2}\sin t) & e^{-3t}(\cos t-\frac{1}{2}\sin t)\\ e^t(\sin t-\frac{1}{2}\cos t) & e^{-3t}(\sin t+\frac{1}{2}\cos t)\end{bmatrix},$$

which is clearly unbounded. □

Let us now turn to exponential stability. The linear system (1) is *(uniformly) exponentially stable* if there exist for every $t_0\in T$ a constant $M>0$, and a decay rate $\omega<0$ which may depend upon t_0 (resp. independent of t_0), such that

$$\|\Phi(t,t_0)\|\le Me^{\omega(t-t_0)}, \quad t\in T_{t_0}. \tag{11}$$

The next theorem is rather surprising. A similar result does not hold for nonlinear systems.

Theorem 3.3.8. *The system* (1) *is uniformly exponentially stable if and only if it is uniformly asymptotically stable.*

Proof: The only if part follows immediately from (11) and Proposition 3.3.2. Conversely suppose that (1) is uniformly asymptotically stable. By Proposition 3.3.2 there exists $\tau\in T$ such that $\|\Phi(t+\tau,t)\|\le 1/2$ for all $t\in T$. Hence using the concatenation property of Φ

$$\|\Phi(t_0+k\tau,t_0)\| \le \|\Phi(t_0+k\tau,t_0+(k-1)\tau)\| \dots \|\Phi(t_0+\tau,t_0)\| \le 2^{-k}.$$

Now suppose $t_0+k\tau \le t < t_0+(k+1)\tau$, $t \in T_{t_0}$, $k \in \mathbb{N}$, then

$$\|\Phi(t,t_0)\| \le \|\Phi(t,t_0+k\tau)\|\|\Phi(t_0+k\tau,t_0)\| \le \|\Phi(t,t_0+k\tau)\|2^{-k}.$$

By Proposition 3.3.1 there exists $M'>0$ such that $\|\Phi(t,t_0+k\tau)\| \le M'$ for all $t \ge t_0+k\tau$, $k \in \mathbb{N}$, and hence

$$\|\Phi(t,t_0)\| \le M'2^{-[(t-t_0)/\tau-1]}, \qquad t \in T_{t_0},\ t_0 \in T.$$

Setting $M=2M'$, $\omega=-(\ln 2)/\tau$ we obtain (11). □

In his doctoral thesis in 1892 Liapunov introduced characteristic numbers associated with the flow generated by the differential equation (1a). They are now known as *Liapunov exponents* and we will be particularly interested in the upper one which characterizes the supreme exponential growth rate of the system. Our definition is applicable to both continuous and discrete time systems (1).

Definition 3.3.9 (Liapunov exponents). If $\Phi(\cdot,\cdot)$ is the evolution operator of (1) and $t_0 \in T$, the upper and lower Liapunov exponents $\overline{\alpha}(\Phi)$, $\underline{\alpha}(\Phi)$ are defined by

$$\begin{aligned}\overline{\alpha}(\Phi) &= \inf\{\omega \in \mathbb{R};\ \exists M_\omega>0\ \forall\, t \in T_{t_0}:\ \ \|\Phi(t,t_0)\| \le M_\omega e^{\omega(t-t_0)}\}\\ \underline{\alpha}(\Phi) &= \sup\{\omega \in \mathbb{R};\ \exists M_\omega>0\ \forall\, t \in T_{t_0}\ \forall x \in \mathbb{K}^n:\ \ \|\Phi(t,t_0)x\| \ge M_\omega e^{\omega(t-t_0)}\|x\|\}\end{aligned}$$

(where we set $\inf\emptyset := \infty$, $\sup\emptyset := -\infty$).

It is easily seen that the two Liapunov exponents do not depend upon t_0 in the continuous time case. In the discrete time case this is also true if $\det A(t) \neq 0$ for all $t \in T$. But, if $\det A(t_1)=0$ for some $t_1 \in T$ then $\det\Phi(t,t_0)=0$ for all (t,t_0) with $t_0 \le t_1 \le t$. So by (3) $\overline{\alpha}(\Phi)=-\infty$ if we choose $t_0 \le t_1$ (as we will always do in this case). Therefore we need not indicate the dependency on t_0 in our notation of the Liapunov exponents.
While exponential stability can be characterized by $\overline{\alpha}(\Phi)<0$ (see the next remark), *uniform* exponential stability can be characterized in terms of the upper *Bohl exponent* introduced by Bohl in 1913.

Definition 3.3.10 (Bohl exponents). If $\Phi(\cdot,\cdot)$ is the evolution operator generated via (1), the upper and lower Bohl exponents $\overline{\beta}(\Phi)$, $\underline{\beta}(\Phi)$ are defined by

$$\begin{aligned}\overline{\beta}(\Phi) &= \inf\{\omega \in \mathbb{R};\ \exists M_\omega\ \forall t_0 \in T\ \forall t \in T_{t_0}:\ \|\Phi(t,t_0)\| \le M_\omega e^{\omega(t-t_0)}\},\\ \underline{\beta}(\Phi) &= \sup\{\omega \in \mathbb{R};\ \exists M_\omega\ \forall t_0 \in T\ \forall t \in T_{t_0}\ \forall x \in \mathbb{K}^n:\ \|\Phi(t,t_0)x\| \ge M_\omega e^{\omega(t-t_0)}\|x\|\}.\end{aligned}$$

Remark 3.3.11. (i) Clearly $\overline{\alpha}(\Phi) \le \overline{\beta}(\Phi)$ and $\underline{\beta}(\Phi) \le \underline{\alpha}(\Phi)$.

(ii) If $\|A(t)\| \le \gamma$, for all $t \in T$ and some $\gamma>0$, it follows from (3) in the discrete time case that $\overline{\beta}(\Phi) \le \ln\gamma$, whereas in the continuous time case we have $\overline{\beta}(\Phi) \le \gamma$ and $\underline{\beta}(\Phi) \ge -\gamma$ by (5).

(iii) Suppose that $\overline{\alpha}(\Phi)<\infty$ (resp. $\overline{\beta}(\Phi)<\infty$), then given $\gamma>\overline{\alpha}(\Phi)$ (resp. $\gamma>\overline{\beta}(\Phi)$) there exists M depending on γ such that $\|\Phi(t,t_0)\| \le Me^{\gamma(t-t_0)}$, $t \in T_{t_0}$ for a given $t_0 \in T$ (resp. $\|\Phi(t,t_0)\| \le Me^{\gamma(t-t_0)}$, $t_0 \in T$, $t \in T_{t_0}$). So we conclude that the system (1) is exponentially stable at time t_0 (resp. uniformly exponentially stable) if and only if $\overline{\alpha}(\Phi)<0$ (resp. $\overline{\beta}(\Phi)<0$).

(iv) If $\tilde{\Phi}(t,t_0) = \Phi(t_0,t)^*$ denotes the evolution operator generated by $-A(t)^*$, then $\underline{\beta}(\Phi) = \overline{\beta}(\tilde{\Phi})$. □

In general the Bohl and Liapunov exponents are not the same as the following scalar example shows.

Example 3.3.12. (Perron). Consider the scalar system

$$\dot{x}(t) = a(t)x(t)\,, \quad \text{where} \quad a(t) = \sin\ln t + \cos\ln t, \ \ t > 0. \tag{12}$$

The corresponding evolution operator is

$$\Phi(t,t_0) = e^{t\sin\ln t - t_0\sin\ln t_0}\,, \quad t \geq t_0 > 0$$

and $\|\Phi(t,1)\| \leq e^t$, $t \geq 1$ so $\overline{\alpha}(\Phi) \leq 1$.
For small $\varepsilon > 0$ let $\ln t_n = 2n\pi + \pi/4 + \varepsilon$, $\ln t_{0n} = 2n\pi + \pi/4$, then

$$t_n \sin\ln t_n - t_{0n}\sin\ln t_{0n} = e^{2n\pi+\pi/4}\left[e^{\varepsilon}\sin(\pi/4+\varepsilon) - \sin\pi/4\right].$$

But for small ε, $e^{\varepsilon}\sin(\pi/4+\varepsilon) - \sin\pi/4 \approx (1+\varepsilon)(1+\varepsilon)/\sqrt{2} - 1/\sqrt{2} \approx \sqrt{2}\,\varepsilon$. Hence given any small $\delta > 0$ there exists $\varepsilon > 0$ such that $e^{\varepsilon}\sin(\pi/4+\varepsilon) - \sin\pi/4 \geq (\sqrt{2}-\delta)(e^{\varepsilon}-1)$. And so for this ε

$$t_n \sin\ln t_n - t_{0n}\sin\ln t_{0n} \geq (\sqrt{2}-\delta)(t_n - t_{0n})\,.$$

But then

$$|\Phi(t_n,t_{0n})| \geq e^{(\sqrt{2}-\delta)(t_n-t_{0n})}\,.$$

Since $t_n - t_{0n} \to \infty$ as $n \to \infty$, this shows $\overline{\beta}(\Phi) \geq \sqrt{2}$. Now $|a(t)| \leq \sqrt{2}$, $t > 0$. Hence $|\Phi(t,t_0)| = |e^{\int_{t_0}^t a(s)ds}| \leq e^{\sqrt{2}(t-t_0)}$, $t \geq t_0 > 0$ and so in fact $\overline{\beta}(\Phi) = \sqrt{2}$. □

Remark 3.3.13. In the continuous time case if $A(\cdot)$ generates $\Phi(\cdot,\cdot)$, then for any $\lambda \in \mathbb{C}$, $A(\cdot) + \lambda I_n$ generates $\Phi_\lambda(t,t_0) = e^{\lambda(t-t_0)}\Phi(t,t_0)$ and

$$\overline{\alpha}(\Phi_\lambda) = \overline{\alpha}(\Phi) + \operatorname{Re}\lambda, \qquad \overline{\beta}(\Phi_\lambda) = \overline{\beta}(\Phi) + \operatorname{Re}\lambda\,. \tag{13}$$

If $a(\cdot)$ is as in the above example and $-\sqrt{2} < \lambda < -1$ we see that $\overline{\alpha}(\Phi_\lambda) < 0$ and $\overline{\beta}(\Phi_\lambda) > 0$ so that all solutions of $\dot{x} = (a(t)+\lambda)x$ decrease exponentially although $\overline{\beta}(\Phi_\lambda) > 0$. □

It is easily verified (see Ex. 5) that for the upper upper Liapunov exponent we have

$$\overline{\alpha}(\Phi) = \limsup_{t\to\infty} \frac{\ln\|\Phi(t,0)\|}{t}\,. \tag{14}$$

The corresponding formula for the Bohl exponent is given in the next proposition.

Proposition 3.3.14. $\overline{\beta}(\Phi) < \infty$ *if and only if*

$$\sup_{t_0,t\in T,\, 0\leq t-t_0\leq 1} \|\Phi(t,t_0)\| < \infty\,, \tag{15}$$

and when this is the case

$$\overline{\beta}(\Phi) = \limsup_{t_0,\, t-t_0\to\infty} \frac{\ln\|\Phi(t,t_0)\|}{t-t_0}\,. \tag{16}$$

Proof: Suppose $\overline{\beta}(\Phi) < \infty$, then choosing $\gamma > \max\{\overline{\beta}(\Phi), 0\}$ there exists $M(\gamma) > 0$ such that

$$\|\Phi(t, t_0)\| \leq M(\gamma) e^{\gamma(t-t_0)}, \quad t_0 \in T, \, t \in T_{t_0}. \tag{17}$$

Hence $\sup_{0 \leq t-t_0 \leq 1} \|\Phi(t,t_0)\| \leq M(\gamma)e^{\gamma} < \infty$. Conversely suppose (15) holds so that $\|\Phi(\tau, \sigma)\| \leq K$ for some $K \geq 1$ and all $\sigma, \tau \in T$, $0 \leq \tau - \sigma \leq 1$. Then for every $t_0 \in T$, $t \in T_{t_0}$ such that $t_0 + (n-1) \leq t < t_0 + n$

$$\|\Phi(t,t_0)\| \leq \|\Phi(t, t_0+n-1)\| \prod_{k=1}^{n-1} \|\Phi(t_0+k, t_0+k-1)\| \leq K^n \leq K e^{(t-t_0)\ln K}. \tag{18}$$

So $\overline{\beta}(\Phi) \leq \ln K$ and this concludes the proof of the equivalence statement.
To prove (16) we suppose $\overline{\beta}(\Phi) < \infty$. Then (17) holds for every $\gamma > \overline{\beta}(\Phi)$ and so

$$\mu = \limsup_{t_0, t-t_0 \to \infty} \frac{\ln \|\Phi(t,t_0)\|}{t-t_0} \leq \limsup_{t_0, t-t_0 \to \infty} \frac{\ln M(\gamma)}{t-t_0} + \gamma = \gamma.$$

Hence $\mu \leq \overline{\beta}(\Phi)$. Conversely, for every $\gamma > \mu$ there exists a time $t_\gamma \in T$ such that

$$\frac{\ln \|\Phi(t,t_0)\|}{t-t_0} \leq \gamma, \quad \text{i.e.} \quad \|\Phi(t,t_0)\| \leq e^{\gamma(t-t_0)}, \quad t_0 \in T_{t_\gamma}, \, t \in T_{t_0+t_\gamma}.$$

By (18)

$$K_\gamma := \sup\{\|\Phi(t,t_0)\|;\, t_0,\, t \in T,\, 0 \leq t - t_0 \leq t_\gamma\} \leq K e^{t_\gamma \ln K} < \infty. \tag{19}$$

So

$$\|\Phi(t,t_0)\| \leq K_\gamma e^{|\gamma| t_\gamma} e^{\gamma(t-t_0)}, \quad t_0 \leq t \leq t_0 + t_\gamma. \tag{20}$$

Therefore

$$\|\Phi(t,t_0)\| \leq N e^{\gamma(t-t_0)}, \quad t_0 \in T_{t_\gamma}, \, t \in T_{t_0},$$

where $N = \max\{1, K_\gamma e^{|\gamma| t_\gamma}\}$. But by (20) this same estimate is also valid for $0 \leq t_0 \leq t \leq t_\gamma$. Finally if $t_0 \leq t_\gamma < t$ we have

$$\|\Phi(t,t_0)\| \leq \|\Phi(t,t_\gamma)\| \|\Phi(t_\gamma, t_0)\| \leq N e^{\gamma(t-t_\gamma)} N e^{\gamma(t_\gamma - t_0)} = N^2 e^{\gamma(t-t_0)}$$

and so there exists M such that $\|\Phi(t,t_0)\| \leq M e^{\gamma(t-t_0)}$, for all $t_0 \in T$, $t \in T_{t_0}$. Thus $\overline{\beta}(\Phi) \leq \mu$ and (16) is proved. □

Note that in the discrete time case (15) holds if and only if $\sup_{t \in T} \|A(t)\| =: \gamma < \infty$ in which case $\overline{\alpha}(\Phi) \leq \overline{\beta}(\Phi) \leq \ln \gamma$.
The following theorem gives an alternative characterization for uniform exponential stability of (1). It is closely related to the Liapunov results which we will develop in Subsection 3.3.4.

Theorem 3.3.15. *Suppose the evolution operator Φ of (1) satisfies $\overline{\beta}(\Phi) < \infty$ then the following statements are equivalent.*

(i) The system (1) is uniformly exponentially stable.

(ii) $\overline{\beta}(\Phi) < 0$.

(iii) For any $p \in (0,\infty)$ there exists a constant c independent of $t_0 \in T$ such that

$$\int_{t_0}^{\infty} \|\Phi(t,t_0)\|^p dt \le c \quad \Big(\textit{resp.} \quad \sum_{t=t_0}^{\infty} \|\Phi(t,t_0)\|^p \le c\Big), \quad t_0 \in T. \tag{21}$$

(iv) For any $p \in (0,\infty)$ there exists a constant c independent of $t_0 \in T$ such that

$$\int_{t_0}^{\infty} \|\Phi(t,t_0)x\|^p dt \le c\|x\|^p \,\Big(\textit{resp.} \sum_{t=t_0}^{\infty} \|\Phi(t,t_0)x\|^p \le c\|x\|^p\Big), \; x \in \mathbb{K}^n, \; t_0 \in T. \tag{22}$$

Proof: The proof is for the continuous time case. (i) $\Leftrightarrow$ (ii) and (iii) $\Rightarrow$ (iv) is clear.
(i) $\Rightarrow$ (iii): Suppose (i) then there exist constants $M > 0, \omega < 0$ independent of $t_0 \in T$ such that $\|\Phi(t,t_0)\| \le Me^{\omega(t-t_0)}$, for all $t_0 \in T$, $t \in T_{t_0}$. Hence (21) holds with $c = M^p/p(-\omega)$.
(iv) $\Rightarrow$ (i): Since $\overline{\beta}(\Phi) < \infty$ there exists $\overline{M}, \overline{\omega} > 0$ independent of t_0 such that $\|\Phi(t,t_0)\| \le \overline{M}e^{\overline{\omega}(t-t_0)}$, for all $t_0 \in T$, $t \in T_{t_0}$. So

$$\begin{aligned} \frac{1-e^{-p\overline{\omega}(t-t_0)}}{p\overline{\omega}}\|\Phi(t,t_0)x\|^p &= \int_{t_0}^{t} e^{-p\overline{\omega}(t-s)}\|\Phi(t,t_0)x\|^p ds \qquad (23)\\ &\le \int_{t_0}^{t} e^{-p\overline{\omega}(t-s)}\|\Phi(t,s)\|^p\|\Phi(s,t_0)x\|^p ds \\ &\le \overline{M}^p \int_{t_0}^{t} \|\Phi(s,t_0)x\|^p ds \le \overline{M}^p c\|x\|^p, \quad t \ge t_0\,. \end{aligned}$$

Hence there exists γ independent of $t_0 \in T$ such that $\|\Phi(t,t_0)\| \le \gamma$, $t \ge t_0$. But then for $t \ge t_0$, $t_0 \in T$

$$(t-t_0)\|\Phi(t,t_0)x\|^p = \int_{t_0}^{t} \|\Phi(t,t_0)x\|^p ds \le \int_{t_0}^{t} \|\Phi(t,s)\|^p\|\Phi(s,t_0)x\|^p ds \le \gamma^p c\|x\|^p.$$

So for $\tau = 2^p\gamma^p c$

$$\|\Phi(t_0+\tau,t_0)\| \le 1/2\,, \quad t_0 \in T\,. \tag{24}$$

Now suppose $t_0 + (n-1)\tau \le t < t_0 + n\tau$, then from (24)

$$\|\Phi(t,t_0)\| \le \|\Phi(t,t_0+(n-1)\tau)\| \prod_{k=1}^{n-1} \|\Phi(t_0+k\tau,t_0+(k-1)\tau)\| \le \frac{\gamma}{2^{n-1}} < 2\gamma e^{-(\ln 2)(t-t_0)/\tau}.$$

Hence $\overline{\beta}(\Phi) < -(\ln 2)/\tau$. The reader is asked to prove the discrete time case in Ex. 23. □

We now consider the effect of time–varying linear coordinate transformations of the form $\tilde{x}(t) = S(t)^{-1}x(t)$ on the system (1), where $S(\cdot) \in PC^1(T;\mathbf{Gl}_n(\mathbb{C}))$ (resp. $S(t) \in \mathbf{Gl}_n(\mathbb{C}), t \in T$). The associated similarity transformation converts the system (1) into

$$\dot{\tilde{x}}(t) = \tilde{A}(t)\tilde{x}(t), \; t \in T, \quad (\text{resp. } \tilde{x}(t+1) = \tilde{A}(t)\tilde{x}(t), \; t \in T) \tag{25}$$

where

$\tilde{A}(t) = S(t)^{-1}A(t)S(t) - S(t)^{-1}\dot{S}(t), \ t \in T, \quad (\text{resp.} \ = S(t+1)^{-1}A(t)S(t), \ t \in T).$

The evolution operator of the system (25) is

$$\tilde{\Phi}(t,s) = S(t)^{-1}\Phi(t,s)S(s), \quad t,s \in T. \tag{26}$$

In order that these transformations preserve stability properties additional assumptions must be imposed.

Definition 3.3.16 (Liapunov and Bohl transformation). A time-varying transformation $S(\cdot) \in PC^1(T; \mathbf{Gl}_n(\mathbb{C}))$ (resp. $S(t) \in \mathbf{Gl}_n(\mathbb{C})$, $t \in T$) is called a *Liapunov transformation* if $S(\cdot)$, $S(\cdot)^{-1}$ and $\dot{S}(\cdot)$ are bounded on T. It is called a *Bohl transformation* if

$$\inf \left\{ \varepsilon \in \mathbb{R};\ \exists M_\varepsilon > 0 \ \forall t,s \in T : \|S(t)^{-1}\| \, \|S(s)\| \le M_\varepsilon e^{\varepsilon|t-s|} \right\} = 0.$$

It is easily seen that the Liapunov transformations on T form a group with respect to pointwise multiplication, and this group of transformations preserves the properties of stability, instability and asymptotic stability. The next proposition shows that the property of exponential stability is invariant with respect to the larger group of Bohl transformations.

Proposition 3.3.17. *The Bohl exponent is invariant with respect to Bohl transformations.*

Proof: Let $\dot{\tilde{x}}(t) = \tilde{A}(t)\tilde{x}(t)$, (resp. $\tilde{x}(t+1) = \tilde{A}(t)\tilde{x}(t)$) be similar to (1) via a Bohl transformation $S(\cdot)$. Since the evolution operator of the transformed equation is given by (26), we have

$$\|\tilde{\Phi}(t,s)\| \le \|S(t)^{-1}\| \|\Phi(t,s)\| \|S(s)\|, \quad t,s \in T.$$

But by Definitions 3.3.10 and 3.3.16, for every $\varepsilon > 0$, there exists a constant M_ε such that

$$\|S(t)^{-1}\| \|S(s)\| \le M_\varepsilon e^{\varepsilon(t-s)}, \quad \|\Phi(t,s)\| \le M_\varepsilon e^{(\beta(\Phi)+\varepsilon)(t-s)}, \quad t \ge s \in T.$$

So $\beta(\tilde{\Phi}) \le \beta(\Phi)$. Using the fact that $S(\cdot)^{-1}$ is also a Bohl transformation we conclude that $\beta(\tilde{\Phi}) = \beta(\Phi)$. □

It is a simple exercise to show that every time-varying linear system (1) can be transformed into the trivial system $\dot{x} = 0$ by a time-varying coordinate transformation. In the context of stability theory it is interesting to know which time-varying systems can be transformed into time–invariant ones via Liapunov or Bohl transformations. According to a result of Liapunov this is always possible for periodic systems.

Proposition 3.3.18. *Suppose the generator $A(\cdot)$ of* (1) *is periodic with period $\tau > 0$, $\tau \in T$: $A(t+\tau) = A(t)$, $t \in T$, and* $\det A(t) \neq 0, t \in T$ *in the discrete time case. Then there exists a Liapunov transformation such that the transformed system* (25) *is time-invariant.*

Proof: Suppose $\Phi(\cdot,\cdot)$ is generated by $A(\cdot)$. Since $A(\cdot)$ is periodic, we have

$$\dot{\Phi}(t+\tau,0) = A(t)\Phi(t+\tau,0),\ t \in T \ \ (\text{resp. } \Phi(t+\tau+1,0) = A(t)\Phi(t+\tau,0),\ t \in T).$$

So there must exist a constant nonsingular matrix V such that $\Phi(t+\tau,0) = \Phi(t,0)V$. Choose $L \in \mathbb{C}^{n\times n}$ such that $e^L = V$ and set $S(t) = \Phi(t,0)e^{-tL/\tau}$, $t \in T$. Then

$$S(t+\tau) = \Phi(t+\tau,0)e^{-(tL/\tau)-L} = \Phi(t,0)e^L e^{-(tL/\tau)-L} = S(t).$$

Hence $S(\cdot)$ is periodic with period τ. In the continuous time case Φ is automatically invertible and in the discrete time case this is a consequence of the assumption that $\det A(t) \neq 0$, $t \in T$. It follows therefore that $S(t)$, $t \in T$ is invertible. Moreover

$$\dot{S}(t) = A(t)\Phi(t,0)e^{-tL/\tau} - \Phi(t,0)e^{-tL/\tau}\tau^{-1}L = A(t)S(t) - S(t)\tau^{-1}L,\ t \geq 0.$$

And in the discrete case

$$S(t+1) = \Phi(t+1,0)e^{-(t+1)L/\tau} = A(t)S(t)e^{-L/\tau},\ t \in T.$$

So the transformed system (25) is given by $\hat{A}(t) = \tau^{-1}L$, $t \in T$, (resp. $= e^{L/\tau}$). Clearly $S(\cdot) \in PC^1(T;\mathbf{Gl}_n(\mathbb{C}))$ (resp. $S(t) \in \mathbf{Gl}_n(\mathbb{C})$, $t \in T$) and the boundedness of $S(\cdot)$, $S(\cdot)^{-1}$, $\dot{S}(\cdot)$ is a consequence of periodicity. Hence S is a Liapunov transformation and this completes the proof. □

3.3.2 Time–Invariant Systems: Spectral Stability Criteria

We consider systems of the form

$$\dot{x}(t) = Ax(t), \quad t \in T, \qquad (\text{resp. } x(t+1) = Ax(t), \quad t \in T) \tag{27}$$

where $A \in \mathbb{K}^{n\times n}$ and $T = \mathbb{R}_+$ (resp. $T = \mathbb{N}$). The following result relates growth properties of the semigroup generated by the matrix $A \in \mathbb{K}^{n\times n}$ to the spectrum of A, $\sigma(A)$.

Lemma 3.3.19. *Given $A \in \mathbb{K}^{n\times n}$ and $\omega \in \mathbb{R}$. If*

$$\alpha(A) = \max\{\operatorname{Re}\lambda\,;\ \lambda \in \sigma(A)\} < \omega,\ (\textit{resp. } \varrho(A) = \max\{|\lambda|\,;\ \lambda \in \sigma(A)\} < e^{\omega}) \tag{28}$$

then there exists M, depending on ω such that

$$\|e^{At}\| \leq Me^{\omega t}, \quad t \in \mathbb{R}_+, \quad (\textit{resp. } \|A^t\| \leq Me^{\omega t}, \quad t \in \mathbb{N}). \tag{29}$$

Proof: The proof will be for the discrete time case. Since the spectral norm and the spectrum of a real linear operator do not change by complexification we may assume $\mathbb{K} = \mathbb{C}$. Let $(z^1,\dots,z^n)$ be a basis of $\mathbb{C}^n$ consisting of generalized eigenvectors z^i of order m_i, corresponding to eigenvalues λ_i of A. Applying Proposition 3.3.1 to the time–invariant evolution operator $\Phi(t) = (A^t e^{-\omega t})$ we see that $(A^t e^{-\omega t})_{t\in\mathbb{N}}$ is bounded if and only if $(A^t e^{-\omega t} z^i)_{t\in\mathbb{N}}$ is bounded for all $i \in \underline{n}$. Now if $\varrho(A) = 0$, then

$A^t = 0$ for t sufficiently large and it follows from (2.2.55) that for $t \geq m_i - 1$ and $\varrho := \varrho(A) \neq 0$

$$\|A^t e^{-\omega t} z^i\| = e^{-\omega t} \| \sum_{\nu=0}^{m_i-1} \lambda_i^{t-\nu} \binom{t}{\nu} (A - \lambda_i I)^\nu z^i \| \leq \left[\varrho e^{-\omega}\right]^t \sum_{\nu=0}^{m_i-1} \binom{t}{\nu} \varrho^{-\nu} \|(A - \lambda_i I)^\nu z^i\|.$$

Since for every $\alpha \in (0,1)$ and every polynomial $p(t) \in \mathbb{K}[t]$ we have $\lim_{t \to \infty} \alpha^t p(t) = 0$, we see that $\|A^t e^{-\omega t} z^i\| \to 0$ as $t \to \infty$ and so there exists $M > 0$ such that $\|A^t e^{-\omega t}\| \leq M$ for all $t \in \mathbb{N}$. This proves (29). □

If $(\Phi(t))$ is the semigroup of operators generated by A (continuous or discrete time), then it is an easy consequence (see Ex. 1) of the above lemma that the Liapunov and Bohl exponents are equal. They are sometimes called the *growth rate* of Φ, denoted by $\omega(A)$ and are given by

$$\omega(A) = \overline{\beta}(\Phi) = \overline{\alpha}(\Phi) = \alpha(A), \quad \left(\omega(A) = \overline{\beta}(\Phi) = \overline{\alpha}(\Phi) = \ln \varrho(A)\right). \tag{30}$$

For time invariant systems stability and uniform stability properties are equivalent and hence as a consequence of Theorem 3.3.8 *asymptotic stability is equivalent to uniform exponential stability.* The following theorem derives necessary and sufficient conditions for the asymptotic stability of the system (27).

Theorem 3.3.20. *The system* (27) *is asymptotically (or, equivalently, exponentially) stable if and only if*

$$\operatorname{Re} \lambda < 0, \qquad (\textit{resp.}\ |\lambda| < 1), \quad \lambda \in \sigma(A). \tag{31}$$

Proof: The proof is for the discrete time case. If (31) holds then $\ln \varrho(A) < 0$ and so by Lemma 3.3.19 there exists $\omega < 0$, and M such that

$$\|A^t x_0\| \leq \|A^t\| \|x_0\| \leq M e^{\omega t} \|x_0\|, \qquad t \in \mathbb{N},\ x^0 \in \mathbb{K}^n.$$

This implies that (27) is exponentially stable. To prove necessity suppose there exists $\lambda \in \sigma(A)$ such that $|\lambda| \geq 1$ and let $z \in \mathbb{C}^n$ be a corresponding eigenvector. Then

$$\|A^t z\| = \|\lambda^t z\| \geq \|z\| \qquad t \in \mathbb{N}$$

and hence (27) is not asymptotically stable. □

Theorem 3.3.21. *The system* (27) *is stable if and only if both of the following conditions hold for all* $\lambda \in \sigma(A)$

(i) $\operatorname{Re} \lambda \leq 0$, *(resp.* $|\lambda| \leq 1$*).*

(ii) If $\operatorname{Re} \lambda = 0$ *(resp.* $|\lambda| = 1$*) then there exist* k_λ *linearly independent eigenvectors, where* k_λ *is the algebraic multiplicity of* λ.

Proof: The proof is for the discrete time case. By Proposition 3.3.1 the origin is stable if and only if all the (generalized) eigenmotions are bounded. This clearly implies $|\lambda| \le 1$ for all $\lambda \in \sigma(A)$. Now suppose there exists $\lambda \in \sigma(A)$ with $|\lambda| = 1$ and a generalized eigenvector z of order $m > 1$ then for $t \ge m-1$

$$A^t z = \lambda^t \sum_{\nu=0}^{m-1} \binom{t}{\nu} \lambda^{-\nu}(A-\lambda I)^\nu z \quad \text{and} \quad \|A^t z\| = \|\sum_{\nu=0}^{m-1} \binom{t}{\nu} \lambda^{-\nu}(A-\lambda I)^\nu z\|.$$

The RHS is a polynomial in t of degree $m-1 \ge 1$ and is therefore unbounded. Thus conditions (i), (ii) are necessary.
Conversely if (i) and (ii) hold there exist generalized eigenvectors of order $m > 1$ only for eigenvalues $\lambda \in \sigma(A)$ with $|\lambda| < 1$. We know already that these eigenmotions tend exponentially to the origin as $t \to \infty$. On the other hand if z is an eigenvector corresponding to $\lambda \in \sigma(A)$, then since $|\lambda| \le 1$

$$\|A^t z\| = \|\lambda^t z\| \le \|z\|.$$

Hence all generalized eigenmotions are bounded and (27) is stable. □

Figure 3.3.1 shows the stability regions for the eigenvalues in the continuous and discrete time case. They are denoted by $\mathbb{C}_-$ and $\mathbb{D}$ respectively.

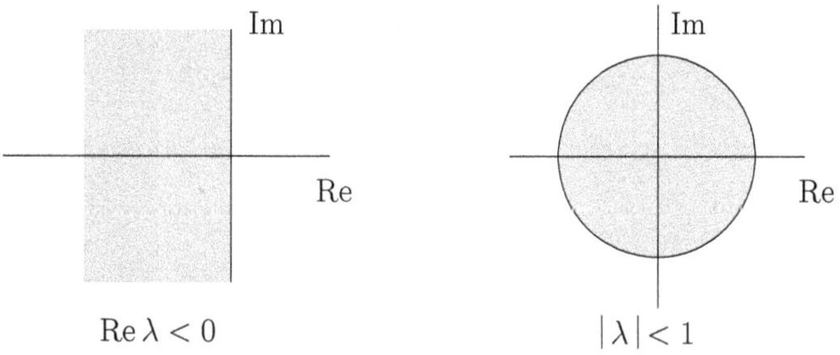

Figure 3.3.1: Stability regions for continuous and discrete time systems

As a consequence of Lemma 2.3.9 and Proposition 2.3.10 we have the following characterization of asymptotic stability in terms of properties of the solutions of the controlled systems

$$\dot{x}(t) = Ax(t) + Bu(t), \ t \in \mathbb{R}_+, \quad x(t+1) = Ax(t) + Bu(t), \ t \in \mathbb{N}. \tag{32}$$

We denote the solutions with $x(0) = x^0 \in \mathbb{K}^n$ by $\varphi(t; x^0, u(\cdot))$, $t \in \mathbb{R}_+$ (resp. $\mathbb{N}$).

Proposition 3.3.22. *The following are equivalent.*

(i) $\sigma(A) \subset \mathbb{C}_-$ *(resp.* $\sigma(A) \subset \mathbb{D}$*).*

(ii) For every $x^0 \in \mathbb{K}^n$, $e^{At}x^0 \to 0$ *(resp.* $A^t x^0 \to 0$*) as* $t \to \infty$.

(iii) If $u(\cdot) \in L^2(\mathbb{R}_+; \mathbb{K}^m)$ *(resp.* $\ell^2(\mathbb{N}; \mathbb{K}^m)$*), then* $\varphi(\cdot; x^0, u(\cdot)) \in L^2(\mathbb{R}_+; \mathbb{K}^n)$ *(resp.* $\ell^2(\mathbb{N}; \mathbb{K}^n)$*) for all* $x^0 \in \mathbb{K}^n$.

Proof: The proof is for the continuous time case. (i) implies (ii) by Theorem 3.3.20. Let $z \in \mathbb{C}^n$ be an eigenvector for the eigenvalue $\lambda \in \sigma(A)$, then $e^{At}z = e^{\lambda t}z$. Choosing $x^0 = z$ in the complex case and $x^0 = \operatorname{Re} z$ or $\operatorname{Im} z$ in the real case we see from (2.2.44) that if (ii) holds then necessarily $\operatorname{Re}\lambda < 0$. So (ii) implies (i). By

Proposition 2.3.10 with $C = I_n$ and $t_0 = 0$, (i) implies (iii). Now suppose $u(\cdot) \equiv 0$ and (iii) holds, then $x(\cdot) = e^{A\cdot}x^0 \in L^2(\mathbb{R}_+; \mathbb{K}^n)$ for all $x^0 \in \mathbb{K}^n$. But $\dot{x}(\cdot) = Ax(\cdot)$ and so $x(\cdot)$ is absolutely continuous and $\dot{x}(\cdot) \in L^2(\mathbb{R}_+; \mathbb{K}^n)$. Applying Lemma 2.3.9 with $p = n$, $y(\cdot) = x(\cdot)$ and $t_0 = 0$ we see that $x(t) = e^{At}x^0 \to 0$ as $t \to \infty$ for all $x^0 \in \mathbb{K}^n$. So (iii) implies (ii). □

Remark 3.3.23. In the previous section (Proposition 3.3.6) we showed that a periodic system with evolution operator Φ is stable (resp. uniformly asymptotically stable) if and only if the associated discrete time system with system matrix $\Phi(\tau, 0)$ is stable (resp. asymptotically stable). The eigenvalues of $\Phi(\tau, 0)$ are called the *characteristic multipliers* of (1). It follows from Theorems 3.3.20 and 3.3.21 that a periodic system (1) is stable (resp. asymptotically stable) if and only if its characteristic multipliers $\mu \in \sigma(\Phi(\tau, 0))$ satisfy conditions (i) (ii) of Theorem 3.3.21 (resp. (31)) in their discrete time versions. In contrast we have seen in Example 3.3.7 that, in general, the stability properties of a periodic time-varying system cannot be determined via the eigenvalues of $A(t)$. □

In the next two examples we illustrate the stability criteria by applying them to second order scalar systems.

Example 3.3.24. Consider the second order differential equation

$$\ddot{\xi}(t) + 2\alpha\dot{\xi}(t) + \beta\xi(t) = 0, \quad t > 0. \tag{33}$$

The matrix A of the corresponding state space system has eigenvalues $\lambda_{1,2} = -\alpha \pm \sqrt{\alpha^2 - \beta}$. So

(a) if $\alpha > 0$, $\beta > 0$, the origin is *exponentially stable*;

(b) if $\alpha > 0$, $\beta = 0$, the origin is *marginally stable* (i.e. stable but not asymptotically stable);

(c) if $\alpha = 0$, $\beta > 0$, the origin is *marginally stable*;

(d) if $\alpha = 0$, $\beta = 0$, there is a generalized eigenvector for the zero eigenvalue and so the origin is *unstable*;

(e) if $\alpha < 0$ or $\beta < 0$, the origin is *unstable*.

The stability chart, i.e. the set of all parameter values $(\alpha, \beta) \in \mathbb{R}^2$ for which the system is asymptotically stable is given by the positive orthant $(0, \infty)^2$. □

Example 3.3.25. Using the approximations

$$\dot{\xi}(t) \approx \frac{\xi(t+\tau) - \xi(t)}{\tau}, \quad \ddot{\xi}(t) \approx \frac{\xi(t+2\tau) - 2\xi(t+\tau) + \xi(t)}{\tau^2}, \quad \tau > 0$$

the differential equation of the previous example gives rise to the difference equation

$$\xi(t+2\tau) - 2(\alpha\tau - 1)\xi(t+\tau) + (1 - 2\alpha\tau + \beta\tau^2)\xi(t) = 0, \qquad t \in \mathbb{N}\tau. \tag{34}$$

We will examine the stability properties of this discrete time system and compare the results with those obtained in the previous example. In order to do this we first obtain results for the general second order difference equation

$$\xi(t+2) + a_1\xi(t+1) + a_0\xi(t) = 0, \quad t \in \mathbb{N}. \tag{35}$$

The eigenvalues of the matrix $A = \begin{bmatrix} 0 & 1 \\ -a_0 & -a_1 \end{bmatrix}$ of the corresponding state space system are

$$\lambda_{1,2} = (1/2)[-a_1 \pm (a_1^2 - 4a_0)^{1/2}].$$

The parameter set (a_0, a_1) for which the system is stable must satisfy

$$\begin{aligned} -1 \leq \quad (1/2)[-a_1 \pm (a_1^2 - 4a_0)^{1/2}] \quad &\leq 1 \text{ if } a_1^2 \geq 4a_0 \\ (1/4)(a_1^2 + (4a_0 - a_1^2)) \quad &\leq 1 \text{ if } a_1^2 < 4a_0. \end{aligned}$$

The first condition is equivalent to $a_1^2 - 4a_0 \leq (2 + a_1)^2$ and $(a_1 - 2)^2 \geq a_1^2 - 4a_0$, i.e.

$$1 + a_1 + a_0 \geq 0 \text{ and } 1 - a_1 + a_0 \geq 0. \tag{36}$$

The second condition is equivalent to

$$1 - a_0 \geq 0 \text{ if } a_1^2 < 4a_0. \tag{37}$$

This leads to the stability chart for (35) shown on the LHS of Figure 3.3.2. The shaded

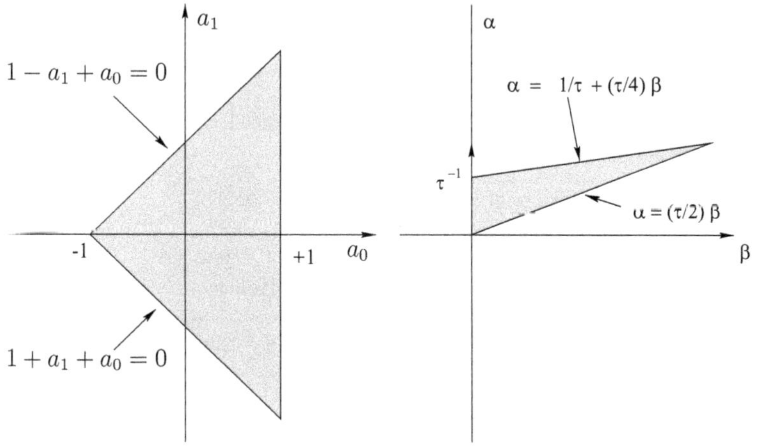

Figure 3.3.2: Stability charts for (35) and (34)

region inside the left triangle represents values of the parameters (a_0, a_1) for which $|\lambda_i| < 1$, $i = 1, 2$ and hence values for which the system is asymptotically stable. Now consider the boundary of the triangle. When $1 - a_1 + a_0 = 0$ (resp. $1 + a_1 + a_0 = 0$) then $\sigma(A) = \{-1, -a_0\}$ (resp. $\{+1, a_0\}$) and so if $a_0 < 1$ the system is marginally stable, and this is also the case if $a_0 = 1$, $|a_1| < 2$. However if $a_0 = 1$, $|a_1| = 2$ there are generalized eigenvectors of order 2 so the system is unstable.

Note that if $a_0 = a_1 = 0$ the system matrix A is nilpotent and so $A^{t+2} = 0$ for $t \in \mathbb{N}$. Thus any initial state is transferred to the origin in finite time. This can never occur in the differentiable case.

For the discretized differential equation (34), we have

$$a_0 = 1 - 2\alpha\tau + \tau^2\beta, \quad a_1 = 2\alpha\tau - 2.$$

Hence $1+a_1+a_0 = \tau^2\beta$, $1-a_1+a_0 = 4-4\alpha\tau+\tau^2\beta$, $1-a_0 = 2\alpha\tau-\tau^2\beta$. So the discretized system will be asymptotically stable if $\beta > 0$, $2\alpha\tau - \tau^2\beta > 0$, $4 - 4\alpha\tau + \tau^2\beta > 0$. The stability chart is shown on the RHS of Figure 3.3.2. Note that for any $\beta > 0$, $\alpha > 0$ there exists τ sufficiently small such that the discretized system is asymptotically stable and as $\tau \to 0$ the shaded region fills up the positive orthant $(0,\infty)^2$. Thus the stability chart for the discretized system gradually approaches the stability region of the differentiable system as $\tau \to 0$. This is not the case for all discretization schemes as we will show in the next subsection. □

We conclude this subsection with a brief discussion of the relationship between spectral stability criteria for continuous and discrete time systems (27) (see Ex. 27). There is a well known rational map transforming the open left half plane $\mathbb{C}_-$ onto the open unit disk $\mathbb{D}$ and vice versa, the so-called Möbius map

$$m(\cdot)\ :\ \lambda \mapsto \frac{\lambda+1}{\lambda-1},\quad \lambda \in \mathbb{C}\setminus\{1\} \tag{38}$$

with inverse $m^{-1}(\cdot) = m(\cdot)$. In particular this map sends 0 to -1, ∞ to 1, -1 to 0. The matrix version of this transformation

$$A \mapsto (A+I)(A-I)^{-1} \tag{39}$$

is well defined on $\{A \in \mathbb{K}^{n\times n};\ 1 \notin \sigma(A)\}$ and is known as the *Cayley transform.*

Proposition 3.3.26. *Given* $A \in \mathbb{K}^{n\times n}$, $1 \notin \sigma(A)$, *let* $\hat{A} = (A+I)(A-I)^{-1}$, *then* $A = (\hat{A}-I)^{-1}(\hat{A}+I)$ *and*

$$\sigma(\hat{A}) = \left\{(\lambda+1)(\lambda-1)^{-1},\ \lambda \in \sigma(A)\right\}. \tag{40}$$

Proof: Suppose $Ax = \lambda x$, $x \in \mathbb{C}^n$, $\lambda \in \mathbb{C}$, $x \neq 0$, then $(A-I)x = (\lambda-1)x$, $(\lambda-1)^{-1}x = (A-I)^{-1}x$ and $(A+I)x = (\lambda+1)x$. Hence

$$\hat{A}x = (\lambda+1)(\lambda-1)^{-1}x$$

and so $(\lambda+1)(\lambda-1)^{-1} \in \sigma(\hat{A})$, i.e. $m(\sigma(A)) \subset \sigma(\hat{A})$. Since $\hat{A}(A-I) = A+I$, we get $(\hat{A}-I)A = \hat{A}+I$ that is $A = (\hat{A}-I)^{-1}(\hat{A}+I)$. So applying the above argument to $\hat{A}$ instead of A and making use of $m^{-1}(\cdot) = m(\cdot)$ we obtain (40). □

As a result of the above proposition and Theorem 3.3.20 we see that the continuous time system $\dot{x} = Ax$ is asymptotically stable if and only if the discrete time system $x(t+1) = (A+I)(A-I)^{-1}x(t)$ is asymptotically stable. For further details and applications of the Cayley transform see Subsection 3.4.6 and Subsection 5.3.7.

3.3.3 Numerical Stability of Discretization Methods

In this subsection we examine the stability of linear multistep discretizations methods as described in Subsection 2.3.1 (see (35))

$$x^\tau(k+1) = x^\tau(k-p) + \tau[b_{-1}f_{k+1} + b_0 f_k + \ldots + b_\nu f_{k-\nu}],\quad p \in \mathbb{N}. \tag{41}$$

We apply this integration formula to the scalar differential equation,

$$\dot{x} = ax \tag{42}$$

(with $a \in \mathbb{R}$, $a\tau b_{-1} \neq 1$), i.e. we set $f_k = ax^\tau(k)$. Substitution in (41) yields the difference equation

$$x^\tau(k+1) = x^\tau(k-p) + a\tau[b_{-1}x^\tau(k+1) + b_0 x^\tau(k) + \ldots + b_\nu x^\tau(k-\nu)].$$

Let us first assume $\nu \geq p$, then introducing $x_1(k) = x^\tau(k-\nu)$, $x_2(k) = x^\tau(k-\nu+1), \ldots, x_{\nu+1}(k) = x^\tau(k)$, we obtain the matrix difference equation

$$x(k+1) = Ax(k)$$

where $x(k) = [x_1(k), \ldots, x_{\nu+1}(k)]^\top \in \mathbb{R}^{\nu+1}$ and

$$A = \begin{bmatrix} 0 & 1 & 0 & \ldots & \ldots & \ldots & 0 \\ 0 & 0 & 1 & \ldots & \ldots & \ldots & 0 \\ \ldots & \ldots & \ldots & \ldots & \ldots & \ldots & \ldots \\ 0 & 0 & 0 & \ldots & \ldots & \ldots & 1 \\ a\tau b_\nu \gamma^{-1} & \ldots & \ldots & \ldots & (1+a\tau b_p)\gamma^{-1} & \ldots & a\tau b_0 \gamma^{-1} \end{bmatrix}, \gamma = 1 - a\tau b_{-1} \neq 0.$$

The characteristic equation of A, after multiplication by γ is

$$(1 - a\tau b_{-1})\lambda^{\nu+1} - \lambda^{\nu-p} - a\tau[b_0\lambda^\nu + \ldots + b_\nu] = 0. \tag{43}$$

If $\tau = 0$, the eigenvalues of A are 0 (with multiplicity $\nu - p$) and the $p+1$ distinct roots $\omega_1, \ldots, \omega_{p+1}$ of $z^{p+1} = 1$. It follows from Corollaries 4.2.4 and 4.2.3 of the next chapter that for small $\tau \geq 0$ the eigenvalues of A can be written in the form $\lambda_1(\tau), \ldots, \lambda_\nu(\tau)$ where the first $p+1$ eigenvalues with $\lambda_i(0) = \omega_i$ are analytic in τ

$$\lambda_i(\tau) = \omega_i + \alpha_i\tau + O(\tau^2), \quad i = 1, 2, \ldots, p+1 \tag{44}$$

and the remaining eigenvalues with $\lambda_i(0) = 0$, $i = p+2, \ldots, \nu+1$ are continuous in τ. Hence if $\tau \geq 0$ is sufficiently small, then

$$|\lambda_i(\tau)| < 1, \quad i = p+2, \ldots, \nu+1.$$

Substituting (44) in (43) and equating terms of order 1 in τ, yields

$$\alpha_i = \frac{a}{p+1}[b_{-1}\omega_i + b_0 + \ldots + b_\nu\omega_i^{-\nu}], \quad i = 1, \ldots, p+1. \tag{45}$$

Hence

$$\lambda_i^{p+1}(\tau) = \omega_i^{p+1} + a\tau[b_{-1}\omega_i^{p+1} + b_0\omega_i^p + \ldots + b_\nu\omega_i^{p-\nu}] + O(\tau^2)$$

and since $\omega_i^{p+1} = 1$,

$$|\lambda_i^{p+1}(\tau)|^2 = 1 + 2\tau a \operatorname{Re}[b_{-1} + b_0\omega_i^p + \ldots + b_\nu\omega_i^{p-\nu}] + O(\tau^2).$$

We say that a particular discretization method is *stable* if, on application to an asymptotically stable scalar differential equation (42), the resulting discrete time system is also asymptotically stable for sufficiently small τ. Thus the discretization method (41) is stable if and only if all the roots of (43) lie in $\mathbb{D}$, for every $a < 0$ and τ sufficiently small $0 < \tau \leq \delta(a)$. So a sufficient condition is

$$\operatorname{Re}[b_{-1} + b_0\omega_i^p + \ldots + b_\nu\omega_i^{p-\nu}] > 0, \quad i = 1, \ldots, p+1, \tag{46}$$

and a necessary condition is

$$\mathrm{Re}[b_{-1} + b_0\omega_i^p + \ldots + b_\nu \omega_i^{p-\nu}] \geq 0, \quad i = 1, \ldots, p+1.$$

In the case of equality higher order approximation of $\lambda_i(\tau)$ must be considered to determine whether or not the discretization method (41) is stable.
The case $p > \nu$ leads to the same conclusions and is slightly easier to analyze since when $\tau = 0$ there are no roots at 0. Now let us apply the results to some of the integration schemes introduced in Section 2.5.

Example 3.3.27. (Euler's method). In this case (see (2.5.25)), $p = 0$, $\nu = 0$, $b_{-1} = 0$, $b_0 = 1$, so (46) holds and Euler's method is stable. □

Example 3.3.28. (Runge-Kutta method). This is a single step method with $p = 0$, $\nu = 0$, $b_{-1} = 0$, $b_0 = 1 + (1/2)\tau a + (1/6)\tau^2 a^2 + (1/24)\tau^3 a^3$ (see Example 2.5.12), so (46) holds and the Runge-Kutta method is stable. □

Example 3.3.29. (Midpoint method). For this method (see Example 2.5.11) $p = 1$, $\nu = 0$, $b_{-1} = 0$, $b_0 = 2$. So $\omega_1 = +1$, $\omega_2 = -1$ and

$$\mathrm{Re}[b_0\omega_1] > 0 \;\text{ but }\; \mathrm{Re}[b_0\omega_2] < 0.$$

Hence the midpoint method is unstable. □

Example 3.3.30. (Adams-Bashforth methods). For these methods (see Example 2.5.13), $p = 0$, $b_{-1} = 0$ and some typical values of b_i, $i = 0, \ldots, \nu$ are given in Table 2.5.11. The stability condition (46) is $b_0 + b_1 + \ldots + b_\nu > 0$. Note that the values of b_i given in Table 2.5.11 all have the property that $\sum_{i=0}^{\nu} b_i = 1$, so the Adams-Bashforth methods are stable. □

Example 3.3.31. (Milne's method). The implicit corrector of Milne's method (see Example 2.5.14) applied to the scalar differential equation (42), yields

$$x^\tau(k+1) = x^\tau(k-1) + (a\tau/3)[x^\tau(k+1) + 4x^\tau(k) + x^\tau(k-1)].$$

Hence $p = 1$, $\nu = 1$, $b_{-1} = 1/3$, $b_0 = 4/3$, $b_1 = 1/3$. So $\omega_1 = 1$, $\omega_2 = -1$ and substitution in the left hand side of (46) gives

$$\mathrm{Re}[b_{-1} + b_0\omega_1 + b_1] = 2 > 0, \quad \text{but} \quad \mathrm{Re}[b_{-1} + b_0\omega_2 + b_1] = -2/3 < 0.$$

So the corrector of Milne's method is unstable and this will be the case for the predictor-corrector algorithm as well. □

These results seem to contradict some of the convergence properties of the above schemes as described in Section 2.5. However, the important thing to remember is that convergence is defined relative to a *finite* interval $[0, b]$ whereas stability is a requirement on the asymptotic behaviour as $t \to \infty$. We illustrate this distinction by the following example.

Example 3.3.32. (Instability of the midpoint rule). Applying the midpoint rule to the scalar differential equation (42) yields the difference equation

$$x^\tau(k+1) = x^\tau(k-1) + 2\tau a x^\tau(k), \quad k \in \mathbb{N} \tag{47}$$

Assume $a < 0$ so that (42) is asymptotically stable. The eigenvalues of the second order system (47) are given by

$$\lambda_1(\tau) = a\tau + \sqrt{1 + a^2\tau^2} = 1 + a\tau + O(\tau^2), \quad \lambda_2(\tau) = a\tau - \sqrt{1 + a^2\tau^2} = -1 + a\tau + O(\tau^2).$$

Every solution of (47) can be represented in the form

$$x^\tau(k) = c_1\lambda_1^k(\tau) + c_2\lambda_2^k(\tau). \tag{48}$$

If $\tau = t/k$, with $t > 0$ fixed arbitrarily, then the first eigenmotion

$$x^\tau(k) = \lambda_1(\tau)^k x_0 = (1 + at/k + O(t^2/k^2))^k x_0 \ \rightarrow\ e^{at}x_0 \quad \text{as} \quad k \rightarrow \infty.$$

Hence, on any compact interval $[0, b]$, this eigenmotion of (47) generated by the initial conditions $x^\tau(0) = x_0$, $x^\tau(1) = \lambda_1(\tau)x_0$ yields a uniform approximation of the eigenmotion $e^{at}x_0$ of (42) generated by $x(0) = x_0$. Moreover the eigenmotion $x^\tau(k) = \lambda_1(\tau)^k x_0$ tends to 0 for $k \rightarrow \infty$, as does $e^{at}x_0$. However, the discretization (47) has a second eigenmotion

$$x^\tau(k) = \lambda_2(\tau)^k x_0 = (-1 + a\tau + O(\tau^2))^k, \quad k \in \mathbb{N}$$

which is an *unbounded* oscillation. This eigenmotion of (47) is called *spurious* or *parasitic* since it does not correspond to a solution of the differential equation (42). Any deviation from the initial conditions $x^\tau(0) = x_0$, $x^\tau(1) = \lambda_1(\tau)x_0$ or any rounding error will excite this spurious eigenmotion and then, for any given $\tau > 0$, this eigenmotion will completely dominate the true solution after some time. This is illustrated in Table 3.3.3 where we apply the midpoint rule to

$$\dot{x}(t) = -x(t), \quad x(0) = 1. \tag{49}$$

We started the algorithm with the exact value $x^\tau(0) = 1$ and an order 2 approximation of

t	TRUE SOL'N	ERRORS
$k\tau$	$x(k\tau)$	
0.0	1.00000000	0.0000000
0.5	0.60653066	-0.0000003
1.0	0.36787944	-0.0000006
1.5	0.22313016	-0.0000011
2.0	0.13533528	-0.0000019
2.5	0.08208500	-0.0000031

t	TRUE SOL'N	ERRORS
$k\tau$	$x(k\tau)$	
5.0	0.00673795	-0.0000371
7.5	0.00055308	-0.0004520
10.0	0.00004540	-0.0055066
12.5	0.00000373	-0.0670841
15.0	0.00000031	-0.8172517
17.5	0.00000003	-9.9561596

Table 3.3.3: Errors of the midpoint rule applied to (49) with $\tau = 0.001$

the corresponding value at time τ : $x^\tau(1) = 1-\tau$. Note that we have a good approximation of the true solution for t in the range $[0, 2.5]$ because of the relatively small stepsize $\tau = 0.001$. However, for values $t \geq 15$ the parasitic oscillations excited by the initial errors and by rounding errors become so strong that any correlation between the true and the "approximate" solutions is lost. □

The previous example illustrates the general problem. Suppose we apply a linear multistep method of the form (41) to an initial value problem $\dot{x} = Ax$, $x(0) = x_0$. If the differential system is n-dimensional and $\nu \geq 1$ or $p \geq 1$ then the dimension of the state space X of the corresponding discrete time system $x(t+1) = A_\tau x(t)$ is higher, namely

$$\dim X = n \cdot (\max\{\nu, p\} + 1). \tag{50}$$

Only n of the $n \cdot (\max\{\nu, p\} + 1)$ eigenvalues of A_τ (counting multiplicities) approximate eigenvalues of A, all the others correspond to parasitic eigenmotions of the discrete time system introduced by the multistep method. Thus a crucial question is whether or not these parasitic eigenmotions are tending to zero with an appropriate decay rate as $t \to \infty$. *Numerically unstable* integration methods of the form (41) generate unstable discrete time systems (27) when applied to certain asymptotically stable differentiable systems (27). Although these integration methods may be very efficient for the solution of initial value problems *on fixed compact intervals* they are not suitable for the approximation of differentiable *systems* by discrete time *systems* (see Section 2.5).

3.3.4 Liapunov Functions for Time-Varying Linear Systems

In this subsection we return to time-varying linear systems of the form (1). The time domain T is either an interval in $\mathbb{R}$ unbounded to the right or an interval in $\mathbb{Z}$ unbounded to the right. For linear systems it is natural to choose (time–varying) quadratic forms $x \mapsto V(t,x) = \langle x, P(t)x\rangle$, $t \in T$, as possible candidates for Liapunov functions. Here we characterize stability properties of (1) in terms of these quadratic Liapunov functions. In contrast to the previous section (where we assumed Liapunov functions to be given) we develop a systematic construction procedure. At the end of the subsection we will see that quadratic Liapunov functions provide a tool for deriving stability properties of a given nonlinear system trajectory from stability properties of the associated linearized model. Thus we will use Liapunov's direct method in order to prove the validity of Liapunov's indirect method.

Throughout the subsection we assume $P(t)$, $t \in T$ is symmetric if $\mathbb{K} = \mathbb{R}$ and Hermitian if $\mathbb{K} = \mathbb{C}$. Moreover in the continuous time case we suppose that $P(\cdot) : T \mapsto \mathcal{H}_n(\mathbb{K})$ is continuous and piecewise continuously differentiable, i.e. $P(\cdot) \in PC^1(T; \mathcal{H}_n(\mathbb{K}))$[1]. We do not assume $P(\cdot)$ to be continuously differentiable since our construction process will only yield piecewise continuously differentiable $P(\cdot)$ if $A(\cdot) \in PC(T; \mathbb{K}^{n\times n})$ has jump points.

Now consider

$$V(t,x) = \langle x, P(t)x\rangle \qquad (t,x) \in T \times \mathbb{K}^n \tag{51}$$

as a candidate for a Liapunov function for the linear system (1). In the continuous time case the derivative of V along the flow of (1a) is defined by

$$\begin{aligned} \dot V(t,x) &= \langle x, \dot P(t)x\rangle + \langle A(t)x, P(t)x\rangle + \langle x, P(t)A(t)x\rangle \\ &= \langle x, (\dot P(t) + A(t)^* P(t) + P(t)A(t))x\rangle, \quad (t,x) \in T \times \mathbb{K}^n \end{aligned} \tag{52}$$

[1]This means that the derivative $\dot P(t)$ exists for all $t \in T \setminus S$ where $S \subset T$ is a subset without accumulation point in $\mathbb{R}$ and the limit $\lim_{t\downarrow s} \dot P(t)$ exists at every $s \in S$. Extending $\dot P(\cdot)$ by $\dot P(s) = \lim_{t\downarrow s} \dot P(t)$ to all of T, we obtain a piecewise continuous and right continuous matrix function $\dot P(\cdot) : T \to \mathcal{H}_n(\mathbb{K})$.

where $\dot{P}(t)$ is defined for all $t \in T$ as in the footnote. In the discrete time case

$$\begin{aligned}\dot{V}(t,x) &= \langle A(t)x, P(t+1)A(t)x\rangle - \langle x, P(t)x\rangle \\ &= \langle x, (A(t)^*P(t+1)A(t) - P(t))x\rangle\,, \quad (t,x) \in T \times \mathbb{K}^n\,. \end{aligned} \tag{53}$$

Suppose we define a matrix $Q(t) \in \mathbb{K}^{n\times n}$, $t \in T$ by

$$\dot{P}(t) + A(t)^*P(t) + P(t)A(t) + Q(t) = 0\,, \quad t \in T \tag{54a}$$

$$A(t)^*P(t+1)A(t) - P(t) + Q(t) = 0\,, \quad t \in T\,. \tag{54b}$$

Then in the continuous time case $Q(\cdot) \in PC(T;\mathcal{H}_n(\mathbb{K}))$ and in the discrete time case $Q(\cdot) = (Q(t))_{t\in T} \in \mathcal{H}_n(\mathbb{K})^T$, i.e. $Q(\cdot)$ is a sequence in $\mathcal{H}_n(\mathbb{K})$ defined on T. In both cases

$$\dot{V}(t,x) = -\langle x, Q(t)x\rangle\,, \quad (t,x) \in T \times \mathbb{K}^n\,. \tag{55}$$

As a counterpart of Theorem 3.2.17 for quadratic Liapunov functions we have

Theorem 3.3.33. *Suppose that $P(\cdot) \in PC^1(T;\mathcal{H}_n(\mathbb{K}))$ and $Q(\cdot) \in PC(T;\mathcal{H}_n(\mathbb{K}))$ (resp. $P(\cdot)$, $Q(\cdot) \in \mathcal{H}_n(\mathbb{K})^T$) satisfy* (54). *If $\alpha_1, \alpha_2, \alpha_3 > 0$, then*

(i) $\forall t \in T : P(t) \succeq \alpha_1 I_n,\ Q(t) \succeq 0 \quad \Rightarrow \quad$ *stability of* (1) *at any time* $t_0 \in T$.

(ii) $\forall t \in T : \alpha_2 I_n \succeq P(t) \succeq \alpha_1 I_n\,,\ Q(t) \succeq 0 \quad \Rightarrow \quad$ *uniform stability of* (1) *on* T.

(iii) $\forall t \in T : \alpha_2 I_n \succeq P(t) \succeq \alpha_1 I_n\,,\ Q(t) \succeq \alpha_3 I_n \Rightarrow$ *uniform asymptotic stability of* (1).

Proof: In the discrete time case the theorem is a specialization of Theorem 3.2.17 using (51) as a Liapunov function. However, for the continuous time case, V will not, in general be a Liapunov function in the sense of Definition 3.2.16 since V may not be continuously differentiable on $T \times \mathbb{K}^n$. But $t \mapsto V(t,x(t)) = \langle x(t), P(t)x(t)\rangle$ is continuous and piecewise continuously differentiable for trajectories $x(\cdot)$ of (1). By (52), for all $t \in T$ where $x(\cdot)$ and $P(\cdot)$ are both differentiable, we have

$$\frac{dV}{dt}(t,x(t)) = \langle \dot{x}(t), P(t)x(t)\rangle + \langle x(t), \dot{P}(t)x(t)\rangle + \langle x(t), P(t)\dot{x}(t)\rangle = \dot{V}(t,x(t)). \tag{56}$$

Hence if the premises in (i) are satisfied, V is a generalized Liapunov function for (1) and then (i) follows from Theorem 3.2.7. In a similar way (ii) and (iii) follow since V is bounded in the sense of (5) (and strictly decreasing along the flow of (1)) if the premises in (ii) (resp. (iii)) hold. □

For quadratic functions the instability Theorem 3.2.22 specializes to the following result.

Theorem 3.3.34. *Suppose that $P(\cdot) \in PC^1(T;\mathcal{H}_n(\mathbb{K}))$ and $Q(\cdot) \in PC(T;\mathcal{H}_n(\mathbb{K}))$ (resp. $P(\cdot), Q(\cdot) \in \mathcal{H}_n(\mathbb{K})^T$) satisfy* (54). *If there exists $(t_0, x^0) \in T \times \mathbb{K}^n$ and positive constants α_3, α_2 such that $\langle x^0, P(t_0)x^0\rangle < 0$ and for all $t \in T_{t_0}$, $x \in \mathbb{K}^n$*

$$\langle x, P(t)x\rangle < 0 \quad \Rightarrow \quad \langle x, Q(t)x\rangle \geq \alpha_3\|x\|^2 \quad \text{and} \quad |\langle x, P(t)x\rangle| \leq \alpha_2\|x\|^2$$

then (1) *is unstable at time* $t_0 \in T$.

The proof is set as Ex. 9.

Example 3.3.35. The damped Mathieu equation is of the form

$$\ddot{y} + 2\zeta\dot{y} + (a - 2r\cos 2t)y = 0\,, \quad t \geq 0$$

where $\zeta > 0$, $a > 0$, $r \in \mathbb{R}$ are constants. If $x_1 = y$, $x_2 = \dot{y}$ we obtain the state space system

$$\begin{bmatrix} \dot{x}_1(t) \\ \dot{x}_2(t) \end{bmatrix} = \begin{bmatrix} 0 & 1 \\ -a + 2r\cos 2t & -2\zeta \end{bmatrix} \begin{bmatrix} x_1(t) \\ x_2(t) \end{bmatrix} =: A(t)x(t)\,, \quad t \geq 0\,. \tag{57}$$

Consider the matrix function

$$P(t) = \begin{bmatrix} 2\rho\zeta^2 + a - 2r\cos 2t & \rho\zeta \\ \rho\zeta & 1 \end{bmatrix}, \quad t \geq 0$$

where ρ is constant. A straight forward calculation yields

$$\dot{P}(t) + A(t)^*P(t) + P(t)A(t) = \begin{bmatrix} -2\zeta\rho(a - 2r\cos 2t) + 4r\sin 2t & 0 \\ 0 & -\zeta(4 - 2\rho) \end{bmatrix}.$$

So $Q(t)$ as defined by (54a) is

$$Q(t) = \begin{bmatrix} 2\zeta\rho(a - 2r\cos 2t) - 4r\sin 2t & 0 \\ 0 & \zeta(4 - 2\rho) \end{bmatrix}, \quad t \geq 0\,.$$

There exist positive constants $\alpha_1, \alpha_2, \alpha_3$ such that $\alpha_2 I_2 \succeq P(t) \succeq \alpha_1 I_2$, $Q(t) \succeq \alpha_3 I_2$, $t \geq 0$

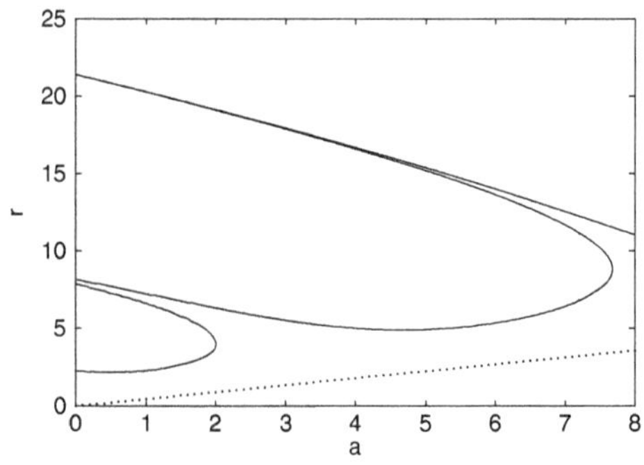

Figure 3.3.4: Stability domain for the Mathieu equation

provided

$$0 < \rho < 2\,, \quad 2\rho\zeta^2 + a - 2r\cos 2t > \rho^2\zeta^2\,, \quad 2\rho\zeta(a - 2r\cos 2t) - 4r\sin 2t > 0\,, \quad t \geq 0\,.$$

And these inequalities will hold if the following time–invariant inequalities are satisfied:

$$0 < \rho < 2\,, \quad \rho(2-\rho)\zeta^2 + a > 2|r|\,, \quad \rho\zeta a > 2|r|(1 + \rho^2\zeta^2)^{1/2}\,. \tag{58}$$

Now suppose $\zeta a > |r|\,(1 + 4\zeta^2)^{1/2}$ then by choosing ρ close to 2 it can be shown that (58) holds and hence by Theorem 3.3.33 the time-varying system (57) is uniformly exponentially stable. For $\zeta = 1$ the stability domain determined by this inequality is shown in Figure 3.3.4 (below the dotted line) together with the actual stability boundaries (below the continuous lines). □

In order to prepare the ground for Liapunov's indirect method we now seek a partial converse to statement (iii) in Theorem 3.3.33. For this the following lemma is useful.

Lemma 3.3.36. *Suppose that $A(\cdot)$ generates a uniformly exponentially stable evolution operator $\Phi(\cdot,\cdot)$. Given a bounded $Q(\cdot) \in PC(T;\mathcal{H}_n(\mathbb{K}))$ (resp. bounded $Q(\cdot) \in \mathcal{H}_n(\mathbb{K})^T$), the only bounded $P(\cdot) \in PC^1(T;\mathcal{H}_n(\mathbb{K}))$ (resp. bounded $P(\cdot) \in \mathcal{H}_n(\mathbb{K})^T$) which solves (54) is*

$$P(t) = \int_t^\infty \Phi(s,t)^* Q(s)\Phi(s,t)ds\,, \quad t \in T, \tag{59a}$$

$$P(t) = \sum_{s=t}^{\infty} \Phi(s,t)^* Q(s)\Phi(s,t)\,, \qquad t \in T\,. \tag{59b}$$

Proof: The proof is for the continuous time case. Suppose $P(\cdot) \in PC^1(T;\mathcal{H}_n(\mathbb{K}))$ is a bounded solution of (54a) then

$$\begin{aligned}\frac{\partial}{\partial s}\left(\Phi(s,t)^*P(s)\Phi(s,t)\right) &= \Phi(s,t)^*\left[\dot P(s) + A(s)^*P(s) + P(s)A(s)\right]\Phi(s,t)\\ &= -\Phi(s,t)^*Q(s)\Phi(s,t)\,, \quad a.e.\ s > t,\ t \in T.\end{aligned} \tag{60}$$

By assumption there exist constants $p, q, M > 0$ and $\omega < 0$ such that $\|P(s)\| \le p$ and

$$\|Q(s)\| \le q\,, \quad \|\Phi(s,t)\| \le Me^{\omega(s-t)}\,, \quad s \ge t,\ t \in T. \tag{61}$$

So we may integrate (60) on $[t,\infty)$, $t \in T$ to obtain (59a). It remains to show that $P(\cdot)$ defined by (59a) is a bounded solution of (54a) on T. Since (61) holds, $P(\cdot)$ is well defined by (59a), Hermitian and bounded on T (see (62a)). Now just as in the proof of Lemma 3.3.4, we have

$$\frac{\partial \Phi(s,t)}{\partial t} = -\Phi(s,t)A(t) \ \text{ for a.e. } s > t.$$

Differentiating the integral in (59a)

$$\begin{aligned}\dot P(t) &= -Q(t) - A(t)^*\int_t^\infty \Phi(s,t)^*Q(s)\Phi(s,t)ds - \int_t^\infty \Phi(s,t)^*Q(s)\Phi(s,t)A(t)ds\\ &= -Q(t) - A(t)^*P(t) - P(t)A(t)\,, \quad a.e.\ t \in T.\end{aligned}$$

This shows that $P(\cdot) \in PC^1(T;\mathcal{H}_n(\mathbb{K}))$ and solves (54a). □

Note that if (61) holds and $P(\cdot)$ is given by (59) then $P(\cdot)$ is bounded by

$$\|P(t)\| \le M^2q\int_t^\infty e^{2\omega(s-t)}ds = M^2q/(-2\omega)\,, \quad t \in T\,, \tag{62a}$$

$$\|P(t)\| \le M^2q\sum_{s=t}^{\infty} e^{2\omega(s-t)} = M^2q/(1-e^{2\omega})\,, \quad t \in T\,. \tag{62b}$$

$Q(t)$ needs not necessarily be positive definite in order to conclude uniform asymptotic stability of (1) via the Liapunov function (51). Suppose that $Q(t) = C(t)^*C(t)$, where $C(\cdot) \in PC(T;\mathbb{K}^{p\times n})$ (resp. $C(\cdot) \in (\mathbb{K}^{p\times n})^T$). We will need an extra assumption which is expressed in terms of the matrices

$$\mathcal{Q}(t,t_0) = \int_{t_0}^{t} \Phi(s,t_0)^* C(s)^* C(s) \Phi(s,t_0) ds, \quad t \in T_{t_0},\ t_0 \in T \tag{63a}$$

$$\mathcal{Q}(t,t_0) = \sum_{s=t_0}^{t-1} \Phi(s,t_0)^* C(s)^* C(s) \Phi(s,t_0), \quad t \in T_{t_0},\ t_0 \in T. \tag{63b}$$

We will say that $(A(\cdot),C(\cdot))$ is *uniformly observable* on T (see Volume II) if there exist constants $\tau > 0, c > 0$ such that

$$\mathcal{Q}(t_0+\tau,t_0) \succeq cI_n, \quad t_0 \in T. \tag{64}$$

Clearly in this case $\mathcal{Q}(t_1,t_0) \succeq cI_n$ for all $t_0,t_1 \in T$ such that $t_1 - t_0 \geq \tau$. In the next theorem we will show that condition (64) (instead of $Q(t) \succeq \alpha_3 I_n$, see Theorem 3.3.33 (iii)) suffices to obtain uniform asymptotic stability of (1) via the Liapunov function (51).

Remark 3.3.37. If $Q(t) = C(t)^*C(t)$ satisfies

$$Q(t) \succeq \alpha_3 I_n,\ t \in T, \tag{65}$$

then in the discrete time case $(A(\cdot),C(\cdot))$ is *uniformly observable* and we may choose the observability time $\tau = 1$. This need not be the case for continuous time systems (see Ex. 22). However, if $\underline{\beta}(A) > -\infty$ and (65) holds, then $(A(\cdot),C(\cdot))$ is *uniformly observable* and the observability time τ can be made arbitrarily small (with $c > 0$ chosen appropriately). In fact, we have, by assumption, the existence of $\varepsilon > 0, \omega < 0$ such that

$$\Phi(t,t_0)^*\Phi(t,t_0) \succeq \varepsilon e^{\omega(t-t_0)} I_n, \quad t \geq t_0,\ t_0 \in T.$$

Hence it follows from (65) that for all $\tau > 0$,

$$\mathcal{Q}(t_0+\tau,t_0) \succeq \alpha_3 \int_{t_0}^{t_0+\tau} \Phi(s,t_0)^*\Phi(s,t_0) ds \succeq c_\tau I_n, \quad t_0 \in T,$$

where $c_\tau = \varepsilon\alpha_3(1-e^{\omega\tau})/(-\omega)$. □

Theorem 3.3.38. *Suppose that* $Q(t) = C(t)^*C(t)$ *for a bounded* $C(\cdot) \in PC(T;\mathbb{K}^{p\times n})$ *(resp. bounded* $C(\cdot) \in (\mathbb{K}^{p\times n})^T$*) and* $(A(\cdot),C(\cdot))$ *is uniformly observable on* T*. Then the following are equivalent.*

(i) $A(\cdot)$ *generates a uniformly exponentially stable evolution operator.*

(ii) There exists a solution $P(\cdot) \in PC^1(T;\mathcal{H}_n(\mathbb{K}))$ *(resp.* $P(\cdot) \in \mathcal{H}_n(\mathbb{K})^T$*) of (54) such that* $\alpha_2 I_n \succeq P(t) \succeq \alpha_1 I_n$*,* $t \in T$*, for some* $\alpha_1,\alpha_2 > 0$*.*

(iii) There exists a bounded positive definite solution $P(\cdot) \in PC^1(T;\mathcal{H}_n(\mathbb{K}))$ *(resp. bounded positive definite solution* $P(\cdot) \in \mathcal{H}_n(\mathbb{K})^T$*) of (54).*

Proof: The proof is for the continuous time case.
(i) $\Rightarrow$ (ii) Suppose that (1a) is uniformly exponentially stable, then by Lemma 3.3.36

$$P(t) = \int_t^\infty \Phi(s,t)^* C(s)^* C(s) \Phi(s,t) ds \quad \succeq 0 \tag{66}$$

satisfies (54a) and hence satisfies (60). An upper bound for $P(t)$ is given by (62a). (60) implies that for $t_1 \geq t_0$, $t_0 \in T$

$$0 \preceq \Phi(t_1,t_0)^* P(t_1) \Phi(t_1,t_0) = P(t_0) - \int_{t_0}^{t_1} \Phi(s,t_0)^* C(s)^* C(s) \Phi(s,t_0) ds \, . \tag{67}$$

By the observability assumption there exists $\tau > 0$ satisfying (64) where $Q(t,t_0)$ is defined by (63a). Hence (ii) follows from

$$P(t_0) \succeq \int_{t_0}^{t_1} \Phi(s,t_0)^* C(s)^* C(s) \Phi(s,t_0) ds \succeq cI_n \, , \quad t_1 \geq t_0 + \tau, \;\; t_0 \in T. \tag{68}$$

(ii) $\Rightarrow$ (iii) is trivial.
(iii) $\Rightarrow$ (i) Suppose (iii) and let $v(t) = \langle x(t), P(t)x(t) \rangle$ where $x(t) = \Phi(t,t_0)x^0$, $(t_0, x^0) \in T \times \mathbb{K}^n$. Then $\dot{v}(t) = -\langle x(t), C(t)^* C(t) x(t) \rangle$ (see (56) and (55)) and integrating from t_0 to $t_1 > t_0$ yields (67) and hence (68). So $\alpha_2 I_n \succeq P(t) \succeq cI_n$, $t \in T$ for some $\alpha_2 > 0$ and $c > 0$ as in (64). Using again the assumption of uniform observability, we have

$$v(t+\tau) - v(t) = -\langle x(t), Q(t+\tau,t)x(t) \rangle < -c\|x(t)\|^2 \leq -(c/\alpha_2)v(t) \, , \quad t \geq t_0, \; t_0 \in T.$$

Setting $\tilde{c} = c/\alpha_2$, we obtain $v(t+\tau) \leq (1-\tilde{c})v(t)$, $t \geq t_0$ and hence $\tilde{c} < 1$ and

$$0 \leq v(t+k\tau) \leq (1-\tilde{c})^k v(t) \, , \qquad k \in \mathbb{N} \, , \;\; t \geq t_0 \, , \;\; t_0 \in T.$$

For every $t = t_0 + k\tau + r$ where $k \in \mathbb{N}$, $r \in [0,\tau)$ we get

$$v(t) \leq (1-\tilde{c})^k v(t_0+r) \leq (1-\tilde{c})^{(t-t_0-\tau)/\tau} \langle x(t_0+r), P(t_0+r)x(t_0+r) \rangle \, .$$

By Theorem 3.3.33 the system (1a) is uniformly stable on T and hence there exists $M > 0$ such that $\|\Phi(t,t_0)x^0\| \leq M\|x^0\|$, $t > t_0$, $t_0 \in T$. So

$$\langle \Phi(t,t_0)x^0, P(t)\Phi(t,t_0)x^0 \rangle = v(t) \leq M^2 \alpha_2 (1-\tilde{c})^{-1} (1-\tilde{c})^{(t-t_0)/\tau} \|x^0\|^2 \, .$$

Now using the lower bound of $P(t)$ and choosing $\omega = (2\tau)^{-1} \ln(1-\tilde{c}) < 0$, we conclude

$$\|\Phi(t,t_0)\| \leq M([\tilde{c}(1-\tilde{c})])^{-1/2} e^{\omega(t-t_0)} \, , \quad t \geq t_0, \;\; t_0 \in T.$$

Hence (1) is uniformly exponentially stable on T. □

We now derive a necessary and sufficient *instability* criterion in terms of quadratic Liapunov functions.

Theorem 3.3.39. *Let $t_0 \in T$ and suppose that $Q(t) = C(t)^* C(t)$, for a bounded $C(\cdot) \in PC(T_{t_0}; \mathbb{K}^{p \times n})$ (resp. bounded $C(\cdot) \in (\mathbb{K}^{p \times n})^{T_{t_0}}$), $(A(\cdot), C(\cdot))$ is uniformly observable on T_{t_0} and there exists a bounded $P(\cdot) \in PC^1(T_{t_0}; \mathcal{H}_n(\mathbb{K}))$ (resp. bounded $P(\cdot) \in \mathcal{H}_n(\mathbb{K})^{T_{t_0}}$) which solves (54). Then the following are equivalent* *(i)* (1) *is not exponentially stable at time* t_0.

(ii) (1) *is unstable at* t_0.

(iii) *There exist* $\tau > 0$, $\omega > 0$, $M_\omega > 0$ *and* $x^0 \in \mathbb{K}^n$, $x^0 \neq 0$ *such that*

$$\|\Phi(t_0 + k\tau, t_0)x^0\| \geq M_\omega e^{k\tau\omega}\|x^0\|, \quad k \in \mathbb{N}.$$

(iv) *There exist* $x^0 \in \mathbb{K}^n$ *such that* $\langle x^0, P(t_0)x^0\rangle < 0$.

Moreover, when this is the case the Bohl exponent $\overline{\beta}(\Phi) > 0$.

Proof: The proof is for the continuous time case.
(iii) $\Rightarrow$ (ii) $\Rightarrow$ (i) is obvious.
(i) $\Rightarrow$ (iv): Suppose $\langle x, P(t)x\rangle \geq 0$ for all $(t, x) \in T_{t_0} \times \mathbb{K}^n$ then by Theorem 3.3.38 $A(\cdot)$ generates an exponentially stable evolution operator at time t_0.
(iv) $\Rightarrow$ (iii): Suppose that $x^0 \in \mathbb{K}^n$ is such that $\langle x^0, P(t_0)x^0\rangle < 0$. Setting $t_1 = t_0 + k\tau$, $k \in \mathbb{N}$ in (67) (where τ satisfies the uniform observability condition (64)) we obtain from (67)

$$\begin{aligned}
\langle x^0, P(t_0)x^0\rangle &- \langle \Phi(t_0 + k\tau, t_0)x^0, P(t_0 + k\tau)\Phi(t_0 + k\tau, t_0)x^0\rangle \\
&= \int_{t_0}^{t_0+k\tau} \|C(s)\Phi(s, t_0)x^0\|^2 ds = \sum_{j=1}^{k} \int_{t_0+(j-1)\tau}^{t_0+j\tau} \|C(s)\Phi(s, t_0)x^0\|^2 ds \\
&= \sum_{j=1}^{k} \int_{t_0+(j-1)\tau}^{t_0+j\tau} \|C(s)\Phi(s, t_0 + (j-1)\tau)\Phi(t_0 + (j-1)\tau, t_0)x^0\|^2 ds \\
&\geq c \sum_{j=1}^{k} \|\Phi(t_0 + (j-1)\tau, t_0)x^0\|^2 . \qquad (69)
\end{aligned}$$

Assume $\langle x^0, P(t_0)x^0\rangle = -\alpha\|x^0\|^2$, $\alpha > 0$, $\|P(t)\| \leq \alpha_2$ for all $t \in T_{t_0}$ and set $r_k = \sum_{j=1}^{k} \|\Phi(t_0 + (j-1)\tau, t_0)x^0\|^2$, then $r_1 = \|x^0\|^2$ and using (69)

$$cr_k + \alpha\|x^0\|^2 \leq -\langle \Phi(t_0 + k\tau, t_0)x^0, P(t_0 + k\tau)\Phi(t_0 + k\tau, t_0)x^0\rangle \leq \alpha_2\|\Phi(t_0 + k\tau, t_0)x^0\|^2.$$

Hence

$$\alpha_2(r_{k+1} - r_k) = \alpha_2\|\Phi(t_0 + k\tau, t_0)x^0\|^2 \geq cr_k + \alpha\|x^0\|^2 \;, \quad r_1 = \|x^0\|^2 . \qquad (70)$$

So $r_{k+1} \geq (1 + c/\alpha_2)r_k + \alpha/\alpha_2\|x^0\|^2$. From which it is easy to see that

$$r_k \geq \left[(1 + c/\alpha_2)^{k-1}(1 + \alpha/c) - \alpha/c\right] \|x^0\|^2 .$$

Inserting this inequality in (70) we obtain

$$\alpha_2\|\Phi(t_0 + k\tau, t_0)x^0\|^2 \geq (1 + c/\alpha_2)^{k-1}(\alpha + c)\|x^0\|^2 = \alpha_2 M^2 e^{2k\omega\tau}\|x^0\|^2 \qquad (71)$$

where $M = ((\alpha+c)/(\alpha_2+c))^{1/2}$, $\omega = (2\tau)^{-1}\ln(1+c/\alpha_2) > 0$. This proves (iii) and $\overline{\beta}(\Phi) \geq \omega > 0$. □

Remark 3.3.40. (i) The system (1) is not uniformly exponentially stable if and only if there exists a $t_0 \in T$ such that one of the conditions (ii)-(iv) is satisfied.
(ii) Since the in the continuous time case the Liapunov exponent $\overline{\alpha}(\Phi)$ is independent of t_0 we must have $\overline{\alpha}(\Phi) > 0$. The same conclusion also holds for the discrete time case provided $\det A(t) \neq 0$ for all $t \in T$ (this will be the case if $\underline{\beta}(\Phi) > -\infty$).
(iii) In comparison with the Stability Theorem 3.3.38 the Instability Theorem 3.3.39 is rather unsatisfactory since it assumes the existence of a bounded solution of (54). On the other hand Theorem 3.3.39 is surprising in that it shows that under its assumptions marginal stability cannot occur. More precisely we obtain as a consequence of Theorem 3.3.39 and Theorem 3.3.38: If $\overline{\beta}(\Phi) = 0$ and $Q(t) = C(t)^*C(t)$, where $C(\cdot) \in PC(T;\mathbb{K}^{p\times n})$ is bounded (resp. $C(\cdot) \in (\mathbb{K}^{p\times n})^T$ is bounded) then either there is no bounded $P(\cdot) \in PC^1(T_{t_0};\mathcal{H}_n(\mathbb{K}))$ (resp. $P(\cdot) \in \mathcal{H}_n(\mathbb{K})^{T_{t_0}}$) solving (54), for any $t_0 \in T$, or $(A(\cdot), C(\cdot))$ is not uniformly observable. □

We now describe *Liapunov's indirect method* of stability analysis which proceeds via linearization. Consider the nonlinear equations (2.27), namely

$$\dot{x}(t) = f(t, x(t)), \quad t \in T \tag{72a}$$
$$x(t+1) = f(t, x(t)), \quad t \in T. \tag{72b}$$

Let $\overline{x} \in X \subset \mathbb{K}^n$ be an equilibrium point of (72) and assume that f satisfies the conditions (A1), (A2) in Subsection 3.2.2. In addition we also require that

$$f(t,x) = A(t)(x - \overline{x}) + h(t, x - \overline{x}) \ , (t,x) \in T \times X \, . \tag{73}$$

where $A(\cdot) \in PC(T;\mathbb{K}^{n\times n})$ (resp. $A(\cdot) \in (\mathbb{K}^{n\times n})^T$) and for any $\varepsilon > 0$ there exists $\delta > 0$, such that

$$\|h(t, x - \overline{x})\| \leq \varepsilon \|x - \overline{x}\|, \quad (t,x) \in T \times B(\overline{x}, \delta) \, . \tag{74}$$

Theorem 3.3.41 (Liapunov's indirect method). *Suppose that f satisfies (73), (74), $A(\cdot)$ generates a uniformly exponentially stable evolution operator and $(A(\cdot), I)$ is uniformly observable. Then the equilibrium point $\overline{x}$ is uniformly exponentially stable for the nonlinear system (72). More precisely for $x^0 \in B(\overline{x}, r)$, $r > 0$ sufficiently small, the solutions $\varphi(t; t_0, x^0)$ of (72) have infinite life time and there exist constants $M > 0, \omega < 0$ such that for all $t_0 \in T$*

$$x^0 \in B(\overline{x}, r) \Longrightarrow \forall t \in T_{t_0} : \ \|\varphi(t; t_0, x^0) - \overline{x}\| \leq M e^{\omega(t - t_0)} \|x^0 - \overline{x}\|. \tag{75}$$

Proof: The proof is for the discrete time case. Let $Q(t) \equiv I_n$, then by Theorem 3.3.38 there exists a solution $P(t) \in \mathcal{H}_n(\mathbb{K})$, $t \in T$ of (54b) with $\alpha_2 I_n \succeq P(t) \succeq \alpha_1 I_n$, $t \in T$, for some $\alpha_1, \alpha_2 > 0$. Consider $V(t,x) = \langle x - \overline{x}, P(t)(x - \overline{x})\rangle$, $(t,x) \in T \times X$. Setting $\Delta x = x - \overline{x}$ we obtain by (73) for every $(t,x) \in T \times X$

$$\begin{aligned}
\dot{V}(t,x) &= V(t+1, f(t,x)) - V(t,x) \\
&= \langle A(t)\Delta x + h(t,\Delta x), P(t+1)(A(t)\Delta x + h(t,\Delta x))\rangle - \langle \Delta x, P(t)\Delta x\rangle \\
&= -\|\Delta x\|^2 + 2\,\mathrm{Re}\langle h(t,\Delta x), P(t+1)A(t)\Delta x\rangle + \langle h(\Delta x, t), P(t+1)h(t,\Delta x)\rangle.
\end{aligned}$$

So

$$\dot{V}(t,x) \le -\|\Delta x\|^2 + 2\|h(t,\Delta x)\| \, \|P(t+1)\| \, \|A(t)\| \, \|\Delta x\| + \|P(t+1)\| \, \|h(t,\Delta x)\|^2.$$

In the discrete time case $\|A(\cdot)\|$ is bounded on T by the uniform stability assumption, so we may choose $\varepsilon > 0$ sufficiently small to obtain

$$1 - 2\alpha_2\|A(t)\|\varepsilon - \alpha_2\varepsilon^2 \ge 1/2, \quad t \in T.$$

By (74), there exists a $\delta > 0$, such that $B(\overline{x},\delta) \subset X$ and

$$\|h(t,\Delta x)\| \le \varepsilon\|\Delta x\| \quad \text{for} \quad (t,x) \in T \times B(\overline{x},\delta).$$

Hence

$$\dot{V}(t,x) \le -(1/2)\|\Delta x\|^2 \quad \text{for} \quad (t,x) \in T \times B(\overline{x},\delta).$$

Setting $D = B(\overline{x},\delta)$ we may apply Corollary 3.2.20 with $p = 2$ to conclude that the equilibrium point $\overline{x}$ is uniformly exponentially stable and there exist constants $M > 0, \omega < 0$ such that (75) holds. □

Remark 3.3.42. (i) The formulation of Theorem 3.3.41 in terms of uniform exponential stability is essential and cannot be replaced by for example asymptotic stability (see *Bellman* (1953) [44] for a counter example).

(ii) By Remark 3.3.37 $(A(\cdot), I)$ is necessarily uniformly observable in the discrete time case and this will also hold in the continuous time case if $\underline{\beta}(\Phi) > -\infty$. □

In order to apply the previous theorems to the stability analysis of a non-constant *trajectory* of the nonlinear system (72) we proceed as described in Subsection 2.1.4. Suppose that $\tilde{x} : T_{t_0} \to X$ is a trajectory of (72) and that $(t, \tilde{x}(t) + x) \in X$ for all $t \in T_{t_0}$, $x \in B(0,\rho)$, $\rho > 0$. Let

$$g(t,x) = f(t,\tilde{x}(t)+x) - f(t,\tilde{x}(t)), \quad (t,x) \in T_{t_0} \times B(0,\rho).$$

Then $\overline{x} = 0$ is an equilibrium point of (72) with f replaced by g and $\overline{x} = 0$ is exponentially stable for this system if and only if $\tilde{x}(\cdot)$ is exponentially stable for (72). Now g will satisfy (73), (74) with $T = T_{t_0}$ (so that we may apply Theorem 3.3.41 to the equilibrium point $\overline{x} = 0$ of the system described by g) if and only if f is uniformly differentiable along the trajectory $\tilde{x}(\cdot)$ in the following sense

$$f(t,x) = f(t,\tilde{x}(t)) + A(t)(x - \tilde{x}(t)) + h(t, x - \tilde{x}(t)), \quad (t,x) \in T_{t_0} \times X\,. \tag{76}$$

where $A(\cdot) \in PC(T_{t_0}; \mathbb{K}^{n\times n})$ (resp. $A(\cdot) \in (\mathbb{K}^{n\times n})^{T_{t_0}}$) and for any $\varepsilon > 0$ there exists $\delta > 0$, such that

$$\|h(t, x - \tilde{x}(t))\| \le \varepsilon\|x - \tilde{x}(t)\|, \quad (t,z) \in T_{t_0} \times B(\tilde{x}(t),\delta)\,. \tag{77}$$

This condition will be satisfied if, for example, f is twice continuously differentiable with respect to x in an ε-neighbourhood of the integral curve $\{(t,\tilde{x}(t)); t \in T_{t_0}\}$, and its second derivative is bounded on this neighbourhood. This follows from the fact that

$$\|f(t,x) - f(t,\tilde{x}(t)) - f'(t,\tilde{x}(t))(x - \tilde{x}(t))\| \le (1/2)\|f''(t,\tilde{x}(t) + \theta(t)(x - \tilde{x}(t)))\|\|x - \tilde{x}(t)\|^2$$

where f' and f'' denote the first and the second derivative of f with respect to x and $\theta(t) \in [0,1]$.
In the following instability theorem we use a quadratic Liapunov function which is associated with a perturbation of the linearization (1) of (72). This widens the applicability of the theorem.

Theorem 3.3.43. *Assume that f satisfies (73), (74) and $t_0 \in T$. For some $r \geq 0$ let $A_r(t) = A(t) - rI_n$ (resp. $r \geq 1$, $A_r(t) = r^{-1}A(t)$), $t \in T$ and suppose the following hold*

(i) $(A_r(\cdot), I_n)$ is uniformly observable on T_{t_0}.

(ii) For $Q(t) \equiv I_n$ there exists a bounded $P_r(\cdot) \in PC^1(T_{t_0}; \mathcal{H}_n(\mathbb{K}))$ (resp. bounded $P_r(\cdot) \in \mathcal{H}_n(\mathbb{K})^{T_{t_0}}$) which solves (54) with $A(t)$ replaced by $A_r(t)$ on T_{t_0}.

Then, if $\dot{x}(t) = A_r(t)x(t)$ (resp. $x(t+1) = A_r(t)x(t)$) is unstable at time t_0, the equilibrium point $\overline{x}$ of the nonlinear system (72) will also be unstable at time t_0.

Proof: The proof is for the continuous time case. Suppose $\dot{x}(t) = A_r(t)x(t)$ is unstable at time t_0. By applying Theorem 3.3.39 with $Q(t) \equiv I_n$ we see that there exists $\tilde{x} \in X$, $\|\tilde{x}\| = 1$ such that $\langle \tilde{x}, P_r(t_0)\tilde{x}\rangle < 0$. Let $V(t,x) = \langle x - \overline{x}, P_r(t)(x - \overline{x})\rangle$ for $(t,x) \in T \times X$. Setting $\Delta x = x - \overline{x}$, the derivative of V along the flow of (72a) is given by

$$\begin{aligned} \dot{V}(t,x) &= \langle \Delta x, \dot{P}_r(t)\Delta x\rangle + \langle A(t)\Delta x + h(t,\Delta x), P_r(t)\Delta x\rangle + \langle \Delta x, P_r(t)(A(t)\Delta x + h(t,\Delta x))\rangle \\ &= -\|\Delta x\|^2 + 2r\langle \Delta x, P_r(t)\Delta x\rangle + 2\,\mathrm{Re}\langle h(t,\Delta x), P_r(t)\Delta x\rangle, \quad (t,x) \in T \times X. \end{aligned}$$

So

$$\dot{V}(t,x) \leq -\|\Delta x\|^2 + 2rV(t,x) + 2\|h(t,\Delta x)\|\,\|P_r(t)\|\,\|\Delta x\|.$$

By assumption there exists $\alpha_2 > 0$ such that $\|P_r(t)\| \leq \alpha_2$ for all $t \in T_{t_0}$. Choose $\varepsilon > 0$ such that $4\varepsilon\alpha_2 < 1$. By (74), there exists a $\delta > 0$, such that $B(\overline{x},\delta) \subset X$ and

$$\|h(t,\Delta x)\| \leq \varepsilon\|\Delta x\| \quad \text{for} \quad (t,x) \in T \times B(\overline{x},\delta).$$

Hence

$$\dot{V}(t,x) - 2rV(t,x) \leq -(1/2)\|\Delta x\|^2 \quad \text{for} \quad (t,x) \in T \times B(\overline{x},\delta). \tag{78}$$

First suppose that $r = 0$. Setting $D = B(\overline{x},\delta)$ and choosing $x^0 = \overline{x} + \rho\tilde{x}$ for any $\rho \in (0,\delta)$ we have an $x^0 \in B(\overline{x},\rho)$ with $V(t_0,x^0) < 0$. Moreover since $\dot{V}(t,x) \leq -(1/2)\|\Delta x\|^2$ and $|V(t,x)| \leq \alpha_2\|\Delta x\|^2$, for all $(t,x) \in T_{t_0} \times B(\overline{x},\delta)$, we may apply the Instability Theorem 3.2.22 to conclude that $\overline{x}$ is unstable at t_0 for (72a) .
Now suppose that $r > 0$ and assume by way of contradiction that $\overline{x}$ is stable for (72a) at time t_0. Then there exists a $\tilde{\delta} \in (0,\delta)$ such that

$$\|x - \overline{x}\| < \tilde{\delta} \Longrightarrow \|\varphi(t;t_0,x) - \overline{x}\| < \delta, \quad t \geq t_0.$$

For every $\rho \in (0,\tilde{\delta})$ we again choose $x^0 = \overline{x} + \rho\tilde{x}$, then $\|x^0 - \overline{x}\| = \rho$ and $V(t_0,x^0) < 0$. By (78) we have

$$\frac{d}{dt}\left[e^{-2r(t-t_0)}V(t,\varphi(t;t_0,x^0))\right] \leq -(1/2)e^{-2r(t-t_0)}\|\varphi(t;t_0,x^0) - \overline{x}\|^2 \leq 0, \quad t \geq t_0.$$

Hence, for $t \geq t_0$

$$V(t, \varphi(t; t_0, x^0) \leq e^{2r(t-t_0)} V(t_0, x^0) = -e^{2r(t-t_0)} |\langle \tilde{x}, P_r(t_0)\tilde{x}\rangle| \|x^0 - \overline{x}\|^2.$$

This contradicts the fact that $V(t, x)$ is bounded on $T \times B(\overline{x}, \delta)$. So $\overline{x}$ is unstable for (72a) at time t_0. □

One might think that if the linear system is unstable at time t_0, there will not exist any bounded solution $P_r(t)$ on T_{t_0} which solves (54) with $A(t)$ replaced by $A_r(t)$ and $Q(t) \equiv I_n$. In part this is suggested by the solution formulas (59) for the uniformly exponentially stable case. Note that even if the right hand side of this formula is well defined we cannot use the resulting $P_r(t)$ in applying the theorem since it is positive definite for all $t \in T_{t_0}$. Let us denote by $\Phi_r(\cdot,\cdot)$ the evolution operator generated by $A_r(\cdot)$, then integrating the equation (60) with the above replacements from t_0 to t, the unique solution with initial value $P_r(t_0)$ is

$$P_r(t) = \Phi_r(t_0, t)^* P_r(t_0) \Phi_r(t_0, t) - \int_{t_0}^{t} \Phi_r(s, t)^* \Phi_r(s, t) ds, \quad t \geq t_0. \tag{79}$$

So in order to apply the theorem we have to seek a non-positive definite $P_r(t_0)$ for which the $P_r(\cdot)$ defined by (79) is bounded on T_{t_0}. Similar considerations apply in the discrete time case, see Ex. 14. We illustrate the continuous time case for the case $r = 0$ in the following simple example, see also Ex. 15.

Example 3.3.44. Consider the scalar system $\dot{x}(t) = a(t)x(t)$, $t \in \mathbb{R}$. First let us assume that $a(t) \equiv a > 0$, $t \in \mathbb{R}$, then the solution given by (59a) is not defined. However by (79) we have $p(t) = e^{-2a(t-t_0)}(p(t_0) + 1/2a) - 1/2a$, $t \geq t_0$. Choosing $p(t_0) = -1/2a$ we conclude from the above theorem that any nonlinear system satisfying (73), (74) for which $\dot{x}(t) = ax(t)$ is the linearization will be unstable at any time t_0.
Now suppose $a(t) = t$, $t \in \mathbb{R}$, then $\Phi(t, t_0) = e^{(t^2 - t_0^2)/2}$ and so again the solution given by (59a) is not defined. But from (79) we get

$$p(t) = e^{t_0^2 - t^2} p(t_0) - \int_{t_0}^{t} e^{s^2 - t^2} ds, \quad t \geq t_0.$$

Hence for $t \geq t_0 \geq 1/2$

$$e^{t_0^2 - t^2} p(t_0) \geq p(t) \geq e^{t_0^2 - t^2} p(t_0) - \int_{t_0}^{t} 2s e^{s^2 - t^2} ds = e^{t_0^2 - t^2} p(t_0) - (1 - e^{t_0^2 - t^2}).$$

So $p(t)$ is bounded for $t \geq t_0 \geq 1/2$ and since we may choose $p(1/2) < 0$ we conclude from the above theorem that any nonlinear system satisfying (73), (74) for which $\dot{x}(t) = tx(t)$ is the linearization will be unstable at any time $t_0 \geq 1/2$. □

3.3.5 Liapunov Functions for Time-Invariant Linear Systems

In this subsection we specialize the results of the previous one to the time–invariant case. The dynamic Liapunov equations (54) then reduce to linear matrix equations for which effective solution procedures are available. This enables us to construct Liapunov functions in an efficient way. Moreover we obtain a more satisfactory

instability criterion (Theorem 3.3.49 (iv)), and the application of Liapunov's indirect method is simplified by the fact that it only requires the differentiability of the right hand side of the nonlinear system at the equilibrium point in question.
In a time–invariant setting it is natural to assume that Q is constant and to require *time–invariant* solutions of (54). Then the dynamic Liapunov equations become static and take the form

$$A^*P + PA + Q = 0 \tag{80a}$$

$$A^*PA - P + Q = 0\,. \tag{80b}$$

In contrast to the dynamic Liapunov equations it is not clear whether these linear matrix equations have solutions. For the case where $\sigma(A) \subset \mathbb{C}_-$ (resp. $\sigma(A) \subset \mathbb{D}$) a solution could be constructed as in Lemma 3.3.36. However this would presuppose asymptotic stability. We need to prove the existence of solutions under more general conditions and with a view to later applications we do this by characterizing the eigenvalues of the Liapunov maps for a generalized version of the equations (80).

Proposition 3.3.45. *Suppose $A \in \mathbb{K}^{n\times n}$, $A_1 \in \mathbb{K}^{n_1\times n_1}$ and let $\mathbf{L}$ (resp. $\mathbf{L}^D$) be the associated generalized Liapunov operator*

$$\mathbf{L} : \mathbb{K}^{n_1\times n} \to \mathbb{K}^{n_1\times n}, \quad X \to \mathbf{L}(X) = A_1X + XA \tag{81a}$$

$$\mathbf{L}^D : \mathbb{K}^{n_1\times n} \to \mathbb{K}^{n_1\times n}, \quad X \to \mathbf{L}^D(X) = A_1XA - X. \tag{81b}$$

Then

$$\sigma(\mathbf{L}) = \{\mu_1 + \mu;\ \mu_1 \in \sigma(A_1),\ \mu \in \sigma(A)\} \tag{82a}$$

$$\sigma(\mathbf{L}^D) = \{\mu_1\mu - 1;\ \mu_1 \in \sigma(A_1),\ \mu \in \sigma(A)\}. \tag{82b}$$

In particular, $\mathbf{L}, \mathbf{L}^D : \mathbb{K}^{n_1\times n} \to \mathbb{K}^{n_1\times n}$ is a linear isomorphism if and only if

$$\mu_1 + \mu \neq 0 \quad (\textit{resp. } \mu_1\mu \neq 1), \qquad \mu_1 \in \sigma(A_1),\ \mu \in \sigma(A). \tag{83}$$

Proof: The proof is for the continuous time case. Suppose that $A_1x^1 = \mu_1x^1, x^1 \in \mathbb{C}^{n_1}, x^1 \neq 0$ and $xA = \mu x$, $x^\top \in \mathbb{C}^n, x \neq 0$. Then for $X = x^1x$, we have

$$\mathbf{L}(X) = A_1X + XA = A_1x^1x + x^1xA = (\mu_1 + \mu)x^1x = (\mu_1 + \mu)X.$$

Hence $\mu_1 + \mu \in \sigma(\mathbf{L})$, i.e. the inclusion $\subset$ in (82a). To prove the converse we transform A to Schur form. In Section 4.5 we will show that for $A \in \mathbb{K}^{n\times n}$, there exists a unitary matrix $U \in \mathbf{U}_n(\mathbb{C})$ such that U^*AU is in upper triangular complex Schur form, namely

$$U^*AU = S = \begin{bmatrix} s_{11} & s_{12} & \cdots & s_{1n} \\ 0 & s_{22} & \cdots & s_{2n} \\ . & 0 & \ldots & . \\ . & . & \ldots & . \\ 0 & 0 & \ldots & s_{nn} \end{bmatrix} \tag{84}$$

where each diagonal element is an eigenvalue of A. Now suppose that λ is an eigenvalue of $\mathbf{L}$ with eigenvector $X \in \mathbb{K}^{n_1\times n}, X \neq 0$. Then $A_1X + XA = \lambda X$ and multiplying on the right by U, we obtain

$$A_1XU + XUU^*AU = A_1XU + XUS = \lambda XU.$$

Defining $Z = [z^1\, z^2\, \dots z^n] := XU$, $z^j \in \mathbb{C}^{n_1}, j \in \underline{n}$, then $Z \neq 0$ and

$$[A_1 + s_{jj} I_{n_1}] z^j = \lambda z^j - \sum_{i=1}^{j-1} s_{ij} z^i, \quad j \in \underline{n}. \tag{85}$$

Since $Z \neq 0$ there exists $j \in \underline{n}$, such that $z^j \neq 0$ and $z^k = 0, k < j$. But then from (85) $[A_1 + s_{jj} I_{n_1}] z^j = \lambda z^j$ and hence $\lambda - s_{jj} \in \sigma(A_1)$ and this completes the proof of (82). Since the linear map $\mathbf{L}$ is a vector space isomorphism if and only if $0 \notin \sigma(\mathbf{L})$ the second assertion follows. □

As a direct consequence of Proposition 3.3.45 the *generalized Liapunov equations*

$$A_1 P + PA + Q = 0, \quad (resp. \ \ A_1 P A - P + Q = 0). \tag{86}$$

have unique solutions $P \in \mathbb{K}^{n_1 \times n}$ for every $Q \in \mathbb{K}^{n_1 \times n}$ if and only if condition (83) holds. In the present context the particular case where $A_1 = A^*$ is of special interest. If $Q = Q^*$ is Hermitian and P is a solution of (86) then P^* is also a solution of (86). Hence if (86) has a unique solution then the solution is necessarily Hermitian. This leads us to introduce the following *Liapunov operator* on the real vector space $\mathcal{H}_n(\mathbb{K})$ of Hermitian $n \times n$ matrices

$$\mathbf{L}_A : \mathcal{H}_n(\mathbb{K}) \to \mathcal{H}_n(\mathbb{K}), \quad X \mapsto A^*X + XA \quad (resp. \, \mathbf{L}_A^D : X \mapsto A^*XA - X). \tag{87}$$

As an immediate consequence of Proposition 3.3.45 we obtain

Corollary 3.3.46. *Suppose $A \in \mathbb{K}^{n \times n}$. The Liapunov operator $\mathbf{L}_A$ (resp. $\mathbf{L}_A^D$) is a linear bijection from $\mathcal{H}_n(\mathbb{K})$ onto itself if and only if*

$$\lambda + \overline{\mu} \neq 0, \quad (resp. \ \ \lambda \overline{\mu} \neq 1) \quad \lambda, \mu \in \sigma(A). \tag{88}$$

In this (and only in this) case the algebraic Liapunov equation (80) *has a unique (Hermitian) solution for every $Q \in \mathcal{H}_n(\mathbb{K})$.*

If $\sigma(A) \subset \mathbb{C}_-$ (resp. $\sigma(A) \subset \mathbb{D}$), then by Lemma 3.3.36 we know that the solution of (80) is given by

$$P = \int_t^\infty e^{A^*(s-t)} Q e^{A(s-t)} ds = \int_0^\infty e^{A^*\rho} Q e^{A\rho} d\rho, \tag{89a}$$

$$P = \sum_{s=t}^\infty A^{*(s-t)} Q A^{s-t} = \sum_{\rho=0}^\infty A^{*\rho} Q A^\rho. \tag{89b}$$

Clearly, if $Q \succ 0$ then P defined by (89) is positive definite. Therefore

Corollary 3.3.47. *Suppose $A \in \mathbb{K}^{n \times n}$ and $\sigma(A) \subset \mathbb{C}_-$ (resp. $\sigma(A) \subset \mathbb{D}$). Then the Liapunov operator $\mathbf{L}_A$ (resp. $\mathbf{L}_A^D$) : $\mathcal{H}_n(\mathbb{K}) \to \mathcal{H}_n(\mathbb{K})$ is invertible and $-\mathbf{L}_A^{-1}$ (resp. $-(\mathbf{L}_A^D)^{-1}$) is a positive operator from the vector space $\mathcal{H}_n(\mathbb{K})$ ordered by $\succeq$ into itself, i.e.*

$$Q \succ 0 \Rightarrow P = -\mathbf{L}_A^{-1}(Q) \succ 0, \quad (resp.. - (\mathbf{L}_A^D)^{-1}(Q) \succ 0, \tag{90}$$

Remark 3.3.48. As a consequence of the next theorem the converse of Corollary 3.3.47 is also true. Hence $-\mathbf{L}_A^{-1}$, (*resp.* $-(\mathbf{L}_A^D)^{-1}$) is a positive operator on $\mathcal{H}_n(\mathbb{K})$ if and only if the associated system $\dot{x} = Ax$ (resp. $x(t+1) = Ax(t)$) is asymptotically stable. This observation shows that there is a close relationship between the stability theory of time–invariant linear systems and the theory of positive operators. □

For time-invariant systems the matrix $\mathcal{Q}(t, t_0)$ defined in (63) takes the form

$$\mathcal{Q}(t,t_0) = \int_{t_0}^{t} e^{A^*(s-t_0)} C^* C e^{A(s-t_0)} ds = \int_0^{t-t_0} e^{A^*\rho} C^* C e^{A\rho} d\rho, \tag{91a}$$

$$\mathcal{Q}(t,t_0) = \sum_{s=t_0}^{t-1} A^{*(s-t_0)} C^* C A^{s-t_0} = \sum_{\rho=0}^{t-t_0-1} A^{*\rho} C^* C A^{\rho}. \tag{91b}$$

Hence the pair (A, C) is uniformly observable if and only if there exists $c > 0, \tau > 0$, such that $\mathcal{Q}(\tau, 0) \geq cI_n$. And it is not difficult to show (cf. Volume II) that this will be the case if and only if (A, C) is *observable* in the sense that

$$\bigcap_{i=1}^{n} \ker CA^{i-1} = \{0\}. \tag{92}$$

As a consequence of these observations the results developed in the previous subsection take a simpler form.

Theorem 3.3.49. *Suppose $Q = C^*C$, where $C \in \mathbb{K}^{p\times n}$.*

(i) If (A, C) is observable, then (27) is asymptotically stable if and only if there exists a solution P of (80) with $P \succ 0$.

(ii) If there exists a solution P of (80) with $P \succeq 0$ and $\ker P \neq \{0\}$, then (A, C) is not observable.

(iii) If there exists a solution P of (80) with $P \succ 0$, then the time-invariant system (27) is stable, and if in fact it is asymptotically stable then (A, C) is necessarily observable.

(iv) Suppose (A, C) is observable and there exists a solution $P \in \mathcal{H}_n(\mathbb{K})$ of (80). Then there exists $x^0 \in \mathbb{K}^n$ with $\langle x^0, Px^0\rangle < 0$ if and only if $\operatorname{Re}\lambda > 0$ (resp. $|\lambda| > 1$) for some $\lambda \in \sigma(A)$.

Proof: The proof is for the continuous time case.
(i) The "if" statement follows from Theorem 3.3.38. Conversely, suppose that (27) is asymptotically stable. Then P defined by (89a) solves (80a) and is positive definite by the observability of (A, C).
(ii) If $P \succeq 0$ is a solution of (80a) then it follows from (67) that

$$e^{A^*t} P e^{At} = P - \int_0^t e^{A^*\rho} C^* C e^{A\rho} d\rho, \quad t \geq 0.$$

Now suppose $x \in \ker P$, $x \neq 0$ then $-\int_0^t \|Ce^{A\rho}x\|^2 d\rho = \langle x, e^{A^*t}Pe^{At}x\rangle \geq 0$ and hence $\langle x, \mathcal{Q}(t,0)x\rangle = \int_0^t \|Ce^{A\rho}x\|^2 d\rho = 0$ for all $t \geq 0$. Thus (A, C) is not observable.

(iii) Suppose that $P \succ 0$ solves (80a) then (27) is stable by Theorem 3.3.33. If additionally $\sigma(A) \subset \mathbb{C}_-$ then $\int_0^\infty e^{A^*\rho}C^*Ce^{A\rho}d\rho = P \succ 0$ and so (A, C) is observable.
(iv) follows from Theorem 3.3.39 and Theorem 3.3.38, see Remark 3.3.40. □

Remark 3.3.50. (i) If $Q \succ 0$ then rank $C = n$ and (92) is automatically satisfied.

(ii) For higher dimensions it is a nontrivial task to solve the linear matrix equation (80). In the next chapter we will describe an algorithm based on the reduction of A to Schur form. An alternative is to make an inspired choice of a $P = P^* \succ 0$ and compute Q from (80). If $Q \succ 0$ ($\succeq 0$) then (27) is asymptotically stable (stable) whereas if this is not the case no conclusion can be drawn.

(iii) If $P \succ 0$ solves (80) with $Q = Q^* \succ 0$ then $V(x) = \langle x, Px\rangle = \langle x, x\rangle_P = \|x\|_P^2$ satisfies

$$\dot{V}(x) = 2\,\mathrm{Re}\langle x, Ax\rangle_P < 0 \quad (\dot{V}(x) = \|Ax\|_P^2 - \|x\|_P^2 < 0), \quad x \in \mathbb{K}^n,\, x \neq 0. \tag{93}$$

So the flow is contracting with respect to the induced norm $\|\cdot\|_P$, i.e. the distance from the origin measured by this norm is continually decreasing along the trajectory.

(iv) If the spectral abscissa $\alpha(A) = 0$ (resp. $\varrho(A) = 1$) and $Q \succ 0$, then there is no solution of (80), see Remark 3.3.40.

(v) Given $r \in \mathbb{R}$ (resp. $r > 0$), $Q = Q^* \succ 0$, let us assume that (88) holds for the matrix $(A - rI)$ (resp. $r^{-1}A$). Then from the above theorem we have that $\mathrm{Re}\,\lambda > r$ (resp. $|\lambda| > r$) for some $\lambda \in \sigma(A)$ if and only if the solution $P_r \in \mathcal{H}_n(\mathbb{K})$ of following equation (94)

$$P(A - rI) + (A - rI)^*P + Q = 0 \tag{94a}$$
$$r^{-2}\, A^*PA - P + Q = 0. \tag{94b}$$

satisfies $\langle x^0, P_r x^0\rangle < 0$ for some $x^0 \in \mathbb{K}^n$. Moreover (88) will hold for all but a finite number of values of r. □

Example 3.3.51. We again consider the linear oscillator studied in Example 3.3.24. For different parameter combinations ($\alpha \in \mathbb{R}, \beta \geq 0$) we will determine the stability properties via the use of Liapunov functions. The Liapunov equation (80a) takes the following form

$$\begin{bmatrix} 0 & 1 \\ -\beta & -2\alpha \end{bmatrix}^* \begin{bmatrix} p_1 & p_2 \\ p_2 & p_3 \end{bmatrix} + \begin{bmatrix} p_1 & p_2 \\ p_2 & p_3 \end{bmatrix} \begin{bmatrix} 0 & 1 \\ -\beta & -2\alpha \end{bmatrix} + \begin{bmatrix} q_1 & q_2 \\ q_2 & q_3 \end{bmatrix} = 0$$

i.e.

$$-2\beta p_2 + q_1 = 0,\ p_1 - 2\alpha p_2 - \beta p_3 + q_2 = 0,\ 2(p_2 - 2\alpha p_3) + q_3 = 0.$$

1. case: $\alpha \neq 0$, $\beta > 0$. In this case we choose $Q = I_2$ and obtain the solution

$$p_1 = \frac{\alpha}{\beta} + \frac{1}{4\alpha}(1+\beta), \quad p_2 = \frac{1}{2\beta}, \quad p_3 = \frac{1+\beta}{4\alpha\beta}.$$

Since $p_1p_3 - p_2^2 > 0$, $P \succ 0$ if and only if $p_1 > 0$ (or $p_3 > 0$). $\langle x, Px\rangle_{\mathbb{R}^2} < 0$ for some $x \in \mathbb{R}^2$ (in fact $-P \succ 0$) if and only if $p_1 < 0$ (or $p_3 < 0$). Thus by Theorem 3.3.49 the system is *asymptotically stable* if $\alpha > 0$, $\beta > 0$ and it is *unstable* if $\alpha < 0$, $\beta > 0$.
If $\alpha = 0$ or $\beta = 0$ there are no solutions of the Liapunov equation when $Q = I_2$, so we examine the modified Liapunov equation (94a).
2. case: $\alpha < 0$, $\beta = 0$. The solution of (94a) with $Q = I_2$ is

$$P_r = \frac{1}{4r(\alpha+r)(2\alpha+r)} \begin{bmatrix} 2(\alpha+r)(2\alpha+r) & 2\alpha+r \\ 2\alpha+r & 2r^2+2\alpha r+1 \end{bmatrix}.$$

Now $2(\alpha+r)(2\alpha+r)(2r^2+2\alpha r+1)-(2\alpha+r)^2 = r(2\alpha+8\alpha^3)+O(r^2)$, so for $\alpha<0$, $r>0$ sufficiently small there exists $x\in\mathbb{R}^2$ such that $\langle x, P_r x\rangle < 0$. Hence by Remark 3.3.50 A has an eigenvalue with $\operatorname{Re}\lambda > r$ for small $r>0$. So there is a $\lambda\in\sigma(A)$ with $\operatorname{Re}\lambda>0$ and hence the system is *unstable*.

If $\alpha\geq 0$, $\beta=0$ it is easily verified from the above formula that $P_r\succ 0$ for $r>0$ and $\langle x^0, P_r x^0\rangle < 0$ for some $x^0\in\mathbb{K}^n$ when $r<0$ is near $r=0$. Hence there is a $\lambda\in\sigma(A)$ with $\operatorname{Re}\lambda=0$. A similar analysis can be carried out for the case $\alpha=0$, $\beta>0$. But as in the case where $\beta=0$, $\alpha\geq 0$ no stability or instability result is obtained (only the existence of $\lambda\in\sigma(A)$ with $\operatorname{Re}\lambda=0$). Thus stability results for these remaining cases cannot be obtained with the choice of $Q=I_2$, even if we use the modified Liapunov equation (94a). In order to proceed using quadratic Liapunov functions we need to make an inspired choice for P (or equivalently Q). The total energy of the oscillator is $\frac{1}{2}(\beta x_1^2+x_2^2)$, so let us consider $P=\begin{bmatrix}\frac{1}{2}\beta & 0\\ 0 & \frac{1}{2}\end{bmatrix}$, with the associated

$$Q=-A^*P-PA=\begin{bmatrix}0 & 0\\ 0 & 2\alpha\end{bmatrix}=\begin{bmatrix}0 & \sqrt{2\alpha}\end{bmatrix}\begin{bmatrix}0\\ \sqrt{2\alpha}\end{bmatrix}=: C^*C. \tag{95}$$

The parameter values we still have to analyze are $\alpha\geq 0$, $\beta=0$ and $\alpha=0$, $\beta\neq 0$, however for these values there is more than one solution of (80a) with Q given by (95).

3. case: $\alpha=0$, $\beta>0$. In this case there are many solutions of (80a)

$$P(\gamma,\delta)=\begin{bmatrix}\beta\delta & \gamma\\ -\gamma & \delta\end{bmatrix},\qquad \gamma,\delta\in\mathbb{R}.$$

which are, in general, non-symmetric. However, $P(0,1)\in\mathcal{H}_2(\mathbb{R})$ is positive definite and so the system is *stable* by Corollary 3.3.46.

4. case: $\beta=0$, $\alpha>0$. In this case there are many symmetric solutions of (80a) with Q given by (95)

$$P(\gamma)=\begin{bmatrix}2\alpha\gamma & \gamma\\ \gamma & \frac{1}{2}+\frac{\gamma}{2\alpha}\end{bmatrix},\qquad \gamma\in\mathbb{R}.$$

Since $P(1)\succ 0$ for $\alpha>0$ we conclude from Theorem 3.3.49 the system is *stable*.

5. case: $\alpha=0$, $\beta=0$. In this case there are again many symmetric solutions of (80a)

$$P(\delta)=\begin{bmatrix}0 & 0\\ 0 & \delta\end{bmatrix},\qquad \delta\in\mathbb{R}$$

and the pair (A,C) is unobservable since $C=0$. We know from Example 3.3.24 that the system is (marginally) unstable in the present case but we cannot infer this result from Theorem 3.3.49.

This example illustrates the usual situation when Liapunov equations are used. Stability or instability can be deduced for most of the parameter values by the choice of $Q=I_n$. However certain combinations of the parameters (associated with the case $\operatorname{Re}\lambda=0$, $\lambda\in\sigma(A)$) require a more subtle analysis. □

We now turn to the time-invariant version of Liapunov's indirect method. Consider the nonlinear equations

$$\dot{x}(t) = f(x(t)),\qquad t\in\mathbb{R}, \tag{96a}$$
$$x(t+1) = f(x(t)),\qquad t\in\mathbb{Z} \tag{96b}$$

where f is Lipschitz continuous (resp. continuous) on an open subset $X \subset \mathbb{K}^n$, $\overline{x} \in X, f(\overline{x}) = 0$ (resp. $f(\overline{x}) = \overline{x}$). In addition, suppose that f is differentiable at $\overline{x}$ and $f'(\overline{x}) = A$, i.e.

$$f(x) = A(x - \overline{x}) + h(x - \overline{x}), \qquad x \in X. \tag{97}$$

and for any $\varepsilon > 0$ there exists $\delta > 0$, such that

$$\|h(x - \overline{x})\| \leq \varepsilon \|x - \overline{x}\|, \qquad x \in B(\overline{x}, \delta). \tag{98}$$

Theorem 3.3.52. *Assume that* (97), (98) *hold for the nonlinear system* (96). *Then*

(i) if $\operatorname{Re}\lambda < 0$ *(*$|\lambda| < 1$*) for all* $\lambda \in \sigma(A)$*, the equilibrium state* $\overline{x}$ *is exponentially stable with respect to the nonlinear system* (96).

(ii) If $\operatorname{Re}\lambda > 0$ *(*$|\lambda| > 1$*) for some* $\lambda \in \sigma(A)$ *then the equilibrium state* $\overline{x}$ *is unstable with respect to the nonlinear system* (96).

Proof: Since (A, I_n) is uniformly observable (i) is an immediate consequence of Theorem 3.3.20 and Theorem 3.3.41. We prove (ii) for the continuous time case leaving the proof for the discrete time case to the reader (Ex. 23). Suppose $\operatorname{Re}\lambda_0 > 0$ for some $\lambda_0 \in \sigma(A)$ and choose $r \in (0, \operatorname{Re}\lambda_0)$ such that (88) holds for $A_r = A - rI_n$. Then (A_r, I_n) is uniformly observable and there exists a solution $P_r \in \mathcal{H}_n(\mathbb{K})$ of (94a). Moreover $\dot{x} = A_r x$ is unstable and so by Theorem 3.3.43 we must have that $\overline{x}$ is unstable for the nonlinear system (96a). □

As an immediate consequence of Theorem 3.3.52 we know that if the equilibrium point $\overline{x}$ of the nonlinear system (96) is unstable then there exists $\lambda \in \sigma(A)$ such that $\operatorname{Re}\lambda \geq 0$ ($|\lambda| \geq 1$), but the linearized system is not necessarily unstable. Conversely if the equilibrium point $\overline{x}$ of the nonlinear system is (asymptotically) stable then necessarily $\operatorname{Re}\lambda \leq 0$ ($|\lambda| \leq 1$), $\lambda \in \sigma(A)$, but we cannot infer that the linearization is (asymptotically) stable. In contrast in the case of *exponential* stability there is a tighter relationship between the behaviour of a nonlinear system near an equilibrium point and its linearization. In order to express this relationship in a succinct way we need the following definition. The solution of (96) with initial state x^0 will be denoted by $\varphi(t; x^0)$, $t \in T(x^0)$.

Definition 3.3.53. Let $r > 0$ be such that $B(\overline{x}, r) \subset X$. The infimum of all $\omega \in \mathbb{R}$ for which there exists $M_\omega \geq 1$ such that

$$x^0 \in B(\overline{x}, r) \Longrightarrow \forall t \in T(x^0) : \|\varphi(t; x^0) - \overline{x}\| \leq M_\omega e^{\omega t} \|x^0 - \overline{x}\| \tag{99}$$

is called the *(upper) growth rate* of the nonlinear system (96) with initial state in $B(\overline{x}, r)$ and is denoted by $\omega(f, \overline{x}, r)$. $\omega(f, \overline{x}) := \lim_{r \searrow 0} \omega(f, \overline{x}, r)$ is said to be the *(upper) growth rate* of (96) *at the equilibrium state* $\overline{x}$.

It follows from the definition that $0 < r_1 < r_2$ implies $\omega(f, \overline{x}, r_1) \leq \omega(f, \overline{x}, r_2)$ and therefore $\omega(f, \overline{x}) = \inf_{r>0} \omega(f, \overline{x}, r)$. By definition $\omega(f, \overline{x}, r) = \infty$ if there does not exist an $M_\omega \geq 1$, $\omega \in \mathbb{R}$ such that (99) holds.

Example 3.3.54. Let $f(x) = Ax, x \in \mathbb{K}^n$ where $A \in \mathbb{K}^{n\times n}$ is given and $\overline{x} = 0$. Then $\omega(f,0,r) = \omega(f,0) = \omega(A)$ for all $r > 0$ where $\omega(A)$ equals the upper Liapunov (or Bohl) coefficient of the semigroup $\Phi(t) = e^{At}$ generated by A, see (30). Hence Definition 3.3.53 generalizes the concept of growth rate as introduced in Subsection 3.3.2 for time-invariant linear systems. □

Theorem 3.3.55. *Assume* (97), (98) *hold for the nonlinear system* (96). *Then the equilibrium point* $\overline{x}$ *is exponentially stable if and only if the linearization at* $\overline{x}$ *is exponentially stable. In this case* $\omega(f,\overline{x}) = \omega(A)$.

Proof: The proof is for the continuous time case, the proof for the discrete time case is set as Ex. 23. Assume $\omega(A) < 0$ and $\beta \in (0, -\omega(A))$. Given $\varepsilon > 0$ choose $\delta > 0$ such that (98) holds and consider the time-varying nonlinear equation

$$\dot{z}(t) = (A+\beta I_n)z(t) + \tilde{h}(t,z(t)), \quad \tilde{h}(t,z) = e^{\beta t}h(e^{-\beta t}z),\ z \in B(0,\delta),\ t \geq 0. \quad (100)$$

Now

$$\|z\| < \delta \Longrightarrow \|\tilde{h}(t,z)\| = e^{\beta t}\|h(e^{-\beta t}z)\| \leq e^{\beta t}\varepsilon\|e^{-\beta t}z\| = \varepsilon\|z\|, \quad t \geq 0. \quad (101)$$

Hence $\tilde{h}(\cdot,\cdot)$ has the property (74) for the pair (ε,δ). Moreover $\sigma(A+\beta I_n) \subset \mathbb{C}_-$ and $(A+\beta I_n, I_n)$ is uniformly observable. So we may apply Theorem 3.3.41 to conclude that there exist positive constants $\tilde{\delta}, \tilde{\varepsilon}, \tilde{M}$ such that

$$\|z(0)\| < \tilde{\delta} \Longrightarrow \|z(t)\| \leq \tilde{M}e^{-\tilde{\varepsilon}t}\|z(0)\|, \quad t \geq 0.$$

Let $x^0 = \overline{x} + z(0)$ and $\varphi(t;x^0) - \overline{x} = e^{-\beta t}z(t)$, $t > 0$, then

$$\|x^0 - \overline{x}\| < \tilde{\delta} \Longrightarrow \|\varphi(t;x^0) - \overline{x}\| \leq \tilde{M}e^{-(\beta+\tilde{\varepsilon})t}\|x^0 - \overline{x}\|, \quad t \geq 0.$$

Moreover for $t > 0$, we have

$$\begin{aligned} \dot{\varphi}(t;x^0) &= -\beta e^{-\beta t}z(t) + e^{-\beta t}[(A+\beta I_n)z(t) + e^{\beta t}h(e^{-\beta t}z(t))] \\ &= A(\varphi(t;x^0) - \overline{x}) + h(\varphi(t;x^0) - \overline{x}). \end{aligned}$$

So $\varphi(t;x^0)$ is the solution of (96a) with initial state x^0. We see, therefore, that $\overline{x}$ is exponentially stable for the system (96a) and its growth rate at $\overline{x}$, $\omega(f,\overline{x}) \leq \omega(A)$. Conversely, assume that $\overline{x}$ is exponentially stable for (96a). Then given $\varepsilon > 0$ and $\omega \in (\omega(f,\overline{x}), 0)$, there exists positive constants δ, M such that (98) holds and

$$\|x^0 - \overline{x}\| < \delta \Longrightarrow \|\varphi(t;x^0) - \overline{x}\| \leq Me^{\omega t}\|x^0 - \overline{x}\|, \quad t \geq 0.$$

Choose $\beta \in (0, -\omega)$ such that $\lambda + \overline{\mu} + 2\beta \neq 0$ for all $\lambda, \mu \in \sigma(A)$ and set $z(t) = e^{\beta t}(\varphi(t;x^0) - \overline{x})$, $t \geq 0$. Then $z(\cdot)$ satisfies (100) with initial state $x^0 - \overline{x}$ and by (101) $\tilde{h}(\cdot,\cdot)$ has the property (74) for the pair (ε,δ). Now if $\lambda_\beta \in \sigma(A+\beta I_n)$, then $\lambda_\beta = \lambda + \beta$ for some $\lambda \in \sigma(A)$. So by the restriction on the choice of β we see that $\lambda_\beta + \overline{\mu}_\beta \neq 0$ for all $\lambda_\beta, \mu_\beta \in \sigma(A+\beta I)$ and hence there exists a solution P of the algebraic Liapunov equation (80a) with A replaced by $A+\beta I_n$ and $Q = I_n$. Finally, since $(A+\beta I_n, I_n)$ is observable, we see that all the conditions of Theorem 3.3.43 for the equation given by (100) are satisfied. But $\|z(t)\| \leq Me^{(\omega+\beta)t}\|z(0)\|, t \geq 0$ and so the the equilibrium point 0 of (100) is exponentially stable. Therefore $A+\beta I_n$ cannot be unstable and $\operatorname{Re}\lambda \leq -\beta$ for all $\lambda \in \sigma(A)$. Thus $\omega(A) \leq \omega(f,\overline{x})$ and this completes the proof. □

Liapunov's indirect method provides a very simple way of determining whether or not an equilibrium state is stable since it relates the nonlinear flow to that of the linearized flow. However it is important to stress that stability or instability of an equilibrium state is a local property and from a practical point of view may give misleading information. For example an equilibrium state may be asymptotically stable but its basin of attraction may be so small that from a practical standpoint one should think of it as being unstable. Similar considerations apply to unstable equilibrium points. Although the construction of Liapunov functions for nonlinear systems may be difficult, the great advantage of Liapunov's direct method is that it provides information about the basin of attraction.

Example 3.3.56. Consider the nonlinear oscillator

$$\ddot{y} + h(y, \dot{y})\dot{y} + g(y) = 0$$

where $g(0) = 0$. Setting $x = [x_1, x_2]^\top = [y, \dot{y}]^\top$, we get the corresponding state space system

$$\dot{x} = \begin{bmatrix} x_2 \\ -g(x_1) - h(x_1, x_2)x_2 \end{bmatrix} := f(x).$$

Since

$$\frac{\partial f}{\partial x}(0) = \begin{bmatrix} 0 & 1 \\ -g'(0) & -h(0,0) \end{bmatrix},$$

the origin will be exponentially stable if and only if $g'(0) > 0$ *and* $h(0,0) > 0$. It will be unstable if either $g'(0) < 0$ *or* $h(0,0) < 0$. □

Example 3.3.57. Let us analyze the stability of an oscillator with nonlinear friction described by the following equation

$$\ddot{\xi} + (2\alpha + \dot{\xi}^2)\dot{\xi} + \beta\xi = 0.$$

Mechanical systems with this equation of motion are used to regulate the angular position ξ of a gyrating mass. The corresponding state space system is

$$\dot{x} = \begin{bmatrix} \dot{x}_1 \\ \dot{x}_2 \end{bmatrix} = \begin{bmatrix} 0 & 1 \\ -\beta & -2\alpha \end{bmatrix} \begin{bmatrix} x_1 \\ x_2 \end{bmatrix} - \begin{bmatrix} 0 \\ x_2^3 \end{bmatrix}, \tag{102}$$

which has one equilibrium state at $(0,0)$. The linearization about this equilibrium state is $\dot{x} = \begin{bmatrix} 0 & 1 \\ -\beta & -2\alpha \end{bmatrix} x$ and we have analyzed the stability of this system in Example 3.3.24. Using these results and Theorem 3.3.52, we are able to conclude that the origin is exponentially stable if $\alpha > 0$, $\beta > 0$ and it is unstable if $\alpha < 0$. If $\alpha = 0$, $\beta > 0$ the origin of the linearized system is a centre and since it is only marginally stable we cannot apply Theorem 3.3.52. In order to obtain information about this case and the basin of attraction when $\alpha \geq 0$, consider the function

$$V(x) = (1/2)(\beta x_1^2 + x_2^2), \quad x \in \mathbb{R}^2.$$

This function associates with any state x, the corresponding total energy of the system. Then $\lim_{\|x\| \to \infty} V(x) = \infty$ and $\dot{V}(x) = -(2\alpha + x_2^2)x_2^2$. The largest invariant subset in $\{x \in \mathbb{R}^2 : \dot{V}(x) = 0\} = \{(x_1, 0); x_1 \in \mathbb{R}\}$ for (102) is $\{(0,0)\}$ when $\alpha \geq 0$, $\beta > 0$. So by Corollary 3.2.29 the origin is asymptotically stable even when $\alpha = 0$, $\beta > 0$. Moreover since every sublevel set $V(x) < \rho$ is bounded the asymptotic stability is global. □

Example 3.3.58. The discrete time system

$$x_1(t+1) = \alpha x_1(t) + x_2^2(t), \quad x_2(t+1) = x_1(t) + \beta x_2(t) \tag{103}$$

has two equilibrium points $\overline{x}^1 = (0,0)$ and $\overline{x}^2 = ((1-\alpha)(1-\beta)^2, (1-\alpha)(1-\beta))$.
The linearized system about $(0,0)$ is given by the matrix $\begin{bmatrix} \alpha & 0 \\ 1 & \beta \end{bmatrix}$ which has eigenvalues α, β. So the equilibrium state $(0,0)$ will be exponentially stable if $|\alpha| < 1$ and $|\beta| < 1$. It will be unstable if $|\alpha|$ or $|\beta|$ is greater than one.
The linearized system about the second equilibrium state is given by the matrix

$$\begin{bmatrix} \alpha & 2(1-\alpha)(1-\beta) \\ 1 & \beta \end{bmatrix}.$$

The characteristic equation is $(\lambda - \alpha)(\lambda - \beta) = 2(1-\alpha)(1-\beta)$. The shaded region in Figure 3.3.5 corresponds to those values of α, β for which this equilibrium state is exponentially stable. The boundaries $\alpha = 1$, $\beta = 1$ are obtained when $\lambda = +1$, the

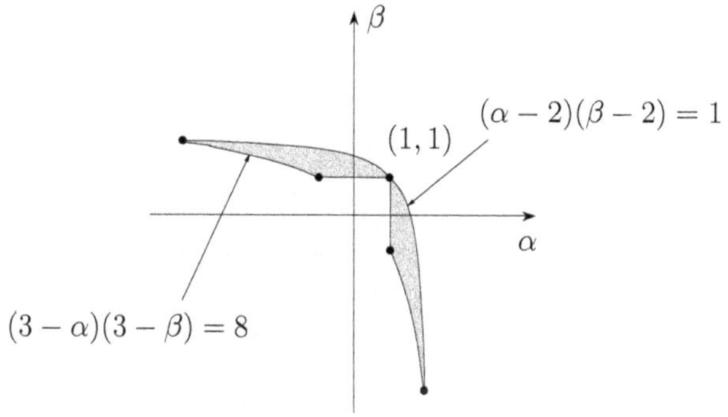

Figure 3.3.5: Stability chart for $\overline{x}^2$ with respect to (103)

boundary $(3-\alpha)(3-\beta) = 8$ is obtained when $\lambda = -1$ and the other part of the boundary is determined by setting $\lambda = e^{\imath\theta}$ with $\cos\theta = (\alpha+\beta)/2$. Note that the first equilibrium point is unstable if the second one is exponentially stable. □

3.3.6 Exercises

1. Prove that the growth rate $\omega(A) = \inf\{\omega \in \mathbb{R}; \exists M > 0 : \|\Phi(t)\| \le Me^{\omega t}\}$ of a continuous (resp. discrete) time semigroup Φ with generator $A \in \mathbb{C}^{n\times n}$ is equal to the spectral abscissa $\alpha(A) = \max_{\lambda\in\sigma(A)} \operatorname{Re}\lambda$ (resp. the logarithm of the spectral radius $\ln \varrho(A) = \ln \max_{\lambda\in\sigma(A)} |\lambda|$).

2. If $A = \begin{bmatrix} 0 & 1 \\ -2 & -2 \end{bmatrix}$ show that

$$e^{At} = e^{-t}\begin{bmatrix} \cos t + \sin t & \sin t \\ -2\sin t & \cos t - \sin t \end{bmatrix}, \quad A^t = (\sqrt{2})^t \begin{bmatrix} \cos\frac{\pi}{4}t - \sin\frac{\pi}{4}t & \sin\frac{\pi}{4}t \\ -2\sin\frac{\pi}{4}t & \sin\frac{\pi}{4}t + \cos\frac{\pi}{4}t \end{bmatrix}.$$

Determine

$$\lim_{t\to\infty} \frac{\ln\|e^{At}\|}{t}, \quad \text{and} \quad \lim_{t\to\infty} \frac{\ln\|A^t\|}{t}.$$

3. Find the continuous time evolution operator generated by $a(t) = t\sin t$ on $T = \mathbb{R}_+$. Show that the upper Liapunov exponent is $+1$, whereas the upper Bohl exponent is not finite.

4. Show the upper Liapunov exponent for a continuous time system (1) is finite if $\sup_{t\in\mathbb{R}_+} \int_t^{t+1} \|A(s)\| ds < \infty$.

5. Show that the Liapunov exponent of (1) is given by

$$\overline{\alpha}(\Phi) = \limsup_{t\to\infty} \frac{\ln\|\Phi(t,0)\|}{t} \quad \text{(where } \ln\|\Phi(t,0)\| = -\infty \text{ if } \|\Phi(t,0)\| = 0\text{)}.$$

6. Prove Theorem 3.3.15 in the discrete time case.

7. Consider

$$A(t) = \begin{bmatrix} -11/2 + (15/2)\sin 12t & (15/2)\cos 12t \\ (15/2)\cos 12t & -11/2 - (15/2)\sin 12t \end{bmatrix}.$$

Show that the system $\dot{x} = A(t)x$ is exponentially stable even though $\sigma(A(t)) = (2, -13)$ for all $t \geq 0$.

8. Prove that every scalar system $\dot{x}(t) = a(t)x(t)$, $t \in \mathbb{R}_+$ (resp. $x(t+1) = a(t)x(t)$, $t \in \mathbb{N}$) which has a finite upper Bohl exponent β, can be transformed via a Bohl transformation $\theta(t) = e^{\int_0^t (a(s)-\beta)ds}$ (resp. $\theta(t) = e^{-\beta t}\prod_{s=0}^{t-1} a(s)$) into the time invariant system $\dot{x}(t) = \beta x(t)$, (resp. $x(t+1) = e^{\beta}x(t)$).

9. Consider the scalar system $\dot{x}(t) = (4t\sin t - 2t)x(t)$, $t > t_0$, $x(t_0) = x_0$, $t_0 \in \mathbb{R}_+$. Prove that the solution is

$$\Phi(t,t_0)x_0 = x_0 \exp(4\sin t - 4t\cos t - t^2 - 4\sin t_0 + 4t_0\cos t_0 + t_0^2).$$

Hence show that the origin is asymptotically stable at any time $t_0 \in \mathbb{R}_+$. Prove that $\Phi((2n+1)\pi, 2n\pi) = \exp((4n+1)\pi(4-\pi))$ for $n \in \mathbb{N}$ and so the origin is not uniformly asymptotically stable.

10. If $a(t) = -(1+t)^{-1}$, $q(t) = 2(1+t)^{-1} - 3(1+t)^{-2}$ for $t \in \mathbb{R}_+$ show that a solution of the Liapunov equation (54a) is $p(t) = 1 - (1+t)^{-1}$. What conclusion can be drawn about the stability properties of the evolution operator generated by $a(\cdot)$ on $\mathbb{R}_+$.

11. Let $A(t) = \begin{bmatrix} -2+\cos t & -\sin t \\ -\sin t & -2-\cos t \end{bmatrix}$, $t \in \mathbb{R}_+$, choose $P(t) \equiv I_2$ and compute $Q(t)$ such that the Liapunov equation (54a) is satisfied. Use this to prove that the evolution operator generated by $A(\cdot)$ is uniformly exponentially stable.

12. Let $A(t) = \begin{bmatrix} 0 & 1 \\ -a(t) & -1/2 \end{bmatrix}$, $t \in \mathbb{N}$, choose $P(t) = \begin{bmatrix} a(t)^2 + 1/4 & 0 \\ 0 & 1 \end{bmatrix}$ and compute $Q(t)$ such that the discrete time Liapunov equation (54b) is satisfied. Hence show that the evolution operator generated by $A(\cdot)$ is uniformly exponentially stable if $|a(t)| < 1/2, t \in \mathbb{N}$.

13. Prove the Instability Theorem 3.3.34

14. Show that the unique solution of (54b) on T_{t_0} with initial state $P(t_0)$ is

$$P(t) = \Phi(t_0,t)^* P(t_0)\Phi(t_0,t) - \sum_{s=t_0}^{t-1} \Phi(s,t)^* Q(s)\Phi(s,t), \quad t \geq t_0.$$

15. Suppose $A(t) = \begin{bmatrix} t & 0 \\ 0 & 0 \end{bmatrix}$, $t \in \mathbb{R}$. Show that for $Q(t) = I_2$ there is not a bounded solution of (54a). However for every $r > 0$ if $A_r(t) = A(t) - rI_2$ there are bounded solutions of (54a) on $T_{1/2+r}$ when $A(t)$ is replaced by $A_r(t)$.

16. Determine conditions on a, b, ρ for which the equilibrium state is asymptotically stable for Goodwin's model of supply and demand considered in Example 1.2.1.

17. Determine whether or not the following matrices correspond to asymptotically stable systems in the continuous time and discrete time cases

$$\text{(a)} \begin{bmatrix} 0 & \frac{1}{2} \\ \frac{1}{2} & 0 \end{bmatrix} \quad \text{(b)} \begin{bmatrix} -1 & 1 \\ 1 & -2 \end{bmatrix} \quad \text{(c)} \begin{bmatrix} 0 & 1 \\ -\frac{1}{8} & -\frac{1}{2} \end{bmatrix}.$$

Verify your conclusions by solving the Liapunov equations with $Q = I_2$.

18. Find the linearized equations of motion about the equilibrium states in Ex. 2.5, 2.6. Determine whether or not these systems are asymptotically stable.

19. Show that the system $\dot{x} = \alpha x^3$ is asymptotically stable if $\alpha < 0$ and unstable if $\alpha > 0$. Note that the linearized system about the origin is marginally stable for all $\alpha \in \mathbb{R}$. This example shows that no conclusions for the stability with respect to the nonlinear system can be drawn from this fact.

20. Consider the discrete time system with matrix

$$A = \begin{bmatrix} 0 & 1 \\ -a_1 & -a_0 \end{bmatrix}.$$

Determine the values of a_0, a_1 for which there is not a unique solution of the Liapunov equation (80b). Solve the Liapunov equation when $Q = I_2$ and hence determine those values of a_0, a_1 for which the system is asymptotically stable, marginally stable or unstable. Solve the modified Liapunov equation (94b). What further conclusions can be drawn? Compare your results with those given in Example 3.3.25.

21. Suppose that the system $\dot{x} = Ax$ is asymptotically stable, where $A \in \mathbb{R}^{n \times n}$. For a step size $\tau > 0$ consider the following discretizations

$$\text{(a)} \quad \frac{x^\tau(t+1) - x^\tau(t)}{\tau} = Ax^\tau(t), \quad \text{(b)} \quad \frac{x^\tau(t+1) - x^\tau(t)}{\tau} = Ax^\tau(t+1), \quad t \in \mathbb{N}.$$

Show that the system in (b) is necessarily asymptotically stable but the system in (a) need not be asymptotically stable.

22. If $\Phi(\cdot,\cdot)$ is generated by $a(t) = -t$, $t \in \mathbb{R}_+$ show that for every $\tau > 0$, $\int_{t_0}^{t_0+\tau} \Phi(s,t_0)^2 ds \leq (2t_0)^{-1}$, $t_0 \geq 0$. This example shows that the pair $(a(\cdot), 1)$ is not uniformly observable.

23. Prove Theorems 3.3.43, 3.3.52 and 3.3.55 for the discrete time case.

24. Consider the system

$$\dot{x}_1 = x_2, \quad \dot{x}_2 = -2x_1 - 3x_2 - h(x_1 + x_2)$$

where $h : \mathbb{R} \to \mathbb{R}$ is continuous $h(0) = 0$ and $xh(x) > 0$ for all $x \neq 0$.

(i) Find a matrix $P \in \mathbb{R}^{n\times n}$ such that

$$PA + A^\top P + 4I = 0 \quad \text{where } A = \begin{bmatrix} 0 & 1 \\ -2 & -3 \end{bmatrix}.$$

(ii) Use the function $V(x) = x^\top Px + \int_0^{x_1+x_2} h(s)\,ds$ to show that the origin is asymptotically stable.

25. Show that for the Lur'e problem (2.58) the function V (2.60) has the property

$$\dot{V}(x,\sigma) = -\langle x, Qx\rangle - rf^2(\sigma) + 2f(\sigma)\langle Pb + (1/2)c, x\rangle$$

where $PA + A^\top P + Q = 0$. Hence prove that if $\sigma(A) \subset \mathbb{C}_-$, $f(0) = 0$, $\sigma \neq 0 \Rightarrow \sigma f(\sigma) > 0$ and

$$r > \langle Pb + (1/2)c, Q^{-1}(Pb + (1/2)c)\rangle,$$

then the origin $x = 0$, $\sigma = 0$ is asymptotically stable.

26. Newton's method for solving the equation $F(x) = 0$, where $F : \mathbb{R}^n \to \mathbb{R}^n$ is differentiable, is the iterative scheme

$$x(t+1) = x(t) - [F'(x(t))]^{-1} F(x(t)) \quad t \in \mathbb{N}$$

where the inverse is assumed to exist.
This method is used to solve the scalar equation $e^{-x} - x = 0$, so that

$$x(t+1) = x(t) + \frac{e^{-x(t)} - x(t)}{e^{-x(t)} + 1}, \quad t \in \mathbb{N}. \tag{104}$$

If $\overline{x}$ is the required unique solution find the linearized equations about $\overline{x}$ and show that $\overline{x}$ is an exponentially stable equilibrium state of (104). Use the function $V(x) = |x - \overline{x}|$ to obtain an estimate for the basin of attraction of this equilibrium state.

27. The second order differential system $\ddot{\xi} + \alpha\dot{\xi} + \beta\xi = 0$ is asymptotically stable if and only if $\alpha > 0$, $\beta > 0$. Use the Cayley transform to obtain necessary and sufficient conditions for the second order difference equation

$$\xi(t+2) + a\,\xi(t+1) + b\,\xi(t) = 0, \quad t \in \mathbb{N}$$

to be asymptotically stable.

28. Case study: A model for a continuous flow stirred tank reactor is given by

$$\begin{aligned} \dot{T} &= a(T_0 - T) + bkCe^{-\alpha/T} \\ \dot{C} &= a(C_0 - C) - kCe^{-\alpha/T} \end{aligned}$$

where C_0, T_0 are the concentration and temperature of the reactant in the influent and C, T are the concentration and temperature of the reactant in the effluent. a, b, α, k are positive constants.

(i) Show that all equilibrium states (C_e, T_e) satisfy

$$1 + \frac{a}{k}e^{\alpha/T_e} = \frac{bC_0}{T_e - T_0}, \qquad \frac{C_e - C_0}{T_e - T_0} = -\frac{1}{b}.$$

(ii) Linearize the equations about an equilibrium state and hence show that an equilibrium state is stable if

$$\frac{C_0}{T_e - T_0} > \frac{\alpha C_e}{T_e^2}.$$

(iii) If $T_0 = 300$, $C_0 = 10$, $a = 2^{-9}$, $b = 30$, $k = 0.5$, $\alpha = 3600 \ln 2$, find three equilibrium states and determine whether or not they are stable. Are there any other equilibrium states?

(iv) Use a computer to obtain a phase portrait of the system around the three equilibrium points.

3.3.7 Notes and References

Many of the results for time-varying linear systems can be found in *Daleckii and Krein* (1974) [118]. The notion of Bohl exponent is due to *Bohl* (1913) [65]. The proof of Theorem 3.3.15 is given in [118] and was proved for the case $p = 2$ in [120]. Our proof is based on that of [115]. Many of the books quoted in Section 3.2 contain results for time-varying systems. For further results on *time-varying* Liapunov transformations, see *Gantmacher* (1959 Vol. 2) [183].

The result that the growth rate of a strongly continuous semigroup is $\sup_{\lambda \in \sigma(A)} \text{Re}\,\lambda$ is known as the *spectrum determined growth condition.* It holds for a large class of strongly continuous semigroups on infinite dimensional Banach spaces. However it is not true in general, see *Zabczyk* (1975) [540] for a counterexample with $\sup_{\lambda \in \sigma(A)} = 0$ yet $\|S(t)\| = e^t$. For a discussion of numerical stability of discretization methods see for example *Stoer and Bulirsch* (1978) [485] and the references in Section 4.5.

The quadratic Liapunov function for linear systems was introduced in Liapunov's original work and many of the results in Subsection 3.3.4 and Subsection 3.3.5 can be found there. A good account can also be found in *Barbashin* (Translation 1970) [33]. Extensions of Liapunov's result which relate the inertia $i(A)$ to the inertia $i(P)$ where $PA + A^*P + Q = 0$, are called *inertia theorems* see *Carlson and Schneider* (1963) [91], *Wimmer* (1975) [531], *Glover* (1984) [188] and *Datta* (1999) [123].

Generalizations of Liapunov's Theorem 3.3.33 to infinite dimensional systems, continuous or discrete time have been obtained by *Datko* (1970) [119] and *Zabczyk* (1974) [539].

In the late 60's determining stability domains via Liapunov functions (or otherwise) was much in vogue and there have been many such attempts for the Mathieu equation. For example *Narenda and Taylor* (1973) [387] obtained the stability domain $\pi a \zeta / 2 > |q|$, $a \gg \zeta^2$, $\zeta \ll 1$.

The linearization result in Subsection 3.3.5 is essentially due to Liapunov and the fact that an equilibrium point is exponentially stable if and only if the linearization at the equilibrium point is exponentially stable can be found in *Zabczyk* (1992) [541].

3.4 Stability Criteria for Polynomials

We have seen that for the asymptotic stability of continuous time systems (3.1a) (resp. discrete time systems (3.1b)) it is required that all the eigenvalues lie in $\mathbb{C}_-$ (resp. $\mathbb{D}$). These spectral stability criteria were already known in the 19th century. However, in the absence of systematic solution procedures for algebraic equations of order $n \geq 5$ and without computers for their approximate solution these spectral criteria could only be verified for lower dimensional systems. It was therefore a problem of fundamental importance, both for mathematical stability theory and its applications, to express the spectral stability criteria by verifiable conditions on the coefficients of the characteristic polynomial. This problem was stated by Maxwell in 1868. In mathematics the analysis of real algebraic equations was one of the driving forces in algebra and analysis in the early decades of the 19th century. Many leading mathematicians contributed to this field and developed methods for determining the number of roots of a polynomial in certain locations of the complex plane (e.g. the real axis, the upper half-plane). Some further historical comments are given in the extended *Notes and References.* The purpose of this section is to present some of the most important methods and results which have been obtained in this field.

In the first two subsections we will deal with analytic methods and results. In Subsection 3.4.1 we obtain stability criteria by the argument principle. Both real and complex polynomials are considered. Important results are the *Hermite-Biehler Theorem* and a recursive stability test due to *Routh* who was the the winner of the Adams Prize mentioned in the introduction to this chapter. In Subsection 3.4.2 we characterize stable polynomials by the *Cauchy index* of an associated real rational function.

The algebraic methods we present in this section are based on quadratic forms. The idea of using quadratic forms for the root location of polynomials is due to *Hermite* (1856) [226]. In Subsection 3.4.3 it is shown that the number of roots of a polynomial in $\mathbb{C}_-$ can be obtained from the *Hermite form.* Moreover we will discuss its relationship with another important quadratic form in this context, the *Bézoutiant.* In Subsection 3.4.4 the *Hankel form* will be introduced. Hankel matrices play an important role, not only in stability analysis, but also in other areas of systems theory such as *realization theory* and *model reduction.* We will prove *Kronecker's Theorem* which is fundamental in realization theory, and show that the Cauchy index of a real rational function can be expressed by the signature of an associated Hankel matrix. These results will then be used to derive Brockett's Theorem on the connected components of the space of real rational functions of order n. In Subsection 3.4.5 we prove the classical stability criteria of *Liénard–Chipart* and *Hurwitz.* Subsection 3.4.6 is dedicated to the discrete time case, i.e. the problem of characterizing those polynomials whose roots are all located in $\mathbb{D}$ (Schur polynomials). Counterparts of the Hermite form, the Bézoutiant and the Hankel form are introduced, the *Schur-Cohn Theorem* is proved and a recursive stability test similar to the Routh test is presented. We conclude the chapter with a unifying framework for obtaining stability criteria with respect to a large class of algebraic stability regions (including arbitrary open half-planes and disks) and derive for these stability regions a Liapunov-type stability criterion.

3.4.1 Stability Criteria and the Argument Principle

In the next five subsections we will derive stability criteria for real and complex polynomials with respect to the open left half plane $\mathbb{C}_-$. A polynomial p is said to be a *Hurwitz polynomial* or*Hurwitz stable* if all the roots s_i of p lie in the open left half plane $\mathbb{C}_-$. We begin in this subsection by applying some methods from complex analysis and we will see that the argument function plays a central role. First we deal with arbitrary complex polynomials and then later derive special stability criteria for real ones. We conclude this subsection with the Routh test for Hurwitz stability and its application to a problem of feedback stabilization first studied by Maxwell.

Complex polynomials

Given any continuous function $f : \imath\mathbb{R} \to \mathbb{C}^* := \mathbb{C} \setminus \{0\}$, we denote by $\Delta_a^b \arg f(\imath\omega)$ the change of the argument of the arc $\gamma : \omega \mapsto f(\imath\omega)$ on $[a, b]$, see Section A.2. The *change of the argument of f along the imaginary axis* is defined by

$$\Delta_{-\infty}^{\infty} \arg f(\imath\omega) := \lim_{k\to\infty} \Delta_{-k}^{k} \arg f(\imath\omega).$$

The following example illustrates how the Hurwitz stability of a linear polynomial can be characterized by its change of the argument along $\imath\mathbb{R}$.

Example 3.4.1. Let $s_1 = \rho_1 + \imath\omega_1 \in \mathbb{C} \setminus \imath\mathbb{R}$. We want to determine the change of the argument of $p(s) = s - s_1$ along the imaginary axis. It is intuitively clear (see Figure 3.4.1) that

$$\Delta_{-\infty}^{\infty} \arg p(\imath\omega) = \begin{cases} \pi & \text{if } \rho_1 < 0 \\ -\pi & \text{if } \rho_1 > 0 \end{cases}.$$

(As ω moves from $-\infty$ to ∞, $\arg p(\imath\omega)$ changes continuously from $-\pi/2$ to $\pi/2$ if $\operatorname{Re} s_1 < 0$

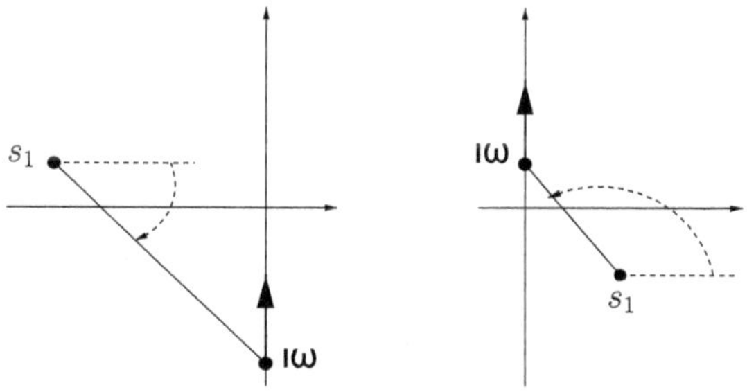

Figure 3.4.1: $\arg(\imath\omega - s_1)$ increases if $\operatorname{Re} s_1 < 0$, decreases if $\operatorname{Re} s_1 > 0$

and from $3\pi/2$ to $\pi/2$ if $\operatorname{Re} s_1 > 0$). We will verify that this agrees with our definition of the *change of the argument* in the Section A.2. Denoting the arc $\omega \mapsto p(\imath\omega)$ on $[-k, k]$ by γ_k we have by (A.2.8)

$$\Delta_{-k}^{k} \arg p(\imath\omega) = \operatorname{Im} \int_{\gamma_k} \frac{ds}{s} = \int_{-k}^{k} \operatorname{Im} \frac{\imath\, d\omega}{\imath\omega - (\rho_1 + \imath\omega_1)} = \int_{-k}^{k} \frac{-\rho_1\, d\omega}{\rho_1^2 + (\omega - \omega_1)^2} = \int_{\frac{k-\omega_1}{\rho_1}}^{-\frac{k+\omega_1}{\rho_1}} \frac{dt}{1+t^2}.$$

Hence

$$\Delta_{-\infty}^{\infty} \arg p(\imath\omega) = \lim_{k\to\infty} \Delta_{-k}^{k} \arg p(\imath\omega) = \int_{\mp\infty}^{\pm\infty} \frac{dt}{1+t^2} = [\arctan t]_{\mp\infty}^{\pm\infty} = \pm\pi$$

according to whether $\rho_1 < 0$ or $\rho_1 > 0$. □

In order to extend this criterion to arbitrary complex polynomials we need the following lemma.

Lemma 3.4.2. *Let $D \subset \mathbb{C}$ be open, $f, g : D \to \mathbb{C}$ holomorphic and $\alpha : [a,b] \to D$ an integration path which does not hit any zero of f or g. Then*

$$\Delta_a^b \arg(fg)(\alpha(t)) = \Delta_a^b \arg f(\alpha(t)) + \Delta_a^b \arg g(\alpha(t)).$$

Proof: The proof follows directly from (A.2.10) since

$$\frac{(fg)'(s)}{(fg)(s)} = \frac{f'(s)g(s) + f(s)g'(s)}{(fg)(s)} = \frac{f'(s)}{f(s)} + \frac{g'(s)}{g(s)}, \quad s \in D,\ (fg)(s) \neq 0.$$

□

The next proposition shows that the number of roots of a polynomial which lie in the left half plane can be determined by the change of the argument of the polynomial along the imaginary axis.

Proposition 3.4.3. *Given a polynomial $p(s) \in \mathbb{C}[s]$ of degree n without zeros on the imaginary axis. Then*

$$\Delta_{-\infty}^{\infty} \arg p(\imath\omega) = (n - 2\nu)\pi$$

where ν is the number of zeros of $p(s)$ in the open right half-plane (taking account of multiplicities). In particular, $p(s)$ is Hurwitz stable if and only if $\Delta_{-\infty}^{\infty} \arg p(\imath\omega) = n\pi$.

Proof: Let $p(s) = a_n \prod_{j=1}^{n}(s - s_j)$. Then by Example 3.4.1 and Lemma 3.4.2

$$\Delta_{-\infty}^{\infty} \arg p(\imath\omega) = \lim_{k\to\infty} \sum_{j=1}^{n} \Delta_{-k}^{k} \arg(\imath\omega - s_j) = (n-\nu)\pi - \nu\pi.$$

□

An alternative proof of this proposition can be given by applying the Argument Principle (see Section A.2) to p on a semicircle in the right half-plane with centre at the origin and radius $r \to \infty$, the reader is asked to prove this in Ex. 1.

Having determined the overall change of the argument of $p(\imath\omega)$ along the imaginary axis, we will now analyze the *rate of change* of the argument. Given a polynomial $p(s) \in \mathbb{C}[s]$, let $\varphi(\omega) = \arg p(\imath\omega)$ be any argument function for the curve $\gamma : \omega \mapsto \gamma(\omega) = p(\imath\omega)$, $\omega \in I$ where $I \subset \mathbb{R}$ is an interval such that p does not have a zero on the segment $\imath I$ of the imaginary axis. Since $\gamma'(\omega) = \left.\frac{dp(s)}{ds}\right|_{s=\imath\omega} \frac{d(\imath\omega)}{d\omega} = \imath p'(\imath\omega)$ we have by (A.2.6)

$$\frac{d \arg p(\imath\omega)}{d\omega} = \frac{d\varphi(\omega)}{d\omega} = \operatorname{Im} \frac{\gamma'(\omega)}{\gamma(\omega)} = \operatorname{Re} \frac{p'(\imath\omega)}{p(\imath\omega)}, \quad \omega \in I. \tag{1}$$

Sometimes it is useful to express this rate of change of the argument directly in terms of the real and imaginary parts of $p(\imath\omega)$. Given any complex polynomial $p(s) = \sum_{k=0}^{n} a_k s^k$ we denote by $\overline{p}(s)$ the polynomial with conjugate complex coefficients: $\overline{p}(s) = \sum_{k=0}^{n} \overline{a_k} s^k$.

Definition 3.4.4. Let $n \in \mathbb{N}$ be fixed. For every $p(s) = \sum_{i=0}^{n} a_i s^i \in \mathbb{C}[s]$, the *Hurwitz-reflection* of p is defined by

$$p^\star(s) = \overline{p}(-s).$$

$p(s) \in \mathbb{C}[s]$ is called *Hurwitz-symmetric* if $p = p^\star$.

For Hurwitz stability, the Hurwitz-symmetric polynomials play the role of real polynomials. In fact, if p is symmetric and $s \in \imath\mathbb{R}$, then $-s = \overline{s}$ and $\overline{p(s)} = \overline{p}(\overline{s}) = \overline{p}(-s) = p^\star(s) = p(s)$, hence every symmetric polynomial is real on the imaginary axis, the boundary of the Hurwitz stability region $\mathbb{C}_-$.
The reflection operator $p \mapsto p^\star$ is an $\mathbb{R}$-linear degree preserving bijection of $\mathbb{C}[s]$ onto itself with $(ap)^\star = \overline{a}p^\star$ for all $a \in \mathbb{C}$ and $(pq)^\star = p^\star q^\star$. Moreover $p \mapsto p^\star$ is involutive, i.e. $(p^\star)^\star = p$. For polynomials of degree $\leq n$ the reflection operator induces the following transformation of the coefficient vectors

$$\star_n : a = (a_0, a_1, \ldots, a_n) \mapsto a^\star = (\overline{a_0}, -\overline{a_1}, ..., (-1)^n \overline{a_n}), \quad a \in \mathbb{C}^{n+1}.$$

Every polynomial $p(s) \in \mathbb{C}[s]$ can be decomposed into a *symmetric part* p_+ and an *antisymmetric part* $\imath p_-$ (satisfying $(\imath p_-)^\star = -\imath p_-$) as follows

$$p(s) = p_+(s) + \imath p_-(s) \text{ where } p_+(s) = (p(s) + p^\star(s))/2, \; p_-(s) = (p(s) - p^\star(s))/(2\imath).$$

The symmetric polynomials p_+, p_- have real values on $\imath\mathbb{R}$ and so the polynomials

$$p_R(s) := (p(\imath s) + p^\star(\imath s))/2 = p_+(\imath s), \quad p_I(s) := (p(\imath s) - p^\star(\imath s))/(2\imath) = p_-(\imath s). \tag{2}$$

are real on $\mathbb{R}$ and hence have real coefficients. p and $p^\star$ can be expressed by these real polynomials as follows

$$p(\imath s) = p_R(s) + \imath p_I(s) \quad \text{and} \quad p^\star(\imath s) = p_R(s) - \imath p_I(s), \qquad s \in \mathbb{C}. \tag{3}$$

For $\omega \in \mathbb{R}$, $p_R(\omega)$ and $p_I(\omega)$ are the real and imaginary parts of $p(\imath\omega)$

$$p_R(\omega) = \operatorname{Re} p(\imath\omega) \quad \text{and} \quad p_I(\omega) = \operatorname{Im} p(\imath\omega), \qquad \omega \in \mathbb{R}. \tag{4}$$

The rate of change of the argument of $p(\imath\omega)$ can now be expressed in terms of the real and imaginary parts of $p(\imath\omega)$ by

$$\frac{d \arg p(\imath\omega)}{d\omega} = \operatorname{Im} \frac{d(p_R(\omega) + \imath p_I(\omega))/d\omega}{p_R(\omega) + \imath p_I(\omega)} = \frac{p_R(\omega)p_I'(\omega) - p_I(\omega)p_R'(\omega)}{p_R(\omega)^2 + p_I(\omega)^2}, \quad \omega \in I. \tag{5}$$

Proposition 3.4.5 (Phase increasing property). *If $p(s) \in \mathbb{C}[s]$ is a complex Hurwitz polynomial of degree $n \geq 1$ then $\frac{d}{d\omega} \arg p(\imath\omega) > 0$ for all $\omega \in \mathbb{R}$, i.e.*

$$p_R(\omega)p_I'(\omega) - p_I(\omega)p_R'(\omega) > 0. \tag{6}$$

Proof: Suppose $p(s) = a_n \prod_{j=1}^{n} (s - s_j)$ and $\omega \in \mathbb{R}$. Then by (1)

$$\frac{d \arg p(\imath\omega)}{d\omega} = \operatorname{Re} \frac{p'(\imath\omega)}{p(\imath\omega)} = \sum_{j=1}^{n} \operatorname{Re}\left(\frac{1}{\imath\omega - s_j}\right).$$

Hence it suffices to show that $\operatorname{Re}\left(\frac{1}{\imath\omega - (\rho_1 + \imath\omega_1)}\right) > 0$ for all $\rho_1, \omega_1 \in \mathbb{R}, \rho_1 < 0$. But this follows from

$$\operatorname{Re} \frac{1}{-\rho_1 + \imath(\omega - \omega_1)} = \operatorname{Re} \frac{-\rho_1 - \imath(\omega - \omega_1)}{\rho_1^2 + (\omega - \omega_1)^2} = \frac{-\rho_1}{\rho_1^2 + (\omega - \omega_1)^2} > 0. \tag{7}$$

Now equation (6) is a direct consequence of (5). □

Real Polynomials

For real polynomials, some of the stability criteria we have derived can be simplified or strengthened. Moreover, we can prove some additional stability criteria which only hold for this case.

A simple, necessary (but not sufficient) criterion for a real polynomial p to be Hurwitz stable is that all its coefficients are non-zero and of the same sign. In particular, if p is a monic real Hurwitz polynomial all its coefficients must be positive.

Proposition 3.4.6. *If a real polynomial $p(s) = \sum_{j=0}^{n} a_j s^j$ of degree n is Hurwitz stable, then*

$$a_i\, a_j > 0, \quad i, j \in \underline{n}.$$

Proof: Suppose $p(s) = a_n \prod_{j=1}^{n}(s - s_j)$ is Hurwitz stable. By hypothesis for each s_j, either s_j is real and $s_j < 0$, or s_j is complex in which case its complex conjugate $\overline{s}_j$ is also a root, and

$$(s - s_j)(s - \overline{s}_j) = s^2 - 2(\operatorname{Re} s_j)\, s + |s_j|^2$$

with $\operatorname{Re} s_j < 0$. Hence all the coefficients of the polynomial $\prod_{j=1}^{n}(s - s_j)$ will be positive, i.e. all coefficients of $p(s) = a_n \prod_{j=1}^{n}(s - s_j)$ are non-zero and have the same sign as a_n. □

For a real polynomial $p(s) \in \mathbb{R}[s]$ we have $p(-\imath\omega) = \overline{p(\imath\omega)}$ and so $\Delta_{-\infty}^{0} \arg p(\imath\omega) = \Delta_0^{\infty} \arg p(\imath\omega)$. Hence the formula in Proposition 3.4.3 can be replaced by

$$\Delta_0^{\infty} \arg p(\imath\omega) = (n - 2\nu)\pi/2$$

and we obtain the following special version of the stability criterion for real polynomials

$$p(s) \in \mathbb{R}[s] \text{ is Hurwitz stable} \Leftrightarrow \Delta_0^{\infty} \arg p(\imath\omega) = n\pi/2. \tag{8}$$

The phase increasing property can be strengthened for *real* Hurwitz polynomials.

Proposition 3.4.7 (Phase increasing property). *If $p(s) \in \mathbb{R}[s]$ is a real Hurwitz polynomial then*

$$\frac{d \arg p(\imath\omega)}{d\omega} \geq \left|\frac{\sin(2 \arg p(\imath\omega))}{2\omega}\right|, \qquad \omega \in \mathbb{R}^*, \tag{9}$$

with equality if $\deg p = 1$ *and strict inequality if* $\deg p \geq 2$.

Proof: First consider a linear polynomial $p(s) = s - \rho_1 \in \mathbb{R}[s]$, $\rho_1 < 0$. Then

$$\left|\frac{\sin(2\arg p(\imath\omega))}{2\omega}\right| = \left|\frac{\sin(\arg p(\imath\omega))\cos(\arg p(\imath\omega))}{\omega}\right| = \frac{\omega}{\omega\sqrt{\rho_1^2+\omega^2}}\frac{-\rho_1}{\sqrt{\rho_1^2+\omega^2}} = \frac{-\rho_1}{\rho_1^2+\omega^2}.$$

Hence, applying (7) with $\omega_1 = 0$, we see that (9) holds with equality.
Now consider $p(s) = (s-\rho_1)(s-\rho_2) = s^2 - (\rho_1+\rho_2)s + \rho_1\rho_2$ with $\rho_1, \rho_2 < 0$. Then

$$\frac{d\arg p(\imath\omega)}{d\omega} = \operatorname{Re}\frac{1}{\imath\omega-\rho_1} + \operatorname{Re}\frac{1}{\imath\omega-\rho_2} = \frac{-\rho_1}{\rho_1^2+\omega^2} + \frac{-\rho_2}{\rho_2^2+\omega^2}.$$

But $2\arg p(\imath\omega) = 2(\alpha+\beta)$ where $\alpha = \arg(\imath\omega - \rho_1)$, $\beta = \arg(\imath\omega-\rho_2)$, hence

$$\begin{aligned}\left|\frac{\sin(2\arg p(\imath\omega))}{2\omega}\right| &\le \frac{|\sin 2\alpha|}{2|\omega|}|\cos 2\beta| + \frac{|\sin 2\beta|}{2|\omega|}|\cos 2\alpha| \\ &= \frac{|\sin\alpha||\cos\alpha|}{|\omega|}|\cos 2\beta| + \frac{|\sin\beta||\cos\beta|}{|\omega|}|\cos 2\alpha| \\ &\le \frac{-\rho_1}{\rho_1^2+\omega^2}|\cos 2\beta| + \frac{-\rho_2}{\rho_2^2+\omega^2}|\cos 2\alpha| \le \frac{d\arg p(\imath\omega)}{d\omega}.\end{aligned}$$

Since $\rho_1, \rho_2 < 0$ we have $0 < \alpha, \beta < \pi/2$ if $\omega > 0$ and $-\pi/2 < \alpha, \beta < 0$ if $\omega < 0$ so that $|\cos 2\alpha|, |\cos 2\beta| < 1$. Thus strict inequality holds in (9) for quadratic polynomials with two real roots.
Finally consider the case where p is of degree 2 and has non-real roots, i.e. $p(s) = (s-(\rho_1+\imath\omega_1))(s-(\rho_1-\imath\omega_1)) = s^2 - 2\rho_1 s + \rho_1^2 + \omega_1^2$ where $\rho_1 < 0$. Then

$$\frac{d\arg p(\imath\omega)}{d\omega} = \operatorname{Re}\frac{p'(\imath\omega)}{p(\imath\omega)} = \operatorname{Re}\frac{2\imath\omega - 2\rho_1}{-\omega^2 - 2\imath\rho_1\omega + \rho_1^2 + \omega_1^2} = \frac{-2\rho_1(\rho_1^2+\omega_1^2+\omega^2)}{(\rho_1^2+\omega_1^2-\omega^2)^2 + 4\rho_1^2\omega^2}.$$

On the other hand, for all $\omega \in \mathbb{R}^*$,

$$\begin{aligned}\left|\frac{\sin(2\arg p(\imath\omega))}{2\omega}\right| &= \left|\frac{\sin(\arg p(\imath\omega))\cos(\arg p(\imath\omega))}{\omega}\right| \\ &= \left|\frac{-2\rho_1\omega}{\omega\sqrt{(\rho_1^2+\omega_1^2-\omega^2)^2+4\rho_1^2\omega^2}}\frac{\rho_1^2+\omega_1^2-\omega^2}{\sqrt{(\rho_1^2+\omega_1^2-\omega^2)^2+4\rho_1^2\omega^2}}\right| \\ &= \left|\frac{-2\rho_1(\rho_1^2+\omega_1^2-\omega^2)}{(\rho_1^2+\omega_1^2-\omega^2)^2+4\rho_1^2\omega^2}\right| < \frac{-2\rho_1(\rho_1^2+\omega_1^2+\omega^2)}{(\rho_1^2+\omega_1^2-\omega^2)^2+4\rho_1^2\omega^2} = \frac{d\arg p(\imath\omega)}{d\omega}.\end{aligned}$$

This proves that strict inequality holds in (9) for all real polynomials of degree 2.
Now suppose that (9) has been proved for all real polynomials of degree $n \ge 1$ and let $p(s) \in \mathbb{R}[s]$ be a polynomial of degree $n+1$. Then there exist $p_1(s) \in \mathbb{R}[s]$ of degree $n-1$ and $p_2(s) \in \mathbb{R}[s]$ of degree 2 such that $p = p_1p_2$. Now let $\omega \in \mathbb{R}^*$, $\alpha = \arg p_1(\imath\omega)$, $\beta = \arg p_2(\imath\omega)$, then by induction

$$\left|\frac{\sin(2\arg p(\imath\omega))}{2\omega}\right| = \left|\frac{\sin 2(\alpha+\beta)}{2\omega}\right| \le \frac{|\sin 2\alpha| + |\sin 2\beta|}{2|\omega|} < \frac{d\arg p_1(\imath\omega)}{d\omega} + \frac{d\arg p_2(\imath\omega)}{d\omega}.$$

Since by Lemma 3.4.2 the latter sum is equal to $\dfrac{d\arg p(\imath\omega)}{d\omega}$, this concludes the proof. □

Remark 3.4.8. Note that in each of the three cases considered in the above proof we have

$$\lim_{\omega\to 0}\left(\frac{d\arg p(\imath\omega)}{d\omega}-\left|\frac{\sin(2\arg p(\imath\omega))}{2\omega}\right|\right)=0$$

and so this equality holds for any $p(s)\in\mathbb{R}[s]$. □

We will now prove the Hermite–Biehler Theorem which expresses the stability criterion (8) for p in terms of its even and odd parts. Given any complex polynomial $p(s)\in\mathbb{C}[s]$ the even and odd parts u, v of p are, by definition, the unique polynomials $u,v\in\mathbb{C}[s]$ such that

$$p(s)=p^e(s^2)+sp^o(s^2)=u(s^2)+sv(s^2). \tag{10}$$

If $\deg p=2m$ is even, then $\deg u=m$ and $\deg v\le m-1$, whereas if $\deg p=2m+1$ is odd then $\deg v=m$ and $\deg u\le m$.
Now suppose that $p(s)\in\mathbb{R}[s]$ is *real*. Then the real and imaginary parts of $p(\imath\omega)$ (see (2)) can be expressed via its even and odd parts as follows:

$$p_R(\omega)=\operatorname{Re}p(\imath\omega)=u(-\omega^2),\quad p_I(\omega)=\operatorname{Im}p(\imath\omega)=\omega v(-\omega^2). \tag{11}$$

As a consequence we obtain the following equivalent formulation of the phase increasing property (9) in terms of u and v. Remember that $p'(s)=\dfrac{dp}{ds}(s)$.

Remark 3.4.9. Let $p(s)\in\mathbb{R}[s]$ and $\varphi(\omega)=\arg p(\imath\omega)$, $\omega\in\mathbb{R}$. Then

$$\begin{aligned}
\frac{d\arg p(\imath\omega)}{d\omega} &= \operatorname{Re}\frac{p'(\imath\omega)}{p(\imath\omega)}=\operatorname{Re}\frac{v(-\omega^2)-2\omega^2v'(-\omega^2)+2\imath\omega u'(-\omega^2)}{u(-\omega^2)+\imath\omega v(-\omega^2)}\\
&= \frac{\left[v(-\omega^2)-2\omega^2v'(-\omega^2)\right]u(-\omega^2)+2\omega^2u'(-\omega^2)v(-\omega^2)}{u(-\omega^2)^2+\omega^2v(-\omega^2)^2},\\
\frac{\sin(2\varphi(\omega))}{2\omega} &= \frac{\sin\varphi(\omega)\cos\varphi(\omega)}{\omega}=\frac{u(-\omega^2)}{\omega\sqrt{u(-\omega^2)^2+\omega^2v(-\omega^2)^2}}\,\frac{\omega v(-\omega^2)}{\sqrt{u(-\omega^2)^2+\omega^2v(-\omega^2)^2}}
\end{aligned}$$

Therefore (9) can be written

$$\frac{v(-\omega^2)u(-\omega^2)+2\omega^2\left[u'(-\omega^2)v(-\omega^2)-v'(-\omega^2)u(-\omega^2)\right]}{u(-\omega^2)^2+\omega^2v(-\omega^2)^2}\ge\left|\frac{v(-\omega^2)u(-\omega^2)}{u(-\omega^2)^2+\omega^2v(-\omega^2)^2}\right|.$$

We conclude that the phase increasing property (9) is equivalent to the fact that $\omega\in\mathbb{R}$, $p(\imath\omega)\neq 0$ implies

$$\begin{aligned}
u'(-\omega^2)v(-\omega^2)-v'(-\omega^2)u(-\omega^2)&\ge 0 &&\text{if } v(-\omega^2)u(-\omega^2)\ge 0\\
\omega^2(u'(-\omega^2)v(-\omega^2)-v'(-\omega^2)u(-\omega^2))&\ge|v(-\omega^2)u(-\omega^2)| &&\text{if } v(-\omega^2)u(-\omega^2)<0,
\end{aligned}$$

with equality if $\deg p=1$ and strict inequality if $\deg p\ge 2$. □

Let $\varphi(\omega)=\arg p(\imath\omega)$ be an argument function for $p(\imath\omega)$, $\omega\in\mathbb{R}_+$ with initial value $\arg p(0)=\arg u(0)\in\{0,\pi\}$. Then we have $p(\imath\omega)=|p(\imath\omega)|e^{\imath\varphi(\omega)}$ and (11) implies the following relationship between the zeros of u,v on the negative real axis and values of the argument function for $\omega>0$

$$u(-\omega^2)=0\ \Leftrightarrow\ \arg p(\imath\omega)\in\pi/2+\mathbb{Z}\pi;\quad v(-\omega^2)=0\ \Leftrightarrow\ \arg p(\imath\omega)\in\mathbb{Z}\pi. \tag{12}$$

In order to state the Hermite–Biehler Theorem the following definition is useful.

Definition 3.4.10. A pair of real polynomials (u, v) is said to be a *positive pair* if the leading coefficients of u and v have the same sign, the roots t_i of u and t'_j of v are all simple, real and negative and satisfy one of the following two *interlacing conditions* where $m = \deg u, \ell = \deg v$

$$m = \ell \quad \text{and} \quad t'_m < t_m < t'_{m-1} < \ldots < t'_1 < t_1 < 0, \tag{13}$$
$$m = \ell + 1 \quad \text{and} \quad t_m < t'_{m-1} < t_{m-1} < \ldots < t'_1 < t_1 < 0. \tag{14}$$

Theorem 3.4.11 (Hermite–Biehler). *A polynomial $p(s) = u(s^2) + sv(s^2)$ with real coefficients is Hurwitz stable if and only if (u, v) is a positive pair.*

Proof: Suppose that p is Hurwitz stable of degree n and let $\varphi(\omega) = \arg p(\imath\omega)$ be the argument function for p on the positive imaginary axis considered above. Then $\varphi(\omega)$ moves monotonically from 0 (resp. π) to $n\pi/2$ (resp. $\pi + n\pi/2$) by Proposition 3.4.7 and (8). For $n = 2m$ even, as ω is moving through $(0, \infty)$ $\varphi(\omega)$ first arrives at an odd multiple of $\pi/2$ then at a multiple of π, then at the next odd multiple of $\pi/2$ and so on until it finally approaches a multiple of π from below. Denoting the corresponding values of ω by

$$0 < \nu_1 < \nu'_1 < \nu_2 < \ldots < \nu'_{m-1} < \nu_m \quad \varphi(\nu_i) \in (\pi/2 + \mathbb{Z}\pi), \quad \varphi(\nu'_i) \in \mathbb{Z}\pi$$

we see from (12) that $t_i = -\nu_i^2$ are zeros of u whilst $t'_i = -(\nu'_i)^2$ are zeros of v and we have $0 > t_1 > t'_1 > \ldots > t_m$. Since $\deg u = m$, all the m zeros of u must be simple, real and negative. Since $\deg v \leq m - 1$ and we have just seen that v has $m - 1$ zeros t'_j, all the zeros of v must also be simple, real and negative and $\deg v = m - 1$. Finally by Proposition 3.4.6 the leading coefficients of u and v have the same sign. Therefore u, v form a positive pair. A similar proof shows that u, v form a positive pair if p has an odd degree.
Conversely assume that (u, v) is a positive pair and suppose, for instance, that $\deg p = 2m$ is even, i.e. (14) holds. Let

$$0 < \nu_1 < \nu'_1 < \nu_2 < \ldots < \nu'_{m-1} < \nu_m, \quad -\nu_i^2 = t_i, \ -(\nu'_j)^2 = t'_j$$

where the t_i, t'_j are the zeros of u and v, respectively, as in (14). By assumption there are no joint zeros of u and v, so that $p(\imath\omega) \neq 0$ for all $\omega \geq 0$. Let ω increase from 0 to ∞ and consider the open frequency intervals $I_j = (\nu'_{j-1}, \nu_j)$ (where $\nu'_0 := 0$) and $I'_j = (\nu_j, \nu'_j)$ (where $\nu'_m := \infty$) for $j \in \underline{m}$. Both $\omega \mapsto u(-\omega^2)$ and $\omega \mapsto v(-\omega^2)$ do not change signs on these intervals so that we can associate with each of these intervals the sign pattern

$$\begin{aligned}(S_j(u), S_j(v)) &= (\operatorname{sign} u(-\omega^2), \operatorname{sign} v(-\omega^2)), \quad \omega \in I_j,\\ (S'_j(u), S'_j(v)) &= (\operatorname{sign} u(-\omega^2), \operatorname{sign} v(-\omega^2)), \quad \omega \in I'_j.\end{aligned}$$

Since the zeros of u (resp. v) are simple, u (resp. v) are changing signs at each of their zeros, respectively. This implies that the sign patterns of the intervals change as follows

$$(S'_j(u), S'_j(v)) = (-S_j(u), S_j(v)), \quad (S_{j+1}(u), S_{j+1}(v)) = (S'_j(u), -S'_j(v)) \tag{15}$$

By assumption $u(t)$ and $v(t)$ have the same sign for large $t > 0$ and hence for all $t > -\nu_1^2$ since both polynomials have no zeros on $(-\nu_1^2, \infty)$. As a consequence we have $S_1(u) = S_1(v) = \pm 1$ i.e. $p(\imath\omega) = u(-\omega^2) + \imath\omega v(-\omega^2)$ is either in the open first or the open third quadrant for $\omega \in I_1$ starting at $\omega = 0$ at either a positive or negative value. $p(\imath\omega)$ hits the imaginary axis at the zeros of u and the real axis at the zeros of v. Moreover, according to the sign change rule (15) if it is positive at $\omega = 0$, as ω increases, it moves from the first to the second (resp. third to the fourth if negative at $\omega = 0$) quadrant at ν_1. Then from the second to the third (resp. from the fourth to the first) quadrant at ν_1' and so on. Hence the change of the argument of $p(\imath\omega)$ is $\pi/2$ on each compact interval $\overline{I_j}$, $j \in \underline{m}$ and on each compact interval $\overline{I_j'}$, $j = 1, \ldots, m-1$. Finally, since $\lim_{\omega\to\infty} p(\imath\omega)/|p(\imath\omega)| \in \mathbb{R}$ and since $p(\imath\nu_m) \in \imath\mathbb{R}^*$, we have $\Delta_{\nu_m}^{\infty} \arg p(\imath\omega) = \pi/2$. This shows that $\Delta_0^{\infty} \arg p(\imath\omega) = n\pi/2$ and hence p is Hurwitz by (8). □

In the following lemma we apply the above theorem to derive an iterative procedure for testing whether or not a polynomial is Hurwitz stable.

Lemma 3.4.12. *Let $p(s) = \sum_{k=0}^{n} a_k s^k = u(s^2) + sv(s^2)$ be a real polynomial of degree $n \geq 2$, $a_n > 0$. Then p is Hurwitz stable if and only if $a_{n-1} > 0$ and*

$$q(s) = a_{n-1}s^{n-1} + (a_{n-2} - \kappa a_{n-3})s^{n-2} + a_{n-3}s^{n-3} + (a_{n-4} - \kappa a_{n-5})s^{n-4} + \ldots$$

with $\kappa = a_n/a_{n-1}$ and $a_{n-k} = 0$ for $k > n$, is Hurwitz stable.

Proof: The decomposition of q into its even and odd parts has the form

$$q(s) = \begin{cases} (u(s^2) - \kappa s^2 v(s^2)) + sv(s^2) & \text{if } n \text{ is even,} \\ u(s^2) + s(v(s^2) - \kappa u(s^2)) & \text{if } n \text{ is odd.} \end{cases}$$

Let us, for instance, consider the case where $n = 2m$ is even. Then p and q have the same odd part v. The even part of q is given by $\tilde{u}(t) = u(t) - \kappa t v(t)$, and so the values of u and $\tilde{u}$ coincide at the zeros of v and at $t = 0$.
Now assume that p is Hurwitz stable. Then $a_{n-1} > 0$ by Proposition 3.4.6, and by the Hermite-Biehler Theorem all the roots of u, v are negative, simple and satisfy the interlacing condition (13). Thus $\tilde{u}(0) = u(0), \tilde{u}(t_1') = u(t_1'), \ldots, \tilde{u}(t_{m-1}') = u(t_{m-1}')$ form an alternating sequence and therefore $\tilde{u}$ has exactly $m-1$ zeros, one in each of the intervals $(t_{m-1}', t_{m-2}'), \ldots, (t_2', t_1'), (t_1', 0)$. Hence all the zeros of $\tilde{u}$ and v are simple, real, negative, and interlacing. Moreover $\tilde{u}(0)v(0) = u(0)v(0) > 0$ and since $\tilde{u}, v$ have no zeros on $\mathbb{R}_+$, the leading coefficients of $\tilde{u}$ and v have the same sign. Thus these two polynomials form a positive pair and q is Hurwitz stable by the Hermite-Biehler Theorem.
Conversely, assume that q is Hurwitz stable and $a_{n-1} > 0$. Then $(\tilde{u}, v)$ is a positive pair and the above arguments show that u has $m-1$ zeros interlacing the $m-1$ zeros of v. By the Hermite-Biehler Theorem it only remains to prove that there is another root of u in $(-\infty, t_{m-1}')$. Since $\deg u = \deg \tilde{u} + 1$ and the leading coefficients of u and $\tilde{u}$, a_n and $a_{n-2} - \kappa a_{n-3}$, are both positive, we have $u(t)\tilde{u}(t) < 0$ for $t \to -\infty$. Since $\tilde{u}(t)$ does not have a zero in $(-\infty, t_{m-1}')$, the sign of $\tilde{u}(t)$ for $t \to -\infty$ must be

equal to the sign of $\tilde{u}(t'_{m-1}) = u(t'_{m-1})$. Hence $u(t)u(t'_{m-1}) < 0$ for $t \to -\infty$ and we conclude that u has a zero in $(-\infty, t'_{m-1})$.
The proof for n odd is similar and is left as an exercise, see Ex. 3. □

This lemma gives rise to the following recursive algorithm for testing whether or not a polynomial $p(s) \in \mathbb{R}[s]$ of degree $n \geq 2$ with $a_n > 0$ is Hurwitz stable.

Algorithm 3.4.13 (Test for Hurwitz stability of real polynomials).

1. Start: $p_0(s) = p(s)$, $i = 0$.
2. Verify that all the coefficients of $p_i(s)$ are positive. If not, p is not Hurwitz. If yes and $\deg p_i = 2$ then p is Hurwitz. If yes and $\deg p_i > 2$, continue.
3. Construct $p_{i+1}(s) = q(s)$ from $p_i(s)$ according to Lemma 3.4.12 with p replaced by p_i, set $i := i + 1$ and go back to 2.

Remark 3.4.14. Algorithm 3.4.13 is equivalent to the *Routh test*[1] for Hurwitz polynomials. Given a real polynomial $p(s) = \sum_{k=0}^{n} a_k s^k$ with $a_n > 0$ the associated *Routh array* has $n + 1$ rows and is given by

$$\begin{array}{cccccccc}
a_n & a_{n-2} & a_{n-4} & a_{n-6} & \cdots & a_{n-2j+2} & \cdots & 0 \\
a_{n-1} & a_{n-3} & a_{n-5} & a_{n-7} & \cdots & a_{n-2j+1} & \cdots & 0 \\
c_{3,1} & c_{3,2} & c_{3,3} & c_{3,4} & \cdots & c_{3,j} & \cdots & 0 \\
c_{4,1} & c_{4,2} & c_{4,3} & c_{4,4} & \cdots & c_{4,j} & \cdots & 0 \\
\vdots & & & & & & & \vdots \\
c_{i,1} & c_{i,2} & c_{i,3} & c_{i,4} & \cdots & c_{i,j} & \cdots & \cdots \\
\vdots & & & & & & & \vdots \\
c_{n+1,1} & c_{n+1,2} & c_{n+1,3} & c_{n+1,4} & \cdots & c_{n+1,j} & \cdots & \cdots
\end{array}$$

where $a_{n-k} := 0$ for $k > n$ and every row of index $i > 2$ is obtained from the preceding two by the following rule.

> From the entries of the row $i - 2$ subtract the corresponding entries in row $i - 1$ multiplied by the number which makes the difference in the first column zero. Delete this 0 entry and shift the row by one entry to the left

$$c_{i,j} = c_{i-2,j+1} - \frac{c_{i-2,1}}{c_{i-1,1}} c_{i-1,j+1}.$$

Routh proved that a necessary and sufficient condition for p to be Hurwitz is that the first column of the Routh array contains only positive elements.
The relationship between the Routh array and the Algorithm 3.4.13 is simple. If e.g. *deg* $p = 2m$ is even, then the coefficients of the even and the odd parts of $p_0(s)$ are given in rows 1 and 2, in this order, starting with the leading coefficients. The coefficients of the odd and the even parts of $p_1(s)$ are given, in this order, in rows 2 and 3, respectively. Those of the even and odd parts of $p_2(s)$ are given in rows 3 and 4, respectively, etc. In particular, we see that all non-zero entries in the Routh array of a Hurwitz polynomial with positive coefficients must be positive. □

[1]E. J. Routh [439] was the winner of the Adams Prize mentioned in the introduction to this chapter.

Example 3.4.15. Consider the cubic polynomial $p(s) = s^3 + a_2 s^2 + a_1 s + a_0$. The associated Routh array is

$$\begin{matrix} 1 & a_1 & 0 & 0 & \dots \\ a_2 & a_0 & 0 & 0 & \dots \\ (a_1 a_2 - a_0)/a_2 & 0 & 0 & 0 & \dots \\ a_0 & 0 & 0 & 0 & \dots \end{matrix} . \tag{16}$$

So $p(s)$ is Hurwitz if and only if $a_2, a_0 > 0$ and $a_1 a_2 - a_0 > 0$. The same result is obtained by Algorithm 3.4.13 after the first step which gives $p_1(s) = a_2 s^2 + (a_1 - a_0/a_2)s + a_0$. Now consider the quartic polynomial $p(s) = s^4 + a_3 s^3 + a_2 s^2 + a_1 s + a_0$. Algorithm 3.4.13 generates the following two polynomials

$$\begin{aligned} p_1(s) &= a_3 s^3 + (a_2 - \kappa_1 a_1)s^2 + a_1 s + a_0, & \kappa_1 &= 1/a_3 \\ p_2(s) &= (a_2 - \kappa_1 a_1)s^2 + (a_1 - \kappa_2 a_0)s + a_0, & \kappa_2 &= a_3^2/(a_3 a_2 - a_1). \end{aligned}$$

Thus p is Hurwitz if and only if

$$a_0, a_1, a_3 > 0, \quad (a_2 - (1/a_3)a_1) > 0, \quad a_1 - a_3^2 a_0/(a_3 a_2 - a_1) > 0.$$

The same result is provided by the Routh test:

$$\begin{matrix} 1 & a_2 & a_0 \\ a_3 & a_1 & 0 \\ a_2 - a_1/a_3 & a_0 & 0 \\ a_1 - a_3^2 a_0/(a_3 a_2 - a_1) & 0 & 0 \\ a_0 & 0 & 0 \end{matrix}$$

□

Example 3.4.16. In one of the first mathematical analyses of feedback control systems, *Maxwell* (1868) in his paper "On Governors" [364] considered the problem of regulating the angular velocity of driving shafts. We will describe his study of Jenkin's governor. It is rather difficult to understand the exact form of this governor from Maxwell's paper and so we have used the schematic given in *Bennett* (1979) [50], see Figure 3.4.2. Experiments were made with the governor by Maxwell, Balfour Stewart and Jenkin in 1863. Unfortunately, no description of the governor from that time has been found, although the governor itself is preserved in the Whipple Museum of Science at Cambridge University. The purpose of the governor is to ensure that deviations of the angular speed of the drive shaft from a nominal value are small. It is basically a friction governor; if the angular speed increases above its nominal value the fly balls move out and the force between the fly balls and the friction ring is increased. This causes the ring to rotate at an increased angular velocity, which in turn has the following effects:

the weight in the damping fluid is raised to provide hydraulic damping,
the toothed worm gear worked by a revolving spiral causes the band brake to tighten on the drive shaft,
an extra torque proportional to the angular deviation of the friction ring from a nominal value is applied to the driving shaft.

The friction ring can rotate in either direction so that if the angular speed decreases below its nominal value the last two effects are reversed. Let

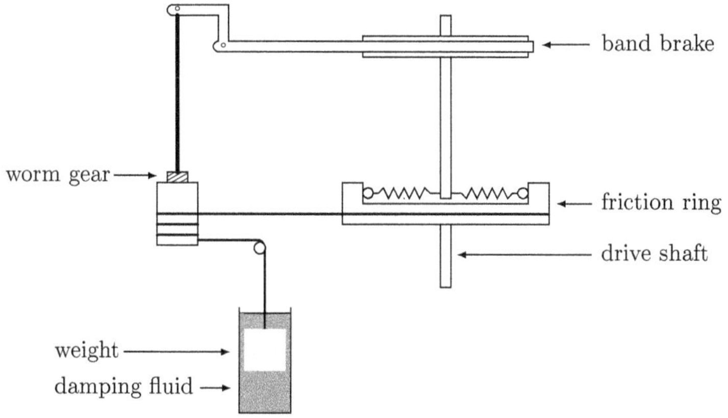

Figure 3.4.2: Jenkin's governor

$$
\begin{aligned}
y &= \text{angular deviation of ring from a nominal value} \\
\frac{dx}{dt} &= \text{angular velocity of shaft} \\
P &= \text{driving torque on the shaft} \\
R &= \text{fixed load torque on the shaft} \\
M &= \text{moment of inertia of the machine} \\
G &= \text{constant relating torque applied to the machine to } y \\
F &= \text{friction coefficient} \\
V_1 &= \text{fixed lowest possible operating velocity} \\
B &= \text{moment of inertia of the ring} \\
Y &= \text{viscous damping coefficient} \\
W &= \text{torque due to weight.}
\end{aligned}
$$

Maxwell used the equations

$$
\begin{aligned}
M\frac{d^2x}{dt^2} &= P - R - F\left(\frac{dx}{dt} - V_1\right) - Gy \\
B\frac{d^2y}{dt^2} &= F\left(\frac{dx}{dt} - V_1\right) - Y\frac{dy}{dt} - W.
\end{aligned} \tag{17}
$$

We will take as state variables

$$x_1 = \frac{dx}{dt}, \quad x_2 = y, \quad x_3 = \frac{dy}{dt}$$

then (17) may be written in the form

$$
\begin{aligned}
M\dot{x}_1 &= P - R - F(x_1 - V_1) - Gx_2 \\
\dot{x}_2 &= x_3 \\
B\dot{x}_3 &= F(x_1 - V_1) - Yx_3 - W.
\end{aligned} \tag{18}
$$

There is an equilibrium state $x_e = (x_{1e}, x_{2e}, x_{3e})$ where

$$x_{1e} = \frac{W}{F} + V_1, \quad x_{2e} = \frac{P - R - W}{G}, \quad x_{3e} = 0.$$

Setting $x = x_e + x'$ and substituting in (18) we obtain the following equations for the perturbation x'

$$\dot{x}' = \begin{bmatrix} \dot{x}'_1 \\ \dot{x}'_2 \\ \dot{x}'_3 \end{bmatrix} = \begin{bmatrix} -F/M & -G/M & 0 \\ 0 & 0 & 1 \\ F/B & 0 & -Y/B \end{bmatrix} \begin{bmatrix} x'_1 \\ x'_2 \\ x'_3 \end{bmatrix}.$$

The eigenvalues of the above matrix satisfy the characteristic equation

$$p(\lambda) = \lambda^3 + \left(\frac{Y}{B} + \frac{F}{M}\right)\lambda^2 + \frac{FY}{MB}\lambda + \frac{FG}{BM} = 0.$$

Hence by applying (16) the polynomial p will be Hurwitz if

$$\left(\frac{Y}{B} + \frac{F}{M}\right)\frac{FY}{MB} - \frac{FG}{BM} > 0$$

or

$$\frac{Y}{B} + \frac{F}{M} > \frac{G}{Y}. \tag{19}$$

Thus the equilibrium state x_e will be asymptotically stable if (19) is satisfied. This is the result obtained by Maxwell from which he concluded *"If it is not fulfilled there will be a dancing motion of the governor which will increase till it is as great as the limits of motion of the governor. To ensure stability the value of Y must be sufficiently great ..."*. □

3.4.2 Characterization of Stability via the Cauchy Index

In this subsection we will characterize complex Hurwitz polynomials $p(s) \in \mathbb{C}[s]$ via the *Cauchy index* of an associated real rational function.

Definition 3.4.17 (Cauchy index). Let $f(s) \in \mathbb{R}(s)$ be a real rational function. The *local Cauchy index* of f at a pole $s_0 \in \mathbb{R}$ is by definition

$$C_{s_0}(f) = \frac{1}{2}\left[\lim_{s \downarrow s_0} \frac{f(s)}{|f(s)|} - \lim_{s \uparrow s_0} \frac{f(s)}{|f(s)|}\right].$$

If $-\infty \le a < b \le \infty$ then the sum of the local Cauchy indices at the poles of f in (a, b) is said to be the *Cauchy index* of f on the interval (a, b) and is denoted by $CI_a^b(f)$. $CI_{-\infty}^{\infty}(f)$ is called the (global) Cauchy index of f.

Intuitively speaking, $CI_a^b(f)$ is the difference in the number of jumps of $f(s)$ from $-\infty$ to $+\infty$ and the number of jumps from $+\infty$ to $-\infty$ as s goes from a to b on the real axis, see Figure 3.4.3.

Remark 3.4.18. All the *finite poles* of f are taken into account by $CI_{-\infty}^{\infty}(f)$. However, the pole at infinity is neglected and as a consequence $CI_{-\infty}^{\infty}(f) = CI_{-\infty}^{\infty}(f + p)$ for every real rational function f and every real polynomial p. In order to take into account the pole at infinity, we set

$$CI(f) = CI_{-\infty}^{\infty}(f) + \frac{1}{2}\left[\lim_{s \to -\infty} \frac{f(s)}{|f(s)|} - \lim_{s \to +\infty} \frac{f(s)}{|f(s)|}\right].$$

Then $CI(f) = CI_{-\infty}^{\infty}(f)$ if and only if the polynomial part of f is of even degree. If it is of odd degree and a_n is its leading coefficient then $CI(f) = CI_{-\infty}^{\infty}(f) - \operatorname{sign} a_n$. □

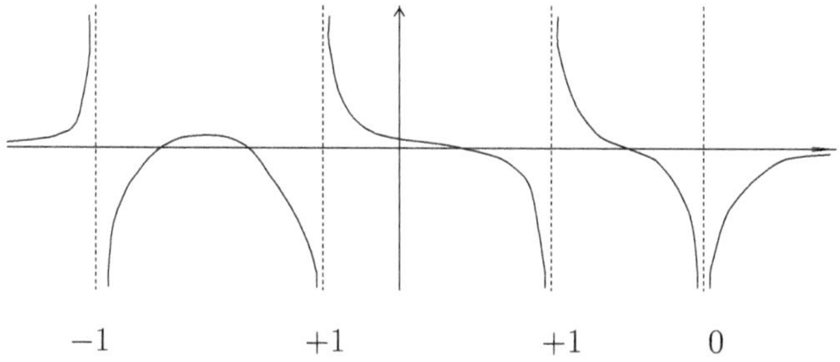

Figure 3.4.3: Local Cauchy indices of a function with $CI_{-\infty}^{\infty}(f) = 1$

Example 3.4.19. By partial fraction decomposition every real rational function can be represented in the form

$$f(s) = \sum_{i=1}^{r} \frac{q_i(s)}{(s - s_i)^{m_i}} + f_0(s)$$

where $s_1, \ldots, s_r$ are the real poles of odd order, the q_i are real polynomials of degree $< m_i$ with $q_i(s_i) \neq 0$ and f_0 has no real poles of odd order. The sign of $q_i(s_i)$ determines the local Cauchy index of f at s_i: $CI_{s_i}(f) = \operatorname{sign} q_i(s_i)$. The global Cauchy index $CI_{-\infty}^{\infty}(f)$ is the number of positive $q_i(s_i)$ minus the number of negative $q_i(s_i)$. For instance, if

$$f_\nu(s) = \sum_{k=-\nu}^{n-\nu} \frac{k}{s+k}, \qquad \nu = 0, 1, \ldots, n,$$

then the Cauchy index $CI_{-\infty}^{\infty}(f_\nu) = n - 2\nu$. In fact it is easy to see that $-n$, $-n+2$, ..., $n-2$, n are the only possible values of the Cauchy index of a real rational function with denominator degree n. □

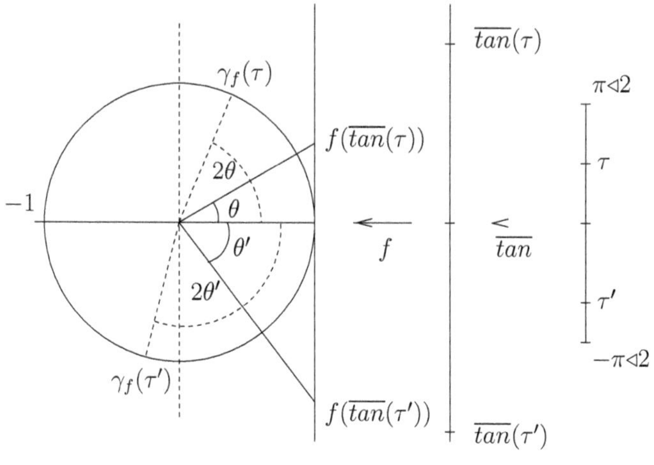

Figure 3.4.4: Construction of γ_f

We will now show that the index $CI(f)$ of a function $f(s) = u(s)/v(s) \in \mathbb{R}(s)$ can be expressed as the winding number of a closed curve γ_f associated with f. To obtain a closed curve we parametrize the compactified real axis $\overline{\mathbb{R}} = [-\infty, \infty]$ by the order preserving homeomorphism

$$\overline{\tan} : [-\pi/2, \pi/2] \to \overline{\mathbb{R}}, \ \ \overline{\tan}(\tau) = \tan\tau, \ \ \tau \in (-\pi/2, \pi/2), \ \ \overline{\tan}(\pm\pi/2) = \pm\infty$$

and let $\theta : \overline{\mathbb{R}} \to [-\pi/2, \pi/2]$ be the inverse of $\overline{\tan}$

$$\theta(t) = \arctan t, \ \ t \in \mathbb{R}, \quad \theta(\pm\infty) = \pm\pi/2. \tag{20}$$

Define the curve $\gamma_f : \ [-\pi/2, \pi/2] \to \mathbb{C}$ by

$$\gamma_f(\tau) = e^{2\imath\theta(f(\overline{\tan}(\tau)))}, \quad \tau \in [-\pi/2, \pi/2], \tag{21}$$

see Figure 3.4.4. γ_f is a (continuous) curve with values in $S^1 = \{s \in \mathbb{C} : \ |s| = 1\}$. If $\deg u \le \deg v$, then the limits $f(\pm\infty) := \lim_{t\to\pm\infty} f(t)$ are equal and $\alpha := f(\pm\infty) \in \mathbb{R}$, otherwise we have $f(\pm\infty) \in \{-\infty, \infty\}$. In the former case the curve starts and returns to the point $s_0 = e^{2\imath \arctan \alpha}$ and in the latter it starts and returns to $s_0 = -1$. So the curve is always closed. The winding number $w(\gamma_f, 0)$ (see Section A.2) counts the number of times that the curve $\gamma_f(\tau)$ crosses -1 in the anticlockwise sense minus the number of crossings in the clockwise sense as τ moves from $-\pi/2$ to $\pi/2$. Now $\gamma_f(\tau_0) = -1$ if and only if $|f(t_0)| = \infty$ for $t_0 = \overline{\tan}(\tau_0)$, i.e. $v(t_0) = 0$ or $|t_0| = \infty$ with $\deg u > \deg v$. $\gamma_f(\tau)$ crosses -1 in the anticlockwise (resp. clockwise) direction at $\tau_0 \in (-\pi/2, \pi/2)$ if and only if $f(t)$ jumps from $+\infty$ to $-\infty$ (resp. $-\infty$ to $+\infty$) at $t_0 = \overline{\tan}(\tau_0)$. Besides these contributions to the winding number from finite t, there will be a contribution of $+1$ (resp. -1) if $f(+\infty) = +\infty, f(-\infty) = -\infty$ (resp. $f(+\infty) = -\infty, f(-\infty) = +\infty$).[2] If however $f(-\infty) = f(+\infty)$ there will be no added contribution. Thus

$$CI(f) = -w(\gamma_f, 0). \tag{22}$$

The graph of γ_f describes a curve on the cylinder $[-\pi/2, \pi/2] \times S^1$. Figure 3.4.5 shows this graph for the rational function f illustrated in Figure 3.4.3.
Since $\gamma_f(-\pi/2) = \gamma_f(+\pi/2)$, we may identify the ends of the cylinder to obtain a

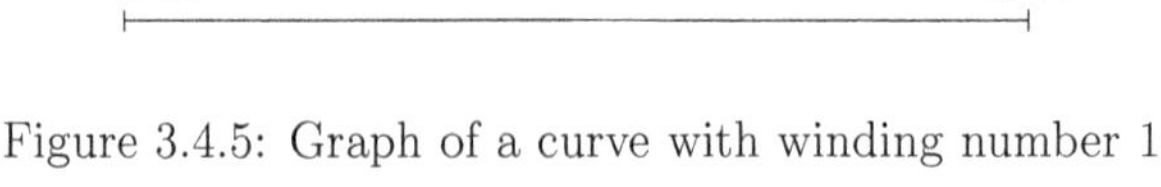

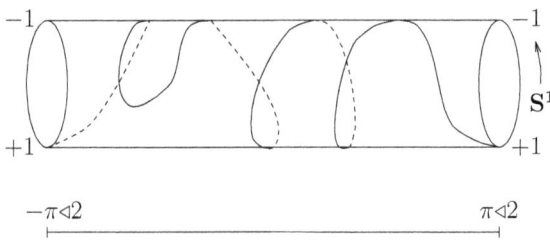

Figure 3.4.5: Graph of a curve with winding number 1

closed curve on a torus. The winding number $w(\gamma_f, 0)$ is just the net number of times the graph winds around the torus in the positive sense as τ increases from $-\pi/2$ to $\pi/2$.

[2]To see this the reader should consider the reparametrized curve $\tau \mapsto e^{2\imath\theta(f(\overline{\tan}(\tau)))}$ for $\tau \in [-\pi/2 + \varepsilon, \pi/2 + \varepsilon]$ at $\tau_0 = \pi/2$ $(0 < \varepsilon \ll 1)$. As τ crosses τ_0 in increasing order the corresponding $t = \tan(\tau)$ jumps from ∞ to $-\infty$.

Remark 3.4.20. For every $t \in \mathbb{R}^*$ we have

$$\theta(-1/t) = \arctan(-1/t) = \arctan(t) + \pi/2 = \theta(t) + \pi/2 \qquad \mod \pi$$

and since $e^{\imath 2\theta(\pm\infty)} = -1$ it follows that $e^{\imath 2\theta(-1/t)} = e^{\imath 2(\theta(t)+\pi/2)} = -e^{\imath 2\theta(t)}$ for all $t \in \mathbb{R} \cup \{-\infty, \infty\}$. Thus, for any $f(s) = u(s)/v(s) \in \mathbb{R}(s)$,

$$\gamma_{-1/f}(\tau) = e^{\imath 2\theta(-1/f(\overline{\tan}(\tau)))} = -e^{\imath 2\theta(f(\overline{\tan}(\tau)))} = -\gamma_f(\tau), \quad \tau \in [-\pi/2, \pi/2].$$

Hence $w(\gamma_{-1/f}, 0) = w(\gamma_f, 0)$ and so

$$CI(v/u) = -CI(-1/f) = w(\gamma_{-1/f}, 0) = w(\gamma_f, 0) = -CI(u/v). \tag{23}$$

In particular if p is a real polynomial then $CI(1/p) = -CI(p) \in \{+1, -1, 0\}$. □

We now express the change of the argument of a polynomial by the Cauchy index of an associated real rational function. Consider a complex polynomial $p(s) = \sum_{j=0}^{n} a_j s^j \in \mathbb{C}[s]$, and let $p_R(\omega)$ and $p_I(\omega)$ be the real and the imaginary parts of $p(\imath\omega)$, see (2). The explicit formulas for these real polynomials are different for the even and odd cases. Suppose $a_j = \alpha_j + \imath\beta_j$, $\alpha_j, \beta_j \in \mathbb{R}$. If n is even then

$$\begin{aligned} p_R(\omega) &= (-1)^{\frac{n}{2}}(\alpha_n\omega^n + \beta_{n-1}\omega^{n-1} - \alpha_{n-2}\omega^{n-2} - \beta_{n-3}\omega^{n-3} + + - - \\ p_I(\omega) &= (-1)^{\frac{n}{2}}(\beta_n\omega^n - \alpha_{n-1}\omega^{n-1} - \beta_{n-2}\omega^{n-2} + \alpha_{n-3}\omega^{n-3} + - - + \end{aligned} \tag{24}$$

and if n is odd then

$$\begin{aligned} p_R(\omega) &= (-1)^{\frac{n+1}{2}}(\beta_n\omega^n - \alpha_{n-1}\omega^{n-1} - \beta_{n-2}\omega^{n-2} + \alpha_{n-3}\omega^{n-3} + - - + \\ p_I(\omega) &= (-1)^{\frac{n-1}{2}}(\alpha_n\omega^n + \beta_{n-1}\omega^{n-1} - \alpha_{n-2}\omega^{n-2} - \beta_{n-3}\omega^{n-3} + + - - \end{aligned} \tag{25}$$

Proposition 3.4.21. *Suppose $p(s) \in \mathbb{C}[s]$ is a polynomial of degree n without zeros on the imaginary axis and $p_R(s), p_I(s) \in \mathbb{R}[s]$ are given by (24) or (25). If ν is the number of zeros of $p(s)$ in the open right half-plane (taking into account multiplicities), then*

$$CI_{-\infty}^{\infty}(f_p) = (n - 2\nu) \quad \text{where } f_p(\omega) = \begin{cases} p_R(\omega)/p_I(\omega) & \text{if } \deg p_R \le \deg p_I \\ -p_I(\omega)/p_R(\omega) & \text{if } \deg p_I < \deg p_R \end{cases} \tag{26}$$

and $\Delta_{-\infty}^{\infty} \arg p(\imath\omega) = CI_{-\infty}^{\infty}(f_p)\pi$. In particular, p is Hurwitz stable if and only if $CI_{-\infty}^{\infty}(f_p) = n$.

Proof: Note that, for all n, the real rational function $f = f_p$ has the form $f(\omega) = u(\omega)/v(\omega)$ where u, v do not have joint real roots (since p has no roots on the imaginary axis). Now assume e.g. that $\deg p_I < \deg p_R$. Let $\varphi(\omega) = \arg p(\imath\omega)$ be an argument function for p on $\imath\mathbb{R}$, hence $\tan\varphi(\omega) = p_I(\omega)/p_R(\omega)$ for all $\omega \in \mathbb{R}$ such that $p_R(\omega) \neq 0$. Define θ by (20) and $\gamma_{-f} = \gamma_{p_I/p_R}$ by (21). Then $\theta(-f(\omega)) = \arctan(p_I(\omega)/p_R(\omega)) = \varphi(\omega) \mod 2\pi$ for all $\omega \in \mathbb{R}$ such that $p_R(\omega) \neq 0$. Thus

$$e^{\imath\theta(-f(\omega))} = e^{\imath\varphi(\omega)} = p(\imath\omega)/|p(\imath\omega)|$$

for all $\omega \in \mathbb{R}$, $p_R(\omega) \neq 0$. Hence

$$\gamma_{-f}(\tau) = e^{2\imath\theta(-f(\tan\tau))} = e^{2\imath\varphi(\tan\tau)} = p(\imath\tan\tau)^2/|p(\imath\tan\tau)|^2$$

for all $\tau \in (-\pi/2, \pi/2)$ satisfying $p_R(\tan\tau) \neq 0$. By continuity it follows that

$$\gamma_{-f}(\tau) = p(\imath\tan\tau)^2/|p(\imath\tan\tau)|^2, \quad \tau \in [-\pi/2, \pi/2],$$

and so $\gamma_{-f}(\tau)$, $\tau \in [-\pi/2, \pi/2]$ is just a reparametrization of $p(\imath\omega)^2/|p(\imath\omega)|^2$, $\omega \in \overline{\mathbb{R}}$. This implies

$$w(\gamma_{-f}, 0)2\pi = \Delta_{-\pi/2}^{\pi/2} \arg\gamma_{-f}(\tau) = \Delta_{-\infty}^{\infty} \arg\frac{p(\imath\omega)^2}{|p(\imath\omega)|^2}$$

whence by (22)

$$\Delta_{-\infty}^{\infty} \arg p(\imath\omega) = w(\gamma_{-f}, 0)\pi = -CI_{-\infty}^{\infty}(-f)\,\pi = CI_{-\infty}^{\infty}(f)\,\pi.$$

Making use of Proposition 3.4.3 this proves the proposition for $\deg p_I < \deg p_R$. The proof is similar in the other case. □

Remark 3.4.22. (i) In Remark 3.4.20 we saw that for every real rational function f we have $CI(1/f) = -CI(f)$. Moreover, the Cauchy index and the extended Cauchy index coincide for proper functions $f(s) \in \mathbb{R}(s)$. Therefore, using the extended Cauchy index formula (26) simplifies to

$$CI(p_R/p_I) = CI(-p_I/p_R) = CI_{-\infty}^{\infty}(f_p) = (n - 2\nu).$$

(ii) Suppose that the leading coefficient of p is real, i.e. $a_n = \alpha_n, \beta_n = 0$. Then $\deg p_R > \deg p_I$ if n is even, and $\deg p_R < \deg p_I$ if n is odd. Hence, using the expressions (24) and (25) for p_R, p_I we obtain the following unified formula for f_p in (26)

$$f_p(\omega) = \frac{\alpha_{n-1}\omega^{n-1} + \beta_{n-2}\omega^{n-2} - \alpha_{n-3}\omega^{n-3} - \beta_{n-4}\omega^{n-4} + + - - \cdots}{\alpha_n\omega^n + \beta_{n-1}\omega^{n-1} - \alpha_{n-2}\omega^{n-2} - \beta_{n-3}\omega^{n-3} + + - - \cdots}$$

If p is a *real* polynomial this formula simplifies further to

$$f_p(\omega) = \frac{\alpha_{n-1}\omega^{n-1} - \alpha_{n-3}\omega^{n-3} + \alpha_{n-5}\omega^{n-5} - + \cdots}{\alpha_n\omega^n - \alpha_{n-2}\omega^{n-2} + \alpha_{n-4}\omega^{n-4} - + \cdots}$$

(iii) Let p be any complex Hurwitz polynomial of degree n and $f(s) = -p_I(s)/p_R(s)$. Since $CI(f) = n$ (see (i)) we must have $n \geq \deg p_R \geq n - 1$. Moreover all the poles of f (roots of p_R) must be real and simple, say t_i, $i = 1, \ldots, n$ or $n - 1$. Let us now consider the real zeros, t_i' of f. First suppose there are n poles, then f is proper and hence $CI_{-\infty}^{\infty}(f) = n$. So there must be at least $n - 1$ real zeros which interlace the poles

$$t_n < t_{n-1}' < t_{n-1} < \ldots < t_2 < t_1' < t_1.$$

If $\deg p_I = n$, there will be one more zero of f and, depending on the sign of α_n/β_n (see (24) and (25)), it will either be greater than t_n or less than t_1.
Now suppose there are only $n - 1$ poles, then $CI(f) = n$ implies $\deg p_I = n$ and $CI_{-\infty}^{\infty}(p_R/p_I) = n$. Hence in this case the $n - 1$ real roots t_i of p_R and the n real roots t_j' of p_I are again all simple and interlaced as follows

$$t_n' < t_{n-1} < \ldots t_2 < t_2' < t_1 < t_1'.$$

Summarizing we see that the roots of p_R and p_I are all simple, real and interlacing (as for real Hurwitz polynomials). Note, however, that in contrast with the real case the roots of p_R and p_I may be nonnegative and the interlacing condition alone is not sufficient for Hurwitz stability (see the following example). A necessary and sufficient condition for Hurwitz stability is given in Corollary 3.4.63. □

Example 3.4.23. Consider the polynomial $p(s) = s^2 + (\alpha_1 + \imath\beta_1)s + \imath\beta_0$, where $\beta_0, \alpha_1, \beta_1 \in \mathbb{R}$. Then for the f_p given by (26), we have

$$f_p(\omega) = -\frac{p_I(\omega)}{p_R(\omega)} = \frac{\alpha_1\omega + \beta_0}{\omega(\omega + \beta_1)}.$$

So if $\beta_1 = 0$, p cannot be Hurwitz since p_R would have a double root. Now suppose $\beta_1 > 0$, then $CI_{-\infty}^{\infty}(f_p) = 2$, if and only if $\beta_0 > 0$ and $-\alpha_1\beta_1 + \beta_0 < 0$. Whereas if $\beta_1 < 0$, then $CI_{-\infty}^{\infty}(f_p) = 2$ if and only if $\beta_0 < 0$ and $-\alpha_1\beta_1 + \beta_0 > 0$. Hence by Proposition 3.4.21 p is Hurwitz if and only if $\alpha_1 > 0$ and $\alpha_1\beta_0\beta_1 > \beta_0^2$.
Suppose $\beta_1 < 0$ and p is Hurwitz, then by Remark 3.4.22, we have the interlacing $0 < -\beta_0/\alpha_1 < -\beta_1$, but note that this condition is not sufficient for p to be Hurwitz (choose β_0 positive, α_1 negative and $\beta_1 < \beta_0/\alpha_1$). □

3.4.3 Hermite Forms and Bézoutiants

In this subsection and the next we will use quadratic forms in order to determine how many roots of a given complex algebraic equation lie in the open left half-plane $\mathbb{C}_-$. We first consider a quadratic form which has been introduced by Hermite in order to separate the roots of algebraic equations with complex coefficients. Then we introduce another quadratic form, the Bézoutiant, and study its relationship with the Hermite form.

Hermite Form

For any complex polynomial $p(s) = \sum_{k=0}^{n} a_k s^k$ with $p^\star$ its Hurwitz-reflection, we have

$$\overline{p}(s)p(w) - \overline{p}^\star(s)p^\star(w) = \overline{p}(s)p(w) - p(-s)\overline{p}(-w) = \sum_{i=0}^{n}\sum_{j=0}^{n} \overline{a_i}a_j(s^i w^j - (-s)^j(-w)^i).$$

For $0 \le i, j \le n$, necessarily $(s^i w^j - (-s)^j(-w)^i) = 0$ if $i = j$, and for $i < j$

$$\begin{aligned} s^i w^j - (-s)^j(-w)^i &= (-1)^i \left[(-s)^i w^j - (-s)^j w^i\right] = (-1)^i(-s)^i w^i \left[w^{j-i} - (-s)^{j-i}\right], \\ &= s^i w^i (s+w)\left[w^{j-i-1} + w^{j-i-2}(-s) + \ldots w(-s)^{j-i-2} + (-s)^{j-i-1}\right]. \end{aligned}$$

Hence there exist $h_{ij} \in \mathbb{C}$, $i, j \in \underline{n}$ such that

$$\overline{p}(s)p(w) - \overline{p}^\star(s)p^\star(w) = (s+w)\sum_{i=1}^{n}\sum_{j=1}^{n} h_{ij}s^{i-1}w^{j-1}.$$

Indeed comparing coefficients of equal powers yields

$$\overline{a_i}a_j - (-1)^{i+j}\overline{a_j}a_i = h_{i,j+1} + h_{i+1,j}, \quad h_{i,0} = h_{0,j} = h_{n+1,j} = h_{i,n+1} = 0.$$

But $-h_{ij} = \sum_{k=1}^{\min\{i,n+1-j\}}(-1)^k(h_{i-k,j+k} + h_{i-k+1,j+k-1})$ and hence

$$h_{ij} = \sum_{k=1}^{\min\{i,n+1-j\}} (-1)^{k-1}\left[\overline{a_{i-k}}a_{j+k-1} - (-1)^{i+j-1}a_{i-k}\overline{a_{j+k-1}}\right]. \tag{27}$$

For any $n \in \mathbb{N}$ we introduce the following notation

$$l(w)^\top = [1, w, w^2, \ldots, w^{n-1}], \quad w \in \mathbb{C}. \tag{28}$$

Definition 3.4.24. Given a complex polynomial $p(s) = \sum_{k=0}^{n} a_k s^k$, the associated *Hermite generating function* is given by

$$h(p; s, w) = \frac{\overline{p}(s)p(w) - \overline{p}^\star(s)p^\star(w)}{s+w} = \sum_{i=1}^{n}\sum_{j=1}^{n} h_{ij} s^{i-1} w^{j-1} = l(s)^\top \mathbf{H}_n(p) l(w). \tag{29}$$

The matrix $\mathbf{H}_n(p) = (h_{ij}) \in \mathbb{C}^{n\times n}$ is called the associated *Hermite matrix* (of order $n \geq \deg p$), and the corresponding bilinear form $(x, y) \mapsto x^*\mathbf{H}_n(p)y$ on $\mathbb{C}^n$ is called the associated *Hermite form.* If $n = \deg p$ we write $\mathbf{H}(p)$ for $\mathbf{H}_n(p)$.

The matrix $\mathbf{H}_n(p)$ is Hermitian since

$$\sum_{i,j=1}^{n} \overline{h_{ij}} s^{i-1} w^{j-1} = \overline{h(p; \overline{s}, \overline{w})} = \frac{p(s)\overline{p}(w) - p^\star(s)\overline{p}^\star(w)}{s+w} = h(p; w, s) = \sum_{i,j=1}^{n} h_{ij} w^{i-1} s^{j-1}.$$

Also since $h_{ij} = 0$ for $i > \deg p$ or $j > \deg p$, we have

$$\operatorname{rank} \mathbf{H}_n(p) = \operatorname{rank} \mathbf{H}(p), \quad \operatorname{sign} \mathbf{H}_n(p) = \operatorname{sign} \mathbf{H}(p), \quad n \geq \deg p. \tag{30}$$

In the following example we use (27) to determine $\mathbf{H}_n(p)$ for $n = 1, 2, 3$.

Example 3.4.25. (a) If $p(s) = a_1 s + a_0$, then $\mathbf{H}_1(p) = (\overline{a_0}a_1 + a_0\overline{a_1})$.

(b) If $p(s) = a_2 s^2 + a_1 s + a_0$, then

$$\mathbf{H}_2(p) = \begin{bmatrix} \overline{a_0}a_1 + \overline{a_1}a_0 & (\overline{a_0}a_2 - a_0\overline{a_2}) \\ -(\overline{a_0}a_2 - a_0\overline{a_2}) & \overline{a_1}a_2 + \overline{a_2}a_1 \end{bmatrix}.$$

(c) If $p(s) = a_3 s^3 + a_2 s^2 + a_1 s + a_0$, a short calculation yields

$$\mathbf{H}_3(p) = \begin{bmatrix} (\overline{a_0}a_1 + a_0\overline{a_1}) & (\overline{a_0}a_2 - a_0\overline{a_2}) & (\overline{a_0}a_3 + a_0\overline{a_3}) \\ -(\overline{a_0}a_2 - a_0\overline{a_2}) & (\overline{a_1}a_2 + a_1\overline{a_2} - \overline{a_0}a_3 - a_0\overline{a_3}) & (\overline{a_1}a_3 - a_1\overline{a_3}) \\ (\overline{a_0}a_3 + a_0\overline{a_3}) & -(\overline{a_1}a_3 - a_1\overline{a_3}) & (\overline{a_2}a_3 + a_2\overline{a_3}) \end{bmatrix}.$$

Note that if $a_3 = 0$ the last row and the last column of this matrix are zero and the remaining entries coincide with the corresponding entries of $\mathbf{H}_2(p)$. □

Given a Hermitian matrix $H \in \mathcal{H}_n(\mathbb{C})$ we denote by $n_+(H)$, $n_0(H)$, $n_-(H)$ the number of its positive, zero and negative eigenvalues, respectively, accounting for multiplicities. The triplet $i(H) = (n_+(H), n_0(H), n_(H))$ is called the *inertia* of H. The *signature* of H, denoted by $\operatorname{sign}(H)$, is by definition the number of its positive eigenvalues minus the number of its negative eigenvalues, $\operatorname{sign}(H) = n_+(H) - n_-(H)$. Two Hermitian matrices $H, K \in \mathcal{H}_n(\mathbb{C})$ are called *congruent* if there exists a non-singular matrix $T \in \mathbf{Gl}_n(\mathbb{C})$ such that $T^*HT = K$. By Sylvester's Law of Inertia (see Theorem A.1.33) two Hermitian matrices in $\mathcal{H}_n(\mathbb{C})$ are congruent if and only if they have the same rank and the same signature.

We will now investigate the rank and the signature of the Hermite matrix $\mathbf{H}_n(p)$. Let $d(s) = (p(s), p^\star(s))$ be the normalized greatest common divisor of $p(s)$ and $p^\star(s)$, so $p(s) = d(s)p_0(s)$, $p^\star(s) = d(s)p_0{}^\star(s)$. Then $p_0(s)$ and $p_0{}^\star(s)$ are coprime, $d^\star(s) = d(s)$ and

$$d(\lambda) = 0 \quad \Leftrightarrow \quad p(\lambda) = 0 \text{ and } p(-\overline{\lambda}) = 0 \quad \Leftrightarrow \quad d(-\overline{\lambda}) = 0. \tag{31}$$

In particular, purely imaginary roots of p are also roots of d, and p_0 has no roots on the imaginary axis.

Lemma 3.4.26. *$\mathbf{H}(p)$ and $\mathbf{H}(p_0)$ have the same invariants*

$$\operatorname{rank}\mathbf{H}(p) = \operatorname{rank}\mathbf{H}(p_0), \quad \operatorname{sign}\mathbf{H}(p) = \operatorname{sign}\mathbf{H}(p_0).$$

Proof: Let $d(s) = s^m + d_{m-1}s^{m-1} + \ldots + d_0 = (p(s), p^\star(s))$ and $n = \deg p$. Since

$$\frac{\overline{p}(s)p(w) - \overline{p}^\star(s)p^\star(w)}{s+w} = \overline{d}(s)d(w)\frac{\overline{p_0}(s)p_0(w) - \overline{p_0}^\star(s){p_0}^\star(w)}{s+w}$$

we have

$$\begin{bmatrix} 1 \\ s \\ \vdots \\ s^{n-1} \end{bmatrix}^\top \mathbf{H}(p) \begin{bmatrix} 1 \\ w \\ \vdots \\ w^{n-1} \end{bmatrix} = \begin{bmatrix} \overline{d}(s) \\ s\overline{d}(s) \\ \vdots \\ s^{n-m-1}\overline{d}(s) \end{bmatrix}^\top \mathbf{H}(p_0) \begin{bmatrix} d(w) \\ wd(w) \\ \vdots \\ w^{n-m-1}d(w) \end{bmatrix} = \begin{bmatrix} 1 \\ s \\ \vdots \\ s^{n-1} \end{bmatrix}^\top D^* H D \begin{bmatrix} 1 \\ w \\ \vdots \\ w^{n-1} \end{bmatrix}$$

where[3]

$$H = \left[\begin{array}{c|c} \mathbf{H}(p_0) & 0 \\ \hline 0 & 0 \end{array}\right]_{n\times n}, \quad D = \left[\begin{array}{cccccccccc} d_0 & d_1 & d_2 & \cdots & \cdots & 1 & 0 & \cdots & \cdots & 0 \\ 0 & d_0 & d_1 & d_2 & \ddots & \ddots & 1 & 0 & \cdots & 0 \\ \vdots & \ddots & \ddots & \ddots & \ddots & \cdots & \cdots & \ddots & \ddots & \vdots \\ 0 & \cdots & 0 & d_0 & d_1 & d_2 & \cdots & \cdots & 1 & 0 \\ 0 & \cdots & \cdots & 0 & d_0 & d_1 & d_2 & \cdots & \cdots & 1 \\ \hline \multicolumn{5}{c|}{I_m} & \multicolumn{5}{c}{0_{m\times(n-m)}} \end{array}\right]_{n\times n}.$$

But two polynomials in $\mathbb{C}[s,w]$ having the same values for all $(s,w) \in \mathbb{C}\times\mathbb{C}$ must have the same coefficients. Hence $\mathbf{H}(p) = D^* H D$, and since D is nonsingular H and $\mathbf{H}(p)$ have the same invariants. This concludes the proof. □

Before enunciating the next lemma we recall the notion of the *resultant* of two polynomials.

Definition 3.4.27. Given two complex polynomials

$$p(s) = \sum_{k=0}^{n} a_k s^k, \qquad q(s) = \sum_{k=0}^{m} b_k s^k, \quad n,\, m \geq 1$$

the associated $(m+n)\times(m+n)$ resultant matrix is

$$R(p,q) = R_{n,m}(p,q) = \left[\begin{array}{cccccccc} a_0 & a_1 & \cdots & a_n & 0 & \cdots & \cdots & 0 \\ 0 & a_0 & a_1 & \cdots & a_n & 0 & \cdots & 0 \\ \vdots & \ddots & \ddots & \ddots & \cdots & \ddots & \ddots & \vdots \\ \vdots & \cdots & \ddots & \ddots & \ddots & \cdots & \ddots & 0 \\ 0 & \cdots & \cdots & 0 & a_0 & a_1 & \cdots & a_n \\ \hline b_0 & b_1 & \cdots & b_{m-1} & b_m & 0 & \cdots & 0 \\ 0 & b_0 & b_1 & \cdots & b_{m-1} & b_m & \ddots & \vdots \\ \vdots & \ddots & \ddots & \ddots & \cdots & \ddots & \ddots & 0 \\ 0 & \cdots & 0 & b_0 & b_1 & \cdots & b_{m-1} & b_m \end{array}\right]_{(m+n)\times(m+n)} \begin{array}{l} \left.\vphantom{\begin{array}{c}a\\a\\a\\a\\a\end{array}}\right\} m \\ \left.\vphantom{\begin{array}{c}a\\a\\a\\a\end{array}}\right\} n \end{array}. \tag{32}$$

[3]The precise column alignment in the matrix D and in next matrix $R(p,q)$ will depend on m and n.

It is well known that $R_{n,m}(p,q)$ is singular if and only if the two polynomials p, q have a common non-trivial factor or $a_n = b_m = 0$ [503, §34].

Lemma 3.4.28. *Suppose $p(s) \in \mathbb{C}[s]$ is such that $p(s)$ and $p^\star(s)$ are coprime. If $p = p_1 p_2$ in $\mathbb{C}[s]$ then*

$$\mathbf{H}(p) = R^* \begin{bmatrix} \mathbf{H}(p_1) & 0 \\ 0 & \mathbf{H}(p_2) \end{bmatrix} R$$

where R is the (nonsingular) resultant of ${p_1}^\star(s)$ and $p_2(s)$. In particular,

$$\operatorname{rank} \mathbf{H}(p) = \operatorname{rank} \mathbf{H}(p_1) + \operatorname{rank} \mathbf{H}(p_2), \quad \operatorname{sign} \mathbf{H}(p) = \operatorname{sign} \mathbf{H}(p_1) + \operatorname{sign} \mathbf{H}(p_2).$$

Proof: Let $n = \deg p$, $\ell = \deg p_1$, $m = \deg p_2$, so that $n = \ell + m$. Since

$$\begin{aligned} &\frac{\overline{p}(s)p(w) - \overline{p}^\star(s)p^\star(w)}{s+w} \\ &= \overline{p_2}(s)p_2(w)\frac{\overline{p_1}(s)p_1(w) - \overline{p_1}^\star(s){p_1}^\star(w)}{s+w} + \overline{p_1}^\star(s){p_1}^\star(w)\frac{\overline{p_2}(s)p_2(w) - \overline{p_2}^\star(s){p_2}^\star(w)}{s+w} \end{aligned}$$

we have

$$\begin{bmatrix} 1 \\ s \\ \vdots \\ s^{n-1} \end{bmatrix}^\top \mathbf{H}(p) \begin{bmatrix} 1 \\ w \\ \vdots \\ w^{n-1} \end{bmatrix} = \begin{bmatrix} \overline{p_2}(s) \\ s\overline{p_2}(s) \\ \vdots \\ s^{\ell-1}\overline{p_2}(s) \end{bmatrix}^\top \mathbf{H}(p_1) \begin{bmatrix} p_2(w) \\ wp_2(w) \\ \vdots \\ w^{\ell-1}p_2(w) \end{bmatrix} + \begin{bmatrix} \overline{p_1}^\star(s) \\ s\overline{p_1}^\star(s) \\ \vdots \\ s^{m-1}\overline{p_1}^\star(s) \end{bmatrix}^\top \mathbf{H}(p_2) \begin{bmatrix} {p_1}^\star(w) \\ w{p_1}^\star(w) \\ \vdots \\ w^{m-1}{p_1}^\star(w) \end{bmatrix}.$$

If

$${p_1}^\star(w) = \sum_{k=0}^{\ell} a_k w^k,\ a_\ell \neq 0, \qquad p_2(w) = \sum_{k=0}^{m} b_k w^k,\ b_m \neq 0,$$

we obtain $\mathbf{H}(p) = F_1^* \mathbf{H}(p_1) F_1 + F_2^* \mathbf{H}(p_2) F_2$ where[4]

$$F_1 = \begin{bmatrix} b_0 & b_1 & \cdots & b_m & 0 & \cdots & 0 \\ 0 & b_0 & b_1 & \cdots & b_m & \ddots & \vdots \\ \vdots & \ddots & \ddots & \ddots & \cdots & \ddots & 0 \\ 0 & \cdots & 0 & b_0 & b_1 & \cdots & b_m \end{bmatrix}_{\ell \times n}, \quad F_2 = \begin{bmatrix} a_0 & a_1 & \cdots & a_\ell & 0 & \cdots & 0 \\ 0 & a_0 & a_1 & \cdots & a_\ell & \ddots & \vdots \\ \vdots & \ddots & \ddots & \ddots & \cdots & \ddots & 0 \\ 0 & \cdots & 0 & a_0 & a_1 & \cdots & a_\ell \end{bmatrix}_{m \times n}.$$

Hence

$$\mathbf{H}(p) = F^* \begin{bmatrix} \mathbf{H}(p_1) & 0 \\ 0 & \mathbf{H}(p_2) \end{bmatrix} F \quad \text{where } F = \begin{bmatrix} F_1 \\ F_2 \end{bmatrix}_{n \times n}.$$

Since F is the resultant of $p_2(s)$ and ${p_1}^\star(s)$ and $({p_1}^\star(s), p_2(s))$ are coprime, F is nonsingular and the results follow. □

The following theorem, due to Hermite, shows how the Hermite form can be used in order to determine the number of roots of a polynomial in the open left and right half-planes (assuming that the polynomial does not have roots on the imaginary axis).

[4]Again the column alignment will depend on ℓ and m.

Theorem 3.4.29 (Hermite). *Let $p(s) \in \mathbb{C}[s]$ be a polynomial of degree n and $\mathbf{H}(p)$ the associated Hermite matrix. If $r = \operatorname{rank} \mathbf{H}(p)$, $\sigma = \operatorname{sign} \mathbf{H}(p)$, then $p(s)$ has $n-r$ roots in common with $p^\star(s)$, $(r+\sigma)/2$ additional roots in the open left half-plane, and $(r-\sigma)/2$ additional roots in the open right half-plane. In particular, p is Hurwitz stable if and only if $\mathbf{H}(p)$ is positive definite.*

Proof: Let $d(s) = (p(s), p^\star(s))$, $m = \deg d$, $p = dp_0$ and $p_0(s) = a_n \prod_{k=1}^{n-m}(s - s_k)$. By induction we obtain from the previous lemma

$$\operatorname{rank} \mathbf{H}(p_0) = \sum_{k=1}^{n-m} \operatorname{rank} \mathbf{H}(s - s_k), \quad \operatorname{sign} \mathbf{H}(p_0) = \sum_{k=1}^{n-m} \operatorname{sign} \mathbf{H}(s - s_k).$$

But $\mathbf{H}(s - s_k) = -(\overline{s_k} + s_k) = -2 \operatorname{Re} s_k$, see Example 3.4.25. Now since p_0 does not have any imaginary roots we obtain from Lemma 3.4.26

$$r = \operatorname{rank} \mathbf{H}(p) = n - m = n_-(p_0) + n_+(p_0), \quad \sigma = \operatorname{sign} \mathbf{H}(p) = n_-(p_0) - n_+(p_0).$$

This proves the first statement of the theorem. The second follows immediately since by (31) p is Hurwitz stable if and only if $p = p_0$ and $n_-(p) = n$. □

Example 3.4.30. Consider again the polynomial studied in Example 3.4.23, $p(s) = s^2 + (\alpha_1 + \imath\beta_1)s + \imath\beta_0$, where $\beta_0, \alpha_1, \beta_1 \in \mathbb{R}$. Using Example 3.4.25, we have

$$\mathbf{H}(p) = \begin{bmatrix} -\imath\beta_0(\alpha_1 + \imath\beta_1) + \imath\beta_0(\alpha_1 - \imath\beta_1) & -2\imath\beta_0 \\ +2\imath\beta_0 & 2\alpha_1 \end{bmatrix} = \begin{bmatrix} 2\beta_0\beta_1 & -2\imath\beta_0 \\ +2\imath\beta_0 & 2\alpha_1 \end{bmatrix}.$$

Hence $\mathbf{H}(p) \succ 0$ if and only if $\alpha_1 > 0$ and $\alpha_1\beta_0\beta_1 > \beta_0^2$. These conditions are precisely the necessary and sufficient conditions for the stability of p obtained in Example 3.4.23. □

Bézoutiants

Given two complex polynomials

$$u(s) = \sum_{k=0}^{n} u_k s^k \quad \text{and} \quad v(s) = \sum_{k=0}^{n} v_k s^k \tag{33}$$

the polynomial in two variables $u(s)v(w) - u(w)v(s)$ is divisible by $s - w$ since

$$u(s)v(w) - u(w)v(s) = \sum_{i=0}^{n}\sum_{j=0}^{n} u_i v_j (s^i w^j - w^i s^j) = \sum_{0 \le i < j \le n} (u_i v_j - v_i u_j)(s^i w^j - w^i s^j)$$

and $s^i w^j - w^i s^j = (w - s)\sum_{k=0}^{j-i-1} s^{i+k} w^{j-k-1}$ for $0 \le i < j \le n$. Therefore there exist uniquely determined $b_{ij} \in \mathbb{C}$ such that

$$\frac{u(s)v(w) - u(w)v(s)}{s - w} = \sum_{i=1}^{n}\sum_{j=1}^{n} b_{ij} s^{i-1} w^{j-1} = l(s)^\top \mathbf{B}_n(u, v) l(w) \tag{34}$$

where $\mathbf{B}_n(u, v) = (b_{ij})_{i,j \in \underline{n}}$. Indeed, multiplying (34) by $s - w$ and comparing the coefficients of equal powers yields

$$u_i v_j - u_j v_i = b_{i,j+1} - b_{i+1,j}, \quad b_{i,0} = b_{0,j} = b_{n+1,j} = b_{i,n+1} = 0. \tag{35}$$

But $-b_{ij} = \sum_{k=1}^{\min\{i,n+1-j\}} b_{i-k,j+k} - b_{i-k+1,j+k-1}$ and hence

$$b_{ij} = \sum_{k=1}^{\min\{i,n+1-j\}} u_{j+k-1}v_{i-k} - u_{i-k}v_{j+k-1}, \quad i,j \in \underline{n}. \tag{36}$$

Clearly $b_{ij} = 0$ for $i, j > \max\{\deg u, \deg v\}$.

Definition 3.4.31. The matrix $\mathbf{B}_n(u,v) = (b_{ij}) \in \mathbb{C}^{n\times n}$ defined by (34) is called the *Bézout matrix* or *Bézoutian* of order n for the polynomial pair (u,v) given by (33) and the associated bilinear form $(x,y) \mapsto x^*\mathbf{B}_n(u,v)y$ is called the *Bézoutiant* of (u,v) on $\mathbb{C}^n$. The Bézout matrix of order $\max\{\deg u, \deg v\}$ for (u,v) is denoted by $\mathbf{B}(u,v)$.

The Bézoutian of order n for (u,v) is defined whenever $\max\{\deg u, \deg v\} \leq n$. By definition and (35) it has the following elementary properties.

Lemma 3.4.32. *Given $u(s), v(s) \in \mathbb{C}[s]$ with $\max\{\deg u, \deg v\} \leq n$, then*

(i) $\mathbf{B}_n(u,v)$ *is symmetric, i.e.* $\mathbf{B}_n(u,v)^\top = \mathbf{B}_n(u,v)$.

(ii) $\mathbf{B}_n(u,v) = -\mathbf{B}_n(v,u), \quad \mathbf{B}_n(u,u) = 0_{n\times n}$.

(iii) $\mathbf{B}_n(u,v)$ *is bilinear in* (u,v).

(iv) $\operatorname{rank}\mathbf{B}_n(u,v) = \operatorname{rank}\mathbf{B}(u,v)$.

(v) $\mathbf{B}_n(u,v) \in \mathbb{R}^{n\times n}$ *if* $u, v \in \mathbb{R}[s]$*, and in this case* $\operatorname{sign}\mathbf{B}_n(u,v) = \operatorname{sign}\mathbf{B}(u,v)$.

The Bézoutians for $n = 1, 2, 3$ are given in the next example.

Example 3.4.33. (a) For $n = 1$:

$$\mathbf{B}_1(u,v) = u_1v_0 - u_0v_1.$$

(b) For $n = 2$:

$$\mathbf{B}_2(u,v) = \begin{bmatrix} u_1v_0 - u_0v_1 & u_2v_0 - u_0v_2 \\ u_2v_0 - u_0v_2 & u_2v_1 - u_1v_2 \end{bmatrix}.$$

(c) For $n = 3$:

$$\mathbf{B}_3(u,v) = \begin{bmatrix} u_1v_0 - u_0v_1 & u_2v_0 - u_0v_2 & u_3v_0 - u_0v_3 \\ u_2v_0 - u_0v_2 & u_2v_1 - u_1v_2 + u_3v_0 - u_0v_3 & u_3v_1 - u_1v_3 \\ u_3v_0 - u_0v_3 & u_3v_1 - u_1v_3 & u_3v_2 - u_2v_3 \end{bmatrix}.$$

□

The following proposition relates the Hermite form of a complex polynomial with the real Bézout matrix of the pair (p_R, p_I) where $p_R(s), p_I(s)$ are the real and imaginary parts of $p(\imath s)$, see (2).

Proposition 3.4.34. *Suppose $p(s) \in \mathbb{C}[s]$ and let $p_R(s), p_I(s) \in \mathbb{R}[s]$ be the real polynomials satisfying $p(\imath s) = p_R(s) + \imath p_I(s)$. Then the Hermite matrix $\mathbf{H}_n(p)$ and the Bézoutian $\mathbf{B}_n(p_I, p_R)$ are congruent*

$$\mathbf{H}_n(p) = 2D\mathbf{B}_n(p_I, p_R)D^*, \quad \textit{where} \quad D = \operatorname{diag}(1, \imath, \ldots, \imath^{n-1}).$$

Proof: Since $p(s) = p_R(-\imath s) + \imath p_I(-\imath s)$ and $\overline{p}(s) = p_R(\imath s) - \imath p_I(\imath s)$, we obtain by straight forward calculations

$$\begin{aligned} l(s)^\top \mathbf{H}_n(p) l(w) &= \frac{\overline{p}(s)p(w) - \overline{p}^\star(s)p^\star(w)}{s+w} \\ &= 2\frac{p_R(\imath s)p_I(-\imath w) - p_R(-\imath w)p_I(\imath s)}{-\imath(s+w)} = 2\frac{p_I(\imath s)p_R(-\imath w) - p_R(\imath s)p_I(-\imath w)}{\imath s - (-\imath w)} \\ &= 2\, l(\imath s)^\top \mathbf{B}_n(p_I, p_R)\, l(-\imath w) = 2l(s)^\top D\mathbf{B}_n(p_I,p_R)D^*\, l(w), \quad s, w \in \mathbb{C}. \end{aligned}$$

□

As a consequence of Theorem 3.4.29 and this proposition we obtain

Corollary 3.4.35. *With the notations of the proposition, suppose* $\deg p = n$ *and let* r *be the rank and* σ *the signature of* $\mathbf{B}_n(p_I, p_R)$. *Then* $p(s)$ *has* $n-r$ *roots in common with* $p^\star(s)$, *and additionally* $(r+\sigma)/2$ *roots in the open left half plane,* $(r-\sigma)/2$ *roots in the open right half plane. In particular,* p *is Hurwitz stable if and only if* $\mathbf{B}_n(p_I, p_R) \succ 0$.

Example 3.4.36. Consider again the polynomial $p(s) = s^2 + (\alpha_1 + \imath\beta_1)s + \imath\beta_0$, where $\beta_0, \alpha_1, \beta_1 \in \mathbb{R}$. Then $p_R(s) = -s^2 - \beta_1 s$, $p_I(s) = \alpha_1 s + \beta_0$ and using the expression given in Example 3.4.33

$$\mathbf{B}_2(p_I, p_R) = \begin{bmatrix} \beta_0\beta_1 & \beta_0 \\ \beta_0 & \alpha_1 \end{bmatrix}.$$

Then $\mathbf{B}_2(p_I, p_R) \succ 0$ if and only if $\alpha_1 > 0$ and $\alpha_1\beta_0\beta_1 > \beta_0^2$. □

For *real* polynomials $p(s)$ analogous results can be obtained in terms of the even and odd parts of $p(s)$.

Proposition 3.4.37. *Suppose* $p(s) = u(s^2) + sv(s^2) \in \mathbb{R}[s]$. *Then the invariants of the Hermite matrix* $\mathbf{H}(p)$ *are equal to the sum of the invariants of* $\mathbf{B}(u,v)$ *and* $\mathbf{B}(\tilde{v}, u)$ *where* $\tilde{v}(t) = tv(t)$

$$\operatorname{rank} \mathbf{H}(p) = \operatorname{rank} \mathbf{B}(u,v) + \operatorname{rank} \mathbf{B}(\tilde{v},u), \quad \operatorname{sign} \mathbf{H}(p) = \operatorname{sign} \mathbf{B}(u,v) + \operatorname{sign} \mathbf{B}(\tilde{v},u). \quad (37)$$

Proof: Choose $n = 2m \geq \deg p$, then $\deg u \leq m$ and $\deg \tilde{v} \leq m$ and we have

$$\begin{aligned} l(s)\mathbf{H}_n(p)\, l(w) &= \frac{p(s)p(w) - p(-s)p(-w)}{s+w} \\ &= 2sw\frac{u(s^2)v(w^2) - v(s^2)u(w^2)}{s^2 - w^2} + 2\frac{s^2v(s^2)u(w^2) - w^2v(w^2)u(s^2)}{s^2 - w^2} \\ &= 2swl(s^2)^\top \mathbf{B}_m(u,v)\, l(w^2) + 2l(s^2)^\top \mathbf{B}_m(\tilde{v},u)\, l(w^2). \end{aligned}$$

Now let P be the $n \times n$-permutation matrix such that

$$[1, s, s^2, \ldots, s^{n-1}]P = [s, s^3, \ldots, s^{2m-1}, 1, s^2, \ldots, s^{2(m-1)}].$$

Then

$$l(s)^\top \mathbf{H}_n(p)\, l(w) = 2l(s)^\top P \begin{bmatrix} \mathbf{B}_m(u,v) & 0 \\ 0 & \mathbf{B}_m(\tilde{v},u) \end{bmatrix} P^\top l(w).$$

From this the equations (37) follow by Lemma 3.4.32 and (30). □

As a consequence of the preceding proposition we obtain the following stability criterion.

Corollary 3.4.38. *Suppose* $p(s) = u(s^2) + sv(s^2) \in \mathbb{R}[s]$, $\tilde{v}(t) = tv(t)$. *Then* p *is Hurwitz stable if and only if* $\mathbf{B}(u, v) \succ 0$ *and* $\mathbf{B}(\tilde{v}, u) \succ 0$.

We illustrate the application of this result in the following example.

Example 3.4.39. Consider the real cubic polynomial $p(s) = a_3s^3 + a_2s^2 + a_1s + a_0$. Then $u(t) = a_2t + a_0$, $v(t) = a_3t + a_1$, $\tilde{v}(t) = a_3t^2 + a_1t$ and using the formulas in Example 3.4.33 we have

$$\mathbf{B}_1(u, v) = a_2a_1 - a_0a_3, \qquad \mathbf{B}_2(\tilde{v}, u) = \begin{bmatrix} a_0a_1 & a_0a_3 \\ a_0a_3 & a_3a_2 \end{bmatrix}.$$

Then $\mathbf{B}_1(u, v) \succ 0$ and $\mathbf{B}_2(\tilde{v}, u) \succ 0$ if and only if

$$a_2a_1 - a_0a_3 > 0, \quad a_0a_1 > 0, \quad a_0a_1a_2a_3 > a_0^2a_3^2.$$

This is equivalent to either the coefficients all being positive or all being negative plus $a_2a_1 - a_0a_3 > 0$ and reduces to the conditions given in Example 3.4.15 for the case where $a_3 = 1$. □

3.4.4 Hankel Matrices and Rational Functions

We now introduce another quadratic form, the *Hankel form*, which plays an important role not only in stability analysis, but also in other areas of systems theory such as *realization theory* and *model reduction* (see Volume II). We will see that this quadratic form has a close relationship with the Bézoutiant and with strictly proper rational functions. In particular, we will see that the Cauchy index of a real rational function can be expressed by the signature of the associated Hankel matrix. We conclude the subsection with a proof of Brockett's Theorem which determines the connected components of the space of real rational functions of fixed degree.
Every rational function $g(s) = c(s)/d(s)$ where $c(s), d(s) \in \mathbb{C}[s]$ can be developed as a power series in s^{-1} (Laurent expansion at ∞, see (A.2.20)),

$$g(s) = \sum_{k=-\nu}^{\infty} g_k s^{-k} = \sum_{k=0}^{\nu} g_{-k} s^k + \sum_{k=1}^{\infty} g_k s^{-k}, \quad (\text{where } \nu \geq 0). \tag{38}$$

The first sum on the RHS is called the *polynomial part* of $g(s)$ and the second one is called its *strictly proper part.* $g(s)$ is strictly proper (i.e. $\deg c < \deg d$) if and only if its polynomial part is zero.

Definition 3.4.40. Given any formal Laurent series of the form (38) with complex coefficients g_k, $k \geq -\nu$, the associated infinite Hankel matrix (resp. Hankel matrix of order $n \geq 1$) is given by $\mathbf{Hk}(g) = (g_{i+j-1})_{i,j \in \mathbb{N}^*}$ (resp. $\mathbf{Hk}_n(g) = (g_{i+j-1})_{i,j \in \underline{n}}$)

$$\mathbf{Hk}(g) = \begin{bmatrix} g_1 & g_2 & g_3 & g_4 & \cdots \\ g_2 & g_3 & g_4 & g_5 & \cdots \\ g_3 & g_4 & g_5 & g_6 & \cdots \\ \vdots & \vdots & \vdots & \vdots & \vdots \end{bmatrix}, \quad \mathbf{Hk}_n(g) = \begin{bmatrix} g_1 & g_2 & \cdots & g_n \\ g_2 & g_3 & \cdots & g_{n+1} \\ \vdots & \vdots & & \vdots \\ g_n & g_{n+1} & \cdots & g_{2n-1} \end{bmatrix}. \tag{39}$$

An infinite complex matrix $\mathbf{Hk} = (h_{ij})_{i,j\in\mathbb{N}^*}$ is called a Hankel matrix if there exists a sequence $(h_k)_{k\in\mathbb{N}^*}$ in $\mathbb{C}$ such that $h_{ij} = h_{i+j-1}$ for $i,j \in \mathbb{N}^*$. Analogously, a matrix $\mathbf{Hk} \in \mathbb{K}^{M\times N}$ is called a Hankel matrix if it is of the form $\mathbf{Hk} = (h_{i+j-1})_{i\in\underline{M},j\in\underline{N}}$.

In the real case there is a close relationship between Hankel matrices and Bézoutians which is explained in the following theorem.

Theorem 3.4.41. *Let $c,d \in \mathbb{R}[s]$ be of the form*

$$c(s) = \sum_{k=0}^{n} c_k s^k; \qquad d(s) = \sum_{k=0}^{n} d_k s^k, \quad d_n \neq 0 \tag{40}$$

and $g(s) = c(s)/d(s)$. Then the associated Bézout and Hankel matrices of order n, $\mathbf{B}_n(d,c)$ and $\mathbf{Hk}_n(g)$ are congruent and hence have the same rank and signature

$$\mathbf{B}_n(d,c) = D^\top \mathbf{Hk}_n(g) D \quad \textit{where} \quad D = \begin{bmatrix} d_1 & d_2 & \cdots & d_{n-2} & d_{n-1} & d_n \\ d_2 & d_3 & \cdots & d_{n-1} & d_n & 0 \\ \vdots & \vdots & \vdots & \vdots & \vdots & \vdots \\ d_{n-1} & d_n & 0 & \cdots & \cdots & 0 \\ d_n & 0 & \cdots & \cdots & \cdots & 0 \end{bmatrix}_{n\times n} = D^\top. \tag{41}$$

Proof: If $\mathbf{B}_n(d,c) = (b_{ij})$ we have by Definition 3.4.31

$$\sum_{i,j=1}^{n} b_{ij} s^{i-1} w^{j-1} = \frac{d(s)c(w) - d(w)c(s)}{s-w} = d(s)d(w)\frac{g(w)-g(s)}{s-w}$$

$$= d(s)d(w)\sum_{i=1}^{\infty} g_i \frac{w^{-i}-s^{-i}}{s-w} = d(s)d(w)\sum_{i=1}^{\infty} g_i \sum_{k,\,l\geq 1,\,k+l-1=i} s^{-k}w^{-l}$$

$$= d(s)d(w)\sum_{k,l=1}^{\infty} g_{k+l-1}s^{-k}w^{-l} = \sum_{k,l=1}^{\infty} g_{k+l-1}(d_n s^{n-k} + \ldots + d_0 s^{-k})(d_n w^{n-l} + \ldots + d_0 w^{-l}).$$

Since there are no negative powers of s, w on the LHS we may omit those on the RHS and so

$$\sum_{i,j=1}^{n} b_{ij}s^{i-1}w^{j-1} = \sum_{k,l=1}^{n} g_{k+l-1}(d_n s^{n-k} + \ldots + d_{k+1}s + d_k)(d_n w^{n-l} + \ldots + d_{l+1}w + d_l).$$

Or, in matrix terms

$$l(s)^\top \mathbf{B}_n(d,c)\, l(w) = l(s)^\top D^\top \mathbf{Hk}_n(g) D\, l(w)$$

where D is given by (41). □

Observe that the matrix D in this theorem can be viewed as a Bézout matrix, $D = \mathbf{B}_n(d,1)$. For later use we note the following immediate consequence of the previous theorem and the properties of Bézoutians given in Lemma 3.4.32(ii): If $g(s) = c(s)/d(s)$, $\deg c = \deg d = n$ then

$$\operatorname{rank} \mathbf{Hk}_n(g) = \operatorname{rank} \mathbf{Hk}_n(1/g) \quad \text{and} \quad \operatorname{sign} \mathbf{Hk}_n(g) = \operatorname{sign} \mathbf{Hk}_n(-1/g). \tag{42}$$

Corollary 3.4.42. *The principal minors from the lower right corner of* $\mathbf{B}_n(d,c)$ *are given by the principal minors of the upper left corner of* $\mathbf{Hk}_n(g)$

$$\det(b_{ij})_{i,j=n-k+1}^{n} = d_n^{2k} \det \mathbf{Hk}_k(g), \quad k \in \underline{n}.$$

Proof: If $(b_{ij})_{i,j=n-k+1}^{n} = \mathbf{B}_k$ equation (41) can be written in the form

$$\begin{bmatrix} M_{11} & M_{12} \\ M_{21} & \mathbf{B}_k \end{bmatrix} = \begin{bmatrix} D_{11}^\top & D_{21}^\top \\ D_{12}^\top & 0 \end{bmatrix} \begin{bmatrix} \mathbf{Hk}_k & H_{12} \\ H_{21} & H_{22} \end{bmatrix} \begin{bmatrix} D_{11} & D_{12} \\ D_{21} & 0 \end{bmatrix},$$

where $D_{11} \in \mathbb{R}^{k\times(n-k)}$, $D_{12} \in \mathbb{R}^{k\times k}$ and the other matrices are of commensurate dimensions. Hence

$$\begin{bmatrix} b_{n-k+1,\,n-k+1} & \cdot & b_{n-k+1,\,n} \\ \cdot & \cdot & \cdot \\ b_{n,\,n-k+1} & \cdot & b_{n,\,n} \end{bmatrix} = \begin{bmatrix} d_{n-k+1} & \cdot & d_n \\ \cdot & \cdot & \cdot \\ d_n & \cdot & 0 \end{bmatrix} \begin{bmatrix} g_1 & \cdot & g_k \\ \cdot & \cdot & \cdot \\ g_k & \cdot & g_{2k-1} \end{bmatrix} \begin{bmatrix} d_{n-k+1} & \cdot & d_n \\ \cdot & \cdot & \cdot \\ d_n & \cdot & 0 \end{bmatrix}.$$

The proof is concluded by taking determinants. □

We will now explain the relationship between Hankel matrices and rational functions in some detail. By definition, two rational functions $g(s), \tilde{g}(s)$ determine the same Hankel matrix if their strictly proper parts coincide. Let $g(s) = c(s)/d(s)$ be a proper rational function where $c(s), d(s) \in \mathbb{K}[s]$ are of the form (40). Then $g(s)$ has a Laurent expansion of the form (38) with $\nu = 0$. The coefficients g_j, $j \in \mathbb{N}$ are uniquely determined by the equation

$$c(s) = d(s)(g_0 + g_1 s^{-1} + g_2 s^{-2} + \ldots). \tag{43}$$

Comparing coefficients of s^{-j}, $j \geq 1$, yields the following recurrence relation

$$0 = d_0 g_j + d_1 g_{j+1} + \ldots + d_{n-1} g_{j+n-1} + d_n g_{j+n}, \quad j \geq 1. \tag{44}$$

And comparing coefficients of s^{n-j}, $0 \leq j \leq n$, yields

$$\begin{aligned} c_n &= d_n g_0 \\ c_{n-1} &= d_{n-1} g_0 + d_n g_1 \\ &\vdots \\ c_0 &= d_0 g_0 + d_1 g_1 + \ldots + d_{n-1} g_{n-1} + d_n g_n. \end{aligned} \tag{45}$$

Together, (44) and (45) are equivalent to (43). The relation (44) is expressed by saying that $d(s) = d_n s^n + d_{n-1} s^{n-1} + \ldots + d_0$ is a *recursive polynomial* for the sequence $(g_1, g_2, g_3, \ldots)$. Given $g_1, \ldots, g_n$, the recursive polynomial determines all the subsequent Laurent coefficients g_k, $k \geq n+1$ of $g(s)$. Such a recursive polynomial of degree $n \geq 1$ exists if and only if the $n+1$-st row (column) of the infinite Hankel $\mathbf{Hk}(g)$ is linearly dependent on the n previous ones. Let rank $\mathbf{Hk}(g)$ denote the dimension of the linear subspace of $\mathbb{K}^{\mathbb{N}^*}$ generated by the rows (or, equivalently, columns) of $\mathbf{Hk}(g)$. If rank $\mathbf{Hk}(g) \leq n$ then there exists $r \leq n$ such that the $r+1$-st row of $\mathbf{Hk}(g)$ depends linearly on the previous r rows, hence there exists a recursive polynomial of degree $r \le n$. We now want to show the converse, i.e. that $\operatorname{rank} \mathbf{Hk}(g) \le n$ if there exists a recursive polynomial of degree n for the sequence $(g_k)_{k\in\mathbb{N}^*}$. Let S denotes the left shift on the sequence space $\mathbb{K}^{\mathbb{N}^*}$, i.e. $S(g_1, g_2, g_3, \dots) = (g_2, g_3, g_4, \dots)$. Then the i-th row (column) of $\mathbf{Hk}(g)$ is given by the sequence $S^{i-1}(g_k)_{k\in\mathbb{N}^*}$ written as an infinite row (resp. column) vector. The recurrence relation (44) is equivalent to saying that the operator $d(S) = \sum_{k=0}^{n} d_k S^k$ annuls the first row (resp. column) of $\mathbf{Hk}(g)$. But then $d(S)$ annuls the sequence $S^{i-1}(g_k)_{k\in\mathbb{N}^*}$, i.e. the $n+i$-th row (column) of $\mathbf{Hk}(g)$ depends linearly on the preceding n rows (resp. columns). Hence $\operatorname{rank} \mathbf{Hk}(g) \le n$. Altogether we have proved

Lemma 3.4.43. *There exists a recursive polynomial of degree $\le n$ for the sequence $(g_k)_{k\in\mathbb{N}^*}$ if and only if* $\operatorname{rank} \mathbf{Hk}(g) \le n$.

Let $\mathrm{Rat}_n(\mathbb{K})$ denote the set of all strictly proper rational functions of degree n with coefficients in $\mathbb{K}$. The elements of $\mathrm{Rat}_n(\mathbb{K})$ are of the form $g(s) = c(s)/d(s) \in \mathbb{K}(s)$ where c, d are coprime polynomials as in (40) with $d_n \ne 0, c_n = 0$. A formal Laurent series (38) is said to be *rational* if $g(s)$ is the Laurent expansion of a rational function $c(s)/d(s)$ at ∞. We write $g(s) \in \mathrm{Rat}_n(\mathbb{K})$ if $g(s) = c(s)/d(s)$ in a neighbourhood of ∞ and $c(s)/d(s) \in \mathrm{Rat}_n(\mathbb{K})$. The following theorem due to *Kronecker* (1881) [320] determines the relationship between infinite Hankel matrices of finite rank and strictly proper rational functions.

Theorem 3.4.44 (Kronecker). *Let $g(s) = \sum_{k=1}^{\infty} g_k s^{-k}$ be any strictly proper formal Laurent series with coefficients in $\mathbb{K}$. Then*

(i) $g(s)$ is rational if and only if there exists a recursive polynomial for its coefficient sequence (g_k).

(ii) $g(s) \in \mathrm{Rat}_n(\mathbb{K})$ if and only if $\operatorname{rank} \mathbf{Hk}(g) = n$. *In this case*

$$\operatorname{rank} \mathbf{Hk}(g) = \operatorname{rank} \mathbf{Hk}_N(g) = n \quad \text{for all} \quad N \ge n.$$

Moreover, if $\mathbb{K} = \mathbb{R}$ then $\operatorname{sign} \mathbf{Hk}_N(g) = \operatorname{sign} \mathbf{Hk}_n(g)$ *for all $N \ge n$.*

Proof: (i) We have seen that if $g(s) = c(s)/d(s)$ is rational then $d(s)$ is a recursive polynomial for $(g_k)_{k\in\mathbb{N}^*}$. Conversely, if $d(s) = \sum_{k=0}^{n} d_k s^k$ is a monic recursive polynomial for (g_k) and the coefficients of c are defined by (45) with $g_0 := 0$ then $g(s) = c(s)/d(s)$.

(ii) Suppose $\operatorname{rank} \mathbf{Hk}(g) = n$. By Lemma 3.4.43 there exists a monic recursive polynomial d of degree n for (g_k) and there does not exist a recursive polynomial of smaller degree. Its coefficients $d_0, \dots, d_{n-1} \in \mathbb{K}$ satisfy the linear equations

$$\begin{bmatrix} g_1 \\ g_2 \\ \vdots \\ g_N \end{bmatrix} d_0 + \begin{bmatrix} g_2 \\ g_3 \\ \vdots \\ g_{N+1} \end{bmatrix} d_1 + \dots + \begin{bmatrix} g_n \\ g_{n+1} \\ \vdots \\ g_{N+n-1} \end{bmatrix} d_{n-1} + \begin{bmatrix} g_{n+1} \\ g_{n+2} \\ \vdots \\ g_{N+n} \end{bmatrix} = 0 \tag{46}$$

for all $N \ge 1$. Defining c by (45) (with $g_0 := 0$) we get $g(s) = c(s)/d(s)$, $\deg c < \deg d$. The pair (c, d) is coprime, since if this were not the case there would exist

a recurrence relation (44) of length $\leq n-1$ among the g_i's, i.e. d would not be a recursive polynomial of minimal degree. Hence $g(s) = c(s)/d(s) \in \mathrm{Rat}_n(\mathbb{K})$.
Now suppose $g(s) = c(s)/d(s) \in \mathrm{Rat}_n(\mathbb{K})$ where d, c are coprime polynomials of the form (40) with $d_n = 1$ and $c_n = 0$. Then by equation (44) the $(n+1)$-st column of the Hankel matrix $\mathbf{Hk}(g)$ is linearly dependent on the preceding n columns. Hence $\operatorname{rank}\mathbf{Hk}(g) \leq n$. But if $r = \operatorname{rank}\mathbf{Hk}(g) < n$ then the r-th column of $\mathbf{Hk}(g)$ would depend linearly on the preceding columns, and there would exist a recursive polynomial $\tilde{d}$ for (g_j) with $\deg\tilde{d} \leq r \leq n-1$. Defining $\tilde{c}$ via (45) (with d replaced by $\tilde{d}$) we would obtain $g(s) = \tilde{c}(s)/\tilde{d}(s)$ and this would contradict the coprimeness of c, d. Hence $\operatorname{rank}\mathbf{Hk}(g) = n$.
Since $\operatorname{rank}\mathbf{Hk}_n(g) \leq \operatorname{rank}\mathbf{Hk}_N(g) \leq \operatorname{rank}\mathbf{Hk}(g) = n$ for all $N \geq n$ it remains to prove $\operatorname{rank}\mathbf{Hk}_n(g) = n$. Now there exists $N \geq n$ such that $\operatorname{rank}\mathbf{Hk}_N(g) = n$. Otherwise the first n columns of $\mathbf{Hk}_N(g)$ would be linearly dependent for all $N \geq n$ and so there would exist a recursive polynomial of degree $< n$ for (g_k), in contradiction with $\operatorname{rank}\mathbf{Hk}(g) = n$. Choose $N \geq n$ such that $\operatorname{rank}\mathbf{Hk}_N(g) = n$ and set

$$D_N = \left[\begin{array}{cccc|cccccc} \multicolumn{4}{c|}{I_n} & \multicolumn{6}{c}{0_{n\times(N-n)}} \\ \hline d_0 & d_1 & \cdots & d_{n-1} & 1 & 0 & 0 & \cdots & & 0 \\ 0 & d_0 & d_1 & \cdots & d_{n-1} & 1 & 0 & \cdots & & 0 \\ 0 & 0 & d_0 & d_1 & \cdots & d_{n-1} & 1 & \ddots & & \vdots \\ \vdots & \ddots & \ddots & \ddots & \ddots & \cdots & \ddots & \ddots & & 0 \\ 0 & \cdots & 0 & 0 & d_0 & d_1 & \cdots & d_{n-1} & & 1 \end{array}\right]_{N\times N} \in \mathbf{Gl}_N(\mathbb{K})$$

where the d_k are the coefficients of $d(s)$. Then using (44)

$$D_N \mathbf{Hk}_N(g) D_N^\top = \begin{bmatrix} \mathbf{Hk}_n(g) & 0 \\ 0 & 0 \end{bmatrix}$$

and therefore $\operatorname{rank}\mathbf{Hk}_n(g) = \operatorname{rank}\mathbf{Hk}_N(g) = n$ (and $\operatorname{sign}\mathbf{Hk}_N(g) = \operatorname{sign}\mathbf{Hk}_n(g)$ in the real case). □

For future use we define $\operatorname{sign}\mathbf{Hk} = \operatorname{sign}\mathbf{Hk}_n$ if $\mathbf{Hk}$ is an infinite real Hankel matrix of finite rank n.

Corollary 3.4.45. *Let $c(s), d(s) \in \mathbb{K}[s]$ be of the form* (40), $\deg d = n$, *and* $g(s) = c(s)/d(s)$. *Then* $\dim\ker\mathbf{Hk}_n(g)$ *is equal to the degree of the greatest common divisor of d and c. In particular, d and c are coprime if and only if* $\det\mathbf{Hk}_n(g) \neq 0$.

Proof: If $c_n \neq 0$, then we may write $c(s)/d(s) = c_n/d_n + \tilde{c}(s)/d(s)$ where $\deg\tilde{c} < \deg d$ and the greatest common divisor of d and c is the same as that of d and $\tilde{c}$. So without restriction of generality we may assume $\deg c < \deg d$. Let $\tilde{n} = n - \deg\tilde{d}(s)$ where $\tilde{d} = (c, d)$ is the greatest common divisor of d, c. Then $g \in \mathrm{Rat}_{\tilde{n}}(\mathbb{K})$, hence $\dim\ker\mathbf{Hk}_n(g) = n - \operatorname{rank}\mathbf{Hk}_n(g) = n - \tilde{n} = \deg\tilde{d}$ by Theorem 3.4.44. □

We illustrate this result in the following example.

Example 3.4.46. Consider $c(s) = s^2 + s$, $d(s) = s^3 + s^2 + 2s + 2$ then equations (44) and (45) take the form

$$\begin{aligned} 2g_1 + 2g_2 + g_3 + g_4 &= 0 \\ 2g_2 + 2g_3 + g_4 + g_5 &= 0 \end{aligned} \quad \text{resp.} \quad \begin{aligned} g_1 &= 1 \\ g_1 + g_2 &= 1, \\ 2g_1 + g_2 + g_3 &= 0 \end{aligned}$$

Hence $g_1 = 1$, $g_2 = 0$, $g_3 = -2$, $g_4 = 0$, $g_5 = 4$ and $\mathbf{Hk}_3(c/d) = \begin{bmatrix} 1 & 0 & -2 \\ 0 & -2 & 0 \\ -2 & 0 & 4 \end{bmatrix}$. So $\dim\ker \mathbf{Hk}_3(c/d) = 1$ and hence the highest common factor of c and d is of degree 1; it is, of course, $\tilde{d}(s) = s + 1$. □

Let Hank($\mathbb{K}$) be the set of all *infinite* Hankel matrices with entries in $\mathbb{K}$, and, for any $n \in \mathbb{N}^*$

$$\mathrm{Hank}(n, \mathbb{K}) = \{\mathbf{Hk} \in \mathrm{Hank}(\mathbb{K}); \mathrm{rank}\, \mathbf{Hk} = n\}, \quad n \in \mathbb{N}.$$

The corresponding counterpart of finite Hankel matrices of rank n is denoted by

$$\mathrm{Hank}_n^*(\mathbb{K}) = \{\mathbf{Hk} \in \mathrm{Hank}_n(\mathbb{K}); \mathrm{rank}\, \mathbf{Hk} = n\} \subset \mathrm{Hank}_n(\mathbb{K})$$

where $\mathrm{Hank}_n(\mathbb{K}) \cong \mathbb{K}^{2n-1}$ denotes the vector space of all $n \times n$ Hankel matrices. Kronecker's Theorem establishes a close relationship between $\mathrm{Rat}_n(\mathbb{K})$, $\mathrm{Hank}(n, \mathbb{K})$ and $\mathrm{Hank}_n^*(\mathbb{K})$. We will now determine this relationship more precisely, endowing the sets with their natural topologies. For many purposes (e.g. parametrization problems, the study of identification algorithms, analysis of uncertain systems, continuity arguments etc.) topologies are needed on sets of rational functions and Hankel matrices. We first introduce a topology on the space $\mathrm{Rat}_n(\mathbb{K})$. For every $g \in \mathrm{Rat}_n(\mathbb{K})$ there exists a unique pair of coprime polynomials d, c of the form (40) such that $g(s) = c(s)/d(s)$ and $d_n = 1, c_n = 0$. Identifying $g(s)$ with its *coefficient vector* $(d_0, \ldots, d_{n-1}; c_0, \ldots c_{n-1}) \in \mathbb{K}^{2n}$ we provide $\mathrm{Rat}_n(\mathbb{K})$ with the topology and Euclidean metrics induced from $\mathbb{K}^{2n}$. Note that with this identification $\mathrm{Rat}_n(\mathbb{K})$ may be considered as an open subset of $\mathbb{K}^{2n}$, obtained by removing from $\mathbb{K}^{2n}$ the closed subset of vectors $(d_0, \ldots, d_{n-1}; c_0, \ldots c_{n-1})$ for which the resultant $R(d, c)$ of the associated polynomials $d(s), c(s)$ has zero determinant, see Definition 3.4.27. Identifying every $\mathbf{Hk} = (h_{i+j-1})_{i,j \in \mathbb{N}^*} \in \mathrm{Hank}(\mathbb{K})$ with the corresponding sequence $(h_k)_{k \in \mathbb{N}^*}$ of its entries, Hank($\mathbb{K}$) is identified with the sequence space $\mathbb{K}^{\mathbb{N}^*}$ and we provide Hank($\mathbb{K}$) with the product topology of $\mathbb{K}^{\mathbb{N}^*}$. With respect to this topology a sequence of Hankel matrices $\mathbf{Hk}^\ell = (h^\ell_{i+j-1})_{i,j \in \mathbb{N}^*}$ converges in Hank($\mathbb{K}$) to $\mathbf{Hk} = (h_{i+j-1})_{i,j \in \mathbb{N}^*}$ if and only if the sequences $h_i^\ell \to h_i$ as $\ell \to \infty$, for all $i \in \mathbb{N}^*$. The subsets $\mathrm{Hank}(n, \mathbb{K}) \subset \mathrm{Hank}(\mathbb{K})$ are provided with the induced topologies. Note that theses subspaces are not closed in Hank($\mathbb{K}$). If $(\mathbf{Hk}^\ell)_{\ell \in \mathbb{N}}$ converges to $\mathbf{Hk}$ in Hank($\mathbb{K}$) and $\mathbf{Hk}^\ell \in \mathrm{Hank}(n, \mathbb{K})$ for all $\ell \in \mathbb{N}$, it may be that $\mathrm{rank}\, \mathbf{Hk} < n$. The space of finite Hankel matrices $\mathrm{Hank}_n(\mathbb{K}) \cong \mathbb{K}^{2n-1}$ is endowed with the usual topology of $\mathbb{K}^{2n-1}$, and with respect to this topology the subset $\mathrm{Hank}_n^*(\mathbb{K})$ is open and dense in $\mathrm{Hank}_n(\mathbb{K})$. The next proposition is a topological version of Kronecker's Theorem.

Proposition 3.4.47. *For every $n \in \mathbb{N}^*$ the following maps are homeomorphisms.*

$$\begin{aligned} \mathtt{H} &: \mathrm{Rat}_n(\mathbb{K}) \to \mathrm{Hank}(n, \mathbb{K}), \quad g \mapsto \mathbf{Hk}(g); \\ \mathtt{H}_n &: \mathrm{Rat}_n(\mathbb{K}) \to \mathrm{Hank}_n^*(\mathbb{K}) \times \mathbb{K}, \quad g \mapsto (\mathbf{Hk}_n(g), g_{2n}); \\ \pi_n &: \mathrm{Hank}(n, \mathbb{K}) \to \mathrm{Hank}_n^*(\mathbb{K}) \times \mathbb{K}, \quad \mathbf{Hk} = (h_{i+j-1}) \mapsto (\mathbf{Hk}_n, h_{2n}). \end{aligned}$$

Proof: By Kronecker's Theorem $\mathbf{Hk}(g) \in \mathrm{Hank}(n, \mathbb{K})$, $\mathbf{Hk}_n(g) \in \mathrm{Hank}_n^*(\mathbb{K})$ for all $g \in \mathrm{Rat}_n(\mathbb{K})$ and $\mathbf{Hk}_n \in \mathrm{Hank}_n^*(\mathbb{K})$ for all $\mathbf{Hk} \in \mathrm{Hank}(n, \mathbb{K})$. Therefore the above mappings are well defined with the specified image spaces. Clearly π_n is continuous. The Laurent coefficients $g_0 = 0, g_1, \ldots, g_n$ of $g(s) = c(s)/d(s) \in \mathrm{Rat}_n(\mathbb{K})$ are determined from the linear equations (43) (with $c_n = 0$) and depend continuously (even rationally) on the coefficients $c_0, \ldots, c_{n-1}$ and $d_0, \ldots, d_{n-1}$. The g_j for $j \geq n+1$ are then determined recursively by (44). Hence $\mathtt{H} : g \to \mathbf{Hk}(g)$ and $\mathtt{H}_n : g \mapsto (\mathbf{Hk}_n, h_{2n})$ are continuous on their domains of definition. For this we identify every finite Hankel matrix with the finite sequence of its entries. This yields

$$\mathrm{Hank}_n^*(\mathbb{K}) \times \mathbb{K} \cong \{(h_1, \ldots, h_{2n}) \in \mathbb{K}^{2n};\ (h_{i+j-1})_{i,j \in \underline{n}} \in \mathrm{Hank}_n^*(\mathbb{K})\}.$$

Let $h = (h_1, \ldots, h_{2n}) = (\mathbf{Hk}_n, h_{2n})$ be any vector in this set and define $g_k = h_k$ for $k = 1, \ldots, 2n$. Then there exists a unique solution $(d_0, \ldots, d_{n-1})$ of the equation (46) with $N = n$. Define g_{n+j}, $j = n+1, \ldots$ by the recursive equation (44) (with $d_n = 1$). Then $(d_0, \ldots, d_{n-1})$ and g_{n+j}, $j = n+1, \ldots$ depend continuously (even rationally) on h. By construction $d(s) = \sum_{k=0}^{n} d_k s^k$ (with $d_n = 1$) satisfies (44) for all $j \geq 1$ and so $c(s)$ defined by (43) is a polynomial such that $g(s) := c(s)/d(s) = \sum_{k=1}^{\infty} g_k s^{-k}$ and hence $\mathbf{Hk}_n(g) = h$. Since $n = \mathrm{rank}\,\mathbf{Hk}_n = \mathrm{rank}\,\mathbf{Hk}_n(g) \leq \mathrm{rank}\,\mathbf{Hk}(g) \leq \deg d = n$ it follows from Kronecker's Theorem that $g \in \mathrm{Rat}_n(\mathbb{K})$. Whence $\mathtt{H}_n$ is surjective, and the coefficient vectors of $c(s), d(s)$ depend continuously on h. On the other hand the rational function $g(s) = c(s)/d(s) = \sum_{k=1}^{\infty} g_k s^{-k} \in \mathrm{Rat}_n(\mathbb{K})$ is uniquely determined by $\mathtt{H}_n(g) = h$, since this equation is equivalent to equations (46) and (43) where $g_i = h_i$, $i = 1, \ldots, 2n$, $g_0 = 0$. This proves that $\mathtt{H}_n$ is a homeomorphism from $\mathrm{Rat}_n(\mathbb{K})$ onto $\mathrm{Hank}_n^*(\mathbb{K}) \times \mathbb{K}$.
The same arguments show that for every $h = (h_1, \ldots, h_{2n}) = (\mathbf{Hk}_n, h_{2n}) \in \mathrm{Hank}_n^*(\mathbb{K}) \times \mathbb{K}$ there exists exactly one $\mathbf{Hk} \in \mathrm{Hank}(n, \mathbb{K})$ such that $\pi_n(\mathbf{Hk}) = (\mathbf{Hk}_n, h_{2n})$, and the entries of $\mathbf{Hk}$ depend continuously on h. Thus π_n is a homeomorphism between $\mathrm{Hank}(n, \mathbb{K})$ and $\mathrm{Hank}_n^*(\mathbb{K}) \times \mathbb{K}$. This concludes the proof since $\pi_{2n} \circ \mathtt{H} = \mathtt{H}_n$. □

Remark 3.4.48. The preceding proof shows that for any given sequence $(h_1, \ldots, h_{2n}) \in \mathbb{K}^{2n}$ with $\mathrm{rank}(h_{i+j-1})_{i,j \in \underline{n}} = n$ there exists a unique infinite sequence $(g_k)_{k \in \mathbb{N}^*}$ satisfying $g_k = h_k$ for $k = 1, \ldots, 2n$ and $\mathrm{rank}\,\mathbf{Hk}(g) = n$. Such a sequence $g = (g_k)$ is called a *singular extension* of $(h_1, \ldots, h_{2n})$. Moreover, $\pi_{2n}^{-1} : (h_1, \ldots, h_{2n}) \mapsto (g_k)_{k \in \mathbb{N}^*}$ is a homeomorphism from $\mathrm{Hank}_n^*(\mathbb{K}) \times \mathbb{K}$ onto $\mathrm{Hank}(n, \mathbb{K})$. □

We will now specialize to *real* rational functions and *real* Hankel matrix. We begin by determining the signature of the *real* Hankel matrix $\mathbf{Hk} = \mathbf{Hk}(g)$ where $g(s) = c(s)/d(s) \in \mathrm{Rat}_n(\mathbb{R})$. Suppose that d is monic and decompose $d(s)$ into irreducible real polynomials

$$d(s) = \prod_{j=1}^{r} (s - \rho_j)^{m_j} \prod_{j=r+1}^{\ell} (s^2 - 2\rho_j s + \rho_j^2 + \omega_j^2)^{m_j}, \quad \rho_j, \omega_j \in \mathbb{R}, \omega_j > 0. \tag{47}$$

$\rho_1, \ldots \rho_r$ are the distinct real roots of d and $\rho_{r+1} \pm \imath\omega_{r+1}, \ldots, \rho_\ell \pm \imath\omega_\ell$ are the distinct pairs of complex conjugate roots of d. As a consequence of the above factorization of d we obtain the following representation of g (e.g. by partial fraction decomposition)

$$g(s) = \sum_{j=1}^{r} \frac{c_j(s)}{(s-\rho_j)^{m_j}} + \sum_{j=r+1}^{\ell} \frac{c_j(s)}{(s^2 - 2\rho_j s + \rho_j^2 + \omega_j^2)^{m_j}}$$

where the $c_j(s)$ are real polynomials with $\deg c_j < m_j$ for $j \in \underline{r}$ and $\deg c_j < 2m_j$ for $j = r+1, \dots, \ell$. Let

$$\mathbf{Hk}_j = \mathbf{Hk}\left(\frac{c_j(s)}{(s-\rho_j)^{m_j}}\right), \; j \in \underline{r}, \;\; \mathbf{Hk}_j = \mathbf{Hk}\left(\frac{c_j(s)}{(s^2 - 2\rho_j s + \rho_j^2 + \omega_j^2)^{m_j}}\right), \; r < j \le \ell \quad (48)$$

be the associated infinite Hankel matrices. Then $\mathbf{Hk}(g) = \mathbf{Hk}_1 + \dots + \mathbf{Hk}_\ell$ and

$$\operatorname{rank} \mathbf{Hk} = n = m_1 + \dots + m_r + 2(m_{r+1} + \dots + m_\ell) = \sum_{j=1}^{\ell} \operatorname{rank} \mathbf{Hk}_j$$

by Kronecker's Theorem. Applying Lemma A.1.32 of Section A.1 it follows that

$$\operatorname{sign} \mathbf{Hk} = \operatorname{sign} \mathbf{Hk}_1 + \dots + \operatorname{sign} \mathbf{Hk}_\ell. \quad (49)$$

So it remains to determine the signatures of the Hankel matrices $\mathbf{Hk}_j$. For this we will use a transformation technique where the rational functions represented by the Hankels are continuously transformed into simpler rational functions for which the signature is easily determined. This technique is based on the following lemma.

Lemma 3.4.49. *If $\mathcal{H} \subset \mathcal{H}_n(\mathbb{C})$ is a connected set of $n \times n$ Hermitian matrices of fixed rank r, then all the matrices in $\mathcal{H}$ have the same signature.*

Proof: For $\nu = -r, \dots, r$ let $\mathcal{H}_\nu = \{H \in \mathcal{H};\, \operatorname{sign}(H) = \nu\}$. If $H \in \mathcal{H}_\nu$ then H has exactly $(r+\nu)/2$ positive eigenvalues, $(r-\nu)/2$ negative eigenvalues and $n-r$ zero eigenvalues (taking into account multiplicities). Since all the matrices in $\mathcal{H}$ have exactly $n-r$ zero eigenvalues, it follows from the continuous dependence of the spectrum $\sigma(H)$ on the entries of H that $\mathcal{H}_\nu$ is an open (possibly empty) subset of $\mathcal{H}$. But these subsets form a partition of $\mathcal{H}$, and so, by connectivity of $\mathcal{H}$ there exists a ν_0 such that $\mathcal{H}_{\nu_0} = \mathcal{H}$. □

We now apply this lemma to the above *real* Hankel matrices by transforming the associated rational functions continuously to more simple ones.

Lemma 3.4.50. *Suppose $\rho \in \mathbb{R}$, $m, n \in \mathbb{N}^*$, $m \le n$ and $c(s) \in \mathbb{R}[s]$, $\deg c < m$, $c(\rho) \neq 0$. Then*

$$\operatorname{sign} \mathbf{Hk}_n\left(\frac{c(s)}{(s-\rho)^m}\right) = \begin{cases} 0 & \text{if } m \text{ even} \\ \operatorname{sign} c(\rho) & \text{if } m \text{ odd.} \end{cases} \quad (50)$$

Proof: Dividing $c(s)$ by $(s-\rho)$ we obtain $c(s) = \tilde{c}(s)(s-\rho) + c(\rho)$ where $\tilde{c}(s) \in \mathbb{R}[s]$. Let

$$c_t(s) = (1-t)\,\tilde{c}(s)(s-\rho) + c(\rho), \quad t \in [0,1].$$

Then $c_0(s) = c(s)$, $c_1(s) = c(\rho)$ and the polynomials $c_t(s)$ and $d(s) = (s-\rho)^m$ are coprime for all $t \in [0,1]$. It follows from Theorem 3.4.44 that $\operatorname{rank} \mathbf{Hk}_n(c_t/d) = m$, and so the preceding Lemma implies that $\operatorname{sign} \mathbf{Hk}_n(c/d) = \operatorname{sign} \mathbf{Hk}_n(c(\rho)/d)$. Now we transform $d(s) = (s-\rho)^m$ continuously to $d_0(s) = s^m$ (replacing ρ by $t\rho$, $t \in [0,1]$) and applying the same argument we obtain

$$\operatorname{sign} \mathbf{Hk}_n(c/d) = \operatorname{sign} \mathbf{Hk}_n(c(\rho)/d) = \operatorname{sign} \mathbf{Hk}_m(c(\rho)/s^m) = \operatorname{sign} \begin{bmatrix} 0 & 0 & \dots & 0 & c(\rho) \\ \vdots & & & \cdot^{\cdot^{\cdot}} & \vdots \\ c(\rho) & 0 & \dots & 0 & 0 \end{bmatrix}_{m\times m}.$$

From this (50) follows, since this latter matrix has two eigenvalues $\pm c(\rho)$, where both eigenvalues have the same multiplicity $m/2$ if m is even, whereas $c(\rho)$ is of multiplicity $(m+1)/2$ and $-c(\rho)$ of multiplicity $(m-1)/2$ if m is odd. □

Lemma 3.4.51. *Suppose $d(s) = (s^2 - 2\rho s + \rho^2 + \omega^2)^m$ where $\rho, \omega \in \mathbb{R}$, $\omega \neq 0$, $m \in \mathbb{N}^*$, and let $c(s) \in \mathbb{R}[s]$, $\deg c < 2m$, $c(\rho \pm \imath\omega) \neq 0$. Then*

$$\operatorname{sign} \mathbf{Hk}_n(c/d) = 0, \quad n \geq 2m.$$

Proof: Dividing $c(s)$ by $(s^2 - 2\rho s + \rho^2 + \omega^2)$ we obtain

$$c(s) = \tilde{c}(s)(s^2 - 2\rho s + \rho^2 + \omega^2) + e_1 s + e_0, \quad \text{where } \tilde{c}(s) \in \mathbb{R}[s],\ (e_1, e_0) \in \mathbb{R}^2 \setminus \{(0,0)\}.$$

For $t \in [0,1]$, we define $d_t(s) = [s^2 - 2(1-t)\rho s + (1-t)^2(\rho^2 + \omega^2)]^m$ and

$$c_t(s) = (1-t)\,\tilde{c}(s)(s^2 - 2(1-t)\rho s + (1-t)^2(\rho^2 + \omega^2)) + (1-t)[t\delta(e_1)s + e_1 s + e_0] + t,$$

where $\delta(e_1) = 0$ if $e_1 \neq 0$ and $\delta(0) = 1$. Then $c_0(s) = c(s)$, $c_1(s) \equiv 1$ and $d_0(s) = d(s)$, $d_1(s) = s^{2m}$. Now since $(1-t)[(t\delta(e_1)s + e_1 s + e_0)] + t$ is a linear real polynomial in s for every $t \in (0,1)$ and therefore has only real roots we see that $c_t(s)$, $d_t(s)$ are coprime for all $t \in [0,1]$. Hence $\operatorname{rank} \mathbf{Hk}_n(c_t/d) = 2m$ for all $t \in [0,1]$ by Theorem 3.4.44, and so by Lemma 3.4.49 $\operatorname{sign} \mathbf{Hk}_n(c/d) = \operatorname{sign} \mathbf{Hk}_n(1/s^{2m})$. Now the result follows from the previous lemma. □

Remark 3.4.52. The proof shows that under the conditions of the preceding lemma $g(s) = c(s)/d(s)$ can be connected by an arc in $\mathrm{Rat}_{2m}(\mathbb{R})$ to $1/s^{2m}$. □

Theorem 3.4.53 (Hermite–Hurwitz). *Let $g(s) = c(s)/d(s)$ be a proper real rational function and consider the decomposition*

$$g(s) = c_n/d_n + \sum_{j=1}^{r} \frac{c_j(s)}{(s-\rho_j)^{m_j}} + \sum_{j=r+1}^{\ell} \frac{c_j(s)}{(s^2 - 2\rho_j s + \rho_j^2 + \omega_j^2)^{m_j}} \tag{51}$$

where $c_j(s)$, $j \in \underline{\ell}$ are real polynomials of degree $< m_j$ for $j \in \underline{r}$ and $< 2m_j$ for $j = r+1, \dots, \ell$, and $\rho_1, \dots \rho_r$ are the distinct real roots and $\rho_{r+1} \pm \imath\omega_{r+1}, \dots, \rho_\ell \pm \imath\omega_\ell$ are the distinct pairs of complex conjugate roots of d. Then

$$\operatorname{sign} \mathbf{Hk}(g) = \sum_{j \in \underline{r},\ m_j \text{ odd}} \operatorname{sign} c_j(\rho_j) = CI_{-\infty}^{\infty}(g). \tag{52}$$

Proof: Define $\mathbf{Hk}_j$, $j \in \underline{\ell}$ by (48). Then, applying the previous two lemmata to the $\mathbf{Hk}_j$ and making use of (49) we get

$$\operatorname{sign} \mathbf{Hk} = \operatorname{sign} \mathbf{Hk}_1 + \ldots + \operatorname{sign} \mathbf{Hk}_\ell = \sum_{j \in \underline{r},\ m_j \text{ odd}} \operatorname{sign} \mathbf{Hk}_j = \sum_{j \in \underline{r},\ m_j \text{ odd}} \operatorname{sign} c_j(\rho_j).$$

The proof is concluded by noting that the RHS is just the Cauchy index of $g(s)$, see Example 3.4.19. □

The following corollary serves as an illustration of how Hankel matrices can be used to count the number of roots of a polynomial located in certain subsets of the complex plane. It is of historical interest since its proof by Jacobi is considered to be the first application of quadratic forms to the investigation of the roots of an algebraic equation, see [318] and *Notes and References*.

Corollary 3.4.54. *Let $p(s)$ be a real non-constant polynomial and $g(s) = p'(s)/p(s)$. Then the number of distinct roots of $p(s)$ is equal to* $\operatorname{rank} \mathbf{Hk}(g)$, *and the number of distinct real roots of $p(s)$ is equal to* $\operatorname{sign} \mathbf{Hk}(g)$.

Proof: Let $s_1, \ldots, s_\ell$ be the distinct roots of p and $p(s) = a_n \prod_{i=0}^{\ell}(s - s_i)^{m_i}$ the corresponding factorization of $p(s)$. Then

$$g(s) = \frac{p'(s)}{p(s)} = \sum_{i=0}^{\ell} \frac{m_i}{s - s_i} \in \operatorname{Rat}_\ell(\mathbb{R}), \tag{53}$$

and it follows from Kronecker's Theorem that $\ell = \operatorname{rank} \mathbf{Hk}(g)$. On the other hand, (53) implies that the Cauchy index of $g(s)$ is equal to the number r of real roots of p (see Example 3.4.19), hence $r = \operatorname{sign} \mathbf{Hk}(g)$ by Theorem 3.4.53. □

The next corollary is of importance for stability analysis since it gives necessary and sufficient conditions for $\mathbf{Hk}_n(g)$ to be positive definite.

Corollary 3.4.55. *Suppose $g(s) = c(s)/d(s)$ is a proper real rational function, $\deg d = n$ and $d(s), c(s)$ are coprime. Then the following conditions are equivalent.*

(i) The Hankel matrix $\mathbf{Hk}_n(c/d)$ is positive definite.

(ii) All the roots t_j, $j \in \underline{n}$ of d are real and simple, and the numbers $c(t_j)/d'(t_j)$, $j \in \underline{n}$ are all positive.

(iii) All the roots t_j of d and t_i' of c are real, simple and interlacing, and $d_n c_{n-1} - c_n d_{n-1} > 0$.

(iv) All the roots t_j of d and t_i' of c are real, simple, interlacing, and

$$\exists t \in \mathbb{R} : \quad d'(t)c(t) - d(t)c'(t) > 0. \tag{54}$$

If one of these conditions is satisfied, then the inequality in (54) holds for all $t \in \mathbb{R}$.

Proof: (i) $\Longrightarrow$ (ii): By the previous theorem we know that $\mathbf{Hk}_n(g)$ is positive definite, i.e. $\operatorname{sign} \mathbf{Hk}(g) = n$, if and only if there are n real roots t_j of d of odd degree, such that $c_j(t_j) > 0$, $j \in \underline{n}$ (see (52)). This means that all the roots of d are real and simple, and $g(s)$ has the representation

$$g(s) = \gamma_0 + \sum_{j=1}^{n} \frac{\gamma_j}{s - t_j}, \qquad t_n < \ldots < t_1 \tag{55}$$

with $\gamma_j > 0$, $j \in \underline{n}$. Multiplying this equality by $d(s) = d_n \prod_{i \in \underline{n}}(s - t_i)$ we obtain

$$c(s) = \gamma_0 d(s) + \sum_{j=1}^{n} \gamma_j d_n \prod_{i \in \underline{n}, i \neq j} (s - t_i) \quad \text{and} \quad c(t_j) = \gamma_j d_n \prod_{i \in \underline{n}, i \neq j} (t_j - t_i) = \gamma_j d'(t_j).$$

Hence $c(t_j)/d'(t_j) = \gamma_j > 0$ for all $j \in \underline{n}$, i.e. (ii) holds.
(ii) $\Longrightarrow$ (iii): Suppose (ii) so that $g(s)$ has the representation (55) with $\gamma_j = c(t_j)/d'(t_j) > 0$ for $j \in \underline{n}$. It follows that

$$\frac{d(t)c'(t) - d'(t)c(t)}{d(t)^2} = g'(t) = -\sum_{j=1}^{n} \frac{\gamma_j}{(t-t_j)^2} < 0, \quad t \in \mathbb{R} \setminus \{t_j; j \in \underline{n}\}, \tag{56}$$

and so $g(t)$ is strictly decreasing on each interval $I \subset \mathbb{R} \setminus \{t_j; j \in \underline{n}\}$. Moreover, since $d(s)/(s-t_j) = d_n \prod_{i \in \underline{n}, i \neq j}(s - t_i) \to d'(t_j) \neq 0$ as $s \to t_j$,

$$d'(t)c(t) - d(t)c'(t) = \sum_{j=1}^{n} \frac{c(t_j)}{d'(t_j)} \frac{d(t)^2}{(t-t_j)^2} > 0, \quad t \in \mathbb{R}. \tag{57}$$

It follows from this inequality that every real zero of $c(\cdot)$ cannot also be a zero of $c'(\cdot)$ and hence must be simple. Now $g(t)$ tends to ∞ (resp. $-\infty$) as t tends from the right (resp. left) towards any pole t_j, $j \in \underline{n}$. So $g(t)$ must have exactly one zero between each pair of neighbouring poles. Thus $\deg c(s) \geq n-1$. If $\deg c(s) = n-1$ then all the roots t'_j of $c(s)$ must be real and can be ordered as follows

$$t_n < t'_{n-1} < t_{n-1} < \cdots < t'_2 < t_2 < t'_1 < t_1.$$

If $\deg c(s) = n$, again each interval between neighbouring poles must contain exactly one simple root of $c(s)$. The n-th root of $c(s)$ is either smaller than t_n (if $\gamma_0 = \lim_{|t| \to \infty} g(t) > 0$) or larger than t_1 (if $\gamma_0 < 0$). This proves the interlacing condition. Now (57) shows that $d'(t)c(t) - d(t)c'(t)$ tends to ∞ as t^{2n-2} for $t \to \infty$. On the other hand, the leading terms of $d'(t)c(t)$ and $d(t)c'(t)$ cancel, so that the leading power of $d'(t)c(t) - d(t)c'(t)$ is t^{2n-2} with coefficient

$$[nd_n c_{n-1} + (n-1)d_{n-1}c_n] - [nc_n d_{n-1} + (n-1)c_{n-1}d_n] = d_n c_{n-1} - c_n d_{n-1}.$$

Therefore it follows that $d_n c_{n-1} - c_n d_{n-1} > 0$.
(iii) $\Longrightarrow$ (iv): This is clear, since $d_n c_{n-1} - c_n d_{n-1} > 0$ implies $d'(t)c(t) - d(t)c'(t)$ for all large t by the previous consideration.
(iv) $\Longrightarrow$ (i): Since all the roots of d are simple and real, $g(s)$ can be represented in the form (55). We will now prove that all the γ_i are of the same sign. In fact, if this were not so, there would exist an $1 \leq i < n$ such that $\gamma_i \gamma_{i+1} < 0$. Assume e.g. $\gamma_i > 0, \gamma_{i+1} < 0$, then $g(t)$ would tend to $-\infty$ as $t \to t_i$ from the left and as $t \to t_{i+1}$ from the right. As a consequence $g(t)$ would either have no zero or a zero of higher multiplicity or at least two zeros in $[t_i, t_{i+1}]$. But the interlacing condition excludes all these possibilities. Hence all the γ_i in (55) have the same sign. But then it follows from (54) and (56) that all γ_i must be positive. □

Remark 3.4.56. Suppose that $g(s) = c(s)/d(s) \in \mathrm{Rat}_n(\mathbb{R})$ with c and d coprime and the roots of d are all real and simple. Then the previous proof shows that sign $\mathbf{Hk}(g) = \pm n$, i.e. all coefficients γ_i in (55) are of the same sign, if and only if the roots of c and d are real, simple and interlacing.
The additional conditions in (iii) and (iv) only serve to ensure that $\mathbf{Hk}_n(g)$ is not negative but positive definite, or, in terms of the rational function g, that $g'(t) < 0$ (and not $g'(t) > 0$) for all $t \in \mathbb{R} \setminus \{t_j; j \in \underline{n}\}$. □

We conclude this subsection by determining the connected components of the space $\mathrm{Rat}_n(\mathbb{R})$ of strictly proper real rational functions of degree n. For this we need the following lemmata.

Lemma 3.4.57. *Let $g_t^i(s) = c_t^i(s)/d_t^i(s)$, $t \in [0,1]$ be two arcs in $\mathrm{Rat}_{n_i}(\mathbb{R})$ with $\deg d_t^i(s) = n_i$ for all $t \in [0,1]$ and suppose that $d_t^1(s)$, $d_t^2(s)$ have no common roots for all $t \in [0,1]$. Then $g_t(s) = g_t^1(s) + g_t^2(s)$, $t \in [0,1]$ is an arc in $\mathrm{Rat}_{n_1+n_2}(\mathbb{R})$.*

Proof: We have

$$g_t(s) = \frac{c_t^1(s)}{d_t^1(s)} + \frac{c_t^2(s)}{d_t^2(s)} = \frac{c_t^1(s)d_t^2(s) + c_t^2(s)d_t^1(s)}{d_t^1(s)d_t^2(s)}.$$

By assumption the greatest normalized divisor $(c_t^i(s), d_t^i(s)) = 1$, $i \in \underline{2}$ and also $(d_t^1(s), d_t^2(s)) = 1$, for each $t \in [0,1]$. Hence if λ is a root of $d_t^1(s)$ then $(c_t^1 d_t^2 + c_t^2 d_t^1)(\lambda) = c_t^1(\lambda)d_t^2(\lambda) \neq 0$. Similarly if $d_t^2(\lambda) = 0$, then $(c_t^1 d_t^2 + c_t^2 d_t^1)(\lambda) = d_t^1(\lambda)c_t^2(\lambda) \neq 0$. Hence $d_1 d_2$ and $c_t^1 d_t^2 + c_t^2 d_t^1$ are coprime so that $g_t \in \mathrm{Rat}_{n_1+n_2}(\mathbb{R})$. Finally it is clear that the coefficients of the polynomials $(c_t^1 d_t^2 + c_t^2 d_t^1)(s)$ and $(d_t^1 d_t^2)(s)$ depend continuously on $t \in [0,1]$ since the coefficients of $c_t^1, c_t^2, d_t^1, d_t^2$ have this property by assumption. □

Lemma 3.4.58. *Suppose $\gamma_1, \gamma_2 \in \mathbb{R}$, $\gamma_1\gamma_2 < 0$ and $\rho_1, \rho_2 \in \mathbb{R}$, $\rho_1 \neq \rho_2$ and let $\rho = (\rho_1 + \rho_2)/2$. Then there exists an arc $g_t(s)$, $t \in [0,1]$ in $\mathrm{Rat}_2(\mathbb{R})$ connecting $g_0(s) = \gamma_1/(s-\rho_1) + \gamma_2/(s-\rho_2)$ with $g_1(s) = \gamma_1(\rho_1 - \rho_2)/(s-\rho)^2$ such that all the poles of g_t, $t \in [0,1]$ are real and lie between ρ_1 and ρ_2.*

Proof: Transforming γ_i continuously to $\operatorname{sign}\gamma_i$ and applying the preceding lemma we may assume that $|\gamma_i| = 1$ and hence $\gamma_1 + \gamma_2 = 0$. Suppose e.g. $\rho_1 < \rho_2$ and define $\rho_i(t) = \rho_i + t(\rho - \rho_i)$, $\gamma_i(t) = \gamma_i/(1-t)$ so that $\rho_1(t) - \rho_2(t) = (1-t)(\rho_1 - \rho_2)$ and $\rho_i(t) \in [\rho_1, \rho_2]$. Now $\gamma_1(t) + \gamma_2(t) = 0$ for $t \in [0,1)$, and hence

$$\begin{aligned}\gamma_1(t)(s-\rho_2(t)) + \gamma_2(t)(s-\rho_1(t)) &= \gamma_1(t)(s-\rho_2(t)) - \gamma_1(t)(s-\rho_1(t)) \\ &= \gamma_1(t)(\rho_1(t) - \rho_2(t)) = \gamma_1(\rho_1 - \rho_2), \quad t \in (0,1).\end{aligned}$$

Since $\rho_1(t) \neq \rho_2(t)$ for all $t \in (0,1)$ we have by Lemma 3.4.57

$$g_t(s) := \frac{\gamma_1(t)}{s-\rho_1(t)} + \frac{\gamma_2(t)}{s-\rho_2(t)} = \frac{\gamma_1(\rho_1-\rho_2)}{(s-\rho_1(t))(s-\rho_2(t))} \in \mathrm{Rat}_2(\mathbb{R}), \quad t \in [0,1).$$

This equation shows that $t \mapsto g_t$ can be extended continuously to $[0,1]$ and describes an arc in $\mathrm{Rat}_2(\mathbb{R})$ connecting $g_0(s)$ with $g_1(s)$. □

The previous lemmata allow us to determine the number of connected components of $\mathrm{Rat}_n(\mathbb{R})$ for $n = 1, 2$.

Example 3.4.59. (a) $\mathrm{Rat}_1(\mathbb{R})$ consists of all rational functions of the form

$$g(s) = c/(s-\rho), \qquad \rho \in \mathbb{R},\ c \in \mathbb{R}^*.$$

Thus $\mathrm{Rat}_1(\mathbb{R})$ is homeomorphic (via the map $g \mapsto (\rho, c)$) to $\mathbb{R} \times \mathbb{R}^*$ and has two connected components. These two connected components are classified by the Cauchy index: $c < 0$

if and only if $CI_{-\infty}^{\infty}g = -1$ and $c > 0$ if and only if $CI_{-\infty}^{\infty}g = 1$.
(b) Each element in $\mathrm{Rat}_2(\mathbb{R})$ has a real partial fraction decomposition of one of the following forms

$$(i)\; g(s) = \frac{\gamma_1}{s-\rho_1} + \frac{\gamma_2}{s-\rho_2},\; \rho_i \in \mathbb{R}, \rho_1 > \rho_2 \quad \text{or} \quad (ii)\; g(s) = \frac{c_1 s + c_0}{(s-\rho)^2 + \omega^2},\; \rho \in \mathbb{R}, \omega \geq 0$$

where $\gamma_1, \gamma_2 \in \mathbb{R}^*$ and $(c_0, c_1) \in \mathbb{R}^2 \setminus \{(0,0)\}$, respectively. The proof of Lemma 3.4.51 shows that in case (ii) $g(s)$ can be connected to $1/s^2$ in $\mathrm{Rat}_2(\mathbb{R})$ (see Remark 3.4.52). By the preceding Lemma the same holds true in case (i) if $\gamma_1\gamma_2 < 0$. Thus every $g \in \mathrm{Rat}_2(\mathbb{R})$ of Cauchy index 0 can be connected to $f_{2,0}(s) := 1/s^2$ by an arc in $\mathrm{Rat}_2(\mathbb{R})$. It remains to consider the case (i) with $\gamma_1\gamma_2 > 0$. Then the numerators γ_i can be continuously deformed to $\operatorname{sign}\gamma_i$ and the poles ρ_1, ρ_2 to -1 and -2, respectively. Therefore, if $g(s) \in \mathrm{Rat}_2(\mathbb{R})$ has Cauchy index 2, i.e. $\operatorname{sign}\gamma_i = 1$, then g can be connected to $f_{2,2} := 1/(s+1) + 1/(s+2)$ by an arc in $\mathrm{Rat}_2(\mathbb{R})$; and if g has Cauchy index -2, i.e. $\operatorname{sign}\gamma_i = -1$, then g can be connected to $f_{2,-2} := -[1/(s+1)+1/(s+2)]$. We will see in a more general context below that two elements of $\mathrm{Rat}_n(\mathbb{R})$ of different Cauchy index cannot be connected by an arc in $\mathrm{Rat}_n(\mathbb{R})$. Thus $\mathrm{Rat}_2(\mathbb{R})$ has exactly 3 connected components which are classified by the Cauchy index. □

The following theorem generalizes the findings of the previous example.

Theorem 3.4.60 (Brockett). *For $n \geq 1$, $\mathrm{Rat}_n(\mathbb{R})$ has $n+1$ connected components given by*

$$\mathrm{Rat}(n,\nu) = \{g \in \mathrm{Rat}_n(\mathbb{R});\; CI_{-\infty}^{\infty}(g) = n - 2\nu\}, \quad \nu = 0, \ldots, n. \tag{58}$$

Proof: Let g_t, $0 \leq t \leq 1$ be an arc in $\mathrm{Rat}_n(\mathbb{R})$ then $\mathcal{H} = \{\mathbf{Hk}_n(g_t); 0 \leq t \leq 1\}$ is a connected set of Hermitian matrices of fixed rank n (by Proposition 3.4.47), and so all the $\mathbf{Hk}_n(g_t)$ have the same signature by Lemma 3.4.49. Hence all g_t, $0 \leq t \leq 1$ have the same Cauchy index by Theorem 3.4.53. Therefore any arc in $\mathrm{Rat}_n(\mathbb{R})$ which intersects $\mathrm{Rat}(n,\nu)$ is contained in $\mathrm{Rat}(n,\nu)$.
It remains to prove that $\mathrm{Rat}(n,\nu)$ is connected. For this it suffices to verify that every $f \in \mathrm{Rat}(n,\nu)$ can be continuously transformed to a fixed rational function $f_{n,\nu} \in \mathrm{Rat}(n,\nu)$. We will show this in several steps.
1. Given any $f(s) \in \mathrm{Rat}(n,\nu)$, let $g(s) = c(s)/d(s) \in \mathrm{Rat}(n,\nu)$ be a function such that the number of distinct real roots minus the number of pairs of complex roots with distinct non-zero imaginary parts is minimal amongst all elements in $\mathrm{Rat}(n,\nu)$ which can be connected to $f(s)$ by an arc in $\mathrm{Rat}(n,\nu)$. $d(s)$ can be factorized as

$$d(s) = \prod_{i=1}^{r}(s-\rho_i)\prod_{j=1}^{h}(s^2 - 2\rho_{r+j}s + \rho_{r+j}^2 + \omega_j^2)$$

where $\rho_1, \ldots, \rho_r$ are distinct real numbers, $\rho_{r+j} \in \mathbb{R}$ and $\omega_j \geq 0$ for $j \in \underline{h}$. Then necessarily $\omega_1, \ldots, \omega_h$ are all positive and mutually distinct. Otherwise, $\mathrm{Rat}_n(\mathbb{R})$ being open and hence locally connected, the number of complex roots with distinct imaginary parts could be increased by small perturbations (replacing ω_j by $\omega_j + \varepsilon_j$ where $\varepsilon_1, \ldots, \varepsilon_h > 0$ are distinct and sufficiently small to ensure $c(s)/d_\varepsilon(s) \in \mathrm{Rat}(n,\nu)$). It follows that $g(s)$ has the partial fraction decomposition

$$g(s) = \sum_{i=1}^{r}\frac{\gamma_i}{(s-\rho_i)} + \sum_{j=1}^{h}\frac{c_{j1}s + c_{j0}}{(s^2 - 2\rho_{r+j}s + \rho_{r+j}^2 + \omega_j^2)}. \tag{59}$$

2. We will now show that the number r of distinct real poles ρ_i of $g(s)$ is equal to the absolute value of the Cauchy index $n-2\nu$ and all the residues γ_i, $i \in \underline{r}$ have the same sign: $\operatorname{sign}\gamma_i = \operatorname{sign}(n-2\nu)$. In fact, since (see Example 3.4.19)

$$n - 2\nu = CI(g) = \sum_{j=1}^{r} \operatorname{sign}\gamma_j \tag{60}$$

it follows that $r \geq |n-2\nu|$, and that $r > |n-2\nu|$ if and only if there are $i,j \in \underline{r}$ such that γ_i and γ_j have different signs. Now suppose $r > |n-2\nu|$. Then there exist $i,j \in \underline{r}$ such that $\gamma_i\gamma_j < 0$ and $\rho_i < \rho_j$ are "neighbours" in the sense that $\rho_k \notin [\rho_i, \rho_j]$ for all $k \in \underline{r} \setminus \{i,j\}$. After suitable renumbering we may assume $i = 1$, $j = 2$. By Lemma 3.4.58 the function $\gamma_1/(s-\rho_1)+\gamma_2/(s-\rho_2)$ can be continuously deformed in $\mathrm{Rat}_2(\mathbb{R})$ to $\gamma_3/(s-\rho)^2$ (where $\rho = (\rho_1+\rho_2)/2$) such that the poles of the deformations always remain in $[\rho_1, \rho_2]$. Now $\gamma_3/(s-\rho)^2$ can be further deformed along a small arc in $\mathrm{Rat}_2(\mathbb{R})$ into $\gamma_3/[(s-\rho)^2+\omega^2]$ where $\omega > 0$ is chosen sufficiently small so that $\omega < \min\{\omega_j; j \in \underline{h}\}$. Let $f_t, 0 \leq t \leq 1$ be an arc in $\mathrm{Rat}_2(\mathbb{R})$ along which $f_0(s) = \gamma_1/(s-\rho_1)+\gamma_2/(s-\rho_2)$ is deformed into $f_1(s) = \gamma_3/[(s-\rho)^2+\omega^2]$. Then it follows from Lemma 3.4.57 that

$$g_t(s) = \sum_{i=3}^{r} \frac{\gamma_i}{(s-\rho_i)} + f_t(s) + \sum_{j=1}^{h} \frac{c_{j1}s + c_{j0}}{(s^2 - 2\rho_{r+j}s + \rho_{r+j}^2 + \omega_j^2)}, \quad t \in [0,1]$$

describes an arc in $\mathrm{Rat}(n,\nu)$ connecting $g_0(s) = g(s)$ with a function $g_1(s) \in \mathrm{Rat}(n,\nu)$ having a smaller number of real poles and more pairs of complex poles with distinct positive imaginary parts. This is a contradiction to the choice of g. Hence $r = |n-2\nu|$ and, as a consequence of (60), $\operatorname{sign}\gamma_i = \operatorname{sign}(n-2\nu)$, $i \in \underline{r}$.

3. It remains to prove that the function $g(s) \in \mathrm{Rat}(n,\nu)$ given by (59) with $r = |n-2\nu|$ and $\operatorname{sign}\gamma_i = \operatorname{sign}(n-2\nu)$, $i \in \underline{r}$ can be connected in $\mathrm{Rat}(n,\nu)$ to a fixed function $f_{n,\nu}(s) \in \mathrm{Rat}(n,\nu)$. Since $r = |n-2\nu|$ we must have $h = \nu$ if $n-2\nu \geq 0$ and $h = n-\nu$ if $n-2\nu < 0$. We choose

$$f_{n,\nu}(s) = \sum_{i=1}^{|n-2\nu|} \frac{\operatorname{sign}(n-2\nu)}{s+i} + \sum_{j=1}^{h} \frac{1}{s^2+j^2}.$$

We can enumerate the ρ_i and ω_j in such a way that $\rho_r < \ldots < \rho_1$ and $\omega_1 < \ldots < \omega_\nu$. Then we can transform $g(s)$ into $f_{n,\nu}(s)$ by moving the complex poles of $g(s)$ horizontally to the imaginary axis, and transforming γ_i to $\operatorname{sign}\gamma_i = \operatorname{sign}(n-2\nu)$ and ρ_i to $-i$ for $i \in \underline{r}$, $c_{j1}s + c_{j0}$ to 1 and ω_j to j^2 for $j \in \underline{h}$. More precisely, let

$$\hat{g}_t(s) = \sum_{i=1}^{|n-2\nu|} \frac{\gamma_i(t)}{s-\rho_i(t)} + \sum_{j=1}^{h} \frac{c_{j1}(t)s + c_{j0}(t)}{s^2 - 2\rho_{r+j}(t)s + \rho_{r+j}(t)^2 + \omega_j(t)^2}, \quad t \in [0,1]$$

where $\gamma_i(t) = (1-t)\gamma_i + t\operatorname{sign}\gamma_i$, $\rho_i(t) = (1-t)\rho_i - ti$ for $i \in \underline{r}$ and

$$\begin{aligned} c_{j1}(t)s + c_{j0}(t) &= (1-t)[(t\delta(c_{j1})s + c_{j1}s + c_{j0})] + t \\ \rho_{r+j}(t) &= (1-t)\rho_{r+j},\ \omega_j(t)^2 = (1-t)\omega_j^2 + tj^2, \quad j \in \underline{h}, \end{aligned}$$

where $\delta(c_{j1}) = 0$ if $c_{j1} \neq 0$ and $\delta(0) = 1$. Then as in the proof of Lemma 3.4.51 it is easily verified that the numerators and denominators in the previous sum are (non-zero and) coprime and the order $\rho_r(t) < \ldots < \rho_1(t)$, $0 < \omega_1(t) < \ldots < \omega_h(t)$ is preserved, whence $\hat{g}_t(s)$ has n simple poles for all $t \in [0,1]$. Since the above coefficient functions are obviously continuous and $\operatorname{sign}\gamma_i(t) = \operatorname{sign}\gamma_i = \operatorname{sign}(n-2\nu)$ for all $t \in [0,1]$, it follows from Lemma 3.4.57 that $\hat{g}_t$, $0 \le t \le 1$ describes an arc in $\mathrm{Rat}(n,\nu)$ connecting $\hat{g}_0(s) = g(s)$ with $\hat{g}_1(s) = f_{n,\nu}(s)$. Summarizing we see that every $f(s) \in \mathrm{Rat}(n,\nu)$ can be connected to $f_{n,\nu}(s)$ by an arc in $\mathrm{Rat}(n,\nu)$. This concludes the proof. □

Remark 3.4.61. Since $\mathrm{Rat}_n(\mathbb{R}) \cong \mathrm{Hank}(n,\mathbb{R}) \cong \mathrm{Hank}_n^*(\mathbb{R}) \times \mathbb{R}$ by Proposition 3.4.47, it follows from Theorem 3.4.53 that $\mathrm{Hank}(n,\mathbb{R})$ and $\mathrm{Hank}_n^*(\mathbb{R})$ each have $n+1$ connected components classified by the signature.
In the complex case, $\mathrm{Rat}_n(\mathbb{C}) \cong \mathrm{Hank}(n,\mathbb{C}) \cong \mathrm{Hank}_n^*(\mathbb{C}) \times \mathbb{C}$ is connected for all $n \ge 1$. This result can be proved along similar lines, but the proof is easier since the realness constraints are absent and all poles can be moved onto the imaginary axis, see Ex. 8. □

3.4.5 Applications to Stability

We now return to the problem of characterizing Hurwitz stable polynomials and derive a variety of classical stability criteria by means of Bézout and Hankel matrices. We have seen in the second subsection that a complex polynomial $p(s)$ of degree n is Hurwitz stable if and only if the associated real rational function $f_p(s)$ (26) has Cauchy index n and, by Theorem 3.4.53, this will be the case if and only if $\mathbf{Hk}_n(f_p) \succ 0$. We will now give a purely algebraic proof of this result by making use of the relationship between Hankel and Bézout matrices.

Theorem 3.4.62. *If $p(s)$ is a complex polynomial of degree n and $f_p(s)$ is the associated rational function* (26) *then p is Hurwitz stable if and only if $\mathbf{Hk}_n(f_p) \succ 0$.*

Proof: Define the real polynomials p_R, p_I by $p(\imath s) = p_R(s) + \imath p_I(s)$, see (2). By Corollary 3.4.35 p is Hurwitz stable if and only if the Bézout matrix $\mathbf{B}_n(p_I, p_R)$ is positive definite. If $\deg p_R \le \deg p_I$ then $\mathbf{B}_n(p_I, p_R)$ is congruent to $\mathbf{Hk}_n(p_R/p_I) = \mathbf{Hk}_n(f_p)$ (see (26)) by Theorem 3.4.41. Thus the theorem holds in this case. If $\deg p_R > \deg p_I$ then (again see (26)) $f_p = -p_I/p_R$, and by Theorem 3.4.41 $\mathbf{B}_n(p_R, p_I)$ is congruent to $\mathbf{Hk}_n(p_I/p_R)$. Hence $\mathbf{Hk}_n(f_p)$ is congruent to $-\mathbf{B}_n(p_R, p_I)$ which is equal to $\mathbf{B}_n(p_I, p_R)$ by Lemma 3.4.32 and this concludes the proof. □

Similarly one can prove a complex version of the Hermite-Biehler Theorem.

Corollary 3.4.63 (Hermite–Biehler, complex version). *Suppose that $p(s) = \sum_{i=0}^{n}(\alpha_i + \imath\beta_i)s^i$ is a complex polynomial of degree n and the real polynomials p_R, p_I are defined by $p(\imath s) = p_R(s) + \imath p_I(s)$. Then $p(s)$ is Hurwitz stable if and only if the following two conditions are satisfied.*

(i) All the roots of p_R, p_I are real, simple and interlacing.

(ii) $\alpha_{n-1}\alpha_n + \beta_{n-1}\beta_n > 0$.

In this case $p_I'(\omega)p_R(\omega) - p_I(\omega)p_R'(\omega) > 0$ for all $\omega \in \mathbb{R}$.

Proof: By the previous theorem it is sufficient to show that $\mathbf{Hk}_n(f_p)$ is positive definite if and only if the two conditions are satisfied. This follows directly from Corollary 3.4.55 and the unified formula for f_p given in Remark 3.4.22. In fact, if p is monic then this formula yields $f_p(s) = c(s)/d(s)$ with $c_n = 0$, $c_{n-1} = \alpha_{n-1}$ and $d_n = \alpha_n = 1$, $d_{n-1} = \beta_{n-1}$, whence $d_n c_{n-1} - c_n d_{n-1} = \alpha_{n-1}$. On the other hand condition (ii) reduces to $\alpha_{n-1} > 0$ in the monic case. Therefore conditions (i) and (ii) combined are equivalent to statement (iii) of Corollary 3.4.55, i.e. to $\mathbf{Hk}_n(f_p) = \mathbf{Hk}_n(c/d) \succ 0$. This proves the theorem for monic p. For the general case it suffices to note that the monic polynomial $a_n^{-1}p$ has the second coefficient $(\alpha_{n-1} + \imath\beta_{n-1})/(\alpha_n + \imath\beta_n)$ whose real part is $(\alpha_{n-1}\alpha_n + \beta_{n-1}\beta_n)/(\alpha_n^2 + \beta_n^2)$. This real part is positive if and only if condition (ii) is satisfied. Finally the last statement of the corollary follows from Corollary 3.4.55. □

Remark 3.4.64. Suppose that only the first condition in the previous corollary is satisfied. Then we know from Remark 3.4.56 that $\mathbf{Hk}_n(f_p)$ is either positive or negative definite. If $p^\star(s) = \overline{p}(-s)$ is the Hurwitz-reflection of p then the associated Hankel matrix is $\mathbf{Hk}_n(f_{p^\star}) = -\mathbf{Hk}_n(f_p)$ since $(p^\star)_R(s) = p_R(s)$ and $(p^\star)_I(s) = -p_I(s)$, see (3). By Theorem 3.4.62 $\mathbf{Hk}_n(f_p) \succ 0$ if and only if p is stable, and $\mathbf{Hk}_n(f_p) \prec 0$ if and only if $\mathbf{Hk}_n(f_{p^\star}) = -\mathbf{Hk}_n(f_p) \succ 0$, i.e. $p^\star$ is stable or, equivalently, p is anti-stable. Summarizing we see that if condition (i) is satisfied then p is either stable or anti-stable. □

We now turn to *real* polynomials and present a more algebraic proof of the real version of the Hermite–Biehler Theorem, which was proved by analytic means in the first subsection.

Corollary 3.4.65 (Hermite–Biehler, real version). *A real polynomial $p(s) = u(s^2) + sv(s^2)$ of degree n is Hurwitz stable if and only if (u, v) is a positive pair. Moreover, if p is Hurwitz stable then*

$$u'(\omega)v(\omega) - u(\omega)v'(\omega) > 0, \quad \omega \in \mathbb{R}. \tag{61}$$

Proof: Suppose that p is Hurwitz and let $\tilde{v}(s) = sv(s)$. Then $\deg v \le \deg u \le \deg \tilde{v}$, and $\mathbf{B}(u, v) \succ 0$, $\mathbf{B}(\tilde{v}, u) \succ 0$ by Corollary 3.4.38. $\mathbf{B}(u, v) \succ 0$ implies $\mathbf{Hk}_m(v/u) \succ 0$ (where $m = \deg u$), hence (61) is satisfied and all the roots of u, v are real, simple and interlacing by Corollary 3.4.55. Since p is Hurwitz, the leading coefficients of $u(s)$ and $v(s)$ have the same sign. In order to prove (13) resp. (14), it only remains to show that the largest roots t_1 of u and t_1' of v satisfy $t_1' < t_1 < 0$. Now (61) implies that the derivative of $v(t)/u(t)$ is negative to the right of t_1. Hence $t_1' > t_1$ would imply that $v(t)/u(t) < 0$ for large t, which is impossible since the leading coefficients of $u(s)$ and $v(s)$ are of the same sign. So $t_1' < t_1$. To prove $t_1 < 0$ we make use of $\mathbf{B}(\tilde{v}, u) \succ 0$, i.e. $\mathbf{Hk}_{\tilde{m}}(u/\tilde{v}) \succ 0$ where $\tilde{m} = \deg \tilde{v}$. From Corollary 3.4.55 we obtain

$$\tilde{v}'(\omega)u(\omega) - \tilde{v}(\omega)u'(\omega) = (v(\omega) + \omega v'(\omega))u(\omega) - \omega v(\omega)u'(\omega) > 0, \quad \omega \in \mathbb{R}.$$

Combining this with (61) we obtain

$$v(\omega)u(\omega) > \omega[u'(\omega)v(\omega) - u(\omega)v'(\omega)] \ge 0, \quad \omega \ge 0.$$

This shows that all the roots of $u(s), v(s)$ are negative.
Conversely, suppose that (u, v) is a positive pair. Then $u(t)$ and $v(t)$ have the same sign to the right of $t_1 < 0$. Thus $u'(t_1)v(t_1) > 0$. Moreover, all the roots of v, u are simple, real, and interlacing. Hence we conclude that $\mathbf{Hk}_m(v/u) \succ 0$ by Corollary 3.4.55 and therefore $\mathbf{B}(u, v) \succ 0$ by Theorem 3.4.41. On the other hand, all the roots of $\tilde{v}$ and u are interlacing, and since the roots are negative it follows that $-\tilde{v}(t_1)u'(t_1) = -t_1 v(t_1)u'(t_1) > 0$. Hence (54) holds at t_1 with $c(t_1) = u(t_1)$ and $d(t_1) = \tilde{v}(t_1)$ and it follows from Corollary 3.4.55 that $\mathbf{Hk}_{\tilde{m}}(u/\tilde{v}) \succ 0$. So $\mathbf{B}(\tilde{v}, u) \succ 0$ and p is Hurwitz by Corollary 3.4.35. □

The proof shows that the condition $\mathbf{B}(\tilde{v}, u) \succ 0$ is only needed to ensure that the simple real roots of u are negative. If we assume that u has only positive or only negative coefficients, then $|u(\omega)| > 0$ for all $\omega \geq 0$ so that all its real roots must be negative. Hence in this case Hurwitz stability is equivalent to $\mathbf{B}(u, v) \succ 0$.

Corollary 3.4.66 (Liénard–Chipart). *A real polynomial $p(s) = u(s^2) + sv(s^2)$ of degree n is Hurwitz stable if and only if all its coefficients a_{2k}, $k = 0, \ldots, [n/2]$ have the same sign and $\mathbf{B}(u, v) \succ 0$ (or, equivalently, $\mathbf{Hk}_n(v/u) \succ 0$).*

In order to apply one of the above stability criteria based on quadratic forms it is necessary to compute the Laurent series for the rational function u/v or the entries of the Bézout matrix and then check for positive definiteness. In order to apply the Hermite–Biehler criterion one must either plot the graphs or compute the roots of the even and odd parts of the polynomial in question. We will now derive the Hurwitz criterion which is expressed directly in terms of the coefficients $a_0, \ldots, a_{n-1}, a_n$ of the real polynomial p.

Lemma 3.4.67. *Let $c(s), d(s) \in \mathbb{C}[s]$ be of the form* (40) *and $g(s) = c(s)/d(s)$. Then the principal minors of $\mathbf{Hk}_n(g)$ satisfy*

$$d_n^{2k} \det \mathbf{Hk}_k(g) = \det M_{2k}, \quad k \in \underline{n}$$

where

$$M_{2k} = \begin{bmatrix} d_n & d_{n-1} & d_{n-2} & \cdots & \cdots & d_{n-2k+1} \\ c_n & c_{n-1} & c_{n-2} & \cdots & \cdots & c_{n-2k+1} \\ 0 & d_n & d_{n-1} & \cdots & \cdots & d_{n-2k+2} \\ 0 & c_n & c_{n-1} & \cdots & \cdots & c_{n-2k+2} \\ \vdots & & & & & \vdots \\ 0 & \cdots & \cdots & 0 \; d_n & \cdots & d_{n-k} \\ 0 & \cdots & \cdots & 0 \; c_n & \cdots & c_{n-k} \end{bmatrix}_{2k \times 2k}, \quad c_{-j} = d_{-j} = 0, \; j > 0. \qquad (62)$$

Proof: By equations (44) and (45) the Laurent coefficients g_k of $g(s)$ satisfy

$$\begin{aligned} c_{n-j} &= d_{n-j}g_0 + d_{n-j+1}g_1 + \cdots d_n g_j, \quad j = 0, \ldots, n \\ 0 &= d_0 g_j + d_1 g_{j+1} + \cdots + d_{n-1}g_{j+n-1} + d_n g_{j+n}, \quad j \geq 1. \end{aligned}$$

From these equations we obtain the following equation between $2k \times 2k$-matrices, $k \in \underline{n}$,

$$\begin{bmatrix} 1 & 0 & 0 & \cdots & 0 \\ g_0 & g_1 & g_2 & \cdots & g_{2k-1} \\ 0 & 1 & 0 & \cdots & 0 \\ 0 & g_0 & g_1 & \cdots & g_{2k-2} \\ \vdots & \ddots & \vdots & \vdots & \vdots \\ 0 & \cdots & 0 & 1 \cdots & 0 \\ 0 & \cdots & 0 & g_0 \cdots & g_k \end{bmatrix} \begin{bmatrix} d_n & d_{n-1} & d_{n-2} & \cdots d_{n-2k+1} \\ 0 & d_n & d_{n-1} & \cdots d_{n-2k+2} \\ 0 & 0 & d_n & \cdots d_{n-2k+3} \\ \vdots & \vdots & \ddots & \ddots \quad \vdots \\ 0 & 0 & \cdots & 0 \quad d_n \end{bmatrix} = \begin{bmatrix} d_n & d_{n-1} & d_{n-2} & \cdots & d_{n-2k+1} \\ c_n & c_{n-1} & c_{n-2} & \cdots & c_{n-2k+1} \\ 0 & d_n & d_{n-1} & \cdots & d_{n-2k+2} \\ 0 & c_n & c_{n-1} & \cdots & c_{n-2k+2} \\ \vdots & \ddots & \vdots & \vdots & \vdots \\ 0 & \cdots & 0 & d_n \cdots & d_{n-k} \\ 0 & \cdots & 0 & c_n \cdots & c_{n-k} \end{bmatrix}.$$

Taking determinants of both sides gives

$$\det M_{2k} = d_n^{2k} \det \begin{bmatrix} 1 & 0 & 0 & \cdots & \cdots & \cdots & 0 \\ g_0 & g_1 & g_2 & \cdots & \cdots & \cdots & g_{2k-1} \\ 0 & 1 & 0 & \cdots & \cdots & \cdots & 0 \\ 0 & g_0 & g_1 & \cdots & \cdots & \cdots & g_{2k-2} \\ \vdots & \ddots & & & & & \vdots \\ 0 & \cdots & 0 & 1 & 0 & \cdots & 0 \\ 0 & \cdots & 0 & g_0 & g_1 & \cdots & g_k \end{bmatrix}, \quad k \in \underline{n}.$$

The right hand determinant can be rearranged by interchanging rows an even number of times so that it is equal to

$$\det \left[\begin{array}{cccccc|ccc} 1 & 0 & 0 & \cdots & & 0 & 0 & \cdots & 0 \\ 0 & 1 & 0 & \cdots & & 0 & 0 & \cdots & 0 \\ 0 & 0 & 1 & \cdots & & 0 & 0 & \cdots & 0 \\ \vdots & \ddots & \ddots & & & \vdots & \vdots & \vdots & \vdots \\ 0 & \cdots & \cdots & 0 & & 1 & 0 & \cdots & 0 \\ \hline 0 & \cdots & \cdots & 0 & & g_0 & g_1 & \cdots & g_k \\ 0 & \cdots & 0 & g_0 & & g_1 & g_2 & \cdots & g_{k+1} \\ \vdots & \vdots & \vdots & \vdots & & \vdots & \vdots & \vdots & \vdots \\ 0 & g_0 & g_1 & \cdots & & g_{k-2} & g_{k-1} & \cdots & g_{2k-2} \\ g_0 & g_1 & g_2 & \cdots & & g_{k-1} & g_k & \cdots & g_{2k-1} \end{array}\right] = \det \begin{bmatrix} g_1 & \cdots & g_k \\ \vdots & \cdots & \vdots \\ g_k & \cdots & g_{2k-1} \end{bmatrix} = \det \mathbf{Hk}_k(g).$$

□

Theorem 3.4.68 (Hurwitz criterion, complex version). *Suppose that $p(s)$ is a complex polynomial of degree n and the real polynomials p_R, p_I are defined by $p(\imath s) = p_R(s) + \imath p_I(s)$ and write $p_R(s) = \sum_{j=0}^{n} c_j s^j$, $p_I(s) = \sum_{j=0}^{n} d_j s^j$. Then p is Hurwitz stable if and only if the even order principal minors of the matrix*

$$M^{\mathbb{C}}(p) = \begin{bmatrix} d_n & d_{n-1} & d_{n-2} & \cdots & \cdot & d_0 & 0 & \cdot & \cdots & 0 \\ c_n & c_{n-1} & c_{n-2} & \cdots & \cdot & c_0 & 0 & \cdot & \cdots & 0 \\ 0 & d_n & d_{n-1} & d_{n-2} & \cdot & \cdots & d_0 & \cdot & \cdots & 0 \\ 0 & c_n & c_{n-1} & c_{n-2} & \cdot & \cdots & c_0 & \cdot & \cdots & 0 \\ \vdots & & & & & & & & & \vdots \\ 0 & \cdots & \cdots & 0 & d_n & d_{n-1} & d_{n-2} & \cdot & \cdots & d_0 \\ 0 & \cdots & \cdots & 0 & c_n & c_{n-1} & c_{n-2} & \cdot & \cdots & c_0 \end{bmatrix}_{2n \times 2n} \in \mathbb{R}^{2n \times 2n} \qquad (63)$$

are positive, i.e.

$$\det\begin{bmatrix} d_n & d_{n-1} \\ c_n & c_{n-1} \end{bmatrix} > 0, \quad \det\begin{bmatrix} d_n & d_{n-1} & d_{n-2} & d_{n-3} \\ c_n & c_{n-1} & c_{n-2} & c_{n-3} \\ 0 & d_n & d_{n-1} & d_{n-2} \\ 0 & c_n & c_{n-1} & c_{n-2} \end{bmatrix} > 0, \; \ldots, \quad \det M^{\mathbb{C}}(p) > 0.$$

Proof: By Corollary 3.4.35 p is Hurwitz stable if and only if $\mathbf{B}(p_I, p_R) \succ 0$. Suppose that $\deg p_I \geq \deg p_R$, then necessarily $d_n \neq 0$ and so by Corollary 3.4.42 and Lemma 3.4.67 the principal minors from the lower right corner of $\mathbf{B}(p_I, p_R)$ coincide with the even order principal minors of $M^{\mathbb{C}}(p)$. Now $\mathbf{B}(p_I, p_R) \succ 0$ if and only if all these minors are positive, and so the theorem follows in this case. If $\deg p_I < \deg p_R$ we consider the polynomial $\tilde{p}(s) = \imath p(s)$ for which $\tilde{p}_R = -p_I$ and $\tilde{p}_I = p_R$, so that $\deg \tilde{p}_I = n > \tilde{p}_R$. By the first step $\tilde{p}$ is Hurwitz stable if and only if the even order principal minors of $M^{\mathbb{C}}(\tilde{p})$ are positive. But $M^{\mathbb{C}}(\tilde{p})$ is obtained from $M^{\mathbb{C}}(p)$ by interchanging the d-rows and the c-rows and multiplying the d-rows by -1. Hence the even order principal minors of $M^{\mathbb{C}}(\tilde{p})$ and $M^{\mathbb{C}}(p)$ coincide and the theorem is proved. □

The matrix $M^{\mathbb{C}}(p)$ in the previous theorem is called the *Hurwitz matrix* of the complex polynomial p. Explicit expressions for c and d in terms of the coefficients of $p(s)$ are given in (24) and (25).

Example 3.4.69. Consider once again the polynomial $p(s) = s^2 + (\alpha_1 + \imath\beta_1)s + \imath\beta_0$, where $\beta_0, \alpha_1, \beta_1 \in \mathbb{R}$. Then as in Example 3.4.36 $p_R(s) = -s^2 - \beta_1 s$, $p_I(s) = \alpha_1 s + \beta_0$ and hence $c_2 = -1$, $c_1 = -\beta_1$, $c_0 = 0$, $d_2 = 0$, $d_1 = \alpha_1$, $d_0 = \beta_0$. So the Hurwitz matrix is

$$M^{\mathbb{C}}(p) = \begin{bmatrix} 0 & \alpha_1 & \beta_0 & 0 \\ -1 & -\beta_1 & 0 & 0 \\ 0 & 0 & \alpha_1 & \beta_0 \\ 0 & -1 & -\beta_1 & 0 \end{bmatrix}.$$

The two even order principal minors are α_1 and $\alpha_1\beta_1\beta_0 - \beta_0^2$ and so we obtain the same stability criteria as in Example 3.4.36. □

For real polynomials we have the following counterpart.

Definition 3.4.70. Given a real polynomial $p(s) = a_n s^n + a_{n-1}s^{n-1} + \cdots + a_0$, $a_n \neq 0$,

$$M^{\mathbb{R}}(p) = \begin{bmatrix} a_{n-1} & a_{n-3} & a_{n-5} & \cdots & a_{n-2n+3} & a_{n-2n+1} \\ a_n & a_{n-2} & a_{n-4} & \cdots & a_{n-2n+4} & a_{n-2n+2} \\ 0 & a_{n-1} & a_{n-3} & \cdots & a_{n-2n+5} & a_{n-2n+3} \\ 0 & a_n & a_{n-2} & \cdots & a_{n-2n+6} & a_{n-2n+4} \\ \vdots & & & & & \vdots \\ 0 & 0 & \cdots & \cdots & a_1 & 0 \\ 0 & 0 & \cdots & \cdots & a_2 & a_0 \end{bmatrix} \in \mathbb{R}^{n\times n}, \quad a_{-k} = 0,\; k > 0 \quad (64)$$

is called the *Hurwitz* matrix of the real polynomial p.

Theorem 3.4.71 (Hurwitz criterion, real version). *A real polynomial* $p(s) = \sum_{j=0}^{n} a_j s^j$ *with* $a_n > 0$ *is Hurwitz stable if and only if all the principal minors* Δ_k, $k \in \underline{n}$ *of the associated Hurwitz matrix* $M^{\mathbb{R}}(p)$ *are positive,*

$$a_{n-1} > 0, \ \det \begin{bmatrix} a_{n-1} & a_{n-3} \\ a_n & a_{n-2} \end{bmatrix} > 0, \ \det \begin{bmatrix} a_{n-1} & a_{n-3} & a_{n-5} \\ a_n & a_{n-2} & a_{n-4} \\ 0 & a_{n-1} & a_{n-3} \end{bmatrix} > 0, \ \ldots, \ \det M^{\mathbb{R}}(p) > 0.$$

Proof: Let $p(s) = u(s^2) + sv(s^2)$. By Corollary 3.4.38 p is Hurwitz stable if and only if $\mathbf{B}(u,v) \succ 0$ and $\mathbf{B}(\tilde{v},u) \succ 0$ where $\tilde{v}(s) = sv(s)$. Suppose e.g. that $n = 2m$ is even, then

$$u(t) = a_n t^m + a_{n-2} t^{m-1} + \cdots + a_0, \quad v(t) = a_{n-1} t^{m-1} + a_{n-3} t^{m-2} + \cdots + a_1.$$

By Corollary 3.4.42 and Lemma 3.4.67, $\mathbf{B}(u,v) \succ 0$ holds if and only if

$$\det \begin{bmatrix} a_n & a_{n-2} & a_{n-4} & \cdots & \cdot & \cdot & \cdots & a_{n-4k+2} \\ 0 & a_{n-1} & a_{n-3} & \cdots & \cdot & \cdot & \cdots & a_{n-4k+3} \\ 0 & a_n & a_{n-2} & \cdots & \cdot & \cdot & \cdots & a_{n-4k+4} \\ 0 & 0 & a_{n-1} & \cdots & \cdot & \cdot & \cdots & a_{n-4k+5} \\ \vdots & & & & & & & \vdots \\ 0 & \cdots & \cdots & 0 & a_n & a_{n-2} & \cdots & a_{n-2k} \\ 0 & \cdots & \cdots & 0 & 0 & a_{n-1} & \cdots & a_{n-2k+1} \end{bmatrix}_{2k \times 2k} = a_n \Delta_{2k-1} > 0, \quad k \in \underline{m}$$

where $a_{-j} = 0, \ j > 0$. On the other hand, we have

$$\tilde{v}(t) = a_{n-1} t^m + a_{n-3} t^{m-1} + \cdots + a_1 t.$$

Hence by Lemma 3.4.67, $\mathbf{B}(\tilde{v},u) \succ 0$ if and only if

$$\det \begin{bmatrix} a_{n-1} & a_{n-3} & a_{n-5} & \cdots & a_{n-4k+1} \\ a_n & a_{n-2} & a_{n-4} & \cdots & a_{n-4k+2} \\ 0 & a_{n-1} & a_{n-3} & \cdots & a_{n-4k+3} \\ 0 & a_n & a_{n-2} & \cdots & a_{n-4k+4} \\ \vdots & & & & \vdots \\ 0 & 0 & \cdots & \cdots & a_{n-2k+1} \\ 0 & 0 & \cdots & \cdots & a_{n-2k} \end{bmatrix}_{2k \times 2k} = \Delta_{2k} > 0, \quad k \in \underline{m}.$$

This concludes the proof for even n. The case where n is odd can be dealt with in a similar way. □

Example 3.4.72. Consider the real cubic polynomial $p(s) = a_3 s^3 + a_2 s^2 + a_1 s + a_0$ as in Example 3.4.39, but now with $a_3 > 0$. Then

$$M^{\mathbb{R}}(p) = \begin{bmatrix} a_2 & a_0 & 0 \\ a_3 & a_1 & 0 \\ 0 & a_2 & a_0 \end{bmatrix}.$$

The minors are a_2, $a_2 a_1 - a_0 a_3$ and $a_0(a_2 a_1 - a_0 a_3)$. So p will be Hurwitz if and only if all the coefficients are positive and $a_2 a_1 - a_0 a_3 > 0$. □

By the criterion of Liénard-Chipart (Corollary 3.4.66) only half of the determinant inequalities in the previous theorem need to be verified if the polynomial p has positive coefficients.

Corollary 3.4.73 (Liénard-Chipart). *A real polynomial $p(s) = \sum_{j=0}^{n} a_j s^j$ with positive coefficients is Hurwitz stable if and only if all the principal minors of odd or even order of the Hurwitz matrix $M^{\mathbb{R}}(p)$ are positive:*

$$\Delta_1 > 0, \Delta_3 > 0, \ldots \quad \text{or} \quad \Delta_2 > 0, \Delta_4 > 0, \ldots .$$

Thus Hurwitz' determinant inequalities are not independent from each other for polynomials with positive coefficients. In particular, if the principal minors of odd order are positive then also the principal minors of even order are positive, and vice versa. We leave the details of the proof to the reader, see Ex. 16.

3.4.6 Schur Polynomials

A polynomial

$$p(z) = a_n z^n + a_{n-1} z^{n-1} + \ldots + a_0 \in \mathbb{C}[z] \tag{65}$$

is said to be a *Schur polynomial* or *Schur stable* if all the roots z_i of p lie in $\mathbb{D}$. Most of the results on Hurwitz polynomials which we have derived in the previous subsections have counterparts for Schur polynomials. In this subsection we will derive and discuss some of them. We begin with counterparts to the analytic stability criteria presented in Subsection 3.4.1.

Analytic criteria for Schur stability

A counterpart of Proposition 3.4.3 can be directly inferred from the argument principle (see Section A.2).

Proposition 3.4.74. *Given a polynomial $p(z) \in \mathbb{C}[z]$ of degree n without roots on the unit circle $\partial\mathbb{D}$, then*

$$\Delta_0^{2\pi} \arg p(e^{\imath\theta}) = \int_0^{2\pi} \frac{p'(e^{\imath\theta})}{p(e^{\imath\theta})} e^{\imath\theta} d\theta = (n-\nu)2\pi$$

where ν is the number of roots of $p(z)$ outside of $\overline{\mathbb{D}}$ (taking into account multiplicities). In particular, $p(z)$ is Schur stable if and only if $\Delta_0^{2\pi} \arg p(e^{\imath\theta}) = n2\pi$.

Proof: By the argument principle the number n_z of zeros of p in $\mathbb{D}$ is

$$n_z = \frac{1}{2\pi i} \int_\Gamma \frac{p'(s)}{p(s)} ds = \frac{1}{2\pi i} \int_{p(\Gamma)} \frac{ds}{s} = w(p(\Gamma), 0)$$

where Γ is the positively oriented unit circle. Since $n_z = n - \nu$, it follows that $\Delta_0^{2\pi} \arg p(e^{\imath\theta}) = (n-\nu)2\pi$ (see (A.2.11)), whence the result. □

In geometric terms this proposition says that the polynomial p is Schur stable if and only if the *frequency plot* $\theta \mapsto p(e^{\imath\theta})$, $\theta \in [0, 2\pi]$ describes a curve which surrounds the origin n times in the anticlockwise direction.

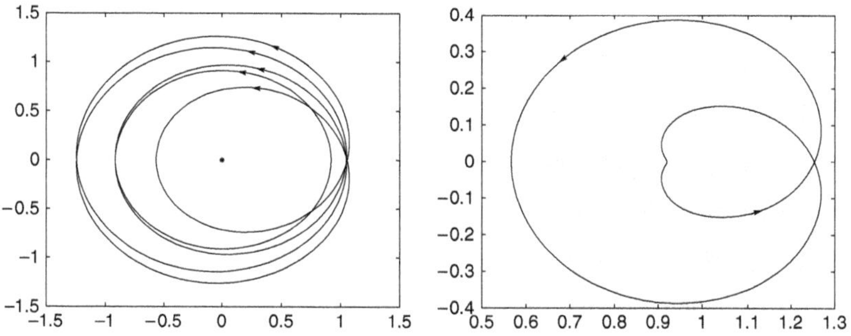

Figure 3.4.6: Frequency plot of a Schur and an anti-Schur polynomial

Example 3.4.75. Consider the Schur polynomial $p(z) = (z+3/4)(z\pm 2\imath/5)(z+3/8\pm\imath/4)$ and the anti-stable Schur-reflection polynomial $p^*(z) = z^5 p(1/z)$. Since $\deg p = 5$ the frequency plot $\omega \mapsto p(\imath\omega)$ encircles the origin anticlockwise five times (see the left hand figure in Figure 3.4.6), whilst the frequency plot $\omega \mapsto p^*(\imath\omega)$ does not encircle the origin (see the right hand figure in Figure 3.4.6). □

It is intuitively clear that, for any $z_0 \in \mathbb{D}$, $\arg(e^{\imath\theta} - z_0)$ is strictly increasing with $\theta \in [0, 2\pi]$. This yields the following counterpart of Proposition 3.4.5.

Proposition 3.4.76 (Phase increasing property). *If $p(z) \in \mathbb{C}[z]$ is a non-constant Schur polynomial then* $\frac{d}{d\theta} \arg p(e^{\imath\theta}) > 0$ *for all* $\theta \in [0, 2\pi)$.

Proof: Let $\gamma(\theta) = p(e^{\imath\theta})$, $\theta \in [0, 2\pi]$ and $p(z) = a_n \prod_{i=1}^n (z - z_i)$. Then by (1)

$$\frac{d}{d\theta} \arg p(e^{\imath\theta}) = \operatorname{Im} \frac{\gamma'(\theta)}{\gamma(\theta)} = \operatorname{Re} \frac{p'(e^{\imath\theta}) e^{\imath\theta}}{p(e^{\imath\theta})} = \sum_{i=1}^n \operatorname{Re} \left(\frac{e^{\imath\theta}}{e^{\imath\theta} - z_i} \right) > 0, \quad \theta \in [0, 2\pi).$$

since $\operatorname{Re}(1/(1-z)) > 0$ for all $z \in \mathbb{D}$. □

In the Hurwitz case we defined the Hurwitz-reflection of a polynomial by reflecting its roots at the imaginary axis, the boundary of the Hurwitz stability domain $\mathbb{C}_-$. Analogously, we now define the Schur-reflection of a polynomial by reflecting its roots at the unit circle, the boundary of the Schur stability domain. Let $\mathbb{K}_n[z]$ denote the $n+1$-dimensional vector space of polynomials $p(z) \in \mathbb{K}[z]$ of degree $\leq n$ and hence representable in the form (65).

Definition 3.4.77. Let $n \in \mathbb{N}^*$ be fixed. For every $p(z) = \sum_{i=0}^n a_i z^i \in \mathbb{C}_n[z]$, the *Schur-reflection*[5] of p is defined by

[5] Since a polynomial p is contained in all spaces $\mathbb{C}_n[z]$ with $n \geq \deg p$ its reflection is not uniquely determined but depends on the chosen n. This dependency on a previously chosen degree bound n distinguishes the Schur-reflection $p \mapsto p^*$ from the Hurwitz-reflection $p \mapsto p^*$ which is defined on the whole polynomial ring $\mathbb{C}[s]$ without degree constraints. In order to avoid this restriction and achieve uniqueness one could allow $n \in \mathbb{N}$ to vary and set $n = \deg p$ in the definition. This would, however, destroy the additivity of the reflection operator $p \mapsto p^*$.

$$p^*(z) = z^n \overline{p}(1/z) = \sum_{i=0}^{n} \overline{a_i} z^{n-i}.$$

$p(z) \in \mathbb{C}[z]$ is called *Schur-symmetric* if $p = p^*$.

In this subsection we will – as a rule – only be concerned with reflections in the sense of the previous definition and use the epithet "Schur-" only when it is necessary for their distinction from Hurwitz-reflections.

Schur reflections induce the following decomposition of any $p \in \mathbb{C}_n[z]$ into its symmetric and antisymmetric parts

$$p(z)=p_+(z)+\imath p_-(z), \quad p_+(z)=(p(z)+p^*(z))/2, \;\; p_-(z)=(p(z)-p^*(z))/(2\imath) \quad (66)$$

Note that with this notation p_+ and p_- are both symmetric, p_+ is the symmetric part and $\imath p_-$ is the antisymmetric part of p.

If z_i is a non-zero root of p then its reflection at the unit circle, $1/\overline{z_i}$, is a root of $p^*(z)$, and the multiplicities of these roots are equal. This follows from the factorizations

$$p(z)=a_m \prod_{i=1}^{m}(z-z_i) \;\Rightarrow\; p^*(z)=z^n\overline{p(1/\overline{z})}=z^n\,\overline{a_m \prod_{i=1}^{m}(1/\overline{z}-z_i)} = \overline{a}_m z^{n-m} \prod_{i=1}^{m}(1-\overline{z_i}z).$$

In particular, if p is of degree n and symmetric then every root z_i of p is non-zero and its reflection $1/\overline{z_i}$ is also a root of p (of the same multiplicity as z_i). Conversely, it can be shown that every polynomial with this property is symmetric after multiplication with a suitable non-zero constant factor, see Ex. 20.

The reflection $p \mapsto p^*$ on $\mathbb{C}_n[z]$ is completely described by the corresponding transformation of the coefficient vectors

$$*_n : (a_0, a_1, \ldots, a_n) \mapsto (\overline{a}_n, \ldots, \overline{a}_1, \overline{a}_0), \quad a \in \mathbb{C}^{n+1}.$$

Besides reflections the use of rotations will also be helpful in the sequel. For any polynomial $p(z) \in \mathbb{C}_n[z]$ define the *rotation* p_α of p by an angle α as follows

$$p_\alpha(z) = e^{-\imath\alpha n/2} p(e^{\imath\alpha} z). \quad (67)$$

Obviously, the roots of p_α are obtained by rotating the roots of p

$$p_\alpha(z_0) = 0 \quad \Leftrightarrow \quad p(e^{\imath\alpha} z_0) = 0. \quad (68)$$

Further elementary properties of reflections and rotations are summarized in the following lemma.

Lemma 3.4.78. *Let* $\alpha, \beta \in \mathbb{R}$, $p \in \mathbb{C}_n[z]$.

(i) *The map* $p \mapsto p^*$ *is an* $\mathbb{R}$*-linear bijection of* $\mathbb{C}_n[z]$ *onto itself with* $(ap)^* = \overline{a}p^*$ *for all* $a \in \mathbb{C}$. *Moreover* $p \mapsto p^*$ *is involutive, i.e.* $(p^*)^* = p$.

(ii) $\deg p^* = n$ *if and only if* $p(0) \neq 0$.

(iii) *If* p *is symmetric then* $\theta \mapsto e^{-\imath\theta n/2} p(e^{\imath\theta})$ *is a real-valued function on* $\mathbb{R}$.

(iv) *The map* $p \mapsto p_\alpha$ *is a degree preserving vector space isomorphism of* $\mathbb{C}_n[z]$ *onto itself, and* $(p_\alpha)_\beta = p_{\alpha+\beta}$.

(v) $(p_\alpha)^* = (p^*)_\alpha$, $(p_\alpha)_+ = (p_+)_\alpha$ *and* $(p_\alpha)_- = (p_-)_\alpha$.

(vi) $p \in \mathbb{C}_n[z]$ *is symmetric if and only if* p_α *is symmetric.*

(vii) *If* d *is a greatest common divisor of* p *and* p^*, *then* d_α *is a greatest common divisor of* p_α *and* $(p_\alpha)^*$.

Proof: (i),(ii), (iv) follow directly from the above formula for the reflection operator $*_n$ and (67).
(iii) If p is symmetric then

$$\overline{e^{-\imath\theta n/2}p(e^{\imath\theta})} = e^{\imath\theta n/2}\overline{p}(e^{-\imath\theta}) = e^{\imath\theta n/2}e^{-\imath\theta n}p^*(e^{\imath\theta}) = e^{-\imath\theta n/2}p(e^{\imath\theta}).$$

(v) The first equation in (v) follows from

$$(p_\alpha)^*(z)=z^n\,\overline{p_\alpha(1/\overline{z})} = z^n e^{\imath\alpha\frac{n}{2}}\,\overline{p(e^{\imath\alpha}/\overline{z})} = z^n e^{\imath\alpha\frac{n}{2}}\,\overline{p}(1/(e^{\imath\alpha}z)) = e^{-\imath\alpha\frac{n}{2}}\,p^*(e^{\imath\alpha}z) = (p^*)_\alpha(z).$$

The two remaining equations in (v) follow from the first one and (66).
(vi) follows directly from (v).
(vii) Suppose $p(z) = d(z)q(z)$ then $p_\alpha(z) = d_\alpha(z)q(e^{\imath\alpha}z)$, hence d_α divides p_α. Combining this with (v), statement (vii) follows. □

An efficient method for deriving algebraic criteria for Schur stability proceeds via the linear fractional transformation $m : s \mapsto (s+1)/(s-1)$ (the Möbius map, see Subsection 3.3.2 and Subsection 5.3.7). We will need the following elementary properties of this map.

Lemma 3.4.79. *The Möbius map* $m : s \mapsto (s+1)/(s-1)$ *is a bijection of* $\mathbb{C}\setminus\{1\}$ *onto itself with inverse* $m^{-1} = m$. *It maps* $\mathbb{C}_-$ *onto* $\mathbb{D}$, *and* $\imath\mathbb{R}$ *onto* $\partial\mathbb{D}\setminus\{1\} = \{e^{\imath\theta}; 0 < \theta < 2\pi\}$. *Moreover the map*

$$\omega \mapsto \theta(\omega) = \arg(m(\imath\omega)) = \arg((\imath\omega+1)/(\imath\omega-1)) \in (0, 2\pi), \quad \omega \in \mathbb{R}$$

is strictly increasing. In particular, if two sequences $(t_i)_{i\in\underline{n}}$ *and* $(t'_j)_{j\in\underline{m}}$ *(where* $m = n$ *or* $m = n-1$*) are interlacing in* $\mathbb{R}$ *then* $(\theta(t_i))_{i\in\underline{n}}$ *and* $(\theta(t'_j))_{j\in\underline{m}}$ *are interlacing in* $(0, 2\pi)$.

The Möbius map induces the following transformation of polynomials.

Definition 3.4.80. The *Möbius transform* of a polynomial $p(z) \in \mathbb{C}_n[z]$ of the form (65) is defined by

$$\tilde{p}(z) = (z-1)^n p\left(\frac{z+1}{z-1}\right) = \sum_{i=0}^{n} a_i(z+1)^i(z-1)^{n-i}. \tag{69}$$

We will obtain results on the location of the roots of $p(\cdot)$ relative to the stability domain $\mathbb{D}$ from the ones that we have obtained in the previous subsections for $\tilde{p}(\cdot)$ relative to $\mathbb{C}_-$. In order to highlight this in the sequel we will usually write $\tilde{p}(s)$ instead of $\tilde{p}(z)$. Some elementary properties of the Möbius transform of polynomials are summarized in the next lemma.

Lemma 3.4.81. *(i) The map $p \mapsto \tilde{p}$ is a vector space isomorphism of $\mathbb{C}_n[z]$ onto itself. Moreover, $p \mapsto \tilde{p}$ is involutive modulo a non-zero constant: $\tilde{\tilde{p}} = 2^n p$.*

(ii) If $\nu \geq 0$ is the maximal integer such that $(z-1)^\nu$ divides $p(z)$ then $\deg \tilde{p} = n - \nu$.

(iii) A polynomial $p \in \mathbb{C}_n[z]$ has k roots in $\mathbb{D}$ (resp. in $\mathbb{D}_+$) if and only if $\tilde{p}$ has k roots in $\mathbb{C}_-$ (resp. in $\mathbb{C}_+$), taking into account multiplicities.

(iv) A polynomial $p \in \mathbb{C}_n[z]$ of degree n with $p(1) \neq 0$ is Schur stable if and only if $\tilde{p}$ is Hurwitz stable.

(v) If $p \in \mathbb{C}_n[z]$ then $\tilde{\overline{p}}(s) = \overline{\tilde{p}}(s)$ and $\widetilde{p^*}(s) = (-1)^n \overline{\tilde{p}}(-s) = (-1)^n (\tilde{p})^\star(s)$. Thus the Möbius transform of a Schur-symmetric polynomial is Hurwitz-symmetric if n is even, and Hurwitz-antisymmetric if n is odd.

(vi) Suppose $\deg p = n$ and $p(1) \neq 0$. Then d is a g.c.d. of p and its Schur-reflection p^* if and only if $\hat{d}(s) = (s-1)^{\deg d} d(m(s)) \in \mathbb{C}_n[s]$ is a g.c.d. of $\tilde{p}(s)$ and its Hurwitz-reflection $(\tilde{p})^\star(s)$.

Proof: (i) The linearity of $p \mapsto \tilde{p}$ follows directly from (69). If $\tilde{p} = 0$ then $p(m(z)) = 0$ for all $z \in \mathbb{C} \setminus \{1\}$ and hence $p = 0$. Thus $p \mapsto \tilde{p}$ is a vector space isomorphism of $\mathbb{C}_n[z]$ onto itself. Moreover

$$\tilde{\tilde{p}}(z) = (z-1)^n \tilde{p}(m(z)) = (z-1)^n (m(z)-1)^n p(m(m(z))) = 2^n p(z).$$

(ii) Suppose $\deg p = m$ and $p(z) = a_m (z-1)^\nu \prod_{i=1}^{m-\nu} (z - z_i)$ where $z_i \neq 1$ for $i \in \underline{m-\nu}$. Then $\deg \tilde{p} = n - \nu$ follows from

$$\begin{aligned}\tilde{p}(s) &= a_m (s-1)^n \left[\frac{s+1}{s-1} - 1\right]^\nu \prod_{i=1}^{m-\nu} \left(\frac{s+1}{s-1} - z_i\right) = a_m (s-1)^{n-m} 2^\nu \prod_{i=1}^{m-\nu} (s+1-z_i(s-1)) \\ &= 2^\nu a_m (s-1)^{n-m} \prod_{i=1}^{m-\nu} (1 - z_i) \prod_{i=1}^{m-\nu} (s - m(z_i)). \qquad (70)\end{aligned}$$

(iii), (iv) Since $m(z_i)$ is in $\mathbb{C}_-$ (resp. $\mathbb{C}_+$) if and only if $z_i \in \mathbb{D}$ (resp. $\mathbb{D}_+$), the statements (iii) and (iv) are direct consequences of the preceding formula.

(v) By definition $\tilde{\overline{p}}(s) = (s-1)^n \overline{p}(m(s)) = \overline{(\overline{s}-1)^n p(m(\overline{s}))} = \overline{\tilde{p}(\overline{s})} = \overline{\tilde{p}}(s)$.
The second equation in (v) follows from

$$(s-1)^n p^*(m(s)) = (s-1)^n (m(s))^n \, \overline{p}\left(\frac{1}{m(s)}\right) = (-1)^n (-s-1)^n \, \overline{p}\left(\frac{-s+1}{-s-1}\right).$$

(vi) Suppose $\deg p = n$ and $p(1) \neq 0$, hence also $\deg \tilde{p} = n$ and $\tilde{p}(1) \neq 0$. If $p(z) = d(z)q(z)$ and $\hat{d} = (s-1)^{\deg d} d(m(s))$, $\hat{q} = (s-1)^{\deg q} q(m(s))$ then $\tilde{p}(s) = \hat{d}(s)\hat{q}(s)$ so that $\hat{d}|\tilde{p}$. Hence, if d is a g.c.d. of p and its Schur reflection p^*, and e is a g.c.d. of $\tilde{p}$ and its Hurwitz reflection $(\tilde{p})^\star = (-1)^n \widetilde{p^*}$ (see (v)), then $\hat{d}$ divides e, and $\hat{e} = (s-1)^{\deg e} e(m(s))$ divides $\tilde{\tilde{p}} = 2^n p$ and $\widetilde{\widetilde{p^*}} = 2^n p^*$, hence $\hat{e}|d$. Since $\deg d = \deg \hat{d} \leq \deg e = \deg \hat{e} \leq \deg d$ (because of $d(1)e(1) \neq 0$, $\hat{d}|e$ and $\hat{e}|d$) it follows that $\hat{d} = ae$ and $\hat{e} = bd$ for some constants $a, b \in \mathbb{C}^*$. □

The following example shows that the Möbius transform $\tilde{p}(s)$ may be Hurwitz stable without $p(z)$ being Schur stable. By (70) and statements (ii), (iii) of the previous Lemma this can only occur if $\deg \tilde{p} < \deg p = n$, hence $p(1) = 0$.

Example 3.4.82. Let $n = 2$ and $p(z) = z(z-1)$, then $\tilde{p}(s) = 2(s+1)$ by (70) is Hurwitz stable whereas $p(z)$ is not Schur stable. □

The previous lemma shows that the Schur stability of a polynomial p can be tested by applying Hurwitz stability criteria to the Möbius transform $\tilde{p}(s)$ defined by (69) with $n = \deg p$. Alternatively it is possible to derive direct tests from the known criteria for continuous time systems by pull-back to the Schur case via the map $p \mapsto \tilde{p}$. We will now follow this route and first establish a counterpart of the Hermite-Biehler Theorem 3.4.63 for complex Schur polynomials.
We say that two finite subsets $\{z_i; i \in \underline{n}\}$ and $\{z'_i, i \in \underline{m}\}$ of $\partial\mathbb{D}$ are *interlacing on the unit circle* if any arc in $\partial\mathbb{D}$ connecting two distinct point of the first set contains at least one point of the second set, and vice versa. It follows immediately from this definition that the interlacing property is preserved if both subsets of $\partial\mathbb{D}$ are rotated by the same angle. Moreover, in contrast with the real line, two interlacing subsets on the unit circle must have the same number of elements. Two sets $\{e^{\imath\theta_k}; k \in \underline{n}\}$ and $\{e^{\imath\theta'_k}, k \in \underline{n}\}$, $\theta_k, \theta'_k \in [0, 2\pi)$ of n elements each are interlacing on the unit circle if and only if the sets $\{\theta_k; k \in \underline{n}\}$ and $\{\theta'_k, k \in \underline{n}\}$ are interlacing in $[0, 2\pi)$.

Theorem 3.4.83 (Schur–Biehler). *A polynomial $p(z)$ of the form (65) with decomposition (66) is a Schur polynomial of degree n if and only if the following two conditions are satisfied.*

(i) $|a_n| > |a_0|$.

(ii) p_+ *and* p_- *each have n distinct simple roots on the unit circle, and these roots are interlacing.*

Proof: Suppose $p(z)$ of the form (65) is a Schur polynomial of degree n with roots $z_1, \dots, z_n$. Then $|a_0/a_n| = \prod_{i=1}^n |z_i| < 1$, whence condition (i). The leading coefficients of p_+, p_- are $(a_n + \overline{a_0})/2$ and $(a_n - \overline{a_0})/(2\imath)$, respectively. Hence both polynomials are of degree n because $|a_n| > |a_0|$. Without restriction of generality we suppose that $p_+(1)p_-(1) \neq 0$. (Otherwise, since the conditions (i),(ii) and the property of Schur stability are invariant with respect to arbitrary rotations of z we could replace $p(z)$ by $p_\alpha(z)$ where $\alpha \in \mathbb{R}$ is such that $(p_\alpha)_+(1)(p_\alpha)_-(1) = e^{-\imath\alpha n} p_+(e^{\imath\alpha})p_-(e^{\imath\alpha}) \neq 0$.) Therefore the Möbius transforms $\widetilde{p_+}$ and $\widetilde{p_-}$ of p_+ and p_- are of degree n. Making use of the symmetry of p_+ and p_- we obtain by Lemma 3.4.81 that

$$\tilde{p}(s) = \widetilde{p_+}(s) + \imath\widetilde{p_-}(s), \quad \widetilde{p_+}(s) = (-1)^n \overline{\widetilde{p_+}}(-s), \quad \widetilde{p_-}(s) = (-1)^n \overline{\widetilde{p_-}}(-s).$$

Now let us assume that n is even. Then we obtain for $s = \imath\omega$ that $\widetilde{p_+}(\imath\omega) = \overline{\widetilde{p_+}}(-\imath\omega) = \overline{\widetilde{p_+}(\imath\omega)}$ and $\widetilde{p_-}(\imath\omega) = \overline{\widetilde{p_-}(\imath\omega)}$, i.e. $\widetilde{p_+}(\imath\omega), \widetilde{p_-}(\imath\omega) \in \mathbb{R}$ for all $\omega \in \mathbb{R}$. Since $\tilde{p}(\imath\omega) = \tilde{p}_R(\omega) + \imath\tilde{p}_I(\omega)$ (see (2)) we see that

$$\tilde{p}_R(\omega) = \widetilde{p_+}(\imath\omega) \quad \text{and} \quad \tilde{p}_I(\omega) = \widetilde{p_-}(\imath\omega), \qquad \omega \in \mathbb{R}. \tag{71}$$

But $\tilde{p}$ is Hurwitz stable by Lemma 3.4.81, and so the n roots t_i of $\tilde{p}_R$ and t'_j of $\tilde{p}_I$ are real, simple and interlacing in $\mathbb{R}$ by the complex version of the Hermite-Biehler Theorem 3.4.63. As a consequence *all* the roots of $p_+(z)$ and $p_-(z)$ are simple and of the form $m(\imath t_k) = e^{\imath\theta_k}, k \in \underline{n}$ and $m(\imath t'_j) = e^{\imath\theta'_j}, j \in \underline{n}$, respectively, where the arguments θ_k and θ'_j are interlacing in $(0, 2\pi)$ by Lemma 3.4.79. (Note that here we have used our assumption that $p_+(1), p_-(1) \neq 0$). This proves condition (ii) because p_+ and p_- have the same number of roots.
Conversely, suppose that the two conditions (i), (ii) are satisfied and – without restriction of generality – $p_+(1)p_-(1) \neq 0$. Then the roots of $\tilde{p}_R$ and $\tilde{p}_I$ must be real, simple and interlacing in $\mathbb{R}$ so that by Remark 3.4.64, $\tilde{p}$ is either Hurwitz stable or anti-stable. As a consequence p is either Schur stable or anti-stable. But if p were anti-stable then $|a_0/a_n| = \prod_{i=1}^n |z_i| > 1$, in contradiction to condition (i). This concludes the proof for the case where n is even.
If n is odd then $\widetilde{p_+}(\imath\omega) = -\overline{\widetilde{p_+}(\imath\omega)}$ and $\widetilde{p_-}(\imath\omega) = -\overline{\widetilde{p_-}(\imath\omega)}$, hence $\widetilde{p_+}(\imath\omega), \widetilde{p_-}(\imath\omega) \in \imath\mathbb{R}$ and

$$\tilde{p}(\imath\omega) = \imath\widetilde{p_-}(\imath\omega) + \imath(-\imath\widetilde{p_+}(\imath\omega)), \quad -\imath\widetilde{p_+}(\imath\omega) = \overline{-\imath\widetilde{p_+}(\imath\omega)}, \quad \imath\widetilde{p_-}(\imath\omega) = \overline{\imath\widetilde{p_-}(\imath\omega)}.$$

Therefore

$$\tilde{p}_R(\omega) = \imath\widetilde{p_-}(\imath\omega), \quad \tilde{p}_I(\omega) = -\imath\widetilde{p_+}(\imath\omega) \tag{72}$$

and the above proof can be applied with p_+ replaced by $\imath p_-$ and p_- replaced by $-\imath p_+$. □

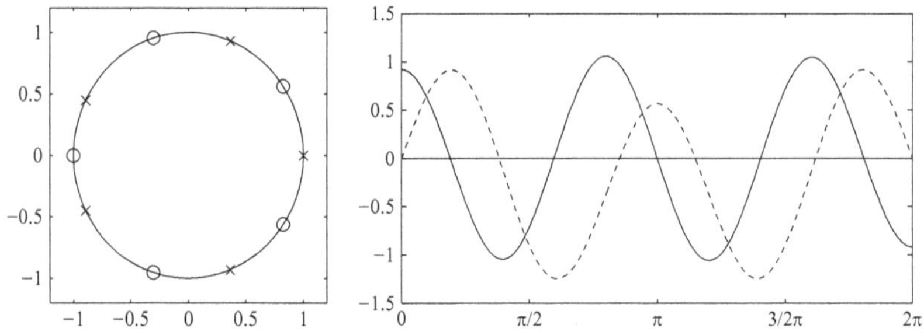

Figure 3.4.7: Roots of p_+, p_- and frequency plots of $e^{-\imath\theta n/2}p_+$ and $e^{-\imath\theta n/2}p_-$

Condition (ii) can be examined by plotting the roots of p_+ and p_- or by plotting the graphs of the functions $\theta \mapsto e^{-\imath\theta n/2}p_+(e^{\imath\theta})$ and $\theta \mapsto e^{-\imath\theta n/2}p_-(e^{\imath\theta})$ on $[0, 2\pi]$ which are real-valued by Lemma 3.4.78 (iii). This is illustrated in Figure 3.4.7 for the polynomial p in Example 3.4.75. The roots of p_- and p_+ are represented by $\times$ and $\circ$, respectively; the graphs of $\theta \mapsto e^{-\imath\theta n/2}p_+(e^{\imath\theta})$ and $\theta \mapsto e^{-\imath\theta n/2}p_-(e^{\imath\theta})$ are represented by continuous and dashed lines, respectively.
The previous proof has shown that condition (ii) is satisfied if and only if p is Schur stable or Schur anti-stable. In the following remark we briefly discuss the real case.

Remark 3.4.84. Suppose p has real coefficients. Then the product of the n-th coefficients of p_+ and $\imath p_-$ is $(a_n^2 - a_0^2)/4$. Hence condition (i) will be satisfied if and only if p_+ and $\imath p_-$ are both of degree n and their leading coefficients are of the same sign. Moreover, we have in this case $p(1) = p^*(1)$, so that 1 is a root of p_-. Since both p_+ and $\imath p_-$ are real polynomials, the roots of p_+, p_- are located symmetrically with respect to the real axis. Therefore, if condition (ii) is satisfied then the zero of p_+p_- on the unit circle with the smallest positive argument must be a zero of p_+. Now since $p^*(-1) = (-1)^n p(-1)$, either $p_+(-1) = 0$ or $p_-(-1) = 0$. In the former case the root of p_+p_- on the unit circle which has the largest argument $< \pi$ must be a root of p_- and in the latter case of p_+ (if condition (ii) is satisfied). In particular, the interlacing condition can be checked by only examining the roots on the upper closed semicircle. □

Hermitian Forms and Schur Polynomials

We will now apply the method of Hermitian forms (developed in Subsections 3.4.3 - 3.4.6 for the Hurwitz case) to Schur polynomials. We begin by introducing the *Schur form* which is a counterpart of the Hermite form studied in Subsection 3.4.3. With every complex polynomial p of the form (65) we associate n pairs of triangular matrices

$$U_m(p) = \begin{bmatrix} \overline{a}_n & \overline{a}_{n-1} & \overline{a}_{n-2} & \cdots & \overline{a}_{n-m+1} \\ 0 & \overline{a}_n & \overline{a}_{n-1} & \cdots & \overline{a}_{n-m+2} \\ \vdots & & & & \vdots \\ 0 & \cdots & 0 & \overline{a}_n & \overline{a}_{n-1} \\ 0 & \cdots & 0 & 0 & \overline{a}_n \end{bmatrix}, \quad L_m(p) = \begin{bmatrix} 0 & \cdots & 0 & 0 & a_0 \\ 0 & \cdots & 0 & a_0 & a_1 \\ \vdots & & & & \vdots \\ 0 & a_0 & a_1 & \cdots & a_{m-2} \\ a_0 & a_1 & a_2 & \cdots & a_{m-1} \end{bmatrix}, \quad m \in \underline{n}. \tag{73}$$

Definition 3.4.85. Given any polynomial p of the form (65), the associated *Schur matrix* of order n is defined by

$$\mathbf{S}_n(p) = U_n(p)^* U_n(p) - L_n(p)^* L_n(p). \tag{74}$$

The bilinear form $(x, y) \mapsto x^* \mathbf{S}_n(p) y$ on $\mathbb{C}^n$ is called the associated *Schur form* and

$$\tilde{h}(p; z, w) = \frac{\overline{p}(z)p(w) - \overline{p}^*(z)p^*(w)}{zw - 1} = \frac{\overline{p}(z)p(w) - z^n p(1/z) w^n \overline{p}(1/w)}{zw - 1} \tag{75}$$

is said to be its *generating function.* If $n = \deg p$ we write $\mathbf{S}(p)$ for $\mathbf{S}_n(p)$.

Note that the formula for the generating function has the same structure as the Hermite generating function (29) with $s+w$ replaced by $zw-1$ and the Hurwitz-reflected polynomial $p^\star(s)$ defined in Definition 3.4.4 replaced by the Schur-reflected polynomial $p^*(z)$ defined by Definition 3.4.77. The next lemma shows that $\tilde{h}(p; z, w)$ does in fact generate the Schur form. We again use the notation $l_m(z)^\top = [1, z, \ldots, z^{m-1}]$.

Lemma 3.4.86. *For any p of the form (65), the entries of the associated Schur matrix $\mathbf{S}_n(p) = (\tilde{h}_{ij})_{i,j \in \underline{n}}$ of order n are given by*

$$\tilde{h}_{ij} = \sum_{k=1}^{\min\{i,j\}} a_{n-i+k}\overline{a}_{n-j+k} - \overline{a}_{i-k}a_{j-k}, \quad i, j \in \underline{n}. \tag{76}$$

The principal submatrices of $\mathbf{S}_n(p)$ are

$$(\tilde{h}_{ij})_{i,j \in \underline{m}} = U_m(p)^* U_m(p) - L_m(p)^* L_m(p), \quad m \in \underline{n}, \tag{77}$$

and the generating function (75) *satisfies*

$$\tilde{h}(p;z,w)=\sum_{i,j=1}^{n}\tilde{h}_{ij}z^{i-1}w^{j-1}=l_n(z)^\top \mathbf{S}_n(p)l_n(w)\in\mathbb{C}[z,w]. \tag{78}$$

Proof: Due to the triangular structures of $U_m(p)$, $L_m(p)$ the k-th coordinate of $U_m(p)l_m(w)$ is given by $\sum_{j=0}^{m-k}\overline{a}_{n-j}w^{k+j-1}$ and the $(m-k+1)$-th coordinate of $L_m(p)l_m(w)$ is given by $\sum_{j=0}^{m-k}a_jw^{k+j-1}$, $j\in\underline{m}$. Moreover

$$l_m(z)^\top U_m(p)^*U_m(p)l_m(w)=\langle U_m(p)l_m(\overline{z}),U_m(p)l_m(w)\rangle$$

and a similar expression holds for $l_m(z)^\top L_m(p)^*L_m(p)l_m(w)$. Therefore

$$l_m(z)^\top(U_m(p)^*U_m(p)-L_m(p)^*L_m(p))l_m(w)=$$

$$=\sum_{k=1}^{m}\left[\left(\sum_{i=0}^{m-k}a_{n-i}z^{k+i-1}\right)\left(\sum_{j=0}^{m-k}\overline{a}_{n-j}w^{k+j-1}\right)-\left(\sum_{i=0}^{m-k}\overline{a}_iz^{k+i-1}\right)\left(\sum_{j=0}^{m-k}a_jw^{k+j-1}\right)\right]$$

$$=\sum_{k=1}^{m}\sum_{i,j=0}^{m-k}(a_{n-i}\overline{a}_{n-j}-\overline{a}_ia_j)z^{k+i-1}w^{k+j-1}=\sum_{\mu,\nu=1}^{m}z^{\mu-1}w^{\nu-1}\sum_{k=1}^{\min\{\mu,\nu\}}a_{n-\mu+k}\overline{a}_{n-\nu+k}-\overline{a}_{\mu-k}a_{\nu-k}$$

where we have set $\mu=k+i$ and $\nu=k+j$. If $m=n$ the LHS of this equality coincides with $l_n(z)^\top\mathbf{S}_n(p)l_n(w)=\sum_{i,j=1}^{n}\tilde{h}_{ij}z^{i-1}w^{j-1}$ by definition (74), and so we obtain (76). Making use of (76) the same equality shows that (77) holds. Thus it only remains to prove (78). Setting $\tilde{h}_{ij}=0$ for $(i,j)\notin\underline{n}\times\underline{n}$ we have

$$(zw-1)\sum_{i,j=1}^{n}\tilde{h}_{ij}z^{i-1}w^{j-1}=\sum_{i,j=1}^{n}\tilde{h}_{ij}z^iw^j-\sum_{i,j=1}^{n}\tilde{h}_{ij}z^{i-1}w^{j-1}=\sum_{i,j=0}^{n}(\tilde{h}_{ij}-\tilde{h}_{i+1,j+1})z^iw^j.$$

(76) yields the following expression for $\tilde{h}_{ij}-\tilde{h}_{i+1,j+1}$ (which also holds in the cases where i or j are either 0 or n)

$$\sum_{k=1}^{\min\{i,j\}}(a_{n-i+k}\overline{a}_{n-j+k}-\overline{a}_{i-k}a_{j-k})-\sum_{k=1}^{\min\{i,j\}+1}(a_{n-i-1+k}\overline{a}_{n-j-1+k}-\overline{a}_{i+1-k}a_{j+1-k})$$

$$=-(a_{n-i}\overline{a}_{n-j}-\overline{a}_ia_j),\ \text{i.e.}\quad \tilde{h}_{ij}-\tilde{h}_{i+1,j+1}=\overline{a}_ia_j-a_{n-i}\overline{a}_{n-j}.$$

Using this equality we obtain

$$\overline{p}(z)p(w)-\overline{p}^*(z)p^*(w)=\left(\sum_{i=0}^{n}\overline{a}_iz^i\right)\left(\sum_{j=0}^{n}a_jw^j\right)-\left(\sum_{i=0}^{n}a_{n-i}z^i\right)\left(\sum_{j=0}^{n}\overline{a}_{n-j}w^j\right)$$

$$=\sum_{i,j=0}^{n}(\overline{a}_ia_j-a_{n-i}\overline{a}_{n-j})z^iw^j=\sum_{i,j=0}^{n}(\tilde{h}_{ij}-\tilde{h}_{i+1,j+1})z^iw^j.$$

Hence

$$(zw-1)\tilde{h}(p;z,w)=\overline{p}(z)p(w)-\overline{p}^*(z)p^*(w)=(zw-1)\sum_{i,j=1}^{n}\tilde{h}_{ij}z^{i-1}w^{j-1}.$$

This concludes the proof. □

Remark 3.4.87. If $p(z)$ is of the form (65) then $p_\alpha(z) = e^{-\imath n\alpha/2}\sum_{k=0}^{n} a_k e^{\imath k\alpha} z^k$ by definition (67). It follows from (76) that the entries $\tilde{h}_{ij}(\alpha)$ of $\mathbf{S}_n(p_\alpha)$ are given by

$$\tilde{h}_{ij}(\alpha)=\sum_{k=1}^{\min\{i,j\}} e^{\imath(n-i+k)\alpha}a_{n-i+k}e^{-\imath(n-j+k)\alpha}\overline{a}_{n-j+k} - e^{-\imath(i-k)\alpha}\overline{a}_{i-k}e^{\imath(j-k)\alpha}a_{j-k} = e^{\imath(j-i)\alpha}\tilde{h}_{ij},$$

hence $\mathbf{S}_n(p_\alpha)=D^{-1}\mathbf{S}_n(p_\alpha)D$ with $D=\operatorname{diag}(e^{\imath\alpha},\dots,e^{\imath n\alpha})$ and so $\operatorname{rank}\mathbf{S}_n(p_\alpha) = \operatorname{rank}\mathbf{S}_n(p)$ and $\operatorname{sign}\mathbf{S}_n(p_\alpha) = \operatorname{sign}\mathbf{S}_n(p)$. □

Example 3.4.88. For $n = 2$ the Schur matrix associated with the polynomial $p(z) = a_2z^2 + a_1z + a_0$ is

$$\mathbf{S}_2(p) = \begin{bmatrix} |a_2|^2 - |a_0|^2 & a_2\overline{a_1} - \overline{a_0}a_1 \\ a_1\overline{a_2} - \overline{a_1}a_0 & |a_2|^2 - |a_0|^2 \end{bmatrix}.$$

If $n = 3$ the Schur matrix for the polynomial $p(z) = a_3z^3 + a_2z^2 + a_1z + a_0$ is

$$\mathbf{S}_3(p) = \begin{bmatrix} |a_3|^2 - |a_0|^2 & a_3\overline{a_2} - \overline{a_0}a_1 & a_3\overline{a_1} - \overline{a_0}a_2 \\ \overline{a_3}a_2 - a_0\overline{a_1} & |a_3|^2 + |a_2|^2 - |a_1|^2 - |a_0|^2 & a_3\overline{a_2} - \overline{a_0}a_1 \\ \overline{a_3}a_1 - a_0\overline{a_2} & \overline{a_3}a_2 - a_0\overline{a_1} & |a_3|^2 - |a_0|^2 \end{bmatrix}.$$

□

The following result is a counterpart of Hermite's Theorem 3.4.29.

Theorem 3.4.89 (Schur–Cohn). *Let $p(z) \in \mathbb{C}[z]$ be a polynomial of degree n and $\mathbf{S}(p)$ be the associated Schur matrix. If $r = \operatorname{rank}\mathbf{S}(p)$, $\sigma = \operatorname{sign}\mathbf{S}(p)$, then $p(z)$ has $n - r$ roots in common with $p^*(z)$, $(r + \sigma)/2$ additional roots in $\mathbb{D}$ and $(r - \sigma)/2$ additional roots in $\mathbb{D}_+$. In particular, p is Schur stable if and only if $\mathbf{S}(p) \succ 0$.*

Proof: If $\tilde{p}$ is the Möbius transform of p then by Definition 3.4.24 and Lemma 3.4.81 the Hermite generating function associated with $\tilde{p}$ can be expressed as follows

$$\frac{\overline{\tilde{p}}(s)\tilde{p}(t) - \tilde{p}(-s)\overline{\tilde{p}}(-t)}{s+t} = 2\frac{(s-1)^n\overline{p}(\frac{s+1}{s-1})(t-1)^n p(\frac{t+1}{t-1}) - (s-1)^n\overline{p}^*(\frac{s+1}{s-1})(t-1)^n p^*(\frac{t+1}{t-1})}{(s+1)(t+1)-(s-1)(t-1)}$$

$$= 2(s-1)^{n-1}(t-1)^{n-1}\frac{\overline{p}(\frac{s+1}{s-1})p(\frac{t+1}{t-1}) - \overline{p}^*(\frac{s+1}{s-1})p^*(\frac{t+1}{t-1})}{\frac{s+1}{s-1}\frac{t+1}{t-1} - 1}$$

where p^* is the Schur-reflection as in Definition 3.4.77. The LHS and the RHS of this equation can be rewritten in terms of the Hermite matrix associated with $\tilde{p}$ and the Schur matrix of p, respectively. Setting $z = (s+1)/(s-1)$, $w = (t+1)/(t-1)$ we obtain

$$l_n(s)^\top \mathbf{H}_n(\tilde{p})l_n(t) = 2(s-1)^{n-1}l_n(z)^\top\mathbf{S}_n(p)l_n(w)(t-1)^{n-1}$$

$$= 2\begin{bmatrix} (s-1)^{n-1} \\ (s-1)^{n-2}(s+1) \\ \vdots \\ (s+1)^{n-1} \end{bmatrix}^\top \mathbf{S}_n(p) \begin{bmatrix} (t-1)^{n-1} \\ (t-1)^{n-1}(t+1) \\ \vdots \\ (t+1)^{n-1} \end{bmatrix} = 2\begin{bmatrix} 1 \\ s \\ \vdots \\ s^{n-1} \end{bmatrix}^\top Q^\top\mathbf{S}_n(\tilde{p})Q \begin{bmatrix} 1 \\ t \\ \vdots \\ t^{n-1} \end{bmatrix}$$

where $Q \in \mathbb{R}^{n\times n}$ is the nonsingular matrix whose rows are the coefficient vectors of the polynomials $(s-1)^{n-1}, (s-1)^{n-2}(s+1), \dots, (s+1)^{n-1}$. We conclude that

$$\mathbf{H}_n(\tilde{p}) = 2Q^\top\mathbf{S}_n(p)Q. \tag{79}$$

Now choose $\alpha \in \mathbb{R}$ such that $p_\alpha(1) = p(e^{\imath\alpha}) \neq 0$. From Remark 3.4.87 and (79) (applied to p_α instead of p) we know that $\operatorname{rank} \mathbf{H}_n(\widetilde{p_\alpha}) = \operatorname{rank} \mathbf{S}_n(p_\alpha) = \operatorname{rank} \mathbf{S}_n(p) = r$ and $\operatorname{sign} \mathbf{H}_n(\widetilde{p_\alpha}) = \operatorname{sign} \mathbf{S}_n(p_\alpha) = \operatorname{sign} \mathbf{S}_n(p) = \sigma$. By Hermite's Theorem 3.4.29 $\widetilde{p_\alpha}$ has $n - r$ roots in common with its Hurwitz-reflection $\widetilde{p_\alpha}^\star$, $(r+\sigma)/2$ additional roots in the open left half-plane, and $(r-\sigma)/2$ additional roots in the open right half-plane. By Lemma 3.4.81 (iii), (vi) this proves the statements of the theorem for p_α instead of p. But then these statements hold for p by (68) and Lemma 3.4.78. □

Example 3.4.90. Consider the polynomial $p(z) = z^2 + (\alpha_1 + \imath\beta_1)z + \imath\beta_0$ where $\alpha_1, \beta_1, \beta_0 \in \mathbb{R}$. By the formula given in Example 3.4.88

$$\mathbf{S}_2(p) = \begin{bmatrix} 1-\beta_0^2 & \alpha_1 - \beta_0\beta_1 + \imath(\alpha_1\beta_0 - \beta_1) \\ \alpha_1 - \beta_0\beta_1 - \imath(\alpha_1\beta_0 - \beta_1) & 1-\beta_0^2 \end{bmatrix}.$$

Hence p is a Schur polynomial if and only if $1-\beta_0^2 > 0$ and $(1-\beta_0^2)^2 > (\alpha_1 - \beta_0\beta_1)^2 + (\alpha_1\beta_0 - \beta_1)^2$. □

Combining Lemma 3.4.86 with the previous theorem we obtain *Jury's criterion* for Schur stability. This criterion is expressed in terms of "inners" of the *Jury* matrix

$$\mathbf{J}_n(p) = \begin{bmatrix} U_n(p) & \overline{L}_n(p) \\ L_n(p) & \overline{U}_n(p) \end{bmatrix} \in \mathbb{C}^{2n\times 2n}. \tag{80}$$

If $M = (m_{ij})$ is any $\ell \times \ell$ matrix, the *inners* of M are the matrices obtained by deleting the first and last rows as well as the first and last columns of M, and then applying the same procedure to the remaining $(\ell-2)\times(\ell-2)$-matrix etc.

Corollary 3.4.91 (Jury). *A polynomial $p(z) \in \mathbb{C}[z]$ of the form* (65) *is a Schur polynomial of degree n if and only if the determinants of the associated Jury matrix* (80) *and all its inners are positive.*

Proof: We write U_m, L_m instead of $U_m(p), L_m(p)$. It follows from the definition of U_m, L_m (73) that the inners of $\mathbf{J}_n(p)$ are of the form $\mathbf{J}_m := \begin{bmatrix} U_m & \overline{L}_m \\ L_m & \overline{U}_m \end{bmatrix}$, $m \in \underline{n}$. It is easily verified that $U_m^* \overline{L}_m$ is of the same structure as L_m, and in particular it is symmetric. Therefore $U_m^*\overline{L}_m = (U_m^*\overline{L}_m)^\top = L_m^*\overline{U}_m$. As a consequence we obtain

$$\begin{bmatrix} U_m^* & -L_m^* \\ 0 & I \end{bmatrix}\begin{bmatrix} U_m & \overline{L}_m \\ L_m & \overline{U}_m \end{bmatrix} = \begin{bmatrix} U_m^*U_m - L_m^*L_m & U_m^*\overline{L}_m - L_m^*\overline{U}_m \\ L_m & \overline{U}_m \end{bmatrix} = \begin{bmatrix} U_m^*U_m - L_m^*L_m & 0 \\ L_m & \overline{U}_m \end{bmatrix}.$$

Now if $p(z) \in \mathbb{C}[z]$ is a Schur polynomial of degree n then $a_n \neq 0$ and hence $\det(U_m^*U_m - L_m^*L_m) = \det \mathbf{J}_m$. But by Lemma 3.4.86, $\det(U_m^*U_m - L_m^*L_m)$ is the m-th principal minor of $\mathbf{S}(p)$, and so $\mathbf{S}(p)$ is positive definite implies $\mathbf{J}_n(p)$ and all its inners have positive determinant. Conversely since $\det \mathbf{J}_2(p) = |a_n|^2 - |a_0|^2 > 0$ we have $a_n \neq 0$ and so again $\det \mathbf{J}_m$ is the m-th principal minor of $\mathbf{S}(p)$. Thus $\mathbf{S}(p) \succ 0$ and so p is Schur stable by Theorem 3.4.89. □

Jury's criterion can be simplified if p is a real polynomial. We describe this result in the following remark, omitting the proof.

Remark 3.4.92. Given a real polynomial $p(z) \in \mathbb{R}[z]$ of the form (65) with $a_n > 0$, let

$$R = \begin{bmatrix} a_n & a_{n-1} & a_{n-2} & \cdots & a_2 \\ 0 & a_n & a_{n-1} & \cdots & a_3 \\ \vdots & & & & \vdots \\ 0 & \cdots & 0 & a_n & a_{n-1} \\ 0 & \cdots & 0 & 0 & a_n \end{bmatrix} \quad S = \begin{bmatrix} 0 & \cdots & 0 & 0 & a_0 \\ 0 & \cdots & 0 & a_0 & a_1 \\ \vdots & & & & \vdots \\ 0 & a_0 & a_1 & \cdots & a_{n-3} \\ a_0 & a_1 & a_2 & \cdots & a_{n-2} \end{bmatrix}. \tag{81}$$

Then it can be shown (see [284]) that p is a Schur polynomial of degree n if and only if

(i) $p(1) > 0, \;\; (-1)^n p(-1) > 0$

(ii) the determinants of $R + S$, $R - S$ and all their inners are positive.

This criterion corresponds to the Liénard-Chipart criterion for Hurwitz polynomials (Corollary 3.4.66). The structure of the matrices $R \pm S$ and their inners is shown below for the case $n = 7$.

$$R \pm S = \begin{bmatrix} a_7 & a_6 & a_5 & a_4 & a_3 & a_2 \pm a_0 \\ 0 & a_7 & a_6 & a_5 & a_4 \pm a_0 & a_3 \pm a_1 \\ 0 & 0 & a_7 & a_6 \pm a_0 & a_5 \pm a_1 & a_4 \pm a_2 \\ 0 & 0 & \pm a_0 & a_7 \pm a_1 & a_6 \pm a_2 & a_5 \pm a_3 \\ 0 & \pm a_0 & \pm a_1 & \pm a_2 & a_7 \pm a_3 & a_6 \pm a_4 \\ \pm a_0 & \pm a_1 & \pm a_2 & \pm a_3 & \pm a_4 & a_7 \pm a_5 \end{bmatrix}.$$

The computation of the determinants is aided by the triangular arrays of zeros in the matrices R and S. □

We will now introduce "discrete time" counterparts to the Bézout and Hankel matrices which we considered in Subsection 3.4.3 for Hurwitz polynomials. In that subsection we saw that the Hermite form of a complex polynomial p can be expressed by the Bézoutiant of the two associated real polynomials p_R and p_I defined by (2). Similarly the Schur form of p can be expressed by a "discrete Bézoutiant" of the two associated symmetric polynomials p_+, p_-. We have

$$\tilde{h}(p; z, w) = l_n(z)^\top \mathbf{S}_n(p) l_n(w) = \frac{\overline{p}(z)p(w) - \overline{p}^*(z)p^*(w)}{zw - 1} =$$

$$\frac{\overline{(p_+ + \imath p_-)}(z)\,(p_+ + \imath p_-)(w) - \overline{(p_+ - \imath p_-)}(z)\,(p_+ - \imath p_-)(w)}{zw - 1} = 2\imath \frac{\overline{p}_+(z)\,p_-(w) - \overline{p}_-(z)\,p_+(w)}{zw - 1}.$$

This equality gives rise to the following definition.

Definition 3.4.93. Given two symmetric polynomials $c(z), d(z)$ of degree n, the Hermitian matrix $\tilde{\mathbf{B}}(c, d)$ satisfying

$$\imath \frac{\overline{c}(z)d(w) - \overline{d}(z)c(w)}{zw - 1} = l_n(z)^\top \tilde{\mathbf{B}}(c, d) l_n(w) \tag{82}$$

is called the *discrete Bézout matrix* for the symmetric pair (c, d), and the corresponding bilinear form $(x, y) \mapsto x^* \tilde{\mathbf{B}}(c, d) y$ is called the *discrete Bézoutiant* of (c, d) on $\mathbb{C}^n$.

The fact that $\tilde{\mathbf{B}}(c, d)$ is well defined by (82) follows from the preceding calculation: setting $p(z) = c(z) + \imath d(z)$ we have $c(z) = p_+(z)$, $d(z) = p_-(z)$ and so

$$\mathbf{S}_n(p) = 2\tilde{\mathbf{B}}(p_+, p_-) = 2\tilde{\mathbf{B}}(c, d), \qquad p(z) = c(z) + \imath d(z),\; c = c^*,\; d = d^*. \tag{83}$$

In particular, it follows from the Schur-Cohn Theorem 3.4.89 that $p(z) = p_+(z) + \imath p_-(z)$ is a Schur polynomial if and only if $\tilde{\mathbf{B}}(p_+, p_-)$ is positive definite.
We now turn to the discrete time counterpart of Hankel matrices, the *Toeplitz matrices*. An infinite complex matrix $T = (t_{ij})_{i,j \in \mathbb{N}^*}$ is called a Toeplitz matrix if there exists a (bi-infinite) sequence $(a_k)_{k \in \mathbb{Z}}$ in $\mathbb{C}$ such that $t_{ij} = a_{j-i}$ for $i, j \in \mathbb{N}^*$.[6] Analogously, a matrix $T \in \mathbb{K}^{M \times N}$ is called a Toeplitz matrix if it is of the form $T = (a_{j-i})_{i \in \underline{M}, j \in \underline{N}}$. With any formal power series $a(z) = \sum_{k \in \mathbb{N}} a_k z^k$ (or polynomial) we associate the upper triangular finite Toeplitz matrices $T_n(a) = (a_{j-i})_{i,j \in \underline{n}}$ where $a_{-k} = 0$ for $k \in \mathbb{N}^*$,

$$T_n(a) = \begin{bmatrix} a_0 & a_1 & a_2 & \cdots & a_{n-1} \\ 0 & a_0 & a_1 & \cdots & a_{n-2} \\ \vdots & \ddots & \ddots & \ddots & \vdots \\ 0 & \cdots & 0 & a_0 & a_1 \\ 0 & \cdots & 0 & 0 & a_0 \end{bmatrix}, \quad n \in \mathbb{N}^*.$$

If $a(z)$, $b(z)$, $c(z) \in \mathbb{K}[[z]]$ and $n \in \mathbb{N}^*$, then it is easily verified that

$$a(z)b(z) = c(z) \quad \Rightarrow \quad T_n(a)T_n(b) = T_n(b)T_n(a) = T_n(c). \tag{84}$$

Definition 3.4.94. Given $n \in \mathbb{N}^*$ and a pair of symmetric polynomials $c(z), d(z)$ of degree n the associated Toeplitz matrix of order n is defined by

$$\mathbf{T}(c,d) = T_n(\gamma) + T_n(\gamma)^* = \begin{bmatrix} \tilde{\gamma}_0 & \gamma_1 & \gamma_2 & \cdots & \gamma_{n-1} \\ \overline{\gamma_1} & \tilde{\gamma}_0 & \gamma_1 & \cdots & \gamma_{n-2} \\ \vdots & \ddots & \ddots & \ddots & \vdots \\ \overline{\gamma_{n-2}} & \cdots & \overline{\gamma_1} & \tilde{\gamma}_0 & \gamma_1 \\ \overline{\gamma_{n-1}} & \cdots & \overline{\gamma_2} & \overline{\gamma_1} & \tilde{\gamma}_0 \end{bmatrix} \tag{85}$$

where γ_i, $i \in \mathbb{N}$ are the Taylor coefficients of $\imath c(z)/d(z) = \sum_{i=0}^{\infty} \gamma_i z^i$ and $\tilde{\gamma}_0 = \gamma_0 + \overline{\gamma_0}$.

The following proposition is a counterpart to Theorem 3.4.41.

Proposition 3.4.95. *Let $c(z), d(z) \in \mathbb{C}[z]$ be symmetric polynomials of the form*

$$c(z) = \sum_{k=0}^{n} c_k z^k, \qquad d(z) = \sum_{k=0}^{n} d_k z^k, \quad d_n, c_n \neq 0. \tag{86}$$

Then the associated Toeplitz and discrete Bézout matrices are congruent,

$$\tilde{\mathbf{B}}(c,d) = T_n(d)^* \mathbf{T}(c,d) T_n(d) \tag{87}$$

In particular, $\tilde{\mathbf{B}}(c,d)$ and $\mathbf{T}(c,d)$ have the same rank and signature.

[6] It is standard notation to associate with a given sequence $(a_k)_{k \in \mathbb{Z}}$ of a formal power series $\sum_{k \in \mathbb{Z}} a_k z^k$ the Toeplitz matrix $(a_{i-j})_{i,j \in \mathbb{N}^*}$ whereas in our definition we associate the transpose of this matrix with the sequence. In the present context our notation is more convenient.

Proof: By the previous definition we have $\imath c(z) = d(z)\sum_{i=0}^{\infty}\gamma_i z^i$ and so by (84) $\imath T_n(c) = T_n(\gamma)T_n(d)$. Therefore

$$T_n(d)^*\left[T_n(\gamma)+T_n(\gamma)^*\right]T_n(d) = \imath\left[T_n(d)^*T_n(c) - T_n(c)^*T_n(d)\right]. \tag{88}$$

Since c is symmetric we have $\overline{c}_i = c_{n-i}$ for $i = 0,\dots,n$ and so

$$T_n(c)^*T_n(d) = \begin{bmatrix} c_n & 0 & \cdots & \cdots & 0 \\ c_{n-1} & c_n & 0 & \cdots & 0 \\ \vdots & \ddots & \ddots & \ddots & \vdots \\ c_2 & c_3 & \cdots & c_n & 0 \\ c_1 & c_2 & \cdots & c_{n-1} & c_n \end{bmatrix} \begin{bmatrix} d_0 & d_1 & d_2 & \dots & d_{n-1} \\ 0 & d_0 & d_1 & \dots & d_{n-2} \\ \vdots & & & & \vdots \\ 0 & \dots & 0 & d_0 & d_1 \\ 0 & \dots & 0 & 0 & d_0 \end{bmatrix}.$$

Hence the (i,j)-entry of $T_n(c)^*T_n(d)$ is $\sum_{k=1}^{\min\{i,j\}} c_{n-i+k}d_{j-k}$, and

$$\left(T_n(d)^*T_n(c) - T_n(c)^*T_n(d)\right)_{ij} = \sum_{k=1}^{\min\{i,j\}} \left(d_{n-i+k}c_{j-k} - c_{n-i+k}d_{j-k}\right).$$

Now let $p(z) = c(z) + \imath d(z) = \sum_{i=0}^{n} a_i z^i$. Then $c(z) = p_+(z)$, $d(z) = p_-(z)$ and thus, for every $\mu,\nu = 0,\dots,n$,

$$d_\mu c_\nu - c_\mu d_\nu = \frac{a_\mu - \overline{a}_{n-\mu}}{2\imath}\frac{a_\nu + \overline{a}_{n-\nu}}{2} - \frac{a_\mu + \overline{a}_{n-\mu}}{2}\frac{a_\nu - \overline{a}_{n-\nu}}{2\imath} = \frac{1}{2\imath}(a_\mu\overline{a}_{n-\nu} - \overline{a}_{n-\mu}a_\nu).$$

Setting $\mu = n-i+k$ and $\nu = j-k$ and substituting the preceding expression into the above formula we get

$$\imath\left(T_n(d)^*T_n(c) - T_n(c)^*T_n(d)\right)_{ij} = \frac{1}{2}\sum_{k=1}^{\min\{i,j\}}\left(a_{n-i+k}\overline{a}_{n-j+k} - \overline{a}_{i-k}a_{j-k}\right) = \frac{1}{2}\tilde{h}_{ij}.$$

This concludes the proof because of (83) and (88). □

As a consequence of this result and the Schur-Cohn Theorem we obtain the following

Corollary 3.4.96. *Let $p(z) \in \mathbb{C}[z]$ be a complex polynomial of degree n and let $\mathbf{T}(p_+,p_-)$ be the associated Toeplitz matrix. Then* rank $\mathbf{T}(p_+,p_-) = n$ *if and only if p_- and p_+ are coprime. In this case* sign $\mathbf{T}(p_+,p_-)$ *is the difference between the numbers of roots of $p(z)$ inside and outside of the unit circle. In particular, p is Schur stable if and only if $\mathbf{T}(p_+,p_-) \succ 0$.*

We have seen that the Cauchy index of a real rational function on the real line can be expressed by the signature of the associated Hankel matrix. We now show that the index of a symmetric rational function (i.e. a quotient of two symmetric polynomials) on the unit circle can be expressed by the signature of the associated Toeplitz matrix. Given two symmetric polynomials $c(z), d(z)$ of degree n, then since both $c(z), d(z)$ are real valued on $\partial\mathbb{D}$ their quotient $f(z) = c(z)/d(z)$ will also be real valued on $\partial\mathbb{D}$. The index of $f(z) = c(z)/d(z)$ along the unit circle (in the positive direction) is defined by

$$CI_0^{2\pi} f(e^{\imath\theta}) = \sum_{i\in\underline{m}} \frac{1}{2}\left[\lim_{\theta\downarrow\theta_i}\frac{f(e^{\imath\theta})}{|f(e^{\imath\theta})|} - \lim_{\theta\uparrow\theta_i}\frac{f(e^{\imath\theta})}{|f(e^{\imath\theta})|}\right]$$

where $e^{\imath\theta_i}$, $i \in \underline{m}$ are the zeros of $d(z)$ on $\partial\mathbb{D}$.

Theorem 3.4.97 (Herglotz). *Given two symmetric polynomials $c(z)$ and $d(z)$ of degree n, the index of $f(z) = c(z)/d(z)$ along the unit circle (in the positive direction) equals the signature of $\mathbf{T}(c,d)$.*

Proof: Let $\alpha \in \mathbb{R}$ be such that $c(e^{\imath\alpha}z)d(e^{\imath\alpha}z) \neq 0$ and set $f_\alpha = c(e^{\imath\alpha}z)/d(e^{\imath\alpha}z)$. It follows from (83), (87) and Remark 3.4.87 that $\mathbf{T}(c_\alpha, d_\alpha)$ is congruent with $\mathbf{T}(c,d)$. Moreover the index of $f_\alpha(z) = f(e^{\imath\alpha}z)$ is equal to the index of f. Therefore it is sufficient to prove the theorem for f_α instead of f. In other words we may assume without restriction of generality that $c(1)d(1) \neq 0$. Now

$$\hat{f} : \omega \mapsto \tilde{f}(\imath\omega) = f(m(\imath\omega)) = \frac{c(m(\imath\omega))}{d(m(\imath\omega))} = \frac{\tilde{c}(\imath\omega)}{\tilde{d}(\imath\omega)}$$

is a real rational function. Since $\theta : \omega \mapsto \theta(\omega) = \arg(m(\imath\omega))$ maps $\mathbb{R}$ onto $(0, 2\pi)$ in a strictly increasing fashion (Lemma 3.4.79) and f has no pole at 1 it follows from the definition and $\hat{f}(\omega) = f(e^{\imath\theta(\omega)})$ that the Cauchy index of $\hat{f}$ on $\mathbb{R}$ is equal to the index of f on the unit circle. From Theorem 3.4.53 we obtain

$$CI_0^{2\pi}(f(e^{\imath\theta})) = CI_{-\infty}^{\infty}(\hat{f}) = \operatorname{sign} \mathbf{Hk}_n(\hat{f}).$$

It remains to prove that $\operatorname{sign} \mathbf{Hk}_n(\hat{f}) = \operatorname{sign} \mathbf{T}(c,d)$. Setting $p(z) = c(z) + \imath d(z)$ we get $\tilde{p}(\imath\omega) = \tilde{c}(\imath\omega) + \imath\tilde{d}(\imath\omega)$. Let us assume that n is even. Then $\tilde{c}(\imath\omega) = \tilde{p}_R(\omega)$ and $\tilde{d}(\imath\omega) = \tilde{p}_I(\omega)$ by (71), whence $\hat{f}(\omega) = \tilde{p}_R(\omega)/\tilde{p}_I(\omega)$. But by Proposition 3.4.34 and Theorem 3.4.41 $\mathbf{Hk}_n(\hat{f}) = \mathbf{Hk}_n(\tilde{p}_R/\tilde{p}_I)$ is congruent to the Hermite matrix $\mathbf{H}_n(\tilde{p})$, and from Proposition 3.4.95, (83) and (79) we know that $\mathbf{T}(c,d)$, $\mathbf{S}(p)$ and $\mathbf{H}_n(\tilde{p})$ are congruent. This proves the theorem in the even case. If n is odd, then by (72)

$$\tilde{p}_R(\omega) = \imath\tilde{d}(\imath\omega), \quad \tilde{p}_I(\omega) = -\imath\tilde{c}(\imath\omega), \quad \tilde{p}_R(\omega)/\tilde{p}_I(\omega) = -\tilde{d}(\imath\omega)/\tilde{c}(\imath\omega) = -1/\hat{f}(\omega).$$

But then $\operatorname{sign} \mathbf{H}_n(\tilde{p}) = \operatorname{sign} \mathbf{Hk}_n(\tilde{p}_R/\tilde{p}_I) = \operatorname{sign} \mathbf{Hk}_n(-1/\hat{f}) = \operatorname{sign} \mathbf{Hk}_n(\hat{f})$ (since $\deg \tilde{c} = n = \deg \tilde{d}$, see (42)), and this concludes the proof. □

We conclude this subsection with a recursive test for Schur stability. The following counterpart of Lemma 3.4.12 is an easy consequence of the Schur-Cohn Theorem.

Corollary 3.4.98. *A polynomial $p(z) \in \mathbb{C}[z]$ of the form (65) is a Schur polynomial of degree n if and only if $|a_n| > |a_0|$ and the polynomial $q(z)$ determined by the equation*

$$zq(z) = \overline{a}_n p(z) - a_0 p^*(z)$$

is a Schur polynomial.

Proof: For any $\lambda \in \mathbb{C}$ we have

$$\begin{aligned}
&(zw-1)\tilde{h}(p+\lambda p^*; z, w) = \overline{(p+\lambda p^*)}(z)(p+\lambda p^*)(w) - \overline{(p+\lambda p^*)}^*(z)(p+\lambda p^*)^*(w) \\
&= \overline{p}(z)p(w) + \overline{\lambda}\overline{p}^*(z)p(w) + \lambda\overline{p}(z)p^*(w) + |\lambda|^2\overline{p}^*(z)p^*(w) - \overline{p}^*(z)p^*(w) \\
&- \lambda\overline{p}(z)p^*(w) - \overline{\lambda}\overline{p}^*(z)p(w) - |\lambda|^2\overline{p}(z)p(w) = (1-|\lambda|^2)\overline{p}(z)p(w) - (1-|\lambda|^2)\overline{p}^*(z)p^*(w) \\
&= (1-|\lambda|^2)(zw-1)\tilde{h}(p; z, w).
\end{aligned}$$

Hence if $\deg(p + \lambda p^*) = n$ and $|\lambda| \neq 1$ we obtain from Theorem 3.4.89 that p and $p + \lambda p^*$ have the same number of zeros in $\mathbb{D}$, taking into account multiplicities. Now let $\lambda = -a_0/\overline{a}_n$. If p is a Schur polynomial then $|a_n|^2 - |a_0|^2 > 0$ by Theorem 3.4.83 (i) and hence $|\lambda| < 1$. Hence $\overline{a}_n(p + \lambda p^*)(z) = zq(z)$ is a Schur polynomial of degree n, i.e. $q(z)$ is a Schur polynomial of degree $n-1$. Conversely, if $|a_n| > |a_0|$ and $q(z)$ is a Schur polynomial then $|\lambda| < 1$, and $p + \lambda p^*$ is a Schur polynomial of degree n, whence p is a Schur polynomial. □

This corollary gives rise to the following recursive algorithm for testing a polynomial $p(z) \in \mathbb{C}[z]$ for Schur stability.

Algorithm 3.4.99 (Test for Schur stability of complex polynomials).

1. Start: $p_0(z) = p(z)$, $i = 0$.
2. If $p_i(z) = \sum_{j=0}^{n-i} a_j^i z^j$ verify that $|a_{n-i}^i| > |a_0^i|$. If not, p is not Schur stable. If yes and $\deg p_i = 1$ then p is Schur stable. If yes and $\deg p_i \geq 2$, continue.
3. Compute $p_{i+1}(z) = z^{-1}[\overline{a}_{n-i}^i p_i(z) - a_0^i p_i^*(z)]$, set $i := i + 1$ and go back to 2.

Example 3.4.100. Consider again the polynomial $p(z) = z^2 + (\alpha_1 + \imath\beta_1)z + \imath\beta_0$ where $\alpha_1, \beta_1, \beta_0 \in \mathbb{R}$. Then $p^*(z) = -\imath\beta_0 z^2 + (\alpha_1 - \imath\beta_1)z + 1$ and

$$p_1(z) = (p(z) - a_0 p^*(z))/z = (1 - \beta_0^2)z + \alpha_1 - \beta_0\beta_1 + \imath(\beta_1 - \alpha_1\beta_0).$$

Hence p is a Schur polynomial if and only if

$$1 - \beta_0^2 > 0, \qquad (\alpha_1 - \beta_0\beta_1)^2 + (\beta_1 - \alpha_1\beta_0)^2 < (1 - \beta_0^2)^2.$$

These are precisely the conditions obtained via the Schur matrix in Example 3.4.90. □

If $p(z)$ is real with coefficients $a_n > a_{n-1} > \ldots > a_0 > 0$ then it is easily verified by induction that all the $p_i(z)$ generated by this algorithm have the same property. Thus we obtain the following sufficient criterion.

Corollary 3.4.101. *Every real polynomial $p(z)$ of the form (65) with coefficients $a_n > a_{n-1} > \ldots > a_0 > 0$ is a Schur polynomial.*

Example 3.4.102. Let $p(z) = z^3 + a_2 z^2 + a_1 z + a_0 \in \mathbb{R}[z]$. Then the preceding algorithm yields the following polynomials,

$$\begin{aligned} p_1(z) &= (p(z) - a_0 p^*(z))/z = (1 - a_0^2)z^2 + (a_2 - a_0 a_1)z + a_1 - a_0 a_2, \\ p_2(z) &= \left[(1 - a_0^2)p_1(z) - (a_1 - a_0 a_2)p_1^*(z)\right]/z \\ &= \left[(1 - a_0^2)^2 - (a_1 - a_0 a_2)^2\right] z + (a_2 - a_0 a_1)(1 - a_0^2 - (a_1 - a_0 a_2)). \end{aligned}$$

Thus $p(z)$ is a Schur polynomial if and only if

$$1 - a_0^2 > |a_1 - a_0 a_2|, \quad (1 - a_0^2)^2 - (a_1 - a_0 a_2)^2 > |(a_2 - a_0 a_1)(1 - a_0^2 - (a_1 - a_2 a_0))|.$$

□

Corollary 3.4.91 shows that the set of Schur polynomials of a given degree can be described by algebraic inequalities on their coefficient vectors. For a monic real polynomial of degree $n = 3$ the inequalities are specified in Example 3.4.102 (see also Ex. 21). In the next chapter we derive some geometric properties of the set of Schur polynomials and determine its convex hull by *linear* inequalities.

Example 3.4.103. (Cohort population model). A cohort population model describes the evolution of the age distribution of a given population in time. Let $x_i(t)$ be the number in the ith age group at time period t, where the groups are indexed from 0 to n, with 0 being the youngest age group. Assume that all age groups are of equal span and that β_i is the constant survival rate of the ith-age group. So

$$x_{i+1}(t+1) = \beta_i x_i(t), \quad i = 0, \ldots, n-1, \quad t \in \mathbb{N} \tag{89}$$

if we take the basic time period to be equal to the span of the age groups. By (89) the age distribution at time period $t = 0, 1, 2, \ldots$ determines the number of individuals in the age groups $1, \ldots, n$ at time $t+1$. The number in the youngest age group at time $(t+1)$ is given by

$$x_0(t+1) = \alpha_0 x_0(t) + \ldots + \alpha_n x_n(t)$$

where $\alpha_i \geq 0$ is the constant birth rate of the ith-age group $i = 0, \ldots, n$.
If we set $x(t) = [x_0(t), \ldots, x_n(t)]^\top$ we obtain the following discrete time system

$$x(t+1) = \begin{bmatrix} \alpha_0 & \alpha_1 & \alpha_2 & \ldots & \alpha_n \\ \beta_0 & 0 & 0 & \ldots & 0 \\ 0 & \beta_1 & 0 & \ldots & 0 \\ \vdots & 0 & 0 & 0 & \vdots \\ 0 & \ldots & 0 & \beta_{n-1} & 0 \end{bmatrix} x(t) \tag{90}$$

with characteristic polynomial

$$z^{n+1} - \alpha_0 z^n - \alpha_1\beta_0 z^{n-1} - \alpha_2\beta_0\beta_1 z^{n-2} \ldots - \alpha_n\beta_0 \ldots - \beta_{n-1}.$$

Here $\beta_0\beta_1 \cdots \beta_k$ can be interpreted as the probability that a new born will reach the kth age group. Now consider the special case that $\alpha_0 = 0$, $\alpha_i = 0$, $i > 2$. So the characteristic polynomial has the form

$$z^{n+1} - \alpha_1\beta_0 z^{n-1} - \alpha_2\beta_0\beta_1 z^{n-2}$$

and the roots are determined by

$$p(z) = z^3 - \alpha_1\beta_0 z - \alpha_2\beta_0\beta_1 \text{ and } z = 0 \text{ (if } n \geq 3).$$

Let $a_1 = -\alpha_1\beta_0$, $a_0 = -\alpha_2\beta_0\beta_1$ and apply the Jury test for real polynomials (see Remark 3.4.92). Since

$$R \pm S = \begin{bmatrix} 1 & 0 \\ 0 & 1 \end{bmatrix} \pm \begin{bmatrix} 0 & a_0 \\ a_0 & a_1 \end{bmatrix} = \begin{bmatrix} 1 & \pm a_0 \\ \pm a_0 & 1 \pm a_1 \end{bmatrix}$$

we obtain the following necessary and sufficient conditions for p to be a Schur polynomial

(i) $1 + a_1 + a_0 > 0$, $1 + a_1 - a_0 > 0$

(ii) $1 + a_1 - a_0^2 > 0$, $1 - a_1 - a_0^2 > 0$.

Condition (ii) is equivalent to $|a_1| < 1 - a_0^2$ and since $a_0 \leq 0$ the first condition in (i) implies the second. Let us assume that $\beta_0 = \beta_1 = 1$, then characteristic polynomial $z^{n-2}p(z)$ is Schur stable if $1 > \alpha_1 + \alpha_2$. So for these parameter values the number of individuals in the age groups will converge to zero and the species will die out. □

3.4.7 Algebraic Stability Domains and Linear Matrix Equations

In this subsection we present a unifying framework for obtaining stability criteria with respect to a large class of algebraic stability regions which include arbitrary half-planes, disks and also many other domains in the complex plane. The approach is based on Hermitian linear matrix equations generalizing Liapunov's equations for the Hurwitz and Schur stability domains. In contrast with the previous subsections the stability criteria are directly applicable to matrices. However they can also be used for polynomials via companion matrices.

Let $q(z_1, z_2) \in \mathbb{C}[z_1, z]$ be a *Hermitian polynomial*, i.e. a polynomial of the form

$$q(z_1, z_2) = \sum_{i=1}^{\ell} \sum_{j=1}^{\ell} c_{ij} z_1^{i-1} z_2^{j-1} = l(z_1)^\top C l(z_2), \quad C = (c_{ij})_{i,j \in \underline{\ell}} \in \mathcal{H}_\ell(\mathbb{C}). \tag{91}$$

where $l(\cdot)$ is given by (28) with $n = \ell$. We will consider algebraic stability domains of the following form

$$\mathbb{D}(q) = \{z \in \mathbb{C}; q(\overline{z}, z) < 0\}. \tag{92}$$

Kalman (1969) [289] posed the following: Which is the largest class of algebraic domains $\mathbb{D}(q)$ for which the property that any polynomial $p \in \mathbb{C}[z]$ of given (but arbitrary) degree n has all its roots in $\mathbb{D}(q)$ can be characterized by a system of rational inequalities in the real and imaginary parts of the coefficients of p? (This system of inequalities will depend on the degree n and on the domain $\mathbb{D}(q)$). Here we obtain a partial answer to this problem.

Given an arbitrary matrix $A \in \mathbb{C}^{n\times n}$, $n \in \mathbb{N}^*$ and a positive definite matrix $Y \in \mathcal{H}_n(\mathbb{C})$, we associate with the Hermitian polynomial q the following *generalized Liapunov equation*

$$\sum_{i=1}^{\ell} \sum_{j=1}^{\ell} c_{ij} (A^*)^{i-1} X A^{j-1} = -Y. \tag{93}$$

Example 3.4.104. (i) Let $q(z_1, z_2) = z_1 + z_2$. Then the associated coefficient matrix C_q, algebraic domain and linear matrix equations are, respectively,

$$C_q = \begin{bmatrix} 0 & 1 \\ 1 & 0 \end{bmatrix}, \quad \mathbb{D}(q) = \mathbb{C}_-, \quad A^*X + XA = -Y.$$

Thus we obtain the classical algebraic Liapunov equation for continuous time systems. More generally, for any half-plane $H = \{z \in \mathbb{C}; \mathrm{Re}(c_1 z) < c_0\}$ (where $c_1 \in \mathbb{C}^*, c_0 \in \mathbb{R}$) we choose $q(z_1, z_2) = \overline{c}_1 z_1 + c_1 z_2 - 2c_0$ to obtain $\mathbb{D}(q) = H$ and $\overline{c_1} A^* X + c_1 X A - 2c_0 X = -Y$.

(ii) Let $q(z_1, z_2) = z_1 z_2 - 1$. Then the corresponding items are

$$C_q = \begin{bmatrix} -1 & 0 \\ 0 & 1 \end{bmatrix}, \quad \mathbb{D}(q) = \mathbb{D}, \quad A^*XA - X = -Y.$$

Hence we obtain the classical Liapunov equation for discrete time systems. To get an arbitrary open disk of center $c \in \mathbb{C}$ and radius $\rho > 0$ as stability domain $\mathbb{D}(q)$ we choose $q(z_1, z_2) = (z_1 - \overline{c})(z - c) - \rho^2$. The corresponding matrix equation is

$$A^*XA - (\rho^2 - |c|^2)X - cA^*X - \overline{c}XA = -Y.$$

(iii) Let $q(z_1, z_2) = 2 - z_1 z_2 (z_1 + z_2)$. Then $q(\overline{z}, z) = 2 - 2|z|^2 \operatorname{Re} z$, hence

$$C_q = \begin{bmatrix} 2 & 0 & 0 \\ 0 & 0 & -1 \\ 0 & -1 & 0 \end{bmatrix}, \quad \mathbb{D}(q) = \{\rho + \imath\omega; \rho(\rho^2 + \omega^2) < 1\}, \quad 2X - (A^*)^2 X A - A^* X A^2 = -Y.$$

$\mathbb{D}(q)$ is neither a half-plane nor a disk. □

We know from Section 3.3 that the classical Liapunov equations have a positive definite solution X for every positive definite Y if and only if $\sigma(A) \subset \mathbb{C}_-$ resp. $\sigma(A) \subset \mathbb{D}$. We will now investigate for which Hermitian polynomials (91) an analogous statement can be proved for the generalized Liapunov equation (93). As a preparation we ask the reader (see Ex. 24) to verify that the spectrum of the *generalized Liapunov operator* associated with A and C

$$\mathbf{L}_{A,C} : \mathcal{H}_n(\mathbb{C}) \to \mathcal{H}_n(\mathbb{C}), \quad X \mapsto \sum_{i=0}^{\ell} \sum_{j=0}^{\ell} c_{ij} \, (A^*)^i X A^j \tag{94}$$

is given by $\sigma(\mathbf{L}_{A,C}) = \{q(\overline{\lambda_1}, \lambda_2); \lambda_1, \lambda_2 \in \sigma(A)\}$. A consequence of this is that $\mathbf{L}_{A,C}$ is an isomorphism on the real vector space $\mathcal{H}_n(\mathbb{C})$ if and only if $q(\overline{\lambda_1}, \lambda_2) \neq 0$ for all $\lambda_1, \lambda_2 \in \sigma(A)$. The following lemma gives a sufficient condition.

Lemma 3.4.105. *Let q be a Hermitian polynomial of the form* (91) *with coefficient matrix C and assume that C has exactly one simple negative eigenvalue, i.e.* $\operatorname{rank} C - \operatorname{sign} C = 2$. *Then $q(\overline{\lambda_1}, \lambda_2) \neq 0$ for all $\lambda_1, \lambda_2 \in \mathbb{D}(q)$.*

Proof: Suppose $\operatorname{rank} C = r$, then C can be written in the form $C = \sum_{k=1}^{r-1} v^k (v^k)^* - v^r (v^r)^*$, where the $v^j \in \mathbb{C}^\ell$ are linearly independent. Thus

$$q(z_1, z_2) = l(z_1)^\top C l(z_2) = \sum_{k=1}^{r-1} q_k(z_1) \overline{q_k}(z_2) - q_r(z_1) \overline{q_r}(z_2) \tag{95}$$

where $(v^k)^\top \in \mathbb{C}^{1 \times \ell}$ is the coefficient vector of q_k. Now let $\lambda_1, \lambda_2 \in \mathbb{D}(q)$. Since by definition of $\mathbb{D}(q)$ we have $q(\overline{\lambda_j}, \lambda_j) < 0$ for $j = 1, 2$ it follows that

$$\sum_{k=1}^{r-1} q_k(\overline{\lambda_j}) \overline{q_k}(\lambda_j) < q_r(\overline{\lambda_j}) \overline{q_r}(\lambda_j), \quad \text{i.e.} \quad \sum_{k=1}^{r-1} |q_k(\overline{\lambda_j})|^2 < |q_r(\overline{\lambda_j})|^2, \quad j = 1, 2.$$

Therefore the assumption that

$$q(\overline{\lambda_1}, \lambda_2) = \sum_{k=1}^{r-1} q_k(\overline{\lambda_1}) \overline{q_k}(\lambda_2) - q_r(\overline{\lambda_1}) \overline{q_r}(\lambda_2) = 0$$

would imply (by the Cauchy-Schwartz inequality)

$$\left| q_r(\overline{\lambda_1}) \overline{q_r}(\lambda_2) \right| = \left| \sum_{k=1}^{r-1} q_k(\overline{\lambda_1}) \overline{q_k}(\lambda_2) \right| \leq \sqrt{\sum_{k=1}^{r-1} |q_k(\overline{\lambda_1})|^2} \sqrt{\sum_{k=1}^{r-1} |\overline{q_k}(\lambda_2)|^2} < |q_r(\overline{\lambda_1})| |\overline{q_r}(\lambda_2)|.$$

This contradiction proves $q(\overline{\lambda_1}, \lambda_2) \neq 0$. □

Example 3.4.106. Let $c = (c_i)_{i \in \underline{\ell}} \in \mathbb{C}^\ell$ and $C = -cc^*$, then the assumption of the above lemma is satisfied. The corresponding Hermitian polynomial is $q(z_1, z_2) = -c(z_1)\overline{c}(z_2)$ where $c(\cdot)$ is the polynomial with coefficient vector $c^\top$. The stability domain is $\mathbb{D}(q) = \{z \in \mathbb{C}; -|\overline{c}(z)| < 0\}$, i.e. the complement of the set of roots of $\overline{c}(\cdot)$ in the complex plane. Given any matrix $A \in \mathbb{C}^{n \times n}$ the associated generalized Liapunov operator (94) is

$$\mathbf{L}_{A,C}(X) : \mathcal{H}_n(\mathbb{C}) \to \mathcal{H}_n(\mathbb{C}), \quad X \mapsto -\sum_{i,j=1}^{\ell} c_i \overline{c_j} (A^*)^{i-1} X A^{j-1} = -[\overline{c}(A)]^* X \overline{c}(A). \quad (96)$$

□

Kharitonov (1981) [302] proved the following theorem.

Theorem 3.4.107. *Suppose q is a Hermitian polynomial of the form* (91) *and the coefficient matrix C has exactly one negative eigenvalue. Then the following conditions are equivalent for arbitrary $A \in \mathbb{C}^{n \times n}$.*

(i) $\sigma(A) \subset \mathbb{D}(q)$.

(ii) There exists $Y \in \mathcal{H}_n(\mathbb{C})$, $Y \succ 0$ such that the generalized Liapunov equation (93) *has a (uniquely determined) positive definite solution.*

(iii) For every $Y \in \mathcal{H}_n(\mathbb{C})$, $Y \succ 0$ (93) *has a (uniquely determined) positive definite solution (i.e. the operator $\mathbf{L}_{A,C} : \mathcal{H}_n(\mathbb{C}) \to \mathcal{H}_n(\mathbb{C})$ is inverse positive).*

Proof: (i) ⇒ (ii). First observe that all the conditions do not change if A is replaced by a similar matrix. Hence we may assume that A is of the form $A = D + N$ where $D = \operatorname{diag}(\lambda_1, \ldots, \lambda_n)$ is diagonal and N is nilpotent and upper triangular. Applying a similarity transformation $S(\varepsilon) = \operatorname{diag}(1, \varepsilon, \ldots, \varepsilon^{n-1})$ we obtain $A(\varepsilon) := S(\varepsilon)^{-1} A S(\varepsilon) = D + \varepsilon N$. For $k \in \mathbb{N}^*$, the k-th power of $A(\varepsilon)$ can be represented in the form $A(\varepsilon)^k = D^k + \varepsilon N(k, \varepsilon)$ where $N(k, \varepsilon)$ is a nilpotent upper triangular matrix which is bounded as $\varepsilon \downarrow 0$ for every $k \in \mathbb{N}^*$. Since $\sigma(A) \subset \mathbb{D}(q)$ we have $q(\overline{\lambda_i}, \lambda_i) \neq 0$, $i \in \underline{n}$ by Lemma 3.4.105. Choose

$$X = -\operatorname{diag}(q(\overline{\lambda_1}, \lambda_1)^{-1}, \ldots, q(\overline{\lambda_n}, \lambda_n)^{-1}) \succ 0.$$

Now $q(\overline{\lambda_k}, \lambda_k)^{-1} \sum_{i,j=1}^{\ell} c_{ij} \overline{\lambda_k}^i \lambda_k^j = 1$ for $k \in \underline{n}$, and there exists a Hermitian matrix $Q(\varepsilon)$ dependent on ε such that $\lim_{\varepsilon \to 0} Q(\varepsilon) = 0$ and

$$\sum_{i,j=1}^{\ell} c_{ij} (A(\varepsilon)^*)^{i-1} X A(\varepsilon)^{j-1} = \sum_{i,j=1}^{\ell} c_{ij} \overline{D}^{i-1} X D^{j-1} + Q(\varepsilon) = -I_n + Q(\varepsilon) \succ 0$$

for $\varepsilon > 0$ sufficiently small. Setting $Y = I_n - Q(\varepsilon)$ we see that there exist $Y \succ 0$ and $X \succ 0$ such that (93) is satisfied. Moreover, since $\sigma(A) \subset \mathbb{D}(q)$, Lemma 3.4.105 implies that the solution X of (93) is uniquely determined by Y.
(ii) ⇒ (i). Suppose $Y \succ 0$ and $X \succ 0$ satisfy (93). For any $\lambda \in \sigma(A)$ let $v \in \mathbb{C}^n$ be an associated eigenvector so that $Av = \lambda v$ and $v^* A^* = \overline{\lambda} v^*$. Then

$$0 > -v^* Y v = v^* \left[\sum_{i,j=1}^{\ell} c_{ij} (A^*)^{i-1} X A^{j-1} \right] v = \sum_{i,j=1}^{\ell} c_{ij} \overline{\lambda}^{i-1} \lambda^{j-1} v^* X v = q(\overline{\lambda}, \lambda) v^* X v.$$

Since $X \succ 0$ we conclude that $q(\overline{\lambda}, \lambda) < 0$ and hence $\lambda \in \mathbb{D}(q)$.
(iii) $\Rightarrow$ (ii) being trivial it only remains to prove (ii) $\Rightarrow$ (iii). Suppose $Y_0 \succ 0$ and $X_0 \succ 0$ satisfy (93). Then $\sigma(A) \subset \mathbb{D}(q)$ by the previous step, and so it follows from Lemma 3.4.105 that the operator $\mathbf{L}_{A,C} : \mathcal{H}_n(\mathbb{C}) \to \mathcal{H}_n(\mathbb{C})$ defined by (94) is an isomorphism. Let $Y_1 \in \mathcal{H}_n(\mathbb{C})$ be positive definite and X_1 the corresponding solution of (93). In order to show that $X_1 \succ 0$ we set

$$Y_t = (1-t)Y_0 + tY_1 \succ 0, \quad X_t = (1-t)X_0 + tX_1, \qquad t \in [0,1]$$

and suppose that there exists an $x \in C^n$, $x \neq 0$ such that $x^* X_1 x \leq 0$. By continuity of $t \mapsto x^* X_t x$ there exists $\tau \in [0,1]$ such that $X_\tau \succeq 0$ and $x^* X_\tau x = 0$. Now let $C = \sum_{k=1}^{r-1} v^k v^{k*} - v^r v^{r*}$ as in the proof of the preceding lemma. We have seen in that proof that $|\overline{q_r}(\lambda)| > 0$ for all $\lambda \in \mathbb{D}(q)$, whence for all $\lambda \in \sigma(A)$. Therefore $\overline{q_r}(A)$ is nonsingular. It follows from the definition of the generalized Liapunov operator $\mathbf{L}_{A,C} : \mathcal{H}_n(\mathbb{C}) \to \mathcal{H}_n(\mathbb{C})$ in (94) that it depends linearly on the coefficient matrix C. Therefore, making use of the formula (96) for $X = X_\tau$

$$\mathbf{L}_{A,C}(X_\tau) = \sum_{k=1}^{r-1} \mathbf{L}_{A,v^k v^{k*}}(X_\tau) - \mathbf{L}_{A,v^r v^{r*}}(X_\tau) = \sum_{k=1}^{r-1} [\overline{q_k}(A)]^* X_\tau \overline{q_k}(A) - [\overline{q_r}(A)]^* X_\tau \overline{q_r}(A).$$

Now choose $x \in \mathbb{C}^n$, $x \neq 0$ such that $X_\tau x = 0$ and $y \in C^n$ such that $\overline{q_r}(A) y = x$. Then, multiplying the equation from the left by y^* and from the right by y we obtain

$$y^* \mathbf{L}_{A,C}(X_\tau) y = \sum_{k=1}^{r-1} y^* [\overline{q_k}(A)]^* X_\tau \overline{q_k}(A) y = -y^* Y_\tau y < 0.$$

This contradicts the fact that $X_\tau \succeq 0$, hence $X_1 \succ 0$. □

Remark 3.4.108. The proof of (ii) $\Rightarrow$ (i) shows that condition (ii) is *sufficient* for $\sigma(A) \subset \mathbb{D}(q)$ without any assumption on the coefficient matrix $C \in \mathcal{H}_\ell(\mathbb{C})$. □

Given a polynomial p, one can use the associated companion matrix for A in order to test the $\mathbb{D}(q)$-stability of p. This is illustrated in the next example.

Example 3.4.109. Let us find the conditions for all the roots of the monic polynomial $p(s) = \sum_{i=0}^{n} a_i s^i$, $a_n = 1$ to lie in $\mathbb{D}(q)$ where $q(z_1, z_2) = 2 - z_1 z(z_1 + z_2)$ is as described in Example 3.4.104 (iii). For this we choose A to be the companion matrix of p and $Y = I_n$. The previous theorem is applicable since the only negative eigenvalue of C is -1. Hence p will have all its roots in $\mathbb{D}(q)$ if and only if the solution X of the linear matrix equation

$$2X - (A^*)^2 X A - A^* X A^2 = -I_n \tag{97}$$

is positive definite. The entries of X are rational functions of the coefficients $a_0, \ldots, a_{n-1}$ of p. Expressing the positive definiteness of X by the positivity of its principal minors we obtain a system of rational inequalities for the coefficients of p which will be satisfied if and only if all the roots of p lie in $\mathbb{D}(q)$. For instance, in the real case with $n = 2$, (97) yields the following equations for the entries $x_1, x_2, x_3 \in \mathbb{R}$ of X (x_2 denotes the off-diagonal entry)

$$\begin{aligned} -2x_1 + 2a_0^2 x_2 - 2a_0^2 x_3 &= 1, \\ -a_0 x_1 + (3a_0 a_1 - 2)x_2 + (a_0^2 - 2a_0 a_1^2)x_3 &= 0, \\ -2a_1 x_1 + (4a_1^2 - 2a_0)x_2 + (2a_0 a_1 - 2a_1^3 - 2)x_3 &= 1. \end{aligned}$$

where (a_0, a_1) is the (real) coefficient vector of p. Solving this linear system we obtain rational expressions for the entries $x_i = x_i(a_0, a_1)$, and so the necessary and sufficient conditions for the $\mathbb{D}(q)$-stability of p are given by

$$x_1(a_0, a_1) > 0, \quad x_1(a_0, a_1)\, x_3(a_0, a_1) - x_2(a_0, a_1)^2 > 0.$$

□

We conclude this subsection with a counterexample which illustrates that, in general, the previous theorem will not hold if C does not have exactly one negative eigenvalue.

Example 3.4.110. Let $q(z_1, z_2) = -9+4(z_1+z_2)-z_1^2 z_2^2$ so that the associated coefficient matrix, algebraic domain and linear matrix equations are, respectively,

$$C_q = \begin{bmatrix} -9 & 4 & 0 \\ 4 & 0 & 0 \\ 0 & 0 & -1 \end{bmatrix}, \ \mathbb{D}(q) = \{\rho + \imath\omega; 8\rho - (\rho^2+\omega^2)^2 < 9\}, \ -9X+4A^*X+4XA-(A^*)^2XA^2 = -Y.$$

C has two positive eigenvalues and so the assumption of Theorem 3.4.107 is not satisfied. Let $A = \operatorname{diag}(1, 2)$ and $Y = \begin{bmatrix} 2 & 1 \\ 1 & 1 \end{bmatrix}$. Then $\sigma(A) \subset \mathbb{D}(q)$, but the solution $X = \begin{bmatrix} 1 & 1 \\ 1 & 1/9 \end{bmatrix}$ is not positive definite. On the other hand if we choose $Y = I_2$ then the solution $X = \operatorname{diag}(1/2, 1/9) \succ 0$. So we see that the conditions (ii) and (iii) of the theorem are not equivalent in this case. □

3.4.8 Exercises

1. Prove Proposition 3.4.3 using the principle of the argument.

2. Real polynomials are characterized (up to multiplication with a non-zero constant factor) by the property that their root sets are invariant with respect to reflection at the real axis (taking into accounting multiplicities). To show that Hurwitz-symmetric polynomials can be characterized in a similar way, prove the following.
(i) If s_i is a non-zero root of a Hurwitz-symmetric polynomial p then its reflection at the imaginary axis, $-\overline{s_i}$, is also a root of p, and the multiplicities of these roots are equal.
(ii) Conversely, every polynomial with this property is symmetric after multiplication with a suitable non-zero constant factor[7].

3. Prove Lemma 3.4.12 for the odd case.

4. Determine the first six terms of the Laurent expansion at ∞ of

$$g(s) = \frac{2s+3}{s^2+3s+2}$$

and check your result using the formulas (44) and (43).

5. Determine the Cauchy index $CI_{-\infty}^{\infty}(g)$, the Hankel matrix $\mathbf{Hk}(g)$ and its signature for each of the following rational functions

$$\begin{aligned} &(\text{i}) \quad g(s) = \frac{2s+3}{s^2+3s+2}, && (\text{ii}) \quad g(s) = \frac{1}{s^2+3s+2} \\ &(\text{iii}) \quad g(s) = \frac{4-2s}{s^2-4s+3}, && (\text{iv}) \quad g(s) = \frac{s^2+s+3}{s^3+s^2+2s+2}. \end{aligned}$$

[7] See Ex. 20 for the Schur case.

6. Let $p(s)$ be a real non-constant polynomial and $g(s) = p'(s)/p(s)$. Prove:
(i) The number of distinct roots of p is equal to codim ker $\mathbf{B}_n(p, p')$.
(ii) The number of distinct real roots of p is equal to sign $\mathbf{B}_n(p, p')$.
(iii) All the roots of p are real if and only if $\mathbf{B}_n(p, p') \succeq 0$.

7. Prove that if the roots of two real polynomials $c(s), d(s)$ are all real, simple, and interlacing then the derivatives $c'(s)$ and $d'(s)$ have the same property.

8. Prove that $\mathrm{Rat}_n(\mathbb{C})$ is connected for all $n \geq 1$.

9. Prove that two real polynomials $c(s), d(s) \in \mathbb{R}[s]$ with $\deg d = m$ and $\deg c = m$ or $m - 1$ form a positive pair if and only if $CI_{-\infty}^{\infty}(c/d) = m$, $CI_{-\infty}^{\infty}(\tilde{c}/d) = -m$ (where $\tilde{c}(s) := sc(s)$), and additionally the leading coefficients of c and d are of equal sign in case $\deg c = \deg d$.

10. Consider the third order system $\dddot{\xi} + a_2\ddot{\xi} + a_1\dot{\xi} + a_0\xi = 0$, where $a_0, a_1, a_2 \in \mathbb{R}$. Determine the *Hermite matrix* $\mathbf{H}_3(p)$ for the associated characteristic polynomial p. If $x = [\xi, \dot{\xi}, \ddot{\xi}]^\top$ and $V(x) = \langle x, \mathbf{H}_3(p)x \rangle$, prove that for the corresponding state space system

$$\dot{V}(x) = -2\langle b, x \rangle^2$$

where $b = [a_0,\, 0,\, a_2]^\top$. Conclude that V is a Liapunov function for the system if and only if the system is asymptotically stable.

11. Show that the set of stable Hurwitz polynomials of fixed degree is, in general, not convex. Give an example of two monic real Hurwitz polynomials $p, q \in \mathbb{R}[s]$ of suitable degree n such that the segment $[p, q] = \{\mu q + (1 - \mu)p; \mu \in [0, 1]\}$ contains a polynomial which is not Hurwitz stable.

12. Let $p(s) = u(s^2) + sv(s^2)$ be a real polynomial of degree $n = 2m$ or $n = 2m + 1$ and $v(s)/u(s) = \sum_{k=0}^{\infty} g_k s^{-k}$. Prove that p is Hurwitz stable if and only if the Hankel matrix $\mathbf{Hk}_m(v/u) = (g_{i+j-1})_{i,j \in \underline{m}}$ is positive definite, the Hankel matrix $\mathbf{Hk}_m(\tilde{v}/u) = (g_{i+j})_{i,j \in \underline{m}}$ is negative definite, and $g_0 > 0$ in case $n = 2m + 1$.

13. Consider the differentiable system

$$\begin{aligned} \dot{x} &= Ax + Bu \\ y &= Cx \end{aligned} \quad \text{where} \quad A = \begin{bmatrix} 0 & 0 & 3 \\ 10 & -3 & 13 \\ 5 & 1 & 6 \end{bmatrix}, \ B = \begin{bmatrix} 1 & 0 \\ 2 & 1 \\ 1 & 1 \end{bmatrix}, \ C = \begin{bmatrix} 1 & 0 & 0.6 \\ 0 & 1 & -0.2 \end{bmatrix}.$$

If static output feedback is applied via the feedback gain matrix $F = -\alpha I_2$ where $\alpha \in \mathbb{R}$ (see Example 2.4.12) prove that the characteristic polynomial of the closed loop system is

$$\lambda^3 + (2.4\alpha - 3)\lambda^2 + (0.2\alpha^2 + 12.2\alpha - 46)\lambda + 0.2\alpha^2 + 14\alpha - 75.$$

Hence show that the closed loop system is asymptotically stable if $\alpha > 0$ is sufficiently large and it can not be asymptotically stable if $\alpha \leq 5$.

14. A continuous time single input single output system with transfer function $g(s) = p(s)/q(s) \in \mathbb{R}(s)$ (p, q coprime) is called *input-output stable* if $q(s)$ is a Hurwitz stable polynomial. Show that the transfer function of a closed loop system with plant $g(s)$ and static linear output feedback $u = -Ky$ is given by $g_K(s) = p(s)/(q + Kp(s))$. Prove that

$$g(s) = \frac{p(s)}{q(s)} = \frac{(s+2)(s+3)}{s^2(s+1)(s+24)(s+30)}.$$

can be input-output stabilized by negative feedback $u = -Ky$, $K > 0$, but that input-output stability is destroyed if the gain $K \gg 0$ is sufficiently large. Determine the set of all static gains $K \in \mathbb{R}$ for which the closed system transfer function $g_K(s)$ is stable (i.e. $p(s) + Kq(s)$ is Hurwitz).

15. A mathematical model for the vibrations of a rotating shaft of mass m, stiffness k and damping c is

$$\ddot{\xi} + 2\Omega i\dot{\xi} - \Omega^2\xi = -(k/m)\xi - (c/m)\dot{\xi}$$

where the displacement of the shaft is given by $\xi = x + iy$ and Ω is the constant angular velocity of the shaft. By using Hermite's Theorem 3.4.29 and the expression for $\mathbf{H}_2(p)$ given in Example 3.4.25 show that the equilibrium state $[0,\, 0]^\top \in \mathbb{C}^2$ of the corresponding state space system is stable provided $\Omega^2 < k/m$.

16. Prove Theorem 3.4.71 and Corollary 3.4.73 for the odd case $n = 2m + 1$. Conclude from the theorem that a quartic polynomial

$$p(s) = a_4s^4 + a_3s^3 + a_2s^2 + a_1s + a_0$$

with positive coefficients is Hurwitz if and only if $\Delta_3 > 0$. Show that the Hurwitz stability of a polynomial of degree $n = 5$ and $n = 6$ can be characterized by the positivity of two principal minors, if the polynomial has positive coefficients.

17. By applying the Liénard-Chipart test Corollary 3.4.66 and using the formula for the Bezoutian $\mathbf{B}_2$ given in Example 3.4.33 derive necessary and sufficient conditions for a real polynomial of degree 5 with positive coefficients to be Hurwitz stable.

18. In Maxwell's paper "On Governors" as well as considering the governor described in Example 3.4.16 he also considered a more complicated governor system described by the equations

$$\begin{aligned} A\ddot{\theta} + X\dot{\theta} + K\dot{\phi} + T\phi + J\psi &= P - R \\ B\dot{\phi} + Y\phi - K\theta &= Q \\ C\ddot{\psi} + Z\dot{\psi} - T\phi &= 0 \end{aligned}$$

where θ, ϕ, ψ are the angles of the main shaft, centrifugal arm and the movable wheel respectively. All of the constants A, X, K, T, J, P, R, B, Y, Q, C and Z are positive. Maxwell studied the stability of the equilibrium state of a corresponding state space system and obtained a fifth degree characteristic polynomial. Prove that this polynomial is

$$s^5 + a_4s^4 + a_3s^3 + a_2 + s^2a_1s + a_0$$

where

$$\begin{aligned} a_0 &= TJK/ABC, \quad a_1 = ZTK/ABC, \quad a_2 = Z(XY + K^2)/ABC + KT/AB, \\ a_3 &= (XY + K^2)/AB + Z(XB + YA)/ABC, \quad a_4 = X/A + Y/B + Z/C. \end{aligned}$$

Show that the conditions given in Ex. 17 for a real polynomial of degree 5 to be Hurwitz stable reduce in the case of the above polynomial to

$$a_2a_1 - a_0a_3 > 0 \quad \text{and} \quad (a_2a_1 - a_0a_3)(a_4a_3 - a_2) > (a_4a_1 - a_0)^2.$$

Hence show that the state space system is not asymptotically stable if

$$Z^2(XY + K^2) + KTCZ < J((XY + K^2)C + Z(XB + YA)).$$

In particular observe that the system will not be asymptotically stable if the damping coefficient Z is sufficiently small. Show, however that if

$$(XY + K^2)(XB + YA) > TKAB$$

then the system will be asymptotically stable for Z sufficiently large.

19. Prove the following result of *Bialas* (1985) [57]: Suppose $p_i(s) = \sum_{j=0}^n a_j^i s^j$, $i = 0, 1$ are two real Hurwitz polynomials with $a_0^i > 0$. Then all polynomials p_μ in the segment $[p_0, p_1] = \{\mu p_1 + (1-\mu)p_0; 0 \le \mu \le 1\}$ are Hurwitz stable if and only if $(M_{n-1}^0)^{-1} M_{n-1}^1$ has no negative eigenvalues, where M_{n-1}^i, $i = 0, 1$ denotes the $(n-1)$-th principal submatrix of the Hurwitz matrix associated with p_i.
(Hint: Compare the proof of Theorem 3.4.71 and make use of Lemma 3.4.49. If the $(n-1)$-th principal minor Δ_{n-1} of a Hurwitz matrix $M^{\mathbb{R}}(p)$ is positive and $a_0 > 0$ then $\det M^{\mathbb{R}}(p) = a_0 \Delta_{n-1}$ is positive, see (64)).

20. Suppose $p(z)$ is a polynomial of degree n with the property that every root z_i of p is non-zero and its reflection at the unit circle, $1/\overline{z}_i$, is a root of p of the same multiplicity as z_i. Prove that $cp(z)$ is Schur-symmetric for a suitable constant $c \in \mathbb{C}^*$.

21. Let $p(z) = a_n z^n + a_{n-1} z^{n-1} + \ldots + a_1 z + a_0 \in \mathbb{R}[z]$ with $a_n > 0$. Prove the following conditions are necessary and sufficient for $p(\cdot)$ to be Schur stable.
(i) in case $n = 2$

$$\text{(a)}\ a_0 + a_1 + a_2 > 0, \quad \text{(b)}\ a_0 - a_1 + a_2 > 0, \quad \text{(c)}\ a_0 < a_2.$$

(ii) in case $n = 3$

$$\text{(a)}\ a_0 + a_1 + a_2 + a_3 > 0, \quad \text{(b)}\ -a_0 + a_1 - a_2 + a_3 > 0, \quad \text{(c)}\ a_3^2 - a_0^2 > |a_0 a_2 - a_1 a_3|.$$

(iii) in case $n = 4$

$$\text{(a)}\ a_0 + a_1 + a_2 + a_3 + a_4 > 0, \quad \text{(b)}\ a_0 - a_1 + a_2 - a_3 + a_4 > 0,$$

$$\text{(c)}\ a_4(a_4^2 + a_0 a_2 - a_1^2 - a_0^2) + a_0(a_1 a_3 - a_0 a_2) > |a_0(a_3^2 - a_0^2) + a_4(a_4 a_0 + a_2 a_4 - a_1 a_3 - a_0 a_2)|.$$

22. Show that the set of monic Schur polynomials of fixed degree is, in general, not convex. Give an example of two monic real Schur polynomials $p, q \in \mathbb{R}[s]$ of suitable degree n such that the segment $[p, q] = \{\mu q + (1-\mu)p; \mu \in [0, 1]\}$ contains a polynomial which is not Schur stable.

23. If $c(z)$ is a Schur-symmetric polynomial of degree n and $f(\theta) = e^{-\imath\theta n/2} c(e^{\imath\theta}), \theta \in \mathbb{R}$, show that f is a trigonometric polynomial of the form

$$f(\theta) = \begin{cases} a_0 + \sum_{k=1}^m (a_k \cos k\theta + b_k \sin k\theta) & \text{if} \quad n = 2m \\ \sum_{k=1}^m (a_k \cos(k - 1/2)\theta + b_k \sin(k - 1/2)\theta) & \text{if} \quad n = 2m - 1. \end{cases}$$

Verify that $f'(\theta) = -\imath e^{-\imath\theta n/2} \hat{c}(e^{\imath\theta})$ where $\hat{c}(z) = (n/2)c(z) - zc'(z)$ is Schur-antisymmetric. Suppose that c and d are symmetric polynomials of degree n and f, g are the associated trigonometric polynomials. Prove that if the roots of c, d (resp. f, g) are all simple, lie on the unit circle (resp. real axis) and are interlacing on the unit circle (resp. the real axis), then the roots of $\hat{c}$ and $\hat{d}$ (resp. f', g') have the same property.

24. If $C \in \mathcal{H}_\ell(\mathbb{C})$, $A \in \mathbb{C}^{n\times n}$, prove that the spectrum of the linear operator

$$\mathbf{L}_{A,C} : \mathcal{H}_n(\mathbb{C}) \to \mathcal{H}_n(\mathbb{C}), \quad X \mapsto \sum_{i=1}^{\ell}\sum_{j=1}^{\ell} c_{ij}(A^*)^{i-1}XA^{j-1}$$

is given by $\sigma(\mathbf{L}_{A,C}) = \{q(\overline{\lambda_1}, \lambda_2); \lambda_1, \lambda_2 \in \sigma(A)\}$ where $q(\cdot,\cdot)$ is given by (91).

25. Case study: The flow of traffic along a motorway is modelled by the partial differential equations

$$\frac{\partial \rho}{\partial t} + \frac{\partial}{\partial x}(\rho v) = 0$$

$$\frac{\partial v}{\partial t} + v\frac{\partial v}{\partial x} = -\alpha\left[v - V(\rho) + \frac{\mu}{\rho}\frac{\partial \rho}{\partial x}\right]$$

where α, μ are positive constants, $V(\cdot) : \mathbb{R}_+ \to \mathbb{R}$ is a given function, $\rho(x,t)$ is the density and $v(x,t)$ is the velocity at distance $x \in [0,\ell]$ along the motorway and time $t \geq 0$. The first equation is the continuity equation and the second states that the acceleration is a function of three terms, the velocity, the density of traffic and the last term $\frac{\mu}{\rho}\frac{\partial \rho}{\partial x}$ represents a model for the way drivers take account of increasing or decreasing density of the traffic ahead. Show that (ρ_0, v_0) is a constant equilibrium state if $v_0 = V(\rho_0)$. It is assumed that the function $V(\cdot)$ is differentiable, has a maximum at zero and decreases to zero at $\rho = \rho_m$ (the value of the density for which the cars are bumper to bumper). For a single lane the maximum density is taken to be 225 vehicles per mile. A model which fitted data for traffic flow through the Lincoln tunnel is $V(\rho) = a\log \rho_m/\rho$ and it was observed that there was a maximum number of cars passing through the tunnel if the velocity was 20 miles per hour.

(i) Show that if $\rho(x,t) = \rho_0 + \rho_1(x,t)$, $v(x,t) = v_0 + v_1(x,t)$ and nonlinear terms in the perturbations ρ_1, v_1 are neglected, then

$$\frac{\partial \rho_1}{\partial t} + v_0\frac{\partial \rho_1}{\partial x} + \rho_0\frac{\partial v_1}{\partial x} = 0$$
$$\frac{\partial v_1}{\partial t} + v_0\frac{\partial v_1}{\partial x} = -\alpha\left[v - V'(\rho_0)\rho_1 + \frac{\mu}{\rho_0}\frac{\partial \rho_1}{\partial x}.\right]$$

(ii) If the perturbations are represented by the real parts of the Fourier series $\rho_1(x,t) = \sum_{k=-\infty}^{\infty}\tilde{\rho}_k(t)e^{\imath k\pi x/\ell}$, $v_1(x,t) = \sum_{k=-\infty}^{\infty}\tilde{v}_k(t)e^{\imath k\pi x/\ell}$, show that the coefficients satisfy

$$\dot{\tilde{\rho}}_k(t) + (\imath v_0 k\pi/\ell)\tilde{\rho}_k(t) + (\imath\rho_0 k\pi/\ell)\tilde{v}_k(t) = 0$$
$$\dot{\tilde{v}}_k(t) + (\imath v_0 k\pi/\ell)\tilde{v}_k(t) + \alpha\left[\tilde{v}_k(t) - V'(\rho_0)\tilde{\rho}_k(t) + (\mu\imath k\pi/(\ell\rho_0))\tilde{\rho}_k(t)\right] = 0.$$

(iii) Prove that the characteristic equation of the the corresponding state space system is

$$\lambda^2 + ((2\imath k\pi/\ell)v_0 + \alpha)\lambda - (k\pi/\ell)^2 v_0^2 + \alpha\mu(k\pi/\ell)^2 + \alpha c_0\imath k\pi/\ell = 0$$

where $c_0 = v_0 + \rho_0 V'(\rho_0)$.

(iv) By using Hermite's Theorem 3.4.29 and the expression for $\mathbf{H}_2(p)$ given in Example 3.4.25 show that this is a Hurwitz polynomial if $\alpha\mu > (v_0 - c_0)^2$ for all values of k.

(v) Calculate the right hand side of this inequality for the above function $V(\cdot)$ and interpret the result, see *Whitham* (1974) [518].

3.4.9 Notes and References

Introductory historical remarks

In 1868 Maxwell posed the mathematical problem of finding conditions under which all the roots of an algebraic equation belong to the open left half of the complex plane. At this time there were about 75,000 steam engines working in England alone, and "large numbers of them were hunting" [356]. Due to changes in engine design (less friction, smaller flywheels) the generated angular velocities of the shaft tended to be unstable: Oscillations about the set angular velocity, caused by load changes did not die out. These technological developments formed the background of Maxwell's problem. He knew that a system behaves in a stable way if all the roots of its characteristic polynomial have negative real parts. But, as Maxwell pointed out in his paper, how this condition could be verified for polynomials of degree > 3 or 4 was unknown. Therefore the pioneering papers of Maxwell and Vyshnegradskii on the stability of controlled systems (like the steam engine) were limited to linear models of low order, see [364], [511], [512], Example 3.4.16 and Ex. 18. However, mathematicians had dealt with the problem of determining the number of roots of algebraic equations in certain locations (on and off the real axis, in half-planes etc.) since the early decades of the nineteenth century. It suffices to mention here the work of Cauchy, Sturm, Jacobi, Borchardt, Cayley, Sylvester and Hermite. In fact, *Hermite* (1853) [226] had already solved Maxwell's problem well before it was stated, but his results were not known outside of world of mathematics. Throughout the nineteenth century the analysis of real algebraic equations was a central theme of mathematics. Excellent surveys, including some historical remarks, are given in the classical paper of *Krein and Naimark* (1939, translation 1981) [318] and in the concluding chapter of *Gantmacher* (1959) [183]. Both sources contain extensive bibliographies where many references on the root location problem can be found. For an exposition of some results of Hermite and Hurwitz in the context of Real Algebra, see *Knebusch* (1989) [310].

It was Maxwell's problem statement which brought the above mathematical developments into contact with the emerging theory of stability and control. In 1877 Routh received the Adams Prize of the University of Cambridge for the best essay on "The Criteria of Dynamic Stability". In [439] he described an algorithm for determining the number of roots of a real polynomial in the right half-plane, based on Sturm's Theorem and the Cauchy index. The corresponding stability test is described in Algorithm 3.4.13 (see also Remark 3.4.14). Two decades later Stodola rediscovered Maxwell's problem in his research on the stability of turbines. He was unaware of Routh's essay and Maxwell's paper, and asked Hurwitz for help in solving the mathematical problem. The determinant inequalities obtained by Hurwitz are nowadays known as the Routh-Hurwitz stability criteria (Subsection 3.4.5). For a description of the early development of stability theory in the context of feedback control, see *Rörentrop* (1971) [436], *Fuller* (1976) [176], and *Bennett* (1979) [50].

Notes and references concerning the subsections

3.4.1 Analytical techniques based on the argument principle and the Cauchy index have been used since the beginning of stability analysis, see [183]. The phase-increasing property of real Hurwitz polynomials has been known in the theory of electrical networks since the sixties (see [68]) and the lower bound (9) was rediscovered in the context of robust stability analysis by *Rantzer* (1992) [427]. In fact Rantzer proved that "convex directions"

$q(s) \in \mathbb{R}[s]$ (for which the stability of p and $p+q$ always implies the stability of the whole segment $[p, p+q]$ of polynomials) are characterized by the reverse inequality to (9). A different characterization can be found in *Kharitonov and Hinrichsen* (1995) [236].
The Hermite-Biehler Theorem 3.4.11 has become an important tool in the stability analysis of polynomials with parametric uncertainty, see *Kharitonov* (1979) [301] and the monograph of *Bhattacharyya et al.* (1995) [56]. Example 3.4.16 is taken from Maxwell's classical paper [364] which has been republished together with other classical papers on stability and control (e.g. Hurwitz' paper [272]) in *Bellman and Kabala (eds.)* 1964 [46], see also the collection of articles edited by *MacFarlane* (1979) [356].

3.4.2 The geometric interpretation of the Cauchy index in Subsection 3.4.2 and its application to the analysis of $\mathrm{Rat}_n(\mathbb{R})$ has been inspired by *Brockett* (1976) [78]. A system theoretic characterization of the Cauchy index is given in *Anderson* (1972) [12].

3.4.3 This and the next subsections on the use of quadratic forms in stability theory have been strongly influenced by the fundamental paper of *Krein and Naimark* (1939) [318] and a lecture series by Kharitonov at the University of Bremen (1993/94). Hermite's Theorem 3.4.29 has been proved in [226] for the case where $p(s)$ and $p^\star(s)$ are coprime. Our proof follows the method of *Liénard and Chipart* (1914) [344] as presented in Kharitonov's lecture series.
Bézout matrices played an important role in the study of root locations and in elimination theory in the 19th century, but have been largely ignored in 20th century mathematics. In systems theory they have been subordinated to Hankel matrices, but their use has been advocated by *Helmke and Fuhrmann* (1989) in [220] where Bézout matrices were applied to construct invariants for static linear output feedback. In comparison with Hankel matrices Bézout matrices have the advantage that they are not constrained by degree restrictions ($\deg c \leq \deg d$), and there are simple explicit formulas for their entries (see (36)) whereas the entries of the Hankel matrix depend in a more complicated manner on the coefficients of the two polynomials. More results about Bézout and Hankel matrices can be found in the textbook of *Fuhrmann* (1996) [173]. This book also contains further applications of quadratic forms in systems theory.

3.4.4 Our proof of the congruence of the Bézout and the associated Hankel matrix (Theorem 3.4.41) follows *Krein and Naimark* (1981) [318]. In the case when the denominator has only simple roots, Theorem 3.4.53 was proved by Sylvester in 1853 and Hermite in 1854, the general result is due to *Kronecker* (1881) [320] and *Hurwitz* (1895) [272]. Topological aspects of this result are discussed in *Byrnes* (1983) [87].
Brockett's theorem was published in 1976, [78] and initiated the investigation of topological problems in linear systems theory. The topology of $\mathrm{Rat}_n(\mathbb{R})$ was studied in more detail by *Segal* (1979) [456]. The problem of parametrizing the connected components of $\mathrm{Rat}_n(\mathbb{R})$ as stated in [78] led to the problem of finding a cell decomposition for $\mathrm{Rat}_n(\mathbb{R})$. Different subdivisions of $\mathrm{Rat}_n(\mathbb{R})$ were proposed by *Fuhrmann and Krishnaprasad* (1986) in [174], via continued fractions. That these subdivisions yield true cell decompositions was established in [238].
The proof of Brockett's Theorem in Subsection 3.4.4 is different from that of [78]. Another proof, based on the cellular subdivision of the space of Hankels of rank $\leq n$ in [237], is given in *Helmke et al.* (1989) [221].
More information about the algebraic theory of Hankel matrices and forms can be found

in *Iohvidov* (1977) [274] and *Heinig and Rost* (1984), [219].

3.4.5 The determinant conditions given in Theorems 3.4.71 were Hurwitz's answer to Stodola's question. Corollary 3.4.66 was first proved by *Liénard and Chipart* (1914) [344]. Surprisingly it took more than half a century to clarify the connection between algebraic stability criteria and Liapunov's equation. In 1962 *Parks* [405] derived the Routh-Hurwitz stability criteria by Liapunov's direct method.

3.4.6 The stability criterion for discrete time systems contained in Theorem 3.4.89 was found by *Schur* (1918) [454] in his analysis of bounded power series on the unit disk. Theorem 3.4.89 itself was first proved by *Cohn* (1922) [107] under the condition that $p(z)$ and $p^*(z)$ are coprime and the general theorem was proved by *Fujiwara* (1926) [175]. The explicit formula (76) for the entries of the Schur form is due to *Wilf* (1959) [522]. The use of symmetric polynomials and Möbius transforms in the treatment of the Schur-Cohn problem is based on *Krein and Naimark* [318]. Theorem 3.4.97, the counterpart of the Hermite-Hurwitz Theorem 3.4.53 was first derived by *Herglotz* (1923) [225] following Hurwitz' method of proof in [272]. The recursive stability test (3.4.99) was found by Schur (1918) [454]. An interesting algebraic derivation of the Schur-Cohn stability criterion via Liapunov's equation for discrete time systems is given in *Kalman* (1965) [288]. In this paper Kalman shows that the Schur-Hermite matrix of a real polynomial p satisfies a Liapunov equation of the form $A^\top XA - X = -cc^\top$ where A is the companion matrix of p and $c \in \mathbb{R}^n$. Additional references are *Marden* (1966) [360] and the book of *Jury* (1982) [284] which presents a unified approach to algebraic stability criteria for discrete and continuous time systems based on the concept of "inners". A proof of Theorem 3.4.91 can be found in [284].
Further information about the algebraic theory of Toeplitz matrices and forms can be found in *Iohvidov* (1977) [274] and *Heinig and Rost* (1984), [219].
The cohort population model is briefly discussed in *Luenberger* (1979) [349].

3.4.7 Several attempts have been made to develop a unifying algebraic theory of stability which is applicable to different stability regions described by algebraic inequalities. As early as the 19th century Hermite studied the problem of determining how many roots of an algebraic equation $p(z) = p(x + \imath y) = 0$ are in a domain of the form $V(x, y) > 0$ where $V(x, y)$ is the imaginary part of some rational function in $z = x + \imath y$, see [226], [318]. An interesting algebraic approach for deriving a general Hermite type criterion for stability regions described by quadratic inequalities has been developed by *Kalman* (1969) [289] for domains described by Hermitian polynomials q of the form (91) with $\operatorname{rank} C = 2$ and $\operatorname{sign} C = 0$. Kalman posed in [289] the "open question" of whether this is the largest class for which the statement $p(z_0) = 0 \implies z_0 \in \mathbb{D}(q)$ can be decided from inequalities employing only rational functions of the coefficients of p. The question was answered in the negative by *Kharitonov* (1981) [302], who showed that a Liapunov type stability test is available for a much larger class of stability regions (see Theorem 3.4.107). Unnoticed in systems theory a more general version of this theorem had already been derived fifteen years earlier by *Schneider* (1965) [451] in an article about positive operators. Thus Kalman's question had already been answered in the negative before it was raised. Subsection 3.4.7 is based on Kharitonov's paper. Further generalized Liapunov criteria for stability regions described by algebraic inequalities can be found in the books of *Jury* (1982) [284] and *Gutman* (1990) [206].

Chapter 4

Perturbation Theory

The aim of this chapter is to study how the root and eigenvalue locations of polynomials and matrices change under perturbations. The chapter is quite a substantial one since we address a number of different issues. First and foremost we consider a variety of perturbation classes, ranging from highly structured perturbations which are determined via a single parameter to unstructured perturbations where all the entries of the matrix or coefficients of the polynomial are subject to independent variation. The size of the perturbations will, in the main, be measured by arbitrary operator norms. Moreover we will develop the theory for both complex and real perturbations which often require quite different approaches.
The first section is concerned with polynomials. We establish some continuity and analyticity results for the roots, then describe the sets of all *Hurwitz* and *Schur* polynomials in coefficient space. We also consider the problem of determining conditions under which all polynomials with real coefficients belonging to prescribed intervals are stable and prove Kharitonov's Theorem. The effect of perturbations on the eigenvalues of matrices is considered in Section 4.2. We first state some simple continuity and analyticity results which follow directly from the results of Section 4.1. Then we assume that the matrix depends analytically on a single parameter and examine the smoothness of eigenvalues, eigenprojections and eigenvectors. Section 4.3 deals with singular values and singular value decompositions which are important tools in the quantitative perturbation analysis of linear systems. Section 4.4 is dedicated to structured perturbations and presents some elements of μ-analysis, both for complex and for real parameter perturbations. We finish the chapter in Section 4.5 with a brief introduction to some numerical issues which are important for Systems Theory, focussing on those aspects which have a relationship with the material of this and the previous chapters.

4.1 Perturbation of Polynomials

In the previous chapter we derived necessary and sufficient conditions for a given polynomial to have all its roots in the open left half plane or inside the open unit disk. Often in applications the coefficients of a polynomial are not known precisely or they depend on physical parameters which may vary between specified bounds. It is important therefore to study the root locations and stability properties for *sets*

or *(parametrized) families* of polynomials. Here we establish some basic results for such problems.
The section is divided into four subsections. In the first one we use function theoretic tools to analyze the dependence of the roots of a polynomial on the coefficient vector. We show that in general this dependence is continuous and for a simple root (multiplicity one) it is analytic. In the second subsection we study *parametrized* polynomials where the coefficient vector is a function of one real or complex parameter. *Critical values* of this parameter are those where the number of distinct roots decreases. We investigate the distribution of critical points and analyze the behaviour of the roots in the vicinity of both critical and non-critical parameter values. In order to obtain quantitative information about root changes under parameter perturbations we introduce the *sensitivity* of a root as a measure of the variation in the root with respect to *infinitesimal* changes in the parameter.
The third subsection is concerned with the sets of *Hurwitz* and *Schur* polynomials, respectively. We establish some elementary topological properties of these sets in the space of coefficient vectors and characterize their boundaries and convex hulls.
In the final subsection we restrict our considerations to Hurwitz polynomials and examine some convexity properties which are useful in studying the effect of *large* parameter variations. The main result will be Kharitonov's Theorem which specifies necessary and sufficient conditions for an n-dimensional interval of polynomials to consist of only Hurwitz ones.

4.1.1 Dependence of the Roots on the Coefficient Vector

Consider the polynomial

$$p(s,a) = a_n s^n + a_{n-1}s^{n-1} + \cdots + a_0 \in \mathbb{C}_n[s] \tag{1}$$

with coefficient vector $a = (a_0, a_1, \ldots, a_n) \subset \mathbb{C}^{n+1}$. We assume that the vector space $\mathbb{C}^{n+1}$ of coefficient vectors is endowed with an arbitrary norm $\|\cdot\|$. Let $n(a) = \deg p(s,a)$ denote the degree of $p(s,a)$. Then $a_{n(a)}$ is called the *leading coefficient* of $p(s,a)$, and we have $a_{n(a)} \neq 0$ and $a_j = 0$ for $j = n(a)+1, \ldots, n$.
By the fundamental theorem of algebra $p(s,a)$ has $n(a)$ complex roots, taking into account multiplicities. We say that a polynomial $p(s,a)$ has (exactly) ℓ *distinct roots* $s_1, \ldots, s_\ell$, $\ell \geq 1$ *with multiplicity* $m_1, \ldots, m_\ell \geq 1$, if and only if $n(a) = \sum_{j=1}^{\ell} m_j$ and $p(s,a)$ admits the factorization

$$p(s,a) = a_{n(a)} \prod_{j=1}^{\ell} (s - s_j)^{m_j} \quad \text{and} \quad s_i \neq s_j \text{ for } i \neq j.$$

We want to analyze the dependency of the roots s_j, $j \in \underline{\ell}$ on the coefficient vector a. Later in this section we will assume that $p(s,a)$ is *monic* of degree n, i.e. $a_n = 1$. But in our first results we do not assume $n(a) = n$. As a consequence, small variations of the coefficient vector may lead to an upward jump in the overall number of roots. The following example illustrates that the "additional" roots generated in this way have large moduli for small variations of the coefficient vector.

Example 4.1.1. Consider the polynomials $p(s,a(z)) = zs^2 - s + z$, $a(z) = (z,-1,z)$ where $z \in \mathbb{C}$ is a parameter. For $z \neq 0$ the polynomial $p(s,a(z))$ has two simple zeros, namely $s_{\pm}(z) = (1 \pm \sqrt{1-4z^2})/(2z) = [1 \pm (1 - 2z^2 + O(z^4))]/2z$. As $z \to 0$, we see that $s_-(z)$ converges to 0 (which is the only root of $p(s, a(0))$) whereas $s_+(z)$ tends to ∞. □

In an intuitive way we may interpret the previous example by saying that the polynomial $p(s, a(0)) = 0s^2 + s + 0$ regarded as an element of $\mathbb{C}_2[s]$ has one zero at the origin and one at infinity, and these zeros attract the zeros of $p(s, a(z))$ as $z \to 0$. A precise statement is given in the next theorem. For any $s_0 \in \mathbb{C}$ we denote the disk of radius $r > 0$ centered at s_0 by $D(s_0, r) = \{s \in \mathbb{C};\ |s - s_0| < r\}$. Whereas by $B(\tilde{a}, \delta)$ we denote the open ball of radius $\delta > 0$ and centre $\tilde{a}$ in the coefficient vector space $\mathbb{C}^{n+1}$, i.e.

$$B(\tilde{a}, \delta) = \{a \in \mathbb{C}^{n+1};\ \|a - \tilde{a}\| < \delta\}.$$

Theorem 4.1.2. *Let $p(s, \tilde{a})$ be a non-constant polynomial of the form (1) having degree $n(\tilde{a}) \le n$ and exactly ℓ distinct roots $\tilde{s}_j$, of multiplicities m_j, $j \in \underline{\ell}$. Then, for any given $\varepsilon > 0$ such that the closed disks $\overline{D(\tilde{s}_j, \varepsilon)}$, $j \in \underline{\ell}$ are mutually disjoint, there exists $\delta(\varepsilon) > 0$ such that for all $a \in B(\tilde{a}, \delta(\varepsilon))$ there are exactly m_j roots of $p(s, a)$ inside the disk $D(\tilde{s}_j, \varepsilon)$ for $j \in \underline{\ell}$ and $n(a) - n(\tilde{a})$ roots outside the disk $D(0, \varepsilon^{-1})$ (taking into account multiplicities).*

Proof: Choose any $\varepsilon > 0$ so that the disks $D_j = D(\tilde{s}_j, \varepsilon)$ and their boundaries $\Gamma_j = \partial D_j$ do not overlap for $j \in \underline{\ell}$. Then $p(s, \tilde{a})$ has exactly one root (of multiplicity m_j) in D_j and $\mu_j := \min_{s \in \Gamma_j} |p(s, \tilde{a})| > 0$. Choose $\delta > 0$ such that

$$a \in B(\tilde{a}, \delta) \quad \Rightarrow \quad \max_{s \in \Gamma_j} |p(s, a) - p(s, \tilde{a})| < \mu_j, \quad j \in \underline{\ell}.$$

Then it follows from Rouché's Theorem A.2.20 that, for any $a \in B(\tilde{a}, \delta)$, the polynomial $p(s, a)$ has exactly m_j roots in D_j accounting for multiplicities. This proves the first assertion in the theorem. We claim that by choosing $\delta(\varepsilon) < \delta$ sufficiently small we can also ensure that the second half holds. Otherwise there would exist a sequence of coefficient vectors $a^k \in B(\tilde{a}, \delta)$ converging to $\tilde{a}$ such that at least one root z_k of $p(\cdot, a^k)$ would lie in $D(0, \varepsilon^{-1}) \setminus \bigcup_{j=1}^n D_j$. By a compactness argument we may assume that the bounded sequence (z_k) is convergent in $\mathbb{C}$. But then the limit $z^0 \in \mathbb{C} \setminus \bigcup_{j=1}^n D_j$ would satisfy $p(z^0, \tilde{a}) = \lim_{k \to \infty} p(z_k, a^k) = 0$ and hence $p(\cdot, \tilde{a})$ would have more than $n(\tilde{a})$ roots (taking into account multiplicities). This contradiction shows that the second assertion also holds. □

We now describe a more specific result which is useful in several areas of systems theory. Real affine one-parameter perturbations of the form $\tilde{a} \rightsquigarrow \tilde{a} + ra$, where $a, \tilde{a}$ are given and $r \ge 0$ is a small real parameter, are considered.

Proposition 4.1.3. *Consider the parametrized polynomial $p(s, a(r)) = p(s, \tilde{a}) + rp(s, a)$ where $r \ge 0$ is a parameter, $p(s, \tilde{a})$ is a polynomial of degree $m < n$ and $p(s, a)$ is a monic polynomial of degree n. Denote by $\tilde{a}_m^{1/(n-m)}$ any fixed $(n-m)$-th root of the leading coefficient $\tilde{a}_m$ of $p(s, \tilde{a})$, and let $\lambda_1, \ldots, \lambda_{n-m}$ be the roots of $s^{n-m} + 1$. Then m of the roots of $p(\cdot, a(r))$ tend to the roots of $p(\cdot, \tilde{a})$ as $r \downarrow 0$, while the remaining $n - m$ roots tend asymptotically to*

$$\mathcal{B}(r) = \{r^{-1/(n-m)}\, \tilde{a}_m^{1/(n-m)} \lambda_j\,;\ j = 1, \ldots, n-m\}, \tag{2}$$

i.e. there are $n-m$ roots $s_{m+1}(r),...,s_n(r)$ of $p(s,a(r))$ such that

$$\lim_{r\downarrow 0} |s_i(r) - r^{-1/(n-m)}\,\tilde{a}_m^{1/(n-m)}\lambda_{i-m}| = 0 \quad \textit{for} \quad i = m+1,...,n.$$

Proof: Suppose that $p(s,\tilde{a})$ has ℓ distinct roots $\tilde{s}_j$ of multiplicities m_j, $j \in \underline{\ell}$. By the previous theorem there exists for every sufficiently small $\varepsilon > 0$ a $\delta(\varepsilon) > 0$ such that the disks $D(\tilde{s}_j,\varepsilon)$ contain exactly m_j roots of $p(s,a(r))$ for $r \in (0,\delta)$, $j \in \underline{\ell}$ (taking account of multiplicities). Thus m of the roots of $p(s,a(r))$ approach the m roots of $p(s,\tilde{a})$ as $r \downarrow 0$.
To prove the second part of the proposition, let $r > 0$ and $\zeta = (r/\tilde{a}_m)^{1/(n-m)}s = \gamma(r)s$ where $\gamma(r) = (r/\tilde{a}_m)^{1/(n-m)}$. Then $a(r) = ra + \tilde{a}$ implies

$$\begin{aligned} p(\gamma(r)^{-1}\zeta, a(r)) &= r\gamma(r)^{-n}\zeta^n + ra_{n-1}\gamma(r)^{-(n-1)}\zeta^{n-1} + \ldots + ra_0 \\ &\quad + \tilde{a}_m\gamma(r)^{-m}\zeta^m + \tilde{a}_{m-1}\gamma(r)^{-(m-1)}\zeta^{m-1} + \ldots + \tilde{a}_0. \end{aligned}$$

Now $\tilde{a}_m\gamma(r)^{-m}/(r\gamma(r)^{-n}) = \tilde{a}_m\gamma(r)^{n-m}/r = 1$ and so

$$\begin{aligned} p(\gamma(r)^{-1}\zeta, a(r)) &= r\gamma(r)^{-n}\left[\zeta^n + \gamma(r)a_{n-1}\zeta^{n-1} + \ldots + \gamma(r)^n a_0 \right. \\ &\quad \left. + \zeta^m + \gamma(r)(\tilde{a}_{m-1}/\tilde{a}_m)\zeta^{m-1} + \ldots + \gamma(r)^m\tilde{a}_0/\tilde{a}_m\right] =: r\gamma(r)^{-n}p_r(\zeta). \end{aligned}$$

But

$$\begin{aligned} p_r(\zeta) &= \zeta^n + \zeta^m + \gamma(r)\left[a_{n-1}\zeta^{n-1} + \gamma(r)a_{n-2}\zeta^{n-2} + \ldots + \gamma(r)^{n-1}a_0 \right. \\ &\quad \left. + (\tilde{a}_{m-1}/\tilde{a}_m)\zeta^{m-1} + \gamma(r)(\tilde{a}_{m-2}/\tilde{a}_m)\zeta^{m-2} + \ldots + \gamma(r)^{m-1}\tilde{a}_0/\tilde{a}_m\right] \end{aligned}$$

and $\lim_{r\downarrow 0}\gamma(r) = 0$. So we obtain from Theorem 4.1.2 that m roots of $p_r(\zeta)$ tend to 0 and the remaining $n-m$ roots tend to the roots of $\zeta^{n-m} + 1$. Since by the above equations, for every $r > 0$ and $\zeta_0 \in \mathbb{C}$, $p_r(\zeta_0) = 0$ if and only if $p(\gamma(r)^{-1}\zeta_0, a(r)) = 0$, the second statement in the proposition follows. □

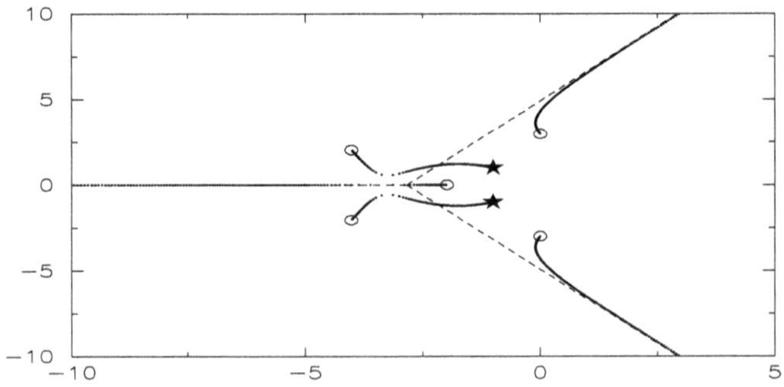

Figure 4.1.1: Butterworth pattern for $\deg p(s,\tilde{a}) = 2$ and $\deg p(s,a) = 5$

Proposition 4.1.3 is illustrated in Figure 4.1.1 where the two roots of the polynomial $p(s,\tilde{a})$ are marked by $\star$, the five roots of the polynomial $p(s,a(1)) = p(s,\tilde{a}) + p(s,a)$ are marked by small circles and the roots of $p(s,a(r)) = p(s,\tilde{a}) + rp(s,a)$ are shown for r decreasing from 1 to 0. The limiting distribution of the far roots is known as a *Butterworth pattern* and has been used in network theory for the design of filters.

It also plays a role in other branches of systems and control theory, see *Notes and References.*
It seems natural to expect that if $\deg p(s,\tilde{a}) = n$ the roots of $p(s,a)$ can be represented as continuous functions of $a \in B(\tilde{a},\delta) \subset \mathbb{C}^{n+1}$ for small $\delta > 0$. However the following example shows that this is not possible in general.

Example 4.1.4. Consider the parametrized monic polynomial $p(s,a(z)) = s^2 - z$ with complex parameter z. The roots of this polynomial are $s_{1,2}(z) = \pm\sqrt{z}$ where for $z = re^{\imath\varphi}$, $0 \le \varphi < 2\pi$, we set $\sqrt{z} = \sqrt{r}e^{\imath\varphi/2}$.
We first consider these roots for parameter values $z \in \mathbb{C}$ close to $\hat{z} = 1$. Let $z = 1 + \delta\, e^{\imath\theta}$, $0 \le \theta < 2\pi$. For small $\delta = |z-1|$ we have the first order approximation

$$\sqrt{z} = \sqrt{1+\delta\, e^{\imath\theta}} \approx \begin{cases} 1 + (\delta/2)\,(\cos\theta + \imath\sin\theta) & \text{if } 0 \le \theta \le \pi, \\ -[1 + (\delta/2)\,(\cos\theta + \imath\sin\theta)] & \text{if } \pi < \theta < 2\pi. \end{cases}$$

Therefore the two roots $s_{1,2}(z) = \pm\sqrt{z} = \pm\sqrt{1+\delta\, e^{\imath\theta}}$ of $p(s,a(z))$ depend *discontinuously* on z. Discontinuities occur at *real* values of $z \ne 1$ (corresponding to $\theta = \pi$ and $\theta = 0$). However a continuous selection of the roots is possible. In fact, for $z = 1 + \delta\, e^{\imath\theta}$ let

$$\tilde{s}_1(z) = \begin{cases} \sqrt{1+\delta\, e^{\imath\theta}} & \text{if } 0 \le \theta \le \pi, \\ -\sqrt{1+\delta\, e^{\imath\theta}} & \text{if } \pi < \theta < 2\pi; \end{cases} \qquad \tilde{s}_2(z) = \begin{cases} -\sqrt{1+\delta\, e^{\imath\theta}}, & \text{if } 0 \le \theta \le \pi, \\ \sqrt{1+\delta\, e^{\imath\theta}}, & \text{if } \pi < \theta < 2\pi. \end{cases}$$

Then $\tilde{s}_1(z)$ and $\tilde{s}_2(z)$ are roots of $p(s,a(z))$ which depend in a continuous way on z in a small neighbourhood of 1.
Now consider the above polynomials close to the parameter value $\hat{z} = 0$. For $z = \delta e^{\imath\varphi}$, $0 \le \varphi < 2\pi$, $0 < \delta \ll 1$, the polynomial $p(s,a(z)) = s^2 - z$ has two simple roots $\pm\sqrt{\delta e^{\imath\varphi}} = \pm\sqrt{\delta}e^{\imath\varphi/2}$. But in this case it is *not* possible to find continuous functions $s_1(\cdot), s_2(\cdot)$ on a small disk $D(0,\varepsilon)$, $\varepsilon > 0$ such that $\{s_1(z), s_2(z)\} = \{\sqrt{z}, -\sqrt{z}\}$ for all $z \in D(0,\varepsilon)$. In fact assume the contrary and let $\delta \in (0,\varepsilon)$ be fixed. Setting $z(\theta) = \delta e^{\imath\theta}$, $0 \le \theta < 2\pi$ and choosing $j \in \{1,2\}$ such that $s_j(z(0)) = \sqrt{\delta}$, then by continuity $s_j(\delta e^{\imath\theta}) = \sqrt{\delta}e^{\imath\theta/2}$ for $\theta \in [0,2\pi)$. But then $s_j(\cdot)$ is discontinuous at $z = \delta = \delta e^{\imath 0} = \delta e^{\imath 2\pi}$. □

From this example we see that, even if $p(s,\tilde{a})$ is a monic polynomial of degree n, it is in general *not* possible to find *continuous* functions $s_j(a)$, $j \in \underline{n}$ on a small neighbourhood of $\tilde{a}$ in $\mathbb{C}^{n+1}$ so that the set of roots of $p(s,a)$ coincides with $\{s_j(a);\, j \in \underline{n}\}$ for $\|a - \tilde{a}\|$ sufficiently small. A problem may occur if $p(s,\tilde{a})$ has multiple roots. Continuous dependence of the roots of a polynomial on the coefficient vector can be established if we consider the whole n-tuple of roots instead of the single roots individually. For any $a = (a_0, \ldots, a_n) \in \mathbb{C}^{n+1}$ with $a_n \ne 0$ let

$$\Lambda(a) = \lfloor s_1(a), \cdots, s_n(a) \rfloor$$

be the *unordered* n-tuple of the roots of $p\,(s,a)$ taking account of multiplicities. We define the *distance*[1] between two unordered n-tuples $\Lambda = \lfloor \lambda_1, \cdots, \lambda_n \rfloor$, $\Lambda' = \lfloor \lambda'_1, \cdots, \lambda'_n \rfloor$ by

$$d(\Lambda, \Lambda') = \min_{\pi \in \Pi_n} \max_{k \in \underline{n}} |\lambda_{\pi(k)} - \lambda'_k|\,, \tag{3}$$

[1] $d(\Lambda, \Lambda')$ is sometimes called "matching distance" in the literature, see [484].

where Π_n is the group of permutations of the set $\underline{n}$. Note that the distances between $(0,0,1)$ and $(0,1,1)$ with respect to this metric is 1 although the two underlying number *sets* are identical. We denote the metric space of all unordered complex n-tuples equipped with the metric (3) by $\mathcal{T}_n(\mathbb{C})$.
Theorem 4.1.2 implies the following corollary which gives a precise meaning to the phrase "the roots of a polynomial depend continuously on its coefficient vector".

Corollary 4.1.5. *The map $a \mapsto \Lambda(a)$ which associates with every coefficient vector $a \in \{x \in \mathbb{C}^{n+1};\, x_n \neq 0\}$ the unordered n-tuple of the roots of $p(s,a)$ is continuous.*

Example 4.1.6. The quadratic polynomial $p(s,a) = s^2 + a_1 s + a_0$ has roots

$$s_1(a) = \frac{-a_1 + \sqrt{a_1^2 - 4a_0}}{2}, \qquad s_2(a) = \frac{-a_1 - \sqrt{a_1^2 - 4a_0}}{2}, \qquad a_0,\, a_1 \in \mathbb{C}$$

where again we set $\sqrt{z} = \sqrt{r}e^{\imath\varphi/2}$ if $z = re^{\imath\varphi}$, $0 \le \varphi < 2\pi$. We have already seen that for the special case $a_1 = 0$, $a_0 = -\delta\, e^{\imath\theta}$ the individual roots $s_j(a)$ are not analytic at $\tilde{a} = 0$ and are not continuous on any ball $B(0,\delta)$, $\delta > 0$. Now let us consider the unordered pair $\Lambda(a) = (s_1(a), s_2(a))$. Suppose $a_1^2 - 4a_0 = r(a)e^{\imath\varphi(a)}$, $0 \le \varphi(a) < 2\pi$, then the only points of discontinuity of $s_j(a)$ occur when $\varphi(a) = 0$. Now if $\varphi(a) > 0$ is close to 0, then $s_1(a) \approx -[a_1 - \sqrt{r(a)}]/2$, $s_2(a) \approx -[a_1 + \sqrt{r(a)}]/2$, whereas if $\varphi(a) < 2\pi$ is close to 2π, $s_1(a) \approx -[a_1 + \sqrt{r(a)}]/2$, $s_2(a) \approx -[a_1 - \sqrt{r(a)}]/2$. So we see that the map $a \mapsto \Lambda(a)$ is continuous at points where $\varphi(a) = 0$. Hence the map $a \mapsto \Lambda(a)$ from $\mathbb{C}^2$ to $\mathcal{T}_2(\mathbb{C})$ is continuous. □

In the following remark we state two important consequences of Corollary 4.1.5.

Remark 4.1.7. Let $\Omega \subset \mathbb{R}^N$ be a parameter set in $\mathbb{R}^N$, $N \ge 1$ and suppose that $a : \omega \to (a_0(\omega), ..., a_n(\omega))$ is a continuous map from Ω to $\mathbb{K}^{n+1}$ such that $a_n(\omega) \neq 0$ for all $\omega \in \Omega$. If $\mathbb{C}_g \subset \mathbb{C}$ is an open subset of the complex plane ("stability region") then the set of all $\omega \in \Omega$ such that $p(s, a(\omega))$ is $\mathbb{C}_g$-stable (i.e. has all its roots in $\mathbb{C}_g$) is open in Ω. If Ω is connected and there exist ω^1, $\omega^2 \in \Omega$ such that $p(s, a(\omega^1))$ is $\mathbb{C}_g$-stable and $p(s, a(\omega^2))$ is not $\mathbb{C}_g$-stable, then there exists $\omega^0 \in \Omega$ such that $p(s, a(\omega^0))$ has all its roots in the closure of $\mathbb{C}_g$ and at least one root on the boundary $\partial\mathbb{C}_g$ of $\mathbb{C}_g$. This latter statement is called the *Boundary Crossing Theorem* in the literature, see Ex. 3. □

We will now extend some of the previous continuity results to *sets of polynomials*. This generalization will be useful later for the spectral analysis of uncertain linear systems. The following notation will be used. For any metric space (X, d) we denote by $\mathcal{K}(X)$ the space of all compact subsets of X provided with the Hausdorff metric

$$d_H(K_1, K_2) = \max\{\max_{x \in K_1} \operatorname{dist}(x, K_2), \max_{x \in K_2} \operatorname{dist}(x, K_1)\} \tag{4}$$

where, as usual, $\operatorname{dist}(x, K) := \min_{y \in K} d(x,y)$ for $x \in X$, $K \in \mathcal{K}(X)$. By this definition we have for all K_1, $K_2 \in \mathcal{K}(X)$, $\varepsilon > 0$

$$d_H(K_1, K_2) < \varepsilon \Leftrightarrow \forall x \in K_1\, \exists y \in K_2 : d(x,y) < \varepsilon \wedge \forall y \in K_2\, \exists x \in K_1 : d(x,y) < \varepsilon. \tag{5}$$

Note that by the definition of the metric on $\mathcal{T}_n(\mathbb{C})$

$$d_H(\{\Lambda\}, \{\Lambda'\}) \le d(\Lambda, \Lambda'), \quad \Lambda, \Lambda' \in \mathcal{T}_n(\mathbb{C})$$

where $\{\Lambda\}$ denotes the *set* of the elements of Λ, i.e. $\{\Lambda\} := \{\lambda_1, \ldots, \lambda_n\}$ for $\Lambda = \lfloor \lambda_1, \ldots, \lambda_n \rfloor$. The following lemma is an easy consequence of the above definitions.

Lemma 4.1.8. *Let* $h : \mathcal{K}(\mathcal{T}_n(\mathbb{C})) \to \mathcal{K}(\mathbb{C})$ *be defined by*

$$h(K) = \bigcup_{\Lambda \in K} \{\Lambda\}, \quad K \in \mathcal{K}(\mathcal{T}_n(\mathbb{C})).$$

Then

$$d_H(h(K), h(K')) \leq d_H(K, K'), \quad K, K' \in \mathcal{K}(\mathcal{T}_n(\mathbb{C})). \tag{6}$$

Proof: Let K, K' be two compact subsets of $\mathcal{T}_n(\mathbb{C})$ and let $\Lambda \in K$ be arbitrary, then we have for every $\lambda \in \{\Lambda\}$

$$\operatorname{dist}(\lambda, h(K')) = \min_{\Lambda' \in K'} \operatorname{dist}(\lambda, \{\Lambda'\}) \leq \min_{\Lambda' \in K'} d(\Lambda, \Lambda') = \operatorname{dist}(\Lambda, K').$$

Hence

$$\max_{\lambda \in h(K)} \operatorname{dist}(\lambda, h(K')) = \max_{\Lambda \in K} \max_{\lambda \in \{\Lambda\}} \operatorname{dist}(\lambda, h(K')) \leq \max_{\Lambda \in K} \operatorname{dist}(\Lambda, K'),$$

and similarly

$$\max_{\lambda' \in h(K')} \operatorname{dist}(\lambda', h(K)) \leq \max_{\Lambda' \in K'} \operatorname{dist}(\Lambda', K).$$

This concludes the proof. □

We will also make use of the following lemma from topology.

Lemma 4.1.9. *Let* X, Y *be metric spaces and* $f \in C(X, Y)$ *be a continuous map. Then the induced map* $K \mapsto f(K)$ *from the metric space* $\mathcal{K}(X)$ *into the metric space* $\mathcal{K}(Y)$ *(both provided with the corresponding Hausdorff metrics) is continuous.*

Proof: Suppose $K_0 \in \mathcal{K}(X)$ and $\varepsilon > 0$. For each $x \in K_0$ there exists $\delta_x = \delta(x, \varepsilon) > 0$ such that $d_Y(f(x), f(y)) < \varepsilon/2$ for all $y \in B(x, \delta_x) = \{w \in X; d_X(w, x) < \delta_x\}$. By compactness there exist finitely many $x_i \in K_0$, $i \in \underline{N}$ such that $K_0 \subset U := \bigcup_{i \in \underline{N}} B(x_i, \delta_i/2)$ where $\delta_i = \delta(x_i, \varepsilon)$. Now choose $0 < \delta < \min_{i \in \underline{N}} \delta_i/2$ and let $K \in \mathcal{K}(X)$ be such that $d_H(K_0, K) < \delta$. Then for every $x \in K_0 \subset U$ there exist $i \in \underline{N}$ and $y \in K$ such that $x \in B(x_i, \delta_i/2)$ and $d(x, y) < \delta$. Hence $x, y \in B(x_i, \delta_i)$ and so

$$\forall x \in K_0 \; \exists y \in K : \; d_Y(f(x), f(y)) \leq d_Y(f(x), f(x_i)) + d_Y(f(x_i), f(y)) < \varepsilon.$$

Similarly

$$\forall y \in K \; \exists x \in K_0 : \; d_Y(f(x), f(y)) \leq d_Y(f(x), f(x_i)) + d_Y(f(x_i), f(y)) < \varepsilon.$$

By (5) the two inequalities together prove $d_H(f(K_0), f(K)) < \varepsilon$. □

In the following proposition we identify a polynomial $p(s, a)$ of degree n with its coefficient vector and consider the space $\mathcal{P}_n(\mathbb{C}) \subset \mathbb{C}_n[s]$ of all complex polynomials of degree n as an (open) subset of $\mathbb{C}^{n+1} \cong \mathbb{C}_n[s]$ with the induced metric.

Proposition 4.1.10. *For every compact subset* $\mathcal{P} \subset \mathcal{P}_n(\mathbb{C})$ *of complex polynomials of degree* n *the associated* root set

$$\mathcal{R}(\mathcal{P}) := \{s \in \mathbb{C}; \exists p \in \mathcal{P} : p(s) = 0\}$$

is compact. Moreover, the maps $\mathcal{P} \mapsto \mathcal{R}(\mathcal{P})$ *and* $\mathcal{P} \mapsto \Lambda(\mathcal{P})$ *from* $\mathcal{K}(\mathcal{P}_n(\mathbb{C}))$ *into* $\mathcal{K}(\mathbb{C})$ *and* $\mathcal{K}(\mathcal{T}_n(\mathbb{C}))$*, respectively, are continuous.*

Proof: Since $\Lambda : \mathcal{P}_n(\mathbb{C}) \to \mathcal{T}_n(\mathbb{C})$ is continuous by Corollary 4.1.5, $\Lambda(\mathcal{P})$ is compact in $\mathcal{T}_n(\mathbb{C})$ for every $\mathcal{P} \in \mathcal{K}(\mathcal{P}_n(\mathbb{C}))$. It follows from Lemma 4.1.9 that $\Lambda : \mathcal{P} \mapsto \Lambda(\mathcal{P})$ is a continuous map from $\mathcal{K}(\mathcal{P}_n(\mathbb{C}))$ into $\mathcal{K}(\mathcal{T}_n(\mathbb{C}))$. But the map $\mathcal{P} \mapsto \mathcal{R}(\mathcal{P})$ is the composition of the map $\mathcal{P} \mapsto \Lambda(\mathcal{P})$ and the map $h : \mathcal{K}(\mathcal{T}_n(\mathbb{C})) \to \mathcal{K}(\mathbb{C})$ defined in Lemma 4.1.8. Hence the proposition follows by application of this lemma. □

In the situation of the proof of Theorem 4.1.2, suppose that $\tilde{s}_j$ is a simple root of $p(s, \tilde{a}) \in \mathbb{C}_n[s]$ for some $j \in \underline{\ell}$, then $D(\tilde{s}_j, \varepsilon)$ will contain exactly one root $s_j(a)$ of $p(s, a)$ for each $a \in B(\tilde{a}, \delta(\varepsilon)) \subset \mathbb{C}^{n+1}$. Applying the Residue Theorem (more precisely, Theorem A.2.20 (ii) with $g(s) = s$) we obtain

$$\frac{1}{2\pi\imath} \int_{\Gamma_j} \frac{s\, p'(s, a)}{p(s, a)}\, ds = s_j(a), \qquad a \in B(\tilde{a}, \delta(\varepsilon))$$

where $p'(s, a) = \frac{d}{ds} p(s, a)$. The LHS is analytic in a and so we have

Theorem 4.1.11. *If $\tilde{s}_j$ is a root of multiplicity one of the polynomial $p(s, \tilde{a})$, then there exist δ, $\varepsilon > 0$ such that for all $a \in B(\tilde{a}, \delta)$ the polynomial $p(s, a)$ has exactly one root $s_j(a)$ (of multiplicity one) in $D(\tilde{s}_j, \varepsilon)$. This root $s_j(a)$ depends analytically on a in $B(\tilde{a}, \delta)$ and satisfies $s_j(\tilde{a}) = \tilde{s}_j$.*

4.1.2 Polynomials with Holomorphic Coefficients

Now suppose the coefficients of the polynomial are continuous functions $a_i(z)$ of *one* real or complex parameter z. As a consequence of Theorem 4.1.2 the unordered n-tuple of roots of $p(s, a(z))$ depends continuously on z. However, the number of distinct roots may change quite irregularly – the splitting and coalescence of roots may take place in a very complicated manner. More specific results can be obtained if the coefficients depend *analytically* on the parameter z so that tools from the complex analytic theory of algebraic functions can be employed. This will be the subject of the present subsection. For the terminology and some elements of Complex Analysis we refer to Section A.2.

Let Ω be a *domain* (connected open subset) in the complex plane and denote by $\mathcal{O}(\Omega)$ the ring of holomorphic, i.e. complex analytic functions on Ω. Since Ω is connected, it follows from the Identity Theorem A.2.9 that a product in $\mathcal{O}(\Omega)$ is zero if and only if one of the factors is zero. Thus $\mathcal{O}(\Omega)$ is an *integral domain*, that is a commutative ring with multiplicative unit[2] and without zero divisors. The quotient field of this integral domain is the *field of meromorphic functions* on Ω, denoted by $\mathcal{M}(\Omega)$. By $\mathcal{O}(\Omega)[s]$ (resp. $\mathcal{M}(\Omega)[s]$) we denote the ring of all polynomials in s

$$p(s, a(z)) = a_n(z)s^n + a_{n-1}(z)s^{n-1} + \cdots + a_0(z) \tag{7}$$

with coefficients $a_i(z)$ in $\mathcal{O}(\Omega)$ (resp. $\mathcal{M}(\Omega)$), $i = 0, \ldots, n$, $n \in \mathbb{N}$. For the convenience of the reader we recall some basic algebraic results concerning polynomials with coefficients in an arbitrary field, see [503, §34], [330].

[2]the constant function 1 on Ω

Remark 4.1.12. Let $\mathbb{K}$ be a field and $\mathbb{K}[s]$ the ring of polynomials in the variable s with coefficients in $\mathbb{K}$. $p(s) \in \mathbb{K}[s]$ is said to be *irreducible* if $\deg p \geq 1$ and p has no proper factor, i.e. there exists no $q \in \mathbb{K}[s]$, $0 < \deg q < \deg p$ such that $q|p$. An irreducible polynomial $p(s) \in \mathbb{K}[s]$ is automatically prime, i.e. if p divides a product $q_1 q_2$ in $\mathbb{K}[s]$ then p divides q_1 or q_2. It follows easily by induction on $n = \deg p$ that an arbitrary polynomial $p(s) \in \mathbb{K}[s]$ admits a factorization of the form

$$p(s) = a_n \prod_{i=1}^{\ell} p_i(s)^{m_i}, \quad m_i \geq 1, \; i \in \underline{\ell}$$

where $a_n \in \mathbb{K}$ is the leading coefficient of $p(s)$ and the factors p_i are monic, irreducible and mutually distinct. Modulo permutation of the p_i this *decomposition of p into its irreducible factors* is uniquely determined.
The *resultant matrix* of two polynomials $p, q \in \mathbb{K}[s]$ is defined in the same way as for complex polynomials. Given $p(s) = \sum_{k=0}^{n} a_k s^k$, $q(s) = \sum_{k=0}^{m} b_k s^k$, $n, m \geq 1$, the resultant matrix $R(p,q) = R_{n,m}(p,q) \in \mathbb{K}^{(m+n)\times(m+n)}$ defined by (3.4.32) is singular if and only if p, q have a common factor of positive degree or $a_m = b_n = 0$. If $\mathbb{K} = \mathbb{C}$ and $\deg p = n, \deg q = m$, $R(p,q)$ vanishes if and only if p and q have a common root.
The *discriminant* $D_n(p)$ of p is by definition the determinant of the $(2n-1)\times(2n-1)$ resultant matrix $R(p,p') = R_{n,n-1}(p,p')$ where $p'(s) = \sum_{k=1}^{n} k a_k s^{k-1}$ is the *derivative* of $p(s)$ with respect to s. The discriminant is a polynomial in the coefficients $a_0, \ldots, a_n$ of p and vanishes exactly at those coefficient vectors $(a_0, \ldots, a_n) \in \mathbb{K}^{n+1}$ for which the polynomials p and p' have a non-trivial common factor or $a_n = 0$. In particular, if $\mathbb{K} = \mathbb{C}$, the discriminant $D_n(p)$ of the complex polynomial $p(s)$ of degree n is zero if and only if p has multiple roots (in which case $p(s)$ and $p'(s)$ have a joint linear factor). □

Definition 4.1.13. Let $p(s, a(z)) \in \mathcal{M}(\Omega)[s]$ be any polynomial with meromorphic coefficients of the form (7) with $a_n(z) \not\equiv 0$. $z_0 \in \Omega$ is said to be a *critical point* or *critical value* of z for p, if the leading coefficient $a_n(z)$ vanishes at z_0 or z_0 is a pole of one of the coefficients $a_i(z)$ or the polynomial $p(s, a(z_0)) \in \mathbb{C}[s]$ has a strictly smaller number of distinct roots than $p(s, a(z))$, for some other $z \in \Omega$. The set of critical points of $p(s, a(z))$ is denoted by C_p.

Outside the set $C_p \subset \Omega$ of critical points, the equation (7) has a *constant* number of distinct roots. The next theorem shows that the critical points are *isolated* in Ω.

Theorem 4.1.14. *Let $p(s) = p(s, a(z)) \in \mathcal{M}(\Omega)[s]$ be a polynomial of the form (7) with $a_n(z) \not\equiv 0$. Then*

(i) the critical set C_p is locally finite in Ω, i.e. for every compact subset $K \subset \Omega$ the set $K \cap C_p$ is finite.

(ii) In every simply connected domain $D \subset \Omega \backslash C_p$ there exist $n = \deg p$ holomorphic functions $s_1(\cdot), \ldots, s_n(\cdot) \in \mathcal{O}(D)$ (not necessarily all distinct), such that

$$p(s, a(z)) = a_n(z) \prod_{i=1}^{n} (s - s_i(z)), \qquad z \in D. \tag{8}$$

(iii) The multiplicity of each root $s_i(z)$ of $p(s, a(z))$ is constant on every simply connected domain $D \subset \Omega \setminus C_p$.

Proof: p can be represented as a product $p = a_n \prod_{i=1}^{\ell} p_i^{m_i}$ where $a_n = a_n(z) \in \mathcal{M}(\Omega)$ is the leading coefficient of p, $m_i \geq 1$ and the $p_i \in \mathcal{M}(\Omega)[s]$, $1 \leq i \leq \ell$ are the irreducible monic factors of p in $\mathcal{M}(\Omega)[s]$, see Remark 4.1.12.
We will first prove assertions (i)-(iii) for irreducible monic polynomials.[3] Let $q(s) = q(s, b(z)) \in \mathcal{M}(\Omega)[s]$ be an irreducible monic polynomial of degree m with coefficient vector $b(z) = (b_0(z), \ldots, b_{m-1}(z), 1)$. The set P_q of all the poles of the coefficients $b_i(z)$ of q is locally finite in Ω. Since $q(s)$ is irreducible, q cannot have a non-trivial common factor with q' in $\mathcal{M}(\Omega)[s]$ and so, applying Remark 4.1.12 to $q \in \mathbb{K}[s]$ where $\mathbb{K} = \mathcal{M}(\Omega)$, we see that the discriminant of q, $\psi_q = D_m(q) \in \mathcal{M}(\Omega)$, must be a *non-zero* meromorphic function on Ω. $\psi_q(z)$ is analytic on $\Omega \setminus P_q$, and for each point $z_0 \in \Omega \setminus P_q$, $\psi_q(z_0)$ is the discriminant of the complex polynomial $q(s, b(z_0)) \in \mathbb{C}[s]$. By Remark 4.1.12 $\psi_q(z_0) = 0$ if and only if $q(s, b(z_0))$ has multiple roots. Let Z_q denote the set of zeros of $\psi_q(z)$ in $\Omega \setminus P_q$. Then Z_q is locally finite in Ω, and $q(s, b(z_0))$ has m distinct simple zeros if $z_0 \in \Omega \setminus (P_q \cup Z_q)$. This proves that the set of critical points of q is given by $C_q = P_q \cup Z_q$ and hence it is locally finite in Ω. Now let D be any simply connected domain in $\Omega \setminus C_q$. For every $z_0 \in D$ there exist by Theorem 4.1.11 a neighbourhood U of z_0 in D and m analytic functions $s_1(\cdot), \cdots, s_m(\cdot) : U \to \mathbb{C}$ such that $q(s, b(z))) = \prod_{i=1}^{m}(s - s_i(z))$ and $s_i(z) \neq s_j(z)$ for $i \neq j$, $z \in U$. It follows from the Monodromy Theorem A.2.24 that this representation of $q(s, b(z))$ as a product of distinct analytic linear factors can be extended to the whole domain D. We see therefore that assertions (i)-(iii) hold for arbitrary irreducible monic polynomials.
We will now prove statements (i)-(iii) for the given polynomial $p(s) = p(s, a(z))$ by applying the above results to its irreducible monic factors p_i, $i \in \underline{\ell}$. Let P_p be the set of poles of the coefficients $a_i(z)$, $i = 0, \ldots, n$ of p in Ω and let Ω_0 be the set of zeros of the leading coefficient $a_n(z)$. Then $P_p \cup \Omega_0$ is locally finite in Ω and contained in C_p by Definition 4.1.13. It is a straightforward exercise (see Ex. 4) to prove that every pole z_0 of one of the coefficients of $p_i(s)$, $i \in \underline{\ell}$ is a pole of one of the coefficients of the product $\prod_{i=1}^{\ell} p_i^{m_i}$, hence belongs to $P_p \cup \Omega_0$. So the union of the pole sets P_{p_i}, $i \in \underline{\ell}$ is contained in C_p. Let $b^i(z)$ be the coefficient vector of p_i, $p_i(s) = p_i(s, b^i(z))$, $n_i = \deg p_i$, $i \in \underline{\ell}$, then p_i has n_i simple roots for $z \in \Omega \setminus C_{p_i}$. For $i, j \in \underline{\ell}$, $i \neq j$ the polynomials p_i, $p_j \in \mathcal{M}(\Omega)[s]$ are coprime and therefore (see Remark 4.1.12) the resultant $\psi_{i,j} = R_{n_i,n_j}(p_i, p_j) \in \mathcal{M}(\Omega)$ must be a *non-zero* meromorphic function on Ω. $\psi_{i,j}(z)$ is analytic on $\Omega \setminus P_p$, and for each point $z_0 \in \Omega \setminus P_p$, $\psi_{i,j}(z_0)$ is the resultant of the complex polynomials $p_i(s, b^i(z_0))$, $p_j(s, b^j(z_0)) \in \mathbb{C}[s]$. By Remark 4.1.12 $\psi_{i,j}(z_0) = 0$ for $z_0 \in \Omega \setminus P_p$ if and only if $p_i(s, b^i(z_0))$ and $p_j(s, b^j(z_0))$ have common roots. The set $Z_{i,j}$ of zeros of $\psi_{i,j}(z)$ in $\Omega \setminus P_p$ is locally finite in Ω. Let Z denote the union of these sets for all $i, j \in \underline{\ell}$, $i \neq j$. If $z_0 \in \Omega \setminus \left(\Omega_0 \cup P_q \cup Z \cup \bigcup_{i \in \underline{\ell}} Z_{p_i}\right)$, the complex polynomial $p(s, a(z_0))$ has $\nu = \sum_{i=1}^{\ell} n_i$ distinct roots, and this is the maximal number of distinct roots of $p(s, a(z))$ for $z \in \Omega \setminus (P_p \cup \Omega_0)$. On the other hand if $z_0 \in Z_{p_i}$, i.e. $p_i(s) = p_i(s, b^i(z_0))$ has strictly less than n_i distinct roots, or if $z_0 \in Z_{i,j}$ for some $i, j \in \underline{\ell}$, $i \neq j$ then $p(s, a(z_0))$ has strictly less than ν distinct

[3]Throughout the proof we make use of the fact that the sets of poles and zeros of a meromorphic function on Ω are locally finite in Ω, see Section A.2.

roots. Thus

$$C_p = P_p \cup \Omega_0 \cup Z \cup \bigcup_{i \in \underline{\ell}} Z_{p_i},$$

so that C_p (as a finite union of locally finite subsets) is locally finite in Ω.
To prove (ii) and (iii) let D be a simply connected subset of $\Omega \setminus C_p$. Since $C_{p_i} = P_{p_i} \cup Z_{p_i} \subset C_p$ for $i \in \underline{\ell}$, all the irreducible factors p_i can be decomposed on D into linear factors with analytical roots. Hence there exist n analytic functions $s_i(\cdot) : D \to \mathbb{C}$ (not necessarily distinct) such that (8) holds. It follows from Definition 4.1.13 that two roots $s_i(z)$, $s_j(z)$ of p which coincide at some point $z \in D \subset \Omega \setminus C_p$ coincide on the whole domain D. This proves (ii) and (iii). □

In the sequel we will focus on *monic* polynomials with *analytic* coefficients on Ω, $p(s, a(z)) \in \mathcal{O}(\Omega)[s]$. Then $z_0 \in \Omega$ is in C_p if and only if the number of roots of $p(s, a(z))$ strictly decreases at $z = z_0$. The following example illustrates what may happen at a critical parameter value $z_0 \in C_p$. In particular it shows that the roots $s_i(z)$ of $p(s, a(z))$ may have a *branch point* at z_0 (see below) but not necessarily so.

Example 4.1.15. Consider the quadratic polynomial $p(s, a(z)) = s^2 + a_1(z)s + a_0(z)$, $z \in \Omega = \mathbb{C}$.
(i) If $a_1(z) = -(1 + 2z)$, $a_0(z) = z + z^2$ then the two roots $s_1(z) = z$, $s_2(z) = 1 + z$ are distinct for all $z \in \mathbb{C}$, so that there are no critical points and $s_1(z)$, $s_2(z)$ are both analytic on $\mathbb{C}$.
(ii) If $a_1(z) = z$, $a_0(z) = 0$, then $s_1(z) = 0$, $s_2(z) = -z$ are analytic functions on $\mathbb{C}$ although there is one critical point at $z = 0$.
(iii) If $a_1(z) = 0$, $a_0(z) = -(1+z^2)$, then $s_{1,2}(z) = \pm(1+z^2)^{1/2}$ are branches of one double valued analytic function with (algebraic) singularities at the two critical points $z = \pm\imath$. □

We now analyze how the roots of a monic polynomial $p(s, a(z)) \in \mathcal{O}(\Omega)[s]$ behave around a critical point $z_0 \in C_p$. For this we rely heavily on results from Complex Analysis and refer to the literature for the required background, see *Notes and References*. Let $D(z_0, r)$, $r > 0$ be a disk such that $\overline{D(z_0, r)} \cap C_p = \{z_0\}$. Suppose that $s_1, \ldots, s_\ell$ are the distinct roots of $p(s, a(z_0))$, with multiplicities $m_1, \ldots, m_\ell$ respectively, and let $\varepsilon > 0$ be sufficiently small so that the closed disks $\overline{D(s_j, \varepsilon)}$, $j \in \underline{\ell}$ are mutually disjoint. By Theorem 4.1.2 there exists $\delta \in (0, r)$ such that $p(s, a(z))$ has exactly m_j roots (accounting for multiplicities) in $D(s_j, \varepsilon)$ for all $z \in D(z_0, \delta)$. Since $D(z_0, \delta) \subset D(z_0, r)$ the number of distinct roots of $p(s, a(z))$ in $D(s_j, \varepsilon)$ will be the same, say κ_j, for all $z \in D^\circ(z_0, \delta) := D(z_0, \delta) \setminus \{z_0\}$. Now consider the punctured disk $D^\circ(z_0, r)$ and the cut disk $D^-(z_0, r) := \{z \in D^\circ(z_0, r); 0 < \arg(z - z_0) < 2\pi\}$ (see Figure 4.1.2). Since $D^-(z_0, r)$ is simply connected, the distinct roots of $p(s, a(z))$ can be represented by analytic functions $s_i(z)$ on $D^-(z_0, r)$ and their multiplicities are constant on $D^-(z_0, r)$ (by the previous theorem). Now for $z \in D^\circ(z_0, \delta) \cap D^-(z_0, r) = D^-(z_0, \delta)$ the root $s_i(z)$ belongs to exactly one of the disks $D(s_j, \varepsilon)$. Grouping the roots on $D^-(z_0, r)$ according to the disk $D(s_j, \varepsilon)$, $j \in \underline{\ell}$ to which they belong when z is restricted to $|z - z_0| < \delta$, we obtain, for each $j \in \underline{\ell}$, κ_j analytic functions $s_{jk}(z)$ on $D^-(z_0, r)$ such that

$$p(s, a(z)) = \prod_{j=1}^{\ell} \prod_{k=1}^{\kappa_j} (s - s_{jk}(z))^{\mu_{jk}}, \quad z \in D^-(z_0, r), \quad \lim_{z \to z_0} s_{jk}(z) = s_j, k \in \underline{\kappa_j}, \quad (9)$$

where $\mu_{jk} \geq 1$ denotes the (constant) multiplicity of the root $s_{jk}(z)$ of $p(s, a(z))$, $z \in D^-(z_0, r)$. In the sequel we identify notationally the analytic functions $s_{jk}(z)$ with the function elements (or power series) they determine at given points in $D^-(z_0, r)$.
The power series $s_{jk}(z)$ can be continued analytically (see Section A.2) along every arc in $D^\circ(z_0, r)$. Now consider the function element $s_{jk}(z)$ for any $j \in \underline{\ell}$, $k \in \underline{\kappa_j}$ in a small neighbourhood of $z_1 = z_0 + \rho e^{-\imath\alpha}$ where $\rho \in (0, r)$, $\alpha \in (0, 2\pi)$ are fixed. Continuing this function element analytically along the circular arc

$$\gamma(t) = z_0 + \rho e^{\imath t}, \quad t \in [-\alpha, 2\pi - \alpha] \tag{10}$$

we obtain a new function element $\tilde{s}_{jk}(z)$ at z_1 for each $j \in \underline{\ell}$, $k \in \underline{\kappa_j}$, (see Figure 4.1.2). The identity theorem implies that the factorization (9) is preserved by analytic continuation, i.e. equality (9) holds with each of the $s_{jk}(z)$ replaced by the corresponding $\tilde{s}_{jk}(z)$. Moreover, continuing any function element $s_{jk}(z)$ at some point in the small cut disk $D^-(z_0, \delta)$ analytically along a circular arc γ_0 in $D^\circ(z_0, \delta)$ the resulting root element $\tilde{s}_{jk}(z)$ remains in $D(s_j, \varepsilon)$ since by construction of δ no continuous root function can leave $D(s_j, \varepsilon)$ along an arc in $D^\circ(z_0, \delta)$. Thus every root function $\tilde{s}_{jk}(z)$ on $D^-(z_0, r)$ obtained by analytic continuation of $s_{jk}(z)$ along the arc γ (10) coincides with one of the analytic roots $s_{jk'}(z)$, $k' \in \underline{\kappa_j}$, and this root has the same multiplicity as $s_{jk}(z)$, i.e. $\mu_{jk} = \mu_{jk'}$ (since multiplicities are preserved by analytic continuation, see Theorem 4.1.14 (iii)).
Summarizing, the map $\pi_{z_0} : s_{jk}(\cdot) \mapsto s_{jk'}(\cdot)$ defines a permutation of the finite set of root functions on $D^-(z_0, r)$, $\{s_{jk}(\cdot);\ j \in \underline{\ell},\ k \in \underline{\kappa_j}\}$, and each $s_{jk}(\cdot)$ belongs to a unique cycle of this permutation. We have just seen that π_{z_0} preserves multiplicities and leaves the sets $\{s_{jk}(\cdot);\ k \in \underline{\kappa_j}\}$ invariant. Suppose that via analytic extension along a circular arc γ, $s_{jk_1}(z)$ produces $s_{jk_2}(z)$, $s_{jk_2}(z)$ produces $s_{jk_3}(z)$ and so on. For some $q \leq \kappa_j$ the analytic extension of $s_{jk_q}(z)$ along γ reproduces $s_{jk_1}(z)$. In this case we say that $s_{jk_1}(z)$ generates a cycle of period q in $D^\circ(z_0, r)$. The cycle $(s_{jk_1}(z), \ldots, s_{jk_q}(z))$ represents a q-valued analytic function $f(z)$ on $D^\circ(r)$ which consists of all function elements obtained by analytic continuation of the power series $s_{jk_1}(z), \cdots, s_{jk_q}(z)$ along arbitrary paths in $D^\circ(r)$. The function elements thus obtained at a point *on* the cut coincides with some $s_{jk_i}(z)$ on the lower half disk and with the corresponding analytic continuation across the cut, $s_{jk_{i+1}}(z)$ (where $k_{q+1} := k_1$), on the upper half disk, (see Figure 4.1.2 where the change in root is

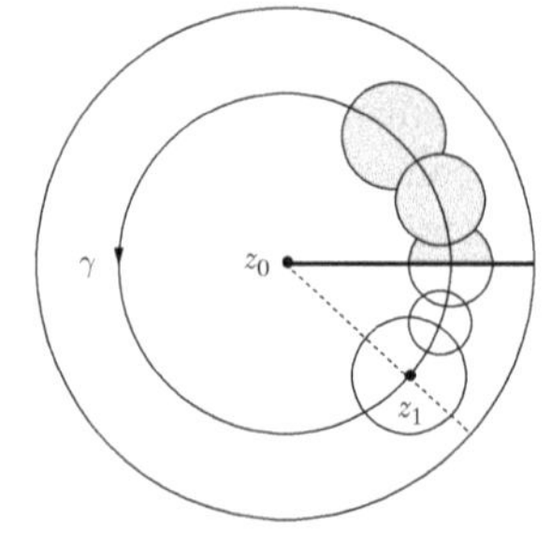

Figure 4.1.2: Analytic continuation defining the permutation π_{z_0}

symbolized by shading). On the other hand, every function element of f at a point

$z_1 \in D^-(z_0, r)$ coincides with the function element determined by some $s_{jk_i}(z)$, $i \in \underline{q}$ at that point. Thus the q-valued function f associates with every point z in the punctured disk $D^\circ(z_0, r)$ a set $f(z)$ of q distinct values of the following form:

$$f(z) = \begin{cases} \{s_{jk_1}(z), \cdots, s_{jk_q}(z)\} & \text{if } z \in D^-(z_0, r), \\ \{\lim_{t\uparrow 2\pi} s_{jk_i}(z_0 + \rho e^{\imath t})\,;\, i \in \underline{q}\} & \text{if } z = z_0 + \rho,\ 0 < \rho < r. \end{cases}$$

By continuity of the roots (Theorem 4.1.2) all the q elements of $f(z)$ converge to s_j as $z \to z_0$. In particular, if $q = 1$, then $z = z_0$ is a removable singularity of f so that f defines an analytic function on the whole disk $D(z_0, r)$. Otherwise z_0 is called a *branch point* of order $q - 1$ of f and the root functions $f_i(z) = s_{jk_i}(z)$, $i \in \underline{q}$ are called the *branches* of f on $D^-(z_0, r)$. Now suppose that $q \geq 2$. We will show that $f(z)$ can be represented by a single-valued analytic function $F(\zeta)$ of $\zeta = (z - z_0)^{1/q}$ near $\zeta_0 = 0$. Let $\tilde{D}^\circ(r) = \{\zeta \in \mathbb{C};\ 0 < |\zeta| < r^{1/q}\}$ and define $\varphi : \tilde{D}^\circ(r) \to D^\circ(z_0, r)$ by $\varphi(\zeta) = z_0 + \zeta^q$. φ maps each sector

$$C_k = \{s \in \tilde{D}^\circ(r); ((k-1)/q)2\pi \leq \arg s < (k/q)2\pi\}, \quad k \in \underline{q}$$

one-to-one onto $D^\circ(z_0, r)$. Now fix any $z_1 = z_0 + \rho e^{-\imath\alpha} \in D^-(z_0, r)$ and choose $\zeta_1 \in \varphi^{-1}(z_1)$ in C_1. The function $\zeta \mapsto s_{jk_1}(\varphi(\zeta))$ is analytic near ζ_1 and hence determines a power series $S_1(\zeta)$ at ζ_1. We can continue $S_1(\zeta)$ along all paths in $\tilde{D}^\circ(r)$ to obtain a (possibly multi-valued) analytic function $F(\zeta)$ on $\tilde{D}^\circ(r)$. In fact $F(\zeta)$ is single-valued. To see this consider the circular path $\tilde{\gamma}$ starting at ζ_1 whose trace is mapped by φ onto the trace of the circular arc γ (10). As ζ travels along $\tilde{\gamma}$ once around 0, $\varphi(\zeta) = z_0 + \zeta^q$ travels q times around z_0 in $D^\circ(z_0, r)$ and $S_1(\zeta) = s_{jk_1}(\varphi(\zeta))$ is analytically continued via $S_2(\zeta) = s_{jk_2}(\varphi(\zeta)), \cdots, S_q(\zeta) = s_{jk_q}(\varphi(\zeta))$, to $S_{q+1}(\zeta) = S_1(\zeta)$. Hence $S_1(\zeta) = s_{jk_1}(\varphi(\zeta))$ remains unchanged by analytic continuation along $\tilde{\gamma}$ and this proves that F is single valued. $F(\zeta)$ coincides with $s_{jk_i}(\varphi(\zeta))$ on the open sector $\operatorname{int} C_i$. As a single-valued analytic function on a punctured disk, $F(\cdot)$ admits a Laurent expansion $F(\zeta) = \sum_{k=-\infty}^{\infty} \alpha_k \zeta^k$ on $\tilde{D}^\circ(r)$ (see Section A.2). Since the roots in $f(z)$ converge to s_j as $z \to z_0$, it follows that $\alpha_k = 0$ for $k < 0$ and $\lim_{\zeta \to 0} F(\zeta) = \alpha_0 = s_j$. Thus $F(\zeta) = s_j + \sum_{k=1}^{\infty} \alpha_k \zeta^k$ is an analytic function on the full disk $D(0, r^{1/q})$ satisfying $F(\zeta) = s_{jk_i}(\varphi(\zeta))$ on the open sectors $\operatorname{int} C_i$, $i \in \underline{q}$. As a consequence we see that the q values of f at $z \in D^\circ(r)$ are given by $F(\zeta_\nu)$, $\nu \in \underline{q}$ where ζ_ν runs through all the q-th roots of $z - z_0$:

$$f(z) = \{F(\zeta_\nu); \zeta_\nu = \rho^{1/q} e^{\nu 2\pi\imath/q} e^{\imath\theta/q},\ \nu \in \underline{q}\}, \quad z = z_0 + \rho e^{\imath\theta} \in D^\circ(z_0, r),\ 0 \leq \theta < 2\pi. \quad (11)$$

Summarizing, we obtain the following

Theorem 4.1.16. *Suppose that $p(s, a(z)) \in \mathcal{O}(\Omega)[s]$ is a monic polynomial of the form (7) with analytic coefficients, $z_0 \in C_p$ is a critical value of z and $D(z_0, r)$ is a disk with centre z_0 such that $D(z_0, r) \cap C_p = \{z_0\}$. Let $(s_1(z), \cdots, s_q(z))$ be any one of the cycles obtained by analytic continuation of the roots of (7) along a circular path around z_0 in $D(z_0, r)$. Then the following hold.*

(i) If $q = 1$, $s_1(z)$ can be continued analytically onto $D(z_0, r)$. If $q \geq 2$, $(s_1(z), \cdots, s_q(z))$ defines by analytic continuation in $D^\circ(z_0, r)$ a q-valued holomorphic function with branch point z_0.

(ii) *The branches of this q-valued analytic function are represented by the Puiseux series*

$$s_\nu(z) = \sum_{k=0}^{\infty} \alpha_k w^{\nu k} (z - z_0)^{k/q}\,, \quad z \in D^-(z_0, r), \;\; \nu = 1, \cdots, q, \tag{12}$$

where $w = e^{2\pi\imath/q}$ *and* $(z - z_0)^{1/q} = \rho^{1/q} e^{\imath\theta/q}$ *for* $z = z_0 + \rho e^{\imath\theta}$, $0 < \theta < 2\pi$.

(iii) *All the roots* $s_\nu(z)$, $\nu \in \underline{q}$ *have the same multiplicities, which are constant throughout* $D^-(z_0, r)$, *and converge to the same root* α_0 *of* $p(s, a(z_0))$ *(the absolute term of the Puiseux series) as* $z \to z_0$.

For later use we note the following.

Remark 4.1.17. (i) Suppose that $s_1, \ldots, s_\ell$ are the distinct roots of $p(s, a(z_0))$, with multiplicities $m_1, \ldots, m_\ell$, and $z_0 \in \mathbb{C}_p$. Let $r > 0$ be such that $D(z_0, r) \cap C_p = \{z_0\}$ so that we have the factorization (9) of $p(s, a(z))$ for $z \in D^-(z_0, r)$. The analytic root branches $s_{jk}(z)$, $k \in \underline{\kappa_j}$ are said to depart from the *centre* s_j by splitting at $z = z_0$. The set of these root elements $s_{jk}(z)$, $k \in \underline{\kappa_j}$ is called the s_j*-group* of roots of $p(s, a(z))$ near $z = z_0$ with centre s_j. In general such a s_j-group consists of several cycles (resp. the associated q-valued analytic functions) with the same centre s_j.
(ii) If f is any q-valued analytic function, $q \geq 2$ in the s_j-group of roots of $p(s, a(z))$ near z_0 and F the associated single valued analytic function constructed above, then (11) implies

$$\{s_j\} \cup \bigcup_{0<|z|<r} f(z) = F(D(0, r^{1/q}))\,.$$

Since F is analytic and non-constant, hence an open map (see Theorem A.2.11), the set $F(D(0, r^{1/q}))$ is an open neighbourhood of s_j for all sufficiently small $r > 0$. □

Example 4.1.18. Consider again Example 4.1.15 (iii) where the roots are $s_{1,2}(z) = \pm(1+z^2)^{1/2}$ and the critical set is $C_p = \{\imath, -\imath\}$. The circular path $\gamma(\theta) = \imath + e^{\imath\theta}$, $\theta \in [0, 2\pi]$ from $z_1 = 1 + \imath$ to z_1 winds once around the critical point $z_0 = \imath$ (see Figure 4.1.3). Since

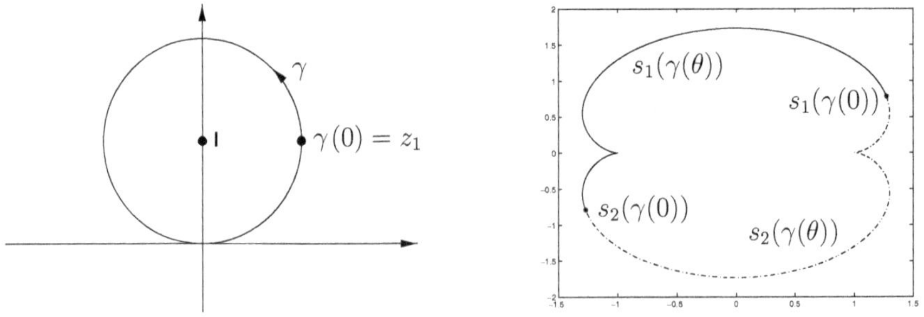

Figure 4.1.3: Interchange of the roots: $s_1(\gamma(2\pi)) = s_2(\gamma(0))$

$$\begin{aligned} s_{1,2}(\gamma(\theta)) = \pm(1 + \gamma(\theta)^2)^{1/2} &= \pm(\gamma(\theta) + \imath)^{1/2}(\gamma(\theta) - \imath)^{1/2} \\ = \pm(2\imath + e^{\imath\theta})^{1/2} e^{\imath\theta/2} &\;, \quad 0 \leq \theta < 2\pi, \end{aligned}$$

we see that the root $s_1(z) = +(1 + z^2)^{1/2}$ produces by analytic continuation along γ the root $s_2(z)$ and vice versa (see Figure 4.1.3). Hence $z_0 = \imath$ is a branch point (of order 1) for the roots of $p(s, a(z)) = s^2 - (1 + z^2)$. Note further that for z near $\imath$, by using the binomial theorem, we have

$$\begin{aligned} s_1(z) &= (1+z^2)^{1/2} = (z+\imath)^{1/2}(z-\imath)^{1/2} = (2\imath)^{1/2}(1+\frac{z-\imath}{2\imath})^{1/2}(z-\imath)^{1/2} \\ &= (2\imath)^{1/2}(z-\imath)^{1/2} + \sum_{h=1}^{\infty} \frac{\frac{1}{2}(\frac{1}{2}-1)\cdots(\frac{1}{2}-h+1)}{h!\,(2\imath)^{(h-\frac{1}{2})}}(z-\imath)^{(h+\frac{1}{2})}. \end{aligned}$$

This is the Puiseux series of $s_1(z)$ at $z_0 = i$. □

If $z_0 \in C_p$ is a branch point for the roots of p, i.e. the permutation π_{z_0} defined by the analytic continuation of the roots along a circular path around z_0 is not the identity, it will not be possible to find *continuous* (single-valued) functions $s_i(z)$, $i = 1, \cdots, n$ representing the complete set of roots $p(s, a(z))$ on a disk around the branch point z_0, see Example 4.1.4. However, a continuous parametrization of the roots is possible if the parameter z is restricted to a *real* interval. In this case it is even sufficient to only assume the continuous dependence of the coefficient vector on $z = r$.

Proposition 4.1.19. *Suppose $I \subset \mathbb{R}$ is an interval and $p(s, a(r)) \in C(I, \mathbb{C})[s]$ is a monic polynomial of the form (7) whose coefficient vector $a(r)$ depends continuously on $r \in I$. Then there exist n (single-valued) continuous functions $s_i(\cdot) : I \to \mathbb{C}$ such that, for each $r \in I$, $\Lambda(a(r)) = \lfloor s_1(r), \cdots, s_n(r) \rfloor$ is the unordered n-tuple of roots of $p(s, a(r))$ taking into account multiplicities.*

The proposition can be proved by induction on n, but since the proof is not instructive and cumbersome (though elementary) we omit it (see *Kato* (1980) [293, Theorem II.5.2].
At a branch point $z = z_0$ of $p(s, a(z)) \in \mathcal{O}(\Omega)[s]$ the distances $|s_i(z) - s_i(z_0)|$ of a root from its value at z_0 will, in general, be of the order $|z - z_0|^{1/q}$ where q is the period of the cycle to which $s_i(z)$ belongs (cf. (12)). If $q \geq 2$ the rate of change of s_i at $z = z_0$ will then be infinitely large. On the other hand, if $z_0 \in \Omega$ is not a branch point of $p(s, a(z))$, all the roots s_j are analytic on some neighbourhood of z_0 (Theorems 4.1.14 and 4.1.16). The derivative $s_j'(z_0) = \frac{ds_j}{dz}(z_0)$ is then called the *sensitivity* of the root s_j at z_0. If $|s_j'(z_0)|$ is large the uncertainty about the location of $s_j(z_0)$ is large when the parameter value z_0 is not known exactly. In the following simple example we derive a formula for the sensitivity of s_j for the case where the coefficient vector $a(z)$ depends affine linearly on z.

Example 4.1.20. Let $p(\cdot)$, $q(\cdot) \in \mathbb{C}[s]$ be two given polynomials with $\deg q < \deg p$ and consider the parametrized family

$$p(s, a(z)) = p(s) + zq(s), \qquad z \in \mathbb{C}. \tag{13}$$

Suppose that $\tilde{s}_j$ is a simple root of $p(\cdot)$ and let $s_j(z)$ be the analytic root of $p(s, a(z))$ in the vicinity of $z_0 = 0$ satisfying $s_j(0) = \tilde{s}_j$ (see Theorem 4.1.11). Differentiation of $p(s_j(z)) + zq(s_j(z)) = 0$ with respect to z at $z = 0$ yields

$$\frac{ds_j}{dz}(0)p'(s_j(0)) + q(s_j(0)) = 0.$$

Hence the sensitivity of the root $\tilde{s}_j$ is given by $s_j'(0) = -\dfrac{q(\tilde{s}_j)}{p'(\tilde{s}_j)}$ and to first order

$$s_j(z) \approx \tilde{s}_j - z\frac{q(\tilde{s}_j)}{p'(\tilde{s}_j)}. \tag{14}$$

□

Formula (14) indicates that the root sensitivity of an affinely perturbed polynomial (13) may become arbitrarily large in the vicinity of a multiple zero z_0 of p, where $p'(z_0) = 0$. But root sensitivities may even be very high for polynomials whose roots are all simple and well-separated. This is illustrated in the next example. Due to the presence of rounding errors the accurate computation of such roots will be difficult.

Example 4.1.21. Consider the Wilkinson polynomial

$$p(s) = (s-1)(s-2)\cdots(s-20) = s^{20} + a_{19}s^{19} + \ldots + a_0$$

and the root $\tilde{s}_{20} = 20$. If we change $a_{19} = -210$ to $a_{19}(1+z)$, so that $q(s) = a_{19}s^{19}$ (see (13)) then the corresponding sensitivity of $\tilde{s}_{20}$ is

$$s_{20}'(0) = -\frac{q(20)}{p'(20)} = \frac{210 \cdot 20^{19}}{19!} \approx 0.9 \cdot 10^{10}.$$

□

4.1.3 The Sets of Hurwitz and Schur Polynomials

In this subsection we consider the sets of *real monic* Hurwitz and Schur polynomials of a given degree n. We establish some elementary topological properties of these sets in the space of coefficient vectors and characterize their boundaries and convex hulls.

Since we only consider *monic* polynomials we omit the leading coefficient from the coefficient vector and identify each real polynomial of the form

$$p(s,a) = s^n + a_{n-1}s^{n-1} + \ldots + a_0 \tag{15}$$

with its coefficient vector $a = (a_0, \ldots, a_{n-1}) \in \mathbb{R}^n$. Let $\mathcal{H}_n$ (resp. $\mathcal{S}_n$) denote the set of coefficient vectors of all the real monic Hurwitz (resp. Schur) polynomials of degree n. By Theorem 4.1.2 the sets $\mathcal{H}_n$ and $\mathcal{S}_n$ are open subsets of $\mathbb{R}^n$, and by the results of Section 3.4 both of them can be described by finite sets of algebraic inequalities. However, it is difficult to get some idea of the geometry of these sets from the inequalities. In the following, we mainly concentrate on the Schur case and leave some of the corresponding development of the Hurwitz case to the exercises (see also *Notes and References*). We first derive some elementary topological properties by studying the boundary $\partial\mathcal{S}_n$ of the set $\mathcal{S}_n$.

Lemma 4.1.22. *The boundary $\partial\mathcal{S}_n$ consists of all those polynomials with roots in $\overline{\mathbb{D}}$ which have at least one root on $\partial\mathbb{D}$.*

Proof: Since the set of roots depends continuously on the coefficient vector, every polynomial $p(s,a)$, $a \in \partial\mathcal{S}_n$ has all its roots in $\overline{\mathbb{D}}$ and at least one root on $\partial\mathbb{D}$. Conversely suppose $p(s,a) = p_1(s,a)\,p_2(s,a)$ where $p_1(s,a) \in \mathbb{R}[s]$ has all its roots in $\mathbb{D}$ and $p_2(s,a) = \prod_{i=1}^k (s - s_i) \in \mathbb{R}[s]$ with $s_i \in \partial\mathbb{D}$ and $k \geq 1$. Then $a \notin \mathcal{S}_n$ and setting $p(s,a(t)) = p_1(s,a)\prod_{i=1}^k (s - ts_i)$ for $t \in [0,1]$ we see that $a(t) \in \mathcal{S}_n$ for $t \in [0,1)$, whence $\lim\limits_{t\to 1} a(t) = a \in \partial\mathcal{S}_n$. □

If a real polynomial $p(s,a) \in \mathbb{R}[s]$ has a root on $\partial\mathbb{D}$, there are two possibilities: either the root is ± 1, or the root is in $\mathbb{C} \setminus \mathbb{R}$ in which case its complex conjugate is also a root. The set of coefficient vectors $a \in \mathbb{R}^n$ with $p(1,a) = 0$ resp. $p(-1,a) = 0$ form two hyperplanes in $\mathbb{R}^n$ denoted by H_1, H_{-1}

$$\begin{aligned} H_1 &= \{a \in \mathbb{R}^n\,;\; 1 + a_{n-1} + a_{n-2} + \cdots + a_0 = 0\}, \\ H_{-1} &= \{a \in \mathbb{R}^n\,;\; 1 - a_{n-1} + a_{n-2} - \cdots + (-1)^n a_0 = 0\}. \end{aligned} \tag{16}$$

Let us now determine the set of coefficient vectors $a \in \mathbb{R}^n$ for which $p(s,a)$ has a pair of complex roots $\alpha \pm \imath\omega \in \partial\mathbb{D}$, $\alpha, \omega \in \mathbb{R}$, $\omega \neq 0$, $\alpha^2 + \omega^2 = 1$. If $p(\alpha \pm \imath\omega, a) = 0$, then $p(s,a)$ is of the form

$$p(s,a) = (s^2 - 2\alpha s + 1)q(s), \quad \text{where } q(s) = s^{n-2} + q_{n-3}s^{n-3} + \cdots + q_0 \in \mathbb{R}[s]. \tag{17}$$

By comparing the coefficients of s^j, $j = 0, \cdots, n-1$ on both sides of the first equation in (17) we see that

$$a_0 = q_0\,, \quad a_1 = q_1 - 2\alpha q_0\,, \quad \text{and } a_i = q_i - 2\alpha q_{i-1} + q_{i-2}, \; i = 2, \ldots, n-1 \tag{18}$$

where $q_{n-1} := 0$. For fixed $\alpha \in (-1,1)$ the coefficient vectors a lie in an $(n-2)$–dimensional linear subspace H_α parametrized by the coefficient vector of the monic polynomial q. Whereas for fixed $q_0, \cdots, q_{n-3}$ the coefficient vector moves along a straight line as α varies between -1 and $+1$. Altogether we get the following decomposition of the boundary of $\mathcal{S}_n$ into the intersections of $\overline{\mathcal{S}_n}$ with two $(n-1)$–dimensional and infinitely many $(n-2)$–dimensional affine subspaces of $\mathbb{R}^n$.

$$\partial\mathcal{S}_n = \overline{\mathcal{S}_n} \cap \left(H_1 \cup H_{-1} \cup \bigcup_{\alpha \in (-1,1)} H_\alpha \right). \tag{19}$$

Example 4.1.23. In this example we determine $\partial\mathcal{S}_n$ for $n = 1, 2, 3$.

(i) $\mathbf{n = 1,\ p(s,a) = s + a_0}$. Then $\mathcal{S}_1 = (-1,1)$, $H_{\pm 1} = \{\pm 1\}$, $\partial\mathcal{S}_1 = \{-1, +1\}$.

(ii) $\mathbf{n = 2,\ p(s,a) = s^2 + a_1 s + a_0}$. In this case $\partial\mathcal{S}_2$ is given by (19) where

$$\begin{aligned} H_1 &= \{a \in \mathbb{R}^2\,;\; 1 + a_1 + a_0 = 0\}, \quad H_{-1} = \{a \in \mathbb{R}^2\,;\; 1 - a_1 + a_0 = 0\} \\ H_\alpha &= \{a \in \mathbb{R}^2\,;\; a_1 = -2\alpha\,,\; a_0 = 1\}, \qquad \alpha \in (-1,1). \end{aligned}$$

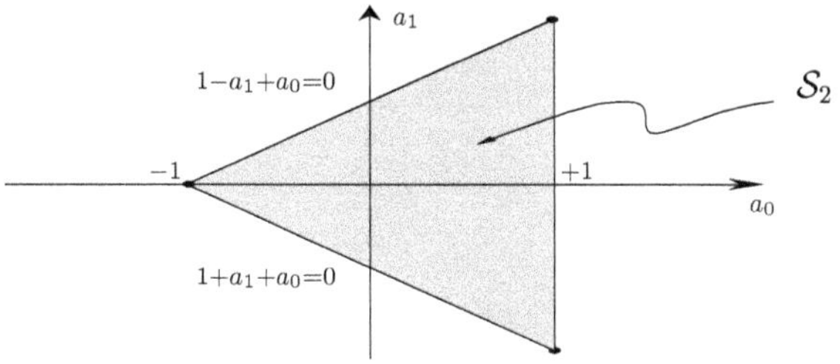

Figure 4.1.4: The set $\mathcal{S}_2$

(iii) $\mathbf{n = 3,\ p(s,a) = s^3 + a_2 s^2 + a_1 s + a_0}$. By (16) and (18) we have

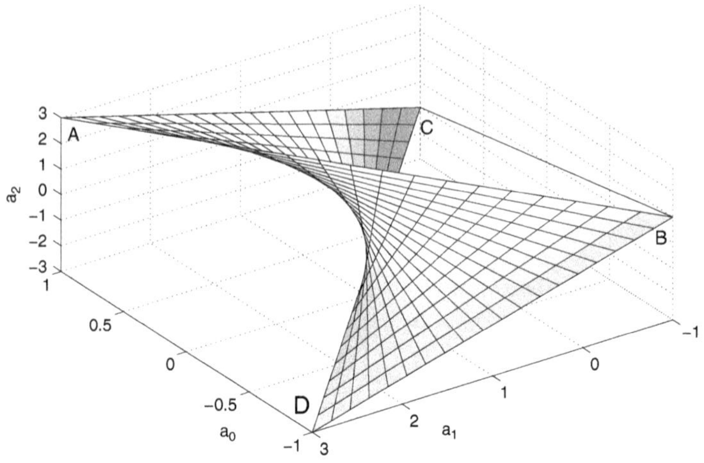

Figure 4.1.5: The set $\mathcal{S}_3$

$$\begin{aligned} H_1 &= \{a \in \mathbb{R}^3;\ 1 + a_2 + a_1 + a_0 = 0\}, \quad H_{-1} = \{a \in \mathbb{R}^3;\ 1 - a_2 + a_1 - a_0 = 0\} \\ H_\alpha &= \{(q_0, 1 - 2\alpha q_0,\ q_0 - 2\alpha) \in \mathbb{R}^3;\ q_0 \in \mathbb{R}\}, \quad \alpha \in (-1,1). \end{aligned}$$

Hence $\bigcup_{\alpha\in(-1,1)} H_\alpha$ is part of the hyperbolic paraboloid described by $a_0a_2 - a_0^2 - a_1 + 1 = 0$, see Figure 4.1.5. The boundary $\partial\mathcal{S}_3$ is composed of the two flat triangles ABC, BCD and the hyperbolic paraboloid spanned between A, B, D, C where $A = (1,3,3)$, $B = (-1,-1,1)$, $C = (+1,-1,-1)$, $D = (-1,3,-3)$. More precisely, $H_1 \cap \partial\mathcal{S}_3$ is just the triangle BCD and $H_{-1} \cap \partial\mathcal{S}_3$ is the triangle ABC. For each $\alpha \in (-1,1)$, $H_\alpha \cap \partial\mathcal{S}_3$ is a straight line joining the point $(1, 1-2\alpha, 1-2\alpha)$ on AC ($q_0 = 1$) to the point $(-1, 1+2\alpha, -1-2\alpha)$ on BD ($q_0 = -1$). □

In the above example, $\mathcal{S}_i$, $i \in \underline{3}$ is the unique bounded connected component of the set $\mathbb{R}^i \setminus \partial\mathcal{S}_i$ $i \in \underline{3}$ obtained by removing the surface $\partial\mathcal{S}_i$ from $\mathbb{R}^i$, and its convex hull is a simplex. We will now generalize these findings to arbitrary n.

Definition 4.1.24. A set $\Omega \subset \mathbb{R}^n$ is contractible to $x^0 \in \Omega$ if there exists a continuous map $F : \Omega \times [0,1] \to \Omega$, such that for all $x \in \Omega$

$$F(x,0) = x\ , \qquad F(x,1) = x^0.$$

Proposition 4.1.25. *The set of Schur polynomials $\mathcal{S}_n$ is contractible to the origin. $\mathbb{R}^n \setminus \partial\mathcal{S}_n$ consists of two connected components of which one is bounded and the other is unbounded. The bounded one coincides with $\mathcal{S}_n$.*

Proof: For any $a \in \mathcal{S}_n$, $t \in [0,1]$ let

$$F(a,t) = ((1-t)^n a_0, (1-t)^{n-1} a_1, \cdots, (1-t)a_{n-1}).$$

Clearly $F(a,0) = a$ and $F(a,1) = 0$ for all $a \in \mathcal{S}_n$. $F(a,t)$ is the coefficient vector of the polynomial

$$\begin{aligned} p^t(s,a) &= s^n + (1-t)a_{n-1}s^{n-1} + \cdots + (1-t)^n a_0 \\ &= (1-t)^n \left([s/(1-t)]^n + a_{n-1}[s/(1-t)]^{n-1} + \cdots + a_0\right), \quad \text{if } t \in [0,1). \end{aligned}$$

So for $t \in [0,1]$, the roots of $p^t(s,a)$ are the same as those of $p^0(s,a) = p(s,a)$ multiplied by $(1-t)$. Hence $F(a,t) \in \mathcal{S}_n$ for all $t \in [0,1]$, $a \in \mathcal{S}_n$ and $F : \mathcal{S}_n \times [0,1] \to \mathcal{S}_n$ is clearly continuous. This proves contractibility and in particular that $\mathcal{S}_n$ is arcwise connected. Moreover, since the coefficients of a polynomial depend continuously (even polynomially) on its roots, $\mathcal{S}_n$ is bounded. Now choose $r > 0$ large enough so that the closure of $\mathcal{S}_n$ is contained in the ball $B(0,r) \subset \mathbb{R}^n$. Since $\mathbb{R}^n \setminus B(0,r)$ is connected, there is only one unbounded connected component C of $\mathbb{R}^n \setminus \partial\mathcal{S}_n$ and this unique unbounded component contains $\mathbb{R}^n \setminus B(0,r)$. Hence $\mathcal{S}_n$ is contained in a bounded connected component C' of $\mathbb{R}^n \setminus \partial\mathcal{S}_n$. But no boundary point of $\mathcal{S}_n$ is contained in C', so $\mathcal{S}_n$ is both open and closed in C', hence $\mathcal{S}_n = C'$. It remains to show that every $a \in \mathbb{R}^n \setminus \partial\mathcal{S}_n$ which is not a coefficient vector of a Schur polynomial is contained in the unbounded connected component C. But then $p(s,a) = \prod_{k=1}^{n} (s - s_k)$ has a root s_j with $|s_j| > 1$ and so all coefficient vectors $a(t)$, $t \geq 1$ defined by $p(s,a(t)) = (s - ts_j)\prod_{k\neq j}(s - s_k)$ are contained in $\mathbb{R}^n \setminus \partial\mathcal{S}_n$. Obviously $\{a(t);\, t > 1\}$ is unbounded and so the curve $t \mapsto a(t)$ will connect a with $C \supset \mathbb{R}^n \setminus B(0,r)$. Therefore $a \in C$, and this concludes the proof. □

The convex hull of $\overline{\mathcal{S}_n}$ has a surprisingly simple form.

Proposition 4.1.26 (Fam–Meditch). *The convex hull of the closure of $\mathcal{S}_n$ is an n-dimensional simplex with vertices given by the coefficient vectors of the polynomials*

$$p_j(s) = (s+1)^j (s-1)^{n-j}, \qquad j = 0, \cdots, n. \tag{20}$$

Proof: By Lemma 3.4.81 every Schur polynomial $\tilde{p}$ can be obtained from a Hurwitz polynomial $p(s,a)$ via the Möbius transformation

$$\begin{aligned} \tilde{p}(s) &= (s-1)^n p\left(\frac{s+1}{s-1}\right) && (21) \\ &= a_n(s+1)^n + a_{n-1}(s+1)^{n-1}(s-1) + \cdots + a_0(s-1)^n = \sum_{j=0}^{n} a_j p_j(s). \end{aligned}$$

Thus every Schur polynomial $\tilde{p}(s)$ is a linear combination of the polynomials $p_j(s)$. Now suppose that $\tilde{p}(s)$ is monic. Then $\sum_{j=0}^{n} a_j = 1$ since all $p_j(s)$ are monic and of degree n. But the coefficients of a Hurwitz polynomial are all of the same sign and so we get $a_j > 0$ for $j = 0, \cdots, n$. Hence every Schur polynomial is a convex combination of the p_j. Since $p_j \in \overline{\mathcal{S}_n}$ for $j = 0, \cdots, n$ it follows that $\operatorname{conv}\overline{\mathcal{S}_n} = \operatorname{conv}\{p_0, \ldots, p_n\}$. Finally $\operatorname{conv}\{p_0, \ldots, p_n\}$ is an n-simplex (i.e. the polynomials $p_0, \ldots, p_n$ are affinely independent) because $\mathcal{S}_n$ has nonempty interior in $\mathbb{R}^n$. □

The above proposition yields a simple necessary *linear* stability criterion. If $p(s,a)$ is a monic polynomial of degree n and if $v^0, \cdots, v^n$ are the coefficient vectors of $p_0, \ldots, p_n$ (20) one solves the equations

$$a = \sum_{j=0}^{n} \alpha_j v^j, \qquad \sum_{j=0}^{n} \alpha_j = 1 \tag{22}$$

for $(\alpha_j) \in \mathbb{R}^{n+1}$, then the necessary condition is that each α_j be positive. The vectors

$\begin{bmatrix} v^0 \\ 1 \end{bmatrix}, \cdots, \begin{bmatrix} v^n \\ 1 \end{bmatrix}$ are linearly independent in $\mathbb{R}^{n+1}$, so there is a unique solution and since the vectors $v^0, \cdots, v^n$ are easy to compute, the necessary condition is readily verified (see Ex. 17). Note however that this stability criterion is far from being sufficient. In fact it can be shown that $\mathrm{vol}(\mathcal{S}_n)/\,\mathrm{vol}(\mathrm{conv}(\overline{\mathcal{S}}_n)) = 2^{-(n^2-n\log n+O(n))/2}$ (see [156]).

Example 4.1.27. $\mathrm{Conv}(\overline{\mathcal{S}}_3)$ is generated by the coefficient vectors of the polynomials $(s+1)^3$, $(s+1)^2(s-1)$, $(s+1)(s-1)^2$, $(s-1)^3$. So its vertices are $A = (1,3,3)$, $B = (-1,-1,1)$, $C = (1,-1,-1)$, $D = (-1,3,-3)$ (see Example 4.1.23 (iii)). □

Making use of the Möbius transformation given by (21) one obtains the following results for the set of Hurwitz polynomials.

Corollary 4.1.28. *The set $\mathcal{H}_n$ of monic Hurwitz polynomials of degree n has the properties.*

(i) The boundary $\partial\mathcal{H}_n$ consists of all coefficient vectors $a \in \mathbb{R}^n$ for which $p(s,a)$ has all its roots in $\overline{\mathbb{C}}_-$ and at least one root on $\imath\mathbb{R}$.

(ii) $\mathcal{H}_n$ is contractible to the coefficient vector of the polynomial $(s+1)^n$, i.e. to the vector $[\binom{n}{1}, \binom{n}{2}, \cdots, 1]$.

(iii) $\mathcal{H}_n$ is a connected component of $\mathbb{R}^n_+ \setminus \partial\mathcal{H}_n$ with boundary $\partial\mathcal{H}_n \subset H'_0 \cup \cup_{\omega>0} H'_\omega$ where

$$H'_0 = \{a \in \mathbb{R}^n_+;\ a_0 = 0\}, \quad H'_\omega = \{a \in \mathbb{R}^n_+;\ (s^2+\omega^2) \text{ divides } p(s,a)\},\ \omega > 0. \quad (23)$$

We leave the proof to the reader (see Ex. 13).
The Möbius map $p(s) \mapsto (s-1)^n p\left(\frac{s+1}{s-1}\right)$ transforms the polynomial $p_j(s)$, defined in (20) into a polynomial of degree j. So it is not clear how one should formulate a counterpart to the Fam–Meditch result for $\mathrm{conv}(\mathcal{H}_n)$. In fact we have

Proposition 4.1.29.

$$\overline{\mathrm{conv}}(\mathcal{H}_n) := \overline{\mathrm{conv}(\mathcal{H}_n)} = \mathbb{R}^n_+. \quad (24)$$

Proof: Consider the Hurwitz polynomials

$$p_j^t(s) = (s+t)^j(s+t^{-1})^{n-j} \quad , \quad t > 0 \quad , j = 0, \cdots, n. \quad (25)$$

For any $a = (a_0, \cdots, a_{n-1}) \in \mathbb{R}^n_+$, let $p^t(s) = \sum_{j=0}^n \alpha_j(t) p_j^t(s)$, where

$$\alpha_j(t) = a_j t^{n-j},\ j = 0, \cdots, n-1, \quad \alpha_n(t) = 1 - (\alpha_0(t) + \ldots + \alpha_{n-1}(t)).$$

Then $\sum_{j=0}^n \alpha_j(t) = 1$ and for t sufficiently small $\alpha_j(t) \geq 0$, $j = 0, \cdots, n$. Hence $p^t \in \mathrm{conv}(\mathcal{H}_n)$ for small t. But

$$p^t(s) = \left(1 - \sum_{j=0}^{n-1} a_j t^{n-j}\right)(s+t)^n + \sum_{j=0}^{n-1} a_j (s+t)^j (ts+1)^{n-j}.$$

Since $p^t(s) \to p(s,a)$ as $t \to 0$, for every $s \in \mathbb{C}$, it follows that $a \in \overline{\mathrm{conv}}(\mathcal{H}_n)$. □

Example 4.1.30. In this example we characterize the sets $\mathcal{H}_n$ and H'_ω, $n = 1, 2, 3$.

(i) $\mathbf{n = 1}$, $\mathbf{p(s, a) = s + a_0}$. Then $\mathcal{H}_1 = \{a_0 \in \mathbb{R} : a_0 > 0\}$ and $H'_\omega = \emptyset$, $\omega > 0$.

(ii) $\mathbf{n = 2}$, $\mathbf{p(s, a) = s^2 + a_1 s + a_0}$. In this case

$$\mathcal{H}_2 = \{a \in \mathbb{R}^2;\ a_1 > 0,\ a_0 > 0\} = \operatorname{int} \mathbb{R}^2_+, \quad H'_\omega = \{a \in \mathbb{R}^2;\ a_1 = 0,\ a_0 = \omega^2\}, \quad \omega > 0.$$

(iii) $\mathbf{n = 3, p(s, a) = s^3 + a_2 s^2 + a_1 s + a_0}$. By the Hermite-Hurwitz Theorem 3.4.53

$$\mathcal{H}_3 = \{a \in \mathbb{R}^3 :\ a_2 > 0,\ a_1 > 0,\ a_0 > 0,\ a_2 a_1 > a_0\}.$$

Moreover, we have

$$\begin{aligned} H'_\omega &= \{a \in \mathbb{R}^3;\ p(s, a) = (s^2 + \omega^2)(s + q_0),\ q_0 \geq 0\} \\ &= \{a \in \mathbb{R}^3;\ a_0 = \omega^2 q_0,\ a_1 = \omega^2,\ a_2 = q_0 \geq 0\}, \quad \omega > 0. \end{aligned}$$

So $\bigcup_{\omega > 0} H'_\omega$ is part of the surface described by $a_1 a_2 = a_0$ in $\mathbb{R}^3_+$. □

4.1.4 Kharitonov's Theorem

In this subsection we will deal with the following problem: "If for a real monic polynomial all we know about the coefficients is that they lie between certain bounds, how can we decide whether it is Hurwitz stable?" In order to answer this question we must find necessary and sufficient conditions for the stability of the whole set of monic polynomials whose coefficients belong to prescribed intervals. Such sets are called *interval polynomials.* We must therefore characterize those interval polynomials which are contained in $\mathcal{H}_n$. This problem was solved by *Kharitonov* (1978) [300] and we will present his result in this subsection.

For the description of interval polynomials it is convenient to introduce the following partial ordering on $\mathbb{R}^n$

$$a = (a_0, \ldots, a_{n-1}) \leq b = (b_0, \ldots, b_{n-1}) \quad :\Longleftrightarrow \quad a_i \leq b_i,\ i = 0, \cdots, n-1.$$

If $\underline{a} \leq \overline{a}$, then $[\underline{a}, \overline{a}]$ denotes the closed n–dimensional interval between $\underline{a}$ and $\overline{a}$, i.e.

$$[\underline{a}, \overline{a}] = \{a \in \mathbb{R}^n;\ \underline{a} \leq a \leq \overline{a}\}.$$

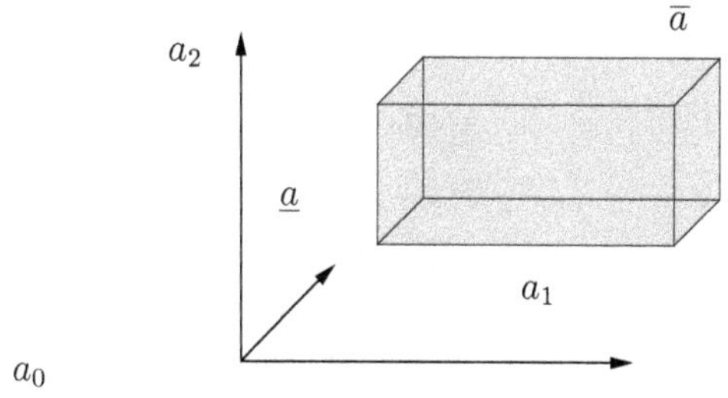

Figure 4.1.6: Interval $[\underline{a}, \overline{a}] \subset \mathbb{R}^3$

With any such interval we associate the *value sets*

$$K(s_0) = \{p\,(s_0, a)\,;\; a \in [\,\underline{a}, \overline{a}\,]\} \subset \mathbb{C}\,, \quad s_0 \in \mathbb{C} \tag{26}$$

which will play a central role in the sequel. In order to describe $K(\imath\omega)$, $\omega \in \mathbb{R}_+$ we need the following four polynomials

$$\left\{ \begin{array}{rcl} v_1(s) & = & \underline{a}_0 + \overline{a}_2 s^2 + \underline{a}_4 s^4 + \cdots\,, \quad v_2(s) = \overline{a}_0 + \underline{a}_2 s^2 + \overline{a}_4 s^4 + \cdots \\ u_1(s) & = & \underline{a}_1 s + \overline{a}_3 s^3 + \underline{a}_5 s^5 + \cdots\,, \quad u_2(s) = \overline{a}_1 s + \underline{a}_3 s^3 + \overline{a}_5 s^5 + \cdots \end{array} \right. \tag{27}$$

Clearly, for $\omega \in \mathbb{R}$, $v_1(\imath\omega)$, $v_2(\imath\omega)$ are real and $u_1(\imath\omega)$, $u_2(\imath\omega)$ are purely imaginary.

Lemma 4.1.31. *For any $\omega \in \mathbb{R}_+$ the value set $K(\imath\omega)$ is a rectangle in the complex plane with vertices $k_{ij}(\imath\omega) = v_i(\imath\omega) + u_j(\imath\omega)$ for $i, j = 1, 2$.*

Proof: For any $a \in [\,\underline{a}, \overline{a}\,]$ we have

$$\begin{array}{rcl} v_1(\imath\omega) \leq & \operatorname{Re} p\,(\imath\omega, a) & \leq v_2(\imath\omega) \\ \imath^{-1} u_1(\imath\omega) \leq & \operatorname{Im} p\,(\imath\omega, a) & \leq \imath^{-1} u_2(\imath\omega), \quad \omega \in \mathbb{R}_+. \end{array}$$

Hence $K(\imath\omega)$ is contained in the rectangle with corners $k_{ij}(\imath\omega)$. To prove the converse it suffices to note that $K(\imath\omega)$ is convex and $k_{ij}(\imath\omega) \in K(\imath\omega)$, $i, j = 1, 2$. □

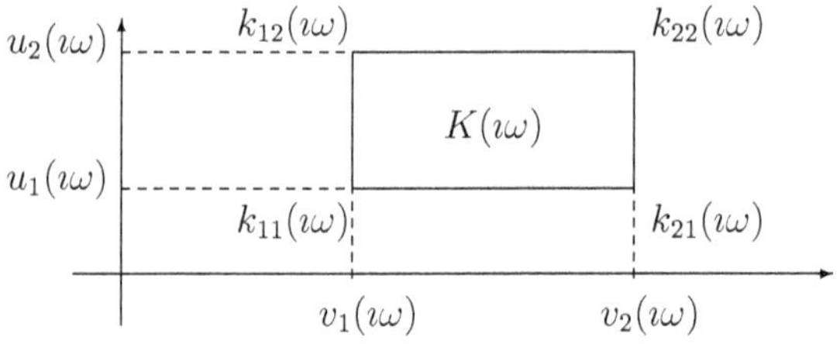

Figure 4.1.7: The value set $K(\imath\omega)$

The four polynomials

$$k_{ij}(s) = v_i(s) + u_j(s)\,, \qquad i, j = 1, 2 \tag{28}$$

are called the *Kharitonov polynomials* associated with $[\,\underline{a}, \overline{a}\,]$. They have coefficient vectors

$$\begin{array}{ll} [{}_{-}{}_{-}{}^{-}{}^{-}] & a^{11} = [\,\underline{a}_0, \underline{a}_1, \overline{a}_2, \overline{a}_3, \underline{a}_4, \underline{a}_5, \cdots\,] \\ [{}_{-}{}^{-}{}^{-}{}_{-}] & a^{12} = [\,\underline{a}_0, \overline{a}_1, \overline{a}_2, \underline{a}_3, \underline{a}_4, \overline{a}_5, \cdots\,] \\ [{}^{-}{}_{-}{}_{-}{}^{-}] & a^{21} = [\overline{a}_0, \underline{a}_1, \underline{a}_2, \overline{a}_3, \overline{a}_4, \underline{a}_5, \cdots\,] \\ [{}^{-}{}^{-}{}_{-}{}_{-}] & a^{22} = [\overline{a}_0, \overline{a}_1, \underline{a}_2, \underline{a}_3, \overline{a}_4, \overline{a}_5, \cdots\,]. \end{array} \tag{29}$$

Figure 4.1.8 illustrates the location of the Kharitonov polynomials in the coefficient space. The shaded area represents the convex set generated by the four Kharitonov polynomials. The evaluation $a \mapsto p\,(\imath\omega, a)$ maps this rectangle affinely onto the value set $K(\imath\omega)$.

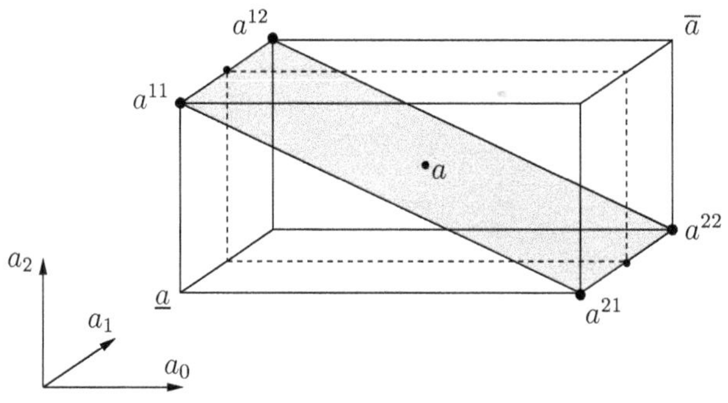

Figure 4.1.8: Kharitonov polynomials in $\mathbb{R}^3$

Theorem 4.1.32 (Kharitonov). *Suppose $\underline{a}$, $\overline{a} \in \mathbb{R}^n$, $\underline{a} \leq \overline{a}$, then all the polynomials $p(s,a)$, $a \in [\underline{a}, \overline{a}]$ are Hurwitz if and only if the associated four Kharitonov polynomials* (29) *are Hurwitz.*

Proof: Necessity is clear. Now assume that the Kharitonov polynomials are Hurwitz, but $[\underline{a}, \overline{a}] \not\subset \mathcal{H}_n$. Since $[\underline{a}, \overline{a}]$ is connected there exists by Corollary 4.1.5 $a \in [\underline{a}, \overline{a}]$ and $\omega \geq 0$ such that $p(\imath\omega, a) = 0$, i.e. $0 \in K(\imath\omega)$, see Remark 4.1.7. By Proposition 3.4.6 $\underline{a}_i > 0$ for $i = 0, \cdots, n-1$, hence $0 \notin K(0)$. Since the four vertices of $K(\imath\omega)$ vary continuously with ω there exists $\tilde{\omega} > 0$ such that 0 lies on the boundary of $K(\imath\tilde{\omega})$. Since no vertex passes through zero there is an edge containing 0 in its interior. Assume for example that this is the bottom edge (see Figure 4.1.9).

Then $k_{11}(\imath\tilde{\omega}) < 0 < k_{21}(\imath\tilde{\omega})$. But by Proposition 3.4.7 $\omega \mapsto \arg k_{ij}(\omega)$ is strictly

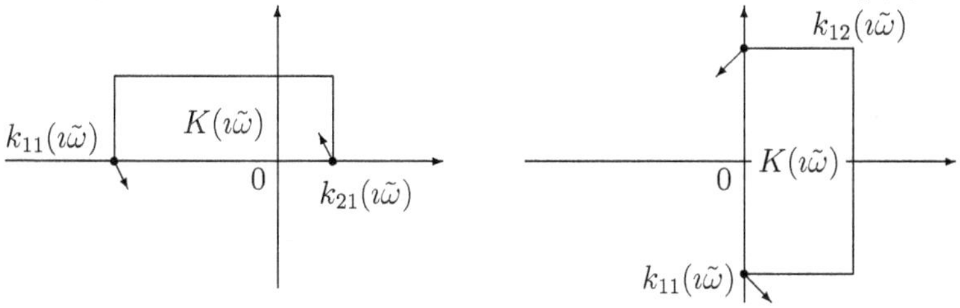

Figure 4.1.9: $0 \in \partial K(\imath\tilde{\omega})$ implies contradiction

increasing. Hence for $\omega > \tilde{\omega}$ sufficiently close to $\tilde{\omega}$ the vertex $k_{11}(\imath\omega)$ will lie in the open third quadrant whilst $k_{21}(\imath\omega)$ lies in the open first quadrant. This contradicts the fact that $\operatorname{Im} k_{21}(\imath\omega) = \operatorname{Im} k_{11}(\imath\omega)$ (by (28)). The cases where 0 lies on one of the other edges are treated analogously (see Figure 4.1.9). □

As a consequence of the above proof and Proposition 3.4.7 we see that the whole value set $K(\imath\omega)$, $0 < \omega < \infty$ is travelling (with increasing ω) counterclockwise through a total angle of $n\pi/2$ and always completely enters one quadrant before crossing into the next.

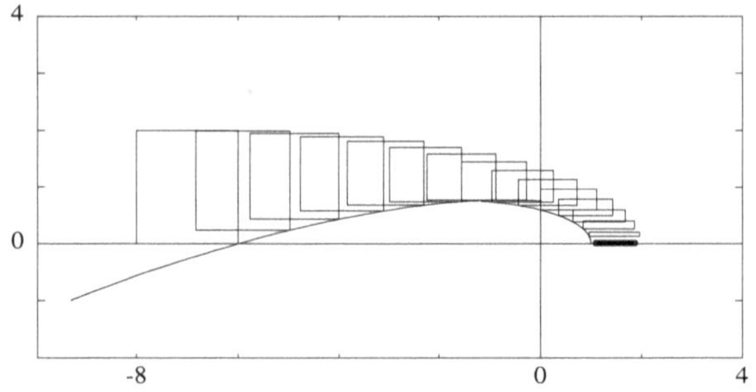

Figure 4.1.10: $\omega \mapsto K(\imath\omega)$

Remark 4.1.33. Kharitonov's Theorem can easily be extended to the non-monic case provided the degree remains invariant over the interval, see Ex. 9. □

Example 4.1.34. Consider the polynomial of degree 3

$$p(s,a) = s^3 + a_2 s^2 + a_1 s + a_0 \quad , \quad a_i > 0 \quad i = 0,1,2.$$

We have seen in Section 3.4 that such a polynomial is Hurwitz if $a_1 a_2 > a_0$. Now suppose that $\underline{a} = (1/2, 2, 2)$, $\overline{a} = (7/2, 4, 4)$, then by Kharitonov's Theorem the box $[\underline{a}, \overline{a}] \subset \mathcal{H}_3$ provided that the points $(1/2, 2, 4)$, $(1/2, 4, 4)$, $(7/2, 2, 2)$, $(7/2, 4, 2)$ all lie in $\mathcal{H}_3$. It is easily seen that this is indeed the case. □

Polynomials which arise from practical problems usually have uncertain coefficients. However it is often known that the coefficient vector lies in some set $\Omega \subset \mathbb{R}^n$. One way of determining whether or not $\Omega \subset \mathcal{H}_n$ is to try to cover it with boxes $[\underline{a}^k, \overline{a}^k] \subset \mathcal{H}_n$, $k \in \underline{N}$. This is illustrated in the next example (see *Notes and References* for other applications of Kharitonov's Theorem).

Example 4.1.35. In Example 3.4.16 we described Maxwell's stability analysis of Jenkin's governor. The system matrix is

$$A = \begin{bmatrix} -F/M & -G/M & 0 \\ 0 & 0 & 1 \\ F/B & 0 & -Y/B \end{bmatrix}$$

where M, B are moments of inertia and G is a torque. We suppose that these quantities are known accurately and for simplicity we assume $M = B = G = 1$. F and Y are friction and viscous damping coefficients and are assumed to be uncertain, both with values in the interval $[3/4, 5/4]$. The characteristic polynomial of A is

$$p(s,a) = s^3 + a_2 s^2 + a_1 s + a_0$$

where $a_2 = F + Y$, $a_1 = FY$, $a_0 = F$. If $(F,Y) \in [3/4, 5/4]^2$ the coefficient vector $a = a(F,Y)$ lies in the 2-dimensional surface $\Omega = \{(F, FY, F+Y); (F,Y) \in [3/4, 5/4]^2\}$ shown in Figure 4.1.11. The vertices of this surface are given by $\underline{a} = a(3/4, 3/4) = (3/4, 9/16, 3/2)$, $b = a(3/4, 5/4) = (3/4, 15/16, 2)$, $c = a(5/4, 3/4) = (5/4, 15/16, 2)$ and $\overline{a} = a(5/4, 5/4) = (5/4, 25/16, 5/2)$. Note that the smallest box containing Ω is $[\underline{a}, \overline{a}]$. In order to apply Kharitonov's Theorem to this box one would require $a^{11} = (3/4, 9/16, 5/2)$, $a^{12} = (3/4, 25/16, 5/2)$, $a^{21} = (5/4, 9/16, 3/2)$ and $a^{22} = (5/4, 25/16, 3/2)$ to be vectors in

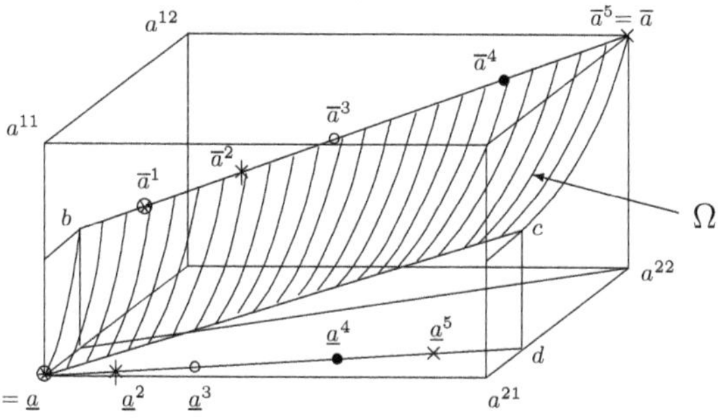

Figure 4.1.11: The set Ω and its covering with 5 Kharitonov boxes

$\mathcal{H}_3$. But $a^{21} \notin \mathcal{H}_3$ and hence $[\underline{a}, \overline{a}] \not\subset \mathcal{H}_3$.
However, we can cover Ω with 5 boxes with opposite vertices denoted by $\otimes$, $*$, $\circ$, $\bullet$, $\times$ as shown in Figure 4.1.11. To simplify the calculations we have chosen the lower vertices on the segment conv$\{\underline{a}, d\}$ where $d = (5/4, 15/16, 3/2)$ is the projection of c onto the bottom face of the box $[\underline{a}, \overline{a}]$. Now any point in conv$\{\underline{a}, d\}$ has the form

$$a(t) = (3/4 + t/2\,, 9/16 + 3t/8\,, 3/2) = (a_0(t), a_1(t), a_2(t)), \quad t \in [0, 1].$$

Since $a_2(t)a_1(t) = (3/2)(9/16 + 3t/8) > 3/4 + t/2 = a_0(t)$, $t \in [0, 1]$ it follows from the Hurwitz criterion (see Example 4.1.34) that $a(t) \in \mathcal{H}_3$ for all $t \in [0, 1]$. Note that by the above definition $a_1(t) = (3/4)\,a_0(t)$. The boxes are generated by setting $\underline{a}^1 = \underline{a}$ and searching along the segment conv$\{b, \overline{a}\}$ for the first point $\tilde{a}$ such that $[\underline{a}^1, \tilde{a}] \not\subset \mathcal{H}_3$. $\overline{a}^1 = (\overline{a}^1_0, \overline{a}^1_1, \overline{a}^1_2)$ is chosen by stepping back slightly from $\tilde{a}$ to ensure $[\underline{a}^1, \overline{a}^1] \subset \mathcal{H}_3$. Then $\underline{a}^2 = (\underline{a}^2_0, \underline{a}^2_1, 3/2)$ is chosen on the segment conv$\{\underline{a}, d\}$ such that $\underline{a}^2_0 = \overline{a}^1_0$, hence $\underline{a}^2_1 = (3/4)\overline{a}^1_0$. Then the process is continued. In this way we can ensure that each box is in $\mathcal{H}_3$ and their union covers Ω. Hence the system is asymptotically stable for all possible parameter values in the interval $[3/4, 5/4]^2$. □

Even though this approach was successful it seems rather artificial and cumbersome. The reason is that in the example the uncertainty was expressed by intervals in the *parameter space*, whereas Kharitonov's Theorem presupposes that the uncertainty is expressed by intervals in the *coefficient space*. If the dimension d of the parameter space is smaller than the degree n of the polynomial, a direct application of the theorem requires one to cover a d-dimensional surface in $\mathbb{R}^n$ by n-dimensional boxes. In general, this will mean that either many thin boxes will have to be used which makes the method cumbersome, or the results will be conservative. For an extension of Kharitonov's result to structured perturbations of the coefficient vector, see *Notes and References*.

4.1.5 Exercises

1. Let $p(s, a_k), a^k \in \mathbb{C}^{n+1}$ be a sequence of (not necessarily monic) polynomials in $\mathbb{C}_n[s]$ and $p(s, a) \in \mathbb{C}_n[s]$. Prove that the following statements are equivalent.

(i) $p(s, a^k) \to p(s, a)$ for all $s \in \mathbb{C}$ as $k \to \infty$.

(ii) There are $n+1$ distinct complex numbers $s_1, \ldots, s_{n+1}$ such that $p(s_j, a^k) \to p(s_j, a)$ for all $j = 1, \ldots, n+1$, as $k \to \infty$.
(iii) $a^k \to a$ in $\mathbb{C}^{n+1}$ as $k \to \infty$.

If the polynomials $p(s, a_k)$, $p(s, a)$ are all monic then (i)-(iii) are equivalent to

(iv) $\Lambda(a^k) \to \Lambda(a)$ as $k \to \infty$, i.e. the unordered n-tuples of the roots of $p(s, a_k)$ converge to the unordered n-tuple of the roots of $p(s, a)$.

2. Let $\mathbb{C}_g \subset \mathbb{C}$ be an open subset of the complex plane and suppose

$$p_k(s) = \sum_{j=0}^{n} a_j^{(k)} s^j, \quad k \in \mathbb{N}$$

is a sequence of $\mathbb{C}_g$-stable polynomials of degree $\leq n$. Prove: If the sequence converges (coefficientwise) to a polynomial $q(s) = \sum_{j=0}^{n} a_j s^j$, then the roots of $q(s)$ are contained in the closure of $\mathbb{C}_g$.

3. Prove the two statements in Remark 4.1.7 and give an example which shows that the Boundary Crossing Theorem does not hold if the constant degree assumption ($a_n(\omega) \neq 0$ for all $\omega \in \Omega$) is dropped. (*Bhattacharyya* [56], pp 34-36.)

4. Let $\Omega \subset \mathbb{C}$ be a domain and $p, q, r \in \mathcal{M}(\Omega)[s]$ monic polynomials of the form

$$p(s) = \sum_{i=0}^{n} a_i(z) s^i, \quad q(s) = \sum_{j=0}^{\ell} b_j(z) s^j, \quad r(s) = \sum_{k=0}^{m} c_k(z) s^k$$

with meromorphic coefficients $a_i(z), b_j(z), c_k(z)$ on Ω, $a_n(z) = b_\ell(z) = c_m(z) \equiv 1$. Prove that if $p = qr$ and z_0 is a pole of one of the coefficients $b_j(z), c_k(z)$ then z_0 is a pole of one of the coefficients $a_i(z)$ (cf. Baumgärtel (1985) [43, A.2.6]).

5. Let $p(s, a(z)) = s^2 + 2(1+z)s + \alpha(1-z) = 0$ where $\alpha \in \mathbb{R}$ is given. Show that for $\alpha \neq 1$ the roots $s_{1,2}(z)$ of $p(s, a(z))$ depend analytically on z for $|z| \ll 1$:

$$s_{1,2}(z) = -1 \pm (1-\alpha)^{1/2} - z\left[1 \mp \frac{2+\alpha}{2(1-\alpha)^{1/2}}\right] \mp z^2 \frac{\alpha(8+\alpha)}{8(1-\alpha)^{3/2}} \cdots, \qquad \alpha \neq 1,$$

whereas, for $\alpha = 1$, we have $s_{1,2}(z) = -1 \pm \sqrt{3} z^{1/2} - z \pm (2\sqrt{3})^{-1} z^{3/2} \ldots$.

6. Determine the sensitivities at $z = 0$ of the roots $s_j(z)$, $j = 1, 2, 3$ of the polynomial

$$p(s, a(z)) = s^3 + (2+z)s^2 - (1-z)s + (z-2).$$

7. Show that if $\underline{a} = (1, 2, 8, 1)$, $\overline{a} = (2, 3, 9, 2)$, then $[\underline{a}, \overline{a}] \subset \mathcal{H}_4$.

8. Show that the polynomial $p(s, a^*) = s^3 + 3s^2 + 12s + 10$ is Hurwitz. Find $\underline{a}$, $\overline{a}$ such that $\underline{a} < a^* < \overline{a}$ and $[\underline{a}, \overline{a}] \subset \mathcal{H}_3$. Calculate $\bar{t} = \sup\{t > 0; [a^* - t(\overline{a} - \underline{a}), a^* + t(\overline{a} - \underline{a})] \subset \mathcal{H}_3$. Find a polynomial in $[a^* - \bar{t}(\overline{a} - \underline{a}), a^* + \bar{t}(\overline{a} - \underline{a})]$ which does not belong to $\mathcal{H}_3$.

9. Prove the following Kharitonov Theorem for non-monic interval polynomials: Suppose $\underline{a}$, $\overline{a} \in \mathbb{R}^{n+1}$, $\underline{a} \leq \overline{a}$ and $\underline{a}_{n+1}\overline{a}_{n+1} > 0$, i.e. the leading coefficients are of the same sign. Then all the polynomials $p(s, a)$, $a \in [\underline{a}, \overline{a}]$ are Hurwitz stable if and only if the associated four Kharitonov polynomials (with coefficients given by (29)) are Hurwitz.

10. If $[\underline{a}, \overline{a}] \subset \mathcal{H}_3$, show that the set of monic polynomials $p\,(s, a)$ where

$$\underline{a}_0 \leq \frac{1-a_0+a_1-a_2}{1+a_0+a_1+a_2} \leq \overline{a}_0, \quad \underline{a}_1 \leq \frac{3-a_2-a_1+3a_0}{1+a_0+a_1+a_2} \leq \overline{a}_1, \quad \underline{a}_2 \leq \frac{3+a_2-a_1-3a_0}{1+a_0+a_1+a_2} \leq \overline{a}_2$$

is Schur stable.

11. For continuous time systems practical considerations often require that the eigenvalues lie to the left of the hyperbola $\omega^2 = v^2 - 1$, $v \leq -1$ in the complex s–plane $s = v + \imath\omega$. This guarantees $1/\sqrt{2}$ damping for second order systems and a real part of all eigenvalues less than -1. Determine the sets $\hat{\mathcal{H}}_2$, $\hat{\mathcal{H}}_3$ which consist of all coefficient vectors of quadratic or cubic polynomials with all roots to the left of the above hyperbola.

12. Consider the discrete time system

$$x(t+2) + (5/6)\,x(t+1) + (1/6)\,x(t) = 0, \quad t \in \mathbb{N}$$

Find the distance $\operatorname{dist}(\overline{a}, \partial\mathcal{S}_2)$ with respect to the ∞–norm where $\overline{a} = [\frac{1}{6}, \frac{5}{6}]$. Hence determine the maximum joint bound on $|\delta_1|$, $|\delta_2|$ guaranteeing that

$$x(t+2) + (5/6+\delta_1)x(t+1) + (1/6+\delta_2)x(t) = 0, \quad t \in \mathbb{N}$$

is asymptotically stable.

13. Prove Corollary 4.1.28 (i), (ii), (iii).

14. If $\tilde{p}(s) = p(s, \tilde{a})$ is the Möbius transform (21) of a monic polynomial $p(s) = p(s, a)$ of degree n then $\tilde{a}_n = 1 + \sum_{j=0}^{n-1} a_j$ is the leading coefficient of $\tilde{p}(s)$ and $\breve{p}(s) = (1 + \sum_{j=0}^{n-1} a_j)^{-1}\tilde{p}(s)$ is the normalized Möbius transform of $p(s)$. Prove that under the transformation $p \mapsto \breve{p}$ the polynomials $p_j^t(s)$, $j = 0, \cdots, n-1$ (see (25)) are transformed into $\breve{p}_j^t(s)$, where

$$\breve{p}_j^t(s) = \frac{[(s+1)+t(s-1)]^j\,[(s+1)t+(s-1)]^{n-j}}{t^{n-j}(1+t)^j + (1+t)^{n-j} - t^{n-j}}, \quad j = 0, \ldots, n-1$$

and hence $\breve{p}_j^t(s) \to p_j(s)$ for $j = 0, \ldots, n-1$ as $t \to 0$ (see (20)). Describe the relationship between the roots of $p_j^t(s)$ and the roots of $p_j(t)$, $j = 0, \cdots, n-1$ via the Möbius transform.

15. Show that

$$\partial\mathcal{H}_4 = \{(0, a_1, a_2, a_3) \in \mathbb{R}_+^4;\ a_2 a_3 \geq a_1\} \cup \{(a_0, a_1, a_2, a_3) \in \mathbb{R}_+^4;\ a_1 a_2 a_3 = a_1^2 + a_0 a_3^2\}$$

16. If $f(s)$ is a polynomial of degree n, then $g = \Gamma f$ defined by

$$g(z) = 2^{-n/2}(z-1)^n f\left(\frac{z+1}{z-1}\right)$$

is called the Γ–*transform* of f. Prove that

(i) g is a polynomial of degree n in z,

(ii) $f = \Gamma g$, that is $f(s) = 2^{-n/2}(s-1)^n g\left(\dfrac{s+1}{s-1}\right)$.

17. Consider the $(n+1) \times (n+1)$ matrix $Q_n = (q_{ij})$ defined by

$$(s+1)^i (s-1)^{n-i} = \sum_{j=0}^{n} q_{ij} s^j \quad , \quad i = 0, 1, \cdots, n.$$

(Note that the row vectors of Q_n after deletion of the last column of Q_n, coincide with the vertices of conv $\bar{\mathcal{S}}_n$). Prove that

(i) $q_{in} = 1$ for $i = 0, \cdots, n$, $q_{0j} = (-1)^{n-j}\binom{n}{j}$ for $j = 0, \cdots, n$.
(ii) $q_{ij} = q_{ij+1} + q_{i-1j} + q_{i-1j-1}$, $i = 1, \cdots, n$, $j = 0, \cdots, n-1$.

(Note that Q_n is completely determined by (i) and (ii)).

(iii) If $f(s) = \sum_{i=0}^{n} a_i s^i$ and $g(z) = \sum_{j=0}^{n} c_j z^j$ is the Γ–transform, then

$$c_j = 2^{-n/2} \sum_{i=0}^{n} a_i q_{ij} \ , \quad a_k = 2^{-n/2} \sum_{j=0}^{n} c_j q_{jk} \ .$$

(iv) $Q_n^2 = 2^n I_{n+1}$ (hence $Q_n^{-1} = 2^{-n} Q_n$).

Determine Q_4 with the aid of (i), (ii).

18. If n is even show that the polynomial map $f_n : \mathcal{S}_1 \times \mathcal{S}_n \to \mathcal{S}_{n+1}$, $(a^{(1)}, a^{(n)}) \mapsto a^{(n+1)}$ defined by

$$a_0^{(n+1)} = a_0^{(1)} a_0^{(n)}, \quad a_1^{(n+1)} = a_0^{(n)} + a_0^{(1)} a_1^{(n)}, \ \ldots, \ a_n^{(n+1)} = a_{n-1}^{(n)} + a_0^{(1)}$$

is surjective but not injective. (Hint: consider the polynomial $p(s, a^{n+1}) = (s + a_0)(s^n + a_{n-1}^n s^{n-1} + \cdots + a_0^n)$). Obtain a similar result for the case where n is odd.

4.1.6 Notes and References

The continuous dependence of the roots of a polynomial on its coefficient vector is proved as an application of Rouché's Theorem in most textbooks on function theory. The study of the roots of a polynomial whose coefficients depend on a *single complex parameter* is a central issue of Complex Analysis, see the chapters on multivalued algebraic functions in classical textbooks such as *Knopp* (1945) [311] and *Ahlfors* (1979) [6]. For a deeper analysis of algebraic functions the concept of a Riemann surface is indispensable, see *Weyl* (1955), *Ahlfors* (1979) [6], *Jones and Singerman* (1986) [279]. A concise review of polynomials with meromorphic coefficients from the viewpoint of Perturbation Theory can be found in the appendix of *Baumgärtel* (1985) [43]. A classical reference for the geometry of the zeros of polynomials is *Marden* (1949) [359].

Newton et al. (1957) [391] discuss the design of filters via the Butterworth pattern. A brief account can be found in the Signal Processing Toolbox of MATLAB. Proposition 4.1.3 also plays a role in *cheap control*, i.e. optimal control problems where the control costs are very small, see *Francis* (1979) [165]. Numerical aspects of root computations are discussed in *Householder* (1970) [268], *Wilkinson* (1965) [524], *Stoer and Bulirsch* (1993) [486], *Golub and Van Loan* (1996) [197]. Root sensitivities have been used for the analysis of stability robustness by *Cruz et al.* (1981) [113]. However, we emphasize that sensitivities give valuable information only for small parameter variations. They can be quite misleading as indices of the robustness of stability, see Section 5.2.

The characterization of the boundary and convex hull of the set of Schur polynomials is due to *Fam and Meditch* (1978) [157]. A systematic description of the relationship between Hurwitz and Schur polynomials (cf. Ex. 16 and Ex. 17) can be found in *Duffin* (1969) [139]. The boundary of the set of Hurwitz polynomials has been analyzed from a geometric point of view by *Levantovskii* (1980) [340], a more transparent proof of his formulas can be found in *Burke et al.* (2004) [83]. The volume of the set of Schur polynomials has been studied and related to the volume of its convex hull by *Fam* (1989) [156]. Proposition 4.1.26 has also been used in the robustness analysis of polynomials and systems, see *Soh et al.* (1985) [470], *Biernacki et al.* (1987) [59]. The interval polynomial problem was first posed by *Faedo* (1953) [155] who obtained some necessary and some sufficient conditions using the Routh-Hurwitz conditions. The elegant result obtained by *Kharitonov* was published in 1978, see [300]. Its significance for control theory was only discovered several years later and generated a flurry of papers presenting new proofs and various generalizations and applications, see the conference proceedings [374] edited by *Milanese et al.* (1988) , the survey *Barmish* (1988) [35], the collection of papers [134] edited by *Dorato and Yedavalli* (1990), and the books by *Barmish (1994)* [36] and *Bhattacharyya et al.* (1995) [56]. Our exposition follows the proof given by *Minichelli et al.* (1989) [376]. As mentioned in Remark 4.1.33 Kharitonov's Theorem can easily be extended to the non-monic case provided the degree remains invariant over the interval for the leading coefficient. *Mori and Kokame* (1992) [383] dealt with the modifications required to extend Kharitonov's Theorem to the case where the degree can fall. A detailed account of the problem of robustness under parametric uncertainty can be found in *Bhattacharyya et al.* (1995) [56]. This book also contains some Kharitonov type results for uncertain polynomials with structured perturbations.

4.2 Perturbation of Matrices

In this section we analyze the effect of perturbations of a matrix on its eigenstructure. There are two basic approaches towards linear perturbation theory. *Classical* perturbation theory has emerged from Mathematical Physics and was developed in an infinite dimensional operator theoretic context. The focus is on qualitative issues like continuity, differentiability and analyticity of eigenvalues and eigenbases. It is mainly concerned with *highly structured* perturbations and most of the results deal with operators depending analytically on *one* complex parameter. *Quantitative perturbation theory* has its origins in Numerical Analysis and is mainly concerned with *unstructured* perturbations of matrices. The aim is to derive tight *bounds* on the variation of eigenvalues and eigenvectors, growth rates, etc. in terms of a norm on the perturbation. On the boundary between these two fields lies *sensitivity analysis* which is concerned with changes of the eigenstructure under infinitesimal parameter variations.

In the first subsection we begin with qualitative aspects of the behaviour of eigenvalues under independent perturbations of all the matrix entries. We use the results of the previous section to discuss the continuity and analyticity of the eigenvalues and also determine the sensitivity of simple eigenvalues with respect to infinitesimal changes of a single parameter.

In the second subsection we study some quantitative issues. Estimates for the change of eigenvalues under *bounded* unstructured parameter variations are derived. These are used to prove Gershgorin's Theorem which provides estimates for the location of the spectrum of a matrix relative to its diagonal entries.

In the third subsection we continue the qualitative studies of the first by investigating the behaviour of eigenprojections and eigenvectors of a holomorphic matrix family $(A(z))_{z\in D}$. A central question will be whether or not the eigenprojections and eigenvectors of $A(z)$ can be expressed as analytic functions of z. The results we obtain will allow us to prove some deeper analyticity results for eigenvalues in the normal case. Finally, if $A(z_0)$ has n different eigenvalues, we obtain an explicit formula for the sensitivity of the eigenvectors of $A(z)$ under small variations of z around z_0.

4.2.1 Continuity and Analyticity of Eigenvalues

As usual we identify a linear operator $A \in \mathcal{L}(\mathbb{C}^n)$ with its matrix representation with respect to the standard basis of $\mathbb{C}^n$. We assume that $\mathbb{C}^{n\times n} \cong \mathcal{L}(\mathbb{C}^n)$ is endowed with an operator norm $\|\cdot\| = \|\cdot\|_{\mathcal{L}(\mathbb{C}^n)}$. For any $A \in \mathbb{C}^{n\times n}$ we denote by $\Lambda(A)$ the unordered n-tuple of the eigenvalues of A taking into account multiplicities. If $\mathcal{A} \subset \mathbb{C}^{n\times n}$ is a set of matrices we set $\Lambda(\mathcal{A}) = \{\Lambda(A);\ A \in \mathcal{A}\}$ and define the *spectrum of* $\mathcal{A}$ by

$$\sigma(\mathcal{A}) = \bigcup_{A\in\mathcal{A}} \sigma(A).$$

Since the coefficients of the characteristic polynomial $\chi_A(s) = \det(sI - A)$ are polynomial functions of the entries of A the following corollaries are immediate consequences of the continuity and analyticity results in the previous section. Corollary 4.1.5 implies

Corollary 4.2.1. *The map $A \mapsto \Lambda(A)$ is continuous on $\mathbb{C}^{n\times n}$, i.e. for any given $A_0 \in \mathbb{C}^{n\times n}$, $\varepsilon > 0$ there exists $\delta > 0$ such that*

$$\|A - A_0\| < \delta \Rightarrow \operatorname{dist}(\Lambda(A), \Lambda(A_0)) < \varepsilon$$

where the distance between n-tuples is defined by (1.3).

By Proposition 4.1.10 and Lemma 4.1.9 we have

Corollary 4.2.2. *For every compact set of matrices $\mathcal{A} \in \mathcal{K}(\mathbb{C}^{n\times n})$ the associated spectrum $\sigma(\mathcal{A})$ is compact and the maps $\mathcal{A} \mapsto \sigma(\mathcal{A})$, $\mathcal{A} \mapsto \Lambda(\mathcal{A})$ from $\mathcal{K}(\mathbb{C}^{n\times n})$ into $\mathcal{K}(\mathbb{C})$ and $\mathcal{K}(\mathcal{T}_n(\mathbb{C}))$, respectively, are continuous (with respect to the Hausdorff metrics).*

Theorem 4.1.11 implies

Corollary 4.2.3. *If $\lambda_0 \in \mathbb{C}$ is a simple eigenvalue of a matrix $A_0 \in \mathbb{C}^{n\times n}$, then for any $\varepsilon > 0$ sufficiently small there exists $\delta = \delta(\varepsilon) > 0$ such that all matrices $A \in B(A_0, \delta)$ have exactly one simple eigenvalue $\lambda(A)$ in $D(\lambda_0, \varepsilon)$, $\lambda(A_0) = \lambda_0$ and $\lambda(A)$ depends analytically on the entries of A in $B(A_0, \delta)$.*

Note that Corollary 4.2.1 does *not* mean that the eigenvalues of $A \in \mathbb{C}^{n\times n}$ can be represented as continuous functions of A (see Example 4.1.6). However, if the entries of A depend continuously on one *real* parameter τ we have the following consequence of Proposition 4.1.19.

Corollary 4.2.4. *Suppose that $I \subset \mathbb{R}$ is an interval and $A(\tau) \in \mathbb{C}^{n\times n}$ depends continuously on $\tau \in I$. Then there exist (single valued) continuous functions $\lambda_i : I \mapsto \mathbb{C}$, $i \in \underline{n}$ such that*

$$\Lambda(A(\tau)) = (\lambda_1(\tau), \dots, \lambda_n(\tau)), \qquad \tau \in I.$$

Remark 4.2.5. If $A(\tau)$ is differentiable in τ on an interval $I \subset \mathbb{R}$ and *$A(\tau)$ is diagonalizable for all $\tau \in I$* then by a theorem of *Kato* (1980) [293, Thm. II.5.6] the $\lambda_i(\cdot)$ in Corollary 4.2.4 can be chosen to be differentiable. □

The following two examples show that diagonalizability of $A(\tau)$ is needed in order to ensure differentiability and that Kato's Theorem *cannot* be generalized to matrices depending on *two* real parameters.

Example 4.2.6. The real analytic matrix

$$A(\tau) = \begin{bmatrix} 0 & 1 \\ \tau & 0 \end{bmatrix}, \qquad \tau \in \mathbb{R}$$

has eigenvalues $\lambda_{\pm}(\tau) = \pm\sqrt{\tau}$ which are continuous on $\mathbb{R}$ but *not* differentiable at $\tau = 0$ (where $A(\tau)$ is not diagonalizable). □

Example 4.2.7. Consider the family of real symmetric (hence diagonalizable) matrices

$$A(\tau_1, \tau_2) = \begin{bmatrix} \tau_1 & \tau_2 \\ \tau_2 & -\tau_1 \end{bmatrix}, \qquad \tau_1, \tau_2 \in \mathbb{R}$$

which is analytic on $\mathbb{R}^2$. The eigenvalues $\lambda_{\pm}(A(\tau_1, \tau_2)) = \pm(\tau_1^2 + \tau_2^2)^{1/2}$ are not differentiable at $(0, 0)$. □

Now suppose that the entries of A depend analytically on one *complex* parameter $z \in \Omega$ where Ω is a domain in $\mathbb{C}$. By the results of Section 4.1, problems with analyticity of the eigenvalues can only arise at exceptional parameter values $z_0 \in \Omega$ for which the number of distinct eigenvalues decreases. $z_0 \in \Omega$ is said to be a critical point for $A(\cdot) = (A(z))_{z\in\Omega}$ if it is a *critical point* for the characteristic polynomial $\chi_{A(z)}(s) = \det(sI - A(z))$ in the sense of Definition 4.1.13. By Theorem 4.1.14 we have

Corollary 4.2.8. *If $A(\cdot) : \Omega \mapsto \mathbb{C}^{n\times n}$ is analytic on a domain $\Omega \subset \mathbb{C}$, then the following hold.*

(i) *The set C_A of critical points of $A(\cdot)$ is locally finite.*

(ii) *In every simply connected domain $D \subset \Omega \setminus C_A$ there exist analytic functions $\lambda_i(\cdot) \in \mathcal{O}(D)$, $i \in \underline{n}$ (not necessarily all distinct) such that*

$$\Lambda(A(z)) = (\lambda_1(z), \ldots, \lambda_n(z)), \quad z \in D.$$

(iii) *The multiplicity of each eigenvalue $\lambda_i(z)$ is constant on every simply connected domain D as in (ii).*

It remains to examine the behaviour of the eigenvalues of $A(z)$ in the neighbourhood of a critical point $z_0 \in C_A$. Let $\lambda_1, \ldots, \lambda_\ell$ be the distinct eigenvalues of $A(z_0)$ with algebraic multiplicities $m_1, \ldots, m_\ell$, $\sum_{j=1}^{\ell} m_j = n$. Suppose that $\Gamma_j, j \in \underline{\ell}$ are positively oriented non-overlapping circles around each λ_j (see Figure 4.2.1), and let $\delta > 0$ be sufficiently small so that $D(z_0, \delta) \cap C_A = \{z_0\}$ and every $A(z)$, $z \in D(z_0, \delta)$ has exactly m_j eigenvalues (accounting for multiplicities) in the disk surrounded by Γ_j (Corollary 4.2.1). The number n_j of distinct eigenvalues of $A(z)$ enclosed by Γ_j is the same throughout the punctured disk $D^\circ(z_0, \delta) := D(z_0, \delta) \setminus \{z_0\}$. We say that the eigenvalue λ_j of $A(z_0)$ *splits* into these n_j eigenvalues as z moves away from z_0. The set of these eigenvalues (or the unordered m_j-tuple of eigenvalues if we want to account for multiplicities) will be called the *λ_j-group of eigenvalues of $A(z)$ near z_0*. In every simply connected domain $D \subset D^\circ(z_0, \delta)$ (e.g. every disk in $D^\circ(z_0, \delta)$) these distinct eigenvalues can be represented by analytic functions $\lambda_{j1}(\cdot), \ldots, \lambda_{jn_j}(\cdot)$ on D and the multiplicity μ_{jk} of each eigenvalue $\lambda_{jk}(z)$ is constant throughout D with $\sum_{k=1}^{n_j} \mu_{jk} = m_j$ (Corollary 4.2.8). As a consequence we obtain the following factorization of the characteristic polynomial of $A(z)$.

$$\chi_{A(z)}(s) = \prod_{j=1}^{\ell} \prod_{k=1}^{n_j} (s - \lambda_{jk}(z))^{\mu_{jk}}, \quad z \in D, \quad \lim_{z\to z_0} \lambda_{jk}(z) = \lambda_j, \; k \in \underline{n_j}, \; j \in \underline{\ell}. \quad (1)$$

A more precise picture of the splitting of λ_j is obtained by analyzing how the $\lambda_{jk}(\cdot) \in \mathcal{O}(D)$ are linked by analytic continuation along arcs in $D^\circ(z_0, \delta)$, see Section 4.1. By analytic continuation along small circles around z_0 the n_j eigenvalues form one or more cycles of the form $(\lambda_{jk_1}(\cdot), \ldots, \lambda_{jk_q}(\cdot))$ where all the eigenvalues in a given cycle have the same constant multiplicity on $D^\circ(z_0, \delta)$. $\lambda_j = \lim_{z\to z_0} \lambda_{jk_i}(z)$ is called *the centre* of these cycles. For all the cycles obtained in this way we obtain the following corollary of Theorem 4.1.16.

Corollary 4.2.9. *Given an analytic matrix function $A(\cdot) : \Omega \mapsto \mathbb{C}^{n\times n}$ on a domain $\Omega \subset \mathbb{C}$, let $D(z_0, r) \subset \Omega$ be a disk with centre $z_0 \in C_A$ such that $D(z_0, r) \cap C_A = \{z_0\}$, and suppose that $(\lambda_1(\cdot), \ldots, \lambda_q(\cdot))$ is one of the cycles obtained by analytic continuation of an eigenvalue element $\lambda_1(z)$ of $A(z)$ along a circular path around z_0 in the punctured disk $D^\circ(z_0, r)$. Then the following hold.*

(i) If $q = 1$, $\lambda_1(\cdot)$ can be continued analytically onto $D(z_0, r)$.

(ii) If $q \geq 2$, $(\lambda_1(z), \ldots, \lambda_q(z))$ defines a q-valued analytic function on $D^\circ(z_0, r)$ with branch point z_0. The branches of this function are represented by a Puiseux series of the form

$$\lambda_\nu(z) = \sum_{k=0}^{\infty} \alpha_k w^{\nu k} (z - z_0)^{k/q}\,, \quad z \in D^-(z_0, r), \;\; \nu = 1, \cdots, q \qquad (2)$$

where $w = e^{2\pi\imath/q}$ and $(z - z_0)^{1/q} = \rho^{1/q} e^{\imath\theta/q}$ for $z = z_0 + \rho e^{\imath\theta}$, $0 < \theta < 2\pi$. In particular $\lim_{z\to z_0} \lambda_\nu(z) = \alpha_0$ for $\nu \in \underline{q}$ where α_0 (the centre of the cycle) is an eigenvalue of $A(z_0)$.

The whole λ_j-group of eigenvalues of $A(z)$ near $z = z_0$ consists of a set of q_i-valued analytic functions f_i, $i = 1, \ldots, h_j$, each one being associated with a cycle $(\lambda_{jk}(\cdot))_{k\in K_i}$ where $q_i = |K_i|$ and $K_i, i \in \underline{h_j}$ forms a partition of $\underline{n_j}$.
Each eigenvalue element in the λ_j-group belongs to exactly one of these cycles. If m_{ji} is the multiplicity of the cycle $(\lambda_{jk}(\cdot))_{k\in K_i}$ (i.e. the algebraic multiplicity μ_{jk} of its elements $\lambda_{jk}(z)$, $k \in K_i$) then $\sum_{i=1}^{h_j} q_i m_{ji} = m_j$ is the algebraic multiplicity of the eigenvalue λ_j of $A(z_0)$. As z moves away from z_0 this eigenvalue splits into the h_j cycles of eigenvalues described by Puiseux series of the form (2) with $\alpha_0 = \lambda_j$. Figure 4.2.1 illustrates this splitting process (as z moves away from z_0 along a given ray $z_0 + tz_1$, $t > 0$).

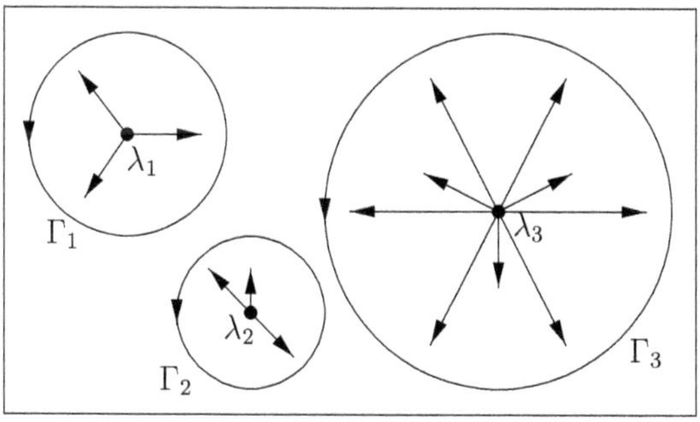

Figure 4.2.1: Splitting of eigenvalues at a critical parameter value z_0

Remark 4.2.10. (i) It may happen that, for a given $j \in \underline{\ell}$, all the cycles with centre λ_j are of period $q = 1$. In this case all the n_j distinct eigenvalues of $A(z)$ enclosed by Γ_j can be represented by analytic functions on the whole disk $D(z_0, \delta)$, and the critical point z_0 will *not* be a branching point for the λ_j-group of eigenvalues of $A(z)$.

(ii) If $(\lambda_1(\cdot), \dots, \lambda_q(\cdot))$ is any cycle of eigenvalues of $A(z)$ near $z = z_0$ and $f(z)$ the associated q-valued analytic function on $D^\circ(z_0, r)$, then the product

$$p_{f(z)}(s) = \prod_{k=1}^{q} (s - \lambda_k(z)) =: p(s, a(z)), \quad 0 < |z - z_0| < r$$

is invariant with respect to analytic continuation along arcs in the punctured disk $D^\circ(z_0, r)$. Therefore the coefficients $a_i(z)$ of this monic polynomial in s can be extended to analytic functions on $D(z_0, r)$. In particular, $a_{n-1}(z)$ and $a_0(z)$, hence the sum and the product of the eigenvalues $\lambda_k(z)$, $k \in \underline{q}$ are analytic on $D(z_0, r)$.

(iii) Given any eigenvalue cycle $(\lambda_1(\cdot), \dots, \lambda_q(\cdot))$ with centre $\lambda \in \sigma(A(z_0))$, the expansion (2) shows that $|\lambda_k(z) - \lambda|$, $k \in \underline{q}$ is, in general, of the order $|z - z_0|^{1/q}$ for small deviations $|z - z_0|$. Therefore, if the cycle's period is $q \geq 2$, the rate of change $|\lambda_k(z) - \lambda|/|z - z_0|$ tends to ∞ as $z \to z_0$, for all $k \in \underline{q}$. Moreover, the larger the period of the cycle, the faster, in general, the associated eigenvalues move away from λ (as illustrated in Figure 4.2.1).

(iv) It may happen that all the eigenvalue elements $\lambda_{jk}(\cdot)$ in a λ_j-group of $A(z)$ at z_0 are constant. In this case λ_j will be an isolated point in $\bigcup_{|z-z_0|<\delta} \sigma(A(z))$, for δ sufficiently small. On the other hand, it follows from Remark 4.1.17 that $\bigcup_{|z-z_0|<\delta} \sigma(A(z))$ is an open neighbourhood of λ_j if the λ_j-group of $A(\cdot)$ near the critical point z_0 contains at least one non-constant analytic function or at least one cycle of period ≥ 2. □

The following example illustrates the previous two corollaries.

Example 4.2.11. (i) Consider $A(z) = \begin{bmatrix} 0 & z \\ z & 0 \end{bmatrix}$, $z \in \mathbb{C}$. $z_0 = 0$ is the only critical point of $A(\cdot)$ and the eigenvalues $\lambda_\pm(z) = \pm z$ are represented by functions which are analytic in the whole complex plane including the critical point. Note that $A(z)$ is normal for all $z \in \mathbb{C}$.

(ii) Let $A(z) = \begin{bmatrix} 0 & 1 \\ z & 0 \end{bmatrix}$. Again $z_0 = 0$ is the only critical parameter value, but here the eigenvalues $\lambda_\pm(z) = \pm z^{1/2}$ constitute one double-valued algebraic function. Contrary to the real case (Example 4.2.6) they cannot be represented by two continuous functions of z in a neighbourhood of $z_0 = 0$. However, in every simply connected domain $D \subset \mathbb{C} \setminus \{0\}$ there exist two analytic functions $\lambda_i(\cdot)$, $i = 1, 2$ such that $\Lambda(A(z)) = (\lambda_1(z), \lambda_2(z))$, $z \in D$.

(iii) Let $A(z) = \begin{bmatrix} 1 & z \\ z & -1 \end{bmatrix}$. The eigenvalues are $\lambda_\pm(z) = \pm(1 + z^2)^{1/2}$ and so the critical set is $C_A = \{\imath, -\imath\}$. Around these critical points the eigenvalues $\lambda_\pm(\cdot)$ are branches of one double-valued algebraic function. The Puiseux series for z near $\imath$ is

$$(2\imath)^{1/2}(z - \imath)^{1/2} + \sum_{k=1}^{\infty} \frac{\frac{1}{2}(\frac{1}{2} - 1) \cdots (\frac{1}{2} - k + 1)}{k!\,(2\imath)^{(k-\frac{1}{2})}} (z - \imath)^{(k+\frac{1}{2})},$$

see Example 4.1.18. □

We have seen in Remark 4.2.10 that, given an eigenvalue element $\lambda(z)$ which generates a cycle of period ≥ 2 and centre λ_j^0 in a small neighbourhood of a critical point z_0 of $A(\cdot)$, the deviation $|\lambda(z) - \lambda_j^0|$ gets infinitely large compared with $|z - z_0|$ as $z \to z_0$. We will now consider the case where an eigenvalue $\lambda(z)$ of $A(z)$ depends in a *differentiable way* on the parameter z at $z = z_0$. Then its *sensitivity* (or rate of change) of $\lambda(z)$ at the point $z = z_0$ is defined by

$$\lambda'(z_0) = \frac{d\lambda}{dz}(z_0)\ .$$

A formula for the sensitivity is given in the next proposition.

Proposition 4.2.12. *Let Ω be an open subset of $\mathbb{R}$ or $\mathbb{C}$ and suppose that $A(\cdot) : \Omega \to \mathbb{C}^{n\times n}$ is differentiable at $z = z_0$ and λ_0 is an eigenvalue of $A(z_0)$ with corresponding left and right eigenvectors w, v such that $w^*v \neq 0$. Suppose $\lambda(\cdot) : \Omega \to \mathbb{C}$, $v(\cdot) : \Omega \to \mathbb{C}^n$ are differentiable at z_0 and satisfy*

$$A(z)v(z) = \lambda(z)v(z)\ , \qquad \lambda(z_0) = \lambda_0\ ,\ v(z_0) = v. \tag{3}$$

Then

$$\lambda'(z_0) = \frac{w^*A'(z_0)v}{w^*v}, \quad \textit{where} \quad A'(z_0) = \frac{dA}{dz}(z_0). \tag{4}$$

Proof: Differentiating the first equation in (3) at z_0 we obtain

$$(\lambda_0 I - A(z_0))v'(z_0) + (\lambda'(z_0)I - A'(z_0))v = 0$$

and multiplying this equation from the left by w^* yields (4). □

If $A(z)$ is analytic at z_0, $\lambda_0 \in \mathbb{C}$ is a simple eigenvalue of $A(z_0)$ and $D(\lambda_0, \varepsilon)$ is a disk whose closure does not contain any other eigenvalue of $A(z_0)$, then for $|z - z_0|$ sufficiently small there exists a unique simple eigenvalue $\lambda(z)$ of $A(z)$ in $D(\lambda_0, \varepsilon)$. $\lambda(z)$ depends analytically on z (Corollary 4.2.3) and since $\lambda(z)I - A(z)$ is of rank $n - 1$ for $|z - z_0|$ sufficiently small, it is easy to see (and will be proved later in a more general context, see Subsection 4.2.3) that there exists an analytic vector $v(z)$ such that (3) is satisfied in a neighbourhood of z_0.
In order to discuss (4) let us suppose $\|v\| = \|w\| = 1$, then

$$|\lambda'(z_0)| \leq \frac{1}{|w^*v|}\|A'(z_0)\|. \tag{5}$$

For this reason $|w^*v|^{-1}$ is sometimes known as the *condition number of the eigenvalue* λ_0. Note that, if w, v are real then w^*v is the cosine of the angle between w and v. If $A(z_0)$ is normal, one may choose $w = v$, see [264, 2.5]. So

$$\lambda'(z_0) \leq \|A'(z_0)\|$$

and this shows that, ceteris paribus, normal matrices have minimal eigenvalue sensitivity.

Example 4.2.13. Let

$$A_0 = \begin{bmatrix} 3 & 4 \\ 4 & -3 \end{bmatrix}, \quad \tilde{A}_0 = \begin{bmatrix} 1 & 1 \\ 0 & 0.01 \end{bmatrix}^{-1} A_0 \begin{bmatrix} 1 & 1 \\ 0 & 0.01 \end{bmatrix}, \quad A_1 = \begin{bmatrix} 0 & 0 \\ 1 & 0 \end{bmatrix},$$

and define $A(\tau) = A_0 + \tau A_1$, $\tilde{A}(\tau) = \tilde{A}_0 + \tau A_1$, $\tau \in \Omega = \mathbb{R}$. A_0 and $\tilde{A}_0$ are similar with eigenvalues ± 5, A_0 is normal. Normalized left and right eigenvectors for $\lambda_0 = 5$ are, respectively,

$$w = v = (-0.8944,\, -0.4472)^\top \quad , \quad \tilde{w} = (0.7053,\, 0.7089)^\top, \quad \tilde{v} = (-0.7,\, 0.7142)^\top$$

with $w^* v = 1$, $\tilde{w}^* \tilde{v} = 0.0126$. Let $\lambda(\tau)$, $\tilde{\lambda}(\tau)$ be the analytic eigenvalues of $A(\tau)$ and $\tilde{A}(\tau)$ in a small neighbourhood of $\tau_0 = 0$ satisfying $\lambda(0) = \tilde{\lambda}(0) = 5$. Then the sensitivities of these eigenvalues at $\tau_0 = 0$ are

$$\lambda'(0) = \frac{w^* A_1 v}{w^* v} = 0.400\,, \quad \tilde{\lambda}'(0) = \frac{\tilde{w}^* A_1 \tilde{v}}{\tilde{w}^* \tilde{v}} = -39.396\,.$$

□

4.2.2 Estimates for Eigenvalues and Growth Rates

We now turn from the qualitative analysis of the previous subsection to the *quantitative analysis* of eigenvalue changes under matrix perturbations. In Chapter 5 we will introduce the set of all eigenvalues which can be obtained from a given matrix via additive (real or complex) perturbations of size less than a given number. We will show how these sets can be characterized and we will see that the computation of the sets from these characterizations can be difficult. Here we derive some easy and useful upper bounds for the change of eigenvalues and growth rates under arbitrary matrix perturbations. The bounds are then used to prove Gershgorin's Theorem.
The estimates we derive will be expressed in terms of the norm of the perturbation matrix, so our results will depend upon the particular norm $\|\cdot\|$ which is chosen on $\mathbb{C}^{n\times n}$. The following lemma holds for a large class of norm including all operator norms. We suppose that $\|\cdot\|$ is *sub-multiplicative*, i.e. $\|XY\| \le \|X\|\,\|Y\|$ for $X, Y \in \mathbb{C}^{n\times n}$.

Lemma 4.2.14. *Let $A \in \mathbb{C}^{n\times n}$, $T \in \mathbf{Gl}_n(\mathbb{C})$, and $\Delta \in \mathbb{C}^{n\times n}$ be arbitrary. Then for every $\mu \in \sigma(A+\Delta) \setminus \sigma(A)$ and every sub-multiplicative norm $\|\cdot\|$ on $\mathbb{C}^{n\times n}$*

$$\|T^{-1}(\mu I_n - A)^{-1} T\|^{-1} \le \|T^{-1} \Delta T\|. \tag{6}$$

Proof: For $\mu \in \sigma(A+\Delta) \setminus \sigma(A)$ we have

$$T^{-1}\left[\mu I_n - (A+\Delta)\right] T = T^{-1}(\mu I_n - A) T \left[I_n - T^{-1}(\mu I_n - A)^{-1} T\,(T^{-1}\Delta T)\right].$$

But the RHS cannot be singular if $\|(T^{-1}(\mu I_n - A)^{-1} T\,(T^{-1}\Delta T)\| < 1$. Hence

$$1 \le \|(T^{-1}(\mu I_n - A)^{-1} T\,(T^{-1}\Delta T)\| \le \|(T^{-1}(\mu I_n - A)^{-1} T\|\,\|T^{-1}\Delta T\| \tag{7}$$

and this proves the lemma. □

Specializing to diagonalizable matrices and a particular class of operator norms we get the following result which is called the lemma of Bauer and Fike in the literature. A norm $\|\cdot\|_{\mathbb{C}^n}$ on $\mathbb{C}^n$ is said to be *absolute* if it satisfies $\|\,|x|\,\|_{\mathbb{C}^n} = \|x\|_{\mathbb{C}^n}$ for all $x \in \mathbb{C}^n$, see Section A.1.

Lemma 4.2.15 (Bauer-Fike). *Let $A \in \mathbb{C}^{n\times n}$ be diagonalizable with $A = TDT^{-1}$, $T \in \mathbf{Gl}_n(\mathbb{C})$, D diagonal and $\Delta \in \mathbb{C}^{n\times n}$ be arbitrary. If $\mu \in \sigma(A+\Delta)$, then*

$$\operatorname{dist}(\mu, \sigma(A)) = \min_{\lambda\in\sigma(A)} |\lambda - \mu| \leq \|T\|\,\|T^{-1}\|\,\|\Delta\| \tag{8}$$

where $\|\cdot\|$ is any operator norm on $\mathbb{C}^{n\times n}$ induced by an absolute norm on $\mathbb{C}^n$.

Proof: Let $\mu \in \sigma(A+\Delta)$. We need only consider the case $\mu \notin \sigma(A) = \sigma(D)$. Since $T^{-1}(\mu I_n - A)^{-1}T = (\mu I_n - T^{-1}AT)^{-1} = (\mu I_n - D)^{-1}$ we obtain from (6)

$$\|\operatorname{diag}((\mu - d_1)^{-1}, ..., (\mu - d_n)^{-1})\|^{-1} = \|(\mu I_n - D)^{-1}\|^{-1} \leq \|T^{-1}\|\,\|\Delta\|\,\|T\|.$$

But $\|\operatorname{diag}((\mu - d_1)^{-1}, ..., (\mu - d_n)^{-1})\| = \max_{i\in\underline{n}} |(\mu - d_i)^{-1}| = [\min_{i\in\underline{n}} |\mu - d_i|]^{-1}$ holds for all operator norms induced by absolute norms (see Theorem A.1.9) and so (8) follows. □

If A is normal, then it is possible to choose T to be unitary, so that $\|T^{-1}\|_{2,2} = \|T^*\|_{2,2} = \|T\|_{2,2} = 1$, and we obtain

Corollary 4.2.16. *Let $A \in \mathbb{C}^{n\times n}$ be normal and $\Delta \in \mathbb{C}^{n\times n}$ be arbitrary. If $\mu \in \sigma(A+\Delta)$ and $\|\cdot\|_{2,2}$ denotes the spectral norm, then*

$$\min_{\lambda\in\sigma(A)} |\lambda - \mu| \leq \|\Delta\|_{2,2}\,. \tag{9}$$

In the general diagonalizable case the column vectors of T may be chosen to form a basis of normalized eigenvectors of A. If the angle between two of these eigenvectors is small (the eigenframe is not well spread out) then $\|T^{-1}\|$ will be large. The estimate (8) suggests that in this case small perturbations of A may result in large variations of the eigenvalues. The following example illustrates this.

Example 4.2.17. The matrix $A(\alpha) = \begin{bmatrix} -1 & \alpha \\ 0 & -2 \end{bmatrix}$, $\alpha \in \mathbb{R}_+$ has normalized eigenvectors $(1\,,\,0)^\top$ and $(\alpha/(1+\alpha^2)^{1/2}\,,\,-1/(1+\alpha^2)^{1/2})^\top$. If $\theta = \theta(\alpha) \in [0, \pi/2]$ is the angle between them, then $\cos\theta = \alpha/(1+\alpha^2)^{1/2}$, $\sin\theta = 1/(1+\alpha^2)^{1/2}$ and for $T = \begin{bmatrix} 1 & \cos\theta \\ 0 & -\sin\theta \end{bmatrix}$ an easy calculation yields $T^{-1} = (\sin\theta)^{-1}\begin{bmatrix} \sin\theta & \cos\theta \\ 0 & -1 \end{bmatrix}$ and

$$T^{-1}A(\alpha)T = \operatorname{diag}(-1, -2), \quad \|T\|_{2,2}^2 = 1 + \cos\theta\,, \quad \|T^{-1}\|_{2,2} = \|T\|_{2,2}/\sin\theta\,.$$

Hence for any $\mu \in \sigma(A(\alpha) + \Delta)$ the estimate (8) (with respect to $\|\cdot\| = \|\cdot\|_{2,2}$) is

$$\operatorname{dist}(\mu, \sigma(A)) \leq \frac{(1+\cos\theta)}{\sin\theta}\|\Delta\|_{2,2} = \left[(1+\alpha^2)^{1/2} + \alpha\right]\|\Delta\|_{2,2} \tag{10}$$

which indicates there may be large variations in the spectrum if θ is small (α large). In fact, consider $\Delta(\varepsilon) = \begin{bmatrix} 0 & 0 \\ \varepsilon & 0 \end{bmatrix}$, $\varepsilon > 0$, then $\mu = [-3 - (1 + 4\varepsilon\alpha)^{1/2}]/2 \in \sigma(A(\alpha) + \Delta(\varepsilon))$ and

$$\operatorname{dist}(\mu, \sigma(A)) = |\mu + 2| = |1 - (1 + 4\varepsilon\alpha)^{1/2}|/2\,. \tag{11}$$

So the spectrum changes considerably under the perturbation $\Delta(\varepsilon)$ if α is large, even for small ε. Note that as $\varepsilon\alpha \to 0$ the RHS of (11) is of order $\varepsilon\alpha$ whereas for large α the RHS of (10) is of order $2\varepsilon\alpha$. Thus, if in the present example $\varepsilon = \alpha^{-2}$ and $\alpha \to \infty$ then the upper bound (8) is "asymptotically tight" modulo a factor of 2. In general, however, this upper bound can be very conservative. □

In Chapter 3 we introduced the growth rate $\omega(A)$ of a continuous time semigroup generated by $A \in \mathbb{C}^{n\times n}$ and showed that $\omega(A) = \sup_{\lambda\in\sigma(A)} \operatorname{Re}\lambda$. If A is diagonalizable, $A = T\operatorname{diag}(\lambda_1, \ldots, \lambda_n)T^{-1}$, we can apply Lemma 4.2.15 to obtain for the growth rate $\omega(A+\Delta)$ of the perturbed matrix $A+\Delta$, $\Delta \in \mathbb{C}^{n\times n}$ the upper bound

$$\omega(A+\Delta) \le \omega(A) + \|T\|\|T^{-1}\|\|\Delta\|.$$

In fact the following stricter perturbation result is valid for arbitrary matrices A.

Proposition 4.2.18. *Suppose $A \in \mathbb{C}^{n\times n}$ and*

$$\|e^{At}\| \le M\, e^{\alpha t}, \quad t \ge 0 \tag{12}$$

where $\alpha \in \mathbb{R}$ and $\|\cdot\|$ is any operator norm on $\mathbb{C}^{n\times n}$. Then for arbitrary $\Delta \in \mathbb{C}^{n\times n}$

$$\|e^{(A+\Delta)t}\| \le M\, e^{(\alpha+M\,\|\Delta\|)\,t}, \quad t \ge 0. \tag{13}$$

Proof: Consider the initial value problem

$$\dot{x} = (A+\Delta)x\,, \quad x(0) = x^0.$$

By the variation-of-constants formula the solution $x(\cdot)$ satisfies

$$x(t) = e^{At}x^0 + \int_0^t e^{A(t-s)}\Delta\, x(s)\, ds, \quad t \ge 0.$$

Hence

$$\|x(t)\|_{\mathbb{C}^n} \le M\, e^{\alpha t}\|x^0\|_{\mathbb{C}^n} + \int_0^t M\, e^{\alpha(t-s)}\|\Delta\|\,\|x(s)\|_{\mathbb{C}^n}\, ds.$$

Let $\gamma(t) = e^{-\alpha t}\|x(t)\|_{\mathbb{C}^n}$, then

$$\gamma(t) \le M\,\|x^0\|_{\mathbb{C}^n} + \int_0^t M\,\|\Delta\|\,\gamma(s)\, ds, \quad t \ge 0.$$

So by Gronwall's Lemma 2.1.18

$$\gamma(t) = e^{-\alpha t}\|e^{(A+\Delta)t}x^0\|_{\mathbb{C}^n} \le Me^{M\,\|\Delta\|t}\,\|x_0\|_{\mathbb{C}^n}, \quad t \ge 0,\ x^0 \in \mathbb{C}^n$$

and this proves (13). □

A corresponding result can also be proved for discrete time systems (see Ex. 9). An immediate corollary of the above proposition is that if the system $\dot{x} = Ax$ is asymptotically stable, so that (12) holds for some M, α with $\alpha < 0$, then the perturbed system $\dot{x} = (A+\Delta)x$ will also be asymptotically stable if

$$\|\Delta\| < |\alpha|/M. \tag{14}$$

Since M in (12) may be decreased by increasing α, one can try to optimize the RHS of (14) with respect to α in order to determine the maximum allowable perturbation bound given by this result (not a particularly easy problem, see Ex. 7). But the optimal bound obtained in this way will, in general, still be conservative. In the next chapter we introduce a stability radius for stable matrices and obtain tight estimates.

As a final result in this subsection we consider a matrix $A = (a_{ij}) \in \mathbb{C}^{n\times n}$ as a perturbation of the diagonal matrix $A_d = \mathrm{diag}(a_{11}, \ldots, a_{nn})$ with the same diagonal as A. This will allow us to obtain approximate information about the location of the spectrum of A relative to its diagonal entries.

Theorem 4.2.19 (Gershgorin). *If $A = (a_{ij}) \in \mathbb{C}^{n\times n}$, then*

$$\sigma(A) \subset \left(\bigcup_{i=1}^{n} \overline{D(a_{ii}, \rho_i)}\right) \cap \left(\bigcup_{j=1}^{n} \overline{D(a_{jj}, \gamma_j)}\right) =: \mathcal{G}_A \tag{15}$$

where the radii of the closed disks $\overline{D(a_{ii}, \rho_i)}$ (resp. $\overline{D(a_{jj}, \gamma_j)}$), $i, j \in \underline{n}$ are given by

$$\rho_i = \sum_{j=1,\, j\neq i}^{n} |a_{ij}| \quad , \quad \left(\text{resp. } \gamma_j = \sum_{i=1,\, i\neq j}^{n} |a_{ij}|\right), \quad i, j \in \underline{n}. \tag{16}$$

Moreover, the number of eigenvalues of A in each connected component of $\mathcal{G}_A$ is equal to the number of a_{ii}'s within this component.

Proof: Let $t \subset [0, 1]$, $A(t) = A_d + t(A - A_d)$. By (7) with $A = A_d$, $\Delta = t(A - A_d)$, $T = I$ and $\|\cdot\| = \|\cdot\|_{\infty,\infty}$ (the operator norm with respect to the ∞-norm on $\mathbb{C}^n$), we obtain for all $\lambda \in \sigma(A(t)) \setminus \sigma(A_d)$

$$1 \le t\|(\lambda I - A_d)^{-1}(A - A_d)\|_{\infty,\infty} = \max_i \frac{t\rho_i}{|\lambda - a_{ii}|}.$$

The last equality holds since $\|B\|_{\infty,\infty} = \max_{i\in\underline{n}} \sum_{j=1}^{n} |b_{ij}|$ for $B = (b_{ij}) \in \mathbb{C}^{n\times n}$, see (A.1.3). With respect to the 1–norm on $\mathbb{C}^n$ we have analogously[1]

$$1 \le t\|(A - A_d)(\lambda I - A_d)^{-1}\|_{1,1} = \max_j \frac{t\gamma_j}{|\lambda - a_{jj}|},$$

for all $\lambda \in \sigma(A(t)) \setminus \sigma(A_d)$. Hence

$$\sigma(A(t)) \subset \left(\bigcup_{i=1}^{n} \overline{D(a_{ii}, t\rho_i)}\right) \cap \left(\bigcup_{j=1}^{n} \overline{D(a_{jj}, t\gamma_j)}\right), \quad t \in [0, 1]. \tag{17}$$

Setting $t = 1$ gives (15). Now suppose that $\mathcal{G}_A = S_1 \dot\cup \cdots \dot\cup S_m$ where the S_i, $i \in \underline{m}$ are the connected components of the Gershgorin set $\mathcal{G}_A$ (maximal connected subsets of $\mathcal{G}_A$). By Corollary 4.2.4 there exist continuous functions $\lambda_i(\cdot)$, $i \in \underline{n}$, on $[0, 1]$ such that $\lambda_i(0) = a_{ii}$ and $\Lambda(A(t)) = (\lambda_1(t), \ldots, \lambda_n(t))$. Since the set on the RHS of (17) is increasing with t, we have $\lambda_i(t) \in S_1 \dot\cup \cdots \dot\cup S_m$ for all $t \in [0, 1]$, $i \in \underline{n}$. But no curve $\lambda_i(t)$, $t \in [0, 1]$ can leave the connected component in which it starts. This concludes the proof. □

[1] making use of (A.1.3) and $1 \le \|(T^{-1}\Delta T)(\mu I_n - D)^{-1}\|$ instead of the first inequality in (7).

Note that the Gershgorin set changes under similarity transformations on A whereas the eigenvalues remain the same. Applying Gershgorin's Theorem to TAT^{-1}, $T \in \mathbf{Gl}_n(\mathbb{C})$ yields additional information about $\sigma(A)$. In particular, one may use scaling transformations $T = \mathrm{diag}(\alpha_1, \ldots, \alpha_n)$ (which leave the diagonal entries of A invariant) to obtain tighter bounds on the location of the eigenvalues, see Ex. 11 and *Notes and References.*

Example 4.2.20. Consider the matrix

$$A = \begin{bmatrix} 6 & 3 & -3 \\ 0 & 2 & 2 \\ -1 & -3 & -7 \end{bmatrix}.$$

Then $\rho_1 = 6$, $\rho_2 = 2$, $\rho_3 = 4$, $\gamma_1 = 1$, $\gamma_2 = 6$, $\gamma_3 = 5$ and the set on the RHS of (15) is given by

$$\begin{aligned} \mathcal{G}_A &= \{\lambda \in \mathbb{C}\,; |\lambda - 6| \le 6 \text{ or } |\lambda - 2| \le 2 \text{ or } |\lambda + 7| \le 4\} \cap \\ &\quad \{\lambda \in \mathbb{C}\,; |\lambda - 6| \le 1 \text{ or } |\lambda - 2| \le 6 \text{ or } |\lambda + 7| \le 5\} \\ &= \{\lambda \in \mathbb{C}\,; |\lambda - 6| \le 6 \text{ and } |\lambda - 2| \le 6\} \cup \{\lambda \in \mathbb{C}\,; |\lambda + 7| \le 4\} \end{aligned}$$

(see Fig. 4.2.2). Applying the scaling transformation $T = \mathrm{diag}(2/3, 2, 2)$, A is transformed

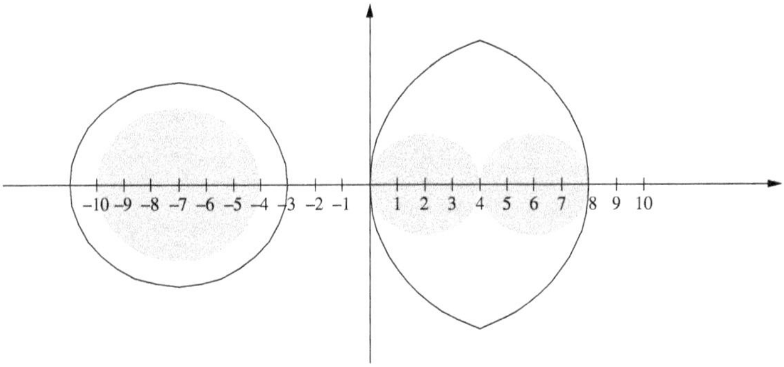

Figure 4.2.2: Gershgorin sets $\mathcal{G}_A$ and $\mathcal{G}_{\tilde{A}}$

to

$$\tilde{A} = TAT^{-1} = \begin{bmatrix} 6 & 1 & -1 \\ 0 & 2 & 2 \\ -3 & -3 & -7 \end{bmatrix}$$

with Gershgorin radii $\tilde{\rho}_1 = 2$, $\tilde{\rho}_2 = 2$, $\tilde{\rho}_3 = 6$, $\tilde{\gamma}_1 = 3$, $\tilde{\gamma}_2 = 4$, $\tilde{\gamma}_3 = 3$ and Gershgorin set

$$\mathcal{G}_{\tilde{A}} = \{\lambda \in \mathbb{C}; |\lambda - 6| \le 2\} \cup \{\lambda \in \mathbb{C}; |\lambda - 2| \le 2\} \cup \{\lambda \in \mathbb{C}; |\lambda + 7| \le 3\} \subset \mathcal{G}_A$$

(the shaded area in Fig. 4.2.2). By Theorem 4.2.19 $\sigma(A) \subset \mathcal{G}_{\tilde{A}}$. It follows from the last statement in Theorem 4.2.19 that each of the three shaded disjoint open disks in Fig. 4.2.2 contains one eigenvalue of A and so the spectrum of A must be real. The actual spectrum of A is $\sigma(A) = \{6.1056,\ 1.4899,\ -6.5956\}$. □

4.2.3 Smoothness of Eigenprojections and Eigenvectors

In this subsection we return to qualitative perturbation theory. We first prove that the sum of the eigenprojections for any λ_j-group of eigenvalues of $A = A_0 + \Delta$ depends analytically on the entries of Δ for $\|\Delta\|$ sufficiently small. We then specialize to matrix families $A(z)$ depending analytically on *one* complex parameter z and study in some detail the problem of whether or not the eigenprojections and eigenvectors of $A(z)$ can be expressed as analytic functions of z. In a similar way to eigenvalues, eigenprojections are analytic at non-critical points. However, at branch points eigenprojections necessarily have poles and thus behave quite differently from eigenvalues. We will see that branching cannot occur at parameter values z_0 for which each neighbourhood contains a point $z \neq z_0$ where $A(z)$ is normal (Rellich's Theorem). We conclude the section by determining the sensitivity of eigenvectors at parameter values z_0 where $A(z_0)$ has n distinct eigenvalues.
Throughout this subsection we assume that $\mathcal{L}(\mathbb{C}^n)$ resp. $\mathbb{C}^{n\times n}$ is provided with a given operator norm $\|\cdot\|$. We begin with some basic results on *resolvents* and *eigenprojections* of single matrices. Let $A \in \mathbb{C}^{n\times n}$, then the operator

$$R(s,A) = (sI - A)^{-1}\,, \quad s \in \rho(A) = \mathbb{C} \setminus \sigma(A) \tag{18}$$

is called the *resolvent* of A at $s \in \rho(A)$, and $\rho(A)$ is called the *resolvent set* of A. $R(s,A)$ commutes with A and this implies via

$$R(s,A) - R(s_0,A) = R(s,A)R(s_0,A)(s_0 I - A) - (sI - A)R(s,A)R(s_0,A),$$

the *resolvent equation*

$$R(s,A) - R(s_0,A) = (s_0 - s)R(s,A)R(s_0,A), \quad s, s_0 \in \rho(A). \tag{19}$$

In particular, $R(s,A)$ and $R(s_0,A)$ commute. Moreover,

$$R(s,A) = R(s_0,A)[I + (s - s_0)R(s_0,A)]^{-1}, \quad s, s_0 \in \rho(A)\,,$$

and we obtain the following absolutely convergent series expansion of $R(s,A)$ at $s_0 \in \rho(A)$.

$$R(s,A) = \sum_{k=0}^{\infty}(s_0 - s)^k (R(s_0,A))^{k+1}, \quad |s - s_0| < \|R(s_0,A)\|^{-1}. \tag{20}$$

Thus $R(s,A)$ is analytic on $\rho(A)$ and its derivatives at $s_0 \in \rho(A)$ are given by

$$R^{(k)}(s_0,A) = (-1)^k\, k!\, (R(s_0,A))^{k+1}, \quad k \in \mathbb{N}^*.$$

For the convenience of the reader we recall some elements from the spectral analysis of A presented in Subsection 2.2.2. Suppose that $\lambda_1, \ldots, \lambda_\ell$ are the distinct eigenvalues of A with algebraic multiplicities $m_1, \ldots, m_\ell$. Consider the spectral decomposition (see Lemma 2.2.4)

$$\mathbb{C}^n = \ker(\lambda_1 I - A)^{m_1} \oplus \cdots \oplus \ker(\lambda_\ell I - A)^{m_\ell} \tag{21}$$

of $\mathbb{C}^n$ into the A-invariant generalized eigenspaces and the corresponding *eigenprojections* of A

$$P_j: \quad \begin{array}{l} \mathbb{C}^n \to \ker(\lambda_j I - A)^{m_j}, \quad j \in \underline{\ell}. \\ x_1 \oplus \cdots \oplus x_\ell \mapsto x_j \end{array} \tag{22}$$

By (2.2.30) the eigenprojections satisfy

$$\sum_{j=1}^{\ell} P_j = I, \quad AP_j = P_j A, \quad P_j P_k = \delta_{jk} P_j, \qquad j,k \in \underline{\ell}. \tag{23}$$

We have $AP_j = P_j A = \lambda_j P_j + N_j$ where $N_j = (A - \lambda_j I)P_j$ is the *eigennilpotent* associated with the eigenvalue λ_j of A. By (2.2.32) the *spectral representation* of A is given by

$$A = A \sum_{j=1}^{\ell} P_j = \sum_{j=1}^{\ell} (\lambda_j P_j + N_j). \tag{24}$$

A is diagonalizable if and only if every eigenvalue of A is semi-simple (i.e. $N_j = 0$ for all $j \in \underline{\ell}$) and in this case

$$A = \sum_{j=1}^{\ell} \lambda_j P_j, \qquad R(s,A) = \sum_{j=1}^{\ell} (s - \lambda_j)^{-1} P_j\,, \quad s \in \rho(A). \tag{25}$$

If A is normal then A is diagonalizable and every eigenprojection P_j is selfadjoint, i.e. $P_j = P_j^*$, with spectral norm $\|P_j\|_{2,2} = 1$.[2] If A is real and symmetric then the eigenvalues λ_j are all real, and it follows from (25) that all the eigenprojections P_j are real (i.e. $P_j x \in \mathbb{R}^n$ for all $x \in \mathbb{R}^n$) and symmetric.

Lemma 4.2.21. *Suppose that $A \in \mathbb{C}^{n\times n}$ has exactly ℓ distinct eigenvalues $\lambda_1, \ldots, \lambda_\ell$, of algebraic multiplicities $m_1, \ldots, m_\ell$ with associated eigenprojections $P_1, \ldots, P_\ell$ and eigennilpotents $N_1, \ldots, N_\ell$. Then we have the partial fraction decomposition,*

$$R(s,A) = \sum_{j=1}^{\ell} \left[(s - \lambda_j)^{-1} P_j + \sum_{k=1}^{m_j - 1} (s - \lambda_j)^{-k-1} N_j^k \right]. \tag{26}$$

If Γ_j is a positively oriented circle in $\rho(A)$ enclosing λ_j but no other eigenvalue of A then

$$P_j = \frac{1}{2\pi\imath} \int_{\Gamma_j} R(s,A)\, ds\,. \tag{27}$$

Proof: If $\tilde{A} = T^{-1}AT$, $T \in \mathbf{Gl}_n(\mathbb{C})$ and the eigenprojections and eigennilpotents of $\tilde{A}$ are denoted by $\tilde{P}_j$, $\tilde{N}_j$, $j \in \underline{\ell}$ then it is easily verified that

$$R(s,\tilde{A}) = T^{-1}R(s,A)T, \quad \tilde{P}_j = T^{-1}P_jT, \quad \tilde{N}_j = T^{-1}N_jT, \quad j \in \underline{\ell}\,. \tag{28}$$

To prove (26) and (27) we may therefore assume that A is in Jordan canonical form

$$A = \bigoplus_{j=1}^{\ell} \bigoplus_{k=1}^{r_j} J(\lambda_j, n_{jk})$$

[2] Recall that the eigenprojections of a normal matrix are orthogonal and orthogonal projections are selfadjoint.

with Jordan blocks $J(\lambda_j, n_{jk}) \in \mathbb{C}^{n_{jk}\times n_{jk}}$, $k = 1, \ldots, r_j$ associated with the eigenvalues λ_j, $j \in \underline{\ell}$. The associated eigenprojections and eigennilpotents are, respectively,

$$P_j = 0_{m_1} \oplus \cdots \oplus 0_{m_{j-1}} \oplus I_{m_j} \oplus 0_{m_{j+1}} \oplus \cdots \oplus 0_{m_\ell} \tag{29}$$

(where 0_{m_j} denotes the $m_j \times m_j$ zero matrix and $m_j = \sum_{k=1}^{r_j} n_{jk}$, $j \in \underline{\ell}$), and

$$N_j = 0_{m_1} \oplus \cdots \oplus 0_{m_{j-1}} \oplus \sum_{k=1}^{r_j} J(0, n_{jk}) \oplus 0_{m_{j+1}} \oplus \cdots \oplus 0_{m_\ell}\,. \tag{30}$$

Now

$$R(s, A) = \bigoplus_{j=1}^{\ell} \bigoplus_{k=1}^{r_j} R(s, J(\lambda_j, n_{jk})). \tag{31}$$

And for any $\lambda \in \mathbb{C}$, $m \in \mathbb{N}$,

$$R(s, J(\lambda, m)) = \begin{bmatrix} s-\lambda & -1 & & 0 \\ & \ddots & \ddots & \\ & & \ddots & -1 \\ 0 & & & s-\lambda \end{bmatrix}^{-1} = \begin{bmatrix} (s-\lambda)^{-1} & (s-\lambda)^{-2} & \cdots & (s-\lambda)^{-m} \\ & \ddots & \ddots & \vdots \\ & & \ddots \quad \ddots & \vdots \\ & & \ddots & (s-\lambda)^{-2} \\ 0 & & & (s-\lambda)^{-1} \end{bmatrix}$$

$$= (s-\lambda)^{-1} I_m + \sum_{k=1}^{m-1} (s-\lambda)^{-k-1} N^k \tag{32}$$

where $N = J(0, m)$. Hence we obtain (26). Moreover, it follows from (32) via the Residue Theorem A.2.19 (applied to each entry of $R(s, J(\lambda_i, n_{ik}))$) that

$$\frac{1}{2\pi\imath} \int_{\Gamma_j} R(s, J(\lambda_i, n_{ik}))\, ds = \begin{cases} 0_{n_{ik}} & \text{if } i \neq j \\ I_{n_{jk}} & \text{if } i = j \end{cases}.$$

But this together with (31) and (29) implies (27). □

Corollary 4.2.22. *Let $A \in \mathbb{C}^{n\times n}$ and Γ be a positively oriented simple closed curve in $\rho(A)$ enclosing the eigenvalues $\lambda_1, \ldots, \lambda_k$ of A and no others, then*

$$\frac{1}{2\pi\imath} \int_{\Gamma} R(s, A)\, ds = \sum_{j=1}^{k} P_j \tag{33}$$

where the P_j are the eigenprojections of A associated with the eigenvalues λ_j, $j \in \underline{k}$.

Proof: By the Residue Theorem A.2.19 we have

$$\frac{1}{2\pi\imath} \int_{\Gamma} R(s, A)\, ds = \sum_{j=1}^{k} \operatorname{Res}(R(s, A), \lambda_j) = \sum_{j=1}^{k} \frac{1}{2\pi\imath} \int_{\Gamma_j} R(s, A)\, ds$$

where Γ_j, $j \in \underline{k}$ are small circles around λ_j as in Lemma 4.2.21. Hence (33) follows from Lemma 4.2.21. □

We now investigate the effect of (small) perturbations on the eigenprojections of a matrix. First recall that if $X \in \mathcal{L}(\mathbb{K}^n)$, $\|X\| < 1$ for some operator norm $\|\cdot\|$ (or, more generally, if $\varrho(X) < 1$), then $I_n - X$ is invertible and its inverse has a square root $(I_n - X)^{-1/2} \in \mathcal{L}(\mathbb{K}^n)$ which is given by the absolutely convergent binomial series

$$(I_n - X)^{-1/2} = I_n + \sum_{k=1}^{\infty} \binom{-1/2}{k} (-X)^k, \text{ where } \binom{\alpha}{k} = \frac{\alpha(\alpha-1)\dots(\alpha-k+1)}{k!}.$$

Hence if $Y \in \mathcal{L}(\mathbb{K}^n)$ commutes with X, then it also commutes with $(I-X)^{-1/2}$. The following proposition will play an important role.

Proposition 4.2.23. *Let $\mathcal{P}$ be the set of pairs of projections $P, Q \in \mathcal{L}(\mathbb{K}^n)$ satisfying $\|P - Q\| < 1$. Then any two projections P, Q with $(P, Q) \in \mathcal{P}$ are similar. The matrix family*

$$U_{P,Q} = [PQ + (I-P)(I-Q)][I - (P-Q)^2]^{-1/2} \in \mathcal{L}(\mathbb{K}^n), \quad (P,Q) \in \mathcal{P} \tag{34}$$

satisfies

$$U_{P,P} = I, \quad U_{P,Q}^{-1} = U_{Q,P}, \quad P = U_{P,Q} Q U_{P,Q}^{-1}. \tag{35}$$

If $\|\cdot\|$ is the spectral norm on $\mathcal{L}(\mathbb{K}^n)$ then $(P^*, Q^*) \in \mathcal{P}$ for all $(P,Q) \in \mathcal{P}$ and

$$U_{P^*,Q^*}^* = U_{Q,P} = U_{P,Q}^{-1}. \tag{36}$$

In particular, if $(P,Q) \in \mathcal{P}$ and $P = P^*$, $Q = Q^*$, then $U_{P,Q}$ is unitary (orthogonal if $\mathbb{K} = \mathbb{R}$).

Proof: Let $R = (P-Q)^2$ and note that

$$PR = P(P-Q)^2 = P(P + Q - PQ - QP) = P - PQP = (P-Q)^2 P = RP.$$

Thus R commutes with P and Q. Since $\|P - Q\| < 1$, $U_{P,Q}$ is well defined by (34) and the first formula in (35) is immediate. Since $(I-R)^{-1/2}$ commutes with P, Q, we have

$$\begin{aligned} &U_{P,Q}[QP + (I-Q)(I-P)][I-R]^{-1/2} \\ &\quad = [PQ + (I-P)(I-Q)][QP + (I-Q)(I-P)][I-R]^{-1} \\ &\quad = [PQP + (I-P)(I-Q)(I-P)][I-R]^{-1} \\ &\quad = [I - Q - P + QP + PQ][I-R]^{-1} = [I - (P-Q)^2][I-R]^{-1} = I. \end{aligned}$$

Hence $U_{P,Q}^{-1} = [QP + (I-Q)(I-P)][I-R]^{-1/2} = U_{Q,P}$. Moreover

$$\begin{aligned} U_{P,Q} Q &= [I-R]^{-1/2}[PQ + (I-P)(I-Q)]Q = [I-R]^{-1/2} PQ \\ &= PQ[I-R]^{-1/2} = P U_{P,Q}. \end{aligned}$$

So P and Q are similar and the other formulas in (35) hold.
Now if $\|\cdot\|$ is the spectral norm and $(P,Q) \in \mathcal{P}$ then $\|P^* - Q^*\| = \|P - Q\| < 1$, hence $(P^*, Q^*) \in \mathcal{P}$ and

$$\begin{aligned} U_{P^*,Q^*}^* &= ([I - R^*]^{-1/2})^* [P^*Q^* + (I-P^*)(I-Q^*)]^* \\ &= [I-R]^{-1/2}[QP + (I-Q)(I-P)] = U_{Q,P}. \end{aligned}$$

This completes the proof. □

In order to investigate the change of eigenprojections under arbitrary perturbations of a matrix we return to Corollary 4.2.1. Suppose that $A_0 \in \mathbb{C}^{n\times n}$ has ℓ distinct eigenvalues $\lambda_1, \dots, \lambda_\ell$ with algebraic multiplicities $m_1, \dots, m_\ell$, $\sum_{j=1}^{\ell} m_j = n$. There exists a $\delta > 0$ and positively oriented non-overlapping circles Γ_j around each λ_j such that for all $\Delta \in \mathbb{C}^{n\times n}$ with $\|\Delta\| < \delta$, Γ_j encloses exactly m_j eigenvalues of $A_0 + \Delta$ taking account of multiplicities. The set (or the unordered m_j-tuple) of these eigenvalues is called the *λ_j-group* of eigenvalues of $A = A_0 + \Delta$ for small $\|\Delta\|$. By Corollary 4.2.22 the operator

$$P_j(A) = \frac{1}{2\pi\imath} \int_{\Gamma_j} R(s, A)\, ds\ , \quad \|A - A_0\| < \delta \tag{37}$$

is equal to the sum of eigenprojections for all the eigenvalues of A lying inside Γ_j. $P_j(A)$ is called the *total projection* for the λ_j-group of eigenvalues of $A = A_0 + \Delta$, $\|\Delta\| < \delta$. $P_j(A_0)$ coincides with the *eigenprojection* for the eigenvalue λ_j of A_0.

Proposition 4.2.24. *Suppose $A_0 \in \mathbb{C}^{n\times n}$ has exactly ℓ distinct eigenvalues $\lambda_1, \dots, \lambda_\ell$, of algebraic multiplicities $m_1, \dots, m_\ell$. Then, for $\|A - A_0\|$ sufficiently small, the total projections $P_j(A)$ (37) depend analytically on the entries of A, $P_j(A)$ and $P_j(A_0)$ are similar and, in particular,* $\operatorname{rank} P_j(A) = \operatorname{rank} P_j(A_0) = m_j$, *$j \in \underline{\ell}$. Moreover if $U_{P_j(A), P_j(A_0)}$ is defined by (34) with $P = P_j(A)$, $Q = P_j(A_0)$, then*

$$P_j(A) = U_{P_j(A),P_j(A_0)}\, P_j(A_0)\, U^{-1}_{P_j(A),P_j(A_0)}, \quad j \in \underline{\ell}, \quad \|A - A_0\| \ll 1.$$

Proof: If $\Delta \in \mathbb{C}^{n\times n}$, $s \in \rho(A_0)$ and $\|\Delta\| < \|R(s, A_0)\|^{-1}$ then $s \in \rho(A)$ for $A = A_0 + \Delta$ and

$$\begin{aligned} R(s, A) &= (sI - (A_0 + \Delta))^{-1} = \left[(I - \Delta(sI - A_0)^{-1})(sI - A_0)\right]^{-1} \\ &= (sI - A_0)^{-1}[I - \Delta R(s, A_0)]^{-1}\,. \end{aligned}$$

Hence

$$R(s, A) = R(s, A_0) \sum_{k=0}^{\infty} [(A - A_0) R(s, A_0)]^k\ , \quad \|A - A_0\| < \|R(s, A_0)\|^{-1}\,. \tag{38}$$

If $\|A - A_0\| < \min_{s\in\Gamma_j} \|R(s, A_0)\|^{-1} =: \delta$ the power series on the RHS of (38) converges uniformly in $s \in \Gamma_j$, so that by (37) the integration of (38) along Γ_j yields

$$P_j(A) = \frac{1}{2\pi\imath} \sum_{k=0}^{\infty} \int_{\Gamma_j} R(s, A_0)[(A - A_0) R(s, A_0)]^k\, ds\,.$$

Thus, for $\|A - A_0\| < \delta$, $P_j(A)$ can be expressed as a power series in the entries of $(A - A_0)$ with coefficients in $\mathbb{C}^{n\times n}$. In particular, for $\|A - A_0\|$ sufficiently small we have $\|P_j(A) - P_j(A_0)\| < 1$. Applying Proposition 4.2.23 completes the proof. □

If $\|A - A_0\|$ is sufficiently small then A has at least as many distinct eigenvalues as A_0. On the other hand every neighbourhood of A_0 in $\mathbb{C}^{n\times n}$ contains matrices with any number of different eigenvalues between $\ell = |\sigma(A_0)|$ and n. The situation greatly simplifies if we consider analytic *one parameter* families of matrices (see Subsection 4.2.1). In this case it follows directly from Proposition 4.2.24 that the total projections depend analytically on z.

Corollary 4.2.25. *Suppose $A(\cdot) : \Omega \to \mathbb{C}^{n\times n}$ is analytic on the domain $\Omega \subset \mathbb{C}$, $z_0 \in \Omega$ and $A_0 = A(z_0)$ has exactly ℓ distinct eigenvalues $\lambda_1, \dots, \lambda_\ell$, of algebraic multiplicities $m_1, \dots, m_\ell$. Then, for $|z - z_0|$ sufficiently small, the total projections $P_j(z) := P_j(A(z))$, defined by (37) with $A = A(z)$, depend analytically on z and are of constant* rank $P_j(z) = m_j$, $j \in \underline{\ell}$.

Moreover, by Corollary 4.2.8 the number $\ell(z) = |\sigma(A(z))|$ of distinct eigenvalues of $A(z)$ is constant ($= \max_{z\in\Omega} |\sigma(A(z))|$) throughout the domain Ω with the exception of isolated critical points $z_0 \in C_A$ at which some of these eigenvalues may coalesce. If $z_0 \in \Omega \setminus C_A$ is an arbitrary non-critical point, $A_0 = A(z_0)$, and Γ_j are non-overlapping circles around the $\ell = \ell(z_0)$ eigenvalues of A_0 then there is exactly one eigenvalue $\lambda_j(z)$ of $A(z)$ lying inside the circle Γ_j for $|z - z_0| \ll 1$. In this case the total projection

$$P_j(z) = \frac{1}{2\pi\imath} \int_{\Gamma_j} R(s, A(z))\, ds = \operatorname{Res}\left(R(s, A(z)), \lambda_j(z)\right) \tag{39}$$

is in fact the *eigenprojection* for $\lambda_j(z)$.[3] Hence we obtain from the preceding corollary that these eigenprojections depend analytically on z in a sufficiently small neighbourhood of z_0. Now suppose that we have a domain $D \subset \Omega \setminus C_A$ and on this set there are ℓ analytic eigenvalue functions $\lambda_j(\cdot) : D \to \mathbb{C}$ such that $\sigma(A(z)) = \{\lambda_1(z), \dots, \lambda_\ell(z)\}$ for $z \in D$. Then the associated eigenprojections $P_j(z) = P_j(A(z))$ (see (22)) are well defined operator (matrix) valued functions on D. Since we have just seen that they are analytic in suitable neighbourhoods of any $z_0 \in D$ we obtain the following result.

Corollary 4.2.26. *Suppose $A(\cdot) : \Omega \to \mathbb{C}^{n\times n}$ is analytic on a domain $\Omega \subset \mathbb{C}$ and assume that there exist ℓ distinct analytic eigenvalue functions $\lambda_j(\cdot) : D \to \mathbb{C}$ on a subdomain D of $\Omega \setminus C_A$ such that $\sigma(A(z)) = \{\lambda_1(z), \dots, \lambda_\ell(z)\}$ for $z \in D$. Then the corresponding eigenprojections $P_j(z)$ are analytic and of constant rank on D. As a consequence, the eigennilpotents $N_j(z) = (A(z) - \lambda_j(z) I_n) P_j(z)$, $j \in \underline{\ell}$ are also analytic on D, and $A(z)$ admits the following analytic spectral representation*

$$A(z) = \sum_{j=1}^{\ell} (\lambda_j(z) P_j(z) + N_j(z)), \quad z \in D. \tag{40}$$

Remark 4.2.27. Note that by Corollary 4.2.8 we know that for any simply connected subdomain $D \subset \Omega \setminus C_A$ there indeed exist ℓ analytic eigenvalue functions $\lambda_j(\cdot) : D \to \mathbb{C}$ such that $\sigma(A(z)) = \{\lambda_1(z), \dots, \lambda_\ell(z)\}$ for $z \in D$ so that the preceding corollary is applicable. $\square$

Example 4.2.28. Consider $A(z) = \begin{bmatrix} 1 & z \\ z & -1 \end{bmatrix}$ on $\mathbb{C}$ (see Example 4.2.11(iii)) with eigenvalues $\lambda_{1,2}(z) = \pm(1 + z^2)^{1/2}$ and critical points at $\pm\imath$. We have seen that these two critical points are branch points of order two for the eigenvalues. By analytic continuation along a small circle around $z_0 = \pm\imath$ the roots form a cycle $(\lambda_1(z), \lambda_2(z))$ of order 2 representing a double-valued algebraic function on a small neighbourhood of z_0.

[3] $\operatorname{Res}(R(s, A(z)), \lambda_j(z))$, the residue of the resolvent $R(s, A(z))$ at $s = \lambda_j(z)$, is defined componentwise by applying Definition A.2.18 to each entry of $R(s, A(z))$.

Now let $D \subset \mathbb{C} \setminus \{\pm\imath\}$ be an arbitrary simply connected domain which does not contain any of these critical points. On D the eigenvalues of $A(z)$ are given by the two analytic functions $\lambda_i(\cdot) : D \to \mathbb{C}$, see Corollary 4.2.8. On the other hand, we have

$$R(s, A(z)) = \frac{1}{s^2 - (1+z^2)} \begin{bmatrix} s+1 & z \\ z & s-1 \end{bmatrix}, \quad s \neq \pm(1+z^2)^{1/2}.$$

To compute the eigenprojections via (39) we integrate the resolvent along small circles around $\lambda_{1,2}(z)$ or, alternatively, calculate the residue of the resolvent at $s = \lambda_{1,2}(z)$, $z \in \mathbb{C} \setminus \{\pm\imath\}$. This yields the following eigenprojections of $A(z)$

$$\begin{aligned} P_1(z) &= \frac{1}{2(1+z^2)^{1/2}} \begin{bmatrix} 1+(1+z^2)^{1/2} & z \\ z & -1+(1+z^2)^{1/2} \end{bmatrix}, \\ P_2(z) &= \frac{1}{2(1+z^2)^{1/2}} \begin{bmatrix} -1+(1+z^2)^{1/2} & -z \\ -z & 1+(1+z^2)^{1/2} \end{bmatrix}, \quad z \in D. \end{aligned}$$

In accordance with Corollary 4.2.26 these projections are represented by analytic matrix functions of constant rank 1 on the simply connected domain $D \subset \mathbb{C} \setminus \{\pm\imath\}$. By analytic continuation along a small circle around the critical points $\pm\imath$ the two eigenprojections form a 2-cycle, i.e. by analytic continuation of $P_1(z)$ along such a cycle we obtain $P_2(z)$ and vice versa. Note further that both $P_1(\cdot)$, $P_2(\cdot)$ have a pole at $\pm\imath$ and so these eigenprojections cannot be extended continuously to $z = \pm\imath$. The total projection $P(z)$ close to the critical point $z_0 = \imath$ is of rank 2 (= the algebraic multiplicity of the eigenvalue 0 of $A(\imath)$, see Corollary 4.2.25), hence $P(z) = I$, $|z - \imath| \ll 1$. An analogous statement holds for the critical point $z = -\imath$. □

Given an analytic eigenvalue $\lambda_j(\cdot)$ of algebraic multiplicity m_j on a subdomain D of $\Omega \setminus C_A$, it is of interest to find for each $z \in D$, m_j generalized eigenvectors $v^{j,k}(z), k = 1, \ldots, m_j$ which form a basis for the generalized eigenspace $\ker(\lambda_j(z)I - A(z))^{m_j}$ and which depend analytically on $z \in D$. The following specialization of Proposition 4.2.23 is our main tool for the construction of analytic (generalized) eigenbases.

Corollary 4.2.29. *Let $P(\cdot)\colon D \to \mathcal{L}(\mathbb{K}^n)$ be an analytic projection-valued function on a connected open set*[4] *$D \subset \mathbb{K}$, and assume $\|P(z) - P(z_0)\| < 1$ for all $z, z_0 \in D$. If*

$$U(z, z_0) = [P(z)P(z_0) + (I - P(z))(I - P(z_0))][I - (P(z) - P(z_0))^2]^{-1/2}, \tag{41}$$

then $U(\cdot,\cdot)$ is analytic on $D \times D$ and, for all $z, z_0 \in D$,

$$U(z_0, z_0) = I, \quad U(z, z_0)^{-1} = U(z_0, z), \quad P(z) = U(z, z_0)P(z_0)U(z, z_0)^{-1}. \tag{42}$$

If for some $z, z_0 \in D$ we have $\overline{z_0}, \overline{z} \in D$ and $P(z)^* = P(\overline{z})$, $P(z_0)^* = P(\overline{z_0})$, then

$$U(\overline{z}, \overline{z_0})^* = U(z, z_0)^{-1}.$$

In particular if $z_0, z \in D$ are real and $P(z), P(z_0)$ are Hermitian (resp. real and symmetric), then $U(z, z_0)$ is unitary (resp. orthogonal).

---

[4]i.e. D is a complex domain if $\mathbb{K} = \mathbb{C}$ and an open interval if $\mathbb{K} = \mathbb{R}$.

Proof: Since $\|P(z) - P(z_0)\| < 1$, for $z, z_0 \in D$, $U(z, z_0)$ is well defined by (41) and analytic on $D \times D$ because $P(\cdot)$ is analytic on D. The proof is completed by applying Proposition 4.2.23 with $P = P(z), Q = P(z_0)$. □

Corollary 4.2.30. *Let $P(\cdot)$ and D be as in Corollary 4.2.29 and $z_0, z \in D$. If $(v^1, \ldots, v^m)$ is a basis for $\operatorname{Im}_{\mathbb{K}} P(z_0) := P(z_0)\mathbb{K}^n$ and $U(z, z_0)$ is defined as in (41), then the vectors $v^i(z) = U(z, z_0)v^i$, $i \in \underline{m}$ form a basis of $\operatorname{Im}_{\mathbb{K}} P(z) \subset \mathbb{K}^n$ depending analytically on $z \in D$.*

Proof: We have

$$P(z)v^i(z) = P(z)U(z, z_0)v^i = U(z, z_0)P(z_0)v^i = U(z, z_0)v^i = v^i(z).$$

So the vectors $v^i(z)$, $i \in \underline{m}$ belong to $\operatorname{Im}_{\mathbb{K}} P(z)$. But $\dim_{\mathbb{K}} \operatorname{Im}_{\mathbb{K}} P(z) = \dim_{\mathbb{K}} \operatorname{Im}_{\mathbb{K}} P(z_0)$ and the transformations $U(z, z_0)$ are invertible, so the vectors $v^i(z)$, $i \in \underline{m}$ are linearly independent and form a basis of $\operatorname{Im}_{\mathbb{K}} P(z)$ for $z \in D$. □

Remark 4.2.31. It is clear from the construction of $U(z, z_0)$ in (41) that if the conditions on the parameter dependent projection are relaxed so that $P(z)$ is e.g. only continuous (resp. differentiable) on D, then the corresponding transformation $U(\cdot, \cdot)$ will be continuous (resp. differentiable) on $D \times D$, and so we obtain a continuous (resp. differentiable) basis of $\operatorname{Im}_{\mathbb{K}} P(z)$ by the preceding corollary. □

Applying Corollary 4.2.30 to the eigenprojections of $A(z)$, analytic generalized eigenvectors can be constructed locally around a non-critical parameter point z_0.

Corollary 4.2.32. *Suppose $A(\cdot) : \Omega \to \mathbb{C}^{n \times n}$ is analytic on a domain $\Omega \subset \mathbb{C}$, C_A is the set of critical points of $A(\cdot)$, $z_0 \in \Omega \setminus C_A$ and λ_j is an eigenvalue of $A(z_0)$ of algebraic multiplicity m_j. Then there exist an analytic function $\lambda_j(\cdot)$ on a suitable disk $D(z_0, r)$, $r > 0$ and m_j analytic functions $v^{j,k}(\cdot) : D(z_0, r) \to \mathbb{C}^n$, $k = 1, \ldots, m_j$ such that $\lambda_j(z)$ is an eigenvalue of algebraic multiplicity m_j of $A(z)$ with $\lambda_j(z_0) = \lambda_j$ and $(v^{j,1}(z), \ldots, v^{j,m_j}(z))$ is a basis of the generalized eigenspace $\ker(\lambda_j(z)I_n - A(z))^{m_j}$ for all $z \in D(z_0, r)$.*

Proof: Since z_0 is non-critical, the λ_j-group of eigenvalues of $A(z)$ near $z = z_0$ consists of only one eigenvalue $\lambda_j(z)$ (of multiplicity m_j), and this eigenvalue of $A(z)$ depends analytically on z in a small disk $D(z_0, \delta)$. The associated total projection $P_j(z)$ defined by (39) is identical with the eigenprojection for $\lambda_j(z)$ and depends analytically on $z \in D(z_0, \delta)$ if $\delta > 0$ is small enough. Now choose $r \in (0, \delta)$ such that $\|P(z) - P(z_0)\| < 1$ for all $z \in D(z_0, r)$ and let $(v^{j,1}, \ldots, v^{j,m_j})$ be a basis of the generalized eigenspace $\ker(\lambda_j I_n - A(z_0))^{m_j}$. If we define $U(z, z_0)$ by (41), then the vectors $v^{j,k}(z) = U(z, z_0)v^{j,k}$, $k \in \underline{m_j}$ form a basis of $\operatorname{Im} P(z) = \ker(\lambda_j(z)I_n - A(z))^{m_j}$ and depend analytically on $z \in D(z_0, r)$ by Corollary 4.2.30. □

The following example shows that if v is an eigenvector of $A(z_0)$, for $z_0 \in \Omega \setminus C_A$ with corresponding eigenvalue λ and $\lambda(z)$ is an eigenvalue of $A(z)$ depending analytically on z in a neighbourhood of $z = z_0$ such that $\lambda(z_0) = \lambda$, then $U(z, z_0)v$ is *not necessarily an eigenvector* corresponding to $\lambda(z)$, for any $z \neq z_0$.

Example 4.2.33. Consider the analytic matrix on the open unit disk

$$A(z) = \begin{bmatrix} z^2(1-z^2)^{-1} & 1 & -z(1-z^2)^{-1} \\ -(1-z^2)^{-1} & 2 & z(1-z^2)^{-1} \\ z(1-z^2)^{-1} & z & -(1-z^2)^{-1} \end{bmatrix}, \quad z \in \mathbb{D}.$$

It is easy to verify that the unordered 3-tuple of eigenvalues of $A(z)$ is constant, $\Lambda(A(z)) = \{1, 1, -1\}$, and even the Jordan form of $A(z)$ is independent of z. The eigenprojection $P(z)$ of $A(z)$ corresponding to $\lambda(z) \equiv 1$ is computed by (39) and the associated transformation $U(z, 0)$ by (41),

$$P(z) = \begin{bmatrix} (1-z^2)^{-1} & 0 & -z(1-z^2)^{-1} \\ 0 & 1 & 0 \\ z(1-z^2)^{-1} & 0 & -z^2(1-z^2)^{-1} \end{bmatrix}, \quad U(z,0) = \begin{bmatrix} (1-z^2)^{-1/2} & 0 & z(1-z^2)^{-1/2} \\ 0 & 1 & 0 \\ z(1-z^2)^{-1/2} & 0 & (1-z^2)^{-1/2} \end{bmatrix}.$$

Consider the eigenvector $v = (1, 1, 0)^\top$ of $A(0)$ corresponding to $\lambda = 1$. Then

$$v(z) = U(z,0)v = ((1-z^2)^{-1/2}, 1, z(1-z^2)^{-1/2})^\top, \quad A(z)v(z) = (1, 2-(1-z^2)^{-1/2}, z)^\top.$$

Thus we see that $v(z)$ is not an eigenvector of $A(z)$ for $z \neq 0$. □

Clearly the situation illustrated in the above example will not occur if the eigenvalue $\lambda(z)$ is semi-simple for all $z \in D$.

Remark 4.2.34. Proposition 4.2.29 is a *local* result since it assumes $\|P(z) - P(z_0)\| < 1$ for $z, z_0 \in D$ and this assumption will in general only be satisfied for small D. Now assume that D is an arbitrary simply connected domain in the complex plane on which $P(z)$ is analytic. In [293, II.4] (see also [43]), *Kato* shows how to construct a transformation $U(z, z_0)$ on $D \times D$ via the solution of differential equations. More precisely he proved: Let $P_1(z), \ldots, P_\ell(z) \in \mathcal{L}(\mathbb{K}^n)$ be projections which are analytic functions of $z \in D$, such that for every $z \in D$

$$\sum_{j=1}^{\ell} P_j(z) = I_n, \quad P_j(z)P_k(z) = \delta_{jk}P_k(z), \quad j,k = 1, \ldots, \ell.$$

Then there is a transformation $U(\cdot,\cdot)$ analytic on $D \times D$ which has the properties given in (42) where the last one is valid for *every* $P_j, j \in \underline{\ell}$, i.e.

$$P_j(z) = U(z, z_0)P_j(z_0)U(z, z_0)^{-1}, \quad z_0, z \in D, \quad j \in \underline{\ell},$$

and in addition $U(\cdot,\cdot)$ satisfies the cocycle condition

$$U(z_2, z_1)U(z_1, z_0) = U(z_2, z_0), \quad z_0, z_1, z_2 \in D.$$

For some hints on the construction, see Ex. 16. The analyticity requirement can be relaxed, but Kato's construction requires at least the differentiability of the projections, see Remark 4.2.31. □

We will now briefly discuss analytic properties of eigenprojections at *critical* parameter values. Suppose that $z_0 \in \Omega$ is a possibly critical point for $A(\cdot)$ and that $A(z_0)$ has exactly ℓ distinct eigenvalues $\lambda_1, \ldots, \lambda_\ell$, of algebraic multiplicities $m_1, \ldots, m_\ell$. Given ℓ non-overlapping positively oriented circles $\Gamma_1, \ldots, \Gamma_\ell$ around these eigenvalues of $A(z_0)$, there exists $\delta > 0$ such that $D(z_0, \delta) \subset \Omega$, the punctured disk

$D^\circ(z_0, \delta) \subset \Omega$ does not contain any critical point of $A(\cdot)$, and for all $z \in D(z_0, \delta)$ each circle Γ_j, $j \in \underline{\ell}$, encloses exactly m_j eigenvalues of $A(z)$, taking account of multiplicities. Let n_j be the number of distinct eigenvalues of $A(z)$ enclosed by Γ_j for $z \in D^\circ(z_0, \delta)$. These eigenvalues constitute the λ_j-group of eigenvalues of $A(z)$ near z_0. By analytic continuation in $D^\circ(z_0, \delta)$ the n_j eigenvalues of the λ_j-group form a number of different cycles $(\lambda_{jk1}(z), \ldots, \lambda_{jkq_{jk}}(z))$, $k = 1, \ldots, h_j$, see Figure 4.2.1. Each cycle $(\lambda_{jk1}(z), \ldots, \lambda_{jkq_{jk}}(z))$ defines a q_{jk}-valued algebraic function f_{jk} on $D^\circ(z_0, \delta)$. The eigenvalues $\lambda_{jki}(z)$, $i \in \underline{q_{jk}}$ of $A(z)$ pertaining to a given cycle are all of the same algebraic multiplicity $m_{jki}(z) = m_{jk}(z)$, and these multiplicities $m_{jk}(z) = m_{jk}$ remain constant throughout the punctured disk $D^\circ(z_0, \delta)$. Altogether $A(z)$ has $\sum_{j=1}^{\ell} n_j = \sum_{j=1}^{\ell} \sum_{k=1}^{h_j} q_{jk}$ distinct eigenvalues for $z \in D^\circ(z_0, \delta)$, and their multiplicities add up to $\sum_{j=1}^{\ell} \sum_{k=1}^{h_j} q_{jk} m_{jk} = n$.
The *total projection* $P_j(z)$ (39) associated with any one of the λ_j-groups

$$\{\lambda_{jki}(z);\ k = 1, \ldots, h_j,\ i = 1, \ldots, q_{jk}\}, \quad j \in \underline{\ell}$$

is the sum of all the eigenprojections corresponding to the n_j eigenvalues of the λ_j-group and depends analytically on z according to Corollary 4.2.25. The behaviour of the *eigenprojections* themselves near a critical point is more complicated. By Corollary 4.2.26 (see also Remark 4.2.27) the eigenprojection of $A(z)$

$$P_{jki}(z) \colon \bigoplus_{j=1}^{\ell} \bigoplus_{k=1}^{h_j} \bigoplus_{i=1}^{q_{jk}} \ker\left(\lambda_{jki}(z) I_n - A(z)\right)^{m_{jki}} \ \to\ \ker\left(\lambda_{jki}(z) I_n - A(z)\right)^{m_{jki}},$$

corresponding to the eigenvalue branch $\lambda_{jki}(z)$ is defined and analytic on the (simply connected) cut disk $D^-(z_0, \delta)$. The same holds for the associated eigennilpotent $N_{jki}(z) = (A(z) - \lambda_{jki}(z) I_n) P_{jki}(z)$. The eigenprojection branches $P_{jki}(z)$ (resp. eigennilpotents branches $N_{jki}(z)$) are linked by analytic continuation in $D^\circ(z_0, \delta)$ if and only if the same holds for the corresponding eigenvalue branches $\lambda_{jki}(z)$. Therefore the operator sum $\sum_{i=1}^{q_{jk}} \lambda_{jki}(z) P_{jki}(z)$ (resp. the sum of eigennilpotents $\sum_{i=1}^{q_{jk}} N_{jki}(z)$) associated with the eigenvalue cycle $(\lambda_{jk1}(z), \ldots, \lambda_{jkq_{jk}}(z))$ is invariant with respect to analytic continuation along circular arcs in $D^\circ(z_0, r)$ and hence can be extended analytically to the punctured disk. $P_j(z) = \sum_{k=1}^{h_j} \sum_{i=1}^{q_{jk}} P_{jki}(z)$, $z \in D^\circ(z_0, r)$ is the total projection associated with the λ_j-group and we have the following spectral representation of $A(z)$

$$A(z) = \sum_{j=1}^{\ell} \sum_{k=1}^{h_j} \sum_{i=1}^{q_{jk}} \left(\lambda_{jki}(z) P_{jki}(z) + N_{jki}(z)\right), \quad z \in D^\circ(z_0, \delta). \tag{43}$$

The eigenprojection branches $P_{jki}(z)$ in (43) are linked by analytic continuation in $D^\circ(z_0, \delta)$ if and only if the same holds for the corresponding eigenvalue branches $\lambda_{jki}(z)$. The proof of the next theorem proceeds from the partial fraction decomposition (26) of $R(s, A(z))$ and can be found e.g. in *Kato* (1980) and *Baumgärtel* (1985), see *Notes and References*.

Theorem 4.2.35. *If $A(\cdot) : \Omega \to \mathbb{C}^{n \times n}$ is an analytic matrix function on a domain $\Omega \subset \mathbb{C}$ the eigenprojections and eigennilpotents of $A(z)$ are branches of analytic functions of $z \in \Omega$ with only algebraic singularities at some (but not necessarily all)*

critical points $z_0 \in \mathbb{C}_A$. The branch points and their orders are the same for the eigenvalues and the eigenprojections.
If $A(z_0) = \sum_{j=1}^{\ell} \lambda_j P_j + N_j$ is the spectral representation of $A(z_0)$ at some (possibly critical) point $z_0 \in \Omega$, there exists $\delta > 0$ such that the spectral representation of $A(z)$ in the punctured disk $D^\circ(z_0, \delta)$ is of the form (43) where h_j is the number of cycles $(\lambda_{jk1}(z), \ldots, \lambda_{jkq_{jk}}(z))$, of order $q_{jk} \geq 1$, in the λ_j-group[5] of eigenvalues of $A(z)$ near z_0 and $(P_{jk1}(z), \ldots, P_{jkq_{jk}}(z))$ resp. $(N_{jk1}(z), \ldots, N_{jkq_{jk}}(z))$ are the corresponding cycles of eigenprojections, resp. eigennilpotents obtained by analytic continuation in $D^\circ(z_0, \delta)$.
If $q_{jk} = 1$, the functions $\lambda_{jk1}(z)$, $P_{jk1}(z)$, $N_{jk1}(z)$ can be continued analytically to the whole disk $D(z_0, \delta)$. If $q_{jk} \geq 2$ the branches of the eigenprojection cycle $(P_{jk1}(z), \ldots, P_{jkq_{jk}}(z))$ near z_0 can be represented by a Laurent-Puiseux series of the form

$$P_{jki}(z) = \sum_{l=-l_{jk}}^{\infty} B_{jkl} \left(w^i (z - z_0)^{1/q}\right)^l, \quad z \in D^-(z_0, \delta) \tag{44}$$

where $w = e^{2\pi i/q_{jk}}$, $l_{jk} \in \mathbb{N}$, $B_{jkl} \in \mathbb{C}^{n\times n}$ for $l \geq -l_{jk}$, and $(z - z_0)^{1/q} = \rho^{1/q} e^{i\theta/q}$ for $z = z_0 + \rho e^{i\theta}$, $0 < \theta < 2\pi$.

Although the eigenvalues and the eigenprojections have common branch points their behaviour near these points are very different. The eigenvalues $\lambda_{jki}(z)$ of a cycle with centre z_0 are continuous at z_0 and as $z \to z_0$ they all converge to the same root λ_j of the polynomial $\det(sI - A(z_0))$, see Corollary 4.2.9. On the other hand we will now show that the eigenprojections $P_{jki}(z)$ necessarily have a *pole* at each branch point.

Theorem 4.2.36 (Butler). *Suppose that $z = z_0$ is a branch point of order $q_{jk} \geq 2$ of an eigenvalue $\lambda_{jki}(z)$ of $A(z)$. Then the Laurent-Puiseux expansion (44) of the associated eigenprojection $P_{jki}(z)$ in powers of $(z - z_0)^{1/q}$ contains negative powers. In particular $\|P_{jki}(z)\| \to \infty$ as $z \to z_0$.*

Proof: Suppose that $P_{jki}(z)$ belongs to the cycle $(P_{jk1}(z), \ldots, P_{jkq_{jk}}(z))$ of eigenprojections. By analytic continuation along a small circle around z_0, $P_{jki}(z)$ is changed to $P_{jk(i+1)}(z)$, $i = 1, \ldots, q_{jk} - 1$ and $P_{jkq_{jk}}(z)$ is changed into $P_{jk1}(z)$. Now assume that the Laurent-Puiseux expansion (44) does not contain negative powers of $(z - z_0)^{1/q}$. Then $P_{jki}(z)$ is continuous at z_0 and it follows that (see (44))

$$B_{jk0} = \lim_{z\to z_0} P_{jk(i+1)}(z) = \lim_{z\to z_0} P_{jki}(z), \qquad (q_{jk} + 1 := 1). \tag{45}$$

But since $P_{jki}(z)P_{jk(i+1)}(z) = 0$ and $P_{jki}(z)P_{jki}(z) = P_{jki}(z)$ by (23), we have $B_{jk0} = B_{jk0}B_{jk0} = 0$. On the other hand $P_{jki}(z)$ is a non-zero projection hence $\|P_{jki}(z)\| \geq 1$. This contradicts $B_{jk0} = 0$ in view of (45). □

We see, therefore, that we cannot expect analyticity of eigenprojections at branch points of an associated eigenvalue. For an illustration, see Example 4.2.28. So it is interesting to ask what properties of $A(z)$ ensure that branch points do not exist for $z \in \Omega$. One such property is given in the following definition.

---
[5] hence $\lim_{z\to z_0} \lambda_{jki}(z) = \lambda_j$ for $j = 1, \ldots h_j$, $i = 1, \ldots, q_{jk}$.

Definition 4.2.37. A continuous matrix function $A(\cdot) : \Omega \to \mathbb{C}^{n\times n}$ is said to satisfy the *normality condition at a point* $z_0 \in \Omega$ if there exists a sequence $(z_i)_{i\in\mathbb{N}^*}$ in $\Omega\backslash\{z_0\}$ converging to z_0 such that the matrices $A(z_i)$ are all normal.

By continuity it follows that under the above condition $A(z_0)$ must also be normal. The next theorem provides an important analyticity result for eigenvalues and eigenprojections.

Theorem 4.2.38 (Rellich). *Let $A(\cdot) : \Omega \to \mathbb{C}^{n\times n}$ be analytic and satisfy the normality condition at $z_0 \in \Omega$. Suppose the spectral representation (24) of $A(z_0)$ is given by $A(z_0) = \sum_{j=1}^{\ell} \lambda_j P_j$ where the distinct eigenvalues λ_j have algebraic multiplicities m_j. Then the following statements hold for sufficiently small disks $D(z_0, \delta) \subset \Omega$ of radius $\delta > 0$ around z_0.*

(i) For each $j \in \underline{\ell}$ there exist $n_j \leq m_j$ distinct analytic functions $\lambda_{j,k}(\cdot) : D(z_0,\delta) \to \mathbb{C}$ such that $\{\lambda_{j,1}(z), \ldots, \lambda_{j,n_j}(z)\}$ is the λ_j-group of eigenvalues of $A(z)$ on $D(z_0, \delta)$. In particular, $\lambda_{j,k}(z_0) = \lambda_j$ for $j \in \underline{\ell}$, $k = 1, \ldots, n_j$.

(ii) For each $j \in \underline{\ell}$ there exist n_j analytic matrix functions $P_{j,k}(\cdot) : D(z_0, \delta) \to \mathbb{C}^{n\times n}$ such that, for every $z \in D^\circ(z_0, \delta)$, $P_{j,k}(z)$ is the eigenprojection of $A(z)$ for the eigenvalue $\lambda_{j,k}(z)$, $k = 1, \ldots, n_j$, and $\sum_{k=1}^{n_j} P_{j,k}(z_0) = P_j$. $P_{j,k}(z)$ is of constant rank on $D(z_0, \delta)$, for every $j \in \underline{\ell}$, $k = 1, \ldots, n_j$.

(iii) The spectral representation of $A(z)$ for $z \in D^\circ(z_0, \delta)$ is given by

$$A(z) = \sum_{j=1}^{\ell} \sum_{k=1}^{n_j} \lambda_{j,k}(z) P_{j,k}(z), \qquad 0 < |z| < \delta. \tag{46}$$

Proof: Let Γ_j, $j \in \underline{\ell}$ be non-overlapping circles around λ_j in $\mathbb{C}$ and let $\delta > 0$ be such that $D(z_0, \delta) \subset \Omega$, $D^\circ(z_0, \delta) \cap C_A = \emptyset$, and each Γ_j encloses exactly m_j eigenvalues of $A(z)$ for $z \in D(z_0, \delta)$ (taking account of multiplicities). Let $(z_i)_{i\in\mathbb{N}^*}$ be a sequence in $D^\circ(z_0, \delta)$ converging to z_0 such that the matrices $A(z_i)$ are all normal. We will first prove (i). If z_0 is not a critical point, then every λ_j-group of eigenvalues of $A(z)$ contains only one eigenvalue ($n_j = 1$) and (i) follows from Corollary 4.2.8. Now suppose $z_0 \in C_A$. For $j \in \underline{\ell}$ let $(\lambda_{j,1}(z), \ldots, \lambda_{j,n_j}(z))$ be the λ_j-group of eigenvalues of $A(z)$ surrounded by Γ_j and $(P_{j,1}(z), \ldots, P_{j,n_j}(z))$ the associated eigenprojections. Since $A(z_i)$ is normal we have $\|P_{j,k}(z_i)\|_{2,2} = 1$ for each eigenprojection $P_{j,k}(z)$, $k \in \underline{n_j}$. Hence by Theorem 4.2.36, z_0 is not a branch point for any $\lambda_{j,k}(z)$ and so the $\lambda_{j,k}(z)$ can be continued analytically onto $D(z_0, \delta)$ by Corollary 4.2.9. Now set $D = D^\circ(z_0, \delta)$. Applying Corollary 4.2.26 the eigenprojections $P_{jk}(z)$ are analytic and of constant rank on D. z_0 is not a branch point for any $\lambda_{j,k}(z)$ and so by Theorem 4.2.35 the associated eigenprojection $P_{j,k}(\cdot)$ and eigennilpotent $N_{j,k}(z) = (A(z) - \lambda_{j,k}(z)I)P_{j,k}(z)$, $j \in \underline{\ell}$, $k = 1, ..., n_j$ can be extended analytically to $D(z_0, \delta)$.[6] But $N_{j,k}(z_i) = 0$ for $i \in \mathbb{N}^*$ since $A(z_i)$ is normal and so $N_{j,k}(z) = 0$ for all $z \in D(z_0, \delta)$ by the Identity Theorem A.2.9 for holomorphic functions. Therefore the spectral representation (24) of $A(z)$, $z \in D$ is given by (46). By continuity the total projections $P_j(z) = \sum_{k=1}^{n_j} P_{j,k}(z)$ (see Corollary 4.2.22) converge to P_j as $z \to z_0$. This shows $\sum_{k=1}^{n_j} P_{j,k}(z_0) = P_j$ and concludes the proof of (ii) and (iii). □

[6]Note, however, that $P_{jk}(z)$ will, in general, not be an eigenprojection for $\lambda_{j,k}(z)$ at $z = z_0$.

A global version of the preceding theorem is presented in the next corollary.

Corollary 4.2.39. *Let $A(\cdot) : \Omega \to \mathbb{C}^{n\times n}$ be analytic, $D \subset \Omega$ a simply connected subdomain and assume that $A(\cdot)$ satisfies the normality condition at each critical point in D. Then there exist analytic functions $\lambda_i : D \to \mathbb{C}$ and analytic projection-valued functions $P_i(\cdot) : D \to \mathbb{C}^{n\times n}$, $i \in \underline{N}$ such that for every non-critical point $z \in D$ the spectral representation of $A(z)$ is given by $A(z) = \sum_{i=1}^{N} \lambda_i(z)P_i(z)$. In particular, $A(z)$ is diagonalizable for all $z \in D$.*

Proof: Let N be the number of distinct eigenvalues of $A(z)$ at the non-critical points of Ω. Given any $z_0 \in D$, there exist (by Theorem 4.2.38) N analytic functions $\lambda_i(\cdot) : D(z_0, \delta) \to \mathbb{C}$ on some small disk $D(z_0, \delta) \subset D$ such that $\sigma(A(z)) = \{\lambda_1(z), \ldots, \lambda_N(z)\}$ for $z \in D(z_0, \delta)$, and N projection-valued analytic functions $P_i(\cdot) : D(z_0, \delta) \to \mathbb{C}^{n\times n}$ such that, for every $z \in D^\circ(z_0, \delta)$, $P_i(z)$ is the eigenprojection of $A(z)$ corresponding to the eigenvalue $\lambda_i(z)$, $i \in \underline{N}$, and the spectral representation of $A(z)$ takes the form $A(z) = \sum_{i=1}^{N} \lambda_i(z)P_i(z)$ for $z \in D^\circ(z_0, \delta)$. Each of the functions $\lambda_i(\cdot)$ and $P_i(\cdot)$ can be continued analytically along any arc in D such that $A(z) = \sum_{i=1}^{N} \lambda_i(z)P_i(z)$ is the spectral representation of $A(z)$ for every noncritical point on the arc. Since D is simply connected this defines, by the Monodromy Theorem A.2.24, N complex analytic functions $\lambda_i(\cdot) : D \to \mathbb{C}$ and N analytic projection-valued functions $P_i(\cdot) : D \to \mathbb{C}^{n\times n}$ which satisfy the corollary.
□

We now return to the problem of constructing an analytic eigenbasis of $A(z)$ and solve it locally under the normality assumption of Rellich's Theorem.

Corollary 4.2.40. *Let $A(\cdot) : \Omega \to \mathbb{C}^{n\times n}$ be analytic and satisfy the normality condition at $z_0 \in \Omega$. Then, for $\delta > 0$ sufficiently small, there exists an analytic matrix function $V(\cdot) : D(z_0, \delta) \to \mathbb{C}^{n\times n}$ such that $V(z) \in \mathbf{Gl}_n(\mathbb{C})$ and $V(z)^{-1}A(z)V(z)$ is diagonal for all $z \in D(z_0, \delta)$.*

Proof: Suppose the same set-up as in the proof of Rellich's Theorem and choose $\delta > 0$ sufficiently small so that additionally $\|P_{j,k}(z) - P_{j,k}(z')\| < 1$ for all $z, z' \in D(z_0, \delta)$, $j \in \underline{\ell}$, $k \in \underline{n_j}$. Then we can apply Proposition 4.2.29 to each analytic projection-valued function $P_{j,k}(z)$ on $D(z_0, \delta)$ and obtain invertible transformations $U_{j,k}(z, z')$ which depend analytically on $(z, z') \in D(z_0, \delta) \times D(z_0, \delta)$. For every $j \in \underline{\ell}, k \in \underline{n_j}$ let $(v^{j,k,1}, \ldots, v^{j,k,m_{jk}})$ be a basis of $\operatorname{Im} P_{j,k}(z_0)$. By Corollary 4.2.30 the vectors $U_{j,k}(z, z_0)v^{j,k,l}$, $l = 1, \ldots, m_{jk}$ form a basis of $\operatorname{Im} P_{j,k}(z)$. Hence by Theorem 4.2.38 the vectors

$$U_{j,k}(z, z_0)v^{j,k,l}, \quad 1 \le j \le \ell,\ 1 \le k \le n_j,\ 1 \le l \le m_{jk}$$

form an eigenbasis of $A(z)$, for each $z \in D(z_0, \delta)$. Choosing these n vectors as columns of $V(z)$ we obtain a diagonal matrix $V(z)^{-1}A(z)V(z)$ whose diagonal entries are the eigenvalues $\lambda_{j,k}(z)$ of $A(z)$. □

Choosing complex parameters in Example 4.2.7 we see that the previous three results do *not* hold for analytic matrix families $A(z_1, z_2)$ depending on *two* parameters. The following example illustrates, that the *normality condition* in Theorem 4.2.38 cannot be replaced by *diagonalizability*.

Example 4.2.41. Consider the affine linear matrix function

$$z \mapsto A(z) = \begin{bmatrix} 0 & z & 0 \\ 0 & 0 & z \\ z & 0 & 1 \end{bmatrix}, \quad z \in \mathbb{C}.$$

The characteristic equation of $A(z)$ is $\lambda^3 - \lambda^2 - z^3 = 0$. For $|z|$ sufficiently small there are three distinct eigenvalues provided $z \neq 0$. Since also $A(0)$ is diagonal, $A(z)$ is diagonalizable in a neighbourhood of $z_0 = 0$. But $z_0 = 0$ is a branch point and the Puiseux series for the eigenvalues at $z_0 = 0$ are of the form

$$\lambda_1(z) = 1 + z^3 + \cdots \quad , \qquad \lambda_{2,3} = \pm\imath z^{3/2} + \cdots \; .$$

Thus local diagonalizability of an analytic matrix family $A(\cdot) : \Omega \to \mathbb{C}^{n\times n}$ at a critical point z_0 does not ensure analyticity of the eigenvalues of $A(z)$ at z_0.
A surprising theorem of Motzkin-Taussky states that if a matrix of the form $A(z) = A_0 + A_1 z$ is diagonalizable for all $z \in \mathbb{C}$ then all eigenvalues of $A(z)$ are affine linear in z, i.e. the eigenvalues of $A(z)$ are of the form $\lambda_j(z) = \lambda_j(A_0) + \eta_j z$ with $\eta_j \in \mathbb{C}$ and the associated eigenprojections $P_j(z)$ on $\mathbb{C}\backslash C_A$ can be extended to analytic functions on $\mathbb{C}$ [293, II.Thm.2.6]. The present example shows that the assumption of *global* diagonalizability is essential for this theorem and cannot be replaced by local diagonalizability at the critical parameter value considered. On the other hand, it follows from the theorem of Motzkin-Taussky that the above matrix $A(z)$ cannot be diagonalizable for all $z \in \mathbb{C}$. In fact one verifies that $A(z)$ is not diagonalizable at the solutions of $z^3 = -4/27$. □

We will now briefly deal with the spectral analysis of matrix families parametrized by a *real* parameter. Suppose $A(\cdot) : I \to \mathbb{C}^{n\times n}$ is analytic on an open interval $I \subset \mathbb{R}$. At every point $\tau_0 \in I$, $A(\cdot)$ can be expanded into a power series $A(\tau) = \sum_{k=0}^{\infty} A_k(\tau - \tau_0)^k$ which is absolutely convergent on a small interval $I(\tau_0, r(\tau_0))$ of radius $r(\tau_0) > 0$ around τ_0. If $r(\tau_0)$ is chosen sufficiently small, the complex analytic extension to the disk $D(z_0, r(\tau_0))$ given by $A(z) = \sum_{k=0}^{\infty} A_k(z - \tau_0)^k$, $z \in D(\tau_0, r(\tau_0))$ does not contain a non-real critical point, i.e. the number of eigenvalues of $A(z)$ at non-real $z \in D(\tau_0, r(\tau_0))$ will be equal to the number of eigenvalues of $A(\tau)$ at non-critical real points $\tau \in I$. The union D of all these disks is a simply connected domain in $\mathbb{C}$ with $I = D \cap \mathbb{R}$. The analytic extension of $A(\cdot)$ to D will again be denoted by $A(\cdot)$. By construction D does not contain any critical point of $A(\cdot)$ off the real axis.

Proposition 4.2.42. *Suppose that $A(\cdot) : I \to \mathbb{C}^{n\times n}$ is analytic on an open interval $I \subset \mathbb{R}$ and ℓ is the maximum number of distinct eigenvalues of $A(\tau)$, $\tau \in I$.*

(i) If I does not contain any critical points of $A(\cdot)$, then there exist analytic functions $\lambda_j : I \to \mathbb{C}$ and $P_j : I \to \mathbb{C}^{n\times n}$, $j \in \underline{\ell}$ such that $\lambda_1(\tau), \ldots, \lambda_\ell(\tau)$ are the distinct eigenvalues of $A(\tau)$ and $P_1(\tau) \ldots, P_\ell(\tau)$ are the corresponding eigenprojections of $A(\tau)$ for all $\tau \in I$.

(ii) Suppose that $A(\cdot)$ satisfies the normality condition at all critical points in I. Then there exist analytic functions $\lambda_j : I \to \mathbb{C}$ and analytic projection-valued functions $P_j : I \to \mathbb{C}^{n\times n}$, $j \in \underline{\ell}$ such that for every non-critical point $\tau \in I$ the spectral representation of $A(\tau)$ is given by $A(\tau) = \sum_{j=1}^{\ell} \lambda_j(\tau) P_j(\tau)$. In particular, $A(\tau)$ is diagonalizable for all $\tau \in I$.

(iii) *Suppose* $A(\tau)^* = A(\tau) \in \mathbb{K}^{n\times n}$ *for all* $\tau \in I$. *Then the previous statement holds with selfadjoint projections* $P_j(\tau) = P_j(\tau)^* \in \mathbb{K}^{n\times n}$. *Moreover, for any* $\tau_0 \in I$, *there exists* $\delta > 0$ *and an analytic orthonormal basis of* $\mathbb{K}^n$ *consisting of eigenvectors of* $A(\cdot)$ *on* $I(\tau_0, \delta)$, *i.e. an analytic matrix function* $V(\cdot) : I(\tau_0, \delta) \to \mathbb{K}^{n\times n}$ *such that* $V(\tau) \in \mathbf{U}_n(\mathbb{K})$ *and* $V(\tau)^{-1}A(\tau)V(\tau)$ *is diagonal for all* $\tau \in I(\tau_0, \delta)$.

Proof: Let $A(\cdot)$ be extended analytically to a simply connected domain $D \subset \mathbb{C}$ with $D \cap \mathbb{R} = I$ as above. If I does not contain any critical points then D does not contain any critical points and so (i) is an immediate consequence of Corollary 4.2.8 and Corollary 4.2.26. (ii) follows directly from Corollary 4.2.39 since D does not contain any non-real critical point and so by assumption $A(\cdot)$ is normal at all its critical points in D.
Now suppose that $A(\tau)$ is Hermitian (i.e. real and symmetric if $\mathbb{K} = \mathbb{R}$) for all $\tau \in I$. Then $A(\tau)$ is normal on I and so the first statement in (iii) follows from (ii). In fact, since the eigenprojections of a Hermitian matrix are selfadjoint, we have $P_j(\tau) = P_j(\tau)^* \in \mathbb{K}^{n\times n}$ at all non-critical points $\tau \in I$ and hence by continuity we obtain $P_j(\tau) = P_j(\tau)^* \in \mathbb{K}^{n\times n}$ for all $\tau \in I$. Similarly it follows that the images $\operatorname{Im}_{\mathbb{K}} P_j(\tau)$, $j \in \underline{\ell}$ (which are the eigenspaces of $A(\tau)$ at non-critical points $\tau \in I$) are mutually orthogonal for all $\tau \in I$. To prove the second statement in (iii) we proceed as in the proof of Corollary 4.2.40. Let $\tau_0 \in I$ and choose $\delta > 0$ sufficiently small so that $\|P_j(\tau) - P_j(\tau')\| < 1$ for all $\tau, \tau' \in I(\tau_0, \delta)$, $j \in \underline{\ell}$. Then we can apply Proposition 4.2.29 to each projection-valued function $P_j(\cdot)$ on $I(\tau_0, \delta)$. Since $P_j(\tau) = P_j(\tau)^*$ we obtain transformations $U_j(\tau, \tau') \in \mathbf{U}_n(\mathbb{K})$ which depend analytically on $(\tau, \tau') \in I(\tau_0, \delta) \times I(\tau_0, \delta)$. For every $j \in \underline{\ell}$, let $(v^{j1}, \ldots, v^{jm_j})$ be an orthonormal basis of $\operatorname{Im}_{\mathbb{K}} P_j(\tau_0)$. By Corollary 4.2.30 the vectors $v^{jk}(\tau) = U_j(\tau, \tau_0)v^{jk}$, $k = 1, \ldots, m_j$ form a basis of $\operatorname{Im}_{\mathbb{K}} P_j(\tau)$ and this basis is orthonormal since the transformations $U_j(\tau, \tau_0)$ are unitary. As we have already seen that the subspaces $\operatorname{Im}_{\mathbb{K}} P_j(\tau)$, $j \in \underline{\ell}$ are mutually orthogonal for all $\tau \in I$, we conclude that the vectors $v^{jk}(\tau)$, $j \in \underline{\ell}$, $k = 1, \ldots, m_j$ form an orthonormal basis of eigenvectors of $A(\tau)$ for all $\tau \in I(\tau_0, \delta)$. This concludes the proof. □

In the next corollary (which will be useful in the next section) we show how to construct locally an analytic unitary (resp. orthogonal) matrix function whose first columns are identical with a given orthonormal family of analytic vectors.

Corollary 4.2.43. *Let* $I \subset \mathbb{R}$ *be an open interval and* $V_1 : I \to \mathbb{K}^{n\times r}$, $r < n$ *an analytic matrix function whose columns form an orthonormal system at each* $\tau \in I$. *Then, for any* $\tau_0 \in I$, *there exists* $\delta > 0$ *and an analytic matrix function* $V_2 : I(\tau_0, \delta) \to \mathbb{K}^{n\times(n-r)}$, *such that* $I(\tau_0, \delta) \subset I$ *and* $V(\tau) = [V_1(\tau)\ V_2(\tau)] \in \mathbf{U}_n(\mathbb{K})$ *for all* $\tau \in I(\tau_0, \delta)$.

Proof: The Hermitian matrix function $A(\cdot) : I \to \mathbb{K}^{n\times n}$ defined by $A(\tau) = V_1(\tau)V_1(\tau)^*$ is analytic on I. Given any $\tau_0 \in I$, there exists by Proposition 4.2.42 an analytic orthonormal basis $(v^{r+1}(\tau), ..., v^n(\tau))$ of the eigenspace $\ker A(\tau) \subset \mathbb{K}^n$ on a sufficiently small interval $I(\tau_0, \delta) \subset I$, $\delta > 0$. Let $V_2(\tau) = [v^{r+1}(\tau)....v^n(\tau)]$, then $[V_1(\tau)\ V_2(\tau)] \in \mathbf{U}_n(\mathbb{K})$ for $\tau \in I(\tau_0, \delta)$. □

Remark 4.2.44. Global versions of Corollary 4.2.30, Proposition 4.2.42 (iii) and Corollary 4.2.43 can be proved by using Kato's construction of globally defined transformation matrices $U(z, z_0)$, see Remark 4.2.34. □

Example 4.2.7 illustrates that the statements of Proposition 4.2.42 do *not* hold for analytic matrix families $A(\tau_1, \tau_2)$ depending on *two real* parameters. The next example shows that the eigenspaces of a real normal matrix family can behave quite irregularly if the assumption of *analyticity* is weakened to *infinite differentiability*. It also illustrates once more the fact that, in general, the eigenspaces of a matrix vary less smoothly than the eigenvalues.

Example 4.2.45. Consider the family of real symmetric matrices

$$A(\tau) = e^{-\frac{1}{\tau^2}} \begin{bmatrix} \cos\frac{2}{\tau} & \sin\frac{2}{\tau} \\ \sin\frac{2}{\tau} & -\cos\frac{2}{\tau} \end{bmatrix}, \ \tau \in \mathbb{R}^*, \quad A(0) = 0_{2\times 2} \,.$$

$A(\cdot)$ is infinitely differentiable on $\mathbb{R}$ and so are the eigenvalues $\lambda_\pm(\tau) = \pm e^{-1/\tau^2}$, $\tau \in \mathbb{R}^*$, $\lambda_\pm(0) = 0$. For $\tau \in \mathbb{R}^*$ the associated eigenspaces are spanned by $(\cos(1/\tau), \sin(1/\tau))^\top$ and $(-\sin(1/\tau), \cos(1/\tau))^\top$, respectively. The corresponding eigenprojections are

$$P_+(\tau) = \begin{bmatrix} \cos^2\frac{1}{\tau} & \cos\frac{1}{\tau}\sin\frac{1}{\tau} \\ \cos\frac{1}{\tau}\sin\frac{1}{\tau} & \sin^2\frac{1}{\tau} \end{bmatrix}, \quad P_-(\tau) = \begin{bmatrix} \sin^2\frac{1}{\tau} & -\cos\frac{1}{\tau}\sin\frac{1}{\tau} \\ -\cos\frac{1}{\tau}\sin\frac{1}{\tau} & \cos^2\frac{1}{\tau} \end{bmatrix}, \ \tau \in \mathbb{R}^*.$$

But these projections cannot be extended to $\tau = 0$ in a continuous fashion. Even more, there does not exist a continuous function $v : [0, \varepsilon) \mapsto \mathbb{R}^2$ for some $\varepsilon > 0$ such that $v(\tau)$ is an eigenvector of $A(\tau)$ for all $\tau \in [0, \varepsilon)$. In fact, suppose e.g. that $A(\tau)v(\tau) = \lambda_+(\tau)v(\tau)$ for $0 \le \tau < \varepsilon$ and let (τ_k) be a positive sequence converging to 0 such that $\cos(1/\tau_{2k}) = 1$ and $\cos(1/\tau_{2k+1}) = 0$, $k \in \mathbb{N}$. Then $v(\tau_{2k}) = \alpha_{2k}(1\ 0)^\top$, $v(\tau_{2k+1}) = \alpha_{2k+1}(0\,,\,1)^\top$, $k \in \mathbb{N}$ for suitable $\alpha_k \in \mathbb{R}$. If $v(\cdot)$ is continuous at 0 then $\lim_{k\to\infty} v(\tau_{2k}) = \lim_{k\to\infty} v(\tau_{2k+1})$ and so (α_k) must converge to 0, whence $v(0) = (0\,,\,0)^\top$ which is not an eigenvector of $A(0)$. □

We conclude this section with a few remarks concerning the *sensitivity* of eigenvectors (for eigenvalues, see Subsection 4.2.1). The eigenvectors of a parametrized matrix $A(z)$ are not uniquely determined. If $v(z)$ is a parametrized eigenvector of $A(z)$, continuously differentiable at $z = z_0$ and $\alpha(z)$ is a complex-valued function, continuously differentiable at $z = z_0$ such that $\alpha(z_0) = 1$. Then the rescaled eigenvector $\tilde{v}(z) = \alpha(z)v(z)$ has sensitivity

$$\tilde{v}'(z_0) = \frac{d\tilde{v}}{dz}(z_0) = \frac{d\alpha}{dz}(z_0)v(z_0) + \alpha(z_0)\frac{dv}{dz}(z_0) = \alpha'(z_0)v(z_0) + v'(z_0)\,. \tag{47}$$

We see therefore that the sensitivity of $v(z)$ at z_0 can be changed arbitrarily in the direction of $v(z_0)$ by scaling.

Proposition 4.2.46. *Let Ω be an open subset of $\mathbb{R}$ or $\mathbb{C}$ and suppose that $A(\cdot) : \Omega \to \mathbb{C}^{n\times n}$ is differentiable and has n distinct eigenvalues λ_j, $j \in \underline{n}$ at $z = z_0$ with corresponding left and right eigenvectors w^j, v^j for $j \in \underline{n}$. For a given $i \in \underline{n}$, let $\lambda_i(\cdot) : \Omega \to \mathbb{C}$, $v^i(\cdot) : \Omega \to \mathbb{C}^n$ be differentiable at z_0 and satisfy*

$$A(z)v^i(z) = \lambda_i(z)v^i(z), \quad \lambda_i(z_0) = \lambda_i, \quad v^i(z_0) = v^i, \quad z \in \Omega. \tag{48}$$

Then

$$\frac{dv^i}{dz}(z_0) = \sum_{j=1}^{n} \beta_{ij} v^j, \quad \textit{where} \quad \beta_{ij} = \frac{w^{j*} A'(z_0) v^i}{(\lambda_i - \lambda_j) w^{j*} v^j}, \quad j \in \underline{n},\ j \neq i. \tag{49}$$

By rescaling $v^i(z)$ *the coefficient* β_{ii} *can be made zero whilst the other coefficients* β_{ij}, $j \in \underline{n}$, $j \neq i$ *are invariant under rescaling of* $v^i(z)$.

Proof: Differentiating the first equation in (48) at $z = z_0$ we have

$$(A(z_0) - \lambda_i I_n)\frac{dv^i}{dz}(z_0) + (A'(z_0) - \lambda_i'(z_0) I_n) v^i = 0\,.$$

Multiplication on the left by w^{j*}, $j \neq i$ gives

$$(\lambda_j - \lambda_i) w^{j*} \frac{dv^i}{dz}(z_0) + w^{j*}(A'(z_0) - \lambda_i'(z_0) I) v^i = 0\,.$$

Now choose $\beta_{ik} \in \mathbb{C}$ such that $\frac{dv^i}{dz}(z_0) = \sum_{k=1}^{n} \beta_{ik} v^k$. Since $w^{j*} v^k = 0$ for $k \neq j$ (because $\lambda_j \neq \lambda_k$), we get $w^{j*} v^j \neq 0$ and

$$(\lambda_j - \lambda_i)\beta_{ij} w^{j*} v^j + w^{j*} A'(z_0) v^i = 0, \quad j \in \underline{n},\ j \neq i.$$

This proves (49). Rescaling $v^i(z)$, i.e. replacing $v^i(z)$ by $\tilde{v}^i(z) = \alpha_i(z) v^i(z)$ where $\alpha_i(z_0) = 1$, we obtain $\frac{d\tilde{v}^i(z)}{dz}(z_0) = (\beta_{ii} + \alpha'(z_0)) v^i + \sum_{j=1, j\neq i}^{n} \beta_{ij} v^j$ from (47). This proves the last part of the proposition. □

If $A(\cdot)$ is analytic and $A(z_0)$ has n distinct eigenvalues then there exist analytic functions $\lambda_i(\cdot)$, $v^i(\cdot)$ satisfying (48) on a neighbourhood of z_0, by Corollary 4.2.3 and Corollary 4.2.26. Expression (49) shows that in this case the sensitivity of the eigenvector $v^i(\cdot)$ at $z = z_0$ is strongly dependent on the separation of $\lambda_i(z_0)$ from the other eigenvalues of $A(z_0)$. In the words of *Golub and Van Loan* (1996) [197], "Eigenvectors associated with nearby eigenvalues are wobbly". This wobbliness anticipates the indeterminacy of eigenvector directions when $\lambda_i = \lambda_j$.

Example 4.2.47. Consider the matrices

$$A(z) = \begin{bmatrix} 10.01 & 0.01 \\ 0 & 9.99 \end{bmatrix} + z \begin{bmatrix} 0 & 0 \\ 1 & 0 \end{bmatrix}, \quad z \in \mathbb{R}$$

at $z_0 = 0$. The eigenvalues of $A_0 = A(0)$ are $\lambda_1 = 10.01$, $\lambda_2 = 9.99$ with corresponding left and right eigenvectors

$$w^1 = \begin{bmatrix} 2 \\ 1 \end{bmatrix}, \quad v^1 = \begin{bmatrix} 1 \\ 0 \end{bmatrix}; \qquad w^2 = \begin{bmatrix} 0 \\ 1 \end{bmatrix}, \quad v^2 = \begin{bmatrix} 1 \\ -2 \end{bmatrix},$$

and sensitivities

$$\lambda_1'(0) = \frac{1}{w^{1*} v^1} w^{1*} A'(0) v^1 = 1/2\,, \quad \lambda_2'(0) = -1/2\,.$$

Since the eigenvalues are close we expect large sensitivities of the eigenvectors. Indeed since $\beta_{12} = w^{2*} A'(0) v^1 / (\lambda_1 - \lambda_2) w^{2*} v^2 = 1/0.02(-2) = \beta_{21}$, we have

$$\frac{dv^1}{dz}(0) = \beta_{12} v^2 = -v^2/0.04 = -25 v^2\,, \quad \frac{dv^2}{dz}(0) = \beta_{21} v^1 = -v^1/0.04 = -25 v^1\,.$$

As a result small changes of the parameter z may strongly affect the eigenvectors whereas the eigenvalues are only slightly perturbed.
For instance, changing $A(0)$ to $A(-0.005) = \begin{bmatrix} 10.01 & 0.01 \\ -0.005 & 9.99 \end{bmatrix}$ we obtain a slightly changed eigenvalue $\lambda_1(-0.005) = 10.0071$, but there is a considerable change in the corresponding normalized right eigenvector $\|v^1(-0.005)\|_2^{-1} v^1(-0.005) = [0.9597\,,\, -0.2811]^\top$. □

4.2.4 Exercises

1. Consider $A(z) = \begin{bmatrix} 1 & \cos z \\ 0.5 & 2 \end{bmatrix}$, $z \in \mathbb{C}$. Find the critical points of $A(\cdot)$ and determine the first three terms of the Puiseux series for the branches of $\Lambda(A(z))$ near the branch point $2\pi/3$.

2. Calculate the sensitivities at $z_0 = 0$ of the eigenvalues of the matrices

$$(i)\quad A(z) = \begin{bmatrix} 6+2z & 8-z \\ 8 & -6+2z \end{bmatrix}, \qquad (ii)\quad A(z) = \begin{bmatrix} 6 & 8-z & 0 \\ 8-z & -6 & 0 \\ 0 & 0 & 10-z \end{bmatrix}, \qquad z \in \mathbb{C}.$$

3. Suppose that $A(\cdot) : \Omega \to \mathbb{C}^{n \times n}$ is continuously differentiable at $z_0 \in \Omega$. Prove that the sensitivity of $e^{A(z)}$ at $z = z_0$ is

$$e^{A(z_0)} \int_0^1 e^{-A(z_0)t} A'(z_0) e^{A(z_0)t}\, dt\,.$$

(Hint: $e^{A(z)t}$ solves the matrix differential equation $\dot{X} = A(z)X$, $X(0) = I$.)

4. Consider

$$A(\tau) = \begin{bmatrix} |\tau|^{1.5} & |\tau|^{1.5} - |\tau|^{2.5}(2 + \sin|\tau|^{-1}) \\ -|\tau|^{1.5} & -|\tau|^{1.5} \end{bmatrix}, \qquad \tau \in \mathbb{R}^*,\ A(0) = 0\,.$$

Show that $A(\cdot)$ is continuously differentiable and diagonalizable for all τ. Calculate the eigenvalues of $A(\tau)$ and show that their derivatives are discontinuous at $\tau = 0$. Hence in Remark 4.2.5 *differentiability* cannot be replaced by *continuous differentiability*.

5. Consider the matrix $A(z)$ of Ex. 2 (ii). Find its eigenvalues $\lambda_i(z)$, $i \in \underline{3}$. Use Corollary 4.2.16 to show that there exist $\mu_i \in \sigma(A(0))$ such that $|\mu_i - \lambda_i(z)| \le |z|$ for $i = 1,2,3$ and all $z \in \mathbb{C}$. Determine the μ_i for each of the $\lambda_i(z)$. Relate your findings to the theorem of Motzkin-Taussky, see Example 4.2.41.

6. Suppose $U^*AU = D + N$ where D is diagonal, U is unitary and $N^p = 0$ for some integer p. If $\Delta \in \mathbb{C}^{n\times n}$, $\mu \in \sigma(A+\Delta)$ show that there exists $\lambda \in \sigma(A)$ such that

$$|\mu - \lambda| \le \max\{\theta, \theta^{\frac{1}{p}}\}$$

where $\theta = \|\Delta\|_{2,2} \sum_{k=0}^{p-1} \|N\|_{2,2}^k$ (see *Golub and Van Loan* (1996) [197, 7.2.1]).

7. If $A = \begin{bmatrix} 0 & 1 \\ -2 & -2 \end{bmatrix}$ show that $\|e^{At}\|_{2,2} = \left|\frac{1}{2}(2 + 5\sin^2 t + [20\sin^2 t + 25\sin^4 t]^{1/2})\right| e^{-t}$. For any $\alpha \in [0,1)$ determine some $M(\alpha) > 0$ such that the RHS of the above expression is bounded by $M(\alpha)e^{-(1-\alpha)t}$. By Proposition 4.2.18 $A + \Delta$, $\Delta \in \mathbb{R}^{2\times 2}$ is a Hurwitz stable matrix if

$$\|\Delta\|_{2,2} < (1-\alpha)/M(\alpha).$$

Show that if $\|\Delta\|_{2,2} < 0.4$ then $A + \Delta$ is Hurwitz stable.

8. Suppose $A \in \mathbb{C}^{n\times n}$ is asymptotically stable. Show that $A+\Delta$, $\Delta \in \mathbb{C}^{n\times n}$ will also be asymptotically stable if

$$\|\Delta\|_{2,2} < \left[\int_0^\infty \|e^{At}\|_{2,2}\, dt\right]^{-1}.$$

9. Suppose $A \in \mathbb{C}^{n\times n}$ and $\|A^t\| \le M\,\alpha^t$ for all $t \in \mathbb{N}$ where $\alpha \ge 0$, $M \ge 1$ and $\|\cdot\|$ is any operator norm. Prove that for any $\Delta \in \mathbb{C}^{n\times n}$

$$\|(A+\Delta)^t\| \le M\,(\alpha + M\,\|\Delta\|)^t, \quad t \in \mathbb{N}.$$

10. Determine the Gershgorin set for the matrix $\begin{bmatrix} 10 & 1 & -1 \\ -2 & 6 & 1 \\ 0 & 1 & 3 \end{bmatrix}$.

11. Applying the scaling transformation $T = \operatorname{diag}(\alpha, 1, 1)$, $\alpha \in \mathbb{R}$ to the matrix $\tilde{A}$ of Example 4.2.20, one obtains

$$A_\alpha = T\tilde{A}T^{-1} = \begin{bmatrix} 6 & \alpha & -\alpha \\ 0 & 2 & 2 \\ -3/\alpha & -3 & -7 \end{bmatrix}.$$

Find conditions on α so that the Gershgorin set $\mathcal{G}_{A_\alpha}$ contains a disjoint disk centered at 6, and prove that there is an eigenvalue of A in such a disk with radius $5-\sqrt{19}$. Carry out similar scaling transformations for the other rows and columns to conclude that there are eigenvalues of A in the disks centered at $2, -7$ with radii $3-\sqrt{3}$, $4-\sqrt{7}$ respectively.

12. If $A = \begin{bmatrix} -2 & -2 & 0 \\ -1 & -3 & 1 \\ 1 & -1 & -3 \end{bmatrix}$ find $\sigma(A)$ and determine the resolvent operator $R(s,A)$. Compute the projections P_j using (27) and verify the equalities in (23), (24).

13. Show that the eigenprojections of $A(z) = \begin{bmatrix} z & 1 \\ 0 & 0 \end{bmatrix}$, $z \in \mathbb{C}$ have a pole at the critical point $z_0 = 0$ (whereas the eigenvalues of $A(z)$ are clearly analytic).

14. Let $A(z) = \begin{bmatrix} 1 & z \\ z & -1 \end{bmatrix}$, $z \in \mathbb{C}$ (see Example 4.2.28) and $z_0 \in \mathbb{C} \setminus \{\imath, -\imath\}$. Following the procedure in the proof of Corollary 4.2.32, construct an analytic eigenbasis for $A(z)$ in a small disk around z_0.

15. If $A(z) = \begin{bmatrix} 1 & z \\ 0 & 2 \end{bmatrix}$, $z \in \mathbb{C}$, calculate the projection $P_1(z)$ corresponding to the eigenvalue $\lambda = 1$ and show that $\|P_1(z) - P_1(z_0)\| < 1$ provided that $z, z_0 \in \mathbb{C}$ satisfy $|z - z_0| < 1$. Observe that $(P_1(z) - P_1(z_0))^2 = 0_2$ and use this to suggest an improvement to Proposition 4.2.29. Prove that the transformation $U(z, z_0)$ given by (41) is $\begin{bmatrix} 1 & z - z_0 \\ 0 & 1 \end{bmatrix}$.

16. This exercise gives some hints for the construction of a transformation $U(z, z_0)$ with the properties given in Remark 4.2.34. Let $P_1(z), \ldots, P_\ell(z)$ be projections which are analytic functions of z on a simply connected domain D, such that for every $z \in D$

$$\sum_{j=1}^{\ell} P_j(z) = I_n, \quad P_j(z)P_k(z) = \delta_{jk}P_k(z), \quad j,k = 1,\ldots,\ell.$$

If $P_j'(z)$ denotes the derivative of $P_j(z)$ with respect to the complex variable z, prove that for all $z \in D$, $j \in \underline{\ell}$

$$P_j'(z)P_j(z)+P_j(z)P_j'(z) = P_j'(z); \; P_j(z)P_j'(z)P_j(z) = 0; \; P_j(z)P_k'(z) = -P_j'(z)P_k(z), \; k \neq j.$$

Let $Q(z) = \frac{1}{2}\sum_{k=1}^{\ell}[P_k'(z)P_k(z) - P_k(z)P_k'(z)]$, $z \in D$. Prove that for $z \in D$, $j \in \underline{\ell}$

$$Q(z)P_j(z) = P_j'(z)P_j(z), \;\; P_j(z)Q(z) = -P_j(z)P_j'(z), \;\; P_j'(z) = Q(z)P_j(z) - P_j(z)Q(z).$$

It is known that for every $z_0 \in D$ there exists a unique solution $U(\cdot, z_0)$ on D of the complex differential matrix equation $X'(z) = Q(z)X(z)$ with initial value $X(z_0) = I$. $U(z, z_0)$ is called the evolution operator of the complex differential equation $X'(z) = Q(z)X(z)$ and has the following properties.

$$U(z_0, z_0) = I, \quad U(z, z_0)^{-1} = U(z_0, z), \quad U(z_1, z)U(z, z_0) = U(z_1, z_0), \;\; z_1, z, z_0 \in D.$$

Prove that $P_j(z)U(z, z_0)$, $j \in \underline{\ell}$ is a solution of the initial value problem

$$X'(z) = Q(z)X(z), \quad X(z_0) = P_j(z_0) \quad z, z_0 \in D.$$

Hence show that $P_j(z)U(z, z_0) = U(z, z_0)P_j(z_0)$, $j \in \underline{\ell}$ and verify that $U(z, z_0)$ has all the properties mentioned in Remark 4.2.34.

17. Calculate the sensitivities of the eigenvectors of $A(z) = \begin{bmatrix} 6+2z & 8-z \\ 8 & -6+2z \end{bmatrix}$ at $z_0 = 0$.

18. Suppose that $A(\cdot) : \Omega \to \mathbb{C}^{n \times n}$ is twice differentiable at $z_0 \in \Omega$ and λ is an eigenvalue of $A(z_0)$ of multiplicity m. Assume there exist $\lambda_i(\cdot) : \Omega \to \mathbb{C}$, $v^i(\cdot) : \Omega \to \mathbb{C}^n$, $w^i(\cdot) : \Omega \to \mathbb{C}^n$, $i \in \underline{m}$, twice differentiable at z_0, satisfying

$$[A(z) - \lambda_i(z)I_n]\, v^i(z) = 0, \quad w^{i*}(z)\,[A(z) - \lambda_i(z)I_n] = 0, \qquad i \in \underline{m}$$

and

$$\lambda_i(z_0) = \lambda, \quad v^i(z_0) = v^i, \quad w^i(z_0) = w^i, \quad w^{i*}v^i \neq 0, \quad i \in \underline{m},$$

where v^i (resp. w^i), $i \in \underline{m}$ are linearly independent right (resp. left) eigenvectors of $A(z_0)$ corresponding to the eigenvalue λ. By (4)

$$\lambda_i'(z_0) = \frac{w^{i*}A'(z_0)v^i}{w^{i*}v^i}, \quad i \in \underline{m}.$$

If in addition $\lambda_{m+1}, \ldots, \lambda_n$ are distinct eigenvalues of $A(z_0)$ with right and left eigenvectors v^i, w^i, $i = m+1, \ldots, n$, and

$$\frac{dv^i}{dz}(z_0) = \sum_{\substack{j=1 \\ j \neq i}}^{n} \beta_{ij}v^j, \qquad \frac{dw^{i*}}{dz}(z_0) = \sum_{\substack{j=1 \\ j \neq i}}^{n} \gamma_{ij}w^{j*}, \qquad i \in \underline{m}.$$

Prove

$$\beta_{ij} = \begin{cases} \dfrac{w^{j*}A'(z_0)v^i}{(\lambda - \lambda_j)w^{j*}v^j}, & i \in \underline{m}, \quad j = m+1, \ldots, n \\[2ex] \dfrac{\frac{1}{2}w^{j*}A''(z_0)v^i + w^{j*}(A'(z_0) - \lambda_i'(z_0)I)\sum_{k=m+1}^{n}\beta_{ik}v^k}{(\lambda_i'(z_0) - \lambda_j'(z_0))w^{j*}v^j}, & 1 \leq i \neq j \leq m \end{cases}$$

$$\gamma_{ij} = \begin{cases} \dfrac{w^{i*}A'(z_0)v^j}{(\lambda - \lambda_j)w^{j*}v^j}, & i \in \underline{m}, \quad j = m+1, \ldots, n \\[2ex] \dfrac{\frac{1}{2}w^{i*}A''(z_0)v^j + \sum_{k=m+1}^{n}\gamma_{ik}w^{k*}(A'(z_0) - \lambda_i'(z_0)I)v^j}{(\lambda_i'(z_0) - \lambda_j'(z_0))w^{j*}v^j}, & 1 \leq i \neq j \leq m. \end{cases}$$

As a consequence we see that for eigenvectors corresponding to eigenvalues of multiplicity greater than one, their sensitivities are not only strongly influenced by the separation of the eigenvalues but also by the *separation of the sensitivities of the eigenvalues.* If $A(0) = I_2$, $A'(0) = \begin{bmatrix} 1 & 0 \\ 0 & .9 \end{bmatrix}$, $A''(0) = \begin{bmatrix} 0 & 1 \\ 1 & 0 \end{bmatrix}$, show that $\lambda_1'(0) = 1$, $\lambda_2'(0) = 0.9$ and $\beta_{12} = \gamma_{12} = 5$.

4.2.5 Notes and References

A very successful method of mathematical analysis (which in some applications may be the only viable approach) is to consider a given system as a slight perturbation of a simpler system for which a complete solution of the problem under consideration is known. In order to derive results for the given system from the simplified model, an appropriate perturbation theory is needed. The perturbation theory of linear operators was initiated by *Rayleigh* and *Schrödinger* (cf. *Sz.–Nagy* (1946), [491]) in their analysis of eigenvalue problems arising from acoustics and quantum mechanics, respectively. Whilst their pioneering work remained mathematically incomplete, the necessary mathematical foundations (convergence proofs etc.) were laid in a series of papers by *Rellich* (1937-42), see the monograph [433]. Motivated by applications in physics most perturbation theory was originally developed in an infinite dimensional (Hilbert space) context. The investigations of Rellich were mainly concerned with real-analytic families of selfadjoint operators where perturbations of isolated eigenvalues of finite algebraic multiplicity were studied. The consideration of complex perturbation parameters allowed the use of function theoretic methods and led to results for non-selfadjoint operators on Hilbert spaces and more generally for bounded and unbounded operators on Banach spaces. *Kato* (1980) [293] gives an excellent wide ranging account of perturbation theory with emphasis on spectral properties of infinite-dimensional linear operators. This standard reference also contains a brief description of the historical development of the theory. A separate concise treatment of finite-dimensional operators based on the exposition in *Kato* (1980) [293] is *Kato* (1982) [294]. A more detailed analysis of the finite dimensional case is given in the monograph of *Baumgärtel* (1985) [43].

Most of our presentation in Subsections 4.2.1 and 4.2.3 is based on *Kato* (1980) [293] and *Baumgärtel* (1985) [43]. The sensitivity of eigenvalues and eigenvectors (Propositions 4.2.12 and 4.2.46) has been a basic tool of robustness analysis in control theory, see *Cruz et al.* (1981) [114]. Sensitivity formulas for matrices $A(z_0)$ with repeated eigenvalues are given in Ex. 18. Formulas for higher derivatives of simple eigenvalues and their eigenprojections can be found in *Kato* (1980) [293, II.2].

The quantitative results presented in Subsection 4.2.2 have their origins in Numerical Analysis. The unavoidable presence of rounding errors in computing eigenvalues and eigenvectors provided a strong incentive for numerical analysts to investigate the effects of matrix perturbations on these computations. As a consequence many textbooks in numerical linear algebra contain estimates for the perturbation of eigenvalues due to errors in the data – we mention in particular *Householder* (1964) [269], *Henrici* (1974) [224], *Stoer and Bulirsch* (1993) [486], *Golub and Van Loan* (1996) [197]. Standard references for the numerical analysis of matrix perturbations are *Wilkinson* (1965) [524] and *Stewart and Sun* (1990) [484] (which contains many further references to the numerical literature), see also *Bhatia* (1987) [55]. Lemma 4.2.15 is due to *Bauer and Fike* (1960) [41]. Generalizations of

Proposition 4.2.18 to infinite dimensional operators and further estimates on the growth of perturbed semigroups of operators are given in *Kato* (1980) [293, IX.2]. *Gershgorin* (1931) [186] established his theorem as a corollary to the fact that a diagonally dominant matrix is nonsingular. The problem of determining what can be achieved by combining scaling transformations with Gershgorin's Theorem has been studied in *Wilkinson* (1965) [524], see also Ex. 11. We will continue our study of Gershgorin type perturbations in Section 5.2.

The integral formula (27) for eigenprojections was first used by *Sz.-Nagy* (1946) [491] and *Kato* (1980) [293] in the context of perturbation theory and is now of fundamental importance in the field. The analytic similarity transformations mentioned in Remark 4.2.34 were developed in connection with the adiabatic theorem in quantum mechanics.

A proof of Theorem 4.2.35 can be found in *Kato* (1980) [293, II.1.5] and in *Baumgärtel* (1985) [43, 3.3]. Butler's Theorem (Theorem 4.2.36) originally appeared in *Butler* (1959) [85]. The normality assumption greatly simplifies the spectral analysis of $A(z)$. Theorem 4.2.38 was proved in *Rellich* (1937) [432]. Further results on normal and selfadjoint perturbations can be found in *Baumgärtel* (1985) [43, 3.5]. An application of Proposition 4.2.42 to the analysis of *singular values* of analytic matrix families will be given in the next section, see Theorem 4.3.17.

4.3 The Singular Value Decomposition

It is well known that the analysis and computation of eigenvalues and eigenvectors is greatly simplified when the matrix is Hermitian. Hermitian matrices have real eigenvalues, orthonormal bases of eigenvectors, and they can be diagonalized by unitary similarity transformations. In the previous section we have seen that perturbation analysis is also greatly simplified for such operators. If a Hermitian matrix[1] depends analytically on a real parameter it can be diagonalized by unitary matrices depending analytically on the parameter, see Proposition 4.2.42 (iii). In this section we will show how these results can be used for the analysis of arbitrary rectangular matrices $G \in \mathbb{K}^{m\times n}$ (and matrix families) by applying them to the associated positive semidefinite Hermitian matrix $G^*G \in \mathcal{H}_n(\mathbb{K})$. The non-negative square roots of eigenvalues of G^*G are called the *singular values* of G and the unitary diagonalization of G^*G leads to the *singular value decomposition* of G. This decomposition has become a powerful tool in many areas of mathematics and is of fundamental importance in Numerical Linear Algebra. In Linear System Theory it has played an increasingly important role since the late seventies, particularly in model reduction and robust control, see *Notes and References.* Some numerical aspects will be discussed in Section 4.5.

We begin with the definition of the singular values of a matrix $G \in \mathbb{K}^{m\times n}$, discuss their geometric significance, and derive a minimax characterization in terms of the associated quadratic form $x \mapsto x^*G^*Gx$ (Courant-Fischer Theorem). In the second subsection we construct the singular value decomposition of G and determine the distance of a matrix from the set of matrices of given lower rank (Schmidt-Mirsky Theorem). We also obtain an elementary but fundamental estimate for the singular values of perturbations of G (Corollary 4.3.11). In the third subsection we consider a matrix $G(\tau)$ which depends analytically on a real parameter τ and derive a modified analytic singular value decomposition of $G(\tau)$ which depends analytically on τ. Finally, in the fourth subsection the relationship between eigenvalues and singular values of a square matrix is discussed.

4.3.1 Singular Values and Singular Vectors

Let $G \in \mathbb{K}^{m\times n}$ be a matrix of rank r. Throughout this section we provide all vector spaces $\mathbb{K}^q$, $q \in \mathbb{N}^*$ with the standard Euclidean inner product $\langle\cdot,\cdot\rangle$ and the associated norm $\|\cdot\|_2 = \langle\cdot,\cdot\rangle^{1/2}$. For short, *the corresponding operator norm (spectral norm)* $\|\cdot\|_{2,2}$ *will be denoted by* $\|\cdot\|$. Since $G^*G \in \mathbb{K}^{n\times n}$ is a positive semidefinite Hermitian matrix of rank r there exists a unitary[2] matrix $V = [v^1, \ldots, v^n] \in \mathbf{U}_n(\mathbb{K})$ such that

$$V^*G^*GV = \Sigma = \operatorname{diag}(\sigma_1^2, \ldots, \sigma_n^2) \tag{1}$$

where the σ_i are the non-negative square roots of the eigenvalues of G^*G ordered by

$$\sigma_1 \geq \sigma_2 \geq \cdots \geq \sigma_r > \sigma_{r+1} = \cdots = \sigma_n = 0. \tag{2}$$

[1] Recall that 'Hermitian' means 'symmetric' if $\mathbb{K} = \mathbb{R}$.

[2] $\mathbf{U}_n(\mathbb{R}) = \mathbf{O}_n$ is the group of orthogonal matrices. Thus 'unitary' means 'orthogonal' if $\mathbb{K} = \mathbb{R}$.

The (right) eigenvectors of G^*G are called *right singular vectors* of G. So the columns of V form an orthonormal basis of $\mathbb{K}^n$ consisting of right singular vectors of G. Moreover the singular vectors $v^1, \ldots, v^r$ form an orthonormal basis of $\operatorname{Im} G^*G = (\ker G)^\perp$ and $v^{r+1}, \ldots, v^n$ form an orthonormal basis of $\ker G$.
The numbers $\sigma_1, \ldots, \sigma_n$ are called the ordered *singular values* of G, and will be denoted by $\sigma_1(G), \ldots, \sigma_n(G)$, respectively. We will always assume they are ordered according to (2). The maximal and minimal singular values of G will be denoted by $\sigma_{\max}(G)$ and $\sigma_{\min}(G)$.
It follows directly from the definition that the singular values of G remain unchanged under unitary transformations $G \mapsto U_1^* G U_2$ $(U_1 \in \mathbf{U}_m(\mathbb{K}),\ U_2 \in \mathbf{U}_n(\mathbb{K}))$. Since $\operatorname{rank} G^* = \operatorname{rank} G = r$ and, for any $\lambda \in \mathbb{C}$,

$$\det(\lambda I_n - G^*G) = 0 \quad \Leftrightarrow \quad \det(\lambda I_m - G\,G^*) = 0\ ,$$

we have

$$\sigma_k(G^*) = \sigma_k(G), \qquad k = 1, \ldots, \min\{m, n\}. \tag{3}$$

A geometric interpretation of the singular values $\sigma_k(G)$, $k \in \underline{n}$ can be given via the following quadratic form on $\mathbb{K}^n$

$$x \mapsto \|Gx\|_2^2 = \langle x, G^*Gx \rangle = \langle V^*x, V^*G^*GVV^*x \rangle = \sum_{j=1}^{r} \sigma_j^2 |(V^*x)_j|^2\,. \tag{4}$$

Here $(V^*x)_j = \langle x, v^j \rangle$ is the j-th coordinate of x with respect to the basis $(v^1, \ldots, v^n)$ of $\mathbb{K}^n$. Now consider the set

$$\mathcal{E}^{-1}(G) := \{x \in \mathbb{K}^n\,;\ \|Gx\| \le 1\} = \{x \in \mathbb{K}^n\,;\ \langle x, G^*Gx \rangle \le 1\}\,. \tag{5}$$

$\mathcal{E}^{-1}(G)$ is the *preimage* of the closed unit ball in $\mathbb{K}^m$ by G, or alternatively the

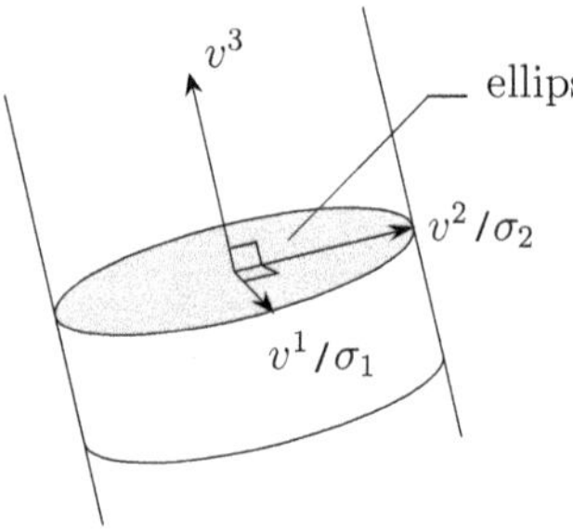

Figure 4.3.1: The cylinder $\mathcal{E}^{-1}(G)$ with $\sigma_3 = 0$

preimage of the closed unit ball in $\mathbb{K}^n$ by the square root of G^*G. This square root is given by

$$(G^*G)^{1/2} = V \operatorname{diag}(\sigma_1, \ldots, \sigma_r, 0, \ldots, 0) V^* \ \in\ \mathbb{K}^{n \times n}. \tag{6}$$

If $r = n$, it follows from (4) that $\mathcal{E}^{-1}(G)$ is an ellipsoid with semi-axes v^j/σ_j, $j \in \underline{n}$,

$$x = \sum_{j=1}^{n} \alpha_j v^j \in \mathcal{E}^{-1}(G) \quad \Leftrightarrow \quad \sum_{j=1}^{n} \sigma_j^2 |\alpha_j|^2 \le 1. \tag{7}$$

If $r < n$, this ellipsoid is degenerate, i.e. $\mathcal{E}^{-1}(G)$ is a 'cylinder' with an ellipsoidal

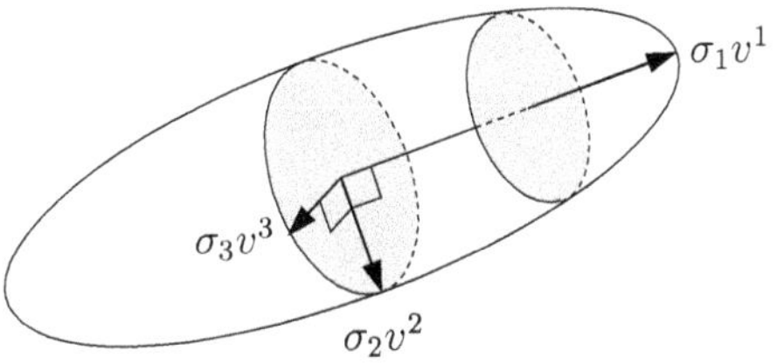

Figure 4.3.2: Ellipsoid $\mathcal{E}(G)$ with $\sigma_1 \geq \sigma_2 \geq \sigma_3 > 0$

base having semi-axes v^j/σ_j, $j \in \underline{r}$, see Figure 4.3.1. An alternative geometric interpretation is obtained by considering the *image* of the unit ball in $\mathbb{K}^n$ by the linear map $(G^*G)^{1/2}$

$$\mathcal{E}(G) := \{(G^*G)^{1/2}x \; ; \; x \in \mathbb{K}^n \; ; \; \|x\|_2 \leq 1\}. \tag{8}$$

By (6), $\mathcal{E}(G)$ is the ellipsoid with semi-axes $\sigma_j v^j$, $j \in \underline{r}$ in $\operatorname{span}\{v^1, \ldots, v^r\} \subset \mathbb{K}^n$, see Figure 4.3.2.

A characterization of the singular values is contained in the following minimax theorem. If E is any linear subspace of the Euclidean space $\mathbb{K}^n$, we denote by $\|G|E\|$ the operator norm of the restriction of G to E and set

$$\sigma^2_{\max}(G|E) = \max_{\substack{x \in E \\ \|x\|_2 = 1}} \langle x, G^*Gx \rangle = \max_{0 \neq x \in E} \frac{\langle x, G^*Gx \rangle}{\|x\|_2^2} = \|G|E\|^2. \tag{9}$$

Theorem 4.3.1 (Minimax Theorem of Courant and Fischer). *For any matrix $G \in \mathbb{K}^{m \times n}$ the ordered singular values $\sigma_k = \sigma_k(G)$ are characterized by*

$$\sigma_k(G) = \min_{\operatorname{codim} E = k-1} \sigma_{\max}(G|E) = \min_{\operatorname{codim} E \leq k-1} \sigma_{\max}(G|E), \quad k \in \underline{n}\,. \tag{10}$$

Proof: Given $k \in \underline{n}$, we have to show

(i) $\sigma_k^2 \leq \sigma^2_{\max}(G|E)$ for any $E \subset \mathbb{K}^n$ with $\operatorname{codim} E \leq k-1$,

(ii) $\sigma_k^2 \geq \sigma^2_{\max}(G|\tilde{E})$ for some $\tilde{E}$ with $\operatorname{codim} \tilde{E} = k-1$.

Let E be any linear subspace of $\mathbb{K}^n$ with $\operatorname{codim} E \leq k-1$. The k–dimensional subspace $\operatorname{span}\{v^1, \ldots, v^k\}$ contains a vector $\hat{x} = \sum_{j=1}^k \hat{\alpha}_j v^j$ of norm $\|\hat{x}\|_2 = 1$ in common with E. By (4) with $x = \hat{x}$, hence $(V^*x)_j = \hat{\alpha}_j$, we have

$$\sigma^2_{\max}(G|E) \geq \langle \hat{x}, G^*G\hat{x} \rangle = \sum_{j=1}^k \sigma_j^2 |\hat{\alpha}_j|^2 \geq \sigma_k^2 \sum_{j=1}^k |\hat{\alpha}_j|^2 = \sigma_k^2 \|\hat{x}\|_2^2\,.$$

This proves (i). To prove (ii), let $E_k = \operatorname{span}\{v^k, \ldots, v^n\}$ so that $\operatorname{codim} E_k = k-1$. Every $x \in E_k$ is of the form $\sum_{j=k}^n \alpha_j v^j$, hence $(V^*x)_j = \langle x, v^j \rangle = 0$ for $1 \leq j \leq k-1$ and so (4) implies

$$\langle \tilde{x}, G^*G\tilde{x} \rangle = \sum_{j=k}^r \sigma_j^2 |(V^*x)_j|^2 \leq \sigma_k^2 \sum_{j=k}^n |\alpha_j|^2 = \sigma_k^2 \|x\|_2^2, \quad x \in E_k.$$

This proves (ii). □

This theorem characterizes the singular values $\sigma_k(G)$, $k \in \underline{n}$ without any reference to the eigenvectors v^j of G^*G. In particular, for $k = 1$, (10) implies that $\sigma_1(G) = \sigma_{\max}(G) = \|G\|$. Moreover, the proof of the above theorem shows that the linear subspaces $E_k = \operatorname{span}\{v^k, \dots, v^n\}$, $k \in \underline{n}$ satisfy

$$\sigma_k = \|G|E_k\| \le \|G|E\|, \qquad E \subset \mathbb{K}^n,\ \operatorname{codim} E \le k-1. \tag{11}$$

An alternative expression of (10) is

$$\sigma_k(G) = \min_{y^1,\dots,y^{k-1}\in\mathbb{K}^n} \max_{\substack{0\ne x\in\mathbb{K}^n \\ x\perp y^1,\dots,y^{k-1}}} \frac{\|Gx\|_2}{\|x\|_2}, \quad k \in \underline{n}. \tag{12}$$

As a direct consequence of the Courant-Fischer Theorem we obtain the following monotonicity property of singular values. If $G \in \mathbb{K}^{m\times n}$, $\tilde{G} \in \mathbb{K}^{\tilde{m}\times n}$ then

$$\Big(\forall x \in \mathbb{K}^n : \|\tilde{G}x\|_2 \le \|Gx\|_2\Big) \;\Rightarrow\; \forall k \in \underline{n} : \sigma_k(\tilde{G}) \le \sigma_k(G). \tag{13}$$

Remark 4.3.2. In an analogous way it is possible to show that the singular values have a *maximum property* in the following sense. If we set for any linear subspace $E \subset \mathbb{K}^n$

$$\sigma^2_{\min}(G|E) := \min_{x\in E,\, \|x\|_2=1} \langle x, G^*Gx\rangle = \min_{0\ne x\in E} \frac{\langle x, G^*Gx\rangle}{\|x\|_2^2}, \tag{14}$$

then for all $k \in \underline{n}$

$$\sigma_k(G) = \max_{\dim E=k} \sigma_{\min}(G|E) = \max_{\dim E\ge k} \sigma_{\min}(G|E) = \max_{y^1,\dots,y^k\in\mathbb{K}^n} \min_{\substack{x\ne 0 \\ x\in\operatorname{span}\{y^1,\dots,y^k\}}} \frac{\|Gx\|_2}{\|x\|_2}. \tag{15}$$

□

Corollary 4.3.3. *Let $G \in \mathbb{K}^{m\times n}$ and $1 \le \nu < n$. If $G_1 \in \mathbb{K}^{m\times\nu}$ is obtained from G by eliminating $n-\nu$ columns of G then*

$$\sigma_{k+n-\nu}(G) \le \sigma_k(G_1) \le \sigma_k(G)\,, \quad 1 \le k \le \nu. \tag{16}$$

Similarly, if $1 \le \mu < m$ and $G_2 \in \mathbb{K}^{\mu\times n}$ is obtained from G by eliminating $m-\mu$ rows of G then

$$\sigma_{k+m-\mu}(G) \le \sigma_k(G_2) \le \sigma_k(G), \quad 1 \le k \le \min\{\mu, \mu+n-m\}. \tag{17}$$

Proof: Let G_1 be obtained by eliminating the columns $j_1, \dots, j_{n-\nu}$ of G and suppose $1 \le k \le \nu$. We denote the standard unit vectors in $\mathbb{K}^n$ by $e^1, \dots, e^n$. Then by (12)

$$\begin{aligned}
\sigma_k(G) &= \min_{y^1,\dots,y^{k-1}\in\mathbb{K}^n} \max_{\substack{0\ne x\in\mathbb{K}^n \\ x\perp y^1,\dots,y^{k-1}}} \frac{\|Gx\|_2}{\|x\|_2} \\
&\ge \min_{y^1,\dots,y^{k-1}\in\mathbb{K}^n} \max_{\substack{0\ne x\in\mathbb{K}^n \\ x\perp y^1,\dots,y^{k-1} \\ x\perp e^{j_1},\dots,e^{j_{n-\nu}}}} \frac{\|Gx\|_2}{\|x\|_2} = \min_{u^1,\dots,u^{k-1}\in\mathbb{K}^\nu} \max_{\substack{0\ne z\in\mathbb{K}^\nu \\ z\perp u^1,\dots,u^{k-1}}} \frac{\|G_1 z\|_2}{\|z\|_2} = \sigma_k(G_1).
\end{aligned}$$

Also by (15)

$$\sigma_{n-\nu+k}(G) = \max_{y^1,\dots,y^{\nu-k}\in\mathbb{K}^n} \min_{\substack{0\neq x\in\mathbb{K}^n \\ x\perp y^1,\dots,y^{\nu-k}}} \frac{\|Gx\|_2}{\|x\|_2}$$

$$\leq \max_{y^1,\dots,y^{\nu-k}\in\mathbb{K}^n} \min_{\substack{0\neq x\in\mathbb{K}^n \\ x\perp y^1,\dots,y^{\nu-k} \\ x\perp e^{j_1},\dots,e^{j_{n-\nu}}}} \frac{\|Gx\|_2}{\|x\|_2} = \max_{u^1,\dots,u^{\nu-k}\in\mathbb{K}^\nu} \min_{\substack{0\neq z\in\mathbb{K}^\nu \\ z\perp u^1,\dots,u^{\nu-k}}} \frac{\|G_1 z\|_2}{\|z\|_2} = \sigma_k(G_1)\,.$$

The second statement of the corollary follows from the first by means of (3). □

As an immediate consequence of this corollary we obtain $\sigma_k(\tilde{G}) \leq \sigma_k(G)$, $k \in \underline{\nu}$ for every submatrix $\tilde{G} \in \mathbb{K}^{\mu\times\nu}$ of $G \in \mathbb{K}^{m\times n}$.

4.3.2 Singular Value Decomposition

The singular values of a matrix $G \in \mathbb{K}^{m\times n}$ are defined via the associated Hermitian matrix G^*G. In the previous section they have been characterized in term of the corresponding quadratic form $x \mapsto \langle x, G^*Gx\rangle$. In this subsection we will relate them directly to G.

Proposition 4.3.4 (Singular Value Decomposition). *Let $G \in \mathbb{K}^{m\times n}$ be a matrix with singular values $\sigma_1 \geq \sigma_2 \geq \cdots \geq \sigma_r > \sigma_{r+1} = \cdots = \sigma_n = 0$ and an associated orthonormal basis $(v^1,\dots,v^n)$ of (right) singular vectors, $V = [v^1,\dots,v^n] \in \mathbf{U}_n(\mathbb{K})$. Then there exists a unitary matrix $W = [w^1,\dots,w^m] \in \mathbf{U}_m(\mathbb{K})$ such that*

$$G = \sum_{i=1}^{r} \sigma_i w^i v^{i*} \tag{18}$$

or, equivalently,

$$G = W\Sigma V^* \quad \text{where} \quad \Sigma = \begin{bmatrix} \Sigma_r & 0 \\ 0 & 0 \end{bmatrix}_{m\times n}, \quad \Sigma_r = \operatorname{diag}(\sigma_1,\dots,\sigma_r). \tag{19}$$

Conversely, if $W \in \mathbf{U}_m(\mathbb{K})$ and $V \in \mathbf{U}_n(\mathbb{K})$ are unitary matrices such that (19) holds, then $\sigma_1,\dots,\sigma_r > 0$ are the non-zero singular values of G and of G^*. Moreover, the columns of V form an orthonormal basis of eigenvectors of G^*G (right singular vectors of G), and the columns w^j of W form an orthonormal basis of eigenvectors of GG^* (right singular vectors of G^*) satisfying $Gv^j = \sigma_j w^j$ and $G^*w^j = \sigma_j v^j$.

Proof: For the first statement, let $W_1 = [w^1,\dots,w^r]$ where

$$w^j = \sigma_j{}^{-1} G v^j, \qquad j \in \underline{r}\,. \tag{20}$$

Then $W_1^*W_1 = \Sigma_r^{-1}[v^1,\dots,v^r]^*\, G^*G\,[v^1,\dots,v^r]\,\Sigma_r^{-1} = I_r$. Augmenting the matrix W_1 by columns $w^{r+1},\dots,w^m$ which form an orthonormal basis of $(\operatorname{Im} G)^\perp = \ker G^*$ we obtain a unitary $m\times m$-matrix W satisfying

$$W \begin{bmatrix} \Sigma_r & 0 \\ 0 & 0 \end{bmatrix}_{m\times n} V^* v^j = W \begin{bmatrix} \Sigma_r & 0 \\ 0 & 0 \end{bmatrix}_{m\times n} e^j = \sigma_j w^j = Gv^j, \quad j \in \underline{n}$$

(because of (20) and $Gv^j = 0$, $j = r+1,\dots,n$). Hence (19) and (18).
The second statement of the proposition follows immediately since (19) implies

$$V^*G^*GV = \operatorname{diag}(\sigma_1^2,\dots,\sigma_r^2,0,\dots,0), \quad W^*GG^*W = \operatorname{diag}(\sigma_1^2,\dots,\sigma_r^2,0,\dots,0)$$

and $Gv^j = \left(\sum_{i=1}^r \sigma_i w^i v^{i*}\right) v^j = \sigma_j w^j$, $G^*w^j = \left(\sum_{i=1}^r \sigma_i v^i w^{i*}\right) w^j = \sigma_j v^j$. □

The factorization (19) of G is called the *singular value decomposition* (SVD) of G. Note that if $\sigma_1 > \sigma_2 > \cdots > \sigma_n$, then Proposition 4.3.4 implies that the columns of V are uniquely determined up to multiplication by scalars $\alpha_i \in \mathbb{K}$ satisfying $|\alpha_i| = 1$. If G is a non-singular $n \times n$-matrix then $\sigma_i > 0$, $i \in \underline{n}$ and (19) implies

$$G^{-1} = (W\Sigma_n V^*)^{-1} = V\Sigma_n^{-1}W^*. \tag{21}$$

So the ordered singular values of G^{-1} are $\sigma_n^{-1} \geq \sigma_{n-1}^{-1} \geq \cdots \geq \sigma_1^{-1}$, and we obtain by applying (10) with $k = 1$ to G^{-1}

$$\sigma_n(G) = \sigma_1(G^{-1})^{-1} = \|G^{-1}\|^{-1}. \tag{22}$$

For later use, the next remark explains the relationship between the SVD of a complex matrix and the SVD of its real form.

Remark 4.3.5. If $G = W\Sigma V^*$ is the singular value decomposition (19) of $G \in \mathbb{C}^{m\times n}$ and $G^{\mathbb{R}}$ is the representation of G in real form, it follows from Lemma A.1.18 that

$$G^{\mathbb{R}} = W^{\mathbb{R}}\,\Sigma^{\mathbb{R}}\,(V^{\mathbb{R}})^\top = \begin{bmatrix} \operatorname{Re} W & -\operatorname{Im} W \\ \operatorname{Im} W & \operatorname{Re} W \end{bmatrix} \begin{bmatrix} \Sigma & 0 \\ 0 & \Sigma \end{bmatrix} \begin{bmatrix} \operatorname{Re} V & -\operatorname{Im} V \\ \operatorname{Im} V & \operatorname{Re} V \end{bmatrix}^\top \tag{23}$$

is the singular value decomposition of $G^{\mathbb{R}} \in \mathbb{R}^{2m\times 2n}$. In particular, we see that $G^{\mathbb{R}}$ has the same singular values as G, but with double multiplicities. □

Proposition 4.3.4 shows how we may define singular values and singular vectors directly in terms of G.

Definition 4.3.6. Let $G \in \mathbb{K}^{m\times n}$ and $\sigma \geq 0$. A normalized pair $(w, v) \in \mathbb{K}^m \times \mathbb{K}^n$, $\|w\| = 1$, $\|v\| = 1$ is called a *singular pair* (or *Schmidt pair*) of G for the *singular value* σ if

$$Gv = \sigma w \qquad \text{and} \qquad G^*w = \sigma v\,. \tag{24}$$

If a pair of non-zero vectors $(w, v) \in \mathbb{K}^m \times \mathbb{K}^n$ satisfies (24), v is called a *right singular vector* and w a *left singular vector* of G for σ.

If a pair of non-zero vectors (w, v) satisfies (24) then $G^*Gv = \sigma^2 v$ and so σ is a singular value and v a right singular vector of G as defined in Subsection 4.3.1. Similarly, w is a right singular vector of G^*. Moreover, (v, w) is a singular pair for G^* if and only if (w, v) is a singular pair for G. The *dyadic decomposition* (18) shows that every matrix $G : \mathbb{K}^n \to \mathbb{K}^m$ of rank r is a linear combination of r rank 1 matrices $w^j v^{j*}$ associated with the singular pairs (w^j, v^j), $j \in \underline{r}$ of G.
Our main reason for introducing singular values is their importance to perturbation theory and model reduction. In fact the $(k+1)$-th singular value of G measures the distance of G from the set of $m \times n$–matrices of rank less than or equal to k (in the operator norm $\|\cdot\|$). We make this precise in the following theorem.

Theorem 4.3.7 (Schmidt-Mirsky Theorem). *Let $G \in \mathbb{K}^{m\times n}$ be a matrix of rank r with singular value decomposition* (18) *and denote by $M_k^{m,n}(\mathbb{K})$, the set*

$$M_k^{m,n}(\mathbb{K}) = \{X \in \mathbb{K}^{m\times n};\ \operatorname{rank} X \leq k\}, \quad k = 0, 1, \ldots, n. \tag{25}$$

Then

$$\sigma_{k+1}(G) = \min_{X \in M_k^{m,n}(\mathbb{K})} \|G - X\| = \|G - G_k\|\,, \quad k = 0, 1, \ldots, n-1 \tag{26}$$

where

$$G_k = \sum_{j=1}^{k} \sigma_j(G)\, w^j v^{j*} \in M_k^{m,n}(\mathbb{K}). \tag{27}$$

Proof: If $X \in M_k^{m,n}(\mathbb{K})$ then $\operatorname{codim} \ker X \le k$, so by the Minimax Theorem 4.3.1

$$\sigma_{k+1}(G) \le \|G|\ker X\| = \max_{0 \ne x \in \ker X} \frac{\|(G - X)x\|_2}{\|x\|_2} \le \|G - X\|\,.$$

On the other hand we have $\operatorname{rank} G_k \le k$ and

$$G - G_k = \sum_{j=k+1}^{r} \sigma_j(G)\, w^j v^{j*}.$$

Hence $\|G - G_k\| = \sigma_{k+1}(G)$ by Proposition 4.3.4 applied to $G - G_k$. □

Remark 4.3.8. Among the many solutions of the minimization problem (26), G_k enjoys the special property that

$$\sigma_\ell(G_k) = \sigma_\ell(G) \quad , \ell \in \underline{k}\,, \tag{28}$$

and hence the distance of G_k to $M_\ell^{m,n}(\mathbb{K})$, $\ell < k$, is the same as the distance of G to $M_\ell^{m,n}(\mathbb{K})$. Moreover $G_\ell = \sum_{j=1}^{\ell} \sigma_j(G) w^j v^{j*}$ is an optimal approximation of rank ℓ to G as well as to G_k. □

If $G \in \mathbb{K}^{n \times n}$ is invertible the above theorem enables us to find the distance of G from the set $M_{n-1}^{n,n}(\mathbb{K})$ of singular matrices.

Corollary 4.3.9. *For any $G \in \mathbf{Gl}_n(\mathbb{K})$, $\sigma_n(G)$ is equal to the distance of G from the set of singular matrices in $\mathbb{K}^{n \times n}$ with respect to the operator norm $\|\cdot\|$*

$$\sigma_n(G) = \min\{\|X\|\,;\ X \in \mathbb{K}^{n \times n}\,,\ \det(G - X) = 0\}. \tag{29}$$

If the singular value decomposition of G is of the form (18), then $G_{n-1} = G - \sigma_n w^n v^{n*}$ is an optimal singular approximation of G.

Example 4.3.10. For fixed $\alpha, \beta \in \mathbb{K}^* = \mathbb{K} \setminus \{0\}$, $\alpha \ne \beta$, consider the matrices

$$G(\gamma) = \begin{bmatrix} \alpha & \gamma \\ 0 & \beta \end{bmatrix}, \quad \gamma \in \mathbb{K}\,.$$

All these matrices are similar to the diagonal matrix $G(0)$ and hence belong to the same similarity class. Obviously $\sigma_1(G(\gamma)) = \|G(\gamma)\| \to \infty$ if $|\gamma| \to \infty$. Applying this to $G(\gamma)^{-1}$ (which has the same form) we get from (22) that $\sigma_2(G(\gamma)) = \|G(\gamma)^{-1}\|^{-1} \to 0$ if $|\gamma| \to \infty$. Thus similarity transformations can change the singular values to make $\sigma_1(G)$ arbitrarily large and $\sigma_2(G)$ arbitrarily small. In particular, by Corollary 4.3.9 the distance between the similarity class of G and the set of singular matrices is zero. This observation extends to arbitrary matrices $G \in \mathbf{Gl}_n(\mathbb{K})$ which are not multiples of I_n, see Ex. 16. □

For the singular values of perturbations of G we have a general estimate analogous to the estimate for eigenvalues in the normal case, see (2.9).

Corollary 4.3.11. *Suppose $G, \Delta \in \mathbb{K}^{m\times n}$ and denote by $\sigma_k(G)$, $\sigma_k(G+\Delta)$, $k \in \underline{n}$ the ordered singular values of G and $G+\Delta$. Then*

$$|\sigma_k(G) - \sigma_k(G+\Delta)| \le \|\Delta\|, \qquad k \in \underline{n}. \tag{30}$$

Proof: By Theorem 4.3.7 for $k = 0, \ldots, n-1$

$$\sigma_{k+1}(G) = \min_{\substack{X\in\mathbb{K}^{m\times n}\\ \operatorname{rank} X\le k}} \|G - X\| \le \min_{\substack{X\in\mathbb{K}^{m\times n}\\ \operatorname{rank} X\le k}} [\|\Delta\| + \|G+\Delta-X\|] = \|\Delta\| + \sigma_{k+1}(G+\Delta).$$

Interchanging the roles of G and $G+\Delta$, we obtain (30). □

The above result explains why, from a numerical point of view, singular values are "safer" to compute than eigenvalues, see Section 4.5.

Example 4.3.12. Consider again the matrices $A(\alpha)$, $\Delta(\varepsilon)$ of Example 4.2.17 and fix $\alpha = 9$. A simple calculation shows that the singular values of $A = A(9)$ are $\sigma_1 = (43+\sqrt{1845})^{1/2} = 9.2711$ and $\sigma_2 = (43-\sqrt{1845})^{1/2} = 0.2157$. The singular values of $A+\Delta(\varepsilon)$ are

$$\left[43 + \varepsilon^2/2 \pm (1845 + 36\varepsilon - 38\varepsilon^2 + \varepsilon^4/4)^{1/2}\right]^{1/2} \approx \left[43 \pm \sqrt{1845} \pm 18\varepsilon/\sqrt{1845}\right]^{1/2}$$

to first order in ε. Hence to first order in ε the change in σ_1 is $9\varepsilon/\sqrt{1845} = 0.213\varepsilon$ and the change in σ_2 is $-9\varepsilon/\sqrt{1845} = -0.209\varepsilon$, confirming (30). The change in the eigenvalues due to a perturbation $\Delta(\varepsilon)$ is more than 40 times larger ($\approx 9\varepsilon$, see Example 4.2.17). □

The following characterizations will be of use in the next section.

Theorem 4.3.13. *Let $G \in \mathbb{K}^{m\times n}$, then for every $k \in \underline{n}$*

$$\begin{aligned} \sigma_k(G) &= \min\{\|\Delta\|;\ \Delta \in \mathbb{K}^{m\times n}, \operatorname{rank}(G-\Delta) < k\} && (31)\\ &= [\inf\{\|\Delta\|;\ \Delta \in \mathbb{K}^{n\times m}, \dim\ker(I_n - \Delta G) \ge k\}]^{-1} && (32)\end{aligned}$$

where, by convention, $\inf\emptyset = \infty$ *and* $\infty^{-1} = 0$.

Proof: (31) follows from Theorem 4.3.7 by setting $G - X = \Delta$.
If $\sigma_k(G) = 0$, i.e. $\operatorname{rank} G < k$, then $\operatorname{rank}(\Delta G) < k$ and $\dim\ker(I_n - \Delta G) < k$ for all $\Delta \in \mathbb{K}^{n\times m}$. So (32) holds in this case. Now suppose $1 \le k \le \operatorname{rank} G$. If $\Delta \in \mathbb{K}^{n\times m}$ and $E := \ker(I_n - \Delta G)$ has dimension $\ge k$, then

$$\|x\|_2 = \|\Delta G x\|_2 \le \|\Delta\|\,\|Gx\|_2, \quad x \in E.$$

Hence $\|\Delta\|\,\sigma_{\min}(G|E) \ge 1$ so that $\sigma_k(G) \ge \|\Delta\|^{-1}$ by (15). On the other hand if (18) is a SVD of G and $\Delta \in \mathbb{K}^{n\times m}$ is defined by

$$\Delta = \sum_{j=1}^{k} \sigma_j(G)^{-1} v^j w^{j*}$$

then $(I_n - \Delta G)v^j = 0$, $j = 1, \ldots, k$ and $\|\Delta\| = \sigma_k(G)^{-1}$. This proves (32). □

Since $\dim\ker(I_n - \Delta G) = \dim\ker(I_m - G\Delta)$, $\Delta \in \mathbb{K}^{n\times m}$ (see Ex. 9), we may replace $(I_n - \Delta G)$ by $(I_m - G\Delta)$ in (32).

4.3.3 Matrices Depending on a Real Parameter

Let us now consider the case where the matrix G depends analytically on a scalar parameter. Since singular values are always real we cannot expect them to depend analytically on a *complex* parameter (every analytic map from a domain $\Omega \subset \mathbb{C}$ into $\mathbb{R}$ is constant). Therefore we consider matrices depending analytically on a *real* parameter τ. In particular, these matrices are Lipschitz continuous in τ and so by Corollary 4.3.11 the ordered singular values $\sigma_i : \tau \to \sigma(G(\tau))$ are Lipschitz continuous. However, they will in general not be analytic. Consider $G(\tau) = \begin{bmatrix} \tau & 0 \end{bmatrix}$, $\tau \in (-1, +1)$. The uppermost singular value is $\sigma_1(G(\tau)) = |\tau|$ and although $G(\cdot)$ is analytic on $(-1, +1)$, this singular value is not. The problem here is the positivity requirement on the singular values. Another problem may be caused by the ordering of the singular values. Figure 4.3.3 shows an example where the uppermost singular value $\sigma_1(G(\tau))$ is not differentiable at the points τ_1, τ_2 where it meets the second singular value $\sigma_2(G(\tau))$. Dispensing with the ordering and positivity requirements we will see in the following that the singular values of $G(\tau)$ *together with their negatives* can be represented by analytic functions. Moreover we will construct an analytic *pseudo-SVD* where the diagonal entries, the "pseudo singular values", are analytic in τ and the set of absolute values of these entries is identical with the set of singular values of $G(\tau)$.
The following lemma provides a basis for our construction. Its proof is straightforward and is set as Ex. 11.

Lemma 4.3.14. *Given $G \in \mathbb{K}^{m\times n}$ of rank r, then $A = \begin{bmatrix} 0 & G \\ G^* & 0 \end{bmatrix} \in \mathbb{K}^{(m+n)\times(m+n)}$ is of rank $2r$ and the unordered $2r$-tuple of non-zero eigenvalues of A is given by*

$$\lfloor \sigma_1(G), ..., \sigma_r(G), -\sigma_1(G), ..., -\sigma_r(G) \rfloor.$$

Suppose $I \subset \mathbb{R}$ is an open interval and $G(\cdot) : I \to \mathbb{K}^{m\times n}$ is an analytic matrix function of *generic rank* $r = \max_{\tau \in I} \operatorname{rank} G(\tau)$. Then $A(\tau) = \begin{bmatrix} 0 & G(\tau) \\ G(\tau)^* & 0 \end{bmatrix} \in \mathcal{H}_{m+n}(\mathbb{K})$ is an analytic family of Hermitian matrices on I of generic rank $2r$. Let $C_G := C_A$ be the set of critical parameter values of $A(\cdot)$, see Section 4.2. Then $C_G \subset I$ is the discrete subset of all $\tau_0 \in I$ where the number of distinct singular values of $G(\tau_0)$ is strictly smaller than the maximal number of distinct singular values of $G(\tau)$ on I.

Lemma 4.3.15. *Suppose $G(\cdot) : I \to \mathbb{K}^{m\times n}$ is an analytic matrix function of generic rank $r = \max_{\tau \in I} \operatorname{rank} G(\tau)$ on an open interval $I \subset \mathbb{R}$ and let $\sigma_j(\tau) = \sigma_j(G(\tau))$, $j \in \underline{n}$ be the ordered singular values of $G(\tau)$, $\tau \in I$. Then*

(i) there exist $2r$ analytic functions $\lambda_j(\cdot) : I \to \mathbb{R}$, $j = 1, \ldots, 2r$ such that

$$\lfloor \lambda_1(\tau), \ldots, \lambda_{2r}(\tau) \rfloor = \lfloor \sigma_1(\tau), ..., \sigma_r(\tau), -\sigma_1(\tau), ..., -\sigma_r(\tau) \rfloor, \quad \tau \in I. \tag{33}$$

(ii) If (33) holds for some analytic $\lambda_j(\cdot) : I \to \mathbb{R}$, $j = 1, \ldots, 2r$ and $J \subset I$ is an open subinterval with $J \cap C_G = \emptyset$ then there exist for each $j \in \underline{r}$ a subindex $1 \le i_j \le 2r$ such that $\sigma_j(\tau) = \lambda_{i_j}(\tau)$ for all $\tau \in J$. In particular, every ordered singular value of $G(\cdot)$ is analytic on J.

Proof: (i) Applying Proposition 4.2.42 (iii) to $A(\tau)=\begin{bmatrix} 0 & G(\tau) \\ G(\tau)^* & 0 \end{bmatrix} \in \mathcal{H}_{m+n}(\mathbb{K})$ we see that there exist $m+n$ analytic functions $\lambda_j(\cdot) : I \to \mathbb{R}$ such that the unordered $(m+n)$-tuple of eigenvalues of $A(\tau)$ is given by $\Lambda(A(\tau)) = \lfloor \lambda_1(\tau), \ldots, \lambda_{m+n}(\tau) \rfloor$ for $\tau \in I$. Since the generic rank of $A(\tau)$ is $2r$ by Lemma 4.3.14, we may order the $\lambda_j(\tau)$ in such a way that $\lambda_j(\tau) \not\equiv 0$ for $j = 1, \ldots, 2r$ and $\lambda_j(\tau) \equiv 0$ for $j = 2r+1, \ldots, m+n$. Then (33) follows from Lemma 4.3.14 (at the critical points $\tau \in C_G$ by continuity).
(ii) Let $J \subset I$ be an open subinterval with $J \cap C_G = \emptyset$. Then the graphs of two distinct non-zero eigenvalue functions $\lambda_i(\cdot), \lambda_j(\cdot)$, $1 \le i, j \le 2r$ do not intersect over the interval J. Hence if $\sigma_j(\tau') = \lambda_{i_j}(\tau')$ for some $\tau' \in J$ then $\sigma_j(\tau) = \lambda_{i_j}(\tau)$ for all $\tau \in J$ by Lemma 4.3.14. □

We now begin with the local construction of an analytic pseudo-SVD of $G(\tau)$. Let $\tau_0 \in I$ be an arbitrary point, possibly critical. By Proposition 4.2.42 (iii) there exist $\delta > 0$, analytic functions $\lambda_j(\cdot) : I(\tau_0, \delta) = (\tau_0 - \delta, \tau_0 + \delta) \to \mathbb{R}$ and an analytic unitary matrix

$$U(\tau) = \begin{bmatrix} u_1^1(\tau) & \cdots & u_1^{m+n}(\tau) \\ u_2^1(\tau) & \cdots & u_2^{m+n}(\tau) \end{bmatrix} \in \mathbf{U}_{m+n}(\mathbb{K}), \qquad \tau \in I(\tau_0, \delta)$$

where $u_1^j(\tau) \in \mathbb{K}^m$, $u_2^j(\tau) \in \mathbb{K}^n$, $j = 1, \ldots, m+n$ are such that

$$A(\tau)U(\tau) = U(\tau)\,\mathrm{diag}\,(\lambda_1(\tau), ..., \lambda_{m+n}(\tau)), \quad \tau \in I(\tau_0, \delta). \tag{34}$$

Choosing $\delta > 0$ sufficiently small we may assume that $I(\tau_0, \delta)$ does not contain any point of $C_G \setminus \{\tau_0\}$. By Lemma 4.3.14 $A(\tau)$ has exactly $2r$ non-zero eigenvalues (taking account of multiplicities) for each $\tau \in I(\tau_0, \delta) \setminus \{\tau_0\}$, the remaining $m + n - 2r$ eigenvalues are identically zero on $I(\tau_0, \delta)$[3]. Reordering the eigenvalue functions $\lambda_j(\cdot)$ and the corresponding columns of $U(\tau)$ appropriately we may assume that $\lambda_{2r+1}(\tau), \ldots, \lambda_{m+n}(\tau)$ are identically zero. We have seen in the proof of Lemma 4.3.15 that the unordered $2r$-tuple of non-zero eigenvalue functions of $A(\tau)$ satisfies the equation (33) on $I(\tau_0, \delta)$. These eigenvalues can be ordered in such a way that the first r of them (taking account of multiplicities), $\lambda_1(\tau), \ldots, \lambda_r(\tau)$, are positive to the left of τ_0 in $I(\tau_0, \delta)$. Since $C_A \cap (I(\tau_0, \delta) \setminus \{\tau_0\}) = \emptyset$ two eigenvalue functions $\lambda_i(\cdot), \lambda_j(\cdot)$, $1 \le i, j \le 2n$ are identically equal if they admit the same value at some point $\tau \in I(\tau_0, \delta) \setminus \{\tau_0\}$. Thus we can order the first r eigenvalue functions in such a way that $\lambda_1(\tau) \ge \ldots \ge \lambda_r(\tau) > 0$ for $\tau \in (\tau_0 - \delta, \tau_0)$. As a consequence we obtain

$$\sigma_j(G(\tau)) = \lambda_j(\tau), \qquad j \in \underline{r}, \;\; \tau \in (\tau_0 - \delta, \tau_0).$$

By (34)

$$\begin{aligned} G(\tau)u_2^j(\tau) &= \lambda_j(\tau)u_1^j(\tau) \\ G(\tau)^*u_1^j(\tau) &= \lambda_j(\tau)u_2^j(\tau), \qquad \tau \in I(\tau_0, \delta), \quad j \in \underline{m+n}. \end{aligned} \tag{35}$$

Hence

$$\begin{aligned} G(\tau)^*G(\tau)u_2^j(\tau) &= \lambda_j(\tau)^2 u_2^j(\tau), \\ G(\tau)G(\tau)^*u_1^j(\tau) &= \lambda_j(\tau)^2 u_1^j(\tau), \qquad \tau \in I(\tau_0, \delta), \quad j \in \underline{m+n}. \end{aligned} \tag{36}$$

[3]If a non-zero eigenvalue function $\lambda_j(\cdot)$ vanishes at $\tau' \in I$ then $\tau' \in C_A$ since at this point the two distinct eigenvalue functions $\lambda_j(\cdot)$ and $-\lambda_j(\cdot)$ of $A(\cdot)$ coalesce.

Lemma 4.3.16. *Let $v^j(\tau) = \sqrt{2}\,u_2^j(\tau)$, $w^j(\tau) = \sqrt{2}\,u_1^j(\tau)$ for $j = 1, \ldots, 2r$. Then the following hold.*

(i) *If $1 \le j \le 2r$ and $\tau \in I(\tau_0, \delta)$, then $\|v^j(\tau)\|_2 = \|w^j(\tau)\|_2 = 1$ and $(w^j(\tau), v^j(\tau))$ is an analytic singular pair of $G(\tau)$ for $\lambda_j(\tau)$ on $I(\tau_0, \delta)$ in the following sense*

$$G(\tau)v^j(\tau) = \lambda_j(\tau)w^j(\tau), \quad G(\tau)^*w^j(\tau) = \lambda_j(\tau)v^j(\tau), \quad \tau \in I(\tau_0, \delta).$$

For each $\tau \in I(\tau_0, \delta)$, either $(w^j(\tau), v^j(\tau))$ or $(w^j(\tau), -v^j(\tau))$ is a singular pair of $G(\tau)$ for the singular value $|\lambda_j(\tau)|$.

(ii) *$\langle v^i(\tau), v^j(\tau)\rangle = \delta_{ij}$, $\langle w^i(\tau), w^j(\tau)\rangle = \delta_{ij}$ for all $1 \le i, j \le r$, $\tau \in I(\tau_0, \delta)$.*

Proof: Let $1 \le j \le 2r$, then $\lambda_j(\tau) \ne 0$ for every $\tau \in I(\tau_0, \delta)$, $\tau \ne \tau_0$. By (35) $u_2^j(\tau) = 0$ (resp. $u_1^j(\tau) = 0$) would imply $u_1^j(\tau) = 0$ (resp. $u_2^j(\tau) = 0$), and since $\|u_2^j(\tau)\|_2^2 + \|u_1^j(\tau)\|_2^2 = 1$ we have $u_2^j(\tau) \ne 0$ and $u_1^j(\tau) \ne 0$ for all $\tau \in I(\tau_0, \delta) \setminus \{\tau_0\}$. By (35) and (36)

$$\lambda_j(\tau)^2 \, \|u_1^j(\tau)\|_2^2 = \|G(\tau)u_2^j(\tau)\|_2^2 = u_2^j(\tau)^* G(\tau)^* G(\tau) u_2^j(\tau) = \lambda_j(\tau)^2 \, \|u_2^j(\tau)\|_2^2.$$

So $\|u_1^j(\tau)\|_2 = \|u_2^j(\tau)\|_2 = 1/\sqrt{2}$ for $\tau \in I(\tau_0, \delta) \setminus \{\tau_0\}$ and hence by continuity for all $\tau \in I(\tau_0, \delta)$. This proves $\|v^j(\tau)\| = \|w^j(\tau)\| = 1$, $\tau \in I(\tau_0, \delta)$, and the remaining statements of (i) are direct consequences of (35). $(w^j(\tau), v^j(\tau))$ is a singular pair of $G(\tau)$ for $\lambda_j(\tau)$ if $\lambda_j(\tau) \ge 0$, whereas $(w^j(\tau), -v^j(\tau))$ is a singular pair for $-\lambda_j(\tau)$ if $\lambda_j(\tau) < 0$.
(ii) It only remains to prove (ii) for $i \ne j$. Let $i \ne j$, $1 \le i, j \le r$ be fixed, $\tau \in I(\tau_0, \delta)$. The columns of $U(\tau)$ are mutually orthogonal, so

$$\langle u_1^j(\tau), u_1^i(\tau)\rangle + \langle u_2^j(\tau), u_2^i(\tau)\rangle = 0. \tag{37}$$

From (35) we have

$$\lambda_i(\tau)\langle u_1^j(\tau), u_1^i(\tau)\rangle = \langle u_1^j(\tau), G(\tau)u_2^i(\tau)\rangle = \langle G(\tau)^* u_1^j(\tau), u_2^i(\tau)\rangle = \lambda_j(\tau)\langle u_2^j(\tau), u_2^i(\tau)\rangle.$$

So by (37)

$$\lambda_i(\tau)\langle u_1^j(\tau), u_1^i(\tau)\rangle = -\lambda_j(\tau)\langle u_1^j(\tau), u_1^i(\tau)\rangle, \text{ i.e. } \quad (\lambda_i(\tau) + \lambda_j(\tau))\langle u_1^j(\tau), u_1^i(\tau)\rangle = 0.$$

If $\lambda_i(\tau) + \lambda_j(\tau) \ne 0$ we obtain $\langle u_1^j(\tau), u_1^i(\tau)\rangle = 0$ and from (37) $\langle u_2^j(\tau), u_2^i(\tau)\rangle = 0$ as well. But $\lambda_i(\tau) + \lambda_j(\tau) \ne 0$ for all but isolated points $\tau \in I(\tau_0, \delta)$ so that (ii) follows for all $\tau \in I(\tau_0, \delta)$ by continuity. □

Lemma 4.3.16 shows that

$$V_1(\tau) = [v^1(\tau), ..., v^r(\tau)], \quad W_1(\tau) = [w^1(\tau), ..., w^r(\tau)], \quad \tau \in I(\tau_0, \delta) \tag{38}$$

define analytic matrix functions with orthonormal columns on $I(\tau_0, \delta)$. By Corollary 4.2.43 there exist analytic matrix functions

$$V_2(\cdot) : I(\tau_0, \delta) \to \mathbb{K}^{n \times (n-r)}, \quad W_2(\cdot) : I(\tau_0, \delta) \to \mathbb{K}^{m \times (m-r)}$$

such that $V(\tau) = [V_1(\tau)\ V_2(\tau)] \in \mathbf{U}_n(\mathbb{K})$ and $W(\tau) = [W_1(\tau)\ W_2(\tau)] \in \mathbf{U}_m(\mathbb{K})$ for all $\tau \in I(\tau_0, \delta)$. For $\tau \in I(\tau_0, \delta) \setminus \{\tau_0\}$ we have $\operatorname{rank} G(\tau)^* G(\tau) = r$ and by

(36) $G(\tau)^*G(\tau)v^j(\tau) = \lambda_j(\tau)^2 v^j(\tau)$ for $j \in \underline{r}$, hence the r-dimensional subspace $\operatorname{Im} G(\tau)^*G(\tau)$ is spanned by the columns of $V_1(\tau)$. It follows that $\ker G(\tau)^*G(\tau) = (\operatorname{Im} G(\tau)^*G(\tau))^\perp = (\operatorname{Im} V_1(\tau))^\perp = \operatorname{Im} V_2(\tau)$ for $\tau \in (I(\tau_0,\delta) \setminus \{\tau_0\})$ and so, by continuity $G(\tau)V_2(\tau) = 0$, $\tau \in I(\tau_0,\delta)$. Similarly $G(\tau)^*W_2(\tau) = 0$ for all $\tau \in I(\tau_0,\delta)$. Thus by (35) we obtain the *analytic pseudo-SVD* : $G(\tau) = W(\tau)\Sigma(\tau)V(\tau)^*$ on $I(\tau_0,\delta)$ where

$$\Sigma(\tau) = W(\tau)^*G(\tau)V(\tau) = \left[\begin{array}{ccccccc} \lambda_1(\tau) & 0 & \cdots & 0 & 0 & \cdots & 0 \\ 0 & \lambda_2(\tau) & \cdots & 0 & 0 & \cdots & 0 \\ \vdots & & \ddots & \vdots & \vdots & & \vdots \\ 0 & \cdots & \cdots & \lambda_r(\tau) & 0 & \cdots & 0 \\ \hline 0 & \cdots & \cdots & 0 & 0 & \cdots & 0 \\ \vdots & & & \vdots & \vdots & & \vdots \\ 0 & \cdots & \cdots & 0 & 0 & \cdots & 0 \end{array}\right]_{m\times n} . \tag{39}$$

We have therefore proved the following theorem.

Theorem 4.3.17. *Suppose $I \subset \mathbb{R}$ is an open interval and $G(\cdot) : I \to \mathbb{K}^{m\times n}$ is an analytic matrix function with $r = \max_{\tau\in I} \operatorname{rank} G(\tau)$. Then for every $\tau_0 \in I$ there exist a neighbourhood $I(\tau_0,\delta)$, $\delta > 0$ and analytic functions $\lambda_j(\cdot) : I(\tau_0,\delta) \to \mathbb{R}$, $j \in \underline{r}$, $V(\cdot) : I(\tau_0,\delta) \to \mathbf{U}_n(\mathbb{K})$, $W(\cdot) : I(\tau_0,\delta) \to \mathbf{U}_m(\mathbb{K})$ such that*

(i) $\lambda_j(\tau) = \sigma_j(G(\tau))$ *for all* $\tau \in (\tau_0 - \delta, \tau_0)$ *and* $j \in \underline{r}$.

(ii) For all $\tau \in I(\tau_0,\delta)$, the unordered r-tuple of (generically) non-zero singular values of $G(\tau)$ is given by $\lfloor |\lambda_1(\tau)|, \ldots, |\lambda_r(\tau)| \rfloor$.

(iii) For all $\tau \in I(\tau_0,\delta)$, we have the analytic pseudo-SVD $G(\tau) = W(\tau)\Sigma(\tau)V(\tau)^*$ where $\Sigma(\tau)$ is the diagonal matrix on the RHS of (39).

Remark 4.3.18. (i) If $\lambda_j(\tau) \neq 0$ for all $\tau \in I(\tau_0,\delta)$, then $\lambda_j(\tau) > 0$ is an (unordered) analytical singular value of $G(\tau)$ on $I(\tau_0,\delta)$ by Lemma 4.3.14. In particular, if $\operatorname{rank} G(\tau) = r$ for all $\tau \in I(\tau_0,\delta)$ then the unordered r-tuple of the non-zero singular values of $G(\tau)$ is given by $\lfloor \lambda_1(\tau), \ldots, \lambda_r(\tau) \rfloor$ for all $\tau \in I(\tau_0,\delta)$. Note, however, that the analytic eigenvalues $\lambda_j(\tau)$ will, in general, *not* be ordered for all $\tau \in I(\tau_0,\delta)$ and, conversely, the ordered non-zero eigenvalues $\sigma_1(G(\tau)) \geq \cdots \geq \sigma_r(G(\tau))$ will, in general, *not* be analytic on I.

(ii) We have noted in Remark 4.2.44 that Proposition 4.2.42 (iii) can be shown to hold globally. Using this, a global version of the previous theorem can be proved, i.e. the analytic functions $\lambda_j(\cdot)$, $V(\cdot)$, $W(\cdot)$ exist on the whole interval I. □

Example 4.3.19. Consider the analytic family of full rank matrices

$$G(\tau) = e^{\imath\tau} I_2 - \begin{bmatrix} 0.1 & -0.3 \\ -0.3 & 0.2 \end{bmatrix} = \begin{bmatrix} \cos\tau - 0.1 + \imath\sin\tau & 0.3 \\ 0.3 & \cos\tau - 0.2 + \imath\sin\tau \end{bmatrix}, \ \tau \in (0, 2\pi).$$

It follows from the global version of Theorem 4.3.17 that there exist two analytic functions $\lambda_1(\tau)$, $\lambda_2(\tau)$ on $(0,2\pi)$ which represent the two (unordered) singular values of $G(\tau)$ for all $\tau \in (0,2\pi)$. The graphs of these functions are shown in Fig. 4.3.3. Clearly the ordered singular values $\sigma_1(\cdot)$ and $\sigma_2(\cdot)$ are not differentiable at the points τ_1, $\tau_2 \in (0,2\pi)$ where the two graphs cross. □

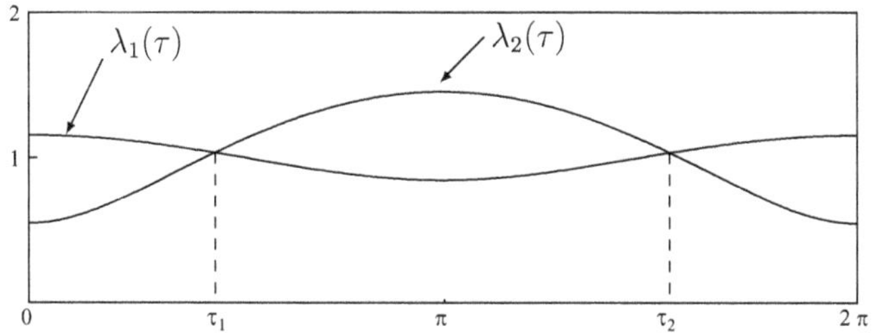

Figure 4.3.3: Singular values of $G(\tau)$, $0 < \tau < 2\pi$

In the next corollary we obtain an equation for the derivative (sensitivity) of $\lambda_j(\cdot)$.

Corollary 4.3.20. *Suppose the conditions of Theorem 4.3.17 and let $v^j(\tau)$, $w^j(\tau)$, $j \in \underline{r}$ denote the first r columns of $V(\tau)$ and $W(\tau)$, respectively. Then*

$$\frac{d\lambda_j(\tau)}{d\tau} = \text{Re}\left(w^j(\tau)^* \frac{dG(\tau)}{d\tau} v^j(\tau)\right), \quad \tau \in I(\tau_0, \delta), \ \ j \in \underline{r}. \tag{40}$$

Proof: (39) implies $\lambda_j(\tau) = w^j(\tau)^* G(\tau) v^j(\tau)$ and by (35) differentiation yields

$$\begin{aligned} \frac{d\lambda_j(\tau)}{d\tau} &= \frac{dw^j(\tau)}{d\tau}^* G(\tau)v^j(\tau) + w^j(\tau)^* \frac{dG(\tau)}{d\tau} v^j(\tau) + w^j(\tau)^* G(\tau) \frac{dv^j(\tau)}{d\tau} \\ &= \lambda_j(\tau) \frac{dw^j(\tau)}{d\tau}^* w^j(\tau) + w^j(\tau)^* \frac{dG(\tau)}{d\tau} v^j(\tau) + \lambda_j(\tau) v^j(\tau)^* \frac{dv^j(\tau)}{d\tau}. \end{aligned}$$

But $w^j(\tau)^* w^j(\tau) = v^j(\tau)^* v^j(\tau) = 1$ and so

$$\text{Re}\left(v^j(\tau)^* \frac{dv^j(\tau)}{d\tau}\right) = \text{Re}\left(\frac{dw^j(\tau)}{d\tau}^* w^j(\tau)\right) = 0.$$

Hence
$$\frac{d\lambda_j(\tau)}{d\tau} = \text{Re}\, \frac{d\lambda_j(\tau)}{d\tau} = \text{Re}\left(w^j(\tau)^* \frac{dG(\tau)}{d\tau} v^j(\tau)\right).$$

□

We conclude this subsection with a result which will be of use in the next section.

Proposition 4.3.21. *Under the conditions of Theorem 4.3.17, suppose that an ordered singular value $\sigma_j(\cdot) \not\equiv 0$ of $G(\cdot)$ has a local extremum (minimum or maximum) at $\hat{\tau} \in I$, then there exists an associated singular pair $(w^j, v^j) \in \mathbb{K}^m \times \mathbb{K}^n$ of $G(\hat{\tau})$ such that*

$$\text{Re}\left(w^{j*} \frac{dG}{d\tau}(\hat{\tau})\, v^j\right) = 0. \tag{41}$$

Proof: Suppose the set-up is that of Theorem 4.3.17 with $\tau_0 = \hat{\tau}$ and let $\sigma_j(\cdot) \not\equiv 0$ be an ordered singular value of $G(\cdot)$. If $\sigma_j(\cdot)$ is analytic on an open subinterval $J \subset I$ containing $\hat{\tau}$ it follows from Theorem 4.3.17 that $\sigma_j(\tau) = \lambda_j(\tau)$ for all $\tau \in J$. Since $\hat{\tau}$ is a local extremum of $\sigma_j(\cdot)$ we have $\frac{d\sigma(\hat{\tau})}{d\tau} = 0$, and so, setting $w^j = w^j(\hat{\tau})$, $v^j = v^j(\hat{\tau})$, equation (41) is a direct consequence of (40).
Now suppose that $\sigma_j(\cdot)$ is not analytic around $\hat{\tau}$, hence $\hat{\tau} \in C_G$ by Lemma 4.3.15, and let $\delta > 0$ be sufficiently small such that $I(\hat{\tau}, \delta) \cap C_G = \{\hat{\tau}\}$. By Theorem 4.3.17

we have $\sigma_j(\tau) = \lambda_j(\tau)$ for $\tau \in (\hat{\tau} - \delta, \hat{\tau})$ and by Lemma 4.3.15 there exists $1 \leq i_j \leq 2r$, $i_j \neq j$ such that $\sigma_j(\tau) = \lambda_{i_j}(\tau)$ for $\tau \in (\hat{\tau}, \hat{\tau} + \delta)$. By continuity $\lambda_j(\hat{\tau}) = \sigma_j(\hat{\tau}) = \lambda_{i_j}(\hat{\tau})$. Let $(w^j(\tau),\, v^j(\tau))$, $(w^{i_j}(\tau),\, v^{i_j}(\tau))$ be analytic singular pairs of $G(\tau)$ for $\lambda_j(\tau)$ and $\lambda_{i_j}(\tau)$ respectively, see Lemma 4.3.16. Then by (40)

$$\frac{d\lambda_j(\tau)}{d\tau} = \operatorname{Re}\left(w^j(\tau)^* \frac{dG(\tau)}{d\tau} v^j(\tau)\right), \quad \frac{d\lambda_{i_j}(\tau)}{d\tau} = \operatorname{Re}\left(w^{i_j}(\tau)^* \frac{dG(\tau)}{d\tau} v^{i_j}(\tau)\right), \quad \tau \in I(\hat{\tau}, \delta).$$

Set $w_\alpha(\tau) = \alpha w^j(\tau) + (1-\alpha^2)^{1/2} w^{i_j}(\tau)$, $v_\alpha(\tau) = \alpha v^j(\tau) + (1-\alpha^2)^{1/2} v^{i_j}(\tau)$ for $\alpha \in [0,1]$, $\tau \in I(\hat{\tau}, \delta)$. Since $j \neq i_j$, $\langle w^j(\tau), w^{i_j}(\tau)\rangle = \langle v^j(\tau), v^{i_j}(\tau)\rangle = 0$ for all $\tau \in I(\hat{\tau}, \delta)$. Hence $(w_\alpha(\hat{\tau}),\, v_\alpha(\hat{\tau}))$ is also a singular pair of $G(\hat{\tau})$ corresponding to $\sigma_j(\hat{\tau})$. Define

$$f(\alpha) = \operatorname{Re} w_\alpha(\hat{\tau})^* \frac{dG}{d\tau}(\hat{\tau})\, v_\alpha(\hat{\tau}).$$

Because $\hat{\tau}$ is a local extremum of $\sigma_j(\cdot)$ and $\sigma_j(\tau) = \lambda_j(\tau)$ on $(\hat{\tau}-\delta, \hat{\tau})$, $\sigma_j(\tau) = \lambda_{i_j}(\tau)$ on $(\hat{\tau}, \hat{\tau}+\delta)$, we have

$$f(1)f(0) = \frac{d\lambda_j}{d\tau}(\hat{\tau})\, \frac{d\lambda_{i_j}}{d\tau}(\hat{\tau}) \leq 0.$$

Hence by continuity there exists $\hat{\alpha} \in [0,1]$ such that $f(\hat{\alpha}) = 0$. Setting $w^j = w_{\hat{\alpha}}(\hat{\tau})$, $v^j = v_{\hat{\alpha}}(\hat{\tau})$ proves the proposition. □

4.3.4 Relations between Eigenvalues and Singular Values

In this subsection we specialize to *square* matrices $G \in \mathbb{K}^{n\times n}$ and discuss the relationship between eigenvalues and singular values of G.
Whereas the eigenvalues of a matrix are similarity invariants, the singular values of a matrix are only invariant with respect to *unitary* similarity transformations. Let $G \in \mathbb{K}^{n\times n}$ have singular values $\sigma_j = \sigma_j(G)$ and eigenvalues $\lambda_j = \lambda_j(G)$ taking account of multiplicity and ordered by

$$|\lambda_1| \geq |\lambda_2| \geq \cdots \geq |\lambda_n|\,. \tag{42}$$

If G has a singular value decomposition $G = W\Sigma V^*$, $\Sigma = \operatorname{diag}(\sigma_1, \ldots, \sigma_n)$, then

$$|\lambda_1 \cdots \lambda_n| = |\det W\Sigma V^*| = \det \Sigma = \sigma_1 \cdots \sigma_n. \tag{43}$$

Now suppose that G is normal. Then there exists a unitary matrix $V \in \mathbf{U}_n(\mathbb{K})$ such that

$$V^*GV = \operatorname{diag}(\lambda_1, \lambda_2, \ldots, \lambda_n).$$

So $V^*G^*GV = (V^*GV)^*V^*GV = \operatorname{diag}(|\lambda_1|^2, \ldots, |\lambda_n|^2)$ and hence

$$\sigma_j = |\lambda_j|, \qquad j \in \underline{n}\,. \tag{44}$$

In particular, if $G \in \mathcal{H}_n(\mathbb{K})$, $G \succeq 0$ then $\sigma_j = \lambda_j \geq 0$, $j \in \underline{n}$. Now let $(v^1, \ldots, v^n)$ be an orthonormal eigenvector basis corresponding to the eigenvalues $\lambda_1 \geq \ldots \geq \lambda_n \geq 0$ and $r = \operatorname{rank}(G)$. If we define w^j by (20) we obtain $w^j = {\sigma_j}^{-1} G v^j = v^j$ for $j = 1, \ldots, r$. So the SVD of G has the form

$$G = \sum_{j=1}^{r} \sigma_j v^j v^{j*} = CC^* \quad \text{where } C = [\sigma_1^{1/2} v^1 \cdots \sigma_r^{1/2} v^r]. \tag{45}$$

Remark 4.3.22. (45) shows that every $G \in \mathcal{H}_n^+(\mathbb{K})$ has a factorization of the form $G = CC^*$ where $C \in \mathbb{K}^{n\times r}$ is a rectangular matrix of full column rank. Setting $B = [\sigma_1^{1/2} v^1 \cdots \sigma_n^{1/2} v^n] = [C \quad 0_{n\times(n-r)}]$ we obtain instead a factorization $G = BB^*$ with a square factor $B \in \mathbb{K}^{n\times n}$. Now B in turn can be factorized in the form $B = LU$ where L is *lower triangular with non-negative diagonal entries* and $U \in \mathbf{U}(\mathbb{K})$ (see Lemma 4.5.12). So we obtain $G = LUU^*L = LL^*$. Such a factorization is called a *Cholesky factorization* of G. If G is positive definite, the Cholesky factor L is uniquely determined, see Ex. 5. □

In general the relationship between eigenvalues and singular values of a square matrix is far less tight than for normal matrices. We have, however, the following estimates.

Proposition 4.3.23 (Weyl). *Let $G \in \mathbb{K}^{n\times n}$ be a square matrix with eigenvalues $\lambda_1, \ldots, \lambda_n$ and singular values $\sigma_1, \ldots, \sigma_n$ ordered by (42) and (2), respectively. Then*

$$|\lambda_1 \cdots \lambda_k| \le \sigma_1 \cdots \sigma_k, \quad k \in \underline{n}, \tag{46}$$

with equality for $k = n$.

Proof: By unitary similarity transformation (see Theorem 4.5.16) G can be brought into upper triangular form (complex Schur form) $T = (t_{ij}) = U^*GU$, $U \in \mathbf{U}_n(\mathbb{C})$ such that the diagonal entries of T are $t_{jj} = \lambda_j$, $j \in \underline{n}$. For any $k \in \underline{n}$, let $U_k \in \mathbb{C}^{n\times k}$ be the submatrix consisting of the first k columns of U. Then $T_k := U_k^* G U_k$ is the upper left $k \times k$-submatrix of T with diagonal entries $\lambda_1, \ldots, \lambda_k$. By Corollary 4.3.3 and the unitary invariance of singular values we have $\sigma_j(T_k) \le \sigma_j(T) = \sigma_j$ for $j = 1, \ldots, k$, hence by application of (43) to T_k

$$|\lambda_1 \cdots \lambda_k| = |\det T_k| = \sigma_1(T_k) \cdots \sigma_k(T_k) \le \sigma_1 \cdots \sigma_k.$$

□

Corollary 4.3.24. *Under the conditions of the preceding proposition*

$$\sigma_n \le |\lambda_n| \le \sigma_n^{\frac{1}{n}} (\sigma_1)^{\frac{n-1}{n}}, \qquad \sigma_1^{\frac{1}{n}} (\sigma_n)^{\frac{n-1}{n}} \le |\lambda_1| \le \sigma_1. \tag{47}$$

Proof: By the previous proposition

$$|\lambda_n|^n \le |\lambda_1||\lambda_2| \cdots |\lambda_n| = \sigma_1\sigma_2 \cdots \sigma_n \le \sigma_1^{n-1}\sigma_n\,.$$

Similarly

$$|\lambda_1|^n \ge |\lambda_1| \cdots |\lambda_n| = \sigma_1\sigma_2 \cdots \sigma_n \ge \sigma_1\sigma_n^{n-1}\,.$$

This proves half of the inequalities in (47). The other half follows from

$$\sigma_1 = \|G\| = \max_{\|x\|_2=1} \|Gx\|_2 \ge |\lambda_1|,$$

$$\sigma_n^{-1} = \|G^{-1}\| = \max_{\|x\|_2=1} \|G^{-1}x\|_2 \ge |\lambda_n|^{-1} \quad \text{if} \quad \sigma_n \ne 0.$$

□

The next proposition shows that the estimate $\sigma_n \le |\lambda_n|$ is tight on every similarity orbit.

Proposition 4.3.25. *For any $G \in \mathbb{C}^{n\times n}$,*

$$\sup_{T\in\mathbf{Gl}_n(\mathbb{C})} \sigma_n(TGT^{-1}) = |\lambda_n(G)|. \tag{48}$$

Proof: By (47) for every $T \in \mathbf{Gl}_n(\mathbb{C})$

$$\sigma_n(TGT^{-1}) \le |\lambda_n(TGT^{-1})| = |\lambda_n(G)|\,.$$

Hence $\le$ holds in (48). Conversely, given any $\varepsilon > 0$ there exists $T_\varepsilon \in \mathbf{Gl}_n(\mathbb{C})$ such that $T_\varepsilon G T_\varepsilon^{-1}$ is in Jordan canonical form but with entries ε instead of 1 on the superdiagonal. Now for diagonal matrices (44) holds and since $\sigma_n(G)$ depends continuously on G by Corollary 4.3.11, we obtain (48). □

Formula (48) has an interesting interpretation. We know that $\sigma_n(TGT^{-1})$ measures the distance of the matrix TGT^{-1} from the set of singular matrices $M_{n-1}^{n,n}(\mathbb{C})$ (see Corollary 4.3.9). Hence $|\lambda_n(G)|$ measures the *supremum of the distances of the matrices in the similarity orbit of G from the set of singular matrices in* $(\mathbb{C}^{n\times n}, \|\cdot\|)$:

$$|\lambda_n(G)| = \sup_{T\in\mathbf{Gl}_n(\mathbb{C})} \operatorname{dist}(TGT^{-1}, M_{n-1}^{n,n}(\mathbb{C})). \tag{49}$$

On the other hand note that (see Example 4.3.10 and Ex. 16)

$$\inf_{T\in\mathbf{Gl}_n(\mathbb{C})} \operatorname{dist}(TGT^{-1}, M_{n-1}^{n,n}(\mathbb{C})) = 0 \tag{50}$$

if G is not a multiple of the identity matrix I_n.

4.3.5 Exercises

1. Find the singular values, right singular vectors and SVD's of the matrices

$$(i)\ \begin{bmatrix} 0 & 1 \\ -1 & 0 \end{bmatrix}, \qquad (ii)\ \begin{bmatrix} 2 & 0 \\ 1 & -1 \end{bmatrix}, \qquad (iii)\ \begin{bmatrix} \imath & 1+\imath \\ 0 & 2\imath \end{bmatrix}.$$

2. Sketch the ellipsoids $\mathcal{E}^{-1}$, $\mathcal{E}$ given by (5) and (8) for the matrices in Ex. 1.

3. Let $G \in \mathbb{K}^{m\times n}$, $m \ge 1$, $n > 1$ and let $\tilde{G}$ be obtained by deleting one column of G. Prove the following interlacing property: If $\sigma_1 \ge \ldots \ge \sigma_n$ (resp. $\tilde\sigma_1 \ge \ldots \ge \tilde\sigma_{n-1}$) are the singular values of G (resp. $\tilde{G}$) then

$$\sigma_1 \ge \tilde\sigma_1 \ge \sigma_2 \ge \tilde\sigma_2 \ge \cdots \ge \tilde\sigma_{n-1} \ge \sigma_n.$$

4. Show that $G = \begin{bmatrix} 2 & 2 & 2 \\ 2 & 4 & 6 \\ 2 & 6 & 10 \end{bmatrix}$ is positive semi-definite and of rank 2. Find a matrix $C \in \mathbb{R}^{3\times 2}$ such that $G = CC^*$.

5. Prove the following results related to Remark 4.3.22.

(i) A Hermitian matrix $H \in \mathcal{H}_n(\mathbb{K})$ is positive semi-definite if and only if it has a Cholesky factorization (i.e. $H = LL^*$ with L lower triangular and non-negative diagonal entries). If H is positive definite, the Cholesky factor L is uniquely determined.

(ii) Let $B, C \in \mathbb{K}^{n\times n}$. Then $BB^* = CC^*$ if and only if there exists a unitary matrix $U \in \mathbf{U}_n(\mathbb{K})$ such that $B = CU$. (Hint: $\ker B^* = \ker C^*$ and $\|B^*x\|_2 = \|C^*x\|_2$ for all $x \in \mathbb{K}^n$.)

(iii) Suppose $B \in \mathbb{K}^{n\times m}$, $C \in \mathbb{K}^{n\times r}$ and $\operatorname{rank} C = r$ (C is of full column rank). Then $BB^* = CC^*$ if and only if there exists a matrix $U \in \mathbb{K}^{r\times m}$ such that $B = CU$ and $UU^* = I_r$.

6. If $A \in \mathbb{K}^{n\times n}$ prove that there exist positive semi-definite matrices P, $\tilde{P} \in \mathcal{H}_n(\mathbb{K})$ and a unitary $U \in \mathbf{U}_n(\mathbb{K})$ such that $A = PU = U\tilde{P}$ (polar decomposition of A). The matrices P, $\tilde{P}$ are uniquely determined by $P = (AA^*)^{1/2}$ and $\tilde{P} = (A^*A)^{1/2}$, see (6). If A is nonsingular then also U is uniquely determined by $U = P^{-1}A$.

7. Suppose $G \in \mathbb{K}^{m\times n}$ has a SVD $G = W\Sigma V^*$ (19) and define $G^\dagger = V\Sigma^+W^* \in \mathbb{K}^{n\times m}$ where $\Sigma^+ \in \mathbb{K}^{n\times m}$ is the transpose of the matrix obtained from Σ by replacing the positive singular values in Σ by their reciprocals. By construction $\operatorname{rank} G = \operatorname{rank} G^\dagger$. Prove:

$$(i)\quad GG^\dagger,\ G^\dagger G \in \mathcal{H}_n(\mathbb{K}), \qquad (ii)\quad GG^\dagger G = G, \qquad (iii)\quad G^\dagger G G^\dagger = G^\dagger.$$

Show that $G^\dagger = G^{-1}$ if $G \in \mathbf{Gl}_n(\mathbb{K})$. The matrix $G^\dagger$ is called the *Moore-Penrose generalized inverse* (or *pseudoinverse*) of G. Show that $G^\dagger$ is uniquely determined by the requirements (i)-(iii).

8. If $\|\cdot\|_F$ denotes the Frobenius norm on $\mathbb{C}^{m\times n}$, prove that $\|G\|_F^2 = \sigma_1(G)^2+\cdots+\sigma_n(G)^2$ for all $G \in \mathbb{C}^{m\times n}$.

9. Let $B \in \mathbb{K}^{n\times m}$, $C \in \mathbb{K}^{m\times n}$. If $x^1,\ldots,x^k \in \mathbb{K}^n$ are linearly independent and span $\ker(I_n - BC)$ prove that $Cx^1,\ldots,Cx^k$ are linearly independent and span $\ker(I_m - CB)$. In particular, $\dim\ker(I_n - BC) = \dim\ker(I_m - CB)$.

10. A norm $\|\cdot\|$ on $\mathbb{C}^{m\times n}$ is called *unitarily invariant* if $\|UGV^*\| = \|G\|$ holds for all $G \in \mathbb{C}^{m\times n}$ and all $U \in \mathbf{U}_m(\mathbb{C}), V \in \mathbf{U}_n(\mathbb{C})$. Prove that if $\|\cdot\|$ is a unitarily invariant norm on $\mathbb{C}^{m\times n}$ then $\|G\|$ can be represented as a function of the singular values $\sigma_i(G)$, $i \in \underline{n}$,

$$\|G\| = \Psi(\sigma_1(G),\ldots,\sigma_n(G)), \quad G \in \mathbb{C}^{m\times n} \tag{51}$$

where $\Psi(\cdot) : \mathbb{R}^n \to \mathbb{R}$ has the following two properties[4]
(i) $x \mapsto \Psi(x)$ defines an absolute norm on $\mathbb{R}^n$.
(ii) $\Psi(\cdot)$ is symmetric, i.e. $\Psi(Px) = \Psi(x)$ for all perturbation matrices P of order n.

11. Prove Lemma 4.3.14.

12. If $G(\tau) \in \mathbb{K}^{m\times n}$, $\tau \in (\tau_0-\delta, \tau_0+\delta)$ is analytic in τ and $\sigma_0 > 0$ is a simple singular value of $G(\tau_0)$, show that there exists an analytic function $\sigma(\cdot)$ on a suitable small neighbourhood J of τ_0 such that $\sigma(\tau_0) = \sigma_0$ and $\sigma(\tau)$ is a simple singular value of $G(\tau)$ for $\tau \in J$. Prove that the sensitivity of $\sigma(\tau)$ at τ_0 is given by $\sigma'(\tau_0) = \operatorname{Re} w^*\, G'(\tau_0)\, v$ where (w, v) is a singular pair of $G(\tau_0)$ for $\sigma(\tau_0)$.

13. Calculate the sensitivities at the origin of the singular values of the matrix $\tilde{A}(\tau)$, $\tau \in \mathbb{R}$ of Example 4.2.13.

14. For $H \in \mathcal{H}_n(\mathbb{K})$ denote by $\lambda_1(H) \geq \ldots \geq \lambda_n(H)$ the eigenvalues of H in decreasing order. Prove that, for each $k \in \underline{n}$, the map $H \mapsto \lambda_k(H)$ is monotonically increasing on $\mathcal{H}_n(\mathbb{K})$ (ordered by $\preceq$).

15. Prove the following generalization of the inequality $\operatorname{Re} z \leq |z|$, $z \in \mathbb{C}$,

$$\lambda_k([A + A^*]/2) \leq \sigma_k(A), \quad k \in \underline{n}, \quad A \in \mathbb{C}^{n\times n},$$

[4] It can be shown that in the case $m = n$ the converse holds true, every function $\Psi(\cdot) : \mathbb{R}^n \to \mathbb{R}$ with the two properties (i),(ii) defines a unitarily invariant norm on $\mathbb{C}^{n\times n}$ via (51), see [484, II.3].

where λ_k denotes the k-th ordered eigenvalue of a Hermitian matrix (as in Ex. 14), and σ_k the k-th ordered singular value, $k \in \underline{n}$.
Hint: Use $x^*(A+A^*)x/2 = \operatorname{Re} x^*Ax \leq \|Ax\|_2$ for $x \in \mathbb{K}^n$, $\|x\|_2 = 1$ and Theorem 4.3.1.

16. If $G \in \mathbb{C}^{n\times n}$ is not a multiple of the identity, prove that

$$\inf_{T\in \mathbf{Gl}_n(\mathbb{C})} \operatorname{dist}(TGT^{-1}, M^{n,n}_{n-1}(\mathbb{C})) = 0.$$

4.3.6 Notes and References

The singular value decomposition of real square matrices goes back to *Beltrami* (1873) [47] and *Jordan* (1874) [280] and has been rediscovered since by various authors. An infinite dimensional analogue of the singular value decomposition was introduced by *Schmidt* (1907) [450]. He was the first to use this decomposition for obtaining optimal approximations of prescribed rank to a given operator. A proof of the SVD for general complex $m \times n$ matrices was given in *Eckart and Young* (1936) [145], but for square complex matrices the result had already been derived over twenty years earlier by *Autonne* (1915) [28]. Singular values of non-selfadjoint operators on Hilbert spaces have been discussed under the name of *s-numbers* in *Gohberg and Krein* (1969) [191]. Counterparts for operators on Banach spaces are studied in *Pietsch* (1987) [412]. This monograph also contains an axiomatic theory of s-numbers. Historical reviews of the early days of the singular value decomposition in Linear Algebra and Functional Analysis can be found in *Horn and Johnson* (1991) [265, 3.0] and *Stewart* (1993) [481], see also the historical remarks in *Pietsch* (1987) [412]. Good modern references for the theory and/or numerics of singular value decompositions in Linear Algebra are *Horn and Johnson* (1991) [265], *Stewart and Sun* (1990) [484] and *Golub and Van Loan* (1996) [197]. More references concerning numerical issues can be found in the *Notes and References* of Section 4.5.
As early as in the 1930s singular value decompositions played an important role in Statistics under the name of "principal component analysis", see *Hotelling* (1933) [266] and (1936) [267]. *Dempster* (1969) [127] gives a geometric treatment of principal component analysis and an overview over its history in multivariate analysis. In Systems Theory singular values became an important tool when the numerical problems of computing the theoretical constructions of geometric control theory were considered in the late seventies, see *Klema and Laub* (1980) [309]. For early applications in model reduction, see *Moore* (1981) [380] and *Glover* (1984) [188], and for early applications in robust stability and control, see *Safonov* (1980) [444] and *Doyle and Stein* (1981) [136].
The Minimax Theorem 4.3.1 is due to *Fischer* (1905) [162]. *Courant* (1920) extended the result to differential operators and therefore the result is often called the Courant-Fischer Theorem. For matrices this result is a special case of Wielandt's Theorem, see [484, IV.4.2]. Theorem 4.3.7 was first established by *Schmidt* (1907) [450] for integral operators with respect to the Schmidt norm. For matrices it was proved in *Eckart and Young* (1936) [145] with respect to the Frobenius norm (the finite dimensional specialization of the Schmidt norm). Later this result was generalized to arbitrary unitarily invariant norms by *Mirsky* (1963) [377]. In the same paper Mirsky extended the perturbation result in Corollary 4.3.11 to unitarily invariant norms. Proposition 4.3.23 was first published in *Weyl* (1949) [517]. More results concerning the relationship between singular values and eigenvalues can be found in *Horn and Johnson* (1991) [265, Ch. 3].

4.4 Structured Perturbations

In the next chapter we will introduce two very useful tools for analyzing spectral properties of dynamical systems with uncertain parameters, *stability radii* and *spectral value sets*. In order to characterize them it will be necessary to calculate, for a given set of perturbations $\mathbf{\Delta} \subset \mathbb{C}^{\ell \times q}$ and a given matrix $G \in \mathbb{C}^{q \times \ell}$, the norm of the smallest perturbation $\Delta \in \mathbf{\Delta}$ which achieves $\Delta G z = z$ for some $z \in \mathbb{C}^{\ell}$, $z \neq 0$. The inverse of the norm of this smallest Δ is called the μ-value of G with respect to the perturbation set $\mathbf{\Delta}$.
Up until now we have considered two types of perturbations to matrices and polynomials. Either the matrices or coefficient vectors of the polynomials have been subjected to arbitrary unstructured perturbations (independent perturbations of the entries/coefficients) or the perturbations are defined in terms of a single real or complex parameter. In many applications parameter perturbations occur which do not fall into either of these categories. So in order to obtain non-conservative estimates we have to build the perturbation structure into our definitions and consequent development. This will be carried out in this section which is subdivided into two subsections.
In the first one we consider general perturbation sets $\mathbf{\Delta} \subset \mathbb{C}^{\ell \times q}$ whose span is provided with an arbitrary operator norm. We define the μ-value of a matrix with respect to these sets of perturbations and derive some elementary properties. In the special case of complex full-block perturbations (i.e. $\mathbf{\Delta} = \mathbb{C}^{\ell \times q}$) the μ-value of a matrix coincides with its operator norm. We then introduce sets of block-diagonal perturbations for which the concept of the μ-value was originally defined, see *Notes and References*, and derive characterizations of the μ-value for these sets. Unfortunately it is usually very difficult to compute the μ-value from these characterizations, for which a global non-convex optimization problem must be solved. So we derive some bounds which can be calculated more easily. In particular we obtain an upper bound which can be computed via the resolution of a *convex* optimization problem. Finally we examine the continuity of the μ-function $G \mapsto \mu_{\mathbf{\Delta}}(G)$.
Most of the results in Subsection 4.4.1 require that $\mathbb{C}\mathbf{\Delta} = \mathbf{\Delta}$ and hence they are not applicable to *real* perturbation sets where $\mathbf{\Delta} \subset \mathbb{R}^{\ell \times q}$. However, we will see in Chapter 5 that the spectral analysis of linear systems with uncertain real parameters leads naturally to such perturbation sets. The problem of determining the μ-value for real perturbation sets is the subject of Subsection 4.4.2. We will not solve this problem in its full generality, but restrict our considerations to *full-block* real perturbations (i.e. $\mathbf{\Delta} = \mathbb{R}^{\ell \times q}$). For the case where $\ell = 1$ or $q = 1$ we obtain formulas which are valid for arbitrary perturbation norms and are comparatively easy to compute. For other values of ℓ and q, we only consider the spectral norm and obtain a characterization which involves a scalar optimization problem for the second singular value of a scaled matrix. subsection 4.4.1

4.4.1 Elements of μ-Analysis

In Section 4.3 we saw that the largest singular value of a matrix $G \in \mathbb{C}^{q \times \ell}$ is characterized by

$$\sigma_{\max}(G) = \left[\inf\{\|\Delta\|;\ \Delta \in \mathbb{C}^{\ell\times q} \text{ and } \det(I_\ell - \Delta G) = 0\}\right]^{-1}$$

where $\|\Delta\|$ is the spectral norm of Δ. We will now define a counterpart to the largest singular value for the case where Δ (viewed as a perturbation) is constrained to a given non-empty subset $\boldsymbol{\Delta} \subset \mathbb{C}^{\ell\times q}$ and its size is not necessarily measured by the spectral norm but by any norm $\|\cdot\|_{\boldsymbol{\Delta}}$ on the subspace $\operatorname{span}_{\mathbb{C}} \boldsymbol{\Delta}$ generated by $\boldsymbol{\Delta}$ over $\mathbb{C}$. This so-called *μ-value* is a flexible tool for dealing with structured uncertainties in the analysis and synthesis of control systems. It was originally introduced for the special case of block-diagonal perturbations (see below) and the spectral norm $\|\Delta\|_{\boldsymbol{\Delta}} = \sigma_{\max}(\Delta)$ [135]. Because of its close relationship with the maximal singular value it was called the *structured singular value.* In this subsection we develop some elements of μ-analysis and derive a number of basic results which are useful for the spectral analysis of uncertain systems studied in Chapter 5.

Definition 4.4.1. Suppose $\boldsymbol{\Delta}$ is any non-empty subset of $\mathbb{C}^{\ell\times q}$ and $\operatorname{span}_{\mathbb{C}} \boldsymbol{\Delta}$ is endowed with a norm $\|\cdot\|_{\boldsymbol{\Delta}}$. Then the μ-value of a matrix $G \in \mathbb{C}^{q\times\ell}$ with respect to the perturbation set $\boldsymbol{\Delta}$ and the norm $\|\cdot\|_{\boldsymbol{\Delta}}$ is defined by

$$\mu_{\boldsymbol{\Delta}}(G) = [\inf\{\|\Delta\|_{\boldsymbol{\Delta}};\ \Delta \in \boldsymbol{\Delta} \text{ and } \det(I_\ell - \Delta G) = 0\}]^{-1}, \quad G \in \mathbb{C}^{q\times\ell}. \tag{1}$$

Here as elsewhere we set $\inf\emptyset = \infty$ and $\infty^{-1} = 0$, so that $\mu_{\boldsymbol{\Delta}}(G) = 0$ if and only if there does not exist $\Delta \in \boldsymbol{\Delta}$ satisfying $\det(I_\ell - \Delta G) = 0$. Throughout the section we tacitly assume that $\boldsymbol{\Delta} \neq \emptyset$.

Remark 4.4.2. In the case where $\boldsymbol{\Delta} \subset \mathbb{R}^{\ell\times q}$ consists only of *real* perturbations, it would seem more natural to define $\mu_{\boldsymbol{\Delta}}(\cdot)$ with respect to a given norm $\|\cdot\|_{\operatorname{span}_{\mathbb{R}}\boldsymbol{\Delta}}$ on the *real* linear subspace generated by $\boldsymbol{\Delta}$,

$$\mu_{\boldsymbol{\Delta}}(G) = \left[\inf\left\{\|\Delta\|_{\operatorname{span}_{\mathbb{R}}\boldsymbol{\Delta}};\ \Delta \in \boldsymbol{\Delta} \text{ and } \det(I_\ell - \Delta G) = 0\right\}\right]^{-1}, \quad G \in \mathbb{C}^{q\times\ell}. \tag{2}$$

This can be made a special case of Definition 4.4.1 by choosing a norm $\|\cdot\|_{\boldsymbol{\Delta}}$ on $\operatorname{span}_{\mathbb{C}} \boldsymbol{\Delta}$ which is *compatible* with $\|\cdot\|_{\operatorname{span}_{\mathbb{R}}\boldsymbol{\Delta}}$ (see Definition A.1.6), i.e.

$$\|\Delta\|_{\boldsymbol{\Delta}} = \|\Delta\|_{\operatorname{span}_{\mathbb{R}}\boldsymbol{\Delta}}, \qquad \Delta \in \operatorname{span}_{\mathbb{R}} \boldsymbol{\Delta} \subset \mathbb{R}^{\ell\times q}.$$

Then $\mu_{\boldsymbol{\Delta}}(G)$ as defined by (1) coincides with $\mu_{\boldsymbol{\Delta}}(G)$ as defined by (2) for all $G \in \mathbb{C}^{q\times\ell}$. For most of this section and nearly all of the next chapter we will assume that $\operatorname{span}_{\mathbb{C}} \boldsymbol{\Delta}$ is provided with a norm $\|\cdot\|_{\boldsymbol{\Delta}}$ which is an *operator norm* with respect to a given pair of norms on $\mathbb{C}^\ell$ and $\mathbb{C}^q$. Again if $\boldsymbol{\Delta} \subset \mathbb{R}^{\ell\times q}$ it would seem more natural to endow $\operatorname{span}_{\mathbb{R}} \boldsymbol{\Delta}$ with an operator norm $\|\cdot\|_{\mathcal{L}(\mathbb{R}^q,\mathbb{R}^\ell)}$ rather than $\|\cdot\|_{\mathcal{L}(\mathbb{C}^q,\mathbb{C}^\ell)}$ and define

$$\mu_{\boldsymbol{\Delta}}(G) = \left[\inf\{\|\Delta\|_{\mathcal{L}(\mathbb{R}^q,\mathbb{R}^\ell)};\ \Delta \in \boldsymbol{\Delta} \text{ and } \det(I_\ell - \Delta G) = 0\}\right]^{-1}, \quad G \in \mathbb{C}^{q\times\ell}. \tag{3}$$

Let $(\|\cdot\|_{\mathbb{R}^\ell}, \|\cdot\|_{\mathbb{R}^q})$ be a given pair of norms on $\mathbb{R}^\ell, \mathbb{R}^q$. A pair of norms $(\|\cdot\|_{\mathbb{C}^\ell}, \|\cdot\|_{\mathbb{C}^q})$ on $\mathbb{C}^\ell, \mathbb{C}^q$ is said to be *compatible* with $(\|\cdot\|_{\mathbb{R}^\ell}, \|\cdot\|_{\mathbb{R}^q})$ if $\|v\|_{\mathbb{C}^\ell} = \|v\|_{\mathbb{R}^\ell}$ for all $v \in \mathbb{R}^\ell$, $\|z\|_{\mathbb{C}^q} = \|z\|_{\mathbb{R}^q}$ for all $z \in \mathbb{R}^q$ and the corresponding operator norms on $\mathbb{R}^{\ell\times q}$ and $\mathbb{R}^{q\times\ell}$ are equal,

$$\|\Delta\|_{\mathcal{L}(\mathbb{C}^q,\mathbb{C}^\ell)} = \|\Delta\|_{\mathcal{L}(\mathbb{R}^q,\mathbb{R}^\ell)},\ \Delta \in \mathbb{R}^{\ell\times q}; \quad \|G\|_{\mathcal{L}(\mathbb{C}^\ell,\mathbb{C}^q)} = \|G\|_{\mathcal{L}(\mathbb{R}^\ell,\mathbb{R}^q)},\ G \in \mathbb{R}^{q\times\ell}.$$

It is shown in Lemma A.1.7 that for any given pair of norms on $\mathbb{R}^\ell, \mathbb{R}^q$ there exists a compatible pair of norms on $\mathbb{C}^\ell, \mathbb{C}^q$. By choosing such a compatible pair of norms and providing $\mathrm{span}_{\mathbb{R}} \mathbf{\Delta}$ with the operator norm $\|\cdot\|_{\mathcal{L}(\mathbb{C}^q,\mathbb{C}^\ell)}$ the μ-value as defined by (1) again coincides with $\mu_{\mathbf{\Delta}}(G)$ as defined by (3) for all $G \in \mathbb{C}^{q\times\ell}$. We see therefore that Definition 4.4.1 also covers the natural definitions for *real* perturbation structures. □

μ-values admit a feedback interpretation as illustrated in Figure 4.4.1. Here we

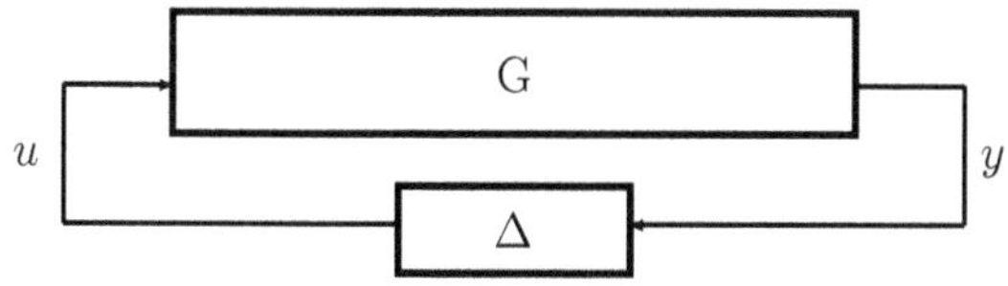

Figure 4.4.1: Feedback interpretation of $\mu_{\mathbf{\Delta}}(G)$

regard $G : \mathbb{C}^\ell \to \mathbb{C}^q$ as the input-output map of a memoryless linear system and $\Delta \in \mathbf{\Delta}$ as an unknown feedback operator. The resulting feedback system $G\square\Delta$ is well-posed (see Section 2.4) if and only if $I_\ell - \Delta G$ is invertible. Thus $\mu_{\mathbf{\Delta}}(G)^{-1}$ is the largest $\rho \in (0,\infty]$ such that $G\square\Delta$ is well-posed for all $\Delta \in \mathbf{\Delta}$ of norm $\|\Delta\|_{\mathbf{\Delta}} < \rho$. The following simple example shows that the calculation of μ may be far from trivial.

Example 4.4.3. Suppose $\mathbf{\Delta}$ has a diagonal structure

$$\mathbf{\Delta} = \{\Delta = \mathrm{diag}\,(\delta_1,\delta_2);\ \delta_1,\delta_2 \in \mathbb{C}\} \quad \text{with norm} \quad \|\Delta\|_{\mathbf{\Delta}} = \max\{|\delta_1|,\ |\delta_2|\}.$$

Consider first the case $G_1 = \begin{bmatrix} 1 & 0 \\ 1 & 1 \end{bmatrix}$, then $I_2 - \Delta G_1 = \begin{bmatrix} 1-\delta_1 & 0 \\ -\delta_2 & 1-\delta_2 \end{bmatrix}$ and so $\mu_{\mathbf{\Delta}}(G_1) = 1$. Now consider $G_2 = \begin{bmatrix} 1 & 1 \\ 1 & 1 \end{bmatrix}$, then

$$\det(I_2 - \Delta G_2) = \det \begin{bmatrix} 1-\delta_1 & -\delta_1 \\ -\delta_2 & 1-\delta_2 \end{bmatrix} = (1-\delta_1)(1-\delta_2) - \delta_1\delta_2 = 1 - \delta_1 - \delta_2.$$

Hence to find $\mu_{\mathbf{\Delta}}(G_2)$ we have to solve the problem (whose optimal value is $\mu_{\mathbf{\Delta}}(G_2)^{-1}$)

$$\text{minimize } \max\{|\delta_1|,\ |\delta_2|\} \quad \text{subject to} \quad 1 - \delta_1 - \delta_2 = 0.$$

The calculation is easy and yields $\mu_{\mathbf{\Delta}}(G) = 2$. Finally consider $G_3 = \begin{bmatrix} 1 & 2\imath \\ 1 & 2 \end{bmatrix}$, then

$$\det(I_2 - \Delta G_3) = \det \begin{bmatrix} 1-\delta_1 & -2\imath\delta_1 \\ -\delta_2 & 1-2\delta_2 \end{bmatrix} = 1 - \delta_1 - 2\delta_2 + 2\delta_1\delta_2(1-\imath).$$

So in order to find $\mu_{\mathbf{\Delta}}(G_3)$ it is necessary to solve the minimization problem

$$\text{minimize } \max\{|\delta_1|,\ |\delta_2|\} \quad \text{subject to} \quad 1 - \delta_1 - 2\delta_2 + 2\delta_1\delta_2(1-\imath) = 0.$$

This time the calculation is not so easy. You are asked to carry it out in Ex. 2. □

The formulas in the next remark are simple but useful consequences of the definition.

Remark 4.4.4. (i) For $G \in \mathbb{C}^{q\times\ell}$, we have

$$\begin{aligned}\mu_{\mathbf{\Delta}}(G) &= \left[\inf\{\|\Delta\|_{\mathbf{\Delta}};\ \Delta\in\mathbf{\Delta} \text{ and } \det\begin{bmatrix} I_\ell & \Delta \\ G & I_q\end{bmatrix}=0\}\right]^{-1},\\ &= \left[\inf\{\|\Delta\|_{\mathbf{\Delta}};\ \Delta\in\mathbf{\Delta} \text{ and } \det(I_q-G\Delta)=0\}\right]^{-1}.\end{aligned}$$

Sometimes these alternative formulas for $\mu_{\mathbf{\Delta}}(G)$ may lead to simpler calculations.
(ii) If $\mathbf{\Delta}$ is a closed subset of $\mathbb{C}^{\ell\times q}$ and $\mu_{\mathbf{\Delta}}(G)>0$ then $\{\Delta\in\mathbf{\Delta};\ \det(I_\ell-\Delta G)=0\}$ is a non-empty closed subset of $\mathbb{C}^{\ell\times q}$ and hence contains a perturbation $\Delta_{\min}$ of minimum norm (which is, in general, not uniquely determined). By definition, $\Delta_{\min}$ satisfies

$$\Delta_{\min}\in\mathbf{\Delta},\quad \det(I_\ell-\Delta_{\min}G)=0,\quad \|\Delta_{\min}\|_{\mathbf{\Delta}}=\mu_{\mathbf{\Delta}}(G)^{-1}. \tag{4}$$

Therefore the "inf" in (1) can be replaced by "min" if $\mu_{\mathbf{\Delta}}(G)>0$ and $\mathbf{\Delta}$ is closed. □

In the next lemma we list some simple properties of the μ-value.

Lemma 4.4.5. *Given any non-void $\mathbf{\Delta}\subset\mathbb{C}^{\ell\times q}$ and a norm $\|\cdot\|_{\mathbf{\Delta}}$ on $\operatorname{span}_{\mathbb{C}}\mathbf{\Delta}$, then*

$$\begin{aligned}\mathcal{A}_{\mathbf{\Delta}} &= \{\alpha\in\mathbb{C}^*;\ \alpha\mathbf{\Delta}=\mathbf{\Delta}\}, && (5)\\ \mathcal{U}_{\mathbf{\Delta}} &= \{U\in\mathbf{Gl}_q(\mathbb{C});\ \mathbf{\Delta}U=\mathbf{\Delta} \text{ and } \|\Delta\|_{\mathbf{\Delta}}=\|\Delta U\|_{\mathbf{\Delta}} \text{ for all } \Delta\in\mathbf{\Delta}\}^1, && (6)\\ \mathcal{C}_{\mathbf{\Delta}} &= \{R\in\mathbf{Gl}_q(\mathbb{C});\ \exists L_R\in\mathbf{Gl}_\ell(\mathbb{C})\ \forall\Delta\in\mathbf{\Delta}: L_R^{-1}\Delta R=\Delta\}^2, && (7)\end{aligned}$$

are subgroups of $\mathbb{C}^*$ and $\mathbf{Gl}_q(\mathbb{C})$, respectively. For all $G\in\mathbb{C}^{q\times\ell}$,

(i) $\mu_{\mathbf{\Delta}}(\alpha G)=|\alpha|\mu_{\mathbf{\Delta}}(G)$ *for all* $\alpha\in\mathcal{A}_{\mathbf{\Delta}}$.

(ii) $\mu_{\mathbf{\Delta}}(UG)=\mu_{\mathbf{\Delta}}(G)$ *for all* $U\in\mathcal{U}_{\mathbf{\Delta}}$.

(iii) $\mu_{\mathbf{\Delta}}(RGL_R^{-1})=\mu_{\mathbf{\Delta}}(G)$ *for all* $R\in\mathcal{C}_{\mathbf{\Delta}}$.

Moreover, if $\|\cdot\|_{\mathbf{\Delta}}$ is an operator norm induced by given norms on $\mathbb{C}^q$ and $\mathbb{C}^\ell$ then

$$\mu_{\mathbf{\Delta}}(G)\le\inf_{R\in\mathcal{C}_{\mathbf{\Delta}}}\|RGL_R^{-1}\|_{\mathcal{L}(\mathbb{C}^\ell,\mathbb{C}^q)}. \tag{8}$$

Proof: The subgroup properties and (i) are easily verified. Statement (ii) follows from the fact that for $U\in\mathcal{U}_{\mathbf{\Delta}}$ we have $\|\Delta\|_{\mathbf{\Delta}}=\|\Delta U^{-1}\|_{\mathbf{\Delta}}$ and

$$(\det(I_\ell-\Delta G)=0 \text{ and } \Delta\in\mathbf{\Delta})\Leftrightarrow(\det(I_\ell-\Delta U^{-1}UG)=0 \text{ and } \Delta U^{-1}\in\mathbf{\Delta}).$$

Similarly, (iii) follows from the fact that for $\Delta\in\mathbf{\Delta}$, $R\in\mathcal{C}_{\mathbf{\Delta}}$

$$\det(I_\ell-\Delta G)=\det(L_R(I_\ell-\Delta G)L_R^{-1})=\det(I_\ell-\Delta RGL_R^{-1}).$$

Finally, if $\|\cdot\|_{\mathbf{\Delta}}=\|\cdot\|_{\mathcal{L}(\mathbb{C}^q,\mathbb{C}^\ell)}$ on $\mathbf{\Delta}$ then for all $\Delta\in\mathbf{\Delta}$

$$\det(I_\ell-\Delta G)=0\Rightarrow\|\Delta\|_{\mathbf{\Delta}}\|G\|_{\mathcal{L}(\mathbb{C}^\ell,\mathbb{C}^q)}\ge1\Rightarrow\|\Delta\|_{\mathbf{\Delta}}^{-1}\le\|G\|_{\mathcal{L}(\mathbb{C}^\ell,\mathbb{C}^q)}.$$

Hence $\mu_{\mathbf{\Delta}}(G)\le\|G\|_{\mathcal{L}(\mathbb{C}^\ell,\mathbb{C}^q)}$ and so (8) is a consequence of (iii). □

---

[2] $\mathbf{\Delta}U:=\{\Delta U;\Delta\in\mathbf{\Delta}\}$.
[2] $\mathcal{C}_{\mathbf{\Delta}}$ is a generalization of the commutant subgroup of a given set $\mathbf{\Delta}\subset\mathbb{C}^{n\times n}$ of square matrices.

Another elementary property of the μ-value is captured in the following monotonicity statement which shows that $\mu_{\mathbf{\Delta}}(G)$ is increased if the perturbation set is increased and the perturbation norm is decreased

$$(\mathbf{\Delta}_1 \subset \mathbf{\Delta}_2 \text{ and } \forall \Delta \in \mathbf{\Delta}_1 \colon \|\Delta\|_{\mathbf{\Delta}_1} \geq \|\Delta\|_{\mathbf{\Delta}_2}) \Rightarrow \mu_{\mathbf{\Delta}_1}(G) \leq \mu_{\mathbf{\Delta}_2}(G), \quad G \in \mathbb{C}^{q \times \ell}. \tag{9}$$

In particular, we obtain for any given perturbation set $\mathbf{\Delta} \subset \mathbb{C}^{\ell \times q}$ and the subset $\mathbf{\Delta}_{\mathbb{R}} = \mathbf{\Delta} \cap \mathbb{R}^{\ell \times q}$ of *real* perturbations in $\mathbf{\Delta}$

$$\mu_{\mathbf{\Delta}_{\mathbb{R}}}(G) \leq \mu_{\mathbf{\Delta}}(G). \tag{10}$$

In the following, $\operatorname{span}_{\mathbb{R}} \mathbf{\Delta}_{\mathbb{R}}$ will always be endowed with the norm induced from $\operatorname{span}_{\mathbb{C}} \mathbf{\Delta}$. The following example illustrates that the inequality in (10) will in general be strict.

Example 4.4.6. Let $\mathbf{\Delta} = \mathbb{C} I_q$ carry the norm $\|z I_q\|_{\mathbf{\Delta}} = |z|$ and consider the matrix $G = \lambda I_q$ where $\lambda \in \mathbb{C} \setminus \mathbb{R}$ is given. Then $\mu_{\mathbf{\Delta}}(G) = |\lambda|$. However, there does not exist a real $\Delta \in \mathbf{\Delta}_{\mathbb{R}} = \mathbb{R} I_q$ such that $\det(I_q - \Delta G) = 0$, whence $\mu_{\mathbf{\Delta}_{\mathbb{R}}}(G) = 0$. □

If $\mathbf{\Delta}$ is a *complex* structure, i.e. $\mathbb{C}\mathbf{\Delta} = \mathbf{\Delta}$, then the μ-value can be characterized in terms of the spectral radius.

Lemma 4.4.7. *Suppose $\mathbf{\Delta}$ is a closed subset of $\mathbb{C}^{\ell \times q}$, $\operatorname{span}_{\mathbb{C}} \mathbf{\Delta}$ is endowed with a norm $\|\cdot\|_{\mathbf{\Delta}}$ and $\mathbb{C}\mathbf{\Delta} = \mathbf{\Delta}$. Then for all $G \in \mathbb{C}^{q \times \ell}$*

$$\mu_{\mathbf{\Delta}}(G) = \max_{\substack{\Delta \in \mathbf{\Delta} \\ \|\Delta\|_{\mathbf{\Delta}} = 1}} \varrho(\Delta G) = \max_{\Delta \in \mathbf{\Delta} \setminus \{0\}} \frac{\varrho(\Delta G)}{\|\Delta\|_{\mathbf{\Delta}}}. \tag{11}$$

If $\mu_{\mathbf{\Delta}}(G) > 0$, we have

$$\mu_{\mathbf{\Delta}}(G)^{-1} = \min_{\substack{\Delta \in \mathbf{\Delta} \\ \varrho(\Delta G) \geq 1}} \|\Delta\|_{\mathbf{\Delta}} = \min_{\substack{\Delta \in \mathbf{\Delta} \\ \varrho(\Delta G) = 1}} \|\Delta\|_{\mathbf{\Delta}}. \tag{12}$$

Proof: Since the spectral radius is positive homogeneous, the second equalities in (11) and (12) are immediate. The first equality in (11) can be verified as follows. If $\varrho(\Delta G) = 0$ for all $\Delta \in \mathbf{\Delta}$ with $\|\Delta\|_{\mathbf{\Delta}} = 1$ then $\sigma(\Delta G) = \{0\}$ for all $\Delta \in \mathbf{\Delta}$ and so (11) holds with $\mu_{\mathbf{\Delta}}(G) = 0$. Now suppose that the maximum in (11) is positive. Since $\Delta \mapsto \varrho(\Delta G)$ is continuous on the non-empty compact set $\{\Delta \in \mathbf{\Delta};\ \|\Delta\|_{\mathbf{\Delta}} = 1\}$ there exists $\Delta_0 \in \mathbf{\Delta}$ such that

$$\|\Delta_0\|_{\mathbf{\Delta}} = 1 \quad \text{and} \quad \varrho_0 := \varrho(\Delta_0 G) = \max_{\substack{\Delta \in \mathbf{\Delta} \\ \|\Delta\|_{\mathbf{\Delta}} = 1}} \varrho(\Delta G).$$

Choosing $\lambda_0 \in \sigma(\Delta_0 G)$ such that $|\lambda_0| = \varrho_0$ we obtain a matrix $\lambda_0^{-1}\Delta_0 \in \mathbf{\Delta}$ satisfying

$$\|\lambda_0^{-1}\Delta_0\|_{\mathbf{\Delta}} = \varrho_0^{-1} \quad \text{and} \quad \det(I_\ell - \lambda_0^{-1}\Delta_0 G) = 0.$$

Thus $\mu_{\mathbf{\Delta}}(G)^{-1} \leq \varrho_0^{-1}$. Conversely, choose $\Delta_{\min} \in \mathbf{\Delta}$ such that (4) holds, then $\varrho(\Delta_{\min} G) \geq 1$ and

$$\mu_{\mathbf{\Delta}}(G)^{-1} = \|\Delta_{\min}\|_{\mathbf{\Delta}} \geq \|\Delta_{\min}\|_{\mathbf{\Delta}} \varrho(\Delta_{\min} G)^{-1} \geq \varrho(\|\Delta_{\min}\|_{\mathbf{\Delta}}^{-1} \Delta_{\min} G)^{-1} \geq \varrho_0^{-1}$$

by the definition of ϱ_0. This concludes the proof of (11).
Now suppose $\mu_{\mathbf{\Delta}}(G) > 0$. Then it follows from (11) and the positive homogeneity of $\varrho(\cdot)$ and $\|\cdot\|$ that

$$\mu_{\mathbf{\Delta}}(G)^{-1} = \min_{\Delta\in\mathbf{\Delta}\setminus\{0\}} \frac{\|\Delta\|_{\mathbf{\Delta}}}{\varrho(\Delta G)} = \min_{\substack{\Delta\in\mathbf{\Delta}\\ \varrho(\Delta G)=1}} \|\Delta\|_{\mathbf{\Delta}}.$$

Thus (12) follows. □

Example 4.4.8. Let $\mathbf{\Delta}$ be the normed perturbation space of Example 4.4.3, $\Delta \in \mathbf{\Delta}$ and $G_2 = \begin{bmatrix} 1 & 1 \\ 1 & 1 \end{bmatrix}$, then $\Delta G_2 = \begin{bmatrix} \delta_1 & \delta_1 \\ \delta_2 & \delta_2 \end{bmatrix}$. Hence $\varrho(\Delta G_2) = |\delta_1 + \delta_2|$ and to calculate $\mu_{\mathbf{\Delta}}(G_2)$ by means of formula (11) we have to solve the optimization problem

$$\text{maximize}\, |\delta_1 + \delta_2| \quad \text{subject to} \quad \max\{|\delta_1|, |\delta_2|\} = 1.$$

We see therefore that in comparison to the direct analysis of Example 4.4.3 the application of Lemma 4.4.7 leads to a reversal of the role of the function to be optimized and the constraint. Clearly $\mu_{\mathbf{\Delta}}(G_2) = 2$ as before. □

The following example illustrates that in general the previous lemma is not valid for *real* perturbation structures.

Example 4.4.9. Let $\mathbf{\Delta} = \mathbb{R}I_2$ (with the norm $\|\delta I_2\|_{\mathbf{\Delta}} = |\delta|$) and $G = \begin{bmatrix} 0 & -1 \\ 1 & 0 \end{bmatrix}$. Then $\mu_{\mathbf{\Delta}}(G) = 0$ but the maximum in (11) is 1. □

Remark 4.4.10. We have seen that the determination of $\mu_{\mathbf{\Delta}}(G)$ for a given matrix and a given disturbance set $\mathbf{\Delta}$ is in general a difficult one. So it is important to have reasonable estimates for it. Lemma 4.4.5 and Lemma 4.4.7 provide upper *and* lower bounds. If $\|\cdot\|_{\mathbf{\Delta}}$ is an operator norm, we have

$$\forall R \in \mathcal{C}_{\mathbf{\Delta}} : \; \mu_{\mathbf{\Delta}}(G) \le \|RGL_R^{-1}\|_{\mathcal{L}(\mathbb{C}^\ell,\mathbb{C}^q)} \tag{13}$$

and in particular $\|G\|_{\mathcal{L}(\mathbb{C}^\ell,\mathbb{C}^q)}$ is an upper bound. Whereas if $\mathbf{\Delta}$ is a closed subset of $\mathbb{C}^{\ell\times q}$ and $\mathbb{C}\mathbf{\Delta} = \mathbf{\Delta}$, then by (11)

$$\Delta \in \mathbf{\Delta},\; \|\Delta\|_{\mathbf{\Delta}} = 1 \quad \Rightarrow \quad \varrho(\Delta G) \le \mu_{\mathbf{\Delta}}(G) \tag{14}$$

and in particular $\varrho(G)$ is a lower bound for $\mu_{\mathbf{\Delta}}(G)$ if $\ell = q$ and $\mathbb{C}I_q \subset \mathbf{\Delta}$. □

In the *full-block case* where $\mathbf{\Delta} = \mathbb{K}^{\ell\times q}$ we write $\mu_{\mathbb{K}}(G)$ for $\mu_{\mathbb{K}^{\ell\times q}}(G)$, i.e.

$$\mu_{\mathbb{K}}(G) = \left[\inf\{\|\Delta\|_{\mathcal{L}(\mathbb{K}^q,\mathbb{K}^\ell)};\; \Delta \in \mathbb{K}^{\ell\times q},\; \det(I_\ell - \Delta G) = 0\}\right]^{-1}, \quad G \in \mathbb{C}^{q\times\ell}. \tag{15}$$

The following proposition extends the result of Theorem 4.3.13 from the spectral norm to arbitrary operator norms and shows that in the full-block case the upper bound in (13) is tight.

Proposition 4.4.11. *Suppose that $\mathbf{\Delta} = \mathbb{K}^{\ell\times q}$ is endowed with an operator norm $\|\cdot\|_{\mathcal{L}(\mathbb{K}^q,\mathbb{K}^\ell)}$, then*

$$\mu_{\mathbb{K}}(G) = \|G\|_{\mathcal{L}(\mathbb{K}^\ell,\mathbb{K}^q)}, \qquad G \in \mathbb{K}^{q\times\ell}. \tag{16}$$

If $G \neq 0$ there exist $u_0 \in \mathbb{K}^\ell, v_0 \in \mathbb{K}^q$ such that

$$v_0^* G u_0 = \|G\|_{\mathcal{L}(\mathbb{K}^\ell,\mathbb{K}^q)}, \qquad \|u_0\|_{\mathbb{K}^\ell} = \|v_0^*\|_{\mathbb{K}^q}^* = 1 \tag{17}$$

(where $\|\cdot\|_{\mathbb{K}^q}^*$ denotes the dual norm of $\|\cdot\|_{\mathbb{K}^q}$ on $\mathbb{K}^{1\times q}$). For each such pair

$$\Delta_{\min} = \|G\|_{\mathcal{L}(\mathbb{K}^\ell,\mathbb{K}^q)}^{-1} u_0 v_0^* = \mu_{\mathbb{K}}(G)^{-1} u_0 v_0^* \tag{18}$$

is of minimal norm amongst all matrices $\Delta \in \mathbb{K}^{\ell\times q}$ satisfying $\det(I_\ell - \Delta G) = 0$.

Proof: Let $G \in \mathbb{K}^{q\times\ell}$ and suppose $G \neq 0$ (otherwise (16) holds trivially). Since

$$\det(I_\ell - \Delta G) = 0 \Longrightarrow \|\Delta\|_{\mathcal{L}(\mathbb{K}^q,\mathbb{K}^\ell)} \|G\|_{\mathcal{L}(\mathbb{K}^\ell,\mathbb{K}^q)} \geq \|\Delta G\|_{\mathcal{L}(\mathbb{K}^\ell,\mathbb{K}^\ell)} \geq 1,$$

we have $\mu_{\mathbb{K}}(G) \leq \|G\|_{\mathcal{L}(\mathbb{K}^\ell,\mathbb{K}^q)}$. To prove the converse inequality choose $u_0 \in \mathbb{K}^\ell$, $\|u_0\|_{\mathbb{K}^\ell} = 1$ such that $\|G\|_{\mathcal{L}(\mathbb{K}^\ell,\mathbb{K}^q)} = \|Gu_0\|_{\mathbb{K}^q}$. By the Hahn-Banach Theorem (see Example A.4.11) there exists a vector $v_0 \in \mathbb{K}^q$ of dual norm $\|v_0^*\|_{\mathbb{K}^q}^* = 1$ such that

$$v_0^* G u_0 = \|G u_0\|_{\mathbb{K}^q} = \|G\|_{\mathcal{L}(\mathbb{K}^\ell,\mathbb{K}^q)}.$$

Now define $\Delta_{\min} = \|G\|_{\mathcal{L}(\mathbb{K}^\ell,\mathbb{K}^q)}^{-1} u_0 v_0^*$. Since $\|u_0 v_0^*\|_{\mathcal{L}(\mathbb{K}^q,\mathbb{K}^\ell)} = 1$ we have

$$\|\Delta_{\min}\|_{\mathcal{L}(\mathbb{K}^q,\mathbb{K}^\ell)} = \|G\|_{\mathcal{L}(\mathbb{K}^\ell,\mathbb{K}^q)}^{-1} \text{ and } \Delta_{\min} G u_0 = u_0.$$

Therefore $I_\ell - \Delta_{\min} G$ is singular and so $\mu_{\mathbb{K}}(G)^{-1} \leq \|\Delta_{\min}\| = \|G\|_{\mathcal{L}(\mathbb{K}^\ell,\mathbb{K}^q)}^{1}$. This concludes the proof. □

It follows from the previous proof that, under the conditions of Proposition 4.4.11, for $G \in \mathbb{K}^{q\times\ell}, G \neq 0$

$$\mu_{\mathbb{K}}(G) = \left[\min\{\|\Delta\|_{\mathcal{L}(\mathbb{K}^q,\mathbb{K}^\ell)};\ \Delta \in \mathbb{K}^{\ell\times q}, \operatorname{rank}\Delta = 1 \text{ and } \det(I_\ell - \Delta G) = 0\}\right]^{-1}.$$

As a consequence we obtain the following corollary which extends the formula (16) to perturbation norms which may not be operator norms.

Corollary 4.4.12. *Suppose that $\mathbf{\Delta} = \mathbb{K}^{\ell\times q}$ is endowed with a norm $\|\cdot\|_{\mathbb{K}^{\ell\times q}}$ which is rank one consistent with an operator norm $\|\cdot\|_{\mathcal{L}(\mathbb{K}^q,\mathbb{K}^\ell)}$, i.e. for all $\Delta \in \mathbb{K}^{\ell\times q}$*

$$\|\Delta\|_{\mathcal{L}(\mathbb{K}^q,\mathbb{K}^\ell)} \leq \|\Delta\|_{\mathbb{K}^{\ell\times q}}, \quad \text{and} \quad \|\Delta\|_{\mathbb{K}^{\ell\times q}} = \|\Delta\|_{\mathcal{L}(\mathbb{K}^q,\mathbb{K}^\ell)} \text{ if } \operatorname{rank}\Delta = 1. \tag{19}$$

Then equality (16) *holds for all $G \in \mathbb{K}^{q\times\ell}$.*

Remark 4.4.13. For $1 \leq p,\, r \leq \infty$ the $(p|r)$-Hölder norm of $\Delta \in \mathbb{C}^{\ell\times q}$ is

$$\|\Delta\|_{p|r} = \left\|(\|\Delta^\top e^1\|_p, \ldots, \|\Delta^\top e^\ell\|_p)^\top\right\|_r \tag{20}$$

where $e^1, \ldots, e^\ell$ are the column vectors of I_ℓ. So if $p, r \in [1,\infty)$ then

$$\|\Delta\|_{p|r} = \left(\sum_{i=1}^{\ell}\left(\sum_{j=1}^{q}|\Delta_{ij}|^p\right)^{r/p}\right)^{1/r}, \quad \Delta = (\Delta_{ij}) \in \mathbb{C}^{\ell\times q}.$$

Otherwise

$$\|\Delta\|_{p|\infty} = \max_{i\in\underline{\ell}} \left[\sum_{j=1}^{q} |\Delta_{ij}|^p\right]^{1/p}, \|\Delta\|_{\infty|r} = \left[\sum_{i=1}^{\ell} [\max_{j\in\underline{q}} |\Delta_{ij}|]^r\right]^{1/r}, \|\Delta\|_{\infty|\infty} = \max_{i\in\underline{\ell}, j\in\underline{q}} |\Delta_{ij}|.$$

These perturbation norms are easy to compute.

When $\mathbb{K}^q$ is normed with an p-norm and $\mathbb{K}^\ell$ is normed with a r-norm, we write $\|\cdot\|_{p,r}$ for the operator norm $\|\cdot\|_{\mathcal{L}(\mathbb{K}^q,\mathbb{K}^\ell)}$. By Lemma A.1.12 the Hölder norm $\|\cdot\|_{p|r}$ is rank one consistent with the operator norm $\|\cdot\|_{\mathcal{L}(\mathbb{K}^q,\mathbb{K}^\ell)}$ induced by the r-norm on $\mathbb{K}^\ell$ and the *dual* p-norm on $\mathbb{K}^q$. That is $\|\cdot\|_{p|r}$ is rank one consistent with $\|\cdot\|_{p^*,r}$, where $1/p + 1/p^* = 1$. Hence (16) holds if the *perturbations* are normed with Hölder norms. Note however, that in this case the corresponding operator norm of G is $\|G\|_{r,p^*}$ which will in general *not* be a Hölder norm. In order to compute most operator norms one needs to solve an optimization problem which may be difficult. But the following operator norms of $G = (g_{ij}) \in \mathbb{K}^{q\times\ell}$ (see (A.1.3)) are computed more easily

$$\|G\|_{1,1} = \max_{j\in\underline{\ell}} \sum_{i=1}^{q} |g_{ij}|, \quad \|G\|_{2,2} = \sigma_{\max}(G), \quad \|G\|_{\infty,\infty} = \max_{i\in\underline{q}} \sum_{j=1}^{\ell} |g_{ij}|.$$

The corresponding Hölder norms on $\mathbf{\Delta}$ are $\|\cdot\|_{\infty|1}$, $\|\cdot\|_{2|2}$ (Frobenius norm) and $\|\cdot\|_{1|\infty}$, respectively. □

Block-diagonal perturbation classes

We will see in the next chapter that model uncertainties are often represented by block-diagonal perturbations. μ-values were originally introduced to deal with such perturbation classes (see *Notes and References*).

Definition 4.4.14. $\mathbf{\Delta} \subset \mathbb{C}^{\ell\times q}$ is said to be a *class of complex block-diagonal perturbations*, if there exist "book-keeping" integers $N \geq 1$, $\ell_i \geq 1$, $q_i \geq 1$ for $i \in \underline{N}$ and a subset $J \subset \underline{N}$ such that $\ell_i = q_i$ for $i \in \underline{N} \setminus J$, $\ell = \sum_{i=1}^N \ell_i$, $q = \sum_{i=1}^N q_i$, and

$$\mathbf{\Delta} = \{\mathrm{diag}\,(\Delta_1, \ldots, \Delta_N); \Delta_i \in \mathbf{\Delta}_i,\ i \in \underline{N}\} \text{ where } \mathbf{\Delta}_i = \begin{cases} \mathbb{C}^{\ell_i\times q_i} & \text{if } \quad i \in J \\ \mathbb{C}I_{q_i} & \text{if } \ i \in \underline{N}\setminus J. \end{cases} \tag{21}$$

The norm on $\mathbf{\Delta}$ is

$$\|\Delta\|_{\mathbf{\Delta}} = \max_{i\in\underline{N}} \|\Delta_i\|_{\mathbf{\Delta}_i}, \quad \Delta = \mathrm{diag}\,(\Delta_1, \ldots, \Delta_N) \in \mathbf{\Delta}$$

where the norms $\|\cdot\|_{\mathbf{\Delta}_i}$ on $\mathbf{\Delta}_i$ coincide with operator norms $\|\cdot\|_{\mathcal{L}(\mathbb{C}^{q_i},\mathbb{C}^{\ell_i})}$ on $\mathbf{\Delta}_i$.

So the i-th block of $\Delta \in \mathbf{\Delta}$ is either an arbitrary complex $\ell_i \times q_i$-matrix (if $i \in J$) or a diagonal matrix with identical diagonal elements (if $i \in \underline{N} \setminus J$). Moreover, $\|\cdot\|_{\mathbf{\Delta}}$ coincides on $\mathbf{\Delta}$ with the operator norm induced e.g. by the vector norms

$$\|(u_i)_{i\in\underline{N}}\|_{\mathbb{C}^\ell} = \left(\sum_{i=1}^{N} \|u_i\|^2_{\mathbb{C}^{\ell_i}}\right)^{1/2}, \quad \|(y_i)_{i\in\underline{N}}\|_{\mathbb{C}^q} = \left(\sum_{i=1}^{N} \|y_i\|^2_{\mathbb{C}^{q_i}}\right)^{1/2}. \tag{22}$$

In fact, there are many other norms on $\mathbb{C}^\ell$ and $\mathbb{C}^q$ which induce the same operator norm on $\mathbf{\Delta}$ (see Ex. 3). Since the upper bound in (8) depends on the specific pair of norms chosen on $\mathbb{C}^\ell$ and $\mathbb{C}^q$ this can be used to tighten it, see Ex. 4.

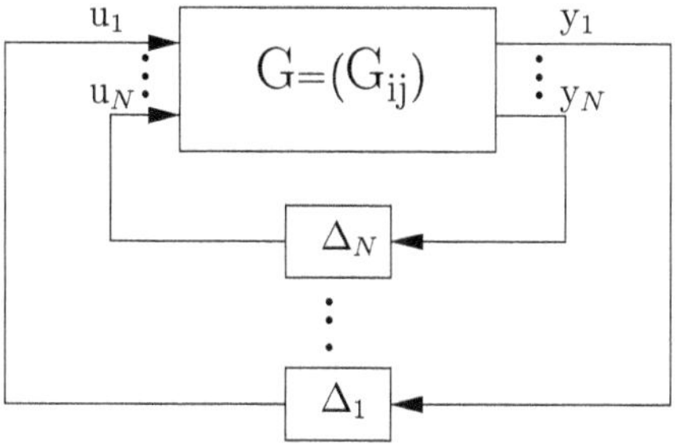

Figure 4.4.2: Multi-loop feedback interpretation of $\mu_{\mathbf{\Delta}}(G)$

The μ-value with respect to a class $\mathbf{\Delta}$ of block-diagonal perturbations admits the following interpretation. Given any matrix $G \in \mathbb{C}^{q\times\ell}$, let G be partitioned according to the decompositions $\mathbb{C}^\ell = \bigoplus_{i=1}^N \mathbb{C}^{\ell_i}$, $\mathbb{C}^q = \bigoplus_{i=1}^N \mathbb{C}^{q_i}$, i.e.

$$G = (G_{ij})_{i,j\in\underline{N}},\ G_{ij} \in \mathbb{C}^{q_i\times\ell_j} \text{ with } Gu = (\textstyle\sum_{j=1}^N G_{ij}u_j)_{i\in\underline{N}},\ u = (u_j)_{j\in\underline{N}} \in \mathbb{C}^\ell. \quad (23)$$

We may interpret $G = (G_{ij})_{i,j\in\underline{N}}$ as an input-output operator of a memoryless linear system with N inputs u_j and N outputs $y_i = \sum_{j=1}^N G_{ij}u_j$. If the inputs and outputs are connected by multi-loop feedback with the feedback operator $\Delta = \operatorname{diag}(\Delta_1,\dots,\Delta_N)$, see Figure 4.4.2, the resulting feedback system $G\Box\Delta$ is well-posed (see Section 2.4) if and only if $I_\ell - \Delta G = I_\ell - (\Delta_i G_{ij})_{i,j\in\underline{N}}$ is invertible. Thus $\mu_{\mathbf{\Delta}}(G)^{-1}$ is the largest $\rho \in [0,\infty]$ such that $G\Box\Delta$ is well-posed for all $\Delta \in \mathbf{\Delta}$ such that $\|\Delta_i\|_{\mathcal{L}(\mathbb{C}^{q_i},\mathbb{C}^{\ell_i})} < \rho$ if $i \in J$ and $\|\Delta_i\| = |\delta_i| < \rho$ if $i \in \underline{N}\setminus J$.
The full-block perturbation class $\mathbf{\Delta} = \mathbb{C}^{\ell\times q}$ which we considered before corresponds to the simple case $N = 1$, $J = \{1\}$. In the following two examples we will consider proper block-diagonal disturbances (with $N > 1$) corresponding to the two extreme cases $J = \underline{N}$ and $J = \emptyset$ in Definition 4.4.14.

Example 4.4.15. (Multi-block perturbations). If $N > 1$ and $J = \underline{N}$ the corresponding perturbation classes have the form

$$\mathbf{\Delta} = \{\operatorname{diag}(\Delta_1,\dots,\Delta_N);\Delta_i \in \mathbb{C}^{\ell_i\times q_i},\ i\in\underline{N}\} \quad (24)$$

where $\sum_{i=1}^N \ell_i = \ell$, $\sum_{i=1}^N q_i = q$. The elements of such a perturbation class are called *multi-block perturbations* with block structure $(\ell_i\times q_i)_{i\in\underline{N}}$. For arbitrary scaling parameters $\gamma=(\gamma_1,\dots,\gamma_N)\in(\mathbb{C}^*)^N$ consider the *scaling transformations* $R_\gamma=\operatorname{diag}(\gamma_1 I_{q_1},\dots,\gamma_N I_{q_N})$, $L_\gamma=\operatorname{diag}(\gamma_1 I_{\ell_1},\dots,\gamma_N I_{\ell_N})$. Then

$$L_\gamma^{-1}\Delta R_\gamma = \Delta, \quad \Delta\in\mathbf{\Delta}$$

and so we obtain from Lemmata 4.4.5 and 4.4.7 that for any $\Delta\in\mathbf{\Delta}$ of norm $\|\Delta\|_{\mathbf{\Delta}} = 1$,

$$\varrho(\Delta G) \le \mu_{\mathbf{\Delta}}(G) \le \inf_{\gamma\in(\mathbb{C}^*)^N} \|R_\gamma G L_\gamma^{-1}\|_{\mathcal{L}(\mathbb{C}^\ell,\mathbb{C}^q)} = \inf_{\gamma\in(0,\infty)^N} \|R_\gamma G L_\gamma^{-1}\|_{\mathcal{L}(\mathbb{C}^\ell,\mathbb{C}^q)} \quad (25)$$

where $\mathbb{C}^\ell, \mathbb{C}^q$ are provided with the norms (22). The last equality results from the fact that L_γ, R_γ are isometries on $\mathbb{C}^\ell$ and $\mathbb{C}^q$ if $|\gamma_i| = 1$ for $i\in\underline{N}$. If $\ell_i = q_i$ for some $i\in\underline{N}$, the index set J in Definition 4.4.14 is not necessarily equal to $\underline{N}$. Hence the perturbation set $\mathbf{\Delta}$ defined by (24) is not the only one having the block structure $(\ell_i\times q_i)_{i\in\underline{N}}$, but it is the largest one. Therefore the upper bound (25) obtained by scaling is valid for all sets of block-diagonal perturbations with this block structure. □

Example 4.4.16. (Diagonal perturbations). If $N > 1$ and $J = \emptyset$ then $\ell_i = q_i$, $i \in \underline{N}$ and the corresponding perturbation class has the form

$$\mathbf{\Delta} = \{\operatorname{diag}(\delta_1 I_{q_1}, \ldots, \delta_N I_{q_N});\ \delta_i \in \mathbb{C},\ i \in \underline{N}.\} \tag{26}$$

Since any operator norm of the identity map is 1 the perturbation norm is given by the ℓ^∞-norm of $\delta = (\delta_1, \ldots, \delta_N)$

$$\|\Delta\|_{\mathbf{\Delta}} = \max_{i \in \underline{N}} |\delta_i| = \|\delta\|_\infty, \quad \Delta = \operatorname{diag}(\delta_1 I_{q_1}, \ldots, \delta_N I_{q_N}) \in \mathbf{\Delta}.$$

Now $R^{-1}\Delta R = \Delta$ for all $R = \operatorname{diag}(R_1, \ldots, R_N)$, $R_i \in \mathbf{Gl}_{q_i}(\mathbb{C})$ and all $\Delta \in \mathbf{\Delta}$. Therefore we obtain from Lemmata 4.4.5 and 4.4.7 that $\varrho(G) \le \mu_{\mathbf{\Delta}}(G)$ and

$$\mu_{\mathbf{\Delta}}(G) \le \inf\{\|RGR^{-1}\|_{\mathcal{L}(\mathbb{C}^\ell, \mathbb{C}^q)};\ R = \operatorname{diag}(R_1, \ldots, R_N),\ R_i \in \mathbf{Gl}_{q_i}(\mathbb{C}) \text{ for } i \in \underline{N}\}. \tag{27}$$

This estimate is in general much tighter than the estimate (25) (which is also applicable in the present case). Consider e.g. the case where $N = 2$ and $G = \operatorname{diag}(G_1, G_2)$, $G_1 \in \mathbb{C}^{q_1 \times q_1}$, $G_2 \in \mathbb{C}^{q_2 \times q_2}$. Then (25) yields the estimate $\mu_{\mathbf{\Delta}}(G) \le \max\{\|G_1\|_{\mathcal{L}(\mathbb{C}^{q_1})}, \|G_2\|_{\mathcal{L}(\mathbb{C}^{q_2})}\}$ whereas (27) yields $\mu_{\mathbf{\Delta}}(G) \le \max\{\varrho(G_1), \varrho(G_2)\}$ [3]. The latter estimate is in fact an equality, hence the estimate yields the precise μ-value in this case, $\mu_{\mathbf{\Delta}}(G) = \max\{\varrho(G_1), \varrho(G_2)\}$ (see Ex. 6). □

Returning to multi-block perturbations of the form (24) we will now see that the upper bound in (25) is tight under a differentiability constraint when the norms on $\mathbb{C}^{\ell_i}$, $\mathbb{C}^{q_i}$, $i \in \underline{N}$ are Euclidean and the norm on the right hand side is the corresponding operator norm. To prove this we need the following lemma.

Lemma 4.4.17. *Let $G = (G_{ij})_{i,j \in \underline{N}}$ be partitioned as in (23), $G_\gamma = R_\gamma G L_\gamma^{-1}$, where $R_\gamma = \operatorname{diag}(\gamma_1 I_{q_1}, \ldots, \gamma_N I_{q_N})$, $L_\gamma = \operatorname{diag}(\gamma_1 I_{\ell_1}, \ldots, \gamma_N I_{\ell_N})$ are the scaling transformations associated with $\gamma \in (0,\infty)^N$. Suppose that in a neighbourhood $B(\hat\gamma, \varepsilon) \subset (0,\infty)^N$ of some $\hat\gamma \in (0,\infty)^N$ we are given a singular value $\sigma(\gamma)$ of G_γ and an associated singular pair $(w(\gamma), v(\gamma))$ partitioned as*

$$w(\gamma) = \begin{bmatrix} w_1(\gamma) \\ \vdots \\ w_N(\gamma) \end{bmatrix} \text{ with } w_j(\gamma) \in \mathbb{K}^{q_j},\quad v(\gamma) = \begin{bmatrix} v_1(\gamma) \\ \vdots \\ v_N(\gamma) \end{bmatrix} \text{ with } v_j(\gamma) \in \mathbb{K}^{\ell_j},\quad \gamma \in B(\hat\gamma, \varepsilon).$$

If, for a given $i \in \underline{N}$, the partial derivatives of $\sigma(\cdot)$, $w(\cdot)$, and $v(\cdot)$ with respect to γ_i exist at $\gamma = \hat\gamma$ then

$$\frac{\partial \sigma}{\partial \gamma_i}(\hat\gamma) = \frac{\sigma(\hat\gamma)}{\hat\gamma_i}\left[\|w_i(\hat\gamma)\|_2^2 - \|v_i(\hat\gamma)\|_2^2\right]. \tag{28}$$

Proof: By definition of a singular pair we have $w(\gamma)^* G_\gamma v(\gamma) = \sigma(\gamma)$ and differentiation yields as in the proof of Corollary 4.3.20

$$\frac{\partial \sigma}{\partial \gamma_i}(\hat\gamma) = \operatorname{Re}\left(w(\hat\gamma)^* \frac{\partial G_{\hat\gamma}}{\partial \gamma_i} v(\hat\gamma)\right)$$

[3] Note that $\min\{\|R_i G_i R_i^{-1}\|_{\mathcal{L}(\mathbb{C}^{q_i})}; R_i \in \mathbf{Gl}_{q_i}(\mathbb{C})\} = \varrho(G_i)$ for i=1,2.

where $\partial G_{\hat{\gamma}}/\partial \gamma_i := \partial G_\gamma/\partial \gamma_i|_{\gamma=\hat{\gamma}}$. Since $G_\gamma = \left(\gamma_j \gamma_k^{-1} G_{jk}\right)_{j,k\in\underline{N}}$ we have

$$\frac{\partial G_{\hat{\gamma}}}{\partial \gamma_i} = \begin{bmatrix} 0 & \cdots & 0 & -\hat{\gamma}_1\hat{\gamma}_i^{-2}G_{1i} & 0 & \cdots & 0 \\ \vdots & \vdots & \vdots & \vdots & \vdots & \vdots & \vdots \\ 0 & \cdots & 0 & -\hat{\gamma}_{i-1}\hat{\gamma}_i^{-2}G_{i-1\,i} & 0 & \cdots & 0 \\ \hat{\gamma}_1^{-1}G_{i1} & \cdots & \hat{\gamma}_{i-1}^{-1}G_{i\,i-1} & 0 & \hat{\gamma}_{i+1}^{-1}G_{i\,i+1} & \cdots & \hat{\gamma}_N^{-1}G_{iN} \\ 0 & \cdots & 0 & -\hat{\gamma}_{i+1}\hat{\gamma}_i^{-2}G_{i+1\,i} & 0 & \cdots & 0 \\ \vdots & \vdots & \vdots & \vdots & \vdots & \vdots & \vdots \\ 0 & \cdots & 0 & -\hat{\gamma}_N\hat{\gamma}_i^{-2}G_{Ni} & 0 & \cdots & 0 \end{bmatrix}$$

$$= \hat{\gamma}_i^{-1}\begin{bmatrix} 0 \\ \vdots \\ I_{q_i} \\ \vdots \\ 0 \end{bmatrix}[0\cdots\ I_{q_i}\ \cdots 0]G_{\hat{\gamma}} - \hat{\gamma}_i^{-1}G_{\hat{\gamma}}\begin{bmatrix} 0 \\ \vdots \\ I_{\ell_i} \\ \vdots \\ 0 \end{bmatrix}[0\cdots\ I_{\ell_i}\ \cdots 0].$$

So

$$\begin{aligned} w(\hat{\gamma})^*\frac{\partial G_{\hat{\gamma}}}{\partial \gamma_i}v(\hat{\gamma}) &= \hat{\gamma}_i^{-1}[w_i^*(\hat{\gamma})\left(G_{\hat{\gamma}}v(\hat{\gamma})\right)_i - \left(w^*(\hat{\gamma})G_{\hat{\gamma}}\right)_i v_i(\hat{\gamma})] \\ &= \sigma(\hat{\gamma})\hat{\gamma}_i^{-1}[w_i^*(\hat{\gamma})w_i(\hat{\gamma}) - v_i^*(\hat{\gamma})v_i(\hat{\gamma})]\ \in \mathbb{R} \end{aligned}$$

by (3.24), and this proves (28). □

Proposition 4.4.18. *Let* $\mathbf{\Delta}$ *be given by (24),* G, G_γ *as in Lemma 4.4.17 and let* $\mathbb{C}^{\ell_i}$, $\mathbb{C}^{q_i}$, $i \in \underline{N}$ *be provided with Euclidean norms. Suppose that for* γ *in a neighbourhood of some* $\hat{\gamma} \in (0,\infty)^N$, $(w_{\max}(\gamma), v_{\max}(\gamma))$ *is a singular pair of* G_γ *for the maximum singular value* $\sigma_{\max}(\gamma) = \sigma_{\max}(G_\gamma)$. *If the partial derivatives of* $\sigma_{\max}(\cdot), w_{\max}(\cdot), v_{\max}(\cdot)$ *exist at* $\hat{\gamma} \in (0,\infty)^N$ *and* $\nabla\sigma_{\max}(\hat{\gamma}) = 0$, *then*

$$\mu_{\mathbf{\Delta}}(G) = \sigma_{\max}(G_{\hat{\gamma}}). \tag{29}$$

Proof: If $\sigma_{\max}(\hat{\gamma}) = 0$, the equality follows immediately from (25). So we may assume $\sigma_{\max}(\hat{\gamma}) > 0$. Since $\nabla\sigma_{\max}(\hat{\gamma}) = 0$ we have by Lemma 4.4.17

$$\frac{\partial\sigma_{\max}(\hat{\gamma})}{\partial\gamma_i} = \frac{\sigma_{\max}(\hat{\gamma})}{\gamma_i}\left[\|w_i(\hat{\gamma})\|_2^2 - \|v_i(\hat{\gamma})\|_2^2\right] = 0, \quad i \in \underline{N}.$$

So $\|w_i(\hat{\gamma})\|_2 = \|v_i(\hat{\gamma})\|_2$ for $i \in \underline{N}$. Let

$$\Delta_i = \sigma_{\max}(\hat{\gamma})^{-1}v_i(\hat{\gamma})w_i(\hat{\gamma})^*/\|w_i(\hat{\gamma})\|_2^2 \quad \text{if } w_i(\hat{\gamma}) \neq 0 \quad \text{and} \quad \Delta_i = 0 \quad \text{otherwise.}$$

Then $\Delta_i w_i(\hat{\gamma}) = \sigma_{\max}(\hat{\gamma})^{-1}v_i(\hat{\gamma})$ for $i \in \underline{N}$, and

$$\|\Delta_i\| = \sigma_{\max}(\hat{\gamma})^{-1}\|v_i(\hat{\gamma})\|_2/\|w_i(\hat{\gamma})\|_2 = \sigma_{\max}(\hat{\gamma})^{-1} \quad \text{if } w_i(\hat{\gamma}) \neq 0.$$

Now for $\Delta = \operatorname{diag}(\Delta_1, \ldots, \Delta_N)$

$$\Delta G v(\hat{\gamma}) = \sigma_{\max}(\hat{\gamma})\Delta w(\hat{\gamma}) = v(\hat{\gamma}).$$

Thus $\sigma_{\max}(\hat{\gamma}) = \|\Delta\|^{-1} \leq \mu_{\mathbf{\Delta}}(G) \leq \|G_{\hat{\gamma}}\| = \sigma_{\max}(\hat{\gamma})$ and (29) is proved. □

Example 4.4.19. Consider the matrix $G = G_3 = \begin{bmatrix} 1 & 2\imath \\ 1 & 2 \end{bmatrix}$ and the same perturbation set as in Example 4.4.3. If $R_\gamma = L_\gamma = \operatorname{diag}(\gamma_1, \gamma_2)$, $\gamma_1 > 0, \gamma_2 > 0$, then

$$G_\gamma := R_\gamma G L_\gamma^{-1} = \begin{bmatrix} 1 & 2\imath\gamma_1\gamma_2^{-1} \\ \gamma_1^{-1}\gamma_2 & 2 \end{bmatrix}.$$

Let $\beta = \gamma_1\gamma_2^{-1}$, then

$$G_\gamma G_\gamma^* = \begin{bmatrix} 1 + 4\beta^2 & \beta^{-1} + 4\imath\beta \\ \beta^{-1} - 4\imath\beta & 4 + \beta^{-2} \end{bmatrix}.$$

So the eigenvalues of $G_\gamma G_\gamma^*$ are the roots of the equation

$$\lambda^2 - (5 + 4\beta^2 + \beta^{-2})\lambda + 8 = 0.$$

In order to find the minimum of the largest eigenvalue with respect to $\beta > 0$ we have to minimize $4\beta^2 + \beta^{-2}$. Clearly the minimum is achieved at $\beta = 1/\sqrt{2}$ and for this value of β the eigenvalues of $G_\gamma G_\gamma^*$ satisfy $\lambda^2 - 9\lambda + 8 = 0$ and hence by (25), $\mu_{\boldsymbol{\Delta}}(G) \le \sqrt{8}$. In fact since there are distinct eigenvalues at $\beta = 1/\sqrt{2}$ (corresponding to $\hat{\gamma} = (1, \sqrt{2})$), $\sigma_{\max}(G_\gamma)$ is analytic in a neighbourhood of $\hat{\gamma}$ and the conditions of Proposition 4.4.18 hold by Corollary 4.3.20. So we actually have $\mu_{\boldsymbol{\Delta}}(G) = \sqrt{8}$. In Remark 4.4.21 we will see that for the type of perturbation structure considered here the upper bound in (25) is tight without requiring the differentiability conditions assumed in Proposition 4.4.18.
The eigenvalues of G are $(3 \pm \sqrt{1 + 8\imath})/2$, i.e. approximately $2.56 + 0.94\imath$ and $0.44 - 0.94\imath$. Hence $\varrho(G) \approx 2.73$, whereas $\mu_{\boldsymbol{\Delta}}(G) \approx 2.83$ and so the lower bound $\varrho(G) \le \mu_{\boldsymbol{\Delta}}(G)$ mentioned in Example 4.4.16 is good but not tight. □

The upper bound in (25) is of special interest since it can be computed by solving a *convex* optimization problem if $\|\cdot\|_{\mathcal{L}(\mathbb{C}^\ell, \mathbb{C}^q)}$ is the spectral norm.

Lemma 4.4.20. *Let* $G = (G_{ij})_{i,j \in \underline{N}} \in \mathbb{C}^{q \times \ell}$, $G_{ij} \in \mathbb{C}^{q_i \times \ell_j}$, $\sum_{j=1}^N \ell_j = \ell$, $\sum_{i=1}^N q_i = q$ *and for* $\alpha = (\alpha_1, \dots, \alpha_N) \in \mathbb{R}^N$,

$$G_\alpha = \operatorname{diag}(e^{\alpha_1} I_{q_1}, \dots, e^{\alpha_N} I_{q_N})\, G \operatorname{diag}(e^{-\alpha_1} I_{\ell_1}, \dots, e^{-\alpha_N} I_{\ell_N}). \tag{30}$$

Then the map $f : \alpha \mapsto \|G_\alpha\| = \sigma_{\max}(G_\alpha)$ *is convex on* $\mathbb{R}^N$.

Proof: Since $f : \mathbb{R}^N \to \mathbb{R}_+$ is continuous it suffices to prove that

$$f(\alpha) + f(\beta) \ge 2f((\alpha + \beta)/2), \qquad \alpha, \beta \in \mathbb{R}^N.$$

Let $\gamma = (\alpha + \beta)/2$ and (v, w) be a singular pair of G_γ for the largest singular value,

$$G_\gamma v = \|G_\gamma\| w, \quad w^* G_\gamma = \|G_\gamma\| v^*,$$

and define L, R by

$$L = \operatorname{diag}(e^{(\beta_1 - \alpha_1)/2} I_{\ell_1}, \dots, e^{(\beta_N - \alpha_N)/2} I_{\ell_N}),\ R = \operatorname{diag}(e^{(\beta_1 - \alpha_1)/2} I_{q_1}, \dots, e^{(\beta_N - \alpha_N)/2} I_{q_N}).$$

Using the fact that $(t + 1/t) \ge 2$ for all $t > 0$ we obtain by simple calculation

$$f(\alpha) + f(\beta) \ge \|G_\alpha + G_\beta\| = \|R^{-1} G_\gamma L + R G_\gamma L^{-1}\| \ge \frac{w^* R\,(R^{-1} G_\gamma L + R G_\gamma L^{-1})\, L v}{\|Rw\|_2\, \|Lv\|_2}$$

$$= \frac{v^* L^2 v + w^* R^2 w}{\|Rw\|_2\, \|Lv\|_2} \|G_\gamma\| = \left[\frac{\|Lv\|_2}{\|Rw\|_2} + \frac{\|Rw\|_2}{\|Lv\|_2}\right] \|G_\gamma\| \ge 2\|G_\gamma\| = 2f(\gamma).$$

This concludes the proof. □

Remark 4.4.21. The upper bound in (25) is in general strictly larger than $\mu_{\mathbf{\Delta}}(G)$. However, we have equality in the multi-perturbation case if $N \leq 3$, see [135]. More precisely, suppose that $\mathbf{\Delta}$ is a class of block-diagonal perturbations with $N_1 = |J|$ square blocks and $N_2 = N - |J|$ diagonal blocks, then it has been proved (see *Notes and References*) that with respect to the spectral norm on $\mathbf{\Delta}$ if $N_1 + 2N_2 \leq 3$, then

$$\mu_{\mathbf{\Delta}}(G) = \inf_{\gamma \in (0,\infty)^N} \sigma_{\max}(L_\gamma G L_\gamma^{-1}), \quad G \in \mathbb{C}^{q\times q}. \tag{31}$$

□

Let $\mathbf{\Delta}$ be an arbitrary class of block-diagonal perturbations (see Definition 4.4.14). From Lemma 4.4.7 we have $\mu_{\mathbf{\Delta}}(G) = \max\{\varrho(\Delta G);\ \Delta \in \mathbf{\Delta},\ \|\Delta\|_{\mathbf{\Delta}} = 1\}$. The next theorem shows that we can replace the set $\mathbf{\Delta}$ on the RHS of this equation with a much smaller set. In order to prove the theorem we need the following lemma.

Lemma 4.4.22. *Let $p(s) = p(s_1, \ldots, s_N) \in \mathbb{C}[s_1, \ldots, s_N]$ be a non-constant complex polynomial in N variables. Then there exists a zero $\hat{z} \in \mathbb{C}^N$ of $p(s)$ such that*

$$\|\hat{z}\|_\infty = \min\{\|z\|_\infty;\ z \in \mathbb{C}^N, p(z) = 0\} =: \delta \quad \text{and} \quad |\hat{z}_i| = \delta,\ i \in \underline{N}.$$

Proof: Let $\hat{z} \in \mathbb{C}^N$ be such that $p(\hat{z}) = 0$, $\|\hat{z}\|_\infty = \delta$ and the index set $\hat{I} := \{i \in \underline{N};\ |\hat{z}_i| = \delta\}$ is maximal amongst all $z \in \mathbb{C}^N$ satisfying $p(z) = 0$ and $\|z\|_\infty = \delta$. Such $\hat{z}$ exists, since the zero set of p is nonempty and closed in $\mathbb{C}^N$. Suppose that $|\hat{z}_k| < \delta$ for some $k \in \underline{N}$. Renumbering the variables if necessary we may assume $k = N$. We can write $p(s)$ in the form

$$p(s_1, \ldots, s_N) = \sum_{j=0}^{d_N} p_j(s_1, \ldots, s_{N-1}) s_N^j.$$

If $p_j(\hat{z}_1, \ldots, \hat{z}_{N-1}) = 0$ for $j = 0, \ldots, d_N$ the vector $z = (\hat{z}_1, \ldots, \hat{z}_{N-1}, \delta)$ satisfies $p(z) = 0$, $\|z\|_\infty = \delta$ and the corresponding index set $I := \{i \in \underline{N};\ |z_i| = \delta\}$ is strictly larger than $\hat{I}$ which is a contradiction. Hence not all $p_j(\hat{z}_1, \ldots, \hat{z}_{N-1})$ are zero. Now consider the non-zero polynomial

$$q(s_N) = p(\hat{z}_1, \ldots, \hat{z}_{N-1}, s_N) = \sum_{j=0}^{d_N} p_j(\hat{z}_1, \ldots, \hat{z}_{N-1}) s_N^j \in \mathbb{C}[s_N].$$

This is a non-constant polynomial of degree, say $d \leq d_N$, which has a zero at $\hat{z}_N$ with $|\hat{z}_N| < \delta$. By Proposition 4.1.2, for $\varepsilon > 0$ sufficiently small, the polynomial

$$p((1-\varepsilon)\hat{z}_1, \ldots, (1-\varepsilon)\hat{z}_{N-1}, s_N) = \sum_{j=0}^{d_N} p_j((1-\varepsilon)\hat{z}_1, \ldots, (1-\varepsilon)\hat{z}_{N-1}) s_N^j \in \mathbb{C}[s_N]$$

has a root z_N of modulus $|z_N| < \delta$. But then $z := ((1-\varepsilon)\hat{z}_1, \ldots, (1-\varepsilon)\hat{z}_{N-1}, z_N)$ is a zero of p satisfying $\|z\|_\infty < \delta$ and this contradicts the minimality of $\|\hat{z}\|_\infty$. Therefore $|\hat{z}_i| = \delta$ for all $i \in \underline{N}$. □

Theorem 4.4.23. *Let $\mathbf{\Delta}$ be a class of multi-block perturbations of the form (24) endowed with a norm as in Definition 4.4.14. Then*

$$\mu_{\mathbf{\Delta}}(G) = \max_{U \in \mathcal{V}_{\mathbf{\Delta}}^\ell, V \in \mathcal{V}_{\mathbf{\Delta}}^q} \varrho(V^* G U) \leq \varrho\left(\left(\|G_{ij}\|_{\mathcal{L}(\mathbb{C}^{\ell_i}, \mathbb{C}^{q_i})}\right)_{i,j \in \underline{N}}\right), \qquad G \in \mathbb{C}^{q\times\ell}, \tag{32}$$

where

$$\begin{aligned} \mathcal{V}_{\mathbf{\Delta}}^\ell &= \{\operatorname{diag}(u_1, \ldots, u_N);\ \forall j \in \underline{N}:\ u_j \in \mathbb{C}^{\ell_j}, \|u_j\|_{\mathbb{C}^{\ell_j}} = 1\} \subset \mathbb{C}^{\ell\times N}, \\ \mathcal{V}_{\mathbf{\Delta}}^q &= \{\operatorname{diag}(v_1, \ldots, v_N);\ \forall j \in \underline{N}:\ v_j \in \mathbb{C}^{q_j}, \|v_j\|_{\mathbb{C}^{q_j}}^* = 1\} \subset \mathbb{C}^{q\times N}. \end{aligned} \tag{33}$$

Proof: Let $U \in \mathcal{V}^\ell_{\mathbf{\Delta}}$, $V \in \mathcal{V}^q_{\mathbf{\Delta}}$

$$\begin{array}{rclllll} U & = & \operatorname{diag}(u_1, \dots, u_N) \in \mathbb{C}^{\ell \times N}, & u_j \in \mathbb{C}^{\ell_j}, & \|u_j\|_{\mathbb{C}^{\ell_j}} = 1, & j \in \underline{N}, \\ V & = & \operatorname{diag}(v_1, \dots, v_N) \in \mathbb{C}^{q \times N}, & v_j \in \mathbb{C}^{q_j}, & \|v_j\|^*_{\mathbb{C}^{q_j}} = 1, & j \in \underline{N}. \end{array} \tag{34}$$

Then $\varrho(V^*GU) = \varrho(UV^*G)$ where $UV^* = \operatorname{diag}(u_1v_1^*, \dots, u_Nv_N^*) \in \mathbf{\Delta}$ and

$$\|UV^*\|_{\mathbf{\Delta}} = \max_{j \in \underline{N}} \|u_jv_j^*\|_{\mathcal{L}(\mathbb{C}^{q_j}, \mathbb{C}^{\ell_j})} = \max_{j \in \underline{N}} \|u_j\|_{\mathbb{C}^{\ell_j}} \|v_j\|^*_{\mathbb{C}^{q_j}} = 1.$$

Hence by Lemma 4.4.7

$$\mu_{\mathbf{\Delta}}(G) = \max_{\substack{\Delta \in \mathbf{\Delta} \\ \|\Delta\|_{\mathbf{\Delta}} = 1}} \varrho(\Delta G) \geq \max_{U \in \mathcal{V}^\ell_{\mathbf{\Delta}}, V \in \mathcal{V}^q_{\mathbf{\Delta}}} \varrho(V^*GU).$$

Conversely, if $\mu_{\mathbf{\Delta}}(G) = 0$, equality follows trivially from the above estimate and so we may suppose $\mu_{\mathbf{\Delta}}(G) > 0$. Let $\Delta_{\min} = \operatorname{diag}(\Delta_1, \dots, \Delta_N) \in \mathbf{\Delta}$ be such that (4) is satisfied. Then there exists $\hat{u} = (\hat{u}_i)_{i \in \underline{N}} \in \mathbb{C}^\ell$, $\hat{u} \neq 0$ such that $(I_\ell - \Delta_{\min}G)\hat{u} = 0$, i.e. $\Delta_i \hat{y}_i = \hat{u}_i$ where the $\hat{y}_i \in \mathbb{C}^{q_i}$ are given by $(\hat{y}_i)_{i \in \underline{N}} = G\hat{u}$. By the Hahn-Banach Theorem (see Example A.4.11) there exists $v_j \in \mathbb{C}^{q_j}$ aligned with $\hat{y}_j$, $j \in \underline{N}$ such that

$$\|v_j\|^*_{\mathbb{C}^{q_j}} = 1, \quad v_j^*\hat{y}_j = \|\hat{y}_j\|_{\mathbb{C}^{q_j}}, \qquad j \in \underline{N}.$$

Choose $u_j^1 \in \mathbb{C}^{\ell_j}$ such that $\|u_j^1\|_{\mathbb{C}^{\ell_j}} = 1$ and $\hat{u}_j = \|\hat{u}_j\|_{\mathbb{C}^{\ell_j}} u_j^1$, $j \in \underline{N}$. Then

$$(r_ju_j^1v_j^*)\hat{y}_j = \hat{u}_j,\ j \in \underline{N} \text{ where } r_j = \|\hat{u}_j\|_{\mathbb{C}^{\ell_j}} / \|\hat{y}_j\|_{\mathbb{C}^{q_j}} \text{ if } \hat{y}_j \neq 0 \text{ and otherwise } r_j = 0.$$

Hence $\det(I - \Delta(r)G) = 0$ for $\Delta(r) = \operatorname{diag}(r_1u_1^1v_1^*, \dots, r_Nu_N^1v_N^*) \in \mathbf{\Delta}$. Since

$$\|\Delta(r)\|_{\mathbf{\Delta}} = \max_{i \in \underline{N}} r_i \leq \max_{i \in \underline{N}} \|\Delta_i\|_{\Delta_i} = \|\Delta_{\min}\|_{\mathbf{\Delta}}$$

it follows from the minimality of $\|\Delta_{\min}\|_{\mathbf{\Delta}}$ that $\delta := \|\Delta_{\min}\|_{\mathbf{\Delta}} = \max_{i \in \underline{N}} r_i$. Now consider the polynomial

$$p(s_1, \dots, s_N) = \det(I - \Delta(s)G) \in \mathbb{C}[s_1, \dots, s_N], \quad \Delta(s) = \operatorname{diag}(s_1u_1^1v_1^*, \dots, s_Nu_N^1v_N^*).$$

$r = (r_1, \dots, r_N)$ is a zero of minimal norm $\|r\|_\infty = \delta$ of $p(s)$. Applying the previous lemma, there exists a zero $\hat{z} = (\hat{z}_1, \dots, \hat{z}_N)$ of $p(s)$ such that $\hat{z}_j \in \mathbb{C}$ and $|\hat{z}_j| = \delta$ for all $j \in \underline{N}$. Now let $u_j = \delta^{-1}\hat{z}_ju_j^1$, $j \in \underline{N}$, v_j as before, and define U, V as in (34). Then $U \in \mathcal{V}^\ell_{\mathbf{\Delta}}$, $V \in \mathcal{V}^q_{\mathbf{\Delta}}$ and $\det(I - \delta UV^*G) = \det(I - \Delta(\hat{z})G) = 0$, and so

$$\varrho(V^*GU) = \varrho(UV^*G) = \delta^{-1}\varrho(\Delta(\hat{z})G)) \geq \delta^{-1} = \mu_{\mathbf{\Delta}}(G).$$

This proves the equality in (32). To prove the inequality note that from the theory of non-negative matrices it is known that (see [183, §2.3])

$$A = (a_{ij}) \in \mathbb{C}^{N \times N},\ B = (b_{ij}) \in \mathbb{R}_+^{N \times N},\ \forall i, j \in \underline{N} : |a_{ij}| \leq b_{ij} \quad \Rightarrow \quad \varrho(A) \leq \varrho(B).$$

Hence

$$\varrho(V^*GU) \leq \varrho\left((|v_i^*G_{ij}u_j|)_{i,j \in \underline{N}} \right) \leq \varrho\left((\|G_{ij}\|_{\mathcal{L}(\mathbb{C}^{\ell_i}, \mathbb{C}^{q_i})})_{i,j \in \underline{N}} \right).$$

This completes the proof. □

Remark 4.4.24. If $\mathbf{\Delta} \subset \mathbb{C}^{\ell\times q}$ is a class of multi-block perturbations of the form (24) and the spaces $\mathbb{C}^{\ell_i\times q_i}$ are endowed with norms $\|\cdot\|_{\mathbf{\Delta}_i} = \|\cdot\|_{\mathbb{C}^{\ell_i\times q_i}}$ which are not operator norms but satisfy condition (19) with respect to operator norms $\|\cdot\|_{\mathcal{L}(\mathbb{C}^{q_i},\mathbb{C}^{\ell_i})}$, $i \in \underline{N}$, then the formula (32) remains valid. □

Proposition 4.4.23 can be viewed as a generalization of Proposition 4.4.11. In fact it implies for the single block case ($N = 1$) that (by the Hahn-Banach Theorem)

$$\mu_{\mathbf{\Delta}}(G) = \max_{\substack{u\in\mathbb{C}^\ell,\, v\in\mathbb{C}^q \\ \|u\|_{\mathbb{C}^\ell}=\|v\|^*_{\mathbb{C}^q}=1}} |v^*Gu| = \|G\|_{\mathcal{L}(\mathbb{C}^\ell,\mathbb{C}^q)}, \qquad G \in \mathbb{C}^{q\times\ell}.$$

The following characterization is applicable to a wide class of perturbation sets, namely block-diagonal structures with prespecified multiplicities of blocks . However, it is assumed that the blocks are square (i.e. $\ell_i = q_i$) and the norm on the perturbation set is the spectral norm.

Proposition 4.4.25. *Suppose that for given integers $m_1, \dots, m_N, q_1, \dots, q_N \geq 1$, $\mathbf{\Delta}$ is the set of block-diagonal matrices of the form*

$$\Delta = \operatorname{diag}(\Delta_1, \dots, \Delta_1; \dots; \Delta_N, \dots, \Delta_N) \in \mathbb{C}^{q\times q},\ q = \sum_{i=1}^{N} m_i q_i, \quad \Delta_i \in \mathbb{C}^{q_i\times q_i}, i \in \underline{N} \tag{35}$$

where the blocks Δ_i are square and repeated m_i times for $i \in \underline{N}$. Then, with respect to the spectral norm on $\mathbf{\Delta}$,

$$\mu_{\mathbf{\Delta}}(G) = \max_{U\in\mathbf{U}_{\mathbf{\Delta}}} \varrho(UG), \qquad G \in \mathbb{C}^{q\times q} \tag{36}$$

where

$$\mathbf{U}_{\mathbf{\Delta}} := \mathbf{\Delta} \cap \mathbf{U}_q(\mathbb{C}) = \{\operatorname{diag}(U_1, \dots, U_1; \dots; U_N, \dots, U_N) \in \mathbf{\Delta};\ U_i \in \mathbf{U}_{q_i}(\mathbb{C}),\ i \in \underline{N}\}.$$

Proof: For every $U \in \mathbf{U}_{\mathbf{\Delta}}$, we have $U \in \mathbf{\Delta}$ and $\|U\|_{\mathbf{\Delta}} = \sigma_{\max}(U) = 1$. Therefore Lemma 4.4.7 implies $\max_{U\in\mathbf{U}_{\mathbf{\Delta}}} \varrho(UG) \leq \mu_{\mathbf{\Delta}}(G)$. If $\mu_{\mathbf{\Delta}}(G) = 0$, equality follows trivially and so we may suppose $\mu_{\mathbf{\Delta}}(G) > 0$. By homogeneity we may assume $\mu_{\mathbf{\Delta}}(G) = 1$. It remains to show that there exists a $U \in \mathbf{U}_{\mathbf{\Delta}}$ such that $\varrho(UG) \geq 1 = \mu_{\mathbf{\Delta}}(G)$. By Lemma 4.4.7 there exists Δ of the form (35) such that $\varrho(\Delta G) = 1$ and $\|\Delta\|_{\mathbf{\Delta}} = 1$. Let $\Delta_i = W_i\Sigma_i V_i^*$ be a singular value decomposition of the block Δ_i, $i \in \underline{N}$, and set

$$W = \operatorname{diag}(W_1, \dots W_1; \dots; W_N, \dots, W_N), \quad V = \operatorname{diag}(V_1, \dots V_1; \dots; V_N, \dots, V_N),$$

$$\Sigma = \operatorname{diag}(\Sigma_1, \dots, \Sigma_1; \dots; \Sigma_N, \dots, \Sigma_N), \quad \operatorname{diag}(\sigma_1, \dots, \sigma_{\bar{q}}) = \operatorname{diag}(\Sigma_1, \Sigma_2, \dots, \Sigma_N).$$

Here the multiplicities of the blocks in W, V, Σ are $m_1, \dots, m_N$ respectively, and $\bar{q} = \sum_{i=1}^N q_i$. Then $\Delta = W\Sigma V^*$ and hence $\sigma = (\sigma_1, \dots, \sigma_{\bar{q}})$ has the ∞-norm $\|\sigma\|_\infty = \|\Delta\|_{\mathbf{\Delta}} = 1$. For any $s = (s_1, \dots, s_{\bar{q}}) \in \mathbb{C}^{\bar{q}}$, let $\Sigma_1(s) = \operatorname{diag}(s_1, \dots, s_{q_1})$, $\Sigma_2(s) = \operatorname{diag}(s_{q_1+1}, \dots s_{q_1+q_2})$, etc. and

$$\Sigma(s) = \operatorname{diag}(\Sigma_1(s), \dots, \Sigma_1(s); \dots; \Sigma_N(s), \dots, \Sigma_N(s)).$$

Now consider the polynomial

$$p(s) = \det(I_q - W\Sigma(s)V^*G) \in \mathbb{C}[s_1, \dots, s_{\bar{q}}].$$

Since $W\Sigma(s)V \in \mathbf{\Delta}$ for all $s \in \mathbb{C}^{\bar{q}}$ and $\|W\Sigma(s)V\|_{\mathbf{\Delta}} = \|s\|_\infty$, σ is a zero of $p(s)$ with minimal ∞-norm. By Lemma 4.4.22 there exists a zero $\hat{z} \in \mathbb{C}^{\bar{q}}$ of $p(s)$ such that $|\hat{z}_i| = 1$, $i = 1, \dots, \bar{q}$. Thus $\Sigma_i(\hat{z}) \in \mathbf{U}_{q_i}$, and $U := W\Sigma(\hat{z})V^* \in \mathbf{U}_{\mathbf{\Delta}}$ satisfies $\det(I - UG) = 0$. So there exists $U \in \mathbf{U}_{\mathbf{\Delta}}$ such that $\varrho(UG) \geq 1 = \mu_{\mathbf{\Delta}}(G)$. □

Unfortunately, the map $U \mapsto \varrho(UG)$ on $\mathbf{U}_{\mathbf{\Delta}}$ may have many local maxima which are not global ones. Therefore optimization algorithms applied to this map will in general only provide lower bounds for $\mu_{\mathbf{\Delta}}(G)$, see *Notes and References.*

Example 4.4.26. Consider again the perturbation set and the matrix $G_3 = \begin{bmatrix} 1 & 2\imath \\ 1 & 2 \end{bmatrix}$ as given in Example 4.4.3. The diagonal perturbations $\Delta \in \mathbf{\Delta}$ are of the form (24) with $\ell_i = q_i = 1$, $i \in \underline{N} = \{1, 2\}$. They are also of the form (35) with $q_1 = q_2 = 1$, $m_1 = m_2 = 1$ and $N = 2$. Therefore, to compute $\mu_{\mathbf{\Delta}}(G_3)$ we may apply both Theorem 4.4.23 and Proposition 4.4.25. Applying the first result we have by (33) $\mathcal{V}^{\ell}_{\mathbf{\Delta}} = \{\mathrm{diag}(u_1, u_2); |u_1| = |u_2| = 1\}$ and $\mathcal{V}^{q}_{\mathbf{\Delta}} = \mathcal{V}^{\ell}_{\mathbf{\Delta}}$, hence by (32)

$$\mu_{\mathbf{\Delta}}(G_3) = \max_{|u_i|=|v_i|=1} \varrho\left(\mathrm{diag}(\overline{v_1}, \overline{v_2}) G_3 \,\mathrm{diag}(u_1, u_2)\right) \leq \varrho\left(\begin{bmatrix} 1 & 2 \\ 1 & 2 \end{bmatrix}\right) = 3.$$

But ϱ is similarity invariant. Moreover, it does not change if the matrix is multiplied by $z \in \mathbf{U}_1(\mathbb{C}) = \{u \in \mathbb{C};\ |u| = 1\}$. Therefore

$$\mu_{\mathbf{\Delta}}(G_3) = \max_{w_1, w_2 \in \mathbf{U}_1(\mathbb{C})} \varrho\left(\mathrm{diag}(w_1, w_2) G_3\right) = \max_{w \in \mathbf{U}_1(\mathbb{C})} \varrho\left(\mathrm{diag}(1, w) G_3\right).$$

Proposition 4.4.25 yields the same result since

$$\mathbf{U}_{\mathbf{\Delta}} = \mathbf{\Delta} \cap \mathbf{U}_1(\mathbb{C}) = \{\mathrm{diag}(u_1, u_2); u_1, u_2 \in \mathbf{U}_1(\mathbb{C})\}$$

and therefore by (36)

$$\mu_{\mathbf{\Delta}}(G_3) = \max_{U \in \mathbf{U}_{\mathbf{\Delta}}} \varrho\left(U G_3\right) = \max_{u \in \mathbf{U}_1(\mathbb{C})} \varrho\left(\mathrm{diag}(1, u) G_3\right).$$

Now let $U = \mathrm{diag}\,(1, u)$, $u \in \mathbb{C}$, $|u| = 1$. Then $UG_3 = \begin{bmatrix} 1 & 2\imath \\ u & 2u \end{bmatrix}$ and the eigenvalues of UG_3 are the roots $s_{1,2}(u)$ of the equation

$$s^2 - (1 + 2u)s + 2u(1 - \imath) = 0.$$

In fact, one can show that the largest absolute value of the roots of this equation is obtained for $u = (4 + 3\imath)/5$. The corresponding root is $\lambda = 2(1 + \imath)$. Hence $\mu_{\mathbf{\Delta}}(G_3) = |\lambda| = \sqrt{8}$ in accordance with the result of Ex. 2. □

We conclude this subsection with a brief discussion of the continuity of the μ-value. The following result holds true for both complex and real perturbation structures.

Lemma 4.4.27. *Let $\mathbf{\Delta}$ be a closed non-empty subset of $\mathbb{C}^{\ell \times q}$ and* $\mathrm{span}_{\mathbb{C}}\, \mathbf{\Delta}$ *be endowed with a norm $\|\cdot\|_{\mathbf{\Delta}}$. Then the map $\mu_{\mathbf{\Delta}}(\cdot) : \mathbb{C}^{q \times \ell} \to \mathbb{R}_+$ is upper semicontinuous, i.e. for every $G_0 \in \mathbb{C}^{q \times \ell}$ and $r > \mu_{\mathbf{\Delta}}(G_0)$ there exists a neighbourhood W of G_0 in $\mathbb{C}^{q \times \ell}$ such that $\mu_{\mathbf{\Delta}}(G) < r$ for all $G \in W$.*

Proof: Given any $G_0 \in \mathbb{C}^{q \times \ell}$ and $r > \mu_{\mathbf{\Delta}}(G_0)$, let

$$\mathbf{\Delta}_r = \{\Delta \in \mathbf{\Delta};\ \|\Delta\|_{\mathbf{\Delta}} \leq r^{-1}\}.$$

By the definition of $\mu_{\mathbf{\Delta}}(G_0)$ we have $\det(I - \Delta G_0) \neq 0$ for all $\Delta \in \mathbf{\Delta}_r$. Since the determinant is continuous, for every $\Delta \in \mathbf{\Delta}_r$, there exist a neighbourhood U_Δ of

Δ in $\mathbf{\Delta}$ and a neighbourhood W_Δ of G_0 in $\mathbb{C}^{q\times\ell}$ such that $\det(I - \tilde{\Delta}G) \neq 0$ for all $\tilde{\Delta} \in U_\Delta$, $G \in W_\Delta$. Since $\mathbf{\Delta}_r$ is compact, finitely many of these neighbourhoods U_{Δ_k}, $k \in \underline{N}$ cover $\mathbf{\Delta}_r$. Thus, setting $W = \bigcap_{k=1}^N W_{\Delta_k}$ we obtain a neighbourhood W of G_0 satisfying $\det(I - \Delta G) \neq 0$ for all $\Delta \in \mathbf{\Delta}_r$, $G \in W$. We conclude that $\mu_{\mathbf{\Delta}}(G) < r$ for all $G \in W$. □

As a consequence of this lemma, $\mu_{\mathbf{\Delta}}(\cdot)$ is continuous at each G_0 satisfying $\mu_{\mathbf{\Delta}}(G_0) = 0$. For complex perturbation classes we obtain the following continuity result from the previous lemma and Lemma 4.4.7.

Proposition 4.4.28. *Suppose $\mathbf{\Delta}$ is a closed non-empty subset of $\mathbb{C}^{\ell\times q}$ such that $\mathbb{C}\mathbf{\Delta} \subset \mathbf{\Delta}$ and $\|\cdot\|_{\mathbf{\Delta}}$ is a norm on* $\operatorname{span}_{\mathbb{C}} \mathbf{\Delta}$. *Then $\mu_{\mathbf{\Delta}}(\cdot) : \mathbb{C}^{q\times\ell} \to \mathbb{R}_+$ is continuous.*

Proof: By Lemma 4.4.7 we have, $\mu_{\mathbf{\Delta}}(G) = \max_{\Delta\in\mathbf{\Delta}, \|\Delta\|_{\mathbf{\Delta}}=1} \varrho(\Delta G)$, for all $G \in \mathbb{C}^{q\times\ell}$. Since the supremum of a family of continuous functions is lower semicontinuous it follows that $\mu_{\mathbf{\Delta}}(\cdot) : \mathbb{C}^{q\times\ell} \to \mathbb{R}_+$ is lower semicontinuous and hence it is continuous by the previous lemma. □

4.4.2 μ-Values for Real Full-Block Perturbations

In the introduction to this section we mentioned that the analysis of spectral variations under *real* parameter perturbations leads to the problem of determining, for a given *complex* matrix $G \in \mathbb{C}^{q\times\ell}$, the norm of the smallest *real* perturbation $\Delta \in \mathbf{\Delta} \subset \mathbb{R}^{\ell\times q}$ for which $\det(I - \Delta G) = 0$. In this subsection we deal with this *real μ-problem* for the full-block case, i.e. $\mathbf{\Delta} = \mathbb{R}^{\ell\times q}$, where we assume that $\mathbb{R}^{\ell\times q}$ is endowed with an operator norm $\|\cdot\|_{\mathcal{L}(\mathbb{R}^q,\mathbb{R}^\ell)}$ induced by a pair of norms on the vector spaces $\mathbb{R}^\ell$ and $\mathbb{R}^q$. For this case we denote the μ-value of $G \in \mathbb{C}^{q\times\ell}$ by $\mu_{\mathbb{R}}(G)$, i.e

$$\mu_{\mathbb{R}}(G) = \left[\inf\{\|\Delta\|_{\mathcal{L}(\mathbb{R}^q,\mathbb{R}^\ell)};\ \Delta \in \mathbb{R}^{\ell\times q} \text{ and } \det(I_\ell - \Delta G) = 0\}\right]^{-1}. \tag{37}$$

In order to determine $\mu_{\mathbb{R}}(G)$ we make use of the representation $G^{\mathbb{R}}$ of G as an $\mathbb{R}$-linear operator (see Subsection A.1.3). Suppose that $G = X + \imath Y$ with $X, Y \in \mathbb{R}^{q\times\ell}$. Then $G^{\mathbb{R}} = \left[\begin{smallmatrix} X & -Y \\ Y & X \end{smallmatrix}\right]$ and by Lemma A.1.18, for every $\Delta \in \mathbb{R}^{\ell\times q}$,

$$\det(I_\ell - \Delta G) = 0 \Leftrightarrow \det(I_\ell - \Delta G)^{\mathbb{R}} = \det\left(I_{2\ell} - \begin{bmatrix} \Delta & 0 \\ 0 & \Delta \end{bmatrix}\begin{bmatrix} X & -Y \\ Y & X \end{bmatrix}\right) = 0. \tag{38}$$

So

$$\mu_{\mathbb{R}}(G) = \mu_{\mathbf{\Delta}}(G^{\mathbb{R}}) \text{ where } \mathbf{\Delta} = \{\operatorname{diag}(\Delta, \Delta);\ \Delta \in \mathbb{R}^{\ell\times q}\}, \tag{39}$$

if we provide $\mathbf{\Delta}$ with the norm $\|\operatorname{diag}(\Delta,\Delta)\|_{\mathbf{\Delta}} = \|\Delta\|_{\mathcal{L}(\mathbb{R}^q,\mathbb{R}^\ell)}$. We see, therefore, that in order to characterize $\mu_{\mathbb{R}}(G)$, we may equally well consider *real* double block perturbations $\operatorname{diag}(\Delta,\Delta) \in \mathbf{\Delta}$ of the real linear operator $G^{\mathbb{R}}$. We find it more convenient to do this.

Most of the results of the previous subsection were developed for complex perturbation structures, i.e. $\mathbb{C}\mathbf{\Delta} = \mathbf{\Delta}$. However some of them are applicable to the present problem. The next lemma summarizes some elementary facts which we will need in the sequel.

Lemma 4.4.29. *Suppose $G \in \mathbb{C}^{q\times\ell}$ is given, $\mu_{\mathbb{R}}(G)$ is defined by* (37) *and* $\mathbf{\Delta}$ *by* (39). *Then*

(i) $\mu_{\mathbb{R}}(G) = \mu_{\mathbf{\Delta}}(G^{\mathbb{R}}) = \mu_{\mathbf{\Delta}^\top}((G^{\mathbb{R}})^\top) = \mu_{\mathbb{R}}(G^*)$ *where* $\mathbf{\Delta}^\top = \{\mathrm{diag}\,(\Delta^\top, \Delta^\top); \Delta \in \mathbb{R}^{\ell\times q}\}$ *is endowed with the norm* $\|\,\mathrm{diag}\,(\Delta^\top,\Delta^\top)\|_{\mathbf{\Delta}^\top} = \|\Delta^\top\|_{\mathcal{L}(\mathbb{R}^\ell,\mathbb{R}^q)}$.

(ii) If $G^{\mathbb{R}}_\gamma = \begin{bmatrix} X & -\gamma Y \\ \gamma^{-1}Y & X\end{bmatrix}$, $\gamma \neq 0$, *then* $\mu_{\mathbf{\Delta}}(G^{\mathbb{R}}) = \mu_{\mathbf{\Delta}}(G^{\mathbb{R}}_\gamma) \le \|G^{\mathbb{R}}_\gamma\|_{\mathcal{L}(\mathbb{R}^{2\ell},\mathbb{R}^{2q})}$.

(iii) The map $\mu_{\mathbb{R}}(\cdot) : \mathbb{C}^{q\times\ell} \to \mathbb{R}_+$ *is upper semicontinuous.*

Proof: (iii) is a specialization of Lemma 4.4.27. (ii) follows from Lemma 4.4.5 (iii) with $R = \begin{bmatrix} \gamma I_q & 0 \\ 0 & I_q\end{bmatrix}$, $L_R = \begin{bmatrix} \gamma I_\ell & 0 \\ 0 & I_\ell\end{bmatrix}$. (i) follows from the definition and (39), since $(G^*)^{\mathbb{R}} = (G^{\mathbb{R}})^\top$, $\|\Delta^\top\|_{\mathcal{L}(\mathbb{R}^\ell,\mathbb{R}^q)} = \|\Delta\|_{\mathcal{L}(\mathbb{R}^q,\mathbb{R}^\ell)}$ and $\det\,(I_q - \Delta^\top G^*) = 0 \Leftrightarrow \det\,(I_\ell - \Delta G) = 0$. □

We first examine the case where $\ell = 1$. Then $G \in \mathbb{C}^{q\times 1}$ is a complex column vector and the matrices Δ are real *row* vectors representing *linear forms* on $\mathbb{R}^q \subset \mathbb{C}^q$, $\Delta \in \mathbb{R}^{1\times q} = (\mathbb{R}^q)^*$. This leads to a considerable reduction in the difficulty of the problem. Given any norm $\|\cdot\|_{\mathbb{R}^q}$ on $\mathbb{R}^q$ the associated operator norm on $\mathbf{\Delta} = \mathbb{R}^{1\times q}$ (with respect to the norm $|\cdot|$ on $\mathbb{R}$) is simply the dual norm of $\|\cdot\|_{\mathbb{R}^q}$,

$$\|\Delta\|_{\mathbf{\Delta}} = \|\Delta\|^*_{\mathbb{R}^q} = \max\{|\Delta X|; X \in \mathbb{R}^q, \|X\|_{\mathbb{R}^q} = 1\}\,.^4 \tag{40}$$

Remark 4.4.30. Of particular interest are the 1, 2, and ∞-norms on $\mathbb{R}^q = \mathbb{R}^{q\times 1}$ and their dual norms on the perturbation set $\mathbf{\Delta} = \mathbb{R}^{1\times q} = (\mathbb{R}^q)^*$. Recall that $\|\cdot\|^*_\infty = \|\cdot\|_1$, $\|\cdot\|^*_2 = \|\cdot\|_2$, and $\|\cdot\|^*_1 = \|\cdot\|_\infty$, see Section A.1. □

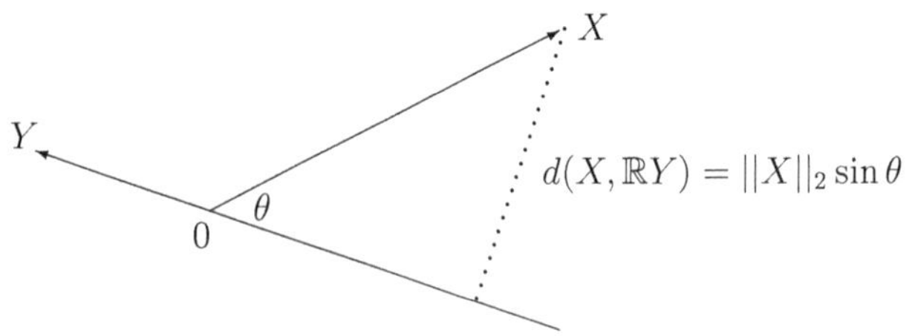

Figure 4.4.3: The distance $\mathrm{dist}(X, \mathbb{R}Y)$ in $(\mathbb{R}^q, \|\cdot\|_2)$ if $Y \neq 0$

Let $\mathrm{dist}\,(X, \mathbb{R}Y)$ denote the distance of the point $X \in \mathbb{R}^q$ from the linear subspace $\mathbb{R}Y = \{\alpha Y; \alpha \in \mathbb{R}\}$ spanned by $Y \in \mathbb{R}^q$ in the normed space $(\mathbb{R}^q, \|\cdot\|_{\mathbb{R}^q})$

$$\mathrm{dist}(X, \mathbb{R}Y) = \min_{\alpha\in\mathbb{R}} \|X - \alpha Y\|_{\mathbb{R}^q}\,. \tag{41}$$

[4] In our discussion of the special case $\ell = 1$ we will use capital letters to denote the elements of $\mathbb{R}^q = \mathbb{R}^{q\times 1}$ considered as the space of $q \times 1$ matrices in which G lives.

In the Euclidean case, illustrated by Figure 4.4.3, it is easy to see that

$$\operatorname{dist}(X,\mathbb{R}Y)^2 = \begin{cases} \|X\|_2^2 - \dfrac{\langle X,Y\rangle^2}{\|Y\|_2^2} & \text{if } Y \neq 0, \\ \|X\|_2^2 & \text{if } Y = 0. \end{cases} \tag{42}$$

We do not have analogous formulas to (42) for the 1 and ∞-norms. However, for these norms one can reduce the optimization problem (41) to a finite search process. The reader is asked to prove this for the ∞-norm in Ex. 7.
The following theorem holds for arbitrary norms on $\mathbb{R}^q$.

Theorem 4.4.31. *Suppose* $\|\cdot\|_{\mathbb{R}^q}$ *is any norm on* $\mathbb{R}^q = \mathbb{R}^{q\times 1}$ *and* $G = X + \imath Y \in \mathbb{C}^{q\times 1}$ *is given, with* $X, Y \in \mathbb{R}^{q\times 1}$. *Then, with respect to the perturbation norm* (40),

$$\mu_{\mathbb{R}}(G) = \mu_{\boldsymbol{\Delta}}(G^{\mathbb{R}}) = \operatorname{dist}(X, \mathbb{R}Y). \tag{43}$$

Proof: By (37) and (38) we have to find a $\Delta \in \mathbb{R}^{1\times q}$ of minimum norm, $\|\Delta\|_{\mathbb{R}^q}^*$, such that the equations

$$\Delta(Xu - Yv) = u, \qquad \Delta(Yu + Xv) = v, \tag{44}$$

admit a non-trivial solution $(u, v) \in \mathbb{R}^2$. But this is equivalent to $\Delta X = 1$ and $\Delta Y = 0$. Suppose that $\Delta \in \mathbb{R}^{1\times q}$ satisfies these two equations, then $\Delta(X - \alpha Y) = 1$ for all $\alpha \in \mathbb{R}$. If $X = \alpha Y$ for some $\alpha \in \mathbb{R}$, we have $\operatorname{dist}(X, \mathbb{R}Y) = \mu_{\mathbb{R}}(G) = 0$, otherwise $\|\Delta\|_{\mathbb{R}^q}^* \geq \|X - \alpha Y\|_{\mathbb{R}^q}^{-1}$ for all $\alpha \in \mathbb{R}$. So

$$\mu_{\mathbb{R}}(G)^{-1} = \min_{\Delta X=1, \Delta Y=0} \|\Delta\|_{\mathbb{R}^q}^* \geq \max_{\alpha\in\mathbb{R}} \|X - \alpha Y\|_{\mathbb{R}^q}^{-1} = [\operatorname{dist}(X, \mathbb{R}Y)]^{-1}$$

and hence $\mu_{\mathbb{R}}(G) \leq \operatorname{dist}(X, \mathbb{R}Y)$. Conversely, by the duality theorem for minimum norm problems (Theorem A.4.12) there exists $z^* \in \mathbb{R}^{1\times q} = (\mathbb{R}^q)^*$ with $\|z^*\|_{\mathbb{R}^q}^* = 1$, such that

$$z^* X = \operatorname{dist}(X, \mathbb{R}Y) \quad \text{and} \quad z^* Y = 0\,.$$

But then $\Delta_0 = [\operatorname{dist}(X, \mathbb{R}Y)]^{-1} z^*$ satisfies (44) for arbitrary $(u, v) \in \mathbb{R}^2$ and has norm $\|\Delta_0\|_{\mathbb{R}^q}^* = [\operatorname{dist}(X, \mathbb{R}Y)]^{-1}$. This proves (43). □

The map $(X, Y) \mapsto \operatorname{dist}(X, \mathbb{R}Y)$ is *not*, in general, continuous at points (X, Y) where $Y = 0$. The geometric reason for this can be seen from Figure 4.4.3 by letting Y tend to zero along a ray which is not orthogonal to X.

Example 4.4.32. Consider $G = (x_1 + \imath y_1,\ x_2 + \imath y_2)^\top$ and first assume $x_1, x_2, y_1, y_2 \neq 0$. Then $X = [x_1\ x_2]^\top$, $Y = [y_1\ y_2]^\top$. So with respect to the 2-norm, we have by (42)

$$\mu_{\mathbb{R}}(G)^2 = \mu_{\boldsymbol{\Delta}}(G^{\mathbb{R}})^2 = x_1^2 + x_2^2 - (x_1 y_1 + x_2 y_2)^2/(y_1^2 + y_2^2) = (x_1 y_2 - x_2 y_1)^2/(y_1^2 + y_2^2).$$

For the ∞-norm on $\boldsymbol{\Delta} = \mathbb{R}^{1\times 2}$ (the dual of the 1-norm on $\mathbb{R}^2$), we have to find the α's which minimize $|x_1 - \alpha y_1| + |x_2 - \alpha y_2|$. This is achieved by $\alpha = x_1/y_1$ or $\alpha = x_2/y_2$ so that in this case

$$\mu_{\mathbb{R}}(G) = \mu_{\boldsymbol{\Delta}}(G^{\mathbb{R}}) = \min\{|(x_2 y_1 - y_2 x_1)/y_1|,\ |(x_2 y_1 - y_2 x_1)/y_2|\}.$$

Now if $y_1 = y_2 = 0$, then with respect to the 2-norm $\mu_{\mathbb{R}}(G)^2 = \|G\|_{2,2}^2 = x_1^2 + x_2^2$ and with respect to the ∞-norm $\mu_{\mathbb{R}}(G) = \|G\|_{1,1} = |x_1| + |x_2|$ (by Proposition 4.4.11). We see therefore that for these two norms the function $G \mapsto \mu_{\mathbb{R}}(G)$ is discontinuous at the non-zero points $G \in \mathbb{R}^{2\times 1} \subset \mathbb{C}^{2\times 1}$. □

In the general case, where G is not a complex vector but a complex matrix ($\ell, q \geq 2$), a computable formula for $\mu_{\mathbb{R}}(G)$ has only been obtained with respect to the *spectral norm* on $\mathbf{\Delta}$. In the following sequence of lemmata and propositions we will derive such a formula. *Throughout the rest of this section we assume that all vector spaces are equipped with Euclidean norms and $\|\cdot\|$ denotes the corresponding operator norm.* We have seen above that $\mu_{\mathbb{R}}(G) = \mu_{\mathbf{\Delta}}(G^{\mathbb{R}})$, where $\mathbf{\Delta}$ is the block-diagonal set of all perturbations $\mathrm{diag}(\Delta, \Delta)$ with $\Delta \in \mathbb{R}^{\ell \times q}$. The scaled matrix $G^{\mathbb{R}}_{\gamma}$ introduced in Lemma 4.4.29 plays a fundamental role in the sequel. By this lemma, for every (non-zero) value of the scaling parameter γ

$$\mu_{\mathbb{R}}(G)^{-1} = \mu_{\mathbf{\Delta}}(G^{\mathbb{R}}_{\gamma})^{-1} = \inf\left\{\|\Delta\|; \Delta \in \mathbb{R}^{\ell \times q},\ \det\left(I_{2\ell} - \begin{bmatrix}\Delta & 0\\ 0 & \Delta\end{bmatrix}\begin{bmatrix} X & -\gamma Y\\ \gamma^{-1}Y & X\end{bmatrix}\right) = 0\right\}. \quad (45)$$

For any $\Delta \in \mathbb{R}^{\ell \times q}$ and $v^1, v^2 \in \mathbb{R}^{\ell}$ we have

$$\begin{bmatrix}\Delta & 0\\ 0 & \Delta\end{bmatrix}\begin{bmatrix} X & -\gamma Y\\ \gamma^{-1}Y & X\end{bmatrix}\begin{bmatrix}v^1\\ v^2\end{bmatrix} = \begin{bmatrix}v^1\\ v^2\end{bmatrix} \ \Rightarrow\ \begin{bmatrix}\Delta & 0\\ 0 & \Delta\end{bmatrix}\begin{bmatrix} X & -\gamma Y\\ \gamma^{-1}Y & X\end{bmatrix}\begin{bmatrix}-\gamma v^2\\ \gamma^{-1}v^1\end{bmatrix} = \begin{bmatrix}-\gamma v^2\\ \gamma^{-1}v^1\end{bmatrix}. \quad (46)$$

If $(v^1, v^2) \neq (0,0)$ then (v^1, v^2) and $(-\gamma v^2, \gamma^{-1}v^1)$ are linearly independent. Hence if $\ker\left(I_{2\ell} - \mathrm{diag}(\Delta,\Delta)\begin{bmatrix} X & -\gamma Y\\ \gamma^{-1}Y & X\end{bmatrix}\right)$ is nontrivial then the dimension of this kernel will be ≥ 2. In view of Theorem 4.3.13, this suggests that a characterization of $\mu_{\mathbb{R}}(G)$ will involve the second singular value. We will see that this is indeed the case. First we obtain an upper bound for $\mu_{\mathbb{R}}(G)$.

Lemma 4.4.33. *If σ_2 denotes the second singular value (see Section 4.3) then*

$$\mu_{\mathbb{R}}(G) = \mu_{\mathbf{\Delta}}(G^{\mathbb{R}}) \leq \inf_{\gamma \neq 0} \sigma_2(G^{\mathbb{R}}_{\gamma}) = \inf_{\gamma \in (0,1]} \sigma_2(G^{\mathbb{R}}_{\gamma}) \ \textit{where}\ G^{\mathbb{R}}_{\gamma} = \begin{bmatrix} X & -\gamma Y\\ \gamma^{-1}Y & X\end{bmatrix}. \quad (47)$$

Proof: By Lemma 4.4.29 (ii) we have $\mu_{\mathbb{R}}(G) = \mu_{\mathbf{\Delta}}(G^{\mathbb{R}}_{\gamma})$, $\gamma \neq 0$ and for $\Delta \in \mathbb{R}^{\ell \times q}$, we get from (46),

$$\det\left(I_{2\ell} - \begin{bmatrix}\Delta & 0\\ 0 & \Delta\end{bmatrix} G^{\mathbb{R}}_{\gamma}\right) = 0 \ \Leftrightarrow\ \dim\ker\left(I_{2\ell} - \begin{bmatrix}\Delta & 0\\ 0 & \Delta\end{bmatrix} G^{\mathbb{R}}_{\gamma}\right) \geq 2.$$

Hence by Theorem 4.3.13, either $\mu_{\mathbf{\Delta}}(G^{\mathbb{R}}_{\gamma}) = \sigma_2(G^{\mathbb{R}}_{\gamma}) = 0$ or $\mu_{\mathbf{\Delta}}(G^{\mathbb{R}}_{\gamma})^{-1} \geq \sigma_2(G^{\mathbb{R}}_{\gamma})^{-1}$. This establishes the inequality in (47). Since

$$\begin{bmatrix} I & 0\\ 0 & -I\end{bmatrix} G^{\mathbb{R}}_{\gamma} \begin{bmatrix} I & 0\\ 0 & -I\end{bmatrix} = G^{\mathbb{R}}_{-\gamma}, \quad \begin{bmatrix} 0 & -I\\ I & 0\end{bmatrix} G^{\mathbb{R}}_{\gamma} \begin{bmatrix} 0 & I\\ -I & 0\end{bmatrix} = G^{\mathbb{R}}_{1/\gamma},$$

the equality in (47) follows because $G^{\mathbb{R}}_{\gamma}$, $G^{\mathbb{R}}_{-\gamma}$ and $G^{\mathbb{R}}_{\gamma^{-1}}$ all have the same singular values. □

Remark 4.4.34. If $G = X + \imath Y \in \mathbb{C}^{q \times \ell}$ and there exist $v^1, v^2 \in \mathbb{R}^{\ell}$ such that $y^1 := Xv^1 - Yv^2$, $y^2 := Yv^1 + Xv^2 \in \mathbb{R}^q$ are linearly independent then there exists $\Delta \in \mathbb{R}^{\ell \times q}$ satisfying

$$\Delta y^1 = v^1 \text{ and } \Delta y_2 = v^2, \ \text{i.e.}\ \begin{bmatrix}\Delta & 0\\ 0 & \Delta\end{bmatrix}\begin{bmatrix} X & -Y\\ Y & X\end{bmatrix}\begin{bmatrix}v^1\\ v^2\end{bmatrix} = \begin{bmatrix}v^1\\ v^2\end{bmatrix}, \ \text{i.e.}\ \Delta G v = v \neq 0, \quad (48)$$

where $v = v^1 + \imath v^2 \in \mathbb{C}^\ell$. Hence $\mu_{\mathbb{R}}(G) \geq \|\Delta\|^{-1} > 0$. Therefore, if $\mu_{\mathbb{R}}(G) = 0$ then $y^1 := Xv^1 - Yv^2$ and $y^2 := Yv^1 + Xv^2 \in \mathbb{R}^q$ are linearly dependent for all $v^1, v^2 \in \mathbb{R}^\ell$ and a similar statement holds for $G^\top$ since $\mu_{\mathbb{R}}(G^\top) = \mu_{\mathbb{R}}(G)$ by Lemma 4.4.29. It follows from this observation that for any $G_0 = X_0 + \imath Y_0 \in \mathbb{C}^{q\times\ell}$ and every $\varepsilon > 0$ there exists $G \in \mathbb{C}^{q\times\ell}$ such that $\|G - G_0\| < \varepsilon$ and $\mu_{\mathbb{R}}(G) > 0$, provided that $\max\{q,\ell\} > 1$. In fact, if $q \geq 2$ and $\mu_{\mathbb{R}}(G_0) = 0$, it suffices to change slightly the first columns of X and Y so that they become linearly independent and the resulting matrix $G_1 = X_1 + \imath Y_1$ satisfies $\|G_1 - G_0\| < \varepsilon$. Then, setting $v^1 = e^1$ (the first standard unit vector of $\mathbb{R}^\ell$) and $v^2 = 0$, the vectors $y^1 := X_1 v^1 - Y_1 v^2$ and $y^2 := Y_1 v^1 + X_1 v^2$ are linearly independent so that $\mu_{\mathbb{R}}(G_1) > 0$. If $q = 1$ but $\ell \geq 2$, we apply the above argument to $G_0^\top$ and obtain the same result by transposition. □

We will show that the inequality in (47) is in fact an equality. In order to do this we have to examine each of the following three possible cases for $\sigma^* = \inf_{\gamma\in(0,1]} \sigma_2(G_\gamma^{\mathbb{R}})$

Case 1 $\sigma^* = \sigma_2(G_{\gamma^*}^{\mathbb{R}})$ for some $\gamma^* \in (0,1)$,

Case 2 $\sigma^* = \sigma_2(G_{\gamma^*}^{\mathbb{R}})$ for $\gamma^* = 1$,

Case 3 $\sigma^* = \liminf_{\gamma\to 0} \sigma_2(G_\gamma^{\mathbb{R}})$.

We will deal with these three cases in a series of lemmata and propositions. First we consider Case 3 and will show that in this case we obtain a formula for $\mu_{\mathbb{R}}(G)$ which does not require the resolution of an optimization problem.

Proposition 4.4.35. *Let $G = X + \imath Y$, $X, Y \in \mathbb{R}^{q\times\ell}$. Then $\sigma^* = \liminf_{\gamma\to 0} \sigma_2(G_\gamma^{\mathbb{R}})$ (case 3) holds if and only if* $\operatorname{rank} Y \leq 1$, *and in this case*

$$\mu_{\mathbb{R}}(G) = \inf_{\gamma\in(0,1]} \sigma_2(G_\gamma^{\mathbb{R}}) = \lim_{\gamma\to 0} \sigma_2(G_\gamma^{\mathbb{R}}) = \max\{\|X|\ker Y\|, \|X^\top|\ker Y^\top\|\}. \tag{49}$$

Proof: By Corollary 4.3.3 we have $\sigma_2(G_\gamma^{\mathbb{R}}) \geq \gamma^{-1}\sigma_2(Y)$. So if $\inf_{\gamma\in(0,1]} \sigma_2(G_\gamma^{\mathbb{R}}) = \liminf_{\gamma\to 0} \sigma_2(G_\gamma^{\mathbb{R}})$, then $\sigma_2(Y) = 0$ and hence $\operatorname{rank} Y \leq 1$.
Conversely, let $\operatorname{rank} Y \leq 1$. If $Y = 0$ then G is real and $G_\gamma^{\mathbb{R}} = \operatorname{diag}(X, X)$ is independent of γ so that(49) follows directly from Proposition 4.4.11 because $\sigma_2(G^{\mathbb{R}}) = \|G\|$, $X|\ker Y = X = G$, $X^\top|\ker Y^\top = X^\top$ and so $\mu_{\mathbb{R}}(G) = \|G\| = \|X|\ker Y\| = \|X^\top|\ker Y^\top\|$. Now suppose $Y \neq 0$ and let $Y = W\Sigma V^\top$ be a singular value decomposition of Y where $W = [w^1 \dots w^q] \in \mathbb{R}^{q\times q}$, $V = [v^1 \dots v^\ell] \in \mathbb{R}^{\ell\times\ell}$ are orthogonal matrices. Setting $W_2 = [w^2 \dots w^q] \in \mathbb{R}^{q\times(q-1)}$, $V_2 = [v^2 \dots v^\ell] \in \mathbb{R}^{\ell\times(\ell-1)}$

$$Y = W\Sigma V^\top = [w^1 \ W_2] \begin{bmatrix} \sigma & 0 \\ 0 & 0 \end{bmatrix}_{q\times\ell} \begin{bmatrix} v^{1\top} \\ V_2^\top \end{bmatrix} = \sigma w^1 v^{1\top} \tag{50}$$

where $\sigma = \|Y\| > 0$ is the only non-zero singular value of Y. The columns of V_2 (resp. W_2) form an orthonormal basis of $\ker Y$ (resp. $\ker Y^\top$) and so

$$\|XV_2\| = \|X|\ker Y\|, \quad \|W_2^\top X\| = \|X^\top W_2\| = \|X^\top|\ker Y^\top\|. \tag{51}$$

Let (w, v) be a singular pair of $W_2^\top X$ for the largest singular value $\sigma_1(W_2^\top X) = \|W_2^\top X\|$ and set $\Delta = \|W_2^\top X\|^{-1} v w^\top W_2^\top \in \mathbb{R}^{\ell\times q}$. Then $\|\Delta\| \leq \|W_2^\top X\|^{-1}$ and since $w^\top W_2^\top X v = \|W_2^\top X\|$ and $W_2^\top Y = 0$ we have

$$\Delta X v = \|W_2^\top X\|^{-1} v w^\top W_2^\top X v = v, \quad \Delta Y v = \|W_2^\top X\|^{-1} v w^\top W_2^\top Y v = 0.$$

Hence $\Delta G v = \Delta X v = v$ and so $\|W_2^\top X\|^{-1} \geq \|\Delta\| \geq \mu_{\mathbb{R}}(G)^{-1}$.
Similarly let $(\tilde{w}, \tilde{v})$ be a singular pair of XV_2 for $\sigma_1(XV_2) = \|XV_2\|$ and set $\tilde{\Delta} = \|XV_2\|^{-1}\tilde{w}\tilde{v}^\top V_2^\top \in \mathbb{R}^{q\times\ell}$. Then $\|\tilde{\Delta}\| \leq \|XV_2\|^{-1}$ and since $\tilde{w}^\top XV_2\tilde{v} = \|XV_2\|$ and $YV_2 = 0$ we have

$$\tilde{\Delta}X^\top \tilde{w} = \|XV_2\|^{-1}\tilde{w}\tilde{v}^\top V_2^\top X^\top \tilde{w} = \tilde{w}, \quad \tilde{\Delta}Y^\top \tilde{w} = \|XV_2\|^{-1}\tilde{w}\tilde{v}^\top V_2^\top Y^\top \tilde{w} = 0.$$

Hence $\tilde{\Delta}G^*\tilde{w} = \tilde{\Delta}(X^\top - \imath Y^\top)\tilde{w} = \tilde{w}$ and so $\|XV_2\|^{-1} \geq \|\tilde{\Delta}\| \geq \mu_{\mathbb{R}}(G^*)^{-1}$. Altogether making use of Lemma 4.4.29 (i) this shows that

$$\max\{\|XV_2\|, \|W_2^\top X\|\} \leq \mu_{\mathbb{R}}(G). \tag{52}$$

We now prove the reverse inequality. The singular value decomposition (50) gives

$$\begin{aligned}\sigma_2(G_\gamma^{\mathbb{R}}) &= \sigma_2\left(\begin{bmatrix} I_q & 0 \\ 0 & w^{1\top} \\ 0 & W_2^\top \end{bmatrix}\begin{bmatrix} X & -\gamma Y \\ \gamma^{-1}Y & X \end{bmatrix}\begin{bmatrix} v^1 & V_2 & 0 \\ 0 & 0 & I_\ell \end{bmatrix}\right) \\ &= \sigma_2\left(\begin{bmatrix} Xv^1 & XV_2 & -\gamma Y \\ \sigma\gamma^{-1} & 0 & w^{1\top}X \\ 0 & 0 & W_2^\top X \end{bmatrix}\right).\end{aligned}$$

Subtracting from this matrix a matrix of rank 1 we obtain by Theorem 4.3.7

$$\begin{aligned}\sigma_2(G_\gamma^{\mathbb{R}}) &\leq \left\|\begin{bmatrix} Xv^1 & XV_2 & -\gamma Y \\ \sigma\gamma^{-1} & 0 & w^{1\top}X \\ 0 & 0 & W_2^\top X \end{bmatrix} - \begin{bmatrix} Xv^1 & 0 & \gamma\sigma^{-1}Xv^1w^{1\top}X \\ \sigma\gamma^{-1} & 0 & w^{1\top}X \\ 0 & 0 & 0 \end{bmatrix}\right\| \\ &= \left\|\begin{bmatrix} 0 & XV_2 & -\gamma(Y + \sigma^{-1}Xv^1w^{1\top}X) \\ 0 & 0 & 0 \\ 0 & 0 & W_2^\top X \end{bmatrix}\right\|.\end{aligned}$$

Hence for $\gamma \to 0$

$$\limsup_{\gamma\to 0} \sigma_2(G_\gamma^{\mathbb{R}}) \leq \left\|\begin{bmatrix} 0 & XV_2 & 0 \\ 0 & 0 & 0 \\ 0 & 0 & W_2^\top X \end{bmatrix}\right\| = \max\{\|XV_2\|, \|W_2^\top X\|\}.$$

Since by Lemma 4.4.33 $\mu_{\mathbb{R}}(G) \leq \inf_{\gamma\in(0,1]} \sigma_2(G_\gamma^{\mathbb{R}}) \leq \liminf_{\gamma\to 0} \sigma_2(G_\gamma^{\mathbb{R}})$ we obtain from (52)

$$\mu_{\mathbb{R}}(G) \leq \liminf_{\gamma\to 0} \sigma_2(G_\gamma^{\mathbb{R}}) \leq \limsup_{\gamma\to 0} \sigma_2(G_\gamma^{\mathbb{R}}) \leq \max\{\|XV_2\|, \|W_2^\top X\|\} \leq \mu_{\mathbb{R}}(G).$$

Thus $\lim_{\gamma\to 0} \sigma_2(G_\gamma^{\mathbb{R}})$ exists and the proof is complete by (51). □

Now let us consider Case 1 where $\sigma^* = \inf_{\gamma\in(0,1]} \sigma_2(G_\gamma^{\mathbb{R}}) = \sigma_2(G_{\gamma^*}^{\mathbb{R}})$ for some $\gamma^* \in (0,1)$. In order to prove equality in (47) we need only consider the case where $\sigma^* > 0$. The following two lemmata concerning singular pairs of $G_\gamma^{\mathbb{R}}$ will be important.

Lemma 4.4.36. *Suppose $\gamma > 0$, $\gamma \neq 1$ and σ is a non-zero singular value of $G_\gamma^{\mathbb{R}}$. Then, for all singular pairs $(w, v) = \left(\left[\begin{smallmatrix} w^1 \\ w^2 \end{smallmatrix}\right], \left[\begin{smallmatrix} v^1 \\ v^2 \end{smallmatrix}\right]\right)$ of $G_\gamma^{\mathbb{R}}$ corresponding to σ, we have $\langle w^1, w^2\rangle = \langle v^1, v^2\rangle$.*

Proof: We have

$$\sigma\begin{bmatrix} w^1 \\ w^2 \end{bmatrix} = \begin{bmatrix} X & -\gamma Y \\ \gamma^{-1}Y & X \end{bmatrix}\begin{bmatrix} v^1 \\ v^2 \end{bmatrix}, \ \sigma\begin{bmatrix} v^1 \\ v^2 \end{bmatrix} = \begin{bmatrix} X^\top & \gamma^{-1}Y^\top \\ -\gamma Y^\top & X^\top \end{bmatrix}\begin{bmatrix} w^1 \\ w^2 \end{bmatrix}. \tag{53}$$

Multiplying the first equality in (53) by $[w^{2\top}\ w^{1\top}]$ and subtracting $[v^{2\top}\ v^{1\top}]$ times the second, yields

$$(\gamma+\gamma^{-1})(\langle w^1, Yv^1\rangle - \langle w^2, Yv^2\rangle) = 2\sigma(\langle w^1, w^2\rangle - \langle v^1, v^2\rangle).$$

Whereas multiplying the first equality in (53) by $[w^{2\top}\ -w^{1\top}]$ and adding $[v^{2\top}\ -v^{1\top}]$ times the second, yields

$$(\gamma-\gamma^{-1})(\langle w^1, Yv^1\rangle - \langle w^2, Yv^2\rangle) = 0.$$

Hence for $\sigma \neq 0$, $\gamma \neq 1$, we have $\langle w^1, w^2\rangle = \langle v^1, v^2\rangle$. □

Our next lemma shows that equality holds in (47) under an extra condition.

Lemma 4.4.37. *Suppose $\hat\sigma = \sigma_2(G^{\mathbb{R}}_{\hat\gamma}) > 0$ for some $\hat\gamma \in (0,1]$. If there exists a singular pair $(w,v) = \left(\left[\begin{smallmatrix} w^1 \\ w^2 \end{smallmatrix}\right], \left[\begin{smallmatrix} v^1 \\ v^2 \end{smallmatrix}\right]\right)$ for $\hat\sigma$ such that*

$$\langle w^1, w^2\rangle = \langle v^1, v^2\rangle, \quad \|w^1\| = \|v^1\|, \quad \|w^2\| = \|v^2\|, \tag{54}$$

then $\hat\gamma$ is a global minimum of $\gamma \mapsto \sigma_2(G^{\mathbb{R}}_{\gamma})$ and $\mu_{\mathbb{R}}(G) = \mu_{\mathbf{\Delta}}(G^{\mathbb{R}}) = \hat\sigma = \sigma^*$.

Proof: If $w^1 = 0$ then necessarily $v^1 = 0$ and $Yv^2 = 0$. In this case, let

$$\Delta = \hat\sigma^{-1}\|w^2\|^{-2}v^2 w^{2\top}. \tag{55}$$

Then $\|\Delta\| = \hat\sigma^{-1}$ and making use of (53) with $\sigma = \hat\sigma$, $\gamma = \hat\gamma$ we get

$$\begin{matrix} \Delta(Xv^1 - \hat\gamma Y v^2) = 0 = v^1, \\ \Delta(\hat\gamma^{-1}Yv^1 + Xv^2) = \hat\sigma\Delta w^2 = v^2 \end{matrix} \quad \text{i.e.} \quad \begin{bmatrix} \Delta & 0 \\ 0 & \Delta \end{bmatrix} G^{\mathbb{R}}_{\gamma} \begin{bmatrix} v^1 \\ v^2 \end{bmatrix} = \begin{bmatrix} v^1 \\ v^2 \end{bmatrix}. \tag{56}$$

So by (45) $\hat\sigma^{-1} = \|\Delta\| \geq \mu_{\mathbf{\Delta}}(G^{\mathbb{R}}_{\hat\gamma})^{-1} = \mu_{\mathbb{R}}(G)^{-1}$. On the other hand we have $\mu_{\mathbb{R}}(G) \leq \sigma^* \leq \hat\sigma$ by Lemma 4.4.33 and this proves $\mu_{\mathbb{R}}(G) = \hat\sigma = \sigma^*$. So $\hat\gamma$ is a global minimum of $\gamma \mapsto \sigma_2(G^{\mathbb{R}}_{\gamma})$. A similar argument shows that the same equality holds if $w^2 = 0$.

Now assume that $w^1 \neq 0$ and $w^2 \neq 0$ are linearly independent and set

$$\Delta = \hat\sigma^{-1}[v^1\ v^2]\begin{bmatrix} \|w^1\|^2 & \langle w^1, w^2\rangle \\ \langle w^1, w^2\rangle & \|w^2\|^2 \end{bmatrix}^{-1}\begin{bmatrix} w^{1\top} \\ w^{2\top} \end{bmatrix}. \tag{57}$$

By linear independence of w^1 and w^2 the inverse in the above expression exists and so Δ is well defined. Now $\Delta[w^1\ w^2] = \hat\sigma^{-1}[v^1\ v^2]$, hence $\hat\sigma\Delta w^j = v^j$, $j = 1,2$ and it follows from (54) that $\hat\sigma\Delta$ defines an isometry from $\operatorname{span}\{w^1, w^2\}$ onto $\operatorname{span}\{v^1, v^2\}$. On the other hand $\hat\sigma\Delta$ vanishes on the orthogonal complement $\operatorname{span}\{w^1, w^2\}^\perp$. Therefore $\|\Delta\| = \hat\sigma^{-1}$ and, applying (53) with $\sigma = \hat\sigma$, $\gamma = \hat\gamma$,

$$\Delta(Xv^1 - \hat\gamma Yv^2) = \hat\sigma\Delta w^1 = v^1, \ \ \Delta(\hat\gamma^{-1}Yv^1 + Xv^2) = \hat\sigma\Delta w^2 = v^2.$$

Hence again the right equation in (56) holds and we get $\hat{\sigma}^{-1} \geq \mu_{\mathbf{\Delta}}(G^{\mathbb{R}}_{\hat{\gamma}})^{-1} = \mu_{\mathbb{R}}(G)^{-1}$ and so $\mu_{\mathbb{R}}(G) = \hat{\sigma} = \sigma^*$ as above.
Finally, if $w^1 \neq 0$, $w^2 \neq 0$ are linearly dependent, then $v^1 \neq 0$, $v^2 \neq 0$ are linearly dependent because

$$0 = \det \begin{bmatrix} \|w^1\|^2 & \langle w^1, w^2 \rangle \\ \langle w^1, w^2 \rangle & \|w^2\|^2 \end{bmatrix} = \det \begin{bmatrix} \|v^1\|^2 & \langle v^1, v^2 \rangle \\ \langle v^1, v^2 \rangle & \|v^2\|^2 \end{bmatrix}.$$

Let $v^2 = \alpha v^1$ and set $\Delta = \hat{\sigma}^{-1}\|w^1\|^{-2} v^1 w^{1\top}$, then $\|\Delta\| = \hat{\sigma}^{-1}$, $\Delta w^1 = \hat{\sigma}^{-1} v^1$, and

$$\Delta w^2 = \hat{\sigma}^{-1}\|w^1\|^{-2} v^1 v^{1\top} v^2 = \hat{\sigma}^{-1}\|w^1\|^{-2} v^1 v^{1\top} \alpha v^1 = \hat{\sigma}^{-1}\alpha v^1 = \hat{\sigma}^{-1} v^2.$$

So again the right equation in (56) holds and we conclude that $\mu_{\mathbb{R}}(G) = \hat{\sigma} = \sigma^*$ as before. □

Our aim, of course, is to show that the conditions (54) required in Lemma 4.4.37 necessarily hold. First we obtain a formula for the derivative with respect to γ of a singular value of $G^{\mathbb{R}}_{\gamma}$.

Lemma 4.4.38. *Let $\eta > 1$ and suppose $\sigma(\gamma)$ is an analytic singular value of $G^{\mathbb{R}}_{\gamma}$ for $\gamma \in (0, \eta)$ and $\left(\begin{bmatrix} w^1(\gamma) \\ w^2(\gamma) \end{bmatrix}, \begin{bmatrix} v^1(\gamma) \\ v^2(\gamma) \end{bmatrix}\right)$ an associated analytic singular pair. Then*

$$\frac{d\sigma}{d\gamma}(\gamma) = \gamma^{-1}\sigma(\gamma)\left(\|w^1(\gamma)\|^2 - \|v^1(\gamma)\|^2\right) = \gamma^{-1}\sigma(\gamma)\left(\|v^2(\gamma)\|^2 - \|w^2(\gamma)\|^2\right). \quad (58)$$

Proof: For $N = 2, G_{11} = X, G_{12} = -Y, G_{21} = Y, G_{22} = X, \gamma_1 = \gamma, \gamma_2 = 1$, the conditions of Lemma 4.4.17 are satisfied. Hence the first equation in (58) follows from (28) and the second from the fact that $\|w^1(\gamma)\|^2 + \|w^2(\gamma)\|^2 = \|v^1(\gamma)\|^2 + \|v^2(\gamma)\|^2 = 1$. □

We want to apply the preceding lemma to $\sigma_2(\gamma) := \sigma_2(G^{\mathbb{R}}_{\gamma})$. Since $G^{\mathbb{R}}_{\gamma}$ is analytic on $(0, 1)$ it follows from Corollary 4.3.11 and Lemma 4.3.15 that this ordered singular value is continuous and piecewise analytic on $(0, 1)$. However, it may not be differentiable on all of $(0, 1)$. Differentiability may be lost at points γ_0 where $\sigma_2(\gamma)$ meets some distinct ordered singular value branch $\sigma(\gamma)$ of $G^{\mathbb{R}}_{\gamma}$. The possible presence of these points is where we encounter the main difficulties.

Lemma 4.4.39. *Suppose $\sigma_2(\gamma)$ has a local extremum at $\hat{\gamma} \in (0, 1)$ and $\sigma_2(\hat{\gamma}) > 0$, then $\hat{\gamma}$ is a global minimum of $\sigma_2(\cdot)$ on $(0, 1)$, $\mu_{\mathbb{R}}(G) = \mu_{\mathbf{\Delta}}(G^{\mathbb{R}}) = \sigma_2(\hat{\gamma}) = \sigma^*$, and there exists a singular pair $(w, v) = \left(\left[\begin{smallmatrix} w^1 \\ w^2 \end{smallmatrix}\right], \left[\begin{smallmatrix} v^1 \\ v^2 \end{smallmatrix}\right]\right)$ of $G^{\mathbb{R}}_{\hat{\gamma}}$ for $\sigma_2(\hat{\gamma})$, such that (54) holds.*

Proof: By Proposition 4.3.21 there exists a singular pair $(w, v) = \left(\left[\begin{smallmatrix} w^1 \\ w^2 \end{smallmatrix}\right], \left[\begin{smallmatrix} v^1 \\ v^2 \end{smallmatrix}\right]\right)$ of $G^{\mathbb{R}}_{\hat{\gamma}}$ for $\sigma_2(\hat{\gamma})$ satisfying $w^\top \frac{dG^{\mathbb{R}}_{\hat{\gamma}}}{d\gamma} v = 0$. Since

$$\frac{dG^{\mathbb{R}}_{\gamma}}{d\gamma} = \begin{bmatrix} 0 & -Y \\ -\gamma^{-2}Y & 0 \end{bmatrix} = \gamma^{-1}\left(\begin{bmatrix} I_q & 0 \\ 0 & 0 \end{bmatrix} G^{\mathbb{R}}_{\gamma} - G^{\mathbb{R}}_{\gamma} \begin{bmatrix} I_\ell & 0 \\ 0 & 0 \end{bmatrix}\right)$$

we have for $\gamma = \hat{\gamma}$

$$0 = w^\top \frac{dG^{\mathbb{R}}_{\hat{\gamma}}}{d\gamma} v = \hat{\gamma}^{-1}\sigma_2(\hat{\gamma})(\|w^1\|^2 - \|v^1\|^2) = \hat{\gamma}^{-1}\sigma_2(\hat{\gamma})(\|v^2\|^2 - \|w^2\|^2).$$

Hence the assertion follows from Lemma 4.4.36 and Lemma 4.4.37. □

It remains to consider Case 2 where $\sigma^* = \inf_{\gamma\in(0,1]} \sigma_2(G^{\mathbb{R}}_\gamma) = \sigma_2(G^{\mathbb{R}}_{\gamma^*}) > 0$ for $\gamma^* = 1$.

Lemma 4.4.40. *Suppose $\sigma^* = \sigma_2(G^{\mathbb{R}}_{\gamma^*}) > 0$ for $\gamma^* = 1$, then*

$$\mu_{\mathbb{R}}(G) = \mu_{\mathbf{\Delta}}(G^{\mathbb{R}}) = \sigma^* = \|G\| = \mu_{\mathbb{C}}(G).$$

Proof: We have seen in Section 4.3 that $G^{\mathbb{R}}_1 = G^{\mathbb{R}}$ has the same singular values as G, but of double multiplicity. First assume that the multiplicity of the largest singular value of G is one. Then $\sigma^* = \sigma_1(G^{\mathbb{R}}) = \sigma_2(G^{\mathbb{R}}) > \sigma_3(G^{\mathbb{R}}_1)$. In this case there exists, by continuity, an open neighbourhood $I \subset (0,2)$ of $\gamma^* = 1$ such that $\sigma_1(G^{\mathbb{R}}_\gamma) \geq \sigma_2(G^{\mathbb{R}}_\gamma) > \sigma_3(G^{\mathbb{R}}_\gamma)$ for $\gamma \in I$ (if $\ell > 1$). Applying Theorem 4.3.17 to the analytic matrix $G^{\mathbb{R}}_\gamma$ at $\gamma^* = 1$ we obtain two analytic functions $\sigma(\cdot), \hat{\sigma}(\cdot) : I(1,\delta) \to (0,\infty)$ on an open neighbourhood $I(1,\delta) \subset I$ of 1 such that $\{\sigma(\gamma), \hat{\sigma}(\gamma)\} = \{\sigma_1(G^{\mathbb{R}}_\gamma), \sigma_2(G^{\mathbb{R}}_\gamma)\}$ for $\gamma \in I(1,\delta)$. It follows that $\sigma(1) = \hat{\sigma}(1) = \sigma^*$ and $\sigma_2(G^{\mathbb{R}}_\gamma) = \min\{\sigma(\gamma), \hat{\sigma}(\gamma)\}$ for $\gamma \in I(1,\delta)$. Now $G^{\mathbb{R}}_\gamma$ and $G^{\mathbb{R}}_{\gamma^{-1}}$ have the same singular values, so $\gamma = 1$ is a minimum of $\sigma_2(G^{\mathbb{R}}_\gamma)$ on $I(1,\delta)$ and we conclude that $\gamma = 1$ is a local minimum of both $\sigma(\gamma)$ and $\hat{\sigma}(\gamma)$ on $I(1,\delta)$. Let $(w(\gamma), v(\gamma)) = \left(\begin{bmatrix} w^1(\gamma) \\ w^2(\gamma) \end{bmatrix}, \begin{bmatrix} v^1(\gamma) \\ v^2(\gamma) \end{bmatrix}\right)$ be an analytic singular pair of $G^{\mathbb{R}}_\gamma$ for the analytic singular value $\sigma(\gamma)$. By Lemma 4.4.36 for $\gamma \neq 1$, we have $\langle w^1(\gamma), w^2(\gamma)\rangle = \langle v^1(\gamma), v^2(\gamma)\rangle$. But then by continuity $\langle w^1(1), w^2(1)\rangle = \langle v^1(1), v^2(1)\rangle$. Also by Lemma 4.4.38, we have

$$0 = \sigma'(1) = \sigma(1)(\|w^1(1)\|^2 - \|v^1(1)\|^2) = \sigma(1)(\|v^2(1)\|^2 - \|w^2(1)\|^2).$$

Hence $\mu_{\mathbb{R}}(G) = \sigma^*$ by Lemma 4.4.37.
Now suppose that the multiplicity of the largest singular value of G is greater than or equal to two and G has the singular value decomposition $G = W\Sigma V^*$ (see (3.19)) where $W \in \mathbf{U}_q(\mathbb{C})$, $V \in \mathbf{U}_\ell(\mathbb{C})$ and $\sigma_1 = \sigma_2 = \|G\|$. Let $v, \tilde{v}$ and $w, \tilde{w}$ be the first two columns of V and W, respectively, and set

$$\begin{bmatrix} w_z \\ v_z \end{bmatrix} := (1+|z|^2)^{-1/2} \begin{bmatrix} w & \hat{w} \\ v & \hat{v} \end{bmatrix} \begin{bmatrix} z \\ 1 \end{bmatrix}, \quad z \in \mathbb{C}.$$

Since $v, \tilde{v}$ (resp. $w, \tilde{w}$) are normalized and orthogonal to each other, (w_z, v_z) is a singular pair of G for $\sigma^* = \|G\|$. Our objective is to choose z, so that $w_z^\top w_z - v_z^\top v_z = 0$. The reason for this is that if $w_z = w_z^1 + \imath w_z^2$, $v_z = v_z^1 + \imath v_z^2$, then

$$\begin{aligned} w_z^\top w_z - v_z^\top v_z &= (w_z^1 + \imath w_z^2)^\top (w_z^1 + \imath w_z^2) - (v_z^1 + \imath v_z^2)^\top (v_z^1 + \imath v_z^2) \\ &= w_z^{1\top} w_z^1 - w_z^{2\top} w_z^2 - v_z^{1\top} v_z^1 + v_z^{2\top} v_z^2 + 2\imath(w_z^{1\top} w_z^2 - v_z^{1\top} v_z^2) \\ &= 2(\|w_z^1\|^2 - \|v_z^1\|^2) + 2\imath(w_z^{1\top} w_z^2 - v_z^{1\top} v_z^2), \end{aligned}$$

since w_z and v_z are normalized. Hence $w_z^\top w_z - v_z^\top v_z = 0$ is equivalent to

$$\|w_z^1\|^2 - \|v_z^1\|^2 = 0, \quad w_z^{1\top} w_z^2 - v_z^{1\top} v_z^2 = 0,$$

which would then show that $\mu_{\mathbb{R}}(G) = \sigma^*$ by Lemma 4.4.37. If $w^\top w - v^\top v = 0$, then there is no need for the construction of w_z, v_z, so we assume $w^\top w - v^\top v \neq 0$. Now

$$w_z^\top w_z - v_z^\top v_z = (1+|z|^2)^{-1}[(w^\top w - v^\top v)z^2 + (w^\top \hat{w} + \hat{w}^\top w - v^\top \hat{v} - \hat{v}^\top v)z + \hat{w}^\top \hat{w} - \hat{v}^\top \hat{v}].$$

There always exists $z \in \mathbb{C}$ such that the RHS of the above expression is zero. The proof is completed by noting that $\sigma^* = \|G^{\mathbb{R}}_1\| = \|G^{\mathbb{R}}\| = \|G\| = \mu_{\mathbb{C}}(G)$. □

Remark 4.4.41. (i) If the infimum of $\gamma \mapsto \sigma_2(G_\gamma^{\mathbb{R}})$ is not achieved at $\gamma = 1$, then there exists $\gamma \in (0,1)$ such that $\sigma_2(G_\gamma^{\mathbb{R}}) < \sigma_2(G_1^{\mathbb{R}}) = \|G\|$, hence $\mu_{\mathbb{R}}(G) < \|G\|$.

(ii) In the second part of the proof of the preceding lemma we did not use the assumption that $\gamma = 1$ is a minimum of $\gamma \mapsto \sigma_2(G_\gamma^{\mathbb{R}})$ but showed that this follows automatically if $\sigma_1(G)$ has multiplicity ≥ 2. Thus we always have $\mu_{\mathbb{R}}(G) = \sigma^* = \|G\| = \mu_{\mathbb{C}}(G)$ if $\sigma_1(G) = \sigma_2(G)$. □

We have now dealt with each of the Cases 1–3, and summarizing we see that equality holds in (47). Actually we have proved a bit more. To state the full result we need the following definition.

Definition 4.4.42. A continuous function $f : (0,1) \to \mathbb{R}_+$ is said to be *unimodal* if any local extremum (maximum or minimum) is a global minimum of f.

We now state the main theorem of this subsection which summarizes the previous results concerning $\mu_{\mathbb{R}}(G)$ (for the general case $\ell, q \geq 1$).

Theorem 4.4.43. *The μ-value of any matrix $G \in \mathbb{C}^{q\times\ell}$ with respect to the perturbation class $\mathbb{R}^{\ell\times q}$ (provided with the spectral norm) is given by*

$$\mu_{\mathbb{R}}(G) = \mu_{\mathbf{\Delta}}(G^{\mathbb{R}}) = \inf_{\gamma\in(0,1]} \sigma_2\left(\begin{bmatrix} X & -\gamma Y \\ \gamma^{-1} Y & X \end{bmatrix}\right) \tag{59}$$

and the function $\gamma \mapsto \sigma_2(G_\gamma^{\mathbb{R}})$ is unimodal on $(0,1)$. Moreover, if $G = X + \imath Y$, $X, Y \in \mathbb{R}^{q\times\ell}$ and $\operatorname{rank} Y \leq 1$, *then*

$$\mu_{\mathbb{R}}(G) = \mu_{\mathbf{\Delta}}(G^{\mathbb{R}}) = \max\{\|X|\ker Y\|, \|X^\top|\ker Y^\top\|\}. \tag{60}$$

Proof: The final statement is a direct consequence of Proposition 4.4.35.
If $\sigma^* = \inf_{\gamma\in(0,1]} \sigma_2(G_\gamma^{\mathbb{R}}) = 0$ formula (59) follows from Lemma 4.4.33. Now suppose $\sigma^* > 0$. If $\sigma^* = \liminf_{\gamma\to 0} \sigma_2(G_\gamma^{\mathbb{R}})$ we can apply Proposition 4.4.35 to conclude that $\operatorname{rank} Y \leq 1$ and (59) holds. Otherwise there exists $\gamma^* \in (0,1]$ such that $\sigma^* = \sigma_2(G_{\gamma^*}^{\mathbb{R}})$ and we can apply Lemmata 4.4.39 and 4.4.40 to obtain (59).
It remains to prove that $\sigma_2(G_\gamma^{\mathbb{R}})$ is unimodal on $(0,1)$. Suppose that $\sigma_2(G_\gamma^{\mathbb{R}})$ has a local extremum at $\gamma_0 \in (0,1)$. If $\sigma_2(G_{\gamma_0}^{\mathbb{R}}) = 0$ then γ_0 is clearly a global minimum of $\sigma_2(G_\gamma^{\mathbb{R}})$. If, however, $\sigma_2(G_{\gamma_0}^{\mathbb{R}}) > 0$ we obtain the same conclusion by applying Lemma 4.4.39. □

As an easy corollary of the previous theorem we obtain the following formula for the distance of an invertible complex matrix from the set of *real* singular matrices, compare Corollary 4.3.9.

Corollary 4.4.44. *Suppose $G \in \mathbb{C}^{n\times n}$ is invertible, then*

$$\inf\{\|\Delta\|;\ \Delta \in \mathbb{R}^{n\times n},\ \det(G-\Delta) = 0\} = [\mu_{\mathbb{R}}(G^{-1})]^{-1} = \sup_{\gamma\in(0,1]} \sigma_{2n-1}(G_\gamma^{\mathbb{R}}). \tag{61}$$

Proof: The first equality follows directly from the definition of $\mu_{\mathbb{R}}(G^{-1})$ since $\det(G-\Delta) = 0 \Leftrightarrow \det(I_n - \Delta G^{-1}) = 0$. To prove the second equality we use $(G^{-1})^{\mathbb{R}} = \left(G^{\mathbb{R}}\right)^{-1}$ (see Lemma A.1.18) and obtain

$$(G^{-1})_\gamma^{\mathbb{R}} = \begin{bmatrix} I_n & 0 \\ 0 & \gamma^{-1} I_n \end{bmatrix} (G^{-1})^{\mathbb{R}} \begin{bmatrix} I_n & 0 \\ 0 & \gamma I_n \end{bmatrix} = \left[\begin{bmatrix} I_n & 0 \\ 0 & \gamma^{-1} I_n \end{bmatrix} G^{\mathbb{R}} \begin{bmatrix} I_n & 0 \\ 0 & \gamma I_n \end{bmatrix}\right]^{-1} = (G_\gamma^{\mathbb{R}})^{-1}.$$

Hence

$$\mu_{\mathbf{\Delta}}((G^{-1})^{\mathbb{R}}) = \inf_{\gamma\in(0,1]} \sigma_2((G^{-1})^{\mathbb{R}}_{\gamma}) = \inf_{\gamma\in(0,1]} \sigma_2((G^{\mathbb{R}}_{\gamma})^{-1}) = \inf_{\gamma\in(0,1]} [\sigma_{2n-1}(G^{\mathbb{R}}_{\gamma})]^{-1}$$

by (3.21). □

In general, numerical algorithms are needed to evaluate the formula (59). There are standard search algorithms such as the golden section search for computing the minima of unimodal functions, which can be employed to determine $\mu_{\mathbb{R}}(G)$ via (59). We illustrate the formula by an example for which the minimization can be carried out using pen and parchment. Further examples of this kind can be found in Ex. 9–13.

Example 4.4.45. Let $M \in \mathbb{R}^{q\times\ell}$, $M \neq 0$ be given and consider $G = G(\theta) = e^{\imath\theta}M$ for an arbitrary but fixed $\theta \in [0, 2\pi]$. Then

$$G^{\mathbb{R}}_{\gamma} = \begin{bmatrix} M\cos\theta & -\gamma M\sin\theta \\ \gamma^{-1}M\sin\theta & M\cos\theta \end{bmatrix} = \begin{bmatrix} \cos\theta & -\gamma\sin\theta \\ \gamma^{-1}\sin\theta & \cos\theta \end{bmatrix} \otimes M, \quad \theta\in[0,2\pi],\ \gamma\in\mathbb{R}^*.$$

Hence, if the ordered singular values of M are σ_i, $i \in \underline{\ell}$, then the unordered 2ℓ-tuple of singular values of $G^{\mathbb{R}}_{\gamma}$ is equal to $\lfloor \sigma_i\,\overline{\sigma}(\gamma), \sigma_i\,\underline{\sigma}(\gamma);\ i\in\underline{\ell}\rfloor$ where $\overline{\sigma}(\gamma)$, $\underline{\sigma}(\gamma)$ are the upper and lower singular values of the 2×2 matrix $\begin{bmatrix} \cos\theta & -\gamma\sin\theta \\ \gamma^{-1}\sin\theta & \cos\theta \end{bmatrix}$ (see Theorem A.1.27). Since the determinant of this matrix is one we must have $\overline{\sigma}(\gamma)\underline{\sigma}(\gamma) = 1$. Now $\sigma_2(G^{\mathbb{R}}_{\gamma}) = \max\{\sigma_1\,\underline{\sigma}(\gamma),\ \sigma_2\,\overline{\sigma}(\gamma)\}$ and there are three cases, depending on the value of θ. If $\theta \in \{0, \pi, 2\pi\}$, then $\overline{\sigma}(\gamma) = \underline{\sigma}(\gamma) = 1$, so $\sigma_2(G^{\mathbb{R}}_{\gamma}) = \sigma_1$ for all $\gamma\in(0,1]$ and consequently $\mu_{\mathbb{R}}(G) = \sigma_1$ by Theorem 4.4.43. If $\theta \in (0,2\pi)\setminus\{\pi\}$ then, by Corollary 4.3.3, $\overline{\sigma}(\gamma)\to\infty$ as $\gamma\to 0$ and since $\overline{\sigma}(\gamma)\,\underline{\sigma}(\gamma) = 1$, we must have $\underline{\sigma}(\gamma)\to 0$ as $\gamma \to 0$. Now assume that $\sigma_2 > 0$. Then $\gamma\mapsto\sigma_2(G^{\mathbb{R}}_{\gamma})$ takes its minimum at some $\gamma^*\in(0,1]$ where $\sigma_1\,\underline{\sigma}(\gamma^*) = \sigma_2\,\overline{\sigma}(\gamma^*)$. Such a γ^* always exists since $\underline{\sigma}(\gamma)\to 0$ and $\overline{\sigma}(\gamma)\to\infty$ as $\gamma\to 0$. But then, since $\overline{\sigma}(\gamma^*)\,\underline{\sigma}(\gamma^*) = 1$, we have $\overline{\sigma}(\gamma^*)^2 = \sigma_1\,\sigma_2^{-1}$. Hence $\sigma_2(G^{\mathbb{R}}_{\gamma^*}) = \sqrt{\sigma_1\,\sigma_2}$. The same formula holds if $\sigma_2 = 0$ because then $\sigma_2(G^{\mathbb{R}}_{\gamma}) = \sigma_1\,\underline{\sigma}(\gamma)\to 0$ as $\gamma\to 0$. Summarizing:

$$\mu_{\mathbb{R}}(G(\theta)) = \begin{cases} \sigma_1 & \text{if } \theta\in\{0,\pi,2\pi\} \\ \sqrt{\sigma_1\,\sigma_2} & \text{otherwise.} \end{cases}$$

In particular, $\mu_{\mathbb{R}}(G(\theta)) = 0$ if and only if rank $M = 1$ and $\theta\notin\{0,\pi,2\pi\}$. Moreover we see that the map $\theta\mapsto\mu_{\mathbb{R}}(G(\theta))$ is discontinuous at $\theta\in\{0,\pi,2\pi\}$ if $\sigma_2 < \sigma_1$. □

Remark 4.4.46. In the overall proof of Theorem 4.4.43 we have distinguished between the following five possibilities for $\sigma^* = \inf_{\gamma\in(0,1]}\sigma_2(G^{\mathbb{R}}_{\gamma})$.

(a) $\sigma^* = \sigma_2(G^{\mathbb{R}}_{\gamma^*})$ where $\gamma^*\in(0,1)$ and $\sigma_2(G^{\mathbb{R}}_{\gamma})$ is smooth at γ^* (Lemma 4.4.38).

(b) $\sigma^* = \sigma_2(G^{\mathbb{R}}_{\gamma^*})$ where $\gamma^*\in(0,1)$ and $\sigma_2(G^{\mathbb{R}}_{\gamma})$ is non-smooth at γ^* (Lemma 4.4.39).

(c) $\sigma^* = \sigma_2(G^{\mathbb{R}}_{\gamma^*})$ where $\gamma^* = 1$ and $\sigma_2(G^{\mathbb{R}}_{1})$ has multiplicity 2 (Lemma 4.4.40).

(d) $\sigma^* = \sigma_2(G^{\mathbb{R}}_{\gamma^*})$ where $\gamma^* = 1$ and $\sigma_2(G^{\mathbb{R}}_{1})$ has multiplicity > 2 (Lemma 4.4.40).

(e) $\sigma^* = \lim_{\gamma\to 0}\sigma_2(G^{\mathbb{R}}_{\gamma})$ (Proposition 4.4.35).

Table 4.4.46 (taken from [424]) illustrates the five possibilities.

Case	G	Singular values of $G_\gamma^{\mathbb{R}}$	$\mu_{\mathbb{R}}(G)$	Minimizing Δ
(a)	$\begin{bmatrix} 4+\imath & 1 \\ -1 & \imath \end{bmatrix}$	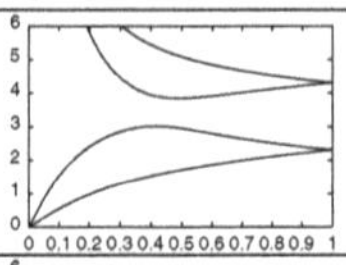	3.80	$\begin{bmatrix} 0.14 & -0.22 \\ 0.22 & 0.14 \end{bmatrix}$
(b)	$\begin{bmatrix} 2+\imath & 1 \\ 1 & 2+\imath \end{bmatrix}$	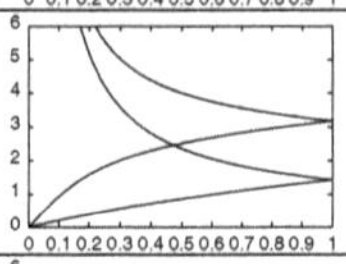	2.45	$\begin{bmatrix} 0.33 & -0.24 \\ 0.24 & 0.33 \end{bmatrix}$
(c)	$\begin{bmatrix} 1+\imath & -1 \\ 1 & 1+\imath \end{bmatrix}$	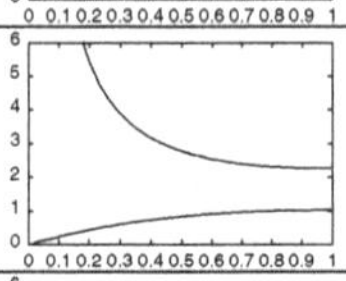	2.24	$\begin{bmatrix} 0.20 & 0.40 \\ -0.40 & 0.20 \end{bmatrix}$
(d)	$\begin{bmatrix} 2+\imath & 0 \\ 0 & 2+\imath \end{bmatrix}$	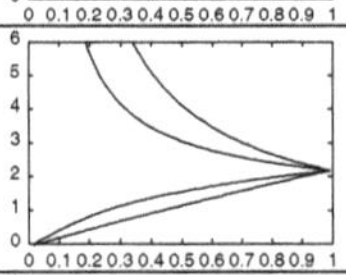	2.24	$\begin{bmatrix} 0.33 & -0.30 \\ 0.30 & 0.33 \end{bmatrix}$
(e)	$\begin{bmatrix} 1+\imath & 2 \\ 0 & 1 \end{bmatrix}$	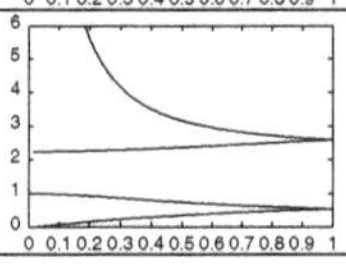	2.24	$\begin{bmatrix} 0.00 & 0.00 \\ 0.40 & 0.20 \end{bmatrix}$

Table 4.4.4: The five different cases

The proofs in the previous lemmata were constructive, i.e. for each of the above possibilities it was shown how a Δ of minimum norm satisfying $\det(I_\ell - \Delta G) = 0$ can, in principle, be determined, see Proposition 4.4.35 for (e) and (55), (57) for the remaining possibilities (a)-(d). □

We have seen above that in contrast to the complex case the real $\mu_{\mathbb{R}}$-value is discontinuous, see Examples 4.4.32, 4.4.45. Note that in both examples the discontinuities occur at *real* matrices G. This is necessarily so, since we will prove that the function $\mu_{\mathbb{R}}(\cdot) : \mathbb{C}^{q\times\ell} \to \mathbb{R}_+$ is continuous at all $G \in \mathbb{C}^{q\times\ell} \setminus \mathbb{R}^{q\times\ell}$. In order to prepare the ground for the proof we need the following lemma.

Lemma 4.4.47. *Suppose that*

$$\Delta_0 U_0 = V_0 \quad \textit{where} \quad U_0 \in \mathbb{R}^{q\times k},\ V_0 \in \mathbb{R}^{\ell\times k} \setminus \{0\},\ \Delta_0 \in \mathbb{R}^{\ell\times q}.$$

If the columns of U_0 are linearly independent, then there exist a neighbourhood $\mathfrak{U}$ of U_0 and an analytic function

$$\Phi : \mathfrak{U} \times \mathbb{R}^{\ell\times k} \to \mathbb{R}^{\ell\times q} \qquad (U,V) \stackrel{\Phi}{\longmapsto} \Delta_{U,V}$$

such that

$$\forall\, (U,V) \in \mathfrak{U} \times \mathbb{R}^{\ell\times k} : \Delta_{U,V} U = V, \quad \Delta_{U_0,V_0} = \Delta_0.$$

Proof: Fix any $U_1 \in \mathbb{R}^{q\times(q-k)}$ such that $[U_0\ U_1] \in \mathbf{Gl}_q(\mathbb{R})$ and set $V_1 = \Delta_0 U_1 \in \mathbb{R}^{\ell\times(q-k)}$ so that $\Delta_0[U_0\quad U_1] = [V_0\quad V_1]$.[5] Hence $\Delta_0 = [V_0\quad V_1][U_0\quad U_1]^{-1}$. Then $\mathcal{U} = \{U \in \mathbb{R}^{q\times k}; \det[U\quad U_1] \neq 0\}$ is an open neighbourhood of U_0 in $\mathbb{R}^{q\times k}$ and the analytic map

$$\Phi : \mathcal{U} \times \mathbb{R}^{\ell\times k} \to \mathbb{R}^{\ell\times q} \qquad \Phi(U,V) = [V\ \ V_1][U\ \ U_1]^{-1} = \Delta_{U,V}$$

has the required properties. □

Theorem 4.4.48. *If $\Delta = \mathbb{R}^{\ell\times q}$ is normed by some operator norm $\|\cdot\| = \|\cdot\|_{\mathcal{L}(\mathbb{R}^q,\mathbb{R}^\ell)}$, the map $\mu_{\mathbb{R}}(\cdot) : \mathbb{C}^{q\times\ell} \to \mathbb{R}_+$ is continuous at all non-real matrices $G \in \mathbb{C}^{q\times\ell} \setminus \mathbb{R}^{q\times\ell}$.*

Proof: By Lemma 4.4.27 $\mu_{\mathbb{R}}(\cdot)$ is upper semicontinuous, therefore continuous at all $G_0 \in \mathbb{C}^{q\times\ell}$ with $\mu_{\mathbb{R}}(G_0) = 0$. It remains to prove that $G \mapsto (\mu_{\mathbb{R}}(G))^{-1}$ is upper semicontinuous at all $G_0 \in \mathbb{C}^{q\times\ell}\setminus\mathbb{R}^{q\times\ell}$ satisfying $\mu_{\mathbb{R}}(G_0) > 0$. For any such G_0 there exist $\Delta_0 \in \mathbb{R}^{\ell\times q}$ and $v^0 \in \mathbb{C}^\ell$, $v^0 \neq 0$ such that $\|\Delta_0\| = (\mu_{\mathbb{R}}(G_0))^{-1}$ and $\Delta_0 G_0 v^0 = v^0$. Taking real and imaginary parts we have

$$\Delta_0 \left[\mathrm{Re}(G_0 v^0)\ \ \mathrm{Im}(G_0 v^0)\right] = [\mathrm{Re}(v^0)\ \ \mathrm{Im}(v^0)], \quad \|\Delta_0\| = (\mu_{\mathbb{R}}(G_0))^{-1}. \tag{62}$$

We have to consider two cases

(i) $\mathrm{Re}(G_0 v^0)$ and $\mathrm{Im}(G_0 v^0)$ are linearly independent.

(ii) $\mathrm{Re}(G_0 v^0)$ and $\mathrm{Im}(G_0 v^0)$ are linearly dependent.

In the first case we can apply Lemma 4.4.47 with

$$U_0 := [\mathrm{Re}(G_0 v^0)\ \ \mathrm{Im}(G_0 v^0)], \qquad V_0 := [\mathrm{Re}(v^0)\ \ \mathrm{Im}(v^0)]$$

to obtain for every $\varepsilon > 0$ a number $\delta > 0$ such that for all $G \in \mathbb{C}^{q\times\ell}$ in a δ-neighbourhood of G_0 there exists a matrix $\Delta \in \mathbb{R}^{\ell\times q}$ in the ε-neighbourhood of Δ_0 satisfying

$$\Delta \left[\mathrm{Re}(G v^0)\ \mathrm{Im}(G v^0)\right] = [\mathrm{Re}(v^0)\ \mathrm{Im}(v^0)] \quad \text{i.e.} \quad \Delta G v^0 = v^0,$$

hence $\mu_{\mathbb{R}}(G)^{-1} \leq \|\Delta\| \leq \|\Delta_0\| + \varepsilon = (\mu_{\mathbb{R}}(G_0))^{-1} + \varepsilon$. This proves the upper semicontinuity of $G \mapsto \mu_{\mathbb{R}}(G)^{-1}$ in a neighbourhood of G_0 in the first case.
Now assume (ii). Then there exist $z \in \mathbb{C}^*$ and a real vector $w^0 \in \mathbb{R}^q \setminus \{0\}$ such that $G_0 v^0 = z w^0$, hence $G_0 v = w^0 \in \mathbb{R}^q$ for $v = z^{-1} v^0$ and $\Delta_0 G_0 v = z^{-1}\Delta_0 G_0 v^0 = v$. It follows that

$$\|\Delta_0\| = \mu_{\mathbb{R}}(G_0)^{-1} \geq \|v\|_{\mathbb{R}^\ell} \|G_0 v\|_{\mathbb{R}^q}^{-1}.$$

But by the Hahn-Banach Theorem there exists $w \in \mathbb{R}^q$ with dual norm $\|w^\top\|_{\mathbb{R}^q}^* = \|G_0 v\|_{\mathbb{R}^q}^{-1}$ such that $w^\top G_0 v = 1$. Setting $\tilde{\Delta}_0 = v w^\top$ we obtain

$$\tilde{\Delta}_0 G_0 v = v w^\top G_0 v = v \quad \text{and} \quad \|\tilde{\Delta}_0\| = \|v\|_{\mathbb{R}^\ell} \|G_0 v\|_{\mathbb{R}^q}^{-1}.$$

[5] U_1 and V_1 are empty matrices if $q = k$.

So $\|\tilde{\Delta}_0\| = \|\Delta_0\|$ and we may assume without restriction of generality that $\Delta_0 = vw^\top$. It then follows that

$$\Delta_0^\top G_0^\top w = w(v^\top G_0^\top w) = w \quad \text{and} \quad \|\Delta_0^\top\| = (\mu_{\mathbb{R}}(G_0^\top))^{-1}$$

where the perturbation norm $\|\cdot\|_{\mathcal{L}(\mathbb{R}^\ell,\mathbb{R}^q)}$ on $\mathbf{\Delta}^\top = \mathbb{R}^{q\times\ell}$ is induced by the same vector norms on $\mathbb{R}^q, \mathbb{R}^\ell$ as the given norm $\|\cdot\| = \|\cdot\|_{\mathcal{L}(\mathbb{R}^q,\mathbb{R}^\ell)}$ on $\mathbf{\Delta}$. Taking real and imaginary parts we see that $\Delta_0^\top$ satisfies (62) with G_0 replaced by $G_0^\top$ and v^0 by w. Again we distinguish between two cases

(i') $\mathrm{Re}(G_0^\top w)$ and $\mathrm{Im}(G_0^\top w)$ are linearly independent.

(ii') $\mathrm{Re}(G_0^\top w)$ and $\mathrm{Im}(G_0^\top w)$ are linearly dependent.

In the first case we obtain as in case (i) that $G^\top \mapsto \mu_{\mathbb{R}}(G^\top)^{-1}$ is upper semicontinuous at $G_0^\top$. Hence $G \mapsto \mu_{\mathbb{R}}(G)^{-1} = \mu_{\mathbb{R}}(G^\top)^{-1}$ is upper semicontinuous at G_0.
Now assume (ii'), then necessarily $\mathrm{Im}(G_0^\top w) = 0$ since $w = \Delta_0^\top G_0^\top w = \Delta_0^\top(\mathrm{Re}\, G_0^\top w + \imath\, \mathrm{Im}\, G_0^\top w) \neq 0$ is real. It follows that $w^\top G_0 u \in \mathbb{R}$ for all $u \in \mathbb{R}^\ell$ and so

$$\Delta(u) := (w^\top G_0 u)^{-1} u w^\top \in \mathbb{R}^{\ell\times q}, \quad \Delta(u) G_0 u = u$$

for all u in a suitable neighbourhood $\mathcal{V} \subset \mathbb{R}^\ell$ of v. Clearly $\Delta(v) = \Delta_0$ and the function $\Delta(\cdot) : \mathcal{V} \to \mathbb{R}^{\ell\times q}$ is continuous. Since G_0 is not real we can find a u arbitrarily close to v such that $\mathrm{Im}(G_0 u) \neq 0$. Then $\mathrm{Re}(G_0 u)$ and $\mathrm{Im}(G_0 u)$ are linearly independent. Otherwise $\mathrm{Re}(G_0 u) = \alpha\, \mathrm{Im}(G_0 u)$ for some $\alpha \in \mathbb{R}$ and then $\Delta(u) G_0 u = u \in \mathbb{R}^\ell$ would imply $\Delta(u)\, \mathrm{Im}(G_0 u) = 0$ and $\Delta(u)\, \mathrm{Re}(G_0 u) = \alpha\, \mathrm{Im}(G_0 u) = 0$. Hence $u = \Delta(u) G_0 u = 0$ which is inconsistent with $\mathrm{Im}(G_0 u) \neq 0$. Now let $\varepsilon > 0$ be arbitrary and choose $u \in \mathcal{V}$ such that $\mathrm{Im}(G_0 u) \neq 0$ and $\|\Delta(u) - \Delta_0\| < \varepsilon/2$. Proceeding as in case (i) (with v^0 replaced by u, Δ_0 by $\Delta(u)$ and ε by $\varepsilon/2$) we find a $\delta > 0$ such that for every G in a δ-neighbourhood of $G_0 \in \mathbb{C}^{q\times}$ there exists a $\Delta \in \mathbb{R}^{\ell\times q}$ in the $\varepsilon/2$-neighbourhood of $\Delta(u)$ such that $\Delta G u = u$. Thus for all G in this δ-neighbourhood

$$\mu_{\mathbb{R}}(G)^{-1} \leq \|\Delta\| \leq \|\Delta(u)\| + \varepsilon/2 < \|\Delta_0\| + \varepsilon = (\mu_{\mathbb{R}}(G_0))^{-1} + \varepsilon.$$

This proves that $G \mapsto (\mu_{\mathbb{R}}(G))^{-1}$ is upper semicontinuous at G_0. □

We conclude the section with the following result proving that $\mu_{\mathbb{R}}(G)(\cdot)$ is even Lipschitz continuous at complex matrices $G \in \mathbb{C}^{q\times\ell}$ whose imaginary part has rank ≥ 2.

Proposition 4.4.49. *Suppose $G_0 = X_0 + \imath Y_0 \in \mathbb{C}^{q\times\ell}$, $X_0, Y_0 \in \mathbb{R}^{q\times\ell}$ and* rank $Y_0 \geq$ *2. Then with respect to the spectral norm on $\mathbf{\Delta} = \mathbb{R}^{\ell\times q}$*

$$|\mu_{\mathbb{R}}(G) - \mu_{\mathbb{R}}(G_0)| \leq \sigma_2(Y_0)^{-1} \mu_{\mathbb{R}}(G_0) \|G - G_0\| \tag{63}$$

for all $G \in \mathbb{C}^{q\times\ell}$ satisfying $\|G - G_0\| < \sigma_2(Y_0)$. Moreover, $\mu_{\mathbb{R}}(\cdot)$ is Lipschitz continuous on $\mathcal{U} := \{X + \imath Y \in \mathbb{C}^{q\times\ell};\ X, Y \in \mathbb{R}^{q\times\ell},$ rank $Y \geq 2\}$.

Proof: Since rank $Y_0 \geq 2$, by Proposition 4.4.35 there exists $\gamma_0 \in (0,1]$ such that $\mu_{\mathbb{R}}(G_0) = \min_{\gamma\in(0,1]} \sigma_2((G_0)^{\mathbb{R}}_\gamma) = \sigma_2((G_0)^{\mathbb{R}}_{\gamma_0})$. Now let $G = X + \imath Y \in \mathbb{C}^{q\times\ell}$ be arbitrary and set $\tilde{G} := G - G_0 = \tilde{X} + \imath\tilde{Y}$. By Corollary 4.3.11,

$$\mu_{\mathbb{R}}(G) = \inf_{\gamma\in(0,1]} \sigma_2(G^{\mathbb{R}}_\gamma) \leq \sigma_2(G^{\mathbb{R}}_{\gamma_0}) \leq \sigma_2((G_0)^{\mathbb{R}}_{\gamma_0}) + \|(G-G_0)^{\mathbb{R}}_{\gamma_0}\| = \mu_{\mathbb{R}}(G_0) + \|\tilde{G}^{\mathbb{R}}_{\gamma_0}\|,$$

and for all $u, v \in \mathbb{R}^\ell$

$$\begin{aligned}\left\|\tilde{G}^{\mathbb{R}}_{\gamma_0}\begin{bmatrix} u \\ \gamma_0^{-1}v\end{bmatrix}\right\|^2 &= \left\|\begin{bmatrix}\tilde{X} & -\gamma_0\tilde{Y} \\ \gamma_0^{-1}\tilde{Y} & \tilde{X}\end{bmatrix}\begin{bmatrix} u \\ \gamma_0^{-1}v\end{bmatrix}\right\|^2 = \left\|\begin{bmatrix}\tilde{X} & -\tilde{Y} \\ \gamma_0^{-1}\tilde{Y} & \gamma_0^{-1}\tilde{X}\end{bmatrix}\begin{bmatrix} u \\ v\end{bmatrix}\right\|^2 \\ &= \|\tilde{X}u - \tilde{Y}v\|^2 + \gamma_0^{-2}\|\tilde{Y}u + \tilde{X}v\|^2 \leq \gamma_0^{-2}[\|\tilde{X}u - \tilde{Y}v\|^2 + \|\tilde{Y}u + \tilde{X}v\|^2] \\ &= \gamma_0^{-2}\left\|\begin{bmatrix}\tilde{X} & -\tilde{Y} \\ \tilde{Y} & \tilde{X}\end{bmatrix}\begin{bmatrix} u \\ v\end{bmatrix}\right\|^2 \leq \gamma_0^{-2}\|\tilde{G}^{\mathbb{R}}\|^2\left[\|u\|^2 + \gamma_0^{-2}\|v\|^2\right],\end{aligned}$$

because $\gamma_0 \leq 1$. Hence $\|\tilde{G}^{\mathbb{R}}_{\gamma_0}\| \leq \gamma_0^{-1}\|\tilde{G}^{\mathbb{R}}\| = \gamma_0^{-1}\|\tilde{G}\|$ and since $\gamma_0^{-1}\sigma_2(Y_0) \leq \sigma_2((G_0)^{\mathbb{R}}_{\gamma_0}) = \mu_{\mathbb{R}}(G_0)$

$$\mu_{\mathbb{R}}(G) - \mu_{\mathbb{R}}(G_0) \leq \gamma_0^{-1}\|\tilde{G}\| \leq \sigma_2(Y_0)^{-1}\mu_{\mathbb{R}}(G_0)\|\tilde{G}\|. \tag{64}$$

Now suppose $\|G - G_0\| = \|\tilde{G}\| < \sigma_2(Y_0)$, then $\|\tilde{Y}\| < \sigma_2(Y_0)$ and we obtain from Corollary 4.3.11 that $\sigma_2(Y) = \sigma_2(Y_0 + \tilde{Y}) > 0$, whence rank $(Y) \geq 2$. So by Proposition 4.4.35 there exists $\gamma_1 \in (0,1]$ such that $\min_{\gamma\in(0,1]} \sigma_2(G^{\mathbb{R}}_\gamma) = \sigma_2(G^{\mathbb{R}}_{\gamma_1})$. Interchanging the roles of $G_0 = G - \tilde{G}$ and G in the above analysis, we obtain the following counterpart of (64)

$$\mu_{\mathbb{R}}(G_0) - \mu_{\mathbb{R}}(G) \leq \gamma_1^{-1}\|\tilde{G}\| \leq \sigma_2(Y)^{-1}\mu_{\mathbb{R}}(G)\|\tilde{G}\|.$$

But $\sigma_2(Y) \geq \sigma_2(Y_0) - \|\tilde{Y}\| \geq \sigma_2(Y_0) - \|\tilde{G}^{\mathbb{R}}\| = \sigma_2(Y_0) - \|\tilde{G}\|$ and so multiplying the above inequality by $\sigma_2(Y_0) - \|\tilde{G}\|$, we get

$$(\sigma_2(Y_0) - \|\tilde{G}\|)(\mu_{\mathbb{R}}(G_0) - \mu_{\mathbb{R}}(G)) \leq \mu_{\mathbb{R}}(G)\|\tilde{G}\|$$

or, equivalently,

$$\sigma_2(Y_0)(\mu_{\mathbb{R}}(G_0) - \mu_{\mathbb{R}}(G)) \leq \mu_{\mathbb{R}}(G)\|\tilde{G}\| + \|\tilde{G}\|(\mu_{\mathbb{R}}(G_0) - \mu_{\mathbb{R}}(G)) = \mu_{\mathbb{R}}(G_0)\|\tilde{G}\|.$$

This together with (64) completes the proof of (63). It remains to prove that every $G_0 \in \mathfrak{U}$ has a neighbourhood on which $\mu_{\mathbb{R}}(\cdot)$ is Lipschitz bounded. By (63) and the continuity of $\sigma_2(\cdot)$ there exists $\delta > 0$ such that for all

$$G = X + \imath Y \in B(G_0, \delta) = \{G \in \mathbb{C}^{q\times\ell};\ \|G - G_0\| < \delta\}$$

we have $\sigma_2(Y) > 0$ and $\sigma_2(Y)^{-1}\mu_{\mathbb{R}}(G) < \sigma_2(Y_0)^{-1}\mu_{\mathbb{R}}(G_0) + 1 =: L$. Now let $G_1, G_2 \in B(G_0, \delta)$ be arbitrary. Applying (63) with G_0 replaced by $G_1 = X_1 + \imath Y_1$ we get

$$|\mu_{\mathbb{R}}(G_2) - \mu_{\mathbb{R}}(G_1)| \leq \sigma_2(Y_1)^{-1}\mu_{\mathbb{R}}(G_1)\|G_2 - G_1\| \leq L\|G_2 - G_1\|.$$

Hence $\mu_{\mathbb{R}}(\cdot)$ is Lipschitz bounded on $B(G_0, \delta)$. $\square$

4.4.3 Exercises

1. If $\mathbf{\Delta} = \mathbb{C}I_q$ is provided with the norm $\|zI_q\|_{\mathbf{\Delta}} = |z|$ show that $\mu_{\mathbf{\Delta}}(G) = \varrho(G)$ for all $G \in \mathbb{C}^{q\times q}$.

2. Suppose $\mathbf{\Delta}$, $\|\cdot\|_{\mathbf{\Delta}}$, G_1, G_2, G_3 are as in Example 4.4.3. Verify that $\varrho(G_1) = \mu_{\mathbf{\Delta}}(G_1)$, $\varrho(G_2) = \mu_{\mathbf{\Delta}}(G_2)$ and determine $\mu_{\mathbf{\Delta}}(G_3)$ by solving the optimization problem in Example 4.4.3. Verify that in this case $\varrho(G_3) < \mu_{\mathbf{\Delta}}(G_3)$.

3. Let $\|\cdot\|_{\mathbb{R}^N}$ be an absolute norm on $\mathbb{R}^N$. Suppose that $\mathbb{C}^\ell = \bigoplus_{i\in\underline{N}} \mathbb{C}^{\ell_i}$ and $\mathbb{C}^q = \bigoplus_{i\in\underline{N}} \mathbb{C}^{q_i}$ are provided with the norms

$$\|(u_i)_{i\in\underline{N}}\|_{\mathbb{C}^\ell} = \|(\|u_i\|_{\mathbb{C}^{\ell_i}})_{i\in\underline{N}}\|_{\mathbb{R}^N}, \quad \|(y_i)_{i\in\underline{N}}\|_{\mathbb{C}^q} = \|(\|y_i\|_{\mathbb{C}^{q_i}})_{i\in\underline{N}}\|_{\mathbb{R}^N}$$

where $\|\cdot\|_{\mathbb{C}^{q_i}}$ and $\|\cdot\|_{\mathbb{C}^{\ell_i}}$ are given norms. Prove that for every $\Delta = \operatorname{diag}(\Delta_1, \ldots, \Delta_N)$ where $\Delta_i \in \mathbb{C}^{\ell_i\times q_i}$, $i \in \underline{N}$, the induced operator norm is given by

$$\|\Delta\|_{\mathcal{L}(\mathbb{C}^q,\mathbb{C}^\ell)} = \max_{i\in\underline{N}} \|\Delta_i\|_{\mathcal{L}(\mathbb{C}^{q_i},\mathbb{C}^{\ell_i})}.$$

4. Suppose $\gamma_i > 0$, $i \in \underline{N}$ and $\mathbf{\Delta}$, $\mathbb{C}^\ell$, $\mathbb{C}^q$ are as in the previous exercise with

$$\|(u_i)_{i\in\underline{N}}\|_{\mathbb{C}^\ell} = \left[\sum_{i=1}^N \gamma_i^2 \|u_i\|^2_{\mathbb{C}^{\ell_i}}\right]^{1/2}, \quad \|(y_i)_{i\in\underline{N}}\|_{\mathbb{C}^q} = \left[\sum_{i=1}^N \gamma_i^2 \|y_i\|^2_{\mathbb{C}^{q_i}}\right]^{1/2}.$$

Use the fact that $\mu_{\mathbf{\Delta}}(G) \le \|G\|_{\mathcal{L}(\mathbb{C}^\ell,\mathbb{C}^q)}$ to infer directly that $\mu_{\mathbf{\Delta}}(G) \le \|R_\gamma G L_\gamma^{-1}\|_{\mathcal{L}(\mathbb{C}^\ell,\mathbb{C}^q)}$, where L_γ, R_γ are as in Example 4.4.15.

5. If $G \in \mathbb{R}^{N\times N}$ is a matrix with non-negative entries and

$$\mathbf{\Delta} = \{\Delta = \operatorname{diag}(\delta_1, \ldots, \delta_N);\ \delta_i \in \mathbb{C},\ i \in \underline{N}\}, \quad \|\Delta\|_{\mathbf{\Delta}} = \max_{i\in\underline{N}} |\delta_i|,$$

prove that $\mu_{\mathbf{\Delta}}(G) = \varrho(G)$.

6. Let $\mathbf{\Delta}$ be a given block-diagonal perturbation class as in Definition 4.4.14. Prove that if $G = \operatorname{diag}(G_1, \ldots, G_N)$, $G_i \in \mathbb{C}^{q_i\times\ell_i}$ then

$$\mu_{\mathbf{\Delta}}(G) = \max\{\max_{j\in J} \|G_j\|_{\mathcal{L}(\mathbb{C}^{\ell_j},\mathbb{C}^{q_j})},\ \max_{i\in\underline{N}\setminus J} \varrho(G_i)\}. \tag{65}$$

7. Let $\mathbb{R}^q$ be provided with the ∞-norm and $X, Y \in \mathbb{R}^q$. In order to compute $\operatorname{dist}(X, \mathbb{R}Y)$ one can proceed as follows: Let $I_0 = \{i \in \underline{q};\ Y_i = 0\}$, $J_i = \{j \in \underline{q} \setminus I_0;\ j < i, (X_j, Y_j) = \pm(X_i, Y_i)\}$ for $i \in \underline{q} \setminus I_0$ and $J_0 = \bigcup_{i\in\underline{q}\setminus I_0} J_i$. Prove that the search for $\alpha \in \mathbb{R}$ in (41) may be restricted to the finite set $R := \{\alpha \in \mathbb{R};\ \exists i, j \in \underline{q} \setminus I_0 \cup J_0 : i \ne j \text{ and } |X_i - \alpha Y_i| = |X_j - \alpha Y_j|\}$. More precisely, prove

(i) $\underline{q} = I_0 \cup J_0$ if and only if $\underline{q} = I_0$.

(ii) If $\underline{q} \setminus (I_0 \cup J_0)$ does not contain more than one element, then $R = \emptyset$ and $\operatorname{dist}(X, \mathbb{R}Y) = \max_{i\in I_0} |X_i|$.

(iii) If $\underline{q} \setminus (I_0 \cup J_0)$ contains more than one element then R is a non-empty finite set and

$$\operatorname{dist}(X, \mathbb{R}Y) = \max\{\max_{i\in I_0} |X_i|,\ \min_{\alpha\in R} \max_{i\in\underline{q}\setminus I_0\cup J_0} |X_i - \alpha Y_i|\}.$$

8. Consider $G = [x_1 + \imath y_1 \ x_2 + \imath y_2]^\top$, where x_1, x_2, y_1, y_2 are real scalars, such that $y_1 \neq 0$. Show that if the perturbation norm (40) is the 1-norm (hence $\|\cdot\|_{\mathbb{R}^2} = \|\cdot\|_\infty$), then

$$\mu_{\mathbb{R}}(G) = \min\{|(x_1 y_2 - y_1 x_2|/|y_1 - y_2|,\ |x_1 y_2 - y_1 x_2|/|y_1 + y_2|\}.$$

In the following exercises the norm is assumed to be the spectral norm.

9. Consider

$$G = \begin{bmatrix} a + \imath b & 0 \\ 0 & \alpha + \imath\beta \end{bmatrix}, \quad a, b, \alpha, \beta \in \mathbb{R},\ a^2 + b^2 > \alpha^2 + \beta^2.$$

By interchanging rows and columns show that $G_\gamma^{\mathbb{R}}$ is congruent to

$$\begin{bmatrix} a & -\gamma b & 0 & 0 \\ \gamma^{-1} b & a & 0 & 0 \\ 0 & 0 & \alpha & -\gamma\beta \\ 0 & 0 & \gamma^{-1}\beta & \alpha \end{bmatrix}.$$

If $\overline{\sigma}(a, b, \gamma)$, $\underline{\sigma}(a, b, \gamma)$, $\overline{\sigma}(\alpha, \beta, \gamma)$, $\underline{\sigma}(\alpha, \beta, \gamma)$ are the upper and lower singular values of

$$\begin{bmatrix} a & -\gamma b \\ \gamma^{-1} b & a \end{bmatrix}, \quad \begin{bmatrix} \alpha & -\gamma\beta \\ \gamma^{-1}\beta & \alpha \end{bmatrix},$$

respectively, show that generically $\mu_{\mathbb{R}}(G) = \underline{\sigma}(a, b, \gamma_0) = \sigma^*$, where γ_0 is chosen so that $\underline{\sigma}(a, b, \gamma_0) = \overline{\sigma}(\alpha, \beta, \gamma_0)$. Hence show that σ^* satisfies

$$b^{-2}[\sigma^{*4} + 2a^2\sigma^{*2} + (a^2 + b^2)^2] = \beta^{-2}[\sigma^{*4} + 2\alpha^2\sigma^{*2} + (\alpha^2 + \beta^2)^2].$$

Discuss some of the nongeneric cases.

10. Find $\mu_{\mathbb{R}}(U)$ if $U \in \mathbb{C}^{q \times q}$ is unitary.

11. Suppose that $M \in \mathbb{R}^{2 \times 2}$ is symmetric with eigenvalues λ_1, λ_2 and let $\omega \in \mathbb{R}$, $\omega \neq 0$. Show that $\mu_{\mathbb{R}}(M + \imath\omega I_2) = (\lambda_1^2 + \lambda_2^2 + \omega^2)^{1/2}$.

12. Prove that $\mu_{\mathbb{R}}(G \oplus G) = \|G\| = \mu_{\mathbb{C}}(G)$.

13. Use Proposition 4.4.35 to prove Theorem 4.4.31 with respect to Euclidean norms.

14. Show that if $G \in \mathbb{C}^{q \times \ell}$ and $R \in \mathbb{R}^{m \times q}$, then

$$\sigma_{\min}(R)\mu_{\mathbb{R}}(G) \leq \mu_{\mathbb{R}}(RG) \leq \sigma_{\max}(R)\mu_{\mathbb{R}}(G).$$

4.4.4 Notes and References

The renaissance of robustness issues in modern control theory began in the seventies with various attempts at generalizing classical notions of gain and phase margins to multivariable systems, see the Special Issue on "Linear Multivariable Control Systems" [445] (1981) of the *IEEE Transactions on Automatic Control.* This development led to multivariable analysis and design techniques based on singular values (see e.g. [25]) and to the emergence of H^∞ Control Theory.

At the start of the eighties it was realized that the available singular value based techniques for determining stability margins or ensuring robust performance gave conservative results

if the perturbations were highly structured. Several authors recognized the importance of block-diagonal perturbations for describing structured uncertainties of interconnected systems where the uncertainty of the overall model is a consequence of uncertainties in its components, see *Safonov* (1978, 1980) in [443], [444] and *Doyle et al.* (1982) [137].
Structured singular values were introduced by *Doyle* (1982) [135] as a tool for analyzing the effect of block-diagonal perturbations. Definition 4.4.1 extends his definition to arbitrary perturbation sets endowed with arbitrary perturbation norms. In this generality, the term "structured singular value" is no longer appropriate since it suggests a relation to Euclidean metrics. This is the reason why we use the term "μ-value". Many authors have contributed to the development of μ-analysis and a comprehensive survey of the state of the theory up to 1993 can be found in *Packard and Doyle* (1993) [402]. This survey also contains a brief historical account and a summary of related work.
The results in Subsection 4.4.1 are mainly extensions of results of *Doyle* (1982) [135] to more general perturbation classes and/or perturbation norms. The observation that a number of these results can be proved for norms which are not operator norms but are bounded below by an operator norm such that condition (19) is satisfied (see Corollary 4.4.12 and Remark 4.4.24) is due to *Op't Hof* (1998) [400]. The importance of absolute norms for the analysis of block-diagonal perturbations has been pointed out in *Hinrichsen and Son* (1998) [257]. Lemma 4.4.20 is due to *Sezginer and Overton* (1990) [458] and Lemma 4.4.22 to *Doyle* (1982) [135]. Theorem 4.4.23 generalizes Theorem 6.4 in [402]. In [135] an analogous characterization of the μ-value has been stated for sets of block-diagonal perturbations with repeated blocks. Proposition 4.4.25 [135] provides the basis for the *power algorithm* which computes lower bounds for the μ-value, see [158], [402]. The upper bound (25) was established for spectral norms by *Safonov* (1978, 1980) in [443], [444], see also *Doyle* (1982) [135], and plays an important role in the numerics of μ-analysis, see e.g. [158], [222]. A proof of the result mentioned in Remark 4.4.21 that for multi-block perturbations with $N \leq 3$ blocks the upper bound (25) is equal to the μ-value can be found in [135], see also Appendix C of *Dullerud and Paganini* (2000) [141]. The analysis of this upper bound and its relationship to the μ-value has been at the centre of theoretical research in μ-analysis. Details of conditions under which equality holds for other block-diagonal perturbation classes and a number of counterexamples are given in [402], see also [141]. The upper bound (65) given in Ex. 5 was derived in [273] for the spectral norm and square diagonal blocks ($\ell_j = q_j$). The fact that the upper bound in (65) always lies above the upper bound in (25) follows from the Balancing Theorem of *Stoer and Witzgall* (1962) [487], see [257]. For the special case of $\ell_j = q_j = 1$, the inequality

$$\inf_{\gamma\in(0,\infty)^N} \left\| \left(\gamma_i G_{ij} \gamma_j^{-1}\right)_{i,j\in\underline{N}} \right\|_{\mathcal{L}(\mathbb{C}^\ell,\mathbb{C}^q)} \leq \varrho\left(\left(\|G_{ij}\|_{\mathcal{L}(\mathbb{C}^{\ell_i},\mathbb{C}^{q_i})}\right)_{i,j\in\underline{N}}\right).$$

can already be found in [444]. Further upper bounds for the μ-value are given in *Chen and Nett* (1992) [97].
The real μ-value studied in the second subsection was motivated by the notions of *real stability radius* and *real spectral value sets* which were introduced by *Hinrichsen and Pritchard* (1986), (1992), [242], [251] and will be studied in detail in the next chapter. In [242] a function was introduced which in essence is the reciprocal of the real μ-value. In (1988) *Hinrichsen and Pritchard* [244] proved the formula given in Theorem 4.4.31 for the case $\ell = 1$ and applied it to obtain stability radii results for polynomials. However the general problem remained an open one for some time. Using tensor product techniques in the context of the stability radius problem, *Qiu and Davison* (1991) [425] obtained several upper bounds for the real μ-value with respect to the spectral norm. In the case where

G is square, they obtained in [426] (1992) the estimate (47) given in Lemma 4.4.33 and conjectured that it was tight. Then in [424] (1993) in conjunction with Bernhardsson, Rantzer, Young and Doyle, they proved the equality. The proof we have given is based on this paper. Their work on the real μ-value motivated Bernhardsson, Rantzer and Qiu to introduce counterparts to singular values for the analysis of *real* perturbations. Analogous to the characterizations (3.32) and (3.31) of $\sigma_j(G)$ they associated with any $G \in \mathbb{C}^{q\times\ell}$ two ordered sequences of non-negative numbers $\tau_1(G) \geq \ldots \geq \tau_\ell(G)$ and $\tilde{\tau}_1(G) \geq \ldots \geq \tilde{\tau}_\ell(G)$ defined by

$$\tau_j(G) = [\inf\{\|\Delta\|;\ \Delta \in \mathbb{R}^{\ell\times q}, \dim\ker(I_n - \Delta G) \geq j\}]^{-1}, \quad j \in \underline{\ell} \tag{66}$$

$$\tilde{\tau}_j(G) = \inf\{\|\Delta\|;\ \Delta \in \mathbb{R}^{\ell\times q},\ \mathrm{rank}\,(G-\Delta) < j\}, \quad j \in \underline{\ell}. \tag{67}$$

The $\tau_j(G)$ (resp. $\tilde{\tau}_j(G)$) are called the *lower (resp. upper) real perturbation values* of G. By the Schmidt-Mirsky Theorem 4.3.7 $\tau_j(G) \leq \sigma_j(G) \leq \tilde{\tau}_j(G)$, but in contrast to the singular values, $\tau_j(G)$ and $\tilde{\tau}_j(G)$ are in general distinct. *Bernhardsson et al.* (1998) [51] gave a full characterization of real perturbation values and discussed their various properties. They showed that

$$\tau_j(G) = \inf_{\gamma\in(0,1]} \sigma_{2j}(G_\gamma^{\mathbb{R}}), \qquad \tilde{\tau}_j(G) = \sup_{\gamma\in(0,1]} \sigma_{2j-1}(G_\gamma^{\mathbb{R}}) \tag{68}$$

where $G_\gamma^{\mathbb{R}}$ is as in (47). Their methods of proof are quite different from the one given in Subsection 4.4.2 and are based on inequalities between an Hermitian form and a symmetric bilinear form associated with G. In [51], they proved the continuity of the real perturbation values on $\mathbb{C}^{q\times\ell} \setminus \mathbb{R}^{q\times\ell}$. This implies the continuity statement of Theorem 4.4.48 for the special case where $\mathbf{\Delta} = \mathbb{R}^{\ell\times q}$ is provided with the spectral norm. The general continuity result is due to *Karow* (2003) [292] and we have followed his proof.

We will see in the next chapter that real perturbations of complex G are important for the spectral analysis of linear systems with uncertain real parameters. It is surprising, therefore, that although μ-values were introduced for complex perturbations circa 1980, the development of real μ-analysis had to wait for about another 10 years. Even today, with the exception of the results presented in Subsection 4.4.2, little is known about $\mu_{\mathbf{\Delta}}$ with respect to arbitrary norms and more general structured perturbation classes $\mathbf{\Delta} \subset \mathbb{R}^{\ell\times q}$. There are a number of interesting unsolved problems in this area which have been open for some time, see the PhD thesis of *Karow* (2003) [292].

4.5 Computational Aspects

Nowadays there are many software packages available for computing solutions to problems encountered in Systems Theory. However one must be cautious, since these packages may give spurious results in applications. To avoid mistakes some knowledge is required about the numerical properties of the algorithms used. In this section we give a brief introduction to a few topics in Numerical Linear Algebra which are of special importance for Systems Theory, focussing on those aspects which have a relationship with the material of this and previous chapters (for more comprehensive accounts, see *Notes and References*).

When solving a problem numerically, errors can occur in the computation because the use of finite arithmetic necessarily means that numbers are rounded. The accumulation of rounding errors may mean that the computed solution is far from the true one. This may be the fault of the particular algorithm used for the resolution of the problem or it may be rooted in the problem itself. Numerical analysts have developed general guidelines in order to consider these questions. They have found it convenient to separate the problem into two parts – one called *conditioning* is problem orientated, and the other called *numerical stability* is algorithm orientated. In the first subsection we introduce the notion of a *condition number* for a given problem and obtain formulas and estimates for the problems of

- solving linear matrix equations,
- solving Liapunov equations,
- determining the eigenvalues of a matrix,
- determining the singular values of a matrix.

In the second subsection we describe *Householder transformations* and show how they can be used to reduce a matrix to Hessenberg and bidiagonal form and to factorize a matrix into a unitary matrix and a triangular one (QR factorization). We also prove that a matrix is unitarily similar to one in *Schur form*.

In the third subsection we explain the details of some important algorithms. Both the transformation to Hessenberg form and the QR factorization are used in an algorithm for computing the eigenvalues of a matrix. The algorithm reduces a matrix to Schur form and is known as Francis' double-shift QR algorithm. We explain the details of this and discuss its simplifications and shortcomings when used to find singular values. We also briefly discuss algorithms for solving linear equations and describe in more detail an algorithm for solving Liapunov equations.

An important consideration in using a particular algorithm is the amount of work the computer is required to carry out in implementing it. This is measured in terms of the number of *flops* that are needed. A flop is a floating point operation such as addition or multiplication of real numbers.[1] The number of flops depends of course on the dimensions of the matrices and when discussing the algorithms we will indicate the approximate number of flops required in the form $\approx Kn^k$ where K, k are constants and n is a typical dimension. We say that an algorithm requires $O(n^k)$ flops if the number of flops increases at the order of n^k as $n \to \infty$.

[1]Note that complex multiplication requires four real multiplications and two real additions. Hence the flop count is six times greater than in the real case.

4.5.1 Condition Numbers

We begin by giving a very general definition of a condition number. Then we derive an explicit formula for the condition number of three specific problems: a) finding solutions of linear matrix equations, b) finding solutions of Liapunov equations and c) computing eigenvalues and singular values. Although these exact formulas are interesting we will see that in the first two cases they suffer the drawback of being given in terms of the unknown solutions of the problems. To get around this we use each formula to find an estimate for the condition numbers which are expressed solely in terms of the given data of the problem.

Suppose that the problem under consideration can be represented abstractly as the evaluation of a continuous function $f : \mathcal{W}_0 \to \mathcal{X}$ at a data point $\overline{w} \in \mathcal{W}_0 \subset \mathcal{W}$. Here $\mathcal{W}$ represents the set of possible problem data, $\mathcal{W}_0$ is the subset of data points $w \in \mathcal{W}$ for which the given problem is solvable, $\mathcal{X}$ is the solution space and f is a solution operator. We assume that $\mathcal{W}$, $\mathcal{X}$ are metric spaces.

In general, it will not be possible to determine $f(\overline{w})$ exactly. Instead a numerical procedure will be carried out which yields, for every $w \in \mathcal{W}_0$, a *computed* value $f^*(w)$ which may be different from $f(w)$.

A problem is said to be *well conditioned* if small changes in the data (assuming perfect computation) do not affect the solution very much. One of the ways of examining whether or not a problem is well conditioned is via a *condition number.* Suppose that $\mathcal{W}_0$ is an open subset of $\mathcal{W}$ and define for sufficiently small $\delta > 0$

$$\begin{aligned} \kappa_\delta(f,\overline{w}) &= \inf\{r > 0\,; f(B_{\mathcal{W}}(\overline{w},\delta)) \subset B_{\mathcal{X}}(f(\overline{w}), r\delta)\}, & (1)\\ \kappa(f,\overline{w}) &= \limsup_{\delta\to 0}\ \kappa_\delta(f,\overline{w}) & (2) \end{aligned}$$

where, for any metric space $(\mathcal{Y}, d_{\mathcal{Y}})$, we denote by $B_{\mathcal{Y}}(y,\delta)$ the open ball in $\mathcal{Y}$ centred at $y \in \mathcal{Y}$ with radius $\delta > 0$. $\kappa_\delta(f,\overline{w})$ (resp. $\kappa(f,\overline{w})$) is called the *δ-condition number* (resp. *condition number*) of f at $\overline{w}$. If for small δ the δ-condition number is not large then the problem is well conditioned. Of course the terms "small" and "large" are subjective and their meaning depends on the concrete application.

There are other definitions of condition numbers. The ones we have given here were introduced by *Rice* (1966) [434] and are widely used. A nice feature is that if $\mathcal{W}$ and $\mathcal{X}$ are Banach spaces, $\mathcal{W}_0$ is an open subset of $\mathcal{W}$ and f is Fréchet differentiable at $\overline{w} \in \mathcal{W}_0$, then

$$\kappa(f,\overline{w}) = \|f'(\overline{w})\|_{\mathcal{L}(\mathcal{W},\mathcal{X})} \tag{3}$$

where $f'(\overline{w})$ is the Fréchet derivative of f at $\overline{w}$, see Definition A.4.16. This makes the computation of κ tractable as we illustrate below in examples where $\mathcal{W}$ and $\mathcal{X}$ are spaces of matrices.

In order to check whether for a given $\overline{w} \in \mathcal{W}_0$ the evaluation $f^*(\overline{w})$ is a good approximation of $f(\overline{w})$ we would need to estimate the errors at every arithmetic operation of, perhaps, a large series of complex calculations. This is usually impossible. It turns out, however, that often there is a $w \in \mathcal{W}_0$ such that $f^*(\overline{w}) = f(w)$. An algorithm f^* for computing f is said to be *numerically stable* if for every $\overline{w} \in \mathcal{W}_0$ there exists $w \in \mathcal{W}_0$ near $\overline{w}$ such that $f^*(\overline{w}) = f(w)$. Recall that a problem is well conditioned if w near $\overline{w}$ implies $f(w)$ near $f(\overline{w})$. Thus good conditioning of the problem *plus* numerical stability of the algorithm ensure that the computed solution $f^*(\overline{w})$ is near the true solution $f(\overline{w})$.

Condition Number for Solving Linear Matrix Equations

Suppose $A \in \mathbb{K}^{n\times n}$, $B \in \mathbb{K}^{n\times m}$, where A is invertible and consider the linear matrix equation $AX = B$ with solution $X = A^{-1}B \in \mathbb{K}^{n\times m}$. Our aim is to obtain a formula for the condition number of this problem. Now since the distance of an invertible matrix $A \in \mathbb{K}^{n\times n}$ from the set of singular ones (with respect to the spectral norm) is $\sigma_{\min}(A)$ we would expect that the condition number should involve $\sigma_{\min}(A)$. We will see that this is indeed the case.

If $M \in \mathbb{K}^{q\times \ell}$ we denote by $\|M\|_F = \left(\sum_{i=1}^{q}\sum_{j=1}^{\ell}|m_{ij}|^2\right)^{1/2}$ its Frobenius norm. For the above linear matrix equation we have the following data set $\mathcal{W}$, solution set $\mathcal{X}$ and solution operator $f_S : \mathcal{W}_0 \to \mathcal{X}$ on $\mathcal{W}_0 = \mathbf{Gl}_n(\mathbb{K}) \times \mathbb{K}^{n\times m}$

$$\mathcal{W} = \mathbb{K}^{n\times n} \times \mathbb{K}^{n\times m}, \quad \mathcal{X} = \mathbb{K}^{n\times m} \text{ and } f_S(w) = A^{-1}B, \; w = (A, B) \in \mathcal{W}_0$$

where we endow the spaces $\mathcal{W}$ and $\mathcal{X}$ with the Frobenius norms

$$\|w\|_{\mathcal{W}} = \|(A,B)\|_{\mathcal{W}} = (\|A\|_F^2 + \|B\|_F^2)^{1/2}, \quad \|X\|_{\mathcal{X}} = \|X\|_F. \tag{4}$$

Besides the ease by which it can be calculated there is another reason for taking the Frobenius norm. If $M \in \mathbb{K}^{q\times\ell}$ we denote by $\operatorname{vec}(M)$ the vector in $\mathbb{K}^{q\ell}$ formed by stacking each column of the matrix M (from the left to the right) one beneath the other (see Subsection A.1.4). Then the Frobenius norm of a matrix $M \in \mathbb{K}^{q\times\ell}$ is the same as the Euclidean norm of $\operatorname{vec}(M) \in \mathbb{K}^{q\ell}$. By (3) the condition number for solving the linear matrix equation is $\kappa(f_S,(A,B)) = \|f_S'(A,B)\|_{\mathcal{L}(\mathcal{W},\mathcal{X})}$ where $f_S'(A,B)$ is the Fréchet derivative of f_S at the point (A,B). In the following proposition we identify $f_S'(A,B) \in \mathcal{L}(\mathcal{W},\mathcal{X})$ with its standard matrix representation in $\mathbb{K}^{nm\times(n^2+nm)}$.

Proposition 4.5.1. *For* $(A,B) \in \mathbf{Gl}_n(\mathbb{K}) \times \mathbb{K}^{n\times m}$ *we have*

$$f_S'(A,B) = \left[-(A^{-1}B)^\top \otimes A^{-1} \;|\; I_m \otimes A^{-1}\right], \tag{5}$$

$$\kappa(f_S,(A,B)) = \|A^{-1}\|_{2,2}\left(\|A^{-1}B\|_{2,2}^2 + 1\right)^{1/2}, \tag{6}$$

where the symbol $\otimes$ *denotes the Kronecker product.*

Proof: For $(\Delta_A, \Delta_B) \in \mathcal{W}$ small enough, we have $(A+\Delta_A, B+\Delta_B) \in \mathcal{W}_0$ and

$$\begin{aligned} f_S(A+\Delta_A, B+\Delta_B) - f_S(A,B) &= (A+\Delta_A)^{-1}(B+\Delta_B) - A^{-1}B \\ &= [I + \sum_{k=1}^{\infty}(-1)^k(A^{-1}\Delta_A)^k]A^{-1}(B+\Delta_B) - A^{-1}B. \end{aligned}$$

Hence

$$f_S'(A,B)(\Delta_A,\Delta_B) = -A^{-1}\Delta_A A^{-1}B + A^{-1}\Delta_B. \tag{7}$$

In order to express the linear operator $f'_S(A,B)$ as a matrix, we represent Δ_A and Δ_B as vectors in $\mathbb{K}^{n^2}$ and $\mathbb{K}^{nm}$ respectively and then use the Kronecker product (see Subsection A.1.4). In Proposition A.1.25 we see that $\operatorname{vec}(M_0XM_1) = (M_1^\top \otimes M_0)\operatorname{vec}(X)$. So from (7) we have

$$\operatorname{vec}\left(f'_S(A,B)(\Delta_A,\Delta_B)\right) = -((A^{-1}B)^\top \otimes A^{-1})\operatorname{vec}(\Delta_A) + (I_m \otimes A^{-1})\operatorname{vec}(\Delta_B).$$

From which (5) follows and

$$\kappa(f_S,(A,B)) = \|f'_S(A,B)\|_{2,2} = \left\|\left[-(A^{-1}B)^\top \otimes A^{-1} \mid I_m \otimes A^{-1}\right]\right\|_{2,2} \qquad (8)$$

where $\|\cdot\|_{2,2}$ denotes the spectral norm. Now (see Subsection A.1.4) for $M \in \mathbb{K}^{q\times\ell}$, $M_1 \in \mathbb{K}^{q_1\times\ell_1}$, $M_2 \in \mathbb{K}^{q_1\times\ell_2}$

$$\begin{aligned}[][M_1\otimes M \mid M_2 \otimes M] &= ([M_1 \mid M_2]\otimes M),\\ \|M_1\otimes M\|_{2,2} = \|M_1\|_{2,2}\|M\|_{2,2}\ , &\qquad \|M_1\otimes M\|_F = \|M_1\|_F\|M\|_F.\end{aligned}$$

Since $\|[M \mid \alpha I_m]\|_{2,2}^2 = \|M\|_{2,2}^2 + |\alpha|^2$ for every $M \in \mathbb{K}^{m\times n}$, $\alpha \in \mathbb{K}$, we have

$$\begin{aligned}\kappa(f_S,(A,B)) &= \left\|\left[-(A^{-1}B)^\top \mid I_m\right]\otimes A^{-1}\right\|_{2,2} = \|A^{-1}\|_{2,2}\left\|\left[-(A^{-1}B)^\top \mid I_m\right]\right\|_{2,2}\\ &= \|A^{-1}\|_{2,2}\left(\|A^{-1}B\|_{2,2}^2+1\right)^{1/2}.\end{aligned}$$

□

When the Frobenius norms of the data $A \in \mathbb{K}^{n\times n}$, $B \in \mathbb{K}^{n\times m}$ and the solution $A^{-1}B \in \mathbb{K}^{n\times m}$ have different orders of magnitude, it is appropriate to consider relative errors instead of absolute errors. If, for instance, a change Δ_A in A results in a change in the solution Δ_X, one compares $\|\Delta_X\|/\|X\|$ with $\|\Delta_A\|/\|A\|$ rather than $\|\Delta_X\|$ with $\|\Delta_A\|$. This leads to the use of the following weighted norms on $\mathcal{W}$ and $\mathcal{X}$

$$\|(w_1,w_2)\|_{\mathcal{W}} = \left(\|w_1\|_F^2/\|A\|_F^2 + \|w_2\|_F^2/\|B\|_F^2\right)^{1/2},\ \|X\|_{\mathcal{X}} = \|X\|_F/\|A^{-1}B\|_F. \quad (9)$$

With respect to these norms the condition number $\kappa_{rel}(f_S,(A,B))$ is given by

$$\kappa_{rel}(f_S,(A,B)) = \|A^{-1}B\|_F^{-1}\left\|\left[(-(A^{-1}B)^\top\otimes A^{-1})\|A\|_F \mid (I_m\otimes A^{-1})\|B\|_F\right]\right\|_{2,2}.$$

Then in a similar way to the unweighted case one gets

$$\kappa_{rel}(f_S,(A,B)) = \|A^{-1}\|_{2,2}\|A^{-1}B\|_F^{-1}\left(\|A^{-1}B\|_{2,2}^2\|A\|_F^2 + \|B\|_F^2\right)^{1/2}. \qquad (10)$$

Both formulas (6) and (10) suffer from the fact that we have to know the norm of the solution $A^{-1}B$ before they can be evaluated. To get around this problem we use the following upper estimate for $\|A^{-1}B\|_{2,2}$ and lower estimate for $\|A^{-1}B\|_F$.

$$\begin{aligned}\|A^{-1}B\|_{2,2} &\le \|A^{-1}\|_{2,2}\|B\|_{2,2}, \qquad \|A^{-1}B\|_{2,2} \le \|A^{-1}B\|_F,\\ \|B\|_F &= \|AA^{-1}B\|_F \le \|A\|_{2,2}\|A^{-1}B\|_F.\end{aligned}$$

Then

$$\begin{aligned}\kappa(f_S,(A,B)) &\le \|A^{-1}\|_{2,2}\left(\|A^{-1}\|_{2,2}^2\|B\|_{2,2}^2+1\right)^{1/2} = \sigma_{\min}(A)^{-2}\left(\|B\|_{2,2}^2+\sigma_{\min}(A)^2\right)^{1/2},\\ \kappa_{rel}(f_S,(A,B)) &\le \|A^{-1}\|_{2,2}\left(\|A\|_F^2+\|A\|_{2,2}^2\right)^{1/2} = \sigma_{\min}(A)^{-1}\left(\|A\|_F^2+\|A\|_{2,2}^2\right)^{1/2}. \quad (11)\end{aligned}$$

The above formulas show the importance of $\sigma_{\min}(A)$ in obtaining an estimate of the condition number for solving linear matrix equations $AX = B$.

Remark 4.5.2. (i) If $B = I_n$ is not subject to perturbations then it is easy to see from the above calculations that the condition number and the relative condition number for inverting a matrix $A \in \mathbf{Gl}_n(\mathbb{K})$ are, respectively,

$$\kappa(f_I, A) = \|A^{-1}\|_{2,2}^2 \quad \text{and} \quad \kappa_{rel}(f_I, A) = \|A^{-1}\|_{2,2}^2 \|A\|_F \|A^{-1}\|_F^{-1} \le \|A\|_F \|A^{-1}\|_F.$$

The reader is asked to prove this in Ex. 1. If for the weighted norm (9) (with $B = I_n, w_2 = 0$) we use the weight $\|A\|_{2,2}^{-1}$ rather than $\|A\|_F^{-1}$, we get $\kappa_{rel}(f_I, A) = \|A^{-1}\|_{2,2}^2 \|A\|_{2,2} \|A^{-1}\|_F^{-1}$ and hence the estimate $\kappa_{rel}(f_I, A) \le \|A\|_{2,2} \|A^{-1}\|_{2,2}$.

(ii) Suppose $x \in \mathbb{K}^n$ is the true solution and y is the computed solution of a linear equation $Ax = b$ with $A \in \mathbb{K}^{n\times n}$ and $0 \ne b \in \mathbb{K}^n$. Let $\|\cdot\|_{\mathbb{K}^n}$ be any norm on $\mathbb{K}^n$ and $\|\cdot\|$ the associated operator norm on $\mathbb{K}^{n\times n}$. If $r = b - Ay$ is the residual vector, then it is easily seen that

$$\frac{\|y - x\|_{\mathbb{K}^n}}{\|x\|_{\mathbb{K}^n}} \le \|A\|\,\|A^{-1}\| \frac{\|r\|_{\mathbb{K}^n}}{\|b\|_{\mathbb{K}^n}},$$

cf. Ex. 3. In the literature, $\kappa(A) = \|A\|\,\|A^{-1}\|$ is called the *condition number of* A. Actually, $\kappa(A)$ is the operator norm of the Fréchet derivative of the map $A \mapsto A^{-1}$ with respect to weighted operator norms on the matrix spaces and is therefore a condition number for inverting A in the sense of (3), see Ex. 2 and *Notes and References*.

(iii) In the above formulas for $\kappa(f_I, A)$, $\kappa_{rel}(f_I, A)$ and $\kappa(A)$ there is a problem in computing $\|A^{-1}\|$ without knowing A^{-1}. For the spectral norm one can utilize $\|A^{-1}\|_{2,2} = \sigma_{\min}(A)^{-1}$. With respect to other norms there are computational procedures for estimating $\|A^{-1}\|$, see *Notes and References*. □

Condition Number for Solving Liapunov Equations

In the following we only consider Liapunov Equations for the *real* continuous time case

$$XA + A^\top X + Q = 0 \tag{12}$$

(see (3.3.80a)) and assume

$$(A, Q) \in \mathcal{W}_1 \times \mathbb{R}^{n\times n}, \quad \mathcal{W}_1 = \{A \in \mathbb{R}^{n\times n};\ \forall \lambda, \mu \in \sigma(A) : \lambda + \overline{\mu} \ne 0\}.$$

We do not require that Q is symmetric. Let $\mathbf{L}_A : \mathbb{R}^{n\times n} \to \mathbb{R}^{n\times n}$ be the Liapunov operator $X \mapsto \mathbf{L}_A(X) = XA + A^\top X$. Then by Theorem 3.3.46 $\mathbf{L}_A$ is invertible and the solution of the Liapunov equation is $-\mathbf{L}_A^{-1}(Q)$. Our aim is to obtain a formula for the corresponding condition number (with respect to perturbations of the data $w = (A, Q)$). For this we define

$$\mathcal{W} = \mathbb{R}^{n\times n} \times \mathbb{R}^{n\times n},\ \mathcal{W}_0 = \mathcal{W}_1 \times \mathbb{R}^{n\times n},\ \mathcal{X} = \mathbb{R}^{n\times n},\ \ f_L(A, Q) = -\mathbf{L}_A^{-1}(Q),\ \ (A, Q) \in \mathcal{W}_0$$

where the matrix spaces $\mathcal{W}$ and $\mathcal{X}$ are again provided with Frobenius norms

$$\|w\|_{\mathcal{W}} = \|(A, Q)\|_{\mathcal{W}} = \left(\|A\|_F^2 + \|Q\|_F^2\right)^{1/2}, \quad \|X\|_{\mathcal{X}} = \|X\|_F.$$

It follows from (3) that the condition number for the Liapunov equation is given by $\kappa(f_L, (A, Q)) = \|f_L'(A, Q)\|_{\mathcal{L}(\mathcal{W},\mathcal{X})}$. To express $\|f_L'(A, Q)\|_{\mathcal{L}(\mathcal{W},\mathcal{X})}$ in terms of the data, we identify $f_L'(A, Q) \in \mathcal{L}(\mathcal{W}, \mathcal{X})$ with its standard matrix representation in $\mathbb{R}^{n^2\times 2n^2}$

and write the Liapunov equation as an ordinary linear equation for vec(X) using Kronecker products. Then (12) takes the form

$$(A^\top \otimes I_n + I_n \otimes A^\top)\mathrm{vec}(X) = -\mathrm{vec}(Q). \tag{13}$$

So one might think that one could apply the previous results on solving linear matrix equations to obtain the condition number for solving the Liapunov equation. However this would mean that arbitrary perturbations of the matrices $A^\top \otimes I_n + I_n \otimes A^\top \in \mathbb{R}^{n^2 \times n^2}$ and $Q \in \mathbb{R}^{n \times n}$ would be taken into account. Whereas we only consider perturbations of $(A, Q) \in \mathbb{R}^{n \times n} \times \mathbb{R}^{n \times n}$.

Proposition 4.5.3. *Suppose $(A, Q) \in \mathcal{W}_0$, $X = \mathbf{L}_A^{-1}(Q)$, and P is a permutation matrix such that $P\,vec(Z) = vec(Z^\top)$ for all $Z \in \mathbb{R}^{n \times n}$. Then*

$$f'_L(A, Q) = -\left(A^\top \otimes I_n + I_n \otimes A^\top\right)^{-1}\left[(I_n \otimes X + (X^\top \otimes I_n)P) \mid I_{n^2}\right], \tag{14}$$

$$\kappa\left(f_L, (A, Q)\right) \le \|\left(A^\top \otimes I_n + I_n \otimes A^\top\right)^{-1}\|_{2,2}\left(1 + \|I_n \otimes X + (X^\top \otimes I_n)P)\|_{2,2}^2\right)^{1/2}. \tag{15}$$

Proof: For $\|\Delta_A\|$ sufficiently small, we have $A + \Delta_A \in \mathcal{W}_1$, so that $f_L(A + \Delta_A, Q + \Delta_Q) =: X + \Delta_X$ is well defined. By the definition of f_L we have

$$(X + \Delta_X)(A + \Delta_A) + (A + \Delta_A)^\top(X + \Delta_X) + Q + \Delta_Q = 0,$$

and so

$$\Delta_X A + A^\top \Delta_X = -\Delta_Q - X\Delta_A - (\Delta_A)^\top X - \Delta_X\Delta_A - \Delta_A^\top\Delta_X.$$

Making use of the Kronecker product we obtain

$$\begin{aligned}&(A^\top \otimes I_n + I_n \otimes A^\top)\mathrm{vec}(\Delta_X) = \\ &\quad -\mathrm{vec}(\Delta_Q) - (I_n \otimes X)\mathrm{vec}(\Delta_A) - (X^\top \otimes I_n)\mathrm{vec}(\Delta_A^\top) - \mathrm{vec}(\Delta_X\Delta_A + \Delta_A^\top\Delta_X).\end{aligned}$$

But $\mathrm{vec}(\Delta_A^\top) = P\mathrm{vec}(\Delta_A)$, so (14) holds and

$$\begin{aligned}\kappa(f_L, (A, Q)) &= \|\left(A^\top \otimes I_n + I_n \otimes A^\top\right)^{-1}\left[I_n \otimes X + (X^\top \otimes I_n)P \mid I_{n^2}\right]\|_{2,2} \\ &\le \|\left(A^\top \otimes I_n + I_n \otimes A^\top\right)^{-1}\|_{2,2}\left(1 + \|I_n \otimes X + (X^\top \otimes I_n)P)\|_{2,2}^2\right)^{1/2}.\end{aligned}$$

□

Since

$$\|(I_n \otimes X + (X^\top \otimes I_n)P)\|_{2,2} \le 2\|X\|_{2,2}, \tag{16}$$

we obtain from (15) the following estimate

$$\kappa(f_L, (A, Q)) \le \|\left(A^\top \otimes I_n + I_n \otimes A^\top\right)^{-1}\|_{2,2}\left(1 + 4\|X\|_{2,2}^2\right)^{1/2}. \tag{17}$$

If instead we endow $\mathcal{W}$ and $\mathcal{X}$ with the weighted Frobenius norms

$$\|(w_1, w_2)\|_{\mathcal{W}} = \left(\|w_1\|_F^2/\|A\|_F^2 + \|w_2\|_F^2/\|Q\|_F^2\right)^{1/2}, \quad \|X\|_{\mathcal{X}} = \|X\|_F/\|\mathbf{L}_A^{-1}(Q)\|_F,$$

then

$$\kappa_{rel}(f_L, (A, Q)) \le \|X\|_F^{-1}\|(A^\top \otimes I_n + I_n \otimes A^\top)^{-1}\|_{2,2}\left[\|Q\|_F^2 + \|I_n \otimes X + (X^\top \otimes I_n)P\|_{2,2}^2\|A\|_F^2\right]^{1/2}.$$

Using the estimate in (16) we obtain

$$\kappa_{rel}(f_L,(A,Q)) \le \|X\|_F^{-1} \|\left(A^\top \otimes I_n + I_n \otimes A^\top\right)^{-1}\|_{2,2} \left(\|Q\|_F^2 + 4\|X\|_{2,2}^2\|A\|_F^2\right)^{1/2}. \quad (18)$$

Again both estimates (17) and (18) suffer from the drawback that the solution $X = \mathbf{L}_A^{-1}(Q)$ is required before they can be computed. Now

$$\|Q\|_F = \|XA + A^\top X\|_F \le 2\|X\|_F\|A\|_{2,2}, \; \|\mathbf{L}_A^{-1}\|_{\mathcal{L}(\mathbb{R}^{n\times n})} = \| \left(A^\top \otimes I_n + I_n \otimes A^\top\right)^{-1} \|_{2,2}$$

where $\mathbb{R}^{n\times n}$ carries the Frobenius norm. Hence

$$\begin{aligned} \kappa\,(f_L,(A,Q)) &\le \|\mathbf{L}_A^{-1}\|_{\mathcal{L}(\mathbb{R}^{n\times n})} \left(1 + 4\|\mathbf{L}_A^{-1}\|^2_{\mathcal{L}(\mathbb{R}^{n\times n})}\|Q\|^2_{2,2}\right)^{1/2}, \\ \kappa_{rel}(f_L,(A,Q)) &\le 2\|\mathbf{L}_A^{-1}\|_{\mathcal{L}(\mathbb{R}^{n\times n})} \left(\|A\|^2_{2,2} + \|A\|_F^2\right)^{1/2}. \end{aligned} \quad (19)$$

The above estimates clearly show, as one would expect, the importance of the term $\|\mathbf{L}_A^{-1}\|_{2,2}$. However one should be cautious since there are examples where the estimates given by (19) can be very bad in comparison with the exact values of $\kappa\,(f_L,(A,Q))$ and $\kappa_{rel}(f_L,(A,Q))$, see *Higham* (2002) [230].

Remark 4.5.4. In the above analysis we have not assumed that Q and hence X are symmetric. Clearly we could have restricted our analysis to the symmetric case, but then we would have been forced to assume that the perturbations Δ_Q are also symmetric. This assumption might be too restrictive. □

Condition Number for Determining the Eigenvalues of a Matrix

We have seen in Corollary 4.2.3 that if λ_0 is a simple eigenvalue of $A_0 \in \mathbb{C}^{n\times n}$, then there exists a neighbourhood $B(A_0,\varepsilon)$, $\varepsilon > 0$ of A_0 in $\mathbb{C}^{n\times n}$ and an analytic function $A \mapsto \lambda(A)$ on $B(A_0,\varepsilon)$ such that $\lambda(A)$ is a simple eigenvalue of A for all $A \in B(A_0,\varepsilon)$ and $\lambda(A_0) = \lambda_0$. In this case we can use Proposition 4.2.12 to determine the condition number of $\lambda(A)$ at A_0. For this we endow $\mathcal{W} = \mathbb{C}^{n\times n}$ with the Frobenius norm and set

$$\mathcal{W}_0 = B(A_0,\varepsilon)\,; \quad \mathcal{X} = \mathbb{C}\,; \quad f_E(A) = \lambda(A), \;\; A \in \mathcal{W}_0.$$

The condition number of f_E at A_0 is $\kappa(f_E, A_0) = \|f'_E(A_0)\|_{\mathcal{L}(\mathcal{W},\mathcal{X})}$. In the following proposition we identify $f'_E(A_0) \in \mathcal{L}(\mathcal{W},\mathcal{X})$ with its matrix representation in $\mathbb{C}^{1\times n^2}$.

Proposition 4.5.5. *Let $A_0 \in \mathbb{C}^{n\times n}$ and assume that for some $\varepsilon > 0$ we are given two functions, $\lambda(\cdot) : B(A_0,\varepsilon) \to \mathbb{C}$ and $v(\cdot) : B(A_0,\varepsilon) \to \mathbb{C}^n \setminus \{0\}$, which are differentiable at A_0 and satisfy*

$$Av(A) = \lambda(A)v(A), \quad A \in B(A_0,\varepsilon).$$

If w is a left eigenvector of A_0 corresponding to the eigenvalue $\lambda_0 = \lambda(A_0)$ and $w^*v \ne 0$ where $v = v(A_0)$, then

$$\begin{aligned} f'_E(A_0) &= (w^*v)^{-1} v^\top \otimes w^*, & (20)\\ \kappa(f_E, A_0) &= |w^*v|^{-1}\|w\|_2\|v\|_2. & (21) \end{aligned}$$

Proof: Given $\Delta \in \mathbb{C}^{n\times n}$, $\Delta \neq 0$, let $\Omega = \{z \in \mathbb{C}; |z| < \varepsilon\|\Delta\|_F^{-1}\}$ and $A(z) = A_0 + z\Delta$, $z \in \Omega$. Then the assumptions of Proposition 4.2.12 hold with $z_0 = 0$. Hence from (2.4)

$$f_E'(A_0) : \Delta \mapsto (w^*v)^{-1}w^*\Delta v.$$

Thus

$$\mathrm{vec}(f_E'(A_0)(\Delta)) = (w^*v)^{-1}v^\top \otimes w^*\mathrm{vec}(\Delta).$$

So (20) holds and $\kappa(f_E, A_0) = |w^*v|^{-1}\|v^\top \otimes w^*\|_2 = |w^*v|^{-1}\|w\|_2\|v\|_2$. □

By Proposition 4.2.24 and Corollary 4.2.3 the assumptions of the previous proposition will hold if λ_0 is a simple eigenvalue of A_0. But if λ_0 is not simple it may not be possible to find a differentiable eigenvalue function $\lambda(\cdot) : B(A_0, \varepsilon) \to \mathbb{C}$ satisfying $\lambda(A_0) = \lambda_0$. In this case the condition number can no longer be expressed via a derivative as in (3). However, our definitions (1), (2) are applicable to the λ_0-group of eigenvalues of $A \in B(A_0, \varepsilon)$ denoted by $\lfloor\lambda_1(A), \ldots, \lambda_{k_0}(A)\rfloor$ (ε sufficiently small). It follows from Remark 4.2.10 (iii) that the condition number of the λ_0-group (considered as a map from $B(A_0, \varepsilon)$ to the metric space of k_0-tuples) is infinite if for some $A_1 \in \mathbb{C}^{n\times n}$ the λ_0-group of $A(z) = A_0 + zA_1$ has a branch point at $z = 0$. Formula (21) shows that under the conditions of the previous proposition the problem of determining the eigenvalue $\lambda_0 = \lambda(A_0)$ will be ill-conditioned if $w^*v \approx 0$. On the other hand, if A_0 is normal we have seen in Section 4.2 that we can choose $w = v$ and in this case Proposition 4.5.5 yields $\kappa(f_E, A_0) = 1$. But $\kappa(f_E, A_0) \geq 1$ for all $A_0 \in \mathbb{C}^{n\times n}$. So of all the matrices satisfying the assumption of the previous proposition the normal ones have the best condition number for determining eigenvalues.

Let $A_0 = B_0^*B_0$, $B_0 \in \mathbb{C}^{m\times n}$, then we have $\kappa(f_E, B_0^*B_0) = 1$ and one might conjecture that the condition number for determining the singular values of B_0 is also 1. We will now see that this indeed the case. Let $\mathbb{C}^{m\times n}$ be provided with the operator norm $\|\cdot\|_{2,2}$, and consider

$$\mathcal{W} = \mathcal{W}_0 = \mathbb{C}^{m\times n}; \quad \mathcal{X} = \mathbb{R}; \quad \sigma_i : \mathcal{W} \to \mathcal{X},\ A \mapsto \sigma_i(A)$$

where σ_i denotes the i-th ordered singular value, $i \in \underline{n}$. From Corollary 4.3.11 we know that for arbitrary $A_0, A \in \mathbb{C}^{m\times n}$ we have

$$|\sigma_i(A) - \sigma_i(A_0)| \leq \|A - A_0\|_{2,2}\,.$$

Thus $\sigma_i(B_{\mathcal{W}}(A_0, \delta)) \subset B_{\mathcal{X}}(\sigma_i(A_0), \delta)$ for all $\delta > 0$. It follows from our definitions (1) and (2) that $\kappa_\delta(\sigma_i, A_0) \leq 1$ for all $\delta > 0$ and so $\kappa(\sigma_i, A_0) \leq 1$. Since the i-th ordered singular value is zero for all $A \in \mathbb{C}^{m\times n}$ if $\min\{m, n\} < i \leq n$, it is only of interest to determine the condition number of the first $\min\{m, n\}$ singular values.

Proposition 4.5.6. *For $i = 1, \ldots, \min\{m, n\}$ and $\delta > 0$, the condition number and the δ-condition number of the i-th ordered singular value of $A_0 \in \mathbb{C}^{m\times n}$ are both equal to 1, i.e. $\kappa(\sigma_i, A_0) = \kappa_\delta(\sigma_i, A_0) = 1$.*

Proof: It only remains to prove that $\kappa_\delta(\sigma_i, A_0) \geq 1$ if $1 \leq i \leq \min\{m, n\}$, and this is left to the reader. □

4.5.2 Matrix Transformations

In this subsection we introduce *Householder transformations* and show how they can be used (i) to reduce a matrix $A \in \mathbb{K}^{n\times n}$ to Hessenberg form, (ii) to reduce a matrix by pre- and post-multiplication to bidiagonal form, and (iii) to factorize A into a unitary (resp. orthogonal) matrix and an upper triangular one. Moreover, we prove that every $A \in \mathbb{C}^{n\times n}$ can be reduced by unitary similarity transformations to upper triangular complex Schur form and every $A \in \mathbb{R}^{n\times n}$ can be reduced by orthogonal similarity transformations to upper quasi-triangular real Schur form. These pseudo-canonical forms are not uniquely determined for a given matrix A, but play a fundamental role in Numerical Linear Algebra and in the numerics of Linear Control. In particular, the eigenvalues of A can be read directly from the diagonal entries (blocks) of the complex (resp. real) Schur form of A. An iterative algorithm for reducing a given matrix $A \in \mathbb{R}^{n\times n}$ to real Schur form will be presented in the next subsection.

Throughout this subsection we only consider unitary (resp. orthogonal) transformations. The reason for this is that unitary transformations do not increase the size of perturbations of A if these are measured by the spectral or the Frobenius norms. Indeed, these two matrix norms are *unitarily invariant* in the sense that

$$\|UAV\|_{2,2} = \|A\|_{2,2}\,, \quad \|UAV\|_F = \|A\|_F, \qquad U, V \in \mathbf{U}_n(\mathbb{K}).$$

Therefore, if A is perturbed to $A + \Delta$ where $\Delta \in \mathbb{K}^{n\times n}$ is arbitrary, the resulting perturbation $\hat{A} \rightsquigarrow \hat{A} + \hat{\Delta} = \hat{A} + U\Delta V$ of the transformed matrix $\hat{A} = UAV$ satisfies $\|\hat{\Delta}\|_{2,2} = \|\Delta\|_{2,2}$ and $\|\hat{\Delta}\|_F = \|\Delta\|_F$. Whereas if $S, T \in \mathbb{K}^{n\times n}$ are not unitary and $\hat{\Delta} = S\Delta T$, then one has $\|\hat{\Delta}\|_{2,2} \le \|S\|_{2,2}\|\Delta\|_{2,2}\|T\|_{2,2}$ and $\|\hat{\Delta}\|_F \le \|S\|_{2,2}\|\Delta\|_F\|T\|_{2,2}$. In particular, for $S = T^{-1}$, $T \in \mathbf{Gl}_n(\mathbb{K})$ we get

$$\|\hat{\Delta}\|_{2,2} \le \|T\|_{2,2}\|\Delta\|_{2,2}\|T^{-1}\|_{2,2}\,, \qquad \|\hat{\Delta}\|_F \le \|T\|_{2,2}\|\Delta\|_F\|T^{-1}\|_{2,2}, \tag{22}$$

indicating that there may be numerical problems if $\kappa(T) = \|T\|_{2,2}\|T^{-1}\|_{2,2}$ is large.

Remark 4.5.7. In order to determine the Jordan form of a given matrix A, general similarity transformations must be used. The above considerations indicate why it may be risky to apply numerical methods which are based on the computation of Jordan normal forms. Any Jordan normal form of A depends discontinuously on A in any neighbourhood of a non-diagonalizable matrix. The $n \times n$ diagonalizable matrices are dense in $\mathbb{R}^{n\times n}$ and therefore changes of some entries in a non-diagonalizable matrix, however small, can radically alter its Jordan block structure. Also the matrix of eigenvectors of a matrix which is nearly non-diagonalizable can be poorly conditioned. For instance, any matrix $T \in \mathbf{Gl}_n(\mathbb{K})$ which diagonalizes $\begin{bmatrix} 1+\varepsilon & 1 \\ 0 & 1-\varepsilon \end{bmatrix}$, $0 < \varepsilon \ll 1$ has condition number $\kappa(T) = \|T\|_{2,2}\|T^{-1}\|_{2,2}$ of order ε^{-1}. □

Householder transformations

Householder transformations are rank one modifications of the identity and take the following form

$$V = I_n - 2vv^* \; \in \mathbb{K}^{n\times n}, \tag{23}$$

where $v \in \mathbb{K}^n$, $\|v\|_2 = 1$. Note that $V = V^*$ and

$$VV^* = I_n - 4vv^* + 4vv^*vv^* = I_n,$$

so V is a unitary involution (orthogonal in the real case). Geometrically $x \mapsto Vx$ describes the reflection at the hyperplane $(\mathbb{K}v)^\perp$ in the Euclidean space $\mathbb{K}^n$. Therefore V is also called a *Householder reflection*, and v is called the associated *Householder vector*.

A basic step in the reduction of a $n \times n$ matrix A to Hessenberg form (or in the construction of a QR factorization of A) consists in transforming a given vector $c \in \mathbb{K}^n$ to a multiple of the first standard unit vector e^1 of the same norm.[2] In the real case if $c \in \mathbb{R}^n \setminus \mathbb{R}e^1$, this can be achieved by *reflecting* the vector c across either the hyperplane H^+ or the hyperplane H^-, where $H^+ = \{x \in \mathbb{R}^n;\ \langle x, c - \|c\|_2 e^1\rangle = 0\}$ and $H^- = \{x \in \mathbb{R}^n;\ \langle x, c + \|c\|_2 e^1\rangle = 0\}$. This yields two possibilities for the Householder vector v, $v = v^+/\|v^+\|_2$ or $v = v^-/\|v^-\|_2$, where $v^+ = c - \|c\|_2 e^1$, $v^- = \|c\|_2 e^1 + c$, see Figure 4.5.1. Mathematically either v^+ or v^- will suffice, but

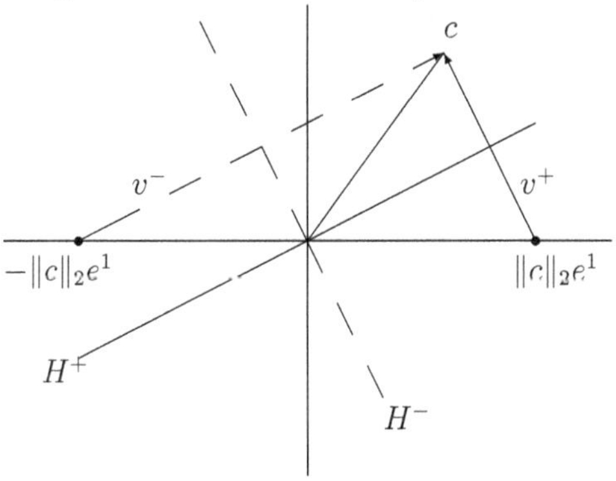

Figure 4.5.1: Householder reflections

from a numerical point of view it is best to choose the one which has the same order of magnitude as c. For, if say v^+ is of much smaller magnitude than c, then the calculation of $v^+/\|v^+\|_2$ may suffer errors. So if c is very close to $\|c\|_2 e^1$ one would choose $v^- \approx 2c$ instead of v^+.

In the complex case a Householder reflection which transforms a non-zero vector $c \in \mathbb{C}^n$ into a multiple of e^1 can be constructed as follows (see Ex. 8). If $c = [c_1, ..., c_n]^\top$ and $c_1 = |c_1|e^{\imath\theta}$, $\theta \in [0, 2\pi)$, define $u = c + e^{\imath\theta}\|c\|_2 e^1$ and $v = u/\|u\|_2$. Then the Householder transformation $V = I_n - 2vv^*$ maps c to $Vc = -e^{\imath\theta}\|c\|_2 e^1$.

Transformation to Hessenberg form

A matrix $H = (h_{ij})_{i,j \in \underline{n}} \in \mathbb{K}^{n \times n}$ is said to be in *Hessenberg form* if $h_{ij} = 0$ for all $i, j \in \underline{n}$ such that $i - j \geq 2$. For any given matrix $A \in \mathbb{K}^{n \times n}$ our objective is

[2] While Householder reflections can be used to introduce zeros at all but the first components of a vector $c \in \mathbb{R}^n$, selected components of c may be zeroed utilizing so-called Givens rotations $G = (g_{ij})$ which differ from the identity by a 2×2 principal submatrix where $g_{ii} = g_{jj} = \cos\theta, g_{ij} = -g_{ji} = -\sin\theta$, $1 \leq i < j \leq n$. These matrices describe a rotation of θ degrees in the (i, j) coordinate plane, see *Notes and References*.

to construct a unitary matrix $U \in \mathbf{U}_n(\mathbb{K})$ such that $H := U^*AU$ is in Hessenberg form. This will be accomplished via $n-2$ Householder transformations. Note that if $n \leq 2$, then A is already in Hessenberg form, so we may assume that $n \geq 3$. Suppose that $k \geq 1$ and after $k-1$ iterations A_k has the form

$$A_k = \left[\begin{array}{ccccc|ccccc} * & * & \cdots & * & * & * & . & . & . & * \\ * & * & & . & . & . & & & & . \\ 0 & * & & . & . & . & & & & . \\ \vdots & \vdots & \ddots & \vdots & \vdots & . & . & . & . & . \\ 0 & 0 & \cdots & * & * & * & . & . & . & * \\ \hline 0 & 0 & \cdots & 0 & \star & * & . & . & . & * \\ \vdots & \vdots & & \vdots & \vdots & \vdots & & & & \vdots \\ 0 & 0 & \cdots & 0 & \star & * & . & . & . & * \end{array}\right] = \left[\begin{array}{ccc|c} & H_k & & F_k \\ \hline 0 & \cdots & 0 \; c^k & C_k \end{array}\right] \tag{24}$$

where $c^k \in \mathbb{K}^{n-k}$, $F_k \in \mathbb{K}^{k\times(n-k)}$, $C_k \in \mathbb{K}^{(n-k)\times(n-k)}$ and $H_k \in \mathbb{K}^{k\times k}$ is in Hessenberg form. The entries of A_k marked by $\star$ are the components of the vector c^k which is to be transformed into a multiple of $e^1 \in \mathbb{K}^{n-k}$ by a suitable Householder transformation. Note that for $k = 1$, (24) takes the form $A_1 := A = \left[\begin{array}{c|c} a_{11} & F_1 \\ \hline c^1 & C_1 \end{array}\right]$ where $a_{11} \in \mathbb{K}$, $c^1 \in \mathbb{K}^{n-1}$, $F_1 \in \mathbb{K}^{1\times(n-1)}$, $C_1 \in \mathbb{K}^{(n-1)\times(n-1)}$. If c^k is a multiple of $e^1 \in \mathbb{K}^{n-k}$, set $v^{k+1} = 0_{n-k}$ and $W_{k+1} = I_n$. Otherwise choose $v^{k+1} \in \mathbb{K}^{n-k}$ such that

$$(I_{n-k} - 2v^{k+1}(v^{k+1})^*)c^k \quad \text{is a multiple of} \quad e^1 \in \mathbb{K}^{n-k} \quad \text{and} \quad \|v^{k+1}\|_2 = 1. \tag{25}$$

Set $V_{k+1} = I_{n-k} - 2v^{k+1}(v^{k+1})^*$ and

$$W_{k+1} = \left[\begin{array}{c|c} I_k & 0 \\ \hline 0 & V_{k+1} \end{array}\right] = \left[\begin{array}{c|c} I_k & 0 \\ \hline 0 & I_{n-k} - 2v^{k+1}(v^{k+1})^* \end{array}\right] \in \mathbf{U}_n(\mathbb{K}). \tag{26}$$

Then define

$$A_{k+1} = W_{k+1}A_kW_{k+1} = \left[\begin{array}{ccc|c} & H_k & & F_kV_{k+1} \\ \hline 0 & \cdots & 0 \; f^k & V_{k+1}C_kV_{k+1} \end{array}\right],$$

where $f^k = V_{k+1}c^k = [*, 0, \cdots, 0]^\top \in \mathbb{K}^{n-k}$. Note that $A_{k+1} = (a_{ij}^{(k+1)})$ is in Hessenberg form up to the $(k+1)$-th column, i.e. $a_{ij}^{(k+1)} = 0$ for $j \leq k+1$, $i-j \geq 2$. Iterating this step we obtain the following lemma.

Lemma 4.5.8. *Given $A = (a_{ij}) \in \mathbb{K}^{n\times n}$, then A can be reduced to Hessenberg form $H = U^*AU$ with $U \in \mathbf{U}_n(\mathbb{K})$. In the following algorithm this is achieved by means of a sequence of $n-2$ Householder transformations.*

1. Init: Set $k = 1$, $A_1 = A$, $U_1 = I_n$.

2. **while** *$(k < n-1)$*
- *2.1 Write A_k in the form (24).*
- *2.2* **If** *c^k is a multiple of $e^1 \in \mathbb{K}^{n-k}$, set $v^{k+1} = 0_{n-k}$ and $W_{k+1} = I_n$.*
- *2.3* **Else** *determine the Householder vector $v^{k+1} \in \mathbb{K}^{n-k}$ and W_{k+1} such that (25) and (26) hold.*
- *2.4 Put $A_{k+1} = W_{k+1}A_kW_{k+1}$, $U_{k+1} = U_kW_{k+1}$.*
- *2.5 Set $k = k+1$.*

3. End: Return $H = A_{n-1}$ and $U = U_{n-1}$.

The above algorithm requires $\approx (14/3)n^3$ flops.

Remark 4.5.9. (i) Note that by construction the first column (resp. first row) of U is the first standard unit vector e^1 of $\mathbb{K}^n$ (resp. its transpose).

(ii) If in the above iteration one of the terms $c^k = 0$, then A_k, will have a block triangular structure $A_k = \begin{bmatrix} H_k & F_k \\ 0 & C_k \end{bmatrix}$ with H_k in Hessenberg form. Continuing with the algorithm the resulting Hessenberg matrix H will also have a block triangular structure with two square blocks on the diagonal which are both in Hessenberg form. When this is the case the Hessenberg matrix H is said to be *reduced.* □

The Hessenberg form is not unique. In fact, if $A \in \mathbb{K}^{2\times 2}$ then A is in Hessenberg form and for any $U \in \mathbf{U}_2(\mathbb{K})$, U^*AU is another Hessenberg form of A. However, we have the following restricted uniqueness result.

Proposition 4.5.10. *Suppose $A \in \mathbb{K}^{n\times n}$ and $U = [u^1, ..., u^n]$, $\tilde{U} = [\tilde{u}^1, ..., \tilde{u}^n] \in \mathbf{U}_n(\mathbb{K})$ are such that $H = (h_{ij}) = U^*AU$ and $G = (g_{ij}) = \tilde{U}^*A\tilde{U}$ are in Hessenberg form. If H is unreduced and $u^1 = e^{\imath\theta_1}\tilde{u}^1$ for some $\theta_1 \in [0, 2\pi)$ then necessarily G is unreduced and $u^j = e^{\imath\theta_j}\tilde{u}^j$ for some $\theta_j \in [0, 2\pi)$, $j = 2, ..., n$. Moreover*

$$G = DHD^* \quad \text{where } D \text{ is of the form} \quad D = \operatorname{diag}(e^{\imath\theta_1}, ..., e^{\imath\theta_n}), \;\; \theta_j \in [0, 2\pi). \quad (27)$$

(If $\mathbb{K} = \mathbb{R}$ we have $\theta_j \in \{0, \pi\}$ for $j \in \underline{n}$ in (27)).

Proof: Let $Q = [q^1, ..., q^n] \in \mathbf{U}_n(\mathbb{K})$ be the matrix defined by $Q = \tilde{U}^*U$. Then

$$GQ = \tilde{U}^*A\tilde{U}\tilde{U}^*U = \tilde{U}^*AU = \tilde{U}^*UU^*AU = QH.$$

Hence the $(j-1)$-th column of GQ, $j = 2, \ldots, n$ is given by

$$Gq^{j-1} = [q^1, ..., q^n]h^{j-1} = h_{j\,j-1}q^j + \sum_{k=1}^{j-1} h_{k\,j-1}q^k.$$

Thus q^j can be represented as a linear combination of $q^1, ..., q^{j-1}$ and Gq^{j-1}. Since $q^1 = e^{\imath\theta_1}e^1$ for some $\theta_1 \in [0, 2\pi)$ ($q^1 = \pm e^1$ in the real case), it follows by induction that $[q^1, ..., q^n]$ is upper triangular. But since Q is unitary it must be diagonal so that $q^j = e^{\imath\theta_j}e^j$ for some $\theta_j \in [0, 2\pi)$, $j \in \underline{n}$ ($q^j = \pm e^j$ in the real case). Now by definition $u^j = \tilde{U}q^j$ and so $u^j = e^{\imath\theta_j}\tilde{u}^j$, $j \in \underline{n}$ ($= \pm e^j$ in the real case). Finally we have $G = QHQ^*$, so (27) follows by setting $D = Q$. □

The essence of the proposition is that if $U^*AU = H$, $\tilde{U}^*A\tilde{U} = G$ with H an unreduced Hessenberg matrix and U and $\tilde{U}$ have the same first column, then H and G are "essentially equal" in the sense that (27) holds.

Transformation to bidiagonal form

In the next subsection we will see that the first step in the Golub-Kahan-Reinsch algorithm for computing the singular values of a matrix is to reduce the matrix to bidiagonal form. A matrix $B = (b_{ij}) \in \mathbb{K}^{m\times n}$ is in bidiagonal form if $b_{ij} = 0$ for all $i \in \underline{m}$, $j \in \underline{n}$ such that $i - j \geq 1$, $j - i \geq 2$. For a given $A \in \mathbb{K}^{m\times n}$ our objective is to find

$P \in \mathbf{U}_m(\mathbb{K})$, $Q \in \mathbf{U}_n(\mathbb{K})$ such that $B := PAQ$ is in bidiagonal form. The algorithm we describe in Lemma 4.5.11 proceeds by alternately creating zeros in the columns and rows by pre- and post-multiplying the matrix A by Householder transformations. Suppose that after $k \leq \min\{m, n\}$ iterations A has been transformed into a matrix of the form

$$A_k = \left[\begin{array}{cccccc|cccccc} * & * & 0 & . & . & 0 & 0 & 0 & . & . & . & 0 \\ 0 & * & * & 0 & . & 0 & 0 & . & . & . & . & 0 \\ 0 & 0 & * & * & . & 0 & . & . & . & . & . & . \\ . & . & . & . & . & . & . & . & . & . & . & . \\ 0 & . & . & 0 & * & * & 0 & 0 & . & . & . & 0 \\ 0 & . & . & . & 0 & * & * & 0 & . & . & . & 0 \\ \hline 0 & . & . & . & . & 0 & \star & . & . & . & . & * \\ \vdots & & & & & \vdots & \vdots & & & & & \vdots \\ 0 & . & . & . & . & 0 & \star & . & . & . & . & * \end{array}\right] = \left[\begin{array}{c|cccc} & 0 & . & . & 0 \\ & . & . & . & . \\ B_k & . & . & . & . \\ & f_{k,k+1} & . & . & 0 \\ \hline 0 & & & C_k & \end{array}\right] \tag{28}$$

where $B_k \in \mathbb{K}^{k \times k}$ is bidiagonal, $f_{k,k+1} \in \mathbb{K}$ and $C_k \in \mathbb{K}^{(m-k)\times(n-k)}$. Note that for $k = 0$ the matrix B_k is void and $A_0 = C_0 = A$. On the other hand if $k = n$, the submatrices on the right hand side of the vertical line are empty and A_n is in bidiagonal form. Similarly, if $k = m$ the submatrices below the horizontal line are empty and A_m is in bidiagonal form. Now suppose $k < \min\{m, n\}$. The entries of A_k marked by $\star$ form a vector (the first column c^k of C_k) which is to be transformed into a multiple of $e^1 \in \mathbb{R}^{m-k}$ by a suitable Householder transformation. If c^k is a multiple of $e^1 \in \mathbb{K}^{m-k}$, set $v^k = 0_{m-k}$.[3] Otherwise choose $v^k \in \mathbb{K}^{m-k}$ such that

$$(I_{m-k} - 2v^k(v^k)^*)c^k \quad \text{is a multiple of} \quad e^1 \in \mathbb{K}^{m-k} \quad \text{and} \quad \|v^k\|_2 = 1. \tag{29}$$

Set

$$W_k = \left[\begin{array}{c|c} I_k & 0 \\ \hline 0 & I_{m-k} - 2v^k(v^k)^* \end{array}\right] \in \mathbf{U}_m(\mathbb{K}). \tag{30}$$

Then define

$$A_{k+1/2} = W_k A_k = \left[\begin{array}{c|cccc} & 0 & . & . & 0 \\ & . & . & . & . \\ B_k & . & . & . & . \\ & f_{k,k+1} & . & . & 0 \\ \hline 0 & & (I_{m-k} - 2v^k(v^k)^*)C_k & & \end{array}\right] = \left[\begin{array}{c|c} B_{k+1} & \begin{array}{c} 0 \\ \hline (f^{k+1})^* \end{array} \\ \hline 0 & \hat{C}_k \end{array}\right] \tag{31}$$

where $B_{k+1} \in \mathbb{K}^{(k+1)\times(k+1)}$ is bidiagonal, $f^{k+1} \in \mathbb{K}^{n-k-1}, \hat{C}_k \in \mathbb{K}^{(m-k-1)\times(n-k-1)}$. If $k+1 = n$ then $A_{k+1/2}$ is bidiagonal. If $k+2 = n$ then $A_{k+1} := A_{k+1/2}$ is of the form (28) with $k+1$ instead of k. If $k+2 < n$ we post-multiply $A_{k+1/2}$ by a Householder transformation to zero out all but the first element of f^{k+1}. Choose $\hat{v}^{k+1} \in \mathbb{K}^{n-k-1}$ so that

$$(I_{n-k-1} - 2\hat{v}^{k+1}\hat{v}^{k+1*})f^{k+1} \text{ is a multiple of } e^1 \in \mathbb{R}^{n-k-1} \text{ and } \|\hat{v}^{k+1}\|_2 = 1. \tag{32}$$

---

[3]Note that this is always the case if $k = m - 1$.

Set

$$\hat{W}_{k+1} = \left[\begin{array}{c|c} I_{k+1} & 0 \\ \hline 0 & I_{n-k-1} - 2\hat{v}^{k+1}(\hat{v}^{k+1})^* \end{array}\right] \in \mathbf{U}_n(\mathbb{K}). \tag{33}$$

Then A_{k+1} defined by

$$A_{k+1} = A_{k+1/2}\hat{W}_{k+1} = \left[\begin{array}{c|c} B_{k+1} & \begin{array}{c} 0 \\ \hline (f^{k+1})^*(I_{n-k-1} - 2\hat{v}^{k+1}(\hat{v}^{k+1})^*) \end{array} \\ \hline 0 & \hat{C}_k(I_{n-k-1} - 2\hat{v}^{k+1}(\hat{v}^{k+1})^*) \end{array}\right], \tag{34}$$

is of the form (28) with $k+1$ instead of k. This leads to the following lemma.

Lemma 4.5.11. *Given $A \in \mathbb{K}^{m\times n}$, $m \geq n$, $m > 1$,*[4] *then A can be reduced to bidiagonal form $B = PAQ$ with $P \in \mathbf{U}_m(\mathbb{K})$, $Q \in \mathbf{U}_n(\mathbb{K})$. This is achieved by means of a sequence of $2n-2$ if $m > n > 1$ or $2n-3$ if $m = n > 1$ Householder transformations as follows:*

1. *Init: Set $A_0 = A$, $P_0 = I_m$, $Q_0 = I_n$, $k = 0$.*
2. **while** *$(k < \min\{n, m-1\})$*
   - 2.1 *Write A_k in the form (28) (with $C_0 = A$ for $k = 0$).*
   - 2.2 **If** *the first column c^k of C_k is a multiple of $e^1 \in \mathbb{K}^{m-k}$, set $v^k = 0_{m-k}$.*
   - 2.3 **Else** *determine the Householder vector $v^k \in \mathbb{K}^{m-k}$ for c^k so that (29) holds.*
   - 2.4 *Put $A_{k+1/2} = W_k A_k, P_{k+1} = W_k P_k$, where W_k is given by (30).*
   - 2.5 **If** *$k < n-2$*
     - 2.5.1 *Write $A_{k+1/2}$ in the form (31).*
     - 2.5.2 **If** *f^{k+1} is a multiple of $e^1 \in \mathbb{K}^{n-k-1}$ set $\hat{v}^{k+1} = 0_{n-k-1}$.*
     - 2.5.3 **Else** *determine the Householder vector $\hat{v}^{k+1} \in \mathbb{K}^{n-k-1}$ for f^{k+1} so that (32) holds.*
     - 2.5.4 *Put $A_{k+1} = A_{k+1/2}\hat{W}_{k+1}$, $Q_{k+1} = Q_k\hat{W}_{k+1}$ where $\hat{W}_{k+1}$ is given by (33).*
   - 2.6 **Else** *Set $A_{k+1} = A_{k+1/2}$, $Q_{k+1} = Q_k$.*
   - 2.7 *Set k=k+1.*
3. *End: Return $B = A_k$, $P = P_k$, $Q = Q_k$.*

The algorithm requires $\approx 4mn^2 - (4/3)n^3$ flops for the computation of B and an additional $\approx 4m^2n - (4/3)n^3$ (resp. $\approx 4n^3$) flops for the computation of P (resp. Q).

QR factorization

For $A \in \mathbb{K}^{n\times n}$ our objective is to determine $Q \in \mathbf{U}_n(\mathbb{K})$ and $R \in \mathbb{K}^{n\times n}$ upper triangular such that $A = QR$. This will be accomplished via $n-1$ Householder

[4] Essentially the same algorithm can also be used for the case $m < n$, but the stopping condition and number of Householder transformations will, of course, be different.

transformations. Let $n \geq 2$ and suppose that after $k \geq 0$ iterations A_k has the form

$$A_k = \left[\begin{array}{cccccc|ccccc} * & * & \cdots & \cdots & * & * & * & \cdot & \cdot & \cdot & * \\ 0 & * & \cdots & \cdots & \cdot & \cdot & \cdot & & & & \cdot \\ \vdots & \ddots & \ddots & \ddots & & \vdots & & & \cdot & & \\ \cdot & \cdot & \cdot & * & & * & \cdot & \cdot & \cdot & \cdot & \cdot \\ 0 & \cdot & \cdots & 0 & & * & * & \cdot & \cdot & \cdot & * \\ \hline 0 & \cdot & \cdots & \cdot & & 0 & \star & * & \cdot & \cdot & * \\ \vdots & & & & & \vdots & \vdots & & & & \vdots \\ 0 & \cdot & \cdots & \cdot & & 0 & \star & * & \cdot & \cdot & * \end{array}\right] = \left[\begin{array}{c|c} R_k & F_k \\ \hline 0 & C_k \end{array}\right] \tag{35}$$

where $F_k \in \mathbb{K}^{k\times(n-k)}$, $C_k \in \mathbb{K}^{(n-k)\times(n-k)}$, and $R_k \in \mathbb{K}^{k\times k}$ is in triangular form. Note that for $k = 0$ the matrices R_k and F_k are void and $A_0 = C_0$. The algorithm proceeds by reducing all but the first components of the first column c^k of C_k to zero via a suitable Householder transformation. If c^k is already in this form set $v^{k+1} = 0_{n-k}$ and $W_{k+1} = I_n$. Otherwise choose $v^{k+1} \in \mathbb{K}^{n-k}$, $\|v^{k+1}\|_2 = 1$ and W_{k+1} so that (25) and (26) hold and set

$$A_{k+1} = W_{k+1}A_k = \left[\begin{array}{c|c} R_k & F_k \\ \hline 0 & (I_{n-k} - 2v^{k+1}v^{k+1\top})C_k \end{array}\right] = \left[\begin{array}{c|c} R_{k+1} & F_{k+1} \\ \hline 0 & C_{k+1} \end{array}\right].$$

Then the $(k+1) \times (k+1)$ principal submatrix of A_{k+1} is in triangular form. Repeating this step for $k = 0, \ldots, n-2$, we obtain the following lemma.

Lemma 4.5.12. *Given $A \in \mathbb{K}^{n\times n}$, $n \geq 2$ then A can be factorized so that $A = QR$ with $Q \in \mathbf{U}_n(\mathbb{K})$ and R upper triangular. This is achieved by means of a sequence of $n-1$ Householder transformations as in the following algorithm.*

1. *Init: Set $k = 0$ and $A_0 = A$, $Q_0 = I_n$.*
2. **while** *$(k < n-1)$*
   - 2.1 *Write A_k in the form (35) ($C_0 = A$ for $k = 0$).*
   - 2.2 **If** *the first column of C_k, c^k is a multiple of $e^1 \in \mathbb{K}^{n-k}$ then set $v^{k+1} = 0_{n-k}$ and $W_{k+1} = I_n$.*
   - 2.3 **Else** *determine the Householder vector $v^{k+1} \in \mathbb{K}^{n-k}$ for c^k and W_{k+1} so that (25) and (26) hold.*
   - 2.4 *Put $A_{k+1} = W_{k+1}A_k$, $Q_{k+1} = W_{k+1}Q_k$ and set $k = k+1$.*
3. *End: Return $R = A_{n-1}$ and $Q = Q^*_{n-1}$.*

In the real case the above algorithm requires $\approx (4/3)n^3$ flops.

Remark 4.5.13. If $A = [a^1, \ldots, a^n] \in \mathbb{K}^{n\times n}$ is nonsingular and we have a QR factorization $A = QR$, then the columns $q^1, \ldots, q^n$ of $Q = AR^{-1}$ form an orthonormal basis of $\mathbb{K}^n$ satisfying $\text{span}\{q^1, \ldots, q^j\} = \text{span}\{a^1, \ldots, a^j\}$ for $j = 1, \ldots, n$. Conversely, a QR factorization of A can be obtained by applying an orthonormalization procedure to the columns of A. Recall that the Gram-Schmidt procedure determines an orthonormal sequence of vectors q^j, $j = 1, \ldots, n$ from a linearly independent sequence of vectors a^j, $j = 1, \ldots, n$ by computing sequentially for $j = 1, \ldots, n$

$$f^j = a^j - \sum_{i=1}^{j-1} \langle a^j, q^i \rangle q^i, \qquad q^j = f^j / \|f^j\|_2.$$

If $Q = [q^1, \dots, q^n]$, $A = [a^1, \dots, a^n]$ and $T = (t_{ij})$ where $t_{ij} = \langle a^j, q^i \rangle$ for $j = 2, \dots, n$, $i = 1, \dots, j-1$, and $t_{ij} = 0$ otherwise, then

$$Q \operatorname{diag}(\|f^1\|_2, \dots, \|f^n\|_2) = [f^1, \dots, f^n] = A - QT.$$

Hence $A = QR$, where $R = \operatorname{diag}(\|f^1\|_2, \dots, \|f^n\|_2) + T$ is upper triangular and $Q \in \mathbf{U}_n(\mathbb{K})$. So if A is nonsingular it is possible to find a QR factorization via the Gram-Schmidt orthonormalization process. Unfortunately this method has poor numerical properties in that there may be a severe loss of orthogonality among the computed q^j. It is possible to rearrange the calculations into what is known as a *modified* Gram-Schmidt algorithm which has certain advantages over the algorithm given in the above lemma. But again it may suffer a loss of orthogonality in Q in comparison with the Householder approach if the condition number $\kappa(A) = \|A\|_{2,2} \|A^{-1}\|_{2,2}$ is large, see *Notes and References*. □

Schur form

Every matrix $A \in \mathbb{C}^{n \times n}$ can be converted into upper triangular form (complex Schur form) by unitary similarity transformations. This is a key result from Numerical Linear Algebra which is also of special importance in the numerics of Linear Systems Theory. There does not exist an exact counterpart for the real case, since a matrix $A \in \mathbb{R}^{n \times n}$ with non-real eigenvalues cannot be transformed into real upper triangular form by any similarity transformations. However, it can be reduced by an orthogonal similarity transformation $U \in \mathbf{O}_n$ into upper *quasi-triangular* form (real Schur form)

$$U^\top A U = S = \begin{bmatrix} S_{11} & S_{12} & \cdots & S_{1\ell} \\ 0 & S_{22} & \cdots & S_{2\ell} \\ \vdots & \vdots & \ddots & \vdots \\ 0 & 0 & \cdots & S_{\ell\ell} \end{bmatrix} \in \mathbb{R}^{n \times n}, \tag{36}$$

where each diagonal block $S_{ii}, i \in \underline{\ell}$ is either a real 1×1 matrix consisting of an eigenvalue of A or a real 2×2 matrix with eigenvalues a complex pair $\lambda, \overline{\lambda} \in \sigma(A)$. In the sequel we will derive these two basic results. Since the proof for the real case is slightly more complicated than for the complex one, we give a detailed proof for the real case, and leave the details for the complex case as an exercise.

Theorem 4.5.14 (Real Schur form). *Suppose $A \in \mathbb{R}^{n \times n}$ and $\lambda_1, \dots, \lambda_n$ is a given ordering of the eigenvalues of A accounting for multiplicities where we assume that none of the complex conjugate pairs of eigenvalues is separated by the numbering. Then there exists an orthogonal matrix $U \in \mathbf{O}_n$ such that $U^\top A U$ has the quasi-triangular form* (36) *where the eigenvalues of the blocks $S_{11}, \dots, S_{\ell\ell}$ occur in the prescribed order.*

Proof: The proof is by induction. Suppose we have already constructed $U_k \in \mathbf{O}_n$ such that

$$U_k^\top A U_k = \left[\begin{array}{c|c} S_k & * \\ \hline 0 & A_{n-k} \end{array} \right], \qquad A_{n-k} \in \mathbb{R}^{(n-k) \times (n-k)}, \tag{37}$$

where S_k is in real Schur form with the k eigenvalues $\lambda_1, \ldots \lambda_k$ of the diagonal blocks occurring in the prescribed order. For $k = 0$, (37) holds with S_0 void, if we set $A_n = A$ and $U_0 = I_n$.

It follows from the assumption that $\lambda_{k+1} \in \sigma(A_{n-k})$. First suppose that λ_{k+1} is a *real* eigenvalue of A_{n-k} with some corresponding normalized eigenvector $x \in \mathbb{R}^{n-k}$, $\|x\|_2 = 1$. Choose $z^i \in \mathbb{R}^{n-k}$, $i = 2, \ldots, k$ so that $\|z^i\|_2 = 1$ and $(x,\, z^2, \ldots, z^{n-k})$ is an orthonormal basis for $\mathbb{R}^{n-k}$, then

$$A_{n-k}\,[x\; z^2\; \ldots\; z^{n-k}] = [x\; z^2\; \ldots\; z^{n-k}] \begin{bmatrix} \lambda_{k+1} & * & \cdots & * \\ 0 & * & \cdots & * \\ \vdots & \vdots & \vdots & \vdots \\ 0 & * & \cdots & * \end{bmatrix}.$$

So if $V_{n-k} = [x\; z^2\; \ldots\; z^{n-k}]$, then V_{n-k} is orthogonal and

$$V_{n-k}^\top A_{n-k} V_{n-k} = \left[\begin{array}{c|c} \lambda_{k+1} & * \\ \hline 0 & A_{n-k-1} \end{array}\right]$$

for some $A_{n-k-1} \in \mathbb{R}^{(n-k-1)\times(n-k-1)}$. Setting $U_{k+1} = U_k \begin{bmatrix} I_k & 0 \\ 0 & V_{n-k} \end{bmatrix}$, we obtain

$$U_{k+1}^\top A U_{k+1} = \left[\begin{array}{c|c} S_k & * \\ \hline 0 & V_{n-k}^\top A_{n-k} V_{n-k} \end{array}\right] = \left[\begin{array}{c|c|c} S_k & * & * \\ \hline 0 & \lambda_{k+1} & * \\ \hline 0 & 0 & A_{n-k-1} \end{array}\right] = \left[\begin{array}{c|c} S_{k+1} & * \\ \hline 0 & A_{n-k-1} \end{array}\right],$$

where S_{k+1} is in real Schur form with the $k+1$ eigenvalues $\lambda_1, \ldots \lambda_{k+1}$ of the diagonal blocks occurring in the prescribed order.

Now suppose that λ_{k+1} is a *non-real* eigenvalue of A_{n-k}, then λ_{k+1} and $\overline{\lambda_{k+1}} = \lambda_{k+2}$ form a pair of complex conjugate eigenvalues of A_{n-k}. If $x, \overline{x} \in \mathbb{C}^{n-k}$ is a pair of corresponding eigenvectors then

$$A_{n-k}\,[\operatorname{Re} x\;\; \operatorname{Im} x] = [\operatorname{Re} x\;\; \operatorname{Im} x] \begin{bmatrix} \operatorname{Re}\lambda_{k+1} & \operatorname{Im}\lambda_{k+1} \\ -\operatorname{Im}\lambda_{k+1} & \operatorname{Re}\lambda_{k+1} \end{bmatrix}.$$

Now choose an orthonormal basis $(v^1,\, v^2)$ of $\operatorname{span}\{\operatorname{Re} x,\, \operatorname{Im} x\} \subset \mathbb{R}^{n-k}$ and let $\alpha, \beta, \gamma, \delta \in \mathbb{R}$ such that

$$\begin{array}{rcl} \operatorname{Re} x & = & \alpha\, v^1 + \beta\, v^2 \\ \operatorname{Im} x & = & \gamma\, v^1 + \delta\, v^2 \end{array}, \quad \text{i.e.} \quad [\operatorname{Re} x\;\; \operatorname{Im} x] = [v^1\;\; v^2] \begin{bmatrix} \alpha & \gamma \\ \beta & \delta \end{bmatrix},$$

then

$$A_{n-k}\,[v^1\, v^2] = A_{n-k}\,[\operatorname{Re} x\;\; \operatorname{Im} x] \begin{bmatrix} \alpha & \gamma \\ \beta & \delta \end{bmatrix}^{-1} = [v^1\, v^2]\, A_{22}$$

where

$$A_{22} = \begin{bmatrix} \alpha & \gamma \\ \beta & \delta \end{bmatrix} \begin{bmatrix} \operatorname{Re}\lambda_{k+1} & \operatorname{Im}\lambda_{k+1} \\ -\operatorname{Im}\lambda_{k+1} & \operatorname{Re}\lambda_{k+1} \end{bmatrix} \begin{bmatrix} \alpha & \gamma \\ \beta & \delta \end{bmatrix}^{-1}.$$

Let $v^3, \ldots, v^{n-k}$ be normalized vectors such that $v^1,\, v^2, \ldots,\, v^{n-k}$ is an orthonormal basis for $\mathbb{R}^{n-k}$, then $V_{n-k} = [v^1\, v^2 \ldots v^{n-k}] \in \mathbf{O}_{n-k}$, and we have

$$A_{n-k} V_{n-k} = V_{n-k} \left[\begin{array}{c|c} A_{22} & * \\ \hline 0 & A_{n-k-2} \end{array}\right], \quad A_{22} \in \mathbb{R}^{2\times 2}, \quad \sigma(A_{22}) = \{\lambda_{k+1},\, \overline{\lambda_{k+1}}\}$$

where $A_{n-k-2} \in \mathbb{R}^{(n-k-2)\times(n-k-2)}$. Setting $U_{k+2} = U_k \begin{bmatrix} I_k & 0 \\ 0 & V_{n-k} \end{bmatrix}$ we get

$$U_{k+2}^\top A U_{k+2} = \left[\begin{array}{c|c} S_k & * \\ \hline 0 & V_{n-k}^\top A_{n-k} V_{n-k} \end{array}\right] = \left[\begin{array}{c|c|c} S_k & * & * \\ \hline 0 & A_{22} & * \\ \hline 0 & 0 & A_{n-k-2} \end{array}\right] = \left[\begin{array}{c|c} S_{k+2} & * \\ \hline 0 & A_{n-k-2} \end{array}\right]$$

where S_{k+2} is in real Schur form with the $k+2$ eigenvalues $\lambda_1, \dots, \lambda_{k+2}$ of the diagonal blocks occurring in the prescribed order. □

Remark 4.5.15. One should note that the 2×2 blocks in (36) are similar, but cannot in general be chosen equal to $\begin{bmatrix} \mathrm{Re}\,\lambda & \mathrm{Im}\,\lambda \\ -\mathrm{Im}\,\lambda & \mathrm{Re}\,\lambda \end{bmatrix}$, $\lambda \in \sigma(A)$. The reader is asked to provide a counterexample in Ex. 9. □

The reduction of $A \in \mathbb{C}^{n\times n}$ by unitary transformations to complex Schur form can be accomplished in exactly the same way as in the first part of the above proof using unitary instead of orthogonal transformations (Ex. 10).

Theorem 4.5.16 (Complex Schur form). *Suppose $A \in \mathbb{C}^{n\times n}$ and $\lambda_1, \dots, \lambda_n$ is a given ordering of the eigenvalues of A accounting for multiplicities. Then there exists a unitary matrix $U \in \mathbf{U}_n(\mathbb{C})$ such that*

$$U^* A U = \begin{bmatrix} \lambda_1 & * & \cdots & * \\ 0 & \lambda_2 & \cdots & * \\ \vdots & \vdots & \ddots & \vdots \\ 0 & 0 & \cdots & \lambda_n \end{bmatrix} \in \mathbb{C}^{n\times n}. \tag{38}$$

As is the case of the Hessenberg form neither the orthogonal matrix U nor the Schur form S constructed in the proof of Theorem 4.5.14 are unique, see Ex. 10. Note however that if $S = U^*AU = \mathrm{diag}(\lambda_1, \dots, \lambda_n) + N$ is a Schur decomposition of A with N strictly upper triangular, then $\|N\|_F$ is independent of the choice of U. In fact, since the Frobenius norm of a matrix is not changed by unitary pre- or post-multiplications, it follows that $\|S\|_F = \|A\|_F$, hence

$$\|N\|_F^2 = \|A\|_F^2 - \sum_{i=1}^{n} |\lambda_i|^2.$$

The expression on the RHS is called the *departure of A from normality* with respect to the Frobenius norm.

4.5.3 Algorithms

In this subsection we give details of algorithms for computing the eigenvalues and singular values of a real matrix and also for computing the solution of a Liapunov equation. The most widely used algorithm for reducing a matrix to Schur form is Francis' double–shift QR algorithm ([166], [167]). This is an iterative algorithm which, after first reducing A to Hessenberg form, performs at each iteration a QR factorization. We describe the essential features of this algorithm and the simplifications that occur when A is symmetric. Computation of the Schur form of A will yield the eigenvalues of A and when applied to the symmetric matrix $A^\top A$ it will yield the singular values of A. However we will see that working with $A^\top A$ can cause discrepancies so we also describe the Golub-Kahan-Reinsch algorithm which computes the singular values directly from a bidiagonal matrix which is unitarily equivalent to A, see Lemma 4.5.11. Finally we describe the Bartels and Stewart algorithm for solving Liapunov equations which in a first step reduces A to Schur form. For the precise coding of the algorithms, see *Notes and References.*

Solution of linear equations

We only make some very brief remarks about algorithms for solving linear equations. If $A \in \mathbb{R}^{n\times n}$, $B \in \mathbb{R}^{n\times m}$ the matrix equation $AX = B$ is usually solved by computing the solutions of m linear equations of the form $Ax^i = b^i$ where $b^i, i \in \underline{m}$ are the columns of B. All the books we recommend in *Notes and References* have extensive sections on algorithms for solving linear equations of the form $Ax = b$, $A \in \mathbb{R}^{n\times n}$, $b \in \mathbb{R}^n$. The algorithms are basically divided into two groups, direct and iterative. Direct methods such as Gaussian elimination compute the exact answer after a finite number of iterations (in the absence of roundoff errors). In general the methods proceed by decomposing $A = LU$ via Householder transformations into a real lower triangular matrix $L = (l_{ij})$ with $l_{ii} = 1$, and a real upper triangular matrix U.[5] Then the triangular equations $Ly = b$, $Ux = y$ are solved. The LU factorization costs $\approx (1/3)n^3$ flops and the solution of the triangular equations $\approx (1/2)n^2$ flops. Iterative methods produce a sequence of vectors (x^k) which converge to the desired solution. Since they use the matrix A only in matrix-vector multiplications they are suitable for the solution of *large* systems of linear equations. Jacobi and Gauss-Seidel iterations and Krylov subspace methods are classical algorithms in this group, see for example *Datta* (1995) [121], *Trefethen and Bau* (1997) [500] and *Demmel* (1997) [126]. They are particularly useful for large *sparse* matrices and typically cost $O(n^2)$ flops.
If the condition number of A, $\kappa(A)$, is large then one would expect difficulties in both direct and iterative methods. Suppose $M \in \mathbb{R}^{n\times n}$ is invertible, then solving $Ax = b$ is equivalent to solving $MAx = Mb$. It may be possible to improve the conditioning of the problem by a suitable choice of M. Finding a good *pre-conditioner* M is usually a difficult task. For a survey, see [500]. A different approach to ill-conditioned or even singular linear equations is based on regularization methods, see *Notes and References.*

The QR Algorithm

The QR algorithm is the most widely used general purpose eigenvalue algorithm. It is an iterative algorithm where for a given $n \times n$ matrix A a sequence (A_k) is constructed which, under mild conditions (see [525]), converges to a Schur form of A. Contrary to the procedure in the proof of Theorem 4.5.14 which constructs such a Schur form in a finite number of steps, the QR algorithm does not presuppose knowledge of the eigenvalues of A. In the following we will describe the main features of the algorithm in real arithmetic. There is a complex arithmetic analogue working

[5] This is always possible if the n principal minors of A do not vanish.

for complex matrices. Let $A \in \mathbb{R}^{n\times n}$ and $n \geq 3$. The purpose of the QR algorithm is to determine $U \in \mathbf{O}_n$ such that $U^\top AU = S$ is in real Schur form. In each iteration there are two main steps.

Step 1: Given $A_k \in \mathbb{R}^{n\times n}$, compute a QR factorization $Q_kR_k = A_k$.

Step 2: Compute $A_{k+1} = R_kQ_k$.

As a consequence $A_{k+1} = R_kQ_k = Q_k^\top A_kQ_k$, so that A_{k+1} and A_k are orthogonally similar. In general each QR factorization costs $\approx (4/3)n^3$ flops. However if A_k is in Hessenberg form the cost is reduced to $\approx 6n^2$ flops. Moreover the Hessenberg form is inherited by A_{k+1}. In fact, if A_k is in *unreduced* Hessenberg form, the first $n-1$ columns of A_k, $a_k^1, \dots, a_k^{n-1}$ must be linearly independent. But since $A_k = Q_kR_k$ with R_k triangular, one has $\text{span}\{a_k^1, ..., a_k^j\} \subset \text{span}\{q_k^1, ..., q_k^j\}$, $j = 1, ..., n-1$ where q_k^j are the columns of Q_k. But $\{a_k^1, ..., a_k^j\}$ is linearly independent and hence $\text{span}\{a_k^1, ..., a_k^j\} = \text{span}\{q_k^1, ..., q_k^j\}$, for $j = 1, ..., n-1$. Thus each column q_k^j of Q_k is a linear combination of $a_k^1, \dots, a_k^j$. Hence Q_k is in Hessenberg form and since $A_{k+1} = R_kQ_k$, so is A_{k+1}. Thus after a single first cost of $\approx (14/3)n^3$ flops required to reduce A to Hessenberg form, the remaining iterations need only $\approx 6n^2$ flops. Steps 1 and 2 are iterated provided each A_k is *unreduced*, $k = 0, 1, \dots$. If at some stage A_k is in reduced form, then the problem is deflated by applying the algorithm to the unreduced Hessenberg submatrices. It is therefore necessary to decide when a subdiagonal term can be neglected. A natural criterion is to choose a small ε and set a subdiagonal term[6] $A_k(j+1, j)$ to zero if $|A_k(j+1,j)| \leq \varepsilon\|A\|_F$. Then the error in computing $A_{k+1} = Q_k^\top A_kQ_k$ assuming perfect computation is $E = Q_k^\top A_k(j+1,j)e^{j+1}e^{j\top}Q_k$ and the relative error is $\|E\|_F/\|A\|_F \leq \varepsilon$.
In order to speed up convergence, shifts are employed, i.e. the above two steps are replaced by

Step 1: Given $A_k \in \mathbb{R}^{n\times n}$ and $\alpha_k \in \mathbb{R}$, compute $Q_k \in \mathbf{O}_n$, R_k triangular so that $Q_kR_k = A_k - \alpha_kI_n$.

Step 2: Compute $A_{k+1} = R_kQ_k + \alpha_kI_n$.

If A_k is in Hessenberg form then so is $A_k - \alpha_kI_n$. Moreover, since

$$A_{k+1} = R_kQ_k + \alpha_kI_n = Q_k^\top(Q_kR_k + \alpha_kI_n)Q_k = Q_k^\top A_kQ_k,$$

we see that once again A_{k+1} and A_k are orthogonally similar. The idea is to try to choose α_k close to a real eigenvalue of A_k. Indeed if α_k is a real eigenvalue then $A_k - \alpha_kI_n$ is singular, so R_k is also singular and hence must have a diagonal element equal to zero. If, in this case, $A_k - \alpha_kI_n$ is in *reduced* Hessenberg form, we may deflate the problem in iteration k and continue with the unreduced Hessenberg submatrices. If, however, $A_k - \alpha_kI_n$ is in *unreduced* Hessenberg form, then as shown above Q_k is in Hessenberg form. Thus R_kQ_k and consequently $A_{k+1} = R_kQ_k + \alpha_kI_n$ are in *reduced* Hessenberg form. So we may deflate the problem in iteration $k+1$ and continue with the unreduced Hessenberg submatrices. In practice, of course, we do not know the eigenvalues of A_k. Indeed they are the very objects that we are

[6] Here we adopt the MATLAB notation for matrix entries. If M is any matrix we denote by $M(i,j)$ the entry of M at position (i,j).

trying to compute. For the moment let us suppose that the block $S_{\ell\ell}$ in the Schur form S of A (36) is the 1×1 matrix consisting of the real eigenvalue λ_n. Then if A_k converges to S we must have $A_k(n, n-1) \to 0$ and $A_k(n,n) \to \lambda_n$. So $\alpha_k = A_k(n,n)$ is a good choice for the shift and in fact will yield local quadratic convergence, see [126].

We are left with the question of how to choose α_k to accelerate the convergence to a *complex* eigenvalue. For this we assume that A_k is in *unreduced* Hessenberg form, since otherwise we may deflate the problem and apply the following considerations to the unreduced diagonal blocks of A_k. If α_k is chosen to be complex then the arithmetic used in Steps 1 and 2 must be complex and this may increase the computational cost by a factor of about 6 over that of real arithmetic. However suppose that for complex α_k the above single shifts are replaced by double shifts as follows:

$$\begin{aligned} Q_k' R_k' &= A_k - \alpha_k I_n, & A_{k+1/2} &= R_k' Q_k' + \alpha_k I_n, \\ Q_k'' R_k'' &= A_{k+1/2} - \overline{\alpha_k} I_n, & A_{k+1} &= R_k'' Q_k'' + \overline{\alpha_k} I_n, \end{aligned} \tag{39}$$

where $Q_k', Q_k'' \in \mathbf{U}_n(\mathbb{C})$ and R_k', R_k'' triangular. Then $A_{k+1/2} = Q_k'^* A_k Q_k'$, $A_{k+1} = Q_k''^* A_{k+1/2} Q_k''$ and hence $A_{k+1} = Q_k''^* Q_k'^* A_k Q_k' Q_k''$. Now since $Q_k'' R_k'' = A_{k+1/2} - \overline{\alpha_k} I_n = R_k' Q_k' + (\alpha_k - \overline{\alpha_k}) I_n$, we have

$$\begin{aligned} Q_k' Q_k'' R_k'' R_k' &= Q_k'(R_k' Q_k' + (\alpha_k - \overline{\alpha_k}) I_n) R_k' = (Q_k' R_k' + (\alpha_k - \overline{\alpha_k}) I_n) Q_k' R_k' \\ &= (A_k - \alpha_k I_n + (\alpha_k - \overline{\alpha_k}) I_n)(A_k - \alpha_k I_n) = (A_k - \overline{\alpha_k} I_n)(A_k - \alpha_k I_n). \end{aligned}$$

So $(Q_k' Q_k'')(R_k'' R_k')$ is a QR decomposition of the *real* Hessenberg matrix $M_k := (A_k - \overline{\alpha_k} I_n)(A_k - \alpha_k I_n)$. This suggests that one should use a real QR decomposition of this matrix, i.e. $M_k = Q_k R_k$ with $Q_k \in \mathbf{O}_n$, R_k real upper triangular, and define $A_{k+1} = Q_k^\top A_k Q_k$. This is the essence of the idea put forward by Francis and is reflected in the epithet "double" in the title of the algorithm.

Unfortunately in order to compute Q_k and R_k one must first compute the real matrix $M_k = A_k^2 - 2(\operatorname{Re}\alpha_k) A_k + |\alpha_k|^2 I_n$ and this step requires $O(n^3)$ flops. In order to circumvent this, one uses the so-called *implicit QR iterations* which require $O(n^2)$ flops. These are based on Proposition 4.5.10 and proceed as follows:

Step 1: Compute the first column $b^k = M_k e^1$ of M_k (see below) and find a Householder transformation $W \in \mathbb{R}^{n\times n}$ such that $W b^k \in \mathbb{R} e^1$.

Step 2: Set $A' = W A_k W$.

Step 3: Reduce A' to Hessenberg form $H' = U^\top A' U$ by an orthogonal transformation $U \in \mathbf{O}_n$, see Lemma 4.5.8.

Step 4: Set $W_k = WU$ and $A_{k+1} = W_k^\top A_k W_k$.

It follows from equation (40) below that $b^k \notin \mathbb{R} e^1$, since A_k is not reduced and so $A_k(2,1) A_k(3,2) \neq 0$. In particular, $b^k \neq 0$ and $W e^1 \in W(\mathbb{R} W b^k) = \mathbb{R} b^k$. Now by construction $U e^1 = e^1$ (see Remark 4.5.9) and hence the first column of W_k is $W_k e^1 = W e^1 \in \mathbb{R} b^k$. But if $M_k = Q_k R_k$ were a real QR factorization of M_k, the first column of the orthogonal matrix Q_k would also be a unit vector in $\mathbb{R} b^k$. Hence $W_k e^1 = \pm Q_k e^1$ and so by Proposition 4.5.10 the constructed W_k is essentially equal to Q_k provided the Hessenberg matrix $A_{k+1} = W_k^\top A_k W_k$ is unreduced. So the calculation of the matrix Q_k can be replaced by the Steps 1-4.

Since A_k is in Hessenberg form, only the first three components of the first column of A_k^2 (and hence of M_k) are possibly non-zero and these are

$$\begin{bmatrix} A_k(1,1) & A_k(1,2) \\ A_k(2,1) & A_k(2,2) \\ 0 & A_k(3,2) \end{bmatrix} \begin{bmatrix} A_k(1,1) \\ A_k(2,1) \end{bmatrix} = \begin{bmatrix} A_k(1,1)^2 + A_k(1,2)A_k(2,1) \\ A_k(2,1)A_k(1,1) + A_k(2,1)A_k(2,2) \\ A_k(2,1)A_k(3,2) \end{bmatrix}.$$

Hence b^k is of the form $b^k = [b^\top, 0_{n-3}^\top]^\top$ where

$$b := \begin{bmatrix} b_1^k \\ b_2^k \\ b_3^k \end{bmatrix} = \begin{bmatrix} A_k(1,1)^2 + A_k(1,2)A_k(2,1) - 2\operatorname{Re}\alpha_k A_k(1,1) + |\alpha_k|^2 \\ A_k(2,1)A_k(1,1) + A_k(2,1)A_k(2,2) - 2\operatorname{Re}\alpha_k A_k(2,1) \\ A_k(2,1)A_k(3,2) \end{bmatrix}. \tag{40}$$

This calculation can be carried out in $O(1)$ flops and means that the matrix W in Step 1 has the form $W = \begin{bmatrix} V & 0 \\ 0 & I_{n-3} \end{bmatrix}$ where $V \in \mathbb{R}^{3\times 3}$ is a Householder transformation such that $Vb = [a, 0, 0]^\top$ for some $a \in \mathbb{R}$. As a consequence the matrix A' has the following structure

$$A' = \left[\begin{array}{c|c} B & C \\ \hline f^\top & \\ \cline{1-1} 0 & H \end{array}\right], \quad B \in \mathbb{R}^{3\times 3},\ C \in \mathbb{R}^{3\times(n-3)},\ f \in \mathbb{R}^3,\ H \in \mathbb{R}^{(n-3)\times(n-3)},$$

where H is in Hessenberg form. This reduces the computational burden in Step 3. The result is that Step 3 can be carried out in $O(n^2)$ flops rather than the $O(n^3)$ flops required to reduce a general matrix to Hessenberg form.

As in the case of a single real shift one would like to choose the complex shift α_k close to a complex eigenvalue of A_k. But the eigenvalues of A_k are not known. Now suppose that the block $S_{\ell\ell}$ in the Schur form of A is a 2×2 matrix which has complex eigenvalues. Then if A_k converges to S, we have $A_k(n-1, n-2) \to 0$ and the 2×2 blocks

$$A_k(\ell) = \begin{bmatrix} A_k(n-1,n-1) & A_k(n-1,n) \\ A_k(n,n-1) & A_k(n,n) \end{bmatrix}$$

converge to $S_{\ell\ell}$ as $k \to \infty$. So it seems a good idea to choose α_k to be an eigenvalue of $A_k(\ell)$ and this choice will in fact yield local quadratic convergence.

Note that in order to compute the vector b^k we only need to know $2\operatorname{Re}\alpha_k$ and $|\alpha_k|^2$, and these are the trace t and determinant d of $A_k(\ell)$, if we choose $\alpha_k \in \sigma(A_k(\ell))$. This allows us to use real arithmetic. We do not need to check that $A_k(\ell)$ has in fact non-real eigenvalues, i.e. $d > t^2/4$, since the double shift algorithm also works if the eigenvalues of $A_k(\ell)$ are real. Alternatively, in this case a single shift may be employed and then it is usual to employ the *Wilkinson shift*, i.e. the real eigenvalue of $A_k(\ell)$ which is closest to $A_k(n,n)$.

After first balancing[7] the matrix A and transforming A into Hessenberg form, the QR algorithm usually starts by choosing shifts based on the 2×2 block in the bottom right hand corner. Deflation occurs at iteration k if $A_k(n-1, n-2)$ is sufficiently

[7] Balancing is briefly discussed in the next subsubsection.

small and when this is the case the process is repeated for the Hessenberg submatrix obtained by removing rows and columns $n-1$ and n. Finally all 2×2 blocks that have real eigenvalues are upper triangularized. In practice the QR algorithm with double shifts almost always converges and on average only two shifts are required before a reduction occurs. However there are small sets of matrices for which this is not the case. When these are discovered, "exceptional shifts" are introduced to patch the algorithm.
We now discuss the application of the QR algorithm to three different problems.

Determining the eigenvalues of a matrix

Since the middle of the 20th century there have been many significant advances in the numerical resolution of the eigenvalue problem, see *Notes and References.* The essential feature of most of these methods is that a given matrix $A \in \mathbb{R}^{n\times n}$ is reduced by orthogonal transformations to real Schur form S.
The power and inverse power are particularly simple methods for approximating eigenvalues and eigenvectors of a matrix. Moreover they have connections to the QR algorithm which indicate why the sequence of iterates of the QR algorithm converges. To explain this let us assume that $A \in \mathbb{R}^{n\times n}$ is diagonalizable with eigenpairs $(\lambda_i,\, x^i)_{i\in\underline{n}}$, $\|x^i\|_2 = 1$ and that A has a unique simple eigenvalue of maximum magnitude

$$|\lambda_1| > |\lambda_2| \geq |\lambda_3| \geq \ldots \geq |\lambda_n|. \tag{41}$$

The power method is particularly suited for finding the dominant eigenvalue λ_1 and an associated eigenvector of A. (To approximate other eigenvalues, one may use shifts, see below.) For every $z^0 = \sum_{i=1}^n \gamma_i x^i \in \mathbb{C}^n$, $\gamma_i \in \mathbb{C}$, we have

$$z^k := A^k z^0 = \sum_{i=1}^n \gamma_i A^k x^i = \sum_{i=1}^n \gamma_i \lambda_i^k x^i = \lambda_1^k \left(\gamma_1 x^1 + \sum_{i=2}^n \gamma_i \left(\frac{\lambda_i}{\lambda_1}\right)^k x^i \right), \qquad k \in \mathbb{N}.$$

Because of (41) the second term in the parenthesis tends to zero since $(\lambda_i/\lambda_1)^k \to 0$ as $k \to 0$. As a consequence we have $z^k \approx \lambda_1^k \gamma_1 x^1$ for large k and if $\gamma_1 \neq 0$ then the associated *Rayleigh quotients*[8] $\rho_k := \langle Az^k, z^k\rangle / \langle z^k, z^k\rangle$ tend to λ_1. The rate of convergence will depend mainly on the ratio $|\lambda_\ell/\lambda_1|$ where ℓ is the smallest index $i \geq 2$ such that $\gamma_i \neq 0$ and convergence will be slow if $|\lambda_\ell/\lambda_1|$ is close to 1.
Because $(\lambda_1^k)_{k\in\mathbb{N}}$ tends to zero or is unbounded if $|\lambda_1| \neq 1$, it is convenient to scale the sequence (z^k). This leads us to consider, for any given $z^0 \in \mathbb{C}^n$ with $\gamma_1 \neq 0$, the normalized sequence (y^k) and the associated sequence of Rayleigh quotients defined by

$$y^0 = z^0/\|z^0\|_2, \quad y^{k+1} = Ay^k/\|Ay^k\|_2, \quad \rho_k = \langle Ay^k, y^k\rangle, \quad k \in \mathbb{N}.$$

It is easily seen by induction that in fact $y^k = z^k/\|z^k\|_2$, $k \in \mathbb{N}$, and therefore the Rayleigh quotients of y^k and z^k coincide. As a consequence we obtain (if $\gamma_1 \neq 0$)

$$|\rho_k - \lambda_1| \to 0, \quad \|y^k - (\lambda_1/|\lambda_1|)^k\, (\gamma_1/|\gamma_1|)\, x^1\|_2 \to 0 \quad \text{as } k \to \infty, \tag{42}$$

i.e. (ρ_k) approximates the dominant eigenvalue of A and the sequence (y^k), although not necessarily convergent, approximates the eigenspace spanned by x^1.

[8]The Rayleigh quotient $\langle Az, z\rangle/\langle z, z\rangle$ of a vector $z \in \mathbb{K}^n$, $z \neq 0$ can also be characterized as the value of $\mu \in \mathbb{K}$ which minimizes $\|Az - \mu z\|_2$, see Ex. 14.

If A is invertible, the power method can be applied to A^{-1}, provided there is a unique and simple eigenvalue of A of smallest magnitude, i.e. $|\lambda_n| < |\lambda_{n-1}|$. To approximate different simple eigenvalues shifts are used. For any $\alpha \in \mathbb{C}$, the inverse power method applied to the shifted matrix $A - \alpha I_n$ and the vector z^0 proceeds as follows

$$y^0 = z^0/\|z^0\|_2,\ (A-\alpha I_n)z^{k+1} = y^k,\ \rho_k = \langle z^{k+1}, y^k\rangle,\ y^{k+1} = z^{k+1}/\|z^{k+1}\|_2,\ k \in \mathbb{N} \quad (43)$$

where the initial vector $z^0 \in \mathbb{K}^n$, $z^0 \neq 0$ is arbitrary. Since the eigenvalues of $(A - \alpha I_n)^{-1}$ are $(\lambda_i - \alpha)^{-1}$, the inverse power method will yield the eigenvalue closest to α, provided this eigenvalue is unique and simple.

Now suppose that we are at the kth iteration in the QR algorithm with single, possibly complex, shifts α_k. For simplicity of notation we write (A, A_1, α, Q, R) instead of $(A_k, A_{k+1}, \alpha_k, Q_k, R_k)$ and set $A_\alpha = A - \alpha I_n$, $Q = [Q_1, q]$, $R^* = [R_1, r_{nn}e^n]$ with $q \in \mathbb{C}^n$, $r_{nn} \in \mathbb{C}$. The QR step can be written in the form

$$\begin{bmatrix} Q_1^* \\ q^* \end{bmatrix} A_\alpha = \begin{bmatrix} R_1^* \\ \overline{r_{nn}} e^{n\top} \end{bmatrix},\ A_1 = Q^* A_\alpha Q + \alpha I_n = \begin{bmatrix} Q_1^* A_\alpha Q_1 + \alpha I_{n-1} & Q_1^* A_\alpha q \\ q^* A_\alpha Q_1 & q^* A_\alpha q + \alpha \end{bmatrix}. \quad (44)$$

If $\alpha \notin \sigma(A)$, then $q = r_{nn}(A_\alpha^*)^{-1}e^n$ and since $\|q\|_2 = 1$ we have $|r_{nn}| = \|(A_\alpha^*)^{-1}e^n\|_2^{-1}$. This shows that the last column q of the matrix Q generated by a QR factorization of A_α is the result of an inverse power iteration (43) with the matrix A_α^* applied to the vector $e^{\imath\theta}e^n$ where $r_{nn} = e^{\imath\theta}|r_{nn}|$, $\theta \in [0, 2\pi)$. So by the above power and inverse power analysis one might expect that q is a good approximation of an eigenvector of A_α^*. To explore this further we again assume that A is diagonalizable, but now $(x^i)_{i\in\underline{n}}$, $\|x^i\|_2 = 1$ are eigenvectors of $A^\top$ with eigenvalues $\overline{\lambda_i}$. Then $A_\alpha^* x^i = (\overline{\lambda_i - \alpha})x^i$, $i \in \underline{n}$ and by (44)

$$q^* A_\alpha = \overline{r_{nn}} e^{n\top} = e^{-\imath\theta}|r_{nn}|e^{n\top} = \frac{e^{-\imath\theta}e^{n\top}}{\|(A_\alpha^*)^{-1}e^n\|_2}.$$

Suppose that $e^n = \sum_{i=1}^n \gamma_i x^i$, $\gamma_i \in \mathbb{C}$, $\gamma_1 \neq 0$, then, if we set $\varepsilon = \lambda_1 - \alpha$,

$$\|(A_\alpha^*)^{-1}e^n\|_2 = \|\gamma_1 \overline{\varepsilon}^{-1} x^1 + \sum_{i=2}^n \gamma_i (\overline{\lambda_i - \alpha})^{-1} x^i\|_2.$$

Hence

$$\|q^* A_\alpha\|_2 = |\varepsilon||\gamma_1|^{-1}|\,\|x^1 + \overline{\varepsilon}\sum_{i=2}^n (\gamma_i/\gamma_1)(\overline{\lambda_i - \alpha})^{-1} x^i\|_2^{-1}$$

Now suppose that λ_1 is a simple eigenvalue of A and that $|\lambda_1 - \alpha| = |\varepsilon|$ is small with $|\varepsilon| \ll \min_{i=2,\dots,n} |\lambda_i - \alpha| = |\lambda_j - \alpha|$. Then there exists a constant C such that $\|\sum_{i=2}^n (\gamma_i/\gamma_1)(\overline{\lambda_i - \alpha})^{-1} x^i\|_2 \leq C/|\lambda_j - \alpha|$. So if ε is sufficiently small we have

$$\|q^* A_\alpha\|_2 \leq |\varepsilon||\gamma_1|^{-1}(1 - |\varepsilon|C/|\lambda_j - \alpha|)^{-1}.$$

We see therefore that for small ε, the vector q is indeed an approximate eigenvector of A_α^* with approximate eigenvalue zero.

Actually the above analysis also suggests a possible shift for $A_{k+1} = A_1$. By the second equality in (44), we have for $\alpha_1 \in \mathbb{C}$

$$\|e^{n\top} A_1 - \alpha_1 e^{n\top}\|_2^2 = \|q^* A_\alpha Q_1\|_2^2 + |\alpha + q^* A_\alpha q - \alpha_1|^2.$$

Since for small ε the first term on the right hand side in the above expression is small, $\alpha_1 = \alpha + q^* A_\alpha q = q^* A q$ is an approximate eigenvalue of A_1 with approximate left eigenvector e^n. The shift $q^* A q$ is called the *Rayleigh quotient shift* because it is the Rayleigh quotient $\langle AQe^n, Qe^n \rangle$. Note that this is the single shift strategy advocated in the first part of the description of the QR algorithm for the real eigenvalues.

Let us make some comments regarding the computation of the Schur form. Since the eigenvalues can be determined from the Schur form its computation will be *ill-conditioned* for certain matrices. Nevertheless it can be shown that Francis' implicit double-shift QR algorithm is numerically stable in that there exist $\hat{U} \in \mathbf{O}_n$, $\Delta \in \mathbb{R}^{n \times n}$ such that the computed matrix $\tilde{S}$ is in real Schur form with

$$\tilde{S} = \hat{U}^\top (A + \Delta) \hat{U}, \tag{45}$$

where $\|\Delta\|_{2,2}/\|A\|_{2,2}$ is a small multiple of machine precision. The computed matrix $\tilde{U}$ is nearly orthogonal and close to $\hat{U}$, but it need not be close to any U satisfying (36). We see from (45), that

$$\|\tilde{S} - \hat{U}^\top A \hat{U}\|_{2,2} = \|\hat{U}^\top \Delta \hat{U}\|_{2,2} = \|\Delta\|_{2,2}. \tag{46}$$

Although (46) is very satisfactory we cannot conclude that the computed eigenvalues are near to those of A because of the possible ill-conditioning of the eigenvalue problem.

If the eigenvectors are also required they can be computed from the eigenvectors of the Schur form $S = U^\top A U$. For example if S has an eigenpair (λ, x), then A will have an eigenpair (λ, Ux). In computing the eigenvectors of S one uses a back-substitution procedure exploiting the quasi-triangular form of S, see *Notes and References*. The extra cost of determining the eigenvectors is $\approx 6n^2$ flops.

The `eig` function in MATLAB chooses from 16 different algorithms for computing eigenvalues depending on the structure of the matrix, whether it is symmetric or non-symmetric, whether eigenvectors are required etc., see *Notes and References*. For the general non-symmetric case after first pre-conditioning A, MATLAB reduces A to Hessenberg form, then uses a QR algorithm to reach Schur form. There are two types of pre-conditioning: (i) one seeks a permutation matrix P so that $PAP^\top$ is nearly triangular or close to real Schur form, (ii) one tries to *balance* the matrix by seeking a diagonal matrix D so that the rows and columns of $D^{-1}AD$ have approximately equal norms. In general, pre-conditioning the matrix improves the accuracy of the QR iterations. It costs $O(n^2)$ flops.

Determining the singular values of a matrix

Given $A \in \mathbb{R}^{m \times n}$, the squares of the singular values of A are the eigenvalues of $A^\top A$. So Francis' double-shift QR algorithm may be applied to the symmetric matrix $A^\top A$ to obtain the singular values of A. Now by Proposition 4.5.6 the condition number for computing singular values σ_i is $\kappa(\sigma_i, A) = 1$ and since we have just seen that Francis' double-shift QR algorithm is numerically stable, their computation by this method will in general be reliable. There are two simplifications when the QR algorithm is applied to symmetric matrices:

- There is no need to consider complex shifts since $\sigma(A^\top A) \subset \mathbb{R}$.
- If $U^\top A^\top AU = H = (h_{ij})$, with H in Hessenberg form, then H is tridiagonal, i.e. $h_{ij} = 0$, $|i-j| > 1$, $i, j = 1, 2, ..., n$.

Moreover, at each iteration symmetry is preserved and since the single shift QR steps $H_k - \alpha_k I_n = Q_k R_k$, $H_{k+1} = R_k Q_k + \alpha_k I_n$, $\alpha_k \in \mathbb{R}$ also preserve the Hessenberg structure, the tridiagonal structure of the H_k is inherited by H_{k+1}. Now the QR factorization of a tridiagonal matrix requires only $O(n)$ flops, so there is an order of magnitude less work than that involved in computing eigenvalues of non-symmetric matrices.

The symmetric structure can also be used in calculating the shifts α_k. Suppose that H_k is in unreduced tridiagonal Hessenberg form

$$H_k = \begin{bmatrix} h_{11} & h_{12} & 0 & \cdots & 0 \\ h_{21} & h_{22} & \ddots & \ddots & \vdots \\ 0 & \ddots & \ddots & \ddots & 0 \\ \vdots & \ddots & \ddots & h_{n-1n-1} & h_{n-1n} \\ 0 & \cdots & 0 & h_{nn-1} & h_{nn} \end{bmatrix}.$$

Wilkinson suggested that one should choose α_k to be the eigenvalue of the bottom 2×2 block which is closest to h_{nn} and he showed in [526] that when this is the case the algorithm is locally cubically convergent for almost all matrices.

Although it would seem that the scheme that we have described here is very satisfactory, one must be cautious. Consider $A = \begin{bmatrix} 1 & 1 \\ \varepsilon & 0 \\ 0 & \varepsilon \end{bmatrix}$, where ε is so small that to machine precision $1+\varepsilon^2$ is the same as 1. Then $A^\top A = \begin{bmatrix} 1+\varepsilon^2 & 1 \\ 1 & 1+\varepsilon^2 \end{bmatrix}$ and the computed singular values would be $(\sqrt{2}, 0)$, leading to the erroneous conclusion that $\operatorname{rank} A = 1$. The discrepancy occurs because we have worked with $A^\top A$, so introducing unnecessarily ε^2. *Golub and Kahan* (1965) [196] have suggested a way around this by working directly with A and not $A^\top A$. Their algorithm involves a variant of implicit QR iterations and below we describe its essential features.

In the first step Householder transformations are used[9] to reduce A to bidiagonal form (see Lemma 4.5.11):

$$B = \begin{bmatrix} d_1 & e_1 & \cdots & . & 0 \\ 0 & d_2 & \ddots & . & 0 \\ \vdots & \vdots & \ddots & \ddots & \vdots \\ 0 & 0 & \cdots & d_{n-1} & e_{n-1} \\ 0 & 0 & \cdots & 0 & d_n \end{bmatrix}.$$

[9] If $B = PAQ$ with $P \in \mathbf{O}_m$, $Q \in \mathbf{O}_n$ then $B^\top B = Q^\top A^\top P^\top PAQ = Q^\top A^\top AQ$, so that A and B have the same singular values.

Here we have taken $m = n$ since the rows of zeros when $m > n$ do not really play a role. We also assume that all the d_i and e_i are non-zero. This is because if $e_i = 0$ the problem can be deflated. And actually this is also the case if $d_i = 0$ since then it is possible to zero the superdiagonal entry e_{i-1} by pre-multiplication by suitable rotations, see *Notes and References*. It follows therefore that the tridiagonal matrix $B^\top B$ is unreduced. We will describe the first iteration of the algorithm and see later that it amounts to a QR factorization . The 2×2 upper left and lower right submatrices of $B^\top B$ are

$$\begin{bmatrix} d_1^2 & d_1 e_1 \\ d_1 e_1 & d_2^2 + e_1^2 \end{bmatrix}, \qquad \begin{bmatrix} d_{n-1}^2 + e_{n-2}^2 & d_{n-1}e_{n-1} \\ d_{n-1}e_{n-1} & d_n^2 + e_{n-1}^2 \end{bmatrix}.$$

Suppose that α is the Wilkinson shift based on the right hand matrix, then the following two steps are carried out.

Step 1: Choose a Householder vector $v \in \mathbb{R}^2$ with $\|v\|_2 = 1$ such that

$$(I_2 - 2vv^\top)\begin{bmatrix} d_1^2 - \alpha \\ d_1 e_1 \end{bmatrix} = \begin{bmatrix} \beta \\ 0 \end{bmatrix} \quad \text{where } \beta \in \mathbb{R}^* \text{ and set } \quad V = \begin{bmatrix} I_2 - 2vv^\top & 0 \\ 0 & I_{n-2} \end{bmatrix}.$$

Step 2: Find $W_1, W_2 \in \mathbf{O}_n$ with the first column of W_2 equal $e^1 \in \mathbb{R}^n$ such that $\hat{B} = W_1^\top B V W_2$ is bidiagonal.

The fact that Step 2 can indeed be carried out will be demonstrated below. Since Householder transformations preserve Euclidean norms we have $\sqrt{(d_1^2 - \alpha)^2 + d_1^2 e_1^2} = |\beta|$. By construction we have $V(B^\top B - \alpha I_n)e^1 = \beta e^1 = \beta W_2 e^1$. Multiplying these equalities by V from the left and using $V^2 = I_n$, we conclude that the first column of $\beta V W_2$ is equal to the first column of $B^\top B - \alpha I_n$. Moreover, $\hat{B}$ being bidiagonal, the matrix $\hat{B}^\top \hat{B} = W_2^\top V^\top B^\top B V W_2$ is tridiagonal and hence in Hessenberg form. Now if $B^\top B - \alpha I_n = QR$ was a QR factorization, then necessarily the first column of Q would be the first column of $B^\top B - \alpha I_n$ divided by its norm $\pm\beta$. Hence the first column of Q would be equal to $\pm V W_2 e^1$. But $B^\top B - \alpha I_n$ is unreduced and this implies, as we have seen in the description of the QR algorithm, that Q and $Q^\top(B^\top B - \alpha I_n)Q = RQ$ are also in Hessenberg form. Hence it follows from Proposition 4.5.10 (with $A = B^\top B - \alpha I_n$, $U = V W_2$ and $\tilde{U} = Q$) that

$$V W_2 = QD \quad \text{where } D \text{ is of the form} \quad D = \operatorname{diag}(\pm 1, ..., \pm 1).$$

But $(QD)(D^{-1}R)$ would be just another QR factorization of $B^\top B - \alpha I_n$ and so determining $V W_2$ and $\hat{B}$ mimics one iteration of the QR algorithm. The purpose of the algorithm is to achieve deflation by zeroing either e_{n-1} or d_{n-1}. By repeated application of the two steps one constructs a sequence of bidiagonal matrices $(\hat{B}_k)$ with entries $d_{n-1}(k)$ and $e_{n-1}(k)$. From the fact that the QR algorithm with Wilkinson shifts almost always converges we can expect that $d_{n-1}(k)e_{n-1}(k)$ converges to zero and so the problem can be deflated. Criteria for smallness within the bidiagonal band are usually of the form

$$|e_i| \le \varepsilon(|d_i| + |d_{i+1}|), \qquad |d_i| \le \varepsilon \|B\|_F,$$

where ε is a small multiple of machine accuracy. After successive deflations B is eventually reduced to diagonal form. Accumulating all the orthogonal matrices one obtains $U, V \in \mathbf{O}_n$ such that $U^\top AV = \operatorname{diag}(\sigma_1, ..., \sigma_n)$. Hence $U \operatorname{diag}(\sigma_1, ..., \sigma_n)V^\top$ is a singular value decomposition of A.

Let us comment on Step 2. This can be carried out by reducing BV to bidiagonal form by the algorithm described in Subsection 4.5.2. However the fact that B is bidiagonal can be used to simplify the algorithm. In fact both the first column of C_k in (28) and the vector f^{k+1} in (31) will have all zero entries with the exception of the first two. As a consequence the first iterations starting from BV develop in the following way

$$BV = \begin{bmatrix} * & * & 0 & . & . & . & 0 \\ + & * & * & 0 & . & . & 0 \\ 0 & 0 & * & * & 0 & . & 0 \\ \vdots & \vdots & & & & & \vdots \\ 0 & 0 & & & & & * \end{bmatrix} \rightsquigarrow (BV)_{1/2} = \left[\begin{array}{c|cccccc} * & * & + & 0 & . & . & 0 \\ \hline 0 & * & * & 0 & . & . & 0 \\ 0 & 0 & * & * & 0 & . & 0 \\ \vdots & \vdots & & & & & \vdots \\ 0 & 0 & & & & & * \end{array}\right] \rightsquigarrow$$

$$(BV)_1 = \left[\begin{array}{c|cccccc} * & * & 0 & . & . & . & 0 \\ \hline 0 & * & * & 0 & . & . & 0 \\ 0 & + & * & * & 0 & . & 0 \\ \vdots & \vdots & & & & & \vdots \\ 0 & 0 & & & & & * \end{array}\right] \rightsquigarrow (BV)_{3/2} = \left[\begin{array}{cc|ccccc} * & * & 0 & . & . & . & 0 \\ 0 & * & * & + & . & . & 0 \\ \hline 0 & 0 & * & * & 0 & . & 0 \\ \vdots & \vdots & & & & & \vdots \\ 0 & 0 & & & & & * \end{array}\right] \rightsquigarrow$$

So the bulge denoted by $+$ moves from the $(2, 1)$ position to the $(1, 3)$ position and then to the $(3, 2)$ position etc. Successive applications chases the bulge down the subdiagonal until the bidiagonal form is reached.

Note that $(BV)_1$ is obtained from $(BV)_{1/2}$ and $(BV)_2$ is obtained from $(BV)_{3/2}$ by Householder transformations of the form

$$\hat{W}_1 = \left[\begin{array}{c|cc|c} 1 & 0 & 0 & 0 \\ \hline 0 & * & * & 0 \\ 0 & * & * & 0 \\ \hline 0 & 0 & 0 & I_{n-3} \end{array}\right], \qquad \hat{W}_2 = \left[\begin{array}{c|cc|c} I_2 & 0 & 0 & 0 \\ \hline 0 & * & * & 0 \\ 0 & * & * & 0 \\ \hline 0 & 0 & 0 & I_{n-4} \end{array}\right]$$

respectively. The first column of both these matrices is e^1 and this will also be the case for the subsequent matrices $\hat{W}_3, ..., \hat{W}_{n-2}$. Hence $W_2 e^1 = \hat{W}_1 ... \hat{W}_{n-2} e^1 = e^1$.

Remark 4.5.17. The books we cite in *Notes and References* give more comprehensive accounts of the algorithm and describe it in terms of Givens rotations rather than Householder transformations. □

Since the computed singular values can be shown to be the singular values of $A + \Delta$, where $\|\Delta\|_{2,2}/\|A\|_{2,2}$ is a modest multiple of machine precision, the algorithm is numerically stable. It is regarded as the most reliable way of, for example, calculating the rank of a matrix. The flop count is dominated by the first reduction to bidiagonal form and for $m \geq n$ is $\approx 4mn^2 - (4/3)n^3$ if only the singular values are computed and $\approx 4mn^2 + 4m^2n + (8/3)n^3$ if U and V are also required.

Solving Liapunov equations

We describe an algorithm due to *Bartels and Stewart* for solving the Liapunov equation

$$A^*P + PA + Q = 0 \tag{47}$$

where it is assumed that (3.3.88) holds, i.e. $\lambda + \overline{\mu} \neq 0$ for all $\lambda, \mu \in \sigma(A)$. Under this condition (47) has a unique solution P. This algorithm works for both the real and the complex case. Since the real case is more complicated than the complex one we will assume $A, Q \in \mathbb{R}^{n\times n}$ and show how a solution of (47) can be obtained in real arithmetic. The Bartels-Stewart algorithm [38] for the solution of a Liapunov equation (47) proceeds via the following three steps (the notation is the same as that in Theorem 4.5.14).

Step 1: Francis' double-shift QR algorithm is used to transform A into real Schur form $S = U^\top AU$ where $U \in \mathbf{O}_n$.

Step 2: Compute $\hat{Q} = U^\top QU$. Then P is a solution of (47) if and only if $\hat{P} = U^\top PU$ solves

$$S^\top \hat{P} + \hat{P}S + \hat{Q} = 0. \tag{48}$$

Step 3: Compute the solution $\hat{P}$ of (48) and return the solution $P = U\hat{P}U^\top$ of (47).

To compute $\hat{P}$ in Step 3 the quasi-triangular structure of S is utilized in the following way. Let

$$\hat{Q} = [\hat{q}^1, \ldots, \hat{q}^n], \quad S = (s_{ij}) = \begin{bmatrix} S_{11} & S_{12} & \cdots & S_{1\ell} \\ 0 & S_{22} & \cdots & S_{2\ell} \\ \vdots & \vdots & \ddots & \vdots \\ 0 & 0 & \cdots & S_{\ell\ell} \end{bmatrix} \in \mathbb{R}^{n\times n}.$$

where each diagonal block $S_{jj}, j \in \underline{\ell}$ is either a real 1×1 matrix consisting of an eigenvalue of A or a real 2×2 matrix with eigenvalues a complex pair $\lambda, \overline{\lambda} \in \sigma(A)$.

Algorithm 4.5.18. This algorithm computes successively the columns $\hat{p}^1, \ldots, \hat{p}^n$ of $\hat{P}$, determining in each iteration either a single column or a pair according to whether the corresponding diagonal block of S has order one or two.

1. Init: Set $k = 1$.
2. **while** $(k \leq n)$
   2.1 **If** $k = n$ or $s_{k+1k} = 0$ (i.e s_{kk} is a real eigenvalue of A)[10] solve the following linear equation for $\hat{p}^k$

$$[S^\top + s_{kk}I_n]\,\hat{p}^k = -\hat{q}^k - \sum_{i=1}^{k-1} s_{ik}\,\hat{p}^i. \tag{49}$$

   Set $k = k + 1$.

   2.2 **Else** (i.e. s_{kk} belongs to a diagonal 2×2 block of S) solve the following linear equation for $[\hat{p}^k\ \hat{p}^{k+1}]$

$$S^\top[\hat{p}^k\ \hat{p}^{k+1}] + [\hat{p}^k\ \hat{p}^{k+1}] \begin{bmatrix} s_{kk} & s_{kk+1} \\ s_{k+1k} & s_{k+1k+1} \end{bmatrix} = -[\hat{q}^k\ \hat{q}^{k+1}] - \sum_{i=1}^{k-1} [s_{ik}\,\hat{p}^i\ \ s_{ik+1}\,\hat{p}^i]. \tag{50}$$

   Set $k = k + 2$.

[10] Note that in this algorithm k can only take values for which s_{kk} is either a real eigenvalue of A or the upper left entry of a 2×2 block S_{jj}. Hence, for these values of k, the entry s_{kk} is a real eigenvalue of A if and only if $k = n$ or $s_{k+1k} = 0$.

3. Set $\hat{P} = [\hat{p}^1, \ldots, \hat{p}^n]$.
4. End: Return $P = U\hat{P}U^\top$.

Note that in Step 2.1 the vectors $\hat{p}^i$, $i = 1, ..., k-1$ have been computed in previous iterations so that (49) has a unique solution $\hat{p}^k$ since $s_{kk} \in \sigma(S) \cap \mathbb{R}$ and hence $-s_{kk} \notin \sigma(S) = \sigma(S^\top)$ by assumption.
Also in Step 2.2 the vectors $\hat{p}^i$, $i = 1, ..., k-1$ have been calculated in previous iterations so that (50) admits a unique solution $[\hat{p}^k\ \hat{p}^{k+1}]$. In order to see this let $T \in \mathbb{C}^{2\times 2}$ be such that

$$T \begin{bmatrix} s_{kk} & s_{kk+1} \\ s_{k+1k} & s_{k+1k+1} \end{bmatrix} T^{-1} = \begin{bmatrix} \lambda_k & 0 \\ 0 & \overline{\lambda_k} \end{bmatrix}.$$

Multiplying (50) on the right by T^{-1} and setting $[\hat{p}^k\ \hat{p}^{k+1}]T^{-1} = [x\ y]$, $x, y \in \mathbb{C}^n$, the left hand side of (50) becomes $[(S^\top + \lambda_k I_n)x,\ (S^\top + \overline{\lambda_k} I_n)y]$. Since $\lambda_k, \overline{\lambda_k} \in \sigma(A)$ it follows by assumption that $-\lambda_k, -\overline{\lambda_k} \notin \sigma(S^\top)$ and so equation (50) has a unique solution. If we rewrite (50) as a vector equation

$$\begin{bmatrix} S^\top + s_{kk} I_n & s_{k+1k} I_n \\ s_{kk+1} I_n & S^\top + s_{k+1k+1} I_n \end{bmatrix} \begin{bmatrix} \hat{p}^k \\ \hat{p}^{k+1} \end{bmatrix} = - \begin{bmatrix} q^k \\ q^{k+1} \end{bmatrix} - \begin{bmatrix} \sum_{i=1}^{k-1} s_{ik}\, \hat{p}^i \\ \sum_{i=1}^{k-1} s_{ik+1}\, \hat{p}^i \end{bmatrix},$$

and reorder the scalar equations of this system via the permutation

$$(1, N+1, 2, N+2, \ldots, N, 2N),$$

a banded system of linear equations is obtained that can be solved in $O(n^2)$ flops.
In Subsection 4.5.1 we derived various estimates of the condition number for the problem of solving a Liapunov equation. However even if these are small, the above algorithm may not give good results. This is because the computation of the Schur form in Step 1 is not necessarily reliable if the eigenvalue problem is ill-conditioned. Nevertheless the algorithm seems to be the most successful one for solving Liapunov equations. It is the basis of `lyap` in MATLAB, although there A is first reduced to *complex* Schur form, see *Notes and References.*

4.5.4 Exercises

1. Consider the problem of inverting a matrix A in the following setting

$$\mathcal{W} = \mathbb{K}^{n\times n}, \quad \mathcal{W}_0 = \mathbf{Gl}_n(\mathbb{K}), \quad \mathcal{X} = \mathbb{K}^{n\times n} \quad \text{and } f_I(w) = A^{-1},\ w \in \mathcal{W}_0$$

where $\mathcal{W}$ and $\mathcal{X}$ are endowed with Frobenius norms. Prove that $\kappa(f_I, A) = \|A^{-1}\|_{2,2}^2$. Prove that with respect to weighted norms $\|w\|_\mathcal{W} = \|w\|_F/\|A\|_F$, $\|X\|_\mathcal{X} = \|X\|_F/\|A^{-1}\|_F$, $\kappa_{rel}(f_I, A) = \|A^{-1}\|_{2,2}^2 \|A\|_F \|A^{-1}\|_F^{-1}$.
Hint: Use the same method as in the proof of Proposition 4.5.1 with $B = I_n$ fixed.
Calculate these condition numbers for the matrices $\begin{bmatrix} 1 & 0 \\ 1 & 1 \end{bmatrix}$, $\begin{bmatrix} 1 & 1 \\ 1 & 1.1 \end{bmatrix}$.

2. Consider the same problem as that in Ex. 1 but where now $\mathcal{W}$ and $\mathcal{X}$ are endowed with the weighted norms $\|w\|_\mathcal{W} = \|w\|/\|A\|$, $\|X\|_\mathcal{X} = \|X\|/\|A^{-1}\|$ with $\|\cdot\|$ a given operator norm on $\mathbb{K}^{n\times n}$. Prove that $\kappa_{rel}(f_I, A) = \kappa(A) := \|A\|\,\|A^{-1}\|$.
Hint: Use $f_I'(A)\Delta_A = -A^{-1}\Delta_A A^{-1}$ to obtain the estimate $\kappa_{rel}(f_I, A) \le \kappa(A)$ and then choose a particular Δ_A to conclude equality.

3. Suppose $A \in \mathbf{Gl}_n(\mathbb{K})$, $Ax = b \in \mathbb{K}^n$ and $(A + \Delta_A)y = b + \Delta_b$ where $\|A^{-1}\|\,\|\Delta_A\| < 1$ and $\|\cdot\|$ is the operator norm induced by the norm $\|\cdot\|_{\mathbb{K}^n}$ on $\mathbb{K}^n$. Prove that

$$\frac{\|y - x\|_{\mathbb{K}^n}}{\|x\|_{\mathbb{K}^n}} \leq \frac{\kappa(A)}{1 - \|A^{-1}\|\,\|\Delta_A\|}\left(\frac{\|\Delta_A\|}{\|A\|} + \frac{\|\Delta_b\|_{\mathbb{K}^n}}{\|b\|_{\mathbb{K}^n}}\right), \quad \kappa(A) := \|A\|\,\|A^{-1}\|.$$

If z is the computed solution, then $r = b - Az$ is called the residual vector. Show that

$$\frac{\|z - x\|_{\mathbb{K}^n}}{\|x\|_{\mathbb{K}^n}} \leq \kappa(A)\frac{\|r\|_{\mathbb{K}^n}}{\|b\|_{\mathbb{K}^n}}.$$

4. Suppose that $A = \begin{bmatrix} 1 & 0 \\ 2 & -1 \end{bmatrix}$. Let $\mathcal{T}$ be the set of regular real matrices T, such that

$$T^{-1}AT = \begin{bmatrix} 1 & 0 \\ 0 & -1 \end{bmatrix}.$$

Prove that the condition number $\kappa(f_I, A) = \|A^{-1}\|_{2,2}^2$ (see Ex. 1) is bounded above by $\|T\|_{2,2}^2\|T^{-1}\|_{2,2}^2$ for any $T \in \mathcal{T}$. Minimize this upper bound over the set $\mathcal{T}$ and compare your answer with $\kappa(f_I, A)$.

5. Suppose $A \in \mathbb{R}^{n\times n}$ has eigenvalues $\lambda_1, ..., \lambda_n$. Prove that

$$\sum_{i=1}^{n} |\lambda_i|^2 = \inf_{\det T \neq 0} \|T^{-1}AT\|_F^2,$$

and show that A is normal if and only if $\sum_{i=1}^{n} |\lambda_i|^2 = \|A\|_F^2$.

6. Calculate the unweighted condition numbers for solving the Liapunov equations, where $Q = I_2$ and

$$(i)\ A = \begin{bmatrix} 0 & 1 \\ -2 & -3 \end{bmatrix}, \quad (ii)\ A = \begin{bmatrix} -1 & 0 \\ 1 & 1.1 \end{bmatrix}.$$

Also calculate the estimate (14) for the two Liapunov equations.

7. Consider the discrete time Liapunov equation

$$A^\top XA - X + Q = 0. \tag{51}$$

Let $\mathbf{L}_A^D : \mathbb{R}^{n\times n} \to \mathbb{R}^{n\times n}$ be the corresponding Liapunov map $X \mapsto \mathbf{L}_A^D(X) = A^\top XA - X$ and

$$\mathcal{W} = \mathbb{R}^{n\times n} \times \mathbb{R}^{n\times n}, \ \mathcal{W}_0 = \mathcal{W}_1^D \times \mathbb{R}^{n\times n}, \ \mathcal{X} = \mathbb{R}^{n\times n}, \ f_L^D(A,Q) = -(\mathbf{L}_A^D)^{-1}(Q), \ (A,Q) \in \mathcal{W}_0$$

where $\mathcal{W}_1^D = \{A \in \mathbb{R}^{n\times n};\ \forall \lambda, \mu \in \sigma(A) : \lambda\overline{\mu} \neq 1\}$. If $(A,Q) \in \mathcal{W}_0$ and the matrix spaces $\mathcal{W}$, $\mathcal{X}$ are provided with the Frobenius norms

$$\|w\|_{\mathcal{W}} = \|(A,Q)\|_{\mathcal{W}} = \left(\|A\|_F^2 + \|Q\|_F^2\right)^{1/2}, \quad \|X\|_{\mathcal{X}} = \|X\|_F,$$

prove that the condition number for solving (51) satisfies

$$\kappa(f_L^D, A, Q) \leq \|[A^\top \otimes A^\top - I_{n^2}]^{-1}\|_{2,2}\,[1 + \|I_n \otimes A^\top X + (A^\top X \otimes I_n)P\|_{2,2}^2]^{1/2},$$

where P is a suitable permutation matrix.

8. Suppose $c = (c_i) \in \mathbb{C}^n$, $c \neq 0$ and let $c_1 = |c_1|e^{\imath\theta}$ with $\theta \in [0, 2\pi)$. Defining $u = c + e^{\imath\theta}\|c\|_2 e^1$ and $v = u/\|u\|_2$, prove that the Householder transformation $V = I_n - 2vv^* \in \mathbf{U}_n(\mathbb{C})$ maps c to $Vc = -e^{\imath\theta}\|c\|_2 e^1$. Show that V can be expressed as follows

$$V = I_n - \alpha uu^* \text{ where } u = (e^{\imath\theta}(|c_1| + \|c\|_2), c_2, \ldots, c_n)^\top, \quad \alpha = [\|c\|_2(\|c\|_2 + |c_1|)]^{-1}.$$

9. Construct a counterexample illustrating Remark 4.5.15.

10. Prove Theorem 4.5.16 and show by a counterexample that the complex and the real Schur forms of a given matrix are, in general, not uniquely determined.

11. Suppose $S = \begin{bmatrix} A & F \\ 0 & C \end{bmatrix}$ is in Schur form. Prove that if $\lambda \neq \mu$, for all $\lambda \in \sigma(A)$, $\mu \in \sigma(C)$ then there is a matrix T of the form $T = \begin{bmatrix} I & R \\ 0 & I \end{bmatrix}$ such that $T^{-1}ST = \begin{bmatrix} A & 0 \\ 0 & C \end{bmatrix}$.

12. Prove that if λ is an eigenvalue of an unreduced upper Hessenberg matrix $H \in \mathbb{R}^{n\times n}$, then its geometric multiplicity is one.

13. Suppose $A \in \mathbb{R}^{n\times n}$ is singular and is in Hessenberg form. Prove that if $A = QR$ is a QR factorization of A, then $R(n, n) = 0$.

14. Given $A \in \mathbb{K}^{n\times n}$ and $v \in \mathbb{K}^n$, $\|v\|_2 = 1$ show that $\mu = \langle Av, v\rangle$ achieves $\min_{\mu\in\mathbb{K}} \|Av - \mu v\|_2^2 = \|Av\|_2^2 - |\langle Av, v\rangle|^2$.

15. Given $A = \begin{bmatrix} 1 & 1 & 1 \\ 0 & 2 & 1 \\ .9999 & -2 & 0 \end{bmatrix}$ use MATLAB to compute (i) a Schur form (ii) a singular value decomposition and (iii) the solution of the Liapunov equation $PA + A^\top P + I = 0$. Find estimates for the condition numbers for determining (i) and (iii) if the matrix spaces involved in these problems are provided with the Frobenius norm.

4.5.5 Notes and References

For an historical account of numerical problems going back to the 16th century, see *Goldstine* (1977) [195]. Details about floating-point arithmetic can be found in *Goldberg* (1991) [192]. The problems caused by the use of finite arithmetic and rounding errors are discussed in most books cited in the third paragraph.

We have used the definition of condition number due to *Rice* (1966) [434]. For a discussion of numerical stability and conditioning see *Wilkinson* (1965) [524], *Stewart* (1973) [480] and *Stoer and Bulirsch* (1993) [486]. Motivation for using $\kappa(A) = \|A\|\,\|A^{-1}\|$ as a condition number for solving linear equations can be found in most of the books we cite here. The fact that for operator norms $\kappa(A) = \|A\|\,\|A^{-1}\|$ is the weighted norm of the Fréchet derivative of the map $A \mapsto A^{-1}$ is proved in *Higham* (2002) [230]. For a discussion on the problem of computing $\|A^{-1}\|$ see Chapter 14 in [230] and the references therein. Important references for eigenvalue problems are *Wilkinson* (1965) [524] and *Golub and Wilkinson* (1976) [198].

In preparing the material for Subsections 4.5.2 and 4.5.3 we have made use of the comprehensive books by *Datta* (1995) [121], *Golub and Van Loan* (1996) [197], *Higham* (2002) [230], *Demmel* (1997) [126], *Trefethen and Bau* (1997) [500], *Stewart* (1998), (2001) [482], [483] and *Higham and Higham* (2002) [229]. The transformation which now takes his name

and many other aspects of Numerical Analysis can be found in the book by *Householder* (1974) [269]. Details of the modified Gram-Schmidt algorithm (MGS) are given in *Golub and Van Loan* (1996) [197]. *Björck* (1967) [62] has shown that MGS produces a computed Q which satisfies $Q^\top Q = I + E_{MGS}$ where $\|E_{MGS}\| \approx \mathtt{eps}\,\kappa(A)$ (for spectral norms) and eps is the machine precision. The corresponding result for the Householder approach yields $Q^\top Q = I + E_H$ where $\|E_H\| \approx \mathtt{eps}$.
The Schur form originally appeared in *Schur* (1909) [453]. The papers *Francis* (1961,1962) [166], [167] describe the double-shift QR algorithm. An alternative algorithm for obtaining a QR factorization is given in *Stewart* (2001) [483]. This first reduces the matrix to Hessenberg form. Then Givens rotations are used to zero the subdiagonal entries at a cost of $O(n^2)$ flops for $n \times n$ Hessenberg matrices. Givens rotations are discussed in most of the textbooks cited above.
Another approach to solving linear equations $AX = B$ when the matrix A is ill-conditioned or even singular is based on *regularization* techniques. Perhaps the most important method is *Tikhonov's regularization* of the pseudo-inverse which utilizes the formula $X_\varepsilon = (A^\top A + \varepsilon^2 I_n)^{-1} A^\top B$. Regularization methods have mostly been discussed in an infinite dimensional context. A standard reference for the numerical solution of finite dimensional linear equations via regularization methods is *Hansen* (1997) [216] and an instructive tutorial survey is *Neumaier* (1998) [390]. Wilkinson shifts and the fact that for symmetric matrices one obtains cubic convergence are discussed in *Wilkinson* (1968) [526]. The algorithm we have given for computing singular values was proposed by *Golub and Kahan* (1965) [196]. If one of the diagonal terms d_i in a bidiagonal matrix is zero it is possible to zero the superdiagonal entry e_{i-1} by pre-multiplication by a sequence of Givens transformations, see e.g. [197]. The back-substitution method for computing the eigenvectors of a matrix in Schur form is discussed e.g in *Stewart* (2001) [483]. The Bartels-Stewart algorithm was originally developed for solving *Sylvester equations* $AX + XB = C$, see *Bartels and Stewart* (1972) [38].
LAPACK contains a vast number of freely available well tested linear algebraic subroutines, for a user's guide see *Anderson et al.* (1999) [14]. As a user's guide to MATLAB we recommend *Higham and Higham* (2002) [229].
Linear Systems Theory has provided a rich source of new problems for Numerical Linear Algebra. The difficulties associated with the Hessenberg form for *systems*, where both the system matrix and either an input or output matrix are simultaneously simplified, have been discussed in *Laub and Linnemann* (1986) [333]. They outline some basic applications of the reduction but also show that it can be extremely sensitive to perturbations. More comprehensive treatments of numerical problems which arise in System Theory can be found e.g. in the books by *Laub et al.* (1994) [334],[335], *Datta* (1999) [122] and *Pichler et al.* (2000) [410]. The European Network of Numerics in Control (NICONET) has developed a subroutine library for Systems and Control Theory which is based on Numerical Linear Algebra routines from LAPACK, see [122]. More details can be found on the web site `http://www.win.tue.nl/niconet/niconet.html`.

Chapter 5

Uncertain Systems

The first step in most applications of mathematics is to determine a mathematical model for the system under investigation. The model may be used in a number of different ways. For example, a mathematical and computational analysis of the model often leads to a better understanding of the real physical system it represents. From a more practical viewpoint the model can be used to make predictions about the future behaviour of the system, or to design algorithms of automatic control which ensure that the system behaves in some desirable fashion. However, in each of these applications it is of fundamental importance to keep in mind that the model is only a model, its behaviour and that of the real system might be quite different. The origins and causes of this possible discrepancy are many and in the systems theory literature are collectively referred to as *model uncertainties*:

- PARAMETER UNCERTAINTY. The model may depend on some physical parameters which are not known precisely.
- IMPERFECT KNOWLEDGE OF THE DYNAMICS. There may be nonlinear and/or time-varying effects which are not known accurately.
- UNKNOWN INPUTS AND NEGLECTED DYNAMICS. A system is usually in dynamic interaction with its environment and it is often not clear where the boundary of the system should be drawn. Uncertainties arise if parts of the real system dynamics are not accounted for in the model and if the inputs to the system from the environment are not accurately known.
- MODEL SIMPLIFICATION. Although an accurate complex model of the real physical system may be available, it is often necessary to simplify this for the purpose of analysis and design. E.g. nonlinearities and time-variations are neglected, infinite dimensional systems are replaced by finite dimensional ones and sometimes further model reduction techniques are used to reduce the dimension of the system.
- DISCRETIZATION AND ROUNDING ERRORS. If simulations are carried out on a computer, discretization methods must be applied and rounding errors are introduced which will lead to unknown nonlinear model perturbations.

Some of the above points can be illustrated by the examples we have discussed in Chapter 1. For instance, the friction and stiffness forces in the mechanical systems of Section 1.3 will usually depend in a nonlinear way on velocity and position, and their linearizations (see (1.3.2)) at uncertain operating points will yield uncertain coefficients c and k. In Example 1.1.3 there may be other predators and prey present but their dynamics and their influence on the evolution of the real predator-prey system have been neglected. The temperature at the ends of the heated rod in Section 1.6 may not be a given constant but may vary because of changes in the temperature of the environment. The pendulums in Section 1.3 have been modelled as rigid bodies whereas, in fact, they may be flexible, requiring an infinite dimensional system for their description.

The development of methods for coping with the problem of model uncertainty is a great challenge for mathematical scientists today. There is a lack of tools for quantifying the effects of model uncertainties. Sometimes this leads to the development of more and more complex models in an attempt to account for all possible relevant phenomena and to reduce the modelling error to a minimum. However this strategy can fail. It may lead to models which are too complex for mathematical analysis so that experimental studies via simulations remain the only source of information, but then the size of the model could create new numerical uncertainties.

In this chapter we will be mainly concerned with developing tools for the spectral analysis of time-invariant linear systems with uncertain parameters. Spectral methods have proved to be very successful in solving many problems in the mathematical sciences. For example, eigenvalues yield information about resonance, stability, rates of growth or decay, and together with eigenvectors provide a means of approximation (e.g. from PDE to ODE). However we have seen in Section 4.2 that their sensitivities may be high and so, in the presence of rounding errors and parameter uncertainties, their computation may give misleading results. In order to take account of such perturbations we use a multi-model approach and assume that the spectral properties of the real system are portrayed with sufficient accuracy by at least one of the models. A nominal model is chosen on the basis of a "best guess" for the parameter vector and then the *uncertain system* is modelled as a *set* of linear systems whose parameter vectors lie in a given ball centred at the nominal parameter vector.

In the previous chapter we derived classical perturbation results which were mostly of a *qualitative* type, namely smoothness properties of eigenvalues and the eigenframe (continuity, analyticity) under small parameter perturbations (often restricted to a single parameter). Here we use the results of Section 4.3 and 4.4 on singular values and the μ-function to develop a *quantitative* theory, namely bounds and quantitative information about the variation of the spectrum under arbitrary parameter perturbations without any restriction on the number of parameters.

Full information about the "spectrum of the uncertain system" is given by its *spectral value sets*. These are the unions of the spectra of all perturbed systems with perturbation of size less than a given number (uncertainty level). The variation of these sets as the uncertainty level changes provides a kind of *spectral portrait* of the nominal system characterizing the behaviour of its spectrum under given sets of perturbations.

Very often, and especially so in control, desired properties of a system can be expressed by constraints on the spectrum. We say that a system is $\mathbb{C}_g$-stable if its spectrum lies in a prescribed open subset $\mathbb{C}_g$ of the complex plane (stability region).

In this chapter we do not deal with the design problem of moving the spectrum of a given system into $\mathbb{C}_g$ by feedback control, but assume that this has been carried out successfully so that the nominal model (of the closed loop system) satisfies the constraints. We then introduce a *stability radius* as a measure of the smallest perturbation for which the perturbed system no longer satisfies the constraints. This is a worst case robustness measure expressed by a single number and provides an efficient tool for assessing the robustness of the stability of a given system.
If an asymptotically stable linear system has a stability radius which is very small compared to the distance of its spectrum from the imaginary axis, this may indicate an unpleasant transient behaviour. In certain directions small initial deviations from the origin will generate large transient deviations before these are reduced to zero in the long term. In the presence of model uncertainties an analysis of the transient behaviour is particularly important and is an essential complement of spectral stability analysis. If the state trajectories of a linearized model move temporarily far away from the origin they may incite neglected nonlinearities which drive the system permanently away from the corresponding equilibrium point. From a practical viewpoint such equilibria are unstable (due to an extremely flat basin of attraction), even though the spectral analysis of the linearized model promises stability, see *Notes and References.* In this chapter we will study the transient behaviour of linear systems in some detail and discuss its relationship with spectral value sets and stability radii.
At the end of the chapter we will extend the analysis to wider perturbation classes accounting for time-varying parameter perturbations, neglected nonlinearities and neglected dynamics. For each of these perturbation classes we introduce a corresponding stability radius and study the relationship between them.
We will now give a brief outline of the material contained in each section.

Section 5.1 Here we define *spectral value sets* and *stability radii* in a very general setting for systems with arbitrary parameter uncertainty. Elementary properties which do not depend on special perturbation structures are derived. We then introduce more specific perturbation structures which will be considered in later sections.

Section 5.2 In this substantial section we define *spectral value sets* with respect to the structured perturbations introduced in Section 5.1 and show that they can be characterized via the μ-function of an associated transfer function. Particular attention is given to complex and real *single block perturbations* where the characterizations are computationally feasible and lead to algorithms for their visualization. We also consider unstructured perturbations for which the spectral value sets have been called *pseudospectra* in the literature, see *Notes and References.*

Section 5.3 Stability radii are introduced and analyzed following essentially the same programme as in Section 5.2. For complex full-block perturbations and the spectral norm the stability radii can be characterized by parametrized Riccati equations or, equivalently, by parametrized Hamiltonian matrices and this leads to an algorithm for their computation.

Section 5.4 By specializing the results of Sections 5.2 and 5.3, we obtain explicit formulas for *root sets* and *stability radii* of polynomials with uncertain coefficients.

Section 5.5 The concept of *(M,β)-stability* is introduced which implies both a satisfactory *transient behaviour* and exponential stability. We derive various estimates for the state trajectories of a stable linear system and will see that spectral value sets and stability radii play an important role in deriving these estimates.

Section 5.6 We introduce *stability radii for time-varying, nonlinear and dynamic perturbations* and show that they are all equal in the *complex* case, whereas they may differ in the *real* case.

Throughout this chapter the main emphasis will be on *analysis*, but we will also deal with the problem of *computing* spectral value sets and stability radii.

5.1 Models of Uncertainty and Tools for their Analysis

In this section we introduce the general notions of *spectral value set* and *stability radius* for arbitrary parametrized sets of time-invariant linear systems. We suppose that a nominal parameter vector is given, and view the other parameter values as deviations from it. The uncertain system is then modelled by the set of systems whose parameter deviations are bounded in norm by a given uncertainty level ("multi-model with norm bounded uncertainty").

Without further assumptions no specific characterizations or algorithms are available in order to determine spectral value sets and stability radii. So, after introducing the general framework and illustrating the new concepts by figures and examples we only derive some basic general properties in the first subsection. More specific perturbation structures will be considered in the second subsection where we introduce *linear fractional representations of parameter uncertainties*. Many parameter uncertainties encountered in control applications can be represented in this way, and we will restrict our studies to these in all but the last section of this chapter. In principle, arbitrary rational parameter dependencies can be represented in linear fractional form. However, for computational purposes further restrictions have to be imposed on the perturbation structure. Uncertain systems, for which spectral value sets and stability radii can be explicitly calculated, will be studied more closely in later sections.

5.1.1 General Definitions and Basic Properties

There are many different ways of modelling uncertain dynamical systems. If e.g. the parameters of a linear system are uncertain but some statistical information is available, one may choose a stochastic linear model where the random entries of the system matrix fluctuate according to some probabilistic law about their given mean values. In this chapter we will, however, only consider deterministic models of uncertainty. More precisely we will study parametrized sets of *time-invariant* linear finite dimensional systems of the following form

$$\dot{x}(t) = A_\omega x(t),\ t \geq 0, \quad \text{or} \quad x(t+1) = A_\omega x(t),\ t \in \mathbb{N} \quad (\omega \in \Omega) \tag{1}$$

where the system matrix $A_\omega \in \mathbb{K}^{n \times n}$ depends on some parameter vector $\omega \in \Omega$ and Ω is a given parameter set which is a subset of a finite dimensional normed vector space $(V, \|\cdot\|)$. The whole parametrized set (1) represents our a priori knowledge about the "real" system we wish to analyze, and although we do not know which parameter value $\omega \in \Omega$ is the right one, the modelling assumption is that at least one of the systems in the set describes the real system with sufficient accuracy. The individual models in (1) may be thought of as either uncontrolled dynamical systems or controlled systems in which the loop has been closed by feedback.

In addition to the model class (1) there is often some a priori knowledge available concerning the range of values of the parameter vector. For example it may be known that the parameter values belong to certain real intervals thus specifying lower and upper limits for the parameters. Here we regard the individual systems in (1) as perturbations of a given *nominal* system with matrix $A = A_{\omega^0}$, which represents the "best" or "most probable" or "averaged" model within the system family. It is assumed that the norm of the deviation of the parameters from the nominal value is bounded by some given level of uncertainty $\delta > 0$ so that a satisfactory multi-model of the "real system" is given by the set of systems with matrices

$$A(\Delta) = A_{\omega^0+\Delta}, \quad \Delta \in \mathbf{\Delta}_0 := \Omega - \omega^0, \ \|\Delta\| < \delta. \tag{2}$$

The vector $\Delta \in \mathbf{\Delta}_0$ represents the *deviation* of the parameter vector $\omega = \omega^0 + \Delta$ from the nominal parameter vector ω^0, and $A(\Delta)$ is the perturbed system matrix associated with this deviation. Identifying the matrix $A_\omega = A_{\omega^0+\Delta}$ with the corresponding continuous or discrete time system in (1), the matrix family $(A(\Delta))_{\Delta \in \mathbf{\Delta}_0, \|\Delta\| < \delta}$ models a (time-invariant, linear, finite dimensional) system with norm bounded uncertainty, centred at the nominal system $A = A(0) = A_{\omega^0}$.

Throughout the section we suppose the following:

Assumption 5.1.1. *$(A(\Delta))_{\Delta \in \mathbf{\Delta}_0}$ is a given continuous matrix family in $\mathbb{K}^{n \times n}$, whose parameter set $\mathbf{\Delta}_0$ is a subset of a finite dimensional normed $\mathbb{K}$-linear space $(V, \|\cdot\|)$. The closure $\mathbf{\Delta} = \overline{\mathbf{\Delta}_0}$ of $\mathbf{\Delta}_0$ in V is starlike with respect to the origin*[1], *and the* well-posedness radius $\boldsymbol{\delta}_0 = \inf\{\|\Delta\|; \Delta \in \mathbf{\Delta} \setminus \mathbf{\Delta}_0\}$ *is strictly positive ($\boldsymbol{\delta}_0 := \infty$ if $\mathbf{\Delta} = \mathbf{\Delta}_0$).*

Let $B(\delta)$, $B_{\mathbf{\Delta}_0}(\delta)$ and $B_{\mathbf{\Delta}}(\delta)$ denote the open balls with radius δ around the origin in V, $\mathbf{\Delta}_0$ and $\mathbf{\Delta}$, respectively. For later use we note that, as a consequence of the previous assumption, the closed balls around the origin in $\mathbf{\Delta}_0$ with radius $\delta < \boldsymbol{\delta}_0$ are starlike and compact:

$$\overline{B_{\mathbf{\Delta}_0}(\delta)} = \overline{B_{\mathbf{\Delta}}(\delta)} = \overline{B(\delta)} \cap \mathbf{\Delta} = \overline{B(\delta)} \cap \mathbf{\Delta}_0, \quad \delta \in (0, \boldsymbol{\delta}_0). \tag{3}$$

In control applications it often occurs (see the next subsection) that the perturbations Δ belong to some linear space $\mathbf{\Delta}$ and the system matrix depends rationally on the parameter deviations such that no entry $a_{ij}(\Delta) = q_{ij}(\Delta)/p_{ij}(\Delta)$ of $A(\Delta)$ has a pole at $\Delta = 0$. In this case the set $\mathbf{\Delta}$ is given first and then $\mathbf{\Delta}_0$ is taken as the set of all the $\Delta \in \mathbf{\Delta}$ for which all the entries have denominators $p_{ij}(\Delta) \neq 0$. Then $\boldsymbol{\delta}_0$ is the largest $\delta > 0$ so that $A(\Delta)$ is well defined for all $\Delta \in \mathbf{\Delta}$, $\|\Delta\| < \delta$. With these definitions the above assumption will be satisfied.

Our aim is to examine the extent to which dynamic properties of the nominal system are preserved under parameter perturbations $\Delta \in \mathbf{\Delta}_0$ of bounded norm. We

[1] i.e. $\Delta \in \mathbf{\Delta} \Rightarrow \alpha\Delta \in \mathbf{\Delta}$ for all $\alpha \in [0, 1]$.

suppose that these properties can be expressed by spectral constraints $\sigma(A) \subset \mathbb{C}_g$ where $\mathbb{C}_g \subset \mathbb{C}$ is a given open subset (the "good part") of the complex plane. $\mathbb{C}_g$ will also be called the "stability region". Then the above problem can be stated as a problem of robust stability: Given a bound $\delta > 0$ on the norm of the parameter deviation Δ, are all the perturbed spectra $\sigma(A(\Delta))$ contained in the prescribed stability region? This leads to the following definition where $A(\cdot)$ denotes the matrix family $(A(\Delta))_{\Delta \in \mathbf{\Delta}_0}$.

Definition 5.1.2. For every $\delta > 0$ the set

$$\sigma_{\mathbf{\Delta}}(A(\cdot);\delta) = \bigcup_{\Delta \in \mathbf{\Delta}_0, \|\Delta\| < \delta} \sigma(A(\Delta))$$

is called the *spectral value set* of $A = A(0)$ *under perturbations of the form* $A \rightsquigarrow A(\Delta)$, $\Delta \in \mathbf{\Delta}_0$, at the uncertainty level δ.

Thus $\sigma_{\mathbf{\Delta}}(A(\cdot);\delta)$ is the set of all complex numbers to which an eigenvalue of A can be moved by a parameter perturbation $\Delta \in \mathbf{\Delta}_0$ of size $< \delta$. The shape of these sets clearly depends on the parameter dependence $\Delta \mapsto A(\Delta)$, the parameter set $\mathbf{\Delta}_0$, the chosen norm $\|\cdot\|$, and the uncertainty level δ.

Remark 5.1.3. We have chosen to write $\sigma_{\mathbf{\Delta}}$ rather than $\sigma_{\mathbf{\Delta}_0}$ since – as mentioned above – the perturbation class $\mathbf{\Delta}$ is usually fixed first and as a second step the set $\mathbf{\Delta}_0$ is specified as the set of all $\Delta \in \mathbf{\Delta}$ for which the matrices $A(\Delta)$ are well defined. □

Methods for the computation of spectral value sets and their visualization on the computer screen are only available for a few special perturbation structures and special parameter sets. Some of these methods will be dealt with in the next section. Here we make use of them in order to compute the spectral value sets of some examples which illustrate the concept.

We have seen in Chapter 1 that rational parameter dependencies often occur in linear models of physical systems. As a simple example we consider the spectral value set of a linear oscillator with uncertain coefficients.

Example 5.1.4. Consider the damped oscillator $m\ddot{\xi} + c\dot{\xi} + k\xi = 0$ where the mass $m = m_0 + \Delta_m$, damping coefficient $c = c_0 + \Delta_c$ and spring coefficient $k = k_0 + \Delta_k$ are uncertain, with $m_0, c_0, k_0 > 0$. Taking $\omega_0 = (m_0, c_0, k_0)$ as the nominal parameter vector and setting $\Delta = (\Delta_m, \Delta_c, \Delta_k)$, the system matrix of the state space model of the uncertain linear oscillator is

$$A(\Delta) = \begin{bmatrix} 0 & 1 \\ -(k_0 + \Delta_k)/(m_0 + \Delta_m) & -(c_0 + \Delta_c)/(m_0 + \Delta_m) \end{bmatrix}.$$

Apparently, for physical reasons, the deviation vector $\Delta = (\Delta_m, \Delta_c, \Delta_k)$ should be restricted so that $m > 0, c, k \geq 0$, i.e.

$$\mathbf{\Delta}_0^{\mathbb{R}} = \{\Delta \in \mathbb{R}^3; \Delta_m > -m_0,\ \Delta_c \geq -c_0,\ \Delta_k \geq -k_0\}. \tag{4}$$

However, we shall see later that there is a case for considering *complex* parameter perturbations, see Example 5.3.14. Mathematically, $A(\Delta)$ is defined for all

$$\Delta \in \mathbf{\Delta}_0^{\mathbb{C}} = \{(\Delta_m, \Delta_c, \Delta_k) \in \mathbb{C}^3;\ \Delta_m \neq -m_0\}. \tag{5}$$

With both parameter sets, $\mathbf{\Delta}_0 = \mathbf{\Delta}_0^{\mathbb{R}}$ and $\mathbf{\Delta}_0 = \mathbf{\Delta}_0^{\mathbb{C}}$, the matrix family $(A(\Delta))_{\Delta \in \mathbf{\Delta}_0}$ satisfies Assumption 5.1.1. In order that the spectral value sets give significant information it is essential that the perturbation norm is chosen in a way which reflects the actual parameter uncertainty. For instance, if it is known a priori that the oscillator parameters m, c, k vary in certain real symmetric intervals about their nominal values, the perturbations should be measured by a suitably scaled ∞-norm

$$(i) \quad \|\Delta\| = \max\{\,\alpha_1\,|\Delta_m|,\, \alpha_2\,|\Delta_c|,\, \alpha_3\,|\Delta_k|\,\}, \quad \Delta = (\Delta_m, \Delta_c, \Delta_k) \in \mathbb{R}^3.$$

Then the open ball of radius $\delta > 0$ about the nominal parameter vector (m_0, c_0, k_0) takes the form of a 3-dimensional interval

$$(m_0 - \delta/\alpha_1, m_0 + \delta/\alpha_1) \times (c_0 - \delta/\alpha_2, c_0 + \delta/\alpha_2) \times (k_0 - \delta/\alpha_3, k_0 + \delta/\alpha_3).$$

Any product of intervals of the three parameters $m, c, k \in (0, \infty)$ may be represented in this way by choosing the nominal parameter values and the weights accordingly.

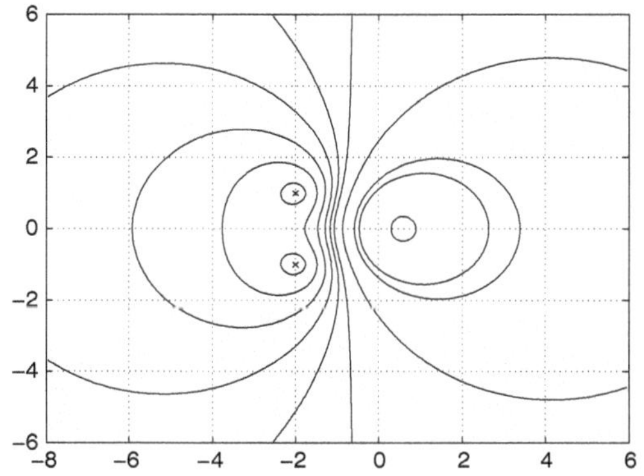

Figure 5.1.1: Evolution of spectral value sets of the oscillator for increasing δ

Similarly, if the uncertainty in the parameters is expressed in terms of their *relative* deviations from the nominal values, adequate norms might be

$$(ii)\ \|\Delta\| = \max\{|\Delta_m|/m_0,\ |\Delta_c|/c_0,\ |\Delta_k|/k_0\},\ (iii)\ \|\Delta\| = \|(|\Delta_m|/m_0,\ |\Delta_c|/c_0,\ |\Delta_k|/k_0)\|_2.$$

By the definition of $\boldsymbol{\delta}_0$ in Assumption 5.1.1, $\boldsymbol{\delta}_0 = \alpha_1 m_0$ for the norm (i) whereas $\boldsymbol{\delta}_0 = 1$ for the norms (ii), (iii) for both parameter sets $\mathbf{\Delta}_0 = \mathbf{\Delta}_0^{\mathbb{R}}$ and $\mathbf{\Delta}_0 = \mathbf{\Delta}_0^{\mathbb{C}}$.
In Figure 5.1.1 we present the spectral value sets of the oscillator with nominal parameters $m_0 = 1, c_0 = 4, k_0 = 5$ for complex perturbations with respect to the norm $\|\cdot\|_\infty$ on $\mathbf{\Delta}$. The eigenvalues of the nominal system are marked by $\times$. The contours represent the boundaries of $\sigma_{\mathbf{\Delta}^{\mathbb{C}}}(A(\cdot);\delta)$ at different uncertainty levels δ, both smaller and larger than the critical value $\boldsymbol{\delta}_0 = 1$. The figure shows, from the left to the right, the spectral value sets for $\delta \in \{0.1, 0.28, 0.46, 0.64, 0.82, 1.0, 1.5, 2.5, 3, 6.0\}$. One observes on the left hand side of the figure that for $\delta = 0.1$ the set consists of two connected components, and each one contains an eigenvalue of $A(0)$. As δ increases the components merge and $\sigma_{\mathbf{\Delta}^{\mathbb{C}}}$ expands until at $\delta = \boldsymbol{\delta}_0 = 1$ the set is unbounded. For $\delta > \boldsymbol{\delta}_0$ the spectral value set consists of that (unbounded) part of the complex plane which is exterior to one of the contours on the right hand side of the figure. The interiors of the nested contours represent regions which are increasingly difficult to reach by eigenvalues of the perturbed system, i.e. the perturbations required for pushing eigenvalues into these regions get larger and larger. □

In Subsection 4.2.3 we derived some estimates for the distance of the spectrum of a perturbed matrix $A(\Delta) = A + \Delta$, $\Delta \in \mathbb{C}^{n\times n}$ from the spectrum of A. These estimates provide upper bounds for the corresponding spectral value sets. In the following example we compare the upper bound provided by the Bauer-Fike Lemma 4.2.14 with the exact result. A similar comparison with the estimate derived from Gershgorin's Theorem will be given in the next section (Example 5.2.13).

Example 5.1.5. In Example 4.2.17 we have applied the Bauer-Fike Lemma to matrices of the form $A = \begin{bmatrix} -1 & \alpha \\ 0 & -2 \end{bmatrix}$ where $\alpha \in \mathbb{R}_+$ is given. For perturbations $A \rightsquigarrow A(\Delta) = A + \Delta$, $\Delta \in \mathbb{C}^{2\times 2}$ we obtained the following estimate

$$\min_{\lambda\in\sigma(A)} |\lambda_\Delta - \lambda| \leq \frac{(1+\cos\theta)}{\sin\theta}\|\Delta\|_{2,2} = \left[(1+\alpha^2)^{1/2} + \alpha\right]\|\Delta\|_{2,2}, \quad \lambda_\Delta \in \sigma(A_\Delta)$$

where $\theta = \theta(\alpha) \in [0, \omega/2]$ is the angle between the two unit eigenvectors $[1\,,\,0]^\top$ and $(1+\alpha^2)^{-1/2}[\alpha\,,\,-1]^\top$ of A. (Note that $\theta(\alpha) \to 0$ as $\alpha \to \infty$). Using this estimate we see that the spectral value set $\sigma_{\mathbb{C}^{2\times 2}}(A(\cdot), 1/2)$ with respect to the spectral norm is contained in the union of the open disks of radius $\left[(1+\alpha^2)^{1/2} + \alpha\right]/2$ around the eigenvalues of A. Figure 5.1.2 compares $\sigma_{\mathbb{C}^{2\times 2}}(A(\cdot), 1/2)$ with this upper bound for two values of the

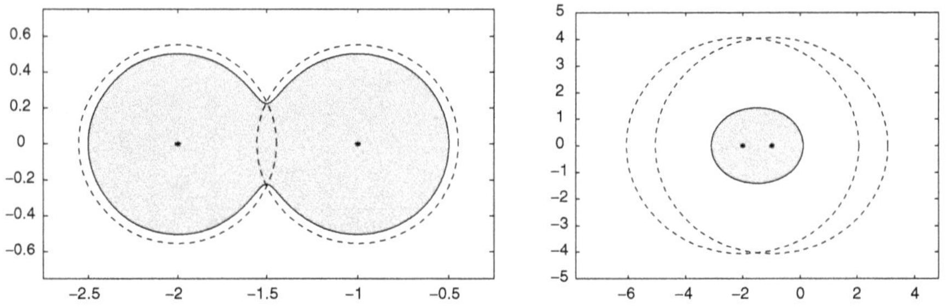

Figure 5.1.2: $\sigma_{\mathbb{C}^{2\times 2}}(A(\cdot), 1/2)$ and the Bauer-Fike bound for $\alpha = 0.1$ and $\alpha = 4$

parameter, $\alpha = 0.1$ and $\alpha = 4$. Note that the bound described by the outer broken lines is tight for $\alpha = 0.1$ (where A is nearly normal) whereas, for $\alpha = 4$, $\sigma_{\mathbb{C}^{2\times 2}}(A(\cdot), 1/2)$ is quite different from that of the upper estimate derived from the Bauer-Fike Lemma. □

The following example and the associated Figure 5.1.3 shows the difference between spectral value sets for *complex* and *real* parameter perturbations. The figure also illustrates how these sets evolve if the level of uncertainty $\delta > 0$ is increasing.

Example 5.1.6. Suppose that the matrix

$$A = \begin{bmatrix} -7 & 5 & 0 & 0 \\ 0 & 0 & 8 & 2 \\ 0 & 0 & 6 & -2 \\ 0 & 0 & 8 & 6 \end{bmatrix} \quad \text{with spectrum} \quad \sigma(A) = \{-7, 0, 6+4i, 6-4i\}$$

is subjected to unstructured perturbations of the form $A \rightsquigarrow A(\Delta) = A + \Delta$ where $\Delta \in \mathbb{C}^{4\times 4}$ or $\Delta \in \mathbb{R}^{4\times 4}$. We denote the corresponding spectral value sets with respect to the spectral norm by $\sigma_{\mathbb{C}}(A;\delta)$ and $\sigma_{\mathbb{R}}(A;\delta)$. These sets are shown in Figure 5.1.3 for the uncertainty levels $\delta = 1, 2, 3$. The spectral value sets for complex and real perturbations

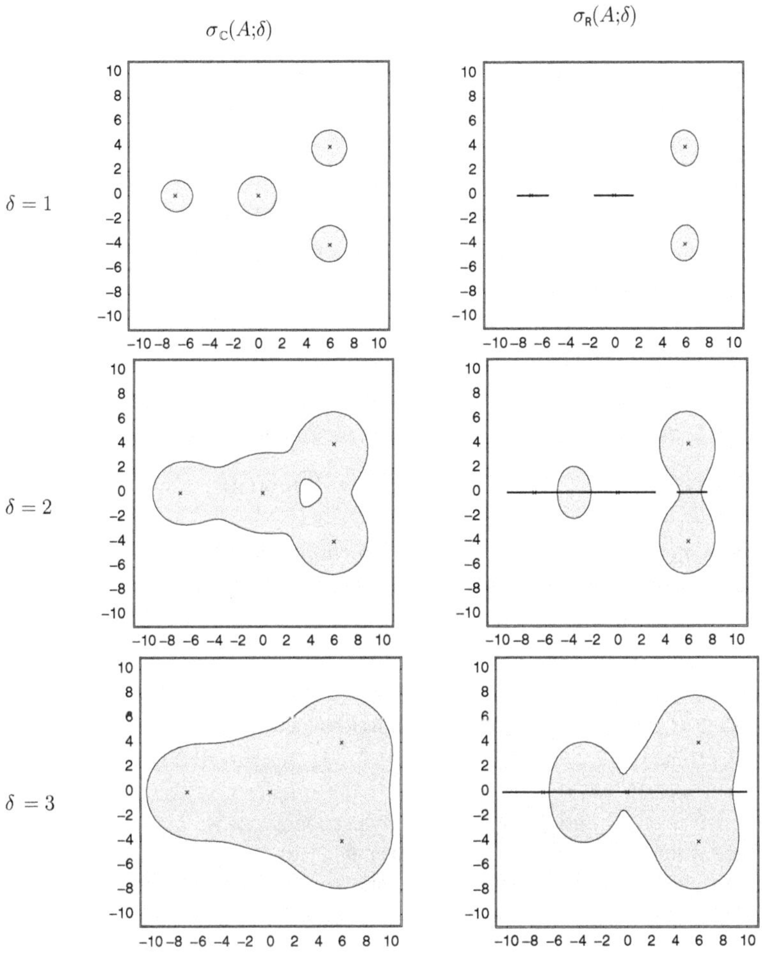

Figure 5.1.3: Evolution of spectral value sets for increasing δ

differ substantially. For $\delta = 1$, $\sigma_{\mathbb{C}}(A;\delta)$ consists of four disks of radius 1 around the eigenvalues of A whereas in the real case the disks around the two real eigenvalues are replaced by real intervals and the other two disks are replaced by sets of an oval form. For $\delta = 2$, $\sigma_{\mathbb{C}}(A;\delta)$ is connected whereas $\sigma_{\mathbb{R}}(A;\delta)$ is not. While all the three sets $\sigma_{\mathbb{C}}(A;\delta)$ are open subsets of $\mathbb{C}$ (the shaded areas inside the contours), none of the three sets $\sigma_{\mathbb{R}}(A;\delta)$ is open because of the real intervals (which form part of these sets) sticking out of the shaded areas. This corresponds to the fact that, under *real* parameter perturbations, the eigenvalues are "more mobile" along the real axis and the real eigenvalues of A may be trapped in $\mathbb{R}$ for small perturbations. Note, however, that by definition $\sigma_{\mathbb{R}}(A;\delta) \subset \sigma_{\mathbb{C}}(A;\delta)$, and so the real intervals sticking out of $\sigma_{\mathbb{R}}(A;\delta)$ are contained in $\sigma_{\mathbb{C}}(A;\delta)$. In fact they bridge the gap between the interior of $\sigma_{\mathbb{R}}(A;\delta)$ and the boundary of $\sigma_{\mathbb{C}}(A;\delta)$. □

The following lemma collects some basic general properties of spectral value sets.

Define $\delta(\cdot) : \mathbb{C} \to [0, \infty]$ by

$$\delta(s) = \inf\{\|\Delta\|; \Delta \in \mathbf{\Delta}_0,\ s \in \sigma(A(\Delta))\}, \quad s \in \mathbb{C} \tag{6}$$

where, as usual, we set $\inf \emptyset = \infty$, whence $\delta(s) = \infty$ for $s \in \mathbb{C} \setminus \left(\bigcup_{\Delta \in \mathbf{\Delta}_0} \sigma(A(\Delta))\right)$.

Lemma 5.1.7. *Suppose Assumption 5.1.1 holds and $A = A(0)$, then*

(i) the spectral value sets $\sigma_{\mathbf{\Delta}}(A(\cdot); \delta)$ are increasing with $\delta \in \mathbb{R}_+$.

(ii) The intersection of all the sets $\sigma_{\mathbf{\Delta}}(A(\cdot); \delta)$, $\delta > 0$ is equal to $\sigma(A)$.

(iii) $\sigma_{\mathbf{\Delta}}(A(\cdot); \delta)$ is a bounded subset of $\mathbb{C}$ for all $\delta \in (0, \boldsymbol{\delta}_0)$.

(iv) If $s_0 \in \sigma_{\mathbf{\Delta}}(A(\cdot); \boldsymbol{\delta}_0)$, then there exists a perturbation $\Delta_0 \in \mathbf{\Delta}_0$ of minimal norm $\|\Delta_0\| = \delta(s_0)$ such that $s_0 \in \sigma(A(\Delta_0))$.

(v) If $\delta \in (0, \boldsymbol{\delta}_0)$, then the closure of $\sigma_{\mathbf{\Delta}}(A(\cdot); \delta)$ in $\mathbb{C}$ is given by

$$\overline{\sigma_{\mathbf{\Delta}}(A(\cdot); \delta)} = \bigcup_{\Delta \in \overline{B_{\mathbf{\Delta}}(\delta)}} \sigma(A(\Delta)) = \bigcap_{\delta' > \delta} \sigma_{\mathbf{\Delta}}(A(\cdot); \delta'). \tag{7}$$

(vi) The closure $\overline{\sigma_{\mathbf{\Delta}}(A(\cdot); \delta)} \in \mathcal{K}(\mathbb{C})$ depends continuously on $\delta \in (0, \boldsymbol{\delta}_0)$ with respect to the Hausdorff metric on $\mathcal{K}(\mathbb{C})$ (the set of compact subsets of $\mathbb{C}$).[2]

Proof: (i) is trivial. (ii) follows from Corollary 4.2.1 since $\Delta \mapsto A(\Delta)$ is continuous.

(iii) For every $\delta \in (0, \boldsymbol{\delta}_0)$, the closed ball $\overline{B_{\mathbf{\Delta}_0}(\delta)} = \overline{B_{\mathbf{\Delta}}(\delta)}$ is compact by (3) and hence $\{A(\Delta);\ \Delta \in \overline{B}_{\mathbf{\Delta}}(\delta)\}$ is a compact set of matrices. Therefore $\bigcup_{\Delta \in \overline{B_{\mathbf{\Delta}}(\delta)}} \sigma(A(\Delta))$ is compact by Corollary 4.2.2 and so $\sigma_{\mathbf{\Delta}}(A(\cdot); \delta) \subset \bigcup_{\Delta \in \overline{B_{\mathbf{\Delta}}(\delta)}} \sigma(A(\Delta))$ is bounded.

(iv) Suppose $s_0 \in \sigma_{\mathbf{\Delta}}(A(\cdot); \boldsymbol{\delta}_0)$. Then there exists a sequence (Δ_k) in $\mathbf{\Delta}_0$ such that $s_0 \in \sigma(A(\Delta_k))$, $\|\Delta_k\| < \boldsymbol{\delta}_0$, and $\|\Delta_k\| \downarrow \delta(s_0)$. By compactness we may assume that the sequence has a limit $\Delta \in \overline{B_{\mathbf{\Delta}}(\delta)}$ for some $\delta < \boldsymbol{\delta}_0$. But then $\|\Delta\| = \delta(s_0)$ by the continuity of the norm, and $s_0 \in \sigma(A(\Delta))$ by the continuity of the spectrum, see Corollary 4.2.1.

(v) From what we have seen in the proof of (iii) it is clear that inclusions $\subset$ hold in (7). Now suppose $s_0 \in \bigcap_{\delta' > \delta} \sigma_{\mathbf{\Delta}}(A(\cdot); \delta')$. By (iv) there exists a perturbation $\Delta_0 \in \mathbf{\Delta}_0$ of minimal norm $\|\Delta_0\| = \delta(s_0)$ such that $s_0 \in \sigma(A(\Delta_0))$. Moreover we have $\|\alpha \Delta_0\| < \delta(s_0) \leq \delta$ for $\alpha \in (0, 1)$ and $\alpha \Delta_0 \in \mathbf{\Delta}_0$ converges towards Δ_0 as $\alpha \to 1$. Applying again Corollary 4.2.1 we see that $\sigma(A(\Delta_0)) \subset \overline{\sigma_{\mathbf{\Delta}}(A(\cdot); \delta)}$. This proves the equalities in (7).

(vi) Since $\overline{B_{\mathbf{\Delta}}(\delta)} = \overline{B(\delta)} \cap \mathbf{\Delta}_0$ is starlike with respect to the origin for $\delta \in (0, \boldsymbol{\delta}_0)$, it is easily verified that the map $\delta \mapsto \overline{B_{\mathbf{\Delta}}(\delta)}$ from $(0, \boldsymbol{\delta}_0)$ into $\mathcal{K}(V)$ is continuous with respect to the Hausdorff metric. Hence, by Corollary 4.2.2 and Lemma 4.1.9, the map

$$\delta \mapsto \sigma\left(\{A(\Delta); \Delta \in \overline{B_{\mathbf{\Delta}}(\delta)}\}\right) = \bigcup_{\Delta \in \overline{B_{\mathbf{\Delta}}(\delta)}} \sigma(A(\Delta)) = \overline{\sigma_{\mathbf{\Delta}}(A(\cdot); \delta)}$$

from $(0, \boldsymbol{\delta}_0)$ into $\mathcal{K}(\mathbb{C})$ is continuous with respect to the Hausdorff metric. □

[2] In Subsection 5.2.1 we will see that the closure of the spectral value sets even depends continuously on the *whole set* of data (A, B, C, D, δ) for $\delta < \boldsymbol{\delta}_0$.

Proposition 5.1.8. *Suppose Assumption 5.1.1 holds. If $A = A(0)$ has m distinct eigenvalues, then $\sigma_{\mathbf{\Delta}}(A(\cdot);\delta)$ has m connected components for $\delta > 0$ sufficiently small. The number $\nu(\delta)$ of connected components of $\sigma_{\mathbf{\Delta}}(A(\cdot);\delta)$ decreases as δ increases in $(0,\boldsymbol{\delta}_0)$, and every connected component of $\sigma_{\mathbf{\Delta}}(A(\cdot);\delta)$ contains at least one eigenvalue of A.*

Proof: Let $\lambda_1,\dots\lambda_m$ be the distinct eigenvalues of $\sigma(A)$ and $\varepsilon > 0$ be such that the disks $D(\lambda_j,\varepsilon)$, $j \in \underline{m}$ are disjoint. Since $\Delta \mapsto A(\Delta)$ is continuous there exists by Corollary 4.2.1 a $\delta > 0$ such that

$$\sigma_{\mathbf{\Delta}}(A(\cdot);\delta) = \bigcup_{\Delta\in\mathbf{\Delta}_0, \|\Delta\|<\delta} \sigma(A(\Delta)) \subset \bigcup_{i=1}^{m} D(\lambda_j,\varepsilon).$$

Since $\sigma_{\mathbf{\Delta}}(A(\cdot);\delta)$ contains $\lambda_1,\dots,\lambda_m$, it has at least m connected components.
Now suppose that $0 < \delta' < \delta < \boldsymbol{\delta}_0$ and C (resp. C') is a connected component of $\sigma_{\mathbf{\Delta}}(A(\cdot);\delta)$ (resp. $\sigma_{\mathbf{\Delta}}(A(\cdot);\delta')$). If $C \cap C' \neq \emptyset$ then $C \cup C'$ is a connected subset of $\sigma_{\mathbf{\Delta}}(A(\cdot);\delta)$, hence $C' \subset C$ since C is a maximal connected subset of $\sigma_{\mathbf{\Delta}}(A(\cdot);\delta)$. For each connected component C' of $\sigma_{\mathbf{\Delta}}(A(\cdot);\delta')$ there exists exactly one connected component of $\sigma_{\mathbf{\Delta}}(A(\cdot);\delta)$ such that $C' \subset C$. It remains to prove that every connected component of $\sigma_{\mathbf{\Delta}}(A(\cdot);\delta)$ contains an eigenvalue of A. This implies that each of the (disjoint) connected components of $\sigma_{\mathbf{\Delta}}(A(\cdot);\delta)$ contains at least one connected component of $\sigma_{\mathbf{\Delta}}(A(\cdot);\delta')$ and thus $\nu(\delta) \le \nu(\delta')$. Now let $\lambda \in C$. Then there exists $\Delta \in \mathbf{\Delta}_0$, $\|\Delta\| < \delta$ such that $\lambda \in \sigma(A(\Delta))$. By Corollary 4.2.4 there exists an arc $\lambda : [0,1] \to \mathbb{C}$ such that $\lambda(t) \in \sigma(A(t\Delta))$ for $t \in [0,1]$. The arc $\lambda([0,1])$ is a connected subset in $\sigma_{\mathbf{\Delta}}(A(\cdot);\delta)$ and $\lambda(1) = \lambda \in C$. This implies $\lambda([0,1]) \subset C$ and so $\lambda(0) \in \sigma(A) \cap C$. □

We say that an eigenvalue $\lambda \in \sigma(A(0))$ is *immovable* under perturbations $A \rightsquigarrow A(\Delta)$ if it is an isolated point of $\sigma_{\mathbf{\Delta}}(A(\cdot);\delta)$ for some $\delta > 0$, i.e. there exists a neighbourhood N of λ such that $\sigma_{\mathbf{\Delta}}(A(\cdot);\delta) \cap N = \{\lambda\}$. Otherwise we say the eigenvalue is *movable*. In Remark 5.1.10 we give an example where the spectral value set consists of both movable and immovable eigenvalues.
By definition (6) the spectral value set at level $\delta < \infty$ can be expressed as a sublevel set of the function $\delta(\cdot) : \mathbb{C} \to [0,\infty]$

$$\sigma_{\mathbf{\Delta}}(A(\cdot);\delta) = \{s \in \mathbb{C}; \delta(s) < \delta\}, \qquad \delta \in (0,\infty).$$

It follows from Lemma 5.1.7 that for $\delta < \boldsymbol{\delta}_0$ the level sets $\{s \in \mathbb{C}; \delta(s) = \delta\}$ coincide with the "frontier" of the spectral value sets in the following sense:

$$\overline{\sigma_{\mathbf{\Delta}}(A(\cdot);\delta)} \setminus \sigma_{\mathbf{\Delta}}(A(\cdot);\delta) = \{s \in \mathbb{C};\ \delta(s) = \delta\}, \quad \delta \in (0,\boldsymbol{\delta}_0).$$

We will see in the next section that the level curves of $\delta(\cdot)$ can be computed for certain perturbation structures.
The frontier of $\sigma_{\mathbf{\Delta}}(A(\cdot);\delta)$ will coincide with its topological boundary if $\sigma_{\mathbf{\Delta}}(A(\cdot);\delta)$ is open. In view of the previous example we cannot expect, in general, a spectral value set to be open if only real parameter perturbations are allowed. For the complex case we have the following result. The proof relies heavily on the method of analytic continuation as discussed in the construction of q-cycles in the previous chapter.

Proposition 5.1.9. *Suppose Assumption 5.1.1 holds and additionally for every $\Delta \in \mathbf{\Delta}_0$ there is an $\varepsilon_\Delta > 0$ such that $z\Delta \in \mathbf{\Delta}_0$ for all $z \in D(1, \varepsilon_\Delta)$ and $z \mapsto A(z\Delta)$ is analytic on the disk $D(1, \varepsilon_\Delta)$. Then $\sigma_{\mathbf{\Delta}}(A(\cdot); \delta) \setminus \sigma(A(0))$ is open for all $\delta \in (0, \boldsymbol{\delta}_0)$.*

Proof: Let $\lambda \in \sigma_{\mathbf{\Delta}}(A(\cdot); \delta) \setminus \sigma(A(0))$ with $\delta \in (0, \boldsymbol{\delta}_0)$, and let $\Delta \in \mathbf{\Delta}_0$ be a minimum norm perturbation such that $\|\Delta\| = \delta(\lambda) < \delta$ and $\lambda \in \sigma(A(\Delta))$ (Lemma 5.1.7). Denote the distinct eigenvalues of $A(\Delta)$ by $\lambda_1, \ldots, \lambda_m$ $(m \le n)$ and choose a radius $r > 0$ sufficiently small so that the disks $D(\lambda_j, r)$, $j \in \underline{m}$ are disjoint. Then choose $\varepsilon \in (0, \varepsilon_\Delta)$ such that $(1+\varepsilon)\|\Delta\| < \delta$ and $\sigma(A(z\Delta)) \subset \bigcup_{j \in \underline{m}} D(\lambda_j, r)$ for all $z \in D(1, \varepsilon)$. Reducing ε further if necessary we may assume that the punctured disk $D^\circ = D(1, \varepsilon) \setminus \{1\}$ does not contain any critical value of the analytic matrix family $A_\Delta(z) := A(z\Delta)$, $z \in D(1, \varepsilon_\Delta)$, see Theorem 4.1.14.
Let $z_0 \in D^\circ$, $|z_0| < 1$. Then $\lambda \notin \sigma(A(z_0\Delta))$, since otherwise $\delta(\lambda) \le |z_0|\,\|\Delta\| < \|\Delta\|$. On the other hand, there exists $\lambda_0 \in \sigma(A(z_0\Delta)) \cap D(\lambda, r)$. In a sufficiently small disk around z_0 the distinct eigenvalues of $A(z\Delta)$ can be expressed by power series $\lambda_i(z)$ (Corollary 4.2.8 (ii)). One of these admits the value λ_0 at z_0, say $\lambda_{i_1}(z_0) = \lambda_0$. By analytic continuation along arcs in D° the power series $\lambda_{i_1}(z)$ generates a cycle $(\lambda_{i_1}(z), \ldots, \lambda_{i_q}(z))$ (represented by a Puiseux series of the form (4.2.2) which belongs to the λ-group of the eigenvalue $\lambda \in \sigma(A(\Delta)) = \sigma(A_\Delta(1))$, i.e. satisfies $\lim_{z \to 1} \lambda_{i_k}(z) = \lambda$ for $k \in \underline{q}$. First suppose that $q = 1$. Then $\lambda_{i_1}(z) \in \sigma(A(z\Delta))$ can be continued analytically to the whole disk $D(1, \varepsilon)$ by Corollary 4.2.9 (i). Since $\lambda_{i_1}(z_0) = \lambda_0 \ne \lambda$, $\lambda_{i_1}(\cdot)$ cannot be constant. But then $z \mapsto \lambda_{i_1}(z)$ is an open mapping and hence $\lambda_{i_1}(D(1, \varepsilon))$ is an open neighbourhood of $\lambda = \lambda_{i_1}(1)$. Since $(1+\varepsilon)\|\Delta\| < \delta$ implies $\|z\Delta\| < \delta$ for all $z \in D(1, \varepsilon)$, it follows that this open neighbourhood is contained in $\sigma_{\mathbf{\Delta}}(A(\cdot); \delta)$.
Now suppose $q \ge 2$. Then the cycle $(\lambda_{i_1}(z), \ldots, \lambda_{i_q}(z))$ defines a q-valued analytic function on D° which can be extended continuously to the whole disk $D(1, \varepsilon)$. The associated value set $\bigcup_{z \in D(1,\varepsilon)} \{\lambda_{i_1}(z), \ldots, \lambda_{i_q}(z)\}$ is open in $\mathbb{C}$ by Remark 4.2.10 and contains λ. Since this set again is contained in $\sigma_{\mathbf{\Delta}}(A(\cdot); \delta)$ (by the same argument as before), the proof is complete. □

Remark 5.1.10. The whole spectral value set $\sigma_{\mathbf{\Delta}}(A(\cdot); \delta)$ will in general not be open since some of the eigenvalues of A may be immovable under the perturbations $A \rightsquigarrow A(\Delta)$. For instance, if $A(\Delta) = \begin{bmatrix} 2 & \Delta_1 \\ 0 & \Delta_2 \end{bmatrix}$ for $\Delta \in \mathbf{\Delta} = \mathbf{\Delta}_0 = \mathbb{C}^2$, then 2 is a fixed eigenvalue of $A(\Delta)$ and, with respect to the ∞-norm on $\mathbb{C}^2$, $\sigma_{\mathbf{\Delta}}(A(\cdot); 1) = \mathbb{D} \cup \{2\}$ is not open, although the assumptions of the previous proposition are clearly satisfied. □

As mentioned at the beginning of this section, spectral value sets are of special interest if one wants to examine whether or not a given $\mathbb{C}_g$-stable system remains $\mathbb{C}_g$-stable under all parameter perturbations of norm below a given bound δ. A natural measure for the robustness of $\mathbb{C}_g$-stability is the supremal value of the bounds $\delta > 0$ which guarantee $\mathbb{C}_g$-stability. This leads to the following definition where we set, as usual, $\inf \emptyset = \infty$.

Definition 5.1.11. Under the conditions of Assumption 5.1.1 suppose that $\mathbb{C}_g \subset \mathbb{C}$ is a given non-trivial open subset of $\mathbb{C}$ and the nominal matrix $A = A(0)$ is $\mathbb{C}_g$-stable. Then

$$r_{\boldsymbol{\Delta}}(A(\cdot);\mathbb{C}_g) = \inf\{\|\Delta\|;\ \Delta \in \boldsymbol{\Delta} \setminus \boldsymbol{\Delta}_0 \ \text{ or } \ (\Delta \in \boldsymbol{\Delta}_0 \text{ and } \sigma(A(\Delta)) \not\subset \mathbb{C}_g)\} \tag{8}$$

is called the $\mathbb{C}_g$-stability radius of A under the perturbations $A \rightsquigarrow A(\Delta)$, $\Delta \in \boldsymbol{\Delta}_0$.

Thus the stability radius is the infimum of the norms of all perturbations $\Delta \in \boldsymbol{\Delta}$ for which $A(\Delta)$ is either not defined or not $\mathbb{C}_g$-stable. If $r_{\boldsymbol{\Delta}}(A(\cdot);\mathbb{C}_g) < \boldsymbol{\delta}_0$ (the well-posedness radius) then only the latter case can occur and the stability radius can be expressed in terms of spectral value sets. In fact, if $\sigma_{\boldsymbol{\Delta}}(A(\cdot);\boldsymbol{\delta}_0) \not\subset \mathbb{C}_g$ and $\mathbb{C}_b := \mathbb{C} \setminus \mathbb{C}_g$ then

$$r_{\boldsymbol{\Delta}}(A(\cdot);\mathbb{C}_g) = \inf\{\delta;\ \sigma_{\boldsymbol{\Delta}}(A(\cdot);\delta) \cap \mathbb{C}_b \neq \emptyset\}. \tag{9}$$

Example 5.1.12. Consider the damped linear oscillator studied in Example 5.1.4 with the real perturbation set $\boldsymbol{\Delta}_0^{\mathbb{R}}$ defined in (4). Then

$$\boldsymbol{\Delta}^{\mathbb{R}} = \overline{\boldsymbol{\Delta}_0^{\mathbb{R}}} = \{\Delta \in \mathbb{R}^3; \Delta_m \geq -m_0, \Delta_c \geq -c_0, \Delta_k \geq -k_0\} \subset \mathbb{R}^3.$$

Let $\mathbb{R}^3$ be provided with the standard Euclidean norm and consider the Hurwitz stability region $\mathbb{C}_g = \mathbb{C}_-$. The perturbed system equations have the state space form

$$\dot{x} = A(\Delta)x = \begin{bmatrix} 0 & 1 \\ -(k_0+\Delta_k)/(m_0+\Delta_m) & -(c_0+\Delta_c)/(m_0+\Delta_m) \end{bmatrix} x.$$

$\Delta = (-m_0, 0, 0)$ is the smallest perturbation in $\boldsymbol{\Delta}^{\mathbb{R}}$ such that $A(\Delta)$ is not defined, and so $\boldsymbol{\delta}_0 = m_0$. On the other hand, if $\Delta \in \boldsymbol{\Delta}^{\mathbb{R}}$, $\Delta_m > -m_0$ the well defined perturbed system $\dot{x} = A(\Delta)x$ will not be Hurwitz stable if and only if $\Delta_c = -c_0$ or $\Delta_k = -k_0$. Hence $r_{\boldsymbol{\Delta}}(A(\cdot);\mathbb{C}_-) = \min\{m_0, c_0, k_0\}$. □

Many more examples will be presented in Section 5.3, where we will also derive characterizations and numerical algorithms for determining stability radii for special perturbation structures. Here we only list some general properties of the stability radius, which are easy consequences of the Definition 5.1.11 and Lemma 5.1.7.

Lemma 5.1.13. *Suppose Assumption 5.1.1 holds and $\sigma(A(0)) \subset \mathbb{C}_g$ where $\mathbb{C}_g$ is a non-trivial open subset of $\mathbb{C}$. Then*

(i) $0 < r_{\boldsymbol{\Delta}}(A(\cdot);\mathbb{C}_g) \leq \boldsymbol{\delta}_0$, *and* $r_{\boldsymbol{\Delta}}(A(\cdot);\mathbb{C}_g) = \boldsymbol{\delta}_0$ *if and only if* $\sigma_{\boldsymbol{\Delta}}(A(\cdot);\boldsymbol{\delta}_0) \subset \mathbb{C}_g$.

(ii) If $r_{\boldsymbol{\Delta}}(A(\cdot);\mathbb{C}_g) < \boldsymbol{\delta}_0$ *there exists a* minimum norm destabilizing perturbation, *i.e.* $\Delta \in \boldsymbol{\Delta}_0$ *such that* $\sigma(A(\Delta)) \not\subset \mathbb{C}_g$ *and* $\|\Delta\| = r_{\boldsymbol{\Delta}}(A(\cdot);\mathbb{C}_g)$.

(iii) $r_{\boldsymbol{\Delta}}(A(\cdot);\mathbb{C}_g) = \infty$ *if and only if* $\boldsymbol{\Delta} = \boldsymbol{\Delta}_0$ *and* $\sigma(A(\Delta)) \subset \mathbb{C}_g$ *for all* $\Delta \in \boldsymbol{\Delta}_0$.

(iv) If the stability region $\mathbb{C}_g$ *increases then* $r_{\boldsymbol{\Delta}}(A(\cdot);\mathbb{C}_g)$ *increases.*

(v) The stability radius remains the same if we replace $\mathbb{C}_g$ *by the complement in* $\mathbb{C}$ *of the boundary* $\partial\mathbb{C}_g = \overline{\mathbb{C}_g} \setminus \mathbb{C}_g$

$$r_{\boldsymbol{\Delta}}(A(\cdot);\mathbb{C}_g) = r_{\boldsymbol{\Delta}}(A(\cdot);\mathbb{C} \setminus \partial\mathbb{C}_g). \tag{10}$$

(vi) If $\boldsymbol{\Delta}_0$ decreases and $\boldsymbol{\delta}_0$ increases, then $r_{\boldsymbol{\Delta}}(A(\cdot);\mathbb{C}_g)$ increases. More precisely, suppose that $\boldsymbol{\Delta}_0' \subset \boldsymbol{\Delta}_0$ satisfies the conditions of Assumption 5.1.1 and $\boldsymbol{\Delta}'$ is the closure of $\boldsymbol{\Delta}_0'$ in V. If $\boldsymbol{\delta}_0' = \inf\{\|\Delta\|; \Delta \in \overline{\boldsymbol{\Delta}_0'} \setminus \boldsymbol{\Delta}_0'\} \geq \boldsymbol{\delta}_0$ then $r_{\boldsymbol{\Delta}'}(A(\cdot)|\boldsymbol{\Delta}_0';\mathbb{C}_g) \geq r_{\boldsymbol{\Delta}}(A(\cdot);\mathbb{C}_g)$ (with respect to the same norm on V).

Proof: Since $\sigma(A(0)) \subset \mathbb{C}_g$, the continuity of the spectrum implies $\sigma(A(\cdot),\delta) \subset \mathbb{C}_g$ for $\delta < \boldsymbol{\delta}_0$ sufficiently small. Therefore $r_{\boldsymbol{\Delta}}(A(\cdot);\mathbb{C}_g) > 0$. The remaining statement in (i) follows directly from Definition 5.1.11.
(ii) If $r_{\boldsymbol{\Delta}}(A(\cdot);\mathbb{C}_g) < \boldsymbol{\delta}_0$ there exists a sequence of $\Delta_k \in \boldsymbol{\Delta}_0$ satisfying $\sigma(A(\Delta_k)) \not\subset \mathbb{C}_g$ and $\|\Delta_k\| \to r_{\boldsymbol{\Delta}}(A(\cdot);\mathbb{C}_g)$ as $k \to \infty$. By a compactness argument we may assume that (Δ_k) is convergent. Then $\Delta := \lim_{k\to\infty} \Delta_k \in \boldsymbol{\Delta}_0$, $\|\Delta\| = r_{\boldsymbol{\Delta}}(A(\cdot);\mathbb{C}_g)$, and $\sigma(A(\Delta)) \not\subset \mathbb{C}_g$ by the continuity of the spectrum and the openness of $\mathbb{C}_g$.
Statements (iii), (iv) and (vi) follow directly from Definition 5.1.11. So it remains to prove (v). Since $\mathbb{C}_g \subset \mathbb{C} \setminus \partial\mathbb{C}_g$, it follows from (iv) that $\leq$ holds in (10). If $r_{\boldsymbol{\Delta}}(A(\cdot);\mathbb{C}_g) = \boldsymbol{\delta}_0$ then equality holds by (i). If $r_{\boldsymbol{\Delta}}(A(\cdot);\mathbb{C}_g) < \boldsymbol{\delta}_0$ then there exists a minimum norm $\Delta \in \boldsymbol{\Delta}_0$ such that $\sigma(A(\Delta)) \not\subset \mathbb{C}_g$ by (ii). Note that $\alpha\Delta \in \boldsymbol{\Delta}_0$ for all $\alpha \in [0,1]$ by Assumption 5.1.1, since $\|\Delta\| = r_{\boldsymbol{\Delta}}(A(\cdot);\mathbb{C}_g) < \boldsymbol{\delta}_0$. By continuity of the spectrum $A(\Delta)$ cannot have any eigenvalue in $\mathbb{C}\setminus\overline{\mathbb{C}_g}$ since otherwise $\sigma(A((1-\varepsilon)\Delta)) \not\subset \mathbb{C}_g$ for $\varepsilon > 0$ sufficiently small and this yields a contradiction. Thus $\sigma(A(\Delta)) \cap \partial\mathbb{C}_g \neq \emptyset$ and this concludes the proof. □

The proof shows that every minimum norm destabilizing perturbation moves at least one eigenvalue onto the boundary $\partial\mathbb{C}_g$. We conclude this subsection with an example illustrating the difference between stability radii of stable real systems for real and for complex perturbations.

Example 5.1.14. Let $\mathbb{C}_g = \mathbb{C}_-$ and consider the matrix family $A(\Delta) = \begin{bmatrix} -1 & 1+\Delta \\ -1-\Delta & -1 \end{bmatrix}$, $\Delta \in \mathbb{R}$ where we provide $\boldsymbol{\Delta} = \boldsymbol{\Delta}_0 = \mathbb{R}$ with the norm $|\cdot|$. Since all the matrices $A(\Delta)$, $\Delta \in \mathbb{R}$ are Hurwitz stable we have $r_{\mathbb{R}}(A(\cdot);\mathbb{C}_-) = \infty$. On the other hand, if we allow complex perturbations $\Delta \in \mathbb{C}$ where $\boldsymbol{\Delta} = \boldsymbol{\Delta}_0 = \mathbb{C}$ is also provided with the norm $|\cdot|$, then $\Delta = -1 - \imath$ is a minimal norm destabilizing perturbation, so $r_{\mathbb{C}}(A(\cdot);\mathbb{C}_-) = \sqrt{2}$. □

5.1.2 Perturbation Structures

At present the computation of spectral value sets is only feasible for very specific perturbation structures. In this subsection we will describe these structures and illustrate by means of examples the type of parameter uncertainty they can portray.

Affine Perturbations

Much of our development will be for affine perturbations of the form

$$A \rightsquigarrow A(\Delta) = A + B\Delta C, \qquad \Delta \in \boldsymbol{\Delta} \tag{11}$$

where $(B,C) \in \mathbb{K}^{n\times\ell} \times \mathbb{K}^{q\times n}$ are given, $\boldsymbol{\Delta} \subset \mathbb{K}^{\ell\times q}$ is a closed convex cone as in Section 4.4 with some operator norm $\|\cdot\|_{\boldsymbol{\Delta}}$ on span $\boldsymbol{\Delta}$. Assumption 5.1.1 is then trivially satisfied with $\boldsymbol{\Delta}_0 = \boldsymbol{\Delta}$ (hence $\boldsymbol{\delta}_0 = \infty$) and $(V, \|\cdot\|) = (\text{span}\,\boldsymbol{\Delta}, \|\cdot\|_{\boldsymbol{\Delta}})$.

The perturbed matrix $A(\Delta)$ can be interpreted as the system matrix of a feedback system obtained by applying *static linear output feedback* $w = \Delta z$ to the system

$$\Sigma: \begin{array}{rcl} \dot{x}(t) &=& Ax(t) + Bw(t),\ t \geq 0 \\ z(t) &=& Cx(t) \end{array} \quad \text{or} \quad \begin{array}{rcl} x(t+1) &=& Ax(t) + Bw(t),\ t \in \mathbb{N} \\ z(t) &=& Cx(t) \end{array} \tag{12}$$

(see Example 2.4.12 and Figure 5.1.4).

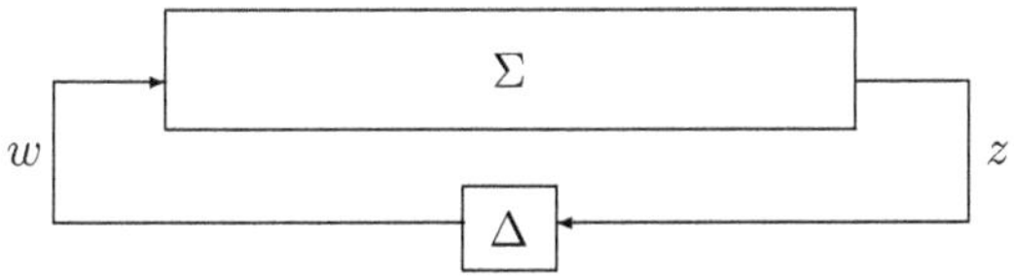

Figure 5.1.4: Feedback interpretation of the perturbed system

Here the matrices B, C are not *given* control and measurement matrices as in Chapter 2 but are *chosen* to reflect the structure and possible scaling of the perturbation of the system matrix A. The input w may be interpreted as a disturbance input coupled to the output z by an unknown feedback matrix Δ. This feedback interpretation is fundamental for the following control theoretic treatment of parameter uncertainties, and is the main reason we represent parameter deviations by matrices instead of vectors, and measure their size by operator norms. In concrete applications the set of all possible perturbed system matrices $\{A(\Delta); \Delta \in \boldsymbol{\Delta}\}$ is determined by our a priori knowledge about the perturbation structure and it is assumed that the "real" system is adequately described by at least one model in this set (multi-model approach).
In general, the model class (11) depends upon the structure matrices B, C and the set $\boldsymbol{\Delta}$ of parameter perturbations, and can be represented in many different ways by choosing B, C and $\boldsymbol{\Delta}$ appropriately. In fact, choosing e.g. $\boldsymbol{\Delta}' = \{B\Delta C;\ \Delta \in \boldsymbol{\Delta}\}$ the set of perturbed models can also be represented in the form $\{A + \Delta';\ \Delta' \in \boldsymbol{\Delta}'\}$ where the corresponding structure matrices are $B' = C' = I_n$.
If $B = C = I_n$ and $\boldsymbol{\Delta} = \mathbb{K}^{n\times n}$, the perturbations of the form

$$A \rightsquigarrow A(\Delta) = A + \Delta,\ \Delta \in \mathbb{K}^{n\times n} \tag{13}$$

are called *unstructured* and the corresponding *unstructured spectral value sets*[3] are denoted by $\sigma_{\mathbb{K}}(A;\delta) := \sigma_{\mathbb{K}^{n\times n}}(A(\cdot);\delta)$.
The flexibility introduced by the matrices B, C can be used in order to reduce the set of parameter perturbations $\boldsymbol{\Delta}$ to a form for which spectral value sets and stability radii can be effectively computed. The simplest type of affine perturbations are the *full-block perturbations* for which $\boldsymbol{\Delta}$ is maximal, i.e. $\boldsymbol{\Delta} = \mathbb{K}^{\ell\times q}$. The associated spectral value sets are denoted by

$$\sigma_{\mathbb{K}}(A;B,C;\delta) = \{\lambda \in \mathbb{C};\ \exists \Delta \in \mathbb{K}^{\ell\times q},\ \|\Delta\|_{\mathbb{K}^{\ell\times q}} < \delta : \lambda \in \sigma(A + B\Delta C)\}. \tag{14}$$

[3]For $\mathbb{K} = \mathbb{C}$ the unstructured spectral value sets $\sigma_{\mathbb{C}}(A;\delta)$ are called *pseudospectra, spectral portraits* or *ε-spectra* of A in the literature (see *Notes and References*).

It often happens that certain submatrices of A do not contain any parameters, since their entries are completely determined by the structure of the system or by the modelling process. By an appropriate choice of B and C, formula (11) with $\mathbf{\Delta} = \mathbb{K}^{\ell \times q}$ can describe parameter uncertainties where all the entries in certain rows and columns of A are uncertain whilst the other entries are precisely known. This is illustrated in the following simple example.

Example 5.1.15. Consider the linear oscillator with mass one, $\ddot{\xi} + c\dot{\xi} + k\xi = 0,\ k,\, c \in \mathbb{R}$, or in state space form

$$\dot{x} = Ax, \qquad A = \begin{bmatrix} 0 & 1 \\ -k & -c \end{bmatrix}. \tag{15}$$

A perturbation of the entries in the first row of A does not make sense since these entries result from the transformation into state space form, they are fixed entries and do not contain any system parameters. Therefore the use of the *unstructured* spectral value set $\sigma_{\mathbb{K}}(A;\delta) = \sigma_{\mathbb{K}}(A; I_2, I_2; \delta)$ would be inappropriate. In order that the perturbations do not affect the first row of A we choose $B = [0\ 1]^\top$. If both the restoring force and friction coefficients, k and c are uncertain, we can take this into account by setting $C = I_2$. Whereas if e.g. only c is uncertain we take $C = [0\ 1]$. In the former case the natural disturbance class is $\mathbf{\Delta} = \mathbb{K}^{1\times 2}$ and in the latter $\mathbf{\Delta} = \mathbb{K}$.
In order to illustrate the difference between spectral value sets with respect to unstructured and structured perturbations, we take $c_0 = 4$, $k_0 = 5$ as nominal parameters, provide all vector spaces with their usual Euclidean norm and consider, for computational reasons, the case of complex perturbations $\mathbb{K} = \mathbb{C}$. Figure 5.1.5 compares the spectral value

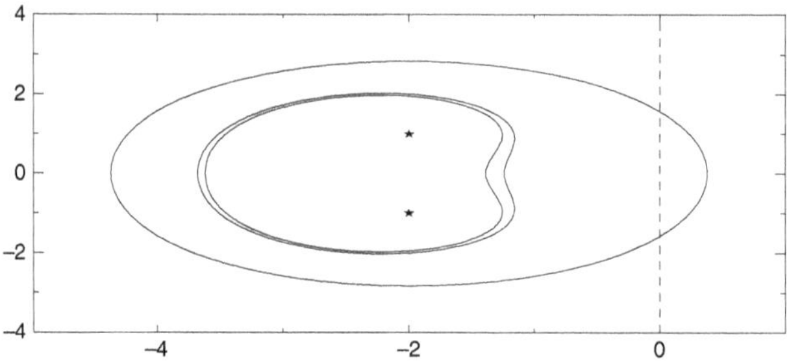

Figure 5.1.5: Spectral value sets $\sigma_{\mathbb{C}}(A; B, [0\ 1]; 1) \subset \sigma_{\mathbb{C}}(A; B, I_2; 1) \subset \sigma_{\mathbb{C}}(A; 1)$

sets $\sigma_{\mathbb{C}}(A; B, I_2; \delta)$ (both physical parameters c, k uncertain), $\sigma_{\mathbb{C}}(A; B, [0\ 1]; \delta)$ (only c is uncertain) and $\sigma_{\mathbb{C}}(A;\delta)$ (unstructured perturbations of A), for $\delta = 1$. Identifying the perturbation $\Delta \in \mathbb{C}^{1\times 2}$ ($\Delta \in \mathbb{C}$) with the unstructured one $\Delta' = B\Delta \in \mathbb{C}^{2\times 2}$ (resp. $\Delta' = B\Delta C \in \mathbb{C}^{2\times 2}$), we obtain a norm preserving embedding of $\mathbb{C}^{1\times 2}$ (resp. $\mathbb{C}$) into $\mathbb{C}^{2\times 2}$, so that necessarily $\sigma_{\mathbb{C}}(A; B, C; \delta) \subset \sigma_{\mathbb{C}}(A; B, I_2; \delta) \subset \sigma_{\mathbb{C}}(A;\delta)$. The figure shows that the unstructured spectral value set can be a misleading indicator of the robustness of stability. In fact, the pseudospectrum $\sigma_{\mathbb{C}}(A;1)$ seems to indicate that the oscillator can be destabilized by perturbations of size < 1 whereas, in reality, the oscillator is robustly stable for all parameter perturbations $\Delta \in \mathbb{C}^{1\times 2}$ of norm < 1, as shown by the spectral value set $\sigma_{\mathbb{C}}(A; B, I_2; 1)$. □

Remark 5.1.16. In general the map $\Delta \mapsto B\Delta C$ from $\mathbf{\Delta}$ into $\mathbb{K}^{n\times n}$ will not be one-to-one so that the matrix perturbation set $\{B\Delta C;\ \Delta \in \mathbf{\Delta}\}$ will be overparametrized.

However, this can often be avoided by choosing B and C carefully. E.g., in the full-block case, one can eliminate the linearly dependent columns in B (resp. rows in C) so that B (resp. C) has full column (resp. row) rank. Then $\Delta \mapsto B\Delta C$ will be one-to-one. □

The matrices B, C determine – together with $\mathbf{\Delta}$ – not only the structure of the perturbations but also the scale of how the overall matrix perturbation $B\Delta C$ depends on the parameter Δ. As in the previous subsection we assume that the uncertainty is *norm bounded*, so that the uncertain system can be described by a system family of the form

$$\dot{x}(t) = (A + B\Delta C)x(t) \quad \text{or} \quad x(t+1) = (A + B\Delta C)x(t), \qquad \Delta \in \mathbf{\Delta}, \ \|\Delta\|_{\mathbf{\Delta}} < \delta.$$

In order to ensure that the bound δ adequately represents the degree of the actual model uncertainty, the matrices B, C must be carefully scaled and the norm $\|\cdot\|_{\mathbf{\Delta}}$ carefully chosen. For instance, if one parameter is quite accurately known whereas another one is very uncertain, their different degrees of uncertainty should be balanced by appropriate scaling of B, C or by the introduction of weighted norms.

Example 5.1.17. In *Franklin et al.* (1986) a mathematical model for the attitude control of a satellite is given. The satellite comprises two wings connected to a central body which supports antennae and sensors. It is required that the antennae, sensors and solar panels are properly orientated and usually this is achieved by pairs of gas jets which can exert torques about each of three perpendicular axes. Once the correct orientations have been obtained, there is a stabilization problem of maintaining them.
As a first step to the general three axis problem of controlling the motion of one of the wings, the motion about a single axis is considered. The wing and the rest of the satellite are modelled as rigid bodies (with moments of inertia J_1, J_2) connected by a rotational spring (with spring constant k) and a rotational viscous damper (with damping constant c). If θ_1, θ_2 are the angular displacement of the wing and the rest of the satellite and τ the control torque exerted on the wing, the equations of motion take the form

$$\begin{aligned} J_1\ddot{\theta}_1 + c(\dot{\theta}_1 - \dot{\theta}_2) + k(\theta_1 - \theta_2) &= \tau \\ J_2\ddot{\theta}_2 + c(\dot{\theta}_2 - \dot{\theta}_1) + k(\theta_2 - \theta_1) &= 0. \end{aligned}$$

Note that this system is a rotational counterpart of the car suspension system discussed in Example 1.3.2 with zero tyre friction and zero spring constants. Let $x = [\theta_2 \ \dot{\theta}_2 \ \theta_1 \ \dot{\theta}_1]^\top$ and $u = \tau/J_1$, then these equations can be written in state space form $\dot{x} = Ax + bu$, where

$$A = \begin{bmatrix} 0 & 1 & 0 & 0 \\ -k/J_2 & -c/J_2 & k/J_2 & c/J_2 \\ 0 & 0 & 0 & 1 \\ k/J_1 & c/J_1 & -k/J_1 & -c/J_1 \end{bmatrix}, \quad b = \begin{bmatrix} 0 \\ 0 \\ 0 \\ 1 \end{bmatrix}. \tag{16}$$

Physical constraints associated with temperature variations imply that the parameters k and c are constrained by

$$0.09 \le k \le 0.4, \qquad 0.04\sqrt{k/10} \le c \le 0.2\sqrt{k/10}. \tag{17}$$

We denote the set of $\omega = (k, c) \in \mathbb{R}^2$ which satisfy the above bounds by Ω. In *Franklin et al.* (1986) the moments of inertia are assumed to be exactly known: $J_1 = 1$, $J_2 = 0.1$ and the problem of designing a feedback control $u = f^\top x$, $f^\top = [f_1, f_2, f_3, f_4] \in \mathbb{R}^{1\times 4}$

which stabilizes the system for all $(k, c) \in \Omega$ is considered. Choosing a nominal $\omega_0 = (k_0, c_0) \in \Omega$, the equations of motion of the uncertain closed loop system $\dot{x} = (A + bf^\top)x$ with $J_1 = 1$, $J_2 = 0.1$ and parameters k, c restricted by (17) can be re-written as

$$\dot{x} = A(k_0, c_0, f)x + P(\Delta_1, \Delta_2)x, \qquad \Delta_1 = k - k_0,\ \Delta_2 = c - c_0$$

where the nominal closed loop system matrix $A(k_0, c_0, f)$ and the overall perturbation matrix $P(\Delta_1, \Delta_2)$ are defined by

$$\begin{aligned} A(k_0, c_0, f) &= \begin{bmatrix} 0 & 1 & 0 & 0 \\ -10k_0 & -10c_0 & 10k_0 & 10c_0 \\ 0 & 0 & 0 & 1 \\ f_1 + k_0 & f_2 + c_0 & f_3 - k_0 & f_4 - c_0 \end{bmatrix}, \\ P(\Delta_1, \Delta_2) &= \begin{bmatrix} 0 & 0 & 0 & 0 \\ -10\Delta_1 & -10\Delta_2 & 10\Delta_1 & 10\Delta_2 \\ 0 & 0 & 0 & 0 \\ \Delta_1 & \Delta_2 & -\Delta_1 & -\Delta_2 \end{bmatrix}. \end{aligned}$$

The perturbation matrices $P(\Delta_1, \Delta_2)$ have a very distinct structure and can be represented as full-block perturbations of the form $B\Delta C$, $\Delta \in \boldsymbol{\Delta} = \mathbb{R}^{1\times 2}$, viz.

$$P(\Delta_1, \Delta_2) = \begin{bmatrix} 0 \\ -10 \\ 0 \\ 1 \end{bmatrix} [\Delta_1,\ \Delta_2] \begin{bmatrix} 1 & 0 & -1 & 0 \\ 0 & 1 & 0 & -1 \end{bmatrix} = B\Delta C, \quad \Delta \in \boldsymbol{\Delta} = \mathbb{R}^{1\times 2}.$$

Let $B_{\boldsymbol{\Delta}}(\delta) = \{\Delta \in \boldsymbol{\Delta};\ \|\Delta\|_{\boldsymbol{\Delta}} < \delta\}$ and $A(\Delta) = A(k_0, c_0, f) + B\Delta C$. Note that only the nominal closed loop system depends upon f whereas the perturbations of the closed loop system remain the same for all possible feedback vectors $f^\top \in \mathbb{R}^{1\times 4}$. Given $(k_0, c_0) \in \Omega$, $f \in \mathbb{R}^4$ such that $\sigma(A(k_0, c_0, f)) \subset \mathbb{C}_-$ one way of testing whether the robustness objective has been met, is to check if there exists a $\delta > 0$ such that

$$\Omega \subset (k_0, c_0) + B_{\boldsymbol{\Delta}}(\delta) \quad \text{and} \quad \sigma_{\boldsymbol{\Delta}}(A(\cdot); \delta) = \bigcup_{\Delta \in B_{\boldsymbol{\Delta}}(\delta)} \sigma(A(k_0, c_0, f) + B\Delta C) \subset \mathbb{C}_-. \quad (18)$$

In order to achieve this, one tries to choose a norm $\|\cdot\|_{\boldsymbol{\Delta}}$ such that $(k_0, c_0) + B_{\boldsymbol{\Delta}}(\delta)$ contains and is a good approximation of Ω for some δ. Now since the maximal deviation of k from

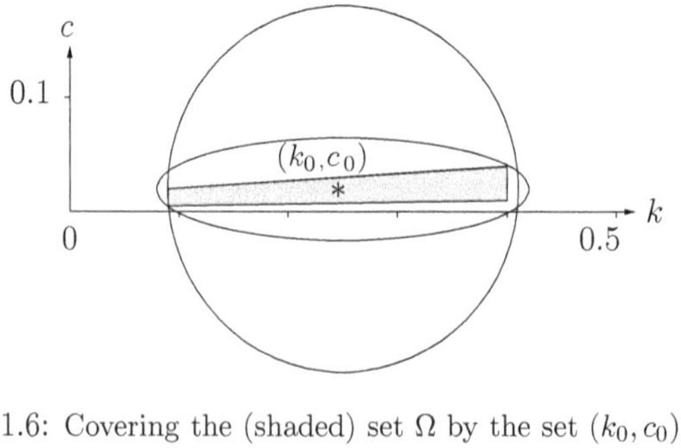

Figure 5.1.6: Covering the (shaded) set Ω by the set $(k_0, c_0) + B_{\boldsymbol{\Delta}}(\delta)$

its nominal value is greater than that of c, one might think of using a weighted perturbation norm $\|\Delta\|_\alpha^2 = \Delta_1^2 + \alpha^2\Delta_2^2$ for some $\alpha > 1$. This is illustrated in Figure 5.1.6 which shows the (shaded) parameter set Ω and the estimates $(k_0, c_0)+B_{\mathbf{\Delta}}(\delta)$ for the standard Euclidean norm and the weighted norm $\|\cdot\|_\alpha$ where $\alpha^2 = 14.67$. Here $(k_0, c_0) = (0.25, 0.02)$ and the $\delta > 0$ is chosen to be minimal so that $(k_0, c_0)+B_{\mathbf{\Delta}}(\delta) \supset \Omega$ ($\delta = 0.1068$ for the unweighted Euclidean norm and $\delta = 0.1684$ for $\|\cdot\|_\alpha$). We see that, for the unweighted Euclidean norm, perturbations yielding much larger values of c than those in Ω must be considered. This suggests that by using the weighted norm there is greater chance of finding a δ such that (18) holds. Of course it might be that there exists a destabilizing Δ in the small part of the set bounded by the ellipse which is not in the set bounded by the circle. So we cannot be sure at this stage that using the weighted norm is better. However we will show that this is indeed the case in Example 5.3.19 where we use a characterization of the stability radius to check, in a succinct way, whether or not a particular feedback control achieves the objective. □

Not every affine matrix perturbation can be represented as a *full-block* perturbation of the form $A \rightsquigarrow A + B\Delta C$ with suitably chosen structure matrices B, C. In fact, if $(B\Delta C)_{ij} = 0$ for all $\Delta \in \mathbb{K}^{\ell\times q}$ then necessarily the i-th row vector of B or j-th column vector of C is zero. Hence if the (i,j)-th element of $A + B\Delta C$ does not depend on Δ then all the elements of the i-th row or j-th column remain unchanged under the perturbation. In particular it is not possible to represent (in this form) affine perturbations of A which affect exclusively the diagonal elements of A. However, every affine perturbation can be represented in the form (11) with $\mathbf{\Delta}$ a linear subspace of diagonal matrices (compare Example 4.4.16). This is shown in the following example.

Example 5.1.18. Consider a general affine matrix valued mapping of the form

$$z = (z_1, \dots, z_N) \mapsto A(z_1, \dots, z_N) = A + \sum_{i=1}^{N} z_i A_i, \quad z \in \mathbb{K}^N$$

where the matrices $A_i \in \mathbb{K}^{n\times n}$ are given. Let $A_i = B_i C_i$ be arbitrary factorizations of the A_i and $B_i \in \mathbb{K}^{n\times \ell_i}$, $C_i \in \mathbb{K}^{\ell_i \times n}$. Then, setting

$$B = [B_1\, B_2, \dots, B_N]; \quad C = [C_1^\top, \dots, C_N^\top]^\top; \quad \Delta(z) = \operatorname{diag}(z_1 I_{\ell_1}, \dots, z_N I_{\ell_N}),\ z \in \mathbb{K}^N$$

we obtain

$$A(z_1, \dots, z_N) = A + B\Delta(z)C, \quad z \in \mathbb{K}^N.$$

Hence the affine perturbation $A \rightsquigarrow A(z_1, \dots, z_N)$ can be represented in the form (11) with $\mathbf{\Delta} = \{\Delta(z);\ z \in \mathbb{K}^N\}$. □

We conclude our discussion of affine matrix perturbations by briefly considering the *interconnection* of systems with affine parameter uncertainties. For this we need subsystems with two (vector) inputs and two (vector) outputs where one input/output pair $(u^{(i)}, y^{(i)})$ is suitable for interconnection whereas the other input/output pair $(w^{(i)}, z^{(i)})$ is used for the feedback representation of uncertainty:

$$\Sigma_i(\Delta_i)\colon \begin{cases} \dot{x}^{(i)} &= A^{(i)}x^{(i)} + B_1^{(i)}w^{(i)} + B_2^{(i)}u^{(i)} \\ z^{(i)} &= C_1^{(i)}x^{(i)} \\ w^{(i)} &= \Delta_i z^{(i)} \\ y^{(i)} &= C_2^{(i)}x^{(i)} \end{cases} \quad \text{i.e.} \quad \begin{cases} \dot{x}^{(i)} = \left(A^{(i)} + B_1^{(i)}\Delta_i C_1^{(i)}\right)x^{(i)} + B_2^{(i)}u^{(i)} \\ y^{(i)} = C_2^{(i)}x^{(i)}. \end{cases}$$

Here the dimensions of $x^{(i)}, w^{(i)}, u^{(i)}, z^{(i)}, y^{(i)}$ are $n_i, \ell_i, m_i, q_i, p_i$, respectively. In the following example we consider the feedback coupling of two uncertain subsystems of this kind.

Example 5.1.19. Consider two uncertain systems $\Sigma_i(\Delta_i)$, $i = 1, 2$ of the above form with full-block uncertainties and associated nominal systems $\Sigma_1 = \Sigma_1(0)$ and $\Sigma_2 = \Sigma_2(0)$. Then the uncertain feedback system $\Sigma(\Delta_1, \Delta_2) = \Sigma_1(\Delta_1) \Box \Sigma_2(\Delta_2)$ obtained by feedback interconnection of $\Sigma_1(\Delta_1)$ and $\Sigma_2(\Delta_2)$ (see Figure 5.1.7 (a)) can equivalently be described

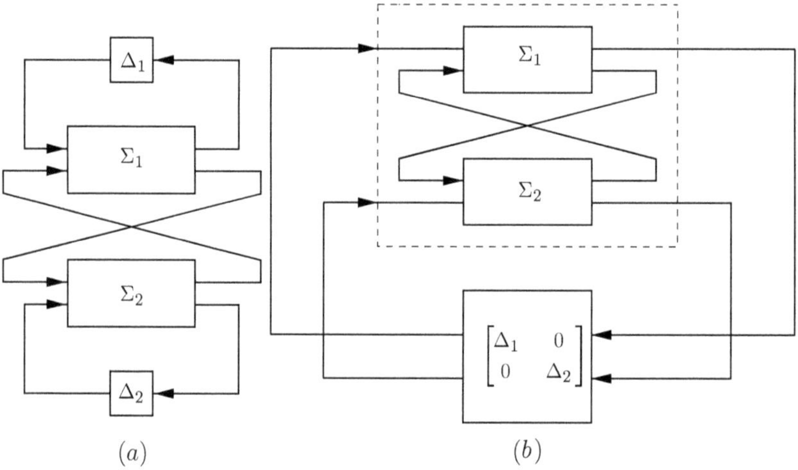

Figure 5.1.7: Feedback interconnection of two uncertain systems: Pulling out the Δs

by applying the perturbation $\text{diag}(\Delta_1, \Delta_2)$ to the nominal feedback system $\Sigma_1 \Box \Sigma_2$ as indicated in Figure 5.1.7 (b). The system matrix $A(\Delta_1, \Delta_2)$ of the uncertain feedback system $\Sigma(\Delta_1, \Delta_2)$ is obtained by setting $u^{(1)} = y^{(2)}$ and $u^{(2)} = y^{(1)}$. So

$$A(\Delta_1, \Delta_2) = \begin{bmatrix} A^{(1)} & B_2^{(1)} C_2^{(2)} \\ B_2^{(2)} C_2^{(1)} & A^{(2)} \end{bmatrix} + \begin{bmatrix} B_1^{(1)} & 0 \\ 0 & B_1^{(2)} \end{bmatrix} \begin{bmatrix} \Delta_1 & 0 \\ 0 & \Delta_2 \end{bmatrix} \begin{bmatrix} C_1^{(1)} & 0 \\ 0 & C_1^{(2)} \end{bmatrix}.$$

Hence the uncertainty of the feedback system can be represented in the form (11) with structure matrices $B = \text{diag}(B_1^{(1)}, B_1^{(2)})$, $C = \text{diag}(C_1^{(1)}, C_1^{(2)})$ and perturbations $\Delta = \text{diag}(\Delta_1, \Delta_2) \in \boldsymbol{\Delta} := \boldsymbol{\Delta}_1 \oplus \boldsymbol{\Delta}_2 = \mathbb{K}^{\ell_1 \times q_1} \oplus \mathbb{K}^{\ell_2 \times q_2}$. □

The previous example illustrates that the full-block uncertainty of two subsystems yields a *block diagonal* perturbation set for their feedback interconnection. The transition from the individual uncertainties of the system components (Figure 5.1.7(a)) to the representation of the resulting uncertainty of the interconnected system in output feedback form (11) is graphically referred to as *"pulling out the Δs"*. We will now see that the same can be done for general networks of uncertain linear systems as considered in Section 2.4, even if not only the subsystems but also the couplings between them are affinely perturbed. This shows the importance of block-diagonal perturbation sets for the analysis of uncertain composite systems.

Example 5.1.20. Consider an arbitrary linear interconnection of N nominal linear systems $(\tilde{A}_i, \tilde{B}_i, \tilde{C}_i) \in \mathbf{L}_{n_i,m_i,p_i}(\mathbb{K})$ as described in Section 2.4. If K_{ij}, $i,j \in \underline{N}$ are the linear coupling matrices between these systems, the system matrix A of the overall interconnected system has the form (see (2.4.32) in Section 2.4)

$$\begin{bmatrix} A_{11} & \dots & A_{1N} \\ \vdots & \vdots & \vdots \\ A_{N1} & \dots & A_{NN} \end{bmatrix} \in \mathbb{K}^{n\times n} \text{ where } \begin{array}{rcll} A_{ii} &=& \tilde{A}_i + \tilde{B}_i K_{ii} \tilde{C}_i \in \mathbb{K}^{n_i\times n_i}, & i \in \underline{N} \\ A_{ij} &=& \tilde{B}_i K_{ij} \tilde{C}_j \in \mathbb{K}^{n_i\times n_j}, & i,j \in \underline{N},\ i \neq j \\ n &=& n_1 + \ldots + n_N. & \end{array} \tag{19}$$

Now suppose that each block A_{ij} is subjected to an affine full block perturbation of the form $A_{ij} \rightsquigarrow A_{ij}(\Delta_{ij}) = A_{ij} + B_{ij}\Delta_{ij}C_{ij}$ where $B_{ij} \in \mathbb{K}^{n_i\times \ell_{ij}}$, $C_{ij} \in \mathbb{K}^{q_{ij}\times n_j}$ are given and the $\Delta_{ij} \in \mathbb{K}^{\ell_{ij}\times q_{ij}}$ are unknown. Let

$$\ell_i = \textstyle\sum_{j=1}^N \ell_{ij}, \quad \ell = \sum_{i=1}^N \ell_i; \quad q_i = \sum_{j=1}^N q_{ij}, \quad q = \sum_{i=1}^N q_i$$

and define the $N \times N$ block matrix $A(\Delta) = (A_{ij}(\Delta_{ij}))_{i,j\in\underline{N}}$ where the parameter vector $\Delta := \mathrm{diag}(\Delta_{ij})_{i,j\in\underline{N}} \in \bigoplus_{i\in\underline{N}} \bigoplus_{j\in\underline{N}} \mathbb{K}^{\ell_{ij}\times q_{ij}}$. Then $A(\Delta) = A + B\Delta C$ where

$$\begin{array}{rcll} B &=& \mathrm{diag}(B_1,\ldots,B_N) \in \mathbb{K}^{n\times \ell}, & B_i = [B_{i1},\ldots,B_{iN}] \in \mathbb{K}^{n_i\times \ell_i}, \quad i \in \underline{N} \\ C &=& [C_1^\top,\ldots,C_N^\top]^\top \in \mathbb{K}^{q\times n}, & C_i = \mathrm{diag}(C_{i1},\ldots,C_{iN}) \in \mathbb{K}^{q_i\times n}, \quad i \in \underline{N} \\ \Delta &=& \multicolumn{2}{l}{\mathrm{diag}(\Delta_{11},\Delta_{12},\ldots,\Delta_{1N};\Delta_{21},\ldots,\Delta_{2N};\ldots;\Delta_{N1},\ldots,\Delta_{NN}) \in \mathbb{K}^{\ell\times q}.} \end{array}$$

Thus the full-block perturbations $A_{ij} \rightsquigarrow A_{ij}(\Delta_{ij}) = A_{ij} + B_{ij}\Delta_{ij}C_{ij}$, $\Delta_{ij} \in \mathbb{K}^{\ell_{ij}\times q_{ij}}$ of the system components result in perturbations of the composite system matrix of the form (11) with block-diagonal perturbation set $\boldsymbol{\Delta} = \{\mathrm{diag}\,(\Delta_{ij})_{i,j\in\underline{N}};\ \forall i,j \in \underline{N} : \Delta_{ij} \in \mathbb{K}^{\ell_{ij}\times q_{ij}}\}$. The perturbation structure is greatly simplified if the B_{ij} only depend on the row index i, i.e. $B_{ij} \equiv B_i \in \mathbb{K}^{n_i\times \tilde{\ell}_i}$, and the C_{ij} only depend on the column index j so that $C_{ij} \equiv C_j \in \mathbb{K}^{\tilde{q}_j\times n_j}$, for all $i,j \in \underline{N}$. Then $A(\Delta)$ can be represented in the form $A + B\Delta C$ with

$$B = \mathrm{diag}(B_1,\ldots,B_N), \;\; C = \mathrm{diag}(C_1,\ldots,C_N), \;\; \Delta = \begin{bmatrix} \Delta_{11} & \dots & \Delta_{1N} \\ \vdots & \vdots & \vdots \\ \Delta_{N1} & \dots & \Delta_{NN} \end{bmatrix} \in \mathbb{K}^{\tilde{\ell}\times\tilde{q}} \tag{20}$$

where $\tilde{\ell} = \sum_{i=1}^N \tilde{\ell}_i$, $\tilde{q} = \sum_{j=1}^N \tilde{q}_j$. So in this case we obtain a full-block representation of the uncertainty of the composite system. A special subcase is obtained if all the subsystems $(\tilde{A}_i, \tilde{B}_i, \tilde{C}_i)$ are certain and only the coupling matrices are perturbed: $K_{ij} \rightsquigarrow K_{ij}(\Delta_{ij}) = K_{ij} + \Delta_{ij}$. Then the resulting uncertainty of the composite system can be described by a full-block perturbation structure as in (20) with $B_i = \tilde{B}_i$, $C_j = \tilde{C}_j$. On the other hand, if only the input matrices $\tilde{B}_i$ or only the output matrices $\tilde{C}_j$ are perturbed to $\tilde{B}_i(\Delta_i) = \tilde{B}_i + \Delta_i$, $\Delta_i \in \mathbb{K}^{n_i\times m_i}$ (resp. $\tilde{C}_j(\Delta_j) = \tilde{C}_j + \Delta_j$, $\Delta_j \in \mathbb{K}^{p_j\times n_j}$) then the resulting overall perturbations of the composite system matrix A are affine and can be represented in block-diagonal form $B\Delta C$, $\Delta = \mathrm{diag}(\Delta_1,\ldots,\Delta_N)$, but in general not by a set of full-block perturbations. We leave it to the reader to determine the corresponding structure matrices B and C. If both the input and output matrices are uncertain, the composite system will no longer depend in an affine way on the parameter perturbations, but in a bilinear one. □

Linear Fractional Perturbations

The parameter dependencies representable by static linear output feedback $w = \Delta z$ can be widened considerably if we introduce a direct feedthrough matrix in the system equations (12). Consider the uncertain feedback system in Figure 5.1.4 where now Σ is of the form

$$\Sigma: \begin{array}{rcl} \dot{x}(t) &=& Ax(t)+Bw(t),\ t\geq 0 \\ z(t) &=& Cx(t)+Dw(t) \end{array} \quad \text{or} \quad \begin{array}{rcl} x(t+1) &=& Ax(t)+Bw(t),\ t\in\mathbb{N} \\ z(t) &=& Cx(t)+Dw(t). \end{array} \tag{21}$$

Here $A, B, C, \mathbf{\Delta}$ are as in (11) and $D \in \mathbb{K}^{q\times\ell}$. The feedback system $\Sigma\square\Delta$ is well-posed if and only if $I_\ell - \Delta D$ is invertible, and in this case we get $w = (I_\ell - \Delta D)^{-1}\Delta C x$ from $w = \Delta z = \Delta(Cx + Dw)$, see (2.4.29) and (2.4.30). Replacing w in (21) by this expression, we see that the feedback system $\Sigma(\Delta) = \Sigma\square\Delta$ has the system matrix

$$A(\Delta) = A + B(I_\ell - \Delta D)^{-1}\Delta C, \qquad \Delta \in \mathbf{\Delta}_0 = \{\Delta \in \mathbf{\Delta}; \det(I_\ell - \Delta D) \neq 0\}. \tag{22}$$

Remark 5.1.21. In complex analysis a map of the form $f : s \mapsto (\alpha + \beta s)/(\gamma + \delta s)$ on the complex plane is called a *linear fractional transformation* if $\det(\alpha\delta - \beta\gamma) \neq 0$. If $\gamma \neq 0$, the map can also be written in the form $f(s) = a + b(1 - sd)^{-1}s$. Comparing with (22) we see that $A(\Delta)$ can be viewed as a linear fractional transformation of Δ with matrix coefficients. More generally, one defines for every partitioned matrix $M = \begin{bmatrix} M_{11} & M_{12} \\ M_{21} & M_{22} \end{bmatrix} \in \mathbb{C}^{(n_1+n_2)\times(m_1+m_2)}$ and any closed convex cone $\mathbf{\Delta} \subset \mathbb{C}^{m_2\times n_2}$ the *linear fractional transformation* with coefficient M on $\mathbf{\Delta}$ by

$$\mathcal{F}(M, \Delta) := M_{11} + M_{12}(I_{m_2} - \Delta M_{22})^{-1}\Delta M_{21} \in \mathbb{C}^{n_1\times m_1}, \quad \Delta \in \mathbf{\Delta}_0 \tag{23}$$

where $\mathbf{\Delta}_0 := \{\Delta \in \mathbf{\Delta}; \det(I_{m_2} - \Delta M_{22}) \neq 0\}$. In terms of this definition the matrix $A(\Delta)$ in (22) is a linear fractional transformation of $\Delta \in \mathbf{\Delta}_0$ with coefficient $\begin{bmatrix} A & B \\ C & D \end{bmatrix}$. □

Thus the incorporation of a direct feedthrough matrix D in (21) allows us to extend the output feedback representation of uncertainty from affine to *linear fractional parameter perturbations*. Whilst this is a very substantial extension – as we shall see soon – a complication arises in comparison to the affine case: the well-posedness problem. The perturbed system matrices $A(\Delta)$ will, in general, not be well defined for *all* $\Delta \in \mathbf{\Delta}$, but only for the Δ in the subset $\mathbf{\Delta}_0$ of *admissible disturbances*

$$\mathbf{\Delta}_0 = \{\Delta \in \mathbf{\Delta};\ \det(I_\ell - \Delta D) \neq 0\}. \tag{24}$$

We have $\mathbf{\Delta} = \overline{\mathbf{\Delta}_0}$. In fact, $\mathbf{\Delta}$ is closed, and for any $\Delta \in \mathbf{\Delta}$ and for all $t < 1$ sufficiently close to 1, we have $t\Delta \in \mathbf{\Delta}_0$ and $\lim_{t\uparrow 1} t\Delta = \Delta$. It follows that Assumption 5.1.1 is satisfied with

$$\boldsymbol{\delta}_0 = \inf\{\|\Delta\|; \Delta \in \mathbf{\Delta} \setminus \mathbf{\Delta}_0\} = \inf\{\|\Delta\|; \Delta \in \mathbf{\Delta},\ \det(I_\ell - \Delta D) = 0\} = \mu_{\mathbf{\Delta}}(D)^{-1}$$

where $\mu_{\mathbf{\Delta}}(D)$ is the μ-value of D with respect to the perturbation set $\mathbf{\Delta}$, as defined in Section 4.4.

Example 5.1.22. We again consider the uncertain linear oscillator of Example 5.1.4. The system matrix of the corresponding state space model is

$$A(\Delta) = \begin{bmatrix} 0 & 1 \\ -(k_0 + \Delta_k)/(m_0 + \Delta_m) & -(c_0 + \Delta_c)/(m_0 + \Delta_m) \end{bmatrix}, \quad \Delta \in \mathbf{\Delta}_0^{\mathbb{C}}$$

and it is easily verified that $A(\Delta)$, $\Delta \in \mathbf{\Delta}_0^{\mathbb{C}}$ can be represented in the form (22) where

$$A = \begin{bmatrix} 0 & 1 \\ -k_0/m_0 & -c_0/m_0 \end{bmatrix}, \; B = \begin{bmatrix} 0 \\ 1 \end{bmatrix}, \; C = -m_0^{-1} \begin{bmatrix} -k_0/m_0 & -c_0/m_0 \\ 0 & 1 \\ 1 & 0 \end{bmatrix},$$

$$D = -m_0^{-1}[1\ 0\ 0]^\top, \qquad \Delta = [\Delta_m\,, \Delta_c\,, \Delta_k] \in \mathbf{\Delta}_0^{\mathbb{C}}.$$

In this representation $\mathbf{\Delta}^{\mathbb{C}} = \mathbb{C}^{1\times 3}$ is a full-block perturbation set. We will see in the next section that this greatly simplifies the computational problem of determining the associated spectral value sets. □

The representation of the uncertainty in the above example by full block linear fractional perturbations can be generalized. In fact in Section 5.4 we will show that every scalar higher order linear differential equation whose coefficient vector is perturbed in an arbitrary affine way can be written as a linear fractionally perturbed system with a full-block perturbation set $\mathbf{\Delta} = \mathbb{K}^{1\times q}$.
It follows from Cramer's formula, $(I_{m_2} - \Delta D)^{-1} = \det(I_{m_2} - \Delta D)^{-1}\mathrm{adj}(I_{m_2} - \Delta D)$, that every linear fractional transformation (23) of Δ is a matrix whose entries depend rationally on the entries of Δ. It can be shown (see *Notes and References*) that the converse is also true, i.e. every matrix depending rationally on a parameter vector $z = (z_1, \ldots, z_N) \in \mathbb{K}^N$

$$R(z) = (q_{ij}(z)/p_{ij}(z))_{i\in\underline{n_1},\, j\in\underline{m_1}}, \quad p_{ij}(z), q_{ij}(z) \in \mathbb{K}[z_1, \ldots, z_N], \;\; p_{ij}(0) \neq 0 \qquad (25)$$

can be represented in linear fractional form (23) by choosing for $\mathbf{\Delta}$ a suitable diagonal perturbation class of the form

$$\mathbf{\Delta} = \{\mathrm{diag}(z_1 I_{d_1}, \ldots, z_N I_{d_N});\; (z_1, \ldots, z_N) \in \mathbb{K}^N\} \subset \mathbb{K}^{m_2\times n_2} = \mathbb{K}^{d\times d}$$

where $(d_1, \ldots, d_N) \in \mathbb{N}^N$ are chosen appropriately, $m_2 = n_2 = d := d_1 + \ldots + d_N$. Making use of the fact that the sum, product and inverse of linear fractional transformations (LFT) can again be represented as linear fractional transformations (Ex. 6, 7, 8), the problem can be reduced to the task of finding an LFT for an arbitrary monomial $\alpha_\nu z_1^{\nu_1} z_2^{\nu_2} \cdots z_N^{\nu_N} E_{ij}$ where $\alpha_\nu \in \mathbb{K}^*$, $\nu = (\nu_1, \ldots, \nu_N) \in \mathbb{N}^N, \nu \neq 0$ and E_{ij} denotes the $p \times m$-matrix whose entries are all zero except for the entry at position (i, j) which is 1. Such LFTs are easily constructed (Ex. 9), but the combination of these LFTs to obtain an LFT representation of the rational matrix $R(z)$ usually leads to an explosion of the dimension $d = d_1 + \ldots + d_N$ of the perturbation matrices Δ . Therefore the above construction procedure is only of theoretical value. It proves the existence of an LFT representation for the given rational matrix (25), but cannot be used in practice. So we will not enter into details of the procedure. There are more efficient construction methods available in the literature, but no general procedures are known to date which produce LFT representations of *minimal* dimension $d = d_1 + \ldots + d_N$ (see *Notes and References*).
We conclude this subsection with an example where the entries of the system matrix depend rationally on three parameters. Whilst an application of the above general procedure would result in a high dimensional linear fractional representation with *diagonal* Δ, we will see that the rational matrix can in fact be represented in a low dimensional linear fractional form by using a *block-diagonal* instead of a diagonal perturbation set.

Example 5.1.23. Let

$$A(z) = A(z_1, z_2, z_3) = \begin{bmatrix} \dfrac{1}{1 - a_1 z_1} & \dfrac{b_2 z_2 + b_3 z_3}{1 - c_2 z_2 - c_3 z_3} \\ 0 & 1 \end{bmatrix}, \quad \begin{array}{rl} (z_1, z_2, z_3) & \in \mathbb{K}^3, \\ a_1 z_1 & \neq 1, \\ c_2 z_2 + c_3 z_3 & \neq 1 \end{array}$$

where $a_1, b_2, b_3, c_2, c_3 \in \mathbb{K}$ are given. Then $A(0) = I_2$ and setting

$$A = A(0), \quad B = \begin{bmatrix} a_1 & b_2 & b_3 \\ 0 & 0 & 0 \end{bmatrix}, \quad C_2 = I_2, \quad D = \begin{bmatrix} a_1 & 0 & 0 \\ 0 & c_2 & c_3 \end{bmatrix}$$

it is easily verified that

$$A(z) = A + B(I_3 - \Delta D)^{-1} \Delta C, \quad \Delta = \begin{bmatrix} z_1 & 0 \\ 0 & z_2 \\ 0 & z_3 \end{bmatrix} \in \mathbf{\Delta}_0 = \{\Delta \in \mathbb{K} \oplus \mathbb{K}^{2\times 1}; \det(I - \Delta D) \neq 0\}.$$

□

5.1.3 Exercises

1. Consider the matrix family A_ω

$$A_\omega = \begin{bmatrix} (3\omega^2 - 2)/(1 - \omega^2) & 1 & -\omega/(1 - \omega^2) \\ -1/(1 - \omega^2) & 0 & \omega/(1 - \omega^2) \\ \omega/(1 - \omega^2) & \omega & (2\omega^2 - 3)/(1 - \omega^2) \end{bmatrix}, \quad \omega \in \Omega = \mathbb{C} \setminus \{-1, 1\},$$

and let $\omega_0 = 0$. Determine $\mathbf{\Delta}_0$, $\mathbf{\Delta}$, $\boldsymbol{\delta}_0$ and the stability radius $r_{\mathbf{\Delta}}(A(\cdot); \mathbb{C}_-)$ for this case. Show that $\sigma_{\mathbf{\Delta}}(A(\cdot); 1) = \{-1, -3\} = \sigma(A_0)$.

2. Consider the matrix $A(\Delta) = \begin{bmatrix} -1 & \cos\Delta \\ 0.5 & -2 \end{bmatrix}$, $\Delta \in \mathbf{\Delta} = \mathbb{R}$. Determine $\sigma_{\mathbf{\Delta}}(A(\cdot); \delta)$ for all $\delta > 0$. Calculate the smallest value $\hat{\delta}$ such that $\sigma_{\mathbf{\Delta}}(A(\cdot); \delta)$ is connected for $\delta > \hat{\delta}$.

3. Suppose

$$A = \begin{bmatrix} a_{11} & a_{12} \\ a_{21} & a_{22} \end{bmatrix}, \quad A(\Delta) = \begin{bmatrix} a_{11} & a_{12} + \Delta_1 \\ a_{21} + \Delta_2 & a_{22} + \Delta_1 + \Delta_2 \end{bmatrix}, \quad (\Delta_1, \Delta_2) \in \mathbb{C}^2.$$

Find B, C such that $A(\Delta) = A + B \operatorname{diag}(\Delta_1, \Delta_2) C$ for all $(\Delta_1, \Delta_2) \in \mathbb{C}^2$. Show that there do *not* exist B, C such that $A(\Delta) = A + B[\Delta_1 \ \ \Delta_2] C$ for all $(\Delta_1, \Delta_2) \in \mathbb{C}^2$.

4. Consider the uncertain system

$$\dot{x} = A(\Delta)x = (1 - \Delta_1 - \Delta_1\Delta_2)^{-1} \begin{bmatrix} 1 - \Delta_1\Delta_2 & \Delta_2^2 \\ \Delta_1^2 & -1 + \Delta_1 + \Delta_2 \end{bmatrix} x.$$

Show that the perturbed system matrix can be represented in the form (22) where

$$A = \begin{bmatrix} 1 & 0 \\ 0 & -1 \end{bmatrix}, \quad B = C = I_2, \quad D = \begin{bmatrix} 1 & 1 \\ 1 & 0 \end{bmatrix}, \quad \Delta = \operatorname{diag}(\Delta_1, \Delta_2).$$

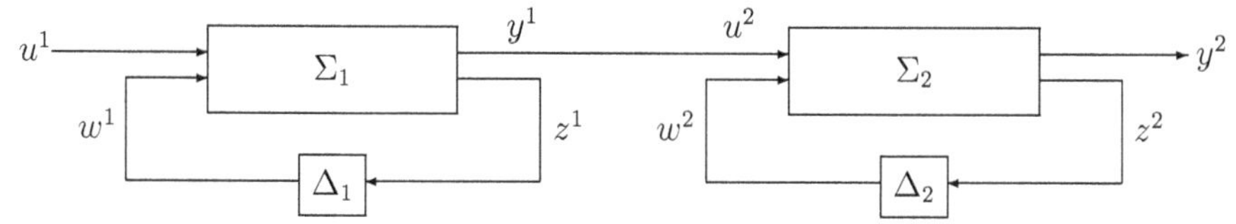

Figure 5.1.8: Series connection of two uncertain systems

5. Consider two uncertain systems connected in series as shown in Figure 5.1.8. Assume that the two systems are of the form

$$\begin{array}{rcl} \dot{x}^i &=& A^i x^i + B_1^i w^i + B_2^i u^i \\ z^i &=& C_1^i x^i + D_{11}^i w^i \\ y^i &=& C_2^i x^i + D_{21}^i w^i \end{array}, \quad i = 1, 2.$$

with uncertain feedback $w^i = \Delta_i z^i$, $\Delta_i \in \boldsymbol{\Delta}_i$, $i = 1, 2$.

Describe the perturbed system matrix $A(\Delta_1, \Delta_2)$ of the series connection in linear fractional form (22) (specify $A, B, C, D, \boldsymbol{\Delta}$).

6. **Sum of LFTs** Consider two linear fractional transformations $\mathcal{F}(M, \Delta_1)$ and $\mathcal{F}(N, \Delta_2)$ where M_{11} and N_{11} have the same format $n_1 \times m_1$. Show that the sum $\mathcal{F}(M, \Delta_1) + \mathcal{F}(N, \Delta_2)$ can be represented as a linear fractional transformation $\mathcal{F}(Q, \Delta)$ with $\Delta = \operatorname{diag}(\Delta_1, \Delta_2)$

$$\mathcal{F}(Q, \Delta) = \mathcal{F}(M, \Delta_1) + \mathcal{F}(N, \Delta_2), \quad Q = \left[\begin{array}{c|cc} M_{11} + N_{11} & M_{12} & N_{12} \\ \hline M_{21} & M_{22} & 0 \\ N_{21} & 0 & N_{22} \end{array}\right].$$

7. **Product of LFTs** Consider two linear fractional transformations $\mathcal{F}(M, \Delta_1)$ and $\mathcal{F}(N, \Delta_2)$ where M_{11} and N_{11} have the formats $n_1 \times m_1$ and $m_1 \times q_1$ respectively. Show that the product $\mathcal{F}(M, \Delta_1)\mathcal{F}(N, \Delta_2)$ can be represented as a linear fractional transformation $\mathcal{F}(Q, \Delta)$ with $\Delta = \operatorname{diag}(\Delta_1, \Delta_2)$

$$\mathcal{F}(Q, \Delta) = \mathcal{F}(M, \Delta_1)\mathcal{F}(N, \Delta_2), \quad Q = \left[\begin{array}{c|cc} M_{11}N_{11} & M_{12} & M_{11}N_{12} \\ \hline M_{21}N_{11} & M_{22} & M_{21}N_{12} \\ N_{21} & 0 & N_{22} \end{array}\right].$$

8. **Inverse of an LFT** Let $M = \begin{bmatrix} M_{11} & M_{12} \\ M_{21} & M_{22} \end{bmatrix}$ and suppose that $\mathcal{F}(M, \Delta)$ is well defined, M_{11} is square and nonsingular. Prove that the inverse of $\mathcal{F}(M, \Delta)$ exists and can be represented as a linear fractional transformation of Δ as follows

$$\mathcal{F}(M, \Delta)^{-1} = \mathcal{F}(Q, \Delta), \quad Q = \begin{bmatrix} M_{11}^{-1} & M_{11}^{-1}M_{12} \\ -M_{21}M_{11}^{-1} & M_{22} - M_{21}M_{11}^{-1}M_{12} \end{bmatrix}.$$

9. Consider the monomial $\alpha_\nu z^\nu E_{ij} = \alpha_\nu z_1^{\nu_1} z_2^{\nu_2} \cdots z_N^{\nu_N} E_{ij}$ with multi-exponent $\nu \in \mathbb{N}^N$, $\nu = (\nu_1, \ldots, \nu_N) \neq 0$, coefficient $\alpha_\nu \in \mathbb{K}^*$ and $E_{ij} = \hat{e}^i \tilde{e}^{j\top}$ where $\hat{e}^i$, $\tilde{e}^j$ denote the standard unit vectors in $\mathbb{K}^p$ and $\mathbb{K}^m$, respectively. Let $\ell \in \underline{N}$ be the largest integer such that $\nu_\ell \neq 0$ and $|\nu| = \nu_1 + \ldots + \nu_\ell$. Define the partitioned matrix M by

$$M_{11} = 0_{p \times m}, \; M_{12} = \alpha_\nu \hat{e}^i (e^1)^\top, \; M_{21} = e^{|\nu|} (\tilde{e}^j)^\top, \; M_{22} = J$$

where $J = J(0,|\nu|)$ denotes the nilpotent Jordan block of order $|\nu|$ (see (2.2.39)) and e^k the k-th standard unit vector in $\mathbb{K}^{|\nu|}$. Verify that

$$\mathcal{F}(M,\Delta) = \alpha_\nu \, \hat{e}^i \, (e^1)^\top \, (I_{|\nu|} - \Delta J)^{-1} \Delta \, e^{|\nu|} \, (\tilde{e}^j)^\top = \alpha_\nu \, z_1^{\nu_1} z_2^{\nu_2} \cdots z_\ell^{\nu_\ell} \, \hat{e}^i \, (\tilde{e}^j)^\top = \alpha_\nu z^\nu E_{ij}$$

for $\Delta = \operatorname{diag}(z_1 I_{\nu_1}, \ldots, z_\ell I_{\nu_\ell})$.
(Hint: Show that
$(e^1)^\top \, (I_{|\nu|} - \Delta J)^{-1} \Delta \, e^{|\nu|} = (e^1)^\top \, \left[I_{|\nu|} + \Delta J + \ldots + (\Delta J)^{|\nu|-1} \right] z_\ell e^{|\nu|} = z_1^{\nu_1} z_2^{\nu_2} \cdots z_\ell^{\nu_\ell}$).

10. Consider the rational matrix

$$R(z) = \begin{bmatrix} 0 & 1/(1+z_1) \\ 1/(1+z_2) & z_2/(1+z_1) \end{bmatrix}.$$

Show that $R(z)$ can be written in the form $q(z)^{-1} P(z)$, where $q(z) = (1+z_1)(1+z_2)$ and

$$P(z) = \begin{bmatrix} 0 & 1 \\ 1 & 0 \end{bmatrix} + \begin{bmatrix} 0 & 0 \\ z_1 & 0 \end{bmatrix} + \begin{bmatrix} 0 & z_2 \\ 0 & 0 \end{bmatrix} + \begin{bmatrix} 0 & 0 \\ 0 & z_1 z_2 \end{bmatrix} + \begin{bmatrix} 0 & 0 \\ 0 & z_2^2 \end{bmatrix}.$$

Show that $P(z) = \mathcal{F}(M, \Delta_1)$ where $\Delta_1 = \operatorname{diag}(z_1, z_2, z_1, z_2, z_2, z_2)$ and

$$M_{11} = \begin{bmatrix} 0 & 1 \\ 1 & 0 \end{bmatrix}, \quad M_{12} = \begin{bmatrix} 0 & 1 & 0 & 0 & 0 & 0 \\ 1 & 0 & 1 & 0 & 1 & 0 \end{bmatrix}, \quad M_{21} = \begin{bmatrix} 1 & 0 \\ 0 & 1 \\ 0 & 0 \\ 0 & 1 \\ 0 & 0 \\ 0 & 1 \end{bmatrix}, \quad M_{22} = \begin{bmatrix} 0 & 0 & 0 & 0 & 0 & 0 \\ 0 & 0 & 0 & 0 & 0 & 0 \\ 0 & 0 & 0 & 1 & 0 & 0 \\ 0 & 0 & 0 & 0 & 0 & 0 \\ 0 & 0 & 0 & 0 & 0 & 1 \\ 0 & 0 & 0 & 0 & 0 & 0 \end{bmatrix}.$$

Prove that $q(z) = \mathcal{F}(N, \Delta_2)$ where $\Delta_2 = \operatorname{diag}(z_1, z_2, z_1, z_2)$ and

$$N_{11} = 1, \quad N_{12} = \begin{bmatrix} 1 & 1 & 1 & 0 \end{bmatrix}, \quad N_{21} = \begin{bmatrix} 1 \\ 1 \\ 0 \\ 1 \end{bmatrix}, \quad N_{22} = \begin{bmatrix} 0 & 0 & 0 & 0 \\ 0 & 0 & 0 & 0 \\ 0 & 0 & 0 & 1 \\ 0 & 0 & 0 & 0 \end{bmatrix}.$$

Hence using the formula given in Ex. 8 for the inverse show that $1/q(z) = \mathcal{F}(Q, \Delta_2)$, where

$$Q_{11} = 1, \quad Q_{12} = \begin{bmatrix} 1 & 1 & 1 & 0 \end{bmatrix}, \quad Q_{21} = -\begin{bmatrix} 1 \\ 1 \\ 0 \\ 1 \end{bmatrix}, \quad Q_{22} = \begin{bmatrix} -1 & -1 & -1 & 0 \\ -1 & -1 & -1 & 0 \\ 0 & 0 & 0 & 1 \\ -1 & -1 & -1 & 0 \end{bmatrix}.$$

Check that this is indeed correct. Use the addition rule given in Ex. 6 to obtain a linear fractional representation of $q(z)^{-1} I_2$. Finally use the product rule given in Ex. 7 to obtain a linear fractional representation of $R(z)$. Check your result.

5.1.4 Notes and References

We have seen that the spectrum of non-normal operators may be highly sensitive to small perturbations. The importance of this fact has been recognized in various fields of mathematics such as Numerical Linear Algebra, Differential Equations, Non-Selfadjoint Operators Theory and Perturbation Theory, see e.g. *Wilkinson* (1963) [523], *Kreiss* (1962)

[317], *Halmos* (1968) [215], *Gohberg and Krein* (1969) [191], *Kato* (1980) [293].
Spectral analysis is a fundamental tool of the mathematical sciences and has been used in many areas of application for more than a century. In recent years awareness has been growing that for certain problems of science and engineering, the predictions based on eigenvalues do not always match the observations, see *Trefethen* (1997) [498]. It was discovered that in these applications the operators involved were highly non-normal with volatile eigenvalues, and this suggested that one should look for a substitute for the spectrum. Since the norm of the resolvent $\|(sI_n - A)^{-1}\|$ tends to infinity as s approaches an eigenvalue of A, it was a natural idea to choose the superlevel sets $\{s \in \mathbb{C}; \|(sI_n - A)^{-1}\| > \delta^{-1}\}$ as a substitute. Due to the growing power of computers it became possible at the beginning of the 1990's to compute and visualize these sets for medium size matrices. It was known that these superlevel sets can alternatively be characterized in terms of the spectra of perturbed operators. In fact we will see in the next section that they coincide with the unstructured spectral value sets in the complex case, $\sigma_{\mathbb{C}}(A; \delta)$. Resolvent based versions of spectral value sets were independently introduced by *Landau* (1975) under the name of "ε-spectrum" [327], by *Godunov et al.* (1990) under the name of "spectral portrait" [190] and by *Trefethen* (1990) [497] under the name of "pseudospectrum". An excellent introduction to the field with ten illustrative examples from various areas of application is *Trefethen* (1997) [498]. This article also contains some historical remarks. More historical remarks and a comprehensive list of applications can be found in the instructive survey article on the computation of pseudospectra by the same author (1999) [499]. Another nice overview of the field is offered on the web site "Pseudospectral Gateway", set up by Embree and Trefethen: `http://web.comlab.ox.ac.uk/projects/pseudospectra/index.html`.
Spectral value sets for unstructured *real* and *structured complex* perturbations were first considered in the context of control theory, see *Hinrichsen and Pritchard* (1991, 1992) [249], [251] and *Hinrichsen and Kelb* (1993) [233]. The general framework presented in Subsection 5.1.1 is new. For more references and comments see the next section. Example 5.1.6 and Figure 5.1.3 are due to *Karow* (2003) [292].
The notion of a stability radius with respect to real and complex, unstructured and structured full-block perturbations was introduced in *Hinrichsen and Pritchard* (1986) [242], [243]. For more references and comments see Section 5.3.
Linear Fractional Transformations were first studied by *Redheffer* (1960) [430]. They have proved to be an important tool in Systems Theory. Their use in modelling uncertainty is discussed in *Packard and Doyle* (1993) [402] and *Zhou et al.* (1995) [546]. These references also give hints of how to represent a rational matrix family in LFT form with $\Delta = \mathrm{diag}(z_1 I_{d_1}, \ldots, z_N I_{d_N})$. A detailed proof can be found in *Op't Hof* (1997) [400]. A systematic graph theoretical method of constructing LFT representations for transfer matrices depending *polynomially* on a finite number of parameters is described in *Zerz* (2000) [545]. This reference also considers the problem of reducing the dimension and shows in particular how to achieve "trim" LFT representations, but trim representations do not guarantee minimality of the dimension, see also *Sugie and Kawanishi* (1995) [490]. The problem of finding a minimal LFT representation (with diagonal Δ) for a given rational matrix in several variables is closely related to the minimal realization of ND-transfer matrices by Roesser type models, see *Roesser* (1975) [435], *Kaczorek* (1985) [285],*Zerz* (2000) [545]. There are no $1D$-like characterizations of minimal realizations for ND systems (see *Zerz* (1999) [544]) and the minimal realization problem is still unsolved for $N \geq 2$.

5.2 Spectral Value Sets

In this section we continue the study of spectral value sets with the aim of deriving explicit and, if possible, computable characterizations of them for different classes of linear fractional perturbation classes $\mathbf{\Delta}$. General results expressed in terms of the μ-function are obtained in the first subsection. These are specialized in the second subsection to *complex* full block perturbations for which computable formulas are obtained. Spectral value sets for *real* full-block perturbations are analyzed in the third subsection. The characterization of these sets is more complicated but still leads to computable formulas. In the final subsection we specialize even further to unstructured perturbations $A \rightsquigarrow A + \Delta$ (pseudospectra) considering both the complex and the real case. Besides presenting the simplified characterizations for the spectral value sets in this case, we also determine bounds for them and analyze the behaviour of pseudospectra along similarity orbits.

5.2.1 General Definitions and Results

We begin by specializing the definition of spectral value sets given in Section 5.1 to linear fractional perturbations and prove some simple properties. The main purpose of this subsection is to characterize spectral value sets in terms of the μ-function. By using the properties derived in Section 4.4, more explicit descriptions and estimates can be obtained with respect to block-diagonal perturbations. As an illustration we apply these characterizations to derive computable formulas for Gershgorin type perturbations of a diagonal matrix (see Subsection 4.2.2). We conclude the subsection with a general theorem on the continuous dependence of spectral value sets on the data.

Throughout the subsection the assumptions are

- $(A, B, C, D) \in \mathbf{L}_{n,\ell,q}(\mathbb{C})$.
- $\mathbf{\Delta} \subset \mathbb{K}^{\ell \times q}$ is a closed convex cone and $\operatorname{span}_{\mathbb{K}} \mathbf{\Delta}$ is provided with a norm $\|\cdot\|_{\mathbf{\Delta}}$ which is an operator norm with respect to a given pair of norms on $\mathbb{K}^{\ell}$, $\mathbb{K}^{q}$. If $\mathbb{K} = \mathbb{R}$, the complex spaces $\mathbb{C}^{\ell}$, $\mathbb{C}^{q}$ are provided with a compatible pair of norms and $\mathbb{C}^{\ell \times q}$, $\mathbb{C}^{q \times \ell}$ with the corresponding operator norms, as explained in Remark 4.4.2.
- We consider perturbations $A \rightsquigarrow A(\Delta)$ where
$$A(\Delta) = A + B(I_\ell - \Delta D)^{-1} \Delta C, \quad \Delta \in \mathbf{\Delta}_0 := \{\Delta \in \mathbf{\Delta};\ \det(I_\ell - \Delta D) \neq 0\}. \quad (1)$$

Then $\overline{\mathbf{\Delta}_0} = \mathbf{\Delta}$ and Assumption 5.1.1 is satisfied with well-posedness radius $\boldsymbol{\delta}_0 = \inf\{\|\Delta\|_{\mathbf{\Delta}};\ \Delta \in \mathbf{\Delta} \setminus \mathbf{\Delta}_0\} = \mu_{\mathbf{\Delta}}(D)^{-1}$. If $D = 0$, the perturbations (1) are affine and $\boldsymbol{\delta}_0 = \infty$. For perturbations of the form (1) Definition 5.1.2 specializes to

Definition 5.2.1. Given a $\delta > 0$, then

$$\sigma_{\mathbf{\Delta}}(A; B, C, D; \delta) = \bigcup_{\Delta \in \mathbf{\Delta}_0,\, \|\Delta\|_{\mathbf{\Delta}} < \delta} \sigma(A(\Delta)) = \bigcup_{\Delta \in \mathbf{\Delta}_0,\, \|\Delta\|_{\mathbf{\Delta}} < \delta} \sigma(A + B(I_\ell - \Delta D)^{-1} \Delta C)$$

is called the *spectral value set* of (A, B, C, D) with respect to the perturbation class $\mathbf{\Delta}$ at the uncertainty level δ. If $D = 0$ the spectral value set is denoted by $\sigma_{\mathbf{\Delta}}(A; B, C; \delta)$.

From the results proved in Section 5.1 for more general types of perturbation, we know that as δ increases from zero, the set $\sigma_{\mathbf{\Delta}}(\delta) = \sigma_{\mathbf{\Delta}}(A; B, C, D; \delta)$ increases and takes different shapes. For small values of δ, $\sigma_{\mathbf{\Delta}}(\delta)$ is a union of disjoint connected components (each one containing exactly one eigenvalue of A) and possibly a set of fixed eigenvalues of A (isolated points of $\sigma_{\mathbf{\Delta}}(\delta)$). As δ increases, the connected components expand and merge, with the number of components decreasing. The rates and ways in which they do this give valuable information about the variability of the spectrum under perturbations of different sizes. From Lemma 5.1.7 we know that the sets $\sigma_{\mathbf{\Delta}}(\delta)$ are *bounded* whenever $\delta < \boldsymbol{\delta}_0 = \mu_{\mathbf{\Delta}}(D)^{-1}$, and their compact closures are given by

$$\overline{\sigma_{\mathbf{\Delta}}(A; B, C, D; \delta)} = \bigcup_{\substack{\Delta \in \mathbf{\Delta} \\ \|\Delta\|_{\mathbf{\Delta}} \leq \delta}} \sigma(A(\Delta)) = \{s \in \mathbb{C}; \exists \Delta \in \mathbf{\Delta}, \|\Delta\|_{\mathbf{\Delta}} \leq \delta : s \in \sigma(A(\Delta))\}. \quad (2)$$

These compact closures depend continuously on $\delta \in [0, \boldsymbol{\delta}_0)$ with respect to the Hausdorff metric on $\mathcal{K}(\mathbb{C})$. Moreover, for *complex* perturbation structures $\mathbf{\Delta}$ (i.e. $\mathbb{C}\mathbf{\Delta} = \mathbf{\Delta}$) we know from Proposition 5.1.9 that $\sigma_{\mathbf{\Delta}}(\delta) \setminus \sigma(A)$ is *open* in $\mathbb{C}$.
Usually we will only consider levels of uncertainty $\delta < \boldsymbol{\delta}_0 = \mu_{\mathbf{\Delta}}(D)^{-1}$ so that all $A(\Delta)$ with $\Delta \in \mathbf{\Delta}$, $\|\Delta\|_{\mathbf{\Delta}} \leq \delta$ are well defined. If $\delta \geq \boldsymbol{\delta}_0$ we can no longer expect $\sigma_{\mathbf{\Delta}}(\delta)$ to be bounded. In fact, if $\delta > \boldsymbol{\delta}_0$, it will be the complement of $\sigma_{\mathbf{\Delta}}(\delta)$ rather than $\sigma_{\mathbf{\Delta}}(\delta)$ which is bounded. This is illustrated in the following example.

Example 5.2.2. Consider the uncertain system

$$\dot{x} = \begin{bmatrix} -(1-\Delta)^{-1} & 0 \\ 0 & 0 \end{bmatrix} x(t) = \begin{bmatrix} -1 - \frac{\Delta}{1-\Delta} & 0 \\ 0 & 0 \end{bmatrix} x(t), \quad \Delta \in \mathbf{\Delta} = \mathbb{C}$$

which can be represented in the form $\dot{x} = A(\Delta)x = (A + B(I_\ell - \Delta D)^{-1}\Delta C)x$ with

$$A = \begin{bmatrix} -1 & 0 \\ 0 & 0 \end{bmatrix}, \quad B = \begin{bmatrix} 1 \\ 0 \end{bmatrix}, \quad C = \begin{bmatrix} -1 & 0 \end{bmatrix}, \quad D = 1.$$

The set of admissible disturbances is given by $\mathbf{\Delta}_0 = \mathbb{C} \setminus \{1\}$, and so the well-posedness radius is $\boldsymbol{\delta}_0 = 1$. Now $s \in \sigma_{\mathbb{C}}(A; B, C, D; \delta)$, provided $s = 0$ or

$$s = -(1-\Delta)^{-1}, \ |\Delta| < \delta, \quad \text{i.e.} \quad (s+1)/s = \Delta, \ |\Delta| < \delta.$$

Therefore

$$\sigma_{\mathbb{C}}(A; B, C, D; \delta) = \{0\} \cup \{s \in \mathbb{C}^*; |s+1|/|s| < \delta\}.$$

$s \in \mathbb{C}^*$ belongs to the boundary of the spectral value set at level $\delta > 0$ if and only if $|s+1|/|s| = \delta$. An easy calculation shows that this is a circle in the complex plane centred at $-(1-\delta^2)^{-1}$ with radius $|\delta/(1-\delta^2)|$. For $\delta \in (0,1)$ this circle is to the left of the line $-(1+\delta)^{-1} + \imath\mathbb{R}$ and as $\delta \to 1$ its centre tends to $-\infty$ and its radius tends to $+\infty$. Moreover for any $s \in \mathbb{C}$ with $\operatorname{Re} s < -1/2$ there exists $\delta < 1$ such that s is in the spectral value set of level δ. So as $\delta \uparrow 1$ the spectral value sets gradually fill up the half plane $\{s \in \mathbb{C}; \operatorname{Re} s < -1/2\}$, see Figure 5.2.1. For $\delta > 1$, the centre $(\delta^2 - 1)^{-1}$ lies in the open right half plane and the associated radius is $\delta/(\delta^2 - 1)$. It is easily verified

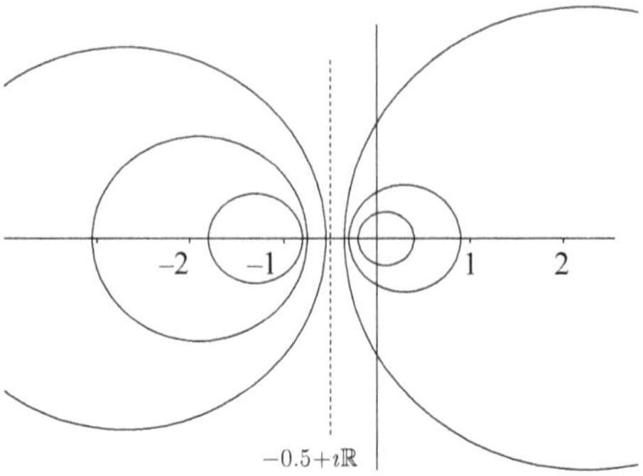

Figure 5.2.1: Spectral value sets for $\delta = 0.4, 0.6, 0.8, 1.2, 2, 3$

that the corresponding spectral value sets (modulo the isolated spectral value $s = 0$) are the *complements* of the closed disks circumscribed by the circles in the right half-plane. Since $(\delta^2-1)^{-1} - \delta(\delta^2-1)^{-1} = -(1+\delta)^{-1}$ and $\delta(\delta^2-1)^{-1} \to \infty$ as $\delta \downarrow 1$ we see that the union of these disks is the open half plane $\operatorname{Re} s > -1/2$. Hence, modulo the point 0, the intersection of all the spectral value sets $\sigma_{\mathbb{C}}(A; B, C, D; \delta) \setminus \{0\}$, $\delta > 1$ is the closed left half-plane $\{s \in \mathbb{C};\ \operatorname{Re} s \leq -1/2\}$. As $\delta > 1$ increases the disks are decreasing and as $\delta \to \infty$ they shrink towards 0 (always containing 0 which belongs to $\sigma_{\mathbb{C}}(A; B, C, D; \delta)$). Summarizing we obtain: For $\delta \in (0, \boldsymbol{\delta}_0) = (0, 1)$, $\sigma_{\mathbb{C}}(\delta)$ is bounded, and for $\delta = 1$, $\sigma_{\mathbb{C}}(1)$ is the open left halfplane $\operatorname{Re} s < -1/2$. For $\delta > 1$, $\sigma_{\mathbb{C}}(\delta)$ is the complement of a disk punctured at 0 whose centre and radius tend to 0 as $\delta \to \infty$. Note that $s = 0$ is an isolated point of $\sigma_{\mathbb{C}}(\delta)$ for all $\delta > 0$. In particular we see that the spectral value sets $\sigma_{\boldsymbol{\Delta}}(A; B, C, D; \delta)$ do *not* necessarily become connected as $\delta \to \infty$. □

We mention two elementary properties of spectral value sets which follow immediately from Definition 5.2.1 and the fact that

$$TAT^{-1} + TB(I_\ell - \Delta D)^{-1}\Delta CT^{-1} = TA(\Delta)T^{-1}, \quad T \in \mathbf{Gl}_n(\mathbb{C}).$$

Lemma 5.2.3. *(i)* Invariance under similarity. *For all $\delta > 0$,*

$$\sigma_{\boldsymbol{\Delta}}(A; B, C, D; \delta) = \sigma_{\boldsymbol{\Delta}}(TAT^{-1}; TB, CT^{-1}, D; \delta), \quad T \in \mathbf{Gl}_n(\mathbb{C}). \tag{3}$$

(ii) Decomposition property. *Given $(A_i, B_i, C_i, D_i) \in \mathbf{L}_{n_i,\ell_i,q_i}(\mathbb{C})$ and perturbation classes $\boldsymbol{\Delta}^i \subset \mathbb{C}^{\ell_i \times q_i}$, $i = 1, 2$, let*

$$A = \begin{bmatrix} A_1 & 0 \\ 0 & A_2 \end{bmatrix}, \quad B = \begin{bmatrix} B_1 & 0 \\ 0 & B_2 \end{bmatrix}, \quad C = \begin{bmatrix} C_1 & 0 \\ 0 & C_2 \end{bmatrix}, \quad D = \begin{bmatrix} D_1 & 0 \\ 0 & D_2 \end{bmatrix},$$

$$\boldsymbol{\Delta} = \{\operatorname{diag}(\Delta_1, \Delta_2);\ \Delta_1 \in \boldsymbol{\Delta}^1,\ \Delta_2 \in \boldsymbol{\Delta}^2\}, \quad \|\operatorname{diag}(\Delta_1, \Delta_2)\|_{\boldsymbol{\Delta}} = \max_{i=1,2} \|\Delta_i\|_{\boldsymbol{\Delta}^i}.$$

Then $\boldsymbol{\Delta}_0 = \{\operatorname{diag}(\Delta_1, \Delta_2); \Delta_1 \in \boldsymbol{\Delta}_0^1,\ \Delta_2 \in \boldsymbol{\Delta}_0^2\}$ (see (1)) and for all $\delta > 0$

$$\sigma_{\boldsymbol{\Delta}}(A; B, C, D; \delta) = \sigma_{\boldsymbol{\Delta}^1}(A_1, B_1, C_1, D_1; \delta) \cup \sigma_{\boldsymbol{\Delta}^2}(A_2, B_2, C_2, D_2; \delta). \tag{4}$$

Remark 5.2.4. (i) Spectral value sets share the property of invariance under similarity transformations with transfer functions. We will see later that the spectral value sets of (A, B, C, D) can be characterized in terms of the transfer function of (A, B, C, D).
(ii) The decomposition property is useful in constructing examples and counter-examples. It shows that every finite union of spectral value sets is again a spectral value set. □

Before dealing with the problem of how spectral value sets can be determined, we illustrate by an example that for some highly non-normal matrices it may be impossible to determine the spectrum itself. This is because the eigenvalues are so sensitive that unavoidable rounding errors render the calculation of the spectrum unreliable, no matter which software is used. Since the notion of a spectral value set is a robust version of the notion of a spectrum, one would expect that spectral value sets can be determined in a more reliable way, if the uncertainty level δ is not too close to the machine precision eps (i.e. $1 + \text{eps}/2$ is indistinguishable from 1). This is illustrated in the following example due to Godunov [189].[1]

Example 5.2.5. (Godunov). Consider the matrices $\tilde{A}$ and $A = L^{-1}\tilde{A}L$ where

$$\tilde{A} = \begin{bmatrix} 1 & 2048 & 256 & 128 & 64 & 32 & 16 \\ 0 & -2 & 1024 & 512 & 256 & 128 & 32 \\ 0 & 0 & 4 & 512 & 1024 & 256 & 64 \\ 0 & 0 & 0 & 0 & 512 & 512 & 128 \\ 0 & 0 & 0 & 0 & -4 & 1024 & 256 \\ 0 & 0 & 0 & 0 & 0 & 2 & 2048 \\ 0 & 0 & 0 & 0 & 0 & 0 & 1 \end{bmatrix}, \quad L = \begin{bmatrix} 1 & 0 & 0 & 0 & 0 & 0 & 0 \\ 0 & 1 & 0 & 0 & 0 & 0 & 0 \\ 1 & 0 & 1 & 0 & 0 & 0 & 0 \\ 0 & 0 & 0 & 1 & 0 & 0 & 0 \\ 0 & 0 & 1 & 0 & 1 & 0 & 0 \\ 1 & 0 & 0 & 0 & 0 & 1 & 0 \\ 0 & 1 & 1 & 0 & 1 & 0 & 1 \end{bmatrix}.$$

Obviously

$$\sigma(A) = \sigma(\tilde{A}) = \{0, 1, 1, \pm 2, \pm 4\}.$$

However, applying standard software for the computation of eigenvalues we obtain the

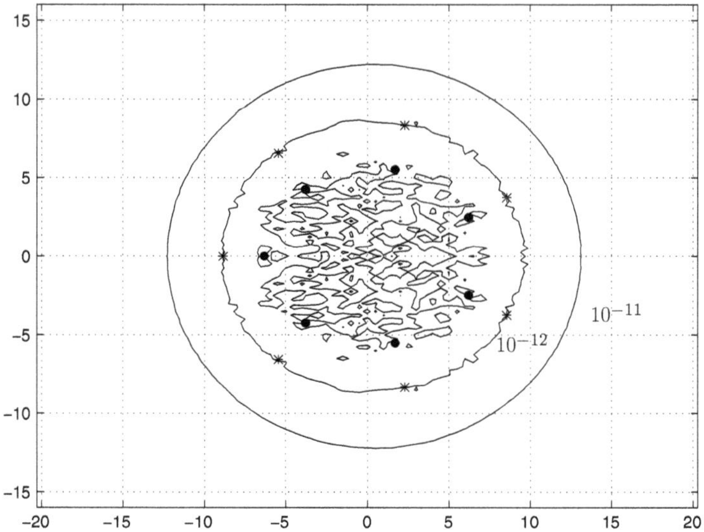

Figure 5.2.2: Computed spectral value sets of the Godunov matrix

following computed spectrum of A

$$\{8.57 \pm 3.73\imath,\ 2.29 \pm 8.33\imath,\ -5.43 \pm 6.56\imath,\ -8.85\}.$$

[1]For a related theoretical result, see Proposition 5.2.48.

Permuting the first two rows and columns by a permutation matrix P and computing the spectrum of $PAP^\top$ we should obtain the same set of eigenvalues but instead we get

$$\{6.21 \pm 2.48\imath,\ 1.70 \pm 5.5\imath,\ -3.78 \pm 6.56\imath,\ -6.29\}.$$

The calculation of the unstructured spectral value sets of A with respect to the spectral norm indicates that the computed values for the spectrum are unreliable. In fact, even for δ-values close to the machine accuracy the computed spectral value set remains connected, whereas theoretically it should separate into as many connected components as there are distinct eigenvalues of A, see Proposition 5.1.8. Figure 5.2.2 shows the boundaries of $\sigma_{\mathbb{C}^{7\times 7}}(A; I_7, I_7; \delta)$ for $\delta = 10^{-11}$, 10^{-12} and 10^{-13}. While the boundaries for the first two values of δ (the two circular contours) appear to be reliable, the highly disconnected contour for the third value of δ ($\approx \|A\|_{1,1}$ eps) reveals that the calculation of this spectral value set is incorrect. The stars and small circles mark the location of the computed eigenvalues of A and $PAP^\top$, respectively. □

We now consider the question of how spectral value sets can be determined. A first idea might be to approximate $\sigma_{\mathbf{\Delta}}(A; B, C, D; \delta)$ by computing $\sigma(A(\Delta))$ for a large number of randomly generated disturbance matrices Δ of norm smaller than δ. But this method is cumbersome, yields only rough approximations and is sometimes misleading. This is illustrated in the following example.

Example 5.2.6. Consider a matrix $A \in \mathbb{R}^{6\times 6}$ in companion form with characteristic polynomial

$$\chi_A(s) = (s + 0.5 \pm 1.2\imath)(s + 1.1 \pm 0.6\imath)(s + 2.2 \pm 1.9\imath).$$

We choose $B = C = I_6$, $D = 0$ and $\mathbf{\Delta} = \mathbb{C}^{6\times 6}$ provided with the spectral norm. Figure 5.2.3 shows the spectral value sets $\sigma_{\mathbf{\Delta}}(A; I, I; \delta)$ for $\delta = 1/2,\ 1/3$.
Two large contours and a very small one form the boundaries of the sets and were ob-

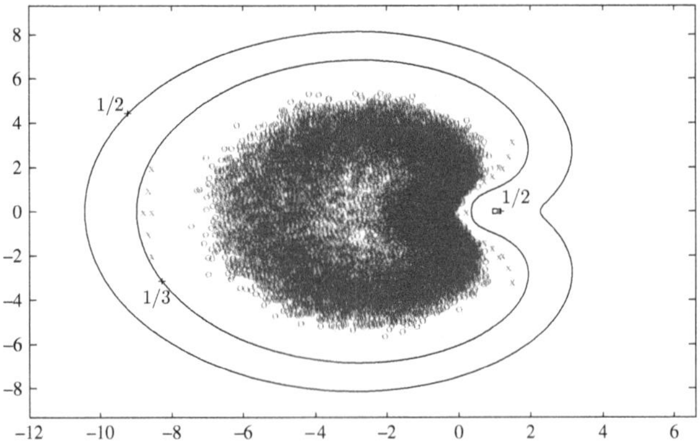

Figure 5.2.3: Spectral value sets and random perturbations

tained by the methods we will describe in this section. The small contour describes a hole in $\sigma_{\mathbf{\Delta}}(A; I, I; 1/2)$. We hoped to get a rough approximation of the set at level $\delta = 1/3$ by plotting the spectra $\sigma(A + \Delta)$, for 5000 pseudo randomly generated $\Delta \in \mathbb{C}^{6\times 6}$ with $\|\Delta\| = 1/3$. Figure 5.2.3 shows the result and reveals that the approximation is quite poor. The reason is that within the sphere of matrices $\Delta \in \mathbb{C}^{6\times 6}$ of norm $\|\Delta\| = 1/3$ the probability of hitting a matrix which moves an eigenvalue of A close to the contour is very low. The perturbations Δ of norm $< 1/3$ which move eigenvalues to the locations "x" are not random but have been constructed via the formulas in Remark 5.2.20. □

Of course, one may argue that if it is very unlikely that a $\Delta \in \mathbf{\Delta}_0$ of norm $< \delta$ will produce eigenvalues near the boundary of the spectral value set, then the set is not a good measure of the way the spectrum is perturbed. This approach has been adopted by some authors who define spectral value sets in probabilistic terms, see *Notes and References.* In contrast, our deterministic approach represents a *worst case analysis.* It ensures that *every* spectral value $\lambda \in \mathbb{C}$ satisfying $\lambda \in \sigma(A(\Delta))$ for some $\Delta \in \mathbf{\Delta}_0$ with norm $\|\Delta\| < \delta$ is accounted for, independent of the fact that the probability of encountering a $\Delta \in \mathbf{\Delta}_0$ with $\|\Delta\| < \delta$ producing an eigenvalue close to λ may be extremely small. Even in the context of a probabilistic investigation a worst case analysis is an indispensable supplementary tool for comparison and control of the results.
Since an experimental approach is not appropriate for a worst case analysis, our first step is to derive a theoretical characterization of spectral value sets. With every quadruple $(A, B, C, D) \in \mathbf{L}_{n,\ell,q}(\mathbb{K})$ we associate the *transfer function*

$$G(s) = D + C(sI_n - A)^{-1}B \in \mathbb{K}(s)^{q\times\ell}, \tag{5}$$

see Subsection 2.3.1. The following lemma is the main vehicle for obtaining our results.

Lemma 5.2.7. *Suppose* $\Delta \in \mathbb{C}^{\ell\times q}, \det(I_\ell - \Delta D) \neq 0$ *and* $s_0 \in \rho(A)$. *Then*

$$s_0 \in \sigma(A(\Delta)) \quad \Leftrightarrow \quad \det\left(I_\ell - \Delta G(s_0)\right) = 0.$$

Proof: Suppose $s_0 \in \sigma(A(\Delta))$ then there exists $x \in \mathbb{C}^n$, $x \neq 0$, such that

$$s_0 x = (A + B(I_\ell - \Delta D)^{-1}\Delta C)x. \tag{6}$$

Since $s_0 \in \rho(A)$, (6) is equivalent to

$$x = (s_0 I_n - A)^{-1}B(I_\ell - \Delta D)^{-1}\Delta C x.$$

Multiplying this equation with $(I_\ell - \Delta D)^{-1}\Delta C$ from the left we obtain

$$w = (I_\ell - \Delta D)^{-1}\Delta C(s_0 I_n - A)^{-1}Bw,$$

where $w = (I_\ell - \Delta D)^{-1}\Delta C x \in \mathbb{C}^\ell$. This, together with $(s_0 I_n - A)^{-1}Bw = x \neq 0$, implies

$$w = \Delta G(s_0)w \quad \text{and} \quad w \neq 0. \tag{7}$$

Conversely if $\det\left(I_\ell - \Delta G(s_0)\right) = 0$, there exists $w \in \mathbb{C}^\ell$ such that (7) holds. Let $x = (s_0 I_n - A)^{-1}Bw$, then $x \in \mathbb{C}^n$, $\Delta D w + \Delta C x = w$ and since $\det(I_\ell - \Delta D) \neq 0$

$$w = (I_\ell - \Delta D)^{-1}\Delta C x \neq 0.$$

So $x \neq 0$ and $(s_0 I_n - A)x = Bw = B(I_\ell - \Delta D)^{-1}\Delta C x$. Hence x satisfies (6) and therefore $s_0 \in \sigma(A(\Delta))$. □

As a corollary to the above lemma we determine the size of the minimal perturbations Δ achieving $s_0 \in \sigma(A(\Delta))$ for a given $s_0 \in \rho(A)$ (cf.(1.6)).

Corollary 5.2.8. *If* $s_0 \in \rho(A)$ *is such that* $\mu_{\mathbf{\Delta}}(D) < \mu_{\mathbf{\Delta}}(G(s_0))$, *then*

$$\delta(s_0) := \min\{\|\Delta\|_{\mathbf{\Delta}};\ \Delta \in \mathbf{\Delta},\ s_0 \in \sigma(A(\Delta))\} = \mu_{\mathbf{\Delta}}(G(s_0))^{-1}. \tag{8}$$

Proof: If $\Delta \in \mathbf{\Delta}$ is such that $s_0 \in \sigma(A(\Delta))$, then necessarily $\|\Delta\|_{\mathbf{\Delta}} \geq \mu_{\mathbf{\Delta}}(G(s_0))^{-1}$ by Lemma 5.2.7. This shows the inequality $\geq$ in (8). To prove the converse we note that since $\mu_{\mathbf{\Delta}}(G(s_0)) > 0$ there exists $\Delta \in \mathbf{\Delta}$ such that $\|\Delta\|_{\mathbf{\Delta}} = \mu_{\mathbf{\Delta}}(G(s_0))^{-1}$ and $\det(I_\ell - \Delta G(s_0)) = 0$ (see Remark 4.4.4 (ii)). But $I_\ell - \Delta D$ is nonsingular since $\mu_{\mathbf{\Delta}}(D) < \mu_{\mathbf{\Delta}}(G(s_0))$. Hence again by Lemma 5.2.7 $s_0 \in \sigma(A(\Delta))$ and this proves (8). □

The above lemma and corollary enable us to characterize spectral value sets in terms of the μ-function.

Theorem 5.2.9. *Suppose* $0 < \delta \leq \mu_{\mathbf{\Delta}}(D)^{-1}$, *then*

$$\sigma_{\mathbf{\Delta}}(A; B, C, D; \delta) = \sigma(A)\,\dot{\cup}\,\{s \in \rho(A);\ \mu_{\mathbf{\Delta}}(G(s)) > \delta^{-1}\}. \tag{9}$$

Proof: Suppose $s_0 \in \sigma(A(\Delta)) \cap \rho(A)$, $\Delta \in \mathbf{\Delta}$ and $\|\Delta\|_{\mathbf{\Delta}} < \delta$, then by Lemma 5.2.7 $\mu_{\mathbf{\Delta}}(G(s_0))^{-1} \leq \|\Delta\|_{\mathbf{\Delta}} < \delta$. Conversely if $s_0 \in \rho(A)$ and $\mu_{\mathbf{\Delta}}(G(s_0)) > \delta^{-1}$, then $\mu_{\mathbf{\Delta}}(G(s_0)) > \mu_{\mathbf{\Delta}}(D)$ and by the above corollary there exists a Δ with $\|\Delta\|_{\mathbf{\Delta}} = \mu_{\mathbf{\Delta}}(G(s_0))^{-1} < \delta$ such that $s_0 \in \sigma(A(\Delta))$. □

The application of formula (9) requires the calculation of the μ-value $\mu_{\mathbf{\Delta}}(G(s))$ for a grid of s-values covering a region in the complex plane where the spectral value set is to be determined. Since we know already that the calculation of the μ-value of a single matrix can be very difficult (depending on the perturbation class) the evaluation of formula (9) will, in general, be an awesome computational task. We briefly discuss this problem for multi-block perturbations in the next subsection where we will make use of the characterizations and estimates for the μ-function obtained in Section 4.4. As an application we then discuss in some detail *off-diagonal* perturbations of diagonal matrices. We will see that these Gershgorin type perturbations can be represented as block-diagonal perturbations, and in this case (9) yields a computable characterization of the associated spectral value sets.

Multi-Block Perturbations

For $N > 1$, $\ell_i \geq 1$, $q_i \geq 1$, $i \in \underline{N}$, $\ell = \ell_1 + \ldots + \ell_N$, $q = q_1 + \ldots + q_N$, we consider perturbations of the form

$$A \rightsquigarrow A(\Delta) = A + \sum_{i=1}^{N} B_i(I_{\ell_i} - \Delta_i D_i)^{-1}\Delta_i C_i, \quad \Delta_i \in \mathbb{C}^{\ell_i \times q_i},\ i \in \underline{N} \tag{10}$$

where $(A, B_i, C_i, D_i) \in \mathbf{L}_{n,\ell_i,q_i}(\mathbb{C}), i \in \underline{N}$ are given. These perturbations can be recast in the form of a linear fractional perturbation (1) with perturbation set $\mathbf{\Delta} = \{\mathrm{diag}\,(\Delta_1, \ldots, \Delta_N);\ \forall i \in \underline{N}:\ \Delta_i \in \mathbb{C}^{\ell_i \times q_i}\}$ and

$$B = [B_1\, B_2\, \ldots\, B_N],\ C = \begin{bmatrix} C_1 \\ C_2 \\ \vdots \\ C_N \end{bmatrix},\ D = \mathrm{diag}\,(D_1, ..., D_N),\ \Delta = \mathrm{diag}\,(\Delta_1, \ldots, \Delta_N).$$

Given any operator norms $\|\cdot\|_i$ on $C^{\ell_i \times q_i}$, $i \in \underline{N}$, we provide $\mathbf{\Delta}$ with the norm

$$\|\Delta\|_{\mathbf{\Delta}} = \max_{i \in \underline{N}} \|\Delta_i\|_i, \quad \Delta = \operatorname{diag}(\Delta_1, \ldots, \Delta_N) \in \mathbf{\Delta}.$$

This is an operator norm with respect to a variety of pairs of norms on $\mathbb{C}^\ell$ and $\mathbb{C}^q$ (see Definition 4.4.14 and Ex. 4.4.3). We fix such a pair and denote the spectral value sets for this perturbation structure by $\sigma_{\mathbb{C}}(A; (B_i, C_i, D_i)_{i \in \underline{N}}; \delta)$ or, for short, $\sigma_{\mathbb{C}}(\delta)$. We know from Lemma 4.4.7 that

$$\mu_{\mathbf{\Delta}}(G) = \max\{\varrho(\Delta G); \Delta \in \mathbf{\Delta},\ \|\Delta\|_{\mathbf{\Delta}} = 1\}, \quad G \in \mathbb{C}^{q \times \ell} \tag{11}$$

where ϱ denotes the spectral radius. Another characterization (see Theorem 4.4.23) is

$$\mu_{\mathbf{\Delta}}(G) = \max_{U \in \mathcal{V}^\ell_{\mathbf{\Delta}}, V \in \mathcal{V}^q_{\mathbf{\Delta}}} \varrho(V^* G U), \quad G \in \mathbb{C}^{q \times \ell}$$

where

$$\begin{aligned} \mathcal{V}^\ell_{\mathbf{\Delta}} &= \{\operatorname{diag}(u_1, \ldots, u_N);\ \forall j \in \underline{N}:\ u_j \in \mathbb{C}^{\ell_j},\ \|u_j\|_{\mathbb{C}^{\ell_j}} = 1\}, \\ \mathcal{V}^q_{\mathbf{\Delta}} &= \{\operatorname{diag}(v_1, \ldots, v_N);\ \forall j \in \underline{N}:\ v_j \in \mathbb{C}^{q_j},\ \|v_j\|^*_{\mathbb{C}^{q_j}} = 1\}. \end{aligned}$$

Combining these characterizations with Theorem 5.2.9, we get

Corollary 5.2.10. *Suppose* $0 < \delta \le \min_{i \in \underline{N}} \|D_i\|^{-1}_{\mathcal{L}(\mathbb{C}^{\ell_i}, \mathbb{C}^{q_i})}$, *then*

$$\begin{aligned} \sigma_{\mathbb{C}}(A; (B_i, C_i, D_i)_{i \in \underline{N}}; \delta) &= \sigma(A) \,\dot\cup\, \{s \in \rho(A); \exists \Delta \in \mathbf{\Delta} : \|\Delta\|_{\mathbf{\Delta}} = 1 \ \wedge\ \varrho(\Delta G(s)) > \delta^{-1}\} \\ &= \sigma(A) \,\dot\cup\, \{s \in \rho(A); \exists U \in \mathcal{V}^\ell_{\mathbf{\Delta}}, V \in \mathcal{V}^q_{\mathbf{\Delta}} : \varrho(V^* G(s) U) > \delta^{-1}\}. \end{aligned} \tag{12}$$

Unfortunately the above expressions are, in general, very difficult to compute. Upper and lower bounds for $\sigma_{\mathbb{C}}(\delta)$ can be derived from available upper and lower estimates for the μ-function, see Section 4.4. If $\Delta \in \mathbf{\Delta}$, $\|\Delta\|_{\mathbf{\Delta}} = 1$, $R_\gamma = \operatorname{diag}(\gamma_1 I_{q_1}, \ldots, \gamma_N I_{q_N})$ and $L_\gamma = \operatorname{diag}(\gamma_1 I_{\ell_1}, \ldots, \gamma_N I_{\ell_N})$, $\gamma_i > 0$, $i \in \underline{N}$, we have by (4.4.25),

$$\varrho(\Delta G) \le \mu_{\mathbf{\Delta}}(G) \le \|R_\gamma G L_\gamma^{-1}\| \tag{13}$$

where the norm $\|\cdot\|$ is the operator norm on $\mathcal{L}(\mathbb{C}^\ell, \mathbb{C}^q)$. We know from (11) that by a judicious choice of Δ the lower bound in (13) can be made tight but we do not have an algorithm to achieve this. With respect to spectral norms we have seen in Lemma 4.4.20 that the minimization of the upper bound in (13) can be reformulated as a convex optimization problem. However, both the maximizing Δ in (12) and the minimizing scaling vector γ for $G(s)$ depend on s, and consequently the problem of computing optimized upper and lower bounds for $\sigma_{\mathbb{C}}(\delta)$ via (13) is still a formidable one. Very rough but readily computable estimates are obtained by choosing a fixed $\Delta \in \mathbf{\Delta}$ with $\|\Delta\|_{\mathbf{\Delta}} = 1$ and a fixed scaling vector $\gamma = (\gamma_1, \ldots, \gamma_N)$, $\gamma_i > 0$ for all $s \in \rho(A)$

$$\{s \in \rho(A); \varrho(\Delta G(s)) > \delta^{-1}\} \subset \sigma_{\mathbb{C}}(\delta) \setminus \sigma(A) \subset \{s \in \rho(A); \|R_\gamma G(s) L_\gamma^{-1}\| > \delta^{-1}\}. \tag{14}$$

Here the scaled transfer matrix $R_\gamma G(s) L_\gamma^{-1} = D + R_\gamma C(sI - A)^{-1} B L_\gamma^{-1}$ can be interpreted as the transfer matrix of the scaled system $(A, B(\gamma), C(\gamma), D)$ where

$$B(\gamma) = B L_\gamma^{-1} = \left[\gamma_1^{-1} B_1, \cdots, \gamma_N^{-1} B_N\right], \quad C(\gamma) = R_\gamma C = \left[\gamma_1 C_1^\top, \cdots, \gamma_N C_N^\top\right]^\top.$$

We will see in the next subsection that, with these scaled matrices, the upper bound in (14) is identical with the spectral value set of A with respect to full-block perturbations of the form $A \rightsquigarrow A + B(\gamma)(I_\ell - \Delta D)^{-1}\Delta C(\gamma)$, $\Delta \in \mathbb{C}^{\ell\times q}$.
For square multi-block perturbations, with prescribed multiplicities of the blocks, we have shown in Proposition 4.4.25 that with respect to spectral norms we have $\mu_{\mathbf{\Delta}}(G) = \max_{U\in\mathbf{U}_{\mathbf{\Delta}}} \varrho(UG)$, where $\mathbf{U}_{\mathbf{\Delta}}$ is the set of block-diagonal unitary matrices defined in Proposition 4.4.25. We can use this result to characterize the spectral value sets for this type of perturbations, but once again the computational problem is severe.

Gershgorin Type Uncertainty

Gershgorin obtained an estimate for the spectrum of a given matrix $A = (a_{ij}) \in \mathbb{C}^{n\times n}$ by regarding it as a perturbation of the associated diagonal matrix $\mathrm{diag}(A) := \mathrm{diag}(a_{11}, \ldots, a_{nn}) \in \mathbb{C}^{n\times n}$ (see Theorem 4.2.19):

$$\sigma(A) \subset \bigcup_{i\in\underline{n}} \overline{D(a_{ii}, \rho_i)}, \quad \text{where} \quad \rho_i = \sum_{j\neq i} |a_{ij}|, \quad i \in \underline{n}.$$

Better estimates can be obtained by using more information about A than its diagonal elements and the off-diagonal row sums ρ_i, see *Notes and References*. Here we change the point of view: Instead of a *given* matrix A we consider a *set of matrices* $A(\Delta)$ obtained by perturbing a fixed diagonal matrix by off-diagonal disturbances of norm $< \delta$; and instead of deriving an estimate for the spectrum of a given matrix, we determine the union of all the spectra of the matrices $A(\Delta)$, $\|\Delta\| < \delta$.
More precisely, let the set of off-diagonal perturbations

$$\mathbf{\Delta} = \{\Delta \in \mathbb{C}^{n\times n};\ \mathrm{diag}\,(\Delta) = 0_{n\times n}\} \tag{15}$$

be provided with the norm

$$\|\Delta\|_{\mathbf{\Delta}} = \max_{i\in\underline{n}} \sum_{j\neq i} |\Delta_{ij}|, \quad \Delta \in \mathbf{\Delta}. \tag{16}$$

Note that $\|\Delta\|_{\mathbf{\Delta}}$ is the operator norm of $\Delta : \mathbb{C}^n \to \mathbb{C}^n$ if $\mathbb{C}^n$ is endowed with the maximum norm (see (A.1.3)). Our aim is to describe the spectral value sets

$$\sigma_{\mathbf{\Delta}}(A;\delta) = \bigcup_{\Delta\in\mathbf{\Delta},\ \|\Delta\|_{\mathbf{\Delta}}<\delta} \sigma(A+\Delta)$$

where $A = \mathrm{diag}\,(a_{11}, ..., a_{nn})$ is assumed to be diagonal[2]. Then $A(\Delta) = A + \Delta$ has the form

$$A(\Delta) := A + \Delta = \begin{bmatrix} a_{11} & \Delta_{12} & \Delta_{13} & \cdot & \cdot & & \cdot & \Delta_{1n} \\ \Delta_{21} & a_{22} & \Delta_{23} & \cdot & \cdot & & \cdot & \Delta_{2n} \\ \cdot & \cdot & \cdot & \cdot & \cdot & & \cdot & \cdot \\ \cdot & \cdot & \cdot & \cdot & \cdot & & \cdot & \cdot \\ \Delta_{n1} & \Delta_{n2} & \cdot & \cdot & \cdot & & \Delta_{nn-1} & a_{nn} \end{bmatrix}, \quad \Delta = (\Delta_{ij}) \in \mathbf{\Delta}.$$

[2]For non-diagonal $A \in \mathbb{C}^{n\times n}$ the problem of determining $\sigma_{\mathbf{\Delta}}(A;\delta)$ is as yet unsolved.

By Gershgorin's Theorem 4.2.19 we have the following upper bound for $\sigma_{\mathbf{\Delta}}(A;\delta)$

$$\sigma_{\mathbf{\Delta}}(A;\delta) \subset \bigcup_{i\in \underline{n}} D(a_{ii},\delta). \tag{17}$$

In order to derive an exact characterization from Theorem 5.2.9 we first represent the perturbation $A \rightsquigarrow A+\Delta, \Delta \in \mathbf{\Delta}$ as a multi-block perturbation and obtain an improved (and in fact tight) upper bound of the associated μ-function via scaling. For every $\Delta = (\Delta_{ij}) \in \mathbf{\Delta}$, let

$$\tilde{\Delta} = \begin{bmatrix} \Delta_{12} & \Delta_{13} & \cdots & \Delta_{1n} & 0 & 0 & \cdots & \cdot & \cdot & \cdots & \cdot & \cdot & \cdot & \cdots & 0 \\ 0 & 0 & \cdots & 0 & \Delta_{21} & \Delta_{23} & \cdots & \Delta_{2n} & 0 & \cdots & \cdot & \cdot & \cdot & \cdots & 0 \\ \cdot & \cdot & \cdots & \cdot & \cdot & \cdot & \cdots & \cdot & \cdot & \cdots & \cdot & \cdot & \cdot & \cdots & \cdot \\ \cdot & \cdot & \cdots & \cdot & \cdot & \cdot & \cdots & \cdot & \cdot & \cdots & \cdot & \cdot & \cdot & \cdots & \cdot \\ 0 & 0 & \cdots & \cdot & \cdot & \cdot & \cdots & \cdot & \cdot & \cdots & 0 & \Delta_{n1} & \Delta_{n2} & \cdots & \Delta_{n\,n-1} \end{bmatrix} \tag{18}$$

Then we have $\tilde{\Delta} \in \mathbb{C}^{n\times n(n-1)}$ and $\Delta = I_n\tilde{\Delta}C$ for all $\Delta \in \mathbf{\Delta}$ if we define

$$C = \begin{bmatrix} C_1 \\ \cdot \\ \cdot \\ C_n \end{bmatrix}_{n(n-1)\times n}, \qquad C_i = \begin{bmatrix} I_{i-1} & 0 & 0 \\ 0 & 0 & I_{n-i} \end{bmatrix}_{(n-1)\times n}, \quad i \in \underline{n} \tag{19}$$

where I_0 is void. Note that $C_i x = [x_1,..,x_{i-1},x_{i+1},..,x_n]^\top = (x_j)_{j\neq i}$ for $x \in \mathbb{C}^n$ and $\tilde{\Delta} : \bigoplus_1^n \mathbb{C}^{n-1} \mapsto \mathbb{C}^n$. Let $\tilde{\mathbf{\Delta}} \subset \mathbb{C}^{n\times n(n-1)}$ be the vector space of all complex matrices $\tilde{\Delta}$ of the form (18) endowed with the operator norm $\|\cdot\|_{\tilde{\mathbf{\Delta}}}$ induced by the ∞-norms on both $\bigoplus_1^n \mathbb{C}^{n-1} = \mathbb{C}^{n(n-1)}$ and $\mathbb{C}^n$. Then by (A.1.3)

$$\|\tilde{\Delta}\|_{\tilde{\mathbf{\Delta}}} = \max_{i\in\underline{n}} \sum_{j\neq i} |\Delta_{ij}| = \|\Delta\|_{\mathbf{\Delta}}, \quad \Delta \in \mathbf{\Delta}.$$

As a consequence, we have for all $\delta > 0$

$$\sigma_{\mathbf{\Delta}}(A;\delta) = \bigcup_{\Delta\in\mathbf{\Delta},\, \|\Delta\|_{\mathbf{\Delta}}<\delta} \sigma(A+\Delta) = \bigcup_{\tilde{\Delta}\in\tilde{\mathbf{\Delta}},\, \|\tilde{\Delta}\|_{\tilde{\mathbf{\Delta}}}<\delta} \sigma(A+\tilde{\Delta}C) = \sigma_{\tilde{\mathbf{\Delta}}}(A; I_n, C; \delta).$$

The transfer function associated with the triple $(A,\, I_n,\, C)$ is

$$G(s) = \begin{bmatrix} C_1 \\ \cdot \\ \cdot \\ C_n \end{bmatrix} \operatorname{diag}\left((s-a_{11})^{-1}, ..., (s-a_{nn})^{-1}\right) \in \mathbb{C}^{n(n-1)\times n}(s). \tag{20}$$

So every row of $G(s)$ has exactly one entry different from zero and this is of the form $(s-a_{ii})^{-1}$. It follows that the operator norm of $G(s)$ with respect to the ∞-norms on $\mathbb{C}^n$ and $\mathbb{C}^{n(n-1)}$ is

$$\|G(s)\|_{\mathcal{L}(\mathbb{C}^n,\mathbb{C}^{n(n-1)})} = \max_{i\in\underline{n}} |s-a_{ii}|^{-1}. \tag{21}$$

After representing the off-diagonal representations in block-diagonal form we can now apply a scaling technique in order get an upper bound for $\mu_{\tilde{\mathbf{\Delta}}}(G(s))$. The proof of the next proposition shows that the optimized upper bound (13) is in fact tight so that we obtain a computable formula for the spectral value set $\sigma_{\mathbf{\Delta}}(A;\delta)$ via Theorem 5.2.9.

Proposition 5.2.11. *Let $A = \mathrm{diag}(a_{11}, \ldots, a_{nn}) \in \mathbb{C}^{n\times n}$ and suppose that the set of off-diagonal perturbations* $\mathbf{\Delta}$ (15) *is provided with the norm* (16). *Then the spectral value set of A with respect to the perturbations $A \rightsquigarrow A(\Delta) = A + \Delta$, $\Delta \in \mathbf{\Delta}$ is*

$$\sigma_{\mathbf{\Delta}}(A;\delta) = \{s \in \mathbb{C};\ \min_{i,j\in\underline{n},\ i\neq j} |s-a_{ii}||s-a_{jj}| < \delta^2\}, \quad \delta > 0. \tag{22}$$

Proof: Scaling by matrices $R_\gamma = \mathrm{diag}\,(\gamma_1 I_{n-1}, \ldots, \gamma_n I_{n-1})$, $L_\gamma = \mathrm{diag}\,(\gamma_1, \ldots, \gamma_n)$ where $\gamma_i > 0$, $i \in \underline{n}$, results in a scaled transfer function matrix

$$G_\gamma(s) = R_\gamma G(s) L_\gamma^{-1} = \begin{bmatrix} \gamma_1 C_1 \\ \cdot \\ \cdot \\ \gamma_n C_n \end{bmatrix} \mathrm{diag}\,(\gamma_1^{-1}(s-a_{11})^{-1}, \ldots, \gamma_n^{-1}(s-a_{nn})^{-1}). \tag{23}$$

Every row of $G_\gamma(s)$ has exactly one entry which is non-zero and this entry is of the form $(\gamma_i/\gamma_j)(s-a_{jj})^{-1}, i \neq j$. Therefore, with respect to the ∞-norms on $\mathbb{C}^n$, $\mathbb{C}^{n(n-1)}$,

$$\|G_\gamma(s)\|_{\mathcal{L}(\mathbb{C}^n, \mathbb{C}^{n(n-1)})} = \max_{i,j\in\underline{n}, i\neq j} \frac{\gamma_i}{\gamma_j}|s-a_{jj}|^{-1}.$$

Given any $s \in \mathbb{C} \setminus \{a_{11}, \ldots, a_{nn}\}$, let $\gamma_i = |s-a_{ii}|^{-1/2}$, $i \in \underline{n}$, then by (13)

$$\mu_{\tilde{\mathbf{\Delta}}}(G(s)) \leq \|G_\gamma(s)\|_{\mathcal{L}(\mathbb{C}^n, \mathbb{C}^{n(n-1)})} = \max_{i,j\in\underline{n}, i\neq j} |s-a_{ii}|^{-1/2}|s-a_{jj}|^{-1/2}. \tag{24}$$

We want to show that the inequality is in fact an equality. Suppose that the maximum on the RHS of (24) is $m(s)$ and this occurs for $i = \hat{\imath}$, $j = \hat{\jmath}$ with $\hat{\imath} < \hat{\jmath}$. Define $\Delta \in \mathbf{\Delta}$ by $\Delta_{\hat{\imath}\hat{\jmath}} = \Delta_{\hat{\jmath}\hat{\imath}} = 1$ and $\Delta_{ij} = 0$ for all other i, j. Then $\|\tilde{\Delta}\|_{\tilde{\mathbf{\Delta}}} = \|\Delta\|_{\mathbf{\Delta}} = 1$ and $\tilde{\Delta}G(s) = \Delta\,\mathrm{diag}\,((s-a_{11})^{-1}, \ldots, (s-a_{nn})^{-1})$ is a matrix with zero entries except the $(\hat{\imath}, \hat{\jmath})$ entry which is $(s-a_{\hat{\jmath}\hat{\jmath}})^{-1}$ and the $(\hat{\jmath}, \hat{\imath})$ entry which is $(s-a_{\hat{\imath}\hat{\imath}})^{-1}$. Hence

$$\varrho(\tilde{\Delta}G(s))^2 = |s-a_{\hat{\imath}\hat{\imath}}|^{-1}|s-a_{\hat{\jmath}\hat{\jmath}}|^{-1} = m(s)^2$$

and so $m(s) \leq \mu_{\tilde{\mathbf{\Delta}}}(G(s))$ by (13). It follows from (24) that $\mu_{\tilde{\mathbf{\Delta}}}(G(s)) = m(s)$. The proof is completed by applying Theorem 5.2.9 to obtain (22). □

Remark 5.2.12. (i) Under the assumptions of the previous proposition we have by (22)

$$\sigma_{\mathbf{\Delta}}(A;\delta) = \bigcup_{i,j\in\underline{n}\ i\neq j} \{s \in \mathbb{C};\ |s-a_{ii}||s-a_{jj}| < \delta^2\}, \quad \delta > 0.$$

The sets $\{s \in \mathbb{C};\ |s-a_{ii}||s-a_{jj}| < \delta^2\}$ are known as *ovals of Cassini* (see Figure 5.2.4 and *Notes and References*).

(ii) Computable formulas can also be derived for other norms on the Gershgorin perturbation class $\mathbf{\Delta}$. In Ex. 3 the reader is guided towards the result for the Hölder norm $\|\Delta\|_{\mathbf{\Delta}} = \|\Delta\|_{\infty|\infty} = \max_{i,j\in\underline{n}, i\neq j} |\Delta_{ij}|$ on $\mathbf{\Delta}$ (see Section A.1). □

Example 5.2.13. Consider $A = \mathrm{diag}\,(-1,-2,-3)$. The spectral value sets $\sigma_{\mathbf{\Delta}}(A;\delta)$, $\delta = 0.45$, 0.5, 0.55, 0.7, 1, $\sqrt{2}$ with respect to the norm (16) on $\mathbf{\Delta}$ are given in the figure on the left. The figure on the right shows the estimates for these sets obtained by using Gershgorin's Theorem as in (17). In both figures the sets are connected for $\delta > 0.5$. Note that the Cassini oval for $\delta = \sqrt{2}$ touches the imaginary axis and hence if the stability region is $\mathbb{C}_-$ then the stability radius $r_{\mathbf{\Delta}} = \sqrt{2}$. From the figure on the right one obtains the conservative estimate $r_{\mathbf{\Delta}} \geq 1$. □

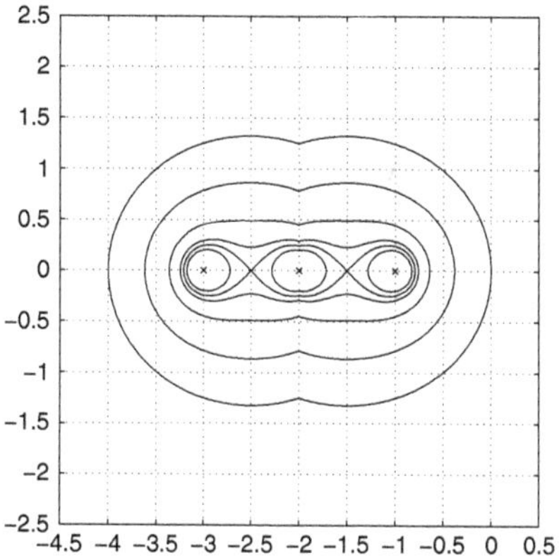

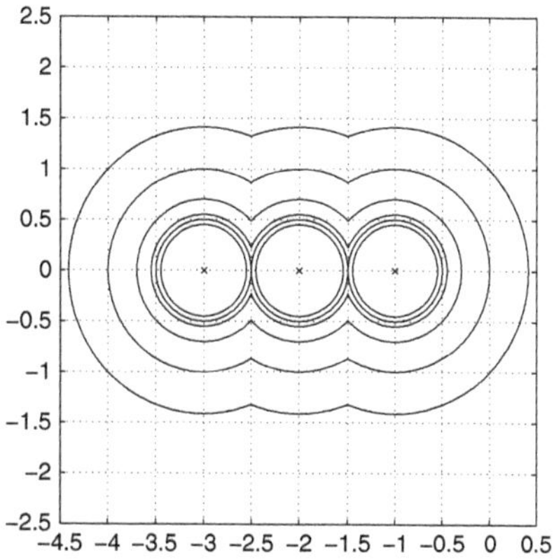

Figure 5.2.4: Spectral value sets and estimates obtained via Gershgorin's Theorem

Dependence on System Data

We now return to the general situation as described by the initial assumptions of this subsection. At the very beginning we mentioned already that the map $\delta \mapsto \overline{\sigma_{\mathbf{\Delta}}(A; B, C, D; \delta)}$ is continuous with respect to the Hausdorff metric. We will now show that the closures of spectral value sets depend continuously on the whole collection of data A, B, C, D, δ. Recall the following definition, see Section 4.1. Given a metric space (Z, d_Z), the Hausdorff metric on the space $\mathcal{K}(Z)$ of compact subsets of Z is defined by

$$d(C_1, C_2) := \max\{\max_{c_1 \in C_1} \operatorname{dist}(c_1, C_2), \max_{c_2 \in C_2} \operatorname{dist}(c_2, C_1)\}, \quad \operatorname{dist}(a, C) := \min_{c \in C} d_Z(a, c).$$

Lemma 5.2.14. *Let (X, d_X), (Y, d_Y), (Z, d_Z) be metric spaces, X locally compact, Y compact, and let the space $\mathcal{K}(Z)$ be provided with the Hausdorff metric. If $f : X \times Y \to Z$ is continuous, then the set-valued map*

$$F : X \to \mathcal{K}(Z), \quad x \mapsto F(x) = f(x, Y) = \{f(x, y);\ y \in Y\}$$

is continuous.

Proof: Let $x^0 \in X$ and C be a compact ball with centre x^0 and radius $r > 0$ in X. Then f is uniformly continuous on the compact set $C \times Y$. For any $\varepsilon > 0$ we find $0 < \delta < r$ such that $d_Z(f(x, y), f(x', y)) < \varepsilon$ for all $y \in Y$ and all $x, x' \in C$ such that $d_X(x, x') < \delta$. But then it follows from the definition of the Hausdorff metric that $d(F(x), F(x^0)) < \varepsilon$ for all $x \in X$ with $d_X(x, x^0) < \delta$. This proves that F is continuous at x^0. □

Theorem 5.2.15. *Suppose $(A^0, B^0, C^0, D^0, \delta^0) \in \mathbf{L}_{n,\ell,q}(\mathbb{C}) \times (0, \infty)$, with $\delta^0 < \mu_{\mathbf{\Delta}}(D^0)^{-1}$ and let $\mathcal{K}(\mathbb{C})$ be provided with the Hausdorff metric. Then the map*

$$(A, B, C, D, \delta) \ \mapsto \ \overline{\sigma_{\mathbf{\Delta}}(A; B, C, D; \delta)} \ \in \mathcal{K}(\mathbb{C})$$

is well defined and continuous on an open neighbourhood of $(A^0, B^0, C^0, D^0, \delta^0)$ in $\mathbf{L}_{n,\ell,q}(\mathbb{C}) \times (0, \infty)$.

Proof: Let $\mathbf{\Delta}_1 = \{\Delta \in \mathbf{\Delta}; \|\Delta\|_{\mathbf{\Delta}} \le 1\}$, then $\mathbf{\Delta}_1$ is compact and by (2)

$$0 < \delta < \mu_{\mathbf{\Delta}}(D)^{-1} \quad \Rightarrow \quad \overline{\sigma_{\mathbf{\Delta}}(A; B, C, D; \delta)} = \bigcup_{\Delta \in \mathbf{\Delta}_1} \sigma(A(\delta\Delta)),$$

where $A(\delta\Delta) = A + B(I_\ell - \delta\Delta D)^{-1}\delta\Delta C$. By the upper semicontinuity of $\mu_{\mathbf{\Delta}}(\cdot)$ (Lemma 4.4.27) there is an open neighbourhood $\mathcal{N}$ of $(A^0, B^0, C^0, D^0, \delta^0)$ in the space $\mathbf{L}_{n,\ell,q}(\mathbb{C}) \times (0, \infty)$ such that $\delta < \mu_{\mathbf{\Delta}}(D)^{-1}$ for all $(A, B, C, D, \delta) \in \mathcal{N}$. Now define $f((A, B, C, D, \delta), \Delta) := A(\delta\Delta)$ for $(A, B, C, D, \delta) \in \mathcal{N}$, $\Delta \in \mathbf{\Delta}_1$. Then $f : \mathcal{N} \times \mathbf{\Delta}_1 \to \mathbb{C}^{n\times n}$ is well defined and continuous on $\mathcal{N} \times \mathbf{\Delta}_1$. Since $\mathcal{N}$ is locally compact and $\mathbf{\Delta}_1$ is compact we conclude from the previous lemma that the map

$$F : \mathcal{N} \to \mathcal{K}(\mathbb{C}^{n\times n}), \quad (A, B, C, D, \delta) \mapsto \{A(\delta\Delta);\ \Delta \in \mathbf{\Delta}_1\}$$

is continuous with respect to the Hausdorff metric on $\mathcal{K}(\mathbb{C}^{n\times n})$. Therefore using Lemma 4.1.9 the map

$$(A, B, C, D, \delta) \mapsto \sigma(F(A, B, C, D, \delta)) = \bigcup_{\Delta \in \mathbf{\Delta}_1} \sigma(A(\delta\Delta)) = \overline{\sigma_{\mathbf{\Delta}}(A; B, C, D; \delta)}$$

is continuous on $\mathcal{N}$. □

5.2.2 Complex Full-Block Perturbations

For complex full-block perturbations Theorem 5.2.9 immediately yields an easily computable criterion for deciding whether any $s \in \mathbb{C}$ belongs to a spectral value set or not. We illustrate this criterion by a simple example and then show how spectral value sets for complex full-block perturbations can be visualized by drawing the level curves ("spectral contours") of the norm of the associated transfer matrix in the complex plane. We conclude the subsection with an illustrative example.
Throughout the subsection we suppose the following

- $(A, B, C, D) \in \mathbf{L}_{n,\ell,q}(\mathbb{C})$.
- The perturbation set is $\mathbf{\Delta} = \mathbb{C}^{\ell\times q}$ provided with a norm $\|\cdot\|_{\mathbf{\Delta}}$ which is an operator norm with respect to a given pair of norms on $\mathbb{C}^\ell$ and $\mathbb{C}^q$.
- The perturbations are of the form
$$A \rightsquigarrow A(\Delta) = A + B(I_\ell - \Delta D)^{-1}\Delta C, \quad \Delta \in \mathbb{C}^{\ell\times q}. \tag{25}$$

The associated spectral value sets are denoted by $\sigma_{\mathbb{C}}(A; B, C, D; \delta)$. The subset $\mathbf{\Delta}_0$ of admissible perturbations is

$$\mathbf{\Delta}_0 = \{\Delta \in \mathbb{C}^{\ell\times q};\ \det(I_\ell - \Delta D) \ne 0\} \quad \text{and hence} \quad \boldsymbol{\delta}_0 = \|D\|^{-1}_{\mathcal{L}(\mathbb{C}^\ell,\mathbb{C}^q)}.$$

Since $\mu_{\mathbf{\Delta}}(G) = \|G\|_{\mathcal{L}(\mathbb{C}^\ell,\mathbb{C}^q)}$ for all $G \in \mathbb{C}^{q\times\ell}$ if $\mathbf{\Delta} = \mathbb{C}^{\ell\times q}$ (see Proposition 4.4.11), the following characterization is an immediate consequence of Theorem 5.2.9.

Theorem 5.2.16. *Suppose* $0 < \delta \le \|D\|^{-1}_{\mathcal{L}(\mathbb{C}^\ell,\mathbb{C}^q)}$, *then*

$$\sigma_{\mathbb{C}}(A; B, C, D; \delta) = \sigma(A)\,\dot{\cup}\,\{s \in \rho(A);\ \|G(s)\|_{\mathcal{L}(\mathbb{C}^\ell,\mathbb{C}^q)} > \delta^{-1}\}. \tag{26}$$

where $G(s) = D + C(sI - A)^{-1}B, \ \ s \in \rho(A)$.

We illustrate this theorem by showing how it may be employed to compute the spectral value sets visualized in Figure 5.1.1.

Example 5.2.17. Consider the perturbed oscillator $(1+\Delta_m)\ddot{\xi}+(4+\Delta_c)\dot{\xi}+(5+\Delta_k)\xi=0$ as described in Example 5.1.4. The perturbed system can be modelled as a linear fractional perturbation with a full-block perturbation set $\mathbf{\Delta}=\mathbb{C}^{1\times 3}$

$$A=\begin{bmatrix} 0 & 1 \\ -5 & -4 \end{bmatrix}, \quad B=\begin{bmatrix} 0 \\ 1 \end{bmatrix}, \quad C=-\begin{bmatrix} -5 & -4 \\ 0 & 1 \\ 1 & 0 \end{bmatrix}, \quad D=\begin{bmatrix} -1 \\ 0 \\ 0 \end{bmatrix},$$

see Example 5.1.22. The associated transfer function is

$$G(s)=-\begin{bmatrix} 1 \\ 0 \\ 0 \end{bmatrix}-(s^2+4s+5)^{-1}\begin{bmatrix} -5 & -4 \\ 0 & 1 \\ 1 & 0 \end{bmatrix}\begin{bmatrix} 1 \\ s \end{bmatrix}=-(s^2+4s+5)^{-1}\begin{bmatrix} s^2 \\ s \\ 1 \end{bmatrix}.$$

If we endow $\mathbf{\Delta}=\mathbb{C}^{1\times 3}$ with the norm $\|\cdot\|_{\mathbf{\Delta}}=\|\cdot\|_\infty$ and $\mathbb{C}^3$ with the corresponding dual norm $\|\cdot\|_1$ on $\mathbb{C}^3$ then

$$\|G(s)\|_1=(1+|s|+|s|^2)/|s^2+4s+5|.$$

So provided $\delta\le\|D\|^{-1}_{\mathcal{L}(\mathbb{C}^\ell,\mathbb{C}^q)}=1$, we have

$$\sigma_{\mathbb{C}}(A;B,C,D;\delta)=\{s\in\mathbb{C};\,(1+|s|+|s|^2)/|s^2+4s+5|>\delta^{-1}\},$$

where we set $|s^2+4s+5|^{-1}=\infty$ at $s=-2\pm\imath$. The spectral value sets for $\delta\in\{0.1, 0.28, 0.46, 0.64, 0.82\}$ are the bounded regions circumscribed by the closed curves on the left hand side of Figure 5.1.1. □

We will now describe how spectral value sets for complex full-block perturbations can be visualized on the computer screen by plotting the level curves of $s\to\|G(s)\|$. For every δ, the level set

$$\mathcal{C}_\delta=\{s\in\rho(A);\ \|G(s)\|_{\mathcal{L}(\mathbb{C}^\ell,\mathbb{C}^q)}=\delta^{-1}\}$$

is called the *spectral contour* of level δ^{-1}. $\mathcal{C}_\delta$ is closed in $\mathbb{C}\setminus\sigma(A)$, but not necessarily in $\mathbb{C}$ (see the following remark). However, $\mathcal{C}_\delta\cup\sigma(A)$ is closed in $\mathbb{C}$.

Remark 5.2.18. (i) In spite of the term "contour" the set $\mathcal{C}_\delta$ will, in general, consist of several curves, some of which may degenerate to isolated points (which must be local minima of $\|G(s)\|$, see Remark 5.2.20 (v) and Example 5.2.21).
(ii) For some values of δ the spectral contour may not be closed. In fact, there may be an isolated eigenvalue $s_0\in\sigma(A)$ which is a removable singularity of $G(s)$ such that $\lim_{s\to s_0}\|G(s)\|_{\mathcal{L}(\mathbb{C}^\ell,\mathbb{C}^q)}=\delta^{-1}$, whereas by definition $s_0\notin\mathcal{C}_\delta$. For instance, if

$$A=\begin{bmatrix} 1 & 0 \\ 0 & 2 \end{bmatrix}, \quad B=I_2,\ C=[1\ 0],\ D=0,\ \mathbf{\Delta}=\mathbb{C}^2,$$

then $G(s)=[(s-1)^{-1}\ 0]$ and the spectral norm of $G(s)$ is $\|G(s)\|=|(s-1)^{-1}|$. Let $\delta=1$ then the eigenvalue $s_0=2$ of A does not lie in $\mathcal{C}_\delta$, but $\lim_{s\to 2}\|G(s)\|=1$ so that $2\in\overline{\mathcal{C}_\delta}\setminus\mathcal{C}_\delta$. □

The next proposition describes how the spectral value set $\sigma_{\mathbb{C}}(A;B,C,D;\delta)$ is determined by the spectral contour $\mathcal{C}_\delta$. The formulations "modulo $\sigma(A)$" in the proposition cannot be avoided since some of the eigenvalues of A may be isolated points in $\sigma_{\mathbb{C}}(A;B,C,D;\delta)$. By $\partial\sigma_{\mathbb{C}}(A;B,C,D;\delta)$ we denote the boundary of $\sigma_{\mathbb{C}}(A;B,C,D;\delta)$ in $\mathbb{C}$.

Proposition 5.2.19. *Let $G(s) = D + C(sI_n - A)^{-1}B$ and $0 < \delta < \|D\|^{-1}_{\mathcal{L}(\mathbb{C}^\ell,\mathbb{C}^q)}$. Then $\sigma_{\mathbb{C}}(A;B,C,D;\delta) \setminus \sigma(A)$ is a bounded open subset of $\mathbb{C}$ and its boundary in $\rho(A)$ is given by*

$$\partial\sigma_{\mathbb{C}}(A;B,C,D;\delta) \setminus \sigma(A) = \mathcal{C}_\delta. \tag{27}$$

Moreover, the following statements hold.

(i) For every $s_0 \in \rho(A)$ with $\|G(s_0)\|_{\mathcal{L}(\mathbb{C}^\ell,\mathbb{C}^q)} = \delta^{-1}$ (i.e. for every $s_0 \in \mathcal{C}_\delta$) there exists a perturbation $\Delta \in \mathbb{C}^{\ell\times q}$ with norm $\|\Delta\| = \delta$ such that $s_0 \in \sigma(A(\Delta))$, and there is no smaller perturbation matrix with this property.

(ii) $\sigma_{\mathbb{C}}(A;B,C,D;\delta)$ is the union of $\sigma(A)$ and of those connected components of $\mathbb{C} \setminus \overline{\mathcal{C}_\delta}$ which contain a point $s_0 \in \rho(A)$ with $\|G(s_0)\| > \delta^{-1}$ (or, equivalently, a pole of $G(s)$).

Proof: For brevity, we write $\sigma_{\mathbb{C}}(\delta) = \sigma_{\mathbb{C}}(A;B,C,D;\delta)$ and use $\|\cdot\|$ to denote the norm $\|\cdot\|_{\mathcal{L}(\mathbb{C}^\ell,\mathbb{C}^q)}$ on $\mathbb{C}^{q\times\ell}$ in this proof. Boundedness and openness follow from (26) in Theorem 5.2.16 since $\|G(\cdot)\|$ is continuous on the resolvent set $\rho(A)$ and $\lim_{|s|\to\infty}\|G(s)\| = \|D\| < \delta^{-1}$. Moreover, the statement in (i) is a direct consequence of Corollary 5.2.8.
The inclusion $\subset$ in (27) follows from the continuity of $s \mapsto \|G(s)\|$ and formula (26). Conversely, let $s_0 \in \rho(A)$, $\|G(s_0)\| = \delta^{-1}$ and suppose $s_0 \notin \partial\sigma_{\mathbb{C}}(\delta)$, hence $s_0 \notin \sigma_{\mathbb{C}}(\delta)$. By (26) this implies that s_0 is a local maximum of $s \mapsto \|G(s)\|$. Choose $y \in \mathbb{C}^q, \|y^*\|^*_{\mathbb{C}^q} = 1, u \in \mathbb{C}^\ell, \|u\|_{\mathbb{C}^\ell} = 1$ such that $|y^*G(s_0)u| = \|G(s_0)\|$ (see Section A.4). The function $f : s \mapsto y^*G(s)u$ is holomorphic on $\rho(A)$ and $s \mapsto |f(s)|$ has a local maximum at s_0. Since $\rho(A)$ is connected, $f(s)$ is constant on $\rho(A)$ (with value δ^{-1}) by the maximum principle (see Theorem A.2.25). But $\lim_{|s|\to\infty}|f(s)| = |y^*Du| < \delta^{-1}$ and so we obtain a contradiction. Thus (27) is proved.
(ii) Let K be a connected component of $\mathbb{C} \setminus \overline{\mathcal{C}_\delta}$ and suppose that there is an $s_0 \in K \setminus \sigma(A)$ satisfying $\|G(s_0)\| > \delta^{-1}$. (Such an s_0 always exists if K contains a pole of $G(s)$). We will show that then $\|G(s)\| > \delta^{-1}$ for every $s \in K \setminus \sigma(A)$, hence $K \subset \sigma_{\mathbb{C}}(\delta)$ by (26). In fact, since K is open and connected in $\mathbb{C}$ there exists, for every $s \in K \setminus \sigma(A)$, an arc $\gamma(t)$, $0 \le t \le 1$ in $K \setminus \sigma(A)$ connecting $s_0 = \gamma(0)$ and $s = \gamma(1)$. Since $t \mapsto \|G(\gamma(t))\|$ is continuous and never takes the value δ^{-1}, it follows that $\|G(s)\| > \delta^{-1}$.
Conversely, assume $s_0 \in \sigma_{\mathbb{C}}(\delta) \setminus \sigma(A)$, hence $\|G(s_0)\| > \delta^{-1}$ by (26). Then $s_0 \notin \overline{\mathcal{C}_\delta}$ and there exists a connected component K of $\mathbb{C} \setminus \overline{\mathcal{C}_\delta}$ such that $s_0 \in K$. It only remains to prove that K contains a pole of $G(s)$. By the above argument it follows that $\|G(s)\| > \delta^{-1}$ for all $s \in K \setminus \sigma(A)$. Now let $P \subset \sigma(A)$ be the set of poles of $G(s)$ and suppose $K \cap P = \emptyset$. Since K is closed in $\mathbb{C} \setminus \overline{\mathcal{C}_\delta}$ we have $\overline{K} \subset K \cup \overline{\mathcal{C}_\delta}$ and so $\overline{K} \cap P = \emptyset$. $G(s)$ can be extended holomorphically to $\mathbb{C} \setminus P \supset \overline{K}$. The

extension $\tilde{G}(s)$ satisfies $\|\tilde{G}(s)\| \geq \delta^{-1} > \|D\|_{\mathcal{L}(\mathbb{C}^\ell,\mathbb{C}^q)}$ on $\overline{K}$ and thus $\overline{K}$ is compact. Let $z \in \overline{K}$ be a maximum of $\|\tilde{G}(s)\|$ on $\overline{K}$. Then $\|\tilde{G}(z)\| > \delta^{-1}$ and $z \in K$. Since K is open in $\mathbb{C}$, z is a local maximum of $\|\tilde{G}(s)\|$. Reasoning as in the first part of the proof we again obtain a contradiction. Hence K must contain a pole of $G(s)$ and the proposition is proved. □

Remark 5.2.20. (i) It follows from the previous theorem and Theorem 5.2.16 that every connected component K of $\mathbb{C} \setminus \overline{\mathcal{C}_\delta}$ is either contained in $\sigma_{\mathbb{C}}(\delta)$ or disjoint from $\sigma_{\mathbb{C}}(\delta) \cap \rho(A)$. $K \subset \sigma_{\mathbb{C}}(\delta)$ (resp. $K \cap \sigma_{\mathbb{C}}(\delta) \cap \rho(A) = \emptyset$) if and only if $\|G(s_0)\| > \delta^{-1}$ (resp. $\|G(s_0)\| < \delta^{-1}$) for an arbitrarily chosen $s_0 \in K \cap \rho(A)$.

(ii) In order to compute a spectral value set $\sigma_{\mathbb{C}}(\delta)$ according to formula (26) the operator norm $\|G(s)\|_{\mathcal{L}(\mathbb{C}^\ell,\mathbb{C}^q)}$ has to be computed for many points $s \in \rho(A)$. It is therefore convenient to choose a perturbation norm $\|\cdot\|$ on $\mathbf{\Delta} = \mathbb{C}^{\ell \times q}$ such that the corresponding operator norm on $\mathbb{C}^{q\times \ell}$ is easily computed, e.g. $\|\Delta\| = \|\Delta\|_{1,1}$, $\|\Delta\| = \|\Delta\|_{\infty,\infty}$ or $\|\Delta\| = \|\Delta\|_{2,2}$. The corresponding norms on $\mathbb{C}^{q\times\ell}$ are $\|G\|_{1,1} = \|G^*\|_{1|\infty}$, $\|G\|_{\infty,\infty} = \|G\|_{1|\infty}$ and $\|G\|_{2,2} = \sigma_{\max}(G)$, respectively, see Remark 4.4.13 and (A.1.3), (A.1.7).

(iii) Given $s_0 \in \rho(A)$, a $\Delta_{\min} \in \mathbb{C}^{\ell\times q}$ of minimum norm such that $s_0 \in \sigma(A(\Delta_{\min}))$ may be constructed as in Proposition 4.4.11: Let $u_0 \in \mathbb{C}^\ell, v_0 \in \mathbb{C}^q$ be such that

$$v_0^* G(s_0) u_0 = \|G(s_0)\|_{\mathcal{L}(\mathbb{C}^\ell,\mathbb{C}^q)}, \qquad \|u_0\|_{\mathbb{C}^\ell} = \|v_0^*\|_{\mathbb{C}^q}^* = 1$$

(where $\|\cdot\|_{\mathbb{C}^q}^*$ is the dual norm of $\|\cdot\|_{\mathbb{C}^q}$). Then for each such pair $\Delta_{\min} = \|G(s_0)\|^{-1}_{\mathcal{L}(\mathbb{C}^\ell,\mathbb{C}^q)} u_0 v_0^*$ achieves $s_0 \in \sigma(A(\Delta_{\min}))$ with $\|\Delta_{\min}\|_{\mathcal{L}(\mathbb{C}^q,\mathbb{C}^\ell)} = \|G(s_0)\|^{-1}_{\mathcal{L}(\mathbb{C}^\ell,\mathbb{C}^q)}$.

(iv) Suppose $\mathbf{\Delta} = \mathbb{C}^{\ell\times q}$ is provided with a norm $\|\cdot\|_{\mathbb{C}^{\ell\times q}}$ which is rank one consistent with an operator norm $\|\cdot\|_{\mathcal{L}(\mathbb{C}^q,\mathbb{C}^\ell)}$ induced by a given pair of norms on $\mathbb{C}^q$ and $\mathbb{C}^\ell$ (see Definition A.1.11). It follows from Corollary 4.4.12 that the spectral value set $\sigma_{\mathbb{C}}(\delta)$ with respect to the perturbation norm $\|\cdot\|_{\mathbb{C}^{\ell\times q}}$ coincides with the spectral value set with respect to the operator norm $\|\cdot\|_{\mathcal{L}(\mathbb{C}^q,\mathbb{C}^\ell)}$ and

$$\sigma_{\mathbb{C}}(\delta) = \bigcup_{\Delta\in\mathbb{C}^{\ell\times q},\, \|\Delta\|_{\mathbb{C}^{\ell\times q}}<\delta} \sigma(A(\Delta)) = \sigma(A) \,\dot{\cup}\, \{s \in \rho(A);\ \|G(s)\|_{\mathcal{L}(\mathbb{C}^\ell,\mathbb{C}^q)} > \delta^{-1}\}.$$

(v) For $\delta < \|D\|^{-1}_{\mathcal{L}(\mathbb{C}^\ell,\mathbb{C}^q)}$, the complement $\mathbb{C}\setminus\sigma_{\mathbb{C}}(\delta)$ has exactly one unbounded connected component.

(vi) The above proof shows that the function $s \mapsto \|G(s)\|_{\mathcal{L}(\mathbb{C}^\ell,\mathbb{C}^q)}$ has no local maxima on $\rho(A)$; it may, however, have strict local minima $\tilde{s}$, see Figure 5.2.5 where the location of such a minimum is indicated by a small square. As a consequence a strictly larger perturbation may be necessary to move an eigenvalue of $A(\Delta)$, to $\tilde{s} \in \mathbb{C}$ than to all the other points in a small disk around $\tilde{s}$. In general, due to the existence of local minima of $\|G(s)\|_{\mathcal{L}(\mathbb{C}^\ell,\mathbb{C}^q)}$, the connected components of $\sigma_{\mathbb{C}}(\delta)$ may not be simply connected, i.e. they may contain holes, for some $\delta > 0$. In many examples the contours $\mathcal{C}_\delta$ contains two disjoint closed curves one of which lies inside the other, see Example 5.2.21. If the smaller one does not contain a pole of $G(s)$, it describes a hole in the corresponding spectral value set.

(vii) Suppose that for some $0 < \delta < \|D\|^{-1}_{\mathcal{L}(\mathbb{C}^\ell,\mathbb{C}^q)}$ we have $\sigma_{\mathbb{C}}(\delta) = \sigma(A)$, i.e. all eigenvalues are fixed for small perturbations. By (26) this is equivalent to $G(s)$ having no poles on $\mathbb{C}$. But $G(s)$ is a proper rational function so that by Liouville's Theorem this can only happen if $G(s)$ is constant, i.e. $G(s) \equiv D$. Thus

$$\exists \delta > 0 : \ \sigma_{\mathbb{C}}(\delta) = \sigma(A) \quad \Leftrightarrow \quad G(s) \equiv D.$$

An equivalent condition is that $C(sI-A)^{-1}B = \sum_{k=1}^{\infty} CA^{k-1}Bs^{-k} \equiv 0$, i.e. $CA^jB = 0$ for all $j \in \mathbb{N}$. In this case we have $\sigma_{\mathbb{C}}(\delta) = \sigma(A)$ for all $\delta \in (0, \|D\|^{-1}_{\mathcal{L}(\mathbb{C}^\ell,\mathbb{C}^q)})$. □

Example 5.2.21. Consider

$$A = \begin{bmatrix} 0 & 1 & 0 & 0 & 0 & 0 \\ 0 & 0 & 1 & 0 & 0 & 0 \\ 0 & 0 & 0 & 1 & 0 & 0 \\ 0 & 0 & 0 & 0 & 1 & 0 \\ 0 & 0 & 0 & 0 & 0 & 1 \\ -1595.48 & -2113.96 & -1361.70 & -518.13 & -122.38 & -15.92 \end{bmatrix}, \quad B = \begin{bmatrix} 0 \\ 0 \\ 0 \\ 0 \\ 0 \\ 1 \end{bmatrix}$$

and $C = I_6$, $D = 0$. So the bottom row of A is subject to perturbations and we suppose these are arbitrary complex numbers and set $\mathbf{\Delta} = \mathbb{C}^{1\times 6}$ with $\|\cdot\|_{\mathbf{\Delta}}$ taken to be the Euclidean norm. A is the companion matrix of a (randomly generated) stable monic polynomial

$$\chi_A(s) = (s + 4.42 \pm 3.83\,\imath)(s + 1.36 \pm 2.39\,\imath)(s + 2.18 \pm 1.19\,\imath).$$

Hence $\sigma(A) \subset \mathbb{C}_-$. The transfer matrix is given by

$$G(s) = C(sI_6 - A)^{-1}B = \frac{1}{\chi_A(s)} \begin{bmatrix} 1 & s & s^2 & s^3 & s^4 & s^5 \end{bmatrix}^\top. \tag{28}$$

We computed the spectral value sets for various values of δ and the results are shown in Figure 5.2.5. The three dimensional portrait in (a) is the graph of the function $\|G(\cdot)\|$

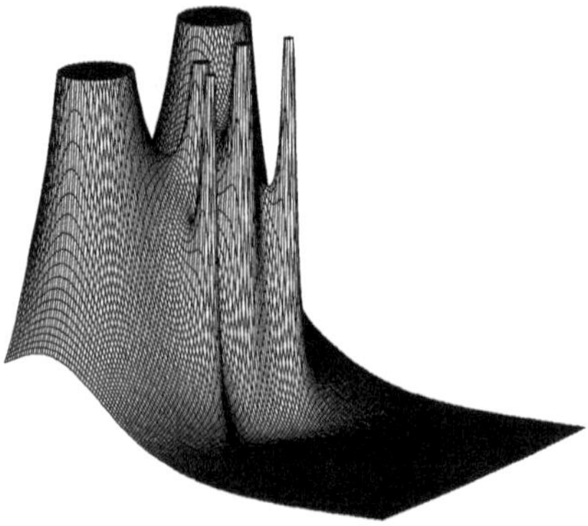

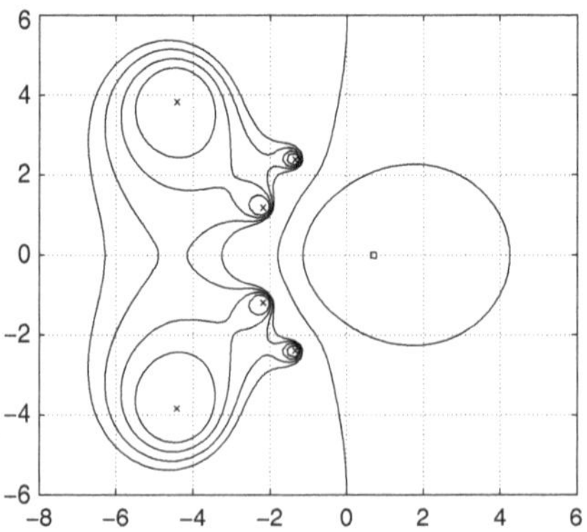

Figure 5.2.5: (a) Graph of $s \mapsto \|G(s)\|$ (b) Spectral contours for specific δ

and the spectral contours shown in (b) are obtained from it via a contour plotter. In (b) the spectral contours have been computed for $\delta = 0.5$, 0.66, 0.83, 1, 8.77, 100. The figures show that the eigenvalues $-4.42 \pm 3.83\imath$ are more sensitive to the perturbations than the

others. This is indicated by the broader peaks around them in (a) and the larger circular contours around them for $\delta = 0.5$ in (b). In fact the four "front line eigenvalues" near the imaginary axis remain stubbornly close to their original location under perturbations of size ≤ 1. At an uncertainty level of about $\delta = 0.83$ the spectral value set is connected and all the eigenvalues "interact". It is only after this stage has been reached, that the spectral value sets begins to approach the imaginary axis. This approach is very slow and the first contour to touch it is at level $\delta = 8.77$. So if the stability region is $\mathbb{C}_-$, then the stability radius $r_{\mathbf{\Delta}} = r_{\mathbb{C}^{1\times 6}} = 8.77$. A similar behaviour of spectral value sets can be observed for many stable matrices and explains why high sensitivity of the eigenvalues can coexist with a considerable robustness of stability.
The closed curve to the right of Figure 5.2.5 (b) belongs to the contour $\mathcal{C}_{100}$. The region circumscribed by this closed curve does not contain a pole of $G(s)$ and hence does not intersect $\sigma_{\mathbb{C}}(A; B, C, D; 100)$. In fact it describes a hole in this spectral value set: $\sigma_{\mathbb{C}}(A; B, C, D; 100)$ contains all points in Figure 5.2.5 (b) except those lying in this hole and its boundary. The small square inside the curve marks the location of a local minimum of $\|G(s)\|$. To push an eigenvalue to this location a disturbance of size $\|\Delta\| \approx 2800$ is required. The outer boundary of the connected set $\sigma_{\mathbb{C}}(A; B, C, D; 100)$ is described by another closed curve contained in $\mathcal{C}_{100}$ which is out of the range of Figure 5.2.5 (b). □

5.2.3 Real Full-Block Perturbations

In the sequel we derive real counterparts to the results of the previous subsection. We first prove two general results applicable to real full-block perturbations provided with an arbitrary operator norm. We then deal separately with the two cases where either $\ell = 1$ and the perturbation norm is arbitrary, or ℓ is arbitrary and the perturbation norm is the spectral norm. We have seen in Section 4.4 that in these two cases the real μ-value $\mu_{\mathbb{R}}$ can actually be computed.
Throughout the subsection we suppose the following

- $(A, B, C, D) \in \mathbf{L}_{n,\ell,q}(\mathbb{R})$.
- The perturbation set is $\mathbf{\Delta} = \mathbb{R}^{\ell\times q}$ provided with a norm $\|\cdot\|$ which is an operator norm with respect to a given pair of norms $(\|\cdot\|_{\mathbb{R}^\ell}, \|\cdot\|_{\mathbb{R}^q})$ on $\mathbb{R}^\ell$ and $\mathbb{R}^q$. Moreover we suppose that $\mathbb{C}^\ell$ and $\mathbb{C}^q$ are provided with a pair of norms $(\|\cdot\|_{\mathbb{C}^\ell}, \|\cdot\|_{\mathbb{C}^q})$, compatible with $(\|\cdot\|_{\mathbb{R}^\ell}, \|\cdot\|_{\mathbb{R}^q})$. (Such a pair of norms always exists, see Lemma A.1.7).
- The perturbations are of the form

$$A \rightsquigarrow A(\Delta) = A + B(I_\ell - \Delta D)^{-1}\Delta C, \quad \Delta \in \mathbb{R}^{\ell\times q}. \tag{29}$$

The associated spectral value sets are denoted by $\sigma_{\mathbb{R}}(A; B, C, D; \delta)$. The subset of admissible perturbations and the corresponding radius of well-posedness are

$$\mathbf{\Delta}_0 = \{\Delta \in \mathbb{R}^{\ell\times q};\ \det(I_\ell - \Delta D) \neq 0\}, \qquad \boldsymbol{\delta}_0 = \|D\|^{-1}_{\mathcal{L}(\mathbb{R}^\ell, \mathbb{R}^q)}.$$

If we allow for *complex* instead of *real* full-block perturbations of the real data we obtain the spectral value set $\sigma_{\mathbb{C}}(A; B, C, D; \delta)$ and since $\|\Delta\|_{\mathcal{L}(\mathbb{R}^q,\mathbb{R}^\ell)} = \|\Delta\|_{\mathcal{L}(\mathbb{C}^q,\mathbb{C}^\ell)}$ for $\Delta \in \mathbb{R}^{\ell\times q}$ we have

$$\sigma_{\mathbb{R}}(A;B,C,D;\delta) \subset \sigma_{\mathbb{C}}(A;B,C,D;\delta). \tag{30}$$

We will see that in general the complex and real sets are quite different with the real ones sometimes exhibiting a somewhat eccentric behaviour. It follows immediately from the definition that for real data the real spectral value sets are symmetric with respect to the real axis, i.e. if $s \in \sigma_{\mathbb{R}}(A;B,C,D;\delta)$, then the complex conjugate $\overline{s} \in \sigma_{\mathbb{R}}(A;B,C,D;\delta)$. So we need only construct the sets in $\{s \in \mathbb{C};\ \mathrm{Im}\, s \geq 0\}$. Recall that for $\mathbf{\Delta} = \mathbb{R}^{\ell \times q}$ the μ-value $\mu_{\mathbf{\Delta}}(G)$ is denoted by $\mu_{\mathbb{R}}(G)$. Therefore Theorem 5.2.9 takes the following form in the present context.

Theorem 5.2.22. *Suppose* $0 < \delta \leq \|D\|^{-1}_{\mathcal{L}(\mathbb{R}^\ell,\mathbb{R}^q)}$, *then*

$$\sigma_{\mathbb{R}}(A;B,C,D;\delta) = \sigma(A) \,\dot{\cup}\, \{s \in \rho(A);\ \mu_{\mathbb{R}}(G(s)) > \delta^{-1}\}, \tag{31}$$

where $G(s) = D + C(sI_n - A)^{-1}B$ *is the transfer function of* (A,B,C,D).

The evaluation of the above formula is complicated by the possible discontinuity of $\mu_{\mathbb{R}}$. We know that $\mu_{\mathbb{R}} : \mathbb{C}^{q\times\ell} \to \mathbb{R}_+$ is continuous at complex $G \in \mathbb{C}^{q\times\ell} \setminus \mathbb{R}^{q\times\ell}$ but it may not be continuous at real $G \in \mathbb{R}^{q\times\ell}$. As a consequence an important role will be played by those points $s \in \rho(A)$ for which $G(s)$ is real.

Definition 5.2.23. Let $G(s) = D + C(sI_n - A)^{-1}B = G_R(s) + \imath G_I(s)$ where $G_R(s), G_I(s) \in \mathbb{R}^{q\times\ell}$ are the real and imaginary parts of $G(s)$ for $s \in \rho(A)$. Then

$$\mathcal{R}_G = \{s \in \rho(A); G_I(s) = 0\} = \{s \in \rho(A); G(s) \in \mathbb{R}^{q\times\ell}\} \tag{32}$$

is called the *realness locus* of G.

Remark 5.2.24. The real and imaginary parts of $G(s)$ are given by the formulas

$$\begin{aligned} G_R(\alpha + \imath\omega) &= D - C(\omega^2 I + (A - \alpha I)^2)^{-1}(A - \alpha I)B \\ G_I(\alpha + \imath\omega) &= -\omega C(\omega^2 I + (A - \alpha I)^2)^{-1}B, \quad \alpha + \imath\omega \in \rho(A). \end{aligned} \tag{33}$$

So the non-real part of the realness locus consists exactly of those $s = \alpha + \imath\omega \in \rho(A) \setminus \mathbb{R}$ for which α, ω satisfy the system of real algebraic equations $C(\omega^2 I + (A - \alpha I)^2)^{-1}B = 0$. □

$\mathcal{R}_G$ always contains the real axis. If $G(s)$ is non-constant then $\mathcal{R}_G$ has empty interior in $\mathbb{C}$. For the scalar case where $G(s) = p(s)/q(s)$, the realness locus is given by $\{s \in \mathbb{C}\,; p(s) = kq(s)$ for some $k \in \mathbb{R}\}$ and in control theory this set is known as the *root locus* of $G(s)$. If there are two non-constant elements $g_1(s)$, $g_2(s)$ of $G(s)$ and the associated root loci curves are in general position then away from the real axis they will only intersect at a finite number of points. So generically if there are three distinct non-constant elements of $G(s)$ the realness locus will be the real axis. Since $\mu_{\mathbb{R}}(G(s)) = \|G(s)\|_{\mathcal{L}(\mathbb{R}^\ell,\mathbb{R}^q)} = \|G(s)\|_{\mathcal{L}(\mathbb{C}^\ell,\mathbb{C}^q)}$ for $s \in \mathcal{R}_G$, we have

$$\begin{aligned} \mathcal{R}_G(\delta) := \sigma_{\mathbb{R}}(A;B,C,D;\delta) \cap \mathcal{R}_G &= \{s \in \mathcal{R}_G;\ \|G(s)\|_{\mathcal{L}(\mathbb{C}^\ell,\mathbb{C}^q)} > \delta^{-1}\} \\ &= \sigma_{\mathbb{C}}(A;B,C,D;\delta) \cap \mathcal{R}_G. \end{aligned} \tag{34}$$

Since $\sigma(A)$ is closed in $\mathbb{C}$ and $\mathcal{R}_G$ is closed in $\rho(A) = \mathbb{C} \setminus \sigma(A)$, the union $\sigma(A) \cup \mathcal{R}_G$ is closed in $\mathbb{C}$. It follows from Theorem 4.4.48 that outside of this closed set the map $\mu_{\mathbb{R}}(G(\cdot)) : s \mapsto \mu_{\mathbb{R}}(G(s))$ is continuous. Hence

$$\sigma_{\mathbb{R}}(A; B, C, D; \delta) \setminus (\sigma(A) \cup \mathcal{R}_G) = \{s \in \rho(A) \setminus \mathcal{R}_G;\ \mu_{\mathbb{R}}(G(s)) > \delta^{-1}\} \tag{35}$$

is open in $\mathbb{C}$. In order to characterize the real spectral value sets in terms of spectral contours as was done in Proposition 5.2.19 for the complex case, we consider the level sets of $\mu_{\mathbb{R}}(G(\cdot)) : \mathbb{C} \setminus (\sigma(A) \cup \mathcal{R}_G) \to \mathbb{R}_+$

$$\mathcal{C}_\delta = \{s \in \mathbb{C} \setminus (\sigma(A) \cup \mathcal{R}_G)\,;\ \mu_{\mathbb{R}}(G(s)) = \delta^{-1}\}. \tag{36}$$

These *spectral contours for the real case* are closed in $\mathbb{C} \setminus (\sigma(A) \cup \mathcal{R}_G)$ by the continuity of $\mu_{\mathbb{R}}(G(s))$ on this set, and as a consequence $\sigma(A) \cup \mathcal{R}_G \cup \mathcal{C}_\delta$ is closed in $\mathbb{C}$. The following proposition is a counterpart to Proposition 5.2.19.

Proposition 5.2.25. *If $0 < \delta < \|D\|^{-1}_{\mathcal{L}(\mathbb{R}^\ell, \mathbb{R}^q)}$, then $\sigma_{\mathbb{R}}(A; B, C, D; \delta) \setminus (\sigma(A) \cup \mathcal{R}_G)$ is a bounded open subset of $\mathbb{C}$, and modulo $\sigma(A) \cup \mathcal{R}_G$ the boundary $\partial\sigma_{\mathbb{R}}(A; B, C, D; \delta)$ of the spectral value set $\sigma_{\mathbb{R}}(A; B, C, D; \delta)$ in $\mathbb{C}$ is given by*

$$\partial\sigma_{\mathbb{R}}(A; B, C, D; \delta) \setminus (\sigma(A) \cup \mathcal{R}_G) = \mathcal{C}_\delta. \tag{37}$$

Moreover, the following two statements hold.

(i) For every $s_0 \in \rho(A)$ with $\mu_{\mathbb{R}}(G(s_0)) = \delta^{-1}$ (in particular for every $s_0 \in \mathcal{C}_\delta$) there exists a perturbation $\Delta \in \mathbb{R}^{\ell \times q}$ of norm $\|\Delta\| = \delta$ such that $s_0 \in \sigma(A(\Delta))$, and there is no smaller disturbance matrix with this property.

(ii) $\sigma_{\mathbb{R}}(A; B, C, D; \delta)$ is the union of $\sigma(A) \cup \mathcal{R}_G(\delta)$ and of those connected components of $\mathbb{C} \setminus \overline{\mathcal{R}_G \cup \mathcal{C}_\delta}$ which contain a point s_0 with $\mu_{\mathbb{R}}(G(s_0)) > \delta^{-1}$. Those components of $\mathbb{C} \setminus \overline{\mathcal{R}_G \cup \mathcal{C}_\delta}$ which contain a point $s_0 \in \rho(A)$ with $\mu_{\mathbb{R}}(G(s_0)) < \delta^{-1}$ are disjoint from $\sigma_{\mathbb{R}}(A; B, C, D; \delta) \setminus \sigma(A)$.

Proof: (i) Suppose $s_0 \in \rho(A)$ and $\mu_{\mathbb{R}}(G(s_0)) = \delta^{-1}$ where $0 < \delta < \|D\|^{-1}_{\mathcal{L}(\mathbb{R}^\ell, \mathbb{R}^q)}$. Then $\mu_{\mathbb{R}}(G(s_0)) > \|D\|_{\mathcal{L}(\mathbb{R}^\ell, \mathbb{R}^q)} = \mu_{\mathbb{R}}(D)$. Hence (i) is a consequence of Corollary 5.2.8.
$\sigma_{\mathbb{R}}(\delta) := \sigma_{\mathbb{R}}(A; B, C, D; \delta)$ is bounded since $\sigma_{\mathbb{C}}(\delta) := \sigma_{\mathbb{C}}(A; B, C, D; \delta)$ is bounded by Proposition 5.2.19 and the inclusion (30) holds. Moreover we have already seen above that the set (35) is open. It remains to prove (37) and (ii). To prove (37), we have to show that for any $s \in \mathbb{C} \setminus (\sigma(A) \cup \mathcal{R}_G)$,

$$s \in \partial\sigma_{\mathbb{R}}(\delta) := \partial\sigma_{\mathbb{R}}(A; B, C, D; \delta) \Leftrightarrow \mu_{\mathbb{R}}(G(s)) = \delta^{-1} \quad (\Leftrightarrow\ s \in \mathcal{C}_\delta\,).$$

Since $\sigma_{\mathbb{R}}(\delta) \setminus (\sigma(A) \cup \mathcal{R}_G)$ is open, no $s \in \partial\sigma_{\mathbb{R}}(\delta) \setminus (\sigma(A) \cup \mathcal{R}_G)$ belongs to $\sigma_{\mathbb{R}}(\delta)$ and so $\mu_{\mathbb{R}}(G(s)) = \delta^{-1}$ follows from Theorem 5.2.22 and the continuity of $\mu_{\mathbb{R}}(G(\cdot))$ on $\mathbb{C} \setminus (\sigma(A) \cup \mathcal{R}_G)$. Conversely, suppose $s_0 \in \mathbb{C} \setminus (\sigma(A) \cup \mathcal{R}_G)$ and $\mu_{\mathbb{R}}(G(s_0)) = \delta^{-1}$. By Theorem 5.2.22 $s_0 \notin \sigma_{\mathbb{R}}(\delta)$ and by (i) there exists $\Delta \in \mathbb{R}^{\ell \times q}$ of norm $\|\Delta\| = \delta$ such that $s_0 \in \sigma(A(\Delta))$. But then every neighbourhood V of s_0 in $\mathbb{C}$ contains an eigenvalue λ of the perturbed matrix $A(t\Delta)$ where $t < 1$ is sufficiently close to 1. So $\lambda \in \sigma(A(t\Delta)) \subset \sigma_{\mathbb{R}}(\delta)$, i.e. $V \cap \sigma_{\mathbb{R}}(\delta) \neq \emptyset$, and therefore $s_0 \in \partial\sigma_{\mathbb{R}}(\delta)$.
(ii) Let K be a connected component of $\mathbb{C} \setminus \overline{\mathcal{R}_G \cup \mathcal{C}_\delta}$ and suppose that there is an

$s_0 \in K \cap \rho(A)$ satisfying $\mu_{\mathbb{R}}(G(s_0)) > \delta^{-1}$. We will show that then $\mu_{\mathbb{R}}(G(s)) > \delta^{-1}$ for every $s \in K \cap \rho(A)$, hence $K \subset \sigma_{\mathbb{R}}(\delta)$ by (31). Since K is open and arc-wise connected, there exists for every $s \in K \cap \rho(A)$ an arc $\gamma(t)$, $0 \le t \le 1$ in $K \cap \rho(A) \subset \rho(A) \setminus \mathcal{R}_G$ connecting $s_0 = \gamma(0)$ with $s = \gamma(1)$. But $t \mapsto \mu_{\mathbb{R}}(G(\gamma(t)))$ is continuous on $[0,1]$ and $\mu_{\mathbb{R}}(G(\gamma(t))) \neq \delta^{-1}$ for all $t \in [0,1]$, since $K \cap \mathcal{C}_\delta = \emptyset$. Hence $\mu_{\mathbb{R}}(G(s)) > \delta^{-1}$ by the intermediate value theorem.
Conversely, assume $s \in \sigma_{\mathbb{R}}(\delta) \setminus (\sigma(A) \cup \mathcal{R}_G(\delta))$. Then $\mu_{\mathbb{R}}(G(s)) > \delta^{-1}$ by (31) and $s \notin \sigma(A) \cup \mathcal{R}_G \cup \mathcal{C}_\delta \supset \overline{\mathcal{R}_G \cup \mathcal{C}_\delta}$, and there exists a connected component K of $\mathbb{C} \setminus \overline{\mathcal{R}_G \cup \mathcal{C}_\delta}$ containing s.
For the last statement of the proposition assume again that K is a connected component of $\mathbb{C} \setminus \overline{\mathcal{R}_G \cup \mathcal{C}_\delta}$ but now suppose there is an $s_0 \in K \cap \rho(A)$ satisfying $\mu_{\mathbb{R}}(G(s_0)) < \delta^{-1}$. By the same argument as above this holds true for every $s \in K \cap \rho(A)$, so $K \cap \sigma_{\mathbb{R}}(\delta) \cap \rho(A) = \emptyset$. □

In order to visualize $\sigma_{\mathbb{R}}(A; B, C, D; \delta)$, one applies the previous proposition and proceeds as follows.

Procedure 5.2.26. *1. Compute $\sigma(A)$ and the realness locus $\mathcal{R}_G$.*

2. *Compute $\mathcal{R}_G(\delta) = \sigma_{\mathbb{R}}(A; B, C, D; \delta) \cap \mathcal{R}_G$. This can be done by intersecting $\mathcal{R}_G$ with the complex spectral value set $\sigma_{\mathbb{C}}(A; B, C, D; \delta)$, see (34).*
3. *Determine the contours $\mathcal{C}_\delta$ so that the set $\overline{\mathcal{R}_G \cup \mathcal{C}_\delta}$ can be visualized.*
4. *Check at any point $s_0 \notin \sigma(A)$ in each connected component K of $\mathbb{C} \setminus \overline{\mathcal{R}_G \cup \mathcal{C}_\delta}$ to see whether $\mu_{\mathbb{R}}(G(s_0)) > \delta^{-1}$ (in which case $K \subset \sigma_{\mathbb{R}}(A; B, C, D; \delta)$) or $\mu_{\mathbb{R}}(G(s_0)) < \delta^{-1}$ (in which case $K \cap \sigma_{\mathbb{R}}(A; B, C, D; \delta) \cap \rho(A) = \emptyset$).*

To carry out this procedure, the essential prerequisite is that the μ-value $\mu_{\mathbb{R}}(G(s))$ can actually be computed. At present this is not feasible for arbitrary operator norms on $\mathbf{\Delta}$ if $\ell, q \ge 2$. Applying the results of Section 4.4 we will now discuss two situations in which $\mu_{\mathbb{R}}(G)$ is computable for arbitrary $G \in \mathbb{C}^{q \times \ell}$.

Real Case $\ell = 1$

Assume $\ell = 1$, $\mathbb{R}^\ell = \mathbb{R}$ is provided with the norm $|\cdot|$, $\mathbb{R}^q$ with an arbitrary norm $\|\cdot\|_{\mathbb{R}^q}$ and $\mathbf{\Delta} = \mathbb{R}^{1\times q} = (\mathbb{R}^q)^*$ with the dual norm $\|\cdot\|_{\mathbb{R}^{1\times q}} = \|\cdot\|^*_{\mathbb{R}^q}$. Note that in this case the transfer matrix $G(s) = D + C(sI_n - A)^{-1}B$ is a q-vector of real proper rational functions. As an immediate consequence of Theorem 5.2.22 and Theorem 4.4.31, we have

Theorem 5.2.27. *Suppose $0 < \delta \le \|D\|^{-1}_{\mathbb{R}^q}$, then*

$$\sigma_{\mathbb{R}}(A; B, C, D; \delta) = \sigma(A) \,\dot{\cup}\, \{s \in \rho(A);\ \mathrm{dist}\,(G_R(s), \mathbb{R}G_I(s)) > \delta^{-1}\}, \qquad (38)$$

where $G(s) = D + C(sI_n - A)^{-1}B = G_R(s) + \imath G_I(s)$, $G_R(s)$, $G_I(s) \in \mathbb{R}^q$, $s \in \rho(A)$ and the distance is taken with respect to the norm $\|\cdot\|_{\mathbb{R}^q}$.

We know that $\mu_{\mathbb{R}}(G(s)) = \mathrm{dist}\,(G_R(s), \mathbb{R}G_I(s))$ is continuous on $\mathbb{C} \setminus (\sigma(A) \cup \mathcal{R}_G)$. Following the Procedure 5.2.26, the visualization of $\sigma_{\mathbb{R}}(A; B, C, D; \delta)$ is based on the partition

$$\sigma_{\mathbb{R}}(A; B, C, D; \delta) = \sigma(A) \,\dot{\cup}\, \mathcal{R}_G(\delta) \,\dot{\cup}\, \{s \in \rho(A) \setminus \mathcal{R}_G;\ \mathrm{dist}\,(G_R(s), \mathbb{R}G_I(s)) > \delta^{-1}\}. \qquad (39)$$

Remark 5.2.28. (i) In the case $q = \ell = 1$, since $\mathbb{R}G_I(s) = \mathbb{R}$ if $s \notin (\sigma(A) \cup \mathcal{R}_G)$, the third term on the RHS of (39) is empty and so

$$\sigma_{\mathbb{R}}(A; B, C, D; \delta) = \sigma(A) \dot{\cup} \mathcal{R}_G(\delta). \tag{40}$$

(ii) Given $s_0 \in \rho(A)$ with $\operatorname{dist}(G_R(s_0), \mathbb{R}G_I(s_0)) > 0$, a minimum norm perturbation $\Delta_{\min} \in \mathbb{R}^{1\times q}$ such that $s_0 \in \sigma(A(\Delta_{\min}))$ may be constructed as in the proof of Theorem 4.4.31: Let $z^* \in \mathbb{R}^{1\times q}$ with $\|z^*\|^*_{\mathbb{R}^q} = 1$ be such that

$$z^* G_R(s_0) = \operatorname{dist}(G_R(s_0), \mathbb{R}G_I(s_0)) \qquad \text{and} \qquad z^* G_I(s_0) = 0\,.$$

Then $\Delta_{\min} = [\operatorname{dist}(G_R(s_0), \mathbb{R}G_I(s_0))]^{-1} z^*$ achieves $s_0 \in \sigma(A(\Delta_{\min}))$ with norm $\|\Delta_{\min}\| = [\operatorname{dist}(G_R(s_0), \mathbb{R}G_I(s_0))]^{-1}$.

(iii) With respect to the 2-norm, we have by (4.4.42)

$$\operatorname{dist}\left(G_R(s), \mathbb{R}G_I(s)\right)^2 = \|G_R(s)\|_2^2 - \frac{\langle G_R(s), G_I(s)\rangle^2}{\|G_I(s)\|_2^2}, \quad s \in \rho(A) \setminus \mathcal{R}_G\,. \tag{41}$$

(iv) Using Corollary 4.4.29 we obtain a similar formula to (39) when $q = 1$ and $\ell \geq 1$ is arbitrary. Let $\mathbb{R}^q = \mathbb{R}$ be provided with the norm $|\cdot|$, $\mathbb{R}^\ell$ with any norm $\|\cdot\|_{\mathbb{R}^\ell}$, $\mathbf{\Delta} = \mathbb{R}^\ell$ with the norm $\|\cdot\|_{\mathbb{R}^\ell}$ and $\mathbb{R}^{1\times\ell}$ with the dual norm $\|\cdot\|^*_{\mathbb{R}^\ell}$, then

$$\sigma_{\mathbb{R}}(A; B, C, D; \delta) = \sigma(A) \dot{\cup} \mathcal{R}_G(\delta) \dot{\cup} \{s \in \rho(A) \setminus \mathcal{R}_G;\ \operatorname{dist}\left(G_R(s), \mathbb{R}G_I(s)\right) > \delta^{-1}\}.$$

where $\mathcal{R}_G(\delta) = \{s \in \mathcal{R}_G;\ \|G(s)\|^*_{\mathbb{R}^\ell} > \delta^{-1}\}$ and the distance is taken with respect to the dual norm $\|\cdot\|^*_{\mathbb{R}^\ell}$ in $(\mathbb{R}^\ell)^* = \mathbb{R}^{1\times\ell}$. □

Example 5.2.29. Consider the nominal oscillator $\ddot{\xi} + 4\dot{\xi} + 5\xi = 0$ of Example 5.2.17, but now we do not consider perturbations to the mass, i.e. the perturbed system is $\ddot{\xi} + (4 + \Delta_c)\dot{\xi} + (5 + \Delta_k)\xi = 0$. This can be modelled as an affine perturbation (1.11) with

$$A = \begin{bmatrix} 0 & 1 \\ -5 & -4 \end{bmatrix}, \quad B = \begin{bmatrix} 0 \\ 1 \end{bmatrix}, \quad C = I_2, \quad D = 0, \quad \Delta = \begin{bmatrix} \Delta_c & \Delta_k \end{bmatrix}.$$

Then $\sigma(A) = \{-2 \pm \imath\}$ and the transfer function is given by

$$G(s) = (s^2 + 4s + 5)^{-1} \begin{bmatrix} 1 \\ s \end{bmatrix}.$$

If $s = \alpha + \imath\omega$ and we write $s^2 + 4s + 5 = q_R(s) + \imath\, q_I(s)$ with $q_R(s), q_I(s) \in \mathbb{R}$, then

$$q_R(\alpha + \imath\omega) = \alpha^2 + 4\alpha + 5 - \omega^2, \quad q_I(\alpha + \imath\omega) = 2\alpha\omega + 4\omega$$

and

$$\begin{aligned} G_R(s) &= [q_R(s)^2 + q_I(s)^2]^{-1} \begin{bmatrix} q_R(s) \\ \alpha q_R(s) + \omega q_I(s) \end{bmatrix}, \quad s = \alpha + \imath\omega \neq -2 \pm \imath \\ G_I(s) &= [q_R(s)^2 + q_I(s)^2]^{-1} \begin{bmatrix} -q_I(s) \\ \omega q_R(s) - \alpha q_I(s) \end{bmatrix}, \quad s = \alpha + \imath\omega \neq -2 \pm \imath. \end{aligned}$$

Hence $\mathcal{R}_G = \mathbb{R}$ and using (41) a short calculation yields that with respect to the 2-norm

$$\operatorname{dist}\left(G_R(s), \mathbb{R}G_I(s)\right)^2 = \omega^2 [q_I(s)^2 + (\omega q_R(s) - \alpha q_I(s))^2]^{-1} \quad s \notin (\mathcal{R}_G \cup \{-2 \pm \imath\}).$$

The contours bounding the real spectral value sets for $\delta = .33,\ 1.5,\ 3.6,\ 4,\ 6$ are shown on the left of Figure 5.2.6. The interval along the real axis belongs to the spectral value set

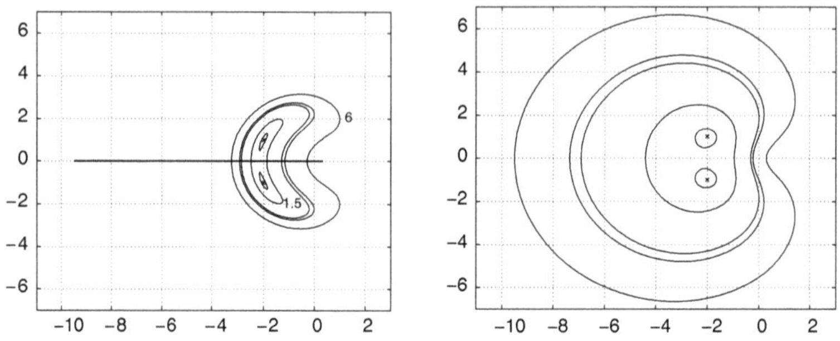

Figure 5.2.6: Real and complex spectral value sets for the 2-norm.

at the uncertainty level $\delta = 6$. For comparison we show the complex contours in the figure on the right for the same values of δ. Note that in accordance with (34) the interval in the left figure is equal to the intersection of $\mathbb{R}$ with $\sigma_{\mathbb{C}}(A; B, C; 6)$. The contour for $\delta = 4$ touches the imaginary axis in the figure on the left, and in the figure on the right this happens for the contour at level $\delta = 3.6$. We see, therefore, that if the stability region is $\mathbb{C}_-$ then the real stability radius is ≈ 4 and the complex one is ≈ 3.6. □

Example 5.2.30. In this example we consider the same system (A, B, C) as in Example 5.2.21 but now we assume the perturbations are real. It follows directly from expression (28) for the associated transfer function that $\mathcal{R}_G = \mathbb{R}$. Figure 5.2.7 shows the real contours with respect to the Euclidean norm for the same values of δ as in Figure 5.2.5, namely $\delta = 0.5,\ 0.66,\ 0.83,\ 1, 8.77,\ 100$. One observes the qualitative and quantitative differences

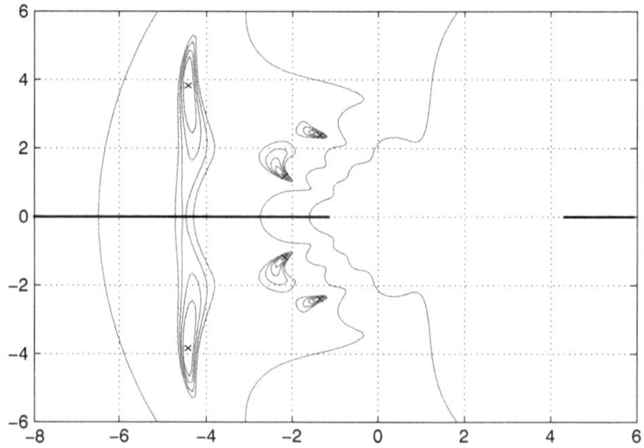

Figure 5.2.7: Real spectral value sets for specific δ

between them and the complex contours of Example 5.2.21. The intervals on the real axis belong to the set $\mathcal{R}_G(100)$. The points located to the right of the rightmost contour, which do not belong to these intervals, do *not* belong to the spectral value set $\sigma_{\mathbb{R}}(A; B, C; 100)$ but to a hole in this set (spiked by the realness locus $\mathcal{R}_G(\delta)$). This hole is much larger than in the complex case, see Figure 5.2.5. It is not possible to move an eigenvalue into this domain with a real perturbation of norm ≤ 100. As in the complex case, the larger contour line surrounding the connected set $\sigma_{\mathbb{R}}(A; B, C; 100)$ is out of the range of Figure 5.2.7. □

General Real Case

We now deal with the general case where $\ell, q \geq 1$ are arbitrary and assume throughout the rest of the subsection that $\mathbf{\Delta} = \mathbb{R}^{\ell \times q}$ and $\mathbb{R}^{q\times\ell}$ as well as their complex counterparts are provided with the spectral norm denoted by $\|\cdot\|$. Then using the characterization of $\mu_{\mathbb{R}}$ in Theorem 4.4.43 we obtain from Theorem 5.2.22

Theorem 5.2.31. *Suppose* $0 < \delta \leq \|D\|^{-1}$, *then*

$$\sigma_{\mathbb{R}}(A;B,C,D;\delta) = \sigma(A) \dot\cup \{s \in \rho(A);\ \inf_{\gamma\in(0,1]} \sigma_2(G_\gamma^{\mathbb{R}}(s)) > \delta^{-1}\} \tag{42}$$

where $G(s) = D + C(sI_n - A)^{-1}B = G_R(s) + \imath G_I(s)$, $G_R(s)$, $G_I(s) \in \mathbb{R}^{q\times\ell}$ *and*

$$G_\gamma^{\mathbb{R}}(s) = \begin{bmatrix} G_R(s) & -\gamma G_I(s) \\ \gamma^{-1}G_I(s) & G_R(s) \end{bmatrix}, \quad s \in \rho(A). \tag{43}$$

Following the Procedure 5.2.26, we compute $\sigma_{\mathbb{R}}(A;B,C,D;\delta)$ according to the partition

$$\sigma_{\mathbb{R}}(A;B,C,D;\delta) = \sigma(A) \dot\cup \mathcal{R}_G(\delta) \dot\cup \{s \in \rho(A) \setminus \mathcal{R}_G;\ \inf_{\gamma\in(0,1]} \sigma_2(G_\gamma^{\mathbb{R}}(s)) > \delta^{-1}\}. \tag{44}$$

In the following example we apply this formula to the case where $\ell = 1$ and re-derive the result in Theorem 5.2.27 with respect to the Euclidean norm.

Example 5.2.32. Suppose $\ell = 1$ and $0 < \delta \leq \|D\|_{\mathbb{R}^q}^{-1}$, then

$$G_\gamma^{\mathbb{R}}(s)^\top G_\gamma^{\mathbb{R}}(s) = \begin{bmatrix} \|G_R(s)\|^2 + \gamma^{-2}\|G_I(s)\|^2 & (\gamma^{-1}-\gamma)\langle G_R(s), G_I(s)\rangle \\ (\gamma^{-1}-\gamma)\langle G_R(s), G_I(s)\rangle & \|G_R(s)\|^2 + \gamma^2\|G_I(s)\|^2 \end{bmatrix}_{2\times 2}.$$

We know from Proposition 4.4.35 that

$$\inf_{\gamma\in(0,1]} \sigma_2(G_\gamma^{\mathbb{R}}(s)) = \lim_{\gamma\to 0} \sigma_2(G_\gamma^{\mathbb{R}}(s)).$$

For a real symmetric matrix $H = \begin{bmatrix} a & b \\ b & c \end{bmatrix}$ the lowest eigenvalue is given by $\lambda_{\min}(H) = [a + c - \sqrt{(a-c)^2 + 4b^2}]/2$. Hence with $f(\gamma) = \sqrt{(\gamma^{-1}+\gamma)^2\|G_I(s)\|^4 + 4\langle G_R(s), G_I(s)\rangle^2}$

$$2\sigma_2(G_\gamma^{\mathbb{R}}(s))^2 = \begin{cases} 2\|G_R(s)\|^2 + (\gamma^2 + \gamma^{-2})\|G_I(s)\|^2 - |\gamma^{-1}-\gamma|\, f(\gamma) & \text{if} \quad s \in \rho(A) \setminus \mathcal{R}_G \\ 2\|G_R(s)\|^2 & \text{if} \quad s \in \mathcal{R}_G. \end{cases}$$

But for $\gamma \downarrow 0$,

$$\begin{aligned}(\gamma^{-2} - \gamma^2)^2\|G_I(s)\|^4 &+ 4(\gamma^{-1}-\gamma)^2\langle G_R(s), G_I(s)\rangle^2 \\ &= \gamma^{-4}\|G_I(s)\|^4 + 4\gamma^{-2}\langle G_R(s), G_I(s)\rangle^2 + O(1) \\ &= \gamma^{-4}\|G_I(s)\|^4 \left[1 + 4\gamma^2\langle G_R(s), G_I(s)\rangle^2/\|G_I(s)\|^4 + O(\gamma^4)\right].\end{aligned}$$

Hence using $(1+x)^{1/2} = 1 + x/2 + O(x^2)$ (for $|x|$ small), we have

$$\begin{aligned}2\sigma_2(G_\gamma^{\mathbb{R}}(s))^2 =& 2\|G_R(s)\|^2 + \gamma^{-2}\|G_I(s)\|^2 - \gamma^{-2}\|G_I(s)\|^2(1 + 2\gamma^2\langle G_R(s), G_I(s)\rangle^2/\|G_I(s)\|^4) \\ &+ O(\gamma^2)\end{aligned}$$

for $s \in \rho(A) \setminus \mathcal{R}_G$. Hence by (41)

$$\lim_{\gamma\to 0} \sigma_2(G_\gamma^{\mathbb{R}}(s))^2 = \|G_R(s)\|^2 - \langle G_R(s), G_I(s)\rangle^2/\|G_I(s)\|^2 = \operatorname{dist}(G_R(s), G_I(s)\mathbb{R})^2.$$

and so

$$\sigma_{\mathbb{R}}(A;B,C,D;\delta) = \sigma(A) \dot\cup \{s \in \rho(A);\ \operatorname{dist}(G_R(s), \mathbb{R}G_I(s)) > \delta^{-1}\}.$$

since $\sigma_2(G_\gamma^{\mathbb{R}}(s)) = \|G_R(s)\| = \operatorname{dist}(G_R(s), \mathbb{R}G_I(s))$ for $s \in \mathcal{R}_G$. □

Remark 5.2.33. Given $s_0 \in \rho(A)$ and $\gamma_0 \in (0,1]$ such that

$$\inf_{\gamma \in (0,1]} \sigma_2(G_\gamma^{\mathbb{R}}(s_0)) = \sigma_2(G_{\gamma_0}^{\mathbb{R}}(s_0)) = \sigma^*.$$

Then a minimum norm $\Delta_{\min}$ satisfying $s_0 \in \sigma(A(\Delta_{\min}))$ may be constructed as in the proof of Lemma 4.4.37. Let $w = \begin{bmatrix} w^1 \\ w^2 \end{bmatrix}$ and $v = \begin{bmatrix} v^1 \\ v^2 \end{bmatrix}$ be left and right singular vectors of $G_{\gamma_0}^{\mathbb{R}}(s_0)$ corresponding to σ^*, such that $w^{1\top} w^2 = v^{1\top} v^2$, $\|w^1\| = \|v^1\|$ and $\|w^2\| = \|v^2\|$. Then if $\|D\| < \sigma^*$ and $w^1 \neq 0$ and $w^2 \neq 0$ are independent,

$$\Delta_{\min} = \sigma^{*-1} [v^1 \; v^2] \begin{bmatrix} \|w^1\|^2 & \langle w^1, w^2 \rangle \\ \langle w^1, w^2 \rangle & \|w^2\|^2 \end{bmatrix}^{-1} \begin{bmatrix} w^{1\top} \\ w^{2\top} \end{bmatrix}$$

achieves $s_0 \in \sigma(A(\Delta_{\min}))$ with norm $\|\Delta_{\min}\| = \sigma^{*-1}$. □

The following example has been especially constructed to obtain a substantial realness locus. It illustrates the relationship between real and complex spectral value sets and the realness locus.

Example 5.2.34. Let $(A_c, B_c, C_c, 0)$, $c \in \mathbb{R}$ be a system with transfer function

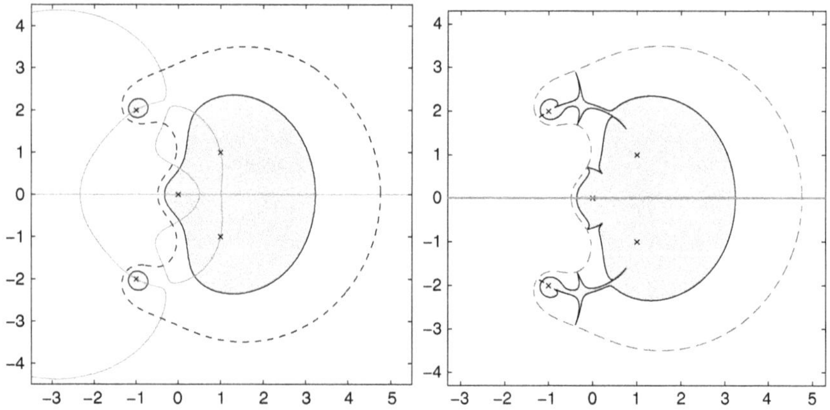

Figure 5.2.8: Realness locus and real and complex spectral value sets

$$G_c(s) = \begin{bmatrix} q(s)/p(s) & 0 \\ 0 & (q(s)+c)/4p(s) \end{bmatrix}, \quad \begin{array}{rcl} p(s) & = & s((s-1)^2+1)((s+1)^2+4)) \\ q(s) & = & ((s+1/2)^2+1)((s+2)^2+1) \end{array}$$

and $\mathbf{\Delta} = \mathbb{R}^{2\times 2}$. The figures have been computed for $\delta = 2$, the one on the left for $c = 0$ and the one on the right for $c = 2$. The crosses mark the eigenvalues of A and the dashed line is the boundary of the complex spectral value set $\sigma_{\mathbb{C}}(A_c; B_c, C_c; 2)$. For $c = 0$ the realness locus consists of the real axis and the curved solid lines which do not circumscribe a shaded area. For $c = 2$ it is reduced to the real axis. The real spectral value set $\sigma_{\mathbb{R}}(A_c; B_c, C_c; 2)$ consists of the shaded area plus (in the case $c = 0$) those parts of the realness locus which lie in the complex spectral value set. □

5.2.4 The Unstructured Case (Pseudospectra)

In this subsection we consider unstructured perturbations $A \rightsquigarrow A + \Delta$, where $\Delta \in \mathbf{\Delta} = \mathbb{K}^{n\times n}$ is arbitrary. So each element of the matrix A is subject to an independent complex or real perturbation. After characterizing the corresponding spectral value sets we will obtain set bounds for them and explore when they are tight. These set bounds are expressed in terms of intervals or disks around the eigenvalues of A. It is well known that the *sensitivities* of the eigenvalues of a matrix may become arbitrarily large under similarity transformations (see Proposition 4.2.12). At the end of the subsection we derive a global version of this local result by analyzing changes of the spectral value sets along similarity orbits in $\mathbb{K}^{n\times n}$.
Unstructured perturbations are a special case of the full-block perturbations considered in the previous two subsections, with corresponding structure matrices $B = C = I_n, D = 0_{n\times n}$. Therefore all the results of those two subsections have specializations to unstructured perturbations. For the sake of brevity we will only give explicit statements for the spectral perturbation norm $\|\cdot\|_{2,2}$ for which more concrete results are available. Throughout this subsection we suppose the following

- $A \in \mathbb{K}^{n\times n}$.
- The perturbation set is $\mathbf{\Delta} = \mathbb{K}^{n\times n}$, the spaces $\mathbb{K}^n$ are provided with the standard Euclidean norms and $\mathbf{\Delta} = \mathbb{K}^{n\times n}$ is provided with the associated operator norm $\|\cdot\| = \|\cdot\|_{2,2}$ (the spectral norm).
- The perturbations are $A \rightsquigarrow A+\Delta$, $\Delta \in \mathbb{K}^{n\times n}$ (whence $\mathbf{\Delta}_0 = \mathbb{K}^{n\times n}$, $\boldsymbol{\delta}_0 = \infty$).

The associated spectral value sets are denoted by

$$\sigma_{\mathbb{K}}(A;\delta) = \sigma_{\mathbb{K}}(A; I_n, I_n, 0_{n\times n};\delta) = \bigcup_{\Delta\in\mathbb{K}^{n\times n},\, \|\Delta\|<\delta} \sigma(A+\Delta).$$

These sets are called (*complex* resp. *real*) *unstructured* spectral value sets or *pseudospectra* of A. The corresponding transfer matrix is $G(s) = (sI_n - A)^{-1}$, the resolvent operator of A. Since $\|\Delta\|_{\mathcal{L}(\mathbb{R}^n)} = \|\Delta\|_{\mathcal{L}(\mathbb{C}^n)}$ for $\Delta \in \mathbb{R}^{n\times n}$ we have

$$\sigma_{\mathbb{R}}(A;\delta) \subset \sigma_{\mathbb{C}}(A;\delta). \tag{45}$$

If $A \in \mathbb{R}^{n\times n}$, the realness locus of $G(s) = (sI - A)^{-1}$ is trivial, i.e. $\mathcal{R}_G = \mathbb{R}$. In fact, $(sI_n - A)^{-1} \in \mathbb{R}^{n\times n}$ if and only if $(sI_n - A) \in \mathbb{R}^{n\times n}$, and this is equivalent to $s \in \mathbb{R}$ if A is real.

Theorem 5.2.35. *Suppose $A \in \mathbb{C}^{n\times n}$ (resp. $A \in \mathbb{R}^{n\times n}$) and $\delta > 0$. Then*

$$\begin{aligned} \sigma_{\mathbb{C}}(A;\delta) &= \{s \in \mathbb{C};\ \sigma_n(sI - A) < \delta\}, & (46)\\ \textit{resp. } \sigma_{\mathbb{R}}(A;\delta) &= \{s \in \mathbb{C};\ \sup_{\gamma\in(0,1]} \sigma_{2n-1}((sI - A)^{\mathbb{R}}_{\gamma}) < \delta\} & (47)\end{aligned}$$

where

$$(sI - A)^{\mathbb{R}}_{\gamma} = \begin{bmatrix} \operatorname{Re}(sI_n - A) & -\gamma \operatorname{Im}(sI_n - A) \\ \gamma^{-1}\operatorname{Im}(sI_n - A) & \operatorname{Re}(sI_n - A) \end{bmatrix} \in \mathbb{R}^{2n\times 2n}, \quad s \in \mathbb{C}. \tag{48}$$

Proof: As an immediate consequence of Theorem 5.2.16 and Theorem 5.2.31, we have for any given $\delta > 0$

$$\begin{aligned}\sigma_{\mathbb{C}}(A;\delta) &= \sigma(A)\,\dot{\cup}\,\{s \in \rho(A);\ \sigma_1((sI-A)^{-1}) > \delta^{-1}\},\\ \sigma_{\mathbb{R}}(A;\delta) &= \sigma(A)\,\dot{\cup}\left\{s \in \rho(A);\ \inf_{\gamma\in(0,1]} \sigma_2\left(((sI-A)^{-1})_\gamma^{\mathbb{R}}\right) > \delta^{-1}\right\}.\end{aligned}$$

Now since $(sI-A)$ is invertible for $s \in \rho(A)$, we have $\sigma_1((sI-A)^{-1}) = (\sigma_n(sI-A))^{-1}$. Moreover $\sigma_n(\lambda I - A) = 0$ for $\lambda \in \sigma(A)$. This proves (46). To prove (47) we use (4.3.21) and obtain for $s \in \rho(A)$ (since $(G^{-1})_\gamma^{\mathbb{R}} = \left(G_\gamma^{\mathbb{R}}\right)^{-1}$ by Lemma A.1.18)

$$\mu_{\mathbb{R}}\left((sI-A)^{-1}\right) = \inf_{\gamma\in(0,1]} \sigma_2\left(((sI-A)^{-1})_\gamma^{\mathbb{R}}\right) = \left[\sup_{\gamma\in(0,1]} \sigma_{2n-1}\left((sI-A)_\gamma^{\mathbb{R}}\right)\right]^{-1}, \quad s \in \rho(A).$$

Moreover, if $\lambda \in \sigma(A)$ then there exists $v = v_1 + \imath v_2 \in \mathbb{C}^n, v \neq 0$ such that $(\lambda I - A)v = 0$. It follows that $(\lambda I - A)_\gamma^{\mathbb{R}}$ annulls both $\begin{bmatrix} v_1 \\ \gamma^{-1} v_2 \end{bmatrix}$ and $\begin{bmatrix} \gamma v_2 \\ -v_1 \end{bmatrix}$ for all $\gamma \in (0,1]$ and this implies $\sigma_{2n-1}((\lambda I - A)_\gamma^{\mathbb{R}}) = 0$, $\gamma \in (0,1]$. Thus (47) holds. □

We illustrate the above formulas by a simple example which can be evaluated analytically. We will use this example later in order to determine the unstructured real spectral value sets for arbitrary normal $A \in \mathbb{R}^{2\times 2}$. As usual we denote by $D(\lambda,\delta)$ resp. $I(\lambda,\delta)$ the open disk (resp. open real interval) with centre λ and radius δ.

Example 5.2.36. Consider the matrix $A = \operatorname{diag}(\lambda_1, \lambda_2)$, $\lambda_1, \lambda_2 \in \mathbb{R}$, and write $s = \alpha + \imath\omega$, $\alpha, \omega \in \mathbb{R}$. Then an easy calculation shows that

$$\sigma_2(sI-A) = \left(\min\{(\alpha-\lambda_1)^2 + \omega^2, (\alpha-\lambda_2)^2+\omega^2\}\right)^{1/2} = \operatorname{dist}(s,\sigma(A)), \quad s = \alpha + \imath\omega \in \mathbb{C}.$$

Hence (46) yields

$$\sigma_{\mathbb{C}}(A;\delta) = \{s \in \mathbb{C}; \operatorname{dist}(s,\sigma(A)) < \delta\} = D(\lambda_1,\delta) \cup D(\lambda_2,\delta). \tag{49}$$

The real case is not quite so simple. We have

$$\begin{aligned}(sI-A)_\gamma^{\mathbb{R}} &= \begin{bmatrix} \alpha-\lambda_1 & 0 & -\gamma\omega & 0 \\ 0 & \alpha-\lambda_2 & 0 & -\gamma\omega \\ \gamma^{-1}\omega & 0 & \alpha-\lambda_1 & 0 \\ 0 & \gamma^{-1}\omega & 0 & \alpha-\lambda_2 \end{bmatrix}\\ &= \begin{bmatrix} 1&0&0&0\\0&0&1&0\\0&1&0&0\\0&0&0&1 \end{bmatrix} \begin{bmatrix} \alpha-\lambda_1 & -\gamma\omega & 0 & 0 \\ \gamma^{-1}\omega & \alpha-\lambda_1 & 0 & 0 \\ 0 & 0 & \alpha-\lambda_2 & -\gamma\omega \\ 0 & 0 & \gamma^{-1}\omega & \alpha-\lambda_2 \end{bmatrix} \begin{bmatrix} 1&0&0&0\\0&0&1&0\\0&1&0&0\\0&0&0&1 \end{bmatrix}.\end{aligned}$$

So the singular values of $(sI-A)_\gamma^{\mathbb{R}}$ are the singular values of $A_1(\gamma,s) = \begin{bmatrix} \alpha-\lambda_1 & -\gamma\omega \\ \gamma^{-1}\omega & \alpha-\lambda_1 \end{bmatrix}$ and $A_2(\gamma,s) = \begin{bmatrix} \alpha-\lambda_2 & -\gamma\omega \\ \gamma^{-1}\omega & \alpha-\lambda_2 \end{bmatrix}$. Let $\sigma_i(\gamma,s,\lambda_j)$, $i = 1,2$ be the ordered singular values of the two matrices $A_j(\gamma,s)$, $j = 1,2$. There are two cases. First suppose $s \in \mathbb{R}$, i.e. $\omega = 0$ and $s = \alpha$, then

$$\sigma_1(\gamma,s,\lambda_j) = \sigma_2(\gamma,s,\lambda_j) = |\alpha - \lambda_j|, \quad \gamma \in (0,1], \ j = 1,2.$$

So $\hat{\sigma}_3(s) := \sup_{\gamma\in(0,1]} \sigma_3((sI-A)^{\mathbb{R}}_\gamma = \min\{|\alpha-\lambda_1|, |\alpha-\lambda_2|\} = \mathrm{dist}(\alpha,\sigma(A))$ and (47) yields

$$\sigma_{\mathbb{R}}(A;\delta)\cap\mathbb{R} = \{\alpha\in\mathbb{R};\ \mathrm{dist}(\alpha,\sigma(A))<\delta\} = I(\lambda_1,\delta)\cup I(\lambda_2,\delta). \tag{50}$$

Now suppose $s=\alpha+\imath\omega\in\mathbb{C}\setminus\mathbb{R}$. An easy calculation shows for $i,j=1,2$

$$\sigma_i(\gamma,s,\lambda_j)^4 - \left[2(\alpha-\lambda_j)^2+\omega^2(\gamma^2+\gamma^{-2})\right]\sigma_i(\gamma,s,\lambda_j)^2 + [(\alpha-\lambda_j)^2+\omega^2]^2 = 0. \tag{51}$$

Note that for $\gamma=1$ the upper and lower singular value of the two matrices $A_j(\gamma,s)$, $j=1,2$ coincide:

$$\sigma_1(1,s,\lambda_j)^2 = \sigma_2(1,s,\lambda_j)^2 = (\alpha-\lambda_j)^2+\omega^2 = |s-\lambda_j|^2.$$

The upper singular values $\sigma_1(\gamma,s,\lambda_j)$ increase strictly monotonically to ∞ as γ goes from 1 to 0, and since the products $\sigma_1(\gamma,s,\lambda_j)\sigma_2(\gamma,s,\lambda_j)$, $j=1,2$ are the determinants of the matrices $A_j(\gamma,s)$ (see Proposition 4.3.23) and these are independent of γ, the lower singular values decrease strictly monotonically to 0 as γ goes from 1 to 0. It follows that in the case $\lambda_1=\lambda_2=\lambda$ the supremum $\hat{\sigma}_3(s) := \sup_{\gamma\in(0,1]}\sigma_3((sI-A)^{\mathbb{R}}_\gamma$ is achieved for $\gamma=1$, so that $\hat{\sigma}_3(s) = \left[(\alpha-\lambda)^2+\omega^2\right]^{-1/2} = |s-\lambda|$. Hence

$$\sigma_{\mathbb{R}}(A;\delta) = D(\lambda,\delta) = \sigma_{\mathbb{C}}(A;\delta) \quad \text{if} \quad \lambda_1=\lambda_2=\lambda. \tag{52}$$

Now suppose $\lambda_1\neq\lambda_2$, then $\hat{\sigma}_3(s) := \sup_{\gamma\in(0,1]}\sigma_3((sI-A)^{\mathbb{R}}_\gamma$ is achieved at a $\hat{\gamma}(s)$ where a lower singular value of one of the matrices meets an upper one of the other matrix. So by (51) $\hat{\sigma}_3(s)$ satisfies

$$\begin{aligned}\omega^2(\hat{\gamma}(s)^2+\hat{\gamma}(s)^{-2})\hat{\sigma}_3(s)^2 &= \hat{\sigma}_3(s)^4 + [(\alpha-\lambda_1)^2+\omega^2]^2 - 2(\alpha-\lambda_1)^2\hat{\sigma}_3(s)^2\\ &= \hat{\sigma}_3(s)^4 + [(\alpha-\lambda_2)^2+\omega^2]^2 - 2(\alpha-\lambda_2)^2\hat{\sigma}_3(s)^2.\end{aligned}$$

Subtracting the two right hand sides from each other and dividing by $(\alpha-\lambda_1)^2-(\alpha-\lambda_2)^2$ yields

$$2\hat{\sigma}_3(s)^2 = (\alpha-\lambda_1)^2+(\alpha-\lambda_2)^2+2\omega^2 = 2\,[\alpha-(\lambda_1+\lambda_2)/2]^2 + 2\,[(\lambda_1-\lambda_2)/2]^2 + 2\omega^2.$$

So

$$\hat{\sigma}_3(s)^2 = [\alpha-(\lambda_1+\lambda_2)/2]^2+\omega^2+[(\lambda_1-\lambda_2)/2]^2 = |s-(\lambda_1+\lambda_2)/2|^2+[(\lambda_1-\lambda_2)/2]^2 \tag{53}$$

for $s=\alpha+\imath\omega$, $\omega\neq0$. It follows from Theorem 5.2.35 that for $\delta\leq|\lambda_1-\lambda_2|/2$ the spectral value set $\sigma_{\mathbb{R}}(A;\delta)$ does not contain non-real $s\in\mathbb{C}$, hence by (50)

$$\delta\leq|\lambda_1-\lambda_2|/2 \quad\Rightarrow\quad \sigma_{\mathbb{R}}(A;\delta) = I(\lambda_1,\delta)\cup I(\lambda_2,\delta). \tag{54}$$

For larger δ, (47) and (50) yield

$$\sigma_{\mathbb{R}}(A;\delta) = I(\lambda_1,\delta)\cup I(\lambda_2,\delta)\cup D\Big(m,(\delta^2-d^2)^{1/2}\Big),\ m=(\lambda_1+\lambda_2)/2,\ d=|\lambda_1-\lambda_2|/2. \tag{55}$$

Note that this formula includes (52) and (54) if we set $D(\lambda,r)=\emptyset$ for $r\in\imath\mathbb{R}$. The formula is illustrated in Figure 5.2.9. The real spectral value set $\sigma_{\mathbb{R}}(A;\delta)$ is the union of the interval $(\lambda_1-\delta,\lambda_2+\delta)$ shown as a thick line and the shaded disk $D(m,(\delta^2-d^2)^{1/2})$. Whereas the complex set $\sigma_{\mathbb{C}}(A;\delta)$ is the union of the disks $D(\lambda_1,\delta)$ and $D(\lambda_2,\delta)$ bounded by the dotted circles. □

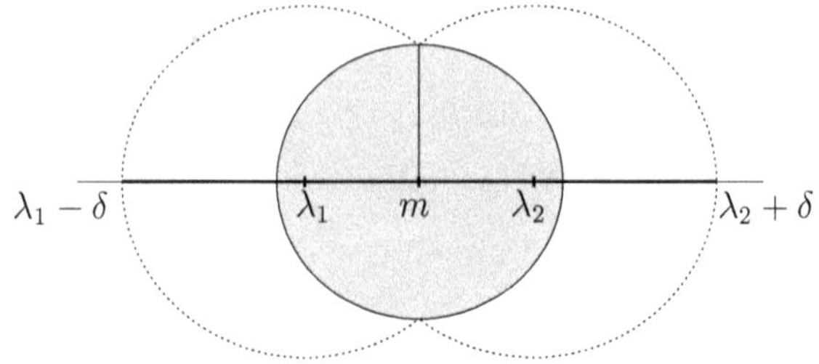

Figure 5.2.9: Illustration of (55)

The difference between the real and the complex unstructured spectral value sets in the previous example can be intuitively explained as follows: In order to move a simple real eigenvalue of A away from the real axis, a *real* perturbation must first produce a collision with another real eigenvalue, after which it can then produce a pair of complex eigenvalues. Therefore, in order to produce a non-real eigenvalue $\lambda \in \mathbb{C} \setminus \mathbb{R}$, a *real* perturbation of much larger size than $\operatorname{dist}(\lambda, \sigma(A))$ may be required if the two simple real eigenvalues λ_1, λ_2 are far apart and λ is lying close to one of the two eigenvalues. This collision mechanism is illustrated in the following example.

Example 5.2.37. Let

$$A = \operatorname{diag}(\lambda_1, \lambda_2) = \begin{bmatrix} 0 & 0 \\ 0 & 4 \end{bmatrix}, \quad \Delta = \begin{bmatrix} a & b \\ c & d \end{bmatrix} \in \mathbb{R}^{2\times 2}$$

and suppose $\lambda := \imath \in \sigma(A + \Delta)$. Then necessarily $\operatorname{trace}(A + \Delta) = a + d + 4 = 0$ so that $\|\Delta\| \geq \max\{|a|, |d|\} \geq 2$ whereas $\operatorname{dist}(\lambda, \sigma(A)) = 1$.

Now consider the spectrum of $A_\gamma = A + \gamma\Delta$ where $\gamma \in [0, 1]$ and $\Delta = \begin{bmatrix} -2 & -\sqrt{5} \\ \sqrt{5} & -2 \end{bmatrix}$. We

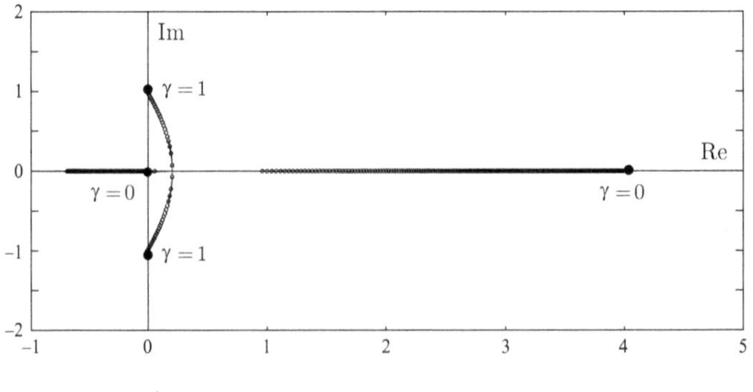

Figure 5.2.10: Eigenvalues of $A_\gamma = A + \gamma\Delta$, $0 \leq \gamma \leq 1$

have $\imath \in \sigma(A + \Delta)$ and by (53)

$$\|\Delta\| = 3 = \left[|\imath - 2|^2 + 2^2\right]^{1/2} = \left[|\imath - (\lambda_1 + \lambda_2)/2|^2 + ((\lambda_2 - \lambda_1)/2)^2\right]^{1/2} = \hat{\sigma}_3(\imath).$$

Since $\hat{\sigma}_3(\imath) = \left[\mu_{\mathbb{R}}\left((\imath I - A)^{-1}\right)\right]^{-1}$, Δ is a minimum norm real perturbation achieving $\imath \in \sigma(A + \Delta)$, see Proposition 5.2.25 (i). The movement of the eigenvalues of A_γ as γ

increases from 0 to 1 is shown in Figure 5.2.10. As γ increases from 0 to 1, the largest eigenvalue of A_γ decreases from 4 while the lowest eigenvalue moves from zero first to the left until its smallest value $2 - 6/\sqrt{5}$ is attained at $\gamma = 4\sqrt{5}/15$. Then the lowest eigenvalue of A_γ turns back and increases until for $\gamma = 2/\sqrt{5}$ the two real eigenvalues collide at $2 - 4/\sqrt{5}$ and split into a pair of complex conjugate eigenvalues ending in $\pm\imath$ for $\gamma = 1$. □

Example 5.2.36 shows that even in a very simple case the evaluation of formula (47) can be complicated and, in fact, in most examples it will be impossible to determine $\sup_{\gamma\in(0,1]} \sigma_{2n-1}((sI - A)_\gamma^{\mathbb{R}})$ analytically. So whilst the formulas in Theorem 5.2.35 are vital for computational purposes, in order to get some analytical estimates for the sets it is often easier to start from the definition of $\sigma_{\mathbb{K}}(A;\delta)$, especially in the real case. Throughout the rest of this subsection we will be particularly interested in comparing the results we obtain for normal matrices with those for non-normal ones. Our first objective is to obtain a *lower* set bound and to determine when it is tight.

For the complex case a lower estimate for the spectral value sets is easily obtained.

Corollary 5.2.38. *If $\sigma(A) = \{\lambda_1, \ldots, \lambda_\ell\}$ then*

$$\bigcup_{j=1}^{\ell} D(\lambda_j, \delta) \subset \sigma_{\mathbb{C}}(A;\delta), \quad \delta > 0. \tag{56}$$

If A is normal then equality holds in (56).

Proof: If $\lambda \in D(\lambda_j, \delta)$ then λ is an eigenvalue of $A + \Delta$ with $\Delta = (\lambda - \lambda_j)I$ and this proves (56). Now equality for normal A follows from the Bauer-Fike lemma, see Corollary 4.2.16. □

Remark 5.2.39. We have seen at the end of Subsection 4.2.1 that the eigenvalues of normal matrices have minimal sensitivity. The previous corollary provides a global version of this result. If $\lambda_1, \ldots, \lambda_\ell$ are any complex numbers, then all complex normal matrices A with $\sigma(A) = \{\lambda_1, \ldots, \lambda_\ell\}$ have exactly the same complex spectral value set at a given uncertainty level. Moreover this set is contained in the complex spectral value set (at the same uncertainty level) of any other matrix which has the same spectrum. In this sense the spectrum of a normal matrix is more *robust* with respect to unstructured parameter perturbations than that of any other matrix. In particular normal matrices have minimal complex spectral value sets in their similarity class. □

To obtain a corresponding bound for the real case is a much more subtle problem and to prepare the ground we will first study the case $n = 2$.

Lemma 5.2.40. *Suppose $A \in \mathbb{R}^{2\times 2}$ and $\lambda = \alpha + \imath\omega \in \mathbb{C}$. Then*

(i) If $\sigma(A) \not\subset \mathbb{R}$ or if $\lambda \in \mathbb{R}$ then there exists $\Delta \in \mathbb{R}^{2\times 2}$ of norm $\|\Delta\| \le \operatorname{dist}(\lambda, \sigma(A))$ such that $\lambda \in \sigma(A + \Delta)$.

(ii) If $\sigma(A) \subset \mathbb{R}$ and $\lambda \notin \mathbb{R}$, then there exists $\Delta \in \mathbb{R}^{2\times 2}$ of norm

$$\|\Delta\| \le \left(|\lambda - m|^2 + d^2\right)^{1/2}, \quad \text{where } m = (\lambda_1 + \lambda_2)/2,\ d = (\lambda_1 - \lambda_2)/2 \tag{57}$$

such that $\lambda \in \sigma(A + \Delta)$.

(iii) If A is normal then the perturbations Δ in (i) (resp. (ii)) are of minimal norm amongst those achieving $\lambda \in \sigma(A+\Delta)$. Moreover, in this case

$$\|\Delta\| = \mathrm{dist}(\lambda, \sigma(A)) \quad \left(\textit{resp. } \|\Delta\| = \left(|\lambda - m|^2 + d^2\right)^{1/2}\right).$$

Proof: It is an easy exercise to show that every matrix $A \in \mathbb{R}^{2\times 2}$ is orthogonally similar to a matrix with equal entries on the diagonal:

$$UAU^\top = \begin{bmatrix} a & b \\ c & a \end{bmatrix} = aI_2 + \begin{bmatrix} 0 & b \\ c & 0 \end{bmatrix}, \quad U \in \mathbf{O}_2(\mathbb{R}).$$

Therefore, since $\lambda \in \sigma(A+\Delta) \Leftrightarrow \lambda \in \sigma(UAU^\top + U\Delta U^\top)$ and $\|U\Delta U^\top\| = \|\Delta\|$, we may assume without restriction of generality that A is of the form $A = \begin{bmatrix} a & b \\ c & a \end{bmatrix}$. Then A has the real eigenvalues $\lambda_{1,2} = a \pm \sqrt{bc}$ if $bc \geq 0$ and the complex eigenvalues $\lambda_{1,2} = a \pm \imath\sqrt{-bc}$ if $bc < 0$.
Now consider perturbations of the form

$$\Delta = \begin{bmatrix} \alpha - a & -\eta \\ \eta & \alpha - a \end{bmatrix} \in \mathbb{R}^{2\times 2}, \quad \eta \in \mathbb{R}. \tag{58}$$

Then $\lambda = \alpha + \imath\omega$, $\omega \geq 0$ is an eigenvalue of $A + \Delta = \begin{bmatrix} \alpha & b-\eta \\ c+\eta & \alpha \end{bmatrix}$ if and only if

$$0 \leq \omega^2 = -(b-\eta)(c+\eta), \quad \text{i.e.} \quad \eta = (b-c)/2 \pm \sqrt{\omega^2 + ((b+c)/2)^2}.$$

Since we want to minimize the norm $\|\Delta\|^2 = (\alpha - a)^2 + \eta^2$, we choose

$$\eta = (b-c)/2 - \mathrm{sign}(b-c)\sqrt{\omega^2 + ((b+c)/2)^2}. \tag{59}$$

(i) First suppose $\sigma(A) \not\subset \mathbb{R}$, i.e. $bc < 0$. Then we have to show that

$$\|\Delta\|^2 = (\alpha - a)^2 + \eta^2 \leq \mathrm{dist}(\lambda, \sigma(A)) = (\alpha - a)^2 + (\omega - \sqrt{-bc})^2$$

i.e.

$$\eta^2 = ((b-c)/2)^2 + \omega^2 + ((b+c)/2)^2 - |b-c|\sqrt{\omega^2 + ((b+c)/2)^2} \leq \omega^2 - bc - 2\omega\sqrt{-bc}.$$

This is equivalent to $(1/2)(b+c)^2 + 2\omega\sqrt{-bc} \leq |b-c|\sqrt{\omega^2 + ((b+c)/2)^2}$, hence to

$$(b+c)^4 + 16\omega^2(-bc) + 8(b+c)^2\omega\sqrt{-bc} \leq (b-c)^2\left(4\omega^2 + (b+c)^2\right),$$

or, equivalently,

$$(b+c)^4 + 4\omega^2\left[-4bc - (b-c)^2\right] \leq (b-c)^2(b+c)^2 - 8(b+c)^2\omega\sqrt{-bc}.$$

Since the bracket is equal to $-(b+c)^2$, this is equivalent to

$$(b+c)^2 - 4\omega^2 \leq (b-c)^2 - 8\omega\sqrt{-bc}, \quad \text{i.e.} \quad 4bc - 4\omega^2 \leq -8\omega\sqrt{-bc}.$$

This last inequality holds since $(\omega - \sqrt{-bc})^2 \geq 0$. This proves (i) for the case $\sigma(A) \not\subset \mathbb{R}$. Now assume $\sigma(A) \subset \mathbb{R}$ and $\lambda \in \mathbb{R}$. To prove (i) for this case it suffices to choose $\Delta = (\lambda - \lambda_1)I_2$ where $\lambda_i \in \sigma(A)$ is chosen such that $|\lambda - \lambda_i| = \operatorname{dist}(\lambda, \sigma(A))$.
(ii) Now suppose $\sigma(A) \subset \mathbb{R}$, i.e. $bc \geq 0$, and $\lambda = \alpha + \imath\omega$, $\omega > 0$. Since $m = a$ and $d = \sqrt{bc}$ it remains to show that

$$\|\Delta\|^2 = (\alpha - a)^2 + \eta^2 \leq (\alpha - a)^2 + \omega^2 + bc,$$

$$\text{i.e.} \quad \eta^2 = (1/2)(b^2 + c^2) + \omega^2 - (1/2)|b - c|\sqrt{4\omega^2 + (b + c)^2} \leq \omega^2 + bc$$

$$\text{i.e.} \quad (b - c)^4 \leq (b - c)^2(4\omega^2 + (b + c)^2), \quad \text{i.e.} \quad (b - c)^2 \leq 4\omega^2 + (b + c)^2.$$

But this inequality holds since $bc \geq 0$.
(iii) Suppose that A is normal. Then it follows directly from Corollary 4.2.16 that the perturbations in (i) are of minimal norm and satisfy $\|\Delta\| = \operatorname{dist}(\lambda, \sigma(A))$. To prove that the perturbations satisfying (57) are necessarily minimal we make use of orthogonal similarity transformations in order to put A into the form $A = \operatorname{diag}(\lambda_1, \lambda_2)$. But then, for any given $\lambda = \alpha + \imath\omega$, $\omega > 0$ we know from (53) and Proposition 5.2.25 (i) that there does not exist a perturbation $\Delta' \in \mathbb{R}^{2\times 2}$ of norm $< \hat{\sigma}_3(\lambda) = (|\lambda - m|^2 + d^2)^{1/2}$ achieving $\lambda \in \sigma(A + \Delta')$. This concludes the proof of (iii). □

Remark 5.2.41. Note that the perturbations constructed in the previous proof are all multiples of isometries with respect to the Euclidean norm. □

As a consequence of the lemma and (55) we obtain the following formulas for $\sigma_{\mathbb{R}}(A;\delta)$ in the 2-dimensional case.

Corollary 5.2.42. *Suppose $A \in \mathbb{R}^{2\times 2}$ has eigenvalues λ_1, λ_2 and $m = (\lambda_1 + \lambda_2)/2$, $d = |\lambda_1 - \lambda_2|/2$. Then, for all $\delta > 0$,*

$$\sigma_{\mathbb{R}}(A;\delta) \supset D(\lambda_1,\delta) \cup D(\lambda_2,\delta) \quad \text{if } \lambda_i \notin \mathbb{R} \tag{60}$$

$$\sigma_{\mathbb{R}}(A;\delta) \supset I(\lambda_1,\delta) \cup I(\lambda_2,\delta) \cup D\left(m, (\delta^2 - d^2)^{1/2}\right) \quad \text{if } \lambda_i \in \mathbb{R}, \tag{61}$$

where $D(\lambda, r) = \emptyset$ for $r \in \imath\mathbb{R}$. The inclusions are equalities if A is normal.

The corollary is illustrated in the following example which shows the four different types of unstructured real spectral value sets of normal matrices $A \in \mathbb{R}^{2\times 2}$.

Example 5.2.43. Consider the normal matrices

$$A_0 = \begin{bmatrix} 1 & 0 \\ 0 & 3.5 \end{bmatrix}, \quad A_1 = \begin{bmatrix} 1 & 0 \\ 0 & 2 \end{bmatrix}, \quad A_2 = \begin{bmatrix} 1 & 0 \\ 0 & 1 \end{bmatrix}, \quad A_3 = \begin{bmatrix} 1 & 1/2 \\ -1/2 & 1 \end{bmatrix}.$$

Figure 5.2.11 shows the spectral value sets of A_i, $i = 0,1,2,3$ for $\delta = 1$. The first three matrices have a real spectrum and the corresponding pictures illustrate the three types of spectral value sets corresponding to this case. A_3 has a pair of complex eigenvalues and so its spectral value set is just the union of two disks. The figures show also the spectra of 3000 perturbed matrices $A_i + \Delta$, where the $\Delta \in \mathbb{R}^{2\times 2}$, $\|\Delta\| = 1$ were produced with the help of a pseudo random generator. □

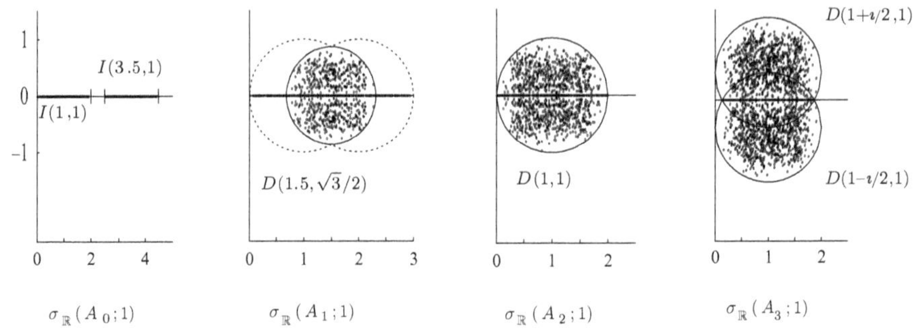

Figure 5.2.11: Spectral value sets of normal matrices

The previous corollary implies that, in the 2-dimensional case, *normal matrices have the smallest real pseudospectra in their similarity class.* It is still an open problem whether this also holds for arbitrary dimensions, see *Notes and References.* The following proposition gives only a partial answer to this question.

Proposition 5.2.44. *Suppose $A \in \mathbb{R}^{n \times n}$ has eigenvalues $\lambda_1, \ldots, \lambda_n$ (taking account of multiplicities), with $\lambda_1 \leq \ldots \leq \lambda_k \in \mathbb{R}$ and $\lambda_{k+1}, \ldots, \lambda_n \notin \mathbb{R}$, and let $\delta > 0$. Then the spectral value set $\sigma_{\mathbb{R}}(A;\delta)$ contains the following sets:*

(i) $I(\lambda_i, \delta) = \{\lambda \in \mathbb{R}; |\lambda - \lambda_i| < \delta\}, \quad i = 1, \ldots, k;$

(ii) $D(\lambda_j, \delta) = \{\lambda \in \mathbb{C}; |\lambda - \lambda_j| < \delta\}, \quad j = k+1, \ldots, n;$

(iii) $D(m_i, (\delta^2 - d_i^2)^{1/2})$ *if* $\lambda_{i+1} - \lambda_i < 2\delta$, $i = 1, \ldots, k-1$ *where* $m_i = (\lambda_i + \lambda_{i+1})/2$ *and* $d_i = (\lambda_{i+1} - \lambda_i)/2$. *(In particular,* $D(\lambda_i, \delta) \subset \sigma_{\mathbb{R}}(A;\delta)$ *if* λ_i *is of multiplicity* ≥ 2*).*

(iv) Assume that A is normal and each simple real eigenvalue $\lambda \in \sigma(A)$ satisfies

$$|\lambda - \mu| \geq 2\delta \quad \text{for all} \quad \mu \in \sigma(A), \quad \mu \neq \lambda. \tag{62}$$

Then

$$\sigma_{\mathbb{R}}(A;\delta) = \bigcup_{\lambda \in \sigma_1} I(\lambda, \delta) \cup \bigcup_{\lambda \in \sigma_2} D(\lambda, \delta) \tag{63}$$

where $\sigma(A) = \sigma_1 \dot{\cup} \sigma_2$ is the partition of $\sigma(A)$ into the set of simple real eigenvalues and the set of eigenvalues which are either not simple or not real.

Proof: (i) If $\lambda \in I(\lambda_i, \delta)$ for some $i \in \underline{k}$ it suffices to set $\Delta = (\lambda - \lambda_i)I$ to get $\lambda \in \sigma(A + \Delta)$.
(ii) Suppose $\lambda \in D(\lambda_j, \delta)$, $\lambda_j \notin \mathbb{R}$. Making use of the Schur canonical form (Theorem 4.5.14) there exists an orthogonal matrix $U \in \mathbf{O}_n$ such that

$$\tilde{A} = U^\top A U = \begin{bmatrix} A_1 & A_3 \\ 0 & A_2 \end{bmatrix} \tag{64}$$

where $A_2 \in \mathbb{R}^{2\times 2}$ has eigenvalues λ_j, $\overline{\lambda_j}$. By Lemma 5.2.40 (i) there exists a matrix $\Delta \in \mathbb{R}^{2\times 2}$ of norm $\|\Delta\| < \delta$ such that $\lambda \in \sigma(A_2 + \Delta)$. Taking

$$A(\Delta) = U\left(\tilde{A} + \begin{bmatrix} 0 & 0 \\ 0 & \Delta \end{bmatrix}\right)U^\top = A + U\begin{bmatrix} 0 & 0 \\ 0 & \Delta \end{bmatrix}U^\top,$$

we obtain $\|A(\Delta) - A\| < \delta$ and $\lambda \in \sigma(A(\Delta))$.
(iii) Suppose $\lambda \in D(m_i, (\delta^2 - d_i^2)^{1/2})$ with $1 \le i \le k-1$. In Theorem 4.5.14 we showed that there exists a Schur canonical form $\tilde{A}$ (64) of A such that $\sigma(A_2) = \{\lambda_i, \lambda_{i+1}\}$. Since $|\lambda - m_i|^2 + d_i^2 < \delta^2$ we can apply Lemma 5.2.40 (ii) and conclude the proof as in (ii).
(iv) By (i) and (ii) it is only necessary to prove the inclusion $\subset$ in (63), under the assumption of (iv). If $s \in \sigma_{\mathbb{R}}(A; \delta)$ then by Corollary 5.2.38 $s \in D(\lambda, \delta)$ for some $\lambda \in \sigma(A)$. If $\lambda \in \sigma_2$, then necessarily s belongs to the RHS of (63). Now suppose that $s \in D(\lambda, \delta)$ for some simple *real* $\lambda \in \sigma(A)$, then we have to show that $s \in \mathbb{R}$. Let $s \in \sigma(A + \Delta)$, $\Delta \in \mathbb{R}^{n \times n}$, $\|\Delta\| < \delta$ and set $A_\varepsilon = A + \varepsilon\Delta$, $\varepsilon \in [0, 1]$. By Corollary 4.2.4 there exist continuous functions $\tilde{\lambda}_j(\cdot) : [0, 1] \to \mathbb{C}$, $j = 1, ..., n$ such that $\sigma(A_\varepsilon) = \{\tilde{\lambda}_1(\varepsilon), ..., \tilde{\lambda}_n(\varepsilon)\}$ for all $\varepsilon \in [0, 1]$ with, say, $\tilde{\lambda}_i(0) = \lambda$. Clearly, $\sigma(A_\varepsilon) \subset \sigma_{\mathbb{R}}(A; \delta) \subset \bigcup_{j=1}^n D(\lambda_j, \delta)$ for all $\varepsilon \in [0, 1]$. Now, by the assumption (62), $D(\lambda, \delta)$ has empty intersection with $D(\mu, \delta)$ for all $\mu \in \sigma(A)$, $\mu \neq \lambda$ and so, by continuity, $\tilde{\lambda}_i(\varepsilon) \in D(\lambda, \delta)$ for all $\varepsilon \in [0, 1]$. By the same reason, no branch $\tilde{\lambda}_j(\varepsilon)$ with $\tilde{\lambda}_j(0) \neq \lambda$ can enter $D(\lambda, \delta)$. But λ is a simple eigenvalue of A and so $\tilde{\lambda}_i(\varepsilon)$ is the only eigenvalue of A_ε in $D(\lambda, \delta)$ taking account of multiplicity. Thus $\tilde{\lambda}_i(\varepsilon) \in \mathbb{R}$ for all $\varepsilon \in [0, 1]$ (otherwise $\tilde{\lambda}_i(\varepsilon) \neq \overline{\tilde{\lambda}_i(\varepsilon)}$ would both be in $D(\lambda, \delta)$) and since $\tilde{\lambda}_i(1) = s$ this concludes the proof. □

Remark 5.2.45. As a consequence of Corollary 5.2.38 we see that the set-valued map $A \mapsto \sigma(A)$ is open on $\mathbb{C}^{n \times n}$ in the sense that $\cup_{A \in \mathcal{U}} \sigma(A)$ is open when $\mathcal{U}$ is an open subset of $\mathbb{C}^{n \times n}$. Whereas by Proposition 5.2.44 we obtain that the same map $A \mapsto \sigma(A)$ is only *conditionally open* on $\mathbb{R}^{n \times n}$, in the sense that $[\cup_{A \in \mathcal{U}} \sigma(A)] \cap \mathbb{R}$ is open in $\mathbb{R}$ and $[\cup_{A \in \mathcal{U}} \sigma(A)] \cap (\mathbb{C} \setminus \mathbb{R})$ is open in $\mathbb{C}$ if $\mathcal{U}$ is an open subset of $\mathbb{R}^{n \times n}$. □

If the condition (62) given in Proposition 5.2.44 (iv) is violated, the real unstructured spectral values sets of normal matrices need not be given by (63) nor can they in general be represented as unions of disks and real intervals as in the 2-dimensional case (Corollary 5.2.42). New shapes appear already for $n = 3$ as illustrated in the next example.

Example 5.2.46. Consider the normal matrices $A_{\alpha,\beta} = \begin{bmatrix} 0 & -\beta & 0 \\ \beta & 0 & 0 \\ 0 & 0 & \alpha \end{bmatrix}$, where $\alpha, \beta \in \mathbb{R}$. In the following figures the the uncertainty level is set at $\delta = 6$. For Figure 5.2.12

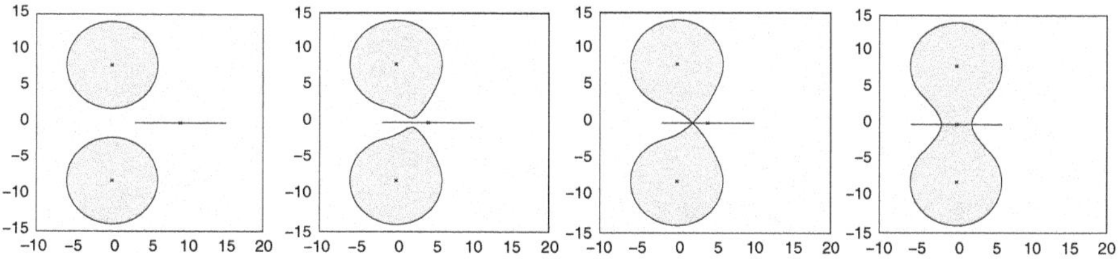

Figure 5.2.12: $\sigma_{\mathbb{R}}(A_{\alpha,8}; 6)$ for $\alpha = 9$, 4.2, 4, 0

$\beta = 8$ and the real unstructured spectral value sets of $A_{\alpha,8}$ have been calculated with

$\alpha = 9$, 4.2, 4, 0. For $\alpha = 9$, the sets are given by (63), but for smaller values of α, the sets around $\pm 8\imath$ seem to be attracted by the interval around the eigenvalue α. This continues until for $\alpha = 4$ the sets touch and then for lower values merge. For Figure 5.2.13 $\alpha = 9$ is kept constant and the real unstructured spectral value sets of $A_{9,\beta}$ have been calculated with $\beta = 8$, 5.8, 5.6, 5.2. For $\beta = 8$, the sets are as in the left hand figure in Figure 5.2.12 and are given by (63). However as β decreases the interval around the eigenvalue $\alpha = 9$ attracts the sets around $\pm\beta\imath$, until they touch. Then as β decreases the sets merge, but a diminishing hole remains temporarily within the connected spectral value set.

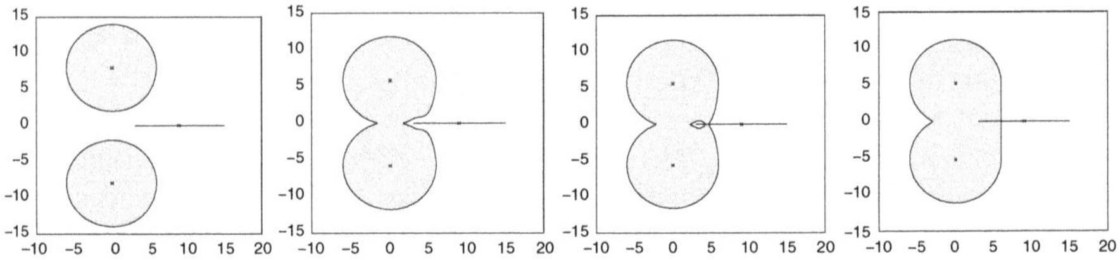

Figure 5.2.13: $\sigma_{\mathbb{R}}(A_{9,\beta}; 6)$ for $\beta = 8$, 5.8, 5.6, 5.2.

□

We conclude this subsection by analyzing how much the unstructured spectral value sets of a given matrix A can be increased by applying similarity transformations to A. We first prove a related lemma which will be used in the next section on stability radii.

Lemma 5.2.47. *Suppose $A \in \mathbb{K}^{n\times n}$, $A \notin \mathbb{K}I$ and $\sigma(A) = \{\lambda_1, \ldots, \lambda_n\}$. Then*

$$\sup_{T\in \mathbf{Gl}_n(\mathbb{K})} \|T^{-1}AT\| = \infty. \tag{65}$$

If $A \in \mathbb{C}^{n\times n}$ there exists a sequence (T_k) in $\mathbf{Gl}_n(\mathbb{C})$ such that

$$\lim_{k\to\infty} T_k^{-1}AT_k = \operatorname{diag}(\lambda_1, \ldots, \lambda_n)\,. \tag{66}$$

If A is real and $\lambda_1, \ldots, \lambda_r \in \mathbb{R}$, $\alpha_1 \pm \imath\beta_1, \cdots, \alpha_h \pm \imath\beta_h \in \mathbb{C}\setminus\mathbb{R}$ $(r + 2h = n)$ are the eigenvalues of A taking account of multiplicities, then there exists a sequence (T_k) in $\mathbf{Gl}_n(\mathbb{R})$ such that

$$\lim_{k\to\infty} T_k^{-1}AT_k = \operatorname{diag}\left(\lambda_1, \ldots, \lambda_r, \begin{bmatrix} \alpha_1 & \beta_1 \\ -\beta_1 & \alpha_1 \end{bmatrix}, \cdots, \begin{bmatrix} \alpha_h & \beta_h \\ -\beta_h & \alpha_h \end{bmatrix}\right). \tag{67}$$

Proof: We prove the lemma for the real case which is slightly more complicated than the complex one. Without restriction we may assume that A is in real Jordan form $A = D+N$ where D is the block-diagonal matrix on the RHS of (67) and N is a nilpotent matrix with zero entries except for possible 1s at the positions $(i, i+1)$, $i = 1, \ldots, r-1$ and possible submatrices I_2 to the right of the complex eigenvalue blocks $\begin{bmatrix} \alpha_j & \beta_j \\ -\beta_j & \alpha_j \end{bmatrix}$, $j = 1, \ldots, h-1$. Let $T(\varepsilon) = \operatorname{diag}(1, \varepsilon, \ldots, \varepsilon^{r-1}, I_2, \varepsilon I_2, \ldots, \varepsilon^{h-1} I_2)$, then

$$T(\varepsilon)^{-1}AT(\varepsilon) = D + T(\varepsilon)^{-1}NT(\varepsilon) = D + \varepsilon N, \quad \varepsilon \in \mathbb{R}. \tag{68}$$

Choosing $T_k = T(1/k)$ proves (67). Moreover, if A is not semi-simple, i.e. $N \neq 0$, then (68) implies (65) by choosing $T = T(\varepsilon)$ with $\varepsilon \to \infty$.
If A is semi-simple, then since $A \neq \mathbb{R}I_n$, A has either two distinct real eigenvalues λ, μ or a pair of complex eigenvalues $\alpha \pm \imath\beta$, $\beta \neq 0$. After a suitable similarity transformation (permutation) we may assume A is of the form

$$A = \operatorname{diag}(\lambda, \mu, A_2) \quad \text{or} \quad A = \operatorname{diag}\left(\begin{bmatrix} \alpha & \beta \\ -\beta & \alpha \end{bmatrix}, A_2\right),$$

where A_2 is of order $n-2$. Let $S(t) = \operatorname{diag}\left(\begin{bmatrix} 1 & t \\ 0 & 1 \end{bmatrix}, I_{n-2}\right)$, so that $S(t)^{-1} = S(-t)$. Then an elementary calculation shows that in both cases

$$\lim_{t\to\infty} \|S(t)^{-1}AS(t)\| = \infty.$$

This completes the proof of (65).
The same proof applies to the complex case ($\mathbb{K} = \mathbb{C}$) with the simplification that in this case D is diagonal. □

In the next proposition we show that for any given δ, however small, the complex pseudospectrum $\sigma_{\mathbb{C}}(A;\delta)$ absorbs every bounded set in $\mathbb{C}$ if A varies along any similarity orbit in $\mathbb{C}^{n\times n}$, $n \geq 2$ which is not a singleton.

Proposition 5.2.48. *Suppose $A \in \mathbb{C}^{n\times n}$, $n \geq 2$, $A \notin \mathbb{C}I_n$, $\delta > 0$, $c \in \mathbb{C}$ and $r > 0$. Then there exists a nonsingular transformation $T \in \mathbf{Gl}_n(\mathbb{C})$ such that*

$$\sigma_{\mathbb{C}}(T^{-1}AT;\delta) \supset D(c,r).$$

Proof: Without restriction of generality we may assume $c = 0$ and that A is of the form

$$A = \begin{bmatrix} A_1 & * \\ 0 & * \end{bmatrix} \quad \text{where } (i)\ A_1 = \begin{bmatrix} \lambda & 0 \\ 0 & \mu \end{bmatrix},\ \lambda \neq \mu, \quad \text{or} \quad (ii)\ A_1 = \begin{bmatrix} \lambda & 1 \\ 0 & \lambda \end{bmatrix}.$$

In the first case we choose $t \geq [(r+|\lambda|)(r+|\mu|)]/(\delta|\lambda-\mu|)$ and for any $s \in \mathbb{C}$

$$S(t) = \operatorname{diag}\left(\begin{bmatrix} 1 & t \\ 0 & 1 \end{bmatrix}, I_{n-2}\right),\ \Delta(s) = \begin{bmatrix} \Delta_1(s) & 0 \\ 0 & 0 \end{bmatrix},\ \Delta_1(s) = \begin{bmatrix} 0 & 0 \\ \frac{(s-\lambda)(s-\mu)}{t(\lambda-\mu)} & 0 \end{bmatrix}.$$

Then

$$S(t)^{-1}AS(t) + \Delta(s) = \begin{bmatrix} A_1(t) + \Delta_1(s) & * \\ 0 & * \end{bmatrix}, \quad A_1(t) + \Delta_1(s) = \begin{bmatrix} \lambda & t(\lambda-\mu) \\ \frac{(s-\lambda)(s-\mu)}{t(\lambda-\mu)} & \mu \end{bmatrix},$$

and so, for all $s \in D(0,r)$,

$$s \in \sigma(A_1(t) + \Delta_1(s)) \subset \sigma\left(S(t)^{-1}AS(t) + \Delta(s)\right), \quad \|\Delta(s)\| < \frac{(r+|\lambda|)(r+|\mu|)}{t|\lambda-\mu|} \leq \delta.$$

Therefore $D(0,r) \subset \sigma_{\mathbb{C}}(T^{-1}AT;\delta)$ for $T = S(t)$ if $t \geq [(r+|\lambda|)(r+|\mu|)]/(\delta|\lambda-\mu|)$.
In the second case we choose $t \geq (r+|\lambda|)^2/\delta$ and

$$S(t) = \operatorname{diag}\left(\begin{bmatrix} 1 & 0 \\ 0 & t \end{bmatrix}, I_{n-2}\right),\ \Delta(s) = \begin{bmatrix} \Delta_1(s) & 0 \\ 0 & 0 \end{bmatrix},\ \Delta_1(s) = \begin{bmatrix} 0 & 0 \\ \frac{(s-\lambda)^2}{t} & 0 \end{bmatrix}.$$

Then

$$S(t)^{-1}AS(t) + \Delta(s) = \begin{bmatrix} A_1(t) + \Delta_1(s) & * \\ 0 & * \end{bmatrix}, \quad A_1(t) + \Delta_1(s) = \begin{bmatrix} \lambda & t \\ \frac{(s-\lambda)^2}{t} & \lambda \end{bmatrix},$$

and so, for all $s \in D(0,r)$,

$$s \in \sigma(A_1(t) + \Delta_1(s)) \subset \sigma(S(t)^{-1}AS(t) + \Delta(s)), \quad \|\Delta(s)\| < \frac{(r + |\lambda|)^2}{t} \leq \delta.$$

Thus again $D(0,r) \subset \sigma_{\mathbb{C}}(T^{-1}AT;\delta)$ for $T = S(t)$ and the proof is complete. □

A similar result also holds in the real case with the added proviso that $n \geq 3$, but the proof is more complicated, see *Notes and References*. The 2-dimensional real case is studied in Ex. 19.
The previous proposition shows that every similarity orbit in $\mathbb{C}^{n\times n}$ which is not reduced to a single element contains matrices whose computed spectrum is arbitrarily uncertain. In fact, choose $\varepsilon > 0$ smaller than half the machine accuracy. Then the computer cannot distinguish between any matrix X and its perturbations $X + \Delta$ with $\|\Delta\| < \varepsilon$. By Proposition 5.2.48, given any matrix $A \in \mathbb{R}^{n\times n}\backslash\mathbb{C}I_n$, there exists, for any $c \in \mathbb{C}$ and $r > 0$, a matrix $\tilde{A} \sim A$ such that the spectra of all the matrices which the computer cannot distinguish from $\tilde{A}$ cover the disk $D(c,r)$.

5.2.5 Exercises

1. Consider

$$A = \begin{bmatrix} 0 & 1 & 0 & 0 \\ 0 & 0 & 1 & 0 \\ 0 & 0 & 0 & 1 \\ -2 & -6 & -8.5 & -5 \end{bmatrix}, \quad B = \begin{bmatrix} 0 \\ 0 \\ 0 \\ 1 \end{bmatrix}, \quad C = \begin{bmatrix} 3.25 & 2 & 1 & 0 \end{bmatrix}, \quad D = 1.$$

Calculate the corresponding scalar transfer function $G(s)$ and visualize $\sigma_{\mathbf{\Delta}}(A;B,C,D;\delta)$ where $\mathbf{\Delta} = \mathbb{C}$ for $\delta = 0.001,\ 0.01,\ 0.1$ (e.g. using the MATLAB-contour plotter for $|G(s)|$).

2. Consider the perturbed first order equation $(1 - \Delta_2)\dot{x} + (1 - \Delta_1)x = 0$ with perturbations $\Delta = (\Delta_1, \Delta_2) \in \mathbf{\Delta} = \mathbb{C}^{1\times 2}$. Show that this can be written in the form $\dot{x} = A(\Delta)x$ where $A(\Delta)$ is of the form (25) and the transfer function of (A,B,C,D) is $G(s) = -(s+1)^{-1}[1\ s]^\top$. Hence show that with respect to the Euclidean norm the boundary of the complex spectral value set $\sigma_{\mathbb{C}}(\delta)$ at level $\delta < 1$ is a circle centred at $-(1-\delta^2)^{-1}$ of radius $(1-\delta^2)^{-1}|\delta|\sqrt{2-\delta^2}$.

3. Suppose the Gershgorin perturbation class $\mathbf{\Delta}$ (15) is provided with the norm

$$\|\Delta\|_{\mathbf{\Delta}} = \max_{i,j \in \underline{n}, i \neq j} |\Delta_{ij}|, \quad \Delta = (\Delta_{ij}) \in \mathbf{\Delta}.$$

Following the same method as Subsection 5.2.1, prove the following counterpart of Proposition 5.2.11: If $A = (a_{ij}) \in \mathbb{C}^{n\times n}$ is diagonal, the spectral value set of A with respect to perturbations of the form $A \rightsquigarrow A(\Delta) = A + \Delta,\ \Delta \in \mathbf{\Delta},\ \|\Delta\|_{\mathbf{\Delta}} < \delta$ is given by

$$\sigma_{\mathbf{\Delta}}(A;\delta) = \sigma(A)\,\dot{\cup}\,\{s \in \rho(A);\ \varrho(P(s)) > \delta^{-1}\},$$

where ϱ denotes the spectral radius and

$$P(s) = \begin{bmatrix} 0 & |s-a_{22}|^{-1} & |s-a_{33}|^{-1} & \cdots & |s-a_{nn}|^{-1} \\ |s-a_{11}|^{-1} & 0 & |s-a_{33}|^{-1} & \cdots & |s-a_{nn}|^{-1} \\ \cdot & \cdot & \cdot & \cdots & \cdot \\ \cdot & \cdot & \cdot & \cdots & \cdot \\ |s-a_{11}|^{-1} & |s-a_{22}|^{-1} & |s-a_{33}|^{-1} & \cdots & 0 \end{bmatrix}, \quad s \in \rho(A). \quad (69)$$

Discuss the relation between this result and Proposition 5.2.11 in the special case $n = 2$.

Hint: Prove the following sequence of statements:

(i) Let $\tilde{\mathbf{\Delta}} \subset \mathbb{C}^{n \times n(n-1)}$ be the set of all $\tilde{\Delta}$ of the form (18) provided with the operator norm $\|\cdot\|_{\tilde{\mathbf{\Delta}}}$ induced by the following Hölder norm on $\mathbb{C}^{n(n-1)}$

$$\|y\| = \max_{i \in \underline{n}} \|y^i\|_1 = \|y\|_{1|\infty}, \quad y = (y^i)_{i \in \underline{n}}, \; y^i \in \mathbb{C}^{n-1}$$

and the ∞-norm on $\mathbb{C}^n$. Show that

$$\|\tilde{\Delta}\|_{\tilde{\mathbf{\Delta}}} = \max_{i,j \in \underline{n}, i \neq j} |\Delta_{ij}| = \|\Delta\|_{\mathbf{\Delta}}, \quad \Delta = (\Delta_{ij}) \in \mathbf{\Delta}.$$

(ii) If C is given by (19) and $G(s)$ by (20), prove that

$$\|G(s)\|_{\mathcal{L}(\mathbb{C}^n, \mathbb{C}^{n(n-1)})} = \max_{i \in \underline{n}} \sum_{j \neq i} |s - a_{jj}|^{-1}, \quad s \in \rho(A)$$

where $\|\cdot\|_{\mathcal{L}(\mathbb{C}^n, \mathbb{C}^{n(n-1)})}$ denotes the operator norm induced by the above vector norms on $\mathbb{C}^n$ and $\mathbb{C}^{n(n-1)}$.

(iii) It is known from Perron-Frobenius theory [183] that $\varrho(P(s))$ is an eigenvalue of the nonnegative matrix $P(s)$ and, since $P(s)$ is irreducible[3], the spectral radius of $P(s)$ is positive and there exists an associated eigenvector $x(s) = (x_i(s)) \in \mathbb{R}^n$ with $x_i(s) > 0$, $i \in \underline{n}$, $s \in \rho(A)$. Let $G_\gamma(s)$ be defined by (23) with $\gamma_i = \varrho(P(s))/x_i(s)$, $i \in \underline{n}$, and prove

$$\mu_{\tilde{\mathbf{\Delta}}}(G(s)) \leq \|G_\gamma(s)\|_{\mathcal{L}(\mathbb{C}^n, \mathbb{C}^{n(n-1)})} = \varrho(P(s)).$$

(iv) Define $\Delta \in \mathbf{\Delta}$ by

$$\Delta_{ij} = \frac{|s - a_{jj}|}{s - a_{jj}}, \quad i, j \in \underline{n}, \; i \neq j, \quad \Delta_{ii} = 0, \; i \in \underline{n}.$$

Show that $\tilde{\Delta} G(s) = \Delta \operatorname{diag}((s-a_{11})^{-1}, ..., (s-a_{nn})^{-1}) = P(s)$, $\|\tilde{\Delta}\|_{\tilde{\mathbf{\Delta}}} = 1$, and conclude the proof.

4. If

$$A = \begin{bmatrix} 1 & 1 \\ -2.5 & -2 \end{bmatrix}, \quad B = \begin{bmatrix} 1 \\ 0 \end{bmatrix}, \quad C = I_2,$$

calculate $\sigma_{\mathbb{C}}(A; B, C; \delta)$ with respect to the spectral norm for $\delta = 2^{-1}$, 1, 2.

---

[3] A nonnegative matrix $M = (m_{ij}) \in \mathbb{R}^{n \times n}$ is said to be irreducible if there does not exist a $n \times n$ permutation matrix P such that $PMP^\top$ has the form $\begin{bmatrix} M_{11} & 0 \\ M_{21} & M_{22} \end{bmatrix}$ where M_{11} and M_{22} are square matrices of order ≥ 1.

5. Calculate the complex spectral value set $\sigma_{\mathbb{C}}(A; B, C; \delta)$ at level $\delta = 1$ for the matrices in Ex. 4 with respect to the 1-norm $\|\Delta\|_1 = |\Delta_1| + |\Delta_2|$ on $\mathbf{\Delta} = \mathbb{C}^{1\times 2}$.

6. If $A = \begin{bmatrix} a_{11} & a_{12} \\ a_{21} & a_{22} \end{bmatrix}$, $B = C = I_2$ show that the complex spectral value sets $\sigma_{\mathbb{C}}(\delta) = \sigma_{\mathbb{C}}(A; I_2, I_2; \delta)$ with respect to the operator norm $\|\cdot\|_{\infty,\infty} = \|\cdot\|_{1|\infty}$ (see (A.13) and (A.1.7)) on $\mathbf{\Delta} = \mathbb{C}^{2\times 2}$ are given by

$$\sigma_{\mathbb{C}}(\delta) = \{s \in \mathbb{C};\ |(s - a_{11})(s - a_{22}) - a_{12}a_{21}| < \delta \max\{|s - a_{11}| + |a_{21}|, |s - a_{22}| + |a_{12}|\}.$$

7. Write a program (e.g. MATLAB code) for visualizing complex spectral value sets for full-block perturbations with respect to the spectral norm. Compute $\|G(s)\|_{2,2} = \|D + C(sI - A)^{-1}B\|_{2,2}$ on a suitable grid in the complex plane and use a contour plotter to visualize the boundary of $\sigma_{\mathbb{C}}(A; B, C, D; \delta)$ (see Proposition 5.2.19).

8. Consider the three matrices

$$A_0 = \begin{bmatrix} 0 & 1 & 0 \\ -1 & 0 & 0 \\ 0 & 0 & 2 \end{bmatrix}, \quad A_1 = \begin{bmatrix} 0 & 1 & 0 \\ 0 & 0 & 1 \\ 2 & -1 & 2 \end{bmatrix}, \quad A_2 = TA_1T^{-1} = \begin{bmatrix} 0 & 4 & 0 \\ 0 & 0 & 4 \\ 1/8 & -1/4 & 2 \end{bmatrix}$$

where $T = \operatorname{diag}(1, 1/4, 1/16)$. A_1 is the companion form of A_0. Visualize the unstructured complex spectral value sets (pseudospectra) of A_0, A_1 and A_2 at the uncertainty level $\delta = 1$ (with respect to the spectral norm on $\mathbf{\Delta} = \mathbb{C}^{3\times 3}$). Why do we necessarily have $\sigma_{\mathbb{C}}(A_0; \delta) \subset \sigma_{\mathbb{C}}(A_i; \delta)$, $i = 1, 2$ for all $\delta > 0$?

9. Suppose $A \in \mathbb{C}^{n\times n}$, $\delta > 0$, and $\mathbf{\Delta} = \mathbb{C}^{n\times n}$ is provided with the spectral norm $\|\cdot\| = \|\cdot\|_{2,2}$. Show that if $\operatorname{diag}(\lambda_1, \ldots, \lambda_n) + N$ is a Schur form of A (N nilpotent, upper triangular), then

$$\sigma_{\mathbb{C}}(A; \delta) \subseteq \bigcup_{i \in \underline{n}} D(\lambda_i, \delta + \|N\|).$$

10. Determine the real spectral value sets with respect to Euclidean norms for the system given in Ex. 2 and $\delta < 1$.

11. Consider
$$A = \begin{bmatrix} 0 & 0 \\ 0 & 0 \end{bmatrix}, \quad B = \begin{bmatrix} 1 & 1 \\ 1 & 1 \end{bmatrix}, \quad C = \begin{bmatrix} 1 & -1 \\ 0 & 1 \end{bmatrix}, \quad D = 0.$$

Show that the real spectral value sets $\sigma_{\mathbb{R}}(A; B, C; \delta)$ with respect to the spectral norm are confined to the real axis for all $\delta > 0$ and are given by

$$\sigma_{\mathbb{R}}(A; B, C; \delta) = \{s \in \mathbb{R};\ |s| < \sqrt{2}\delta\} = I(0, \sqrt{2}\delta).$$

12. Use Remark 5.2.24 and (40) to determine $\sigma_{\mathbb{R}}(A; B, C, D; \delta)$ where the matrices are given in Ex. 1 and $\delta = 0.01, 0.1$. Alternatively, determine $\sigma_{\mathbb{R}}(A; B, C, D; \delta)$ by computing the relevant part of the root locus of (p, q) where $p(s)/q(s)$ is the transfer function of the system (A, B, C, D), and compare the results on the computer screen.

13. Write a program (e.g. MATLAB code) for visualizing real spectral value sets for full-block perturbations with respect to the spectral norm in the case $\ell = 1$, i.e. $G(s)$ is a column vector (see Subsection 5.2.4). Compute $\sigma_{\mathbb{R}}(A; B, C, D; \delta) \cap \mathbb{R}$ separately and visualize it by increased linewidth. Use formula (41) for computing $\operatorname{dist}(G_R(s), \mathbb{R}G_I(s))^2$ on a suitable grid in $\mathbb{C} \setminus \mathbb{R}$ and use a contour plotter to visualize the boundary $\partial\sigma_{\mathbb{R}}(A; B, C, D; \delta) \setminus (\sigma(A) \cup \mathbb{R})$ (see Proposition 5.2.25).

14. Use the program developed in Ex. 13 to visualize the sets $\sigma_{\mathbb{R}}(A; B, C; \delta)$ where the matrices and δ are given as in Ex. 4.

15. Write a program (e.g. MATLAB code) for visualizing real spectral value sets for full-block perturbations with respect to the spectral norm in the general case $q, \ell \geq 1$, i.e. $G(s)$ is a $q \times \ell$-matrix, assuming that the realness locus is equal to the real axis. Compute $\sigma_{\mathbb{R}}(A; B, C, D; \delta) \cap \mathbb{R}$ separately and visualize it by increased linewidth. In order to apply Theorem 5.2.31 write a subroutine for computing $\mu_{\mathbb{R}}(G(s)) = \inf_{\gamma \in (0,1]} \sigma_2(G_\gamma^{\mathbb{R}}(s))$ for given $s \in \mathbb{C} \setminus \mathbb{R}$ employing a suitable minimization algorithm. Using this subroutine compute $\mu_{\mathbb{R}}(G(s)$ on an appropriate grid in $\mathbb{C} \setminus \mathbb{R}$ and use a contour plotter to visualize the relative boundary $\partial \sigma_{\mathbb{R}}(A; B, C, D; \delta) \setminus (\sigma(A) \cup \mathbb{R})$ (see Proposition 5.2.25).

16. Use the program developed in Ex. 15 to visualize the unstructured real spectral value sets $\sigma_{\mathbb{R}}(A_i; \delta)$ where the matrices A_i and δ are given as in Ex. 8. Compare these sets with the corresponding complex counterparts obtained in Ex. 8.

17. Let $\lambda_1, \lambda_2 \in \mathbb{C}$ be given and consider the matrices $A(\alpha) = \begin{bmatrix} \lambda_1 & \alpha \\ 0 & \lambda_2 \end{bmatrix}$ for $\alpha > 0$ (which are all similar). Prove that for all $\delta > 0$

$$(a)\ \alpha_1 < \alpha_2 \Longrightarrow \sigma_{\mathbb{C}}(A(\alpha_1); \delta) \subset \sigma_{\mathbb{C}}(A(\alpha_2); \delta), \quad (b) \bigcup_{\alpha>0} \sigma_{\mathbb{C}}(A(\alpha); \delta) = \mathbb{C}.$$

18. Suppose $\lambda_1, \lambda_2 \in \mathbb{R}$ and $A(\alpha)$ is defined as in Ex. 17. Determine $\bigcup_{\alpha>0} \sigma_{\mathbb{R}}(A(\alpha); \delta)$. Illustrate your findings by visualizing $\sigma_{\mathbb{R}}(A(\alpha); 1)$ for $\lambda_1 = -3, \lambda_2 = -1$ and $\alpha = 10,\ 50,\ 100$, making use of the program developed in Ex. 15.

19. Suppose $A \in \mathbb{R}^{2\times 2}$, $A \notin \mathbb{R} I_2$ and $\delta > 0$.
(i) Prove that for arbitrary $T \in \mathbf{Gl}_2(\mathbb{R})$ and $\Delta \in \mathbb{R}^{2\times 2}$, $\|\Delta\| < \delta$

$$s \in \sigma(T^{-1}AT + \Delta) \setminus \mathbb{R} \quad \Rightarrow \quad |\operatorname{Re} s - \operatorname{trace} A/2| < \delta.$$

Thus $\sigma_{\mathbb{R}}(T^{-1}AT; \delta) \setminus \mathbb{R}$ is contained in the vertical strip $\{s \in \mathbb{C}; |\operatorname{Re} s - \operatorname{trace} A/2| < \delta\}$ for all $T \in \mathbf{Gl}_2(\mathbb{R})$. Compare this upper bound with the tight lower bound for $\sigma_{\mathbb{R}}(T^{-1}AT; \delta)$ as T varies through $\mathbf{Gl}_2(\mathbb{R})$, see Corollary 5.2.42.
(ii) Prove that if $\sigma(A) = \{\lambda, \mu\} \subset \mathbb{R}$

$$\bigcup_{T \in \mathbf{Gl}_2(\mathbb{R})} \sigma_{\mathbb{R}}(T^{-1}AT; \delta) = \mathbb{R} \cup \{s \in \mathbb{C}; |\operatorname{Re} s - (\lambda + \mu)/2| < \delta\}.$$

(iii) Determine $\bigcup_{T \in \mathbf{Gl}_2(\mathbb{R})} \sigma_{\mathbb{R}}(T^{-1}AT; \delta)$ if $\sigma(A) \not\subset \mathbb{R}$, cf. Example 2.9 in [251].

5.2.6 Notes and References

For survey articles, general remarks, applications and historical comments concerning pseudospectra and spectral value sets, see the *Notes and References* of the previous section. Here we only give more specific references related to the material of this section.
A slightly more restricted version of the continuity Theorem 5.2.15 can be found in *Karow* (2000) [291]. This reference also indicates how one can show, using methods from elimination theory (theorem of Tarski-Seidenberg), that the μ-function is a semi-algebraic function and hence spectral value sets are semi-algebraic subsets of $\mathbb{C}$. As a consequence

the boundaries of spectral value sets are piecewise analytic. For more details see [292]. Probabilistic versions of spectral value sets have been studied by *Barmish and Lagoa* (1997) [37], *Tempo et al.* (1997) [493], *Bai et al.* (1998) [29], and *Lagoa et al.* (1998) [324]. Example 5.2.5 is due to *Godunov* (1992) [189]. *Chaitin-Chatelin and Frayssé* (1996) [96] have made use of pseudospectra in rounding error analysis.
Generalizations and refinements of Gershgorin's Theorem have been obtained by, for example, Ostrowski, Brauer and Brualdi (see Chapter 6 in *Horn and Johnson* (1990) [264]). In particular, new inclusion regions have been discovered that are guaranteed to include the eigenvalues of a given matrix. *Brauer* (1947) [75] introduced *Cassini's ovals* in this context. Using his theorem it is possible to show that the RHS of (22) is an upper bound for the spectral value $\sigma_{\mathbf{\Delta}}(A(\cdot);\delta)$. *Brualdi* (1982) [79] used graph theoretic means for a refinement of Brauer's result. More details and references to the original literature can be found in his paper. Our μ-approach to spectral value sets for Gershgorin type perturbations appears to be new (see also Ex. 3). An extension of this method to more general off-diagonal perturbation structures (in the spirit of *Brualdi* (1982) [79]) can be found in *Karow* (2003) [292]. The basic facts from the Perron-Frobenius theory of non-negative matrices on which it is based can be found in Chapter 8 of *Horn and Johnson* (1990) [264] and Chapter XIII of *Gantmacher* (1959) [183].
The characterization of spectral value sets in terms of spectral contours as given in Propositions 5.2.19 and 5.2.25 is based on the papers [233] and [234]. There are efficient algorithms to compute spectral value sets in the complex finite dimensional case, see [178] and the survey article of *Trefethen* (1999) [499].
Real pseudospectra were first considered (under the name of spectral value sets) in *Hinrichsen and Pritchard* (1991) [249]. The partial characterization of real pseudospectra of normal matrices in Proposition 5.2.44 was taken from this paper. A complete solution of the problem has been given by *Karow* (2003) [292]. Examples 5.2.34, 5.2.46 and Figures 5.2.8, 5.2.12, 5.2.13 are due to him. A real counterpart to Proposition 5.2.48 (for $n \geq 3$) can be found in *Hinrichsen and Pritchard* (1992) [251].
In most applications pseudospectra have been investigated for matrices which are approximations to infinite dimensional operators. A central area of application is the stability analysis of fluid flows, see e.g. *Reddy et al.* (1993) [429], *Trefethen et al.* (1993) [501]. A series of examples is presented in *Trefethen* (1997, 1999). Whilst there is a rapidly growing number of applications, there are not many convergence results available which show that the pseudospectra of the finite dimensional approximations in fact converge to the pseudospectra of the corresponding infinite dimensional operators. Convergence results have been obtained for Toeplitz and convolution operators, see *Landau* (1977) [328], *Reichel and Trefethen* (1992) [431], *Böttcher* (1994) [70], *Böttcher et al.* (1997) [71]. The textbook of *Böttcher and Silbermann* (1999) [72] is an excellent introduction to this material.
The notion of a spectral value set has been extended to infinite dimensions in *Gallestey et al.* (2000) [180]. In contrast with pseudospectra this notion allows one to consider spectral value sets for unbounded perturbations of closed linear operators by making use of unbounded structure operators B, C. The corresponding approximation problem has been addressed in *Gallestey* (1998) [179].

5.3 Stability Radii

In the previous section we studied the variation of the spectrum under linear fractional perturbations of size less than a given uncertainty level. Here we consider the same classes of perturbations, but introduce a new feature, a stability region $\mathbb{C}_g \subset \mathbb{C}$, which represents a prescribed set for the location of the system's spectrum. For example, if the system is continuous in time we might set $\mathbb{C}_g = \mathbb{C}_-$, or for discrete time $\mathbb{C}_g = \mathbb{D}$, in order to ensure stability. If, additionally, other system properties are required like, for instance, a minimum decay rate or a minimum damping ratio, these may be reflected by taking $\mathbb{C}_g$ to be a suitable subset of $\mathbb{C}_-$ or $\mathbb{D}$.

In applications it often suffices to know that a $\mathbb{C}_g$-stable system is able to tolerate perturbations below a given size without loosing the property of $\mathbb{C}_g$-stability. Then the detailed pictures given by the spectral value sets are not necessary and it is enough to verify that the (more easily computable) stability radius is larger than the expected level of the perturbations. The stability radius is a *worst case* measure of robustness. It measures the size of the smallest perturbation for which the perturbed system is either not well-posed or does not have spectrum in $\mathbb{C}_g$. If the engineer is not confident that parameter uncertainties will be below this value, it may be possible to increase the radius by feedback control. In Volume II we will address the corresponding *synthesis* problem. In this section we concentrate on the *analysis* of stability radii.

In Section 5.1 we defined the stability radius as a measure of robustness of $\mathbb{C}_g$-stability for very general classes of perturbations. In this section we specialize to linear fractional perturbations $A \rightsquigarrow A(\Delta)$ and obtain more concrete results. The section is divided into seven subsections, starting from the general situation and gradually moving to more specialized ones. In the first, general characterizations are given for linear fractional perturbations in terms of the μ-function. As an application we show that these characterizations lead to computable formulas for the special case of diagonal matrices subject to off-diagonal perturbations. These results yield a substantial refinement of Gershgorin's Theorem (see Section 4.2).

In the next two subsections we consider full-block perturbations and obtain computable formulas for both the complex and the real stability radius. In the fourth subsection we specialize even further and consider the *complex* stability radius for full-block perturbation structures with respect to the spectral norm. We show that in this case the radius can be characterized via Riccati equations and Hamiltonian matrices and give details of an algorithm for computing the stability radius which is quadratically convergent.

Formulae for stability radii with respect to unstructured perturbations are determined in Subsection 5.3.5. We obtain bounds for the real radius and explore when they are tight. We also examine the effect on the radii of similarity transformations of A and show that under similarity the stability radii come arbitrarily close to zero, indicating the need for a carefully chosen coordinate system if the stability radius is to be a useful measure of robustness.

In Subsection 5.3.6 we examine whether or not the stability radii depend continuously on the data. Finally in the last subsection we describe the effect on stability radii of Cayley transformations of the data, which, in particular, allows us to obtain discrete time results from continuous time ones and vice versa.

5.3.1 General Definitions and Results

In this subsection we consider general linear fractional matrix perturbations where the set $\mathbf{\Delta}$ of parameter deviations is only required to satisfy the minimal conditions specified at the beginning of the previous section. The main objective will be to characterize the stability radius under these general conditions. The results are easy consequences of their counterparts for spectral value sets in Subsection 5.2.1. They require the maximization of the μ-value of the associated transfer matrix on the boundary of the stability region. Whilst these maximization problems are hard to solve in general, we will see that they yield computable formulas for the stability radius with respect to special perturbation structures. As an example we will determine the stability radius of arbitrary diagonal matrices with respect to Gershgorin type perturbations in both the real and the complex case.
Throughout this subsection we suppose that the following assumptions are satisfied.

- $\mathbb{C}_g \neq \emptyset$ is an open subset of $\mathbb{C}$ and $\mathbb{C}_b := \mathbb{C} \setminus \mathbb{C}_g \neq \emptyset$.
- $(A, B, C, D) \in \mathbf{L}_{n,\ell,q}(\mathbb{K})$ is $\mathbb{C}_g$-stable, i.e. $\sigma(A) \subset \mathbb{C}_g$.
- $\mathbf{\Delta} \subset \mathbb{K}^{\ell \times q}$ is a closed convex cone and $\operatorname{span}_{\mathbb{K}} \mathbf{\Delta}$ is provided with a norm $\|\cdot\|_{\mathbf{\Delta}}$ which is an operator norm with respect to a given pair of norms on $\mathbb{K}^\ell$ and $\mathbb{K}^q$. If $\mathbb{K} = \mathbb{R}$, the complex spaces $\mathbb{C}^\ell$, $\mathbb{C}^q$ are provided with a compatible pair of norms and $\mathbb{C}^{\ell \times q}$, $\mathbb{C}^{q \times \ell}$ with the corresponding operator norms, as explained in Remark 4.4.2.
- The perturbations $A \rightsquigarrow A(\Delta)$ are given by

$$A(\Delta) = A + B(I_\ell - \Delta D)^{-1}\Delta C, \quad \Delta \in \mathbf{\Delta}_0 := \{\Delta \in \mathbf{\Delta};\, \det(I_\ell - \Delta D) \neq 0\}. \quad (1)$$

For these linear fractional perturbations Definition 5.1.11 takes the following form.

Definition 5.3.1. The stability radius of A with respect to perturbations of the form (1) and the stability region $\mathbb{C}_g$ is defined by

$$r_{\mathbf{\Delta}}(A; B, C, D; \mathbb{C}_g) = \inf\{\|\Delta\|_{\mathbf{\Delta}}; \Delta \in \mathbf{\Delta},\, \det(I_\ell - \Delta D) = 0 \text{ or } \sigma(A(\Delta)) \not\subset \mathbb{C}_g\} \quad (2)$$

where as usual $\inf \emptyset := \infty$.[1]

Here in contrast with Section 5.2 (where we usually assumed $\delta < \boldsymbol{\delta}_0$) the well-posedness question comes into play. Since

$$\boldsymbol{\delta}_0 = \inf\{\|\Delta\|_{\mathbf{\Delta}}; \Delta \in \mathbf{\Delta},\, \det(I_\ell - \Delta D) = 0\} = \mu_{\mathbf{\Delta}}(D)^{-1}$$

we always have

$$r_{\mathbf{\Delta}}(A; B, C, D; \mathbb{C}_g) \leq \mu_{\mathbf{\Delta}}(D)^{-1}. \quad (3)$$

[1] So $r_{\mathbf{\Delta}} = \infty$ if and only if $\det(I_\ell - \Delta D) \neq 0$ and $\sigma(A(\Delta)) \subset \mathbb{C}_g$ for all $\Delta \in \mathbf{\Delta}$.

In particular, if $r_{\mathbf{\Delta}} := r_{\mathbf{\Delta}}(A; B, C, D; \mathbb{C}_g) \geq \boldsymbol{\delta}_0$, then $r_{\mathbf{\Delta}} = \boldsymbol{\delta}_0$ and the stability radius is a measure of how robust the well-posedness of the system is to perturbations lying in the class $\mathbf{\Delta}$. On the other hand if $r_{\mathbf{\Delta}} < \boldsymbol{\delta}_0$ the stability radius is a measure of how robust the $\mathbb{C}_g$–stability of A is to perturbations of the form (1). In this case it can be expressed in terms of spectral value sets as follows:

$$r_{\mathbf{\Delta}}(A; B, C, D; \mathbb{C}_g) = \inf\{\delta \in (0, \infty);\, \sigma_{\mathbf{\Delta}}(A; B, C, D; \delta) \not\subset \mathbb{C}_g\} \quad \text{if } r_{\mathbf{\Delta}} < \boldsymbol{\delta}_0. \tag{4}$$

This characterization is illustrated in Example 5.2.21 and Figure 5.2.5 where $\mathbf{\Delta} = \mathbb{C}^{1\times 6}$ and $D = 0$, so $\boldsymbol{\delta}_0 = \infty$. If $\mathbb{C}_g = \mathbb{C}_-$ we see that at a value of about $\delta = 8.77$ the spectral value sets first touch the imaginary axis and so $r_{\mathbf{\Delta}} \approx 8.77$.

Remark 5.3.2. Using Lemma 5.1.13 which has been proved for even more general perturbation classes, we have the following elementary properties of the stability radius.

(i) $r_{\mathbf{\Delta}} > 0$ and if $r_{\mathbf{\Delta}} < \boldsymbol{\delta}_0$, then there exists a minimum norm destabilizing perturbation, i.e. a perturbation $\Delta_{\min} \in \mathbf{\Delta}_0$ such that $\sigma(A(\Delta_{\min})) \not\subset \mathbb{C}_g$ and $\|\Delta_{\min}\|_{\mathbf{\Delta}} = r_{\mathbf{\Delta}}$. Note that by continuity of the spectrum $\sigma(A(\Delta_{\min})) \subset \overline{\mathbb{C}_g}$ and so $\sigma(A(\Delta_{\min})) \cap \partial\mathbb{C}_g \neq \emptyset$.

(ii) If $r_{\mathbf{\Delta}} \geq \boldsymbol{\delta}_0$, then either $r_{\mathbf{\Delta}} = \boldsymbol{\delta}_0 = \infty$ or there exists $\Delta \in \mathbf{\Delta}$ of norm $\|\Delta\|_{\mathbf{\Delta}} = \boldsymbol{\delta}_0 = r_{\mathbf{\Delta}}$ such that $\det(I_\ell - \Delta D) = 0$.

(iii) If $\mathbf{\Delta} \subset \tilde{\mathbf{\Delta}}$ and $\|\Delta\|_{\mathbf{\Delta}} \geq \|\Delta\|_{\tilde{\mathbf{\Delta}}}$ for all $\Delta \in \mathbf{\Delta}$ then $r_{\mathbf{\Delta}} \geq r_{\tilde{\mathbf{\Delta}}}$.

(iv) The stability radius remains the same if we replace $\mathbb{C}_g$ by the complement in $\mathbb{C}$ of its boundary $\partial\mathbb{C}_g$,

$$r_{\mathbf{\Delta}}(A; B, C, D; \mathbb{C}_g) = r_{\mathbf{\Delta}}(A; B, C, D; \mathbb{C} \setminus \partial\mathbb{C}_g). \tag{5}$$

□

We are now in a position to prove a general characterization of the stability radius under the above assumptions. As usual we set $\alpha/\beta := \infty$ if $\alpha > 0$, $\beta = 0$, and, in particular, $0^{-1} := \infty$.

Theorem 5.3.3. *The stability radius $r_{\mathbf{\Delta}} := r_{\mathbf{\Delta}}(A; B, C, D; \mathbb{C}_g)$ of A is given by*

$$r_{\mathbf{\Delta}} = \left[\max\left\{\mu_{\mathbf{\Delta}}(D),\, \sup_{s\in\partial\mathbb{C}_g} \mu_{\mathbf{\Delta}}(G(s))\right\}\right]^{-1} = \min\{\mu_{\mathbf{\Delta}}(D)^{-1},\, \inf_{s\in\partial\mathbb{C}_g} \mu_{\mathbf{\Delta}}(G(s))^{-1}\}, \tag{6}$$

where $G(s) = D + C(sI_n - A)^{-1}B$ is the transfer function associated with (A, B, C, D).

Proof: Recall that $r_{\mathbf{\Delta}} \leq \mu_{\mathbf{\Delta}}(D)^{-1}$, and if $s_0 \in \rho(A)$ is such that $\mu_{\mathbf{\Delta}}(D) < \mu_{\mathbf{\Delta}}(G(s_0))$, then by Corollary 5.2.8

$$\min\{\|\Delta\|_{\mathbf{\Delta}};\, \Delta \in \mathbf{\Delta},\, s_0 \in \sigma(A(\Delta))\} = \mu_{\mathbf{\Delta}}(G(s_0))^{-1}. \tag{7}$$

Hence $r_{\mathbf{\Delta}} \leq \inf_{s\in\mathbb{C}_b} \mu_{\mathbf{\Delta}}(G(s))^{-1} \leq \inf_{s\in\partial\mathbb{C}_g} \mu_{\mathbf{\Delta}}(G(s))^{-1}$ and so $\leq$ holds in (6). If $\mu_{\mathbf{\Delta}}(D)^{-1} \leq r_{\mathbf{\Delta}} < \infty$ we have seen above that $r_{\mathbf{\Delta}} = \mu_{\mathbf{\Delta}}(D)^{-1}$ and so equality holds in (6) in this case.
Now suppose $r_{\mathbf{\Delta}} < \mu_{\mathbf{\Delta}}(D)^{-1}$. Then $r_{\mathbf{\Delta}} < \infty$ and by Remark 5.3.2 (i) there exists $\Delta \in \mathbf{\Delta}$ of norm $\|\Delta\|_{\mathbf{\Delta}} = r_{\mathbf{\Delta}}$ and $s_0 \in \partial\mathbb{C}_g$ such that $s_0 \in \sigma(A(\Delta))$. By (7) it follows that $\mu_{\mathbf{\Delta}}(D)^{-1} > r_{\mathbf{\Delta}} = \|\Delta\|_{\mathbf{\Delta}} \geq \mu_{\mathbf{\Delta}}(G(s_0))^{-1}$, and this proves equality in (6). □

By (6) the stability radius $r_{\mathbf{\Delta}}$ is infinite if and only if $\|D\|_{\mathbf{\Delta}} = \mu_{\mathbf{\Delta}}(G(s)) = 0$ for all $s \in \partial\mathbb{C}_g$. Even in the case where $\partial\mathbb{C}_g$ is unbounded one cannot automatically dispense with the first term on the RHS of (6) without extra consideration (although $\lim_{|s|\to\infty} G(s) = D$). This is because of possible discontinuities of $\mu_{\mathbf{\Delta}}(\cdot)$, see Remark 5.3.17 (i) and Example 5.3.18.

Remark 5.3.4. The previous proof shows that

$$r_{\mathbf{\Delta}} = \min\left\{\mu_{\mathbf{\Delta}}(D)^{-1},\ \inf_{s\in\mathbb{C}_b} \mu_{\mathbf{\Delta}}(G(s))^{-1}\right\} = \left[\max\left\{\mu_{\mathbf{\Delta}}(D),\ \sup_{s\in\mathbb{C}_b} \mu_{\mathbf{\Delta}}(G(s))\right\}\right]^{-1}. \tag{8}$$

Theorem 5.3.3 implies that the function $s \mapsto \mu(s) = \max\{\mu_{\mathbf{\Delta}}(D), \mu_{\mathbf{\Delta}}(G(s))\}$ satisfies a maximum principle on $\rho(A)$ in the sense that, for every open subset $\Omega \neq \emptyset$ of $\mathbb{C}$ with $\overline{\Omega} \subset \rho(A)$ we have

$$\sup_{s\in\overline{\Omega}} \mu(s) = \sup_{s\in\partial\Omega} \mu(s). \tag{9}$$

In fact, setting $\mathbb{C}_g := \mathbb{C}\setminus\overline{\Omega}$, hence $\mathbb{C}_b = \overline{\Omega}$, $\partial\mathbb{C}_g = \partial\Omega$, the matrix A is $\mathbb{C}_g$-stable and (9) follows from (6) and (8). □

The concept of $\mathbb{C}_g$–stability allows us to develop a unified framework for studying robust stability of continuous and discrete time systems, for which we have the classical stability regions,

$$\mathbb{C}_g = \mathbb{C}_- := \{s \in \mathbb{C};\ \operatorname{Re} s < 0\} \quad \text{and} \quad \mathbb{C}_g = \mathbb{D} := \{s \in \mathbb{C};\ |s| < 1\}.$$

We denote the respective stability radii by $r^-_{\mathbf{\Delta}}(A;B,C,D)$ and $r^1_{\mathbf{\Delta}}(A;B,C,D)$.

Example 5.3.5. The perturbed linear oscillator $(1-\Delta)\ddot{\xi} + 2(1-\Delta)\dot{\xi} + (2-3\Delta)\xi = 0$, $\Delta \in \mathbb{R}$ can be written in state space form $\dot{x} = A(\Delta)x$ with $A(\Delta)$ given by (1) and

$$A = \begin{bmatrix} 0 & 1 \\ -2 & -2 \end{bmatrix}, \quad B = \begin{bmatrix} 0 \\ 1 \end{bmatrix}, \quad C = [1\ \ 0], \quad D = 1, \quad \mathbf{\Delta} = \mathbb{R}.$$

Then $\boldsymbol{\delta}_0 = \mu_{\mathbb{R}}(D)^{-1} = 1$ and $G(s) = 1 + (s^2+2s+2)^{-1}$. Choose $\mathbb{C}_g = \mathbb{C}_-$, then $\partial\mathbb{C}_g = \imath\mathbb{R}$. But if $\omega \neq 0$ then $G(\imath\omega) \in \mathbb{C}\setminus\mathbb{R}$ and so $\mu_{\mathbb{R}}(G(\imath\omega)) = 0$. Hence $\sup_{\omega\in\mathbb{R}} \mu_{\mathbb{R}}(G(\imath\omega)) = \mu_{\mathbb{R}}(G(0)) = 3/2$ and it follows from (6) that $r^-_{\mathbb{R}}(A;B,C,D) = \min\{1, 2/3\} = 2/3$. The same result is obtained by considering the real spectral value sets $\sigma_{\mathbb{R}}(A;B,C,D;\delta)$, $\delta > 0$ and applying the characterization (4). In fact, one easily verifies that $\sigma_{\mathbb{R}}(A;B,C,D;\delta) \not\subset \mathbb{C}_-$ if and only if $\delta > 2/3$. □

Multi-Block Perturbations

As in the previous section it is possible to use the estimates for the μ-function derived in Section 4.4 to obtain estimates for stability radii with respect to various types of block diagonal perturbations. Here we only consider multi-block perturbations ($J = \underline{N}$) and for these we only present the estimate which is most readily computed. We will see in the next subsection that the stability radius $r_{\mathbf{\Delta}}(A;B,C,D;\mathbb{C}_g)$ for *complex full-block* perturbations can be characterized in terms of the norm of the transfer function and this leads to a computable formula. Using the scaling method as in Section 5.2 we obtain an upper estimate for $r_{\mathbf{\Delta}}(A;B,C,D;\mathbb{C}_g)$ in terms of

such computable stability radii.
We adopt the same notation and assumptions as in the corresponding part of Section 5.2. The perturbations are of the form (2.10):

$$A \rightsquigarrow A(\Delta) = A + \sum_{i=1}^{N} B_i (I_{\ell_i} - \Delta_i D_i)^{-1} \Delta_i C_i, \quad \Delta_i \in \mathbb{C}^{\ell_i \times q_i}.$$

The corresponding complex stability radius is denoted by $r_{\mathbb{C}}(A; (B_i, C_i, D_i)_{i \in \underline{N}}; \mathbb{C}_g)$. As a consequence of (2.14) and (6), we have

$$r_{\mathbb{C}}(A; (B_i, C_i, D_i)_{i \in \underline{N}}; \mathbb{C}_g) \geq \sup_{\gamma \in (0,\infty)^N} r_{\mathbb{C}}(A; B(\gamma), C(\gamma), D; \mathbb{C}_g) \tag{10}$$

where $r_{\mathbb{C}}(A; B(\gamma), C(\gamma), D; \mathbb{C}_g)$ denotes the stability radius with respect to complex full-block perturbations of the form $A \rightsquigarrow A(\Delta) = A + B(\gamma)(I_\ell - \Delta D)^{-1} \Delta C(\gamma)$, $\Delta \in \mathbb{C}^{\ell \times q}$ with

$$C(\gamma) = \left[\gamma_1 C_1^\top, \cdots, \gamma_N C_N^\top\right]^\top, \; B(\gamma) = \left[\gamma_1^{-1} B_1, \cdots, \gamma_N^{-1} B_N\right], \; D = \operatorname{diag}(D_1, ..., D_N).$$

As an extended example where the above estimate is tight we will now consider off-diagonal perturbations of a diagonal matrix.

Gershgorin Type Uncertainty

In contrast with Subsection 5.2.1 we study both real and complex Gershgorin type perturbations and derive explicit formulas for the corresponding stability radii with respect to two perturbation norms, one of which was already considered in Subsection 5.2.1 whilst the other was studied in Ex. 5.2.3. The stability radii will be determined for both the classical stability regions, $\mathbb{C}_g = \mathbb{C}_-$ and $\mathbb{C}_g = \mathbb{D}$.
The assumptions are

- $A = \operatorname{diag}(a_{11}, ..., a_{nn})$ where $a_{ii} \in \mathbb{C}_g$ for all $i \in \underline{n}$.
- $\mathbf{\Delta}_{\mathbb{K}} = \{\Delta \in \mathbb{K}^{n \times n};\ \operatorname{diag}(\Delta) = 0_{n \times n}\}$, with perturbation norms (see Section A.1)

$$\|\Delta\|_{\mathbf{\Delta}_{\mathbb{K}}} = \|\Delta\|_{\infty,\infty} = \max_{i \in \underline{n}} \sum_{j \neq i} |\delta_{ij}| = \|\Delta\|_{1|\infty}, \tag{11}$$

$$\|\Delta\|_{\mathbf{\Delta}_{\mathbb{K}}} = \|\Delta\|_{\infty|\infty} = \max_{i,j \in \underline{n},\, i \neq j} |\delta_{ij}|. \tag{12}$$

- The perturbations are given by $A \rightsquigarrow A(\Delta) = A + \Delta, \quad \Delta \in \mathbf{\Delta}_{\mathbb{K}}$.

In Section 5.2 we showed how the spectral value set problem for complex off-diagonal perturbations can be transformed into a spectral value set problem with multi-block perturbation structure and this enabled us to characterize the sets. We will now employ the same method in order to determine the stability radii

$$r_{\mathbf{\Delta}_{\mathbb{K}}}(A; \mathbb{C}_g) = \inf\{\|\Delta\|_{\mathbf{\Delta}_{\mathbb{K}}}; \Delta \in \mathbf{\Delta}_{\mathbb{K}},\ \sigma(A + \Delta) \not\subset \mathbb{C}_g\} \tag{13}$$

for both the real and the complex case. We begin with the stability region $\mathbb{C}_g = \mathbb{C}_-$.

Proposition 5.3.6. *Suppose* $A = \operatorname{diag}(a_{11}, ..., a_{nn})$ *with* $a_{ii} \in \mathbb{C}_-$ *for* $i \in \underline{n}$, *then*

(i) for the perturbation norm (11) $(\|\cdot\|_{(\infty,\infty)} = \|\cdot\|_{(1|\infty)})$, *we have*

$$r_{\mathbf{\Delta}_\mathbb{C}}(A;\mathbb{C}_-) = \min_{\omega\in\mathbb{R}} \min_{i,j\in\underline{n},\, i\neq j} (|\imath\omega - a_{ii}|\,|\imath\omega - a_{jj}|)^{1/2} \geq \min_{i,j\in\underline{n},\, i\neq j} (\operatorname{Re} a_{ii} \operatorname{Re} a_{jj})^{1/2},$$

and if the a_{ii}, $i\in\underline{n}$ *are all real,*

$$r_{\mathbf{\Delta}_\mathbb{R}}(A;\mathbb{C}_-) = r_{\mathbf{\Delta}_\mathbb{C}}(A;\mathbb{C}_-) = \min_{i,j\in\underline{n},\, i\neq j} (a_{ii}a_{jj})^{1/2}. \tag{14}$$

(ii) For the perturbation norm (12) $(\|\cdot\|_{(\infty|\infty)})$, *let*

$$P(s) = \begin{bmatrix} 0 & |s-a_{22}|^{-1} & |s-a_{33}|^{-1} & \cdots & |s-a_{nn}|^{-1} \\ |s-a_{11}|^{-1} & 0 & |s-a_{33}|^{-1} & \cdots & |s-a_{nn}|^{-1} \\ \cdot & \cdot & \cdot & \cdots & \cdot \\ \cdot & \cdot & \cdot & \cdots & \cdot \\ |s-a_{11}|^{-1} & |s-a_{22}|^{-1} & |s-a_{33}|^{-1} & \cdots & 0 \end{bmatrix}. \tag{15}$$

Then

$$r_{\mathbf{\Delta}_\mathbb{C}}(A;\mathbb{C}_-) = \min_{\omega\in\mathbb{R}} \varrho(P(\imath\omega))^{-1} \geq r_{\mathbf{\Delta}_\mathbb{R}}(\operatorname{Re} A;\mathbb{C}_-) \tag{16}$$

where ϱ *denotes the spectral radius. If the* a_{ii} *are real for* $i\in\underline{n}$,

$$r_{\mathbf{\Delta}_\mathbb{R}}(A;\mathbb{C}_-) = r_{\mathbf{\Delta}_\mathbb{C}}(A;\mathbb{C}_-) = \varrho(P(0))^{-1}. \tag{17}$$

Proof: We only prove (ii) and for this we make use of Ex. 5.2.3. The proof of (i) follows the same line making use of Proposition 5.2.11 instead. First we consider the complex case. It follows from the characterization of the spectral value sets in Ex. 5.2.3 that $\imath\omega \in \sigma_{\mathbf{\Delta}_\mathbb{C}}(A;\delta)$ if and only if $\varrho(P(\imath\omega)) > \delta^{-1}$. By (4) this implies the equality in (16).[2] Now

$$|\imath\omega - a_{ii}|^{-1} = ((\omega - \operatorname{Im} a_{ii})^2 + (\operatorname{Re} a_{ii})^2)^{-1/2} \quad \text{hence} \quad \max_{\omega\in\mathbb{R}} |\imath\omega - a_{ii}|^{-1} = |\operatorname{Re} a_{ii}|^{-1}$$

and the maximum occurs at $\omega = \operatorname{Im} a_{ii}$ for $i\in\underline{n}$. Let P_- denote the nonnegative matrix obtained from $P(\imath\omega)$ by replacing the entries $|\imath\omega - a_{ii}|^{-1}$ by $|\operatorname{Re} a_{ii}|^{-1}$. Then $0 \leq P(\imath\omega) \leq P_-$ for all $\omega\in\mathbb{R}$ where we write $(a_{ij}) \leq (b_{ij})$ if $a_{ij} \leq b_{ij}$ for all i,j. It follows that $\varrho(P(\imath\omega)) \leq \varrho(P_-)$ and so $r_{\mathbf{\Delta}_\mathbb{C}} \geq \varrho(P_-)^{-1}$.[3] Note that P_- coincides with $P(0)$ if we replace A by $\operatorname{Re} A$ in (15). Hence the inequality in (16) follows once we have proved (17).

Now assume that A is real. Then P_- coincides with $P(0)$ and so $r_{\mathbf{\Delta}_\mathbb{C}}(A;\mathbb{C}_-) = \varrho(P(0))^{-1}$ by the equality in (16). Since $r_{\mathbf{\Delta}_\mathbb{R}} \geq r_{\mathbf{\Delta}_\mathbb{C}}$ it suffices to construct a destabilizing $\Delta \in \mathbf{\Delta}_\mathbb{R}$ of norm $\varrho(P(0))^{-1}$ to complete the proof of (17). For this we apply Lemma 5.2.7 with $B = C = I_n$, $D = 0_{n\times n}$ whence $G(s) = (sI_n - A)^{-1} = \operatorname{diag}((s-a_{11})^{-1},\ldots,(s-a_{nn})^{-1})$. Define $\Delta^- = (\Delta^-_{ij}) \in \mathbf{\Delta}_\mathbb{R}$ by $\Delta^-_{ij} = \varrho(P(0))^{-1}$ for $i,j\in\underline{n}$, $i\neq j$. Then $\|\Delta^-\|_{\infty|\infty} = \varrho(P(0))^{-1}$ and since $|a_{ii}| = -a_{ii}$ we have $\Delta^- \operatorname{diag}(-a_{11}^{-1},\ldots,-a_{nn}^{-1})) = \varrho(P(0))^{-1}P(0)$ and so

$$\det(I_n - \Delta^- G(0)) = \det(I_n - \varrho(P(0))^{-1}P(0)) = 0.$$

By Lemma 5.2.7 this shows $0 \in \sigma(A+\Delta^-)$ and so Δ^- is a destabilizing real perturbation of norm $\varrho(P(0))^{-1}$. This completes the proof. □

[2] The minimum exists because $\lim_{|\omega|\to\infty} \varrho(P(\imath\omega))^{-1} = \infty$.

[3] It is known from the theory of nonnegative matrices that $0 \leq X \leq Y$ implies $\varrho(X) \leq \varrho(Y)$.

Example 5.3.7. We consider the same uncertain system as that in Example 5.2.13, namely $A = \operatorname{diag}(-1,-2,-3)$. From (14) we obtain that, with respect to the $(\infty|1)$-perturbation norm, $r_{\mathbf{\Delta}_{\mathbb{R}}}(A;\mathbb{C}_-) = r_{\mathbf{\Delta}_{\mathbb{C}}}(A;\mathbb{C}_-) = \min\{\sqrt{2},\sqrt{3},\sqrt{6}\} = \sqrt{2}$. On the other hand

$$P(0) = \begin{bmatrix} 0 & 1/2 & 1/3 \\ 1 & 0 & 1/3 \\ 1 & 1/2 & 0 \end{bmatrix}.$$

$P(0)$ has only real eigenvalues of which the one with maximum absolute value ≈ 1.14 and so with respect to the $(\infty|\infty)$-perturbation norm $r_{\mathbf{\Delta}_{\mathbb{R}}}(A;\mathbb{C}_-) = r_{\mathbf{\Delta}_{\mathbb{C}}}(A;\mathbb{C}_-) \approx 0.88$. □

The reader is asked to prove the following discrete time counterpart of Proposition 5.3.6 (i) in Ex. 1.

Proposition 5.3.8. *Suppose $A = \operatorname{diag}(a_{11},\ldots,a_{nn})$ with $|a_{ii}| < 1$ for $i \in \underline{n}$, then for the perturbation norm* (11), *we have*

$$r_{\mathbf{\Delta}_{\mathbb{C}}}(A;\mathbb{D}) = \min_{\theta\in[0,2\pi]} \min_{i,j\in\underline{n},\ i\neq j} (|e^{\imath\theta}-a_{ii}|\,|e^{\imath\theta}-a_{jj}|)^{1/2} \geq \min_{i,j\in\underline{n},\ i\neq j} [(1-|a_{ii}|)(1-|a_{jj}|)]^{1/2}.$$

If the a_{ii}, $i \in \underline{n}$ are all real,

$$r_{\mathbf{\Delta}_{\mathbb{R}}}(A;\mathbb{D}) = r_{\mathbf{\Delta}_{\mathbb{C}}}(A;\mathbb{D}) = \min_{i,j\in\underline{n},\ i\neq j} \min\{[(1-a_{ii})(1-a_{jj})]^{1/2}, [(1+a_{ii})(1+a_{jj})]^{1/2}\}. \quad (18)$$

5.3.2 Complex Full-Block Perturbations

We consider complex full-block perturbations of a complex system and derive a computable formula for the corresponding stability radius with respect to arbitrary operator norms on $\mathbf{\Delta} = \mathbb{C}^{\ell\times q}$. By means of an example we discuss the significance of complex perturbations of a real system.
The assumptions are

- $\mathbb{C}_g \neq \emptyset$ is an open subset of $\mathbb{C}$ and $\mathbb{C}_b := \mathbb{C}\setminus\mathbb{C}_g \neq \emptyset$.
- $(A,B,C,D) \in \mathbf{L}_{n,\ell,q}(\mathbb{C})$ with $\sigma(A) \subset \mathbb{C}_g$.
- $\mathbf{\Delta} = \mathbb{C}^{\ell\times q}$ is provided with a norm $\|\cdot\|$ which is an operator norm with respect to a given pair of norms on $\mathbb{C}^\ell$ and $\mathbb{C}^q$.
- The perturbations are of the form

$$A \rightsquigarrow A(\Delta) = A + B(I_\ell - \Delta D)^{-1}\Delta C, \quad \Delta \in \mathbb{C}^{\ell\times q}. \quad (19)$$

We denote the corresponding stability radius by $r_{\mathbb{C}}(A;B,C,D;\mathbb{C}_g)$ (if $D = 0$ by $r_{\mathbb{C}}(A;B,C;\mathbb{C}_g)$) and refer to it as the *complex stability radius* of A with respect to perturbations of the form (19). For the stability regions $\mathbb{C}_-$ and $\mathbb{D}$, the complex stability radii are denoted by $r_{\mathbb{C}}^-(A;B,C,D)$ and $r_{\mathbb{C}}^1(A;B,C,D)$. We have seen in Section 4.4 that for complex full-block perturbations $\mu_{\mathbb{C}}(G) := \mu_{\mathbb{C}^{\ell\times q}}(G) = \|G\|_{\mathcal{L}(\mathbb{C}^\ell,\mathbb{C}^q)}$ for all $G \in \mathbb{C}^{q\times\ell}$. So Proposition 5.3.3 directly yields the following characterization.

Theorem 5.3.9. *The complex stability radius of A with respect to perturbations of the form* (19) *is given by*

$$r_{\mathbb{C}}(A;B,C,D;\mathbb{C}_g) = \left[\max\left\{\sup_{s\in\partial\mathbb{C}_g} \|G(s)\|_{\mathcal{L}(\mathbb{C}^\ell,\mathbb{C}^q)}, \|D\|_{\mathcal{L}(\mathbb{C}^\ell,\mathbb{C}^q)}\right\}\right]^{-1}. \quad (20)$$

Remark 5.3.10. If $\partial\mathbb{C}_g$ is unbounded then $\lim_{|s|\to\infty}\|G(s)\|_{\mathcal{L}(\mathbb{C}^\ell,\mathbb{C}^q)} = \|D\|_{\mathcal{L}(\mathbb{C}^\ell,\mathbb{C}^q)}$ implies $\sup_{s\in\partial\mathbb{C}_g}\|G(s)\|_{\mathcal{L}(\mathbb{C}^\ell,\mathbb{C}^q)} \geq \|D\|_{\mathcal{L}(\mathbb{C}^\ell,\mathbb{C}^q)}$, whence

$$r_{\mathbb{C}}(A;B,C,D;\mathbb{C}_g) = \inf_{s\in\partial\mathbb{C}_g}\|G(s)\|^{-1}_{\mathcal{L}(\mathbb{C}^\ell,\mathbb{C}^q)}. \tag{21}$$

Now suppose that $\Omega := \mathbb{C}\setminus\overline{\mathbb{C}_g}$ is non-empty and unbounded, i.e. $\mathbb{C}_b$ has an unbounded interior. This holds e.g. for the classical stability regions $\mathbb{C}_g = \mathbb{C}_-$ and $\mathbb{C}_g = \mathbb{D}$. Then $G(\cdot)$ is holomorphic on Ω, bounded and continuous on the closure $\overline{\Omega}\subset\mathbb{C}_b$. $\|G(\cdot)\|$ is subharmonic (see Section A.2) on Ω and continuous on $\overline{\Omega}$. It follows from an extended version of the maximum principle (see Theorem A.2.27 and the comments which surround it) that

$$\sup_{s\in\Omega}\|G(s)\|_{\mathcal{L}(\mathbb{C}^\ell,\mathbb{C}^q)} = \sup_{s\in\partial\Omega}\|G(s)\|_{\mathcal{L}(\mathbb{C}^\ell,\mathbb{C}^q)}$$

and so

$$\|D\|_{\mathcal{L}(\mathbb{C}^\ell,\mathbb{C}^q)} = \lim_{\substack{|s|\to\infty\\ s\in\Omega}}\|G(s)\|_{\mathcal{L}(\mathbb{C}^\ell,\mathbb{C}^q)} \leq \sup_{s\in\partial\Omega}\|G(s)\|_{\mathcal{L}(\mathbb{C}^\ell,\mathbb{C}^q)} \leq \sup_{s\in\partial\mathbb{C}_g}\|G(s)\|_{\mathcal{L}(\mathbb{C}^\ell,\mathbb{C}^q)}$$

since $\partial\Omega\subset\partial\mathbb{C}_g$. As a consequence formula (20) can be simplified to (21) in this case as well. In particular, we have for $r_{\mathbb{C}}^- = r_{\mathbb{C}}^-(A;B,C,D)$ and $r_{\mathbb{C}}^1 = r_{\mathbb{C}}^1(A;B,C,D)$,

$$r_{\mathbb{C}}^- = \inf_{\omega\in\mathbb{R}}\|G(\imath\omega)\|^{-1}_{\mathcal{L}(\mathbb{C}^\ell,\mathbb{C}^q)} = \|G\|^{-1}_{H^\infty(\mathbb{C}_+;\mathbb{C}^{q\times\ell})} \tag{22}$$

$$r_{\mathbb{C}}^1 = \min_{\theta\in[0,2\pi]}\|G(e^{\imath\theta})\|^{-1}_{\mathcal{L}(\mathbb{C}^\ell,\mathbb{C}^q)} = \|G\|^{-1}_{H^\infty(\mathbb{D}_+;\mathbb{C}^{q\times\ell})}. \tag{23}$$

□

Although a formula like (22) seems at first sight to be quite satisfactory, the com-

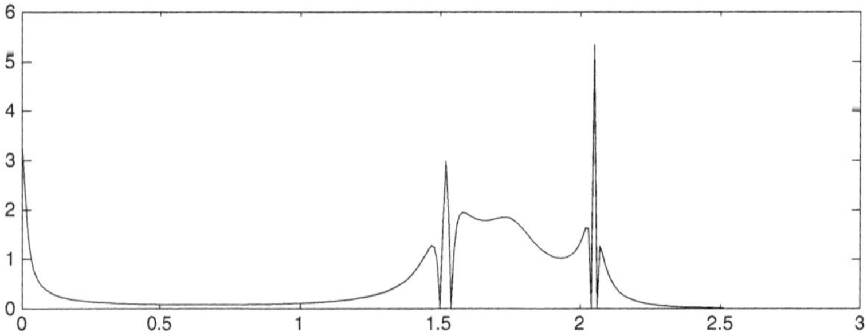

Figure 5.3.1: A computed plot $\omega\to\|G(\imath\omega)\|$

putation of the RHS requires the *global maximization* of the real valued function $\omega\longrightarrow\|G(\imath\omega)\|$ on $\mathbb{R}$, which may have many local minima, spikes etc., see Figure 5.3.1. Solving this global optimization problem is not a trivial numerical task. In the next subsection we will derive, for the special case of the spectral norm, other characterizations of $r_{\mathbb{C}}^-$ and show how they lead to an efficient algorithm for the computation of the complex stability radius.

Remark 5.3.11. Suppose there exists $s_0\in\partial\mathbb{C}_g$ such that

$$\sup_{s\in\partial\mathbb{C}_g}\|G(s)\|_{\mathcal{L}(\mathbb{C}^\ell,\mathbb{C}^q)} = \|G(s_0)\|_{\mathcal{L}(\mathbb{C}^\ell,\mathbb{C}^q)} > \|D\|_{\mathcal{L}(\mathbb{C}^\ell,\mathbb{C}^q)},$$

then a minimum norm destabilizing $\Delta_{\min}$ of rank 1 may be constructed using the formula in Proposition 4.4.11:

$$\Delta_{\min} = \|G(s_0)\|^{-1}_{\mathcal{L}(\mathbb{C}^\ell,\mathbb{C}^q)} u_0 v_0^*, \text{ where } v_0^* G(s_0) u_0 = \|G(s_0)\|_{\mathcal{L}(\mathbb{C}^\ell,\mathbb{C}^q)}, \ \|u_0\|_{\mathbb{C}^\ell} = \|v_0\|^*_{\mathbb{C}^q} = 1.$$

If A, B, C, D and $G(s_0)$ are real, the vectors u_0, v_0 can be chosen to be real so that the minimum norm complex destabilizing perturbation $\Delta_{\min}$ is real. In this case, the complex stability radius $r_{\mathbb{C}}(A; B, C, D; \mathbb{C}_g)$ coincides with the real stability radius $r_{\mathbb{R}}(A; B, C, D; \mathbb{C}_g) := r_{\mathbb{R}^{\ell\times q}}(A; B, C, D; \mathbb{C}_g)$ which will be studied in the next subsection. □

In the scalar case $\ell = q = 1$, if we set $G(\pm\infty) = D$, then $\omega \mapsto G(\imath\omega)$, $\omega \in \overline{\mathbb{R}}$ resp.

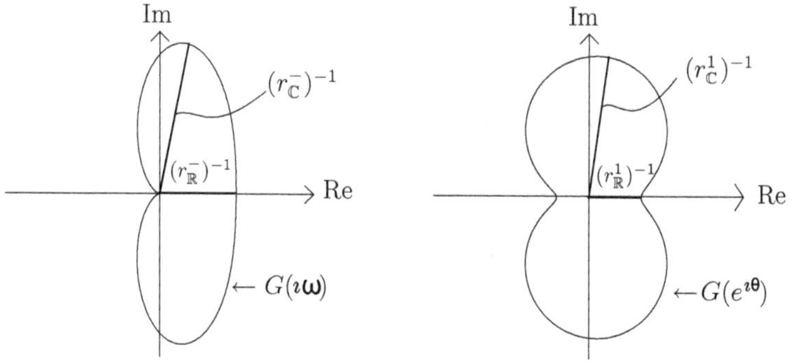

Figure 5.3.2: Complex and real stability radius of a continuous resp. discrete time system (with $D = 0$) illustrated by their Nyquist plots

$\theta \mapsto G(e^{\imath\theta})$, $\theta \in [0, 2\pi]$ describe closed curves Γ in the complex plane (the so-called Nyquist plot). The inverse of the complex stability radius is the maximum distance of a point on these curves from the origin, see Figure 5.3.2. We shall see later that if $r_{\mathbb{R}} < |D|^{-1}$ the real stability radius is the inverse of the maximum distance of a point in $\Gamma \cap \mathbb{R}$ from the origin, see Remark 5.3.17 (iv).

Example 5.3.12. Consider the perturbed oscillator $\ddot{\xi} + (4 + \Delta_1)\dot{\xi} + (5 + \Delta_2)\xi = 0$. By Example 5.2.29 the associated state space system and transfer function are given by

$$A = \begin{bmatrix} 0 & 1 \\ -5 & -4 \end{bmatrix}, \quad B = \begin{bmatrix} 0 \\ 1 \end{bmatrix}, \quad C = I_2, \quad D = 0, \quad G(s) = (s^2 + 4s + 5)^{-1} \begin{bmatrix} 1 \\ s \end{bmatrix}.$$

We choose $\mathbb{C}_g = \mathbb{C}_-$ and illustrate the effect of taking different perturbations norms by considering both the 2-norm and the ∞-norm on the perturbation space $\mathbf{\Delta} = \mathbb{C}^{1\times 2}$. The corresponding norms on $\mathbb{C}^2$ are the 2-norm and the 1-norm, respectively. Since

$$\|G(s)\|_2 = (1 + |s|^2)^{1/2} |s^2 + 4s + 5|^{-1}, \quad \|G(s)\|_1 = (1 + |s|)|s^2 + 4s + 5|^{-1},$$

we have

$$\|G(\imath\omega)\|_2^2 = [(5 - \omega^2)^2 + 16\omega^2]^{-1}(1 + \omega^2), \quad \|G(\imath\omega)\|_1^2 = [(5 - \omega^2)^2 + 16\omega^2]^{-1}(1 + |\omega|)^2.$$

For the 2-norm, a short calculation shows that the maximizing $\omega = 1.863$ and hence $r_{\mathbb{C}}^-(A; B, C) = 3.598$ (see Figure 5.2.6). For the 1-norm of $G(\imath\omega)$ a slightly more difficult

calculation yields the maximizing $\omega = 1.617$ and $r_{\mathbb{C}}^{-}(A; B, C) = 2.634$. Since $\|\Delta\|_2 \geq \|\Delta\|_\infty$ for $\Delta \in \mathbb{C}^{1\times 2}$, the stability radius with respect to the 2-norm must be $\geq$ that for the ∞-norm.
The complex oscillator $\ddot{\xi} + a_1\dot{\xi} + a_0\xi = 0$ will be stable if and only if the polynomial $p(s) = s^2 + a_1 s + a_0$ is Hurwitz and we have seen in Theorem 3.4.29 that this will be the case if and only if the Hermite matrix $\mathbf{H}_2(p) \succ 0$. By Example 3.4.25 we have

$$\mathbf{H}_2(p) = \begin{bmatrix} \overline{a_0}a_1 + \overline{a_1}a_0 & \overline{a_0} - a_0 \\ a_0 - \overline{a_0} & \overline{a_1} + a_1 \end{bmatrix}.$$

The above stability radii are the distances in $\mathbb{C}^{1\times 2}$ of the coefficient vector $(5, 4)$ from the set of coefficient vectors of complex non-Hurwitz polynomials, i.e. from the set $\{a \in \mathbb{C}^{1\times 2}; \mathbf{H}_2(p(\cdot, a)) \not\succ 0\}$ (with respect to the corresponding norms). It is possible to determine the stability radii in this way, but the calculations are quite difficult and more complicated than the ones we have presented above.
In contrast, the distance from the real non-Hurwitz polynomials is directly seen to be 4 since for real coefficients $p(s) = s^2 + a_1 s + a_0$ is Hurwitz if and only if $a_0, a_1 > 0$. □

As a consequence of the characterization (20), the *complex* stability radius has the following remarkable property. Recall that for any two matrices $A_i \in \mathbb{C}^{m_i \times n_i}$, $i = 1, 2$ the matrix $A_1 \oplus A_2$ denotes the block-diagonal matrix $\operatorname{diag}(A_1, A_2) \in \mathbb{C}^{(m_1+m_2)\times(n_1+n_2)}$.

Corollary 5.3.13. (Decomposition Property). *Let* $(A_i, B_i, C_i, D_i) \in \mathbf{L}_{n_i,\ell_i,q_i}(\mathbb{C})$, $\sigma(A_i) \subset \mathbb{C}_g$ *for* $i = 1, 2$ *and*

$$A = A_1 \oplus A_2, \quad B = B_1 \oplus B_2, \quad C = C_1 \oplus C_2, \quad D = D_1 \oplus D_2.$$

Suppose that $\|\cdot\|_{\mathbb{R}^2}$ *is an absolute norm on* $\mathbb{R}^2$ *and* $\mathbb{C}^{\ell_1+\ell_2} = \mathbb{C}^{\ell_1} \oplus \mathbb{C}^{\ell_2}$, $\mathbb{C}^{q_1+q_2} = \mathbb{C}^{q_1} \oplus \mathbb{C}^{q_2}$ *are provided with the norms*

$$\|(u_1, u_2)\|_{\mathbb{C}^{\ell_1+\ell_2}} = \|(\|u_1\|_{\mathbb{C}^{\ell_1}}, \|u_2\|_{\mathbb{C}^{\ell_2}})\|_{\mathbb{R}^2}, \quad \|(y_1, y_2)\|_{\mathbb{C}^{q_1+q_2}} = \|(\|y_1\|_{\mathbb{C}^{q_1}}, \|y_2\|_{\mathbb{C}^{q_2}})\|_{\mathbb{R}^2}$$

where $\|\cdot\|_{\mathbb{C}^{\ell_i}}$, $\|\cdot\|_{\mathbb{C}^{q_i}}$ *are given norms on* $\mathbb{C}^{\ell_i}$ *and* $\mathbb{C}^{q_i}$, *respectively,* $i = 1, 2$. *Then*

$$r_{\mathbb{C}}(A; B, C, D; \mathbb{C}_g) = \min\left\{r_{\mathbb{C}}(A_1; B_1, C_1, D_1; \mathbb{C}_g), r_{\mathbb{C}}(A_2; B_2, C_2, D_2; \mathbb{C}_g)\right\}. \quad (24)$$

Proof: If $G(s)$ and $G_i(s)$ denote the transfer matrices of the systems (A, B, C, D) and (A_i, B_i, C_i, D_i), $i = 1, 2$ respectively, then $G(s) = G_1(s) \oplus G_2(s)$ and thus $\|G(s)\| = \max_{i=1,2} \|G_i(s)\|$ for $s \in \partial\mathbb{C}_g$ (see Ex. 4.4.3). So (24) follows from (20) and $\|D\| = \max_{i=1,2} \|D_i\|$. □

It should be noted that the perturbation class for the LHS of (24) is $\mathbb{C}^{(\ell_1+\ell_2)\times(q_1+q_2)}$ and not $\mathbb{C}^{\ell_1\times q_1} \oplus \mathbb{C}^{\ell_2\times q_2}$ (for which the result would be trivial). We see, rather surprisingly, that the system (A, B, C, D) cannot be destabilized by a perturbation which is smaller than one which destabilizes one of the two subsystems. In particular the corollary says that two $\mathbb{C}_g$–stable state space systems (A_1, B_1, C_1, D_1), (A_2, B_2, C_2, D_2) cannot be destabilized by coupling them together with interconnection matrices (see Subsection 2.4.2) whose norms are strictly smaller than both complex stability radii. An analogous result does *not* hold for the *real* stability radius. In fact in the next example we will see that for any $\delta > 0$ it is possible to find two identical oscillators whose real stability radius is 1 which can be destabilized via a coupling matrix of norm less than δ. This destabilizability is indicated by the complex stability radius $r_{\mathbb{C}} < \delta$ which illustrates why the *complex* stability radius may be of interest for the robustness analysis of *real* systems.

Example 5.3.14. Consider a linear oscillator with given damping $2\,\alpha > 0$ and perturbed restoring force parameter of nominal value 1

$$\ddot{\xi} + 2\,\alpha\,\dot{\xi} + (1+\Delta)\xi = 0. \tag{25}$$

The associated system in state space form is given by $\dot{x}(t) = (A(\alpha) + B\Delta C)x(t)$ where

$$A(\alpha) = \begin{bmatrix} 0 & 1 \\ -1 & -2\alpha \end{bmatrix}, \quad B = \begin{bmatrix} 0 \\ 1 \end{bmatrix}, \quad C = \begin{bmatrix} 1 & 0 \end{bmatrix}.$$

For real perturbation, (25) is asymptotically stable if and only if the characteristic polynomial $s^2 + 2\alpha s + (1+\Delta)$ is Hurwitz, i.e. $1+\Delta > 0$. Hence $\Delta = -1$ is the *real* perturbation of minimum norm that destabilizes the oscillator (25). It follows that the real stability radius $r_{\mathbb{R}}^-(A(\alpha); B, C) = 1$ is independent of the damping, no matter how small.
On the other hand $G(\imath\omega) = C(\imath\omega I - A(\alpha))^{-1}B = (1-\omega^2+\imath 2\alpha\omega)^{-1}$ and hence

$$|G(\imath\omega)|^2 = \left[(\omega^2-1)^2 + 4\alpha^2\omega^2\right]^{-1} = \left[(\omega^2-1+2\alpha^2)^2 + 1 - (2\alpha^2-1)^2\right]^{-1}.$$

If $\alpha \geq 1/\sqrt{2}$ this expression is maximized at $\omega = 0$ so that $r_{\mathbb{C}}^-(A(\alpha); B, C) = |G(0)|^{-1} = 1$ equals the real stability radius.
Now suppose $\alpha < 1/\sqrt{2}$. Then the above expression for $|G(\imath\omega)|^2$ is maximized at $\omega(\alpha) = \sqrt{1-2\alpha^2} > 0$ and so the complex stability radius is given by

$$r_{\mathbb{C}}^-(A(\alpha); B, C) = |G(\imath\omega(\alpha))|^{-1} = 2\alpha\sqrt{1-\alpha^2}, \quad \alpha < 1/\sqrt{2},$$

and hence depends on α if $\alpha < 1/\sqrt{2}$. We conclude that for small $\alpha > 0$ there exists a small complex perturbation $\Delta \in \mathbb{C}$, $|\Delta| = 2\alpha\sqrt{1-\alpha^2} \approx 2\alpha$ such that the system (25) is not asymptotically stable. At first sight this may seem surprising since for real Δ, $|\Delta| < 1$ ensures asymptotic stability. The reason is best explained, perhaps, by considering a purely imaginary perturbation $\Delta = \imath\beta$, $\beta \in \mathbb{R}$. Setting $\xi = \xi_R + \imath\xi_I$, ξ_R, $\xi_I \in \mathbb{R}$ in (25) and separating the real and imaginary parts, we obtain

$$\begin{aligned} \ddot{\xi}_R + 2\,\alpha\,\dot{\xi}_R + \xi_R - \beta\xi_I &= 0 \\ \ddot{\xi}_I + 2\,\alpha\,\dot{\xi}_I + \xi_I + \beta\xi_R &= 0\,. \end{aligned} \tag{26}$$

We see that the perturbation $\Delta = \imath\beta$ introduces a coupling between a pair of identical real oscillators (25). The eigenvalues of the state space system corresponding to the coupled systems (26) are $-[\alpha \pm (\alpha^2 - 1 \pm \imath\beta)^{1/2}]$. If $|\beta| > 2\alpha$, an easy calculation yields that the real part of one pair of these eigenvalues is positive and hence the coupled system (26) is unstable.
This result illustrates that two real oscillators with real stability radius $r_{\mathbb{R}} = 1$ and small complex stability radius $\approx 2\alpha$, $0 < \alpha \ll 1$ may be destabilized by introducing a small real coupling of the order 2α between them if their damping coefficients are sufficiently small. In view of Corollary 5.3.13 this points out an important difference between the real and the complex stability radius, suggesting that the latter may be the more appropriate measure of robust stability in the context of interconnected systems. The example also shows that the imaginary part of a complex perturbation can be interpreted in real terms as a coupling between two identical systems. □

The result in the above example can be generalized. In fact the reader is asked in Ex. 6 to prove the following consequence of Corollary 5.3.13.

Corollary 5.3.15. *Suppose* $(A, B, C, D) \in \mathbf{L}_{n,\ell,q}(\mathbb{R})$, $\sigma(A) \subset \mathbb{C}_g$, *then with respect to spectral norms on* $\mathbb{C}^{\ell \times q}$ *resp.* $\mathbb{R}^{2\ell \times 2q}$

$$r_{\mathbb{C}}(A; B, C, D; \mathbb{C}_g) = r_{\mathbb{R}}(A \oplus A; B \oplus B, C \oplus C, D \oplus D; \mathbb{C}_g) \tag{27}$$

where $r_{\mathbb{R}}$ *denotes the stability radius with respect to real full block perturbations.*

The previous corollary gives a first interpretation of the complex stability radius in real terms. In Section 5.6 the significance of complex stability radii for the robustness analysis of real systems will be explored further.

5.3.3 Real Full-Block Perturbations

In this subsection we consider real full-block perturbations of a real system and derive a computable formula for the corresponding stability radius. Following the characterizations of the real μ-function, $\mu_{\mathbb{R}}$ in Section 4.4, we discuss separately the cases $\ell = 1$ and $\ell > 1$. In the first case we obtain a computable formula for arbitrary operator norms on $\mathbf{\Delta} = \mathbb{R}^{\ell \times q}$, in the second case we only obtain a formula valid for the spectral norm.
The standing assumptions are

- $\mathbb{C}_g \neq \emptyset$ is an open subset of $\mathbb{C}$ and $\mathbb{C}_b := \mathbb{C} \setminus \mathbb{C}_g \neq \emptyset$.
- $(A, B, C, D) \in \mathbf{L}_{n,\ell,q}(\mathbb{R})$ with $\sigma(A) \subset \mathbb{C}_g$.
- $\mathbf{\Delta} = \mathbb{R}^{\ell \times q}$ is provided with a norm $\|\cdot\|$ which is an operator norm with respect to a given pair of norms on $\mathbb{R}^{\ell}$ and $\mathbb{R}^q$. Moreover we suppose that $\mathbb{C}^{\ell}$ and $\mathbb{C}^q$ are provided with a pair of norms $(\|\cdot\|_{\mathbb{C}^{\ell}}, \|\cdot\|_{\mathbb{C}^q})$, compatible with $(\|\cdot\|_{\mathbb{R}^{\ell}}, \|\cdot\|_{\mathbb{R}^q})$. (Such a pair of norms always exists, see Lemma A.1.7).
- The perturbations are given by

$$A \rightsquigarrow A(\Delta) = A + B(I_\ell - \Delta D)^{-1}\Delta C, \quad \Delta \in \mathbf{\Delta} = \mathbb{R}^{\ell \times q}. \tag{28}$$

We denote the corresponding stability radius by $r_{\mathbb{R}}(A; B, C, D; \mathbb{C}_g)$ (if $D = 0$ by $r_{\mathbb{R}}(A; B, C; \mathbb{C}_g)$) and refer to it as the *real stability radius* of A with respect to perturbations of the form (28). By Theorem 5.3.3 we have

$$r_{\mathbb{R}} = \left[\max\left\{\mu_{\mathbb{R}}(D), \sup_{s \in \partial\mathbb{C}_g} \mu_{\mathbb{R}}(G(s))\right\}\right]^{-1} = \min\{\mu_{\mathbb{R}}(D)^{-1}, \inf_{s \in \partial\mathbb{C}_g} \mu_{\mathbb{R}}(G(s))^{-1}\}. \tag{29}$$

For the stability regions $\mathbb{C}_-$ and $\mathbb{D}$, the real stability radii are called $r_{\mathbb{R}}^-(A; B, C, D)$ and $r_{\mathbb{R}}^1(A; B, C, D)$, respectively.
Since by assumption $\|\Delta\|_{\mathcal{L}(\mathbb{R}^q, \mathbb{R}^{\ell})} = \|\Delta\|_{\mathcal{L}(\mathbb{C}^q, \mathbb{C}^{\ell})}$ for $\Delta \in \mathbb{R}^{\ell \times q}$ it follows that

$$r_{\mathbb{R}}(A; B, C, D; \mathbb{C}_g) \geq r_{\mathbb{C}}(A; B, C, D; \mathbb{C}_g). \tag{30}$$

Example 5.3.14 has shown that, in general, the two radii are different and their ratio $r_{\mathbb{R}}/r_{\mathbb{C}}$ can be arbitrarily large (see also Ex. 5).

Case $\ell = 1$

Suppose $\mathbb{R}^\ell = \mathbb{R}^1$ is provided with the norm $|\cdot|$ and $\|\cdot\|_{\mathbb{R}^q}$ is an arbitrary norm on $\mathbb{R}^q$. Then the induced operator norms on $\mathbb{R}^{q\times 1} = \mathbb{R}^q$ and on $\mathbf{\Delta} = \mathbb{R}^{1\times q}$ are the vector norm $\|\cdot\|_{\mathcal{L}(\mathbb{R},\mathbb{R}^q)} = \|\cdot\|_{\mathbb{R}^q}$ and the dual norm $\|\cdot\|_{\mathcal{L}(\mathbb{R}^q,\mathbb{R})} = \|\cdot\|^*_{\mathbb{R}^q}$, respectively. As an immediate consequence of Theorem 4.4.31 and (29), we have

Theorem 5.3.16. *Suppose $\ell = 1$, $\|\cdot\|_{\mathbb{R}^q}$ is an arbitrary norm on $\mathbb{R}^q$ and $\mathbf{\Delta} = \mathbb{R}^{1\times q}$ is provided with the dual norm $\|\cdot\|^*_{\mathbb{R}^q}$. Then the real stability radius of A with respect to perturbations of the form (28) is given by*

$$r_{\mathbb{R}}(A;B,C,D;\mathbb{C}_g) = \left[\max\left\{\sup_{s\in\partial\mathbb{C}_g} \operatorname{dist}\left(G_R(s),\mathbb{R}G_I(s)\right), \|D\|_{\mathbb{R}^q}\right\}\right]^{-1}, \tag{31}$$

where $G(s) = D + C(sI_n - A)^{-1}B = G_R(s) + \imath G_I(s)$ with $G_R(s)$, $G_I(s) \in \mathbb{R}^q$ and $\operatorname{dist}(G_R(s),\mathbb{R}G_I(s))$ *is taken with respect to the vector norm $\|\cdot\|_{\mathbb{R}^q}$ on $\mathbb{R}^q$.*
In particular

$$r_{\mathbb{R}}^-(A;B,C,D) = \left[\max\left\{\sup_{\omega\in\mathbb{R}} \operatorname{dist}(G_R(\imath\omega),\mathbb{R}G_I(\imath\omega)), \|D\|_{\mathbb{R}^q}\right\}\right]^{-1}, \tag{32}$$

$$r_{\mathbb{R}}^1(A;B,C,D) = \left[\max\left\{\max_{\theta\in[0,2\pi]} \operatorname{dist}(G_R(e^{\imath\theta}),\mathbb{R}G_I(e^{\imath\theta})), \|D\|_{\mathbb{R}^q}\right\}\right]^{-1}. \tag{33}$$

We have seen in Section 4.4 that the map $s \mapsto \mu_{\mathbb{R}}(G(s)) = \operatorname{dist}(G_R(s), G_I(s)\mathbb{R})$ is upper semicontinuous so that the maximum in (33) exists (see Proposition A.2.1), but it may be discontinuous at the zeros of $G_I(s)$, so it is best to compute $r_{\mathbb{R}}$ via the formula

$$r_{\mathbb{R}}(A;B,C,D;\mathbb{C}_g) = \min\left\{\|D\|^{-1}_{\mathbb{R}^q}, \left[\sup_{s\in\partial\mathbb{C}_g\cap\mathcal{R}_G}\|G(s)\|\right]^{-1}, \left[\sup_{s\in\partial\mathbb{C}_g\setminus\mathcal{R}_G}\operatorname{dist}(G_R(s),\mathbb{R}G_I(s))\right]^{-1}\right\}$$

where $\mathcal{R}_G = \{s \in \mathbb{C}\setminus\sigma(A) : G_I(s) = 0\} \supset \mathbb{R}$ is the realness locus of $G(s)$.

Remark 5.3.17. (i) Unlike the results in (22) and (23), we cannot remove $\|D\|_{\mathbb{R}^q}$ in (32) and (33) without further consideration. Although we have $\lim_{|\omega|\to\infty} G_R(\imath\omega) = D$ and $\lim_{|\omega|\to\infty} G_I(\imath\omega) = 0_q$, it may be that $\operatorname{dist}(G_R(\imath\omega),\mathbb{R}G_I(\imath\omega))$ does not converge to $\operatorname{dist}(D,0_q) = \|D\|_{\mathbb{R}^q}$ as $|\omega| \to \infty$. This possible discontinuity at ∞ will be illustrated in Example 5.3.18.
(ii) Suppose there exists $s_0 \in \partial\mathbb{C}_g$ such that

$$\sup_{s\in\partial\mathbb{C}_g} \operatorname{dist}\left(G_R(s),\mathbb{R}G_I(s)\right) = \operatorname{dist}\left(G_R(s_0),\mathbb{R}G_I(s_0)\right) > \|D\|,$$

then a minimum norm destabilizing perturbation $\Delta_{\min}$ can be constructed as in the proof of Theorem 4.4.31: Let $z^* \in \mathbb{R}^{1\times q}$ with $\|z^*\|^*_{\mathbb{R}^q} = 1$ be such that

$$z^*G_R(s_0) = \operatorname{dist}(G_R(s_0),\mathbb{R}G_I(s_0)) \qquad \text{and} \qquad z^*G_I(s_0) = 0\,.$$

Then $\Delta_{\min} = [\operatorname{dist}(G_R(s_0),\mathbb{R}G_I(s_0)]^{-1}z^*$ is a minimum norm destabilizing perturbation.
(iii) Using Corollary 4.4.29 we obtain a similar formula to (31) for the case $q = 1$ and $\ell \geq 1$. Let $\mathbb{R}^q = \mathbb{R}$ be provided with the norm $|\cdot|$, $\mathbb{R}^\ell$ with any norm $\|\cdot\|_{\mathbb{R}^\ell}$, $\mathbf{\Delta} = \mathbb{R}^{\ell\times 1}$ with the norm $\|\cdot\|_{\mathbb{R}^\ell}$ and $\mathbb{R}^{1\times\ell}$ with the dual norm $\|\cdot\|^*_{\mathbb{R}^\ell}$, then

$$r_{\mathbb{R}}(A;B,C,D;\mathbb{C}_g) = \left[\max\left\{\sup_{s\in\partial\mathbb{C}_g} \operatorname{dist}(G_R(s),\mathbb{R}G_I(s)), \|D\|^*_{\mathbb{R}^\ell}\right\}\right]^{-1},$$

where the distance is taken with respect to $\|\cdot\|^*_{\mathbb{R}^\ell}$ on $\mathbb{R}^{1\times\ell}$.
(iv) In the case $q=\ell=1$, we have $\operatorname{dist}(G_R(s),\mathbb{R}G_I(s))=0$ if $G_I(s))\neq 0$, and therefore

$$r_{\mathbb{R}}(A;B,C,D;\mathbb{C}_g) = \left[\max\left\{\sup_{s\in\partial\mathbb{C}_g\cap\mathcal{R}_G} |G(s)|, |D|\right\}\right]^{-1}. \tag{34}$$

We will show in the next section that if $\mathbb{C}_g=\mathbb{C}_-$, A is Hurwitz stable and $G(s)$ is not constant, the set $\mathcal{R}_G\cap\imath\mathbb{R}$ consists of at most n elements. In this case $r^-_{\mathbb{R}}$ can be determined from (34) via a search over a finite set, see Figure 5.3.2.
(v) For the stability region $\mathbb{C}_g=\mathbb{C}_-$, we have

$$G_R(\imath\omega) = D - C(\omega^2 I + A^2)^{-1}AB\,, \quad G_I(\imath\omega) = -\omega C(\omega^2 I + A^2)^{-1}B, \qquad \omega\in\mathbb{R},$$

$$\mathcal{R}_G\cap\imath\mathbb{R} = \{\imath\omega\in\imath\mathbb{R};\ \omega=0 \ \text{ or } \ C(\omega^2 I_n + A^2)^{-1}B = 0\}\,,$$

and with respect to Euclidean norms

$$r^-_{\mathbb{R}} = \min\{[\sup_{\omega\in\mathbb{R},\imath\omega\in\mathcal{R}_G} \|G(\imath\omega)\|]^{-1}, [\sup_{\omega\in\mathbb{R},\imath\omega\notin\mathcal{R}_G} \operatorname{dist}(G_R(\imath\omega),G_I(\imath\omega)\mathbb{R})]^{-1}, \|D\|^{-1}_{\mathbb{R}^q}\}, \tag{35}$$

where

$$\operatorname{dist}\left(G_R(\imath\omega),\mathbb{R}G_I(\imath\omega)\right)^2 = \|G_R(\imath\omega)\|_2^2 - \frac{\langle G_R(\imath\omega),G_I(\imath\omega)\rangle^2}{\|G_I(\imath\omega)\|_2^2}, \qquad \omega\in\mathbb{R},\ \imath\omega\notin\mathcal{R}_G.$$

For the case $q>1$ we have mentioned in Section 5.2 that if at least two entries of $G(s)$ are non-constant then generically, away from the real axis, the realness loci of these entries will only intersect at a finite number of points. Hence generically $\mathcal{R}_G\cap\imath\mathbb{R}=\{0\}$. □

Example 5.3.18. We consider the same nominal oscillator as that in Example 5.3.12, namely $\ddot\xi + 4\dot\xi + 5\xi = 0$ and assume that the perturbed system is

$$(1-\Delta_2)\ddot\xi + (4-5\Delta_2)\dot\xi + (5-\Delta_1-5\Delta_2)\xi = 0.$$

So as in Example 5.1.15 the mass, damping and spring constants are all perturbed, but this time the perturbations are not independent. The perturbed system may be written in the form $\dot x = A(\Delta)x$ where $A(\Delta)$ is given by (28) with $A=\begin{bmatrix}0 & 1\\ -5 & -4\end{bmatrix}$, $D=B=[0\ \ 1]^\top$, $C=I_2$, and $\Delta=[\Delta_1,\ \Delta_2]\in\boldsymbol{\Delta}=\mathbb{R}^{1\times 2}$. Accordingly we have

$$G(\imath\omega) = \begin{bmatrix}0\\1\end{bmatrix} + [(5-\omega^2)+4\imath\omega]^{-1}\begin{bmatrix}1\\ \imath\omega\end{bmatrix}.$$

With respect to the Euclidean norm,

$$\|G(\imath\omega)\|_2^2 = 1 + [(5-\omega^2)^2+16\omega^2]^{-1}(1+9\omega^2).$$

A short calculation shows that the maximizing $\omega=2.196$ and hence $r^-_{\mathbb{C}}(A;B,C,D) = 0.797$. For the real case we have

$$[G_R(\imath\omega)\quad G_I(\imath\omega)] = [(5-\omega^2)^2+16\omega^2]^{-1}\begin{bmatrix}5-\omega^2 & -4\omega\\ (5-\omega^2)^2+20\omega^2 & \omega(5-\omega^2)\end{bmatrix}.$$

Hence $G_I(\imath\omega) = 0$ only for $\omega = 0$, and then $\|G_R(0)\| = \sqrt{26}/5$. If $\omega \neq 0$, a calculation yields

$$d(\omega)^2 := \operatorname{dist}(G_R(\imath\omega), \mathbb{R}G_I(\imath\omega))^2 = \|G_R(\imath\omega)\|_2^2 - \frac{\langle G_R(\imath\omega), G_I(\imath\omega)\rangle^2}{\|G_I(\imath\omega)\|_2^2} = \frac{25}{16+(5-\omega^2)^2}.$$

So $d(\omega) \to 0$ as $\omega \to \infty$, whereas $\|D\| = 1$. This illustrates the discontinuity of $d(\cdot)$ at $\omega = \infty$, see Remark 5.3.17 (i). The maximum of $d(\omega)$ occurs when $\omega = \sqrt{5}$ and hence $r_{\mathbb{R}}^-(A; B, C, D) = \min\{5/\sqrt{26}, 4/5, 1\} = 4/5$. □

In our second example for the case $\ell = 1$, we return to the satellite problem described in Section 5.1 and illustrate how state feedback may be used to achieve robust stability in the presence of parameter uncertainty.

Example 5.3.19. For the satellite problem of Example 5.1.17 we assume $J_1 = 1$, $J_2 = 0.1$. The uncontrolled system is not asymptotically stable for any values of (c, k) in the region Ω defined by (1.17) and the control objective is to find a state feedback law $u = f^\top x$ which stabilizes the system for all parameters in this uncertainty region. In the following we show how stability radii can be used to examine whether this objective has been achieved. Consider the state feedback control $u = f_0^\top x := -[0.595,\ 0.275,\ 1.32,\ 1.66]x$. Choosing a nominal parameter pair (k_0, c_0) in the region Ω leads as in Example 5.1.17 to the equation of the perturbed feedback system

$$\Sigma_\Delta: \qquad \dot{x} = A(\Delta)x = [A(c_0, k_0, f_0) + B\Delta C]x, \quad \Delta \in \mathbf{\Delta} = \mathbb{R}^{1\times 2}$$

where

$$A(k_0, c_0, f_0) = \begin{bmatrix} 0 & 1 & 0 & 0 \\ -10k_0 & -10c_0 & 10k_0 & 10c_0 \\ 0 & 0 & 0 & 1 \\ -0.595 + k_0 & -0.275 + c_0 & -1.32 - k_0 & -1.66 - c_0 \end{bmatrix}, \quad B = \begin{bmatrix} 0 \\ -10 \\ 0 \\ 1 \end{bmatrix},$$

$$C = \begin{bmatrix} 1 & 0 & -1 & 0 \\ 0 & 1 & 0 & -1 \end{bmatrix}, \quad \Delta = [k - k_0, c - c_0].$$

For given k_0, c_0 satisfying $\sigma(A(c_0, k_0, f_0)) \subset \mathbb{C}_-$, the system Σ_Δ will be asymptotically stable for all $k, c \in \mathbb{R}$ in the interior of the circle

$$(x - k_0)^2 + (y - c_0)^2 = r(k_0, c_0, f_0)^2$$

where $r(k_0, c_0, f_0) = r_{\mathbb{R}}^-(A(c_0, k_0, f_0); B, C)$ is the real stability radius with respect to the 2-norm. The following table gives $r(k_0, c_0, f_0)$ (computed via (35)) for different values of k_0, c_0. The stability of the closed loop system is guaranteed for all parameter values in Ω since it is contained within the union of the seven disks. We conclude therefore

k_0	0.1	0.15	0.2	0.25	0.3	0.345	0.385
c_0	0.017	0.019	0.02	0.02	0.02	0.02	0.02
$r(k_0, c_0, f_0)$	0.035	0.035	0.034	0.032	0.03	0.029	0.028

Table 5.3.3: Stability radii for different values of k_0, c_0

that the feedback control $u = f_0^\top x$ achieves the above control objective. However the procedure of covering Ω with disks is cumbersome and requires the calculation of seven stability radii. It is not possible to cover the region with a single disk, but we have seen

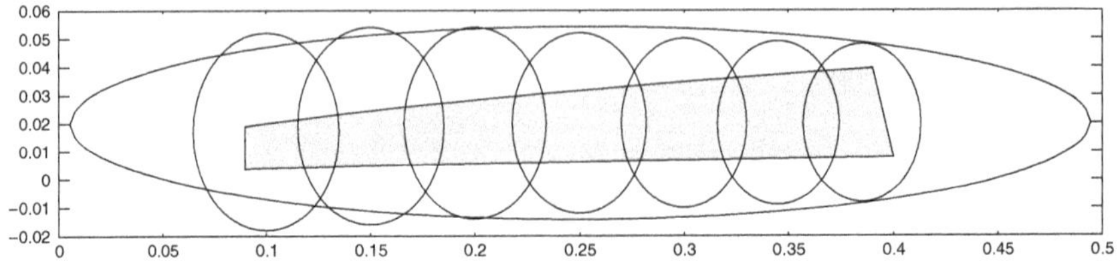

Figure 5.3.4: Covering the region Ω

in Example 5.1.17 that it may be better to use a weighted norm $\|\Delta\|_\alpha^2 = \Delta_1^2 + \alpha^2\Delta_2^2$. Denoting the corresponding radius by $r_\alpha = r_\alpha(k_0, c_0, f_0)$ the system is asymptotically stable for all $(k, c) \in \mathbb{R}^2$ in the interior of the ellipse

$$(x - k_0)^2 + \alpha^2(y - c_0)^2 = r_\alpha^2 \,.$$

For $k_0 = 0.25$, $c_0 = 0.02$, $\alpha = 7.14$ we computed $r_\alpha(k_0, c_0, f_0) = 0.2445$ and this single ellipse covers the whole set Ω. □

Real Case $\ell, \mathbf{q} > 1$

For the general case $\mathbf{\Delta} = \mathbb{R}^{\ell\times q}$ we assume that $\mathbb{R}^{\ell\times q}$ *is provided with the spectral norm.* Then using (29) and the characterization of $\mu_\mathbb{R}$ in Theorem 4.4.43, we have

Theorem 5.3.20. *Suppose $\mathbf{\Delta} = \mathbb{R}^{\ell\times q}$ is provided with the spectral norm $\|\cdot\|_{2,2} = \sigma_1(\cdot)$. Then the real stability radius of A with respect to real full-block perturbations (28) is given by*

$$r_\mathbb{R}(A; B, C, D; \mathbb{C}_g) = \left[\max\left\{\sup_{s\in\partial\mathbb{C}_g}\ \inf_{\gamma\in(0,1]} \sigma_2(G_\gamma^\mathbb{R}(s)), \|D\|_{2,2}\right\}\right]^{-1}, \tag{36}$$

where $G(s) = C(sI_n - A)^{-1}B + D = G_R(s) + \imath G_I(s)$, $G_R(s)$, $G_I(s) \in \mathbb{R}^{q\times\ell}$ and

$$G_\gamma^\mathbb{R}(s) = \begin{bmatrix} G_R(s) & -\gamma G_I(s) \\ \gamma^{-1}G_I(s) & G_R(s)\end{bmatrix}, \quad s \in \rho(A), \quad \gamma \in (0,1]. \tag{37}$$

In particular[4]

$$r_\mathbb{R}^-(A; B, C, D) = \left[\max\left\{\sup_{\omega\in\mathbb{R}}\ \inf_{\gamma\in(0,1]} \sigma_2(G_\gamma^\mathbb{R}(\imath\omega)), \|D\|_{2,2}\right\}\right]^{-1}, \tag{38}$$

$$r_\mathbb{R}^1(A; B, C, D) = \left[\max\left\{\max_{\theta\in[0,2\pi]}\ \inf_{\gamma\in(0,1]} \sigma_2(G_\gamma^\mathbb{R}(e^{\imath\theta})), \|D\|_{2,2}\right\}\right]^{-1}. \tag{39}$$

Remark 5.3.21. (i) Again because of possible discontinuities on the realness locus of the map $s \to \mu_\mathbb{R}(G(s)) = \inf_{\gamma\in(0,1]}\sigma_2(G_\gamma^\mathbb{R}(s))$ one should compute $r_\mathbb{R}$ via

$$r_\mathbb{R}(A; B, C, D; \mathbb{C}_g) = \min\left\{\|D\|_{2,2}^{-1}, \left[\sup_{s\in\partial\mathbb{C}_g\cap\mathcal{R}_G}\|G(s)\|\right]^{-1}, \left[\sup_{s\in\partial\mathbb{C}_g\setminus\mathcal{R}_G}\ \inf_{\gamma\in(0,1]}\sigma_2(G_\gamma^\mathbb{R}(s))\right]^{-1}\right\}.$$

[4]The maximum in (39) exists since the map $s \mapsto \mu_\mathbb{R}(G(s)) = \inf_{\gamma\in(0,1]}\sigma_2(G_\gamma^\mathbb{R}(s))$ is upper semicontinuous (see Lemma 4.4.27 and Proposition A.2.1).

(ii) Since $\lim_{|\omega|\to\infty} G_\gamma^{\mathbb{R}}(\imath\omega) = \left[\begin{smallmatrix} D & 0 \\ 0 & D \end{smallmatrix}\right]$, we have $\lim_{|\omega|\to\infty} \sigma_2(G_\gamma^{\mathbb{R}}(\imath\omega)) = \|D\|_{2,2}$ for $\gamma \in (0,1]$. But again because of the possible discontinuity of $\omega \mapsto \mu_{\mathbb{R}}(G(\imath\omega))$ at ∞ it may be that $\lim_{|\omega|\to\infty} \inf_{\gamma\in(0,1]} \sigma_2(G_\gamma^{\mathbb{R}}(\imath\omega)) \neq \|D\|_{2,2}$, see Example 5.3.18.

(iii) Suppose $s_0 \in \partial\mathbb{C}_g$, $\gamma_0 \in (0,1]$ is a saddle point of $f(\gamma, s) = \sigma_2(G_\gamma^{\mathbb{R}}(s))$ such that

$$\sup_{s\in\partial\mathbb{C}_g} \inf_{\gamma\in(0,1]} \sigma_2(G_\gamma^{\mathbb{R}}(s)) = \sigma_2(G_{\gamma_0}^{\mathbb{R}}(s_0)) > \|D\|_{2,2}.$$

Let $w = \begin{bmatrix} w^1 \\ w^2 \end{bmatrix}$ and $v = \begin{bmatrix} v^1 \\ v^2 \end{bmatrix}$ be left and right singular vectors of $G_{\gamma_0}^{\mathbb{R}}(s_0)$ corresponding to $\sigma_2(G_{\gamma_0}^{\mathbb{R}}(s_0))$, such that $w^{1\top}w^2 = v^{1\top}v^2$, $\|w^1\| = \|v^1\|$ and $\|w^2\| = \|v^2\|$, then we may construct a minimum norm destabilizing perturbation as in Lemma 4.4.37. For example, if $w^1 \neq 0$ and $w^2 \neq 0$ are linearly independent and

$$\Delta_{\min} = [\sigma_2(G_{\gamma_0}^{\mathbb{R}}(s_0))]^{-1}[v^1\ v^2] \begin{bmatrix} \|w^1\|^2 & \langle w^1, w^2\rangle \\ \langle w^1, w^2\rangle & \|w^2\|^2 \end{bmatrix}^{-1} \begin{bmatrix} w^{1\top} \\ w^{2\top} \end{bmatrix}, \tag{40}$$

then $s_0 \in \sigma(A(\Delta_{\min}))$ and $\|\Delta_{\min}\| = r_{\mathbb{R}}(A; B, C, D; \mathbb{C}_g)$. □

In general, even for simple low order systems, the real stability radius is difficult to compute by hand via formula (36). Before trying this for a given example, it is advisable to examine first whether $r_{\mathbb{R}} = r_{\mathbb{C}}$. We have seen in Remark 5.3.11 that this holds, if there exists $s_0 \in \partial\mathbb{C}_g$ such that $G(s_0)$ is real and

$$\max_{s\in\partial\mathbb{C}_g} \|G(s)\|_{\mathcal{L}(\mathbb{C}^\ell,\mathbb{C}^q)} = \|G(s_0)\|_{\mathcal{L}(\mathbb{C}^\ell,\mathbb{C}^q)} > \|D\|_{\mathcal{L}(\mathbb{C}^\ell,\mathbb{C}^q)}.$$

The next example shows that for perturbed positive systems with nonnegative structure matrices the real stability radius always coincides with the complex one and can be determined via a simple formula.

Example 5.3.22. A system $\dot{x} = Ax$, $A \in \mathbb{R}^{n\times n}$ is said to be *positive* if the non-negative orthant is invariant under its flow, i.e. $e^{At} \geq 0$ for all $t \geq 0$. $\dot{x} = Ax$ is positive if and only if A is a Metzler matrix, that is all the off-diagonal entries of A are non-negative. It can be shown that every Metzler matrix satisfies the following entrywise inequality for $\lambda \in \mathbb{C}$

$$\operatorname{Re}\lambda > \alpha(A) \quad \Rightarrow \quad |(\lambda I_n - A)^{-1}| \leq ((\operatorname{Re}\lambda)I_n - A)^{-1}. \tag{41}$$

In particular, $(\alpha I_n - A)^{-1}$ is non-negative for $\alpha > \alpha(A)$. Now assume that A is a $\mathbb{C}_-$-stable Metzler matrix and the structure matrices are non-negative, $B \in \mathbb{R}_+^{n\times\ell}$, $C \in \mathbb{R}_+^{q\times n}$. By (41) the associated transfer matrix $G(s) = C(sI_n - A)^{-1}B$ satisfies

$$|G(\imath\omega)| \leq C|(\imath\omega I_n - A)^{-1}|B \leq -CA^{-1}B = G(0), \quad \omega \in \mathbb{R}$$

and hence, for arbitrary absolute norms on $\mathbb{C}^\ell$, $\mathbb{C}^q$

$$\|G(\imath\omega)\|_{\mathcal{L}(\mathbb{C}^\ell,\mathbb{C}^q)} \leq \|CA^{-1}B\|_{\mathcal{L}(\mathbb{C}^\ell,\mathbb{C}^q)}, \quad \omega \in \mathbb{R}.$$

Thus $\|G(0)\|_{\mathcal{L}(\mathbb{C}^\ell,\mathbb{C}^q)} = \max_{\omega\in\mathbb{R}} \|G(\imath\omega)\|_{\mathcal{L}(\mathbb{C}^\ell,\mathbb{C}^q)}$ and it follows from Remark 5.3.11 that

$$r_{\mathbb{R}}(A; B, C; \mathbb{C}_-) = r_{\mathbb{C}}(A; B, C; \mathbb{C}_-) = \|CA^{-1}B\|^{-1}.$$

□

5.3.4 Hamiltonian Characterization of the Complex Stability Radius

In this subsection we concentrate on complex full-block perturbations of continuous time systems and the corresponding complex stability radius $r_{\mathbb{C}}^{-} = r_{\mathbb{C}}^{-}(A; B, C, D)$, as measured by the *spectral norm*. For this perturbation norm, $r_{\mathbb{C}}^{-}$ admits several interesting characterizations. The first is in terms of the norm of the input-output maps associated with (A, B, C, D) (see (2.3.22) and (2.3.33)), the second is derived from an associated optimal control problem and the third is in terms of Hamiltonian matrices. This latter characterization forms the basis of an algorithm for the computation of $r_{\mathbb{C}}^{-}$. Although we only consider the continuous time case, we will make some comments regarding the slightly more complicated discrete time counterpart in *Notes and References*.

Throughout the subsection we assume

- $(A, B, C, D) \in \mathbf{L}_{n,\ell,q}(\mathbb{C})$ with $\sigma(A) \subset \mathbb{C}_g = \mathbb{C}_-$.
- The vector spaces $\mathbb{C}^q, \mathbb{C}^\ell, \mathbb{C}^n$ are provided with their standard Euclidean norms and the matrix spaces $\mathbf{\Delta} = \mathbb{C}^{\ell\times q}$ and $\mathbb{C}^{q\times\ell}$ are endowed with the corresponding operator norms denoted by $\|\cdot\|$.
- The perturbations are of the form

$$A \rightsquigarrow A(\Delta) = A + B(I_\ell - \Delta D)^{-1}\Delta C, \quad \Delta \in \mathbb{C}^{\ell\times q}.$$

By Theorem 5.3.9 the corresponding complex stability radius is

$$r_{\mathbb{C}}^{-}(A; B, C, D) = \left[\sup_{\omega\in\mathbb{R}} \|G(\imath\omega)\|\right]^{-1} \tag{42}$$

where $G(s) = D + C(sI_n - A)^{-1}B$ is the transfer matrix of the continuous time system

$$\Sigma : \quad \dot{x}(t) = Ax(t) + Bw(t), \;\; z(t) = Cx(t) + Dw(t), \qquad t \in \mathbb{R}. \tag{43}$$

Making use of the results in Subsection 2.3.3 we will now derive a characterization of $r_{\mathbb{C}}^{-}$ in terms of the input-output operator of this system. The output $z(\cdot; w)$ of Σ produced by the (perturbation) input $w(\cdot) \in L^2(\mathbb{R}; \mathbb{C}^\ell)$ is given by

$$z(t; w) = Dw(t) + \int_{-\infty}^{t} Ce^{A(t-\tau)}Bw(\tau)\, d\tau\,, \qquad t \in \mathbb{R}.$$

Since $\sigma(A) \subset \mathbb{C}_-$ we have $z(\cdot, w) \in L^2(\mathbb{R}; \mathbb{C}^q)$ for all $w(\cdot) \in L^2(\mathbb{R}; \mathbb{C}^\ell)$ by Proposition 2.3.15 so that the input-output operator

$$\begin{array}{rrcl} \mathbb{L}: & L^2(\mathbb{R}; \mathbb{C}^\ell) & \to & L^2(\mathbb{R}; \mathbb{C}^q) \\ & (\mathbb{L}w)(t) & = & Dw(t) + \int_{-\infty}^{t} Ce^{A(t-s)}Bw(s)\, ds\,, \quad t \in \mathbb{R} \end{array} \tag{44}$$

is well defined (see Subsection 2.3.1). The norm of this linear operator is by definition

$$\|\mathbb{L}\| = \sup_{w\in L^2(\mathbb{R};\mathbb{C}^\ell),\, w\neq 0} \frac{\|(\mathbb{L}w)(\cdot)\|_{L^2(\mathbb{R};\mathbb{C}^q)}}{\|w(\cdot)\|_{L^2(\mathbb{R};\mathbb{C}^\ell)}}.$$

We have shown in Subsection 2.2.4 that the input-output behaviour of Σ can equivalently be described in frequency domain by means of the transfer matrix $G(s)$. If $\tilde{w}(\cdot)$ is the Fourier-Plancherel transform of $w(\cdot) \in L^2(\mathbb{R};\mathbb{C}^\ell)$, the Fourier-Plancherel transform $\tilde{z}(\cdot)$ of $z(\cdot;w) \in L^2(\mathbb{R};\mathbb{C}^q)$ is determined by

$$\tilde{z}(\omega) = G(\imath\omega)\tilde{w}(\omega), \quad \text{a. e. } \omega \in \mathbb{R}.$$

The norms of the input-output operator $\mathbb{L}$ and of the transfer matrix $G(s)$ (regarded as an element of $H^\infty(\mathbb{C}_+;\mathbb{C}^{q\times\ell})$, see Definition A.3.44) are equal by Theorem 2.3.28.

$$\|\mathbb{L}\|_{\mathcal{L}(L^2(\mathbb{R};\mathbb{C}^\ell),L^2(\mathbb{R};\mathbb{C}^q))} = \sup_{\omega\in\mathbb{R}} \|D + C(\imath\omega I_n - A)^{-1}B\|_{\mathbb{C}^{q\times\ell}} = \|G\|_{H^\infty}. \tag{45}$$

Here and in the rest of the subsection we use the abbreviation H^∞ for $H^\infty(\mathbb{C}_+;\mathbb{C}^{q\times\ell})$. If Σ is at rest at time $t_0 = 0$ ($x(0) = 0$) the future input-output behaviour of Σ is described by the restriction $\mathbb{L}_+$ of the input-output operator $\mathbb{L}$ to the embedded subspace $L^2(\mathbb{R}_+,\mathbb{C}^\ell)$ of $L^2(\mathbb{R};\mathbb{C}^\ell)$. $\mathbb{L}_+$ is defined by

$$\begin{array}{rcl} \mathbb{L}_+ : L^2(\mathbb{R}_+;\mathbb{C}^\ell) & \to & L^2(\mathbb{R}_+;\mathbb{C}^q) \\ (\mathbb{L}_+w)(t) & = & Dw(t) + \int_0^t Ce^{A(t-s)}Bw(s)\,ds\,, \quad t \in \mathbb{R}_+, \end{array} \tag{46}$$

see (2.3.22). $z(\cdot) = (\mathbb{L}_+w)(\cdot)$ is the output signal of Σ produced by the input signal $w(\cdot) \in L^2(\mathbb{R}_+;\mathbb{C}^\ell)$ when the system Σ is initially at rest. By Theorem 2.3.28 $\mathbb{L}_+$ has the same norm as $\mathbb{L}$ and so we obtain from (42) the following characterization of the complex stability radius in time domain.

Proposition 5.3.23. *If $\mathbb{L}$ as in (44) and $\mathbb{L}_+$ as in (46) are the input-output maps associated with the continuous time system (43), then*

$$r_{\mathbb{C}}^-(A;B,C,D) = \|G\|_{H^\infty}^{-1} = \|\mathbb{L}\|^{-1} = \|\mathbb{L}_+\|^{-1}. \tag{47}$$

It follows that

$$1/r_{\mathbb{C}}^- = \inf\{\gamma;\ \forall w \in L^2(\mathbb{R}_+;\mathbb{C}^\ell) : \gamma\|w\|_{L^2} - \|\mathbb{L}_+w\|_{L^2} \geq 0\}. \tag{48}$$

This characterization is closely related to the following optimal control problem.

Problem 5.3.24. *Let $\gamma \geq 0$. For every $x^0 \in \mathbb{C}^n$ find a perturbation input $w_{opt}(\cdot) \in L^2(\mathbb{R}_+;\mathbb{C}^\ell)$ such that w_{opt} minimizes the cost functional*

$$J_\gamma(x^0,w) = \int_0^\infty \left[\gamma^2\|w(t)\|_{\mathbb{C}^\ell}^2 - \|z(t)\|_{\mathbb{C}^q}^2\right] dt \tag{49}$$

amongst all $w(\cdot) \in L^2(\mathbb{R}_+;\mathbb{C}^\ell)$, where $z(t) = z(t;x^0,w(\cdot))$ is determined by

$$\begin{array}{rcl} \dot{x}(t) & = & Ax(t) + Bw(t), \quad t \in \mathbb{R}_+, \quad x(0) = x^0 \\ z(t) & = & Cx(t) + Dw(t). \end{array} \tag{50}$$

The cost $J_\gamma(x^0,w)$ is well defined for all $x^0 \in \mathbb{C}^n$ and $w(\cdot) \in L^2(\mathbb{R}_+;\mathbb{C}^\ell)$ since $z(\cdot\,;x^0,w(\cdot)) \in L^2(\mathbb{R}_+;\mathbb{C}^q)$ by Proposition 2.3.10. For $x^0 = 0$ it follows from (46) that the corresponding output $z(\cdot) = z(\cdot\,;0,w(\cdot))$ is given by $z(\cdot) = (\mathbb{L}_+w)(\cdot)$ and so

$$J_\gamma(0,w) = \gamma^2\|w\|_{L^2}^2 - \|\mathbb{L}_+w\|_{L^2}^2.$$

Hence by (48) $(r_{\mathbb{C}}^-)^{-1}$ is the minimal value of γ for which the cost $J_\gamma(0,w)$ is non-negative for all inputs $w(\cdot) \in L^2(\mathbb{R}_+;\mathbb{C}^\ell)$. In other words, Problem 5.3.24 admits a solution for $x^0 = 0$ (with minimal costs $\inf_{w\in L^2} J_\gamma(0,w) = 0$) if and only if $\gamma \geq (r_{\mathbb{C}}^-)^{-1}$.[5]
Problems of the above kind will be analyzed in detail in Volume II where we will see that the solution is obtained by solving a quadratic matrix equation, called the *algebraic Riccati equation* (parametrized by γ)

$$A^*P + PA - C^*C - (B^*P - D^*C)^*(\gamma^2 I_n - D^*D)^{-1}(B^*P - D^*C) = 0. \quad (51)$$

A solution P of (51) is said to be *stabilizing* if

$$\sigma(A - B(\gamma^2 I_n - D^*D)^{-1}(B^*P - D^*C)) \subset \mathbb{C}_-.$$

For later use we note the following result which will be proved in the second volume.

Theorem 5.3.25. *Let $A \in \mathbb{C}^{n\times n}$ be $\mathbb{C}_-$-stable. Then the following statements are equivalent for $\gamma > \|D\|$.*

(i) For every $x^0 \in \mathbb{C}^n$, Problem 5.3.24 is solvable.

(ii) There exists a stabilizing solution $P \in \mathcal{H}_n(\mathbb{C})$ of (51).

(iii) $\|D + C(\imath\omega I_n - A)^{-1}B\| < \gamma$ for all $\omega \in \mathbb{R}$.

If (i) holds, Problem 5.3.24 is solved by the optimal input (in feedback form)

$$w_{opt}(t) = -(\gamma^2 I_n - D^*D)^{-1}(B^*P - D^*C)x_{opt}(t)$$

where

$$\dot{x}_{opt}(t) = \left[A - B(\gamma^2 I_n - D^*D)^{-1}(B^*P - D^*C)\right] x_{opt}(t), \quad x_{opt}(0) = x^0.$$

The minimal costs are given by $J_\gamma(x^0, w_{opt}) = \langle x^0, Px^0\rangle_{\mathbb{C}^n}$.

As a consequence of Theorems 5.3.9 and 5.3.25 we obtain the following characterization of the complex stability radius in terms of (51).

Corollary 5.3.26. *Suppose $\gamma > \|D\|$, then (51) has a stabilizing Hermitian solution (symmetric in the case of real data) if and only if $\gamma > 1/r_{\mathbb{C}}^-(A;B,C,D)$.*

Remark 5.3.27. (i) If $1/r_{\mathbb{C}}^-(A;B,C,D) = \|G\|_{H^\infty(\mathbb{C}_+;\mathbb{C}^{q\times\ell})} = \|\mathbb{L}_+\| > \|D\|$ the preceding characterization can be extended to include the minimal admissible value of the parameter γ:

$$\text{(51) has a Hermitian solution if and only if } \gamma \geq 1/r_{\mathbb{C}}^-(A;B,C,D).$$

For $\gamma = 1/r_{\mathbb{C}}^-(A;B,C,D)$ there will not, however, exist a *stabilizing* Hermitian solution of (51), but only one which achieves $\sigma(A - B(\gamma^2 I_n - D^*D)^{-1}(B^*P - D^*C)) \subset \overline{\mathbb{C}_-}$. Moreover, it can be proved (see Volume II) that a Hermitian solution with this spectral property is automatically the largest Hermitian solution of (51) (in the sense of the ordering $\preceq$ of Hermitian matrices) and hence unique.
(ii) It is usually not possible to solve the system of coupled quadratic equations (51) analytically. However, there is an extensive literature on the numerical solution of algebraic Riccati equations, see *Notes and References.* □

[5] Note that $\inf_{w\in L^2} J_\gamma(0,w) = -\infty$ if there exists $w \in L^2(\mathbb{R}_+;\mathbb{C}^\ell)$ such that $J_\gamma(0,w) < 0$. In this case Problem 5.3.24 has no solution for $x^0 = 0$.

We will make use of the above Riccati type characterizations of $r_{\mathbb{C}}^-$ in Section 5.6 where we deal with non-parametric model uncertainty. Here we use a closely related characterization in terms of Hamiltonian matrices in order to compute $r_{\mathbb{C}}^-$. With any system (A, B, C, D) and $\gamma > 0$ we associate the parametrized *Hamiltonian matrix*

$$H_\gamma = H_\gamma(A,B,C,D) = \begin{bmatrix} A+BR_\gamma^{-1}D^*C & -BR_\gamma^{-1}B^* \\ \gamma^2 C^*\hat{R}_\gamma^{-1}C & -(A+BR_\gamma^{-1}D^*C)^* \end{bmatrix}, \; \gamma^2 \notin \sigma(D^*D), \quad (52)$$

where $R_\gamma = \gamma^2 I_\ell - D^*D$ and $\hat{R}_\gamma = \gamma^2 I_q - DD^*$.

Definition 5.3.28. A matrix $H \in \mathbb{C}^{2n\times 2n}$ is said to be *Hamiltonian* if

$$JH = (JH)^* \quad \text{where } J = \begin{bmatrix} 0 & I_n \\ -I_n & 0 \end{bmatrix}. \quad (53)$$

The following explicit characterization shows that H_γ (52) is in fact Hamiltonian and has a spectrum which is symmetric with respect to the imaginary axis. We omit the (elementary) proof.

Proposition 5.3.29. *(i) A matrix* $H = \begin{bmatrix} H_{11} & H_{12} \\ H_{21} & H_{22} \end{bmatrix}$ *with* $H_{ij} \in \mathbb{C}^{n\times n}$, $i,j = 1,2$ *is Hamiltonian if and only if*

$$H_{11}^* = -H_{22}, \quad H_{12}^* = H_{12}, \quad H_{21}^* = H_{21}. \quad (54)$$

(ii) If H is Hamiltonian, its spectrum is symmetric with respect to the imaginary axis, i.e. $\lambda \in \sigma(H) \Leftrightarrow -\bar{\lambda} \in \sigma(H)$, and the algebraic/geometric multiplicities of the eigenvalues λ and $-\bar{\lambda}$ of H coincide. Moreover every imaginary eigenvalue of H has multiplicity ≥ 2.

The next lemma is fundamental for a Hamiltonian characterization of the complex stability radius. For later use we prove it without the assumption $\sigma(A) \subset \mathbb{C}_-$.

Lemma 5.3.30. *Suppose $\gamma > 0$, $\gamma^2 \notin \sigma(D^*D)$, $\omega \in \mathbb{R}$ and $\imath\mathbb{R} \subset \rho(A)$, then*

$$\imath\omega \in \sigma(H_\gamma) \iff \gamma^2 \in \sigma(G(\imath\omega)^*G(\imath\omega)). \quad (55)$$

Proof: We will make use of the results for determinants contained in Lemma A.1.13. By (A.1.8) we have $\gamma^2 \notin \sigma(DD^*)$. For ease of notation we set $A_\omega = \imath\omega I_n - A$. Then A_ω is invertible, $\imath\omega I_n + A^* = -A_\omega^*$ and $R_\gamma^{-1}D^* = D^*\hat{R}_\gamma^{-1}$. It follows that

$$\det(\imath\omega I_{2n} - H_\gamma) = \det\begin{bmatrix} A_\omega - BR_\gamma^{-1}D^*C & BR_\gamma^{-1}B^* \\ -\gamma^2 C^*\hat{R}_\gamma^{-1}C & -A_\omega^* + C^*DR_\gamma^{-1}B^* \end{bmatrix}$$

$$= (-1)^n \det(A_\omega^* A_\omega)\det\left(I_{2n} + \begin{bmatrix} -A_\omega^{-1}BR_\gamma^{-1}D^* & A_\omega^{-1}BR_\gamma^{-1} \\ \gamma^2(A_\omega^*)^{-1}C^*\hat{R}_\gamma^{-1} & -(A_\omega^*)^{-1}C^*DR_\gamma^{-1} \end{bmatrix}\begin{bmatrix} C & 0 \\ 0 & B^* \end{bmatrix}\right)$$

$$= (-1)^n \det(A_\omega^* A_\omega)\det\left(I_{q+\ell} + \begin{bmatrix} -(G(\imath\omega)-D)D^*\hat{R}_\gamma^{-1} & (G(\imath\omega)-D)R_\gamma^{-1} \\ \gamma^2(G(\imath\omega)-D)^*\hat{R}_\gamma^{-1} & -(G(\imath\omega)-D)^*DR_\gamma^{-1} \end{bmatrix}\right)$$

$$= (-1)^n \det(A_\omega^* A_\omega)(\det R_\gamma)^{-1}(\det \hat{R}_\gamma)^{-1}\det\begin{bmatrix} \gamma^2 I_q - G(\imath\omega)D^* & G(\imath\omega)-D \\ \gamma^2(G(\imath\omega)-D)^* & \gamma^2 I_\ell - G(\imath\omega)^*D \end{bmatrix}$$

$$= (-1)^n \det(A_\omega^* A_\omega)(\det R_\gamma)^{-1}(\det \hat{R}_\gamma)^{-1}\det\left(\begin{bmatrix} I_q & -G(\imath\omega) \\ G(\imath\omega)^* & -\gamma^2 I_\ell \end{bmatrix}\begin{bmatrix} \gamma^2 I_q & -D \\ D^* & -I_\ell \end{bmatrix}\right)$$

$$= (-1)^n \det(A_\omega^* A_\omega)(\det \hat{R}_\gamma)^{-1}\det(\gamma^2 I_\ell - G(\imath\omega)^*G(\imath\omega)).$$

This last equality follows by the Schur complement formula given in Lemma A.1.17. Since $\det(A_\omega^* A_\omega) \neq 0$, we obtain (55). □

(55) implies that for all $\omega \in \mathbb{R}$, $\gamma > 0$, $\gamma^2 \notin \sigma(D^*D)$

$$\imath\omega \in \sigma(H_\gamma) \iff \gamma = \sigma_j(G(\imath\omega)) \quad \text{for some} \quad j \in \underline{\ell}\,. \tag{56}$$

This relationship between the eigenvalues of H_γ and the singular values of $G(\imath\omega)$ is illustrated in Example 5.3.35. As a consequence of Lemma 5.3.30 we obtain the following characterization of $r_{\mathbb{C}}^-$ in terms of the spectrum of H_γ.

Proposition 5.3.31. *If $\gamma > \|D\|$ and $H_\gamma(A,B,C,D)$ is the Hamiltonian matrix (52) then*

$$\gamma \leq \|G\|_{H^\infty} \iff \gamma \leq \left(r_{\mathbb{C}}^-(A,B,C,D)\right)^{-1} \iff \sigma(H_\gamma) \cap \imath\mathbb{R} \neq \emptyset\,. \tag{57}$$

Moreover if $r_{\mathbb{C}}^- = r_{\mathbb{C}}^-(A,B,C,D) < \|D\|^{-1}$, then

$$\imath\omega_0 \in \sigma(H_{(r_{\mathbb{C}}^-)^{-1}}) \iff \|G(\imath\omega_0)\| = \max_{\omega\in\mathbb{R}} \|G(\imath\omega)\|\,. \tag{58}$$

Proof: By (22) it only remains to prove the second equivalence in (57). Since $\sigma(A) \subset \mathbb{C}_-$ and $R_\gamma^{-1} \to 0$ as $\gamma \to \infty$, we have $\sigma(H_\gamma) \cap \imath\mathbb{R} = \emptyset$ for large γ by continuity of the spectrum (see Corollary 4.2.1). Now suppose $\imath\omega_0 \in \sigma(H_\gamma)$, then by (55),

$$\gamma \leq \|G(\imath\omega_0)\| \leq \sup_{\omega\in\mathbb{R}} \|G(\imath\omega)\| = r_{\mathbb{C}}^{-1}\,.$$

Conversely if $0 < \|D\| < \gamma \leq \sup_{\omega\in\mathbb{R}} \|G(\imath\omega)\|$, then since $\lim_{\omega\to\infty} \|G(\imath\omega)\| = \|D\|$, there exists $\omega_0 \in \mathbb{R}$ such that $\gamma = \|G(\imath\omega_0)\|$. It then follows from (55) that $\imath\omega_0 \in \sigma(H_\gamma)$. This proves (57). Finally (58) follows by setting $\gamma = (r_{\mathbb{C}}^-)^{-1} = \sup_{\omega\in\mathbb{R}} \|G(\imath\omega)\|$ in (55). □

By (22) we have $r_{\mathbb{C}}^-(A,B,C,D) \leq \|D\|^{-1}$ and if $r_{\mathbb{C}}^-(A,B,C,D) < \|D\|^{-1}$ then by (57) and (58)

$$r_{\mathbb{C}}^-(A,B,C,D) = \min\{\gamma^{-1}; \gamma > \|D\| \text{ and } \sigma(H_\gamma) \cap \imath\mathbb{R} \neq \emptyset\}. \tag{59}$$

So we obtain the following graphical characterization of the stability radius $r_{\mathbb{C}}^-$: Draw the eigenloci of H_γ as γ decreases from ∞ to $\|D\|$. If $\sigma(H_\gamma) \cap \imath\mathbb{R} = \emptyset$ for all $\gamma > \|D\|$ then $r_{\mathbb{C}}^- = \|D\|^{-1}$. Otherwise let $\hat{\gamma}$ be the value of γ for which one (or more) of the eigenloci hits the imaginary axis for the first time. Then $r_{\mathbb{C}}^- = \hat{\gamma}^{-1}$.

Remark 5.3.32. It follows from Proposition 4.2.42 that the eigenvalues of H_γ can be represented as continuous piecewise analytic functions of the real parameter γ ("eigenloci") on $(\|D\|, \infty)$ and the ordered singular values $\sigma_j(\omega)$, of $G(\imath\omega)$ are continuous piecewise analytic functions of the frequency ω ("singular loci") on $\mathbb{R}$. The behaviour of these loci with decreasing γ is illustrated in Example 5.3.35.
Once $r_{\mathbb{C}}^- < \|D\|^{-1}$ has been determined, formula (58) yields the set of all global maxima of $\omega \to \|G(\imath\omega)\|$. Since every imaginary eigenvalue $\imath\omega_0 \in \sigma(H_\gamma)$ is of multiplicity ≥ 2, the set $\{\omega \in \mathbb{R}; \imath\omega \in \sigma(H_\gamma)\}$ has not more than n elements for each $\gamma > 0$. In particular, it follows from (58) that there are at most n global maxima of the function $\omega \to \|G(\imath\omega)\|$. □

The characterization (59) provides the basis of an algorithm for computing the complex stability radius $r_{\mathbb{C}}^{-}$, or equivalently $\|G\|_{H^\infty}$.

Algorithm 5.3.33. Assume $G(s) = D + C(sI - A)^{-1}B \not\equiv 0$ so that $\|G\|_{H^\infty} > 0$ and choose a tolerance level $\varepsilon \geq 0$. The algorithm starts from a lower bound γ_0 of $\|G\|_{H^\infty}$ and if $\varepsilon > 0$ it produces a strictly increasing sequence of lower bounds, exiting at a γ_k such that $\gamma_k \leq \|G\|_{H^\infty} \leq \gamma_k(1+\varepsilon)$.

1. Init: Set $\gamma_0 = \max\{\|D\|, \|G(0)\|\}$ and $k = 0$.
2. **loop**
   - 2.1 Set $\gamma_k^\varepsilon = (1+\varepsilon)\gamma_k$.
   - 2.2 Determine the set $\Omega_k = \{\omega \in \mathbb{R}; \imath\omega \in \sigma(H_{\gamma_k^\varepsilon})\}$.
   - 2.3 **If** $\Omega_k = \emptyset$ then **leave** loop.
   - 2.4 Determine $\omega_1^{(k)} < \omega_2^{(k)} < \ldots < \omega_{r_k}^{(k)}$ in Ω_k, such that $\{\omega_1^{(k)}, \ldots, \omega_{r_k}^{(k)}\} = \{\omega \in \Omega_k; \|G(\imath\omega)\| = \gamma_k^\varepsilon\}$.
   - 2.5 Determine the set $\mathbf{I}^{(k)} = \{\mathbf{I}_j^{(k)};\ j \in J_k\}$ of all intervals of the form $\mathbf{I}_j^{(k)} = (\omega_j^{(k)}, \omega_{j+1}^{(k)})$, $j \in \{1, \ldots, r_k\}$, such that $\|G(\imath\omega)\| > \gamma_k^\varepsilon$ for all (equivalently, for some) $\omega \in \mathbf{I}_j^{(k)}$.
   - 2.6 **If** there are no such intervals ($J_k = \emptyset$) then **leave** loop.
   - 2.7 For each interval $\mathbf{I}_j^{(k)}$ set $\mu_j^{(k)} = (\omega_j^{(k)} + \omega_{j+1}^{(k)})/2$, $j \in J_k$ and compute $\gamma_{k+1} = \max_{j \in J_k} \|G(\imath\,\mu_j^{(k)})\|$. Set $k = k+1$ and go to Step 2.1.
3. End: Return γ_k.

If $\|G\|_{H^\infty} \geq (1+\varepsilon)\max\{\|D\|, \|G(0)\|\}$ the iterative part of the algorithm comes into operation. Clearly $\gamma_{k+1} \geq (1+\varepsilon)\gamma_k$ and it has been shown, for real data with $\varepsilon = 0$, that $\gamma_k \nearrow \|G\|_{H^\infty}$ quadratically as $k \to \infty$, i.e. there exists $\alpha > 0$ such that

$$0 \leq \|G\|_{H^\infty} - \gamma_{k+1} \leq \alpha\,(\|G\|_{H^\infty} - \gamma_k)^2,$$

see *Notes and References.* If the loop is left in Step 2.3 at iteration $\hat{k}$ then necessarily $\gamma_{\hat{k}} \leq \|G\|_{H^\infty} < \gamma_{\hat{k}}^\varepsilon$, whereas if the loop is left in Step 2.6, then $\|G\|_{H^\infty} = \gamma_{\hat{k}}^\varepsilon$. Figure 5.3.5 illustrates the last two steps of the loop.

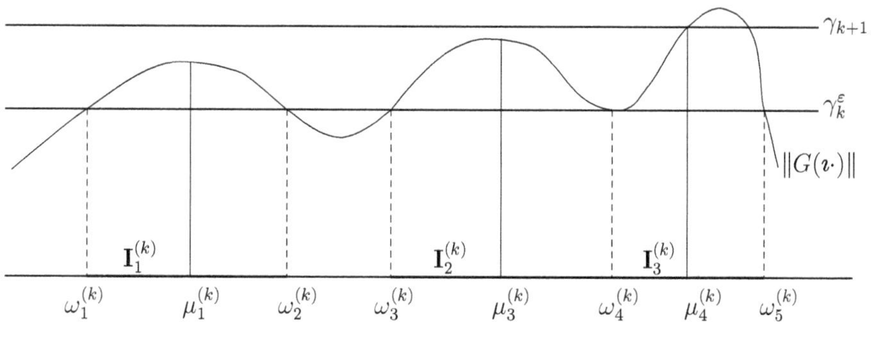

Figure 5.3.5: Steps 4 and 5 of the algorithm

Remark 5.3.34. Below we comment on the various steps of the algorithm.
Init In [80] the authors suggest an initial value γ_0 based on $\|D\|$ and the evaluation of $\|G(\imath\omega)\|$ at two different values of ω, one of which should be $\omega = 0$ and the other determined by the poles of the transfer function.
Step 2.2 If $\gamma_k^\varepsilon \le \|G\|_{H^\infty}$, $\gamma_k^\varepsilon > \|D\|$, then $\Omega_k \neq \emptyset$ by Proposition 5.3.31. By Remark 5.3.32 Ω_k has at most n elements. The set Ω_k is determined by an eigenvalue algorithm such as Francis' double-shift QR algorithm (see Section 4.5) and there may be computational difficulties. Another problem is how one decides whether or not H_γ has purely imaginary eigenvalues. H_γ may have this property but, due to rounding errors, all the computed eigenvalues of H_γ may have non-zero real parts. Therefore it is usual to specify a tolerance level and to include the imaginary parts of eigenvalues in Ω_k if their real parts have a modulus less than this tolerance level.
Step 2.4 This step is carried out by determining the largest singular value of $G(\imath\omega)$, $\omega \in \Omega_k$ and hence the computations are numerically stable. If $\Omega_k \neq \emptyset$ and $\gamma_k^\varepsilon > \|D\|$ there exists $\omega \in \mathbb{R}$ such that $\|G(\imath\omega)\| = \gamma_k^\varepsilon$ (whence $1 \le r_k \le |\Omega_k| \le n$). This follows from the intermediate value theorem because $\lim_{|\omega|\to\infty} G(\imath\omega) = D$ and $\|G(\imath\omega)\| \ge \gamma_k^\varepsilon$ for every $\omega \in \Omega_k$. By (56) we have $\gamma_k^\varepsilon < \|G(\imath\omega)\|$ for all $\omega \in \Omega_k \setminus \{\omega_1^{(k)}, ..., \omega_{r_k}^{(k)}\}$. Since the aim is to maximize $\|G(\imath\omega)\|$ this suggests that one should use *every* $\omega \in \Omega_k$ in order to determine the set of intervals $\mathbf{I}^{(k)}$. Indeed the algorithm proposed in [81] does exactly this and updates the γ_k by computing the value of $\|G(\imath\omega)\|$ at the midpoint of neighbouring points in Ω_k. Although this requires more computing time at each iteration, it may lead to faster convergence.
Of course it is very unlikely that the computed value $\|G(\imath\omega)\|$ will be exactly equal to γ_k^ε for any $\omega \in \Omega_k$, so again one uses a tolerance level at this step and includes those $\omega \in \Omega_k$ for which $\|G(\imath\omega)\| \le \gamma_k^\varepsilon(1+\varepsilon)$. The use of such a tolerance level leads to an algorithm which is intermediate between the two extremes of either using all the frequencies in Ω_k (see [80]) or using only the frequencies $\omega^{(k)}$ as recommended in [73]. Such an intermediate strategy may speed up the convergence in certain cases without leading to an excessive increase of the number of intervals which have to be examined in Step 2.4. For example, in Figure 5.3.5 it looks as if $\|G(\imath\omega_4^{(k)})\| = \gamma_k^\varepsilon$. However suppose that in fact $\|G(\imath\omega_4^{(k)})\| = \gamma_k^\varepsilon + \alpha$ for some small $\alpha > 0$ and $\omega_4^{(k)}$ is excluded in accordance with the algorithm in [73]. Then we see that the next value of γ_{k+1} would be smaller than the one shown (the interval $\mathbf{I}_2^{(k)}$ would become $(\omega_3^{(k)}, \omega_5^{(k)})$).
Step 2.5 It follows from (56) and the intermediate value theorem that $(\omega_j^{(k)}, \omega_{j+1}^{(k)}) \in \mathbf{I}^{(k)}$ if there exists $\omega \in (\omega_j^{(k)}, \omega_{j+1}^{(k)})$ such that $\|G(\imath\omega)\| > \gamma_k^\varepsilon$. The intervals in $\mathbf{I}^{(k)}$ can be found as follows: If there exists $\omega \in \Omega_k$ with $\omega \in (\omega_j^{(k)}, \omega_{j+1}^{(k)})$ then $(\omega_j^{(k)}, \omega_{j+1}^{(k)}) \in \mathbf{I}^{(k)}$. Else check whether $\|G(\imath(\omega_j^{(k)} + \omega_{j+1}^{(k)})/2)\| > \gamma_k^\varepsilon$. When this holds, $(\omega_j^{(k)}, \omega_{j+1}^{(k)})$ is an interval in $\mathbf{I}^{(k)}$. Otherwise it is not. □

If (A, B, C, D) are real and $\imath\omega \in \sigma(H_\gamma)$, then we also have $-\imath\omega \in \sigma(H_\gamma)$. Now suppose that $\max_{\omega\in\mathbb{R}} \|G(\imath\omega)\| = \|G(0)\|$, then 0 will be the midpoint of some $\mathbf{I} \in \mathbf{I}^{(0)}$. Hence the algorithm will converge in the first iteration, even if the initial value $\gamma_0 \in [\|D\|, \|G\|_{H^\infty}]$ was not based on $\omega = 0$.

Example 5.3.35. Consider unstructured perturbations $A \rightsquigarrow A + \Delta$ of the stable system $\dot{x} = Ax$ where $A = \begin{bmatrix} 0 & 1 & 3 \\ -10 & -2 & 0 \\ 0 & 0 & -1.5 \end{bmatrix}$. The associated transfer matrix and Hamiltonian have the form $G(s) = (sI_3 - A)^{-1}$, $H_\gamma = \begin{bmatrix} A & -\gamma^{-2}I_3 \\ I_3 & -A^* \end{bmatrix}$. The eigenloci of H_γ as γ decreases from ∞ and the three singular value loci of $G(\imath\omega)$ are shown in Figure 5.3.7. The eigenloci start at the encircled eigenvalues of A and $-A^*$ with $\gamma = \infty$ and end where the first two pairs of conjugate eigenvalues meet the imaginary axis for $\gamma = (r_{\mathbb{C}}^-)^{-1}$. Table 5.3.6 shows the convergence of the above algorithm to $r_{\mathbb{C}}^- = \|G(\imath\omega_0)\|^{-1} = 2.489^{-1} = 0.4018$, with $\omega_0 = 2.707$. The algorithm was initialized by setting $\gamma_0 = \|G(0)\| = 2.358$ and, since A is real, only non-negative values of ω were considered. There was only one interval $\mathbf{I}_{j_k}^{(k)}$ in each iteration (i.e. $J_k = \{j_k\}$, $k = 1, 2, 3$).

γ_k	$\omega_j^{(k)}$	$\mathbf{I}_j^{(k)}$	$\mu_j^{(k)}$
$\gamma_0 = 2.358$	$\omega_1^{(1)} = 0,\ \omega_2^{(1)} = 2.197,\ \omega_3^{(1)} = 3.076$	$\mathbf{I}_2^{(1)} = [2.197, 3.076]$	$\mu_2^{(1)} = 2.636$
$\gamma_1 = 2.484$	$\omega_1^{(2)} = 2.636,\ \omega_2^{(2)} = 2.774$	$\mathbf{I}_1^{(2)} = [2.636, 2.774]$	$\mu_1^{(2)} = 2.705$
$\gamma_2 = 2.489$	$\omega_1^{(3)} = 2.705,\ \omega_2^{(3)} = 2.709$	$\mathbf{I}_1^{(3)} = [2.705, 2.709]$	$\mu_1^{(3)} = 2.707$

Table 5.3.6: Convergence of the algorithm

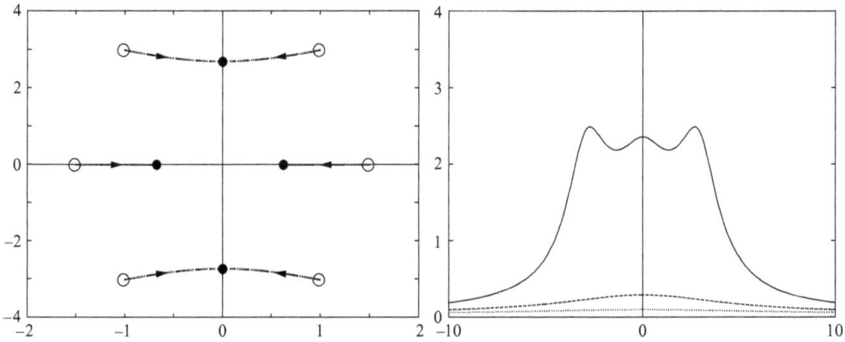

Figure 5.3.7: Eigenloci of H_γ and singular values of $G(\imath\omega)$ □

5.3.5 The Unstructured Case

In this subsection we consider unstructured perturbations $A \rightsquigarrow A + \Delta$ where $\Delta \in \mathbf{\Delta} = \mathbb{K}^{n\times n}$ is arbitrary. After characterizing the stability radii for this special case we will obtain bounds for them and explore when they are tight. One of the upper bounds is expressed in terms of the distance of $\sigma(A)$ to the boundary of the stability region $\partial\mathbb{C}_g$. We will see that under similarity transformations the stability radii come arbitrarily close to this distance and also arbitrarily close to *zero*.
The basic assumptions in the subsection are

- $\mathbb{C}_g \neq \emptyset$ is an open subset of $\mathbb{C}$ and $\mathbb{C}_b := \mathbb{C} \setminus \mathbb{C}_g \neq \emptyset$.
- $A \in \mathbb{K}^{n\times n}$ and $\sigma(A) \subset \mathbb{C}_g$.
- $\mathbf{\Delta} = \mathbb{K}^{n\times n}$ is provided with the spectral norm denoted by $\|\cdot\|$.[6]

[6]Some of the ensuing results have obvious generalizations to arbitrary operator norms.

- The perturbations are of the form

$$A \rightsquigarrow A(\Delta) = A + \Delta, \quad \Delta \in \mathbb{K}^{n\times n}.$$

Note that in this unstructured case $D = 0_{n\times n}$ (so that the well-posedness problem does not play a role) and the associated transfer matrix is the resolvent of A, $G(s) = (sI_n - A)^{-1}$, $s \in \rho(A)$. The corresponding stability radius is called the *unstructured stability radius* of A and is given by

$$d_{\mathbb{K}}(A;\mathbb{C}_g) := r_{\mathbb{K}}(A; I_n, I_n; \mathbb{C}_g) = \inf\{\|\Delta\|; \Delta \in \mathbb{K}^{n\times n},\ \sigma(A+\Delta)\cap\mathbb{C}_b \neq \emptyset\}. \quad (60)$$

If we denote the set of $\mathbb{C}_g$-unstable matrices in $\mathbb{K}^{n\times n}$ by $\mathcal{U}_n(\mathbb{K};\mathbb{C}_g)$, i.e.

$$\mathcal{U}_n(\mathbb{K};\mathbb{C}_g) = \{X \in \mathbb{K}^{n\times n}; \sigma(X)\cap\mathbb{C}_b \neq \emptyset\},$$

we see that $d_{\mathbb{K}}$ is the *distance*, within the normed matrix space $(\mathbb{K}^{n\times n}, \|\cdot\|)$, between A and the set $\mathcal{U}_n(\mathbb{K};\mathbb{C}_g)$. Since $\mathcal{U}_n(\mathbb{K};\mathbb{C}_g)$ is non-empty and closed in $\mathbb{K}^{n\times n}$ (by continuity of the spectrum) there exists a matrix $X \in \mathcal{U}_n(\mathbb{K};\mathbb{C}_g)$ such that

$$\|A - X\| = d_{\mathbb{K}}(A;\mathbb{C}_g) = \operatorname{dist}(A, \mathcal{U}_n(\mathbb{K};\mathbb{C}_g)).$$

Therefore the "inf" in (60) can be replaced by "min". As a consequence of the triangle inequality we have the following estimate

$$|d_{\mathbb{K}}(A;\mathbb{C}_g) - d_{\mathbb{K}}(\tilde{A};\mathbb{C}_g)| \leq \|A - \tilde{A}\|\,. \quad (61)$$

In particular, the function $d_{\mathbb{K}} : A \mapsto d_{\mathbb{K}}(A;\mathbb{C}_g)$ satisfies a global Lipschitz condition on $\mathbb{K}^{n\times n}$. In Figure 5.3.8 we illustrate the difference between the structured and

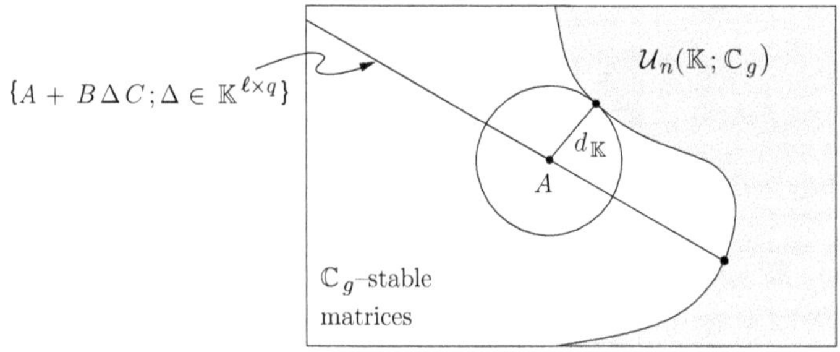

Figure 5.3.8: Unstructured and structured stability radius in $\mathbb{K}^{n\times n}$

unstructured stability radius in the metric space $\mathbb{K}^{n\times n}$. Note, however, that the structured stability radius $r_{\mathbb{K}}(A; B, C; \mathbb{C}_g)$ is not the distance in $\mathbb{K}^{n\times n}$ from A to the closest system in $\mathcal{U}_n(\mathbb{K};\mathbb{C}_g) \cap \{A + B\Delta C; \Delta \in \mathbb{K}^{\ell\times q}\}$ but the distance of the set of destabilizing perturbations from the zero matrix in $\mathbb{K}^{\ell\times q}$.

Theorem 5.3.36. *(i) If $A \in \mathbb{C}^{n\times n}$ is $\mathbb{C}_g$-stable then*

$$d_{\mathbb{C}}(A;\mathbb{C}_g) = \min_{s\in\partial\mathbb{C}_g} \sigma_n(sI_n - A). \quad (62)$$

(ii) If $A \in \mathbb{R}^{n\times n}$ is $\mathbb{C}_g$-stable then

$$d_{\mathbb{R}}(A;\mathbb{C}_g) = \min_{s\in\partial\mathbb{C}_g} \sup_{\gamma\in(0,1]} \sigma_{2n-1}((sI_n - A)^{\mathbb{R}}_{\gamma}), \tag{63}$$

where $(sI_n - A)^{\mathbb{R}}_{\gamma} = \begin{bmatrix} \alpha I_n - A & -\gamma\omega I_n \\ \gamma^{-1}\omega I_n & \alpha I_n - A \end{bmatrix}$ *for* $s = \alpha + \imath\omega \in \mathbb{C}$, $\gamma \in (0,1]$.

Proof: As an immediate consequence of Theorems 5.3.9 and 5.3.3

$$d_{\mathbb{C}}(A;\mathbb{C}_g) = \left[\max_{s\in\partial\mathbb{C}_g} \sigma_1((sI_n - A)^{-1})\right]^{-1}, \quad d_{\mathbb{R}}(A;\mathbb{C}_g) = \left[\max_{s\in\partial\mathbb{C}_g} \mu_{\mathbb{R}}((sI_n - A)^{-1})\right]^{-1}.$$

But $1/\sigma_1((sI_n - A)^{-1}) = \sigma_n(sI_n - A)$, $s \in \rho(A)$ and by (4.3.21)

$$\mu_{\mathbb{R}}((sI_n - A)^{-1}) = [\sup_{\gamma\in(0,1]} \sigma_{2n-1}((sI_n - A)^{\mathbb{R}}_{\gamma})]^{-1}, \quad s \in \rho(A).$$

Hence (62) and (63) hold. □

The distance $d_{\mathbb{C}}$ is relatively easy to compute, whereas the computation of the distance $d_{\mathbb{R}}$ is more difficult. So we explore the possibility of determining bounds for $d_{\mathbb{R}}$ for the classical stability regions $\mathbb{C}_g = \mathbb{C}_-$ and $\mathbb{D}$. Writing $d^-_{\mathbb{K}}(A) = d_{\mathbb{K}}(A;\mathbb{C}_-)$, $d^1_{\mathbb{K}}(A) = d_{\mathbb{K}}(A;\mathbb{D})$, we have by (62) and Corollary 4.3.9 that for $A \in \mathbb{R}^{n\times n}$

$$\min_{\omega\in\mathbb{R}} \sigma_n(\imath\omega I_n - A) = d^-_{\mathbb{C}}(A) \le d^-_{\mathbb{R}}(A) \le \sigma_n(A) \tag{64a}$$

$$\min_{\theta\in[0,2\pi]} \sigma_n(e^{\imath\theta} I_n - A) = d^1_{\mathbb{C}}(A) \le d^1_{\mathbb{R}}(A) \le \min\{\sigma_n(I_n - A), \sigma_n(I_n + A)\}. \tag{64b}$$

Unfortunately these estimates can all be very bad. In fact, as in the structured case, the quotients $d^-_{\mathbb{R}}(A)/d^-_{\mathbb{C}}(A)$, $d^1_{\mathbb{R}}(A)/d^1_{\mathbb{C}}(A)$ can be arbitrarily large (see Ex. 11 and Example 5.3.14). This is also true for the quotient of the upper bound $\sigma_n(A)$ and $d^-_{\mathbb{R}}$ in (64a) as we now illustrate.

Example 5.3.37. Consider the normal matrices $A_\beta = \begin{bmatrix} -1 & -\beta \\ \beta & -1 \end{bmatrix}$, $\beta \in \mathbb{R}$. We will show in Proposition 5.3.38 that $d^-_{\mathbb{C}}(A_\beta) = d^-_{\mathbb{R}}(A_\beta) = \mathrm{dist}(\sigma(A_\beta), \imath\mathbb{R})$. Now $\sigma(A_\beta) = \{-1 \pm \imath\beta\}$ and hence $d^-_{\mathbb{C}}(A_\beta) = d^-_{\mathbb{R}}(A_\beta) = 1$ for all $\beta \in \mathbb{R}$. But $\sigma_2(A_\beta) = \sqrt{1+\beta^2}$ and so the quotient $\sigma_2(A_\beta)/d^-_{\mathbb{R}}(A_\beta)$ is unbounded as $\beta \to \infty$. □

One might think that the distance, $\mathrm{dist}(\sigma(A), \partial\mathbb{C}_g)$, of the spectrum of A from the boundary of the stability region (see Figure 5.3.9) is a robustness indicator for the stability of A. So it is interesting to explore the relationship between this distance and $d_{\mathbb{C}}(A;\mathbb{C}_g)$, $d_{\mathbb{R}}(A;\mathbb{C}_g)$. We do this by applying Proposition 5.2.44 and using the fact that unitary (orthogonal) similarity transformations do not change the distances in $\mathbb{K}^{n\times n}$.

$$d_{\mathbb{K}}(A;\mathbb{C}_g) = d_{\mathbb{K}}(UAU^*;\mathbb{C}_g), \qquad A \in \mathbb{K}^{n\times n}, \ U \in \mathbf{U}_n(\mathbb{K}).$$

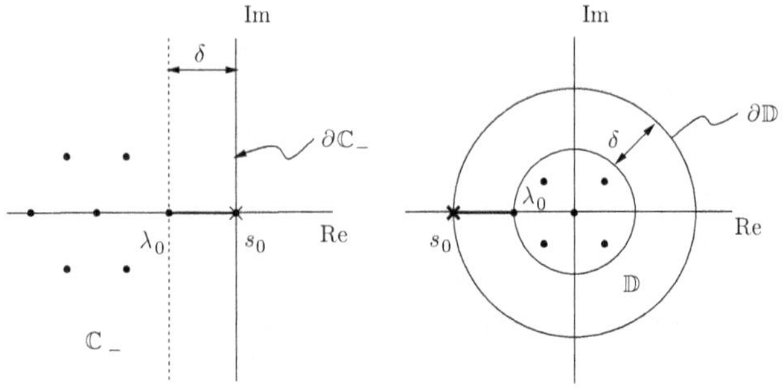

Figure 5.3.9: $\delta = \text{dist}(\sigma(A), \partial\mathbb{C}_g) = |\lambda_0 - s_0|$ for $\mathbb{C}_g = \mathbb{C}_-$, $\mathbb{C}_g = \mathbb{D}$

Proposition 5.3.38. *(i) If $A \in \mathbb{C}^{n\times n}$, then*

$$d_{\mathbb{C}}(A; \mathbb{C}_g) \le \text{dist}(\sigma(A), \partial\mathbb{C}_g) \tag{65}$$

and equality holds if A is normal.
(ii) If $A \in \mathbb{R}^{n\times n}$ and one of the following assumptions is satisfied

$$\begin{aligned} \text{dist}(\sigma(A), \partial\mathbb{C}_g) &= \text{dist}(\sigma(A) \setminus \mathbb{R}, \partial\mathbb{C}_g) \quad \textit{or} \\ \text{dist}(\sigma(A), \partial\mathbb{C}_g) &= \text{dist}(\sigma(A) \cap \mathbb{R}, \partial\mathbb{C}_g \cap \mathbb{R}), \end{aligned} \tag{66}$$

then

$$d_{\mathbb{C}}(A; \mathbb{C}_g) \le d_{\mathbb{R}}(A; \mathbb{C}_g) \le \text{dist}(\sigma(A), \partial\mathbb{C}_g) \tag{67}$$

and again equalities hold if A is normal.

Proof: (i) There exists $(\lambda_0, s_0) \in \sigma(A) \times \partial\mathbb{C}_g$ such that $|\lambda_0 - s_0| = \text{dist}(\sigma(A), \partial\mathbb{C}_g)$, so $\sigma(A + (s_0 - \lambda_0)I_n) \cap \partial\mathbb{C}_g \neq \emptyset$ and hence (65) holds. If A is normal there exists a unitary matrix U such that $UAU^* = \text{diag}(\lambda_1, \ldots, \lambda_n)$. For each $s \in \partial\mathbb{C}_g$

$$\sigma_n(sI_n - A) = \sigma_n(sI_n - \text{diag}(\lambda_1, \ldots, \lambda_n)) = \min_{i\in\underline{n}} |s - \lambda_i| = \text{dist}(\sigma(A), s).$$

Hence it follows from (62) that equality holds in (65) when A is normal.
(ii) Let $A \in \mathbb{R}^{n\times n}$ and $\delta := \text{dist}(\sigma(A), \partial\mathbb{C}_g)$. If the first assumption in (66) is satisfied there is a $\lambda \in \sigma(A)$ such that $\text{Im}\,\lambda \neq 0$ and $\text{dist}(\lambda, \partial\mathbb{C}_g) = \delta$. By Proposition 5.2.44, there exists $\Delta \in \mathbb{R}^{n\times n}$ such that $\|\Delta\| \le \delta$ and $\sigma(A+\Delta) \not\subset \mathbb{C}_g$, hence $d_{\mathbb{R}}(A; \mathbb{C}_g) \le \delta$. Now suppose the second assumption in (66) is satisfied. Then there exist $s_0 \in \partial\mathbb{C}_g \cap \mathbb{R}$, $\lambda_0 \in \sigma(A) \cap \mathbb{R}$ such that $|\lambda_0 - s_0| = \delta$. It follows that $\sigma(A + (s_0 - \lambda_0 I_n)) \cap \mathbb{C}_b \neq \emptyset$ and hence again $d_{\mathbb{R}}(A; \mathbb{C}_g) \le \delta$. This proves (67), since $d_{\mathbb{C}}(A; \mathbb{C}_g) \le d_{\mathbb{R}}(A; \mathbb{C}_g)$. If A is normal, equality holds in (65) by (i) and so equality must hold in (ii). □

Assumption (66) is satisfied for all $A \in \mathbb{R}^{n\times n}$ if $\mathbb{C}_g = \mathbb{C}_-$ or $\mathbb{C}_g = \mathbb{D}$, see Figure 5.3.9. The assumption (66) excludes spectral locations as depicted in Figure 5.3.10 where the minimal distance $\text{dist}(\lambda, \partial\mathbb{C}_g) = \delta$ is only achieved for real $\lambda \in \sigma(A)$ which lie closer to non-real points on $\partial\mathbb{C}_g$ than to real points on this boundary. In Exs. 16, 17 it is shown that Assumption (66) cannot be dispensed with. By Proposition 5.3.38 $\text{dist}(\sigma(A), \partial\mathbb{C}_g)$ is a good robustness measure for normal matrices, but it may be very misleading for non-normal ones as the following example illustrates.

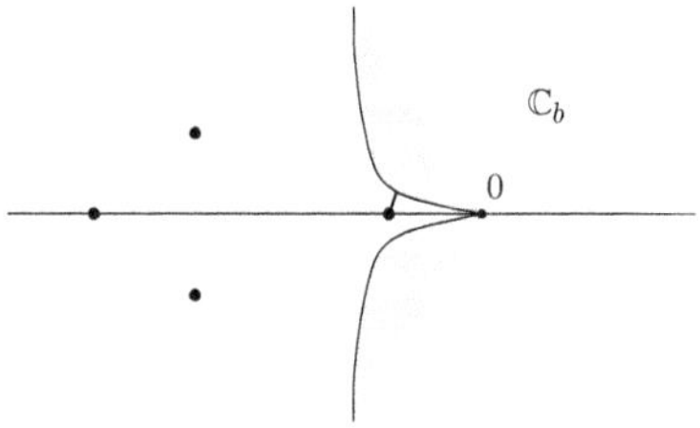

Figure 5.3.10: Assumption (66) not satisfied

Example 5.3.39. Let

$$A_k = \begin{bmatrix} -k & k^4 \\ 0 & -k^2 \end{bmatrix}, \qquad \Delta_k = \begin{bmatrix} 0 & 0 \\ k^{-1} & 0 \end{bmatrix}, \qquad k \in \mathbb{N}.$$

Then $\det(A_k + \Delta_k) = 0$. Hence $d_{\mathbb{R}}^{-}(A_k) \leq k^{-1} \to 0$, whereas $\operatorname{dist}(\sigma(A_k), \imath\mathbb{R}) \to \infty$ as $k \to \infty$. □

Non-unitary similarity transformations $A \mapsto TAT^{-1}$, $T \in \mathbf{Gl}_n(\mathbb{K})$ will in general change the distance from instability. Hence in concrete applications it is essential to carefully choose and scale the coordinate system of the state space in order to be sure that $d_{\mathbb{K}}(A; \mathbb{C}_g)$ is a meaningful indicator of robustness.
We will show that under similarity the stability radii may be made arbitrarily small and be increased arbitrarily close to $\operatorname{dist}(\sigma(A), \partial\mathbb{C}_g)$ but no further.

Proposition 5.3.40. *Suppose $A \in \mathbb{K}^{n\times n}$, $A \notin \mathbb{K}I_n$, then*

$$\sup_{T\in\mathbf{Gl}_n(\mathbb{K})} d_{\mathbb{K}}(TAT^{-1}; \mathbb{C}_g) = \operatorname{dist}(\sigma(A), \partial\mathbb{C}_g), \tag{68}$$

$$\inf_{T\in\mathbf{Gl}_n(\mathbb{K})} d_{\mathbb{K}}(TAT^{-1}; \mathbb{C}_g) = 0 \tag{69}$$

provided that, in the real case, the conditions (66) *for* (68) *(resp. $\partial\mathbb{C}_g \cap \mathbb{R} \neq \emptyset$ for* (69)*) are satisfied.*

Proof: By Proposition 5.3.38 and the assumption (66) we get the inequality $\leq$ in (68) for $\mathbb{K} = \mathbb{R}$ and $\mathbb{C}$. By Lemma 5.2.47, for any given $\varepsilon > 0$, there exists a $T \in \mathbf{Gl}_n(\mathbb{K})$ and a normal matrix $A_d \in \mathbb{K}^{n\times n}$ with $\sigma(A_d) = \sigma(A)$ such that $\|TAT^{-1} - A_d\| < \varepsilon$. Hence, by Proposition 5.3.38 and the triangle inequality

$$d_{\mathbb{K}}(TAT^{-1}; \mathbb{C}_g) \geq d_{\mathbb{K}}(A_d; \mathbb{C}_g) - \varepsilon = \operatorname{dist}(\sigma(A), \partial\mathbb{C}_g) - \varepsilon .$$

This proves (68). To prove (69) we use (62) and (63). For any $s_0 \in \partial\mathbb{C}_g$ (resp. $s_0 \in \partial\mathbb{C}_g \cap \mathbb{R}$ for $\mathbb{K} = \mathbb{R}$) and any $T \in \mathbf{Gl}_n(\mathbb{K})$ we have

$$d_{\mathbb{K}}(TAT^{-1}; \mathbb{C}_g) \leq \sigma_n(s_0 I_n - TAT^{-1})$$

by Corollary 4.3.9. But $\sigma_n(s_0 I_n - TAT^{-1}) = \|T(s_0 I_n - A)^{-1}T^{-1}\|^{-1}$ and so (69) follows from Lemma 5.2.47. □

In geometric terms (69) implies that every similarity orbit in $\mathbb{K}^{n\times n}$ which is not reduced to a singleton comes arbitrarily close to the set $\mathcal{U}_n(\mathbb{K}; \mathbb{C}_g)$ of $\mathbb{C}_g$–unstable matrices.

5.3.6 Dependence on System Data

Our aim in this subsection is to study how the stability radius $r_{\mathbf{\Delta}}(A; B, C, D; \mathbb{C}_g)$ depends on the data (A, B, C, D). We will carry this out under the general assumptions of the first subsection, namely

- $\mathbb{C}_g \neq \emptyset$ is an open subset of $\mathbb{C}$ and $\mathbb{C}_b := \mathbb{C} \setminus \mathbb{C}_g \neq \emptyset$.
- $(A, B, C, D) \in \mathcal{S}_{n,\ell,q}(\mathbb{K}; \mathbb{C}_g)$ where
$$\mathcal{S}_{n,\ell,q}(\mathbb{K}; \mathbb{C}_g) = \{(A, B, C, D) \in \mathbf{L}_{n,\ell,q}(\mathbb{K}); \sigma(A) \subset \mathbb{C}_g\}$$
is provided with the topology induced from $\mathbf{L}_{n,\ell,q}(\mathbb{K})$.
- $\mathbf{\Delta} \subset \mathbb{K}^{\ell \times q}$ is a closed convex cone and $\operatorname{span}_{\mathbb{K}} \mathbf{\Delta}$ is provided with a norm $\|\cdot\|_{\mathbf{\Delta}}$ which is an operator norm with respect to a given pair of norms on $\mathbb{K}^\ell$, $\mathbb{K}^q$.
- The perturbations are given by $A \rightsquigarrow A(\Delta) = A + B(I_\ell - \Delta D)^{-1}\Delta C$, $\Delta \in \mathbf{\Delta}_0$ where $\mathbf{\Delta}_0 = \{\Delta \in \mathbf{\Delta}; \det(I_\ell - \Delta D) \neq 0\}$.

We know from (61) that the unstructured stability radius $d_{\mathbb{K}}(A; \mathbb{C}_g)$ is continuous on $\mathbb{K}^{n \times n}$. However there is no reason to expect the *structured* stability radii, $r_{\mathbf{\Delta}}(A; B, C, D; \mathbb{C}_g)$, to be continuous either as a function of A (with B, C, D fixed) or as a function of the structure matrices B, C, D with A fixed, see Figure 5.3.11. In the case $r_{\mathbf{\Delta}} < \mu_{\mathbf{\Delta}}(D)^{-1}$ discontinuity should be expected at quadruples (A_0, B_0, C_0, D_0) for which the set of perturbed systems $\{A_0 + B_0(I_\ell - \Delta D_0)^{-1}\Delta C_0; \Delta \in \mathbf{\Delta}_0\}$ intersects the set $\mathcal{U}_n(\mathbb{K}; \mathbb{C}_g)$ tangentially at all points $A_0 + B_0(I_\ell - \Delta_0 D_0)^{-1}\Delta_0 C_0$ where $\Delta_0 \in \mathbf{\Delta}_0$ is any minimum norm destabilizing perturbation, see Figure 5.3.11 (which illustrates the case for $D = 0$).

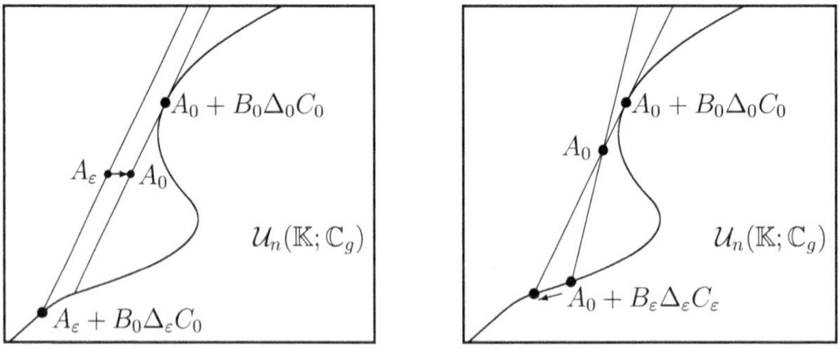

(i) Perturbed system matrix (ii) Perturbed structure matrices

Figure 5.3.11: Discontinuity of structured stability radii

The next proposition shows that the stability radius does not suddenly *decrease* as a result of small perturbations of the data.

Proposition 5.3.41. *The map*

$$r_{\mathbf{\Delta}}(\cdot\,; \mathbb{C}_g) : \mathcal{S}_{n,\ell,q}(\mathbb{K}; \mathbb{C}_g) \to \mathbb{R}_+, \quad (A, B, C, D) \mapsto r_{\mathbf{\Delta}}(A; B, C, D; \mathbb{C}_g)$$

is lower semicontinuous, i.e. if $(A_0, B_0, C_0, D_0) \in \mathcal{S}_{n,\ell,q}(\mathbb{K}; \mathbb{C}_g)$ *and* $\alpha \in \mathbb{R}$ *satisfies* $\alpha < r_{\mathbf{\Delta}}(A_0; B_0, C_0, D_0; \mathbb{C}_g)$*, then there exists a neighbourhood* Ω *of* (A_0, B_0, C_0, D_0) *in* $\mathcal{S}_{n,\ell,q}(\mathbb{K}; \mathbb{C}_g)$ *such that* $\alpha < r_{\mathbf{\Delta}}(A; B, C, D; \mathbb{C}_g)$ *for all* $(A, B, C, D) \in \Omega$.

Proof: Let $r_{\mathbf{\Delta}}(A_0; B_0, C_0, D_0; \mathbb{C}_g) > \alpha$ and suppose there exists a sequence of systems $(A_k, B_k, C_k, D_k) \in \mathcal{S}_{n,\ell,q}(\mathbb{K}; \mathbb{C}_g)$ converging to (A_0, B_0, C_0, D_0) such that $r_{\mathbf{\Delta}}(A_k, B_k, C_k, D_k; \mathbb{C}_g) \leq \alpha$ for all $k \in \mathbb{N}$. Then, for every $k \in \mathbb{N}$ there exists a perturbation $\Delta_k \in \mathbf{\Delta}$, satisfying $\|\Delta_k\|_{\mathbf{\Delta}} \leq \alpha$ and either $\det(I_\ell - \Delta_k D_k) = 0$ and $\|\Delta_k\| = \mu_{\mathbf{\Delta}}(D_k)^{-1}$ or

$$\|\Delta_k\| < \mu_{\mathbf{\Delta}}(D_k)^{-1} \text{ and } \sigma(A_k + B_k(I_\ell - \Delta_k D_k)^{-1}\Delta_k C_k) \cap \mathbb{C}_b \neq \emptyset.$$

Passing over to a subsequence if necessary we may assume (Δ_k) converges to a matrix $\Delta_0 \in \mathbf{\Delta}$. Then $\|\Delta_0\|_{\mathbf{\Delta}} \leq \alpha$ and either $\det(I_\ell - \Delta_0 D_0) = 0$ or $\det(I_\ell - \Delta_0 D_0) \neq 0$ and $\sigma(A_0 + B_0(I_\ell - \Delta_0 D_0)^{-1}\Delta_0 C_0) \cap \mathbb{C}_b \neq \emptyset$ (by the continuity of the spectrum). Hence $r_{\mathbf{\Delta}}(A_0; B_0, C_0, D_0; \mathbb{C}_g) \leq \alpha$ by definition (2) and so yields a contradiction. □

As a consequence of Proposition 4.4.28 and Theorem 5.3.3 the stability radius depends continuously on the data for complex perturbation structures.

Proposition 5.3.42. *Suppose* $\mathbb{C}\mathbf{\Delta} = \mathbf{\Delta}$*, then the map*

$$r_{\mathbf{\Delta}}(\,\cdot\,; \mathbb{C}_g) : \mathcal{S}_{n,\ell,q}(\mathbb{C}; \mathbb{C}_g) \to \mathbb{R}_+, \quad (A, B, C, D) \mapsto r_{\mathbf{\Delta}}(A; B, C, D; \mathbb{C}_g)$$

is continuous.

Proof: By Proposition 5.3.41 it only remains to prove that $r_{\mathbf{\Delta}}(\,\cdot\,; \mathbb{C}_g)$ is upper semicontinuous. Let $(A_0, B_0, C_0, D_0) \in \mathcal{S}_{n,\ell,q}(\mathbb{C}; \mathbb{C}_g)$ and suppose that $\alpha \in \mathbb{R}$ and $r_{\mathbf{\Delta}}(A_0; B_0, C_0, D_0; \mathbb{C}_g) < \alpha$. By Theorem 5.3.3

$$r_{\mathbf{\Delta}}(A_0; B_0, C_0, D_0; \mathbb{C}_g) = \min\{\inf_{s \in \partial\mathbb{C}_g} \mu_{\mathbf{\Delta}}(G_0(s))^{-1}, \mu_{\mathbf{\Delta}}(D_0)^{-1}\} \tag{70}$$

where $G_0(s) = D_0 + C_0(sI_n - A_0)^{-1}B_0$. If $\inf_{s \in \partial\mathbb{C}_g} \mu_{\mathbf{\Delta}}(G_0(s))^{-1} < \alpha$, then there exists an $s_0 \in \partial\mathbb{C}_g$ such that $\mu_{\mathbf{\Delta}}(G_0(s_0))^{-1} < \alpha$ and hence by the continuity of $\mu_{\mathbf{\Delta}}$ on $\mathbb{C}^{q \times \ell}$ and the continuity of the map $(A, B, C, D) \mapsto G(s_0)$, there exists a neighbourhood Ω of (A_0, B_0, C_0, D_0) in $\mathcal{S}_{n,\ell,q}(\mathbb{K}; \mathbb{C}_g)$ such that $\mu_{\mathbf{\Delta}}(G(s_0))^{-1} < \alpha$ for all $G(s_0) = D + C(s_0 I_n - A)^{-1}B$, $(A, B, C, D) \in \Omega$. By (70) this implies $r_{\mathbf{\Delta}}(A; B, C, D; \mathbb{C}_g) < \alpha$ for all (A, B, C, D) in this neighbourhood. On the other hand if $\mu_{\mathbf{\Delta}}(D_0)^{-1} < \alpha$, again by the continuity of $\mu_{\mathbf{\Delta}}$ (Proposition 4.4.28), we have $\mu_{\mathbf{\Delta}}(D)^{-1} < \alpha$ and hence $r_{\mathbf{\Delta}}(A; B, C, D; \mathbb{C}_g) < \alpha$ for all $(A, B, C, D) \in \mathcal{S}_{n,\ell,q}(\mathbb{C}; \mathbb{C}_g)$ with D sufficiently close to D_0. Thus $r_{\mathbf{\Delta}}(\cdot; \mathbb{C}_g)$ is upper semicontinuous and this together with Proposition 5.3.41 completes the proof. □

The following definition is needed to give a sufficient condition for the real stability radius to be continuous at a given quadruple (A, B, C, D).

Definition 5.3.43. Given $(A, B, C, D) \in \mathcal{S}_{n,\ell,q}(\mathbb{K}; \mathbb{C}_g)$, a destabilizing perturbation matrix $\Delta_0 \in \mathbf{\Delta}_0$ is said to be *strongly destabilizing* for (A, B, C, D) (with respect to $\mathbf{\Delta}$ and $\mathbb{C}_g$), if for every $\varepsilon > 0$ there exists $\Delta \in \mathbf{\Delta}_0$, $\|\Delta - \Delta_0\|_{\mathbf{\Delta}} < \varepsilon$ such that $\sigma(A + B(I_\ell - \Delta D)^{-1}\Delta C) \cap \operatorname{int}\mathbb{C}_b \neq \emptyset$ where $\operatorname{int}\mathbb{C}_b$ denotes the interior of $\mathbb{C}_b$.

By definition there does not exist any strongly destabilizing perturbation if $\operatorname{int}\mathbb{C}_b = \emptyset$. But for all stability regions of practical interest we have $\mathbb{C}_b = \overline{\operatorname{int}\mathbb{C}_b}$. The following proposition applies to both real and complex stability radii.

Proposition 5.3.44. *The stability radius, $r_{\mathbf{\Delta}}(\,\cdot\,;\mathbb{C}_g)$ is continuous at every quadruple $(A_0, B_0, C_0, D_0) \in \mathcal{S}_{n,\ell,q}(\mathbb{K};\mathbb{C}_g)$ for which there exists a strongly destabilizing perturbation $\Delta_0 \in \mathbf{\Delta}_0$ of minimum norm $\|\Delta_0\|_{\mathbf{\Delta}} = r_{\mathbf{\Delta}}(A_0; B_0, C_0, D_0; \mathbb{C}_g)$.*

Proof: It only remains to prove that $r_{\mathbf{\Delta}}(\,\cdot\,;\mathbb{C}_g)$ is upper semicontinuous at the quadruple (A_0, B_0, C_0, D_0). If this were not the case there would exist $\varepsilon > 0$ such that each neighbourhood of (A_0, B_0, C_0, D_0) in $\mathcal{S}_{n,\ell,q}(\mathbb{K};\mathbb{C}_g)$ contains a quadruple (A, B, C, D) satisfying $r_{\mathbf{\Delta}}(A; B, C, D; \mathbb{C}_g) \geq r_{\mathbf{\Delta}}(A_0; B_0, C_0, D_0; \mathbb{C}_g) + \varepsilon$. But there exists $\Delta \in \mathbf{\Delta}$ with $\|\Delta - \Delta_0\|_{\mathbf{\Delta}} < \varepsilon$ such that $\det(I_\ell - \Delta D_0) \neq 0$ and $\sigma(A_0 + B_0(I_\ell - \Delta D_0)^{-1}\Delta C_0) \cap \operatorname{int}\,\mathbb{C}_b \neq \emptyset$. Hence there exists a neighbourhood Ω of (A_0, B_0, C_0, D_0), such that $I_\ell - \Delta D$ is invertible and $\sigma(A + B(I_\ell - \Delta D)^{-1}\Delta C) \cap \operatorname{int}\mathbb{C}_b \neq \emptyset$ for all $(A, B, C, D) \in \Omega$. It follows that $r_{\mathbf{\Delta}}(A; B, C, D; \mathbb{C}_g) \leq \|\Delta\|_{\mathbf{\Delta}} < \|\Delta_0\|_{\mathbf{\Delta}} + \varepsilon = r_{\mathbf{\Delta}}(A_0; B_0, C_0, D_0; \mathbb{C}_g) + \varepsilon$ for all $(A, B, C, D) \in \Omega$ and hence a contradiction. □

It is easily deduced from the openness of the spectral value sets $\sigma_{\mathbf{\Delta}}(A; B, C, D; \delta) \setminus \sigma(A)$ for complex perturbation structures $\mathbf{\Delta}$, that every destabilizing $\Delta_0 \in \mathbf{\Delta}_0$ is strongly destabilizing with respect to $\mathbf{\Delta}$. Hence Proposition 5.3.42 can be viewed as a corollary of the preceding result. The following example illustrates that a perturbation $\Delta_0 \in \mathbb{R}^{\ell\times q}$ can be destabilizing without being strongly destabilizing with respect to $\mathbf{\Delta} = \mathbb{R}^{\ell\times q}$.

Example 5.3.45. Consider the Hurwitz stable quadruple defined by

$$A_0 = \begin{bmatrix} 0 & 1 & 0 & 0 \\ 0 & 0 & 1 & 0 \\ 0 & 0 & 0 & 1 \\ -1 & -4 & -6 & -4 \end{bmatrix}, \quad B_0 = \begin{bmatrix} 0 \\ 0 \\ 0 \\ 1 \end{bmatrix}, \quad C_0 = [\ 17 \quad 64 \quad 97 \quad 0\ ],\ D_0 = 0$$

and let $\mathbf{\Delta} = \mathbb{R}$ $(q = \ell = 1)$. The associated (scalar) transfer function is $G_0(s) = c(s)/p(s)$ where $p(s) = (s+1)^4$, $c(s) = 97s^2 + 64s + 17$. From (34) we have

$$r_{\mathbb{R}}(A_0; B_0, C_0; \mathbb{C}_-) = (\max\{|G_0(\imath\omega)|; \omega \in \mathbb{R},\ G_0(\imath\omega) \in \mathbb{R}\})^{-1}.$$

Now $G_0(\imath\omega)$ is real if and only if $\omega = 0$ or $\omega = \pm 1/3$, and the corresponding values of $|G_0(\imath\omega)|$ are 17 and 18. Thus $r_{\mathbb{R}}(A_0; B_0, C_0; \mathbb{C}_-) = 1/18$. Using the Routh–Hurwitz test it can be verified that the perturbed system matrix $A_0 + B_0\Delta C_0$ is $\mathbb{C}_-$-stable for all $\Delta \in (-1/17, 1/17)$ with $\Delta \neq \Delta_0 = 1/18$, hence the affine subspace of perturbed systems, $\{A_0 + B_0\Delta C_0; \Delta \in \mathbb{R}\}$, touches the set of $\mathbb{C}_-$-unstable matrices tangentially at $A_0 + B_0\Delta_0 C_0$, i.e. $\Delta_0 = 1/18$ is not strongly destabilizing. This is illustrated by Figure 5.3.45 which shows the graph of the spectral abscissa $\alpha(A_0 + B_0\Delta C_0)$ as a function of the perturbation $\Delta \in [0.05, 0.06]$.
Now change C_0 to $C_\varepsilon = [\ 17 \quad (64 - 4\varepsilon) \quad 97 \quad 0\ ]$ with the corresponding transfer function $G_\varepsilon(s) = c_\varepsilon(s)/p(s)$. For $0 < \varepsilon \ll 1$ one can show that $G_\varepsilon(\imath\omega)$ is real if and only if $\omega = 0$. Since $|G_\varepsilon(0)| = 17$, we have

$$r_{\mathbb{R}}(A_0, B_0, C_\varepsilon; \mathbb{C}_-) = 1/17, \quad \text{for all } 0 < \varepsilon \ll 1.$$

Similarly, one can show that $r_{\mathbb{R}}(A_\varepsilon, B_0, C_0; \mathbb{C}_-) = 1/17$ for all small $\varepsilon > 0$ where A_ε is obtained from A_0 by subtracting ε from the second entry in the last row, see Ex. 18. □

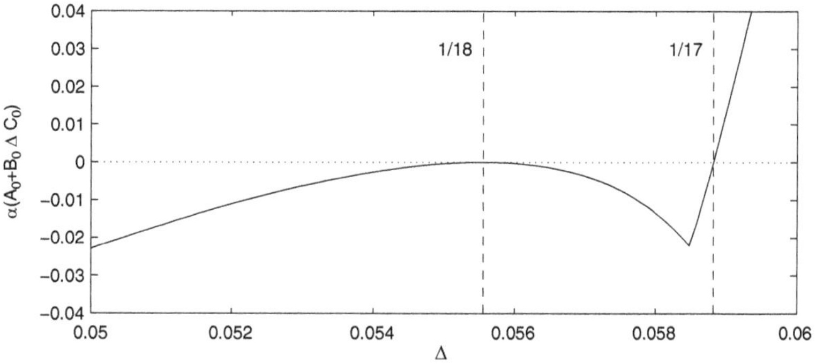

Figure 5.3.12: Graph of $\Delta \mapsto \alpha(A_0 + B_0\Delta C_0)$

This example confirms the above analysis and shows that the real stability radius may jump upwards at quadruples (A_0, B_0, C_0, D_0) for which the smallest destabilizing perturbations are not strongly destabilizing. Another sufficient criterion for the continuity of $r_{\mathbb{R}}(\cdot; \mathbb{C}_g)$ at a given quadruple (A_0, B_0, C_0, D_0) can be derived from Theorem 4.4.48 together with Proposition 5.3.41, see Ex. 21.
It may seem paradoxical that the real stability radius, introduced to measure robustness, may itself be highly sensitive to perturbations of the data. However, discontinuities will only occur when the structure matrices B, C are themselves perturbed or A is perturbed in a way which is not compatible with the given structure. This is in harmony with the inherent logic of structured uncertainty: If the system is subjected to perturbations which are incompatible with the given structure then the corresponding structured stability radius is not a reliable robustness measure. From this point of view one could say that it is not the possible *discontinuity* of the real stability radius which is surprising but the *continuity* of the complex one (see Proposition 5.3.42). *If the perturbation structure is not exactly known, the uncertainty is unstructured* and consequently the unstructured stability radius should be used.

5.3.7 Stability Radii and the Cayley Transformation

In this subsection we assume the general framework is the same as that in the previous one. In Section 3.3 we introduced the Möbius map

$$m : \mathbb{C} \setminus \{1\} \to \mathbb{C} \setminus \{1\}, \quad m(s) = \frac{s+1}{s-1}, \tag{71}$$

which maps $\mathbb{C} \setminus \{1\}$ homeomorphically into itself, with inverse $m^{-1}(s) = m(s)$. This homeomorphism maps $\mathbb{C}_-$ onto $\mathbb{D}$ and vice-versa. We then went on to introduce the Cayley transformation of a matrix and showed that it defined a bijection from the set of matrices $A \in \mathbb{C}^{n \times n}$ with $1 \notin \sigma(A)$ onto itself. In particular we related Hurwitz stability with Schur stability via the Cayley transform. Here our objective is to relate stability radii of Hurwitz stable matrices with stability radii of Schur stable matrices. In order to do this we must also transform the structure matrices B, C, D.

Definition 5.3.46. For every $(A,B,C,D) \in \mathcal{S}_{n,\ell,q}(\mathbb{K};\mathbb{C}\setminus\{1\})$ the *Cayley transform* $\Gamma(A,B,C,D) \in \mathcal{S}_{n,\ell,q}(\mathbb{K};\mathbb{C}\setminus\{1\})$ is defined by

$$\Gamma(A,B,C,D) = ((A+I)(A-I)^{-1}, -\sqrt{2}(A-I)^{-1}B, \sqrt{2}C(A-I)^{-1}, D-C(A-I)^{-1}B).$$

The continuous time Liapunov equation associated with the system (A,B,C,D) is $XA+A^*X+C^*C=0$. For the Cayley transformed data $\Gamma(A,B,C,D)$ this equation takes the form

$$X(A+I)(A-I)^{-1} + (A^*-I)^{-1}(A^*+I)X + 2(A^*-I)^{-1}C^*C(A-I)^{-1} = 0.$$

Multiplying on the left with (A^*-I) and on the right by $(A-I)$ yields

$$(A^*-I)X(A+I) + (A^*+I)X(A-I) + 2C^*C = 0$$

which simplifies, perhaps not surprisingly, to the discrete time Liapunov equation $A^*XA - X + C^*C = 0$. Thus the continuous time Liapunov equation for $\Gamma(A,B,C,D)$ is the discrete time Liapunov equation for (A,B,C,D). The following lemma collects some further basic properties of the Cayley transform.

Lemma 5.3.47. *For each* $(A,B,C,D) \in \mathcal{S}_{n,\ell,q}(\mathbb{K},\mathbb{C}\setminus\{1\})$, *let* $(\tilde{A},\tilde{B},\tilde{C},\tilde{D}) = \Gamma(A,B,C,D)$, *then*

(i) $\sigma(\tilde{A}) = m(\sigma(A))$.

(ii) Γ *maps* $\mathcal{S}_{n,\ell,q}(\mathbb{K},\mathbb{C}\setminus\{1\})$ *homeomorphically onto itself and* $\Gamma^{-1} = \Gamma$.

(iii) If $\mathbb{C}_g \subset \mathbb{C}\setminus\{1\}$ *is open, then* $m(\mathbb{C}_g) \subset \mathbb{C}\setminus\{1\}$ *is open and* Γ *maps* $\mathcal{S}_{n,\ell,q}(\mathbb{K},\mathbb{C}_g)$ *onto* $\mathcal{S}_{n,\ell,q}(\mathbb{K},m(\mathbb{C}_g))$.

(iv) If $G(s)$, $\tilde{G}(s)$ *are the transfer functions of* (A,B,C,D) *and* $(\tilde{A},\tilde{B},\tilde{C},\tilde{D})$, *then*

$$G(s) = \tilde{G}(m(s)), \qquad s \in \mathbb{C}\setminus(\sigma(A)\cup\{1\}). \tag{72}$$

Proof: (i) follows directly from Proposition A.1.15, but can also be verified by elementary calculations.
(ii) It follows from (i) that Γ maps $\mathcal{S}_{n,\ell,q}(\mathbb{K},\mathbb{C}\setminus\{1\})$ onto itself. By definition,

$$[\tilde{A}-I]^{-1} = [(A+I)(A-I)^{-1} - I]^{-1} = (A-I)[(A+I)-(A-I)]^{-1} = (A-I)/2.$$

Hence

$$\begin{aligned}
(\tilde{A}+I)(\tilde{A}-I)^{-1} &= [(A+I)(A-I)^{-1}+I](A-I)/2 = A,\\
-\sqrt{2}(\tilde{A}-I)^{-1}\tilde{B} &= -(1/\sqrt{2})(A-I)(-\sqrt{2}(A-I)^{-1}B) = B,\\
\sqrt{2}\tilde{C}(\tilde{A}-I)^{-1} &= 2C(A-I)^{-1}(A-I)/2 = C,\\
\tilde{D}-\tilde{C}(\tilde{A}-I)^{-1}\tilde{B} &= D - C(A-I)^{-1}B + C(A-I)^{-1}B = D.
\end{aligned}$$

So $\Gamma(\tilde{A},\tilde{B},\tilde{C},\tilde{D}) = (A,B,C,D)$ and Γ is a bijection from $\mathcal{S}_{n,\ell,q}(\mathbb{K},\mathbb{C}\setminus\{1\})$ onto itself with inverse $\Gamma^{-1} = \Gamma$. Since Γ is continuous it is a homeomorphism.
(iii) follows directly from (i) and (ii) and the fact that m defined by (71) is a

homeomorphism.
(iv) If $s \in \mathbb{C} \setminus (\sigma(A) \cup \{1\})$, then $m(s) \in \mathbb{C} \setminus (\sigma(\tilde{A}) \cup \{1\})$ by (i), so $\tilde{G}(m(s))$ is well defined and (72) follows from

$$\begin{aligned}
\tilde{G}(m(s)) &= \tilde{D} + \tilde{C}[(s+1)(s-1)^{-1}I - \tilde{A}]^{-1}\tilde{B} \\
&= D - C(A-I)^{-1}B - 2C(A-I)^{-1}[(s+1)(s-1)^{-1}I - (A+I)(A-I)^{-1}]^{-1}(A-I)^{-1}B \\
&= D + C\{-I - 2[(s+1)(s-1)^{-1}(A-I) - (A+I)]^{-1}\}(A-I)^{-1}B \\
&= D + C\{-I - 2(s-1)[(s+1)(A-I) - (s-1)(A+I)]^{-1}\}(A-I)^{-1}B \\
&= D + C\{-I + (s-1)[sI - A]^{-1})\}(A-I)^{-1}B \\
&= D + C(sI-A)^{-1}[-(sI-A) + (s-1)I](A-I)^{-1}B \\
&= D + C[sI-A]^{-1}B = G(s).
\end{aligned}$$

□

Our objective is to derive formulas for stability radii with respect to the stability region $m(\mathbb{C}_g)$ from formulas with respect to $\mathbb{C}_g$. Note however that, if we start with data $(A, B, C, 0)$, then the transformed data will, in general, have $\tilde{D} \neq 0$. Hence formulas for stability radii with respect to affine perturbations will give rise to formulas for stability radii with respect to special linear fractional ones.
In the following we will assume that $\mathbb{C}_g$ is an open subset of $\mathbb{C} \setminus \{1\}$. Then $m(\mathbb{C}_g)$ is open and we have the partition $\mathbb{C} \setminus \{1\} = m(\mathbb{C}_g) \,\dot{\cup}\, m(\mathbb{C}_b \setminus \{1\})$. Moreover, since m is a homeomorphism from $\mathbb{C} \setminus \{1\}$ onto itself, we have

$$s \in \partial\mathbb{C}_g \setminus \{1\} \iff m(s) \in \partial m(\mathbb{C}_g) \setminus \{1\}. \tag{73}$$

Now $\lim_{|s|\to\infty} m(s) = 1$ and $\lim_{s\to 1} |m(s)| = \infty$ and so it follows that

$$\mathbb{C}_g \text{ is unbounded} \iff 1 \in \partial(m(\mathbb{C}_g)), \quad m(\mathbb{C}_g) \text{ is unbounded} \iff 1 \in \partial\mathbb{C}_g.$$

In order to discuss stability radii for the transformed data we consider separately the cases where $\mathbb{C}_g$ is bounded/unbounded and where $m(\mathbb{C}_g)$ is bounded/unbounded. We set $\tilde{\mathbb{C}}_g = m(\mathbb{C}_g)$ and denote the stability radius for the transformed data with respect to the perturbation class $\mathbf{\Delta}$, by $r_{\mathbf{\Delta}}(\Gamma(A, B, C, D); \tilde{\mathbb{C}}_g)$.

Theorem 5.3.48. *Suppose $\mathbb{C}_g \subset \mathbb{C} \setminus \{1\}$, $(A, B, C, D) \in \mathcal{S}_{n,\ell,q}(\mathbb{K}, \mathbb{C}_g)$ and let $(\tilde{A}, \tilde{B}, \tilde{C}, \tilde{D}) = \Gamma(A, B, C, D)$.*
(i) If $1 \in \partial(m(\mathbb{C}_g))$ and $1 \in \partial\mathbb{C}_g$ (i.e. $\mathbb{C}_g$ and $m(\mathbb{C}_g)$ both unbounded), then

$$r_{\mathbf{\Delta}}(A; B, C, D; \mathbb{C}_g) = r_{\mathbf{\Delta}}(\tilde{A}, \tilde{B}, \tilde{C}, \tilde{D}; \tilde{\mathbb{C}}_g).$$

(ii) If $1 \notin \partial m(\mathbb{C}_g)$ and $1 \in \partial\mathbb{C}_g$ (i.e. $\mathbb{C}_g$ bounded, $m(\mathbb{C}_g)$ unbounded), then

$$r_{\mathbf{\Delta}}(A; B, C, D; \mathbb{C}_g) = \min\{\mu_{\mathbf{\Delta}}(D)^{-1}, r_{\mathbf{\Delta}}(\tilde{A}, \tilde{B}, \tilde{C}, \tilde{D}; \tilde{\mathbb{C}}_g)\}.$$

(iii) If $1 \in \partial m(\mathbb{C}_g)$ and $1 \notin \partial\mathbb{C}_g$ (i.e. $\mathbb{C}_g$ unbounded, $m(\mathbb{C}_g)$ bounded), then

$$\min\{\mu_{\mathbf{\Delta}}(\tilde{D})^{-1}, r_{\mathbf{\Delta}}(A; B, C, D; \mathbb{C}_g)\} = r_{\mathbf{\Delta}}(\tilde{A}, \tilde{B}, \tilde{C}, \tilde{D}; \tilde{\mathbb{C}}_g).$$

(iv) If $1 \notin \partial m(\mathbb{C}_g)$ and $1 \notin \partial\mathbb{C}_g$ (i.e. $\mathbb{C}_g$ and $m(\mathbb{C}_g)$ both bounded), then

$$\min\{\mu_{\mathbf{\Delta}}(\tilde{D})^{-1}, r_{\mathbf{\Delta}}(A; B, C, D; \mathbb{C}_g)\} = \min\{\mu_{\mathbf{\Delta}}(D)^{-1}, r_{\mathbf{\Delta}}(\tilde{A}, \tilde{B}, \tilde{C}, \tilde{D}; \tilde{\mathbb{C}}_g)\}.$$

Proof: If $G(s), \tilde{G}(s)$ are the transfer functions of (A, B, C, D) and $(\tilde{A}, \tilde{B}, \tilde{C}, \tilde{D})$, then $\tilde{D} = G(1)$, $D = \tilde{G}(1)$ by Definition 5.3.46.
(i) By Lemma 5.3.47 (iv) and (73)

$$\{G(s);\ s \in \partial\mathbb{C}_g \setminus \{1\}\} = \{\tilde{G}(s);\ s \in \partial m(\mathbb{C}_g) \setminus \{1\}\}.$$

Hence, by Proposition 5.3.3,

$$\begin{aligned} r_{\mathbf{\Delta}}(A; B, C, D; \mathbb{C}_g)^{-1} &= \max\{\sup_{s\in\partial\mathbb{C}_g} \mu_{\mathbf{\Delta}}(G(s)),\, \mu_{\mathbf{\Delta}}(D)\} \\ &= \max\{\sup_{s\in\partial\mathbb{C}_g\setminus\{1\}} \mu_{\mathbf{\Delta}}(G(s)),\, \mu_{\mathbf{\Delta}}(G(1)),\, \mu_{\mathbf{\Delta}}(D)\} \\ &= \max\{\sup_{s\in\partial m(\mathbb{C}_g)\setminus\{1\}} \mu_{\mathbf{\Delta}}(\tilde{G}(s)),\, \mu_{\mathbf{\Delta}}(\tilde{D}),\, \mu_{\mathbf{\Delta}}(\tilde{G}(1))\} \\ &= \max\{\sup_{s\in\partial m(\mathbb{C}_g)} \mu_{\mathbf{\Delta}}(\tilde{G}(s)),\, \mu_{\mathbf{\Delta}}(\tilde{D})\} = r_{\mathbf{\Delta}}(\tilde{A}, \tilde{B}, \tilde{C}, \tilde{D}; \tilde{\mathbb{C}}_g)^{-1}. \end{aligned}$$

(ii) Since $1 \notin \partial m(\mathbb{C}_g)$ we have $m(\partial\mathbb{C}_g \setminus \{1\}) = \partial m(\mathbb{C}_g)$. Hence by Lemma 5.3.47 (iv)

$$\{G(s);\ s \in \partial\mathbb{C}_g \setminus \{1\}\} = \{\tilde{G}(s);\ s \in \partial m(\mathbb{C}_g)\}.$$

So

$$\begin{aligned} r_{\mathbf{\Delta}}(A; B, C, D; \mathbb{C}_g)^{-1} &= \max\{\sup_{s\in\partial\mathbb{C}_g\setminus\{1\}} \mu_{\mathbf{\Delta}}(G(s)),\, \mu_{\mathbf{\Delta}}(G(1)),\, \mu_{\mathbf{\Delta}}(D)\} \\ &= \max\{\sup_{s\in\partial m(\mathbb{C}_g)} \mu_{\mathbf{\Delta}}(\tilde{G}(s)),\, \mu_{\mathbf{\Delta}}(\tilde{D}),\, \mu_{\mathbf{\Delta}}(D)\} \\ &= \max\{r_{\mathbf{\Delta}}(\tilde{A}, \tilde{B}, \tilde{C}, \tilde{D}; \tilde{\mathbb{C}}_g)^{-1},\, \mu_{\mathbf{\Delta}}(D)\}. \end{aligned}$$

(iii) follows by applying (ii) to $m(\mathbb{C}_g)$ instead of $\mathbb{C}_g$ and making use of $\Gamma = \Gamma^{-1}$.
(iv) This is proved in a similar way to (i) and (ii). We leave the details to the reader (Ex. 22). □

Since $m(\mathbb{C}_-) = \mathbb{D}$, $m(\mathbb{D}) = \mathbb{C}_-$ and $1 \notin \partial\mathbb{C}_-$, $1 \in \partial\mathbb{D}$, we may apply Theorem 5.3.48 (iii) for $\mathbb{C}_g = \mathbb{C}_-$ to obtain

$$r^1_{\mathbf{\Delta}}(\tilde{A}, \tilde{B}, \tilde{C}, \tilde{D}) = \min\left\{[\mu_{\mathbf{\Delta}}(D - C(A-I)^{-1}B)]^{-1}, r^-_{\mathbf{\Delta}}(A; B, C, D)\right\}.$$

This formula enables us to calculate the discrete time radius $r^1_{\mathbb{C}}$ via the algorithm described in Subsection 5.3.4 for the computation of the continuous time radius $r^-_{\mathbb{C}}$, see *Notes and References.*

Example 5.3.49. Suppose $A = \begin{bmatrix} 0 & 1 \\ -5 & -4 \end{bmatrix}$, $D = B = [0\ 1]^\top$, $C = I_2$. Then

$$\tilde{A} = \frac{1}{5}\begin{bmatrix} 0 & -1 \\ 5 & 4 \end{bmatrix}, \quad \tilde{B} = \frac{\sqrt{2}}{10}\begin{bmatrix} 1 \\ 1 \end{bmatrix}, \quad \tilde{C} = \frac{\sqrt{2}}{10}\begin{bmatrix} -5 & -1 \\ 5 & -1 \end{bmatrix}, \quad \tilde{D} = \frac{1}{10}\begin{bmatrix} 1 \\ 11 \end{bmatrix}.$$

We have seen in Example 5.3.18 that with respect to Euclidean norms, for $\mathbf{\Delta} = \mathbb{C}^2$ we have $r^-_{\mathbb{C}}(A, B, C, D) = 0.797$ and for $\mathbf{\Delta} = \mathbb{R}^2$ we have $r^-_{\mathbb{R}}(A, B, C, D) = 4/5$. Now since $\|\tilde{D}\| = \sqrt{122}/10$, we find

$$\begin{aligned} r^1_{\mathbb{C}}(\tilde{A}, \tilde{B}, \tilde{C}, \tilde{D}) &= \min\{10/\sqrt{122},\, r^-_{\mathbb{C}}(A, B, C, D)\} = \min\{10/\sqrt{122},\, 0.797\} = 0.797, \\ r^1_{\mathbb{R}}(\tilde{A}, \tilde{B}, \tilde{C}, \tilde{D}) &= \min\{10/\sqrt{122},\, r^-_{\mathbb{R}}(A, B, C, D)\} = \min\{10/\sqrt{122},\, 4/5\} = 4/5. \end{aligned}$$

□

5.3.8 Exercises

1. Prove Proposition 5.3.8. If $A = \mathrm{diag}\,(0,\ -1/2,\ 1/3)$, determine $r_{\mathbf{\Delta}_\mathbb{C}}(A, \mathbb{D})$ with respect to the perturbation norm $\|\cdot\|_{\infty,\infty} = \|\cdot\|_{1|\infty}$, see (11).

2. If the perturbation norm $\|\cdot\|_{\infty,\infty} = \|\cdot\|_{1|\infty}$ in Proposition 5.3.8 is replaced by the perturbation norm $\|\cdot\|_{\infty|\infty}$ (see (12)), prove that

$$r_{\mathbf{\Delta}_\mathbb{C}}(A;\mathbb{D}) = \min_{\theta\in[0,2\pi]} \varrho(P(e^{\imath\theta}))^{-1} \geq r_{\mathbf{\Delta}_\mathbb{R}}(|A|;\mathbb{D}) = \varrho(P_1)^{-1}, \tag{74}$$

where $|A| := (|a_{ij}|)$ and P_1 is the matrix obtained from $P(s)$ (defined by (15)) by replacing $|s - a_{ii}|^{-1}$ with $(1 - |a_{ii}|)^{-1}$. Calculate $r_{\mathbf{\Delta}_\mathbb{C}}(A, \mathbb{D})$ with respect to the perturbation norm $\|\cdot\|_{\infty|\infty}$ for $A = \mathrm{diag}\,(0,\ -1/2,\ 1/3)$.

In Ex. 3—20 all vector spaces $\mathbb{K}^n$ are provided with Euclidean norms and the matrix spaces with spectral norms.

3. If

$$A = \begin{bmatrix} 1 & 1 \\ -2.5 & -2 \end{bmatrix},\ B = \begin{bmatrix} 1 \\ 0 \end{bmatrix},\ C = I_2, \quad \mathbb{C}_g = \{\lambda \in \mathbb{C};\ \lambda = \alpha + \imath\omega,\ \alpha < -\omega^2\},$$

find $r_\mathbb{C}(A; B, C; \mathbb{C}_g)$ and $r_\mathbb{R}(A; B, C; \mathbb{C}_g)$.

4. Calculate $r_\mathbb{C}^-(A; B, C)$, $r_\mathbb{R}^-(A; B, C)$, $r_\mathbb{C}^1(A; B, C)$ and $r_\mathbb{R}^1(A; B, C)$ for the matrices in Ex. 3. Use the figures generated in Ex. 5.2.4 to verify your results. In each case find minimum norm destabilizing disturbance matrices.

5. Consider the system

$$A = \begin{bmatrix} 0 & 1 \\ -1 & -\varepsilon \end{bmatrix}, \quad B = \begin{bmatrix} 0 \\ -\varepsilon \end{bmatrix}, \quad C = [1,\ 0], \quad D = \varepsilon/2, \quad 0 < \varepsilon < \sqrt{2}.$$

Prove that $r_\mathbb{R}^-(A; B, C, D) = 2/\varepsilon$ and $r_\mathbb{C}^-(A; B, C, D) = (1 + \varepsilon^2/4)^{-1/2}$.

6. Prove Corollary 5.3.15.
(Hint: Let $\Delta = \Delta_1 + \imath\Delta_2$, $\Delta_1, \Delta_2 \in \mathbb{R}^{\ell\times q}$ be such that

$$\|\Delta\| = r_\mathbb{C}(A; B, C, D; \mathbb{C}_g), \quad \sigma(A + B\Delta C) \cap \mathbb{C}_b \neq \emptyset.$$

Consider

$$A(\Delta) = \begin{bmatrix} A & 0 \\ 0 & A \end{bmatrix} + \begin{bmatrix} B & 0 \\ 0 & B \end{bmatrix} \left(I_{\ell_1+\ell_2} - \begin{bmatrix} \Delta_1 & -\Delta_2 \\ \Delta_2 & \Delta_1 \end{bmatrix} \begin{bmatrix} D & 0 \\ 0 & D \end{bmatrix} \right)^{-1} \begin{bmatrix} \Delta_1 & -\Delta_2 \\ \Delta_2 & \Delta_1 \end{bmatrix} \begin{bmatrix} C & 0 \\ 0 & C \end{bmatrix}$$

and show that $\sigma(A + B\Delta C) \subset \sigma(A(\Delta))$.

7. Calculate $d_\mathbb{C}^-(A)$ and $d_\mathbb{R}^-(A)$ for $A = \begin{bmatrix} 0 & 1 \\ -2 & -3 \end{bmatrix}$.

8. Suppose $A \in \mathbb{C}^{n\times n}$, $\sigma(A) \subset \mathbb{C}_-$, and $P = \int_0^\infty e^{A^*t} e^{At} dt$. Show that

$$\|P\| \geq (2 d_\mathbb{C}^-(A))^{-1}.$$

(Hint: Consider $\dot{x} = (A + \Delta)x$ where Δ is a minimum norm destabilizing perturbation. Use an argument based on the fact that for $V(x) = \langle x, Px\rangle$, we have $\dot{V}(x) = -\|x\|^2 + 2\langle Px, \Delta x\rangle$ along the flow of the perturbed system).

9. If $A \in \mathbb{R}^{2\times 2}$, prove that

$$d_{\mathbb{R}}^{-}(A) = \min\{|\operatorname{trace} A/2|, \sigma_2(A)\}.$$

10. If $A \in \mathbb{R}^{2\times 2}$ has ordered singular values $\sigma_1 \geq \sigma_2$ and $A = U \operatorname{diag}(\sigma_1, \sigma_2)V^\top$ is a singular value decomposition of A, with $V^\top U = \begin{bmatrix} \cos\theta & -\sin\theta \\ \sin\theta & \cos\theta \end{bmatrix}$, show that

$$d_{\mathbb{R}}(A) = \begin{cases} \sigma_2 & \text{if } \sigma_1/\sigma_2 \geq 2|\cos\theta|^{-1} - 1 \\ (1/2)(\sigma_1+\sigma_2)|\cos\theta| & \text{otherwise}. \end{cases}$$

$$d_{\mathbb{C}}(A) = \begin{cases} \sigma_2 & \text{if } \sigma_1/\sigma_2 \geq 2|\cos\theta|^{-2} - 1 \\ [\sigma_1\sigma_2 - (1/4)(\sigma_1+\sigma_2)^2\,|\cos\theta|^2]^{1/2}|\cot\theta| & \text{otherwise}. \end{cases}$$

(Hint: Use the fact that $d_{\mathbb{K}}(A) = d_{\mathbb{K}}(\operatorname{diag}(\sigma_1,\sigma_2)V^\top U)$ and for the real case use Ex. 9, see [239]).

11. Deduce from Ex. 10 that for every $q \in (0,1]$, there exists a matrix $A \in \mathbb{R}^{2\times 2}$, $\sigma(A) \subset \mathbb{C}_-$, such that $d_{\mathbb{C}}^{-}(A)/d_{\mathbb{R}}^{-}(A) = q$.

12. Consider the Ostrowski matrix $A \in \mathbb{R}^{n\times n}$ and the matrix $B \in \mathbb{R}^{n\times n}$:

$$A = \begin{bmatrix} -1 & 1 & 1 & . & . & 1 \\ 0 & -1 & 1 & . & . & 1 \\ \vdots & \vdots & \vdots & \vdots & \vdots & \vdots \\ 0 & 0 & 0 & . & . & -1 \end{bmatrix}, \qquad B = \begin{bmatrix} 0 & 0 & 0 & . & . & 0 \\ 0 & 0 & 0 & . & . & 0 \\ \vdots & \vdots & \vdots & \vdots & \vdots & \vdots \\ 1 & 1 & 1 & . & . & 1 \end{bmatrix}.$$

Show that $\det[A + \varepsilon B] = 0$, for $\varepsilon = 2^{-n+1}$. Hence prove that $d_{\mathbb{R}}^{-}(A) \leq \sqrt{n}\,2^{-n+1}$ with respect to the spectral norm.

13. Show that if $A \in \mathbb{R}^{n\times n}$, then

$$d_{\mathbb{C}}^{-}(A)^2 = \min_{\omega\in\mathbb{R}} \min_{\substack{x,y\in\mathbb{R}^n \\ \|x\|^2+\|y\|^2=1}} \{\|Ax\|^2 + \|Ay\|^2 + \omega^2 + 2\omega\langle (A - A^\top)y, x\rangle\}.$$

Hence prove

$$d_{\mathbb{C}}^{-}(A)^2 = \min_{\substack{x,y\in\mathbb{R}^n \\ \|x\|^2+\|y\|^2=1}} \{\|Ax\|^2 + \|Ay\|^2 - \langle (A - A^\top)y, x\rangle^2\}.$$

By choosing first x, y to maximize $\langle (A-A^\top)y, x\rangle^2$, then to minimize $\|Ax\|^2 + \|Ay\|^2$, show that

$$\sigma_n(A)^2 - (1/4)\|A - A^\top\|_2^2 \leq d_{\mathbb{C}}^{-}(A)^2 \leq \|A\|_2^2 - (1/4)\|A - A^\top\|_2^2.$$

14. Suppose $A, \Delta \in \mathbb{C}^{n\times n}$, $\sigma(A) \subset \mathbb{C}_-$,. If $\|\Delta\| < d := d_{\mathbb{C}}^{-}(A)$, prove that

$$\frac{[\int_0^\infty \|e^{(A+\Delta)t} - e^{At}\|^2 dt]^{1/2}}{[\int_0^\infty \|e^{At}\|^2 dt]^{1/2}} = \frac{\|e^{(A+\Delta)\cdot} - e^{A\cdot}\|_{L^2(\mathbb{R}_+;\mathbb{C}^{n\times n})}}{\|e^{A\cdot}\|_{L^2(\mathbb{R}_+;\mathbb{C}^{n\times n})}} \leq \frac{\|\Delta\|}{d - \|\Delta\|}.$$

(Hint: Show that for $\dot{x} = (A+\Delta)x$, $x(0) = x^0$, $\dot{z} = Az$, $z(0) = x^0$, we have

$$x(t) - z(t) = \int_0^t e^{A(t-s)}\Delta(x(s) - z(s))ds + \int_0^t e^{A(t-s)}\Delta z(s)ds,$$

and use the convolution inequality).

15. For any given $\alpha, \varepsilon > 0$, Ex. 10 shows that there exists a 2×2 real matrix A_0 such that $d_{\mathbb{R}}^-(A_0) > \alpha + \varepsilon$ and $d_{\mathbb{C}}^-(A) < \varepsilon$. Using the hint in Ex. 6 show that there exist real 2×2 matrices X, Y such that

$$\sigma\left(\begin{bmatrix} A_0 + X & -Y \\ Y & A_0 + X \end{bmatrix}\right) \cap \mathbb{C}_+ \neq \emptyset, \qquad \left\|\begin{bmatrix} X & -Y \\ Y & X \end{bmatrix}\right\| < \varepsilon.$$

Set $A_1 = A_2 = A_0 + X$ and prove that $d_{\mathbb{R}}^-(A_i) > \alpha$, $i = 1, 2$ yet the two systems $\dot{x}_i = A_i x_i$ can be destabilized by interconnecting them via coupling gains $K_{12} = -K_{21} = Y$, with $\|Y\| < \varepsilon$.

16. Let $\mathbb{C}_g = \{\lambda \in \mathbb{C} : |\operatorname{Im} \lambda| < |\operatorname{Re} \lambda|\}$ and $\mathbb{C}_b = \mathbb{C} \setminus \mathbb{C}_g$. For $A = \operatorname{diag}(-1, -3)$, show that $\operatorname{dist}(\sigma(A), \partial\mathbb{C}_g) = 1/\sqrt{2}$, but $d_{\mathbb{R}}(A, \mathbb{C}_g) = 1$.

17. Show that Assumption (66) is always satisfied if $\mathbb{C}_g$ has the following property

$$\forall r \in \mathbb{R} \cap \mathbb{C}_g : \quad \operatorname{dist}(r, \partial\mathbb{C}_g) = \operatorname{dist}(r, \partial\mathbb{C}_g \cap \mathbb{R}) .$$

Suppose $A = \begin{bmatrix} 0 & 0 \\ 0 & 4 \end{bmatrix}$ and $\mathbb{C}_g = \mathbb{C} \setminus \{\imath\}$. Prove that Assumption (66) is not satisfied and $d_{\mathbb{R}}(A, \mathbb{C}_g) = 3$, whilst $d(\sigma(A), \partial\mathbb{C}_g) = 1$).

18. Prove the last statement in Example 5.3.45.

19. Consider the system

$$A_0 = \begin{bmatrix} 0 & 1 & 0 \\ 0 & 0 & 1 \\ -1 & -3 & -3 \end{bmatrix}, \quad B_0 = \begin{bmatrix} 0 \\ 0 \\ 1 \end{bmatrix}, \quad C_0 = [6 - 4\sqrt{2},\ 1,\ 1], \quad D_0 = 1.$$

Calculate $r_{\mathbb{R}}^-(A_0; B_0, C_0, D_0)$. If $C_\varepsilon = [6 - 4\sqrt{2},\ 1 - \varepsilon,\ 1]$, $0 < \varepsilon < 4/9 - 4(1 - \sqrt{2})^2/3$, calculate $r_{\mathbb{R}}^-(A_0; B_0, C_\varepsilon, D_0)$ and show that $\lim_{\varepsilon \to 0} r_{\mathbb{R}}^-(A_0; B_0, C_\varepsilon, D_0) = 1 + (6 - 4\sqrt{2}) \neq r_{\mathbb{R}}^-(A_0; B_0, C_0, D_0)$.

20. Use the results of Ex. 5 and the Cayley transform to obtain $r_{\mathbb{R}}^1(A; B, C, D)$ and $r_{\mathbb{C}}^1(A; B, C, D)$ for

$$A = \frac{1}{2+\varepsilon}\begin{bmatrix} -\varepsilon & -2 \\ 2 & \varepsilon \end{bmatrix}, \ B = \frac{-\sqrt{2}}{2+\varepsilon}\begin{bmatrix} \varepsilon \\ \varepsilon \end{bmatrix}, \ C^\top = \frac{-\sqrt{2}}{2+\varepsilon}\begin{bmatrix} 1+\varepsilon \\ 1 \end{bmatrix}, \ D = \varepsilon^2/(4+2\varepsilon),$$

where $0 < \varepsilon < \sqrt{2}$.

21. Suppose $\mathbf{\Delta} = \mathbb{R}^{\ell \times q}$, $(A_0, B_0, C_0, D_0) \in \mathcal{S}_{n,\ell,q}(\mathbb{R}; \mathbb{C}_g)$ and $G_0(s) = D_0 + C_0(sI - A_0)^{-1}B_0$. Prove: If $\mu_{\mathbb{R}}(G_0(s))$, $s \in \partial\mathbb{C}_g$ takes its maximum at some $s_0 \in \partial\mathbb{C}_g$ such that $G(s_0)$ is not real, then the map

$$r_{\mathbb{R}}(\,\cdot\,; \mathbb{C}_g) : \mathcal{S}_{n,\ell,q}(\mathbb{R}; \mathbb{C}_g) \to \mathbb{R}_+, \qquad (A, B, C, D) \mapsto r_{\mathbb{R}}(A; B, C, D; \mathbb{C}_g)$$

is continuous at (A_0, B_0, C_0, D_0).

22. Prove Theorem 5.3.48 (iv).

5.3.9 Notes and References

The notion of a stability radius with respect to full-block affine perturbations was first introduced by *Hinrichsen and Pritchard* (1986) [242], although the idea behind it can be found in many different fields, see e.g. *Rudin* (1973) [440]. The distance from instability has been analyzed by *Van Loan* (1985) [505] and *Hinrichsen and Pritchard* (1986) [242]. The complex full-block case has a well developed theory, see the survey by *Hinrichsen and Pritchard* (1990) [248]. It also extends to *time-varying linear systems*, see *Hinrichsen et al.* (1989) [232], to *infinite dimensional time-invariant and time-varying linear systems*, see *Pritchard and Townley* (1989) [420], *Hinrichsen and Pritchard* (1994) [253], *Jacob* (1995) [277], [276] and to *stochastic systems*, see *Bouhtouri and Pritchard* (1992) [149], *Morozan* (1995) [384], *Hinrichsen and Pritchard* (1996) [254], *Bouhtouri et al.* (2000) [148].
Linear fractional transformations have been an important tool in system theory for many years. It is surprising therefore that stability radii for linear fractional perturbations have only recently been developed. Our main reference for this is the diploma thesis by *Op't Hof* (1998) [400].
The formula for the real full-block stability radius in the case q or $\ell = 1$ was first given by *Hinrichsen and Pritchard* (1988) [245] and was used to obtain stability radii for polynomials, see *Hinrichsen and Pritchard* (1992) [252] and the next section. The characterization of the general real full-block stability radius with respect to the spectral norm is of more recent vintage. *Qiu and Davison* (1991), [425] obtained several lower bounds and in (1992), [426], for the case where $G(s)$ is square, they obtained the formula (36) given in Theorem 5.3.20 as a lower bound and conjectured that it was tight. Then in (1993), [423] in conjunction with Bernhardson, Rantzer, Young and Doyle, they proved the formula for the μ-function and hence the formula (36). Example 5.3.22 is based on *Hinrichsen and Son* (1998) [256].
A standard reference for the theory of algebraic Riccati equations is *Lancaster and Rodman* (1995) [326]. For the numerical solution of algebraic Riccati equations, see *Mehrmann* (1991) [372] and *Sima* (1996) [466]. A variety of solution algorithms for these equations can be downloaded from the web site `http://www.win.tue.nl/niconet`.
Algorithms for computing the complex stability radius have been developed by several authors, see e.g. *Hinrichsen et al.* (1989) [235], *Bruinsma and Steinbuch* (1990) [81], and *Boyd and Balakrishnan* (1990) [73]. The one we have described was proposed in [73] and is a refinement of the algorithm described in [81], [80]. The quadratic convergence of this algorithm was also proved in [73]. A fast algorithm for computing the real stability radius has been developed by *Sreedham et al.* (1996) [478].
The continuity results and the notion of strongly destabilizing perturbations for the case where $D = 0$ were first given in *Hinrichsen and Pritchard* (1990) [247]. The result for the real full-block case is based on the continuity result for the μ-function, see Section 4.4.
The subsection on transformation of the data is based on *Op't Hof* (1998) [400].
Discrete time results analogous to the characterization of the complex stability radius via the Hamiltonian have been developed by *Hinrichsen and Son* (1991) [255] in terms of symplectic matrix pencils. An algorithm for the computation of $r^1_{\mathbb{C}}$ which is based on these results has been given by *Schwiedernoch* (1991) [455] and in his thesis it is compared with the algorithm of Subsection 5.3.3 applied to the Cayley transformed data.

5.4 Root Sets and Stability Radii of Polynomials

The stability properties of uncertain higher order differential (resp. difference) equations of the form

$$a_n(\Delta)\xi^{(n)}(t) + a_{n-1}(\Delta)\xi^{(n-1)}(t) + \ldots + a_0(\Delta)\xi(t) = 0, \quad \Delta \in \mathbf{\Delta}$$

where $\xi^{(k)}(t) = \frac{d^k\xi}{dt^k}(t)$ (resp. $\xi^{(k)}(t) = \xi(t+k)$), $k \in \underline{n}$ can be determined from the location of the roots of the uncertain polynomial

$$p(s, a(\Delta)) = a_n(\Delta)s^n + a_{n-1}(\Delta)s^{n-1} + \ldots + a_0(\Delta), \quad \Delta \in \mathbf{\Delta}.$$

In Section 4.1 we mainly dealt with continuity and smoothness properties of the roots of a polynomial with coefficient vector depending analytically on a single complex parameter. In this section we will consider uncertain polynomials whose coefficient vectors depend affinely on an arbitrary number of complex or real parameters. We assume that a nominal coefficient vector is given and consider the coefficients of the affine family of polynomials as (additive linear) perturbations of the nominal coefficients. For these uncertain polynomials we will investigate analogous problems to those in the previous two sections. First we will be interested in the set of roots of all the perturbed polynomials with perturbations of norm less than a given uncertainty level $\delta > 0$. These sets are called "root sets" of the uncertain polynomial. Then we assume that the nominal polynomial has all its roots in a prescribed stability region $\mathbb{C}_g \subset \mathbb{C}$ and determine the smallest level of δ ("stability radius") for which either the degree of a perturbed polynomial is less than n or its root set is no longer contained in the given stability region.
We begin in the first subsection by showing how the two problems can be reformulated in terms of matrices so that the results of Section 5.2 and Section 5.3 can be applied. Then we employ these results to obtain general formulas for the root sets and stability radii of uncertain polynomials (with respect to arbitrary perturbation norms and arbitrary stability regions). In the two subsequent subsections we specialize these results to Schur and Hurwitz polynomials with respect to special norms and perturbation structures, dealing separately with complex and with real perturbations. In particular we will show in Subsection 5.4.3 how Kharitonov's Theorem can be used to determine the unstructured stability radius of a real Hurwitz polynomial for arbitrarily scaled ∞-norms.

5.4.1 General Formulas

In this subsection we study root sets and stability radii of given nominal polynomials under arbitrary linear perturbations of the coefficient vector. Every family of polynomials $p(s, a(\Delta))$, $\Delta \in \mathbb{K}^{1\times q}$ whose coefficient vector $a(\Delta) = a(\Delta_1, \ldots, \Delta_q)$ depends affinely on q parameters $\Delta_i \in \mathbb{K}$ can be written in the form $p(s, a + \Delta C))$, $\Delta \in \mathbb{K}^{1\times q}$ for a suitable matrix $C \in \mathbb{K}^{q\times(n+1)}$. Here we interpret a as the *nominal coefficient vector* and Δ as the *vector of parameter deviations* (*deviation vector*). The matrix C determines the structure of the perturbation $a \rightsquigarrow a + \Delta C$ and is called the *structure matrix*. Our basic assumptions are

- $a = [a_0, \ldots, a_n] \in \mathbb{K}^{1\times(n+1)}$, $a_n \neq 0$, $C = [c^0, \ldots, c^n] = (c_{ij})_{i\in\underline{q}, j=0,\ldots,n} \in \mathbb{K}^{q\times(n+1)}$ where $n, q \in \mathbb{N}^*$.[1] In particular, the nominal polynomial
$$p(s,a) = a_n s^n + a_{n-1}s^{n-1} + \cdots + a_0, \tag{1}$$
is of degree $n \geq 1$.

- The coefficients of $p(s,a)$ are subjected to perturbations of the form
$$a_j \rightsquigarrow a_j(\Delta) = a_j + \Delta c^j = a_j + \sum_{i=1}^{q} \Delta_i c_{ij},\ j = 0, \ldots, n \text{ or } a \rightsquigarrow a(\Delta) = a + \Delta C \tag{2}$$
where $\Delta = [\Delta_1, \ldots, \Delta_q] \in \mathbb{K}^{1\times q}$.

- $\boldsymbol{\Delta} = \mathbb{K}^{1\times q}$ is provided with an arbitrary norm $\|\cdot\|_{\boldsymbol{\Delta}}$ and the vector space $\mathbb{K}^q$ is endowed with the dual norm which we denote by $\|\cdot\|_{\mathbb{K}^q}$. So $\|\cdot\|_{\boldsymbol{\Delta}}$ is the dual norm of $\|\cdot\|_{\mathbb{K}^q}$ and hence $\|\cdot\|_{\boldsymbol{\Delta}} = \|\cdot\|_{\mathcal{L}(\mathbb{K}^q,\mathbb{K})} = \|\cdot\|^*_{\mathbb{K}^q}$.

- $\mathbb{C}_g$ is an open subset of $\mathbb{C}$ with $\mathbb{C} = \mathbb{C}_g \dot\cup \mathbb{C}_b$, $\mathbb{C}_g, \mathbb{C}_b \neq \emptyset$ and $p(s,a)$ is $\mathbb{C}_g$-stable, i.e. $\mathcal{R}(a) \subset \mathbb{C}_g$ where $\mathcal{R}(a)$ denotes the set of all the roots of $p(s,a)$.

The latter assumption is only needed for stability radius considerations.

Remark 5.4.1. If the deviation vectors are measured by the p-norm, $1 \leq p \leq \infty$
$$\|\Delta\|_p = \left[\sum_{i=1}^{q} |\Delta_i|^p\right]^{1/p}, \quad 1 \leq p < \infty, \quad \|\Delta\|_\infty = \max_{i\in\underline{q}} |\Delta_i|,$$
the associated norm on $\mathbb{K}^q$ is the p^*-norm where p^* is the conjugate exponent satisfying $1/p + 1/p^* = 1$, see Section A.1. □

The *root sets* of an uncertain polynomial are defined in an analogous way to the definition of spectral value sets for uncertain matrices.

Definition 5.4.2. Given an uncertainty level $\delta > 0$, the corresponding *root set* of the uncertain polynomial $p(\cdot, a(\Delta)) = p(\cdot, a + \Delta C)$, $\Delta \in \mathbb{K}^{1\times q}$ is
$$\mathcal{R}_{\mathbb{K}}(a; C; \delta) = \{s \in \mathbb{C};\ \exists \Delta \in \mathbb{K}^{1\times q} : \|\Delta\|^*_{\mathbb{K}^q} < \delta \text{ and } p(s, a(\Delta)) = 0\}. \tag{3}$$

If $\mathbb{R}^{1\times q}$ is provided with the norm induced from $\mathbb{C}^{1\times q}$ it is clear that
$$\mathcal{R}_{\mathbb{R}}(a; C; \delta) \subset \mathcal{R}_{\mathbb{C}}(a; C; \delta), \quad \delta > 0. \tag{4}$$

We now describe some special perturbation structures which will be considered in the sequel. Many of the results of this section will be developed for the *monic case* where the leading coefficient of the nominal and the perturbed polynomials is fixed at $a_n(\Delta) \equiv 1$. This special case is accounted for by taking $c^n = 0$. Another important special case is that of *unstructured perturbations* where all coefficients are perturbed independently and on the same scale. Then the perturbed polynomial is
$$p(s, a(\Delta)) = (a_n + \Delta_{n+1})s^n + (a_{n-1} + \Delta_n)s^{n-1} + \ldots + (a_0 + \Delta_1), \quad \Delta \in \mathbb{K}^{1\times(n+1)},$$

[1] Note that throughout this section the $n+1$ columns of $C \in \mathbb{K}^{q\times(n+1)}$ are indexed by $0, \ldots, n$ (in parallel to the indexing of the coefficient vector $a = [a_0, \ldots, a_n] \in \mathbb{K}^{1\times(n+1)}$).

or in the monic case

$$p(s, a(\Delta)) = s^n + (a_{n-1} + \Delta_n)s^{n-1} + \ldots + (a_0 + \Delta_1), \quad \Delta \in \mathbb{K}^{1\times n}.$$

Unstructured perturbations can be represented in the form (2) by setting $q = n+1$, $C = I_{n+1}$ and in the monic case $q = n$, $C = [I_n, 0_n] \in \mathbb{R}^{n\times(n+1)}$. We denote the corresponding root set by $\mathcal{R}_{\mathbb{K}}(a;\delta)$ and it will be clear from the context whether or not this refers to the general unstructured case or the monic unstructured case. Another special case of interest is obtained when one coefficient a_j is perturbed while the other coefficients remain unchanged. This case is represented by the structure matrix $C = e^{j\top} \in \mathbb{R}^{1\times(n+1)}$ where e^j, $j = 0, 1, \ldots, n$ are the standard unit vectors in $\mathbb{R}^{n+1}$.[2] The corresponding root set $\mathcal{R}_{\mathbb{K}}(a; e^{j\top}; \delta)$ is the set of all roots of the polynomials

$$p(s, a(\Delta)) = a_n s^n + a_{n-1}s^{n-1} + \cdots + (a_j + \Delta)s^j + \cdots + a_0, \quad \Delta \in \mathbb{K},\ |\Delta| < \delta.$$

With any polynomial (1) of degree n we associate the monic polynomial $p(s, a/a_n)$ and the corresponding companion matrix

$$A = \begin{bmatrix} 0 & 1 & 0 & \cdots & 0 \\ 0 & 0 & 1 & \cdots & 0 \\ \vdots & \vdots & & \ddots & \vdots \\ 0 & 0 & & & 1 \\ -a_0/a_n & -a_1/a_n & \cdots & \cdots & -a_{n-1}/a_n \end{bmatrix}_{n\times n} . \tag{5}$$

Then $\mathcal{R}(a) = \sigma(A)$. Our aim is to find structure matrices $B, \tilde{C}, D$ such that $\mathcal{R}(a(\Delta)) = \sigma(A(\Delta))$ for all $\Delta \in \mathbb{K}^{1\times q}$ for which $A(\Delta)$ is well defined by (3.1). We have seen in Proposition 4.1.3 that if a perturbation Δ decreases the degree of $p(s,a)$ to $m = \deg p(s, a(\Delta)) < n$ then $n - m$ roots of $p(s, a((1+\varepsilon)\Delta))$ tend to ∞ for $\varepsilon \searrow 0$ in the form of a Butterworth pattern. To exclude this possibility and ensure that $A(\Delta)$ is well defined by (3.1) we only consider perturbations of norm

$$\|\Delta\|^*_{\mathbb{K}^q} < |a_n| \|c^n\|^{-1}_{\mathbb{K}^q} \tag{6}$$

where $c^n = (c_{in})_{i\in \underline{q}}$ is the last column of C. This norm bound guarantees that the degree of p is not decreased by the perturbation, i.e. deg $p(s, a(\Delta)) = n$ for all $\Delta \in \mathbb{K}^{1\times q}$ satisfying (6). Note that in the monic case we have $c^n = 0$ so that the RHS of (6) is infinite and all perturbations are admissible.[3]

Lemma 5.4.3. *Let A be defined by* (5) *and $B \in \mathbb{K}^n, \tilde{C} \in \mathbb{K}^{q\times n}, D \in \mathbb{K}^q$ by*

$$B = [0, \ldots, 0, 1]^\top, \quad D = -a_n^{-1} c^n, \quad \tilde{C} = -a_n^{-1}[c^0 + a_0 D, \ldots, c^{n-1} + a_{n-1}D]. \tag{7}$$

Then $\|D\|_{\mathbb{K}^q} = |a_n|^{-1}\|c^n\|_{\mathbb{K}^q}$ and for all $\Delta \in \mathbb{K}^{1\times q}$ satisfying (6)*, we have*

$$\mathcal{R}(a(\Delta)) = \sigma(A(\Delta)) \text{ where } A(\Delta) = A + B(1 - \Delta D)^{-1}\Delta\tilde{C}. \tag{8}$$

So $\mathcal{R}_{\mathbb{K}}(a; C; \delta) = \sigma_{\mathbb{K}}(A; B, \tilde{C}, D; \delta)$ for all $\delta < |a_n| \|c^n\|^{-1}_{\mathbb{K}^q}$.

---

[2] In contrast to our usual notation (see the Glossary) we denote – throughout this section – by e^j the $j+1$-st column of I_{n+1}, $j = 0, 1, \ldots, n$.

[3] Recall that we set $\alpha/\beta := \infty$ if $\alpha > 0$, $\beta = 0$.

Proof: It follows from (7) and (8) that $A(\Delta)$ is a matrix in companion form whose last row is obtained by adding to the last row of A given by (5) the row vector

$$(1-\Delta D)^{-1}\Delta\tilde{C} = -(1+a_n^{-1}\Delta c^n)^{-1}a_n^{-1}[\Delta c^0 - a_0 a_n^{-1}\Delta c^n, \ldots, \Delta c^{n-1} - a_{n-1}a_n^{-1}\Delta c^n].$$

But $(1+a_n^{-1}\Delta c^n)^{-1}a_n^{-1} = (a_n+\Delta c^n)^{-1} = (a_n(\Delta))^{-1}$ and so $A(\Delta)$ has the form

$$A(\Delta) = \begin{bmatrix} 0 & 1 & \cdots & 0 \\ \vdots & & \ddots & \vdots \\ 0 & & & 1 \\ -\tilde{a}_0(\Delta)/a_n(\Delta) & \cdots & & -\tilde{a}_{n-1}(\Delta)/a_n(\Delta) \end{bmatrix}$$

where

$$\begin{aligned} -\tilde{a}_j(\Delta) &= -a_j a_n^{-1} a_n(\Delta) - (\Delta c^j - a_j a_n^{-1}\Delta c^n) \\ &= a_n^{-1}[-a_j(a_n+\Delta c^n) - a_n(\Delta c^j - a_n^{-1}a_j\Delta c^n)] \\ &= -(a_j+\Delta c^j) = -a_j(\Delta), \quad j = 0,1,\ldots,n-1. \end{aligned}$$

So $A(\Delta)$ is the companion matrix of the monic polynomial $p(s, a_n(\Delta)^{-1}a(\Delta))$ and this proves (8). The last statement is a consequence of Definition 5.2.1 and the fact that $\|\Delta\|_{\boldsymbol{\Delta}} = \|\Delta\|^*_{\mathbb{K}^q}$ is the operator norm of Δ as a linear map from $(\mathbb{K}^q, \|\cdot\|_{\mathbb{K}^q})$ to $(\mathbb{K}, |\cdot|)$. □

By the previous lemma the root sets of arbitrary affine families of polynomials are special cases of spectral value sets so that all the results from Sections 5.1 and 5.2 – in particular those for the case $\ell = 1$ – have their counterparts in the polynomial context. Note however, that even in the monic case ($c^n = 0$), *unstructured* perturbations of the polynomial $p(s,a)$ correspond to *structured* perturbations of the matrix A given by (5).

Theorem 5.4.4. *Suppose that the nominal polynomial $p(s,a)$ is subjected to perturbations of the form (2), let $\delta < |a_n|\|c^n\|^{-1}_{\mathbb{K}^q}$ and set*

$$G(s) = -p(s,a)^{-1}C(s) \ \ \textit{where}\ \ C(s) = C\begin{bmatrix} 1 \\ s \\ \vdots \\ s^n \end{bmatrix} = \begin{bmatrix} c_1(s) \\ \vdots \\ c_q(s) \end{bmatrix} = \begin{bmatrix} \sum_{j=0}^n c_{1j}s^j \\ \vdots \\ \sum_{j=0}^n c_{qj}s^j \end{bmatrix}. \tag{9}$$

(i) If $\mathbb{K} = \mathbb{C}$ then

$$\mathcal{R}_{\mathbb{C}}(a; C; \delta) = \mathcal{R}(a) \cup \{s \in \mathbb{C}\setminus\mathcal{R}(a);\ \|G(s)\|_{\mathbb{C}^q} > \delta^{-1}\}. \tag{10}$$

(ii) If $\mathbb{K} = \mathbb{R}$ then

$$\mathcal{R}_{\mathbb{R}}(a; C; \delta) = \mathcal{R}(a) \cup \{s \in \mathbb{C}\setminus\mathcal{R}(a);\ \mathrm{dist}(G_R(s), \mathbb{R}G_I(s)) > \delta^{-1}\} \tag{11}$$

where $G_R(s)$, $G_I(s)$ are the real and imaginary parts of $G(s)$ and dist *is measured with respect to $\|\cdot\|_{\mathbb{R}^q}$.*

Proof: Let $(A, B, \tilde{C}, D)$ be as defined in Lemma 5.4.3 and $s \in \mathbb{C} \setminus \mathcal{R}(a)$. In order to prove that $G(s)$ defined by (9) is the transfer matrix associated with $(A, B, \tilde{C}, D)$, define $x = [x_1, \ldots, x_n]^\top$ by $(sI - A)x = B$, i.e. $x = (sI - A)^{-1}B$. Then

$$\begin{array}{rcl} sx_1 - x_2 \qquad\qquad\qquad\qquad & = & 0 \\ \ddots \qquad\qquad & & \vdots \\ sx_{n-1} - x_n & = & 0 \\ sx_n + a_0/a_n x_1 + a_1/a_n x_2 + \cdots + a_{n-1}/a_n x_n & = & 1. \end{array}$$

So $x_j = s^{j-1}x_1$ for $j \in \underline{n}$ and $p(s, a/a_n)\, x_1 = 1$ for $s \notin \mathcal{R}(a)$, whence $x = p(s, a/a_n)^{-1} \left[1, s \cdots, s^{n-1}\right]^\top$. Thus the transfer function of $(A, B, \tilde{C}, D)$ is

$$\begin{aligned} D+&\tilde{C}(sI_n - A)^{-1}B \\ &= D - a_n^{-1} p(s, a/a_n)^{-1} [c^0 + a_0 D, \ldots, c^{n-1} + a_{n-1}D][1, s, s^2, \ldots, s^{n-1}]^\top \\ &= p(s,a)^{-1}\{D[p(s,a) - a_{n-1}s^{n-1} \ldots - a_0] - [c^0, c^1, \ldots, c^{n-1}][1, s, s^2, \ldots, s^{n-1}]^\top\} \\ &= -p(s,a)^{-1}\{c^n s^n + [c^0, c^1, \ldots, c^{n-1}][1, s, s^2, \ldots, s^{n-1}]^\top\} = G(s). \end{aligned}$$

The well-posedness radius of $(A, B, \tilde{C}, D)$ is $\boldsymbol{\delta}_0 = \|D\|_{\mathbb{K}^q}^{-1} = |a_n| \|c^n\|_{\mathbb{K}^q}^{-1}$. Hence using Lemma 5.4.3, (10) follows from Theorem 5.2.16 and (11) from Theorem 5.2.27. □

In the monic case the previous theorem is applicable for all $\delta > 0$ and the formula (9) for the transfer function $G(s)$ specializes to

$$G(s) = -p(s,a)^{-1}C(s), \;\; C(s) = [c_1(s), \ldots, c_q(s)]^\top \in \mathbb{K}^q[s], \;\; c_i(s) = \sum_{j=0}^{n-1} c_{ij}s^j. \quad (12)$$

So in this case $G(s)$ is a strictly proper rational vector, whereas in the general case it is only proper rational. In the following example we illustrate the non-monic case where the leading coefficient is perturbed.

Example 5.4.5. Consider the perturbed linear polynomial

$$p(s, a(\Delta)) = (1 + \Delta_2)s + 1 + \Delta_1$$

where the nominal coefficient vector $a = [a_0, a_1] = [1, 1]$ is subjected to unstructured perturbations $a \rightsquigarrow a + \Delta$, i.e. $C = I_2$. By (9) $G(s) = -(s+1)^{-1}[1\ s]^\top$. Hence given an uncertainty level $\delta < a_1 \|c^1\|^{-1} = 1$ and Euclidean norms on $\mathbb{K}^q = \mathbb{K}^2$ and $\boldsymbol{\Delta} = \mathbb{K}^{1\times 2}$, the root set of $p(s,a) = s + 1$ with respect to unstructured complex perturbations is

$$\mathcal{R}_{\mathbb{C}}(a;\delta) = \{s \in \mathbb{C}; (|s|^2 + 1)^{-1}|s+1|^2 < \delta^2\},$$

which is the open disk in $\mathbb{C}$ centred at $-(1-\delta^2)^{-1}$ with radius $(1-\delta^2)^{-1}\sqrt{2\delta^2 - \delta^4}$. This result is easily verified by direct calculation, since by definition

$$\mathcal{R}_{\mathbb{C}}(a;\delta) = \left\{ -\frac{1+\Delta_1}{1+\Delta_2};\ (\Delta_1, \Delta_2) \in \mathbb{C}^2,\ |\Delta_1|^2 + |\Delta_2|^2 < \delta^2 \right\}.$$

Restricting Δ_1, Δ_2 to the reals we see that for $\delta < 1$, $\mathcal{R}_{\mathbb{R}}(a;\delta)$ is the open interval in $\mathbb{R}$ centred at $-(1-\delta^2)^{-1}$ with radius $(1-\delta^2)^{-1}\sqrt{2\delta^2 - \delta^4}$. In Ex. 5 the reader is asked to prove this by using the characterization of $\mathcal{R}_{\mathbb{R}}(a;\delta)$ as given in formula (11). □

We will now derive formulas for *stability radii* of polynomials under arbitrary complex and real affine perturbations. The stability radius is a measure of the size of the smallest perturbation Δ for which $p(s, a(\Delta))$ is either no longer $\mathbb{C}_g$–stable or of degree less than n.

Definition 5.4.6. The stability radius of the polynomial (1) under perturbations of the form (2) is

$$r_{\mathbb{K}}(a; C; \mathbb{C}_g) = \inf\{\|\Delta\|^*_{\mathbb{K}^q};\, \Delta \in \mathbb{K}^{1\times q}, \deg p(s, a(\Delta)) < n \text{ or } \exists s \in \mathbb{C}_b : p(s, a(\Delta)) = 0\}.$$

Clearly, if $p(s, a)$ is $\mathbb{C}_g$-stable and $\mathbb{R}^{1\times q}$ is provided with the norm induced by $\|\cdot\|^*_{\mathbb{C}^q}$ then

$$0 < r_{\mathbb{C}}(a; C; \mathbb{C}_g) \leq r_{\mathbb{R}}(a; C; \mathbb{C}_g).$$

The most important choices of $\mathbb{C}_g$ are $\mathbb{C}_-$ and $\mathbb{D}$, the cases of Hurwitz and Schur polynomials. We denote the corresponding stability radii by $r^-_{\mathbb{K}}(a; C)$ and $r^1_{\mathbb{K}}(a; C)$, respectively. For unstructured perturbations the radius will be denoted by $d_{\mathbb{K}}(a; \mathbb{C}_g)$, and by $d^-_{\mathbb{K}}(a)$ (resp. $d^1_{\mathbb{K}}(a)$) for the Hurwitz (resp. Schur) case. We will use the same notation for the non-monic ($C = I_{n+1}$) and the monic case ($C = [I_n, 0_n]$), but will mention it explicitly if the monic case is considered. For the one-parameter case where only one coefficient a_j is perturbed while the other coefficients remain unchanged we denote the radii by $r_{\mathbb{K}}(a; e^{j\top}; \mathbb{C}_g)$ and for the Hurwitz (resp. Schur) case by $r^-_{\mathbb{K}}(a; e^{j\top})$ (resp. $r^1_{\mathbb{K}}(a; e^{j\top})$).
$d_{\mathbb{K}} = d_{\mathbb{K}}(a, \mathbb{C}_g)$ represents the distance of a $\mathbb{C}_g$-stable polynomial from the set of $\mathbb{C}_g$-unstable polynomials in the coefficient space $\mathbb{K}^{n+1}$ (resp. $\mathbb{K}^n$ in the monic case).[4] Depending on the norm $\|\cdot\|^*_{\mathbb{K}^q}$, $q = n+1$, ($q = n$ in the monic case), the maximal open ball $\{a + \Delta;\, \Delta \in \mathbb{K}^{1\times q}, \|\Delta\|^*_{\mathbb{K}^q} < d_{\mathbb{K}}\}$ of $\mathbb{C}_g$-stable polynomials about the central (nominal) polynomial $p(s, a)$ has different geometric forms, see Figure 5.4.1 for an illustration in the monic case. The following theorem presents explicit general

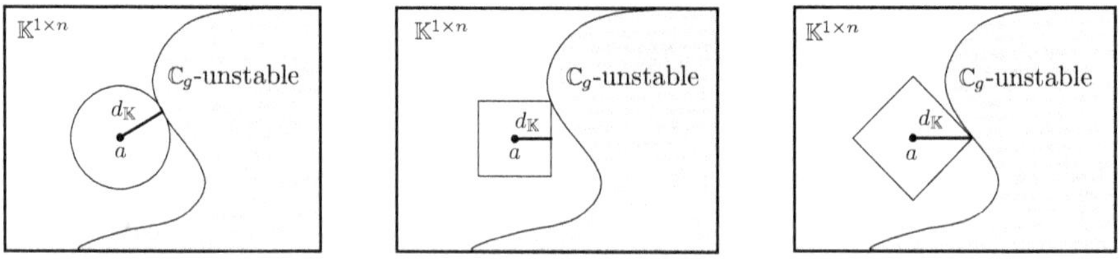

(a) $\|\cdot\|^*_{\mathbb{K}^n} = \|\cdot\|_2$ (b) $\|\cdot\|^*_{\mathbb{K}^n} = \|\cdot\|_\infty$ (c) $\|\cdot\|^*_{\mathbb{K}^n} = \|\cdot\|_1$

Figure 5.4.1: Distances from the set of $\mathbb{C}_g$-unstable monic polynomials in $\mathbb{K}^{1\times n}$

formulas for the complex and real stability radii of arbitrary $\mathbb{C}_g$-stable polynomials under arbitrary linear perturbations of the coefficient vector.

[4] As in previous chapters we identify polynomials $p(s, a)$ of degree n with their coefficient vector $a = [a_0, \ldots, a_n]$. In the monic case where $a_n = 1$ we will sometimes omit the leading coefficient (which is not perturbed) and call $[a_0, \ldots, a_{n-1}]$ the coefficient vector of $p(s, a)$ and denote it also by a if there is no risk of confusion.

Theorem 5.4.7. *Suppose $p(s,a)$ is a $\mathbb{C}_g$-stable polynomial subjected to perturbations of the form (2), and let $G(s)$ be given by (9).*

(i) If $\mathbb{K}=\mathbb{C}$, then

$$r_{\mathbb{C}}(a;C;\mathbb{C}_g)=\min\left\{|a_n|\|c^n\|_{\mathbb{C}^q}^{-1},\ \left[\sup_{s\in\partial\mathbb{C}_g}\|G(s)\|_{\mathbb{C}^q}\right]^{-1}\right\}. \tag{13}$$

(ii) If $\mathbb{K}=\mathbb{R}$, then

$$r_{\mathbb{R}}(a;C;\mathbb{C}_g)=\min\left\{|a_n|\|c^n\|_{\mathbb{R}^q}^{-1},\ \left[\sup_{s\in\partial\mathbb{C}_g}\operatorname{dist}\left(G_R(s),\mathbb{R}G_I(s)\right)\right]^{-1}\right\}, \tag{14}$$

where $G_R(s)$, $G_I(s)$ are the real and imaginary parts of $G(s)$ and the distance dist *is measured with respect to $\|\cdot\|_{\mathbb{R}^q}$.*

Proof: Let A be defined by (5), $B,\tilde{C}$ and D by (7) and $r_{\mathbb{K}}(A;B,\tilde{C},D;\mathbb{C}_g)$ the associated stability radius defined in the previous section. Then $\|D\|_{\mathbb{K}^q}=|a_n|^{-1}\|c^n\|_{\mathbb{K}^q}$. Moreover since $a_n(\Delta)=a_n+\Delta c^n=0$ if and only if $\det(1-\Delta D)=0$ it follows from the definition of $r_{\mathbb{K}}(A;B,\tilde{C},D;\mathbb{C}_g)$ and (8) that

$$r_{\mathbb{K}}(a;C;\mathbb{C}_g)=r_{\mathbb{K}}(A;B,\tilde{C},D;\mathbb{C}_g).$$

So (13) and (14) are special cases of (3.20) and (3.31) respectively. □

Remark 5.4.8. (i) If $\partial\mathbb{C}_g$ is unbounded then $\lim_{|s|\to\infty}\|G(s)\|=|a_n|^{-1}\|c^n\|_{\mathbb{C}^q}$ implies $\sup_{s\in\partial\mathbb{C}_g}\|G(s)\|_{\mathbb{C}^q}\geq|a_n|^{-1}\|c^n\|_{\mathbb{C}^q}$ whence by (13)

$$r_{\mathbb{C}}(a;C;\mathbb{C}_g)=\left[\sup_{s\in\partial\mathbb{C}_g}\|G(s)\|_{\mathbb{C}^q}\right]^{-1}. \tag{15}$$

Also in the case where $\Omega:=\mathbb{C}\setminus\overline{\mathbb{C}_g}$ is non-empty and unbounded, formula (13) can be simplified to (15), see Remark 5.3.10. Note however that (15) will in general not hold if $\mathbb{C}_b$ is bounded. This is illustrated in Example 5.4.9 (ii).

(ii) In the *monic case* $G(s)$ is given by (12) and so

$$r_{\mathbb{C}}(a;C;\mathbb{C}_g)=\left[\max_{s\in\partial\mathbb{C}_g}\|G(s)\|_{\mathbb{C}^q}\right]^{-1},\quad r_{\mathbb{R}}(a;C;\mathbb{C}_g)=\left[\max_{s\in\partial\mathbb{C}_g}\operatorname{dist}\left(G_R(s),\mathbb{R}G_I(s)\right)\right]^{-1}. \tag{16}$$

Note that in the monic case we are allowed to replace the sup in the formulas of Theorem 5.4.7 by max since $G(s)$ is strictly proper rational. □

Example 5.4.9. (i) Consider again the perturbed polynomial studied in Example 5.4.5

$$p(s,a(\Delta))=(1+\Delta_2)s+(1+\Delta_1).$$

Then for Euclidean norms $\|G(\imath\omega)\|_2=\sqrt{1+\omega^2}/|\imath\omega+1|\equiv 1$ for all $\omega\in\mathbb{R}$ and $|a_1|/\|c^1\|_2=1$, so $d_{\mathbb{C}}^-(a)=1$. For the real stability radius it is obvious that $d_{\mathbb{R}}^-(a)=1$. The reader is asked to show how this can be derived from (14) in Ex. 5.

(ii) To illustrate the last statement in Remark 5.4.8 (i) let $\mathbb{C}_g=\mathbb{C}\setminus\overline{\mathbb{D}}$ and consider the perturbed polynomial

$$p(s,a(\Delta))=(1+\Delta_2)s+(5+\Delta_1),\quad \text{i.e.}\quad p(s,a)=s+5,\ \ C=[c^0\,,\,c^1]=I_2.$$

Then $p(s,a) = s+5$ is $\mathbb{C}_g$-stable, $G(s) = (s+5)^{-1}[1,\, s]^\top$ and $\|c^1\|_2/|a_1| = 1$. But on $\partial\mathbb{C}_g = \partial\mathbb{D}$ we have

$$\max_{\theta\in[0,2\pi]} \|G(e^{\imath\theta})\|_2 = \max_{\theta\in[0,2\pi]} \sqrt{2}/|e^{\imath\theta}+5| = \sqrt{2}/4\,.$$

So by (13)

$$d_{\mathbb{C}}(a,\mathbb{C}_g) = r_{\mathbb{C}}(a;C;\mathbb{C}_g) = \min\{1, 4/\sqrt{2}\} = 1 < 4/\sqrt{2} = \left[\max_{\theta\in[0,2\pi]} \|G(e^{\imath\theta})\|_2\right]^{-1}.$$

□

If $r_{\mathbb{K}}(a;C;\mathbb{C}_g) < \infty$ there exists a perturbation of norm $r_{\mathbb{K}}(a;C;\mathbb{C}_g)$ which either decreases the degree or destabilizes the given $\mathbb{C}_g$-stable polynomial $p(s,a)$, $a \in \mathbb{K}^{1\times(n+1)}$. We conclude this subsection with three remarks on how to construct such minimum norm destabilizing or degree decreasing perturbations with respect to the 1-, 2-, and ∞-norms on $\mathbb{K}^{1\times q}$.

Remark 5.4.10. If $r_{\mathbb{K}}(a;C;\mathbb{C}_g) = |a_n|\|c^n\|_{\mathbb{K}^q}^{-1} < \infty$ a minimum norm degree decreasing $\Delta \in \mathbb{K}^{1\times q}$ must satisfy the following alignment condition

$$\Delta c^n = -a_n \quad \text{and} \quad \|\Delta\|_{\mathbf{\Delta}} = \|\Delta\|_{\mathbb{K}^q}^* = |a_n|\,\|c^n\|_{\mathbb{K}^q}^{-1}. \tag{17}$$

Such a Δ always exists by the Hahn-Banach Theorem, see Example A.4.11. If $|c_{kn}| = \max_{j\in\underline{q}} |c_{jn}| = \|c^n\|_\infty$, it is easy to check that with respect to the 1-, 2-, and ∞-norms on $\mathbb{K}^{1\times q}$ the following Δ's satisfy (17)

$$\begin{array}{rrcl} \text{1-norm}: & \Delta & = & -a_n\,\|c^n\|_\infty^{-1}\,[0,\ldots,0,\bar{c}_{kn}/|\bar{c}_{kn}|,0,\ldots,0], \\ \text{2-norm}: & \Delta & = & -a_n\,\|c^n\|_2^{-2}\,[\bar{c}_{1n},\ldots,\bar{c}_{qn}], \\ \infty\text{-norm}: & \Delta & = & -a_n\,\|c^n\|_1^{-1}\,[\bar{c}_{1n}/|\bar{c}_{1n}|,\ldots,\bar{c}_{qn}/|\bar{c}_{qn}|]. \end{array} \tag{18}$$

Here $^-$ denotes complex conjugation and we set $z/|z| := 0$ if $z = 0$. □

If $r_{\mathbb{K}}(a;C;\mathbb{C}_g) < |a_n|\|c^n\|_{\mathbb{K}^q}^{-1}$ there exists a destabilizing perturbation $\Delta \in \mathbb{K}^{1\times q}$ of norm $r_{\mathbb{K}}(a;C;\mathbb{C}_g)$. In the following two remarks we show how to construct such minimum norm destabilizing perturbations for the complex and real cases.

Remark 5.4.11. Suppose that $a \in \mathbb{C}^{1\times(n+1)}$, $r_{\mathbb{C}}(a;C;\mathbb{C}_g) < |a_n|\|c^n\|_{\mathbb{C}^q}^{-1}$ and that $s_0 \in \partial\mathbb{C}_g$ maximizes $\|G(s)\|_{\mathbb{C}^q}$ on $\partial\mathbb{C}_g$. Then $\Delta \in \mathbb{C}^{1\times q}$ (see Proposition 4.4.11) will be a minimum norm destabilizing perturbation if it satisfies

$$\Delta G(s_0) = 1, \quad \|\Delta\|_{\mathbb{C}^q}^* = \|G(s_0)\|_{\mathbb{C}^q}^{-1}. \tag{19}$$

If $G(s_0) = [\gamma_1,\ldots,\gamma_q]^\top$, $|\gamma_k| = \max_{j\in\underline{q}} |\gamma_j|$ the following formulas yield minimum norm destabilizing perturbations with respect to the 1-, 2-, or ∞-norms on $\mathbb{C}^{1\times q}$, respectively

$$\begin{array}{rcl} \text{1-norm}: \Delta & = & \|G(s_0)\|_\infty^{-1}\,[0,\ldots,0,\bar{\gamma}_k/|\bar{\gamma}_k|,0,\ldots,0], \\ \text{2-norm}: \Delta & = & \|G(s_0)\|_2^{-2}\,[\bar{\gamma}_1,\ldots,\bar{\gamma}_q], \\ \infty\text{-norm}: \Delta & = & \|G(s_0)\|_1^{-1}\,[\bar{\gamma}_1/|\bar{\gamma}_1|,\ldots,\bar{\gamma}_q/|\bar{\gamma}_q|]. \end{array} \tag{20}$$

□

Remark 5.4.12. Suppose that $\mathbb{K} = \mathbb{R}$, $r_{\mathbb{R}}(a;C;\mathbb{C}_g) < |a_n|\|c^n\|_{\mathbb{R}^q}^{-1}$ and $s_0 \in \partial\mathbb{C}_g$ maximizes the distance between $G_R(s)$ and $\mathbb{R}G_I(s)$ on $\partial\mathbb{C}_g$

$$\text{dist}(G_R(s_0), \mathbb{R}G_I(s_0)) = \max_{s\in\partial\mathbb{C}_g} \text{dist}(G_R(s), \mathbb{R}G_I(s)) = r_{\mathbb{R}}(a;C;\mathbb{C}_g)^{-1}.$$

If $G_I(s_0) = 0$ then it suffices to choose any Δ aligned with $G_R(s_0) = G(s_0)$ to obtain a minimum norm destabilizing perturbation.

$$\Delta G_R(s_0) = 1, \quad \|\Delta\|_{\mathbb{R}^q}^* = \|G(s_0)\|_{\mathbb{R}^q}^{-1}.$$

$G(s_0)$ being real, the formulas given in (20) will yield minimum norm destabilizing real perturbations with respect to the 1-, 2- and ∞-norms on $\mathbb{R}^{1\times q}$. Now suppose $G_I(s_0) \neq 0$, then $G_R(s_0)$ and $G_I(s_0)$ are linearly independent since otherwise $\operatorname{dist}(G_R(s_0), \mathbb{R}G_I(s_0)) = 0$, i.e. $r_{\mathbb{R}}(a; C; \mathbb{C}_g) = \infty$ which contradicts $r_{\mathbb{R}}(a; C; \mathbb{C}_g) < |a_n| \|c^n\|_{\mathbb{R}^q}^{-1}$. By Remark 5.3.17 (ii) a minimum norm destabilizing perturbation Δ satisfies

$$\Delta G_I(s_0) = 0\,, \quad \Delta G_R(s_0) = 1\,, \quad \|\Delta\|_{\mathbb{R}^q}^* \|G_R(s_0) - \hat{\alpha} G_I(s_0)\|_{\mathbb{R}^q} = 1 \tag{21}$$

where $\hat{\alpha} \in \mathbb{R}$ is such that $\|G_R(s_0) - \hat{\alpha} G_I(s_0)\|_{\mathbb{R}^q} = \operatorname{dist}(G_R(s_0), \mathbb{R}G_I(s_0))$. Geometrically, the set of all $\Delta \in \mathbb{R}^{1\times q}$ satisfying (21) is the intersection of the $(q-2)$–dimensional affine subspace determined by the first two equations and the sphere with centre 0 and radius $r_{\mathbb{R}}(a; C; \mathbb{C}_g) = \operatorname{dist}(G_R(s_0), \mathbb{R}G_I(s_0))^{-1}$ in $(\mathbb{R}^{1\times q}, \|\cdot\|_{\mathbb{R}^q}^*)$. If s_0 and $\hat{\alpha}$ are known, a point Δ in this intersection can be determined. The reader is asked to prove the explicit expressions for such Δ with respect to the 1-, 2- and ∞-norms given in Ex. 9. □

In the next two subsections we will use the formulas in (10)–(14) to obtain more explicit results in a number of interesting special cases.

5.4.2 Complex Perturbation Structures

Throughout this subsection we will consider complex perturbations and obtain characterizations for root sets and stability radii with respect to two special perturbation structures, viz. unstructured perturbations and single parameter linear perturbations of the coefficient vector. Otherwise the basic assumptions are the same as those in the previous subsection with $\mathbb{K} = \mathbb{C}$, but we only consider the cases where the perturbation space $\mathbb{C}^{1\times q}$ is provided with a p-norm for $p = 1, 2, \infty$.
We first consider *unstructured* perturbations. In this case $q = n+1$, $C = I_{n+1}$ and (9) yields

$$G(s) = -p(s,a)^{-1}[1, s, \ldots, s^{n-1}, s^n]^\top. \tag{22}$$

The following corollary is a consequence of (13).

Corollary 5.4.13. *Suppose $\mathbb{C}^{1\times(n+1)}$ is provided with the 1-, 2- or ∞-norm and $G(s)$ is defined by (22), then the root sets of $p(s,a)$ under unstructured complex perturbations of size smaller than $\delta \in (0, |a_n|)$ are given by*

$$\begin{aligned}
\text{1-norm}: \mathcal{R}_{\mathbb{C}}(a;\delta) &= \mathcal{R}(a)\dot{\cup}\{s \in \mathbb{C}\setminus\mathcal{R}(a);\ \|G(s)\|_\infty > \delta^{-1}\} \\
&= \left\{s \in \mathbb{C};\ \min_{0\le j\le n} \frac{|p(s,a)|}{|s^j|} < \delta\right\}, \\
\text{2-norm}: \mathcal{R}_{\mathbb{C}}(a;\delta) &= \mathcal{R}(a)\dot{\cup}\{s \in \mathbb{C}\setminus\mathcal{R}(a);\ \|G(s)\|_2 > \delta^{-1}\} \\
&= \left\{s \in \mathbb{C};\ \frac{|p(s,a)|}{(1+|s|^2+\cdots+|s|^{2n-2}+|s|^{2n})^{1/2}} < \delta\right\}, \\
\infty\text{-norm}: \mathcal{R}_{\mathbb{C}}(a;\delta) &= \mathcal{R}(a)\dot{\cup}\{s \in \mathbb{C}\setminus\mathcal{R}(a);\ \|G(s)\|_1 > \delta^{-1}\} \\
&= \left\{s \in \mathbb{C};\ \frac{|p(s,a)|}{(1+|s|+\cdots+|s|^{n-1}+|s|^n)} < \delta\right\}.
\end{aligned}$$

In the monic case, where $G(s)$ is given by (12), there is no restriction on δ and for the 1-norm the minimization is for $j = 0, \ldots, n-1$, for the 2-norm the term $|s|^{2n}$ is absent and for the ∞-norm the term $|s|^n$ is absent.
Let us now consider the *complex stability radius* of *Schur* and *Hurwitz* polynomials under unstructured perturbations. Choosing $\mathbb{C}_g = \mathbb{C}_-$ or $\mathbb{C}_g = \mathbb{D}$ in (15) we have

$$d^1_{\mathbb{C}}(a) = \left[\max_{\theta\in[0,2\pi]} \|G(e^{\imath\theta})\|_{\mathbb{C}^q}\right]^{-1}, \quad d^-_{\mathbb{C}}(a) = \left[\sup_{\omega\in\mathbb{R}} \|G(\imath\omega)\|_{\mathbb{C}^q}\right]^{-1}$$

where $G(s)$ is given by (22). In particular we obtain

Corollary 5.4.14. *Suppose $p(\cdot\,,a)$ is a Schur polynomial. Then its distance from the set of non-Schur polynomials in the coefficient space $(\mathbb{C}^{1\times(n+1)}, \|\cdot\|_p)$, $p = 1, 2, \infty$ is given by*

$$\begin{aligned} \text{1-norm}: d^1_{\mathbb{C}}(a) &= \min_{\theta\in[0,2\pi]} |p(e^{\imath\theta}, a)|, \\ \text{2-norm}: d^1_{\mathbb{C}}(a) &= \min_{\theta\in[0,2\pi]} \frac{|p(e^{\imath\theta}, a)|}{\sqrt{n+1}}, \\ \infty\text{-norm}: d^1_{\mathbb{C}}(a) &= \min_{\theta\in[0,2\pi]} \frac{|p(e^{\imath\theta}, a)|}{n+1}. \end{aligned}$$

So the ratios between the distances $d^1_{\mathbb{C}}(a)$ with respect to the above norms are given by $1 : \sqrt{n+1} : n+1$, independently of the particular Schur polynomial $p(s,a)$.

Corollary 5.4.15. *Suppose $p(\cdot\,,a)$ is a Hurwitz polynomial. Then its distance from the set of non-Hurwitz polynomials in the coefficient space $(\mathbb{C}^{1\times(n+1)}, \|\cdot\|_p)$, $p = 1, 2, \infty$ is given by*

$$\begin{aligned} \text{1-norm}: \quad d^-_{\mathbb{C}}(a) &= \inf_{\omega\in\mathbb{R}} \min_{0\le j\le n} \frac{|p(\imath\omega, a)|}{|\omega|^j}, \\ \text{2-norm}: \quad d^-_{\mathbb{C}}(a) &= \inf_{\omega\in\mathbb{R}} \frac{|p(\imath\omega, a)|}{(1+\omega^2+\cdots+\omega^{2n-2}+\omega^{2n})^{1/2}}, \\ \infty\text{-norm}: \quad d^-_{\mathbb{C}}(a) &= \inf_{\omega\in\mathbb{R}} \frac{|p(\imath\omega, a)|}{(1+|\omega|+\cdots+|\omega|^{n-1}+|\omega|^n)}. \end{aligned}$$

There are obvious adjustments in the above formulas for the monic case: For monic Schur polynomials $n+1$ is replaced by n in the formulas for $d^1_{\mathbb{C}}(a)$ and for monic Hurwitz polynomials n is replaced by $n-1$ in the formulas for $d^-_{\mathbb{C}}(a)$.

Example 5.4.16. Consider the monic polynomial $p(s,a) = s^2 + a_1 s + a_0$ where $a = [a_0, a_1] \in \mathbb{R}^2$. We will first calculate the root set $\mathcal{R}_{\mathbb{C}}(a;\delta)$ for the monic case with respect to the 2-norm. Since

$$p(re^{\imath\theta}, a) = r^2\cos 2\theta + a_1 r\cos\theta + a_0 + \imath(r^2\sin 2\theta + a_1 r\sin\theta), \quad \theta\in[0,2\pi],$$

a straightforward calculation yields

$$\begin{aligned} |p(re^{\imath\theta}, a)|^2 &= (r^2\cos 2\theta + a_1 r\cos\theta + a_0)^2 + (r^2\sin 2\theta + a_1 r\sin\theta)^2 \\ &= a_1^2 r^2 + (r^2 - a_0)^2 + 2a_1 r(r^2 + a_0)\cos\theta + 4a_0 r^2\cos^2\theta \\ &= \begin{cases} r^4 + 2a_1 r^3\cos\theta + a_1^2 r^2 & \text{if } a_0 = 0 \\ 4a_0\left[r\cos\theta + \dfrac{a_1(r^2+a_0)}{4a_0}\right]^2 + \dfrac{(r^2-a_0)^2}{4a_0}[4a_0 - a_1^2] & \text{if } a_0 \neq 0\,. \end{cases} \end{aligned}$$

Hence $\mathcal{R}_{\mathbb{C}}(a;\delta)$ is the set of all $s = re^{\imath\theta}$, $(r,\theta) \in \mathbb{R}_+ \times [0, 2\pi]$, such that

$$\begin{cases} (1+r^2)\,\delta^2 > r^4 + 2a_1 r^3 \cos\theta + a_1^2 r^2 & \text{if} \quad a_0 = 0\,, \\ (1+r^2)\,\delta^2 > 4a_0 \left[r\cos\theta + \dfrac{a_1(r^2+a_0)}{4a_0} \right]^2 + \dfrac{(r^2-a_0)^2}{4a_0}[4a_0 - a_1^2] & \text{if} \quad a_0 \neq 0\,. \end{cases}$$

Using these formulas for the root sets explicit formulas for the unstructured stability radii $d_{\mathbb{C}}^1(a)$ and $d_{\mathbb{C}}^-(a)$ in the monic case can be derived. We illustrate this for the Schur case, see Ex. 2 for the Hurwitz case. Recall that the set of real monic Schur polynomials of degree n is denoted by $\mathcal{S}_n$. To determine $d_{\mathbb{C}}^1(a)$ for $a \in \mathcal{S}_2$ we set $r = 1$ in the above formulas for $|p(re^{\imath\theta}, a)|$ and minimize the RHS of the two inequalities with respect to θ. If $a_0 \neq 0$ this means minimizing or maximizing the first term on the RHS of the second inequality depending on whether $a_0 > 0$ or $a_0 < 0$. For example if $a_0 > 0$, $|a_1|(1+a_0) < 4a_0$ one chooses $\theta \in [0, 2\pi]$ such that $\cos\theta + \dfrac{a_1(1+a_0)}{4a_0} = 0$. Otherwise $\cos\theta = +1$ or -1. After some calculations one finds that for any $a \in \mathcal{S}_2$

$$d_{\mathbb{C}}^1(a) = \begin{cases} \dfrac{1-a_0}{\sqrt{2}} \left[1 - \dfrac{a_1^2}{4a_0} \right]^{1/2} & \text{if} \quad |a_1|(1+a_0) < 4a_0, \ a_0 > 0\,, \\ \dfrac{1-|a_1|+a_0}{\sqrt{2}} & \text{if} \quad |a_1|(1+a_0) \geq 4a_0. \end{cases}$$

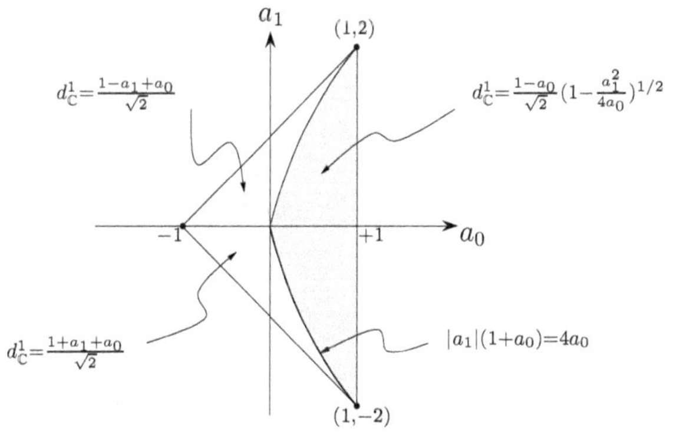

Figure 5.4.2: The distances $d_{\mathbb{C}}^1(a)$, $d_{\mathbb{R}}^1(a)$ for $a \in \mathcal{S}_2$

This result is illustrated in Figure 5.4.2 where the distances of $a \in \mathcal{S}_2$ from the set of complex resp. real non-Schur coefficient vectors are compared. The triangle is the set of real monic Schur polynomials of degree 2. Hence the distance $d_{\mathbb{R}}^1(a)$ of $a \in \mathcal{S}_2$ from the set of real non-Schur coefficient vectors is easily determined: It is just the Euclidean distance of a to the boundary of the triangle. An elementary geometric consideration shows that

$$d_{\mathbb{R}}^1(a) = \min\left\{ \frac{1+a_1+a_0}{\sqrt{2}}, \frac{1-a_1+a_0}{\sqrt{2}}, 1-a_0 \right\}, \quad a \in \mathcal{S}_2.$$

In the shaded area the first formula for $d_{\mathbb{C}}^1(a)$ applies, so that here

$$2(d_{\mathbb{C}}^1(a))^2 = \frac{(1-a_0)^2}{4a_0}(4a_0 - a_1^2) \leq \frac{(1-a_0)^2}{4a_0}(4a_0 - a_1^2) + 4a_0\left(|a_1|\frac{1+a_0}{4a_0} - 1\right)^2 = (1-|a_1|+a_0)^2.$$

Hence $d^1_{\mathbb{C}}(a) < d^1_{\mathbb{R}}(a)$ in this part of $\mathcal{S}_2$. On the other hand in the non-shaded area of $\mathcal{S}_2$ the latter two formulas for $d^1_{\mathbb{C}}(a)$ apply and so $d^1_{\mathbb{C}}(a) = d^1_{\mathbb{R}}(a)$ in this part of $\mathcal{S}_2$. □

The distances $d^-_{\mathbb{C}}(a)$ and $d^1_{\mathbb{C}}(a)$ yield adequate robustness measures if all the coefficients $p(\cdot\,, a)$ are subject to independent perturbations of equal weight. On the other hand it is interesting to consider the case where the coefficients of $p(\cdot\,, a)$ all depend on one and the same unknown $\Delta \in \mathbb{C}$ (i.e. $q = 1$, $C = [c_0, \ldots, c_n] \in \mathbb{C}^{1\times(n+1)}$). This case has received special attention in Numerical Analysis and Perturbation Theory (see Subsection 4.4.1).

Corollary 5.4.17. *Suppose that $p(s,a)$ is subjected to perturbations of the form $a \rightsquigarrow a + \Delta c$, $\Delta \in \mathbb{C}$ where $c = [c_0, \ldots, c_n] \in \mathbb{C}^{1\times(n+1)}$, then*

$$\mathcal{R}_{\mathbb{C}}(a;c;\delta) = \left\{ s \in \mathbb{C};\ \frac{|p(s,a)|}{|c(s)|} < \delta \right\}, \quad \delta < \frac{|a_n|}{|c_n|}, \tag{23}$$

and if $p(s,a)$ is $\mathbb{C}_g$-stable,

$$r_{\mathbb{C}}(a;c;\mathbb{C}_g) = \min\left\{ \frac{|a_n|}{|c_n|},\ \inf_{s\in\partial\mathbb{C}_g} \frac{|p(s,a)|}{|c(s)|} \right\} \tag{24}$$

where $c(s) = c_n s^n + \cdots + c_1 s + c_0$.

For the monic case ($c_n = 0$), (24) implies $r_{\mathbb{C}}(a;c;\mathbb{C}_g) = \min_{s\in\partial\mathbb{C}_g} |p(s,a)|/|c(s)|$. Of particular interest is the case where only one coefficient is perturbed, i.e. $c = e^{j\top} \in \mathbb{R}^{1\times(n+1)}$ and $c(s) = s^j$. Then, for $j = 0, \ldots, n$,

$$\mathcal{R}_{\mathbb{C}}(a;e^{j\top};\delta) = \left\{ s \in \mathbb{C};\ \frac{|p(s,a)|}{|s^j|} < \delta \right\}, \quad \text{(for } j = n \text{ we require } \delta < |a_n|\text{)}.$$

If $p(\cdot\,, a)$ is a monic Hurwitz or Schur polynomial, then

$$r^-_{\mathbb{C}}(a;e^{j\top}) = \min_{\omega\in\mathbb{R}} \frac{|p(\imath\omega,a)|}{|\omega|^j},\ r^1_{\mathbb{C}}(a;e^{j\top}) = \min_{\theta\in[0,2\pi]} |p(e^{\imath\theta},a)| = d^1_{\mathbb{C}}(a),\ j=0,\ldots,n-1 \tag{25}$$

where $d^1_{\mathbb{C}}(a)$ (monic case) is taken with respect to the 1-norm. Note that in the Schur case the stability radii $r^1_{\mathbb{C}}(a, e^{j\top})$, $j = 0, \ldots, n-1$ are independent of j.
In Example 4.1.20 we considered the parametrized family $p(s, a(z)) = p(s) + zq(s)$, $z \in \mathbb{C}$ and showed that the sensitivity of a simple root $s_j(z)$ of $p(s) + zq(s)$ at $z = 0$ is $\xi_j = -q(s_j(0))/p'(s_j(0))$. Using this result one can obtain good estimates for $s_j(z)$ for *small* values of $|z|$. In contrast the root sets $\mathcal{R}_{\mathbb{C}}(a;C;\delta)$ and the stability radius $r_{\mathbb{C}}(a;C;\mathbb{C}_g)$ can provide information about the root variations for *large* parameter changes. We illustrate these two approaches by means of a monic Hurwitz polynomial which is known to be "ill-conditioned" in a Numerical Analysis sense (cf. [26], [523]), see also Section 4.5.

Example 5.4.18. Consider the (Hurwitz stable) Wilkinson polynomial

$$\begin{aligned} p(s) = p(s,a) &= (s+1)(s+2)\cdots(s+7) \\ &= s^7 + 28s^6 + 322s^5 + 1960s^4 + 6769s^3 + 13132s^2 + 13068s + 5040\,. \end{aligned}$$

Suppose that only the coefficient $a_6 = 28$ is perturbed, i.e. $q(s) = s^6$ and $p(s, a(z)) = p(s) + zs^6$. The corresponding sensitivities of the roots $s_j(0) = -j$, $j = 1, \ldots, 7$ are

$$\xi_j = -\frac{j^6}{\prod_{i\neq j}^{7}(-j+i)} = (-1)^j \frac{j^6}{(j-1)!(7-j)!}.$$

For example $\xi_1 = -1/720$, $\xi_7 \approx -163.4$.
In order to examine the effect of "large" variations of the coefficient $a_6 = 28$ we computed $\mathcal{R}_{\mathbb{C}}(a; e^{6\top}; \delta)$ for various values of δ. The sets are illustrated in Figure 5.4.3 for $\delta =$

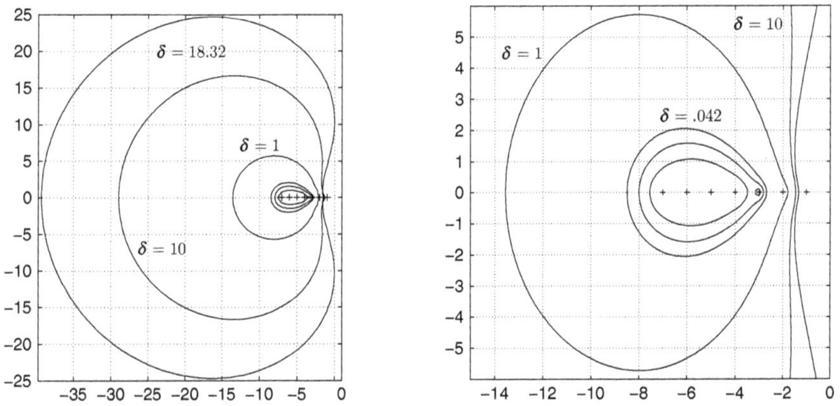

Figure 5.4.3: Root sets $\mathcal{R}_{\mathbb{C}}(a; e^{6\top}, \delta)$ for the Wilkinson polynomial

$18.32, 10, 1, .042, .0192$ and $.0069$, where the figure on the right is a zoom in of the one on the left. As indicated by the low sensitivity $\xi_1 = -1/720$ the root nearest the imaginary axis is quite stubborn with respect to perturbations of a_6. It hardly moves at all so that no contour line is visible around $s_1 = -1$ even for relatively large perturbations. Moreover, the figures show that although for small δ the root sets around $s_3, \ldots\ldots s_7$ are expanding fast in accordance with the sensitivity measures, the expansion slows down for larger δ so that, for example, when $\delta = 1$ the expansion to the left is less than $1/10$ of what the sensitivity $\xi_7 \approx -163.4$ would predict by linear extrapolation. The discrepancy between local sensitivity and global robustness analysis becomes even more evident when we consider the complex stability radius with respect to the Hurwitz partition $\mathbb{C}_g = \mathbb{C}_-$. In fact linear extrapolation using sensitivity indicates that the root $s_7(z)$ should hit the imaginary axis for $z = -7/163.4 \approx -.0428$. But $r_{\mathbb{C}}^-(a, e^{6\top}) = 18.3194$. Thus the robustness of stability of $p(s, a)$ with respect to a perturbation of a_6 is about 400 times larger than sensitivity indicates. □

5.4.3 Real Perturbation Structures

In this subsection we carry out essentially the same programme as that of the previous one but now the data and the perturbations are assumed to be real

$$a = [a_0, \ldots, a_n] \in \mathbb{R}^{1\times(n+1)},\ a_n \neq 0; \quad C = [c^0, \ldots, c^n] \in \mathbb{R}^{q\times(n+1)}; \quad \mathbf{\Delta} = \mathbb{R}^{1\times q}$$

where $n, q \in \mathbb{N}^*$. We first assume that $\mathbb{R}^q$ is provided with an arbitrary norm $\|\cdot\|_{\mathbb{R}^q}$ and $\mathbf{\Delta} = \mathbb{R}^{1\times q}$ with the corresponding dual norm $\|\cdot\|_{\mathbb{R}^q}^*$. Later we will consider the case where $\|\cdot\|_{\mathbb{R}^q}^*$ is the ∞-norm and relate the corresponding stability radius problem to Kharitonov's Theorem. Further results for other norms can be found in the exercises.

By $G_R(s)$, $G_I(s)$ we denote the real and imaginary parts of $G(s)$ given in (9). We have seen in Section 4.4 that the map $s \mapsto \operatorname{dist}(G_R(s), G_I(s)\mathbb{R})$ may not be continuous at the zeros of $G_I(s)$ and so as in Subsections 5.2.2 and 5.3.3 we introduce the realness locus

$$\mathcal{R}_G = \{s \in \mathbb{C} \setminus \mathcal{R}(a); G_I(s) = 0\} \supset \mathbb{R}. \tag{26}$$

Note that on the realness locus $G_R(s) = G(s)$. By (11) we obtain

$$\begin{aligned}\mathcal{R}_{\mathbb{R}}(a; C; \delta) &= \mathcal{R}(a) \,\dot\cup\, \{s \in \mathcal{R}_G;\ \|G(s)\|_{\mathbb{R}^q} > \delta^{-1}\} \\ &\quad \dot\cup\, \{s \in \mathbb{C} \setminus (\mathcal{R}_G \cup \mathcal{R}(a));\ \operatorname{dist}(G_R(s), \mathbb{R}G_I(s)) > \delta^{-1}\},\ \ \delta < |a_n|\,\|c^n\|_{\mathbb{R}^q}^{-1}.\end{aligned} \tag{27}$$

If $p(s,a)$ is $\mathbb{C}_g$-stable then by (14)

$$r_{\mathbb{R}}(a; C; \mathbb{C}_g) = \min\left\{\frac{|a_n|}{\|c^n\|_{\mathbb{R}^q}},\ \inf_{s \in \mathcal{R}_G \cap \partial\mathbb{C}_g} \|G(s)\|_{\mathbb{R}^q}^{-1},\ \inf_{s \in \partial\mathbb{C}_g \setminus \mathcal{R}_G} \operatorname{dist}(G_R(s), \mathbb{R}G_I(s))^{-1}\right\}. \tag{28}$$

Remark 5.4.19. If $p(s,a)$ is $\mathbb{C}_g$-stable then $\mathcal{R}_G \cap \partial\mathbb{C}_g$ is closed and so the continuous function $s \mapsto \|G(s)\|_{\mathbb{R}^q}$ has a maximum on $\mathcal{R}_G \cap \partial\mathbb{C}_g$ if this set is non-empty and $G(s)$ is strictly proper. Therefore the first inf in (28) may be replaced by a min in the monic case ($c^n = 0$) if the realness locus of $G(s)$ intersects the boundary of the stability region. □

We now derive more concrete versions of (27) and (28) for special perturbation structures $C \in \mathbb{R}^{q\times(n+1)}$ and begin with the one parameter case ($\Delta \in \mathbb{R}$, $q = 1$, $C = c = [c_0, \ldots, c_n]$, $\|\cdot\|_{\mathbb{R}^q}^* = |\cdot|$). Then

$$G(s) = G_R(s) + \imath G_I(s) = -\frac{c(s)}{p(s,a)}, \quad c(s) = \sum_{j=0}^{n} c_j s^j .$$

In this case the last term on the RHS of (27) and (28) can be omitted (since $\mathbb{R}G_I(s) = \mathbb{R}$ if $G_I(s) \neq 0$), so we have the following real counterpart to Corollary 5.4.17.

Corollary 5.4.20. *Suppose that $p(s,a)$ is subjected to perturbations of the form $a \rightsquigarrow a + \Delta c$, $\Delta \in \mathbb{R}$ where $c = [c_0, \ldots, c_n] \in \mathbb{R}^{1\times(n+1)}$, then*

$$\mathcal{R}_{\mathbb{R}}(a; c; \delta) = \mathcal{R}(a) \,\dot\cup\, \left\{s \in \mathcal{R}_G;\ \frac{|p(s,a)|}{|c(s)|} < \delta\right\}, \quad \delta < \frac{|a_n|}{|c_n|}. \tag{29}$$

If $p(s,a)$ is $\mathbb{C}_g$-stable, then

$$r_{\mathbb{R}}(a; c; \mathbb{C}_g) = \min\left\{\frac{|a_n|}{|c_n|},\ \inf\left\{\frac{|p(s,a)|}{|c(s)|};\ s \in \mathcal{R}_G \cap \partial\mathbb{C}_g\right\}\right\}. \tag{30}$$

By Remark 5.4.19 if $\mathcal{R}_G \cap \partial\mathbb{C}_g$ is non-empty (e.g. if $\mathbb{R} \cap \partial\mathbb{C}_g \neq \emptyset$) we have *in the monic case*

$$r_{\mathbb{R}}(a; c; \mathbb{C}_g) = \min\{|p(s,a)|/|c(s)|;\ s \in \mathcal{R}_G \cap \partial\mathbb{C}_g\}. \tag{31}$$

The following example shows that it is possible that $\mathcal{R}_G \cap \partial\mathbb{C}_g = \emptyset$.

Example 5.4.21. Consider the perturbed monic polynomial

$$p(s, a(\Delta)) = s + 1 + \Delta, \quad \Delta \in \mathbb{R}$$

(i.e. $n = 1, q = 1$, $a = [1\,,\,1]$, $c = [1\,,\,0]$) and let $\mathbb{C}_g = \{s \in \mathbb{C};\, |\operatorname{Im} s| < 1\}$. Then all the perturbed polynomials are $\mathbb{C}_g$-stable so that $r_{\mathbb{R}}(a; c; \mathbb{C}_g) = \infty$. In fact we have $G(s) = 1/(s+1)$ and $\mathcal{R}_G \cap \partial\mathbb{C}_g = \mathbb{R} \cap \partial\mathbb{C}_g = \emptyset$. □

The RHS of (30) gives rise to an efficient procedure for the computation of $r_{\mathbb{R}}(a; c; \mathbb{C}_g)$. We illustrate this for the Hurwitz case. In a first step we decompose $c(s)$ and $p(s, a)$ into their even and odd parts

$$c(s) = c^e(s^2) + sc^o(s^2)\,, \quad p(s, a) = p^e(s^2) + sp^o(s^2) \tag{32}$$

where $c^e(s^2)$, $c^o(s^2)$, $p^e(s^2)$, $p^o(s^2)$ are real polynomials in s^2 of degree $\leq n/2$ (see Section 3.4 where it follows from Proposition 3.4.6 that $p^e(0)$ and $p^o(0)$ are both non-zero if p is Hurwitz). Since

$$G(\imath\omega) = -\frac{c^e(-\omega^2) + \imath\omega c^o(-\omega^2)}{p^e(-\omega^2) + \imath\omega p^o(-\omega^2)}$$

we have

$$\begin{aligned} G_R(\imath\omega) &= -\frac{c^e(-\omega^2)p^e(-\omega^2) + \omega^2 c^o(-\omega^2)p^o(\omega^2)}{p^e(-\omega^2)^2 + \omega^2 p^o(-\omega^2)^2}, \\ G_I(\imath\omega) &= -\frac{\omega[c^o(-\omega^2)p^e(-\omega^2) - c^e(-\omega^2)p^o(-\omega^2)]}{p^e(-\omega^2)^2 + \omega^2 p^o(-\omega^2)^2}, \qquad \omega \in \mathbb{R}. \end{aligned}$$

So $G_I(\imath\omega) = 0$ is equivalent to

$$\omega = 0 \quad \text{or} \quad [c^o(-\omega^2)p^e(-\omega^2) - c^e(-\omega^2)p^o(-\omega^2)] = 0 \tag{33}$$

where the expression in the bracket is a real polynomial of degree $< n$ in ω^2. It follows that for $\omega \in \mathbb{R}$, $\omega \neq 0$ and $G(\imath\omega) \in \mathbb{R}$, we have

$$G(\imath\omega) = -\frac{c^e(-\omega^2)p^e(-\omega^2) + \imath\omega c^e(-\omega^2)p^o(-\omega^2)}{p^e(-\omega^2)[p^e(-\omega^2) + \imath\omega p^o(-\omega^2)]} = -\frac{c^e(-\omega^2)}{p^e(-\omega^2)} = -\frac{c^o(-\omega^2)}{p^o(-\omega^2)}. \tag{34}$$

Now suppose that $G(s)$ is not constant. Then the polynomial in (33) cannot be identically zero in ω^2 since otherwise by (34) and the identity theorem of complex analysis $G(s) = -c^e(s^2)/p^e(s^2)$. But $G(s)$ is proper rational and not constant, so there is at least one zero $\lambda \in \mathbb{C}$ of p^e which is not cancelled by a zero of c^e, i.e. $G(s)$ has poles at $s_\pm = \pm(\lambda)^{1/2}$. This contradicts the assumption that the poles of $G(s)$ lie in $\mathbb{C}_-$. We conclude, therefore, that the non-empty set $\mathcal{R}_G^+ := \mathcal{R}_G \cap \imath\mathbb{R}_+$, i.e.

$$\mathcal{R}_G^+ = \{\omega \in \mathbb{R}_+; G_I(\imath\omega) = 0\} = \{\omega \geq 0; \omega\,[c^o(-\omega^2)p^e(-\omega^2) - c^e(-\omega^2)p^o(-\omega^2)] = 0\} \tag{35}$$

has at most n elements. To calculate $r_{\mathbb{R}}^-(a; c)$ the main step is to compute $\mathcal{R}_G^+$ by determining the positive solutions of (33). Then it only remains to compute the minimum of a set of at most n real numbers

$$r_{\mathbb{R}}^-(a; c) = \min\left\{\frac{|a_n|}{|c_n|}, \min_{\omega \in \mathcal{R}_G^+} \frac{|p^e(-\omega^2)|}{|c^e(-\omega^2)|}\right\}. \tag{36}$$

The following specialization of Corollary 5.4.20 deals with the problem of determining the size of the smallest real perturbation of a *single* coefficient which destabilizes a given monic Hurwitz polynomial, i.e. $c = e^{j\top}$, $j = 0, \ldots, n-1$. See Ex. 4 for the Schur case.

Corollary 5.4.22. *Suppose that $p(s,a)$ is a monic real Hurwitz polynomial and $p^e(\cdot)$, $p^o(\cdot)$ are as in (32). Then for $j = 0, \ldots, n-1$*

$$r_{\mathbb{R}}^-(a; e^{j\top}) = \begin{cases} \min\left\{|p^o(-\omega^2)/\omega^{j-1}|\,; \omega \geq 0,\ \omega p^e(-\omega^2) = 0\right\}, & j \text{ odd}, \\ \min\left\{|p^e(-\omega^2)/\omega^{j}|\,; \omega \geq 0,\ \omega p^o(-\omega^2) = 0\right\}, & j \text{ even}, \end{cases} \tag{37}$$

where $0^0 := 1$.

Proof: We prove the odd case. Since in this case $c(s) = s^j = sc^o(s^2)$, hence $c^e(-\omega^2) \equiv 0$ and $c^o(-\omega^2) = (-1)^{(j-1)/2}\omega^{j-1}$, we have by (35) $\mathcal{R}_G^+ = \{\omega \in \mathbb{R}_+;\ \omega = 0 \text{ or } p^e(-\omega^2) = 0\}$ and $|G(\imath\omega)| = |\omega^{j-1}/p^o(-\omega^2)|$ for $\omega \in \mathcal{R}_G^+$. This proves (37) for the odd case. □

Many intricacies of stability theory are due to the fact that stability is a non-convex property. If $p(\cdot\,,a)$ and $p(\cdot\,,b)$ are $\mathbb{C}_g$-stable, the roots of the convex combinations

$$p(\cdot\,, \gamma a + (1-\gamma)b)\,, \qquad 0 \leq \gamma \leq 1 \tag{38}$$

need not be $\mathbb{C}_g$-stable. It is natural to ask what conditions must be imposed in order to guarantee that all the polynomials in the segment (38) between $p(\cdot\,,a)$ and $p(\cdot\,,b)$ are in fact $\mathbb{C}_g$-stable.

Corollary 5.4.23. *Given $a,\ b \in \mathbb{R}^{n+1}$ with $a_n b_n > 0$, then all the polynomials in (38) have their roots in $\mathbb{C}_g$ if and only if*

$$p\left(\cdot\,, (a+b)/2\right) \text{ has all its roots in } \mathbb{C}_g \quad \text{and} \quad \inf_{s \in \mathcal{R}_G \cap \partial\mathbb{C}_g} |G(s)|^{-1} > 1 \tag{39}$$

where $G(s) = G_R(s) + \imath G_I(s) = \dfrac{p(s,(a-b)/2)}{p(s,(a+b)/2)}$.

Proof: The polynomials in (38) have all their roots in $\mathbb{C}_g$ if and only if this holds for all polynomials of the form

$$p(\cdot\,, (a+b)/2 - \Delta_0(a-b)/2)\,, \quad \Delta_0 \in [-1,1]\,.$$

But this is equivalent to

$$r_{\mathbb{R}}((a+b)/2; (a-b)/2; \mathbb{C}_g) > 1.$$

Hence Corollary 5.4.23 follows from Corollary 5.4.20 since $|a_n + b_n||a_n - b_n|^{-1} > 1$ because $a_n b_n > 0$. □

We now consider the following *scaled perturbations* of *monic* polynomials

$$a = [a_0, \ldots, a_{n-1}, 1] \rightsquigarrow a(\Delta) = [a_0 + \Delta_1 c_0, \ldots, a_{n-1} + \Delta_n c_{n-1}, 1] \tag{40}$$

where $\Delta = [\Delta_1, \ldots, \Delta_n] \in \mathbb{R}^{1\times n}$. This is a special case of (2) where $q = n$ and $C \in \mathbb{R}^{n\times(n+1)}$ has nonnegative entries $c_{ii} = c_i$ for $i = 0, \ldots, n-1$ and $c_{ij} = 0$ for all other i, j. Such matrices are called scaling matrices for the monic case. We will only consider the case where the size of perturbations is measured by the ∞-norm since in this case an interesting relationship with Kharitonov's Theorem can be established.

Proposition 5.4.24. *Suppose that $p(s,a)$, $a \in \mathbb{R}^{1\times(n+1)}$ is a monic Hurwitz polynomial, $C = [\mathrm{diag}\,(c_0, c_1, \ldots, c_{n-1}), 0_n] \in \mathbb{R}^{n\times(n+1)}$, $c_i \geq 0, i = 0, \ldots, n-1$ and*

$$\begin{aligned}\tilde{c} &= [-c_0, -c_1, c_2, c_3, -c_4, -c_5, \ldots, \pm c_{n-1}, 0] \in \mathbb{R}^{1\times(n+1)},\\ \hat{c} &= [-c_0, c_1, c_2, -c_3, -c_4, c_5, \ldots, \pm c_{n-1}, 0] \in \mathbb{R}^{1\times(n+1)}.\end{aligned}$$

Then, with respect to the ∞-norm on $\mathbf{\Delta} = \mathbb{R}^{1\times n}$,

$$r_{\mathbb{R}}^{-}(a; C) = \min\{r_{\mathbb{R}}^{-}(a; \tilde{c}),\ r_{\mathbb{R}}^{-}(a; \hat{c})\}. \tag{41}$$

Proof: For $\Delta = [-\Delta_0, -\Delta_0, \Delta_0, \Delta_0, -\Delta_0, \ldots] \in \mathbb{R}^{1\times n}$, $\Delta_0 \in \mathbb{R}$ we have

$$a + \Delta C = [a_0 - \Delta_0 c_0, a_1 - \Delta_0 c_1, a_2 + \Delta_0 c_2, a_3 + \Delta_0 c_3, \ldots, 1] = a + \Delta_0 \tilde{c}.$$

Similarly for $\Delta = [-\Delta_0, \Delta_0, \Delta_0, -\Delta_0, -\Delta_0, \ldots]$ we have $a + \Delta C = a + \Delta_0 \hat{c}$. Since in both cases $\|\Delta\|_\infty = |\Delta_0|$, the inequality $\leq$ in (41) follows.
To prove the converse inequality, let $r = r_{\mathbb{R}}^{-}(a; C)$ and $c = [c_0, c_1, \ldots, c_{n-1}, 0]$. By (40) we have

$$\{a + \Delta C; \Delta \in \mathbb{R}^{1\times n}, \|\Delta\|_\infty \leq r\} = [\underline{a}, \overline{a}] \ \text{ where } \underline{a} = a - rc,\ \overline{a} = a + rc, \tag{42}$$

and so the interval $[\underline{a}, \overline{a}]$ contains the coefficient vector of an unstable polynomial. Moreover, $a + \Delta_0 \tilde{c}$ and $a + \Delta_0 \hat{c}$, $\Delta_0 = \pm r$ are the coefficient vectors of the four Kharitonov polynomials for the interval $[\underline{a}, \overline{a}] = [a - rc, a + rc]$. By Kharitonov's Theorem at least one of these is unstable and so $\min\{r_{\mathbb{R}}^{-}(a; \tilde{c}), r_{\mathbb{R}}^{-}(a; \hat{c})\} \leq r$. Hence (41) follows. □

The computation of $r_{\mathbb{R}}^{-}(a; C)$ with respect to the ∞-norm is considerably simplified by the application of the formula (41) since $r_{\mathbb{R}}^{-}(a; \tilde{c})$, $r_{\mathbb{R}}^{-}(a; \hat{c})$ can be computed via (35) and (36).
Proceeding as in the previous proof we can use stability radii to determine conditions for a multi-dimensional interval of monic polynomials to have roots in arbitrary stability domains.

Proposition 5.4.25. *Given $\underline{a}, \overline{a} \in \mathbb{R}^{1\times(n+1)}$ with $\underline{a} \leq \overline{a}$ and $\underline{a}_n = \overline{a}_n = 1$, the interval $[\underline{a}, \overline{a}]$, consists of monic polynomials with roots in $\mathbb{C}_g$ if and only if the central polynomial $p(s,a)$ where $a = (\underline{a} + \overline{a})/2$ is $\mathbb{C}_g$-stable and with respect to the ∞-norm $r_{\mathbb{R}}(a; C; \mathbb{C}_g) > 1$ where*

$$C = \left[\mathrm{diag}\ \left((\overline{a}_0 - \underline{a}_0)/2, \ldots, (\overline{a}_{n-1} - \underline{a}_{n-1})/2\right)\ , 0_n\right].$$

Proof: As in (42) (with $r = 1$) we have $[\underline{a}, \overline{a}] = \{a + \Delta C; \Delta \in \mathbb{R}^{1\times n}, \|\Delta\|_\infty \leq 1\}$ and this proves the proposition. □

The radius $r_{\mathbb{R}}(a; C; \mathbb{C}_g)$ in Proposition 5.4.25 is the minimal factor by which the interval $[\underline{a}, \overline{a}]$ with centre a must be blown up before one of the monic polynomials in the interval has a root in $\mathbb{C}_b$.

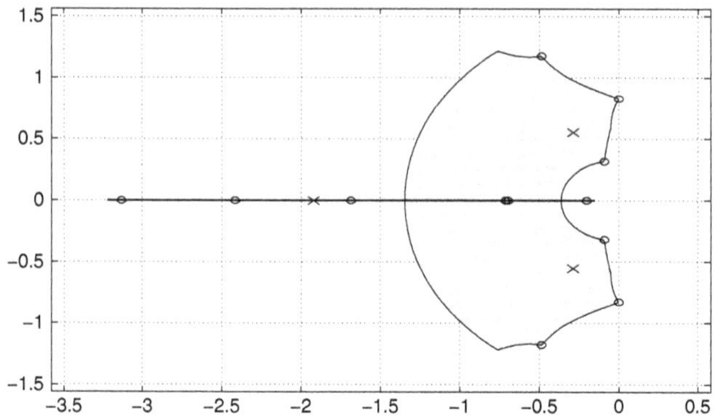

Figure 5.4.4: The root set $\mathcal{R}_{\mathbb{R}}(a; C; 1.63)$ with respect to the ∞-norm

Example 5.4.26. Consider the monic Hurwitz polynomial $p(s,a)$ of degree $n = 3$ with coefficient vector $a = [3/4,\ 3/2,\ 5/2,\ 1]$ subjected to perturbations $a \rightsquigarrow a + \Delta C$ where $C = [\mathrm{diag}\,(1/4,\ 1/2,\ 1/2),\ 0_3] \in \mathbb{R}^{3\times 4}$ and $\Delta \in \mathbb{R}^{1\times 3}$. The root set $\mathcal{R}_{\mathbb{R}}(a; C; \delta_1)$ with respect to the ∞-norm on $\mathbb{R}^{1\times 3}$ is shown for the value $\delta_1 = 1.63$ in Figure 5.4.4. The crosses are the roots of the nominal polynomial $p(s,a)$, and the interval $(-3.25, -.15)$ which is represented by a thick line in the figure is the real part of the root set. We see from the figure that $r_{\mathbb{R}}^-(a; C)$ with respect to the ∞-norm is approximately 1.63. If we set $c = (1/4,\ 1/2,\ 1/2,\ 0)$ then by (42), for any $\delta > 0$, $\{a + \Delta C; \Delta \in \mathbb{R}^{1\times 3}, \|\Delta\|_\infty \le \delta\} = [\underline{a}^\delta, \overline{a}^\delta]$ where

$$\begin{aligned}\underline{a}^\delta &= a - \delta c = [(3-\delta)/4, (3-\delta)/2, (5-\delta)/2, 1]\,,\\ \overline{a}^\delta &= a + \delta c = [(3+\delta)/4, (3+\delta)/2, (5+\delta)/2, 1]\,.\end{aligned}$$

The most critical Kharitonov polynomial for the interval $[\underline{a}^\delta, \overline{a}^\delta]$ has the coefficient vector $[(3+\delta)/4,\ (3-\delta)/2,\ (5-\delta)/2,\ 1]$ and by the Hurwitz criterion (see Example 3.4.72) this is $\mathbb{C}_-$-stable if and only if

$$3 > \delta \;\text{ and }\; (3-\delta)(5-\delta) > (3+\delta).$$

The second condition requires $\delta^2 - 9\delta + 12 > 0$ and together with the first one yields $\delta < (9 - \sqrt{33})/2 \approx 1.63 = \delta_1$, confirming the above computation. The roots of the critical Kharitonov polynomial when $\delta = \delta_1$ are $\pm 0.8283i, -1.6861$ two of which lie on the intersection of the boundary of $\mathcal{R}_{\mathbb{R}}(a; C; 1.63)$ with the imaginary axis. The 12 roots of the 4 Kharitonov polynomials for the interval $[\underline{a}^{\delta_1}, \overline{a}^{\delta_1}]$ are marked with circles in the figure (the circles around two real roots in the interval $[-1, -0.5]$ overlap). □

As an application of Proposition 5.4.24 we derive a formula for the unstructured stability radius $d_{\mathbb{R}}^-(a)$ of a Hurwitz polynomial with respect to the ∞-norm in the monic case. Formulas with respect to the 1- and 2-norms are set as Ex. 10 and 11.

Corollary 5.4.27. *If $p(\cdot, a)$ is a monic Hurwitz polynomial and p^e, p^o are as in (32), then with respect to the ∞-norm on $\mathbb{R}^{1\times n}$ for the monic case*

$$d_{\mathbb{R}}^-(a) = \min_{\omega\in\Omega_0} \frac{|p^e(-\omega^2)|}{1 + \omega^2 + \cdots + \omega^{n-2}}, \qquad n \text{ even} \tag{43}$$

where $\Omega_0 = \{0\} \cup \{\omega \in \mathbb{R}_+;\ |p^e(-\omega^2)| = |p^o(-\omega^2)|\}$,

$$d_{\mathbb{R}}^-(a) = \min_{\omega\in\Omega_1} \frac{|p^e(-\omega^2)|}{1+\omega^2+\cdots+\omega^{n-1}}, \qquad n \text{ odd} \tag{44}$$

where $\Omega_1 = \{0\} \cup \{\omega \in \mathbb{R}_+;\ (1+\omega^2+\cdots\omega^{n-3})|p^e(-\omega^2)| = (1+\omega^2+\cdots\omega^{n-1})|p^o(-\omega^2)|\}$.

Proof: Since $C = [I_n, 0_n]$, the corresponding vectors $\tilde{c}, \hat{c}$ (see Proposition 5.4.24) are

$$\tilde{c} = [-1,-1,+1,+1,-1,-1,\cdots,0], \quad \hat{c} = [-1,+1,+1,-1,-1,\cdots,0].$$

Suppose n is odd and as in (32) $\tilde{c}(s) = \tilde{c}^e(s^2) + s\tilde{c}^o(s^2)$, $\hat{c}(s) = \hat{c}^e(s^2) + s\hat{c}^o(s^2)$, then

$$\tilde{c}^e(-\omega^2) = -(1+\omega^2+\ldots+\omega^{n-1}) = \hat{c}^e(-\omega^2), \quad \tilde{c}^o(-\omega^2) = -(1+\omega^2+\ldots+\omega^{n-3}) = -\hat{c}^o(-\omega^2).$$

Hence by (35) and (36)

$$r_{\mathbb{R}}^-(a;\tilde{c}) = \min_{\omega\in\Omega_1'} \left| \frac{p^e(-\omega^2)}{1+\omega^2+\cdots+\omega^{n-1}} \right|$$

where

$$\Omega_1' = \{0\} \cup \{\omega \in \mathbb{R}_+; (1+\omega^2+\cdots+\omega^{n-3})p^e(-\omega^2) = (1+\omega^2+\cdots+\omega^{n-1})p^o(-\omega^2)\}$$

and

$$r_{\mathbb{R}}^-(a;\hat{c}) = \min_{\omega\in\Omega_1''} \left| \frac{p^e(-\omega^2)}{1+\omega^2+\cdots+\omega^{n-1}} \right|$$

where

$$\Omega_1'' = \{0\} \cup \{\omega \in \mathbb{R}_+; (1+\omega^2+\cdots+\omega^{n-3})p^e(-\omega^2) = -(1+\omega^2+\cdots+\omega^{n-1})p^o(-\omega^2)\}.$$

(44) follows from (41) since $\Omega_1 = \Omega_1' \cup \Omega_1''$. A similar proof can be given for the even case. □

We conclude this section with an example illustrating the above corollary.

Example 5.4.28. Consider the Hurwitz polynomial $p(s,a) = s^4 + 5s^3 + 8s^2 + 8s + 3$. Then

$$p^e(-\omega^2) = 3 - 8\omega^2 + \omega^4, \qquad p^o(-\omega^2) = 8 - 5\omega^2.$$

Hence the set Ω_0 in (43) is

$$\Omega_0 = \{\omega \in \mathbb{R}_+;\ \omega^4 - 3\omega^2 - 5 = 0 \quad \text{or} \quad \omega^4 - 13\omega^2 + 11 = 0\}.$$

A simple calculation gives

$$\Omega_0 = \left\{ \sqrt{(3+\sqrt{29})/2}, \sqrt{(13\pm 5\sqrt{5})/2} \right\} = \{2.0476\,,\ 3.4771\,,\ 0.95385\}\,.$$

$\omega_0 = \sqrt{0.9098}$ minimizes $|p^e(\omega^2)|/(1+\omega^2) = |\omega^4 - 8\omega^2 + 3|\,|1+\omega^2|^{-1}$ on Ω_0 with value $|p^e(\omega_0^2)|/(1+\omega_0^2) = 1.807$. So by (43) with respect to the ∞–norm, we have

$$d_{\mathbb{R}}^-(a) = \min\{3\,,\ 1.807\} = 1.807$$

(for the monic case). □

5.4.4 Exercises

1. If $p(s,a) = s^2 + s$, use Corollary 5.4.13 to show that the root set $\mathcal{R}_{\mathbb{C}}(a;\delta)$ of the polynomial $p(\cdot, a)$ with respect to the 1-norm in the monic case is the set of all $s = re^{\imath\theta}$, $(r,\theta) \in \mathbb{R}_+ \times [0, 2\pi]$, such that (compare Example 5.4.16)

$$\min\{r^4 + 2r^3 \cos\theta + r^2,\ r^2 + 2r\cos\theta + 1\} < \delta^2.$$

2. Use Corollary 5.4.15 to prove that for a Hurwitz polynomial $p(s,a) = s^2 + a_1 s + a_0$ in the monic case with respect to the 2-norm

$$d_{\mathbb{C}}^-(a) = \begin{cases} a_0 & \text{if } (1+a_0)^2 \le 1 + a_1^2, \\ \left[a_1^2 - 2a_0 - 2 + 2((1+a_0)^2 - a_1^2)^{1/2}\right]^{1/2} & \text{if } (1+a_0)^2 > 1 + a_1^2. \end{cases}$$

Compare your results with the real distance $d_{\mathbb{R}}^-(a)$ (in the monic case) and produce a similar figure to Figure 5.4.2 for the Hurwitz case.

3. If $a = [1,\ 2,\ 1]$, $c = [0,\ 1,\ 0]$ and $\delta < 1$, show that

$$\mathcal{R}_{\mathbb{R}}(a;c;\delta) = \{-r;\ 0 \le r^{-1}(r-1)^2 < \delta\} \cup \{e^{\imath\theta};\ 2 + 2\cos\theta < \delta\}.$$

4. Prove that if $p(s,a)$ is a real Schur polynomial, then

$$r_{\mathbb{R}}^{1}(a; e^{j\top}) = \min_{\theta\in\Omega} |p(e^{\imath\theta}, a)|, \quad j = 0, \ldots, n-1$$

where $\Omega = \{\theta \in [0, 2\pi];\ \mathrm{Im}[e^{-\imath j\theta} p(e^{\imath\theta}, a)] = 0\}$. (Compare this with the formula for the complex case, see (25)).

5. Consider the data in Example 5.4.5 and prove the results stated in the example for the real case. In particular, show that

$$[G_R(re^{\imath\theta}),\ G_I(re^{\imath\theta})] = -(1 + r^2 + 2r\cos\theta)^{-1} \left[\begin{bmatrix} (1 + r\cos\theta) \\ r\cos\theta + r^2 \end{bmatrix}, \begin{bmatrix} -r\sin\theta \\ r\sin\theta \end{bmatrix}\right],$$

$$\begin{aligned} \mathrm{dist}\left(G_R(re^{\imath\theta}), \mathbb{R}G_I(re^{\imath\theta})\right) &= 1/\sqrt{2} && \text{if } \theta \ne 0, \pi,\ r \ne 0, \\ &= 1 && \text{if } r = 0, \\ &= (1+r)^{-1}\sqrt{r^2+1} && \text{if } \theta = 0, \\ &= |1-r|^{-1}\sqrt{r^2+1} && \text{if } \theta = \pi. \end{aligned}$$

Hence prove

$$\mathcal{R}_{\mathbb{R}}(a;\delta) = \{re^{\imath\pi} \in \mathbb{C};\ (r^2+1)^{-1}|r-1|^2 < \delta^2\}.$$

Note the discontinuity of dist at the points where $G_I(s) = 0$. Prove that

$$\mathrm{dist}\left(G_R(\imath\omega), \mathbb{R}G_I(\imath\omega)\right) \le \mathrm{dist}\left(G_R(0), \mathbb{R}G_I(0)\right) = 1, \quad \omega \in \mathbb{R}$$

and hence verify that $d_{\mathbb{R}}^-(a) = 1$.

6. While Kharitonov's Theorem provides a powerful method for determining the stability of boxes of polynomials, it yields conservative results when applied to determine the stability of segments of polynomials. Use Kharitonov's Theorem to find the largest value of $\delta > 0$ such that the open segment

$$p(s, a(\Delta)) = s^3 + 2s^2 + 4s + 2 - \Delta_0(s^2 + 2s + 1), \quad \Delta_0 \in (-\delta, \delta)$$

is Hurwitz stable by examining the boxes $[a - \alpha c, a + \alpha c]$, where $a = [2,\ 4,\ 2,\ 1]$, $c = [1,\ 2,\ 1,\ 0]$, $\alpha > 0$. Then calculate $r_{\mathbb{R}}^-(a, c)$ and compare the results.

7. Use (28) to prove that for a real Schur polynomial $p(s,a) = s^2 + a_1 s + a_0$ with respect to the 2-norm in the monic case (compare Example 5.4.16), we have

$$d_{\mathbb{R}}^1(a) = \min\left\{1 - a_0,\ (1 + a_1 + a_0)/\sqrt{2},\ (1 - a_1 + a_0)/\sqrt{2}\right\}.$$

8. If $a = [1,\ 2,\ 1,\ 1]$, $b = [13,\ 10,\ 5,\ 1]$, use Corollary 5.4.23 to prove that the convex combination $\gamma p(s,a) + (1-\gamma)p(s,b)$ is Hurwitz stable for all $\gamma \in [0,1]$. Show that this does not hold if b is replaced by $\tilde{b} = [29,\ 10,\ 5,\ 1]$ although $p(s,\tilde{b})$ is stable.

9. Under the assumptions of Remark 5.4.12 suppose that $s_0 \in \partial\mathbb{C}_g$ maximizes the distance $\operatorname{dist}(G_R(s), \mathbb{R}G_I(s))$ on $\partial\mathbb{C}_g$ (for the 1-, 2- and ∞-norm, respectively)

$$\operatorname{dist}(G_R(s_0), \mathbb{R}G_I(s_0)) = \|G_R(s_0) - \hat{\alpha}G_I(s_0)\|_{\mathbb{R}^q} = \max_{s\in\partial\mathbb{C}_g} \operatorname{dist}(G_R(s), \mathbb{R}G_I(s))$$

and $G_R(s_0) - \hat{\alpha}G_I(s_0) = \gamma = [\gamma_1, \ldots, \gamma_q]^\top$. Prove that minimum norm destabilizing perturbations $\Delta = [\Delta_1, \ldots, \Delta_q]$ with respect to the 1-, 2- and ∞-norm in the monic case are

<u>1–norm</u> $\Delta_i = 0,\ \ i \in \underline{q} \setminus \mathcal{J},\ \mathcal{J} = \{j \in \underline{q} : |\gamma_j| = \|\gamma\|_\infty\}$; $\operatorname{sign}\Delta_j = \operatorname{sign}\gamma_j,\ \ j \in \mathcal{J}$; $\sum_{j\in\mathcal{J}} |\Delta_j| = \|\gamma\|_\infty^{-1}$ and the Δ_i, $i \in \underline{q} \setminus \mathcal{J}$ are chosen to satisfy the first condition in (21).

<u>2–norm</u>: $\Delta_i = \gamma_i / \|\gamma\|_2^2$, $i \in \underline{q}$;

<u>∞–norm</u>: $\Delta_k = \|\gamma\|_1^{-1} \operatorname{sign}\gamma_k$, $k \subset \mathcal{K} - \{k \in \underline{q}; \gamma_k \neq 0\}$ and the Δ_i, $i \in \underline{q} \setminus \mathcal{K}$ are chosen to satisfy the first condition in (21).

10. Prove that in the monic case, if $p(\cdot, a)$ is a Hurwitz polynomial then with respect to the 1-norm

$$d_{\mathbb{R}}^-(a) = \min\{a_0,\ \min_{\omega\in\mathbb{R}_+} f(\omega)\},$$

where

$$f(\omega) = \begin{cases} \max\{|p^o(-\omega^2) - p^e(-\omega^2)|, |p^o(-\omega^2) + p^e(-\omega^2)|\}, & |\omega| \le 1, \\ \max\{|\omega|^{2-n}|p^o(-\omega^2) - p^e(-\omega^2)|, |\omega|^{2-n}|p^o(-\omega^2) + p^e(-\omega^2)|\}, & |\omega| > 1,\ n \text{ even}, \\ \max\{|\omega|^{1-n}|p^o(-\omega^2) - \omega^2 p^e(-\omega^2)|, |\omega|^{1-n}|p^o(-\omega^2) + \omega^2 p^e(-\omega^2)|\}, & |\omega| > 1,\ n \text{ odd}. \end{cases}$$

11. Prove that in the monic case, if $p(\cdot, a)$ is a Hurwitz polynomial then with respect to the 2-norm

$$d_{\mathbb{R}}^-(a) = \begin{cases} \min\left\{a_0,\ \min_{\omega\in\mathbb{R}_+}\left[\dfrac{p^e(-\omega^2)^2 + p^o(-\omega^2)^2}{1 + \omega^4 + \ldots + \omega^{2n-4}}\right]^{1/2}\right\}, & n \text{ even}, \\ \min\left\{a_0, \min_{\omega\in\mathbb{R}_+}\left[\dfrac{p^e(-\omega^2)^2(1+\ldots+\omega^{2n-6}) + p^o(-\omega^2)^2(1+\ldots+\omega^{2n-2})}{(1 + \omega^4 + \ldots + \omega^{2n-6})(1 + \omega^4 + \ldots + \omega^{2n-2})}\right]^{1/2}\right\}, & n \text{ odd}. \end{cases}$$

12. Suppose $p(s,a)$ is a real monic polynomial such that $\mathcal{R}(a) \subset \mathbb{C} \setminus \mathbb{R}$. If d is the distance with respect to the Euclidean norm of $p(s,a)$ to the set of real monic polynomials which have at least one real root, prove that

$$d = \min_{r\in\mathbb{R}} \frac{|p(r,a)|}{(1 + r^2 + \ldots + r^{2(n-1)})^{1/2}}.$$

(Hint: Take $\mathbb{C}_g = \mathbb{C} \setminus \mathbb{R}$). In the quadratic case show that

$$d = \min_{r \in R} \frac{|r^2 + a_1 r + a_0|}{(1+r^2)^{1/2}} \quad \text{where } R \text{ is the set of real roots of } \; s^3 + (2-a_0)s + a_1 = 0.$$

In particular if $a_0 = 2$, $a_1^2 < 8$ show that $d = (2 - |a_1|^{2/3})(1 + |a_1|^{2/3})^{1/2}$.

13. If $p(\cdot\,, a)$, $a \in \mathbb{R}^{n+1}$ is a monic Hurwitz polynomial, n even and

$$\tilde{a} = [a_0 + \imath a_1, 0, a_2 + \imath a_3, 0, \ldots, a_{n-2} + \imath a_{n-1}, 0, 1]$$

$$C = \begin{bmatrix} 1 & 0 & 0 & 0 & 0 & \cdot & \cdot & \cdot & 0 & 0 \\ 0 & 0 & 1 & 0 & 0 & \cdot & \cdot & \cdot & 0 & 0 \\ 0 & 0 & 0 & 0 & 1 & \cdot & \cdot & \cdot & 0 & 0 \\ \cdot & \cdot & \cdot & \cdot & \cdot & \cdot & \cdot & \cdot & \cdot & \cdot \\ 0 & 0 & 0 & 0 & 0 & \cdot & \cdot & 1 & 0 & 0 \end{bmatrix}_{\frac{n}{2} \times (n+1)}.$$

Prove that with respect to the Euclidean norm in the monic case (cf. [246])

$$d_{\mathbb{R}}^{-}(a) = \min\{a_0, r_{\mathbb{C}}(\tilde{a}; C; \mathbb{C} \setminus \imath\mathbb{R})\}.$$

5.4.5 Notes and References

The robustness analysis of polynomials has received a good deal of attention over the last 25 years. Much of this work is described in the books by *Ackermann* (1993) [4], *Barmish* (1994) [36] and *Bhattacharyya et al.* (1995) [56]. In these books the authors acknowledge the importance that Kharitonov's Theorem (see Section 4.1) brought to the field. Kharitonov's Theorem determines sufficient stability conditions for intervals of polynomials via an analysis of existing stability criteria for single polynomials. Extensions and system theoretic applications can be found in the above books and are based on the works of e.g. *Bose* (1985) [67], *Hollot and Bartlett* (1986) [261], *Bartlett et al.* (1988) [39]. In the latter paper it is shown that for examining the stability of an arbitrary polytope of polynomials it suffices to check the edges (*Edge Theorem*).
Other stability results for sets of polynomials have been derived from the Nyquist stability criterion, see *Yeung* (1983) [535], *Białas and Garloff* (1985) [58], *Argoun* (1987) [17], or from other sufficient stability criteria, see *Bose et al.* (1986) [69], *Anderson et al.* (1987) [13].
Soh et al. (1985) [470] characterized the distance of a Schur (or Hurwitz) polynomial from the set of non-Schur (resp. non-Hurwitz) polynomials in coefficient space. Extensions of their formulas to other stability domains and to structured perturbations can be found in *Soh et al.* (1987) [471] and *Biernacki et al.* (1987) [59]. Mathematically these results are based on Theorem 4.1.26.
A different approach to the robust stability analysis of polynomials is based on optimization ideas. A class of special subsets in the coefficient space (e.g. ellipsoids or polytopes) are parametrized by a positive real number, and the supremum of this number for which the corresponding subset consists only of Hurwitz (or Schur) polynomials, is determined. In the context of Kharitonov's Theorem, optimization problems have been considered by *Barmish* (1984) [34], *Białas and Garloff* (1985) [58], *Rantzer* (1992) [427], *Kharitonov and Tempo* (1994) [305]. The approach that we have adopted in this section is similar in that the subset of polynomials is parametrized by a single number, but the results are based on stability radii for matrices rather than Kharitonov's Theorem, see (1992) [252]. *Tsypkin*

and Polyak (1991) [502] considered the problem of determining the Hurwitz stability of a ball of polynomials specified via weighted ℓ_p norms. Their solution was graphical in terms of the so-called Tsypkin-Polyak loci.
Robustness analysis has been extended to quasipolynomials by *Kharitonov* (1991) [303] and the results have been used to obtain robust stability criteria for delay equations, see *Kharitonov and Zhabko* (1994) [306]. Robustness analysis has also been extended to polynomials with probabilistic uncertainty, see the *Notes and References* of Section 5.2.
A polynomial p, $\deg p < n$ is called a convex direction for the set $\mathcal{H}_n$ of real Hurwitz polynomials of degree n if $\mathcal{H}_n$ is convex in the direction $\mathbb{R}_+ p$, i.e. for all $q \in \mathcal{H}_n$, $\mu > 0$

$$q + \mu p \in \mathcal{H}_n \quad \Longrightarrow \quad \text{the segment } [q, q + \mu p] = \{q + \alpha\mu p; \alpha \in [0,1]\} \subset \mathcal{H}_n.$$

Convex directions have been investigated by *Rantzer* (1992) [427] and by *Kharitonov and Hinrichsen* (1997) [304]. For a survey on convex directions, see *Atanassova et al.* (1997) [23].

5.5 Transient Behaviour

Trajectories of an asymptotically stable linear system may temporarily move a long way from the origin before approaching it as $t \to \infty$. Such transient behaviour is often exhibited by highly non-normal systems. From a practical point of view, if the "state excursions" are very large the stable system actually behaves like an unstable one. Moreover, if the system is obtained by linearization of a nonlinear system around an equilibrium point, the large transients of the linear part may incite the nonlinearities to drive the system permanently far away from the equilibrium point. In such cases the practical instability of the equilibrium point is reflected by an extreme thinness of its domain of attraction in some directions of the state space. In fluid dynamics the interaction between large transient motions of the linearization and its nonlinear perturbations has recently been put forward as an explanation for observed instabilities of flows which are inconsistent with the results of spectral stability analysis, see *Notes and References.*
In the first subsection we introduce a new concept of stability ((M, β)-stability) which combines information about the decay rate and the transient behaviour of a system. As a quantitative index of the transient amplification of initial state perturbations the notion of a transient bound for a given exponential rate is introduced and the interplay between the bound and the rate is discussed. In particular we characterize those rates for which the bound is 1. Estimates for the transient bound are obtained via the distance of A from the set of normal matrices. In the second subsection the concept of a contraction semigroup is discussed and it is shown how transient bounds can be estimated via Liapunov norms. These are auxiliary norms with respect to which a given system is contractive. In the third subsection we discuss the relationship between the transient bound, stability radii and spectral value sets. Finally, in the last subsection perturbation bounds are presented which ensure (M, β)-stability of an uncertain system.
We will only consider the continuous time case, but set some of the discrete time results as exercises.

5.5.1 Transient Bounds and Initial Growth Rate

Throughout this subsection it is assumed that $A \in \mathbb{K}^{n\times n}$, $\mathbb{K}^n$ is provided with an arbitrary fixed norm $\|\cdot\|$ and $\mathbb{K}^{n\times n}$ with the corresponding operator norm which we also denote by $\|\cdot\|$. We begin by introducing the concept of (M, β)-stability.

Definition 5.5.1. Given $M \geq 1$ and $\beta \in \mathbb{R}$, the system $\dot{x} = Ax$ is said to be (M, β)-stable if its solutions $\varphi(t; x^0) = e^{At}x^0$ satisfy

$$\|\varphi(t; x^0)\| \leq Me^{\beta t}\|x^0\|, \quad x^0 \in \mathbb{K}^n, \quad t \geq 0. \tag{1}$$

$\dot{x} = Ax$ is said to be strictly (M, β)-stable if there exists $\varepsilon > 0$ such that (1) holds with $\beta - \varepsilon$ instead of β.

The inequality (1) is satisfied if and only if $\|e^{At}\| \leq Me^{\beta t}$ for all $t \geq 0$. In this case we also say that the semigroup $(e^{At})_{t\geq 0}$ or the matrix A are (M, β)-stable. In most applications interest will be focussed on the case where $\beta \leq 0$. Then (M, β)-stability

guarantees both a specific decay rate (given by $-\beta \geq 0$) and a specific bound on the transient behaviour (given by M). We have included the possibility $\beta > 0$ in our definition since it may be that in some applications the system is unstable and the growth rate needs to be bounded.
If $\dot{x} = Ax$ is (M, β)-stable then necessarily the spectral abscissa $\alpha(A) \leq \beta$ and $A_\beta := A - \beta I_n$ is stable. If it is strictly (M, β)-stable then $\alpha(A) < \beta$ and A_β is asymptotically stable. Conversely, for every $\beta > \alpha(A)$, A_β is exponentially stable and so there *exists* a constant M (depending on β), such that (1) holds. Hence the system is strictly (M, β)-stable for *some* $M \geq 1$. Similarly, if A_β is stable, then the system $\dot{x} = Ax$ is (M, β)-stable for *some* $M \geq 1$.
We know that the eigenvalues of A determine the long term behaviour of the system in the sense that the *growth rate* of $(e^{At})_{t \geq 0}$ is equal to the spectral abscissa $\alpha(A)$, see (3.3.30)

$$\lim_{t \to \infty} \frac{\ln \|e^{At}\|}{t} = \alpha(A). \tag{2}$$

However no conclusion can be drawn from the spectrum of A about the transient behaviour of $\dot{x} = Ax$. It cannot be decided upon the basis of its eigenvalues alone whether or not (1) is satisfied for prescribed $M \geq 1$, $\beta \in \mathbb{R}$.
Another basic difference between the stability concepts considered in Chapter 3 and the notion of (M, β)-stability is that the former do *not* depend upon the specific norm on $\mathbb{K}^n$, whereas the set of pairs (M, β) satisfying (1) and consequently the concept of (M, β)-stability and strict (M, β)-stability depend on the given norm.
We will now briefly discuss some elementary topological properties of the set of (M, β)-stable systems $\dot{x} = Ax$. For this we make use of the following lemma.

Lemma 5.5.2. *Suppose $M > 1$. Then $\dot{x} = Ax$ is strictly (M, β)-stable if and only if $\alpha(A) < \beta$ and $\|e^{At}\| < Me^{\beta t}$ holds for all $t > 0$.*

Proof: The necessity of the condition is obvious. To prove sufficiency, assume $\alpha(A) < \beta$ and $\|e^{At}\| < Me^{\beta t}$ for all $t > 0$. Since $\|e^{At}\| = 1$ for $t = 0$ and $M > 1$, we must have $\|e^{At}\| < Me^{\beta t}$ for all $t \geq 0$. Choose $\delta > 0$ such that $\alpha(A) + 2\delta < \beta$. Then there exists $\tilde{M} \geq 1$ such that $\|e^{At}\| \leq \tilde{M} e^{(\beta - 2\delta)t}$ for all $t > 0$. Let t_0 be such that $\tilde{M} e^{-\delta t} < M$ for $t \geq t_0$. By assumption we have $\sup_{t \geq 0} \|e^{At}\| e^{-\beta t} = \max_{t \geq 0} \|e^{At}\| e^{-\beta t} < M$ and hence there exists $\varepsilon \in (0, \delta)$ sufficiently small so that $e^{\varepsilon t} \|e^{At}\| e^{-\beta t} < M$ for $t \in [0, t_0]$. Since $e^{\varepsilon t} \|e^{At}\| \leq \tilde{M} e^{(\beta - \delta)t} < Me^{\beta t}$ for $t \geq t_0$ we obtain $e^{\varepsilon t} \|e^{At}\| < Me^{\beta t}$ for all $t \geq 0$ and this concludes the proof. □

We will see in Example 5.5.27 that Lemma 5.5.2 does not hold in the case $M = 1$.

Proposition 5.5.3. *For any $M \geq 1, \beta \in \mathbb{R}$, the set $\mathcal{S}_n(\mathbb{K}; M, \beta)$ of all $A \in \mathbb{K}^{n \times n}$ generating (M, β)-stable semigroups is closed in $\mathbb{K}^{n \times n}$. The interior of $\mathcal{S}_n(\mathbb{K}; M, \beta)$ consists of all the matrices $A \in \mathbb{K}^{n \times n}$ generating strictly (M, β)-stable semigroups*

$$\operatorname{int} \mathcal{S}_n(\mathbb{K}; M, \beta) = \bigcup_{\beta' < \beta} \mathcal{S}_n(\mathbb{K}; M, \beta'). \tag{3}$$

Proof: If $\lim_{k \to \infty} A_k = A$ in $\mathbb{K}^{n \times n}$ then $\lim_{k \to \infty} \|e^{A_k t}\| = \|e^{At}\|$ for every $t \geq 0$ and so it follows from Definition 5.5.1 that $\mathcal{S}_n(\mathbb{K}; M, \beta)$ is closed in $\mathbb{K}^{n \times n}$. If $A \in \mathcal{S}_n(\mathbb{K}; M, \beta)$

is not strictly (M,β)-stable, then $A+\varepsilon I_n \notin \mathcal{S}_n(\mathbb{K};M,\beta)$ for all $\varepsilon>0$. Hence A belongs to the boundary of $\mathcal{S}_n(\mathbb{K};M,\beta)$ in $\mathbb{K}^{n\times n}$. Now suppose that $A\in\mathcal{S}_n(\mathbb{K};M,\beta)$ is strictly (M,β)-stable, then there exists $\varepsilon>0$ such that $\|e^{At}\|\le Me^{(\beta-\varepsilon)t}$ for all $t\ge 0$, hence $A\in\mathcal{S}_n(\mathbb{K};M,\beta-\varepsilon)$. Applying Proposition 4.2.18 we see that for all $\Delta\in\mathbb{K}^{n\times n}$

$$\|e^{(A+\Delta)t}\|\le M\,e^{(\beta-\varepsilon+M\,\|\Delta\|)\,t}, \quad t\ge 0.$$

Therefore $A+\Delta\in\mathcal{S}_n(\mathbb{K};M,\beta)$ for all $\Delta\in\mathbb{K}^{n\times n}$ satisfying $\|\Delta\|\le\varepsilon/M$. So A is an interior point of $\mathcal{S}_n(\mathbb{K};M,\beta)$. The same inequality shows that $\mathcal{S}_n(\mathbb{K};M,\beta')\subset\operatorname{int}\mathcal{S}_n(\mathbb{K};M,\beta)$ for $\beta'<\beta$, and this concludes the proof. □

The previous proposition shows that there is an analogy in the relationship between stable and asymptotically stable systems and between (M,β)-stable and strictly (M,β)-stable systems. In fact, the set of asymptotically stable systems is the interior of the set of stable systems and the set of strictly (M,β)-stable systems is the interior of the set of (M,β)-stable systems. Note however, that the set of stable systems is not closed in $\mathbb{K}^{n\times n}$.

Definition 5.5.4. For every $\beta\ge\alpha(A)$ the *transient bound* of $(e^{At})_{t\ge 0}$ for the exponential rate β is defined to be

$$M_\beta(A)=\inf\{M\in\mathbb{R};\ \forall t\ge 0:\ \|e^{At}\|\le Me^{\beta t}\}=\sup_{t\ge 0}\|e^{(A-\beta I_n)t}\| \tag{4}$$

(where as usual $\inf\emptyset=\infty$).

It is clear that

$$\beta'\ge\beta\ge\alpha(A)\implies 1\le M_{\beta'}\le M_\beta. \tag{5}$$

If $\beta>\alpha(A)$, then $M_\beta(A)<\infty$ and the 'inf' can be replaced by 'min' in (4). However (see the next example) it is possible that $M_\beta(A)\to\infty$ as $\beta\to\alpha(A)$ (hence it may be that $M_\beta(A)=\infty$ for $\beta=\alpha(A)$). This will happen if and only if the Jordan canonical form of A contains blocks of order ≥ 2 corresponding to eigenvalues $\lambda\in\sigma(A)$ with $\operatorname{Re}\lambda=\alpha(A)$.
Obviously $\alpha(A)\le\beta$ if and only if $\alpha(A-\beta I_n)\le 0$ and it follows from (4), that

$$M_\beta(A)=M_0(A_\beta)\quad\text{where}\quad A_\beta=A-\beta I_n,\qquad \beta\ge\alpha(A). \tag{6}$$

Suppose $\dot{x}=Ax$ is stable. Then $M_0(A)$ is finite and $M_0(A)$ is the maximal factor by which the size of any initial deviation $x^0\in\mathbb{K}^n$ from the equilibrium $\bar{x}=0$ is amplified in the course of the trajectory $e^{At}x^0$, $t\ge 0$. Therefore $M_0(A)$ is called the *transient amplification factor* of the system.
Now if $\beta>\alpha(A)$, then $\lim_{t\to\infty}\|e^{A_\beta t}\|=0$ and so there exists $t_\beta\ge 0$ (not necessarily unique) such that

$$\|e^{At_\beta}\|=M_\beta(A)\,e^{\beta t_\beta}.$$

Therefore $\alpha(A)<\beta<\beta'$ implies

$$M_\beta(A)e^{\beta t_\beta}=\|e^{At_\beta}\|\le M_{\beta'}(A)e^{\beta' t_\beta}=\|e^{At_{\beta'}}\|e^{\beta'(t_\beta-t_{\beta'})}\le M_\beta(A)e^{\beta t_{\beta'}}e^{\beta'(t_\beta-t_{\beta'})}$$

and hence

$$M_{\beta'}(A)e^{(\beta'-\beta)t_{\beta'}} \le M_\beta(A) \le M_{\beta'}(A)e^{(\beta'-\beta)t_\beta}, \quad \alpha(A) < \beta < \beta'. \tag{7}$$

In particular if $M_{\beta'}(A) > 1$, then since $t_{\beta'} > 0$, we obtain

$$\alpha(A) < \beta < \beta' \implies M_{\beta'}(A) < M_\beta(A) \text{ and } t_{\beta'} \le t_\beta. \tag{8}$$

Proposition 5.5.5. *For any given* $\beta \in \mathbb{R}$ *the map*

$$M_\beta : \mathcal{S}_n(\mathbb{K};\beta) \to [1,\infty], \quad A \mapsto M_\beta(A) \quad \textit{where } \mathcal{S}_n(\mathbb{K};\beta) = \{A \in \mathbb{K}^{n\times n}; \alpha(A) \le \beta\}$$

is lower semicontinuous on $\mathcal{S}_n(\mathbb{K};\beta)$ *and continuous on* $\operatorname{int}\mathcal{S}_n(\mathbb{K};\beta) = \{A \in \mathbb{K}^{n\times n}; \alpha(A) < \beta\}$.

Proof: Let $A \in \mathcal{S}_n(\mathbb{K};\beta)$. Suppose there exist $M < M_\beta(A)$ and a sequence (A_k) in $\mathcal{S}_n(\mathbb{K};\beta)$ converging to A such that $M_\beta(A_k) \le M$ for all $k \in \mathbb{N}$. Then

$$\|e^{At}\| = \lim_{k\to\infty} \|e^{A_k t}\| \le Me^{\beta t}, \quad t \ge 0,$$

which contradicts the definition of $M_\beta(A)$. This proves the lower semicontinuity of M_β. Now let $A \in \operatorname{int}\mathcal{S}_n(\mathbb{K};\beta)$ so that $\alpha(A) < \beta$, and suppose $M > M_\beta(A)$. By Lemma 5.5.2 there exists $\varepsilon > 0$ such that $\|e^{At}\| \le Me^{(\beta-\varepsilon)t}$ for all $t \ge 0$. From Proposition 4.2.18 we conclude that $\|e^{(A+\Delta)t}\| \le M\,e^{\beta t}$ for all $t \ge 0$ and all $\Delta \in \mathbb{K}^{n\times n}$, $\|\Delta\| \le \varepsilon/M$. This proves that $M_\beta(\cdot)$ is upper semicontinuous at A. □

In the following example we analyze in some detail the dependency of the transient bound $M_0(A)$ on the off-diagonal entry of an upper triangular real 2×2 matrix.

Example 5.5.6. Suppose $A = \begin{bmatrix} a & c \\ 0 & b \end{bmatrix}$, with $a, b, c \in \mathbb{R}$, $a < 0$, $b < 0$. Then its matrix exponential is given by

$$e^{At} = \begin{bmatrix} e^{at} & c(a-b)^{-1}(e^{at}-e^{bt}) \\ 0 & e^{bt} \end{bmatrix}, \; a \ne b, \qquad e^{At} = \begin{bmatrix} 1 & ct \\ 0 & 1 \end{bmatrix} e^{at}, \; a = b.$$

For the spectral norm Figure 5.5.1 shows the function $t \mapsto \|e^{At}\|$ for $a = -.6, b = -1$ and various values of c. One sees that the transient bound $M_0(A)$ increases as c increases and the time at which the maximum is achieved is almost constant (≈ 1.28 for $c = 4$ as predicted by the formula for $t_{\max}$ below). The state trajectories starting at $x^0 = [0,1]^\top$ are shown in the figures on the right for $c = 8$ and 24, for which $M_0(A) = 3.76$ and 11.2 respectively. The large transient motions are clearly visible. The straight lines with arrows represent contracting eigenmotions of the system. The angle between them is $\arctan(0.4/c)$, so that as c increases the angle is reduced, i.e. the eigenvectors become more aligned. We know from Section 4.2 that this is an indicator that the spectrum of a matrix is highly sensitive to perturbations. In Proposition 5.5.12 we will see that this is also an indicator of large transient bounds.
For $b \ne a$, we have

$$(a-b)^{-1}(e^{at}-e^{bt}) \ge \max\{-b^{-1}e^{at}, -a^{-1}e^{bt}\}, \quad t \ge (a-b)^{-1}\ln(b/a).$$

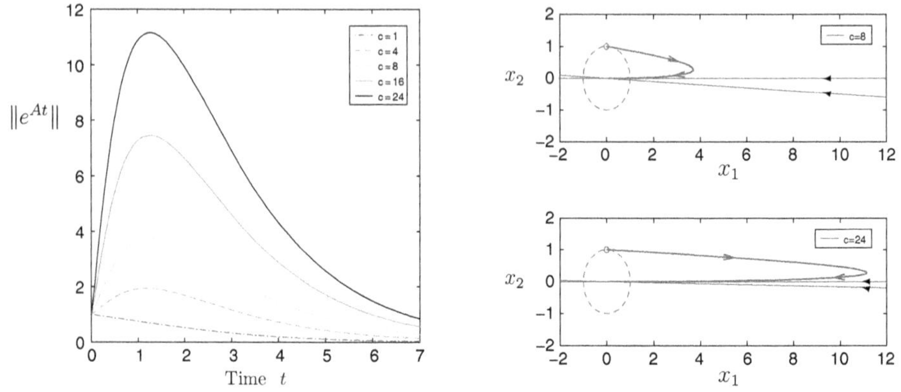

Figure 5.5.1: The function $t \to \|e^{At}\|$ for various c, and state trajectories for $c = 8, 24$.

Hence for large $|c|$

$$\|e^{At}\| \approx |c|\,(a-b)^{-1}(e^{at} - e^{bt}), \quad t \geq (a-b)^{-1}\ln(b/a).$$

The RHS is maximized for $t_{\max} = (a-b)^{-1}\ln(b/a)$ and substitution yields, for large $|c|$,

$$M_0(A) \approx -\frac{|c|}{a}\left[\frac{b}{a}\right]^{\frac{b}{a-b}} \text{ and hence by (6) } M_\beta(A) \approx -\frac{|c|}{a-\beta}\left[\frac{b-\beta}{a-\beta}\right]^{\frac{b-\beta}{a-b}}, \quad \beta > \max\{a,b\}.$$

For $a = b$ and large $|c|$ we have $\|e^{At}\| \approx |c|te^{at}$, $t \geq -a^{-1}$. The maximum on the RHS occurs at $t_{\max} = -a^{-1}$. So for large $|c|$

$$M_0(A) \approx -|c|(ae)^{-1} \quad \text{and} \quad M_\beta(A) \approx -|c|((a-\beta)e)^{-1}.$$

Note that if $a > b$, $M_\beta(A) \to |c|(a-b)^{-1}$ as $\beta \searrow a$, whereas if $a = b$, $M_\beta(A) \to \infty$ as $\beta \searrow a$.

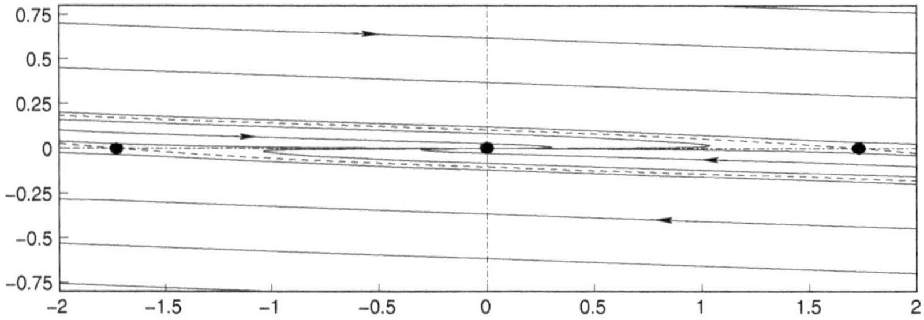

Figure 5.5.2: Phase portrait for (9)

Now consider the nonlinear equation

$$\dot{x}_1 = -0.6x_1 + 24x_2 + x_1^3, \quad \dot{x}_2 = -x_2. \tag{9}$$

There are three equilibrium points at $[0,0]^\top$ and $[\pm\sqrt{0.6}, 0]^\top$. The former is stable and the latter ones are unstable. The phase-portrait is shown in Figure 5.5.2. The linearization of (9) at the equilibrium point $[0,0]^\top$ has the system matrix $A = \begin{bmatrix} -0.6 & 24 \\ 0 & -1 \end{bmatrix}$. One might expect that the domain of attraction of this equilibrium point will be "thin" in directions which generate large transient excursions of the linearized system. This is illustrated in Figure 5.5.2 where we see that small deviations in the x_2 direction from the origin generate unbounded trajectories. The boundary of the basin of attraction of the origin is depicted by the two dashed lines. □

To obtain good upper bounds for $\|e^{At}\|$, $t \geq 0$ (which are often required in e.g. stability analysis), one would like to have the exponential rate β as small as possible (usually as negative as possible) and at the same time have a small transient bound $M_\beta(A)$. However, (5) indicates that there may be a trade off between the two objectives. Nevertheless we will see that in some cases there is an "ideal" transient behaviour in the sense that both optimal bounds can be achieved, i.e. $\beta = \alpha(A)$ and $M_\beta(A) = 1$. Before describing these special cases, we first determine conditions on β and A so that $M_\beta(A) = 1$. Then, later we will fix β and determine estimates for $M_\beta(A)$.

Initial Growth Rate

We can always achieve the optimal transient bound $M_\beta(A) = 1$ by choosing β sufficiently large, since

$$\beta \geq \|A\| \quad \Longrightarrow \quad \|e^{At}\| \leq e^{\|A\|\,t} \leq e^{\beta t}, \quad t \geq 0.$$

It is of interest to know which is the smallest $\beta \in \mathbb{R}$ such that $\|e^{At}\| \leq e^{\beta t}$, $t \geq 0$.

Definition 5.5.7. For any $A \in \mathbb{C}^{n\times n}$ the *initial growth rate* of e^{At} is defined by

$$\nu(A) = \min\{\beta \in \mathbb{R};\ \forall t \geq 0:\ \|e^{At}\| \leq e^{\beta t}\}. \tag{10}$$

This terminology is explained by the following characterization of $\nu(A)$. Here we denote the right derivative by $\frac{d^+}{dt}$.

Proposition 5.5.8. *For any* $A \in \mathbb{C}^{n\times n}$

$$\nu(A) = \frac{d^+}{dt} \ln \|e^{At}\|\Big|_{t=0} = \frac{d^+}{dt}\|e^{At}\|\Big|_{t=0} = \lim_{h\searrow 0} \frac{\|I_n + hA\| - 1}{h} = \lim_{\tau\nearrow\infty} \|A + \tau I_n\| - \tau, \tag{11}$$

where the last two expressions are monotonically decreasing with $h \searrow 0$, *resp.* $\tau \nearrow \infty$.

Proof: Since the function $f : h \mapsto \|I_n + hA\|$ is convex on $\mathbb{R}$ the difference quotient $(\|I_n + Ah\| - 1)/h$ is monotonically decreasing with $h \searrow 0$. But for $h \geq 0$

$$1 = \|I_n + hA - hA\| \leq \|I_n + hA\| + h\|A\|,$$

and so $-\|A\| \leq (\|I_n + hA\| - 1)/h$. This shows that the limit in the last but one expression in (11) exists and the convergence to it is monotone.
For any given $\varepsilon > 0$, there exists $\delta > 0$ such that

$$\big|(\|e^{Ah}\| - 1)/h - (\|I_n + hA\| - 1)/h\big| \leq \|e^{Ah} - I_n - hA\|/h < \varepsilon, \quad 0 < h < \delta.$$

Hence the right derivative of $\|e^{At}\|$ exists at $t=0$ and

$$\tfrac{d^+}{dt}\|e^{At}\|\big|_{t=0} = \lim_{h\searrow 0}\frac{\|e^{Ah}\|-1}{h} = \lim_{h\searrow 0}\frac{\|I_n+hA\|-1}{h} = \lim_{\tau\to\infty}\|A+\tau I_n\|-\tau$$

where the last equality follows by setting $h^{-1}=\tau$. Also by the chain rule

$$\tfrac{d^+}{dt}\ln\|e^{At}\|\big|_{t=0} = \left(\|e^{At}\|\big|_{t=0}\right)^{-1}\tfrac{d^+}{dt}\|e^{At}\|\big|_{t=0} = \tfrac{d^+}{dt}\|e^{At}\|\big|_{t=0}.$$

This establishes the equality of the last four expressions in (11).
Finally, note that for any $\delta>0$, since $\|e^{At}\|\le\|e^{At/k}\|^k$ for $k\in\mathbb{N}^*$, we have

$$\forall t>0: \|e^{At}\|\le e^{\beta t} \Leftrightarrow \forall t\in(0,\delta]: \|e^{At}\|\le e^{\beta t} \Leftrightarrow \forall t\in(0,\delta]: \frac{\|e^{At}\|-1}{t}\le\frac{e^{\beta t}-1}{t}.$$

Hence $\|e^{At}\|\le e^{\beta t}$, $t\ge 0$ implies $\tfrac{d^+}{dt}\|e^{At}\|\big|_{t=0}\le\beta$ and so $\tfrac{d^+}{dt}\|e^{At}\|\big|_{t=0}\le\nu(A)$. Conversely, suppose that $\beta=\tfrac{d^+}{dt}\|e^{At}\|\big|_{t=0}$, then given any $\varepsilon>0$, there exists $\delta>0$ such that

$$\|e^{At}\|\le 1+(\beta+\varepsilon)t\le e^{(\beta+\varepsilon)t}, \quad t\in(0,\delta].$$

This implies $\|e^{At}\|\le e^{(\beta+\varepsilon)t}$ for all $t\ge 0$. So the first equality in (11) holds and this completes the proof. □

The proposition shows that $\nu(A)$ can be interpreted as a directional derivative of the norm function evaluated at I_n in the direction of A. In particular, the initial growth rate is determined by the values of $\|e^{At}\|$ on an arbitrarily small interval $[0,\varepsilon]$, $\varepsilon>0$. The first equation in (11) explains why $\nu(A)$ is called *logarithmic derivative* by some authors. Below we list some properties of the initial growth rate.

Lemma 5.5.9. *For all $A,B\in\mathbb{C}^{n\times n}$ we have*

(*i*) $-\|A\|\le-\nu(-A)\le\alpha(A)\le\nu(A)\le\|A\|$, (*ii*) $e^{-\nu(-A)t}\le\|e^{At}\|\le e^{\nu(A)t}$, $t\ge 0$,
(*iii*) $\nu(A+zI_n)=\nu(A)+\operatorname{Re}z$, $z\in\mathbb{C}$, (*iv*) $\nu(\gamma A)=\gamma\nu(A)$, $\gamma>0$,
(*v*) $\operatorname{Re}\lambda\in[-\nu(-A),\nu(A)]$ *for all* $\lambda\in\sigma(A)$, (*vi*) $\nu:\mathbb{C}^{n\times n}\to\mathbb{R}$ *is convex,*
(*vii*) $\max\{\nu(A)-\nu(-B),-\nu(-A)+\nu(B)\}\le\nu(A+B)\le\nu(A)+\nu(B)$.

Proof: (ii), (iii) and (iv) are immediate from the definition of $\nu(\cdot)$. Suppose $Av=\lambda v$ with $\lambda\in\mathbb{C}$, $\|v\|=1$. Then

$$e^{\operatorname{Re}\lambda t}=\|e^{At}v\|\le\|e^{At}\|\le e^{\nu(A)t} \quad\text{and}\quad e^{-\operatorname{Re}\lambda t}=\|e^{-At}v\|\le\|e^{-At}\|\le e^{\nu(-A)t}$$

hence $-\nu(-A)\le\operatorname{Re}\lambda\le\nu(A)$. This proves (i) and (v). Since

$$\frac{\|I_n+h(A+B)\|-1}{h}=\frac{\|2I_n+2h(A+B)\|-2}{2h}\le\frac{\|I_n+2hA\|-1}{2h}+\frac{\|I_n+2hB\|-1}{2h}.$$

we obtain $\nu(A+B)\le\nu(A)+\nu(B)$. Replacing A by $A+B$ and B by $-B$, yields $\nu(A)\le\nu(A+B)+\nu(-B)$ or $\nu(A)-\nu(-B)\le\nu(A+B)$. By symmetry we have $\nu(B)-\nu(-A)\le\nu(A+B)$ and this establishes (vii). Finally, convexity is a consequence of (iv) and (vii). □

Although the initial growth rate has been called the *logarithmic norm* in the literature (see *Notes and References*), it is neither a norm nor a semi-norm on $\mathbb{K}^{n\times n}$. Negative values are possible and the ν-function is sign-sensitive: $\nu(A)$ is in general different from $\nu(-A)$. In Ex. 6 it is shown how the initial growth rate can be used to obtain upper and lower estimates on the solutions of time-varying linear equations.

Remark 5.5.10. In contrast with the (long term) growth rate $\alpha(A)$ the initial growth rate $\nu(A)$ depends upon the specific norm $\|\cdot\|$. The function $t \mapsto \ln\|e^{At}\|/t$, which may be interpreted as the mean exponential growth rate of $\|e^{At}\|$ on the interval $[0,t]$, is not monotonically decreasing. However, one can show the following relationship between the growth rates:

$$\nu(A) = \sup_{t>0}\frac{\ln\|e^{At}\|}{t} = \lim_{t\searrow 0}\frac{\ln\|e^{At}\|}{t} \geq \frac{\ln\|e^{At}\|}{t} \geq \lim_{t\to\infty}\frac{\ln\|e^{At}\|}{t} = \inf_{t>0}\frac{\ln\|e^{At}\|}{t} = \alpha(A),\ t>0.$$

□

$\nu(A)$ is easily computable for the 1-, 2- and ∞-norms. In the following lemma we compare formulas for the corresponding operator norms with those for $\nu(A)$.

Lemma 5.5.11. *Let $A = (a_{ij}) \in \mathbb{C}^{n\times n}$, then*

$$\|\cdot\| = \|\cdot\|_{1,1}: \qquad \|A\| = \max_{j\in\underline{n}}\sum_i |a_{ij}|, \qquad \nu(A) = \max_{j\in\underline{n}}\left(\operatorname{Re} a_{jj} + \sum_{i\neq j}|a_{ij}|\right),$$

$$\|\cdot\| = \|\cdot\|_{2,2}: \qquad \|A\| = \sqrt{\lambda_{\max}(A^*A)}, \qquad \nu(A) = \lambda_{\max}(A+A^*)/2,$$

$$\|\cdot\| = \|\cdot\|_{\infty,\infty}: \qquad \|A\| = \max_{i\in\underline{n}}\sum_j |a_{ij}|, \qquad \nu(A) = \max_{i\in\underline{n}}\left(\operatorname{Re} a_{ii} + \sum_{j\neq i}|a_{ij}|\right).$$

Proof: The formulas for the operator norms are well known, see (A.1.3). Here we will only derive the formula for the initial growth rate with respect to the ∞-norm. The reader is asked to prove the formula for the 1-norm in Ex. 3. For the 2-norm the formula will follow directly from Corollary 5.5.26.
To prove the result for the norm $\|\cdot\| = \|\cdot\|_{\infty,\infty}$ we first note that for every $\lambda \in \mathbb{C}$

$$\lim_{\tau\to\infty}(|\lambda+\tau| - \tau) = \operatorname{Re}\lambda, \quad \lambda \in \mathbb{C}. \tag{12}$$

In fact, since $(1+x)^{1/2} \leq 1 + x/2$ for $x \in \mathbb{R}, |x| < 1$ we get for $\tau > 0$ sufficiently large

$$\operatorname{Re}\lambda \leq |\lambda+\tau| - \tau = \tau\left(\sqrt{1 + \frac{2\operatorname{Re}\lambda}{\tau} + \frac{|\lambda|^2}{\tau^2}} - 1\right) \leq \tau\left(\frac{\operatorname{Re}\lambda}{\tau} + \frac{|\lambda|^2}{2\tau^2}\right),$$

whence (12). Applying (12) to the diagonal entries of A the formula for $\nu(A)$ (with respect to $\|\cdot\| = \|\cdot\|_\infty$) follows from

$$\begin{aligned}\nu(A) &= \lim_{\tau\to\infty}\|A + \tau I_n\| - \tau = \lim_{\tau\to\infty}\max_{i\in\underline{n}}\left(|\tau + a_{ii}| + \sum_{j\neq i}|a_{ij}|\right) - \tau \\ &= \max_{i\in\underline{n}}\left(\lim_{\tau\to\infty}(|\tau + a_{ii}| - \tau) + \sum_{j\neq i}|a_{ij}|\right).\end{aligned}$$

□

Using this lemma the initial growth rate is easy to determine with respect to the 1- and ∞-norms. Moreover, in these cases two matrices $A, B \in \mathbb{C}^{n\times n}$ have the same initial growth rate if the absolute values of their off-diagonal entries and the real parts of their diagonal entries coincide. This is a rare property which is not shared by the spectral abscissa, the stability radius and the transient bounds.
As a special consequence of the previous lemma one obtains a simple criterion for real matrices to generate a contraction semigroup with respect to the ∞-norm. $A \in \mathbb{R}^{n\times n}$ has this property if and only if A is diagonally dominant with non-positive diagonal entries, see Ex. 13.
The relationship between the initial growth rate and the transient bounds is, in general, not very close. However, it follows directly from the definition that $M_\beta(A) = 1$ for all $\beta \geq \nu(A)$. On the other hand, if $\alpha(A) < \beta < \nu(A)$ one has from (7) (with $\beta' = \nu(A)$) only that

$$1 \leq M_\beta(A) \leq e^{(\nu(A)-\beta)t_\beta}.$$

Here t_β may be replaced by any $\tilde{t}$ for which it is known that $\|e^{A_\beta t}\|$ takes its maximum on $\mathbb{R}_+$ in the interval $[0, \tilde{t}\,]$.
If $\nu(A) = \alpha(A)$ the system $\dot{x} = Ax$ has an "ideal" transient behaviour in the sense that it satisfies the exponential estimate (1) with both minimal exponential rate $\beta = \alpha(A)$ and minimal transient bound $M_\beta(A) = 1$,

$$\|e^{At}\| \leq e^{\alpha(A)t}. \tag{13}$$

For the spectral norm on $\mathbb{C}^{n\times n}$ all normal matrices enjoy this property.

Proposition 5.5.12. *Suppose that $A \in \mathbb{C}^{n\times n}$ is diagonalizable and $A = SDS^{-1}$ with diagonal D for some $S \in \mathbf{Gl}_n(\mathbb{C})$. If $\|\cdot\|$ is an absolute norm on $\mathbb{C}^n$, then*

$$\|e^{At}\| \leq \kappa(S)\, e^{\alpha(A)t}, \qquad t \geq 0, \qquad \kappa(S) = \|S\|\,\|S^{-1}\|. \tag{14}$$

In particular, if A is normal then A satisfies (13) *with respect to the spectral norm and $\nu(A) = \alpha(A)$.*

Proof: (14) follows from

$$\|e^{At}\| = \|Se^{Dt}S^{-1}\| \leq \|S\|\,\|S^{-1}\|\,\|e^{Dt}\|$$

and the fact that the operator norm of a diagonal matrix with respect to an absolute vector norm is equal to the maximal absolute value of its diagonal entries (Theorem A.1.9). The second statement of the proposition follows because for normal A, the diagonalizing S can be chosen to be unitary, and unitary matrices have spectral norm equal to 1. □

Comparing Proposition 5.5.12 and Corollary 5.2.38 we notice the close relationship between transient behaviour of a nominal system and its robustness under (unstructured, complex) perturbations. If $A \in \mathbb{C}^{n\times n}$ is normal then it has the smallest pseudospectrum and the smallest transient bounds $M_\beta = 1$, $\beta \geq \alpha(A)$ amongst all the matrices of its similarity class.
If A is not normal, one may hope to obtain an estimate for $\nu(A) - \alpha(A)$ in terms

of some measure of non-normality of A. To conclude the subsection we will briefly discuss this problem. Through the rest of this subsection we assume $\|\cdot\|$ is the spectral norm on $\mathbb{C}^{n\times n}$. In the literature various measures have been used to quantify the non-normality of a matrix. The most obvious one is the distance of the matrix from the set $\mathcal{N}_n(\mathbb{C})$ of complex normal $n \times n$ matrices

$$\operatorname{dist}(A, \mathcal{N}_n(\mathbb{C})) = \min_{X \in \mathcal{N}_n(\mathbb{C})} \|A - X\|_{2,2}. \tag{15}$$

Another index of non-normality is defined via the *Schur decomposition* $U^*AU = D_U + N_U$ of A where $U \in \mathbf{U}_n(\mathbb{C})$, D_U is diagonal and N_U is upper triangular with zeros on the diagonal (and hence nilpotent). Amongst all Schur decompositions of A there is one for which the spectral norm $\|N_U\|_{2,2}$ is minimal[1] and this norm is called the *departure* of A from normality (with respect to the spectral norm, see [223])

$$\operatorname{dep}(A) = \min\{\|N_U\|_{2,2}; U \in \mathbf{U}_n(\mathbb{C}),\ U^*AU = D_U + N_U \text{ Schur decomposition}\}. \tag{16}$$

Note that, although the nilpotent part N_U of a Schur decomposition $U^*AU = D_U + N_U$ varies with U, the *Frobenius norm* $\|N_U\|_F$ of N_U does not depend on U since

$$\|N_U\|_F^2 = \|A\|_F^2 - \sum_{i \in \underline{n}} |\lambda_i|^2 \quad \text{where} \quad \Lambda(A) = \lfloor \lambda_1, \ldots, \lambda_n \rfloor.$$

As a consequence of Proposition 5.5.12 and Lemma 5.5.9 we have

Corollary 5.5.13. *If $A \in \mathbb{C}^{n\times n}$ then with respect to the spectral norm,*

$$\nu(A) \le \alpha(X) + \nu(A - X) \le \alpha(X) + \|A - X\|, \quad X \in \mathcal{N}_n(\mathbb{C}). \tag{17}$$

Moreover

$$\nu(A) \le \alpha(A) + \operatorname{dep}(A). \tag{18}$$

Proof: Let $X \in \mathcal{N}_n(\mathbb{C})$. By Lemma 5.5.9 (vii)

$$\nu(A) = \nu(X + A - X) \le \nu(X) + \nu(A - X).$$

So the first inequality in (17) follows from Proposition 5.5.12 and the second from Lemma 5.5.9 (i). The inequality (18) follows by setting $X = D_{\hat{U}}$ where $\hat{U} \in \mathbf{U}_n(\mathbb{C})$ achieves the minimum in (16). $\square$

Example 5.5.14. Consider A as in Example 5.5.6 and let $X = \begin{bmatrix} -0.6 & c/2 \\ -c/2 & -0.6 \end{bmatrix}$, then $X \in \mathcal{N}_2(\mathbb{C})$, $\alpha(X) = -0.6$ and $A - X = \begin{bmatrix} 0 & c/2 \\ c/2 & -0.4 \end{bmatrix} \in \mathcal{N}_2(\mathbb{C})$. Hence

$$\nu(A - X) = \alpha(A - X) = -0.2 + (0.04 + c^2/4)^{1/2}$$

and so by the corollary we obtain the estimate $\nu(A) \le -0.8 + (0.04 + c^2/4)^{1/2}$, which is tight for all $c \in \mathbb{R}$, see Example 5.5.27 (ii). Note that in this case X is not a closest

[1] That such a minimum exists follows by a compactness argument making use of the fact that $\mathbf{U}_n(\mathbb{C})$ is compact, see Ex. 15.

normal matrix although it provides via (17) an optimal estimate for $\nu(A)$. It can be shown that, amongst all normal real matrices, the matrix $\hat{X} = \begin{bmatrix} -0.8 & c/2 \\ -c/2 & -0.8 \end{bmatrix}$ has a minimal distance from A. Now $\|A - \hat{X}\|_{2,2} = [0.04 + c^2/4]^{1/2}$ and $\alpha(\hat{X}) = -0.8$, so the second inequality in (17) yields again the estimate $\nu(A) \leq -0.8 + (0.04 + c^2/4)^{1/2}$ which is tight. On the other hand, since $\alpha(A) = -0.6$ and $\operatorname{dep}(A) = |c|$, (18) yields the inferior estimate $\nu(A) \leq -0.6 + |c|$. □

5.5.2 Contractions and Estimates of the Transient Bound

A system $\dot{x} = Ax$ combines ideal transient behaviour ($M_0(A) = 1$) with (asymptotic) stability if $\nu(A) \leq 0$ (resp. $\nu(A) < 0$). Systems with these properties are called *contractions.* In this subsection we will first characterize these systems for a given norm and then show how, for non-contractive systems, upper estimates of $M_0(A)$ can be derived via *Liapunov norms.* These are auxiliary norms with respect to which a given system is contractive. If one uses Hilbert space norms this leads to Liapunov equations and yields a constructive method for obtaining upper estimates. As in the previous subsection we will assume throughout that $\mathbb{K}^n$ is provided with an arbitrary, but fixed norm $\|\cdot\|$, and $\mathbb{K}^{n\times n}$ with the corresponding operator norm which we also denote by $\|\cdot\|$.

Definition 5.5.15. $\left(e^{At}\right)_{t\geq 0}$ is said to be a contraction semigroup if $\nu(A) \leq 0$, i.e.

$$\|e^{At}\| \leq 1 \text{ for all } t > 0. \tag{19}$$

It is said to be a *strict contraction semigroup* if $\nu(A) < 0$, i.e. $\|e^{At}\| \leq e^{-\varepsilon t}$, $t \geq 0$ for some $\varepsilon > 0$. It is called a *strong contraction semigroup* if $\|e^{At}\| < 1$ for all $t > 0$.

A generates a contraction semigroup if and only if the closed unit ball $\overline{B(0,1)} = \{x \in \mathbb{K}^n; \|x\| \leq 1\}$ is invariant under the flow of $\dot{x} = Ax$, i.e. $e^{At}\overline{B(0,1)} \subset \overline{B(0,1)}$ for all $t > 0$. Similarly, A generates a strong contraction semigroup if and only if e^{At} maps the closed unit ball into the open unit ball for all $t > 0$.

Remark 5.5.16. Given any $\beta \in \mathbb{R}$ we will call (e^{At}) a *β-contraction semigroup* if $\|e^{At}\| \leq e^{\beta t}$ for all $t \geq 0$. A semigroup $\left(e^{At}\right)$ is a (strict) contraction semigroup if and only if it is a β-contraction semigroup for $\beta = 0$ (resp. for some $\beta < 0$). For any $\beta \in \mathbb{R}$, A generates a β-contraction semigroup if and only if $\nu(A) \leq \beta$. Since this condition defines a closed set and ν is convex by Lemma 5.5.9, the set of all $A \in \mathbb{K}^{n\times n}$ generating a β-contraction semigroup forms a closed convex subset of $\mathbb{K}^{n\times n}$. □

If A generates a contraction semigroup then it is necessarily stable. Moreover, (19) implies that $t \to \|e^{At}x^0\|$ is monotonically decreasing for all $x^0 \in \mathbb{C}^n$. If additionally $\gamma = \|e^{A\tau}\| < 1$ for some $\tau > 0$ then $\|e^{At}\| \leq \|e^{Ak\tau}\| \leq \gamma^k$ for all $t \geq k\tau$. In particular, every strong contraction semigroup is asymptotically stable. By the same argument, if A generates a contraction semigroup which is not asymptotically stable then necessarily $\|e^{At}\| = 1$ for all $t \geq 0$. Clearly, every strict contraction semigroup is a strong contraction semigroup. However not every strong contraction semigroup is strict, see Example 5.5.27. If A is asymptotically stable and generates a contraction semigroup one might expect that it is necessarily strong. For this to be the case one must rule out the possibility that for some $\tau > 0$ we have $\|e^{At}\| = 1$, $t \in [0, \tau]$. With respect to the spectral norm this would imply $\lambda_{\max}(e^{A^*t}e^{At}) = 1$, $t \in [0, \tau]$ and so $\det[I_n - e^{A^*t}e^{At}] = 0$, $t \in [0, \tau]$. But the LHS of this equality is analytic in t and hence must equal 0 for all $t \geq 0$ which contradicts the assumption of asymptotic stability. In Ex. 11 conditions on the norm $\|\cdot\|$ are specified under which every asymptotically stable contraction semigroup is a *strong* contraction. The fact that this is not so in general is explained in the following remark and illustrated in the subsequent example.

Remark 5.5.17. Suppose $A \in \mathbb{K}^{n\times n}$ is asymptotically stable with $\nu(A) > 0$ and let

$$\|\!|x|\!\|_A := \max_{s\geq 0} \|e^{As}x\|, \quad x \in \mathbb{K}^n. \tag{20}$$

Then $\|\!|\cdot|\!\|_A$ defines a norm on $\mathbb{K}^n$ and A generates a contraction semigroup with respect to the induced operator norm. Moreover there exists a $\tau > 0$ such that $\|\!|e^{At}|\!\|_A = 1$ for $t \in [0, \tau]$. The reader is asked to prove these statements in Ex. 9. □

Example 5.5.18. Consider the real linear system

$$\Sigma: \qquad \dot{x} = Ax, \quad \text{where } A = \begin{pmatrix} -1 & -1 \\ 1 & -1 \end{pmatrix}.$$

Examining the trajectories starting on the boundary of the unit square, on can show that A generates a strong contraction semigroup with respect to the maximum norm $\|\cdot\|_\infty$. In Ex. 10 the reader is asked to prove this and the following observations. Consider the

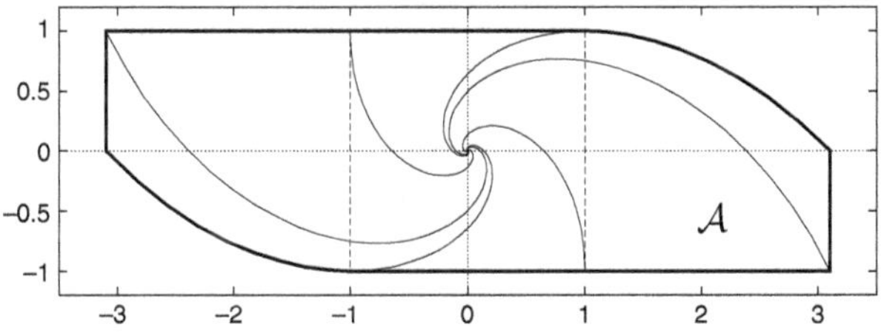

Figure 5.5.3: Two trajectories with $\|e^{At}x\| = 1$, $t \in [0, \pi/4]$.

rectangle $\mathcal{R} := [-\alpha, \alpha] \times [1, 1]$ where $\alpha = \sqrt{2}e^{\pi/4} > 1$. It is easily verified that the solution starting at $[\alpha, 0]^\top$ is given by $\alpha e^{-t}[\cos t, \sin t]^\top$ and that this curve remains entirely inside the box $\mathcal{R}$, only touching the border $\partial\mathcal{R}$ in $[1, 1]^\top$. If the area in the box above the curve segment given by $t \in [0, \pi/4]$ and the area below its central symmetric counterpart in the lower left corner are clipped away, the remaining area $\mathcal{A}$ is invariant with respect to the flow of Σ, see Figure 5.5.3. $\mathcal{A}$ is a convex and symmetric neighbourhood of the origin. Hence the corresponding Minkowski functional, (see Definition A.4.13) $p_{\mathcal{A}}(x) = \inf\{\gamma > 0,\ \gamma^{-1}x \in \mathcal{A}\}$ is a norm on the state space $\mathbb{R}^2$. The thick lines in Figure 5.5.3 describe the boundary of the unit ball with respect to this norm. By construction we have $p_{\mathcal{A}}(e^{At}) = 1$, $t \in [0, \pi/4]$ where $p_{\mathcal{A}}(e^{At})$ is the operator norm of e^{At} with respect to the norm $p_{\mathcal{A}}(\cdot)$ on $\mathbb{R}^2$. In particular, we have $\nu(A) = 0$ with respect to this norm although the corresponding system is asymptotically stable. Note that the norm $p_{\mathcal{A}}(\cdot)$ coincides, for this example, with the norm $\|\!|x|\!\|_A$ constructed in the previous remark (starting from the norm $\|\cdot\|$ whose unit ball is $\mathcal{R}$). □

By Definitions 5.5.7 and 5.5.15, A generates a (strict) contraction semigroup if and only if $\nu(A) \le 0$ (resp. $\nu(A) < 0$). Strong contraction semigroups, however, cannot be characterized via their initial growth rates. In Example 5.5.27 (iii) we will encounter a strong contraction semigroup for which $\nu(A) = 0$.
In a Hilbert space, contraction properties of a semigroup can be directly expressed by the dissipativity of its generator. This characterization can be generalized to arbitrary normed spaces X by introducing a *semi-scalar product* on X. The existence of a semi-scalar product follows directly from the Hahn-Banach Theorem. We will only consider the finite dimensional case.

Lemma 5.5.19. *Suppose $\|\cdot\|$ is a norm on $\mathbb{K}^n$, then there exists a scalar valued function $(x,y) \to [x,y]$ on $\mathbb{K}^n \times \mathbb{K}^n$ such that for all $x, y, z \in \mathbb{K}^n$, $\lambda \in \mathbb{K}$*

$$[x+y,z] = [x,z] + [y,z], \quad [\lambda x, y] = \lambda[x,y],$$
$$[x,x] = \|x\|^2, \quad |\,[x,y]\,| \le \|x\|\|y\|.$$

$[\cdot,\cdot]$ is called a semi-scalar product *on the normed linear space* $(\mathbb{K}^n, \|\cdot\|)$.

Proof: By the Hahn-Banach Theorem A.4.10 for every $y \in \mathbb{K}^n$ there exists a (not necessarily unique) vector $f_y \in \mathbb{K}^{1\times n}$ of dual norm $\|f_y\|^* = \|y\|$ such that f_y is aligned with y, i.e. $f_y y = \|y\|^2$. Clearly $[x,y] = f_y x$ defines a semi-scalar product for $\|\cdot\|$. □

Note that semi-scalar products are linear and hence continuous in the first argument. Example 5.5.21 below shows that they need not be continuous in the second argument.

Definition 5.5.20. A matrix $A \in \mathbb{K}^{n\times n}$ is *dissipative* (resp. *strictly dissipative*) with respect to a semi-scalar product $[\cdot,\cdot]$ on $(\mathbb{K}^n, \|\cdot\|)$ if

$$\mathrm{Re}[Ax,x] \le 0 \quad (\text{resp. } \mathrm{Re}[Ax,x] \le -\varepsilon\|x\|^2 \text{ for some } \varepsilon > 0) \quad \text{for all} \quad x \in \mathbb{K}^n.$$

Just as the concept of a contraction semigroup depends on the norm $\|\cdot\|$ on $\mathbb{K}^n$ so does the property of dissipativity. However we will see in Theorem 5.5.23 that dissipativity does not depend on the specific semi-scalar product $[\cdot,\cdot]$ that is chosen on $(\mathbb{K}^n, \|\cdot\|)$. We will now illustrate these concepts by an example of a matrix which is strictly dissipative for the 2-norm, but not dissipative for the 1-norm.

Example 5.5.21. Let $\mathbb{R}^2$ be normed by the 1-norm, then the dual norm is the ∞-norm and for any vector $y \in \mathbb{R}^2$ the linear form

$$f_y : x \mapsto f_y(x) = \|y\|_1 \left(x_1 y_1/|y_1| + x_2 y_2/|y_2|\right), \quad \text{where } y_i/|y_i| := 0 \text{ if } y_i = 0$$

is aligned with y since $f_y(y) = \|y\|_1(y_1^2/|y_1| + y_2^2/|y_2|) = \|y\|_1^2$ and $\|f_y\|^* = \|y\|_1$. So $[x,y] = f_y(x)$ is a semi-scalar product for $\|\cdot\|_1$ on $\mathbb{R}^2$. Note that $y \mapsto [x,y]$ is not continuous at $y = (1,0)^\top$ if we choose e.g. $x = (0,1)^\top$. In fact, setting $y_\varepsilon = (1,\varepsilon)^\top$ we have $y_\varepsilon \to y$ and $[x, y_\varepsilon] = (1+\varepsilon) \to 1$ if $\varepsilon \searrow 0$, but $[x,y] = 0$.

Now consider the matrix $A = \begin{bmatrix} a_{11} & a_{12} \\ a_{21} & a_{22} \end{bmatrix}$, then A will be dissipative with respect to the semi-scalar product $[\cdot,\cdot]$ if

$$\|x\|_1^{-1}[Ax,x] = \frac{(a_{11}x_1 + a_{12}x_2)x_1}{|x_1|} + \frac{(a_{21}x_1 + a_{22}x_2)x_2}{|x_2|} \leq 0 \quad \text{for all} \quad \begin{bmatrix} x_1 \\ x_2 \end{bmatrix} \in \mathbb{R}^2.$$

The conditions on $A = (a_{ij}) \in \mathbb{R}^{2\times 2}$ for the above to hold are given in Ex. 13 and are not the same as in the Euclidean case. To see this consider $A = \begin{bmatrix} -1 & 2 \\ 2 & -9 \end{bmatrix}$, then $A \prec 0$ and so A is strictly dissipative with respect to the Euclidean inner product on $\mathbb{R}^2$, but

$$\|x\|_1^{-1}[Ax,x] = \frac{(-x_1 + 2x_2)x_1}{|x_1|} + \frac{(2x_1 - 9x_2)x_2}{|x_2|}$$

and for $x_1 = 8/9, x_2 = 1/9$, we find $\|x\|_1^{-1}[Ax,x] = +1/9$. □

Throughout the rest of this section we will make use of the well known connections between a semigroup and its Laplace transform, the resolvent operator associated with its generator, see Example A.3.22.

Lemma 5.5.22. *Suppose that $A \in \mathbb{C}^{n\times n}$ and $R(s,A) = (sI_n - A)^{-1}, s \in \rho(A)$ is its resolvent operator, then*

$$\begin{aligned} R(s,A) &= \int_0^\infty e^{-st}e^{At}dt, \qquad \operatorname{Re} s > \alpha(A), \\ e^{At} &= \lim_{m\to\infty}(I_n - m^{-1}tA)^{-m} = \frac{1}{2\pi\imath}\int_\Gamma e^{st}R(s,A)ds, \quad t \geq 0, \end{aligned} \tag{21}$$

where Γ is any positively-oriented, piecewise smooth, simple, closed curve enclosing $\sigma(A)$.

We have the following theorem which also holds in an infinite dimensional setting, see *Notes and References*.

Theorem 5.5.23. *Let $A \in \mathbb{K}^{n\times n}$ and $\|\cdot\|$ be a norm on $\mathbb{K}^n$, then A generates a (strict) contraction semigroup on $\mathbb{K}^n$ with respect to the norm $\|\cdot\|$ if and only if A is dissipative (resp. strictly dissipative) with respect to some (or, equivalently, every) semi-scalar product on $\mathbb{K}^n$ for $\|\cdot\|$.*

Proof: Let $\lambda > 0$ and assume A is dissipative for some semi-scalar product $[\cdot,\cdot]$ for $\|\cdot\|$. Then for any $x \in \mathbb{K}^n$, we have

$$\lambda\|x\|^2 = \lambda[x,x] \leq \operatorname{Re}(\lambda[x,x] - [Ax,x]) = \operatorname{Re}[(\lambda I - A)x,x] \leq \|(\lambda I - A)x\|\|x\|.$$

Hence $\lambda\|x\| \leq \|(\lambda I - A)x\|$ for all $x \in \mathbb{K}^n$ and thus $\lambda I - A$ is invertible with $\|(\lambda I - A)^{-1}\| \leq \lambda^{-1}$, or equivalently $\|(I - \lambda^{-1}A)^{-1}\| \leq 1$. So $\|e^{At}\| \leq 1$ by applying the second formula in (21) with $\lambda = m^{-1}t$.
Conversely assume that A generates a contraction semigroup, then for any semi-scalar inner product $[\cdot,\cdot]$ for $\|\cdot\|$

$$\operatorname{Re}[e^{At}x - x, x] = \operatorname{Re}[e^{At}x,x] - \|x\|^2 \leq \|e^{At}x\|\|x\| - \|x\|^2 \leq 0.$$

Hence by the continuity of $v \mapsto [v, x]$ we get $\text{Re}[Ax, x] = \lim_{t\to 0} \text{Re}\{t^{-1}[e^{At}x - x, x]\} \le 0$. A is strictly dissipative if and only if $A + \varepsilon I_n$ is dissipative for some $\varepsilon > 0$. Now $e^{(A+\varepsilon I_n)t} = e^{At}e^{\varepsilon t}$ and so the strict case is a consequence of applying the contraction result to $A + \varepsilon I_n$ since $\text{Re}[(A + \varepsilon I_n)x, x] = \text{Re}[Ax, x] + \varepsilon\|x\|^2$. □

Corollary 5.5.24. *$A \in \mathbb{K}^{n\times n}$ generates a contraction semigroup with respect to a given norm $\|\cdot\|$ on $\mathbb{K}^n$ if and only if its resolvent satisfies*

$$\|(\lambda I - A)^{-1}\| \le \lambda^{-1}, \quad \lambda > 0. \tag{22}$$

Proof: The result follows from Theorem 5.5.23 and the first part of its proof. □

Corollary 5.5.25. *Let $A \in \mathbb{K}^{n\times n}$, then with respect to any semi-scalar product $[\cdot,\cdot]$ on $\mathbb{K}^n$ for $\|\cdot\|$,*

$$\nu(A) = \sup_{x\neq 0} \frac{\text{Re}[Ax, x]}{\|x\|^2}. \tag{23}$$

Proof: By definition $\nu(A)$ is the smallest $\beta \in \mathbb{R}$ such that $\left(e^{(A-\beta I_n)t}\right)_{t\ge 0}$ is a contraction semigroup. By Theorem 5.5.23 $\left(e^{(A-\beta I_n)t}\right)_{t\ge 0}$ is a contraction semigroup if and only if

$$0 \ge \text{Re}[Ax, x] - \text{Re}[\beta x, x] = \text{Re}[Ax, x] - \beta\|x\|^2, \quad x \in \mathbb{K}^n$$

or, equivalently $\beta \ge \sup_{x\neq 0} \text{Re}[Ax, x]/\|x\|^2$. This proves (23). □

In specializing the previous result to the Hilbert space case we obtain a computable formula for $\nu(A)$ in the next corollary. Moreover we obtain a characterization of $\nu(A)$ in terms of the *numerical range* or *field of values* of A defined by

$$\Theta(A) = \{\langle Ax, x\rangle;\ \|x\| = 1\}. \tag{24}$$

By a theorem of Hausdorff [218] $\Theta(A)$ is a *convex* compact subset of the complex plane containing the spectrum of A, see *Notes and References.*

Corollary 5.5.26. *Suppose that $\mathbb{K}^n$ is provided with an arbitrary inner product $\langle\cdot,\cdot\rangle$ and $\|x\| = \langle x, x\rangle^{1/2}$ is the associated norm, then for any $A \in \mathbb{K}^{n\times n}$*

$$\nu(A) = \sup \text{Re}\,\Theta(A) = \lambda_{\max}(A + A^*)/2 = \min\{\nu \in \mathbb{R};\ A + A^* \preceq 2\nu I_n\}, \tag{25}$$

where A^* is the adjoint of A and $\Theta(A)$ is the numerical range, both with respect to the inner product $\langle\cdot,\cdot\rangle$. In particular A generates a (strict) contraction semigroup if and only if $A + A^* \preceq 0$ (resp. $A + A^* \prec 0$).

Proof: Since $\text{Re}\langle Ax, x\rangle = [\langle Ax, x\rangle + \langle x, Ax\rangle]/2 = \langle (A + A^*)x, x\rangle/2$ for all $x \in \mathbb{K}^n$ and $\nu(A) = \sup_{x\neq 0} \text{Re}\langle Ax, x\rangle/\|x\|^2$ by (23), it suffices to take the maximum over $\{x \in \mathbb{C}^n; \|x\| = 1\}$ and recall that $\lambda_{\max} = \max_{\|x\|=1}\langle Hx, x\rangle$ (see (A.4.17)) and $H \preceq \nu I_n \Leftrightarrow \lambda_{\max}(H) \le \nu$ for every Hermitian matrix H. □

Example 5.5.27. (i) Consider the matrix $A = \begin{bmatrix} -1 & 1 \\ -1 & -2 \end{bmatrix}$, then with respect to the usual Euclidean inner product $A+A^* = \mathrm{diag}\,(-2,-4)$. Hence $\nu(A) = -1$ and $\|e^{At}\| \le e^{-t}$, $t \ge 0$. Note that A is not normal and $\alpha(A) = -3/2$.
(ii) Now consider the matrix A of Example 5.5.6 with $a = -0.6$, $b = -1$, then

$$A + A^* - 2\nu I_2 = \begin{bmatrix} -1.2 - 2\nu & c \\ c & -2 - 2\nu \end{bmatrix}.$$

So $A + A^* - 2\nu I_2 \preceq 0$ if and only if $1.2 + 2\nu \ge 0$ and $(1.2 + 2\nu)(2 + 2\nu) \ge c^2$. Hence $\nu(A) = -0.8 + (0.04 + c^2/4)^{1/2}$ confirming that the first two estimates in Example 5.5.14 are tight.
(iii) Consider the matrix $A = \begin{bmatrix} -1 & 2 \\ 0 & -1 \end{bmatrix}$, then $A+A^* = \begin{bmatrix} -2 & 2 \\ 2 & -2 \end{bmatrix} \preceq 0$. So A is dissipative but not strictly dissipative. Now

$$e^{At} = e^{-t} \begin{bmatrix} 1 & 2t \\ 0 & 1 \end{bmatrix}$$

and hence with respect to the spectral norm a short calculation yields

$$\|e^{At}\| = e^{-t} \left[1 + 2t^2 + 2t\sqrt{1+t^2}\right]^{1/2} = e^{-t}(t + \sqrt{1+t^2}).$$

Since we know that this expression is ≤ 1 and cannot be identically equal to 1 on any interval $[0,\tau]$, $\tau > 0$, we must have $\|e^{At}\| < 1$, $t > 0$. So A generates a strong contraction semigroup although it is not strictly dissipative and hence does not generate a strict contraction semigroup. □

Estimates for Transient Bounds via Liapunov Norms

We now explore the possibility of obtaining estimates for transient bounds via Liapunov norms. In the 2-norm case this leads to the problem of finding well conditioned solutions of strict Liapunov inequalities.

Definition 5.5.28. Given $A \in \mathbb{K}^{n\times n}$ a norm $p(\cdot)$ on $\mathbb{K}^n$ is called a *(strict) Liapunov norm* for a system $\dot{x} = Ax$ if A generates a (strict) contraction semigroup with respect to p. More generally, given any $\beta \in \mathbb{R}$, p is called a *β-Liapunov norm* for $\dot{x} = Ax$ if A generates a β-contraction semigroup with respect to p, i.e.

$$p(e^{At}x^0) \le e^{\beta t} p(x^0), \quad x^0 \in \mathbb{K}^n,\ t \ge 0.$$

By Definition 3.2.6 a (strict) Liapunov norm for the system $\dot{x} = Ax$ defines a global (strict) generalized Liapunov function for the flow of $\dot{x} = Ax$.

Definition 5.5.29. Suppose $p(\cdot)$ is any norm on $\mathbb{K}^n$. Then the eccentricity of p with respect to $\|\cdot\|$ is given by

$$\mathrm{ecc}(p) = \mathrm{ecc}(p, \|\cdot\|) = \frac{\max_{\|x\|=1} p(x)}{\min_{\|x\|=1} p(x)}. \tag{26}$$

A simple calculation shows that $\mathrm{ecc}(p, \|\cdot\|) = \mathrm{ecc}(\|\cdot\|, p)$.

Remark 5.5.30. If $W \in \mathbb{K}^{n\times n}$ is nonsingular, the eccentricity of the norm $p(x) = \|Wx\|$ on $\mathbb{K}^n$ equals the condition number of W with respect to the norm $\|\cdot\|$, $\operatorname{ecc}(p) = \kappa(W) = \|W\|\,\|W^{-1}\|$. In particular, if $\|\cdot\| = \|\cdot\|_2$ then $\operatorname{ecc}(p) = \sigma_{\max}(W)/\sigma_{\min}(W)$. □

If the eccentricity of p and the initial growth rate $\nu_p(A)$ of $\dot{x} = Ax$ with respect to p are known, the following lemma can be applied in order to obtain an exponential estimate for $\|e^{At}\|$. We denote by $p(T)$ the operator norm of a matrix $T \in \mathbb{K}^{n\times n}$ with respect to the norm p on $\mathbb{K}^n$.

Lemma 5.5.31. *Suppose $p(\cdot)$ is a norm on $\mathbb{K}^n$ and $\nu_p(A)$ is the initial growth rate of e^{At} with respect to p and $\operatorname{ecc}(p)$ is the eccentricity of p with respect to $\|\cdot\|$, then*

$$\|e^{At}\| \leq \operatorname{ecc}(p)\, e^{\nu_p(A)t}, \quad t \geq 0. \tag{27}$$

Proof: For all $y \in \mathbb{K}^n$, $y \neq 0$ we have

$$\min_{\|x\|=1} p(x) \leq p\left(\frac{y}{\|y\|}\right) \leq \max_{\|x\|=1} p(x), \qquad \frac{p(y)}{\max_{\|x\|=1} p(x)} \leq \|y\| \leq \frac{p(y)}{\min_{\|x\|=1} p(x)}. \tag{28}$$

This implies for every $T \in \mathbb{K}^{n\times n}$

$$\|T\| = \max_{\|x\|\neq 0} \frac{\|Tx\|}{\|x\|} \leq \max_{\|x\|\neq 0} \left(\frac{p(Tx)}{\min_{\|z\|=1} p(z)}\right)\left(\frac{p(x)}{\max_{\|z\|=1} p(z)}\right)^{-1} = (\operatorname{ecc} p)p(T).$$

Since $p(e^{At}) \leq e^{\nu_p(A)t}$ for all $t \geq 0$, setting $T = e^{At}$ gives the desired result. □

It follows from this lemma that if

$$\operatorname{ecc}(p) \leq M, \quad \nu_p(A) \leq \beta \tag{29}$$

for some $M \geq 1$, $\beta \in \mathbb{R}$ then $\dot{x} = Ax$ is (M,β)-stable. The converse also holds true.

Proposition 5.5.32. *If $\dot{x} = Ax$ is (M,β)-stable (with respect to $\|\cdot\|$) then there exists a β-Liapunov norm $p(\cdot)$ on $\mathbb{K}^n$ such that (29) holds.*

Proof: Consider the norm on $\mathbb{K}^n$ defined by $p(x) = \sup_{t\geq 0} e^{-\beta t}\|e^{At}x\|$, $x \in \mathbb{K}^n$. By (M,β)-stability this norm is well defined, see Ex. 9 and we have

$$\|x\| \leq p(x) \leq e^{-\beta t} M e^{\beta t}\|x\| = M\|x\|, \quad x \in \mathbb{K}^n.$$

It follows from (26) that $\operatorname{ecc}(p) \leq M$. Moreover, since by definition

$$e^{-\beta\tau} p(e^{A\tau}x) = e^{-\beta\tau} \sup_{t\geq 0} e^{-\beta t}\|e^{At}e^{A\tau}x\| = \sup_{t\geq 0} e^{-\beta(t+\tau)}\|e^{A(t+\tau)}x\| \leq p(x), \quad \tau \geq 0$$

we have $p(e^{A\tau}x) \leq e^{\beta\tau}p(x)$, $\tau \geq 0$ for all $x \in \mathbb{K}^n$ and so $\nu_p(A) \leq \beta$. □

Although this proposition shows that the upper estimate (27) can be made arbitrarily tight, there is no safe and efficient numerical procedure available for constructing Liapunov norms whose eccentricity comes close to the optimum $M_\beta(A)$. Note that in order to compute $\mathrm{ecc}(p)$ for the Liapunov norm defined in the previous proof one needs to compute $\sup_{t\geq 0} e^{-\beta t}\|e^{At}\|$ which is just $M_\beta(A)$ and so it would be simpler to determine this value directly rather than via a construction of the tight Liapunov norm.

However, if $\mathbb{K}^n$ is provided with the 2-norm, suboptimal estimates can be obtained by a solution of Liapunov equations. To explain this, we assume in the rest of this subsection that the given norm is $\|x\| = \langle x, x\rangle^{1/2}$, $x \in \mathbb{K}^n$ where $\langle \cdot\,, \cdot\rangle$ is the usual Euclidean inner product. Now choose any $P \in \mathcal{H}_n(\mathbb{K}), P \succ 0$ and define another inner product $\langle \cdot, \cdot \rangle_P$ with associated norm $\|\cdot\|_P = \langle \cdot, \cdot\rangle_P^{1/2}$ on $\mathbb{K}^n$ by

$$\langle x,\, y\rangle_P = \langle Px, y\rangle, \quad \|x\|_P = \langle Px, x\rangle^{1/2} = \|P^{1/2}x\|_2, \qquad x, y \in \mathbb{K}^n,$$

see Proposition A.4.28. We have seen that if A generates a contraction (resp. strong contraction) semigroup then A is stable (resp. asymptotically stable). In the sequel we derive a partial converse to this result. First note that

$$2\,\mathrm{Re}\langle Ax, x\rangle_P = 2\,\mathrm{Re}\langle PAx, x\rangle = \langle PAx, x\rangle + \langle x, PAx\rangle = \langle (PA + A^*P)x, x\rangle$$

where $A^* = \overline{A}^\top$ is the adjoint of A with respect to the standard inner product. It therefore follows from Corollary 5.5.26 that A generates a (strict) contraction semigroup on $\mathbb{K}^n$ with respect to the norm $\|\cdot\|_P$ if and only if

$$PA + A^*P \preceq 0 \qquad (\text{resp. } PA + A^*P \prec 0).$$

Proposition 5.5.33. *Let $A \in \mathbb{K}^{n\times n}$, $Q \in \mathcal{H}_n^+(\mathbb{K})$, $\beta \in \mathbb{R}$ and suppose that $P \in \mathcal{H}_n^+(\mathbb{K})$, $P \succ 0$ solves*

$$PA + A^*P + Q - 2\beta P = 0. \tag{30}$$

Then

$$\|e^{At}\| \leq [\sigma_{\max}(P)/\sigma_{\min}(P)]^{1/2}\, e^{\beta t}, \quad t \geq 0. \tag{31}$$

In particular, for every matrix A with spectral abscissa $\alpha(A) < \beta \leq 0$ there exists an inner product $\langle \cdot, \cdot\rangle_P$ on $\mathbb{K}^n$ with respect to which A is strictly dissipative and generates a strict contraction semigroup with initial growth rate $\nu_P(A) < \beta$.

Proof: Setting $A_\beta = A - \beta I_n$ the equation (30) can be rewritten in the form $A_\beta^* P + PA_\beta + Q = 0$. From $A_\beta^* P + PA_\beta \preceq 0$ we obtain that A_β generates a contraction semigroup on $\mathbb{K}^n$ with respect to the norm $\|\cdot\|_P$. Hence $\nu_P(A_\beta) = \nu_P(A) - \beta \leq 0$, i.e. $\nu_P(A) \leq \beta$. Now apply Lemma 5.5.31 to the norm $p(x) = \|x\|_P = \|P^{1/2}x\|_2$. By Remark 5.5.30 we have $\mathrm{ecc}(p) = \sigma_{\max}(P^{1/2})/\sigma_{\min}(P^{1/2}) = (\sigma_{\max}(P)/\sigma_{\min}(P))^{1/2}$. Therefore (31) follows from (27).

The second statement of the proposition follows by choosing $Q \succ 0$. In fact, since $\alpha(A) < \beta \leq 0$ we have $\sigma(A_\beta) \subset \mathbb{C}_-$ and so we know from Theorem 3.3.49 that there exists a unique solution $P = P_\beta(Q) \succ 0$ of (30). Since $A_\beta^* P + PA_\beta = -Q \prec 0$ we conclude that A_β (and hence A) generates a strict contraction semigroup on $\mathbb{K}^n$ with respect to the norm $\|\cdot\|_P$ and $\nu_P(A_\beta) = \nu_P(A) - \beta < 0$. □

Given a solution $P \succ 0$ of $PA + A^*P \preceq 0$, one is interested in minimizing $\beta \in \mathbb{R}$ such that (31) holds. In the following corollary we determine the smallest $\beta \leq 0$ such that $\|\cdot\|_P$ is a β-Liapunov norm for A.

Corollary 5.5.34. *If $P \succ 0$ solves $PA + A^*P = -Q \preceq 0$ then* (31) *holds with*

$$\beta := -\max\{\alpha \in \mathbb{R}_+; Q - 2\alpha P \succeq 0\} = -\min_{x \neq 0}\langle Qx, x\rangle / \langle 2Px, x\rangle = \nu_P(A) \qquad (32)$$

where $\nu_P(A)$ denotes the initial growth rate of A with respect to the norm $\|\cdot\|_P$.

Proof: The second equality is easily verified. Since

$$-\langle Qx, x\rangle / \langle 2Px, x\rangle = \langle (PA + A^*P)x, x\rangle / \langle 2Px, x\rangle = \operatorname{Re}\langle PAx, x\rangle / \langle Px, x\rangle$$

for $x \in \mathbb{K}^n, x \neq 0$, the third equality follows from (23). If β is defined by (32) then

$$PA + A^*P - 2\beta P = -(Q + 2\beta P) \preceq 0.$$

Thus (31) follows from the previous proposition. □

On the other hand, given any $\beta > \alpha(A)$ one may be interested in finding a solution $P \succ 0$ of $PA + A^*P \preceq 0$ such that $\kappa(P) = \sigma_{\max}(P)/\sigma_{\min}(P) = \lambda_{\max}(P)/\lambda_{\min}(P)$ is minimized. This leads to the following problem (where we assume without loss of generality that $\beta = 0$).

Problem 5.5.35. *Given a matrix $A \in \mathbb{C}^{n \times n}$ with $\sigma(A) \subset \mathbb{C}_-$, determine $Q \in \mathcal{H}_n^+(\mathbb{K})$ such that for the unique solution $P = P(Q)$ of the Liapunov equation*

$$PA + A^*P + Q = 0 \qquad (33)$$

the condition number $\kappa(P) = \sigma_{\max}(P)/\sigma_{\min}(P)$ is minimized.

This problem appears to be still an open one (see *Notes and References*) and we will only derive an existence result. We need the following

Lemma 5.5.36. *Suppose $P_1, P_2 \in \mathcal{H}_n(\mathbb{K})$, $P_1 \succ 0$, $P_2 \succ 0$. Then*

$$\kappa(P_2) < \kappa(P_1) \quad \Longrightarrow \quad \kappa(P_2 + P_1) < \kappa(P_1). \qquad (34)$$

Proof: Since $\sigma_{\max}(P_1 + P_2) \leq \sigma_{\max}(P_1) + \sigma_{\max}(P_2)$ and $\sigma_{\min}(P_1 + P_2) \geq \sigma_{\min}(P_1) + \sigma_{\min}(P_2)$, we have

$$\kappa(P_2 + P_1) = \frac{\sigma_{\max}(P_2 + P_1)}{\sigma_{\min}(P_2 + P_1)} \leq \frac{\sigma_{\max}(P_2) + \sigma_{\max}(P_1)}{\sigma_{\min}(P_2) + \sigma_{\min}(P_1)}.$$

But because $\kappa(P_2) < \kappa(P_1)$, we obtain $\sigma_{\max}(P_2)\sigma_{\min}(P_1) < \sigma_{\max}(P_1)\sigma_{\min}(P_2)$ and thus

$$(\sigma_{\max}(P_2) + \sigma_{\max}(P_1))\sigma_{\min}(P_1) < (\sigma_{\min}(P_2) + \sigma_{\min}(P_1))\sigma_{\max}(P_1).$$

Hence

$$\kappa(P_2 + P_1) \leq \frac{\sigma_{\max}(P_2) + \sigma_{\max}(P_1)}{\sigma_{\min}(P_2) + \sigma_{\min}(P_1)} < \frac{\sigma_{\max}(P_1)}{\sigma_{\min}(P_1)} = \kappa(P_1).$$

□

The following proposition shows that an optimal Q always exists and it is necessarily singular if A is not dissipative.

Proposition 5.5.37. *Let $A \in \mathbb{C}^{n\times n}$ be such that $\sigma(A) \subset \mathbb{C}_-$. Then there exists $\hat{Q} \in \mathcal{H}_n^+(\mathbb{K})$ with $(A,\hat{Q})$ observable such that the corresponding solution $\hat{P}$ of (33) has a minimal condition number $\hat{\kappa} = \kappa(\hat{P})$ amongst all $P \in \mathcal{H}_n^+(\mathbb{K})$, $P \neq 0$ satisfying $A^*P + PA \preceq 0$. Moreover if $\hat{\kappa} > 1$, then* $\operatorname{rank} \hat{Q} < n$.

Proof: Let $P(Q)$ be the solution of (33) for any given $Q \succeq 0$. Then since $\sigma(A) \subset \mathbb{C}_-$, we have by (3.3.89a)

$$P(Q) = \int_0^\infty e^{A^*t}Qe^{At}dt \succeq 0, \quad \text{and} \quad \|P(Q)\| \le \|Q\| \int_0^\infty \|e^{At}\|^2 dt < \infty.$$

Since $\kappa(\alpha P) = \kappa(P)$ for all $\alpha > 0$ we may restrict Q to the compact set

$$\mathcal{Q} = \{Q \in \mathcal{H}_n(\mathbb{K})\,;\, Q \succeq 0, \|Q\| = 1\}.$$

By definition $\kappa(P(Q)) = \infty$ if $P(Q) \neq 0$ is singular. Let $(Q_j)_{j\in\mathbb{N}}$ be a minimizing sequence in $\mathcal{Q}$ such that $\kappa(P(Q_{j+1})) \le \kappa(P(Q_j)) < \infty$, $j \in \mathbb{N}$ and

$$\lim_{j\to\infty} \kappa(P(Q_j)) = \inf\{\kappa(P(Q)); Q \in \mathcal{Q}\} =: \hat{\kappa}.$$

By compactness of $\mathcal{Q}$ we may assume that $(Q_j)_{j\in\mathbb{N}}$ is convergent. Now since $Q \mapsto P(Q)$ is continuous on $\mathcal{Q}$, $\min\{\sigma_{\max}(P(Q)); Q \in \mathcal{Q}\}$ exists and is positive. Hence it follows from the boundedness of $(\kappa(P(Q_j)))_{j\in\mathbb{N}}$ that there exists an $\varepsilon > 0$ such that $\sigma_{\min}(P(Q_j)) \ge \varepsilon$ for all $j \in \mathbb{N}$. As a consequence the limit $\hat{Q} = \lim_{j\to\infty} Q_j$ is a minimum of $Q \to \kappa(P(Q))$ on $\mathcal{Q}$:

$$\kappa(P(\hat{Q})) = \sigma_{\max}(\lim_{j\to\infty} P(Q_j))/\sigma_{\min}(\lim_{j\to\infty} P(Q_j)) = \lim_{j\to\infty} \kappa(P(Q_j)) = \hat{\kappa}.$$

Moreover $(A,\hat{Q})$ is observable since $P(\hat{Q}) \succ 0$, see Theorem 3.3.49. This proves the existence result. Finally assume that $\hat{\kappa} > 1$, but $\hat{Q} \succ 0$. Choose $\tau > 0$ such that $Q_\tau := \hat{Q} - \tau(A + A^*) \succeq 0$. Then $P(Q_\tau) = P(\hat{Q}) + \tau I_n$. But $\kappa(\tau I_n) = 1 < \hat{\kappa} = \kappa(P(\hat{Q}))$ and this implies by Lemma 5.5.36 that $\kappa(P(Q_\tau)) = \kappa(P(\hat{Q}) + \tau I_n) < \kappa(P(\hat{Q})) = \hat{\kappa}$. Therefore $\hat{Q}$ must be singular if $\hat{\kappa} > 1$. □

We illustrate this result and the estimate (31) by the following simple example.

Example 5.5.38. We consider the same A as in Example 5.5.6 with $a = -0.6$, $b = -1$, $c = 1.2$ and $\beta = -0.2$. Then $A_\beta = A - \beta I_2 = \begin{bmatrix} -0.4 & 1.2 \\ 0 & -0.8 \end{bmatrix}$ and $(A_\beta + A_\beta^*)/2$ is not negative semidefinite since $0.4 \times 0.8 < 0.6^2$. By Corollary 5.5.26 A_β is not dissipative (with respect to the standard inner product on $\mathbb{R}^2$). Let $Q(\alpha) = \operatorname{diag}(1,\alpha)$ and $P(Q(\alpha)) = \begin{bmatrix} p_1 & p_2 \\ p_2 & p_3 \end{bmatrix}$ solve (33), then $p_1 = p_2 = 5/4$ and $p_3 = 15/8 + 5\alpha/8$. Hence

$$16\,\sigma_{\max}(P_{Q(\alpha)}) = 5[5+\alpha+\sqrt{(1+\alpha)^2+16}], \quad 16\,\sigma_{\min}(P(Q(\alpha))) = 5[5+\alpha-\sqrt{(1+\alpha)^2+16}]$$

and

$$\kappa(P(Q(\alpha))) = \sigma_{\max}(P_{Q(\alpha)})/\sigma_{\min}(P_{Q(\alpha)}) = [5+\alpha+\sqrt{(1+\alpha)^2+16}]/[5+\alpha-\sqrt{(1+\alpha)^2+16}].$$

$\tilde{\alpha} = 3$ minimizes the above expression yielding $\kappa(P(Q(\tilde{\alpha}))) = 3+2\sqrt{2}$. For this value of α we have $Q(\tilde{\alpha}) = \operatorname{diag}(1,3)$, $P(Q(\tilde{\alpha}))) = \begin{bmatrix} 5/4 & 5/4 \\ 5/4 & 15/4 \end{bmatrix}$ and by (31) we obtain the estimate $\|e^{At}\| \leq [3+2\sqrt{2}]^{1/2}e^{-0.2t} = (1+\sqrt{2})e^{-0.2t} = 2.14e^{-0.2t}$.
Since $\tilde{Q} := Q(\tilde{\alpha}) \succ 0$ and $\kappa(P(\tilde{Q})) > 1$, we may improve this estimate via the construction carried out at the end of the proof of Proposition 5.5.37. Now

$$\tilde{Q} - \tau(A + A^*) = \begin{bmatrix} 1 & 0 \\ 0 & 3 \end{bmatrix} - \tau \begin{bmatrix} -0.8 & 1.2 \\ 1.2 & -1.6 \end{bmatrix} = \begin{bmatrix} 1+0.8\tau & -1.2\tau \\ -1.2\tau & 3+1.6\tau \end{bmatrix}.$$

A short calculation shows that $\tilde{Q} - \tau(A + A^*) \prec 0$ for $\tau < 25.73$ and is singular for $\tau = 25.73$. Now $\kappa(P_{\tilde{Q}} + 25.73I_2) = 1.134$. This is a considerable improvement and yields the estimate $\|e^{At}\| \leq 1.065e^{-0.2t}$. It is a good approximation of the best estimate $\|e^{At}\| \leq 1.05728e^{-0.2t}$ which can be obtained by optimizing Q. □

It is known (see *Notes and References*) that even the *optimal* bound obtainable for $M_\beta(A)$, $\beta > \alpha(A)$ via *quadratic* norms may be quite conservative. We conclude this subsection with a result which shows that tight estimates can be obtained by utilizing solutions of *differential* Liapunov equations.

Proposition 5.5.39. *Suppose $\mathbb{K}^{n\times n}$ is provided with the spectral norm, $A \in \mathbb{K}^{n\times n}$, $\beta > \alpha(A)$ and $A_\beta = A - \beta I_n$. If $P_\beta(\cdot) : \mathbb{R}_+ \to \mathcal{H}_n(\mathbb{K})$ is any continuously differentiable matrix function satisfying*

$$\dot{P}_\beta(t) - A_\beta^* P_\beta(t) - P_\beta(t)A_\beta \succeq 0, \quad t > 0 \quad \text{and} \quad P_\beta(0) \succ 0, \tag{35}$$

then

$$M_\beta(A)^2 \leq \sup_{t\geq 0} \sigma_{\max}(P_\beta(t))/\sigma_{\min}(P_\beta(0)). \tag{36}$$

Moreover, if the inequality $\succeq$ in (35) is replaced by an equation and $P_\beta(\cdot)$ is the solution of this differential equation with $P_\beta(0) = I_n$, then $M_\beta(A)^2 = \sup_{t\geq 0} \sigma_{\max}(P_\beta(t))$.

Proof: Suppose $P_\beta(\cdot)$ solves (35) and let

$$Q(t) = \dot{P}_\beta(t) - A_\beta^* P_\beta(t) - P_\beta(t)A_\beta, \quad t > 0$$

then $Q(t) \in \mathcal{H}_n^+(\mathbb{K})$ for $t \geq 0$ and

$$\begin{aligned} \frac{d}{ds}\left[e^{-A_\beta^* s}P_\beta(s)e^{-A_\beta s}\right] &= e^{-A_\beta^* s}\left[-A_\beta^* P_\beta(s) - P_\beta(s)A_\beta + A_\beta^* P_\beta(s) + P_\beta(s)A_\beta + Q(s)\right]e^{-A_\beta s} \\ &= e^{-A_\beta^* s}Q(s)e^{-A_\beta s}, \quad s \geq 0. \end{aligned}$$

Integrating from 0 to t, yields

$$e^{-A_\beta^* t}P_\beta(t)e^{-A_\beta t} - P_\beta(0) = \int_0^t e^{-A_\beta^* s}Q(s)e^{-A_\beta s}ds.$$

Hence

$$P_\beta(t) = e^{A_\beta^* t} P_\beta(0) e^{A_\beta t} + \int_0^t e^{A_\beta^*(t-s)} Q(s) e^{A_\beta(t-s)} ds. \tag{37}$$

So if $P_\beta(\cdot)$ satisfies (35), then $P_\beta(t) \succeq e^{A_\beta^* t} P_\beta(0) e^{A_\beta t} \succeq \sigma_{\min}(P_\beta(0)) e^{A_\beta^* t} e^{A_\beta t}$. Hence (36) follows from

$$\sigma_{\max}(P_\beta(t)) \geq \|e^{A_\beta^* t} P_\beta(0) e^{A_\beta t}\| \geq \sigma_{\min}(P_\beta(0)) \|e^{A_\beta t}\|^2 = \sigma_{\min}(P_\beta(0)) e^{-2\beta t} \|e^{At}\|^2.$$

For $P_\beta(0) = I_n$, the solution of the equation in (35) is $P_\beta(t) = e^{A_\beta^* t} e^{A_\beta t}$ and so

$$\sup_{t\geq 0} \sigma_{\max}(P_\beta(t)) = \sup_{t \geq 0} \|e^{A_\beta t}\| = M_\beta(A)^2. \qquad \square$$

If $\beta > \alpha(A)$ then by (37) $\sigma_{\max}(P_\beta(t))$ will be uniformly bounded for $t \geq 0$ and the smallest bound is obtained for $Q = 0$. This suggests that we should have restricted our considerations to the equality in (35). We have chosen not to do so because this precludes the possibility of constant solutions. In its present form the proposition is applicable to constant solutions so that (31) can be viewed as a special case of (36).

5.5.3 Spectral Value Sets and Transient Behaviour

If $A \in \mathbb{K}^{n\times n}$ is an asymptotically stable matrix ($\alpha(A) < 0$), it is plausible to conjecture that systems with large transient motions are somehow close to instability. By this we mean that some small perturbation Δ of A will move some of the eigenvalues of A from the open left to the closed right half-plane. Information about this possibility is contained in the unstructured stability radius $d_{\mathbb{C}}^-(A)$ and the pseudospectra $\sigma_{\mathbb{C}}(A;\delta), \delta > 0$. Therefore we expect a close relationship between pseudospectra, stability radius and transient behaviour which we will now investigate.
The following proposition shows that there is a relationship between the contraction property $\nu(A) < 0$, the pseudospectra of A and its distance from normality, $\operatorname{dist}(A, \mathcal{N}_n(\mathbb{K}))$. For every $\gamma \in \mathbb{R}$, we denote by $\mathbb{C}_\gamma$ the open left-half plane $\{s \in \mathbb{C}; \operatorname{Re} s < \gamma\}$.

Proposition 5.5.40. *Suppose $A \in \mathbb{C}^{n\times n}$, $\mathbb{C}^{n\times n}$ carries the spectral norm $\|\cdot\|$, and*

$$\overline{\sigma_{\mathbb{C}}(A;\delta)} \subset \mathbb{C}_{-\delta} \tag{38}$$

for some given $\delta > 0$. Then $\nu(A) < 0$ if $\operatorname{dist}(A, \mathcal{N}_n(\mathbb{C})) \leq \delta$.

Proof: Let $X \in \mathcal{N}_n(\mathbb{C})$ be such that $\|A - X\| \leq \delta$. Then by (38) we have $\alpha(X) < -\delta$ and hence by (17), $\nu(A) \leq \alpha(X) + \|A - X\| < 0$. $\square$

Remark 5.5.41. The largest $\hat{\delta}$ such that the inclusion (38) holds for all $\delta < \hat{\delta}$ satisfies $\hat{\delta} = d_{\mathbb{C}}^-(A + \hat{\delta} I_n)$ and can be determined from this equation (see Ex. 16). $\square$

We will now apply Lemma 5.5.22 to obtain estimates for $M_0(A)$. The following theorem is a "continuous time version" of a theorem due to Kreiss and improved by Spijker, see *Notes and References.* After the proof we will show how the result can be reformulated in terms of stability radii and spectral value sets.

Theorem 5.5.42 (Kreiss-Spijker). *Let $\|\cdot\|$ be any operator norm on $\mathbb{K}^{n\times n}$ and $A \in \mathbb{K}^{n\times n}$ be such that*

$$R(A) := \sup_{\operatorname{Re} s>0} \operatorname{Re} s \, \|(sI_n - A)^{-1}\| < \infty. \tag{39}$$

Then $\dot{x} = Ax$ is stable and

$$R(A) \le M_0(A) = \sup_{t\ge 0} \|e^{At}\| \le e \cdot n \, R(A). \tag{40}$$

Proof: Since $R(A) < \infty$ we have $\sigma(A) \subset \overline{\mathbb{C}_-}$. So by (21)

$$(sI_n - A)^{-1} = \int_0^\infty e^{-st} e^{At} dt, \quad \operatorname{Re} s > 0,$$

and we obtain the first inequality in (40) by taking norms

$$\|(sI_n - A)^{-1}\| \le \sup_{t\ge 0} \|e^{At}\| \int_0^\infty e^{-t \operatorname{Re} s} \, dt = M_0(A) \, (\operatorname{Re} s)^{-1}, \quad \operatorname{Re} s > 0. \tag{41}$$

Given any $\gamma > 0$ and $r > \gamma$, let $\Gamma(r, \gamma)$ denote the Bromwich contour which consists

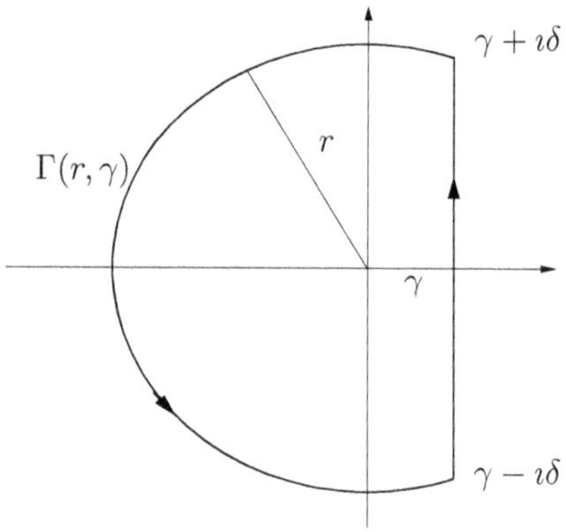

Figure 5.5.4: Bromwich contour

of a straight line from $\gamma - \imath\delta$ to $\gamma + \imath\delta$ where $\delta = \sqrt{r^2 - \gamma^2}$, and the positively oriented arc $\Gamma(r, \gamma)$ connecting $\gamma + \imath\delta$ with $\gamma - \imath\delta$ along the circle with radius r around the origin, see Figure 5.5.4. Now consider any $r > 0$ sufficiently large so that all the eigenvalues of A lie inside the contour (this is possible since $\sigma(A) \subset \overline{\mathbb{C}_-}$). Given any $t > 0$ choose $y \in \mathbb{K}^n$, $\|y^*\|^*_{\mathbb{K}^n} = 1$, $x \in \mathbb{K}^n$, $\|x\|_{\mathbb{K}^n} = 1$ such that $|y^* e^{At} x| = \|e^{At}\|$, then from (21)

$$y^* e^{At} x = \frac{1}{2\pi\imath} \int_{\Gamma(r,\gamma)} e^{st} g(s) ds + \frac{1}{2\pi\imath} \int_{\gamma-\imath\delta}^{\gamma+\imath\delta} e^{st} g(s) ds,$$

where $g(s) = y^*(sI_n - A)^{-1} x$. By partial integration we obtain for all r sufficiently large

$$y^* e^{At} x = -\frac{1}{2\pi\imath t} \int_{\Gamma(r,\gamma)} e^{st} g'(s) ds - \frac{1}{2\pi\imath t} \int_{\gamma-\imath\delta}^{\gamma+\imath\delta} e^{st} g'(s) ds,$$

where $g' = dg/ds$. Since g is strictly proper rational there exists a constant $M > 0$ such that $|g'(s)| \le M/r^2$ on $\Gamma(r,\gamma)$ for all sufficiently large $r \gg 0$, and so

$$\left|\int_{\Gamma(r,\gamma)} e^{st} g'(s)\, ds\right| \le \frac{M}{r^2}\int_{\Gamma(r,\gamma)} e^{(\mathrm{Re}\, s)\, t}\, |ds| \le \frac{M}{r^2} e^{\gamma t} 2\pi r \to 0 \quad \text{as } r \to \infty$$

($\gamma > 0, t > 0$ arbitrary but fixed). It follows that for all $t, \gamma > 0$

$$y^* e^{At} x = -\frac{1}{2\pi \imath t} \lim_{r\to\infty} \left[\int_{\Gamma(r,\gamma)} e^{st} g'(s) ds + \int_{\gamma-\imath\delta}^{\gamma+\imath\delta} e^{st} g'(s) ds\right] = -\frac{1}{2\pi\imath t}\int_{\gamma-\imath\infty}^{\gamma+\imath\infty} e^{st} g'(s) ds.$$

Thus, setting $\gamma = t^{-1}$, we obtain

$$\|e^{At}\| = |y^* e^{At} x| \le \frac{1}{2\pi t}\int_{\gamma-\imath\infty}^{\gamma+\imath\infty} e^{\gamma t} |g'(s)|\, |ds| = \frac{e}{2\pi t}\int_{-\infty}^{\infty} |g'(t^{-1} + \imath\omega)|\, d\omega, \quad t > 0.$$

Now apply the continuous time version of Spijker's Lemma A.2.21 to the strictly proper rational function $g(s)$ and note that the denominator of $g(s)$ is of degree $\le n$, then

$$\int_{t^{-1}-\imath\infty}^{t^{-1}+\imath\infty} |g'(s)|\, |ds| \le 2\pi n \sup_{\omega\in\mathbb{R}} |g(t^{-1} + \imath\omega)| \le 2\pi n \sup_{\omega\in\mathbb{R}} \|((t^{-1} + \imath\omega) I_n - A)^{-1}\|, \quad t > 0.$$

Therefore

$$\sup_{t>0} \|e^{At}\| \le \sup_{t>0} \frac{e\cdot n}{t} \sup_{\omega\in\mathbb{R}} \|((t^{-1} + \imath\omega) I_n - A)^{-1}\| = e\cdot n \sup_{\mathrm{Re}\, s>0} \mathrm{Re}\, s\, \|(sI_n - A)^{-1}\|.$$

Since $\sup_{t>0} \|e^{At}\| < \infty$ implies stability of $\dot{x} = Ax$, this completes the proof. □

If $\dot{x} = Ax$ is stable (but not necessarily asymptotically stable), then $M_0(A) < \infty$ by Proposition 3.3.1 and it follows from (41) that $R(A) < \infty$. On the other hand, we always have $R(A) \ge 1$ since $\alpha\|(\alpha I_n - A)^{-1}\| = \|(I_n - \alpha^{-1}A)^{-1}\| \to 1$ as $\alpha \to \infty$. The above theorem establishes bounds for $M_0(A)$. While the upper and the lower bounds clearly depend on the matrix $A \in \mathbb{K}^{n\times n}$, their quotient ($= e \cdot n$) is independent of A. Note that the factor grows linearly with the system's dimension and it has been shown that the factor $e\cdot n$ in (40) cannot be reduced, if the inequality is required to hold for all stable matrices A of arbitrary order, see *Notes and References.* As a direct consequence of the Kreiss-Spijker Theorem we obtain estimates for all $M_\beta(A)$, $\beta > \alpha(A)$.

Corollary 5.5.43. *Let $\|\cdot\|$ be any operator norm on $\mathbb{K}^{n\times n}$, $A \in \mathbb{K}^{n\times n}$ and $\beta > \alpha(A)$. Then*

$$R_\beta(A) \le M_\beta(A) \le e\cdot n\, R_\beta(A). \tag{42}$$

where $R_\beta(A) = \sup_{\mathrm{Re}\, s>\beta} (\mathrm{Re}\, s - \beta)\, \|(sI_n - A)^{-1}\| < \infty$.

Proof: Since $\beta > \alpha(A)$ we have $\sigma(A_\beta) \subset \mathbb{C}_-$ and hence $R(A_\beta) < \infty$ for $A_\beta = A - \beta I_n$. Now

$$R(A_\beta) = \sup_{\mathrm{Re}\, s>0} \mathrm{Re}\, s\, \|((s+\beta)I_n - A)^{-1}\| = \sup_{\mathrm{Re}\, s>\beta} (\mathrm{Re}\, s - \beta)\, \|(sI_n - A)^{-1}\| = R_\beta(A).$$

So (42) follows by applying Theorem 5.5.42 with A replaced by A_β and noting that by (6), $M_0(A_\beta) = M_\beta(A)$. □

Remark 5.5.44. The Hille-Yosida Theorem states that a necessary and sufficient condition for a closed linear operator A with dense domain in a Banach space X to generate a strongly continuous semigroup is that there exist real numbers M, β such that for all s with $\operatorname{Re} s > \beta$, we have $s \in \rho(A)$ and

$$\|(sI - A)^{-m}\| \leq \frac{M}{(\operatorname{Re} s - \beta)^m}, \quad m \in \mathbb{N}^*.$$

Moreover this condition is equivalent to $\|e^{At}\| \leq Me^{\beta t}$, $t \geq 0$.
On the other hand if $\|(sI - A)^{-1}\| \leq M/(\operatorname{Re} s - \beta)$ holds for all $\operatorname{Re} s > \beta$, i.e. $R_\beta(A) \leq M$, then the previous corollary only yields that $\|e^{At}\| \leq e{\cdot}nMe^{\beta t}$, $t \geq 0$. So by restricting to a single power of the resolvent one obtains an inferior estimate which cannot be generalized to infinite dimensions. □

The following example illustrates that, with respect to the spectral norm, the lower bound in (40) is achieved by all normal matrices.

Example 5.5.45. Suppose $A \in \mathbb{K}^{n \times n}$ is normal, then we know from Proposition 5.5.12 that with respect to the spectral norm $M_\beta(A) = 1$ for all $\beta \geq \alpha(A)$. Now by (3.64a)

$$R_\beta(A) = \sup_{\alpha > \beta} \sup_{\omega \in \mathbb{R}} (\alpha - \beta) \, \|((\alpha + \imath\omega)I_n - A)^{-1}\| = \sup_{\alpha > \beta} (\alpha - \beta)/d_{\mathbb{C}}^-(A_\alpha)$$

where $d_{\mathbb{C}}^-(A_\alpha)$ denotes the distance of $A_\alpha = A - \alpha I_n$ from instability. But Proposition 5.3.38 implies that $d_{\mathbb{C}}^-(A_\alpha) = -\alpha(A_\alpha) = \alpha - \alpha(A)$ for $\alpha > \alpha(A)$. Hence for all $\beta \geq \alpha(A)$

$$\sup_{\alpha > \beta}(\alpha - \beta)/d_{\mathbb{C}}^-(A_\alpha) = \sup_{\alpha > \beta}(\alpha - \beta)/(\alpha - \alpha(A)) = \lim_{\alpha \to \infty} (\alpha - \beta)/(\alpha - \alpha(A)) = 1.$$

So (42) takes the form $R_\beta(A) = 1 = M_\beta(A) \leq e{\cdot}n$. □

In some applications we may be only interested in the transient behaviour of the system in certain directions, e.g. if a perturbation is known to affect only certain coordinates of the state vector. This can be taken into account by introducing structure matrices $(B, C) \in \mathbb{K}^{n \times \ell} \times \mathbb{K}^{q \times n}$ and considering $Ce^{At}B$. If $\dot{x} = Ax$ is stable the proof of Theorem 5.5.42 can easily be extended to yield

$$R(A; B, C) \leq \sup_{t \geq 0} \|Ce^{At}B\| \leq e{\cdot}n \, R(A; B, C), \tag{43}$$

where $R(A; B, C) = \sup_{\operatorname{Re} s > 0} \operatorname{Re} s \, \|C(sI_n - A)^{-1}B\|$. This formula for $R(A; B, C)$ is reminiscent of the formula for the complex stability radius. Note that $\sigma(A_\alpha) \subset \mathbb{C}_-$ for $\alpha > 0$, since $\dot{x} = Ax$ is assumed to be stable. Making use of (3.22) $R(A; B, C)$ can be expressed in terms of this radius as follows

$$R(A; B, C) = \sup_{\alpha > 0} \alpha \sup_{\omega \in \mathbb{R}} \|C((\alpha + \imath\omega)I_n - A)^{-1}B\| = \sup_{\alpha > 0} \alpha / r_{\mathbb{C}}^-(A_\alpha; B, C),$$

so that for every stable system $\dot{x} = Ax$

$$\sup_{\alpha > 0} \alpha / r_{\mathbb{C}}^-(A_\alpha; B, C) \;\leq\; \sup_{t \geq 0} \|Ce^{At}B\| \;\leq\; e{\cdot}n \sup_{\alpha > 0} \alpha / r_{\mathbb{C}}^-(A_\alpha; B, C). \tag{44}$$

Thus, if for some $\alpha > 0$ the stability radius of (A_α, B, C) is small compared with α then we can expect large values of $\|Ce^{At}B\|$ for some $t > 0$.

More generally, if $A \in \mathbb{K}^{n\times n}$ is arbitrary and we set $R_\beta(A;B,C) := R(A_\beta;B,C)$ for any $\beta > \alpha(A)$, we have

$$R_\beta(A;B,C) = \sup_{\mathrm{Re}\, s>\beta} (\mathrm{Re}\, s - \beta)\, \|C(sI_n - A)^{-1}B\| = \sup_{\alpha>\beta} (\alpha-\beta)/r_{\mathbb{C}}^{-}(A_\alpha;B,C),$$

and so

$$\sup_{\alpha>\beta} (\alpha-\beta)/r_{\mathbb{C}}^{-}(A_\alpha;B,C) \le \sup_{t\ge 0} \|Ce^{(A-\beta I_n)t}B\| \le e{\cdot}n \sup_{\alpha>\beta} (\alpha-\beta)/r_{\mathbb{C}}^{-}(A_\alpha;B,C). \quad (45)$$

We will now interpret the lower bound $R(A;B,C)$ in terms of spectral value sets. For this we introduce the following

Definition 5.5.46. Given $(A,B,C) \in \mathbb{K}^{n\times n} \times \mathbb{K}^{n\times \ell} \times \mathbb{K}^{q\times n}$, the *$\delta$-spectral abscissa* of A under perturbations of the form $A \rightsquigarrow A(\Delta) = A + B\Delta C$ is given by

$$\alpha_\delta(A;B,C) = \sup\{\mathrm{Re}\, s\,;\; s \in \sigma_{\mathbb{C}}(A;B,C;\delta)\}.$$

In the unstructured case ($B=C=I_n$) the δ-spectral abscissa is denoted by $\alpha_\delta(A)$.

By Proposition 5.2.19 we have for $\delta > 0$ and $s \in \rho(A)$

$$\|C(sI_n - A)^{-1}B\| = \delta^{-1} \quad \Leftrightarrow \quad s \in \partial\sigma_{\mathbb{C}}(A;B,C;\delta).$$

Hence, setting $G(s) = C(sI_n - A)^{-1}B$ and assuming $G(s) \not\equiv 0$, $\sigma(A) \subset \overline{\mathbb{C}_-}$ we get

$$R(A;B,C) = \sup_{\mathrm{Re}\, s>0} \mathrm{Re}\, s\, \|G(s)\| = \sup_{\delta>0} \sup_{\substack{\mathrm{Re}\, s>0\\ \|G(s)\|=\delta^{-1}}} \mathrm{Re}\, s\, \|G(s)\| = \sup_{\delta>0} \alpha_\delta(A;B,C)/\delta,$$

where $\sup\emptyset := 0$. Here we have used that for every $\delta > 0$ satisfying $\alpha_\delta(A;B,C) > 0$

$$\alpha_\delta(A;B,C) = \sup\{\mathrm{Re}\, s;\; s \in \partial\sigma_{\mathbb{C}}(A;B,C;\delta)\} = \sup\{\mathrm{Re}\, s;\; s \in \mathbb{C}_+,\; \|G(s)\| = \delta^{-1}\}$$

by (2.27). We conclude that

$$\sup_{\delta>0} \alpha_\delta(A;B,C)/\delta \;\le\; \sup_{t\ge 0} \|Ce^{At}B\| \;\le\; e{\cdot}n \sup_{\delta>0} \alpha_\delta(A;B,C)/\delta. \quad (46)$$

More generally, since $\alpha_\delta(A_\beta;B,C) = \alpha_\delta(A;B,C) - \beta$ we have for $\beta > \alpha(A)$

$$\sup_{\delta>0} (\alpha_\delta(A;B,C) - \beta)/\delta \;\le\; \sup_{t\ge 0} \|Ce^{A_\beta t}B\| \;\le\; e{\cdot}n \sup_{\delta>0} (\alpha_\delta(A;B,C) - \beta)/\delta. \quad (47)$$

In particular, we obtain for the unstructured case if $\beta > \alpha(A)$

$$\sup_{\delta>0} (\alpha_\delta(A) - \beta)/\delta \;\le\; M_\beta(A) = \sup_{t\ge 0} \|e^{A_\beta t}\| \;\le\; e{\cdot}n \sup_{\delta>0} (\alpha_\delta(A) - \beta)/\delta. \quad (48)$$

This formula, together with (45) gives a precise meaning to the intuitive reasoning at the beginning of this subsection: If A is Hurwitz stable and if for small δ the spectral value set $\sigma_{\mathbb{C}}(A;\delta)$ moves deep into the right half plane (see Figure 5.5.5) then since $\sup_{t\ge 0} \|e^{At}\| \ge \sup_{\delta>0} \alpha_\delta(A)/\delta \gg 0$ some trajectories of the system $\dot{x} = Ax$ will make large transient excursions. This means that small initial deviations of the system's state from the equilibrium $\bar{x} = 0$ will be largely amplified during transient motion.

Example 5.5.47. Consider the following matrix A in real Schur form with spectrum $\sigma(A) = \{-1 \pm 10\imath, -1 \pm 20\imath, -1, -1 \pm 25\imath\}$.

$$A = \begin{bmatrix} -1 & -100 & 0 & -150 & 0 & 200 & -1000 \\ 1 & -1 & 1 & -10 & 25 & 11 & -200 \\ 0 & 0 & -1 & 400 & -30 & 0 & 250 \\ 0 & 0 & -1 & -1 & 5 & 5 & 200 \\ 0 & 0 & 0 & 0 & -1 & -2 & 30 \\ 0 & 0 & 0 & 0 & 0 & -1 & -625 \\ 0 & 0 & 0 & 0 & 0 & 1 & -1 \end{bmatrix}.$$

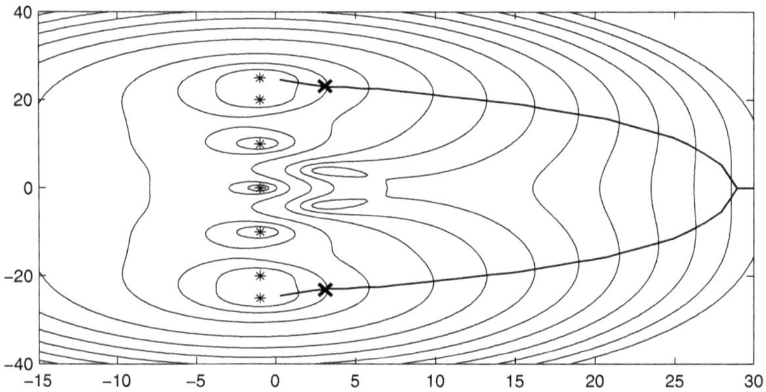

Figure 5.5.5: Spectral contours and front locus

In Figure 5.5.5 we have plotted the spectral contours $\mathcal{C}_\delta = \{s \in \rho(A); \|(sI - A)^{-1}\| = \delta^{-1}\}$ for $\delta = 0.01, 0.02, 0.04, 0.08, 0.12, 0.16, ..., 0.4$. The spectral value sets move quickly into the right half plane for $\delta > d_{\mathbb{C}}^{-}(A) = 0.004$, so one would expect large transient excursions. Denote by $S(\delta)$ the set of points on the contour $\mathcal{C}_\delta$ with largest real part. The real parts of these points are equal to the δ-pseudospectral abscissae of A. The set

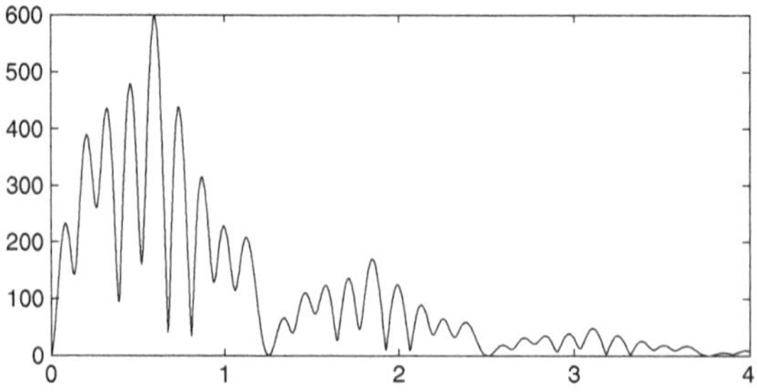

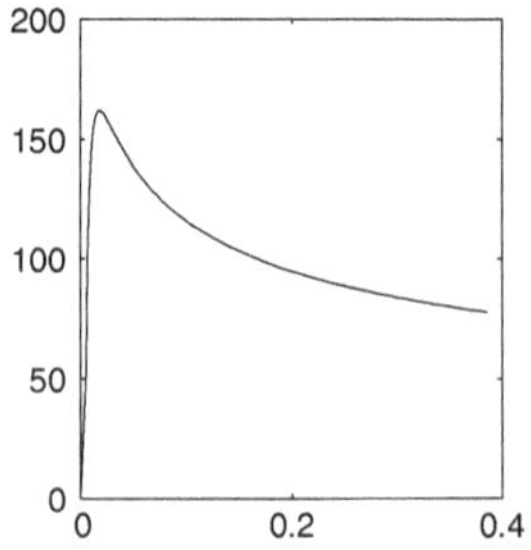

Figure 5.5.6: Graph of $t \mapsto \|e^{At}\|$ and graph of $\delta \mapsto \alpha_\delta(A)/\delta$

$\mathcal{F} = \cup_{\delta>0} S(\delta)$ will be called the *front locus* of A. This is plotted with thick lines for $0 < \mathrm{Re}\, s < 30$ in Figure 5.5.5. We see for small δ that $S(\delta)$ consists of two points which follow with increasing $\delta > 0$ parabolic-like paths until they become united for $\delta \approx 0.365$. From this value of δ onwards the front locus lies on the real axis. The quotient $\alpha_\delta(A)/\delta$

takes its maximum value $\alpha_{\hat{\delta}}(A)/\hat{\delta} \approx 162.3$ at $\hat{\delta} = 0.0181$. The corresponding points $\hat{s}_{\pm} = 2.9308 \pm 23.2i$ of the contour $\mathcal{C}_{\hat{\delta}}$ with largest real part are indicated by two crosses in Figure 5.5.5. At these points the function $\operatorname{Re} s \,\|(sI - A)^{-1}\|$ attains its maximum on $\mathbb{C}_+$, i.e. $R(A) = \operatorname{Re} \hat{s}_{\pm}\|(\hat{s}_{\pm}I - A)^{-1}\| \approx 162.3$. This value is to be compared with $M_0(A) = \sup_{t\geq 0} \|\exp(At)\| \approx 598.45$. In Figure 5.5.6 the graph of $t \mapsto \|e^{At}\|$ is shown alongside the graph of $\delta \mapsto \alpha_\delta(A)/\delta$. The Kreiss-Spijker Theorem compares the maxima of these two graphs and tells us that $M_0(A)$ belongs to the interval $[162.3, 3088.3]$ which is not a particularly good estimate in this case. □

5.5.4 Robustness of (M, β)-Stability

In this subsection we examine the robustness of (M, β)-stability under full-block affine perturbations. Throughout the subsection the assumptions are

- $M \geq 1$, $\beta \in \mathbb{R}$ are given.
- $(A, B, C) \in \mathbf{L}_{n,\ell,q}(\mathbb{K})$ and the nominal system $\dot{x} = Ax$ is (M, β)-stable.
- The vector spaces $\mathbb{K}^n$, $\mathbb{K}^\ell$, $\mathbb{K}^q$ are provided with fixed norms (all denoted by $\|\cdot\|$) and the operators between them are provided with the corresponding operator norms (also denoted by $\|\cdot\|$).
- The perturbations of the system matrix A are of the form

$$A \rightsquigarrow A(\Delta) = (A + B\Delta C), \quad \Delta \in \mathbb{K}^{\ell \times q}. \tag{49}$$

Definition 5.5.48. The (M, β)-stability radius of A under perturbations of the form (49) is defined by

$$r_{\mathbb{K}}(A; B, C; M, \beta) = \inf \{\|\Delta\|; \Delta \in \mathbb{K}^{\ell \times q}, \exists t > 0 : \|e^{(A+B\Delta C)t}\| > Me^{\beta t}\}.$$

It follows from this definition and the fact that $\mathcal{S}_n(\mathbb{K}; M, \beta)$ is closed that every system $\dot{x} = A(\Delta)x$ with $\|\Delta\| \leq r_{\mathbb{K}}$ is (M, β)-stable. Proposition 5.5.3 implies that $r_{\mathbb{K}}(A; B, C; M, \beta) > 0$ if A is strictly (M, β)-stable. In the unstructured case ($B = C = I_n$) the (M, β)-stability radius $r_{\mathbb{K}}(A; I_n, I_n; M, \beta)$ is the distance of A from the open set of non-(M, β)-stable systems in $\mathbb{K}^{n \times n}$ and is denoted by $d_{\mathbb{K}}^-(A; M, \beta)$. In the special case where $M = 1$ this distance can be expressed via the initial growth rate.

Proposition 5.5.49. *Suppose that $A \in \mathbb{K}^{n \times n}$ generates a β-contraction semigroup, i.e. $\|e^{At}\| \leq e^{\beta t}$ for all $t \geq 0$. Then the distance of A from the set $\mathcal{X}_\beta \subset \mathbb{K}^{n \times n}$ of $n \times n$-matrices which do not generate a β-contraction semigroup is given by*

$$d_{\mathbb{K}}^-(A; 1, \beta) := \operatorname{dist}(A, \mathcal{X}_\beta) = \beta - \nu(A). \tag{50}$$

Proof: Suppose A generates a β-contraction semigroup and $X \in \mathcal{X}_\beta$. Then $A_\beta := A - \beta I_n$ generates a contraction semigroup whilst $X_\beta := X - \beta I_n$ does not, and so by Lemma 5.5.9 (vii) and (i)

$$0 < \nu(X_\beta) \leq \nu(A_\beta) + \nu(X_\beta - A_\beta) \leq \nu(A_\beta) + \|A - X\|.$$

Hence $d_{\mathbb{K}}^{-}(A;1,\beta) = \operatorname{dist}(A, \mathfrak{X}_\beta) \geq -\nu(A_\beta) = -\nu(A) + \beta$. Now for any $\varepsilon > 0$, we have by Lemma 5.5.9 (iii)

$$\nu(A + (-\nu(A) + \beta + \varepsilon)I_n) = \nu(A) + (-\nu(A) + \beta + \varepsilon) = \beta + \varepsilon.$$

So $A + (-\nu(A) + \beta + \varepsilon)I_n \in \mathfrak{X}_\beta$ and $\|(-\nu(A) + \beta + \varepsilon)I_n\| = -\nu(A) + \beta + \varepsilon$ and this proves $\operatorname{dist}(A, \mathfrak{X}_\beta) \leq -\nu(A) + \beta$. □

Specializing the above result to Euclidean norms and $\beta = 0$, the distance between any generator A of a contraction semigroup from the set of generators of non-contractive semigroups is given by (see Corollary 5.5.26)

$$d_{\mathbb{K}}^{-}(A;1,0) = -\lambda_{\max}(A + A^*)/2. \tag{51}$$

We will now examine robustness aspects of (M,β)-stability by means of Liapunov norms. We have seen in Subsection 5.5.2 that a single system $\dot{x} = Ax$ is (M,β)-stable if and only if there exists a β-Liapunov norm p for $\dot{x} = Ax$ on $\mathbb{K}^n$ such that (29) holds. Our aim is to extend this result to *sets* of systems. A direct generalization does not work: If there exists a *common β-Liapunov norm* satisfying (29) for all A in a given set $\mathcal{A} \subset \mathbb{K}^{n\times n}$, then each system $\dot{x} = Ax$, $A \in \mathcal{A}$ is (M,β)-stable, but the converse is not true, see Ex. 19. Instead of considering each time-invariant linear system $\dot{x} = Ax$, $A \in \mathcal{A}$ separately, we have to consider the *differential inclusion*

$$\dot{x}(t) \in \mathcal{A}x(t), \quad t \geq 0. \tag{52}$$

An absolutely continuous function $\varphi(\cdot) : \mathbb{R}_+ \to \mathbb{K}^n$ is said to be a solution of (52) if $\dot{\varphi}(t) \in \mathcal{A}\varphi(t)$ for almost all $t \geq 0$. In other words for almost all $t \in \mathbb{R}_+$ there exists $A(t) \in \mathcal{A}$ such that $\dot{\varphi}(t) = A(t)\varphi(t)$.[2]

Proposition 5.5.50. *Suppose p is a joint β-Liapunov norm on $\mathbb{K}^n$ for the set $\mathcal{A} \subset \mathbb{K}^{n\times n}$ satisfying $\operatorname{ecc}(p) \leq M$ and $\nu_p(\mathcal{A}) := \sup_{A\in\mathcal{A}} \nu_p(A) \leq \beta$. Then*

$$\|\varphi(t)\| \leq M\, e^{\beta t}\|\varphi(0)\|, \quad t \geq 0 \tag{53}$$

for all solutions $\varphi(\cdot)$ of (52). Conversely, if (53) holds for all solutions of (52) then there exists a joint β-Liapunov norm p for $\mathcal{A}$ with $\operatorname{ecc}(p) \leq M$ and $\nu_p(\mathcal{A}) \leq \beta$.

Proof: Note that φ is a solution of the differential inclusion (52) if and only if $t \mapsto e^{-\beta t}\varphi(t)$ solves the differential inclusion $\dot{\varphi}(t) \in (\mathcal{A} - \beta I_n)\varphi(t)$ and by Lemma 5.5.9 we have $\nu_p(\mathcal{A} - \beta I_n) = \nu_p(\mathcal{A}) - \beta$ for any norm p on $\mathbb{K}^n$. Hence, without any restriction, we may assume that $\beta = 0$.
First suppose p is a joint Liapunov norm on $\mathbb{K}^n$ for the set $\mathcal{A} \subset \mathbb{K}^{n\times n}$ satisfying $\operatorname{ecc}(p) \leq M$ and $\nu_p(\mathcal{A}) \leq 0$. Let $\varphi(\cdot) : \mathbb{R}_+ \to \mathbb{K}^n$ be a solution of (52) on $\mathbb{R}_+$. Then there exist $A(t) \in \mathcal{A}$, $t \geq 0$ such that $t \mapsto A(t)$ is measurable and $\dot{\varphi}(t) = A(t)\varphi(t)$ for almost all $t \geq 0$. Since the norm $p(\cdot) : \mathbb{K}^n \to \mathbb{K}^n$ satisfies a global Lipschitz condition, the composed function $p(\varphi(\cdot))$ is absolutely continuous. Therefore $p(\varphi(\cdot))$

[2] It is known that the matrices $A(t)$ can be chosen in such a way that $t \mapsto A(t)$ is measurable.

is almost everywhere differentiable and since $\varphi(t+h) = \varphi(t) + hA(t)\varphi(t) + r(t,h)$ with $\lim_{h\to 0} r(t,h)/h = 0$ for a.e. $t \geq 0$ we have by (11)

$$\begin{aligned}\tfrac{d^+}{dt}p(\varphi(t)) &= \lim_{h\searrow 0}\left[p(\varphi(t+h)) - p(\varphi(t))\right]/h = \lim_{h\searrow 0}\left[p([I_n + hA(t)]\varphi(t)) - p(\varphi(t))\right]/h \\ &\leq \lim_{h\searrow 0}\left[p(I_n + hA(t)) - 1\right]p(\varphi(t))/h \leq \nu_p(A(t))p(\varphi(t)) \leq 0, \quad \text{a.e. } t \geq 0.\end{aligned}$$

Integrating from 0 to t we obtain $p(\varphi(t)) - p(\varphi(0)) \leq 0$ for $t \geq 0$ and so making use of (28) $\|\varphi(t)\| \leq \operatorname{ecc}(p)\|\varphi(0)\| \leq M\,\|\varphi(0)\|$, $t \geq 0$ whence (53) for $\beta = 0$. Conversely, suppose that (53) holds with $\beta = 0$ for all solutions of (52). For any $x^0 \in \mathbb{K}^n$ let $S(x^0)$ be the set of all solutions $\varphi(\cdot)$ of (52) with $\varphi(0) = x^0$, and define

$$p(x) = \sup_{\varphi\in S(x)}\sup_{t\geq 0}\|\varphi(t)\|, \quad x \in \mathbb{K}^n. \tag{54}$$

It is easily verified that p is a norm on $\mathbb{K}^n$ and $\|x\| \leq p(x) \leq M\|x\|$ for $x \in \mathbb{K}^n$, see Ex. 18. For every $A \in \mathcal{A}$, $\tau \geq 0$, $x^0 \in \mathbb{K}^n$, and $\varphi \in S(e^{A\tau}x^0)$ let $\psi_\varphi : \mathbb{R}_+ \to \mathbb{K}^n$ be defined by

$$\psi_\varphi(t) = e^{At}x^0,\ t \in [0,\tau] \quad \text{and} \quad \psi_\varphi(t) = \varphi(t-\tau),\ t \geq \tau.$$

Then $\psi_\varphi \in S(x^0)$ and $\sup_{t\geq 0}\|\varphi(t)\| \leq \sup_{t\geq 0}\|\psi_\varphi(t)\|$. Therefore

$$p(e^{A\tau}x^0) = \sup_{\varphi\in S(e^{A\tau}x^0)}\sup_{t\geq 0}\|\varphi(t)\| \leq \sup_{\psi\in S(x^0)}\sup_{t\geq 0}\|\psi(t)\| = p(x^0), \quad \tau \geq 0.$$

This proves $\nu_p(A) \leq 0$ for all $A \in \mathcal{A}$. Finally, since $\|x\| \leq p(x) \leq M\|x\|$ for $x \in \mathbb{K}^n$ it follows from Definition 5.5.29 that $\operatorname{ecc}(p) \leq M$. □

The differential inclusion (52) is called (M,β)-stable if (53) holds for all solutions $\varphi(\cdot)$ of (52). Proposition 5.5.50 shows that there exists a joint β-Liapunov norm satisfying (29) for all $A \in \mathcal{A}$ if and only if the differential inclusion (52) is (M,β)-stable. Since for any norm p the function $\nu_p : \mathbb{K}^{n\times n} \to \mathbb{R}$ is convex by Lemma 5.5.9, we obtain from Proposition 5.5.50 the following

Corollary 5.5.51. *The differential inclusion* (52) *is* (M,β)*-stable if and only if the differential inclusion* $\dot{x}(t) \in (\operatorname{conv}\mathcal{A})x(t)$ *is* (M,β)*-stable.*

Remark 5.5.52. In the next section we will consider more general classes of perturbations, in particular time-varying ones. Suppose that $\mathbf{\Delta} \subset \mathbb{K}^{\ell\times p}$ is closed, $0 \in \mathbf{\Delta}$, $\|\cdot\|$ is an operator norm on $\mathbb{K}^{\ell\times p}$ and $A(\cdot) : \Delta \mapsto A(\Delta)$ is a continuous map from $\mathbf{\Delta}$ to $\mathbb{K}^{n\times n}$. We assume that $\dot{x} = A_0x$ with $A_0 = A(0)$ is (M,β)-stable. Let $\mathcal{A}(r) = \{A(\Delta); \Delta \in \mathbf{\Delta}, \|\Delta\| \leq r\}$ for $r \in \mathbb{R}_+$. Then the (M,β)-stability radius of $A(\cdot)$ with respect to *time-varying* perturbations is defined by

$$r_{\mathbf{\Delta},t} = r_{\mathbf{\Delta},t}(A(\cdot); M,\beta) = \sup\{r \in \mathbb{R}_+; \dot{x} \in \mathcal{A}(r)x \text{ is } (M,\beta)\text{-stable}\}. \tag{55}$$

So, for all time-varying perturbations $\Delta(\cdot) \in L^\infty(\mathbb{R}_+, \mathbf{\Delta})$ with $\|\Delta(\cdot)\|_{L^\infty} < r_{\mathbf{\Delta},t}(A(\cdot); M,\beta)$, the solutions φ of

$$\dot{x}(t) = A(\Delta(t))x(t), \quad t \geq 0 \tag{56}$$

satisfy (53). On the other hand, for every $\varepsilon > 0$ there exists a perturbation $\Delta(\cdot) \in L^\infty(\mathbb{R}_+, \mathbf{\Delta})$ with $\|\Delta(\cdot)\|_{L^\infty} < r_{\mathbf{\Delta},t} + \varepsilon$ and a solution φ of (56) such that $\|\varphi(t)\| > M\, e^{\beta t}\|\varphi(0)\|$ for some $t > 0$.
Proposition 5.5.50 implies that, for every $r \in \mathbb{R}_+$, $r < r_{\mathbf{\Delta},t}$ there exists a joint Liapunov norm p for the set $\mathcal{A}(r)$ satisfying $\operatorname{ecc}(p) \le M$ and $\sup_{A \in \mathcal{A}(r)} \nu_p(A) \le \beta$. As mentioned above, an analogous result does not necessarily hold for $\mathcal{A}(r_{\mathbf{\Delta}})$ where

$$r_{\mathbf{\Delta}} = r_{\mathbf{\Delta}}(A(\cdot); M, \beta) = \sup\{r \in \mathbb{R}_+;\ \dot{x} = A(\Delta)x \text{ is } (M,\beta)\text{-stable for all } A(\Delta) \in \mathcal{A}(r)\}.$$

Although by definition all the systems $\dot{x} = A(\Delta)x$, $\Delta \in \mathbf{\Delta}$, $\|\Delta\| \le r$ are (M,β)-stable if $r < r_{\mathbf{\Delta}}$, there may not exist a joint Liapunov norm p for these sets of systems satisfying $\operatorname{ecc}(p) \le M$ and $\sup_{A \in \mathcal{A}(r)} \nu_p(A) \le \beta$. □

While Proposition 5.5.50 is satisfactory from a theoretical point of view, it is unclear how a joint β-Liapunov norm p for the set $\mathcal{A} \subset \mathbb{K}^{n\times n}$ could be constructed in order to obtain estimates of the form (53). We will now show that for the spectral norm estimates of the (M,β)-stability radius of A can be obtained from solutions of differential and algebraic Riccati equations.

Proposition 5.5.53. *Suppose there exist* $P^o \in \mathcal{H}_n(\mathbb{K})$, $Q \in \mathcal{H}_q(\mathbb{K})$, $R \in \mathcal{H}_\ell(\mathbb{K})$, $P^o \succ 0$, $Q \succ 0$, $R \succ 0$ *such that*

$$\dot{P} - A^*P - PA + 2\beta P - C^*QC - PBRB^*P = 0, \quad P(0) = P^o \tag{57}$$

has a solution on $\mathbb{R}_+$ *which satisfies*

$$\sigma_{\max}(P(t))/\sigma_{\min}(P^o) \le M^2, \quad t \ge 0. \tag{58}$$

Then with respect to Euclidean norms $r_{\mathbb{C}}(A; B, C; M, \beta) \ge (\sigma_{\min}(Q)\,\sigma_{\min}(R))^{1/2}$.

Proof: If P satisfies (57) on $\mathbb{R}_+$, then for $\Delta \in \mathbb{C}^{\ell\times q}$

$$\begin{aligned}
&\dot{P} - (A - \beta I_n + B\Delta C)^*P - P(A - \beta I_n + B\Delta C)\\
&\quad = C^*QC + PBRB^*P - (B\Delta C)^*P - PB\Delta C\\
&\quad = (B^*P - R^{-1}\Delta C)^*R(B^*P - R^{-1}\Delta C) + C^*(Q - \Delta^*R^{-1}\Delta)C.
\end{aligned}$$

Now assume that $\|\Delta\|^2 \le \sigma_{\min}(Q)\sigma_{\min}(R)$, then $Q \succeq \sigma_{\min}(R)^{-1}\Delta^*\Delta \succeq \Delta^*R^{-1}\Delta$, hence for all $t \ge 0$

$$V(t) := (B^*P(t) - R^{-1}\Delta C)^*R(B^*P(t) - R^{-1}\Delta C) + C^*(Q - \Delta^*R^{-1}\Delta)C \succeq 0.$$

In the same way that (37) was established, we have

$$P(t) = e^{(A(\Delta)^* - \beta I_n)t}P^o e^{(A(\Delta) - \beta I_n)t} + \int_0^t e^{(A(\Delta)^* - \beta I_n)(t-s)}V(s)e^{(A(\Delta) - \beta I_n)(t-s)}ds.$$

Therefore $e^{A(\Delta)^*t}P^o e^{A(\Delta)t} \preceq e^{2\beta t}P(t)$ and (58) implies

$$\|e^{A(\Delta)t}\|^2 \le e^{2\beta t}\sigma_{\max}(P(t))/\sigma_{\min}(P^o) \le M^2 e^{2\beta t}, \qquad t \ge 0.$$

So $\dot{x} = A(\Delta)x$ is (M,β)-stable for all $\Delta \in \mathbb{K}^{\ell\times q}$, $\|\Delta\|^2 \le \sigma_{\min}(Q)\sigma_{\min}(R)$. □

The usefulness of the lower bound for $r_{\mathbb{C}}(A;B,C;M,\beta)$ obtained from this proposition depends on a judicious choice of $Q \succ 0$, $R \succ 0$ and $P^o \succ 0$. The simplest choices are $Q = \alpha_Q I_n$, $R = \alpha_R I_n$, $P^o = I_n$, where $\alpha_Q > 0$ and $\alpha_R > 0$ have to be selected carefully, see Example 5.5.56.

Remark 5.5.54. If $P = P^\circ$ is a constant solution of (57), i.e.

$$A^*P + PA - 2\beta P + C^*QC + PBRB^*P = 0, \tag{59}$$

then the proof of the previous proposition shows that

$$(A - \beta I_n + B\Delta C)^*P + P(A - \beta I_n + B\Delta C) \preceq 0.$$

Hence $\|x\|_P = \langle Px, x\rangle^{1/2}$ defines a joint β-Liapunov norm for all the perturbed systems $\dot{x} = A(\Delta)x$ where $\|\Delta\|^2 \leq \sigma_{\min}(Q)\sigma_{\min}(R)$. □

We conclude this section by deriving a bound on time-varying, nonlinear perturbations which ensures global (M,β)-stability of the perturbed systems. Consider the time-varying nonlinear equation

$$\dot{x} = Ax + B\Delta(t, Cx), \quad x(0) = x^0, \tag{60}$$

where $\Delta : \mathbb{R}_+ \times \mathbb{K}^q \mapsto \mathbb{K}^\ell$, $(t,z) \mapsto \Delta(t,z)$ is continuous in $(t,z) \in \mathbb{R}_+ \times \mathbb{K}^q$, locally Lipschitz with respect to z and satisfies $\Delta(t,0) = 0$ for all $t \geq 0$. The following proposition extends Proposition 5.5.53 to perturbations of the RHS of the form $Ax \rightsquigarrow Ax + B\Delta(t, Cx)$.

Proposition 5.5.55. *Under the assumptions of Proposition 5.5.53 suppose that*

$$\|\Delta(t,z)\|_2 \leq (\sigma_{\min}(Q)\sigma_{\min}(R))^{1/2}\|z\|_2, \quad (t,z) \in \mathbb{R}_+ \times \mathbb{K}^q.$$

Then for every $x^0 \in \mathbb{K}^n$, there exists a unique solution $x(\cdot, x^0)$ of (60) on $\mathbb{R}_+$ and

$$\|x(t,x^0)\|_2 \leq Me^{\beta t}\|x^0\|_2, \text{ for all } t \geq 0.$$

Proof: The existence of a unique solution $x(\cdot) = x(\cdot, x^0)$ of (60) on $\mathbb{R}_+$ follows from Theorem 2.1.14 and Proposition 2.1.19. By assumption there exists a Hermitian solution $P(\cdot)$ of the Riccati equation (57) on $\mathbb{R}_+$ for which we have

$$\begin{aligned}
&\frac{d}{ds}\langle x(t-s), P(s)x(t-s)\rangle \\
&= \langle x(t-s), \dot{P}(s)x(t-s)\rangle - 2\,\mathrm{Re}\left\langle Ax(t-s) + B\Delta(t-s, Cx(t-s)), P(s)x(t-s)\right\rangle \\
&= -\,2\beta\langle x(t-s), P(s)x(t-s)\rangle + \langle Cx(t-s), QCx(t-s)\rangle + \\
&\quad + \langle B^*P(s)x(t-s), RB^*P(s)x(t-s)\rangle - 2\,\mathrm{Re}\langle \Delta(t-s, Cx(t-s)), B^*P(s)x(t-s)\rangle \\
&= \langle Cx(t-s), QCx(t-s)\rangle - \left\langle R^{-1}\Delta(t-s, Cx(t-s)), \Delta(t-s, Cx(t-s))\right\rangle + \\
&+\left\langle B^*P(s)x(t-s) - R^{-1}\Delta(t-s, Cx(t-s)), R\left[B^*P(s)x(t-s) - R^{-1}\Delta(t-s, Cx(t-s))\right]\right\rangle \\
&\quad - 2\beta\langle x(t-s), P(s)x(t-s)\rangle.
\end{aligned}$$

Now suppose that $\sigma_{\min}(Q)\sigma_{\min}(R)\|z\|_2^2 \geq \|\Delta(t,z)\|_2^2$, $(t,z) \in \mathbb{R}_+ \times \mathbb{K}^q$, then setting $z := Cx(t-s)$ and $u = \Delta(t-s,z)$ we obtain $\|u\|_2^2 = \|\Delta(t-s,z)\|_2^2 \leq \sigma_{\min}(Q)\sigma_{\min}(R)\|z\|_2^2$ and

$$\frac{d}{ds}\langle x(t-s), P(s)x(t-s)\rangle + 2\beta\langle x(t-s), P(s)x(t-s)\rangle \geq \langle z, Qz\rangle - \langle R^{-1}u, u\rangle \geq 0.$$

After multiplying by $e^{2\beta s}$, this yields

$$\frac{d}{ds}\left[e^{2\beta s}\langle x(t-s), P(s)x(t-s)\rangle\right] \geq 0$$

and integrating from 0 to t

$$\sigma_{\max}(P(t))e^{2\beta t}\|x^0\|_2^2 \geq e^{2\beta t}\langle x^0, P(t)x^0\rangle \geq \langle x(t), P^o x(t)\rangle \geq \sigma_{\min}(P^o)\|x(t)\|_2^2.$$

Thus under the conditions of Proposition 5.5.53, $\|x(t)\|_2 \leq Me^{\beta t}\|x^0\|_2$, $t \geq 0$. □

The sufficient condition given in Propositions 5.5.53 and 5.5.55 can be effectively used even in the borderline case where $M = 1$ (in which case P^o must necessarily be a multiple of the identity matrix by (58)). This is illustrated in the following example.

Example 5.5.56. Suppose A is a normal matrix, $A = U^* \operatorname{diag}(\lambda_1, \lambda_2, \ldots, \lambda_n)\,U$, with U unitary, $\operatorname{Re}\lambda_i < 0$, $i \in \underline{n}$ and $B = C = I_n$. Let $P^o = I_n$, $Q = \alpha^2 I_n$, $R = \alpha^2 I_n$. A matrix of the form $\hat{P}(t) = UP(t)U^* = \operatorname{diag}(p_1(t), p_2(t), \ldots, p_n(t))$ solves (57) if and only if its diagonal entries solve the following set of n decoupled scalar differential Riccati equations:

$$\dot{p_i} - (\lambda_i + \lambda_i^* - 2\beta)p_i - \alpha^2 - \alpha^2 p_i^2 = 0, \quad p_i(0) = 1, \; i \in \underline{n}.$$

Let $\gamma_i = -(\lambda_i + \lambda_i^* - 2\beta)/2$, $\gamma_1 \leq \gamma_2 \leq \ldots \leq \gamma_n$ and suppose β is such that $\gamma_1 > 0$. If $\alpha^2 = \gamma_1$, then $p_1(t) \equiv 1$ and $p_i(t) \leq 1$ for all $t \geq 0$ and $i \in \underline{n}$. So $1 \geq \|\hat{P}(t)\| = \|P(t)\|$, $t \geq 0$. Thus, with respect to Euclidean norms, $r_{\mathbb{K}}(A; I, I; 1, \beta) \geq \alpha^2 = \gamma_1 = -\left[\lambda_{\max}(A + A^*) - 2\beta\right]/2$. In particular, if $\sigma(A) \subset \mathbb{C}_-$ then $r_{\mathbb{K}}(A; I, I; 1, 0) \geq -\lambda_{\max}(A + A^*)/2 = \nu(A)$. (By Proposition 5.5.49 we know that in fact the equality holds).
Now suppose that $\Delta : \mathbb{R}_+ \times \mathbb{K}^n \mapsto \mathbb{K}^n$ is continuous in $(t,x) \in \mathbb{R}_+ \times \mathbb{K}^n$, locally Lipschitz with respect to x and satisfies $\Delta(t,0) = 0$ for all $t \geq 0$. Then by Proposition 5.5.55 if

$$1/2\,\left[\lambda_{\max}(A + A^*) - 2\beta\right] < 0 \quad \text{and} \quad \|\Delta(t,x)\|_2 \leq -1/2\,\left[\lambda_{\max}(A + A^*) - 2\beta\right]\|x\|_2,$$

then the solutions $\varphi(\cdot)$ of the time-varying nonlinear equation $\dot{x} = Ax + \Delta(t,x)$ satisfy $\|\varphi(t)\|_2 \leq e^{\beta t}\|\varphi(0)\|_2$ for all $t \geq 0$. □

5.5.5 Exercises

If not specified otherwise, the vector norms in the following exercises are arbitrary and the matrix norms are the corresponding operator norms. Both are denoted by $\|\cdot\|$.

1. Prove that for all $A \in \mathbb{C}^{n \times n}$

$$\sigma(A) \subset \mathbb{C}_- \iff \exists t > 0 : \|e^{At}\| < 1.$$

2. If $A = \begin{bmatrix} -5 & 3 & 1 \\ 0 & -1 & 1 \\ 1 & 2 & 1 \end{bmatrix}$, find $\nu(A)$ with respect to the 1-, 2- and ∞-norms and use Lemma 5.5.9 to compute intervals in $\mathbb{R}$ containing the real parts of the eigenvalues of A.

3. Prove that if $A = (a_{ij}) \in \mathbb{C}^{n\times n}$, then with respect to the 1-norm

$$\nu(A) = \max_{j\in \underline{n}} \big(\operatorname{Re} a_{jj} + \sum_{i\in \underline{n}\setminus\{j\}} |a_{ij}|\big).$$

4. If $\nu_p(A)$ denotes the initial growth rate of $A \in \mathbb{C}^{n\times n}$ with respect to the p-norm for $1 \le p \le \infty$, prove

$$\nu_2(A) \le (\nu_1(A) + \nu_\infty(A))/2.$$

(Hint: Use Gershgorin's Theorem 4.2.19.)

5. Suppose $\|\cdot\|$ and $\|\cdot\|^*$ are dual vector norms on $\mathbb{C}^n$. Show that the initial growth rate of $A \in \mathbb{C}^{n\times n}$ with respect to the norm $\|\cdot\|$ is given by

$$\nu(A) = \max_{\|x\|=1} \max_{\substack{\|y\|^*=1 \\ \langle x,y\rangle=1}} \operatorname{Re}\langle Ax, y\rangle.$$

If $\nu^*(A)$ denotes the initial growth rate with respect to the dual norm, prove that $\nu^*(A) = \nu(A^*)$ and if $\nu_p(A)$ is the initial growth rate with respect to the p-norm, then $\nu_2(A) \le (\nu(A) + \nu^*(A))/2$ and so, in particular,

$$\nu_2(A) \le (\nu_p(A) + \nu_{p^*}(A))/2, \quad 1 < p, p^* \le \infty,\ 1/p + 1/p^* = 1.$$

6. Suppose $T \subset \mathbb{R}$ is an interval and $A(\cdot) : T \to \mathbb{C}^{n\times n}$ a continuous matrix function. Prove that for any $t_0 \in T$, $x^0 \in \mathbb{C}^n$ the solution $x(\cdot) = x(\cdot; t_0, x^0)$ of the initial value problem $\dot{x} = A(t)x$, $x(t_0) = x^0$ satisfies

$$\exp\left[\int_{t_0}^t -\nu(-A(s))ds\right] \|x^0\| \le \|x(t)\| \le \exp\left[\int_{t_0}^t \nu(A(s))ds\right] \|x^0\|, \quad t \in T, t \ge t_0.$$

Assuming $\sup T = \infty$, conclude that the system $\dot{x} = A(t)x$ is stable at time t_0 if there exists a constant $c = c(t_0) > 0$ such that $\int_{t_0}^t \nu(A(s))ds \le c$ for all $t \ge t_0$ and it is asymptotically stable at time t_0 if $\int_{t_0}^t \nu(A(s))ds \to -\infty$ as $t \to \infty$, see [111], [508].

7. If $A \in \mathbb{K}^{n\times n}$ generates a contraction semigroup such that $\|e^{A\tau}\| = 1$ for some $\tau > 0$ show that there exists an $x \in \mathbb{K}^n$ such that $\|e^{At}x\| = 1$, $t \in [0, \tau]$.

8. If $A = \begin{bmatrix} -1 & 4 \\ 0 & -2 \end{bmatrix}$, find a norm on $\mathbb{R}^2$ with respect to which A generates a contraction semigroup.

9. If $\|\cdot\|$ is a norm on $\mathbb{K}^n$ and $\sup_{t\ge 0} \|e^{At}\| < \infty$, prove that $\|\!|\cdot|\!\|_A$ defined by

$$\|\!|x|\!\|_A = \sup_{t\ge 0} \|e^{At}x\|, \quad x \in \mathbb{K}^n$$

is a norm on $\mathbb{K}^n$ with respect to which A generates a contraction semigroup. If $\sigma(A) \subset \mathbb{C}_-$ and $\sup_{t\ge 0} \|e^{At}\| > 1$ prove there exist $x \in \mathbb{K}^n$ and $\tau > 0$ such that the trajectory $e^{At}x$ remains on the unit sphere $\{x \in \mathbb{K}^n;\ \|\!|x|\!\|_A = 1\}$ for all $t \in [0, \tau]$, see Fig. 5.5.3.

10. Prove the statements in Example 5.5.18.

11. Suppose that a norm $\|\cdot\|$ on $\mathbb{K}^n$ has the following property:

If $x(t) \in \mathbb{K}^n \setminus \{0\}$ is analytic on an interval $I \subset \mathbb{R}$, then there exists a $y^*(t) \in \mathbb{K}^{1\times n}$ which is also analytic on I such that $\|y^*(t)\|^* = \|x(t)\|$ and $y^*(t)x(t) = \|x(t)\|^2$ for all $t \in I$ where $\|\cdot\|^*$ is the dual norm.

Prove that if A is asymptotically stable and generates a contraction semigroup on $(\mathbb{K}^n, \|\cdot\|)$, then the semigroup must be a strong contraction. (Hint: Make use of Ex. 7.)

12. Show that $[\cdot,\cdot]$ defined by

$$[x,y] = \begin{cases} x_1 y_1 & \text{if } |y_1| > |y_2| \\ x_2 y_2 & \text{if } |y_2| \geq |y_1| \end{cases}, \qquad x, y \in \mathbb{R}^2$$

is a semi-scalar product on $\mathbb{R}^2$ with respect to the ∞-norm. Hence prove that the real matrix $A = \begin{bmatrix} a_{11} & a_{12} \\ a_{21} & a_{22} \end{bmatrix}$ generates a contraction semigroup on $\mathbb{R}^2$ with respect to the ∞-norm if and only if $a_{11} + |a_{12}| \leq 0$, $a_{22} + |a_{21}| \leq 0$.

13. Prove that $A = (a_{ij}) \in \mathbb{C}^{n\times n}$ generates a contraction semigroup with respect to the ∞-norm if and only if $\sum_{j\neq i} |a_{ij}| \leq -\operatorname{Re} a_{ii}$ for all $i \in \underline{n}$, i.e. if and only if $\operatorname{Re} a_{ii} \leq 0$, $i \in \underline{n}$ and the matrix obtained from A by replacing a_{ii} by $\operatorname{Re} a_{ii}$ is diagonally dominant.

14. Show that if $A = \begin{bmatrix} a & b \\ c & a \end{bmatrix} \in \mathbb{C}^{2\times 2}$, a closest normal matrix with respect to the spectral norm is $\hat{X} = \begin{bmatrix} a & (b-c)/2 \\ -(b-c)/2 & a \end{bmatrix}$, and $\|A - \hat{X}\|_{2,2} = |b+c|/2$. Show that $\nu(A) < 0$ if $a < 0$ and $2|a| > |b+c|$. So in this case Corollary 5.5.13 gives a tight result.

15. Given any $A \in \mathbb{K}^{n\times n}$ prove that there exists $\hat{U} \in \mathbf{U}_n(\mathbb{C})$ such that $\hat{U}^* A \hat{U} = D_{\hat{U}} + N_{\hat{U}}$ is in upper Schur form, $D_{\hat{U}}$ is diagonal, $N_{\hat{U}}$ is upper triangular with zeros on the diagonal and $\|N_{\hat{U}}\|_{2,2} = \operatorname{dep}(A)$ where $\operatorname{dep}(A)$ is defined by (16).

16. Show that (38) holds if and only if

$$\max_{\omega\in\mathbb{R}} \|((-\delta + \imath\omega)I_n - A)^{-1}\| < \delta^{-1}.$$

Hence prove the statement in Remark 5.5.41.

17. If $A = \begin{bmatrix} -1 & 6 \\ 0 & -1 \end{bmatrix}$ and $A_\alpha = A - \alpha I_2$, $\alpha \in \mathbb{R}$ show that for the spectral norm

$$d_{\mathbb{C}}^-(A_\alpha)^2 = 18 + (1+\alpha)^2 - 6[9 + (1+\alpha)^2]^{1/2}, \quad \alpha > 0.$$

Prove that $d_{\mathbb{C}}^-(A_\alpha)/\alpha$ is minimized at $\alpha = \hat{\alpha} = 5/4$ and so $R(A) = \hat{\alpha}/d_{\mathbb{C}}^-(A_{\hat{\alpha}}) = 5/3$ where $R(A)$ is the Kreiss constant defined in (39).
For given $\delta > 0$ show that $\alpha + \imath\omega \in \partial\sigma_{\mathbb{C}}(A,\delta)$, $\alpha \in \mathbb{R}$ provided

$$\delta^2 = 18 + (1+\alpha)^2 + \omega^2 - 6\left[9 + ((1+\alpha)^2 + \omega^2)\right]^{1/2}.$$

Hence show that $\alpha_\delta(A) = \sqrt{\delta^2 + 6\delta} - 1$, provided $\delta > \sqrt{10} - 3$. Use this result to confirm that $R(A) = 5/3$. Prove that

$$e^{At} = \begin{bmatrix} 1 & -6t \\ 0 & 1 \end{bmatrix} e^{-t}, \quad t \geq 0.$$

and hence $\|e^{At}\|^2 = [1 + 18t^2 + 6t\sqrt{9t^2+1}]\, e^{-2t}$, i.e. $\|e^{At}\| = e^{-t}(\sqrt{9t^2+1} + 3t)$.
Show that $t \approx 0.89$ maximizes the right hand side and this yields $M_0(A) \approx 5.35$. Whereas the estimates in (40) are $5/3 \leq M_0(A) \leq 10e/3 = 9.06$.

18. Suppose $\mathcal{A} \subset \mathbb{K}^{n\times n}$ and all solutions φ of the differential inclusion $\dot{x}(t) \in \mathcal{A}x(t)$ satisfy $\|\varphi(t)\| \le M\|\varphi(0)\|$, $t \ge 0$. For any $x^0 \in \mathbb{K}^n$ let $S(x^0)$ be the set of all solutions $\varphi(\cdot)$ of $\dot{x}(t) \in \mathcal{A}x(t)$ with $\varphi(0) = x^0$, and define

$$p(x) = \sup_{\varphi\in S(x)} \sup_{t\ge 0} \|\varphi(t)\|, \quad x \in \mathbb{K}^n.$$

Prove that p is a norm on $\mathbb{K}^n$ and $\|x\| \le p(x) \le M\|x\|$ for $x \in \mathbb{K}^n$.

19. Consider $A = \begin{bmatrix} -1 & 4 \\ 0 & -1 \end{bmatrix}$. Prove that $A+A^*$ is unstable. Explain why it is not possible to find $P \succ 0$, $Q_i \succ 0$, $i = 1, 2$ such that

$$PA + A^*P + Q_1 = 0, \qquad AP + PA^* + Q_2 = 0.$$

Let $\beta \in (-1, 0)$ and $M = M_\beta(A) = M_\beta(A^*)$. Then this exercise shows that although both A and A^* are (M, β)-stable, there does not exist a joint β-Liapunov function for them.

20. If

$$A = \begin{bmatrix} 0 & 1 \\ -1 & -1 \end{bmatrix}, \quad B = \begin{bmatrix} 0 \\ 1 \end{bmatrix}, \quad C = [0,\ 1],$$

show that $P(t) \equiv I_2$ is a constant solution of (57) with $R = 1$, $Q = 1$, $\beta = 0$. Hence show that $r_{\mathbb{K}}(A; B, C; 1, 0) \ge 1$ (see Definition 5.5.48). Prove that this estimate is tight, i.e. $r_{\mathbb{K}}(A; B, C; 1, 0) = 1$.

21. Consider the nonlinear scalar initial value problem

$$\ddot{\xi} + \dot{\xi} + \xi + \Delta(\xi) = 0, \quad \xi(0) = \xi_0,\ \dot{\xi}(0) = \xi_1$$

where $\xi_0, \xi_1 \in \mathbb{R}$, $\Delta : \mathbb{R} \to \mathbb{R}$ is locally Lipschitz and satisfies $|\Delta(\xi)| \le |\xi|$, $\xi \in \mathbb{R}$. If $\xi(t)$ solves this initial value problem, use Ex. 20 to show that $\dot{\xi}(t)^2 + \xi(t)^2 \le \xi_0^2 + \xi_1^2$ for all $t \ge 0$.

In the next four exercises the reader is asked to prove some results about the transient behaviour of discrete time semigroups $(A^t)_{t\in\mathbb{N}}$ where $A \in \mathbb{K}^{n\times n}$ is given. For simplicity we assume that $\mathbb{K}^{n\times n}$ is endowed with the spectral norm $\|\cdot\|$ and set

$$\begin{aligned} M_\beta(A) &= \min\{M \in \mathbb{R}_+;\ \forall t \in \mathbb{N} : \|A^t\| \le M\beta^t\}, \quad \beta > \varrho(A) \\ \nu(A) &= \inf\{r \in \mathbb{R}_+;\ \forall t \in \mathbb{N} : \|A^t\| \le r^t\}. \end{aligned}$$

22. Prove that $M_\beta(A) = M_1(\beta^{-1}A)$ for $\beta > 0$ and $\nu(A) = \lambda_{\max}(AA^*)^{1/2} = \|A\|$. If $A = \begin{bmatrix} 1 & 1 \\ 0 & 1 \end{bmatrix}$ prove that $A^t = \begin{bmatrix} 1 & t \\ 0 & 1 \end{bmatrix}$, $t \in \mathbb{N}$ and determine $\varrho(A), \nu(A), M_\beta(A)$ for $\beta > \varrho(A)$.

23. Suppose $\beta > \varrho(A)$. Prove that any Hermitian solution $P_\beta(\cdot)$ of the difference inequality

$$P(t+1) - \beta^{-2}A^*P(t)A \succeq 0, \quad t \in \mathbb{N} \tag{61}$$

satisfies $\sup_{t\in\mathbb{N}} \overline{\sigma}(P_\beta(t))/\sigma_{\min}(P_\beta(0)) \ge M_\beta(A)^2$. Prove also that the solution $P_\beta(\cdot)$ of the initial value problem

$$P(t+1) = \beta^{-2}A^*P(t)A, \quad t \in \mathbb{N}; \quad P(0) = I_n$$

satisfies $\sup_{t\in\mathbb{N}} \overline{\sigma}(P_\beta(t)) = M_\beta(A)^2$.

24. Prove a counterpart of the Kreiss-Spijker Theorem for the discrete time case, namely if

$$R(A) = \sup_{|z|>1} (|z| - 1)\|(zI_n - A)^{-1}\| < \infty,$$

then

$$R(A) \leq \sup_{t \in \mathbb{N}} \|A^t\| \leq e \cdot n\, R(A).$$

(Hint: For the first inequality you may wish to use the Neumann series

$$(zI_n - A)^{-1} = z^{-1}I_n + z^{-2}A + z^{-3}A^2 + ..., \qquad |z| > \varrho(A).$$

And for the second Cauchy's integral representation

$$A^t = \frac{1}{2\pi \imath} \int_\Gamma z^t (zI_n - A)^{-1} dz, \quad t \in \mathbb{N},$$

where Γ is a positively oriented, piecewise smooth, simple, closed curve enclosing $\sigma(A)$, see Example A.3.9).

25. If $\varrho_\delta(A) = \sup\{|z|; z \in \sigma_{\mathbb{C}}(A;\delta)\}$ is the *δ-spectral radius* of A then prove that for $R(A)$ as in the previous exercise we have

$$R(A) = \sup_{r>1} (1 - r^{-1})/d^1_{\mathbb{C}}(r^{-1}A) = \sup_{\delta>0} (\varrho_\delta(A) - 1)/\delta.$$

26. For the spectral norm show that

$$\|A^t\| \leq \sum_{i=0}^{n-1} \binom{t}{i} [\varrho(A)]^{t-i} [\mathrm{dep}\,(A)]^i = [\varrho(A)]^{t-n} \sum_{i=0}^{n-1} \binom{t}{i} [\varrho(A)]^{n-i} [\mathrm{dep}\,(A)]^i, \quad t \in \mathbb{N},$$

where dep (A) is the departure of A from normality defined in (16), see [223].

5.5.6 Notes and References

An awareness of the importance of transient behaviour in fluid dynamics goes back many years. Indeed more than one hundred years ago *Thomson* (1887) [494] recognized that before their eventual decay there was a large transient growth of disturbances in Poiseuille flows. But it is only recently that *Boberg and Brosa* (1988) [64], *Gustavsson* (1991) [205], *Reddy and Henningson* (1993) [428], *Butler and Farrell* (1992) [86] found the extent of this growth. Their results are startling since amplifications by many hundreds can occur, see *Trefethen* (1997) [498]. The coupling between large transient motions of the linear part and nonlinearities has been put forward as an explanation for the onset of turbulence at Reynolds numbers which differ significantly from the critical Reynolds numbers obtained from linear stability analysis, see *Reddy et al.* (1993) [429], *Trefethen et al.* (1993) [501]. Another motive for studying the transient behaviour of linear systems comes from the stability analysis of discretization methods for ordinary and partial differential equations. *Ważewski* (1949) [514] used scalar functions $m(\cdot)$, $M(\cdot)$ satisfying $m(t)I \preceq \frac{1}{2}(A(t) + A^*(t)) \preceq M(t)I$ in order to derive exponential estimates for the solutions of time-varying linear systems $\dot{x} = A(t)x$. Motivated by a study of error estimates for numerical integration of ordinary differential equations *Dahlquist* (1959) [117] and *Lozinskii* (1958) [348] independently introduced the *initial growth rate* (11) using the terms *logarithmic derivative* and *logarithmic norm*. For further properties and applications see *Ström* (1975)

[488] and *Desoer and Vidyasagar* (1975) [130] (where $\nu(A)$ is called *matrix measure*). In the paper of *Mohler and Van Loan* (1978) [378] on "nineteen dubious ways of computing the exponential" the initial growth rate was used to obtain sensitivity estimates, see also *Mohler and Van Loan* (2003) [379]. *Henrici* (1962) [223] defined the departure from normality with respect to arbitrary norms and obtained the estimate given in Ex. 26. The results of Henrici, Lozinskii, and Ważewski have been extended to time-varying and infinite dimensional systems by *Gil'* (1998) [187]. This monograph also contains a chapter on the Aizerman conjecture.
Bauer (1962) [40] linked the initial growth rate to the numerical range. The convexity of the numerical range was first proved by *Hausdorff* (1919) [218], see also *Toeplitz* (1918) [496]. A detailed survey on the numerical range is given in Chapter 1 of *Horn and Johnson* (1991) [265].
Lumer (1961) [350] introduced semi-scalar products and *Lumer and Phillips* (1961) [351] proved Theorem 5.5.23, see also *Yosida* (1974) [538].
Definition 5.5.4, Example 5.5.6 and Propositions 5.5.37, 5.5.39 have been taken from *Hinrichsen et al.* (2001) [240]. Problem 5.5.35 has been studied by *Khusainov et al.* (1984) [308]. In this reference it is also shown that that the *optimal* bound obtainable for the transient bounds via quadratic Liapunov norms may be quite conservative. That Problem 5.5.35 may be recast as a semidefinite program with LMI constraints can be seen from *Boyd et al.* (1994) [74]. A number of different estimates for $\|e^{At}\|$ are compared in a recent paper of *Veselic'* (2003) [507].
In some control applications it is desirable to use state or output feedback in order to obtain both a reasonable decay rate and a satisfactory transient behaviour of the closed loop system. This leads to the problem of finding feedback laws which, for prescribed constants M and β, achieve (M,β)-stability of the controlled system. Some results on this (M,β)-stabilization problem can be found in *Hinrichsen et al.* (2001) [240] (2002) [241].
For material on spectral value sets see the *Notes and References* of Section 5.2. Kreiss' original theorem was proved for the discrete time case, i.e. for matrix powers instead of matrix exponentials, see *Kreiss* (1962) [317]. It is one of the fundamental results available for establishing numerical stability of discretization methods, see *Hairer and Wanner* (1991) [211] and *Dorsselaer* (1993) [504]. The upper Kreiss bound was improved by many authors until *LeVeque and Trefethen* (1984) [341] obtained $2\,e{\cdot}n$ as the constant and conjectured that it could be $e{\cdot}n$. This was finally proved to be the case by *Spijker* (1991) [476]. A nice account of the development can be found in *Wegert and Trefethen* (1994) [515]. An instructive paper on the sharpness of the upper Kreiss estimate (in the original discrete time case) is *Spijker et al.* (2002) [477].
For an introduction to the theory of differential inclusions see *Smirnov* (2002) [468]. This book also contains some historical comments, a brief survey of the literature and some applications. A short introduction to differential inclusions in a control theoretic context is given in the last chapter of *Clarke et al.* (1998) [104]. The exponential growth of linear differential inclusions has been thoroughly investigated by *Wirth* (2005) [533]. In his work, the concept of *eccentricity* and special joint Liapunov norms (called *Barabanov norms*) play a central role. These norms were introduced by *Barabanov* (1988) to determine the Liapunov exponents of differential inclusions, see [32]. The literature on Liapunov exponents is of key importance for the determination of stability radii with respect to time-varying parameter perturbations (to be studied in the next section).

5.6 More General Perturbation Classes

So far, we have dealt mainly with *affine parameter uncertainties* of time-invariant linear systems. Other model uncertainties were not considered and consequently not only the nominal models but also the perturbed ones were time-invariant and linear. From our discussion in the introduction to this chapter we know that parameter uncertainties are not the only source of model uncertainties. In applications, time-invariant linear models are often obtained by linearizing a nonlinear system around an equilibrium point or by neglecting time-varying effects. In other cases they are the result of model reduction, i.e. the approximation of high or infinite dimensional plants by lower dimensional models (neglected dynamics). In these cases the "real" system can be viewed as a nonlinear, time-varying or dynamic perturbation of the simplified time-invariant linear model. We conclude this chapter by introducing *stability radii* with respect to these wider perturbation classes and analyze some of their properties. It will be of particular interest to investigate whether or not the real and complex stability radii studied in Section 5.3 are misleading indicators of stability in the presence of more general perturbations of the system equations.
Our basic approach will be, as in the previous sections, to assume that the nominal system is a time-invariant linear model and that the perturbations can be represented in *output feedback form.* However now, we will allow feedback operators that are (i) nonlinear, (ii) time-varying linear or (iii) dynamic. As a consequence a number of new features appear:

- Spectral analysis and the concept of $\mathbb{C}_g$-stability are no longer applicable. Instead the analysis must take place in the time domain with the resulting need for a separate discussion of continuous and discrete time systems. Here we only consider continuous time systems.
- In the case of nonlinearities we have to discuss the stability of a particular given equilibrium state since, in contrast with the linear case, different equilibria may exhibit different stability properties. Moreover equilibrium points may change under nonlinear perturbations.
- There are different stability concepts, e.g. uniform, exponential, asymptotic, globally asymptotic, etc. and each of these may give rise to different stability radii. Here we focus on global asymptotic stability and make occasional comments on other possibilities.

In the first subsection we describe the three perturbation classes (i)-(iii) in more detail and illustrate their scope by examples. Complex and real stability radii are introduced for each of the three perturbation classes in the second subsection. Whereas the real stability radii are in general quite different for each of the classes, we will discover the surprising fact that the complex stability radii with respect to the three perturbation classes are all equal to $r_{\mathbb{C}}^-$. Moreover, in the case of real data, the real stability radius with respect to *dynamic* perturbations is in fact equal to $r_{\mathbb{C}}^-$. In the final subsection we restrict our considerations to time-invariant nonlinear perturbations. Using the Riccati characterization of $r_{\mathbb{C}}^-$ we show that it is possible to find a single Liapunov function for a maximal ball of perturbed systems, both linear and nonlinear. This allows us to prove that the *Aizerman conjecture*, which is known to be false in its original version (i.e. for the real case) is true over the field of complex numbers.

5.6.1 The Perturbation Classes

In this subsection we describe in detail the three classes of perturbations to be considered: (i) nonlinear, (ii) linear but time-varying, and (iii) dynamic. We will see that these perturbations form $\mathbb{K}$-linear spaces with compatible norms. The scope of the widest perturbation class (iii) is illustrated by two examples. The first one shows that a nonlinear delay equation with time-varying delay can be viewed as a dynamically perturbed time-invariant linear system, the other shows that neglected dynamics can be represented as a dynamic perturbation. In the case of perturbations (i) and (ii), the existence of solutions for the perturbed equations follows from Carathéodory's Theorem 2.1.14. However this is not the case for the dynamic perturbation class and so a global existence theorem for this class will be proved. We conclude the subsection by describing the stability concepts underlying the new stability radii.
For the sake of simplification we set $D = 0$. Throughout the whole section we assume the following:

- $\Sigma = (A, B, C) \in \mathbf{L}_{n,\ell,q}(\mathbb{K})$. All vector spaces are provided with Euclidean norms $\|\cdot\| = \|\cdot\|_2$ and all matrix spaces are endowed with the corresponding operator norms (spectral norms) which will also be denoted by $\|\cdot\|$. The nominal model
$$\dot{x}(t) = Ax(t), \quad t \in \mathbb{R}_+ \tag{1}$$
is asymptotically stable. The time domain is $T = \mathbb{R}_+$.

It is assumed that the perturbations of this model can be represented in output feedback form as illustrated in Figure 5.6.1, i.e. the perturbed system Σ_Δ is obtained by applying the uncertain output feedback $w = \Delta z$ to the system $\Sigma = (A, B, C)$. In

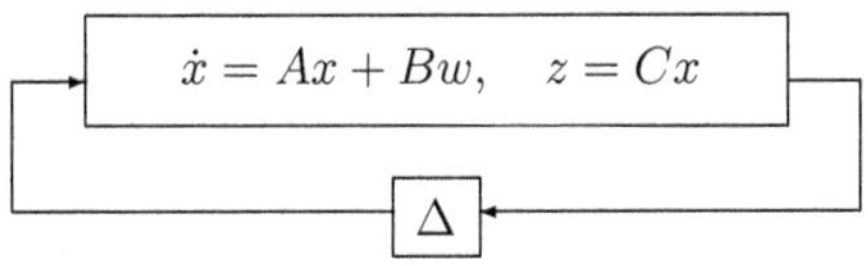

Figure 5.6.1: Feedback interpretation of the perturbed system Σ_Δ

the previous sections we considered the special case where the unknown operator Δ is memoryless, time-invariant, linear, i.e. $w(t) = \Delta z(t)$, $\Delta \in \mathbb{K}^{\ell \times q}$. Here we consider three wider classes of feedback operators $\Delta : z(\cdot) \mapsto w(\cdot)$. The first two are again *memoryless* (i.e. the output of the feedback operator at time t only depends on the input at time t) whereas the third class is dynamic (i.e. the output at time t depends on the input values during the whole past interval $[0, t]$). In this latter case Δ does not operate on single values of the output signal $z(\cdot)$ but on the whole function. More precisely, we assume that the unknown feedback $w(\cdot) = \Delta(z(\cdot))(\cdot)$ (representing the perturbation) is defined by a *causal* operator $\Delta : L^2(\mathbb{R}_+; \mathbb{K}^q) \to L^2(\mathbb{R}_+; \mathbb{K}^\ell)$. The causality of Δ means that for all $z(\cdot)$, $\tilde{z}(\cdot) \in L^2(\mathbb{R}_+; \mathbb{K}^q)$ and every $t > 0$

$$z(\tau) = \tilde{z}(\tau) \text{ for a.e. } \tau \in [0, t] \quad \Longrightarrow \quad \Delta(z(\cdot))(\tau) = \Delta(\tilde{z}(\cdot))(\tau) \text{ for a.e. } \tau \in [0, t].$$

In order to distinguish between the elements of the three perturbation classes we use different symbols, namely N for time-invariant nonlinearities, $\Delta(\cdot)$ for time-varying linear feedback gains and $\mathcal{N}$ for (possibly nonlinear) causal feedback operators. The three perturbation sets are defined as follows.

(i) **Nonlinear memoryless time-invariant feedback operators.** Let $\mathcal{P}_n(\mathbb{K})$ be the set of all nonlinearities $N : \mathbb{K}^q \mapsto \mathbb{K}^\ell$, differentiable at the origin with $N(0) = 0$, satisfying a global Lipschitz condition, i.e. there exists $\gamma \geq 0$ such that $\|N(z) - N(\tilde{z})\|_{\mathbb{K}^\ell} \leq \gamma \|z - \tilde{z}\|_{\mathbb{K}^q}$ for all $z, \tilde{z} \in \mathbb{K}^q$. For this class the perturbed system equations are of the form

$$\Sigma_N : \dot{x}(t) = Ax(t) + BN(Cx(t)), \quad t \geq 0, \qquad N \in \mathcal{P}_n(\mathbb{K}). \tag{2}$$

(ii) **Time-varying memoryless linear feedback operators.** Let $\mathcal{P}_t(\mathbb{K})$ be the set $L^\infty(\mathbb{R}_+, \mathbb{K}^{\ell \times q})$ of all essentially bounded time-varying feedback matrices. For this class the perturbed system equations are of the form

$$\Sigma_{\Delta(\cdot)} : \dot{x}(t) = Ax(t) + B\Delta(t)Cx(t), \quad \text{a.e. } t \geq 0, \qquad \Delta(\cdot) \in \mathcal{P}_t(\mathbb{K}). \tag{3}$$

(iii) **Dynamic feedback operators.** Let $\mathcal{P}_d(\mathbb{K})$ be the set of all causal L^2-stable input-output operators $\mathcal{N} : L^2(\mathbb{R}_+; \mathbb{K}^q) \mapsto L^2(\mathbb{R}_+; \mathbb{K}^\ell)$ with $\mathcal{N}(0) = 0$ which are of finite Lipschitz gain, i.e. there exists $\gamma \geq 0$ such that for all $z(\cdot), \tilde{z}(\cdot) \in L^2(\mathbb{R}_+; \mathbb{K}^q)$

$$\|\mathcal{N}(z(\cdot))(\cdot) - \mathcal{N}(\tilde{z}(\cdot))(\cdot)\|_{L^2(\mathbb{R}_+;\mathbb{K}^\ell)} \leq \gamma \|z(\cdot) - \tilde{z}(\cdot)\|_{L^2(\mathbb{R}_+;\mathbb{K}^q)}.$$

For this class the perturbed system equations are of the form

$$\Sigma_{\mathcal{N}} : \quad \dot{x}(t) = Ax(t) + B\mathcal{N}(Cx(\cdot))(t), \quad \text{a.e. } t \geq 0, \qquad \mathcal{N} \in \mathcal{P}_d(\mathbb{K}). \tag{4}$$

Remark 5.6.1. (i) Although the possibility of multiple equilibrium points is not excluded for the nonlinear system (2) , it is assumed that the equilibrium point under investigation, $\bar{x} = 0$, is preserved under perturbations (because of $N(0) = 0$).
(ii) In most applications, nonlinearities will not satisfy global Lipschitz conditions on $\mathbb{K}^q$. Then a local analysis is necessary which we will address briefly in the final subsection. □

In the rest of this subsection if there is no risk of confusion we use the abbreviation $L^2(T)$ or even L^2 for $L^2(T; \mathbb{K}^k)$ (where $T \subset \mathbb{R}$ is any interval and $k \geq 1$).
The sets of perturbations $\mathcal{P}_n(\mathbb{K})$, $\mathcal{P}_t(\mathbb{K})$, $\mathcal{P}_d(\mathbb{K})$ are $\mathbb{K}$–linear spaces and are provided with the following norms.

$$\begin{aligned}
\|N\|_n &= \inf \{\gamma \geq 0; \forall z, \tilde{z} \in \mathbb{K}^q : \|N(z) - N(\tilde{z})\|_{\mathbb{K}^\ell} \leq \gamma \|z - \tilde{z}\|_{\mathbb{K}^q}\}, \\
\|\Delta\|_t &= \operatorname{ess\,sup}\{\|\Delta(t)\|; \, t \in \mathbb{R}_+\}, \\
\|\mathcal{N}\|_d &= \inf \left\{\gamma \geq 0; \forall z(\cdot), \tilde{z}(\cdot) \in L^2(\mathbb{R}_+, \mathbb{K}^q) \colon \|\mathcal{N}(z) - \mathcal{N}(\tilde{z})\|_{L^2} \leq \gamma \|z - \tilde{z}\|_{L^2}\right\}.
\end{aligned}$$

Remark 5.6.2. If $\mathbb{G}_+ : L^2(\mathbb{R}_+; \mathbb{K}^q) \to L^2(\mathbb{R}_+; \mathbb{K}^\ell)$ is a causal bounded linear operator then $\mathbb{G}_+ \in \mathcal{P}_d(\mathbb{K})$ and $\|\mathbb{G}_+\|_d$ is the operator norm of $\mathbb{G}_+$. □

There is an obvious norm preserving embedding of the space $\mathbb{K}^{\ell\times q}$ of time-invariant linear perturbations into the normed spaces $\mathcal{P}_n(\mathbb{K})$, $\mathcal{P}_t(\mathbb{K})$:

$$\mathbb{K}^{\ell\times q} \subset \mathcal{P}_n(\mathbb{K}), \qquad \mathbb{K}^{\ell\times q} \subset \mathcal{P}_t(\mathbb{K}). \tag{5}$$

Similarly, there are *norm preserving inclusions*

$$\mathcal{P}_n(\mathbb{K}) \subset \mathcal{P}_d(\mathbb{K}), \qquad \mathcal{P}_t(\mathbb{K}) \subset \mathcal{P}_d(\mathbb{K}). \tag{6}$$

The second inclusion is obtained by identifying any time varying $\Delta(\cdot) \in \mathcal{P}_t(\mathbb{K})$ with the multiplication operator $\mathcal{N}_\Delta$ defined by $\mathcal{N}_\Delta(z(\cdot))(t) = \Delta(t)z(t)$. We leave it to the reader to prove that $\mathcal{N}_\Delta \in \mathcal{P}_d(\mathbb{K})$ and has the norm $\|\mathcal{N}_\Delta\|_d = \|\Delta\|_t$. For the first inclusion, $N \in \mathcal{P}_n(\mathbb{K})$ is identified with $\mathcal{N}_N$ defined by $\mathcal{N}_N(z(\cdot))(t) := N(z(t))$ for arbitrary $z(\cdot) \in L^2(\mathbb{R}_+;\mathbb{K}^q)$. To see that $\mathcal{N}_N \in \mathcal{P}_d(\mathbb{K})$, note that for arbitrary $z(\cdot)$, $\tilde{z}(\cdot) \in L^2(\mathbb{R}_+;\mathbb{K}^q)$, $\mathcal{N}_N(z(\cdot))(\cdot)$, $\mathcal{N}_N(\tilde{z}(\cdot))(\cdot)$ are measurable and since

$$\|\mathcal{N}_N(z(\cdot)) - \mathcal{N}_N(\tilde{z}(\cdot))\|_{L^2}^2 = \int_0^\infty \|N(z(t)) - N(\tilde{z}(t))\|_{\mathbb{K}^\ell}^2\, dt \le \|N\|_n^2 \int_0^\infty \|z(t) - \tilde{z}(t)\|_{\mathbb{K}^q}^2\, dt$$

it follows that $\mathcal{N}_N : L^2(\mathbb{R}_+;\mathbb{K}^q) \mapsto L^2(\mathbb{R}_+;\mathbb{K}^\ell)$ is of finite Lipschitz gain and $\|\mathcal{N}_N\|_d \le \|N\|_n$. Also if $z(\tau) = \tilde{z}(\tau)$ for a.e. $\tau \in [0,t]$, $t > 0$, then

$$\mathcal{N}_N(z(\cdot))(\tau) = N(z(\tau)) = N(\tilde{z}(\tau)) = \mathcal{N}_N(\tilde{z}(\cdot))(\tau) \text{ for a.e. } \tau \in [0,t]. \tag{7}$$

So $\mathcal{N}_N$ is causal and hence $\mathcal{P}_n(\mathbb{K}) \subset \mathcal{P}_d(\mathbb{K})$. It remains to show that $\|\mathcal{N}_N\|_d \ge \|N\|_n$. For any $\varepsilon > 0$, there are z, $\tilde{z} \in \mathbb{K}^q$ such that $\|N(z) - N(\tilde{z})\|_{\mathbb{K}^\ell} \ge (\|N\|_n - \varepsilon)\|z - \tilde{z}\|_{\mathbb{K}^q}$. Let $I \subset \mathbb{R}_+$ be an interval of finite length and define $z(t) = z$, $\tilde{z}(t) = \tilde{z}$ for $t \in I$, $z(t) = \tilde{z}(t) = 0$, $t \in \mathbb{R}_+ \setminus I$. Then

$$\begin{aligned} \|\mathcal{N}_N(z(\cdot)) - \mathcal{N}_N(\tilde{z}(\cdot))\|_{L^2(\mathbb{R}_+;\mathbb{K}^\ell)}^2 &= \int_I \|N(z) - N(\tilde{z})\|_{\mathbb{K}^\ell}^2\, dt \ge \\ &\ge (\|N\|_n - \varepsilon)^2 \int_I \|z - \tilde{z}\|_{\mathbb{K}^q}^2\, dt = (\|N\|_n - \varepsilon)^2 \|z(\cdot) - \tilde{z}(\cdot)\|_{L^2(\mathbb{R}_+;\mathbb{K}^q)}^2. \end{aligned}$$

Hence $\|\mathcal{N}_N\|_d = \|N\|_n$.

Remark 5.6.3. Let $\mathcal{P}_{nt}(\mathbb{K})$ be the set of *time-varying* Lipschitzian nonlinearities $N(z,t)$ with $N(0,t) = 0$, $t \ge 0$ which are measurable in (z,t) and satisfy

$$\|N(z,t) - N(\tilde{z},t)\| \le \gamma(t)\|z - \tilde{z}\|, \quad z, \tilde{z} \in \mathbb{K}^q,\ t \ge 0$$

for some $\gamma(\cdot) \in L^\infty(\mathbb{R}_+,\mathbb{R}_+)$. $\mathcal{P}_{nt}(\mathbb{K})$ is a $\mathbb{K}$-linear space and is provided with the norm

$$\|N\|_{nt} = \inf\{\|\gamma(\cdot)\|_{L^\infty};\gamma(\cdot) \in L^\infty, \forall z, \tilde{z} \in \mathbb{K}^q\ \forall t \in \mathbb{R}_+ : \|N(z,t) - N(\tilde{z},t)\|_{\mathbb{K}^\ell} \le \gamma(t)\|z - \tilde{z}\|_{\mathbb{K}^q}\}.$$

Identifying every $N \in \mathcal{P}_{nt}(\mathbb{K})$ with the operator $\mathcal{N} : L^2(\mathbb{R}_+;\mathbb{K}^q) \to L^2(\mathbb{R}_+;\mathbb{K}^\ell)$ defined by $\mathcal{N}(z(\cdot))(t) = N(z(t),t)$, $t \ge 0$ for $z(\cdot) \in L^2(\mathbb{R}_+;\mathbb{K}^q)$ the linear space $\mathcal{P}_{nt}(\mathbb{K})$ is embedded isomorphically as a linear subspace in $\mathcal{P}_d(\mathbb{K})$. One can prove as in the time-invariant case that this embedding is also norm-preserving, $\|\mathcal{N}\|_d = \|N\|_{nt}$. □

A basic stability requirement is that solutions exist on intervals unbounded to the right (no finite escape time, see Section 3.2). By Proposition 2.1.19 for any $N \in \mathcal{P}_n(\mathbb{K})$, $\Delta(\cdot) \in \mathcal{P}_t(\mathbb{K})$ and initial data $(t_0, x^0) \in \mathbb{R}_+ \times \mathbb{K}^n$ there exist unique solutions $x(\cdot) = x(\cdot\,; t_0, x^0)$ of Σ_N, $\Sigma_{\Delta(\cdot)}$, respectively, on $[t_0, \infty)$ satisfying $x(t_0) = x^0$. For dynamic perturbations the existence problem is more complicated. This is not surprising since $\mathcal{P}_d(\mathbb{K})$ is a very large set of perturbations and contains operators which give rise to different classes of infinite dimensional systems, e.g. delay equations, integro-differential equations and other systems with possibly unbounded memory (see Example 5.6.5).
Since $\mathcal{N}(z(\cdot))(t)$ depends not only on $z(t)$ but the whole "past" $z(\cdot)|_{[0,t]}$ of $z(\cdot)$, the initial value problem for $\Sigma_{\mathcal{N}}$ must include initial functions.

Definition 5.6.4. Let $\mathcal{N} \in \mathcal{P}_d(\mathbb{K})$, $(t_0, x^0, \phi) \in \mathbb{R}_+ \times \mathbb{K}^n \times L^2(0, t_0; \mathbb{K}^q)$ be given and $I = [t_0, t_1)$, $t_0 < t_1 \leq \infty$. A function $x(\cdot) : I \mapsto \mathbb{K}^n$ is said to be a solution of $\Sigma_{\mathcal{N}}$ on I with initial data (t_0, x^0, ϕ) if it satisfies the following conditions:

(i) $x(\cdot)$ is absolutely continuous on I,
(ii) $x(\cdot)$ satisfies $\quad \dot{x}(t) = Ax(t) + B\mathcal{N}(z_\phi(\cdot))(t) \quad$ a.e. $t \in I$, where

$$z_\phi(t) = \begin{cases} \phi(t) & \text{if } 0 \leq t < t_0 \\ Cx(t) & \text{if } t \in I \\ 0, & \text{otherwise} \end{cases}, \qquad t \in \mathbb{R}_+, \tag{8}$$

(iii) $x(t_0) = x^0$.

Note that by causality $\mathcal{N}(Cx(\cdot))(\cdot)$ is determined a.e. on $[0, t]$ by ϕ and $x(\cdot)|_{[t_0,t]}$ for every $t \in I$. If $\mathcal{N}$ is memoryless, i.e. $\mathcal{N}(z(\cdot))(t) = N(t, z(t))$ for some time-varying nonlinearity $N \in \mathcal{P}_{nt}$ then the function ϕ can be omitted from the initial data.
In the following examples we consider two different types of dynamical perturbations. The first example shows how a nonlinear delay system with time-varying delay can be represented as a perturbed system of the form $\Sigma_{\mathcal{N}}$ with an appropriate $\mathcal{N} \in \mathcal{P}_d(\mathbb{K})$. In the second, we show how neglected dynamics can be accounted for by introducing a suitable dynamic perturbation $\mathcal{N} \in \mathcal{P}_d(\mathbb{K})$.

Example 5.6.5. Suppose $h(\cdot) : \mathbb{R}_+ \to \mathbb{R}_+$ is a continuously differentiable function such that for some $\varepsilon > 0$

$$h'(t) \leq 1 - \varepsilon, \quad t \geq 0,$$

and consider the nonlinear system with time-varying (and possibly unbounded) delay of the form

$$\dot{x}(t) = Ax(t) + BN(Cx(t - h(t))), \quad t \geq t_0 \tag{9}$$

where $N \in \mathcal{P}_n(\mathbb{K})$. For instance, we may have $h(t) \equiv h$ (constant delay) or we may have $h(t) = t/2$ (unbounded delay as $t \to \infty$). We want to represent this system in the form (4). Since the initial functions of $\Sigma_{\mathcal{N}}$ are defined on intervals of the form $[0, t_0]$ (and not on intervals $[-h, 0]$ as e.g. in Examples 2.1.25 and 2.3.20) we need to choose the initial time t_0 sufficiently large so that the initial delayed argument $t_0 - h(t_0)$ is non-negative. By assumption $\alpha(t) := t - h(t)$ has a positive derivative $\alpha'(t) \geq \varepsilon$ and so there exists a unique $\tau_0 \geq 0$ such that $\alpha(\tau_0) = 0$. For all later times we have $\alpha(t) = t - h(t) > 0$. Hence we may choose the initial time t_0 arbitrarily in $[\tau_0, \infty)$. A function $x(\cdot) : I \to \mathbb{K}^n$ on an interval $I = [t_0, t_1)$, $t_0 < t_1 \leq \infty$ is said to be a solution of the retarded differential equation

(9) with initial data $(t_0, x^0, \psi) \in [\tau_0, \infty) \times \mathbb{K}^n \times L^2(t_0 - h(t_0), t_0; \mathbb{K}^n)$ if it is absolutely continuous on I, and satisfies

$$\begin{aligned} \dot{x}^{\psi}(t) &= Ax^{\psi}(t) + BN(Cx^{\psi}(t - h(t))), \quad \text{a.e. } t \in I \\ x^{\psi}(t_0) &= x^0 \end{aligned}$$

where $x^{\psi} : [t_0 - h(t_0), t_1) \to \mathbb{K}^n$ is the following extension of $x(\cdot)$ to $[t_0 - h(t_0), t_1)$

$$x^{\psi}(t) = \begin{cases} \psi(t) & \text{if } t \in [t_0 - h(t_0), t_0) \\ x(t) & \text{if } t \geq t_0 \end{cases}, \qquad t \in [t_0 - h(t_0), t_1). \tag{10}$$

Now define the operator $\mathcal{N} : L^2(\mathbb{R}_+; \mathbb{K}^q) \to L^2(\mathbb{R}_+; \mathbb{K}^\ell)$ by

$$\mathcal{N}(z(\cdot))(t) = \begin{cases} 0 & \text{if } 0 \leq t < \tau_0 \\ N(z(t - h(t))) & \text{if } t \geq \tau_0 \end{cases}, \qquad z(\cdot) \in L^2(\mathbb{R}_+; \mathbb{K}^q). \tag{11}$$

To show that $\mathcal{N} \in \mathcal{P}_d(\mathbb{K})$ we make use of $\alpha(t) = t - h(t)$ as a (strictly increasing, continuously differentiable) time transformation. $\alpha(\cdot)$ maps $[\tau_0, \infty)$ onto $\mathbb{R}_+$. Denoting the inverse of $\alpha(\cdot)$ by $\beta(\cdot)$ we obtain for arbitrary $z(\cdot)$, $\tilde{z}(\cdot) \in L^2(\mathbb{R}_+; \mathbb{K}^q)$

$$\begin{aligned} \|\mathcal{N}(z(\cdot)) - \mathcal{N}(\tilde{z}(\cdot))\|^2_{L^2(\mathbb{R}_+;\mathbb{K}^\ell)} &= \int_{\tau_0}^{\infty} \|N(z(\alpha(t))) - N(\tilde{z}(\alpha(t)))\|^2_{\mathbb{K}^\ell}\, dt \\ &\leq \|N\|_n^2 \int_{\tau_0}^{\infty} \|z(\alpha(t)) - \tilde{z}(\alpha(t))\|^2_{\mathbb{K}^q}\, dt \\ &= \|N\|_n^2 \int_0^{\infty} \|z(\alpha) - \tilde{z}(\alpha)\|^2_{\mathbb{K}^q} \frac{1}{\alpha'(\beta(\alpha))} d\alpha \\ &\leq \|N\|_n^2\, \varepsilon^{-1} \|z - \tilde{z}\|^2_{L^2(\mathbb{R}_+;\mathbb{K}^q)}. \end{aligned} \tag{12}$$

In particular, if $z(t) = \tilde{z}(t)$ almost everywhere, then $\mathcal{N}(z(\cdot))(t) = \mathcal{N}(\tilde{z}(\cdot))(t)$ almost everywhere and so $\mathcal{N}$ is well defined on $L^2(\mathbb{R}_+; \mathbb{K}^q)$ by (11). Moreover choosing $\tilde{z}(t) \equiv 0$ we see that $\mathcal{N}$ maps $L^2(\mathbb{R}_+; \mathbb{K}^q)$ into $L^2(\mathbb{R}_+; \mathbb{K}^\ell)$. Since it has finite Lipschitz gain by (12) and is causal by definition, we obtain $\mathcal{N} \in \mathcal{P}_d(\mathbb{K})$ and $\|\mathcal{N}\|_d \leq \varepsilon^{-1/2} \|N\|_n$.
Now suppose that $x(\cdot) : I \to \mathbb{K}^n$ is a solution of (9) on $I = [t_0, t_1)$ with initial data $(t_0, x^0, \psi) \in [\tau_0, \infty) \times \mathbb{K}^n \times L^2(t_0 - h(t_0), t_0; \mathbb{K}^n)$ and define $z_\phi(\cdot) \in L^2(\mathbb{R}_+; \mathbb{K}^q)$ by (8) with

$$\phi(t) = \begin{cases} 0 & \text{if } t \in [0, t_0 - h(t_0)) \\ C\psi(t) & \text{if } t \in [t_0 - h(t_0), t_0) \end{cases}. \tag{13}$$

Then $z_\phi(t) = Cx^{\psi}(t)$ for $t \in (t_0 - h(t_0), t_1)$ and hence $\mathcal{N}(z_\phi(\cdot))(t) = N(Cx^{\psi}(t - h(t)))$ for $t \in I$ since $t_0 \geq \tau_0$. Applying Definition 5.6.4 we see that $x(\cdot)$ is a solution of $\Sigma_{\mathcal{N}}$ on I. Conversely, if $x(\cdot)$ is a solution of $\Sigma_{\mathcal{N}}$ on I with initial data $(t_0, x^0, \phi) \in [\tau_0, \infty) \times \mathbb{K}^n \times L^2(0, t_0; \mathbb{K}^q)$ and the initial function ϕ is of the form (13) with $\psi \in L^2(t_0 - h(t_0), t_0; \mathbb{K}^n)$ then $x(\cdot)$ is a solution of (9) on I with initial condition (t_0, x^0, ψ). □

Example 5.6.6. Consider a system

$$\begin{bmatrix} \dot{x}_1(t) \\ \dot{x}_2(t) \end{bmatrix} = \begin{bmatrix} A_{11} & A_{12} \\ A_{21} & A_{22} \end{bmatrix} \begin{bmatrix} x_1(t) \\ x_2(t) \end{bmatrix}, \qquad t \geq 0 \tag{14}$$

where $(A_{11}, A_{12}, A_{21}, A_{22}) \in \mathbb{K}^{n_1\times n_1} \times \mathbb{K}^{n_1\times n_2} \times \mathbb{K}^{n_2\times n_1} \times \mathbb{K}^{n_2\times n_2}$, $\sigma(A_{22}) \subset \mathbb{C}_-$. Suppose that $\dot{x}_1(t) = A_{11}x_1(t)$ is taken as a simplified model for the system (14), neglecting a part of the stable dynamics of the overall system. Let $(A, B, C) = (A_{11}, A_{12}, A_{21})$ and

$$\mathcal{N}(z(\cdot))(t) = \int_0^t e^{A_{22}(t-s)}z(s)ds, \quad t \in \mathbb{R}_+, \qquad z(\cdot) \in L^2(\mathbb{R}_+; \mathbb{K}^{n_2}). \tag{15}$$

Since $\sigma(A_{22}) \subset \mathbb{C}_-$ it follows from (3.45) that (15) defines a bounded linear operator $\mathcal{N} : L^2(\mathbb{R}_+; \mathbb{K}^{n_2}) \to L^2(\mathbb{R}_+; \mathbb{K}^{n_2})$ with operator norm

$$\|\mathcal{N}\| = \max_{\omega\in\mathbb{R}} \|(\imath\omega I_{n_2} - A_{22})^{-1}\|.$$

Since $\mathcal{N}$ is linear we have $\|\mathcal{N}\| = \|\mathcal{N}\|_d$ (see Remark 5.6.2). We want to represent the overall system (14) as a dynamic perturbation of the reduced model $\dot{x}_1 = A_{11}x_1$, see Figure 5.6.2. Clearly $\mathcal{N}$ is causal and so $\mathcal{N} \in \mathcal{P}_d(\mathbb{K})$. Let $x_1(\cdot)$ be a trajectory of the

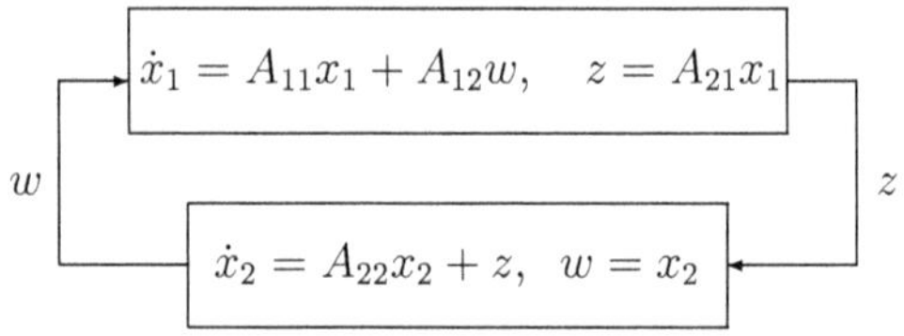

Figure 5.6.2: Feedback representation of neglected dynamics

system $\Sigma_{\mathcal{N}}$ with initial data $(t_0, x_1^0, \phi) \in (0,\infty) \times \mathbb{K}^{n_1} \times L^2(0, t_0; \mathbb{K}^{n_2})$ where $t_0 > 0$ is fixed. By Definition 5.6.4 and (15) this means

$$\begin{aligned} \dot{x}_1(t) &= A_{11}x_1(t) + A_{12}\int_{t_0}^t e^{A_{22}(t-s)}A_{21}x_1(s)ds + A_{12}\int_0^{t_0} e^{A_{22}(t-s)}\phi(s)ds, \ t \geq t_0 \\ x_1(t_0) &= x_1^0. \end{aligned} \tag{16}$$

Now define $x_2(\cdot) : [t_0, \infty) \to \mathbb{K}^{n_2}$ by

$$x_2(t) = e^{A_{22}t}\left[\int_{t_0}^t e^{-A_{22}s}A_{21}x_1(s)ds + \int_0^{t_0} e^{-A_{22}s}\phi(s)ds\right], \quad t \geq t_0. \tag{17}$$

Then

$$\dot{x}_2(t) = A_{22}x_2(t) + A_{21}x_1(t), \quad t \geq t_0, \quad x_2(t_0) = x_2^0 \tag{18}$$

where

$$x_2^0 = \int_0^{t_0} e^{A_{22}(t_0-s)}\phi(s)ds. \tag{19}$$

So every trajectory $x_1(\cdot)$ of $\Sigma_{\mathcal{N}}$ with initial data (t_0, x_1^0, ϕ) defines via (17) a function $x_2(\cdot) : [t_0, \infty) \to \mathbb{K}^{n_2}$ such that $(x_1(t), x_2(t))$ satisfies the overall system (14) with $(x_1(t_0), x_2(t_0)) = (x_1^0, x_2^0)$.

Conversely, let $(x_1(t), x_2(t))$ be a solution of (14) with $(x_1(t_0), x_2(t_0)) = (x_1^0, x_2^0) \in \mathbb{K}^{n_1} \times \mathbb{K}^{n_2}$. There exists $\phi \in L^2(0, t_0; \mathbb{K}^{n_2})$ such that (19) holds. In fact it suffices to choose

$$\phi(s) = e^{A_{22}^*(t_0-s)}y \quad \text{where} \quad \left(\int_0^{t_0} e^{A_{22}\tau}e^{A_{22}^*\tau}d\tau\right)y = x_2^0 \tag{20}$$

(which is always possible since $t_0 > 0$ and the integral defines a positive definite matrix). As a consequence $x_2(\cdot)$ satisfies (17) and so (16) holds. But then, as we have seen above, $x_1(\cdot)$ is a trajectory of $\Sigma_{\mathcal{N}}$ with initial condition (t_0, x_1^0, ϕ). □

For solutions of the perturbed system equations $\Sigma_{\mathcal{N}}$ (4) we have the following existence and uniqueness theorem, together with some estimates which we will use later in the stability analysis of these systems.

Theorem 5.6.7. *Suppose $\mathcal{N} \in \mathcal{P}_d(\mathbb{K})$, $\|\mathcal{N}\|_d \le \gamma < r_{\mathbb{C}}^-(A;B,C)$, $(t_0, x^0, \phi) \in \mathbb{R}_+ \times \mathbb{K}^n \times L^2(0,t_0;\mathbb{K}^q)$, then there exists a unique solution $x(\cdot) = x(\cdot;t_0,x^0,\phi)$ of $\Sigma_{\mathcal{N}}$ on $[t_0,\infty)$ with initial data (t_0,x^0,ϕ). Moreover there exist constants K_1, K_2 depending only on A, B, C, γ such that for all initial data $(t_0,x^0,\phi) \in \mathbb{R}_+ \times \mathbb{K}^n \times L^2(0,t_0;\mathbb{K}^q)$*

$$\begin{aligned} \|x(\cdot;t_0,x^0,\phi)\|_{L^2(t_0,\infty;\mathbb{K}^n)} &\le K_1\left(\|x^0\| + \|\phi\|_{L^2(0,t_0;\mathbb{K}^q)}\right) && (21)\\ \|x(t;t_0,x^0,\phi)\|_{\mathbb{K}^n} &\le K_2\left(\|x^0\| + \|\phi\|_{L^2(0,t_0;\mathbb{K}^q)}\right), \quad t \ge t_0 && (22)\end{aligned}$$

and $x(t;t_0,x^0,\phi) \to 0$ as $t \to \infty$.

Proof: Let $\mathbb{L}_+ : L^2(\mathbb{R}_+;\mathbb{K}^\ell) \to L^2(\mathbb{R}_+;\mathbb{K}^q)$ be the input-output operator of the system $(A,B,C,0)$ as defined by (3.46), see also (2.3.22). Suppose $(t_0,x^0,\phi) \in \mathbb{R}_+ \times \mathbb{K}^n \times L^2(0,t_0;\mathbb{K}^q)$ is given and assume that $\mathcal{N} \in \mathcal{P}_d(\mathbb{K})$ satisfies $\|\mathbb{L}_+\,\mathcal{N}\|_d \le c$ for some constant $c < 1$. Note that this assumption is automatically satisfied if $\|\mathcal{N}\|_d \le \gamma < r_{\mathbb{C}}^-(A;B,C)$ since $r_{\mathbb{C}}^-(A;B,C) = \|\mathbb{L}_+\|^{-1}$ by Corollary 5.3.23 and so

$$\|\mathbb{L}_+\mathcal{N}\|_d \le \|\mathbb{L}_+\|_d\,\|\mathcal{N}\|_d \le r_{\mathbb{C}}^-(A;B,C)^{-1}\gamma =: c < 1.$$

Consider the operator $\Psi : L^2(t_0,\infty;\mathbb{K}^q) \to L^2(t_0,\infty;\mathbb{K}^q)$ defined by

$$\Psi(z(\cdot))(t) = Ce^{A(t-t_0)}x^0 + C\int_{t_0}^{t} e^{A(t-s)}B\mathcal{N}(z_\phi(\cdot))(s)\,ds,\ t \ge t_0,\ z(\cdot) \in L^2(t_0,\infty;\mathbb{K}^q) \quad (23)$$

where $z_\phi(\tau) = z(\tau)$ for $\tau \ge t_0$ and $z_\phi(\tau) = \phi(\tau)$ for $\tau \in (0,t_0)$. Then

$$\|z_\phi\|_{L^2(\mathbb{R}_+)} = \left(\|\phi\|^2_{L^2(0,t_0)} + \|z\|^2_{L^2(t_0,\infty)}\right)^{1/2} \le \|\phi\|_{L^2(0,t_0)} + \|z\|_{L^2(t_0,\infty)} \quad (24)$$

and

$$\begin{aligned}\Psi(z(\cdot))(t) &= Ce^{A(t-t_0)}x^0 + \mathbb{L}_+\mathcal{N}(z_\phi(\cdot))(t) - Ce^{A(t-t_0)}\int_0^{t_0} e^{A(t_0-s)}B\mathcal{N}(z_\phi(\cdot))(s)\,ds\\ &= z^0(t) + \mathbb{L}_+\mathcal{N}(z_\phi(\cdot))(t), && (25)\end{aligned}$$

where

$$z^0(t) = Ce^{A(t-t_0)}\left(x^0 - \int_0^{t_0} e^{A(t_0-s)}B\mathcal{N}(z_\phi(\cdot))(s)\,ds\right), \quad t \ge t_0. \quad (26)$$

Clearly $z^0(\cdot) \in L^2(t_0,\infty;\mathbb{K}^q)$ and if $z(\cdot)$, $\tilde{z}(\cdot) \in L^2(t_0,\infty;\mathbb{K}^q)$ then by causality $\mathcal{N}(z_\phi(\cdot))(t) = \mathcal{N}(\tilde{z}_\phi(\cdot))(t)$ for a.e. $t \in [0,t_0]$. Hence with norms in $L^2 = L^2(t_0,\infty;\mathbb{K}^q)$

$$\|\Psi(z(\cdot)) - \Psi(\tilde{z}(\cdot))\|_{L^2} = \|\mathbb{L}_+\mathcal{N}(z_\phi(\cdot)) - \mathbb{L}_+\mathcal{N}(\tilde{z}_\phi(\cdot))\|_{L^2} \le c\|z - \tilde{z}\|_{L^2}\,.$$

So Ψ is a contraction on $L^2(t_0,\infty;\mathbb{K}^q)$. Let $z(\cdot)$ be the (unique) fixed point of Ψ and consider $x(\cdot) : [t_0,\infty) \to \mathbb{K}^n$ defined by

$$x(t) = e^{A(t-t_0)}x^0 + \int_{t_0}^{t} e^{A(t-s)}B\mathcal{N}(z_\phi(\cdot))(s)\,ds\,, \quad t \ge t_0. \quad (27)$$

Then $x(\cdot)$ is absolutely continuous on $[t_0,\infty)$, $x(t_0) = x^0$ and

$$Cx(t) = (\Psi(z(\cdot))(t) = z(t), \qquad \text{a.e.} \quad t \geq t_0 .$$

So $x(\cdot)$ solves the initial value problem formulated in Definition 5.6.4. For any other solution $\tilde{x}(\cdot)$ of this initial value problem, the function $\tilde{z}(t) = C\tilde{x}(t)$, $t \geq t_0$ is also a fixed point of Ψ and hence $\tilde{z}(t) = z(t)$, a.e. $t \geq t_0$. But then $\tilde{z}_\phi = z_\phi$ and thus $x(t) = \tilde{x}(t)$, $t \geq t_0$.
For the remainder of the proof define $\phi_0(\cdot) \in L^2(0,\infty;\mathbb{K}^q)$ by $\phi_0(t) = \phi(t)$ for $t \in [0,t_0]$ and $\phi_0(t) = 0$ for $t > t_0$. Then by the Cauchy-Schwarz inequality (A.3.20) and the causality of $\mathcal{N}$ there exists a constant $\alpha > 0$ such that

$$\int_0^{t_0} \|e^{A(t_0-s)} B\mathcal{N}(z_\phi)(s)\|\, ds \leq \alpha\Big(\int_0^{t_0} \|\mathcal{N}(z_\phi)(s)\|^2\, ds\Big)^{1/2} = \alpha\Big(\int_0^{t_0} \|\mathcal{N}(\phi_0)(s)\|^2\, ds\Big)^{1/2}$$
$$\leq \alpha\|\mathcal{N}(\phi_0)\|_{L^2(0,\infty;\mathbb{K}^\ell)} \leq \alpha\|\mathcal{N}\|_d\, \|\phi\|_{L^2(0,t_0;\mathbb{K}^q)}$$

for any $z(\cdot) \in L^2(t_0,\infty;\mathbb{K}^q)$. Since $\sigma(A) \subset \mathbb{C}_-$ it follows from (24), (25) and (26) that there exists a constant K such that the fixed point $z(\cdot)$ of Ψ satisfies

$$\|z\|_{L^2(t_0,\infty)} = \|\Psi z\|_{L^2(t_0,\infty)} \leq \|z^0\|_{L^2(t_0,\infty)} + \|\mathbb{L}_+\mathcal{N}(z_\phi)\|_{L^2(t_0,\infty)}$$
$$\leq K\left(\|x^0\| + \alpha\|\mathcal{N}\|_d\, \|\phi\|_{L^2(0,t_0)}\right) + c\left(\|\phi\|_{L^2(0,t_0)} + \|z\|_{L^2(t_0,\infty)}\right) .$$

Now $0 < c < 1$ and so there exists a constant K_3 such that

$$\|z(\cdot)\|_{L^2(t_0,\infty;\mathbb{K}^q)} \;\leq\; K_3\left(\|x^0\| + \|\phi\|_{L^2(0,t_0;\mathbb{K}^q)}\right) . \tag{28}$$

By (24) and (28) $w(\cdot) := \mathcal{N}(z_\phi)(\cdot) \in L^2(\mathbb{R}_+;\mathbb{K}^\ell)$ satisfies

$$\|w\|_{L^2(\mathbb{R}_+)} \;\leq\; \|\mathcal{N}\|_d(\|\phi\|_{L^2(0,t_0)} + \|z\|_{L^2(t_0,\infty)}) \leq K_3'\|\mathcal{N}\|_d\left(\|x^0\| + \|\phi\|_{L^2(0,t_0)}\right) . \tag{29}$$

where $K_3' = K_3 + 1$. Since $\|e^{At}\| \leq Me^{-\omega t}$ for some $M \geq 1$, $\omega > 0$ by the convolution inequality (A.3.24)

$$\|x(\cdot)\|_{L^2(t_0,\infty;\mathbb{K}^n)} \leq (M/\sqrt{2\omega})\|x^0\| + \|e^{A\cdot}B\|_{L^1(\mathbb{R}_+;\mathbb{K}^{n\times\ell})}\|w(\cdot)\|_{L^2(t_0,\infty;\mathbb{K}^\ell)}$$

which together with (29) proves (21) for a sufficiently large K_1. (22) and the fact that $x(t;t_0,x^0,\phi) \to 0$ as $t \to \infty$ follows from Proposition 2.3.10 (with $C = I_n$). □

Remark 5.6.8. (i) The previous proof shows that if $\mathcal{N} \in \mathcal{P}_d(\mathbb{K})$, $\|\mathbb{L}_+\mathcal{N}\|_d \leq c$ for some $c < 1$ and $(t_0, x^0, \phi) \in \mathbb{R}_+ \times \mathbb{K}^n \times L^2(0,t_0;\mathbb{K}^q)$ then there exists a unique solution $x(\cdot)$ of $\Sigma_\mathcal{N}$ on $[t_0,\infty)$ with initial data (t_0,x^0,ϕ) which satisfies the inequality (22).
Note that if $\mathcal{N} \in \mathcal{P}_d(\mathbb{K})$ is a linear convolution operator defined by an integrable convolution kernel $\mathcal{K}(\cdot)$ (see Section 2.3) then the transfer matrix corresponding to the input output operator $\mathbb{L}_+\mathcal{N} : L^2(0,\infty;\mathbb{K}^q) \to L^2(0,\infty;\mathbb{K}^q)$ is given by $C(sI_n - A)^{-1}BK(s)$ where $K(s)$ is the Laplace transform of $\mathcal{K}(t)$. In this case $\|\mathcal{N}\|_d = \|K(\cdot)\|_{H^\infty}$ by Theorem 2.3.28 and the assumption $\|\mathbb{L}_+\mathcal{N}\|_d < 1$ is equivalent to

$$\max_{\omega\in\mathbb{R}} \|C(\imath\omega I_n - A)^{-1}BK(\imath\omega)\| < 1.$$

(ii) Theorem 5.6.7 has been stated here in a simplified form which is needed later in the context of the robust stability problems studied in this section. However, existence and uniqueness can be proved without the stability assumption $\sigma(A) \subset \mathbb{C}_-$ and also for arbitrary $\mathcal{N} \in \mathcal{P}_d(\mathbb{K})$. Clearly in this case the inequality (22) must be modified to allow for exponential growth, see *Notes and References.*

(iii) The theorem can be extended to a wider perturbation class where in the definition of $\mathcal{P}_d(\mathbb{K})$ the global Lipschitz condition is replaced by the weaker condition that $\mathcal{N}$ is Lipschitz continuous (i.e. locally Lipschitz) and of *finite gain*, i.e.

$$\|\mathcal{N}(z(\cdot))\|_{L^2(\mathbb{R}_+;\mathbb{K}^\ell)} \leq \gamma \|z\|_{L^2(\mathbb{R}_+;\mathbb{K}^q)}, \qquad z(\cdot) \in L^2(\mathbb{R}_+;\mathbb{K}^q) \tag{30}$$

for some constant $\gamma > 0$. Let $\tilde{\mathcal{P}}_d(\mathbb{K})$ denote this class of perturbations with norm $\|\mathcal{N}\|_{\tilde{\mathcal{P}}_d} =$ the minimal gain γ such that (30) holds. Then the previous theorem and also the following propositions can be proved with $\mathcal{P}_d(\mathbb{K})$ replaced by $\tilde{\mathcal{P}}_d(\mathbb{K})$. However the proofs are more complicated. These results are stronger in two respects: the perturbation class is wider and the norm is weaker, i.e. $\|\mathcal{N}\|_{\tilde{\mathcal{P}}_d} \leq \|\mathcal{N}\|_{\mathcal{P}_d}$, see *Notes and References.* □

The stability concept we use in this section is that of global asymptotic stability. The definition is the standard one for Σ_N, $\Sigma_{\Delta(t)}$ (see Definition 3.1.6). However since we have not developed a state space description of $\Sigma_\mathcal{N}$, our general definitions in Chapter 3 are not applicable. Instead we give a "phase space" definition viewing the solutions of $\Sigma_\mathcal{N}$ as trajectories in $\mathbb{K}^n$. In the definition we make use of the fact that for every $(t_0, x^0, \phi) \in \mathbb{R}_+ \times \mathbb{K}^n \times L^2(0, t_0; \mathbb{K}^q)$ there exists a unique solution $x(\cdot) = x(\cdot; t_0, x^0, \phi)$ of $\Sigma_\mathcal{N}$ on $[t_0, \infty)$ with initial data (t_0, x^0, ϕ), see Remark 5.6.8 (ii).

Definition 5.6.9. $\Sigma_\mathcal{N}$ is said to be *globally asymptotically stable* (g.a.s.) if it satisfies the following two conditions

(i) The origin $\overline{x} = 0$ is *stable* for $\Sigma_\mathcal{N}$, i.e. for every $t_0 \geq 0$ and $\varepsilon > 0$ there exists $\delta = \delta(t_0, \varepsilon)$, such that

$$\|x^0\| < \delta\,, \quad \|\phi\|_{L^2(0,t_0;\mathbb{K}^q)} < \delta \implies \forall t \geq t_0 : \|x(t; t_0, x^0, \phi)\| < \varepsilon.$$

(ii) The origin $\overline{x} = 0$ is *globally attractive*, i.e. for all $(t_0, x^0, \phi) \in \mathbb{R}_+ \times \mathbb{K}^n \times L^2(0, t_0; \mathbb{K}^q)$ we have $\lim_{t\to\infty} x(t; t_0, x^0, \phi) = 0$.

We say that a perturbation $\mathcal{N} \in \mathcal{P}_d(\mathbb{K})$ *destabilizes* the system $\Sigma = (A, B, C)$ if $\Sigma_\mathcal{N}$ is not g.a.s..

Remark 5.6.10. (i) For memoryless linear time-invariant perturbations global asymptotic stability is equivalent to global exponential stability, but this is not the case for the more general classes of perturbations considered here. It is rather difficult to work with a stability concept requiring exponential decay for the perturbation class $\mathcal{P}_d(\mathbb{K})$. However in Subsection 5.6.3 we will obtain some results for the class $\mathcal{P}_n(\mathbb{K})$.

(ii) Whilst *global* asymptotic stability is a desirable property, most nonlinear perturbations in applications will not preserve it, e.g. if the nonlinearly perturbed system Σ_N possesses more than one equilibrium state. When this is the case one is interested in conditions which ensure local asymptotic stability with an appropriate domain of attraction around the given equilibrium point. We will say something about this in Subsection 5.6.3. □

Example 5.6.11. Consider again the system (14) and $\Sigma_\mathcal{N}$ as defined in Example 5.6.6. We have seen there that if $x_1(\cdot)$ is a trajectory of $\Sigma_\mathcal{N}$ with initial data $(t_0, x^0, \phi) \in (0, \infty) \times \mathbb{K}^{n_1} \times L^2(0, t_0; \mathbb{K}^q)$ then there exists a function $x_2(\cdot) : [t_0, \infty) \to \mathbb{K}^{n_2}$ such that $(x_1(\cdot), x_2(\cdot))$

is a solution of (14) with $x_1(t_0) = x_1^0$, $x_2(t_0) = x_2^0$ where by (19) $\|x_2^0\| \le K_1 \|\phi\|_{L^2}$ for some constant K_1. Conversely, if $(x_1(\cdot), x_2(\cdot))$ is a solution of (14) with $x_1(t_0) = x_1^0$, $x_2(t_0) = x_2^0$ then $x_1(\cdot)$ is a solution of $\Sigma_{\mathcal{N}}$ with initial data (t_0, x^0, ϕ) where $\phi \in L^2(0, t_0; \mathbb{K}^q)$ is defined by (20) and satisfies $\|\phi\|_{L^2} \le K_2 \|x_2^0\|$ for some constant $K_2 > 0$. From these two facts it follows that the overall system (14) is stable if and only if $\Sigma_{\mathcal{N}}$ is stable in the sense of Definition 5.6.9 (i). Moreover, if (14) is g.a.s. then every trajectory $x_1(t)$ of $\Sigma_{\mathcal{N}}$ being the first component of a solution of (14) tends to zero as $t \to \infty$. Conversely, if $\Sigma_{\mathcal{N}}$ is g.a.s. then (14) must be g.a.s. since otherwise (14) would be stable but not asymptotically stable and hence possess a non-zero periodic eigenmotion $(x_1(\cdot), x_2(\cdot))$ corresponding to a purely imaginary eigenvalue. But this implies a contradiction. In fact the first component $x_1(\cdot)$ is a trajectory of $\Sigma_{\mathcal{N}}$, hence must be identically zero if periodic because $\Sigma_{\mathcal{N}}$ is assumed to be asymptotically stable. But then the second component would be a non-zero periodic solution of $\dot{x}_2 = A_{22}x_2$, which is impossible since by assumption $\sigma(A_{22}) \subset \mathbb{C}_-$. □

5.6.2 Stability Radii

In this subsection we introduce both real and complex stability radii for the three perturbation classes described above and explore the relationships between them.

Definition 5.6.12. The stability radius of A with respect to the perturbation structure (B, C) and perturbation class $\mathcal{P}_d(\mathbb{K})$ is

$$r_{\mathbb{K},d}^{-}(A; B, C) = \inf\{\|\mathcal{N}\|_d;\ \mathcal{N} \in \mathcal{P}_d(\mathbb{K}) \text{ and } \Sigma_{\mathcal{N}} \text{ is not g.a.s.}\}\ .$$

Stability radii with respect to the perturbation classes $\mathcal{P}_n(\mathbb{K})$, $\mathcal{P}_t(\mathbb{K})$ are defined in an analogous way; they are denoted by $r_{\mathbb{K},n}^{-}(A; B, C)$ and $r_{\mathbb{K},t}^{-}(A; B, C)$, respectively.[1]

Proposition 5.6.13.

$$r_{\mathbb{K}}^{-}(A; B, C) \ge r_{\mathbb{K},n}^{-}(A; B, C) \ge r_{\mathbb{K},t}^{-}(A; B, C) \ge r_{\mathbb{K},d}^{-}(A; B, C)\ . \qquad (31)$$

Proof: In view of the norm preserving inclusions (5), (6) it is only necessary to prove $r_{\mathbb{K},n}^{-} \ge r_{\mathbb{K},t}^{-}$. Assume $N \in \mathcal{P}_n(\mathbb{K})$, $\|N\|_n < r_{\mathbb{K},t}^{-}$, then $\|N'(0)\| \le \|N\|_n < r_{\mathbb{K}}^{-}$ and so the linearization of Σ_N at $\overline{x} = 0$, $\dot{x} = (A + BN'(0)C)x$, is asymptotically stable. It follows from Liapunov's linearization Theorem 3.3.52 that the origin is asymptotically stable for Σ_N. Now suppose $x(\cdot) : \mathbb{R}_+ \to \mathbb{K}^n$ is an arbitrary solution of Σ_N and $z(t) = Cx(t)$. Define $\Delta \in \mathcal{P}_t(\mathbb{K})$ by

$$\Delta(t) = \begin{cases} N(z(t))z^*(t)/\|z(t)\|^2 & \text{if } z(t) \neq 0, \\ 0 & \text{otherwise,} \end{cases} \qquad t \in \mathbb{R}_+.$$

Then $x(\cdot)$ is a trajectory of $\Sigma_{\Delta(t)}$. Since $\|\Delta\|_t \le \|N\|_n < r_{\mathbb{K},t}^{-}$ we conclude that $x(t) \to 0$ as $t \to \infty$. Hence Σ_N is g.a.s.. □

In fact equalities hold in (31) if $\mathbb{K} = \mathbb{C}$. To see this note that if $\mathcal{N} \in \mathcal{P}_d(\mathbb{C})$, $\|\mathcal{N}\|_d < r_{\mathbb{C}}^{-}$ then by Theorem 5.6.7 $\Sigma_{\mathcal{N}}$ is g.a.s.. Hence $r_{\mathbb{C},d}^{-} \ge r_{\mathbb{C}}^{-}$ and so we obtain by (31)

[1] Here $^{-}$ is not used as an indicator of spectral constraints but to distinguish continuous time stability radii from discrete time ones.

Theorem 5.6.14.

$$r^-_{\mathbb{C}}(A;B,C) = r^-_{\mathbb{C},n}(A;B,C) = r^-_{\mathbb{C},t}(A;B,C) = r^-_{\mathbb{C},d}(A;B,C)\ . \tag{32}$$

As a consequence we see that the *complex* stability radius has the remarkable property that it is invariant under the above extensions of the perturbation class $\mathbb{C}^{\ell\times q}$.

Example 5.6.15. Consider the system (9) with time-varying delay described in Example 5.6.5 and assume $N \in \mathcal{P}_n(\mathbb{K})$, $\|N\|_n < \varepsilon^{1/2} r^-_{\mathbb{C}}(A;B,C)$ and $h'(t) \le 1-\varepsilon$ for $t \ge 0$. If $x(\cdot)$ solves the delay system with initial data $(t_0, x^0, \psi) \in [\tau_0, \infty) \times \mathbb{K}^n \times L^2(t_0 - h(t_0), t_0; \mathbb{K}^n)$ we have seen that $x(\cdot)$ is a solution of (4) with initial data (t_0, x^0, ϕ) where $\mathcal{N}$ is defined by (11) and ϕ is defined by (13). Since $\|\mathcal{N}\|_d \le \varepsilon^{-1/2}\|N\|_n < r^-_{\mathbb{C}}(A;B,C)$ by (12) it follows from Theorem 5.6.14 that the delay system is globally asymptotically stable for all $t_0 \ge \tau_0$. In particular if $h(t) \equiv h > 0$ is an arbitrary constant, then (9) will be g.a.s. for all nonlinearities $N \in \mathcal{P}_n(\mathbb{K})$ satisfying $\|N\|_n < r^-_{\mathbb{C}}(A;B,C)$. □

Example 5.6.16. Consider the system (14) in Example 5.6.6, where $\sigma(A_{11}) \subset \mathbb{C}_-$, and assume that

$$\sigma(A_{22}) \subset \mathbb{C}_- \quad \text{and} \quad \max_{\omega\in\mathbb{R}} \|A_{21}(\imath\omega I_{n_1} - A_{11})^{-1}A_{12}\| < \min_{\omega\in\mathbb{R}} \sigma_{\min}(\imath\omega I_{n_2} - A_{22}). \tag{33}$$

By (5.3.45) the linear operator $\mathcal{N} : L^2(\mathbb{R}_+;\mathbb{K}^{n_2}) \to L^2(\mathbb{R}_+;\mathbb{K}^{n_2})$ as defined by (15) has norm $\|\mathcal{N}\|_d = \max_{\omega\in\mathbb{R}} \|(\imath\omega I_{n_2} - A_{22})^{-1}\|$. Hence by Theorem 5.6.14, $\Sigma_{\mathcal{N}}$ is g.a.s. if

$$\max_{\omega\in\mathbb{R}} \|(\imath\omega I_{n_2} - A_{22})^{-1}\| < r^-_{\mathbb{C}}(A_{11};A_{12},A_{21}) = \left[\max_{\omega\in\mathbb{R}} \|A_{21}(\imath\omega I_{n_1} - A_{11})^{-1}A_{12}\|\right]^{-1}.$$

In conclusion we see from Examples 5.6.6 and 5.6.11 that the overall system (14) is asymptotically stable if the system with neglected dynamics $\dot{x}_1 = A_{11}x_1$ is asymptotically stable and the assumption (33) is satisfied. □

Remark 5.6.17. It follows from the previous theorem that equalities hold in (31) in the real case *if* $r^-_{\mathbb{R}} = r^-_{\mathbb{C}}$. In particular, if A, B, C are real and $\max\{\|G(\imath\omega)\|; \omega \in \mathbb{R}\} = \|G(0)\|$ holds for $G(s) = C(sI_n - A)^{-1}$ then by Remark 5.3.11 and Theorem 5.3.9

$$r^-_{\mathbb{R}}(A;B,C) = r^-_{\mathbb{R},n}(A;B,C) = r^-_{\mathbb{R},t}(A;B,C) = r^-_{\mathbb{R},d}(A;B,C) = r^-_{\mathbb{C}}(A;B,C) = \|G(0)\|^{-1}.$$

□

However, if $r^-_{\mathbb{R}} > r^-_{\mathbb{C}}$, the stability radius with respect to time-varying perturbations, $r^-_{\mathbb{R},t}$, may lie anywhere between $r^-_{\mathbb{C}}$ and $r^-_{\mathbb{R}}$. We illustrate this by the next example.

Example 5.6.18. Consider the linear oscillator

$$\ddot{y}(t) + 2\alpha\dot{y}(t) + (1+\Delta(t))y(t) = 0\ ,\quad t \ge 0 \tag{34}$$

where $\alpha > 0$ is given and $\Delta : \mathbb{R}_+ \to \mathbb{R}$ represents an unknown time-varying perturbation of the restoring force. The associated state space system can be written in the form $\dot{x}(t) = (A(\alpha) + B\Delta(t)C)x(t)$ where

$$A(\alpha) = \begin{bmatrix} 0 & 1 \\ -1 & -2\alpha \end{bmatrix},\quad B = \begin{bmatrix} 0 \\ 1 \end{bmatrix},\quad C = \begin{bmatrix} 1 & 0 \end{bmatrix}.$$

We know from Example 5.3.14 that the real stability radius $r_{\mathbb{R}}^{-}(A(\alpha); B, C)$ equals 1 independently of the damping coefficient $\alpha > 0$ whereas the complex stability radius $r_{\mathbb{C}}^{-}(A(\alpha); B, C)$ equals 1 only for $\alpha \geq 1/\sqrt{2}$ and satisfies $r_{\mathbb{C}}^{-}(A(\alpha); B, C) = 2\alpha\sqrt{1-\alpha^2}$ for $0 < \alpha < 1/\sqrt{2}$. In particular, $r_{\mathbb{C}}^{-}(A(\alpha); B, C) \sim 2\alpha$ tends to zero as $\alpha \searrow 0$.
For a given matrix interval $[A^-, A^+]$ in $\mathbb{R}^{2\times 2}$, *Gonzalez* (1991) [199] has derived necessary and sufficient conditions under which all the time-varying systems $\dot{x}(t) = A(t)x(t)$ with measurable (or piecewise constant) $A(\cdot) : \mathbb{R}_+ \to [A^-, A^+]$ are asymptotically stable. These conditions enable us to decide, for any $\rho > 0$, if the perturbed oscillator (34) is asymptotically stable for all measurable functions $\Delta : \mathbb{R}_+ \to [-\rho, \rho]$. The stability radius $r_{\mathbb{R},t}^{-}(A(\alpha), B, C)$ is the supreme value of these ρ and can be determined by solving a set of three nonlinear scalar equations, see Ex. 2 and *Notes and References.* We omit the details and only present the result. Figure 5.6.3 shows the three graphs of

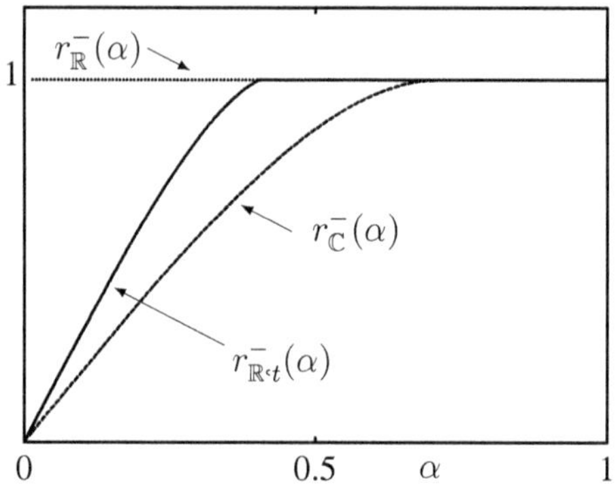

Figure 5.6.3: Stability radii $r_{\mathbb{C}}^{-}$, $r_{\mathbb{R},t}^{-}$ and $r_{\mathbb{R}}^{-}$ of the oscillator

$r_{\mathbb{R}}^{-}(\alpha) := r_{\mathbb{R}}^{-}(A(\alpha); B, C) \equiv 1$, $r_{\mathbb{C}}^{-}(\alpha) := r_{\mathbb{C}}^{-}(A(\alpha); B, C)$ and $r_{\mathbb{R},t}^{-}(\alpha) := r_{\mathbb{R},t}^{-}(A(\alpha), B, C)$ as functions of $\alpha \in (0, 1]$. By (31) and Theorem 5.6.14 the graph of $r_{\mathbb{R},t}^{-}(\alpha)$ must lie between the graphs of $r_{\mathbb{C}}^{-}(\alpha)$ and $r_{\mathbb{R}}^{-}(\alpha)$. Since the latter two graphs coincide for $\alpha \geq 1/\sqrt{2}$, $r_{\mathbb{R},t}^{-}(\alpha)$ must coincide with $r_{\mathbb{R}}^{-}(\alpha)$ for $\alpha \geq 1/\sqrt{2} \approx 0.7071$. In fact, one can show that $r_{\mathbb{R},t}^{-}(\alpha) = r_{\mathbb{R}}^{-}(\alpha) \equiv 1$ for $\alpha \geq \alpha_1$ where $\alpha_1 \approx 0.405$, see Figure 5.6.3.
In particular we see that if α is sufficiently small, (34) can be destabilized by a time-varying real disturbance $\Delta(t)$ of arbitrarily small L_∞-norm whereas there does not exist a constant real parameter perturbation Δ of size $|\Delta| < 1$ which destabilizes the system. It can be shown that $r_{\mathbb{R},t}^{-}(\alpha) \sim \pi\alpha$ as $\alpha \searrow 0$, hence $r_{\mathbb{R},t}^{-}(\alpha)/r_{\mathbb{C}}^{-}(\alpha)$ is bounded, see Ex. 2. □

A good deal of analysis has been carried out on $r_{\mathbb{R},t}^{-}$ but its computation is still a difficult problem even for modest dimensions n, see *Notes and References.* Much less is known about $r_{\mathbb{R},n}^{-}$. For both $r_{\mathbb{R},n}^{-}$ and $r_{\mathbb{R},t}^{-}$ there appear to be no general necessary and sufficient conditions under which they coincide with either $r_{\mathbb{R}}^{-}$ or $r_{\mathbb{C}}^{-}$. Also, to our knowledge there are no better general estimates available for both of them than $r_{\mathbb{C}}^{-} \leq r_{\mathbb{R},n}^{-} \leq r_{\mathbb{R}}^{-}$ and $r_{\mathbb{C}}^{-} \leq r_{\mathbb{R},t}^{-} \leq r_{\mathbb{R}}^{-}$. In contrast we have the following simple characterization of $r_{\mathbb{R},d}^{-}$.

Definition 5.6.19. A state space system $(\hat{A}, \hat{B}, \hat{C}, \hat{D}) \in \mathbf{L}_{\hat{n},\hat{\ell},\hat{q}}(\mathbb{K})$ is said to be a *multivariable oscillator* of dimension $\hat{n}$ if its transfer matrix is of the form

$$\hat{D} + \hat{C}(sI - \hat{A})^{-1}\hat{B} = \frac{1}{p(s)}[K_0 + K_1 s + K_2 s^2]$$

where the $K_i \in \mathbb{K}^{\hat{q}\times\hat{\ell}}$ are constant matrices and $p(s)$ is a non-zero polynomial of $\deg p \leq 2$.[2]

We will show that $r^-_{\mathbb{R},d} = r^-_{\mathbb{C}}$. The idea of the proof is to interpolate a complex minimum norm destabilizing perturbation matrix Δ by the transfer matrix of a stable *real* multivariable oscillator at a critical frequency where $\|C(\imath\omega I - A)^{-1}B\|$ attains its maximum.

Theorem 5.6.20. *For any real triple $(A,B,C) \in \mathbf{L}_{n,\ell,q}(\mathbb{K})$, $\sigma(A) \subset \mathbb{C}_-$ and $\varepsilon > 0$ there exists a stable real multivariable oscillator with input-output operator $\widehat{\mathbb{L}}$ for which the dynamic system $\Sigma_{\widehat{\mathbb{L}}}$ of the form (4) is not globally asymptotically stable and $\|\widehat{\mathbb{L}}\|_d \leq r^-_{\mathbb{C}}(A;B,C) + \varepsilon$. In particular*

$$r^-_{\mathbb{R},d}(A;B,C) = r^-_{\mathbb{C}}(A;B,C). \tag{35}$$

Proof: It follows from Theorem 5.6.14 that $r^-_{\mathbb{C}} = r^-_{\mathbb{C},d} \leq r^-_{\mathbb{R},d}$. To prove the converse inequality suppose $\Delta = \Delta_1 + \imath\Delta_2 \in \mathbb{C}^{\ell\times q}$, where $\Delta_1,\ \Delta_2 \in \mathbb{R}^{\ell\times q}$ are such that Δ is a minimum norm complex destabilizing perturbation of A with respect to the structure (B,C). Hence $\|\Delta\| = r^-_{\mathbb{C}}(A;B,C)$ and $\imath\mathbb{R} \cap \sigma(A+B\Delta C) \neq \emptyset$. Replacing Δ by $\overline{\Delta}$ if necessary, we may assume that $\imath\omega_0 \in \sigma(A+B\Delta C)$ for some $\omega_0 \geq 0$. If $\omega_0 = 0$ then $r^-_{\mathbb{C}} = r^-_{\mathbb{R}}$ and there exists a *real* destabilizing $\Delta \in \mathbb{R}^{\ell\times q}$ of norm $r^-_{\mathbb{C}}$. In this case no dynamics is needed in the feedback loop: It suffices to choose a zero-dimensional oscillator where $\hat{A}, \hat{B}, \hat{C}$ are void and $\hat{D} = \Delta$. Thus it remains to consider the case where $\omega_0 > 0$. By Lemma 5.2.7 we have

$$\det\left[I_\ell - \Delta C(\imath\omega_0 I_n - A)^{-1}B\right] = 0. \tag{36}$$

We want to construct a transfer matrix $\hat{\Delta}(s)$ of a stable real multivariable oscillator such that $\hat{\Delta}(\imath\omega_0) = \Delta$ and $\|\hat{\Delta}\|_d = \max_{\omega\in\mathbb{R}} \|\hat{\Delta}(\imath\omega)\|$ is close to $\|\Delta\|$. Consider

$$\hat{\Delta}(s) = \frac{\gamma^{-1}\Delta_2 s^2 + \omega_0\gamma^{-1}\Delta_1 s}{s^2 + \omega_0\gamma^{-1}s + \omega_0^2} \tag{37}$$

where $\gamma > 1$ is a parameter which will be chosen later to ensure that $\|\hat{\Delta}\|_d$ is close to $\|\Delta\| = r^-_{\mathbb{C}}$. An elementary calculation shows that $\hat{\Delta}(\imath\omega_0) = \Delta$ for all $\gamma > 1$ and $\hat{\Delta}(s)$ is the transfer matrix of the real multivariable oscillator $(\hat{A}, \hat{B}, \hat{C}, \hat{D})$ given by

$$\hat{A} = \begin{bmatrix} 0 & 1 \\ -\omega_0^2 & -\omega_0\gamma^{-1} \end{bmatrix} \otimes I_q = \begin{bmatrix} 0_{q\times q} & I_q \\ -\omega_0^2 I_q & -\omega_0\gamma^{-1}I_q \end{bmatrix}, \quad \hat{B} = \begin{bmatrix} 0 \\ 1 \end{bmatrix} \otimes I_q = \begin{bmatrix} 0_{q\times q} \\ I_q \end{bmatrix},$$
$$\hat{C} = \left[\ -\gamma^{-1}\omega_0^2\Delta_2\,,\ \ \gamma^{-1}\omega_0(\Delta_1 - \gamma^{-1}\Delta_2)\ \right], \quad \hat{D} = \gamma^{-1}\Delta_2\,.$$

Since $\det(sI_{2q} - \hat{A}) = (s^2 + \omega_0\gamma^{-1}s + \omega_0^2)^q$ by Proposition A.1.22 (k) , we have $\sigma(\hat{A}) \subset \mathbb{C}_-$. To see that the associated input-output operator $\widehat{\mathbb{L}} : L^2(\mathbb{R}_+;\mathbb{R}^q) \to L^2(\mathbb{R}_+;\mathbb{R}^\ell)$ defined by

$$\widehat{\mathbb{L}}(z(\cdot))(t) = \hat{D}z(t) + \hat{C}\int_0^t e^{\hat{A}(t-s)}\hat{B}z(s)\,ds\,, \quad z(\cdot) \in L^2(\mathbb{R}_+;\mathbb{R}^q)$$

[2] If $\deg p < 2$ then necessarily $K_i = 0$ for $\deg p < i \leq 2$.

destabilizes (A,B,C) we first prove that the feedback coupling of (A,B,C) with $(\hat{A},\hat{B},\hat{C},\hat{D})$ is not asymptotically stable. We have

$$\begin{aligned}
&\det\left(sI_{n+\hat{n}}-\begin{bmatrix} A+B\hat{D}C & B\hat{C} \\ \hat{B}C & \hat{A}\end{bmatrix}\right)\\
&=\det(sI_{\hat{n}}-\hat{A})\det[sI_n-(A+B\hat{D}C)-B\hat{C}(sI_{\hat{n}}-\hat{A})^{-1}\hat{B}C]\\
&=\det(sI_{\hat{n}}-\hat{A})\det\left(\left[I_n-B\hat{D}C(sI_n-A)^{-1}-B\hat{C}(sI_{\hat{n}}-\hat{A})^{-1}\hat{B}C(sI_n-A)^{-1}\right][sI_n-A]\right)\\
&=\det(sI_{\hat{n}}-\hat{A})\det(sI_n-A)\det\left[I_n-\hat{D}C(sI_n-A)^{-1}B-\hat{C}(sI_{\hat{n}}-\hat{A})^{-1}\hat{B}C(sI_n-A)^{-1}B\right]\\
&=\det(sI_{\hat{n}}-\hat{A})\det(sI_n-A)\det\left[I_n-\hat{\Delta}(s)C(sI_n-A)^{-1}B\right].
\end{aligned}$$

By (36) $\det[I_n-\hat{\Delta}(\imath\omega_0)C(\imath\omega_0 I_n-A)^{-1}B]=0$ and so there exists a non-zero real periodic solution $x(\cdot)$ of the feedback system

$$\begin{bmatrix}\dot{x}_1(t)\\ \dot{x}_2(t)\end{bmatrix}=\begin{bmatrix} A+B\hat{D}C & B\hat{C} \\ \hat{B}C & \hat{A}\end{bmatrix}\begin{bmatrix}x_1(t)\\ x_2(t)\end{bmatrix}. \tag{38}$$

The first component $x_1(\cdot)$ must be non-zero, since otherwise $x_2(\cdot)$ would be a non-zero periodic function satisfying $\dot{x}_2=\hat{A}x_2$ which is in contradiction with $\sigma(\hat{A})\subset\mathbb{C}_-$. Let $t_0>0$ be fixed and choose $\phi\in L^2(0,t_0,\mathbb{R}^{2q})$ such that

$$\int_0^{t_0} e^{\hat{A}(t_0-s)}\hat{B}\phi(s)ds=x_2(t_0). \tag{39}$$

That this is possible follows from the controllability of the multivariable oscillator $(\hat{A},\hat{B},\hat{C},\hat{D})$ (the reachability matrix $[\hat{B}\ \ \hat{A}\hat{B}]$ has full rank $\hat{n}=2q$, cf. Volume II). Proceeding as in Example 5.6.6 one can show that the non-zero periodic function $x_1(\cdot)$ is equal to the trajectory of the system $\Sigma_{\widehat{\mathbb{L}}}$ with initial data $(t_0,x_1(t_0),\phi)$. This proves that $\Sigma_{\widehat{\mathbb{L}}}$ is not g.a.s., and so the input-output operator $\widehat{\mathbb{L}}$ of the oscillator $(\hat{A},\hat{B},\hat{C},\hat{D})$ destabilizes (A,B,C). It follows that $\|\widehat{\mathbb{L}}\|_d\geq r_{\mathbb{C}}^-(A;B,C)$ by Theorem 5.6.14. Given any $\varepsilon>0$, it remains to prove that there exists $\gamma>1$ such that $\|\widehat{\mathbb{L}}\|_d\leq r_{\mathbb{C}}^-(A,B,C)+\varepsilon$. Now by (3.45)

$$\|\widehat{\mathbb{L}}\|_d^2=\max_{\omega\in\mathbb{R}}\|\hat{\Delta}(\imath\omega)\|^2 .$$

For any $z\in\mathbb{C}^q$, $\|z\|=1$ if $\eta:=\omega/\omega_0$, we have by (37)

$$\begin{aligned}
\|\hat{\Delta}(\imath\omega)z\|^2 &= \frac{\gamma^{-2}\omega^2\|(\omega_0\Delta_1+\imath\omega\Delta_2)z\|^2}{|(\omega_0^2-\omega^2)+\imath\omega\omega_0\gamma^{-1}|^2}=\frac{\|(\Delta_1+\imath\eta\Delta_2)z\|^2}{1+(\eta-\eta^{-1})^2\gamma^2}\\
&=\frac{\|\Delta_1 z\|^2+\eta^2\|\Delta_2 z\|^2-\imath\eta\left(\langle\Delta_1 z,\Delta_2 z\rangle-\langle\Delta_2 z,\Delta_1 z\rangle\right)}{1+(\eta-\eta^{-1})^2\gamma^2}\\
&=\frac{\|\Delta_1 z\|^2+\eta^2\|\Delta_2 z\|^2+\eta\left(\|\Delta z\|^2-\|\Delta_1 z\|^2-\|\Delta_2 z\|^2\right)}{1+(\eta-\eta^{-1})^2\gamma^2}.
\end{aligned}$$

Since $\hat{\Delta}(s)$ is a *real* transfer matrix we have $\|\hat{\Delta}(\imath\omega)\|=\|\hat{\Delta}(-\imath\omega)\|$ and so it suffices to consider $\eta>0$, i.e. $\omega>0$ (since $\omega_0>0$). Then because $\|\Delta z\|\leq r_{\mathbb{C}}^-\|z\|$, we have

$$\|\hat{\Delta}(\imath\omega)z\|^2\leq\frac{\eta(r_{\mathbb{C}}^-)^2\|z\|^2+(1-\eta)\|\Delta_1 z\|^2+\eta(\eta-1)\|\Delta_2 z\|^2}{1+(\eta-\eta^{-1})^2\gamma^2}. \tag{40}$$

Now choosing $y \in \mathbb{R}^q$, $\|y\| = 1$ such that $\|\Delta_1 y\| = \|\Delta_1\|$, we get

$$\|\Delta_1\|^2 \leq \|\Delta_1 y\|^2 + \|\Delta_2 y\|^2 = \|\Delta y\|^2 \leq (r_{\mathbb{C}}^-)^2.$$

Similarly we have $\|\Delta_2\|^2 \leq (r_{\mathbb{C}}^-)^2$. If $0 < \eta \leq 1$

$$\|\hat{\Delta}(\imath\omega)\|^2 = \sup_{z \in \mathbb{C}^q, \|z\|=1} \|\hat{\Delta}(\imath\omega)z\|^2 \leq \frac{\eta (r_{\mathbb{C}}^-)^2 + (1-\eta)\|\Delta_1\|^2}{1 + (\eta - \eta^{-1})^2\gamma^2} \leq \frac{(r_{\mathbb{C}}^-)^2}{1 + (\eta - \eta^{-1})^2\gamma^2}.$$

Hence $\|\hat{\Delta}(\imath\omega)\|^2 \leq (r_{\mathbb{C}}^-)^2$ if $\eta \leq 1$. On the other hand if $\eta > 1$, then

$$\|\hat{\Delta}(\imath\omega)\|^2 \leq \frac{\eta (r_{\mathbb{C}}^-)^2 + \eta(\eta - 1)\|\Delta_2\|^2}{1 + (\eta - \eta^{-1})^2\gamma^2} \leq \frac{\eta^2 (r_{\mathbb{C}}^-)^2}{1 + (\eta - \eta^{-1})^2\gamma^2} =: f(\eta)\,.$$

It is easy to see that for $\gamma > 1$ $f(\eta)$ is maximized at $\hat{\eta} = \left(\dfrac{2\gamma^2}{2\gamma^2 - 1}\right)^{1/2} > 1$ and so

$$\|\widehat{\mathbb{L}}\|_d^2 = \max_{\omega \in \mathbb{R}_+} \|\hat{\Delta}(\imath\omega)\|^2 \leq f(\hat{\eta}) = \frac{4\gamma^2}{4\gamma^2 - 1}(r_{\mathbb{C}}^-)^2$$

if $\gamma > 1$. Choosing γ sufficiently large completes the proof. □

Remark 5.6.21. We have seen in Example 5.3.14 that for the family of linear oscillators (3.25) with damping coefficient 2α the quotient $r_{\mathbb{R}}^-/r_{\mathbb{C}}^-$ tends to ∞ as $\alpha \to 0$. Thus the quotient $r_{\mathbb{R}}^-/r_{\mathbb{R},d}^-$ is unbounded by (35). It follows from Ex. 2 and Ex. 3 that the quotients $r_{\mathbb{R}}^-/r_{\mathbb{R},t}^-$ and $r_{\mathbb{R},n}^-/r_{\mathbb{R},t}^-$ are unbounded, too. At present it is an open question whether or not the quotient $r_{\mathbb{R},t}^-/r_{\mathbb{C}}^-$ is bounded, see *Notes and References.* □

As a consequence of the above theorem and Theorem 5.6.14 one should take $r_{\mathbb{C}}^-$ rather than $r_{\mathbb{R}}^-$ as a robustness measure whenever neglected dynamics play an important role.

5.6.3 The Aizerman Conjecture

In this subsection we only consider *time-invariant nonlinearities.* The zero equilibrium state of $\dot{x} = Ax$ is said to be *absolutely stable* relative to a set of nonlinear perturbations $\mathcal{P}$ if Σ_N is globally asymptotically stable for all $N \in \mathcal{P}$. The origin of this notion can be traced to the work of *Lur'e* where it was introduced to cope with uncertainty in the implementation of feedback controls, see *Notes and References.* For example, the precise behaviour of amplifiers, resistors, etc. may not be known, e.g. due to saturation effects. Of course the main difficulty is to determine a set $\mathcal{P}$ which reflects the practical constraints. In 1948 Aizerman made the following conjecture for the case $\mathbb{K} = \mathbb{R}$, $\ell = q = 1$, see Figure 5.6.4.

Aizerman's conjecture: *Suppose* $(A, b, c) \in \mathbb{L}_{n,1,1}(\mathbb{R})$ *and the feedback systems*

$$\dot{x} = Ax + bu\,, \quad y = c^\top x\,, \quad u = ky$$

are asymptotically stable for all k, $k_1 < k < k_2$. *Then the origin is a globally asymptotically stable equilibrium point of the nonlinear feedback systems*

$$\dot{x} = Ax + bu\,, \quad y = c^\top x\,, \quad u = N(y), \qquad N \in \mathcal{P}$$

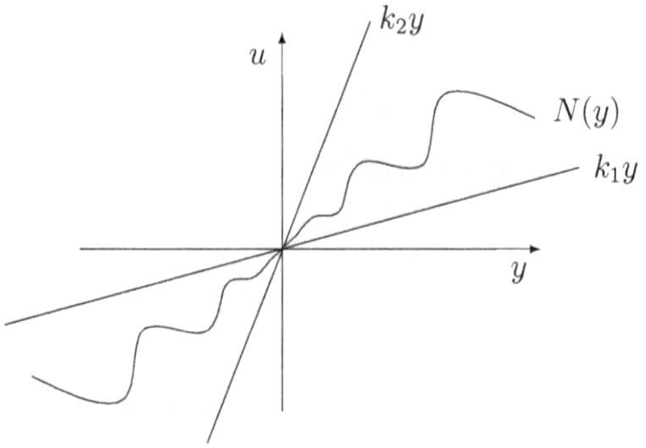

Figure 5.6.4: The Aizerman conjecture: sector conditions

where $\mathcal{P} := \{N \in \mathcal{C}(\mathbb{R},\mathbb{R});\, N$ *locally Lipschitz and* $\forall y \in \mathbb{R}^*: k_1y^2 < yN(y) < k_2y^2\}$. $\mathcal{P}$ consists of all locally Lipschitzian nonlinearities $N : \mathbb{R} \to \mathbb{R}$ whose graphs lie in the sector between the lines $u = k_1y$ and $u = k_2y$, see Figure 5.6.4. In particular $N(0) = 0$. Note that the above set $\mathcal{P}$ can also be described by

$$\mathcal{P} = \{N \in \mathcal{C}(\mathbb{R},\mathbb{R});\ N \text{ locally Lipschitz and } \forall y \in \mathbb{R}^* : |N(y) - k_0y| < r|y|\}\,.$$

where $k_0 = (k_1+k_2)/2$, $r = (k_2-k_1)/2$. Now let $\tilde{A} = A + k_0bc^\top$, $\tilde{N}(y) = N(y) - k_0y$, then an alternative formulation of Aizerman's conjecture is:

Suppose the linear systems

$$\dot{x} = \tilde{A}x + bu \quad , \quad y = c^\top x \quad , \quad u = ky$$

are asymptotically stable for all $k \in \mathbb{R}$, $|k| < r$. *Then the origin is a globally asymptotically stable equilibrium point of all the nonlinear feedback systems*

$$\dot{x} = \tilde{A}x + bu\,, \quad y = c^\top x\,, \quad u = \tilde{N}(y), \qquad \tilde{N} \in \tilde{\mathcal{P}}$$

where $\tilde{\mathcal{P}} = \left\{\tilde{N} \in \mathcal{C}(\mathbb{R},\mathbb{R});\ \tilde{N} \text{ locally Lipschitz and } \forall y \in \mathbb{R}^* : |\tilde{N}(y)| < r|y|\right\}$.

Thus a natural generalization of the conjecture to multivariable systems over both the real and the complex fields is the following

Multivariable version of Aizerman's conjecture: *Suppose* $\sigma(A) \subset \mathbb{C}_-$. *For any* $\delta > 0$, *the origin is a globally asymptotically stable equilibrium point for all the nonlinear feedback systems*

$$\Sigma_N : \quad \dot{x} = Ax + BN(Cx) \tag{41}$$

where $N : \mathbb{K}^q \mapsto \mathbb{K}^\ell$ *is locally Lipschitz and satisfies*

$$\|N(z)\| < \delta\|z\|\,, \quad z \in \mathbb{K}^q\,,\ z \neq 0 \tag{42}$$

if and only if all the linear systems

$$\Sigma_\Delta : \qquad \dot{x} = Ax + B\Delta Cx\,, \quad \Delta \in \mathbb{K}^{\ell\times q}\,, \quad \|\Delta\| < \delta$$

are asymptotically stable.

It is well known that Aizerman's conjecture does not hold over the field of real numbers, see [527]. In fact there are examples which show that $r_{\mathbb{R}}^{-}/r_{\mathbb{R},n}^{-}$ can become arbitrarily large, see [248]. In contrast, using the characterization of the complex stability radius given in Theorem 5.6.14, we have the following

Theorem 5.6.22. *The Aizerman conjecture holds true over the field of complex numbers.*

Proof: The "if" part of the conjecture is obvious. Conversely suppose that the linear systems Σ_Δ are asymptotically stable for all $\Delta \in \mathbb{C}^{\ell\times q}$, $\|\Delta\| < \delta$. Then necessarily $\delta \le r_{\mathbb{C}}^{-}(A;B,C)$. By Remark 5.3.27 there exists a Hermitian solution P of the Riccati equation (3.51) with $\gamma = \delta^{-1}$ and $D = 0$, i.e. of

$$PA + A^*P - C^*C - \delta^2 PBB^*P = 0.$$

Multiplying this equation by δ^2 and replacing $\delta^2 P$ by P there exists a Hermitian solution P of

$$PA + A^*P - \delta^2 C^*C - PBB^*P = 0. \tag{43}$$

Since P also satisfies

$$P = -\int_0^\infty e^{A^*t}[\delta^2 C^*C + PBB^*P]e^{At}dt, \tag{44}$$

(see (3.3.89a)) we must have $P \preceq 0$. Multiplying (43) from the right by $x \in \mathbb{C}^n$ and from the left by x^* we see that $\ker P \subset \ker C$. Suppose $N \in \mathcal{P}_n(\mathbb{C})$ satisfies (42) and consider the derivative of $V(x) = -\langle x, Px\rangle$ along the solutions of the nonlinear system (41). We have

$$\begin{aligned}\dot V(x) &= -\langle Ax + BN(z), Px\rangle - \langle x, P(Ax + BN(z))\rangle\\ &= -\delta^2\|z\|^2 - \|B^*Px\|^2 - 2\operatorname{Re}\langle B^*Px, N(z)\rangle\\ &= -\|B^*Px + N(z)\|^2 - [\delta^2\|z\|^2 - \|N(z)\|^2]\end{aligned}$$

where $z = Cx$. Hence $\dot V(x) \le 0$ for all $x \in \mathbb{C}^n$, and by (42) $\dot V(x) = 0$ only if $z = Cx = 0$.

Let $X_1 = (\ker P)^\perp$, $X_2 = \ker P$ and $\pi_1 : \mathbb{C}^n = X_1 \oplus X_2 \to X_1$ be the orthogonal projection onto X_1. Then the restriction of the quadratic form $x \mapsto V(x)$ to X_1 is positive definite. Hence there exists $\varepsilon > 0$ such that $V(x_1) \ge \varepsilon\|x_1\|^2$ for all $x_1 \in X_1$. Given any $x^0 \in \mathbb{C}^n$ there exists by Proposition 2.1.19 a solution $x(\cdot) = x(\cdot, x^0)$ of the nonlinear differential equation (41) on $\mathbb{R}_+$ with initial state $x(0) = x^0$. Let $x_1(t) = \pi_1(x(t))$. Since $V(x(t)) = -\langle x(t), Px(t)\rangle = -\langle x_1(t), Px_1(t)\rangle = V(x_1(t))$ is not increasing, we have

$$\varepsilon\|x_1(t)\|^2 \le V(x_1(t)) \le V(x_1(0)) = V(x^0) \le \|P\|\|x^0\|^2.$$

Hence $x_1(t)$ is bounded and there exists a constant $c > 0$ such that

$$\|z(t)\| = \|Cx(t)\| = \|Cx_1(t)\| \le c\|x^0\|, \quad t \ge 0. \tag{45}$$

But since $\|e^{At}\| \le Me^{-\omega t}, t \ge 0$ for suitable $M, \omega > 0$, we have by (42) and (45)

$$\|x(t)\| \le Me^{-\omega t}\|x^0\| + \int_0^t Me^{-\omega(t-s)}\|B\|\,\|N(z(s))\|\,ds \le \tilde{c}\|x^0\|, \quad t \ge 0 \tag{46}$$

for a suitable constant $\tilde{c} > 0$. Hence the origin is stable for Σ_N. It remains to prove for all $x^0 \in \mathbb{C}^n$ that $x(t) = x(t, x_0) \to 0$ as $t \to \infty$. By (42) and (46), we have for any $t_0 \ge 0$

$$\|x(t)\| \le Me^{-\omega(t-t_0)}\|x(t_0)\| + \hat{c}\|z\|_{L^\infty(t_0,\infty;\mathbb{K}^q)}, \quad t \ge t_0 \tag{47}$$

with a suitable constant $\hat{c} > 0$ independent of t_0. Applying LaSalle's Invariance Principle (Proposition 3.2.28) we conclude that $x(t) \to S$ as $t \to \infty$ where S is the largest invariant set in $\dot{V}^{-1}(0)$. Since $\dot{V}^{-1}(0) \subset \ker C$ we get $z(t) \to 0$ as $t \to \infty$. Hence choosing t_0 in (47) sufficiently large we see that $x(t) \to 0$ as $t \to \infty$. □

Remark 5.6.23. Theorem 5.6.14 implies that the origin is absolutely stable with respect to the set $\mathcal{P}_0$, where

$$\mathcal{P}_0 = \{N \in \mathcal{P}_n(\mathbb{C});\ \|N\|_n < r_{\mathbb{C}}^-(A; B, C)\}. \tag{48}$$

In contrast Theorem 5.6.22 shows that the origin is absolutely stable relative to the set $\mathcal{P}_{\mathbb{C}}$ where

$$\mathcal{P}_{\mathbb{C}} = \{N\,; N : \mathbb{C}^q \mapsto \mathbb{C}^\ell \text{ is locally Lipschitz, } \|N(z)\| < r_{\mathbb{C}}^-\|z\|,\ z \in \mathbb{C}^q,\ z \ne 0\}. \tag{49}$$

This is a considerable improvement on (48) not only because the Lipschitz bound is replaced by a finite gain condition, but also because $\|N\|_n < r_{\mathbb{C}}^-(A; B, C)$ implies $\|N(z)\| \le \delta\|z\|$ for some $\delta < r_{\mathbb{C}}^-$ which is more restrictive than $N \in \mathcal{P}_{\mathbb{C}}$. □

Remark 5.6.24. Suppose that the pair (A, C) is observable, see Subsection 3.3.5. Then it follows from (44) that every Hermitian solution P of (43) is negative definite. Hence the above proof shows that $V(x) = -\langle x, Px\rangle$ is a *joint global quadratic Liapunov function* for all systems Σ_N where $N \in \mathcal{P}_{\mathbb{C}}$ (49). In particular, $V(x)$ is a *joint Liapunov function* for *all* the linearly perturbed systems

$$\Sigma_\Delta : \quad \dot{x} = Ax + B\Delta Cx\,, \quad \Delta \in \mathbb{C}^{\ell\times q}, \quad \|\Delta\| < r_{\mathbb{C}}^-\,.$$

Now if $\delta > r_{\mathbb{C}}^-$ there is no joint Liapunov function for all perturbed systems Σ_Δ with $\Delta \in \mathbb{C}^{\ell\times q}$, $\|\Delta\| < \delta$. So one could say that $V(x)$ is a quadratic Liapunov function for $\dot{x} = Ax$ of *maximal robustness* with respect to perturbations of the form $A \rightsquigarrow A(\Delta) = A + B\Delta C$, $\Delta \in \mathbb{C}^{\ell\times q}$. □

If a given nonlinearity N does not satisfy $\|N(z)\| < r_{\mathbb{C}}^-\|z\|$ for all $z \in \mathbb{K}^q$, $z \ne 0$ or is not of finite gain at all, it may be possible that the above methods can be applied *locally* to obtain *local* stability results with guaranteed domains of attraction for all systems Σ_N where N satisfies $\|N(z)\| < r_{\mathbb{C}}^-\|z\|$, $z \ne 0$ locally. For this, one proceeds as follows. If Ω is an open neighbourhood of 0 in $\mathbb{K}^q$ and $0 < \delta \le r_{\mathbb{C}}^-(A; B, C)$ we consider the perturbation set

$$\mathcal{P}_{\mathbb{K}}(\Omega, \delta) = \{N\,; N : \Omega \to \mathbb{K}^\ell \text{ is locally Lipschitz, } \|N(z)\| < \delta\|z\|,\ z \in \Omega\backslash\{0\}\}. \tag{50}$$

For $N \in \mathcal{P}_{\mathbb{K}}(\Omega, \delta)$ the RHS of (41) is defined on the preimage $C^{-1}(\Omega)$ of Ω by C. We say that a neighbourhood $D \subset C^{-1}(\Omega)$ of 0 is a *guaranteed domain of attraction* of the origin for all $N \in \mathcal{P}_{\mathbb{K}}(\Omega, \delta)$ if $x_N(\cdot, x^0)$ exists on $\mathbb{R}_+$ and $x_N(t, x^0)$ tends to zero as $t \to \infty$ for all trajectories $x_N(\cdot, x^0)$ of Σ_N (41) starting in $x^0 \in D$ where $x^0 \in D$ and $N \in \mathcal{P}_{\mathbb{K}}(\Omega, \delta)$ are arbitrary. Note that if the origin is an equilibrium state of a differentiable system and N represents the nonlinear residual after linearization at 0, then N will belong to $\mathcal{P}_{\mathbb{K}}(\Omega, \delta)$ for any given $0 < \delta \leq r_{\mathbb{C}}^-(A; B, C)$ if Ω is reduced appropriately.

Theorem 5.6.25. *Suppose that Ω is an open neighbourhood of the origin in $\mathbb{K}^q$, $\delta \leq r_{\mathbb{C}}^-(A; B, C)$ and P is one of the Hermitian solutions of (43). Assume that the following implication holds for a given $\rho > 0$ and arbitrary $x \in \mathbb{K}^n$*

$$-\langle x, Px \rangle < \rho \quad \Longrightarrow \quad Cx \in \Omega, \tag{51}$$

i.e. the ellipsoid $D_\rho = \{x \in \mathbb{K}^n; -\langle x, Px \rangle < \rho\}$ lies in the preimage of Ω by C. Then the origin is an asymptotically stable equilibrium point of (41) with an invariant guaranteed domain of attraction D_ρ for all $N \in \mathcal{P}_{\mathbb{K}}(\Omega, \delta)$.

Proof: Let $N \in \mathcal{P}_{\mathbb{K}}(\Omega, \delta)$. If we define $V(x) = -\langle x, Px \rangle$, $x \in \mathbb{K}^n$, then just as in the proof of Theorem 5.6.22, we have for the derivative of V along the trajectories of Σ_N (41)

$$\dot{V}(x) = -\|B^* Px + N(z)\|^2 - [\delta^2 \|z\|^2 - \|N(z)\|^2], \quad x \in D_\rho \tag{52}$$

where $z = Cx$. Hence $\dot{V}(x) \leq 0$ for $x \in D_\rho$. We want to show that D_ρ is invariant under the flow of Σ_N. For this we restrict the RHS of (41) to D_ρ. Let $x^0 \in D_\rho$ and suppose that the solution $x(t) = x(t, x^0)$ of (41) has a finite maximal existence interval $[0, t_+(x^0))$. The RHS of (41) restricted to D_ρ is affinely bounded because $N \in \mathcal{P}_{\mathbb{K}}(\Omega, \delta)$ and therefore, applying Proposition 2.1.19 to Σ_N on D_ρ, we see that $x(t)$ must be bounded. But by (52)

$$\overline{\{x(t); t \in [0, t_+(x^0))\}} \subset \{x \in \mathbb{K}^n; V(x) \leq V(x^0)\} \subset D_\rho.$$

So by Lemma 3.2.14 $x(\cdot)$ cannot leave D_ρ in finite time and we conclude that $x(t, x^0)$ exists for all $t \geq 0$ and remains in D_ρ. Moreover we see from (52) and (50) that $Cx = 0$ for every $x \in D_\rho$ with $\dot{V}(x) = 0$. From here onwards we can proceed in exactly the same way as in the proof of Theorem 5.6.22 to conclude that the origin is stable for Σ_N and $x(t, x^0) \to 0$ for all $x^0 \in D_\rho$. □

We have seen that $\ker P \subset \ker C$ for every Hermitian solution P of (43). Hence there always exists a $\rho > 0$ such that (51) holds. But for each Hermitian solution P, one is interested in finding a $\rho > 0$ satisfying (51) such that the invariant guaranteed domain of attraction D_ρ is *large*. We illustrate this by an example where we will also see that in general there are no set inclusion properties between the domains generated via different solutions of (43). Thus, in order to find a more comprehensive guaranteed domain of attraction, it may be a reasonable strategy to maximize the ρ (under the constraint provided by (51)) for different Hermitian solutions of (43).

Example 5.6.26. Consider nonlinear oscillators of the form

$$\Sigma_N : \qquad \ddot{\xi} + 2\dot{\xi} + \xi = N(\xi), \tag{53}$$

where $N : \Omega \to \mathbb{R}$ is locally Lipschitz on $\Omega = (-\omega, \omega)$. Introducing the state $x = [\xi, \dot{\xi}]^\top$, (53) can be rewritten in the form Σ_N (41) with

$$A = \begin{bmatrix} 0 & 1 \\ -1 & -2 \end{bmatrix}, \quad B = \begin{bmatrix} 0 \\ 1 \end{bmatrix}, \quad C = [1 \ 0].$$

It is not difficult to show that $r_{\mathbb{C}}^{-}(A; B, C) = 1$ and there are two solutions of the Riccati equation (43) for $\delta = 1$. The largest and the smallest solutions of (43) for the above data are given by

$$P_+ = -\begin{bmatrix} 2 & 1 \\ 1 & 2 - \sqrt{2} \end{bmatrix} \quad \text{and} \quad P_- = -\begin{bmatrix} 2 & 1 \\ 1 & 2 + \sqrt{2} \end{bmatrix},$$

respectively. The nonlinearity N belongs to $\mathcal{P}_{\mathbb{R}}(\Omega, \delta)$ with $\delta = 1$ if and only if

$$|N(\xi)| < |\xi| \quad \text{provided} \quad \xi \in (-\omega, \omega), \ \xi \neq 0. \tag{54}$$

For example, if $\omega = 1$ then $N \in \mathcal{P}_{\mathbb{K}}(\Omega, \delta)$ for all $N(\xi) = \xi^p$, $p \geq 1$. Since for $x \in \mathbb{R}^2$

$$-\langle x, P_+ x\rangle = 2x_1^2 + 2x_1x_2 + (2-\sqrt{2})x_2^2 = (2-1/(2-\sqrt{2}))x_1^2 + (2-\sqrt{2})[x_2 + x_1/(2-\sqrt{2})]^2$$

implication (51) holds for $P = P_+$ if and only if $\rho \leq (2-1/(2-\sqrt{2}))\omega^2 = (1-\sqrt{2}/2)\omega^2$. We conclude therefore that if we choose $P = P_+$ the largest guaranteed domain of attraction of the origin for all $N \in \mathcal{P}_{\mathbb{R}}(\Omega, \delta)$ obtainable from Theorem 5.6.25 is

$$D_{\rho_+} = \{[x_1, x_2]^\top \in \mathbb{R}^2; 2x_1^2 + 2x_1x_2 + (2-\sqrt{2})x_2^2 < \rho_+\}, \quad \rho_+ = (1-\sqrt{2}/2)\omega^2.$$

Carrying out a similar analysis using P_- instead of P_+ shows that

$$D_{\rho_-} = \{[x_1, x_2]^\top \in \mathbb{R}^2; 2x_1^2 + 2x_1x_2 + (2+\sqrt{2})x_2^2 < \rho_-\}, \quad \rho_- = (1+\sqrt{2}/2)\omega^2$$

is also a common domain of attraction for all $N \in \mathcal{P}_{\mathbb{R}}(\Omega, \delta)$ and is the largest one

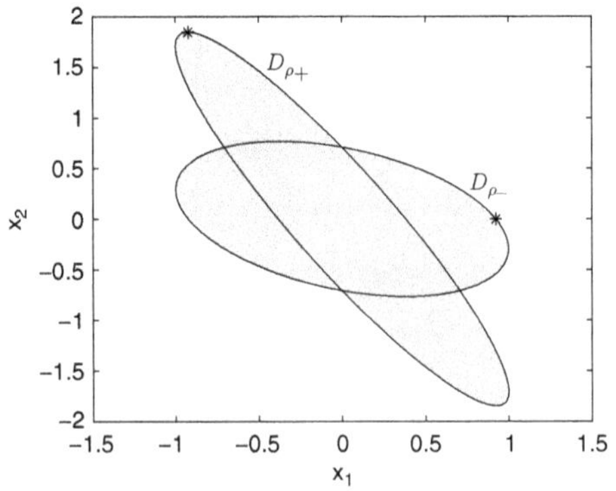

Figure 5.6.5: Two guaranteed domains of attractions obtained via Theorem 5.6.25

obtainable via Theorem 5.6.25 if we choose $P = P_-$. Note that $((\sqrt{2+\sqrt{2}}/2)\,\omega, 0)$ lies on ∂D_{ρ_-} but not in $\overline{D_{\rho_+}}$ and that $((-\sqrt{2+\sqrt{2}}/2)\,\omega, \sqrt{2+\sqrt{2}}\,\omega)$ lies in ∂D_{ρ_+} but not in $\overline{D_{\rho_-}}$, see Figure 5.6.5. For $\omega = 1$ these two points and the two elliptic domains D_{ρ_+}, D_{ρ_-} are shown in Figure 5.6.5. The union $D_{\rho_+} \cup D_{\rho_-}$ is a guaranteed domain of attraction of the origin for all nonlinear oscillators Σ_N where $N \in \mathcal{P}_{\mathbb{K}}(\Omega, 1)$, $\Omega = (-1, 1)$. □

If in the above theorem we assume the more restrictive condition that $\delta < r_{\mathbb{C}}^{-}(A; B, C)$, then we can strengthen asymptotic stability to exponential stability. By the continuity of $r_{\mathbb{C}}^{-}(\cdot)$ (see Proposition 5.3.42) if $\delta < r_{\mathbb{C}}^{-}(A; B, C)$ there exists $\alpha > 0$, such that $\sigma(A + \alpha I_n) \subset \mathbb{C}_-$ and

$$\delta \leq r_{\mathbb{C}}^{-}(A + \alpha I_n; B, C). \tag{55}$$

Theorem 5.6.27. *Suppose Ω is an open neigbourhood of the origin in $\mathbb{K}^q$ which is starlike with respect to 0 and $\delta < r_{\mathbb{C}}^{-}(A; B, C)$. Let $\alpha > 0$ be such that (55) holds and P_α be one of the Hermitian solutions of*

$$P(A + \alpha I_n) + (A + \alpha I_n)^* P - \delta^2 C^* C - PBB^* P = 0. \tag{56}$$

If for a given $\rho > 0$ and all $x \in \mathbb{K}^n$

$$-\langle x, P_\alpha x \rangle < \rho \quad \Longrightarrow \quad Cx \in \Omega, \tag{57}$$

then the origin is an exponentially stable equilibrium point of Σ_N (41) for all $N \in \mathcal{P}_{\mathbb{K}}(\Omega, \delta)$ (50). Moreover there exists $M \geq 1$ such that for all $N \in \mathcal{P}_{\mathbb{K}}(\Omega, \delta)$

$$\|x_N(t, x^0)\| \leq M e^{-\alpha t} \|x^0\|, \; t \geq 0, \quad x^0 \in D_\rho^\alpha = \{x \in \mathbb{K}^n; -\langle x, P_\alpha x \rangle < \rho\}$$

where $x_N(t, x^0)$ denotes the trajectory of Σ_N with initial state x^0.

Proof: Let $N \in \mathcal{P}_{\mathbb{K}}(\Omega, \delta)$ and define $N_\alpha(z, t) = e^{\alpha t} N(e^{-\alpha t} z)$ for $t \geq 0$, $z \in \Omega$. Then N_α is well defined since Ω is starlike, and by (57) we have for all $t \geq 0$

$$x \in D_\rho^\alpha, \; Cx \neq 0 \implies Cx \in \Omega \text{ and } \|N_\alpha(Cx, t)\| < e^{\alpha t} \delta \|e^{-\alpha t} Cx\| = \delta \|Cx\|. \tag{58}$$

Consider the time-varying nonlinear system

$$\Sigma_{N_\alpha}: \quad \dot{x}(t) = A_\alpha x(t) + B N_\alpha(Cx(t), t), \quad t \geq 0, \quad A_\alpha := A + \alpha I_n. \tag{59}$$

Define $V(x) = -\langle x, P_\alpha x \rangle$, $x \in D_\rho^\alpha$, then along the flow of (59), we have

$$\dot{V}(x) = -\|B^* P_\alpha x + N_\alpha(Cx, t)\|^2 - [\delta^2 \|Cx\|^2 - \|N_\alpha(Cx, t)\|^2], \quad x \in D_\rho^\alpha, \quad t \geq 0.$$

Hence $\dot{V}(x) \leq 0$ for all $x \in D_\rho^\alpha$. Now let us restrict the RHS of the differential equation (59) to $\mathbb{R}_+ \times D_\rho^\alpha$ and suppose that $x_{N_\alpha}(t) = x_{N_\alpha}(t, x^0)$ is any solution of the restricted differential equation (59) with initial state $x_{N_\alpha}(0) = x^0 \in D_\rho^\alpha$ and maximal existence interval $[0, t_+(x^0))$ in $\mathbb{R}_+$. The RHS of (59) is affinely bounded on $\mathbb{R}_+ \times D_\rho^\alpha$ because of (58). Hence, applying Proposition 2.1.19 to Σ_{N_α} on $\mathbb{R}_+ \times D_\rho^\alpha$, we see that $x_{N_\alpha}(t)$, $t \in [0, t_+(x^0))$ is bounded if $t_+(x^0) < \infty$. Since $\dot{V}(x) \leq 0$ for $x \in D_\rho^\alpha$, it follows that $\dot{V}(x_{N_\alpha}(t)) \leq 0$ for $t \in [0, t_+(x^0))$, hence

$$\overline{\{x_{N_\alpha}((t); t \in [0, t_+(x^0))\}} \subset \{x \in \mathbb{K}^n; V(x) \leq V(x^0)\} \subset D_\rho^\alpha.$$

So $t_+(x^0) < \infty$ would yield a contradiction to Lemma 3.2.14. Therefore $x_{N_\alpha}(t, x^0)$ exists and remains in D_ρ^α for all $t \geq 0$, $x^0 \in D_\rho^\alpha$. Proceeding now as in the proof of Theorem 5.6.22 (following the arguments which lead to equation (45) with A

replaced by A_α, $N(z)$ by $N_\alpha(z,t)$) we see that there exists a constant c_α independent of $N \in \mathcal{P}_{\mathbb{K}}(\Omega, \delta)$ such that

$$\|z_{N_\alpha}(\cdot)\|_{L^\infty(0,\infty;\mathbb{K}^q)} \le c_\alpha \|x^0\|, \qquad x^0 \in D_\rho^\alpha \tag{60}$$

where $z_{N_\alpha}(t) = Cx_{N_\alpha}(t, x^0)$ (the reader is asked to prove this in Ex. 8). But

$$x_{N_\alpha}(t) = e^{A_\alpha t}x^0 + \int_0^t e^{A_\alpha(t-s)}BN_\alpha(z_{N_\alpha}(s), s)\,ds\ ,\ t \ge 0,$$

and since $\|e^{A_\alpha t}\| \le M_\alpha e^{-\beta t}$, $t \ge 0$ for suitable M_α, $\beta > 0$, it follows that there exists a constant c'_α such that for all $x^0 \in D_\rho^\alpha$

$$\|x_{N_\alpha}(t)\| \le M_\alpha e^{-\beta t}\|x^0\| + c'_\alpha \|z_{N_\alpha}(\cdot)\|_{L^\infty(0,\infty)} \le M_\alpha e^{-\beta t}\|x^0\| + c'_\alpha c_\alpha \|x^0\|, \quad t \ge 0.$$

Hence $\|x_{N_\alpha}(t)\| \le M\|x^0\|$, $t \ge 0$ for all $x^0 \in D_\rho^\alpha$ where M is a suitable constant only depending upon A, B, α and δ. Let $x(t) = e^{-\alpha t}x_{N_\alpha}(t, x^0)$, where $x^0 \in D_\rho^\alpha$ is arbitrary. Then $Cx(t) \in \Omega$, $t \ge 0$ and

$$\begin{aligned} \dot{x}(t) &= -\alpha x(t) + e^{-\alpha t}\dot{x}_{N_\alpha}(t, x^0) = Ax(t) + e^{-\alpha t}BN_\alpha(Cx_{N_\alpha}(t, x^0), t) \\ &= Ax(t) + BN(Cx(t)), \quad t \ge 0. \end{aligned}$$

So $x(\cdot)$ is the solution of (41) with $x(0) = x^0$ and $\|x(t)\| \le Me^{-\alpha t}\|x^0\|$. This completes the proof. □

Remark 5.6.28. (i) Note that in Theorem 5.6.25 (resp. Theorem 5.6.27) $\ker P \subset D_\rho$ (resp. $\ker P_\alpha \subset D_\rho^\alpha$) so that D_ρ (resp. D_ρ^α) will be unbounded if P (resp. P_α) is not negative definite.

(ii) The significance of the results in Theorem 5.6.25 (resp. Theorem 5.6.27) is not that it yields a lower estimate for the basin of attraction of the origin for a *single* nonlinear system Σ_N, but yields a guaranteed domain of attraction and uniform exponential estimate for *all* systems Σ_N for which the nonlinearity N is in $\mathcal{P}_{\mathbb{K}}(\Omega, \delta)$ with $\delta \le r_{\mathbb{C}}^-$ (resp. $< r_{\mathbb{C}}^-$). If P in Theorem 5.6.25 (resp. Theorem 5.6.27) is nonsingular then $V(x) = -\langle x, Px \rangle$ defines a joint Liapunov function for all the nonlinear systems Σ_N, $N \in \mathcal{P}_{\mathbb{K}}(\Omega, \delta)$. □

Example 5.6.29. Consider the nonlinear oscillator $\ddot{\xi}+\dot{\xi}+\xi = N(\xi)$ where as in Example 5.6.26 $N : \Omega \to \mathbb{R}$ is locally Lipschitz and $\Omega = (-\omega, \omega)$. This can be rewritten in the state space form Σ_N (41) with

$$A = \begin{bmatrix} 0 & 1 \\ -1 & -1 \end{bmatrix}, \quad B = \begin{bmatrix} 0 \\ 1 \end{bmatrix}, \quad C = [1\ \ 0].$$

It follows from Example 5.3.14 that $r_{\mathbb{C}}^-(A; B, C) = \sqrt{3}/2$. The Riccati equation (43) with $\delta = \sqrt{3}/2$ has the unique solution $P = -\begin{bmatrix} 1 & 0.5 \\ 0.5 & 1 \end{bmatrix}$. Proceeding as in Example 5.6.26 we find that the ellipsoid $D = \{[x_1, x_2]^\top \in \mathbb{R}^2; x_1^2+x_1x_2+x_2^2 < 3\omega^2/4\}$ is a guaranteed domain of attraction of the origin with respect to all the systems Σ_N (41) where $N \in \mathcal{P}_{\mathbb{R}}(\Omega, \delta)$, i.e.

$$|N(\xi)| < (\sqrt{3}/2)|\xi| \quad \text{provided} \quad \xi \in (-\omega, \omega),\ \xi \ne 0, \tag{61}$$

and it is the largest one that can be obtained from Theorem 5.6.25. For instance, if $N(\xi) = \xi^2$ we may choose $\omega = \sqrt{3}/2$ to secure (61).
If $\alpha = 1/4$ a short calculation yields $r_{\mathbb{C}}^{-}(A + \alpha I_2; B, C) = \sqrt{3}/4$ and the only Hermitian solution of (56) for $\delta_\alpha = \sqrt{3}/4$ is $P_\alpha = -\begin{bmatrix} 0.5 & 0.25 \\ 0.25 & 0.5 \end{bmatrix} = P/2$. Now consider the set $\mathcal{P}_{\mathbb{R}}(\Omega_\alpha, \delta_\alpha)$ of all nonlinearities N satisfying

$$|N(\xi)| < (\sqrt{3}/4)|\xi| \quad \text{provided} \quad \xi \in \Omega_\alpha = (-\omega_\alpha, \omega_\alpha),\ \xi \neq 0. \tag{62}$$

Then we find that $D^\alpha = \{[x_1, x_2]^\top \in \mathbb{R}^2; x_1^2 + x_1 x_2 + x_2^2 < 3\omega_\alpha^2/4\}$ is a guaranteed domain of attraction of the origin for all nonlinearities satisfying (62) and there exists a constant $M > 0$ such that every solution $x_N(t, x^0)$ of Σ_N (41) satisfies $\|x_N(t, x^0)\| \leq M e^{-\alpha t}$, $t \geq 0$ for all $x^0 \in D^\alpha$. □

We conclude this section with a corollary of Theorem 5.6.27 which provides a sufficient condition for global exponential stability.

Corollary 5.6.30. *Suppose that $\delta < r_{\mathbb{C}}^{-}(A; B, C)$ then the origin is a globally exponentially stable equilibrium point of (41) for all locally Lipschitz $N : \mathbb{K}^q \to \mathbb{K}^\ell$ satisfying (42).*

Proof: Choose $\alpha > 0$ sufficiently small so that (55) holds and set $\Omega = \mathbb{K}^q$. Then (57) is satisfied for all $\rho > 0$ and any Hermitian solution P_α of (56). Since for any $x^0 \in \mathbb{K}^n$ there exists ρ such that $x^0 \in D_\rho^\alpha$, the result follows from Theorem 5.6.27. □

5.6.4 Exercises

1. Consider the system $\dot{x} = \begin{bmatrix} a_{11} & a_{12} \\ a_{21} & a_{22} \end{bmatrix} x$ with $a_{ij} \in \mathbb{R}$ and $a_{11} < 0$, $a_{22} < 0$ as a special case of the system (14) (with $n_1 = n_2 = 1$) studied in Example 5.6.6. Show that the estimate (33) given in Example 5.6.16 reduces to $|a_{21}a_{12}| < a_{11}a_{22}$ and is therefore a tight condition for $\sigma(A) \subset \mathbb{C}_-$.

2. Consider the time-varying oscillator

$$\ddot{\xi}(t) + 2\alpha\dot{\xi}(t) + (1 + \Delta(t))\xi(t) = 0 \ , \quad t \geq 0$$

where $0 < \alpha \leq 1$ is given and $\Delta(\cdot) \in L^\infty(\mathbb{R}; \mathbb{R})$ represents an unknown time-varying perturbation of the restoring force. Suppose $\xi(0) = -1$, $\dot{\xi}(0) = 0$ and $\Delta(t) = \delta > 0$ for $0 \leq t \leq t_1$, where t_1 is the smallest value of t for which $\xi(t) = 0$. If $\alpha_1 = 1 + \delta$, prove that t_1 is uniquely determined by

$$\cos(\alpha_1 - \alpha^2)^{1/2} t_1 = -\alpha/\sqrt{\alpha_1} \ , \quad \frac{\pi}{2\sqrt{\alpha_1 - \alpha^2}} \leq t_1 \leq \frac{\pi}{\sqrt{\alpha_1 - \alpha^2}}$$

and $\dot{\xi}(t_1) = \sqrt{\alpha_1} e^{-\alpha t_1}$. Now suppose $\Delta(t) = -\delta > 0$ for $t_1 < t \leq t_1 + t_2$, where t_2 is the first time that $\dot{\xi}(t_1 + t) = 0$. If $\alpha^2 < \alpha_2 = 1 - \delta$, $\delta < 1$ prove that t_2 is uniquely determined by

$$\cos(\alpha_2 - \alpha^2)^{1/2} t_2 = \alpha/\sqrt{\alpha_2} \ , \quad 0 \leq t_2 \leq \frac{\pi}{2\sqrt{(\alpha_2 - \alpha^2)}}$$

and $\xi(t_1 + t_2) = \sqrt{\alpha_1/\alpha_2} e^{-\alpha(t_1 + t_2)}$. Show that if α and δ are sufficiently small, then $t_1 \approx t_2 \approx \pi/2$ and $\xi(t_1 + t_2) > 1$ provided $\delta > \pi\alpha$. Explain why this shows that $r_{\mathbb{R},t}^{-}(\alpha) := r_{\mathbb{R},t}^{-}(A(\alpha); B, C) \leq \pi\alpha$ where

$$A(\alpha) = \begin{bmatrix} 0 & 1 \\ -1 & -2\alpha \end{bmatrix}, \quad B = \begin{bmatrix} 0 \\ 1 \end{bmatrix}, \quad C = \begin{bmatrix} 1, & 0 \end{bmatrix}. \tag{63}$$

3. Consider the nonlinear oscillator

$$\ddot{\xi}(t) + 2\alpha\dot{\xi}(t) + \xi(t) + N(\xi(t)) = 0 \ , \quad t \geq 0$$

where $\alpha > 0$, $N : \mathbb{R} \to \mathbb{R}$ is locally Lipschitz and satisfies $|N(\xi)| < |\xi|$, $\xi \in \mathbb{R}$, $\xi \neq 0$. Use the Liapunov function

$$V(\xi, \dot{\xi}) = \xi^2 + \dot{\xi}^2 + 2\int_0^{\xi} N(s)ds + 2\alpha\xi\dot{\xi} + 2\alpha^2\xi^2$$

to prove that the origin of the corresponding state space system is globally asymptotically stable. Hence show that $r^-_{\mathbb{R},n}(A(\alpha); B, C) = 1$ for all values of $\alpha > 0$, if $(A(\alpha), B, C)$ is defined by (63). Determine the limit of $r^-_{\mathbb{R},t}(A(\alpha); B, C)/r^-_{\mathbb{R},n}(A(\alpha); B, C)$ as $\alpha \searrow 0$ (compare Ex. 2).

4. Consider the uncertain system

$$\Sigma_{\Delta,h}: \quad \dot{x}(t) = Ax(t) + B\Delta Cx(t-h), \;\; A = \begin{bmatrix} 0 & 1 \\ -1 & -2 \end{bmatrix}, \; B = \begin{bmatrix} 0 \\ 1 \end{bmatrix}, \; C = [1, 1]$$

where $h > 0$ is given and $\Delta \in \mathbb{K}$ is unknown. As in Example 5.6.5 define a dynamic perturbation $\mathcal{N} : L^2(\mathbb{R}_+; \mathbb{K}) \to L^2(\mathbb{R}_+; \mathbb{K})$ by

$$\mathcal{N}(z(\cdot))(t) = \begin{cases} 0 & 0 \leq t < h \\ \Delta z(t-h) & t \geq h \end{cases}, \qquad z(\cdot) \in L^2(\mathbb{R}_+; \mathbb{K}).$$

Given $t_0 = h$, $x^0 \in \mathbb{K}^2$, $\psi \in L^2(0, t_0; \mathbb{K}^2)$, the solution of the delay equation $\Sigma_{\Delta,h}$ with initial data (t_0, x^0, ψ) coincides with the solution of $\Sigma_{\mathcal{N}}$ with initial data $(t_0, x^0, C\psi)$, where

$$\Sigma_{\mathcal{N}}: \quad \dot{x}(t) = Ax(t) + B\mathcal{N}(Cx(\cdot))(t), \quad t \geq t_0.$$

Prove that $\|\mathcal{N}\|_d = |\Delta|$. The real and complex stability radii of $\dot{x}(t) = Ax(t)$ with respect to the delayed perturbations of the form $B\Delta Cx(t-h)$, $\Delta \in \mathbb{K}$ are defined by

$$r^-_{\mathbb{K},h}(A; B, C) = \inf\{|\Delta|;\ \Delta \in \mathbb{K},\ \Sigma_{\Delta,h} \text{ is not g.a.s.}\}.$$

Prove $r^-_{\mathbb{C},h}(A; B, C) = r^-_{\mathbb{C}}(A; B, C)$ for all $h > 0$ and

$$\lim_{h \to 0} r^-_{\mathbb{R},h}(A; B, C) = r^-_{\mathbb{R}}(A; B, C), \qquad \lim_{h \to \infty} r^-_{\mathbb{R},h}(A; B, C) = r^-_{\mathbb{C}}(A; B, C).$$

5. In this example we guide the reader to prove that $r^1_{\mathbb{R},d}(A; B, C) = r^1_{\mathbb{C}}(A; B, C)$ for discrete time systems. Suppose $\Delta = wz^*$, $w \in \mathbb{C}^\ell$, $z \in \mathbb{C}^q$ destabilizes the stable real discrete time system (A, B, C) and is of *minimum norm* :

$$\sigma(A + B\Delta C) \not\subset \mathbb{D} \quad \text{and} \quad \|\Delta\| = \left[\max_{\theta \in [0, 2\pi)} \|G(e^{\imath\theta})\|\right]^{-1} = \|G(e^{\imath\theta_0})\|^{-1} = r^1_{\mathbb{C}}(A, B, C),$$

where $G(s) = C(sI_n - A)^{-1}B$. Explain why we may assume $\theta_0 \in (0, \pi)$. Let $w = [w_1, ..., w_\ell]^\top$ and choose $\theta_j \in [\pi, 2\pi)$, $\tilde{w}_j \in \mathbb{R}$ such that $w_j = \tilde{w}_j e^{\imath\theta_j}$, $j = 1, \ldots, \ell$. Let $a_j = (e^{\imath\theta_0} - e^{\imath\theta_j})/(1 - e^{\imath(\theta_j + \theta_0)})$, $j \in \underline{\ell}$ and set

$$w_j(s) = \begin{cases} w_j & \text{if } w_j \in \mathbb{R}, \\ \tilde{w}_j/s & \text{if } \theta_j = -\theta_0, \\ \tilde{w}_j(s - a_j)/(1 - a_j s) & \text{otherwise}, \end{cases} \qquad s \in \mathbb{C}.$$

Prove that $a_j \in \mathbb{R}$, $|a_j| > 1$ and

$$w_j(e^{\imath\theta_0}) = w_j, \quad |w_j(e^{\imath\theta})| = |w_j| \text{ for all } \theta \in [0, 2\pi).$$

Let $w(s) = [w_1(s), \ldots, w_\ell(s)]^\top$, and construct the real, proper rational vector $z(s) = [z_1(s), \ldots, z_q(s)]^\top$, corresponding to the vector z, in the same way. Prove that $\Delta(s) = w(s)z(s)^*$ destabilizes (A, B, C) and satisfies

$$\|\Delta(e^{\imath\theta})\| = r_{\mathbb{C}}^1 \quad \text{for all } \theta \in [0, 2\pi).$$

6. Show that the Aizermann conjecture does not hold for arbitrary *time-varying linear* perturbations, neither for $\mathbb{K} = \mathbb{R}$ nor for $\mathbb{K} = \mathbb{C}$. More precisely, show that if $\Delta(\cdot) \in \mathcal{P}_t(\mathbb{R})$ satisfies $\|\Delta(t)\| < r_{\mathbb{C}}^-(A, D, E)$, $t \in \mathbb{R}_+$, it does not necessarily follow that $\Sigma_{\Delta(\cdot)}$ (3) is asymptotically stable.
(Hint: Consider the perturbed scalar system $\dot{x}(t) = -x(t) + \Delta(t)x(t)$ with $\Delta(t) = 1 - e^{-t}$.)

7. This exercise shows that the Aizermann conjecture for $\mathbb{K} = \mathbb{C}$ cannot be extended to *time-varying* nonlinearities satisfying

$$\sup_{t \in \mathbb{R}_+} \|N(z, t)\| < r_{\mathbb{C}}^- \|z\|, \quad z \in \mathbb{C}^q, \quad z \neq 0.$$

Consider the nominal system $\dot{x} = -x$ and let $\tilde{x}(t) = e^{e^{-t}}$ so that $\tilde{x}(0) = e$ and $\tilde{x}(t)$ decreases monotonically to 1 as $t \to \infty$. Define for every $t \in \mathbb{R}_+$, $x \in \mathbb{C}$

$$N(x, t) = \begin{cases} \dfrac{(1 - \ln|x|)x}{1 + (t + \ln\ln|x|)^2} & \text{if} \quad |x| \in (1, e] \\ 0 & \text{otherwise} \end{cases}.$$

Show that $N \in \mathcal{P}_{nt}(\mathbb{C})$ (see Remark 5.6.3) and $\sup_{t \in \mathbb{R}_+} |N(x, t)| < |x|$, $x \in \mathbb{C}$, $x \neq 0$. Prove that $N(\tilde{x}(t), t) = (1 - e^{-t})\tilde{x}(t)$, $t \geq 0$, and show that the system $\dot{x}(t) = -x(t) + N(x(t), t)$ is not g.a.s..

8. Prove that (60) holds (see the proof of Theorem 5.6.27).

9. Suppose the conditions in Theorem 5.6.25 hold with $\delta = r_{\mathbb{C}}^-(A; B, C)$ and in addition $N \in \mathcal{P}_{\mathbb{K}}(\Omega, \delta)$ is differentiable at the origin with derivative $N'(0)$ satisfying $\|N'(0)\| < r_{\mathbb{C}}^-(A; B, C)$. Prove:

(i) Every solution $x(t, x^0)$ of Σ_N with initial state $x^0 \in D_\rho$ decays exponentially.

(ii) For any $\tilde{\rho} < \rho$ there exists an $\alpha > 0$ and $M > 0$ (both depending on $\tilde{\rho}$ and N but not on x^0) such that $\|x(t, x^0)\| \leq Me^{-\alpha t}\|x^0\|$, $t \geq 0$, $x^0 \in D_{\tilde{\rho}}$.

(Hint: For (i) use Theorem 3.3.52 and for (ii) show that there exists $\gamma < r_{\mathbb{C}}^-$ such that $\|Nz\| \leq \gamma\|z\|$ for all $z \in CD_{\tilde{\rho}}$ and apply Theorem 5.6.27 with $\Omega = CD_{\tilde{\rho}}$).

5.6.5 Notes and References

The stability of feedback systems, with a given time-invariant linear system in the forward path and an *uncertain nonlinearity* in the feedback path, is a classic problem in control and has led to the concept of absolute stability (see below). In the classical context the uncertain memoryless feedback operator described a nonlinear control mechanism for which

a precise mathematical model was not known e.g. a servo-motor of constant speed with deadzones, see *Lur'e* (1951) [352]. In our context the uncertain feedback operator is just a convenient way of describing various types of perturbations.
This section is based on the paper by *Hinrichsen and Pritchard* (1992) [250] where both discrete and continuous time systems are considered. This paper also contains a proof of the more general existence and uniqueness result for systems of the form $\Sigma_{\mathcal{N}}$ with arbitrary $\mathcal{N} \in \tilde{\mathcal{P}}_d(\mathbb{K})$ (see Remark 5.6.8 (iii)). For existence and uniqueness results without stability assumptions (in an infinite dimensional setting), see *Jacob* (1995) [277].
Whereas not very much is known about the stability radius of time-invariant linear systems with respect to time-invariant nonlinear perturbations, the theory of stability radii under time-varying parametric uncertainties has made substantial progress over the last decade. One important source of information is provided by the spectral analysis of flows on vector bundles and in particular the theory of Liapunov exponents of bilinear control systems, see the comprehensive monograph of *Colonius and Kliemann* (2000) [108] which contains applications to various stability radii problems. A closely related field which provides efficient tools for the analysis of $r^-_{\mathbb{R},t}$ is the theory of linear differential inclusions. Here we refer to the forthcoming work of *Wirth* (2005) [533] which introduces the concept of extremal norms for characterizing the exponential growth of the semigroup associated with a linear inclusion. Roughly speaking these norms generalize the concept of a Liapunov norm (see Section 5.5) to linear inclusions.
For 2-dimensional interval systems *Gonzalez* (1991) [199] developed a method of constructing a time-varying system matrix with values in a given matrix interval which achieves the largest growth rate. An application of his results to the perturbed oscillator in Example 5.6.18 can be found in [250].
A numerical method for computing the exponential growth rate of bilinear control systems has been developed by Grüne, see Appendix D in [108] and the rate of convergence of the algorithm was analyzed in *Grüne and Wirth* (2000) [204]. The method is based on the solution of a Hamilton-Jacobi equation associated with a discounted infinite horizon optimal control problem and can also be used in order to compute the stability radius $r^-_{\mathbb{R},t}$ for systems of modest dimension, see *Wirth* (2005) [533].
For an early paper in the spirit of Theorem 5.6.20 see *Desoer and Chan* (1975) [129].
The basic idea of *absolute stability* was introduced by *Lur'e and Postnikov* (1945) [353]. Research on the absolute stability problem generated a wealth of important results in nonlinear feedback theory, see the early monographs of *Aizerman and Gantmacher* (1963) [8] and *Lefschetz* (1963) [337]. In 1949 *Aizerman* [7] put forward the conjecture which now bears his name. *Kalman* (1957) [286] formulated a more restrictive conjecture which required the nonlinearity N to be differentiable with derivative $N'(y) \in [k_1, k_2]$, $y \in \mathbb{R}$. The discovery of counterexamples to these conjectures showed that one cannot hope to reduce the stability analysis of nonlinear feedback systems Σ_N to the analysis of a set of linear systems, see the books of *Hahn* (1963) [208], *J. L. Willems* (1970) [530] and *J. C. Willems* (1971) [527]. A detailed analysis of the counterexample in [527] shows that for each $\delta \in (0,1]$ there exist stable systems (A, B, C) such that $r^-_{\mathbb{R},n}(A,B,C) \leq \delta\, r^-_{\mathbb{R}}(A,B,C)$, see [250]. However, by strengthening the conditions on the linear system, it is possible to define a restricted class of systems for which the Aizerman conjecture holds true over the reals. Perhaps the most notable result in this direction is the *Popov* criterion (1962) [418] which yields a (conservative) sufficient condition for the global asymptotic stability

of Σ_N. Popov's paper initiated a new frequency-response approach towards the stability theory of nonlinear feedback systems. Another important result in this vein is the *circle criterion* originally proved by *Sandberg* (1964) [447] and *Zames* (1966) [543]. In contrast to Popov's criterion the circle criterion is also applicable to time-varying nonlinearities. The question "How conservative is the circle criterion" has been discussed by *Megretski* (1998) in the open problem book [63]. It is closely related to the open problem (mentioned in Remark 5.6.21) of whether or not the quotient $r_{\mathbb{R},t}^{-}/r_{\mathbb{C}}^{-}$ is bounded. For instructive surveys on the early developments in the stability theory of nonlinear feedback systems and for reprints of some of the classical papers, see the IEEE volume on *Frequency-Response Methods in Control Systems* edited by *MacFarlane* (1979). In [250] it is shown that the complex version of the Aizerman conjecture can be extended to certain classes of time-varying nonlinearities. However some constraints are needed, see Ex. 6.

Popov has shown that his results include those obtainable via the use of Liapunov functions consisting of a quadratic form plus an integral of the nonlinearity. *Yakubovich* (1962) [534] established the converse result that, if the Popov criterion is satisfied with some additional constraints then there exists a Liapunov function consisting of a quadratic form plus an integral of the nonlinear term. Recently, significant progress has been made in the construction of Liapunov functions for perturbed nonlinear differential equations, see e.g. *Lin et al.* (1996) [345]. In the proofs of these converse theorems for uncertain nonlinear systems, the concept of *input-to-state stability* due to Sontag and several variations of this concept play a central role, see *Sontag and Wang* (1996) [474] and the recent monograph of *Grüne* (2002) [203].

As early as the 1960s *Zubov* (1964) [547] derived a method of constructing a Liapunov function which characterizes the complete basin of attraction of a given asymptotically stable equilibrium. This method is based on a partial differential equation called the *Zubov equation*. Recently *Camilli et al.* (2000) [88] generalized Zubov's method to perturbed systems. They introduced the concept of robust domains of attraction and developed a method of computing them via the viscosity solution of a suitable modification of Zubov's equation, see [89], [203] and [533].

Appendix

In this appendix we collect some definitions and results from

- Linear Algebra
- Complex Analysis
- Convolutions and Transforms
- Operator Theory

which are used in the text. We assume, however, that the reader is familiar with the basic elements of these fields. Some of the results reported in the following sections are not used in this volume, but have been included with a view towards Volume II.

A.1 Linear Algebra

Most of the results in this section can be found in *Horn and Johnson* (1999) [264], *Stewart and Sun* (1990) [484] or *Horn and Johnson* (1991) [265]. We give proofs of those results for which we were not able to find references.

A.1.1 Norms of Vectors and Matrices

In this subsection we give some definitions and results on vector and matrix norms which are needed throughout the book and, in particular, in Chapters 4 and 5.

Definition A.1.1. For $p \in [1, \infty]$ the p-norm of $x \in \mathbb{C}^n$ is defined by

$$\|x\|_p = (|x_1|^p + |x_2|^p + \ldots + |x_n|^p)^{1/p}, 1 \le p < \infty, \quad \|x\|_\infty = \max\{|x_1|, |x_2|, \ldots, |x_n|\}.$$

If X is a normed space the dual space is denoted by X^*, see Section A.4. The dual space of $\mathbb{K}^n$ can be identified with $\mathbb{K}^{1\times n}$ by associating with the linear form $f \in (\mathbb{K}^n)^*$ the row vector $[f_1, f_2, \ldots, f_n] = [f(e^1), f(e^2), \ldots, f(e^n)]$ where $(e^1, \ldots, e^n)$ is the standard basis of $\mathbb{K}^n$. Note that $[f(e^1), f(e^2), \ldots, f(e^n)]$ is just the matrix representation of f with respect to the standard basis of $\mathbb{K}^n$.

Definition A.1.2. If $\|\cdot\|_{\mathbb{K}^n}$ is a norm on $\mathbb{K}^n$, the dual norm on the dual space $(\mathbb{K}^n)^*$ is

$$\|f\|^*_{\mathbb{K}^n} = \max\{|f_1x_1 + \ldots + f_nx_n|;\ \|x\|_{\mathbb{K}^n} = 1\}, \quad f = [f_1, f_2, \ldots, f_n] \in \mathbb{K}^{1\times n}. \tag{1}$$

For $p \in [1, \infty]$ the dual norm of a p-norm is $\|\cdot\|^*_p = \|\cdot\|_{p^*}$, where $p^* \in [1, \infty]$ is the *conjugate exponent* defined by $1/p + 1/p^* = 1$. In particular if $p = 1$ then $p^* = \infty$.

Example A.1.3. For $p = 2$, the norm $\|\cdot\|_2$ on $\mathbb{C}^n$ is induced by the inner product

$$\langle x, y\rangle_{\mathbb{C}^n} = y^*x = \overline{y_1}x_1 + \overline{y_2}x_2 + \ldots + \overline{y_n}x_n, \quad x, y \in \mathbb{C}^n. \tag{2}$$

In this case we have $p^* = 2$ and $(\mathbb{C}^n)^*$ can be identified with $\mathbb{C}^n$ by associating with any $y \in \mathbb{C}^n$ the linear form $f_y : x \mapsto y^* x$ on $\mathbb{C}^n$. The map $y \mapsto f_y$ is a conjugate linear bijection from $\mathbb{C}^n$ onto $(\mathbb{C}^n)^*$ ($f_{y+z} = f_y + f_z$, $f_{\alpha y} = \overline{\alpha} f_y$ for $y, z \in \mathbb{C}^n, \alpha \in \mathbb{C}$) and preserves the norms: $\|f_y\|_2^* = \|y\|_2$. □

For every $p \in [1, \infty]$ we have:

$$|\langle x, y\rangle| \leq \sum_{i=1}^{n} |x_i|\,|y_i| \leq \|x\|_p \|y\|_{p*}, \quad x, y \in \mathbb{C}^n.$$

This is *Hölder's inequality* and generalizes the *Cauchy-Schwarz inequality*

$$|\langle x, y\rangle| \leq \sum_{i=1}^{n} |x_i|\,|y_i| \leq \|x\|_2 \|y\|_2, \quad x, y \in \mathbb{C}^n.$$

Definition A.1.4. If $\|\cdot\|_{\mathbb{K}^n}$ and $\|\cdot\|_{\mathbb{K}^m}$ are given norms on $\mathbb{K}^n$ and $\mathbb{K}^m$, then the corresponding *operator norm* of any matrix $B \in \mathbb{K}^{n \times m}$ is defined by

$$\|B\|_{\mathcal{L}(\mathbb{K}^m, \mathbb{K}^n)} := \max\{\|Bx\|_{\mathbb{K}^n};\ x \in \mathbb{K}^m,\ \|x\|_{\mathbb{K}^m} = 1\}.$$

The operator norm of $B \in \mathbb{K}^{n \times n}$ corresponding to the norm $\|\cdot\|_{\mathbb{K}^n}$ on $\mathbb{K}^n$ is denoted by $\|B\|_{\mathcal{L}(\mathbb{K}^n)}$.

When $\mathbb{K}^m$ is normed with a p-norm and $\mathbb{K}^n$ is normed with an r-norm, $1 \leq p, r \leq \infty$, we write $\|B\|_{\mathcal{L}(\mathbb{K}^m, \mathbb{K}^n)} = \|B\|_{p,r}$. If $B = (b_{ij}) \in \mathbb{K}^{n \times m}$, we have

$$\|B\|_{1,1} = \max_{j \in \underline{m}} \sum_{i=1}^{n} |b_{ij}|, \quad \|B\|_{2,2} = \sigma_{\max}(B), \quad \|B\|_{\infty,\infty} = \max_{i \in \underline{n}} \sum_{j=1}^{m} |b_{ij}|. \tag{3}$$

If $\|\cdot\|_{\mathbb{C}^n}$ and $\|\cdot\|_{\mathbb{C}^m}$ are given norms on $\mathbb{C}^n$,$\mathbb{C}^m$ and $\mathbb{R}^n$, $\mathbb{R}^m$ (considered as $\mathbb{R}$-linear subspaces of $\mathbb{C}^n$,$\mathbb{C}^m$) are provided with the induced norms $\|\cdot\|_{\mathbb{R}^n}$ and $\|\cdot\|_{\mathbb{R}^m}$, then

$$\|B\|_{\mathcal{L}(\mathbb{R}^m, \mathbb{R}^n)} \leq \|B\|_{\mathcal{L}(\mathbb{C}^m, \mathbb{C}^n)}, \qquad B \in \mathbb{R}^{n \times m}. \tag{4}$$

The notation $\|\cdot\|_{p,r}$ does not reveal whether the underlying spaces are real or complex. We will use the convention that if it is not stated otherwise the underlying spaces are complex. It is important to make this distinction since the inequality in (4) may be strict.

Example A.1.5. Consider the matrix $B = \begin{bmatrix} 1 & -1 \\ 1 & 1 \end{bmatrix}$ as a map from $(\mathbb{K}^2, \|\cdot\|_\infty)$ into $(\mathbb{K}^2, \|\cdot\|_1)$. Then $\|Bx\|_1 = |x_1 - x_2| + |x_1 + x_2|$ for $x \in \mathbb{C}^2$, and for $x_1 = 1, x_2 = \imath$ we have $\|x\|_\infty = 1$, $\|Bx\|_1 = 2\sqrt{2}$, hence $\|B\|_{\infty,1} \geq 2\sqrt{2}$ in the complex case. But in the real case $\max\{\|Bx\|_1;\ x \in \mathbb{R}^2, \|x\|_\infty = 1\} = 2$. □

The above considerations motivate the following definition.

Definition A.1.6. (i) A norm $\|\cdot\|_{\mathbb{C}^n}$ on $\mathbb{C}^n$ is said to be compatible with a given norm $\|\cdot\|_{\mathbb{R}^n}$ on $\mathbb{R}^n$ if $\|x\|_{\mathbb{C}^n} = \|x\|_{\mathbb{R}^n}$ for all $x \in \mathbb{R}^n$.

(ii) Let $(\|\cdot\|_{\mathbb{R}^m}, \|\cdot\|_{\mathbb{R}^n})$ be a given pair of norms on $\mathbb{R}^m, \mathbb{R}^n$. A pair of norms $(\|\cdot\|_{\mathbb{C}^m}, \|\cdot\|_{\mathbb{C}^n})$ on $\mathbb{C}^m, \mathbb{C}^n$ is said to be *compatible* with $(\|\cdot\|_{\mathbb{R}^m}, \|\cdot\|_{\mathbb{R}^n})$ if $\|\cdot\|_{\mathbb{C}^n}$ is compatible with $\|\cdot\|_{\mathbb{R}^n}$, $\|\cdot\|_{\mathbb{C}^m}$ is compatible with $\|\cdot\|_{\mathbb{R}^m}$ and the corresponding operator norms on $\mathbb{C}^{m \times n}$, $\mathbb{C}^{n \times m}$ are compatible with those on $\mathbb{R}^{m \times n}$, $\mathbb{R}^{n \times m}$ in the following sense:

$$\|B\|_{\mathcal{L}(\mathbb{C}^m, \mathbb{C}^n)} = \|B\|_{\mathcal{L}(\mathbb{R}^m, \mathbb{R}^n)},\ B \in \mathbb{R}^{n \times m}; \quad \|C\|_{\mathcal{L}(\mathbb{C}^n, \mathbb{C}^m)} = \|C\|_{\mathcal{L}(\mathbb{R}^n, \mathbb{R}^m)},\ C \in \mathbb{R}^{m \times n}.$$

It follows from the formulas in (3) that if $\mathbb{R}^m, \mathbb{R}^n$, $\mathbb{C}^m, \mathbb{C}^n$ are jointly normed with either the 1-, 2- or ∞-norms, then the norm pairs $(\|\cdot\|_{\mathbb{R}^m}, \|\cdot\|_{\mathbb{R}^n})$ and $(\|\cdot\|_{\mathbb{C}^m}, \|\cdot\|_{\mathbb{C}^n})$ will be compatible. The next lemma shows that for any given pair of norms on $\mathbb{R}^m, \mathbb{R}^n$ there always exists a compatible pair on $\mathbb{C}^m, \mathbb{C}^n$.

Lemma A.1.7. *Let $(\|\cdot\|_{\mathbb{R}^m}, \|\cdot\|_{\mathbb{R}^n})$ be a given pair of norms on $\mathbb{R}^m, \mathbb{R}^n$. Then the pair of norms $(\|\cdot\|_{\mathbb{C}^m}, \|\cdot\|_{\mathbb{C}^n})$ defined by*

$$\|x\|_{\mathbb{C}^m} = \sup_{0\le\theta\le 2\pi} \|\cos\theta \operatorname{Re} x + \sin\theta \operatorname{Im} x\|_{\mathbb{R}^m}, \quad \|y\|_{\mathbb{C}^n} = \sup_{0\le\theta\le 2\pi} \|\cos\theta \operatorname{Re} y + \sin\theta \operatorname{Im} y\|_{\mathbb{R}^n}$$

for $x \in \mathbb{C}^m$, $y \in \mathbb{C}^n$, is a compatible pair of norms on $\mathbb{C}^m, \mathbb{C}^n$. In particular, the norm $\|\cdot\|_{\mathbb{C}^n}$ on $\mathbb{C}^n$ is compatible with the norm $\|\cdot\|_{\mathbb{R}^n}$ on $\mathbb{R}^n$.

Proof: We first show that $\|\cdot\|_{\mathbb{C}^m}$ is in fact a norm on $\mathbb{C}^m$ which induces the norm $\|\cdot\|_{\mathbb{R}^m}$ on $\mathbb{R}^m$. The same proof applies to $\|\cdot\|_{\mathbb{C}^n}$.

(i) $\|x\|_{\mathbb{C}^m} = 0 \Leftrightarrow x = 0$ is clear from the definition (choose $\theta = 0, \pi/2$).

(ii) The triangle inequality follows directly from the definition.

(iii) $\|\alpha x\|_{\mathbb{C}^m} = |\alpha|\|x\|_{\mathbb{C}^m}$ for all $\alpha \in \mathbb{C}$, $x \in \mathbb{C}^m$. This is clear for real $\alpha \ge 0$. Hence it remains only to show it for $\alpha = e^{\imath\varphi}$. But this follows from the equality

$$\|x\|_{\mathbb{C}^m} = \sup_{0\le\theta\le 2\pi} \|\cos\theta \operatorname{Re} x + \sin\theta \operatorname{Im} x\|_{\mathbb{R}^m} = \sup_{0\le\theta\le 2\pi} \|\operatorname{Re}(e^{-\imath\theta}x)\|_{\mathbb{R}^m}, \ x \in \mathbb{C}^m$$

by applying it to $e^{\imath\varphi}x$ and x.

(iv) For $x \in \mathbb{R}^m$ we have $\|x\|_{\mathbb{C}^m} = \sup_{0\le\theta\le 2\pi} \|\cos\theta\, x\|_{\mathbb{R}^m} = \|x\|_{\mathbb{R}^m}$.

It remains to prove the reverse inequality to the one in (4) for every $B \in \mathbb{R}^{n\times m}$. By definition we have for every $x \in \mathbb{C}^m$

$$\begin{aligned}\|Bx\|_{\mathbb{C}^n} &= \sup_{0\le\theta\le 2\pi} \|\cos\theta \operatorname{Re} Bx + \sin\theta \operatorname{Im} Bx\|_{\mathbb{R}^n} = \sup_{0\le\theta\le 2\pi} \|B(\cos\theta \operatorname{Re} x + \sin\theta \operatorname{Im} x)\|_{\mathbb{R}^n} \\ &\le \|B\|_{\mathcal{L}(\mathbb{R}^m,\mathbb{R}^n)}\|x\|_{\mathbb{C}^m}.\end{aligned}$$

Hence $\|B\|_{\mathcal{L}(\mathbb{R}^m,\mathbb{R}^n)} \ge \|B\|_{\mathcal{L}(\mathbb{C}^m,\mathbb{C}^n)}$. □

Definition A.1.8. If $x = (x_i) \in \mathbb{K}^n$, we set $|x| = (|x_i|)$ and say that $|x| \le |y|$ if $|x_i| \le |y_i|$ for all $i = 1, 2, \ldots, n$. A norm on $\mathbb{K}^n$ is said to be

(i) *monotone* if $|x| \le |y|$ implies $\|x\| \le \|y\|$ for all $x, y \in \mathbb{K}^n$,

(ii) *absolute* if $\|x\| = \|\,|x|\,\|$ for all $x \in \mathbb{K}^n$.

By definition all p-norms, $1 \le p \le \infty$ are absolute.

Theorem A.1.9. *Let $\|\cdot\|$ be a norm on $\mathbb{K}^n$, and $\|\cdot\|_{\mathcal{L}(\mathbb{K}^n)}$ the corresponding operator norm it induces on $\mathbb{K}^{n\times n}$. Then the following are equivalent*

(i) $\|\cdot\|$ *is a monotone norm on* $\mathbb{K}^n$.

(ii) $\|\cdot\|$ *is an absolute norm on* $\mathbb{K}^n$.

(iii) For every diagonal matrix $D = \operatorname{diag}(d_1, d_2, \ldots, d_n) \in \mathbb{K}^{n\times n}$ *we have* $\|D\|_{\mathcal{L}(\mathbb{K}^n)} = \max\{|d_1|, |d_2|, \ldots, |d_n|\}$.

Definition A.1.10. For $1 \le p,\, r \le \infty$ the $(p|r)$-Hölder norm of norm $B \in \mathbb{C}^{n\times m}$ is

$$\|B\|_{p|r} = \left\|(\|B^\top e^1\|_p, \ldots, \|B^\top e^n\|_p)^\top\right\|_r, \tag{5}$$

where $e^1, \ldots, e^n$ are the column vectors of I_n. So, if $p, r \in [1, \infty)$, then

$$\|B\|_{p|r} = \left(\sum_{i=1}^{n}\left(\sum_{j=1}^{m}|b_{ij}|^p\right)^{r/p}\right)^{1/r}, \quad B=(b_{ij})\in\mathbb{C}^{n\times m}, \tag{6}$$

and otherwise

$$\|B\|_{p|\infty}=\max_{i\in\underline{n}}\left[\sum_{j=1}^{m}|b_{ij}|^p\right]^{1/p}, \ \|B\|_{\infty|r}=\left[\sum_{i=1}^{n}\left(\max_{j\in\underline{m}}|b_{ij}|\right)^r\right]^{1/r}, \ \|B\|_{\infty|\infty}=\max_{i\in\underline{n},j\in\underline{m}}|b_{ij}|. \tag{7}$$

If $p=r$ the norms $\|\cdot\|_{p|p}$ are just the p-norms on matrices in $\mathbb{C}^{n\times m}$ regarded as vectors in $\mathbb{C}^{nm}$, i.e. $\|B\|_{p|p}=\|\mathrm{vec}\,(B)\|_p$. Here $\mathrm{vec}\,(B)\in\mathbb{C}^{nm}$ is formed by stacking each column of the matrix B (from the left to the right) one beneath the other, see Definition A.1.23. For example if $n=m=2$

$$\mathrm{vec}\,(B)=[b_{11}\ b_{21}\ b_{12}\ b_{22}]^\top.$$

In particular, $\|\cdot\|_F=\|\cdot\|_{2|2}$ is the Frobenius norm. This norm is associated with the following inner product on $\mathbb{K}^{n\times m}$

$$\langle X,Y\rangle=\mathrm{trace}(XY^*), \quad X,Y\in\mathbb{K}^{n\times m}$$

Note that for $n>1$ the Frobenius norm is not the operator norm $\|\cdot\|_{2,2}$. From (3) and (7) we see that the Hölder norm $\|B\|_{1|\infty}$ is the operator norm $\|B\|_{\infty,\infty}$ induced by the ∞-norms on $\mathbb{K}^n$ and $\mathbb{K}^m$, and $\|B^\top\|_{1|\infty}$ is the operator norm $\|B\|_{1,1}$ induced by the 1-norms on $\mathbb{K}^n$ and $\mathbb{K}^m$. But, in general, $\|\cdot\|_{p|r}$ is not an operator norm.

Definition A.1.11. A norm $\|\cdot\|_{\mathbb{K}^{n\times m}}$ on $\mathbb{K}^{n\times m}$ is said to be *rank one consistent* with an operator norm $\|\cdot\|_{\mathcal{L}(\mathbb{K}^m,\mathbb{K}^n)}$ if for all $B\in\mathbb{K}^{n\times m}$,

$$\|B\|_{\mathcal{L}(\mathbb{K}^m,\mathbb{K}^n)}\le\|B\|_{\mathbb{K}^{n\times m}}, \quad \text{and} \quad \|B\|_{\mathbb{K}^{n\times m}}=\|B\|_{\mathcal{L}(\mathbb{K}^m,\mathbb{K}^n)} \text{ if } \mathrm{rank}\,B=1.$$

Lemma A.1.12. *For arbitrary $p,r\in[1,\infty]$, the Hölder norm $\|\cdot\|_{p|r}$ is rank one consistent with the operator norm $\|\cdot\|_{\mathcal{L}(\mathbb{K}^m,\mathbb{K}^n)}$ induced by the dual p-norm on $\mathbb{K}^m$ and the r-norm on $\mathbb{K}^n$. That is $\|\cdot\|_{p|r}$ is rank one consistent with $\|\cdot\|_{p^*,r}$, where $1/p+1/p^*=1$.*

Proof: For $B=(b_{ij})\in\mathbb{K}^{n\times m}$, we denote the ith row by $b^{i\top}$, $i\in\underline{n}$. Since by Hölder's inequality $|y^\top x|\le\|y\|_p\|x\|_{p^*}$ for $p\in[1,\infty]$, we have for $x\in\mathbb{K}^m$

$$\begin{aligned}\|Bx\|_r^r &= \sum_{i=1}^{n}|b^{i\top}x|^r\le\sum_{i=1}^{n}\|b^i\|_p^r\|x\|_{p^*}^r=\|B\|_{p|r}^r\|x\|_{p^*}^r, \quad 1\le r<\infty\\ \|Bx\|_\infty &= \max_{i\in\underline{n}}|b^{i\top}x|\le\max_{i\in\underline{n}}\|b^i\|_p\|x\|_{p^*}=\|B\|_{p|\infty}\|x\|_{p^*}.\end{aligned}$$

Hence $\|B\|_{p^*,r}\le\|B\|_{p|r}$ for all $r,p\in[1,\infty]$.
Now suppose that $B=dc^\top$, $d\in\mathbb{K}^n$, $c\in\mathbb{K}^m$, then clearly $\|B\|_{p|r}=\|d\|_r\|c\|_p$. If $1\le p<\infty$ define $x_j=\overline{c_j}|c_j|^{p-2}$ if $c_j\ne0$ and $x_j=0$ if $c_j=0$. Then

$$\|x\|_{p^*}^{p^*}=\sum_{j=1}^{m}|c_j|^{p^*}|c_j|^{(p-2)p^*}=\sum_{j=1}^{m}|c_j|^{(p-1)p^*}=\|c\|_p^p.$$

Hence $\|x\|_{p^*}=\|c\|_p^{p/p^*}=\|c\|_p^{p-1}$ and

$$|c^\top x|=\|c\|_p^p=\|c\|_p\|x\|_{p^*}, \quad \|Bx\|_r=\|d\|_r\|c\|_p\|x\|_{p^*}.$$

So $\|B\|_{p^*,r}\ge\|d\|_r\|c\|_p=\|B\|_{p|r}\ge\|B\|_{p^*,r}$. If $|c_k|=\max_{j\in\underline{m}}|c_j|$, by choosing $x_k=1$ and $x_j=0$, $j\ne k$, a similar simpler proof goes through for the case $p=\infty$. □

The following table lists the operator norms which are rank one consistent to some Hölder norms

Hölder norm	operator norm	Hölder norm	operator norm
$\|\cdot\|_{1\|\infty}$	$\|\cdot\|_{\infty,\infty}$	$\|\cdot\|_{1\|1}$	$\|\cdot\|_{\infty,1}$
$\|\cdot\|_{2\|2}$	$\|\cdot\|_{2,2}$	$\|\cdot\|_{\infty\|\infty}$	$\|\cdot\|_{1,\infty}$

Table A.1.1: Rank one consistent norms

A.1.2 Spectra and Determinants

For any $A \in \mathbb{C}^{n\times n}$ the characteristic polynomial is denoted by $\chi_A(s)$, the spectrum by $\sigma(A)$, i.e.

$$\chi_A(s) = \det(sI_n - A) \quad \text{and} \quad \sigma(A) = \{\lambda \in \mathbb{C};\ \chi_A(\lambda) = 0\},$$

and the resolvent set by $\rho(A) = \mathbb{C} \setminus \sigma(A)$.

Lemma A.1.13. *Suppose $M \in \mathbb{C}^{n\times m}$, $N \in \mathbb{C}^{m\times n}$ and $m \le n$. Then*

$$\det(I_n - MN) = \det(I_m - NM) \quad \textit{and} \quad \chi_{MN}(s) = s^{n-m}\chi_{NM}(s). \tag{8}$$

In particular, MN and NM have the same non-zero eigenvalues (taking account of multiplicities).

Definition A.1.14. The *inertia* of $A \in \mathbb{C}^{n\times n}$ is the triple $i(A) = (n_+(A), n_0(A), n_-(A))$ where, accounting for multiplicities,

$$\begin{aligned} n_+(A) &= \text{number of eigenvalues of } A \text{ with } \operatorname{Re}\lambda > 0,\\ n_0(A) &= \text{number of eigenvalues of } A \text{ on the imaginary axis},\\ n_-(A) &= \text{number of eigenvalues of } A \text{ with } \operatorname{Re}\lambda < 0. \end{aligned}$$

If f is an entire function and $f(s) = \sum_{k\in\mathbb{N}} a_k s^k$ its representation by a globally convergent power series (see Section A.2), then $f(A) \in \mathbb{C}^{n\times n}$ is defined for every $A \in \mathbb{C}^{n\times n}$ by the absolutely convergent power series in $\mathbb{C}^{n\times n}$

$$f(A) = \sum_{k\in\mathbb{N}} a_k A^k, \qquad A \in \mathbb{C}^{n\times n}. \tag{9}$$

We have the following elementary version of the Spectral Mapping Theorem.

Proposition A.1.15. *If f is an entire function and $A \in \mathbb{C}^{n\times n}$ then*

(i) $f(A)$ is invertible if and only if $f(\lambda) \neq 0$ for all $\lambda \in \sigma(A)$.

(ii) $\sigma(f(A)) = f(\sigma(A))$.

An analogous statement holds for any rational function of a matrix A.

Proposition A.1.16. *Suppose $f(s) = p(s)/q(s)$, $p, q \in \mathbb{C}[s]$ is a complex rational function and $A \in \mathbb{C}^{n\times n}$ is a matrix whose spectrum does not contain any root of q. Then $q(A)$ is invertible and $f(A) := p(A)q(A)^{-1}$ has the spectrum $\sigma(f(A)) = f(\sigma(A))$.*

We conclude this subsection with the Schur complement formula for the determinant of a 2×2 block matrix with square diagonal blocks.

Lemma A.1.17. *Suppose* $A = \begin{bmatrix} M & N \\ P & Q \end{bmatrix} \in \mathbb{C}^{(q+\ell)\times(q+\ell)}$ *and* $\det M \neq 0$. *Then*

$$\det A = \det M \det(Q - PM^{-1}N).$$

$Q - PM^{-1}N$ is called the *Schur complement* of M in A. If $\det Q \neq 0$, the Schur complement of Q in A is $M - NQ^{-1}P$, and the corresponding formula for the determinant is $\det A = \det Q \det(M - NQ^{-1}P)$.

A.1.3 Real Representation of Complex Matrices

If $G \in \mathbb{C}^{q\times\ell}$ and $G = X + \imath Y$ with $X, Y \in \mathbb{R}^{q\times\ell}$ then the representation of G in real form is given by

$$G^{\mathbb{R}} = \begin{bmatrix} X & -Y \\ Y & X \end{bmatrix} \in \mathbb{R}^{2q\times 2\ell}.$$

$G^{\mathbb{R}}$ represents the $\mathbb{R}$-linear map $G : u \mapsto Gu$ with respect to the standard bases $(e^1, \dots, e^\ell; \imath e^1, \dots, \imath e^\ell)$ and $(e^1, \dots, e^q; \imath e^1, \dots, \imath e^q)$ of the real vector spaces $\mathbb{C}^\ell \cong \mathbb{R}^\ell \times \mathbb{R}^\ell$ and $\mathbb{C}^q \cong \mathbb{R}^q \times \mathbb{R}^q$, respectively.

Lemma A.1.18. *The map* $G \mapsto G^{\mathbb{R}}$ *from* $\mathbb{C}^{q\times\ell}$ *to* $\mathbb{R}^{2q\times 2\ell}$ *has the following properties.*

(a) $G \mapsto G^{\mathbb{R}}$ *is an* $\mathbb{R}$*-linear isomorphism from* $\mathbb{C}^{q\times\ell}$ *into* $\mathbb{R}^{2q\times 2\ell}$.

(b) $(GH)^{\mathbb{R}} = G^{\mathbb{R}} H^{\mathbb{R}}$ *for* $G \in \mathbb{C}^{q\times n}$, $H \in \mathbb{C}^{n\times\ell}$.

(c) If $G \in \mathbb{C}^{n\times n}$ *is invertible then so is* $G^{\mathbb{R}} \in \mathbb{R}^{2n\times 2n}$ *and* $\left(G^{-1}\right)^{\mathbb{R}} = \left(G^{\mathbb{R}}\right)^{-1}$.

(d) $(G^*)^{\mathbb{R}} = (G^{\mathbb{R}})^\top$ *for all* $G \in \mathbb{C}^{q\times\ell}$.

(e) $G \mapsto G^{\mathbb{R}}$ *maps unitary matrices* $G \in \mathbf{U}_n(\mathbb{C})$ *into orthogonal matrices* $G^{\mathbb{R}} \in \mathbf{O}_{2n}$.

In particular, $G \mapsto G^{\mathbb{R}}$ defines an injective ring homomorphism from $\mathbb{C}^{n\times n}$ into $\mathbb{R}^{2n\times 2n}$, and injective group homomorphisms from $\mathbf{Gl}_n(\mathbb{C})$ into $\mathbf{Gl}_{2n}(\mathbb{R})$ and from $\mathbf{U}_n(\mathbb{C})$ into $\mathbf{O}_{2n}$.

A.1.4 Direct Sums and Kronecker Products

In this subsection we first define the direct sum of subspaces and matrices and then briefly review the main properties of Kronecker products. More details can be found in [265, Ch. 4].

Definition A.1.19. Suppose V_1 and V_2 are linear subspaces of a vector space V. Then V is said to be the direct sum of V_1 and V_2 (written $V = V_1 \oplus V_2$) if $V_1 \cap V_2 = \{0\}$ and every element of V can be expressed as the sum of an element of V_1 and an element of V_2.

The direct sum of two linear maps is defined as follows.

Definition A.1.20. Suppose $A_i : X_i \to Y_i$, $i = 1, 2$ are linear maps between vector spaces, then $A_1 \oplus A_2 : X_1 \times X_2 \to Y_1 \times Y_2$ is defined by $(A_1 \oplus A_2) \begin{bmatrix} x_1 \\ x_2 \end{bmatrix} = \begin{bmatrix} A_1 x_1 \\ A_2 x_2 \end{bmatrix}$ for $x_1 \in X_1$, $x_2 \in X_2$.

Applying this definition to matrices $A_1 \in \mathbb{K}^{n_1 \times m_1}$, $A_2 \in \mathbb{K}^{n_2 \times m_2}$ (identified with the corresponding linear maps) the resulting direct sum has a block-diagonal matrix representation

$$(A_1 \oplus A_2) = \operatorname{diag}(A_1, A_2) \in \mathbb{K}^{(n_1+n_2)\times(m_1+m_2)}.$$

The *Kronecker product* of two matrices is defined as follows.

Definition A.1.21. If $A \in \mathbb{K}^{m\times n}$, $A = (a_{ij})$, $B \in \mathbb{K}^{p\times q}$, the Kronecker product of A and B is the block-diagonal matrix $A \otimes B = (a_{ij}B)_{i\in\underline{m}, j\in\underline{n}} \in \mathbb{K}^{mp\times nq}$.

For example if $m = n = 2$

$$A \otimes B = \left[\begin{array}{c|c} a_{11}B & a_{12}B \\ \hline a_{21}B & a_{22}B \end{array}\right].$$

Proposition A.1.22. *The Kronecker product has the following properties.*

(a) $(\alpha A) \otimes B = A \otimes (\alpha B) = \alpha(A \otimes B)$, $\quad \alpha \in \mathbb{K}$, $A \in \mathbb{K}^{m\times n}$, $B \in \mathbb{K}^{p\times q}$.

(b) $(A \otimes B)^\top = A^\top \otimes B^\top$, $\quad A \in \mathbb{K}^{m\times n}$, $B \in \mathbb{K}^{p\times q}$.

(c) $(A \otimes B)^* = A^* \otimes B^*$, $\quad A \in \mathbb{C}^{m\times n}$, $B \in \mathbb{C}^{p\times q}$.

(d) $(A \otimes B) \otimes C = A \otimes (B \otimes C)$, $\quad A \in \mathbb{K}^{m\times n}$, $B \in \mathbb{K}^{p\times q}$, $C \in \mathbb{K}^{r\times s}$.

(e) $(A + B) \otimes C = A \otimes C + B \otimes C$, $\quad A \in \mathbb{K}^{m\times n}$, $B \in \mathbb{K}^{m\times n}$, $C \in \mathbb{K}^{r\times s}$.

(f) $A \otimes (B + C) = A \otimes B + A \otimes C$, $\quad A \in \mathbb{K}^{m\times n}$, $B \in \mathbb{K}^{p\times q}$, $C \in \mathbb{K}^{p\times q}$.

(g) $A \otimes B = 0$ *if and only if* $A = 0$ *or* $B = 0$, $\quad A \in \mathbb{K}^{m\times n}$, $B \in \mathbb{K}^{p\times q}$.

(h) $(A \otimes B)(C \otimes D) = AC \otimes BD$, $\quad A \in \mathbb{K}^{m\times n}$, $B \in \mathbb{K}^{p\times q}$, $C \in \mathbb{K}^{n\times s}$, $D \in \mathbb{K}^{q\times r}$.

(i) *If* $A \in \mathbb{K}^{n\times n}$ *and* $B \in \mathbb{K}^{q\times q}$ *are nonsingular, then* $(A \otimes B)^{-1} = A^{-1} \otimes B^{-1}$.

(j) $[B \otimes A \,|\, C \otimes A] = ([B \,|\, C] \otimes A)$ *and there exists a permutation matrix* Q *such that* $[A \otimes B \,|\, A \otimes C] = (A \otimes [B \,|\, C])Q$ *for all* $A \in \mathbb{K}^{m\times n}$, $B \in \mathbb{K}^{p\times q}$, $C \in \mathbb{K}^{p\times s}$.

(k) *If* $A \in \mathbb{K}^{n\times n}$ *and* $B \in \mathbb{K}^{q\times q}$ *then* $\det(A \otimes B) = (\det A)^q(\det B)^n = \det(B \otimes A)$.

In the study of matrix equations it is sometimes useful to consider matrices in $\mathbb{K}^{m\times n}$ as vectors by ordering their entries in a convenient way.

Definition A.1.23. With each matrix $A = (a_{ij}) \in \mathbb{K}^{m\times n}$ we associate the vector $\operatorname{vec}(A) \in \mathbb{K}^{mn}$ defined by

$$\operatorname{vec}(A) = [a_{11}, a_{21}, \ldots, a_{n1}, a_{12}, a_{22}, \ldots, a_{n2}, \ldots, a_{1m}, a_{2m}, \ldots, a_{nm}]^\top.$$

It follows that $\|\operatorname{vec}(A)\|_2 = \|A\|_F$ where $\|\cdot\|_F$ is the Frobenius norm.

Proposition A.1.24. *For every* $m, n \in \mathbb{N}^*$ *there exists a unique* $mn \times mn$ *permutation matrix* $P(m,n)$ *such that* $\operatorname{vec}(X^\top) = P(m,n)\operatorname{vec}(X)$ *for all* $X \in \mathbb{K}^{m\times n}$. *With these permutation matrices the Kronecker products* $A \otimes B$ *and* $B \otimes A$ *are related by*

$$B \otimes A = P(m,p)^\top (A \otimes B) P(n,q), \qquad A \in \mathbb{K}^{m\times n},\ B \in \mathbb{K}^{p\times q}.$$

In particular, if $m = n$, $p = q$ *then* $A \otimes B$ *and* $B \otimes A$ *are similar via the permutation* $P(m,p)$.

The following proposition is useful for converting a linear matrix equation for X into a vector equation for $\mathrm{vec}(X)$.

Proposition A.1.25. *Suppose $A \in \mathbb{K}^{m\times n}$, $X \in \mathbb{K}^{n\times q}$ and $C \in \mathbb{K}^{q\times s}$, then*

$$\mathrm{vec}\,(AXC) = (C^\top \otimes A)\mathrm{vec}\,(X).$$

By this proposition we see that for $A \in \mathbb{K}^{n\times n}$, $B \in \mathbb{K}^{m\times m}$, $C \in \mathbb{K}^{n\times m}$, $X \in \mathbb{K}^{n\times m}$ the equation $AX + XB = C$ can be re-written as $(I_m \otimes A + B^\top \otimes I_n)\mathrm{vec}\,(X) = \mathrm{vec}\,(C)$. The eigenvalues of the Kronecker product of two square matrices are determined as follows.

Theorem A.1.26. *Suppose $A \in \mathbb{K}^{n\times n}$ and $B \in \mathbb{K}^{q\times q}$. If $\lambda \in \sigma(A)$ and $x \in \mathbb{C}^n$ is a corresponding eigenvector, and if $\mu \in \sigma(B)$ and $y \in \mathbb{C}^q$ is a corresponding eigenvector, then $\lambda\mu \in \sigma(A\otimes B)$ with corresponding eigenvector $x\otimes y \in \mathbb{C}^{nq}$. Every eigenvalue of $A\otimes B$ arises as such a product of eigenvalues of A and B. If $\sigma(A) = \{\lambda_1, \dots, \lambda_n\}$, $\sigma(B) = \{\mu_1, \dots, \mu_q\}$, then $\sigma(A \otimes B) = \{\lambda_i\mu_j;\ i = 1, \dots, n,\ j = 1, \dots, q\}$ (taking account of multiplicities). In particular, $\sigma(A \otimes B) = \sigma(B \otimes A)$ and* $\mathrm{trace}(A \otimes B) = \mathrm{trace}(B \otimes A) = \mathrm{trace}(A)\ \mathrm{trace}(B)$.

The following result about the singular value decomposition of a Kronecker product is an immediate consequence of this theorem and Proposition A.1.22 (h).

Theorem A.1.27. *Suppose $A \in \mathbb{C}^{m\times n}$ and $B \in \mathbb{C}^{p\times q}$ have singular value decompositions $A = W_1\Sigma_1V_1^*$ and $B = W_2\Sigma_2V_2^*$, then $A \otimes B = (W_1 \otimes W_2)(\Sigma_1 \otimes \Sigma_2)(V_1 \otimes V_2)^*$. The non-zero singular values of $A \otimes B$ are $\sigma_i(A)\sigma_j(B)$, $i = 1, \dots, \mathrm{rank}(A)$, $j = 1, \dots, \mathrm{rank}(B)$ taking account of multiplicities. In particular, $A \otimes B$ and $B \otimes A$ have the same non-zero singular values and* $\mathrm{rank}(A \otimes B) = \mathrm{rank}(B \otimes A) = \mathrm{rank}(A)\ \mathrm{rank}(B)$.

As a direct consequence we obtain

$$\|A \otimes B\|_{2,2} = \sigma_{\max}(A \otimes B) = \sigma_{\max}(A)\sigma_{\max}(B) = \|A\|_{2,2}\|B\|_{2,2}.$$

Similarly we have by Theorem A.1.26 and Proposition A.1.22 (h) for the Frobenius norm

$$\|A \otimes B\|_F = \mathrm{trace}[(A \otimes B)\,(A \otimes B)^*] = \mathrm{trace}\,(AA^* \otimes BB^*) = \|A\|_F\|B\|_F.$$

A.1.5 Hermitian Matrices

Here we give a review of some basic properties of Hermitian matrices and their order relation $\succeq$.

Definition A.1.28. For any $A = (a_{ij}) \in \mathbb{C}^{n\times n}$ let $A^* := \overline{A}^\top = (\overline{a_{ji}})$ be its complex conjugate transpose. The matrix A is said to be *Hermitian* if $A^* = A$. It is said to be *skew-Hermitian* if $A^* = -A$.

We note the following.

- The set $\mathcal{H}_n(\mathbb{K})$ of Hermitian matrices in $\mathbb{K}^{n\times n}$ form a *real* vector space, i.e if $A, B \in \mathcal{H}_n(\mathbb{K})$, then $\alpha A + \beta B \in \mathcal{H}_n(\mathbb{K})$ for all $\alpha, \beta \in \mathbb{R}$.
- A real matrix $A \in \mathbb{R}^{n\times n}$ is Hermitian if and only if it is symmetric, i.e. $A = A^\top$.
- If $A \in \mathbb{C}^{n\times n}$ then $A+A^*$, AA^*, A^*A are all Hermitian and $A-A^*$ is skew-Hermitian.
- $A \in \mathbb{C}^{n\times n}$ is Hermitian if and only if the associated quadratic form $Q(x) = x^*Ax$ is real on $\mathbb{C}^n$.

Hermitian matrices have special spectral properties.

Theorem A.1.29. *A matrix* $A \in \mathbb{C}^{n\times n}$ *is Hermitian if and only if there is a unitary matrix* $U \in \mathbf{U}_n(\mathbb{C})$ *and a real diagonal matrix* Λ *such that* $A = U\Lambda U^*$. *A is real and Hermitian (i.e., real symmetric) if and only if there is an orthogonal matrix* $U \in \mathbf{O}_n$ *and a real diagonal matrix* Λ *such that* $A = U\Lambda U^\top$.

As a consequence of this theorem if $A \in \mathcal{H}_n(\mathbb{K})$, we have

$$\|A\|_{2,2} = \max_{\|x\|=1} |\langle Ax, x\rangle|, \quad \text{and} \quad \lambda_{\max}(A) = \max_{\|x\|=1} \langle Ax, x\rangle, \tag{10}$$

where $\langle\cdot,\cdot\rangle$ is the usual inner product on $\mathbb{K}^n$.
$A \in \mathcal{H}_n(\mathbb{K})$ is said to be *positive definite* $(A \succ 0)$ (resp. *positive semi-definite* $(A \succeq 0)$) if

$$\langle Ax, x\rangle = \langle x, Ax\rangle > 0 \quad (\text{resp.} \geq 0)\,, \qquad x \in \mathbb{K}^n,\ x \neq 0.$$

The set $\mathcal{H}_n^+(\mathbb{K})$ of positive semi-definite matrices on $\mathbb{K}^n$ is a pointed convex cone, i.e.

$$\alpha\mathcal{H}_n^+(\mathbb{K}) \subset \mathcal{H}_n^+(\mathbb{K}),\ \alpha > 0,\ \mathcal{H}_n^+(\mathbb{K}) + \mathcal{H}_n^+(\mathbb{K}) \subset \mathcal{H}_n^+(\mathbb{K}),\ \mathcal{H}_n^+(\mathbb{K}) \cap (-\mathcal{H}_n^+(\mathbb{K})) = 0.$$

The associated order on the vector space $\mathcal{H}_n(\mathbb{K})$ is defined by $A \succeq B \Leftrightarrow A - B \succeq 0$. We write $A \succ B$ if $A, B \in \mathcal{H}_n(\mathbb{K})$ and $A - B \succ 0$.

Theorem A.1.30. *The following are equivalent*

(a) $A \succ 0$ *(resp.* $A \succeq 0$*),*
(b) $\lambda > 0$ *(resp.* $\lambda \geq 0$*) for all* $\lambda \in \sigma(A)$*,*
(c) *the leading principal minors of A are positive (resp. all the principal minors of A are non-negative),*
(d) *there exists a unique matrix* $B \succ 0$ *(resp.* $B \succeq 0$*) such that* $B^2 = A$ *(we write* $B = A^{1/2}$*).*

Since the eigenvalues of an Hermitian matrix are all real they are either positive, negative or zero. Specializing Definition A.1.14 to Hermitian matrices we get

Definition A.1.31. The *inertia* of a Hermitian matrix $A \in \mathcal{H}_n(\mathbb{K})$ is the ordered triple $i(A) = (n_+(A), n_0(A), n_-(A))$ where $n_+(A), n_0(A), n_-(A)$ denote the number of positive, zero and negative eigenvalues of A, respectively, taking account of multiplicities.
$\operatorname{sign}(A) := n_+(A) - n_-(A)$ is called the *signature* of A.

Notice that the rank of a Hermitian matrix is $n_+(A) + n_-(A)$ and so the inertia of a Hermitian matrix A is uniquely determined by $\operatorname{rank}(A)$ and $\operatorname{sign}(A)$. We have the following lemma.

Lemma A.1.32. *For* $A, B \in \mathcal{H}_n(\mathbb{K})$ *we have*

$$\operatorname{rank}(A + B) = \operatorname{rank} A + \operatorname{rank} B \quad \Rightarrow \quad \operatorname{sign}(A + B) = \operatorname{sign}(A) + \operatorname{sign}(B).$$

Two matrices $A, B \in \mathbb{C}^{n\times n}$ are said to be *congruent* if there exists a nonsingular matrix $S \in \mathbf{Gl}_n(\mathbb{C})$ such that $B = SAS^*$. One can show that two real symmetric matrices $A, B \in \mathcal{H}(\mathbb{R})$ are congruent if and only if there exists an $S \in \mathbf{Gl}_n(\mathbb{R})$ such that $B = SAS^\top$.

Theorem A.1.33 (Sylvester's Law of Inertia). *Two Hermitian matrices* $A, B \in \mathcal{H}_n(\mathbb{K})$ *are congruent if and only if they have the same inertia* $i(A) = i(B) = (n_+, n_0, n_-)$*. In this case they are both congruent to* $I_{n_+} \oplus 0_{n_0\times n_0} \oplus I_{n_-}$ *(via a matrix* $S \in \mathbf{Gl}_n(\mathbb{K})$*).*

A.2 Complex Analysis

In this section we summarize some basic concepts and results from Complex Analysis. Excellent textbooks in this field are *Ahlfors* (1979) [6], *Cartan* (1995) [92], *Conway* (1978, 1995) [109], [110], *Fischer and Lieb* (1992) [163], *Narasimhan* (1985) [386] and *Rudin* (1987) [441]. A comprehensive introduction which contains many variants and extensions of classical results and instructive comments on the literature is *Burckel* (1979) [82].

A.2.1 Topological Preliminaries

Suppose X is a metric space or, more generally, a topological Hausdorff space. For $S \subset X$ we denote by $\operatorname{int} S$ the *interior* of S (the largest open subset of X contained in S) and by $\overline{S}$ the *closure* of S in X (smallest closed subset of X containing S). $S \subset X$ is called *dense* in X if $\overline{S} = X$. A point $a \in X$ is said to be an *accumulation point* of $S \subset X$ if every neighbourhood of a contains infinitely many elements of S. S is said to be a *discrete* (or *locally finite*) subset of X if every $x \in X$ has a neighbourhood which contains at most finitely many points of S, i.e. S has no accumulation point in X. A discrete subset of X is always closed in X. s_0 is said to be an *isolated point* of S if there exists a neighbourhood V of s_0 in X such that $V \cap S = \{s_0\}$. A closed subset is discrete if and only if it consists of isolated points.
A function $u : X \to \overline{\mathbb{R}} := [-\infty, \infty]$ is *upper semicontinuous* if, for every $a \in \mathbb{R}$, the set $\{x \in X; u(x) < a\}$ is an open subset of X. Similarly, a function $u : X \to \overline{\mathbb{R}}$ is *lower semicontinuous* if, for every $a \in \mathbb{R}$, the set $\{x \in X; u(x) > a\}$ is open. If $(u_i)_{i\in I}$ is a family of upper (resp. lower) semicontinuous functions then $\inf_{i\in I} u_i$ (resp. $\sup_{i\in I} u_i$) is upper (resp. lower) semicontinuous.

Proposition A.2.1. *If K is a non-empty compact subset of X and u is an upper (resp. lower) semicontinuous function, then there exists an $x^0 \in K$ such that $u(x^0) \geq u(x)$ (resp. $u(x^0) \leq u(x)$) for all $x \in K$.*

An *arc* or *curve* in X is a continuous map γ from a compact interval $[\alpha, \beta] \subset \mathbb{R}$ into X. $[\alpha, \beta]$ is called the *parameter set* and $\gamma([\alpha, \beta]) = \{\gamma(t);\ \alpha \leq t \leq \beta\}$ the *trace* of γ. $\gamma(\alpha)$ is called the *initial point* and $\gamma(\beta)$ the *final point* of γ. γ is said to be a *closed curve* if $\gamma(\alpha) = \gamma(\beta)$. X is said to be *arcwise connected* if, for any two points $x_0, x_1 \in X$, there exists an arc in X with initial point x_0 and final point x_1. Two arcs γ_0, γ_1 in X with joint parameter interval $[\alpha, \beta]$ and $\gamma_0(\alpha) = \gamma_1(\alpha) =: a$, $\gamma_0(\beta) = \gamma_1(\beta) =: b$ are said to be *homotopic* in X, if one can be continuously transformed into the other leaving the initial and final points fixed. More precisely, they are homotopic if there exists a continuous function $h : [\alpha, \beta] \times [0, 1] \to X$ such that $h(t, 0) = \gamma_0(t)$, $h(t, 1) = \gamma_1(t)$, $t \in [\alpha, \beta]$ and $h(\alpha, r) = a$, $h(\beta, r) = b$ for all $r \in [0, 1]$.
If $\Omega \subset \mathbb{C}$ is open, every $x \in \Omega$ is contained in a maximal open and arcwise connected subset of Ω. These subsets are called the *connected components* of Ω. Any two connected components of Ω are disjoint and Ω is the union of these components. $\Omega \subset \mathbb{C}$ is called a *domain* if it is open and arcwise connected.
$U \subset \mathbb{C}$ is said to be a *neighbourhood of* ∞ if it contains a set of the form $\{s \in \mathbb{C}; |s| > r\}$, $r > 0$. If $\gamma : [\alpha, \beta] \to \mathbb{C}$ is any arc then its trace is contained in some disk in $\mathbb{C}$ and so $\mathbb{C} \setminus \gamma([\alpha, \beta])$ has exactly one unbounded connected component (which is a neighbourhood of ∞). A closed curve $\gamma : [\alpha, \beta] \to \mathbb{C}$ is called *simple* if it is one-to-one on $[\alpha, \beta)$. A simple closed curve is called a *Jordan curve*. A fundamental result on the topology of the complex plane is the following.

Theorem A.2.2 (Jordan Curve Theorem). *If $\gamma : [\alpha, \beta] \to \mathbb{C}$ is a Jordan curve then $\mathbb{C} \setminus \gamma([\alpha, \beta])$ consists of two connected components, one bounded and one unbounded. $\gamma([\alpha, \beta])$ is the common boundary of both components.*

A.2.2 Path Integrals

Every $s \in \mathbb{C}^*$ has a polar representation $s = re^{i\phi}$ where $r = |s| > 0$ and $\phi \in \mathbb{R}$. In this case we call $\arg s := \phi$ an *argument* of s. Such an argument is uniquely determined modulo a multiple of 2π. We will now associate with every arc in $\mathbb{C}^*$ a continuous argument function. For the next proposition, see e.g. [82, IV.2].

Proposition A.2.3. *Given an arbitrary interval $I \subset \mathbb{R}$ and a continuous function $\gamma : I \to \mathbb{C}^*$, there exists a continuous function $\phi : I \to \mathbb{R}$ such that*

$$\gamma(t) = |\gamma(t)|e^{\imath\phi(t)} = e^{\ln|\gamma(t)|+\imath\phi(t)}, \quad t \in I. \tag{1}$$

Moreover ϕ is differentiable at each point $t \in I$ where γ is differentiable.

Since the exponential function is a homomorphism of the additive group $\mathbb{C}$ onto the multiplicative group $\mathbb{C}^*$ with kernel $2\pi\imath\mathbb{Z}$, the continuous function ϕ is uniquely determined by (1) up to an additive constant $2\pi k$, $k \in \mathbb{Z}$.

Definition A.2.4. Given an arbitrary interval $I \subset \mathbb{R}$ and a continuous function $\gamma : I \to \mathbb{C}^*$, any continuous function $\phi : I \to \mathbb{R}$ satisfying (1) is called an *argument function* or *continuous argument* of γ. In this case we write $\arg\gamma(\cdot) = \phi(\cdot)$. If $I = [\alpha, \beta]$, the *net change of the argument* of $\gamma(t)$ as t moves from α to β is given by

$$\Delta_\alpha^\beta \gamma(t) = \phi(\beta) - \phi(\alpha). \tag{2}$$

If $\gamma : [\alpha, \beta] \to \mathbb{C}$ is a closed curve and $a \in \mathbb{C} \setminus \gamma([\alpha, \beta])$ then the *winding number* of the point a with respect to the closed curve γ is defined by

$$w(\gamma, a) = (2\pi)^{-1}(\psi(\beta) - \psi(\alpha)) = (2\pi)^{-1}\,\Delta_\alpha^\beta(\gamma(t) - a) \tag{3}$$

where ψ is any argument function of the closed curve $t \mapsto \gamma(t) - a$.

The winding number $w(\gamma, a)$ is well defined by (3) and does not depend upon the particular argument function ψ chosen in (3). Moreover it is integer valued. The winding number measures the net number of encirclements the curve γ makes around the point a (in the

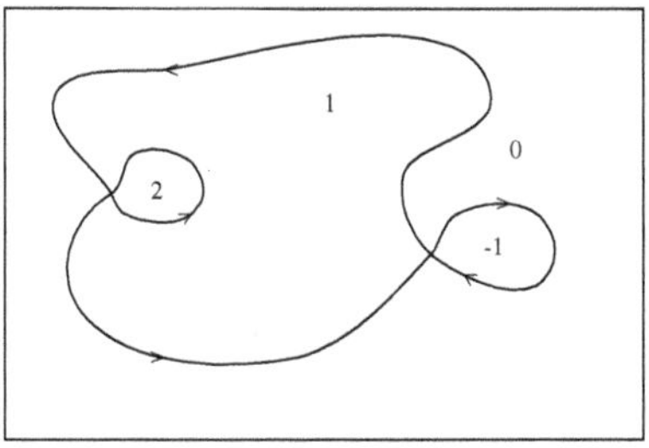

Figure A.2.2: Value of $w(\gamma, .)$ in various connected components of $\mathbb{C} \setminus \gamma([a, b])$

positive, i.e. anticlockwise sense). If, for example, $\gamma(t) = a + e^{2\pi ikt}$, $0 \le t \le 1$ then

$w(\gamma, a) = k$. For every closed curve γ, $w(\gamma, a)$ depends continuously upon $a \in \mathbb{C} \setminus \gamma([\alpha, \beta])$ and as a consequence $w(\gamma, a)$ is constant on the connected components of $\mathbb{C} \setminus \gamma([\alpha, \beta])$ (see Figure A.2.2). It vanishes on the unbounded connected component. If γ is a Jordan curve then $w(\gamma, a) = \pm 1$ if a belongs to the bounded component. We say that a simple closed curve γ *borders the domain* D *in the positive (anticlockwise) sense* if D is the bounded connected component of $\mathbb{C} \setminus \gamma([a, b])$ and $w(\gamma, a) = 1$ for all $a \in D$.

An arc $\gamma : [\alpha, \beta] \to \mathbb{C}$ is said to be *piecewise continuously differentiable* (or *piecewise smooth*) if there are real numbers $t_0 = \alpha < t_1 < \ldots < t_N = \beta$ such that γ is continuously differentiable on each subinterval $[t_{i-1}, t_i]$[3]. An *integration path* or simply a *path* in a subset $\Omega \subset \mathbb{C}$ is a piecewise smooth arc $\gamma : [\alpha, \beta] \to \mathbb{C}$ with $\gamma([a, b]) \subset \Omega$. If γ is an integration path and $f : \gamma([\alpha, \beta]) \to \mathbb{C}$ is continuous then the integral of f over γ is defined by

$$\int_\gamma f(s)ds = \int_\alpha^\beta f(\gamma(t))\dot{\gamma}(t)dt. \tag{4}$$

This *path integral* is invariant under a change of parameter. If ψ is a continuously differentiable one-to-one mapping of an interval $[\tilde{\alpha}, \tilde{\beta}]$ onto $[\alpha, \beta]$, such that $\psi(\tilde{\alpha}) = \alpha$, $\psi(\tilde{\beta}) = \beta$, and if we define the path $\tilde{\gamma} : [\tilde{\alpha}, \tilde{\beta}] \to \mathbb{C}$ by $\tilde{\gamma} := \gamma \circ \psi$, then

$$\int_\gamma f(s)ds = \int_{\tilde{\gamma}} f(s)ds.$$

If $\gamma : [\alpha, \beta] \to \mathbb{C}$ is subdivided into a finite number of subpaths $\gamma_i : [r_{i-1}, r_i] \to \mathbb{C}$, $i \in \underline{n}$ where $r_0 = \alpha < r_1 < \ldots < r_n = \beta$ and $\gamma_i = \gamma|[r_{i-1}, r_i]$, $i \in \underline{n}$ we write symbolically $\gamma = \gamma_1 + \ldots + \gamma_n$. The path integral is additive with respect to this operation

$$\int_\gamma f(s)ds = \int_{\gamma_1} f(s)ds + \ldots + \int_{\gamma_n} f(s)ds.$$

If γ^{-1} is the opposite path $t \mapsto \gamma(\alpha + \beta - t)$, $t \in [\alpha, \beta]$, then

$$\int_{\gamma^{-1}} f(s)ds = -\int_\gamma f(s)ds.$$

A different type of path integral, which does not depend on the orientation of γ, is obtained by integrating f with respect to the path length

$$\int_\gamma f(s)\,|ds| := \int_\alpha^\beta f(\gamma(t))\,|\dot{\gamma}(t)|\,dt, \quad f \in C(\gamma([\alpha, \beta]), \mathbb{C}). \tag{5}$$

(5) is called the *integral of* f *over* γ *with respect to the path length* and satisfies

$$\int_\gamma f(s)|ds| = \int_{\gamma^{-1}} f(s)|ds| \quad \text{and} \quad \left|\int_\gamma f(s)\,ds\right| \leq \int_\gamma |f(s)|\,|ds|.$$

We will now derive an integral expression for the change of the argument along an arbitrary path $\gamma : [\alpha, \beta] \to \mathbb{C}^*$. Let $\phi : [\alpha, \beta] \to \mathbb{R}$ be an argument function of γ, $\phi(t) = \arg \gamma(t)$. It follows from (1) that for all $t \in [\alpha, \beta]$ where γ is differentiable (hence for all but finitely many $t \in [\alpha, \beta]$)

$$\frac{\dot{\gamma}(t)}{\gamma(t)} = \frac{d}{dt}\left[\ln|\gamma(t)| + \imath\phi(t)\right], \quad \dot{\phi}(t) = \operatorname{Im}\frac{\dot{\gamma}(t)}{\gamma(t)}. \tag{6}$$

---

[3] γ is continuously differentiable on $[t_{i-1}, t_i]$ if γ is continuously differentiable on (t_{i-1}, t_i), has a right resp. left derivative at t_{i-1} resp. t_i and $\frac{d^+\gamma}{dt}(t_{i-1}) = \lim_{t \searrow t_{i-1}} \dot{\gamma}(t)$, $\frac{d^-\gamma}{dt}(t_i) = \lim_{t \nearrow t_i} \dot{\gamma}(t)$.

So

$$\int_\gamma \frac{ds}{s} = \int_\alpha^\beta \frac{\dot\gamma(t)}{\gamma(t)} dt = \ln|\gamma(\beta)| - \ln|\gamma(\alpha)| + \imath[\phi(\beta) - \phi(\alpha)]. \tag{7}$$

Thus the *change of the argument* of $\gamma(t)$ as t moves from α to β is given by

$$\Delta_\alpha^\beta \arg\gamma(t) = \phi(\beta) - \phi(\alpha) = \operatorname{Im}\int_\alpha^\beta \frac{\dot\gamma(t)}{\gamma(t)} dt = \operatorname{Im}\int_\gamma \frac{ds}{s}. \tag{8}$$

Now suppose that $\gamma : [\alpha,\beta] \to \mathbb{C}$ is any closed path (piecewise smooth) in $\mathbb{C}$, $a \in \mathbb{C} \setminus \gamma([\alpha,\beta])$ and apply the above considerations to the closed path $\gamma - a : t \mapsto \gamma(t) - a$ with parameter interval $[\alpha,\beta]$. If ψ is a continuous argument for $\gamma - a$, then (3) and (7) imply that the *winding number* of the point a with respect to the closed curve γ is given by

$$w(\gamma, a) = \frac{\psi(\beta) - \psi(\alpha)}{2\pi} = \frac{1}{2\pi\imath}\int_\alpha^\beta \frac{\dot\gamma(t)}{\gamma(t) - a} dt = \frac{1}{2\pi\imath}\int_{\gamma - a} \frac{ds}{s}. \tag{9}$$

A.2.3 Holomorphic Functions

Let Ω be an open subset of $\mathbb{C}$. A function $f : \Omega \to \mathbb{C}$ is said to be complex differentiable at $s_0 \in \Omega$ with (complex) derivative $f'(s_0)$ if

$$f'(s_0) = \lim_{s \to s_0} \frac{f(s) - f(s_0)}{s - s_0}.$$

$f : \Omega \to \mathbb{C}$ is said to be *holomorphic* on Ω if it is complex differentiable at each point $s_0 \in \Omega$. The set of holomorphic functions on Ω is a commutative unitary ring with respect to pointwise addition and multiplication and is denoted by $\mathcal{O}(\Omega)$. If $f \in \mathcal{O}(\mathbb{C})$ then f is said to be an *entire* function.

If $f \in \mathcal{O}(\Omega)$ and $\gamma : [\alpha,\beta] \to \Omega$ is an integration path which does not hit any zero of f, then (8) applied to $\gamma_f = f \circ \gamma : t \mapsto f(\gamma(t))$ yields

$$\Delta_\alpha^\beta \arg\gamma_f(t) = \operatorname{Im}\int_\alpha^\beta \frac{f'(\gamma(t))\dot\gamma(t)}{f(\gamma(t))} dt = \operatorname{Im}\int_\gamma \frac{f'(s)}{f(s)} ds. \tag{10}$$

$\Delta_\alpha^\beta \arg\gamma_f(t)$ is called the *change of the argument* of f along the arc $\gamma(t)$, $\alpha \le t \le \beta$. If additionally γ is a closed path and $a \in \mathbb{C} \setminus f(\gamma([\alpha,\beta]))$, then by (3) and (9) the winding number of $\gamma_f = f \circ \gamma$ with respect to the point a is

$$w(f \circ \gamma, a) = \frac{1}{2\pi\imath}\int_{\gamma_f - a} \frac{ds}{s} = \frac{1}{2\pi\imath}\int_\alpha^\beta \frac{f'(\gamma(t))\dot\gamma(t)}{f(\gamma(t)) - a} dt = \frac{1}{2\pi\imath}\int_\gamma \frac{f'(s)}{f(s) - a} ds. \tag{11}$$

The following theorem is of fundamental importance in Complex Analysis.

Theorem A.2.5 (Cauchy's Integral Theorem and Integral Formula). *Let $\Omega \subset \mathbb{C}$ be open and $\gamma : [\alpha,\beta] \to \Omega$ a closed curve which does not surround any point outside Ω (i.e. $w(\gamma, a) = 0$ for all $a \in \mathbb{C} \setminus \Omega$). If $f : \Omega \to \mathbb{C}$ is holomorphic on Ω then*

$$\int_\gamma f(s) ds = 0. \tag{12}$$

Moreover, for every $z \in \Omega \setminus \gamma([\alpha,\beta])$,

$$w(\gamma, z) f(z) = \frac{1}{2\pi\imath}\int_\gamma \frac{f(s)}{s - z} ds. \tag{13}$$

It follows that if γ_0 and γ_1 are two homotopic paths in Ω and $f : \Omega \to \mathbb{C}$ is holomorphic then

$$\int_{\gamma_0} f(s)ds = \int_{\gamma_1} f(s)ds. \tag{14}$$

For any $a \in \mathbb{C}$ and $r > 0$ let $D(a,r)$ denote the open disk of centre a and radius r. A function $f : \Omega \to \mathbb{C}$ is said to be *(complex) analytic* if it is representable by power series in Ω, i.e. on every disk $D(a,r) \subset \Omega$ it can be expanded as an absolutely convergent power series

$$f(s) = \sum_{k=0}^{\infty} c_k(s-a)^k, \quad s \in D(a,r). \tag{15}$$

Every analytic function f on Ω is holomorphic and has holomorphic derivatives of arbitrary order. The coefficients c_k in (15) are uniquely determined and given by $c_k = f^{(k)}(a)/k!$, $k \in \mathbb{N}$. Conversely, it follows from Cauchy's Integral Formula (13) that every $f \in \mathcal{O}(\Omega)$ can be represented by a power series in Ω. Therefore a function $f : \Omega \to \mathbb{C}$ is holomorphic on Ω if and only if it is analytic on Ω. In particular, every $f \in \mathcal{O}(\Omega)$ has holomorphic derivatives of arbitrary order. Taking derivatives in (13), Cauchy's Integral Formula implies the following integral formulas for the derivatives of f

$$w(\gamma,z)f^{(k)}(z) = \frac{k!}{2\pi\imath}\int_\gamma \frac{f(s)}{(s-z)^{k+1}}ds, \quad z \in \Omega \setminus \gamma([\alpha,\beta]), \quad k \in \mathbb{N}. \tag{16}$$

Suppose $F \in \mathcal{O}(\Omega)$ and $F' = f$ then it follows from the Fundamental Theorem of Calculus that for every path $\gamma : [\alpha,\beta] \to \Omega$

$$\int_\gamma f(s)ds = F(\gamma(\beta)) - F(\gamma(\alpha)). \tag{17}$$

In this case equation (12) holds for every closed path γ in Ω. The next theorem states a converse result.

Theorem A.2.6. *Suppose $f : \Omega \to \mathbb{C}$ is continuous and* (12) *holds for every closed path in Ω. Then there exists a function $F \in \mathcal{O}(\Omega)$ such that $F' = f$. In particular $f \in \mathcal{O}(\Omega)$.*

If Ω is convex, it suffices in the previous theorem that (12) holds for the positively oriented boundaries $\gamma = \partial\Delta$ of all triangles $\Delta = \mathrm{conv}\{a,b,c\} \subset \Omega$. By localization this fact implies the following useful converse of Cauchy's Integral Theorem on arbitrary domains $\Omega \subset \mathbb{C}$.

Theorem A.2.7 (Morera). *Suppose $f : \Omega \to \mathbb{C}$ is continuous and $\int_{\partial\Delta} f(s)ds = 0$ for every triangle $\Delta = \mathrm{conv}\{a,b,c\} \subset \Omega$. Then f is holomorphic on Ω.*

Theorems A.2.5 and A.2.7 imply that the set $\mathcal{O}(\Omega)$ is closed in $\mathcal{C}(\Omega,\mathbb{C})$ with respect to the topology of uniform convergence on compact subsets of Ω. Applying (16) as well, we obtain the following result.

Corollary A.2.8 (Weierstrass). *Suppose $f_k \in \mathcal{O}(\Omega)$ for $k \in \mathbb{N}$ and $f_k \to f$ uniformly on compact subsets of Ω as $k \to \infty$. Then $f \in \mathcal{O}(\Omega)$ and $f_k^{(j)} \to f^{(j)}$ uniformly on compact sets $K \subset \Omega$, for every $j \in \mathbb{N}^*$.*

The following theorem is often used for proving the equality of holomorphic functions.

Theorem A.2.9 (Identity Theorem). *If f, g are holomorphic on a domain Ω and $f(z) = g(z)$ for all z in a non-discrete subset of Ω then $f(s) = g(s)$ for all $s \in \Omega$.*

Corollary A.2.10. *If $\Omega \subset \mathbb{C}$ is a domain then the commutative ring $\mathcal{O}(\Omega)$ has no zero-divisors and is therefore an integral domain.*

As a consequence of the Identity Theorem the *zero set* $Z_f = \{z \in \Omega; f(z) = 0\}$ of any non-zero function $f \in \mathcal{O}(\Omega)$ is a discrete subset of Ω. Each compact subset of Ω contains only finitely many zeros of f. For every $z \in Z_f$ there is a unique $m = m(z) \in \mathbb{N}^*$ such that

$$f(s) = (s - z)^m g(s)$$

where $g \in \mathcal{O}(\Omega)$ and $g(z) \neq 0$. m is called the *order* or *multiplicity* of the zero z of f. $z \in \Omega$ is a zero of multiplicity m if and only if $f(z) = 0, \ldots, f^{(m-1)}(z) = 0$ and $f^{(m)}(z) \neq 0$ or, equivalently if, in a sufficiently small disk around z the function f can be represented by a power series of the form

$$f(s) = \sum_{k=m}^{\infty} a_k (s - z)^k, \quad a_m \neq 0.$$

Theorem A.2.11 (Open Mapping Theorem). *If Ω is a domain and $f \in \mathcal{O}(\Omega)$ is not constant then $f(\Omega)$ is a domain.*

A.2.4 Isolated Singularities

For any $a \in \mathbb{C}$ we denote the *punctured disk of centre a and radius $r > 0$* by

$$D^\circ(a, r) = \{s \in \mathbb{C};\ 0 < |s - a| < r\}.$$

Definition A.2.12. $f \in \mathcal{O}(\Omega)$ is said to have an *isolated singularity* at a point $a \in \mathbb{C} \setminus \Omega$ if $D^\circ(a, r) \subset \Omega$ for some $r > 0$. The singularity is said to be *removable* if f can be extended to a function which is holomorphic on the whole disk $D(a, r)$.

Theorem A.2.13. *If $a \in \mathbb{C}$ and f is a function which is holomorphic and bounded on some punctured disk $D^\circ(a, r)$, $r > 0$, then f has a removable singularity at a.*

Isolated singularities are classified by the following result.

Theorem A.2.14. *If $a \in \mathbb{C}$ and f is holomorphic on some punctured disk $D^\circ(a, r)$, $r > 0$, then one of the following three cases must occur*

(i) f has a removable singularity at a.

(ii) There are $c_1, \ldots, c_m \in \mathbb{C}$, $m \geq 1$, $c_m \neq 0$ such that $f(s) - \sum_{k=1}^{m} c_k/(s - a)^k$ has a removable singularity at a.

(iii) For every $\rho \in (0, r]$ the image $f(D^\circ(a, \rho))$ is dense in $\mathbb{C}$.

If (ii) holds, f is said to have a *pole of order (or multiplicity)* m at a. In this case $|f(s)| \to \infty$ as $s \to a$. If (iii) holds, f is said to have an *essential singularity* at a. In this case there exists, for every $z \in \mathbb{C}$, a sequence $(s_k)_{k \in \mathbb{N}}$ in $D^\circ(a, r)$ converging to a such that $\lim_{k \to \infty} f(s_k) = z$.

Definition A.2.15. A *meromorphic* function on an open set $\Omega \subset \mathbb{C}$ is a holomorphic function $f : \Omega \setminus P_f \to \mathbb{C}$ where P_f is a discrete subset of Ω consisting of poles of f. The set of meromorphic functions on Ω is denoted by $\mathcal{M}(\Omega)$.

For any domain $\Omega \subset \mathbb{C}$ the set $\mathcal{M}(\Omega)$ provided with the operations of pointwise addition and multiplication is a field.

Theorem A.2.16. *If $\Omega \subset \mathbb{C}$ is a domain and $f \in \mathcal{M}(\Omega)$ then there exist $g, h \in \mathcal{O}(\Omega)$ such that $f = g/h$.*

Theorem A.2.17 (Laurent expansion). *Let $0 \leq r < R \leq \infty$, $s_0 \in \mathbb{C}$ and let f be holomorphic on the annulus $U = \{s \in \mathbb{C}\,; r < |s - s_0| < R\}$. Then there exist (uniquely determined) coefficients $a_k \in \mathbb{C}$, $k \in \mathbb{Z}$ such that*

$$f(s) = \sum_{k=-\infty}^{\infty} a_k (s - s_0)^k, \quad s \in U. \tag{18}$$

The series on the RHS of (18) is absolutely convergent and converges to f uniformly on compact subsets of U. The Laurent coefficients a_k are determined by

$$a_k = \frac{1}{2\pi\imath} \int_{\Gamma_\rho(s_0)} \frac{f(s)}{(s - s_0)^{k+1}}\, ds = \frac{1}{2\pi} \int_{-\pi}^{\pi} \frac{f(s_0 + \rho e^{\imath\theta})}{\rho^k e^{\imath k\theta}}\, d\theta, \quad k \in \mathbb{Z} \tag{19}$$

where $\rho \in (r, R)$ is arbitrary and $\Gamma_\rho(s_0) : \theta \mapsto s_0 + \rho e^{\imath\theta}$, $\theta \in [-\pi, \pi]$ is the positively oriented circle of radius ρ about s_0.

In the important special case where $r = 0$, the annulus U reduces to the punctured disk $D^\circ(s_0, R)$. Then (18) is called the *Laurent expansion* of f at s_0. In Linear Systems Theory the inverse special case where $R = \infty$ also plays an important role. Suppose that $f(s)$ is holomorphic on $\{s \in \mathbb{C}; |s| > r\}$ for some $r > 0$, then $f(s)$ can be represented in this neighbourhood of ∞ by the locally uniformly convergent series

$$f(s) = \sum_{k=-\infty}^{\infty} a_k s^{-k}, \quad |s| > r; \qquad a_k = \frac{\rho^k}{2\pi} \int_{-\pi}^{\pi} f(\rho e^{\imath\theta}) e^{\imath k\theta}\, d\theta, \quad k \in \mathbb{Z} \tag{20}$$

where $\rho > r$ is arbitrary. (20) is called the *Laurent expansion of f at infinity.* Note that f has a Laurent expansion of the form

$$f(s) = \sum_{k=0}^{\infty} a_k s^{-k}, \quad |s| > r \tag{21}$$

if and only if f is bounded in some neighbourhood of ∞. In this case $\lim_{|s|\to\infty} f(s) = a_0$.

Definition A.2.18. If f is holomorphic on a punctured disk $D^\circ(s_0, R)$ and $\rho \in (0, R)$, then

$$\operatorname{Res}(f, s_0) := \frac{1}{2\pi\imath} \int_{\Gamma_\rho(s_0)} f(s)\, ds = \frac{\rho}{2\pi} \int_0^{2\pi} f(s_0 + \rho e^{\imath\theta}) e^{\imath\theta} d\theta$$

is called the *residue* of f at s_0.

If (18) is the Laurent expansion of f at s_0, then $\operatorname{Res}(f, s_0) = a_{-1}$. If s_0 is a pole of first order of f then $\operatorname{Res}(f, s_0) = \lim_{s\to s_0}(s - s_0) f(s)$.

Theorem A.2.19 (Residue Theorem). *Suppose $\Omega \subset \mathbb{C}$ is open, S_0 a discrete subset of Ω and $f : \Omega \setminus S_0 \to \mathbb{C}$ is holomorphic. Then for any closed curve $\gamma : [\alpha, \beta] \to \Omega \setminus S_0$ satisfying $w(\gamma, a) = 0$ for all $a \in \mathbb{C} \setminus \Omega$,*

$$\frac{1}{2\pi\imath} \int_\gamma f(s)\, ds = \sum_{s_0 \in S_0} \operatorname{Res}(f, s_0)\, w(\gamma, s_0)$$

and this sum only contains finitely many non-zero terms.

The following theorem lists some important consequences of the Residue Theorem.

Theorem A.2.20. *Let Ω be a domain in $\mathbb{C}$ and $\gamma : [\alpha, \beta] \to \Omega$ a simple closed curve in Ω bordering a bounded domain $D \subset \overline{D} \subset \Omega$ in the anticlockwise sense and satisfying $w(\gamma, a) = 0$ for all $a \in \mathbb{C} \setminus \Omega$.*

(i) **(Principle of the Argument)** . *If f is a meromorphic function on Ω having no poles or zeros on the trace of γ, then*

$$\frac{1}{2\pi i} \int_\gamma \frac{f'(s)}{f(s)} ds = n_z(f, D) - n_p(f, D) \tag{22}$$

where $n_z(f, D)$ and $n_p(f, D)$ are the numbers of zeros and poles of f in D counted according to their multiplicities.

(ii) If z_i, $i \in I$ and p_j, $j \in J$ are the zeros and poles of f in D, of order $m(z_i)$ and $m(p_j)$ respectively, and $g \in \mathcal{O}(\Omega)$ then[4]

$$\frac{1}{2\pi i} \int_\gamma \frac{f'(s)}{f(s)} g(s) ds = \sum_{i \in I} m(z_i)\, g(z_i) - \sum_{j \in J} m(p_j)\, g(p_j). \tag{23}$$

(iii) **(Rouché's Theorem)**. *If f, g are meromorphic on Ω without poles on the trace $\gamma([\alpha, \beta])$ and $|f(s) - g(s)| < |f(s)|$ for all $s \in \gamma([\alpha, \beta])$ then*

$$n_z(f, D) - n_p(f, D) = n_z(g, D) - n_p(g, D).$$

Since

$$\frac{1}{2\pi i} \int_\gamma \frac{f'(s)}{f(s)} ds = \frac{1}{2\pi i} \int_{f \circ \gamma} \frac{ds}{s} = w(f \circ \gamma, 0) \tag{24}$$

the number of zeros minus the number of poles of f in D is just the winding number of the closed curve $f \circ \gamma$ around the origin. This is illustrated in Figure A.2.4 for a function f with two zeros and three poles in the circular domain D.

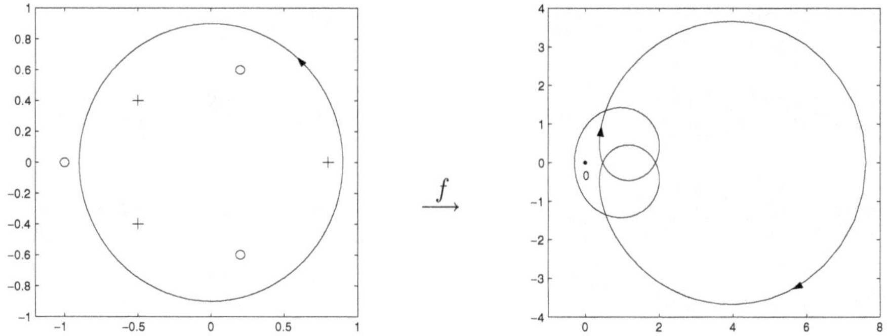

Figure A.2.4: Principle of the Argument

We conclude this subsection with a "continuous time version" of a lemma due to Spijker [476] which we will need in Section 5.5, see also [504].

[4] Since $\overline{D} \subset \Omega$ is compact, D contains only finitely many zeros and poles of f so that the sums in (23) are finite.

Lemma A.2.21. *Let $g(s) = p(s)/q(s)$ where $p, q \in \mathbb{C}[s]$ are coprime complex polynomials and $\deg p \leq \deg q = n$. If $\alpha \in \mathbb{R}$ and g has no poles on the trace of $\Gamma_\alpha : \omega \mapsto \alpha + \imath\omega$, $\omega \in \mathbb{R}$ then*

$$\int_{\Gamma_\alpha} |g'(s)|\,|ds| = \int_{-\infty}^{\infty} |g'(\alpha + \imath\omega)|\,d\omega \leq 2\pi n \sup_{\operatorname{Re} s = \alpha} |g(s)|,$$

where $g' = \frac{d}{ds} g$ is the complex derivative of g.[5]

Proof: Let $\mu(s) = [(\alpha+1)s+(\alpha-1)]/(s+1)$ be the linear fractional transformation, which maps $\partial\mathbb{D} \setminus \{-1\}$ onto $\Gamma_\alpha(\mathbb{R}) = \alpha + \imath\mathbb{R}$, and define $\zeta : [-\pi, \pi] \to \partial\mathbb{D}$ by $\zeta(t) = e^{\imath t}$. Then $\gamma : t \mapsto \gamma(t) = (\mu \circ \zeta)(t)$, $t \in (-\pi, \pi)$ maps $(-\pi, \pi)$ onto $\alpha + \imath\mathbb{R}$. We write $\gamma(t) = \alpha + \imath\beta(t)$. By straightforward calculations one shows that $\lim_{t \searrow -\pi} \beta(t) = -\infty$, $\lim_{t \nearrow \pi} \beta(t) = \infty$ and $\beta'(t) = -\imath\gamma'(t) > 0$ for all $t \in (-\pi, \pi)$. For every $\varepsilon \in (0, \pi)$, let $a_\varepsilon = -\pi + \varepsilon$, $b_\varepsilon = \pi - \varepsilon$. Then

$$I(\varepsilon) := \int_{\beta(a_\varepsilon)}^{\beta(b_\varepsilon)} |g'(\alpha + \imath\omega)|\,d\omega = \int_{a_\varepsilon}^{b_\varepsilon} |g'(\alpha + \imath\beta(t))|\,\beta'(t)\,dt = \int_{a_\varepsilon}^{b_\varepsilon} |g'(\gamma(t))|\,|\gamma'(t)|\,dt$$

and

$$\int_{a_\varepsilon}^{b_\varepsilon} |g'(\gamma(t))|\,|\gamma'(t)|\,dt = \int_{a_\varepsilon}^{b_\varepsilon} |g'(\mu(\zeta(t)))|\,|\mu'(\zeta(t))|\,|\zeta'(t)|\,dt = \int_{a_\varepsilon}^{b_\varepsilon} |(g \circ \mu)'(\zeta(t))|\,|\zeta'(t)|\,dt.$$

Since g has no pole on $\Gamma_\alpha(\mathbb{R}) = \alpha + \imath\mathbb{R}$ and g is proper, $g \circ \mu$ is a rational function of degree $\leq n$ which has no poles on $\partial\mathbb{D}$. Applying Spijker's Lemma [476] to the rational function $g \circ \mu$ we obtain the following inequality

$$\int_{-\pi}^{\pi} |(g \circ \mu)'(\zeta(t))|\,|\zeta'(t)|\,dt = \int_{\zeta} |(g \circ \mu)'(s)|\,|ds| \leq 2\pi n \max_{s \in \partial\mathbb{D}} |g \circ \mu(s)| = 2\pi n \sup_{s \in \Gamma_\alpha(\mathbb{R})} |g(s)|.$$

On the other hand,

$$\int_{-\infty}^{\infty} |g'(\alpha+\imath\omega)|\,d\omega = \lim_{\varepsilon \to 0} I(\varepsilon) = \lim_{\varepsilon \to 0} \int_{a_\varepsilon}^{b_\varepsilon} |(g\circ\mu)'(\zeta(t))|\,|\zeta'(t)|\,dt = \int_{-\pi}^{\pi} |(g\circ\mu)'(\zeta(t))|\,|\zeta'(t)|\,dt$$

and this concludes the proof. □

A.2.5 Analytic Continuation

A domain $\Omega \subset \mathbb{C}$ is called *simply connected* if, for every closed curve γ in Ω, $w(\gamma, a) = 0$ for all $a \in \mathbb{C} \setminus \Omega$. Intuitively speaking, a domain Ω is simply connected if it does not contain holes. In this case Cauchy's Theorem is applicable to every closed path in Ω. Every convex domain in $\mathbb{C}$ is simply connected. A punctured disk $D^\circ(a, r)$ is not simply connected but a *cut disk* $D^-(a, r) = \{s \in D(a, r); 0 < \arg(s - a) < 2\pi\}$ is.

Theorem A.2.22. *A domain $\Omega \subset \mathbb{C}$ is simply connected if and only if one of the following equivalent conditions is satisfied.*

(i) *For every $f \in \mathcal{O}(\Omega)$ and closed path γ in Ω we have $\int_\gamma f(s)ds = 0$.*

(ii) *Any two arcs in Ω with the same initial and final points are homotopic in Ω.*

(iii) *Ω is homeomorphic to the open unit disk $\mathbb{D}$.*

(iv) *For every $f \in \mathcal{O}(\Omega)$ there exists $F \in \mathcal{O}(\Omega)$ such that $F' = f$.*

(v) *If $f \in \mathcal{O}(\Omega)$ has no zeros on Ω then there exists a holomorphic logarithm of f on Ω, i.e. a holomorphic function $g : \Omega \to \mathbb{C}$ such that $f(s) = e^{g(s)}$ for all $s \in \Omega$.*

---

[5]Since $g'(s) = (p'(s)q(s) - p(s)q'(s))/q(s)^2$ and $\deg(p'q - pq') \leq \deg q^2 - 2$ the function $|g'(s)|$ is integrable over Γ_α.

A holomorphic function f on a domain Ω is said to possess an analytic continuation across a boundary point $b \in \partial\Omega$ if there exists a disk $D = D(b, r)$, $r > 0$ and a holomorphic function $g \in \mathcal{O}(D)$ such that $f(s) = g(s)$ for all s in some non-void open set $U \subset \Omega \cap D$. $b \in \partial\Omega$ is said to be a *singular point* of f if f does not possess an analytic continuation across b. Suppose that the power series

$$f(s) = \sum_{k=0}^{\infty} c_k (s-a)^k$$

has the radius of convergence $R = \left[\limsup_{k\to\infty} |c_k|^{1/k}\right]^{-1} > 0$, then the boundary of the open disk $D(a, R)$ contains a singular point of f.

Definition A.2.23. A *function element* is a pair (f, D) consisting of an open disk D and a holomorphic function $f \in \mathcal{O}(D)$. Two function elements (f_0, D_0) and (f_1, D_1) are *direct continuations* of each other (notation: $(f_0, D_0) \sim (f_1, D_1)$) if $D_0 \cap D_1 \neq \emptyset$ and $f_0(s) = f_1(s)$ for all $s \in D_0 \cap D_1$.
Given an arc $\gamma : [\alpha, \beta] \to \mathbb{C}$, if there are numbers $t_1 = \alpha < t_2 < \ldots < t_n = \beta$ and a sequence of function elements $(f_i, D_i), i \in \underline{n}$ such that $\gamma(\alpha)$ is the centre of D_1, $\gamma(\beta)$ the centre of D_n, $\gamma([t_i, t_{i+1}]) \subset D_i$ and $(f_i, D_i) \sim (f_{i+1}, D_{i+1})$ for $i = 1, \ldots, n-1$, then (f_n, D_n) is said to be an *analytic continuation of* (f_1, D_1) *along the arc* γ.

If f is holomorphic on a domain Ω then f defines by restriction to any disk $D_0 = D(s_0, r_0) \subset \Omega$ a unique function element $(f|D_0, D_0)$. This function element is completely described by a convergent power series centred at s_0 and is often identified with it. If the particular disk D_0 is of no interest we briefly speak of the function element of f at s_0 and mean by this any function element $(f|D, D)$ with $D = D(s_0, r)$ and $r > 0$ sufficiently small.

Theorem A.2.24 (Monodromy Theorem). *Suppose Ω is a simply connected domain, and (f_0, D_0) is a function element on a disk $D_0 \subset \Omega$ which can be analytically continued along every arc in Ω starting at the centre of D_0, then there exists $f \in \mathcal{O}(\Omega)$ such that $f|D_0 = f_0$.*

A.2.6 Maximum Principle and Subharmonic Functions

The following classical maximum principle plays an important role in Complex Analysis.

Theorem A.2.25 (Maximum Principle for Holomorphic Functions). *Suppose f is a holomorphic function on a domain $\Omega \subset \mathbb{C}$. If $|f|$ has a local maximum at some point $z_0 \in \Omega$ then f is constant on Ω. If Ω is bounded and f can be extended continuously to the closure $\overline{\Omega}$, then*

$$\max_{s\in\overline{\Omega}} |f(s)| = \max_{s\in\partial\Omega} |f(s)|. \tag{25}$$

Definition A.2.26. Suppose Ω is an open set in $\mathbb{C}$ and $u : \Omega \to \mathbb{R}$ is upper semi-continuous, then u is said to be *subharmonic* on Ω if, for all, $a \in \Omega$, there is $r_a > 0$ such that $\overline{D(a, r_a)} \subset \Omega$ and

$$u(a) \leq \frac{1}{2\pi} \int_0^{2\pi} u(a + re^{i\theta})\, d\theta, \quad 0 < r < r_a. \tag{26}$$

If $u : \Omega \to \mathbb{R}$ is continuous and both u and $-u$ are subharmonic, i.e. equality holds in (26), then u is *harmonic*.

A continuous function $u : \Omega \to \mathbb{C}$ is harmonic if and only if $\Delta u = 0$ where Δ denotes the Laplacian. It is well known that the real and imaginary parts of a holomorphic function f on Ω are harmonic whilst $|f|$ is subharmonic on Ω. Similarly, if the matrix-valued function $G : \Omega \to \mathbb{C}^{p\times m}$ is holomorphic on Ω (i.e. all the entries g_{ij} of $G(s)$ depend holomorphically on $s \in \Omega$) and $\|\cdot\|$ is any operator norm on $\mathbb{C}^{p\times m}$, then $s \mapsto \|G(s)\|$ is a subharmonic function on Ω. In fact, if $\|\cdot\|$ is induced by the vector norms $\|\cdot\|_{\mathbb{C}^m}$ and $\|\cdot\|_{\mathbb{C}^p}$ on $\mathbb{C}^m$ and $\mathbb{C}^p$, respectively, then by the Hahn-Banach Theorem, see Example A.4.11

$$\|G(s)\| = \sup\{\, |y^*G(s)u|; u \in \mathbb{C}^m,\ \|u\|_{\mathbb{C}^m} = 1,\ y \in \mathbb{C}^p, \|y^*\|^*_{\mathbb{C}^p} = 1\}$$

where $\|\cdot\|^*_{\mathbb{C}^p}$ is the dual norm on $(\mathbb{C}^p)^*$. This proves the subharmonicity of $s \mapsto \|G(s)\|$ on Ω since the supremum of a locally bounded family of subharmonic functions is subharmonic if it is upper semicontinuous [110, 19.4].
A local maximum principle similar to Theorem A.2.25 holds for harmonic but not for subharmonic functions. However, we have the following result.

Theorem A.2.27 (Maximum Principle for Subharmonic Functions). *Suppose u is a subharmonic function on a bounded domain Ω in $\mathbb{C}$ with boundary $\partial\Omega$. If u attains its global maximum in Ω, i.e. $u(s_0) = \sup_{s\in\Omega} u(s)$ for some $s_0 \in \Omega$, then u is constant on Ω. If u is not constant then*

$$u(s) < \sup_{w\in\partial\Omega}\ \limsup_{s\in\Omega,\, s\to w} u(s), \qquad s \in \Omega.$$

We also make use of the following extended version which applies to *unbounded* open sets.

Theorem A.2.28 (Extended Maximum Principle). *Suppose Ω is a proper open subset of $\mathbb{C}$ with boundary $\partial\Omega$ and $u : \Omega \to \mathbb{R}$ is subharmonic and bounded above on Ω, then*

$$\sup_{s\in\Omega} u(s) = \sup_{w\in\partial\Omega}\ \limsup_{s\in\Omega,\, s\to w} u(s). \tag{27}$$

Proof: Since the boundary of each connected component of Ω lies in $\partial\Omega$ it suffices to prove the theorem for connected Ω. But then the result is a direct consequence of a theorem of Phragmén and Lindelöf [82, Thm. 7.15]. □

If in this theorem the subharmonic function u is continuous and can be continuously extended to $\overline{\Omega}$ then (27) can be simplified to

$$\sup_{s\in\Omega} u(s) = \sup_{w\in\partial\Omega} u(w).$$

A.3 Convolutions and Transforms

In this section we introduce some important sequence and function spaces which serve as signal spaces in systems theory. We discuss the convolution of sequences and functions, and present some basic results on Laplace, Fourier and **z**-transforms. A textbook which covers much of the material of this section, written from an engineering point of view and aimed at undergraduates is *Kwakernaak and Sivan* (1991) [322]. More mathematical books are cited in the subsections for which they are relevant. Throughout the section we will assume that $\mathbb{K}^n$, $\mathbb{K}^m$ are provided with norms $\|\cdot\|_{\mathbb{K}^n}$, $\|\cdot\|_{\mathbb{K}^m}$ and $\mathbb{K}^{n\times m}$ is endowed with the corresponding operator norm.

A.3.1 Sequences: Convolution and z-Transforms

For the material in this subsection, see e.g. *Cartan* (1995) [92], *Kalman et al.* (1969) [290] and *Ogata* (1987) [396].

Let $T \subset \mathbb{Z}$ be given. If V is a vector space we denote by V^T the vector space of V-valued sequences with indices in T. The elements of V^T are denoted by $v(\cdot)$ or by $v = (v(t))_{t\in T}$. The vector space of all scalar valued sequences with index set T is denoted by $\mathbb{K}^T$.

Definition A.3.1. Let $T \subset \mathbb{Z}$ be given and $(X, \|\cdot\|_X)$ a Banach space. For $p \in [1,\infty]$ the p-norm of a sequence $x(\cdot) : T \to X$ is defined by

$$\|x(\cdot)\|_p = \left[\sum_{t\in T} \|x(t)\|_X^p\right]^{1/p}, \; 1 \le p < \infty, \qquad \|x(\cdot)\|_\infty = \sup_{t\in T} \|x(t)\|_X. \tag{1}$$

The vector space of all $x(\cdot) \in X^T$ with finite p-norm is denoted by $\ell^p(T;X)$. We have

$$1 \le p < r \le \infty \quad \Rightarrow \quad \ell^p(T;X) \subset \ell^r(T;X). \tag{2}$$

$\ell^p(T;X)$ provided with the norm $\|\cdot\|_p$ is a Banach space. In the case $T = \{1,...,n\}$, $X = \mathbb{K}$ the space $\ell^p(T;X)$ is just the space $\mathbb{K}^n$ provided with the p-norm, $\|\cdot\|_\infty$ is the maximum norm and $\|\cdot\|_1$ is the sum norm on $\mathbb{K}^n$.

Now suppose that $(X, \langle\cdot,\cdot\rangle_X)$ is a Hilbert space. An inner product on $\ell^2(T;X)$ is defined by

$$\langle x(\cdot), y(\cdot)\rangle = \sum_{t\in T} \langle x(t), y(t)\rangle_X, \qquad x(\cdot),\, y(\cdot) \in \ell^2(T;X).$$

The fact that the above inner product in $\ell^2(T;X)$ is well defined is a consequence of the Cauchy-Schwarz inequality in $\ell^2(T;X)$:

$$\left|\sum_{t\in T} \langle x(t), y(t)\rangle_X\right| \le \sum_{t\in T} |\langle x(t), y(t)\rangle_X| \le \|x(\cdot)\|_2 \|y(\cdot)\|_2, \qquad x(\cdot), y(\cdot) \in \ell^2(T;X). \tag{3}$$

Equality holds in (3) if and only if $x(\cdot)$, $y(\cdot)$ are linearly dependent. Provided with the above inner product $\ell^2(T;X)$ is a Hilbert space. In the case $T = \{1,...,n\}$, $X = \mathbb{K}$ the space $\ell^2(T;X)$ is just the space $\mathbb{K}^n$ and $\langle\cdot,\cdot\rangle$ is the usual inner product on $\mathbb{K}^n$, $\|\cdot\|_2$ is the associated norm on the Euclidean space $\mathbb{K}^n$.

The Cauchy-Schwarz inequality in $\ell^2(T;X)$ is a special case of Hölder's inequality. This states that if $p,q,r \in [1,\infty]$, $1/p + 1/q = 1/r$, $x(\cdot) \in \ell^p(T;X)$ and $y(\cdot) \in \ell^q(T;X)$, then $(\langle x(t), y(t)\rangle_X)_{t\in T} \in \ell^r(T;\mathbb{C})$ and

$$\|(\langle x(t), y(t)\rangle_X)_{t\in T}\|_{\ell^r(T;\mathbb{C})} \le \|x(\cdot)\|_{\ell^p(T;X)} \|y(\cdot)\|_{\ell^q(T;X)}. \tag{4}$$

In the special case where $r = 1$ and $1/p + 1/q = 1$, the number $p^* := q = p/(p-1)$ is called the *conjugate exponent* of p.
For $1 \leq p \leq \infty$ the *forward* shift operator S and the *backward* shift operator S^* on $\ell^p(\mathbb{Z}; \mathbb{K}^m)$, $m \in \mathbb{N}^*$ are defined by

$$(Su)(t) = u(t-1), \quad (S^*u)(t) = u(t+1), \ t \in \mathbb{Z}, \quad u(\cdot) \in \ell^p(\mathbb{Z}; \mathbb{K}^m). \tag{5}$$

Note that for $p = 2$, if $\ell^2(\mathbb{Z}; \mathbb{K}^m)$ is provided with its standard inner product, S^* is the Hilbert space adjoint of S, i.e. $\langle Su(\cdot), v(\cdot)\rangle = \langle u(\cdot), S^*v(\cdot)\rangle$ for all $u(\cdot), v(\cdot) \in \ell^2(\mathbb{Z}; \mathbb{K}^m)$, see Definition A.4.23.

Definition A.3.2. For $g(\cdot), u(\cdot) \in \ell^1(\mathbb{Z}; \mathbb{K})$ the convolution $(g * u)(\cdot)$ is defined by

$$(g * u)(t) = \sum_{s=-\infty}^{\infty} g(t-s)u(s), \qquad t \in \mathbb{Z}.$$

The series is in fact summable so that $(g*u)(t)$ is well defined for all $t \in \mathbb{Z}$. Moreover, as a consequence of the convolution inequality (see Proposition A.3.3), $(g*u)(\cdot) \in \ell^1(\mathbb{Z}; \mathbb{K})$ and $\|(g * u)(\cdot)\|_1 \leq \|g(\cdot)\|_1 \|u(\cdot)\|_1$ for all $g(\cdot), u(\cdot) \in \ell^1(\mathbb{Z}; \mathbb{K})$. So the Banach space $\ell^1(\mathbb{Z}; \mathbb{K})$ provided with the multiplication $*$ is a commutative *Banach algebra*. The unit impulse e defined by $e(0) = 1$ and $e(t) = 0$ for $t \in \mathbb{Z}^*$ is an identity of this algebra.
Extending Definition A.3.2 the convolution of a sequence of matrices $\mathcal{G}(\cdot) \in \ell^1(\mathbb{Z}; \mathbb{K}^{n\times m})$ and a sequence of vectors $u(\cdot) \in \ell^p(\mathbb{Z}; \mathbb{K}^m)$, $1 \leq p \leq \infty$, is defined by

$$(\mathcal{G} * u)(t) = \left(\sum_{j=1}^{m} (\mathcal{G}_{ij} * u_j)(t)\right)_{i \in \underline{n}} = \sum_{s=-\infty}^{\infty} \mathcal{G}(t-s)u(s), \quad t \in \mathbb{Z}. \tag{6}$$

The series in (6) converges absolutely for every $t \in \mathbb{Z}$ and every $u(\cdot) \in \ell^p(\mathbb{Z}; \mathbb{K}^m) \subset \ell^\infty(\mathbb{Z}; \mathbb{K}^m)$, $1 \leq p \leq \infty$. The following *convolution inequality* plays an important role in the analysis of input-output systems.

Proposition A.3.3 (Convolution inequality). *Suppose* $\mathcal{G}(\cdot) \in \ell^1(\mathbb{Z}; \mathbb{K}^{n\times m})$ *and* $u(\cdot) \in \ell^p(\mathbb{Z}; \mathbb{K}^m)$, $1 \leq p \leq \infty$, *then* $(\mathcal{G} * u)(\cdot) \in \ell^p(\mathbb{Z}; \mathbb{K}^n)$ *and*

$$\|(\mathcal{G} * u)(\cdot)\|_{\ell^p(\mathbb{Z};\mathbb{K}^n)} \leq \|\mathcal{G}(\cdot)\|_{\ell^1(\mathbb{Z};\mathbb{K}^{n\times m})} \|u(\cdot)\|_{\ell^p(\mathbb{Z};\mathbb{K}^m)}. \tag{7}$$

We will now restrict the time domain from $\mathbb{Z}$ to $\mathbb{N}$. This allows us to define the convolution of sequences in a purely algebraic way, without any convergence requirements.

Definition A.3.4. For $g(\cdot), u(\cdot) \in \mathbb{K}^\mathbb{N}$ the convolution $(g * u)(\cdot) \in \mathbb{K}^\mathbb{N}$ is defined by

$$(g * u)(t) = \sum_{s=0}^{t} g(t-s)u(s), \qquad t \in \mathbb{N}.$$

The vector space $\mathbb{K}^\mathbb{N}$ provided with multiplication $*$ is a commutative algebra over $\mathbb{K}$ with identity $e = (1, 0, 0, ...)$. The convolution of a sequence of matrices $\mathcal{G}(\cdot) \in (\mathbb{K}^{n\times m})^\mathbb{N}$ and a sequence of vectors $u(\cdot) \in (\mathbb{K}^m)^\mathbb{N}$ is defined similarly by

$$(\mathcal{G} * u)(t) = \left(\sum_{j=1}^{m} (\mathcal{G}_{ij} * u_j)(t)\right)_{i \in \underline{n}} = \sum_{s=0}^{t} \mathcal{G}(t-s)u(s), \quad t \in \mathbb{N}. \tag{8}$$

On $(\mathbb{K}^m)^\mathbb{N}$ the *forward* shift operator S and *backward* shift operator S^* are defined by

$$S(u(0), u(1), ...) = (0, u(0), u(1), ...), \ S^*(u(0), u(1), ...) = (u(1), u(2), ...), \ u(\cdot) \in (\mathbb{K}^m)^\mathbb{N}.$$

Again, if $\mathbb{K}^m$ is provided with its standard inner product and we restrict the shift operators to $\ell^2(\mathbb{N}; \mathbb{K}^m)$, then S^* is the Hilbert space adjoint of S, i.e. $\langle Su(\cdot), v(\cdot)\rangle = \langle u(\cdot), S^*v(\cdot)\rangle$ for all $u(\cdot), v(\cdot) \in \ell^2(\mathbb{N}; \mathbb{K}^m)$.

z-Transform

Associated with any sequence $u(\cdot) \in \mathbb{K}^{\mathbb{N}}$ is the *formal power series* $\sum_{t=0}^{\infty} u(t)z^{-t}$ in z^{-1}. Addition, scalar multiplication and multiplication of formal power series are defined by

$$\begin{aligned}\sum_{t=0}^{\infty} u(t)z^{-t} + \sum_{t=0}^{\infty} v(t)z^{-t} &= \sum_{t=0}^{\infty}(u(t)+v(t))z^{-t}, \quad \alpha\sum_{t=0}^{\infty} u(t)z^{-t} = \sum_{t=0}^{\infty} \alpha u(t)z^{-t}\\ \sum_{t=0}^{\infty} u(t)z^{-t}\sum_{t=0}^{\infty} v(t)z^{-t} &= \sum_{t=0}^{\infty} w(t)z^{-t} \quad \text{where} \quad w(t) = \sum_{j,k\in\mathbb{N}, j+k=t} u(j)v(k).\end{aligned}$$

The set of all formal power series with coefficients in $\mathbb{K}$ is denoted by $\mathbb{K}[[z^{-1}]]$. Provided with the above operations $\mathbb{K}[[z^{-1}]]$ is a commutative algebra over $\mathbb{K}$ with identity $e(z) = \sum_{t=0}^{\infty} e(t)z^{-t} \in \mathbb{K}$ (where $(e(t))_{t\in\mathbb{N}}$ is the unit impulse).

Definition A.3.5. Suppose that $u(\cdot) \in \mathbb{K}^{\mathbb{N}}$, then the **z**-*transform* of u is the formal power series

$$\hat{u}(z) = (\mathcal{Z}u)(z) := \sum_{t=0}^{\infty} u(t)z^{-t}. \tag{9}$$

The **z**-transform is an isomorphism of the algebra $\mathbb{K}^{\mathbb{N}}$ onto the algebra $\mathbb{K}[[z^{-1}]]$ and via this isomorphism the convolution of sequences in $\mathbb{K}^{\mathbb{N}}$ corresponds to the multiplication of formal power series in $\mathbb{K}[[z^{-1}]]$.
One defines the **z**-transform of a vector sequence $(u(t))_{t\in\mathbb{N}} \in (\mathbb{K}^m)^{\mathbb{N}}$ and matrix sequence $(\mathcal{G}(t))_{t\in\mathbb{N}} \in (\mathbb{K}^{n\times m})^{\mathbb{N}}$ by taking the **z**-transform of each component, e.g. if $u(t) = (u_i(t))_{i\in\underline{m}}$, then $(\mathcal{Z}u)(z) = ((\mathcal{Z}u_i)(z))_{i\in\underline{m}}$. The vector spaces of these formal power series are denoted by $\mathbb{K}^m[[z^{-1}]]$ and $\mathbb{K}^{n\times m}[[z^{-1}]]$, respectively.

Proposition A.3.6. *The* **z**-*transform has the following properties.*

(i) If S and S^* are the forward and backward shift operators, then

$$(\mathcal{Z}Su)(z) = z^{-1}(\mathcal{Z}u)(z), \qquad (\mathcal{Z}S^*u)(z) = z((\mathcal{Z}u)(z) - u(0)), \quad u(\cdot) \in (\mathbb{K}^m)^{\mathbb{N}}.$$

(ii) If $\mathcal{G}(\cdot) \in (\mathbb{K}^{n\times m})^{\mathbb{N}}, u(\cdot) \in (\mathbb{K}^m)^{\mathbb{N}}$ and $y(t) = (\mathcal{G} * u)(t)$, $t \in \mathbb{N}$, then

$$\hat{y}(z) = G(z)\hat{u}(z) \;\; \textit{where} \;\; G(z) = (\mathcal{Z}\mathcal{G})(z),\; \hat{u}(z) = (\mathcal{Z}u)(z),\; \hat{y}(z) = (\mathcal{Z}y)(z). \tag{10}$$

It is often useful to view a **z**-transform $\hat{u}(\cdot)$ as a complex analytic function on its domain of convergence. For any $u(\cdot) \in (\mathbb{K}^m)^{\mathbb{N}}$, $\gamma > 0$ define $u_\gamma(t) = u(t)\gamma^{-t}$, $t \in \mathbb{N}$ and set

$$\mathcal{S}_\gamma(\mathbb{K}^m) = \{u(\cdot) \in (\mathbb{K}^m)^{\mathbb{N}};\; u_\gamma(\cdot) \in \ell^1(\mathbb{N};\mathbb{K}^m)\}, \tag{11}$$

$$\mathbb{D}_\gamma^+ := \{z \in \mathbb{C}; |z| > \gamma\}, \quad \mathbb{D}_+ = \{z \in \mathbb{C}; |z| > 1\}. \tag{12}$$

If $u(\cdot) \in \mathcal{S}_\gamma(\mathbb{K}^m)$, the series on the RHS of (9) is uniformly absolutely convergent for all $z \in \overline{\mathbb{D}_\gamma^+}$, since

$$\sum_{t=0}^{\infty} \|u(t)z^{-t}\|_{\mathbb{K}^m} \le \sum_{t=0}^{\infty} \|u_\gamma(t)\|_{\mathbb{K}^m}(|z|/\gamma)^{-t} \le \|u_\gamma(\cdot)\|_1, \qquad |z| \ge \gamma. \tag{13}$$

For every $u(\cdot) \in \mathcal{S}_\gamma(\mathbb{K}^m)$ the **z**-transform defines a continuous function $\hat{u}(\cdot)$ on $\overline{\mathbb{D}_\gamma^+}$ which is bounded by (13) and analytic on $\mathbb{D}_\gamma^+$. The series on the RHS of (9) is the Laurent expansion of $\hat{u}(\cdot)$ at ∞, see Subsection A.2.4. We will use the same symbol $\hat{u}(z)$ to denote the *formal power series* and the associated *complex analytic function*. Note that

$u(t),\ t\in\mathbb{N}$	$\hat{u}(z)$	$z\in\mathbb{D}_{\gamma}^{+},\ \gamma>$
$e(t)$	1	0
1	$z/(z-1)$	1
t	$z/(z-1)^2$	1
$t^n,\ n\in\mathbb{N}^*$	$\lim_{a\to 0}\frac{\partial^n}{\partial a^n}\left(\frac{(-1)^n z}{z-e^{-a}}\right)$	1
$e^{-\lambda t},\ \lambda\in\mathbb{C}$	$z/(z-e^{-\lambda})$	$e^{-\operatorname{Re}\lambda}$
$te^{-\lambda t},\ \lambda\in\mathbb{C}$	$ze^{-\lambda}/(z-e^{-\lambda})^2$	$e^{-\operatorname{Re}\lambda}$
$\sin at,\ a\in\mathbb{R}$	$z\sin a/(z^2-2z\cos a+1)$	1
$\cos at,\ a\in\mathbb{R}$	$z(z-\cos a)/(z^2-2z\cos a+1)$	1

Table A.3.1: **z**-transforms

if $u(\cdot)\in\mathcal{S}_{\gamma}(\mathbb{R}^m)$ then $\overline{\hat{u}(z)}=\hat{u}(\overline{z})$, $|z|\geq\gamma$. Some elementary **z**-transforms, together with their domains of analyticity $\mathbb{D}_{\gamma}^{+}$, are presented in Table A.3.1.
If $u(\cdot)\in\mathcal{S}_{\gamma}(\mathbb{K}^m)$, $\mathcal{G}(\cdot)\in\mathcal{S}_{\gamma}(\mathbb{K}^{n\times m})$ and $y(\cdot)=(\mathcal{G}*u)(\cdot)$, then $y_{\gamma}=\mathcal{G}_{\gamma}*u_{\gamma}$ because

$$y_{\gamma}(t)=\gamma^{-t}\sum_{s=0}^{t}\mathcal{G}(t-s)u(s)=\sum_{s=0}^{t}\gamma^{-(t-s)}\mathcal{G}(t-s)\gamma^{-s}u(s)=\sum_{s=0}^{t}\mathcal{G}_{\gamma}(t-s)u_{\gamma}(s),\ t\in\mathbb{N}.$$

Hence (7) implies

$$\|y_{\gamma}(\cdot)\|_{\ell^1(\mathbb{N};\mathbb{K}^n)}\leq\|\mathcal{G}_{\gamma}(\cdot)\|_{\ell^1(\mathbb{N};\mathbb{K}^{n\times m})}\|u_{\gamma}(\cdot)\|_{\ell^1(\mathbb{N};\mathbb{K}^m)}. \tag{14}$$

So the **z**-transform $\hat{y}(\cdot)$ is analytic on $\mathbb{D}_{\gamma}^{+}$ and we obtain from Proposition A.3.6 the following result.

Theorem A.3.7. *Let $\gamma>0$ and suppose $u(\cdot)\in\mathcal{S}_{\gamma}(\mathbb{K}^m)$, $\mathcal{G}(\cdot)\in\mathcal{S}_{\gamma}(\mathbb{K}^{n\times m})$. If $y(\cdot)=(\mathcal{G}*u)(\cdot)$, and $G(z)=(\mathcal{Z}\mathcal{G})(z)$, then $y(\cdot)\in\mathcal{S}_{\gamma}(\mathbb{K}^n)$ and*

$$\hat{y}(z)=(\mathcal{Z}(\mathcal{G}*u))(z)=G(z)\hat{u}(z),\qquad |z|\geq\gamma. \tag{15}$$

(7) is an equality between formal power series whereas (15) is an equality between functions. We have the following inversion theorem for the **z**-transform, see (A.2.20).

Theorem A.3.8. *Suppose that $u(\cdot)\in\mathcal{S}_{\gamma}(\mathbb{K}^m)$ for some $\gamma>0$ and $r>\gamma$. If $\hat{u}(z)=(\mathcal{Z}u)(z)$ and Γ_r is the positively oriented circle of radius r around $z=0$, we have the inversion formula*

$$u(t)=(\mathcal{Z}^{-1}\hat{u})(t)=\frac{1}{2\pi\imath}\int_{\Gamma_r}z^{t-1}\hat{u}(z)dz=\frac{r^t}{2\pi}\int_{-\pi}^{\pi}e^{\imath t\theta}\hat{u}(re^{\imath\theta})d\theta,\qquad t\in\mathbb{N}. \tag{16}$$

*In particular, two sequences $u(\cdot),v(\cdot)\in\mathcal{S}_{\gamma}(\mathbb{K}^m)$ are equal if their **z**-transforms are equal on $\mathbb{D}_{\gamma}^{+}$.*

Example A.3.9. Let $A\in\mathbb{K}^{n\times n}$ be a matrix with spectral radius $\varrho(A)$ and let $\gamma>\varrho(A)$ be arbitrary. Choosing any $\tilde{\gamma}\in\mathbb{R}$ such that $\varrho(A)<\tilde{\gamma}<\gamma$ we obtain from Lemma 3.3.19 that there exists $M>0$ such that $\|A^t\|_{\mathcal{L}(\mathbb{K}^n)}\leq M\tilde{\gamma}^t$, $t\in\mathbb{N}$. Hence $\|\gamma^{-t}A^t\|_{\mathcal{L}(\mathbb{K}^n)}\leq M(\tilde{\gamma}/\gamma)^t$, $t\in\mathbb{N}$ and so $(A^t)_{t\in\mathbb{N}}\in\mathcal{S}_{\gamma}(\mathbb{K}^{n\times n})$. We conclude that

$$\left(\mathcal{Z}\left((A^t)_{t\in\mathbb{N}}\right)\right)(z) = \sum_{t=0}^{\infty} A^t z^{-t} = z(zI_n - A)^{-1}, \quad |z| > \varrho(A)$$

is analytic on $D^+_{\varrho(A)}$. In fact the resolvent $(zI_n - A)^{-1}$ is defined and analytic on the whole resolvent set $\rho(A) = \mathbb{C} \setminus \sigma(A) \supset \mathbb{D}^+_{\varrho(A)}$. It follows that if $(A, B, C, D) \in \mathbf{L}_{n,m,p}(\mathbb{K})$ and $\mathcal{G}$ is the impulse response of the corresponding discrete time system, i.e.

$$\mathcal{G} = (\mathcal{G}(t))_{t\in\mathbb{N}}, \quad \mathcal{G}(0) = D, \quad \mathcal{G}(t) = CA^{t-1}B \quad \text{for} \quad t \in \mathbb{N}^*,$$

then $\mathcal{G} \in \mathcal{S}_\gamma(\mathbb{K}^{p\times m})$ for every $\gamma > \varrho(A)$ and the associated transfer matrix

$$G(z) = (\mathcal{Z}\mathcal{G}(\cdot))(z) = D + \sum_{t=1}^{\infty} CA^{t-1}Bz^{-t} = D + C(zI_n - A)^{-1}B, \qquad z \in \rho(A) \quad (17)$$

is analytic on $\rho(A)$. Hence we conclude from the inversion theorem that

$$\mathcal{G}(t) = \frac{1}{2\pi\imath}\int_{\Gamma_r} z^{t-1}(D + C(zI_n - A)^{-1}B)dz, \quad t \in \mathbb{N} \quad (18)$$

for all $r > \varrho(A)$. □

A.3.2 Lebesgue Spaces, Convolution of Functions, Laplace Transforms

In this subsection we assume that the reader is familiar with the elements of Lebesgue integration theory. For a brief introduction to abstract integration theory, Lebesgue integration and L^p-spaces we refer to *Rudin* (1987) [441]. More comprehensive accounts can be found in e.g. *Kolmogorov and Fomin* (1957) [312] and *Bauer* (2001) [42]. Standard references on the Laplace transform are *Widder* (1966) [521] and *Doetsch* (1974) [133], see also *Körner* (1988) [314]. Material on convolutions of functions and measures can be found in [441], [42] and *Desoer and Vidyasagar* (1975) [130].

Definition A.3.10. Let $T \subset \mathbb{R}$ be an interval and $(X, \|\cdot\|_X)$ a Banach space. For $p \in [1, \infty]$ the p-norm of a Lebesgue measurable function $x(\cdot) : T \to X$ is given by

$$\|x(\cdot)\|_p = \left[\int_T \|x(t)\|_X^p dt\right]^{1/p}, \qquad \|x(\cdot)\|_\infty = \inf\{\alpha; \|x(t)\|_X \leq \alpha, \text{ a.e. } t \in T\}.$$

Recall that the set $\mathcal{L}^p(T;X)$ of all Lebesgue measurable functions $x(\cdot) : T \to X$ with finite p-norm is a linear space, but it is not a normed linear space, since $\|x(\cdot)\|_p = 0$ only implies that $x(\cdot)$ is a *zero-function*, i.e. $x(t) = 0$ for almost all $t \in T$. In order to make it into a normed space it is necessary to consider equivalence classes of functions $[x(\cdot)]$ where $[x(\cdot)]$ is the class of functions which equal $x(\cdot)$ almost everywhere. These equivalence classes form a linear space and $\|[x(\cdot)]\|_p = \|x(\cdot)\|_p$ defines a norm on it. The resulting normed linear space is a Banach space and is denoted by $L^p(T;X)$. Throughout the text we will not distinguish between a function $x(\cdot) \in \mathcal{L}^p(T;X)$ and the corresponding equivalence class $[x(\cdot)] \in L^p(T;X)$. Any function $x(\cdot)$ defined almost everywhere on T is identified with its trivial extension to all of T. Note that if $\lambda(T) < \infty$ the following implication holds

$$p < r \quad \Longrightarrow \quad L^r(T;X) \subset L^p(T;X), \qquad (19)$$

but this is *not* so if $\lambda(T) = \infty$.
The following proposition establishes a relationship between the absolute convergence of a series of functions in $L^p(T;\mathbb{C}^m)$ and its pointwise convergence.

Proposition A.3.11. *Let $(v_k(\cdot))_{k\in\mathbb{N}}$ be a sequence in $L^p(T;\mathbb{C}^m)$ such that $\sum_{k=0}^{\infty}\|v_k(\cdot)\|_p < \infty$. Then the series $\sum_{k=0}^{\infty} v_k(t)$ is absolutely convergent for almost every $t\in T$. If we set $v(t)=\sum_{k=0}^{\infty} v_k(t)$ for all $t\in T$ for which the series converges and set $v(t)=0$ elsewhere, then the series $\sum_{k=0}^{\infty} v_k(\cdot)$ converges to $v(\cdot)$ in $L^p(T;\mathbb{C}^n)$ and*

$$\|v(\cdot)-\sum_{k=0}^{N} v_k(\cdot)\|_p \le \sum_{k=N+1}^{\infty} \|v_k(\cdot)\|_p.$$

A measurable function $x(\cdot): T\to X$ is said to be locally p-integrable if $\int_{t_1}^{t_2}\|x(t)\|_X^p dt < \infty$ for arbitrary $t_1,t_2\in T$, $t_1<t_2$. The vector space of all such functions modulo zero functions is denoted by $L^p_{\text{loc}}(T;X)$.
If $(X,\langle\cdot,\cdot\rangle_X)$ is a Hilbert space, an inner product on $L^2(T;X)$ is defined by

$$\langle x(\cdot),y(\cdot)\rangle = \int_T \langle x(t),y(t)\rangle_X\, dt, \qquad x(\cdot),y(\cdot)\in L^2(T;X).$$

The fact that the above inner product in $L^2(T;X)$ is well defined is a consequence of the Cauchy-Schwarz inequality on $L^2(T;X)$

$$\left|\int_T \langle x(t),y(t)\rangle_X dt\right| \le \int_T |\langle x(t),y(t)\rangle_X|\, dt \le \|x(\cdot)\|_2\|y(\cdot)\|_2, \quad x(\cdot),y(\cdot)\in L^2(T;X). \tag{20}$$

Equality holds in (20) if and only if $x(\cdot)$, $y(\cdot)$ are linearly dependent. Provided with the above inner product, $L^2(T;X)$ is a Hilbert space.
The Cauchy-Schwarz inequality is a special case of Hölder's inequality. This states that if $p,q,r\in[1,\infty]$, $1/p+1/q=1/r$, $x(\cdot)\in L^p(T;X)$ and $y(\cdot)\in L^q(T;X)$, then the function $\langle x(\cdot),y(\cdot)\rangle_X : t\mapsto \langle x(t),y(t)\rangle_X$ is r-integrable and

$$\|\langle x(\cdot),y(\cdot)\rangle_X\|_{L^r(T;\mathbb{C})} \le \|x(\cdot)\|_{L^p(T;X)}\|y(\cdot)\|_{L^q(T;X)}. \tag{21}$$

Definition A.3.12. Let $I\subset\mathbb{R}$ be an interval. A function $f: I\to\mathbb{K}$ is said to be *absolutely continuous* on I if for every $\varepsilon>0$, there exists a $\delta>0$, such that for every finite collection $\{(t_k,t'_k)\}_{k\in\underline{n}}$ of non-overlapping open subintervals of I

$$\sum_{k=1}^{n}(t'_k-t_k)<\delta \quad\Longrightarrow\quad \sum_{k=1}^{n}|f(t'_k)-f(t_k)|<\varepsilon.$$

The fundamental theorem of calculus for Lebesgue measurable functions states that a function $f:[a,b]\to\mathbb{K}$ has the form

$$f(t)=f(a)+\int_a^t \varphi(s)ds, \quad t\in[a,b] \tag{22}$$

for some integrable $\varphi(\cdot)$ on $[a,b]$ if and only if f is absolutely continuous on $[a,b]$. In this case f is continuous on $[a,b]$ and almost everywhere differentiable with derivative $f'(t)=\varphi(t)$ on (a,b).

Definition A.3.13. For $g(\cdot),u(\cdot)\in L^1(\mathbb{R};\mathbb{K})$ the convolution $(g*u)(\cdot)$ is defined by

$$(g*u)(t)=\int_{-\infty}^{\infty} g(t-s)u(s)\,ds, \quad \text{a.e.} \quad t\in\mathbb{R}.$$

The integral in the above equation exists almost everywhere and defines an integrable function on $\mathbb{R}$. The vector space $L^1(\mathbb{R};\mathbb{K})$ provided with the multiplication $*$ is in fact a commutative algebra over $\mathbb{K}$. It is without an identity, since there is no function $e(\cdot) \in L^1(\mathbb{R};\mathbb{K})$ such that $g * e = e * g = g$ for every $g \in L^1(\mathbb{R},\mathbb{K})$. The convolution of a matrix function $\mathcal{G}(\cdot) \in L^1(\mathbb{R};\mathbb{K}^{n\times m})$ and a vector function $u(\cdot) \in L^1(\mathbb{R};\mathbb{K}^m)$ is defined by

$$(\mathcal{G} * u)(t) = \left(\sum_{j=1}^{m}(\mathcal{G}_{ij} * u_j)(t)\right)_{i\in\underline{n}} = \int_{-\infty}^{\infty} \mathcal{G}(t-s)u(s)\,ds\,,\ a.e.\ \ t \in \mathbb{R}. \tag{23}$$

In the continuous time context the convolution inequality takes the following form.

Proposition A.3.14 (Convolution inequality). *Suppose* $\mathcal{G}(\cdot) \in L^1(\mathbb{R};\mathbb{K}^{n\times m})$ *and* $u(\cdot) \in L^p(\mathbb{R};\mathbb{K}^m)$, $1 \le p \le \infty$ *then* $y(\cdot) = (\mathcal{G} * u)(\cdot) \in L^p(\mathbb{R};\mathbb{K}^n)$ *and*

$$\|y(\cdot)\|_{L^p(\mathbb{R};\mathbb{K}^n)} = \|(\mathcal{G} * u)(\cdot)\|_{L^p(\mathbb{R};\mathbb{K}^n)} \le \|\mathcal{G}(\cdot)\|_{L^1(\mathbb{R};\mathbb{K}^{n\times m})}\, \|u(\cdot)\|_{L^p(\mathbb{R};\mathbb{K}^m)}. \tag{24}$$

In particular, if $g(\cdot), u(\cdot) \in L^1(\mathbb{R};\mathbb{K})$ then $\|(g * u)(\cdot)\|_{L^1(\mathbb{R};\mathbb{K})} \le \|g(\cdot)\|_{L^1(\mathbb{R};\mathbb{K})}\|u(\cdot)\|_{L^1(\mathbb{R};\mathbb{K})}$ and so $L^1(\mathbb{R};\mathbb{K})$ is a commutative Banach algebra.
We now restrict the time domain from $\mathbb{R}$ to $\mathbb{R}_+$ and this allows us to define the convolution of locally integrable functions on $\mathbb{R}_+$. This is particularly useful in Systems Theory where often input functions are considered which are not integrable over $\mathbb{R}_+$.

Definition A.3.15. If $g(\cdot), u(\cdot) \in L^1_{loc}(\mathbb{R}_+;\mathbb{K})$ the convolution $(g * u)(\cdot) : \mathbb{R}_+ \to \mathbb{K}$ is defined by

$$(g * u)(t) = \int_0^t g(t-s)u(s)\,ds, \quad a.e.\ t \in \mathbb{R}_+.$$

The integral in the above equation exists almost everywhere and defines a locally integrable function on $\mathbb{R}_+$. The vector space $L^1_{loc}(\mathbb{R}_+;\mathbb{K})$ provided with the multiplication $*$ is in fact a commutative algebra over $\mathbb{K}$ (without identity). The convolution of a matrix function $\mathcal{G}(\cdot) \in L^1_{loc}(\mathbb{R}_+;\mathbb{K}^{n\times m})$ and a vector function $u(\cdot) \in L^1_{loc}(\mathbb{R}_+;\mathbb{K}^m)$ is defined by

$$(\mathcal{G} * u)(t) = \left(\sum_{j=1}^{m}(\mathcal{G}_{ij} * u_j)(t)\right)_{i\in\underline{n}} = \int_0^t \mathcal{G}(t-s)u(s)\,ds, \quad \text{a.e. } t \in \mathbb{R}_+. \tag{25}$$

Remark A.3.16. At some points in the text we need to convolve a function and a measure. In order to avoid integrability conditions we restrict our considerations to the time domain $T = \mathbb{R}_+$. A positive Borel measure μ on $\mathbb{R}_+$ is said to be *finite* if $\|\mu\| := \mu(\mathbb{R}_+) < \infty$ and it is said to be *locally finite* if $\mu(J) < \infty$ for all compact intervals $J \subset \mathbb{R}_+$. For every $a \in \mathbb{R}_+$ we denote by δ_a the *Dirac impulse* or *unit mass* at a. This is a finite measure on $\mathbb{R}_+$ defined on the σ-algebra $\mathcal{B}$ of Borelian subsets of $\mathbb{R}_+$ by $\delta_a(E) = 1$ if $a \in E$ and $\delta_a(E) = 0$ if $a \notin E$, $E \in \mathcal{B}$. We are particularly interested in measures of the form

$$\mu = \sum_{i=0}^{\infty} g_i \delta_{t_i} \tag{26}$$

where $g = (g_i) \in \ell^1(\mathbb{N};\mathbb{R})$, $g_i > 0$, $0 \le t_1 < t_2 < \ldots$ with $t_i \to \infty$ as $i \to \infty$. Then μ is a finite measure on $\mathbb{R}_+$ with norm $\|\mu\| = \|g\|_{\ell^1(\mathbb{N};\mathbb{R})}$.
Now suppose $u(\cdot) \in L^1_{loc}(\mathbb{R}_+;\mathbb{K})$ and μ is a locally finite measure on $\mathbb{R}_+$, then the convolution of μ and u is defined by

$$(\mu * u)(t) = (u * \mu)(t) = \int_0^t u(t-s)\mu(ds), \quad \text{a.e.} \quad t \in \mathbb{R}_+.$$

The integral in the above equation exists almost everywhere and defines a locally integrable function on $\mathbb{R}_+$. If $u(\cdot) : \mathbb{R}_+ \to \mathbb{K}$ is continuous and μ is the measure defined in (26), then $(\mu * u)(t)$ is defined for every $t \in \mathbb{R}_+$, is piecewise continuous, and we have

$$(\mu * u)(t) = \sum_{i=1}^{k} g_i u(t - t_i), \quad t \in \mathbb{R}_+$$

where k is the largest integer such that $t_k \leq t$.
If $u \in L^1(\mathbb{R}_+; \mathbb{K})$ and μ is a finite Borel measure we have the convolution inequality

$$\|\mu * u\|_{L^1(\mathbb{R}_+;\mathbb{K})} \leq \|\mu\| \, \|u\|_{L^1(\mathbb{R}_+;\mathbb{K})}.$$

□

Laplace Transform

For $u(\cdot) \in L^1_{loc}(\mathbb{R}_+; \mathbb{K}^m)$, $\alpha \in \mathbb{R}$ we define $u_\alpha(\cdot) : t \to e^{-\alpha t} u(t)$ on $\mathbb{R}_+$ and set

$$\mathcal{E}_\alpha(\mathbb{K}^m) = \{u(\cdot) \in L^1_{loc}(\mathbb{R}_+; \mathbb{K}^m);\ u_\alpha(\cdot) \in L^1(\mathbb{R}_+; \mathbb{K}^m)\}, \tag{27}$$

$$\mathbb{C}_\alpha^+ = \{s \in \mathbb{C}; \operatorname{Re} s > \alpha\}, \quad \mathbb{C}_+ = \{s \in \mathbb{C}; \operatorname{Re} s > 0\}. \tag{28}$$

If $u(\cdot) \in \mathcal{E}_\alpha(\mathbb{K})$ then $t \to u(t)e^{-st}$ is integrable on $\mathbb{R}_+$ for all $s \in \overline{\mathbb{C}_\alpha^+}$.

Definition A.3.17. Suppose $u(\cdot) \in \mathcal{E}_\alpha(\mathbb{K})$. Then the *Laplace transform* of $u(\cdot)$ is defined on $\overline{\mathbb{C}_\alpha^+}$ by

$$\hat{u}(s) = (\mathcal{L}\, u)(s) := \int_0^\infty u(t)e^{-st}dt, \qquad \operatorname{Re} s \geq \alpha.$$

The Laplace transform $\hat{u}(\cdot)$ is continuous on $\overline{\mathbb{C}_\alpha^+}$, analytic on $\mathbb{C}_\alpha^+$ and bounded because

$$|\hat{u}(s)| \leq \|u_\alpha(\cdot)\|_{L^1(\mathbb{R}_+;\mathbb{K})}, \quad s \in \overline{\mathbb{C}_\alpha^+}. \tag{29}$$

If there is no risk of confusion, any analytic extension of $\hat{u}(\cdot)$ to a complex domain $\Omega \supset \mathbb{C}_\alpha^+$ will be denoted by the same symbol. If $u(\cdot)$ takes its values in $\mathbb{R}$, then $\overline{\hat{u}(s)} = \hat{u}(\overline{s})$. The Laplace transform of vectors and matrices with entries in $\mathcal{E}_\alpha(\mathbb{K})$ for some $\alpha \in \mathbb{R}$ are defined by taking the Laplace transform componentwise.

Remark A.3.18. Definition A.3.17 can be extended to positive Borel measures on $\mathbb{R}_+$ as follows. Assume that μ is a positive Borel measure on $\mathbb{R}_+$ satisfying $\int_0^\infty e^{-\alpha t}\mu(dt) < \infty$ for some $\alpha \in \mathbb{R}$. Then the Laplace transform of μ is defined by

$$\hat{\mu}(s) = (\mathcal{L}\, \mu)(s) = \int_0^\infty e^{-st}\mu(dt), \quad \operatorname{Re} s \geq \alpha.$$

$\hat{\mu}(s)$ is continuous on $\overline{\mathbb{C}_\alpha^+}$, analytic on $\mathbb{C}_\alpha^+$ and bounded since $|\hat{\mu}(s)| \leq \int_0^\infty e^{-\alpha t}\mu(dt)$. In particular, if $\mu(\cdot)$ is the measure defined in (26) with $(g_i)_{i\in\mathbb{N}} \in \ell^1(\mathbb{N}; \mathbb{R})$, then

$$\hat{\mu}(s) = \sum_{i=0}^{\infty} g_i e^{-st_i}, \quad \operatorname{Re} s \geq 0$$

is analytic on $\mathbb{C}_+$, bounded and continuous on $\overline{\mathbb{C}_+}$. □

We have the following inversion and uniqueness result for the Laplace transform of functions.

Theorem A.3.19. *(i) Suppose that* $u(\cdot) \in \mathcal{E}_\alpha(\mathbb{K}^m)$ *for some* $\alpha \in \mathbb{R}$ *and let* $\hat{u}(s) = (\mathcal{L}u)(s)$, $s \in \overline{\mathbb{C}_\alpha^+}$. *Then for fixed* $\beta > \alpha$, *the function* $\hat{u}(\beta + \imath\,\cdot) \in L^1(\mathbb{R};\mathbb{C}^m)$ *and we have the inversion formula*

$$u(t) = (\mathcal{L}^{-1}\hat{u})(t) = \frac{1}{2\pi\imath}\int_{\beta-\imath\infty}^{\beta+\imath\infty} e^{st}\hat{u}(s)ds = \frac{e^{\beta t}}{2\pi}\int_{-\infty}^{\infty} e^{\imath\omega t}\hat{u}(\beta+\imath\omega)d\omega, \quad a.e.\ t \in \mathbb{R}_+. \quad (30)$$

If $u(\cdot)$ *is continuous then* (30) *holds for all* $t \in \mathbb{R}_+$.

(ii) If $u(\cdot), v(\cdot) \in \mathcal{E}_\alpha(\mathbb{K}^m)$ *and* $(\mathcal{L}u)(s) = (\mathcal{L}v)(s)$, $s \in \mathbb{C}_\alpha^+$ *then* $u(t) = v(t)$, *a.e.* $t \in \mathbb{R}_+$.

The following proposition lists some important properties of the Laplace transformation. For $\tau > 0$ let S_τ be the forward shift operator on $L^1_{loc}(\mathbb{R}_+;\mathbb{K}^m)$ defined by

$$(S_\tau u)(t) = u(t-\tau), \quad t \geq \tau \quad \text{and} \quad (S_\tau u)(t) = 0, \quad t \in [0,\tau).$$

Proposition A.3.20. *Suppose* $a, b \in \mathbb{C}$, $\alpha \in \mathbb{R}$ *and* $u(\cdot), v(\cdot) \in \mathcal{E}_\alpha(\mathbb{K}^m)$, $\hat{u}(s) = (\mathcal{L}u)(s)$. *Then*

(i) $\mathcal{L}(au+bv)(s) = a\mathcal{L}(u)(s) + b\mathcal{L}(v)(s), \quad \operatorname{Re} s \geq \alpha.$

(ii) If $c \in \mathbb{C}$ *and* $u_c(t) = e^{-ct}u(t)$ *then* $(\mathcal{L}u_c)(s) = \hat{u}(s+c), \quad \operatorname{Re} s \geq \alpha - \operatorname{Re} c.$

(iii) If $\tau > 0$ *then* $(S_\tau u)(\cdot) \in \mathcal{E}_\alpha(\mathbb{K}^m)$ *and*

$$(\mathcal{L}S_\tau u)(s) = e^{-\tau s}\hat{u}(s), \quad \operatorname{Re} s \geq \alpha.$$

(iv) If $c > 0$, *then* $(\mathcal{L}u(ct))(s) = c^{-1}\hat{u}(s/c), \quad \operatorname{Re} s \geq c\alpha.$

(v) If $u(\cdot) : \mathbb{R}_+ \to \mathbb{K}^m$ *is* k *times differentiable on* $\mathbb{R}_+$ *(with right-hand derivatives* $u^{(j)}(0)$ *at* 0*) and* $u^{(j)}(\cdot) \in \mathcal{E}_\alpha(\mathbb{K}^m)$ *for* $j = 0,\ldots,k$, *then*

$$(\mathcal{L}u^{(k)})(s) = s^k\hat{u}(s) - s^{k-1}u(0) - s^{k-2}u'(0) - \cdots - u^{(k-1)}(0), \qquad \operatorname{Re} s \geq \alpha.$$

(vi) Let $w(\cdot) : t \mapsto t\,u(t)$, $t \in \mathbb{R}_+$, *then* $(\mathcal{L}w)(s) = -(\hat{u})'(s)$, $\operatorname{Re} s > \alpha$.

$u(t),\ t \geq 0$	$\hat{u}(s)$	$s \in \overline{\mathbb{C}_\alpha^+},\ \alpha >$
$\delta(t)$	1	$-\infty$
$e^{-\lambda t},\ \lambda \in \mathbb{C}$	$1/(s+\lambda)$	$-\operatorname{Re}\lambda$
$e^{-\lambda t}t^n/n!,\ \lambda \in \mathbb{C}, n \in \mathbb{N}$	$1/(s+\lambda)^{n+1}$	$-\operatorname{Re}\lambda$
$\sin bt,\ b \in \mathbb{R}$	$b/(s^2+b^2)$	0
$\cos bt,\ b \in \mathbb{R}$	$s/(s^2+b^2)$	0
$t\sin bt,\ b \in \mathbb{R}$	$2bs/(s^2+b^2)^2$	0
$t\cos bt,\ b \in \mathbb{R}$	$(s^2-b^2)/(s^2+b^2)^2$	0
$e^{-at}\sin bt,\ a, b \in \mathbb{R}$	$b/[(s+a)^2+b^2]$	a
$e^{-at}\cos bt,\ a, b \in \mathbb{R}$	$(s-a)/[(s+a)^2+b^2]$	a

Table A.3.2: Laplace transforms

Some elementary Laplace transforms, together with their domains of analyticity $\mathbb{C}_\alpha^+$, are given in Table A.3.2. Using partial fraction decomposition the inverse Laplace transform of arbitrary proper rational functions can be obtained from the results in this table making use of the above properties of the Laplace transformation.

The next theorem provides a counterpart to Theorem A.3.7.

Theorem A.3.21. *Suppose $\alpha \in \mathbb{R}$ is given, $u(\cdot) \in \mathcal{E}_\alpha(\mathbb{K}^m)$ and $\mathcal{G}(\cdot) \in \mathcal{E}_\alpha(\mathbb{K}^{n\times m})$ with Laplace transform $G(s) = (\mathcal{L}\,\mathcal{G})(s)$, $s \in \overline{\mathbb{C}_\alpha^+}$. Then $y(\cdot) = (\mathcal{G} * u)(\cdot) \in \mathcal{E}_\alpha(\mathbb{K}^n)$, and*

$$\hat{y}(s) = (\mathcal{L}\,(\mathcal{G} * u))(s) = G(s)\hat{u}(s), \qquad \operatorname{Re} s \geq \alpha. \tag{31}$$

The fact that $y(\cdot) \in \mathcal{E}_\alpha(\mathbb{K}^n)$ in the above theorem follows from the convolution inequality

$$\|y_\alpha(\cdot)\|_{L^1(\mathbb{R}_+;\mathbb{K}^n)} = \|(\mathcal{G}_\alpha * u_\alpha)(\cdot)\|_{L^1(\mathbb{R}_+;\mathbb{K}^n)} \leq \|\mathcal{G}_\alpha(\cdot)\|_{L^1(\mathbb{R}_+;\mathbb{K}^{n\times m})}\|u_\alpha(\cdot)\|_{L^1(\mathbb{R}_+;\mathbb{K}^m)}. \tag{32}$$

Example A.3.22. Suppose $A \in \mathbb{K}^{n\times n}$, $B \in \mathbb{K}^{n\times m}$, $C \in \mathbb{K}^{p\times n}$, the spectral abscissa of A is $\alpha(A)$ and $\alpha > \alpha(A)$ is arbitrary. Choosing any $\tilde{\alpha} \in \mathbb{R}$ such that $\alpha(A) < \tilde{\alpha} < \alpha$ we obtain from Lemma 3.3.19 that there exists $M > 0$ such that $\|e^{At}\|_{\mathcal{L}(\mathbb{K}^n)} \leq Me^{\tilde{\alpha}t}$ for all $t \in \mathbb{R}_+$. Hence $\|e^{-\alpha t}e^{At}\|_{\mathcal{L}(\mathbb{K}^n)} \leq Me^{-(\alpha-\tilde{\alpha})t}$ and so $\mathcal{G}(\cdot) : t \mapsto Ce^{At}B$ belongs to $\mathcal{E}_\alpha(\mathbb{K}^{p\times m})$ for every $\alpha > \alpha(A)$. We conclude that the Laplace transform of $\mathcal{G}$ is well-defined and analytic on $\mathbb{C}^+_{\alpha(A)}$. Since $\lim_{t\to\infty} e^{(A-sI_n)t} = 0_{n\times n}$ for $\operatorname{Re} s > \alpha(A)$ we get

$$(\mathcal{L}\,\mathcal{G})(s) = \int_0^\infty Ce^{-st}e^{At}Bdt = \int_0^\infty \tfrac{d}{dt}\left[C(A - sI_n)^{-1}e^{(A-sI_n)t}B\right]dt = C(sI_n - A)^{-1}B \tag{33}$$

for all $s \in \mathbb{C}^+_{\alpha(A)}$. In fact $G(s) := (\mathcal{L}\,\mathcal{G})(s) = C(sI_n - A)^{-1}B$ can be extended analytically to the whole resolvent set $\rho(A)$. Moreover, by the inversion formula

$$Ce^{At}B = \frac{1}{2\pi\imath}\int_{\beta-\imath\infty}^{\beta+\imath\infty} e^{st}C(sI_n - A)^{-1}Bds \quad \text{for all } t \in \mathbb{R}_+, \quad \beta > \alpha(A). \tag{34}$$

□

A.3.3 Fourier Series and Fourier Transforms

In this subsection we present the definitions and some basic results on Fourier series and Fourier transforms. Good references are *Kolmogorov and Fomin* (1957) [312], *Goldberg* (1965) [193], *Katznelson* (1968) [297], *Rudin* (1987) [441], *Folland* (1992) [164] and *Körner* (1988) [314]. It is assumed that all finite dimensional vector spaces $\mathbb{K}^m$ are equipped with the Euclidean norm $\|\cdot\|_{\mathbb{K}^m}$ induced by the usual inner product $\langle\cdot,\cdot\rangle_{\mathbb{K}^m}$.

Fourier Series

Let $l > 0$. A function $u(\cdot)$ on $\mathbb{R}$ is $2l$-periodic if $u(t + 2l) = u(t)$ for all $t \in \mathbb{R}$. Such a function is completely determined by its values on the interval $[-l, l]$. Conversely, every function $u(\cdot)$ on $[-l, l]$ satisfying $u(-l) = u(l)$ has a unique $2l$-periodic extension to $\mathbb{R}$. In the following we will not distinguish notationally between a function $u(\cdot) \in L^1(-l, l; \mathbb{C}^m)$ and its $2l$-periodic extension.[6] Under certain conditions a $2l$-periodic function can be represented as a superposition of harmonic oscillations.

Definition A.3.23. Given $l > 0$ and $u(\cdot) \in L^1(-l, l; \mathbb{C}^m)$, the *Fourier coefficients* of $u(\cdot)$ are defined by

$$u_k = \frac{1}{2l}\int_{-l}^{l} u(\theta)e^{-\imath k\pi\theta/l}d\theta, \qquad k \in \mathbb{Z} \tag{35}$$

and the associated *Fourier series* is

$$\sum_{k=-\infty}^{\infty} u_k e^{\imath k\pi\theta/l}, \quad \theta \in \mathbb{R}. \tag{36}$$

[6]Since $u(\cdot)$ may be altered on a subset of measure zero, we may assume $u(-l) = u(l)$.

Note that if $u(\cdot)$ takes its values in $\mathbb{R}^m$, then $\overline{u_k} = u_{-k}$ for $k \in \mathbb{Z}$.
The two-sided sequence of Fourier coefficients $(u_k)_{k\in\mathbb{Z}}$ needs not be summable and so the series (37) may not converge for some $\theta \in [-l, l]$. In fact Kolmogorov has shown that there exists a function $u(\cdot) \in L^1(-l, l; \mathbb{C})$ whose Fourier series diverges everywhere. If, however, $(u_k)_{k\in\mathbb{Z}}$ is summable in $\mathbb{C}^m$, then the series

$$S(u,\theta) = \sum_{k=-\infty}^{\infty} u_k e^{\imath k\pi\theta/l}, \quad \theta \in \mathbb{R} \tag{37}$$

defines a continuous $2l$-periodic function $S(u,\cdot) : \mathbb{R} \to \mathbb{C}^m$.

Theorem A.3.24. *Suppose that $u(\cdot) \in L^1(-l, l; \mathbb{C}^m)$, $l > 0$ and $(u_k)_{k\in\mathbb{Z}}$ is the sequence of its Fourier coefficients defined by (35). Then*

(i) $\|u_k\|_{\mathbb{C}^m} \leq (2l)^{-1}\|u(\cdot)\|_{L^1(-l,l;\mathbb{C}^m)}$ *for all* $k \in \mathbb{Z}$ *and* $\lim_{|k|\to\infty} u_k = 0$.

(ii) If $(u_k)_{k\in\mathbb{Z}} \in \ell^1(\mathbb{Z}; \mathbb{C}^m)$ then the Fourier series $S(u,\theta)$ of $u(\cdot)$ defined by (37) is absolutely convergent, uniformly in $\theta \in \mathbb{R}$, and

$$u(\theta) = \sum_{k=-\infty}^{\infty} u_k e^{\imath k\pi\theta/l}, \quad a.e.\ \theta \in [-l, l].$$

If, additionally, $u(\cdot)$ is continuous then this equality holds everywhere on $[-l, l]$.

(iii) If $u(\cdot)$ is piecewise continuously differentiable on $[-l, l]$, then the symmetric partial sums

$$S_N(u,\theta) = \sum_{k=-N}^{N} u_k e^{\imath k\pi\theta/l}, \quad \theta \in [-l, l], \quad N \in \mathbb{N} \tag{38}$$

converge pointwise to $\frac{1}{2}[u(\theta+) + u(\theta-)]$ as $N \to \infty$.[7]

(iv) The Cesáro means of the symmetric partial sums $S_k(u,\theta)$, $k \in \mathbb{N}$

$$\sigma_N(u,\theta) = \frac{1}{N+1}\left(S_0(u,\theta) + \cdots + S_N(u,\theta)\right) = \sum_{k=-N}^{N}\left(1 - \frac{|k|}{N+1}\right) u_k e^{\imath k\pi\theta/l}, \quad \theta \in [-l, l]$$

converge to $u(\cdot)$ in $L^1(-l, l; \mathbb{C}^m)$ as $N \to \infty$. If $u(\cdot)$ is continuous and $u(-l) = u(l)$, then $\sigma_N(u,\theta)$ converges to $u(\theta)$ uniformly on $[-l, l]$.

The second result in (i) is called Riemann's Lemma, (iv) is known as Fejér's Theorem. Statement (iii) indicates that the pointwise convergence of the Fourier series is connected with smoothness properties of $u(\cdot) \in L^1(-l, l; \mathbb{C}^m)$. We have the following regularity result.

Theorem A.3.25. *Suppose the same situation as in the previous theorem and let $n \in \mathbb{N}^*$.*

(i) If $u(\cdot)$ can be extended to an n-times differentiable $2l$-periodic function on $\mathbb{R}$ such that $u^{(n)}(\cdot)$ is integrable on $[-l, l]$ then

$$\|u_k\|_{\mathbb{C}^m} \leq (1/2l) \min_{0\leq j\leq n} \frac{\|u^{(j)}(\cdot)\|_{L^1(-l,l;\mathbb{C}^m)}}{|k|^j}, \quad k \in \mathbb{Z}.$$

(ii) If $\sum_{k=-\infty}^{\infty} |k|^n \|u_k\|_{\mathbb{C}^m} < \infty$ then $u(\cdot)$ can be extended to an n-times continuously differentiable function on $\mathbb{R}$.

[7] By definition, $u(l+) := u(-l+)$ and $u(-l-) := u(l-)$.

The pointwise convergence of Fourier series (i.e. of the symmetric partial sums $S_N(u,\theta)$) is a complicated problem and some questions in this area are still open. This is in sharp contrast with the L^2-theory of Fourier series which is simple and elegant due to the fact that the exponential functions $\psi_k(\cdot) = (\sqrt{2l})^{-1}e^{\imath k\pi(\cdot)/l}$, $k \in \mathbb{Z}$ form an orthonormal basis of $L^2(-l,l;\mathbb{C})$, see Example A.4.6. It follows from Theorem A.4.4 that

$$u(\cdot) = \sum_{k\in\mathbb{Z}} \langle u(\cdot), \psi_k(\cdot)\rangle_{L^2((-l,l;\mathbb{C})}\, \psi_k(\cdot) = \sum_{k\in\mathbb{Z}} u_k e^{\imath k\pi(\cdot)/l}, \quad u(\cdot) \in L^2(-l,l;\mathbb{C}).$$

More details are given in the next theorem.

Theorem A.3.26. *Suppose $u(\cdot) \in L^2(-l,l;\mathbb{C}^m)$ and $(u_k)_{k\in\mathbb{Z}}$ is the sequence of Fourier coefficients defined by* (35). *Then*

(i) $(u_k)_{k\in\mathbb{Z}} \in \ell^2(\mathbb{Z};\mathbb{C}^m)$, *and* $u(\cdot) \in L^2(-l,l;\mathbb{R}^m)$ *if and only if* $\overline{u_k} = u_{-k}$, $k \in \mathbb{Z}$.

(ii) The Fourier series (37) *converges (absolutely) to* $u(\cdot)$ *in* $L^2(-l,l;\mathbb{C}^m)$.

(iii) $\|u\|_{L^2(-l,l;\mathbb{K}^m)} = \sqrt{2l}\, \|(u_k)_{k\in\mathbb{Z}}\|_{\ell^2(\mathbb{Z};\mathbb{C}^m)}$.

(iv) If $u(\cdot), v(\cdot) \in L^2(-l,l;\mathbb{K}^m)$, *then*

$$\int_{-l}^{l} \langle u(\theta), v(\theta)\rangle_{\mathbb{K}^m}\, d\theta = 2l \sum_{k=-\infty}^{\infty} \langle u_k, v_k\rangle_{\mathbb{C}^m}.$$

(iv) Given any sequence $(a_k)_{k\in\mathbb{Z}} \in \ell^2(\mathbb{Z};\mathbb{C}^m)$ *there exists a unique* $u(\cdot) \in L^2(-l,l;\mathbb{C}^m)$ *such that* $a_k = u_k$ *for all* $k \in \mathbb{Z}$.

It follows from this theorem that the map $u(\cdot) \mapsto \sqrt{2l}\,(u_k)_{k\in\mathbb{Z}}$ is a Hilbert space isomorphism from $L^2(-l,l;\mathbb{C}^m)$ onto $\ell^2(\mathbb{Z};\mathbb{C}^m)$.

Fourier Transforms

We first define the Fourier transform of L^1 functions.

Definition A.3.27. Suppose $u(\cdot) \in L^1(\mathbb{R};\mathbb{C}^m)$, then the *Fourier transform* of $u(\cdot)$ is the function $\tilde{u} : \mathbb{R} \to \mathbb{C}^m$ defined by

$$\tilde{u}(\omega) = (\mathcal{F}\, u)(\omega) := \int_{-\infty}^{\infty} u(t)e^{-\imath\omega t}dt, \qquad \omega \in \mathbb{R}. \tag{39}$$

Since $\|u(t)e^{-\imath\omega t}\|_{\mathbb{C}^m} = \|u(t)\|_{\mathbb{C}^m}$, the function $\tilde{u}(\cdot) : \omega \to \tilde{u}(\omega)$ is well defined on $\mathbb{R}$. Note that if u takes its values in $\mathbb{R}^m$, then $\overline{\tilde{u}(\omega)} = \tilde{u}(-\omega)$, $\omega \in \mathbb{R}$. Some basic properties of the Fourier transform are similar to those of the Laplace transform. We list a few of these in the following proposition.

Proposition A.3.28. *Suppose* $u(\cdot) \in L^1(\mathbb{R};\mathbb{C}^m)$, $a > 0$, $\tau \in \mathbb{R}$ *and* $\tilde{u}(\cdot) = (\mathcal{F}\, u)(\cdot)$. *Then*

(i) $\tilde{u}(\cdot) : \mathbb{R} \to \mathbb{C}^m$ *is uniformly continuous and bounded on* $\mathbb{R}$, $\|\tilde{u}(\omega)\|_{\mathbb{C}^m} \le \|u(\cdot)\|_{L^1(\mathbb{R};\mathbb{C}^m)}$ *for all* $\omega \in \mathbb{R}$, *and* $\lim_{|\omega|\to\infty} \tilde{u}(\omega) = 0_m$.

(ii) $(\mathcal{F}\, u(\cdot - \tau))(\omega) = e^{-\imath\tau\omega}\tilde{u}(\omega)$, $\omega \in \mathbb{R}$ *and* $(\mathcal{F} e^{\imath\tau(\cdot)}\, u(\cdot))(\omega) = \tilde{u}(\omega - \tau)$, $\omega \in \mathbb{R}$.

(iii) $(\mathcal{F}\, u(a\,\cdot))(\omega) = a^{-1}\tilde{u}(\omega/a)$, $\omega \in \mathbb{R}$.

(iv) If $u(\cdot) : \mathbb{R} \to \mathbb{C}^m$ *is absolutely continuous on every finite interval in* $\mathbb{R}$ *and* $\dot{u}(\cdot) \in L^1(\mathbb{R};\mathbb{C}^m)$, *then* $(\mathcal{F}\, \dot{u})(\omega) = \imath\omega\, \tilde{u}(\omega)$ *for all* $\omega \in \mathbb{R}$.

(v) *If* $v(\cdot): t \mapsto tu(t)$, $t \in \mathbb{R}$ *is in* $L^1(\mathbb{R};\mathbb{C}^m)$, *then* $(\mathcal{F}v)(\omega) = \imath(\mathcal{F}u)'(\omega)$, $\omega \in \mathbb{R}$.

(vii) *If* $\mathcal{G}(\cdot) \in L^1(\mathbb{R};\mathbb{K}^{n\times m})$, $\tilde{\mathcal{G}}(\cdot) = (\mathcal{F}\mathcal{G})(\cdot)$ *and* $y(\cdot) = (\mathcal{G} * u)(\cdot)$ *then*

$$\tilde{y}(\omega) = (\mathcal{F}(\mathcal{G} * u))(\omega) = \tilde{\mathcal{G}}(\omega)\tilde{u}(\omega), \quad \omega \in \mathbb{R}.$$

(vi) *If* $v^1(t) = u(t)$, $t \geq 0$, $v^2(t) = u(-t)$, $t \geq 0$ *and* $(\mathcal{L}v^1)(\cdot), (\mathcal{L}v^2)(\cdot)$ *are the Laplace transforms of* $v^1(\cdot), v^2(\cdot) \in L^1(\mathbb{R}_+;\mathbb{C}^m)$, *then*

$$(\mathcal{F}u)(\omega) = (\mathcal{L}v^1)(\imath\omega) + (\mathcal{L}v^2)(-\imath\omega), \qquad \omega \in \mathbb{R}. \tag{40}$$

In particular if $u(\cdot)$ *vanishes on* $(-\infty, 0)$

$$(\mathcal{F}u)(\omega) = (\mathcal{L}u|_{\mathbb{R}_+})(\imath\omega), \qquad \omega \in \mathbb{R}. \tag{41}$$

Tables of Fourier transforms can be obtained from tables of Laplace transforms via (40). The last result in (i) is usually called Riemann's Lemma. We have the following inversion theorem.

Theorem A.3.29. *(i) Suppose* $u(\cdot) \in L^1(\mathbb{R};\mathbb{C}^m)$ *and set*

$$u_N(t) = \frac{1}{2\pi}\int_{-N}^{N} (1 - |\omega|/N)\,\tilde{u}(\omega)e^{\imath\omega t}d\omega, \quad t \in \mathbb{R},$$

then $u_N(\cdot) \to u(\cdot)$ *in* $L^1(\mathbb{R};\mathbb{C}^m)$ *as* $N \to \infty$. *In particular, if two* L^1*-functions have the same Fourier transform then they are equal almost everywhere.*

(ii) If $\tilde{u}(\cdot) = (\mathcal{F}u)(\cdot) \in L^1(\mathbb{R};\mathbb{C}^m)$, *then* $(\mathcal{F}\tilde{u})(-\,\cdot): t \mapsto (\mathcal{F}\tilde{u})(-t) = \int_{-\infty}^{\infty}\tilde{u}(\omega)e^{\imath\omega t}d\omega$ *is uniformly continuous and bounded on* $\mathbb{R}$ *and*

$$u(t) = (\mathcal{F}^{-1}\tilde{u})(t) = \frac{1}{2\pi}\int_{-\infty}^{\infty}\tilde{u}(\omega)e^{\imath\omega t}d\omega = \frac{1}{2\pi}(\mathcal{F}\tilde{u})(-t), \quad \text{a.e. } t \in \mathbb{R}. \tag{42}$$

If, additionally, $u(\cdot)$ *is continuous then this inversion formula holds for all* $t \in \mathbb{R}$.

There is a regularity result similar to Theorem A.3.25.

Theorem A.3.30. *Suppose* $u(\cdot) \in L^1(\mathbb{R};\mathbb{C}^m)$ *and let* $\tilde{u}(\cdot) = (\mathcal{F}u)(\cdot)$.

(i) *If* $u(\cdot)$ *is* $(n-1)$*-times differentiable on* $\mathbb{R}$, $n \geq 1$, $u^{(n-1)}(\cdot)$ *is absolutely continuous on every finite interval in* $\mathbb{R}$ *and* $u^{(1)}, \ldots, u^{(n)} \in L^1(\mathbb{R};\mathbb{C}^m)$ *then*

$$|\omega|^n\,\|\tilde{u}(\omega)\|_{\mathbb{C}^m} \to 0 \qquad \text{as } |\omega| \to \infty.$$

(ii) *On the other hand, if* $\tilde{u}(\cdot)$ *and the functions* $\omega \mapsto \omega^k\tilde{u}(\omega)$ *belong to* $L^1(\mathbb{R};\mathbb{C}^m)$ *for* $k = 1, \ldots, n$, *then* $u(\cdot)$ *is* n*-times differentiable on* $\mathbb{R}$.

As a corollary to this theorem and Theorem A.3.29, we have

Corollary A.3.31. *Suppose* $u(\cdot): \mathbb{R} \to \mathbb{C}^m$ *is twice differentiable on* $\mathbb{R}$ *with* $u(\cdot), u'(\cdot)$, $u''(\cdot) \in L^1(\mathbb{R};\mathbb{C}^m)$, *then* $\tilde{u}(\cdot) \in L^1(\mathbb{R};\mathbb{C}^m)$ *and*

$$u(t) = (\mathcal{F}^{-1}\tilde{u})(t) = \frac{1}{2\pi}\int_{-\infty}^{\infty}\tilde{u}(\omega)e^{\imath\omega t}d\omega, \qquad t \in \mathbb{R}.$$

Remark A.3.32. Definition A.3.27 can be extended to positive measures on $\mathbb{R}$ as follows. Assume that μ is a finite Borel measure on $\mathbb{R}$, see [297, VI.2]. Then the Fourier-Stieltjes transform of μ is defined by

$$\tilde{\mu}(\omega) = (\mathcal{F}\,\mu)(\omega) = \int_{-\infty}^{\infty} e^{-\imath\omega t}\mu(dt), \quad \omega \in \mathbb{R}.$$

$\tilde{\mu}(\omega)$ is continuous and bounded on $\mathbb{R}$ with $|\tilde{\mu}(\omega)| \leq \|\mu\|$ for all $\omega \in \mathbb{R}$. In particular, if μ is the following finite measure carried by a discrete subset $\{t_k; k \in \mathbb{Z}\}$ of $\mathbb{R}$

$$\mu = \sum_{k\in\mathbb{Z}} v_k\, \delta_{t_k} \quad \text{where } (v_k)_{k\in\mathbb{Z}} \in \ell^1(\mathbb{Z};\mathbb{R}),$$

then

$$\tilde{\mu}(\omega) = \sum_{k=-\infty}^{\infty} v_k e^{-\imath\omega t_k}, \quad \omega \in \mathbb{R}.$$

□

Fourier-Plancherel Transforms

In order to obtain a counterpart to Theorem A.3.26 we need to define the Fourier transform for L^2 functions. There is an initial difficulty to be overcome since $L^2(\mathbb{R};\mathbb{C}^m) \not\subset L^1(\mathbb{R};\mathbb{C}^m)$. The following theorem indicates how this problem is surmounted.

Theorem A.3.33 (Plancherel). *Suppose $u(\cdot) \in L^2(\mathbb{R};\mathbb{C}^m)$ and define for $N \in \mathbb{N}$*

$$\tilde{u}_N(\omega) = \int_{-N}^{N} u(t)e^{-\imath\omega t}dt, \quad \omega \in \mathbb{R}.$$

Then $\tilde{u}_N(\cdot)$ converges in $L^2(\mathbb{R};\mathbb{C}^m)$ to a limit $\tilde{u}(\cdot)$ as $N \to \infty$ ($\tilde{u}(\cdot)$ is called the *Fourier-Plancherel transform* of $u(\cdot)$ and is denoted by $(\mathcal{F}u)(\cdot)$)[8]. *If we set*

$$u_N(t) = \frac{1}{2\pi}\int_{-N}^{N} \tilde{u}(\omega)e^{\imath\omega t}d\omega, \qquad t \in \mathbb{R}$$

then $u_N(\cdot)$ converges to $u(\cdot)$ in $L^2(\mathbb{R};\mathbb{C}^m)$ for $N \to \infty$. Moreover $(\sqrt{2\pi})^{-1}\mathcal{F}$ is a Hilbert space isomorphism from $L^2(\mathbb{R};\mathbb{C}^m)$ onto $L^2(\mathbb{R};\mathbb{C}^m)$. In particular,

$$\sqrt{2\pi}\|u(\cdot)\|_{L^2(\mathbb{R};\mathbb{K}^m)} = \|\tilde{u}(\cdot)\|_{L^2(\mathbb{R};\mathbb{C}^m)}, \quad u(\cdot) \in L^2(\mathbb{R};\mathbb{K}^m), \tag{43}$$

$$\int_{-\infty}^{\infty} \langle u(t), v(t)\rangle_{\mathbb{K}^m} dt = \frac{1}{2\pi}\int_{-\infty}^{\infty} \langle \tilde{u}(\omega), \tilde{v}(\omega)\rangle_{\mathbb{C}^m} d\omega, \quad u(\cdot), v(\cdot) \in L^2(\mathbb{R};\mathbb{K}^m). \tag{44}$$

Note that for $u(\cdot) \in L^1(\mathbb{R};\mathbb{C}^m)$, the Fourier-Plancherel transform $\tilde{u}(\cdot)$ is defined unambiguously for every $\omega \in \mathbb{R}$, whereas for $u(\cdot) \in L^2(\mathbb{R};\mathbb{C}^m)$ the Fourier-Plancherel transform $\tilde{u}(\cdot)$ is only determined as an element of $L^2(\mathbb{R};\mathbb{C}^m)$, i.e. almost everywhere.

Example A.3.34. For any $a > 0$, let $1_{[-a,a]}(\cdot)$ be the indicator function of the interval $[-a, a]$, i.e. $1_{[-a,a]}(\omega) = 1$ if $|\omega| \leq a$ and $1_{[-a,a]}(\omega) = 0$ if $|\omega| > a$. Then $v(\cdot) = \pi\, 1_{[-a,a]}(\cdot) \in L^2(\mathbb{R};\mathbb{C})$ and so there exists $u(\cdot) \in L^2(\mathbb{R};\mathbb{C})$ such that $\tilde{u}(\cdot) = (\mathcal{F}u)(\cdot) = v(\cdot)$. By the previous theorem we have $u(\cdot) = (\mathcal{F}^{-1}v)(\cdot) = \lim_{N\to\infty} u_N(\cdot)$ in $L^2(\mathbb{R};\mathbb{C})$ where, for $N \geq a$,

$$u_N(t) = \frac{1}{2\pi}\int_{-N}^{N} \tilde{u}(\omega)e^{\imath\omega t}d\omega = \frac{1}{2\pi}\int_{-a}^{a} \pi e^{\imath\omega t}d\omega = \frac{e^{\imath at} - e^{-\imath at}}{2\imath t} = \frac{\sin at}{t}, \qquad \text{a.e. } t \in \mathbb{R}.$$

[8] This does not lead to an inconsistency, since the Fourier-Plancherel transform of any $u(\cdot) \in L^1(\mathbb{R};\mathbb{C}^m) \cap L^2(\mathbb{R};\mathbb{C}^m)$ coincides with the Fourier transform of $u(\cdot)$

Hence $u(t) = (\mathcal{F}^{-1}v)(t) = (\sin at)/t = a \operatorname{sinc}(at)$ almost everywhere[9], i.e.

$$(\mathcal{F} \operatorname{sinc}(a\,\cdot))(\omega) = (\pi/a)\, 1_{[-a,a]}(\omega), \quad \text{a.e. } \omega \in \mathbb{R}. \tag{45}$$

□

In the following proposition we list some particular properties of the Fourier-Plancherel transform which are needed in Chapter 2.

Proposition A.3.35. *Suppose $u(\cdot) \in L^2(\mathbb{R};\mathbb{K}^m)$, $a > 0$, $\tau \in \mathbb{R}$ and let $\tilde{u}(\cdot) = (\mathcal{F}u)(\cdot)$ be the Fourier-Plancherel transform of $u(\cdot)$. Then*

(i) $u(\cdot) \in L^2(\mathbb{R};\mathbb{R}^m)$ if and only if $\overline{\tilde{u}(\omega)} = \tilde{u}(-\omega)$ for a.e. $\omega \in \mathbb{R}$.

(ii) $(\mathcal{F}\, u(\cdot - \tau))(\omega) = e^{-\imath\tau\omega}\tilde{u}(\omega)$, a.e. $\omega \in \mathbb{R}$.

(iii) If $\mathcal{G}(\cdot) \in L^1(\mathbb{R};\mathbb{K}^{n\times m})$ and $y(\cdot) = (\mathcal{G} * u)(\cdot)$ is the convolution on $\mathbb{R}$ defined by (23), then $y(\cdot) \in L^2(\mathbb{R};\mathbb{K}^n)$ and the Fourier-Plancherel transform of $y(\cdot)$ satisfies

$$\tilde{y}(\omega) = (\mathcal{F}(\mathcal{G} * u))(\omega) = \tilde{\mathcal{G}}(\omega)\tilde{u}(\omega), \quad \text{a.e. } \omega \in \mathbb{R}. \tag{46}$$

Moreover

$$\|\tilde{y}(\cdot)\|_{L^2(\mathbb{R};\mathbb{C}^n)} \le \|\tilde{\mathcal{G}}(\cdot)\|_{L^\infty(\mathbb{R};\mathbb{C}^{n\times m})}\, \|\tilde{u}(\cdot)\|_{L^2(\mathbb{R};\mathbb{C}^m)}. \tag{47}$$

Discrete Fourier Transform

If, for a given $l > 0$, we identify any two-sided sequence $(u_k)_{k\in\mathbb{Z}} \in \ell^1(\mathbb{Z};\mathbb{K})$ with the discrete measure $\mu = \sum_{k\in\mathbb{Z}} u_k\, \delta_{k\pi/l}$ and apply the Fourier transform to μ as in Remark A.3.32, we are led to the following definition.

Definition A.3.36. Suppose $u(\cdot) = (u_k)_{k\in\mathbb{Z}} \in \ell^1(\mathbb{Z};\mathbb{C}^m)$ and $l > 0$, then $\tilde{u}(\cdot) : \mathbb{R} \to \mathbb{C}^m$ defined by

$$\tilde{u}(\theta) = \sum_{k=-\infty}^{\infty} u_k e^{-\imath k\pi\theta/l}, \quad \theta \in \mathbb{R} \tag{48}$$

is called the *discrete $2l$-periodic Fourier transform* of the sequence $u(\cdot)$ and is denoted by $(\mathcal{F}_D\, u)(\cdot)$.

The series on the RHS of (48) is uniformly absolutely summable in θ and hence $\tilde{u}(\theta)$ is well defined for each $\theta \in \mathbb{R}$. Note that if $(u_k)_{k\in\mathbb{Z}} \in \ell^1(\mathbb{Z};\mathbb{C}^m)$ is a sequence of Fourier coefficients of $u(\cdot) \in L^1(-l,l;\mathbb{C}^m)$ as in (35), then $\tilde{u}(-\theta)$ is the sum of the Fourier series (37) for $\theta \in \mathbb{R}$ and thus $\tilde{u}(-\theta) = u(\theta)$ for a.e. $\theta \in [-l,l]$ by Theorem A.3.24.
The discrete Fourier transform has the following properties.

Proposition A.3.37. *Suppose $(u_k)_{k\in\mathbb{Z}} \in \ell^1(\mathbb{Z};\mathbb{C}^m)$, $l > 0$ and $\tilde{u}(\cdot)$ is the discrete Fourier transform of $(u_k)_{k\in\mathbb{Z}}$ defined by (48). Then*

(i) $\tilde{u}(\cdot) : \mathbb{R} \to \mathbb{C}^m$ is continuous and $2l$-periodic.

(ii) If S is the unit forward shift defined in (5) and $v(\cdot) = (Su)(\cdot)$, then

$$\tilde{v}(\theta) = e^{-\imath\pi\theta/l}\tilde{u}(\theta), \quad \theta \in \mathbb{R}. \tag{49}$$

[9] For a definition of the sinc function, see (2.5.9).

(iii) If we define $v^1(\cdot), v^2(\cdot) \in \ell^1(\mathbb{N};\mathbb{C}^m)$ *by* $v^1_k = u_k$ *for* $k \in \mathbb{N}$ *and* $v^2_0 = 0_m$, $v^2_k = u_{-k}$ *for* $k \in \mathbb{N}^*$ *then the* $\mathcal{Z}$*-transforms* $(\mathcal{Z}\, v^i)(\cdot)$, $i = 1,2$ *are continuous on* $\overline{D_+}$ *and*

$$(\mathcal{F}_D\, u)(\theta) = (\mathcal{Z}\, v^1)(e^{\imath\pi\theta/l}) + (\mathcal{Z}\, v^2)(e^{-\imath\pi\theta/l}), \quad \theta \in \mathbb{R}. \tag{50}$$

In particular if $u(\cdot)$ *vanishes on* $\mathbb{Z} \setminus \mathbb{N}$

$$(\mathcal{F}_D\, u)(\theta) = (\mathcal{Z}\, u|_{\mathbb{N}})(e^{\imath\pi\theta/l}), \quad \theta \in \mathbb{R}. \tag{51}$$

Tables of discrete Fourier transforms can be obtained from tables of **z**-transforms via (50). Definition A.3.36 is not directly applicable to arbitrary square summable sequences because $\ell^2(\mathbb{Z};\mathbb{C}^m)$ is not contained in $\ell^1(\mathbb{Z};\mathbb{C}^m)$. However, since the normalized exponential functions $\psi_k(\cdot) = (\sqrt{2l})^{-1}\, e^{\imath k\pi(\cdot)/l}$, $k \in \mathbb{Z}$ form an orthonormal basis of $L^2(-l,l;\mathbb{C})$, the series on the RHS of (48) is absolutely summable in $L^2(-l,l;\mathbb{C}^m)$ for every sequence $u(\cdot) \in \ell^2(\mathbb{Z};\mathbb{C}^m)$, see Proposition A.4.3. The sum $\tilde{u}(\cdot) \in L^2(-l,l;\mathbb{C}^m)$ of this series in $L^2(-l,l;\mathbb{C}^m)$ is called the *discrete Fourier transform* of $u(\cdot) \in \ell^2(\mathbb{Z};\mathbb{C}^m)$ and is again denoted by $(\mathcal{F}_D\, u)(\cdot)$. Note that for $u(\cdot) \in \ell^1(\mathbb{Z};\mathbb{C}^m)$ the discrete $2l$-periodic Fourier transform $\tilde{u}(\cdot)$ is a pointwise determined continuous function on $\mathbb{R}$ whereas for $u(\cdot) \in \ell^2(\mathbb{Z};\mathbb{C}^m)$, $\tilde{u}(\cdot)$ is a function in $L^2(-l,l;\mathbb{C}^m)$ and the corresponding $2l$-periodic function on $\mathbb{R}$ (identified with $\tilde{u}(\cdot)$) is only determined modulo zero functions.

Remark A.3.38. The discrete Fourier transform $\mathcal{F}_D : \ell^2(\mathbb{Z};\mathbb{C}^m) \to L^2(-l,l;\mathbb{C}^m)$ is the discrete time counterpart of the Fourier-Plancherel transform $\mathcal{F} : L^2(-l,l;\mathbb{C}^m) \to L^2(-l,l;\mathbb{C}^m)$. In a similar way to the normalized Fourier-Plancherel transform the normalized discrete Fourier transform $(2l)^{-1/2}\mathcal{F}_D$ is a Hilbert space isomorphism. □

In the next proposition we present a discrete time counterpart of Proposition A.3.35.

Proposition A.3.39. *Suppose* $\mathcal{G}(\cdot) \in \ell^1(\mathbb{Z};\mathbb{K}^{n\times m})$, $u(\cdot) \in \ell^2(\mathbb{Z};\mathbb{K}^m)$, *and* $y(\cdot) = (\mathcal{G} * u)(\cdot)$ *is their convolution defined by* (6). *Then* $y(\cdot) \in \ell^2(\mathbb{Z};\mathbb{K}^n)$ *and, for any* $l > 0$, *we have the following equality between the associated discrete* $2l$*-periodic Fourier transforms*

$$\tilde{y}(\theta) = (\mathcal{F}_D(\mathcal{G} * u))(\theta) = \tilde{\mathcal{G}}(\theta)\tilde{u}(\theta), \quad a.e.\ \theta \in \mathbb{R}. \tag{52}$$

Moreover,

$$\|\tilde{y}(\cdot)\|_{L^2(-l,l;\mathbb{C}^n)} \le \|\tilde{\mathcal{G}}(\cdot)\|_{L^\infty(-l,l;\mathbb{C}^{n\times m})}\, \|\tilde{u}(\cdot)\|_{L^2(-l,l;\mathbb{C}^m)}\,. \tag{53}$$

A.3.4 Hardy Spaces

In this subsection we briefly summarize some basic facts about Hardy spaces. These spaces provide a natural setting for the study of analytic functions on $\mathbb{D}^+_\gamma$, $\gamma > 0$ (resp. $\mathbb{C}^+_\alpha$, $\alpha \in \mathbb{R}$) which converge in some sense to a boundary function on the circle $\partial\mathbb{D}^+_\gamma = \{s \in \mathbb{C};\, |s| = \gamma\}$ (resp. the vertical line $\partial\mathbb{C}^+_\alpha = \alpha + \imath\mathbb{R}$). Such functions are obtained e.g. by taking the **z**-transform of sequences in $\mathcal{S}_\gamma(\mathbb{K}^m)$ (resp. Laplace transform of functions in $\mathcal{E}_\alpha(\mathbb{K})$). For material on Hardy spaces, see *Hoffman* (1962) [260], *Duren* (1970) [144] and *Cima and Ross* (2000) [100]. An early application of Hardy spaces to problems in infinite dimensional system theory can be found in *Fuhrmann* (1981) [172].
All finite dimensional vector spaces $\mathbb{K}^m$ are equipped with the Euclidean norm $\|\cdot\|_{\mathbb{K}^m}$ associated with the usual inner product $\langle\cdot,\cdot\rangle_{\mathbb{K}^m}$ on $\mathbb{K}^m$.

Definition A.3.40. For $1 \le p \le \infty$, $\gamma > 0$ we denote by $H^p(\mathbb{D}_\gamma^+; \mathbb{C}^m)$ the space of all analytic functions $u(\cdot)$ on $\mathbb{D}_\gamma^+$ with values in $\mathbb{C}^m$ satisfying $\|u(\cdot)\|_{H^p(\mathbb{D}_\gamma^+;\mathbb{C}^m)} < \infty$ where

$$\|u(\cdot)\|_{H^p(\mathbb{D}_\gamma^+;\mathbb{C}^m)} = \begin{cases} \sup_{r>\gamma} \left(\int_{-\pi}^{\pi} \|u(re^{\imath\theta})\|_{\mathbb{C}^m}^p d\theta \right)^{1/p} & \text{if } 1 \le p < \infty, \\ \sup_{z \in \mathbb{D}_\gamma^+} \|u(z)\|_{\mathbb{C}^m} & \text{if } p = \infty. \end{cases} \tag{54}$$

It is known that (54) defines norms on the vector spaces $H^p(\mathbb{D}_\gamma^+; \mathbb{C}^m)$ and provided with these norms they are Banach spaces. In order to explore the relationship between the functions in $H^p(\mathbb{D}_\gamma^+; \mathbb{C}^m)$ and the associated boundary functions on $\partial\mathbb{D}_\gamma^+$, let σ be the Lebesgue measure on $\partial\mathbb{D}_\gamma^+$ (defined as the image of the Lebesgue measure on $[-\pi, \pi]$ via the map $\theta \mapsto \gamma e^{\imath\theta}$) and let $L^p(\partial\mathbb{D}_\gamma^+; \mathbb{C}^m)$ be the corresponding L^p-spaces, $1 \le p \le \infty$ endowed with their usual norms so that e.g. for $u(\cdot) \in L^p(\partial\mathbb{D}_\gamma^+; \mathbb{C}^m)$ and $1 \le p < \infty$

$$\|u(\cdot)\|_{L^p(\partial\mathbb{D}_\gamma^+;\mathbb{C}^m)} = \left(\int_{\partial\mathbb{D}_\gamma^+} \|u(z)\|_{\mathbb{C}^m}^p \, d\sigma(z) \right)^{1/p} = \left(\int_{-\pi}^{\pi} \|u(\gamma e^{\imath\theta})\|_{\mathbb{C}^m}^p \, d\theta \right)^{1/p}.$$

The map $u(\cdot) \mapsto u(\gamma e^{\imath\cdot})$ yields a Banach space isomorphism between $L^p(\partial\mathbb{D}_\gamma^+; \mathbb{C}^m)$ and $L^p(-\pi, \pi; \mathbb{C}^m)$. We have the following proposition.

Proposition A.3.41. *Let $1 \le p \le \infty$ and $\gamma > 0$. For every $u(\cdot) \in H^p(\mathbb{D}_\gamma^+; \mathbb{C}^m)$ the pointwise limit $u^0(\gamma e^{\imath\theta}) = \lim_{r\downarrow\gamma} u(re^{\imath\theta})$ exists for a.e. $\theta \in [-\pi, \pi]$, and the boundary function of $u(\cdot)$ defined by*

$$u^0(\cdot) : \partial\mathbb{D}_\gamma^+ \to \mathbb{C}^m, \quad \gamma e^{\imath\theta} \mapsto u^0(\gamma e^{\imath\theta}) = \lim_{r\downarrow\gamma} u(re^{\imath\theta}), \quad \theta \in [-\pi, \pi]$$

satisfies $u^0(\cdot) \in L^p(\partial\mathbb{D}_\gamma^+; \mathbb{C}^m)$. Moreover, the map $u(\cdot) \mapsto u^0(\cdot)$ is a linear isometry from $H^p(\mathbb{D}_\gamma^+; \mathbb{C}^m)$ onto a closed linear subspace of $L^p(\partial\mathbb{D}_\gamma^+; \mathbb{C}^m)$. In particular,

$$\|u(\cdot)\|_{H^p(\mathbb{D}_\gamma^+;\mathbb{C}^m)} = \|u^0(\cdot)\|_{L^p(\partial\mathbb{D}_\gamma^+;\mathbb{C}^m)}, \quad u(\cdot) \in H^p(\mathbb{D}_\gamma^+; \mathbb{C}^m). \tag{55}$$

The closed linear subspace of $L^p(\partial\mathbb{D}_\gamma^+; \mathbb{C}^m)$ composed of all the boundary functions $u^0(\cdot)$, $u(\cdot) \in H^p(\mathbb{D}_\gamma^+; \mathbb{C}^m)$ is denoted by $H^p(\partial\mathbb{D}_\gamma^+; \mathbb{C}^m)$ and endowed with the induced L^p-norm. For the case $p = 2$ and $\gamma = 1$ we have the following complete characterization of those functions $u(\cdot) \in L^2(\partial\mathbb{D}; \mathbb{C}^m)$ which belong to $H^2(\partial\mathbb{D}; \mathbb{C}^m)$.

Theorem A.3.42 (F. and M. Riesz). *A function $u(\cdot) \in L^2(\partial\mathbb{D}; \mathbb{C}^m)$ is in $H^2(\partial\mathbb{D}; \mathbb{C}^m)$ if and only if the Fourier coefficients $v_k \in \ell^2(\mathbb{Z}; \mathbb{C}^m)$ of the function $v(\cdot) : \theta \mapsto u(e^{\imath\theta})$ on $[-\pi, \pi]$ vanish for $k \in \mathbb{Z} \setminus \mathbb{N}$.*

The above theorem is a discrete version of the Paley-Wiener Theorem A.3.47 (see below). In fact it is valid for all $p \in [1, \infty]$, but we only need the result for $p = 2$.
We have seen in Subsection A.3.1 that the **z**-transform $\hat{u}(\cdot) = (\mathcal{Z}u)(\cdot)$ of every $u(\cdot) \in \mathcal{S}_\gamma(\mathbb{K}^m)$ is a continuous function on $\overline{\mathbb{D}_\gamma^+}$ which is bounded and analytic on $\mathbb{D}_\gamma^+$. Hence $\mathcal{Z}$ maps $\mathcal{S}_\gamma(\mathbb{C}^m)$ into $H^\infty(\mathbb{D}_\gamma^+; \mathbb{C}^m)$ for $\gamma > 0$. In particular, $\mathcal{Z}$ maps $\ell^1(\mathbb{N}; \mathbb{C}^m) \subset \ell^p(\mathbb{N}; \mathbb{C}^m)$ into $H^\infty(\mathbb{D}_+; \mathbb{C}^m) \subset H^p(\mathbb{D}_+; \mathbb{C}^m)$, $1 \le p \le \infty$. For $p = 2$ we have

Theorem A.3.43. $(\sqrt{2\pi})^{-1}\mathcal{Z}$ *is a linear isometry from $\ell^2(\mathbb{N}; \mathbb{C}^m)$ onto $H^2(\mathbb{D}_+; \mathbb{C}^m)$.*

The Hardy spaces for the domains $\mathbb{C}_\alpha^+$, $\alpha \in \mathbb{R}$ are defined as follows.

Definition A.3.44. For $1 \le p \le \infty$ and $\alpha \in \mathbb{R}$ denote by $H^p(\mathbb{C}_\alpha^+;\mathbb{C}^m)$ the space of all analytic functions $u(\cdot)$ on $\mathbb{C}_\alpha^+$ with values in $\mathbb{C}^m$ satisfying $\|u(\cdot)\|_{H^p(\mathbb{C}_\alpha^+;\mathbb{C}^m)} < \infty$ where

$$\|u(\cdot)\|_{H^p(\mathbb{C}_\alpha^+;\mathbb{C}^m)} = \begin{cases} \sup_{\beta>\alpha}\left(\int_{-\infty}^{\infty}\|u(\beta+\imath\omega)\|_{\mathbb{C}^m}^p d\omega\right)^{1/p} & \text{if } 1 \le p < \infty\,, \\ \sup_{s\in\mathbb{C}_\alpha^+}\|u(s)\|_{\mathbb{C}^m} < \infty & \text{if } p=\infty\,. \end{cases} \quad (56)$$

It is known that (56) defines norms on the vector spaces $H^p(\mathbb{C}_\alpha^+;\mathbb{C}^m)$, $1 \le p \le \infty$ and provided with these norms they are Banach spaces. In the following proposition $L^p(\partial\mathbb{C}_\alpha^+;\mathbb{C}^m)$ is the L^p-space with respect to the Lebesgue measure on the boundary $\partial\mathbb{C}_\alpha^+ = \alpha + \imath\mathbb{R}$ (image of the Lebesgue measure on $\mathbb{R}$ by the map $\omega \mapsto \alpha + \imath\omega$).

Proposition A.3.45. *Let $1 \le p \le \infty$ and $\alpha \in \mathbb{R}$. For every $u(\cdot) \in H^p(\mathbb{C}_\alpha^+;\mathbb{C}^m)$ the pointwise limit $u^0(\alpha+\imath\omega) = \lim_{\beta\downarrow\alpha} u(\beta+\imath\omega)$ exists for a.e. $\omega \in \mathbb{R}$, and the boundary function of $u(\cdot)$ defined by*

$$u^0(\cdot) : \partial\mathbb{C}_\alpha^+ = \alpha + \imath\mathbb{R} \to \mathbb{C}^m, \quad \omega \mapsto u^0(\alpha+\imath\omega) = \lim_{\beta\downarrow\alpha} u(\beta+\imath\omega),$$

satisfies $u^0(\cdot) \in L^p(\partial\mathbb{C}_\alpha^+;\mathbb{C}^m)$. Moreover, the map $u(\cdot) \mapsto u^0(\cdot)$ is a linear isometry from $H^p(\mathbb{C}_\alpha^+;\mathbb{C}^m)$ onto a closed linear subspace of $L^p(\partial\mathbb{C}_\alpha^+;\mathbb{C}^m)$. In particular, for any $1 \le p \le \infty$,

$$\|u(\cdot)\|_{H^p(\mathbb{C}_\alpha^+;\mathbb{C}^m)} = \|u^0(\cdot)\|_{L^p(\partial\mathbb{C}_\alpha^+;\mathbb{C}^m)}, \quad u(\cdot) \in H^p(\mathbb{C}_\alpha^+;\mathbb{C}^m). \quad (57)$$

Given $1 \le p \le \infty$, we denote the vector space of all the boundary functions $u^0(\cdot)$, $u(\cdot) \in H^p(\mathbb{C}_\alpha^+;\mathbb{C}^m)$ (modulo zero functions) by $H^p(\partial\mathbb{C}_\alpha^+;\mathbb{C}^m)$, and provide this linear space with the norm induced from $L^p(\partial\mathbb{C}_\alpha^+;\mathbb{C}^m)$. With respect to this norm, $H^p(\partial\mathbb{C}_\alpha^+;\mathbb{C}^m)$ is a Banach space. It follows from Definition A.3.17 and (29) that the Laplace transformation $\mathcal{L}$ maps $\mathcal{E}_\alpha(\mathbb{C})$ into $H^\infty(\mathbb{C}_\alpha^+;\mathbb{C}^m)$, in particular $\mathcal{L}$ maps $L^1(\mathbb{R}_+;\mathbb{C}^m)$ into $H^\infty(\mathbb{C}_+;\mathbb{C}^m)$. For the case $\alpha = 0$, $p = 2$ the following theorem gives a complete characterization of those functions $u(\cdot) \in L^2(\imath\mathbb{R};\mathbb{C}^m)$ which lie in $H^2(\imath\mathbb{R};\mathbb{C}^m)$.

Theorem A.3.46 (Paley-Wiener). *A function $u(\cdot) \in L^2(\imath\mathbb{R};\mathbb{C}^m)$ belongs to $H^2(\imath\mathbb{R};\mathbb{C}^m)$ if and only if the Fourier-Plancherel transform $\tilde{v}(\cdot) \in L^2(\mathbb{R};\mathbb{C}^m)$ of the function $v(\cdot) : t \mapsto u(\imath t)$ vanishes almost everywhere on $(-\infty, 0)$.*

Again this theorem is valid for all $p \in [1,\infty]$. The following theorem is the continuous time counterpart of Theorem A.3.43.

Theorem A.3.47. *The normalized Laplace transform $\left(\sqrt{2\pi}\right)^{-1}\mathcal{L}$ is a linear isometry from $L^2(\mathbb{R}_+;\mathbb{C}^m)$ onto $H^2(\mathbb{C}_+;\mathbb{C}^m)$.*

A.4 Linear Operators and Linear Forms

This section is divided into four subsections. In a first subsection we briefly recall some definitions and results concerning the summability of series in Banach or Hilbert spaces, in the second we consider linear operators on Banach spaces, in the third we introduce some additional definitions and theorems for linear operators on Hilbert spaces and in the fourth we describe some elements of spectral analysis. Good references for the material contained in the section are *Dunford and Schwartz* (1958 and 1963) [142], [143], *Rudin* (1973) [440], *Kato* (1980) [293], *Naylor and Sell* (1971) [388], and *Kreyszig* (1978) [319].

A.4.1 Summability and Generalized Fourier Series

Definition A.4.1. A family $(x_i)_{i\in I}$ of elements of a normed space $(X, \|\cdot\|)$ over $\mathbb{K}$ is said to be *summable* if there exists $x \in X$ and for every $\varepsilon > 0$ there exists a finite subset $J_0 \subset I$ such that for every finite subset J of I

$$J_0 \subset J \quad \Rightarrow \quad \|x - \sum_{i\in J} x_i\| < \varepsilon.$$

In this case x is called the sum of $(x_i)_{i\in I}$ and we write $x = \sum_{i\in I} x_i$.

The sum x of a summable family $(x_i)_{i\in I}$ is uniquely determined. If $(X,\|\cdot\|_X)$ and $(Y,\|\cdot\|_Y)$ are normed linear spaces, $A : X \to Y$ is a bounded linear operator (see the next subsection), and $(x_i)_{i\in I}$ is a summable family of elements of X with sum x, then $(Ax_i)_{i\in I}$ is a summable family of elements of Y with sum $y = Ax$.

Proposition A.4.2. *Let $(X, \|\cdot\|)$ be a Banach space.*

1. *A family $(x_i)_{i\in I}$ of elements of X is summable if and only if for every $\varepsilon > 0$ there exists $J_0 \subset I$ such that for every finite $J \subset I$*

$$J \cap J_0 = \emptyset \quad \Rightarrow \quad \|\sum_{i\in J} x_i\| < \varepsilon.$$

2. *$(x_i)_{i\in I}$ is summable with sum $x \in X$ if and only if either the set $I_0 = \{i \in I; x_i \neq 0\}$ is finite and $\sum_{i\in I_0} x_i = x$, or the set is countable and $\lim_{N\to\infty} \sum_{k=0}^{N} x_{\iota(k)} = x$ for every bijection $\iota : \mathbb{N} \to I_0$.*

If $(x_i)_{i\in I}$ is a family of elements of a Banach space $(X, \|\cdot\|)$ and $(\|x_i\|)_{i\in I}$ is summable in $\mathbb{R}$, then $(x_i)_{i\in I}$ is summable in X. In this case we say that $(x_i)_{i\in I}$ is *absolutely summable.* Thus in any Banach space X absolute summability implies summability. The converse statement holds true if X is finite dimensional, but does not necessarily hold if X is infinite dimensional. In a Hilbert space, we have

Proposition A.4.3. *A family $(x_i)_{i\in I}$ of mutually orthogonal elements of a Hilbert space $(X, \langle\cdot,\cdot\rangle)$ is summable if and only if $(\|x_i\|^2)_{i\in I}$ is summable in $\mathbb{R}$. In this case*

$$\|\sum_{i\in I} x_i\|^2 = \sum_{i\in I} \|x_i\|^2. \tag{1}$$

A family $(x_i)_{i\in I}$ in a Hilbert space is said to be *orthonormal* if the x_i are mutually orthogonal and of norm $\|x_i\| = 1$, $i \in I$.

Theorem A.4.4. *For any orthonormal family $(x_i)_{i\in I}$ in a Hilbert space X the following statements are equivalent.*

(i) *If* $x \in X$ *is orthogonal to every* x_i, $i \in I$ *then* $x = 0$.

(ii) *Every* $x \in X$ *is the sum of its generalized Fourier series*

$$x = \sum_{i \in I} \langle x, x_i \rangle \, x_i \, . \tag{2}$$

(iii) *For all* $x, y \in X$

$$\langle x, y \rangle = \sum_{i \in I} \langle x, x_i \rangle \langle x_i, y \rangle \, . \tag{3}$$

(iv) *The generalized Parseval equation holds for every* $x \in X$

$$\|x\|^2 = \sum_{i \in I} |\langle x, x_i \rangle|^2. \tag{4}$$

Definition A.4.5. An orthonormal family $(x_i)_{i \in I}$ in a Hilbert space X is said to be an *orthonormal basis* (or *Hilbert basis*) of X if it satisfies the equivalent conditions of the previous theorem.

Example A.4.6. Consider the Hilbert space $X = L^2(-l, l; \mathbb{C})$ provided with the inner product

$$\langle u(\cdot), v(\cdot) \rangle_X = \int_{-l}^{l} u(\theta) \, \overline{v(\theta)} \, d\theta, \quad u(\cdot), v(\cdot) \in L^2(-l, l; \mathbb{C}). \tag{5}$$

Then the family $(\psi_k(\cdot))_{k \in \mathbb{Z}}$ of functions $\psi_k : \theta \mapsto (\sqrt{2l})^{-1} e^{\imath k \pi \theta / l}$ on $[-l, l]$ form an orthonormal basis of $(X, \langle \cdot, \cdot \rangle_X)$. In Subsection 2.5.2, in order to simplify some of the formulas, we endow $X = L^2(-l, l; \mathbb{C})$ with the normalized inner product

$$\langle u(\cdot), v(\cdot) \rangle_X = \frac{1}{2l} \int_{-l}^{l} u(\theta) \, \overline{v(\theta)} \, d\theta, \quad u(\cdot), v(\cdot) \in L^2(-l, l; \mathbb{C}).$$

With respect to this inner product the functions $\psi_k : \theta \mapsto e^{\imath k \pi \theta / l}$, $k \in \mathbb{Z}$ form an orthonormal basis of $L^2(-l, l; \mathbb{C})$. □

A.4.2 Linear Operators on Banach Spaces

Let X, Y be normed linear spaces over $\mathbb{K}$. The corresponding norms are denoted by $\|\cdot\|_X$ and $\|\cdot\|_Y$, respectively. A linear operator $A : X \to Y$ is continuous if and only if it is bounded, i.e. its *operator norm*

$$\|A\| = \|A\|_{\mathcal{L}(X,Y)} = \sup_{x \neq 0} \frac{\|Ax\|_Y}{\|x\|_X} = \sup_{\|x\|_X \leq 1} \|Ax\|_Y = \sup_{\|x\|_X = 1} \|Ax\|_Y \tag{6}$$

is finite. The vector space of all bounded linear operators $A : X \to Y$ is denoted by $\mathcal{L}(X, Y)$. Recall that for $\alpha \in \mathbb{K}$, $A, B \in \mathcal{L}(X, Y)$ the maps $\alpha A \in \mathcal{L}(X, Y)$ and $A + B \in \mathcal{L}(X, Y)$ are defined pointwise,

$$(\alpha A)x = \alpha Ax, \quad (A + B)x = Ax + Bx, \qquad x \in X.$$

We have the following theorem.

Theorem A.4.7. *If* X, Y *are normed linear spaces, then* (6) *defines a norm on* $\mathcal{L}(X, Y)$. *If* Y *is a Banach space then so is* $\mathcal{L}(X, Y)$.

The normed linear space of all bounded linear operators from X to itself is denoted by $\mathcal{L}(X)$. In $\mathcal{L}(X)$ the product of two operators A_1, A_2 is defined by $(A_1A_2)x = A_1(A_2x)$, $x \in X$. If X is a Banach space then $\mathcal{L}(X)$ is a Banach algebra (complete normed algebra). The inverse of a bijective linear operator $A \in \mathcal{L}(X,Y)$ is not necessarily bounded. But it is so if X and Y are Banach spaces.

Theorem A.4.8 (Open Mapping Theorem of S. Banach). *If X, Y are Banach spaces and $A \in \mathcal{L}(X,Y)$ is a bijective linear operator. Then the inverse linear operator $A^{-1} : Y \to X$ is bounded, $A^{-1} \in \mathcal{L}(Y,X)$.*

If X is a normed space, the dual space X^* (i.e. the linear space of all bounded linear functionals on X) is provided with the *dual norm:*

$$\|f\|_{X^*} = \|f\|_{\mathcal{L}(X,\mathbb{K})} = \sup_{\|x\|_X \le 1} |f(x)| = \sup_{\|x\|_X = 1} |f(x)|. \tag{7}$$

X^* is a Banach space by Theorem A.4.7.
The following theorem is used to extend continuous linear functionals from a linear subspace of X to the whole space without changing its norm.

Definition A.4.9. A function $p : X \to \mathbb{R}$ defined on a linear space X over $\mathbb{K}$ is called a *semi–norm* on X if it satisfies the following conditions

$$\begin{aligned} p(x+y) &\le p(x) + p(y), \quad \text{for all} \quad x,\, y \in X \\ p(\alpha x) &= |\alpha| p(x) \quad \text{for all} \quad \alpha \in \mathbb{K},\, x \in X. \end{aligned}$$

Theorem A.4.10 (Hahn – Banach). *Let X be a linear space over $\mathbb{K}$ and p a semi–norm on X. If f is a linear functional on a $\mathbb{K}$–linear subspace $V \subset X$ satisfying $|f(x)| \le p(x)$ for $x \in V$ then there exists a $\mathbb{K}$–linear extension F of f on X satisfying $|F(x)| \le p(x)$ for all $x \in X$.*

Example A.4.11. Suppose X is a normed linear space, $x^0 \in X$, $x^0 \ne 0$, $p(x) = \|x\|_X$, $V = \text{span}\{x^0\}$ and $f(v) = \alpha\|x^0\|_X$ for $v \in V$, $v = \alpha x^0$, $\alpha \in \mathbb{K}$. Then $|f(v)| = \|v\|_X = p(v)$ for $v \in V$. Hence by the Hahn-Banach Theorem there exists $F \in X^*$ such that $|F(x)| \le \|x\|_X$ for all $x \in X$. Hence $\|F\|_{X^*} \le 1$, but since $|F(v)| = \|v\|_X$ we actually have $\|F\|_{X^*} = 1$. In particular we see that given $x^0 \in X$, $x^0 \ne 0$, there exists an aligned[10] linear form $F \in X^*$ such that $F(x^0) = \|x^0\|_X$ and $\|F\|_{X^*} = 1$. □

Another result following from the Hahn-Banach Theorem is the following *duality theorem for minimum norm problems.* If V is a linear subspace of a normed space X we denote by $V^\perp$ the *orthogonal complement* of V defined by

$$V^\perp = \{f \in X^*;\ \forall v \in V : f(v) = 0\} \subset X^*.$$

Theorem A.4.12. *Let $(X, \|\cdot\|_X)$ be a real normed linear space, $x^0 \in X$ and $V \subset X$ a linear subspace of X. Then*

$$d := \inf_{v \in V} \|x^0 - v\|_X = \max_{f \in V^\perp, \|f\|_{X^*} = 1} f(x^0) \tag{8}$$

where the maximum on the right is achieved for some $f^0 \in V^\perp$ with $\|f^0\|_{X^*} = 1$. If the infimum on the left is achieved for some $v^0 \in V$ then f^0 is aligned with $x^0 - v^0$, i.e. $f^0(x^0 - v^0) = \|x^0 - v^0\|_X = d$.

[10] $x_0 \in X$ and $F \in X^*$ are called *aligned* if $F(x^0) = \|F\|_{X^*}\|x^0\|_X$.

Definition A.4.13. Let K be a convex neigbourhood of the origin in a normed space X over $\mathbb{K}$ and suppose that K is balanced, i.e. $\alpha x \in K$ for all $x \in K$ and all $\alpha \in \mathbb{K}, |\alpha| \leq 1$. Then the functional

$$p_K(x) = \inf\{r \in \mathbb{R}; r > 0,\ r^{-1}x \in K\}, \quad x \in X$$

is called the *Minkowski functional* of K.

If K is the (open or closed) unit ball in $(X, \|\cdot\|)$ then $p_K(\cdot) = \|\cdot\|$. In general we have

Proposition A.4.14. *Let K be a balanced convex neigbourhood of the origin in a normed space X over $\mathbb{K}$. Then the Minkowski functional of K is a continuous semi–norm on X and one has*

$$\overline{K} = \{x \in X;\ p_K(x) \leq 1\}, \quad \operatorname{int} K = \{x \in X;\ p_K(x) < 1\}.$$

If K is bounded in X, then p_K is a norm on X (equivalent to the given norm on X).

Contrary to the finite dimensional case different types of convergence must be considered for infinite dimensional operators.

Definition A.4.15. Let X, Y be normed linear spaces and $A, A_k \in \mathfrak{L}(X,Y)$, $k \in \mathbb{N}$. The sequence (A_k) is said to converge *uniformly* (or *in norm*) to A if $\|A_k - A\|_{\mathfrak{L}(X,Y)} \to 0$ as $k \to \infty$. (A_k) is said to conrverge *strongly* to A if $\|A_k x - Ax\|_Y \to 0$ as $k \to \infty$ for all $x \in X$. And (A_k) is said to converge *weakly* to A if $|f(A_k x) - f(Ax)| \to 0$ as $k \to \infty$ for all $x \in X$ and all $f \in Y^*$, the dual space of Y.

Uniform convergence implies strong convergence and strong convergence implies weak convergence.

Definition A.4.16. Suppose X, Y are Banach spaces and $\Omega \subset X$ is open. A map $f : \Omega \to Y$ is said to be Fréchet differentiable at $x^0 \in \Omega$ if there exists a bounded linear operator $A \in \mathfrak{L}(X,Y)$ such that

$$\lim_{\|h\|_X \to 0} \frac{\|f(x^0 + h) - f(x^0) - Ah\|_Y}{\|h\|_X} = 0. \tag{9}$$

In this case A is said to be the *Fréchet derivative*[11] of f at x^0 and is denoted by $f'(x^0)$.

A generalization of the semigroup $(e^{At})_{t \in \mathbb{R}_+}$ to infinite dimensional spaces is given in the following

Definition A.4.17. A *strongly continuous semigroup* on a Banach space X is an operator-valued function $\Phi(\cdot) : \mathbb{R}_+ \longrightarrow \mathfrak{L}(X)$ with the following properties:

$$\Phi(t+s) = \Phi(t)\Phi(s),\ t, s \geq 0; \quad \Phi(0) = I_X; \quad \forall x \in X : \|\Phi(t)x - x\|_X \to 0 \text{ as } t \searrow 0.$$

It can be shown that $\Phi(t)x$ is continuous for all $t > 0$, $x \in X$.

Definition A.4.18. If X, Y are Banach spaces, a linear operator $A \in \mathfrak{L}(X,Y)$ is said to be *compact* if it maps bounded subsets of X into relatively compact subsets of Y (subsets of Y whose closures are compact). The vector space of compact linear operators from X to Y will be denoted by $\mathfrak{C}(X,Y)$ ($\mathfrak{C}(X)$ if $Y = X$).

[11] A is uniquely determined by (9).

Clearly any composition of a compact linear operator and a bounded linear operator is compact. For compact modifications of the identity the Fredholm alternative holds: If $A \in \mathfrak{C}(X, Y)$, either the equation $x + Ax = 0$ has a nontrivial solution or the equation $x + Ax = u$ has a uniquely determined solution for every $u \in X$. The solution depends continuously on the RHS. In other words,

Theorem A.4.19 (Fredholm Alternative). *Suppose that X is a Banach space and $A \in \mathfrak{L}(X)$ is compact. Then $I_X + A$ is injective if and only if it is surjective, and in this case $I_X + A$ is invertible in $\mathfrak{L}(X)$.*

Sometimes we need to consider linear operators between Banach spaces X and Y which are not defined for all vectors $x \in X$ (e.g. the differential operator $\frac{d}{dt}$ on the space $C(\mathbb{R}, \mathbb{R})$ of continuous real functions on $\mathbb{R}$). Let $D \subset X$ be a linear subspace of X and A a linear map from D to Y, then A is said to be an (unbounded) linear operator from X to Y. D Is called the *domain of definition*, or simply the *domain* of A and is denoted by $\mathcal{D}(A)$. Of particular importance are the unbounded operators which are closed in the sense of the following definition.

Definition A.4.20. A linear operator $A : \mathcal{D}(A) \subset X \to Y$ is called a *closed linear operator from X to Y* if its graph

$$\text{Graph}(A) = \{(x, Ax);\, x \in \mathcal{D}(A)\} \subset X \times Y \tag{10}$$

is closed in $X \times Y$. The space of closed operators from X to Y will be denoted by $\mathcal{C}(X, Y)$ ($\mathcal{C}(X)$ if $Y = X$).

Theorem A.4.21 (Closed Graph Theorem of S. Banach). *If X, Y are Banach spaces, every $A \in \mathcal{C}(X, Y)$ with domain $\mathcal{D}(A) = X$ is bounded.*

Definition A.4.22. The *infinitesimal generator* of a strongly continuous semigroup $\Phi(t)$ on a Banach space X is an operator $A : \mathcal{D}(A) \longrightarrow X$ such that

$$Ax = \lim_{t \searrow 0} (\Phi(t)x - x)/t\, x, \qquad x \in \mathcal{D}(A),$$

where $\mathcal{D}(A)$ is the set of elements in X for which the limit exists.

It can be shown that $\mathcal{D}(A)$ is dense in X and if $x \in \mathcal{D}(A)$, then $\Phi(t)x \in \mathcal{D}(A)$ for all $t \geq 0$. Moreover $\frac{d}{dt}\Phi(t)x = \Phi(t)Ax = A\Phi(t)x$, $x \in \mathcal{D}(A)$.

A.4.3 Linear Operators on Hilbert Spaces

In this subsection we assume that all the underlying spaces are $\mathbb{K}$-Hilbert spaces. The inner product in a Hilbert space X will be denoted by $\langle \cdot, \cdot \rangle_X$.

Definition A.4.23. Let X, Y be Hilbert spaces and $A : \mathcal{D}(A) \mapsto Y$ be a linear operator whose domain $\mathcal{D}(A) \subset X$ is dense in X. Then the *adjoint operator* (*Hilbert space adjoint*) of A is the unique linear operator $A^* : \mathcal{D}(A^*) \subset Y \to X$ satisfying

$$\langle Ax, y \rangle_Y = \langle x, A^* y \rangle_X\,, \qquad x \in \mathcal{D}(A),\ y \in \mathcal{D}(A^*) \tag{11}$$

where $\mathcal{D}(A^*) = \{y \in Y; \exists z \in X : \langle Ax, y \rangle_Y = \langle x, z \rangle_X \text{ for all } x \in \mathcal{D}(A)\}$.

If $A \in \mathcal{C}(X,Y)$ is densely defined, then $\mathcal{D}(A^*)$ is dense in Y, $A^* \in \mathcal{C}(Y,X)$ and $A^{**} = A$. If $A \in \mathcal{L}(X,Y)$ then $\mathcal{D}(A^*) = Y$, $A^* \in \mathcal{L}(Y,X)$ and $\|A\|_{\mathcal{L}(X,Y)} = \|A^*\|_{\mathcal{L}(Y,X)}$.
We make use of the following range–kernel duality between a closed linear operator and its dual.

Theorem A.4.24 (Closed Range Theorem of S. Banach). *If X, Y are Hilbert spaces and $A : \mathcal{D}(A) \mapsto Y$ is a closed linear operator with dense domain, then the following statements are equivalent.*

(i) $\operatorname{Im} A$ *is closed in* Y.

(ii) $\operatorname{Im} A^*$ *is closed in* X.

(iii) $\operatorname{Im} A = \ker(A^*)^\perp = \{y \in Y;\ \langle y, z\rangle_Y = 0 \text{ for all } z \in \ker(A^*)\}$.

(iv) $\operatorname{Im} A^* = \ker(A)^\perp = \{x \in X;\ \langle z, x\rangle_X = 0 \text{ for all } z \in \ker(A)\}$.

Example A.4.25. Let $S : x = (x_0, x_1, x_2, ...) \mapsto Sx = (0, x_0, x_1, x_2, ...)$ be the forward shift operator on $X = \ell^2(\mathbb{N}; \mathbb{K})$. Then for $y = (y_0, y_1, y_2, ...) \in X$

$$\langle Sx, y\rangle_X = 0 + x_0\overline{y_1} + x_1\overline{y_2} + x_2\overline{y_3} + \cdots = \langle x, S^*y\rangle_X$$

where $S^* : y = (y_0, y_1, y_2, \ldots) \mapsto S^*y = (y_1, y_2, y_3, \ldots)$ is the backward shift operator on $\ell^2(\mathbb{N}; \mathbb{K})$. Let $(e^j)_{j\in\mathbb{N}}$ be the standard orthonormal basis of $\ell^2(\mathbb{N}; \mathbb{K})$, i.e. $e^j_k = \delta_{jk}$, $k \in \mathbb{N}$ for each $j \in \mathbb{N}$. Clearly $\ker S = \{0\}$, $\operatorname{Im} S^* = X$, $\ker S^* = \operatorname{span}\{e^0\}$, $\operatorname{Im} S = \operatorname{span}\{e^1, e^2, e^3, \ldots\}$. So we see that (i)–(iv) of Theorem A.4.24 are satisfied. □

An operator $A \in \mathcal{L}(X)$ is said to be *selfadjoint* or *Hermitian* if $A = A^*$ and *normal* if $AA^* = A^*A$. Every selfadjoint operator $A \in \mathcal{L}(X)$ is normal. The set of all selfadjoint operators $A \in \mathcal{L}(X)$ is a real vector space and will be denoted by $\mathcal{H}(X)$.

Theorem A.4.26. *If $A \in \mathcal{H}(X)$ then*

$$\|A\|_{\mathcal{L}(X)} = \sup_{\|x\|\le 1} |\langle Ax, x\rangle_X|. \tag{12}$$

$A \in \mathcal{H}(X)$ is said to be *positive semi-definite*

$$\langle Ax, x\rangle_X = \langle x, Ax\rangle_X \ge 0\,, \qquad x \in X.$$

The set $\mathcal{H}^+(X)$ of positive semi-definite operators on X is a pointed convex cone, i.e.

$$\alpha\mathcal{H}^+(X) \subset \mathcal{H}^+(X),\ \alpha > 0,\ \mathcal{H}^+(X) + \mathcal{H}^+(X) \subset \mathcal{H}^+(X),\ \text{and}\ \mathcal{H}^+(X) \cap (-\mathcal{H}^+(X)) = 0.$$

The associated order on the real vector space $\mathcal{H}(X)$ is defined by $A \succeq B \Leftrightarrow A - B \succeq 0$. Every decreasing sequence in this vector space which is bounded below has a pointwise limit in $\mathcal{H}(X)$. More precisely,

Proposition A.4.27. *Suppose (A_k) is a decreasing sequence in $\mathcal{H}(X)$ which is bounded below, i.e. for all $k \in \mathbb{N}$, $x \in X$ we have $\langle A_k x, x\rangle_X \ge \langle A_{k+1}x, x\rangle_X$ and there exists $B \in \mathcal{H}(X)$ such that $A_k \succeq B$. Then there exists $A \in \mathcal{H}(X)$ such that $\lim_{k\to\infty}\langle A_k x, x\rangle_X = \langle Ax, x\rangle_X$, for all $x \in X$.*

Positive semi-definite operators have a square root.

Proposition A.4.28. *Suppose that $A \in \mathcal{H}^+(X)$ then there exists a unique $B \in \mathcal{H}^+(X)$ such that $B^2 = A$. B is called the square root of A and denoted by $B = A^{1/2}$. A is invertible in $\mathcal{L}(X)$ if and only if B is invertible in $\mathcal{L}(X)$.*

As a consequence $A \in \mathcal{H}^+(X)$ is invertible in $\mathcal{L}(X)$ if and only if there exists $\varepsilon > 0$ such that

$$\langle Ax, x\rangle_X = \|Bx\|_X^2 \geq \varepsilon^2 \|x\|_X^2 \qquad x \in X. \tag{13}$$

Occasionally we will need the following modification of the previous proposition.

Corollary A.4.29. *If X, Y are Hilbert spaces and $B \in \mathcal{L}(X, Y)$ then $A = B^*B \in \mathcal{H}^+(X)$. Conversely, if $A \in \mathcal{H}^+(X)$ then there exist a Hilbert space Y and $B \in \mathcal{L}(X, Y)$ such that $A = B^*B$. If $\operatorname{rank} A = q$ then one can choose $Y = \mathbb{K}^q$.*

A.4.4 Spectral Theory

In this subsection we first define the spectrum and resolvent of an operator $A \in \mathcal{C}(X)$ where X is a Banach space. Then we examine the simplifications when firstly $A \in \mathcal{L}(X)$ and secondly $A \in \mathfrak{C}(X)$ and conclude with a spectral theorem for closed linear operators on Hilbert spaces with compact normal resolvents. The Banach and Hilbert spaces in this subsection are assumed to be complex.

Definition A.4.30. (i) Let X be a Banach space and $A \in \mathcal{C}(X)$. The set of $s \in \mathbb{C}$ such that $sI_X - A : \mathcal{D}(A) \to X$ is bijective with bounded inverse

$$R(s, A) := (sI_X - A)^{-1} \in \mathcal{L}(X),$$

is called the *resolvent set* of A and is denoted by $\rho(A)$. The operator-valued function $R(\cdot, A) : s \mapsto R(s, A)$ on $\rho(A)$ is called the *resolvent* of A.

(ii) The complement of the resolvent set, $\sigma(A) = \mathbb{C} \setminus \rho(A)$, is called the *spectrum* of A. $\lambda \in \mathbb{C}$ is said to be an *eigenvalue* and a non-zero $x \in \mathcal{D}(A)$ an *eigenvector* of A if $Ax = \lambda x$. The collection of all eigenvalues of A is called the *point spectrum* and denoted by $\sigma_p(A)$.

The *continuous spectrum* of A, denoted by $\sigma_c(A)$, is the set of all $\lambda \in \mathbb{C}$ such that $\lambda I_X - A$ has its range dense in X, is one-to-one, but does not have a bounded inverse.

The *residual spectrum* of A, denoted by $\sigma_r(A)$, is the set of all $\lambda \in \mathbb{C}$ such that $\lambda I_X - A$ is one-to-one but does not have its range dense in X.

By definition

$$\sigma(A) = \sigma_p(A) \,\dot\cup\, \sigma_c(A) \,\dot\cup\, \sigma_r(A).$$

$\sigma(A)$ is always closed, but for unbounded $A \in \mathcal{C}(X)$ it is possible that $\sigma(A)$ is empty or covers the whole complex plane. Correspondingly, $\rho(A)$ is always open in $\mathbb{C}$, but may be empty. The resolvent operator satisfies the *resolvent equation*

$$R(\lambda_1, A) - R(\lambda_2, A) = (\lambda_2 - \lambda_1) R(\lambda_1, A) R(\lambda_2, A), \qquad \lambda_1, \lambda_2 \in \rho(A). \tag{14}$$

In particular $R(\lambda_1, A)$ and $R(\lambda_2, A)$ commute. For λ, $\lambda_0 \in \rho(A)$, we have

$$R(\lambda, A) = R(\lambda_0, A)[I_X - (\lambda_0 - \lambda)R(\lambda_0, A)]^{-1} = \sum_{k=0}^{\infty} (\lambda_0 - \lambda)^k R(\lambda_0, A)^{k+1}$$

and the series converges in norm if $|\lambda_0 - \lambda| < \|R(\lambda_0, A)\|_{\mathcal{L}(X)}^{-1}$. Thus $R(\lambda, A)$ is analytic on the open (but not necessarily connected) set $\rho(A)$. We will now deal with the case where A is bounded.

Definition A.4.31. The *spectral radius* of $A \in \mathfrak{L}(X)$ is given by

$$\varrho(A) = \sup\{|\lambda|; \lambda \in \sigma(A).\}$$

The next theorem shows that if $A \in \mathfrak{L}(X)$, then neither $\rho(A)$ nor $\sigma(A)$ are empty. Moreover, it implies that in the previous definition the "sup" may be replaced by "max".

Theorem A.4.32. *Let X be a Banach space and $A \in \mathfrak{L}(X)$, then the following hold.*

(i) The spectrum $\sigma(A)$ is compact and nonempty.

(ii) The spectral radius is determined by

$$\varrho(A) = \lim_{k\to\infty} \|A^k\|^{1/k}_{\mathfrak{L}(X)} = \inf_{k\geq 1} \|A^k\|^{1/k}_{\mathfrak{L}(X)}. \tag{15}$$

(iii) If $A, B \in \mathfrak{L}(X)$, then $\varrho(AB) = \varrho(BA)$.

(iv) If A is normal then $\varrho(A) = \|A\|_{\mathfrak{L}(X)}$.

(v) The resolvent operator of A has the following Laurent expansion at infinity

$$R(s, A) = \sum_{k=1}^{\infty} s^{-k} A^{k-1}, \quad |s| > \varrho(A) \tag{16}$$

where the series converges in norm.

Example A.4.33. Consider the forward and backward shift operators S and S^* on $X = \ell^2(\mathbb{N}; \mathbb{C})$ as in Example A.4.25. Then

$$(\lambda I_X - S)x = (\lambda x_0, \lambda x_1 - x_0, \lambda x_2 - x_1, ...).$$

So for all $\lambda \in \mathbb{C}$, $(\lambda I_X - S)x = 0$ implies $x = 0$ and hence $\sigma_p(S) = \emptyset$. Since $\|S^k x\|_X = \|x\|_X$ for all $x \in X$ and $k \in \mathbb{N}$, the spectral radius formula (15) shows that $\varrho(S) = 1$ and hence $\sigma(S) \subset \overline{\mathbb{D}}$ where $\mathbb{D} = \{\lambda \in \mathbb{C}; |\lambda| < 1\}$ is the open unit disk. Now let $x \in X$ and $y = (\lambda I_X - S)x$, i.e. $y_k = \lambda x_k - x_{k-1}$, $k \geq 1$, $y_0 = \lambda x_0$, then

$$\textstyle\sum_{k=0}^{N} \lambda^k y_k = \lambda x_0 + \lambda(\lambda x_1 - x_0) + \cdots + \lambda^N(\lambda x_N - x_{N-1}) = \lambda^{N+1} x_N.$$

For $|\lambda| \leq 1$, $\lambda^{N+1} x_N \to 0$ as $N \to \infty$ and hence $\sum_{k=0}^{\infty} \lambda^k y_k = 0$. Thus when $|\lambda| < 1$ the range of $\lambda I_X - S$ is orthogonal to $x_\lambda = (1, \overline{\lambda}, \overline{\lambda}^2, ...) \in X$ and hence cannot be dense in X. This proves that $\mathbb{D} \subset \sigma_r(S)$. It is shown in [388] that $\partial\mathbb{D} = \{\lambda \in \mathbb{C}; |\lambda| = 1\} \subset \sigma_c(S)$ and so

$$\sigma(S) = \overline{\mathbb{D}}, \quad \sigma_p(S) = \emptyset, \quad \sigma_c(S) = \partial\mathbb{D}, \quad \sigma_r(S) = \mathbb{D}.$$

For the backward shift operator S^*

$$(\lambda I_X - S^*)x = (\lambda x_0 - x_1, \lambda x_1 - x_2, \lambda x_2 - x_3, ...).$$

So every λ with $|\lambda| < 1$ is an eigenvalue of S^* with associated eigenvector $(1, \lambda, \lambda^2, \lambda^3, \dots) \in X$. Now $\|S^{*k} x\|_X \leq \|x\|_X$ and $\|S^{*k} S^k x\|_X = \|S^k x\|_X$ for $x \in \ell^2(\mathbb{N}; \mathbb{C})$ and $k \in \mathbb{N}$. So $\|S^{*k}\|_{\mathfrak{L}(X)} = 1$ for all $k \in \mathbb{N}$ and hence $\varrho(S^*) = 1$ by (15). Again it can be shown (see [388]) that $\partial\mathbb{D} \subset \sigma_c(S)$ and so

$$\sigma(S^*) = \overline{\mathbb{D}}, \quad \sigma_p(S^*) = \mathbb{D}, \quad \sigma_r(S^*) = \emptyset, \quad \sigma_c(S^*) = \partial\mathbb{D}.$$

□

For any $A \in \mathfrak{L}(X)$, we have $\sigma(A^*) = \{\overline{\lambda} \in \mathbb{C};\ \lambda \in \sigma(A)\}$, but as the above example shows this does not mean that if $\lambda \in \sigma(A)$ is an eigenvalue of A, then $\overline{\lambda}$ is an eigenvalue of A^*. If $A \in \mathfrak{L}(X)$ is selfadjoint then $\sigma(A) \subset \mathbb{R}$.
In the case where A is compact we have the following theorem.

Theorem A.4.34. *Suppose X is a Banach space and $A \in \mathfrak{C}(X)$. Then $\sigma(A)$ is a countable set with no non-zero accumulation point and each non-zero $\lambda \in \sigma(A)$ is an eigenvalue of A with finite multiplicity.*

Most operators which arise in physics are not compact, but often they have compact resolvents. For such operators we have

Theorem A.4.35. *Suppose X is a Banach space, $A \in \mathcal{C}(X)$ is such that $\rho(A) \neq \emptyset$ and $R(\lambda, A)$ is compact for some $s_0 \in \rho(A)$. Then $\sigma(A)$ consists entirely of isolated eigenvalues of A with finite multiplicities, and $R(s, A)$ is compact for every $s \in \rho(A)$.*

For the case where X is a Hilbert space with inner product $\langle \cdot, \cdot \rangle$ we have the following spectral theorems. They show that compact normal operators on a Hilbert space have a spectral decomposition analogous to normal matrices.

Theorem A.4.36. *Suppose X is a Hilbert space and $A \in \mathfrak{C}(X)$ is normal . Then there exists an orthonormal basis of eigenvectors (v^k) of A with associated eigenvalues λ_k such that for ull $x \subset X$*

$$Ax = \sum_k \lambda_k \langle x, v^k \rangle\, v^k.$$

If $A \in \mathfrak{C}(X)$ is selfadjoint then at least one of the two numbers $\pm\|A\|$ is an eigenvalue of A. If A has a maximal (resp. minimal) eigenvalue, these are given by

$$\lambda_{\max}(A) = \max_{\|x\|=1} \langle Ax, x \rangle \quad (\textit{resp. } \lambda_{\min}(A) = \min_{\|x\|=1} \langle Ax, x \rangle). \tag{17}$$

One can also express an unbounded operator as a sum of eigen-projections if its resolvent is compact and normal.

Theorem A.4.37. *Suppose X is a Hilbert space, $A \in \mathcal{C}(X)$ is such that $\rho(A) \neq \emptyset$ and $R(\lambda, A)$ is compact and normal for some $\lambda \in \rho(A)$. Then there exists an orthonormal basis of eigenvectors (v^k) of A with associated eigenvalues λ_k such that for every $x \in \mathcal{D}(A)$*

$$Ax = \sum_k \lambda_k \langle x, v^k \rangle\, v^k$$

and $\mathcal{D}(A) = \{x \in X;\ \sum_k |\lambda_k|^2 |\langle x, v^k \rangle|^2 < \infty\}$. Moreover if $\sup_k \operatorname{Re} \lambda_k < \infty$ and we define

$$\Phi(t)x = \sum_k e^{\lambda_k t} \langle x, v^k \rangle\, v^k, \quad t \geq 0, \ \ x \in X$$

then $\Phi(t)$ is a strongly continuous semigroup.

We see that under the conditions of the above theorem, given any $x \in \mathcal{D}(A)$, the function $x(\cdot) : t \mapsto \Phi(t)x$ is the (unique) solution of the initial value problem

$$\dot{x}(t) = Ax(t), \quad t > 0, \quad x(0) = x.$$

Bibliography

[1] R. A. Abraham and J. E. Marsden. *Foundations of Mechanics.* Benjamin/Cummings, Reading, MA, 1978.

[2] J. Ackermann. Entwurf durch Polvorgabe. *Regelungstechnik*, 6, 7:137–179, 209–215, 1977.

[3] J. Ackermann. *Sampled-data Control Systems, Analysis and Synthesis, Robust System Design.* Communication and Control Engineering Series. Springer-Verlag, Berlin, 1985.

[4] J. Ackermann. *Robust Control. Systems with Uncertain Physical Parameters.* Springer-Verlag, London, 1993.

[5] R. P. Agarwal. *Difference Equations and Inequalities.* Number 155 in Pure and Applied Mathematics. Marcel Dekker, New York, 1992.

[6] L. V. Ahlfors. *Complex Analysis. An Introduction to the Theory of Analytic Functions of One Complex Variable.* McGraw-Hill, New York, 3rd edition, 1979.

[7] M. A. Aizerman. On a problem concerning the stability "in the large" of dynamical systems. *Uspekhi Mat. Nauk*, 4:187–188, 1949.

[8] M. A. Aizerman and M. A. Gantmacher. *Absolute Stability of Controlled Systems.* Izd. Akad. Nauk SSSR, 1963. (In Russian).

[9] R. G. D. Allen. *Mathematical Economics.* Macmillan, London, 2nd edition, 1966.

[10] K. T. Alligood, T. D. Sauer, and J. A. Yorke. *Chaos: An Introduction to Dynamical Systems.* Springer-Verlag, New York, 1997.

[11] H. Amann. *Ordinary Differential Equations: An Introduction to Nonlinear Analysis*, volume 13 of *De Gruyter Studies in Mathematics.* De Gruyter, Berlin - New York, 1990.

[12] B. D. O. Anderson. On the computation of the Cauchy index. *Quarterly of Applied Mathematics*, 29:577–581, 1972.

[13] B. D. O. Anderson, E. I. Jury, and M. Mansour. On robust Hurwitz polynomials. *IEEE Trans. Automat. Contr.*, AC-32:909–913, 1987.

[14] E. Anderson, Z. Bai, C. Bischof, S. Blackford, J. Demmel, J. Dongarra, J. DuCroz, A. Greenbaum, S. Hammarling, A. McKenney, and D. Sorensen. *LAPACK Users' Guide*, volume 9 of *Software - Environments - Tools.* SIAM Publications, Philadelphia, PA, 3rd edition, 1999.

[15] S. S. Antman. The simple pendulum is not so simple. *SIAM Review*, 40(4):927–930, 1998.

[16] M. A. Arbib. *Algebraic Theory of Machines, Languages, and Semigroups.* Academic Press, New York-London, 1968.

[17] M. B. Argoun. Stability of a Hurwitz polynomial under coefficient perturbations: Necessary and sufficient conditions. *Int. J. Control*, 45:739–744, 1987.

[18] V. I. Arnold. *Mathematical Methods of Classical Mechanics.* Number 60 in Graduate Texts in Mathematics. Springer-Verlag, New York-Heidelberg-Berlin, 1978.

[19] V. I. Arnold. *Ordinary Differential Equations.* MIT Press, Cambridge, MA, 1978.

[20] V. I. Arnold. *Geometrical Methods in the Theory of Ordinary Differential Equations.* Number 250 in Grundlehren der mathematischen Wissenschaften. Springer-Verlag, New York-Heidelberg-Berlin, 1983.

[21] K. J. Arrow and F. H. Hahn. *General Competitive Analysis.* Holden-Day, San Francisco, 1971.

[22] D. K. Arrowsmith and C. M. Place. *An Introduction to Dynamical Systems.* Cambridge University Press, Cambridge, 1990.

[23] L. Atanassova, D. Hinrichsen, and V. L. Kharitonov. Convex directions for stable polynomials and quasipolynomials: A survey of recent results. In L. Dugard and E. I. Verriest, editors, *Stability and Control of Time-Delay Systems*, number 228 in Lecture Notes in Control and Information Sciences, pages 72–91. Springer-Verlag, London, 1998.

[24] M. Athans. On large scale systems and decentralized control. *IEEE Trans. Automat. Contr.*, AC-23(2), 1978.

[25] M. Athans and M. G. Safonov. Gain and phase margin for multiloop LQG regulators. In *Proc. 15th IEEE Conf. Decis. Control*, pages 361–368, Clearwater Beach, Florida, 1976.

[26] K. E. Atkinson. *An Introduction to Numerical Analysis.* John Wiley & Sons, New York, 2nd edition, 1989.

[27] B. Aulbach and T. Wanner. Integral manifolds for Carathéodory type differential equations in Banach spaces. In B. Aulbach and F. Colonius, editors, *Six Lectures on Dynamical Systems*, pages 45–119. World Scientific, Singapore, 1996.

[28] L. Autonne. Sur les matrices hypohermitiennes et sur les matrices unitaires. *Ann. Univ. Lyon, Nouvelle Série I*, 38:1–77, 1915.

[29] E. W. Bai, R. Tempo, and M. Fu. Worst-case properties of the uniform distribution and randomized algorithms for robustness analysis. *Math. Control Signals Syst.*, 11(3):183–196, 1998.

[30] N. T. J. Bailey. *The Mathematical Theory of Epidemics.* Hafner, New York, 1957.

[31] N. T. J. Bailey. *The Mathematical Theory of Infectious Diseases and its Applications.* Charles Griffin & Company, London - High Wycombe, 2nd edition, 1975.

[32] N. E. Barabanov. Absolute characteristic exponent of a class of linear nonstationary systems of differential equations. *Siberian Mathematical Journal*, 29(4):521–530, 1988.

[33] E. A. Barbašin. *Introduction to the Theory of Stability.* Wolters-Noordhoff Series of Monographs and Textbooks on Pure and Applied Mathematics. Wolters-Noordhoff, Groningen, 1970.

[34] B. R. Barmish. Invariance of the strict Hurwitz property for polynomials with perturbed coefficients. *IEEE Trans. Automat. Contr.*, AC-29:935–936, 1984.

[35] B. R. Barmish. New tools for robustness analysis. In *Proc. 27th IEEE Conf. Decis. Control, Austin, Texas*, pages 1–6, 1988.

[36] B. R. Barmish. *New Tools for Robustness of Linear Systems.* Macmillan Publishing Co., New York, 1994.

[37] B. R. Barmish and C. M. Lagoa. The uniform distribution: A rigorous justification for its use in robustness analysis. *Math. Control Signals Syst.*, 10(3):203–222, 1997.

[38] R. H. Bartels and G. W. Stewart. Algorithm 432: The solution of the matrix equation $AX + XB = C$. *Comm. ACM*, 15(9):820–826, 1972.

[39] A. C. Bartlett, C. V. Hollot, and H. Lin. Root location of an entire polytope of polynomials: It suffices to check the edges. *Math. Control Signals Syst.*, 1:61–71, 1988.

[40] F. L. Bauer. On the field of values subordinate to a norm. *Numer. Math.*, 4:103–113, 1962.

[41] F. L. Bauer and C. T. Fike. Norms and exclusion theorems. *Numer. Math.*, 2:137–141, 1960.

[42] H. Bauer. *Measure and Integration Theory*, volume 26 of *de Gruyter Studies in Mathematics*. de Gruyter, Berlin, 2001.

[43] H. Baumgärtel. *Analytic Perturbation Theory for Matrices and Operators*. Number 15 in Operator Theory: Advances and Applications. Birkhäuser, Basel, 1985.

[44] R. Bellman. *Stability Theory of Differential Equations*. McGraw-Hill, New York, 1953.

[45] R. Bellman and K. L. Cooke. *Differential-Difference Equations*. Number 6 in Mathematics in Science and Engineering. Academic Press, New York, 1963.

[46] R. Bellman and R. Kalaba, editors. *Selected Papers on Mathematical Trends in Control Theory*. Dover, New York, 1964.

[47] E. Beltrami. Sulle funzioni bilineari. *Giornale Math.*, 11:98–106, 1873.

[48] J. J. Benedetto. Irregular sampling and frames. In C. K. Chui, editor, *Wavelets: A Tutorial in Theory and Applications*, volume 2 of *Wavelet Anal. Appl.*, pages 445–507. Academic Press, San Diego, CA, 1992.

[49] S. Benedetto, E. Biglieri, and V. Castellani. *Digital Transmission Theory*. Prentice Hall, Englewood Cliffs, NJ, 1987.

[50] S. Bennett. *A History of Control Engineering 1800–1930*. IEE Control Engineering Series 8. Peter Peregrinus Ltd., Stevenage, UK, 1979.

[51] B. Bernhardsson, A. Rantzer, and L. Qiu. Real perturbation values and real quadratic forms in a complex vector space. *Lin. Alg. Appl.*, 270:131–154, 1998.

[52] J. E. Bertram and R. E. Kalman. Control systems analysis and design via the second method of Lyapunov. *Trans. ASME J. Basic Eng.*, 82:371–392, 1960.

[53] F. J. Beutler. Error-free recovery of signals from irregularly spaced samples. *SIAM Review*, 8:328–335, 1966.

[54] N. P. Bhatia and G. P. Szegö. *Stability Theory of Dynamical Systems*. Number 161 in Grundlehren der mathematischen Wissenschaften. Springer-Verlag, Berlin, New York, Heidelberg, 1970.

[55] R. Bhatia. *Perturbation Bounds for Matrix Eigenvalues*. Number 162 in Pitman Research Notes in Mathematics. John Wiley & Sons, New York, 1987.

[56] S. P. Bhattacharyya, H. Chapellat, and L. H. Keel. *Robust Control. The Parametric Approach*. Prentice-Hall, Upper Saddle River, NJ, 1995.

[57] S. Białas. A necessary and sufficient condition for the stability of interval matrices. *Int. J. Control*, 37:717–722, 1983.

[58] S. Białas and J. Garloff. Stability of the strict Hurwitz property for polynomials with perturbed coefficients. *IEEE Trans. Automat. Contr.*, AC-30:935–936, 1985.

[59] R. M. Biernacki, H. Hwang, and S. P. Bhattacharyya. Robust stability with structured real parameter perturbations. *IEEE Trans. Automat. Contr.*, AC-32(6):495–506, 1987.

[60] G. Birkhoff and T. C. Bartee. *Modern Applied Algebra.* McGraw-Hill, New York, 1970.

[61] G. D. Birkhoff. Stability and the equations of dynamics. *American Journal of Mathematics*, 49:1–38, 1927.

[62] Å. Björck. Solving linear least squares problems by Gram-Schmidt orthogonalization. *BIT*, 7:1–21, 1967.

[63] V. D. Blondel, E. D. Sontag, M. Vidyasagar, and J. C. Willems, editors. *Open Problems in Mathematical Systems and Control Theory.* Springer-Verlag, London, 1999.

[64] L. Boberg and U. Brosa. Onset of turbulence in a pipe. *Z. Naturforsch.*, 43a:697–726, 1988.

[65] P. Bohl. Über Differentialgleichungen. *Journal für Reine und Angewandte Mathematik*, 144:284–313, 1913.

[66] G. Boole. *An Investigation of the Laws of Thought, on Which are Founded the Mathematical Theories of Logic and Probability.* Dover, 1973. Reprint of 1854.

[67] N. K. Bose. A system-theoretic approach to stability of sets of polynomials. *Contemp. Math.*, 47:25–34, 1985.

[68] N. K. Bose. Argument conditions for Hurwitz and Schur polynomials from network theory. *IEEE Trans. Automat. Contr.*, AC-39:345–346, 1994.

[69] N. K. Bose, E. I. Jury, and E. Zeheb. On robust Hurwitz and Schur polynomials. In *Proc. 25th IEEE Conf. Decis. Control*, pages 739–744, Athens, 1986.

[70] A. Böttcher. Pseudospectra and singular values of large convolution operators. *J. Integral Equations and Applications*, 6:267–301, 1994.

[71] A. Böttcher, S. Grudsky, and B. Silbermann. Norms of inverses, spectra, and pseudospectra of large truncated Wiener-Hopf operators and Toeplitz matrices. *New York J. Math.*, 3:1–31, 1997.

[72] A. Böttcher and B. Silbermann. *Introduction to Large Truncated Toeplitz Matrices.* Springer-Verlag, New York, 1999.

[73] S. Boyd and V. Balakrishnan. A regularity result for the singular values of a transfer matrix and a quadratically convergent algorithm for computing its L^∞-norm. *Syst. Control Lett.*, 15(1):1–7, 1990.

[74] S. Boyd, L. E. Ghaoui, E. Feron, and V. Balakrishnan. *Linear Matrix Inequalities in Systems and Control Theory*, volume 15 of *Studies in Applied Mathematics.* SIAM Publications, Philadelphia, PA, 1994.

[75] A. Brauer. Limits for the characteristic roots of a matrix II. *Duke Math. J.*, 14:21–26, 1947.

[76] R. W. Brockett. The status of stability theory for deterministic systems. *IEEE Trans. Automat. Contr.*, AC-11:596–606, 1966.

[77] R. W. Brockett. *Finite Dimensional Linear Systems.* John Wiley & Sons, New York, 1970.

[78] R. W. Brockett. Some geometric questions in the theory of linear systems. *IEEE Trans. Automat. Contr.*, AC-21:449–455, 1976.

[79] R. A. Brualdi. Matrices, eigenvalues, and directed graphs. *Linear Multilinear Algebra*, 11(2):143–165, 1982.

[80] N. A. Bruinsma and M. Steinbuch. A fast algorithm to compute the H^∞-norm of a transfer function matrix. *Syst. Control Lett.*, 14:287–293, 1990.

[81] N. A. Bruinsma and M. Steinbuch. H^∞-norm computation using a Hamiltonian matrix. *Selected Topics in Identification, Modelling and Control*, 1:45–50, 1990.

[82] R. B. Burckel. *An Introduction to Classical Complex Analysis*, volume 64 of *Mathematische Reihe*. Birkhäuser, Basel, 1979.

[83] J. Burke, A. Lewis, and M. Overton. Variational analysis of the abscissa mapping for polynomials via the Gauss-Lucas theorem. *J. Global Opt.*, 28(3–4), 2004.

[84] T. D. Burton. *Introduction to Dynamic Systems Analysis*. McGraw-Hill, Singapore, 1994.

[85] J. B. Butler. Perturbation series for eigenvalues of analytic non-symmetric operators. *Arch. Math.*, 10:21–27, 1959.

[86] K. M. Butler and B. F. Farrell. Three dimensional optimal perturbations in viscous shear flow. *Phys. Fluids A*, 4:1637–1650, 1992.

[87] C. I. Byrnes. On a theorem of Hermite and Hurwitz. *Lin. Alg. Appl.*, 50:61–101, 1983.

[88] F. Camilli, L. Grüne, and F. Wirth. A regularization of Zubov's equation for robust domains of attraction. In A. Isidori et al., editors, *Nonlinear Control in the Year 2000*, volume 258 of *Lecture Notes in Control and Information Sciences*, pages 277–290. Springer-Verlag, Berlin, 2000.

[89] F. Camilli, L. Grüne, and F. Wirth. A generalization of Zubov's method to perturbed systems. *SIAM J. Control Optim.*, 40(2):496–515, 2001.

[90] A. B. Carlson. *Communication Systems*. McGraw-Hill, New York, 3rd edition, 1986.

[91] D. H. Carlson and H. Schneider. Inertia theorems for matrices: the semi-definite case. *J. Math. Anal. Appl.*, 6:430–446, 1963.

[92] H. Cartan. *Elementary Theory of Analytic Functions of One or Several Complex Variables*. Dover, New York, 1995.

[93] H. Cavendish. An account of some attempts to imitate the effects of the torpedo by electricity. *Philos. Trans. Roy. Soc. London*, 66:196–225, 1776.

[94] A. Cayley. The Newton-Fourier imaginary problem. *American Journal of Mathematics*, II(92), 1879.

[95] A. Cayley. Sur les racines d'une équation algébrique. *C.R. Acad. Sci., Paris, Ser. I*, 110:215–218, 1890.

[96] F. Chaitin-Chatelin and V. Frayssé. *Lectures on Finite Precision Computations*. SIAM Publications, Philadelphia, 1996.

[97] J. Chen and C. N. Nett. Bounds on generalized structured singular values via the Perron root of matrix majorants. *Syst. Control Lett.*, 19:439–449, 1992.

[98] N. G. Chetaev. *The Stability of Motion*. Pergamon Press, 1961.

[99] F. Chorlton. *Textbook of Dynamics*. Ellis Horwood, Chichester, 2nd edition, 1983.

[100] J. A. Cima and W. T. Ross. *The backward shift on the Hardy space*, volume 79 of *Mathematical Surveys and Monographs*. American Mathematical Society, Providence, RI, 2000.

[101] C. W. Clark. *Mathematical Bioeconomics: The Optimal Management of Renewable Resources*. Wiley-Interscience, New York, 1976.

[102] C. W. Clark. *Bioeconomic Modelling and Fisheries Management.* Wiley-Interscience, New York, 1985.

[103] R. N. Clark. *Control System Dynamics.* Cambridge University Press, Cambridge, 1995.

[104] F. H. Clarke, Y. S. Ledyaev, R. J. Stern, and P. R. Wolenski. *Nonsmooth Analysis and Control Theory*, volume 178 of *Graduate Texts in Mathematics.* Springer, New York, 1998.

[105] C. M. Close and D. K. Frederick. *Modeling and Analysis of Dynamic Systems.* J. Wiley, New York, 2nd edition, 1995.

[106] B. Cohen, editor. *Benjamin Franklin's Experiments.* Harvard University Press, 1941.

[107] A. Cohn. Über die Anzahl der Wurzeln einer algebraischen Gleichung in einem Kreise. *Math. Zeitschrift*, 14:110–148, 1922.

[108] F. Colonius and W. Kliemann. *The Dynamics of Control.* Birkhäuser, Boston, 2000.

[109] J. B. Conway. *Functions of One Complex Variable.* Springer-Verlag, New York, 2nd edition, 1978.

[110] J. B. Conway. *Functions of One Complex Variable. II.* Springer-Verlag, New York, 1995.

[111] W. A. Coppel. *Stability and Asymptotic Behavior of Differential Equations.* D. C. Heath and Company, Boston, 1965.

[112] R. Courant and D. Hilbert. *Methods of Mathematical Physics.* Interscience, New York, 1953.

[113] J. B. Cruz Jr., J. S. Freudenberg, and D. P. Looze. A relationship between sensitivity and stability of multivariable feedback systems. *IEEE Trans. Automat. Contr.*, AC-26:66–74, 1981.

[114] J. B. Cruz Jr., D. P. Looze, and W. R. Perkins. Sensitivity analysis of non-linear feedback-systems. *J. Franklin Inst.*, 312(3–4):199–215, 1981.

[115] R. F. Curtain and A. J. Pritchard. *Infinite-Dimensional Linear Systems Theory.* Number 8 in Lecture Notes in Control and Information Sciences. Springer-Verlag, Berlin, 1978.

[116] R. F. Curtain and H. J. Zwart. *An Introduction to Infinite-Dimensional Linear Systems Theory.* Number 21 in Texts in Applied Mathematics. Springer-Verlag, New York, 1995.

[117] G. Dahlquist. Stability and error bounds in the numerical integration of ordinary differential equations. *Transactions Royal Inst. of Technology*, 130, 1959.

[118] J. L. Daleckiĭ and M. G. Kreĭn. *Stability of Solutions of Differential Equations in Banach Spaces.* Number 43 in Translations of Mathematical Monographs. American Mathematical Society, Providence, RI, 1974.

[119] R. Datko. Extending a theorem of A. M. Liapunov to Hilbert space. *J. Math. Anal. Appl.*, 32:610–616, 1970.

[120] R. Datko. Uniform asymptotic stability of evolutionary processes in a Banach space. *SIAM J. Math. Anal.*, 3:428–445, 1972.

[121] B. N. Datta. *Numerical Linear Algebra and Applications.* Brooks/Cole Publishing Co., Pacific Grove, CA, 1995.

[122] B. N. Datta, editor. *Applied and Computational Control, Signals, and Circuits*, volume 1. Birkhäuser, Basel, 1999.

[123] B. N. Datta. Stability and inertia. *Lin. Alg. Appl.*, 302–303:563–600, 1999.

[124] C. A. de Coulomb. Première Mémoire sur l'Électricité et Magnétisme. *Histoire de l'Académie Royale Sciences*, 569, 1785.

[125] D. F. Delchamps. *State Space and Input–Output Linear Systems.* Springer-Verlag, New York, 1988.

[126] J. W. Demmel. *Applied Numerical Linear Algebra.* SIAM Publications, Philadelphia, 1997.

[127] A. P. Dempster. *Elements of Continuous Multivariate Analysis.* Addison-Wesley, Reading, MA, 1969.

[128] J. T. Desaguliers. Some thoughts and experiments concerning electricity. *Philos. Trans. Roy. Soc. London*, 41:186–210, 1739.

[129] C. A. Desoer and W. S. Chan. The feedback interconnection of lumped linear time-invariant systems. *J. Franklin Inst.*, 300:335–351, 1975.

[130] C. A. Desoer and M. Vidyasagar. *Feedback Systems: Input-Output Properties.* Academic Press, New York, 1975.

[131] R. L. Devaney. *An Introduction to Chaotic Dynamical Systems.* Addison-Wesley, Redwood City, CA, 2nd edition, 1989.

[132] J. Dieudonné. *Treatise of Modern Analysis*, volume 2. Academic Press, New York, 1970.

[133] G. Doetsch. *Introduction to the Theory and Application of the Laplace Transformation.* Springer-Verlag, Berlin, 1974.

[134] P. Dorato and R. K. Yedavalli, editors. *Recent Advances in Robust Control.* IEEE Press, New York, 1990.

[135] J. C. Doyle. Analysis of feedback systems with structured uncertainties. *IEE Proc., Part D*, 129:242–250, 1982.

[136] J. C. Doyle and G. Stein. Multivariable feedback design: concepts for a classical/modern synthesis. *IEEE Trans. Automat. Contr.*, AC-26:4–16, 1981.

[137] J. C. Doyle, J. E. Wall, and G. Stein. Performance and robustness analysis for structured uncertainty. In *Proc. 21st IEEE Conf. Decis. Control*, pages 629–636, 1982.

[138] R. D. Driver. *Ordinary and Delay Differential Equations.* Number 20 in Applied Mathematical Sciences. Springer-Verlag, New York, 1977.

[139] R. J. Duffin. Algorithms for classical stability problems. *SIAM Review*, 11:196–213, 1969.

[140] G. E. Dullerud. *Control of Uncertain Sampled-Data Systems.* Systems and Control: Foundations and Applications. Birkhäuser, Basel, 1996.

[141] G. E. Dullerud and F. Paganini. *A Course in Robust Control Theory*, volume 36 of *Texts in Applied Mathematics.* Springer-Verlag, New York, 2000.

[142] N. Dunford and J. T. Schwartz. *Linear Operators, Part I: General Theory.* Interscience, New York, 1958.

[143] N. Dunford and J. T. Schwartz. *Linear Operators, Part II: Spectral Theory.* Interscience, New York, 1963.

[144] P. L. Duren. *Theory of H^p Spaces*, volume 38 of *Pure and Applied Mathematics.* Academic Press, New York and London, 1970.

[145] C. Eckart and G. Young. The approximation of one matrix by another of lower rank. *Psychometrika*, 1:211–218, 1936.

[146] L. Edelstein-Keshet. *Mathematical Models in Biology.* Mathematics Series. Random House, New York, 1988.

[147] S. Eilenberg. *Automata, Languages and Machines.* Number 59-A in Pure and Applied Mathematics. Academic Press, New York, London, 1974.

[148] A. El Bouhtouri, D. Hinrichsen, and A. J. Pritchard. Stability radii of discrete-time stochastic systems with respect to blockdiagonal perturbations. *Automatica*, 36:1033–1040, 2000.

[149] A. El Bouhtouri and A. J. Pritchard. Stability radii of linear systems with respect to stochastic perturbations. *Syst. Control Lett.*, 19:29–33, 1992.

[150] O. I. Elgerd. *Control System Theory.* McGraw-Hill Electrical and Electronic Engineering Series. McGraw-Hill, New York, 1967.

[151] R. S. Elliott. *Electromagnetics: History, Theory and Applications.* IEEE Press, 1993.

[152] K.-J. Engel and R. Nagel. *One-Parameter Semigroups for Linear Evolution Equations.* Number 194 in Graduate Texts in Mathematics. Springer-Verlag, Berlin, 2000.

[153] N. D. Evans and A. J. Pritchard. A control theoretic approach to containing the spread of rabies. *IMA J. Math. Appl. Med. Biol.*, 18(1):1–23, 2001.

[154] E. D. Fabricius. *Modern Digital Design and Switching Theory.* CRC Press, Boca Raton, Florida, 1992.

[155] S. Faedo. Un nuovo problema di stabilità per le equazioni algebriche a coefficienti reali. *Ann. Scuola Norm. Sup. Pisa, Sci. Fis. Mat., III. Ser. 7*, pages 53–63, 1953.

[156] A. T. Fam. The volume of the coefficient space stability domain of monic polynomials. In *Proc. IEEE ISCAS*, pages 1780–1783, 1989.

[157] A. T. Fam and J. S. Meditch. A canonical parameter space for linear systems design. *IEEE Trans. Automat. Contr.*, AC-23:454–458, 1978.

[158] M. K. H. Fan and A. L. Tits. Characterization and efficient computation of the structured singular value. *IEEE Trans. Automat. Contr.*, AC-31:734–743, 1986.

[159] M. Faraday. Faraday's diary, being entries in his laboratory notebook in 1821. Bell and Sons Ltd. London, 1932.

[160] P. Fatou. Sur les équations fonctionelles. *Bulletin de la Societé Mathématique de France*, 47:161-271, 48:33–94 and 208–314, 1919, 1920.

[161] R. P. Feynman, R. B. Leighton, and M. Sands. *The Feynman Lectures on Physics*, volume 1. Addison-Wesley, Reading, MA, 1975.

[162] E. Fischer. Über quadratische Formen mit reellen Koeffizienten. *Monatsheft für Mathematik und Physik*, 16:234–249, 1905.

[163] W. Fischer and I. Lieb. *Funktionentheorie.* Vieweg, 6th edition, 1992.

[164] G. B. Folland. *Fourier Analysis and its Applications.* Wadsworth & Brooks/Cole, Pacific Grove, CA, 1992.

[165] B. A. Francis. The optimal linear-quadratic time-invariant regulator with cheap control. *IEEE Trans. Automat. Contr.*, AC-24(4):616–621, 1979.

[166] J. G. F. Francis. The QR transformation. A unitary analogue to the LR transformation, part I. *Comput. J.*, 4:265–272, 1961.

[167] J. G. F. Francis. The QR transformation, part II. *Comput. J.*, 5:332–345, 1962.

[168] G. F. Franklin, J. D. Powell, and A. Emami-Naeini. *Feedback Control of Dynamic Systems.* Addison-Wesley, Reading, MA, 1986.

[169] G. F. Franklin, J. D. Powell, and M. L. Workman. *Digital Control of Dynamic Systems.* Addison-Wesley, Reading, MA, 3rd edition, 1998.

[170] G. F. Franklin and J. R. Ragazzini. *Sampled-Data Control Systems.* McGraw-Hill, New York, 1958.

[171] A. Freeman. *English Translation of J. B. Fourier's 'Théorie Analytique de la Chaleur'.* Dover Publications, New York, 1955.

[172] P. A. Fuhrmann. *Linear Systems and Operators in Hilbert space.* McGraw-Hill, New York, 1981.

[173] P. A. Fuhrmann. *A Polynomial Approach to Linear Algebra.* Springer-Verlag, 1996.

[174] P. A. Fuhrmann and P. S. Krishnaprasad. Towards a cell decomposition for rational functions. *IMA Journal of Mathematical Control & Information*, 3:137–150, 1986.

[175] M. Fujiwara. Über die algebraischen Gleichungen, deren Wurzeln in einem Kreise oder in einer Halbebene liegen. *Math. Zeitschrift*, 24:161–169, 1926.

[176] A. T. Fuller. The early development of control theory. *Trans. ASME, J. Dyn. Syst. Meas. Contr.*, 89:109–118, 224–235, 1976.

[177] K. Furuta, H. Kajiwara, and R. Kosuge. Digital control of a double inverted pendulum on an inclined rail. *Int. J. Control*, 32(5):907–924, 1980.

[178] E. Gallestey. Computing spectral value sets using the subharmonicity of the norm of rational matrices. *BIT*, 38(1):22–33, 1998.

[179] E. Gallestey. *Theory and Numerics of Spectral Value Sets.* PhD thesis, University of Bremen, Germany, October 1998.

[180] E. Gallestey, D. Hinrichsen, and A. J. Pritchard. Spectral value sets of closed linear operators. *Proc. R. Soc. Lond. Ser. A*, 456:1397–1418, 2000.

[181] G. Gandolfo. *Economic Dynamics: Methods and Models.* North-Holland, Amsterdam, New York, Oxford, 2nd edition, 1980.

[182] F. R. Gantmacher. *The Theory of Matrices (Vol. I).* Chelsea, New York, 1959.

[183] F. R. Gantmacher. *The Theory of Matrices (Vol. II).* Chelsea, New York, 1959.

[184] F. R. Gantmacher. *Lectures in Analytical Mechanics.* Mir Publishers, Moscow, 1975.

[185] G. F. Gause. *The Struggle for Existence.* Dover Publications, New York, 1964.

[186] S. A. Gershgorin. Über die Abgrenzung der Eigenwerte einer Matrix. *Izvestia Akad. Nauk SSSR, Ser. Fis-Mat.*, 6:749–754, 1931.

[187] M. I. Gil'. *Stability of Finite and Infinite Dimensional Systems.* Kluwer Academic Publishers, Boston/Dordrecht/London, 1998.

[188] K. Glover. All optimal Hankel-norm approximations of linear multivariable systems and their L_∞-error bounds. *Int. J. Control*, 39(6):1115–1193, 1984.

[189] S. K. Godunov. Spectral portraits of matrices and criteria of spectrum dichotomy. In L. Atanassova et al., editors, *Computer arithmetic and enclosure methods. Proceedings of the 3rd International IMACS-GAMM Symposium on Computer Arithmetic and Scientific Computing (SCAN-91)*, pages 25–35, Amsterdam, 1992. North-Holland.

[190] S. K. Godunov, O. P. Kiriljuk, and W. I. Kostin. Spectral portraits of matrices. Technical Report 3, Inst. of Math. Acad. Sci. USSR, Soviet Union, 1990.

[191] I. C. Gohberg and M. G. Kreĭn. *Introduction to the Theory of Linear Nonselfadjoint Operators.* Number 18 in Transl. Math. Monogr. American Mathematical Society, Providence, RI, 1969.

[192] D. Goldberg. What every computer scientist should know about floating-point arithmetic. *ACM Computing Surveys*, 23(1):5–48, 1991.

[193] R. R. Goldberg. *Fourier Transforms.* Cambridge University Press, Cambridge, 1965.

[194] H. Goldstein. *Classical Mechanics.* Addison-Wesley, Reading, MA, 2nd edition, 1980.

[195] H. H. Goldstine. *A History of Numerical Analysis from the 16th through the 19th Century.* Springer-Verlag, New York-Heidelberg-Berlin, 1977.

[196] G. Golub and W. Kahan. Calculating the singular values and pseudo-inverse of a matrix. *SIAM J. Numer. Anal.*, 2:205–224, 1965.

[197] G. H. Golub and C. F. van Loan. *Matrix Computations.* Number 3 in Johns Hopkins series in the mathematical sciences. Johns Hopkins University Press, Baltimore, MD, 3rd edition, 1996.

[198] G. H. Golub and J. H. Wilkinson. Ill-conditioned eigensystems and the computation of the Jordan canonical form. *SIAM Review*, 18:578–619, 1976.

[199] H. González. Estabilidad absoluta de sistemas de segundo orden. In *Proc. Primero Simposio Acerca del Desarollo de la Matematica*, La Habana, Cuba, 1991.

[200] R. M. Goodwin. Dynamic coupling with especial reference to markets having production lags. *Econometrica*, 15:181–204, 1947.

[201] S. Gray. Several experiments concerning electricity. *Philos. Trans. Roy. Soc. London*, 37:18–44, 1731.

[202] D. Green. *Modern Logic Design.* Electronic Systems Engineering Series. Addison-Wesley, Wokingham, UK, 1986.

[203] L. Grüne. *Asymptotic Behavior of Dynamical and Control Systems under Perturbation and Discretization*, volume 1783 of *Lecture Notes in Mathematics.* Springer-Verlag, Berlin, 2002.

[204] L. Grüne and F. Wirth. On the rate of convergence of infinite horizon discounted optimal value functions. *Nonlinear Anal., Real World Appl.*, 1:499–515, 2000.

[205] L. H. Gustavsson. Energy growth of three-dimensional disturbances in plane Poiseuille flow. *J. Fluid Mech.*, 224:241–260, 1991.

[206] S. Gutman. *Root Clustering in Parameter Space.* Number 141 in Lecture Notes in Control and Information Sciences. Springer-Verlag, Berlin, 1990.

[207] W. Hahn. Über die Anwendung der Methode von Liapunov auf Differenzengleichungen. *Mathematische Annalen*, 136:430–441, 1958.

[208] W. Hahn. *Theory and Application of Liapunov's Direct Method.* Prentice Hall, Englewood Cliffs, NJ, 1963.

[209] W. Hahn. *Stability of Motion.* Springer-Verlag, Berlin, 1967.

[210] E. Hairer, S. P. Nørsett, and G. Wanner. *Solving Ordinary Differential Equations I, Nonstiff Problems*, volume 8 of *Springer Series in Computational Mathematics.* Springer-Verlag, Berlin, 2nd edition, 1993.

[211] E. Hairer and G. Wanner. *Solving ordinary differential equations II: Stiff and differential-algebraic problems*, volume 14 of *Springer Series in Computational Mathematics.* Springer-Verlag, Berlin, 2nd edition, 1996.

[212] J. K. Hale. Dynamical systems and stability. *Journal of the Institute of Mathematics and its Applications*, 26:39–59, 1969.

[213] J. K. Hale. *Theory of Functional Differential Equations.* Number 3 in Applied Mathematical Sciences. Springer-Verlag, New York, 2nd edition, 1977.

[214] J. K. Hale. *Ordinary Differential Equations*, volume XXI of *Pure and Applied Mathematics.* Wiley-Interscience, New York, 2nd edition, 1980.

[215] P. Halmos. Quasitriangular operators. *Acta Sci. Math. (Szeged.)*, 29:283–293, 1968.

[216] P. C. Hansen. *Rank-Deficient and Discrete Ill-Posed Problems. Numerical Aspects of Linear Inversion*, volume 4 of *SIAM Monographs on Mathematical Modeling and Computation*. SIAM Publications, Philadelphia, PA, 1997.

[217] P. Hartman. *Ordinary Differential Equations*. Birkhäuser, Basel, 2nd edition, 1982.

[218] F. Hausdorff. Der Wertvorrat einer Bilinearform. *Math. Zeitschrift*, 3:314–316, 1919.

[219] G. Heinig and K. Rost. *Algebraic Methods for Toeplitz-like Matrices and Operators*. Birkhäuser, Basel, 1984.

[220] U. Helmke and P. A. Fuhrmann. Bezoutians. *Lin. Alg. Appl.*, 122–124:1039–1097, 1989.

[221] U. Helmke, D. Hinrichsen, and W. Manthey. A cell decomposition of the space of real Hankels of rank $\leq n$ and some applications. *Lin. Alg. Appl.*, 122–124:331–355, 1989.

[222] J. W. Helton. A numerical method for computing the structured singular value. *Syst. Control Lett.*, 10:21–26, 1988.

[223] P. Henrici. Bounds for iterates, inverses, spectral variation and fields of values of non-normal matrices. *Numer. Math.*, 4:24–40, 1962.

[224] P. Henrici. *Applied and Computational Complex Analysis*, volume 1 and 2. John Wiley & Sons, New York-London-Sydney-Toronto, 1974 and 1977.

[225] G. Herglotz. Über die Wurzelanzahl algebraischer Gleichungen innerhalb und auf dem Einheitskreis. *Math. Zeitschrift*, 19:26–34, 1923.

[226] C. Hermite. Sur le nombre de racines d'une équation algébrique comprise entre des limites données. *Journal für Reine und Angewandte Mathematik*, 52:39–51, 1856.

[227] J. R. Hicks. *A Contribution to the Theory of the Trade Cycle*. Oxford University Press, 1950.

[228] J. R. Higgins. *Sampling Theory in Fourier and Signal Analysis: Foundations*. Clarendon Press, Oxford, 1996.

[229] D. J. Higham and N. J. Higham. *MATLAB Guide*. SIAM Publications, Philadelphia, PA, 2002.

[230] N. J. Higham. *Accuracy and Stability of Numerical Algorithms*. SIAM Publications, Philadelphia, PA, 2nd edition, 2002.

[231] E. Hille and R. S. Phillips. *Functional Analysis and Semigroups*. American Mathematical Society, Providence, RI, 1957.

[232] D. Hinrichsen, A. Ilchmann, and A. J. Pritchard. Robustness of stability of time-varying linear systems. *J. Diff. Eqns.*, 82(2):219–250, 1989.

[233] D. Hinrichsen and B. Kelb. Spectral value sets: A graphical tool for robustness analysis. *Syst. Control Lett.*, 21:127–136, 1993.

[234] D. Hinrichsen and B. Kelb. Stability radii and spectral value sets for real matrix perturbations. In U. Helmke, R. Mennicken, and J. Saurer, editors, *Systems and Networks: Mathematical Theory and Applications. Volume II: Invited and Contributed Papers*, pages 217–220. Akademie-Verlag, Berlin, 1994.

[235] D. Hinrichsen, B. Kelb, and A. Linnemann. An algorithm for the computation of the structured complex stability radius. *Automatica*, 25:771–775, 1989.

[236] D. Hinrichsen and V. L. Kharitonov. Stability of polynomials with conic uncertainties. *Math. of Control, Signals and Systems*, 8:97–117, 1995.

[237] D. Hinrichsen and W. Manthey. The Bruhat parametrization of infinite real Hankel matrices of rank $\leq$ n. In *Proc. 25th IEEE Conf. Decis. Control*, volume II, pages 527–529, Athens, 1986.

[238] D. Hinrichsen and W. Manthey. On a cell decomposition for Hankel matrices and rational functions. *Journal für Reine und Angewandte Mathematik*, 451:15–50, 1994.

[239] D. Hinrichsen and M. Motscha. Optimization problems in the robustness analysis of linear state space systems. In A. Gomez, M. A. Jimenez, and G. Lopez, editors, *Approximation and Optimization*, number 1354 in Lecture Notes in Mathematics, pages 54–78. Springer-Verlag, Berlin, 1988.

[240] D. Hinrichsen, E. Plischke, and A. J. Pritchard. Liapunov and Riccati equations for practical stability. In *Proc. 6th European Control Conf. 2001, Porto, Portugal*, pages 2883–2888, 2001. Paper no. 8485 (CDROM).

[241] D. Hinrichsen, E. Plischke, and F. Wirth. State feedback stabilization with guaranteed transient bounds. In *Proc. MTNS-2002*, Notre Dame, Indiana, 2002. Paper no. 2132 (CDROM).

[242] D. Hinrichsen and A. J. Pritchard. Stability radii of linear systems. *Syst. Control Lett.*, 7:1–10, 1986.

[243] D. Hinrichsen and A. J. Pritchard. Stability radius for structured perturbations and the algebraic Riccati equation. *Syst. Control Lett.*, 8:105–113, 1986.

[244] D. Hinrichsen and A. J. Pritchard. On the robustness of root locations of polynomials under complex and real perturbations. In *Proc. 27th IEEE Conf. Decis. Control, Austin, Texas*, pages 1410–1414, 1988.

[245] D. Hinrichsen and A. J. Pritchard. On the robustness of root locations of polynomials under complex and real perturbations. Report 147, Control Theory Centre, University of Warwick, 1988.

[246] D. Hinrichsen and A. J. Pritchard. An application of state space methods to obtain explicit formulae for robustness measures of polynomials. In M. Milanese et al., editors, *Robustness in Identification and Control*, Proc. Internat. Workshop Torino 1988, pages 183–206. Plenum Press, 1989.

[247] D. Hinrichsen and A. J. Pritchard. A note on some differences between real and complex stability radii. *Syst. Control Lett.*, 14:401–409, 1990.

[248] D. Hinrichsen and A. J. Pritchard. Real and complex stability radii: A survey. In D. Hinrichsen and B. Mårtensson, editors, *Control of Uncertain Systems*, pages 119–162. Birkhäuser, Basel, 1990.

[249] D. Hinrichsen and A. J. Pritchard. On the robustness of stable discrete time linear systems. In *New Trends in Systems Theory (Proc. Conf. Genova, 1990)*, pages 393–400. Birkhäuser, Basel, 1991.

[250] D. Hinrichsen and A. J. Pritchard. Destabilization by output feedback. *Differ. Integral Equ.*, 5(2):357–386, 1992.

[251] D. Hinrichsen and A. J. Pritchard. On spectral variations under bounded real matrix perturbations. *Numer. Math.*, 60:509–524, 1992.

[252] D. Hinrichsen and A. J. Pritchard. Robustness measures for linear systems with application to stability radii of Hurwitz and Schur polynomials. *Int. J. Control*, 55(4):809–844, 1992.

[253] D. Hinrichsen and A. J. Pritchard. Robust stability of linear evolution operators on Banach spaces. *SIAM J. Control Optim.*, 32:1503–1541, 1994.

[254] D. Hinrichsen and A. J. Pritchard. Stability radii of systems with stochastic uncertainty and their optimization by output feedback. *SIAM J. Control Optim.*, 34:1972–1998, 1996.

[255] D. Hinrichsen and N. K. Son. Stability radii of linear discrete-time systems and symplectic pencils. *Int. J. Robust & Nonlinear Control*, 1:79–97, 1991.

[256] D. Hinrichsen and N. K. Son. μ-analysis and robust stability of positive linear systems. *Appl. Math. and Comp. Sci.*, 8:253–268, 1998.

[257] D. Hinrichsen and N. K. Son. Stability radii of positive discrete-time systems under affine parameter perturbations. *Int. J. Robust & Nonlinear Control*, 8:1169–1188, 1998.

[258] W. M. Hirsch and S. Smale. *Differential Equations, Dynamical Systems and Linear Algebra.* Academic Press, New York, 1974.

[259] J. Hofbauer and K. Sigmund. *The Theory of Evolution and Dynamical Systems: Mathematical Aspects of Selection*, volume 7 of *London Mathematical Society Student Texts.* Cambridge University Press, Cambridge, 1988.

[260] K. Hoffman. *Banach Spaces of Analytic Functions.* Prentice Hall, London, 1962.

[261] C. V. Hollot and A. C. Bartlett. Some discrete-time counterparts to Kharitonov's stability criterion for uncertain systems. *IEEE Trans. Automat. Contr.*, AC-31:355–356, 1986.

[262] F. Hoppensteadt. *Mathematical Methods of Population Biology.* Number 4 in Cambridge Studies in Mathematical Biology. Cambridge University Press, Cambridge, 1982.

[263] F. C. Hoppensteadt and C. S. Peskin. *Mathematics in Medicine and the Life Sciences.* Springer-Verlag, New York, 1992.

[264] R. A. Horn and C. R. Johnson. *Matrix Analysis.* Cambridge University Press, Cambridge, 1990.

[265] R. A. Horn and C. R. Johnson. *Topics in Matrix Analysis.* Cambridge University Press, Cambridge, 1991.

[266] H. Hotelling. Analysis of a complex of statistical variables into principle components. *J. Educ. Psych.*, 24:417–441,498–520, 1933.

[267] H. Hotelling. Simplified calculation of principal components. *Psychometrika*, 1:27–35, 1936.

[268] A. S. Householder. *The Numerical Treatment of a Single Nonlinear Equation.* McGraw-Hill, New York, 1970.

[269] A. S. Householder. *The Theory of Matrices in Numerical Analysis.* Dover Publications, New York, 2nd edition, 1975.

[270] D. A. Huffman. The synthesis of sequential switching circuits. *J. Franklin Inst.*, 257:161–190, 275–303, 1954.

[271] E. V. Huntington. Sets of independent postulates for the algebra of logic. *Trans. Am. Math. Soc.*, 5:288–309, 1904.

[272] A. Hurwitz. Über die Bedingungen, unter welchen eine Gleichung nur Wurzeln mit negativen reellen Teilen besitzt. *Mathematische Annalen*, 46:273–284, 1895.

[273] D. C. Hyland and E. G. Collins Jr. An M-matrix and majorant approach to robust stability and performance analysis for systems with structured uncertainty. *IEEE Trans. Automat. Contr.*, 34:699–710, 1989.

[274] I. S. Iohvidov. *Hankel and Toeplitz Matrices and Forms: Algebraic Theory.* Birkhäuser, Boston-Basel-Stuttgart, 1977.

[275] A. Isidori. *Nonlinear Control Systems.* Communication and Control Engineering Series. Springer-Verlag, New York, 2nd edition, 1989.

[276] B. Jacob. Stability radius for evolution operators with respect to dynamical perturbations. In *Proc. 3rd European Control Conf. 1995*, pages 3298–3303, Roma, 1995.

[277] B. Jacob. *Time-Varying Infinite Dimensional State-Space Systems.* PhD thesis, Universität Bremen, Bremen, 1995.

[278] D. E. Johnson, J. R. Johnson, and J. L. Hilburn. *Electric Circuit Analysis.* Prentice-Hall, Englewood Cliffs, NJ, 2nd edition, 1992.

[279] G. A. Jones and D. Singerman. *Complex Functions: An Algebraic and Geometric Viewpoint.* Cambridge University Press, Cambridge, 1987.

[280] C. Jordan. Mémoire sur les formes bilinéares. *Journal de Mathématiques Pures et Appliquées Deuxième Série*, 19:35–54, 1874.

[281] J. P. Joule. On the heat evolved by metallic conductors of electricity. Phil. Mag., 19:260–265, 1841.

[282] G. Julia. Mémoire sur l'itération des fonctions rationelles. *Journal de Mathématiques Pures et Appliques*, 8.1:47–245, 1918.

[283] E. I. Jury. *Sampled Data Control Systems.* John Wiley & Sons, New York, London, 1958.

[284] E. I. Jury. *Inners and Stability of Dynamic Systems.* Robert E. Krieger Publishing Company, Malabar, FL, 2nd edition, 1982.

[285] T. Kaczorek. *Two-Dimensional Linear Systems*, volume 68 of *Lecture Notes in Control and Information Sciences.* Springer-Verlag, Berlin, 1985.

[286] R. E. Kalman. Nonlinear aspects of sampled-data control-systems. In *Proc. Sympos. Nonlinear Circuit Analysis*, volume 6, pages 273–313, 1957.

[287] R. E. Kalman. Mathematical description of linear dynamical systems. *SIAM J. Control Optim.*, 1:152–192, 1963.

[288] R. E. Kalman. On the Hermite-Fujiwara theorem in stability theory. *Quarterly of Applied Mathematics*, 23:279–282, 1965.

[289] R. E. Kalman. Algebraic characterization of polynomials whose zeros lie in algebraic domains. *Proc. Nat. Acad. Sci. U.S.A.*, 664:818–823, 1969.

[290] R. E. Kalman, P. L. Falb, and M. A. Arbib. *Topics in Mathematical System Theory.* McGraw-Hill, New York, 1969.

[291] M. Karow. Spectral value sets for structured matrix perturbations. In *Proc. MTNS-2000*, Perpignan, France, 2000. (CDRom).

[292] M. Karow. *Geometry of Spectral Value Sets.* PhD thesis, University of Bremen, Germany, 2003.

[293] T. Kato. *Perturbation Theory for Linear Operators.* Springer-Verlag, Heidelberg, 2nd edition, 1980.

[294] T. Kato. *A Short Introduction to Perturbation Theory for Linear Operators.* Springer-Verlag, New York-Heidelberg-Berlin, 1982.

[295] A. Katok and B. Hasselblatt. *Introduction to the Modern Theory of Dynamical Systems*, volume 54 of *Encyclopedia of Mathematics and its Applications.* Cambridge University Press, Cambridge, 1995.

[296] R. H. Katz. *Contemporary Logic Design.* Benjamin/Cummings, Redwood City, CA, 1994.

[297] Y. Katznelson. *An Introduction to Harmonic Analysis.* Wiley, 1968.

[298] W. G. Kelley and A. C. Peterson. *Difference Equations. An Introduction with Applications.* Harcourt/Academic Press, San Diego, CA, 2nd edition, 2001.

[299] H. K. Khalil. *Nonlinear Systems.* Prentice-Hall, Upper Saddle River, NJ, 2nd edition, 1996.

[300] V. L. Kharitonov. The asymptotic stability of the equilibrium state of a family of systems of linear differential equations. *Diff. Uravn.*, 14(11):2086–2088, 1978. (in Russian).

[301] V. L. Kharitonov. Asymptotic stability of an equilibrium position of a family of systems of linear differential equation. *J. Diff. Eqns.*, 14:1483–1485, 1979.

[302] V. L. Kharitonov. Distribution of the roots of the characteristic polynomial of an autonomous system. *Avtomatica i Telemekhanica*, 5:42–47, 1981.

[303] V. L. Kharitonov. Interval stability of quasipolynomials. In S. P. Bhattacharyya and L. H. Keel, editors, *Control of Uncertain Systems*, pages 439–446. CRC Press, Littleton, MA, 1991.

[304] V. L. Kharitonov and D. Hinrichsen. On convex directions for stable polynomials. *Autom. Rem. Control*, 58(3):394–402, 1997.

[305] V. L. Kharitonov and R. Tempo. On the stability of a weighted diamond of real polynomials. *Syst. Control Lett.*, 22:5–7, 1994.

[306] V. L. Kharitonov and A. P. Zhabko. Robust stability of time-delay systems. *IEEE Trans. Automat. Contr.*, 39:2388–2397, 1994.

[307] B. Khoussainov and A. Nerode. *Automata Theory and its Applications*, volume 21 of *Progress in Computer Science and Applied Logic.* Birkhäuser, Boston, MA, 2001.

[308] D. Y. Khusainov, Y. A. Komarov, and Y. A. Yun'kova. Constructing optimal Lyapunov functions for linear differential equations. *Sov. Autom. Control*, 17 (6):80–83, 1984.

[309] V. C. Klema and A. J. Laub. The singular value decomposition: Its computation and some applications. *IEEE Trans. Automat. Contr.*, AC-25:164–176, 1980.

[310] M. Knebusch and C. Sandberg. *Einführung in die reelle Algebra.* Fried. Vieweg & Sohn, Braunschweig/Wiesbaden, 1989.

[311] K. Knopp. *Theory of Functions, volume I and II.* Dover, New York, 1945 and 1947.

[312] A. Kolmogorov and S. Fomin. *Elements of the Theory of Functions and Functional Analysis I/II.* Greylock Press, Rochester, NY, 1957/61. Reprinted in a single volume by Dover Publications, New York, 1999.

[313] A. N. Kolmogorov. Sulla teoria di Volterra della lotta per l'esistenza. *Giorn. Instituto Ital. Attuari*, 7:74–80, 1936.

[314] T. W. Körner. *Fourier Analysis.* Cambridge University Press, Cambridge, 1988.

[315] V. A. Kostitzin. *Symbiose, Parasitisme et Évolution.* Étude mathématique. Hermann, Paris, 1934.

[316] N. N. Krasovskii. *Stability of Motion.* Stanford University Press, 1963.

[317] H.-O. Kreiss. Über die Stabilitätsdefinition für Differenzengleichungen die partielle Differentialgleichungen approximieren. *BIT*, 2:153–181, 1962.

[318] M. G. Kreĭn and M. A. Naimark. The method of symmetric and Hermitan forms in the theory of the separation of the roots of algebraic equations. *Linear and Multilinear Algebra*, 10:265–308, 1981.

[319] E. Kreyszig. *Introductory Functional Analysis with Applications.* John Wiley & Sons, London, 1978.

[320] L. Kronecker. Zur Theorie der Elimination einer Variablen aus zwei algebraischen Gleichungen. *Monatsber. der Königl. Preuss. Akad. der Wiss., Berlin*, pages 535–600, 1881. See also Werke, Vol. II, Chelsea, New York, 1968.

[321] B. C. Kuo. *Digital Control Systems.* HRW series in Electrical and Computer Engineering. Holt, Rinehart & Winston, New York, 2nd edition, 1980.

[322] H. Kwakernaak and R. Sivan. *Modern Signals and Systems.* Prentice Hall, Englewood Cliffs, NJ, 1991.

[323] H. Kwakernaak and H. Westdijk. Regulability of a multiple inverted pendulum system. *Control Theory and Advanced Technology*, 1:1–9, 1995.

[324] C. M. Lagoa, P. S. Shcherbakov, and B. R. Barmish. Probabilistic enhancement of classical robustness margins: The unirectangularity concept. *Syst. Control Lett.*, 35:31–43, 1998.

[325] J. L. Lagrange. *Analytical Mechanics*, volume 191 of *Boston Studies in the Philosophy of Science.* Kluwer Academic Publishers, Dordrecht, 1997. Translated from the 1811 French original. Edited by Auguste Boissonnade and Victor N. Vagliente.

[326] P. Lancaster and L. Rodman. *Algebraic Riccati Equations.* Clarendon Press, Oxford, 1995.

[327] H. J. Landau. On Szegö's eigenvalue distribution theory and non-Hermitian kernels. *J. Anal. Math.*, 28:335–357, 1975.

[328] H. J. Landau. The notion of approximate eigenvalues applied to an integral equation of laser theory. *Q. Appl. Math.*, 35:165–172, 1977.

[329] L. D. Landau and E. M. Lifshitz. *Course of Theoretical Physics. Vol. 1: Mechanics.* Pergamon Press, Oxford, 3rd edition, 1976.

[330] S. Lang. *Algebra.* Addison-Wesley, Reading, MA, 1965.

[331] J. P. LaSalle and Z. Artstein. *The Stability of Dynamical Systems.* Number 25 in Regional Conference in Applied Mathematics. SIAM Publications, Philadelphia, PA, 1976.

[332] J. P. LaSalle and S. Lefschetz. *Stability by Liapunovs's Direct Method.* Academic Press, New York, 1961.

[333] A. J. Laub and A. Linnemann. Hessenberg and Hessenberg/triangular forms in linear system theory. *Int. J. Control*, 44(6):1523–1547, 1986.

[334] A. J. Laub, R. V. Patel, and P. M. van Dooren, editors. *Numerical Linear Algebra Techniques for Systems and Control.* IEEE Press, New York, 1994.

[335] A. J. Laub, R. V. Patel, and P. M. van Dooren. Numerical and computational issues in linear control and systems theory. In W. S. Levine, editor, *Control system fundamentals*, chapter 21, pages 403–418. CRC Press, Boca Raton, FL, 2000.

[336] E. B. Lee and L. Markus. *Foundations of Optimal Control Theory.* John Wiley & Sons, New York, 1967.

[337] S. Lefschetz. *Differential Equations: Geometric Theory*, volume VI of *Pure and Applied Mathematics.* Interscience, New York-London, 2nd edition, 1963.

[338] C. T. Leondes, editor. *Advances in Industrial Systems*, volume 37 of *Control and Dynamic Systems*. Academic Press, San Diego, CA, 1990.

[339] C. T. Leondes, editor. *Advances in Large Scale Systems Dynamics and Control*, volume 36 of *Control and Dynamic Systems*. Academic Press, San Diego, CA, 1990.

[340] L. V. Levantovskii. The boundary of a set of stable matrices. *Russian Mathematical Surveys*, 35:249–250, 1980.

[341] R. J. LeVeque and L. N. Trefethen. On the resolvent condition in the Kreiss matrix theorem. *BIT*, 24:584–591, 1984.

[342] W. S. Levine, editor. *The Control Handbook*. IEEE Press, Piscataway, NJ, 1996.

[343] T. Li and J. A. Yorke. Period three implies chaos. *Amer. Math. Monthly*, 82:985–992, 1975.

[344] A. Liénard and M. Chipart. Sur le signe de la partie réelle des racines d'une equation algebrique. *J. Math. Pures Appl.*, 10:291–346, 1914.

[345] Y. Lin, E. D. Sontag, and Y. Wang. A smooth converse Lyapunov theorem for robust stability. *SIAM J. Control Optim.*, 34(1):124–160, 1996.

[346] F. N. Lorenz. Deterministic nonperiodic flow. *J. Atmospheric Sci.*, 20:130–141, 1963.

[347] A. Lotka. *Elements of Mathematical Biology*. Dover, New York, 1956.

[348] S. M. Lozinskiĭ. Error estimates for numerical integration of ordinary differential equations. *Izv. Vysš. Učebn. Zaved Matematika*, 5:52–90, 1958. Errata 5:222, 1959.

[349] D. G. Luenberger. *Introduction to Dynamic Systems: Theory, Models and Applications*. John Wiley & Sons, New York, 1979.

[350] G. Lumer. Semi-inner product spaces. *Trans. Amer. Math. Soc.*, 100:29–43, 1961.

[351] G. Lumer and R. S. Phillips. Dissipative operators in a Banach space. *Pacific J. Math.*, 11:679–698, 1961.

[352] A. I. Lur'e. *Some Nonlinear Problems from the Theory of Automatic Control*. Gostekhizdat, Moscow, 1951. (Russian), German translation: Akademie-Verlag, Berlin, 1957.

[353] A. I. Lur'e and V. N. Postnikov. On the theory of the stability of controlled systems. *Prikl. Mat. i Mech. IX*, 5, 1945.

[354] A. M. Lyapunov. Problème général de la stabilité du mouvement. *Ann. Fac. Sci. Toulouse*, 9:203–474, 1907. Translation of the original paper published in 1893 in *Comm. Soc. Math. Kharkow* and reprinted as Vol. 17 in *Ann. Math Studies*, Princeton Univerity Press, Princeton, NJ, 1949.

[355] A. G. J. MacFarlane. *Dynamical System Models*. Harrap, London, 1970.

[356] A. G. J. MacFarlane. The development of frequency-response methods in automatic control. *IEEE Trans. Automat. Contr.*, AC-24:250–265, 1979. Reprinted in: A. G. J. MacFarlane (editor), *Frequency-Response Methods in Control Systems*, IEEE Press, New York, 1979, 1–16.

[357] W. Maletinski, M. F. Senning, and F. Wiederkehr. Observer based control of a double pendulum. In *Proc. 8th Triennial World Congress of IFAC*, Oxford, 1982. Pergamon.

[358] T. R. Malthus. *An essay on the principle of population, as it affects the future improvement of society, with remarks on the speculations of Mr. Godwin, M. Condorcet, and other writers*. Macmillan, London, 1926. Reprint of the original book, London 1798.

[359] M. Marden. *The Geometry of the Zeros of a Polynomial in a Complex Variable.* American Mathematical Society, New York, 1949.

[360] M. Marden. *Geometry of Polynomials.* American Mathematical Society, Providence RI, 1966.

[361] J. E. Marsden and T. S. Ratiu. *Introduction to Mechanics and Symmetry. A Basic Exposition of Classical Mechanical Systems*, volume 17 of *Texts in Applied Mathematics.* Springer-Verlag, New York, 2nd edition, 1999.

[362] J. E. Marsden and A. J. Tromba. *Vector Calculus.* W. H. Freeman, New York, 4th edition, 1996.

[363] J. C. Maxwell. A dynamical theory of the electromagnetic field. *Philos. Trans. Roy. Soc. London*, 155(450), 1865.

[364] J. C. Maxwell. On governors. *Proc. R. Soc. Lond.*, 16:270–283, 1868.

[365] J. C. Maxwell, editor. *The Scientific Papers of the Honourable Henry Cavendish, revised by J. Larmor.* Cambridge University Press, Cambridge, 1921.

[366] R. M. May. *Stability and Complexity in Model Ecosystems.* Princeton University Press, Princeton, NJ, 2nd edition, 1974.

[367] R. M. May. Simple mathematical models with very complicated dynamics. *Nature*, 261:459–467, 1976.

[368] R. M. May. *Theoretical Ecology: Principles and Applications.* Blackwell, Oxford, 2nd edition, 1981.

[369] N. H. McClamroch. *State Models of Dynamic Systems.* Springer-Verlag, New York-Heidelberg-Berlin, 1980.

[370] C. Mead and L. Conway. *Introduction to VLSI Design.* Addison-Wesley, Reading, MA, 1980.

[371] G. H. Mealy. A method for synthesizing sequential circuits. *Bell System Tech. J.*, 34:1045–1079, 1955.

[372] V. Mehrmann. *The Autonomous Linear Quadratic Control Problem, Theory and Numerical Solution.* Number 163 in Lecture Notes in Control and Information Sciences. Springer-Verlag, Heidelberg, 1991.

[373] J. Michell. *A Treatise of Artificial Magnets.* London, 1750.

[374] M. Milanese, R. Tempo, and A. Vicino, editors. *Robustness in Identification and Control.* Plenum Press, New York, 1989.

[375] R. K. Miller and A. N. Michel. *Ordinary Differential Equations.* Academic Press, New York, 1982.

[376] R. J. Minnichelli, J. J. Anagnost, and C. A. Desoer. An elementary proof of Kharitonov's stability theorem with extensions. *IEEE Trans. Automat. Contr.*, AC-34:995–998, 1989.

[377] L. Mirsky. Results and problems in the theory of doubly-stochastic matrices. *Z. Wahrsch. verw. Gebiete*, 1:319–334, 1963.

[378] C. B. Moler and C. F. van Loan. Nineteen dubious ways to compute the exponential of a matrix. *SIAM Review*, 20(4):801–837, 1978.

[379] C. B. Moler and C. F. van Loan. Nineteen dubious ways to compute the exponential of a matrix – twenty-five years later. *SIAM Review*, 45(1):3–49, 2003.

[380] B. C. Moore. Principal component analysis in linear systems: controllability, observability, and model reduction. *IEEE Trans. Automat. Contr.*, AC-26:17–32, 1981.

[381] E. F. Moore. *Sequential Machines: Selected Papers.* Addison-Wesley, Reading, MA, 1964.

[382] S. Mori, H. Nishihara, and K. Furuta. Control of unstable mechanical system – control of pendulum. *Int. J. Control*, 23:673–692, 1976.

[383] T. Mori and H. Kokame. Stability of interval polynomials with vanishing extreme coefficients. In *Recent Advances in Mathematical Theory of Systems, Control Networks and Signal Processing I.*, pages 409–414. Mita Press, Tokyo, 1992.

[384] T. Morozan. Stability radii for some stochastic differential equations. *Stochastics and Stochastics Reports*, 54:281–291, 1995.

[385] J. D. Murray. *Mathematical Biology*, volume 19 of *Biomathematics*. Springer-Verlag, Berlin, 1993.

[386] R. Narasimhan. *Complex Analysis in One Variable.* Birkhäuser, 1985.

[387] K. S. Narenda and J. H. Taylor. *Frequency Criteria for Absolute Stability.* Academic Press, New York, 1973.

[388] A. W. Naylor and G. R. Sell. *Linear Operator Theory in Engineering and Science.* Holt, Rinehart & Winston, New York, 1971.

[389] V. V. Nemytskii and V. V. Stepanov. *Qualitative Theory of Differential Equations.* Princeton University Press, London, 1960.

[390] A. Neumaier. Solving ill-conditioned and singular linear systems: A tutorial on regularization. *SIAM Review*, 40(3):636–666, 1998.

[391] C. G. Newton Jr., L. A. Gould, and J. F. Kaiser. *Analytical Design of Linear Feedback Controls.* John Wiley & Sons, New York, 1957.

[392] J. Nielsen and J. Villadsen. *Bioreaction Engineering Principles.* Kluwer Academic/Plenum Press, New York, 1994.

[393] H. Nijmeijer and A. J. van der Schaft. *Nonlinear Dynamical Control Systems.* Springer-Verlag, Berlin-Heidelberg-New York, 1990.

[394] H. Nyquist. Certain topics in telegraph transmission theory. *Trans. Amer. Inst. Elec. Engineering*, 47:617–644, 1928.

[395] H. Nyquist. Regeneration theory. *Bell Syst. Techn. J.*, 11:126–147, 1932. Reprinted in R. Bellman and R. Kalaba (editors), *Selected Papers on Mathematical Trends in Control Theory*, Dover, New York, 1964, and in A. G. J. MacFarlane (editor), *Frequency-response Methods in Control Systems*, IEEE Press, New York, 1979.

[396] K. Ogata. *Discrete-time Control Systems.* Prentice Hall, New York, 1987.

[397] K. Ogata. *System Dynamics.* Prentice Hall, Englewood Cliffs, NJ, 3rd edition, 1998.

[398] G. S. Ohm. Determination of the laws whereby metals conduct contact electricity. *J. Chemie und Physik (Schweigger's Journal)*, 46(137), 1826.

[399] A. V. Oppenheim, A. S. Willsky, and S. H. Nawab. *Signals and Systems.* Prentice Hall, Upper Saddle River, NJ, 2nd edition, 1997.

[400] F. Op't Hof. Stabilitätsradien von Matrizen unter gebrochen-linearen Störungen. Master's thesis, Universität Bremen, November 1997.

[401] M. N. Ozisik. *Heat Conduction.* John Wiley & Sons, New York, 1993.

[402] A. Packard and J. C. Doyle. The complex structured singular value. *Automatica*, 29(1):71–109, 1993.

[403] R. E. A. C. Paley and N. Wiener. *Fourier Transforms in the Complex Domain.* Number XIX in A.M.S. Colloquium Publications. American Mathematical Society, New York, 1934.

[404] T. Palis and W. de Melo. *Geometric Theory of Dynamical Systems.* Springer-Verlag, New York, 1982.

[405] P. C. Parks. A new proof of the Routh-Hurwitz stability criterion using the second method of Liapunov. *Proc. Cambridge Philos. Soc.*, 58:694–702, 1962.

[406] A. Pazy. *Semigroups of Linear Operators and Applications to Partial Differential Equations.* Springer-Verlag, New York, 1983.

[407] H. O. Peitgen, D. Saupe, and F. von Haeseler. Cayley's problem and Julia sets. *Mathematical Intelligencer*, 6(2):11–20, 1984.

[408] L. Perko. *Differential Equations and Dynamical Systems.* Number 7 in Texts in Applied Mathematics. Springer-Verlag, New York, 3rd edition, 2001.

[409] A. W. Phillips. Stabilisation policy in a closed economy. *Economic Journal*, 64:290–323, 1954.

[410] F. Pichler, R. Moreno-Díaz, and P. Kopacek, editors. *Computer aided systems theory – EUROCAST '99*, volume 1798 of *Lecture Notes in Computer Science.* Springer-Verlag, Berlin, 2000.

[411] E. C. Pielou. *Mathematical Ecology.* Wiley-Interscience, New York, 2nd edition, 1977.

[412] A. Pietsch. *Eigenvalues and s-Numbers.* Cambridge University Press, Cambridge, 1987.

[413] R. Pintelon and J. Schoukens. *System Identification: A Frequency Domain Approach.* John Wiley & Sons, 2001.

[414] S. J. Plimpton and W. E. Lawton. A very accurate test of Coulomb's law of force between charges. *Phys. Rev. A*, pages 1066–1077, 1936.

[415] H. Poincaré. *Les Méthodes Nouvelles de la Méchanique Céleste.* Gauthier-Villars, Paris, 1892, 1899. (Reprint: Dover, New York, 1960).

[416] J. W. Polderman and J. C. Willems. *Introduction to Mathematical Systems Theory. A Behavioral Approach.* Springer-Verlag, New York, 1997.

[417] J. Polking, A. Boggess, and D. Arnold. *Differential Equations.* Prentice Hall, Upper Saddle River, NJ, 2001.

[418] V. M. Popov. Absolute stability of nonlinear systems of automatic control. *Autom. Remote Control*, 22:857–875, 1962.

[419] D. Prätzel-Wolters. *Feedback morphisms between linear systems — a unified approach to state space systems, transfer functions and system matrices.* PhD thesis, Institut für Dynamische Systeme, Report No. 39, Universität Bremen, Germany, 1981.

[420] A. J. Pritchard and S. Townley. Robustness of linear systems. *J. Diff. Eqns.*, 77(2):254–286, 1989.

[421] A. J. Pritchard and J. Zabczyk. Stability and stabilizability of infinite dimensional systems. *SIAM Review*, 23:25–52, 1983.

[422] F. P. Prosser and D. E. Winkel. *The Art of Digital Design, an Introduction to Top-down Design.* Prentice Hall, Englewood Cliffs, NJ, 2nd edition, 1987.

[423] L. Qiu, B. Bernhardsson, A. Rantzer, E. J. Davison, P. M. Young, and J. C. Doyle. On the real structured stability radius. In *Proceedings of the 12th IFAC World Congress*, volume 8, pages 71–78, 1993.

[424] L. Qiu, B. Bernhardsson, A. Rantzer, E. J. Davison, P. M. Young, and J. C. Doyle. A formula for computation of the real stability radius. *Automatica*, 31(6):879–890, 1995.

[425] L. Qiu and E. J. Davison. The stability robustness determination of state space models with real unstructured perturbations. *Math. Control Signals Syst.*, 4(3):247–267, 1991.

[426] L. Qiu and E. J. Davison. Bounds on the real stability radius. In *Robustness of Dynamic Systems with Parameter Uncertainties, Proc. 2nd Workshop, Ascona/Switz.*, pages 139–145. Birkhäuser, Basel, 1992.

[427] A. Rantzer. Stability conditions for polytopes of polynomials. *IEEE Trans. Automat. Contr.*, AC-37:79–89, 1992.

[428] S. C. Reddy and D. S. Henningson. Energy growth in viscous channel flows. *J. Fluid Mech.*, 252:209–238, 1993.

[429] S. C. Reddy, P. J. Schmid, and D. S. Henningson. Pseudospectra of the Orr-Sommerfeld operator. *SIAM J. Appl. Math.*, 53:15–47, 1993.

[430] R. Redheffer. On a certain linear fractional transformation. *J. Math. Phys.*, 39:269–286, 1960.

[431] L. Reichel and L. N. Trefethen. Eigenvalues and pseudo-eigenvalues of Toeplitz matrices. *Lin. Alg. Appl.*, 162/164:153–185, 1992.

[432] F. Rellich. Störungstheorie der Spektralzerlegung, I. *Mathematische Annalen*, 113:600–619, 1937.

[433] F. Rellich. *Perturbation Theory of Eigenvalue Problems.* Gordon and Breach, New York-London-Paris, 1969.

[434] J. R. Rice. A theory of condition. *SIAM J. Numer. Anal.*, 3:287–310, 1966.

[435] R. P. Roesser. A discrete state-space model for linear image processing. *IEEE Trans. Automat. Contr.*, AC–20:1–10, 1975.

[436] K. Rörentrop. *Entwicklung der Modernen Regelungstechnik.* R. Oldenbourg, München, 1971.

[437] C. H. Roth. *Fundamentals of Logic Design.* West Publ. Co, St. Paul, MN, 3rd edition, 1985.

[438] N. Rouche and J. Mawhin. *Ordinary Differential Equations.* Surveys and Reference Works in Mathematics. Pitman, London, 1980.

[439] E. J. Routh. *A Treatise on the Stability of a Given State of Motion.* Taylor and Francis, London, 1975. Reprint.

[440] W. Rudin. *Functional Analysis.* McGraw-Hill, New York, 1973.

[441] W. Rudin. *Real and Complex Analysis.* McGraw-Hill, New York, 3rd edition, 1987.

[442] W. J. Rugh. *Linear System Theory.* Information and System Sciences Series. Prentice Hall, NJ, 1993.

[443] M. G. Safonov. Tight bounds on the response of multivariable systems with component uncertainty. In *Allerton Conference on Communication, Control and Computing*, pages 451–460, 1978.

[444] M. G. Safonov. *Stability and Robustness of Multivariable Feedback Systems.* MIT Press, Cambridge, MA, 1980.

[445] M. K. Sain, editor. *Special Issue on Linear Multivariable Control Systems*, volume AC-26 of *IEEE Trans. Automat. Contr.*, 1981.

[446] P. A. Samuelson. A synthesis of the principle of acceleration and the multiplier. *Journal of Political Economy*, 47:786–797, 1939.

[447] I. W. Sandberg. A frequency domain condition for stability of feedback systems containing a single time-varying nonlinear element. *Bell Syst. Tech. J.*, 43:1601–1608, 1964.

[448] S. Sastry. *Nonlinear Systems. Analysis, Stability, and Control*, volume 10 of *Interdisciplinary Applied Mathematics*. Springer-Verlag, New York, 1999.

[449] M. B. Schaefer. Some aspects of the dynamics of populations important to the management of commercial marine fisheries. *Inter-Amer. Trop. Tuna Comm*, 1:25–56, 1954.

[450] E. Schmidt. Zur Theorie der linearen und nichtlinearen Integralgleichungen. *Mathematische Annalen*, 63:433–476, 1907.

[451] H. Schneider. Positive operators and an inertia theorem. *Numer. Math.*, 7:11–17, 1965.

[452] D. G. Schultz and J. E. Gibson. The variable gradient method for generating Liapunov functions. *Trans. AIEE*, 81(II):203–210, 1962.

[453] I. Schur. Über die charakteristischen Wurzeln einer linearen Substitution mit einer Anwendung auf die Theorie der Integralgleichungen. *Mathematische Annalen*, 66:488–510, 1909.

[454] I. Schur. Über Potenzreihen, die im Innern des Einheitskreises beschränkt sind. *Journal für Reine und Angewandte Mathematik*, 148:122–145, 1918. (In German).

[455] L. Schwiedernoch. Zur Theorie und Berechnung des Stabilitätsradius zeitdiskreter linearer Systeme. Master's thesis, University of Bremen, Germany, 1991.

[456] G. B. Segal. The topology of spaces of rational functions. *Acta Math.*, 143:39–72, 1979.

[457] G. R. Sell. *Topological Dynamics and Ordinary Differential Equations*, volume 33 of *Annals of Mathematics Studies*. Van Nostrand Rheinhold, London, 1971.

[458] R. Sezginer and M. Overton. The largest singular value of $e^X A_0 e^{-X}$ is convex on convex sets of commuting matrices. *IEEE Trans. Automat. Contr.*, 35:229–230, 1990.

[459] C. E. Shannon. Symbolic analysis of relay and switching circuits. *Trans. AIEE*, 57:713–723, 1938.

[460] C. E. Shannon. A mathematical theory of communication. *Bell Syst. Tech. J.*, 27:379–423 and 623–656, 1948.

[461] C. E. Shannon and J. McCarthy, editors. *Automata Studies*, volume 34 of *Annals of Mathematics Studies*. Princeton University Press, Princeton, NJ, 1956.

[462] J. L. Shearer, A. T. Murphy, and H. H. Richardson. *Introduction to System Dynamics*. Addison-Wesley, Reading, MA, 1971.

[463] J. A. Shercliff. *Vector Fields: Vector Analysis Developed Through its Application to Engineering*. Cambridge University Press, Cambridge, 1977.

[464] M. Shub. Stability and genericity for diffeomorphisms. In Peixoto, editor, *Dynamical Systems*, pages 493–514. Academic Press, New York, 1973.

[465] D. D. Šiljak. *Decentralized Control of Complex Systems*, volume 184 of *Mathematics in Science and Engineering*. Academic Press, Boston, MA, 1991.

[466] V. Sima. *Algorithms for Linear-Quadratic Optimization*. Marcel Dekker, New York, 1996.

[467] S. Smale. Differentiable dynamic systems. *Bull. Amer. Math. Soc.*, 73:747–817, 1967.

[468] G. V. Smirnov. *Introduction to the Theory of Differential Inclusions.* American Mathematical Society, Providence, RI, 2002.

[469] S. L. Sobolev. *Partial Differential Equations in Mathematical Physics.* Pergamon, 1964.

[470] C. B. Soh, C. S. Berger, and K. P. Dabke. On the stability properties of polynomials with perturbed coefficients. *IEEE Trans. Automat. Contr.*, AC-30:1033–1036, 1985.

[471] C. B. Soh, C. S. Berger, and K. P. Dabke. Addendum to 'On the stability properties of polynomials with perturbed coefficients'. *IEEE Trans. Automat. Contr.*, AC-32:239–240, 1987.

[472] E. D. Sontag. *Mathematical Control Theory, Deterministic Finite Dimensional Systems.* Springer-Verlag, New York, 2nd edition, 1998.

[473] E. D. Sontag. The ISS philosophy as a unifying framework for stability-like behavior. In A. Isidori, F. D. Lamnabhi-Lagarrigue, and W. Respondek, editors, *Nonlinear Control in the Year 2000*, volume 258 of *Lecture Notes in Control and Information Sciences*, pages 443–468. Springer-Verlag, Berlin, 2000.

[474] E. D. Sontag and Y. Wang. New characterizations of input to state stability. *IEEE Trans. Automat. Contr.*, 41:1283–1294, 1996.

[475] C. Sparrow. *The Lorenz Equations: Bifurcations, Chaos, and Strange Attractors.* Number 41 in Applied Mathematical Sciences. Springer-Verlag, 1982.

[476] M. N. Spijker. On a conjecture by LeVeque and Trefethen related to the Kreiss matrix theorem. *BIT*, 31:551–555, 1991.

[477] M. N. Spijker, S. Tracogna, and B. D. Welfert. About the sharpness of the stability estimates in the Kreiss matrix theorem. *Math. Comp.*, 72(242):697–713, 2002.

[478] J. Sreedhar, P. M. van Dooren, and A. L. Tits. A fast algorithm to compute the real structured stability radius. In R. Jeltsch and M. Mansour, editors, *Stability theory*, volume 121 of *Internat. Ser. Numer. Math.*, pages 219–230. Birkhäuser, Basel, 1996. International Conference "Centennial Hurwitz on Stability Theory", Ascona, 1995.

[479] F. Stenger. *Numerical Methods Based on Sinc and Analytic Functions.* Springer Series in Computational Mathematics. 20. Springer-Verlag, New York, 1993.

[480] G. W. Stewart. *Introduction to Matrix Computations.* Academic Press, New York-London, 1973.

[481] G. W. Stewart. On the early history of the singular value decomposition. *SIAM Review*, 35(4):551–566, 1993.

[482] G. W. Stewart. *Matrix Algorithms. Vol. I: Basic Decompositions.* SIAM Publications, Philadelphia, PA, 1998.

[483] G. W. Stewart. *Matrix Algorithms. Vol. II: Eigensystems.* SIAM Publications, Philadelphia, PA, 2001.

[484] G. W. Stewart and J. Sun. *Matrix Perturbation Theory.* Computer Science and Scientific Computing. Academic Press, Boston, MA, 1990.

[485] J. Stoer and R. Bulirsch. *Einführung in die Numerische Mathematik II.* Springer-Verlag, Berlin, Heidelberg, New York, 2nd edition, 1978.

[486] J. Stoer and R. Bulirsch. *Introduction to Numerical Analysis.* Springer-Verlag, New York, 2nd edition, 1993.

[487] J. Stoer and C. Witzgall. Transformations by diagonal matrices in a normed space. *Numer. Math.*, 4:158–171, 1962.

[488] T. Ström. On logarithmic norms. *SIAM J. Numer. Anal.*, 12(5):741–753, 1975.

[489] F. M. Sudo and J. R. Ziegler. *The Golden Age of Theoretical Ecology, 1923-1940. A Collection of Works by Volterra, Kostitzin, Lotka, and Kolmogoroff.* Number 22 in Lecture Notes in Biomathematics. Springer-Verlag, Berlin, New York, 1978.

[490] T. Sugie and M. Kawanishi. μ analysis/synthesis based on exact expressions of physical parameter variations and its application. In *Proc. 3rd European Control Conf. 1995*, pages 159–164, 1995.

[491] B. Sz.-Nagy. Perturbations des transformations autoadjointes dans l'espace de Hilbert. *Comm. Math. Helv.*, 19:347–366, 1947.

[492] G. I. Taylor. Stability of a viscous liquid contained between two rotating cylinders. *Phil. Trans. A*, 223:289–343, 1923.

[493] R. Tempo, E. W. Bai, and F. Dabbene. Probabilistic robustness analysis: Explicit bounds for the minimum number of samples. *Syst. Control Lett.*, 30:237–242, 1997.

[494] W. Thomson. Stability of fluid motion-rectilinear motion of viscous fluid between two parallel plates. *Philosophical Magazine*, 24, 1887.

[495] K. Thulasiraman and M. N. S. Swamy. *Graphs: Theory and Algorithms.* John Wiley & Sons, New York, 1992.

[496] O. Toeplitz. Das algebraische Analogon zu einem Satze von Fejér. *Math. Zeitschrift*, 2:187–197, 1918.

[497] L. N. Trefethen. Approximation theory and numerical linear algebra. In J. C. Mason and M. G. Cox, editors, *Algorithms for Approximation II*, pages 336–360. Chapman and Hall, London, 1990.

[498] L. N. Trefethen. Pseudospectra of linear operators. *SIAM Review*, 39(3):383–406, 1997.

[499] L. N. Trefethen. Computation of pseudospectra. *Acta Numerica*, 8:247–295, 1999.

[500] L. N. Trefethen and D. Bau, III. *Numerical Linear Algebra.* SIAM Publications, Philadelphia, PA, 1997.

[501] L. N. Trefethen, A. E. Trefethen, S. C. Reddy, and T. A. Driscol. Hydrodynamic stability without eigenvalues. *Science*, 261:578–584, 1993.

[502] Y. Z. Tsypkin and B. T. Polyak. Frequency domain criteria for l_p-robust stability of continuous linear systems. *IEEE Trans. Automat. Contr.*, AC-36:1464–1469, 1991.

[503] B. L. van der Waerden. *Algebra. Volume I.* Springer-Verlag, New York, 1991. Transl. of the 7th German ed.

[504] J. van Dorsselaer, J. Kraaijevanger, and M. Spijker. Linear stability analysis in the numerical solution of initial value problems. *Acta Numerica*, pages 199–237, 1993.

[505] C. van Loan. How near is a stable matrix to an unstable matrix? *Contemp. Math.*, 47:465–478, 1985.

[506] P. F. Verhulst. Notice sur la loi que la population suit dans son accroissement. *Corr. Math. et Phys.*, 10:113–121, 1838.

[507] K. Veselić. Bounds for exponentially stable semigroups. *Lin. Alg. Appl.*, 358:195–217, 2003.

[508] M. Vidyasagar. *Nonlinear Systems Analysis.* Prentice Hall, Englewood Cliffs, NJ, 2nd edition, 1993.

[509] V. Volterra. *Variazioni e Fluttuazioni del Numero d'Individui in Specie Animali Conviventi.* R. Comitato Talassografico Italiano, Memoria 131, 1927.

[510] V. Volterra. *Leçons sur la Théorie Mathématique de la Lutte pour la Vie.* Gauthiers-Villars, Paris, 1931.

[511] J. Vyshnegradskii. Sur la Théorie Générale des Régulateurs. *C.R. Acad. Sci., Paris, Ser. I*, 83:318–321, 1876.

[512] J. Vyshnegradskii. Über direkt wirkende Regulatoren. *Der Civilingenieur*, 23:95–132, 1877.

[513] J. F. Wakerly. *Digital Design Principles and Practices.* Prentice Hall, Englewood Cliffs, NJ, 1990.

[514] T. Ważewski. Sur la limitation des intégrales des systèmes d'équations différentielles linéaires ordinaires. *Stud. Math.*, 10:48–59, 1948.

[515] E. Wegert and L. N. Trefethen. From the Buffon needle problem to the Kreiss matrix theorem. *Amer. Math. Monthly*, 101:132–139, 1994.

[516] P. E. Wellstead. *Introduction to Physical System Modelling.* Academic Press, London, 1979.

[517] H. Weyl. Inequalities between the two kinds of eigenvalues of a linear transformation. *Proc. Nat. Acad. Sci. USA*, 35:408–411, 1949.

[518] G. B. Whitham. *Linear and Nonlinear Waves.* Pure and Applied Mathematics. John Wiley & Sons, New York, 1974.

[519] E. T. Whittaker. On the functions which are represented by the expansion of the interpolation theory. *Proc. Royal Soc. Edinburgh*, 35:181–194, 1915.

[520] E. T. Whittaker. *A Treatise on the Analytical Dynamics of Particles and Rigid Bodies.* Cambridge University Press, Cambridge, 4th edition, 1970.

[521] D. V. Widder. *The Laplace-Transform.* Princeton University Press, Princeton, NJ, 1966.

[522] H. Wilf. A stability criterion for numerical integration. *J. Assoc. Comput. Mach.*, 6:363–365, 1959.

[523] J. H. Wilkinson. *Rounding Errors in Algebraic Processes.* Prentice Hall, Englewood Cliffs, NJ, 1963.

[524] J. H. Wilkinson. *The Algebraic Eigenvalue Problem.* Clarendon Press, Oxford, 1965.

[525] J. H. Wilkinson. Convergence of the LR, QR and related algorithms. *Computer Journal*, 8:77–84, 1965.

[526] J. H. Wilkinson. Global convergence of tridiagonal QR algorithm with origin shifts. *Lin. Alg. Appl.*, 1:409–420, 1968.

[527] J. C. Willems. *The Analysis of Feedback Systems.* Number 62 in Research Monograph. MIT Press, Cambridge, MA, 1971.

[528] J. C. Willems. The generation of Lyapunov functions for input-output stable systems. *SIAM J. Cont.*, 9:105–133, 1971.

[529] J. C. Willems. Paradigms and puzzles in the theory of dynamical systems. *IEEE Trans. Automat. Contr.*, AC-36(3):259–294, 1991.

[530] J. L. Willems. *Stability Theory of Dynamical Systems.* Studies in Dynamical Systems. Thomas Nelson and Sons, London, 1970.

[531] H. K. Wimmer. Generalizations of theorems of Lyapunov and Stein. *Lin. Alg. Appl.*, 10:139–146, 1975.

[532] F. Wirth. The generalized spectral radius and extremal norms. *Lin. Alg. Appl.*, 342:17–40, 2002.

[533] F. Wirth. *Stability theory of perturbed systems: Joint spectral radii and stability radii.* Lecture Notes in Mathematics. Springer-Verlag, Berlin, 2005. To appear.

[534] V. A. Yakubovich. The solution of certain matrix inequalities in automatic control theory. *Soviet Math. Dokl*, 3:620–623, 1962. Translation from Dokl. Akad. Nauk SSSR 143.

[535] K. S. Yeung. Linear system stability under parameter uncertainties. *Int. J. Control*, 38:459–464, 1983.

[536] T. Yoshizawa. Stability of sets and perturbed systems. *Funkcialay Ecvacioj*, 5:31–69, 1963.

[537] T. Yoshizawa. *The Stability Theory by Liapunov's Second Method.* Mathematical Society of Japan, Tokyo, 1966.

[538] K. Yosida. *Functional Analysis.* Springer-Verlag, Berlin, 4th edition, 1974.

[539] J. Zabczyk. Remarks on the control of discrete-time distributed parameter systems. *SIAM J. Cont.*, 12(4):721–735, 1974.

[540] J. Zabczyk. A note on C_0-semigroups. *Bull. Acad. Pol. Sci. Ser. Sci. Mat.*, 23(6):895–898, 1975.

[541] J. Zabczyk. *Mathematical Control Theory: An Introduction.* Systems and Control: Foundations and Applications. Birkhäuser, Boston, 1992.

[542] L. A. Zadeh and C. A. Desoer. *Linear System Theory — The State Space Approach.* McGraw-Hill, New York, 1963.

[543] G. Zames. On the input-output stability of nonlinear time-varying feedback systems, parts i and ii. *IEEE Trans. Automat. Contr.*, 11:228–238,465–476, 1966.

[544] E. Zerz. LFT representations of parametrized polynomial systems. *IEEE Trans. Circuits Syst. I Fund. Theory Appl.*, 46(3):410–416, 1999.

[545] E. Zerz. *Topics in Multidimensional Linear Systems Theory*, volume 256 of *Lecture Notes in Control and Information Sciences.* Springer-Verlag, London, 2000.

[546] K. Zhou, J. C. Doyle, and K. Glover. *Robust and Optimal Control.* Prentice Hall, Upper Saddle River, NJ, 1996.

[547] V. I. Zubov. *The Methods of A. M. Lyapunov and Their Applications.* Noordhoff, Groningen, 1964.

Glossary

Standard mathematical symbols

$\mathbb{N}$, $\mathbb{N}^*$	set of natural numbers including and excluding 0
$\mathbb{R}$, $\mathbb{C}$, $\mathbb{R}^*$, $\mathbb{C}^*$	field of real and complex numbers including and excluding 0
$\mathbb{K}$	field, either $\mathbb{R}$ or $\mathbb{C}$ unless explicitly stated otherwise
δ_{jk}	Kronecker symbol $\delta_{jk} = 1$ for $j = k$, $\delta_{jk} = 0$ for $j \neq k$
$\underline{n}$	the set of natural numbers $1, 2, ..., n$
$\binom{\alpha}{k}$	binomial coefficient for $\alpha \in \mathbb{R}$, $k \in \mathbb{N}$, $\alpha(\alpha-1)...(\alpha-k+1)/k!$
$\mathbb{R}_+$	non-negative real numbers, $\{x \in \mathbb{R};\ x > 0\}$
$\mathbb{Z}$, $\mathbb{Z}_p$	ring of integers, ring of integers modulo p
$\mathbb{B}$	Boolean algebra $\mathbb{B} = \{0, 1\}$ with the operations $\wedge$ and $\vee$
X^n	n-th power of the set X, $\{(x_1, \ldots, x_n);\ \forall i \in \underline{n} : x_i \in X\}$
$\mathbb{K}^n$	space of n-vectors with entries in $\mathbb{K}$
$\mathbb{R}^n_+$	positive orthant, $\{(x_1, ..., x_n) \in \mathbb{R}^n; \forall i \in \underline{n} :\ x_i \in \mathbb{R}_+\}$
$a \leq b$	elementwise comparison of vectors $a, b \in \mathbb{R}^n$, $b - a \in \mathbb{R}^n_+$
$\operatorname{Re} z$, $\operatorname{Im} z$	real and imaginary parts of a complex number/vector $z \in \mathbb{C}^n$
$\mathbb{K}^{n \times m}$	space of $n \times m$ matrices with entries in $\mathbb{K}$
0	zero scalar, vector, or matrix.
0_n, $0_{m \times n}$	zero vector in $\mathbb{K}^n$, zero matrix in $\mathbb{K}^{m \times n}$
$I\ (I_n)$	the identity matrix (in $\mathbb{K}^{n \times n}$)
$e^1, ..., e^n$	the column vectors of I_n, standard basis of $\mathbb{K}^n$
$\operatorname{diag}(\alpha_1, ..., \alpha_n)$	diagonal matrix in $\mathbb{K}^{n \times n}$ with diagonal entries $\alpha_1, \ldots, \alpha_n \in \mathbb{K}$
$\operatorname{span}_{\mathbb{K}} S$	$\mathbb{K}$-linear subspace generated by a subset S of a vector space
$\operatorname{conv}(S)$	convex hull of a set S in a vector space
$\operatorname{trace} A$	trace of a matrix $A = (a_{ij}) \in \mathbb{K}^{n \times n}$, $\sum_{i=1}^n a_{ii}$
$\operatorname{im} A$, $\ker A$	image and kernel of a matrix A
$\operatorname{rank} A$	rank of a matrix A
$A^\top$, $\overline{A}$, A^*	transpose, complex conjugate and adjoint of a matrix A
$\operatorname{adj} A$	adjugate of a matrix A
$A^{\mathbb{R}}$	real form of a complex matrix $A \in \mathbb{C}^{m \times n}$: 465, 720
$\sigma(A)$, $\rho(A)$	spectrum and resolvent set of a matrix or linear operator A: 759
$\chi_A(s)$	characteristic polynomial of a matrix A, $\det(sI_n - A)$: 719
$\varrho(A)$	spectral radius of a matrix A, $\max\{\lvert\lambda\rvert; \lambda \in \sigma(A)\}$
$\alpha(A)$	spectral abscissa of a matrix A, $\max\{\operatorname{Re} \lambda; \lambda \in \sigma(A)\}$
$i(A)$	inertia of A, $(n_+(A), n_0(A), n_-(A))$: 719
$R(s, A)$	resolvent operator of a matrix or linear operator, $(sI - A)^{-1}$

$P_j(A)$, $N_j(A)$	eigenprojection and eigennilpotent of a matrix A: 410
$J(\lambda, m)$	Jordan block of order m with an eigenvalue λ: 106
$\sigma_i(A)$	i^{th} ordered singular value of $A \in \mathbb{C}^{m\times n}$, $\sigma_1 \geq \cdots \geq \sigma_n$: 431
$\sigma_{\max}(A)$, $\sigma_{\min}(A)$	maximal and minimal singular values of A
$V_1 \oplus V_2$	direct sum of two subspaces V_1 and V_2 of a vector space: 105, 720
$A \oplus B$	direct sum of two matrices A and B: 107, 720
$A \otimes B$	Kronecker product of $A \in \mathbb{K}^{m\times n}$ and $B \in \mathbb{K}^{p\times q}$: 721
$\mathrm{vec}(A)$	vector of stacked columns of $A \in \mathbb{K}^{m\times n}$: 721
$\kappa(A)$	condition number for inverting a matrix: 488
$\mathbf{Gl}_n(\mathbb{K})$	group of invertible $n \times n$ matrices over the field $\mathbb{K}$
$\mathbf{U}_n(\mathbb{K})$, $\mathbf{O}_n$	group of unitary matrices in $\mathbb{K}^{n\times n}$, orthogonal matrices in $\mathbb{R}^{n\times n}$
$\mathcal{H}_n(\mathbb{K})$	real vector space of Hermitian matrices in $\mathbb{K}^{n\times n}$
$\mathcal{H}_n^+(\mathbb{K})$	convex cone of positive semi-definite matrices in $\mathcal{H}_n(\mathbb{K})$
$\mathrm{sign}(H)$	$n_+(H) - n_-(H)$, signature of $H \in \mathcal{H}_n(\mathbb{K})$: 314, 723
$A \succeq 0$, $A \succ 0$	$A \in \mathcal{H}_n(\mathbb{K})$ is positive semi-definite/positive definite: 723
$p(s, a)$	polynomial with coefficient vector a: 370
$\deg p$	degree of a polynomial p
$\mathbb{K}[s]$	algebra of polynomials in s with coefficients in any field $\mathbb{K}$
$\mathbb{K}_n[s]$	space of polynomials $p(s) \in \mathbb{K}[s]$ of degree $\leq n$
$\mathbb{K}(s)$	field of rational functions in s with coefficients in $\mathbb{K}$
$\mathbb{K}^n[[s]]$	algebra of all formal power series in s with coefficients in $\mathbb{K}^n$
$R(p, q)$	resultant of two polynomials $p, q \in \mathbb{K}[s]$: 315
$\overline{S}$, $\mathrm{int}\ S$, ∂S	closure, interior and boundary of a subset S of a topological space
$d(x, y)$	distance between two points x and y in a metric space
$\mathrm{dist}(x, S)$	distance of a point x from a set S
$B(x, \delta)$	open ball in a metric space X with centre $x \in X$ and radius δ
$D(x, \delta)$, $I(x, \delta)$	open disk in $\mathbb{C}$ and open interval in $\mathbb{R}$ with centre x and radius δ
1_S	indicator function of $S \subset X$, $1_S(x) = 1$ if $x \in S$, $1_S(x) = 0$ if $x \in X \setminus S$
$\Vert x\Vert_F$	Frobenius norm of $x \in \mathbb{K}^n$: 486
$\Vert x\Vert_p$	p-norm of $x \in \mathbb{K}^n$: 715
$\Vert \cdot \Vert_X$	norm of a normed space X
$\langle \cdot, \cdot \rangle_X$	inner product on a Hilbert space X
I_X	identity operator on a vector space X
X^*	dual space of a normed linear space X: 755
$\Vert \cdot \Vert_X^*$	dual norm on X^*: 755
A^*	(Hilbert space) adjoint operator of A: 757
X^T	set of all maps $f : T \to X$
$\mathcal{C}(X, Y)$	space of continuous maps between metric spaces X and Y
$\mathcal{C}^m(T; X)$	space of m times continuously differentiable maps $x : T \to X$
$\mathcal{L}(X)$, $\mathfrak{C}(X)$	space of bounded/compact linear operators on X
$\mathcal{L}(X, Y)$	space of bounded linear operators from X to Y: 754
$\mathcal{C}(X, Y)$	space of closed linear operators from X to Y: 757
$\mathcal{D}(A)$	domain of an unbounded linear operator A: 757
$PC(T; X)$	space of piecewise continuous functions $x : T \to X$: 83
$PC^1(T; X)$	$\{x \in PC(T; X); \dot{x} \in PC(T; X)\}$
$L^p(T; X)$	space of p-integrable functions $x : T \to X$: 739
$L^p_{\mathrm{loc}}(T; X)$	space of locally p-integrable functions $x : T \to X$: 740
$L^p(a, b; X)$	$L^p([a, b]; X)$
$\ell^p(T; X)$	space of p-summable sequences $x : T \to X$: 735

Chapter 2

T, X	time domain and state space	74, 76
$U, \mathcal{U}$	input value and input function spaces	75
$Y, \mathcal{Y}$	output value and output function spaces	75
$\varphi(\cdot)$	state transition map, $\varphi : \mathcal{D}_\varphi \to X$	76
$\mathcal{D}_\varphi$	domain of definition of $\varphi(\cdot)$, $\mathcal{D}_\varphi \subset T^2 \times X \times \mathcal{U}$	77
Σ	dynamical system $(T, U, \mathcal{U}, X, Y, \varphi, \eta)$	77
$\eta(\cdot)$	output map $\eta : T \times X \times U \to Y$	77
$T_{t_0, x^0, u(\cdot)}$	life span of $\varphi(\cdot; t_0, x^0, u(\cdot))$	77
T_{t_0}	$\{t \in T;\ t \geq t_0\}$	78
$T^2_\geq$	$\{(t, t_0) \in T^2;\ t \geq t_0\}$	78
ψ	next state function ψ of an automaton (U, X, Y, ψ, η)	79
S_τ	forward $(\tau > 0)$ and backward $(\tau < 0)$ shift operators	89
$\Phi(t, t_0)$	linear evolution operator, $\varphi(t; t_0, \cdot; 0_\mathcal{U})$	101
$(\Phi(t))$	semigroup, usually (e^{At}) or (A^t)	101
$\Theta(t, t_0)$	input-to-state map, $\varphi(t; t_0, 0_X; \cdot) : u(\cdot) \mapsto x(t)$	101
$(g * u)(\cdot)$	convolution of sequences or functions	125, 126
$\delta_{t_0}(\cdot)$	Dirac impulse at $t_0 \in \mathbb{R}$	128
$\mathbb{L}_+, \mathbb{L}$	input-output operators, time domain $\mathbb{R}_+$, $\mathbb{R}$ (or $\mathbb{N}$, $\mathbb{Z}$)	131
$\mathbb{C}_-, \mathbb{C}_+$	$\{z \in \mathbb{C} : \mathrm{Re}(z) < 0\}$, $\{z \in \mathbb{C} : \mathrm{Re}(z) > 0\}$	132, 742
$\mathbb{D}, \mathbb{D}_+$	$\{z \in \mathbb{C}; \lvert z\rvert < 1\}$, $\{z \in \mathbb{C}; \lvert z\rvert > 1\}$	132, 737
$\mathbb{G}_+, \mathbb{G}$	input-output operators, $\mathcal{U}_+ \to \mathcal{Y}_+, \mathcal{U} \to \mathcal{Y}$	136
$\tilde{u}(\omega) = (\mathcal{F}u)(\omega)$	Fourier transform of a function $u : \mathbb{R} \to \mathbb{C}^m$	138, 748
$\hat{u}(z) = (\mathcal{Z} u)(z)$	$\mathbf{z}$-transform of a sequence $u = (u_k)_{k \in \mathbb{N}}$	139, 737
$\hat{u}(s) = (\mathcal{L}u)(s)$	Laplace transform of a function $u : \mathbb{R}_+ \to \mathbb{C}^m$	139, 742
$\tilde{u}(\theta) = (\mathcal{F}_D u)(\theta)$	discrete Fourier transform of a sequence $u = (u_k)_{k \in \mathbb{Z}}$	139, 746
$\mathcal{E}_\alpha(\mathbb{K}^m)$	$\{u(\cdot) \in L^1_{loc}(\mathbb{R}_+; \mathbb{K}^m);\ u(\cdot)e^{-\alpha \cdot} \in L^1(\mathbb{R}_+; \mathbb{K}^m)\}$, $\alpha \in \mathbb{R}$	139, 742
$\mathcal{S}_\gamma(\mathbb{K}^m)$	$\{u(\cdot) \in (\mathbb{K}^m)^\mathbb{N};\ u(\cdot)\gamma^{-\cdot} \in \ell^1(\mathbb{N}; \mathbb{K}^m)\}$, $\gamma > 0$	139, 737
$G(s)$	transfer function	140
$H^p(\mathbb{C}_+; \mathbb{C}^m)$	Hardy space on $\mathbb{C}_+$	147, 752
$H^p(\mathbb{D}_+; \mathbb{C}^m)$	Hardy space on $\mathbb{D}_+$	148, 751
$\mathbf{L}_{n,\ell,q}(\mathbb{K})$	$\{(A, B, C, D); A \in \mathbb{K}^{n \times n}, B \in \mathbb{K}^{n \times \ell}, C \in \mathbb{K}^{q \times n}, D \in \mathbb{K}^{q \times \ell}\}$	155
Σ/Σ_1	quotient system of Σ by Σ_1	157
$\Sigma_1 \oplus \Sigma_2$	direct sum of systems Σ_1 and Σ_2	159
$\Sigma_1 \Box \Sigma_2$	feedback connection of systems Σ_1 and Σ_2	162
$\mathrm{sinc}(z)$	sinc function	173

Chapter 3

$\mathcal{F} = (T, X, \varphi)$	local flow, time domain T, state space X, transition map φ	196
$\varphi(\cdot\,; t_0, x^0)$	trajectory initialized at (t_0, x^0)	196
$T_{t_0}(x^0)$	domain of definition (life span) of $\varphi(\cdot\,; t_0, x^0)$	196
$t_+(t_0, x^0)$	$\sup T_{t_0}(x^0)$	196
$X_\infty(t_0)$	initial states at t_0 with infinite life span	196
$t_+(x^0)$, $T(x^0)$	$t_+(0, x^0)$, $T_0(x^0)$, in the time-invariant case	197
$\varphi(t; x^0)$, X_∞	$\varphi(t; 0, x^0)$, $X_\infty(0)$ in the time-invariant case	197
$\omega(x)$	forward limit set of $\varphi(\cdot; x)$	202
$\mathcal{O}(x^0)$	orbit initialized at x^0, $\{\varphi(t; x^0);\ t \in T(x^0)\}$	202
$\mathcal{A}(t_0, \Omega)$	basin of attraction of an attractor Ω at time t_0	211
$\mathcal{K}$, $\mathcal{K}_\infty$, $\mathcal{LK}$	classes of comparison functions	219

$\overline{\alpha}(\Phi)$, $\underline{\alpha}(\Phi)$	upper/lower Liapunov exponent, evolution operator Φ	258
$\overline{\beta}(\Phi)$, $\underline{\beta}(\Phi)$	upper/lower Bohl exponent, evolution operator Φ	258
$\omega(A)$	growth rate of the semigroup generated by A	264
$m(\cdot)$	Möbius transformation, $m(s) = \frac{s+1}{s-1}$	268
$\Delta_a^b \arg \gamma(t)$	change of argument of $\gamma(t)$ as t goes from a to b	297, 727
$p_R(s)$, $p_I(s)$	real and imaginary parts of $p(\imath s) = p_R(s) + \imath p_I(s)$	299
$p^\star(s)$	Hurwitz-reflection of a polynomial $p(s) \in \mathbb{C}[s]$, $\bar{p}(-s)$	299
$p_+(s)$	symmetric part of p, Hurwitz case, $(p(s)+p^\star(s))/2$	299
$p_-(s)$	antisymmetric part of p, Hurwitz case, $(p(s)-p^\star(s))/(2\imath)$	299
$p^e(s^2)$, $p^o(s^2)$	even and odd part of a polynomial $p(s)=p^e(s^2)+sp^o(s^2)$	302, 639
$CI_a^b(f)$	Cauchy index of $f(s) \in \mathbb{R}(s)$ on the interval (a,b)	308
$CI(f)$	global Cauchy index of a rational function $f(s) \in \mathbb{R}(s)$	308
$w(\gamma, s_0)$	winding number of a closed curve γ about $s_0 \in \mathbb{C}$	310, 727
$\mathbf{H}_n(p)$	Hermite matrix of a polynomial p, of order $n \geq \deg p$	314
$\mathbf{B}_n(u,v)$	Bézout matrix of a polynomial pair (u,v), of order n	317
$\mathbf{B}(u,v)$	Bézout matrix of order $\max\{\deg u, \deg v\}$	318
$\mathbf{Hk}(g)$	infinite Hankel matrix of a rational function $g(s) \in \mathbb{C}(s)$	321
$\mathbf{Hk}_n(g)$	upper left $n \times n$ matrix of $\mathbf{Hk}(g)$	321
$\mathrm{Rat}_n(\mathbb{K})$	set of strictly proper rational functions of degree n	323
$\mathrm{Hank}(\mathbb{K})$	set of infinite Hankel matrices with entries in $\mathbb{K}$	325
$\mathrm{Hank}(n,\mathbb{K})$	set of Hankel matrices in $\mathrm{Hank}(\mathbb{K})$ of rank n	325
$\mathrm{Hank}_n(\mathbb{K})$	set of Hankel matrices in $\mathbb{K}^{n\times n}$	325
$\mathrm{Hank}_n^*(\mathbb{K})$	set of Hankel matrices in $\mathrm{Hank}_n(\mathbb{K})$ of full rank n	325
$\mathrm{Rat}(n,\nu)$	set of $g \in \mathrm{Rat}_n(\mathbb{R})$ with Cauchy index $CI(g) = n - 2\nu$	332
$M^{\mathbb{K}}(p)$	Hurwitz matrix of a polynomial $p(s) \in \mathbb{K}[s]$	337
$p^*(z)$	Schur-reflection of a polynomial $p(z) \in \mathbb{C}_n[z]$, $z^n \bar{p}(z^{-1})$	341
$p_+(z)$	symmetric part of p, Schur case, $(p(z)+p^*(z))/2$	342
$p_-(z)$	antisymmetric part of p, Schur case, $(p(z)-p^*(z))/(2\imath)$	342
$\tilde{p}(z)$	Möbius transform of $p(z)$, $(z-1)^n p((z+1)/(z-1))$	343
$\mathbf{S}_n(p)$	Schur matrix of a polynomial p, of order $n \geq \deg p$	347
$\mathbf{S}(p)$	Schur matrix of a polynomial p, of order $n = \deg p$	347
$\mathbf{J}_n(p)$	Jury matrix of a polynomial p, of order $2n$	350
$\tilde{\mathbf{B}}(c,d)$	discrete Bézout matrix of a polynomial pair (c,d)	351
$\mathbf{T}(c,d)$	Toeplitz matrix of a polynomial pair (c,d)	352

Chapter 4

$\Lambda(a)$	unordered n-tuple of roots of $p(s,a)$	373
$d(\Lambda, \Lambda')$	distance between two unordered n-tuples	373
$\mathcal{K}(X)$	space of compact subsets of a metric space X	374
$d_H(K_1, K_2)$	distance between $K_1, K_2 \in \mathcal{K}(X)$ in the Hausdorff metric	374
$\mathcal{O}(\Omega)$	ring of complex analytic functions on a domain Ω	376
$\mathcal{M}(\Omega)$	field of meromorphic functions on Ω	376
$\mathcal{O}(\Omega)[s]$	ring of polynomials with coefficients in $\mathcal{O}(\Omega)$	376
$\mathcal{M}(\Omega)[s]$	ring of polynomials with coefficients in $\mathcal{M}(\Omega)$	376
C_p	set of critical points of $p(s,a(z)) \in \mathcal{M}(\Omega)[s]$	377
$D^\circ(z,\delta)$	the punctured disk $D(z,\delta) \setminus \{z\}$	379
$D^-(z_0,r)$	the cut disk $\{z \in D^\circ(z_0,r); 0 < \arg(z - z_0) < 2\pi\}$	379
$\mathcal{H}_n$	set of coefficient vectors of real monic Hurwitz polynomials	384

$\mathcal{S}_n$	set of coefficient vectors of real monic Schur polynomials	384
$[\underline{a}, \overline{a}]$	closed n-dimensional interval between $\underline{a}$ and $\overline{a}$	389
$\Lambda(A)$	unordered n-tuple of eigenvalues of A	398
$\Lambda(\mathcal{A})$	$\{\Lambda(A);\ A \in \mathcal{A}\}$ where $\mathcal{A}$ is a set of matrices	398
$\sigma(\mathcal{A})$	$\bigcup_{A\in\mathcal{A}} \sigma(A)$ where $\mathcal{A}$ is a set of matrices	398
C_A	set of critical points of an analytic matrix family	400
$\mathbf{\Delta}, \mathbf{\Delta}_{\mathbb{R}}$	perturbation class in $\mathbb{C}^{\ell\times q}$, $\mathbf{\Delta}_{\mathbb{R}} = \mathbf{\Delta} \cap \mathbb{R}^{\ell\times q}$	450, 453
$\|\cdot\|_{\mathbf{\Delta}}$	norm on the subspace $\mathrm{span}_{\mathbb{K}}\,\mathbf{\Delta}$	450
$\mu_{\mathbf{\Delta}}(G)$	μ-value of $G \in \mathbb{C}^{q\times\ell}$ for the perturbation class $\mathbf{\Delta}$	450
$\mu_{\mathbb{K}}(G)$	μ-value of $G \in \mathbb{C}^{q\times\ell}$ for the perturbation class $\mathbb{K}^{\ell\times q}$	454
$\|\cdot\|_{p\|r}$	Hölder norm	455, 717
$\|\cdot\|_{p,r}$	operator norm of a map from $(\mathbb{K}^m, \|\cdot\|_p)$ to $(\mathbb{K}^n, \|\cdot\|_r)$	456, 716
$\mathbf{L}_A, \mathbf{L}_A^D$	continuous and discrete time Liapunov operators	488, 514

Chapter 5

$\mathbf{\Delta}_0$	perturbations $\Delta \in \mathbf{\Delta}$ for which $A(\Delta)$ is well defined	521
$\boldsymbol{\delta}_0$	well-posedness radius	521
$\mathbb{C}_g, \mathbb{C}_b$	$\mathbb{C}_g$ open set in $\mathbb{C}$, $\mathbb{C}_b = \mathbb{C}\backslash\mathbb{C}_g$, stability/instability regions	522, 529
$\sigma_{\mathbf{\Delta}}(A(\cdot);\delta)$	spectral value set, perturbation structure $\mathbf{\Delta}$, level δ	522
$r_{\mathbf{\Delta}}(A(\cdot);\mathbb{C}_g)$	$\mathbb{C}_g$-stability radius of $A(0)$, perturbation structure $\mathbf{\Delta}$	529
$\sigma_{\mathbb{K}}(A;B,C;\delta)$	spectral value set, full block perturbations, level δ	531
$\mathcal{F}(M,\Delta)$	linear fractional transformation	538
$\sigma_{\mathbf{\Delta}}(A;B,C,D;\delta)$	spectral value set, linear fractional perturbations, level δ	544
$\sigma_{\mathbb{K}}(A;B,C,D;\delta)$	$\sigma_{\mathbf{\Delta}}(A;B,C,D;\delta)$, $\mathbf{\Delta} = \mathbb{K}^{\ell\times q}$	556, 567
$\mathcal{C}_\delta$	spectral contour, level δ^{-1}	557
$\mathcal{R}_G$	realness locus	562
$\mathcal{R}_G(\delta)$	$\sigma_{\mathbb{R}}(A;B,C,D;\delta) \cap \mathcal{R}_G$	562
$G_R(s)$, $G_I(s)$	real and complex part of $G(s)$	562
$\sigma_{\mathbb{K}}(A;\delta)$	unstructured spectral value set (pseudospectrum), level δ	569
$r_\Delta(A;B,C,D;\mathbb{C}_g)$	$\mathbb{C}_g$-stability radius, linear fractional perturbations	586
$r_\Delta^-(A;B,C,D)$	stability radius for $\mathbb{C}_g = \mathbb{C}_-$	588
$r_\Delta^1(A;B,C,D)$	stability radius for $\mathbb{C}_g = \mathbb{D}$	588
$r_{\mathbb{K}}(A;B,C,D;\mathbb{C}_g)$	$r_\Delta(A;B,C,D;\mathbb{C}_g)$ with $\mathbf{\Delta}=\mathbb{K}^{\ell\times q}$	591, 596
$H_\gamma(A,B,C,D)$	Hamiltonian matrix of (A,B,C,D), parameter γ	605
$d_{\mathbb{K}}(A;\mathbb{C}_g)$	distance of A from $\mathbb{C}_g$-instability	610
$\mathcal{U}_n(\mathbb{K};\mathbb{C}_g)$	set of $\mathbb{C}_g$-unstable matrices, $\{X \in \mathbb{K}^{n\times n};\ \sigma(X)\cap\mathbb{C}_b \neq \emptyset\}$	610
$d_{\mathbb{K}}^-(A)$, $d_{\mathbb{K}}^1(A)$	$d_{\mathbb{K}}(A,\mathbb{C}_g)$ with $\mathbb{C}_g = \mathbb{C}_-$ and $\mathbb{C}_g = \mathbb{D}$	611
$\mathcal{S}_{n,\ell,q}(\mathbb{K};\mathbb{C}_g)$	$\{(A,B,C,D) \in \mathbf{L}_{n,\ell,q}(\mathbb{K});\ \sigma(A) \subset \mathbb{C}_g\}$	614
$\Gamma(A,B,C,D)$	Cayley transform of (A,B,C,D)	618
$\mathcal{R}(a)$	root set of a polynomial with coefficient vector a	626
$\mathcal{R}_{\mathbb{K}}(a;C;\delta)$	root set, structure matrix C, level δ	626
$r_{\mathbb{K}}(a;C;\mathbb{C}_g)$	$\mathbb{C}_g$-stability radius of $p(s,a)$, structure matrix C	630
$r_{\mathbb{K}}^-(a;C)$	structured stability radius, Hurwitz polynomial	630
$r_{\mathbb{K}}^1(a;C)$	structured stability radius, Schur polynomial	630
$d_{\mathbb{K}}(a;\mathbb{C}_g)$	unstructured $\mathbb{C}_g$-stability radius of $p(s,a)$	630
$d_{\mathbb{K}}^-(a)$, $d_{\mathbb{K}}^1(a)$	distance of Hurwitz, Schur polynomials to instability	630
$\mathcal{R}_{\mathbb{K}}(a;\delta)$	root set, unstructured perturbations, level δ	633

$\mathcal{S}_n(\mathbb{K}; M, \beta)$	set of $A \in \mathbb{K}^{n\times n}$ generating (M,β)-stable semigroups	649
$M_\beta(A)$	transient bound of $A \in \mathbb{K}^{n\times n}$ for exponential rate β	650
$\mathcal{S}_n(\mathbb{K}; \beta)$	$\{A \in \mathbb{K}^{n\times n};\ \alpha(A) \le \beta\}$	651
$\nu(A)$	initial growth rate, $\min\{\beta \in \mathbb{R};\ \forall t \ge 0 : \|e^{At}\| \le e^{\beta t}\}$	653
$\mathcal{N}_n(\mathbb{K})$	set of normal matrices in $\mathbb{K}^{n\times n}$	657
$\operatorname{dep}(A)$	departure of A from normality	657
$[\cdot,\cdot]$	semi-scalar product on a normed space	660
$\Theta(A)$	numerical range of A, $\{\langle Ax, x\rangle;\ \|x\| = 1\}$	662
$\operatorname{ecc}(p)$	eccentricity of a norm $p(\cdot)$ on $\mathbb{K}^n$ w.r.t. a norm $\|\cdot\|$	663
$R(A)$	Kreiss constant, $\sup_{\operatorname{Re} s>0} \operatorname{Re} s\, \|(sI_n - A)^{-1}\|$	670
$\alpha_\delta(A; B, C)$	δ-spectral abscissa, $\sup\{\operatorname{Re} s;\ s \in \sigma_{\mathbb{C}}(A; B, C; \delta)\}$	673
$r_{\mathbb{K}}(A; B, C; M, \beta)$	(M,β)-stability radius	675
$d_{\mathbb{K}}^{-}(A; M, \beta)$	unstructured (M,β)-stability radius, $r_{\mathbb{K}}(A; I_n, I_n; M, \beta)$	675
$r_{\boldsymbol{\Delta},t}(A(\cdot); M, \beta)$	(M,β)-stability radius, time-varying perturbations	677
$\mathcal{P}_n(\mathbb{K})$	vector space of nonlinear perturbations	688
$\mathcal{P}_t(\mathbb{K})$	vector space of linear time-varying perturbations	688
$\mathcal{P}_d(\mathbb{K})$	vector space of dynamic perturbations	688
$\|\cdot\|_n$, $\|\cdot\|_t$, $\|\cdot\|_d$	norms on perturbation spaces $\mathcal{P}_n$, $\mathcal{P}_t$, $\mathcal{P}_d$	688
$\mathcal{P}_{nt}(\mathbb{K})$	vector space of nonlinear time-varying perturbations	689
$r_{\mathbb{K},t}^{-}(A; B, C)$	stability radius, time-varying perturbations	696
$r_{\mathbb{K},n}^{-}(A; B, C)$	stability radius, nonlinear perturbations	696
$r_{\mathbb{K},d}^{-}(A; B, C)$	stability radius, dynamic perturbations	696

Abbreviations

RHS, LHS	right/left hand side
SVD	singular value decomposition
siso	single input single output (system)

Index

A/D-converter, 168
Absolutely continuous, 83, 740
Aizerman's conjecture, 701, 712
  multivariable, 702
  theorem, 703
Algebraic stability domains, 357
Algorithms
  for complex stability radius, 607
  for eigenvalues, 506–508
    power methods, 506
  for Liapunov equations
    Bartels-Stewart, 512
  for singular values, 508–511
    Golub-Kahan-Reinsch, 509
  Francis' double-shift QR, 502–506
Ampère's law, 42
Analytic continuation, 380, 400, 402, 733
Arc, 297, 380, 724
Argument, 297, 725
  change of, 297, 725, 727
  principle of, 297, 731
  rate of change, 298, 299
Argument function, 298, 725
Attractive
  closed set, 211
  equilibrium point, 200
  globally, 200, 695
Attractor, 211
  asymptotically stable, 227, 240
Automaton, 79, 80, 95
  –, *see also* Digital systems
  next-state function, 79
Automobile suspension system, 16, 533

Bandwidth, 173
Basin of attraction, 223, 225, 232
  guaranteed domain of attraction, 705
  of an attractor, 211
  of an equilibrium point, 200
Bauer-Fike Lemma, 404, 524
Bézout matrix, 317–320
  discrete, 351
Bidiagonal form, 495
  algorithm for reducing to, 497
Birkhoff's Recurrence Theorem, 209
Bohl exponent, 258, 259, 262, 278
Bohl transformation, 262
Boundary Crossing Theorem, 374
Branch point, 381, 401, 419
  order, 401, 419
Brockett's Theorem, 332
Bromwich contour, 670
Butler's Theorem, 419
Butterworth pattern, 372

Carathéodory conditions, 84
Carathéodory's Theorem, 84
Cart-pendulum system, 23, 33, 165
Cassini ovals, 554, 584
Cauchy index, 308–312
Cauchy's Integral Theorem, 727
Cauchy-Schwarz inequality
  for functions, 134, 740
  for sequences, 174, 735
  for vectors, 716
Causality, 77, 135, 687
Cayley transform, 268, 618
Cayley's problem, 211
Characteristic multipliers, 266
Cholesky factorization, 445, 446
Closed curve, 310, 724
  simple, 724
Closed Graph Theorem, 757
Closed Range Theorem, 758
Cobweb model, 8
Cocycle property, 77, 196, 417
Cohort population model, 356
Comparison functions
  of class $\mathcal{K}$, 219, 248
  of class $\mathcal{K}_\infty$, 219
  of class $\mathcal{LK}$, 219, 248
Condition number, 666, 667
  for determining eigenvalues, 490
  for determining singular values, 491

for solving Liapunov equations, 488
for solving linear equations, 486
Configuration space, 29, 30
Congruent, 314, 318, 321, 723
Conjugate exponent, 715
Connected
arcwise, 387, 564, 724
simply, 377, 559, 732
Connected components, 212, 527, 545, 558–559, 563–564, 724
of $\mathrm{Rat}_n(\mathbb{R})$, 332
Constraint, 27
holonomic, 29
Continuity equation, 43
Contraction, *see* Semigroup
Convergence
in norm, uniform, strong, weak, 756
Convex combinations of polynomials, 640
Convex direction, 366, 647
Convolution
of functions, 126, 740
of sequences, 125, 736
Convolution inequality
for functions, 133, 741
for sequences, 149, 736
Convolution kernel, 127, 136
Convolution system, 136
transfer function, 147–151
Coulomb's law, 40
Coupling matrices, 160, 161
Courant-Fischer Minimax Theorem, 433
Critical point
of a parametrized matrix, 400
of a parametrized polynomial, 377

D/A-converter, 168
Decreasing along a trajectory, 221
strictly, 221
Delay system, 90, 141, 710
time-varying delay, 690
Departure from normality, 501, 657
Diagonalizable, 399, 405, 423, 723
Differential inclusion, 676
Digital systems, 56–68
finite state machines, 56
latches and flip–flops, 62
parity check machine, 58, 95
realization, 69
shift register, 80
three bit counter, 67
memoryless, 56, 80
half adder, full adder, 57
logic gates, 59, 80
Dirac impulse, 128, 741
Direct sum
of matrices, 594, 720
of subspaces, vector spaces, 105, 720
of systems, 159, 594
Directed graph, 50, 164–166
cut-set, 51
cycle, edge, vertex, 50
incidence map, 50
strongly connected, 51
Discretization methods
Adams-Bashforth, 184
stability, 270
Euler, 178
stability, 270
explicit, implicit, 182
Heun, 182
midpoint, 183
stability, 270
Milne, 185
stability, 270
multi-step, single-step, 182
of order p, 185
predictor-corrector, 182
Runge-Kutta, 183
stability, 270
Discriminant, 377
Dissipative, 660, 661
strictly, 660, 661
Distance from instability, 609, 669, 672
bounds, 611, 612
characterizations, 610
of a normal matrix, 612
of polynomials, 630, 634, 642
under similarity transformations, 613
Distance from non-dissipativity, 675
Distance from normality, 657, 669
Distance from singularity, 437, 474
Domain, 376, 724
Dyadic decomposition, 143
Dynamical system, 77
complete, 78
differentiable, 83
finite, 90
finite dimensional, 90

- infinite dimensional, 90, 115, 690
- inputs, outputs, 74
- linear, 91
- memoryless, 56, 80
- output map, 77
- recursive, 88
- reversible, 79
- state, 75
- state transition map, 76
- time–invariant, 89, 104

Eccentricity, 663, 676
Eigenbasis
- analyticity, 416, 421
  - single real parameter, 423

Eigenloci, 606
Eigenmode, *see* Eigenmotion
Eigenmotion
- complex, 106, 113
  - generalized, 105, 113
- real, 107, 114
  - generalized, 107

Eigennilpotents, 105, 410
- analyticity, 414
- behaviour near a critical point, 419

Eigenprojections, 105, 410
- analyticity, 414, 420, 421
  - single real parameter, 422
- behaviour near a critical point, 419
- total projection, 413
  - analyticity, 413

Eigenvalues, 759
- algebraic, geometric multiplicity, 104
- algorithm for determining, 506
- analyticity, 399, 420, 421
  - single real parameter, 422
- behaviour near a critical point, 401
- continuity, 399
  - single real parameter, 399
- differentiability, 425, 428
  - single real parameter, 399
- λ-group of eigenvalues, 400, 413, 420
- n-tuple of eigenvalues, 399
- sensitivity, 403

Eigenvectors, 759
- analyticity, 415
- differentiability, 425, 428
- generalized, 105
- sensitivity, 424, 428

Electrical circuit, 39–50, 94, 95, 122
Energy
- dissipation, 31
- kinetic, 28, 30
- potential, 28, 31
- thermal, 70

Entire function, 173, 727
Equilibrium state, 87, 199, 214
Ergodic set, 211
Event space, 77
Evolution operator, 101, 254, 428

Fam–Meditch Theorem, 387
Faraday's law, 43
Feedback
- dynamic output, 161
- representation of model uncertainties, 531, 536, 687
- static output, 162
- static state, 162
- well-posedness condition, 164

Fejér's Theorem, 745
Fibonacci sequence, 115
Finite escape time, 84, 85, 199, 230, 690
Fisheries model, 98
Flip-flop, 62, 80
Flow
- differentiable, 81
- global, 196, 206
- local, 196
- time-invariant, 196

Focus, stable and unstable, 111
Fourier coefficients, 744
Fourier series, 171, 745–746
- generalized, 754

Fourier transform, 172, 746
- discrete, 139, 749
- inverse transform, 174, 747

Fourier's law, 71
Fourier-Plancherel transform, 172, 748
Fréchet differentiable, 485, 756
Frequency response, 145
Front locus, 674
Function element, 380, 733

Gain response, 145
Generalized coordinates, 30
Generator of a semigroup, 120, 757
Gershgorin set, 407
Gershgorin type uncertainty, 552, 580, 589

Gershgorin's Theorem, 398, 407, 552
Godunov matrix, 547
Goodwin's model, 9, 293
Gram-Schmidt orthogonalization, 498
Gronwall's Lemma, 86
    generalized, 85
Growth rate
    at an equilibrium point, 288, 289
    of a semigroup, 264, 406, 649

Hahn-Banach Theorem, 632, 660, 755
Hamilton's equations, 33
Hamilton's principle, 32
Hamiltonian, 33, 217, 246
Hamiltonian matrix, 605
    of a system, 605
Hankel form, 320
Hankel matrix, 320–330
    Kronecker's Theorem, 323
    singular extension, 326
Hardy spaces
    on $\mathbb{C}_+$, $\mathbb{D}_+$, 147
    on $\mathbb{C}^+_\alpha$, $\mathbb{D}^+_\gamma$, 750, 751
Hausdorff metric, 374, 545, 555
Heat conduction equation, 115
Herglotz' Theorem, 354
Hermite form, 313, 314
Hermite generating function, 314
Hermite matrix of a polynomial, 314–317
Hermite's Theorem, 317
Hermite–Biehler Theorem, 303, 334, 335
Hermite–Hurwitz Theorem, 328
Hermitian matrix, 722
    inertia, 314
    positive definite, 276, 317, 723
    positive semi-definite, 446, 723
    signature, 314, 723
    spectral theorem, 723
Hessenberg form, 493
    algorithm for reducing to, 495
    reduced, 495
Hilbert space, 116, 753
Hille-Yosida Theorem, 672
Hold, 168, 170
    k^{th}–order, 170
Hold equivalent system, 178
Hölder norm, 455, 554, 718
Hölder's inequality
    for functions, 740
    for sequences, 735
    for vectors, 716
Holomorphic function, 376, 727
Hooke's law, 14
Householder reflection, 493
Householder transformation, 492–493
Householder vector, 493
Hurwitz matrix
    of a complex polynomial, 338
    of a real polynomial, 338, 340

Identity Theorem, 376, 420, 728
Impulse response, 126, 128
Inertia
    of a complex matrix, 295, 719
    of a Hermitian matrix, 723
    of an Hermitian matrix, 314
    Sylvester's Law of Inertia, 723
Initial growth rate, 653, 662, 675, 676
    of normal systems, 656
Inners of a matrix, 350
Input-output operator
    on $\mathbb{N}$, 124, 148
    on $\mathbb{R}$, 134, 150, 602
    on $\mathbb{R}_+$, 131, 148, 603
    on $\mathbb{Z}$, 150
Input-output system, 135
    –, *see also* Convolution system
    linear, 136
    L^q-stable, 132, 137
    time-invariant, 136
Input-state map, 119, 124
Input-to-state map, 101
Interconnection
    feedback connection, 161
    of linear systems, 154
    of uncertain systems, 535
    parallel connection, 160
    series connection, 160
Interlacing condition
    on the real line, 303
    on the unit circle, 345
Interval polynomial, 389, 641
Invariant set, 197, 213, 237, 238
    weakly, 197, 237
Inverted pendulum, 22, 35, 110, 176

Jenkin's governor, 306
Joint Liapunov function, 704, 708
    joint Liapunov norm, 676
Jordan Curve Theorem, 725

Julia set, 210
Jury matrix, 350
Jury's Stability Criterion, 350

Kharitonov polynomials, 390, 641
Kharitonov's Theorem, 391
Kirchhoff's laws, 52
Kreiss-Spijker Theorem, 670
Kronecker product, 486, 489, 720
Kronecker symbol, 143
Kronecker's Theorem, 323
  topological version, 325

Lagrange's equations, 31
Lagrangian, 31
Laplace transform, 139, 742
  inverse, 142, 742
Laplace's equation, 41
Laplacian, 41, 71
LaSalle's invariance principle
  for flows, 227
  for time-invariant flows, 238
Laurent series, 320, 730
Laurent-Puiseux series, 419
Legendre transformation, 33, 38
Liapunov equation
  algebraic, 283, 665
  differential, difference, 273, 668
  generalized, 357, 359
Liapunov exponent, 258, 259
Liapunov function
  for flows, 222
  for time-invariant systems, 236
  for time-varying systems, 231
  joint, *see* Joint Liapunov function
  of maximal robustness, 704
  quadratic, for linear systems
    time-invariant, 282–290
    time-varying, 272–282
Liapunov operator, 284, 488
  generalized, 283, 358
Liapunov transformation, 262
Liapunov's direct method, 217
  instability theorems
    for flows, 226
    time-invariant linear, 285
    time-invariant nonlinear, 245
    time-varying linear, 273, 277
    time-varying nonlinear, 234
  stability theorems
    for flows, 223, 227, 248
    time-invariant linear, 285
    time-invariant nonlinear, 236
    time-varying linear, 273, 276
    time-varying nonlinear, 232, 233
Liapunov's indirect method, 253
  exponential stability theorem, 289
  instability theorem
    time-invariant, 288
    time-varying, 281
  stability theorem
    time-invariant, 288, 696
    time-varying, 279
Liénard's equation, 250
Liénard–Chipart Theorem, 336, 340
Limit cycle, 203
  stable, 204
Limit point, 202
Limit set, 202–204, 208, 209, 214
Linear fractional transformation, 538
  sum, product, inverse, 541
Linear operator
  adjoint, 662, 757
  bounded, 124, 756
  closed, 756, 761
  compact, 756, 761
  normal, 573, 758, 761
  selfadjoint, 758
  unbounded, 120, 757
Linear system, 100
  decomposition principle, 100
  difference, 103
  differentiable, 101
  isomorphism, 155
  morphism, 155
  similar, 35, 156
  superposition principle, 100
  time-invariant, 102
  time-varying, 101
Linearization, 92
  along a trajectory, 280
  around an equilibrium point, 279, 288
Lipschitz
  continuous, locally Lipschitz, 235, 680, 695
  gain, 688, 689
  global Lipschitz condition, 688, 695
Logarithmic norm, *see* Initial growth rate
Logistic growth model, 3, 98, 215

Lorentz force, 44
Lorenz attractor, 212, 241
Lotka-Volterra model, *see* Predator-prey system
Lumped parameter models, 14

Markov parameter, 125
Mass-spring system, 13–17, 94, 157
Mathieu's equation, 274
Maximum principle, 588, 592
 extended, 734
 for holomorphic functions, 733
 for subharmonic functions, 734
Maxwell's equations, 43
Meromorphic function, 376, 377, 729
Method of steps, 90
Minimal set, 203, 209
Minkowski functional, 659, 756
Möbius map, 268, 343, 617
Möbius transform
 of a polynomial, 343, 387
Model uncertainties, 517
 –, *see also* Perturbations of systems
Momentum
 angular, 18, 19
 linear, 17
Monodromy Theorem, 733
Morera's Theorem, 728
Multi-model, 518
Multiplication operator
 on $\mathbb{C}_+$, $\mathbb{D}_+$, 148
 on $\imath\mathbb{R}$, $\partial\mathbb{D}$, 149
μ-value, 450, 550, 586
 block-diagonal perturbations, 456
 bounds
  lower, 454
  upper, 452
 complex structures
  characterization, 453
  continuity, 465
 diagonal perturbations
  upper bound, 458
 full-block perturbations, complex
  characterization, 454
  of minimum norm, 454
 full-block perturbations, real
  characterization q or $\ell = 1$, 467
  characterization $q, \ell \geq 2$, 474
  continuity, 477
  Lipschitz continuity, 478
  upper semicontinuity, 466
 multi-block perturbations
  characterization, 459, 461, 463
  upper bound, 457, 460
 properties, 452
 upper semicontinuity, 464

Neglected dynamics, 690, 697
Network
 combinational switching, 56
 electrical, 50–55
 sequential switching, 56, 80
Newton's second and third laws, 13
Non-wandering set, 207, 208
Norm
 absolute, 404, 717
 compatible, 450, 716
 dual, 455, 466, 715, 755
 Frobenius, 456, 718
 Hölder, 455, 554, 717
 Liapunov, 663
  β-Liapunov, 663, 676
  strict, 663
 monotone, 717
 of input-output operator, 602
 operator norm, 716
 p-norm of a function, 739
 p-norm of a sequence, 715
 p-norm of a vector, 735
 rank one consistent, 455, 463, 559, 718
 sub-multiplicative, 404
Normality of a matrix family at a point, 420–422
Numerical range, 662
Numerical stability, 268
Nyquist plot, 145, 593

Observable, uniform, 276, 285
Ohm's law, 39
Open Mapping Theorem, 528, 755
Orbit, 78
Orthonormal, 423, 753
 basis, 117, 432, 754
Oscillations, parasitic, 185, 271
Oscillator
 linear, 16, 108, 129
 multivariable, 698, 699
 nonlinear, 111, 250, 290, 706
 perturbed

spectral value sets, 522, 529, 557
stability radii, 593, 697
Ostrowski matrix, 622

Parity check machine, 58, 95
Pendulum, 25, 82
stability analysis, 241
Periodic points, 214
Periodic system, 262, 266
criteria for stability, 256
Perron's example, 259
Perturbation norm, 450, 456, 688
Perturbations of polynomials
complex structured, 628
complex unstructured, 633
minimum norm complex, 632
minimum norm real, 633, 645
of one coefficient, 627
real structured, 637
real unstructured, 642
scaled unstructured, 630, 640
unstructured, 626
Perturbations of systems
affine, 530
block-diagonal, 456, 536
diagonal, 458
multi-block, 457, 550, 588
specified block multiplicities, 463
complex structure, 453
full-block, 454
complex, 556, 591
real, 561, 596
unstructured, 569, 609
linear fractional, 537, 544, 586
minimum norm, 699
complex, 558, 559, 593
real, 563, 565, 568, 597, 601
neglected dynamics, 690, 697
nonparametric
dynamic, 688
linear time-varying, 688, 709
nonlinear time-invariant, 688
nonlinear time-varying, 689
off-diagonal, 552, 580, 589
Phase increasing property
Hurwitz, 299
real case, 300
Schur, 341
Phase portrait, 82, 111, 243, 652
Phase response, 145
Plancherel's Theorem, 172, 748
Poincaré's Recurrence Theorem, 208
Poincaré-Bendixson Theorem, 203
Poisson stability, 208
Poisson's equation, 41
Polar plot, 145, 146
Pole, 146, 308, 521, 729
Polynomial
–, *see also* Roots of a polynomial
even and odd parts, 302, 639
Hermitian, 357
Hurwitz, 297, 384
irreducible, 377
leading coefficient, 370
monic, 370
positive pair, 303
real and imaginary parts, 311
recursive, 322
reflection of
Hurwitz polynomial, 299
Schur polynomial, 342
rotation of, 342
Schur, 340, 384
symmetric
Hurwitz polynomial, 299
Schur polynomial, 342
with analytic coefficients, 376
Positive definite
function, 220
away from a closed set, 219
matrix, *see* Hermitian matrix
Positive system, 247, 601
Predator-prey system, 4–6, 243, 518
Proper rational function
strictly, 141, 320
Pseudospectra, 531, 569, 669
characterization, 569
for complex normal matrices, 573
for real normal matrices, 576
lower bounds
complex case, 573
real case, 576
under similarity transformations, 579, 583
Puiseux series, 382, 401, 528
Pulse amplitude modulation, 169

QR Algorithm, 502

QR Factorization, 497
Quotient system, 157

Rational dependence on parameters, 521, 539
Rational systems in the plane, 209
Real perturbation values
  lower, 483
  upper, 483
Real representation of complex matrices, 465, 600, 720
Realness locus, 562, 568, 597, 600, 638
Recurrence, 206
Recurrent point, 209
Rellich's Theorem, 420
Residue Theorem, 411, 730
Resolvent
  equation, 409, 759
  partial fraction decomposition, 410
Resolvent operator, 409, 661, 759
  compact, 761
Resolvent set, 409, 759
Resultant, 315, 377
Riccati equation
  algebraic, 604, 679, 703
    stabilizing solution, 604
  differential, 678
Riemann's Lemma, 138, 745, 747
Root locus, 562, 582
Root set of an uncertain polynomial, 626
  complex perturbations
    scalar, 636
    structured, 628
    unstructured, 633
  real perturbations
    interval polynomial, 642
    scalar, 638
    structured, 628, 638
Roots of a polynomial, 370
  analyticity, 370, 376
  continuity, 370, 374
  n-tuple of roots, 373, 374
  sensitivity, 370, 383, 636
Rouché's Theorem, 371, 731
Routh array, 305
Routh Test, 305

Saddle point, 111
Sample rate, 169, 192
Sampled system, 175
Sampler, 168
Sampling period, 169
Sampling Theorem, 173
Samuelson-Hicks multiplier-accelerator, 10
Satellite model, 93
Scaling parameter, 468, 551
Scaling transformation, 457, 458, 551, 554, 588
Schaefer's model, 3
Schmidt pair, 436
Schmidt-Mirsky Theorem, 436
Schur complement formula, 606, 719
Schur form of a matrix, 499–501
  algorithm for reducing to, 502
  complex, 501
  real, 499
Schur generating function, 347
Schur matrix of a polynomial, 347
Schur Test, 355
Schur–Biehler Theorem, 345
Schur–Cohn Theorem, 349
Semi-norm, 655, 755
Semi-scalar product, 660
Semi-simple, 410
Semicontinuous
  lower, 465, 651, 724
  upper, 464, 466, 651, 724
Semigroup, 101, 119
  contraction, 658, 661
    β-contraction, 658, 663
    strict, 658, 661, 663
    strong, 658
  generator, 120, 757
  strongly continuous, 120, 672, 756, 761
Shift operator, 88, 132, 137
  for functions, 743
  for sequences, 736, 760
Singular value decomposition, 435
  of Kronecker product, 722
  pseudo, 439
    analyticity, 442
Singular value loci, 609
Singular values, 431
  algorithm for determining, 508
  continuity, 439
  pseudo, 439
Singular vectors, 432, 436
  pair, 436

Singularity of a complex function
  isolated, removable, essential, 729
Spectral abscissa, 263, 649
  δ-spectral abscissa, 673
Spectral contour
  complex perturbations, 557
  real perturbations, 563
Spectral Decomposition Lemma, 105
Spectral Mapping Theorem, 719
Spectral radius, 453, 760
  δ-spectral radius, 684
Spectral representation, 105, 410, 418
  analyticity, 414, 420, 421
    single real parameter, 422
Spectral value set, 522, 545
  bounds, 551
  characterization, 550, 551
  connected components, 545, 558–559, 563–564
  continuity, 555
  full-block complex perturbations, 524, 673
    boundary of, 558
    characterization, 556
  full-block real perturbations, 524, 561
    boundary of, 563
    chacterization for $\ell = 1$, 564
    chacterization for spectral norm, 567
    general characterization, 562
  of a direct sum of systems, 546
  off-diagonal perturbations, 554, 580
  under similarity transformations, 546
  unstructured, *see* Pseudospectra
Spectrum, 759
  point, residual, continuous, 759
Spijker's Lemma, 671, 731
Splitting of an eigenvalue, 400, 401, 573
Stability criteria for polynomials
  Hurwitz polynomials
    Bézout matrix, 319, 320
    Cauchy index, 311
    change of argument, complex polynomial, 298
    change of argument, real polynomial, 300
    Hankel matrix, 334
    Hermite matrix, 317
    Hurwitz matrix, complex case, 337
    Hurwitz matrix, real case, 339
    Liénard–Chipart Theorem, 336
    recursive test, real polynomial, 305
    Routh test, 305
  Schur polynomials
    change of argument, 340
    recursive test, 355
    Schur matrix, 349
    Schur–Biehler Theorem, 345
    Schur–Cohn Theorem, 349
    Toeplitz matrix, 353
Stability criteria for systems
  –, *see also* Liapunov's indirect method
  –, *see also* Liapunov's direct method
  time-invariant linear systems
    algebraic stability domains, 357, 359
    spectral criteria, 264
    via algebraic Liapunov equations, 285
  time-varying linear systems, 260
    via differential Liapunov equations, 273, 276
Stability definitions
  absolute stability, 701
  asymptotic stability
    of a trajectory, 199
    of an equilibrium point, 200
  $\mathbb{C}_g$-stability, 522
  exponential stability, 707
    for linear systems, 257
    of nonlinear systems, 233
  global asymptotic stability, 695
    of nonlinear systems, 200, 238
  global exponential stability, 709
  L^q-stability, 132
  marginal stability, 218
  stability
    of an attractor, 211
    of a trajectory, 199
    of an equilibrium point, 200
  uniform asymptotic stability
    of a trajectory, 199
  uniform stability
    of a trajectory, 199
  (M, β)-stability, 648, 664
    of differential inclusions, 677
    strict, 648
Stability radius of polynomials, 630
  complex perturbations
    of one coefficient, 636
    scalar, 636

structured, 630
unstructured, 634
real perturbations
of one coefficient, 640, 644
scalar, 638
scaled unstructured, 640
structured, 630, 638
unstructured, 642, 645
Stability radius of systems, 529, 586
bound for multi-block perturbations, 589
characterization, 587
continuity
complex perturbations, 615
lower semicontinuity, 614
real perturbations, 616, 623
full-block complex perturbations, 591
algorithm for computing, 607
characterization, 591
characterization for spectral norm, 602–606
continuous time, 592
discrete time, 592
full-block real perturbations, 596
characterization for $\ell = 1$, 597
general characterization, 600
of a direct sum of systems, 594, 596
off-diagonal perturbations
continuous time, 589
discrete time, 591
under dynamic perturbations, 696
real case, 699
under nonlinear perturbations, 697
under time-varying perturbations, 697
under unstructured perturbations, *see* Distance from instability
(M, β)-stability radius, 675, 678
time-varying perturbations, 677
Stability region, 265, 522, 529, 585
algebraic, 357
Starlike set, 521, 707
State, *see* Dynamical system
of a delay system, 91
of a shift register, 80
State transition map, 76
Steady state response, 144
Step response, 126
Strongly destabilizing perturbation, 616
Structured singular value, 450
Subharmonic function, 592, 733
Sublevel set, 220, 237
attractive, 240
Subsystem, 156
Summability, 140, 753
absolute, 753
Supply and demand curves, 8
Sylvester's Law of Inertia, 314, 723
System embedding, 156
System graph, 165
System projection, 157

Toeplitz matrix, 352–354
Transfer function, 140–142, 549, 587
dyadic decomposition, 143
experimental identification, 144
of feedback interconnection, 162
of parallel interconnection, 161
of series interconnection, 160
Transfer matrix, *see* Transfer function
Transient amplification factor, 650
Transient behaviour, 648
ideal, 656
Transient bound
of a linear system, 650, 669, 671
normal system matrix, 656, 672
structured, 672, 673
time-varying nonlinear perturbations, 679, 680
Tree, 51
spanning, 51
Trolley example, 15

Unimodal, 474
Unit impulse, 125

Value set, 390
Van der Pol equation, 244
Verhulst model, *see* Logistic growth model
Volume preserving flow, 207

Wandering point, 206
Well-posedness radius, 521, 538, 544, 556, 561, 587, 629
Weyl's Theorem, 445
Wilkinson polynomial, 384, 636
Wilkinson shift, 509
Winding number, 310, 725, 727

z-transform, 139, 737

Texts in Applied Mathematics

(continued after page ii)

31. *Brémaud*: Markov Chains: Gibbs Fields, Monte Carlo Simulation, and Queues.
32. *Durran*: Numerical Methods for Wave Equations in Geophysical Fluids Dynamics.
33. *Thomas*: Numerical Partial Differential Equations: Conservation Laws and Elliptic Equations.
34. *Chicone*: Ordinary Differential Equations with Applications.
35. *Kevorkian*: Partial Differential Equations: Analytical Solution Techniques, 2nd ed.
36. *Dullerud/Paganini*: A Course in Robust Control Theory: A Convex Approach.
37. *Quarteroni/Sacco/Saleri*: Numerical Mathematics.
38. *Gallier*: Geometric Methods and Applications: For Computer Science and Engineering.
39. *Atkinson/Han*: Theoretical Numerical Analysis: A Functional Analysis Framework, 2nd ed.
40. *Brauer/Castillo-Chávez*: Mathematical Models in Population Biology and Epidemiology.
41. *Davies*: Integral Transforms and Their Applications, 3rd ed.
42. *Deuflhard/Bornemann*: Scientific Computing with Ordinary Differential Equations.
43. *Deuflhard/Hohmann*: Numerical Analysis in Modern Scientific Computing: An Introduction, 2nd ed.
44. *Knabner/Angermann*: Numerical Methods for Elliptic and Parabolic Partial Differential Equations.
45. *Larsson/Thomée*: Partial Differential Equations with Numerical Methods.
46. *Pedregal*: Introduction to Optimization.
47. *Ockendon/Ockendon*: Waves and Compressible Flow.
48. *Hinrichsen/Pritchard*: Mathematical Systems Theory I.
49. *Bullo/Lewis*: Geometric Control of Mechanical Systems: Modeling, Analysis, and Design for Simple Mechanical Control Systems.
50. *Verhulst*: Methods and Applications of Singular Perturbations: Boundary Layers and Multiple Timescale Dynamics.
51. *Bondeson/Rylander/Ingelström*: Computational Electromagnetics.
52. *Holmes*: Introduction to Numerical Methods in Differential Equations.
53. *Pavliotis/Stuart*: Multiscale Methods: Averaging and Homogenization.
54. *Hesthaven/Warburton*: Nodal Discontinuous Galerkin Methods.
55. *Allaire/Kaber*: Numerical Linear Algebra.
56. *Mark H. Holmes*: Introduction to the Foundations of Applied Mathematics.

GPSR Compliance
The European Union's (EU) General Product Safety Regulation (GPSR) is a set of rules that requires consumer products to be safe and our obligations to ensure this.

If you have any concerns about our products, you can contact us on

ProductSafety@springernature.com

In case Publisher is established outside the EU, the EU authorized representative is:

Springer Nature Customer Service Center GmbH
Europaplatz 3
69115 Heidelberg, Germany

www.ingramcontent.com/pod-product-compliance
Lightning Source LLC
Chambersburg PA
CBHW071613100426
42873CB00003B/35
9783540441250